Abstracts

28th International Geological Congress

Washington, D.C. USA
July 9-19, 1989

Abstracts are presented in alphabetical order. A detailed directory of the presenters in this volume is provided at the back of the book.

A number of abstracts were received too late to be included with the major alphabetical listing. Many of these late abstracts may be found in a separate section at the back of volume 3.

S

SABIR, H. M., BRGM Geological Mission, Jeddah, Saudi Arabia

Metallogenic Studies Applied to Mineral Exploration: 15 Years of Saudi Arabian Experience

Many mineral deposits of varied economic importance have recently been discovered in the Precambrian Shield of the Kingdom of Saudi Arabia. These discoveries, coupled with the presence of a great volume of mineralized core material, have given impetus to the initiation of an extensive metallogenic research program in 1974 that provides basic knowledge related to the specific types of mineralization, parageneses, mineral content, and chemical parameters.

Progress achieve in applied metallogenesis during the last 15 years is presented, with typical examples from two of the most important and widely recognized geologic environmental hosting mineral deposits in the Kingdom, namely (1) volcanosedimentary successions and (2) plutonic complexes.

(1) The volcanosedimentary rocks contain two distinct but closely related mineralization types: (a) Polymetallic sulfide deposits are marked by a complex mineralogical association mainly comprising pyrite, chalcopyrite, sphalerite and pyrrhotite, together with minor galena, arsenopyrite, cubanite, mackinawite, and marcasite. The trace minerals are numerous and varied and include: sulfosalts of copper-lead-antimony-silver, tellurides of gold, silver, lead, bismuth, nickel, iron and mercury, native metals of silver, gold and bismuth, together with magnetite, molybdenite and cassiterite. This global mineralogic composition is characteristic of volcanogenic massive sulfide deposits associated with island-arc volcanism recognized in several parts of the world. (b) A newly-discovered disseminated gold mineralization hosted by brecciated and silicified, saccharoidal quartzitic rocks in proximity to gossanous facies, where gold is closely associated with quartz, goethite and hematite.

(2) Plutonic complexes host a wide variety of rare-element mineralization, and which may be recognized: (a) Gold deposits with a complex mineral assemblage comprising pyrite, arsenopyrite, sphalerite, chalcopyrite, gold, galena, stannite, together with native bismuth and lead-bismuth-silver sulfosalts. (b) Tungsten-tin deposits, the paragenesis of which is marked by an association of wolframite, cassiterite, stannite, molybdenite, pyrite, chalcopyrite, galena, native bismuth and lead-bismuth sulfides. (c) Silver deposits containing varied proportions of sphalerite, pyrite and chalcopyrite, together with acanthite, polybasite, pyrargyrite, tetrahedrite, native silver and stromeyerite.

Comprising an integral part of the Directorate General of Mineral Resources' work program, these studies have led to a better understanding of the process of ore formation and to a definition of certain guidelines that have helped in orienting mineral exploration programs in the Kingdom of Saudi Arabia.

SABURO, A., Toyo University, Asaka Saitama, Japan, K. NORIHIKO, National Institute of Industrial Health, Kawasaki, Japan; and I. TOSHIO, University of Tokyo Nakano, Tokyo, Japan

Sedimentary History and Chemical Characteristics of Clay Minerals in Drilled Cores of ODP Leg 116

Three closely spaced drill core samples were recovered from the equatorial Indian Ocean about 800 km south of Sri Lanka in ODP, Leg 116. These bottomed in Miocene age. The purpose of this study is to clarify sedimentary history and chemical characteristics of clay minerals in the drilled core samples since the Himalayan Mountains uplifted 17 million years ago. Seventy-eight samples of the three drilled cores were collected for clay mineral analyses by means of XRD and ATEM. Results obtained show that values of the average content of individual clay mineral species in sediment samples of the same age at each site at very similar but show distinguishable features in different geological ages. Detrital clay minerals such as illite and chlorite deposited more abundantly than kaolinite and smectite from early Miocene to middle Miocene. This means that the Himalayan uplift was vigorous at least until middle Miocene. The denudation by the sea-level fluctuations during Miocene time must also have played a role in supplying detrital clay minerals. In upper Miocene time, illite shows the highest concentration but chlorite content decreases relatively in place of increasing kaolinite content. This suggests the weakness of the Himalayan uplift and the sea-level changes while chemical weathering for kaolinite formation prevailed inland. It had been reinforced more in Pliocene time so that kaolinite showed the highest concentration at the distal part of the Bengal Fan. This must be concordant with the faintness of the Himalayan uplift. In Pleistocene time, the vigorous Himalayan uplift is characterized by illite-rich sediments in compensation for the kaolinite that prevailed in Pliocene time. The eustatic sea-level changes must have also played a significant role in supplying illite to the equatorial Indian Ocean. In Holocene time, the sedimentary history of clay mineral species at the distal part of the Bengal Fan has greatly changed. Smectite shows the highest concentration in place of illite or kaolinite, the predominant clay minerals in previous geological ages. Increasing smectite concentration suggests the Himalayan uplift to have been stable after Pleistocene time. Otherwise, smectite may have been exclusively supplied from the Afanasy Nitikin Seamounts close to ODP Leg 116 site.

Of the clay minerals identified, chemical compositions of smectite and chlorite have variable characteristics. Chemical characteristics of smectite in samples from Miocene to Pliocene time show it to be a di-octahedral Fe-rich type, ranging from $(Si_{3.38}Al_{0.62})_{4.00}$ $(Al_{0.70}Mg_{0.20}Fe^{3+}{}_{1.12})_{2.02}$ $(Ca_{0.02}K_{0.39}Na_{0.32})_{0.73}$ $O_{10}(OH)_2$ to $(Si_{3.34}Al_{0.66})_{4.00}$ $(Al_{1.01}Mg_{0.18}Fe^{3+}{}_{0.83})_{2.02}$ $(Ca_{0.16}K_{0.20}Na_{0.22})_{0.59}$ $O_{10}(OH)_2$. In Pleistocene samples, beidelite is the predominant species: $(Si_{3.46}Al_{0.54})_{4.00}$ $(Al_{1.34}Mg_{0.21}Fe^{3+}{}_{0.37})_{1.82}$ $(Ca_{0.14}K_{0.12}Na_{0.84})_{1.10}$ $O_{10}(OH)_2$. Chlorite in samples of both Miocene and Pleistocene ages belongs to the di-octahedral Fe-Mg type, ranging from $(Si,Al)_{4.00}$ $(Al,Mg)_{3.26}$ $Fe^{3+}{}_{1.47}$ $O_{10}(OH)_8$ to $(Si,Al)_{4.00}$ $(Al,Mg)_{2.14}$ $Fe^{3+}{}_{2.59}$ $O_{10}(OH)_8$. Tri-octahedral chlorite is not as prevalent as the di-octahedral type. The genesis of clay minerals in this study has not been clarified. However, Fe-rich smectite in Miocene samples might have partly originated from the neighboring seamounts such as Afanasay Nitikin.

SACCHI, R., Universitá di Torino, Torino, Italy

Pan-African Rejuvenation

This paper presents a discussion and a reinterpretation of some aspects of the so-called Pan-African event. Early in the sixties, it was recognized that many age determinations (essentially K-Ar and Rb-Sr on single mineral phases) clustered around 500 Ma throughout Africa. It was also felt that orogenic developments of the same age were embarrassingly lacking, with rare exceptions (Katanga and Damara). The exceptions grew in number in the years that followed: collision-type orogeny with addition of juvenile crust is now known in the Arabian–Nubian Shield, in the Pharusian Belt of West Africa, and, more debatably, in parts of the Mozambique Belt. But little seems to have happened elsewhere in "Pan-African" time, apart from scattered emplacement of granites. It should be noted, however, that this abnormal age frequency is by no means evident when methods with a less labile memory, namely Rb-Sr whole rock isochron and U-Pb on zircons, are used.

For this reason, the Pan-African event (so named by Kennedy, 1964) has remained in a limbo of "tectonothermal events"; it has never really established itself as a fully fledged orogeny. As Almond (1984) stresses, "the importance of an event, which, among other things, set or reset radiometric ages over an area of 20 million km^2 in Africa alone cannot be disputed." Let us not forget that 20 million km^2 is two-thirds of Africa, and that the same picture, not unknown in Australia, is widespread in Antarctica and

South America. "The Pan-African s.s. age province," as Almond (1984) concludes, "is very much more extensive than any pervasive tectonism that can reasonably be related to it."

We would here point out that 500 Ma ago the African craton and the (then) adjoining Gondwana continents underwent a probably unique combination of events. A very large eustatic rise (Vail et al, 1977) followed a period of widespread peneplanation and brought about flooding of the order of 50% of the emergent land in a time of generalized continental pull apart (cit. p. 94) and of steep geothermal gradient. It is suggested that this combination triggered large-scale hydrothermal circuits and was responsible for widespread radiometric rejuvenation.

We also suggest that it may conceivably have a bearing on another aspect of the Pan-African history. Martin and Bonin (1976) analyze the case of water becoming available in a late stage during the cooling of an initially dry magma, and reach the conclusion that such an event might trigger the emplacement of minor bodies of granite. They start from the factual observation that in an environment of extensional tectonics, the association of hypersolvous (i.e., perthite-bearing) granite with later subsolvous granite (bearing the two feldspars as discrete grains) is common, and argue thus: "We propose that a late influx of water, possibly of meteoric origin, 'soaked' the cooling hypersolvous granite once crystallized, inducing local near complete fusion and remobilization of the early granite. No reheating need be implied as the 'dry' solidus is even higher than the 'wet' liquidus." This reasoning is supported by an experimental phase diagram for the system albite–K-feldspar–water at 1.25 and at 5 Kbars of $P(H_2O) = P(total)$. Increasing the water partial pressure both lowers the solidus and raises the solvus curves (with respect to T) so as to annihilate the stability field of sanidine. A model calculation of the thermodynamic properties of the alkali feldspar-H_2O system could throw some light on the matter and show to what extent Martin and Bonin's hypothesis is viable. This we have attempted, but the relevant data unfortunately appear to be lacking, particularly for the water/silicate melt interaction.

Clearly, our neo-Neptunist view is not free from problems. One way can be conceived whereby the hypothesis could be upgraded to the rank of theory. This involves verifying the predictions. For the hypothesis to hold true, a number of features of the Pan-African, rejuvenated basement are expected to behave accordingly, e.g., (1) granite feldspar mineralogy, (2) oxygen isotope composition, (3) radiometric age pattern, and (4) areal distribution of the radiometric rejuvenation and its relation to the Cambrian geography. A study of the literature has shown that the present data for each of these topics are inadequate. Consequently, our idea cannot yet be taken further than the stage of a heuristic model.

REFERENCES

Almond, D.C., 1984, The concepts of "Pan-African episode" and "Mozambique Belt" in relation to the geology of E and NE Africa: Bull. Fac. Earth Sci. Abdulaziz Univ., v. 6, p. 71–87.

Kennedy, W.Q., 1964, The structural differentiation of Africa in the Pan-African (+500 m.y.) tectonic episode: Univ. Leeds Res. Inst. African Geol. Ann. Rept., v. 8, p. 48–49.

Martin, R.F., and B. Bonin, 1976, Water and magma genesis: the association hypersolvous granite-subsolvous granite: Canadian Mineralogy, v. 14, p. 228–237.

Vail, P.R., R.M. Mitchum, Jr., and S. Thompson III, 1977, Seismic stratigraphy and global changes of sea level, Part 4: Global cycles of relative changes of sea level, *in* C.E. Payton, ed., Seismic stratigraphy—application to hydrocarbon exploration: AAPG Memoir 26, p. 83–97.

SACKS, L. A., and J. S. HERMAN, University of Virginia, Charlottesville, VA, A. L. VELA, Universidad Complutense, Madrid, Spain, and L. F. KONIKOW, U. S. Geological Survey, Reston, VA

Hydrologic and Geochemical Dynamics between Groundwater and Interdunal Lakes in Doñana National Park, Spain

Several shallow, interdunal lakes occur in Doñana National Park, an important European wetland located in southwestern Spain. Of three lakes occurring within a 1.5 km^2 area, one is intermittent and two are perennial. Their salinities differ by more than an order of magnitude, with the least saline lake located upgradient and the most saline lake located downgradient along the regional groundwater flow path. The lakes are closely interrelated with the groundwater system, and the occurrence of distinct wet and dry seasons may result in seasonally dynamic groundwater–lake interactions. In July and December of 1985, water samples were collected from the lakes and shallow groundwater below the perimeters of the lakes, and potentiometric heads were measured along the shores of the lakes.

This study investigated the effects of seasonally dynamic groundwater flow on the hydrologic and solute budgets of the lakes using a modeling approach. In order to better understand the controls on lake salinities, the influence of the local and regional groundwater flow systems on the lakes was examined. Furthermore, chemical data collected from the lakes and the shallow groundwater were interpreted to elucidate chemical processes occurring as groundwater seeps into and out of the lakes.

A two-dimensional solute-transport model was modified to allow for groundwater interaction with the lakes. The model was used to simulate transient flow and solute-transport patterns near the lakes, as well as seasonal changes in lake salinity, surface area, and stage. Results indicted that seasonal changes in recharge give rise to dynamic hydrologic and solute interactions between the groundwater and the lakes. Water table mounding during the wet season leads to local flow reversals near the lakes. These flow reversals, in turn, result in changes in the locations of groundwater seepage into the lakes and the hydrologic budgets of the lakes.

The responses of the three lakes, however, vary depending upon the hydrogeologic setting of each lake within the regional flow system. The least saline lake is a flow-through lake, where dilute groundwater seeps into the lake on its upgradient end, is concentrated by evaporation, and seeps into the groundwater on its downgradient end. These solutes are then transported along the regional flow path in the aquifer toward the more saline lakes. The flow system around the higher salinity, permanent lake is characterized by seasonal groundwater flow reversals on its downgradient end. These reversals lead to complex seepage patterns and a higher solute residence time in the lake. In addition, higher salinity groundwater enters the lake because of the seasonal reversals and because of hydrologic connections with the upgradient flow-through lake. The intermittent lake has a highly variable solute concentration, depending upon the time of year. Thus, its salinity is influenced more strongly by the wetting and drying cycle than the position of the lake in the regional flow system.

While evaporation controls the concentration of conservative solutes in the groundwater–lake system, geochemical and biochemical reactions occurring both in the lake sediments and in the lakes control the concentration of nonconservative solutes. The depletion of ions in the lake waters relative to groundwater inflow can be attributed to calcite precipitation, biological uptake, and sulfate reduction in the lake sediments. Likely reactions occurring in the lake sediments. Likely reactions occurring in the lake sediments as water seeps out of the lake include calcite and silica dissolution, cation exchange, and further sulfate reduction. The rewetting of the intermittent lake is accompanied by the addition of conservative solutes stored within the lake sediments, the dissolution of calcite, and the dissolution of sulfate from the dried lake sediments.

The chemical composition of the lake waters are influenced by evaporative concentration, seasonal variations in groundwater seepage patterns, and chemical reactions occurring within the lake water and the sediments. The composition of the shallow groundwater near the lakes is a reflection of seepage patterns to and from the lakes. Seasonal reversals in seepage patterns, however, can result in complicated groundwater compositions near a lake that can vary considerably over the course of the year. Thus, understanding the transient groundwater flow patterns around the lakes enables a more accurate interpretation of the hydrologic and chemical relationships between the lakes and the groundwater.

SADOONI, F.N., Oil Exploration Co., Baghdad, Iraq

Suggested Scheme for Diagenesis of Evaporites

Evaporites display a wide variety of diagenetic features both megascopic and microscopic. The adoption of the sabkha theory in the assessment of the origin of evaporites during the sixties greatly influenced evaporitic sediment studies. Megascopic features in outcrops and well cores have been the subject of wide and careful studies due to the weight given to them by this theory. Many important works of such type were carried out on various "fossil" sabkhas all over the world and especially in the Middle East.

The present study, however, depends on microscopic examination of evaporitic sediments from the Gotnia, Dhiban, Jeribe, and the Lower Fars formations in Iraq that range in age from Triassic to Middle Miocene and in depth from surface to more than 3000 m.

It is concluded that the sabkha theory suffers from at least two main shortcomings, which are summarized below:

1. Great emphasis has been put on diagenetic evaporites on the expense of primary evaporites; this has led to false interpretation in some cases.

2. Much weight has been put on the importance of certain megascopic features such as nodular anhydrite, enterolithic folding, etc., with almost complete negligence of the microscopic features that play an extremely important role in a more precise understanding and thus in the interpretation of the origin and history of these sediments.

It is thought, therefore, that there is an urgent need to establish an organized scheme for the diagenesis of evaporites similar in some respects to the diagenetic system of carbonates, taking into consideration, however, the definitive characters of evaporites and their relation with carbonates and other sediments. The similarity between the two systems is based on the fact that both evaporites and carbonates are similar in many respects, as evaporites could be generally compared to the chemically precipitated lime mudstone.

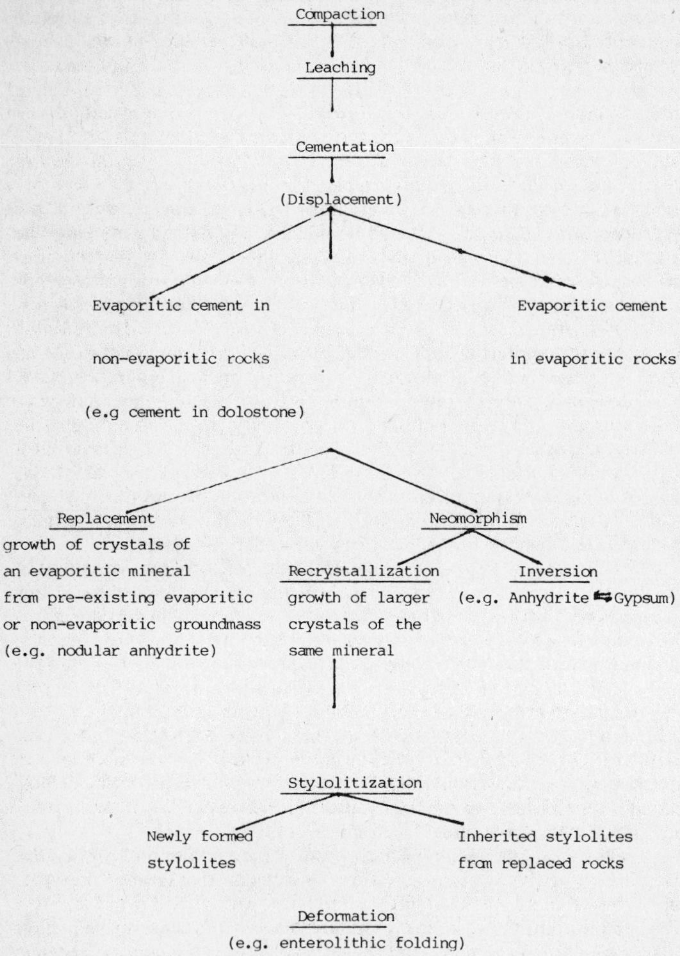

Figure 1.

SAET, Y.E., and E.K. BURENKOV, Ministry of Geology of the USSR, Moscow, USSR

Geological and Geochemical Problems and Methods Applied for Studying Urbanization Areas

Ore minerals are primary sources for metals that are involved in technogenesis. All further methods of ore transformation lead to wide dissemination of chemical elements in the environment. Dissemination follows natural migration flows (air, water, biogenic), thereby bringing about additional loads both on the flows proper and on the contacting natural bodies. Therefore, environmental pollution is a geochemical process that follows general laws of element migration established for hypergenesis zones and natural landscapes.

The intensity of chemical element supply by anthropogenic pollution sources can be assessed through comparison with extreme natural geochemical conditions observed in ore deposits. In the process of natural denudation, ore fields yield tens to hundreds of tons of metal per year, whereas a nearby big city supplies to the environment thousands of tons of zinc, copper, chrome, and lead; hundreds of tons of nickel, tin, and tungsten; and tens of tons of cadmium and molybdenum, as inferred from a comparison analysis.

Applied geochemical methods used in prospecting for mineral deposits have proved most effective in studying pollution and evaluating its results. The applied geochemical investigations aimed as solving present-day ecological-geochemical problems evolve along four lines: (1) geochemical assessment of the main sources of pollution and the intensity of loading from sources in the environment, qualitative and quantitative assessment of solid domestic wastes, dust-gas outbursts, effluents and chemical products used in agriculture, forestry and water economy as pollution sources, and at the same time as potential secondary mineral raw materials; (2) identification of major qualitative and quantitative parameters and formative conditions of technogenic and agrogenic aureoles and stray fluxes; (3) evaluation of the intensity of biogeochemical and biological reactions of living organisms in pollution zones; and (4) practical application of geochemical data for working out nature conservation measures.

Identified geochemical and biogeochemical correlations in the distribution of chemical elements between environmental natural components are empirical statistical models that are helpful in compiling maps and differentiating urbanized areas by levels of pollution.

The experience gained in geochemical mapping of environments that take in pollution products (soils, snow, bottomset beds) shows that in all cases the pollution sources are accompanied by anomalous contents of chemical elements in natural objects. Therefore, express geochemical mapping without any prolonged stationary observations required for studying dynamically unstable environmental components such as air or water permits us to identify pollution sources and zones exposed to the pollution and outline areas that require assessment from hygienic and ecological points of view.

Ecological-geochemical techniques developed in the Soviet Union for studying environments enable us to solve a large scope of scientific and practical tasks. It has become an important tool that helps provide information for many nature conservation projects.

SAG, L., Geographical Research Institute of the Hungarian Academy of Sciences, Budapest, Hungary, and G. MAJOROS, Mecsek Ore Mining Enterprise, Pécs, Hungary

Structure and Development of Basement of Pannonian Basin in Central Europe and Buildup of Its Hercynian and Older Structural Stages

The so-called Carpathian Basin, which is an inter-arc basin in the southeastern part of Central Europe between the Carpathian arc, the Eastern and Southern Alps, and the Dinarids, comprises the territory that was historical Hungary. Within this area, there are several basins separated by mountain groups and chains. By the term Pannonian Basin we now mean the greatest subbasin of the area, which involves the whole area of present Hungary. This Basin is bordered on the north by the Western Carpathians, on the east by the Bihar Mountains (Muntii Apuseni), on the south by the Inner Dinarides and the Vardar zone, and on the west by Eastern (and partly Southern) Alpine units. Although several mountainous ranges are outcropping in the area, these can be considered as inselbergs, and the greater part is a basin filled with Neogene and Quaternary sediments. Fortunately, since World War II–thanks to increasing explorational activity–more than 10,000 boreholes were deepened, aiming at the exploration of hydrocarbons, ores, coal, nonmetallics, and water, with a depth of several hundred to several thousand meters. The detailed study of the logs, cores, and cuttings resulted in an extensive knowledge of the basement of the sedimentary basins and in recognition of the connections of the surface and subsurface occurrences of the similar sequences.

As a result of this widespread explorational and drilling activity, the establishment of basement-structure reconstructions started in the early 1950s. Several different structural zones and units can be distinguished. Within each structural zone vertical structural units, i.e., structural stages, can be differentiated.

The main structural units of the Hercynian and older basement of the Pannonian Basin are (see Figure 1): (1) The *Lower East Alpine* nappe system (Unterostalpin); (2) The *Upper East Alpine* system (continuation of the *"Paleozoic of Graz"*); (3) The range of the *Hungarian Uplands* ("Mittelgebirge"); this mountain range, together with its subsided wings, is rather heterogenous, analyzing its structure either in a northwest-southeast or in a southwest-northeast direction. Therefore, we have to distinguish at least two zones from each other: (3/a) the *Bakonyides* in the northwest and (3/b) the *Bükkides* in the southeast; (4) the *Central Hungarian* (semi-) *autochthon Zone,* which involves the Mecsek Mountains in southern Transdanubia in the west and the Bihar Mountains (Muntii Apuseni) in the east, together with the basement of the middle part of the Great Hungarian Plain (Alföld); (5) The *Codru* nappe system, which includes—besides the surface and subsurface formations in Romania—the southern part of the basement of the Hungarian Alföld, the Villány Mountains in Transdanubia and the Paleozoic of the mountains of northern Croatia.

The detailed characterization of the older formations of these structural units could be performed only by presenting and analyzing the most important and characteristic stratigraphic columns within the units based on the sequences known from the mountain ranges as well as from the boreholes, respectively. We shall try this in the full text of the paper; here we give only a very short summary of the structural stages within each zone.

In the Lower East Alpine unit rock series of Proterozoic (Baikalian?), Caledonian, and Hercynian origin (or deposition) can be found. The rocks are usually polymetamorphic; a Caledonian, Hercynian (or Variscan), and Alpine phase can be distinguished. The main rock types are metapelites-metasammites, acidic gneisses, and amphibolites. The thermal gradient of the pre-Hercynian metamorphism is lower than 34° C/km (greenschist facies), and of the Alpine is lower than 34° C/km (greenschist facies).

The continuation of the "Paleozoic of Graz" is a part of the Upper East Alpine system. The sequence represents the sediments and magmatic rocks of the Caledonian and Hercynian cycles, from the Upper Ordovician to the Lower Carboniferous (Upper Viséan). The rocks are mostly slates with diabase, with platform carbonates and Carboniferous pelagic sediments on the upper parts. Hercynian and older metamorphism is absent; only some parts were effected by a weak Alpine metamorphism.

The zone of the Hungarian Uplands can be divided into two different ranges. The Bakonyides are considered as the continuation of the Drauzug. On the southern-southeastern part there is a zone of thrusted metamorphic belts; the initial rocks are of Caledonian or even older origin (deposition), the metamorphism is Hercynian; medium (amphibolite) and low (greenschist) grade. This zone corresponds to the Quartzphyllite of the Alps and is separated tectonically from the pre-Permian series of the Uplands ("Mittelgebirge" = mountains of medium height) itself. Here a mostly metapelitic–metapsammitic series can be found with acidic and basic volcanic products and with carbonate sediments on the top of the sequence. The age is considered to be Ordovician–Lower Carboniferous (Viséan). The low and mostly very low grade metamorphism is Hercynian. The Upper Hercynian structural substage is represented by a postorogenic conglomerate (molasse) and granitic igneous activity. The Middle–Upper Permian sediments are the initial members of the thick Alpine sedimentary series. The Bükkides are mostly similar to the former sequences covering about the same time span. The lithological difference is in the greater abundance of carbonate rocks and the lack of volcanics. The Hercynian metamorphism is absent; only a weak Alpine metamorphism is known, as in the Paleozoic of Graz.

In the Central Hungarian autochthon zone, the following structural stages are known: Proterozoic series with pre-Baikalian (Assyntian) metamorphism of medium grade, with ortho- and paragneisses, amphibolites, micaschists, a Baikalian metamorphic sequence of low grade, with synorogenic granites, rocks of Caledonian and Hercynian origin with a low-grade Hercynian metamorphism; Carboniferous–Permian, mostly clastic sedimentary deposits, covered by Permo-Triassic series.

In the Codru nappe system the oldest rocks are of Proterozoic origin and of progressive metamorphism (maybe pre-Baikalian and Baikalian) with micaschists, granitoid intrusions, and intensive folding. The Hercynian series is built up of slates, conglomerates, greywackes, quartzites, limestones with spilite, and diabase; the age expands until the middle of the Carboniferous. The deposition of a mostly clastic Middle–Upper Permian–Lower Triassic (and younger) sequence terminates the Hercynian cycle.

The basement of the whole basin is divided by the Mid-Hungarian Structural Zone. The northwestern zones are parts of the Alpine system and have southern, African affinity, while the southeastern zones are coming from a northern domain and can be correlated with Hercynian Europe (the higher Permian, Triassic, and Jurassic series have a germanotype character). These two main structural units reached their present-day geographical position during the Oligocene and Early Miocene, through several hundred kilometer long horizontal and rotational movements. The main structural line between these units is characterized by horizontal movements, and the Rába structural zone, too, while the structural boundaries between the other units are mostly thrust zones.

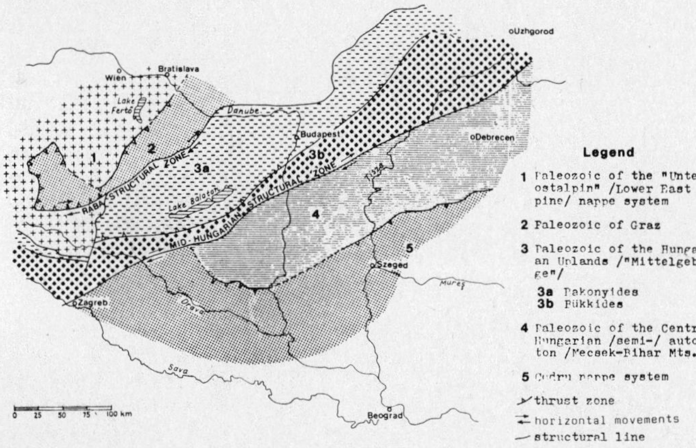

Figure 1—Major structural units of the Hercynian and older basement of the Pannonian Basin.

SAG, L., Hungarian Academy of Sciences, Budapest, Hungary, and G. SZILI, Central Office of Geology, Budapest, Hungary, and R. I. VANDORFI, National Petroleum Gas Trust, Budapest, Hungary

The Geological Principles, History of Development and Economic Role of Hydrocarbon Production in Hungary

Hungary is situated in the Carpathian basin in the southeastern part of Central Europe. Within this interarc basin we can distinguish some subbasins, among which the greatest and most important is the so-called Pannonian Basin. This basin, between the Western Carpathians, the Bihar Mts., the Dinarides and the Eastern and southern Alps, involves the whole area of present Hungary and is registered as one of the hydrocarbon-bearing basins of Europe. In the Hungarian part 77 thousand square km of the country's area (93,000 km^2 are suitable for hydrocarbon accumulation. While the young sedimentary cover is almost homogeneous, although different in thickness, the basement can be divided into several zones; these form two entirely different units, separated by the Mid-Hungarian Structural Zone (Line). Overviewing the different formations of the basin areas, the following oil and gas bearing horizons ("reservoir rocks") can be mentioned: Upper Miocene (Pannonian s.l. or Pontian and Pannonian) rocks as Pontian sands-sandstones, Pannonian sandstones, marls and basal (conglomerate) horizons; Middle Miocene limestones, conglomerates, breccias etc., Oligocene (Lower) clastic sediments; Paleogene-Upper Cretaceous flysch, Upper Cretaceous limestones, marls; Jurassic carbonate rocks; Triassic carbonates (limestone-dolomite) and clastics and the Proterozoic metamorphic basement in the southern part of the Aolföld (Great Plain) area. In 1964 30% of the petroleum and 65% of the gas reserves was found in the Upper Miocene (Pannonian-Pontiac) rocks and 68-17% in Mesozoic rocks, but in 1979 this changed completely to a ratio of 55%-78% in Upper Miocene and 25-1% in Mesozoic, respectively.

Traces of hydrocarbons were known in the past, before the 20th century. In historical Hungary (the Carpathian Basin) several hydrocarbon occurrences were found before 1918 (Kissarmas in Transylvania, 1909, gas; Ebell-now Slovakia-in 1913, Bujavica-Croatia-in 1918). The country was among the first in the world where the state monopoly of hydrocarbon exploration was codified by Parliament in 1911. After World War I, the renewed activity did not result in discoveries til 1937.

Since the Danube splits the country to two parts: Transdanubia on the west and the "Alföld" (Great Plain) on the east, present Hungary is traditionally divided to these units. On the former area prospecting and exploration was carried out with the help of western (Anglo-Saxon) capital, while on the latter (Alföld) the State Treasury monitored explorational activity; later they had to make way for German capital too. In Transdanubia first the Hungarian Oil Syndicate Ltd. (a subsidiary of Anglo-Persian Oil Co.) deepened continued the activities. On the Great Plain area the Treasury could not find oil, but thermal and gaseous water was discovered at Hajduszoboszlo near Debrecan. North of the Alföld at the foothills of Bükk Mts, the small oil field of Bükkszék was discovered; before World War II the Hungarian-German Mineral Oil Works Ltd. (MANAT) was involved, too.

Really modern oil and gas exploration started in the 1930s and the first period of this can be counted from 1933 til 1944. The EUROGASCO founded MAORT in 1938 (Hungarian-American Oil Industrial Corp.), and this company controlled the oil industrial til the nationalization in 1949. The further result of this successful beginning was the Lovászi field in 1940. MANAT found on the Alföld area the small occurrences of Korösszegapfi, Tótkmolós and Ferencszállás. After the war the second period (1945-1956) was very troubled. The nationalization, the political (preconceptional, as it was later cleared) process at court against MAORT, the discovery of Nagylengyel (Zala basin) and Mezokeresztes (foothill area of Bükk Mts.) oil fields and the 1956 political events were the milestones of this time interval. The third period lasted from 1957 til about 1971. The characteristics of this interval are: boom and climax. The results: Pusztaföldvár, Hajdúszoboszló (great gas occurrence), Ullés, Szank, Algyó (far the greatest Hungarian oil and gas field), Asotthalom, Ferencszallas, Ortahaza. The fourth recent period: exploration among more and more complicated natural conditions since 1972. The most important discoveries were Szeged, Kiskunhalas, NE, Sarkadkeresztur, the deep horizon at Ulles.

The oil geologists in Hungary have been dealing since the late 1950s with the elaboration of a subbasin-system within the great basin, serving the goals of hydrocarbon prospecting and exploration. We give here an up-to-date listing of these:

1. Western Kisalfold (Little Plain) neogene subbasin (= sb).
2. Eastern Kisalfold neogene sb
3. Zala-Southern Balaton coast sb (center: Nagykanizsa)
4. Drava-valley sb
5. Central-Danubene neogene sb
6. Kiskunsag neogene sb (Kiskunhalas)
7. Szeged neogene sb

8. Northern Alfold (Great Plain) neogene sb (Jaszbereny)
9. Nagykunsag neogene sb
10. Bekes neogene sb (Oroshaza)
11. Hajdusag-Nyirseg neogene sb (Debrecen)
12. Bihar neogene sb
13. Northern Paleogene basin (Eger)

The most productive area the Szeged (7), Zala (3), Bekes (10) and the Nagykunsag (9) and Kiskunsag (6) subbasins. In the Nationa Hydrocarbon Balance cca. 125 occurrences are registered by the Central Office of Geology (KFH) and the National Petroleum and Gas Industrial Trust (OKGT). Among these far the most important is the Algyo Field, which is situated a few km ENE from the city of Szeged. In Tables 1 and 2 we list the parameters of the 12 most important oil and gas fields of Hungary.

As mentioned above, oil was discovered in 1938 in Hungary. The sudden boom of Hungarian oil production—beside the traditional East European producers, the USSR and Romania was a surprise. During the war the production peak was 0.8 Mt/year. In the postwar period around 1950 the country represented a good average level in Europe. Nowadays Hungary was preceded by a number of countries even in Eastern Europe, but her place in world ranking is still about the fiftieth, e.g., 49, in 1987.

Hungarian oil explorational geophysics is in the European forefront and the country's oil refining industry is better than the average level.

NAME	Year of discovery	Sub-basin	Trap type layer-like	Trap type massive	Age and lithology of reservoir rocks	Initial re-coverable reserves Mt	Cumulative production
						as of 1. I. 1987	
1. Algyő	1965	7	+	+	N_1^3 S+SS, BH; PR WF MB	30,5	18,6
2. Nagylengyel(Barabásszeg-Szilvágy)	1951	3	+	+	N_1^2 SS; K_2LS-SS; T_3D	24,0	19,0
3. Lovászi	1940	3	+		N_1^3 SS	6,5	6,4
4. Budafa-Kiscsehi	1937	3	+		N_1^3 SS	5,8	5,3
5. Szeged	1972	7	+	+	N_1^3 M; N_1^2 SS+CB; T_2D; T_1BH; PR MB	3,9	2,4
6. Szank (-SzankW)	1964	6	+	+	N_1^3 SS; N_1^2 BH; PR WF-MB	2,4	2,1
7. Pusztaföldvár	1958	10	+	+	N_1^3 S+SS, BH	2,3	2,1
8. Szeghalom	1979	9	+	+	N_1^3 BH; N_1^2 LS; PR MB	2,0	0
9. Kiskundorozsma	1971	7		+	N_1^2 CB; PR WF MB	1,8	0,2
10. Kiskunhalas NE	1974	6		+	N_1^2 LS, BH; K_1LS; T_2 D; PR WF MB	1,7	1,0
11. Demjén	1954	13	+		OL SS	1,3	1,2
12. Ruzsa	1978	7	+	+	N_1^3SS+BH; N_1^2 SS+CB	1,1	0

Explanation to both tables:
Ages: PR=Proterozoic; T=Triassic; J=Jurassic; K=Cretaceous; OL=Oligocene; N_1^2=Middle Miocene; N_1^3=Upper Miocene, Pannonian and Pontian
Rock types: S=sands; SS=sandstone; BH=basal horizon; CB=conglomerate and breccia; LS=limestone; D=dolomite; M=marl; F=flysch; MB=metamorphic basement; WF=weathered and fractured

Table 1—The 12 most important fields of Hungary.

NAME	Year of discovery	Sub-basin	Trap type layer-like	Trap type massive	Age and lithology of reservoir rocks	Initial re-coverable reserves Gm^3	Cumulative production
						as of 1. I. 1987	
1. Algyő	1965	7	+	+	N_1^3 S+SS, BH; PR WF MB	85,6	45,1
2. Hajdúszoboszló	1959	11	+	+	N_1^3 S+SS; KF; N_1^2 LS	25,6	22,6
3. Pusztaföldvár	1958	10	+	+	N_1^3 S+SS, BH	13,1	9,8
4. Üllés (deep horizon)	1974	7	+	+	N_1^3 SS, BH; N_1^2 CB; T_2 D; PR WF MB	12,6	2,8
5. Szeghalom	1979	9	+	+	N_1^3 S+BH	11,8	0
6. Szank-(SzankW)	1964	6	+	+	N_1^3 SS; N_1^2BH; PR WF MB	9,5	8,5
7. Nagykörü-(Tisza-püspöki-Surján)	1964	9	+		N_1^3 SS	9,1	0,5
8. Sarkadkeresztúr	1976	12	+	+	N_1^3SS; PR WF MB	6,3	3,5
9. Lovászi	1940	3	+		N_1^3 SS	4,3	4,3
10. Endrőd III.	1975	9	+		N_1^3 SS	3,8	1,0
11. Zsana-N	1979	6		+	N_1^2 LS; JM	3,7	2,7
12 Ferencszállás	1969	7	+	+	N_1^3 SS+BH	3,4	2,6

Table 2—The 12 most important natural gas occurrences of Hungary.

SAGEMAN, B. B., and E. G. KAUFFMAN, University of Colorado, Boulder, CO

Primary Taphonomic Processes in Black Shales

Blackshale facies representing oxygen-depleted paleoenvironments provide a unique opportunity to study the primary mechanisms and levels of taphonomic change that are normally altered or bypassed in active oxygenated waters. The taphonomic processes related to current and wave reworking, intense bioturbation, bottom scavenging, and rapid aerobic bacterial decay are highly attenuated in the low-energy, low-oxygen benthic environments of black shale deposition, as is early biogenic diagenesis. These conditions commonly result in excellent preservation of hard and soft parts, skeletal articulation, and abnormally high representation of pelagic and planktonic organisms not characteristic of more oxygenated, actively circulating aquatic facies. Many lines of evidence, however, support the concept that taphonomic destruction does proceed, more slowly and through more primary processes, in black shale facies/environments. This observation, and the exceptional preservation of black shale biotas, allows recognition of a detailed taphonomic gradient reflecting primary processes not normally preserved in the fossil record. This gradient extends from anaerobic processes of slow decay and dissolution, through a dysaerobic gradient involving higher levels of decay, selective scavenging at the sediment-water interface, and (with more oxygen) in shallow infaunal settings, selective bottom-feeding, and limited bioturbation. This gradient is illustrated in Figure 1 and applies equally to both introduced and resident taxa. Sources of organic remains in low-oxygen benthic environments include (1) settling plankton and nekton (dead or alive), (2) introduced shallow-water organisms (e.g., via turbidites) (dead or alive), (3) resident low-oxygen adapted taxa including bacterial chemosymbionts, and (4) event communities established on the bottom as a result of short-term modifications of the benthic zone through overturn, mixing, storm/tsunami events, rapid stratification, etc. No matter what the origin, organic remains that are transported to or live on the bottom of oxygen-depleted basins are then subjected to this taphonomic gradient regulated by oxygen levels and biological processes reflecting these levels. If all other factors are held constant, this results in a series of taphofacies that reflect the oxygen/biological process gradient except for retaining the signal of taphonomic modification of introduced organisms during death and transport. The gradient is, however, commonly perturbed by short-term events that may change the taphonomic field, more often toward the more oxygenated end of the gradient than toward the more anaerobic end. These perturbational events can result in mixed taphonomic signals within the same strata.

Black Shale Taphonomy

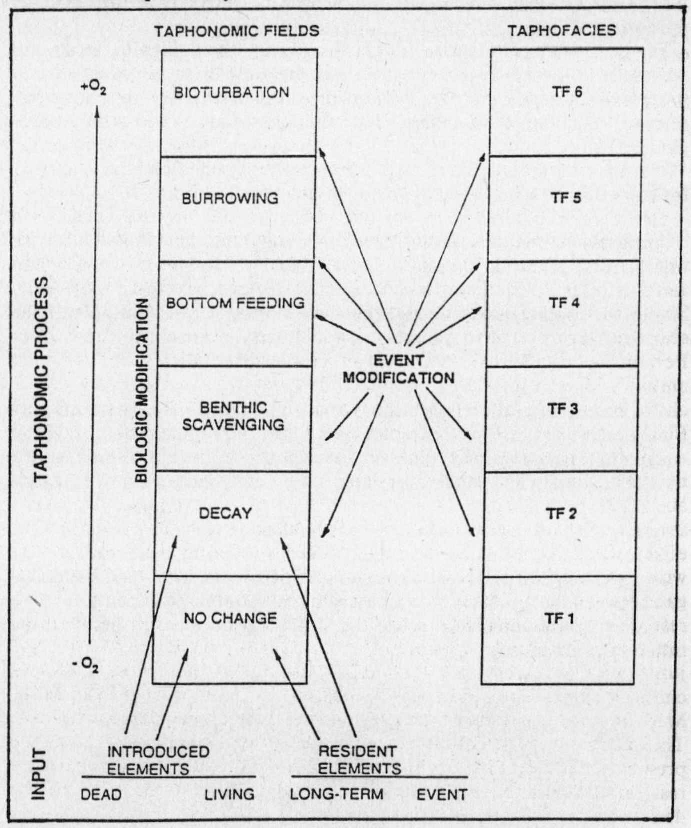

SAGER, W.W., and T.W.C. HILDE, Texas A & M University, College Station, TX, and D.W. HANDSCHUMACHER, NORDA, Stennis Space Center, MS

Mesozoic Tectonic Evolution of Pacific Plate

The early history of the Pacific plate was traced by the mapping of magnetic isochrons formed by seafloor spreading. The Pacific plate seems to have evolved from a small plate that formed near the Phoenix–Farallon–Izanagi triple junction about 180–188 Ma at about 17 °N, 160 °E in present coordinates. Its history was punctuated by several significant plate boundary reorganizations that had significant effects on its evolution. The first was at M21 time (153 Ma) and coincided with a 24° clockwise reorientation of the Pacific–Izanagi ridge. This reorganization was accompanied by ridge propagation and ridge jumps. As a result, the Pacific–Izanagi–Farallon triple junction probably degenerated into two or three triple junctions. The possible co-existence of several triple junctions and differences in spreading rates measured on the Japanese isochrons suggests that a microplate, the Shatsky plate, existed from M21–M15 time in the vicinity of the present west flank of the Shatsky Plateau. The Pacific–Shatsky–Farallon triple junction migrated northeast along the present trace of the Shatsky Plateau. At M15 (142 Ma) time, it apparently resumed its identity as the Pacific–Izanagi–Farallon junction with the demise of the Shatsky microplate.

M15 time also saw the formation of another microplate, the Magellan plate, at the Pacific–Phoenix–Farallon triple junction at the eastern tip of the Pacific plate. This microplate evidently evolved independently for about 10 Ma until it was assimilated by the Pacific plate at M11–M10 time.

At M11–M10 time (132 Ma), another plate boundary reorganization occurred, this one evidenced by the 15° clockwise shift of the Pacific–Farallon ridge. A magnetic bight off the northeast tip of the Shatsky Plateau constrains the location and configuration (RRR) of the Pacific–Izanagi–Farallon triple junction at this time. The trace of the junction from M21 to M10 lies beneath the Shatsky Plateau. A puzzling observation is that geometric constraints do not allow the triple junction to have remained in either of the allowable stable configurations (RRR or RFF) for the entire time in between.

As the Pacific plate continued to grow, the Pacific–Izanagi–Farallon triple junction migrated northeast whereas the Pacific–Phoenix–Farallon junction moved east. At the beginning of the Cretaceous Quiet Period, the former must have been located a few degrees west of Nintoku Seamount in the Emperor chain whereas the latter must have been just to the west of the Line Islands near the equator. Without isochrons as a guide, the subsequent evolution of the Pacific is unclear, but central Pacific topography and fracture zones suggest that there was another reorganization not long after the beginning of the Quiet Period, perhaps at about 110–100 Ma.

SAGER, W.W., and T.W.C. HILDE, Texas A & M University, College Station, TX, and D.W. HANDSCHUMACHER, NORDA, Stennis Space Center, MS

Magnetic Isochrons and Mesozoic Tectonic Evolution of Pacific Plate

The early history of the Pacific plate was traced by the mapping of magnetic isochrons formed by seafloor spreading. We reanalyzed isochrons in the region east of the Japan as southeast, southwest, and northwest of the Shatsky Plateau. Among these magnetic anomalies are a series located on seafloor previously thought to have been formed during the Jurassic Quiet Period. We believe that these are pre-Cretaceous isochrons older than anomaly M29. Additionally, we assimilated the results of other recent isochron analyses from the regions to the north and northeast of the Shatsky Plateau and in the Central Pacific Basin.

The Pacific plate seems to have evolved from a small plate that formed near the Phoenix–Farallon–Izanagi triple junction about 180–188 Ma at about 17°N, 160°E in present coordinates. Its history was punctuated by several significant plate boundary reorganizations that had significant effects on its evolution. The first was at M21 time (153 Ma) and coincided with a 24° clockwise reorientation of the Pacific–Izanagi ridge. This reorganization was accompanied by ridge propagation and ridge jumps. As a result, the Pacific–Izanagi–Farallon triple junction probably degenerated into two or three triple junctions. The possible co-existence of several triple junctions and differences in spreading rates measured on the Japanese isochrons suggests that a microplate, the Shatsky plate, existed from M21–M15 time in the vicinity of the present west flank of the Shatsky Plateau. The Pacific–Shatsky–Farallon triple junction migrated northeast along the present trace of the Shatsky Plateau. At M15 (142 Ma) time, it apparently resumed its identity as the Pacific–Izanagi–Farallon junction with the demise of the Shatsky microplate.

M15 time also saw the formation of another microplate, the Magellan plate, at the Pacific–Phoenix–Farallon triple junction at the eastern tip of the Pacific plate. This microplate evidently evolved independently for about 10 Ma until it was assimilated by the Pacific plate at M11–M10 time.

At M11–M10 time (132 Ma) another plate boundary reorganization occurred, this one evidenced by the 15° clockwise shift of the Pacific–Farallon ridge. A magnetic bight off the northeast tip of the Shatsky Plateau constrains the location and configuration (RRR) of the Pacific–Izanagi–Farallon triple junction at this time. The trace of the junction from M21 to M10 lies beneath the Shatsky Plateau. A puzzling observation is that geometric constraints do not allow the triple junction to have remained in either of the two allowable stable configurations (RRR or RFF) for the entire time in between.

As the Pacific plate continued to grow, the Pacific–Izanagi–Farallon triple junction migrated northeast whereas the Pacific–Phoenix–Farallon junction moved east. At the beginning of the Cretaceous Quiet Period, the former must have been located a few degrees west of Nintoku Seamount in the Emperor chain whereas the latter must have been just to the west of the Line Islands near the equator. Without isochrons as a guide, the subsequent evolution of the Pacific is unclear, but central Pacific topography and fracture zones suggest that there was another reorganization not long after the beginning of the Quiet Period, perhaps at about 110–100 Ma.

SAGREDO, J., M. PEINADO, J.L. BRANDLE, R. ARENAS, and E. ANCOCHEA, Universidad Complutense de Madrid, Madrid, Spain

Middle Cambrian Volcanic Rocks in Ossa Morena Zone (Iberian Massif): Petrological and Geochemical Features

In the southwestern part of the Iberian Massif (southwest Spain), i.e., the Ossa Morena Zone, a volcanic event of Middle Cambrian age erupted different materials, which are interlayerd with clastic shaly and psammitic shelf sediments.

The volcanic rocks outcrop discontinuously in the Zafra region, where basic, intermediate, and acid units are closely related. They were folded and metamorphosed under epizonal conditions during the Hercynian Orogeny. Original volcanic structures are preserved at places, in spite of the strong hydrothermal transformations shown by the volcanic rocks.

The basic unit is found in the lower part of the middle Cambrian sequence. It displays shallow submarine eruptive features, including close-packed pillow-lavas and hyaloclastites. The pillows contain vesicles filled with chlorite and calcite-ankerite. The volcanic rocks show porphyritic textures with preserved plagioclase phenocrysts. The original mafic minerals have been destroyed during the postvolcanic alteration. Their geochemical features [Nb/Y-Zr/TiO_2, Figure 1; ΣREE = 138–230; (La/Lu)cn = 7.1–9.6; Eu/Sm = 0.31–0.37] indicate an equivalence with alkaline basalts and trachybasalts (Winchester and Floyd, 1977). Their Zr-Ti-Y (Pearce and Cann, 1973) and Zr/Y-Zr (Pearce and Norry, 1979) ratios support an equivalence with intraplate basalts.

The intermediate unit is mainly volcaniclastic although some massive eruptions are also found. The pyroclastic rocks contain a groundmass variably transformed into carbonates. The lithologies of this unit are porphyritic, with plagioclase, K-feldspar, and rare quartz and biotite phenocrysts, embedded in a fluidal groundmass with microlitic feldspars and opaque minerals. At places the fragments of some volcaniclastic rocks show vitreous groundmasses. Geochemical features of the intermediate rocks suggest that the contents of some major elements have been strongly modified during postmagmatic alteration (high alkali content, low CaO and MgO, and high Fe_2O_3/FeO ratio). Nevertheless, the content of other major and trace elements do not seem to have been greatly modified. These contents, along with the Nb/Y-Zr/TiO_2 ratios (Figure 1), are characteristic of trachyandesites. Other geochemical features (ΣREE = 212–230; (La/Lu)cn = 4–9.3; Eu/Sm = 0.2–0.32] are also typical of intermediate rocks related with alkaline basalts. TiO_2-Zr ratios are compatible with an intraplate environment.

The acid unit is mostly pyroclastic, with common ignimbritic structures; massive rocks, likely subvolcanic, can also be found. Both clastic and massive rocks are rhyolitic in composition. They have quartz and alkali feldspar phenocrysts in a groundmass ranging from vitreous (with widespread spherulitic textures) to holocrystalline, this one consisting of tiny feldspar microlites with a flow orientation. Taking into account the SiO_2 content of these rocks and the Nb/Y-Zr/Ti ratios (Figure 1), they can be classified either as rhyolites or comendites–pantellerites. Other geochemical features [ΣREE = 452–869; (La/Lu)cn = 4.4–10.8; Eu/Sm = 0.13–0.23] suggest that they are felsic lavas related to alkaline basalts. The pattern of selected trace elements normalized to the ORG composition (Pearce et al, 1984) allow us to classify them as effusive equivalents of intraplate granitic rocks. High Rb, Th, and Nb contents are interpreted as indicative of a mantle source enriched in incompatible elements. Furthermore, some features point to some extent of crustal contamination: low Nb and high Sm contents as opposed, respectively, to Rb-Th and Zr-Y contents, along with a linear correlation between pairs of residual elements.

The REE patterns of the three units (Figure 2) are parallel, showing similar LREE/HREE ratios. The ΣREE increases continuously from the basic to the acid unit, with a moderate Eu anomaly. These features suggest that they are cogenetic, plagioclase, or plagioclase + alkali feldspar crystal fractionation being the most likely mechanism involved in the process.

All these rocks thus constitute a moderately alkaline intraplate series, differentiation being mainly controlled by crystal fractionation. A crustal contamination was likely involved in the most evolved rocks. Taking into account all these data, along with the geological features of the Zafra region, the most probable geodynamic setting for this volcanism is a passive continental margin during a period of moderate crustal rifting in the Middle Cambrian.

REFERENCES

Pearch, J.A., and J.R. Cann, 1973, Earth and Planetary Science Letters, v. 19, p. 290–300.

——— M.J. Norry, 1979, Contributions to Mineralogy and Petrology, v.

69, p. 33–47.
Winchester, J.A., and P.A. Floyd, 1977, Chemical Geology, v. 20, p. 325–343.

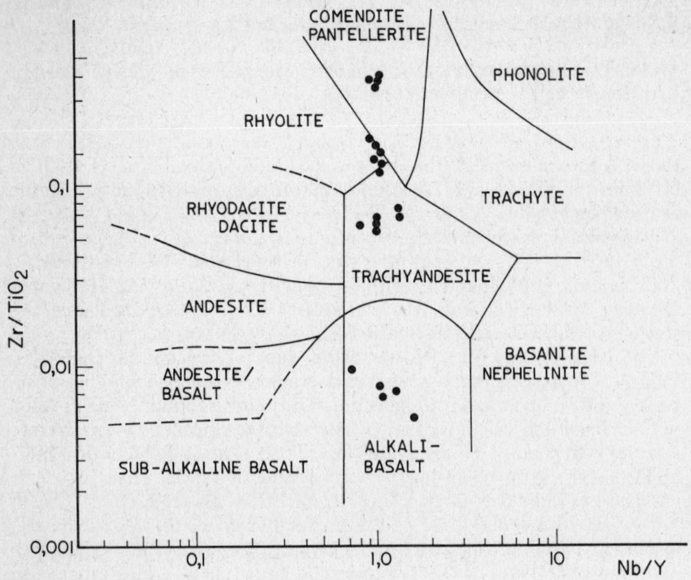

Figure 1.

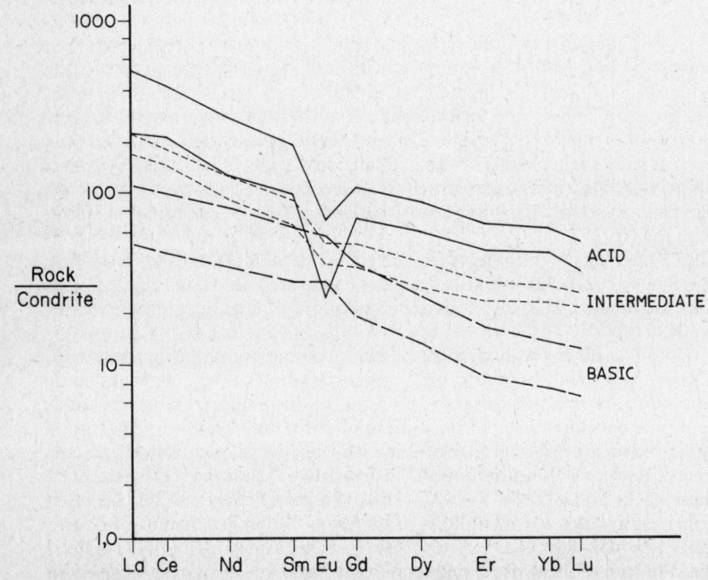

Figure 2.

SAHA, D., Indian Statistical Institute, Calcutta, India

Compressional Structures in Shallow Platform-Miogeoclinal Sequence within Godavari Valley Proterozoic, South India: Remnants of Collisional Suture, Aulacogen, or Intracratonic Orogen?

The *Bhandara* and *Eastern Dharwar* cratons of Peninsular India are separated by middle Proterozoic (Pakhal suite) and younger sediments of Godavari Valley (Rogers, 1986). Although most other rift valley lineaments in Indian continental crust follow sites of middle Proterozoic orogenic belts, the Godavari Rift has so far not been known to be associated with any compressional structure. However, on the basis of a marked difference in sedimentary and magmatic suites of *Bhandara* and *Eastern Dharwar* cratons, a case for a collisional suture along the Godavari Rift has been made (Rogers, 1986). The lithofacies association and structure of a sequence of rocks exposed in a narrow belt (Somanpalli–Albaka belt) bordering the *Bhandara* craton is discussed in relation to the issue raised above. The case study is to be viewed in conjunction with the hypothesis that formation of cratonic basins follows major collisional orogenic events (Klein, 1987).

The Somanpalli–Albaka terrane with a strike-length of over 100 km consists of a tectonosedimentary welt with parallel strips of peritidal limestone–lime marl with breccia pipes; ash beds–mass flow arenites with subordinate porphyritic flows; tidal–intertidal shale–quartz arenite-stromatolitic limestone with oolitic interbeds; cherty limestone–greywacke turbidite; banded iron formation; and cherty argillite–black chert. The overall association is miogeoclinal. In this terrane, the conventional layer cake stratigraphy is grossly defied by the presence of tectonic dislocations, which are steep at current erosion level, between strikingly dissimilar lithofacies association. The presence of small regional folds with variable vergence, strong axial planar fabric, small thrust sequences, and strong shear zone fabrics are apparently products of a contractional regime. However, local inheritance from basin margin faults of the rifting stage is not totally excluded.

The pyroclastics and volcanic flows in a fold-thrust belt as above may be interpreted as products of explosive volcanism in a convergent margin setting with offscraped ocean floor sediments (the greywacke turbidites; cherty argillite–black chert–limestone) juxtaposed against nearshore continental margin sediments (the peritidal carbonate-siliciclast belt). On closer scrutiny, however, there is no sign of oceanic crust in this tectonosedimentary welt. The relative volume of volcanic flows is also small. What, then, are the alternative tectonic models to explain the basin infill and its structure?

An aulacogen? The deformation and metamorphism at the southern end of the southwestern belt of Proterozoic rocks in Godavari trough bordering *Eastern Dharwar* craton has been interpreted as an overprint of the Eastern Ghats orogeny (Rogers, 1986), thus raising the possibility of connecting the middle Proterozoic Eastern Ghats mobile belt with the Godavari trough in a triple junction. However, known Proterozoic aulacogens, e.g., Athapuscow aulacogen in North America, tend to converge toward the craton interior. There is no suggestion that sedimentation in the Godavari trough was more widespread at the southern end as compared to the northern end. Moreover, in one reassembly of Indian ocean continents to a predrift pattern, the northern margin of Western Australia is fitted with the southeast coast of the Indian Peninsula, bringing the Lower Proterozoic Hammersley basin in line with the Godavari trough (Crawford, 1969). Although there are doubts about its age, the Bailadila Group iron ore deposits occur to the north of and within a few tens of kilometers of the Godavari trough–Bhandara craton margin. Thus, proposing a parallel in development of iron formations of intracratonic Hammersley basin and those of Godavari trough and adjoining areas, a case can be made for lower Proterozoic tectonic and sedimentation history of the latter. This has to be independent of the development of middle Proterozoic Eastern Ghats mobile belt.

As a third alternative, let us consider a case for mantle diapirism, subsidence through gravitational instability, and subsequent folding and faulting as a result of basin collapse consequent upon a detachment and sinking of a dense lithospheric slab (Martin and Paroda, 1977). Any direct evidence for mantle diapirism is lacking in the fold-fault belt of the Godavari trough bordering the Bhandara craton. However, the breccia bodies within the carbonate platform provide some clues here. Not only do these bodies tend to fall on a line bordering a tectonic dislocation between ash beds–volcanic flows–greywacke turbidites, these are also sites of granitoid veining as well as widespread soda metasomatism resulting in disseminations or segregations of nepheline crystals. The volcanic eruptions, widespread disruption of shallow-water carbonate formations, and their metasomatism are interpreted here as a higher order crustal mobility having their agent provocateur in mantle diapirism under a continental lithosphere. Following this, mechanical subsidence through normal faulting led to the formation of a deeper ensialic trough. Slope instabilities related to basin marginal faults and eruptive pulses catapulted the turbidite sedimentation; the cherty argillites–black cherts probably formed in deeper parts of the basin or during quiescence between turbidite inputs. At the present erosion level, plutonic bodies are rare in the Proterozoic belts of Godavari trough; however, the structural complexity and lithofacies association of the terrane described above have parallels in the intracratonic branch of the Damara Orogen in southwest Africa. The dynamics of the latter is thought to be explained better by concepts based on gravitational instability.

REFERENCES

Crawford, A.R., 1969, India, Ceylon and Pakistan: new age data and comparisons with Australia: Nature, v. 223, p. 380–384.
Klein, G. deV., 1987, Origin of cratonic basins: Geology, v. 15, p. 1094–1098.
Martin, H., and H. Paroda, 1977, The intracratonic branch of Damara Orogen in Southwest Africa. I. Discussion of geodynamic models: Precambrian Research, v. 5, p. 311–338.
Rogers, J.J.W., 1986, The Dharwar craton and the assembly of Peninsular India; Journal of Geology, v. 94, p. 129–143.

SAHAGIAN, D., Lamont Doherty Geological Observatory at Columbia University, Palisades, NY

Quantification of Past Sea Levels Using Shoreline Sediments on Stable Frame of Reference

Sea level is the relation between the elevation of the global sea surface and the elevation of the earth's crustal surface. There is copious evidence that sea level has changed significantly throughout geologic time. Measurement of past sea levels is problematic for two classes of reasons, however. The first is the problem of identifying elevation markers to quantify ancient sea levels, and the second is due to the fact that parts of the crust as well as the sea surface have experienced variations in elevation, such that the question arises: "Which part of the crustal surface should be used as a reference frame against which to measure past sea levels?"

A useful sea-level elevation marker can be found in shoreline sediments deposited during marine transgressions and highstands (Sahagian, 1987). The present elevation of these sediments on a stable reference frame is a measure of sea level at the time of their deposition. Shoreline sediments are relatively easy to identify in the stratigraphic record. They are commonly transgressive sandstones (such as the Dakota Sandstone of the United States), and are between shallow-marine sediments and low elevation terrestrial deposits of the same age.

Whereas the problem of identifying a sea-level marker is relatively simple, that of defining a stable reference frame for measurement is more significant. Ad hoc sea-level reference frames have been used in the past associated with a variety of techniques of sea-level measurement. Those that have used sequence stratigraphy have primarily depended on a passive margin reference frame, the source of the bulk of the oil industry's sea-level data. However, it is difficult to distinguish sea-level changes from epeirogeny, since passive margin subsidence curves must be carefully calculated, so that the departures from expected subsidence history indicated by onlap-offlap sequences can be attributed to sea-level changes. Another approach to sea-level measurement involves hypsometry, which utilizes the present-day cross-sectional profiles of the continents and the percentages of present continental area flooded in the past to infer sea-level changes as well as epeirogenic motions. The reference frame implicitly used in this method is that of the average elevation of the continental marginal regions affected by past marine transgressions. Although global in distribution, this reference frame is directly affected by a passive margin subsidence, orogeny, and other tectonic processes that cannot be filtered out or quantified using hypsometric analyses.

A more reliable reference frame can be found in undeformed continental interiors. A large region of the Russian Platform is covered by flatlying upper Cenomanian (mid-Cretaceous) marine sediments that rest at a uniform elevation of 200 m (Nikonov and Romanov, 1966). If these were strictly shoreline sediments, they would indicate a Cenomanian sea level of 200 m above present. These upper Cenomanian sediments were not deposited at Cenomanian sea level, however. At the beginning of the Cenomanian, shallow-marine and deltaic sediments (assumed to reflect water depths of up to 50 m) were being deposited on the Russian Platform, but by the end of the Cenomanian, deeper water facies (limestone, including chalk) were accumulating (Vinogradov, 1968). The widespread transition from sandstone to chalk is interpreted to be reflective of a water depth increase of between 50 and 100 m, although paleowater depths are difficult to estimate on the basis of chalk deposition and have been the topic of considerable discussion. Ziegler et al (1985) estimated the sea-level rise within the Cenomanian to be 80 m based on a similar facies change in North America. If there was appreciable water depth at the beginning of the Cenomanian, and if late Cenomanian water depths on the Russian Platform were estimated to be 100 m (Naidin, 1960), rather than the 50 m minimum, the calculated difference between Cenomanian and present sea levels would be 270 m. Thus, it can be concluded that eustatic sea level was higher than the level at which the sediments were deposited (200 m above present sea level) by an amount equal to the water depth (50–100 m) less a correction for water loading (15–30 m), such that sea level at the end of the Cenomanian must have been 235–270 m above present. Thus, the present elevation of the Russian Platform sediments can be completely attributed to post-Cenomanian sea-level change.

Since the undeformed part of the Russian Platform is so extensive, the region must be regarded as a reliable frame of reference for measuring sea-level changes, despite uncertainties in water depth estimation (Sahagian, in press). However, it may be desirable to define a sea-level reference frame not on any single locality, but to base it on all available level and apparently undeformed continental areas. The 235–270 m value for sea-level change is similar to the earlier calculations based on a locality in Minnesota on the North American craton of 270 m, and with the elevation of Cenomanian deposits in two stable areas in Africa of 250–300 m (Sahagian, 1987, 1988). Since 235 m is taken to be a minimum for post-Cenomanian sea-level change, and 270 m is considered a maximum, as argued above, the actual eustatic sea level change is probably at some intermediate value, and 250 m will be chosen arbitrarily as the magnitude of post-Cenomanian eustatic sea-level change.

The quantification of sea-level changes based on a stable Eurasian reference frame makes it possible to use sea level as an elevation marker for the measurement of epeirogenic motions of other continental areas that have been less stable than the Russian Platform. As such, this area of Eurasia may be considered to be "normal" and can be used as a standard against which to compare other regions in terms of epeirogeny as well as sea level, heat flow, and thermal structure.

An important implication of the existence of undisturbed Cenomanian sediments over such a large region in Eurasia is that the underlying mantle has been quiescent, or at least has not developed any dynamic or thermal anomalies since the Cenomanian. This is not to say that the mantle beneath the Eurasian lithosphere is physically and thermally homogeneous. The uniformity of the Cenomanian surface simply suggests that there have been no net *changes* in the mantle-lithosphere interaction since the Cenomanian. This is in sharp contrast to the implications of the deformed Cenomanian surfaces in Africa and North America. It must be concluded that generalities regarding the mantle and its effects on the continental lithosphere cannot be made, as each continent is affected very differently by the underlying mantle. However, since a large region in Eurasia remains undeformed since the Mid-Cretaceous, it can be used as a frame of reference against which to measure eustatic sea-level changes.

REFERENCES

Naidin, D.P., 1960, The stratigraphy of the Upper Cretaceous of the Russian Platform: Stockholm Contributions to Geology, v. 6, p. 39–61.

Nikonov, M., and L. Romanov, 1966, Geologicesk Carta CCCP, 1:7 500 000, Moscow.

Sahagian, D.L., 1987, Epeirogeny and eustatic sea level changes as inferred from Cretaceous shoreline deposits: applications to the central and western U.S.: Journal of Geophysical Research, v. 92, p. 4895–4904.

——— 1988, Epeirogenic motions of Africa as inferred from Cretaceous shoreline deposits: Tectonics, v. 7, p. 125–138.

——— in press, Epeirogeny of Europe and western Asia: Cretaceous Research.

Vinogradov, A.P. 1968, Atlas of lithological-paleogeographical maps of the U.S.S.R., v. 3—Triassic, Jurassic and Cretaceous: Moscow, Akademia Nauk S.S.S.R.

Ziegler, A.M., D.B. Rowley, A.L. Lottes, D.L. Sahagian, M.H. Hulver, and T.C. Gierlowski, 1985, Paleogeographic interpretation: with an example from the Mid-Cretaceous: Annual Review of Earth and Planetary Science, v. 13, p. 385–425.

SAHU, K.C., Indian Institute of Technology, Bombay, India

Urban Flux Transfer: Case for Bombay City

A simple calculation shows that modern man by way of his activities moves material fluxes on earth surface on the order of 10^{16} g/y, equivalent to any geological processes such as erosion or crust-mantle formation. The culture of man becomes intense and focused where he congregates in large number. In every respect, these congregations—cities, metropolises, or future megalopolises—are anomalies in an otherwise homogeneous domain of "earth-people" system and therefore subject to environmental degradation.

Cities are sinks of lithogenic, anthropogenic, and sociogenic fluxes. Lithogenic fluxes are developed by natural erosion but accelerated by large-scale land development for citiculture. Anthropogenic fluxes are represented by movement of a large amount of various types of materials—primary, secondary, and processed goods—in and out of the city, producing a conspicuous change in composition of land, water, and air of the crustal segment occupied by the city. Sociogenic fluxes are represented by migration of people from hinterland resulting in classical "mutual poisoning," a mere "sociosostatic game" in the "earth-people" system.

The island city of Bombay once consisted of seven islands developed as a natural harbor because of its bays, carved out by impact of Arabian Sea on intercalated basalt and tuffs. The basaltic rocks provide ideal construction materials for development of an azoic "concrete jungle" and reclamation work for a land-hungry citiculture in the south resulting in enhanced beach erosion in the north. Rapid changes in land forms by way of removal of construction materials from existing basaltic ridges in the north and creation of the new citiscape in the south on massive scales are violent processes unmatched by nature's mass transfers for isostatic balance. Besides these, the Panvel flexure in Deccan Trap close to the east, Indian plate edge on the west, a chain of oozing hot springs in the north, and the recent stampede of oil extraction from the adjacent soft offshore sediments are thought-provoking for micro-seismotectonic stability of the island.

In a polluted air–water–land system, the standard formulae of rock mechanics for assured natural strength of rocks or terrain are likely to be on low key and the technological assurance of stability is less predictable and therefore less reliable. Thus, the signs of subsidence in the Bandra tunneling operation in a western suburb and corrosion to near collapse of the 10 year old trans-Thana Creek bridge, connecting the island city to the mainland in the east are but symbolic in nature. One of the artificial reservoirs, Khardi, storing water for city supply has already indicated reservoir seismicity.

A host of north-south communication channels—railways, roads, aqueducts, sewerage, electrical lines, etc.,—together with extensive ground asphalting in the city to the south have of late resulted in increasing intensity of flooding.

At least 5000 truckloads of materials daily enter the city by road. A total of 2000 "train trips" at full capacity move men and materials up and down the city daily, and so do 66,000 flights and 4600 ships with 2.6 million tons of cargo per year. Each year 2600 MLD of water runs into the city and about 0.22 million people come to settle, adding to the 2–3% of the native growth. The Municipal Corporation transfers 260 MGD of sewage and 3500 tons of garbage, the former discharged into the sea only to be thrown back to the shore and then later piled up in the central region to contaminate the solum. An analysis of material movement in and out of Bombay by surface, sea, and air routes with a modest leakage of 1% in the utility cycle and their discharge into air, water, and land during handling, processing, and consumption indicate conspicuous changes in geochemical characters of the top biospheric layer of the island city. Samples of surface soil analyses show concentration of 35–75 ppm Pb, 110–230 ppm Zn, 3–4 ppm Cd, 80–260 ppm Cu, etc.

The subtropical hot humid climate weathers the basaltic rocks to smectite-rich solum that on washout, reaches the bays and creeks forming extensive mudflats around the city. Leakage of the materials stated above, including sewage, industrial effluents, leachates from municipal dumps, and atmospheric fallouts (about 2000 tons per day) ultimately reach the mudflat, which becomes a repository or sink of all the toxic pollutants. Preliminary surveys indicate toxic metal concentration in the mud to the extent of 150 ppm Pb, 270 ppm Zn, 5 ppm Cd, 250 ppm Cu, 10 ppm Hg, etc. The mud-bank being the "protein bank" (fishing ground) of a large section of city people, the pollutants find an entry into human life through classic food chain.

SAINT-BLANQUAT, M. DE, and M. BRUNEL, Université de Montpellier et U.A., Montpellier, France

Ductile Strain and Metamorphism in Extensional Shear-Zone From French Pyrenees (Saint-Barthelemy Massif)

The Saint-Barthelemy Massif is located north of the North Prenean Fault. It was structured during the Hercynian orogeny and was later thrust over the surrounding Mesozoic sediments during the Pyrenean orogeny at Eocene time.

The present study is concerned with the internal structure of the Massif and focuses on the post-Hercynian–pre-Pyrenean period during which extensional tectonics have been detected.

One the most prominent features of the Massif is the Main Mylonite Band (MMB) (Passchier, 1982). The MMB separates the hanging wall made of Paleozoic rocks (from bottom to top: migmatites, Cambro-Silurian micaschists, and Devono-Carboniferous metasedimentary series) from the footwall made of gneisses affected by a granulitic metamorphism dated at 300 Ma.

The MMB strikes N1100 °E and dips 200 °N; it is about a hundred meters thick. It consists of several superposed bands. Each one involves mylonites, ultramylonites (centimeters to meters thick), basic levels, and folded and mylonitized carbonates containing fragments of the surrounding quartz mylonite. The deformation intensity increases from the underlying gneisses to the top of the shear-zone, ending with a several meter thick ultramylonite band. Strain drastically decreases in the migmatites where only retrometamorphic thin shear-zones may be observed. The foliation wears a strong biotite–muscovite north–south-stretching lineation. This syntectonic mineral yields Mesozoic age (130–100 Ma by Ar-Ar dating; Costa and Maluski, 1988).

The shear criteria analysis at different observation scales shows a south-directed movement of the hanging wall block.

The MMB and the associated structures cross-cut all the Hercynien and late-Hercynian granitic bodies.

The main results of the petrographic study are: (1) An HT-LP granulitic metamorphism (quartz–K feldspar–garnet–cordierite–sillimanite–o. pyroxene assemblages) has affected the basement. Textural observations indicate static blastesis of the metamorphic minerals. (2) From this previous assemblages a retrometamorphic evolution is observed (amphibolite to greenschist facies conditions). It is related to a shearing deformation involving development of the gneiss foliation. (3) The MMB deformation is achieved under greenschist facies conditions during this retrometamorphic evolution. Strain becomes more and more localized within a more ductile zone, probably related to weakness of carbonate levels. (4) Finally, we observed the development of a quartz–chlorite–albite–epidote assemblage contemporaneous with a brittle extensional deformation.

If we restore the structures and microstructures to their initial position, we notice that the MMB was horizontal or gently dipping southwards before the rotation due to Pyrenean thrusting and was, consequently, a ductile normal shear-zone.

As a conclusion, this petro-structural analysis of the MMB and surrounding rocks has shown the gradual uprising of the lower crust related to extensional shearing.

In the regional Pyrenean context, this tectono-metamorphic process may be interpreted as: (1) collapse of a Hercynian thickened crust by an extensional shearing (driven by internal forces); and (2) extensional tectonics of Mesozoic age in relation with the opening of the Bay of Biscay (driven by external plate motion).

SALEM, R., Consulting Geologist, Fort Worth, Texas, USA, and H. OZBAHCECI, A. UNGOR and M. ISBILIR, Turkish Petroleum Corporation, Ankara, Turkey

Raman/Bati-Raman Oil Fields, Southeast Turkey

The Raman/Bati-Raman oil fields are located south of the Miocene thrust belt that crosses southeast Turkey from the east to west. The productive areas occupy two segments on the large east-west trending structure. The two segments are bounded on the south by high angle reverse faults with the down throw to the south. The two segments are separated from each other by a system of cross faults with the down throw to the west.

The Raman Field constitutes the eastern productive segment. It was discovered in 1948 based on surface geologic mapping. The Raman Field produces 19 Gravity oil from the Mardin Carbonates (Turonian-Cenomanian) from the crestal part of the structure. It also produces oil from the crest and flanks of the same structure from the Garzan Reefoid Carbonates (Maestrichtian).

The Bati-Raman Field constitutes the western productive area. It was discovered in 1961 and produces 12 Gravity oil only from the Garzan Formation from the northern flanks of the structure. At least three oil water contacts are recognized from the Raman/Bati-Raman oil fields. Both fields have produced a cumulative amount in excess of 78 million barrels of oil. They have a combined estimated reserve in excess of 2 billion barrels of oil.

SALER, A., Nuclear Power Plant Krško, Vrbina, Yugoslavia

General Approach and Site Selection Criteria for Low- and Intermediate-Level Radioactive Waste Disposal in Yugoslavia

Industrial land and intermediate radioactive wastes are continuously generated in Yugoslavia since the first and only Yugoslav nuclear power plant (NPP) in Krsko has started operation. During the period 1982–1987, 6230 drums (or about 1246 m^3) of solidified low and intermediate radioactive wastes have been generated. Since NPP Krsko interim storage is almost full, the final radioactive waste disposal, i.e., the construction of a radioactive waste repository in Yugoslavia, has to be ready as soon as possible. Therefore, the electric power production organizations of two Yugoslav republics—Slovenia and Croatia—that have been investors of NPP Krsko, established in 1986 at NPP Krsko site the Radioactive Waste Repository Department. The task of the Department is to coordinate three main groups of activities: radioactive waste disposal site selection, design activities (both of them include safety assessment), and license requirements procedure. Site selection activities include geotechnical (geology and similar fields of activities), bio-ecological, social-geographic, and regional planning investigations. According to the time schedule, the radioactive waste disposal repository is planned to be finished in the 1992.

The site selection activities in Croatia (56,538 km^2) and Slovenia (20,251 km^2) consist of five stages (three of them are already finished) respecting relevant multidiscipline (geotechnical, bio-ecological, social-geographic, and regional planning) criteria. In this article, only geotechnical criteria will be considered.

In the first stage, the site screening of both republic territories (with total area of 76,789 km^2) was performed. The evaluation of the area has been carried out with respect to these weighted criteria: lithology, tectonics, seismicity, hydrogeology, pedology, demography, land use, and transportation. According to this evaluation, five macrolocations of 200–400 km^2 each have been selected by overlapping through each criterion positively assessed areas. Each of them include 450–1100 m high mountain regions, formed by igneous (granite, basalt, tuff) and metamorphic (gneiss, amphibolite, schists) rocks. Within these potential macrolocations, respective to the mentioned criteria and considering available data, in the next stage nine so-called closer macrolocation areas (with 40–50 km^2 per unit area) have been defined. The last already accomplished stage was selection of dozen most suitable microlocations (with 2–4 km^2 per unit area) situated in any one of existing closer areas. As a result, microlocations were listed according to their range (i.e., suitability degree). Finally, during the next stage all needed detailed field investigations will be carried out at the 2-4 best ranged microlocations in order to define the final repository site within any one of included microlocations. The site area is planned to be 15–20 hectares large.

All geotechnical investigations at selected closer macrolocation areas have been accomplished by Yugoslav companies (INA-PROJECT, Zagreb, and Geološki zavod, Ljubljana). Corresponding documentation was reviewed by experts in Zagreb and Ljubljana.

These investigations were carried out according to IAEA recommendations and consultations with an expert team from Bechtel (USA). As the decision on final repository type has not yet been made (both a tunnel type and a shallow ground repository are alternatives), criteria lists were created for both variants. First, all needed exclusionary criteria (that had to be satisfied at all for each suitable location) were evaluated. In such a way, defined microlocations have been compared according to corresponding weighted criteria. Numerical evaluation, i.e., points given to each microlocation according to defined weighted criteria, was performed in such a way that every considered discipline receives a corresponding weight factor. The value of that factor varied due to importance of particular discipline in the process of repository site selection. The total amount of weight factors was 100. Weight factors were distributed by considered disciplines as follows: hydrogeology 30, tectonics/seismicity 25, rock properties 15, site geometry 10, engineering geomorphology/geology 10, hydrology 5, and human activities 5. In every one of listed disciplines/criteria there is one or more subcriteria (i.e., "subweight factors") and their sum is equal to weight factor of particular discipline/criterion. Each weight factor (or "subweight factor") is to be multiplied with corresponding quality factor (within the span 1–5) according to determined quality of particular terrain property. The sum of all multiplication products of weight- and quality factors represents the rank of particular microlocation.

Further activities, listed in the site characterization plan, are referred to field investigations at four most suitable (i.e., best ranked) microlocations in the area of interest (two of them in Croatia and two in Slovenia). In this way, all relevant properties of considered microlocations will be studied in detail. So, in the last stage, the final repository site will be defined at the best of those four microlocations.

Field investigations will comprise the following geotechnical disciplines: (1) engineering geomorphology/geology, (2) lithology, (3) tectonics, (4) seismology, (5) hydrogeology, (6) groundwater chemistry, (7) groundwater hydraulics and radionuclide dispersion model, (8) hydrology (surface waters), (9) pedology, (10) physical and chemical properties of host rock, (11) geophysics, and (12) rock mechanics.

SALINAS, E., B. BEAUDOIN, I. COJAN, D. MERCIER, Ecole des Mines de Paris, Fontainebleu, France

Sedimentary Dynamics in French Coal-Measures Restituted by Decompaction and Litho/Bio-Facies Analysis

In the French North-Coal Measures Basin, the total thickness is about 2500 m (Namurian-Westphalian D). We have selected an interval of 200 m (Westphalian A) of fluvio-deltaic sequences between two ephemeral marine ingressions, known by 100 boreholes over 10km2.

Three approaches have been carried out:

–simulation of compaction processes for decompaction and restitution of primary geometries;

–determination of faunal and floral associations by satisfied methods;

–characterization of bio- and litho-facies for sedimentary dynamics restitutions.

The effects of compaction have been after recognized long time ago within coal-bearing sequences (Ashley, 1907, see in Beaudoin et al, 1988). The compaction ratio ($\varrho h_0 h$ with h_0 being the original thickness) may be directly measured or inferred from porosity versus depth evolution. Three basic facies have been distinguished: sandstone considered as uncompactable (first approximation), shale (ϱ up to 3) and coal (ϱ up to 7). Compaction ratio for intermediate facies are interpolated (Figure 1). The use of these compaction laws allows to decompact each section under the chosen level: a direct application is better correlations.

Floral and faunal associations were established by correspondence analysis on a set of 4,000 fossiliferous levels (9,000 mentions, 144 species). It permits the individualization of 6 associations (Figure 2); *limnic fauna, Lycpods and spores, Calamites, Cordaites, A and B* (Mercier et al, 1987). Lithofacies and previous associations are not independent. Cross correlation (Figure 3) leads to the definition of litho- and bio-facies.

On the bases, we developed the analysis of the sedimentary piles along the time: decompacted logs are used for correlations and sedimentary bodies geometry resitutions; maps (true isopachs, litho-facies and bio-facies) are drawn (Figure 4) which help for environmental interpretations. The next step is the determination of the role of tectonics (subsidence and compaction of facies distributions and migrations).

REFERENCES

Beaudoin, B., Cojan, I., Fries G., Maillart, J., Parize O., Pinault, M., Pinoteau B., Truyol V., 1988, Mesure directe de la compaction dans les sediments in Genèse et évolution des bassins sédimentaires. Notes et Mémoires, n. 21, chap. 11, Total C.F.P. Paris.

Mercier, D., Cojan, I., Beaudoin, B., Salinas Zuniga, E., 1987, Apport des associations fooristiques dans la caractérization des paléoenvironnements sédimentaires (Bassin du Nord-Pas de Calais), Ann. Soc. Géol. Nord, CVI, 155-161, September, 1987, Lille.

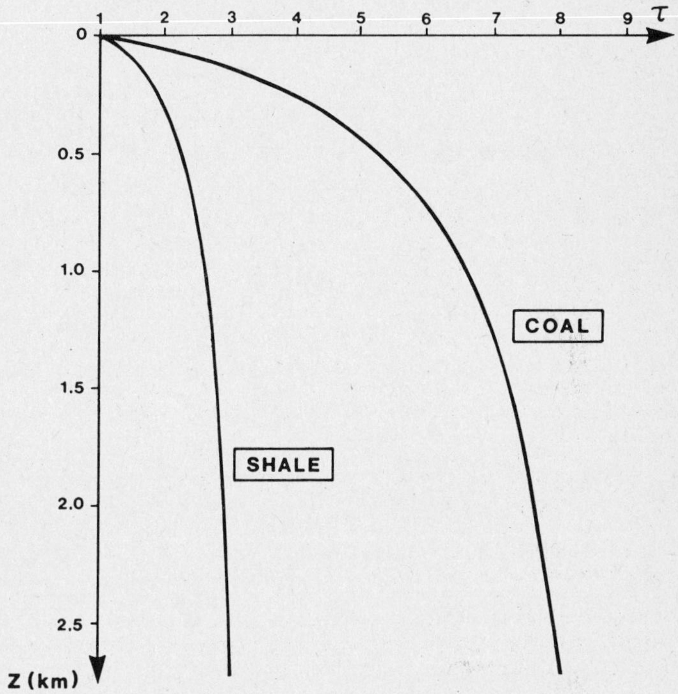

Figure 1—Compaction ratio (ϱ) versus burial depth (z).

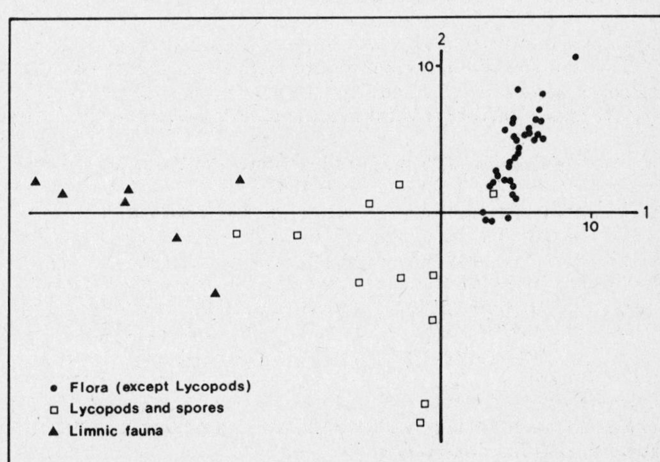

Figure 2—Projections of variables on axes 1 and 2, showing floral and faunal associations.

	LIMN	SPORE	LYCO	CALAM	CORD	A	B
SAND			* *	*	*		
SILT				* *	*		*
SHALE	* *	* *				* *	
COAL		*					

Figure 3—Relations between litho- and bio-facies.

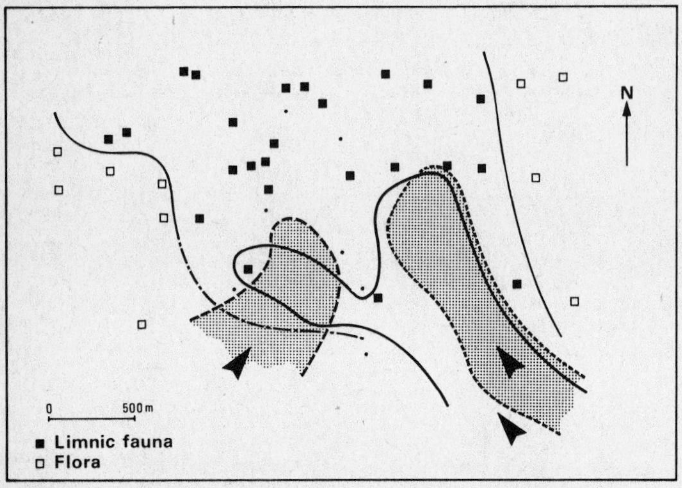

Figure 4—Restitution of sedimentary dynamics of sandstone bodies over a pool.

SALJE, E., University of Cambridge, Cambridge, United Kingdom

Time-Dependent Ordering Processes in Minerals

A new universal rate law has recently been derived for the description of kinetic processes in minerals (Salje, 1988). It explains most of the macroscopic properties of minerals in nonequilibrium states, such as inhomogeneous ordering, microstructures, and time-dependent order parameters. First experimental results on Na-feldspar, Mg-cordierite, and omphacite confirm the theoretical predictions; their individual rate laws of Al, Si ordering are discussed.

REFERENCE

Salje, E., 1988, Kinetic rate laws as derived from order parameter theory: Phys. Chem. Min., v. 15, p. 336–348 and v. 16, in press.

SALJE, E., University of Cambridge, Cambridge, United Kingdom

Kinetic Rate Laws and Their Application to Al, Si Ordering in Na-Feldspar

The kinetic behavior of Al,Si ordering in Na-feldspar at 1080 °C has been determined using Hard Mode Infrared Spectroscopy (HMIS) and high-resolution X-ray powder diffraction techniques. The order parameter Q_{od}, which describes the degree of Al,Si ordering, evolves as

$$Q_{od} = v_o e^{\Delta Ga/k/t}\, \partial G/\partial Q_{od}$$

where ΔG_a is the free energy of activation and G is the Landau potential of Na-feldspar. Fluctuations lead to broad distributions of Q_{od} for most non-equilibrium states with the tweed pattern as the dominant microstructure. Experimental details and analytical descriptions are presented.

SALJE, E., University of Cambridge,, Cambridge, United Kingdom

Hard Mode Infrared Spectroscopy and Determination of Thermodynamic Order Parameters

Thermodynamic and structural properties of minerals are derived from changes of live profiles of infrared absorption signals. Simple correlations between the relative changes of the absorption frequencies ω, the intensities of the absorption signal I, and the thermodynamic order parameter Q have been derived:

$$\Delta(\omega^2) \alpha Q^m$$
$$\Delta I \alpha Q^2$$

with $m = 2$ in most applications, besides crystals with trigonal symmetry. The method is in close analogy with Hard Mode Raman Spectroscopy (HMRS) as recently reviewed by Bismayer (1988). Experimental details and applications in mineral physics are discussed.

REFERENCE

Bismayer, U., 1988, New developments in Raman Spectroscopy on structural phase transitions, *in* Physical properties and thermodynamic behaviour of minerals: Reidel, ASI series n. 225, p. 143–184.

SALMAN, A.G., and S.B. FELITZIN, Academy of Sciences of the USSR, Moscow, USSR

Uppermost Crust of Apennine Peninsula: Correlation between Seismicity and Thermal Regime

On the global scale, a relationship between seismicity and heat flow field is observed. The seismicity and the heat flow value appear to be different manifestations of a single process, i.e., mantle convection. We shall not discuss this relationship, which is the subject of separate investigations, but will confine ourselves only to the connection between the thermal and seismic regimes of the crust within individual regions. The assumption is that the crustal thermal regime along with the tension field is one of the main factors controlling deformation and destruction of rocks.

The best region for studying seismicity and thermal regime of the crust is one where seismic activity is high and geothermal parameters are viable. In this respect, the crust of the Apennine peninsula is the best choice. The following parameters have been estimated for this region (Sicily excluded), which is subdivided into blocks 50 × 50 km with 81 elementary cells: mean temperature at a depth of 2000 m – $\hat{T}$; the number of earthquakes in the magnitude range from 4 to 4.5, from 4.5 to 5.0, and from 5.0 to 6.0, etc.

The regionalization of the Apennine peninsula has been made using the R-model factor analysis and is based on the quantitative ratio of thermal to seismic regimes of the upper crust. The regionalization shows that the coldest and the hottest provinces have the highest seismicity. Figure 1 demonstrates our observation that the number of earthquakes is a temperature-dependent factor; i.e., the number of earthquakes decreases with the crustal temperature increase in low-temperature environments. This observation is supported by statistical calculations. Further increase in temperature enhances seismicity. The reverse dependence of seismicity on crustal temperature has been established for seismic provinces in Turkmenia.

The following explanation of the phenomenon can be offered for discussion.

A decrease in the number of earthquakes with magnitude 4–6 in the low-temperature province suggests the similarity in the destruction processes and kinetic mechanism of solid body destruction across the entire crust. In a simple case, the increment of the energy ΔW during the deformation of rocks for the Maxwell's body is expressed as

$$\Delta W = \sigma \epsilon t \text{ where } \epsilon = \text{const} \quad (1)$$

where σ is the existing tension, ϵ is the deformation rate, t is the time; ΔW is proportional to the time starting from the moment the load is applied and to the moment of destruction, i.e., to the lasting time. According to the kinetic theory (equation 2)

$$\tau = \tau_o \exp\,(U_o - \gamma s)/KT \quad (2)$$

the lasting time of the material τ under loading σ decreases with the temperature T increase; in other words, the increment of the mechanical energy decreases with the temperature increase. Assuming that the energy released during earthquake is proportional to the increment of energy ΔW, we get the reverse relationship of the seismic indicators (the number of earthquakes of different energetic categories) on the temperature. This kind of reverse relationship is typical for the uppermost crust of the Apennine peninsula.

The direct dependence of the number of earthquakes on the temperature at a depth of 2000 m reflects the well-known fact of the positive connection between the seismicity and heat flow in tectono-active regions, such as, for example, the Baikal rift zone.

The regionalization of the uppermost crust of the Apennine peninsula on the durability and tectonic activity is based on this result.

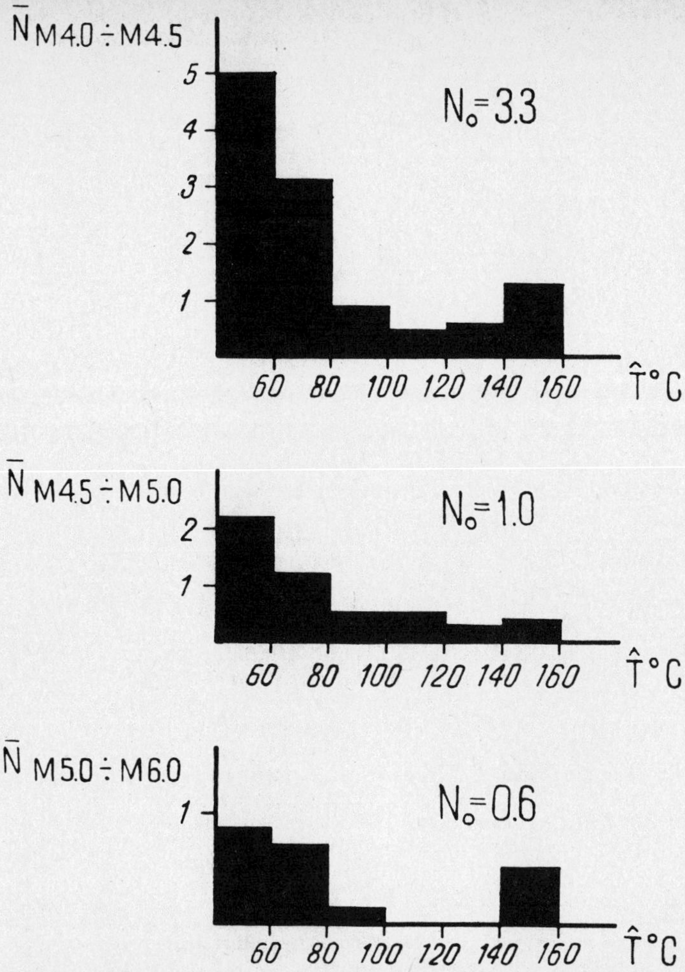

Figure 1—Dependence of an average number of the earthquakes in the crustal blocks 50 × 50 km on the value of the average temperature at the depth 2000 m for the Apennine peninsula. N_o, average value of a number of the earthquakes in the single block for the whole territory.

SALMINEN, R., Geological Survey of Finland, Espoo, Finland

Combining Geochemical Data of Till and Lake Sediments, An Experimental Study from Finland

A study has been carried out to investigate the effects of combining the results of different types of regional surveys out in Filand. The whole country (337,000 km^2 has been mapped using till samples at a density of 1/300 km^2. A small area (80,000 km^2 has also been mapped using lake sediment sampling at a density of 1/4 km^2. The till samples were analyzed by ICP after both a total leach(HF + H_3BO_3) and a partial leach (Aqua Regia); and the lake sediment samples were analyzed by AAS after a partial leach (6M HCI).

The results for Cu, Ni and Mn are discussed. Firstly, a subset of the lake sediment data at the sampling density, i.e., 1/300 km^2, as the till survey was prepared. This subset was then substituted for the till data on the Finnish wide survey and anomaly maps prepared on the basis of the new combined data set.

The main features of the anomaly patterns for Ni and Cu remain the same compared with the national till survey. However, because the range of Mn in the lake sediments is wide in comparison with that exhibited by the tills, a misleading anomaly pattern is generated with the combined raw Mn data set. A normalization procedure was therefore employed for the Mn data from both surveys, the combination of these normalized data yields reasonable maps that reveal similar patterns in comparison with the national till survey.

SALTERS, V.J.M., and S.R. HART, Massachusetts Institute of Technology, Cambridge, MA, and G. PANTO, Hungarian Academy of Sciences, Budapest, Hungary

Characteristics of Upper Mantle Beneath Carpathians, Hungary, as Inferred from Alkaline and Calcalkaline Volcanism and Lherzolite Nodules

Medium to high K calcalkaline (CA) volcanics are remnants of an Eocene to Miocene volcanic arc, which is related to the subduction of the European plate under the Pannonian plate fragment. The CA volcanism was followed by alkali basalts (AB) in the Quaternary. The alkalic volcanism is partly contemporaneous with and partly postdates the opening of the Pannonian basin. Although the AB and CA range in $^{87}Sr/^{86}Sr$ from 0.703265 to 0.71125 and in $^{143}Nd/^{144}Nd$ from 0.51221 to 0.51303 the volcanics form a narrow array on an Sr-Nd isotope correlation diagram, despite the large variation in time, location, and composition. Lherzolite inclusions in the alkali basalts exhibit a restricted range in $^{87}Sr/^{86}Sr$, 0.7027–0.70348, while they range in $^{143}Nd/^{144}Nd$ from 0.512373 to 0.513638 ϵ_{Nd} = 5.6–+19.8.

The Pb isotope variation of the volcanics is restricted in $^{207}Pb/^{204}Pb$ and $^{208}Pb/^{204}Pb$ 15.61–15.72 and 38.99–39.20, respectively, while ranging in $^{206}Pb/^{204}Pb$ from 18.69 to 19.42. The alkali basalts lie within the field of ocean-ridge basalts while the CA have higher $^{207}Pb/^{204}Pb$ and $^{208}Pb/^{204}Pb$ for a given $^{206}Pb/^{204}Pb$ leading to positive Δ7/4Pb and Δ8/4Pb. The large variation in $^{206}Pb/^{204}Pb$ in the AB indicates that at least two components contributed to the Ab volcanism. The lherzolite nodules in the ABs rule out contamination of the Abs during ascent to the surface. Thus the presence of two components in the ABs has to be caused by source mixing. One of the two components is a "real" mantle component with $^{87}Sr/^{86}Sr$ = 0.70326, $^{143}Nd/^{144}Nd$ = 0.51303, and $^{206}Pb/^{204}Pb$ = 19.4. The second component has $^{87}Sr/^{86}Sr$ and $^{143}Nd/^{144}Nd$ similar to the other mantle component, but has a lower $^{206}Pb/^{204}Pb$ (= 18.6–18.8). These isotopic characteristics are similar to the isotopic characteristics of the kaersutite vein in the spinel lherzolite.

The spinel lherzolites have trace element characteristics that prohibit their being the source or having been in equilibrium with the AB, and host ABs and lherzolites are unrelated (see Figure 1).

The CA volcanics have low Nb, Ta, Zr/Zr* and Ti/Ti* as is typical for subduction-related volcanics. The silicic end of the CA compositional range exhibits high $^{87}Sr/^{86}Sr$ and low $^{143}Nd/^{144}Nd$ indicating that crustal contamination was involved in the evolution of the CA series. Based on the High Field Strength Elements (HFSE) and the REE, the Ca are not related to the AB in a simple way. However, the trace element characteristics of the lherzolites are similar to that of the CA (see Figure 1). Melting of the lherzolites, brought to the surface by the AB, will result in volcanics with HFSE-REE characteristics similar to basic CA volcanics. It is proposed that this "calcalkaline" mantle component is refractory in major elements and resides high in the upper mantle. In a subduction-type setting, a component from the subducted slab, rich in alkaline elements, is able to flush the refractory mantle and lower the solidus, enabling melting of this "calcalkaline" component.

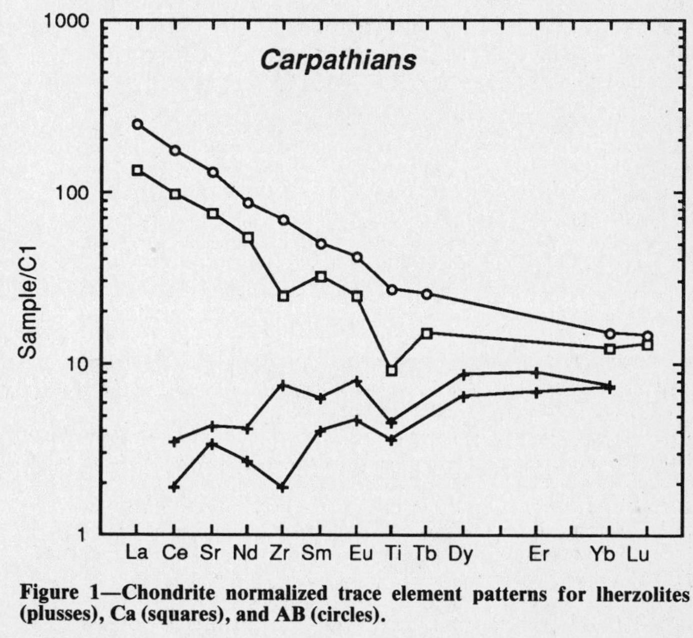

Figure 1—Chondrite normalized trace element patterns for lherzolites (plusses), Ca (squares), and AB (circles).

SALVADOR, A., The University of Texas at Austin, Austin, TX

Gulf of Mexico Basin

The Gulf of Mexico Basin is a relatively simple, roughly circular structural basin filled with 10 to 15 km of sedimentary rocks ranging in age from Late Triassic to Holocene. Paleozoic and older rocks are known from a few outcrop areas and from only modest penetrations by a limited number of wells around the rim of the basin. They are considered to from the "basement" of the basin and little is known about their stratigraphic or tectonic history.

The central part of the basin is occupied by the Gulf of Mexico, which covers an area of more than 1,500,000 km^2, and attains water depths of 3750 km (12,303 f) in its abyssal plain. The floor of the Gulf of Mexico rises steeply to the east and south along the Florida and Campeche escarpments, which form the gulfward limits of the shallow, underwater parts of the Florida and Yucatan platforms. Elsewhere, the floor of the Gulf rises more gently toward the coast through generally well defined rise, slope, and continental shelf. The Gulf of Mexico is surrounded to the north and west by a low coastal plain ranging in width from less than 50 km in east central Mexico to more than 550 km in the central part of the U. S. Gulf coastal plain, in the states of Louisiana, Mississippi, and Arkansas.

The "basement" in the central, deepest part of the basin has been estimated to be at a depth of about 12 to 16 km below sea level. Geophysical surveys indicate that this deep central part of the basin is underlain by an oceanic type crust, whereas most of the U. S. Gulf coastal plain, the Florida and Yucatan platforms, and the remainder of Mexico are underlain by continental crust. "Transitional" crust, inferred to represent "rifted" or "stretched" continental crust, occurs between them under the northern and southern parts of the Gulf of Mexico. Along the western and eastern limits of the deeper, central part of the Gulf, the boundary between oceanic and continental crust is sharper, without a significant intervening belt of "transitional" crust.

Information concerning the Upper Triassic, the Lower Jurassic, and most of the Middle Jurassic limited. The earliest stratigraphic interval of which more is known is the uppermost Middle Jurassic, represented over most of the Gulf of Mexico Basin by extensive salt deposits. The plastic behavior of this salt interval under the load of the overlying sediments has resulted in the formation of many diapiric salt domes, pillows, and ridges throughout the basin, some of which are still active.

Overlying the salt, or the oceanic "basement" in the central part of the basin, is a very thick sequence of sediments, deposited for the most part in marine environments. The thickest sections are found along the north flank of the basin where up to 15 km of sediments, mainly Tertiary clastics, have prograded into the basin.

The Upper Jurassic, Cretaceous, and Cenozoic sedimentary section of the Gulf of Mexico Basin was deposited under remarkably stable tectonic conditions—the persistent subsidence of the basin was only modified by the local deformation of the Jurassic salt and by growth-faulting bordering major depocenters. As a result of this tectonic stability the environments of deposition, the main avenues of terrigenous clastic influx, and, consequently, the lithologic composition of the sedimentary sequence persisted, with minor modifications from the Late Jurassic to the present. In a broad sense, three distinctive "lithofacies provinces" can be recognized in the sedimentary sequence of the Gulf of Mexico Basin:

1. To the east and southeast, over the Florida and Yucatan platforms, the sedimentary section is entirely composed of carbonates and lesser amounts of evaporites.

2. To the southwest and west, in the area of the Isthmus of Tehuantepec and in east central Mexico, carbonates and fine-grained terrigenous clastics predominate. Coarser-grained clastics are much less common and generally restricted to local areas.

3. To the northwest and north, from northeastern Mexico to Alabama, coarse-grained terrigenous clastics are much more prevalent, indicating the presence of important streams draining the continental interior and depositing their load along the shores of the Gulf of Mexico of the time.

The present Gulf of Mexico Basin is believed to have originated in Late Triassic time as the result of rifting within the North American Plate at the time it began to crack and drift away from the African and South American plates. Rifting probably continued through Lower and Middle Jurassic time with the formation of "stretched" or "transitional" continental crust throughout the central part of the basin. Intermittent advance of the sea into the basin from the west during late Middle Jurassic time resulted in the formation of the extensive salt deposits. It has been suggested that the main rifting episode, which resulted in the southward separation of the Yucatan block from the North American Plate and the formation of oceanic crust in the central part of the basin, took place during the Middle or early Late Jurassic time, after the formation of the salt deposits.

Since Late Jurassic time, the basin has been a stable geologic province characterized by the persistent subsidence of its central part, probably due at first to thermal cooling and later to sediment loading as the basin filled with thick prograding clastic wedges along its northwestern and northern margins particularly during the Cenozoic. To the east and south, the stable Florida and Yucatan platforms were not covered by the sea until the beginning of Cretaceous time; since then the formation of platform carbonates has characterized the geologic history of these two stable areas. Most of the basin was rimmed during the Early Cretaceous by carbonate platforms and its western flank was involved during the latest Cretaceous and early Tertiary in a compressive deformation episode—the Laramide Orogeny—which created the Sierra Madre Oriental of eastern Mexico.

The posters presented illustrate the present configuration of the Gulf of Mexico Basin and the history of its development.

SAMANI, B. A., Atomic Energy Organization of Iran, Tehran, Iran

Metallogenesis and Ore Provinces of Iran: In The Contexts of Plate Tectonics

Iran as a part of Alpine-Hymalayan mobile belt forms a solid bridge between Afro-Arabia and Eurasia. On the basis of structural evolution, magmatism and metallogenesis, it could be subdivided into; Talesh-Qarahdagh, Orumieh-Takab, Alborz-Binalud, Kopet Dagh, Central Kavir, Sabzevar, BaftNaiin, Yazd, Tabas, Sanandaj-Sirjan, Lut, East Iran, Makran, and Zagros mobile belts (zones) and stable blocks. The late Precambrian orogeny accompanied by metamorphism, and magmatism, affected the whole basement of the area and created continental crust. Several blocks (e.g. Orumieh-Takab, Central Kavir, Yazd, Tabas and Lut) were located partly between the belts, partly within their frames. These are massifs of the old sialic crust which played the role of micro-continents in deep-sea basin of oceanic type. On the basis of time and space distribution of different type ore formation, and magmatism, several tectonic settings such as; intercontinental rift, ocean rise, Andean type magmatic, island arc and interarc basin, continent-continent collision and post collision volcanic zones were identified. Metallogenesis and magmatism evolution from early stage oceanic to orogenic plateau province occurred and different metallogenic districts and ore zones were made. The main ore forming epochs were as follow:

–Pan-African intercontinental riftogenic type Fe, Mn, Pb, Zn, apatite, phosphate, evaporites, and metasomatic-hydrothermal type Fe, U, Th, REE, sulphides, rare metals.

–Middle-Upper Triassic carbonate hosted Alpian type Pb-Zn-F_2Ca-SO,Ba, and Cimmerid's bauxite.

–Mesozoic oceanic massive sulphide, podiform chromite, Mn, Cu, Co and etc., Lower Cretaceous Mississippi type Pb-Zn-Ba-(Ag, Cu), Cimmerian and Early Alpine skarn-hydrothermal type Fe, Pb, Zn, Cu, Mo, Sn, W, Au, S, etc.

–Mid-gate Alpine (Tertiary) island arc, inter-arc, Andean type, collision and post collision volcanic type deposits of Fe, Mn, Cu, Mo, Pb, Zn, Au, Ag, rare, metals, U, Th, Sb, Hg, As, Co, Bi, Ni, alunite, turquoise, rock salt, etc.

Different groups of deposits are characterized by platform, geosyncline, foLded zones, and region of autonomous tectonic-magmatic reacticization elements. According to the plate tectonic modelling, not only the peculiarities of known deposits could be explained, but also new favourable metallogenic districts of Au, Ag, Hg, Sb, U, rare metals and rare earth elements for further exploration are presented.

SAMANI, B. A., Atomic Energy Organization of Iran, Tehran, Iran

Uranium and Thorium Mineralizations Related to the Pan-African Magmatism and Alkaline Metasomatism in Central Iran

The Central part of Precambrian Bafq-Poshte-Badam block in Central Dran which is considered as a Pan-African taphrogenic zone is occupied by Pyroclastites, sandstones, carbonates, acid to basic lava flows, evaporites, ironstones and pyroxenite to syenitic, granotoids, and carbonatite intrusions. Alkaline metasomatism and hydrothermal alterations strongly developed on the Peripheral zones of alkaline massives, and produced associated mineralizations such as iron, uranium, thorium, vanadium, molybdenum, manganese, apatite, and rare earth elements (La, Ce, Y). The magmatic activities and deformations were occurred during the Pan-African tectono-thermal event.

At least three main genetic: "(I) Carbonatite, (II) metasomatite, and (III) Post metasomatite, hydrothermal vein types" Ore forming of thorium and uranium are recognized.

The Carbonatite types mainly consist of thoriferous ferrocarbonatite dykes and veins with high content of Th, U, REE, and apatite.

Most of the radioactive mineralized metasomatic bodies have domal shapes and extension. Hydrothermal alterations (metasomatism) occurred in the pyroclastic rocks and produced amphibole, amphibole-albite, Chlorite-Phlogopite and Feldspar (albite) metasomatites. Uranium, and uranium-thorium mineralizations are know in amphibole, phlogopite and Carbonate metasomatites. Uranium on disperse forms occurs in ilmenite, magnetite, hematite and locations of silicification and sericitization. According to U-Pb radioisotope age determination the ore forming stage is 540.7 million years old. Post metasomatite hydrothermal vein type uranium ore forming mainly occurs in association with magnetite-molybdenite, serpentine-magnetite.

Uranium transported by alkaline solutions and deposited on the reductory environments, such as magnetite horizons and $Fe^{2/3}2$ bearing rock units. According to U-Pb radioisotope age determination, the ore forming stage is 513.6 m.y. old.

SAMANIEGO, C. M., and L. L. CHYI, University of Akron, Akron, OH

Trace Element Distribution in Keystone Vein of Antamok Gold Deposits in Luzon, Northern Philippines and Their Genetic Implications

The deposits lie in a major vein system that trends in a northwest-southeast fashion. The mineral assemblage is typical of epithermal type deposits and is composed of quartz, calcite, clay minerals, plagioclase, anhydrite, minor tellurides, pyrite, and other base metal sufides together with trace amounts of native gold. The host rocks consist predominantly of metamorphosed Cretaceous to Paleogene volcanic and sedimentary units of island-arc derivation. Rock types are typically andesitic and dioritic in nature. Alteration assemblages are dominated by chlorite and to a lesser extent, sericite. Studies show that the deposit is of Late Tertiary age and may have formed during Pleistocene time.

The Keystone vein is approximately 6 m in width, and filled predominantly with quartz and calcite. The vein has an average gold content of 5.5 gm/t. It is almost vertical and extends 440 m downward. The mineralization was considered to be formed by circulating hydrothermal fluids of dominantly meteoritic origin.

Twenty-one samples were collected along the vertical vein from 10 different mine levels. The paragenetical order of the ore minerals is divided into three stages. Stage one marks the deposition of gray quartz, pyrite and native gold. Stage two continues with pyrite mineralization and a change from gray to white quartz accompanied by rhodonite which is closely associated with increasing gold contents. Minor base metal sulfides were formed at this time. Stage three is characterized by the influx of massive anhydrite and calcite veins that are generally barren of gold.

The distribution of Au in sulfide minerals has been examined by SEM imaging with the microprobe. Most of the gold contained in the ore is carried by disseminated native gold. However, significant amounts of Au is found in chalcopyrite and sphalerite, presumably in solid solution. The SEM images have uniform dot distribution and the phases appear to be homogenous under the microscope.

Five samples were selected for the study of trace element partition between coexisting sulfide and non-sulfide phases. It appears that Au is selectively enriched in sulfide phases versus non- sulfide phases by approximately 22 times. Au-Ag ratios are also higher in the sulfide phases as compared to the non-sulfide fraction. Au appears to be more chalcophilic than Ag.

The total Au-Ag ratio of the ores generally decreases with depth from values of 10 to about 1 in approximately a 200 m change in elevation. In contrast to this, however, the Au-Ag ratio in the discrete gold particles increases with depth from a value of 4.7 to 18.2 for the same elevation change. The tendency of Ag in forming complexes and subsequent loss of these complexes may have some bearing on the lowered Au-Ag ratio in ore moving upward.

Rubidium contents are too low to be measured in sulfide phases. In non-sulfide phases, the ratios vary from 230 to 380. K-Rb ratio of common terrestrial rocks rarely fall outside the limits of 160 and 300, with an average of about 230. Mafic rocks have values that tend toward and beyond the upper limit of this ratio range. Possible mantle material are known to have ratios much higher than 300. The K-Rb ratio characteristics suggest that the deposits may have been derived from ophiolitic fragments.

REE were also analyzed by instrumental neutron activation analyses. LREE are found to concentrate in the ore by a factor of up to 10 as compared to chondritic abundances. HREE are found to be depleted by a factor also close to 10 as compared to chondritic abundances. Between sulfide and non-sulfide phases, Eu is found to concentrate in the sulfide phases. Lower REE content of these ores as compared to other crustal materials is in keeping with the suggestion from K-Rb characteristics that the deposit may be related to ophiolitic fragments genetically. Enrichment of LREE suggests additional differentiation to produce the ore with possible contamination of crustal materials.

SAMIR, A., Zagazig University, Zagazig, Egypt

Groundwater Exploration in Arid Regions

This paper presents some examples of geoelectric surveys applied to hydrogeologic investigations in some desert areas in Egypt. Under such arid conditions, the field curves are usually contaminated with noise and are, in many cases, inadequate for quantitative analysis. The purpose of processing is to try to overcome the influence of the disturbing factors and to utilize all the useful information provided by the field data. Computational facilities help in developing programs for smoothing noisy data without loss of information as well as in the physical inversion of curves into layer parameters using the well known filter theory. The problem of smoothing is recently presented where the Schlumberger sounding curves can be converted into pole-pole system and interpretation is carried out in the two-pole domain. Another procedure is to retransform once more into the Schlumberger domain. Another comparison is presented using various techniques of smoothing beside a known geologic section of a discovered producing well.

For a particular hydrogeological setting and as long as the geology does not change significantly, a type resistivity sounding can be computed. The resistivity range of the encountered geological units has to be determined from previous or parameteric measurements. This procedure may help in establishing the resolving power or limitation of the method for a particular unit. In addition, the principle of equivalence or suppression can be sufficiently considered.

The results are presented in the form of 2-D sections illustrating the distribution of the various electrostratigraphic horizons in the surveyed area. Correlation is made with the available geologic drilling information.

SANCHEZ CELA, V., J. FERNANDEZ, L.F. AUQUE, B. GUARAS, M.P. LAPUENTE, and J.M. TENA, Universidad de Zaragoza, Zaragoza, Spain

Proposed Mechanism for Generation of Basic-Igneous Rocks in Upper Continental Crust

In the Spanish Pyrenees and Iberian Mountains there exist abundant outcrops, generally of small sizes, of basic igneous rocks, curiously always associated with Triassic–Liassic sedimentary materials, mainly evaporitic Keuper marls. The most common type of igneous rocks present are defined as ophites or dolerites, although others also exist, such as various types of gabbros, monzonites, syenites, and even granitic rocks and ultramafics. In some places and associated with many of the previous rocks appear "granulitic-hornfeld" rocks with high T-P minerals (spinel, sillimanite, orthopyroxene, etc.) formed in relation to high dynamic-thermal conditions in punctual-lineal zones.

In this paper, a petrogenetic mechanism is proposed for the generation of basic rocks (ophites, dolerites, gabbros, etc.), associated with Triassic–Liassic stratigraphic levels in the upper crust. In such a "mechanism" the basic and the associated intermediate-acidic rocks and even some ultramafics, are related to transformation processes, which in some cases could reach the melting stage, between suitable sedimentary rocks, mainly Keuper materials, and "sialic" allochthonous chemical elements at different physical and chemical conditions.

Although many of these "ophites–dolerites" were affected by later structural episodes, some of them conserve the features of the original petrogenetic process. In such a case, the geological field data—gradual transition between wall rocks and the igneous ones—complemented by abundant petrographic and geochemical studies appear to indicate that many mineralogical and chemical components of the igneous rocks have an evident sedimentary origin.

To understand this petrogenetic model, the following points should be taken into account:

1. Existence of structurally and compositionally suitable sedimentary materials to be transformed when they are affected by physical and chemical changes.

2. Mobilization of active sialic elements related to the presence of a faulting sialic basement in thickening crustal zones.

3. Possible petrogenetic processes when such sedimentary materials are affected by allochthonous active chemical elements at moderate-high T-P conditions.

4. Geological and petrological information, including experimental petrology, on the possible mineralogical phases that can be formed in various physiochemical environments between suitable marly-evaporitic materials and various allochthonous chemical elements, mainly mobile sialic elements.

5. Formation of suitable domed structures that, at punctual top places, allow the existence of high T-P conditions because of the impermeable-

refractory behavior of the carbonate rocks and the first generation of igneous rocks. This causes the strength of the petrogenetic processes to increase with depth.

6. Formation of high T-P conditions at shallow crustal levels. High P conditions are created because of the constraining conditions of the chemical fluid phase. As a result, the petrogenetic processes can persist at high temperatures in a solid state because of the relatively high pressure of the fluids. High T conditions result from thermal phenomena related to sialic thickening and thermal energy released at faulting zones by the transformation of mechanical energy into heat. The differential thermal behavior of the various materials within the domed Triassic–Liassic levels may also contribute to the development of thermal storage zones.

Considering the above arguments, it is not very difficult to understand "why" and "where" this petrogenetic mechanism, which forms basic igneous and related rocks, takes place.

On the other hand, the existence of various chemical-physical environments during this petrogenetic mechanism condition the formation of paragenesis from undersaturated at high T and moderate P to oversaturated at high–moderate T and low P. In this way, rocks with minerals such as orthopyroxene, forsterite, spinel, etc., and others with clinopyroxene, amphibole, biotite, feldspars, etc., can be formed by transformations of chloritic-illitic minerals with the participation of carbonate-evaporitic components.

If, during the formation of rocks with undersaturated-oversaturated minerals at high temperatures and relatively high pressures (e.g., 4–6 Kbars), a rapid decompression phenomenon takes place, for example, because of the formation of fractures, through which fluids can escape, partial or total fusion can take place and extrusive volcanics can be formed.

The above petrogenetic grounds and considerations, applied in this case to Triassic–Liassic ophitic-doleritic rocks of supposed mantle provenance, could be applied to other basic and even ultrabasic rocks present in many stratigraphic levels of the crust. For this reason, more information on the evolution of the basic igneous rocks throughout geological times and also on the meaning of their frequent association with carbonate and marly evaporitic materials is necessary.

SANCHEZ CELA, V., M.P. LAPUENTE, L.F. AUQUE, J. FERNANDEZ, and B. GUARAS, Universidad de Zaragoza, Zaragoza, Spain

Dacitic and High T-P Metamorphic Rocks Related to Sialic-Thickening and Shear-Friction Zones

The Spanish dacitic rocks are distributed in various geological environments and date from the Paleozoic to the Tertiary. Many of these volcanics exhibit metamorphic enclaves with corundum, spinel, sillimanite, and also isolated crystals, mainly garnets.

In this paper, the origin of dacitic rocks and their metamorphic enclaves is related to the anatexis and the transformation of suitable wall rocks at various T-P conditions under a high chemical activity, associated with sialic-thickening phenomena at faulting-shear-friction zones. The crustal thickening phenomena is related to the origin of sialic matter from the mantle.

This sialic-thickening phenomenon, as well as the formation of "juvenile granitic masses," involves the generation of great energy which, in many cases, is released through deep faults. These faults are mainly formed in the structural ruptures of thickening zones (borders of domes). In such fault zones, as well as the "primary" energy liberated in the transformation of the upper mantle matter into sialic matter, a "secondary" energy is also formed. This energy, mechanical in nature, is created by the relative displacement of sialic blocks, and thus transformed into heat.

In these domed fault zones, high pressures can also be reached under suitable structural conditions, for example, when these active faults are fossilized by an "impermeable" sedimentary cover. In such fault zones, many active and mobile chemical elements are released, from the wall rocks and from deeper sources.

In Spain, the materials in the domed fault zones associated with dacitic rocks are lutites, greywackes, shales, and schists, mainly formed of illitic and chloritic minerals, with other detrital minerals (Fe-oxides, quartz, etc.) and chemical-evaporitic components ($CO_3^{3/4}$, $SO_4^{3/4}$, etc.).

Geological, petrological, and geochemical data indicate that first the metamorphic rocks (enclaves) and later the dacitic rocks were formed in three T-P petrogenetic stages. In the first stage, at high T-P conditions, high transformation metamorphic processes take place with the formation of high-grade metamorphic rocks, with corundum, spinel, fibrolite-sillimanite, etc. In a subsequent stage, partial decompression can occur. This "transitional" metamorphic-volcanic episode is characterized by the presence of new "igneous" minerals (e.g., plagioclase, biotite, orthopyroxene, garnet, etc.) together with partial-melting phenomena, predecessor of the anatectic-volcanic process. At the same time, new prograde-retrograde metamorphic minerals are formed. Later, in a more decompressive stage, a more general anatexis takes place, giving rise to the volcanic phenomenon. This, according to their chemistry, fluid pressure, etc., will create intrusive, extrusive, or fragmentary-pyroclastic volcanics that, in many cases, enclose metamorphic rocks or isolated crystals (e.g., garnets) created in previous petrogenetic stages. Even within igneous crystals (e.g., plagioclases) and also in the groundmass it is possible to observe restitic metamorphic minerals (e.g., spinels, sillimanite, etc.). Here the crystallization of new igneous minerals (plagioclases, biotites, etc.) from a previous melting stage is evident.

The final composition of these volcanics will depend on the degree of partial melting, the contribution of new matter from deeper zones (e.g., mantle), differentiation mechanisms (e.g., fractional crystallization, etc.), and perhaps also contamination by wall rocks at superficial zones. All of these phenomena will take place in shear-friction zones associated with thickening-sialic phenomena, at relatively shallow crustal levels.

SANCHEZ, J. R., R. SEGURA, R. GARCIA, E. MILIAN, J. FERNANDEZ, C. PERERA, Centro des Investigacions del Petroleo, La Habana, Cuba

Subsurface Stratigraphy of Cuba Based on Deep Wells

Subsurface stratigraphic information based on deep wells drilled in Cuba during late years is expressed, which gives evidence of the complex structural setting of the island showed on the Tectonic Map scale 1:500,000 published in 1985. The Northern Cuban Basin includes different zones that were tectonically displaced by overthrusting, in which several tectonostratigraphic units can be established. Other basins "epieugeosynclines" correspond to the Upper Cretaceous Tertiary basins called western, central and eastern subdivided in minor ones.

Northern Cuban Basin. Guaniguanico, La Esperanza and Bahia Honda Tectonostratigraphic Units were established on the western part of Cuba. At Guaniguanico the Pinar 1 (5200 m) well was drilled discovering several allochthonous carbonatic-sequences, ranging from the Tithonian to the Albian stage, overlying a polymictioclay oliostrome of Campanian-Maestrichtian age or younger, which in turn overlies the autochthonous ? carbonate rocks from the Kimeridgian to the Albian stage, dislocated by the faults.

In the Esperanza Tectonostratigraphic Unit, Dimas 1 (5505 m) well is represented by a thick sequence of allochthonous terrigenous-carbonate flyschoid sediments, from the Upper Jurassic to the Lower Cretaceous (Neocomian) that overlies another olistostrome of Paleocene age, not completely drilled. Mariel 1 (3204 m) well was drilled on Bahia Honda Tectonostratigraphic Unit and is essentially characterized by the ophiolitic sequence as a melange, together with the rocks of the Cretaceous island-arc forming nappes over the continental margin, which comprises carbonatic-terrigenous sediments of Upper Cretaceous age.

In directions to the east the northern basin includes Habana-Matanzas and Varadero-Corralillo oil and gas bearing regions. In both, numerous wells have been drilled being Boca de Jaruco 3 (4054 m) and Littoral 1 (3435 m) two of the most representatives. They are characterized by the following structural stages which correspond to the Placetas and Camajuani Tectonostratigraphic Units: *parautochthonous* comprising carbonatic-terrigenous sediments from the Oxfordian ? to the early Lower Eocene. The last includes a polymictic olistostrome which underlies the *allochthonous* structural stage, that consists of carbonatic-clay rocks of Aptian-Turonian age in its lower part, followed by the ophiolitic melange, and the top consists of overthrusted terrigenous-carbonitic sediments of Campanian-Maestrichtian age. The *neoautochthonous* structural stage includes carbonatic sediments from the late Lower Eocene to the Neogene.

The Remedios Tectonostratigraphic Unit lies north of the Camajuani Unit and is well exposed in central and eastern Cuba. The inner part of the unit corresponds to the Colorados 1 (4,060 m) well being the most prominent characteristic the presence of calcirudites and calcarenites of Middle and Upper Cretaceous age, reaching a thickness of 1000 m. The outer part is well representative in Camagüey 3 (2264 m) and Gibara 1 (4527 m) wells, which penetrated a bank-type limestone sequence with well developed dolomite beds from the Neocomian to the Maestrichtian stage.

The Cayo Coco Unit is the northernmost tectonostratigraphic unit recognizied in Cuba. The last well drilled was Gloria 1 (4001.6 m). The Upper Jurassic-Lower Cretaceous shallow water sediment are followed by more normal marine beds (pelagic) from the Aptian to the Cenomanian stage, which indicates in that time the existence of a marine depression located at the periphery of the Bahamas platform.

Upper Cretaceous-Tertiary basin. These basins are located on the Zaza Tectonostratigraphic Unit that comprises the volcanic-arc rocks. In surface these rocks are mixed tectonically with the ophiolitic sequence. Three basins are considered: western basin is the biggest. In the Candelaria 1 (4384 m) well a carbonate sequence of 2050 m thickness was penetrated in the Neogene mainly Miocene age, which is the maximum thickness reported in Cuba. The underlying sequence consists of terrigenous-carbonatic sediments from the Oligocene to the Campanian stage, ending with rests of the volcanic-arc erosion.

In the Vegas 1 (5030 m) well the sequence has variation in thickness in the Tertiary sediments but the underlying deposits involve tuffs of the volcanic-arc and the upper part of the ophiolitic complex. The central basin in that most drilled area in Cuba. The deepest well was Jatabonico 78 (4336 m) that penetrated island arc volcanic material (lava and pyroclastics) from the Albian to the Pre-Campanian age, interbedded non volcanic sediments; all these rocks overlying the ophiolitic sequence (tholeiitic basalts, gabbros and serpentinized ultramafites). In the lower portion metamorphic schists with affinities to those of the Escambray Unit were found. Some depressed areas in the basin reached greater thickness, at least 1000 m, in the Upper Cretaceous (Catalina wells) either in the Tertiary (Cometa 1) well.

Eastern basin: the best stratigraphic information in this basin was obtained from the Granma 1 (3017 m) well. Good cores were recovered from carbonatic-terrigenous sequences from the Pliocene to the Lower-Middle Eocene. The well reached the upper part of the paleogenic volcanic-arc of eastern Cuba.

Data based on deep wells drilled in Cuba are important and must be considered for paleotectonic reconstructions in all Caribbean regions.

SANDOWSKI G. R., Universidade de São Paulo, São Paulo, Brazil

Cubatãq Fault System in Southeastern Brazil: Example of Megafault

A net of mylonite belts with an extension of more than 1000 km borders the Santos Basin on its continental frame. It is represented by an originally Late Precambrian to Early Paleozoic post inversion transcurrent fault system composed by nylonitic belts with different local names that isolate tectonic lenses or fault blocks. The main linear discontinuity is the Cubatã-Alé Paraiba –Lancinha fault zone. Along this structure a whole and complete transition of brittle to ductile shear deformations are observed affecting upper crustal levels to the southwest and deeper to the northwest. While in the upper levels tectonic breccias and stratigraphic discontinuities together with strong stretching and flattening of certain rock types are dominant, in the lower levels expressive stretching, ductile dragging of regional structures across several kilo meters transversally to the main deformation zone, and concomitant flattening and parallelization of previous folds are the most evident features. The deep mylonites show stretching parameters of X:Y:Z above 35:1:1, typical quartz c-axes girdles sheath folds and recrystallization.

The faults were reactivated differently during the Phanerozoic. Chevron folding affected the mylonites in some locals when the rocks were already in higher levels, and Paleozoic sedimentation was partially controlled by this fault system in some places of the Parana Basin during certain periods. Tertiary continental rifting as later activations along these faults conditioned the formation of the Serra do Mar Rift System. Some recent kinematic analyses allow us to admit that the faults have been subjected to a compressional regime during Late Tertiary to Quaternary times.

SANDRONE, R., Politecnico corso Duca degli Abruzzi, Torino, Italy

Geology of the Italian High-quality Cosmetic Talc

The Italian high-quality cosmetic talc is produced at the Fontane mine near Pinerolo (Piedmont, Italy). This mine is the only active survivor of a series of workings along a mineralized horizon forming part of the terranes of the Dora-Maira Massif.

This massif is a pennidic unit of continental crust formed of an upper element consisting of pre-Carboniferous metapelites, marbles and metabasites and a lower one consisting of metasediments of probably (Permo)-Carboniferous age; both elements contain metagranitoid with granitic to tonalitic chemistry. All these lithologies have clearly undergone metamorphic re-equilibration into greenschist facies of Lepontine age, although early-Alpine high-pressure paragenesis and occasional relics of a pre-Alpine metamorphic event are preserved locally.

The talc occurs in the pre-Carboniferous basement and forms a conformable, sheetlike body.

The Fontane lithological series (from below upwards) consists of marbles (interbedded on the regional scale in the basement micaschists), followed by micaschists, Augengneiss and more micaschists enclosing the mineralized body.

On an average this body trends N25 °W and dips west-southwestward at an angle of 30°. Its workable thickness is usually 1-3 m but is sometimes more than 10 m. Two foldings phases with subhorizontal (or slightly inclined) N-S and N25 °W trending axes are responsible for the variations in thickness and attitude of the mineralization.

The talc is normally very pure, although accessory minerals are by no means uncommon. The mineralized body also contains predominantly carbonatic or siliceous "xenoliths" of various sizes and usually ellipsoid.

The genesis of the talc can probably be attributed to regional metamorphism, very likely in Hercynian time, of an original sepiolitic horizon.

SANDY, M.R., University of Dayton, Dayton, OH

Jurassic and Cretaceous Brachiopods as Indicators of Tethys–Atlantic–Pacific Mesozoic Gateways

Jurassic and Cretaceous brachiopod faunas from northwest and southern Europe and the Soviet Union are abundant and diverse at certain horizons. However, it is the lesser known (and often unpublished) brachiopod occurrences from other continental blocks of Pangaea, subsequently Laurasia and Gondwanaland, that are keys in interpreting Tethys–Atlantic–Pacific gateways for Mesozoic brachiopods. This paper is a synthesis incorporating new data and new interpretations. Similarity coefficients, diversity, and abundance data will be discussed during the presentation.

Opening of the Atlantic Ocean: The opening of the Atlantic Ocean had a direct effect on brachiopod connections along Tethys and into the Pacific. During the opening of the Atlantic Cretaceous, brachiopods show evidence of migration via four major routes: (1) the Hispanic Corridor (European Tethys) to the eastern seaboard of the Pacific; (2) from eastern Tethys via eastern Africa to the South Atlantic; (3) south along the opening South Atlantic to the Antarctic Peninsula; (4) northward to East Greenland (Ager and Walley, 1977). For routes 1-3 direction of movement may prove to have been in either direction.

Transatlantic brachiopods and the Hispanic Corridor: The Atlantic Ocean has been expanding in a predominantly east-west direction from the Jurassic to the present day. The terebratulid brachiopod *Flabellothyris* from the Middle Jurassic of Mexico, Morocco, and India (Ager and Walley, 1977) and *Rhynchonella s.s.* from the Late Jurassic of Mexico, northwest Europe, and the Russian Platform (Ager, 1986) provide examples of the Hispanic Corridor linking early trans-Atlantic faunas.

The number of generic links between the trans-Atlantic brachiopod faunas rose during the Early Cretaceous from the Jurassic as free communication along the Hispanic Corridor increased. From the Valanginian of Mexico the brachiopod genera *Sellithyris, Cyrtothyris* and *Colinella* have been recorded (Owen, 1981). *Sellithyris* and *Cyrtothyris* are known from the Early Cretaceous of northwest Europe and *Colinella* from the Hauterivian of Morocco. The record of a single specimen of the terebratulid *Musculina sanctaecrucis* (Catzigras) from the Cretaceous of South America (British Museum [Natural History], London) probably also supports this Hauterivian connection. Early Cretaceous brachiopods from Colombia have affinities with genera originally described from Europe, i.e., *Gemmarcula* and *Oblongarcula* (National Museum of Natural History, Paris). These examples show the continuing influence of the Hispanic Corridor on brachiopod distributions through western Tethys to the eastern Pacific margin during the Early Cretaceous.

Toward the end of the Early Cretaceous, there are still links at the generic level between American and European brachiopod faunas that point toward the continued existence of the Hispanic Corridor or at least trans-Atlantic connections. Albian brachiopods from Arizona include the rhynchonellid *Cyclothyris* (also recorded from the Aptian–Albian of Texas) and the terebratellids *Psilothyris* and *Gemmarcula. Cyclothyris, Psilothyris,* and *Gemmarcula* (the latter probably present on Vancouver Island, Canada; Owen, 1981; *Cyclothyris* and *Kingena* may also be present in this region) appear to be widespread genera, and alternative routes and distributions should not be dismissed as the Atlantic Ocean increased in width. The Hispanic Corridor was too wide for free faunal dispersal although Arctic links may have still been maintained.

East Pacific fauna and possible Boreal links: The rhynchonellid *Ptilorhynchia* has been recorded from the Middle and Late Jurassic of California and British Columbia, respectively. It is also known from the Late Jurassic to Early Cretaceous of Northern Siberia and the Early Cretaceous of the Barents Sea (Sandy, in preparation). *Ptilorhynchia* has also been recorded from the Aptian of Alexander Island, Antarctic Peninsula, and possibly the Turonian–Coniacian of James Ross Island, Antarctic Peninsula (Sandy, in preparation). A northern high-latitude, eastern Pacific seaboard through to southern high-latitude distribution is emerging for the genus. *Proteorhynchia* from the Early Cretaceous of Mexico may be an endemic off-shoot from *Ptilorhynchia.*

The occurrence of the enigmatic rhynchonellid *Peregrinella* in the Early Cretaceous of Europe, Tibet, and California has been cited as possible evidence in support of an expanding earth or seamount "hopping" across the Pacific (Ager, 1986). The Californian occurrence was at a paleolatitude greater than 40°N. Records of *Peregrinella* are scarce, and the genus appears to have had a wider distribution than purely "Tethyan." During the Late Jurassic and Early Cretaceous, molluscan faunas from California display mixed faunal affinities (Tethyan ammonites and Boreal belemnites) showing that this was a region of faunal overlap.

Another rhynchonellid *Lacunosella schucherti* (Stanton) has also been recorded from the Early Cretaceous of California. The genus *Lacunosella* has a Tethyan distribution in Europe. However, *Lacunosella schucherti* (Stanton) has ligate folding and may be a long-looped terebratellid with affinities to *Rugitela* (USNM, Smithsonian Institution, Washington, DC). The genus *Rugitela* has a predominantly Boreal to Jura distribution in Europe. A terebratellid brachiopod, probably referable to the genus *Rugitela,* has recently been collected from Albian strata of the Svedrup Basin from Prince Patrick Island, Canadian Arctic Islands (Canadian Geological Survey, Ottawa). If both occurrences are confirmed as belonging to *Rugitela,* it would indicate trans-Arctic links with faunas from Europe and the Russian Platform.

Tethys–Indian Ocean–South Atlantic Ocean: Brachiopod faunas from the Albian of southern Africa and Madagascar have links at the subfamilial and generic level with brachiopods described from the Antarctic Peninsula (Sandy, in preparation). The southern Atlantic Ocean was therefore not too wide to prevent migration of faunal elements between Africa and Antarctica at about this time. These faunas have links with Tethys via eastern Africa (northern Zululand and Madagascar).

The genus *Kingena* has a widespread distribution (e.g., Europe, India, Madagascar, Western Australia, and Antarctica) and may have spread east through Tethys via eastern Africa to the Antarctic. Genera such as *Kingena* with pandemic distributions may have had wide environmental tolerance ranges or longer larval stages than endemic genera.

Opening of the South Atlantic Ocean: The presence of "European" Cretaceous brachiopod genera (e.g., *Rectithyris, "Capillithyris," Cretirhynchia*[?]) from James Ross Island, Antarctica, suggests that there may have been a direct link between European and Antarctic brachiopod faunas via the opening South Atlantic Ocean possibly by the late Albian and certainly by the late Turonian (Sandy, in preparation).

Opening of the North Atlantic: The presence of "typically Tethyan" brachiopods in the Valanginian of East Greenland has been interpreted as an indication of the development of an early Gulf Stream in the North Atlantic Ocean (Ager and Walley, 1977). Other more widespread brachiopods are also found in these northern mid-to high paleolatitudes, for example the terebratulid *Moutonithyris moutoniana* (d'Orbigny) (BM[NH], also references in Ager and Walley, 1977).

Acknowledgment: I thank Dr. E.F. Owen (British Museum [Natural History], London) for his valuable comments on an earlier draft of this abstract.

REFERENCES

Ager, D.V., 1986, Migrating fossils, moving plates and an expanding earth: Modern Geology, v. 10, p. 377–390.

——— and C.D. Walley, 1977, Mesozoic brachiopod migrations and the opening of the North Atlantic: Palaeogeography, Palaeoclimatology, Palaeoecology, v. 21, p. 85–99.

Middlemiss, F.A., 1984, Distribution of Lower Cretaceous brachiopods and its relation to climate, *in* P. Brenchley, ed., Fossils and climate: John Wiley and Sons Ltd., p. 165–170.

Owen, E.F., 1981, Distribution of some Mesozoic brachiopods in North America, *in* J. Gray, A.J. Boucot, and W.B.N. Berry, eds., Communities of the past, Hutchinson Ross, Stroudsburg, Pennsylvania, p. 297–309.

SANGSTER, D. F.,, G. S. NOWLAN, C. R. BARNES, and M. W. HITCH, Geological Survey of Canada, Ottawa, Canada, D. F. STRONG, Memorial University of Newfoundland, St. John's, Canada, and C. M. SAUNDERS, Newfound Department of Mines, St. John's, Canada

Are Mississippi Valley Type (MVT) Deposits Thermal Anomalies?

Certain genetic models for MVT deposits are predicated on the assumption that ore fluids were hotter than the surrounding country rock at the time of ore deposition. This assumption has been most commonly based on burial depths and temperatures derived from reconstruction(s) of the stratigraphy at an assumed time of mineralization. Because the latter is seldom known for MVT deposits, burial temperatures derived for country rocks during MVT mineralization are therefore of questionable reliability.

We report here on initial attempts to determine maximum temperatures reached by country rocks surrounding several MVT deposits by compiling Color Alteration Indices (CAI) of conodonts in the host carbonates. In all cases (1) CAI values from carbonate samples within ore zones were indistinguishable from those taken up to 20 km from the ore; (2) within the accuracy of the method, temperatures determined from CAI data correlated well with homogenization temperatures of fluid inclusions in main stage sphalerite in the ores. Thus it would appear that, at least for the districts studied to date, fluids responsible for MVT deposits were at, or near, thermal equilibrium with their host rocks at the time of mineralization.

A detailed study was made in an area of Lower Paleozoic carbonates in western Newfoundland containing several dozen MVT occurrences in addition to one operating mine. Many of these are in carbonates yielding CAI values of 2–3 (90°–200°C), agreeing well with homogenization temperatures in sphalerite. Several others, however, with similar homogenization temperatures, occur in rocks with CAI values as high as 5 (360°–550°C), suggesting *negative* thermal anomalies of at least 300°C. Discrepancies of this sense and magnitude may indicate mineralization, at least in this region, took place only after the host rocks were buried, heated, and then returned to a near surface position.

SAPPAL, K. K., Curtin University of Technology, Perth, Australia

Geology and Petrology of Coal from Permian Intracratonic Basin, Western Australia

The Early to Late Permian intracratonic Collie Basin of Western Australia is situated 150 km south-southeast of Perth and 27 km east of the Darling Fault, and it is the major north–south-trending tectonic feature of Western Australia (Fig. 1). The basin is bilobate and it is subdivided into three unequal sub-basins, namely Cardiff, Muja, and Shotts.

The three sub-basins cover an area of approximately 230 km^2 within a basement complex of Archean granite and gneiss. In broad terms the stratigraphy of the Collie Basin is simple and consists of three sedimentary formations overlying Archean basement. At the base the earliest Permian (Sakmarian) Stockton Formation, is overlain conformably by the Early to Late Permian Collie Coal Measures, which is overlain unconformably by the Nakina Formation of Cretaceous to Tertiary age. The sequence is similar in all three sub-basins, except that the uppermost Permian Coal Measures are absent from the Shotts Sub-Basin. The southwestern flank of each sub-basin is steeper than the northeastern flank, and contains thicker deposition of Permian sediments. The Cardiff Sub-Basin is the deepest and contains at least 1100 m of Permian sediments. The Muja and Shotts Sub-Basins contain 850 m and 580 m of Permian sediments respectively.

Sedimentation in the Collie Basin began in early Permian times with the deposition of basal tillites followed by glaciofluvial sediments, which together comprise the Stockton Formation. This unit is conformably overlain by the Collie Coal Measures which includes three major coal bearing members. The coal seams of the lowermost Ewington Member can be broadly correlated over the entire basin, however, the seams of the upper two members, the Collieburn and Cardiff Members of the Cardiff Sub-Basin and the Premier and Muja Members of the Shotts and Muja Sub-Basins are considered to be contemporaneous but cannot be correlated. The Collie Coal Measures form a succession of coal seams, sandstones, siltstones, shales, clays, grits, and conglomerates. The coal measures are overlain unconformably over the entire basin by a thin Tertiary sequence of the Nakina Formation. It is composed of poorly cemented sands, clays, and grits of lacustrine and/or fluviatile origin.

The Collie Coal is finely banded band the dominant lithotypes of the coal are dull, dull banded, banded types, with minor bright and bright banded types. The maceral composition of the individual seams is variable, and the dominant maceral groups are vitrinite and inertinite, and the exinite and mineral matter contents are low. The maximum reflectance (RO% MAX) of vitrinite lies between 0.43% to 0.49%, within the range of subbituminous rank of coal. The dominant component of mineral matter is clay, which is usually associated with maceral groups inertinite and vitrinite. The ratio of structured (fusinite and semifusinite) to detrital inertinite (inertodetrinite and micrinite) is described as the "semifusinite" ratio. The semifusinite ratio for the coal is low to medium, due to higher content of inertodetrinite.

The coal types and maceral composition of coal from the Collie Basin are related to the position of the water table in the basin, and on the basis of the abundance of banded, dull banded, and dull coal types, the water table postulated in the basin was mostly fluctuating with occasional low water levels, which accounts for minor fusain bands in the coal. The pattern of sedimentation in the basin was essentially braided fluvial with fluvio lacustrine and peat forming environments during periods of low sedimentation. The vitrinite reflects formation of peat from a woody vegetation within relatively stable and high ground water regimes. The higher content of inertodetrinite in the coal perhaps represents transported material from the marginal areas where greater fluctuate in groundwater conditions would occur.

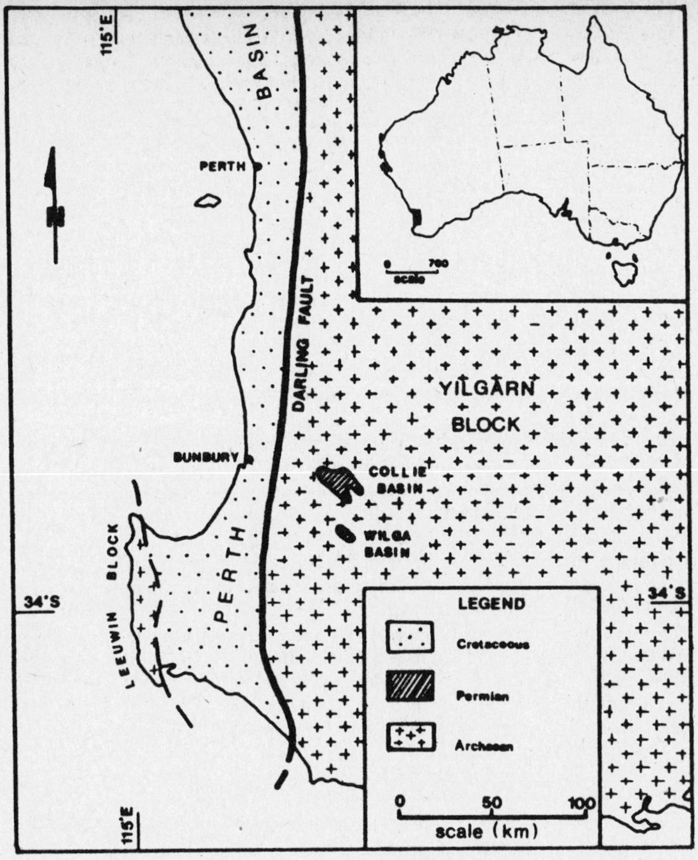

Figure 1—Regional setting and locality plan, Collie Basin.

SARDAROV, S., and K. KAZIEV, Academy of Sciences of the USSR, Daghestan Branch, Makhachkala, USSR

Thermal History of Fault in Sedimentation Basin

An area of stable and prolonged sinking of the earth's crust during which sedimentary rocks are formed is called a sedimentation basin. The sinking could be either the result of sediment accumulation due to isostasy or a cause thereof. An energy loss for warming-up the accumulating sediments results either way. This loss is made up partly by the sinking of foundation into more-heated regions of the earth's crust. Simultaneously, in the process of diagenesis void waters are squeezed out generating a vertical heat-and-mass transfer and mechanogeneous heat sources are created. The process described has been calculated in the rigorous mathematical formulation for a two-layer model with a movable upper boundary with a condition of conjugation at foundation-sediment line and a special condition at the lower boundary with parameters typical of the East-Manytch sagging, Near-Caspian-Kubin depression zone, and Saatli super-deep hole. The thermomechanical effects caused by disturbance of equilibrium of a compacting medium are disregarded in the problem, though in general they could lead to serious errors in determination of temperature distribution under study (Sharp, 1976). The simulation results show that a partial compensation of thermal losses is possible owing to the sinking, which reaches 15–20 mW m^{-2} and overall losses are to be limited by 20–30 mW m^{-2}. The statistical analysis of experimental measurements of the heat flow allows us to estimate its modal value for the East-Manytch sagging at 32 mW m^{-2} and for the whole region, 55.4 mW m^{-2} (Sardarov et al, 1984). The difference agrees well with theory. Still, the process described fails to explain the entire structure of geothermal fields of sedimentation basins.

Their thermal condition is greatly affected by faults along which the foundation blocks may shift with respect to each other. The physical modeling (Sherman et al, 1983) shows that in this case a thick zone of jointing A is formed in the stratum of young plastic sediments. It is located either vertically (shear, normal fault) or in the inclined position (shove fault). In the case of normal and shove faults, this zone is complicated by small sections B and C of near-surface decompaction. On the whole, an area of dynamic influence of the fault (ADIF) is formed in which the thermal and physical characteristics vary strongly from point to point. The laboratory measurements of the cores and in situ studies have shown that for the similar lithological differences the spread may reach 50% and is of random nature. During tectonic activations, thermal sources of comparatively small capacity are created. The mathematical simulation based on the solution of heat transfer equation in randomly inhomogeneous medium has led us to the conclusion that the dispersion of heat flow over the ADIF zone is considerably higher than beyond. The average value of heat flow during tectonic activations is higher and at the platform stage is 20 and more percent less than in the nondisturbed area. The latter fact is confirmed by the analysis of geothermal data on the North American platform, West Siberian platform with the Ural plicate zone, Australian platform, and other regions.

The ADIF and the fault itself represent permeable zones, which is confirmed by anomal flows of helium. The sedimentation and infiltration waters, the flow of fluids of hypogene origin coming up the fault stimulate the convectional heat transfer. Our feeling is that heat flow above the fault should increase by overlapping the above-mentioned mechanisms. However, the detailed geothermal survey has shown that this is at least a nonuniversal phenomenon and that more often the geothermal field is of a mosaic structure. The results on the hydrodynamic instability allow us to explain this fact by convectional self-organization in the saturated permeable medium. It is possible to derive an equation for temperature in the hydrodynamically active ADIF from the set of equations for mass, energy, and momentum (Sardarov and Savina, 1984). For the majority of real faults in sedimentation basins its solution confirms the necessity of formation of convectional cells with a period along the fault extension equal to $2S(S^{2-}H\ L^{-1})^2$, where H is width of the ADIF zone, $L \geq H$ is thickness of the superfractural stratum of young sediments, and $S \geq 2$ is a whole number. The heat flows above the faults in the region of the Karpin swell, Uytash fault in Daghestan, were studied in detail for testing. The measurements were conducted in specially drilled holes 50–100 m deep. The thermal field represented a regulated set of zones of increased and reduced values of geothermal flows.

The systematized model of thermal history of the fault in sedimentation basin has been found useful in working out new methods of petroleum and gas prospecting in tectonically active areas, which was confirmed during their approbation.

REFERENCES

Sardarov, S.S., Jr., and E.V. Savina, 1984, Geothermal field of fault: Doklady Akademii Nauk SSSR, v. 276, n. 5, p. 1091–1094.

———, A.V. Savin, and M.G. Pashuk, 1984, Normal and anomal geothermal fields and their connection with hierarchy of geological bodies: Doklady Akademii Nauk SSSR, v. 275, n. 5, p. 1084–1087.

Sharp, J.M., 1976, Momentum and energy balance equations for compacting sediments: Mathematical Geology, v. 8, n. 3, p. 305–322.

Sherman, S.I., S.A. Berknov, and V.Y. Buddo, 1983, Areas of fault dynamic influence: Nauka Publishers, Siberian Branch, 110 p.

SARDAROV, SARD, and MIKHAIL PASHUK, USSR Academy of Science, Daghestan Branch, Makhachkala, USSR

Stable Oscillations of Gas Emission in Seismic Areas

There are certain cases of a sharp change in gas emission before earthquakes, eruptions, and other dramatic changes in the state of geological medium. However, the oscillations in degassing intensity may have other causes. A geochemical system in tectonically active areas is an open one. Temporal oscillations of its parameters may be determined by external action of initial conditions. These are called "trivial structures." Of the utmost interest are temporal structures formed in the system when a certain parameter determining it exceeds the critical value, and the existence of which bears no connection with the structure of boundary conditions. Such "dissipative" structures represent a qualitatively new kind of organization and exist far from equilibrium due to influx of energy and matter (Prigozhin, 1980). The existence of trivial structures of geochemical fields is a known fact. The analysis of systemic observations of natural gas composition in free jets in the Khibin workings (Hitarov et al, 1979), and in gas and thermal water fields in Daghestan (Sardarov et al, 1982) show the variation of chemical composition with a periodicity of 14 days, 24 hours, and 12 hours that correlate with tidal variations of thermal well yield and are the result of tidal deformations in the earth's crust. The task of singling out other structures from many trivial ones against the background of strong noises necessitated the use of precision methods of analyzing quite a long series of observations.

To estimate the spectral density of temporal series of in-field observations on a computer, a number of algorithms were employed: the maximazation of spectrum entropy in Burg modification (Burg, 1967), autoregression estimate, a number of Fourier transform modifications: the modified Krylov method, Gibbs factors method (Serebrennikov and Pervozvanny, 1965), Fourier analysis with the use of approximation by parabolic splines.

Finally, a unified synthetic algorithm was worked out, which was tested on models adequate to the natural process (Sardarov and Pashuk, 1986).

Analyzed by this method were the results of experimental observations of helium, hydrogen, nitrogen, and methane concentrations in thermal water. The measurements were carried out on a chromatograph positioned right at the well-head for 1014 hours with 1 hour intervals. The following average bulk concentrations were obtained: nitrogen, 57.7; hydrogen, 0.05; methane, 24.7; helium, 0.03 vol%. The seismic situation was calm during the whole observation period. The behavior of autocorrelation functions shows clear 24 h periodicity of all the gases; besides, nitrogen and methane demonstrate a strikingly coherent behavior. The hydrogen oscillations are less stable, its cross-correlation with other gases also being weak. This is associated with the fact that in this case hydrogen is mainly generated as a result of secondary processes in the shaft of the well, e.g., due to reaction of hydrogen sulfide with the metal of casing columns. Such phenomena are well known. The cross-correlation of oscillations of He, N_2, and CH_4 suggests a community of processes leading to formation of the structures despite the difference in genesis and chemical properties. The detailed study enables us to explain the harmonics with periods of 12, 24, and 26 hours by the action of lunar and solar tide and thermal tide because the similar oscillations were also detected in the yields of the flowing wells. Harmonics with periods of 8, 10, 15, 20, 27, 29, and 34 hours manifesting themselves with all the gases and stable for 1014 hours are associated with dynamic degassing.

Several models leading to a stable discrete spectrum may be suggested: pressure auto-oscillations in fluid during filtration into the well; spatial-and-temporal structures in the process of dissolving-to-degassing-to-dissolving, and so on. They all are both likely and possibly far from reality. We tried to find objective evidence that the structures revealed are dissipative ones. The distinctive feature of dissipative structures is the increase of fluctuation level in the evolution of the system toward the point of qualitative restructuring (Burg, 1967). Between the two instabilities the development of the system is governed by deterministic laws. In the vicinity of these points, a substantial role is played by the oscillations having a big amplitude and propagating to large distances. Presuming that an earthquake is associated with the approach to the instability state, we are proceeding from the fact that in this case the fluctuations are to increase and the structure is to change (Chelidze, 1984). The analysis of observations of N_2 and CH_4 for 3 years at two stations belonging to the Geology Institute, Makhachkala, and located in two different seismic regions in Daghestan confirmed this assumption. The correlation coefficient is quite high and reaches 0.93. The seismic activity was estimated by the formula $e^M R^{-1}$, where M is earthquake magnitude, and R is distance from epicenter to geochemical station. The level of fluctuations of geochemical parameters was estimated as σS^{-1}, where σ is root-mean-square value, and S is average value over the interval of σ determination.

Thus, the seismic complication is consequently related to bifurcation of the geochemical field. In the vicinity of the instability point, the formation of a new geochemical temporal structure is preceded by the growth of oscillations. And this shows that the nontrivial structures are of dissipative nature.

REFERENCES

Burg, J., 1967, Maximum entropy spectral analysis: Paper presented at the 37th Meeting Soc. Explorat. Geophys., Oklahoma City, October 31, 1967.

Chelidze, T.L., 1984, Gerlands Beitr. Geophys., Leipzig, b. 93, n. 3, S. 202.

Hitarov, A.I., A.I. Kravtsov, G.I. Voitov et al, 1979, Sov. Geologia, n. 2, p. 62.

Prigozhin, E., 1980, Uspehki Fizitcheskikh Nauk, v. 131, n. 2, p. 185.

Sardarov, S.S., S.S. Sardarov, Jr., and A.B. Mantikov, 1982, Doklady Akademii Nauk SSSR, v. 262, n. 6, p. 1351.

Sardarov, S.S., Jr., and M.G. Pashuk, 1986, Stable temporal structures in geochemical fields, Geokhimia, n. 9, p. 1329.

Serebrennikov, M.G., and A.A. Pervozvanny, 1965, Detection of hidden periodicities: Moscow, Nauka Publishers, 244 p.

SARFATI, J., Center Geologique et Geophysique, Montpellier, France

Stromatolites in Archean and Proterozoic Record: An Example of Biosedimentation.

Stromatolites are organo-sedimentary structures which can be traced from the Early Archean to present-day depositional settings. The way they build a laminated boundstone, by trapping and binding of particles and in situ precipitation of carbonate is now widely accepted but the fact that they are organo-sedimentary structures is not. The emphasis now put on biosedimentation, a complex process, appears very promising to improve stromatolite understanding and trace early evolution of the sedimentation.

Such long lasting models as stromatolites, offering modern counterparts to the ancient, are very attractive. Unfortunately the discoveries of some stromatolite characters, in the Recent have been applied as over-simplified models to ancient stromatolites interpretation. The main feature of Recent stromatolites, used as modern strict analogs to the ancient stromatolites is the rough morphology of the build-up. Comparisons have been made without careful studies of the ancient stromatolite conducing to minimize their other characteristics.

Advances in the study of Archean and Proterozoic sedimentary basins, using stromatolites and sedimentary facies together, have revealed the incredible diversity of stromatolite shapes and environmental setting. Researchers in the field of microbiology, organic chemistry. isotopic fractionation from recent microbial mats communities have shown the incredible complexity of their metabolic activity in response to environmental changes. The simple analogies proposed in the past and still presented that are based on a simplistic linkage of some features cannot be sustained.

Four examples, from Lower and Upper Proterozoic stromatolites which have been studied with such attention will be discussed here. Two are giant stromatolites; one built by columnar forms (Little Dal formation, McKenzie Mountains, Canada, Aitken, 1981) end the other from , mainly flat laminae in low relief domes (Schisto-calcaire formation, Congo, Western Africa, Bertrand-Sarfati end Milandou, 1988). The other two examples are biostromes of columnar stromatolites, displaying a great variety of columnar shapes. One is also from Little Dal and the other from Atar Group, West African Craton (Bertrand-Sarfati and Moussine-Pouchkine. 1988). All these stromatolites have been interpreted as growing in subtidal, low to relatively low energy settings. Three are located on the higher part of a ramp-model sedimentation basin, one more peculiar is attributed to an almost completely flat ramp, a cratonic area of sedimentation. These four sedimentary basins seem to differ by the amount of sediment input. At this point of the study, it is impossible to give an environmental reason for explaining the fact that they display such a diversity in stromatolites gross morphology (bioherm or biostromes), column shape (parallel ramified oblical etc.), and microstructure. To get further in the study of sedimentation, using sophisticated methods (isotopes, organic biomarkers etc.), a precise understanding of stromatolite features and their relations to sedimentation will be needed. One must look carefully at all the stromatolite and sediment features around or within the build-up as well as facies analysis of the encasing rocks, in order to try to come back to the recent examples without prejudices and then try to interpret the ancient by the knowledge of mechanisms from the Recent.

Recent studies have outlined the role of microbial mats in the precipitation of carbonate particles (Monty, 1986 ; Trichet et al., 1988 ; Hahn, 1988 ; Krumbein et al., 1988) as homogeneous micrites or round biospheres both with radiating or concentric layers. This point may be of great interest in the study of mechanisms of growth in ancient stromatolites in that it may permit discrimination between the microbial micrite and the reworked micrite, at laminae scale.

Some conclusions can be drawn from that review: (1) The discoveries made through the study of the Recent must be used as clues to understand the ancient not as shackling models that result in forced modern analogs. (2) The role of microbial communities must be emphasized in the stromatolite boundstone features. It is more or less admitted that microstructures in ancient stromatolites may reflect the best the microbial activity. The columnar shape of the stromatolites may also reflects the microbial community growth habit. Furthermore, all features of stromatolites remain under control of both the environment conditions and the biological activity of microbial communities. (3) When trying to go further in the knowledge of environmental conditions by using more sophisticated methods, one must rely on the careful study of stromatolites and their very proximate environment. It is the only way to improve the Proterozoic stratigraphy.

Stromatolite characteristics are the result of a cooperation, or a competition between two different but both essential processes: one is a sedimentary process, the other a biological process. We cannot go further in understanding the way life appears and respect this fundamental fact.

References

Aitken. J. D., 1981. Stratigraphy and sedimentology of the Upper Proterozoic Little Dal Group, Mackenzie Mountains, Northwest territories. in "Proterozoic Basins of Canada," Campbell editor, Geological Survey of Canada, pap. 81-10, pp. 47-71.

Bertrand-Sarfati, J., and Moussine-Pouchkine, A., 1988. Is oratonic sedimentation consistent with available models ? An example from the Upper Proterozoic of the West African craton. in "Aspects of Proterozoic sedimentary geology," J. Jackson editor, Sedimentary Geology, 58, pp 255-276.

Bertrand-Sarfati, J. and Milandou, R., 1988. Sedimentation on a carbonate ramp, related to an Upper Proterozoic intracratonic rift : the "Schisto-calcaire" (Niari, Congo). Geology, in press.

Höhn, A., 1988. Calcium carbonate deposition associated with laminated algal mats and sediments. Lagoa Vermelha, R.J., Brazil (Abst.). Terra Cognita. 8, p. 226.

Krumbein, W.E., Dyer, B.D., Gasiewitz, A., Gerdes, G., Brochwicz-

Lewinski, W. and Schellen-Huber, H.J., 1988. Microbial matsiridium-events-fossil biospheres-an overview of geophysiology (Abst.). Terra Cognita, 8, p. 227.

Monty, Cl., 1986. Microbial dolomites. 12th International Sedimentol. Congress, Canberra, Abst., p. 215. Trichet. J., Defarge, C. and Chambers, L.A., 1988. Kopara from Tuamotu Archipelago: a possible model for interpreting fossil stromatolites (Abst.) Terra Cognita, 8, p. 229.

SARKAR, S. C., Jadavpur University, Calcutta, India

C 17. Evolution of Ore-Forming Processes Through Geologic Time: The Indian Scenario

Some broad changes in orogenetic processes through geologic time is a common observation. In any fresh effort to bring out a generalized picture, it is necessary to piece together the regional or even the local pictures. In that context it may be worthwhile to sketch the Indian scenario as it appears today. For brevity and relevance, the discussion will be limited to gold, uranium, iron, manganese, lead, zinc, copper, tin and tungsten.

Major gold deposits of India occur in differentially evolved komatiite-tholeiite (Kolar, 800 t Au) and tholeiitic rocks (Hutti and Ramagiri) of the Archean (2.7 Ga) greenstone belts of South India. Ore mineralization took place in ductile-brittle shear zones from metamorphogenic H_2O–CO_2 ore fluids in which Au must have been in S/S—As complexes (1). Weak mineralization of the metal is reported from the contemporaneous 'Dharwar type' and older 'Sargur type' greenstone belts. Archean (> 3.2 Ga) Kunderkocha deposit in South Singhbhum, eastern India, is also small. The picture more or less conforms with the one proposed by Groves et al, (2) for the Archean deposits elsewhere. Small deposits of gold occur in the early Proterozoic north Singhbhum mobile belt and mid-Proterozoic Sakoli Group rocks in central India where volcanic rocks are intimately related to the ore zone.

Although lean uranium mineralization is recorded from the quartz-pebble conglomerate at the bottom of the Dharwar Supergroup (3 Ga), substantial mineralization developed as veins in the metamorphic tectonites of the Singhbhum Cu-U belt during 1.6-1.7 Ga (1,3) temporally closed to the first appearance of large vein deposits of uranium in Canada, Australia and China. Low temperature (200 °C) hydrothermal vein deposit occurs within S-type granitic rocks at Dhabi, central India (800 ± 20 Ma) (4).

BIF-related, broadly Lake Superior type, moderately large iron ore deposits of Gorumahisani-Sulaipat-Badampahar belt appear older than 3.12 Ga old granite intrusive into it (5). Paleontological studies suggest a Proterozoic age for the Jamda-Koira valley ores also Lake Superior type. Boiladilla ore could also be early Proterozoic (6,7). Archean Bababudan ores in South India are Algoma type and ores of Hospet-Bellary-Sandur belt broadly resemble Lake Superior type. Relatively small, amphibolite-granulite facies metamorphosed, BIF bodies in the deep Archean crust of South India appear to have been deposited in the shallow basins at the continent margin (1,8). MnO up to 7.6% in these BIFs occur in silicates only. The only Mn-ore deposit in India, the Archean age of which not generally doubted till today, is that of Sandur in South India. The best developed Mn-deposits of India are in the Proterozoic of central and eastern India. Most of these Mn deposits are not closely associated with volcanic rocks. It is not easy to locate the source of Fe and Mn for the Indian deposits. It could be ocean waters, aided and abetted by volcanism in cases. Lesser concentration of manganese relative to iron into deposits at any time could be explained by their differential distribution in the crust, their unequal oxygen-demand and the supply of oxygen. The development of sedimentary Fe-ore deposits in the Archean suggests that the then O_2-content in the atmosphere (photolytic origin?), though much lower than PAL, must have been sufficient to oxidize iron for its precipitation into moderately large deposits.

India does not have a Ni (–Cu) sulfide deposit, unlike its Gondwana days partners, Australia and South Africa. Due to mantle heterogeneity ? Neither any Cu-Zn (with Au and Ag) type Archean massive sulfide deposit, common in Canadian greenstone belts is reported. A reason could be poor differentiation in the Indian greenstones. REE studies (9, 10) have shown that even amongst differentiated volcanic associations the ore-bearing ones possibly have a different genetic history. Rather a relatively small Cu-sulfide deposit at Aladahalli and Pb-Zn-Cu deposit at Mamandur, highly metamorphosed along with the host sediments are reported from South India. Volcanogenic Archean Cu-sulfide deposits at Kalyadi and Chitradurga are also small (1).

Moderately large (> 100 mt, 1.4% Cu) volcanogenic Cu-sulfide deposit developed along the Singhbhum Cu-belt in the early Proterozoic (2.3 Ga). Almost at the same time (2.3 Ga) developed a Vein type Cu-sulfide deposit (100 mt. 1.23% Cu) in tonalitic granodiorite rock ('I type', 'magnetite series') at Malanjkhand, central India. Stratabound, principally sediment-hosted, Cu-deposits along the Khetri belt, Zn (–Cu) deposit at Rampura Agucha, Zn-Pb deposit along the Zawar belt and the Zn-Pb-Cu deposit at Rajpura-Daribo, all in Rajasthan, NW India, developed during 1.6-1.8 Ga.Pb in the ores is mainly radiogenic (11).

Tungsten as scheelite developed in high $a_{Ca}2+$ zones right from the Archean (Kolar and Hutti gold deposits) to the late Proterozoic-early Phanerozoic (carbonate rocks in the Himalaya). But as wolframite developed only with middle to late Proterozoic granitic intrusives at Degana and Sirohi (Rajasthan), Agargaon (Madhya Pradesh) and Chhendapathar and Porapahar (West Bengal). Sn (–Nb, Ta) mineralization in granitic pegmatities (Lower/Middle Proterozoic) in Baster and Koraput, Eastern India, and A-type granite (735-750 Ma) rocks at Tosham, Hariyana (12) are what are expected in an evolved crust.

SARKAR, S. C., Jadavpur University, Calcutta, India

Plate Tectonics and Precambrian Metallogeny: The Concept and the Constraints

There is a conceivable urge amongst the earth scientists to explore the application of the plate tectonic concept for the explanation of the Precambrian geology and metallogeny. But they are baffled by the fact that the observations supporting the concept are so numerous for the Phanerozoic tectogens and these are comparatively so few in most cases of the Precambrians. How to proceed in such situations? The author makes a paradigmatic discussion with respect to two important Proterozoic metallotects of India, the Aravalli-Delhi belt(s) of the Rajasthan, NW India and the North Singhbhum belt of Eastern India. Both are located at "continent margins."

Rajasthan. An intracratonic thermal plume-generated rift-basin developed in the late Archean in Bhilwara on the Banded Gneissic Complex (BGC) basement (3.5 Ga). Sediments, derived dominantly from the felsic country rocks and some from ocean-floor type basic rocks in the rift, filled the basin and subsequently deformed and metamorphosed into amphibolite-granulite facies. In the Aravalli-Jharol belt the dolomite with stromatolites and phosphorites represent shelf facies and the pelites in the Jharol area, a deep water facies, the two together constituting passive continental margin deposits. The Delhi belt with a substantive development of arenaceous, argillaceous and carbonate sediments in general and low-K tholeiitic basic volcanics in the southern part and 'ophiolite suite' at Phulad, has been suggested to be an island arc with a foreland to the west and the plate movement towards east.

Strata-bound Rajpura-Dariba, Rampura-Agucha and Pur-Banera Zn-Pb-Cu deposits formed in the Bhilwara basin. The Zawar belt Zn-Pb deposits developed in a fault-bonded basin on the passive continental margin. The Basantgarh-Golia Zn-Pb-Cu deposits in the southern part of Delhi belt are volcanogenic, and could be arc-related while the tectonic settings of Khetri and Saladipura deposits in northern Delhi is still obscure. Available Pb-isotope data suggest metallization in most of the above cases took place during 1.6-1.8 Ga ago. In Basatgarh-Golia it was about 1.1 Ga (1). The W-mineralization at Degana and Sirohi and Sn-mineralization at Tosham are related to granitic rocks (735 Ma).

A better appreciation of the plate tectonic concept in this context will need clarification of (1) How sedimentary (-diagnetic)/volcanogenic base metal deposition in the Delhi Supergroup could be spread over a period of about 700 Ma (1800-1100 Ma) and granitic activities spread over 1 billion yeras (1700-735 Ma) ? An episodic development in an otherwise long continuous process? (2) The metamorphic history in the Basantgarh-Golia section where the geochemistry of the basic volcanics suggests a volcanic arc, shows a clockwise P-T-t trend (4) while in the north, where the volcanics are insignificant, it is reverse.

Singhbhum. North Singhbhum mobile belt in Eastern India is a 200 km long, almost E-W trending belt, about 100 km at its widest. The original age of some of its early basic volcanic rocks (Dhanjori) is dated to be 2.3 Ga and the peak of tectono-thermal evolution, 1.6-1.7 Ga. Lithotectonic domains, south to north are: Singhbhum Granite proto-continent (3.2 Ga), dominantly tholeiitic Dhanjori Formation, shallow water psammopelitic sediments of the Chaibasa Formation, high MgO (18-20%) komatiitic tholeitite-dominant Dalma volcanics, deep water sediments, and then the Chhotonagpur Granite Complex (1-1.6 Ga). Metamorphism is Barrovian type. A couple of Plate tectonic models have been suggested but none appears viable on the strength of geological information presently available. As a result the well known Cu-U minralization(s) and a couple of small ones are still generally explained independently of the plate tectonic models, suggested for the Phanerozoic tectogens.

SARKAR, S., and K. BOSE PRADIP, Jadavpur University, Calcutta, India

Evolutionary Record of Epicontinental Basin: Proterozoic Penganga Group, India

Penganga Group, one of the several packages of Proterozoic formations in intracratonic basins, separated from each other by highs in granitic basement in Godavari Rift valley system, India, is little known for its sedimentation aspect. Its intra- and interformational successions record progressive marine transgression. Direct emplacement of shelf sands with clear fingerprints of alongshore and tidal currents as well as waves is possibly due to eustatic rise in sea level.

Vertical superposition of these shelf sands by micritic limestone formed well below the wave base reflect a sharp rise in relative sea level. Lateral juxtaposition of the shelf siliciclastics with thick basinal limestone, observed in drill cores, implies large-scale differential subsidence of two crustal blocks on the onset of rifting, induced in all probability by thermal stress. Subsequent progressive rise in relative sea level was, however, comparatively much slower. Upward change, rather rapid, from pink to grey limestone accompanied by sharp rise in the content of organic carbon attests to shifting of depositional loci across the pycnocline. Gradual transition, further up, from grey to black limestone in conjunction with enrichment of sediments in pyrite and the gradual passage of the limestone into red clay rich in volcanic fines and striped with millimeter-thick lighter carbonate turbidites mirror ongoing transgression of the sea.

Abrupt rise in relative sea level concomitant with onset of rifting not only terminated terrigenous supply but also caused paleogeographic shifting from shelf to deep starved basin. Dissolution seams omnipresent throughout the limestone sequence are invariably present also within autoclasts of synsedimentary mass flow deposits. This implies generation of dissolution of carbonate sediments on or near the seafloor. Deposition is likely to have taken place within the lysocline zone, and that is corroborated by increasing silica content up the limestone sequence. Further support comes from abundant evidence of pressure welding between limeclasts within the conglomerates, including the matrix-supported ones, implying delayed cementation in pore water relatively undersaturated with calcium and carbonate ions owning to high pressure of carbon dioxide and/or low temperature.

Eustatic rise in sea level might have played a role in the long run, but tectonics were, indeed, the overriding factor for evolution of the Penganga basin. Although subsidence of the basin was evidently considerable during initial rifting, the basin extended in subsequent phases dominantly in horizontal dimension. This is clearly attested by innumerable listric faults of various scales in the limestone sequence. Translation of strata took place consistently northward. In the direction of translation, the initially steep slip planes rapidly turn bedding parallel and in faults of shorter lateral extent, transformation from normal to low angle reverse nature is evident (Figure 1). The reverse faults are accompanied by generally north-vergent asymmetric and box folds. The striated slip planes are characteristically covered by sheets of chert and manganese. Along major fault planes, dominantly lamina-parallel manganese mineralization, possibly through hydrothermal replacement, generated downslip wedging ore bodies of considerable economic prospect. Such major faults often induced mass failure resulting in downcurrent wedging conglomeratic sedimentary prisms in the proximal parts and muddy turbidites in the distal. Northward orientation of steepest part of the faults, striations on them, northward wedging of scree deposits, and orientation/imbrication of clasts imply extension of the basin in north-south direction. The listric faults, irrespective of their scale, are hardly expected to affect the basement. Omnipresent truncation surfaces and thinning of beds on them in the limestone sequence elicit syndepositional tectonics and dilation cracks in chert. Common evidence of fault block rotation and local concentration of antithetic faults evoke its extensional mode.

Lateral extension of the rifted basin concomitant with progressive relative rise in sea level, in conspicuous absence of pillow lavas and plutonic dykes mirrors, in essence, an aulacogen pattern of basin evolution. Downdip, Pengenga Group abuts against a limestone replete with detached stromatolite columns with preferred leaning along a poorly exposed fault. Neither the stratigraphic affinity of the limestone nor the nature of the fault is yet known. Overprint of postdepositional disturbance and compression in forms of faults of four distinct phases and multigeneration stylolites is also detectable within the limestone sequence in the Penganga Group. Nonetheless, the course of terminal orogeny reverting the trend in Penganga basin evolution is still a matter of speculation.

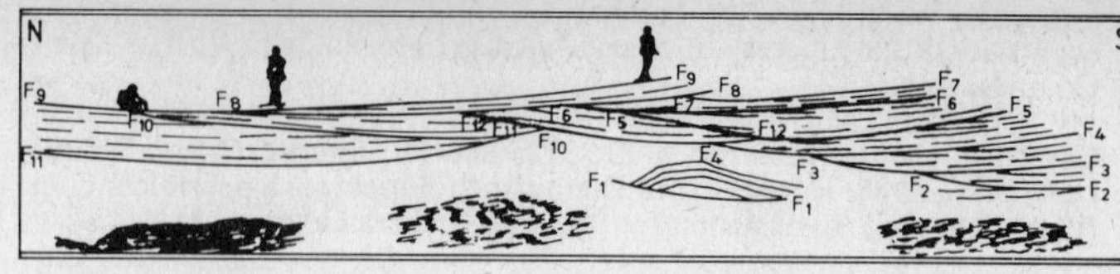

Figure 1—Traces of successive fault planes, steeper and normal at south and reverse at north. Mark F-F lines and fold on F_1-F_1.

SARMA, V. V. J., Andhra University, Visakhapatnam, India

Baseline Groundwater Quality Studies in Tammileru Drainage Basin, India

An analysis of the geohydrochemical data obtained from Tammileru drainage basin, located on the eastern ghats of India, has resulted in providing not only the baseline data under normal conditions but also the changes in the groundwater quality due to normal and artificial interferences. A total of 887 water samples were collected from existing open and bore wells during the pre and post monsoon periods over a three year duration. The samples were analyzed for their chemical ionic characteristics and the data is processed using a stochastic approach. The influence of geomorphic and geohydrological features as well as the regional geological features are taken into consideration in arriving at the baseline data. The data is standardized either in the form of characteristic ionic concentrations or ionic ratios. It is shown that the results can be used as a tool to identify the subsurface geological formations and also alienate the residual anomalous zones caused due to natural or artificial activity of local origin. The method obviates the need for strenuous subsurface studies.

SARMAH, S.K., Gauhati University, Assam, India

Seismicity and Seismic Risk in Northeast India

The northeastern part of India, bounded by 22°–30°N latitude and 89°–98°E longitude, is seismically very active. Historical documents indicate the occurrence of destructive earthquakes in the region in the years 1548, 1596, 1642, 1663, and 1696–1714. In modern times, destructive earthquakes have occurred in 1869, 1875, 1897 ($M = 8.7^+$), 1918 ($M = 7.6$), 1930 ($M = 7.1$), and 1950 ($M = 8.7$) in the region. Since 1969, occurrences of high-magnitude earthquakes in the region are gradually diminishing.

A study of the distribution of all available earthquake epicenters in the region shows that the distribution is not uniform throughout the region. Most of the epicenters do not seem to be associated with known faults of the region, nor do they follow the major lineaments of the region. Khattri et al, (1987) have identified a narrow region between the epicenters of the 1897 and 1950 earthquakes where the pericentral density as well as the rate of occurrence of earthquakes is low. They termed it as "Assam Gap." However, the whole region shows some small elliptical gaps, at the boundary of which the epicenters lie and within which no earthquake is found to occur. On the basis of earthquake occurrence, the whole region can be divided into six tectonic blocks: Eastern Himalaya, Naga Hills and Patkai Synclinorium, Arakan–Yoma ranges, Shillong plateau and Mikir Hills, Brahmaputra valley, and Bengal Basin including Surma valley. The first four are highly active, and the last two are the least active blocks of the region.

Statistical calculation of the return period for various magnitude earthquakes have been made for the region under consideration by using magnitude-frequence relation of Gutenberg and Richter and extreme value method of Gumbel (Type I and Type III distribution). On the basis of the available earthquake catalogs of the region up to 1984, it is not possible to predict the average return periods of very large earthquakes with certainty. However, the estimate methods tend to indicate that most probable largest magnitude annual earthquake that may occur in the region is close to 6. The most probable maximum magnitude earthquake that may occur in a 50 year period is of magnitude 8 or greater.

By using the data from the earthquakes recorded by local seismic networks, the temporal variation of the compressional wave velocity (V_p) and shear wave velocity (V_s) ratio have been studied for earthquakes of magnitude less than 6. It has been found that before each relatively high magnitude earthquake, the velocity ratio undergoes diminution by a maximum of about 10–40% of its normal value, which is again regained just before the earthquake. The relationship between the duration of the anomaly (T) of compressional and shear wave velocity ratio, and magnitude (M) of the earthquake is found to be $\log T = 0.49M - 0.72$, for the region. Here T is in days. The relation is similar to that suggested by Whitcomb et al (1973). This shows that dilatancy occurs before the earthquakes of this region. The extrapolation of the results to the high-magnitude range of 8.0–8.9 gives V_p/V_s anomaly duration period ranging from 4.5 to 12.5 years. This seems to agree with the seismic quiescence period of about 10 years before the earthquake of 1950 (M = 8.7), which occurred in this region. Gradual decrease of earthquakes with magnitude greater than 6 since about 1969 in the region, observed reduction of P wave velocities of the earthquakes recorded at Shillong observatory in some directions of the region by Gupta et al (1980) since about 1969, and gradual reduction of lower magnitude earthquakes in recent times probably indicate that the region is currently undergoing dilatancy—a preparatory stage for the occurrence of a big earthquake.

These observations, coupled with the results of the statistical studies of return period of high-magnitude earthquakes, indicate the possibility of occurrence of a high-magnitude earthquake in the region in the near future.

By utilizing all available earthquake data for magnitudes greater than 4 from 1897 to 1984 for the region, seismic risk in terms of probabilistic estimate of maximum ground acceleration to be expected has been studied by dividing the region into a number of elementary sources. From distribution of earthquakes in these elementary sources, the distribution of acceleration is computed for each grid center of $1/2^0 \times 1/2^0$ area of the region. Probabilistic acceleration contours with 25, 50 and 100 year return periods are then drawn. The results indicate that the Shillong plateau, the eastern part of the Eastern Himalaya, and the Indian–Burma border are high-risk regions, whereas Brahmaputra valley and Bengal Basin including Surma valley are relatively low-risk regions.

REFERENCES

Gupta, H.K., and V.P. Singh, 1980, Teleseismic P wave residual investigation at Shillong, India: Tectonophysics, v. 66, n. 4, P. 19.

Khattri, K.N., 1987, Great earthquakes, seismicity gap and potential for great earthquake disaster along the Himalayan Plate Boundary: Tectonophysics, v. 138, p. 79–92.

Whitcomb, J.H., J.D. Garmany, and D.L. Anderson, 1973, Earthquake prediction: variation of seismic velocity before the San Francisco Earthquake: Science, v. 180, p. 632–635.

SARMAN, R., and D.F. PALMER Kent State University, Kent, OH

Application of Radio Frequency Method to Study Dielectric Behavior of Rocks under Uniaxial Compressive Stress

The main purpose of this research is to provide experimental foundation for developing a stress estimating radar (SER) that can be used to locate highly stressed zones prior to the structural failure. The basis of the experiment is the measurement of the stress dependence of the dielectric properties of rocks.

Most solids behave as conductors at low frequence alternating current and as dielectrics at high frequencies (Houck, 1984). At high frequencies, the dielectric constant decreases with increasing frequency until about 1 MHz, beyond which it becomes constant (Sen, 1981; Zheluder, 1971). Laboratory studies to determine the effect of stress changes on the dielectric constant of rocks and minerals include those of Parkhomenko (1967) and Decker and Zhao (1983). These studies concentrated on the behavior of materials under very high temperature and pressure conditions similar to those encountered at great depths in the earth. It is observed that as pressure increases, the ratio Ep/Eo (where Ep and Eo are the dielectric constants at pressure P and at atmospheric pressure, respectively) increases sharply at low pressure and then stabilizes at higher pressures.

In an effort to evaluate the dependence of dielectric constant on state of stress, we have measured the transmission of radio waves through limestone and sandstone at variable loading. The present study was conducted using square wave modulated electromagnetic signals from a Hewlitt-Packard sweep oscillator operating at 19–22 GHz frequency range. The signal was transmitted using wave guides attached to aluminum plates that were flush with the sample. Samples were shielded with aluminum foil to limit external sources of radiation. Both transmitted and reflected signals were recorded using a Hewlitt-Packard scalar network analyzer (HP8756A). Measurements were guided and the data were processed and stored using a Hewlitt-Packard computer.

Samples were stressed under uniaxial compression and measurements were made at several predetermined load increments reaching beyond the microfracture state of the rocks. The load level at which the microfractures developed was indicated by a slight drop in the rate of loading shown on the compression machine. Tests were conducted on samples of massive limestone and thinly bedded Berea sandstone. Compressive strength of the samples was determined using the point load test. Results of this test show that compressive strength of limestone is about 88 MPa ($\approx$13000 psi) and that of the sandstone is about 24.8 MPa ($\approx$3600 psi). Reproducibility was assessed by making multiple measurements at predetermined load intervals while first incrementally increasing the load and then while decreasing the load. The results show that the experimental error was small (5%) compared to the changes in the dielectric constant due to the overall increase in stress. The procedure also showed that within the elastic limit of the sample, the electrical properties were independent of the loading sequence. Since the reflected signal did not change significantly in response to changes in stress, only the transmitted signal was used for analysis.

The experimental results show the effects of the constructive and destructive resonance generated by standing waves within the sample. The output signal received is the net result of the resonance. The curves shown in Figure 1 give the logarithmic difference (Ep/Eo) between the transmitted signal received at different loads and the reference signal at 100 lbs load. Limestone was stressed to a maximum of 44 MPa (40,000 lbs load), which is about 50% of its uniaxial compressive strength. Microfracturing was developed at about 41 MPa ($\approx$37200 lbs), which is approximately 46% of its uniaxial strength. The curves on the left show that there is a gradual increase in Ep/Eo until 35,000 lbs load (38.5 MPa) and then the ratio increases suddenly with the development of microfractures. A similar trend of increase in Ep/Eo is observed in the case of Berea sandstone (right-hand curves) where the microfracturing was developed at 21,234 lbs. (17.4 MPa, about 70% of the compressive strength). This is indicated by a sharp increase in the amplitudes of curves at 22,000 lbs (18 MPa), and then the amplitude becomes independent of the stress increase even when the load is close to the peak stress condition. Both sets of curves indicate a gradual shift in the location of peak toward higher frequencies before microfracturing. After microfracturing, the peaks shift toward the lower frequency. This indicates a gradual phase shift along with the amplitude, caused by the changes in stress conditions.

Changes in the transmission characteristics of EM radiation measured here indicate a significant change in dielectric constant at 50–70% of the peak strength. These results offer real promise for the development of a ground-penetrating radar system that can be used to measure changes in stress conditions in near-surface materials such as unstable slopes and embankments and underground structure.

REFERENCES

Decker, D.L., and Y.X., Zhao, 1983, Changes in the $BaTiO_3$ phase transition to 40 KBar, *in* High pressure in science and technology, Part 1: New York, North-Holland, p. 179–182.

Houck, R.T., 1984, Subsurface imaging with ground penetrating radar: Unpublished Ph.D. Dissertation, Pennsylvania State University, 172 p.

Parkhomenko, E.I., 1967, Electrical properties of rocks: New York, Plenum Press, 314 p.

Sen, P.N., 1981, Relation of certain geometrical fractures to the dielectric constant of rocks: Geophysics, v. 46, n. 12, p. 1714–1720.

Zheluder, I.S., 1971, Physics of crystalline dielectrics: New York, Plenum Press, v. 2, 620 p.

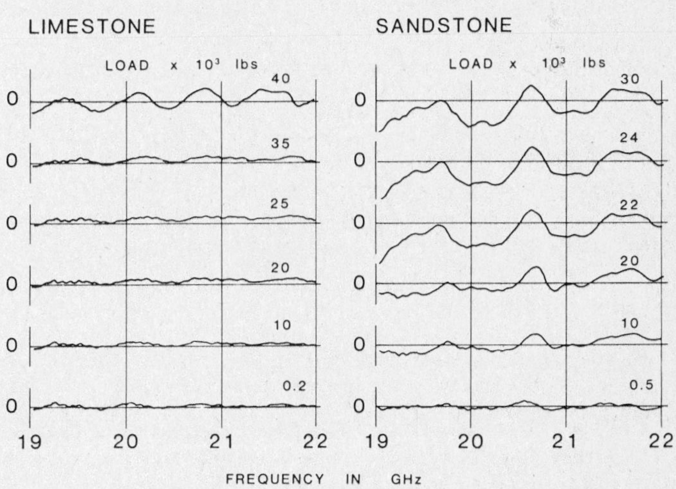

Figure 1—Logarithmic difference between the transmitted signals at different load levels and at 100 lbs load. Left: massive limestone; right: thinly bedded Berea sandstone.

SARNTHEIN, M., and R. TIEDEMANN, Uiversität Kiel, Kiel, Federal Republic of Germany

Sea Level and Temperature Fluctuations During Last 5 M.Y.: Oxygen Isotope Records from Eastern Atlantic

In recent years, several $\delta^{18}O$ records of benthic and planktonic foraminifera were produced from the mid- and low-latitude eastern Atlantic, which covers parts or the whole of the Pliocene (e.g., Shackleton et al, 1984; Keigwin, 1986; Sarnthein and Tiedemann, in press; Curry and Miller, in press). The records form a vertical paleo-oceanographic transect from the sea surface to 2300–4750 m water depth. Their sampling density provides a time resolution, which is better than 7–12 k.y. per datum and comes close to that of the Quaternary.

Based on a comparison of records from different water masses, the signals of temperature and global ice-volume fluctuations are tentatively distinguished in the isotopic curves. We surmise that the major long-term $\delta^{18}O$ events at 2.0, 2.6/2.65, 3.2/3.25 Ma, and possibly at 4.0/4.1 and prior to 4.7 Ma are linked to ice-volume and sea-level fluctuations. A climatic "Golden-Age" marked the time span 4.0–3.25 Ma, when the range of short-term benthic $\delta^{18}O$ fluctuations was smaller than 0.5–0.7‰ and the average $\delta^{18}O$ level was about 0.25‰ lighter than during the Holocene. Hence, the sea level was accordingly higher for an extended time span. Subsequently, the range of short-term variability increased up to 1.0‰. At 2.65–2.0 Ma it went up to 1.25‰ and in rare cases (2.4 Ma) up to 1.6‰. This value almost corresponds to the range of Quaternary glaciations. The time span 1.97–1.85 Ma formed a short-term exception from the long-term trend in showing only small-scale fluctuations of up to 0.6‰ that ranged slightly below the Holocene level.

The high time resolution of several records permitted time series analyses of the short-term cyclicity, which indicates the well-known periodicities linked to orbital forcing such as during the Quaternary (e.g., Imbrie et al, 1984).

REFERENCES

Curry, W.B., and R.C. Miller, in press, Oxygen and carbon isotope variation in Pliocene benthic foraminifera of the equatorial Atlantic, *in* W.F. Ruddiman, M. Sarnthein, J. Baldauf, et al, eds., Proceedings Initial Report of ODP-Leg 108, Part B: U.S. Government Printing Office.

Imbrie, J., J.D. Hays, D.G. Martinson, et al, 1984, The orbital theory of Pleistocene climate: support from a revised chronology of the marine $\delta^{18}O$ record, *in* A. Berger and J. Imbrie, eds., Milankovitch and climate: D. Reidel Publ., p. 269–305.

Keigwin, L., 1986, Pliocene stable-isotope record of Deep Sea Drilling Project Site 606: sequential events of ^{18}O enrichment at 3.1 Ma, *in* W.F. Ruddiman, R.B. Kidd, E. Thomas, et al, eds., Initial Reports of DSDP, v. 94: U.S. Government Printing Office, p. 911–920.

Sarnthein, M., and R. Tiedemann, in press, High resolution stable isotope stratigraphy of the last 3.3 million years, ODP Sites 658 and 659 off Northwest Africa, *in* W.F. Ruddiman, M. Sarnthein, J. Baldauf, et al, eds., Proceedings Initial Report of ODP-Leg 108, Part B: U.S. Government Printing Office.

Shackleton, N.J., and the Leg 81 Shipboard Scientific Party, 1984, Oxygen isotope calibration of the onset of ice rafting and history of glaciation in the North Atlantic Region: Nature, v. 307, p. 620–623.

SAROGLU, F., MTA Genel Müdürlüğü, Jeoloji Etüdleri Dairesi, Ankara, Turkey, and Y. YILMAZ, ITÜ Maden Fakültesi, Jeoloji Bölümü. Tesvikiye, Istanbul, Turkey

Example of Triple Junction of Intersecting Faults in Turkey: Karliova Region

The North Anatolian Fault (NAF) and the East Anatolian Fault (EAF) are the two major faults in Turkey. They are active strike-slip faults crossing Anatolia in east-west and northeast-southwest directions. The North Anatolian Fault is a right-lateral strike-slip fault and is approximately 1100 km long. However, the East Anatolian Fault is a left-lateral strike-slip fault with a length of 650 km. These faults intersect near Karliova in Eastern Turkey, so this part of the country will be referred as the Karliova region in this study.

The rock units exposed in the Karliova region are various metamorphics of Paleozoic–Lower Mesozoic; an ophiolitic melange named the Kazan formation and marine clastics referred to as the Keleres formation, both in Oligocene age; the Adilcevaz limestone of reef origin and Lower Miocene in age; the Solhan and Karliova volcanics; the Zirnak formation composed of volcano-sedimentary rocks, all in Upper Miocene-Pliocene age; the Quaternary Bingöl volcanics; and terrestrial deposits of the Boran and Anzar formations. The beginning of Neotectonic period in Eastern Turkey is assumed to be Middle Miocene, and it is characterized by a compressional regime that is still active. The main structural features of the Karliova region are strike-slip faults; the NAF, which trends N65°W, and the EAF, which trends N50°E in this region, are well-defined. However, they are not single faults but are composed of parallel fault segments that develop a fault zone for each. A right-lateral offset of 7.5 km for NAF and a 17 km left-lateral offset for EAF are measured. Evidence indicates that the intersecting of NAF and EAF postdates the Zirnak formation of Lower Miocene age and predates the Quaternary Boran formation.

The EAF is not observed continuing to the north of the intersecting point. But the NAF is continuous to the east for about 30 km, so a triple junction is developed by these faults to the east of Karliova. Besides these two major faults, there are several minor strike-slip faults of various types and sizes in this region. To the west of the intersecting point inside the acute angle of the NAF and EAF, the minor strike-slip faults also have normal fault components. However, outside of this area to the east minor strike-slip faults have thrust components. All these minor strike-slip faults appear to be controlled by the NAF and EAF and are the elements of a conjugate fault system. Because of the movements of NAF and EAF and their being the boundary structures, a lithospheric plate that is called the Anatolian plate and moves westward is developed between them.

The morphologic development of the Karliova region is controlled by active tectonics. Almost all morphological features of the strike-slip faults are observed. Extension is present in the wedge defined by NAF and EAF to the west of the intersection. Outside of this area, compression is present. Therefore, the Karliova basin, which lies in the western wedge, is extending. There are Quaternary alluvial fans developing on the northern and southern boundaries of the Karliova basin. Similar fans outside of the western wedge are also present. The younging direction of these faults is from west to east. The extension of the Karliova basin, which is approximately 18 km long, is assumed to have begun in the late Pliocene–early Quaternary. The geometry and the amount of the sedimentation in the Karliova basin are mostly controlled by the amount and the direction of the movement of the Anatolian plate.

SARTI, M., Universitá della Calabria, Cosenza, Italy, M. BOSELLINI, Universitá di Ferrara, Ferrara, Italy, and E.L. WINTERER, Scripps Institution of Oceanography, La Jolla, CA

Basin Geometry and Architecture of Alpine Passive Margin: Implications of Mechanisms of Rifting

Structural styles in the Southern Alps reflect the mechanisms that extended the crust to form a Jurassic passive margin. The Southern Alps consist of an array of half-grabens, each with a west-tilted floor and bounded on the west by an east-dipping master fault. Antithetic faults divide each half-graben into a western depocentral zone and an eastern, ramping marginal plateau, shallower and less subsident. The half-grabens from west to east are: M. Nudo–Arbostora, M. Generoso–Albenza, West Sebino–Botticino, East Sebino–Trento, and Belluno–Friuli. Typical antithetic faults, commonly with further Cenozoic displacement, include the Ballino and Carnian Prealps fault systems. Timing of half-graben formation shows an eastward younging age progression, suggesting that extension propagated stepwise into unextended crust, over much of early and middle Liassic time.

The continent-facing dip of faults in the Southern Alps is the reverse of the dip predicted by simple symmetrical extension models with only ductile extension in the lower lithosphere (McKenzie, 1978), but accords with models in which low-angle shearing cuts through the entire lithosphere (Wernicke, 1985; Lister et al, 1986) and results in structural asymmetry of conjugate margins. We suggest that the polarity of the observed shallow structures is the consequence of a gently east-dipping master shear cutting through the lithosphere at depth beneath the Brianconnais–Southern Alps margin system.

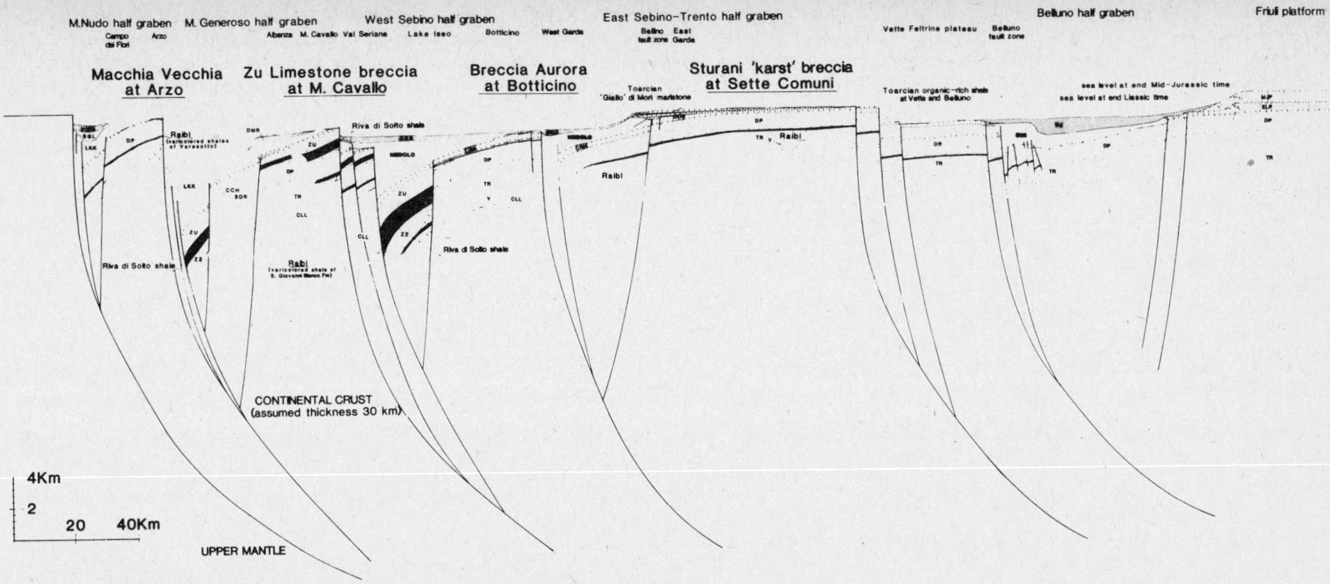

Figure 1—Palinspastic cross-section of the Southern Alps rift basins and marginal high. Vertical exaggeration is 10 times.

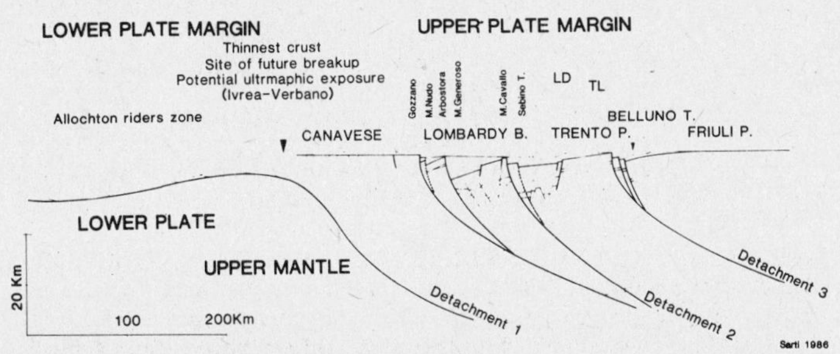

Figure 2—Lithospheric shearing model applied to the Southern Alps margin. Uneven crustal thickness producing multiple rifts across the margin is accounted for by the multiple detachment model. Left portion of the scheme represents the situation in the Brianconnais. Multiple detachment system dipping toward the continent of Apulia accounts for the polarity of normal faults and tilted blocks that face the ocean. The Canavese "outer high" and the Trento plateau zone are zones of thickest crust and latest extension. TL, Triassic–Liassic boundary, that is base of the Liassic rift-stage sequence; LD, Late Liassic seafloor topography.

SASADA, M., Geological Survey of Japan, Tsukuba, Japan

Some Problems of Application of Fluid Inclusion Microthermometry to Geothermal Explorations

Fluid inclusion microthermometry has been applied to geothermal explorations so far in more than 40 geothermal fields of 11 countries. Fluid inclusions in hydrothermal minerals reflect the thermal history of geothermal fields, but the thermal records of fluid inclusions are not always complete. Homogenization temperatures (Th) should correspond to the formation temperatures of geothermal reservoir where the hydrothermal minerals are precipitated, if the pressure correction for trapping temperature is negligible. Actual Th of fluid inclusions in hydrothermal minerals represents the reservoir temperatures in most cases at very active geothermal fields where the borehole temperatures are close to a boiling point curve for water (Tagushi, 1982; Hedenquist, 1983). The pressure-corrected Th may also represent the temperature of aquifer of the cooling reservoir. Th or pressure-corrected Th are, however, not accordant to the formation temperatures, when the geothermal reservoir undergoes the following thermal histories after trapping of the fluid inclusions: (1) hydrothermal veins completely filled with minerals on cooling, (2) transition from hot water-dominated system to vapor-dominated system, and (3) recent temperature increases. The comparison between Th and the final temperature recovery of boreholes gives more accurate thermal history of geothermal reservoirs.

SASOWSKY, I. D., and W. B. WHITE, The Pennsylvania State University, University Park, PA

Evolutionary Pathways of Valley Longitudinal Profiles in Karst

Field observation has shown that valley longitudinal profiles in karst can be characterized on the basis of the presence or absence of slope reversals along the valley thalweg. Slope reversals are reaches where the thalweg has an aspect reverse from the overall trend of the valley. They occur where a sinkhole or swallet is present in the valley floor.

The three categories of profile type are: (1) smooth profiles have no slope reversals apparent at the chosen scale; (2) ragged profiles have slope reversals apparent at the chosen scale; and (3) no discernible thalweg—in some areas, particularly those without surrounding clastic uplands, it is not possible to define a coherent surface valley, and therefore impossible to characterize a profile.

In the simplest sense the difference between smooth and ragged profiles can be accounted for by the predomination of either surface of subsurface erosional processes. By this it is meant that if surface erosional processes predominate that the profile will be smooth because lowering of the stream bed will outstrip subsurface erosion. Alternatively, if subsurface erosion predominates along a certain reach, the surface stream will be captured, leaving swallets and sinkholes caused by underdraining of the valley.

From this simple basis, one can, by evaluating controlling variables in the drainage system, determine numerous pathways leading to observed valley forms. The primary controlling factors are the ratio of high discharge to low discharge occurrences of the surface stream (called "exceedance probability" by Smart, 1988), and the ratio of sediment supply to sediment transporting capability of the karst conduit system. Simply stated: If the surface stream is frequently flowing with significant erosive power, the surface erosion will allow a smooth profile to be maintained. If the surface stream frequently has dry reaches, due either to lack of high flow regimes or efficient underdraining by the conduit, then subsurface erosion may predominate yielding a ragged profile.

The four main variables, expressed as ratios above, are in fact controlled by numerous variables themselves. Some of these are: Bedrock type of both the karst and any uplands, sediment type, climate, basin size, vegetation, structural features (both macro and micro), and relief.

As the controlling parameters are both spatially and temporally variable, the profile type may vary through time in response to both external and internal conditions. The two types of profiles cannot be though of as different "stages" in a Davisian sense, but rather as responses to differing boundary conditions.

REFERENCES

Smart, C. C., 1988, A deductive model of karst evolution based on hydrological probability: Earth Surface Processes and Landforms, v. 13, p. 271–288.

SASSA, K., Kyoto University, Kyoto, Japan

Initiation Process and Moving Mechanism of Debris Flows

"Debris flows" may be used as a collective term for a broad spectrum of phenomena observed in the gravity-induced mass movement of debris according to Chen (1987). However, this definition may include landslides. To distinguish debris flows from landslides and debris avalanches, the author likes to describe debris flows as the gravity-induced movement of debris being saturated as a whole and mobilized as a whole. "Saturated/ mobilized as a whole" does not mean a 100% degree of saturation/ mobilization. It means that saturation and mobilization extend over the whole body. Usually, landslides are partly saturated and partly mobilized. Motion of the well-mobilized but not wholly saturated mass may be called debris (soil) avalanches or slope failures, but those are included in the wider meaning of landslides.

Takahashi (1978) proposed that the initiation of debris flows was caused by the sudden appearance of surface flow having a certain depth. Failure of a natural dam may cause this kind of surface flow. But it does not seem to be so often in the head part of torrent. Monitoring of debris flows in volcanic torrents in Japan has revealed that debris flows are caused when the short-term (10–20 minutes) rainfall intensity exceeds a critical value. The author investigated such debris flow torrents and noticed the existence of many tales, which deposit at the critical angle without pore-water pressure. Therefore, when the rainfall intensity exceeds the permeable capacity of the ground below the talus, a groundwater flow appears above the ground. Then pore-water pressure is generated inside the talus, which should result in the sliding of the talus. The sliding talus rides on the torrent deposit and usually causes the undrained loading and the undrained shear of the torrent deposit. Then, sometimes the torrent deposit is liquefied and changes to a debris flow; sometimes the torrent deposit is sheared and starts to move and, gradually being mobilized, changes to a debris flow; sometimes it moves without the whole mobilization and stabilizes again as a landslide inside the torrent. This process seems to be one of the most probable process of the initiation of debris flows. The critical condition of liquefaction and also undrained shear of the torrent deposit was presented (Sassa, 1986b).

Sassa (1984) has developed a new high-speed ring shear apparatus for the study of shear resistance of debris flow during motion. The test results (Sassa, 1986a) showed that shear resistance due to viscosity of pore water may be usually negligible in comparison with frictional resistance in real disaster-causing debris flows (flow depths are in order of meters), though it may not be negligible in a channel experiment and the Bagnold experiment where the normal stress is only 0.05–2 gf/cm^2. The measured friction angle during motion remained constant in the shear speed of 0.01 cm/s to 100 cm/s, i.e., the increase of speed did not cause the decrease of friction angle. So, the apparent low friction angle of debris flow should result from a high pore-water pressure. A high pore-water pressure may be caused by two mechanisms, (1) Natural mixing of saturated debris during motion inside the debris flow mass causes floatation of relatively fine particles and results in a high density of pore water and a high pore-water pressure. This was confirmed by the experimental results (Sassa, 1985, 1988) that the measured pore-water pressure increased with the mixing velocity of saturated debris inside the double cylinder cell. (2) When debris flows move on the saturated torrent deposit, a high pore-water pressure is caused inside the torrent deposit by the undrained loading and the undrained shear. The pore pressure generation can be estimated by the pore pressure parameters proposed by Sassa (1985, 1988) and the loaded stresses. When pore pressure (1) is greater than (2), the major shear of debris flow is inside the flowing mass. When pore pressure (2) is greater than (1), the major shear of debris flow is inside the torrent deposit. In this case, the debris flow moves, scraping (eroding) the torrent deposit.

REFERENCES

Sassa, K., 1986a, The mechanism of unsaturated debris flows: Annals of DPRI, Kyoto University, n. 29 B-1, p. 315–329 (in Japanese).

——— 1986b), The mechanism of debris flows and the forest effect on their prevention: Proceedings of 18th IUFRO (Int. Union of Forest Res. Organizations) World Congress, Div. 1, v. 1, p. 227–238.

——— 1988, Geotechnical model for the motion of landslides: Proceedings of 5th International Symposium on Landslides, v. 1, p. 37–55.

SASSANO, G. P., Concordia University, Montreal, Quebec, Canada

Genesis of Pyritic Columnar Aggregates and Associated Framboidal Pyrite from the Acton Vale Sector, Quebec Appalachians

In addition to the well-established early-diagenetic mode of formation of framboidal pyrite in mudstone, its late-diagenetic formation is likely in limestones. This concept stems from the apparent absence of framboidal pyrite from Recent carbonate sediments and its occasional presence in carbonate rocks. During deep burial, ferrous iron is commonly available and reduced sulfur is at least occasionally available as demonstrated by the cementation history of limestone, the change in composition with depth of burial of authigenic-carbonates in mudstones and by sulfide-producing processes in zones of hydrocarbon generation. Breakdown and partial dissolution of iron-bearing compounds, mainly in mudstones, at relatively high temperatures constant with deep burial, constitute the major source of dissolved ferrous iron which migrates during compaction into more permeable lithologies such as limestones. Thermochemical reduction of sulfate, mainly in limestones, near the zone of hydrocarbon generation, is probably the major source of HLS; it tends to be removed from associated hydrocarbons by dissolution in migrating waters. Mixing of aqueous iron and sulfide-bearing solutions might lead to the precipitation of iron sulfide; drops of oil trapped in these solutions would concentrate the precipitate which would then assume the spheroidal form of the oil drop and thus attain that typical of framboidal pyrite. Early-syndiagenetic framboidal pyrite in siltstones and shales and late-diagenetic framboidal pyrite in limestones occur side by side in the Cambro-Ordovician sequence of the Acton-Vale quarry (Quebec Appalachians).

The geology of the area has been documented by Lesperance (1979) while Sassano and Procyshyn (1985, 1986) studied the mineralogy of the metallic and gangue minerals occurring in the quarry. The pit itself, about 500 m by 1000 m, has walls that expose a massive, poorly bedded crystalline limestone with an apparent thickness of at least 100 m, including its drill indicated subsurface extent. The southwestern wall is in unconformable contact with a polymictic conglomerate, and is overlain by a finely banded sequence of siltstones and black shales. Mafic metavolcanics cut the lithologies which are also intruded by poorly exposed cross-cutting alkaline microgabbros. In turn, the entire sequence, with the exception of the microgabbros, is cut by several generations of calcite and calcite-quartz veins and segregations. Framboidal pyrite, finely disseminated subhedral pyrite and aggregates of these are present in all sedimentary rocks. Pyrite is most abundant in the siltstone-shale sequence but occurs as well in the limestones being irregularly distributed throughout. In 1988, Sassano and Procyshyn recognized seven generations of sediment-hosted pyrite, at least two of which include framboidal pyrite. The early syndiagenetic framboidal pyrite, is present in a siltstone-shale sequence and in limestones; it is considerably more abundant in the former and in the contiguous limestones than in those lithologies remote from the siltstone-shale sequence. The later carbonate-hosted framboidal pyrite, although morphologically similar to that in siltstones and shales, differs that it displays incipient replacement of chalcopyrite. A third variety of pyrite is present in a unique occurrence of columnar aggregates of upward-growths. Such columnar aggregates composed of pyrite microcrysts, formed along incipient fractures and "vent-like" conduits set within veins filled with calcite, have to our knowledge, not been reported from elsewhere with the exception of Hudson's pyrite stalactites (1982) and Larter et al.'s pyrite chimneys from the Ballynoe barite deposit (1981).

The author ascribes the formation of the columnar aggregates of pyrite and their cogenetic pyrite framboids to a genuine hydrothermal event spatially related to the post-tectonic emplacement of an intrusive and extrusive suite of igneous rocks of the Acton-Vale sector. It is argued that the metal brines derived from the expulsion of metal-rich formation waters, during compaction, were contaminated and set in motion by magmatic solutions emanating from the rising plutons of post-Taconic age.

SASSI, F.P., University of Padova, Padova, Italy

Correlation of Pre-Hercynian and Hercynian Events in Alpine-Mediterranean Belt: Final Results of IGCP Project No. 5

Numerous records of the pre-Hercynian and Hercynian events are still recorded in the AMB (which includes all Alpine chains from Spain-Morocco to Caucasus). However, their reconstruction, sometimes their detection, and in any case their deciphering are difficult due to the strong, multistage, thermal, and deformational reworking of Alpine age.

The description of the pre-Permian records basing on standard criteria, their long-distance correlation, and attempts at their interpretation were the goals of IGCP Project No. 5, in which about 200 geoscientists from 17 countries participated.

These records can be classified in three main groups:

1. *Records of Late Proterozoic processes:* In the eastern part of the AMB, they are represented by situations (sedimentation, metamorphism, magmatism) referred, according to some interpretations, to the Dalslandian and/or Baikalian events; however, some of these situations are not yet well understood, and their chronological frame is not yet well established. In the western part of the AMB, they only consist of few radiometric ages of old zircons incorporated within younger rocks; they only indicate Pro-

terozoic petrogenetic events the products of which were completely eroded and/or recycled.

2. *Records of pre-Hercynian Paleozoic processes:* They are relatively widespread, and mainly consist of:

- foliation-producing, garnet-amphibolite facies, regional metamorphism, and related migmatites (about 460 Ma)
- tectonic activity, recorded in the regional foliation and in micro- to mesoscopic folds
- post-tectonic granitoid intrusions (440–420 Ma)
- post-tectonic acidic volcanism (Upper Ordovician)
- local evidence of a pre-Hercynian uplifting with erosion of the Ordovician granitoids
- sedimentation sequences of various time span in various localities, globally ranging from Cambrian to Carboniferous.

Local metaultrabasites and eclogites are also to be mentioned; their age is unknown, but it is certainly pre-Hercynian.

3. *Records of the Hercynian event:* They are ubiquitous. A two-stage metamorphism (about 350 and 320 Ma), tectonic activity including nappe emplacement, mainly acidic post-tectonic plutonism and volcanism, have been carefully described, particularly in the western AMB. The Paleozoic sedimentation, its fossil content, and its paleoenvironment meaning have also been studied, and represented in special "stratigraphic correlations forms," which make the interregional correlations easier.

All above results (here outlined in a rough way due to the large disproportion between the available pages and the amount of data and size of the area) are presented in the volumes listed below (which can be obtained on request:

- a special volume of Miner. Slovaka (1988), which presents a set of regional papers
- a special volume of Mem. Soc. Geol. Ital. (1989), which shows all Stratigraphic Correlation Forms and their explanatory texts
- seven issues of newsletters

In addition, a booklet (1989) listing all the publications related to IGCP No. 5 (~1000 papers) is also available.

Among the many final results worthy to be reported, let us point out only two of them.

1. The existence in the AMB of an Ordovician thermotectonic event is now not matter of discussion but of geodynamic interpretation. Some authors called it "Caledonian," because it developed in the same time range in which the Caledonian orogeny took place in the Caledonides. However, it is worthy to stress that it also took place in the same time range in which the Pan-African processes largely rejuvenated the African craton; and that probably it can find a more suitable interpretation in the ambit of the Pan-African event.

2. The Hercynian event displays, at least in parts of the AMB, some peculiarities in comparison to the models established in the Mesoeurope. In particular, the post-tectonic magmatism displays geochemical features that suggest significant mantle contributions. Furthermore, the two-stage development of the Hercynian metamorphism recognized in some areas of the AMB, and the possible local existence in it of early-Hercynian eclogites, could suggest new models for the Hercynian event within the AMB.

SASSI, F.P., and C. MAZZOLI, University of Padova, Padova, Italy, C.V. GUIDOTTI, University of Maine, Orono, ME, and J.G. BLENCOE, Oak Ridge National Lab, Oak Ridge, TN

Compositional Control of Cell Volume along Muscovite-Paragonite Pseudobinary Join

The *a*, *b*, and *c* dimensions of muscovite (Mu) and paragonite (Pg) are a function of the Na/(Na + K) ratio (XII substitutions) and replacement of Al^{VI} by Mg, Fe^{+2}, and Fe^{+3} (VI substitutions). Consequently, cell volumes of Mu and Pg reflect the cumulative effects of these substitutions on *a*, *b*, and *c*. The degree to which XII and VI substitutions affects each of *a*, *b*, and *c* differs within a given mica and between the two end members considered herein. The differences arise due to structural and/or crystallochemical factors.

New data plus data from the literature (144 Mu and 24 Pg) enables quantification of most of the effects of the XII and VI substitutions of *a*, *b*, and $1/2c\ \sin\beta$ for both Mu and Pg.

For Mu the following effects have been recognized:

1. *a* decreases as Na content increases, but increases as the $\Sigma(Mg + Fe_t)$ increases. Observed changes of *a* are proportionate with concomitant changes of *b*. However, absolute changes of *b* are larger than those shown by *a*.

2. The maximum decrease of *b* is about 0.034 Å in the Na-richest Mu, corresponding to a shortening of ~0.07% per 1% increase of 100 Na/(Na + K). However, *b* increases by 0.05% per 0.1 increase in the $\Sigma(Mg + Fe_t)$.

3. $1/2c\ \sin\beta$ displays a significant decrease for both XII and VI substitutions: ~0.03% per 1% increase of 100 Na/(Na + K) and ~0.07% per 0.1 increase of the $\Sigma(Mg + FE_t)$.

As with Mu, *a*, *b*, and *c* of Pg are affected by the XII and VI substitutions, but the significantly smaller database and the smaller extent of the VI substitutions in Pg precludes a quantitative assessment.

Available data indicate: 1. *a* and *b* of Pg behave qualitatively in a manner similar to that of Mu in response to VI substitutions.

2. $1/2c\ \sin\beta$ increases for decreasing Na content (~0.005% per 1% decrease of the 100 Na/(Na + K)); and decreases as the $\Sigma(Mg + Fe_t)$ increases (roughly, ~0.4% to 0.6% per 0.1 of $\Sigma(Mg + Fe_t)$.

The net results on unit cell volumes (*V*) of the above variations can be summarized as follows:

1. For Mu the *V* is only modestly affected by the VI substitutions, increasing only modestly due to concomitant decrease of $1/2c\ \sin\beta$ and increase of *a* and *b*. It is strongly affected by the XII substitutions. Increasing Na content produces a significant *V* decrease as a cumulative response to decreases of *a* and *b* and $1/2c\ \sin\beta$.

2. For Pg, *V* is also controlled largely by the XII substitutions, increasing K content resulting in a significant *V* increase due to increases of *a*, *b*, and $1/2c\ \sin\beta$. Because VI substitutions are very minor in Pg, they have undetectable effects on *V*.

Available data on the *V* of Mu and Pg as a function of Na/(Na + K) ratio along the pseudobinary join can be fit to parabolic curves having the equations (in cubic Å);

(A) (for Mu) $V = 935.400 - 0.254\,Na - 0.007\,Na^2$

(B) (for synth. Pg) $V = 1090.805 - 3.760\,Na + 0.016\,Na^2$

Curve A is concave downward, indicating a positive ΔV_{mix}. Curve B appears to be concave upward; however, consideration of: (1) the small data base, (2) the effects of possible analytical errors, and (3) the only small concavity of the curve, also permits interpreting the Pg data in terms of a linear fit (i.e., $\Delta V_{mix} = 0$).

Observations discussed above lend themselves to fairly straightforward interpretations based on well-understood crystal structure aspects (e.g., Radoslovich, 1962; Güven, 1971). Of particular interest, Güven showed that VI substitution decreases markedly the α rotation. This results in enlarged XII sites. Guidotti (1984) suggested that such an enlargement enables the alkali ions to sink deeper into the tetrahedral rings, thereby causing the decrease of $1/2\ \sin\beta$ that is observed in Mg, Fe-rich Mu.

Petrologic and crystallochemical implications that follow from the above include:

1. Increase of pressure should move the Mu limb of the Mu-Pg solvus to more K-rich compositions. The Pg limb may move similarly or remain stationary. Such motions will bear on using the Mu-Pg solvus for geothermometry.

2. Guidotti (1984) suggested that in Mu, this interrelationships among $\Sigma(Mg + Fe_t)$ content, α rotation, and size of the XII sites might explain the low Na contents typical of Mg, Fe-rich Mu: the larger XII sites being less able to hold in the small Na ion. Reversal of this argument for Pg may explain the small amounts of Mg and Fe typically found in Pg. Pg commonly has a large α angle (near 16°) and this may be in part a requirement to hold in the small Na ion. This requirement might preclude substitution of significant Mg or Fe for Al^{VI}, as it would cause angle α to decrease, thereby increasing the size of the XII site to the extent that the Na:O bond lengths would become unstable.

REFERENCES

Guidotti, C.V., 1984, Micas in metamorphic rocks, *in* S.W. Bailey, ed., Micas: Min. Soc. Amer., Reviews in Mineralogy, v. 13, p. 357–468.

Güven, N., 1971, The crystal structures of $2M_1$ phengite and $2M_1$ muscovite: Zeit. f. Kristallog, v. 134, p. 196–212.

Radoslovich, E.W., 1962, The cell dimensions and symmetry of layer-lattice silicates: II. Regression relations: American Mineralogist, v. 47, p. 617–636.

SATHYAMURTHY, K., Geological Survey of India, Hyderabad, India

Hydrogeology of Deccan Traps

Deccan traps, the most extensive tholetitic fluid basalt eruptions of Cretaceous-Eocene age, occupying about 5,000,000 sq. km area in the western and central parts of Peninsular India has always been a fascinating subject. These stratified lava flows measure up to 1700 m in the exposed western part while it thins out to few meters only in the margins. While it is dominantly 'Pahoehoe' type in the western part, it is 'aa' type in the south, eastern and southeastern parts. The thickness of the individual flows may be 20, 30 or even 50 m.

The Deccan traps overlie the Archaean metamorphics over a large part of the area. In the south and southwest it overlies the Precambrian Bhima and Kaledgi group of sediments. Along its northern boundary it overlies the Vindhyans while in the eastern margins it overlies the Archaeans and Gondwana Group of rocks.

Apart from the cooling joints of the individual flows a system of fractures and shears of varied intensity cut across a number of flows. Major faults are rare except in the Satpura and Gujarat areas and in the offshore areas near the coast of Bombay.

In general in any trap flow two distinct horizons the lower massive and upper vesicular can be distinguished. The massive part of the flow is greenish to dark gray in color, hard and compact in nature. The vesicular part is generally dark grey to pinkish or purplish with vesicles often filled with secondary minerals or at places empty. The lava flows are also associated with volcanic products such as tuffs, breccias, ash beds and sedimentary deposits (intertrappeans) formed during the interval between the flows. A discontinuous redbole layer less than a meter thick is also invariably present in between the massive and vesicular flows believed to be an alteration product of weathering. Black cotton soils ranging in thickness from 3 to 8 m cover the major part of the Deccan trap terrain and alluvial materials covering the traps are restricted to areas adjacent to river courses.

Only during the last two decades or so the Deccan traps have been subjected to systematic groundwater surveys. The interesting feature of the different basaltic units is the contrast in their water bearing properties. The massive traps with their fracture porosities, the minutely interconnected and partly filled vesicles of the vesicular traps with the overlying redbole beds acting as aquicludes and the intertrappen sediments with their primary porosities have a decisive role to play in determining the groundwater regimen of different regions. The weathering of basalts results in widening of joints which again increases the secondary porosity, while partial weathering of vesicular unit gives rise to porous granular zones on account of presence of zeolites, agate, quartz, etc.

Groundwater occurs under both water table and confined conditions in the Deccan lava flows. The near surface weathered and jointed flows of massive trap units and vesicular traps together constitutes the water table aquifers which are tapped mainly by dugwells. These units locally show both vertical and lateral variations in the degree of weathering and porosity which naturally plays a dominant role in the success or failure of wells. These observations are also reflected in the variation of Transmissivity values in various directions thereby indicating that the Deccan basalts are hydrologically anisotropic and heterogeneous. The deeper aquifers generally tapped by dug cum borewells and tubewells, can be expected to be under confined condition provided they are disposed favorably to receive the recharge. The intertrappean sediments of limited areal extent are not much of significance except for local water supply under favorable conditions. The entire succession of lava flows present a multiaquifer system with productive as well as less productive zones. The yields of wells range from 5 to 10 klph with specific capacity of the aquifers varying from 10 to 400 lpm/m of drawdown.

The Deccan traps by and large fall into three domains with its distinct hydrological characteristics, based on the permeability data of the water table aquifers in the Bhima, Godavari and Wardha river basins. The statistical analysis of the data has revealed a decrease in permeability range as one proceeds from the Bhima basin (8-92 klpd/sqm low rain fall area) through the Godaveri basin (3-40 klpd/sq. m. assured rain fall area) to the high rain fall area of Wardha basin (1 to 18 klpd/sq. m).

In general the waters in the Deccan traps fall under excellent class as their sodium absorption ratio is less than 10. However, in the area underlain by silty alluvial deposits the quality in the majority of cases is not suitable for irrigation in view of their high TDS and chlorides. This item included in the poster session is intended to display some illustrative examples of the Deccan trap aquifer systems.

SATO, H. Hiroshima University, Hiroshima, Japan

Redox States of Magmas Estimated by Mg-Fe Partitioning Between Plagioclase and Magma

Longhi et al (1976) suggested that Mg-Fe partitioning between plagioclase and magma can be employed to estimate fO_2 of plagioclase–phyric magmas, based on the difference of the partition coefficients of terrestrial and lunar basalts. In the present study, we investigated the effect of fO_2 on the partition relationship of Mg and Fe between plagioclase and silicate liquid at 1 atm in some basaltic systems, and found that the Mg-Fe partition relation could be a good oxygen fugacity meter for most plagioclase–phyric basalts.

Two relatively fresh ocean-floor basalts of Hole 504B, ODP Leg 111, and two basaltic rocks of 1986 eruption products of Izu-Oshima Volcano, Japan, were used as starting materials. The ODP basalts contain several percents of phenocrysts of plagioclase, olivine, and/or augite, whereas, the Izu-Oshima lavas are nearly aphyric with sporadic phenocrysts of plagioclase, augite, and/or orthopyroxene. Compositions of plagioclase in a representative ODP basalt are shown in Figure 1.

1 atm experiments were performed using the wire-loop method (Presnall and Brenner, 1974). We used a vertical quench furnace of Siliconit with SiC heating elements. The furnace atmosphere was controlled by mixing CO_2 and H_2 gases, and its oxygen fugacity was calibrated against the reaction; $Ni + 1/2O2 = NiO$. The furnace temperature was calibrated against the melting point of gold and diopside at the sample sites.

Plagioclase appeared as liquidus phases in all the samples with liquidus temperatures ranging from 1250 °C to 1190 °C. Because of the very low diffusivity of elements in plagioclase crystal, reequilibration experiments or reversal experiments were not successful to measure equilibrium partition coefficients of Mg and Fe between plagioclase and liquid. We obtained apparent partition coefficients of Mg and Fe by analyzing coexisting glass and plagioclase, which have grown from liquid 30–70 °C below the liquidus temperatures. Figure 1 shows the compositions of both experimentally grown plagioclases and natural plagioclases of ODP samples. MgO concentration in plagioclase is rather constant in wide range of fO_2, whereas the FeO content varies from 0.2 to 0.3 wt% at $\log fO_2 = -11.5$ to ~3 wt% at $\log fO_2 = -0.7$ (Figure 1), suggesting much different distribution coefficients for ferrous and ferric iron between plagioclase and liquid. The simple weight ratio distribution coefficients of Mg and Fe between plagioclase and liquid are shown in Figure 2, together with the distribution coefficients observed in the natural samples (phenocryst-bulk rock). It is noted that the distribution coefficients of Mg of the natural systems are distinctly lower than those of the experimental systems. This discrepancy may be due to much larger departure from equilibrium distribution in the experimental systems, in which larger degree of supercooling conditions prevailed. The disequilibrium effects can be evaluated by using the equation of Burton et al (1953); $D' = D/[D + (1-D)\exp(-Vd/R)]$, where D' and D are apparent and true distribution coefficients, V, d, and R are growth rate, thickness of diffusion layer, and diffusion coefficient of an element under consideration, respectively. Using available data, the departure from equilibrium distribution as represented by D'/D are calculated to be 1.12, 1.16, and 1.09 for Mg, Fe^{2+}, and Fe^{3+}, respectively. Therefore, when we consider the exchange partition coefficient of Mg and Fe between plagioclase and liquid, K = (Mg/Fe)plag/(Mg/Fe)liquid, the disequilibrium effects would be much less than in the case of simple ratio distribution coefficients. Figure 3 demonstrates that the Mg-Fe partition coefficients strongly depend on logarithm of oxygen fugacity. Rather uniform exchange partition coefficients in the range of $\log fO_2 = -6$ to 0 (Figure 3) may be due to probable decomposition reaction of ferric-rich plagioclase under oxidized conditions. These plagioclases are accompanied by abundant magnetite crystals. The partition coefficients observed in natural ODP basalts indicate the redox conditions of these basalts are near the one defined by fayalite–quartz–magnetite assemblage. Similar experimental work on basaltic rocks of Izu-Oshima 1986 eruption products showed 0-1 logarithm unit more oxidized condition of crystallization of plagioclase for these arc volcanic rocks. The present method, Mg-Fe exchange partitioning between

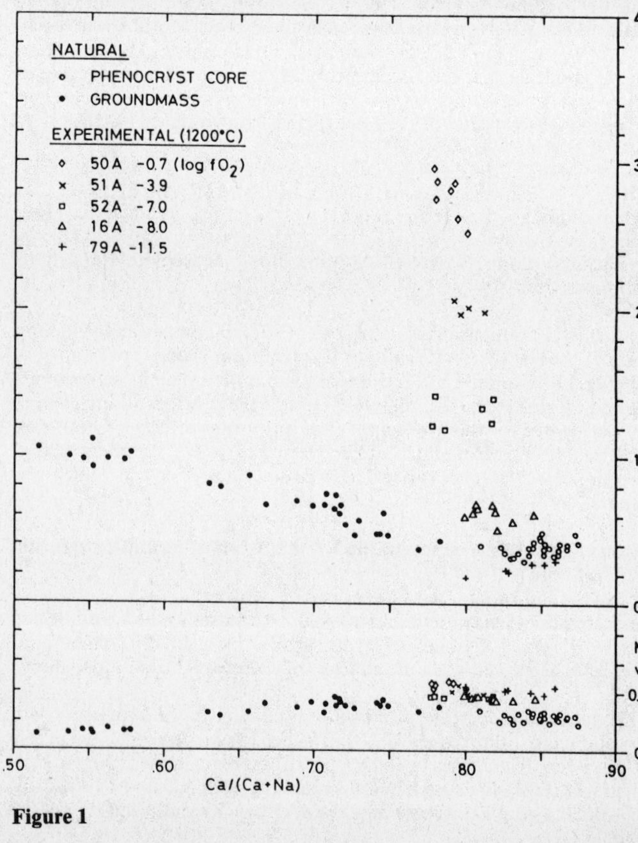

Figure 1

phenocryst plagioclase and bulk rock composition, gives us redox informations of the intratelluric phenocryst crystallization stage of magma evolution, whereas the method of Christie (1986, EPSL, v. 79, p. 397), which is after Kilinc et al (1983), records redox conditions of erupted magmas. The present method of estimating redox states of magmas by Mg-Fe exchange partitioning between plagioclase and liquid have wide applicability because of the widespread occurrences of plagioclase phenocrysts in volcanic rocks and the apparently low susceptibility of the Mg and Fe contents in basalt and plagioclase to alteration.

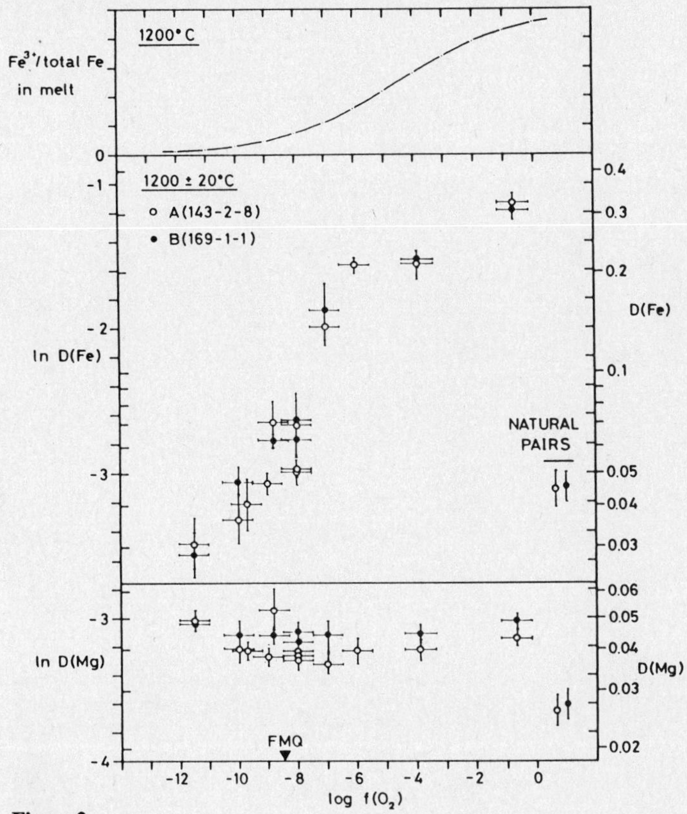

Figure 2

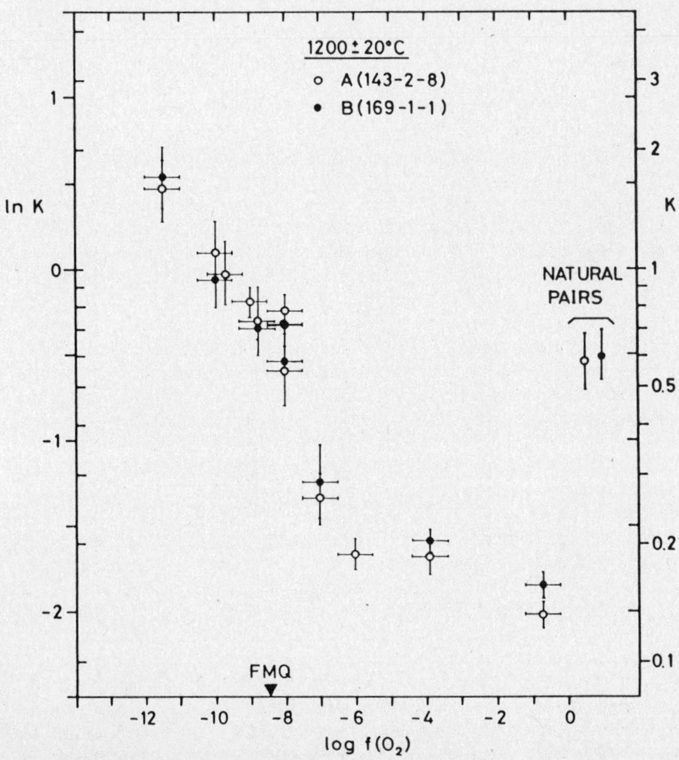

Figure 3

SATO, H., University of Alberta, Edmonton, Canada, and I.S. SACKS, Carnegie Institution of Washington, Washington, DC

Magma Generation in Midocean Ridge and Island Arc: Constraints from Seismic Measurements in Peridotite at High Pressure and Temperature

Knowledge of temperature and degree of partial melt in the upper mantle yields important constraints on magma generation in the earth. In order to determine actual temperatures in magmatic zones, it is necessary to perform measurements under controlled laboratory conditions in a way that these can be related to conditions in the earth. Seismic velocity and attenuation determined in the laboratory at high pressures and temperatures can be used to constrain thermal structure in the earth. Seismic velocity and attenuation determined in the laboratory at high pressures and temperatures can be used to constrain thermal structure in the earth. Peridotite is widely considered to be the dominant component of the uppermost mantle. The measurements carried out above the solidus temperature of a dry peridotite can be directly compared with the seismic structure in partially molten regions. Because extensive seismic studies have determined velocity and Q structures as a function of depth and plate age, both vertical and lateral temperature profiles may be determined from laboratory seismic results. The laboratory results show that the pressure dependence of seismic velocity and Q is well accounted for by the pressure dependence of the melting (solidus) temperature (Tm) of the peridotite, i.e., both seismic velocity and Q depend on the homologous temperature (T/Tm). This observation allows us to extrapolate laboratory results to higher pressures (greater depths) with some confidence, requiring only a knowledge of the solidus as a function of pressure. We compare the laboratory results with seismic velocity and Q structures, and derive temperature and fraction of the partial melt in the upper mantle. Generally, laboratory velocity and Q data give a consistent thermal structure.

At the midocean ridges, the oceanic plates spread apart and new seafloor is created. The comparison between laboratory and seismic results yields a large amount of partial melt (≥ vol%) and a higher temperature than the dry peridotite solidus in 0–5 m.y. asthenosphere under the Iceland Plateau and beneath the active marginal basin of the east Philippine Sea. Partial melting is not expected in the asthenosphere older than ~5 m.y. under the Iceland Plateau and older than ~40 m.y. under the Pacific Ocean. Even though the volume fraction of partial melt is not so large (≤ vol%) under the Pacific plate, the melting zone may extend to a distance of above 2000 km from the ridge. If substantial partial melt occurs far from the spreading ridge, some melt may penetrate the lithosphere, forming seamounts. This may explain the large number of seamounts observed on the Pacific plate. The temperatures of the asthenosphere beneath the Iceland Plateau derived from seismic Q values are 1290, 1040, and 980 °C at 45 km depth for 0–5, 5–10, and 10–20 m.y. regions, respectively. We show that there is a good agreement between temperatures derived from seismic studies and heat flow data. Since the laboratory measurements were made on a dry peridotite and we employ a dry peridotite solidus for determining the geotherm, the consistency in temperature suggests that the oceanic upper mantle (remote from subduction zones) is fairly dry. It appears that the spreading process in midocean ridge causes asthenosphere material to ascent from greater depths where it was at higher temperature. This gives rise to pressure release melting for the source magma of midocean ridge basalts.

The consistency in temperatures derived from the laboratory seismic data and heat flow beneath the Pacific indicates that the laboratory results may be applicable to other areas of the upper mantle. Here we examine the thermal structure in a subduction zone. There are two general classes of models for magma generation in subduction zones: melting of the subducted slab, or magma derivation from the asthenosphere wedge overlying the subducted slab. By comparing laboratory results with seismic observations in Japanese Island, we determine that temperatures are generally below the solidus of the dry peridotite (T < 0.93 Tm). Both velocity and Q data indicate no general partial melting beneath Japan under a volatile free condition. However, active island-arc volcanisms suggest the presence of partial melt. If volatiles (probably water) are present, the melting temperature of the peridotite is greatly reduced and partial melt can be produced. The results of our study suggest melt due to fluxing with water probably from the subducted lithosphere. The results also indicate that melting of the slab or the sediments is not required for magma generation beneath the island arc.

SATO, M., U.S. Geological Survey, Reston, VA

Anomalous Hydrogen Emission Along Active Faults and Implications for Earthquake-Triggering Mechanisms

Earthquakes originate in stress loading, but they can be triggered when (A) the shear stress exceeds the existing high frictional resistance of the fault zone, or (B) the frictional resistance of the fault zone falls below the existing low shear stress level. Case A is the more usual explanation, whereas case B has been discussed in published literature primarily in terms of hydrofracturing induced by reservoir filling or injection in deep wells.

The patterns of anomalous H_2 emission that preceded five M > 5 earthquakes recorded in California and those described by investigators in China, Japan, and the USSR do not seem to support case A. The major difficulties are that (1) anomalous H_2 emission increased rather abruptly and often peaked before the correlated earthquake, instead of steadily increasing with increasing stress and peaking co-seismically at maximum stress; (2) epicentral distances from the monitoring sites showing anomalies often exceeded hundreds of kilometers and crossed other active faults and crustal block boundaries, where shear stress would have been dissipated; (3) although the magnitude of the anomalies tended to be larger for larger and closer earthquakes at a given site when the site showed anomalies, the combination of the sites that showed anomalies varied with each earthquake; and (4) sometimes the precursory time was shortest at the site closest to the epicenter.

To explain the observed behavior of H_2 along active faults on the basis of processes that can generate free H-2 abundantly and also can trigger earthquakes, I propose a tectonochemical mechanism for case B that is applicable to earthquakes that occur at convergent plate boundaries and possibly, with a little modification, to those that occur in intraplate seismic regions where horizontal compression of the crust is in progress.

According to the scenario of plate tectonics, crustal blocks are passively transported by convecting mantle. When the free movement of a crustal block is impeded by collision or friction with another crustal block, displacements should occur within the crust in the direction of the mantle movement and in planes parallel to the interface between the moving mantle and the impeded crustal block. Displacement-induced stresses would follow this pattern. In the case of an obliquely converging plate boundary such as the San Andreas fault system in central and northern California, this oblique convergence would cause compressive stress normal to the fault and shear stress in the fault strike direction in proportion to the relative dimensions of the respective vectors. This simple elastic model is disturbed, however, by vertical inhomogeneity of crustal rocks and differing nature of water-rock interactions. In the greenschist-facies environment that prevails at depths of about 10–20 km, rocks would readily react with water, but the resulting changes in physical properties would be very different depending on whether rocks are saturated with silica (sialic rocks) or are undersaturated (mafic rocks). Sialic rocks (which are common in the upper continental crust) would become more rigid owning to silica cementing and/or recrystallization, whereas mafic rocks (dominant in oceanic crust and probably common in deeper continental crust) would become more ductile due to the formation of weak, hydrated minerals. Extreme examples of the latter case are serpentinites, which form by hydration of ultramafic rocks such as dunite in a greenschist-facies environment. Cowan and Mansfield (1970) estimated the shear strength of serpentinites in New Idria, California, at 1 bar. The sharp brittle-ductile (or viscous) transition that occurs at about 12 km near the San Andreas fault is probably caused by the sialic-mafic lithologic transition as originally suggested by Scholz et al (1969) on the basis of the seismic reflection and refraction profiling by Stewart (1968). In addition to making ultramafic rocks very weak, hydration reactions in these rocks also generate abundant hydrogen; 1cm^3 of H_2, simultaneously producing magnetite.

The ductile hydrated mafic layer decouples the brittle sialic crust from the moving mantle by allowing schistose slip, unless the frictional resistance of the fault plane is substantially reduced. Horizontal convergence of two plates requires that vertical components be added in the crustal movements. Folding or thrusting of sialic crust and subduction of dense mafic crust are well-known tectonic processes that meet the requirement. An additional process indicated by geologic observation is the squeezing up of mobile serpentinites into weak fault zones to relieve the compressive stress arising from plate convergence. "Cold" (meaning nonmagmatic temperatures) intrusions of serpentinites and partly serpentinized ultramafic rocks along fault zones are commonly observed in many tectonic settings in the world. Serpentinites are denser than most upper-crustal host rocks, so such intrusions require compressive stress at the "source," i.e., in the mafic crust.

When compressive stress increases in a segment of the lower crust, it is transmitted over a wide area via the ductile layer and exerts upward-directed stresses to the base of the brittle upper crust. Where mobile serpentinites are available because of groundwater penetration and the ensuing hydration of ultramafic rocks, serpentinites can be squeezed into weak places in the brittle crust (most likely unhealed fault zones) to relieve the compressive stress in the ductile layer. The segment of a fault zone above such an intrusion bulges and dilates, releasing hydrogen and steam, the latter released from serpentinites under a reduced confining pressure. This situation may be *somewhat* analogous to the intrusion of igneous dikes or domes, although the temperature is much lower and the viscosity may be greater. Heavier gases such as radon and carbon dioxide may be carried by the steam flow until the steam condenses within several kilometers of the surface. Buoyant and mobile H_2 rushes out to the surface and may locally be exhausted soon after the dilation of the fault zone, as is often observed, unless the intrusion is massive. Local fault creep may also occur as the compressive stress normal to the fault is temporarily reduced by the bulging near the surface and/or by wedging near the base of the brittle crust in addition to the increased fluid pressure. When a wide enough span of the fault zone is thus wedged and lubricated, shear stress will be concentrated at locations where fault creep has been retarded due to relatively intense cementing and/or lack of lubrication in the lower part of the brittle crust (seismic asperities). Because no data are available as to the behavior of H_2 at an epicenter, I am uncertain as to whether a serpentinite intrusion within the hypocentral area is necessary to trigger a damaging earthquake. Concentration of shear stress by the lubrication/wedging of the surrounding areas alone may be able to trigger an earthquake.

This tectonochemical mechanism may reasonably explain the observed behavior of H_2, which is difficult to explain on the basis of conventional high-shear-stress physical models. H_2 anomalies may be widespread, even across fault-block boundaries, when the compressive stress in the ductile crust is large, and such anomalies may migrate toward the potential epicenter with time. Also they may occur suddenly when a particular segment is dilated, peaking out before the correlatable earthquake. The suggested mechanism can also explain reported premonitory geophysical changes in or near an active fault, such as bulging, water level changes, decreasing electrical resistivity, and rising temperatures, without invoking high-shear-stress required for the dilatation hypothesis.

REFERENCES

Cowan, D.S., and C.F. Mansfield, 1970, Serpentine flow in Joaquin Ridges, Southern Coastal Ranges, California: GSA Bulletin, v. 81, p. 2615–2628.

Scholz, C.H., M. Wyss, and S.W. Smith, 1969, Seismic and aseismic slip in the San Andreas fault: Journal of Geophysical Research, v. 74, p. 2049–2069.

Stewart, S.W., 1968, Preliminary comparison of seismic travel times and inferred crustal structure adjacent to the San Andreas fault in the Diablo and Gabilan Ranges of central California: W.R. Dickinson and A. Grantz, eds., Proceedings of Conference on Geological problems of San Andreas Fault System: Stanford Univ. Publ. Geol. Sci., v. 11, p. 218–230.

SATO, M., U.S. Geological Survey, Reston, VA

Redox Evolution Model for Earth and Moon Based on Thermochemical Relations of C-H-O and Fe-Bearing Phases

In understanding planetary redox evolution, the Earth and the Moon together present a challenge. The Fe° core of the Earth constitutes 31 wt% of the Earth's mass, whereas the existence of a core in the Moon is uncertain, and, if it exists, it may even be Fe-Ni-S (Brett, 1973). Fe^{3+} is normally present in terrestrial basalts (unoxidized by air), whereas Fe° is almost ubiquitous in lunar basalts. Thus, the bulk Earth is more reduced than the bulk Moon, but terrestrial basalts are more oxidized than lunar basalts.

To explain the marked difference in redox state of basaltic rocks between the two planets, I proposed (Sato, 1978) that the oxygen fugacity (fO_2) of basaltic magma was controlled by elemental carbon and oxidized carbon species in the source region in both planets and that the marked difference in the redox state of basaltic magma arose in the crust because of differences in H_2O content. This hypothesis assumes that refractory elemental carbon was left in the mantle in both planets after planet-wide melting (possibly by the mechanism of Cameron, 1985) and core formation. Graphite or diamond is found in mantle rocks of the Earth. The existence of some carbon in the lunar interior was indicated by Apollo 17 orange glass (Sato, 1979) as described below. The hypothesis is based on the highly pressure-dependent characteristics of the CCO buffer. Relative to those of iron-bearing assemblages, the fO_2 of this buffer increases with pressure and decreases with temperature. At magmatic temperatures, C is not strongly reducing in the source region of basaltic magma (between the WM and IW buffers), but it can reduce Fe^{2+} to Fe° near the surface if the magma is H_2O free.

Lunar rocks formed from H_2O-free magmas; no evidence has been found to the contrary. The Moon probably could not retain light, volatile H_2O during the initial planet-wide melting because of its small mass and hence weak gravitational field. Of lunar rocks, the highest fO_2 (above the IW buffer) was exhibited by Apollo 17 orange glass, which probably represents a basaltic magma quenched under pressure in a CO-driven pyroclastic eruption. This glass was self-reduced and exhibited fO_2 typical of lunar lavas when reheated above 1120 °C at 1 bar. The self-reducing pattern was duplicated by similarly heating a synthetic basaltic glass with a minute amount of graphite (Sato, 1979). The Fe° content in lunar mare basalts is typically about 0.1%, requiring a few hundred ppm C to form by carbon reduction. Once Fe° thus forms, the fO_2 of the magma is buffered by the Fe°-Fe^{2+} (in melt) assemblage. This scenario of the fO_2 control by the CCO buffer of anhydrous melt produced in a small planetary body probably also applies to parent bodies of some achondrites (Brett and Sato, 1984). Under very low pressures and at high temperatures, C could reduce silicates even to a solution of Fe°-Si° found in highly reduced enstatite chondrites (Ringwood, 1966).

The presence of H_2O in the magma source region of the more massive Earth and, hence, in the magma modifies the above simple carbon reduction pattern. With decreasing pressure during magmatic ascent, H_2O reacts with C to form H_2, CO_2, and some CO (very little under pressure) and, thus, retards the reduction by C in a closed environment. Near the surface where the pressure is less than about 1 Kbar, extremely mobile, buoyant H_2 begins to escape and C is completely oxidized. To oxidize 500 ppm C to CO_2, only 0.15 wt% H_2O is required. Continued escape of H_2 generated by thermal dissociation of H_2O results in the oxidation of Fe^{2+} to Fe^{3+}. Fractional crystallization of ferrous silicates also helps to oxidize the magma by concentrating Fe^{3+} into residual melt. The progressive oxidation of hydrous magma eventually makes magnetite stable and raises SO_2 fugacity. Both the crystallization of magnetite and the degassing of SO_2 have a reducing effect on the melt. Also, H_2fugacity diminishes as fO_2 increases, slowing down the H_2 loss. Hence, the fO_2 of terrestrial basaltic magma is balanced by the combination of these factors and tends to settle near those of the FMQ and NNO buffers in the near-surface environment, unless it is further oxidized by post-photosynthetic atmosphere.

The above scenario of the control of the redox state of the magma by the counteracting roles of C and H_2O also supports the thesis by Kennedy (1948) that high fO_2 induces early crystallization of iron oxides and thus suppresses iron-enrichment trend during crystallization differentiation. Because fractional crystallization takes place at a depth where magma can cool slowly and degassing is restricted, the balance between C and H_2O becomes critical. In a given reservoir setting, H_2 loss is easier if fO_2 is higher. The H_2O-free lunar basalts show an extreme iron-enrichment trend (Papike et al, 1976). Relatively H_2O-poor tholeiitic basalts show a moderately strong iron-enrichment trend. An example of this is the Skaergaard intrusion, for which a closed-system crystallization model failed to account for about one-third of Fe^{3+} in the upper zone, and, thus, some H_2 loss was suggested (Wager and Deer, 1939). Such tholeiitic layered intrusions show evidence of C being present in the lower zones (Sato and Valenza, 1980). Being richer in H_2O, alkaline basalts and calcalkaline basalts are likely to be more oxidized due to higher H_2 loss, regardless of the source of origin, and tend to show a subdued iron-enrichment trend due to higher fO_2, as suggested by Osborn (1959) in a slightly different context. A good correlation between the H_2O content and the SO_4^{2-}/S_2total ratio (an index of redox state) of such basalts was shown by Sakai, et al (1984).

Fluid inclusions in rocks of mantle origin are almost pure CO_2 (Roedder, 1965) and indicate dominance of CO_2 among gaseous species in the Earth's mantle. Graphite or diamond is also found in these rocks, whereas carbonates are rarely found. If we assume that the CCO buffer controlled the fO_2 of the whole mantle at the end of global melting and vigorous outgassing, the redox gradient within the mantle can be inferred on the basis of thermodynamic data and high-pressure molar volume data of iron-bearing phases and carbon phases. The law of barometric equilibrium dictates that for a gas not to move in radial directions, its fugacity must increase with depth according to the relation: $\ln(f_2/f_1) = M \cdot g \cdot (z_2 - z_1)/RT$, where f_1 and $f_2$are fugacities of the gas at depths z_1 and z_2, respectively, M is the molecular weight of the gas, g the gravitational acceleration, R the gas constant, and T the absolute temperature. Because g and T vary with depth, we must integrate the relation by small increments (e.g., 10 km) from the surface to the core–mantle boundary (CMB). Considering that the primitive atmosphere of the hot Earth was probably dense (no ocean nor limestone), I assumed that the fO_2 required for degassing near the surface was 1 Kbar and that fO_2 in the mantle obeyed the barometric relation with this value. The fO_2 of the CCO buffer thus computed agreed well with the fO_2 of the CMB computed as in Sato (1984), indicating that C could have reduced Fe^{2+} to Fe°, itself being oxidized to CO_2, in the core region, because the temperature effect dominated the pressure effect. The streaming up of the CO_2 toward the surface, which could have occurred more easily than sinking of iron, would have been an efficient mechanism of bringing up oxygen from the core region to the upper mantle. If any Fe° was left in the mantle, the CO_2 would have oxidized it to Fe^{2+} while being reduced to C.

The redox relation between the CCO buffer and the iron-bearing buffers is highly dependent on the P-T balance (geotherm). I infer, therefore, that, at a given depth in the lower crust and the mantle, the rocks are more reduced along an oceanic geotherm than along a continental geotherm.

REFERENCES

Brett, R., 1973, Geochimica et Cosmochimica Acta, v. 37, p. 165–170.
——— and M. Sato, 1984, Geochimica et Cosmochimica Acta, v. 48, p. 111–120.
Cameron, A.G.W., 1985, Icarus, v. 62, p. 319–327.
Kennedy, G.C., 1948, American Journal of Science, v. 246, p. 529–549.
Osborn, E.F., 1959, American Journal of Science, v. 257, p. 609–647.
Papike, J.J., et al, 1976, Review of Geophysical Space Phys., v. 14, p. 475–540.
Ringwood, A.E., 1966, Geochimica et Cosmochimica Acta, v. 30, p. 41–104.
Roedder, E., 1965, American Mineralogist, v. 50, p. 736–782.
Sakai, H., et al, 1984, Geochimica et Cosmochimica Acta, v. 48, p. 2433–2441.
Sato, M., 1978, Geophysical Research Letters, v. 5, p. 447–449.
——— 1979, Proceedings of 10th Lunar Planetary Science Conference, p. 311–325.
——— 1984, Proceedings of 27th International Geological Congress, v. 11, p. 405–433, Amsterdam, VNU Sci. Press.
——— and M. Valenza, 1980, American Journal of Science, v. 280-A, p. 134–158.
Wager, L.R., and W.A. Deer, 1939, Medd. Gronland, v. 105, p. 1–352.

SATO, T., University of Tsukuba, Ibaraki, Japan

Geotectonics and Mineral Resources of East and Southeast Asia

East and southeast Asia is a complex assemblage of tectonic units of different origin. Two major types are distinguished in the tectonic units; one is the continental mass of Eurasia and the other is the island arc systems bordering its eastern margin.

The Eurasian continent is not a homogeneous land mass but is composed of "paleoplates" of different constitution, which have successively come to be juxtaposed with each other, since probably Late Precambrian time. Siberian, North China and South China (including Yangtze platform) are major paleoplates. These paleoplates are themselves made up of Precambrian core continents and Paleozoic-Mesozoic fold belts.

The latter belts are generally accretionary fold belts, disposed between the core continents. They comprise within themselves small continental blocks such as Qaidam, Tibet, Kontum massifs besides many others. A number of accretionary fold belts can be individualized in Eurasia, of which the age of activity ranges from early Paleozoic to Cenozoic. Unlike the European continent, Mesozoic accretion phases are much more important in East Asia, particularly in the south of China.

The internal structure of the major paleoplates is, however, not well understood. Older structures are often much obliterated by superposition of later deformations, both ductile and brittle. Disintegration of the continents always followed the accretion. Even at present, fragmentation of Eurasia is going on, as a result of the collision of Indian Peninsula.

The eastern margin of the core continent is bordered by island arc systems. The presently active systems, East Japan, Izu-Bonin, Mariana, Philippines, Sunda and Banda arcs, among others, are constructed at the continental plate margin next to the oceanic plates of the Pacific, Philippines, and India-Australia. Generally subduction occurs towards the continent, except for the arc of Taiwan-West Luzon which is towards the Philippine Sea.

Late Mesozoic arc systems, like SW Japan, Palawan-NE Borneo, Sulawesi and Burmese-Malayan fold belts, survived alongside or within the present arc systems, where later deformation has not complete masked the original structure. Important mineral resources are located in these fold belts. Their geologic structure awaits further investigation.

Better understanding of geotectonic structure is the prerequisite for systematic mineral exploration. To promote such organized exploration in East and Southeast Asia, the publication of a series of geotectonic maps (8 sheets illustrated in Figure 1) is now being undertaken, as one of the CCOP/CPC Projects. An international workshop composed of specialists coming from 9 countries concerned has been organized for coordination of drafting and editing. The leading philosophy of the compilation is to keep the information as factual as possible, which may be used as the basic data for systematic exploration as well as for possible tectonic interpretation.

SATO, T., University of Tsukuba, Ibaraki, Japan

Mesozoic Paleogeography, Seaways, and Ocean Connections of Japan and Eastern Asia

The triangular geometry of Tethys Ocean is generally adopted in every paleogeographic reconstruction. Its eastern mouth is assumed to open largely toward the Pacific, to allow free connection between two oceans. The paleogeography of the eastern Laurasian continent has not been, however, depicted in detail. A tentative reconstruction of the Late Jurassic paleogeography, among others, is shown in Figure 1. Constraints for determining the paleopositions of continents and oceans in Mesozoic East Asia are as follows.

1. Until Triassic time, the Siberian platform was not united to the continental blocks of China but was bounded on its southern side by the narrow ocean of Paleotethys. The sea spread over Japan and surrounding areas at that time and was connected directly to Paleotethys. This is also strongly supported by the wide distribution of common ammonite genera.

2. From Jurassic onward, Paleotethys was closed and an integrated Eurasian continent was formed. The continental blocks of China and southeast Asia came to be juxtaposed with the Siberian platform. Neotethys has newly opened to the south of the Tibetan blocks. A narrow "Tanggula Sea" can be assumed on the southern margin of this continent. These seaways can be presumed also by ammonite occurrences.

3. The present marginal seas behind the West Pacific arcs did not exist in Mesozoic. The geographic location of these future arcs is not exactly determined, but it should be along the eastern margin of Eurasian continent. The Mesozoic rocks found at present in these arcs, Japan, southwestern Philippines, Borneo, and West Sulawesi, for instance, are generally marine sediments, except for nonmarine parts of Upper Cretaceous.

4. In the northwest Pacific, blocks of Okhotsk, Omolon, and Chukotka existed off the main Siberian continent. The sea fringing the eastern coast of Eurasia connected with the "Boreal" ocean through the seaways between the mainland and these islands. Ammonites of both "Boreal" and "Tethys" origin coexisted in some horizons of the Japanese Jurassic, suggesting that Japan in Jurassic time was at the site of junction of both oceanic realms.

5. Mesozoic ammonites of East Asia show generally world-wide distribution. Some genera constitute, however, endemic faunas, indicating an individualized Pacific province. On the other hand, genera found in the northern Gondwana coast are not uncommon. A few genera occur in common in both South America and Japan.

6. Bivalve faunas form a separate biogeographic province in East Asia. In Tethys ocean, Maorian and Ethiopian provinces can be set up for Jurassic bivalves. This resulted probably from ecological segregation, but not from the interruption of seaways between East Asia and Tethys oceans.

7. Land plants of the Japanese Mesozoic take part of the Chinese land floras, suggesting the close geographic position to the Chinese continent.

8. Paleomagnetic studies indicate that some of the East Asian Mesozoic terranes were located in the equatorial region, but these data are contradictory often with the above biogeographic information.

Figure 1—Tentative paleogeographic reconstruction of East Asia in Late Jurassic. IC, Indochina block; S-G, Songpang-Garze fold belt of Triassic accretion; SL, Songliao block.

SAUNDERS, R.S., Jet Propulsion Laboratory, California Institute of Technology, Pasadena, CA

Magellan Venus Mapping Mission: Science Objectives

The Magellan mission is scheduled to be the first planetary mission to be launched by the resumed Shuttle operations. The launch is scheduled for late April 1989. The type IV trajectory results in an August 1990 arrival for the 17 day in-orbit checkout and 243 day mission. Placed in a 3.15 hour near-polar orbit, Magellan's synthetic aperture radar system will produce an image swath from the north pole to about 70°S latitude on each orbit. This image swath is 20 km wide and 16,000 km in length. Pixels will be 75 m square. Extended mission operations will provide for filling of gaps in the image maps and multiple incidence angle coverage. Image resolution will be 120 m in azimuth and 118 to 300 m in range. Incidence angle is variable from approximately 52° near periapsis to 19° over the north pole. The images will be made into mosaics at full resolution, covering about 5° square regions, and a series of smaller scale mosaics covering larger regions. The 243 day mission will return over 3 Gbytes of image data.

Because of the broad similarities between Earth and Venus, Venus mapping remains an important priority for planetary exploration. In particular, Venus mapping will help to further our understanding of the interiors and tectonic mechanisms of Earth-like planets. Analysis of U.S. Pioneer Venus Orbiter data and data from the several Soviet Venera missions by Magellan investigators and others is providing a wide range of hypotheses for the origin and support of Venusian topography. These include spreading, analogous to that of the Earth's midocean ridge systems, migrating hot-spot tectonics, as in the Hawaiian–Emperor volcanic chain, and ancient crustal blocks that would resemble the oldest cratons on Earth. These hypotheses are testable with Magellan synthetic aperture radar images, and new gravity data to be obtained. By allowing us to address such fundamental questions, Magellan is expected to provide an important bridge between terrestrial geology and planetary science.

SAUZAY, E., H. TEIL, M. VANNIER, and L. ZANONE, Ecole Nationale Supérieure des Mines de Paris, Fontainebleau, France

Geological Software for Microcomputers: Automatic Data-Processing System and Mineral Prospecting Teaching Tool

An ever increasing number of geologists are turning to microcomputers for their data-processing requirements and for use in teaching because of the relatively low-cost and easier accessibility. There is a wide availability of hardware, but the choice of software is limited and may not be well adapted to a geologist's requirements. Using the experience gained over 20 years in geological data processing and teaching, the research geologists at the IGM have developed software that can be used in documentary, research, or teaching environments in geological organizations. An integrated, comprehensive data processing system, GEONIX, has been developed for the simultaneous treatment of documentary, technical, and scientific data, as well as a sophisticated simulation tool, CLAIM, for computer- assisted instruction into the techniques of mineral prospecting.

Both software packages have been developed on microcomputers under the increasingly popular UNIX operating system, taking full advantage of its development tools. This system was chosen because of its portability from micros to main-frames, and because it works in a multiuser and multitasking environment. Modular, open-ended programs have been created, allowing special applications to be easily incorporated.

Geological data are particularly difficult to process because of the extreme complexity of the concepts handled. A simple concept such as that of a "vein" in a mineral deposit may have many different parameters associated with it (thickness, paragenesis, age, or host- rock), and several of these associations may occur in one deposit. This structured aspect of data therefore needs to be represented in a data record describing the deposit. Similarly, relationships between geological files are very complex, and cannot be predetermined in a system. GEONIX can deal with these relationships and allows the user freedom in expressing them. It offers flexibility for defining the full-screen input record, so that the user can include free-text, numerical data, predefined names, x-y tables, or structured data in the same record.

Geological applications are distinguished by the necessity to have sufficiently important mathematical, statistical, and graphic procedures. A great emphasis has therefore been placed on these procedures in GEONIX, which has its own mathematical language enabling the user to perform calculations or logical operations on user-chosen fields from the main field or a preselected subfile. This language is also available for defining the variables to be plotted on one-, two-, or three-dimensional diagrams (in normal or high resolution) for subsequent screen viewing or printing. The resulting diagrams may be memorized and recalled when required for report editing or image mixing. Integrated also are a digitizer interface and statistical methods such as principal component analysis and discriminant analysis.

The simulation game CLAIM, which involves the interaction of a geological model, a set of rules, and teaching materials, is used to teach students and professionals attending refresher courses the techniques of mineral exploration. Using geochemical data and stream sediment, topographical, and aerial geological maps, the students have to plan an exploration strategy, control the exploration budget, discover the structure and geology of the area represented by the model, outline probable mineralization, and evaluate the ore reserves. To achieve these objectives, they choose the borehole locations, the type of drilling machine, and the drilling depth, taking continually into account the financial aspects. The computer controls the running of the game, printing out the detailed borehole logs, controlling the budget and drilling equipment, and enforcing the rules of the game to simulate an actual exploration project. The computer-generated geological model is multidisciplinary, involving stratigraphy, sedimentology, structural geology, tectonics, climatic conditions, metallogeny, mineralogy, and petrology, with complex mathematical functions being used to express certain characteristics.

GEONIX is in use at various geological or mining surveys, and at the Centre de Géologie Généérale et Minière for developing an African mineral deposit data bank as well as, for example, thermodynamical or geochemical files. A nonmicro version of the mineral simulation game has run for many years at the Centre and also at the University of Lausanne, and the CLAIM model containing several gold and copper deposits has been developed recently to take advantage of the interactive graphical features offered by microcomputers (Fig. 1).

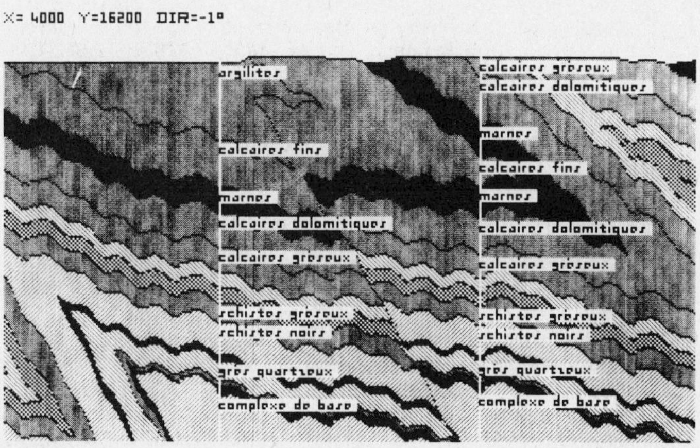

Figure 1—Computer-generated cross-section representing 1.6 km of the CLAIM model (20 × 30 km).

SAVERIKKO, M., Geological Survey of Finland, Espoo, Finland

Komatiitic Explosive Volcanism—Special Evidence for an Archaean Craton

Pyroclastic komatiites are found in most Precambrian shields but they constitute only a minor volcanic feature in ultramafic rock piles. The exception is a komatiite zone of distinct explosive origin in the Archaean granite—greenstone terran of the Fennoscandian (Baltic) Shield; the explosions, which were predominantly magmatic in nature, took place in three volcanic cycles, altaining regional dimensions during the last cycle along with accelerated mantle-activated rifting within the cratonic environment, in northern Finland, Norway and USSR. The stratigraphic order of the Lapland Greenstone Belt is the opposite of the classic Archaean greenstone-belt sequence as terrigeneous metasediments are in excess in the Lower and Middle Lapponian and komatiites do not prevail until in the Upper Lapponian.

The Lapponian komatiitic eruption phases were associated with the Cwenan disastrophism at 3.0-2.5 Ga. The first phase, with few exposures, refers to the initial volcanism, which resulted in an eruptive outburst when the komatiitic greenstones discharge onto the Saamian (3.5-3.0 Ga) sialic crust. A subsequent cratonic sedimentation of quartzite-carbonate-slate association preceded a bimodal volcanism which included minor komatiitic eruptions, too: the rare pyroclasticity of the komatiites is attributed to water indicating explosions mainly in a terrestrial environment. As a result of cratonic rifting, these volcanic complexes of the stable platform were superimposed by a sequence of arkose-slate-quartzite association. Tectonic quiescency, demonstrated by thin euxinic-exhalative strata, separated the following mantle-activated riftal period characterized by the main explosive komatiite centres. The alignment of the pyroclastic komatiite complexes is indicative of a mantle hot-line. The Paleoenvironment seems to have been the subaerial-subaqueous border of a subsiding depositional basin filled with spilitic greenstones.

Vulnerability to explosive eruptions was promoted by the presence of komatiitic rock species differing slightly from the conventional classification. Perioditic komatiites (>30 wt % MgO) are coarse-grained olivine-pyroxene cumulates with random volcanic features or with almost nonexistent pyroclastics. Pyroxene periodotitic komatiites (18-30 wt % MgO) are present as fine-grained tremolite-chlorite rock often displaying structures of ejecta and fragmentary lava flows with or without basal cumulates. Komatiitic basalts (9-18 wt % MgO) are now actinolite-hornblende (-plagioclase) rocks of compact lavas and infrequent ejecta. All these rocks produced volcaniclastic detritus.

The main, i.e., third eruption phase occurred in isolated centers, around which the komatiite complexes are mainly pyroxene peridotitic komatiites responsive to magmatic shattering. Only one complex (at Sattasvaara) forms a gradual series including all three rock types. With the exception of the first flood eruption of peridotitic komatiite, the magma poured out in Hawaiian to Strombolian-type eruptions along with increasing ultrabasicity connected with the spreading of komatiitic volcanism; the terminal flows were a crystal mush of peridotitic komatiite. As a comagmatic suite they may manifest compositional zonation of the magma chamber, which may be due to fractional crystallization.

Pahoehoe flows of komatiitic basalt which erupted in the earlier stage contain amygdules with static-vesicles and flow-vesicles as further proof of low viscosity. The weakly developed vesicularity in the fragmentary flows of pyroxene peridotitic komatiite which discharged in the later stage may be attributed to the originally low content of volatiles in these lavas, since the discharge was usually followed by enormous expansions liberating gases only later; the block lavas together with the voluminous pyroclastics largely of vitric tuff or lapilli tuff (also present in the form of a cinder cone) implicate highly viscous magma.

The formation of a volcanic glass depends on the cooling rate and viscosity of a stable silicate liquid (Fisher and Schmincke 1984, p. 76). Because of the low volatile content and small number of preexisting solid fragments, i.e. cumulus minerals and accidental or accessory ejecta, the viscosity of the pyroxene peridotitic komatiite may have been caused primarily by the "low" eruption temperature. This could explain the absence of the spinifex textures, which occur in the komatiites of the earlier Lapponian volcanic cycles. Low-temperature lavas may not have a high enough cooling rate and degree of supercooling, both of which, according to Donaldson (1982), are necessary for the development of spinifex texture.

The "low" temperature of the ultramafic magma may result from loss of heat in a mantle-derived material during its slow ascent through the thick cratonic crust. A radial swarm of aulacogens and crustal fractures reveals that the linear mantle upwelling with continental dimensions which is partly visible in the pyroclastic komatiite zone involved domal uplift. Therefore, fractional differentiation in a magma chamber near the surface may also be feasible. A shallow crustal level is suggested by the concentration of volatiles in the products of earlier eruptions that is, in the upper part of the magma chamber, and by the advance cooling which increased the viscosity of residual magma extruded later.

Owing to the high viscosity of the magma, the cratonic paleoenvironment may require the explosive komatiitic volcanism to have been a regional phenomenon. Then the explosions should have been of magmatic origin rather than a phreatic or preatomagmatic outburst.

SAVOSTIN, L., and S.S. DRACHEV, Academy of Sciences of the USSR, Moscow, USSR, J. BONIN, and B. KRESS, Institute de Physique du Globe de Strasbourg, Strasbourg, France, and S. MERKURIEV, A.M. KARASIK, and N. SOCHEVANOVA, Academy of Sciences of the USSR, Leningrad, USSR

Evolution of Arctic Basin and Continental Rift System of Northeastern Asia: Late Mesozoic–Cenozoic Geodynamics

The present-day structure of the region under study is generally controlled by the interaction between the two large lithospheric plates: Eurasian and North American. The boundary between them is distinctly delineated by the Arctic seismic belt, associated with the Mohns, Knipovich, and Gakkel Ridges and, within northeastern Asia, with the Chersky Range.

A study of the present motion of lithospheric plates in the Eurasian Basin has been made using data on fault plane solutions, linear magnetic anomalies, and bottom topography as well as geological evidence from northeastern Asia.

New data concerned with relative motions of North American, Eurasian, and some other plates of the Arctic region were obtained as the result of analysis of low-level aeromagnetic data recorded in the western Arctic Basin by USSR scientists.

The kinematic parameters of the past lithospheric motion have been determined by fitting the paired isochrons represented by the two sequences of the magnetic anomalies: (21–32[34] in the Makarov Basin and 1–24 in the Eurasian Basin) and have been used in the set of joint Soviet-French paleogeographic maps.

Comparison of the parameters of rotation between the lithospheric plates obtained in the Arctic Basin with those earlier determined in the North Atlantic (Pitman and Talwani, 1972) and in the Norwegian Greenland Sea (Talwani and Eldholm, 1977) has shown that the history of the Eurasian Basin opening is separated by anomaly 12 in two stages. From the initiation of the active spreading (about 56 m.y. ago) until 36 m.y. ago the pole of opening of the Eurasian Basin differed significantly from those of the Norwegian Greenland Sea and the North Atlantic, suggesting the existence of some other lithospheric plates in the western part of the region. It is concluded that in addition to the well-known Greenland plate, there must have existed another plate, the West European, which could partly accommodate the interaction between the Eurasian and North American plates. The boundary between the West European and Eurasian plates presumably followed the Jan Mayen fracture zone and Poland–Danish furrow.

From the Late Oligocene to Middle Miocene (13-6 magnetic anomalies) sharp reorganization of plate motion in the region began. The Euler pole of rotation moved to Movosibirsky islands, which made the compression regime predominant on the territory. Some Eocene layers were eroded, and sandstone–gravellite formations were deposited. On the west (Laptev sea and Novosibirsky islands) compression led to fold-thrust deformations of Paleocene–Eocene layers and local interruptions in sedimentary rock deposition. On the east (Chaunsko–Kolymia region), the development of grabens continued. Some transgressions are fixed there. Geodynamics of the region were determined by right-lateral shift.

At the beginning of the Late Miocene (6 magnetic anomaly), the rotation pole moved to the northern foothills of the Chersky range and then increasingly southward. Modern Momsky rift system began formation in this time; the oldest deposits related to this configuration of the Eurasian/North American plate boundary (Late Miocene) are known in grabens of Buor-Haia bay and Polousny range. The rift advancement to the southeast is emphasized by appropriate rejuvenation of its sedimentary deposits. This stage is characterized by formation of the great archuative raise and connected with it lowering foundation of the East Siberia with deposition within it continuous cover of friable continental deposits. Perhaps in the Quaternary one more "jump" of the pole of rotation to the north not fixed in the ocean magnetic anomalies took place; it is now situated in the Buor-Haia bay region. This results in modern kinematics of the Momsky rift; the largest part of earthquakes here have a compression mechanism.

SAVOSTIN, L.A., and A.M. HEIPHETZ Academy of Sciences of the USSR, Moscow, USSR

Paleomagnetism and Tectonics of Olutorsky Terrane

Abundant data collected lately have contributed greatly to the idea that the Koryak highland is a collage of allochthonous terranes (Fujita and Hewberry, 1982; Zonenshain et al, 1987). However, many problems concerning the limits of the terranes, their inner structure, geological evolution, and processes of their accretion to the Eurasian active margin remain unsolved. The objective of this topic is to consider results of the first paleomagnetic investigation of the Olutorsky terrane—the best presently known terrane in the Southern Koruak range.

A major thrust-fault boundary dipping south-southwestward—the Vatyna megathrust—has been documented between the Olutorksy terrane and the Upper Cretaceous Ukelayat flysch to the north (Mitrofanov, 1977; Alexandrov et al, 1980). Southern and eastern boundaries are hidden by the Bering sea, so their position is uncertain. The inner structure of the terrane is very complex. It is an original paleoaccretional prism composed of at least three large tectonic nappes (Astrachantzev et al, 1987; Kazimirov et al, 1987). The upper part of sequence in every nappe is composed of Maestrichtian–Paleocene island-arc tuff, lava, sandstone, mudstone, and chert. They overlap various oceanic lithologies of Albian–Campanian age. Within the lowermost nappe, island-arc rocks cover basalts, supposed to be formed at seamounts.

In the middle nappe they cover pillow basalts with interlayers and lenses of jaspers, and clay jaspers seem to be formed within the short distance from a seamount. In the uppermost nappe island-arc rocks cover abyssal jaspers; several thin flows of abyssal pillow-basalts were detected. All these nappes were warped into complexly folded, sometimes recumbent folds by later tectonism. At least two stages of deformation are recognized (Astrachantzev et al, 1987). The first stage resulted in amalgamation of the terrane during subduction, and the second occurred during the collision between the island arc and Eurasian margin.

To determine the original position of the Olutorsky terrane, we studied paleomagnetically both island-arc mudstone and fine tuff (Maestrichtian) and seamount basalts (Campanian). Special attention was paid to the fold axis plunge, cleavage, and reheating.

After the demagnetization (reheating and alternating field), we succeeded in finding prefold components (Table 1).

The characteristic magnetization shows significant poleward motion both of seamount and island-arc rocks. The latitudinal anomaly for Campanian seamount basalts ($39.4^{\circ -11.1 \circ +8.7 \circ}$) suppose their origination in the central part of the Pacific, possibly near the Hawaiian hot spot. And island-arc rocks latitudinal anomalies ($18.2^{\circ -14.6 \circ +11.7 \circ}$ for Apuka and $23.6^{\circ -13.2 \circ +10.6 \circ}$ for bay of Machevna) show that the island arc existed in the northwestern Pacific at the end of Cretaceous a great distance from the Eurasian continent.

Age	locality	nappe	D	I	K_s	K_s/Kg	α_{95}
Maastrichtian	Upper river Apuka	uppermost	321.4°	70.8°	13.4	12.6	10.8°
	Bay of Machevna[1]	lowermost	338.6°	66.5°	-	-	10.4°
Campanian	Bay of Iavevin[1]	lowermost	297.1°	51.3°	-	-	11.7°

1/ Demagnetisation circles technies (Halls, 1976) was used to determine the direction.

SAWAF, T.,, D. AL-SAAD, M. AL-OTRI, and A. GEBRAN, Syrian Petroleum Company, Ministry of Petroleum and Mineral Resources, Damascus, Syrian Arab Republic, and M. BARAZANGI, J. BEST, T. CHAIMOV, and J. McBRIDE, INSTOC, Cornell University, Ithaca, NY

Seismic Reflection Profiling across Intracratonic Palmyra Fold and Thrust Belt and Surrounding Arabian Platform, Central Syria: Transect under Auspices of Global Geoscience Transects (GGT) Project

More than 2000 km of seismic reflection data correlated to eight exploration wells along with detailed geologic maps and gravity and magnetic data are used to constrain the subsurface structure along an approximately 100 km wide corridor that traverses central Syria in a southeast-northwest direction for a distance of about 500 km from the Iraqi border in the southeast to the Turkish border in the northwest. This transect in central Syria is the first of three such transects that represent the Syrian Arab Republic's contribution to the Global Geoscience Transects (GGT) Project. Three major tectonic elements of central Syria are crossed from south to north along the central transect: the Rutba uplift, the northeast sector of the Palmyra fold and thrust belt, and the Aleppo plateau. Preliminary results are:

1. Paleozoic isopachs indicate that the Rutba uplift was an early Paleozoic depocenter.
2. Mesozoic stratigraphic sequences dramatically thicken into the Palmyrides from the adjacent Paleozoic platforms consistent with the hypothesis of the belt as a site of early Mesozoic rifting.
3. Four prominent regional reflectors beneath the Rutba uplift represent major Paleozoic unconformities that dip gradually toward and disappear beneath the Palmyrides. The deepest subhorizontal reflector (6–7 km) is a pervasive bright reflector that may represent an Infracambrian feature and/or an Alpine decollement. This reflector appears again beneath the Aleppo plateau at a depth of about 5 km.
4. The regional structure of the deformed Palmyra belt consists of a complex of faulted anticlines and synclines plunging northeast beneath the flat-lying Cenozoic sediments in the Euphrates valley. The outer flanks of the deformed belt are marked by smaller asymmetric anticlines with an outward vergence (i.e., northwest and southeast) giving the belt a rough structural symmetry.
5. Late Cretaceous to Neogene deformation is sharply restricted to the narrow width of the belt (≈ 100 km) and appears to have reactivated antecedent structure. Folds and faults mapped from the seismic data often offset Quaternary sediments indicating continued reactivation. The intense deformation of the Palmyra belt remarkably contrasts with the relatively undeformed Phanerozoic strata of the adjacent platforms to the north and south.

In summary, the new data and results are consistent and strengthen the hypothesis that the Palmyra fold and thrust belt began its evolution as a Triassic failed rift (i.e., aulacogen) that was later inverted and complexly deformed by late Mesozoic and Cenozoic transpression.

SBORSHCHIKOV, I.M., Institute of Oceanology, Moscow, USSR

Microplates of Tethys: Kinematics and Correlation with Motion of Main Plates

In the scheme of development of the Tethys ocean, which was reduced to the Alpine folded system, can be noted displacement of several microplates. Some researchers believe that these "median" blocks represent fragments of the passive margin of Gondwana and that their moving to the north was correlated with the closing of the northern branch of the ocean (Paleotethys) and the opening of the southern branch (Neotethys). However, the real situation was more complex than that suggested by this scheme. For example, the Iran and Afghan blocks (which sometimes are considered as parts of a united Kimmerian microcontinent) have not advanced simultaneously. The Iranian block was jointed with Eurasia in latest Triassic, whereas the Central Afghan block was separated from it up to Late Cretaceous. The microplates behaved very differently during closing of Mesozoic Tethys. The Apoulia block, for example, rotated accordingly with displacement of the African plate relative to Eurasia. The motions of the Menderes and Kirshekhir blocks were produced by kinematics of the main plates, but Lesser Caucasus, Armenian, and Lut blocks did not have such dependence.

Several blocks were moving in the latitudinal direction (along Tethys) and such motions are not indicated by paleomagnetic measurements. On the other hand, it is supposed that large-scale wrench-fault displacements also caused rotation of the microplates. Dzirula massiv on the Caucasus rotated anticlockwise in late Triassic, whereas distinct sinistral displacement between Gondwana and Eurasia took place during Jurassic. Kinematic analyses also show that rotation of Iberia (Cretaceous time) was accompanied by spreading in the Biscay to the north and by stretching in the other margin to the south. This situation can't be explained by the simple "ball-bearing" effect.

During the history of Tethys, tremendous volumes of oceanic crust disappeared, but at the same time new back-arc basins originated in several regions. At the present time, there are three regions between Gibraltar and Pamir where subduction of the oceanic lithosphere is continuing now: the Calabrian, Hellinic, and Makran zones. The Calabrian arc is connected with the opening of Tyrrhenian sea. It is considered that the origin of this basin results from the break-up of Calabrian blocks and the movement Beniof zone in consequence of eastward asthenospheric flows.

The final stage of closing of Tethys is the stage of collision and of reorientation of the continental fragments in conditions of strong compression. The main difficulty for the analysis of recent kinematics is represented by the lack of space and by possible intraplate deformations. The comparison of geological data from terrestrial and oceanic regions indicated a reducing of microplates of about 30%. Such a process in thick continental lithosphere is practically impossible. There are numerous geophysical data about the geological stratification of lithosphere and the presence of ductile layer near the crust–mantle boundary. It is assumed that during collision this layer can be affected by a decollement surface. The upper, crustal plates are not as rigid as the underlying lithosphere and can be deformed.

In a global scale, interactions between main plates are dependent. In regional reconstructions, we must take into account independent displacements of individual blocks (microplates) showing no direct connections with motion of macroplates. In collision situations "dermal" tectonics of thin slab of crust are possible.

SCANLON, K.M., U.S. Geological Survey, Woods Hole, MA, and D.G. MASSON, Institute of Oceanographic Sciences, Wormley, United Kingdom

Sedimentary Processes in Deep Ocean Areas of Puerto Rico and Virgin Islands

Puerto Rico and the Virgin Islands lie along the plate boundary between the North American and the Caribbean plates. Until Eocene time, this plate boundary was characterized by north-south convergence, which resulted in subduction of the North American plate beneath the Caribbean plate and the formation of the Puerto Rico Trench. Since the end of the Eocene, the relative plate motion has been left-lateral slip plus a small component of convergence. As a result of this tectonic history, Puerto Rico and the Virgin Islands are surrounded by a variety of marine sedimentary environments.

To explore the deep-water (>200 m) environments, we conducted a GLORIA long-range sidescan-sonar survey. Overlapping sidescan imagery surrounding Puerto Rico and the Virgin Islands and extending about 200 km both north and south of the northern Caribbean plate boundary was obtained. The data have been processed by computer to correct distortions that result from slant-range geometry and changes in ship's speed and to remove striping and speckle noise. The processed data have been digitally mosaicked to create a single overview of the entire 2000,000 km^2 area. This mosaic has revealed new information about the morphology and seafloor character of the insular slopes, trenches, and deep ocean basins that surround Puerto Rico and the Virgin Islands.

The GLORIA mosaic shows numerous canyons cutting the steep insular slopes surrounding the islands. We attribute the high acoustic backscatter exhibited by the canyon floors and walls to outcrops of the Tertiary limestones into which the canyons are cut. The canyons are narrow (100–500 m), closely spaced (generally less than 4 km apart), and deeply incised (as deep as 400 m) on the upper slopes. Many canyons merge, become wider, and are less deeply incised on the middle and the lower slopes. Most begin near the shelf edge and can be traced downslope continuously for 30–50 km. Canyons north of Puerto Rico end at the base of the slope in the fore-arc region of the Puerto Rico Trench, 50 km south of and 1500 m above the floor of the trench. Sediment transported down the canyons apparently does not reach the trench floor. Most canyons south of Puerto Rico end within a deformation front along the northern side of the Muertos Trough. Two large downslope-facing amphitheater-shaped scarps, 30 and 60 km across, were revealed by the GLORIA sonographs on the slope north of Puerto Rico. The larger scarp is 1500 m high and represents the removal of as much as 1250 km^3 of sedimentary section. A feature such as this, if formed in a single event, would pose serious tsunamic danger to coastal communities and warrants further study.

The fore-arc region between the Puerto Rico Trench and the base of the insular slope consists of a series of subparallel east–west-trending ridges and intervening basins. The basins, which are partially filled with flat-lying sediments, can be identified on the GLORIA mosaic as irregularly shaped areas of low backscatter. The flat-lying sediments on the floor of the Puerto Rico Trench demonstrate similar backscatter properties.

North of the Puerto Rico Trench, the Greater Antilles Outer Ridge, an 1800 km long and 110–220 km wide sediment drift, rises several hundred meters above the abyssal plain. The GLORIA mosaic reveals an area of anomalously high backscatter over part of the southeastern section of the drift. Dredged samples and bottom photographs indicate that the high backscatter is probably caused by a field of manganese nodules. Seismic-reflection profiles show that the acoustically transparent sediments, which constitute the Greater Antilles Outer Ridge, are thinner in this region. This indicates either a low sedimentation rate or active erosion, either of which would favor manganese nodule development.

SCHAEBEN, H., and S. AUERBACH, University of Bonn, Bonn, Federal Republic of Germany

Computer-Aided Geometric Design in Geosciences: III. Practical Application to Three-Dimensional Geologic Modeling and Digital Geological Maps

When conventionally constructed by hand, geological maps and cross-sections provide static two-dimensional images of the geologist's spatial model—his/her concepts of where rock bodies are, how they are arranged, what they consist of, and how they originated and developed through time. The model is an interpretation that should be consistent with available data and with expectations based on background knowledge of the environment that formed and deformed the rock bodies (Loudon, 1986, personal communication).

In the past, much effort has been spent on writing software packages for graphic data presentation, e.g., contour mapping packages, which the geologist can interpret to prepare his/her map. However, these presentations, whatever sophisticated algorithms have been used, never match the geologist's background knowledge completely. Even worse, these presentation graphics do not provide a truly three-dimensional model nor do they provide helpful means for the actual constructing process. But this is exactly what is needed: digital representation of the spatial model that enables the geologist to express his/her interpretations and codify his/her background knowledge within the computer model, most favorably interactively using three-dimensional graphic display and efficient means to locally control and modify the geometry of the model. Thus, methods of computer-aided geometric design (cagd) have been introduced into geology (Schaeben, in press). Concerning geometry, this model is internally consistent due to the truly three-dimensional constructing process. Once the geological spatial model has been built, geological maps or cross-sections can be easily generated or updated automatically by combining with a digital terrain model.

What will be modeled is a geologic solid consisting of distinct units, to be distinguished by geological parameters. Thus, the geologic solid is essentially structured by surfaces that separate these distinct geological units, e.g., the joined surface of two distinct layers in sedimentary rocks. Since a direct approach to modeling the geologic solid or equivalently the family of surfaces as a whole, though desirable, (e.g., when interior points of layer are endowed with material properties, say hydraulic conductivity for modeling groundwater flow or solute transport), appears to be too complex for the time being, it seems reasonable to model its internal surfaces successively. Due to the ways of sampling geological information as compared to the task at hand, the purpose of modeling a geological surface gives rise to a rather pathological mathematical problem that some consider a nonmathematical problem altogether (cf. Gold, 1980; Vinken, 1986).

The problem becomes evident if we think of the configuration and the content of the available numerical, i.e., quantitative or quantifiable, data versus likely features of geological surfaces. The sparseness and erratic regional distribution of data by comparison with the features to be observed amount to a rather technical problem; the geological background information which includes (1) previous knowledge, e.g., about some structural features of different geological surfaces (faults, fractures as opposed to folds); (2) basic knowledge, e.g., concerning the major tectonic elements; (3) basic ideas, e.g., concerning the chronological order of geological processes that contributed to the situation to be observed today; (4) fundamental ideas concerning geodynamical processes in general, i.e., paradigms like plate tectonics; and which are all at the geologist's hand when constructing a map but cannot be quantified immediately if at all poses the essential problem of mathematically modeling of geological surfaces: How to deal with soft information. And common features of geological surfaces such as multiplicity requiring an appropriate parameterization, discontinuities (usually of known location), and constraints on the set of surfaces constituting the geologic solid do not relax the problem but indeed show the necessity of using the soft information in any semiautomatic modeling procedure.

Using advanced practical methods of interpolation/approximation (Gmelig Meyling, 1986) and extending it for purposes of interactive geometric design (Schaeben and Auerbach, in press), we summarize the procedure proposed here as follows: (1) Find an initial representation of an individual geological surface as good as possible with respect to the fact that we are usually not able to provide all the geological knowledge as input data for the specific algorithm used and that consequently the representation will fail to completely meet the geologist's idea; (2) do carefully some further design work on the initial surface approximation that will preserve the initial geometry where it is required, i.e., use interactive means for local changes where needed to meet all of the geologist's objectives; (3) apply this two-stage construction successively to all geological surfaces involved in the spatial model of the geologic solid. It is especially the second stage where cagd methods are utilized, i.e., where the control vertices (Schaeben and Auerbach, in press) are manipulated by the expert geologist until the surface meets all his/her ideas. The final set of control vertices may be stored as a representation of the accepted surface model. Now we may proceed to the next surface in the solid. If constraints on neighboring surfaces are fairly simple, e.g., almost parallel (constant thickness of layer), or a simple trend in the thickness that can be reasonably expressed with a bivariate function, then this function may be applied to the first set of control vertices to yield an initial guess of the second surface immediately. Otherwise, this two-stage procedure should be applied successively to all surfaces, taking care that none of the constraints of geological consistency must be violated. In some unfortunate instances, this may result in starting all over again with the first surface to achieve a consistent model. Thus, the process described here is one of interactively learning and increasing understanding. It will still require a lot of interaction of the expert geologist. But we explicitly aimed not at automatization in geological cartography but at computer-aided geologic design, the acronym (CAGD) of which coincides with computer-aided geometric design. This should not really bother us because that is basically what at least the second stage boils down to.

REFERENCES

Gmelig Meyling, R.H.J., 1986, Polynomial spline approximation in two variables: Academisch Proefschrift, Universiteit van Amsterdam.

Gold, C., 1980, Geological mapping by computer, *in* D.R.F. Taylor, ed., The computer in contemporary cartography: New York, J. Wiley and Sons, p. 151–190.

Schaeben, H., in press, Improving the geological significance of computed surfaces by cagd methods (Einführung von Methoden des rechnergestützten geometrischen Konstruierens in die Geologie): Geol. Jb., v. A104.

———, and S. Auerbach, in press, Computer aided geometric design in geosciences: I. Methodology: Polynomial splines on triangles: This volume.

Vinken, R., 1986, Digital geoscientific maps: priority program of the German society for the advancement of scientific research: Mathematical Geology, v. 18, p. 237–246.

SCHAEBEN, H., and S. AUERBACH, University of Bonn, Bonn, Federal Republic of Germany

Computer-Aided Geometric Design in Geosciences: I. Methodology: Polynomial Splines on Triangles

Essentials of multivariate spline theory have been developed by Dahmen (1986), Dahmen and Micchelli (1982a,b), Höllig (1986a,b), and Micchelli (1979); they have been adapted and elaborated for the purposes of practical approximation in two dimensions and gave rise to two-dimensional polynomial b-splines on triangles (Gmelig Meyling, 1986). Preserving almost all favorable properties of tensor product b-splines, they provide a much greater flexibility to approximate fast-changing or even discontinuous functions.

Furthermore, triangulation allows adjusting to the areal distribution of the locations of the data in the domain. To avoid any possible misunderstanding right from the beginning, we would like to emphasize here that the vertices of the triangles initially define the knots of the spline function and generally do not coincide with the places where the data are given. This fact constitutes one of the basic differences from other schemes of interpolation/approximation based on triangulation that take the location of the data for vertices of the triangles, and should be kept in mind throughout. We would also like to remind the reader of the fact that with splines the knots play the essential role much as the polynomial degree does with polynomials. Splines may be thought of as piecewise polynomials, pieced together obeying given constraints concerning the smoothness in the joints, the locations of which are the knots.

Following Schoenberg's geometrically inclined volumetric definition of b-splines, it is cited from Böhm (1983) that a general b-spline may be defined as the "shadow" of a simplex of m-dimensional space onto an n-dimensional space, $n < m$. More exactly, the shadow is a projection or function: the "loss of intensity" of light passing through the simplex. The "usual" shadow is the support of the b-spline. One can prove that this b-spline is a piecewise polynomial of degree $r = m - n$. Projecting a three-dimensional simplex, i.e., a tetrahedron, onto one-dimensional space, i.e., a straight line, yields the classical univariate quadratic b-spline; projecting onto a two-dimensional plane defines a bivariate linear b-spline. Note that the projections of the vertices of the simplex give the knots of the b-spline. Thus, the "shadow" depends on the knots and also on the volume of the simplex.

To summarize, a multivariate b-spline is a locally supported nonnegative piecewise smooth polynomial defined by its corresponding set of knots. Their distinct advantage over tensor-product b-splines implying a grid partition of the domain is their flexibility in approximation, e.g., the use of varying knot density and local control over the smoothness properties.

More specifically, we employ piecewise quadratic b-splines in two variables to obtain a C1-smooth surface. After triangulation of the domain (Auerbach and Schaeben, in press) we associate two additional points with each vertex of a given triangle. Using the combinatorial recipe by Dahmen and Micchelli, (1982a) we obtain six quadratic b-splines for every triangle. Each of them is supported on the convex hull of five knots selected from the nine according to the very rule, and C1-smooth provided that the five knots are in general position, i.e., every three of them span a nondegenerated triangle. The entire triangular mesh with T being the number of triangles generates $6*T$ quadratic b-splines that are linearly independent, thus span all quadratics over the domain (Dahmen and Micchelli, 1982b).

If we consider the vertices of the triangulation as threefold knots, i.e., if we allow the two additional points to coincide with the original vertex they are associated with, the bivariate quadratic b-splines turn into the well-known bivariate quadratic polynomials in Bernstein-Bezier form (Dahmen, 1986; Farin, 1986). Bernstein-Bezier polynomials constitute a basis for quadratic polynomials with respect to each individual triangle; however, their linear combination over the entire domain does not result in a C1-smooth surface. Thus, we might be led to think of b-splines as of smoothed versions of Bernstein-Bezier polynomials with respect to the entire domain. Next, we deduce from the degenerate Bernstein-Bezier situation rules how to place the additional points associated with each vertex to establish knot configurations that allow the modeling of discontinuities in the function itself or its derivatives. We find that four collinear knots out of the set of five defining an individual quadratic b-spline generate a discontinuity in the surface along the line they constitute, and that analogously three collinear knots generate a discontinuity in the first derivatives.

As with all computer-aided geometric design (cagd) applications of Bezier-polynomials or b-splines, the fundamental idea is to visualize the coefficients of their linear combination geometrically as control points and exploit their favorable features. These control points are not in the surface, but the spline-surface mimics the overall behavior of the two-dimensional polygon constituted of the control points. More specifically, these control points allow us to locally manipulate the surface that was initially determined to approximate the data. Moving a single control point in any direction causes a local change of the surface to smoothly follow this movement. Unfortunately, with b-splines on triangles the relationship between surface and control points is not as direct as for Bernstein-Bezier polynomials or tensor product b-splines because in our case the "convex hull" property is lost.

Nevertheless, this formalism provides the means needed to construct geologic maps semiautomatically, because the expert geologist got an easy-to-use geometric approach to insert his/her additional geological knowledge, which is otherwise hard to quantify into the computer-aided construction process that now becomes an interactive one of progressive understanding of the geologic situation to be studied (Schaeben and Auerbach, in press).

REFERENCES

Auerbach, S., and H. Schaeben, in press, Computer aided geometric design in geosciences: II. Practical application to digital elevation models and topographic maps: This volume.

Böhm, W., 1983, Subdividing multivariate splines: CAD, v. 15, p. 345–352.

Dahmen, W., 1986, Bernstein-Bezier representation of polynomial surfaces: Vorlesungsreihe SFB 72, n. 31, Universität Bonn.

——— and C.A. Micchelli, 1982a, Multivariate splines—a new constructive approach, *in* R.E. Barnhill and W. Böhm, eds., Surfaces in computer aided geometric design: Proceedings of the conference held at the Mathematical Research Institute at Oberwolfach, West Germany, 25–30 April 1982, Amsterdam, North Holland Publishing Company, p. 191–215.

——— ——— 1982b, On the linear independence of multivariate b-splines. I. Triangulations of simploids: SIAM, Journal Numer. Anal., v. 19, p. 992–1012.

Farin, G., 1986, Triangular Bernstein-Bezier patches: Computer Aided Geometric Design, v. 3, p. 83–127.

Gmelig Meyling, R.H.J., 1986, Polynomial spline approximation in two variables: Academisch Proefschrift, Universiteit van Amsterdam.

Höllig, K., 1986a, Box splines: Computer Sciences Technical Report no. 640, University of Wisconsin, Madison, 30 p.

——— 1986b, Multivariate splines: Lecture notes, AMS short course series, New Orleans.

Micchelli, C.A., 1979, On a numerically efficient method for computing multivariate B-spline, *in* W. Schempp, and K. Zeller, eds., Multivariate approximation theory, Basel, Birkhäuser, p. 211–248.

Schaeben, H., and S. Auerbach, in press, Computer aided geometric design in geosciences: III. Practical application to three dimensional geologic modeling and digital geological maps: This volume.

SCHAEBEN, H., University of Bonn, Bonn, Federal Republic of Germany, and H. SIEMES, Technical University of Aachen, Aachen, Federal Republic of Germany

Benefits of Use of Explicit Mathematical Models: II. Practical Applications of Entropy Optimization in Texture Goniometry

Pursuing the approach of finite series expansion methods to the texture problem of reproducing the orientation distribution function (odf) from pole distribution functions (pdf) measured in X-ray or neutron diffraction experiments, the mathematical model of entropy maximization is applied to pole figure data and numerical results are presented. It is shown that this model as well as its results compare favorably with other methods of odf reproduction.

From theoretical consideration (Schaeben, in press a,b) as well as from practical applications (Schaeben et al, in press) including odf reproduction from a single complete or incomplete pdf, and from several complete or (initially not normalized) incomplete pdfs of cubic or trigonal/hexagonal materials, a selection of which will be presented in this talk, it can be concluded that the technical properties and practical advantages of this new method are that:

1. it immediately applies to complete or incomplete pdfs as well as
2. to simultaneous processing of several pdfs of different crystal forms,
3. it detects faked pdfs or the erroneous mixture of pdfs of different crystal forms from different samples,
4. it provides an adjustable relaxation parameter to account for non-negligible experimental errors in the measured pdfs,
5. the necessary nonnegativity of the odf to be reproduced is a truly constitutive element,
6. possible zero components in the pdf are exactly and efficiently accounted for; a uniform pdf gives a uniform odf,
7. it provides the most safely balanced solution with respect to the inevitable compromise that must be made between spatial resolution and the accuracy of amplitude measurements,
8. it is numerically straightforward and stable,
9. it requires storage for only the vector of iterated approximate solutions, which makes parallel processing an attractive possibility, and
10. it uses a property of the odf itself as a stopping rule for the iteration, which is the size of the k + 1st increment of the entropy in accordance with the rate of convergence of the algorithm; stopping rules in all other methods of odf reproduction are based on the residual or simple functions of it, thus referring to properties of the recalculated pdfs. Due to the general ambiguity of the texture problem, they do not provide sufficient criteria for the physical goodness of the reproduced odf.

Comparison with other discrete methods of odf reproduction, e.g., Imhof (1987), Matthies and Vinel (1982), Pawlik and Pospiech (1987), and Williams (1968), reveals that they formally resemble some elements of the method proposed here. However, all of them are lacking any explicitly given mathematical model, and neither do they provide any theoretical justification as for example entropy optimization. For a sort of justification, they merely offer to replace arithmetic by geometric or harmonic means. From a mathematical point of view, they all have a strong heuristic flavor in common. This situation should be avoided in mathematical geology (Schaeben, 1988).

REFERENCES

Imhof, J., 1987, *in* H.J. Bunge, ed., Theoretical methods of texture analysis: DGM, Oberursel, p. 79–88.

Matthies, S., and G. Vinel, 1982, Phys. Stat. Sol. (b), v. 112, p. K111–K120.

Pawlik, K., and J. Pospiech, 1987, *in* H.J. Bunge, ed., p. 127–139.

Schaeben, H., 1988, Geol. Rdschau., v. 77, p. 591–607.

——— in press a, Journal of Applied Physics.

——— in press b, This volume.

——— H. Siemes, and S. Auerbach, in press, Phys., Stat., Sol. (b).

Williams, R.O., 1968, Journal of Applied Physics, v. 39, p. 4329–4335.

SCHAEBEN, H., University of Bonn, Bonn, Federal Republic of Germany

Benefits of Use of Explicit Mathematical Models: I. Methodology of Entropy Optimization in Texture Goniometry

Pursuing the approach of finite series expansion to the texture problem of reproducing the orientation distribution function (odf) from its pole distribution functions (pdf) measured in X-ray or neutron diffraction experiments, the mathematical model of entropy maximization is introduced. The entropy maximizing solution yields the odf with the smallest information content consistent with the experimental data, thus avoiding artificial textural components for which there is no evidence in the measured pdfs, especially "ghosts" caused by the specific properties of the diffraction experiment.

The solution of the inverse problem of texture goniometry presented here uses finite series expansion, explicitly relates to a well-established body of mathematical and information theory including Bayesian probability, and is given by an algorithm that allows efficient programming (Schaeben in press; Schaeben et al, in press).

The concept of entropy optimization principally works as follows. A real valued random variable Z takes on numerical values z within a given interval C_i with probabilities $0 \leq p_i \leq 1$. Then

$$W(Z \epsilon C_i) = p_i, i = 1,..,n$$
$$\Sigma_{i-1}^{n} p_i = 1$$

The information content with respect to a possible result of an experiment that probes the ensemble ("draw a ball from the urn") depends on the distribution function, or more specifically on its form. In case of a uniform distribution, the lack of a priori knowledge is complete: Each possible result of an experiment has the same a priori probability; uncertainty is the larger the larger the number *n* of possible results. In case of a well-pronounced unimodal distribution, the lack of knowledge is relaxed: Some results are much more likely or probable than others. In case this distribution degenerates to Dirac's δ-distribution, the knowledge is complete again and we are back to the (almost) deterministic situation.

The information theoretical concept of entropy provides a quantitative measure of the information inherent in, resp the uncertainty represented by the probability distribution. It is maximum if the distribution is uniform and smaller if not. The entropy of a discrete random variable respectively of its discrete probability distribution is formally defined as

$$S(p_1,\ldots,p_n) = -\Sigma_{i-1}^{n} p_i \ln p_i$$

Now, in case of a given, i.e., known, distribution, its information content quantified by its entropy may be calculated immediately. In another case, the discrete probabilities may be thought of as unknowns given by constraints in form of a system of linear equations

$$Ap = q \tag{1}$$

with some given (m × n) matrix A and some experimentally given values q_i, i = 1,...,m. If the matrix A is of full rank, the system has the unique solution

$$p = A^{-1}q \quad (2)$$

and the information of this distribution quantified by its entropy may again be calculated immediately. If the matrix is of rank deficiency, the system has several different solutions, in fact infinitely many, of which we may choose a specific one by an additional mathematical model assumption. Thus, we may well choose the solution with maximum entropy. Fortunately, there exists an iterative algorithm that converges toward the entropy optimizing solution of the system (1) in case it is rank deficient and that is numerically straightforward and stable (Censor, 1983).

This situation applies to the problem of texture goniometry, resp. truly quantitative texture analysis.

Adapting the mathematical model of entropy optimization to the purposes of texture goniometry, the texture problem is completely restated in a discrete setting and solved.

Applications of the mathematical model of entropy optimization to pole figure inversion and their numerical results will be presented in a companion communication (Schaeben and Siemes, in press), and it will be shown that it actually yields satisfying numerical results at justifiable computational costs.

The infinite series expansion or harmonic method (Bunge and Morris, 1982) of texture analysis proved in practice sufficient to calculate all macroscopic physical properties of an aggregate that share the symmetry properties of the experimental pole figures, which applies for almost all macroscopic properties of any interest in material sciences. It suits also the purpose of calculating the odf from individual crystal measurements because in this case all the problems around the lack of information in experimental pole figures vanish. However, it proved insufficient for detailed quantitative texture studies, particularly if noncentrosymmetric properties are to be considered, one of which is plasticity. Plasticity not only depends on a given direction but on the full orientation. Adapting iterative techniques (1) to incorporate the required nonnegativity of the odf to be reproduced into the harmonic method and (2) to make it capable of processing incomplete pdfs, it loses all its theoretical fascination, i.e., its mathematical elegance as well as its straightforward computability.

For the specific purposes of geological texture studies, the maximum entropy model solution that is nonnegative and maximally uniform in the sense of information content seems to be adequate, as it provides the safest choice in the sense of the inevitable compromise that must be made in all tomographical problems between spatial resolution and the accuracy of amplitude measurements. More specifically, all solutions of the texture problem are consistent with the data but differ in resolution and noise. Mathematically, there is no reason to favor one solution over any other, but maximum entropy provides the most safely balanced solution.

REFERENCES

Bunge, H.J., and P.R. Morris, 1982, Texture analysis in materials science: London, Butterworths.

Censor, Y., 1983, Proceedings of IEEE, v. 71, p. 409.

Schaeben, H., in press, Journal of Applied Physics.

———, and H. Siemes, in press, This volume.

———, ———, and S. Auerbach, in press, Phys. Stat. Sol. (b).

SCHAEFFER, M.F., Duke Power Company, Charlotte, NC, F.J. HAGYE, Law Engineering Inc., Greenville, SC, R.E. BRYANT, Duke Power Company, Salem, SC, and L.E. McAULIFFE, Westminster, SC

Detailed Studies at Bad Creek Project, Northwestern South Carolina, USA: Identification of Geologic Factors and Their Effects on Construction and Plant Operation

The Bad Creek Project, presently under construction in the Blue Ridge Geologic Province of northwestern South Carolina, is a 1000 megawatt hydroelectric pumped storage station that will consist of four reversible pump turbines in a powerhouse 200 m underground. The underground structures consist of approximately 4000 m of tunnels, 400 m of vertical shafts, and a 132 × 22.5 × 50 m powerhouse. The aboveground structures consist of two large dams and a saddledike (earth-rock construction) that will form the upper reservoir and the intake and discharge structures.

Initial site geologic studies provided a basis for the original design and construction scheme but could not define or anticipate all geologic factors that may influence the final design, construction, and operation of the plant. The studies underway are designed to detect unfavorable geologic conditions not anticipated during design and to document the conditions encountered. The permanent aboveground geologic records provide valuable information for later use in interpreting any possible postconstruction problems (such as excessive seepage, differential settlement, local instability, etc.) and in planning remedial action if necessary. The records will be used to interpret postconstruction foundation instrumentation data. The underground studies assist in the selection of instruments or other methods to study rock mass behavior around the excavations, provide a record of special excavation areas, identify water-bearing fractures and weathered zones, and may be used to correlate any possible tunnel lining problems with mapped geologic features. The primary documents are detailed geologic maps of the dam foundations, intake and discharge excavations, tunnels, shafts, and powerhouse excavation. The features shown on the maps include rock types (soils), foliation, joints and fracture zones, faults and shear zones, and other significant geologic features.

The detailed mapping has delineated several aspects of the deformational history of the site. The understanding of this history is essential for the interpretation of geologic features that may influence the plant construction and operation. The site is underlain by lithologies of the Toxaway Gneiss (granitic gneiss and biotite-hornblende schist) and Tallulah Falls Formation (metagraywacke, mica schist, and amphibolite). The orientation of the dominant foliation is very consistent (N35–50E; 28–41SE) throughout the site area. The earliest set of folds are isolated, intrafolial fragments that indicate the earliest foliation (S_1) has been transposed by the development of the dominant foliation (S_2) related to isoclinal folding (second set of folds). The third set of folds are isoclinal to open, fold S_2, and have a variably developed secondary foliation (S_3). S_2 and S_3 are folded by later open fold events. Shear zones are present and generally parallel S_2. Along some of the shears, brecciation has occurred after shearing and has been healed by greenschist mineralization. The predominant joint sets in the reservoir area are oriented N77E; 82NW, N42E; 74NW, and N47W; 88SW. In the underground works, the predominant joint set varies between N70W and N70E with steep north and south dips, and a weakly developed set strikes N45W with steep southwest dips. All joint sets have some degree of mineralization, but the northeast and east-west sets contain a greater percentage of mineralized joints. Fault zones are associated with the joints and have offsets ranging from <1 cm to >4 m. Breccia is present along some fault planes and consist of rock, quartz/feldspar, and vein quartz fragments in a fine-grained matrix of chlorite–epidote. All fault planes are mineralized by various combinations of greenschist minerals. This faulting correlates with the brittle deformation in the shear zones.

The geologic studies are directly implemented into the ongoing construction and design of the plant. Geologic factors are identified and their possible impacts on construction and future plant operation are analyzed. Based on these studies, remedial action is planned and implemented to minimize possible effects. The aboveground geologic studies during construction to date have been used to: (1) determine orientation and angle of grout holes for the formal and exploratory grout curtain along the dam centerlines and a portion of the reservoir rim, (2) identify geologic factors affecting final preparation for dam foundations, (3) identify and project a major water-bearing fault zone approximately perpendicular to the centerline of the Main Dam and the design and implementation of a supplementary grouting program to minimzie seepage, (4) identify and define the extent of a zone of blocky rock related to minor down-dip movement on the right abutment of the Main Dam and plan the treatment of the area to insure a stable foundation and prevent seepage and excessive water pressure gradients at the foundation-core contact, (5) identify areas in rock cuts that require rock bolts for long-term stability in the intake/discharge excavations, and (6) plan remedial action to stabilize an old landslide above the discharge structure.

The underground geologic studies have been used to: (1) predict the location of significant geologic features along proposed tunnel alignments, (2) relocate the portal of a bypass tunnel from a potentially unstable to a stable area, (3) project and predict the location of water-bearing features in the penstock, powerhouse, and draft tube areas, (4) locate drainage holes in the penstock area to intercept known and potential water-bearing features, (5) select the location of piezometers for monitoring water pressure in the rock mass surrounding the penstocks, (6) determine the location and orientation of grout holes in the penstock tunnels to minimize water seepage, (7) redesign the powerhouse drainage system to intercept known and potential water-bearing features, (8) select areas for installation of radial drains to intercept groundwater behind the concrete lining in the vertical access shaft, and (9) identify areas requiring additional ground support.

The detailed studies have resulted in modification of the initial design and construction schemes to incorporate the actual geologic conditions encountered. The identification of geologic factors during the construction of the Bad Creek Project and the implementation of remedial programs to minimize their effects helps to ensure the long-term safety and reliability of the plant from a geologic perspective.

SCHAFER, A., University of Bonn, Bonn, Federal Republic of Germany

Intermontane Saar-Nahe Basin (West Germany): Variscan Structure and Permo-Carboniferous Molasse Sedimentation

The Saar–Nahe Basin is a Permo-Carboniferous molasse basin within the Variscan orogenic belt (Schäfer. 1986).

The basin is partially exposed at an area of 100 × 40 km just south of the Rheinisches Schiefergebirge in the Saxothuringian Zone (see Figure 1: SNB). The total basin extension, however, below a thick Mesozoic cover, is approximately 300 × 100 km (Donsimoni, 1981). It is confined by the ancient Variscan sutures that divide Central Europe in several structural blocks (Behr and Heinrichs, 1987; Anderle, 1987; Schwaab, 1987).

In that basin, 4.1 km Upper Carboniferous and 1.6 km Lower Rotliegend grey fluvio-lacustrine deposits and additional 1.8 km Upper Rotliegend red alluvial deposits together with various types of volcanites were accumulated (as is shown by the overall stratigraphic profile compiled from three wells; Schäfer, in press).

The depositional environments of Upper Carboniferous to Lower Rotliegend are alluvial along the northwestern margin of the basin, but braided and meandering fluvial, deltaic, and lacustrine within the basin (Schäfer and Stapf, 1978; Schäfer and Sneh, 1983; Schäfer, 1986; Boy, 1987). Marine ingressions are not encountered.

Significant amounts of coal were developed only in the Westfalian: cwA-cwD middle part (Kneuper and Schönenberg, 1964; Engle, 1986; Kerp and Fichter, 1985). Together with coal, volcanic tuffs (so-called tonsteins) proved important for stratigraphy (Hess and Lippolt, 1986; Burger and Billig, 1987).

Paleocurrent data indicate a sediment transport during Westfalian from the northern Rhenohercynian (Rheinisches Schiefergebirge). The paleocurrents supplied lithic arenites (Zimmerle, 1976) to a depocenter south of the presently exposed part of the Saar–Nahe Basin. However, from Stefanian onward the sediments were delivered from a Moldanubian source (Schwarzwald, Vosges, and Massif Central). The depocenter of the basin progressively shifted northeastward and approached the Rheinisches Schiefergebirge. Arkosic arenites with an increasing content of total feldspar and of K-feldspar were deposited. This tendency was even further enhanced at the beginning of Upper Rotliegend (Schäfer, in press).

Tectonics at the end of Westfalian induced uplift and gentle folding. Later peneplenization of these beds occurred with up to 1500 m erosion in places. Stefanian and Lower Rotliegend above followed discordantly.

At the beginning of Upper Rotliegend crustal shortening formed anticlinal structures together with overthrusting (Engel, 1985). Syngenetic magmatism produced rhyolitic intrusives and tuffs as well as lati-andesitic lavas (Eckhardt, 1979). Both structural events and a more arid climate had consequences for the Permo-Carboniferous geomorphology of the basin and favored an alluvial style of sedimentation (Stapf, 1982; Sneh, 1988).

REFERENCES

Anderle, H.J., 1987, The evolution of the South Hunsrück and Taunus borderzone: Tectonophysics, v. 137, p. 101–114.

Behr, H.J., and T. Heinrichs, 1987, Geological interpretation of DEKORP 2-S: a deep seismic reflection profile across the Saxothuringian and possible implications for the Late Variscan structural evolution of Central Europe: Tectonophysics, v. 142, p. 173–202.

Boy, J.A., 1987, Die Tetrapoden-Lokalitäten des saarpfälzischen Rotliegenden (?Ober-Karbon–Unterperm; SW-Deutschland) und die Biostratigraphie der Rotliegend- Tetrapoden: Mainzer geowiss. Mitt., v. 16, p. 31–65.

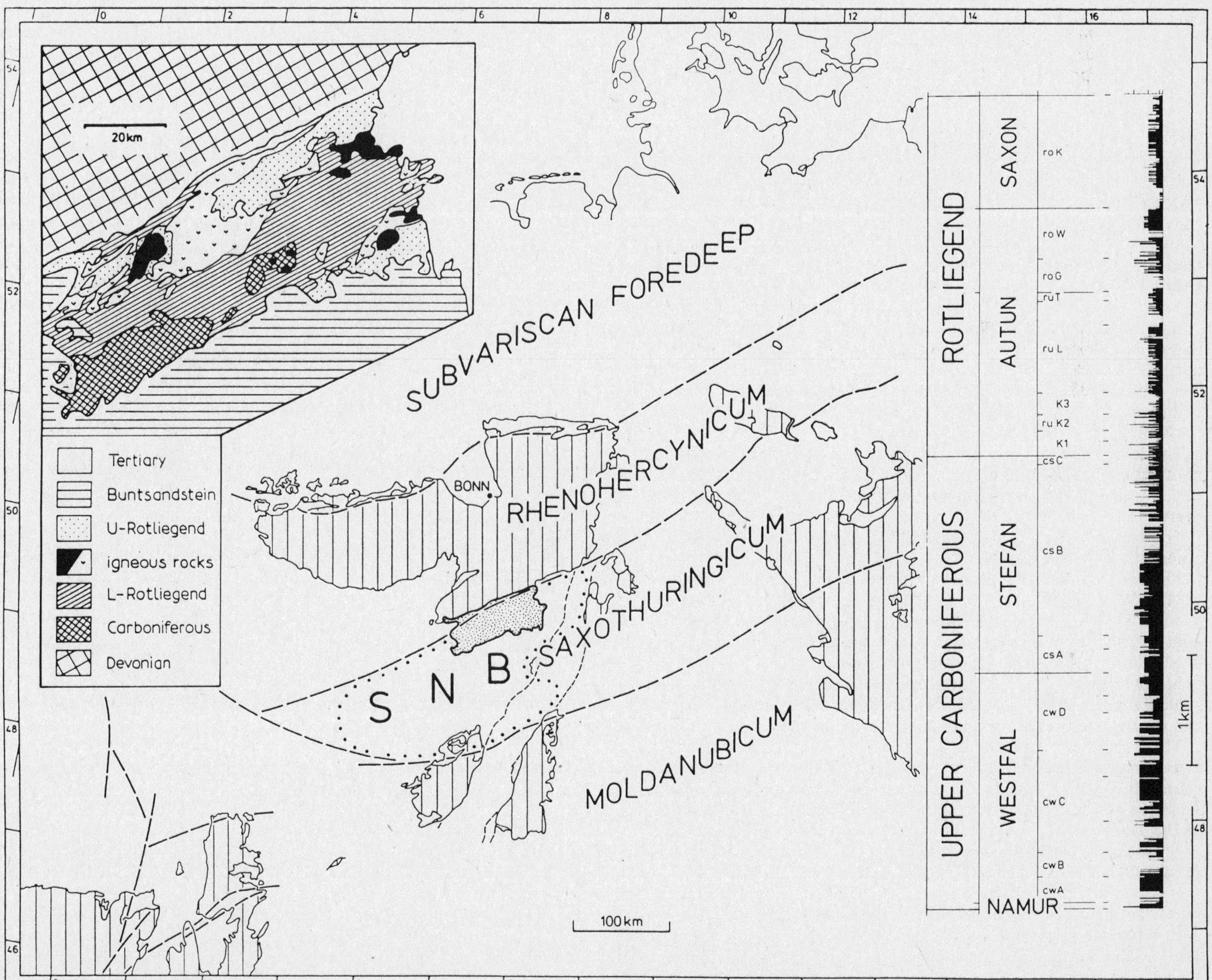

Figure 1.

Burger, K., and M. Billig, 1987, Vulkanogene Glasscherbenrelikte im Kohlentonstein 3 des Saar-Lothringer Ober-karbons: Z. dt. geol. Ges., v. 138, p. 103–119.
Donsimoni, M., 1981, Le bassin houiller lorrain: BRGM, v. 117, 102 p.
Eckhardt, F.J., 1979, Der permische Vulkanismus Mitteleuropas: Geol. Jb., v. D-35, p. 3–84.
Engel, H., 1985, Zur Tektogenese des Saarbrücker Hauptsattels und der Südlichen Randüberschiebung, *in* G. Drozdzewski, et al, Beiträge zue Tiefentektonik westdeutscher Steinkohlenlagerstätten, p. 217–236.
——— 1986, Palynologie stefanischer Flöze des Saarlandes und der übergang vom Karbon zum Perm: Cour. Forsch.-Inst. Senckenberg, v. 86, p. 113–124.
Hess, J.C. and H.J. Lippolt, 1986, ^{40}Ar/^{39}Ar ages of tonstein and tuff sanidines: new calibration points for the improvement of the Upper Carboniferous time scale: Chemical Geology, v. 59, p. 143–154.
Kerp, H., and J. Fichter, 1985, Die Makrofloren des saarpfälzischen Rotliegenden (?Ober-Karbon–Unterperm; SW-Deutschland): Mainzer geowiss. Mitt., v. 14, p. 159–286.
Kneuper, G., and R. Schönenberg, 1964, Neue Ergebnisse zur Frage des Werdeganges der Saar-Nahe-Senk und der Entwicklung des Saarbrücker Steinkohlengebirges: Congr. Int. Strat. et de Geol. du Carbonifere, Paris 1963, p. 505–511.
Schäfer, A., 1986, Die Sedimente des Oberkarbon und Unterrotliegenden im Saar-Nahe-Becken: Mainzer geowiss. Mitt., v. 15, p. 239–365.
——— in press, Variscan molasse in the Saar-Nahe Basin (W-Germany), Upper Carboniferous and Lower Permian: Geol. Rundschau.
——— and A. Sneh, 1983, Lower Rotliegend fluvio-lacustrine sequences in the Saar-Nahe Basin: Geol. Rundschau, v. 72, p. 1135–1145.
——— and K.R.G. Stapf, 1978, Permian Saar-Nahe Basin and Recent Lake Constance (Germany): two environments of lacustrine algal carbonates, *in* A. Matter and M.E. Tucker, eds., Modern and ancient lakes: Spec. Publs. Int. Ass. Sediment., v. 2, p. 83–107.
Schwab, K., 1987, Compression and right-lateral strike-slip movement at the Southern Hunsrück Borderfault (Southwest Germany): Tectonophysics, v. 137, p. 115–126.
Sneh, A., 1988, Permian dune patterns in Northwestern Europe challenged: Journal of Sedimentary Petrology, v. 58, p. 44–51.
Stapf, K.R.G., 1982, Schwemmfächer-und Playa-Sedimente im Ober-Rotliegenden des Saar-Nahe-Beckens (Permokarbon, SW-Deutschland). Ein überblick über Faziesanalyse und Faziesmodell: Mitt. Pollichia, v. 70, p. 7–64.
Zimmerle, W., 1976, Die Tiefbohrung Saar 1: Petrographische Beschreibund und Deutung der erbohrten Schichten: Geol. Jb., v. A-27, p. 91–305.

SCHALAMUK, I., M. A. LEGUIZAMON, and M. E. TERUGGI, Universidad Nacional de La Plata, La Plata, Argentina

New Greenstone Belt from San Miguel, of the South Precambrian, Paraguay

The greenstone belt is located in the south Precambrian of Paraguay, that covers an area of about 2000 km^2. The greenstone belt is situated 5 km north far from San Miguel, in the eastern area of Paraguay. It is a narrow belt of 5-6 km long and tens of m of wide, lineal shaped in E-W direction within the precambrian sialic basement. Two well defined shear zones are the boundaries of the belt and the basement.

The lowest unit of the greenstone is a bimodal sequence formed by komatiitic ultramafic flows with typical microspinifere structure and perioditic serpentinizated komatiites, which alternate with thin levels of cherts.

The komatiite rock content about 25-26 percent of MgO, % SiO_2, 3000 ppm of Cr y 2000 ppm of Ni.

There is a tabular deposits of talc in the ultramafic rock, with direction of NE-SW dipping subvertically to 80° NE. It has 550 meters long and 5.5 m wide. The ultramafic rocks (komatiites) and talcons (schists and talc of high purity) are cut by micropegmatities, aplitees and thin vein of quartz. The country rock is a characteristic pink gneiss.

Structurally the rocks of the greenstone belt are disposed in isoclinal folding dipping 75-80° toward north, possibly is may be a very tight synclinoria with vertical faults that limits infolede sequences.

The mineral association and the well preserved primary structures show a very low grade metamorphism (greenschist facies) some of the structures, especially in the ultramafic flows, have developed orientated cataclastics facies, as well as irregular breccias in the zones close to the shear which limits the basement.

The siolic basement pregreen rock does not show too much tectonic complexity. Only shows a gneissity parallel to trend of the greenstone belt.

It would indicate perhaps a new arrangement by tectonic action of the internal structure in the compressive effects generated by the belt.

In Table 1 and Figure 3 present chemical analyses and molecular proportion ratio diagrams show their Komatiitic nature rocks. The Paraguayan Komatiite greenstone belt might represent the suture of collision of small continental plates; in which case it would be a remnant of an early oceanic crust.

PRECAMBRICO SUR. Según A. KANZLER, 1987 (simplif. y modif.)

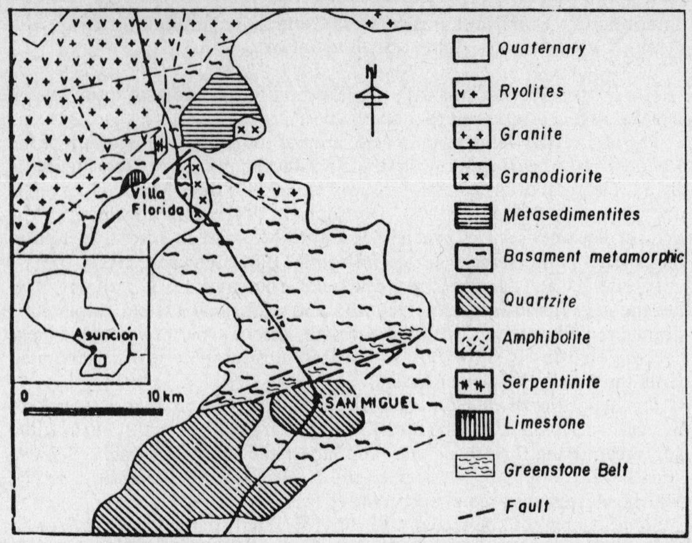

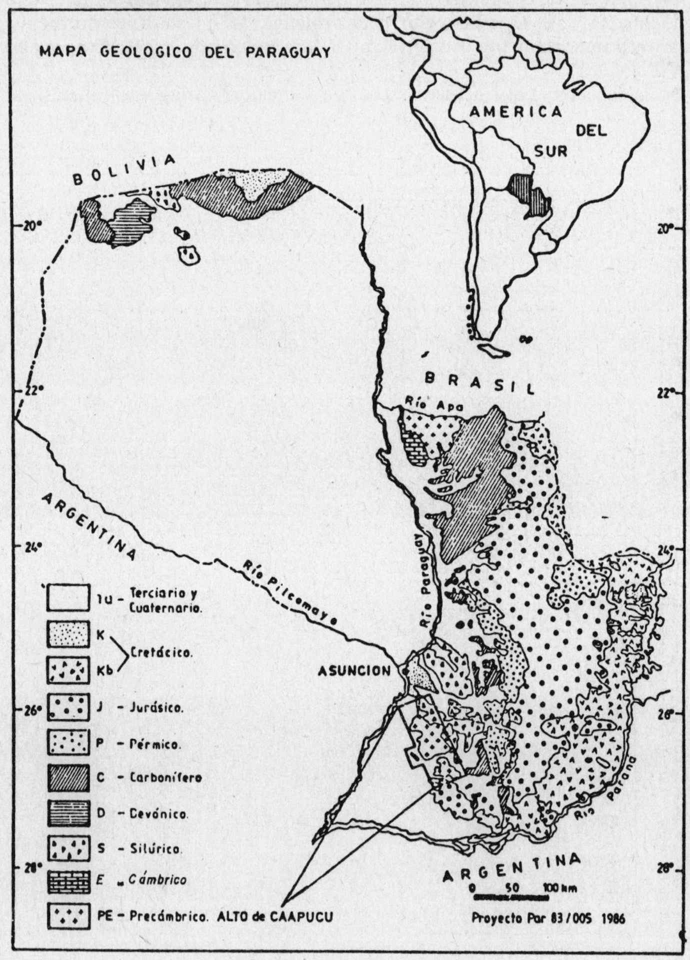

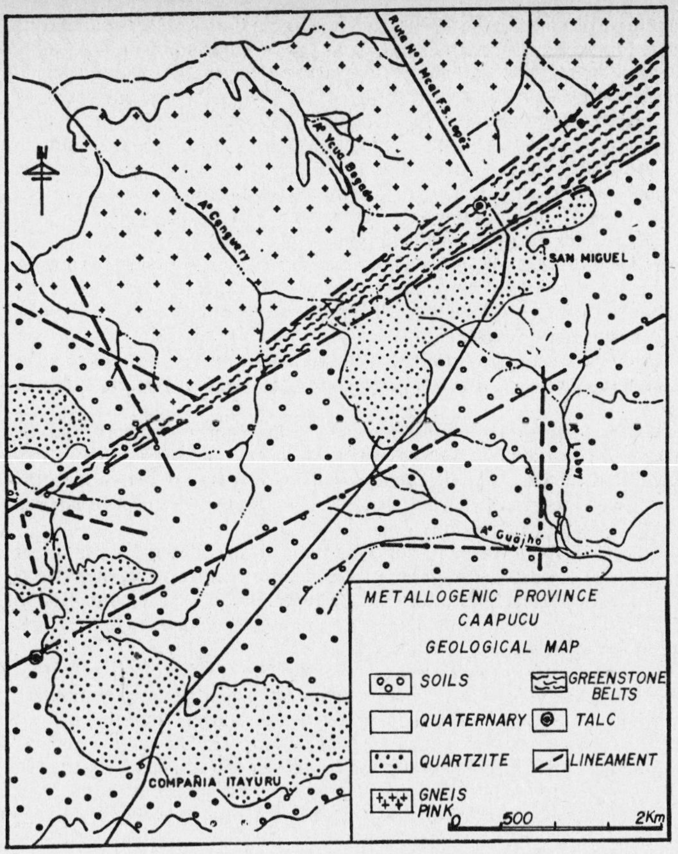

Table 1

CHEMICAL ANALYSES OF PARAGUAYAN KOMATIITES

Whole-rock analyses

	83.	56.	58.	57.	81.	23.
SiO2	45.80	44.50	46.10	45.80	46.20	44.80
Al2O3	7.70	10.20	9.80	9.70	10.30	9.90
TiO2	0.85	0.10	0.60	0.70	0.90	1.40
FeO+Fe2O3	9.80	8.70	9.40	8.80	9.50	9.50
MgO	26.90	25.80	26.10	25.90	25.10	26.50
CaO	7.85	8.40	7.15	6.85	6.20	6.20
K2O	0.60	0.40	0.35	0.75	0.45	0.35
NaO	0.30	0.35	0.40	0.60	0.55	0.40
P2O5	-	0.10	0.07	-	0.06	0.08
Cr	-	3.300	3.000	-	3.000	1.900
Ni	-	1.800	1.500	-	2.500	1.800
Cu	35	41	22	28	32	35
Y	14	12	17	18	14	16
Zr	43	40	39	44	53	55

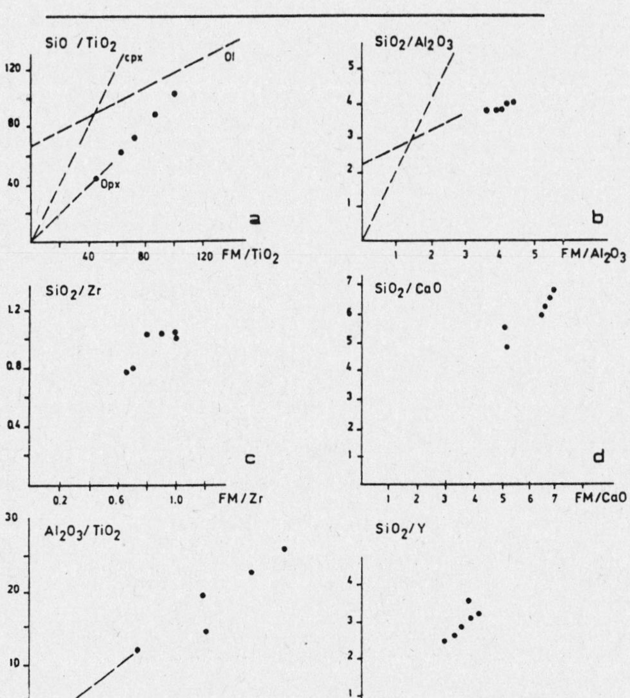

Fig. 3 Molecular proportion ratio (MPR) for major and trace elements. Dashed lines in diagrams(a) and (b) are clinopyroxene (Cpx), olivine (Ol) and ortopyroxene (Opx). In diagram (e) dashed line is ortopyroxene. (Opx).

SCHEEPERS, D. J., and D. H. CORNELL, University of Stellenbosch, South Africa

Host Rock Protolithology of Metamorphosed Ore Deposits

Models for the origin of metamorphosed syngenetic ore deposits depend on interpretations of the original rock types in which the ore is enclosed. Thus, while the conditions and tectonic setting of metamorphism are determined from the sequence of mineral parageneses, an equally important task for the economic petrologist is the derivation of protolithology.

This work aims to establish a practical, objective, and general set of criteria by which the protolith of samples in any metamorphic facies can be derived. It is based on our experience of massive sulfide deposits of the 1200 Ma Namaqua–Natal Structural Province of South Africa, which are found in highly deformed amphibolite to granulite grade host rocks. Computer programs are being developed to do the calculations.

It is assumed that metamorphism has been *isochemical,* apart from water and CO_2. The products of ore forming hydrothermal processes are included in our protolith classification.

The scheme requires the following information:

1. Representative mineral mode, *or* major element analysis.
2. Modal proportion of carbonate minerals *or* CO_2 analysis.
3. Textural information such as the scale and mineralogical expression of banding.
4. Some trace element and rare earth element criteria may be used for refinement.

The following definitions are used for the major protolith groups. They are distinguished using the criteria given, and a further distinction of genetically different subgroups is also made.

Metapelite originally had >20% (by mass) clay minerals. It plots in the A-An-F triangle of ACF diagrams, and also in the A′-20K-20F triangle of A′KF diagrams, and has <80% quartz plus feldspar. The products of sericitic alteration plot as metapelite but often have high Mg/Fe ratios and birdwing rare earth profiles.

Basite has a major element composition similar to that of basalt, although it need not have an igneous protolith. On ACF it plots in the Anorthite-Diopside-F triangle. *Orthobasite,* of igneous origin in generally unlayered or has only <10mm plagioclase rich (segregation) layers, and may show relict igneous textures. *Parabasite,* of sedimentary origin (silicate- carbonate mixes) often has >10 mm thick layers with parageneses which define a C-wards vector in ACF. Further distinctions are based on rare earth profiles and trace elements such as Ni and Cr. The tectonic setting of orthobasites may be revealed by the careful use of geochemical discriminant and MORB-normalized spidergrams.

Ultrabasite originally contained >80% dark minerals. It plots within the tremolite-F-garnet triangle on ACF, and within 20% of F on A′KF. *Metaperidotite,* of igneous or mantle origin, plots below the An-F line on ACF, and normally has >500 ppm Cr and Ni, and flat rare earth profiles. The products of *chloritic alteration* may plot above the An-F line, and often have heterogeneous rare earth profiles.

Iron formation originally contained >30% iron oxide, sulfide, or carbonate minerals. It may still do so, or may contain predominant iron end-member silicates. It plots in the triangle F-50A- 50C, and has Fe/Mg ratio >5. A <10 mm color banding indicates chemical sedimentary origin. Such rocks with >10% Mn may represent the top of hydrothermal deposits. Minor elements like Cu, Pb, and Zn may discriminate between massive sulfides related to localized igneous hydrothermal or to regional sedimentary processes.

Marble, the metamorphosed equivalent of limestone, dolomite, or (exceptionally) carbonatite, contains >50% carbonate minerals. Metacarbonatites usually contain >100 ppm Sr and light rare earths, while sedimentary marbles have <10 ppm of these elements.

Calcsilicate originally contained >30% carbonate minerals, now partly or completely (depending on grade) reacted with silicates to form Ca and Mg-bearing minerals. It plots in the anorthite-diopside-C triangle of ACF. The common >10 mm layering comprises parageneses which when plotted in ACF may define vectors from C towards the silicate protolith end-member (metapelite, orthobasite).

Quartzofeldspathite or leucogneiss contains >80% quartz and feldspar (or feldspathoid). >40% quartz indicates *paragneiss,* (>80% metaquartzite), of sedimentary origin. Leucogneisses with <40% quartz are either igneous *orthogneiss* or paragneiss, and may be distinguished by textural and chemical homogeneity tests. *Alkali orthogneiss* may contain nepheline and other feldspathoids. Zircon morphology studies may distinguish extrusive, plutonic, and sedimentary gneisses.

When a rock has been assigned to one or more categories using the above criteria, an attempt is made to prove the credibility of the interpretation. This is done by modeling the major element composition of the rock using mixes of mineral analyses considered typical of the protolith. Only if an adequate math is achieved is the interpretation regarded as valid.

Protolithology interpretations using this scheme have enabled us to characterize the Prieska Mine Zn-Cu deposit as a proximal volcanogenic massive sulfide surrounded by former sericitic and chloritic alteration zones, lying on a metadacite and covered by quartzofeldspathic and basaltic volcanics and their sedimentary derivatives. This deposit was previously described by Middelton (1975) as an unmetamorphosed volcanogenic deposit, and by Wagener (1980) as a sedimentary basin deposit. We interpret the Annex deposit 5 km south of Prieska Mine, and the Areachap Zn mine 200 km north, to represent distal volcanic deposits with transported chloritic hydrothermalites set in reworked volcanic sediments had basaltic volcanics. The copper occurrences at Boksputs in the middle of the belt are set in a sequence of orthobasites interpreted as metabasalts, in which variable amounts of carbonate were interlayered. This gave rise to calcsilicate bands, at first suggesting that the sequence comprised parabasites. The protolithological assemblage established for the mid-Proterozoic Areachap Group in which all these deposits occur, leads us to propose ore deposition in a rifted back-arc basin environment.

SCHEIBNER, E., New South Wales Geological Survey, Sydney, Australia

Tectonics and Some Aspects of Metallogeny of Southwest Pacific

The Circum-Pacific Tectonic Maps are part of an integrated series of maps showing the distribution of known hydrocarbon and mineral resources in relation to the geology, tectonics and lithospheric dynamics of the Pacific region. The tectonic maps distinguish oceanic and continental crustal domains, the passive margin interface between these domains, and active plate margins. With the search for hydrocarbons in mind, there is further subdivision into basement complexes, deemed to be too deformed to contain hydrocarbons, and cover sequences. Symbols in red mark active plate boundaries and colored patterns show tectonic units at active plate margins. The tectonic development of oceanic crust is shown by episodes of seafloor spreading which appear to correlate with the rift and drift sequences at passive continental margins and episodes of tectonic activity at active plate margins. Color coding shows the age of deformation of basement complexes, transitional tectonic (molassic) and reactivation sequences, and the inception of platform basins. Colored patterns show the age of inception of continental-margin (rift and drift) sequences.

It is generally recognized that tectonic setting is one of the determining factors in the formation of metal deposits (e.g. Cox et al., 1986). The clarity with which such relationships can be defined varies from well to hypothetical. For example, the distribution of Mn nodules in present oceanic basins has been mapped. The black smokers at spreading centers are a good model for the metallogeny of massive sulfides in an oceanic crustal environment, while the origin of many epigenetic deposits is contentious. In similar way the usefulness of tectonic maps in the study of metallogeny and mineral exploration varies. They can be used in the direct search deposits related to oceanic lithospheric rocks, as the distribution of these rocks in orogenic belts is shown. The various active plate margin tectonic units, especially volcanic arcs, are the subject of the latest gold rush, besides being the subject of the search for polymetallic deposits. It appears that lamprophyres, especially shoshonitic rocks, are the ultimate source of gold in the crust (Rock and Groves, 1988). Shoshonites, besides characterizing mature volcanic arcs, occur in an extensional setting in volcanic arcs (Stern et al., 1988). The Southwest Quadrant Circum-Pacific Tectonic Map depicts such extensional settings and thus could be used in the rationalization of exploration and metallogenic studies. Various extensional features, especially rifts and volcanotectonic depressions, are the sites of polymetallic syn- and epigenetic deposits, with epithermal gold deposits creating the most excitement in explorers.

Recently structures similar to metamorphic core complexes were described from the D'Entrecasteaux Islands (Davies and Warren, 1988), in an area of active westward propagating rifting of the Woodlark Basin. The presence of similar structures elsewhere in the Southwest Pacific can be expected. The metallogenic importance of core complexes, and more specifically of the extensional processes which are so well understood in the Cordielleran region of the United States, has not as yet been appreciated in the southwest Pacific region.

Experience from the Northeast Quadrant of the Circum-Pacific shows that besides extensional settings, contractional structures, i.e., the various sutures between tectonostratigraphic terranes, deserve special attention in gold exploration. Some more metallogenic problems related to tectonic settings will be discussed in the talk.

A question can rightly be put: in what way are the Circum-Pacific Tectonic Maps useful for the study of metallogeny? The scale of these maps is a drawback in their usefulness in practical exploration, but they give a unique synoptic view of the tectonics of over half of the earth's surface, and they can be directly integrated with other sets of data, including the distribution of major known deposits. These tectonic maps should be useful in the design of long-term strategies in exploration, and especially in the testing of some of the conceptual models (cf. Mitchell and Garson, 1981; Sawkins, 1984) that attempt to explain the genesis of metal deposits.

REFERENCES

Cox, D. P., and D. A. Singer, eds., 1986, Mineral Deposits Models: U. S. Geol. Surv. Bull. 1963, 379 p.

Davies, H. L., and R. G. Warren, 1988, Origin of eclogite-bearing, domed, layered metamorphic complexes ("core complexes") in the D'Entrecasteaux Islands, Papua New Guinea: Tectonics, v. 7, p. 1–21.

Mitchell, A. H. G., and M. S. Garson, 1981, Mineral deposits and global tectonic setting: London, Academic Press, 405 p.

Rock, N. M. S., and D. I. Groves, 1988, Can lamprophyres resolve the genetic controversy over mesothermal gold deposits?: Geology, v. 16, 538–541.

Stern, R. J., S. H. Bloomer, Ping-Nan Lin, E. Ito, and J. Morris, 1988, Shoshonitic magmas in nascent arcs: New evidence from submarine volcanoes in the northern Marianas: Geology, v. 16, p. 426–430.

Sawkins, F. J., 1984, Metal deposits in relation to plate tectonics: Berlin, Springer Verlag, 325 p.

SCHEIBNER, E., New South Wales Geological Survey, Sydney, Australia

Geoscience Transect Broken Hill to Sydney, New South Wales, Australia

This geoscience transect follows the earlier "Geodynamics" profile #4 through the Tasmanides (cf. Rickard et al., 1983). The present transect conforms with the requirements of the Global Geoscience Transect Program in that it attempts to show the composition and structure of the entire crust. Unfortunately a deep seismic reflection survey along the transect is in the distant future and so the crustal interpretations are based on fragmentary crustal seismic data, the projection of deep seismic data from the adjoining region to the north (cf. Finlayson, this volume), gravity, aeromagnetic and magnetotelluric data, the composition of lower crustal and mantle xenoliths in young intraplate igneous rocks, the results of new geologic mapping, and the latest structural geologic and tectonic studies. In brief, what is presented is a working hypothesis based on the work of earth scientists at universities, in the exploration industry, and in state and federal public service research organizations.

The Tasman Line (Fig. 1) separates the orthotectonic Tasman Fold Belt System (TFBS) or Tasmanides in the east from the Early to Middle Proterozoic Australian craton in the west and its associated epicratonic basins and paratectonic fold belts. The composite, Late Proterozoic to Mesozoic, orogenic TFBS comprises five fold belts: Kanmantoo (KFB), Thomson (TFB), Lachlan (LFB), Hodgkinson–Broken River (HBFB) and New England (NEFB), and the molasic Bowen–Sydney Basin.

The transect starts in the west, in the paratectonic Adelaide Fold Belt (AFB), specifically the Broken Hill block, an internal massif. This internal massif was affected (retrograde metamorphism and emplacement of various igneous bodies) during the Early Paleozoic when both the paratectonic AFB, to which it is the basement, and the orthotectonic KFB were formed.

Further east the transect traverses the Early Paleozoic KFB in the region where it is complicated by the cross cutting Darling River Lineament, an ancient transform structure. In the region south of the transect, beneath the Cainozoic Murray Basin, geophysical data indicate westward thrusting of structural zones in the KFB. The two coincidental Bouguer gravity and aeromagnetic ridges (Mt. Wright–Pathaway and Stavely–Lake Wintlow) represent remnants of two Cambrian volcanic arcs, or the same arc doubled up by strike-slip displacement, or perhaps a split arc. Latest structural studies in South Australia indicate westward thrusting of the continental margin complexes (Kanmantoo Supergroup) probably due to collision of the Mt. Wright–Pathway arc during the Early Paleozoic Delamerian Orogeny, and a foreland fold and thrust belt structure developed at the same time in the AFB. Synkinematic granitic intrusive activity was typical for the zone of strongest deformation, with post-kinematic igneous activity affecting a wider region. During the Ordovician A- and M-type intrusives were associated with extension and limited terrane dispersal.

Further east is the LFB (cf. Scheibner, 1987), in itself a complex orogenic belt whose Early Paleozoic history culminated in a Late Ordovician to Early Silurian collisional belt during the Benambran Orogeny. The Ordovician to Early Silurian collisional belt during the Benambran Orogeny. The Ordovician volcanic arc, built upon a micro-continent, collided along the Gilmore Suture with the back-arc basin complex in the west. During the subsequent Silurian terrane dispersal (strike-slip displacement, rifting and basin formation), the area east of the Gilmore Suture was affected. During the latest Silurian Bindi–Bowning Orogeny, the area immediately east of the Gilmore Suture was deformed and basins inverted. The areas further east were not affected until Middle Devonian time. During the Early Devonian, rifting and basin formation occurred to the west of the Gilmore Suture.

A working hypothesis depicted on the transect suggests that the Gilmore Suture was reactivated and functioned as a deep detachment fault during the Early Devonian extension. Lithospheric thinning along this detachment fault during the Early Devonian extension. Lithospheric thinning along this detachment enabled upwelling of the asthenosphere and derivation of M-type Alaskan-type intrusions. In the zones of strong upper crustal (brittle) extension within the Devonian Darling Basin, dense igneous rocks forming sills were emplaced into the ductile lower crust partly to compensate and balance the upper crustal extension. The geometry of the major crustal units were little disturbed during subsequent deformation and therefore the relatively positive gravity ridges in the Darling Depression are related to rifted zones. Strong lower crustal reflectors, possibly such sills, occur in some basinal areas further north (cf. Finlayson, this volume).

The eastern part of the transect depicts only the cover relationship of the latest Carboniferous to Triassic Sydney Basin with the LFB. The fore-deep thrust relationship of the NEFB with the Sydney Basin occurs further north and is the subject of another transect. However, the crustal structure beneath the Sydney Basin is interesting. A distinctive longitudinal Bouguer gravity ridge can be followed from the southern end of the Sydney Basin for over 1200 km all the way into the Bowen Basin in Queensland. This gravity ridge is interpreted as the site where the latest Carboniferous–Early Permian rifting occurred, and the length to width ratio of over 24 excludes a pull-apart origin. This ratio points to a cause by plate interaction, possibly at an active plate margin at that time in the east. Published with permission of the Secretary, New South Wales Department of Mineral Resources.

REFERENCES

Rickard, M. J., B. A. Duff, R. J. Korsch, E. J. Heidecker, E. Scheibner, and N. J. Turner, 1983, Tasman Orogen, Australia: Amer. Geophys. Union Geodynamics Series n. 10, p. 5–19.

Scheibner, E., 1987, Paleozoic tectonic development of eastern Australia in relation to the Pacific region: Amer. Geophys. Union Geodynamics Series n. 18, p. 133–165.

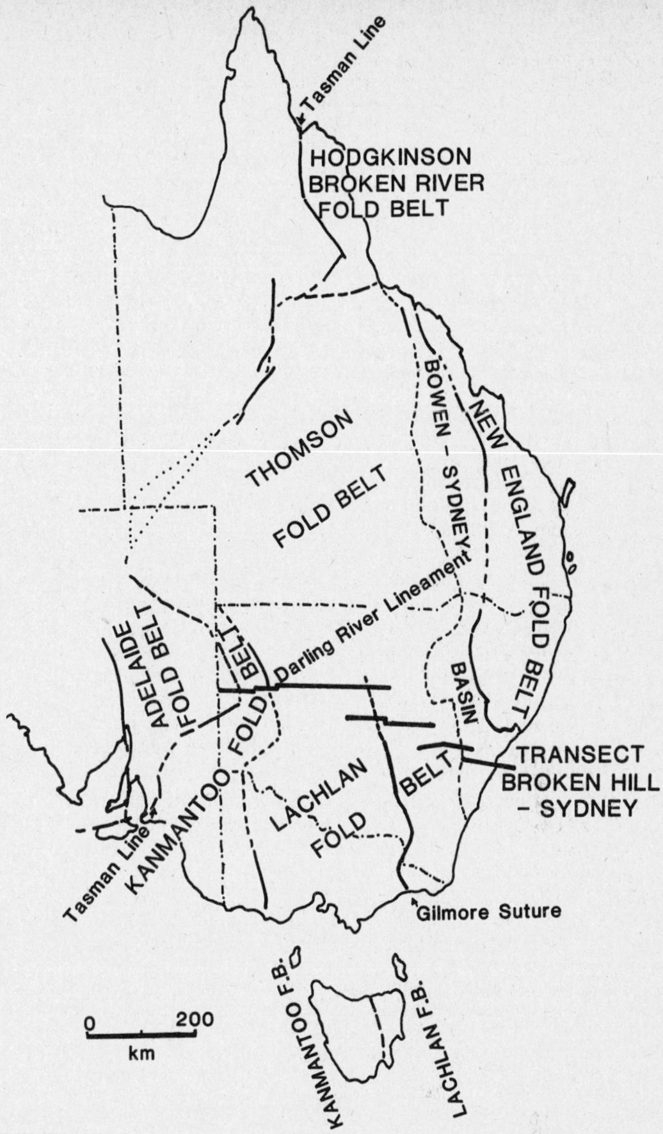

Figure 1—Position of the transect (heavy line). Fold belts east of the Tasman Line constitute the Tasman Fold Belt System.

SCHERER, W., N. JORDAN and O. COLMENARES, INTEVEP S.A., Los Teques, Venezuela

The Maracaibo Coal Basin—Western Venezuela, South America

The Maracaibo basin is a well known hydrocarbon province in Western Venezuela with several giant oilfields producing since the early nineteen hundreds. Explorations for mineral resources in the Andean foothills bordering this basin during the last 200 years have shown the presence of important coal seams over a major part of this basin. The recent development of two coal districts: Guasare in the northwest and Lobatera-Santo Domingo in the southwest corners of this province, as well as the increasing exploration of the Urumaco coal district in the northeast, indicate the need to establish a regional geologic evaluation of the Maracaibo coal basin, Figure 1. The purpose of this paper is to present the coal distribution and geologic characteristics of the major coal forming cycles on a sedimentary tectonic frame of basin evolution and to show pertinent geologic and mining characteristics of the producing coal mines.

High volatile bituminous coals and lignites are found in two sedimentary sequences of the Maracaibo basin. The regressive phase of Cycle I, the Cretaceous-Paleocene-lower Eocene sequence, deposited widespread coal beds in the deltaic formations of the incipiente Andean drainage in Western Venezuela and Eastern Colombia. The Los Cuervos and Marcelina formations of Paleocene to lower Eocene age are the main coal bearing formations in this sequence, Figure 2. Their thickness varies between 300-550 m, generally increasing to the north.

Both the transgressive and regressive phases of Cycle II, the upper

Eocene-Miocene Andean sequence originated widespread lignite beds in the Maracaibo and Eastern Venezuela or Oficina basin. In the Maracaibo coal basins, the Carbonera and Cerro Pelado formations of Miocene age are the main coal bearing formations in this sequence. Their thickness varies between 380-1000 m, again increasing to the north. The Oficina, Quebradón and Naricual formations of Oligo-Miocene age are the coal bearing formations in the Eastern Venezuela or Oficina basin.

The intimate association of coal seams and petroleum reservoirs in the Oficina basin and the Orinoco Oil Belt has been reported in the literature for 30 years (1) (2). To a lesser degree this same association can be observed in the subsurface of the Maracaibo basin.

The number of coal seams and coal cycles varies according to sedimentary-tectonic conditions in any individual locality of the basin. A maximum of 35 seams are reported in Cycle I in the Santo Domingo coal district, of these 19 are economically exploitable having a thickness of 0,8 m or more. The Guasare district has 22 well defined coal seams varying in thickness between 0.5 to 12 m. The important seam N° 4 has an average thickness of 12 m in the Paso Diablo Mine.

The Cycle II sequence produced more than 80 individual, regionally correlatable coal cycles in the Oficina basin under exceptionally favorable geologic conditions. The same interval in the Maracaibo basin shows up to 25 coal seams, 8 of them with a thickness of more than 0.8 m.

The most remarkable characteristic of the Western Venezuelan coals is their large variability in maceral and microlithotype composition. Depositional environments vary from strictly terrestrial producing humic coal to lacustrine environments producing sapropelic coals and lignites. Redeposition of pre-existing coal seams in paralic marine environments has been reported from unstable shelf areas. This variability reflects varying sources of detrital organic matter, different organic depositional facies and diverging diagenetic histories. Regional understanding of these conditions is attempted in a series of compositional, trend and facies maps. Markov chain analysis is used to quantitatively describe the conditions of coal facies development through geologic time on a macroscopic bed to bed scale for each sequence. Organic petrography on microscopic scale using microlithotype analysis unravels some of the more puzzling regional changes in coal quality, and relates present day organic facies (3) (4) to ancient organic depositional environments.

Characterization of organic depositional environments, paleoecological changes and plant communities existent at the time of deposition of each major cycle is based on qualitative and quantitative attributes of palynomorph assemblages.

The local geologic conditions of the three main coal districts in this basin, i.e. Guasare, Lobatera and Santo domingo, Figure 3, are shown together with the coal quality (6). Heat capacity available reserve data and industrial uses of these coals are presented in Table I.

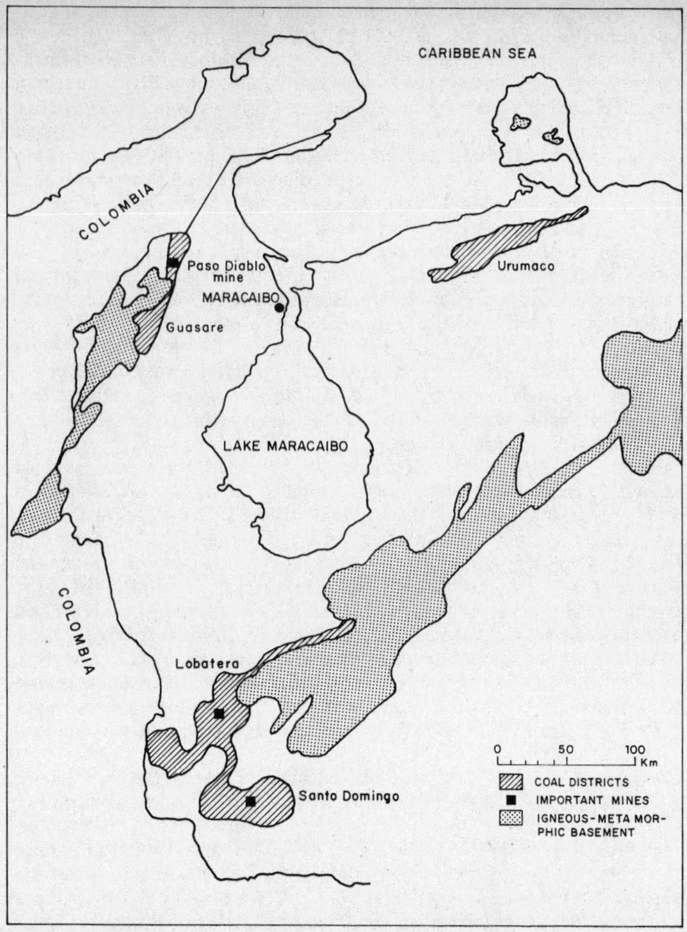

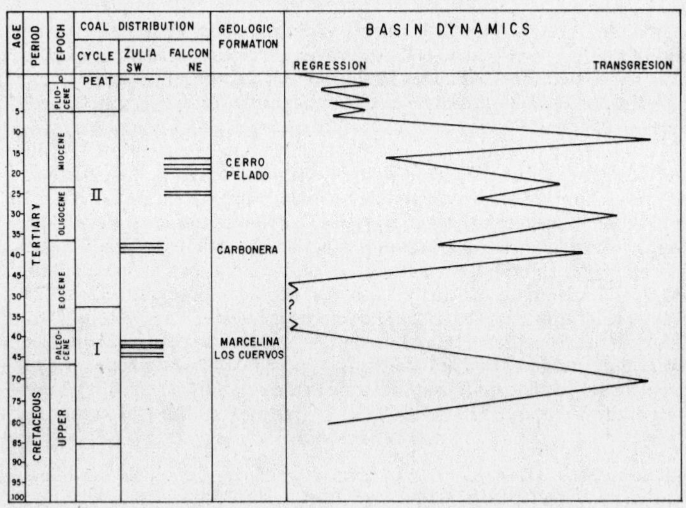

REFERENCES

Banks, L.M., 1959, Oil-Coal Association in Central Anztegui, Venezuela. AAPG Bull. v. 43, No. 8, p. 1998-2003.

Scherer, W. and Jordan, N., 1985. Consideraciones sobre Origen, Distribucin y Usos de los Carbones en la Faja Petrolfera del Orinoco, in Memorias VI Conqreso Geologico Venezolano. Vol. VI, p. 3547-3584.

Jordan, N. 1988. Organic Facies, Diagenesis and Sedimentation of Peats on the Eastern Coast of Lake Maracaibo. Western Venezuela. Rev. Tec. Intevep. V. 8 No. 1.

Jordan, N. 1988. Ciclotemas del Terciario y Cuaternario de Venezuela. Jornadas 50 Aniversario Escuela de Geologa, Minas y Geofsica. Caracas, Rev. Geos No. 29.

Ascanio, G., Daz, M., Camacho, F. and Venegas, G., 1983, Report of the Coal Commission, E. Prato Coordinator, Ministry of Energy and Mines, Venezuela, 78 p.

Lopez, V.M. and Salsamendi, M.K., 1978, El Carbon Mineral. Ministerio de Energa y Minas, Venezuela, 175 p.

Sociedad Venezolana de Geologos y Carbozulia, 1988, Memorias II Jornadas Geologicas del Carbon, Maracaibo, 267 p.

IEA (International Energy Agency) COAL RESEARCH-1983, London. Concise Guide to world coalfields, 308 p.

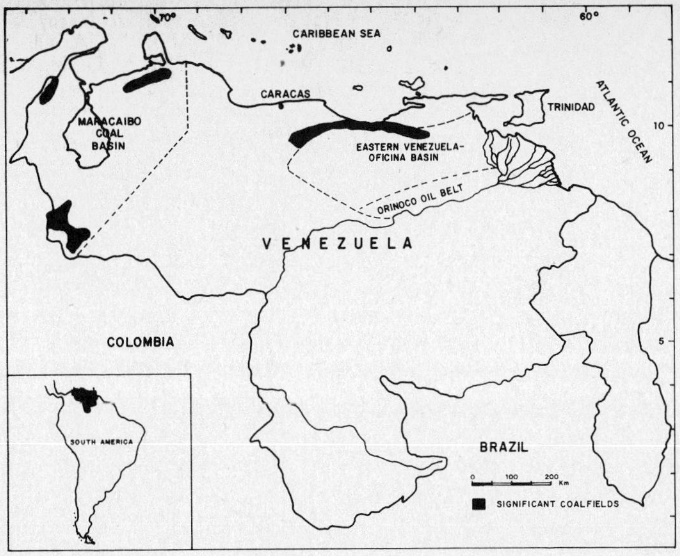

TABLE I - MARACAIBO COAL BASIN

COAL DISTRICT	Accesible Coal [8] or Proven Reserves [5] x 10^6 Ton	Heat Capacity Kcal/Kg	Uses
GUASARE	1117	6800 - 7100	thermoelectric heating
LOBATERA	40	6300	coke
SANTO DOMINGO	300	5300 - 7300	thermoelectric heating chemical
URUMACO	46	N.A.	N.A.
MARACAIBO BASIN	1503		

SCHEUING, D. F., Lamont-Doherty Geological Observatory at Columbia University, Palisades, NY, and R. D. JACOBI, State University of New York at Buffalo, Amherst, NY

Significance of Geochemical Abundances in Matrices of Melanges: Study from Central Newfoundland

The geochemical signature of a melange matrix provides information on the origin of the melange as well as providing a basis for regional correlations. The island of Newfoundland is well known for its excellent exposures of melange units. These melanges have played key roles in the classic plate tectonic reconstruction of the Newfoundland Appalachians (e.g., Bird and Dewey, 1970). Through the use of detailed structural and sedimentological analyses we have come to some understanding of the strain history in individual melanges, but a few outstanding problems still remain: (1) the source terrane and age of the melange matrix, and (2) the interrelationship among the melanges, i.e., do various melanges exposed in orogenic belts represent vestiges of a single decollement, or does each represent a discrete shear zone? Our research shows that these questions can be addressed through geochemical fingerprinting of the melange matrix.

The Dunnage melange of north central Newfoundland is thought by some to mark an Iapetus suture (Bird and Dewey, 1970), as is the Carmanville melange to the east (Pajari et al., 1979). The plate tectonic setting of the Cobbs Arm melange (located north of the Dunnage) is least understood. Karlstrom et al. (1982) proposed that the Dunnage Zone was tectonically transported westward during the early Acadian orogeny and that the slip zone underlying the Dunnage Zone was a melange horizon now represented by the various melanges in central Newfoundland.

The geochemistry of the Dunnage melange closely resembles that of the New Bay Formation, which lies to the west within Notre Dame Bay. Significant geochemical differences are observed between Dunnage argillites and the pre-Caradocian Davidsville Group shales to the east, as well as other black shales of Caradocian age, such as the Dark Hole Formation to the north, and the Luscombe Formation and Lawrence Harbor Shale to the southwest. The geochemical similarity with the New Bay and dissimilarity with other pre-Caradocian shales supports earlier suggestions that the source for the Dunnage melange was the New Bay Formation. Independent structural and sedimentological data suggest the initial disruption in the Dunnage melange resulted from slumping towards the east. Geochemical discrimination diagrams, including SiO_2 vs. Mn, SiO_2 vs. Al_2O_3, Mn vs. TiO_2, and (K_2O/Na_2O) vs. SiO_2, conclusively show that the matrix of the Dunnage was derived partly from a volcanic arc. However, comparisons between the analyses of the Dunnage matrix to argillaceous sediments from modern fore-arc and back-arc settings does not discriminate a specific paleotectonic setting.

Comparison of the geochemistry of Dunnage matrix argillites with that of the Carmanville melange reveal wide differences in elemental abundances, suggesting that the two melanges had different sources. Karlstrom et al. (1982) and Pajari et al. (1979) both infer from structural and sedimentological evidence that the Dunnage and Carmanville melanges are very close related. If both melanges formed in a convergence zone and are of approximately the same age, then their distinct sources suggest that the broadly coeval melanges do not mark a single fault zone along which the Dunnage Zone was transported relatively great distances. Rather, the sampling of different sources in different melanges suggests an imbricate thrust zone, or possibly a duplex, similar to those found in fore-arc or retro-arc thrusting (assuming the master faults were originally subhorizontal with dip- slip motion). If the faults were originally steeply dipping with strike-slip motion, the different sources for the melanges imply the juxtaposition of crustal blocks by multiple fault zones, such as in southern California.

REFERENCES

Bird, J. M., and J. F. Dewey, 1970, Lithosphere plate-continental margin tectonics and the evolution of the Appalachian orogen: Geol. Soc. of Amer. Bull., v. 81, p. 1031-1060.

Karlstrom, K. E., B. A. van der Pluijm, 1982, Structural interpretation of the eastern Notre Dame Bay area, Newfoundland: regional post-Middle Silurian thrusting and asymmetrical folding: Can. Jour. Earth Sci., v. 19, p. 2325-2341.

Pajari, G. E., R. K. Pickerill, and K. L. Currie, 1979, The nature, origin, and significance of the Carmanville ophiolitic melange, northeastern Newfoundland, Can. Jour. Earth Sci., v. 16, p. 1439-1451.

SCHIDLOWSKI, M., and H. GORZAWSKI, Max-Planck-Institut fur Chemie, Mainz, Federal Republic of Germany, and I. DOR, Hebrew University of Jerusalem, Jerusalem, Israel

Isotopically Heavy Biomass from Microbial Mats: Predictor Variables from Experimental Hypersaline Ponds

Stromatolite-type microbial ecosystems from coastal brine pools of the Sinai Peninsula were shown to be substantially enriched in ^{13}C (Schidlowski et al, 1984, 1985) and to represent the heaviest organic carbon as yet encountered in the biosphere ($\delta^{13}C_{org}$ = -5.4 ± 1.1 ‰ [PDB]). Ineffective discrimination against ^{13}C by the mat-forming microbenthos had been explained in terms of reduced CO_2-solubilities in heliothermal brines and the concomitant establishment of a diffusion-limited assimilatory pathway in which the large isotope fractionations of the RuBP carboxylase reaction were basically suppressed.

To quantify inferred relationships between $_7d^{13}C_{org}$ and various environmental parameters, isotope analyses were performed on the microbial biomass of experimental hypersaline ponds for which the basic hydrological (temperature, pH, ionic strength) and microbiological (community composition, productivity rates) background data were known. Recordings of $\delta^{13}C_{org}$ as a function of salinity and the light- and temperature-related productivity gradient along the slope of the ponds have shown that the impact of productivity rates of $\delta^{13}C_{org}$ values is decidedly more pronounced than that of ionic strength. Expanding on our previously proposed concept for the generation of "superheavy" organic matter, we conclude that the exceptional enrichment of ^{13}C in the biomass of these microbial ecosystems is primarily a response to high rates of productivity sustained under conditions of eutrophication (notably in sabkha environments). Increased CO_2-consumption rates and salinity-induced low CO_2 supplies are likely to combine in exerting excessive demands on the local (aquatic) carbon dioxide pools of these and related habitats. Under such conditions, the trend to utilize heavy carbon is proportionally enhanced since high productivity rates cannot be sustained in these environments except by a largely indiscriminate metabolization of all CO_2 supplies available, giving rise to exceptionally heavy organic matter. This is consonant with our previous conclusion (Schidlowski et al, 1984, 1985) that the formation of unusually heavy biomass is primarily a habitat-specific adaption to inadequate CO_2 supplies.

REFERENCES

Schidlowski, M., U. Matzigkeit, and W.E. Krumbein, 1984, Naturwissenschaften, v. 71, p. 303–308.

——— W.G. Mook, and W.E. Krumbein, 1985, *in* G.F. Friedman and W.E. Krumbein, eds., Hypersaline ecosystems, Berlin, Springer, p. 381–401.

SCHLAGER, W., Free University, Amsterdam, Netherlands

Carbonate Platforms vs. Siliciclastic Systems in Sequence

Construction of sea-level curves from sequence stratigraphy is complicated by the fact that different depositional systems respond differently to changes of sea level and that factors other than sea level may control sequence geometry and sequence boundaries. These points are examined by comparing siliciclastic systems and carbonate platforms.

It is a well-established fact that in the Pleistocene, siliciclastic sediment supply to the deep sea was at its maximum during glacial lowstands of sea level. Pleistocene carbonate platforms were exactly in antiphase to this rhythm. They produced and exported most sediment during interglacial highstands when the platforms were flooded ("highstand shedding," Figure 1). Highstand shedding is best documented for the Bahama Banks, but occurs also in other Caribbean platforms as well as in the Indian Ocean. Highstand turbidites also differ in composition from their lowstand counterparts as they contain more ooids, pellets, and grapestones—grains that require flooded bank tops for their formation. Turbidite abundance and composition together provide a faithful record of the Pleistocene sea-level cycles that is not easily erased by diagenesis and should be looked for elsewhere in the geologica record. Geometrically, platforms respond to sea level by forming highstand turbidite wedges and lowstand drapes of pelagic sediment—the opposite of siliciclastic systems.

Drowning unconformities are an example for sequence boundaries caused by lithologica change. Flanks of carbonate platforms are generally steeper than siliciclastic slopes (Schlager and Camber, 1986). When the platforms are drowned and buried by siliciclastics, an unconformity ensues because the clastics are unable to assume the steep carbonate slope angle or because they are shed from other directions than the carbonates. Examples of drowning unconformities include the mid-Cretaceous unconformity in the Gulf of Mexico as well as unconformities on the Early Cretaceous platforms off eastern North America and off western Afric. In the African example, we reinterpret the Valanginian lowstand of Vail et al (1977) as a drowning unconformity, based on observations on carbonate slope angles as well as on evidence for platform drowning in the carbonate turbidites.

Highstand shedding and drowning unconformities illustrate that not all depositional systems respond alike to changes in sea level and that sequence boundaries may be caused by lithologica change. These stratigraphic turning points need not be related to sea level. In a very general way, sequence boundaries can be viewed as changes in the pattern of sediment input and dispersal in a basin. Sea-level fluctuations are one way to induce such changes, but tectonic movements and environmental change represent important alternatives that compete with sea level for control on sequence stratigraphy.

REFERENCES

Droxler, A. W. and W. Schlager, 1985, Glacial versus interglacial sedimentation rates and turbidite frequency in the Bahamas: Geology, v. 13, p. 799–802.

Schlager, W., and O. Camber, 1986, Submarine slope angles, drowning unconformities, and self-erosion of limestone escarpments: Geology, v. 14, p. 762–765.

Vail, P. R., R. M. Mitchum, R. G. Todd, J. M. Widmier, S. Thompson, J. B. Sangree, J. N. Bubb, and W. G. Hatlelid, 1977, Seismic stratigraphy and global changes of sea level, *in* C. E. Payton, ed., Seismic stratigraphy—applications to hydrocarbon exploration: AAPG Memoir 26, p. 49–212.

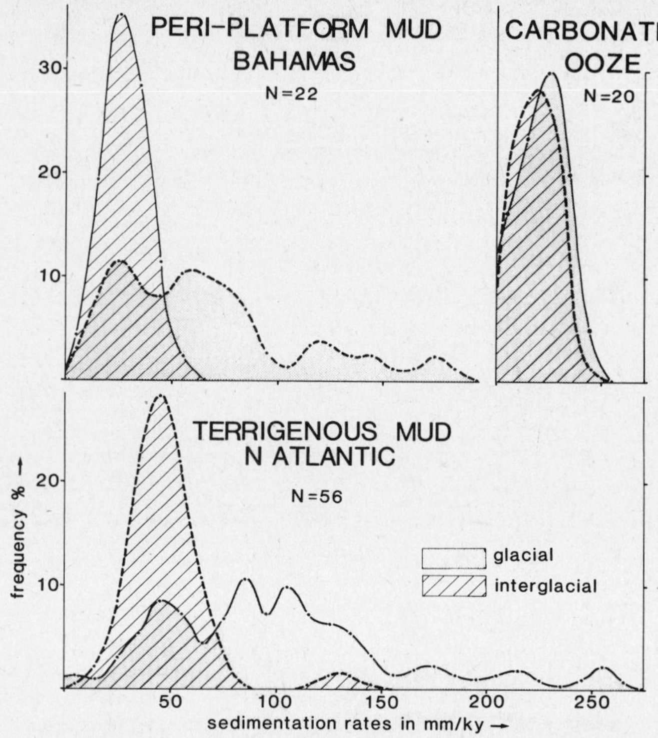

Figure 1—Frequency distribution of glacial and interglacial sedimentation rates for three different dee-sea sediments. In terrigenous muds from the continental rise, rates are low and stable during glacials, and high and variable (turbidite-controlled) during interglacials. In peri-platform muds from the flanks of the Bahama Banks, the pattern is reversed, conforming to the principle of highstand shedding of carbonate platforms. In pelagic carbonate ooze, rates change little between glacial and intergalcial periods. Data from Droxler and Schlager (1985).

SCHLAGER, W., Free University, Amsterdam, Netherlands

Cretaceous Carbonate Platforms, Oceanic Anoxia and Sea Level

Coral reefs and carbonate platforms are highly sensitive monitors of the marine and atmospheric environment that complement the dee-sea record in many ways. The cretaceous is a time of peculiar alternation of spectacular growth and abrupt demise of carbonate platforms. Growth was particularly extensive in the central Pacific and along the young margins of Tethys and North Atlantic, where some platforms grew in excess of 100 m/Ma and rose up to 4 km above the ocean floor. Pronounced episodes of crisis and demise occurred in the Valanginian, the Aptian, the late Albian, and the late Cenomanian. At lease some of these crises can be shown to have been global in nature and caused by reef growth being outpaced by the relative rise in sea level. Burial by siliciclastic sediments usually follows the drowning with some delay; platform termination by clastic influx is the exception. It is very unlikely that the drowning events were caused solely by rapid rises of sea level, because documented and inferred rates of eustatic sea-level change for the period are one order of magnitude lower than the growth potential of healthy platforms. We suggest that changes in the ocean environment reduced the growth potential of platforms and so led to the mass drownings. In this context, it is intriguing that platform crises tend to coincide with oceanic anoxic events in the pelagic realm (Figure 1). Open-ocean overturn may be the causal link between drowning and anoxia (Hallock and Schlager, 1986). Overturn would have created episodic upwelling that flooded the reefs with nutrients, caused blooms of heterotrophic organisms, and thus damaged the autotrophic, carbonate-secreting benthos—an effect amply demonstrated by modern sewage outfalls on reefs. The coincidence of platform crises and oceanic oxygen deficiency also holds for other parts of the geologic record, notably the late Devonian (Frasnian/Famennian reef extinction) and for the Early Jurassic (Toarcian), when Tethyan platforms had to give way to deeper-water black shales.

The world wide drowning of platforms had most diverse implications for the global sediment record. We will discuss two of them. (1) Drowning seems to have shifted the locus of calcium carbonate fixation in the ocean from shallow-water benthos to nannoplankton. In the Late Jurassic and earliest Cretaceous, carbonate platforms rimming the young Central Atlantic were tying up calcium at the rate of 12% of the present annual river supply to the ocean. The elimination, in the Valanginian through Cenomanian, of this platform belt and similar ones in the Tethyan Ocean closed a significant sink in the calcium carbonate cycle. The reduction of shallow-water carbonates coincides with a rapid increase of planktonic carbonate deposition in the world ocean. Even though cause and effect in this relationship remain unclear, the coincidence in time is remarkable and the end effect is that of a shift from the shelf reservoir to the deep-sea reservoir of calcium carbonate.

(2) Another effect of platform drowning is the formation of pronounced unconformities and sequence boundaries because of change in sediment input and dispersal in the basin. These drowning unconformities resemble lowstand unconformities in geometry but relate to lithologic change during rises or highstands of sea level.

REFERENCES

Hallock, P., and W. Schlager, 1986, Nutrient excess and the demise of coral reefs and carbonate platforms: Palaios, v. 1, p. 389–398.

Jenkyns, H. C., 1980, Cretaceous anoxic events: from continents to oceans: Journal of the Geological Society of London, v. 137, p. 171–188.

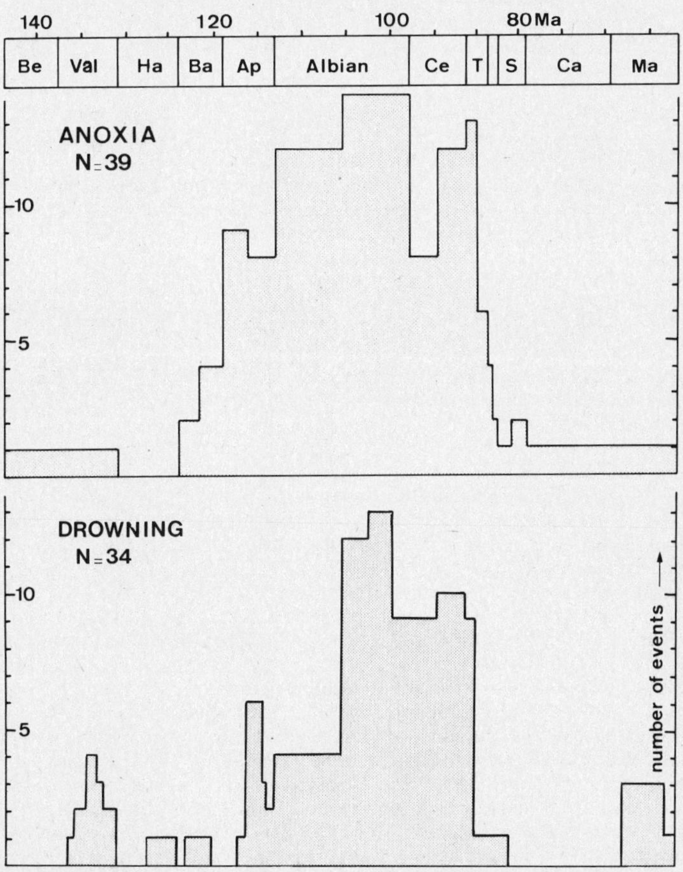

Figure 1—Frequency distribution through time of oceanic anoxic events (top) and platform drowning (bottom) in the Cretaceous. Anoxic events after Jenkyns (1980).

SCHLEE, J. S., and J. M. ROBB, U. S. Geological Survey, Woods Hole, MA

Sedimentary Domains on Mid-Atlantic Continental Rise

A detailed GLORIA sidescan survey (125,000 km^2) of the continental slope and rise off New York, New Jersey, Delaware, Maryland, and Virginia has revealed a complex association of depositional features and transport channels on the upper continental rise; the relatively flat upper rise deepens from 2000 to 4000 m over a distance of 150 km. The upper rise has four large sedimentary domains or areas of distinctive acoustic pattern: large area of slump and debris flows off the Virginia–North Carolina coast, three triangular gather areas where sediment transport paths converge, large irregular areas of sedimentary drape, and several conspicuous channelways. Most of these subdivision boundaries are curt across the isobaths, and the mosaic appears to show the cumulative effect of Pleistocene–Holocene downslope gravitational processes.

The slump-debris flow area is a massive tongue-shaped mass that is up to 40 km wide and that extends 120 km seaward. It appears as a streaked light grey area on the mosaic. The varied sonic expression on the mosaic is interpreted to represent a complex of slumps, slides, and debris flows. Seismic profiles in the shallower areas show closely spaced diffuse hyperbolae over a moderate relief of 300 m. A few profiles show a double bottom over an older topography of low relief. Three subparallel scarps are present upslope from the slide complex on the continental slope.

The three gather areas (Norfold, Baltimore–Wilmington, and Accomac) are convergent dispersal ways at the base of the slope where transport paths of sediment from many submarine canyons are combined into major channelways that cross the continental rise. These channels provide major dispersal routes to the lower rise and abyssal plains beyond. High-resolution profiles across the gather areas show multilayered drape and zones of hyperbolic returns where channels ways are cut.

Areas of hemipelagic drape lie between the gather areas and appear as a uniform dark gray on the mosaic; in some parts, the sediment drape is finely gullied and occurs adjacent to major channelways. High-resolution seismic profiles are marked by gently rolling, closely spaced hyperbolae that change laterally to a multilayered signature; however, over most of the area, the multilayered signature prevails. Toward the seaward part of the drape areas, a few narrow channels trend down slope toward the lower rise.

Channelways traverse the rise in broadly curving arcs within which are narrow active relatively straight channels. Some of the channelway systems, like the Hudson, are fairly direct pathways to the lower rise with only a few meanders. Others, at the apex of gather areas (Baltimore–Wilmington), are up to 9 km wide and extend toward the lower rise in broad arcs that are well defined topographically. These channelways are obvious drainage ways for sediment traversing the rise from the "headwaters" of the slope to the abyssal plain.

The GLORIA mosaic shows active patterns of sediment transport on the upper rise. We infer that the transport is probably Pleistocene–Holocene, mainly because the sources of active sediment input are presently far removed from the shelf edge; and most of the present-day sediment delivered to the shelf is trapped in the many coastal estuaries (Meade, 1969) along the Atlantic coast. Yet many sampling and coring studies show that the rise is veneered over large areas by a layer of varying thickness of silty clay, which the GLORIA signal can see into to reveal Pleistocene sedimentary patterns beneath.

Most patterns of sediment transport on the sonographs show movement toward the abyssal plain. Linear streaks of debris flows and the morphology of channelways all indicate the importance of downslope transport sediment in the recent past. The upper continental rise of the greater New York Bight has been seen as a site for deposition of pelagic sediment from gather areas and through channelways.

The subtleties of shading and texture of the GLORIA mosaic are sufficiently varied so that any future efforts to distinguish the many major sedimentary provinces will benefit from locating sample locations on the mosaic.

REFERENCE

Meade, R. H., 1969, Landward transport of bottom sediment in estuaries of the Atlantic Coastal Plain: Journal of Sedimentary Petrology, v. 39, n. 1, p. 222–234.

SCHLEICHER, H., and J. KELLER, Mineralogisches-Petrographisches Institut, Freiburg, Federal Republic of Germany, and U. KRAMM, Münster, Federal Republic of Germany

Sr, Nd, and Pb Isotope Studies on Alkaline Volcanics and Carbonatites from Kaiserstuhl, Federal Republic of Germany

The Miocene alkaline volcano of the Kaiserstuhl belongs to the Cenozoic volcanic province of Central Europe and is situated in the southern part of the Rhinegraben rift valley. Geophysical evidence reveals updoming of the crust–mantle boundary in this region, possibly as the result of a mantle plume.

The volcanic and subvolcanic rocks of the Kaiserstuhl complex include a great variety of rocks from the tephritic and phonolitic families (e.g., leucite tephrites, essexites, shonkinites, theralites, phonolites, ledmorites, tinguaites, hauynophyres). Olivine nephelinites and melilitites as well as basanites (limburgites) are also of importance. In the subvolcanic center carbonatites occur in form of stocks (sövites) and numerous dikes (alvikites). Also extrusive carbonatitic volcanism (lapilli tuffs) is reported.

In geochemical terms, a subdivision in K-resp. Na-dominant series seems to be possible. The Sr isotopic data strongly support these results. $^{87}Sr/^{86}Sr$ ratios of primary mantle melts (olivine nephelinites, melilitites, and nepheline-basanites, the main representatives of the Na-series) range from 0.7032 to 0.7040 ($^{143}Nd/^{144}Nd$ = 0.5129–0.51285), supporting the concept of a metasomatic event > 100 Ma ago. On the other hand, different degrees of crustal contamination can be shown for rocks of the tephritic-essexitic as well as the phonolithic families ($^{87}Sr/^{86}Sr$ = 0.7039–0.7051; $^{143}Nd/^{144}Nd$ = 0.51274–0.51284). The detailed interpretation gives evidence for a further metasomatic overprinting, related to the upwelling of the mantle diapir beneath the Kaiserstuhl.

The carbonatitic rocks of the Kaiserstuhl are characterized by a very uniform Sr isotopic composition (mean = 0.70365 ± 2). A genetic linkage between carbonatites and ultramafic diatreme breccias as well as melts from the turjaitic magma family (bergalites) is supported by both Nd and Sr isotopic data.

SCHLEUSENER, J. L., and H. L. BARNES, The Pennsylvania State University, University Park, PA, and S. E. DRUMMOND and D. A. PALMER, Oak Ridge National Laboratory, Oak Ridge, TN

Stability of Acetate in Hydrothermal Solutions

Recent calculations for Pb and Zn complexes with Clminus and HSminus yield solubilities which are below the minimum necessary for transport of Pb and Zn in Mississippi Valley Type ore deposits. This has led to speculation on the role that organic ligands may play in these deposits. For example, acetate (CH_3COO^-) has been found in basin solutions such as petroleum brines at concentrations as high as 10,000 ppm. For acetate to be an effective ligand in the hydrothermal transport of ore metals, it must possess sufficient thermal stability to survive at least one pass through the system.

The stability of acetate depends on the rate of the following simplified decomposition of decarboxylation reactions:

$$CH_3COOH_{(aq)} \rightarrow CH_4 + CO_2$$

$$CH_3COO^{minus}{}_{(aq)} + H_2O \rightarrow CH_4 + HCO_{3-}$$

Previous work in stainless steel and titanium vessels has shown the above reactions to be heterogeneously catalyzed such that the rate of acetate decomposition depends on the solution temperature, pH, and the nature of contacted solid surfaces. The purpose of our research has been to determine the effects that geologically relevant catalytic surfaces such as quartz, calcite, pyrite, hematite, and montmorillonite have on decarboxylation kinetics and therefore on the stability of acetate.

One molar acetic acid solutions were loaded along with crushed and sieved mineral samples into titanium pressure vessels. Sodium acetate was substituted when using calcite because calcite is soluble in acetic acid. The vessels were heated in rocking tube furnaces to 335 °C and 355 °C. Solution samples were analyzed for acetic acid/acetate by both potentiometric titration and ion exchange chromatography. Gas phase samples were collected at temperature at the termination of each experiment and analyzed for composition using mass spectrometry. The following first order rate constants, enthalpies and entropies of activation, and half-lives were derived from the experimental data:

Mineral	$k_{335°C}$ (s^{-1})	$k_{355°C}$ (s^{-1})
Acetic Acid Blank	$8.94x10^{-9}$	$2.54x10^{-8}$
Quartz	$1.51x10^{-8}$	$5.41x10^{-8}$
Ca-Mont	$1.44x10^{-8}$	$8.03x10^{-8}$
Pyrite	$1.19x10^{-7}$	$1.83x10^{-7}$

Na Acetate Blank at 350°C $k = 2.34x10^{-7}\ s^{-1}$

Calcite at 350°C $k = 4.36x10^{-7}\ s^{-1}$

Mineral	$\Delta H^‡$ (kJ/mol)	$\Delta S^‡$ (J/mol)	$t_{1/2}$ 100°C (yrs)
Quartz	197	-76.5	$2.20x10^{11}$
Pyrite	62.0	-281	$1.21x10^{3}$
Ca-Mont	234	-15.2	$2.97x10^{12}$

Quartz, calcite, and Ca-montmorillonite were found to have a negligible catalytic effect on the decarboxylation reaction. The low ΔH, low ΔS, and short half-life for the reaction in the presence of pyrite demonstrates that pyrite is an effective catalyst. In order to test the dependence of the reaction on the amount of solid surface available, acetic acid decarboxylation experiments were conducted at 355 °C with different pyrite surface areas. The first order rate constant varied linearly with the log of (surface area to solution mass ratio) proving as expected, that the reaction is dependent on surface area.

Several surfaces did not yield data from which a first order rate constant could be derived. The experiment for hematite at 355 °C had an extremely high initial rate which decreased quickly with time and, at the conclusion of the experiment, the hematite had been completely converted to magnetite. The experiments for an iron-bearing montmorillonite (3.1% Fe as Fe_2O_3) showed a moderate initial rate which decreased to a rate similar to that for the iron-free montmorillonite after several hundred hours. A possible explanation for this behavior is that the decarboxylation reaction mechanism involves an electron transfer step requiring the availability of an electron accepting species such as Fe^{3+}. The resulting conversion of Fe^{3+} to Fe^{2+} aids the conversion of hematite to magnetite and decreases the ability of montmorillonite to act as a decarboxylation catalyst because Fe^{3+} is depleted.

The stability of acetate in basin solutions thus far has been shown to depend upon temperature; pH; and the availability of effective catalytic surfaces such as pyrite, hematite, and iron- bearing montmorillonite. The durability of acetate in a hydrothermal system, and its potential as an effective ligand for metals transport, depends greatly on the nature of host lithofacies and the residence time of the solution therein.

SCHMIDT, H., Universität Mainz, West Germany

Sequences and Systems Tracts in Fore-arc Coastal Embayments of Central America

The southern Central American landbridge emerged from the collision zone between the Cocos and the Caribbean plates. During Albian and Campanian time, this intraoceanic suture formed as a result of thickening of the oceanic plate due to the Caribbean sill event that gave rise to a plateau of primitive tholeiites. From the Lower to Mid-Campanian time, the Caribbean plate margin was shortened by telescoping. Subsequently during Maestrichtian to Eocene time, an andesitic island arc formed behind this tholeiitic ridge. The Oligocene to Present time sees this island arc gradually transforming into a continental landbridge. A series of short-lived sedimentary basins formed on the Pacific fore-arc side.

Shallow-marine strata that formed within such basins crop out along the Pacific coast of southwestern Nicaragua, Costa Rica, and western Panama. They mirror the complex interaction of (1) changes of sea level, (2) subsidence, (3) tectonic uplift, and (4) variable sediment supply. Changes of sea level and tectonic displacement control the architecture and geometry of the depositional sequences. The facies framework of system tracts, in addition, is controlled by sediment supply.

Sequential stratigraphic correlations within these embayments indicate two depositional sequences that are separated by tectonically enhanced unconformities. These correspond to the TA4/TBI and TB2/TB3 boundaries of second-order supercycles (Haq et al, 1988).

The first depositional sequence covers the time between the Uppermost Oligocene and the Middle Miocene. The basal part of the sequence consists of fine-grained shelf sediments of a lowstand systems tract. The turn from transgressive toward regressive conditions corresponds to a gradual coarsening and finally to progradation of deltaic sediments. A ravinement surface, which was produced by erosional shoreface retreat during a rise of relative sea level, forms the boundary between the lowstand systems tract and a transgressive systems tract. This systems tracts is made up by nearshore sediments and fan delta deposits. In Central Costa Rica, fluvial sediments overlie the marine succession. This depositional sequence shows a general shallow-upward trend, which reflects the gradual decrease in water depth due to filling up of the embayment and local uplift.

The second depositional sequence developed during the Uppermost Miocene and the Pliocene. In comparison to the former troughs, these embayments were smaller and ephemeral. The juxtaposition of fully marine and estuarine-marine facies types, as well as coarse-grained cliff-fall sediments and fan delta deposits within the depositional systems record mainly tectonic control of sedimentation within these basins.

SCHMIDT, P. W., CSIRO Division of Exploration Geoscience, North Ryde, Australia, and Z. X. LI, Macquarie University, North Ryde, Australia

Gondwana Paleomagnetic Data in Relation to Paleozoic Distribution of Continents

The Paleozoic paleomagnetic data of Gondwanaland has been reviewed in the context of continental reconstructions. In the seventies, three markedly different APWP interpretations were put forward, but little advance has been made with this problem until the last few years. There are now a number of new pole positions available from Australia and Africa for the Paleozoic, many coming from studies utilizing field tests. These new poles are from: (1) the late Middle–early Upper Devonian Comerong Volcanics, NSW (Schmidt et al, 1986); (2) the late Silurian–early Devonian Snowy River Volcanics NSW (Schmidt et al, 1987); (3) the late Devonian Canning Basin limestones, WA (Hurley and Van der Voo, 1987); (4) the early–mid-Silurian Air Ring Complex, Nigeria (Hargraves et al, 1987); (5) the late Devonian–earliest Carboniferous Hervey Group, NSW (Li et al, 1988); (6) the late Devonian Worange Point Formation, NSW (Thrupp et al, 1988); and (7) the late Permian Dundee Ryodacite, NSW (Lackie, 1988).

Using only those poles that are classified A and B, according to the scheme of Briden and Duff (1981), a more robust APWP has been constructed for Gondwanaland. This has been used to reconstruct the Silurian to Permian events that lead to the formation of Pangaea. There is a perception at large that such quantitative reconstructions are intrinsically sound. However, it is important to remember the assumptions that underpin these methods. While the Axial Geocentric Dipole (AGD) hypothesis is of paramount importance to paleomagnetism, it is nevertheless widely acknowledged that zonal non-dipole terms have been significant during the Tertiary. This gives rise to an error in paleolatitude estimates. In addition, because of the indeterminancy of longitude, paleomagnetic reconstructions are nonunique, allowing continents to be placed at any relative longitude as long as they do not overlap irreconcilably. For instance, it is possible to close the apparent ocean shown between North American and Gondwanaland in the early–mid-Silurian, to open an ocean for the Silurian–Devonian, or to close the mid–late Devonian ocean. A further degree of uncertainty is incurred through the confidence limits on the pole positions.

The early to mid-Carboniferous model shows alternative Pangaea A2 and B positions for North America permitted by the data. However, it is difficult to envisage how the continents could evolve from B2 into a Wegnerian arrangement before breakup in the Jurassic. The A2 position is preferable. The slight overlap of A2 could be simply due to pole error and non-dipole effects.

Ironically, reconstructions for the Permo-Carboniferous and Permo-Triassic (periods for which poles are subject to less uncertainty) appear to be the least acceptable geologically. The large degree of overlap is inexplicable in terms of the confidence limits on the pole position (see alternative Gondwanaland displaced south) and indicates that one assumption, at least, has been seriously violated. These discrepancies have been noted in the past, and many workers have suggested that non-dipole effects are responsible. It is noted that the offset is in the same sense as that in the Tertiary, i.e., far-sided north poles and near-sided south poles, and deleteriously affects high latitude sites more so than low-latitude sites by virtue of the dipole equation. This consistency also suggests that the non-dipole term may be stable and, to some extent, predictable.

REFERENCES

Briden, J.C., and B. A. Duff, 1981, Pre-Carboniferous paleomagnetism of Europe north of the Alpine orogenic Belt, *in* M. W. MeElhinny and D. A. Valencio, eds., Paleoreconstructions of the continents: Geodynamics Series, v. 2, p. 137–149.

Hargraves, R. B., E. M. Dawson, and F. B. Van Houten, 1987, Palaeomagnetism and age of mid-Palaeozoic ring complexes in Niger, West Africa, and tectonic implications: Geophysical Journal of the Royal Astronomical Society, v. 90, p. 705–729.

Jurley, N. F., and R. Van der Voo, 1987, Paleomagnetism of Canning Basin reefal limestones—new constraints on the Gondwana apparent polar wander path: Abstract 12th Int. Sed. Congress, Canberra, Australia, p. 149.

Lackie, M., 1988, The palaeomagnetism and magnetic fabric of the Late Permian Dundee Rhyodacite, New England, SW: The New England Orogen, University of New England, Armidale, NSW.

Li, Z. X., P. W. Schmidt, and B. J. J. Embleton, 1988, Paleomagnetism of the Hervey group, Central New South Wales and its tectonic implications: Tectonics, v. 7, p. 351–367.

Schmidt, P. W., B. J. J. Embleton, T. J Cudahy, and C. McA Powell, 1986, Prefolding and premegakinking magnetizations from the Devonian Comerong Volcanics, new South Wales, Australia, and their bearing on the Gondwana Pole Path: Tectonics, v. 5, p. 135–150.

——— and H. C. Palmer, 1987, Pre- and post-folding magnetizations from the Devonian Snowly River Volcanics and Buchan Caves Limestones, Victoria: Geophysical Journal of the Royal Astronomical Society, v. 91, p. 155–170.

Thrupp, G. A., D. V. Kent, and P. W. Schmidt, 1988, Preliminary paleomagnetic results from Late Devonian redbeds of southeast Australia: AGU Spring meeting, San Francisco.

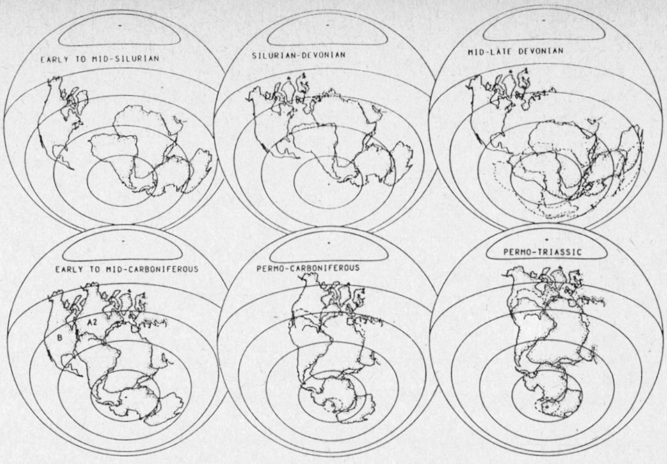

SCHMIDT, S. T., Universitaät Heidelberg, Heidelberg, West Germany

Alteration under Conditions of Burial Metamorphism in North Shore Volcanic Group Mineralogical and Geochemical Zonation

The North Shore Volcanic Group (NSVG) in northeastern Minnesota forms part of the Upper Precambrian Keweenawan plateau lava province. The NSVG lavas mainly consist of subaerial olivine tholeiites and transitional basalts (Green, 1983). Eruption occurred mainly from fissures in a relatively low relief craton environment. The lava flows are highly amygdaloidal at the top and bottom. Alteration has affected the lava flows, has filled former vesicles and fractures, and has altered igneous minerals and intercalated fluviatile sediments.

On the basis of the alteration minerals, including minerals in amygdules and veins, a metamorphic zonation is observed in the southern limb of the NSVG between Duluth (D) and Tofte (T). The following zones are recognized from the top (near Tofte) to the bottom (near Duluth) of the sequence: (1) thomsonite–escolecite–smectite facies; (2) heulandite–stilbite–mordenite facies; (3) laumontite–chlorite facies; (4) laumontite–prehnite–chlorite facies; (5) epidote–chlorite–pumpellyite facies; and (6) beginning epidote–actinolite–chlorite facies. The clear facies succession is disturbed by a recurrence of a higher metamorphic facies in the upper part of the sequence.

In addition to the zonation of alteration minerals, a compositional zonation of the newly formed albite replacing igneous Ca-rich plagioclase and the newly formed phyllosilicates replacing mafic igneous components is observed *within the sequence* and *within the single flow* (Figure 1). In the upper parts of the sequence, albite shows compositions far from that of ideal albite. In lower parts of the sequence, i.e., in flows with higher metamorphic facies the composition of albite lies between 95 and 99 mol% Ab. In addition, albite in flow tops and bottoms shows compositions that are always closer to a stoichiometric albite than albite from close to the massive flow interiors. The front of albitization within the single flow continuously approaches the massive flow interiors with increasing metamorphic grade. Phyllosilicates (mainly smectites and chlorites) in the upper part of the sequence display high SiO_2-MgO values whereas in the lower part of the sequence the values for Al_2O_3 and FeO increase.

Stable isotopes ($\delta^{18}O$, $\delta^{13}C$) on calcite and quartz are consistent with the observed zonation of alteration minerals becoming lighter with greater burial depth.

Whole-rock major and trace element chemical compositions have been determined on profiles across selected flows of the different metamorphic facies. Mainly the LIL-elements show a special mobility behavior and were redistributed depending on metamorphic facies and sampling position within the single flow.

The alteration history can be summarized in the following way:

1. Diagenetic stage: After deposition, and probably simultaneously to cooling of the lava flows, oxidation of Fe^{2+} to Fe^{3+} and initial hydration near the surface occurred.

2. Stage of low-temperature alteration: Mobilization of elements resulted in deposition of smecites and illites followed by Si-rich calcite and zeolites.

3. Metamorphic stage: Flows in stratigraphically deeper levels underwent metamorphism from laumontite–chlorite to beginning greenschist facies. Albitization of Ca-rich plagioclase is a major process.

The three main aspects of the alteration should be stressed: large fluid flow, different water-rock ratios resulting in different parageneses, and the

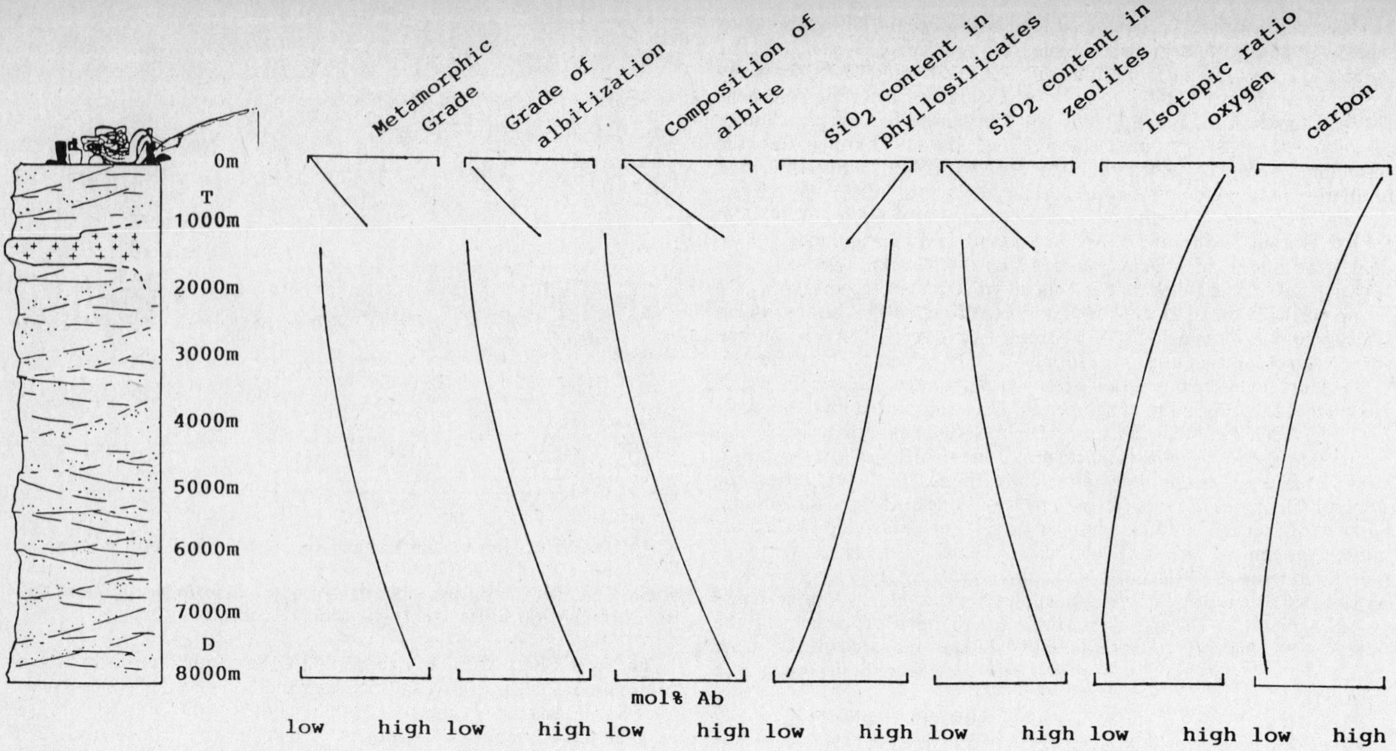

Figure 1—Main features of alteration in the North Shore Volcanic Group, Minnesota.

lack of equilibrium especially in stratigraphically upper parts of the sequence.

REFERENCE

Green, J. C., 1983, Geologic and geochemical evidence for the nature and development of the Middle Proterozoic (Keweenanwan) Midcontinent Rift of North America: Tectonophysics, v. 94, p. 413–437.

SCHMINCKE, H. U. and M. SIMON-NEUSER, Ruhr Universität Bochum, Bochum, Federal Republic of Germany

Evolution of Volcanism Associated with Aseismic Ridges as Reflected in Volcaniclastic Sediments

The primary composition, volcanic evolution, and alteration history of two aseismic ridges of the South Atlantic, in part with subaerial activity, are evaluated from mass flow and fallout derived volcaniclastic rocks of DSDP cores. The drill sites on either side of the MAR are located on 74–76 Ma old crust of the South American plate (Site 357 and 516F) and 70–72 Ma (Sites 525A and 528) resp. about 64 Ma old crust (Site 359) of the South African plate. The main parts of the aseismic ridges Rio Grande Rise (RGR) and Walvis Ridge (WR) have a common origin at an anomalous segment of the MAR (e.g., Barker, 1983), which may have lasted for about 30 Ma up to the Cretaceous–Tertiary boundary.

The volcaniclastic sediments exhibit two different types of magmatism (spreading center and intraplate type, respectively), both related to the aseismic ridges. They are mainly characterized by their position within the sequence of strata, their depositional features, and their primary composition.

The primary composition of the volcaniclastic sediments is inferred by the concentrations and ratios of elements that are insensitive to the low temperature subseafloor alteration (mainly Zr and Nb). Thus, four chemical groups with specific Zr/Nb-ratios can be distinguished: tholeiitic (Zr/Nb > 12), enriched tholeiitic (5–7), alkalic (4–6), and highly alkalic (<3).

The so-called basement volcaniclastics (BVC) are genetically related to the final phases of spreading center volcanism that results in an intercalation of the volcaniclastics in a sequence of basalts and autochthonous sediments. In time, the primary composition of the magmas shifts from tholeiitic (> 74 Ma, Site 516F, RGR) to enriched tholeiitic (72–65 Ma, Sites 525A and 528, WFR).

Intraplate volcaniclastics (IPVC), representing the second type of magmatism, appear far above the oceanic crust within the series of marine sediments and are primarily of alkalic (Site 359, WFR) and highly alkalic composition (Sites 357 and 516F, RGR). Both types comprise monomict as well as polymict deposits that are generally dominated by vitric tuffs with various amounts of crystals, xenoliths, and nonvolvanic components (mostly of biogenic origin). They are of mafic to slightly differentiated or felsic composition, the IPVC showing the higher degrees of differentiation. This is reflected by the concentrations of immobile elements and by the type of volcanic particles as well. Despite the pronounced alteration, various degrees of differentiation can be distinguished even in vitric deposits free of co-magmatic crystals. Here, shape and vesicularity of the vitric particles and in some cases even alteration products are diagnostic.

The mafic to slightly differentiated volcaniclastic deposits are mainly poor in crystals. Co-magmatic crystals included in vitric particles and less occurring as single clasts are plagioclase, olivine, and clinopyroxene with plagioclase dominating the BVC and olivine prevailing in the IPVC. In the IPVC, chromite occurs as a characteristic accessory phase. Felsic volcaniclastics of both occurrences may be either poor or very rich in crystals (alkali-feldspar). The bimodal occurrence of crystal-poor vitric tuffs on one and crystal tuffs on the other hand shows the effective sorting of products of highly explosive eruptions by the combination of subaerial and subaquatic processes. Amphibole and biotite are minor phases in the IPVC while in the BVC only amphibole occurs as an accessory phase. This reflects the enriched tholeiitic character of the BVC in contrast to the alkalic nature of the BVC.

The various stages of volcanic evolution generally parallel the magmatic differentiation. They range from effusive to weakly explosive and highly explosive eruptions and are best developed in the IPVC of the Rio Grande Rise. The inferred fragmentation processes show a general decrease in water depth superposed to at least two cycles of magmatic differentiation. Rough estimations of the erupted volume show that this shoaling is not only due to the growth of a complex of seamounts and volcanic islands centered upon the preexisting submarine ridge but also to an uplift of the RGR itself. This uplift may be caused by reheating and additional intrusion of sills (Staudigel and Schmincke, 1984) associated with the Middle Eocene volcanic activity.

REFERENCES

Barker, P. F., 1983, Tectonic evolution and subsidence history of the Rio Grande Rise, *in* P. F. Barker, R. L. Carlson, and D. A. Johnson, et al, eds., Initial Reports of the Deep Sea Drilling Project, v. 72, p. 953–976.

Staudigel, H., and H. U. Schmincke, 1984, The Pliocene seamount series of La Palma/Canary Islands: Journal of Geophysical Research, v. 89, p. 11195–11212.

SCHMITZ, B., Lawrence Berkeley Laboratory, Berkeley CA

Recent Formation of Barite and Celestite in Weathering Cretaceous–Tertiary Boundary Clays

In air-exposed, strongly weathered Cretaceous–Tertiary (K-T) boundary clay at Stevns Klint in Denmark abundant small barite concretions are found. The concretions are in general 0.1–2 mm in diameter and are readily distinguishable due to their white color. The barite concretions are confined mainly to the basal red layer of the clay (Layer S5; see microstratigraphy in Schmitz, in press), where they occur together with extremely abundant (10% by weight), sometimes hollow, goethite spherules (0.1–1 mm in diameter). Often the barite is found intergrown with goetithe.

An explanation to the origin of the barite and the goethite spherules follows from studies of relatively unweathered clay excavated at depths of about 1 m in the Stevns Klint exposure. In such clay, spherules are mostly massive and made up of cryptocrystalline pyrite instead of goethite. A thin rim of goethite, however, may cover the pyrite. More weathered pyrite spherules at intermediate depths in the exposure have thicker goethite rims. In unweathered clay, no barite concretions have been found. Apparently the barite in weathered clay has formed in connection with sulfate saturation of pore waters during recent oxidation of pyrite. The process taking place can be represented stoichiometrically as follows:

$$FeS_2 + 2H_2O + 4O_2 \rightarrow FeO(OH) + 2SO_4^{2-} + 3H^+ + e^- \quad (1)$$

$$Ba^{2+} + SO_4^{2-} \rightarrow BaSO_4$$

An interesting aspect of the barite formation is the fast rate with which it takes place. Due to the rapid erosion of the sea cliffs at Stevns Klint, it can be ascertained that the barite concretions have formed during the last 20–100 years.

In addition to small barite concretions weathered, K–T boundary clay at Stevns Klint contains abundant celestite ($SrSO_4$). The celestite is fibrous and is typically associated with the weathered fractions of large pyrite concretions (up to 5 cm in diameter), which also are common in the basal clay. Also in this case supersaturation of pore waters with sulfate during weathering explains the origin of the mineral.

Comparative studies of unweathered and weathered K-T boundary clay at Stevns Klint are of great importance for understanding the mineralogy and geochemistry of other K-T boundary clays that are found in deeply weathered exposures (such as Gubbio in Italy, Caravaca in Spain, and Woodside Creek in New Zealand). Not only at Stevns Klint, but also at Caravaca and Woodside Creek, the boundary clay is made up of a basal (about 0.5 cm thick) goethite-rich layer below a dark or black organic-rich zone. Considering the great lithostratigraphic similarity with Stevns Klint, it seems likely that also at the two other sites, the red layers represent weathered pyrite horizons. However, there are also other ways of reaching this conclusion. For example, the strongly weathered red layer at caravaca contains abundant sulfate minerals such as gypsum (Schmitz, in press) and celestite (Izett, 1987). Pyrite, however, is relatively rare, whereas goethite makes up about 15–20% of the layer. By analogy with the red layer at Stevns Klint, the sulfate minerals and the goethite may to a large extent have formed due to pyrite weathering.

Additional evidence for the importance of pyrite weathering in K-T boundary clays comes from studies of ubiquitous goethite spherules recovered from other marine K-T boundary clays. Many of these are hollow and have infillings of sulfate minerals such as barite and gypsum (Bohor, 1988), indicating a pyrite precursor.

Pyrite weathering may also explain other features of the basal red layers of the marine K-T boundary clays. It is known from the research area of ore geology that Gossan formations (strongly weathered sulfide deposits) contain very high concentrations of elements such as As and Sb. In the red layers of marine K-T boundary clays all over the world As and Sb are strongly enriched (see, e.g., Schmitz, in press). The world wide distributed As and Sb anomalies are sometimes used as an argument for a volcanic rather than an asteroidal origin of K-T boundary Ir enrichments. However, an explanation for the enrichments that is in much better agreement with actual mineralogical features of the clays is that the elements have accumulated partly as chalcophile elements during diagenetic pyrite precipitation and partly due to absorption in goethite lattices during weathering. In general, the As content of marine K-T boundary red clay is higher in strongly weathered than in less weathered material.

More attention has to be directed toward an understanding of how recent weathering processes affect the marine K-T boundary clays that are found in continental exposures. Far too much large-scale modeling of the causes of the K/T boundary extinctions have neglected the effects of these processes.

REFERENCES

Bohor, B. F., 1988, Shocked quartz and more: impact signatures in K-T boundary clays and claystones: (Abstract) Global Catastrophes in Earth History, LPI Contribution n. 673, p. 17–18.

Izett, G. A., 1987, Authigenic "spherules" in K-T boundary sediments at Caravaca, Spain, and Raton Basin, Colorado and New Mexico, may not be impact derived: GSA Bulletin, v. 99, p. 78–86.

Schmitz, B., in press, The origin of microlayering in worldwide distributed Ir-rich marine K-T boundary clays: Geology.

SCHNABEL, S., G. TIETZ, and A. KRÖNER, Universität Mainz, Mainz, Federal Republic of Germany

Zircon Morphology as Aid to Establish Protolith Origin and Metamorphic History in High-Grade Rocks

We show that zircon morphology examined by scanning electron microscopy can be used to deduce the origin and metamorphic history of high-grade metamorphic rocks. Previous work was largely restricted to the typification of magmatic rocks by length-to-width ratios. Our examples come from the late Proterozoic granulite terrain of Sri Lanka where both igneous and sedimentary rocks have been affected by multiphase deformation and metamorphism so that the original rock types are difficult to identify in the field. We compare the zircons from these rocks with zircons of unmetamorphosed Jurassic granites from Nigeria and identify the following features in igneous, sedimentary, and metamorphic rocks:

1. Primary zircons derived from cooling of magmatic rocks are characterized by crystals with sharp edges. It is easy to identify the prisms and pyramidal ends by Miller indices. These grains fit exactly into the scheme of Pupin and Turco (1971), which details the main prism and pyramid combinations of primary magmatic zircons. We observe magmatic zonation, a distinct growth pattern in successive layers.

2. Detrital grains are characterized by smooth surfaces rounded by mechanical abrasion. The extent of abrasion is inferred to be related to the duration of the transport process. Initially, the euhedral grains roll around their prisms, first wearing out the faces along the c-axes. During further transport by rolling, the pyramidal ends are also affected by abrasion until the grains become more and more spherical. For such grains with advanced abrasion, crystallographic orientation is no longer obvious. Such abrasion can occur during more than one sedimentary cycle. In general, only zircons with no crystal defects are able to withstand long sedimentary transport. This resistance to abrasion is based on different degrees of radiation damage, i.e, grains with high concentrations of U and/or Th develop cracks resulting from radioactive decay, that causes tensions in the crystals.

3. The morphology of zircons changed by chemical processes during metamorphism is significantly different from that of mechanically abraded grains. Mechanical attack during sedimentary transport preferentially abrades the long prismatic faces, whereas chemical attack first affects the pyramidal ends. In the latter case, dissolution starts by rounding the edges formed by pyramidal faces, while long faces and the edges between them remain unaffected. In many cases, we found secondary growth of zircon over the corroded parts of the crystal. The most likely cause of this growth is a change in the chemistry of the surrounding rock fluid. During this secondary growth, additional crystal faces are formed that do not fit into the system of Pupin and Turco (1972).

We can define principally two different types of new growth of crystal faces during metamorphism: In type 1, secondary growth occurs on magmatically formed primary crystal faces, while in type 2 new crystal faces are formed on the crystal edges. We have also observed rare amorphous accretion of $ZrSiO_4$ on long prismatic surfaces but, strangely, not on pyramidal ends although much more energy is needed to accrue zircon material on long faces than on pyramidal ends. The type 2 new crystal faces may be the results of repeated growth and subsequent dissolution that may finally lead to rather complex, multifaceted crystals.

Another variation of metamorphic growth are so-called agglutinated grains where two or more crystals have grown together, most probably by $ZrSiO_4$. In cases of very high concentrations of Zr in the host rock, a coating of many small grains may encrust the surfaces of the host crystal.

In general, our results indicate that zircon morphology reacts more strongly to changes in the host rock than was previously supposed. The grain shape and the chemistry of the newly formed zircon material is determined by the following factors: corrosion, sedimentation, new growth of crystal faces, agglutination, and formation of small-crystal coating. We therefore conclude that it is possible to distinguish primary igneous zircons from thos affected by sedimentary transport and/or metamorphism. The morphological appearance of the crystal faces definitely shows whether or not these grains have been affected by a sedimentary cycle. The zircons showing rounded edges of prisms but well-preserved pyramidal ends can only be explained by mechanical attack, i.e., by a sedimentary process. Zircons in high-grade metamorphic rocks are characterized by chemical corrosion that leaves a typical grain shape with damaged pyramidal ends and grooves, and by new growth producing many small crystal faces with high Miller indices.

The identification of different types of secondary zircon growth during metamorphism has important consequences for single zircon dating: such growth occurs not only around easily identified old cores but also along previously damaged crystal faces, and it may then escape detection during optical microscopy.

REFERENCE

Pupin, J. P., and G. Turco, 1972, Une typologie originale du zircon accessoire: Soc. fr. Minéral. Cristallogr. Bull., v. 95, p. 348-359.

SCHNEER, C. J., Newfields, NH

Idea of Time

In the apprehension of time, Man is alone among all organic species. There is evidence, in the form of the Blanchard bone tool, of the human linkage of time with counting as long ago as 30,000 B.P. Marshack has distinguished 29 sets of marks engraved on the tool in a serpentine line, each set at a different time and with a different tool. Each mark represents a day, with the twists and turns of the line corresponding exactly to the changes in phase of the moon. Here in embryonic form, is the common sense of time, Newton's "...relative, apparent and common time,...some sensible, external measure of duration by the means of motion, such as an hour, a day, a month, a year," which was nevertheless an approximation of "Absolute, true and mathematical time [that] of itself, and from its own nature, flows equably and without relation to anything external."

The unique distinction between geology and the other sciences is in the central role played by time. The attempts by classical and Renaissance savants to couple their understanding of natural processes with the systematic observation of nature led to the concept of a history of the earth and the "deep time" to accommodate it. Leonardo and Robert Hooke are two examples. In visualizing fossils as the "hieroglyphics" to be deciphered to make possible another downward extension of history parallel to the efforts of his contemporaries to establish a history for Egypt, Hooke spurred the processes culminating in the development of the geological scale of time. A truly empirical scale, independent of the progress of physics of evolutionary biology for that matter, the geological time scale-geological column is the ultimate invariant. Was Einstein, in identifying the contradiction between Newton's absolute and relative time, influenced at all by the concept of geological time? The new mobilism, latest development in the reordering of the fundamentals of geological science, rests firmly on the geomagnetic time scale, an extension of the classical geological time scale.

SCHNEIDER, J. A., University of Chicago, Chicago, IL

Teleost Fish and Retreat of Stalked Crinoids

The disappearance of stalked crinoids from shallow water during the Cretaceous and Early Tertiary ecological reorganization of shallow-marine communities has previously been recognized (Meyer and Macurda, 1977; Vermeij, 1977). This reorganization was a consequence of the diversification of predators such as bony fish (teleosts) and crustaceans, and a corresponding decline in many sessile benthic groups (brachiopods, bryozoans, and echinoderms) that had been dominant since the Paleozoic. Recent investigations (Meyer, 1985; Schneider, 1987), in press) have uncovered evidence that teleost fish prey upon crinoids, overturning long-held notions that crinoids are not subject to predation. In this study, global diversity patterns of marine teleost fish were compared with those of the five major orders of articulate crinoids: isocrinids, comatulids, millericrinids, cyrtocrinids, and bourgueticrinids. Additionally, time-environment diagrams (Bottjer and Jablonski, in press) have been constructed for each order of articulate crinoids to trace the bathymetric distribution of crinoids from the Triassic to the Recent.

The unstalked, mobile comatulids account for approximately 570 of 650 Recent crinoid species and occur from littoral to bathyal depths. All other Recent crinoids are sessile and are restricted to depths greater than 100 m, where durophagous teleosts are much less common.

Crinoids, like other echinoderms, are capable of regenerating lost or injured body parts. Isocrinid and comatulid crinoids evolved the ability to autotomize their arms, which minimizes injury-related trauma. Arm autotomy in these crinoids is related to syzygies, which are immovable ligamentary brachial articulations. Comatulids have more syzygies autotomize more readily than those of isocrinids. Other articulate crinoids lack syzygies. Oji (1985) separated isocrinids into an "old" group and a "new" group based on the type of articulation in the primibrachials. New isocrinids have primibrachial syzygies, whereas they are lacking in old isocrinids. Syzygies are found in the more distal portions of the arms of all isocrinids, but occur more frequently in new isocrinids, so new isocrinids have more sites of autotomy near the central soft body mass than do old isocrinids.

Teleost fish originated in the Late Triassic but did not begin diversifying until the Cenomanian when they evolved their modern durophagous jaw apparatus. There were fewer than 25 teleost genera in the middle of the Cretaceous; there were more than 400 genera by the start of the Tertiary.

Isocrinids, represented by about 30 Recent species, attained approximately their present diversity in the Jurassic, and as previously shown (Bottjer and Jablonski, in press) retreated from shallow water during the teleost diversification (Cenomanian to Eocene). However, this retreat was accompanied by only a minor global diversity decline in the Paleogene, attributable to a poor fossil record of deep-water isocrinids. Old isocrinids, with fewer trauma-reducing syzygies, retreated earlier from the Cenomanian to the Maestrichtian, whereas the more predation-resistant new isocrinids retreated from the Campanian to the Chattian. Recent comatulids and isocrinids co-exist, and there is no evidence of competition between members of these two groups. Isocrinids have the shallowest overall depth distribution of stalked crinoids, and their ability to autotomize their arms may be a factor in their shallower bathymetric distribution.

Millericrinids and cyrtocrinids each reached their peak diversity in the Late Jurassic, then declined in diversity and retreated into deeper water during the Early Cretaceous, before the Cenomanian teleost diversification, suggesting that predatory fish were not responsible for the dwindling of those groups. Competition with comatulids was not a likely factor in the retreat of these stalked crinoids, for the millericrinid and crytocrinid diversifications were contemporaneous with the diversification of comatulids. There are only three Recent monospecific genera of crytocrinids, and Recent millericrinids are comprised of five genera containing eight species. The depauperate diversity and distribution may be attributed to their lack of syzygies and therefore lack of ability to withstand predation pressure. However, the Early Cretaceous decline in diversity of millericrinids and cyrtocrinids antedates the Cenomanian rise of durophagous teleosts, and if these predators were not a factor in the decline and retreat of these crinoids, then other factors may presently control the diversity and distribution of cyrtocrinids and millericrinids.

Bourgueticrinids originated in the Turonian, after the initial diversification of teleosts. Bourgueticrinid diversity has steadily increased and is now greater than that of isocrinids, although the bourgueticrinid overall bathymetric distribution is deeper. Bourgueticrinids did not retreat until the Eocene. Their later retreat is probably just a function of their late origination.

In addition to mobility and the most highly developed xyzygies, many comatulids have semicryptic habits, biochemical defenses not found in stalked crinoids, swimming ability. and spike-like oral pinnules to protect the visceral mass. These characteristics apparently make comatulids the most predation-resistent crinoids, and they are the only articulates that did not undergo environmental retreat. The evidence suggests that comatulids have been better able than the isocrinids, and possibly all other articulates, to withstand the predation pressure in shallow water since the diversification of the teleosts.

REFERENCES

Bottjer, D. J., and D. Jablonski, in press, Palaios.

Meyer, D. L., 1985, Paleobiology, v. 11, p. 154-164.

———, and D. B. Macurda, 1977, Paleobiology, v. 3, p. 74-82.

Oji, T., 1985, Palaeontology, v. 28, p. 629-642.

Schneider, J. A., 1987, GSA Abstracts with Program, v. 19, p. 833-834.

——— in press, Proceedings of the Sixth International Echinoderm Conference, Victoria: Rotterdam, A. A. Balkema.

Vermeij, G. J., 1977, Paleobiology, v. 3, p. 245-258.

SCHOLL, D. W., A. J. STEVENSON, T. L. VALLIER, H. F. RYAN, and E. L. GEIST, U. S. Geological Survey, Menlo Park, CA

Aleutian Arc-Trench System: Perspective of Ocean Margin Evolution Controlled by Regional Changes in Plate-Boundary Conditions

The Aleutian Arc forms the northwestern rim of the Pacific Basin (Figure 1). Striking westward from Alaska, the arc is a mostly submerged 2200 km long and 150 km wide mountain range separating the deep-water Bering Sea from the abyssal depths of the Aleuian Trench and north Pacific Basin. The rock sequence of the arc documents a three-phases evolution, each phase, it is argued, a response to changes in plate-boundary conditions.

Initial arc growth—the first phase—occurred in early to middle Eocene time (50–55 Ma) at a newly established intraplate convergence zone. Arc magmatism at the Aleutian subduction zone was caused by the seaward buckling of a northwestward-moving oceanic plate (Kula) of probable

Early Cretaceous age (Figures 2A, 2B). Landward of the rising arc, the trapped oceanic terrane of Kula crust accreted to the continental margin connecting Alaska and Siberia (Figure 2C). The natal Aleutian Arc may have extended from Alaska to Kamchatka, but as shown on Figure 2B, probably only as far westward as the present Shirshov Ridge, which is either a remnant sector of the arc or possibly a coeval or older oceanic plateau or hotspot-generated volcanic massif.

Geologically, the significant consequence of terrane accretion and offshore arc growth was the addition of a deep water basin—the Aleutian Basin—to the North American plate (Figure 2C). During the past 50 m.y., the infilling of this basin by thickening sequences of sedimentary and arc-volcanic rocks has been building new continental crust (Figures 2C, 3B, and 4B).

The medial phase of arc evolution began at about 42 Ma and lasted until near the end of the Miocene. At 42 Ma (late Eocene), the Kula plate of the north Pacific welded to the Pacific plate and the convergence direction at the Aleutian Trench rotated from northward to the northwest, which was, and remains, oblique to its curving trend (Figure 3A). In the Oligocene and Miocene, increasingly younger and bathymetrically higher ocean crust formed at the extinct Kuly–Pacific spreading ridge entered the Aleutian subduction zone (Figure 3B).

These second-phase plate-boundary events slowed are magmatism, which favored the emplacement of epizonal plutons and thus elevation of the arc crest, as did the underthrusting of an increasingly shallower lower plate. Shear stresses arising from oblique convergence structurally disrupted the arc's igneous massif, forming summit basins that filled with volcaniclastic debris. Some of this detritus spilled seaward and mantled the arc's volcanically quiescent flanks. In the Miocene, increasingly oblique plate boundary conditions may have induced arc-rifting to form the Komandorsky Basin, thus abandoning Shirshov Ridge as a remnant fragment of the original Aleutian Arc (Figure 4A).

Toward the end of the Miocene, the Aleutian arc-trench system entered its third or modern evolutionary phase. At roughly 5 Ma, the rate of convergence increased, the arc began to collide with the extinct Kula–Pacific spreading ridge, and, along the landward trench wall, a thick accretionary wedge formed in response to rapid trench sedimentation sustained by glaciated Alaskan drainages (Figures 4A, 4B). These plate-boundary changes kindled the eruption of the modern chain of arc volcanoes and induced regional subsidence as older crust increasingly underthrust the arc. In the fore-arc, deep-water basins formed within arc-basement rocks in response to the structural emplacement of an underlying accretionary wedge and the lateral or strike-slip movement of arc-basement terranes. Possibly owing to the underthrusting of the extinct spreading ridge and the accretionary wedge, stronger intraplate coupling accelerated arc fragmentation. Tectonic blocks of the arc massif moved rapidly westward, some rotating clockwise and creating large and new summit basins in their wakes (Figures 4A, 4B). Terranes of the arc massif collided with Kamchatka, either accreting to or underthrusting the peninsula.

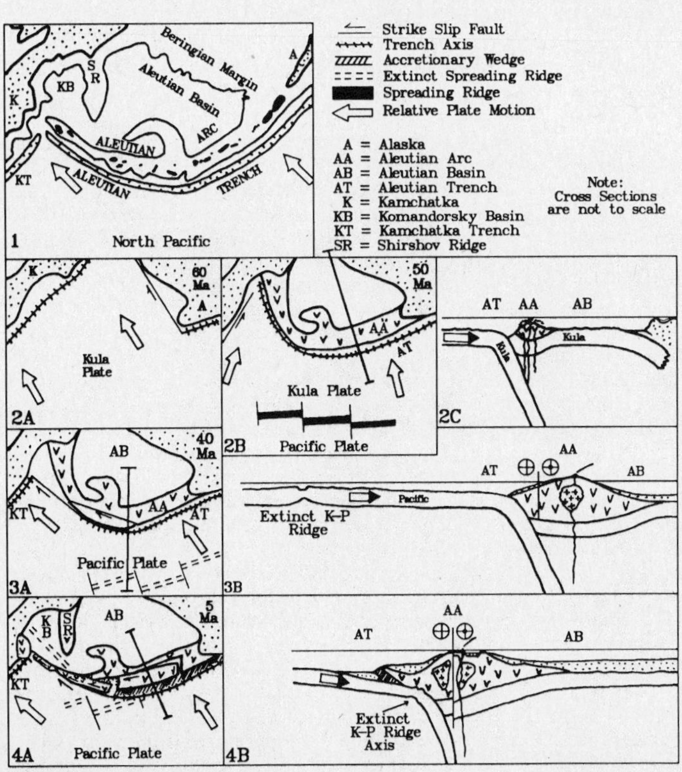

Figure 1

A geologic perspective gained by observing plate-tectonic controls on the evolutionary history of the Aleutian arc-trench system is that regional changes in plate-boundary conditions can initiate episodes of either continental growth, through the formation and infilling of stable ocean-margin basins, or continental wastage, by crustal fragmentation and dispersal of ocean-margin terranes to other plate boundaries.

SCHOT, P. P., University of Utrecht, Utrecht, the Netherlands

Groundwater Systems Analysis and Hydrochemistry of Wetlands in Central Netherlands

The Vecht river plain is about 30 km^2 and is located in the Central Netherlands. It is the natural discharge zone of a Pleistocene sand aquifer some 150–200m thick. Under the river plain, the aquifer is semiconfined by Holocene clay and peat deposits 1–5m thick. Groundwater in the aquifer is recharged in a sandy hill ridge bordering the river plain.

During the last century, most of the marshy river plain has been drained and turned into agricultural land. In the remaining area, wetlands are present of which several have been designated as nature reserved. The wetlands function as breeding and forage places for waterfowl and also contain important species-rich vegetation types. The plant species of these vegetations depend on seepage of mesotrophic groundwater from the Pleistocene aquifer.

The supply of mesotrophic groundwater has become increasingly threatened. Withdrawal of groundwater in the recharge area for drinking water and industrial water supply has decreased the amount of seepage in the river plain. Moreover, human activities in urban areas in the recharge area pose a threat to the quality of the recharge water. In the river plain, the wetlands have become hydrologically elevated with respect to their surroundings due to drainage of the agricultural areas. This has induced output of water from the wetlands through infiltration. Decreased seepage from the recharge area along with drainage of the river plain have resulted in water deficits in the wetlands during dry summer periods. These deficits are compensated by suppletion of external surface water. The suppletion water is, however, of different chemical composition and has negative effects on the occurrence of mesotrophic plant species in the wetlands.

Management directed toward preservation and restoration of mesotrophic wetland ecosystems must take into account the role of groundwater. To formulate adequate measures, a hydrological framework is needed. For this purpose, a regional groundwater systems analysis has been carried out. A groundwater system may be defined as a coherent unit of groundwater and earth materials in space and time, natural or influenced by man. Groundwater systems connect recharge areas with discharge areas. The following methods were employed in the groundwater systems analysis:

1. Two-dimensional computer simulation of vertical groundwater flow. The vertical groundwater flow pattern is of particular importance as it shows the connection between recharge areas and discharge areas. Equipotential and flow lines were calculated and plotted for a number of inhomogeneous, anisotropic sections of the subsoil.
2. Isotope analysis of groundwater and surface water. Oxygen-18 and deuterium were used to trace infiltration of surface water and mixing of waters of different origin. Tritium and carbon-14 were used to determine groundwater age. Carbon-13 provided information on hydrochemical processes.
3. Chemical analysis of groundwater and surface water. Some 6000 analyses with up to 18 chemical parameters were used to determine the spatial distribution of solutes in groundwater. A number of chemical parameters were used as tracers of groundwater origin and flow direction. Principal component factor analysis and cluster analysis were used to study the mutual relationships between the chemical variables and between the analyses. Hydrochemical processes were deduced from changes in solute concentrations in the direction of flow. Special emphasis was put on groundwater quality as it is a principal factor for the occurrence of plant species in wetland ecosystems.

The results of the separate methods were integrated to define groundwater systems. The chemical and isotope tracers proved to be essential to correct the results of the computer model.

The analyses showed a large number of groundwater systems are present. In the hill ridge, two main systems exist. Part of the recharge area has a natural discharge in a belt along the foot of the hill ridge. In this zone important wetland ecosystems are found. The remaining part of the recharge area comprises an artificial system in which groundwater discharges through the wells of the public drinking water supply. In the river plain a cascade of groundwater systems is found. The artificially controlled surface water levels in the wetlands and agricultural polders become ever lower in the direction of the Vecht river. This has lead to the development of numerous infiltration-discharge systems from one polder to another.

Groundwater quality is mainly controlled by the following processes: (1) mineralization of peat deposits in the river plain; (2) reduction of nitrate and sulfate and solution of Fe^{2+} and Mn^{2+} under anaerobic conditions; (3) solution of calcite; (4) infiltration of polluted recharge water in the hill ridge and of polluted surface water in the river plain; and (5) cation-exchange following intrusion of brackish water in parts of the river plain.

Water quality of a groundwater system is primarily determined by processes in the recharge area. Groundwater composition observed in wetland ecosystems depends on the position of the ecosystem with respect to the hierarchy of groundwater systems.

SCHREIBER, H. D., Virginia Military Institute, Lexington, VA

Redox State of Variable Valence Elements in Equilibrium Models of Basaltic Magmas

Introduction: The mineralogy of the elemental distribution within a basalt depend significantly on the redox state of the crystallizing magma. This redox state is governed by internal parameters (bulk composition and concentrations of active redox components) as well as external parameters (prevailing oxygen fugacity and melt temperature). A complete understanding of how each of these variables affects the melt's redox state will aid in the interpretation of the chemistry of basalts and of the evolution of magmas.

Objective. The objective of this study is to delineate the factors that control the redox state of multivalent elements in a typical silicate melt. An alkali borosilicate composition (Table 1) was employed to provide a simple model of basaltic magmas.

Oxidation-reduction equilibrium: A simplified chemical equation for oxidation-reduction equilibria in silicate liquids is:

$$4M^{m+} + 2nO^{2-} \leftrightarrow 4M^{(m-n)+} + nO_2 \quad (1)$$

where M is the multivalent element of charge m+ in its oxidized state and of (m-n)+ in its reduced state. The ions do not exist as discrete entities but as solvated complexes within the melt.

Oxygen fugacity: The dependence of the redox state of the melt on the oxygen fugacity can be expressed as,

$$\log X_M = (n/4)(-\log fO_2) + E'_M \quad (2)$$

at constant temperature and composition. X_M is the ratio of the reduced to oxidized ion concentrations for M; n is the number of electrons transferred to establish the equilibrium; fO_2 is the oxygen fugacity; and E_M' is the reference reduction potential for M and is a measure of the ease of reduction of M^{m+}. A series of reference reduction potentials, provided in Table 1, has been determined for numerous redox couples in the model composition.

Bulk composition: The composition of the melt controls the oxide ion activity and thus, according to equation 1, the direction of the redox equilibrium. However, a change from the model composition has been generally found to shift all individual equilibria in the same direction to about the same degree. That is, the electromotive force series established in Table 1 is applicable to all compositions upon the offset in each E_M' value by a constant amount.

Redox components: If many variable valence elements are dissolved simultaneously in a silicate melt, internal electron exchange reactions can occur. For example,

$$sM^{m+} + nR^{(r-s)+} \leftrightarrow sM^{(m-n)+} + nR^{r+} \quad (3)$$

represents the mutual interaction of multivalent elements M and R. Even if a particular redox state, M^{m+}, is stabilized in a melt at a particular temperature and oxygen fugacity, the presence of a sufficiently strong reducing agent can result in the reaction to form $M^{(m-n)+}$. Experiments in the model composition have shown that these oxidation-reduction reactions are analogous to such reactions in water. When in the oxidized form, thos redox couples at the top of the electromotive force series (Table 1) will oxidize the reduced form of the elements below it in the table. When more than two redox components are present in the melt, one element assumes the role of a redox buffer, in that it is preferentially oxidized or reduced by the other components in accordance with its position in Table 1. In geological situations, the ferrous/ferric couple usually acts as the redox buffer because of its central position in Table 1 and because iron typically exists in higher concentration than any other redox component. Thus, the ferrous/ferric couple minimizes the presence of oxidized and reduced states of other multivalent elements over a wide range of intermediate oxygen fugacities.

Temperature: High temperatures are synonymous with reducing conditions. This effect of temperature can be quantitatively expressed by:

$$\ln X_M = (\Delta H/RT) + c \quad (4)$$

where ΔH is the enthalpy of reduction for equation 1, R is the ideal gas constant, T is the temperature in degrees absolute, and c is a constant. The enthalpy of reduction can, thus, be used to describe the dependence of the redox ration, X_M, on temperature. The ΔH values, shown in Table 1 for the model composition, have been determined to be applicable to other bulk compositions.

Discussion: The experimental values compiled in Table 1 for a model composition allow the estimation of the concentrations of the oxidized and reduced forms of variable valence elements under any set of conditions. Only the position of the Fe^{2+}/Fe^{3+} redox couple (its E_M' value) has to be experimentally determined in this new composition to calibrate the series of reference reduction potentials to that system. A knowledge of the redox states present in a particular magma then allows (1) an appraisal of the prevailing conditions during the production of the magma, and (2) a basis for possible evolutionary pathways for the resulting crystalline basalts.

Redox Couple	E_M'	ΔH_{red}	Redox Couple	E_M'	ΔH_{red}
$Ni^{3+} \rightarrow Ni^{2+}$	1.7		$^*Fe^{3+} \rightarrow Fe^{2+}$	-2.0	41±10
$Co^{3+} \rightarrow Co^{2+}$	1.4		$U^{5+} \rightarrow U^{4+}$	-2.2	
$Au^{3+} \rightarrow Au^0$	>0.8		$Cu^+ \rightarrow Cu^0$	-3.3	
$Mn^{3+} \rightarrow Mn^{2+}$	0.8	13±3	$Cr^{3+} \rightarrow Cr^{2+}$	-3.4	
$Ag^+ \rightarrow Ag^0$	0.5		$Mo^{6+} \rightarrow Mo^{5+}$	-3.8	
$Se^{6+} \rightarrow Se^{4+}$	0.3		$V^{4+} \rightarrow V^{3+}$	-4.0	
$Se^{4+} \rightarrow Se^0$	0.2		$Eu^{3+} \rightarrow Eu^{2+}$	-4.3	
$O^2 \rightarrow O^{2-}$	0.0		$Ti^{4+} \rightarrow Ti^{3+}$	-5.0	35±5
$Ce^{4+} \rightarrow Ce^{3+}$	-0.1	11±3	$Ni^{2+} \rightarrow Ni^0$	-5.3	
$Cr^{6+} \rightarrow Cr^{3+}$	-0.3	10±4	$Sn^{4+} \rightarrow Sn^{2+}$	-5.5	
$Sb^{5+} \rightarrow Sb^{3+}$	-0.3		$Se^0 \rightarrow Se^{2-}$	-5.7	
$Cu^{2+} \rightarrow Cu^+$	-0.8	42±10	$Co^{2+} \rightarrow Co^0$	-6.0	
$As^{5+} \rightarrow As^{3+}$	-1.4		$Fe^{2+} \rightarrow Fe^0$	-6.3	
$U^{6+} \rightarrow U^{5+}$	-1.5	41±13	$^*Fe^{2+} \rightarrow Fe^0$	-7.2	80±10
$Fe^{3+} \rightarrow Fe^{2+}$	-1.7	58±25	$Mo^{5+} \rightarrow Mo^0$	-15.8	
$V^{5+} \rightarrow V^{4+}$	-1.9		$SO_4^{2-} \rightarrow S^{2-}$	-19.2	188±30

Table 1—A series of reference reduction potentials (E_M', in relative units) and enthalpies of reduction (ΔH_{red}, in Kcal/mole of redox element) for multivalent elements in an alkali borosilicate composition (57.9 wt% SiO_2, 1.0% TiO_2, 0.5% ZrO_2, 14.7% B_2O_3, 0.5% La_2O_3, 2.0% MgO, 17.7% Na_2O, and 5.7% Li_2O) at 1150 °C. The nominal concentration of the multivalent component is 1 wt%, except where noted by *(10 wt%).

SCHREYER, W., Ruhr-Universität Bochum, Bochum, West Germany

Ultrahigh-Pressure Metamorphism of Magnesian Pelites in Experiment and Nature

Since the discovery of the mineral yoderite in a talc-kyanite schist from Tanzania (McKie, Mineral Magazine, v. 32, 1959), a number of metamorphic rocks with similar chemistry have been detected in basement areas all over the world covering a wide range of metamorphic grades, although none of the later findings contained yoderite. On account of the talc + kyanite pair Schreyer (Earth and Planetary Science Letters, v. 43, 1977) used the term whiteschist for these rocks in distinction to greenschists characterized by the alternative assemblage Mg-chlorite + quartz.

The chemical compositions of whiteschists fall largely (up to 98%) into the model system $MgO\text{-}Al_2O_3\text{-}SiO_2\text{-}H_2O$ (MASH0, but there may also be additional K_2O (up to 4%). Occasionally, boron is strongly enriched. Since rocks with this chemistry occur as mudstones in Triassic evaporites in Algeria, they are considered to originate as evaporitic peliets.

Experimental investigations in the MASH system can be used directly to deduce the metamorphic reactions of the initial chlorite + quartz assemblage, which remains stable at 6 Kbars water pressure to a temperature maximum of 620 °C (Massonne, Journal of Metamorphic Geology, in press). At lower pressures, chlorite + quartz react to form cordierite + talc + H_2O at temperatures decreasing to 550 °C at 1 Kbar. This assemblage was found by Reinhardt (Journal of Metamorphic Geology, v. 5, 1987) in low-pressure evaporitive metapelites of Queensland, Australia. At water pressures between 6 and 18 Kbars the pair chlorite + quartz breaks down to the typical whiteschist assemblage kyanite + talc at temperatures decreasing from 620 ° to 530 °C with increasing pressure. The considerable extension and the negative slope of this reaction curve imply that the talc + kyanite field can be reached along linear geotherms from about 30 °C/km down to as little as 8 °C/km, which is a typical gradient in subduction zones. At even lower gradients, talc + kyanite can be generated at the expense of Mg-carpholite + quartz between 18 Kbars, 530 °C, and about 38 Kbars, 560 °C. Above about 38–45 Kbars and temperatures of 550°–750 °C, talc + kyanite react to an assemblage carrying the new synthetic phase MgMgAl-pumpellyite (Shcreyer, Mineral Magazine, v. 52, 1988). Beyond 800 °C the field is limited by the thermal stability of talc.

A subdivision of the huge whiteschist field can only be accomplished by means of the presence of additional phases. While talc-kyanite may coexist with quartz or coesite nearly over the whole PT-range, restrictions come in because of the extended stability of chlorite in silica-poor whiteschist. Such whiteschists react between about 18 and 31 Kbars, 750°–720 °C, to form pyrope at the expense of talc + kyanite + chlorite. This reaction can be seen from the textures of the sensational pyrope rock from the Dora Maira Massif, Western Alps (Chopin, Contributions to Mineralogy and Petrology, v. 86, 1984), which yields a minimum pressure for the PT-path of this rock during subduction.

Above some 28 Kbars, 750°–800 °C, the talc + kyanite field is terminated by the reaction to form pyrope + SiO_2, below about 30 Kbars as quartz, above as coesite. In the Dora Maira rock (Chopin, 1984) this reaction has just been reached, and thus, pyrope and coesite coexist with talc and kyanite. These are the highest PT-conditions yet reported for a magnesian metapelite.

Pure Mg-staurolite cannot coexist with a silica phase and is thus confined to silica-deficient bulk compositions. Chopin (Philosophical Transactions of the Royal Society of London, v. A321, 1987) found small crystals of nearly end-member composition as inclusions within pyrope. This is in agreement with its high-pressure, high-temperature stability field (> 14 Kbars; > 700 °C; Schreyer, 1988), but the exact reaction relationships at Dora Maira are not clear as yet.

The Dora Maira pyrope-coesite rock exhibits all the chemical characteristics as reported from whiteschists. As in the Afghanistan rock (Schreyer and Abraham, Contributions to Mineralogy and Petrology, v. 59, 1979), boron is represented by Mg-tourmaline forming relic inclusions in pyrope. The element Ti, P, and Zr are contained in the intriguing new high-pressure mineral ellenbergerite (Chopin, et al, Contributions to Mineralogy and Petrology, v. 92, 1986), which is presently being studied experimentally. The common Ti-end-member ellenbergerite has a lower pressure stability limit of about 28 Kbars at 650 °C and forms at the expense of the assemblage talc + kyanite + chlorite + rutile (Chopin, Shcreyer, Baller, unpublished data), in general agreement with theoretical deductions of Chopin (GSA Memoir, 164, 1986).

Potassium in the magnesian metapelites forms the dioctahedral mica phengite that coexists with all other minerals mentioned. Thus, PT-conditions are still in the stability field of talc + phengite (Massonne and Schreyer, European Journal of Mineralogy, v. 1, 1989; Massone, this symposium). The Si content of the phengite can be used for barometry and also yields pressures in excess of 30 Kbars for the Dora Maira pyrope rock.

The typical whiteschist chemistry is also encountered in some granulites such as an enstatite-sapphirine-quartz rock from Antarctica (Grew, American Mineralogist, v. 65, 1980). It is interesting to speculate about the PTt-path of metamorphism applying to this rock. Was it formed along a normal progressive path, or could it actually have been at ultrahigh pressures exhibiting a pyrope-bearing assemblage? The Antarctica assemblage could be formed more or less isothermally simply by pressure release from a rock like that from Dora Maira.

The multitude of different crystalline phases of the KMASH system with characteristic stability and compatibility ranges make magnesian metapelites an ideal rock composition to study deep subduction and subsequent obduction processes by means of changes in their mineral assemblages and compositions.

SCHROEDER, J. H., Technische Universitaet Berlin, Berlin, West Germany

Rudists as Indicators of Porosity Development in Cretaceous Limestones from Xamur, North Somalia

The Mesozoic–Cenozoic sedimentary sequence of the North Somalia Basin around Bosaso contains several rudist-bearing strata among the carbonate layers of the Tisje Formation that interfinger with clastic deposits of the Yesomma Formation. In the outcrops at the northern escarpment of the Al Mado Mountains west of the village of Xamur, the rudists, almost exclusively of the species *Hippurites inaequicostatus*, are associated with corals, mostly single colonies. Rudists and corals are found in growth positions; they form no mounds or structures, but occur in loose and irregular clusters within layers of biosparitic limestones containing mainly benthic foraminifers, pelecypods, and echinoids.

In view of the diagenetic variations frequently observed in multicomponent carbonate rocks (Schroeder, 1988), this study focused on the skeletons of the above-mentioned rudist species to explore their indicative potential with respect to the porosity development of the host rock. Morphologically, the right valve of this pelecypod forms a slightly bent cone of up to 24 cm height and 12 cm diameter; the left valve forms an obtuse conical lid of up to few cm height. The first alteration affecting these skeletons was the encrustation by organisms, probably curstose coralline algae. Simultaneously and slightly later, the skeletons were bored by pelecypods, sponges, worms (?), and endolithic microbes. The borings as well as intraskeletal pores were filled by internal sediment, frequently with geopetal structures. Irregular rims of early cement of either distinct individual needles or composite spherulites are indicated by relics within neomorphic calcite; by analogy with observation in Recent carbonates, these were presumably marine aragonite cements. Neomorphism of this first cement and relatively sparse growth of sparitic calcite cement are attributed to early freshwater influence.

The next process traced was fracturing of skeletons and previous cements that probably occurred in the course of early burial. Fractures of this first generation as well as many other (older) pores were then filled by sparitic Fe-calcite cement characterized by its crystal size of up to several millimeters, but its complex crystal habit involving a combination of rhombohedral and scalenohedral faces, and by crystal growth zones varying in iron content. This cement is pervasive and volumetrically the most important in the rudists; it is related to early burial. Next followed an ankerite cement characterized by saddle-shaped crystals, curved cleavage planes, and sweeping extinction. This cement is comparable to saddle dolomite, which is generally ascribed to temperatures of several tens to more than 100 °C thus to moderate burial depths, an interpretation corroborated in this case by more than 1000 m overlying sedimentary rocks of the stratigraphic column.

Another fracture generation followed; these fractures were filled or lined by dolomite cements, which in addition, grew on Fe-calcite or ankerite cements in some older remaining pores. finally, the ankerite cement in some cases was invaded by Fe-oxide minerals and in the process transformed into finely crystalline carbonate; this alteration affected entire patches of this cement, entire cement crystals, or only the outer portions of cement crystals. Fe-oxide minerals also are found as cements in the second generation of fractures. The formation of these Fe-oxide minerals is ascribed either to oxidizing conditions at depth due to the passage of corresponding solutions or to surface weathering.

In this case study, individual rudist skeletons are sites of the dynamic interplay of constructive and destructive processes known from Recent and fossil reefs (Schroeder and Zankl, 1974). The net effect of the genetic processes is that a given skeleton is no homogeneous entity, but a highly heterogenous piece of rock. During diagenesis, different stages are reflected by various cements of distinct mineralogy and/or chemical composition; cement characteristics indicate the conditions of the respective diagenetic environment and their fluctuations. Intermittent fracture generations are "dated" by subsequent cement filling(s). Very much as the results of genetic processes, those of diagenetic processes, i.e., various cements and fractures, are all observed in individual skeletons, but are neither evenly distributed within them, nor between different skeletons of the same unit: The spatial diversity is considerable.

Comparison with host rock and with coexisting coral skeletons shows that rudists keep a more complete record of diagenetic processes than other components, probably due to their large and varied intraskeletal pores. Therefore, they can hardly serve as indicators of porosity and permeability available in a given carbonate unit at respective diagenetic stages. However, these rudists are very good indicators of the processes effective in the porosity development of a carbonate rock and, in the context of basin analysis, they can be used for cement stratigraphy and the establishment of successive paleohydrological environments and gradients.

REFERENCES

Schroeder, J. H., 1988, Spatial variations in the porosity development of carbonate sediments and rocks: Facies, v. 18, p. 181–204.

——— and H. Zankl, 1974, Dynamic reef formation: a sedimentological concept based on studies of Recent Bermuda dn Bahama reefs: Proceedings of IInd International Symposium Coral Reefs, Brisbane 1973, v. 2, p. 413–428.

SCHROEDER, N., Roskilde University Center, Roskilde, Denmark

Reducing Environmental Impact of Mining and Waste Dumping in Shallow Seas

A research program is proposed that will develop methods to reduce the environmental impact from mining and waste dumping in shallow seas.

Through the years, an intensified use of land materials has resulted in increasing conflicts between sand and gravel exploitation and other interests—use of groundwater, conservation, etc.

The growing interest in exploiting the potentials of the seabed could reduce or eliminate many of the problems of land-based mining. Waste deposits in old sites and waste materials from coal burning seems to be growing problems (see Figure 1).

Increasing exploitation of marine resources depends, however, on the controlling environmental hazards.

Today, offshore materials are recognized as of minor importance, although offshore materials are first-grade building materials. Also it is often seen that dumping of dredging spoil from harbors and fairways can devastate large areas of the seafloor.

It is hoped that detailed geological/geophysical mapping and site planning combined with precision navigational aids (SYLEDIS-system) and the use of Push/TUC Barge system as shown in Figure 2 may reduce the impact on the biosphere to one or two orders of magnitude.

For example, the North Sea basin has a long geological history of subsidence and almost uninterrupted sedimentation. In recent years, the knowledge of the postglacial history of the area has reached a level that will make it possible to locate sites which in the future will be very safe sinks for pollutants. It may even be possible to find areas with submerged beach deposits, usable for gravel mining, where restoration of the seafloor could include disposal of waste from coal burning.

In order to use such areas for waste disposal, steps must be taken to ensure that the waste can be transported and dumped with minimum effect on the biosphere. Therefore, an investigation of the necessity of solidification (pelletation) of the waste products prior to disposal must be undertaken.

The aim of the project as a whole is to provide a base for extended use of sea-based area potentials and mineral resources through:

1. Tests of new technics (SYLEDIS, Push TUC Barge system, etc.) to reduce environmental impact of material handling at sea.
2. Working out geological/geophysical methods and standards of site planning at sea.
3. Working out methods and standards for proper handling and disposal of spoils and wastes.
4. Working out methods for control of sites, insuring that standards are followed.

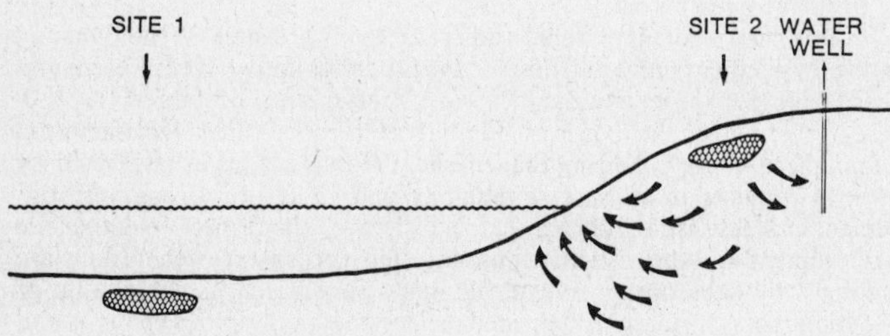

Figure 1—Depositing of waste products. Site 1. Burial in the seafloor sediments will keep the waste products out of the biosphere for so long that there will be no radical alteration to our natural environment. Site 2. A coastal deposit, which is the alternative, will—because of precipitation and resulting groundwater movements—deliver waste products to the biosphere relatively quickly either through the water supply or through seepage to the sea.

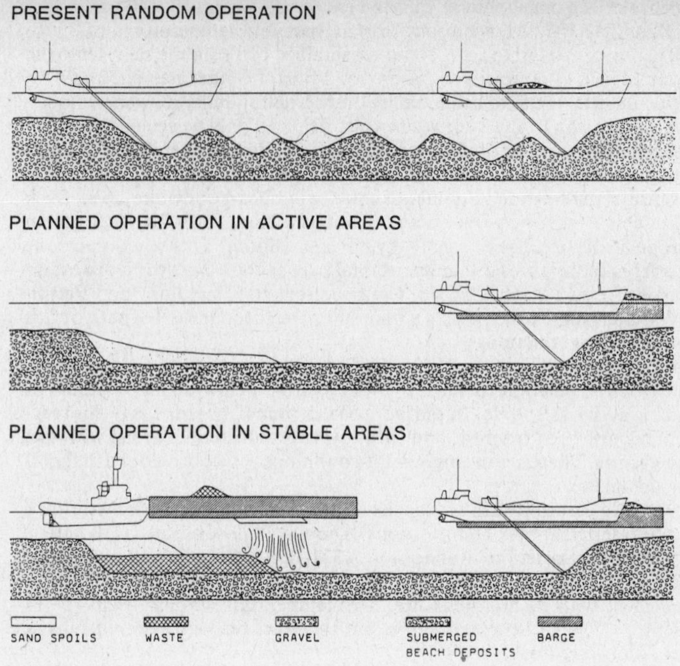

Figure 2.

SCHUBERT, G., University of California, Los Angeles, CA

Numerical Modeling of Mantle Dynamics

Supercomputers have made it possible to construct numerical models of mantle convection that incorporate many of the fundamental characteristics of the structure and dynamics of the actual mantle. Prior to supercomputers, numerical models of mantle convection were limited in their realism and were usually two-dimensional in character with simplified equations of state and rheological behaviors. Realistic models of mantle convection require numerically intensive computations because of the rich complexity of the phenomenon. Mantle convection is fully three-dimensional, occurs in a spherical shell, and is highly nonlinear. The mantle is a compressible medium with a viscosity that depends nonlinearly on temperature, pressure, and stress. The radial structure of mantle convection is complicated by possible changes in chemistry, phase, rheology, and thermal properties at a depth of about 670 km. Mantle convection is inherently time-dependent, as manifest in the temporal variability of plate tectonics, the surface expression of mantle dynamics. Changes in the pattern of ridges and trenches reveal a temporal variability in mantle convection on a time scale of a hundred million years. Variability in hot spot activity demonstrates that times as small as about a million years characterize temporal changes in mantle convection. Mantle convection occurs on a variety of spatial scales, from many thousands of kilometers down to several kilometers. The diameters of hot spots and the thicknesses of boundary layers such as the lithosphere and the D'-layer reveal the importance of 10 to 100 km scales in mantle convection. The nonlinearity, complex material behavior, and wide range of temporal and spatial scales all combine to make numerical modeling of mantle convection a formidable task.

Numerical calculations of nonlinear, compressible, fully three-dimensional, spherical convection have been carried out on a Cray supercomputer. The results show that upwelling plumes and downwelling sheets are a fundamental mode of mantle convection. The cylindrical upwellings are surrounded by the linear downwellings that form a polygonal network. Numerical models of mantle boundary layers and plumes reveal the inherent unsteadiness of these features and suggest that solitary waves may propagate along plumes from the core–mantle boundary to the surface, causing variability in hot-spot activity.

SCHUBERT, G., University of California, Los Angeles, CA

Evolution of Mars' Interior

The thermal evolution of Mars is governed by subsolidus mantle convection beneath a thick lithosphere. Models of the interior evolution are developed by parameterizing the mantle convective heat transport in terms of mantle viscosity, superadiabatic temperature rise across the mantle, concentration of mantle heat sources, and other thermal parameters of the mantle. Geological, geophysical, and geochemical observations of the composition and structure of Mars' interior and of the timing of major events in Mars' evolution such as core, crust, and atmosphere formation are used to constrain the model computations. Numerical calculations of fully three-dimensional, spherical, compressible convection in a shell the size of the Martian mantle are carried out to explore plausible patterns of Martian mantle convection and to relate convective features such as plumes to surface features such as Tharsis.

SCHULZ, K. J., U. S. Geological Survey, Reston, VA, and P. K. SIMS, U. S. Geological Survey, Denver, CO

Tectonics and Metallogeny of the Penokean Orogen, Lake Superior Region, USA

The Penokean orogen is a major Early Proterozoic tectonic belt along the southern margin of the Archean Superior craton. The orogen consists of a northern fold and thrust terrane composed of an epicratonic sedimentary and lesser volcanic assemblage overlying Archean basement and southern island-arc terranes, the Wisconsin magmatic terranes. The northern and southern portions of the orogen are separated by the east–west-trending Niagara fault zone. Recent studies of lithologic assemblages in the Penokean orogen provide important new constraints on the geologic and tectonic evolution of the orogen and help establish a basis for predictive metallogeny.

The epicratonic rocks of the northern terrane consist of quartzite, dolomite, iron formation, graywacke, slate, tholeiitic basalt and minor rhyolite. Stratigraphic and sedimentalogical features suggest that the sedimentary rocks record the progression from a shallow water stable shelf environment to restricted rift basins to a deep water passive margin environment (Larue and Sloss, 1980). The associated volcanic rocks have chemical characteristics typical of within-plate, rift-related magma types (Ueng et al., 1988). A comparison with modern analogs indicates that these epicratonic and related volcanic rocks record the evolution of an Early Proterozoic rifted continental margin. The major Lake Superior type iron formations of the northern terrane probably were deposited during the rift–basin stage of evolution, at least in part, contemporaneously with volcanic activity. Also, some iron formation may have been deposited in a foreland basin environment during the later collision related deformation of the rifted continental margin (Hoffman, 1987).

The southern boundary of the northern epicratonic terrane is the Niagara fault zone. Ophiolitic ultramafic and basaltic rocks, compositionally similar to those of the Troodos and Oman ophiolites, are found along, and to the south of, the fault zone (Schulz, 1987). The presence of the ophiolitic rocks shows that the Niagara fault zone marks a collisional suture zone.

The Wisconsin magmatic terranes to the south of the Niagara fault zone (suture zone) consist of two terranes—a northern terrane that lacks evidence of older basement rocks and a southern terrane that has localized Archean rocks. Early Proterozoic rocks in the terranes consist of widespread calcalkaline and tholeiitic volcanic sequences ranging in age from 1880 to 1860 Ma and an areally more restricted calcalkaline suite that formed between about 1845 and 1835 Ma. The volcanic sequences are compositionally similar to recent island arc suites. Magmatic rocks in the northern terrane show a north to south enrichment in K, Th, and other large ion lithophile elements. Synvolcanic granitoid bodies are predominantly calcic to calcalkaline but include rocks of alkali–calcic affinity. The Wisconsin magmatic terranes are interpreted to be arc system that collided with the continental margin of the Superior craton, ca 1850 Ma. The collision (The Penokean orogeny) resulted in recumbant folding and thrusting of the epicratonic rocks of the northern continental margin terrane (Klasner, 1988).

The only known volcanogenic massive sulfide deposits in the region occur within the northern subterrane of the Wisconsin magmatic terrane (Sims, 1987). These deposits, consisting principally of copper and zinc sulfides, are restricted to bimodal, calcalkaline volcanic units that are compositionally distinct from other calcalkaline suites within the terrane. The massive sulfide hosting volcanic rocks, characterized by basalt–basaltic andesite and rhyolite, high Al_2O_3, and distinctive rare earth and other trace element patterns, may have formed in a within-arc rift. These distinctive compositions are particularly useful for massive sufide exploration within the Wisconsin magmatic terrane.

Late orogenic, postorogenic, and younger (1760 Ma and 1500–14,700 Ma) anorogenic intrusions and lesser amounts of volcanic rocks are locally abundant within the Penokean orogen. Compositionally distinct syn to postcollisional granite plutons occur near the Niagara fault zone, whereas younger anorogenic plutons are particularly abundant in the southern terrane of the Wisconsin magmatic terranes. Minor amounts of molybdenum mineralization are associated with some syn to postcollisional granites, whereas some anorogenic granites are anomalous in Sn, Ta, Nb, U, and Th. In particular, a recently discovered 1733 Ma alkali feldspar granite in the northern continental margin terrane of the Penokean orogen is compositionally similar to rare metal rich granites of Nigeria and the Arabian Shield. This compositional similarity suggests the potential for Sn-W (Ta-Nb) mineralization in portions of the Penokean orogen.

REFERENCES

Hoffman, P. F., 1987, Early Proterozoic foredeeps, foredeep magmatism, and Superior-type iron-formations of the Canâdian Shield: *in* A. Kroner, ed., Proterozoic lithospheric evolution: American Geophysical Union Geodynamics Series, v. 17, p. 85–98.

Klasner, J. S., R. W. Ojakangas, K. J. Schulz, and G. L. Laberge, 1988, Widespread evidence for Early Proterozoic overthrusting in the Penokean orogen of northern Michigan: Geological Association of Canada, Program with Abstracts, v. 13, p. A67.

Larue, D. K., and L. L. Sloss, 1980, Early Proterozoic sedimentary basins of the Lake Superior region: GSA Bulletin, pt. 1, v. 91, p. 450–452.

Schulz, K. J., 1987, An Early Proterozoic ophiolite in the Penokean Orogen: Geological Association of Canada Program with Abstracts, v. 12, p. 87.

Sims, P. K., 1987, Metallogeny of Archean and Proterozoic terranes in the Great Lakes region—A brief overview: U. S. Geological Survey Bulletin 1694-E, p. 55–74.

Ueng, W. C., T. P. Fox, D. K. Larue, and J. t. Wilband, 1988, Geochemistry and petrogenesis of the Early Proterozoic Hemlock volcanic rocks and the Kiernan sills, southern Lake Superior region: Canadian Journal of Earth Sciences, v. 25, p. 528–546.

SCHULZE, D. J., Queen's University, Kingston, Canada

Ultramafic Nodules from Kaalvallei Kimberlite, O.F.S., South Africa: Evidence for Thermal Perturbation of Lithosphere during Megacryst Formation

The ultramafic xenolith suite of the Kaalvallei kimberlite, located near Virginia, O.F.S., South Africa, contains members of most mantle rock types, including spinel peridotites, coarse and deformed garnet peiodotites, Cr-poor silicate and ilmenite megacrysts, ecologites, and alkremites. Coarse garnet peridotites equilibrated between 825 °C, 33 Kbars, and 1000 °C, and 47 Kbars, approximating a steady-state subcontinental geothermal gradient. Deformed garnet periodotites range from 1075 °C, 48 Kbars, to 1435 °C, 68 Kbars, the lower end of which falls on the steady-state geotherm, but the upper reaches of which are hundreds of degrees above a shield geotherm. Spinel peridotites equilibrated at temperature below 900 °C.

Ecologites include Cr-pyrope-bearing clinopyroxenites (Cr_2O_3 to 2.5 wt% in garnet) and relatively Ca-poor ecologites (Ca/(Ca + Mg + Fe) in garnet < 0.2), the most Fe-rich (Mg/(Mg + Fe) (*mg*) = .494 in garnet) of which contains diamond. Unlike most other mantle eclogite suites, eclogites containing Ca-Fe rich garnets are absent, and the suite is dominated by a group with garnets rich in Mg and Ca (Ca/(Ca + Mg + Fe) from 0.24 to 0.47, Fe/(Ca + Mg + Fe) < 0.2). An exceptionally magnesian member of this group [garnet with Mg/(Mg + Fe) = 0.919, cpx with 0.3 wt% FeO] contains primary Mg-Al spinel [*mg* = 0.937, Cr/(Cr + Al) = 0.050]. Alkremite garnets also plot in this group, near Ca/Mg/Fe-.29/.53/.18, but are not considered to be related to the eclogites. Eclogite equilibration temperatures (calculated at 50 Kbars) range from 1090 °C to 1325 °C, with most examples below 1200 °C. The spinel-bearing eclogite equilibration temperature is 860 °C. Such a temperature distribution suggests that the eclogites coexisted with the garnet peridotites that equilibrated on the geothermal gradient.

Cr-poor megacrysts span a wide compositional range. Mg/(Mg + Fe) in garnet, for example, ranges from 0.809 to 0.667 (with a gap from 0.77 to 0.73), and exhibits a positive correlation with Cr_2O_3 (1.2 to < 0.1 wt%). Variation of TiO_2 with *mg* resembles that from Monastery Mine megacrysts, in which, as *mg* declines, Ti first increases and then, with the appearence of ilmenite, decreases. The expected TiO_2 "peak" in the Kaalvallei suite, however, occurs in the *mg* gap. The correlations of Ti and Cr with *mg* are both expected results of fractional crystallization. Ca/(Ca + Mg) in clinopyroxene ranges from 0.288 to 0.471, equivalent to temperatures in the range 1470–930 °C, but correlation with mg is complex. At high temperatures, *mg* is constant (~0.87 from 1470 to 1300 °C). One intermediate cpx occurs at 1270 °C, mg = 0.846, and a negative correlation of temperature and *mg* occurs in a lower temperature, Fe-rich group, ranging from approximately *mg* = 0.825, 1240 °C to *mg* = 0.847, 930 °C. Aligned sulfide blebs occur in some of the more magnesian garnets and a clinopyroxene from the Mg-rich group, due to early separation of an immiscible sulfide magma.

All silicate–ilmenite intergrowths occur in the Fe-rich (low-temperature) group, including graphic intergrowths of olivine–ilmenite and clinopyroxene–ilmenite. Orthopyroxenes (*mg* = 0.832–0.864) and olivines (*mg* = 0.815–0.847), with or without ilmenite, belong to the Fe-rich group.

Orthopyroxenes equilibrated in the small temperature range 1160–1230 °C at 55 ± 1 Kbars, which is no more than 50° above the steady-state geothermal gradient. Monomineralic ilmenites range from 4.0 to 10.4 wt% MgO, with Cr_2O_3 < 0.6 wt%, and thos intergrown with silicates are restricted to the range MgO-8.6–9.7 wt%, Cr_2O_3 < 0.8 wt%.

If the Mg-rich and Fe-rich groups are comagmatic, which is suggested by the lack of a discontinuity in correlation between $Cr_2O_3$and *mg* across the garnet *mg* gap, precipitation of orthopyroxene or olivine may have caused the shift to more Fe-rich compositions at approximately 1300 °C. Precipitation of ilmenite could have caused the shift to more magnesian compositions with decreasing temperature in the Fe-rich silicate group.

Two important nodules provide a physical link between the megacrysts and the garnet peridotites. Each nodule consists of a Cr-pyrope core (one of which contains orthopyroxene and serpentinized olivine) armored by Ti-rich pyrope, the latter compositionally indistinguishable from individual Cr-poor garnet megacrysts. The contacts between the Cr-rich garnets (7.8 wt% Cr_2O_3 in one example and 10.3 wt% Cr_2O_3 in the other) and the Ti-pyropes (0.5 wt% Cr_2O_3) are marked by necklaces of ilmenite blebs and, in one nodule, clinopyroxene, and in the other clinopyroxene + orthopyroxene + olivine. Orthopyroxene included in the Cr-pyrope core yields equilibration conditions of ~1060 °C, 50.5 Kbars, the temperature of which is identical to that calculated for clinopyroxene in the contact assemblage. In the other nodule, the contact assemblage clinopyroxene + orthopyroxene (+ olivine) yields equilibration conditions of 1090 °C, 51.6 Kbars, and both estimates fall on the steady-state geothermal gradient.

The geochemical trends within the megacryst suite are evidence for an origin for this suite through fractional crystallization during cooling. Further evidence for disequilibrium exists in the mantled pyrope xenocrysts, which show that the magma that precipitated the megacrysts was not in chemical equilibrium with the surrounding peridotite wall rock. Furthermore, the high-temperature, Mg-rich group of megacrysts, the equilibration temperatures of which overlap those of the highest temperature deformed garnet peridotites, are compositionally distinct from the peridotite minerals, again suggesting disequilibrium.

Estimates of pressures and temperatures of equilibration indicate that peridotites and megacrysts coexisted over a range of >500 °C. It is proposed that the parent magma of the megacrysts, with a liquidus temperature of ≥1470 °C intruded garnet peridotite host rock at a depth of approximately 220 km (pressure ≅68 Kbars), heating the wall rocks from an ambient temperature near 1250 °C to magmatic temperatures. The Fe-rich megacryst group, representing an evolved equivalent of the original magma, similarly affected the host rocks, although the magmatic temperatures were not much above those of the wall rocks at lower pressures. No effects were observed that suggested interaction between eclogites and the megacryst magma.

SCHUMANN, H. H., U.S. Geological Survey, Phoenix, AZ

Geophysical Investigations of Earth Fissures Caused by Land Subsidence Near Apache Junction, Arizona

A pioneering program of surface- and borehole-geophysical surveys was conducted to investigate the hydrogeologic conditions that may control the formation of earth fissures in the alluvial sediments that constitute the principal groundwater reservoir in south-central Arizona. Earth fissures and land subsidence in the alluvial basins results from groundwater overdraft. Large water-level declines have caused more than 4.5 m of land subsidence, which has produced earth-fissure zones as much as 15 km long in parts of southern Arizona. Differential land subsidence and earth fissures have caused millions of dollars in damage to engineering structures, have adversely affected urban and rural land use, and continue to pose serious environmental hazards. water-level declines in excess of 100 m indicate a large-scale depletion of groundwater in the alluvial aquifer near Apache Junction, Arizona. Land-subsidence rates as great as 60 mm/y have been measured. New earth fissures have been observed and old fissures have extended in the study area.

A program of borehole-geophysical surveys, seismic reflection and seismic refraction profiling, and gravity surveys was developed to investigate earth fissures observed in the alluvial sediments near the route of the Central Arizona Project aqueduct that passes through the study area. Surface-geophysical surveys revealed anomalies that were interpreted to be irregularities in the surface of the basement rock. These irregularities occur beneath existing earth fissures at depths as much as 200 m below the land surface. A series of deep test holes was drilled to calibrate the surface geophysical surveys. Second-order level surveys indicated that differential land subsidence was occurring across the geophysical anomalies. Differential land subsidence produced tensional stresses that caused the fissures in the alluvial sediments overlying the irregularities in the surface of the buried basement rock.

SCHWAB, W. C., U.S. Geological Survey, Woods Hole, MA, E. UCHUPI, D. A. GALLO, R. D. BALLARD, and T. K. DETTWEILER, Woods Hole Oceanographic Institution, Woods Hole, MA, R. T. HOLCOMB, U.S. Geological Survey, Seattle, WA, and S. E. HARRIS, Marine Imaging Systems, Pocasset, MA

High-Resolution Mapping of Segments of Neovolcanic Zone of Mohn's Ridge and East Pacific Rise Using Argo Imaging System

The Argo imaging system, which carries a charge-couple device (CCD) electronic still camera, three silicon-intensified target (SIT) TV cameras, a 35 mm photographic still camera, and a 100 kHz sidescan sonar system, allows marine geologists to rapidly map hectare- to square kilometer-scale areas of the seafloor in high resolution. SIT cameras and film have been used together in complement, but photographic film is not well matched in sensitivity to the SIT camera. The typical range for the Argo 35 mm still camera is 5–10 m, whereas the SIT cameras collect high-resolution digital imagery routinely at 10–20 m above the seafloor. To overcome this limitation, an underwater electronic still camera has been developed and integrated into the Argo system. The electronic still camera represents a major advance in high-quality underwater visual images that have large area coverage. Using scientific-grade large-format CCD arrays, the electronic still camera makes high-resolution (centimeter scale) digital images at long ranges (commonly 10–20 m) with real-time image display and processing capability.

Argo was used along with previously collected Seabeam bathymetry data to map the distribution of volcanic flow types and structural setting (detailed bathymetry, fracture patterns, faults, etc.) of a segment of the neovolcanic zone of the East Pacific Rise from latitudes 10°19 'N to 11°53 'N. Similarly, Argo and the SeaMARC 1B sidescan sonar system were used to explore for the presence of hydrothermal activity along a segment of the neovolcanic zone of Mohn's Ridge in the Norwegian-Greenland Sea, from longitudes 6°00 'W to 0°05 'E. Both the Seabeam bathymetry and the SeaMARC sonographs provided base maps for the higher resolution Argo investigations.

Three areas of Mohn's Ridge were investigated. The Mohn's Ridge neovolcanic zone displays considerable relief and is characterized by an axial graben 1 km deep and 15 km wide. The area between 5°30 'W and 5°25 'W (about 10 Km^2) has a bioturbated rippled sediment cover of 100%. Volcanic ridges 100–200 m high, which are composed mostly of pillow flows, comprise the other two areas, from 3°40 'W to 3°00 'W (about 330 Km^2) and from 1°05 'W to 0°05 'E (about 400 km^2). These pillowed ridges are aligned oblique to the strike of the neovolcanic zone. The summits of some pillowed ridges are capped by volcanic sheet flows. The Argo imagery indicates that the area between 1°05 'W and 0°05 'E is an area of hydrothermal activity. These indications include minor temperature anomalies, sediment suspected to be of hydrothermal origin, sediment elutriation structures along fissures, and large colonies of sponges. The SeaMARC sonographs depicted a water-column density anomaly in the same area that is though to represent a hydrothermal plume.

The crest of the East Pacific Rise between the Clipperton and the Orozco Fracture Zones consists of a chain of subparallel topographic highs 2–10 km wide, 200–400 m high, and 40–140 km long. The East Pacific Rise study area includes a continuous ridge segment that extends from the Clipperton Fracture Zone (10°18 'N) to approximately 11°53 'N, where it forms the western limb of overlapping spreading centers, and the southern terminus of another ridge segment that forms the eastern limb of the overlapping spreading centers. The neovolcanic zone in this area of the East Pacific Rise displays considerable topographic variation in cross-section and contains a narrow summit graben 10–100 m wide and 20 m deep. The imagery permits mapping of three volcanic rocks units: (I) fresh, glassy surfaced sheet flows, (II) partly glassy to dull pillow basalts and sheet flows, and (III) pillow basalts having a sediment cover of 15–20%. Evidence of past hydrothermal activity, including sulfide stacks and hydrothermal staining, was found in the first two units and in some places along fractures in the third. With a few exceptions, active vents were found only in unit I.

Our observations along the crest of the East Pacific Rise from the Clipperton Fracture Zone to the western limb of the 11°45 'N overlapping spreading centers suggest that an initial phase of magmatic activity, which emplaced unit III pillow flows (>6 Ka), was followed first by a period of volcanic quiescence and then by a second phase of activity (about 500 years ago) that deposited sheet flows of Units I and Ii atop the pillows. The mapping also suggests that although there is morphologic overlap between the eastern and western limbs of the 11°45 'N overlapping spreading centers, there is no overlap in the active volcanic zones. The imagery shows that recent volcanic activity occurred only on the eastern limb; the western limb possesses only older unit II and III flows.

SCHWARCZ, H. P., McMaster, University, Hamilton, Canada and R. GUÜN, Subdepartment of Quaternary Research, Cambridge, United Kingdom

ESR Dating of Prehistoric Archeological Sites

Electron spin resonance (ESR) spectroscopy allows us to detect unparried electrons that have been trapped in crystals as a result of environmental radiation. This method is applicable to hydroxyapatite of tooth enamel, calcitic speleothem, and heated flint, all of which are found commonly in age-critical situations in archeological sites. The range of the ESR method is from a few k.y. up to about 2 m.y. Precision over most of this range is ± 10–15%. The method is based on the following assumptions: (1) the lifetime of trapped charges (electrons or holes) is large compared with the age of the sample; (2) the environmental radiation dose rate has remained constant through time, or has changed in a known manner; (3) the amplitude of the natural ESR signal can be measured above background; (4) the sample is not so old that either all available traps have been filled with charges or the sample has reached a steady state (rate of discharge of traps = rate of filling). The age, T, is obtained from solution of the following equation:

$$A = \int_0^T D(t)\,dt$$

where A is the acquired dose (in grays) and $D(t)$ is the dose rate as a function of time. To find A, the response of the ESR signal to additional doses of artificial gamma rays is determined and the equivalent dose of the natural ESR signal is inferred. Normally, the response curve is almost linear up to doses much larger than the natural dose.

For samples such as tooth enamel, which receive a large part of their radiation from uranium (U) trapped in the tooth itself, the age critically depends on the history of U uptake by the tooth. Appropriate models of U uptake must be used. We have shown that simultaneous measurement of U-series disequilibrium (^{230}Th/^{234}U ratio) and A from ESR analysis allows us to fit a simple exponential U-uptake model that gives consistent data for replicate samples from single sites (Grün et al, in press).

U-series dates on tooth enamel have so far been used to determine the age of important archeological and hominid sites in the Old World ranging up to 300 k.y., including La Micoque, Bilzingsleben, Pontnewydd, Djebel Irhoud, Monte Circeo, Skhul, Qafzeh, and the Neanderthal skeleton at Kebara (Schwarcz and Grün, in press). At several of these sites, we have shown good agreement with two other methods of dating; U-series (on speleothem); and TL on burnt flint (Valladas et al, 1988). Teeth from the australopithecine site of Sterkfontein, South Africa, give preliminary ages of 2.3 ± 0.6 m.y.; the large error is because we could not measure the dose rate in the environment of the teeth; new samples for which in situ dose rates are known are being analyzed.

While the ESR method is less precise than some other methods (^{30}Ar/^{40}Ar, U-series by mass spectrometry) it is applicable to a very wide range of archeological sites. For example, we are investigating the dating of calcite deposits form a "sandwich" with the paint in the middle. ESR dates of the calcite would put limits on the time of painting.

Acknowledgments: This research was supported by grants to HPS from the Natural Sciences and Research Council of Canada.

REFERENCES

Grün, R., H. P. Schwarcz, and S. Zymela, 1987, ESR dating of tooth enamel: Canadian Journal of Earth Science, v. 24, p. 1022–1037.

——— J. Chadam, and H. P. Schwarcz, in press, ESR dating of tooth enamel: coupled correction for U-uptake and U-series disequilibrium: Nuclear Tracks.

Schwarcz, H. P. and R. Grün, in Press, ESR dating of level L2/3 at Lar Micoque (Dordogne), France: excavations of Debenath and Rigaud: Geoarchaeology.

——— ——— A. G. Latham, D. Mania, and K. Brunnacker, 1988, New evidence for the age of the Bilzingsleben archaeological site: Archaeometry, v. 30, p. 5–17.

——— ——— B. Vandermeersch, O. Bar-Yosef, H. Valladas, and E. Tchernov, in press, ESR dates for the hominid burial site of Qafzeh in Israel: Journal of Human Evolution.

Valladas, H., J. L. Joron, G. Valladas, B. Arensburg, O. Bar-Yosef, A. Belfer-Cohen, P. Goldberg, H. Laville, L. Meignen, Y. Rak, E. Tchernov, A. M. Tillier, and B. Vandermeersch, 1987, Thermoluminescence dates for the Neanderthal burial site at Kebara in Israel: Nature, v. 330, p. 159–160.

SCHWARTZ, F. W., University of Alberta, Edmonton, Canada, and L. SMITH, University of British Columbia, Vancouver, Canada

Continuum Approach for Modeling Mass Transport in Fractured Media

A continuum approach has been developed that can represent anisotropic dispersion in a network of fractures. The technique requires that statistics describing the fracture network and the character of particle transport be collected from a discrete network model of a fracture system. This discrete model is, in practice, a small piece of a much larger system. These statistics are used to control the pattern of transport in a larger continuum and enable us to mimic in the larger system the particle transport that is observed in the smaller system.

Testing of the model has verified the modeling concept. We are able to duplicate patterns of anisotropic dispersion predicted by de Josselin de Jong's model for equidistant fractures of infinite length. Comparisons of the two procedures were conducted for a variety of different orientations of this simple network with respect to the hydraulic gradient. The model was also reasonably successful in simulating the pattern of dispersion in the discrete network from which the transport statistics were collected. In this case, the two sets of simulation results were not exactly the same. The discrepancy is due to the fact that transport in the discrete network is approaching but has not yet encompassed a large enough piece of the network to produce the "average" behavior obtained with the continuum model. Examination of the breakthrough curves in the case of the discrete model shows that the spreading is often non-Gaussian. For the continuum model, spreading is Gaussian along nearly the entire distance of travel. Thus, at comparable scales the two models give different results. This outcome is due to averaging that develops because in the continuum model only statistical features of the particle transport are preserved. We cannot account for the local variability that can develop in the discrete model.

Indications are that de Josselin de Jong's model for estimating dispersivities for simple networks cannot be extended generally to represent more complex networks. However, a geometry based on orthogonal networks is probably the most severe test in evaluating the approach because of the pathological behavior in the calculation when the network is aligned parallel and perpendicular to the mean hydraulic gradient. Thus, further work to examine the possibility of extending de Josselin de Jong's model appears to be warranted.

The simple application of the continuum model presented here shows how complicated dispersion can be in a fracture network. The factors that contribute to spreading in the major and minor principal directions change as the orientation of the network is changed relative to the mean hydraulic gradient. Spreading in the direction of the principal axis is mainly a result of the intrinsic geometric properties of the network and the variation in fracture lengths between intersections. The only exception is when the mean hydraulic gradient is oriented close to the attitude of one of the fracture sets.

Spreading in the direction of the minor principal axis is controlled mainly by variability in velocity when flow is not oriented closely to one of the fracture sets. When gradients and fractures ae coincident, variability related to the d_{ij} is the main cause of dispersion.

An important feature of our modeling approach is that dispersion is accounted for by the variability in flow conditions instead of an empirical parameter such as dispersivity. The approach sidesteps the calculation of an anisotropic dispersion tensor. Additional work is required to fully develop the continuum approach as a practical modeling tool. There is an initial requirement for a more general discrete-network model, which can account for at least an order of magnitude larger number of intersections. Being able to include a larger number of nodes would reduce the inherent variability in the network and partical statistics that are caused by the small sample size. This extension would also reduce the subjective "tinkering" that is now required with the distributions. Another important requirement for not only this but other continuum models is to account for diffusion into the matrix. This process is typically not considered in modeling yet appears to have important implications for contaminant attenuation.

SCHWARZ, G., H. J. GÖTZE, and P. J. WIGGER, Free University of Berlin, Berlin, West Germany

Structure of Earth's Crust and Upper Mantle in Central Andes (20°–26°S) as Inferred from Combined Geophysical Studies

The Andes represent one of the oldest examples of plate tectonics. In the classical view, the mountain belt began to rise when the Nasca plate began to submerge under western South America. But in fact, the database to confirm this model has been poor for years and different explanations for the origin of the Andean mountains must be established.

We have been investigating the structure of the Central Andean crust and upper mantle for more than six years by means of gravity surveys, magnetotelluric and geomagnetic depth sounds, seismological, and seismic refraction studies. In an east-west segment between 20° and 26°S, traversing the Central Andes from the Pacific Ocean to the Brazilian shield (see Figure 1), more than 3500 gravity data were measured, about 100 sites were occupied to determine the electrical resistivity of the crust, catalog as well as own data were used for earthquake analysis, and a net of reversed refraction profiles covering representative units was observed with shot points on land as well as offshore. In the following paragraphs, we will summarize the geophysical results obtained up to now, going from west to east.

In the *Coastal Cordillera*, we observe a gravity high with local maxima, centered along the coastline. Seismic velocity reaches values of about 6.7 km/s—a typical value for the lower crust at a depth of about 6–8 km. Thus, seismics as well as gravity point to material of high density positively compensated at shallow depth. Indeed, lower crustal rocks are exposed in the Coastal Cordillera. We conclude that the continental margin is uplifted by subduction processes and the old upper crust is eroded. A very clear seismic discontinuity with velocities of 8.3 km/s was observed at depths between 38 and 43 km. A variation of crustal structure in north-south direction is indicated. Fault systems striking northwest-southwest have very low resistivities, thus indicating recent tectonic activity. Earthquakes occur at shallow depth.

The *Pre-Cordillera* is characterized by regional high values of gravity due to compressional structures and isostatically uprising blocks. Crustal thickness was calculated to about 60–65 km; seismic refraction data on a nonreversed profile yield about 70 km for crustal thickness. A clear seismic discontinuity at about 30 km depth might indicate a Jurassic Paleo-Moho. This enormous thick lower crust (40 km, mean velocity 6.7 km/s) might be generated mainly by accretion processes. The area of the Pre-Cordillera at about 22.5°S shows a maximum of released seismic energy from about 100 km depth. Electrical resistivity is found to have normal crustal values.

The *Western Cordillera* shows the absolute minimum of the Bouguer gravity of about −450 mgal, corresponding to a maximum of crustal thickness of about 70 km. Local minima in gravity are caused by structures within shallow depth. Electrical resistivity is decreasing to extreme low values of less than 1 ohm m at about 8–10 km depth; the lower bound must be at depth greater than 30 km leading to a total conductance greater than that of the Pacific Ocean. These extreme values may be explained by unsolidified acid intrusions, a result that seems to be confirmed by reduced seismic velocities, attenuation, and scattering of seismic waves in this area.

The *Altiplano* in Bolivia is underlain—at 40–50 km depth—by very low resistive material. We interpret these low values as being due to the partially melted state. Gravity shows alternating highs and lows superposing the still existing large minimum, which yield a crustal thickness of 70 km. The increase of gravity toward the east is very strong.

At the western border of the *Eastern Cordillera*, a low resistive layer is found at 20 km depth, and the total conductance increases strongly toward the east bordering an uprising, highly resistive Pre-Cambrian basement near Tarija. The very low resistivities may be connected to shear zones in the upper crust, as well as pointing to an increased mobilization of these crustal structures. Gravity increases to values of about −300 to −250 mgal.

The *Subandean Belt* and *Chaco* display thin-skinned tectonics. Crustal thickness calculated from gravity data amounts to about 50 km. The sedimentary cover above a highly resistive three-dimensional basement shows low resistivities. Tectonic activity is manifested in the occurrence of shallow earthquakes.

This brief summary of our geophysical investigations shows the complexity of the Central Andean crust and upper mantle structure but points as well to a better insight into the ongoing processes of Andean mountain building.

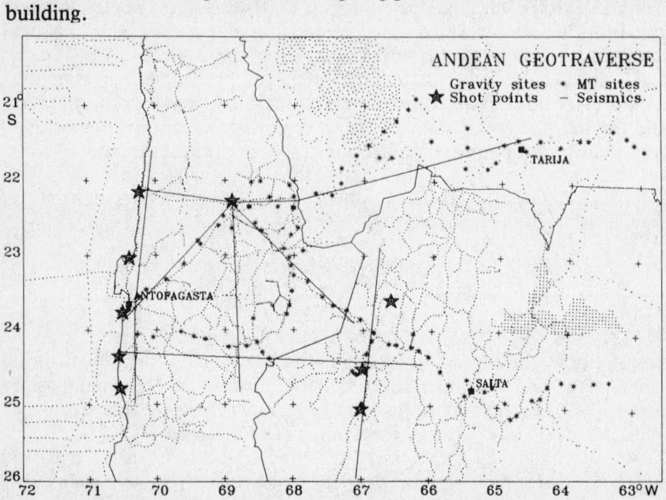

SCHWEIZER, T., T. BECHSTÄDT, A. HAGEMEISTER, Geologisches Institut der Universität Freiburg, Freibur, West Germany

Terrigenous/Carbonate Raibl Group (Upper Triassic, Eastern Alps): Record of Sea-Level Changes in Carnian

Marine transgressions and regressions are controlled by two main parameters: the relative rise of sea level (as a combination of subsidence and eustatic sea-level changes) and the sedimentation rate (mainly terrigenous influx or carbonate production). According to Vail and others (1977), during relative rise of sea level the amount of terrigenous input determines whether transgressive, regressive, or stationary facies belts are produced. Coarser grained terrigenous clastics encroach the coast during transgressional periods. A change from carbonate sequences to terrigenous sands should be indicative of a regressional period. The model of Vail and other (1977) is based on a vertical point of view. Lateral transport and reworking of clastics onto platform and shelf areas are not considered. The middle and late Carnian Raibl group of the Eastern Alps is a good example of these processes.

The Raibl group is under- and overlain by thick carbonate platform sequences (the Wetterstein limestone and the Dachsteinkalk/Hauptdolomit). The three third-order cycle of the Raibl group (each approximately 5–100 m thick) consist of terrigenous clastics, overlain by a (partly evaporitic) carbonate sequence. Facies and fauna indicate a mostly inter- to supratidal environment of the carbonates, subtidal, sometimes intertidal conditions during sedimentation of the terrigenous clastics. The terrigenous clastic beds are interpreted as more open marine, the carbonates as more restricted marine environments.

Subcycles of fourth order are shallowing upward; they allow a detailed lithostratigraphic subdivision and correlation from nearshore to offshore. Subcycles within terrigenous clastics consist of the following facies types (from base to top): (A) black shales, partly bioturbated with ammonites; (B) sandstones, at the base strongly bioturbated (*Diplocraterion*), at the top partly clean, either even parallel laminated or cross-bedded; (c) marly beds; (D) cross-bedded oolites with a highly diverse shallow marine fauna.

The characteristic sandstones (unit B) within these fourth order cycles are interpreted as submarine bars (sand ridges); they were deposited under increasing energy conditions toward the top.

The subcycles within the carbonate sequences start with peioid wackestones/packstones (with few forams and some megalodonts), overlain by blindstones and finally evaporitic collapse breccias or flat pebbles.

During a regressive phase, a restricted carbonate platform was built up, situated close to a mainland. Only in nearshore positions can the input of minor shales be observed within the carbonates. Coarser clastics were absent; they were deposited along the coastline, mainly in deltaic areas, and bypassed the platform. To move sediment from the shoreline to the shelf area involves breaking through the "littoral energy fence" (Swift, 1967). This can be accomplished by a variety of processes, including river mouth and shoreface bypassing and storm surge turbidity flows. In the sequence described, shoreface bypassing seems to be one of the main mechanisms for the redistribution of sand onto the carbonate platform. During the rapid transgressions, the coastal and deltaic clastics became "palimpsest sands"; they were reworked, transported, and piled up to form flat sandbodies (compare Tillman and Siemers, 1984, and Johnson and Baldwin, 1986).

According to Haq and others (1986), two eustatic transgressive/regressive cycles have to be assumed for all of the Carnian. The Raibl sequence of the Eastern Alps indicates three transgressive events in the middle and upper Carnian, each of them culminating within the terrigenous clastics of the third-order cycles.

REFERENCES

Haq, B. U., J. Hardenbol, and P. R. Vail, 1986, Chronology of fluctuating sea levels since the Triassic: Science, v. 235, p. 1156–1167.

Johnson, H. D. and C. T. Baldwin, 1986, Shallow siliciclastic seas; in H. G. Reading, Sedimentary environments and facies: Oxford, Blackwell Scientific Publ., p. 229–282.

Swift, D. J. W., 1976, Continental shelf sedimentation, in D. J. Stanley and D. J. P. Swift, eds., Marine sediment transport and environmental management: New York, John Wiley & Sons; p. 311–350.

Tillman, R. W. and C. T. Siemers, 1984, Siliciclastic shelf sediments: Society of Economic Paleontologists Mineralogists, Special Publications, v. 34, p. 1–251.

Vail, P. R., R. M. Mitchum, S. Thompson, III, 1977, Seismic stratigraphy and global changes of sea level. Part 3: AAPG Memoir 26, p. 63–81.

SCOON, R. N. and H. V. EALES, Rhodes University, Grahamstown, South Africa

Platinum-Group Element Mineralization in Critical Zone of Bushveld Complex: Distribution and Mineralization Types

Introduction: "Primary platinum-group element (PGE) mineralization, which is concentrated at the base of most cyclic units within the Critical Zone of The Bushveld Complex, is scavenged from new influxes of primitive magma concommittantly with the development of ultramafic cumulates. Chromitite layers are envisages as the fundamental concentration mechanism, with early-formed base-metal sulfides (BMSs) locally playing a supportive role. Postcumulus processes may redistribute the primary mineralization, but usually only over vertical heights of a few centimeters. In other words, the PGE mineralization is essentially concentrated by syngenetic processes at the crystal–liquid interface and is not related to mechanisms that Barnes and Campbell (1988) neatly summarized as "uppers" (i.e., deposition from upward-migrating intercumulus liquids or hydrothermal fluids) or "downers" (i.e., accumulation of PGE-rich sulfides that have achieved a high (R-value). These (deliberately) provocative statements, which are based on a comprehensive knowledge of field relationships rather than accepted geochemical theories, are directed at stimulating a rethink on the mechanisms that concentrated the PGEs within the Critical Zone of the Bushveld Complex.

Current Models: The "hydrothermal" model (see MacDonald, 1987) has gained little acceptance among South African-based geologists, largely because it fails to explain key issues such as the regionally contiguous nature of the "reefs," the consistency of grade × thickness values, the remarkable stratabound nature of the mineralization, and the absence of mineralized veins. These features, particularly the regional field relationships, must imply some form of synmagmatic control. The high R-value model of Campbell et al (1983), which is currently widely accepted (MacDonald, 1987), addresses many of these features and is an elegant explanation of how an immiscible sulfide liquid can become enriched in PGEs. However, this model has four principal flaws when applied unreservedly to PGE-rich cumulates in the Critical Zone of the Bushveld Complex: (1) the remarkable association of anomalous concentrations of PGEs with chromitite layers (probably all of the chromitite layers in the Critical Zone are enriched in PGEs at their base) has been largely ignored; (2) only a minor proportion of the PGE-rich cumulates are associated with anomalous concentration of BMSs; accordingly, the role of immiscible sulfide liquids as collectors is overemphasized; (3) the behavior of the six individual PGEs, as well as mineralogical data, define (albeit poorly understood) predictable trends that cannot be related entirely to concentration by sulfides; and (4) for the BMSs to achieve a high R-value each new influx of magma must mix with the supernatent liquid column (by entering the chamber as a jet or plume), a model which is in direct conflict with how we perceive the cyclic units of the Critical Zone to have formed (this latter point is obviously a biased objection).

Development of the Critical Zone: The Rhodes University school of thought argues that the upper Critical Zone has largely developed by numerous influxes of primitive magma (preserved as cyclic units) that have been injected into the chamber as hot, dense basal flows (inter alia Eales et al, 1987, 1988; Scoon and de Klerk, 1987). Successive injections led to thermal erosion of the floor, causing "dimpling" and "potholing" on a local scale, and elimination of entire noritic-anorthositic cumulates on a regional scale. Each cyclic unit is attributed to a new influx of magma; accordingly, lateral stratigraphic variations may be related to facies changes and both truncation and beheading of the uppermost parts of earlier units. Each cyclic unit in the upper Critical Zone may reflect two periods of "mixing"—the lower, ultramafic members (chromitite, harzburgite, pyroxenite) are derivatives of contamination between the new influx and the eroded floor material [mixing event (a)], whereas the upper, leucocratic members (norite, anorthosite) crystallized during progressive hybridization of the fractionated residua of the contaminated new influx with the supernatant liquid column [mixing event (b)]. Mixing event (a) accounts for the robust basal chromitite layers, and event (b) explains, first, the juxtaposition of harzburgite and anorthosite layers without the intervening pyroxenite–norite cumulates and, second, the thin "reaction" chromitite stringers found in the central parts of units. Cryclic units in the lower Critical Zone probably formed from a similar scenario, except that new influxes truncated the developing unit prior to reaching mixing event (b), and mixing event (a) involved ultramafic floor rocks rather than norite–anorthosite cumulates.

Note that we attribute all of the cyclic units in the upper Critical Zone to one magma lineage ("U-type" of Irvine and Sharpe, 1986), with the exception of the Merensky and Bastard units that reflect the initial stages of the partial melting event evolving to a higher R_0 magma (culminating in the "A-type" magma in the Main Zone. Thus, the bulk of the mixing processes described above are attributed to mixing of primitive and evolved U-type magmas. However, mixing event (a) at the base of the Merensky unit occurred between a primitive "U/A-type" and cumulates that were derived from "U-type" magma lineage.

Distribution of the Mineralization: PGE mineralization, and PGE and BMS mineralization that occurs in juxtaposition, is essentially restricted to ultramafic cumulates that occur at the base of cyclic units. Accordingly, primary (or cumulus) mineralizing events are probably related to mixing event (a), and must have occurred prior to mixing event (b), such that any immiscible sulfide droplets could not have equilibrated with the supernatant liquid column. Furthermore, streaming of influxes of magma across the floor of the chamber, continuously feeding new, primitive magma to the developing cyclic unit, obviates mass-balance considerations (that have constrained much of our thinking to date).

Mineralization Types: Six mineralization types are recognized within the Critical Zone of the Busveld Complex; of these, types 1–3 are primary and types 4–6 are postcumulus.

1. *Chromitite layers associated with ultramafic cumulates*: These are essentially restricted to the lower Critical Zone (e.g., LG 1–7; Mg 1–2) and all contain anomalous (yet subeconomic) PGE mineralization. This mineralization type is laterally persistent, is intimately stratabound within chromitite layers (and associated stringers), and contains background values of BMSs (characterized by [Pt*1000]/Cu ratios of 10–20).

2. *Chromitite layers associated with norite-anorthosite cumulates*: These are restricted to the upper Critical Zone (e.g., MG 3–4, UG 1–3, chromitite stringers associated with the Pseudoreef, Merensky, and Bastard units) and contain anomalous to high-grade PGE mineralization. This mineralization type is also stratabound, contains minor to background values of BMSs, and reports [Pt*1000]/Cu ratios of up to 100.

3. *Sulfide-rich ultramafic cumulates*: This mineralization associated with chromitite, harzburgite, and pyroxenite cumulates (e.g., Merensky and Bastard). It exhibits considerable lateral variation (in grade, mineralogy, host-rock lithology), is stratabound, but exhibits minor discordancies, and reports [Pt*1000]/Cu ratios of close to unity.

4. *Pegmatoidal layers*: The enigmatic coarse-grained harzburgites or pyroxenites (of similar composition to their cumulate hosts) that occur as poorly defined layers may, if associated with specific localities where primary mineralization is present, be mineralized (e.g., Merensky and UG-2 pegmatoids).

5. *Replaced "reefs"*: Discordant bodies of iron-rich ultramafic pegmatite (which are of entirely different composition to their cumulate hosts) that replace cumulate packages where primary mineralization occurs may contain PGEs (e.g., the replaced Merensky Reef).

6. *Discordant platiniferous pipes*: Four composite pipe-like bodies of magnesian- and iron-rich dunite, which contain PGE-rich cores, occur in the Critical Zone of the Bushveld Complex (e.g., Onverwacht).

Origin: The postcumulus mineralization-types are all attributed to upward-migrating fluids that have leached cumulates enriched in PGEs; however, the bulk of the PGEs within types 4 and 5 were redeposited within a few centimeters of their primary location. Proponents of the hydrothermal model for the Merensky Reef often cite the pipe-like bodies as evidence for the concentration of PGEs within late-stage fluids—we would argue the reverse, i.e., PGEs were leached from fluids migrating through cumulates enriched in primary mineralization. Accordingly, we emphasize that cumulus processes were largely responsible for concentrating the PGEs within the Critical Zone of the Bushveld Complex. The recognition of two types of "chromitite control" is fundamental: type 1 is duplicated by chromite-rich layers, pods, etc., of (probably) all layered, ophiolite, and Alaskan-type complexes, whereas type 2 is unique to the Bushveld (note that the upper Critical Zone is a package of rocks that is not duplicated in other layered complexes; e.g., in Bushveld-terms, the Stillwater chromitites lies within the lower Critical Zone, and the J-M Reef within the Main Zone). The mechanisms that concentrated the PGEs within the chromitite layers is not known, although we favor nucleation of discrete platinum-group minerals (PGMs) in the absence of an immiscible sulfide liquid. Evidence of the early-crystallizing nature of the PGMs (in these circumstances) is found in the occurrence of laurite enclosed within chromite grains.

Conclusions: We attribute the mineralization within the UG-2 Reef, which is relatively both persistent and consistent, to a single event (type 2); it reflects primary magmatic controls. A study of the UG-2 footwall pegmatoid, which has locally inherited PGEs from the chromitite (mineralization type 4), has implications for the origin of the Merensky pegmatoid. Furthermore, study of the mineralized chromitite layers of the Critical Zone, which we interpret as evidence of a fundamental magmatic control on the PGEs, may also provide the key to understanding the enigmatic Merensky Reef. Our preliminary thoughts on the Merensky Reef are that it has formed from a combination of processes involving chromitite control, sulfide collection, and postcumulus overprinting (mineralization types 2, 3, and 4)—clearly it represents the first stage of major sulfur saturation within the Bushveld chamber, an event that can be related to the influx of a new magma-type. Finally, the geochemical behavior of the PGEs within the upper Critical Zone of the Bushveld Complex may be unique; scenarios in which nonchromitite concentration mechanisms dominate will almost certainly be applicable to other PGE deposits.

REFERENCES

Barnes, S., and I. H. Campbell, 1988.

Campbell, I. H., A. J. Naldrett, and S. J. Barnes, 1983, A model for the origin of platinum-rich sulfide horizons in the Bushveld and Stillwater Complexes: Journal of Petrology, v. 24, p. 133–165.

Eales, H. V., M. Field, W. J. deKlerk, and R. N. Scoon, 1988, Regional trends of chemical variation and thermal erosion in the upper critical zone, western Bushveld Complex: Mineralogical Magazine, v. 52, p. 63–79.

——— J. S. Marsh, A. A. Mitchell, W. J. de Klerk, F. J. Kruger, and M. Field, 1986, Some geochemical constraints upon models for the crystallization of the upper critical-main zone interval, northwestern Bushveld Complex: Mineralogical Magazine, v. 50, p. 567–582.

Macdonald, A. J., 1987, Ore deposit models 12. The platinum-group element deposits: classification and genesis: Geoscience Canada, v. 14, p. 155–166.

Scoon, R. N., and W. J. deKlerk, 1987, The relationship of olivine cumulates and mineralization to cyclic units in part of the upper Critical Zone of the Western Bushveld Complex: Canadian Mineralogist, v. 25, p. 51–77.

SCOTESE, C. R., Chairman, Working Group 2B, International Lithosphere Program, Shell Oil, Houston, TX, and M. I. ROSS, Rice University, Houston, TX

Phanerozoic Plate Tectonic Reconstructions: Preliminary Results of PALEOMAP Project

Fifty plate tectonic reconstructions are presented illustrating the movements of the continents and the formation of the ocean basins from the latest Precambrian (600 Ma) to the present. In addition, 20 paleogeographic reconstructions have been produced illustrating the relative distribution of land, sea, and mountains during the Phanerozoic. Speculative maps illustrating the probable positions of the continents 50, 100, and 150 m.y. in the future are also presented. These maps represent a synthesis of regional plate tectonic models produced by an international panel of geologists and marine geophysists. The goal of the PALEOMAP project is to produce an atlas and an electronic folio that illustrates the evolution of the continents and ocean basins during the last 600 m.y.

The Paleozoic plate tectonic reconstructions are based on a combination of paleomagnetic, biogeographic, and paleoclimatic information. The Paleozoic maps illustrate the progressive closure of Early Paleozoic oceans (i.e., Iapetus) and the formation of the western half of Pangaea by Late Paleozoic times. During the Early Paleozoic, North America moved south from an equatorial position (Cambrian) to a subtropical position (Early Devonian). During the same time interval, Gondwana moved northward across the South Pole. In the Middle Ordovician, the south pole was located in Northwest Africa. During the Silurian the pole moved southwest across the Amazon Basin and into southeastern Argentina by the Middle Devonian. Starting in the Early Devonian, North America began to collide with the northern fringe of Gondwana and moved with Gondwana northward across the Equator (Carboniferous–Permian). By the Middle Paleozoic, the North and South China platforms had rifted away from eastern Gondwana. The presence of tillites in Burma suggests that much of Southeast Asia, however, may not have separated from eastern Gondwana until after the Carboniferous.

The Mesozoic and Cenozoic reconstructions are based on a global database compiled by Working Group 2B that contains digital representations of linear magnetic anomalies, tectonic features of the ocean floor, and fracture zone trends derived from satellite altimetry. The Mesozoic and Cenozoic maps illustrate the assembly of Asia during the Early Mesozoic, the progressive breakup of Pangaea during the Jurassic and Cretaceous, and the formation of the modern ocean basins during the Late Cretaceous and Cenozoic. The hot spots and mantle tomography have been used to produce a reference frame for Early Cretaceous and younger plate reassemblies. Prior to the Early Cretaceous (120 Ma), paleomagnetic data was used to orient the reconstructions with respect to the earth's rotation axis.

Maps illustrating the possible configuration of the plates in the future have been produced by extrapolating present plate motions forward in time. According to this method, during the next 50 m.y. the Atlantic and Circum-Antarctic oceans will continue to widen, while the Pacific Ocean contracts. Baja California, as expected, will move northward along the western coast of North America eventually accreting to southern Alaska. Simple forward extrapolation, however, cannot be used beyond 50 m.y. because future continental collisions, e.g., Africa and Europe (closing the Mediterranean) and Australia and Southeast Asia (collapsing Indonesia and the Philippines), will dramatically alter plate motions. Maps for 100 and 150 m.y. have been produced using the following assumptions: (1) subduction will begin along western shores of the Atlantic as a result of the propagation of the Puerto Rican and Scotian trenches, (2) the Tethyan subduction zone will jump south to a position in the Central Indian Ocean, and (3) the East African Rift may develop into a small ocean basin.

SCOTESE, C. R., Chairman, Working Group 2B, International Lithosphere Program, Shell Development Company, Houston, TX

Mesozoic Plate Tectonic Reconstructions: Global Constraints on Mesozoic Seaways

The two most important factors controlling the general pattern of oceanic circulation and the formation of interoceanic seaways during the Mesozoic have been the plate tectonic development of the ocean basins and the history of continental collisions. Ten plate tectonic reconstructions are presented illustrating the assembly of the Pangaea during the Early Mesozoic and its subsequent breakup and dispersal during the Late Mesozoic. The data used to produce these maps were compiled as part of the PALEOMAP Project, Working Group 2B, International Lithosphere Program. Paleogeographic reconstructions for the Induan, Norian, and Pliensbachian illustrate the assembly of Asia as a result of the progressive collision and accretion of Southeast Asia, South China, and North China along the southern margin of Sibera. The suturing of these continents began in the Late Triassic (Red River suture between Southeast Asia–South China; Qinling suture between South China and North China) and was not complete until the earliest Cretaceous (closure of Stanovoy suture between North China and Siberia). Jurassic and Early Cretaceous paleogeographic reconstructions (Callovian, Volgian, Valanginian, and Aptian) illustrate the breakup of the western half of Pangaea and the formation of the Atlantic and Indian Oceans. During the Middle Jurassic, the central North Atlantic opened as Africa separated from North America. At about the same time, in the southern hemisphere, Gondwana began to break apart. In the Late Jurassic, a rift separated Western Gondwana (Africa/South America) from Eastern Gondwana (India/Antarctica Australia) opening the Mozambique and Somali Basins. During the Early Cretaceous, rifts propagated into the South Atlantic separating South America from Africa and into the Indian Ocean resulting in the separation of India from Australia/Antarctica. Late Cretaceous paleogeographic reconstructions (Cenomanian, Coniacian, and Maestrichtian) illustrate the final phases of the breakup of Pangaea. During the mid-Cretaceous, India rifted away from Madagascar and began its rapid journey northward. During Late Cretaceous, the Tasman Sea opened as New Zealand rifted away from Australia and western Antarctica. It was also during the Late Cretaceous that Australia slowly began to separate from eastern Antarctica, and the northern continent of Laurasia began to break up as the result of seafloor spreading in the Labrador Sea and in the vicinity of the Rockall Plateau.

The motion of the major plates also played an indirect role in the formation of interoceanic seaways by controlling the development of the complex, multiplate "buffer zones" that exist between the major plates. These buffer zones, such as the Gulf of Mexico, Caribbean Sea, Mediterranean Sea, Scotia Sea, and seaways of Southeast Asia represent the "closure error" that results from subtle differences in the relative motions of the major plates. These buffer zones tend to form at convergent plate boundaries and have plate geometries that are rarely stable for long periods of time. Due to their instability, these buffer zones often act as "triggers" or "gateways" that can modify the general pattern of oceanic circulation over relatively short time intervals.

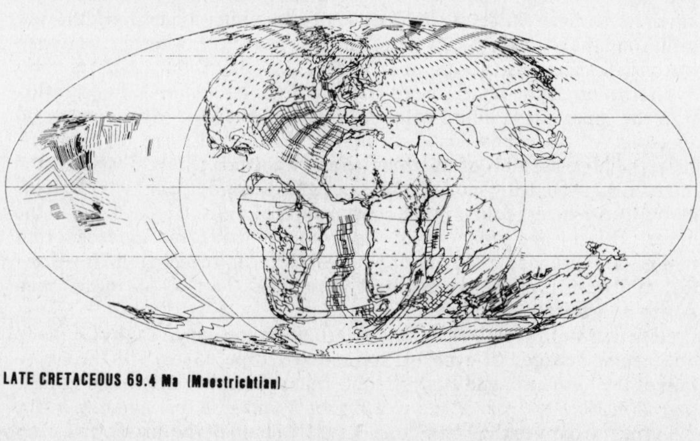

LATE CRETACEOUS 69.4 Ma (Maestrichtian)

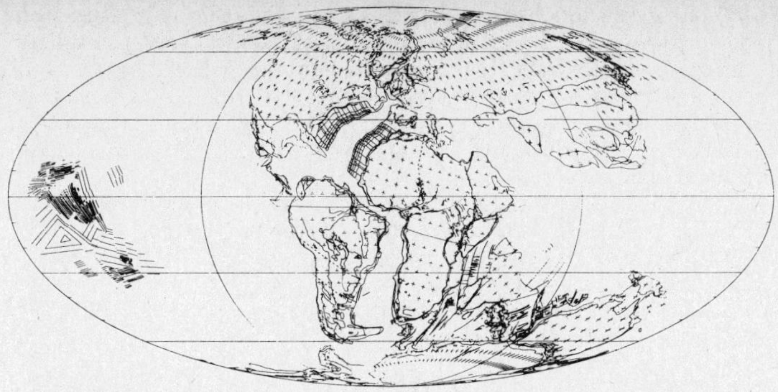

LATE CRETACEOUS 94.0 Ma (Cenomanian)

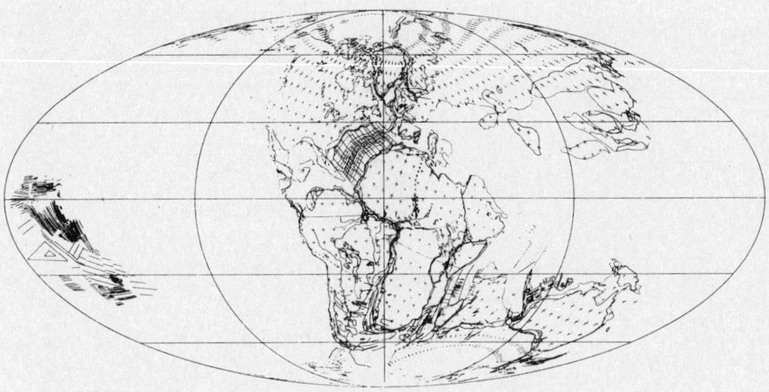

EARLY CRETACEOUS 118.7 Ma (Aptian)

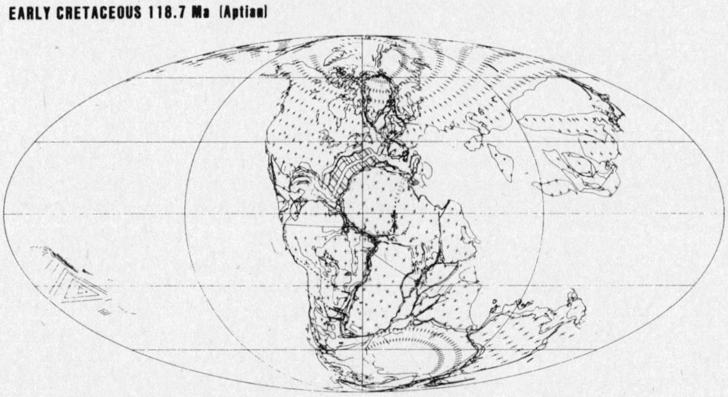

LATE JURASSIC 152.2 Ma (Volgian)

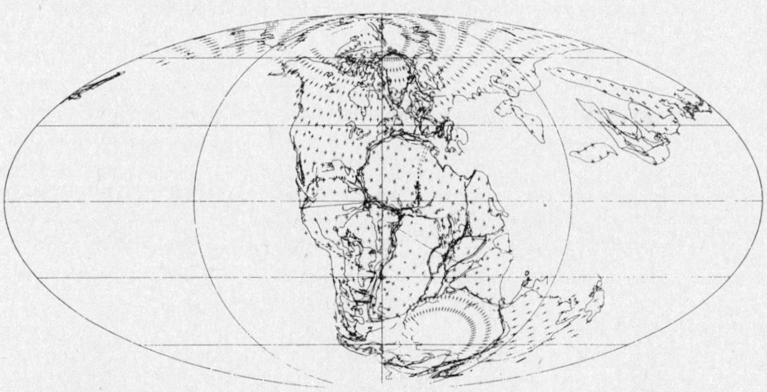

EARLY JURASSIC 195.0 Ma (Pliensbachian)

SCOTT SMITH, B. H., North Vancouver, Canada

Lamproites in India

The Majhgawan diamond mine and nearby Hinota pip, Madhya Pradesh are classified here as olivine lamproites and not kimberlites. These bodies, therefore, comprise a newly recognized province of diamondiferous olivine lamproites which are extremely rare worldwide. Other intrusions in India have been proposed as lamproites (e.g., Paul and Sarkar, 1984; Bergman and Baker, 1984; Reddy, 1987). Information for relevant localities is reviewed (Table 1).

Rocks from *Majhgawan* and *Hinota* are lapilli tuffs. The juvenile lapilli are composed of olivine macrocrysts and phenocrysts set in glassy groundmasses containing mica, vesicles, apatite and altered perovskite. The petrographic classification of these rocks is problematic because of their glassy nature and paucity in primary minerals. Various features, however, are atypical of kimberlites and more typical of lamproites. These include the occurrence of the glass, in particular of scoriaceous juvenile lapilli, the complex shapes of the olivine macrocrysts (probably imposed morphology) and phenocrysts (crystal aggregates) and polysynthetic twinning in the phlogopite. Also phlogopite compositions (high TiO_2 contents—5.3–7.3 wt%) and whole rock compositions (e.g., high Ba—3634–27,300 ppm) are more similar to lamproites than kimberlites. These rocks are very similar to those found at Ellendale and Argyle (Western Australia), Prairie Creek (Arkansas) and Kapamba (Zambia). The apparent paucity of mantle-derived xenoliths and xenocrysts at Majhgawan is also similar to that noted in the other lamproites. These data suggest therefore that Majhgawan and Hinota are not kimberlites but should rather be classified as lamproites. The samples examined are hence classified mostly as glassy olivine lamproite lapilli tuffs although some autolithic breccias may be present.

The *Wajrakarur–Lattavaram* bodies (Andra Pradesh) are composed mostly of mica-bearing monticellite kimberlite. Contamination by late stage xenolith digestion produces pectolite-, clinopyroxene- bearing kimberlite. Texturally, the bodies are predominantly hypabyssal but pelletal tuffisitic kimberlite breccia (diatreme- facies) is present in the largest body. They probably represent the root zones of substantially eroded diatremes. Pipes 1, 3, 4, and 6 are typical kimberlites and contain some diamond and relatively common mantle-derived xenoliths and xenocrysts. Pies 2 and 5 display some petrographic features (e.g., paucity of olivine macrocrysts, probable melilite pseudomorphs) which are not typical of, but do not preclude their classification as, kimberlites. Pipes 2 and 5 also appear to be devoid of diamond, mantle-derived xenoliths and xenocrysts. These differences are not sufficient to suggest that these two pipes represent a different rock type such as lamproite. They are, rather, more extreme varieties of kimberlite.

Lamprophyre dykes both at *Chelima* and in the *Gondwana Coalfields* could include lamproites. Information for *Angor, Banda,* and *Jungel* suggests that they are not kimberlites or lamproites. Insufficient information is available for the intrusions at *Zangamrajupalle, Maddur,* and *Warangal* to comment further but it is doubtful whether they are kimberlites or lamproites.

It is noteworthy that the only confirmed lamproites (Majhgawan) and kimberlites (Wajrakarur–Lattavaram) constitute the only known primary sources of diamond in India. They are Proterozoic and similar in age to other lamproites, kimberlites and related rocks worldwide. The tectonic setting of these provinces may also be similar to those elsewhere. The Wajrakarur–Lattavaram kimberlites occur well within Dharwar craton, while it appears that the lamproites (Majhgawan–Hinota) may occur toward the margin of the Aravalli craton and might even be associated with a rift. Although there are numerous published papers on the "kimberlitic" bodies of India, this review shows the great need for new, detailed petrological studies of many of these localities.

Acknowledgments: The author is indebted to R. B. Hargraves, F. R. Boyd, P. Kresten, and S. C. Bergman for donating samples without which this study would not have been possible.

REFERENCES

Bergman, S. C., and N. R. Baker, 1984, A new look at the Proterozoic dykes from Chelima, Andhra Pradesh, India: diamondiferous lamproite?: GSA Abstracts, v. 16, p. 444.

Paul, D. K., and A. Sarkar, 1984, Petrogenesis of some Indian lamprophyres, *in* Proceedings of SYMPET, Jaipur 1981: Geol. Surv. India Special Publ. v. 12, p. 45–54.

Reddy, T. A. K., 1987, Kimberlite and lamproite rocks of Vajrakarur area, Andhra Pradesh: Geol. Soc. India Journal v. 30, p. 1–12.

SCOTT, A. C., University of London, Egham, United Kingdom

Observations on Nature and Origin of Fusain

Evidence for a wildfire origin of fusain is examined. Charcoal produced by natural fires and artificial charring resembles fossil fusain both in physical size and shape and in optical reflectance. Artificially charred leaves and naturally burnt wood are illustrated using the SEM. Fusain is shown to be composed mainly of fusinite and semifusinite. Reflectance studies on modern charcoals suggest that much semifusinite may be produced by pyrolysis. The ecological implications of a wildfire origin of fusain are discussed and it is concluded that fires have been a feature of terrestrial ecosystems from at least the Late Devonian.

SCOTT, R. W., Amoco Production Co., Tulsa, OK

Early-Mid-Cretaceous Sea-Level Changes and Evolution of Tethyan Reef Communities

The replacement of corals by rudists in Early Cretaceous reefal communities spanned a 30 m.y. period when sea level rose and drowned continental shelves. During this time, corals formed communities in the deeper parts of reefs and rudists occupied the shallow, high-energy habitats. By Aptial time, rudists dominated reefs that fringed interior shelf basins and corals formed reefs with rudists at the open shelf margins. By late Albian, coral communities were greatly limited in areal extent, presumably because of complex environmental changes and cycles of organic productivity. Most Late Cretaceous reefs were dominated by rudist communities having a less complex trophic structure.

Two important events of eustatic sea-level rise are represented by unconformities separating carbonate depositional sequences on the Arabian platform that correlate with sequence boundaries on the Gulf Coast platform. Graphic correlation techniques test the synchroneity of these events. A composite standard time scale dates these sea-level rises at 115.8 Ma and 94.6 Ma; a third, intra-Albian event at 104.3 Ma is present in many places and may also be eustatic. Associated with these sea-level rises were apparent changes in ocean water chemistry as evidenced by changes in isotopes and trace elements, where diagenetic effects can be discounted. During this time, the climate became more humid and atmospheric CO_2 increased. The concomitant environmental changes in the oceanic conditions presumably stressed the deeper coral communities on reefs. The emergence of rudists as reef contributors had a profound effect on Late Cretaceous depositional conditions and the development of hydrocarbon reservoirs.

SCRIBANO, V., Istituto di Scienze della Terra dell'Università, Catania, Italy

Insights Into the Nature of Lower Lithosphere Beneath South-Eastern Sicily (Italy) from the Study of Xenoliths in Volcanic Rocks

The Hyblean Plateau (south-eastern Sicily, Italy), is part of the Pelagian block, that is the northern foreland of African Plate. Hyblean Plateau consists of a thick Meso-Cenozoic carbonatic sequence with interbedded volcanic horizons. NNE-SSW and NE-SW trending fault systems, which have controlled the volcanic activity throughout the time, constitute the dominant tectonic feature of the region (e.g. Lentini, et al, 1986).

Apart from some Miocenic tuff-breccia pipes and associated pyroclastic-flow deposits, most of the Hyblean volcanic rocks consist of fissural-type lava flows, basalts (with alkaline and tholeiitic affinity) constitute the dominant lithotypes. Those volcanic rocks belonging to the alkaline suite, carry a number of deep-seated xenoliths, whose study—though still in a preliminary stage has provided some insights into the nature of the Hyblean lower lithosphere (Scribano, 1988; 1987a, b). It must be emphasized that nodules occurring into the above-mentioned tuff-breccia pipes, represent pristine lower-lithosphere material, since the "diatremic-type" emplacement is a very rapid volcanic process: nodule-bearing magma would, in fact, traverse the thickness of the continental crust in less than 15 minutes, preserving xenoliths from heavy reactions with host lava.

Hyblean ULTRAMAFIC NODULES consist of peridotites and pyroxenites. the former are spinel-bearing harzburgite, with porphyroclastic to protogranular texture, and spinel-lherzolites, with typical protogranular texture. Glassy patches, due to minor partial melting episodes, are ubiquitous in all our peridotites. Pyroxenites consist of several types of websterites, which may be split into two main groups: Cr-diopside-bearing and Al-augite-bearing, coarse unstrained textures dominate for all the above-mentioned pyroxenites, except for some Al-augite websterites with well-developed foliation. Hyblean ultramafic nodules equilibrated within the spinel-pyrolite field, as the lack of garnet and plagioclase suggests. Thus they represent fragments of the upper mantle beneath south-eastern Sicily: in particular, websterites derive from the (recrystallization of magmatic liquids intruded as dikes into the peridotite levels. Fractional partial melting of the peridotites and fractional crystallization of primary mafic liquids may explain the origin of the different websterite types.

PLAGIOCLASE-BEARING NODULES are also ubiquitous in the Hyblean volcanic rocks. These nodules mostly consist of spinel-bearing two-pyroxene granulites, which may be split into two sub-types. The first (G1) displays medium to coarse polygonal texture, with An (45-47) plagioclase, fassaite, Al-rich enstatitic-bronzite and Al-spine. The second granulite typic (G2) has finer grained texture than the first, mineral chemistry also showing a significant differences: the plagioclase is more rich in anorthite, the pyroxenes and spinel display lower Al and higher Fe content than those of the G1-type granulites. Both G1 and G2-type granulites equilibrated under static conditions within the spinel-gabbro field, that is in the lower crust.

Modal proportions between plagioclase and pyroxenes widely vary from nodule to nodule from both G1 and G2 granulitic types: plagioclase-bearing websteritic granulite represents the malonocratic end-term, a modally "leuconorite" the leucocratic one, the dominant types being mesocratic. Above-mentioned modal variations reflect upon the whole-nodule chemistry, which on the average is "basaltic," with mildly subalkaline (G2) to transitional (G1) affinity.

Finally, it is worth of note the rare occurrence, in the Hyblean volcanic rocks, of massif-type anorthosite and granitoid nodules. It is therefore hypothesized that the Hyblean deep-crust consists of stratiform bodies of mafic granulites (probably derived from the static recrystallization of layered "gabbroic" intrusions) with lenses of anorthosites and granitoids, representing relics of an older sialic crust.

REFERENCES

Lentini, F., Grasso, M., Carbone S., 1987, Introduzione alla geologia della Sicilia, Univ. de Catania, Ist. de Scienze della Terra, Catania, 59 p.

Scribano, V., 1986, The harzburgite xenoliths in a Quaternary basinitoid lava near Scordia (Hyblean Plateau), Rend. SIMP, 41, 245-256.

——— 1987a, The ultramafic and mafic nodule suite in a tuff-breccia pipe from Cozzo Molino (Hyblean Plateau, SE Sicily), Rend. SIMP 42, 203-217.

——— 1987b, Origin of websterite nodules from some alkaline volcanic rocks of Hyblean Plateau (southeastern Sicily). Periodico de Mineral. (in press).

SEABER, P. R., Illinois State Geological Survey, Champaign, IL, and J. V. BRAHANA, U.S. Geological Survey, Nashville, TN

Application of Regional Aquifer Coefficients to Evaluate Hydrogeologic Controls on Permeability and Porosity

Hydrogeologic processes and controls that affect the permeability and porosity of rocks are numerous, with interrelations that are complex and poorly quantified (Brahana et al, in press); Seaber et al, in press). This paper describes the rationale and preliminary application of a methodology that generates relevant regional aquifer coefficients from the few quantitative data that are readily available.

The methodology is based on the technique of Seaber and Hollyday (1965, 1966a, 1966b) wherein specific capacity data are compiled and statistically evaluated. The application of the original method of Seaber and Hollyday has been expanded from one hydrogeologic control (lithology), to three controls (lithology, structure, and topography). Specific capacity data were chosen because they commonly are easy and inexpensive to obtain and they are generally reliable. Because these data represent points only, it is necessary to evaluate many to define the range of regional response. By comparing point and regional data to known, quantifiable hydrogeologic parameters statistically, it is possible to extend understanding about measurable hydraulic effects that each hydrogeologic parameter may control.

Application of the method requires the determination of two aquifer coefficients, aquifer capability and aquifer variability. Calculation of the aquifer coefficients involves the following steps. The specific capacity values in liters per second per meter of drawdown (L/s/m) for each formation are arranged in order of magnitude, plotted on log-normal probability paper, a straight line is fitted to the data (reported data trend), and the trend is adjusted to a common pumping duration of 180 days (Figure 1a). The median and standard deviation of the specific capacities are determined from the log-normal distribution. The median value defines the coefficient of aquifer capability and the standard deviation defines the coefficient of aquifer variability.

In the example illustrated in Figure 1, the coefficients are determined only for lithology; in extended application, the coefficients may be related to any hydrogeologic variable that may be quantified. Generalized results are based on an analysis of specific capacity data from 56 formations that were classified into five major lithologic types for the Susquehanna River basin (Seaber and Hollyday, 1965, 1966a, 1966b). Aquifer coefficients are graphically represented on Figure 1b. A large median value indicates high transmissivity, and a large standard deviation indicates great range in local transmissivities. Aquifer capability, expressed as L/s/m, is directly related to the regional permeability and porosity. It is a measure of the volume of groundwater that a rock unit will transmit in a specified time interval. Aquifer variability, expressed in log cycles, is a measure of the heterogeneity and anisotropy of the formation. It is felt that aquifer variability is a valuable indicator of the scale at which regional trends in permeability should be assessed.

Major lithologic types form five distinct groupings in Figure 1b. For the most part, the range of coefficients for relatively pure lithologic types does not overlap within the Susquehanna River Basin. Assessment of other river basins throughout North America shows that this general trend holds true,

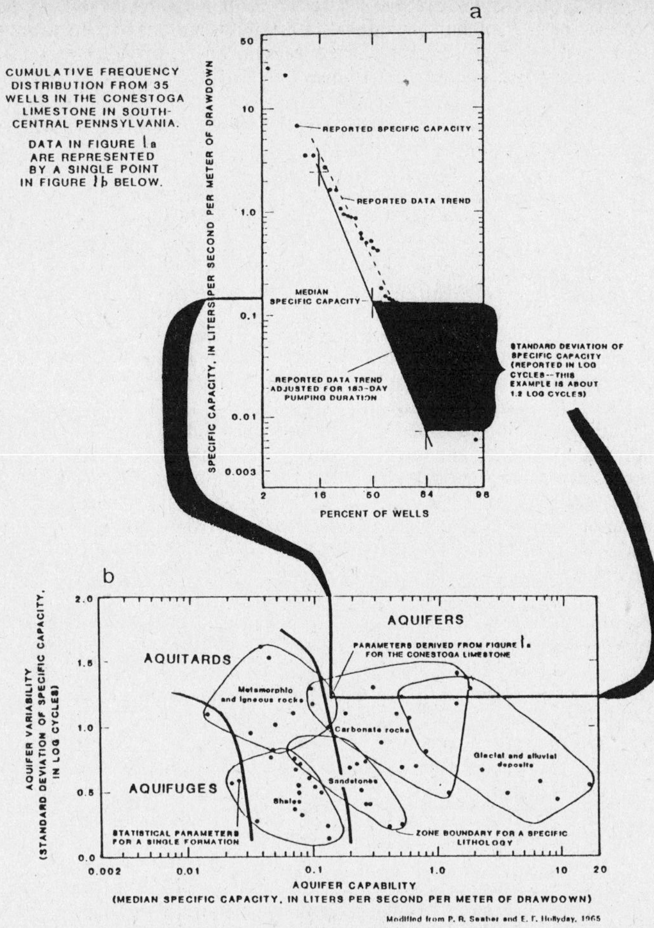

Figure 1—Relation between lithology and statistical measures of specific capacity for 56 formations in the Susquehanna River basin.

with the exception of extrusive igneous rocks and carbonates. Extrusive igneous rocks and carbonate rocks encompass such a broad diversity of permeability types (and evolution) that lithologic boundaries are not as uniquely defined for these as they are for other rocks. Some carbonate rock coefficient overlap can reflect the inclusion of metamorphic and sandy carbonates, and some can reflect the fact that carbonate rocks encompass a wider range of lithologic types (chalk, marble, lithographic limestone, coquina, dolomite, etc.) than do other categories, which are essentially single lithologies (sandstone, shale). These considerations notwithstanding, lithologic control of aquifer coefficients is surprisingly consistent throughout North America. In general, carbonate and metamorphic rocks have standard deviations greater than 0.75 log cycles, and sandstones and shales have standard deviations less than 0.75 log cycles. Carbonates and sandstones generally have medians greater than 0.20 L/s/m, and shales and metamorphic rocks have medians less than 0.20 L/s/m. Glacial deposits have medians greater than 0.70 L/s/m, but a wide range in standard deviations. Interbedded lithologies have intermediate coefficients depending on the heterogeneity of the mapped unit.

Observations of aquifer coefficients with respect to structure and topography indicate that other major hydrogeologic parameters may have significant control within specific basins, but that on regional evaluation, the effect of lithology dominates. Variations introduced by structure and topography (as well as other processes and controls not considered by this paper) indicate that the interrelation between all hydrogeologic factors and permeability and porosity is complex. The method described herein can quantify and simplify the interrelations and improve understanding about the hydrogeologic controls on permeability and porosity.

REFERENCES

Brahana, J. V., J. Thrailkill, T. Freeman, and W. C. Ward, in press, Carbonate Rocks, *in* W. Back, J. S. Rosenshein, and P. R. Seaber, eds., Hydrogeology: Boulder, Colorado, Geological Society of America, The Geology of North America, v. 0-2.

Seaber, P. R., and E. F. Hollyday, 1965, An appraisal of the ground-water resources of the lower Susquehanna River basin: U.S. Geological Survey Open-File Report, 75 p.

——— ——— 1966a, Statistical analysis of the regional aquifers (abs): GSA Abstracts with Programs, p. 196.

——— ——— 1966b, An appraisal of the ground-water resources of the Juniata River basins: U.S. Geological Survey Open-File Report, 125 p.

——— J. V. Brahana, and E. F. Hollyday, in press, Region 20, Appalachian Plateaus and Valley and Ridge, *in* W. Back, J. S. Rosenshein, and P. R. Seaber, eds., Hydrogeology: Boulder, Colorado, Geological Society of America, The Geology of North America, v. 0-2.

SEARLE, M. P., University of Leicester, Leicester, United Kingdom

Mesozoic Paleogeography of Tethys between Oman and Burma Seaways and Ocean Connections

The Mesozoic paleogeography of Tethys in the Middle East–Central Asia region can be approximately reconstructed by unraveling the stratigraphy and restoring the structure of the major suture zones and thrust sheets that separate the Arabian and Indian plates to the south from the collage of plates that make up Eurasia to the north.

Oman: A series of oceanic Tethyan thrust sheets have been emplaced southwestwards onto the passive margin of the Arabian plate in the Oman Mountains. The Sumeini, Hawasina, and Haybi complex rocks structurally overlie Mesozoic shelf sediments and late Cretaceous fore-deep deposits and have been overthrust by the massive Semail ophiolite complex—a 14 km thick oceanic crust and mantle sequence radiometrically and biostratigraphically dated as Cenomanian (97–93 Ma). The allochthonous thrust sheets restore to a minimum width of at least 250 km between the autochthonous Cretaceous shelf edge and the northeast-dipping subduction zone along the base of the Semail ophiolite, with its inverted, high-temperature, medium-pressure (max. P-T 850°C, 4–5 Kbars) metamorphic sole. The lower thrust sheets comprise Triassic to mid-Cretaceous outer shelf-slope deposits (Sumeini Group) overlain by deeper basin sediments, including turbidites and radiolarian cherts (Hawasina complex). The higher thrust sheets include isolated Permian and Triassic exotic limestones, Triassic alkaline and tholeiitic within-plate volcanics, and sedimentary and tectonic melanges (Haybi complex). The emplacement of these thrust sheets spanned ~25 Ma (93–68 Ma) and terminated in the Maestrichtian. Unlike the Zagros and Himalayan sutures, the closure of Tethys was not completed in the Oman region, and heat flow data shows that the Gulf of Oman is presently underlain by Cretaceous ocean crust, a relic of Neotethys.

Pakistan: Mesozoic oceanic sediments and ophiolites separate the Indian plate from the Lut, Afghan, and Helmand blocks along both the Sistan suture and the sinistral Chaman strike-slip fault system. Restoration of the precollision paleogeography in the Hindu Kush–Karakoram–Kohistan region is extremely complex, but two major Mesozoic seaways are recognized along the Main Mantle Thrust (MMT) and the Shyok Suture Zone (SSZ) bounding the south and north margins of the Kohistan island arc-microplate in northern Pakistan. The SSZ closed in the mid-Cretaceous; the MMT has been reactivated as a breakback thrust, is folded around the huge Nanga Parbat syntaxis, and is analogous to the Indus suture zone (ISZ).

Western Himalaya: The Indus suture zone of Ladakh contains Mesozoic oceanic sediments (Lamayuru complex), Permian and Triassic exotic limestone blocks, Triassic alkali volcanics, serpentinite melanges, and blueschists. These Tethyan oceanic rocks occur in a 20–30 km wide belt separating the Aptian–Eocene Ladakh batholith to the north from the northern continental passive margin sediments of the Indian plate along the Zanskar Range to the south. The ISZ can be traced for 2500 km north of the Himalaya from Baltistan and Ladakh to south Tibet and Burma and closed during the mid-Eocene at ca. 50 Ma (anomaly 22 in the Indian ocean). The Kohistan–Dras island arc within Tethys is only present in the western Himalaya. East of Ladakh only one major Tethyan oceanic seaway existed between the Indian plate and Tibet during the late Jurassic–Cretaceous.

Tibet: The ISZ continues eastwards from Ladakh across southern Tibet. Oceanic rocks in the ISZ are time equivalent but of deeper water facies to the passive margin Mesozoic sediments on the Indian plate. Paleomagnetic data from Tibet indicate that at least 2000 km of Tethyan seaway separated India from Tibet at the end of the Cretaceous. Approximately 550 km of shortening can be demonstrated from restoration of balanced cross-sections across the western Himalaya. The Banggong suture separating the Lhasa block to the south and the Qiantang block to the north probably represents a Mesozoic Tethyan seaway separating north and south Tibet. Southward-directed ophiolite obduction in the Jurassic was followed by early Cretaceous closure and uplift. Southern Tibet and the Karakoram were uplifted during the late Cretaceous and became an Andean-type margin bounding the northern margin of Tethys.

Burma: The major suture zones of the Himalaya and Tibet follow the northeastern margin of the Indian plate around the Namche Barwa (eastern syntaxis of the Himalaya southwards into Burma). Late Triassic collision of the western southeast Asia block with the Indo-China block was followed by eastward subduction and subsequent accretion of the Burma volcanic arc during Jurassic–early Cretaceous time. Fore-arc accretionary sediments and ophiolites are exposed in the Indo-Burman Ranges (Arakan–Yoma belt).

SEBRIER, M., J. L., MERCIER, J. CABRERA, J. MACHARE, J. F. DUMONT, and D. HUAMAN, Université de Paris Sud, Orsay, France

State of Stress Above Subduction Zone: Example of Central Andes (Central and Southern Peru)

The state of stress of a continental lithosphere situated above a subduction zone is analyzed in the northern part of the Central Andes (Peru). The Central Andes are characterized by the highest Cordilleran topography, by a conspicuous oroclinal bend (Bolivian orocline), and by a segmented Benioff geometry. The north-northwest-south-southwest- to northwest-southeast-trending Peruvian Andes allow us to compare the stress pattern in an overriding plate situated above either a 30° dipping slab (southern Peru) or a flat slab (central Peru). Stress pattern has been determined using both kinematics of Recent or active faults observed in the field and available focal mechanisms of shallow earthquakes.

South Peruvian Andes are characterized by an active volcanic arc, a 300 km wide mean topography that is over 4 km, and a narrow (100 km wide) foreland. Available gravimetric data suggest that this high plateau (High Andes) is essentially compensated by a nearly 70 km thick crust under the Western Cordillera and the Altiplano while the Eastern Cordillera is slightly undercompensated. Conversely, both the fore-arc (Pacific Lowlands) and the foreland (sub-Andean Lowlands) appear overcompensated. In the High Andes and Pacific Lowlands, Quaternary and active faults are normal. The amount of stretching is small, being of the order of 1% during the last 1-2 Ma. In the sub-Andes, folds and reverse faults affecting Neogene and early Quaternary deposits indicate that back-arc thrusting prevails in the eastern Andean foothills. Data from structural analysis and from earthquake focal mechanisms indicate that the High Andes and Pacific Lowlands are characterized by north–south-trending extension and bounded by two zones of east–west-trending compression: the sub-Andes to the east and the contact between the Nazca and South American plates to the west. This state of stress may be interpreted at a large wavelength as an effect of high topography. In this model the vertical stress σzz increases with the topography and $\sigma Hmax$ is considered as fairly constant, trending east-west roughly parallel to the convergence. On both edges of the Andes, tectonics being compressional, σzz is $\sigma 3$ and $\sigma Hmax$ is $\sigma 1$; in the High Andes σzz becomes $\sigma 1$, then $\sigma Hmax$ is $\sigma 2$, allowing $\sigma Hmin$ trending north-south to be $\sigma 3$. The predominately extensional tectonics observed along the Pacific Coast may be due to a topographical effect of the nearby deep Peru–Chile trench.

Central Peruvian Andes show an extinct volcanic arc, a narrower (150 km wide) and slightly lower (4 km high) topography, and wider (nearly 250 km wide) sub-Andean Lowlands. Notwithstanding such different characteristics, the state of stress in central Peru displays many similarities with that analyzed in southern Peru (i.e., north–south-trending extension in the Western Cordillera and along the Pacific Coast as well as east–west-trending compression on both sides of the Andes) and thus may be also interpreted as an effect of high topography. However, a conspicuous difference occurs: compressional strike-slip tectonics is observed in the Eastern Cordillera of central Peru; this is characterized by east–west-trending compression and north–south-trending extension. This intermediate case, where the vertical stress σzz is $\sigma 2$, seems to be explained by a stronger coupling between the two plates, a consequence of the flat subduction, and by a lower mean elevation of the Andes. Thus the $\sigma Hmax$ value should be higher in central Peru than in southern Peru causing a more widespread distribution of compressional tectonics. Finally, the present-day state of stress in the Peruvian Andes does not appear strongly modified either by the subduction of submarine topography (i.e., aseismic Nazca ridge) or by the development of a fore-arc sliver due to the occurrence of oblique convergence in southern Peru.

From Eocene to Present, the Andean tectonic evolution shows an alternation of three types of stress regime. The first is characterized by compression trending roughly parallel to the convergence. The second type is characterized by extension orthogonal to the convergence and is illustrated by the present-day state of stress. The third type is characterized by extension roughly orthogonal to the Andean margin (i.e., to the Peru–Chile trench). These stress regimes result from the interaction between the boundary forces that are related to the convergence and the body forces that are due to lateral variations of the lithospheric mass (i.e., topography). The analysis of the Pliocene and Quaternary stress pattern suggests that boundary forces might undergo stronger variations than body forces indicating thus that the Andean stress pattern is probably more subject to modifications of $\sigma Hmax$ than σzz values. These modifications of $\sigma Hmax$ value should be due either to oceanic slab ruptures under its own weight or to variations of the Andean convergence rate.

SECCOMBE, P. K., and M. N. HICKS, The University of Newcastle, New South Wales, Australia

Isotopic and Geochemical Modeling of Early Metamorphic Turbidite-Hosted Gold Mineralization at Hill End, New South Wales, Australia

Gold-bearing quartz veins in the Hawkins Hill area of the Hill End goldfield, New South Wales, comprise bedding–parallel vein sets that are hosted by a multiply deformed Late Silurian slate–metagraywacke turbidite sequence. The veins are localized close to the crest of the Hill End Anticline and formed in response to slip mechanisms operating during the early stages of Early Carboniferous greenschist facies metamorphism. Tight to ptygmatic folds and boudinage are characteristic features of the veins. Vein emplacement is accompanied by a narrow zone of chlorite and white mica wallrock alteration and more extensive carbonate (calcite, ferroan-magnesium calcite, and ankerite).

P-T estimates of 2.9 Kbars and 420 °C during vein emplacement and metamorphism are obtained from the calcite, ferroan- magnesian calcite geothermometer and the Si content of white mica.

Fluid inclusions from vein quartz are commonly three phase $CO_2(1)$–$CO_2(g)$–$H_2O(1)$ types with variable L-V ratios. High CO_2 contents and probable additional CH_4 are argued from depressed CO_2 melting temperatures (x = -57.1 °C) and slightly elevated CO_2- hydrate melting temperatures (x = 10.1 °C) when compared with the pure CO_2–H_2O system. Ice melting temperatures (x = ~1.6 °C) indicate a maximum salinity of 2.5 wt% NaCl equivalent. Homogenization temperatures for both liquid- and vapor-rich inclusions are similar (x = 343 °C; range = 260° to 361 °C; n = 133). The ore forming fluid is likely to have been trapped as two unmixed low salinity fluids (one CO_2-rich, the other H_2O-rich) close to the solvus in the H_2O–CO_2 (NaCl, CH_4) system. Homogenization temperatures therefore may be close to the true trapping temperatures.

A preliminary sulfur isotope study of the vein sulfides suggests that the underlying turbidite sequence has acted as a source of sulfur ($\delta^{34}S$ for vein sulfides: (x = 6.7 per mil; range - -2.8 to 17.8 per mil; n = 7).

The timing of vein formation and the lack of an obvious igneous intrusive source suggests that gold, as for sulfur, may have been derived from the sedimentary sequence. However, an alternative model may be viable, based on the extraction of gold from a probable underlying shoshonitic basaltic-andesite sequence representing a rifted Ordovician island arc. An isotopic study of trace lead in gold and associated sulfides is underway to discriminate between these two possible sources of gold. Deposition of gold at Hill End has taken place in apparent response to redox changes in the ore fluids resulting from CO_2-consuming wallrock alteration reactions and CO_2 and H_2O loss through transient pressure drops during the crack-seal cycle of vein formation.

SECCOMBE, P. K., The University of Newcastle, New South Wales, Australia, and B. A. BRILL, The University of Melbourne, Parkville, Australia

Fluid Inclusion and S, O, H, and C Isotopic Evidence for Metamorphic Cu, Zn, Pb, and Au Ore Formation at Cobar, New South Wales, Australia

Cu-Zn-Pb, Cu-Au, and Zn-Pb vein and discordant massive deposits in the Cobar region, New South Wales, are hosted by multiply deformed Early Devonian turbidites. Microthermometry and laser Raman spectroscopy of inclusion fluids from vein quartz and calcite associated with all styles of mineralization indicate three principal inclusion types: (1) low salinity (less than 2 wt% NaCl equivalent), aqueous, mixed CO_2 and CH_4 type (low total CO_2 content; CO_2:CH_4 ratios in the gas phase range from 9:1 to 7:3), (2) CH_4-rich type, and (3) low salinity, aqueous type.

Homogenization temperatures (T_h) and inclusion types are similar for individual deposits. Type 1 and 2 inclusions from the CSA Cu-Zn-Pb vein deposit yield an average T_h of 305 °C for quartz and 273 °C for calcite. Type 1 inclusions for the central mineralized area (principal deposits are New Cobar, Chesney, Occidental, Queen Bee, Great Cobar) indicate a range in average T_h from 215 °C to 325 °C. Type 1 inclusions in quartz from the massive Zn-Pb Elura deposit average 320 °C. Average T_h measurements on type 3 inclusions in barren quartz from the central area and from nonmineralized syntectonic extension veins adjacent to the Elura deposit range from 146 °C to 190 °C.

Type 1 fluids are considered to have been trapped close to the solvus of the H_2O–CO_2–CH_4–(NaCl) system and hence homogenization temperatures approximate trapping temperatures. Type 2 inclusions are associated with Pb-Zn mineralization and late chlorite shears in the CSA Mine and may represent a separate, low CO_2 ore fluid trapped late in the paragenesis. T_h data for Type 3 inclusions require a temperature correction of approximately +160°C, which corresponds to the metamorphic load of 2.5–3Kbars and indicates a temperature range from 300°C to 350°C for metamorphic fluids present during sub-vertical extension of the Cobar orebodies.

Sulphur isotopic compositions for the major deposits (Elura: range 4.7 to 12.6 per mil, average 12.6 per mil; CSA, range to 2.7 per mil, average 7.5 per mil) are similar to those from minor or barren mineralization (Central area deposits: range 3.8 to 11.2 per mil, average per mil). $\delta^{34}S$ data for sulfides from the magnetite dominated Cu-Au orebodies (Great Cobar, Occidental) are similar to those from magnetite-free mineralization and suggest a high pH ore fluid for these two ore deposits. Sulfur isotopic disequilibria is characteristic of the vein deposits, however, in the massive pipe-like Elura orebody. Coexisting sphalerite and galena in particular, are equilibrated isotopically at a calculated late metamorphic temperature of 275°C.

A general depletion of approximately 1–2 per mil is shown for $\delta^{18}O$ and $\delta^{13}C$ for minerals and whole rocks in the zone of alteration and mineralization at the CSA Mine. Oxygen isotope values for quartz from mineralized veins range from +10.4 to +12.4$^o/_{oo}$ (vs. V-SMOW) in contrast with metamorphic quartz veins outside the alteration halo (range +11.2 to +13.3$^o/_{oo}$. Whole rock $\delta^{18}O$ values show a range of 16.8 to +7.6$^o/_{oo}$ in the alteration zone and +8.9 to +9.7$^o/_{oo}$ in unaltered metasiltstones. Oxygen and carbon isotope signatures of calcite from the mineralized zone ($\delta^{18}O_{V\text{-}SMOW}$ = +9.5 to +9.6$^o/_{oo}$, $\delta^{13}C_{V\text{-}SMOW}$ = -8.4 to -8.7$^o/_{oo}$) are depleted isotopically with reference to metamorphic vein calcite outside the alteration halo ($\delta^{18}O$ = +11.3 to +12.4$^o/_{oo}$, $\delta^{13}C$ = -0.5 to $^o/_{oo}$).

Hydrogen-isotope data from altered whole rock samples in the CSA mine alteration zone (chlorite is the major OH-bearing phase) yield $\delta D_{V\text{-}SMOW}$ values from -51 to -61$^o/_{oo}$ whereas unaltered whole rock samples (muscovite is the major OH-bearing phase) show $\delta D_{V\text{-}SMOW}$ values from -42 to -44$^o/_{oo}$. The calculated isotopic composition of the CSA ore fluid is $\delta^{18}O_{V\text{-}SMOW}$ = +3.7 to +5.5$^o/_{oo}$. $\delta^{13}C_{V\text{-}PDB}$ = -6.3 to -6.7$^o/_{oo}$ and the measured isotopic composition of fluid inclusions is $\delta D_{V\text{-}SMOW}$ = -42 to -44$^o/_{oo}$. The calculated isotopic composition of the regional metamorphic fluid is $\delta^{18}O_{V\text{-}SMOW}$ = +5.1 to +6.8$^o/_{oo}$, $\delta^{13}C_{V\text{-}PDB}$ = -4.4 to -6.7$^o/_{oo}$ and $\delta D_{V\text{-}SMOW}$ = -33 to -38$^o/_{oo}$.

The fluid inclusion and isotopic studies strongly support structural, textural, and mineral chemical evidence that indicate a metamorphic origin for all of the Cobar orebodies. Metal and sulfur appear to have been derived from the Cobar Supergroup sedimentary pile during early metamorphic dewatering. Fluid focusing has been achieved for many of the deposits by regional and local structural and permeability controls. Differences in metal content of individual orebodies in the Cobar region cannot be attributed to differences in ore fluid source, temperature or salinity, or major variations in the source rock composition. Rather, Cu-Au orebodies are likely to have formed from low salinity, neutral to alkaline ore fluids wherein Cu and Au transport took place by bisulfide complexing. Cu-Zn-Pb and Zn- Pb orebodies formed from more acid ore fluids than those apparently responsible for Cu-Au mineralization. Transport of the base metals in the acid, low salinity, $CO_2(CH_4)$-rich ore fluids was likely to have been achieved by chloride or hydroxy carbonate complexing.

SEEGER, C. R., Western Kentucky University, Bowling Green, KY

Cryptoexplosion Structures in United States

The the XVI International Geological Congress, Walter H. Bucher (1933) presented a paper entitled "Cryptovolcanic structures in the United States." Six such structures were described, maps were presented, and other structural and topographic features were noted. They are listed here in Table 1.

Bucher compared the structures to known geological features of other kinds, and concluded thus (p. 1080): "I believe, then that all American cryptovolcanic structures represent special phases of the ascent of basic magmas into the central plateau region of the United States, an assumption that is supported by the scattered occurrences of basic dikes.

Although the starting point for Bucher's application of "cryptovolcanic" in name and theory (Steinheim Basin, West Germany) and all of the above structures (Table 1) would not be considered by most geologists to represent probable impact or remains of craters of the type called central peak or central uplift, Bucher's work was very important in recognizing the commonality of these features, classifying them together, and calling them to the attention of succeeding generations of geologists. They are now called "cryptoexplosion structures."

Indeed, his description of the structural type is still apt (p. 1055): characterized by a nearly circular outline; a central uplift with intense structural derangement; and a marginal, ring-shaped depression with irregular and local faulting . . . intensely disordered structure and local brecciation of the central, uplifted portion; the presence . . . of peculiar fracture patterns, the "shatter cones"; and the presence and character of folding in the marginal zone.

The purpose of this paper is to review work subsequent to Bucher on his "cryptovolcanic" structures (Table 1) and to add to the list the cryptoexplosion structures in Table 2, seven in number, and to provide some brief information about them. Locations are shown on Figure 1.

Most of these structures are shatter coned (*), which is considered a definitive criterion of impact origin. The ones that are not and the reasons for inclusion follow, along with other notes. Many of them also have been found to have shock metamorphic features in quartz, or other minerals; monomict, polymict and mega-block breccias (sometimes named with different terms), and there are scattered reports of coesite in the literature:

2. Jeptha Knob: breccias (as above; iridium anomaly in uppermost breccias; undeformed Silurian caprock above deformed Ordovician; gravity and magnetic geophysical work, and limited drilling show shallow depth of deformation, no basement counterpart.

5. Upheaval Dome: recently reported shatter cones (E. M. Shoemaker).

9. Kilmichael: recently reported (and still somewhat tentative) quartz shock lamellae (V. L. Sharpton); seismic sections show deformation dies out below surface deformations.

10. Manson: although buried, is the largest; has quartz shock lamellae; is likely to have drilling exploration because of Cretaceous–Tertiary boundary age.

13. Versailles: although small, probably represents an exposure of a level under the floor of an impact; geophysics (gravity, magnetics) show feature is shallow, no basement counterpart; possible pair with Jeptha Knob.

Similar structures with shatter cones but completely buried are Des Plaines, Illinois; Glasford, Illinois; Glover Buff, Wisconsin; and Red Wing Creek, North Dakota.

Complete references are not possible, but Baldwin (1963), French and Short (1968), and Silver and Schultz (1982) are general works with historic importance and further information on these structures.

REFERENCES

Baldwin, R. B., 1963, The measure of the moon: Chicago, Illinois, University of Chicago Press, 488 p.

Bucher, W. H., 1933, Cryptovolcanic structures in the United States (with discussion): International Geological Congress, 16th, Washington, D.C., 1933, Report, v. 2, p. 1055–1084.

French, B. M., and N. M. Short, eds., 1968, Shock metamorphism of natural materials: Baltimore, Mono Book Corp., 644 p.

Silver, L. T., and P. H. Schultz, eds., 1982, Geological implications of impacts of large asteroids and comets on the earth: GSA Special Paper 190, 528 p.

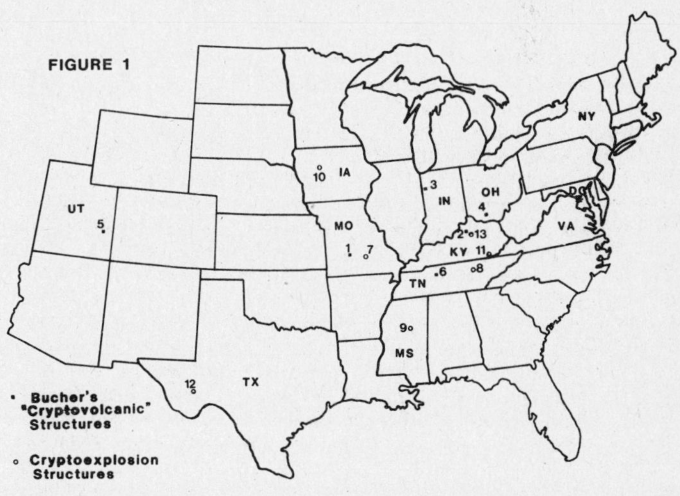

Figure 1.

Table 1—Bucher's "Cryptovolcanic Structures."

	NAME	LATITUDE	LONGITUDE	DIAMETER	AGE
				km	m.y.
1.	Decaturville, Missouri*	37 54'N	092 43'W	6	<300
2.	Jeptha Knob, Kentucky	38 11'N	085 07'W	4.3	438 ± 50
3.	Kentland, Indiana*	40 45'N	087 24'W	13	300
4.	Serpent Mound, Ohio*	39 02'N	083 24'W	6.4	300
5.	Upheaval Dome, Utah*	38 28'N	109 57'W	5.2	66
6.	Wells Creek, Tennessee*	36 23'N	087 40'W	14	200 ± 100

Table 2—Additional Cryptoexplosion Structures.

	NAME	LATITUDE	LONGITUDE	DIAMETER	AGE
				km	m.y.
7.	Crooked Creek, Missouri*	37 50'N	091 23'W	5.6	320 ± 80
8.	Flynn Creek, Tennessee*	36 16'N	085 37'W	3.8	360 ± 20
9.	Kilmichael, Mississippi	33 03'N	089 33'W	9.6	<40
10.	Manson, Iowa, U.S.A.	42 35'N	094 31'W	35	<70
11.	Middlesboro, Kentucky*	36 37'N	083 44'W	6	300
12.	Sierra Madera, Texas*	30 36'N	102 55'W	13	100
13.	Versailles, Kentucky	38 05'N	084 41'W	2	<458

SEGERO, A. S., Geoloconsultants Network, Nairobi, Kenya

Kimberlite Potential in Western Kenya Greenstone Belt

Extensive countrywide investigation of kimberlite/diamonds was undertaken in 1960's and 70's by the Mineral Exploration Company (Kenya) Ltd. utilizing stream loam and barrage samples. No major kimberlite pipes were discovered during the programme. The ground follow-up of airborne electromagnetic/magnetic anomalies in the Western Kenya Archaean greenstone belt (1978-1981) employing integrated and interdisciplinary techniques involving geophysics, geochemistry and geology successfully revealed the presence of concealed kimberlite pipes.

Geophysical methods included Horizontal Loop, Vertical Loop EM and magnetics while both stream sediment and grid soil geochemistry were employed. Three kimberlite pipes confirmed through drilling are well documented Segero (1981, 85 and 86), Ito et al (1981 and 83). Petrographical studies together with microprobe analysis, X-Ray diffraction, DTA-TG-DTG, IR were carried out on YIB-1, YIB-3 and YA-17 kimberlite with coordinates 34°11.2'E and 0°4.7'S 34°14.0'E, 0°3.1'S, 34°30.9'E, 0°9.5'N. X-ray powder diffraction patterns have been utilised to distinguish natrolite thomsonite, cancrinite and sodalite in the YA-17 xenoliths (Figure 3).

They contain macrocrysts of olivine, perovoskite, amphibole, pyroxene, chromite, garnet in the matrix of olivine, phlogopite, serpentine and calcite. Both zoned and unzoned phlogopites were found with some chemical differences. The rim part is richer in $C_{r2}O_3$ and poorer in TiO_2, and FeO. Calcic pyrope-almandine garnet in YiB-1 and diopsitic pyroxene in YA-17 were identified.

Volatile rich nature of the kimberlites is expressed in H_2O, CO_2 and P_2O_5 levels. In keeping with recent development in definition and classification of kimberlites the present study has indicated that some of the Western Kenya kimberlites are kimberlites sensu-stricto in the context of Clement et al, and Scott Smith et al (1984). For comparison with other kimberlites Al_2O_3 – MgO – FeO (total) diagram for YIB-1 kimberlites has been plotted on that of South Africa, Lesotho, Arkansas and Greenland. The diagram indicates that YIB-1 kimberlite from Western Kenya is richer in MgO and poorer in Al_2O_3 and FeO. Based on petrochemical characteristics a diamond bearing degree of Western Kenya kimberlites is plotted on (Na_2O + K_2O + Al_2O_3) – MgO – Fe) total diagram Chong (1981) Figure 2. YIB-1 kimberlite plot in diamond-rich and diamond-poor fields and YA-17 in diamond-free field. However samples from only one drill hole each sunk on the kimberlite pipe about one kilometer in diameter may not give a representative cross section.

Based on field and laboratory investigation the writer has delineated a kimberlite province, in the Western Kenya greenstone belt north of the Winam Gulf. (Segero 1985 and 86). Future investigation of kimberlites may take two trends: to search for more kimberlite pipes and find economic possibilities of those kimberlites which have been confirmed. Airborne and ground investigations have indicated clusters of probable kimberlites in the kimberlite province. For the proposed program to be effective exploration methods will be modified and refined as recommended in ground follow-up Segero (1981 and 86) to suite the present geologic environment, and petrographical studies together with petrochemical and laboratory analyses on the lines indicated by Ito et al 1981 and 83 will be necessary for ascertaining the kimberlites. The Western Kenya kimberlites appear to have a favorable geological environment in the Tanzania craton in which Mwadui Diamond Mines occur. In addition occurrences of diamonds were reported in the area.

SEGUIN, M. K., and J. FRYDECKI, Laval University, Quebec, Canada

Gravimetric Analysis of Charlevoix Intraplate Seismic Zone, Canada

The study area is bounded by latitudes 69°50'–70°50'W and latitudes 46°50'–48°00'N. Some 900 data points are distributed over this area on both sides of the St. Lawrence River. A non-classical method making use of a mobile window filter having a variable dimension is the approach selected to define the regional gravity map of the Charlevoix area which is located in the central granulite terrain of the Grenville structural province of the Precambrian Shield. The area located to the south of the St. Lawrence River is part of the northern Appalachians. The mean weighed density obtained with a correlation: Free Air anomaly vs. altitude is 3.1 mg/m^{-3} and this value is used for the computation of the Bouguer gravity contour map. In view of the uneven space distribution of the data points, a 2 km × 2 km mobile window filter is chosen to further improve the content of the Bouguer anomaly map. Using differing mobile window sizes, the subtraction of a Bouguer anomaly data set obtained with a mobile window filter of a specific size from that obtained with a filter of another size leads to a residual anomaly map which characterizes the geological signature contained between two specific levels at depth. In this fashion, shallow (1–4 km), intermediate (4–10 km), and deeper (10–30 km) geological objects within the earth's crust can be extracted.

North–south and northeast–southwest-trending faults are well defined on the residual gravity maps imaging shallow and intermediate sources but attenuate and fade out at the bottom of the earth's crust ($\approx$ 30 km). On the south shore of the St.. Lawrence River, a linear northeast-trending negative anomaly is observed only on the residual anomalies of intermediate and deep sources suggesting a deep seated (Precambrian) source for this anomaly. The south central part of the area is characterized by a highly heterogeneous triangular area depicted on all the residual gravity maps. The cluster of heterogeneous bodies is most clearly observed at the intermediate level (4–10 km). This large number of contiguous bodies of widely varying densities surely reflects some mechanical instability in this sector of the earth's crust. Thus, a gravity analysis of the La Malbaie sector leads to a tectonic model characteristic of a largely heterogeneous upper sector of the earth's crust. This interpretation indicates that intraplate readjustments along gravity faults represent a more probable cause of the high seismicity of this region than postglacial rebound or the occurrence of the Charlevoix meteoritic impact. A three-dimensional gravity model (mean corrected density and independent depth extent) of some of the individual object (e.g., St. Urbain anorthosite) is achieved.

SEIDEL, E., Mineralogisch-Petrographisches Institut der Univrsität zu Köln, Köln, Federal Republic of Germany, and T. THEYE, Ruhr-Universität, Bochum, Federal Republic of Germany

Prograde High P/T Metamorphism of Sediments in External Hellenides of Peloponnesus and Crete, Greece

A late Oligocene–early Miocene high-pressure metamorphic belt can be traced over a distance of about 600 km along the external Hellenides from the northern Peloponnesus to eastern Crete. It consists of two tectonic units, namely the Plattenkalk Series and the Phyllite–Quartzite (PQ) Unit.

The PQ unit has a Permo-Triassic volcanosedimentary protolith. Occurrences of magnesio-riebeckite/crossite (eastern Crete) and glaucophane (western Crete, Peloponnesus), aegirine-augite (eastern Crete), and omphacite (western Crete) as well as lawsonite (Crete, Peloponnesus) in basic to intermediate metavolcanites prove the high-pressure type of metamorphism. Regional differences in the mineralogical composition of equivalent rocks in the PQ Unit indicate a gradation in P-T conditions of metamorphism from eastern Crete (5–7 Kbars, 250–350°C) via western Crete (7–9 Kbars, 300–450°C).

In the course of prograde metamorphism, characteristic minerals and mineral assemblages appear in the metasediments of the PQ unit. Oxidized *Fe-Al-rich metapelites* with high Mg/Fe^{2+} ratios commonly contain the di/trioctahedral chlorite mineral sudoite in eastern Crete. In rocks with low X_{Mg}, the place of sudoite is taken by chlorite + pyrophyllite, suggesting the equilibrium:

$$\text{sudoite + quartz = chlorite + pyrophyllite} + H_2O. \quad (1)$$

The first appearance of rock-forming Fe-Mg-carpholite is due to progression of the reaction:

$$\text{pyrophyllite} + \text{chlorite} + H_2O = \text{carpholite} + \text{quartz}. \quad (2)$$

Except in the vicinity of quartz–calcite segregations (channel-ways of fluid), there was generally not enough water available for this hydration reaction to go to completion. Thus, the parageneses chlorite–pyrophyllite or chlorite–pyrophyllite–carpholite (varying positions of the 3-phase triangle in AFM-diagrams) persisted at reduced water activity.

Passing from eastern to central Crete, chlorite + pyrophyllite disappears and chloritoid occurs, suggesting the solid–solid reaction:

$$\text{chlorite} + \text{pyrophyllite} = \text{chloritoid} + \text{carpholite} + \text{quartz}. \quad (3)$$

In AFM-diagrams, the resulting 3-phase triangles chloritoid–carpholite–chlorite and chloritoid–carpholite–pyrophyllite vary in position, indicating local control of a_{H2O}. Only the compositions of the limiting assemblages with the most Fe-rich minerals are fixed corresponding to a water activity of pure H_2O. With increasing grade of metamorphism (western Crete, Peloponnesus) these critical parageneses become richer in Mg by the simplified reaction:

$$\text{carpholite} = \text{chloritoid} + \text{quartz} + H_2O. \quad (4)$$

Carpholite is present at high water activities only. Otherwise the reaction:

$$\text{chlorite} + \text{pyrophyllite} = \text{chloritoid} + \text{quartz} + H_2O \quad (5)$$

takes place.

The highest grade of metamorphism is reflected by the appearance of almandine-rich garnet in metapelites of the Peloponnesus.

Diagnostic minerals are also present in *Ca-Al-rich metapelites* and *impure marbles.* Sudoite occurs in eastern Crete in company with dolomite and calcite, while lawsonite-bearing aragonite marbles and metapelites are frequent in western Crete. Textures indicate the formation of lawsonite by means of the reaction:

$$\text{sudoite} + \text{calcite} + \text{quartz} = \text{lawsonite} + \text{dolomite} \quad (6)$$

In the Peloponnesus lawsonite in marbles is replaced by zoisite or clinozoisite/epidote. A possible reaction is:

$$\text{lawsonite} + \text{calcite} = \text{zoisite} + CO_2 + H_2O \quad (7)$$

Further minerals of diagnostic value are found in *Na–Al-rich metasediments.* In eastern and central Crete they contain chlorite + albite. Passing to the west blue amphibole appears according to the reaction:

$$\text{chlorite} + \text{albite} + \text{quartz} = \text{glaucophane} + \text{paragonite} + H_2O. \quad (8)$$

Glaucophane schists produced this way are rare in westeren Crete but widespread in the Peloponnesus.

Chloritoid-glaucophane schists formed by the reaction:

$$\text{chlorite} + \text{paragonite} + \text{quartz} = \text{glaucophane} + \text{chloritoid} + H_2O \quad (9)$$

are typical for the Peloponnesus, whereas chlorite + paragonite is still stable in the metapelites of western Crete.

Na–pyroxene-bearing metasediments of the Peloponnesus represent the most advanced stage of prograde metamorphic evolution. Here, aegirine-jadeite + chloritoid was formed under relatively high oxygen fugacity according to the simplified reaction:

$$\text{glaucophane} + \text{paragonite} = \text{Na-pyroxene} + \text{chloritoid} + \text{quartz} + H_2O. \quad (10)$$

In conclusion, sediments hitherto disregarded as suitable indicators of blueschist metamorphism may allow an adequate characterization of low-grade high P/T metamorphism. A necessary condition for the formation of diagnostic assemblages with sudoite, Fe-Mg-carpholite, and chloritoid is the enrichment of alumina and iron in pelitic and psammitic rocks of high maturity, which demands a special paleogeographic environment. Further minerals of diagnostic value occur in metasediments containing sodium (glaucophane, Na-pyroxene) or calcium (lawsonite, zoisite) in addition.

REFERENCES

Seidel, E., H. Kreuzer, and W. Harre, 1982, A late Oligocene/early Miocene high-pressure belt in the external Hellenides: Geol. Jb., v. E23, p. 165–206.

Theye, T., 1988, Austeigende Hochdruckmetamorphose in Sedimenten der Phyllit-Quarzit-Einheit Kretas und des Peloponnes: Thesis, TU Braunschweig.

——— and E. Seidel, 1987, Low-grade high P/T metamorphism of pelitic rocks—evidence of water deficiency: Terra Cognita, v. 7, p. 288.

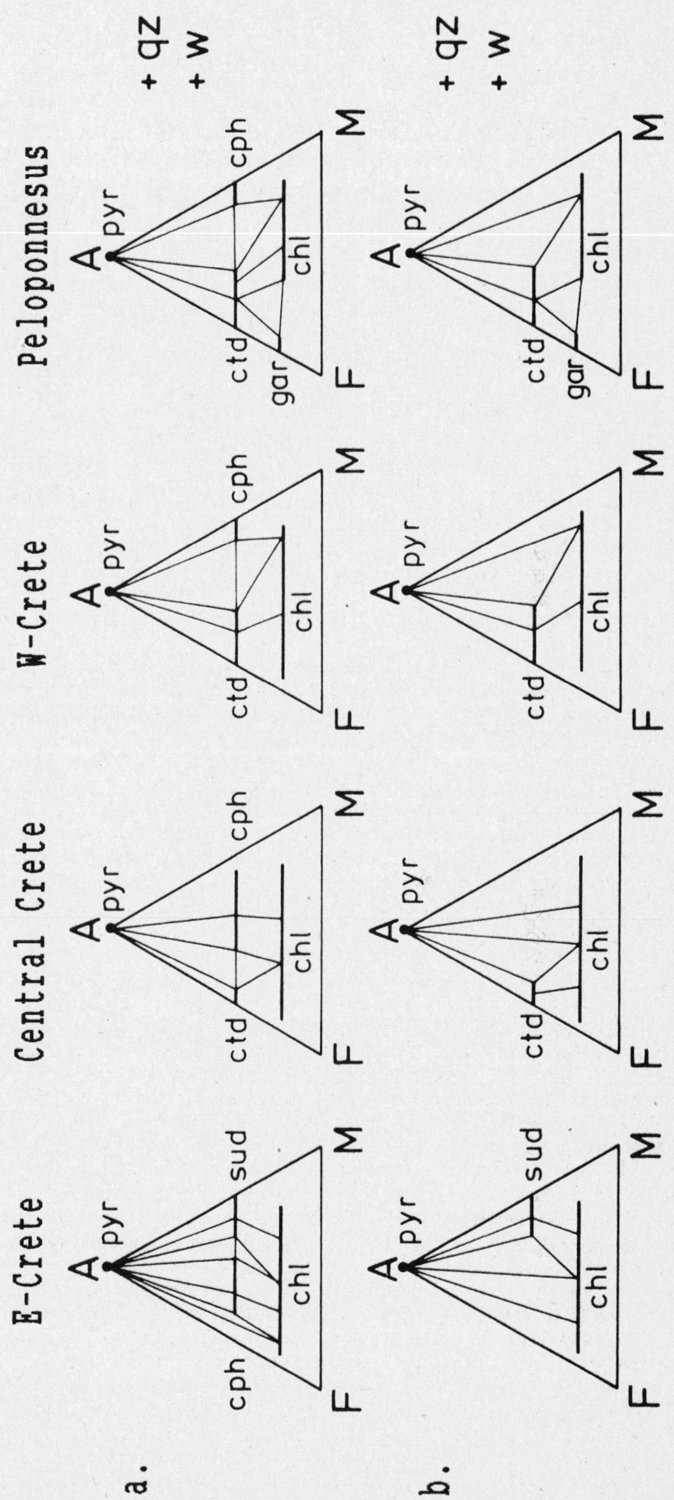

Figure 1—AFM-diagrams illustrating the prograde high P/T metamorphism of Fe-Al-rich sediments in the PQ Unit from eastern Crete to the Peloponnesus: a, high water activity; b, low water activity; chl, chlorite; cph, Fe-Mg-carpholite; ctd, chloritoid; gar, garnet; pyr, pyrophyllite; qz, quartz; sud, sudoite; w, H_2O.

SEIFERT, A., H. BRAUSE, J. RASCHER, K. UNGER, P. REINHARDT, and H. J. KAMPS, VEB K Geologische Forschung und Erkundung Halle, Halle, German Democratic Republic

Geology of Brown Coal District and Niederlausitz in German Democratic Republic

Brown coal is the main source for energy production in the German Democratic Republic. It supplies 75% of the demand for electric energy and for processing (briquetting, coking, gasification). It exploitation started as early as 100 years ago. Now it is in a range of 320 million tons per year. One of the largest brown coal districts is the Miocene one of Niederlausitz in the southeastern part of the GDR.

The resources of the first Miocene seam are nearly exhausted here; mainly the second is mined, the fourth is in exploration and the third Miocene and Oligocene seam Calau are unminable.

The rank of brown coal of the second Miocene seam ranges from fuel coal to coking coal, its heat value (dry) from 22.2 to 23.5 MJ, its ash content (dry) from 6 to 13%, its moisture from 57 to 59%, and its seam thickness from 10 to 12 m.

General aspects: The Oligocene and Miocene brown coal formation was synchronous with the alpidic orogenesis and the seafloor spreading of the North Atlantic Ocean. North- and east-oriented migrations of the labile basement of Central Europe were caused by both. Periods of compression alternated with longer times of isostatic subsidence and sedimentation. Additionally the trends of thickness and facies of sediments were controlled by a block system in the basement of the brown coal district of Niederlausitz, uplifting, subsiding, collapsing, turning, or spreading.

Stratigraphy: Despite the lithostratigraphic units of the Oligocene and Miocene, continental to brackish sediments in Niederlausitz can only partly be correlated with the marine units in the Tethys; their lithostratigraphic subdivision is very detailed. Its base is the geophysical logging of thousands of boreholes being logged for the past 30 years here. Widespread brown coal seams are lithostratigraphic guide horizons connected with ingressions or transgressions and related to world-wide tectonic evidence. Contrarily, their coal qualities are determined by local factors such as groundwater level and moor facies.

Seam petrography and coal quality: A seam consists of several cycles of moor/swamp facies, closely linked to the swamp environment and its plant associations. Each cycle starts with an increasing groundwater table and finishes with decreasing water level. These hydrological phases are connected with typical swamp plant associations. An ideal cycle begins with forest and bush swamps (coniferous wood, swamp facies K, and angiospermous bush, swamp facies A) and reed marshes (glumiflorous facies G). The regression phase leads at first to increasingly oligotrophic conditions, then to forest swamps (pinus, palms, facies P), and then to oligothrophic ombrogenous moss bogs (Marcoduria, facies M) characterized by roots within the seam. At the foot of the cycle Alnus-Liquidambar wood associations can be developed (ferns, facies F). Thus, an ideal cycle involves the facies sequence (F), K, A, G, P, M. These facies are macroscopically recognizable and determinable in routine investigations of drill cores. Such observations allow us to reconstruct the paleogeographic development of a seam. Moreover, due to the close relationship between swamp facies and the main coal quality parameters, the coal quality can be directly determined from the drill log: K- and G/P-facies characterize fuel coals, A- and M-facies yield coals for briquetting and processing to coke.

Structure: The second Lower Miocene seam is situated in the north of Niederlausitz at sea level and rises to the south to + 150 to + 180 m above sea level, caused by considerable widespread subsidence and uplifting since Lower Miocene, modified by local block movements. Additional suberosion in the time and after the seam formation is seldom the reason for greater thickness of seam or troughlike subsidences. Halocinetic effects occur locally, too.

Horizontal tectonic movements caused the collapse of asymmetric formed grabens with slight rotation tendency. Regional shear movements led to block faulting followed by volcanism. Counterclockwise rotation of the basement blocks is assumed. This hypothesis is supported by recent tension measurements and seismic observations.

The entire district was covered by inland ice in the Pleistocene. Wide destruction zones, narrow and deep grooves, intensively folded or imbricated seam structures and gravity-caused plastic structures were formed, affecting the recent distribution of seams, their exploitation, and the quality of run-of-mine coal. All brown coal is mined in open pits. Mine sites are restored to fertilized regions for agriculture or forestry and to forests and lakes for recreation.

REFERENCES

Rascher, J., and A. Seifert, 1985, Zur Anwendung kohlepetrographischer und paläobotanischer Untersuchungsverfahren in der Braunkohlenerkindung: Z. Angew, Geol., Berl, v. 31, n. 4, p. 85–91.

Schneider, W., 1978, Zu einigen Gesetzmäigkeiten der faziellen Entwicklung im 2. Lausitzer Flöz: Z. Angew. Geol., Berlin, v. 24, n. 3, p. 125–130.

Sontag, E., and M. Suss, 1973, Die Anwendung der quantitativen mikropetrographischen Analyse (ZMA) zur Kennzeichnung von Braunkohler hinsichtlich ihrer effektiven Verwertung: Z. Geol. Wiss., v. 1, n. 12, p. 1565–1589.

Vulpius, R., and K. -H. Neubert, 1986, Zum Einflu der Tektonik auf die Bildung und stoffliche Differenzierung des 2. Lausitzer (miozänen) Flözes: WTI, Berlin, v. 27, A, n. 3, p. 40–51.

SEILACHER, A., University of Tuebingen, Tuebinger, Federal Republic of Germany

Oyster Beds: Biological and Taphonomic Response to Storm-Dominated Regimes

Many oysters are adapted to live on quite, muddy bottoms with low sedimentation rates. Such conditions prevail, in terms of biological life spans, near storm wave base, particularly during transition from a regressive to a transgressive phase. Decreasing sedimentation rates within such intervals are reflected by the regular succession of mud-sticking and reclining oyster morphotypes. At the taphonomic times scale, however, catastrophic winnowing and reworking by rare storm events become the dominating factor. It leads to the gradual accumulation of oyster beds and mounds that may reach tens of meters in thickness. In the Upper Cretaceous of Jordan, huge oyster mounds, or "tells," are associated with rich phosphate deposits. These are found in depressions between the mounds and reflect a similar winnowing effect—only that the remnant particles were formed during the quiet intervals not by biological shell growth, but by diagenetic pellet formation under low-oxygen conditions. The occurrence of similar oyster beds in shallow-marine basins of Jurassic, Cretaceous, and Tertiary age and possibly even in East African lake deposits underlines the general significance of the phenomenon.

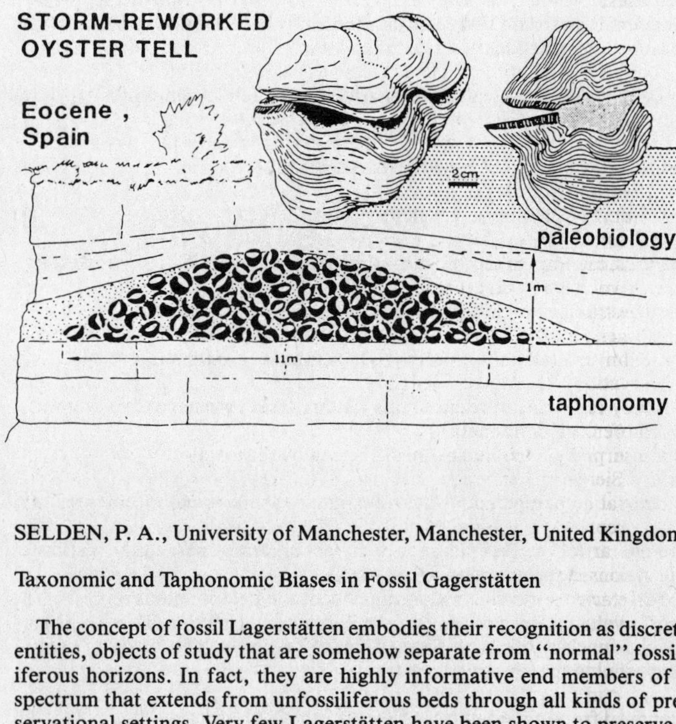

SELDEN, P. A., University of Manchester, Manchester, United Kingdom

Taxonomic and Taphonomic Biases in Fossil Gagerstätten

The concept of fossil Lagerstätten embodies their recognition as discrete entities, objects of study that are somehow separate from "normal" fossiliferous horizons. In fact, they are highly informative end members of a spectrum that extends from unfossiliferous beds through all kinds of preservational settings. Very few Lagerstätten have been shown to preserve a biota truly representative of that which was alive at the time. The Burgess Shale is one such (Conway Morris, 1985), but nearly all Konservat-Lagerstätten are biased in some way. This study is an attempt to define the nature of these distortions so that they can be predicted, thus ensuring clear understanding of the limitations of know Lagerstätten and providing the ability to seek new exceptional biotas in favorable sedimentary environments. The record of the arthropod phylum Chelicerata is used as the database.

Chelicerates range from the Cambrian to the Recent. At present they outnumber all other terrestrial animal species except insects and were the top carnivores in many Paleozoic ecosystems. Table 1 gives an outline fossil record of the Chelicerata; the bulk of the information, at least for the terrestrial groups, comes from a few Lagerstätten. Note the taxonomic bias: Some of these exceptional biotas are acclaimed because they preserve che-

Table 1—Fossil record of the Chelicerata.

Cenozoic:	Neogene Schizomida, Aranese, and Palpigradi. Oligocene Baltic amber fauna of Opiliones, Pseudoscorpiones, Uropygi, Amblypygi, Araneae, and mites.
Mesozoic:	Cretaceous Araneae (Canadian amber; Botswana; lithographic limestone, Spain), Opiliones, mites. Jurassic Palpidradi?, mite, Araneae (USSR). Triassic Araneae (France).
Paleozoic:	Stephanian Trigonotarbi, mite. Westphalian Opiliones, Anthracomarti, Solifugae, Trigonotarbi, Uropygi, Amblypygi, Araneae, Phalangiotarbi, Ricinulei, mites. Namurian Ricinulei. Dinantian Opiliones and mites (Scotland).
	Devonian Pseudoscorpiones (Gilboa), Trigonotarbi (Gilboa, Rhynie, Aiken), Araneae? (Rhynie), mites (Gilboa, Rhynie).
	Ordovician–Permian Eurypterida
Silurian–Recent Scorpiones	
Cambrian–Recent Xiphosura	

licerates, animals that otherwise are rarely encountered in the fossil record.

Eurypterids are the most numerous Paleozoic chelicerates and are usually found in "eurypterid beds." These (e.g., Lesmahagow and Hagshaw Hills in Scotland, Bertie Waterlime in New York) show an atypical marine fauna in association, often including fish. There is an ecological bias at work, but would the eurpterids and fish also have lived in the "normal" marine habitats of the same age? Is there a combination of ecological and taxonomic (i.e., constructional morphological) factors that must interact to produce preservation? The celebrated dolomitic limestones of Saaremaa, Estonia, and Gotland, Sweden, which yield the exquisitely preserved eurypterid *Baltoeurypterus tetragonophthalmus* (Selden, 1981), are included by Conway Morris (1985) in his listing of principal Lagerstätten. Yet other species of eurypterids, and the xiphosurans *Pseudoniscus* and *Bunodes*, found in the same beds are less well preserved (Eldredge, 1974).

Scorpions are rare in the middle Paleozoic, but some occur in the eurypterid beds mentioned above. It is becoming increasingly obvious that the remains of these animals are actually quite common in particular lithologies (e.g., shaly coals) in the upper Paleozoic, in the absence of those of other arthropods. The reason put forward by Bartram, et al (1987) was the unusual hyaline exocuticle known from Recent scorpions being preferentially preserved. This layer may occur in eurypterids, too. This aspect of preservational bias in fossil Lagerstätten is the subject of the contribution by Jeram in this symposium.

Terrestrial arachnids have a much lower chance of burial than aquatic animals, and their record is almost entirely due to exceptional preservation. The Devonian Rhynie Chert of Scotland, Gilboa mudstones of New York, and Westphalian Mazon Creek nodules of Illinois are the important arachnid Lagerstätten. The presence of these arachnids adds weight to their recognition as Lagerstätten, and in the case of Gilboa there may be preservational bias in their favor as well. Curiously, Mesozoic arachnids are very rare. Their virtual absence from Mesozoic Lagerstätten such as Solnhofen, which contains numerous examples of terrestrial forms, is especially surprising. Five spider specimens are known from a similar lithology in the Sierra de Montsech, northeast Spain (Lower Cretaceous). Jeram (personal communication) has suggested that this may be because these biotas represent obrution deposits: Rapid burial may not aid the preservation of arthropod cuticle as much as postdepositional physicochemical conditions in the sediment. Later arachnid-bearing biotas are the ambers of the Cretaceous and Oligocene.

The value of fossil Lagerstätten lies in the information that they provide on the functional morphology and geological history of extinct animals that would otherwise be poorly known. While there is a taxonomic bias inherent in the definition of Lagerstätten, more important to recognize is the input from constructional morphology (e.g., scorpion hyaline exocuticle). With their nonmineralized cuticles, chelicerates are nevertheless not "soft-bodied" (cf. graptolites and palynomorphs) and are abundant in many sediments not otherwise classified as Lagerstätten. As more becomes known about the preservability of organic cuticles, more rigorous definition of what constitutes an extraordinary occurrence will become necessary.

REFERENCES

Bartram, K. M., A. J. Jeram, and P. A. Selden, 1987, Arthropod cuticles in coal: Journal of Geological Society of London, v. 144, p. 513–517.

Conway Morris, S., 1986, The community structure of the Middle Cambrian Phyllopod Bed (Burgess Shale): Palaeontology, v. 29, p. 423–467.

Eldredge, N., 1974, Revision of the suborder Synziphosurina (Chilicerata, Merostomata), with remarks on merostome phylogeny: Am. Mus. Novit., v. 2543, p. 1–41.

Selden, P. A., 1981, Functional morphology of the prosoma of *Baltoeurypterus ptetragonophthalmus* (Fischer) (Chelicerata: Eurypterida): Transactions of Royal Society of Edinburgh Earth Sciences, v. 72, p. 9–48.

SEMPERE, J. C., University of Washington, Seattle, WA, and G. M. PURDY and H. SCHOUTEN, Woods Hole Oceanographic Institution, Woods Hole, MA

Segmentation of Mid-Atlantic Ridge Between Kane and Atlantis Fracture Zones: Results of SeaBeam Investigation

Contrary to the East Pacific Rise, where the segmentation of the spreading center has been demonstrated, the segmentation of the Mid-Atlantic Ridge has only been postulated from surface magnetic field data and the extremely limited multibean bathymetric coverage available. The lack of substantial, high-resolution, bathymetric data along the strike of the Mid-Atlantic Ridge has been a serious restriction in the development of models of lithospheric accretion along slow spreading centers.

During the fall of 1988, we conducted a detailed geophysical investigation of the Mid-Atlantic Ridge between the Kane and Atlantis Fracture Zones using the SeaBeam multibeam echo-sounder. For the first time, complete, continuous coverage of a significant portion of the slow-spreading Mid-Atlantic Ridge was obtained. Our SeaBeam survey allows us to define the nature and scale of the segmentation of accretionary processes along the Mid-Atlantic Ridge, as well as the variability of volcanic and tectonic activity along the spreading center. The data also allow us to study the nature of the ridge axis discontinuities that offset the spreading center along its strike and to compare their structure to that of ridge axis discontinuities encountered along intermediate and fast-spreading centers. Results of this study, combined with existing seismic, magnetic, and geochemical data, will provide fundamental constraints for our understanding of the accretionary processes along the slow-spreading Mid-Atlantic Ridge.

SEMPERE, J.C., and J. PAMER, University of Washington, Seattle, WA, and J. P. MORGAN, M.I.T., Cambridge, MA, D. CHRISTIE, Oregon State University, Corvallis, OR, and A. SHOR, University of Hawaii, Honolulu, HI

Seafloor Spreading within Australian–Antarctic Discordance

The Australian–Antarctic Discordance (AAD) is a region of very rugged bathymetry located along the Southeast Indian Ridge (SEIR) between 120 °E and 128 °E. The AAD is centered within a 2000 km wide depression in which the seafloor can be up to 1000 m deeper than expected from empirical age-depth relationships in other ocean basins. It is a major geochemical boundary along the SEIR as well as a major morphotectonic boundary. Klein et al (1988) have shown that the very narrow and abrupt transition between Indian Ocean and Pacific Ocean isotopic compositions occurs within the AAD. Weissel and Hayes (1974) have suggested that the Australian–Antarctic Discordance is a region of asthenospheric downflow beneath the midocean ridge system. The presence of upper mantle that is colder than usual, a corollary of this hypothesis, is confirmed by seismic studies (Forsyth et al, 1987).

In January and February of 1988, we conducted a detailed geophysical petrologic study within and east of the Australian–Antarctic Discordance, using the swath mapping capability of the SeaMARC II sidescan sonar and bathymetric system. The plate boundary within the AAD consists of 5 segments bounded by transform offsets (Figure 1). It is highly crenulated and can be broken into 3 north-offset (B1, B3, B5) and 2 south-offset segments (B2, B4). Comprehensive aeromagnetic coverage (Bogt et al, 1984) indicates that except for the 3 north-offset segments, the SEIR has spread asymmetrically, the northern flank being faster by about 5 mm/y. Asymmetric spreading is, therefore, at least partly responsible for the isolation of the 3 northern segments from the rest of the plate boundary. Time constraints limited our survey to the 3 easternmost segments (B3, B4, B5) and to their bounding transforms. The geochemical boundary between Indian Ocean and Pacific Ocean isotopic compositions occurs between segments B3 and B5 (Klein et al, 1988).

Our Sea MARC II study indicates that the 2 north-offset segments surveyed are characterized by rift valleys, which are more typical of slower spreading centers (the full spreading rate in the AAD is about 7 cm/y). Segment B5 is associated with a 400–1200 m deep, 11–13 km wide rift valley. It can be divided into 2 subsegments separated by a nontransform discontinuity (offset ~ 3 km) that corresponds to a large depression along the spreading axis. A large volcano in the easternmost portion of segment B5 appears to be the youngest member of a chain that extends to the south. Segment B3 is similar to B5 and corresponds to a 1200–1500 m deep, 8–10 km wide rift valley. It can be broken into 3 subsegments bounded by nontransform discontinuities (offsets ≈ 3 and 5 km). In contrast, B4 is highly unusual, exhibiting morphotectonic characteristics very different from those of the north-offset segments. Near its intersection with the B4/B5 transform, it corresponds to a narrow graben 3 km wide and 1000 m deep. The spreading center is then offset to the north by ≈ 4.5 km where it is associated with a shallow (2700 m), 3–7 km wide ridge. The western end of this ridge rotates to the north in a transform parallel direction and decreases progressively in relief.

The spreading segments we surveyed are offset by transform faults with highly variable characteristics. Segments B4 and B5 are separated by a 120 km long, left-stepping transform fault. The morphology of this offset is similar to that of slow slipping transforms (Fox and Gallo, 1984), even though crustal accretion occurs at a rate of ≈ 7 cm/y. There is no well-defined transform fault linking the two offset spreading segments B4 and B3. Instead, there appears to be a wide transform region (13–18 km wide) in which the transform-parallel portion of segment B4 is found. The morphotectonic characteristics of the B2/B3 transform offset are similar to those thought to be typical of intermediate to fast-slipping transforms (Fox and Gallo, 1984). Two relay zones (15 and 35 km long) are present along the portion of the transform we surveyed. Each of these small spreading segments corresponds to a 1000–1500 m deep, 4–7 km wide graben.

In summary, the spreading segments and transform offsets along the SEIR within the Australian–Antarctic Discordance exhibit a wide range of morphotectonic characteristics, often associated with midocean ridges with different spreading rates. The results of our SeaMARC II study suggest highly variable, locally controlled mechanisms of crustal accretion and provide additional insights into the nature and evolution of the AAD.

REFERENCES

Forsyth, D.W., R. L. Ehrenbard, and S. Chapin, 1987, Anomalous upper mantle beneath the Australian-Antarctic Discordance: Earth and Planetary Science Letters, v. 84, p. 471–478.

Fox, P. J., and D. G. Gallo, 1984, A tectonic model for ridge-transform-ridge plate tectonics: implications for the structure of oceanic lithosphere: Tectonophysics, v. 104, p. 205–242.

Klein, E.M., C. H. Langmuir, A. Zindler, H. Staudigel, and B. Hamelin, 1988, Isotopic evidence of a mantle convection boundary at the Australian-Antarctic Discordance: Nature, v. 333, p. 623–629.

Vogt, P. R., N. Z. Cherkis, and G. A. Morgan, 1984, Project Investigator-I: evolution of the Australia-Antarctic Discordance from a detailed aeromagnetic study, *in*R. L. Oliver, P. R. James, and J. B. Jago, eds., Antarctic Earth Science, Proceedings 4th International Symposium on Antarctic Earth Science, Australian Academy of Science, Canberra.

Weissel, J. K., and D. E. Hayes, 1974, The Australian-Antarctic Discordance: new results and implications: Journal of Geophysical Research, v. 197, p. 2579–2587.

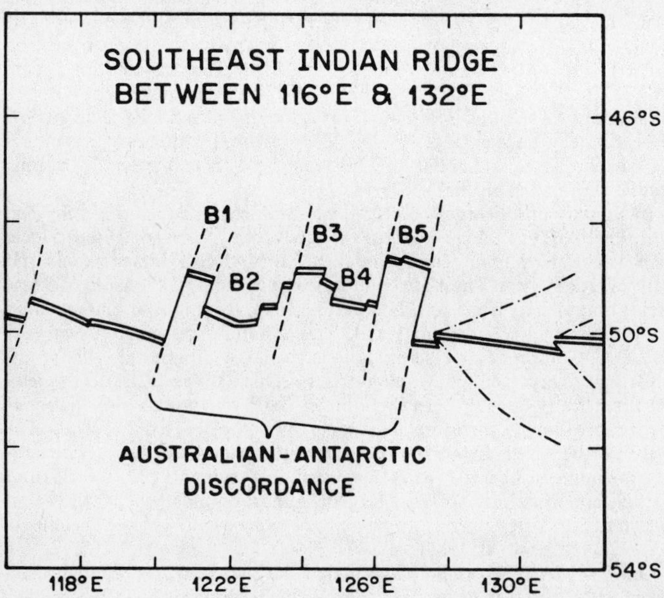

Figure 1— The plate boundary within the Australian–Antarctic Discordance (from Vogt et al, 1984).

SEMPERE, T., and P. BABY, Convenio ORSTOM-YPFB, Santa Cruz, Bolivia, G. Huc accenterail, ORSTOM, La Paz, Bolivia, and J. OLLER, Yacimientos Petrolíferos Fiscales Bolivianos, Santa Cruz, Bolivia

Geological Structure and Tectonic History of Bolivian Orocline

The Bolivian orocline results from the autoindentation of South America by part of its western margin in an area occupied by a thick, northwest-trending, initially 800 km wide, Paleozoic basin mostly filled with shales, which provided the main decollements used by the observed great-scale propagation of thin-skinned thrusting toward the craton. North and South of this Paleozoic basin, the shallower Precambrian basements acted as buttresses resisting somehow rigidly the advance of deformation, resulting in distinct, partly thick-skinned, tectonic styles.

Subduction of the Nazca plate below South America seems to have only been the motor of the continental margin deformation, of which the actual structural features were mainly determined by heterogeneities of the upper plate, probably influencing in return the geometry of the subducted one. However, major tectonic crises, resulting in apparent abrupt changes in the modalities of deformation and magmatism, were triggered by changes in convergence direction and/or velocity, since they nearly coincide in time. Present data obtained in the orocline allow us to locate the last two major events at 10–11 Ma and 26–28 Ma.

Phanerozoic basin reconstructions and structural mapping in the Bolivian orocline area (12 °–24 °S) have impelled the definition of fault-bounded tectonostratigraphic domains, clearly shortened and displaced in relation to each other during the Cenozoic Andean orogeny. These domains are quite distinct from the traditional physiographic units that have constantly grounded any geologic model of the Central Andes until now. We do not call them "terranes" because they all have been linked to one another at least since Upper Paleozoic times and are not exotic to South America.

To describe the orocline general structure, the tectonostratigraphic domains can be grouped in two major structural belts. South of 14 °S, the very complex western belt began to be wedged beneath the Precambrian basement and/or Paleozoic cover of the eastern belt in Upper Oligocene times, thus triggering the eastward propagation of tectonic imbrication that has resulted during the Neogene in individualization of the four tectonostratigraphic domains now recognized in this belt. The characteristic asymmetry of the orocline is due to differential propagation of the deformation. In the Vilcabamba "hinge zone" and northwest-trending branch, propagation of thrusts has been relatively impeded by rapidly northeast-tapering of Paleozoic units, and/or Carboniferous to Traissic deformational and magmatic heterogeneities. This resulted in considerable tectonic wedging, backthrusting, and vertical piling-up, from which the Cordillera Real inherited some of the highest elevations in the Andes. The very subsident Northern Altiplano basin formed in response to the Cordillera Real "backthrust belt," and also to major strike-slip motions that deeply affected the whole Altiplano, in part probably reactivating ancient fault zones. In contrast, the north-south branch formed in the area of maximum thickness of flat-lying Paleozoic units, which led to easier propagation of shortening, wider thrust belts and lesser altitudes. Accommodation of deformation along the southwestern margin of the Paleozoic basin (21 °–24 °S) is expressed by common northeast-trending right-lateral strike-slip faulting. This compensates the limited thrust motion of the vilcabamba "hinge zone" and suggests a gross anticlockwise rotation of the eastern belt of the orocline, in good agreement with the published paleomagnetic data.

Timing of deformation in the orocline is approximated by dating the functioning of the associated foreland basins, which discontinuously shifted in time from west to east. Two of them, of Late Jurassic–Early Cretaceous and Upper Cretaceous–Paleocene ages, respectively, are known in the western belt, whereas the other two locate in the eastern belt, covering the Eocene–Middle Oligocene and Late Oligocene–Present time spans. The Eocene–Middle Oligocene foreland basin, which located in the Bolivian Altiplano and Cordillera Oriental, evolved during the 26–28 Ma tectonic crisis to become the Altiplanic intermontane basin, while eastward progression of thrusting built the present Cordillera Oriental and triggered the individualization of the easternmost foreland basin. Relative tectonic quiescence seems to have occurred during the 17–20 Ma interval in most part of Bolivia. Tectonism and magmatism, however, resumed at 15–17 Ma, notably increased during the 10–11 Ma crisis until ca. 6 Ma and have lingered on until now.

Crustal thickening occurs particularly below the Altiplano–Puna plateau. Its origin should be south in (1) the underthrusting of the western belt crust and deformed cover beneath the Precambrian basement and Paleozoic rocks of the eastern belt, and (2) probably complex duplexing of both brittle and ductile crusts of the eastern belt, which would be responsible for the crustal fusion evidenced in the easternmost, ore-bearing, Neogene magmatic belt.

The Bolivian orocline enlightens the processes that deform a continental margin submitted to subduction. Its main characteristics show that at all scales continental heterogeneities have imposed some important controls upon the actual three-dimensional structures. A model of indenters-and-buttresses tectonics is proposed to explain most of them, and the whole orocline is one large-scale example of it.

Conclusions reached between 12° and 24°S can be extended to the 6°–32°S interval. A detailed study of the geology of the orocline is to be presented in a forthcoming paper.

SEMPERE, T., Convenio ORSTOM-YPFB, Santa Cruz, Bolivia

Paleozoic Evolution of Central Andes (10–26°S)

The "Paleozoic" history of the present Central Andes (10–26°S) records the evolution of part of the western Gondwanaland active margin from Late Cambrian to Early Triassic Times. Most outcrop zones locate in what was then a northwest-trending back-arc basin (Andean Peru, Bolivia), and several areas (coastal Southern Peru, Northern Chile, northwestern Argentina) show magmatic arc rocks or fore-arc deposits that also define a roughly northwest strike for the Paleozoic margin.

Important tensional episodes occurred in the Late Cambrian, Ashgillian–Early Llandoverian, and Pennsylvanian. The main periods of Paleozoic compressive deformation in the back-arc basin are Llandeilian–Caradocian ("Caledonian"), Late Famennian–Visean ("Hercynian"), and Middle to Late Permian ("Gondwanian").

The Paleozoic back-arc basin individualized in Late Cambrian times starting from a rift, nearly perpendicular to the margin, that propagated into the continent with a north-northeast trend. Its evolution led in Early Ordovician times to the formation of ophiolitic rocks in northwestern Argentina. In southern Bolivia, tensional processes are only marked along the same north-northeast trend by numerous spilitic sills and dykes, the frequency of which decreases northwards. Approximately at 20°S, the basin axis progressively turns west to take a northwest trend and shows little syn-sedimentary magmatism from there on.

This development created a paleogeography somehow similar to the modern Gulf of California. Extension lasted till Llanvirnian times (maximum areal transgression). During the Llandeilian–Caradocian period, the north–northeast-striking ophiolite-floored "straits" cut in the Precambrian continental margin closed, triggering the functioning of the northwest-trending Peru–Bolivian back-arc as a marine foreland basin, where very thick shallow-water deposits (4000 m) are known for this time interval. Intense deformation in the "straights" area produced structures with cleavage and local metemorphism, dying out in a northeast direction, and the southwest edge of the continent was then intruded by numerous granitoids.

This "Caledonian" crisis was followed in Bolivia by an abrupt deepening of the back-arc basin in the Ashgillian–Early Llandoverian, as recorded by deposition of euxinites and debris-flows sharply above the Caradocian shallow-marine beds. Some clasts show evidence of contemporary glaciation in an unknown area. From Late Llandoverian to Early Famennian, the back-arc basin progressively filled under unconstrained conditions, showing onlap in a northeast direction.

The Late Famennian–Visean tectonics partly coincide in time with a Late Famennian–Tournaisian glaciation. Within the back-arc basin, the "Hercynian" deformed area only occupies a northwest-trending belt clearly located north of the "Caledonian" one. Deformation is maximum in Peru and decreases into Bolivian territory. Where more intense, it is mainly characterized by asymmetrical to recumbent folds with associated cleavage, incipient metamorphism and, very locally, medium, to high-grade metamorphism and anatexis. Most structures are southwest-verging, which suggests northeast-dipping tectonic wedging. Deformation is probably of thin-skinned type in Peru, laterally changes to thick-skinned type in western Andean Bolivia, and dies out eastwards. Parts of the back-arc basin went on functioning during the orogeny, receiving the erosion products of the uplifted areas, but emersion seems to have been general in the Namurian.

The southwestern continental edge recorded a magmatic peak near the Devonia–Carboniferous boundary. Fore-arc turbiditic sediments were complexly deformed within accretionary wedges, suggesting a northeast-dipping subduction. On the proto-Pacific margin, deformation lasted until Pennsylvanian times and may have been partly caused or complicated by accretion of exotic terranes, such as the Mejillones terrane and possibly others now located west of the present shoreline. Apparently as a result of accretion, the magmatic are shifted to the west in the Late Pennsylvanian, and numerous granitoids were emplaced.

The eastern part of the back-arc basin recorded, in a syntectonic setting, the sea-level falls and mostly resedimented lowstand deposits linked to the Gondwana glaciations, until Latest Pennsylvanian when a climatic warming occurred. Transgressive carbonates deposited on both sides of the Late Pennsylvania–Early Permian magmatic arc. The back-arc marine basin, rich in warm-water fauna, was then bounded to the east (65°W) by eolian and fluvial sands.

The "Gondwanian" tectonic crises started in Mid-Permian times, only affecting an apparently narrow belt in Perus and westernmost Bolivia, geographically located within the "Hercynian" belt. As a consequence the depositional area rapidly migrated eastwards. Shallow-water chert-rich carbonates were mainly deposited until Early Triassic times, when the basin gradually emerged (siliceous stromatolites). During the Late Pennsylvanian–Permian period, the magmatic arc was the site of abundant volcanism and plutonism, and of locally thick lacustrine sedimentation.

Intensities of "Hercynian" and "Gondwanian" deformations in the back-arc basin both show a significant decrease from Peru into Bolivia, which points to a probable strong crustal difference between the two areas. Present lack of outcropping rocks older than Jurassic along the coast of Central Peru might be linked to it.

The Middle to Late Triassic general extension is considered to mark the beginning of the "Andean" evolution, which culminated with the building of the present Andes Cordillera. it must be stressed that the geometry of pre-Cenozoic basins and structures has been deeply modified during this orogeny. The "Andean" overprint on the Paleozoic geologic record thus makes somewhat difficult any precise understanding of the pre-Andean history of the western margin of Gondwanaland, at least at the present time.

SEN GUPTA, B. K., and J. J. GALLUZZO, Louisiana State University, Baton Rouge, LA, R. B. LESCHBER, Glasgow College of Technology, Glasgow, United Kingdom, and M. L. GABY, Sun Exploration and Production Company, Dallas, TX

Late Quaternary Benthic Foraminifera and Water Masses of Caribbean Deep Basins

The distribution of benthic foraminiferal assemblages in the cold water sphere of Caribbean basins is related to the spreading of the Caribbean Bottom Water (CBW) and the Subantarctic Intermediate Water (SA/W) in depths greater than 1000 m. CBW has a variable component of well-oxygenated North Atlantic Deep Water (NADW) whose influx into the Caribbean is at the 1800 m deep Anegada-Jungfern Passages and the 1600 m deep Windward Passage (Figure 1). Samples from numerous stations in the Venezuela, Grenada, and Colombia Basins and the Cayman Trough (Figure 1) demonstrate a link between the NADW fauna and the CBW fauna, noticeably in the abundance of Cibicides wuellerstorfi. This species is a dominant member of the high-diversity assemblage (Shannon-Wiener function > 3) in the deep Venezuela Basin (> 4000 m), which has a direct influx of NADW, and even in the southwestern corner of the Grenada Basin, at depths near 3000 m, at its Venezuela Basin bottom- water connection. In the western and central Caribbean Sea, it constitutes a significant proportion of the fauna below 2500 m in the Cayman Trough, but shows a reduced abundance in comparable depths of the Colombia Basin. The most pronounced difference between a typical NADW assemblage and a CBW assemblage in all of the deep basins is in the proportion of Epistominella exigua-high in the former and very low in the latter.

The SA/W fauna is developed to a much greater water depth (almost 2000 m) in the Grenada Basin than in the other basins, because a large volume of the poorly oxygenated, relatively low salinity SA/W enters this relatively small basin from the Atlantic. The assemblage, between 1000 and

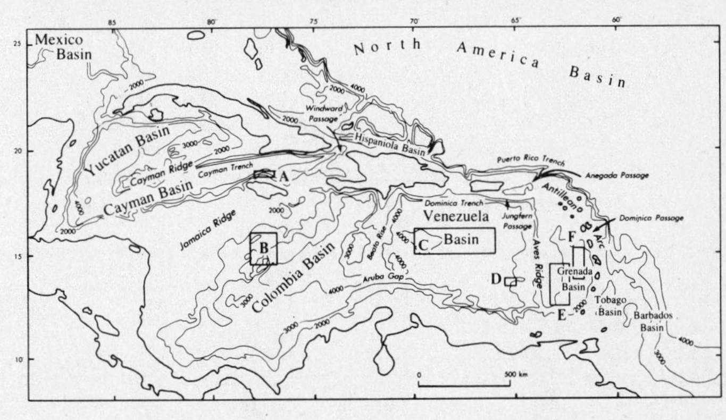

Figure 1—Map of Caribbean Sea, with the principal basins and passages; isobaths in meters. The sampling areas are shown as boxes A-F. High abundances of Cibicides wuellerstorfi (> 20% of assemblage) are associated with CBW in areas A,C,D, and E. The most typical SA/W assemblage is present in areas E and F.

1700 m, is dominated by Bulimina aculeata, Uvigerina peregrina, U. auberiana, and Osangularia culter. The fauna in similar depths of the Cayman Trough and Colombia Basin, however, is distinctly different, no stenobathic species showing a clear dominance.

Principal Components Analyses of the faunal data from many deeper (> 3000 m) sites clearly indicate a major transition at the Pleistocene-Holocene (Ericson Y-Z) boundary. The reappearance of Cibicides wuellerstorfi at this boundary coincides with that of the planktonic marker Globorotalia menardii and indicates a relatively pronounced influx of NADW over the deeper Caribbean sills about 10,500 years ago. In the typical SA/W assemblage, the faunal change at this boundary, although present, is recognized only with difficulty.

SEN Subhasis, Central Fuel Research Institute, Nagpur, India

Autochthonous and Allochthonous Modes of Coal Seam Formation—A Critical Assessment

Studies on one of the major aspects of coal genesis, pertaining to the deposition of peat forming vegetal debris, is largely based on coal measure fossils, such as tree trunks, stigmaria, concretions as well as the nature of underclay or seat earth composing the floor of the coal seams. Such studies, carried out for a protracted period by geologists and paleobotanists have generally given rise to two sharply contrasting views on the modes of deposition of vegetal debris, namely, autochthonous or in situ accumulation, involving accumulation of vegetal matter in the place of their growth and allochthonous or drift process essentially requiring transportation of the plant debris before their deposition in a depression away from their place of growth.

The author considers that the concept of deposition of vegetal debris to form peat bogs representing the earliest stage of coal formation—right from the stage of uprooting trees or falling of branches and leaves to the forest floor to the formation of basic infrastructure of banding of coal, need a thorough revision in the light of knowledge accrue on physical and chemical properties of the coal forming ingredients, their subjection to environmental conditions, transportation and deposition in the form of sediments.

The present study incorporates a critical assessment and comparison of the two distinct processes of coal formation which remained a matter of speculation and controversy for a protracted period.

Teichmüller (1962) reconstructed depositional condition of coal, essentially of autochthonous type, in which zonal distribution of various lithotypes have been suggested depending on the condition of water table. It can be pointed out that under such a set up origin of a sharp banding in coal seam cannot be explained. The present author (Sen, 1982) has, however, explained the nature and origin of banded structure of coal seam—essentially a sedimentological feature which has formed by drift or allochthonous process of deposition of both chemical and clastic sediments and their sorting and subsequent induration in a sedimentary basin.

The protagonists of authochthonous origin subscribe their views mainly based on evidences of upright tree trunks, stigmaria and nature of underclay or seat earth. All the major coal deposits of the world are associated with sedimentary strata, frequently showing cyclic process of sedimentation giving rise to characteristic cylothem, which, however, tend to rule out the autochthonous origin of major coal deposits.

A critical analysis of coal deposits, associated with upright tree trunks and stigmaria, confirm that the so-called evidences are not present in the form of carbonaceous or coaly substances, although these are intimately associated with the coal seams usually in a vertical deposition. The petrified wood fossils, in the form of upright tree trunks, stigmaria or concretions evidently do not represent the same generation of coal forming vegetation, though frequently they may belong to and represent the identical species because of small time gap. Further, it is indeed inconceivable that upright stems of tree would remain erect and survive the process of peatification involving extensive bacterial decay and prolonged chemical action. It can be argued that on the basis of evidences of petrified roots or tree trunks, accumulation of coal by autochthonous process is incompatible since petrification of wood evidently preceded the deposition of vegetal debris and their subsequent peatification.

It has been stressed that occurrence of rootlets or small and fragmented parts of tree trunks in coalified condition or in the form of pyrofusinite do not necessarily guarantee autochthonous process of a coal seam formation in subsequent periods. The same argument would generally hold good to preclude underclay or seat earth as evidences of autochthony.

The present author considers that the plant fragments composed of cellulose of lignin bearing tissues on falling a forest floor would be acted upon by various biochemical and chemical agents owing to which the cellulose derivatives would be segregated and form a chemical sediment that would be deposited in a depression in a solution phase as chemical sediment. The residual lignin-enriched cellular tissues, mingled with variable proportions of clay and sticky plant ingredients such as spore, cuticle, wax and resin would form clastic sediment and deposit in a basin forming durainous lithotypes, sometimes grading into argillaceous units of sedimentary deposits.

Evidences support that the basin or depression of vegetal accumulation were originally constituted low laying forest floors, frequently ravaged by flood prior to vegetal deposition. It has been conceived that flood water caused temporary concealment of some of the forest trees of low laying areas by sands which consequently suffered petrification owing to replacement of wood tissues by silica-rich solution. In a subsequent stage, the loose sand covering the devastated forest was duned exposing the silicified tree trunks and stigmaria over which a new generation of vegetal debris of allochthonous affinity were accumulated.

REFERENCES

Sen, Subhasis, 1982, Origin and evolution of layered structure in coal seam, Quart. Jour. Geol. Min. Met. Soe. India, v. 54, u 3 and 4, p. 18-27.

Teichmuller, M., 1962, Die genese der kohle, c.r.4, Congr. Intern. Strat. Geol. Carbonifere, Heerlen, 1958 3, Maastrich, p. 699-722.

SEN, D. K., GSI, Calcutta, India

On Techniques of X-Ray Microanalysis of Indian Bauxite from East Coast

Independent grains from three bauxite samples from the east coast of India, collected by Sri P. I. Raman, were studied by an MS46 microprobe X-ray microanalyzer. The main objective of the investigation was to examine the general belief of unsuitability of microanalysis for bauxites because of their softness and porosity. Special precautions were taken in order to address these difficulties. The samples were sent to BRGM, FRance, for sample preparation. Microolitic/oelitic textures have been the main subject of study. Amon the frequently occurring phases the pure gibbsite phase has been identified and the goethite phase is always found to be associated with a reasonably high concentration of gibbsite and a small amount of rutile. The other phase examined is ilmenite with good amount of rutile and small amount of gibbsite. Each analysis was based on an average of 10 point scanning. The uniformity in each of the observations and the fitting of the modal pattern indicates success of good quantitative analyses of bauxites by adopting special procedure for sample preparation.

Grainwise phase distribution pattern revealed by the analysis is shown in Table 1.

Table 1.

Mineral %	Composition of the Grain									
	1	2	3	4	5	6	7	8	9	10
Goethite	-	-	73.47	-	63.49	-	50.25	60.32	-	-
Gibbsite	3.39	100.00	22.48	100.00	33.31	1.67	47.49	35.39	1.12	100.00
Ilmenite	72.00	-	-	-	-	73.20	-	-	68.37	-
Rutile	24.60	-	4.06	-	3.20	25.13	2.26	3.79	30.00	-

SEN, S. and A. BANDYOPADHYAY, Calcutta University, Calcutta, India

Structure Style of Delhi (Ajmer–Pindwara) Fold Belt, Rajasthan, and Time Relation of Metamorphism and Deformation

The area lies within tectonic zone 3 of the Ajmer–Pindwara fold belt (Sen, 1981) in the Aravalli Range, Northwestern Indian Shield. It is occupied wholly by the Delhi Supergroup (Heron, 1953) and to its east is in contact with the Archaean BGC country.

The metasedimentaries of the Delhi Supergroup, comprising quartzite, metapelites and psammopelites, cal-granulites, calc-gneisses, amphibolites and marbles, and contemporaneous intermediate or basic metavolcanics, are regionally folded into a tight overturned synform by F_2, the Jaltakhera (25°57′N:74°18′E) fold with a north-south or north-northeast–south-southwest-trending axial trace. East of Kashipura (26°11′N:74°29′E) this axial trace swerved east into a north-facing antiformal bend midway between Lir(26°13′N:74°34′E) and Shamgarh (26°07′N:74°26′E) antiformal hingers, which describe the last regional fold F_3. An earlier set of folds, F_1, is represented by minor isoclinal rootless intrafolial folds whose well-formed axial plane schistosity S_1 is parallel to the bedding S_0, except at the hinges, and serve as the form surface of all later folds. A regional F_1 fold was mapped in the field. It is rotated along the hinge of F_2 at

Jaltakhera. The F_3 fold is nonplanar, noncylindrical, as evident from the unsystematic variation of the plunge of the major Liri–Shamgarh antiform, measured at different structural levels. The eastern limbs of the Shamgarh antiform and the synform east of Liri are cut off by a narrow (300–500 m) shear zone that separates the Delhi sequence from the older Archaean rocks to its southeast. The synform has been systematically broken up by a set of synthetic faults that trend at an angle of more than 45° with that of the shear zone.

Measurement of the megascopic and mesoscopic folds show the fold to be of flexural slip origin with a superposed flattening that apparently increases southeastward. These flattened folds fall close to or within the field of class 1C.

The metasedimentary-metavolcanic group is intruded toward the late stages of folding by ultramafic rocks (post-F_1 but pre-F_2) and granitic rocks (pre- to post-F_3). Peak of metamorphism seems to have been attained after the close of the F_2 phase, but prior to the F_3 folding. The F_3 folds cause disruption or bending of all such minerals as garnet, staurolite, and kyanite. This indicates a time gap before the initiation of the last fold generation.

SEN, S., Central Fuel Research Institute, Nagpur, India

Critical Analysis of Concepts of Plate Tectonics and Expansion of Earth

The various theories put forward to explain fragmentation and segregation of crustal blocks suffer from a serious flaw, namely, the rigid condition of the mantle that would resist widespread mobilistic phenomena. In the model of plate tectonics it has, however, been conjectured that the mantle on a long time scale would behave like a fluid body.

A recent study (Sen, 1986) posits that the condition of the mantle of the primordial earth was indeed in a mobile state owing to incorporation of ocean-forming water under ultra-high pressure condition, which has been deduced from the model of the oceanless condensed earth as conceived by Hilgenberg (1933) by reducing the radius of the earth by two- thirds of its present value. This view is based on experimental studies of silicate rocks under hydrothermal and high-pressure condition confirming considerable increase in mobility of such rocks owing to alteration of melting point (Roy and Tuttle, 1961), which greatly simulate the condition of the primordial mantle rocks of the condensed oceanless earth. This vital deduction, in contrast to the conjecture of plate tectonics, is based on experimental evidence and supports widespread mobilistic phenomena such as expansion of the earth.

It can be further conceived that the expansion of the planet caused development of cracks over the sialic crust through which, with progress of expansion, extensive molten mantle rocks disgorged forming ocean basins and releasing a large quantum of ocean-forming water from the mantle. It is the release of water from the mantle that caused the formation of oceans as well as gradually turned the mantle material into a rigid mass. A stable and dimensionally unaltered earth with ocean water already segregated from the mantle in the primordial stage of the earth, as conceived in the concept of plate tectonics, leaves no scope for considering sufficiently fluid or mobile condition of the mantle in which plate movements can be contemplated.

The earth expansion theory essentially considers a uniform pattern of sialic envelope over the primordial mantle - a reasonably justified and expected condition in the primal stage of the planet that is in fair agreement with the condition of the outermost layer of most other planets. Continental drift and plate tectonics, on the other hand, suggest that in the primordial stage the sialic land masses were united as a single continent surrounded by a solitary ocean—a condition that cannot be reasonably accepted, especially in the initial and juvenile stages of the planet. It is indeed most rational to consider that after formation of the earth with a solid crust, it was covered by an envelope of fairly uniform nature rather than segregating into two discrete contiguous units of continent and ocean.

The earth expansion theory further contemplates that if the radius of the earth could be reduced to two-thirds of its present value, all the continental landmasses would adjust themselves in a perfectly snug-fit manner, covering the entire part of the oceanless globe. This is highly significant since in the condensed globe the continents are not only appropriately adjusted solving the crustal jigsaw puzzle but they also cover the entire part of the condensed globe. In doing so the total surface area of the condensed globe exactly balances the total surface area of the detached continental fragments together with epicontinental seas. Such a vast array of significant matching incidences cannot be explained as chance events, but must have a meaningful bearing on the earth's dimensional aspect—evidently pointing to the expansion of the earth.

The occurrence of midoceanic ridges girdling some continental fragments in a parallel manner cannot be appropriately explained by transgression of the continental plates that would, on the contrary, obliterate such remarkable parallelism. In case of earth expansion theory occurrence of such expansion cracks or midoceanic ridges are expected through which basaltic magma would extrude and spread on either side keeping a close balance with the process of expansion and development of oceans.

A further evidence decidedly favoring the expansion in preference to plate transgression is the fitting of the detached continental masses in an imperfect manner over the present crustal surface. It is quite significant to note that all such arrangements are invariably marked by gaps and overlaps. This constraint can easily be avoided by inducing a little curvature or bending over the detached units, which evidently means that these fragments were originally arranged over a condensed globe.

Comparatively younger age of the ocean basins with the youngest rocks occurring in the vicinity of midoceanic ridges can be rationally explained by earth expansion theory.

Considering the geometric problem involving movements in a sphere it has been shown by Rajarajan (1977) that the plate tectonics hypothesis is not in conformity with the mathematical concept of velocity of motion and structural analysis on a curved surface. According to him, the mechanism of slow three-dimensional expansion and consequent radial increase in the earth's surface adequately explains many crustal features. In a recent work, Weijermars (1986) has also pointed out that slow global expansion can appropriately explain the development of surface features of the globe.

REFERENCES

Hilgenberg, O.C., 1933, Vom wachsenden Eedball: Berlin Selbstverlag.

Rajarajan, K., 1977, Review of plate tectonic hypothesis and continental drift in relation to Gondwanaland, in B. Laskar and C.S. Raja-Rao, eds., 4th International Gondwana Symposium, Calcutta, v. 2, p. 751-755, Hindusthan Publishing Corporation, New Delhi, 1979.

Roy, R., and O.F. Tuttle, 1961, Investigation under hydrothermal conditions, in L.H. Ahrens, K. Rankama, and S.K. Runcorn, eds., Physics and Chemistry of the Earth, v. 1, p. 138-180.

Sen, S., 1986, A new concept on the earth's interior, geomagnetism and crustal fragmentation: Proceedings, International Symposium on Neotectonics in South Asia, Survey of India, Dehra Dun, p. 83-94.

Weijermars, R., 1986, Slow but not fast global expansion may explain surface dichotomy of Earth: Phys. Earth Planet. Inter., v. 43, p. 67-89.

SEN, S., Central Fuel Research Institute, Nagpur, India

Structure and Dynamics of Earth's Deep Interior: Relevance of Reverse Gravity

Serious discrepancies have been noted in the prevalent concept of three-dimensional aspect and distribution of temperature, pressure, and gravity in the deep interior of the earth.

The trend of gravity in the interior of the earth has so far been considered to be directed inwardly, giving rise to zero gravity at the center. At that point, instead of weightlessness or absence of pressure, enormous pressure and temperature have been deduced by extending the crustal values that seem incongruous. This serious lacuna can, however, be solved by considering the presence of an additional force of reverse gravity in the deep interior of the earth caused by the attraction of the mass of superincumbent layers (Sen, 1986). Existence of such a force in nature, acting in a direction opposite to gravity, has recently been suggested by Fischbach et al, (1986). The present author considers that the trend and magnitude of gravity in the planet's deep interior are decided by the resultant of normal gravity and reverse gravity. While over the crustal surface owing to absence of effective superincumbent mass the reverse gravity would be zero, at the center of the earth where the normal gravity is zero, the resultant gravitational force would be directed toward the crustal surface, thereby inducing a low- pressure zone around the core.

In contrast to the prevalent concept, it has been considered that the solid inner core, composed of iron, is a permanent magnet suspended freely in a nonsolid geosphere. Besides the solid nature of the core and magnetic behavior of the earth, the view is strongly supported by the structure and texture of iron meteorites, especially by the presence of Neumall lines in them that are destroyed at 800 °C.

It has also been contemplated that toward the middle portion of the mantle, owing to convergence of pressures from opposite directions generated by normal and reverse gravity, a highly compressed zone would be developed that would be periodically disturbed by the gravitational pull induced by the moon. The friction thus produced owing to interaction of gravitational forces at the relevant surface would perpetually generate heat energy that would escape to the crustal surface as the terrestrial source of thermal energy.

It has been suggested by Hilgenberg (1933) that if the radius of the earth could be reduced to two-thirds of its present length, all the continental fragments would be adjusted in a perfectly snug-fit manner. The present author

(Sen, 1986) has pointed out that since the condensed globe was devoid of oceans, the primordial mantle must have been considerably fluid and expansile owing to incorporation of ocean-forming water under ultra-high pressure. This vital clue has been based on experimental studies of silicate rocks confirming depression of melting point under hydrothermal and high- pressure condition (Roy and Tuttle, 1961).

In the interior of the earth, occurrence of a major fluid zone at great depth sandwiched between two solid geospheres of rocky outer shell and metallic inner core poses major constraints that also can be solved by the earth expansion theory. It has been suggested that in the condensed earth, the semifluid mantle and the solid core were juxtaposed and the nonsolid geosphere, matching remarkably with the magnitude of expansion in regard to thickness, was developed in subsequent periods owing to the earth's expansion.

The expansion of the planet appears to have taken place owing to gravitational attraction from extraterrestrial planetary bodies causing rupture of the core-mantle conjunction, swelling or upliftment of the semifluid mantle, development of expansion cracks over the solid sialic envelope, and segregation of the fragmented crustal units. Owing to further expansion, effected in a gradual manner, a void zone along the disrupted core-mantle conjunction was developed. Concurrently with the expansion, the zone was filled up with finer particles and volatiles, derived from the adjoining surfaces, thereby developing into a major geosphere of nonsolid or pseudofluid characteristics.

It can also be conceived that through the expansion cracks over the crustal surface a large quantum of water emerged associated with widespread volcanism and degassing of volatiles forming ocean bodies. In the innermost part of the mantle cracks may also develop through which volatile products are likely to escape to the pseudofluid geosphere. During the younger geological periods, it is evidenced that the mantle itself became sufficiently rigid owing to desiccation and, hence, incapable for further widespread expansion.

Following the basic principles of physics, it can be inferred that owing to the earth's rotation, the solid core formed an ellipsoid with its major axis originally coinciding with the axis of rotation while the semifluid mantle tended to bulge along the original equatorial region.

REFERENCES

Fishbach, E., et al, 1986, The Fifth Force: Reprint, Proceedings XXIII International Conference on High Energy Physics, Berkeley, California, U.S. Department of Energy, p. 1-11.

Hilgenberg O.C., 1933, Vom Wachsenden Erdball: Berlin, Geissmann and Bartsch, p. 56.

Roy, R., and O.F. Tuttle, 1961, Investigation under hydrothermal conditions, in L.H. Ahrens, K. Rankama, and S.K. Runcorn, eds., Physics and Chemistry of the earth (2nd ed), v. 1, p. 138-180.

Sen, S., 1986, A new concept on the earth's interior, geomagnetism and crustal fragmentation: Proceedings International Symposium on Neotectonics in South Asia, Survey of India, Dehradum, p. 83-94.

SEN, S., and R. DAS, Calcutta University, Calcutta, India

Eastern Boundary Shear Zone of Central Aravalli Range, India: Its Tectonic Significance

In the Central Aravalli sector between parallels 26°45 'N and 25°15 'N, the contact between the youngest Proterozoic fold belt comprising Delhi Supergroup rocks and the Archaean BGC (Banded Gneissic Complex) country is a narrow shear zone averaging 500 m wide.

The shear zone shows all the rock types of the Delhis with some occasional slices of BGC. There is some difference in composition in the late phase intrusives from thos of Delhi rocks. Presence of a thin banded association of harzburgite, pyroxinite, hornblendite, scaly serpentinite, metabasics (amphibolites), granitoids including plagiogranites, and a late dyke of olivine picrite resembles an ophiolite sequence. There is no true high-pressure mineral in any of the assemblages, but the presence of stilpnomelane in hornblendite, of a light blue colored hornblende in amphibolites or calsilicate rocks, and the preponderance of zoisite in the last two groups oint to a moderately high pressure environment.

It is a dominantly ductile shear zone. Its strain history is different from that of the ofld belt of which it forms an integral part. In the field, the shear zone is seen to consist of a number of specific zones of variable width and extent that are highly sheared into mylonites and ultramylonites. They occur in curved trends forming a conjugate set cutting at small angles, and they occur in an asymmetric anastomosing intersecting pattern. The enclosed think elongated lenses exhibit considerable flattening seen in the early and contemporary structures. In the individual shear domains a strong non-coaxial strain is proved by the S-C mylonites, which show a gradual convergence of the S and C planes with progressive strain, ribbon quartz, rotated prophyroclasts, asymmetric fold trains in mylonites, fold axes rotated on the axial surface of the folds, structures like mic-fish, and asymmetric pull-apart structures, recrystalized tails of porphyroclasts, and pressure shadows. But the overall pattern of the anastomosing shear domains cannot be explained by a purely non-coaxial deformation. Under coaxial strain, conjugate shear zones or C bands in mylonites develop symmetrically and remain symmetrical with progressive strain. In the present case, the pattern of the individual shear zones and the flattening of the shear lenses indicate a deformation within the "flattening field" of Gapais et al (1987).

The pattern is more pronounced close to the horizontal surface. There was thus a coaxial element acting roughly along a horizontal plane that was superimposed on the non-coaxial flow.

Narrow zones of thin banded ultramafic and associated rocks described as ophiolites occur nearly parallel to the regional shear zone. They locally cut across the anastomosing mylonite bands and are invariably bordered by a later generation ultramylonite. Within each such narrow zone, the rocks occur in repeated partial or full sequence, separated by ultramylonites. These are interpreted as imbricate thrust slices that have been emplaced piggy-back on the two bounding thrusts of the packets. The thrusted nature of the packets is evident from the contrasting sense of shear, sinistral, and dextral, respectively, in the northwest and the southeast walls, on nearly vertical sections looking north.

The petrographic associations and the overall map feature of tectonic mixing of rock units indicate that the belt is a subduction-related melange as proposed earlier by Sen. The reconstructed tectonic model, helped by the aspects of structural features, their time sequence, and their symmetry, differs from that of Sen in some details but follows his arguments with regard to the broad outline of the model and its basic derivation from the plate tectonic paradigm.

The contribution of a Proterozoic subduction zone in influencing the specific characters of the belt seem well established. On this premise, the pure shear part of the stress history is ascribed to an oblique approach rather than a head-on collision of the two involved lithospheric blocks; the northwestern carrying the Delhi fold belt on its back, the other to the southwest, bearing the older BGC country. It explains a flattening by transpression with a concomittant non-coaxial strain producing the described features. This is also evident from the pattern and symmetry of the shear zones suggesting a bulk sinistral shear. Part of the squeezed, deformed, and folded prism of rocks of tectonic zone 3 was being dragged down by the subducting underthrusting oceanic crust below the BGC country, with a concomittant emplacement by piggy-back thrusting of slices of oceanic crust on composite thrust sheets.

SENDEROV, E. E., Vernadsky Institute of Geochemistry and Analytical Chemistry, Moscow, USSR, and P. P. HEARN, Jr., U.S. Geological Survey, Reston, VA

X-Ray Structural State Analyses of Secondary K- Feldspars

Numerous occurrences of sediment-hosted secondary K-feldspar, variously described as authigenic, epigenetic, or hydrothermal, have been reported in the geological literature during the past 75 years. For many years, many such occurrences were assumed to be either syngenetic or early diagenetic in origin. However, recent $^{40}Ar/^{39}Ar$ and K/Ar analyses, together with petrographic and fluid inclusion studies, strongly suggest that many occurrences of abundant secondary K-feldspar were formed by the migration of sedimentary brines mobilized by tectonic activity. These findings have focused new attention of the chemistry and crystal structure of secondary K-feldspars, both with regard to its usefulness as a chemical indicator, and also with regard to the behavior of argon within the lattice.

We report here the unit cell parameters for a suite of 16 samples of K-feldspar, including authigenic overgrowths and their detrital cores, metasomatized clays, and homogenous hydrothermal vein and vug filings. Purified separates of authigenic overgrowths were produced by a combination of air- abrasion and gravity-settling techniques. Other samples were either analyzed in bulk (metasomatized clays) or hand picked to remove contaminant phases (hydrothermal vein and vug-filling samples).

The most distinctive characteristic of these feldspar samples is the relative imperfection of their crystal structures. The presence of extremely small crystalline particles and domains, the variable structural state of different domains, and the simultaneous existence of triclinic and monoclinic phases in the same samples results in broadening of primary X-ray reflections and lowers the precision of the unit-cell determinations.

While all of the samples of secondary K-feldspar were close to end-member orthoclase in composition, the degree of ordering was highly variable, ranging from sanidine to maximum microcline. Most samples of overgrowths were mixtures of well-ordered microcline and orthoclase (both phases having $t_1o + t_1m \geq 0.9$). The simultaneous presence of two phases indicates the inversion of one phase into the other; the relative absence of intermediate microcline in the mixtures suggests that this transition is rather abrupt. The structural character of detrital cores appears to have little influence on that of subsequently formed overgrowths.

Secondary K-feldspars form under physiochemical conditions where maximum microcline should be the stable phase. Consequently, the occurrence of secondary sanidine, orthoclase, microcline + orthoclase, together with monomineralic microcline can only be explained by widely variable kinetic conditions during crystallization. It is difficult to imagine that ordering could occur in the solid state at the relatively low temperatures (< 200 °C) of formation for secondary K-feldspar. Consequently, the existence of mixtures of disordered secondary microcline and orthoclase suggests a relatively long period of crystallization, whereby more ordered atomic layers are formed initially and then are closed off from solutions by subsequent crystal growth. Thus, fluctuation of the various physicochemical factors affecting the speed of crystallization will ultimately determine the degree of ordering.

Sample	Location	Formation	Host Lithology	Host Age	Type	Phase Composition
1	Scotland	Fucoid Beds	dolomitic argillite	Cam.	overgrowth	orthoclase + microcline
2	"	"	"	"	detrital core	microcline + minor orthoclase
3	Vermont	Winooski	dolomite	Cam.	overgrowth	microcline
4	"	"	"	"	detrital core	microcline
5	Tennessee	Maynardville	dolomite	Cam.	overgrowth	microcline + minor orthoclase
6	"	"	"	"	detrital core	orthoclase + minor microcline
7	"	"	"	"	altered clay	orthoclase + minor microcline
8	Missouri	Bonneterre	dolomite	Cam.	overgrowth	orthoclase
9	"	"	"	"	detrital core	microcline + minor orthoclase
10	Texas	Hickory	sandstone	Cam.	overgrowth	orthoclase
11	"	"	"	"	detrital core	microcline + minor orthoclase
12	Arkansas	Womble	shale	Ord.	vein	orthoclase
13	Pennsylvania	Beekmantown	dolomite	Ord.	vein	microcline
14	Zambia	Mwashia	dolomite	Precam.	discrete crystals	microcline + orthoclase
15	Wisconsin	Deicke	bentonite	Ord.	altered tuff	sanidine
16	Siberia	Yukhtinskiy	mudstone	Jur.	vug-fill	sanidine

SENGOR, A. M. C., and Y. YILMAZ, İTÜ Maden Fakûltesi, Jeoloji Boľûmû, Teçsvikiye, Istanbul, Turkey

Strike-Slip Faulting in Orogenic Processes: Significance and Example from Tethysides

The presence and importance of strike-slip faulting parallel with the trend of orogenic belts has been noted early in the twentieth century, and plate tectonics has provided a rationale for its place. Despite this, cross-sectional thinking dominated early plate-tectonic models of orogeny and methods have been developed to construct balanced cross- sections across entire orogens so as to deduce deep structure. In recent years, sideways motion along orogenic belts has again become popular, mainly because of the longitudinal motion of numerous crustal fragments along the North American Cordillera, called "terranes."

The purpose of this paper is to document the late Paleozoic-Cainozoic evolution of the Middle Eastern Tethysides, in which there has been major strike-slip motion along the trend of the orogen of both senses and in various orientations. The main tenet of the model developed here is to study the Pontide/Dzirula/Adzharia-Trialeti/ Artvin-Karabagh/Sanandaj- Sirjan zones, collectively a north-northeast facing Paleo- tethyan arc (the "Podataksasi zone"), located originally to the west-northwest of the present northwest and central Iran. This arc began to rift from Gondwanaland toward the end of the Permian, thus opening the Zagros/Oman branch of Neotethys. Following the accretion of the Podataksasi zone plus the northwest and central Iranian fragments as a unified monolith (part of the Cimmerian Continent) to Laurasia, strike-slip faulting began disrupting it internally. The origin of this strike-slip faulting remains unclear. The Podataksasi zone began moving left-laterally, thus ripping a central Iranian microcontinent away from northwest Iran and opening the central Iranian minioceans (Nain-Baft, Sabzevar, Sistan). Continuing motion of the Podtaksasi zone rotated the central Iranian microcontinent nearly 130° counterclockwise and brought the Sanandaj-Sirjan zone to its south and southwest. This multiblock mosaic of Iran was further disrupted during the latest Mesozoic compressional episodes by extensive strike-slip faulting, and, by the medial Cainozoic, the present gross geometry of Iran had formed.

In late Mesozoic time, the easternmost extension of the Apulian platform (the Bitlis and Pûtürge massifs) had collided with eastern Arabia and was subsequently strike- slipped northwards while opening the Maden marginal basin complex. This basin was shut in the Eocene by collision and was accreted to Eurasia. In the middle-late Miocene, Arabia collided both with Iran and with Anatolia and generated the presently active tectonic escape regime pushing a considerable chink of Turkey westwards and redisrupting Iran internally. The best-known products of this last episode are the North and East Anatolian, Great Kavir, and the Main Zagros strike-slip systems.

Although the collision-related and subduction-related strike-slip systems are easy to understand, the origin of the very large offset strike-slip systems, such as the one that displaced the Podataksasi zone for nearly 200 km during the course of the Mesozoic, are more difficult to comprehend. They imply abrupt changes in the motion of plates larger than simple fore-arc slivers and may suggest a control of buoyant objects on sizeable plates.

An important implication of this study is that cross- sectional area in orogenic belts is not conserved during any stage of orogeny. Consequently, balanced cross-sections across entire orogens cannot be drawn unless one has established that material was not lost from the cross- sections. Our contention, derived from our study of the Middle Eastern Tethysides, is that material is almost always lost from cross-sections and that in places strike-slip faulting may dominate orogenic processes.

In order to see whether this is so on a still larger scale, we have looked at the entire Tethyside orogenic complex from the Atlantic Ocean to the Pacific. We considered motion only along sutures. Our analysis has revealed that over 60% (by length) of the Tethyside sutures had considerable strike-slip motion on them. More than 25% had more than 1000 km of motion and over 50% had more than a few hundred kilometers. At least 15% of them had more than one major episode of motion. Many had a number of episodes with opposing senses.

We conclude that modeling of orogenic belts must be based on careful and detailed field observations, simply because mass calculations cannot be made on assumptions of mass conservancy in certain volumes. We underline the role of paleobiogeographic and paleomagnetic studies in supplementing classical field geological studies to establish orogen- parallel strike-slip faulting of various kinds and in various structural levels.

SENGOR, A. M. C., and N. GORUR, ITU Maden Fakultesi, Jeoliji Bolumu, Tesvikiye, Istanbul, Turkey

Tethyan Oceanic Connections From Mediterranean to Southeast Asia

Tethys or the Tethyan domain denotes a large, triangular, eastward-gaping oceanic embayment of the Pahnagea together with its continental margins and smaller continental fragments that populated the oceanic floor. Paleotethys was the original gap that formed when the late Paleozoic (Hecynide in the Americas and western Europe/northwest Africa and Altaid in Asia, Fig. 1) continental collisions gathered Pangaea in late Carboniferous-Permian time. In late Permian time, a continental strip was rifted off the northern margin of Gondwanaland between the Balkans and Thailand. It can be shown that in many places (e.g., Greece, Turkey, Iran, the Pamirs, Thailand), this rifting event was caused by back-arc spreading over a Gondwania-ward dipping subduction zone. In the Mediterranean and the Middle East this rifting opened, from west to east, the Pindos/Karakaya marginal basin (now sutures 15 and II, Fig. 1), the Zagros/Oman branch of Neotethys (25), and Neotethyan oceans south and southwest of the Bajgan/Deyadar/Deh Salm arc fragments in Iran. Farther east, this activity first opened the Waser/Rushan-Pshart/Banggong Co-Nu Jiang ocean (IX, X, XXIV), and rifted the Farah (1), northern Central Pamir(s), and western Qangtangs (s''), and also possibly the Sibmasu (s''') blocks from Gondwanaland. These fragments probably collectively formed a thin and long continent, the Cimmerian Continent. Coevally in the east and somewhat later westwards (middle Triassic in the Mediterranean), the Kirsehir(h), Apulia(e), northwest Itan (j), the Lut(k), Helmand (m), Lhasa (t), and the Burmese (also t) blocks also detached from Gondowanaland along the Neotethyan sutures (18, 19, 20, 25, 27, 28, 29, 30, 31, 32). One continental block, the Mount Victoria Land (not shown on Fig. 1, located just to the south of 32), remained attached to the Exmouth plateau. This block rifted off the Australian continental margin in Callovian time.

The continental blocks farther east in the Cimmerids collage, now forming much of China, had been also detached from Gondwanaland, but during the middle Paleozoic. Their agglomeration around the Altaid nucleus of Asia began in Carboniferous-Permian time and lasted until the latest Triassic. The seaways that existed in China in Mesozoic time largely sat along the foreland flexural troughs associated with the sutures of these latest Paleozoic-Triassic zones of collision. These molasse basins localized much coal deposition during the Triassic-Jurassic interval. By contrast, evaporites were widespread behind the northern Paleotethyan sutures (VII, VIII, XI, XIII, XIX, XIV, XVI), possibly because of the rain shadow generated by the suture mountains.

The closure of the Paleotethyan sutures west of China (and partly in China, in northern Tibet) was dominantly an early Jurassic affair. Only in northern Turkey was the closure delayed until the middle Jurassic, except for the Karakaya ocean that shut in the late Triassic. In Turkey and westwards, newer branches of Neotethys (9, 11, 12, 13, 14, 17) opened in Sinemurian–Hettangian time, simultaneously with the opening of the Atlantic Ocean-related Tethyan basins (2, 3, 4, 5, 6, 7, 8), thus establishing the initial equatorial oceanic connections. At this time, the equatorial oceanic circulation may have been maintained along both the juvenile Neotethyan oceans and along the remnant Paleotethyan dependencies (IX, X, XXIV). This accounts for the amazing similarities among the sedimentary successions of the various Tethyan basins at large. These similarities have been misinterpreted frequently in terms of single basin models (e.g., in Turkey, the "internal" versus "external" interpretations of the Antalya and the Cypriot nappes).

During much of the Jurassic, quiet carbonate platforms surrounded by deep oceanic basins have dominated the Tethyan scenery except in the north. The Helman and the Lhasa parts of the Cimmerian continent collided with the rest of Asia in early Cretaceous time and sealed off all of the Paleotethyan remnants, thus limiting deep circulation only to Neotethyn oceans. Shallow-marine connections survived atop the amalgamated blocks and gave rise, in the middle Cretaceous, to the widespread carbonate rocks that reach from the Mediterranean area to Burma.

In late Cretaceous time, extensive ophiolite obduction onto some of the "free-floating" continental blocks (e.g., Kirsehir, h) but mainly to the Neotethyan southern margin (from Turkey to Oman, possibly beyond) took place creating extensive marginal fore-deep basins. This was in places accompanied by island arc-continent collisions both along the south Tethyan margin (e.g., southeastern Turkey) and along the north Tethyan margin (e.g., Kohistan/Ladakah). These local collisions did not have a major effect on the oceanic circulation.

The main Tethyan influence on oceanic circulation came after the Mesozoic, by the beginning major collisions in the Tethys (Paleocene-Eocene and Miocene-Pliocene) that may have influenced the formation of the present ice caps.

Figure 1—Simplified tectonic map of Eurasia showing areas of consolidation and distribution of Tethyside sutures.

SENGUPTA, S., Jadavpur University, Calcutta, India

High-Temperature Mylonitization in Association with Regional Folding, and Migmatization in the Schirmacher Hills, Queen Maud Land, East Antarctica

The Precambrian basement rocks in the Schirmacher Hills, situated between 70° 44′33″S and 70°46′30″S latitudes and 11°22′04″E and 11°54′00″E longitudes, show several generations of folding, development of shear zones, and two distinct metamorphic-migmatitic events. The rocks of this region have undergone an initial granulite-facies metamorphism closely followed by charnockitization. Very coarse hypersthene- and garnet-bearing pegmatites are emplaced as parallel bands within the proxene-granulites and convert the whole rock into a migmatite. This event was followed by a diapthoresis under amphibolite-facies conditions and regional granitization leading to the development of the majority of the quartzofeldspathic gneisses of the area.

Some of the porphyroblasts of a tightly crenulated Si, marked by included needles of sillimanite. These early crenulations have been entirely obliterated in the matrix of the rock, where a well-defined gneissosity is marked by parallel orientation of sillaminite grains and very long grains of garnet that grew parallel to the crenulation cleavage in the matrix. The crenulated Si (*) and the later-formed external foliation (S_1) mark out the two phases of an early deformation (D_1), when the rocks were still in the granulite facies of metamorphism. The banding in the pyroxene-granulite and the migmatitic-banding in the charnockitic rocks is parallel to S_1. During the second deformation D_2, S_1 in all the rocks was intensely deformed, with the development of two sets of broadly coaxial isoclinal folds (F_{2A} and F_{2B}) with the axis having a southwesterly pitch of around 60° on the moderately dipping dominant foliation (S_2) of the gneisses. S_2, the dominant foliation of the area, is axial planar to the F_2 group of folds. The D_2 deformation was followed by two other generations of folding, F_3 and F_4, of which F_3 is coaxial to F_2, but with steep north-south striking axial surfaces, while 4 folds have steep, east-west trending axial surfaces with subhorizontal axes.

The area is dissected by numerous shear zones ranging from the microscopic-scale to a few meters in width. There are mainly three types of shear zones: (1) ductile shear zones at low angle to the axial surfaces of F_2 folds with mylonitic foliation essentially parallel to the dominant gneissic foliation S_2 and mylonitic lineation subparallel to the F_2 axes, (2) steep or subvertical ductile shear zones trending north-northeast and east and at high angles to the F_2 axial surfaces or S_2, (3) discrete shear-surfaces cutting across all the earlier structures. The ductile shear zones anastomose around somewhat less deformed lozenge-shaped pods. The mylonitic banding in the intensely sheared rocks is often folded and a new axial planar foliation occasionally cuts across the hinge zones of the folds. This new foliation usually shows all the microscopic characters of a mylonitic foliation.

There has been extensive granitization synchronous with D_2 deformation and with the development of the ductile shear zones. Pegmatitic veins with diffuse borders are profusely emplaced along the shear zones and the pegmatites themselves are mylonitized to different extents. The ductile shear zones were often intersected by cross-cutting veins of pegmatites indicating thereby that granitic activity continued after the cessation of ductile shearing. Discrete bodies of pegmatites were emplaced also along the discrete shear fractures that were the last to develop.

The major part of ductile shearing must have taken place when the rocks were still under amphibolite facies conditions. In the sheared calc-gneisses, large megacrysts of diopside often show a core-and-mantle structure with small equant grains of diopside recrystallization from the strongly strained peripheral parts of the porphyrocrysts. In some of the ultramylonites, very small euhedral crystals of garnet are found to neocrystallize from the fine-grained groundmass. The Fe-Mg partitioning of coexisting neocrystallized garnet and biotite indicate that the temperature of mylonitization was around 50 °C.

The microstructure of the quartzofeldspathic mylonite also indicates high temperature of mylonitization. These mylonites are characterized by the presence of long unstrained or weakly strained ribbons of quartz alternating with feldspar- rich bands of small equant recrystallized grains of feldspars. There are a few megacrysts of feldspar occurring as strongly elongate porphyoclasts. The occurrence of such unstrained quartz ribbons, many times larger than the recrystallized feldspar grains in the groundmass, is characteristic of high-temperature mylonitization. TEM studies show that the unbound dislocation densities of ribbons are very low. It is likely that the quartz ribbons developed by metadynamic grain growth with crystallization outlasting deformation. Postmylonitic feldspathization has caused the development of ovoid or globular, completely unstrained feldspar grains within a very fine grained mylonitic groundmass.

SENO, T., University of Tokyo, Tokyo, Japan

Evolution of Marginal Basins of Western Pacific

Marginal basins of the world are concentrated in the western Pacific. Most of them are back-arc basins of Tertiary age. It is known that back-arc spreading has been episodic; ages of the back-arc basin formation of the western Pacific can be categorized into three periods: Paleocene-Eocene, Oligocene- middle Miocene, and late Miocene-Present.

There have been a number of theories that tried to explain the back-arc basin formation; they are, for example, secondary flow beneath an island arc, anchored slab hypothesis, trench roll-back, and buckling of the slab. However, none of the theories seems to have succeeded in explaining all the above features.

I propose here a new idea for the origin of back-arc basins, in which back-arc spreading and midocean ridge spreading is not discriminated. Instead, a new classification, i.e., active spreading and passive spreading, is introduced (Figure 1). In active spreading, there is subduction toward the spreading center; the mid-Atlantic ridge and most of the back- arc basins of the western Pacific are of this type. In passive spreading, there is subduction away from the spreading center; the East Pacific Rise and the southeast Indian Ocean ridge are of this type.

We can notice that the active spreading centers at present are located in the low-order geoid-high region and the passive ones are, on the contrary, in the geoid-low region. I examine this relationship for the geological past by the reconstruction of the marginal basins of the western Pacific by rotating major continents with respect to hot spots (Engebretson et al, 1985, GSA Special Paper 206). It is found that almost all the back-arc basins were located in the geoid-high region when they were formed.

Hager et al, (1985, Nature, v. 313., p. 541) showed that the low-order geoid high corresponds to the low velocity in the lower mantle revealed by the recent seismic tomographic studies (e.g., Dziewonski, 1984, JGR, v. 89, p. 5929). Thus, the above correspondence between the geoid high and the active spreading suggests that the origin of the back-arc spreading may be deep-seated in the lower mantle or in the core-mantle boundary. This inference is further reinforced by the initiation timings of active spreading. They have a periodicity of about 30 Ma and can be correlated well with the 30 Ma period of the mass extinction and the high frequence of the magnetic reversals (e.g., Raup, 1984, Nature, v. 314, p. 341). The instability of the D layer above the core-mantle boundary, which may have a time constant of tens of millions of years (e.g., Olson et al, 1987, Nature, v. 327, p. 409), can produce the high rate of heat conduction across the core-mantle boundary (Loper and McCartney, 1986, GRL, v. 13, p. 1525); this would produce the higher rate of the lower mantle convection, which may in turn trigger active spreading. This mechanism can explain why back-arc basins are located only in the western Pacific and also why there are no back-arc basins of Mesozoic or older age. This is because the back-arc spreading occurred only when the subduction zones surrounding the major continents entered the high-geoid region of the western Pacific after the splitting of Pangaea in the Mesozoic.

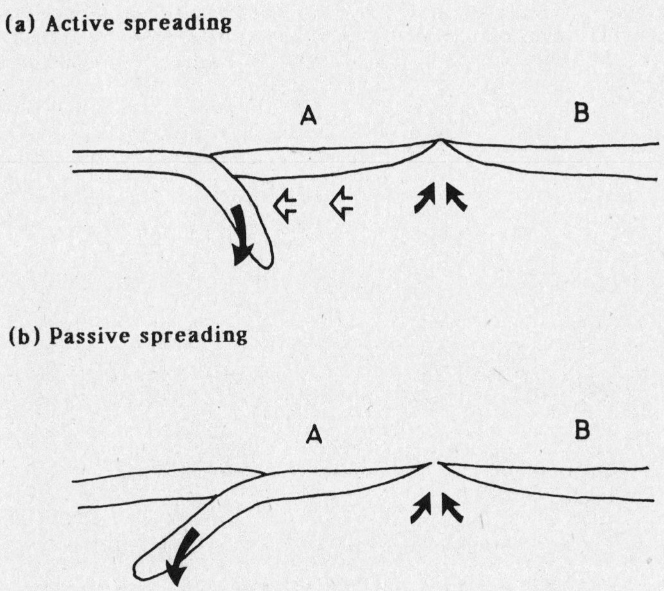

Figure 1. Two modes of seafloor spreading.

SENO, T., University of Tokyo, Tokyo, Japan, S. SEKIGUCHI, National Research Center for Disaster Prevention, Tskuba, Japan, and E. T. PETERSON, Mobil R&P, Dallas, TX

Pacific Plate Motions Calculated from Reconstruction of Subduction Zones and Ages of Plate In Western Pacific

We calculate Pacific plate motions for the past 50 m.y. on the basis of the location of subduction/collision zones and mid-ocean ridges and the age distribution of the Pacific plate. The geographical locations of the subduction zones are reconstructed by the rotation of major continents with respect to hot spots (Engbretson et al, 1985); the western Pacific marginal basins are closed at appropriate times when they opened. We identify subduction/collision on each segment of the boundary based on the geological data.

Midocean ridges are reconstructed based on the plate kinematic data of the Pacific ocean (Engbretson et al, 1985) and the magnetic anomaly data. For the early Tertiary, we also use the North New Guinea plate hypothesis (Seno, 1984), which maintains that another plate existed during the early Tertiary in the South Pacific, marked by the ridge-transform fault system from the Pacific plate. This ridge system is assumed to have been annihilated beneath the West Phillipine Basin and the Australian continent around 43 Ma.

The age distributions of the Pacific plate are estimated from the Pacific plate motions with respect to hot spots and the magnetic anomaly patterns. As the time dates back, there appears a vast oceanic area in the western Pacific for which no direct age data are available. We estimate the age for this area from the timings of ridge subduction at the western Pacific subduction zones and interpolation with the available magnetic anomaly data. The North New Guinea plate hypothesis is also utilized for the age reconstruction.

The plate driving forces, i.e., slab pull and ridge push forces, are calculated from the age of the plate at each segment of the subduction boundaries. The collision force of 200 bars is assigned to the collision zones normal to their strike. Summing up these forces along the boundaries, we obtain a torque vector of the Pacific plate. This torque is balanced with the slab resistance forces or mantle drag forces that are assumed to be proportional to the absolute velocity of the plate. Two types of slab resistance are conceived: One is the slab edge resistance in which the resistance forces is simply proportional to the plate velocity, and another is the slab side resistance in which the resistance force is proportional to the plate velocity multiplied by the slab length. The torque balance provides an angular vector of the absolute Pacific plate motion for each case of the resistance force. The reconstructed Pacific plate motions in this manner produce significant changes in pole location and can explain the observed fluctuated trend of the Emperor-Hawaiian chain.

REFERENCES

Engebreston, D.C., A. Cox, and R. G. Gordon, 1985, Relative motions between oceanic and continental plates in the Pacific basin: GSA Special Paper 206, 59 p.

Seno, T., 1984, Was there a North New Guinea plate?: Rep. Geol. Surv. Jpn, v. 263, p. 29-42.

SENO, T., University of Tokyo, Tokyo, Japan, S. SEKIGUCHI, National Research Center Center for Disaster Prevention, Tskuba, Japan, and E. T. PETERSON, Mobil R&P, Dallas, TX

Evolution of Subduction Zones and Marginal Basins of Western Pacific: Implications for Pacific Plate Motion

We present the reconstruction of subduction zones and marginal basins of the western Pacific for the past 50 m.y. Major continents are rotated with respect to hot spots using the parameters listed in Engbretson et al (1985). Marginal basins are closed at appropriate times, results from studies of each marginal basin are used for this reconstruction. We identify subduction/collision on each segment of the consuming boundary based on the geological data.

We also reconstruct the location of the midocean ridges based on the plate kinematic and magnetic anomaly data of the Pacific Ocean; the Pacific plate motions with respect to hot spots by Engbretson et al (1985) are used as the plate kinematic data. For the early Tertiary, we use the North New Guinea plate hypothesis (Seno, 1984), which maintains that another plate existed during the early Tertiary in the south Pacific, marked by the ridge-transform fault system from the Pacific plate. This ridge system is assumed to have been annihilated beneath the West Phillipine Basin and the Australian continent around 43 Ma.

We calculate the Pacific plate absolute motions as an application of the above reconstruction. The plate driving forces, i.e., slab pull and ridge push forces, are calculated at each segment of the subduction boundaries. Because the plate driving force is dependent on the age of the subducting plate, the age distributions of the Pacific plate are estimated from the magnetic anomaly patterns. As the time dates back, there appears a vast oceanic area in the western Pacific for which no direct age data are available. We estimate the age for this area by the timings of ridge subduction at the western Pacific subduction zones and interpolation with the available magnetic anomaly data. The North New Guinea plate hypothesis is also utilized for the age reconstruction.

The collision force of 200 bars is assigned to the collision zones normal to their strike. Summing up these forces along the boundaries, we obtain a torque vector of the Pacific plate. This torque is balanced with the slab resistance forces or mantle drag forces that are assumed to be proportional to the absolute velocity of the plate. Two types of slab resistance are conceived: One is the slab edge resistance in which the resistance force is simply proportional to the plate velocity, and another is the slab side resistance in which the resistance force is proportional to the plate velocity multiplied by the slab length. The torque balance provides an angular velocity vector of the absolute Pacific plate motion for each case of the resistance force. The reconstructed Pacific plate motions in this manner produce significant changes in pole location and can explain the observed fluctuated trend of the Emperor-Hawaiian chain.

REFERENCES

Engebretson, D.C., A. Cox, and R.G. Gordon, 1985, Relative motions between oceanic and continental plates in the Pacific basin: GSA Special Paper 206, 59 p.

Seno, T., 1986, Was there a North New Guinea plate?: Rep. Geol. Surv. Jpn, v. 263, p. 29-42.

SENSKE, D. A., and J. W. HEAD, Brown University, Providence, RI

Western Eisila Regio, Venus: Geology and Origin of Equatorial Highland

Introduction: Venera 15/16 imaging of the northern high latitudes of Venus, north of 30 °N, shows that the northern highland of Ishtar Terra is characterized by broad zones of compression (Barsukov et al, 1986; Crumpler et al, 1986; Vorder Breugge, 1987). In contrast, highlands in the equatorial region of the planet (e.g., Beta Regio, Aphrodite Terra) appear to be associated with zones of extension (Stofan et al, in press; Head and Crumpler, 1987). In order to compare the processes forming highlands in the equatorial region with those in the northern high latitudes, the morphology, geology, and geophysical characteristics of a number of highland features in the Venus equatorial region have been studied (Senske and Head, 1988a). In this study, we examine in detail the geologic and geophysical characteristics of the equatorial highland of Western Eisila Regio. We first describe the morphology and general structure of this highland using Pioneer Venus (PV) altimetry data. The local geologic history along with patterns of tectonic deformation are then examined from a radar unit map produced from the PV radar image and structural mapping from Venera 15/16 imaging. In addition, analysis of line-of-sight (LOS) gravity data is carried out in order to estimate an apparent depth of compensation. Finally, we use the geologic and geophysical information to assess formational models for this region and compare them with a formational model for the highland of Ishtar Terra.

Structure and morphology: Western Eisila Regio is the western most of a chain of equatorial Upland Rises characterized by broad radar-dark domal topographic rises, individual peaks, and extensive volcanism (Senske and Head, 1988b). This highland centered at 20 °N, 355 ° between Sedna Planitia and Guenevere Planitia is characterized by (1) an elongated topographic rise approximately 2400 km wide striking N 60 ° and (2) two peaks, Sif Mons and Gula Mons located on the crest of the topographic rise that rise 2.2 km and 4.0 km above a datum of 6051.0 km, respectively (Figure 1).

PV radar unit map: Using the PV radar image, seven radar units have been defined on the basis of radar brightness and texture and have been used to map Western Eisila Regio. This map shows the broad topographic rise of Western Eisila to be characterized by units ranging from Very dark in the east, showing characteristics similar to the surrounding plains, to Mottled bright in the west. Mottled bright units are associated with the summit and flanks of Sif Mons. The flanking Mottled bright deposits extend radially (averaging 315 km) in all directions from the summit, are lobate, and suggest the presence of a complex system of lava flows superimposed on the broad high (Senske and Head, 1988b).

The radar units mapped on Gula Mons, located 740 km to the east of Sif Mons, show properties different from those mapped on Sif Mons. The summit is characterized by a Very bright unit with flanking Dark and Very dark Units confined to areas of lower topography forming a pattern radial from a region just below the summit of Gula Mons (18 °N, 358 °). These deposits form a cross-cutting relation with adjacent Mottled bright and Mottled dark units associated with Sif Mons, suggesting that the Very dark units in this area are relatively young and are associated with a system of flows originating from flank eruptions. Gula Mons is therefore interpreted to be younger than Sif Mons.

Venera 15/16 imaging: Venera 15/16 radar imaging (Barsukov et al, 1986) overlaps the far north and northeast corner of our study area (Figure 1). A number of subparallel ridges are located in the overlap region in lowlying plains near the flanks of the broad topographic rise between latitudes of 25 °N and 40 °N and longitudes 2 °E and 10 °E. The ridges show a relatively uniform spacing of 20 km and parallel the north- northwest trend of high topography along the eastern edge of Western Eisila Regio suggesting they are structurally controlled. Between latitudes of 35 °N and 40 °N and longitudes of 5 °E and 10 °E, the ridge pattern becomes chaotic. This area corresponds closely to the point where the general trend of high topography changes from a north-northwest direction to east-west. The ridges closest to the flanks of the topographic rise do not have radar shadows, suggesting that they are scarps, possibly associated with normal faulting. Further to the east, radar shadows are present and may be the result of ridge formation due to the rotation of blocks along low angle normal faults.

LOS gravity: Examination of line-of-sight (LOS) gravity data shows Western Eisila Regio to be characterized by a maximum positive anomaly of + 35 mgal at a spacecraft altitude of 150 km. An analysis of gravity due to topography at spacecraft altitude (Sjogren et al, 1983) predicts an anomaly of + 40 mgal. Using this information, assuming that the gravity can be modeled by a single wavelength (λ = 3200 km), and incorporating the method of Esposito et al, (1982), we calculate an apparent depth of compensation for Western Eisila Regio of approximately 1000 km.

Formational models: On the basis of our observation, we propose two models to explain the formation of Western Eisila Regio: (1) construction and crustal thickening due to volcanic loading in the absence of a thermal anomaly and (2) doming due to a thermal anomaly with secondary volcanism forming Sif Mons and Gula Mons. The presence of Sif Mons and Gula Mons on the crest of a topographic rise suggests a formational mechanism by construction and crustal thickening. The flows radiating from Sif Mons, however, are of limited extent and appear superimposed on the domal rise, suggesting that they represent a thin veneer of material and that constructional volcanism is localized, arguing against a thick crust built up of volcanic deposits. Mapping shows that the broad topographic rise exhibits properties similar to material forming the surrounding plains. This suggests a formational mechanism to be associated with normal faulting form a pattern consistent with models of updoming under regional extension as proposed by Withjack and Scheiner (1982). The estimated apparent depth of compensation can be interpreted in terms of representing a crustal root or dynamic support of topography. The interpretation of the ridges forming in association with normal faulting related to doming, the presence of plains material similar to adjacent regions on the domal rise, and the interpretation of volcanic deposits as being a thin veneer, along with the deep apparent compensation argue for a model of formation related to doming due to a thermal uplift. Similar models have been proposed for Beta Regio (Esposito et al, 1982; Stofan et al, in press), Bell Regio (Basilevsky and Janle, 1987), and Atla Regio (Senske and Head, 1988a). In a global context, the equatorial region of Venus appears to be the site of mantle upwelling and extension forming domal rises along with being the site of possible crustal accretion (Head and Crumpler, 1987). The highland of Ishtar Terra in the north polar region, on the other hand, is characterized by large-scale

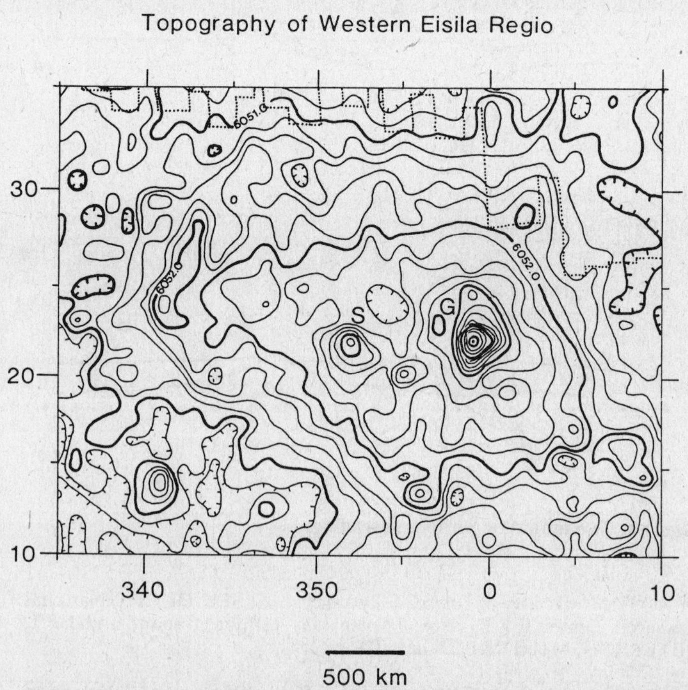

Figure 1. Pioneer Venus topography of Western Eisila Regio. Data is contoured at 200 m intervals. Two peaks, Sif Mons (S) and Gula Mons (G), interpreted to be volcanoes are located on the crest of a broad topographic rise. The dotted line marks the southern extent of Venera 15/16 image coverage.

compression, forming high topography through thrust faulting and crustal thickening (Crumpler et al, 1986; Vorder Breugge, (1987).

REFERENCES

Barsukov, V.L., et al, 1986, JGR, v. 91, p. D378.
Basilevsky, A.T., and P. Janle, 1987, Astron. Vestnik., v. 21, p. 109.
Crumpler, L., et al, 1986, Geology, v. 14, p. 1031.
Esposito, P.B. et al, 1982, Icarus, v. 51, p. 448.
Head, J., and L. Crumpler, 1987, Science, v. 238, p. 1280.
Senske, D.A., and J.W. Head, 1988a, LPSC XIX, p. 1061.
Senske, D.A., and J.W. Head, 1988b, LPSC XIX, p. 1063.
Sjogren, W.L., et al, 1983, JGR, v. 88, p. 1119.
Stofan, E.R., et al, in press, GSA Bull.
Vorder Bruegge, R., 1987, MS Thesis.
Withjack, M.D., and C. Scheiner, 1982, AAPG Bulletin, v. 66, p. 302.

SEYED-EMAMI, K., University of Tehran, Tehran, Iran

Stratigraphy and Structural Evolution of Iran during Mesozoic

The geologic evolution of Iran during the Mesozoic is characterized by four tectonosedimentary cycles, bounded by major tectonic movements. Each cycle begins with an extensive transgression and ends by the next following movement.

The Lower and Middle Triassic of Iran consists mainly of shallow-water sediments. Exceptions are the Triassic of Nakhlak and Agh-Darband. Because of the Early Cimmerian movements around the boundary of Carnian to Norian, the previously Iranian territory is divided into two parts, i.e., southwest and Central/North Iran (formation of the Neotethys). From this point, the two basins show completely different lithological, faunal, and structural developments. Whereas in Southwest Iran the more or less calm and continuous sedimentation lasts until the end of the Mesozoic (only with slight epeirogenic movements), in Central and North Iran the main Alpine phases (partly with magmatism and even metamorphism) are present. Here, because of the same movements, is an abrupt change from carbonatic to thick, molassic sediments (Shemshak Formation). The Shemshak Formation transgresses with a more or less pronounced unconformity on Triassic or even older formations and is bounded again by a major movement in the Middle Jurassic. The mid-Cimmerian (Lutian) movement intensively affected North Alborz, Koppeh-Dagh, and East Central Iran. It is also of utmost importance for the formation and further evolution of the same basins. Following the mid-Cimmerian movements, there is a wide-reaching transgression, almost everywhere in North and Central Iran, in Late Bajocian. During the rest of the Middle and Upper Jurassic calm, continuous and more or less carbonatic sedimentation existed in North and Northwest Iran. But in Central Iran, many differential basins, with thick and mainly clastic sediments, were formed.

Around the boundary of Jurassic and Cretaceous, there is again an important movement (Late Cimmerian) in North and Central Iran. Because of the following uplift, Neocomian sediments are either missing or represented by red, clastic, and partly evaporitic sediments. Continuous sedimentation from Jurassic to Cretaceous (with pelagic sediments) is only developed along narrow troughs in the northern Alborz (along the "North Iran Suture") and along the Great Kavir Fault.

An extensive transgression follows in Late Barremian and Aptian. Thick (Orbitolina-) limestones of this age unconformably rest on Jurassic or older formations in many parts of North and Central Iran. This is probably the most extensive transgression in the Mesozoic of Iran, where a shallow and uniform sea covered the entire Iranian territory. The Upper Cretaceous of North and Central Iran is again signified by a big transgression following the Austrian movements around the boundary of Albian to Cenomanian. On the whole, the Late Cretaceous in North and Central Iran is characterized by tectonic unrest and the formation of many differentiated basins, which are strongly folded at the end of the Cretaceous (Laramide).

SHAFIQULLAH, M., P. E. DAMON, and G. H. DAVIS, University of Arizona, Tucson, AZ

Determination of Timing of Structural Movements Through K-Ar Dating of Fault Gouge and Cataclasite

Establishing timing of fault movements is critical in understanding the complex tectonic history of an area. In the absence of isotopic dating, the age of tectonic events can only be estimated from the age and correlation of displaced units and overlying strata. It becomes especially difficult to date events involving continental deposits where regional time markers so common in the marine record are absent. We are developing a technique that may be of significant practical importance to structural geologists, regional tectonists, and economic geologists. It involves dating submicron illitic clays that form in fault gouge. Our study area is in the Basin and Range tectonic province of Arizona where fault gouge is commonly preserved in the arid setting.

The insight for this research emerged from research on K-Ar analyses of clays in sedimentary. It has been established that K-Ar dating of marine bentonitic shales yield the time of diagenesis and not the time of deposition of the clays. Detrital or recycled illite from the same environments yields older hybrid dates. Although it is difficult to achieve complete separation of diagenetic and detrital illites from argillaceous rocks such as shales and mudstones, it is clear that the proportion of diagenetic illite increases with decreasing grain size of the clay fraction. At the beginning of our work, we speculated that if K-Ar dating of illite from the finest fractions of marine shales yields timing of diagenesis, then K-Ar dating of illite from the finest fractions of fault gouge might yield timing of fault movement and not the age of the rock that was comminuted.

Illite has an activation energy of 38.6 Kcal/mole and the potential loss of radiogenic argon from

0.1 micron size illite heated to 125 °C for 150 Ma is about 5% (Glassman, 1987). Based on our work, we are beginning to conclude that submicron illite that forms in a fault gouge retains all the radiogenic argon produced in the gouge since its formation. If this is substantiated through further work, the use of illite K-Ar dating of fault gouge clay would provide a powerful method of determining fault movement chronology.

We have applied our hypothesis to the dating of movement on the Catalina Fault, a major detachment fault that bounds the Catalina-Rincon Metamorphic Core Complex near Tucson, Arizona. Preliminary work on submicron illitic clays from thin gouge zones exposed from place to place along the Catalina Fault yield concordant K-Ar dates of approximately 21 Ma. The ages are perfectly consistent with geologic constraints on the timing.

Three gouge samples and one ultracataclasite sample were analyzed. The gouge samples were collected along parts of the fault zone that separate a mylonitized granitic footwall from a variety of hanging wall upper plate rocks: Lower Cretaceous Bisbee Group sedimentary rocks at site 1 (see Table 1); Permian limestones at site 2; and Precambrian Rincon Valley granite at site 3. The cataclasite collected at site 4 is derived from mylonitized lower plate Precambrian granite. Although K and Ar contents vary considerably in the fault gouge samples, the calculated K-Ar dates fro the products of brittle deformation are concordant at about 21 Ma.

Table 1

Site	Upper Plate	$K	Ar pm	Date	
1.	Clastic sediments	3.053	111.8	21.0 ±	0.5
2.	Limestone	3.956	144.7	21.0	0.5
3.	Crystalline rocks	5.041	183.7	20.9	0.4
4.	Ultracataclasite	2.881	107.3	21.4	0.5

We realize that the submicron illite samples dated may contain minor amounts of detrital components-mica, feldspar, and recycled illite, and hence the calculated dates may be a maximum. The dates, in theory, will come in closer and closer to the time of faulting as the component of extraneous detrital component becomes smaller and smaller. If, however, a detrital component is present, a correction must be applied based on knowledge of the level of contamination. The consistency of the K-Ar dates for sites 1-4 along the Catalina Fault suggest that the extraneous component is negligible. We are evaluating this interpretation through analysis of the composition of the fault gouge samples using electron microscope and X-ray techniques.

Although limitations on the use of fine-grained illite/smectite in K-Ar dating must still be assessed, Aronson and Lee (1986) and Glassman et al (1989) have demonstrated the effectiveness of the method in the study of burial and diagenesis of sediments and petroleum maturation. If results are as promising as our preliminary data suggest, K-Ar dating of fault gouge may become an important, standard dating technique for evaluating the chronology of tectonic events involving fault displacements.

REFERENCES

Aronson, J.L., and M. Lee, 1986, K/Ar systematics of bentonite and shale in a contact metamorphic zone, Cerrios, New Mexico: Clays and Clay Minerals, v. 34, p. 483-487.
Glassman, J.R., 1987, Argon diffusion in illite during diagenesis: how good is the K/Ar clock?: Abstracts, 24th Annual Meeting, The Clay Mineral Society, Socorro, New Mexico, p. 60.

SHAHIN, A. N., Gulf of Suez Petroleum Company, Cairo, Egypt

Application of Quantitative Modeling to Petroleum Exploration in the Northern Western Desert, Egypt

The Western Desert (WD) of Egypt occupies a vast area of 270,000 sq. mi. of an intracratonic basin in an unstable shelf region where basinal areas are widespread. The northern part of the WD has about 110,000 mi^3 of prospective sedimentary rocks, ranging in age from Paleozoic to Tertiary, in four main basins: Abu Gharadig (AG), Alamein, Matruh-Shushan, and Faghur (Figure 1).

The aim of this study is to make use of quantitative modeling to provide insights to the distribution of potential petroleum reserves among the prospective basins. This would imply better prediction of prospective structures or stratigraphic horizons favorable for oil accumulation.

Method: Predictions based on existing fields distribution and size or absolute statistics, are not favored because of possible reserve appreciation to discovered fields or personal judgment. In this work, a combination of both deterministic methods (based on analytical data) and probabilistic methods (analytical data treated statistically for normal distribution) were employed:

Geodynamic modeling which incorporates, beside the subsidence history, the diagenetic effect of heat flow, uplifts and magnitude of erosion at unconformities (Lopatin, 1971; Wapes, 1980, 1985). Maturation measurements (vitrinite reflectance ‰) and the successive steps of hydrocarbon generation were considered to achieve a match between observed and modeled maturity. This procedure allowed the delineation of the mature "oil-expelling" basins. The time of migration within each basin was related to the main tectonic events in the region in order to establish preservation deadlines for the potential oil accumulations.

The Geochemical material balance method (Tissot and Welte, 1978; Waples, 1985) has the advantage of covering most of the key genetic factors of oil and gas occurrence, e.g., the pyrolysis-derived genetic potential of source rock (in ppm), the respective rock and oil densities, the level of maturation reached and volume of the effective source rocks resulted from the modeling. Each of these variables was then exposed to quantitative risk analysis employing Monte Carlo simulation technique (Newendorp, 1975), and amounts of oil generated in each basin were calculated (Table 1). Based on the expulsion and migration efficiency factors suggested by Momper (1987), Barker and Dickey (1984), and Waples (1985), and the recovery factors established in the AG basin (Shahin et al 1986, 1987), the amounts of undiscovered recoverable oil and gas were estimated. For easy comparison among basins, the exploration performance parameters (Table 1) were normalized in relation to the sedimentary volume of the respective basin (Rapoport and Grender, 1986).

Abu Gharadig (AG) Basin: Half of the exploratory wells in the WD were drilled in the AG basin, resulting in eight fields and two discoveries (Figure 1). The basin has the highest drilling density in the WD, 37 ft/mi^3 with a discovered density of 8,000 BOE/mi^3. The basin is multi hydrocarbon sourced by the Cretaceous Khoman Fm, Abu Roash F and G Mbrs and Bahariya Fm; the Jurassic Safa Mbr of the Khatatba Fm; and possibly the Devonian Zeitun Fm and the Carboniferous Dhiffah Fm. The basin is estimated to hold 10.4 billion BOE of recoverable reserves, about 35 times the proven reserves (Table 1).

Alamein Basin: This basin feeds four fields and one discovery, including the first commercial oil field in the WD, Alamein Field (1977). Although the basin ranks last in its sedimentary volume it has the highest discovery density in the WD (14,000 BOE/mi^3) and the highest exploration response (625 BOE/ft). The Jurassic Safa Mbr of the Khatatba Fm is the source of hydrocarbons. The estimated recoverable reserves are 7×10^9 BOE, about 40 times the proven reserves (173×10^3 BOE). Hydrocarbon migration in Alamein and eastern AG basins is suggested to coincide with, or post-dates, the Laramide structural trap-forming episode. The western AG basin, despite the possible preservation risk for its earlier maturation, its potential became beyond question after the discovery of Badr El-Din (BED) Field (Figure 1). Although the "Syrian Arc" system is most prominent in the WD (Abdine, 1984), the BED discovery implies that the Laramide-aged structures, contrary to earlier beliefs, were not always disturbed due to subsequent overprinting by the Alpine tectonic episode the "Syrian Arc" of the Oligo-Miocene. The latest significant event, the Late Miocene "Messinian" salinity crisis, was a gentle event of regression and erosion, though of high magnitude emergence.

Matruh-Shushan Basin: This basin holds about the same volume of sediments as the AG basin, though less explored. The drilling density is 23 ft/mi^3. The discovery density is only 1000 BOE/mi^3 compared to 8×10^3 and 14×10^3 in Alamein and AG basins respectively. This basin is estimated to host 23×10^9 BOE about half of the WD calculated recoverable reserves. The last five years witnessed an undeniable success in this basin with five new fields and three discoveries. The Salam Field (Figure 1), the first in the WD to produce from the Jurassic, represents a turning-point in the WD prospectivity. The Jurassic Safa Mbr source rocks hold enormous potential in all basins. However, it may not be mature to generate hydrocarbons in the Paleozoic-aged Faghur basin due to shallow burial. A low risk is attached to the preservation of hydrocarbon accumulations as oil migration post-dates the severe Laramide tectonics. Only in the vicinity of Umbarka field, a greater risk is involved as maturation and migration were earlier (pre-Cenomanian) for its high heat flow. However, Umbarka Field itself is an instance of old structural trap (Late Jurassic "Cimmerian") modified by later successive events with no significant leakage.

Faghur basin: This basin covers a mature areal extent of 8,000 mi^2 trending NW-SE and extends into the Libyan territory to the west. To date, the Paleozoic in this basin has not been tested. This can be attributed to lack of exploration as the drilling density is only 3 ft/mi^3, the lowest in the WD. Organically rich intervals are indicated within the oil-prone Devonian Zeiton Fm (3,500 ppm) and the Carboniferous gas-prone Dhiffah Fm (3,000 ppm). Peak-oil generation is suggested to have occurred the earliest in the region. Such timing would imply a degree of preservation uncertainty. Nevertheless, as concluded from BED discovery in AG basin, the Umbarka Field in the Matruh-Shushan basin, is another instance of old-but-preserved oil. Umbarka is in part a Paleozoic trap, and may yet prove possible entrapment of Paleozoic origin hydrocarbons. Faghur basin is estimated to contain recoverable reserves of 2×10^9 BOE of Paleozoic oil and gas.

The predictions presented in this work at such regional scale, although tentative, they can be utilized to improve estimation of exploration risk and prospect assessments in the Western Desert.

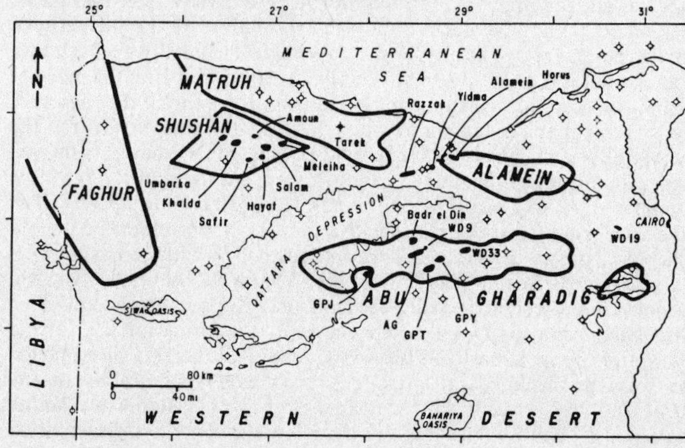

Figure 1—Fields and hydrocarbon generating basins in the northern Western Desert, Egypt.

WESTERN DESERT BASINS	ABU GHARADIG	ALAMEIN	MATROUH-SHUSHAN	FAGHUR	Total
Number of Fields	8	4	7	0	19
Wells drilled	120	36	87	10	253
Feet drilled (10^3ft)	1368	277	870	70	2585
Sediment volume($10^3 mi^3$)	37	12	38	23	110
Source rocks age	K, Jr, Pz (?)	Jr	Jr, Pz (?)	Pz	
Generated HC (10^9BOE)	150	95	315	27	587
Proven Rec. Res. (10^6BOE)	296	173	37+	0	506
Estimated Rec. Res. (10^9BOE)	10	7	23	2	42
Drilling density (ft/mi^3)	37	23	23	3	
Discovery density(10^3 BOE/mi^3)	8	14	1	0	
Response (BOE/ft)	216	625	43+	0	
Estimated / Proven Res.	35	40	620	?	

Table 1—Hydrocarbon potential of basins in the northern Western Desert. BOE (barrels oil equivalent), HC (hydrocarbons), Rec. (recoverable), Res. (reserves), K (Cretaceous), Jr (Jurassic), Pz (Paleozoic).

SHANGYOU, NIE, University of Chicago, Chicago, IL

Paleozoic Orientations of Chinese Microcontinents Based on Climatically Sensitive Sediments and Paleomagnetic Data

The distribution of climatically sensitive sediments, such as sillites, coal, evaporites, dolomites, pure limestone, and reefs provides independent paleolatitudinal controls for plate reconstruction. Such first-order constraints are especially important for regions where the paleomagnetic data are limited, as is the case for the Chinese blocks in the Paleozoic. A set of preliminary plate orientations are presented based primarily on climate indicators. I assume that the ancient atmospheric and oceanic circulation patterns operated as today. These patterns result from the equator to pole temperature gradient and the rotation of the earth. Complicating factors, such as size and distribution of continents that can alter zonal pattern of climatic belts, are ignored.

A brief summary of the distribution of climatically sensitive sediments is given in Table 1 for the three major blocks of North China, Tarim, and South China. Presence of typical tropical and/or subtropical sediments, such as evaporites, dolomites, and pure limestone for all three blocks suggests that they occupied relatively low latitudinal positions in the Cambrian and Ordovician and could have been part of the same major continent, such as gondwana. during the Silurian and Devonian, the South China Block, as indicated by continuous distribution of limestone, dolomite, and reefs, appears to have maintained its low latitudinal position, while Tarim is dominated by clastic sequences that are much more difficult to interpret in terms of their latitudinal association. Little can be said about the North China Block from the Late Ordovician to Early Carboniferous since this period coincides with the major disconformity that is characteristic of the whole of the microcontinent as usually defined.

Beginning with the Middle Carboniferous, coal occurs extensively on the North China Block, representing a moist environment. These coals contain the Cathaysian flora, which has been interpreted to have inhabited an equatorial rainforest, based on such factors as leaf morphology and the absence of seasonal growth rings. The extent of the coal formation decreased in the Permian and was replaced by evaporitic sequences in the early Triassic, suggesting that the North China Block traveled from the equator to the subtropical dry zone in this time span. For south China, the most widespread coal formation occurred in the late Permian, which is best explained as equatorial, since it contains the Cathaysian flora that coexisted with well-developed carbonate and dolomitic sequences. There is no coal formation on the Tarim Block for the Carboniferous and Early Permian probably because it did not have much land area and was instead dominated by shallow-marine limestone, implying also a relatively low latitude position.

As an example, Figure 1 gives a series of orientations for the South China Block with both paleoclimatic and paleomagnetic data presented. the two sets of data agree reasonably well, both indicating a relatively low latitudinal position for the South China Block during the entire Paleozoic. As shown, paleomagnetic data is only available for the Early Cambrian, Middle Silurian, Late Carboniferous, and Late Permian. Moreover, the reliability of these results is subject to conflicting opinions because of problems such as remagnetization and tectonic distortion. The nearly continuous distribution of climatically sensitive sediments therefore gives better control on the first-order plate positions of south China. For comparison, two early Mesozoic intervals have also been included.

The orientation of these microcontinents is mainly determined using paleomagnetic data. Spatial distribution patterns of climatically sensitive sediments on a single plate can also provide useful information for plate orientation. For example, an orderly zonal distribution of evaporite/dolomite/limestone/coal would indicate a change from dry to wet environment, such as from the subtropics to the equator. An interesting example is given by coal distribution in the Permian of south China. The northeast-southwest (present day) trending belt of coal beds in the eastern part of south China was established earlier in the stratigraphic succession than in the west. A reasonable explantion is that the eastern part of south China crossed the equator earlier, giving an orientation that agrees also with paleomagnetic data.

SHANKS, W. C., III,, U. S. Geological Survey, Reston, VA, and R. A. ZIERENBERG, U. S. Geological Survey, Menlo Park, CA

Stable Isotope Studies of Sulfide Deposits and Hydrothermally Altered Sediments, Escanaba Trough, Gorda Ridge

Escanaba Trough comprises the southern portion of the Gorda Ridge, located about 200–300 km off the coast of northern California in the northeast Pacific Ocean. Up to 500 m of turbidites and hemipelagic sediments cover the southern 90 km of the slow- spreading Gorda Ridge. Massive sulfide deposits have been discovered associated with volcanic edifices which penetrate the sediment cover to near seafloor positions. Investigations utilizing deeply towed bottom photography, dredging and coring, and submersible dives in *DSRV ALVIN* have defined two major areas of sulfide accumulation centered about 40°N latitude, 127°30'W longitude and 40°45' N latitude, 127°30'W longitude.

Sulfide deposits are characterized by large mounds (up to 20 by 30 by 50 m) and small chimneys dominantly of phyrhotite. Hydrothermal fluids at 220 °C have been sampled from one site in the northern area, but most of the deposits presently are inactive. Some chimney and mound samples are enriched in isocubanite, sphalerite, galena, barite, and anhydrite. In addition, hydrothermally altered near-surface sediments have been recovered by gravity coring and dredging. Such samples are characterized by sulfide-talc breccia, veinlets of sulfide and thermogenic hydrocarbons cross-cutting slightly altered sediments, and turbidites and hemipelagic sediments pervasively altered to near-pure chlorite.

Oxygen and sulfur isotope studies have been employed to better understand the conditions and mechanisms of formation of sulfide, sulfate, and hydrothermal silicate minerals. Bulk sediment oxygen isotope values ($\delta^{18}O_{smow}$) cluster into two groups. Unaltered turbidites and hemipelagic sediments have $\delta^{18}O$ values of 8.9 to 14.3‰. Pervasively altered chloritic sediments and talc-rich sediments have $\delta^{18}O$ values from 0.2 to 3.7‰, indicating high temperature hydrothermal alteration of these surficial (less than 3 m depth) sediments. Intermediate between these two groups at 5.8‰ is a turbidite sample with pyrrhotite–petroleum veins and secondary smectite overgrowth on quartz. The most intensely altered samples are the chlorite-rich sediments with approximately 25 wt% MgO, which are nearly monomineralic chlorite. Hydrothermal fluids from other mid-ocean ridge vent sites have $\delta^{18}O$ values of 0 to 1‰ indicating their derivation from deep circulating seawater. If similar fluids formed the Escanaba Trough chloritic deposits with values as low as 0.2‰, then chlorite–water isotopic fractionation requires temperatures in the 300 to 350 °C range. Alternatively, Escanaba vent fluids may be ^{18}O enriched due to deep reactions in the sediment column. For example, a 2‰ hydrothermal fluid could form the chlorite deposits at 220 °C. The latter lower temperature process seems more likely because subsurface mixing of ambient seawater is required to provide the necessary Mg.

Sulfur isotope values ($\delta^{34}S_{cdt}$) of sulfide minerals from mounds and chimneys range from 2 to 12‰. Sulfide minerals from veinlets in sediments range from 2 to 7‰. Sulfate minerals, including barite from veinlets and anhydrite from chimneys, all have $\delta^{34}S$ values from 19.7 to 21.5‰, close to or slightly higher than normal seawater sulfate $\delta^{34}S$ values.

Sulfur isotope values of the Escanaba sulfide minerals are very ^{34}S enriched compared to other seafloor hydrothermal systems from unsedimented mid-ocean ridges. For example, deposits at 21 °N, East Pacific Rise, and southern Juan de Fuca Ridge have $\delta^{34}S$ values ranging from 0.9 to 5.5‰. Thus the δS values of Escanaba sulfides indicate possible derivation of sulfur from the following sources: (1) basaltic sulfide with $\delta^{34}S$ values near 0 per mil, (2) ^{34}S enriched sulfide from hydrothermal reduction of seawater sulfate with basalt in the deep subsurface, and (3) ^{34}S enriched sulfide form hydrothermal reduction of seawater sulfate during formation of thermogenic petroleum during fluid ascent through the sediments. Source (3) is of dominant importance in forming the ^{34}S enriched sulfides of the Escanaba deposits, accounting for 20–60% of the sulfide in the deposits.

In summary, the Escanaba Trough sulfide deposits form as a result of hydrothermal seawater circulation driven by heat from volcanic edifices that penetrate the sediment blanket which overlies the ridge crest. Oxygen and sulfur isotope studies, and geochemical studies of altered sediments, indicate near surface entrainment of ambient seawater to account for Mg and SO_4 contributions.

SHANNON, R. D., and M. A. SUBRAMANIAN, E. I. du Pont de Nemours & Co., Inc., Wilmington, DE, A. N. MARIANO, Carlisle, MA, and G. R. ROSSMAN, California Institute of Technology, Pasadena, CA

Mineral Dielectric Constants and Oxide Additivity Rule

The Clausius-Mosotti equation relates dielectric constants to molecular polarizabilities:

$$\alpha_D = 3/(4\pi) \cdot V_m \cdot (\kappa - 1)/(\kappa + 2)$$

where Vm = molar volume and κ = dielectric constant at 10^5–10^{10}Hz and leads to the oxide additivity rule, where, for example, $\alpha_D(MgAl_2O_4)$ = $\alpha_D(MgO)$ + $\alpha_D(Al_2O_3)$. The additivity rule, although approximately valid for isotropic crystals, has not been adequately tested for most minerals and in particular for noncubic crystals.

In this study we have accurately determined the 1 MHz dielectric constants of BeO (synthetic), CaO (synthetic), Be_2SiO_4 (phenacite), $BeAl_2O_4$:Cr (synthetic), $MgAl_2O_4$ (spinel), Mg_2SiO_4 (forsterite), Be_2BO_3OH (hambergite), $CaB_2Si_2O_8$ (danburite), $Y_3Al_5O_{12}$ (synthetic), $Ca_3Al_2Si_3O_{12}$ (grossular), and $Mg_2Al_4Si_5O_{18}$ (cordierite) using the two-terminal method with edge corrections. With these data, the Y_2O_3, and quartz, and estimated polarizabilities of B_2O_3 and ice, compound or mineral polarizabilities calculated from dielectric constants using the Clausius-Mosotti equation and from the oxide additivity rule are compared. Table 1 shows polarizabilities calculated according to the two methods agree to within 0.5%.

Compound	$\langle\kappa\rangle$	Predicted α_T (oxide rule)	Measured* α_T(exp)	Δ%
$BeAl_2O_4$	8.93	9.84	9.87	0.3
Be_2SiO_4	6.21	9.31	9.35	0.5
$MgAl_2O_4$	8.18	10.96	11.01	0.5
Mg_2SiO_4	7.00	11.53	11.56	0.2

*Using the Clausius-Mosotti relationship

Table 1—Comparison of observed and predicted single crystal dielectric polarizabilities.

SHARMA, B. L., R. UPADHYAYA, S. M. ALI, N. K.CHAUHAN, and H. BHU, University of Rajastan, Udaipur, India

Deformation Pattern as Deduced from Early Proterozoic Conglomerate Pebbles of Aravalli Supergroup, Udaipur Rajastan (India)

Phyllitic and arkosic conglomerate horizons of the Proterozoic Aravelli rocks of Udaipur region (Rajastan, India) have been studied for regional strain analysis. The depositional setting of pebbles and matrices of these conglomerates have been deformed by varying amount of strains due to subsequent folding. The variation in the pebble-matrix deformation is used here to calculate the finite strain.

These two conglomerate horizons are only a few kilometers apart and represent different stratigraphic positions but the same structural domain. The structural analysis (120 km2) suggests two dominant and penetrative episodes of deformation.

A total of 482 pebbles from arkosic matrix and 433 pebbles from phyllitic matrix were collected on a grid pattern. Quartzite and vein quartz pebbles are dominant. However, substantial numbers of pebbles of mafic rocks, gneissic rocks, carbonate rocks, tourmaline, and grit are also present, and their data have been included in the strain analysis. Pebble size varies from 7.5 mm to 440 mm. Shape is usually triaxial ellipsoid.

For strain analyses the two conglomerates have been divided into various sectors depending upon dominance of (1) pebble size and shape, (2) pebble lithology, and (3) regional structural setting. Hence, the strain analysis considers (1) matrix, (2) pebble lithology, (3) pebble size, (4) pebble shape, and (5) structural setting of the sectors. Pebble concentration and pebble contacts (with pebbles and matrix) have also affected the strain behavior.

The X-Y plane ($X > Y > Z$) of the pebbles is usually parallel to the bedding schistosity, axial planar to the first folding. The long axes of the pebbles are uniformly parallel to each other and parallel to the down dip first-fold axis forming the first tectonic lineation.

The important observations a as follows:

Uniformly unimodel histograms with high modal frequencies for X/Y and Y/Z for all pebbles of arkosic matrix are consistent with undeformed planar or nonrandom imbricate fabric of the pebbles. This is supported by $R_f\phi$ plots that give an asymmetric distribution of ellipses implying original preferred orientation of the clasts. Unimodality also suggests that the strain suffered during deformations may be of the same type. In the Phyllitic matrix, the modes of histogram are not uniform. The constant shift in the modal values of axial ratios suggests an involvement of rotational component in the deformation.

Axian rations X/Y and Y/Z are strongly positively correlated for most of the lithologies indicating a strain of pure shear type in which Y-axis appears to remain unchanged during deformation. Comparatively smaller values of "r" for axial lengths suggest that larger clasts might not have deformed homogeneously with the matrix.

Plots of the three ratios of the semiaxes of the pebbles and their best fit lines always have finite intercepts indicating nonspherical original shapes of the pebbles. The slopes of the lines of best fit in the same set of diagrams are suggestive of different shears experienced by the two conglomerates: dominantly pure shear type by the arkosic and simple shear type by the phyllitic conglomerates.

Axial ratios of the total finite strain ellipsoids and values of natural octahedral unit shear suggest that the strain experienced by the arkosic conglomerate belongs to the line of plane strain. In phyllitic conglomerate, the strain recorded by different pebbles shows considerable variation.

Calculated viscosity ratios for pebble-pebble and matrix-pebble combinations show considerable variation from locality to locality. It may be due to varying physical conditions at each locality during deformation.

Flinn plots for arkosic conglomerate show arithmetic average, of initial axial ratio (I), final axial ratio (f), and calculated axial ratio for the strain ellipsoid (S), mostly falling in the field of apparent flattening.

Conclusions:

1. Initially the pebbles were nonspherical and their setting was nonrandom. Increasing elongation of the pebbles seems to be associated with increasing parallelism of the preferred orientation of the long axes. This supports the special coincidence of sedimentary and tectonic fabric producing rather abnormally shaped clasts.

2. The dominant first stage of strain was pure shear with very feeble rotational component in the incompetent clasts and matrix.

3. "Fluctuation" calculations have suggested that the second stage strain was coaxial with the first strain. It was a simple shear component resulting in superposed flattening parallel to the differential movements in the pebble-nonpebble stratum (logarithmic Flinn Plots).

4. In response to the strain, pebble lithologies have behaved differently in different environments (of different matrix, different pebble concentration, and pebble-pebble and pebble martix contacts).

5. Plots ϵ_s (natural strain) and μ (Lode's parameter) on the strain plane confirm that the deformation paths are different for different lithologies. The pebbles in the arkosic conglomerate record more strain than the matrix, whereas in phyllitic conglomerate matrix records more strain than the pebbles.6. The evolution of strain pattern is shown in the following figure. The location on the F2 limb is the structural domain covered in the present investigation.

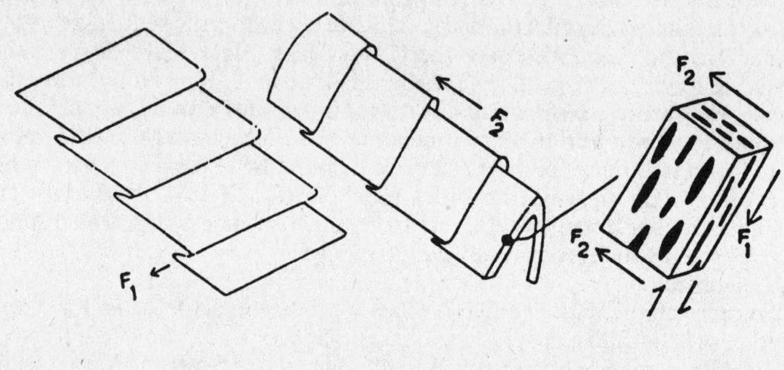

SHARMA, K. K., Wadia Institute of Himalayan Geology, Dehra Dun, India

Tectonic Evolution of Northwestern Himalaya: Fission Track Age Constraint

A large number of fission track (FT) ages determined on the coexisting mineral pairs (zircon, sphene, epidote, apatite) by the author and his coworkers during the last decade from the northwestern Himalaya between Doda-Kishtwar (Lesser Himalaya) and Ladakh (Transhimalaya) in the west to the easternmost part of the Kumaun Lesser Himalaya (Figure 1) have been synthesized in the light of the available information, the uplift rates for various segments calculated, and their possibile constraint on the timing of various thermotectonics events in the northwestern Himalaya. The evolution of the Himalaya is attributed to the compression caused in the leading edge of the northward-moving Indian plate before and after its collision with Asia.

For this paper, the Himalaya is broadly divided into three zones each separated from the other by a thrust (Figure 1): (1) Indus-Tsangpo Suture Zone, (2) Salkhala-Vaikrita-Tibetan slab crystallines (Tethyan Crystallines) and their sedimentary pile (Cambrian to Cretaceous Tethyan rocks), and (3) Jutogh-Munsiari Crystallines (Jutogh-Munsiari and Chail-Ramgarh rocks) pushed southward to rest on its sedimentary cover and now seen as nappes and klippen. The two crystalline basements (i.e., Tethyan basement and Lesser Himalayan basement) are separated by a prominent thrust MCT-Vaikrita (V) that marks an important boundary between the Higher and the Lesser Himalaya. Recent fossil finds from the Lesser Himalayan sedimentaries (Azmi, 1983; Tewari, 1984; Kumar et al, 1987) supported by sedimentation records from Tethyan and Lesser Himalayan basins and the observations of the author from the northwestern Himalaya suggest that the two crystallines were brought closer during Late Proterozoic and possibly uplifted to some extent during Early Paleozoic, thus restricting the deposition of the younger (Krol Belt) sedimentaries to the southern part of the Lesser Himalayan basin.

FT ages on coexisting mineral pairs along with mineral age data by K-Ar, Rb-Sr techniques wherever available bring out the following thermotectonic history for the northwestern Himalaya.

1. The oldest ages (82-80 Ma) of zircon in the Lesser Himalayan Crystallines pushed farthest away from the Central Crystalline Axis (CCA) and now exposed around Chor and Dudhatoli suggest that these areas possibly preserve the signature of the first phase of uplift in the Himalaya during Upper Cretaceous. Earlier reported K-Ar age of 98 Ma for the muscovite from Chor supports FT age data. FT zircon ages of 72 Ma from Dalhousie,

60 Ma from Mandi, and sphene age of 69 Ma from Chor preserve the records of subsequent uplift and cooling suffered by the Lesser Himalayan crystallines during Late Mesozoic-Early Tertiary. The uplift of the CCA during Upper Cretaceous brought about changes in sedimentation in Tethyan basin and transgression in the Lesser Himalaya.

2. FT ages from the Dras island arc in relation to mineral ages by other techniques suggest that the arc got rapidly uplifted around 60 ; 2 Ma and 38 ; 2 Ma due to the two-stage collision of the arc with the approaching continental margins (Sharma, 1987).

3. Soon after the first-stage collision of the Dras island arc with the leading edge of the Indian plate, the Tethyan basement started uplift around 60 Ma ago around Doda-Thathari in Jammu region and around Leopargial in Satluj valley. The doming of the basement resulted in separation of Chamba basin from Kashmir basin and Zanskar-Spiti basin from the Garhwal-Kumaun basin (Figure 1). Incidentally, it may be mentioned that these domal uplifts trend northeast-southwest and the earlier workers have reported northeastward plunging tight isoclinal folds (Aravalli Trend) particularly in Kishtwar ad Satluj valley. These areas are also characterized by abrupt bending of MBT and lithotectonic units and possibly point to the presence of buried ridges of the Indian Shield.

The effects of India-Asia collisions are also recorded by the fission track clocks in resistant minerals of the Lesser Himalayan crystallines exposed in the central axis as well as in the klippen and the nappes.

4. The fission track chronometers suggest uplift in the Himalaya during Early Miocene (25-20 Ma) and Middle Miocene (17-15 Ma). The Tethyan basement in lower Swat-Hazara, Doda-Thathari, Rohtang regions, and in many other parts uplifted at a very fast rate (0.8 mm/y to 1 mm/y) during this period along MCT (V) in comparison to slow rate (0.3-0.4 mm/y) of uplift in the Lesser Himalaya, thus resulting in topographic break between the Lesser and Higher Himalaya. The Middle Miocene uplift in the Lesser Himalaya seems to be due to large-scale translation along thrusts that caused frictional heat (more than 300 °C) locally along these thrusts, thus completely resetting the FT clock even in resistant minerals around 16 ; 2 Ma. It appears the compression caused by the northward movement of India during Middle Miocene was largely consumed in the translation of the thrust sheets along gently dipping planes. Renewed movements along some of the thrust planes have been recorded by low temperature (100 °C) chronometer (apatite) around 10 and 5 Ma.

5. Another phase of compression built up in the Indus Suture Zon, Tethyan, and Lesser Himalayan basements is evident from rapid uplift (more than 1 mm/y) of Wangtu dome, Rampur-Larji window, Kishtwar window, Nanga Parbat massif, and Gilgit. This uplift phase, unlike the Middle Miocene thrust-related uplift, seems to be due to the compression in the basement that pushed up the sedimentary cover rocks to pierce through the crystalline thrust sheet overriding the sedimentaries. The cross-folding of thrust sheets in the Lesser Himalaya is also attributed to this compression. It is interesting to note that this zone of compression initiated during Late Miocene (10 Ma) and continuing during Pliocene-Pleistocene is more or less parallel to Karakorum-Gandak fault and needs careful analysis of the data.

The significant contribution of the FT age data from the northwestern

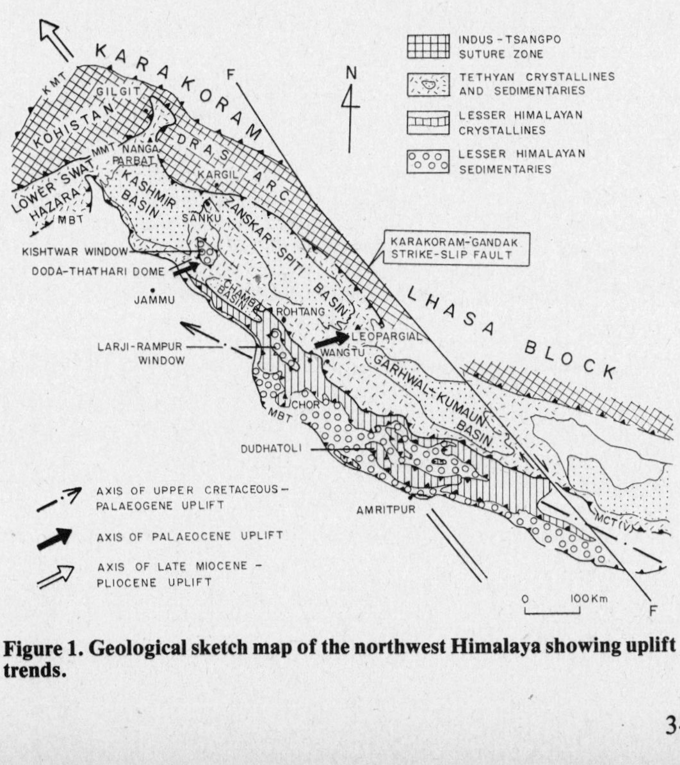

Figure 1. Geological sketch map of the northwest Himalaya showing uplift trends.

Himalaya has been the recognition of the Upper Cretaceous uplift in the crystalline basements before the collision of India and Asia. During northward movement of India, its leading edge under the weight of the Tethyan sediments (more than 10 km thick) induced compression in the crystalline basements causing uplift and conspicuous changes in the sedimentation history of the Tethyan and Lesser Himalaya basins. This conclusion has further implications suggesting that the passive margin of the Indian plate underwent thermotectonic activity much before its collision with the active margin.

REFERENCES

Azmi, R.J., 1983, Microfauna and age of the Lower Tal phosphorite of Mussoorie Syncline, Garwal Lesser Himalaya: Him. Geol., v. 11, p. 373-409.

Kumar, G., A. Joshi, and V. Mathur, 1987, Redlichiid Trilobites from the Tal Formation, Lesser Himalaya, India: Current Science, v. 56, n. 13, p. 659-663.

Sharma, K.K., 1987, Crustal growth and two-stage India-Eurasia collision in Ladakh: Tectonophysics, v. 134, p. 17-28.

Tewari, V.C., 1984, Stromatolites and Precambrian-Lower Cambrian biostratigraphy of the Lesser Himalaya, India: Proceedings of Vth Indian Geophytological Conference, Lucknow, 1983, Special Publication, p. 71-97.

SHARMA, P., and S. K. GUPTA, Physical Research Laboratory, Ahmedabad, India

Tritium Tracer Study of Soil Moisture Movement

Data of isotopic tracer studies involving tritium, deuterium and oxygen are used to describe the fate of a component of precipitation that eventually becomes part of groundwater through the process of infiltration and deep percolation.

The isotopic signature of precipitation is first modified by a variable amount of evapotranspiration and run-off loss from different rainfall events in a given rainy season. The extent of this modification can be estimated using an evapotranspiration-run off model described.

The isotopic labeling of the infiltrated water is further modified during the precess of movement and distribution within the unsaturated zone. The soil moisture movement can be conceptually described as a process involving pulsed drainage of soil moisture (in excess of field capacity) through a series of interconnected expandable cells.

The isotopic modification of the fraction of participation reaching the water table after evapotranspiration, run-off and movement through the unsaturated zone can be further visualized in terms of a dispersion-advection model to estimate flow characteristics of aquifers using tracers.

The attempt in the paper is to show that application of tritium and other isotopic tracers in arid/semi-arid regions of north-western India has not only led us to estimate vitally needed groundwater recharge parameter but has also enabled us to understand the physical processes taking place within the unsaturated zone.

Based on the results of controlled tritium tagging experiments, we estimate that on average about 14% of the precipitation annually recharges groundwaters in alluvial parts of Sabarmati basin in Gujarat State. Even in the Thar desert of Western Rajasthan significant amount of rainwater (6-14%) infiltrates to recharge shallow aquifers. Comparison of data from Sabarmati basin with similar data from alluvial parts of Indo-Gangetic plains in northern India having similar soil characteristics and total annual rainfall reveals the significant role of winter precipitation in groundwater recharge. Similar studies in Mahi Right Bank Canal (MRBC) command area enabled quantification of the seepage from irrigated fields.

SHARMA, R. S., Banaras Hindu University, Varanasi, India

P-T Path for High-Pressure Granulites from Sand Mata, Rajasthan, Northwestern Indian Shield

The Sand Mata granulite suite forms a part of the Arahean gneiss complex in Rajasthan, northwestern peninsular India. Its dominant rock is the pelitic granulite having garnet, sillimanite, kyanite, biotite, cordierite, hypersthene, quartz, and feldspar in various combination. Compositional banding with garnet leptynite or garnet-quartz-feldspar rock is seen on varying scales. Another rock type of the suite is the basic granulite containing the assemblage of garnet, diopside, hypersthene, plagioclase, and minor quartz. These granulite types have been introduced by norite dykes of uncertain age. They are characterized by blastophitic texture and at places are olivine bearing. The granulite suite is oval-shaped and surrounded by a gneiss-schist-granite complex called the Banded Gneissic Complex. Their contact is marked by a ductile shear zone with steep down-dip lineation.

The basic granulite is characterized by the corona of garnet-diopside-quartz at the interface of hypersthene and plagioclase. Random orientation of the corona minerals indicates their recrystallization in static conditions of the granulite facies metamorphism. That the garnet-diopside-quartz association has not formed during cooling is evidenced by their P-T estimates and by the development of garnet in the norite dykes which, owing to their blastophitic texture, seem to have intruded during waning stages of the granulite metamorphism. The assemblages in the pelitic granulite are characterized by overprinting of kyanite by sillimanite which, in turn, is followed by the growth of second-generation kyanite, mostly in the form of short needles.

Estimates of temperature conditions by updated calibrations give values that cluster about 850 °C and 650 °C for the basic assemblages and 650 °C ± 50 °C for the pelitic assemblages. These two temperature concentrations of 850 ° and 650 °C are interpreted as climactic and reequilibration (blocking) temperatures, respectively, for these granulites. Application of different geobarometers to the granulite assemblages yields pressures that have concentrations in the vicinity of 10 ± 2 Kbars and 5 ± 1 Kbars. These concentrations of P at about 8-11 Kbars and near 5 Kbars, with almost no record of intermediate values, are taken to indicate that the rocks were suddenly transported from deeper levels and emplaced to shallower depths (~5 Kbars), where frozen-in equilibrium was attained in the constituent phases of the granulites. This interpretation finds support from the occurrence of the peripheral shear zone, stated earlier. Furthermore, estimates in the pelitic granulite give appreciably lower values of pressure for the garnet-core composition than for the garnet-rim composition, using both hydrous equilibria (involving cordierite) and anhydrous equilibria involving garnet-plagioclase-Al_2SiO_5-quartz. This feature perhaps suggests that there was loading during cooling of the Sand Mata granulites.

Based on the above P-T estimates and textural criteria for the crystallization sequence of the Al-silicate polymorphs (kyanite → sillimanite → kyanite), the derived P-T path (Figure 1) for the Sand Mata granulites is found to have an anticlockwise sense and hence reverse to that characterizing the collision zone, thereby placing considerable constraints on the tectonometamorphic evolution of the region in the northwest part of the Indian Shield.

REFERENCES

Holdaway, M.J., 1971, American Journal of Science, v. 271, p. 97-131.

Green, D.H., and A.E. Ringwood, 1967, Geochimica et Cosmochimica Acta, v. 31, p. 767-833.

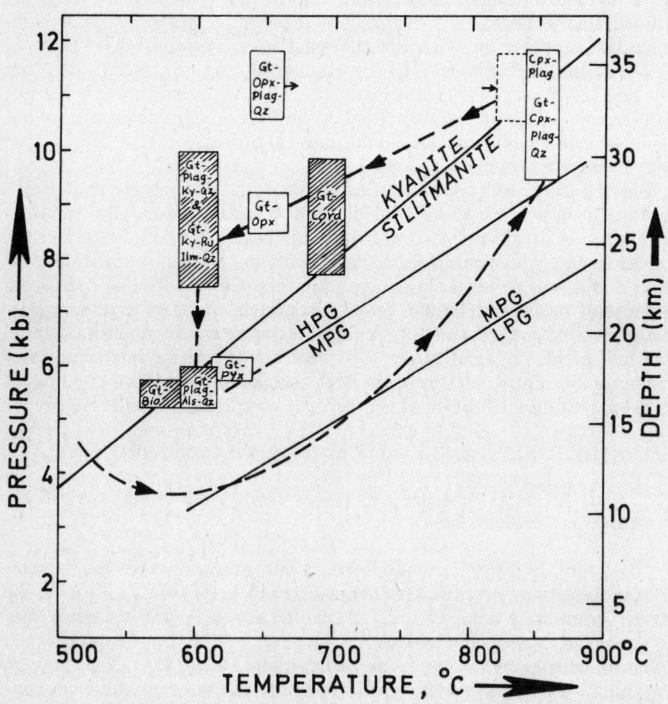

Figure 1. A synoptic diagram showing P-T values and tectonic path of the Sand Mata granulites, depicted in relation to the kyanite-sillimanite univariant curve (Holdaway, 1971) and the boundaries of high-pressure granulite (HPG), medium-pressure granulite (MPG), and low-pressure granulite (LPG) after Green and Ringwood (1967). The P-T boxes for the pelitic granulite are ruled and for the basic granulite are unruled.

SHARP, J. M., JR., The University of Texas, Austin, TX

Stratigraphic, Geomorphic, and Structural Controls of Edwards Aquifer, Texas, USA

The Edwards aquifer of Texas is the principal source of water for almost two million people and is a superlative example of how various geologic processes control an aquifer's evolution as well as its present areal extent, water quality, and hydraulic properties. Important processes include deposition in varying sedimentary environments, early diagenesis, tectonic activity, karstification and fluvial erosion, late diagenesis, and modern geologic processes both within the aquifer itself and in the adjoining Gulf of Mexico Basin.

The Edwards aquifer comprises a narrow (less than 64 km) belt which stretches some 400 km from a groundwater divide near the Rio Grande to the northeast, where it becomes a minor aquifer because of thinning and sedimentary facies changes. Equivalent carbonate rocks form the Edwards-Trinity Plateau aquifer, connected to the Edwards aquifer proper only at its western extremities. Uplift, faulting, and subsequent erosion have largely separated the two aquifers. The Edwards aquifer fringes the present day Llano Uplift. Miocene faulting and subsequent erosion control the location of the unconfined and confined zones of the aquifer. The outcrop's northern-western edge delineates one hydraulic boundary, except in the westernmost areas; the southern-eastern hydraulic boundary is the "badwater line" where dissolved solids increase to over 100 mg/L.

The carbonate rocks making up the Edwards aquifer were deposited during the Cretaceous under generally shallow marine conditions, but changes in depositional environments created varying lithologies. In areas of relative stability, the San Marcos Arch and the Central Texas Platform, formations were thinner and more dolomitized. Early diagenetic effects include dolomitization, micritization, and selective leaching. Flanking these more stable areas were the Maverick Basin, the North Texas-Tyler Basin, and the Karnes and Atascosa troughs in which the units are shalier and much thicker, especially in the Maverick Basin. Bioherms are present in aquifer rocks and extensive reef trends border the larger basins—the Devils River Reef around the Maverick Basin and the Stuart City trend located at the Cretaceous shelf margin. Evaporitic conditions prevailed periodically. Gypsum is widely disseminated, but also concentrated in laterally extensive evaporite members.

Regression and erosion followed deposition of the Edwards Group formations. Over 100 feet of erosion is estimated, with karstification concentrated in zones of greater paleo-elevation. Gulfward and downdip from the aquifer, solutioned zones have produced oil and gas. Concurrent leaching of evaporites also occurred which created numerous brecciated and collapse features in certain aquifer formations (the Person, McKnight, and Segovia formations). The aquifer's interstitial porosity and permeability trends were largely established by these Cretaceous solutioning processes.

Erosional and karstification were followed by transgression and deposition of Washita Group units, the lowest of which also comprises the uppermost Edwards aquifer. Continued deposition produced the overlying, very low permeability Del Rio Clay and other clastic units. The aquifer is confined both above and below, except in outcrop or in zones of significant faulting. Toward the Gulf of Mexico, thick clastic sediments overlie the equivalent formations. Finally, numerous small volcanic plugs penetrated these formations during the Late Cretaceous. On their periphery, these provide zones of fracture enhanced permeability often associated with anomalous water chemistries.

The continued development of the Gulf Coast Basin is associated with Miocene uplift and faulting. High-angle normal faults, dipping predominantly toward the Gulf, offset the aquifer. The faulting is most intensive over the San Marcos Arch. These faults and numerous near-vertical joints roughly parallel the Edwards aquifer outcrop. Simultaneously, fluvial erosion, continuing to the present, has downcut through and isolated the Edwards aquifer from equivalent units in the Edwards Plateau. A few large streams, flowing normal to strike, incised deeply and created points of discharge. Because of the extreme anisotropy of the carbonates, flow changed from a relatively slow downdip sense of motion to a more open system flowing rapidly almost parallel to strike along faults, joints, and leached horizons. The discharge was and still is concentrated at a few large springs near incised major streams. Infiltration of losing streams flowing over unconfined outcrop areas accounts for approximately 85% of total recharge.

Groundwater still dissolves the carbonates, enhancing secondary permeability and the degree of anisotropy. Karstification has been reactivated, particularly in unconfined zones. In some cases, old sinkholes have been rejuvenated. Modern diagenetic processes within the aquifer include dedolomitization, formation of subsurface travertines, and continued leaching of disseminated evaporite minerals.

The badwater zone is created by several processes—continued dissolution of disseminated evaporite minerals, leakage along faults from underlying and overlying units, and the slow updip flow of brines from the compacting Gulf Coast basin (including Edwards aquifer formations). Brine movement is associated with hydrocarbon seeps in aquifer units. The varying importance of these processes creates a systematic change in the hydrochemical facies along the badwater line. Flow conditions within the badwater zone are relatively stagnant and, although in hydraulic connection with the freshwater zone, the transition from fresh to brackish water is often very abrupt. This transition and the change of hydrochemical facies along strike are strongly controlled by faulting.

Finally, the aquifer is under intensive development, which may cause dewatering of portions of the unconfined system, possible updip migration of the badwater line, contamination problems, and diminution of recreationally important springs. Several artificial recharge projects and water resource management plans have been proposed.

SHATA, A. A., Desert Institute, Mataria, Cairo, Egypt

Contribution to the Hydrogeology of the Quaternary Alluvial Aquifer System in the Nile Valley and Delta

The Nile Valley in Egypt, having a north-south length of about 1100 km, an average width varying between 10 km and 100 km, and an average depth in excess of 200 m, is subdivided on hydrogeological basis into three sectors. From south to north, these sectors comprise:

-The High Dam Lake or the Nubian Sector;

-The Nile Gorge or the Upper Egyptian Sector, which includes the Faiyum Depression;

-The Nile Delta or the Lower Egyptian Sector;

The High Dam Lake Sector (± 5000 km^2), is dominantly occupied by dark colored sandstone strata, which are commonly referred to as the Nubian Sandstone Complex (Mesozoic-Paleozoic). These strata act as the main aquifer and constitute only a portion of the major regional aquifer in North Africa. This aquifer, with recorded piezometric levels reaching 100 m $^+$MSL, is in direct contact with the alluvial aquifer system in a small portion of the Nile Gorge to the north of Aswan, but it then disappears below the ground surface as a result of the regional northward dip. An indirect contact between both aquifers is, however, suspected through faulting.

In Upper Egypt, the Nile Valley is morphologically a gorge, filled with Quaternary alluvium, having a maximum reported thickness of about 300 m. The early members of this Quaternary section were deposited by the Prenile in Late Pleistocene times ($\pm 300{,}000$ BP to 200,000 BP) from source rocks located to the south of Egypt, and are dominated by graded coarse sand and gravel. These are followed by coarse gravel and sand and were deposited by a number of pre-existing channels during a great pluvial period of rains, that affected the high land to the east of the Nile in Middle Pleistocene times ($\pm 200{,}000$ BP to 120,000 BP). It is suspected that these deposits have taken, originally, the form of alluvial cones and were developed at the outlets of the main channels, which cut the cliffs bounding the Nile Gorge particularly on the eastern side. The last phase of the Quaternary alluvial section, developed essentially into a silt-clay cap, was deposited by the present Nile during the flood seasons in Holocne times ($\pm 25{,}000$ BP) rocks are essentially the volcanic exposures of the Ethupian Plateau. The Quaternary alluvium is underlain by Pliocene plastic clays of shallow marine origin and is, in places, in contact with the Eocene carbonate rocks, dominantly as a result of faulting. Faulting and also folding, play both a conspicuous role in determining the thickness of the alluvial deposits in the Nile Valley and the regional picture is a marked thinning in the direction of Aswan to the south and in the direction of Cairo to the north.

The aquifer system in the Quaternary alluvium of Upper Egypt is developed into two distinct types, a semi-confined type (leaky aquifer) and an unconfined type (phreatic aquifer) and the Pliocene clay beds at the base act as an aquiclude. In almost all previous work, this system is generally assumed not to be in hydraulic connection with the surrounding karstified carbonate rocks and the underlying Nubian sandstone aquifer system. Recent research work in the Nile Valley is expected to give more evidences, which will lead to disprove this assumption. The extension of the semi-confined aquifer system is determined by the geographical distribution of the Holocene silty clay layer (aquitard), which in other words dominates the cultivated old land in the Nile Valley in Upper Egypt (60 per cent of the valley area). The unconfined aquifer system (phreatic), stretches on the outer sides of the cultivated land area and continues irregularly into the down stream portions of dry streams (wadis) now connected to the valley. The River Nile, which behaves in a similar way to the longitudinal rivers in the northern hemisphere in favouring an active action along the eastern portion of the valley, cuts into the aquifer and acts either as an effluent stream or an influent stream.

Generally speaking, the aquifer types, phreatic or semiconfined are highly productive, with a hydraulic conductivity in the range of 40 to 50 m/day and a transmissivity of 5000 to 15,000 m^2/day. The recharge to the alluvial aquifer system is continuously through infiltration from irrigation water in the cultivated lands, through the irrigation canals, locally through the Nile channel and occasionally through the dry streams (wadi) during the rainy storms (once every seven or ten years). The regional flow direction is northward from 80 m + MSL near Aswan to 20 m + MSL near Cairo. The natural discharge of the groundwater takes place mainly through the River Nile itself (± 1.7 Billion m^{3}y) and through evaporation processes both within the Nile Valley itself and within El Fayoum Depression. That depression lies a short distance to the west of the valley and is connected to it by a narrow passage in the surrounding hills. The lowest part of that depression is occupied by a shallow brackish lake (Birket Qarun) of 200 km^2 in area, situated at about 44 m-MSL. Groundwater is also discharged artificially through abstraction from wells, established both for irrigation and for municipal purposes (± 60 million m^{3}y). The quality of the abstracted water is generally good with salinities less than 1500 ppm.

The Nile Delta (Lower Egypt) reservoir represents the northern portion of the Quaternary alluvial aquifer system. It is a main taphrogeosynclinal basin, where the sedimentary section is in excess of 5000 meters. The top portion of that section is dominated by alluvium, having a thickness varying from less than 100 m in the vicinity of Cairo to more than 1000 m along the Mediterranean Coast.

Similar to the Nile Valley in Upper Egypt, the Quaternary aquifer system is distinct into different types and in view of the development of more clay facies along the coastal zone almost complete confinement is manifested. The aquifer is also underlain by the Pliocene plastic clays, which act as an aquiclude. The aquifer is not only restricted to the deltaic portion, but it continues eastward into Sinai in the area to the north of the Bitter Lakes. It continues also westward to Wadi El Natrun, where it gets in contact with the Moghra Aquifer (Oligo-miocene fluviomarine deposits). The transmissivity of the Quaternary aquifer varies between 10,000 and 20,000 m^2/d, the horizontal hydraulic conductivity amounts to 75 m/d, the vertical hydraulic conductivity amounts to 25 m/d and the effective porosity is in the range of 20 per cent. The groundwater level descends from 20 $_+$MSL at Cairo to close to sea level along the Mediterranean Coast.

The aquifer is essentially recharged by infilteration from irrigation water and its net work through the silty-clay cap. It is also recharged in part through lateral seepages from the same aquifer in the Nile Valley at the junction located in the vicinity of Cairo. The total available groundwater storage in the Nile Delta is of the order of 300 Billion m^3 and the present value of groundwater extraction amounts to 1.5 Billion m^{3}y. The aquifer is naturally discharged through some reaches of the Nile branches, through Wadi El Natrun-Qattar Depressions and through upward leakage along the Mediterranean Zone.

The data available about the geometry of the alluvial aquifer system in both the Nile Valley and Delta, the hydraulic parameters, the boundary conditions and eventually about the regional hydrogeological features have permitted the establishment of mathematical models for the aquifer simulation. The results obtained are continuously updated in view of the availability of more information about the hydrogeology of the aquifer system.

SHATA, A. A., Desert Institute, Mataria, Cairo, Egypt

Aspects of the Hydrogeology of the Nubian Sandstone Aquifer System in North AFrica and in Arabia

The great desert of the old world, dominating North Africa and its extension into Arabia, is underlain by some of the well known large deep-lying aquifers. These are recently described as extensive water mines containing huge groundwater reserves (in excess of 500×10^{12}m^3. The geometry of such aquifers is, generally, determined by the regional geological structure, where it takes the shape of broad geosynclinal basins. In such basinal areas, the strata of hydrogeological interest are dominated by sandstone facies (Paleozoic-Mesozoic), and are given different local or regional names including the Nubian Sandstone Complex, the Intercalary Continental Complex, the Continental Terminal Complex, the Wasie-Tabuq Complex, etc. The thickness of that sandstone complex varies from less than 500 m to more than 3000 m, and it is underlain by hard rocks (Pre-Cambrian Basement Complex), which slope regionally northward and eastward, but with some local and regional structural complications. In spite of such complications, the sandstone section forms two major hydrogeological systems, a western system and an eastern system, with some local connection across North Sinai.

Within each of these two major systems, the aquifers are distinct into a number of multi-layered artesian basins, where the huge groundwater storage reserves were accumulated, essentially during the Pluvials of the Quaternary. Such reserves are, therefore, considered as fossil reserves and although they are known to be practically non-newable, an effective communication between one basin and the other, and also between some basins and the rainy areas, particularly in Africa and Southeast Arabia is undoubted. However, in view of the slow motion of water in the sandstone complex, which is in the order of 140 m/y, the magnitude of the present recharge, especially when compared to the storage reserves, can be considered negligible.

The thick carbonate rocks (mainly Upper Cretaceous and Eocene), which overlie the sandstone aquifer system, display both active and inherited karstified features and are occasionally recharged by upward leakage from the underlying sandstone aquifer system, generally containing high pressure water.

The discharged water of the Nubian Sandstone aquifer system, either naturally or artificially, taking place principally from storage, accounts for the gradual drop in the hydrostatic pressure and for the development of scattered cones of depressions. The areal extent and the magnitude of such depressions depend on the volume of the water abstracted and also on the hydrodynamics of the local basin or bains. In this regards attention can especially be directed to lessons learned from the New Valley Project in the Western Desert of Egypt, which can be taken as a case study.

Because the great storage reserves of the groundwater in the Nubian Sandstone aquifer system are situated in some of the extremely arid regions of the world and represent accordingly the only and or the main available source of water, they have taken increasing importance in the socio-economical development of the desert communities. Large scale development of such groundwater reserves is underway in Maroco ($\pm$20 Billion m^3y), Algeria (20 Billion m^3y), Tunisia (1.2 Billion m^3y), Libya (20 Billion m^3y), Egypt (1.5 Billion m^3y), Saudi Arabia (3.0 Billion m^3y), etc. More than 75% of the water abstracted in such countries is from storage. Similar cases are proven in other arid and subarid areas of the world, including in particular Australia and California in U.S.A.

At present there is a continued debate about the beneficial development of this type of water, where a continued drop in the groundwater pressure is manifested and which may be in excess of 100 m within the coming 50 years. Although it is expected that the advanced technological achievements, shall make the withdrawal of water from such depths economical with advent of the coming century either by use of the traditional energy resources or the non-renewable energy resources, there is vital need to:

-adopt adequate strategies for the development of such non renewable groundwater reserves either in situ or through conveyance techniques to areas away from the abstraction sites.

-stimulate action to integrate technical and organizational efforts among countries both in North Africa and is Arabia in order to establish, through their water policies, a rigid control on such regional aquifer systems.

The lessons learned from the previous experiences of the International Hydrological Program sponsored by UNESCO, and also of the regional organizations like the African Unity and the Arab League will be of supreme importance when formulating the proposed, strategies and when taking a regional action.

SHATA, A. A., Desert Institute, Cairo, Egypt

Geology and Land Utilization With Special Emphasis on Urbanization

In the International Geological Congress held in New Delhi in 1964, the Bureau of the International Association of Engineering Geologists was elected. The main objectives of that association were and still are to add more to the knowledge about the role of earth materials in the socioeconomic development of communities, particularly with regard to the standard of living, the improvement of transportation and communication facilities, housing, urban development, and development of utilities. During the discussions in that congress, the Bureau agreed to classify the aspects of engineering geology into five principal groups, with certain essential subdivisions: (1) naturally occurring materials for construction and industry; (2) relations between geology and engineering structures; (3) prospection techniques; (4) specialized maps used in engineering geology; and (5) documentation, including case histories.

Four years later, UNESCO in cooperation with the Australian Commonwealth Scientific Organization (CSIRO) organized in 1968 the symposium on land evaluation or terrain evaluation. The subject was identified as the assessment of the suitability of land for any possible uses to which it may be put by man, which includes urban development. The term land encompassed, as outlined by the symposium, not only all natural features of the earth's surface, but also all man-made features. In that symposium the "landscape approach," which called for the definition of a part of the landscape with common appearance, was favored and recommended as an adequate system for land classification of large areas of land. The "parametric approach," where individuals were identified objectively from one or more parameters that vary together, was proposed for smaller parts of that large area. Attention was then directed to two important topics: (1) the increasing use of mathematical models of natural process and human management systems, particularly for small areas, as a means of improving predictions and assessing their likely errors, clarifying concepts, defining what data are most relevant to a particular process or activity, and anticipating possibly unforeseen effects of changes in land use; (2) the need for more competence in planning land use and land management, particularly for developing countries, where there is a general low-level technical knowledge, lack of technical resources, and lack of means to avoid wasteful mistakes and delays.

In most developing countries, although the attitude is to avoid or to minimize using sophisticated methodologies to evaluate the land and the potential resources, a reasonable performance is at present attained by a new generation of scientists who have entered the field during the past two decades and who have adequate background in applied earth sciences and physical geography.

The present paper discusses in some detail the lessons learned about the application of geology and geomorphology to urban development in some countries in North Africa and Asia. Emphasis is given to four main topics: (1) the location and type of community, (2) the availability of water resources, (3) the availability of building materials, and (4) the infrastructure.

Cairo City, which is the largest metropolis of Africa and which will reach a population of more than 10,000,000 with the advent of the twenty-first century, is taken as a case study. In that cosmopolitan city, 2500 years old, which occupies an area of about 500 km^2, the land-use situation is as follows: (1) the urban portion, covering an area of about 300 km^2 (60% of the total area); (2) the agricultural portion, covering an area of about 50 km^2 (10% of the total area); (3) the wetland portion, covering an area of about 25 km^2 (5% of the total area); and (4) the quarrying and wasteland portion or the desert, covering an area of about 125 km^2 (25% of the total area).

SHAW, C. A., U.S. Geological Survey, Lakewood, CA, and W. W. HAY, University of Colorado, Boulder, CO

Recent Advances in Mass-Balanced Paleogeographic Modeling

The mass-balanced paleogeographic map program (BPAM) reconstructs the paleotopography and paleobathymetry of a source-sink system based on mass-balance principles (isotasy) and sediment erosion-deposition cycles. The mass-balanced paleogeographic maps are quantitative reconstructions of the surface of the earth at times in the past. The program considers (1) changes in elevation and bathymetry due to bulk sediment transfer, (2) compaction of sediment, (3) isostatic adjustments related to transfer of mass, (4) sea-level changes, (5) subsidence of ocean crust with age, (6) subsidence of passive continental margins with age, (7) uplift and subsidence within the continental interior, (8) changes in the paleodrainage system, and (9) variable uplift rates within the drainage basin.

The output consists of (1) a summary file showing the value of key variables in the modeling process, (2) paleotopographic and paleobathymetric map data, (3) reconstructed uplift/subsidence maps, (4) data showing the thickness of sediment eroded from each grid square during the time interval considered, (5) map of the calculated thickness of oceanic and attenuated continental crust (where applicable), (6) map of the average total sediment thickness in each grid square, and (7) hypsographic information on the reconstructed drainage basin.

Previous mass-balanced modeling programs assumed a closed sediment source-link system. Therefore, the amount of uplift (subsidence) in a drainage basin was related to the ratio of calculated total sediment yield from erosion of the reconstructed topography to the mass of sediment existing in the basin (Shaw, 1987; Shaw and Hay, in press; Hay et al, in press). In reality, however, sediment is usually supplied to a basin from several drainage basins. By establishing a maximum allowable rate of uplift (subsidence) quantitative constraints are imposed on the mass of sediment that may be eroded from a given source area and on the relative amount of sediment that may have been transported out of the basin proper. For example, since the opening of the North Atlantic, the Appalachians have undergone primarily topographic decay (uplift rate 0 m m.y.-1). Balanced paleogeographic reconstructions using this information show that nearly 85% of the mass in the western North Atlantic basin and continental margin was derived from sources other than the eastern flank of the Appalachians.

The method employed to reconstruct paleobathymetry and paleotopography has also been refined to more accurately represent current isostatic theory. The process of isostasy requires that at some depth in the earth the weight of the overlying material is the same. The depth of compensation (Td) is the sum of the thickness of the overlying water (Tw), sediment (Ts), crust (Tc), and mantle (Tm) (equation 1).

$$T_d = T_w + T_s + T_c + T_m \quad (1)$$

The pressure at the depth of compensation (Pd) is assumed to be a constant (k) both spatially and temporally and is represented by the general formula:

$$T_w\rho_w + T_s\rho_s + T_c\rho_c + T_m\rho_m = P_d \quad (2)$$

For the present-day earth, only the thickness of crust (T_c) and mantle (T_m) and the density of mantle (ρ_m) are not known. However, we can make some reasonable estimates as to the average values of both T_c and ρ_m from seismic, gravity, and petrographic information that allows us to calculate the present-day value of P_d. Because the density of the mantle beneath oceanic crust increases with age due to cooling, the pressure exerted by the mantle also increases with age and, therefore, distance from the spreading center. The age-dependent density of the mantle underlying ocean crust can be calculated by rearranging equation 2 and substituting for Tm in the denominator from equation 1. The thickness of the sediment is known from chronostratigraphic information, and the average depth is estimated from seafloor subsidence curves developed by Sclater and others.

After estimating the density of the mantle with age, we are left with two equations (1 and 2) and two unknowns (Tm and Tc). Rearranging equation 2 to solve for the thickness of the mantle and substituting that number into equation 1, the thickness of the oceanic or attenuated continental crust can be calculated. The thickness of crust is assumed to remain constant with time and so is calculated only once at the beginning of the model run.

The reconstructed paleobathymetry is then calculated by rearranging equation 2 to solve for the thickness of water (depth), which is dependent on the age of the underlying ocean crust (mantle density), thickness of ocean or attenuated continental crust, and thickness of sediment cover:

$$\text{Depth} = (k - 100\rho_m + T_c\rho_m + T_s\rho_m - T_c\rho_c - T_s\rho_s)/(\rho_w - \rho_m)$$

Paleotopography is reconstructed using identical principles and a slight modification to the above equation. Because the depth of water is zero when reconstructing elevation, the topography is:

$$\text{Altitude} = (k - T_c\rho_c - T_s\rho_s)/\rho_m$$

These and other smaller refinements in the mass-balanced modeling program allow more accurate reconstructions of the evolution of sedimentary basins and their source areas that are constrained by geologic information and modern earth-forming principles.

SHAW, C. A., U.S. Geological Survey, Lakewood, Colorado, and M. A. LESLIE, University of Colorado, Boulder Colorado

Interdependent Nature of Paleoclimatic, Paleohydrologic, and Mass Balance Modeling

Balanced paleogeographic modeling (BPAM) reconstructs paleotopography and paleobathymetry in an open sediment source-sink system by considering (1) erosion and deposition cycles, (2) thermal subsidence of ocean crust with age, (3) sea level changes, (4) sediment compaction, (5) sediment recycling, (6) subsidence of passive continental margins with age, (7) changes in porosity and pore fluids, (8) changes in paleodrainage systems, (9) variable rates of continental uplift, and (10) isostasy. Standard BPAM output consists of maps showing (1) reconstructed paleotopography and paleobathymetry, (2) estimates of continental uplift, (3) mass of sediment eroded from areas in the drainage basin per unit time, (4) hypsographic information, and (5) summary data concerning the average area, elevation, and sediment mass associated with denudation of the source area. The output from such modeling provides necessary quantitative input for paleoclimatic, paleoceanographic, and paleohydrologic reconstructions.

The global climate can be generalized as a function of primarily four large scale factors; (1) latitude, (2) the distribution of oceans and land masses, (3) the distribution of mountains, and (4) orbital variations. Although many other factors are important in climate modeling to the first order the above four elements generally represent the major processes. Therefore, the accuracy of paleoclimatic reconstructions is dependent upon the quality of data describing paleogeography and paleotopography and is constrained by geologic and biologic data. The position of large land masses through time both with respect to one another and, to a varying degree of accuracy, with respect to latitude, is reasonably well known and has been documented in numerous paleogeographic studies. More recently, global analysis of accreted terranes has refined the classical view of global palegeography (Wilson et al.) and has helped provide an excellent quantitative base for the distribution of land and oceans through time.

In contrast, although the distribution of ancient mountain ranges and highlands may be inferred from stratigraphic and/or structural evidence, a quantitative reconstruction of orography is rarely available. The effect of a mountain range on circulation patterns is a function of its orientation to the air flow and maximum elevations (i.e. the amount of convergence and the associated change in velocity). The distribution of elevations within the orogenic belt further controls the interference of the trough on the leeward side of the topographic ridge. Circulation models run with a linear mountain belt normal to air flow and divide of equal elevation show atmospheric circulation with parallel flow patterns (Figure 1a). In reality, variations in summit elevations produce non-parallel circulation patterns caused by differing degrees of convergence and vorticity (figure 1b). Although general trends and relative heights of mountain ranges have previously substituted in general circulation models for the lack of numerical paleotopographic maps, BPAM allows quantitative reconstructions of ancient topography that are essential for refining paleoclimatic and global circulation models.

The hydrologic conditions in a sediment source-sink system are primarily a function of topography, rock type, and climate. The accuracy of paleohydrologic reconstructions, therefore, is inherently linked to the reliability of paleotopographic, paleogeologic, and paleoclimatic models. Topographic gradients and the hypsometry of the drainage basin create spatial differences in the hydraulic head that result in transient surface and subsurface flow conditions over time. Surface flow conditions are governed by runoff, groundwater discharge, elevation, and slope and are represented in the erosion-deposition cycle by the mass-age distribution of lithologies eroded from the source area and deposited in the basin. The pattern of sediment distribution in the basin is related to the distribution of elevations in the source area. BPAM reconstructions of the size and relief of a watershed provide paleohydrologic models with quantitative boundary conditions that define necessary areal and volumetric limits, as well as paleodrainage patterns in the watershed through time. Subsurface flow conditions are a function of the porosity and permeability of the transmissive layer. BPAM reconstructions of the precompaction porosity of lithologic units in the subsurface provide paleohydrologic models with quantitative estimates of the compaction history and original thickness of sediments in a basin thereby imposing limits on the size of the transmissive layers and on the volume of the groundwater system.

The hydrologic cycle is measured by its water budget which in turn is a function of the orography of its drainage basin, geology of the source-sink system, and climate. The present relationship between climate and the hydrologic cycle emphasizes that paleohydrologic models must incorporate precise paleoclimatic reconstructions which in turn should incorporate precise paleotopographic reconstructions. Paleohydrology is also directly related to changes in watershed relief, ground slopes, and the porosity of both rocks in the drainage basin and the transmissive layer and, therefore, is critically dependent upon paleotopographic and paleogeologic reconstructions. Marrying BPAM reconstructions of paleotopography and paleobathymetry with paleoclimatic and paleohydrologic models provides a critical component of topography that has not previously been well quantified as input to other 3D models.

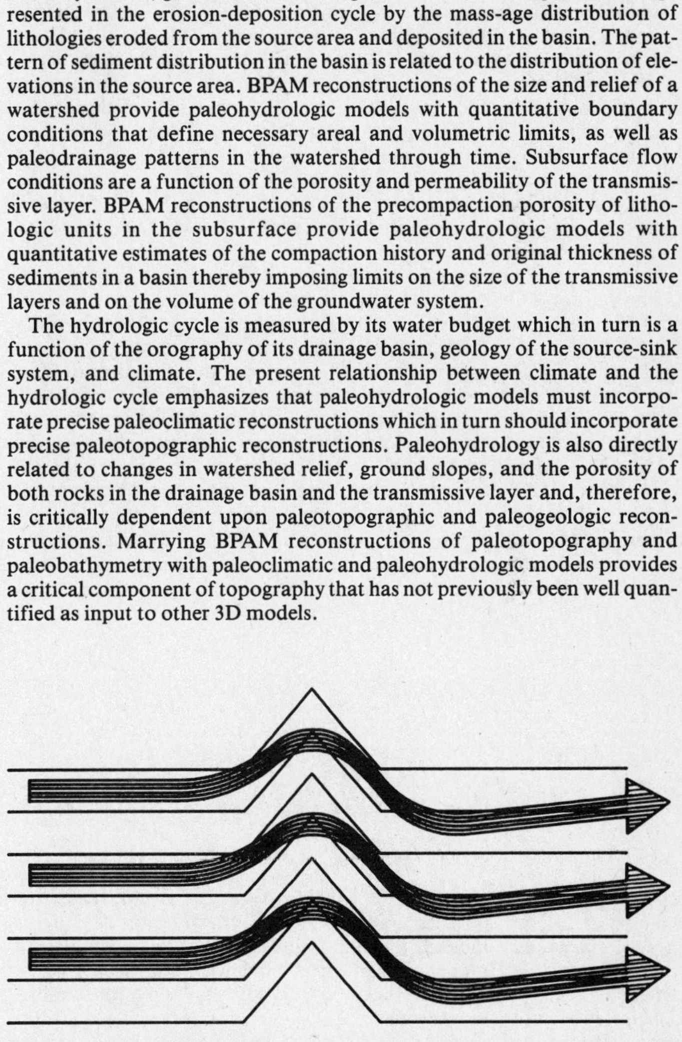

Figure 1a—Schematic showing parallel patterns over hypothetical mountain range with identical peak elevations.

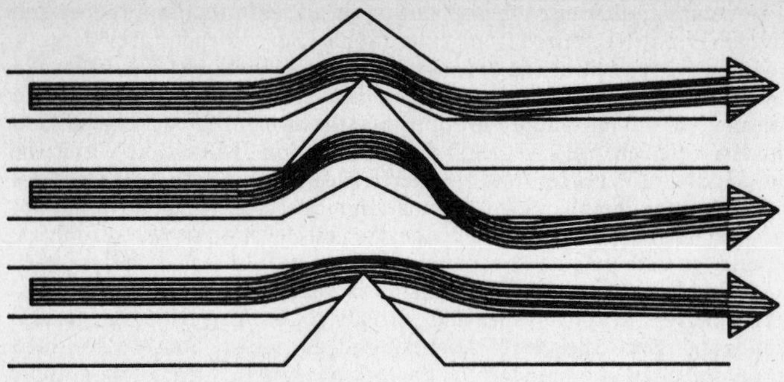

Figure 1b—Schematic showing unparallel flow patterns over mountain range with irregular topography.

SHAW, H. R., and B. CHOUET, U.S. Geological Survey, Menlo Park, CA

Hierarchical Volume Budget of Fracture-Related Magma Transport in Hawaii

Summations of tremor data by Shaw and Chouet (1988) for regions beneath Kilauea, Mauna Loa, and Loihi volcanoes give 1.60 $\times$ 10^4 min for deep tremor (30–60 km depth) and 0.94 $\times$ 10^4 min of intermediate-depth tremor (5–15 km depth from 1 January 1962 through 31 December 1983; the depth interval 15–30 km had little detectable tremor activity during this time. Average vertical tremor gradients are 24.2 min y^{-1} km^{-1} (30–60 km), 42.7 min yr^{-1} km^{-1} (5–15 km), and 21.0 min y^{-1} km^{-1} (5–60 km). When the quiet interval is combined with the deep and intermediate intervals, the value 21 min y^{-1} km^{-1} satisfies the 5–60 km average as well as the 5–25 and 25–60 km averages. We have taken this as a steady-state reference frame for vertical distribution of tremor rates.

For the sake of creating a budget, we define a characteristic tremor event as that portion of a tremor seismogram (typically ranging from 1 to 100 min) that is equivalent in time to a long-period earthquake. Then the numbers of characteristic extensional fracture events of the tremor process are obtained by dividing tremor durations by an event duration time of 20 s, typical of long-period seismograms (Koyanagi et al, 1987; Chouet, 1988). These 20-s events are analogous to pump strokes or pulses in the magma transport rate, and a summation of all strokes over all episodes gives the magma displacement by tremor-related fracture. Averaging over characteristic times and volumes defines the mean concentrations, mass fluxes, and the drift velocities of magma transport. Aseismic extensional fracture is then included by introducing a factor of $1/k$ (below). Taking 1 km as a maximum for crack length (a few hundred meters is more consistent with estimates of tremor source lengths), the steady-state gradient thus gives 63 detachable events (magma pulses) per year per 1 km layer (i.e., 21 min/20 s).

Long-period events and tremor are clustered in both the vertical and the horizontal directions (Koyanagi et al, 1987). A mean-field model permits us to write idealized mass balances as functions of event counts and to test departures from the continuum state. Balances are then written for a total supply rate (intrusive + extrusive) of 0.1 km^3 y^{-1}, i.e.,

$$V_o = (k\phi\Delta S)/n, \qquad (km^3) \qquad (1)$$

and

$$\nu = 0.1/(\phi\Delta S), \qquad (km\ yr^{-1}) \qquad (2)$$

where V_o is the volume equivalent to a 20 s event, ΔS is the area across which vertical transport occurs, ϕ is the volume fraction of magma involved in fracture-induced transport within a 1 km layer (we do not include static or intergranular magma in this balance), n is the number of detectable events per layer (average = 63 y^{-1}), k is the ratio of the number of detected events to the total number of virtual events (seismic + aseismic) required to satisfy the magma supply rate, and υ is the mean upward transport velocity (dz/dt). Data for dike dimensions and source mechanisms of tremor (Shaw and Chouet, 1988) are used to assign $V_o = 2 \times 10^{-5}$ km^3 (2 $\times$ 10^4 m^3). To satisfy both equations for other parameters, we have included an aseismic component of transport. Making k = 0.05, we satisfy V_o and n for reasonable ϕ and ΔS and for ratios of tremor moments to magma transport rates (Aki and Koyanagi, 1981). Given V_o, k, and n, we solve directly for $\phi\Delta S$ ($\sim$ 0.025 km^2) and υ ca $\sim$4 km/y). The relationships between magma fractions and domain sizes have been expressed in terms of logϕ vs. log R_d, where R_d is the radius of a sphere equivalent to a specific domain volume; e.g., individual tracks, the estimated average volume of long-period event hypocenters, conduit-like paths of Ryan (1988), and the total subvolcanic lithosphere.

Results demonstrate that a model in which magma-filled cracks are clustered hierarchially is consistent with magma transport rates that are proportional to tremor rates (asesimic and seismic transport are closely coupled). Magma fractions increase within envelopes of decreasing R_d according to a power law in which the exponent depends on assumptions concerning the size scale of a completely different molten region, where ϕ = 1. There are two general cases for this limiting size scale: (A) size is referred to the locus of a single magma-filled crack or crack cluster with a volume between 2 $\times$ 10^4 and 4 $\times$ 10^5 m^3; (B) size is referred to the locus of a chamber with a volume of about 1 km^3. Exponents are converted to fractal scaling dimensions, D, based on definitions given by dr Vaucouleurs (1970) and Mandelbrot (1982, p. 85). We have obtained D = 1.5 $\pm$ .3 for limit (A), and D = 0.3 $\pm$ .3 for limit (B). Interestingly, scaling is nearly the same as that of de Vaucoleura (1970) for the clustering of galaxies (limit A) and for point-like distributions of stars (limit B). This is not surprising, however, if both stem from dissipative (convergent) nonlinear dynamical processes (cf. Shaw, 1987; 1988).

REFERENCES

Aki, K., and R. Koyanagi, 1981, Deep volcanic tremor and magma ascent mechanism under Kilauea, Hawaii: Journal of Geophysical Research, v. 86, p. 7095–7109.

Chouet, B., 1988, Resonance of a fluid-driven crack: radiation properties and implications for the source of long-period events and harmonic tremor: Journal of Geophysical Research, v. 93, p. 4375–4400.

De Vaucouleurs, G., 1970, The case for a hierarchial cosmology: Science, v. 167, p. 1203–1213.

Koyanagi, R. Y., B. Chouet, and K. Aki, 1987, Origin of volcanic tremor in Hawaii. Part I. Data from the Hawaiian Volcano Observatory, *in* R. W. Decker, T. L. Wright, and P. H. Stauffer, eds., Volcanism in Hawaii: U. S. Geological Survey Professional Paper, v. 1350, p. 122–125.

Mandelbrot, B. B., 1982, The fractural geometry of nature: San Francisco, W. H. Freeman, 460 p.

Ryan, M. P., 1988, The mechanics and three-dimensional internal structure of active magmatic systems: Kilauea, Hawaii: Journal of Geophysical Research, v. 93, p. 4213–4248.

Shaw, H. R., 1987, The periodic structure of the natural record, and nonlinear dynamics: Eos v. 68, p. 1651–1665.

——— 1988, Terrestrial–cosmological correlations in evolutionary processes: U S. Geological Survey Open-File Report 88–43, 85 p.

——— and B. Chouet, 1988, Application of nonlinear dynamics to the history of seismic tremor at Kilauea Volcano, Hawaii: U. S. Geological Survey Open-File Report 88–539, 78 p.

SHCHEGLOV, A. D., Ministry of Geology of the USSR, Bolshaya, Gruzinskaya

Ore Deposits–Indicators of Areas of Tectonomagmatic Activation

One of the most important general features of ore deposits distribution in the earth's crust, which has a role of a geological law, is the association of certain groups of deposits with definite tectonic regimes and structures which characterize them.

Areas of tectonomagmatic activation–nongeosynclinal mobile structures of the earth's crust represent an independent category of regional (planetary) structures of the same rank as geosynclines and platforms, which show up most distinctly during the continental stage of crustal evolution on a rigid consolidated basement (substrate) of fold areas, platforms and large median masses of different ages. Examples of tectonomagmatic activation areas are: Basin and Range Province in the Western United States, Southwest China, Brazil and Aldan Shield, East Africa and other regions.

Areas of tectonomagmatic activation display a number of features, among them a specific complex of endogenic deposits, which are typical only for given structures. These deposits serve as peculiar indicators of the regime of tectonomagmatic activation.

In the evolution of activation areas two independent stages can be clearly distinguished, each of them corresponding to definite superimposed structures. During the first stage, the superimposed gentle troughs commonly appear, filled by continental volcanic-detrital formations. They are formed in connection with major faults, which impart a block structure to the substrate, on which these structures are emplaced. Commonly, such troughs have an isometric shape. The formation of such structures is associated with shows of fracture, often subvolcanic intrusions of different composition, accompanied by diverse mineralization.

The second stage of activation is noted for the appearance of superimposed terrigenous basins, i.e. embryonal rifts, filled by coarse clastic continental coal-bearing deposits. These structures are commonly rimmed by major zones of long-lived tectonic faults. Sometimes these are depressions extending along faults, separated by transverse uplifts, composed of basement rocks; in these cases, garlands of basins appear. The most contrasting, extreme expression of the second activation stage are young continental rift zones.

Areas of tectonomagmatic activation are, as a rule, major ore provinces and areas, where each stage has a specific complex of ore deposits. The most diverse spectrum of deposits appears during the first stage. There are tin, tungsten, molybdenum, gold ore, polymetallic, copper, uranium and rare earth (carbonatite) deposits. They are characterized by multistage shows, development of moderate-temperature productive stages; in rare metal deposits, the emplacement of independent sulfide ores; superimposed on tin, tungsten and molybdenum proper earlier mineral associations; phenomena of horizontal and reverse vertical zonation.

During the second stage, epithermal deposits are formed practically in all the regions; they are represented by fluorite, barite, manganese, less frequently antimony, iron ore, cinnabar, uranium and tungsten ore formation.

In a general case, this group of deposits is characterized by the following successive series of their shows: fluorite + gold + antimonite + ferberite (scheelite) + cinnabar.

The analyses of the specific features of shows of certain elements in the deposits of activation areas demonstrates that there are indicator-elements, typical of endogenic ore processes, associated with the development of these structures. The most characteristic indicator element is fluorine, which forms large concentrations in the deposits of activated areas. It is usually connected with barium, manganese, less frequently lead in single, complex deposits.

Another indicator-element is manganese. In the deposits of the first stage it is fixed in hübnerite, triplite, rhodonite, rhodochosite, manganocalcite, less frequently, manganese oxides; in epithermal deposits of the second stage, of utmost significance are manganese oxides and hydroxides along with a limited development of other manganese minerals. In some activation areas, deposits of the second stage display exotic combination of elements: antimony + tungsten + cinnabar; tungsten + chromium.

The most characteristic indicator deposit of tectonomagmatic activation areas, among them continental rifts, are fluorite deposits. They are typical of many regions and primarily the Western United States (Basin and Range Province) and Western Transbaikal area. Three groups are distinguished among fluorite deposits: a, fluorite and, less frequently, barite-fluorite deposits, in most of which fluorite is closely associated with chalcedony-like quartz; b, sulfide-fluorite deposits (vein and stratiform); c, fluorite-bearing carbonates.

The common feature of the considered group of fluorite deposits is their close connection with zones of regional faults, which are particularly evident in areas with the ancient consolidated basement. Areas of development of fluorite deposits, as a rule, coincide with the areas of shows of activation processes, in whose evolution two stages are distinguished. In rare instances, in fluorite-bearing provinces, these stages are not clearly expressed or are represented in the "embryonal" development (Illinois and Kentucky states).

The endogenic deposits in activation areas are characterized by the presence of liquid and hard bitumens, which, to a certain degree, indicates not only their belonging to these structures, but also outlines links between ore and oil deposits.

In many regions, areas of tectonomagmatic activation are surrounded by telethermal polymetallic and fluorite deposits of a stratiform type, which allows to presume the connection of these deposits with development of activation areas, of which they are a peculiar reflection.

The presence of fluorite deposits in the regions, where tectonomagmatic activation processes are not revealed, should point to the development of such processes and a necessity of a comprehensive additional analysis of the evolution history of the given territory, as well as the possibility of discovering the deposits of other genetic groups, which characterize areas of tectono-magmatic activation, in these regions. Therefore, even the presence of ore shows and small fluorite deposits of the above groups, particularly in the regions, which represent ancient stabilized structures, is of great significance for their general predictive evaluation.

SHCHEGOLEV, I. N., Voronezh State University, USSR

The World Pre-Cambrian Iron-Formations: Structure, Genesis, Evolution

Early stages of earth evolution are characterized by considerable duration of rock-forming and ore-forming processes, specific paleogeographic and paleotectonic conditions, peculiarity of geochemical and metamorphic processes. All these factors exerted substantial influence over originality of the conditions of formation of Pre-Cambrian thicknesses and definite ore-formation types timed to them. Iron takes the leading place among them. Productive horizons of the named element have spreading in Pre-Cambrian sections of ancient platforms of all the continents. These horizons occupy vast spaces (hundreds of km), are characterized by considerable thicknesses (hundreds of m), and possess an enormous industrial accumulation (many milliards of tons).

Comparative analyses if sections of Pre-Cambrian shields and massifs allowed to establish quite definite stratigraphic levels with iron-ore beds and horizons. They are often found among Archean and Lower Proterozoic thicknesses. Iron-formations of the same age have a lot of similar features with all world iron-ore provinces: Voronezh massif (Kursk Magnetic Anomaly), Ukrainian, Western Australian, Canadian, Indian, China-Korean, Brazilian, Rhodesian, Transvaalian and other shields (Table 1).

Geological age	Geochronological age (million years)	Regions: Southern parts of Russian platform: Voronezh massif (Kursk Magnetic Anomaly)	Southern parts of Russian platform: Ukrainian shield	African platform: Rhodesian, Transvaalian shields	Indian platform	China-Korean platform	Western part of Australian platform	Canadian shield (Lake Superior region)	Guianian, Western and Central Brazilian shields
Upper parts of Lower Proterozoic	1650±50	Oscol	Frunzenskaja	Pretoria	Upper Dharvar	Huto	Wyloo	Baraga	Piracicaba
Lower Proterozoic	1900±100	Kursk	Krivoy Rog	Ghaap, Black-Reef	Medium Dharvar	Upper Utai	Hamersley, Forteskew	Menominee, Chokoley	Itabira, Carasa
Upper Archean	2700±100	Mikhailov	Konksko-Verhovzevskaja	Swaziland	Lower Dharvar	Lower Utai	Kulgardie-Kalgoorlie	Keewatin	Rio Dos Velijs, Pastobarama
Lower Archean	3000±100	Oboyan	Centralno-Priasov, Aulskaja, Dnestrovo-Bugskaja	Granite-Gneiss	Indostan Gneiss	Taishan Gneiss	Granite-Gneiss	Pre-Keewatin	Itamaca
	3500±100								

The names of series, complexes, systems are in accordance with the native ones.

Table 1—Interregional stratigraphic correlation of Lower Pre-Cambrian sections of the main world iron-ore provinces.

We see quite another picture while comparing different time iron-formations even within same region. According to the conditions of formation they are characterized by considerable distinctions, peculiarity of structure, chemical and mineralogical composition and productivity. Confrontation of leading features of iron-formations from older to younger ones reveals irreversible alterations in them.

Thus Lower Archean iron-formations of all the world platforms have on the whole magnetite composition, coarse-grained texture, banded or indistinct banded structure, small thickness, low extent, poor productivity, even high-grade metamorphism, with geochronology data more than 3000 million years. Among Upper Archean greenstone belts the productivity of iron-formations are characterized by clear banded structure and medium-grained texture. The degree of metamorphism of them is middle. In most cases iron-formations are bound up with basic and intermediate volcanic rocks. The time of forming of these iron-formations is between 3000 and 2700 m.y. ago. There are enormous iron deposits among Lower Proterozoic rocks. These iron-formations have thinly banded structure, fine-grained textures. Magnetite and hematite are the main ore-forming minerals of them. These productive horizons are regionally metamorphosed in degree of green slate, and occupy middle parts of Lower Proterozoic section. Iron-formations are occurred as a rule in slates and seldom in carbonate rocks. These ore bodies form very consistent horizons with the strike of about hundreds of km and thicknesses about hundreds of m. The sedimentation of these iron-formations took place between 2700 and 1900 m.y. ago. There are ferruginous beds situated on higher Lower Proterozoic stratigraphic level. They differ from the previous ones as they sedimented as clastic products of weathering of jaspilites and older iron ores. The time of forming of these iron-formations is between 1900-1650 m.y. back.

The evolution of Lower PreCambrian may be observed by volcanic and intrusive rocks as well as tectonic processes, degree of metamorphism, etc.

Comparing Pre-Cambrian rock sections with iron-formations of continental platforms has permitted to detect many similarities in their structure. Large stratigraphic units (series, complexes or systems) as a result are separated from each other by stratigraphic or structural unconformities, by the presence of old crusts of weathering. These units are characterized by peculiarities of forming. Iron-formations and their enclosing rocks have similar but not identical features of structures. Large scale of sedimentary and volcanic rocks and many other features confirm the changeability of their conditions of forming.

SHCHERBAK, N. P., E. V. BIBIKOVA, and M. A. SEMIKHATOV, Inst. of Geochem. and Anal. Chem., USSR Academy of Science, Moscow, USSR

Geochronological Scale of the USSR Precambrian

General stratigraphic scale of the USSR Precambrian (SSP) is based on chronostratigraphic principle and rests on the natural sequence of selected type rock complexes. The chronostratigraphic approach provides material essence for the established subdivisions, enables to get their paleontological, radiometric, paleomagnetic, geochemical, etc. characteristics and to broaden methodical base of correlation.

Geochronological scale of the USSR Precambrian is derived from the SSP and proved by U-Pb and Sm-Nd datings in the lower part (over 1650 Ma) and by U-Pb, Rb-Sr and partly K-Ar datings in the upper.

An important role in periodization of Precambrian events is played by distinguishing large tectono-magmatic cycles whose beginning is marked by initial volcanism and sedimentation in the newly formed structural framework, and termination by crust-forming events. Contrary to the adopted routine, it is suggested to define major stratigraphic boundaries by the beginning and not by the termination of such cycles.

The structure of the USSR Archean prove its two-member subdivision. The Lower-Upper Archean boundary is defined by origination an older-than-3.3-3.4 Ga tonalite-gneiss basement of greenstone belts dated at 3000 Ma on the Baltic and Aldan and at 3200 Ma on the Ukrainian Shield. This settles the boundary at 3200 ± 100 Ma (see the Table). Late Archean magmatism terminated on these shields 2650-2600 Ma ago.

The onset of Lower Proterozoic volcanism and sedimentation in North Eurasia is marked by the development of a new structural pattern independent of the preexisting one. Basal horizons of the Lower Proterozoic in Karelian sections regarded as the type ones for the USSR and in the Ukranian ones are dated at 2500-2450 Ma. This defines the Archean-Proterozoic boundary age as 2500 ± 50 Ma.

The two-member subdivision of the Proterozoic accepted in the SSP is proved unique for the Proterozoic crust-forming significance of the pre-Riphean events and geological unity of the North Eurasia Riphean. The Lower-Upper Proterozoic boundary (the Lower Riphean boundary) is defined by riftogenesis and basins origination on the epi-Svecofennian cratons; it coincides with important events in the paleontological record. Its isotopic age estimated at 1650 ± 20 Ma is proved by data on termination of East European platform basement cratonization (1670-1700 Ma) and on volcanics at the base of the Riphean type in the Urals (1635 ± 30 Ma).

Two general subdivisions can be discriminated within the Lower Proterozoic. Their boundary (1970 ± 20 Ma) in type Baltic Shield sections is defined by the base of the Svecofennian rock sequence (Kalevian base).

The Upper Proterozoic in the USSR is subdivided into three-member Riphean and two-member Vendian. Their boundaries are defined by geological events in selected type areas and by some changes in the organic world which are fixed in the succession of microfossil and stromatolite assemblages and in the Upper Vendian by Metazoa appearance.

The Lower-Middle Riphean boundary is dated at 1350 ± 20 Ma by the mutually consistent Rb-Sr and zircon U-Pb data on basal Middle Riphean volcanics and post-Lower Riphean granites. Estimation of the Middle-Upper Riphean boundary age (1000 ± 50 Ma) is supported by K-Ar data only. The age of the Riphean-Vendian (650 ± 20 Ma) and Lower-Upper Vendian (620 ± 15 Ma) boundaries is defined by mutually consistent datings on the pre-Vendian (U-Pb) and Vendian (Rb-Sr) granites and Vendian sedimentary geochronometers (Rb-Sr, K-Ar).

The age of the Archean-Proterozoic boundary and the Lower Proterozoic time-range in the proposed version of the SSP corresponds to those accepted in the chronometric Precambrian scale of the International Subcommission on Precambrian stratigraphy. Peculiarities of the late Precambrian geological history of North Eurasia reflected in the SSP stipulated the differences of the relevant parts of the above scales.

General subdivisions				Isotopic age, Ma	Types (underline) and typical representatives			
					Baltic Shield (Eastern part)	Ukrainian Shield	Urals and E. Europ. platform	Aldan Shield
Proterozoic	Upper	Vendian	Upper	620±15			Valday Gr. Volyn' Gr.	Yudoma Gr.
Proterozoic	Upper	Vendian	Low	650±20			Vilchany Gr.	
Proterozoic	Upper	Riphean	Upper	1000±50			Karatau Gr.	Uy Gr. Lakhanda Gr.
Proterozoic	Upper	Riphean	Middle	1350±20			Yurmata Gr.	Kerpyl' Gr. Aimchan Gr.
Proterozoic	Upper	Riphean	Lower	1650±20		Tolkachy Gr.	Burzyan Gr.	Uchur Gr.
Proterozoic	Lower	PR_1^2		1970±20	Vepsian, Kalevian & other Svecofennian supracrustals	Ovruch Gr. Pugachevka Gr. Klyosovo Gr.		Ulkan-Uyan Gr. Akitkan Gr.
Proterozoic	Lower	PR_1^1		2500±50	Ludikovian Jatulian Sariolian Sumian	Frunze Gr. Krivoy Rog Gr.	Taratash Gr.	Udokan Gr.
Archean	Upper			3200±100	Lopian	Konka-Verkhovtsevo Gr.		Olondo Gr.
Archean	Lower				Vodlozero gneisses	Aul Gr. Novopavlovka Gr.		Kurulta Gr.

Table 1—Geochronological scale of the USSR Precambrian.

SHEARER, C. K., J. J. PAPIKE, and S. B. SIMON, Institute for the Study of Mineral Deposits, SDSM&T, Rapid City, SD, N. SHIMIZU, Massachusetts Institute of Technology, Cambridge, MA, and H. YURIMOTO and S. SUENO, University of Tsukuba, Ibaraki, Japan

Ion Microprobe Studies of Trace Elements in Lunar "Volcanic" Glass Beads and Comparisons to Mare Basalts

Homogeneous, mafic lunar glasses are believed to be volcanic in origin (Ridley, et al, 1973; Delano, 1986). The major element geochemistry of these glass beads indicates that unlike mare basalts, which have undergone significant near-surface fractionation and possible crystal accumulation, the volcanic glasses are closer to liquid compositions and in many cases represent less fractionated magmas (Delano, 1986). Because of those important attributes, we have initiated an ambitious program to obtain ion microprobe data for trace elements in the lunar volcanic glass groups identified by Delano (1986). Trace element evaluation of the petrogenetic connection between mare basalts and the picritic volcanic glasses is difficult due to the rarity and small size of individual glass beads (less than 100μm). Only in rare occurrences (A-15 green glass; A-17 orange glass) have trace element analyses of picritic glass beads been performed. With the introduction and advance of trace element microbeam technologies (e.g., secondary ion mass spectrometry, SIMS), the analysis of minute lunar glass beads (10-100 μm) is possible (Simon et al, in press), and is not restricted to glass types present in high concentrations. Here we focus on the trace element characteristics of the picritic glass beads at the Apollo 14 landing site, their relation to mare basalts, and the nature of their source regions within the lunar mantle.

Glass beads were identified and documented in selected polished thin sections of A-14 regolite breccias using an electron microprobe. This allowed us to place the volcanic glasses within the chemical classification scheme as defined by Delano (1986). These chemically documented glass beads were then analyzed with Cameca IMS-3f ion microprobes operated by the MIT-Brown-Harvard Consortium and the University of Tsukuba. Apollo 14 Green B, Green A, and VLT glasses are typically LREE enriched with Sm/Eu = 6-14. The total REE abundances of Green A and VLT overlap (Ce = 30 to 70 × chondrite) whereas the REE abundances in the more MgO-rich Green B are substantially lower (Ce = 12 to 14 × chondrite). Zirconium (66-508 ppm) and Ba/Sr (> 1.4) in these glasses are distinguishably different from low-Ti mare basalts at other sites but are similar to those of basalts sampled at the Apollo 14 site. The high-Ti (orange and black) picritic glasses exhibit a flat to slightly positive LREE slope and a negative HREE slope. The orange glass has a higher total REE abundance (Ce = 85 to 120 × chonrite) and a slightly larger negative Eu anomaly (Sm/Eu = 4-7) than the black glass (Ce = 55 to 90 × chondrite; Sm/Eu = 3-5). Relative to Apollo 17 orange glass (74220), the Apollo 14 high Ti picritic glasses are enriched in REE, LREE/HREE, Y, V, Zr, Sr, Ba, and Ba/Sr and are similar in the alkali elements (Li, Rb), Co, and Sc.

Genetic linkages among the picritic glasses, between the picritic glasses and the aluminous basalts at the Apollo 14 site, and between the picritic glasses and other mare basalts can be evaluated using the liquid line of descent and crystallization sequence constructed from Longhi (1987), distribution coefficients in the literature, and trace element characteristics of the Apollo 14 picritic glasses documented in this study. Liquid lines of descent suggest the black, orange, Green A, and VLT-Green B are not related by low-pressure crystallization processes. Modeling of the trace element abundances of basaltic liquids along the liquid lines of descent using the different picritic glasses indicates (1) calculated basaltic liquids derived from orange or black glasses have Sr, Ba, and REE concentrations considerably different from most high Ti basalts; (2) it is possible that either the Green A or VLT glass compositions may be derived by crystallization of olivine from a liquid of Green B glass composition; (3) Apollo 17 VLT basalts are not derived from liquids represented by Apollo 14 VLT glasses; (4) picritic volcanic glasses sampled at the Apollo 14 site are not related by low-pressure crystallization processes to produce the Apollo 14 aluminous mare basalts; and (5) if the glasses sampled at the Apollo 14 site represent unmodified primary basaltic liquids, the basalts derived from these picritic parents have not been recognized.

Although the above modeling indicates low-pressure crystallization processes do not petrogenetically link the different basaltic liquid compositions represented by the picritic glasses and basalts at the Apollo 14 site, many of the basaltic compositions share similar trace element signatures (e.g., Ba/Sr, Sm/Eu, REE enrichment, LREE enrichment). This suggests an intrinsic character of the crust-mantle at the Apollo 14 site that imparts this signature to many of the basalts produced. This signature has been attributed to the assimilation of a KREEP component (Shervais et al, 1985; Dickinson et al, 1985) or partial melting of a hybrid mantle source with varying proportions of trapped or mixed evolved liquid (Hughes et al, 1988).

REFERENCES

Delano, J. W., 1986, Pristine lunar glasses: criteria, data and implications: Proceedings of 16th Lunar Planetary Science Conference, p. D201-D213.

Dickinson, T., et al, 1985, Apollo 14 aluminous mare basalts and their possible relationship to KREEP: Proceedings of the 15th Lunar Planetary Science Conference, p. C365-C374.

Hughes, S. S., et al, 1988, Chemistry of 74220 orange mare volcanic glass; implications for their magmatic source composition: Lunar and Planetary Science XIX, p. 517-518.

Longhi, J., 1987, On the connection between mare basalts and picritic volcanic glasses: Proceedings of 17th Lunar Planetary Science Conference, p. E349-E360.

Ridley, W. I., et al, 1973, Apollo 15 green glass: Phys. Earth Planet. Interiors, v. 7, p. 133-136.

Shervais, J. W., et al, 1985, Apollo 14 mare basalts: petrology and geochemistry of clasts from consortium breccia 14321: Proceedings of 15th Lunar Planetary Science Conference, p. C375-C395.

Simon, S. B., et al, in press, Petrology of Apollo 14 regolith breccias and ion microprobe studies of glass beads: Proceedings of 19th Lunar Planetary Science Conference.

SHEDLOCK, R. J., and P. J. PHILLIPS, U. S. Geological Survey, Towson, MD, D. A. WILCOX, U. S. Fish and Wildlife Service, Ann Arbor, MI, and T. A. THOMPSON, Indiana Geological Survey, Indiana University, Bloomington, IN

Importance of Both Regional and Local Patterns of Flow and Chemistry in Hydrogeologic Investigations of Freshwater Wetlands

Most scientific investigations of freshwater wetlands in the literature have been conducted by biologists and ecologists. However, wetlands are created and modified by a variety of biologic, hydrologic, and geologic processes. Because of the interaction of these processes, investigations of wetlands are best conducted using a multidisciplinary approach (Wilcox, 1988). In recent literature, hydrogeologic conditions have been related to wetland vegetation patterns (Siegel and Glaser, 1987; Wilcox et al., 1986; LaBaugh et al., 1987). These studies show that relationships between wetlands and groundwater systems are complex and wetlands are influenced strongly by local and regional groundwater flow systems. These and other studies also demonstrate that spatial variations in water chemistry in wetlands have important ecological implications and are often related to the groundwater flow systems.

Regional and local investigations were conducted at Indiana Dunes, on the south shore of Lake Michigan. Indiana Dunes has several well preserved dune–beach complexes that were deposited inland from the modern lakeshore when lake levels were higher. The lowlands within and between the dune–beach complexes are wetlands. The entire area is underlain by a complex aquifer system comprised of unconsolidated glacial, lacustrine, and aeolian deposits. Analysis of hydrogeologic and hydrochemical data from a regional network of test holes, cores, and wells shows that the wetlands are influenced by regional, intermediate, and local groundwater flow systems. The interior sections of the Great Marsh, the largest wetland between the dune complexes, is a discharge zone for intermediate and regional flow systems. These regional and intermediate flow systems are recharged in glacial end moraines landward of the dune–beach complexes and discharge to the Great Marsh by upward leakage through glacial tills that underlie the marsh sediments. The margins of the Great Marsh and nearly all of the other wetlands at Indiana Dunes are influenced by local groundwater flow systems that are recharged by precipitation.

Recharge–discharge relations in these local flow systems were investigated using several years of data from a series of closely spaced continuous recording wells along profiles through the wetlands and adjacent dunes. Cores of the near surface sediments were also taken at each of the well sites. Fluctuations in the water table and changes in water chemistry along these profiles showed that recharge is locally variable in time and space and seems to be concentrated at the dune–wetland margins. Transient water table mounds and seepage facies were observed at the dune–wetland margins after significant rainfall and snowmelt events. The location and duration of the water table mounds depends not only on the frequency, duration, and intensity of the rainfall and snowmelt events, but also on the stratigraphy of the near surface sediments, antecedent soil moisture, and the thickness of the unsaturated zone.

Relations between freshwater wetlands and regional and local hydrogeologic processes have also been investigated in the central upland of the Delmarva Peninsula, in northern Delaware. As at Indiana Dunes, changes in water table profiles through the wetlands and adjacent uplands show that recharge is quite variable in time and space. However, in contrast to Indiana Dunes, which is near a regional discharge zone, the wetlands in the central upland of the Delmarva Peninsula are located in a major recharge area for the regional aquifer system in the Atlantic Coastal Plain. The transient changes in the water table profile in this area may influence groundwater flow paths from the surficial aquifer to the underlying confined aquifers.

This last point suggests that regional hydrogeologic investigations of wetland settings have implications beyond the hydrology of individual wetlands. Before human development, the types of landscapes represented by wetland settings were once more common. The patterns of groundwater flow and chemistry in these settings probably prevailed over larger areas before manmade changes in drainage, land use, and land cover. Because they represent conditions before such anthropogenic changes, wetland settings may have important paleohydrologic implications for regional aquifer systems. Wetland settings more accurately reflect patterns of shallow groundwater flow and chemistry during the recharge of waters that are now much deeper in the groundwater flow system. Knowledge of hydrogeologic and hydrochemical patterns in wetlands may provide clues to the early stages of the hydrochemical evolution of these older waters.

REFERENCES

LaBaugh, J. W., T. C. Winter, V. A. Adomaitis, and G. A. Swanson, 1987, Hydrology and chemistry of selected prairie wetlands in the Cottonwood Lake area, Stutsman County, North Dakota 1979–1982: U. S. Geological Survey Professional Paper 1431, 26 p.

Siegel, D. I., and P. H. Glaser, 1987, Ground-water flow in a bog-fen complex, Lost River Peatland, northern Minnesota: Journal of Ecology, v. 75, p. 743–754.

Wilcox, D. A., 1988, The necessity of interdisciplinary research in wetland ecology: the Cowles Bog example, *in* D. A. Wilcox, ed., Interdisciplinary Approaches to Wetlands Research: East Lansing, Michigan State University Press, p. 1–10.

———, R. J. Shedlock, and W. H. Hendrickson, 1986, Hydrology, water chemistry and ecological relations in the raised mound of Cowles Bog: Journal of Ecology, v. 74, p. 1103–1117.

SHEMIAKIN, V. M., USSR Academy of Sciences, Leningrad, USSR

Charnockites in the Precambrian Geological History

Granulite-charnockitic complexes are abundant throughout Precambrian terranes of the world, being produced in the course of 2.5 Ga Earth's history at several stages of various duration and importance to form continental earth's crust.

The major and most protracted stage of charnockite formation to produce areal granulite-gneiss complexes occurred in the range of 3.1-2.6 Ga. These charnockites are of global distribution embracing great areas within the cratons of India, Australia, Africa, Greenland, Canada, South America, Antarctica, China, USSR and also occur in the basements of post-Archean mobile belts.

Onward from the Proterozoic till the Vendian including several episodes of charnockite formation are revealed: Svecofennian (1950-1650 Ma), Freiser-Musgrave (1400-1200 Ma), Grenville (1100-900 Ma), Panafrican (650-600 Ma).

The granulite-charnockite complexes differ in their tectonic position. Areal granulites and charnockites are confined to ancient nuclei of the continents; i.e., Archaean cratons. Their structural pattern is defined by multiple folding, i.e., by a combination of several generations of isoclinal, tight and open folds. These folded structures form gentle domal uplifts of isometric or ovoid shapes. Areal granulite metamorphism occurred in the regimes of moderately high geothermal gradients. However, a lateral zoning of thermodynamic regimes of metamorphism occurred in early stages of early Precambrian lithosphere resulting from initial tectonic differentiation of continental crust. Heterogeneous tectonic setting and associated endogeneous regimes of deep petrogenesis are characteristic of the Archean in the Aldan and Baltic Shields, Yilgarn Block, craton of Karnataka, etc. This might be due to basification of preexisting continental crust at the expense of the introduction of great ultramafic and mafic igneous masses in to the lower crust and the deep subsidence of these crustal segments as compared to adjacent sialic blocks which results in the coexistence of low-grade (plunged blocks) and high-grade (uplifted blocks) regimes. Besides a steeper geothermal gradient might be related to uprising deep magmatic melts to the base of crust that provides an additional higher heat flow.

Emplacement of abundant charnockites associated with the areal granulite metamorphism has no counterparts in subsequent geological periods thus suggesting a specific structural-lithologic pattern of the Archean earth's crust.

Post-Archean granulite-charnockite complexes are restricted to mobile linear belts only. The belts show tectonic contacts with the consolidated crustal segments, the sets of linear isoclinal folds being dominant structures in the belts. Their geology is defined by a combination of structural-lithologic basement complexes and younger volcano-terrigeneous rocks with repeated, different age and composition of igneous rocks belonging to the stage of formation of the belt proper. The linear granulite-charnockite complexes unlike areal ones were formed in contrast thermodynamic environments. Their most distinctive feature is the connection with the massifs of anorthosites and gabbro-anorthosites, rare with gabbro-dioritic and fayalite granitic intrusions.

Formation of the linear granulite-charnockite complexes can be explained by some causes. First, granitization of the lower crust to generate great chambers of acid melts uprising due to density inversion. Besides, intense heat flows may be linked to deep (60-70 km) high aluminous andesite-basaltic magmas as a source of anorthositic assemblages. Vertical uprise of mantle asthenolits to produce positive thermal anomaly is also possible cause. As a favorable factor, the initiation and development of deep-seated (up to mantle) faults can be referred to form linear high thermal zones.

SHEMIAKIN, V. M., USSR Academy of Sciences, Leningrad, USSR

Charnockitoid Suites and Their Geochemistry with Reference to Early Precambrian Charnockites of the USSR

Charnockitoids can be referred to as one of the oldest granite formations in the earth's history whose mode of occurrence and succession of formation are variable. However, irrespective of endogeneous regimes giving rise to early Precambrian geostructures, they show a regular evolutionary trend: a particular endogeneous cycle is characterized by a specific charnockitoid suite.

The early stages of mobile belts reveal pre-migmatitic (I) essentially sodic charnockitoids of igneous origin. They are represented by sheeted, phacolite-and laccolith-like intrusions comformable with the enclosing rocks. Charnocktie-migmatites (II) derive from rocks of various compositions and genesis during regional granulite migmatization accompanied by intense folding. They show areal distribution, variable morphology, polygenesis, compositional variations of leucosome and its constituent minerals. Most abundant are postmigmatic autochthonous charnockitoids (III) arisen from granitization of various rocks during regional ultrametamorphism under conditions of granulite facies. The charnockitoids constitute the massifs of irregular shape up to hundreds square km, the tracts of peculiar shape and linear zones ranging in thickness and length. These bodies exhibit mainly diffuse contacts with the source rocks. The autochthonous rocks can contain various relict basement rocks exhibiting various reworking. The charnockitoids are characterized by highly heterogeneous mineral and lithologic composition and range in fabrics due to variable features of the source rocks and subsequent intermittent granitization. Postmigmatitic allochthonous charnockitoids (IV) are terminal products of the charnockite suites. They are rather scarce and form minor intrusions through the massifs up to 5000-6000 sq. km are known in Antarctica. The allochthonous charnockitoids are typically intrusive: they show the zones of eruptive breccia and apophyses into the country rocks, contain xenoliths of the country rocks, thermally affect the latter and cross-cut their structures.

The I-IV suites form a single evolutionary series and are related to areal granulite metamorphism which occurred in the 3.1-2.7 Ga range within the Aldan, Baltic and Ukrainian Shields.

The charnockitoids show a single set of trace elements whose abundances vary in different type rocks. The sets of trace elements in the charnockitoids are quite different from those found in the rocks of close SiO_2-content (granite-tonalites). Geochemical pattern of the charnockitoids is defined by two contrast concentrations of elements: higher amounts of V, Cr, Co, Ni, Sr, Sc, and lower amounts of lithophile elements which are in normal granitoids (1.5-6.0 times) those of the charnockitoids. Chalcophile elements and Zr amounts do not differ in both granitoids types.

All the charnockitoid types have excess (1.5 times clarks) Sc, Co and Ni abundances which in general define their geochemical pattern. Nevertheless, every charnockitoid suite shows a characteristic assemblage of trace elements: I-Cr, V, Sr; II-V, Cr, Cu; III-Sr, V, Cu, i.e. earth-alkaline-siderophile, chalcophile-siderophile, earth-alkaline and earth-alkaline-siderochalkophile, respectively.

On the other hand, the charnockitoid geochemical pattern is also determined by insufficient (smaller than 1.5 factor of clark) trace element abundances which are common either for all the charnockitoids or for a particular suite. Such common elements are Ba, Yb, Rb, Y, Th, U, Li, Be and are specific for: I-Zr, Cu, Ga, Nb, Pb, La; II-Zr, Ga, Pb, La, Nb; III-V, Pb, Cu, La, Cr, Nb; IV-Cr, Ga.

It is a combination of these trace elements of contrast concentrations that can serve a geochemical indicator to distinguish a type of charnockitoid suite, i.e., a particular suite is characterized by a distinctive set of trace elements showing definite qualitative and quantitative features.

SHEN Feng, XUE Yuxuan, LIU Tengyao, CHEN Weixiong, FENG Weiheng, and ZHANG Yunyi, China General Corporation of Nuclear Industry, Beijing, China

Application and Processing of Geological Data with Microcomputer System

Multifunctional uranium geological information system with microcomputer: A geological information system with multifunctions, comprehensive processing ability, multioutputs, quick inquiry, and easy use has been set up. It has various interfaces with special functions using multisubdirectories, multilanguages, and multiinput and output modes including screen, printer, digitizer, and plotter.

Uranium geological information system can be divided into various subsystems, such as resource, exploration, management, and research reports. Function selection is realized via "menu" mode for each subsystem.

The DBase III language is used as a basic language in various databases with more than 100 fields.

Original data are processed mathematically with plotting by using BASIC and Fortran languages.

Calculated results can be transferred into tables of Office system.

Thus, a more suitable geological information system has been realized by using microcomputer including its full storage and peripheral space.

Apart from functions such as database maintaining and reference inquiry, there are many other functions such as entry (including validation), update, data collection, overview of main information, table output, file output, schematic printing map (by screen or printer), and plot output (Figure 1 and Table 1).

Some cases in data processing: A northwest-oriented uranium ore belt has been defined by using uranium evaluation parameters in region A (Figure 2).

Application of fuzzy set theory to gold target selection: Four categories of targets for gold prospecting in region B are outlined by calculating the grade of membership for gold mineralization (Table 2).

Some favorable uraniferous areas in massive C are predicted by using complex parameter correlation method. The complex parameters for 83 subunits in massive C are calculated (Table 3).

Three favorable units, 17, 28, and 30, are outlined in Table 3. Among them: Unit 28 has a known uranium deposit; unit 17 has a prospect; and unit 30 has also a prospect that was proved by trenches.

The regional gamma-spectrometric data are processed by using characteristic parameters "Paleo-uranium abundance" (GU), "activated uranium" (FU), and "relative migration coefficient" (B). Some favorable zones for uranium prospecting in region D are predicted by identification of regional uranium source bodies and studies on mobilization, migration, deposition, and concentration of uranium (Figure 3).

Four favorable uraniferous areas are predicted by using both ISOMIX method and complex parameter correlation method in region E (Figure 4).

NO.	massific	SiO_2 %	W_1	TiO_2 %	W_2	Fe_2O_3 %	W_3	MgO %	W_4	CaO %	W_5	K_2O %	W_6	P.	grades of membership for Au mineralization	T.
1	B1	69.40	0.15	0.46	0.23	1.89	0.19	1.94	0.27	0.42	0.03	5.02	0.13	66.41	$(AuCu)_{0.7}(U)_{0.3}$	I
2	B2	61.01	0.10	1.12	0.32	0.06	0.00	2.84	0.23	4.94	0.22	3.00	0.12	33.05	$(Cu, Au)_{1.0}$	I
3	B2	50.57	0.06	2.28	0.36	0.79	0.02	5.54	0.24	7.29	0.18	1.50	0.13	23.05	$(Cu, Au)_{1.0}$	I
4	B2	59.93	0.11	0.50	0.15	0.01	0.00	2.98	0.26	8.12	0.39	4.30	0.09	34.74	$(Cu, Au)_{1.0}$	I
5	B2	53.84	0.05	2.56	0.33	4.48	0.12	6.90	0.25	5.91	0.12	1.36	0.12	22.62	$(Cu, Au)_{1.0}$	I
6	B2	61.45	0.11	0.85	0.28	2.28	0.15	2.02	0.19	2.79	0.15	3.74	0.12	42.19	$(Cu, Au)_{1.0}$	I
7	B2	51.16	0.06	2.10	0.29	5.82	0.16	4.55	0.17	7.14	0.15	1.07	0.17	27.59	$(Cu, Au)_{1.0}$	I
8	B3	69.66	0.16	0.45	0.24	1.01	0.11	1.06	0.16	1.95	0.17	4.57	0.15	72.06	$(Au, Cu)_{0.6}(U)_{0.4}$	I
9	B3	74.71	0.16	0.80	0.47	0.58	0.07	0.34	0.06	0.95	0.09	5.20	0.15	83.44	$(Au, Cu)_{0.2}(U)_{0.8}$	II
10	B3	63.90	0.10	0.40	0.12	2.84	0.13	0.05	0.00	1.84	0.09	0.74	0.52	77.66	$(Au, Cu)_{0.4}(U)_{0.6}$	II
11	B3	63.54	0.13	0.55	0.22	0.97	0.08	2.01	0.22	3.54	0.22	4.1	0.13	51.39	$(Cu)_{0.9}Fe)_{0.1}$	III
12	B3	69.67	0.09	0.8	0.26	3.31	0.22	2.78	0.25	2.01	0.10	5.21	0.08	45.89	$(Cu, Au)_{0.8}(Cu, Fe)_{0.2}$	I
13	B4	48.81	0.04	1.29	0.13	2.15	0.04	8.14	0.22	9.83	0.15	0.30	0.42	5.60	$(Cu, Au)_{1.0}$	I
14	B4	47.85	0.04	1.30	0.11	1.77	0.03	8.22	0.19	9.95	0.13	0.22	0.50	5.08	$(Cu, Au)_{1.0}$	I
15	B5	75.80	0.31	0.10	0.12	0.13	0.03	0.64	0.21	0.36	0.07	5.56	0.27	165.67	$(Mo)_{1.0}$	N
16	B5	75.40	0.30	0.10	0.11	0.44	0.10	0.51	0.16	0.29	0.05	5.08	0.28	155.08	$(Mo)_{1.0}$	N
17	B5	62.20	0.14	0.60	0.26	4.54	0.40	0.90	0.11	0.55	0.04	12.54	0.04	58.71	$(Au, Cu)_{1.0}$	I
18	B5	69.72	0.23	0.52	0.20	1.52	0.24	0.60	0.13	0.69	0.09	8.82	0.11	107.77	$(Mo)_{1.0}$	N
19	B5	70.18	0.16	0.39	0.21	1.57	0.17	0.84	0.13	1.99	0.17	4.1	0.17	73.59	$(Au, Cu)_{0.5}(U)_{0.5}$	II

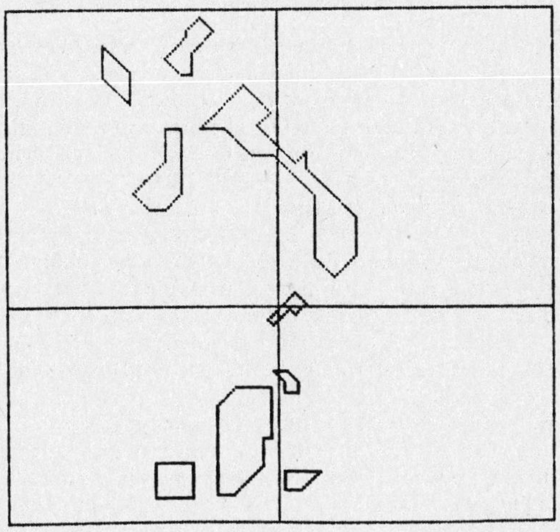

		−0.39/−0.40 ⑭	−0.16/−0.29 ⑮	−0.53/−0.62 ⑯	5.23/5.10 ⑰	−0.21/−0.16 ⑱
−0.09/ −0.16 ⑲	−0.31/ −0.34 ⑳	1.78/ 1.56 ㉑	−0.12/−0.13 ㉒	0.71/ 0.63 ㉓	1.76/ 1.70 ㉔	−0.29/−0.26 ㉕
−0.51/ −0.69 ㉖	−0.14/ −0.20 ㉗	3.88/ 3.82 ㉘	0.67/ 0.63 ㉙	4.48/ 4.37 ㉚	1.10/ 1.08 ㉛	−0.45/−0.58 ㉜
		−0.48/−0.54 ㉝	−0.33/−0.40 ㉞	−0.04/−0.03 ㉟	−0.34/−0.46 ㊱	−0.20/−0.11 ㊲

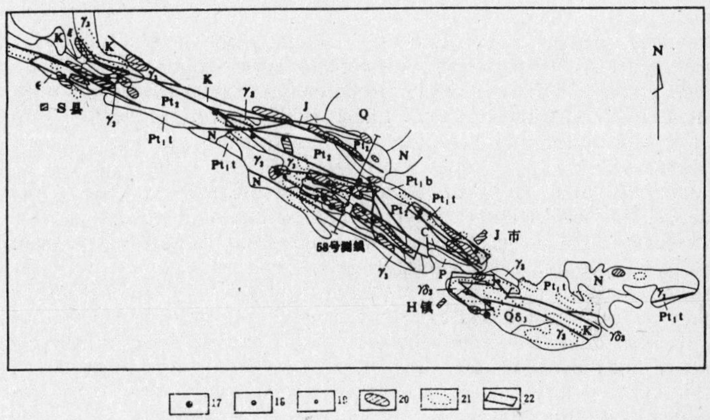

NO. Record	Project	Scale	Quantity in plan	Quantity in fact
5,000,001	Total	1:50000--1:10000	533	
	Gamma survey	1:10000--1:25000	729	
	Gamma-spectrometric survey	1:50000--1:10000	533	
		1:10000--1:25000	729	

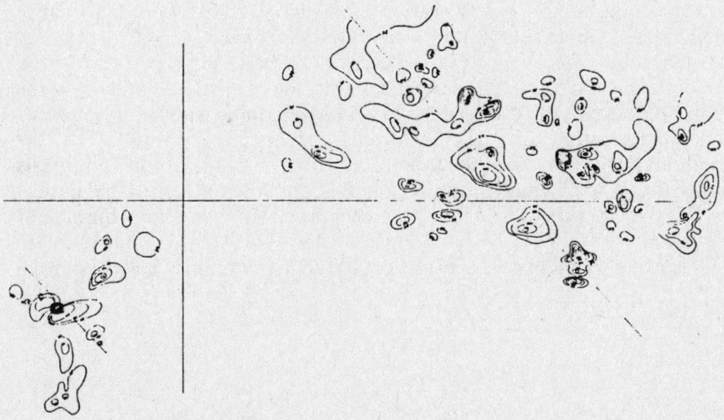

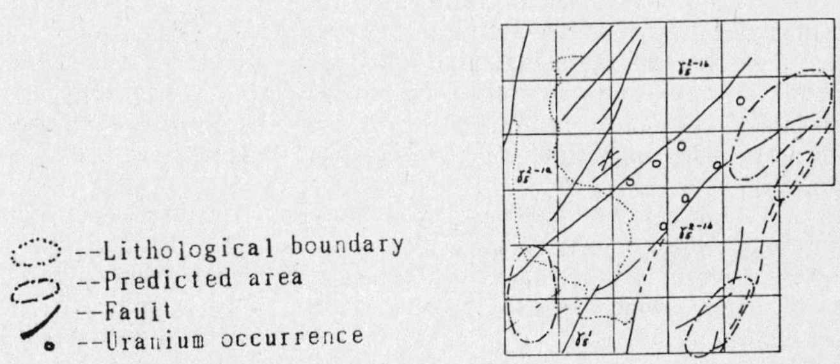

SHEN Zhaoli, XU Shaozhuo, SUN Yong, WANG Yanxin, China University of Geosciences, Beijing, China

Paleohydrogeological Analysis of Some Ore Deposits in Sedimentary Basins of Platform Regions

In recent years, we have made paleohydrogeological research on several ore deposits in China. Our basic working procedure of paleohydrogeological analysis of ore deposits in sedimentary basins of platform areas is as follows: (1) determination of paleohydrogeological periods; (2) restoration of the hydrodynamic field of ancient water-bearing systems, only to locate the ancient discharge area which is the favorable site for ore deposition; (3) superposition of the ancient discharge area with the range of geochemical barrier, to locate the most favorable sites for ore deposition; (4) evaluation of the role of ancient groundwaters in ore-forming process; and (5) effective prediction of prospective ore locations after a synthetic analysis of the available data and application of geomathematical methods. Given below are two examples of our research.

Huangmei Siderite Deposit. As a typical example of stratabound ore deposits in platform regions, the Huangmei siderite deposit is found to have no genetic relationship with magmatic hydrothermal activity. The ore composition is simple, which includes the earlier, gray siderite formed during diagenesis and the later, massive or vein-like yellow siderite formed during catagenesis. Paleohydrodynamic analysis shows that the groundwater cycle in the region studied is "elisional" in nature from the Huanglong age (C_2) to late Triassic (T_3). For this time span, Huangmei was the discharge area of regional groundwater system, the movement of which resulted in the deposition of gray and massive yellow siderite during the IndoChinese movement. Later, as the Yanshanian movement intensified, the region became continental, with a wide-spread fracturing which made infiltrational water enter into the groundwater cycle. The result is a mixing between the infiltrational and the sedimentary water, and the formation of the vein-like yellow siderite.

The ore-forming solution of siderite was mainly sedimentary waster changed during diagenesis and catagenesis. Fluid inclusion study proves that the ore-forming fluid for yellow siderite is weakly alkaline $CaCl_2$-type water.

The prospecting criteria for similar siderite deposits can be concluded as follows: (1) the transition zone between an ancient continent and a sedimentary basin; (2) the region close to a deep fault; (3) the existence of thick shale layers (especially those containing plentiful organic matter and iron) near carbonate strata's distribution; (4) the existence of heat sources near the region.

Tonglushan Copper-Iron Deposit. The famous Tonglushan Cu-Fe deposit in eastern Hubei is typical contact-metasomatic ore deposit. Regional paleohydrogeological analysis shows that, from late Triassic to early Jurassic, the Echendomashoan anticline rose up, and most of the region entered into an infiltrational stage of groundwater cycle. During the time of the intrusion and emplacement of the Tonglushan magma, the Jiangqiao-xialu fault, which passed through the ore field, was highly active, and the sedimentation was still going on in the west of the region. So the Jiangqiao-xialu fault zone was the mixing and discharge sites of sedimentary, infiltrational water and gas-solution released by the cooling magma. The three kinds of water together constituted the ore-forming fluid.

Calculations show that the hydrothermal gas-solution released during the crystallization of orthoclase and quartz was rich in copper, while that released during the crystallization of hornblende and biotite was rich in iron. This may explain why iron is earlier than copper in the ore. The high contents of Fe- and Li_+ in fluid inclusions from skarn and ore minerals indicate the addition of fluid released by magma into the ore-forming solution.

The dolomites of the 4th, 5th and 7th members of the Daye group, which are closely related to ore formation, commonly contain pseudomorphic crystals of gypsum and salt karst breccia. So the sedimentary water in them should be saline, which is proved by the existence of daughter mineral in the fluid inclusions, and rich in Cl^-, Na^+, (K^+), Ca^{2+}, Mg^{2+} as well as SO_4^{2-} and HCO_3^-. This solution not only took part in the hydrothermal process but also in the ore formation period of the Cu-Fe deposit.

The future prospecting activity in the region should be concentrated on those sites where occurred both fractured zone and a contact zone.

SHEN Zhaoli, ZHONG Zuoxin, and WANG Min, China University of Geosciences, Beijing, China

On Acid Rain in China

About 90% of acid rain with pH < 5.6 arises in the south area of the Yangtze River, especially in the cities of Chongqing, Sichuan Province, and Guiyang, Guizhou Province, in China. The average pH values of acid rain in the two cities are 4.14 and 4.02, respectively. Both of the cities frequencies of acid rain are above 80%. Acid rain seldom occurs in the cities of northern China, on the other hand. The annual average pH values of precipitation in Beijing in 1981–1983 were all above 7. There is acid rain in only a few northern cities. Qingdao is a typical example. In 1981–1983, its average annual pH values of precipitation, were 6.4, 4.98, and 4.13, respectively, and the annual frequencies of acid rain were 16.7%, 98.3%, and 87.5%. The ionic concentrations in the acid rainwater generally occur as follows: $SO_4^{2-} > Cl^- > NO_3^- > F^-$, $Ca^{2+} > NH_4^+ > Na^+ > Mg^{2+} > K^+$. The acidity chiefly comes from sulfuric acid. The question is why there is little acid rain in some places, especially in the northern cities, even though the sulfur dioxide contamination is serious. The explanation is this: The formation of acid rain is related not only to SO_2 air pollution, but also to the nature of the atmospheric particles. The situations of Beijing (non-acid rain region of north China) and Guiyang (serious acid rain region of south China) are an example. The average initial pH values of the atmospheric particles in the two cities are 8.2 (Beijing) and 5.72 (Guiyang). The former is obviously greater than the latter. So the buffering capacity with respect to acid rain in Guiyang should be much lower than that in Beijing. This could be the main reason why the Guiyang region has the most serious acid rain in China. The formation of acid rain is connected closely to the geographical environment. Generally, the soil of the northern area is alkaline while the soil of the southern area is acidic. For instance, the pH of Beijing soil is 7–8, but the pH of Guizhou soil is no higher than 5.5 Two-thirds of the total atmospheric particles come from the dust in the wind in north China. However, only one-third come from wind-borne dust in south China. This explains why the pH value of the atmospheric particles in the south is lower than that in the north. In the small city of Ji'an in the south, industry is not developed, and concentrations of gaseous pollutants are very low. Sulfur dioxide and nitrogen oxide are only 0.013–0.034 mg/m^3 and 0.013–0.031 mg/m^3 respectively. But the frequency of acid rain is as high as 84%. The soil of this city is red clay with a pH range of from 4.5 to 5.5. Owing to wind conditions, the atmospheric particle concentration is rather high, and equaling 0.577–0.612 mg/m^3. Therefore, the formation of acid rain is evidently related to the atmospheric particles coming from the red clay. The washout of the acid particles by rainwater in the atmosphere is the major cause of acid rain. Acid rain is not caused completely by industrial air pollution. This typical example shows that the weathered products at the earth's surface have a significant influence on the formation of acid rain.

The research project entitled "The Influence of Acid Rain on the Ecological System" has been carried out for the past 3–4 years in the provinces of Sichuan, Gangxi, Zhejiang, and Guizhou in south China. Following are the results of this project: (1) Above 65% of the branches and twigs in the tested plants are acutely harmed when the rain's pH value is 2.0–3.0. (2) As the rainwater comes through the forest crown cover, most ionic concentrations in the water increase and some kinds of plants are directly influenced: The chloroplast is injured, the nutrient transmissibility of the branches decreases, the root system is undeveloped, and nutrients in the crown cover are lost. (3) Direct influence on crops doesn't seem so serious: Wheat, soybeans, and Chinese cabbage are hurt only when the rain's pH < 3.0; likewise for rice plants when the pH < 2.0. (4) Soils are more directly influenced: With the rain's pH < 4.0, the amount of leaching of ions greatly increases. The concentration of exchangeable aluminum becomes much higher, and exchangeable calcium is leached out. The soils will at last inevitably become acidic. (5) Different soils have different adsorption characteristics to SO_4^{2-}; the tawny soil and the grey-brown soil obey the Temkin equation; the Quaternary red soil is in keeping with the linear equation; and the yellow soil, the crimson soil, and the forest soil conform to the Freundich equation. The SO_4^{2-} concentration in the leaching solution is controlled by the dynamic equilibrium of SO_4^{2-} in the soil and in the rainwater. It is the mobility of SO_4^{2-} that results in the leaching of a large number of the cations.

The influence of acid rain on ancient buildings including grottoes built in sandstone, granite, and limestone, the influence on the total ecological system are all being studied.

SHEN Zhaoli, and ZHONG Zuoxin, China University of Geoscience, Beijing, China

On Water Quality Deterioration and Acid Rain in China

The problems on endemic disease caused by the relative enrichment and poorness of chemical elements in the different geochemical environment are very serious in the vast territory of China. The groundwater with high fluorine concentrations covers a large area and can be found almost everywhere in the north regions of China. About 50 million people drink this kind of groundwater in which the highest fluorine concentration is 40 mg/L. As to the environment problems caused by human activities, the problems are more serious. The water resources have been already polluted to a high degree. The monitoring of 55 surface water supplying sources in 42 cities of China indicates that 58.5% of those water sources exceed NH-N standard, 23.4% the NO_3-N, 33.3% the volatile phenol, and 41.9% the suspended matter. The pollution range of the surface water in China is above 80%. A lot of sewage, feces and toxic industrial waste liquid are discharged into the 3600 KM part of the Yangtze river, the biggest river in China, from Dukou in the upperstream to Shanghai in the downstream. Other rivers neither cannot avoid this disaster. Along the beautiful Songhua river, because of the disposal of the waste water containing Hg from the three big chemical factories in Jilin province into the river, the amount of Hg accumulating in the bodies of the people who drink the river water reaches 16.5 mg, greatly exceeding the limitation of 10 mg of Hg per 50 kg body weight.

From the investigation of groundwater quality, 21 cities are in the poor quality range among the 27 investigated cities. Groundwater are major drinking supply sources in some cities such as Beijing, but the concentrations of nitrate, sulfate and total dissolved solid in the groundwater increase gradually year after year and greatly exceed the domestic standard of drinking water. Owing to the lacking of water resources, several million people in cities have to drink poor quality water.

About 90% of acid rain with pH < 5.6 arises in the south area of the Yangtze river, especially in Chongqing City of Sichuan province and Giuyang city of Guizhou province. The average pH values of acid rain in the two cities are 4.14 and 4.02 respectively. The two cities' frequencies of acid rain are above 80%. On the contrary, acid rain seldom happens in the northern cities of China. The annual average pH values of precipitation in Beijing in 1981-1983 are all above 7. There is acid rain only in a few northern cities. Qindao city is a typical example of this kind. Its average annual pH values of precipitation in 1981-1983 are 6.4, 4.98 and 4.13 respectively, which are reduced with years. But the frequencies of acid rain in 1981-1983 are 16.7%, 98.3% and 87.5% respectively which increase year by year. The ionic concentrations in the acid rain water in China are arranged in the order: $SO^{2-}{}_4 > Cl^- > NO_3 > F^-, Ca^{2+} > NH^+{}_4 > Na^+ > Mg^{2+} > K^+$. The acidity chiefly comes from sulfuric acid. The question is that in some places especially in the northern cities, although the sulfur dioxide contamination is serious, there are few acid rains. It is explained as this: the formation of acid rain is related not only to the air pollution of sulfur dioxide, but also the nature of the atmospheric particles. The situations of Beijing (non-acid rain region of the Northern China) and Giyang (serious acid rain region of South China) have been compared as an example. The average initial pH values of the atmospheric particles in the two cities are 8.20 (Beijing) and 5.72 (Guiyang). The former is obviously bigger than the latter. The buffering capacity to acid rain in Guiyang should be greatly lower than that in beijing. Maybe this is the main reason why Guiyang is the most serious region of acid rain in China. The formation of acid rain is connected closely to the geographical environment. Generally, the soil of north area is alkaline and the soil of south area appears acid in China. For instance, the pH of Beijing soil equals 7-8, but the pH of Guizhou soil is not higher than 5.5. The 2/3 of the total atmospheric particles come from the dust by the wind in the north China. However, it is only 1/3 in the south China. This is one of the reasons why the pH value of the atmospheric particles in the south is lower than that in the north. There is a little city Ji'an (in the south) which industry is not developed. Its concentration of gaseous pollutants is very slow. The concentrations of sulfur dioxide and nitrogen oxide are only 0.013-0,.034 mg/m^2 and 0.013-0.031 mg/m^2 respectively. But its frequency of acid rain is as high as 84%. The soil of this city is red clay which pH value ranges from 4.5 to 5.5. Owing to high frequency of the wind, the atmospheric particles concentration is rather high and equals 0.577-0.612 mg/m^2. So the formation of acid rain is obviously related to the atmospheric particles coming from the red clay. The washout of acid partices by the rain water in the atmosphere may be the major cause of acid rain. Acid rain is not caused completely by the industrial air pollution. This typical example shows that the weathered products have influence significantly on the formation of acid rain.

SHENGLIN, L., Q. SUJUAN, and S. BIN, Department of Earth Sciences, Nanjing University, Beijing, China

Engineering Geological Characteristics of Expansive Soils in China

Expansive soils are widely distributed in China. Its damages to buildings have aroused engineers grave concern. Based on author's researches on expansive soils in Guangxi, Hubei, Henan, Anhui provinces, etc., some problems are discussed in this paper in several sides as follow:

1. Geological formation and geomorphological features. Although there are various formations of expansive soils, they all have close relations with origin of materials and depositional environment and exist obvious regional features. Their common features are that all expansive soils have underwent similar diagenesis of secondary dehydration and compaction. Expansive soils develop generally on terrace and rolling hill, where many gullies and shallow landslides can often be seen.

2. Expansive soils in China are natural high dispersive clays with clay content greater than 30% apart from "red clay" of Guangxi province. Mineral compositions of expansive clays are mainly illite, montmorillonite, kaolinite, chlorite and their mixlayer minerals, which are substantial basis of swelling and shrinkage of expansive soils (Table 1).

3. Physico-chemical properties of expansive soils are active. Cation exchange content (CEC) is generally greater than 200 μeq/g. The expansive soils with main exchanged cation Na, Mg, are greater in swell-shrinking side (Table 2).

4. Expansive soils contain special microstructures. Parallel ranks of flake clay particles form particles intergrowth in domains and turbulent microstructures, which are of great advantage to swelling and shrinkage of soils due to fissures of intelayers absorbing and losing water easily.

5. On the basis of discussions above, the relations among swell-shrinking property, composition, structure are analysed and swell-shrinking mechanics are studied in detail in this paper.

6. Finally, the paper raises noticeable questions in engineering investigation and tests of expansive soils.

SHERIDAN, R. E., Rutgers University, New Brunswick, NJ

U.S. Atlantic Passive Margin

The U.S. Atlantic continental margin is the archetype passive margin formed by the rifting, breakup, and spreading apart of three continents. The rift stage crust with deeply subsided half-grabens underlies the continental side of the province, adjacent to the transitional, rifted, thinned, and intruded crust under the deep Mesozoic–Cenozoic sedimentary basins. The boundary between the transitional crust and true oceanic crust is marked by the East Coast magnetic anomaly (ECMA). Off Florida, this boundary is farther seaward and marked by the Blake Spur magnetic anomaly (BSMA). Draped over these three different basements (continental–transitional–oceanic) are the drift stage sedimentary rocks of middle Jurassic to Holocene age. Five major basins dominate the structure off the United States: Georges Bank Basin, Baltimore Canyon Trough, Carolina Trough, Blake Plateau Basin, and South Florida–Bahamas Basin. Sedimentary thicknesses in these basins reach 12–14 km. The drift stage histories of these basins are similar, with declining rates of tectonic and thermal subsidence allowing sediment infill punctuated by sea-level fluctuations and strandline migration back and forth across the continental shelf and coastal plain. Variations have included the deepening since the late Cretaceous of the broad Blake Plateau caused by cessation of carbonate deposition, sediment starvation, and erosion by the Gulf Stream. The isolated Bahama carbonate banks and intervening channels evolved similarly since the middle to late Cretaceous. Timing of the breakup of the U.S. Atlantic margin is in middle Jurassic close to the breakup age of the Gulf of Mexico. Both areas underwent a major spreading-center shift in a few million years after breakup, leaving an extinct spreading center and oceanic crust along the U.S. passive margin north of Florida, and the unusually wide transitional crust under the Blake–Bahamas region.

SHI Dehong, The Academy of Geology Sciences of China, Beijing, China

Groundwater Resources and Management of Semi-Arid Plain

Groundwater resources: The area under study is located in 35°00'–40°15'N latitude and 114°20'–119°15'E longitude; that is, the famous great plain of North China. It covers an area of 138,000 km^2 with an elevation less than 100 m.

North China is mainly a basin of Cenozoic deposits with the 400–600 m thick Quaternary system. It presents a typical premountain artesian basin. Three belts from the piedmont toward the sea coast are: (1) the piedmont alluvial–diluvial plain, (2) the central alluvial–lakance plain, and (3) the coastal marine plain.

From the center of plain to the sea coast, the upper part of artesian basins forms a saline water layer with total dissolved solids more than 2 g/l. A few freshwater lenses float on the top, 20–60 m thick and 10–30 km in diameter.

Samplings from a depth of less than 100 m at alluvial fan system reveal high contents of tritium, indicating that this water was recharged during the last few decades.

The distribution and character of water resources—Based on the water balance theory, the natural resources of groundwater are represented by average amount of annual recharge. Available resources are estimated by considering the factors of economy and reasonableness. The total available resources including surface water are $364 \times 10^8 \, m^3$. The per capita of water resource os only one-sixth of that for whole country. Furthermore, water-logging and drought, and saline and alkaline conditions coexist in some locations.

As the demands for water increase, especially in the developed regions, water deficits have to be satisfied by overpumping of groundwater.

Optimum groundwater exploitation—Current groundwater exploitation is not being conducted in a rational way in North China. Some abuses have to be controlled. Three research achievements in this area are significant: (1) controlling groundwater table in an optimum situation, (2) setting up a long-term steady water supply system with ground storing of water resources, and (3) transforming and using less brackish water with local conditions.

Scientific management of water resources: Currently, large-scale water resource exploitation has become a problem that requires dealing with a wide range of environmental ecologies and social economies. Water resources system analysis and mathematical models have been undertaken in the alluvial plain of Hutuohe river and Shijiazhuang municipality.

The management model of Hutuohe alluvial plain—Based on studies by the French hydrogeologist G. Castany, a concept model for the groundwater basin with an area of 10,000 km^2 in Hutuohe alluvial plain centered by the city of Shijiazhuang was set up. On the basis of hydrogeological generalization, two-dimensional unsteady flow models have been built up with nonleakage recharge. Using synthetic series of rainfall and runoff, water system predictions are in progress. We adopted the embedding method to realize the coupling of groundwater model and linear programming model for seeking an optimum idea of irrigational water assignation with objective of maximum economic benefit. Its advantage is that when the objective of economy reaches a maximum, the state of water exploitation will be in an optimum state. Our model is a large linear program with 3808 variables and 1708 constraints. Optimization will result in demands on groundwater for agriculture 16–27% less than that in the plan, and output will increase 13–21%. Meanwhile, the trend of sharp decline in the water table could be controlled.

The management model of Shijiazhuang—On the basis of forecasting of groundwater quantity and quality in the cone of depression covering an area of about 500 km^2, where the city of Shijiazhuang is located, an optimum management model of groundwater resources has been established. A response matrix method was adopted here to realize the simulation and optimization model. Two schemes, A and B, are suggested as following: Scheme A with the objective of total fall-depth of water level at all cells in the area going to minimum, readjusts pumping yield to postpone the trends of water table decline and polluters migrating toward the cone area. Scheme B's objectives are the satisfaction of growing demands and control over the development of depression code as well. That is total recovered value of level rise at all cells reaching maximum. Optimization has been considered to determine when, and how much artificial recharge should be done. In making a comparative analysis between the two schemes of A and B, practical opinions have been expressed.

SHI Yaolin, and ZHU Yuanging, Academia, Sinica, Beijing, China

Modeling of Thermal Evolution of Tibet Plateau

Tectonic history of the Tibet Plateau has been revealed by increasing data. Paleomagnetic data show the chronology of continental collision, which occurred at 50 Ma at a convergent rate of 10 cm/y, reduced to 5 cm/y at 45 Ma, and has continued ever since. Igneous and metamorphic rocks disclose the past thermal state. Metamorphic isogrades show apparent symmetry above and below the MCT. Two belts of young granites in southern Tibet are found to have crustal sources. Heat flow measurements directly show the present thermal state. High heat flows south to the Indus-Zangbo suture and active hydrothermal activities suggest recent intrusion of magma into upper crust. Seismic data give hints on the present thermal state at depth. A low velocity zone is suggested by reflection, surface wave, and body wave studies and may be related to partial melting. With these constraints, we carried out a time-dependent finite element modeling on the thermal evolution of the Tibet Plateau since the collision. In contrast to some previous one-dimensional studies, (e.g., Molnar et al, 1984) or two-dimensional studies without consideration of advection (e.g., Jaupart and Provost, 1985; Pinet and Jaupart, 1987), our modeling is two-dimensional and includes effect of advective heat transfer.

It is found that advective heat transfer plays the most important role in the thermal evolution. It has several implications.

1. Initial temperatures before the collision (50 Ma) are not important for successive development of thermal structure except for the first 10 m.y. Thermal state is then dominantly controlled by advective effect due to underthrust. Therefore, uncertainties in initial conditions do not affect our modeling of later events.

2. If the underthrust of the India plate beneath the Tibet Plateau has remained active, though perhaps episodically, the underthrusting of Indian crust can cool down the Tibet Plateau; mantle heat flow from below is screened and radioactive heating is not enough to heat up the crust to partial melting, with thermal parameters chosen similar to some previous work (e.g., Jaupart, 1987). Furthermore, the melting heat of granite is about 100 times greater than the heat capacity of crustal rocks, thus generation of magma requires not only heating up to reach the melting temperature, but also a substantial amount of heat for melting. A plausible way to provide heat sufficiently and rapidly for melting appears to be frictional heating. Frictional heating at adequate magnitude of shear stress may lead to partial melting, while melting may reduce shear stresses on faults; thus, frictional heating and melting may be a self-regulating process. Such characteristics are included in our modeling by introducing a temperature-dependent frictional coefficient.

3. Because the thermal structures are dominantly determined by advective process characterized by tectonic parameters, thermal constraints, can be used to differentiate different tectonic models. Models of compressional thickening, crustal underthrusting, and crustal stacking were compared, and effects of modeling parameters were examined. The effect of an accretionary wedge of low thermal conductivity is also included. Compressional thickening cannot explain the melting in upper crust. Underplating of Indian crust as a unit beneath the Tibetian crust (e.g., Powell and Conaghan, 1973) will result in frictional heating along the whole fault that reaches a depth pf about 35 km and affects temperatures in the whole hanging wall. Melting would occur all across the plateau rather than near the MCT. Stacking of underthrust slices near the collision zone with a decollement beneath the Tibet Plateau at the bottom of the thickened crust (60–70 km of depth) may explain the generation of granites and metamorphic characteristics, and thus appears to be the favorite model. Although observational and experimental data are not enough to impose a strong constraint, the two models predict different heat flow over the Tibet Plateau, and the prediction can be tested in future exploration.

SHI Yaolin, Academia Sinica, Beijing, China

Two-Dimensional Modeling of P-T-t Paths of Metamorphic Rocks at Stacked Thrust Terrains

P-T paths of metamorphic rocks has been intensively studied (e.g. England and Thompson, 1984, Davy and Gillet, 1986) for one-dimensional cases and under the assumption of instantaneously emplacement, however, the two-dimensionality and finite time for thrusting process turns to be important for the thermal evolution of thrust terrains (Shi and Wang, 1987). In this study, we extend our study to stacked thrust terrains, and we include frictional heating in our study. The geotherms at stacked thrust terrains strongly depend on the thrust kinematics including the velocity of thrusting, the duration of thrusting, fault geometry and the sequence of thrust events.

For the stacked underthrust sequence, when the cool plate underthrusts the hot hangingwall, a reversal in the profile of temperature versus depth will occur, although it is not as simple as the saw-toothed profile assumed in one-dimensional models. Low underthrust rate and long duration of thrust process favorite thermal relaxation, resulting removal of the reversal and decrease in temperature of the hangwall. For stacking of many thin slices, the characteristic time of thermal relaxation for each slice is smaller than that of a simple two-unit thrust, but the duration of the stacking is usually much longer than that of the simple thrust if the velocity of thrust are similar. Therefore, the reversal is less apparent or even disappear in case of stacking of many thrust slices, and the hangingwall will have lower temperatures than those of the simple thrust. However, if the thrusting process has proceeded a time long enough, quasi-steady state geotherm can be reached. In this case, either the stacked thrust slices or in the simple thrust terrain, the geotherm will be the same if they have the same tectonic parameters of active faulting. Spatially, temperatures are lower near the root of the thrust than those at the advancing edge of the underthrusted slice.

During uplift and erosion after thrusting, rocks in the thrusted terrains subject to two competitive effects: heating from below and cooling due to approaching the surface by denudation. Thus, the temperatures of the uplifted rocks may increase before they finally decrease to the surface temperature. The P-T-t paths are calculated for our 2-dimensional models of stacked thrust slices. Such increase in temperature during uplift is more obvious for two-unit simple thrust than for stacked thrust slices. Spatial differences exist for stacked slices of the same number. Varying with depth, the increase in temperature with uplift is more distinctive for deeper rocks at the footwall than the shallow rocks at the hangingwall; at shallower depth, the increase may disappear at all. Laterally, such increase in temperature with uplift is more distinctive near the root of the thrust than at the front edge of the underthrusted slice. At the same lateral position near the root of the thrust, the increase in temperature with uplift is also affected by the duration of the thrust. For longer duration of thrust, deeper rocks in the footwall show less increase in temperature with uplift, while the shallower rocks in the hangingwall show more distinctive increase in temperature with uplift. Thus, the P-T-t paths are not only sensitive to thermal parameters and denudation rate as revealed by previous one-dimensional model. They are also sensitive to tectonic parameters of stacking of thrust and the spatial location of rocks. Stacking of underthrust slices and of overthrust slices show different characteristics.

Frictional heating depends on the magnitude of shear stresses on the fault. Partial melting may occur if shear stresses are great enough, however, generation of magma can reduce friction and prevent further increase in temperature. With such temperature dependent frictional behavior, we modeled the thermal structure of the stacked thrust terrain. The most distinctive feature for this model is that the fault plane reaches the high temperature of partial melting and temperature of the whole hanging wall is much warmer than that of models without friction. Reversal in temperature profile still occur, but the minimum appears beneath the fault plane in the footwall instead of at the fault as in models without friction. Although the overall temperature is higher, temperature increase during uplift is less distinctive for deep rocks in the footwall; and no such increase at all for the P-T paths of rocks in the hangingwall.

In summary, P-T-t paths for rocks in stacked thrust terrains are complex and depend on thermal and tectonic parameters. Understanding of the regional tectonics, collection of an array of P-T-t path data and combined with two-dimensional modeling may provide better comprehension to this complicated problem.

SHIKAZONO, Naotatsu, University of Tokyo, Tokyo, Japan and MASATAKA Nakata, Tokyo Gakugei University, Tokyo, Japan

Mineralogic, Geochemical and Geological Characteristics of Se and Fe-Bearing Epithermal Gold Deposits in Japan

Introduction. A large number of epithermal gold deposits occur in Tertiary-Quaternary volcanic terranes in Japan. These deposits are generally accompanied by Se and Te mineralizations. Se-rich deposits are Te free and Cu poor, while Te-rich deposits are associated with small amounts of Se and Cu. In this paper, the characteristics of these Se-bearing and Te-bearing deposits are summarized.

Geologic Characteristics. Se-bearing and Te-bearing deposits are distributed in Tertiary-Quaternary volcanic region; northeast and southwest Hokkaido, middle part of Honshu, southern part of Kyushu. Se-bearing deposits are considerably more abundant than Te-bearing deposits. However, Te-bearing deposits are distributed in the regions similar to the Se-bearing deposits. Sometimes, Te mineralization is found in Se-bearing deposits, though Te minerals do not coexist with Se minerals. Te-bearing deposits are smaller in terms of tonnage than Se-bearing deposits. All Se-bearing and most Te-bearing deposits are vein in form. However, some Te-bearing deposits are massive and stratabound. Dominant host rocks for both types of deposits are volcanic rocks (andesite, dacite). Sometimes Se-bearing deposits are hosted by sedimentary rocks (shale, mudstone).

Mineralogic Characteristics. Major opaque, gangue and alteration minerals from Se-bearing and Te-bearing deposits are as follows:

Se-bearing deposits. Opaque minerals: Se-bearing argentite, polybasite, naumannite, aguilarite, Ag-rich electrum, pyrargyrite, sphalerite (FeS:1-5 wt%), galena, tetrahedrite, chalcopyrite, pyrite. Gaugue minerals: quartz, adularia, sericite/smectite, chlorite/smectite, calcite, rhodochrosite. Alteration minerals: chlorite, adularia, quartz, epidote, chlorite/smectite, sericite/smectite, zeolites, carbones.

Te-bearing deposits. Opaque minerals: hessite, petzite, native Te, sylvanite, Au-rich electrum, chalcopyrite, pyrite, marcasite, tetrahedrite, enargite, bismuthinite, sphaletite (FeS: ~ 1 wt %), galena. Gangue minerals: quartz (very fine grained), barite, sericite, adularia. Alteration minerals: quartz, sericite, kaolinite, alunite, smectite, adularia.

Geochemical Characteristics. $S^{34}S$ values of sulfide sulfur from Se-bearing deposits range widely from -7 to +5%. However, the most data range from 0 to +3%. $S^{34}S$ values of sulfide sulfur from Te-bearing deposits are narrow in a range mostly from 0 to +3%. These data indicate that $S^{34}S$ values for the Se-bearing and Te-bearing deposits are nearly same. $s^{34}s$ values of sulfate sulfur (barite) from Te-bearing deposits are +18 to +289%

Fluid inclusion filling temperatures for the stages of Se and Te mineralizations are 210-250 °C and 230-270 °C, respectively, though the data no Te-bearing deposits are scarce. No filling temperature data on massive Te-bearing deposits are obtained.

Chemical and geologic environments of formation. Based on mineral assemblages, chemical compositions of minerals (FeS content of sphalerite, Ag content of electrum), filling temperatures, possible fs_2, fo_2, pH and temperature regions can be deduced. Roughly, fs_2 and fo_2 for Te-bearing deposits are higher than those for Se-bearing deposits. Alteration and gangue mineral assemblages suggest that pH of ore fluids responsible for Te-bearing deposits may be lower than that for Se-bearing deposits.

High $S^{34}S$ values of barite, alteration assemblages, and relatively low pH and high fs_2 and fo_2 of ore fluids for Te-bearing deposits suggest that SO_2 gas of volcanic origin was involved in the ore fluids responsible for Te mineralization. Te-bearing deposits are mostly hosted by volcanic rocks, while Se-bearing deposits are sometimes by sedimentary rocks. These geologic data suggest that the Te-bearing deposits have formed at the position closer to the volcanic center than Se-bearing deposits.

Heald et al, 1987, divided volcanic hosted epithermal deposits into two types primarily based on vein and alteration mineralogies: acid-sulfate and adularia-sericite types. They considered that (1) acid-sulfate-type deposits form in the upper core of volcanic domes and acid-sulfate waters evolve from the interaction of meteoric waters with magmatic volatiles, and (2) adularia-sericite type deposits form in an extensive meteoric lateral flow regime. According to their classification, some Te-bearing deposits in Japan fall in the acid-sulfate type category, but others in adularia-sericite type. All Se-bearing deposits can be classified as adularia-sericite type.

REFERENCES

Heald, P., N. K. Foley and D. O. Hayba, 1987, Comparative anatomy of volcanic-hosted epithermal deposits; acid-sulfate and adularia-sericite types. Econ. Geol. 82, 1-26.

SHIMAMOTO, T., Hiroshima University, Hiroshima, Japan

Rheological Framework for Island-Arc Tectonics

Based on rheological properties of rocks and on the estimated temperature distributions beneath northeastern Honshu, Japan, an island arc is shown to be a unique place on the earth; that is, (1) the lithosphere is very thin (about 20 km thick at the most) and undergoes large deformation, (2) the assumption of rigid plate is not applicable, and (3) the island arc as a whole may act as a buffer zone to accommodate the relative plate motion. After a brief summary of major rheological properties of rocks, this paper will present the strength profile across northeastern Japan and will discuss characteristics of island-arc tectonics. The Japanese Islands are perhaps the most ideal places for such studies, for the tectonics and thermal state have been thoroughly studied there.

Strength profiles of an island arc across northeastern Honshu (Figure 1) have revealed the following features of an island-arc lithosphere. (1) The island-arc lithosphere behind the volcanic front (VF) is very thin and consists only of the upper part of the crust. The lithosphere in the uppermost mantle does not exist there, so that the island-arc lithosphere is much weaker than the continental lithosphere (cf. Bird, 1978; Ranalli and Murphy, 1987). (2) The triangular region beneath the outer part of the arc (ACT in Figure 1) is strong because the temperature is low there due to the subduction of a cold slab. This portion is the "fore-arc wedge," and an island arc is reinforced by it. (3) The lithosphere beneath the Japan Sea is much thicker than that beneath Honshu (Figure 1a), mainly because the crust is oceanic and thin there. (4) The lower crustal asthenosphere extends somewhat into the lithosphere underneath point I and the aseismic front, AF (Figure 1a).

An island arc is thus by far the weakest part of a continental plate (cf. Vink et al, 1984), and marginal seas such as the Japan Sea must have formed where the lithosphere is weakest. Deformability of the thin island-arc lithosphere can be seen clearly by the development of a great many active faults throughout the Japanese Islands (Research Group for Active Faults, 1980). Numerous active faults are developed where the lithosphere is thin (from point A toward the Japan Sea), but few are developed above the thick part of the fore-arc wedge, i.e., around AD (Figure 1). The rheological structure also supports Fujita's (1983) suggestion that the mountain ranges in the Japanese Islands formed as a result of upper crustal buckling with wavelengths of 80–100 km. That the thin island-arc lithosphere undergoes large deformation also manifests in the large, geodetically-determined crustal strain rate (Satoh, 1978). The strain rate exceeds 10^{-7}/y in many parts of the Japanese Islands, and this rate could be high enough for an island arc to accommodate, while deforming by itself, a considerable fraction of the relative plate motion. Earthquakes occur roughly in the lithosphere shown in Figure 1. Thus, the tectonics and the occurrence of earthquakes in an island arc are strongly influenced by the rheological structures of the lithosphere.

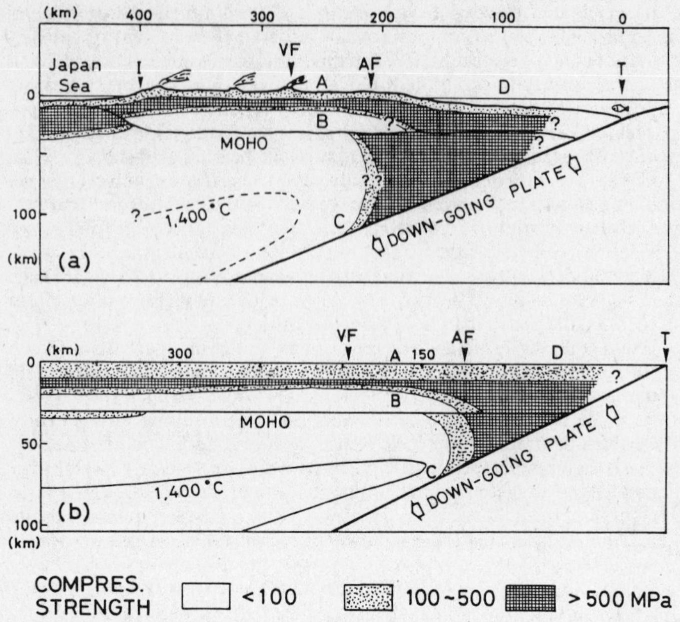

Figure 1—Strength profiles of an island arc across northeastern Honshu, Japan, in the horizontal stress field forming thrusts. The compressive strength, defined as the maximum differential stress sustainable by rocks, was estimated from the frictional strength of rocks (Byerlee, 1978) and from the flow laws of rocks assuming the strain rate of 10^{14}/s. Flow properties of crustal and mantle rocks were represented, respectively, by the flow laws of anorthothite (Shelton, 1981) and dry olivine (Kohlstedt and Goetze, 1974). (a) and (b) were constructed by using temperature distributions beneath the island arc estimated petrologically (Tatsumi et al, 1983) and by computer simulation (Honda, 1982), respectively. The simulation was performed to model the heat flow and magma genesis in the area. The Moho is drawn in accordance with the seismologically determined crustal thickness in (a), whereas its depth is assumed to be constant in (b).

SHIMAMOTO, T., Hiroshima University, Hiroshima, Japan

Fault and Plate Boundary Models

The establishment of realistic fault and plate boundary models is important not only to model large earthquakes, but also to quantitatively estimate the forces acting on plates across their boundaries and thereby evaluate the driving forces of plates. This paper will compare existing fault models and will argue that so-called S-C mylonites are the crucial rocks to examine the deformation mechanisms in the intermediate regime, between the brittle and ductile (or plastic) regimes that are poorly understood at the present time.

A simple model is that of Sibson (1977, 1982), which subdivides a fault zone into brittle and quasi-plastic regimes (AB and BC, respectively, in Figure 1a). Strehlau's (1986) model is basically a two-component model in which Sibson's strength profile is applied to each component (Figure 1b). In his transitional regime (EF in Figure 1b), one component (e.g., quartz) exhibits fully plastic deformation, whereas the other component (e.g., plagioclase) deforms totally in the brittle manner. Such simple models are not consistent with the results from recent shearing experiments on halite (Shimamoto, 1986; Shimamoto and Logan, 1986; Hiraga and Shimamoto, 1987), which have revealed the existence of a wide intermediate regime.

Models of Scholz (1988) and Shimamoto (1988, in press) are intended to incorporate with this intermediate regime (IJ and MN in Figs. 1c and 1d). They both propose that plastic (but not fully plastic) deformation begins to occur already in the intermediate regime to form mylonites. Scholz argues that the potentially unstable fault motion disappears once the plastic deformation begins to occur. Thus, the seismic depth does not extend into the intermediate regime, although he suggests that the rupture extends downwards at the time of a large earthquake (heavy dashed line in Figure 1c). In the halite experiments, however, well-defined stick-slip was recognized well within the intermediate regime where deformation is predominantly plastic. Hence the seismic depth is extended into this regime in Shimamoto's model (Figure 1d, heavy solid line).

S-C mylonites are characterized by mylonite foliation, by grain-size refinement with strain concentration, and by shear surfaces cutting mylonite foliation. Halite experiments have revealed that all of these features are also characteristics of the deformation in the intermediate regime. In particular, stick-slip was recognized only when the deformation became localized at thin and nearly plane and continuous zones, so that the development of the shear surfaces in S-C mylonites should have distinct mechanical implications. The intermediate regime, where the strength is close to maximum, is the primary candidate for the genesis of S-C mylonites.

The seismic depth revealed through halite experiments (Figure 1d, heavy solid line) seems to be consistent with the seismicity along subducting plate boundaries (Shimamoto, 1985). However, Hobbs et al, (1986) argued theoretically that it could extend even deeper (shaded line in Figure 1d), and the exact limit of the seismic depth is still controversial. It is also a future task to extend the single-component model in Figure 1d to multicomponent behaviors.

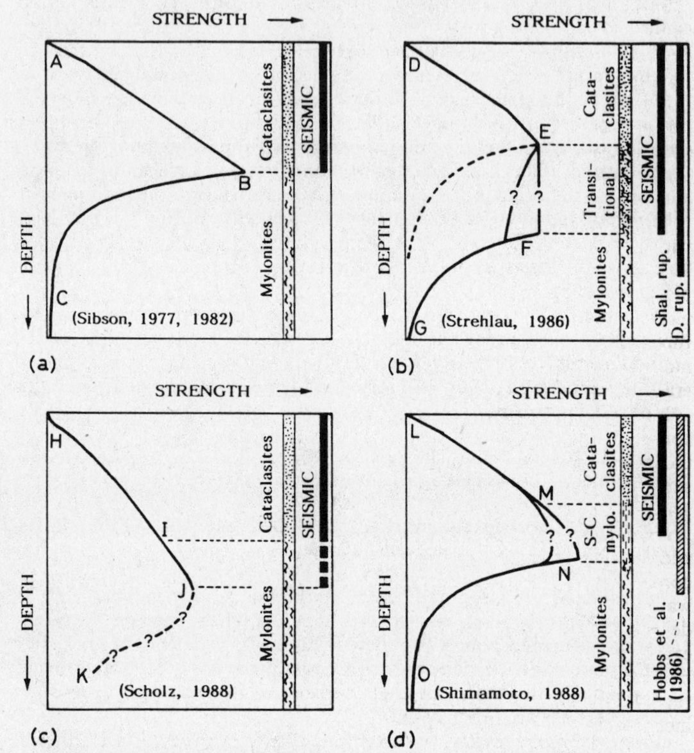

Figure 1—Schematic diagrams illustrating fault-zone models of (a) Sibson (1977, 1982), (b) Strehlau (1986), (c) Scholz (1988), and (d) Shimamoto (1988, in press). Strength profiles under a given stress field are shown by solid lines together with corresponding dominant fault rocks and the presumed seismic depth. The two ranges for the seismic depth in (b) indicate Strehlau's shallow and deep rupture models. Seismic depth after Hobbs et al (1986) is shown by a shaded line in (d) for comparison. The initial rising and final falling portions of the strength profiles were drawn based, respectively, on Byerlee's (1978) empirical equation for the frictional strength of many rocks and on the flow laws of rocks. Note that the scales for the vertical axes in the diagrams are not the same.

SHIMP, R. J., and R. A. RAPAPORT, Procter and Gamble Company, Cincinnati, Ohio, R. M. VENTULLO, University of Dayton, Dayton, Ohio, and W. D. ROBERTSON, J. A. CHERRY, and E. A. SUDICKY, University of Waterloo, Waterloo, Ontario, Canada

Biodegradation and Fate of Chemicals in Septic Tank Systems and Adjoining Aquifers

Approximately 30% of the domestic wastewater produced in the United States enters on-site treatment systems (e.g., septic tanks). These systems typically discharge to subsurface drainage or "tile" fields where chemicals are treated by adsorption and biodegradation. Biodegradation constitutes the most important mechanism by which a chemical is removed, through transformation to simple endproducts (e.g., carbon dioxide) or microbial biomass.

We have evaluated the fate and transport of chemicals in septic tank systems at a field site in Ontario, Canada. The objectives of these studies were:

1. To characterize biodegradation activity and sorption of chemicals in the septic tank tile field and adjoining aquifer.

2. To examine the chemical/physical/microbiological nature of the site.

3. To determine the concentration of consumer product chemicals in the aquifer, in order to validate mathematical models for predicting the transport of chemicals in aquifers near septic tanks.

This presentation focuses on microbial processes at the site, particularly the influence of the septic tank effluent discharge on the biodegradation of consumer product chemicals.

Materials and methods: The field site is located over a shallow unconfined sand and gravel aquifer which is 3–3.5 m thick and is overlain by a 2–2.5 m thick vadose zone. Ground water velocity in the aquifer is approximately 20–40 m/year. A 12 year old septic system serving a family of four discharges to the site at a depth of approximately 0.6 m.

The biodegradation of the following ^{14}C-radiolabelled chemicals was studied: a linear alkylbenzene sulfonate (LAS), nitrilotriacetic acid (NTA), a linear alcohol ethoxylate (LAE), ethylene glycol, cellulose, methyl parathion, and several substituted phenols.

Sodium measurements in the groundwater were used to track the location of an effluent plume from the septic tank. Soil samples used in the biodegradation assays were taken along a linear transect at the plume's "center."

Soil and groundwater samples were collected from locations within, downgradient and upgradient of the tile field. Groundwater samples were collected with Teflon piezometers placed at different depths in the aquifer. Soil samples were collected by coring vertically through the vadose zone and aquifer.

Biodegradation assays were conducted on replicate soil samples dosed with 25–100 ppb of the radiolabelled test chemical and incubated at 15–20 °C. A paper wick containing 1.5N KOH was suspended above each sample to trap CO_2. At selected intervals the wick was removed and assayed for mineralization of the chemical to $^{14}CO_2$ by liquid scintillation counting. Incubations lasted for up to 100 days. CO_2 data were fitted to a first order model which provided an estimate of the extent of mineralization, the mineralization half life, and the biodegradation lag time.

Microbiological parameters, including plate counts, total bacterial counts, coliform counts, enzyme activity, and growth rates (^3H-thymidine incorporation) were also measured in the soil samples. Both water and soil samples were analyzed for specific chemicals (e.g., LAS), as well as an array of general constituents (e.g., organic carbon, particle size).

Results and discussion: Biodegradation of the test chemicals varied widely with respect to location along the transect and depth within each soil core. However, several trends emerged. Generally, biodegradation activity was highest in the near-surface soil samples (< 1 m), and was lowest in the deepest aquifer samples. In samples showing active biodegradation, the extent of mineralization was typically 50–60% for any of the test chemicals. Based on studies in the literature, such extents of mineralization likely represent complete biodegradation. Remaining chemical was presumably incorporated into microbial biomass.

The biodegradation of LAS and NTA, which are components of laundry detergents, was generally highest in soil samples collected from within the septic tank tile field. Mineralization half lives were typically less than 1 day for NTA and 10 days for LAS. Relative to the water velocity in the aquifer, both these half lives are very short. At sites downgradient of the field, LAS and NTA biodegradation activity diminished, suggesting that pre-exposure to septic tank effluent was important for biodegradation. Biodegradation was also limited at the upgradient control site. However, this observation was not universal. For example. LAE (a non-ionic surfactant) and ethylene glycol biodegraded in nearly all samples.

The hypothesis that pre-exposure to some chemicals was a prerequisite for high biodegradation activity was supported by analytical data. For example, while approximately 10,000 ppb of LAS was measured in the septic tank effluent, levels were reduced to 10–50 ppb immediately below the tile field, and were undetectable at both the downgradient and upgradient sites. Thus downgradient and upgradient microbial communities were probably not exposed to enough LAS to support significant biodegradation activity.

Soil microbial characteristics were generally consistent with the biodegradation data. For example, highest enzyme activities and growth rates were observed in surface soils and in the immediate vicinity of the tile field.

Conclusion: This study has demonstrated the importance of biodegradation processes in on-site treatment systems, and the interplay of biodegradation with the hydrogeology, physical/chemical properties, and microbiology of subsurface environments. The program has also provided a valuable demonstration of the value of interdisciplinary research in groundwater studies.

SHIPLEY, T. H., University of Texas, Austin, TX

Accretion, Subduction, and Erosion along Middle America Trench

The Middle America Trench system extends 3000 km along the continental margin from Mexico to Panama. It is dominantly a sediment-starved, pelagic trench with convergence between 6 and 9 cm/y mostly normal to the trench. The subducting oceanic crust is of Miocene to Eocene age with a 200–400 m cover of pelagic carbonates overlain by hemipelagic sediments of similar thickness. Trench axis fill is variable but limited to a few locations where submarine canyons supply material from the shelf. This trench system exhibits evidence for accretion (of both sediment and crustal rocks), underplating, and erosion. The trench is divided into at least three provinces: north of the Tehuantepec Ridge, from the Tehuantepec Ridge to Nicaragua, and from Costa Rica southeastward.

Drilling and geophysical investigations offshore southern Mexico (north of the Tehuantepec Ridge) reveal a late Miocene to recent history of sediment offscraping defined by a zone of landward-dipping seismic reflections composing the middle and lower continental margin slope. Drilling in the middle and upper slope demonstrates continued uplift unrelated to the sediment offscraping at the toe. This uplift is the main evidence of an accretionary mechanism occurring at depth.

Drilling and geophysical investigations offshore Guatemala reveal completely different mechanisms active during convergence. Before drilling, it was predicted that an offscraped section would be sampled, but drilling demonstrated that the arcward trench slope is composed of Cretaceous mafic oceanic crustal rocks. The mafic rocks were emplaced prior to the deposition of the slope cover beginning in early Eocene. Stratigraphic and structural relationships developed from the extensive drilling program suggest that most of the oceanic plate and its overlying sediments have been subducted throughout the Neogene. Thus, a pre-Eocene period of crustal thickening has been followed by a period in which most of the converging plate and overlying sediments are being subducted. The complex history of vertical movement recorded in the slope cover and at the seaward edge of the fore-arc may be evidence for some sort of crustal erosion processes active at times in this same area.

The area south of Nicaragua represents a third province. It is defined by the absence of a large, thickly sedimented fore-arc basin. The Nicoya Peninsula, a largely ophiolitic block of Mesozoic age, is along trend with the fore-arc basin. Whether it is the same as the basement of the forearc to the northwest or is the uplifted trench slope is a matter of speculation. The Cocos Ridge being subducted to the south subtly influences the tectonics and geography of the area. The continental margin has had extensive geophysical investigations, but little drilling. There is ample evidence for a small offscraped section at the base of slope and underthrusting of the crust and most of the overlying sediment at least several kilometers beneath the arcward slope. Unlike either Mexico or Guatemala, the accretionary wedge contains many subhorizontal reflections that are mappable in a three-dimensional seismic survey of a portion of the margin. These seismic reflection images define surfaces up to 5 × 10 km that probably are low-angle thrust surfaces. These thrusts effectively cause thickening of the wedge at very shallow subbottom depths. The slope cover is extensively faulted with high-angle normal faults, possibly a result of continued accretion and uplift beneath the wedge.

The Middle America Trench exhibits rather differing history and processes of margin growth. Such variation along a single simple trench system illustrates the problems of understanding the processes involved in margin growth and modification. The different local histories of the margin system probably have much to do with the details of what actually occurs at the base of the wedge both at the toe and at slightly deeper structural levels. The processes are largely controlled by effective stresses and mechanical strengths of rocks, which are determined by the convergence rate, exact sediment physical properties, and water content.

SHISHKIN, M. A., Paleontological Institute, Moscow, USSR, and V. G. OCHEV, Saratov State University, Saratov, USSR

Spatial Differentiation of Early Triassic Tetrapod Fauna

The history of Triassic land vertebrates is shown to include three successive epochs: the proterosuchian, kannemeyeroidean, and dinosaurian (Ochev and Shishkin, 1984, 1988). The earliest and middle ones are typified by the regional faunal sequence of East Europe (Shishkin and Ochev, 1967, 1985). There the proterosuchuan epoch encompasses the neorhachitome and *Parotosuchus* faunas, the former of which spans, in turn, three groupings. The lowermost one corresponds to the Induan of Greenland by the occurrence of the labyrinthodont genera *Tupilakosaurus* and *Luzocephalus*. The middle grouping is dominated bybenthosuchids and is equated on that account with the Lower Olenekian of the Soviet Far East. The upper grouping is equal to the top of the same substage. The late Scythian age of the *Parotosuchus* fauna is evidenced by occurrence of its guide labyrinthodont genera in the Upper Olenekian (Spathian) estuarine and marine deposits of the Caspian Depression and the Mangyshlak Peninsula.

The kannemeyeroidean epoch also comprises two faunas in East Europe, the *Eryosuchus* and *Mastodonsaurus,* the latter being directly comparable with the fauna of Lettenkohle in Central Europe, which is surely Ladinian in age.

On correlating the Triassic tetrapod sequences over the world, a comparison of the regional faunal events that proceeded during the proterosuchian epoch presents the most difficult problem. The first question to be answered is whether the boundaries between the proterosuchian and the kannemeyeroidean epochs in Gondwana and Laurasia may be considered roughly synchronous. In the Cis-Urals, that level corresponds to the *Parotosuchus-Erysuchus* faunal replacement, i.e., the Lower-Middle Triassic boundary. On the other hand, in South Africa, the same biotic change falls on a transition between the *Lystrosaurus* and *Cynognathus* Zones, both being usually assigned to the Early Triassic. This contradiction implies that either the kannemeyeroidean epoch had begun earlier in Gondwana or the age of the *Cynognathus* Zone should be reevaluated.

We accept the second solution. The idea of the Scythian age of the *Cynognathus* Zone was advocated mainly on the basis of its amphibian fauna, which was believed to include, among others, *Parotosuchus* and *Batrachosuchus,* as well as trematosaurid and rhytidosteid members. However, the African species of "*Parotosuchus*" are markedly advanced over their alleged European congeners. Similarly, *Batrachosuchus,* which is known elsewhere from the Middle Triassic of Zambia, appears to be a descendant of *Batrachosuchoides* from the Upper Olenekian of East Europe. The only rhytidosteid genus previously assigned to the *Cynognathus* assemblage is shown to come from the *Lystrosaurus* Zone. Now the importance of the trematosaurid findings as evidence for the Scythian age of the *Cynognathus* Zone seems weakened due to the allocation of the Upper Triassic genus *Hyperokynodon* to Trematosauridae. As concerns the evidence from reptilian components of the *Cynognathus* Zone, the current idea of the close relationship between the African proterosuchian *Erythrosuchus* and the Eurpoean genus Garjainia (belonging to the *Parotosuchus* fauna) cannot be accepted (Ochev, 1980). To sum up, nothing seems to preclude the placement of the *Cynognathus* Zone in the Middle Triassic.

In this connection, a question arises about the time range of the *Lystrosaurus* Zone, which is currently correlated with the European neorhachitome fauna. Coupled with preceding conclusions, this correlation would imply the lack of the equivalent of the *Parotosuchus* fauna in the Gondwana faunal sequences. But this later viewpoint seems hardly acceptable. In South Africa, the guide amphibians of the *Lystrosaurus* Zone (lydekkerinids, uranocentrodontids) are accompanied by the advanced capitosauroid (mastodonsaurid) *Kestrosaurus.* Similarly, in Australia (Arcadia Formation) lydekkerinids are found together with the true *Parotosuchus.* These facts provide strong evidence for correlating the *Lystrosaurus* fauna with the whole range of the Scythian.

Co-existence of the lydekkerinids and the advanced capitosaurids in Gondwana presents a marked difference from what is known for the Northern hemisphere. The lydekkierinids do not outlast the Induan and become fully displaced by the capitosaurids and trematosaurids (including benthosuchids) as early as the level of the middle grouping of the neorhachitome fauna. On the other hand, both latter groups are quite rare in the most southerly regions of Gondwana (cf. Cosgriff, 1984), the first trematosaurids appearing there at the *Parotosuchus* level (Australia) or even later (South Africa). This seems to indicate the Laurasian origin of both and their rather late spread into the southern areas.

REFERENCES

Cosgriff, J., 1984, The temnospondyl labyrinthodonts of the earliest Triassic: J. Vert. Paleontol., v. 4, n. 1, p. 30-46.

Ochev, V.G., 1980, On *Erythrosuchus (Garjainia)* primus Otschev: Voprosy geologyi Yuzhnogo Urala i Povolzh'a, Vyp. 22, p. 3-22 (in Russian).

and M.A. Shishkin, 1984, Principles of global correlation of the continental Triassic based on tetrapods: Abstracts of the 27 MGC, v. 1, p. 147.

——— 1988, The global correlation of the continental Triassic based on tetrapods: Izvestia Akad. Nauk SSSR, ser. geol., n. 2, p. 3-15 (in Russian).

——— ——— Shishkin, M.A., and V.G. Ochev, 1967, The land vertebrate fauna as a basis of stratification of the continental Triassic sediments of the USSR: Stratigraph. i paleontolog. mezozoyskich i paleogen-neogenov. kontinenta ln.otlozhen.Aziatsk.chastyi SSSR. L:Nauka, p. 74-82 (in Russian).

——— 1985, The significance of the land vertebrates for the stratigraphy of the Triassic of the East-European Platform: Triasovye otlozhenia Vostochno-Evropeyskoy Platformy. Saratov, Saratovyskyi Universitet, p. 28-43 (in Russian).

SHIVAKUMAR, B. S., and S. CHANNABASSAPPA, Bangalore University, Bangalore, India

Facies Concept: Case Study of BIFs of Bababudan Hill Ranges of Karnataka, India

The syndepositional aspects of banded iron formations (BIFs) like facies is a difficult task to be inferred particularly if they are metamorphosed. This needs a careful analysis of the present mineralogy, texture of BIFs, and the field relations, which has been done following James (1954), French (1968), Klein (1973, 1978, 1983), Floran and Paplike (1978), Gole (1981), Mel'nik (1982), Miyano (1987), etc. Sufficient thermodynamic and geochemical support is drawn in favor of the fluctuating and instability in the depositional environments.

Bababudan hill ranges are a prominent part of Western Ghats in the southern part of the Indian Peninsula. BIFs of the Muliangiri Formation in the Bababudan Group (Chadwick et al, 1985) constitute one of the best exposed lithotypes of the late Aracean volcanosedimentary assemblages of the entire peninsula and help understand several aspects including basin analysis, stratigraphy, structure, and metamorphism. James's (1954) concept of facies, when applied here, has revealed that "fluctuating nature" is more prevalent than well defined clear out depositional environments. Although two broad zones (as in Figure 1)-v mixed oxide type and oxide type (following Laajoki, 1977, and Trendall, 1983) of chemogenic deposition, each characterized by different primary mineralogy are envisaged, there is the frequent intermixing of the two in many places indicating fluctuating environment and instability during deposition.

Based on mineralogical studies of BIFs, the following set of primary minerals is inferred. For the older mixed oxide types, a complex combination of chert, fine magnetite, and carbonates such as ferrodolomite/ankerite and siderite, and for the younger oxide types a simpler combination of chert/jaspar and amorphous hydroxide of Fe^{3+} as primary minerals have been proposed. Consequently, the two types broadly reflect reducing and oxidizing environments, respectively. These reducing

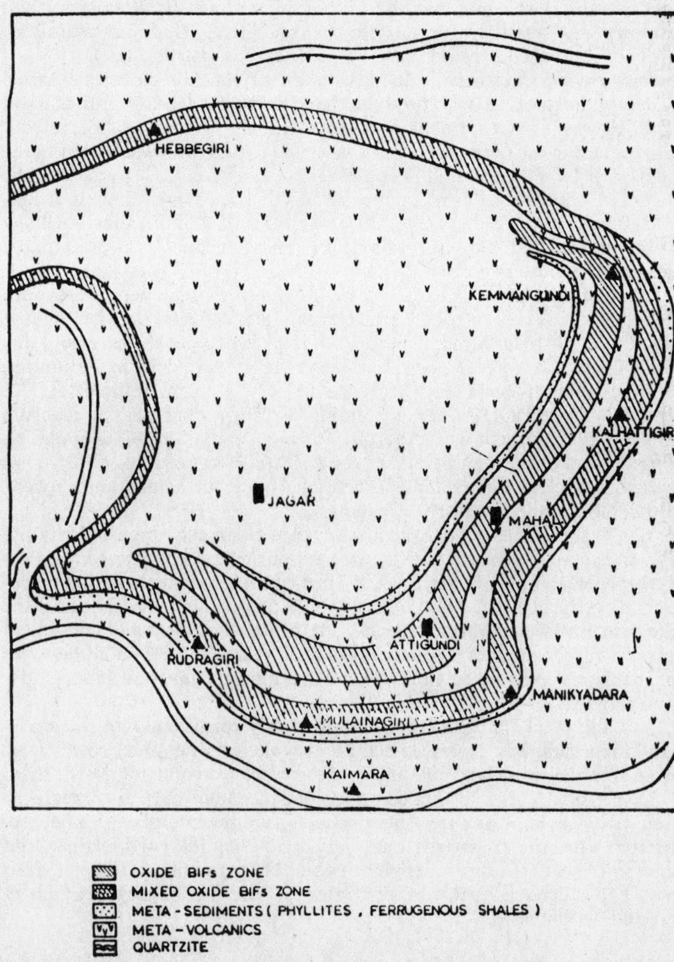

MAP No. 1 SCHEMATIC MAP SHOWING BROAD ZONATION IN BIFs DEPOSITION

and oxidizing environments were largely disturbed by factors such as submarine volcanism, influx of clastics, seasonal fluctuations, photoprecipitation, organic activity, etc., resulting particularly in abrupt changes in Eh and pH conditions. This eventually caused fluctuation in the environment itself, resulting in the formation of different minerals and hence intermixing of mixed oxide type with oxide types and vice versa. Field and geochemical relations also support this inference.

While the development of these two major types took place during deposition, the subsequent postdepositional changes were largely responsible for giving rise to other subtypes. The mineralogical transformation in mixed oxide types is more complex than in oxide types, and it is attempted to understand this aspect. For example, important metamorphic reactions for the formation of common minerals such as grunerite, actinolite, and calcite are as follows (Klein, 1973):

1. $7\ Ca(Fe,Mg)(CO_3)_2 + 8SiO_2 + H_2O\ (Fe,Mg)_7Si_8O_{22}\ (OH) + 7\ CaCO_3 + 7CO_2$
ferrodolomite quartz grunerite calcite

2. $8\ (Fe,Mg)\ CO_3 + 8SiO_2 + H_2O\ (Fe,Mg)\ _7Si_8O_{22}\ (OH)_2 + 7\ CO_2$
siderite grunerite

3. $5\ Ca\ (F3,Mg)\ (CO_3)_2 + 8SiO_2 + H_2O\ Ca_2\ (Fe,Mg)_5Sip_8O_{22}\ (OH)_2 + 3CaCO_3 + 7CO_2$
ferrodolomite quartz actinolite calcite

In conclusion, the overall mineralogy of BIFs of Bababudans is shown in the paragenetic diagram (Table 1).

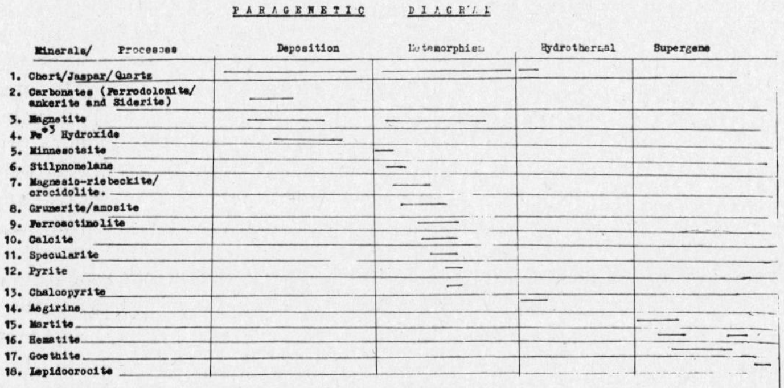

SHIXING, Z., Tianjin Institute of Geology and Mineral Resources, Chinese Academy of Geological Sciences, Tianjin, China

On Evolution of Stromatolite-Building Microorganisms

In China, a lot of stromatolite-building microorganisms have been found from the Proterozoic stromatolites, which occur in the different horizons of the Proterozoic sequences and include the following stromatolite taxa: *Stratifera*, *Gymnosolen*, *Linella*, *Baicalia*, *Scyphus*, *Pseudogymnosolen*, *Jacutophton*, *Conophyton*, and so on.

The Early Proterozoic microorganisms are predominated by quite thin filaments (5 μm in width of sheath or cell), such as *Gunflintia minuta*, *Cyanema minor*, *Eomycetopsis filiformis*, and a part of bacteria (1 μm in width) etc., and the solitary coccoids, such as *Huroniospora* sp., *H. psilata*, *Leptoteichos*, and some *Arcritarcha* sp.

The Middle Proterozoic microorganisms are represented mainly by those filaments with rather wide sheath or cell (5–10 μm), such as *Siphonophycus inornatum*, *Siphonophucus* sp., *Palaeolyngbya maxima*, *Oscillatoriopsis* sp., and so on, and those coccoids in different colonies, such as *Eoentophysalis telcherensis*, *Palaeoanacystis vulgaris*, and *Coniunctiophucus conglobatus*, etc.

The Late Proterozoic stromatolite-building microorganisms differed from microorganisms mentioned above by rather complex morphology and structure. For example, a number of filaments can develop branches and some spore-like bodies at their terminal (*Monostichus cirriformis*, *Polystichus xifengensis*, and *Palaeoreteella multilateralis*, etc.).

These data suggest that the stromatolite-building microorganisms in the Proterozoic age were developed from simple to complex, though their evolution through time is rather slow.

SHNYUKOV, E. F., Institute of Geological Sciences, Kiev, USSR

Mud Volcanism and Ore-Formation

Mud volcanism is spread widely on the earth. There are known hundreds of mud volcanos in the southeast Caucasus, Turkmenia, Kerchensko-Tamansky district, Iran, California, Romania, Sakhalin, southeast Asia, Mexico, Trinidad and others. The mechanism of mud volcano action is defined by anomaly high pressures of fuel gases in bowels of the anticlinal structures, where the gases break loose usually along the tectonic faults from the depth up to 7-9 km.

In geological history the mud volcanism is registered by time from the Silurian to the Recent, but in our country, from the Cretaceous to the present day. The scale of mud volcanism geological activity is huge, and it is expressed by break-looses of millions and even trillions of cubic meters of fuel gases, tens of billions of tons of clay rock and a lot of water too. Along side with combustible gases, Hg, J, Br, and other chemical elements are broken loose too. Sometimes they form the economic concentrations. When mud volcanism appears in the seawater environment, then the mud volcano calderas formed by blowout materials can be filled with the basinal sediments and mud breccia resulted into the pseudotectonic settlement structures. Classical locality of their developments it the Kerchensko-Tamansky districts, where aforesaid structures received local designation such as "caved synclines." The author and P. I. Naumenko discovered the development of the Neogene oolitic iron-ore sediments in the column of many of these synclines. The biggest iron-ore deposits of such type occur in the Kerch Peninsula. Amongst from them there are the Novosyolovskoye (about 100 million tons) and Kamenskoye (about 300 million tons) fields. Many ore shows had been discovered in the areas considered earlier as non-perspective in the Inside-Parpatchian part of the Kerch Peninsula (e.g. the Uzunlarskoye, Repyevskoye, Atchinskoye and Batalnenskoye areas), and in the old ore-bearing districts of the outer-Parpatchian part of Kerch Peninsula (e.g. the Baksinkoye, Bulganakskoye and Kezenskoye fields, etc.). The iron-ore deposits and ore-shows of the Kerch Peninsula "caved synclines" are characterized by the peculiar lithologic and geochemical composition, great thickness and small area. So, the ore-bearing clastic material from 0.5 to 3 mm in size prevails in the composition of iron ores of "caved synclines," where the clay pebble is also present. the specific weight of the oolites of iron ores (only several percent) is somewhat decreased as compared to the original Kerch deposits: the iron contents of oolites is 34, 5-37-41%; but the manganese-somewhat higher (0.47-2.73%), phosphorus-0.45-2.16%, and arsenic-0.08-0.13%. The ore-bearing proven territory of "caved synclines" is usually small and does not exceed 5 sq. km (the Baksinskoye deposit), and as a rule is considerably lesser. The thickness of ore deposits is bigger as compared with the Kerch typical ore deposits and reaches 92 m (the Novoselovskoye deposit), but it is usually from 20 to 30 m. The development of ore deposits takes place only under the conditions of Neogene mud volcano activity (the Kimmerian) in the ore-bearing basin with the accumulation of ore chemogenic and clastic material in lagoons originated by the mud-volcano nature depressions. In the diagenesis process the ore-deposits or underwent by the mineralogical changes. As a rule, the upper 10-15 m of ore deposits are oxidized by the erosion processes. The total iron ore reserves exceed 550 million tons in the "caved synclines." The useful minerals are also the limestone, sands and clays of "caved synclines," and the sulfur shows produced by mud volcanism.

The investigations mud volcanism are not only the regional success of geologists in the Kerchensko-Tamansky area, but these works can play a positive role in the comparative study in some other mud-volcanic regions too.

SHORT, N. M., Bloomsburg University, Bloomsburg, PA

Geomorphic Expression of Accreted Terranes: Klamath Mountains, Oregon

Several concepts underlying current models of the development of accreted or tectonostratigraphic terranes emerged from studies of crustal units making up the Klamath Mountains of southwestern Oregon and northwestern California. More than 20 individual terranes emplaced between Late Paleozoic and Late Cretaceous have been described for this region. These terranes are identified largely from notable differences in lithology, nature of structural boundaries, and, for some, paleomagnetic evidence. Although many workers there and in terrane assemblages elsewhere note distinctive differences in topography and landform expression between terranes in juxtaposition, little systematic analysis of geomorphic variations among these assemblages has been reported. The purpose of this study of the Klamath terranes in Oregon are to: (1) define and describe, qualitatively and quantitatively, the geomorphic characteristics of known (previously mapped) terranes, (2) determine whether terranes within an assemblage have geomorphic signatures sufficiently different to permit their recognition in aerial/space imagery, and perhaps also to better delineate boundaries between them, and (3) assess geomorphic information about terranes as a clue to modes of emplacement, especially wherever landform geometry can be related to structural indicators. The study involves applications of morphometry to conventional and digital topographic databases and, whenever useful, Landsat/SPOT imagery.

Inspection of a Landsat Thematic Mapper 2-image mosaic covering most of the Oregon-California Klamaths reveals visually a collage of diverse surfaces distinguished by obvious differences in stream densities, and in principal ridge orientations, and degree of upland dissection. When a terrane map, shown as black line boundaries drawn on a transparent overlay, is superimposed on the mosaic, the clusters of different appearing topographies broadly coincide with the mapped terranes. Judges qualitatively, the larger terranes are distinguishable by their terrain expression.

The distribution of maximum elevations and relief within entire (or parts of) terranes was calculated from two different data sets. Elevation data were extracted from 15' topographic sheets, sampled on a 1 mi2 (section) grid, then grouped into respective terranes present, and analyzed statistically by the t-test and others. Significant differences were found between many terrane pairs in juxtaposition both in terms of maximum elevation and relief. A typical example occurs near Canyonville, Oregon, where the mean maximum elevation values for the Yolla Bolly and Rogue Valley terranes differ by more than 530 ft (significant at the 0.99 level). However, variations within large individual terranes, sampled over a sequence of quadrangles, may be even greater. Generally similar results, suggesting significant differences, were obtained by computer analysis of U.S.G.S. Digital Elevation Model (DEM) data (150 m cell size) partitioned into entire terranes whose boundaries were separately digitized and registered to the DEM set. Differences in relief between proximate terrane pairs commonly were also significant, but less sensitive than elevation differences.

However, the effect of regional slope gradients (there are two highest areas in the Oregon Klamaths), even where these terrane elevation comparisons were made at the quadrangle scale, can account for some of the apparently significant differences. To compensate for slope effect, trend surface analysis was applied to the DEM data sets covering the entire Oregon Klamaths and to several of the analyzed quadrangle sheets. First- to fourth-order trend surfaces were calculated, from which residuals (local departures in elevation from the general surface) were then contour-plotted both for elevation and relief. A fitted terrane boundaries map served to highlight visually where patterns of positive or negative residuals correlate with terranes. In general, the distribution of residuals lacks strong correspondence with specific terranes, with certain interesting exceptions.

Hypsometric plots of frequency distribution of elevations within terranes (larger ones subdivided into 2-3 subunits) result in a wide range of curves for the 18 terrane units considered. Although distribution patterns are complex and prove difficult to interpret as systematic variations, the more resistant (more metamorphosed) terrane lithologies generally have larger hypsometric integral values (suggesting less extensive downward erosion) and greater fractions of their total elevation range are associated with higher elevations. For the entire Oregon Klamaths, a strong peak on a composite curve for elevations between about 1.3 and 1.65 km is distinctive. Currently, power spectrum analysis of elevations and slopes as functions of variable baseline spacings is underway, from which will be derived fractal dimension data and a related measure of surface roughness.

Stream and ridge densities and orientations measured from topographic maps disclose provocative structural information. Ridge densities vary from 5.4 (Elk terrane) to 2.7 (Snow Camp terrane) km/km2. The regional ridge orientations center on a N20° bearing but most frequent orientations vary between individual terranes. The oldest (inland) terraces show two weak maxima for ridges following north-northwest and northeast trends; the Yolla Bolly-Rogue Valley-Smith River terranes favor ridge orientations to the north-northeast; most coastal terranes have conspicuous west-northwest to northwest ridge trends. Speculatively, these ridge orientations may represent a response to control by streams whose orientations in turn follow structural weaknesses imposed by shifting subduction zone geometry.

Geomorphic measurements made from Landsat imagery are strongly biased by sun azimuth/angle shadowing. Analysis of SPOT stereo imagery largely removes this bias, as demonstrated by comparing ridge orientation plots made from Landsat, SPOT, a topographic map of the same area, and DEM-based shaded relief images generated under various illumination directions.

Two conclusions are emphasized: (1) some morphometric measures effectively separate some terranes, and (2) space imagery must be stereoscopic if it substitutes for topographic maps as well as a regional geomorphic database.

SHREEDHAR, S. K., Govt. College, Dharmsala, India

Geochemical and Tectonic Status of Central Crystalline Rocks of Wangtu Area, District Kinnaur, Himachal Pradesh, India

A very interesting complex of gneissic and schistose rocks, belonging to the Central Crystalline Zone, exposed in the southern part of the Higher Himalaya from Wahgtu to Karcham and beyond in the Sutlej Valley, has been investigated during the present study.

Stoluczka (1866) was the first to report the geology of Wangtu and Spiti areas. In 1866, Mallet followed him by reporting about mineral wealth of this area. Hayden (1904) in his classic memoir furnished a monumental work on Spiti and parts of Bushar and Rupshu areas. Berthelsen (1951) recognized the occurrence of "Wangtu Dome" and "Rampur Dome." Vaikrita Thrust was recognized by Sharma (1976) to designate the overlying rocks as Vaikrita Group and the underlying as Jutoghs. Tewari et al, (1978) gave a brief account of lithounits of the area around Baspa and Sutlej valleys. Bhargava (1982) outlined the regional geology of a vast area between Kulu, Rampur, and Karcham. Bassi and Chopra (1983) accounted for the geology of the Kinnaur Himalayas. Shreedhar (1986) and Kakar (1988) worked on the geology and tectonics of the Central Crystalline Rocks of the Higher Himalaya.

The subject matter of the present communication includes major and trace element geochemistry of these rocks and their structural and tectonic position in relation to some well-established parameters. On the basis of lithology, chemical heterogeneities, rock association, and tectonic set-up, the rocks exposed have been divided into two principal groups; i.e., Wangtu Gneissitic Complex and Karcham Group. Both these litho-units have a significant lithological variation and are separated by a thrust contact known as the Karcham Thrust (Shreedhar, 1986). The Wangtu Gneissic Complex is characterized by dominant feldspar (occurring as plates and also as porphyroblasts) bearing gneisses interbanded with sillimanite-bearing schist rocks, whereas the Karcham Group is characterized by sillimanite and garnet-bearing schistose rocks. The presence of sillimanite indicates that these rocks have suffered high-grade metamorphism.

To work out the chemical heterogeneities and also the changes that took place during and after crystallization, ten representative rock samples have been analyzed chemically following the rapid method of silicate analysis (Shapiro and Brannock, 1962). These rock samples have also been analyzed for trace element (Cu, Co, Ni, Cr, Pb, Zn, Sr) data using an atomic absorption spectrophotometer. The trace and major element data accruing from these analyses have been plotted in various variation diagrams so as to understand the chemical nature of these rocks. The tectonic history of these rocks has also been deciphered on the basis of structural, stratigraphic, and radiometric data comparable with rocks in the other parts of the Himalaya (Pandey, 1981; Kwatra et al, 1984). On the basis of investigations carried out so far, it is understood that the Wangtu Gneissic Complex forms the oldest basement exposed in this part of the Himalaya and is of deep crustal origin.

REFERENCES

Bassi, U.K., and S. Chopra, 1983, A contribution to the geology of Kinnaur Himalaya, Himachal Pradesh: Indian Journal of Earth Science, v. 20, p. 96-99.

Berthelsen, A., 1951, A geological section through the Himalayas: Medd. Densk. Geol. Forening, v. 12, p. 102-104.

Bhargava, O.N., 1982, The tectonic windows of lesser Himalaya: Himalayan Geology, v. 10, p. 135-155.

Hayden, H.H., 1904, The geology of Spiti with parts of Bushar and Rupshu: Memoirs of the Geological Survey of India, v. 36, p. 1-129.

Kakar, R.K., 1988, Geology and tectonic setting of Central Crystalline Rocks of southern part of Higher Himalchal Himalaya: Journal of the Geological Society of India, v. 31, n. 2, p. 243-250.

Kwatra, S.K., V.B. Bhanot, and R.K. Kakar, 1984, New Rb/Sr radiometric data on the granitic rocks of western Himalaya and its geological interpretation: National Seminar on Himalayan Crystallines, Metamorphics and Structures, Abstract 13, 21 p.

Mallet, F.R., 1866, On the gypsum of lower Spiti with a list of minerals collected in the Himalayas: Memoirs of the Geological Survey of India, v. 5, p. 153-173.

Pandey, B.K., 1981, Rb-Sr geochronological studies of some granitic and gneissic rocks of Kumaun and Garhwal Himalaya, U.P., India: Unpublished Ph.D. thesis, Panjab University, Chandigarh, p. 1-186.

Shapiro, L., and W.W. Brannock, 1962, Rapid analysis of silicate, carbonate and phosphate rocks: Bulletin of the Geological Survey, v. 1114A, p. 1-56.

Sharma, K.K., 1976, A contribution to the geology of Sutlej Valley, Kinnaur, Himachal Pradesh, India: Colloques internationaux du CNRS, Paris, France. Ecologie et geologie de Himalaya 268, p. 369-378.

Shreedhar, S.K., 1986, Study of the Central Crystalline rocks occurring in the area east of Wangtu, district Kinnaur, Himachal Pradesh, India: Unpublished Ph.D. thesis, Punjab University, Chandigarh, p. 1-157.

Stoliczka, F., 1866, Geological section across the Himalayan mountains from Wangtu bridge on the river Satluj to Sungdo on the Indus, with an account of the known fossils from that district: Memoirs of the Geological Survey of India, v. 5, p. 1-153.
Tewari, A.P., R.K. Gaur, and S.S. Ameta, 1978, A note on the geology of a part of Kinnaur district, Himachal Pradesh: Himalayan Geology, v. 8, p. 574-582.

SHRINGARPURE, D. M., University of Baroda, Baroda, India

Prograding Deltaic Environmental Controls on the Trace Fossil Distribution Patterns in the Mesozoic Rocks of Kutch-Western India

The Middle Jurassic to Lower Cretaceous Sedimentary deposits of Kutch (West India) exhibit a marked heterogeneity in their rock types with shales, burrowed and laminated siltstones, burrowed sandstones and through cross-bedded channel fill sandstones occurring within 15-20 m thick sequences. The diversity of rock types and their abrupt and lateral transitions between them reflect a mosaic of shifting depositional regimes which typifies a prograding deltaic environment. Spectacular suits of trace fossils in response to the changes are developed in these rocks and include: Arenicolites, Diplocraterion, Skolithos, Cylindrichnus, Spirophyton, Rosselia, Phycodes, Thallasinoides, Planolites, Paleophycos, Rhizocorallium, Zoophycos. Paleocopodichnus, Chondrites, Ophiomorpha, Enteropneust burrows, Gyrochorts, Munesteria, Taenidium, Teichichnus, Scolicia, Nersites, Helminothopsis, Crossopodia, Walcottia, Gyrophyllites, bolonia, Paleodicton, Scalerituba, Spongeliomorpha, Scoyena, Lanicodichnus, Holothurian trails and some unidentified traces. The traces occur in various combinations in the subenvironments: (1) Delta Plain-Distributory and tidal channels, flood basins (swamps, marshes, lakes, bays); and destructive delta margin sands (beaches). (2) Delta Front, with coalesced distribution mouth bar crests and associated distal bars and, (3) The Prodelta, associated with slow rate of deposition in a low energy deep water environment.

Six ichnofacies and ten ichnocoenoses interpreted in terms of their tropic structure indicate useful Paleobathymetric and Paleoenvironmental information. Trace fossils at Omission surfaces further help to understand process intensity, frequency of events, and modification of the sediments by the organisms. Two events of storm deposition can be revealed from the physical and ichnological studies. The biogenic record further provides information on the population strategies ('r' selection opportunistic vs. 'k' selection equilibrium) adopted by the trace making organisms.

SHRODER Jr., J.F., University of Nebraska at Omaha, Omaha, NE

Slope Failure: Extent and Economic Significance in Afghanistan and Pakistan

The complex geology of Afghanistan and Pakistan is a mosaic of sutured and uplifted crustal fragments whose widely varying rock types, great relief, and high seismicity cause many slope failures. Rockfalls and rock slides are among the most common failure types in the Hindu Kush and western Himalayas; one large sackung failure has also been recognized. Debris falls and debris slides from the many deposits of rubble and regolith are common; catastrophic debris flows are most frequent during monsoon rains. Earth falls, earth slides, and earth flows are least common but do occur in loess and clay-rich saprolites. Large complex slump and flow landslips occur in unstable sedimentary rock in a few places in northern Afghanistan. Many of these different landslips have repeatedly destroyed highways, homes, and water works, as well as damming rivers and causing later break-out floods. Slope-failure hazards are an important constraint on development in the two countries.

SHUKOWSKY, W., and M. S. M. MANTOVANI, São Paulo, Brazil, and P. V. ZALAN, Petrobras/Depex/Sebint, Rio de Janeiro, Brazil, and A. C. B. C. de VASCONCELLOS, IG/USP, São Paulo, Brazil

Major Structures of the "Santa Catarina" Transect

Number one among the Brazilian transects, the Santa Catarina (T1) crosses the southernmost part of the Brazilian territory, traveling from the Atlantic coast to the Argentinian border, in approximately east-west direction.

In its eastermost portion, the precambrian terranes of the Dom Feliciano Belt structure outcrop for about 170 km, being westward progressively covered by the sediments and lavaflows of the Parana basin. All the western portion of the transect crosses the Parana sineclysis.

The main regional geophysical data along the transect, available at the moment, are gravity and aeromagnetic surveys. Boreholes drilled for petroleum exploration provide some control of the sediments and lava sequences of the topography of the basement underneath the Parana basin. No deep seismic soundings are available in the area.

Geological data are well detailed in the outcropping portion of the Dom Feliciano belt, and the available reports from the petroleum companies provide valuable informations on the basin structures.

Models of the major structures along T1 are presented based mainly on the regional gravity data, refined by all the other available geophysical, geological and tectonic informations.

SHUNDAO, Tan, China National Nonferrous Metals Industry Corporation, Beijing, China

Important Aspects of Regional Control Facts of Stratabound Deposits in East China Continent

Most ore deposits mined by CNNC are stratabound deposited distributed in East China Continent. These stratabound deposits formed in the post-Proterozoic cratonic environment are provided with some general characteristics. The regional control facts to be important for evaluation are following:

1) Most of the stratabound deposits found in a large area tend to concentrate in a certain stratigraphic unit. For example, in Center–South China region most PbZn, Sb, Hg, W, Sn, Cu stratabound deposits, including the famous Xikuangshan Sb deposit, the huge Dachang cassiterite sulfide deposit, the large Fankou AgPbZn deposit and other PbZn deposits and Shizhuyuan, fluorite–cassiterite–seheelite deposit occur in a same stratigraphic interval, middle and upper Devonian. Also in West Hunan–East Guizhou Hg ore belt, almost all of the stratabound Hg deposits occur in the middle and upper Cambrian. But in East Sichuan region, all of the Hg deposits occur in the upper Proterozoic.

This kind of feature, i.e., stratabound deposits tend to concentrate in a same stratigraphic interval, usually exist in East China Continent and also exist in West–South China, such as Dianzhong—the central region of Yunnan.

It is obvious that there is a favorable stratigraphic interval abound with stratabound deposits exist in a certain area.

2) The most important host rock are carbonite rocks and the orebody occur in transform zone from clastic rock to carbonite rock, but prefer to carbonite rock.

These favorable carbonite rocks usually immediately overlaying on thick aluminosilicatic rocks in regional stratigraphic column. The most favorable host rock is the first carbonite rock series from the bottom to the upper of a certain regional stratigraphic column.

3) In most instance the aluminosilicatic rock underlaying the host carbonite rock series consists of terrigenous sediments with some volcanic materials. These sediments are rich in debris, nodissolved silicate, carbon, iron oxide. In some region these aluminosilicatic rocks characterized by terrestrial intermediate—acidic volcanic rocks. The total thickness of these underlaying rocks can reach more than several thousand meters.

4) Silicolites, evaporites, manganese and iron oxide rich sediments, carboniferous sediments and pyrite rich beds always are the accompanying members of host carbonite rock both in region and stratigraphic section.

5) It is likely that there are some connects between the metallogenetic elements concentrated in overlaying host rock and the constitute of the underlaying aluminosilicatic rock.

SIAL, A. N., and G. MARIANO, Federal University of Pernambuco. Recife, Brazil, J. McMURRY, Texas Tech University, Lubbock, TX, and L. E. LONG, University of Texas, Austin, TX

Isotopic Variation among Late Proterozoic–Cambrian, Potassic Calcalkaline, Coarsely Porphyritic Granitoids of Northeast Brazil

At least 80 widely separated, coarsely porphyritic granitoid plutons were intruded in northeastern Brazil during the Brasiliano (= Pan-African) Orogeny, the last major tectonic event to have affected that region. Some of these plutons, termed "Itaporanga-type" granitoids (Almeida et al., 1971), are found next to the margin or intruded Proterozoic metasediments of the Central Structural Domain (Sial, 1987). They have conspicuous tabular alkali feldspar megacrysts in a dark matrix of plagioclase, biotite, hornblende, microcline, and quartz, with minor amounts of clinopyroxene, sphene, allanite, zircon, and (rarely) Fe-Ti oxides. Megacrysts accentuate foliation textures and suggest that the plutons were intruded diapirically as viscous, crystal-rich mushes, commonly along or near strike-slip fault contacts. In some portions of the plutons, the Itaporanga-type intrusions are further characterized texturally by commingling of the more common porphyritic granodiorite to quartz monzonite with darker, equigranular, K-rich diorite to quartz diorite; this commingling is found on the scale of centimeters and of meters. The physical evidence for commingling in these plutons suggests that some of the porphyritic granitoids may themselves be products of a mixing relationship. Small, fine-grained, elliptical, occasionally cuspate-like, mafic enclaves are ubiquitous in the prophyritic granitoid and often chemically and mineralogically similar to the dark intrusives of the commingled rocks.

Detailed chemical and isotopic studies of three of the Itaporanga-type plutons (the type locality, near Itaporanga, Paraíba; the Bodocó pluton, in western Pernambuco; and the Monte das Gameleiras pluton, on the Rio Grande do Norte–Paraíba border) and a reconnaissance study of three other ones (the Fazenda Nova in eastern Pernambuco; the Serra da Lagoinha, central Paraíba, and Acari, Rio Grande do Norte) reveal that the intrusions are broadly similar but with some important differences. The SiO_2 content of the porphyritic rocks ranges from about 60 to about 72 wt% in the plutons analyzed. (Abundance and size of alkali feldspar megacrysts are variable on outcrop scale and to some extent result in a range of chemical values for samples of the same rock type.) Values of K_2O are in general equal to or greater than Na_2O. The dark colored intrusive rocks, in contrast, have SiO_2 in the range of about 50–58 wt% and have values of CaO greater than either K_2O or Na_2O. The porphyritic rocks are Ba and Sr-enriched and exhibit highly fractionated, similar REE patterns, lacking Eu anomaly.

Quartz separates, less susceptible than whole rocks or other common minerals to post-magmatic isotopic exchange, yield the most reliable oxygen isotope data for the intrusions. Despite contrasting ranges of whole rock values, $\delta^{18}O$ values of quartz separates from the Bodocó and the Monte das Gameleiras plutons are similar to each other and appear to be homogeneous on the scale of the intrusion. Values for quartz separates from the Bodocó pluton range from +9.3 to +9.8 per mil (corresponding whole rock values: +6.8 to +7.2 per mil). The Monte das Gameleiras quartz separates, with $\delta^{18}O$ values of +9.6 and +9.7 per mil, are bracketed by the Bodocó samples, despite a very different range of whole rock values from +7.0 to +9.2 per mil. In contrast, the Itaporanga pluton has elevated oxygen isotope values for quartz separates of +10.0 to +11.4 per mil, and a corresponding range of whole rock values of from +7.5 to +9.2 per mil. The Serra da Lagoinha pluton shows whole rock $\delta^{18}O$ from +6.8 to +8.53 and quartz around +9.;5 per mil, while the Acari massif exhibits values in the range +6.3 to +7.5 per mil. The present oxygen isotope data are compatible with an I-type source for these plutons with some metasedimentary component at variable proportion.

The behavior of Rb and Sr isotopes in the Itaporanga-type granitoids is more variable than that of oxygen isotopes, perhaps reflecting a heterogeneity of source region with respect to these elements, complicated by possible mixing relations and by the difficulty in separating enclaves form large blocks processed for isotope chemistry. Whole rock isochrons for the Itaporanga and Fazenda Nova plutons yielded ages around 630 Ma while the Acari (Jardim de Sá et al., 1987) and Bodocó plutons appear to be considerably younger (around 550 Ma). An isochron for the Monte das Gameleiras pluton, further north, yielded an age of 511 /pmn25 Ma. The initial ^{87}Sr-^{86}Sr ratios for the Itaporanga, Bodocó, and Fazenda Nova agree around 0.7060 while the Acari and Monte das Gameleiras plutons exhibit higher values (0.7075 and 0.7099, respectively). If these dates represent reliable ages of emplacement, they imply that the distinctive Itaporanga-type lithology was produced by Brasiliano orogeny tectonothermal activity spanning about 100 m.y., mainly around 630 and 550 Ma ago, probably corresponding to the time of uplift and cooling of the Pan-African I and onset of the Pan-African II orogens, recognized in West Africa (Dallmeyer et al., 1987).

REFERENCES

Almeida, F. F. M., O. H. Leonardos, Jr., and J. Valença, 1971, Review on granitic rocks of northeast South America: IUGS Committee for the Study of Geological Documentation, 41 p.

Dallmeyer, R. D., M. Villeuneuve, 1987, $^{40}Ar/^{39}Ar$ mineral age record of polyphase tectonothermal evolution in the southern Mauritanide orogen, southeastern Senegal: GSA Bulletin, v. 98, p. 602–611.

Jardim de Sá, E. F., M. H. F. Macedo, J. M. Legrand, I. McReath, A. C. Galindo, and J. M. Sá, 1987, Proterozoic granitoids in a polycyclic setting: The Serido Region, NE Brazil: Intern. Symp. on Granites and Associated Mineralization (ISGAM), Salvador, Brazil, p. 103–109.

Sial, A. N., 1987, Granitic rocks of northeast Brazil: Intern. Symp. on Granites and Associated Mineralization (ISGAM), Salvador, Brazil, p. 61–69.

SIAL, A. N., and V. P. FERREIRA, UFPE, Recife, Brazil, J.A. WHITNEY and D. B. WENNER, University of Georgia, Athens, GA, A. SASAKI, University of Ibaraki, Ibaraki, Japan, and L. E. LONG, University of Texas, Austin, TX

LILE 18O- and 34S-Enriched Mantle Beneath Northeast Brazil: Evidences From Shoshonitic to Ultrapotassic Plutonic Rocks

Around 580 Ma ago, numerous silica-saturated and oversaturated peralkaline magmas were intruded into Precambrian gneiss-migmatite basement and the 20,000 km2 southwest-northeast-trending Cachoeirinha-Salgueiro Fold Belt in the stats of Pernambuoco and Paraíba, Northeast Brazil. Nepheline-normative plutons consist of sodic-potassic to ultrapotassic syenites and two swarms of 250 dikes each, aligned along the southern boundary of the Fold Belt. To the north, plutons of the main syenitoid line grade into quartz monzonite with shoshonitic affinities (Sial et al, in press). Oval, streaky to globular alkali-pyroxenite inclusions and synplutonic dikes of the same composition attest to an ultrapotassic pyroxenite-syenite association.

Silica-saturated plutons are very high in K, Sr, and Ba, high in P and Ti, and low in Zr and Nb. MORB-normalized concentration diagrams show positive Ce, Y, and Sm anomalies and negative Nb and Ti anomalies. Alkali-pyroxenite inclusions are high in K, P, Ba, Sr, and REE contents. MORB-normalized concentration diagrams for shoshonitic compositions show negative P, Ti, and Nb anomalies and positive ones for Zr and Rb. Both groups show LREE-enriched and HREE-depleted patterns with variable Eu anomalies. The pyroxenite inclusions display LREE at 240-900 times chondritic abundances, and HREE at about 20 times only.

O, S, and Sr isotopes: Nearly 100 isotope compositions have been determined on whole rocks and separated minerals from this province. The Triunfo syenite, the largest ultrapotassic pluton, exhibits a broad range of whole-rock $\delta^{18}O$ (+1.75 to +9.58%$_{SMOW}$) suggestive of alteration. Magmatic differentiation cannot use this much variation; furthermore, most $\delta^{18}O$-depleted rocks lie adjacent to fault zones. Low $\delta^{18}O$ in the whole rock (as low as -1.21%) is due to 18O-depleted feldspar. Clinopyroxenes are highly resistant to effects of postmagmatic overprint. The range of $\delta^{18}O$ in clinopyroxene (ferraugite to aegirine-augite) is very narrow: + 7.77 to +8.25% in 9 of 11 samples. If fractionation were +0.25% between whole rock and clinopyroxene at magmatic temperatures (Taylor et al, 1984), then primary $\delta^{18}O$ in the Triunfo magma was $\approx$ +8.2%. Values of whole-rock $\delta^{18}O$ over the entire province deduced by this method of "pyroxene correction" range from +6.6 to +10.4%. These values resemble $\delta^{18}O$ in rocks of the classic Roman ultrapotassic province, Italy, which are considered to have a source in mantle that has been metasomatically enriched in large-ion lithophile elements (LILEs). Most petrologic data from the Triunfo batholith lie on an extension of the Vesuvius-Philegrean and Sabatini trends, and data from other rocks in the syenitoid line plot on the Vico trend of the Roman province (Taylor et al, 1976).

Contamination by fragments of high-$\delta^{18}O$ biotite schist, though present, is not sufficient to modify the primary oxygen isotope composition. Blue amphibole, whether primary or a product of altered clinopyroxene, has higher $\delta^{18}O$ (typically > +10%). Elevated $\delta^{18}O$ in rocks with shoshonitic affinities is probably due to a more pronounced interaction with crustal rocks.

$\delta^{34}S$ in ultrapotassic syenites in the main syenitoid line ranges from +6.6 to 12.3%, the highest values being in pyroxenite and aegirine-syenite from the Triunfo batholith. A Rb-Sr isochron for this batholith provides an age of 583;12 Ma and Ri = 0.7097;0.0002. High initial $_{87}Sr/_{86}Sr$ is consistent with high $\delta^{18}O$ and $\delta^{34}S$ values, and with similar data for other ultrapotassic provinces around the world.

Discussion: The shoshonitic to ultrapotassic rocks lie upon sigmoidal fault zones between the Patos-Aurora and Pernambuco transcurrent lineaments and are probably a consequence of pull-apart tectonics. Chemically, they are similar to rocks in Italy, Australia, East Africa, and Antarctica, which are believed to have a LILE-enriched mantle source (Taylor et al, 1984).

Veined or metasomatized mantle like that proposed as a source for Bouvet or the Walvis Ridge (South Atlantic Ocean) is inferred to lie beneath the Mesozoic Potigua Basin, Rio Grande do Norte state (Fodor et al, 1986). Mesozoic tholeiitic diabases and Tertiary alkali basalts in Rio Grande do Norte and Paraíba show very high K, Ba, Sr, and P (Sial, 1974) as expected from a LILE-enriched mantle source. Similar compositions pertain to the peralkaline Itubia syenite dike (11 x 150 km) in northern Bahia, of Transamazonian (=Eburnean) age, and to a Precambrian biotite pyroxenite-potassic syenite association at Shaki, western Nigeria (Oyawoye, 1976). LILE-enriched mantle beneath Northeast Brazil may have existed since Proterozoic or even since Archean times, produced by successive enrichments or one major event. This must be a rather shallow feature, within the lithosphere, or else it would have been destroyed by mantle convection.

Further detailed Pb, Sr, O, and S isotope investigation is needed to specify the physical extent and age or origin of this mantle anomaly.

REFERENCES

Fodor, R.V., and E.H. McKee, 1986, Tertiary rocks from offshore Northeastern Brazil: geochemistry and K-Ar ages: An. Acad. Bras. Ci., v. 58, n. 2, p. 233-241.

Oyawoye, M.O., 1967, The petrology of a potassic syenite and its associated biotite pyroxenite at Shaki, western Nigeria: Contributions to Mineralogy and Petrology, v. 16, p. 115-138.

Sial, A.N., 1974, Petrology and tectonic significance of the post-Paleozoic basaltic rocks of Northeast Brazil: Ph.D. dissertation, University of California at Davis, 403 p.

and Ferreira, V.P., in press, Brasiliano age peralkaline plutonic rocks of the central structural domain, Northeast Brazil: Rend. della Soc. Italiana di Mineral. e Petrol.

Taylor, H.P., and B. Turi, 1976, High-18O igneous rocks from the Tuscan magmatic province, Italy: Contribution to Mineralogy and Petrology, v. 55, p. 33-54.

and A. Cundari, 1984, 18O/16O and chemical relationships inm K-rich volcanic rocks from Australia, East Africa, Antarctica, and San Venanzo-Cupaello, Italy: Earth and Planetary Science Letters, v. 69, p. 263-276.

SIBRAVA, V., UNESCO, Paris, France

Geoscience Development and Transfer of Technology Needs

Rapid development of technology and the needs of society are having an important impact on the development of earth sciences both in industrial countries and in the developing world. The economic and technological evolution and the shift of industrial priorities are reflected in the changes in priorities related to geological research, the exploration of raw materials and other geological professions. We are witnessing—at least for the time being—a decreasing interest in the exploitation of some traditional raw materials and an increasing need for new materials used in new technological processing. There is a growing importance for some geological branches such as hydrology and environmental geology; as far as basic research is concerned a trend towards interdisciplinary studies can be noted as is also reflected in some international scientific programs. Considerable attention is being given to the investigation of natural hazards leading to the development of broad international programs such as the International Decade for Natural Disaster Reduction which is now in a preparatory stage and which will be implemented within the framework of the United Nations.

These economic changes and technological developments are affecting different regions in varying ways. In some industrial countries they are reflected in structural changes of geological institutions resulting in some cases in a strong reduction of economic geology and mineral exploration, in particular with regard to environmental protection. Nevertheless, mineral exploration still remains the first objective of geological research in many—probably most—countries in particular of the third world. But, even here, we may observe a shift of interest to other kinds of materials than those in the past, i.e., materials necessary for construction activities, agriculture, etc.

Under these circumstances, what are the possibilities, the means and the objectives for the transfer of technological needs to the developing countries? Basic needs represent the development of human resources and of scientific knowledge. Both are necessary conditions for the research and the efficient utilization of equipment and technology that will always be available at different levels. The development of human resources must include the necessary flexibility for the changing needs of society under specific economic conditions, further adaptability to an appropriate environment taking into consideration a realistic view of science, industry and governmental needs. Apart from traditional disciplines, geologists, both in developed and developing countries, should be trained for the preparation and formulation of technical counseling for policy makers and managers.

Most of the developing countries are hampered in several areas with regard to their technological and human resources development: a lack of financial resources and, in particular, the non-convertibility of their currencies represents a serious obstacle–not only for the purchase of necessary equipment but also for the acquisition of basic scientific magazines and publications, including those dealing directly with the geology of their country or region. In many cases, geologists who have received advanced training in highly sophisticated laboratories in industrialized countries are deprived of the possibility of continuing their research in their countries. The main obstacle is the lack of equipment in the laboratories or simply a lack of facilities for field work. Therefore, the building up of an infrastructure constitutes one of the prerogatives for the transfer of technology. On the other hand, some institutes and laboratories which have very modern equipment are far from being used efficiently due to a lack of sufficiently-trained manpower.

The introduction of new methods into geological research has created in some cases an undesirable and adverse effect. Geologists, particularly in regions with difficult working conditions in the field, tend to sit in their offices and rely entirely on these methods for resolving geological problems in their region, quite neglecting the work and direct observation in the field. The use of highly sophisticated and advanced methods is of importance but must be carried out in conjunction with ground observations.

The rapid development in the Earth Sciences requires continuous training at the post-university level and the steady development of international contacts and cooperation. It should be recommended that recycling courses be organized for geologists working in industry as well as in the Government service.

Postgraduate training for developing countries is usually organized either on a national level or under bilateral agreements between developing and industrialized countries or else by international organizations and institutions. Internationally organized training activities, both in the universities or by "training on the job" helps to create links between the countries, regions and continents. The bilateral approach has often limited possibilities, either due to political or language restrictions. Thus, for example, in spite of numerous training and research activities in Africa, little contacts have been developed between French-and English-speaking countries. Crossing the border between these linguistic zones means crossing the border between different stratigraphic, lithologic and tectonic concepts. Not much has been done so far for the cooperation and exchange of scientists from developing countries working on two continents e.g. in Africa and South America for the study of comparable or identical geological formations. A South-South cooperation in geology has still to be further developed.

In order to bridge the geology of different regions and continents, Unesco has launched regional projects and training activities as well as international scientific programmes. In Africa, an attempt has been made to compare geological terminology that has been used in a different way in French and English linguistic zones representing so far the main obstacle for efficient correlation studies.

The general schema of postgraduate training and scientific activities may be outlined as follows : training (training courses and "on the job" training; follow-up research work in local institutes and laboratories; scholarships — if necessary — in advanced institutes and laboratories in industrial countries; transfer of knowledge back to local institutes; participation in exchange programmes; participation in international scientific programmes and feed-back to local institutions.

The changes in the trends of development in the Earth Sciences will surely modify the modalities of the transfer of technology needs and education and training requirements for the future. To keep the role of Geosciences at the level which corresponds to the development of modern technology and to the changing needs of society is one of the high priorities for coming years.

SIBUET, J. C., Ifremer, Plouzané, France, J. DYMENT, Institut de Physique du Globe, Strausbourg, France, and C. BOIS and B. PINET, Institut Françcais du Pétrole, Rueil Malmaison, France

Crustal Structure of Celtic Sea (Southwestern Britain): Constraints on Formation of Continental Basins

A free-air anomaly map of the Celtic Sea and the Western Approaches has been established from all available gravimetric data. Bouguer anomalies on land have been added. The so-called N60° Caledonian grain played a major role in the Variscan orogeny and the formation of the Permo-Triassic sedimentary basins. The fundamental new point that derives from this gravimetric map is the existence of a closely spaced pattern of N105° features that we regard as late Carboniferous transpression structures. In the eastern part of the Celtic Sea, thrusts are orientated N105° West of 6°W longitude, folds and thrusts are trending rather N60°, following the previous Caledonian grain. There, the N105° lineaments may be regarded as dextral shear structures with a possible compressive component. The Variscan shortening of the Celtic domain, located between the Variscan Front and the Lizard suture, resulted from the displacement, along the N150° direction, of the Channel domain with respect to the foreland domain located north of the Variscan Front, the N105° and N135° features being possibly active diachronously later than the N60° ones. Then, both the N60° and N105° trends controlled the formation of the Permo-Triassic basins, the N105° directions being transfer faults.

A two-dimensional gravity modeling of the SWAT profiles in the Celtic Sea shows that the base of the reflective lower crust is the Moho discontinuity everywhere in the Celtic Sea except below the Saint George Channel basin. The lower crust does not change significantly in depth and thickness beneath the Permo-Triassic basins. None of the available models simply satisfies such conditions. They all require at least a modification of the geometry of the lower crust after the formation of the Celtic Sea basins. Nevertheless, if the simple shear model is clearly ruled out mainly from structural arguments, the uniform stretching and at a lesser degree the thermomechanical models seem to be more relevant models if two conditions are fulfilled: (1) the preservation of crustal seismic features formed before the creation of basins must be explained, and (2) a significant portion of the volume of the stretched lower crust (about 40% of this volume for the North Celtic Sea basin) must be added below and perhaps above the lower crust by lateral flow or at the expenses of the upper mantle.

SIBUET, J. C., Ifremer, Plouzané, France

Discussion of Models of Formation of Continental Margins: Example of Starved Northeast Atlantic Passive Margins

Models of formation and evolution of passive margins have been proposed, modified, and tested over the starved passive continental margins of the Northeast Atlantic Ocean where numerous detailed geological and geophysical data have been acquired.

The starved margin of Galicia Bank is a pure tensional margin characterized by a series of north-south tilted fault blocks formed during the lower Cretaceous. Near the continent-ocean boundary, periodotites have been drilled (leg ODP 103). In the lower portion of the margin, the S reflector lies below the tilted fault blocks. In fact, this S reflector is a transition zone about 0.4-0.6 s.d.t.t. thick (about 1.5 km). The superficial extension has been measured from the tilted fault block geometry as a function of depth. These values are compatible with the pure stretching model for the upper part of the continental margin where the S reflector is absent. In the lower part of the margin, above the S reflector, measured values of extension cannot be explained by the stretching model. Nevertheless, tilted blocks above the S reflector are at least partly constituted of syndrift sediments and have not recorded the total amount of extension that has affected the continental crust. Consequently, we cannot eliminate a formation of the whole Galicia margin by stretching. In this hypothesis, the S reflector is an incoherent surface that appeared when the rotation of blocks could not accommodate to stretching values exceeding 2.8-3.0. This type of margin can be considered as the final evolution of a symmetrical intracontinental basin formed by uniform stretching.

On the contrary, for the Northern Bay of Biscay margin, Le Pichon and Barbier (1987) have demonstrated that a decollement surface crossing the whole upper crust down to the brittle-ductile limit has been active since the beginning of rifting. Normal faults along which tilted blocks have been displaced would be antithetic faults. This type of margin can be considered as the final evolution of asymmetrical basin in which a decollement surface is active since the beginning of extension.

In both hypotheses, the point where the S reflector disappears corresponds to the brittle-ductile transition. Near this point, the S reflector corresponds to the brittle-ductile limit in the stretching model and is very close to this limit in Le Pichon and Barbier's model. We propose a model of formation of passive margins by uniform stretching in which a series of preexisting crustal discontinuities could be used as decollement surfaces at any stage of the formation of the margins.

REFERENCES

Le Pichon, X., and F. Barbier, 1987, Passive margin formation by low-angle faulting within the upper crust: the northern bay of Biscay margin: Tectonics, v. 6, p. 133-150.

SIBUET, J.C., J.P. FOUCHER, and J.M. AUZENDE, Ifremer, Plouzané, France

Structure of Back-Arc Basins from Recent Observations in Okinawa Trough, Lau, and North Fiji Basins

We discuss the structure of the back-arc basins from recent observations in three of these: the Okinawa Trough, the Lau, and North Fiji basins.

The Okinawa Trough basin (OTB) is a back-arc basin being opened within the East Chinese continental lithosphere behind the Ryukyu trench-arc system. Initial continental rifting occurred after a major early Miocene change in the motion of the Phillipine plate with respect to Eurasia and ceased during the Pliocene. A second rifting phase started at about 2 Ma at the Plio-Pleistocene boundary and is still active. The formation of the OTB has proceeded to a more advanced stage in its middle and southern parts than it has farther north. Whereas continental lithosphere extension occurs without associated volcanism in the northern OTB, back-arc oceanic spreading started recently in the southern and middle OTB. Several en-echelon, in some cases overlapping, linear basaltic ridges characterize the young oceanic domain. These back-arc volcanic ridges are observed south of Okinawa Island, whereas active-arc volcanoes occur only north of it. The transition between active-arc and back-arc volcanism has been referred to as the volcanic arc-rift migration phenomenon (VAMP). Arc and back-arc basin volcanism seems to be associated in such a manner that spreading tends to migrate simultaneously with a cessation in volcanic activity along the arc (Sibuet et al, 1987).

The Lau basin (LB), an actively spreading back-arc basin located between the Lau and Tonga ridges behind the Tonga trench, was created at about 2.8 Ma by longitudinal splitting of the initial Tonga arc. Though the present and past geometries of the accretionary ridge system remain poorly established, the formation of the LB clearly resulted from several different tectonic phases. In the western portion of the Southern LB, the Valu Fa ridge, a 200 km long, linear active volcanic structure parallel to the Tofua volcanic arc at a distance within 40 km from the arc, may be the southern segment of the present accretionary ridge system. A strong reflector at 1.8-1.6 s underneath the ridge has been interpreted as the top of a 2-3 km wide crustal magma chamber at a subbottom depth of about 3.5 km. Along the northern extension of the Valu Fa ridge, a secondary volcanic chain occupies the axis of the sedimentary trough between the ridge and the Tofua arc and marks the initiation of an en-echelon relay volcanic system from the Tofua arc to the Valu Fa ridge, whereas the continuation of the Valu Fa ridge to the accretionary ridge system in the central LB has not been documented (Foucher et al, 1988).

The North Fiji basin (NFB) is delineated by the Vitiaz trench to the north, the Vanatu Island arc to the southwest, the Hunter Fracture Zone to the southeast, and the Fiji islands to the east. It was created by oceanic back-arc spreading behind the Vanatu (New Hebrides) arc. Its formation can be summarized as follows: (1) from 10 to 8 Ma, reversal of the polarity of the subduction from southward along the Vitiaz zone to northeastward beneath the Vanatu-New Hebrides Arc as a consequence of the collision of the Onting Java plateau near the Salomon arc, (2) from 8 to 3 Ma, beginning of the accretion along an axis initially roughly parallel to the Vitiaz Zone; clockwise rotation of the Vanuatu, (3) from 3 to 0.7 Ma, spreading jump in the central NFB from N150 to a north-south direction at the same time as the beginning of spreading in the LB and possible cessation of the spreading north of the Fiji Islands, (4) from 0.7 Ma to Present, local modification of the spreading geometry in the central NFB with migration of the triple junction. The spreading centers are slightly offset by overlapping spreading centers or small offset fracture zones (Auzende et al, 1988).

The three studied back-arc basins are at different stages of evolution: from continental rifting to oceanic accretion in the OTB; young oceanic spreading in the LB; mature midoceanic type spreading in the NFB. These different stages are reflected in the trace element and isotopic geochemistry of dredged volcanic rocks. Overall structural characteristics suggest a fast succession of tectonic phases associated with jumps of spreading axes probably linked to changes in the subduction rates or geometries of plate convergence; also, transform faults are not observed in the OTB nor in the LB but occur in the NFB. Linear volcanic trends or chains of sediments develop within the back-arc basins with directions either parallel (LB), oblique (OTB) or more or less perpendicular (Bonin) to the arc chains.

REFERENCES

Auzende, J. M., et al, 1988, Seafloor spreading in the North Fiji basin (SW Pacific): Tectonophysics, v. 146, p. 317-352.

Foucher, J. P., et al, 1988, La ride volcanique de Valu Fa dans le bassin de Lau méridional (SW Pacifique): C.R. Acad. Sci. Paris, v. 307, p. 609-616.

Sibuet, J. C., et al, 1987, Back arc extension in the Okinawa trough: Journal of Geophysical Research, v. 92, p. 14041-14063.

SICK, M., University of Tubingen, Tubingen, Federal Republic of Germany, and E. STEINBERG, Escuela Centroamericana de Geologia, San Jose, Costa Rica

Nicoya Complex (Costa Rica): Geotectonic Relevance and Geological Evolution of Jurassic–Cretaceous Ophiolitic Complex in Relation to Early Development of Caribbean Crust—New Perceptions based on Paleomagnetic, Sedimentological, and Biostratigraphic Data

The Nicoya Complex on the west coast of Costa Rica and Panama represents a complete ophiolitic sequence with an allochthonous and autochthonous genetic evolution.

Older genetic models are critically reviewed and modified through new paleomagnetic, sedimentological, and biostratigraphic data, thus confirming a larger subdivision into two different genetic units with a completely different geotectonic setting.

A lower unit comprises the ultrabasic basement of the complex, typical MORB-basalts indicating a true oceanic crust and radiolaritic sediments of Lower Jurassic to Santonian ages. Paleomagnetic data indicate that this unit was built in an Equator-near ridge system and transported through seafloor-spreading to its actual position. During transportation it was rotated toward the north.

The upper unit comprises mainly basalts, indicating a primitive island-arc vulcanism. Associated sediments are documented through basaltic breccia and hemipelagic bituminaceous cherts deposited from the Middle to Upper Cretaceous. Paleomagnetic data confirm a more or less autochthonous build-up of this unit.

According to sedimentological and biostratigraphical data, three main sedimentary environments belonging to two different basins can be described for the Nicoya Complex. This is supported by the paleomagnetic data.

Two main sedimentation cycles can be described, one lasting from the Lower Jurassic to the Middle Cretaceous, the other represented by the bituminaceous cherts, which show a direct relation to the world-wide described Cretaceous Anoxic Event. Both basins present discriminating characteristics that allow a direct correlation with the early geotectonic evolution of the Central American Landbridge and the Caribbean Crust.

Biostratigraphic data based on radiolarian assemblages permit a more concise subdivision of the siliceous strata, thus yielding a precise timescale for the larger regional geologic events. Further on they allow a correlation with international zonations.

SIDOROV, A. A., Northwestern Research Institute, Magaden, USSR

Ore Formation Evolutionary Ranges (As Exemplified Within the Pacific Provinces)

An analysis into the Phanerozoic ore formations within the Pacific provinces has shown that they all in their diversity may be reduced to a certain number of ranges (Table 1). Each range is headed with the formation of the most complex deposits (magmatic, porphyry, sulfide disseminated ores, etc.) and is ended in the ore formations of the lode and rather often monometallic deposits (antimony, mercury). The range model is quite explicable from the point of view of the rocks selective fusion hypothesis and the regularities of the matter pT differentiation within the hydrothermal systems. Generally, the ore formation ranges are complex multistage structures; the data on the well known deposits within the North America and the Soviet Northeast copper-porphyry provinces serve as the basis to elaborate the developmental scheme for a rather noncomplicated ore formation range. The ore formation processes being considered the formations heading these ranges are of a peculiar interest, but not the ordinary ("simple") ones, which developed more or less similarly according to the Emmons zoning. We propose to name these initial ore formations as the base ones (Sidorov, Tomson, 1987; Sidorov, 1987). We distinguish them as the groups of deposits of vein-disseminated (porphyry, disseminated sulfide ores and the like) or massive sulfide ores firstly peculiarized by the greatest diversity of mineral assemblages within a certain region of ore occurrence and, secondly, evidencing a prolonged and repeatedly renewed deposition of the mineral matter during a single or several tectonic-magmatic stages. Generally, the inherent features of the deep (deep crust, mantle) ore matter sources are characteristic of these formations. Essentially, the identification of the base ore formations and their ranges is underlain by an extensive knowledge on the processes of ore matter deposition. This implies the complex deposits originating the whole ore formations range down to the simplest possibly differentiated lodes of the same range. The data on the present hydrothermal activity do not contradict this model of ore formation processes. Thus, according to S. I. Naboko (1980), in various places of the present hydrothermal system composition is correlative with the primary ore generating elements interrelations in a stable and uniform way; the elements are differentiated mainly on the way of migration.

The ore formation range is being developed, while the ore deposition processes are repeated, the factor of time (duration) primarily manifests itself in that each subsequent redeposition of ores promotes their general differentiation (fractionation) even despite the possible paragenesis impoverishment and the ore matter partial dissemination (Schneiderchen, 1957). The ore formation range may be simplified (reduced) or complicated but its essence remains the same. Certainly, it does not concern the cases of various ore formation ranges juxtaposition.

On the basis of the whole matter, the principles of the ore formations and their ranges systematization as follows are proposed (Table 2).

Table 1. Phanerozoic ore formation ranges (as exemplified within the Pacific provinces).

Base ore formations	Range ore formations	Range evolution environment	Regions
I	II	III	IV
I. Chrome	Platinum (accessory), gold-telluride, gold-albite-analcime, mercury	Syngenetic to ultrabasite (dunite-harzburgite) and epigenetic hydrothermal mean-temperatured	New Caledonia, California
II. Copper-nickel	Gold-silver-platinum, gold-silver-telluride, mercury	Syngenetic to ultrabasite, basite near-bottom intrusions and epigenetic hydrothermal high- and mean-temperatured	Cordillera (the Canadian southwest)
III. Ferruginous quartzite	Magnetite-skarn, titanium, manganese	Sedimentary, magmatic, sedimentary-hydrothermal high- and mean-temperatured	Primorje (the Soviet Far East)
IV. A group of massive sulfide and stratiform disseminated sulfide ores	Copper, polymetallic, (skarn-polymetallic), five-element gold-sulfide disseminated ores, gold-sulfide-quartz, gold-quartz, barite, gold-silver, mercury	Submarine hydrothermal-sedimentary and epigenetic hydrothermal mean- and low-temperatured; plutonogenous and metamorphogenous hydrothermal mean-temperatured	Japan (Green Tuff Province), North America (Mexico)
V. A group of disseminated sulfide ores	Gold-sulfide disseminated ores, gold-sulfide-quartz, gold-silver, wolfram, tin-polymetallic, tin-silver, antimony, mercury, etc.	Hydrothermal-sedimentary, orogenous volcanogenous-plutonogenous syngenetic and epigenetic hydrothermal high-, mean- and low-temperatured	Perivolcanic zones of the East Asia volcanogenic belts. Nevada, North America
VI. Copper-porphyry	Polymetallic, gold-silver, antimony, mercury	Orogenous volcanogenous-plutonogenous hydrothermal high-, mean- and low-temperatured	North and South America (Bingham, Chili belt)
VII. Tin-porphyry	Tin (cassiterite-silicate-sulfide), tin-silver, tin-polymetallic, gold-silver, gold-rare metal (gold-sulfide-quartz), antimony	Orogenous volcanogenous-plutonogenous hydrothermal high-, mean- and low-temperatured	Bolivia (Potosi), the Soviet Far East

Table 2. Ore formations systematic scheme.

Ranges of formations	GROUPS OF FORMATIONS										
	BASE			SIMPLE							
	Ultrabasite and basite fusions	Fluid-hydrothermal	Sedimentary hydrothermal-sedimentary	Hydrothermal							
				Deep		Intermediate			Shallow		
				high-temperatured	mean-temp.	high-temp.	mean-temp.	low-temp.	high-temp.	mean-temp.	low-temp.
I. Chrome							+	+		+	
II. Copper-nickel								+		+	
III.			Ferruginous quartzite			+	+	+	+	+	+
IV.			A group of massive and stratiform sulfides			+	+	+	+	+	+
V.		A group of sulfide (disseminated ores)		+	+	+	+	+	+	+	+
VI.		A group of porphyry				+	+	+	+	+	+

\+ - various ore deposits groups of formations identified and probable

SIEGEL, D. I., Syracuse University, Syracuse, NY

Groundwater Flow in Mires: Hydrobiologic System

Peat-forming wetlands, called mires, cover large portions of high latitude North American and Eurasian landscapes formerly covered by Pleistocene age lakes and marginal seas. The origin and controls over the major vegetation types in the mires, bogs, and fens are still uncertain despite several decades of study. Bogs consist of mosses (e.g., *Sphagnum*) and other acid tolerant plants, and have surface water with pH <4.5 and dissolved inorganic solids content <10 mg/L; fens largely consist of reeds and sedges (e.g., *Carex*) that grow in surface water characterized by pH >6.0 and dissolved inorganic solids content up to hundreds of mg/L.

Bogs with convex surface profiles (termed "raised bogs") have been long assumed to be isolated from regional groundwater systems, and thus only receive nutrients and other solutes from precipitation. In contrast, fens are assumed to be groundwater discharge areas, even where the entire mire landscape is covered by peat and mineral soil outcrops that could act as groundwater recharge areas are absent. This paper reviews the results of recent research that challenges these long-held assumptions. First, theoretical computer models of groundwater flow in mires suggest that water table mounds under raised bogs recharge (replenish) *local* groundwater systems which then discharge mineralized to adjacent fens (Fig. 1a). A mixture of only several parts of this groundwater discharge to 20 parts of acidic bog runoff is theoretically sufficient to raise the surface water pH at the bog margin to values above which bog vegetation cannot effectively grow. Significant discharge driven by the water table at raised bogs is reasonable, especially considering that the hydraulic conductivity of deep decomposed peat may be as high as $>10^{-4}$ cm/s, orders of magnitude greater than previously assumed.

Peat porewater chemistry is consistent with the model that raised bogs are groundwater recharge areas. For example, gradients in concentrations of dissolved metals in bogs of southeastern Alaska show that groundwater movement downward through the peat toward the mineral soil probably inhibits chemical transport upward by diffusion.

Field studies also show that the vertical directions of groundwater flow in bogs and fens can reverse where local scale flow systems interact with regional scale systems, and that the trace metal content of bog peat may not reflect the metal loadings at the time of peat burial. Groundwater flow reversals have notably been documented in northern Minnesota where regional discharge of mineralized groundwater occurs under large raised bogs (Fig. 1b). The regional discharge began about 1000 years ago and has caused fen vegetation to replace bog vegetation. The trace metal content of the bog peat strikingly shows the effects of metal transport by groundwater throughout the peat column. Similarly, hydraulic head measurements and computer simulations of groundwater flow show that an Alaskan patterned fen also can either recharge groundwater systems of discharge to them, depending on the scale of the flow system considered.

The combination of theoretical and field studies demonstrate that (1) groundwater flow in mires is best viewed as a part of a "hydrobiologic system," where water table mounds under large raised bogs generate local groundwater systems that chemically can limit the extent of lateral bog growth, (2) the distribution of major vegetation and surface water chemistry of *both* bogs and fens can be mechanistically related to groundwater flow systems at different scales, and (3) the complex interrelationships between mire vegetation and water chemistry are effectively studied within the context of the hydrogeologic systems approach.

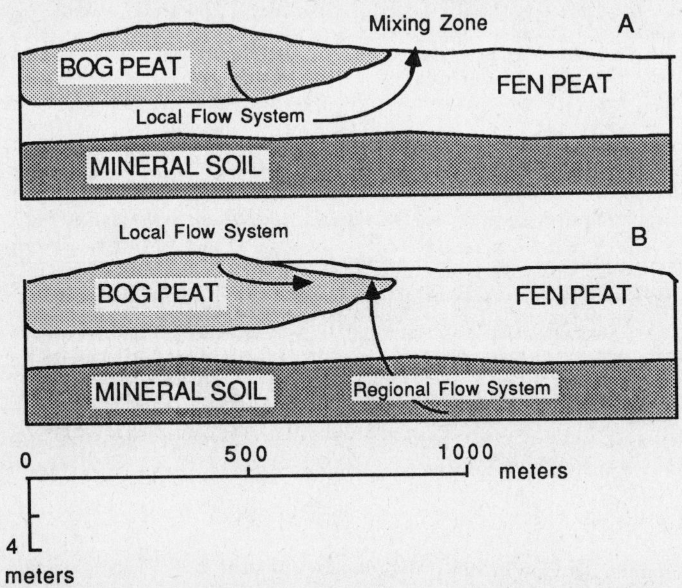

Figure 1—Cartoons showing how: (A) water table mounds under raised bogs produce local groundwater flow systems that discharge to bog margins, (B) regional groundwater discharge can reverse groundwater flow directions under a raised bog.

SIGNOR, P. W., and J. F. MOUNT, University of California, Davis, CA, and White Mountain Research Station, Bishop, CA

Lithological Controls on Stratigraphic Distribution Of Early Cambrian Organisms

The primary source of data bearing on the great Cambrian radiation of Metazoa is the fossil record. Other sources, such as gene sequencing, can provide important evidence about the timing of the radiation and the phylogenetic relationships of early clades, but the fossil record is the source of virtually all information about biotic events of the time. The temporal resolution available from the fossil record cannot be matched through other approaches. It is well known that a variety of processes combine to shape that record, including evolution, biogeographic distributions, diagenesis, and facies effects. But the role such factors play in forming the fossil record of the earliest animals has not been well understood.

In the southwestern Great Basin of North America, the fossil record of the metazoan radiation event is strongly influenced by regional patterns of Early Cambrian sedimentation. Most fossil occurrences are reported from strata that accumulated in shallow, subtidal siliciclastic and carbonate mud-dominated environments below fairweather wave base and above storm wave base. Statistical tests comparing the volumes of each lithofacies and their associated faunas indicates a strong positive correlation between the facies and faunas. Therefore, the processes controlling the deposition of different facies also control the nature of the faunas present in the local fossil record. These processes include episodic changes in the rate of eustatic sea-level rise during the Early Cambrian and local tectonic events.

Comparison with other regions in western North America (Sonora, Mexico, and the Canadian Rockies) corroborate the fossil/lithofacies associations documented in the Great Basin. In the Great Basin and in Sonora, Mexico, three episodes of diversification are evident in the fossil record. Superficially, it appears that this pattern reflects common evolutionary events. But recently refined biostratigraphic correlations demonstrate that the apparent diversifications are not synchronous. Both seem to reflect local paleoenvironmental conditions conductive to the formation and preservation of favorable lithofacies, and not evolutionary events.

The facies controls on the stratigraphic distribution of Lower Cambrian fossils in southwestern Great Basin bear on a number of important problems associated with the Proterozoic-Cambrian transition and the radiation of Metazoa. The lithostratigraphic-specific nature of virtually all taxonomic groups indicates that the accuracy of correlations between sections of contrasting facies will be poor. The timing of the appearance of taxa in various regions will probably reflect facies controls and not evolutionary events. Perceived biogeographic similarities and differences might be a function of facies controls, and not ancient provinces. In sum, the trace and body fossil successions at all stratigraphic sections are, in large part, an artifact of the stratigraphic occurrence of lithofacies.

Associations of taxa and lithofacies also bear on current ideas about the loci of evolutionary innovations. Previous workers have suggested that Phanerozoic adaptive innovations (= the origination of new higher taxa) and their assembly into new communities takes place in nearshore environments. Others have suggested that innovations during the Early Cambrian were progressively offshore, in response to fluctuations in habitat area. The data assembled here suggest that innovations occur within a limited range of mud-dominated subtidal habitats. Most innovations occur where organisms are diverse and abundant.

SIGOLO, J. B., and B. BOULAGÉ, Institute de Giosciencias da Universidade de Sao Paulo, Sao Paulo, Brazil

Pediment Bauxite Deposits of Passa Quatro Alkaline Massif (MG), Brazil: Their Morphological Evolution and Related Surface

The Passa Quatro alkaline massif presents an average altitude of 2000 m and is connected with a pedimentary surface where the average altitude is 900 m.

In this massif, two different types of bauxite deposits can be considered: summital and slope. This paper presents the geological and geomorphological features of the second type.

Slope bauxites are located on alkaline rocks (nepheline syenite, syenites, and phonolites) and of Precambrian basement constituted mainly by biotite and biotite-hornblende gneisses, migmatites, quartz-biotite schists, and quartzite. These bauxites are related to pediments whose materials originated from the alkaline massif.

The piedmont deposits, predominant in the south of the massif, are related to three morphological units forming distinct topographic surfaces. As a whole, these units form a large 5° tilted plane whose upper extremity to the north is anchored on the alkaline massif fringe and that slopes southwards toward the Paraiba do Sul rift bottom.

The geological and geomorphological history of these bauxites has been influenced by the Paraiba do Sul rift formation. It is possible to establish an interrelation between the bauxitization of the alkaline rocks and the tectonic opening of the rift.

The global reserved have been estimated at about 15 million tons, of which 5.5 million correspond to the piedmont deposits. This type of bauxite has average chemical contents of 54% Al_2O_3, 1.9% SiO_2, and 4.2% Fe_2O_3.

SIKKA, S. N., Oil and Natural Gas Commission, New Delhi, India

Quantitative Hydrocarbon Resource Assessment of Krishna-Godavari Basin, India

The paper deals with estimation of hydrocarbon resources of the basin. Krishna-Godavari basin occupies southern part of east coast of India adjoining Bay of Bengal. The basin has an area of about 45,000 sq. km. out of which 20,000 sq. km. is onland and 25,000 sq. km. is in offshore up to 1000 m isobath line.

The basin came into being as a pull apart basin following rifting along eastern continental margin of Indian craton in Early Mesozoic.

Precambrian igneous and metamorphic complexes form the basement. The earliest known sedimentary sequence is of Late Permian. This is followed unconformably by Early Cretaceous sediments. Tertiary and Pleistocene deposits including Decan Trap (Paleocene and Early Eocene) lie over the Cretaceous deposits unconformably. The exposed section has sedimentary thickness of about 4380 m.

The hydrocarbon resources of the basin are assessed considering play analysis and probabilistic model. The plays identified are of Cretaceous, Early Tertiary and Late Tertiary based on geological analysis of the basin. Discovery probability for each of the plays is estimated separately by risk analysis and is assessed as 16%, 22.5% and 24% for Cretaceous, Early Tertiary and Late Tertiary plays, respectively.

Hydrocarbon resources are estimated by probabilistic model and Monte Carlo simulation for each of the plays considering discovery probability and prospects so far identified in the mapped area. Since most of the mapped prospects are of structural type, provision is made for likely accumulations in stratigraphic as well as combination traps. Similarly hydrocarbon resources for the unmapped area are also estimated by extrapolations.

Results of resource estimation indicate hydrocarbon resources for the basin as 257 MMt at 90% probability; 726 MMt as the mean value; 1347 MMt at 10% probability an 2158 MMt at 5% probability.

Conditional pool size distribution curves indicate range of individual pool sizes from 3.00 to 28 MMt, with mean pool size of 6.50 MMt for Late Tertiary play. In case of Early Tertiary play, the individual pool sizes may range from 5.00 to 32.00 MMt with an average pool size of about 10.00 MMt. For the Cretaceous play, pool sizes range from 3.00 to 24.00 MMt with an average pool size of 5.00 MMt.

The quantitative hydrocarbon resource assessment of the basin would help in decision making and formulation of exploration strategy for the basin.

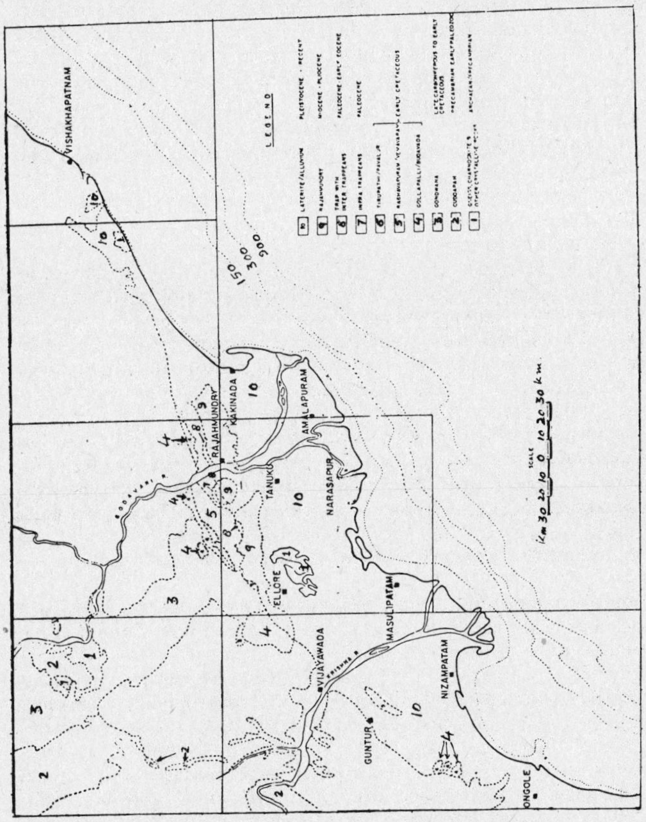

Figure 1—Geological map of Krishna Godavari Basin.

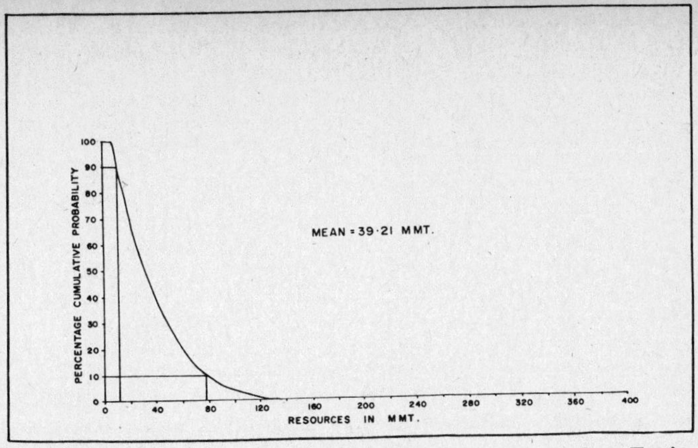

Figure 2—Cumulative probability distribution of resources Late Tertiary play, K. G. Basin.

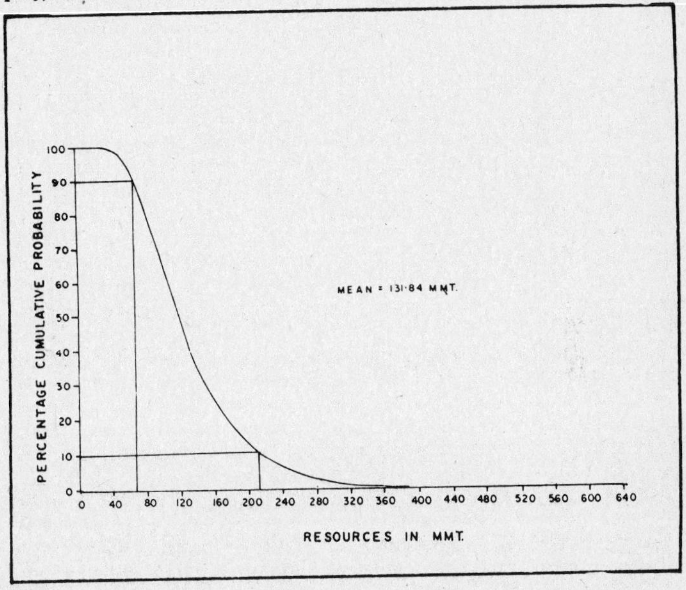

Figure 3—Cumulative probability distribution of resources for Early Tertiary play, K. G. Basin.

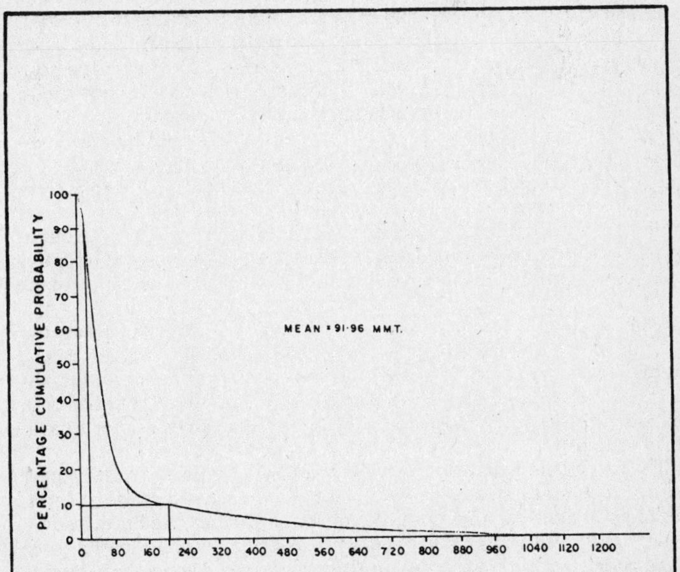

Figure 4—Cumulative probability distribution of resources, Cretaceous play, K. G. Basin.

SILVA, A. A. K., and M. SZIKSZAY, Universidade de São Paulo, São Paulo, Brazil and U. PFISTERER and H.-P. BLUME, Institut fr̈ Pflanzenernährung und Bodenkunde, Kiel, West Germany

Hydrogeologic and Hydrochemical Study of the Unsaturated Zone at the Experimental stations at Cajati-Jacupiranga, Brazil

The study of water in the unsaturated zone became in recent years more and more important for the hydrogeologists, due to its applications in studies of hydrochemical prospection and of groundwater pollution. In Brazil only one study of this type has been done yet, by two of the authors of this paper (Szikszay et al, 1986, 1987, 1988) in allochthonous sediments. In our case, the unsaturated zone is autochthonous, consisting of laterite weathering products, typical formations for tropical and subtropical climate.

As well known, the greatest part of groundwater originates from rain water, infiltrated through the unsaturated zone. During this process the chemical composition of the water is in constant modification, due to interactions between the water and the mineralogical constituents of rock and soil.

Until now, studies in Brazil on the chemical composition of groundwater, have consisted of observations, without taking into consideration the variations of the chemical composition of water during percolation through soil, specially under tropical and subtropical climatic conditions. These variations are the result of processes such as, dissolution/precipitation, adsorption, ion exchange and oxidation-reduction.

In order to study the movement and chemical evolution of water through the unsaturated zone till the groundwater, two Experimental Stations were installed at Cajati-Jacupiranga, State of São Paulo, Brazil, located 250 km SW from São Paulo, in the Ribeira Valley.

As mentioned, the unsaturated zone, in our specific case, consists of laterite weathering products with thicknesses exceeding 40 m. This weathering layer covers the Jacupiranga Alkaline Complex composed of ultramafic, alkaline, peralkaline and carbonatitic rocks, mainly dunites and jacupirangites.

The study of lateritic formations constitutes one of the most important topics in the earth sciences of the tropical and subtropical regions in the world, including Brazil. They may contain exploitable mineral resources accumulated through processes of lateritization as, iron, aluminum, nickel, manganese, chromium and cobalt deposits, being therefore of great importance from the point of view of economic geology.

Unfortunately there is a lack of knowledge as to the genesis of laterites and their transformations under the influence of climatic, geomorphologic, hydrogeologic, hydrochemical and geologic factors, causing difficulties in mineral prospection and exploitation of ore of lateritic regions (Friedrich, et al, 1987).

The aim of this study is the investigation of the hydrogeologic and hydrochemical aspects related both with the behavior of water in the unsaturated zone and with the lateritic formations in the subtropical climate of Brazil.

The Experimental Stations are installed on the area of a private farm that produces mainly bananas and tea. The first one is settled in a hilltop position, among the banana-trees, while the second one is installed downward on a flat plateau amid the primary, untouched subtropical forest.

Each Station is constituted of tensiometers to establish the water tensions, Eh-electrodes to determine the redoxpotentials and porous cups to extract the soil solution, all installed at 20, 50, 100, 200, 400 and 600 cm depth. Water samples were collected from the porous cups every 14 days over a 12-month period, to characterize the unsaturated zone. Groundwater samples were collected from 3 piezometers and a spring. Rain water samples were collected from a rain collector installed near the station in the forest.

There were determined "in situ" the following parameters: temperature, pH, Eh, electrical conductivity, alkalinity, dissolved CO_2 and dissolved O_2. In the laboratory there were measured the contents of major and minor ions, metals, heavy metals, trace elements and C_{org}. An attempt has been made to estimate groundwater recharge and moisture movement using artificial tritium as a tracer.

The results show that the chemical composition of water in the upper part of the unsaturated zone, with higher total dissolved solids, is influenced by organic combinations, the presence of metals and nitrification processes, as well as, by the climate, whereas that in the lower part is more affected by local lithology, due to the presence of dissolved weathering products. This is further demonstrated by the three-dimensional computer treatment of each parameter (PLOT/SURF program), which shows important chemical variations in water with depth in the unsaturated zone. The variations are less pronounced over time, given the more or less uniform conditions of a subtropical climate through the year.

Another aspect that shows the importance of this study is that the results will support new investigations of weathering processes and in the formation of new minerals in tropical and subtropical climate. This study can be of help at hydrochemical prospection for exploitable mineral resources in the laterite. As to the soil sciences, the results are explaining the acidulation and therefore the weathering of the soils in humid climate.

References

Friedrich, G., A. Marker, M. Kanig and A. Germann, 1987, Mineral prospecting and geological mapping in laterite covered areas of Brazil, Final Report, BMFT RG 8301 5, p. 358, Aachen.

Szikszay, M., A. J. Consoni, N. Guiger, et al, Instalacao de uma Estacao Experimental para estudos da dinamica e da evolucão quimica da água na zona não saturada, Rev. Aguas Subt., v. 10, p. 5-20, São Paulo.

——— 1987, Estudo preliminar da hidrodinamica na zona não saturada da Estacão Experimental (Cidade Universitária, São paulo), Rev. Aguas Subt., v. 11, p. 33-62, São Paulo.

——— Estudo hidrogeológico e hidroquimico na zona não saturada, Anais 50 Congr. Bras. Aguas Subt., 14-16, set. 88, p. 133-143, São Paulo.

——— 1988, Evolution of the chemical composition of water passing through the unsaturated zone till the groundwater (Experimental Station—University of São Paulo, Brazil), J. of Hydrol., in print.

SILVA, Z. C. G., Departamento de Geologia, Universidade de Lisboa, Lisboa, Portugal

Anorthosite of Southwest Angola: Preliminary Petrochemical Study Aided by Statistical Interpretation of Major and Trace Elements

The interpretation of geological data on Angola as a result of field work developed by the Geological Survey and private companies operating in that country during a few decades made possible the publication of the geological map in 1:1,000,000 scale in 1984 (Carvalho, 1984). From that general map it became possible to delimitate the area of occurrence of the gabbro–anorthosite Somplex in Southern Angola. The ensuing use of satellite imagery emphasized some structural aspects of the Complex, already known in different areas of the intrusion.

Recent investigation performed by the author at the University of Lisbon on a selected sampling from the total area of outcrop in Angola and collected by Simpson and others (Simpson, 1970) permits the understanding of some aspects mainly related to the layered or massive character of the intrusion, the behavior of major and trace elements throughout the body and is reported hereafter.

The gabbro–anorthosite Complex of Angola is known to be one of the largest anorthosite bodies in the world and covers an estimated area of about 18,000 km^2 in Angola and Namibia.

Petrographic studies were conducted in a few hundred rock samples including all petrographic types occurring in the intrusion: anorthosites, anorthositic gabbros, norites and troctolites, all displaying a typical cumulate texture. The nature of these rocks lead the author to suggest that the exposed portion of the Complex corresponds to the middle zone of the intrusion.

Fifty-five samples from the total collection were selected for detailed petrographic study, microprobe and chemical analyses. These rocks only cover the main body. Country rocks are known to be mainly Precambrian gneisses and migmatites which were not considered for such detailed study presently.

Chemical composition of these rocks and their mineralogy have been recognized as being similar to those of the main and critical zones of the Bushveld Complex (Silva, 1987).

The intrusion displays rhythmic layering in certain areas in the northern part of exposure and detailed mineralogy of the rocks from that area evidences its cryptic layered character, as observed on plagioclases and olivines, the main cumulus phases of the rocks of the Complex. There is a gradational decrease in the calcium content of plagioclases from west to east of the intrusion, with the composition varying from An_{80} to An_{48}. Similarly, the magnesium content of olivines varies from Fo_{79} to Fo_{55}.

Figure 1 is an outline of the area of exposure of the Complex in Angola and the line YY' is a limit arbitrated by the author, based on observation of mineralogical and chemical changes in the rocks as mentioned above. Line XX' is a rough limit set to register the appearance of a distinct type of anorthosite, light colored and massive. From that line to the south, the rhythmic layered character of the intrusion is less conspicuous, although its cryptic layering can still be detected.

Multivariate statistical methods were used to aid the interpretation of the whole rock chemistry obtained on major and trace element analyses of the 55 samples of the main body. When the total samples and the 23 variables (SiO_2, TiO_2, Al_2O_3, Fe_2O_3, FeO_2, MnO, MgO, CaO, Na_2O, K_2O, P_2O_5, Rb, Sr, Y, Zr, Nb, V, Cr, Ni, Cu, Zn, Ba, Th) are considered, clusters consisting of groups of samples either from the west (area B of Fig. 1) or from the east (area A of fig. 1) are formed (Silva, 1987). This type of clustering confirms and illustrates well the observations already mentioned regarding chemical and mineralogical variations from west to east in the intrusion. Cluster dendrogram based on the correlation matrix of the 23 variables

(Fig. 2) emphasizes the inter-element associations and their meaning. It shows that Fe″-Mn, Mg-Ni-(Cr), Si-Na-(Sr), K-Ba have strong correlation, reflecting the proportions of the main mineral phases, olivine and plagioclase, in the rocks.

These results supplement a preliminary interpretation of the chemical evolution of the gabbro–anorthosite Complex, in which the gradual chemical variation in composition, becoming enriched in Si, Na and Sr, K and Ba to the east and decreasing in Mg, Ni and Cr in the same direction, suggest that the top of the intrusion is to the east. They are not sufficient yet to set a petrogenetic model for the Complex, or to indicate the single or multi-pulse nature of the intrusion, but they do show the consistency of data regarding differences in composition of the rocks from the West to the East and that both olivine and plagioclase cumulus phases control the chemistry of the intrusion.

REFERENCES

Garvalho, H., 1984, Geologia de Angola, 1: 000,000 scale. Lab. Nac. Investigaçaõ Cientifica Tropical. Lisboa, Portugal.

Silva, Z. C. G., 1987, Estudo petroquimico do Complexo gabro-anortositico de Angola. PhD. Thesis, Univ. Lisboa, Portugal.

Simpson, E. S. W., 1970, The anorthosite of southern Angola: A review of present data. In Clifford, N. N. & Gass, I. G. (ed)-African Magmatism and Tectonics. Oliver & Boyd, Edinburg, p. 89–96.

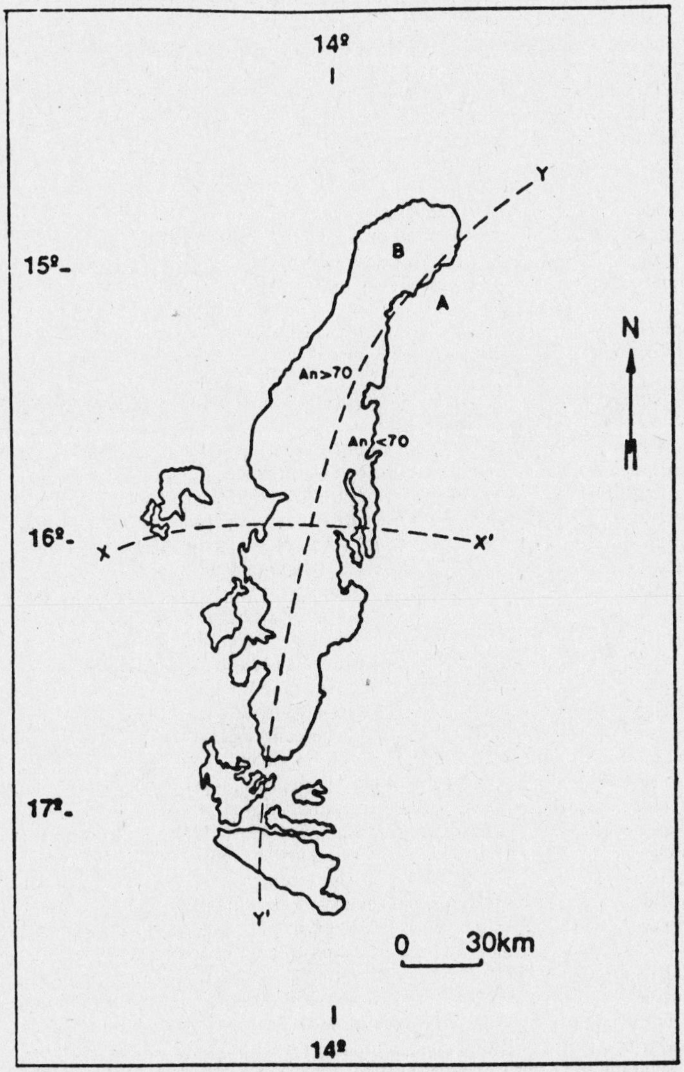

Figure 1—Outline of the exposed area of the gabbro-anorthosite Complex of Angola. YY′ and XX′ are suggested limits to illustrate chemical and mineralogical changes in rock composition.

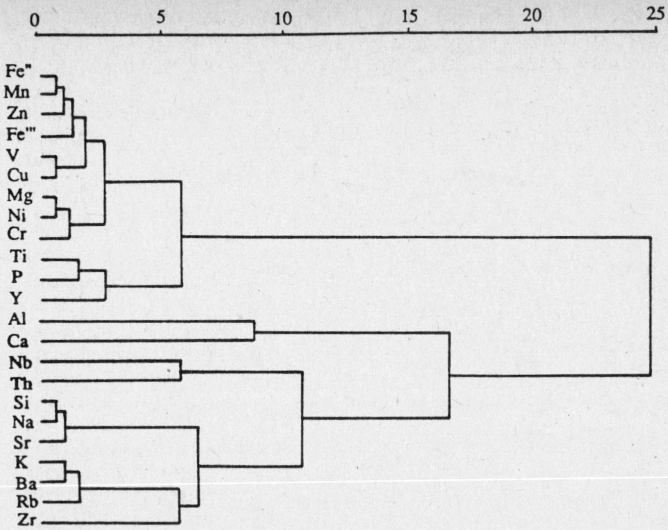

Figure 2—R-mode cluster dendrogram constructed from the correlation coefficient matrix of the 23 variables. (data are from chemical analyses of the 55 rock samples)

SILVER, E. A., University of California, Santa Cruz, CA

Dynamics of Two Arc-Continent Collisions: Panama and Papua New Guinea

Arc-Continent collisions are responsible for significant orogenic activity in northeastern South America and Panama (SAP) and in Papua New Guinea (PNG). Both show major bending of the colliding arc, with cross-arc strike-slip zones accommodating some of the bending, a scissors-like propagating point of initial collision, widespread compressional and strike-slip deformation of the foreland, and maximum horizontal stresses oriented normal to the foreland thrust belts. The latter is well documented in PNG. Both collisions have large fans ponding terrigeneous sediment just beyond the collision point. Both have major ophiolites emplaced within a zone of accreted tectonostratigraphic terranes, and both have well-developed foreland fold and thrust belts. Petroleum deposits occur in both fold and thrust belts.

Despite their great similarities, major differences occur. In the SAP collision, thrusting was dominant beneath the continent, and the Andes represent a major continental subduction. In PNG, some subduction may have occurred beneath the continent, but the dominant subduction is beneath the colliding arc (New Britain). This difference has apparently resulted in attempted subduction of the colliding Panama arc beneath South America, with broad traction created beneath the continental plate. In contrast, the extension of the New Britain arc in the Finisterre range of PNG has overridden the continent. A second difference between the collisions is the tectonics of the ocean side of the collision. In SAP, convergence occurs along the entire Pacific margin of the collision, and this long zone of shear aids in the development of great strike-slip faults in the orogen. Oceanward of the PNG collision, however, the Manus basin is rapidly opening, and thus normal convergence dominates the orogen. Farther west in Irian Jaya, where no extension occurs in the offshore, large strike-slip earthquakes are common.

A third difference is that the collision of Panama with South America has a component of motion parallel to the direction in which the collision point is migrating, whereas the collision of New Britain with PNG has a component of motion opposite to that direction. This difference may provide additional torque on the PNG arc, explain the large offset of New Britain and New Ireland, and be responsible for the propagation of rifting at the tip of the Papua Peninsula. The direction of propagation of the rift tip is opposite to that of the collision tip.

Although some of these differences are of major importance for modern tectonic settings, such differences may prove difficult to discern in ancient orogenic belts. Careful attention to terrane analysis and age dating are key factors in deciphering the ancient tectonic patterns in complex collision zones.

SILVER, L. T., California Institute of Technology, Pasadena, CA

Time, Space, and Source Regions of Cretaceous Cordilleran Batholiths of the U. S.

The Cretaceous batholiths of the western United States were emplaced across a diversity of lithospheric environments in a remarkably systematic set of plate interactions. All of the great units, Peninsular Ranges, Mojave–Salinia, Sierra Nevada, Idaho batholiths appear to have been initiated in a convergent plate tectonic cycle which began in the late Jurassic or earliest Cretaceous and continued into the Paleogene. Throughout the U. S. Cordillera a major change in the style of the plate interactions and magma generation appeared at about 105 /pmn2 Ma. Another significant break occurred at about 75–80 Ma. Each of these changes are reflected in time–space, volume, and chemistry, but the nature of these changes appears to reflect the peculiar local lithospheric setting of each segment.

Spatial and volumetric analysis of the changing loci of magmatism requires removal of the commonly underestimated Cenozoic tectonic overlay which was imposed throughout the Cordilleran batholiths (basin–range extension, California transform faults, metamorphic core complex development, low angle faulting of both extensional and overthrust character). Synplutonic tectonism seen through this overlay appears to have differed in style and degree also, as a function of the setting of each segment.

The most striking character of these batholiths are the remarkable systematic internal gradients in petrological and geochemical characteristics. These gradients offer powerful insights into the relative contributions of different source regions and the mechanisms for magma generation, transport, and emplacement. Proposals that all or more of the batholiths may have formed a single continuous magmatic arc (e.g., Hamilton, 1969) have received support from subsequent tectonic studies in the Cordillera. However, the tectonic and lithospheric context for magmatism differed along the strike length of the batholithic chain. Striking radiogenic and stable isotopic gradients and inflections in these gradients are now well established in all of the great units. They are expressed on many scales, from within small plutons to local plutonic complexes to entire batholiths. They seem to imply formation on, off, and across the boundaries of the earlier continental lithosphere. Pb, Sr, Nd, and O gradients each supply distinctive insights. Major and trace element gradients produce further independent insights. Portions of the chain were clearly primitive oceanic island arcs, developed on oceanic lithosphere. Even discontinuities in the ages of different segments of host oceanic lithosphere appears recorded. These arcs transgressed adjacent continental lithosphere both longitudinally and in transverse time–space migrations. Lateral mixing does not appear important, but vertical mixing seems to have played an extensive role. The third dimension of the source regions seems to have extended deep into heterogeneous continental and oceanic lithospheres, providing some basis for estimation of "juvenile" mantle contributions. Model calculations suggest that mantle sources were very significant on the western margins of all of the batholiths and far into the interior of some of them. Subducted slab contributions do not appear volumetrically important. Thus, these batholiths constitute important elements of new continental crust formation as well as reconstituted older lithosphere.

SILVER, L. T., E. W. JAMES, and S. COTKIN, California Institute of Technology, Pasadena, CA

Some Geological and Geochemical Results of Cajon Pass Scientific Drillhole, Southern California

Situated on the southwestern margin of the Mojave Desert, California, in the vicinity of the great San Andreas fault, the Cajon Pass Scientific Drillhole penetrates 3.5 km into the flank of a large regional basement antiform that reflects the recent transpressional environment along this zone. The drillhole was produced by a combination of conventional rotary and downhole motor drilling with spot coring sampling approximately 3.5% of the total depth. A synthesis of the encountered geologic column has been prepared from cuttings, core, and geophysical and geochemical logs. The primary scientific objectives of the hole were to explore the fundamental relation between the heat flow and stresses along the great fault at seismogenic depths. A significant portion of the geological and geochemical effort was directed toward investigations that provided essential context for the primary experiment. This included the acquisition of information on bulk rock mineral assemblages, chemical compositions, density, fabrics and structures, permeability, heat productivity, fractures, and recent geologic history. The concurrent utilization of direct sampling and downhole geophysical logging provided unusual opportunities for direct testing and calibration of several methods for in situ characterization of the rock column.

A second important scientific objective was the exploration in the third dimension of a critical part of the margin of the North American plate in this part of southern California. This margin has been active for over 250 Ma. At various times it has been the site for magmatic arc development and other dynamics of a convergent margin. At other times, it has been the site of intense intraplate tectonic activity (e.g., metamorphic core complexes). The complexity of the geologic record has never been adequately deciphered. The results from the scientific studies in the well and on the material extracted from the well are producing important new data bearing on the regional geologic history as well as on the recent influence of the transform margin.

The yield of new geological and geochemical insights attained requires the reexamination of previous regional tectonic and petrologic models. The tectonic setting of this experiment is in a late Mesozoic crystalline complex dominated by syntectonic intermediate and felsic intrusions. A thin (500 m) late Cenozoic cover of continental sedimentary basin facies provides stratigraphic and sedimentary control for the latter part of a complex tectonic history.

With the exception of some 250 m of sillamanitic paragneiss and quartzites, the "basement" is comprised of a spectrum of igneous rock compositions ranging from gabbros to true granites. Although all of the rocks analyzed isotopically so far appear to be Cretaceous, they range in primary age from 75 to more than 130 m.y. All rocks are foliated to some degree. Most of the column is markedly gneissic with low angle foliations. Mobilizates, ductile structures, and fabrics are commonly overprinted by cataclasis. Distinct breaks in the column are manifested by brittle fault zones and pervasive alteration assemblages featuring the zeolites laumontite and stilbite. Significant discontinuities in rock ages and radiogenic isotope signatures together with the rock fabrics and structures suggest there are major previously unrecognized low-angle (?) faults intersected by the well. At least one of these faults can be correlated with a strong seismic reflector observed in vertical seismic profiling. The zeolitization phenomena appears to represent a development associated with proximity to the San Andreas fault and may have significant implications for physical properties of the San Andreas fault system elsewhere. The variations in the age and isotopic signatures observed in the well suggest a previously unrecognized late Cretaceous orogenic event and the possibility of fault dislocations of great regional significance.

SILVER, P. G., and R. W. CARLSON, Carnegie Institution, Washington, DC, and P. OLSON, The Johns Hopkins University, Baltimore, MD

Penetrative Convection: Deep Slabs in Chemically Stratified Mantle

We propose a model for mantle convection characterized by slab penetration into the lower mantle and a ~2% compositionally induced density contrast between upper and lower mantle. Referred to as penetrative convection (Silver et al, 1988), this style of convection appears possible when compositional and thermal buoyancy are of the same order. The basic departure from standard two-layer or whole-mantle models, which are dominated by compositional and thermal buoyancy, respectively, is that subducted slabs do not provoke mantle-wide overturn. Rather, the density contrast across the transition zone, while not preventing slab penetration, serves to (1) minimize entrained flow and (2) cause slabs to return to the upper mantle after thermal equilibration (about .2 Ga), because of the intrinsic positive buoyancy of slab material in the lower mantle. The transition zone flux thus consists solely of the descending and "recently" returning slabs. Dynamically significant stratification consequently may survive for billions of years with such a mechanism.

This model has properties that traditionally have been attributed to both whole-mantle and two-layer convection. It is "whole" mantle in that there is significant mass flux between upper and lower mantle and no large thermal boundary layer at the top of the lower mantle. However, because of the density stratification, the mantle will appear geochemically as two distinct reservoirs. The upper mantle is involved in plate-tectonic recycling and consists of fragments of old oceanic lithosphere imbedded in a matrix that is the source of MORB. The lower mantle has not been processed by plate-tectonics and is only rarely sampled as minor entrained material accompanying plumes of slab material ascending from the lower mantle to form major hot sopts. While all the dynamical and geophysical consequences of this model have yet to be established, its potential success in reconciling geophysical and geochemical constraints argues for careful evaluation. Several diagnostic tests of the model will be discussed.

REFERENCES

Silver, P.G., P. Olson, and R.W. Carlson, 1988, Deep slabs, geochemical heterogeneity and the large-scale structure of mantle convection: investigation of an enduring paradox: Annual Review of Earth and Planetary Science, v. 16, p. 477-541.

SILVER, P. G., Carnegie Institution, Washington, D.C., and W. W. CHAN, Teledyne Geotech Alexandria Lab, Alexandria, VA

Constraints on Continental Structure and Evolution from Seismic Anisotropy

Evidence is presented for azimuthal anisotropy in the subcontinental upper mantle from nine stations in North American and Europe. The measurements were obtained from observations of shear-wave splitting in the core phase SKS from broad-band stations from North America and Europe (Figure 1). Since SKS is radially polarized in an isotropic earth, the split shear wave thus appears as transversely polarized energy. For each recording station we use a systematic search procedure to determine the azimuth (clockwise from north) ϕ of the fast polarization direction and the delay time between split arrivals δt for each phase. We search for the parameters that, when used to correct the anisotropy, most completely remove the energy on the transverse component. The resulting parameters represent integral averages of the anisotropy from the CMB to the surface. Several lines of evidence, however, suggest that the anisotropy is localized in the subcontinental upper mantle below the recording station. We obtain a range of δt from 0.6 to 1.75 s. δt is a measure of the magnitude of the anisotropy and can be related to layer thickness L by the approximate relation $L \approx \delta t \beta_o / \epsilon$ where β is the shear velocity and $\epsilon(<<1)$ is the fractional difference in velocity between the fast and slow polarization directions. For 4% anisotropy, a value inferred from natural occurring peridotites, the range of δt translates to a range of mantle thicknesses from 70 to 200 km.

This anisotropy, undoubtedly the result of the preferred orientation of mantle minerals such as olivine, appears to reflect fossil strain associated with the last dominant orogenic episode in the region, in that there is a close correspondence in most cases between the dominant geologic fabric and the fast polarization direction. The station RSON in the Canadian Shield gives the largest anisotropy, 1.75 s, which is one of the largest values of splitting every obtained. The splitting appears to be due to 2.5–2.7 b.y. old "fossil" anisotropy localized in the top 200 km of the upper mantle. This mantle fabric, apparently caused by a series of large-scale deformational episodes at the end of the Archean, has moved coherently with the surface ever since.

Thus, as a geologic tool, seismic anisotropy has the potential for addressing two aspects of continental deformation that up to now have been difficult to constrain. Firstly, we may now begin to study the history of mantle deformation. The results from the Canadian Shield suggest that we may be able to detect "fossil" anisotropy for even the oldest continental masses, some 3.6 b.y. old. Secondly, it may be possible to constrain the nature of deformation at depth that may be associated with present day tectonics. Thus we may begin to address the question of the depth-extent of events such as the collision between India and Eurasia.

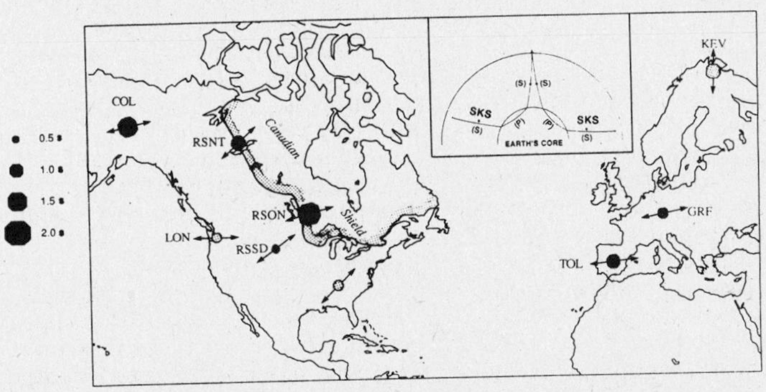

Figure 1—[From Silver and Chan (1988)] Inset: the path of SKS through the earth. On map, the arrow gives the past polarization direction and the size of the closed symbol gives the delay time δt. Shaded symbols are for stations with only one measurement and should be considered preliminary.

SIMKIN, T., Smithsonian Institution, Washington, DC

Subaerial Volcanism in Space and Time: Evidence from Historic Eruptions and Holocene Volcanoes

This introductory talk is based on our Smithsonian datafile of Holocene volcanism. It emphasizes the spatial distribution of Holocene volcanoes and the frequency/magnitude relationship in historic volcanism.

The world's known subaerial volcanoes are strongly concentrated into a relatively small number of linear belts. These account for >94% of known historic eruptions and total 32,000 km in length, but they cover less than 0.6% of the earth's surface. Most are adjacent to deep oceanic trenches, and it is clear that subduction of one plate beneath another causes most of the volcanism that we see. The complementary tectonic process of rifting takes place largely on the deep ocean floor and is not witnessed by humans, but accounts for an estimated three-quarters of the new lava reaching the earth's surface each year (Crisp, 1984, JVGR, v. 20, p. 177-211). Rift volcanism dominates the planet and, although most of it poses no threat to humans, this dominance must be remembered for global perspective.

Like earthquakes, the frequency of eruptions decreases with increasing size—there are many small ones, fewer medium-sized ones, and very few large ones. Figure 1 borrows from the seismologists' use of frequency-magnitude plots, but taking the VEI (volcanic explosivity index), which closely parallels the volume of erupted tephra, as the measure of eruptive magnitude. The number of eruptions recorded over the last 200 years is plotted for each VEI class. We have had one VEI (Tambora, 1815) in the last 200 years, 5 ≥ VEI 6, 13 ≥ VEI 5 (Mt. St. Helens size), and so forth. The data for VEI 2 are based on the last 10 rather than 200 years, since it is the smaller VEI 2 events that are least likely to have been fully reported in the nineteenth century. Eruptions the size of Mt. St. Helens 1980 (10^9 m^3, or 1 km^3) occur perhaps once a decade, and those like Krakatau 1883 (≥10 km^3) two or three times a century. The historic record, however, fails to prepare us for the much larger eruptions of the past. The Toba eruptions, only 75,000 years ago, were 1000 times larger than the Mt. St. Helens eruption, and even larger eruptions are known from the geologic record.

Changes in volcanism with time are difficult to demonstrate historically. The historic record is built by humans, rather than automatic instruments, and socioeconomic influences are clearly shown. Two of the most striking apparent drops in global volcanism coincide with World Wars I and II, when people (including editors) were preoccupied with other things. Many eruptions were probably witnessed during those difficult times, but reports do not survive in the scientific literature. While we see no persuasive historical evidence for global pulsations in volcanism, there are several suggestions of regional groupings of eruptions concentrated in time. Charles Darwin called attention to one of these along the Andean front in 1835 (following the Concepcion earthquake), and another example is the twin disasters of Mont Pelee and Soufriers St. Vincent, only hours apart in 1902 but separated by 165 km along the Caribbean plate margin.

One of the certain contributors to large death tolls in major eruptions is the fact that the repose interval is very much longer than the historic record in many parts of the world. Most eruptions of low explosivity follow repose intervals of only 1 to 10 years, while the more explosive events, of VEI ≥ 5, follow repose intervals measured to hundreds of thousands of years. In regions with short historic records, the human population is commonly unprepared for such events and the results are often tragic. Table 1 lists the 14 most explosive eruptions in the last two centuries-those with a VEI ≥ 5 (or ≥ 1 km^3 of tephra). It is significant that fully 11 of the 14 were the first historic eruptions at that volcano (and for others, like Kralatau, earlier activity was not widely known). Attention is commonly focused on volcanoes that have produced history's longest-running and best-documented eruptions, but we must not neglect the nonfamous, thickly vegetated volcanoes that have had no historic activity.

The following papers in this symposium will provide more detailed discussions of specific regions—all with long historic records—illustrating the major tectonic settings of volcanism: subduction zones (Japan), spreading centers (Iceland), and hot spots (Hawaii). Longer-term views will discuss volcanism in the Cenozoic: explosive volcanism on the North Pacific rim, intraplate continental volcanism of eastern Australia, and the world's most abundant (and least known) form of volcanism—that on the seafloor. Additional papers will provide lessons from Italy, the world's longest records of historic volcanism, and from the common noneruptive activity that is so important in understanding the eruptions that may (or may not) follow.

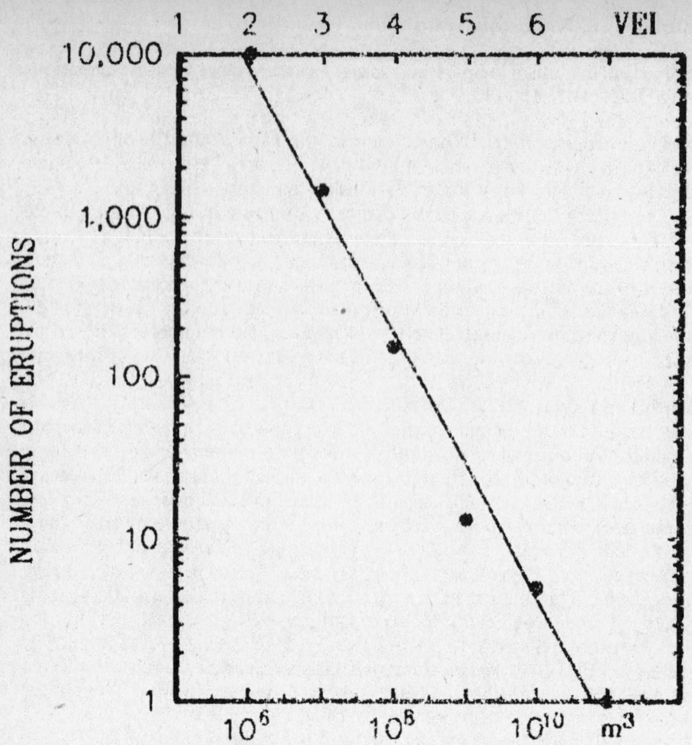

EXPLOSIVE ERUPTIONS >1 km^3

Year	Volcano	1st?	Deaths
1982*	El Chichon	yes	2,000
1980	St Helens	no	66
1956	Bezymianny	yes	0
1932	Cerro Azul (Quizapu)	no	0
1912	Novarupta (Katmai)	yes	2
1907	Ksudach	yes	0
1902	Santa Maria	yes	5,000
1886	Tarawera	yes	150
1883	Krakatau	no	36,417
1875	Askja	yes	0
1854	Sheveluch	yes	0
1835	Cosiguina	yes	10?
1822	Galunggung	yes	4,011
1815	Tambora	yes	92,000

SIMMONS, M. D., U.S. Army Corps of Engineers, Nashville, TN

Harlan, Kentucky, Diversion Tunnels

As a part of a local flood protection scheme, the Nashville District of the U.S. Army Corps of Engineers is planning to divert the full flow of Clover Fork of the Cumberland River around the central business district of the city of Harlan, Kentucky. This is to be accomplished by constructing four tunnels side by side through a meander ridge that has a maximum overburden above the tunnels of approximately 325 ft and averages about 220 ft. The inverted "U" configured tunnels are 32 ft high, 34 ft wide, and in excess of 2000 ft in length. The rock through which they are to be constructed is Pennsylvanian-aged siltstone with an average compressive strength of about 7500 psi and containing over 57% quartz. As a part of the exploratory program, horizontal borings were made from each portal location, each being more than 1000 fr in length. Most of the rock mass explored has RQD values in excess on 95% with zones in excess of 300 ft in the horizontal holes being devoid of all discontinuities. Construction is to be accomplished with a road header mining machine with rock support being furnished by systematic rock bolting. Four inches of steel fiber reinforced shotcrete will be applied as the permanent lining for the tunnels.

SIMO, A., University of Wisconsin, Madison, WI

Carbonate Facies and Platoform Evolution of Upper Cretaceous Tremp Basin, South-Central Pyrenees, Spain

Tremp Basin Upper Cretaceous stratigraphic architecture shows moderate structural deformation and is at a scale comparable to seismic-reflection profiling. Detailed field mapping is the foundation for facies description, facies interpretation, and packaging of depositional facies into depositional sequences. A comprehensive understanding of the processes controlling the sedimentation is achieved by combining the field data with stratigraphy sequence techniques and subsidence modeling.

The Upper Cretaceous depositional sequences of the Tremp Basin in the Pyrenees are a result of a combination of "global rhythms" and tectonics. Five sequences bounded by sequence boundaries are recognized. The three oldest sequences are carbonate-rich, and each sequence exhibits: (1) moderated coastal onlap, with each successively younger carbonate platform backstepping the former; and (2) shelfward encroachment of the basin resulting in submarine slope erosion and increased water depth in the basin. The fourth sequence is the thickest and laterally most extensive with a major shelfward shift of the carbonate platform. The fourth sequence is divided into a siliciclastic-rich lower subsequence and an upper carbonate-rich subsequence. Depositional sequences one through four can be subdivided into a volumetrically small transgressive lower unit and a volumetrically larger regressive upper unit. The fifth depositional sequence is predominantly siliciclastic-rich and shows an overall regression in response to regional thrusting, uplift, and the late Cretaceous sea-level still stand.

Sequence boundaries overmost of the area are concordant surfaces where a sharp transition from shallower to deeper facies occurs and resemble platform drowning surfaces. These surfaces can be traced: shelfward into subaerial surfaces, and basinward into submarine erosional truncation surfaces.

An ideal facies-tract of the fourth youngest carbonate-dominated sequences reveals five depositional facies: carbonate platform facies (lagoonal and shoal-reef complex or bank-margin facies), upper slope facies (breccias, calcarenites, and fine-grained sediments), outer shelf facies (nodular limestones and marls), lower slope facies (burrowed well-laminated limestones with many slumps features). and basin facies (well-bedded marls/shales and calcareous mudstones). Carbonate platform facies-tract changes from the lower and oldest sequence (which comprises carbonate platform, upper slope and basin facies) to the upper and youngest sequence (which comprises the five depositional facies described).

The geometries of the Upper Cretaceous depositional sequences and facies-tracts were influenced by both tectonics and eustacy. Measured field sections and computer modeling have been used to calculate the amount of subsidence caused by sediment load and tectonic basement subsidence in the inner-shelf, midshelf, and basin positions. The thermal subsidence curves match nicely with the coastal onlap curve and the backstepping geometry of the Upper Cretaceous depositional sequences. The more moderate coastal onlap of the three oldest sequences can be explained by slow subsidence rates and erosion on the inner-shelf. The major shelfward shift in coastal onlap of about 60 km at the base of the fourth depositional sequence correlates to the onset of subsidence. The backstepping of each successively younger carbonate platform (sequence 1 through 4) and shelfward encroachment of the basin are explained by: (1) the contrast of subsidence rate on the shelf versus the basin (three times higher in the basin than on the shelf), and (2) by the shelfward increase of subsidence rate with time. Tectonics played a role in the formation of sequence boundaries, including enhancement of the boundaries expression (listric normal faulting and uplift), or diminishment of the boundaries expression during times of rapid subsidence rates.

Calculations and modeling show that tectonic subsidence creates space for potential sediment accumulation but does not account for the total accumulated sediment thickness. The additional space needed (coastal onlap curve) is thus attributed to sea-level movements. The coastal onlap curve was calculated subtracting the predicted thermal subsidence and estimated paleowater depth from the sediment thickness (thickness corrected for compaction from the first coastal onlap horizon of each sequence to the top of the sequence). Magnitudes up to 6-10 cm/100 y of relative sea-level rise in the Tremp Basin are compatible with Cretaceous eustatic movements. Relative sea-level changes are also recorded in the sequence boundaries. Eustatic lowering of sea level exposed the inner-shelf but did not appear to have exposed the shelf-edge. However sea-level drops may have triggered erosion in the basin by increasing current activity and instability of sediments.

Platform progradation occurred during times of slow subsidence and highstand of sea level. Platform aggradation occurred during times of moderate subsidence and sea-level rise. Platform growth ceased when relative sea-level subaerially exposed platform top. Carbonate sediment producers were mostly low-magnesium skeletal, rudist, coral, and few ooids. During platform progradation, platform sediments were mostly massive coarse-grained skeletal grainstone-packstone and reef framestone, and slope sediments were dominated by breccias and calcarenites derived from the platform. During platform aggradation, platform sediments were shallowing-upward lagoonal mud-rich/rudist reef facies and massive coarse-grained facies, and slope sediments were fine-grained limestones and marls derived from the platform and water column.

SIMON, A. A., Empresa de Geología de Pinar del Río, Santa Lucía, Pinar del Río, Cuba

Metallogenetic Evolution of Cuban Island Arc

The geotectonic evolution of the Cuban Archipelago, as part of the Greater Antilles, has been divided into three distinct phases: the passive continental margin, the island arc and the contemporaneous stages.

The first stage is marked by the deposition of continental erosional products at the southern margin of the North American continent during most of the Jurassic. At Callovian–Early Oxfordian that zone becomes a convergent plate boundary, where the Proto-Caribbean oceanic crust is northward subducted beneath the North American plate, beginning the island arc stage. Following the classical pattern of island arc evolution, three volcanic arcs succeed each other in activity, through the opening of interarc basins. Thus, the Septentrional volcanic arc is active up to Neocomian, whereas since Tithonian a back-arc basin develops at the Northern side.

The San Diego de los Baños interarc basin begins its opening at Aptian. The arc volcanic activity is then shifted to the South, leaving a remnant inactive arc to the North. The Median volcanic arc keeps active until Santonian-Early Campanian, when a new split takes part and the Meridional volcanic arc migrates southward, through the opening of the Cauto inter-arc basin. Meanwhile, closure of the San Diego de los Baños inter-arc basin between Late Maastrichtian and Early Eocene provokes collision of the Median and Meridional remnant volcanic arcs, followed by the northward obduction of the former over the latter. The volcanic activity in the Meridional arc finishes in Middle Eocene, after which the contemporaneous stage begins.

In this paper an attempt is made to set a clear fit between geotectonic and metallogenetic units of the same order. Therefore, the North Antillean subduction system is considered as the North Antillean metallogenetic province. Each volcanic arc/back-arc complex represents a metallogenetic subprovince, with two metallogenetic zones, corresponding one to the volcanic arc and the other to the back-arc basin. Secondary metallogenetic factors, as the kind of metamorphic, magmatic or sedimentary processes involved, have conditioned the appearance of smaller subdivisions.

1. *Callovian–Neocomian metallogenetic substage: the Cuyaguateje subprovince.* Cu-Zn-Pb-Ba sediment-hosted submarine-exhalative deposits were formed at the top of the Liassic (?)Lower Oxfordian San Cayetano Formation, simultaneously with the beginning of subduction. These deposits are bed- or lens-shaped, with a Cu-rich "stringer" ore zone. Among them, the Matahambre, Santa Lucia and Castellanos are the most important. The possibly significant contribution of crustal erosional products at the initial subduction process is recorded by their relatively radiogenic lead.

As a result of back-arc spreading at the Esperanza marginal basin, the Besshi-type (pyrite–chalcopyrite) deposits of the Union–Hierro vulcanogenic massive sulfide ore district were originated, in connection with diabase intrusions in a calcareous–terrigenous unit. The still important continental sedimentary afflux obscures the offiolitic nature of the crust.

2. *Aptian–Lower Campanian metallogenetic substage: the Zaza subprovince.* During this substage a considerable volcanic activity is present in the region. Its products are widespread, although the effusive cover has been partially eroded. The McKinley–Tunas metallogenetic zone comprises ore deposits originated in connection with the arc vulcanism and plutonism. Vein Au-Ag (Delita), W (Lela) and Au-Mo (Guaimaro), Fe skarn (Magarabomba, Candelaria), porphyry copper (Arimao) and Kuroki-type (Antonio) ore deposits figure among its most characteristic.

At the same time, back arc spreading at the San Diego do los Baños interarc basin produces the typical offiolitic complex ore deposits: podiform chromites, in close relation with basic and ultrabasic intrusives (MgO-rich, as Caledonia and Casimba, and Al_2O_3-rich, as Merceditas and Cayo Guam), and Cyprus-type vulcanogenic massive sulphide deposits included in the basaltic–sedimentary layer (Jucaro and buenavista).

3. *Early Campanian to Middle Eocene: The Cauto subprovince.* The latest episode of back-arc opening and southward volcanic arc migration took part when the Meridional arc (Cayman Ridge–Sierra Maestra) splits from the Median arc.

Pb-Zn-Cu Kuroko-type (El Infierno), vein Cu-Mo (Buey Cabon, Isabelita), Cu (El Cobre), Ag (La Nicolasa) and barite (El Cedron), magnetite-skarn (Santiago, Sexta) and Mn-vulcanogenic stratiform (Charco Redondo) deposits are present in the Sierra Maestra volcanic arc metallogenetical zone.

The adjoining inter-arc basin is covered by thick post-tectonic sediments.

At the contemporaneous stage hypergenic processes have played the most important role. Large Ni lateritic deposits (Moa, Pinares de Mayari) formed over certain serpentinitic massifs, and oxidation of older sulfide deposits was especially significative at Hierro-Mantua and El Cobre.

The most prominent metallogenetic subdivisions of the Cuban island arc form east-west oriented parallel belts, increasingly eroded from South to North. A conspicuous polyciclicity of the ore forming process is observed: rift- and arc-related deposits were formed at every substage of metallogenetic evolution.

The validity of the Principle of Isomorphism is exemplified by the presence of vulcanogenic–sedimentary cupriferous deposits in the Esperanza and San Diego de los Baños metallogenetic zones, the magnetite-skarn deposits of McKinley–Tunas and Sierra Maestra zones, the Pb, Zn, Cu and barite deposits of Sabalo–Sierra Morena and Sierra Maestra zones, etc.

The increasing weight of magmatic or endogenous factors in the ore forming process characterizes the temporal metallogenetical differentiation: sedimentary–exhalative processes dominated the Jurassic, whereas vulcanogenic, hydrothermal and skarn ores were typical for latter periods.

A great involvement of pelagic material was probably reflected in increasing metallogenetic potential and ore variability.

An important conclusion is that a plate tectonics approach is not only able to explain the geotectonic evolution of the territory, but also to bring light on the complex and often so poorly understood ore forming evolutionary processes.

SIMON, A. A., L. A. GARCÍA and J. A. BÁRZANA, Empresa de Geología de Pinar del Río, Santa Lucía, Pinar del Río, Cuba

Jurassic Metallogenesis in Greater Antilles: Matahambre–Santa Lucía Ore District, Western Cuba

The Matahambre–Santa Lucía ore district, located in Northwestern Pinar del Río, Cuba, comprises numerous Cu, Pb and Zn sediment-hosted massive sulfide occurrences. Among them, the Matahambre (Cu-Pb-Zn), Santa Lucía (Pb-Zn-Ba) and Castellanos (Pb-Zn-Ba-Cu-Au) ore deposits should be mentioned. They are clustered in two parallel, elongated, southwest–northeast trending ore fields: Matahambre–Nieves (SE) and Santa Lucía–Castellanos (NW).

The sulfide mineralization is contained within the upper part of a thick deltaic sequence of Liassic(?) to Lower Oxfordian age, the San Cayetano Formation. The coarse clastic sedimentation characteristic to most of San Cayetano is followed by the deposition of carbonaceous black shales and minor siltstones, limestones, dolomites and gypsum beds (the Castellanos Member). Synchronous volcanic activity has not been clearly demonstrated, although some altered tuffitic material seems to be present within the mineralized beds. The whole San Cayetano has suffered a very slight regional metamorphism, manifested by widespread chloritization and sericitization.

Several overthrusted, northwest-dipping tectonic units alternate, and late steeply dipping, northwest–southeast and northeast–southwest normal and inverse faults complicate the structural framework.

The pelitic-calcareous Castellanos Member hosts a stratiform pyrite–sphalerite–galena–(barite) mineralization. Ore bodies consist of massive sulfide beds and lenses, practically placed within only one mineralized horizon at the bottom of the Castellanos sequence. They reach 1050 m long, over 400 m deep and 60 m thick (19 m thick in average) at Santa Lucía, and 700 m long, over 300 m deep and 85 m thick (20 m thick in average) at Castellanos. Vein and stockwork chalcopyrite–pyrite–(pyrrhotine) mineralization, predominant at Matahambre, is located within the underlying psammitic strata.

The stratiform mineralization usually consists of very fine-grained, rhythmically arranged, laterally continuous siderite–pyrite–sphalerite–galena laminae, 0.55 mm thick, at Santa Lucía, whereas in the vicinity of the cross-cutting ores, at Castellanos, Nieves and Matahambre (Cuerpo 70), large epigenetic metasomatism and recrystallization (pyrite–galena–barite) have obscured the primary ore bedding.

The cross-cutting footwall mineralization mainly consists of segregations, veinlets and veins of chalcopyrite, pyrite, and sometimes pyrrhotine (at Matahambre and Nieves ore deposits), with quartz and minor siderite in brecciated and partly silicified sandstones and siltstones. Ore accumulations are generally linear in shape, and they extend at Matahambre, the better example, over 1500 m down along the paleo-strike, and over 500 m across the profile, with up to 10 m thick.

Average grades are shown in Table No. 1. It can be observed that copper, as in Mt. Isa and Rammelsberg, becomes a significant component.

Sulphur isotopic studies in the massive sulphide ores have shown that $\delta^{34}S$ ranges within large limits, from -30 ‰ to +12.7 ‰ at Castellanos and from -35 ‰ to 17.6 ‰ at Santa Lucía, although extreme negative values predominate, between -20 ‰ and -24 ‰ at Castellanos, and between -28 ‰ and -32 ‰ at Santa Lucía. An almost permanent deviation from the equilibrium condition is observed ($\delta^{34}S_{py} < \delta^{34}S_{sp} < \delta S_{gal}$). The only exception is the pyrite–barite pair in the epigenetic baritic ore at Castellanos, for which ΔSO^{2-}_4-py is +30 ‰. The $\delta^{34}S_{BaSO4}$ is +20 ‰, similar to sea sulfate values.

Lead isotopes ratios from different parts of the Matahambre–Santa Lucía district have revealed surprising similarities, being grouped around the following values:

Pb^{206}/Pb^{204} = 18.73, Pb^{207}/Pb^{204} = 15.67, and Pb^{208}/Pb^{204} = 38.70.

In the $^{207}Pb/^{204}Pb$ versus $^{206}Pb/^{204}Pb$ diagram, those values plot close but above the Plumbotectonics "orogene" evolution curve, possibly reflecting a significant contribution of sialic crustal sediments.

Wall-rock alterations are only manifested by silicification and minor carbonatization (siderite) in the stockwork and veins proximity.

A sinsedimentary genesis is accepted for the sediment-hosted Matahambre–Santa Lucía district ore deposits. Feeder zones are represented by cross-cutting foot-wall mineralization. Other typical features are the metal and mineralogical zonation, the finely banded textures of the stratiform ores aside from the feeder zones, the stratigraphical and lithological controls, the sulfur isotopic composition, etc.

Low energy sedimentary–exhalative process began at the Upper Jurassic as a result of the early stages of subduction at the North American–Caribbean convecting plate margin. Large Paleogene northward overthrust displacements resulted in the extreme fragmentation of the original deposit, now dislocated within several tectonic nappes.

The Matahambre–Santa Lucía district ore deposits rank among the youngest and least metamorphosed world examples of the sediment-hosted massive sulfides class.

Table No. 1.- Average grades of the most important ore deposits at the Matahambre-Santa Lucía ore district.

Ore Deposits	Cu (%)	Pb (%)	Zn (%)	S (%)	$BaSO_4$ (%)
Santa Lucia					
a)	–	2,0	6,2	25	2,5
Castellanos					
a)	–	3,2	6,3	17	7,7
b)	1,9	–	–	29	–
Matahambre					
a)	–	1,7	4,8	*	*
b)	5,3	–	–	*	*
Nieves					
a)	–	1,8	4,6	26	*
b)	0,2	–	–	27	–

Where:
a) Stratiform ores; b) cross-cutting foot-wall mineralization;
*) Not available.

SIMONS, B., Universität Kiel, Kiel, Federal Republic of Germany

Temperature-Dependence of Point-Defect-Concentration in Fayalite

The redox state of the upper mantle not only is reflected in intercrystalline but also intracrystalline equilibria, as site occupancy, order-disorder phenomena, etc. Point defects play a major role and dominate transport properties at high temperatures.

In the present study, the temperature-dependence of point defects in fayalite has been determined.

Nakamura and Schmalzried (1983) established the following defect reactions to describe the relevant majority defects:

(1) $1/s)_2 + Fe^{*}_{Fe} + Si^{*}_{Si} + 1/2Fe_2SiO_4 = Fe'_{Si} + Fe^{\cdot}_{Fe} + 3/2\,SiO_2$
(2) $6Fe^{*}_{Fe} + SiO_2 + O_2 = 2V''_{Fe} + 4Fe^{\cdot}_{Fe} + Fe_2SiO_4$
(3) $Fe'_{Si} + Fe^{\cdot}_{Fe} = (Fe'Fe^{\cdot})$.

Defining the deviation of the silicon/cation ratio by

$$\bar{\xi} = \xi - 1/3 \text{ with } \xi = n_{si}/(n_{si} + n_{Fe})$$

and the anion/cation ratio by

$$\bar{\eta} = \eta - 4/3 \text{ with } \eta = n_o/(n_{si} + n_{Fe})$$

the complete analysis results in (Nakamura and Schmalzried, 1983):

$$\bar{\eta} = A \cdot fO_2^{1/6} \text{ and } \bar{\xi} = 1/4\bar{\eta} - B \cdot fO_2^{1/2},$$

where $A = 8/9\,[K_{2/16}{}^{1/6}$ and $B = 1/3\,\bar{K} \cdot K_2^{-3/2}$ with $\bar{K} = K_2^{3/2} \cdot K_1 \cdot K_3$ being the equilibrium constants of the above defect reactions ($K_1 = K_2$, $K_2 = K_3$, $K_3 = K_{11}$, $\bar{K} = K_{19}$ in Nakamura and Schmalzried's paper).

Nakamira and Schmalzried determined the weight change of a fayalite with constant Si/Fe according equation:

$$\delta = (2)^{3/10} \cdot (3)^{-9/10} \cdot \bar{K}^{1/10} \cdot fO_2^{1/5}$$

In the present study, the temperature-dependence of these equilibrium constants was determined by EMF measurements.

One result is of primary importance: For a defined fayalite, the oxygen fugacity does not change with temperature within the stability field of that fayalite, which may not be confused with the fO_2 – T space between the quartz/fayalite/magnetite and quartz/fayalite/iron buffer curves. Beyond a certain fO_2 – T a destabilization of the system resulted and fayalite changed its Si/Fe ration. This condition is equivalent to an irreversible weight change as observed in Nakamura and Schmalzried's experiments and marks the limiting conditions for the stability of a fayalite with discrete defect concentration.

By the present study, these conditions were monitored in the temperature range 900–1130 °C. From the temperature independence of the oxygen fugacity of a fayalite with a given defect concentration within its stability field and the fO_2 – T limit of that field K and K2 are evaluated. The results are:

$$\log \bar{K} = -9.227 \text{ and } \log K_2 = -2.794 - 6376/T$$

and thus $\log A = -0.7175 - 1063/T$ and $\log B = -5.5131 + 9564/T$.

The above defect reactions are formulated with $a_{sio_2} = 1$ and may be transformed to different standard states. Finally an internally consistent set of 16 equations results (Simons, 1986) and the $\xi - \eta - f_{O2}$ – T hyperspace of fayalite is calculated.

In Figures 1a–c, two parameters are plotted in the temperature range 1130 °C–950 °C, where the upper "sabre-shaped" field represents the 1130 °C stability field and the lower one that at 900 °C. These fields are limited by the activities of the components specified. Both fields are interconnected by 2 lines fixed by the buffer conditions indicated. In Figure 2, lines with constant anion/cation ratio are plotted. Within the field between the heavy lines the condition $\bar{\xi} = 0$ is fulfilled.

The remarkable feature is that a high-temperature fayalite has no low-temperature analogon as can be deduced from Figures 1a–c. Figure 2 shows that a fayalite of a given defect concentration "leaves" its stability field by a change in temperature of somewhat more than 100 °C; it is destabilised as observed in the present EMF-measurements. Various exsolution features can be discussed based on the thermodynamical treatment presented.

REFERENCES

Nakamura, A., and H. Schmalzried, 1983, On the nonstoichiometry and point defects of olivine: Phys. Chem. Minerals, v. 10, p. 27–37.

Simons, B., 1986, Temperatur-und Druckäbhangigkeit der Fehlstellenkonzentration der Olivine und Magnesiowüsite: Habilitation Thesis, Universität Kiel.

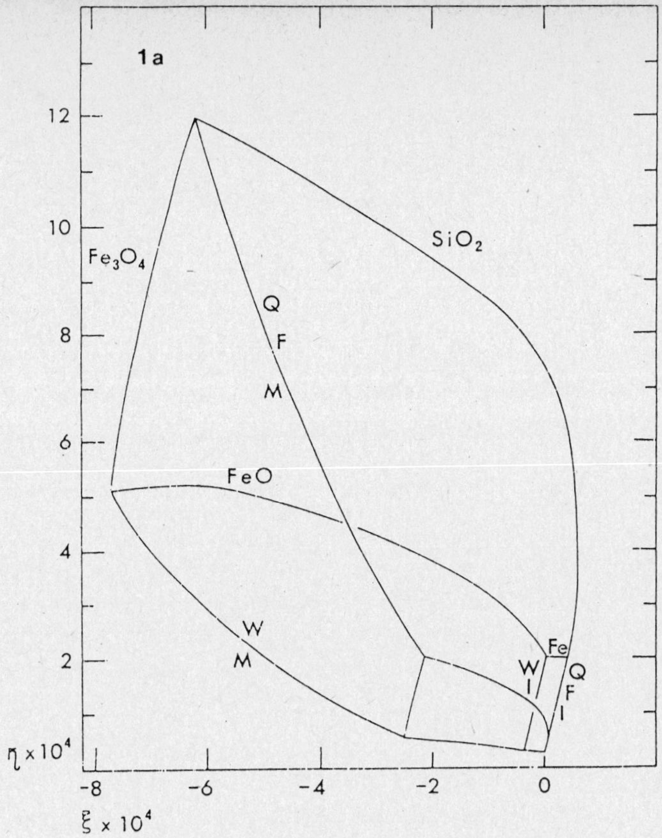

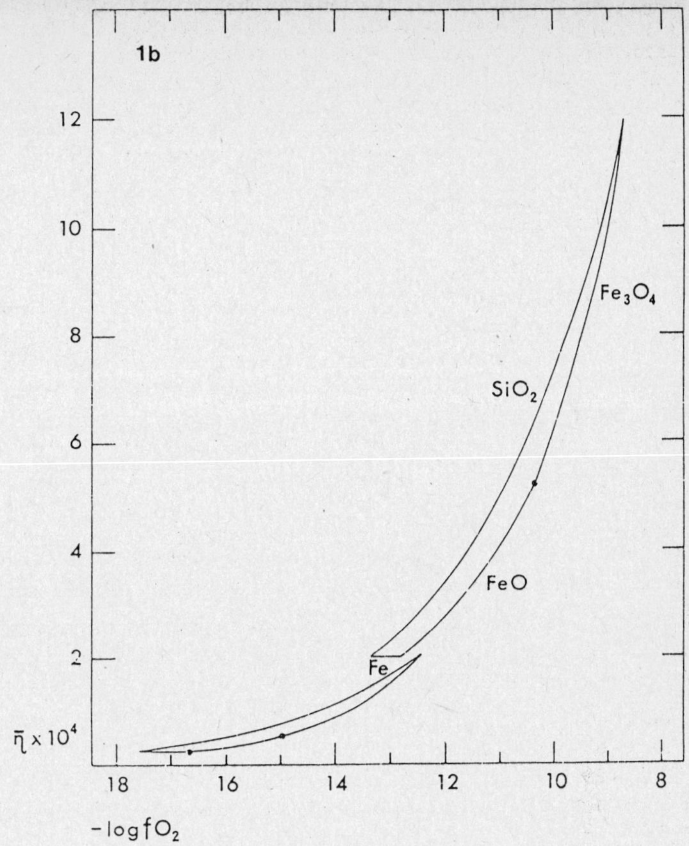

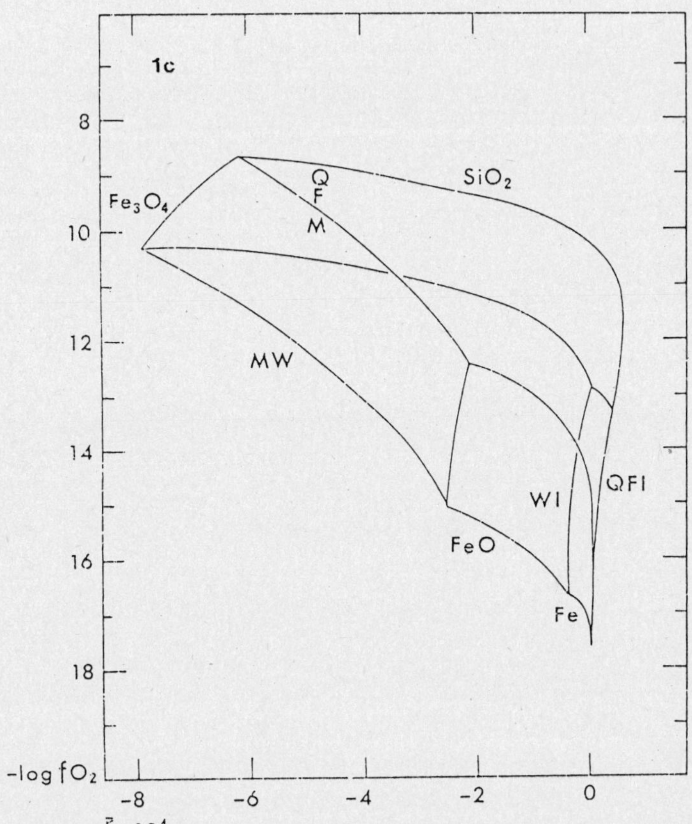

Figure 1

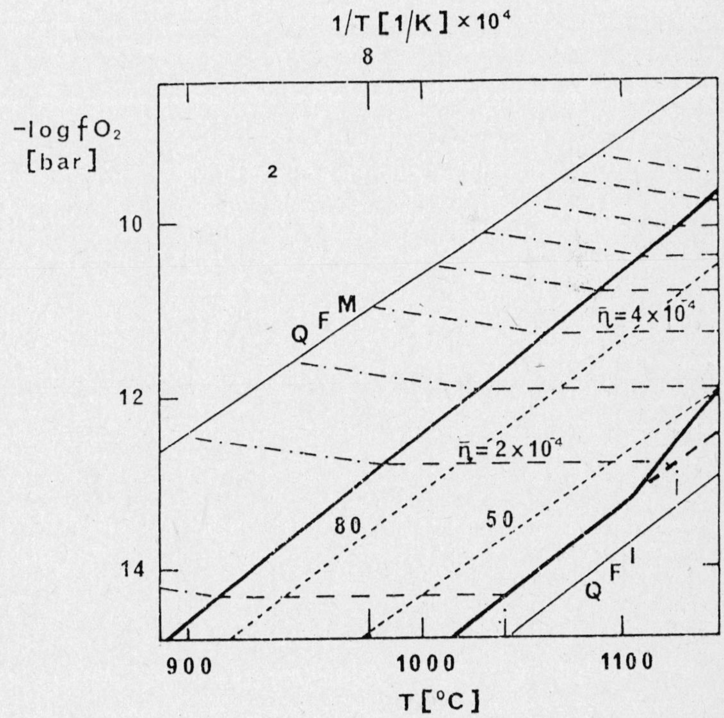

Figure 2

SIMPSON, C., The Johns Hopkins University, Baltimore, MD

Finite and Incremental Strain Fabrics in Ductile Shear Zones

The mechanisms of fabric formation in quartzofeldspathic ductile shear zones vary markedly over a narrow range of deformation conditions. Narrow, semibrittle mylonite and ultramylonite zones in lower greenschist facies rocks commonly exhibit single finite strain-related foliations in outcrop but may also show incremental strain facies in thin section. Broad zones of two-foliation (S-C) mylonites are more typically found in middle greenschist to lowest amphibolite facies rocks. Differences in the relative strengths and grain sizes of dynamically recrystallizing mineral phases and their prograde and retrograde reaction products are of prime importance in determining the types of fabrics developed.

Unaltered feldspar is stronger than quartz under lower greenschist facies conditions and often contains quartz-filled extension cracks indicating significant local material transfer. Pressure solution surfaces are more difficult to detect but may be manifest as undulating mica-rich zones. Oriented fiber growth in crack-seal veinlets may impart an abnormally high degree of lattice preferred orientation (LPO) during subsequent low strain deformation. The retrograde reaction of feldspar to fine-grained white mica and quartz produces material that is weaker than the original igneous quartz and permits ductile deformation zones to surround the relatively rigid quartz grains. In the temperature range of ~250°–350°C, increased mobility of dislocations and grain boundaries in quartz allows the steady state development of lattice and grain shape fabrics whose symmetries are related to the incremental strain ellipsoid. Neocrystallized and recrystallized mica grains and prophyroblasts ("mica fish") also show a comparable symmetry to that of the incremental strain axes. Where a strong rotational strain component is present, these grain shape fabrics form a secondary foliation that is oblique to the bulk flow plane.

With the onset of dynamic recrystallization in feldspar at temperatures above ~450°C, small recrystallized feldspar grains are swept away from the mantles of more rigid clasts to form distinctive feldspar-rich bands. These bands are only a few grains wide in lower temperature, higher stress zones but widen as subgrain and recrystallized grainsize increase with higher temperatures and lower stresses. Grain shape fabrics are rarely found in recrystallized feldspars but may be preserved where strain rates were high with respect to recrystallization rates. Quartz grains show considerable recovery, but grain shape fabrics are still common with ribbon-like aggregates. These aggregates often have a greater volume than original igneous grain aggregates due to coalescence of phases with similar viscosities during deformation. The replacement of alkali feldspar by plagioclase and quartz myrmekitic intergrowths occurs along feldspar grain margins that face the incremental shortening direction. These intergrowths are therefore useful as kinematic indicators, and their fine-grained, equigranular nature serves to further localize strain. Myrmekite formation is thus a high-temperature reaction-softening process. Ribbon-like mica grain aggregates result from the propagation of kink bands through large mica porphyroblasts and concomitant cataclastic flow and/or recrystallization along kink band boundaries.

During the initial stages of S-C mylonite development from a homogeneous granitic rock, alternating quartz-rich and quartz-poor bands form a single (S) foliation that tracks the finite strain axes throughout the subsequent deformation. The secondary shear (C) planes show a considerable grain-size reduction, which suggests that they develop in response to increases in the incremental strain rate. Similarities in mineral chemistries and deformation mechanisms to those in adjacent S planes suggest that the C planes did not act as local fluid conduits.

Above ~550°C, quartz, mica, and feldspar grains are fully recovered and no longer preserve asymmetrical grain shape fabrics related to incremental strain, but LPOs are strongly developed in all these minerals. S-C mylonites are uncommon at this metamorphic grade although a statistical analysis of mica and amphibole grain orientations may indicate the presence of a secondary foliation whose symmetry can be used to deduce shear sense.

Deformation of an incompletely crystallized pluton often produces strongly linear, weakly planar fabrics defined by aligned euhedral feldspar megacrysts and by mafic minerals in elongate aggregates. Within these aggregates, individual mica or amphibole grains are unoriented. These high-temperature fabrics are often irregular in both intensity and distribution and the distinction among incremental and finite strain fabrics has not yet been determined. Large, very slightly deformed and irregular-shaped grains of quartz commonly fill interstices between feldspars. If LPOs are detectable, they indicate deformation on slip systems consistent with very high homologous temperature and very low normalized stress deformation. Feldspars preserve igneous textures and original compositional zoning. They may form myrmekite at their margins due to differential volume changes during dynamic or static cooling of the pluton. If static, the myrmekite shows no preference as to nucleation site geometry, but in the dynamic case, its location reflects the ambient stress field.

Determination of incremental versus finite strain-related fabrics is essential for any kinematic analysis of shear zone rocks. Porphyroblasts that grow during deformation will reflect different geometries of inclusion trails according to which fabric they overgrow. Misinterpretation of an incremental strain fabric for a finite strain fabric may in extreme cases result in a misinterpretation of the overall sense of shear across the zone.

SIMPSON, E. L., Kutztown University, Kutztown PA, and K. A. ERIKSSON, Viriginia Polytechnic Institute and State University, Blacksburg, VA

Depositional Processes and Sea Level Control on Sequence and Parasequence Development in Early Cambrian Hampton and Erwin Formations of Virginia, USA: Example of Early History of Passive Margin

The Chilhowee Group in southern and central Virginia consists of the Unicoi, Hampton, and Erwin Formations. The late Proterozoic to early Cambrian Unicoi Formation overlies rift-generated Catoctin Formation greenstones and Grenville-age basement in central Virginia and Grenville-age basement and rift-generated bimodal volcanics and glaciogenic sediments of the Mt. Rogers Formation in southwestern Virginia. The lower Unicoi Formation comprises alluvial feldspathic sandstones and conglomerates and basalts deposited in a rift setting, whereas the upper Unicoi Formation consists of tide- and wave-influenced, shallow-marine shelf sandstones reflecting the incipient development of an east-facing passive margin bordering the Iapetus Ocean. The early Cambrian Hampton and Erwin Formations record siliciclastic shoreline and shelf sedimentation along this passive margin. The overlying Shady Dolomite reflects the evolution of the passive margin from a ramp to rimmed shelf geometry. During the middle to late Ordovician Taconic Orogeny, the passive margin was deformed into a foreland fold-and-thrust belt.

Occurrence of both the Hampton and Erwin Formations on a number of thrust sheets permits an approximate across-strike reconstruction of the Chilhowee passive margin. The Hampton and Erwin Formations represent an overall offlap package consisting of up to five transgressive-progradational sequences bounded by unconformities and their correlative conformities. Sequences vary systematically between thrust sheets with replacement of shallow-by deeper-water depositional environments from west to east.

Two depositional systems are represented in the Hampton and Erwin Formations: tidal flat (TF) and storm-dominated shelf (SDS); the latter consists of four facies: outer shelf facies (OSF), distal inner shelf facies (DISF), proximal inner shelf facies (PISF), and shoreface facies (SF). OSF include black mudstone and thin to medium beds of sandstone, siltstone, and mudstone that resemble a wide range of Bouma beds. Sedimentation in the outer shelf was dominated by geostrophic flows generated by storms and suspension settling of mud under fair-weather conditions. Thin- to medium-bedded, medium- to fine-grained sandstones, siltstones, and mudstones, along with minor glauconitic beds, compose the DISF. The presence of microhummocky cross-stratification, form-discordant wave ripples, and climbing wave-ripple laminations indicate that DISF resulted from storm-wave reworking below fair-weather wavebase. PISF are dominated by massive, thick-bedded sandstones capped by straight-crested megaripples or hummocks. Hummocky cross-stratified sandstones are minor. Massive beds are considered to be primary and the result of rapid sedimentation from suspension during storms with destruction of primary structures by shearing. Wave periods reconstructed from straight-crested megaripples vary from 7.5 to 13.1 s. Maximum calculated water depths vary from 50 to 66 m. PISF consist of detritus transported from the nearshore and reworked in situ by combined flow currents generated by the complex interaction of storm-surge, ebb-return flow, and waning storm waves. SF comprise very thick-bedded, coarse-grained sandstones with intense bioturbation. Trough and tabular-tangential cross-bedding are present. Skolithos linearis is the only trace fossil preserved. SF sediments display evidence of reworking by longshore currents generated by shoaling fair-weather or waning storm waves. The tidal-flat depositional system is represented by thin- to medium-bedded, coarse-grained sandstone with minor mudstone. Erosional contacts separate the TFF from underlying and overlying SF. Exhumed bedforms include current ripples, wave ripples, and Type I and Type II megaripples. Subaerial exposure is reflected by run-off microdeltas, rills, and oversteeped lee sides of megaripples. Sedimentary structure types coupled with paleocurrent data indicate that tidal-flat sedimentation was dominated by flood currents and late-stage ebb runoff.

Depositional systems and facies are arranged in 65–225 m thick sequences that contain 0.30–6.5 m thick parasequences. Sequences are recognized on the basis of changes in depositional environments, whereas the parasequences are delimited by abrupt changes in bed thickness and types of sedimentary structures. Parasequence thickness decreases form SF to OSF. The Erwin Formation in western thrust sheets contains sequences that consists of the folowing vertical transitions: SF → unconformity → TF unconformity → SF → PISF and SF → PISF → SF. In eastern thrust sheets, the Erwin Formation contains the following vertical transitions: DISF → PISF; and OSF → DISF → PISF. The Hampton exhibits OSF → PISF → SF transitions in western thrust sheets, whereas eastern thrust sheets contain only OSF.

Depositional systems and their constituent facies are associated in system tracts that record fluctuations in relative sea level. Transgressive system Tracts developed during initial rise in sea level and consist of: (1) thin DISF with detrital glauconite, and OSF above conformable sequence boundaries in eastern thrust sheets, and (2) TF followed by SF → PISF above unconformities in western thrust sheets. Associated with maximum rates of transgression, black mudstones (condensed sections) accumulated in distal inner-shelf and outer-shelf settings. Highstand System Tracts reflect a decrease in rate of sea-level rise and represented by (1) OSF → DISF → PISFI and DISF → PISF in eastern thrust sheets, and (2) PISF → SF and OSF → PISF → SF in western thrust sheets.

The Hampton and Erwin Formations reflect outbuilding of the Chilhowee passive margin during regression resulting from a decrease in rate of thermal subsidence. Sequences record lower-order sea-level changes superimposed in the highstand systems tracts.

SIMPSON, P. R., J. HURDLEY, and J. A. PLANT, Geochemistry and Hydrogeology Directorate, London, United Kingdom, and G. C. LALOR, and H. ROBOTHAM, University of the West Indies, Kingston, Jamaica

Geochemistry for Agromineral Studies in the Caribbean Region

A joint research program is presently underway to develop and apply regional geochemical studies to agromineral investigations in the Caribbean Region. Orientation studies for a Regional Geochemical Survey of CARICOM Countries in the Caribbean Region have been undertaken in Jamaica based on six sites which are widely distributed throughout the island and which have been selected to represent, as far as possible, the widest range of geological and environmental conditions which occur in the Caribbean Region. The primary objective of the orientation study is the design of a geochemical sample type or types which can be selected systematically throughout the region, in natural drainage systems where surface drainage systems are generally poorly developed. The sample type selected during this study will be required in sufficient abundance at each sampling site for analysis by a wide range of geochemical methods and to form the basis of a definitive sample library. Furthermore, the sample type selected will be designed to optimize the potential geochemical sensitivity of the combined sampling and analytical system in order to provide an effective basis for Regional Geochemical Mapping in which results from individual sites can be presented with a high degree of confidence in their reproducibility. The control samples collected will therefore be used to prepare standards to monitor seasonal/temporal fluctuations in trace element composition and for calibration and re-calibration of the geochemical database against international standards. Reliable maps generated from this data would indicate the distribution and relative abundance of elements of agricultural (e.g., Ca, Mg, K, P, S, B, Mo, Se, Fe, Zn, Ni, Co), environmental (e.g., Al, Pb, Hg) and economic significance (e.g., Au, PGE); the latter to avoid sterilization of resources by inappropriate development. Sampling of the six sites described in this report has been completed on schedule and the laboratory preparation and analysis has commenced at the Centre for Nuclear Sciences. A full geochemical analysis of these samplings will be undertaken to provide data on which to base a geochemical sampling strategy designed for the Caribbean region.

SIMSEK, S., Mineral Research and Exploration General Directorate (MTA), Energy Department, Ankara, Turkey

Evolution of Büyük Menderes Graben and its Geothermal Energy Possibilities

The regional grabens of Western Anatolia of Turkey have been developed due to neotectonic activity from Miocene until now. East-west aligned B. Menderes, K. Menderes, Gediz, Bakirçay, Edremit, and Simav Grabens have been formed as a result of the doming uplift and mainly north-south tensional forces. The grabens are generally asymmetric but symmetric in some places. A typical step faulting system was determined at the graben flanks. The thickness of the Neogene and Quaternary deposits in the grabens reached up to 2500 m. Geological findings and interpretations have been supported by geophysical, geochemical survey, and drilling data in the grabens.

In this paper, tectonic and geothermal models will be discussed on the basis of geological and geophysical (gravity, resistivity, seismic), geochemical (radon measurement, etc.), shallow (100–200 m) and deep drilling (285–2398 m) data that were realized for geothermal energy exploration in the B. Menderes Graben. Detailed investigation of the graben will be useful for the exploration of other grabens in the region.

The basement rocks in the region are Paleozoic metamorphic rocks of the Menderes Massif. These rocks are composed of mainly gneisses, schists, marble, and quartzite units. The Neogene and Quaternary deposits overlie the basement with an angular unconformity.

The B. Menderes Graben that includes the Aydin nad Denizli provinces is located south of the Menderes Massif. The length of the graben is 200 km, the width is 5–40 km and it has a continuity in the Aegean Sea.

In the region, crust thinning and magma approach have developed in places together with graben formations. Belts suitable for the geothermal systems have developed especially along dislocations or intersection zones of faults. The slip along the northern flank of the faults where the graben is effective is more than the southern flank (approximately a total of 3000 m). Many hot springs and fumaroles are located along these lines. However, there is thermal activity also along the graben faults on the southern flank near Buharkent (Çubukdagi) where the graben is partially symmetric.

The geothermal areas in the region are located especially along the main graben fault where the basement units crossed to young sediments and syntetic and antitetic faults that occurred simultaneously with the main fault.

The hard and brittle lithologies have gained secondary permeability due to active graben tectonism in the region. The first geothermal reservoir in Denizli is limestone (Pliocene) and conglomerates (Miocene) in Aydin. On the other hand, the second reservoir, composed of marble–quartzite unit of the basement, can be used economically.

The geothermal fluid rising along the main faults of the grabens in the field is collected in the permeable zones and moved to the middle parts of the grabens.

These important geothermal fields have been discovered as a result of geothermal exploration carried out by the Mineral Research and Exploration General Directorate of Turkey (MTA) along the B. Menderes Graben. These fields and their reservoir temperatures are as follows: Denizli–Kizidere (212 °C), Aydin–Germencik (232 °C), and Aydin–Salavatli (171 °C).

Various utilizations, especially electricity production, have been initiated and planned in these fields. Explorations are still being carried out in the B. Menderes Graben and other grabens according to the model established in these fields.

SINGH, Dr. A. K., South T. T. Najar, Bhopal, India

Narmadasagar and Sardarsarovar Dams: Precursor of Magnitude >7 Earthquakes

The Narmada, last wild river of India, has been chosen for a river valley development scheme, under which 3000 small, 350 medium and 35 large dams are to be constructed. Two dams, namely the Narmadasagar and Sardarsarovar, will be largest in Asia on completion after twenty years at a twin cost of Rs. 15,000 Crore only. Round the globe, dams are known to act as a conduit to initiate seismicity, as the percolating water destabilises the existing equillibrium between the strength of rocks, pore-pressure and effective pressure.

The Sardarsarovar is to be constructed exclusively on Deccan Trap basalts and the Narmadasagar on quartzite, with a major part of reservoir on Deccan Trap basalts. The basalts have shown a remarkable tendency to drastically alter their mechanical properties on introduction of water (Gupta, 1976). The basalts also bear the unique quality of transmitting earthquake energy waves through them with maximum ease, of all the known rocks. This is why earthquakes have occurred throughout India on dams constructed on Deccan Trap basalts. The latest case being the Bhatsa dam (7 km from Khardi, Thane district, Maharashtra) where on 17 July, 1983, an earthquake of magnitude 4 and then on 15 September, 1983, another earthquake of magnitude 4.8 occurred. On the LANDSAT imagery the lineament patterns of the Bhatsa area, the Narmadasagar area and the Sardarsarovar area have shown striking identicalities, thus strongly pointing out the oncome of seismicity.

On the Indian Peninsular Shield the Narmada is recognized as a distinct seismic zone, as numerous earthquakes of magnitude 5 or more had occurred in the region (Umesh Chandre, 1977), which is unusual as shields are considered stable areas. The latest earthquake in this zone occurred in November, 1985. Geologically, the Narmada being an aborted rift system, is undoubtedly the weakest zone of the Indian Peninsula, along which both the blocks have moved frequently in geological past. Seismicity tends to rejuvenate and creep in pre-existing zones of weakness (Sykes, 1979). From plate tectonic considerations, the Narmada is an arm of the Cambay triple junction (other arms being Panvel, Girnal and Cambay) and in triple junction seismicity is a usual phenomenon. It explains why a very small dam, Koyna, exhibited the maximum known reservoir induced seismicity of 6.3 magnitude, since it is situated on the Panvel arm of the Cambay triple junction. The Koyna dam (incidently again on Deccan Trap basalts) experienced the earthquake of 6.3 magnitude on 10 September, 1967, claiming 200 lives and leaving 1500 injured, 10,000 homeless and uprooted Koyna town.

Realizing the importance of reservoir induced seismicity, UNESCO formed a SPAWLR (Seismic Phenomenon Associated With Large Reservoirs) Group in 1971, which in its concluding report furnished guidelines for detailed seismic monitoring of the areas prior to dam construction. ICOLD has also recommended stepwise systematic seismic monitoring prior to large dam constructions. Unfortunately, not a single specified study, above mentioned, has been carried out in case of these two dams.

The truth that these dams will be extremely prone to seismicity is further substantiated by the fact that the Sukta dam (just 85 km from the Narmadasagar dam) though built on quartzite, is constantly demonstrating seismicity since 1978. Moreover, the Narmada zone is a typical case for showing "fossil-plate" seismicity, a term coined by the author (Singh, 1984). The provision of unusually high seismic coefficients for Narmadasagar and Sardarsarovar dams has been considered a safety measure against earthquakes here. Had the Narmada been an isolated seismic zone, the strategy would have succeeded to certain extent. But the Narmada is connected to the Owen Fracture Zone, which in turn is linked to spreading center rendering the provisions futile, as being a part of global structure the movements cannot be checked. Moreover, providing seismic coefficients bears its inherent limitations, seismicity being a dynamic force, whereas in the laboratory the seismic coefficient values are determined after carrying experimentation on static force, equivalent to the dynamic forces. To cap it all, both the dams are in dangerously close proximity.

Very high levels will be stored in both the dams and a correlation between water levels and earthquake frequency has already been established in case of Koyna, further accentuating the risk of seismicity in both these dams.

Thus, on the basis of the construction of these dams in seismically active zone, geologically weak zone, on the Deccan Trap basalts, very high water levels to be maintained and their close proximity—all point out that after completion they will introduce a reservoir induced seismicity of magnitude 7 or more. Madhya Preadesh, which has a dubious distinction in claiming world's worst industrial disaster (Bhopal gas tragedy) should be ready to accept the honor of being the region of worst reservoir induced seismic event, after completion of these two dams.

REFERENCES

Gupta, H. K., and B. K. Rastogi, 1976, Dams and Earthquakes. Elsevier Publication, p. 236.

Sykes, L. P., 1979, Intra plate seismicity, reactivation of pre-existing zones of weakness, alkali magmatism and other magmatism, Rev. Geophy. Space Phy., v. 16, p. 621–688.

Singh, A. K., 1984, Fossil plate seismicity: origin, implication and evidence. Symp. "International Association of Seismology and Physics of the Earth's Interior" (Abstracts Volume), p. 76.

Umesh Chandra, 1977, Earthquakes of Peninsular India—A seismotectonic study, Bull. Seism. Soc. Am., v. 67 (5), p. 1387–1413.

SINGH, Dr. A. K., South T. T. Najar, Bhopal, India

An Evolutionary Plate Tectonic Model of Indian Shield

Shields are the mosaics of protocontinents fused together. It was earlier proposed that the three protocontinents, namely the Aravalli, Dharwar and Singbhum, merged along the Narmada–Son–Mahanadi lineament giving rise to the Indian Shield (Naqvi et. al., 1974). However, the catechism is:

1. There is no compelling evidence to believe the unification of these protocontinents merely through nucleation, accretion and merger.
2. It fails to explain the identicalities of granite-greenstone terrane and high grade terrane to modern plate boundaries assemblages.
3. Accretion is a very slow process in itself whereas the crustal growth of the Indian Shield was attained in a relatively very short span (about 2.5 Ga ago) of time, and remain more or less the same since then.

On the other hand, it is also proposed (Radhakrishna, 1983) that in all probability the Archaean crust formed a single unit which was later subjected to tapherogeny and collision. He has neither elaborated his statement nor has he supplemented it with field evidence to strengthen his views. Thus there is a need to revise the present status of knowledge regarding the evolution of the Indian Shield.

Based on several criteria like volcanism, ophiolite suites, metamorphic zonations, presence of komatiites, occurrence of deep sea pelagic sediments and melanges, occurrence of flysch and molasse sediments, heatflow and geothermal gradients, polarity of tectonic and magmatic manifestations, asymmetry in the thrusting motion and the folding, asymmetry of the thrusting polarity in the distribution of metallogenic ores, geochronology of tectono-magmatic sequence, photogeologic lineaments and remotely sensed data, structural evidence, seismic data, stratigraphic correlation and provenance studies; the existence of imbricate sutures is revealed in the Aravalli Orogen, Rajasthan and in the Dharwar Craton, Karnataka (Singh, 1987) as a result of complex Precambrian plate tectonic processes. Similar studies for Singbhum Craton are showing encouraging results.

The Vindhyan and Marwar sedimentary basins represent the southeast and northwest molasse of the Aravalli Orogen respectively. Similarly the Cuddapah basin represents the molasse corresponding to the Dharwars.

The Deccan Trap volcanicity is genetically linked with the distansional phase in the continental crust of India. The Indian plate come into existence after the collision of the Aravalli, Dharwar and the Singbhum protocontinents. Though these three protocontinents became one unit (the Indian proto-plate), configuration of their Eular Poles was in different directions, as they were originated separately. The Aravalli Eular Pole was in the northwest, the Dharwar Eular Pole was in the south and the Singbhum Eular Pole was in the northeast. Due to this divergent orientation of rotational Eular Poles, the junction between the three protocontinents became a major stress field. The Indian proto-plate started its movements towards the Eurasian plate at about 75 ma ago. This movement caused the initiation of reactivation in these until now inactive three Eular Poles. As the movement of the Indian proto-plate continued, bit by bit, the three Eular Poles got accelerated slowly. Since the three Eular Poles were pushing the Indian proto-plate in their different directions, it caused the crust to open in a very large number of deep-seated fissures and crevices about 66–60 ma directly linked with upper mantle, the mantle matter immediately flooded the surface in the form of the Deccan Trap lava flows. It is testified by the fact that throughout the Deccan Traps the parent source is proved to be peridotite/eclogite/picrite magma, which is the major component of the heterogenous upper mantle. Moreover, the boundaries of the Aravalli, Dharwar and Singbhum meet each other underneath the Deccan Traps paying credence to above hypothesis.

The encounter of the Indian plate with the Eurasian plate resulted in formation of the Gigantic Himalayas between the two. This slowed down the movement of the Indian plate considerably which is evident from the fading out of magnetic anomalies in the Indian Ocean. The closing of the remaining portion of the Tethys Sea took place during the Eocene period. Once all the oceanic crust beneath the Tethys Sea was consumed, further subduction (a-subduction) was not possible. This resulted in interruption in the mantle convection cycle and hence in its stress pattern. In response to it, a new spreading center trending northwest was created. A new island arc system, the Burmese–Indonesian arc was thus formed to consume the Indian plate for completion of the convection cycle.

Thus the evolution of the Indian Shield was complete.

REFERENCES

Naqvi, S. M., V. Divakar Rao, and Hari Narain, 1974, The protocontinental growth of the Indian Shield and the antiquity of its rift valleys, Precamb. Res., v. 1, p. 345–398.

Radhakrishna, B. P., 1983, Archaean granite greenstone terranes of the South Indian Shield, Mem. Geol. Soc. Ind., v. 4, p. 1–46.

Singh, A. K., 1987, Precambrian plate tectonics of the Aravalli Orogen and LANDSAT interpretations of the Great Boundary Fault, Rajasthan, by remote sensing techniques. Unpublished Ph.D. thesis, p. 238.

SINGH, I. B., Lucknow University, Lucknow, India

Sedimentation Pattern in Gangetic Plain, India: Active Foreland Basin

The Gangetic Plain is part of the active foreland basin (peripheral-type) developed on the underthrusting Indian Plate, in response to the thrust-fold belt loading in the Himalaya. It is part of the Indo–Gangetic Plain and extends between Delhi ridge in the west and Rajmahal ridge in the east, occupying an area of about 250,000 km^2 between longitude 77 °E and 88 °E and latitude 24 °N and 30 °N, and about 350 km wide in a north-south direction. During thrust-fold loading tectonics in the Himalaya, the Son–Narmada lineament much to the south of the foreland basin was reactivated, causing uplift of the Bundelkhand–Vindhyan Plateau and development of northerly slope. The rate of subsidence of the old, rigid, and cold crust of Indian shield was also low and sediment input by rivers high, so that no marine transgressions of Neogene–Quaternary time could enter into this foreland basin.

Gangetic Plains contain major rivers originating in the Himalaya, some rivers draining the Bundelkhand–Vindhyan Plateau in the south, and a few small rivers originating within the alluvial plain. The rivers in the northern Gangetic Plain flow southwards following the regional south-sloping planation surface formed by expanding-contracting alluvial fans. The Central Gangetic Plain shows rivers flowing in east and southeast directions following the easterly sloping regional planation surface on high plateau, and active lineaments and faults developed in the alluvium in response to Himalayan tectonics. These syndepositional faults are normal faults, showing contrasting sediment thicknesses and facies development on either side of the major river valleys oriented northwest-southeast. The southern Gangetic Plain shows rivers flowing in north to northeast directions following the tectonic lineaments of the basement rocks. The rivers are mostly entrenched, indicating uplift of the region as peripheral bulge of the foreland basin.

Geomorphologically, the most important feature of the Gangetic Plain is its east-sloping regional planation, plateau surface (referred to as T_2 surface). This surface shows common occurrence of calcrete horizons, drying lakes and ponds, and abandoned channel belts. The present-day active channels of this surface are small underfit streams carrying very fine sand showing tight meanders, occupying earlier river valleys of large loose meanders. Some of the drying lakes contain regionally extensive deposits of shell layers of about 1 m thickness. In the northern Gangetic Plain, a south-sloping geomorphic surface is present, formed by the activity of the alluvial fans. Present-day belt of alluvial fans is only 20–30 km wide; but there is evidence of 100 km or more wide alluvial fans in the past. The regional plateau surface is broken by a few major streams showing rather broad (30–40 km) river valleys (T_1 surface), occupied by entrenched narrow present-day active channels (T_0 surface), only a few km wide. T_1 surface show large meander scars, while the rivers of T_0-surface of the same river valley are either braided type or show smaller meander waves. Underfit streams, abandoned channel belts, and large meander scars in broad abandoned river valleys indicate one or more periods of much more humid climate in the past than today. All these geomorphic surfaces of the Gangetic Plain are aggradational, where active net deposition is taking place, though at variable rates. Though absolute dates are not yet available, it is argued that age of formation of T_2 surface is 120 Ka, T_1 surface is 25–30 Ka, and T_2 surface is 15 Ka. All these surfaces have received a significant sediment cover during terminal Pleistocene and Holocene. Radiocarbon date of a shell bed in an abandoned lake indicate deposition of about 3 m in the past 8.5 Ka.

Study of subsurface sediments suggests a strongly asymmetrical fill, thinning out southwards. Subsurface sequence in northern Gangetic Plain is mostly made up of gravels, medium to coarse sand altering with muddy sediments, and rare calcrete-bearing clays. Thickness of the sediment fill is 2000 m or more, the sediments derived exclusively from Himalaya. In the central Gangetic Plain, the most dominant lithology is clay with calcrete with thin layers of fine sand, medium sand, and gravelferous horizons in deeper levels. The sediments are derived exclusively from Himalaya, and thickness is 500–1000 m. In the southern Gangetic Plain, basement is only a few tens of meters deep. Just above the basement is a sequence of arkosic gravel-coarse sand, derived from the Peninsular region. This succession is followed by a sequence of clay with calcrete horizons and a few medium to fine sand layers, all derived from the Himalaya. This implies that the Gangetic Plain is expanding southwards in response to the thrust fold loading in the Himalaya. The southern derived arkosic material makes a thin sequence just above the basement, followed by a thick succession of Himalayan-derived lithic sands and clays. Mountainward shift of the alluvial fan belt and a fining-upward sediment cycle in the topmost tens of meters of Gangetic Plain deposits suggests that during terminal Pleistocene–Holocene the sedimentation in the Gangetic Plain was controlled by the fluvial processes and climatic fluctuations, with a subdued role of tectonics in the Himalaya. It is argued that within a tectonically active foreland basin several tens of meters to 100 m thick sediment deposits can be formed exclusively by fluvial processes during periods of tectonic quiescences.

SINGH, I. B., Department of Geology, Lucknow University, Lucknow, India

Facies Architecture of Karewa Deposits, Kashmir: Lacustrine Intermontane Basin

Kashmir valley is about 140 km long and 40 km wide intermontane valley, occupied mainly by a 1300 m thick succession of terrigenous clastic sediments. The Karewa basin developed during collision of Indian and Asian Plates sometime during Middle Miocene and Early Pliocene, and aligned northwest–southeast direction. During sedimentation and also afterwards, southwest flank of the valley (Pir Panjal Range) has risen more than the northeast flank of the valley; consequently better and extensive exposures are present in the southwest part of the valley. Stratigraphically, the sedimentary sequence is divided into a lower part—Hirpur Formation—and an upper part—Nagum Formation.

A vast but shallow lake came into existence right from the initial stages of Karewa basin sedimentation. The lake was mostly 5–10 m deep, maximum may be 30 m, and fed by rivers at various points coming from the surrounding mountainous area and making lake deltas.

The major lithofacies of the Hirpur Formation are grey, non-laminated clay (deposited in the deeper part of the lake), rhythmically laminated sand and mud (deposited under wave-effected shallower part of the lake), sheet sand layer (deposits of delta front), lenticular sand bodies (deposits of lake beach), lignitic mud (deposits of lake–swamp), conglomeratic sand (deposits of braid stream), and thick conglomerates (alluvial fan deposits). The vertical sequences are mostly progradational and/or related to (a) lake water changes of short duration, (b) delta lake switching or change in sand supply, (c) major sequences of lake shrinkage related to tectonics in source area. The lake of Hirpur Formation time possessed a marginal belt of swamps, followed lakeward by zone of lake beach. The deltas were of bird-foot type; the climate was warm, humid, suitable for the preservation of organic matter.

There are two major and several minor tectonic events in Pir Panjal range during deposition of the Lower Karewa sediments. The second major tectonic event caused not only the uplift in the source region, but also uplift of the southwestern part of the Karewa basin, making the initial large lake to shrink and occupy only the northeastern part of the valley. This is coupled by increase in the sediment input and the increased aridity in the climate.

The major lithofacies of the Nagum Formation are yellowish non-laminated mud (deposits of the deeper part of the lake), yellowish mud with rootlets (deposits of vegetated shallow lake bottoms), sand–mud rhythmites (deposits of the shallower wave-effected lake bottom), thick sand bodies (Gilbert-type delta sand bodies), conglomerates (braided stream deposits). The vertical sequences are mostly progradational related to (a) fluctuations in sediment supply, (b) lateral migration of the Gilbert-type delta, (c) changes in the lake size. The lake of the Nagum Formation time was only 1–2 m deep with maximum depth around 10 m, with a marginal belt of lake beach but no marginal swamps.

Filling up of this lake of Nagum Formation and uplift of the Kashmir valley led to the obliteration of any major lake system in the Kashmir valley during early to middle Pleistocene. On the uplifted valley floor sediments (Karewa lake deposit) minor river channels, ponds eolian activity deposited a cover of loamy sediments during Middle Pleistocene–Holocene, sometimes referred to as loess deposits.

SINGH, K.P., Punjab State Council for Science and Technology, Chandigarh, India

Combined Use of Surface and Groundwater—An Economical Solution for the Reclamation of Waterlogged and Saline Areas in Southwest Punjab, India

The area studied lies in southwest Punjab State in India and forms a part of Indo-Gangetic alluvial plains. The area is occupied by Quaternary alluvium comprising clay, silt, and sand. The upper aquifers are phreatic in nature, whereas the deeper ones are under confined—semiconfined conditions. The thickness of the alluvium is about 400 m in the central part. With the advent of canal irrigation in Punjab, a new component of recharge through seepage from unlined canals was imposed on the groundwater system. This led to the increase in storage and rise in water levels since the inception of canal irrigation system. The continuous rise of water levels at an average rate of 0.25 m/yr from 1895 to 1950, and from the year 1950 at the rate of 0.50 m/y has created in some zones groundwater with depth to groundwater between 0.25 m and 3.0 m below ground level, consequently causing waterlogging (water table within 2 m depth). The area under waterlogging increased from 63,090 hectares in 1978 to more than 4,49,239 hectares in the year 1987 and is likely to increase if no remedial measures are taken.

The quality of groundwater in the area varies from fresh to highly saline/alkaline groundwater with electrical conductivity ranging between 500 and 12,000 micromhos/cm at 25 °C. In some areas, circulation of groundwater beneath the canals has been effective in depressing or flushing the more mineralized groundwater, creating lenses of freshwater, which are mainly calcium-magnesium-bicarbonate type as compared with deeper groundwater which is of sodium chloride and sodium sulfate types, besides having more total dissolved solids. A hypothesis combining the processes of solution and evaporation appears to best explain the observed gradual mineralization of groundwater down gradient and away from sources of recharge and the distribution pattern of saline groundwater zones in Punjab.

In the present paper, it is emphasized that tubewell reclamation methods should be adopted by switching irrigation practices from surface to groundwater and combined use of moderately mineralized groundwater with the freshwater be made because such measures are more economical to control waterlogging and salinity as compared with other conventional methods like construction of surface drains, etc., which are presently adopted as measures to reclaim the waterlogged area. Paleochannels of river Stulej located in highly mineralized groundwater and identified on the basis of LANDSAT data and field investigations offer excellent sites for planning tubewell schemes, as groundwater from such channels is of very low mineralization. Combined use of surface and groundwater for salt-resistant crops should be made, along with other measures like lining canals and afforestation. Similar measures can be adopted for reclamation of waterlogged areas having a similar hydrogeological setup in the South Asian countries where waterlogging and salinity irrigated tracts is a common problem.

SINGH, R. N., National Geophysical Research Institute, Hyderabad, India

Modeling Thermal Subsidence of Lithosphere Undergoing Complex Stretching History and Its Implication in Sedimentary Basin Formation

McKenzie's analytical model of the thermal structure of the lithosphere undergoing uniform instantaneous extension (McKenzie, 1978) has been extensively used in the literature to interpret the subsidence history of the sedimentary basins. This model has been analytically extended to include the effects of finite duration of extension (Jarvis and McKenzie, 1980), differential stretching of the crustal and upper mantle parts of the lithosphere (Sclater et al, 1980; Royden and Keen, 1980; Hellinger and Sclater, 1983), and lateral variation in the extension (Beaumont et al, 1983; Cochran, 1983). These analytical models, because of their conceptual clarity and easy applicability, have been used to model a large class of sedimentary basins with varying degrees of success. In general terms, the modeling problem reduces to solving an advection-diffusion partial differential equation of the temperature for the prescribed form of the lithospheric deformation. For some simple forms of deformation, the analytical solutions can be obtained, but for realistic forms it is difficult to obtain the analytical solutions. To take care of these situations involving complex spatio-temporal nature of lithospheric extension, we present here an approximate analytical model. The governing differential equation is averaged over the depth of the lithosphere to get an integral heat balance equation. We then assume a polynomial expression for the depth dependence of temperature field with time varying coefficients and substitute this expression in the integral heat balance equation to obtain an ordinary differential equation for the transient part of the solution. Thus, the solution is reduced to quadrature that can be easily performed. Numerical results for thermal subsidence obtained from this general solution compare favorably with the exact solution for finite duration of extension as obtained in Jarvis and McKenzie (1980). Further, we have extended this approach for random transient variation in the stretching history. Applicability of these results to some actual geological problems will be discussed.

REFERENCES

Beaumont, C., C. E. Keen, and R. Boutilier, 1982, On the evolution of rifted continental margins: comparison of models and observations for the Nova Scotian margin: Geophysical Journal of the Royal Astronomical Society, v. 70, p. 667–715.

Cochran, J. R., 1983, Effects of finite rifting times on the development of sedimentary basins: Earth and Planetary Science Letters, v. 66, p. 289–302.

Hellinger, S. J., and J. G. Sclater, 1983, Some comments on two layer extensional models for the evolution of sedimentary basins: Journal of Geophysical Research, v. 88, n. B10, p. 8251–8269.

Jarvis, G. T., and D. P. McKenzie, 1980, Sedimentary basin formation with finite extension rates: Earth and Planetary Science Letters, v. 48, p. 42–52.

McKenzie, D. P., 1978, Some remarks on the development of sedimentary basins: Earth and Planetary Science Letters, v. 40, p. 25–32.

Royden, L., and C. E. Keen, 1980, Rifting process and thermal evolution of the continental margin of eastern Canada determined from subsidence curves: Earth and Planetary Science Letters, v. 51, p. 343–361.

Sclater, J. G., L. Royden, F. Horvath, B. C. Burchfiel, S. Semken, and L. Stegana, 1980, The formation of the intra-Carpathian basins as determined from subsidence data: Earth and Planetary Science Letters, v. 51, p. 139–162.

SINGH, R. N., and S. N. RAI, National Geophysical Research Institute, Hyderabad, India

On Modeling Geometric Form of Volcanic Constructs

The shapes of volcanoes, aseismic ridges, and seamounts are characteristic features that can be analyzed to constrain the mechanism of volcanic emplacement and the mechanical behavior of the material constituting these volcanic constructs. For modeling their shapes, a reasonable model of the material is that of a porous medium, and for those volcanoes with thicknesses much smaller than their horizontal extents, the method of small perturbations can be used. Combining these two approximations, the evolution of the outer surface is described by the nonlinear Boussinesq equation that has been extensively studied in the hydrogeology. Lacey et al (1981) and Angevine et al (1984) have obtained similarity solution of this equation and showed that the solution explains the general features of several volcanic constructs. In this paper, an approximate integral balance approach has been used to develop a generalized model for the shape of volcanic constructs for the time-dependent rate of magma supply that could be deterministic or random. Here the governing equation is first averaged over the entire length of the volcanic construct to obtain an integral balance equation. Then a polynomial expression in horizontal coordinate with transient coefficients is assumed for the surface profile. Next, we substitute this expression in the integral balance equation and solve the resulting first-order ordinary differential equation by quadrature to obtain the form of the surface profile. This approximate solution is found to be very close to earlier result for which exact solution is available. The applicability of these results to actual geological situations would be discussed.

REFERENCES

Angevine, C. L., D. L. Turcotte, and J. R. Ockendon, 1984, Geometrical form of aseismic ridges, volcanoes and seamounts: Journal of Geophysical Research, v. 89, n. B13, p. 11287–11292.

Lacey, A., J. R. Ockendon, and D. L. Turcotte, 1981, On the geometrical form of volcanoes: Earth and Planetary Science Letters, v. 54, p. 139–143.

SINGH, R.P., Banaras Hindu University, Varanasi, India

Geochemistry of the Regionally Metamorphosed Metapelites Overprinted by Thermal Metamorphism at Mahwaria, District Mirzapur (U.P.) India

It has been found that the entire chemical changes are relative, and the process as a whole has remained isochemical under closed system. Whatever chemical constituents seem to be added during thermal metamorphism have suffered losses during the regional period. Inside the aureole SiO_2 + Al_2O_3 increase nearer the contact, the increase having been reflected in the appearance of large and abundant andaluste crystals. However, it has been calculated that this increase is relative and there has not been any overall addition or subtraction of these constituents so that the order of Si, AlO_4 tetrahedra have been maintained. The losses of A^{+3} have been compensated by gains of Si^{+4} and vice versa on the basis of coupled ionic substitution criteria in which the electrostatic balance of the system is maintained approximately.

Among the other chemical constituents that show increase at the contact are Na_2O, K_2O, MgO, and FeO. The constituent that shows a progressive decrease toward the contact is Fe_2O_3. The rocks from outside the aureole denoting no thermal effect and having suffered only regional metamorphism, the chemical changes comprise an initial decrease of SiO_2 + Al_2O_3 in the lower grades but an eventual increase with increase in metamorphism. Na_2O, on the other hand shows an initial decrease followed by a progressive increase in the higher grades, but eventually again a decrease. K_2O shows, like Na_2O, an initial decrease, followed by a progressive increase. FeO and MgO behave antipathically in the initial stages of progressive metamorphism. MgO shows a progressive increase in the earlier stages, while FeO shows a decrease. In the next stage MgO again increases. FeO at this stage shows a substantive increase, though in relation to the least grade rocks, this increase is not more. The FeO/MgO ratio in the aureole rocks shows, in general, an increase that is quite substantial at the immediate contact. On progress of regional metamorphism, the FeO/MgO ratio shows, in general, a decrease. The most important conclusion is that in both the metamorphic events there has been no net addition or subtraction of any of the chemical constituents, and all the apparent gains and losses are only relative, the overall process thus showing an isochemical behavior.

SINGHAL, B.B.S, and D.C. SINGHAL, University of Roorke, Roorke, India

Evaluation of Aquifer Parameters and Well Characteristics in Fractured Rock Formations of Karnataka, India

The Karnataka state in South India consists of a variety of hard rock formations, vis., granites, charnockits, basalts, limestones, and low- to medium-grade crystalline rocks. The processes of weathering, fracturing, and solution activity have imparted secondary porosity to varying extents. The state suffers from occasional droughts every two to three years.

The groundwater flow characteristics of fractured formations have been investigated by Boulton, Streltsova, Gringarten (1982) and others.

Pumping tests were conducted on wells tapping different geological formations to evaluate formation and well characteristics. The pumping test data show varying trends, which have been compared with similar trends from other areas. The double porosity model suggested by Streltsova (1978) for heterogeneous confined aquifers appears to give satisfactory results. The transmissivity and storativity values of fractures and blocks have been computed for different geological formations. The transmissivity of the fractures is much higher for all types of formations as compared with that of blocks. The fracture transmissivity in granites, charnockites, and limestones is usually higher (115 to 250 m^2/d) than in greywackes and schists (15 to 50 m^2/d). The storativity (S_1) of fractures is smaller (10^{-3} to 10^{-6}) than that of blocks (S_2) (10^{-2} to 10^{-5}) on account of lower values of bulk modulii of elasticity of fractures as compared with that of blocks. The pumping tests also indicate at few places the reverse time-drawdown and reverse distance-drawdown trends.

Specific capacity and productivity (specific capacity index) of the dug wells tapping the shallow aquifers and that of the borewells tapping the deeper fractures were computed. Specific capacity values of dug wells, tapping weathered shallow fractured horizons in laterites, granitic gneiss/granites, and charnockites are in the range of 39.5 to 88.9 lpmin/m, while specific capacity index (specific capacity per unit surface area) is in the range of 0.8 to 1.7 lpmin/m/m^2. The well tapping laterites and granitic gneiss show higher values of specific capacity and specific capacity index (85 lpmin/m and 1.5 lpmin/m/m^2, respectively) and those in charnockites have lower values on the order of 39.5 lpmin/m and 0.8 lpmin/m/m^2, respectively. The specific capacity values of bore wells tapping deeper fractures in greywackes, basalt, laterite, granitic gneiss/granite, chlorite schist, and charnockites fall in the range of 1.3 to 639.3 lpmin/m. Basalt and limestones have higher values than other rock types.

The computation of spacing for bore wells has been carried out using two approaches, i.e., (a) the radius of influence approach, and (b) from the recharge considerations (Table 1). The former approach suggests spacing of 110 to 382 m, while the recharge approach indicates a range of 190 to 417 m with density of wells varying from 6 to 35 wells/km^2. In the radius of influence approach, spacing is calculated on the basis of pumping test data, assuming homogeneous and isotropic media, which is not the case under existing field conditions in hard rock terrains. Therefore, the recharge approach is likely to give more judicious estimate of the well density, as it takes both draft and recharge into consideration.

It may be advisable to use both the radius of influence approach and the recharge method to complement each other. The former approach can be employed to decide the minimum spacing between adjacent wells, whereas the recharge method is recommended for evaluating the maximum density of wells in an area. The recommended minimum spacing between irrigation bore wells may vary between 130 and 275 m, keeping in view the scope of further groundwater development. The overall well density may be in the range of 5–9 per sq km in areas having low to medium rainfall, and 25–35 per sq km in coastal areas having high rainfall.

The three main types of groundwater structures in the state are dug wells, dug-cum-bore wells, and bore wells. Dug-cum-bore wells are constructed by extension drilling in the bottom of dug wells. It is found from the present studies that bore wells are preferable to dug wells and dug-cum-bore wells in terms of cost of groundwater as well as total annual benefits (Figure 1). The average working hours for bore well/dug-cum-bore well has been assumed to be 8 hr/day, while for dug wells it is 4 hr/day. However, dug-cum-bore wells may also be considered economical wherever these are constructed as revitalization structures made by extension drilling in existing dug wells. The success rate of bore wells greatly increases with proper geological and geophysical investigations.

REFERENCES

Gringarten, A.C., 1982, Flow test evaluation of fractured reservoirs, in T.N. Narasimhan, ed., Recent trends in hydrogeology: Geological Society of America Special Paper 189, p. 237-263.

Streltsova-Adams, T.D., 1978, Well hydraulics in heterogeneous aquifer formation, in V.T. Chow, ed., Advances in hydrosciences: v. 11, p. 357-423.

Table 1. Recommended well spacings.

Formation	Radius of Influence Approach Well spacing (m)	Recharge Approach Well spacing (m)	Recharge Approach Well density (per/km^2)
Greywackes	110-136	224-448	6 to 25
Basalt and related rocks	170-256	370-417	7 to 9
Limestones	126	415	7
Gneiss/granite/charnockites	234-382	191-492	5 to 35
Laterites (coastal areas)	130-275	190-225	25 to 35

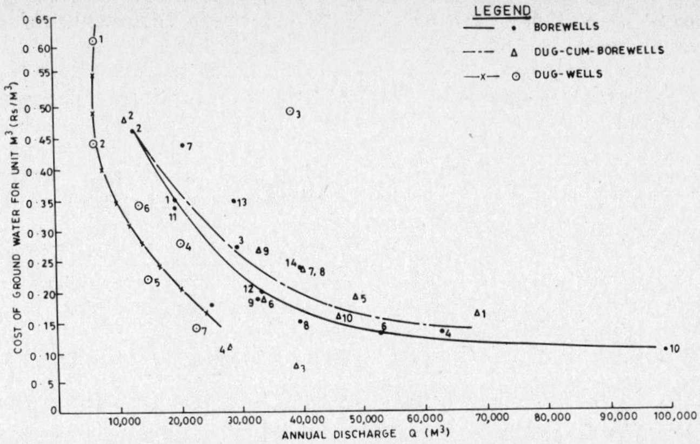

Figure 1—Plot of cost per unit volume of groundwater vs. annual discharge for different groundwater structures.

SINHA, A. K., University of Rajasthan, Jaipur, India, and A. K. TEJWANI, Jaipur, India

Geological Aspect of Urbanization: Jaipur, (India), Case Study

Urbanization is by far the most important social transformation of our times. For developing urban areas, geological aspects such as surficial feature, bedrocks, groundwater table, slope of ground, stability of ground, etc. require careful consideration. The growth capability of any urban center is totally dependent on these geoenvironmental attributes.

Ignorance on the part of urban planners about the geological attributes of the environment in shaping and influencing the life of the urban sector has led to burial of quite a few cities under volcanic debris, collapse of total urban establishments due to seismicity, siltation of drainage and reservoirs due to excessive erosion, floods, shift in the courses of rivers, landslides, subsidence, avalanche, and sand reactivation. Therefore, geoscience has a major role to play in the formulation of long-term urban plans.

Jaipur, the capital of Rajasthan (India), is one of the few planned cities of the eighteenth century in the world. The walled city covers an area of 6 km^2. It was planned for a lac population that has now surpassed 12 lacs (1981 census). It has naturally spilled over beyond the peripheries of the walled city and moved in different directions taking into little or no consideration the character of the various geological formations and processes operative in the region.

Geoenvironment attributes are present here in terms of pre-Cambrian and Quaternary formation, distinct geomorphic units, and water and mineral resources. The major rock types are schist, gneiss, quartzite, marble, and phyllite belonging to pre-Cambrian formation, which are very much tectonically deformed. These rocks are mostly covered by the mantle of sand and alluvium of Quaternary, of which seven distinct units have been recognized in this area. Geomorphologically, the Jaipur region has been divided into three geomorphic units: the recent alluvial plains, the Delhi super group, and the Aravelli super group. The Delhi super group is more resistant than the Aravalli super group; it encircles the city of Jaipur from the north, northeast, and east, preventing the spread of urbanization in those directions. Numerous geomorphic features such as gorges, springs, sand dunes, and longitudinal valleys are present. Drainage is simple but tectonically controlled by intensive gullying. The main drainage of the area is formed by the Amanisha, Jhalana nadi, and Dhund. Several small streams and gulleys join the above rivers most of which originate by headward erosion. The drainage map indicates that the trellis drainage pattern has developed in many areas. Groundwater aquifers belong to both pre-Cambrian and Quaternary. Pre-Cambrian aquifers are jointed quartzites and gneissic rocks. The Quaternary alluvium form aquifers only in the river beds at more or less shallow depth. Groundwater of Jaipur and surrounding area is generally good in quality as revealed during the groundwater inventory.

The presence of sandy flat, undulating topography referred to by many as stabilized sand dunes situated in synclinal trough of Delhi super group probably was the cause for establishment of the walled city of Jaipur by the former king of Jaipur. The remnant hill of the Aravalli mountain range made of Delhi super group surrounding Jaipur from north, northeast, and east—must have acted on one hand as a defense barrier safeguarding Jaipur from the enemy invasion whereas on the other hand, it prohibited the advance of urbanization in these directions.

Quartzite belonging to the Delhi super group is being extensively quarried as building material for constructional activity in this urban zone.

The almost flat, undulating alluvium topography extending in a rough fan shape from north, northeast to south, southwest, and west much beyond the walled city is becoming a site of newer human establishment. This region is also the zone of shrinking internal drainage system. Land cover maps prepared after Landsat MSS and TM data products show intense land transformation and intermingling and overlapping of various land uses in the Jaipur urban zone.

Many of the environmental problems that the city faces today have stemmed from the phenomenon of rapid urbanization. Based on field observation, mapping, and study of relevant background information, the main spheres of environmental hazard are mobilization of sand dunes and their reactivation, extensive quarrying on the hill faces on the urban flank, intensive gullying and badland erosion, congestion of drainage, choking of streams, waterlogging and floods, and construction on relatively loose dune formation.

In the interest of preserving the ecological balance of the urban complex of Jaipur, it is suggested that the dune areas be protected by grass and shrub plantation. Constructional activities should be stopped forthwith on dunal tract. Shrinking internal drainage reaches including its flood plains and spill areas should be protected from urban encroachment by resorting to afforestation immediately. Quarrying activities should be totally banned, particularly on the hill faces on the urban flank. Areas susceptible to large-scale gully erosion should be conserved by proper soil conservation measures and planting of grasses. Proper land use needs to be adopted, keeping in view the natural land capability factors.

SINHA, A. K., Wadia Institute of Himalayan Geology, Dehradun, India

Tectonic Framework of Northwest Sector of Himalaya: Comparative Transects through Kulu–Leh and Karakoram Highway

The northwestern section of the Himalayan Orogen encompass a fascinating area on either sides of Nanga Parbat–Harmosh massif presents challenging tectonic complications. Sine the publication of the classical work of Wadia on the North-Western Sytaxial belt, much more geological research has been done on both sides of the massif, owing to road building activities and expeditional works. The Siwaliks–Tertiary complexes are abutting against the Chail Thrust sheet along the Main Boundary Thrust, marked by activity of basic-rock emplacement of Precambrian to Tertiary age. The granitic emplacement of Caledonian orogenic mass dated by Rb/Sr isotopic method remained controversial due to paucity of data to explain the cutting of Himalayan fabric. Long tongues of Chail and Crystalline Thrsu cover the entire southern zone except an exposed Riphean carbonate-quartzite tectonic window deeply cut by the gorge of the Beas rivers. The Main Axial Zone (MAZ) embodies the "Root zone" of thrust-sheets of intensively metamorphosed rocks. Occupying the uppermost position in topography demarcates the major tectonic feature between Southern and Northern Zones of Himalaya. The northern fringe of Vaikrita Complex, constituting the high-grade metamorphosed Central Crystalline, is tectonically separated by the Tethyan Thrust (TT). This tectonically separated Precambrian Phanerozoic sequence represents a synclinorium recording epeirogenic upheaval at different epochs but pronounced at Variscan and Austrian (Sangchamalla) phase. In contrast to Central Malla Johar, the north-northwest Lahaul–Spiti basin shows a couple of conglomeratic horizons between Cambrian–Ordovician and Silurian–Devonian boundaries. The Devonian and Permo–Carboniferous sequences have been thrust from North to South with an average displacement amplitude of 5–10 km. Regional metamorphism taking place in the lower part of the Paleozoic without any break supports the basic idea in Himalayan metamorphism that this alone should not be regarded as the age index of rocks. The flysch-graben represents trench type sediments. Great Himalayan Brahmputra–Indus Suture (GHS) plays a significant role, embodying ophiolitic melange of mantle origin and representing a continent-continent collision zone. An initial collision age of about 50 m.y. has been estimated with an average convergence rate of 55 mm/y. Continuity of orogen to the north through Karakoram presents a zone of multiple suturing and contraction phenomenon in this part of the earth's crust. In the evolutionary process, the Nanga Parbat–Harmosh–Pamir Knot has played a significant role, being bounded by regional Chaman and Karakoram Faults.

SINHA, A. K., D. K. MISRA, and S. K. PAUL, Wadia Institute of Himalayan Geology, Dehradun, India

Tectonostratigraphic Set-Up and Structural Data from Higher Northwestern Himalayan Zone and Their Implication in Tectonic Evolution of Himalaya

Four main tectonostratigraphic units—Larji–Rampur window group, Chail Group, Jutogh Group, and Vaiktira Group—have been recognized along the Satluj and Beas valleys of Himachal Pradesh. Larji–Rampur window Group consists of quartzarenite and penecontemporaneous basic volcanics together with gneissic Bandal Granite (1840 m.y.) of Rampur Formation tectonically underlain by stromatolotic limestone of Shali Formation. The Chail Group comprises predominantly phyllites, quartzite, mylonitized gneiss, and intrusive body of 500 m.y. dated Mandi Granite. The Chail Group is overthrust by Jutogh Group, which consists of medium-grade metamorphics, foliated micaceous quartzite, and fine-grained banded gneiss. High-grade metamorphics of Central Crystallines (Vaikrita Group) of the Higher Himalayan Zone overlay the Jutogh Group along a tectonic contact. The Vaikrita Group is made up of sillimanite–kyanite–garnet schist, coarse-grained biotite-rich-banded gneiss with tourmaline, and foliated micaceous quartzite. The tectonic contact between the Chail Group and the Larji–Rampur window Group is interpreted as Main Central Thrust, on the basis of regional tectonic analysis. In the north of Rohtant Pass along the valley of Chandra River, the Main Axial Zone (MAZ) represents the root-zone of the Higher Crystalline nappes below the younger phanerozoic deformed sediments. The low-grade metasedimentaries with 500 m.y. dated Jaspa–Granite called the Haimanta Group represent the basal part of the Phanerozoic–Cambrian to Cretaceous Tethys Himalayan sediments.

Systematic structural data have been collected from regions of Satluj, Beas, Chandra, and Spiti valleys in the Higher zone of Himalchal Himalaya including crystalline and Tethyan domains. In the crystalline rocks, the pre-Himalayan F_1 fold hinge lines plunging toward the east-northeast are noticeable, significantly obliterated by the D_2 deformation. The F_{2a} fold hinge line plunges toward northwest-west-northwest Trending northwest-southwest are recline folds that are again coaxially folded (broad open F_{2b} folds) superimposed by F_3 folds trending again northwest-southeast while its fold hinge lines plunge toward north-northwest to north-northeast. The rocks of the Tethyan sequence exhibit only two generations of folding—F_1 tight to isoclinal folds, trending northwest-southeast, are prominently developed throughout the area while F_2 tight folds are feebly developed and plunge toward north-northwest to north-northeast. There is a marked tectonic contact between Vaikrita Crystalline Group rocks against the Tethyan sediments. These two domains developed separately, and are divided by the Main Axial Zone of Himalaya in the south and north.

SINHA, A. S., Central Mine Planning and Design Institute, Bihar, India

Patterns of Coal Seam Deposition in Peninsular Indian Gondwana

More than fifty Lower Gondwana Coalfields (CFs) are strung out in four definitive alignments in the Peninsular India. These are mostly elongated synclinal basins with one boundary depositional and the other faulted against basement Pre-Cambrian metamorphics. Some CFs show basinal structure with centripetal dips; some other CFs have more than one subbasinal structures. The different CFs contain from one to more than thirty coal seams of thickness from less than 1 m to more than 140 m and Mineral Matter Content (MMC) varying from less than 10% to more than 50%.

Vast thickness and MMC data of seams of different CFs are available from exploratory boreholes and mines. Trends in the distribution of seam thickness and MMC in blocks of different CFs have been identified in exploration reports and elsewhere (Ghosh, 1987). A general picture, however, has not emerged so far because, perhaps, of the difficulty of assigning a common reference line or point for the study of the multitude of coal seams with differing geographical, structural, stratigraphical and quality parameters. In this work, the centers of the structural basins or subbasins, showing at least three-quarters closure on surface and borehole evidence, have been taken as reference points and the thickness and MMC variations in each basin have been studied as functions of distance from these reference points.

The Indian coal seams are highly banded, reflecting changes in depositional parameters from one play to the other; even a thick coal ply may represent varying depositional conditions over a length of time. Therefore, only the thinnest but well defined plies of coal seams have separate proximate analyses for selected for study to offset the effect of marginal depositional conditions.

Seventeen plies from six CFs, spread over 1200 kms from Raniganj CF in the east to Umrer CF in the west, with average thickness from 0.5 m to 1.1 m were studied. The raw thickness of a ply in each borehole was transformed by dividing it by its average thickness in the basin ($Xt = Xi/X$) to standardize different thickness ranges. The ash content of the ply was taken as a measure of its MMC, by the equation MMC $\doteq 1.1 \times$ Ash % (Ind. Stnd. 770, 1977).

The plots of the transformed thickness of the plies in boreholes against the distance of the boreholes from the center mostly indicate thinning of the plies away from the center of the basin. Similarly, the plots of the ash content of the plies in borehole samples against the distance of the boreholes from the center mostly indicate an increase in the ash content away from the center of the basin. Finally, the plots of transformed thickness against ash content almost always indicate an increase in ash with the thinning of the ply.

The average rates of change transformed thickness and ash at specified distances for different plies were averaged by regression. Similarly, the average rates of change of transformed thickness with ash content at specified ash levels from different plies were averaged by regression. These regression equations were integrated to generate general models for the thickness-distance, ash-distance and thickness-ash variations. The correlation coefficients of the general models with actual data from most of the plies were found to be close to the correlation coefficients of regression amongst actual data of individual plies.

The results imply that the coal ply deposition generally followed the pattern of gravity fractionated sedimentation in subsiding basins with mineral matter decreasing and coaly material increasing towards the center of the basin. The study also implies that the broad pattern of the subsiding basins of deposition remained unchanged over most of the period of formation of coal seams. It is also implied that the structure of these basins was not affected by post-Damuda (Permian) movements.

It is proposed to test the models further with full seam data. The study may help in locating thicker and qualitatively better reaches of coal seams.

SINHA, A.K., University of Rajasthan, Jaipur, India

Geology of Phosphate-Bearing Rocks of Delhi Super Group, Northeast Rajasthan, India

The early Proterozoic (2500-200 m.y.) tectonic activities, which resulted in the formation of the Aravalli trough in southwest Rajasthan, shifted to northeast Rajasthan during the middle Proterozoic (1800-1400 m.y.) and caused development of fault-bounded troughs, namely Bayana and Alwar. It is believed that the Alwar and Bayana basins were evolved in five stages of subsidence and faulting, The fourth stage marked the cessation of Alwar sedimentation while the fifth phase was responsible for large-scale submergence and concommittant volcanism which generated a conducive environment for phosphogenesis during the Ajabgarh sedimentation.

The deposition, mode of occurrence and structural imprints noted within different members/formations of the Alwar and Ajabagarh groups (of the Delhi super group) indicate at least three time gaps in deposition, These are between the Alwar and Ajabgarh group, between the Kushalgarh and Sariska formation (lower Ajabgarh group), between the Thanagazi and Bharkol formation (upper Ajabgarh group respectively.

Very coarse clastics and minor quantities of fine clastics (quartzite and phyllite of Alwar group) were deposited unconformably over a carbonate and granitic shelf along the shallow coastal/shoreline, The syn-to-post Alwar group tectonic episodes caused development of rift basins, viz. Bayana, Alwar, Sariska-Achrol, Baraud and Khetri, identified by a characteristic tectonosedimentary feature, having restricted connection with sea, which where submerged later owing to large scale subsidence/downwarping, The seawater upwelled and accumulated within restricted basins where chemical and fine terrigenous sediments were deposited, giving rise to the calcareous and phosphogenic sequence of the Kushalgarh formation.

Further uplift followed by erosion resulting in the formation of piedmont basins within continental shelf having restricted connection with sea has been tentatively suggested. Deposition within such basins is characterized by intercalated ferruginous breccia/grit, quartzite, chert, carbonphyllite, with or without stromatolites corresponding to the Sariska formation.

The bands and lenses of variable dimensions of layered and nonlayered phosphorite and phosphate-bearing sandy clay stone are stratigraphically and lithologically controlled. These occur within transition zone comprising kaolinite-sericite-quartz schist between Kushalgarh limestone and interlayered ferruginous breccia, gray-quartzite stromatolite-bearing quartzite, chert, and carbon-phyllite (Sariska formation) grouped under the lower phosphogenic Ajabgarh sequence of the Delhi super group.

The phosphorite is amorphous to equigranular and is fine grained, The phosphate minerals are collophane (A Crybto-crystalline carbonate fluor-apatite), Fluor-apatite, hydroxyl-apatite, strontium-apatite, Na-apatite and possibly francolite. The gangues are quartz clay minerals and iron-ores in major quantity, where in calcite and chlorite occur in traces. The cementing material is phosphatic, siliceous, and mobile ferruginous matter.

Geochemical analyses of seven groups of samples each representing the average analysis of ten samples were taken up. The study reveals that CaO, P_2O_5, F, CO_2, and NaO appear to form phosphate minerals, while SiO_2 and Al_2O_3 form quartz and clay minerals. The average P_2O_5 varies from 20 to 28%. Y and REE as a group is enriched in phosphorite as compared with those of shale, sandstone, and carbonates which are generally the host rocks of phosphorite. Results have been discussed and correlated with the study of various important Indian, as well as, foreign occurrences.

The phosphorite of the area is of medium grade and may be kept under silica-based category, which may be amenable to beneficiation to upgrade the ore to fertilizer and chemical industries grade.

SINHA, R. K., A. K. SINHA, S. G. KRISHNA, V. B. SRIVASTAVA, and S. K. LAHIRI, Geological Survey of India, Jaipur, India

Land System Evolution of Singrauli Coal Field: An Appraisal

The landscape study of about 1000 km^2 in the Singrauli Coal Field with the aid of remote sensing products and aerial photographs and adopting the approach for terrain evaluation followed by Commonwealth Scientific Industrial Research Organization, Australia, lead to the identification of two land systems that show a recurrent pattern of genetically linked land units. The scene thus assembled has been described as:

1. Plateau-Pediplain land system:

a. Climate: Average rainfall of 101 cm/annum.

b. Geology: Shale, sandstone, siltstone, clay, coal, boulder of Talchir Barakar, Raniganj, and Barren Measure Formation. Barakar and Raniganj Formations are coal bearing and are traversed by predominantly northwest-southeast, west-northwest-east-southeast, east-west, northeast-southwest trending lineaments.

c. Landscape: Four levels of plateau ranging in altitude from 340 to 520 m exist, flanked in the north by Bijul pediplain and to the west, south, and east by Waidhan pediplain. The general slope toward north and northeast lies in the 0°–5° slope class, and the scarp relief difference between levels of plateau is 40–100 m.

The plateaus are characterized by benches, which mark the outcrop position of coal seams. The other characteristic features are escarpments, cuestas, mesas, and buttes and deep narrow valleys/gorges with youthful streams that exhibit subtrellis and subparallel drainage arrangement. The plateau top shows development of filled valleys.

The pediplain has an altitude of around 260–340 m and slope class of 0°–2° and 2°–5° and is characterized by older floodplain, relict streams, and buttes. The system is drained by Muher, Matwani, Balia, Kachan, and Mayar streams.

d. Soil: Yellow silty and fine sandy soil with rich ferruginous lenses.

e. Vegetation: Open mixed jungle/cultivated crops.

f. Relief: 40–160 m.

g. Altitude: 260–520 m.

h. Area: 687 km^2.

i. Dominant Process: Fluvial, fluvio-denudational represented by rill and gully erosion and anthropogenic.

2. Ridge-Valley land system:

a. Climate: Average rainfall 101 cm/annum.

b. Geology: Schist, phyllite, quartzite, and gneisses.

c. Landscape: Structural ridges in Archaean metamorphics with sharp narrow crests, moderate dip slope to steep back slope, aligned in a general west-northwest–east-southeast to east-west direction. Structural ridges in Bijawars with rounded crests, moderate dip slope, and steep back slope aligned in a general northwest-southeast to east-west direction. Ridges with an altitude of 320–380 m above m.s.l. are interspersed with narrow linear structural valleys and exhibit a positive relief of 80–170 m with the surrounding undulating structural surface, which shows a characteristic development of valley flats. The overall dominant computed slope class is 5°–35°. This system is drained by Bijul, Mamuar, and Murcha with their tributaries, which collectively exhibit a trellis to subrectangular drainage arrangement.

d. Soils: Clayey and sandy silt, often gritty, rich in ferruginous concretions and occasional kankars.

e. Vegetation: Fairly dense mixed jungle, dry cultivation to open scrub.

f. Relief: 80–170 m.

g. Altitude: 320 m.

h. Area: 313 km^2.

i. Dominant Processes: Denudational, fluvio-denudationaland fluvial.

The streams draining the two land systems form the drainage net of on river system, which exhibits an overall state of vertical incision. A comprehensive evaluation of assemblages and disposition of land element and parameters of drainage and slope morphometry projects the following summarized account of the evolutionary history of the landscape under review:

1. The Archaean metamorphics and the Bijwar metasedimentaries withstood the impact of a series of earth movements that acted upon them through geological times and found their expressions as structural ridges, structural valleys, and undulating structural surface. These land units are aligned in a general west-northwest–east-southeast to east-northeast–west-southwest direction, though northwest-southeast and northeast-southwest alignments were also observed. The alignment of structural valleys and its

continuous extension is suggestive of its close relation with Son–Narmada lineament. Denudation and fluvial processes acted upon and left profound impressions as low slopes of the hills, filled valleys, valley flats, and benches, but structural fabric continued to be the dominant factor in the configuration and disposition of land elements and units developed in this area. Hence, the land system thus developed has been designated as land system of structural origin.

2. Gondwana sediments laid over metamorphics in a trough faulted basin were uplifted with the basement rocks as a plateau, which was subjected to various exegenic processes such as denudational, fluvio-denudational, and fluvial.

3. Slope morphometric studies and drainage morphometric studies indicate that the evolution of land forms took place through scarp retreat and lengthening of wash plains. The study also suggested that fluvial processes and fluvio-denudational processes in combination with tectonic impulses were the principal tools in sculpturing of the four levels of plateaus, benches, and pediplains over the Gondwana sedimentaries.

4. The occurrence of numerous buttes of the Barakar Formation over the Waidhan pediplain is additional evidence to suggest that there was a bulk removal of the overlying sediments by processes as envisaged above. The presence of mesa, butte, and benches over the different levels of plateau is yet more evidence to suggest the same contention. Moreover, the very dominant role of fluvial erosion in carving the major land units and elements is very distinctly imprinted over the Waidhan Pediplain as relict fluvial impressions.

5. The small thickness of alluvium over the Waidhan Pediplain is suggestive of the fact that the streams, most of the time, have been in a degradational state and engaged in rapid removal. The role of tectonic impulses in bringing about an accelerated removal is deduced emphatically by drainage morphometric parameters.

6. The land units have experienced rejuvenation from time to time in the geological past as can be inferred from anomalies in the drainage morphometric parameters. In addition, the same is also reflected by the higher elevation of undulating structural surface than the general elevation of the adjoining plateau.

7. A comparison of Murcha subbasin with Muher and Baliya suggests that though the drainage density, relief ratio of Murch subbasin is higher, the hypsometric integral and TSI are also high. It therefore not only suggests rejuvenation activity along the major weak plain but also possibly that at least a part of the Murcha subbasin was covered in the past by rocks of Gondwana Super Group, overlying the upthrown block of older metamorphic and Bijawar rocks.

The impact of anthropogenic processes is rampant in the form of mining, damming, and excavation. The mining process has resulted in ushering in landforms such as mine dump heaps, benches, and mine pits, which would continue to gain larger aerial extent until the entire coal is taken out and the present plateau system totally degraded. The induced processes would produce derangement of existing drainage patterns, development of zone of rill and gully erosion, damming of the valleys of Matwani and Baliya streams, and subsequent formation of lakes. The continued development of rills and gullies over the dump material lying on the valley side may culminate in the generation of piping in the barrier between such lakes and the lower reaches of these streams, ultimately leading to the breaching of the lake. Some of the mine pits may remain there to be later called lakes. The study of 1968 aerial photographs shows the mine dump head occupying an approximate area of 0.5 km^2 over the plateau. Subsequent field observation shows an excavation rate of 23 million m^3/annum of mined material; a large area (over 1.5 km^2) is presently covered by dumps, rising to about 50 m above the general plateau level. The modified land units and the drainage would enhance the sediment load to the GBPS reservoir when in a state of flash flood.

SINTON, J. M., Hawaii Institute of Geophysics, Honolulu, HI

Magma Chamber Processes at Slow- and Fast-Spreading Midocean Ridges

Because the thickness of oceanic crust formed by normal seafloor spreading processes is essentially constant over the more than 20 × variation in spreading rates observed at midocean ridges, the magma supply to these ridges must vary primarily as a function of spreading rate. This relationship allows a number of predictions to be made with respect to magma chamber geometries and magmatic processes at slow- and fast-spreading ridges. For example, magma chambers should be larger and shallower beneath fast ridges, and regions of high magma supply should correlate with the presence of a restricted range of erupted lava compositions as a consequence of the large, well-mixed magma chambers predicted to occur there. Conversely, slow-spreading ridges, and their low magma supply, should be characterized by the presence of a wide range of lava compositions erupted from small, discontinuous, variably fractionated magma chambers. These predictions are inconsistent with new data from fast-spreading parts of the East Pacific Rise, which confirm earlier speculations that lavas erupted at high-spreading rates are widely variable in composition, but generally more fractionated than those erupted from slow-spreading ridges. Furthermore, the best evidence for magma mixing processes, observed as complexly zoned phenocrysts and disequiibrium relations between phenocrysts and host liquids, is from slow-spreading ridge lavas, whereas fast-spreading ridge lava compositional variations are generally consistent with crystal fractionation models. These characteristics can be accounted for by a model that allows composite magma chambers, in which a melt-dominated reservoir segregates at the tops of crystal-rich magma chambers, to form beneath ridges with high magma supply. At lower magma supplies, melt almost never segregates into regions larger than the volumes of single eruptions, and erupted lavas from slow-spreading ridges represent the composition of a lower reservoir dominated by crystal mush.

The magma chamber beneath the EPR near 13 °N has been images by seismic methods; it is composite and generally similar to that proposed by Sleep (1975). It consists of a shallow, narrow melt-rich region, underlain by a much broader region of hot rock, probably with a small amount of interstitial melt. These observations and a correlation of lava chemistry from ridges with slow, intermediate, fast, and superfast (> 150 mm/y) spreading rates permit the following model for magma chamber geometries with magma supply. At low supply, magma chambers (defined as the region bounded by isotherms appropriate for basalt solidi) are narrow and relatively deep and dominantly consist of a crystal mush with a small amount of interstitial melt. Such chambers will be difficult to detect by all but the most elaborate of seismic experiments. With progressively increasing magma supply, magma chambers become broader and slightly shallower; at supplies appropriate for normal ridges spreading at rates of about 40–50 mm/y magma chambers become composite with small melt-dominated regions segregating at the tops. Ridges tapping these small melt-dominated reservoirs have the lowest average MG #'s, even when propagating rift effects, which dominate the intermediate spreading Juan de Fuca and Galapagos ridges, are removed. At increasing magma supply rates, comparable to those of fast- and superfast-spreading ridges, the melt-dominated reservoir increases in size and continuity along-axis. Commonly observed magma mixing effects mainly reflect processes occurring in the lower crystal-rich reservoirs that are probably steady-state features beneath all spreading ridges. Because slow-spreading ridges tap these reservoirs more or less directly, lavas erupted there show a wide range of complex relations between host liquids and included crystals, which can record long and complicated histories in the crystal-rich magma chambers. Fractionation effects are best observed when eruptions tap melt-dominated reservoirs at intermediate- and fast-spreading ridges. Maximum differentiation is observed when the melt-rich reservoirs are small, such as in the vicinity of propagating rifts and near offsets at intermediate- and fast-spreading ridges.

Samples from the Galapagos 95 °W propagating rift provide a test of this model. The youngest axial samples from this region define a bimodal population with respect to degree of differentiation and crystallinity. High Mg samples are rich in phenocrysts and probably were erupted from crystal-rich reservoirs. Nearly aphyric ferrobasalts represent eruptions from small melt-dominated reservoirs; the distribution of lava compositions is consistent with the progressive development of composite magma chambers with distance behind the propagating rift tip, with some modification close to rift offsets.

Consideration of processes operating in composite magma chambers has implications for the mechanisms of fractionation of MORB magmas as well as for magma mixing processes. Fractionation occurring during upward migration through crystal-dominated reservoirs can at least qualitatively account for some previously enigmatic features of differentiated MORB suites, including "over-enrichment" in the more incompatible elements, and the restriction to a small subset of observed primitive compositions that can be successfully modeled as parents to differentiated lava suites. Although many of the geochemical implications of this model await rigorous quantitative tests, the model proposed here appears to explain why slow-spreading ridges, with their incumbent low magma supply, dominantly erupt a narrow range of undifferentiated lavas, why magma mixing characteristics are most evident in lavas erupted from slow-spreading ridges, why fast-spreading ridges, with mush higher magma supply characteristics, dominantly erupt a wide range of differentiated compositions, and why bimodal lava populations occur in the vicinity of at least some propagating rifts.

SISKOV, G. D., University "Kl. Ohridski," Sofia, Bulgaria

Coal-Bearing Strata in Bulgaria

The Bulgarian coal deposits are of Paleozoic, Mesozoic, and Cenozoic age in different tectonic regions. Depending on the coal rank, the coals are classified as lignite, matt, and bright brown coal, bituminous coal, and anthracite, according to the Bulgarian National Standard. The potential coal resources are about 8,000 million tons. Of the total tonnage, 77% is lignite, 6% brown coal, 17% bituminous coal, and anthracite is less than 1%.

The coal-bearing strata are related to five maxima of coal genesis covering the Upper Paleozoic, Lower Jurassic, Upper Cretaceous, Paleogene, and Neogene, which are distributed in 12 coal provinces.

The oldest coal-bearing strata of Upper Carboniferous age (Namurian, Westphalian, Stephanian) are established in West Sredna Gora (Svoge) and the very east part of the Moesian Platform in North Bulgaria (Dobrudza). By coalification rank, the coal is determined as anthracite (Svoge) and bituminous coal (Dobrudza).

Coal genesis during the Permian is minimal. Only the Belogradcik deposit of bituminous coal is referred to. It is in the Fore Balkan tectonic zone.

The coal deposits of Lower Jurassic age (Liassic, Dogger) are situated in West Sredna Gora, Fore Balkan, and West and Middle Balkan. By rank, the coals are bituminous.

The Balkan coal deposit and some small fields in West Sredna Gora are referred to the Upper Cretaceous (Cenomanian and Turonian). Only the Balkan deposit is exploited, being a source of coking coal in Bulgaria. The Cretaceous coals are bituminous.

The Paleogene coal deposits are concentrated in the area of East Sredna Gora (Burgas, Nikolaevo), West Sredna Gora (Pernik), Kraishtide lineament zone (Bobovdol, Pirin, Suhostrel), and Rhodope Massif (Smoljan, Pcelarovo). The coals are matt and bright brown. Only Suhostrel, Pirin, and Smolja coals are bituminous.

During the Neogene, the coal genesis begins from the Lower Miocene with a maximum in the Upper Miocenen–Pliocene. The coal deposits are concentrated in the West Sredna Gora (Sofa, Cukurova, Aldomirovzi, Stanjanzi), in the East Sredna Gora (East and West Mariza, Elhovo), in the Kraistide zone (Kjustendil, Oranovo), in the Rila-Rhodope Massif (Razlog, Goze Delcev), and in the Moesian Platform (Lom). The coals are lignite and matt brown by rank.

The Bulgarian coal deposits are formed mainly under continental and subcontinental conditions in small grabens and depressions. The Paleozoic and Mesozoic coals originated on the external distal margins of deltas (Dobrudga, Balkan). For most deposits, the thickness of coal-bearing strata is small, the sediments are molasse, while the coal seams are a small number, thin to middle thick, and limited overburdens. The coals have high ash and sulfur content, with the exception of Dobrudga deposit. The molasse sediments are formed during the Variscan and Alpine orogeny.

The coal-bearing strata are connected with swampy localities in foredeep, inner and intramountain depressions and backlands. Weak dynamometamorphic changes are established in the Svoge and the Balkan deposits as well as contact-metamorphic effects in the Dobrudga and the Smoljan ones.

The diversity of coal rank is caused by the nonidentical geological conditions in the Bulgarian territory in time and space, the influence of geodynamic development of the tectonic regions in the Balkan microcontinent, where coal-bearing strata of different thicknesses are formed. For example, the coal-bearing lower molasse of the Svoge deposit is formed during Namurian A to Westphalian C between the arches in the intramountain depression. The coal-bearing strata has been intensively folded many times from Variscan to Late Alpine tectonic cycles.

The Bulgarian coals are mainly of humolithe type. Saprohumolithes, sapropelithes, and gagates are established very rarely. The humolithes form compact to banded coal seams, which are from pure and impure coals and middlings.

On the basis of petrological investigations, the coals from the lignite deposits are divided into three groups according to the huminite amount (organic matter basis): (1) over 95%, (2) 90–95%, and (3) below 90%. The brown coals exhibit analogical petrographical composition.

A regular reduction of the petrographic composition of the bituminous coals is established with the increases of rank. The coals from the Dobrudga deposit are with a variegated maceral composition, while the Balkan coals have a monotonous petrographic composition. In the anthracite, only macerals from vitrinite and inertinite groups take part.

The mineral admixture amount in the whole coal deposits is very high. They are mainly represented by clay minerals and pyrite.

SISODIA, M. S., and D. S. CHAUHAN, University of Jodhpur, Jodhpur, India

The Role of Magnesium Ions During the Formation of Precambrian Aravalli Stromatolitic Phosphorites of Udaipur, Rajasthan, India

The stromatolites occurring around Udaipur city are constituted of carbonate and/or phosphate. The amount of phosphate in these stromatolites is present in varying proportions such that from traces, quite often it constitutes their major framework. This variation in the composition of stromatolites is possibly a reflection of a change in environment, i.e., subtidal to supratidal, under which they developed. Simultaneously, the amount of CaO and MgO representing the carbonate phase of stromatolites is also present in varying proportions. The Ca/Mg ratio, which appears to have been controlled by microorganisms, consequently played an important role in the process of phosphate formation/fixation. It is inferred that the high Mg content of the stromatolites inhibited the process of phosphate formation.

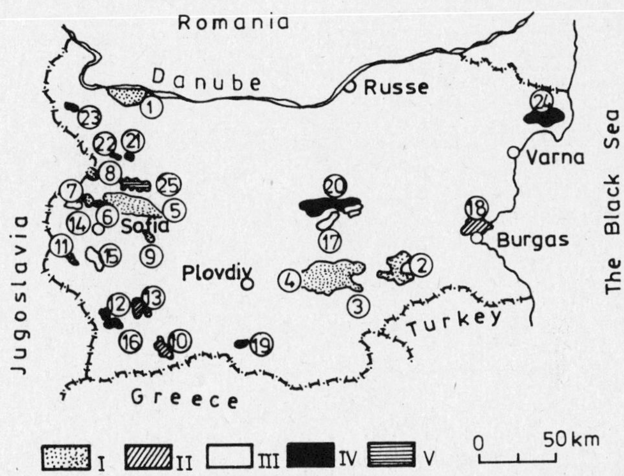

SITTON, D., U. S. Forest Service, Medford, OR

Slope Stability and Hazard Zonation of Enlarged Mt. Ashland Ski Development Area, Jackson County, Oregon, USA

An area proposed for additional ski development on Mt. Ashland was studied to determine the engineering geology and slope stability characteristics. Mt. Ashland is located 17 air miles southeast of Medford, Oregon. The current ski area has 11 ski runs developed on the upper slopes of the 7500 foot mountain. Four more ski runs and two chair lifts are planned for the area. The ski area serves primarily southern Oregon and northern California skiers. The ski area is located within the city of Ashland Watershed. Ashland's drinking water comes from a reservoir which is situated less than 5 miles downstream from the proposed ski area.

Mt. Ashland is located within the Klamath Mountain Geologic Province. This portion of the province is made up entirely of plutonic rocks. The granitics encompassing Mt. Ashland are composed of two episodes of Jurassic age intrusions. The plutonic rock body was exposed after millions of years of uplift and erosion of overlying metamorphic rocks. Five varieties of granitic rocks were mapped in the area including: diorite, tonalite, syenodiorite, granodiorite, and quartz monzonite. Contacts between the rock types are gradational with exception of syenodiorite outcrops on the upper slopes. In many areas the rock outcrops are very weathered. As a result of weathering, much of the rock is partially to completely decomposed. Soils are composed of silty sand and sandy loam. The soil is noncohesive and tends to erode easily. These soils comprise the failure prone mantle over weathered bedrock.

During field mapping numerous areas of instability were located. Unstable terrain is characterized by: alder glades, severe erosion, landslide scarps, gutted channels, over-steepened stream banks and wet deposits. Several landslides mapped are currently active. Some failures are still largely unvegetated with raw soils on their scarps. Active landslides (activity level 1 on accompanying map) have occurred within the past 10–15 years. Many landslides may have been triggered during an intense storm that occurred in 1974.

Alder glades are groundwater exit points, many of which begin at similar elevations throughout the project area. These sites are interpreted as groundwater conduits which have subsequently collapsed and sediments eroded away. Water was brought to the surface when mass movement occurred as a result of groundwater seepage forces in the sapping zone near the origin of groundwater flows. The combination of weak tractive forces, strong seepage and gravity forces, and increased pore pressure on the soils has caused failures to progress upslope. The evolution of groundwater sapping failures to alder glade formation is also evidenced by landslide scarps found just above the beginning of most alder glades. Springs begin to issue at the base of landslide scarps.

Water has been brought to the surface downstream from these zones via piping holes during peak groundwater flows. The piping channels enlarge over time and partially collapse. Eventually more of the pipe collapses creating gully features. Groundwater pipes are also found underneath landslide deposits well downslope from the zones described above.

Creek channels and swales are deeply incised down to bedrock from surface erosion and debris landsliding. These gutted channels are located both in alder glades and through timbered terrain. Sheet, gully, and dry ravel are all erosion processes taking place to develop and deepen these channels after landslides originated them.

A slope stability hazard zonation map was created to categorize the ground within areas proposed for further ski developments. Hazard zonation is a method used by the Rogue River National Forest which assigns risk zones to terrain considered for management. The Mt. Ashland hazard zonation map is used in conjunction with the one-inch to 500 feet scale slope stability map. The hazard zonation map shows four different hazard zones. Each hazard zone evaluates the risk of landsliding associated with developing ski runs and lifts in each zone. Estimated risk percentages for zones increase significantly from natural to developed conditions.

Hazard zone 1 represents the highest risk terrain and is classified as unstable. Hazard zone 2 is not as unstable as zone 1, but is still sensitive (i.e., subactive and potentially active landslides). Hazard zone 2 is classified as low stability (high risk) terrain. Hazard zone 3 represents dry, somewhat rocky soils with a few dormant landslides. Zone 3 is classified as moderately stable. The most stable sites are located in hazard zone 4. It reflects the rocky ridges and relatively flat, smooth slopes.

Risk percentages are used to show how development increases the probability of activating landslides within each zone. The highest risk percentages are for developed conditions in hazard zone 1 and the lowest are in hazard zone 4 for natural conditions. Potential impacts that may be expected from developing the ground within hazard level 1 is higher in terms of total cubic yards of sediment to streams than the other hazard zones. The potential impacts are calculated from past landslide volume totals for each hazard zone. During EIS meetings potential impacts were given in size and volume expected from landslides. The time frame used for potential risks of natural landsliding conditions is from 0–50 years. A shorter time frame is used for developed conditions to indicate that ski development will tend to accelerate landslide frequency.

SIZE, W. B., Emory University, Atlanta, GA

Anatexis and Control of Melt Extraction by Ductile Shearing within Proximal Granitic Gneiss: Mount Arabia Migmatite, Southern Appalachians

The Lower Paleozoic Inner Piedmont crystalline terrane of the Southern Appalachians in Georgia contains extensive regions of granitic gneiss with associated migmatite displaying varying degrees of small scale brittle-ductile deformation. In this region, the Mount Arabia migmatite, a phase of the Lithonia Gneiss, shows a close spatial relationship between zones of closely spaced, intersecting, en echelon ductile shears and neosome formation. The migmatite exhibits marked competence contrasts and structural heterogeneity on the macroscale, which appear to be related to the degree of shearing and melting of the gneiss. Over an area of 20 to 50 m^2, gneissic zones of higher melt proportion include contorted banding, buckle folding, biotite schlieren and pods of nebulite. The ductile shear zones are interpreted as having served as interconnected pathways for the extraction of melt. Melt accumulated where these pathways coalesced, forming small "melting centers" or protodikes. The neosome displays discordant, intrusive structures in areas where melt fractions approach 30% of the migmatite.

Incipient in situ melting of the granitic gneiss is evidenced in the modal similarity between the granite neosome (qtz 41%, micro 27%, plag 28%, biot 0.4%) and the paleosome (qtz 50%, micro 25%, plag 17%, biot 3%) in these melting centers. Trace element differences between the neosome and the proximal granitic gneiss paleosome is accounted for by the lower proportions of biotite, zircon, sphene, apatite, monazite, and allanite in the neosome.

SKEHAN, J. W., S. J., Weston Observatory, Weston, MA, and A. PIQUE, Université Louis Pasteur de Strasbourg, Strasbourg, France

Comparative Evolution of Avalon Zone of New England and of Panafrican Zone of West Africa from Proterozoic through Late Paleozoic Time

Composite stratigraphic columns of the Avalon zone of southeastern New England and of the Panafrican zone of West Africa are presented in Figure 1.

The Eburnean basement (P I Group of Choubert) of high metamorphic grade, is exposed in the WA craton and in Anti-Atlas inliers. It is overlain by shallow shelf deposits (P II Gp.) from which quartzarenites and stromalitic limestone olistostromes (Blackstone Gp. of NE) were derived (Bailey et al, in press) during crustal distention at 788 Ma (Clauer, 1976). Oceanic crust was produced in that phase in the Anti-Atlas (Bou Azzer ophiolite) (LeBlanc, 1975) and gabbro and basalts are also represented in NE (Middlesex Fells V.S.). Pan-African I and Avalonian orogenies resulted in closure of oceanic and thinned-crustal basins with associated metamorphism about 685 Ma. The counterparts of the P II–III sequences of WA and the Panafrican II angular unconformity have not been recognized in NE. The remaining Upper Proterozoic rocks in both WA and NE are characterized by thick, coarse to fine clastic units containing volcanics (Ouarzazate-Tanalt Group and purple shales of the "Lie-de-vin" Gp.-P III, and Mattapan V.S. and Boston Bay Gp., respectively) and 650–575 Ma late- to postorogenic calcalkaline plutonic-volcanic suites present both in WA (Tasulecht Gr.) and in NE (Esmond P.S.) The calcareous Adoudounian Gp. of WA may be represented in NE by the top olistromal unit of the Newport Group. The Adoudounian is cut by alkaline plutonic rocks at 534 Ma.

Lower Cambrian rocks of WA consist of transgressive Archeocyathid limestones (Upper Adoudounian) and fine clastic sediments; those of NE in places have a basal conglomeratic quartzite (Hoppin Quartzite) going up into green and maroon slates with thin hyolite- and trilobite-bearing limestones (Weymouth Fm.), The Middle Cambrian alternating greywackes and siltstones of the Anti-Atlas carry Paradoxides trilobites as is also the case with the Braintree Slate of Boston. The Middle Cambrian of southeast Rhode Island also carries the Paradoxides fauna together with the distinctive *Badulesia tenera* Hartt trilobite having affinities with Morocco and Spain (Skehan et al, 1978). The NE shallow-water Cambrian is thus similar to that of WA in that both contain the Acado–Baltic fauna but differ in that the sequence is thin in NE and thick in WA (8000 m) where volcanic rocks are developed. The Upper Cambrian in NE is not recognized and in Morocco has been partially removed by erosion. The Ordovician in NE has only recently been recognized by acritarchs in the distal turbidites of the Dutch Island Harbor Formation, Rhode Island; in WA, a shallow-water, upward-coarsening clastic sequence, including glacial deposits, forms the Ordovician succession. The Upper Cambrian and Lower Ordovician conglomerates, arkoses, and turbidites suggest a deep sea environment.

During Silurian, Devonian, and Mississippian time, the evolution of these two regions differs markedly. Two episodes of distention are represented in NE by alkaline plutonism and volcanism in Ordovician–Silurian (Quincy P.S.) and in Devonian time (Cape Ann Gr., Scituate P.S.). A thick succession of graptolitic blue shales and intercalated limestone beds in the Silurian represents a transgression due to melting of the Upper Ordovician ice cap. Devonian sedimentary rocks consist of both marine and continental rocks, with Middle Devonian representing the maximum extent of the transgression. The sedimentation ended with Eovariscan folding at the end of Devonian. In Morocco, after the Famennian transgression, blocks were rising and basins subsiding. In the Sidi Bettache basin, several thousand meters of detrital sediments were deposited and alkaline to tholeitic magmas rose (Pique and Michard, in press). Beginning in Late Visean (Middle Mississippian), that basin and the foreland basin of the Eovariscan belt closed progressively. In NE a thick, abundantly fossiliferous sequence of Pennsylvanian coal-bearing, coarse to fine sediments, was deposited in intermontane basins. Several phases of regional metamorphism are recognized and granites (Narragansett Pier P.S.) were produced during the strong Allegheny collision near 272 Ma.

Namurian through Stephanian and Lower Permian aged coal-bearing beds were deposited in WA. In Morocco the last Variscan deformation, including plutonism, occurred before Westphalian C (Late Pennsylvanian), which is represented by limnic redbeds and thin coal sequences.

Thus, the Proterozoic and Paleozoic stratigraphic and tectonic evolution of NE and WA is remarkably similar from Proterozoic through Ordovician time. Their evolution is partially dissimilar from Silurian through Mississippian time with rifting and production of alkaline plutonic rocks taking place earlier in NE and during Mississippian time in Morocco. Both NE and WA have strong similarities during Pennsylvanian sedimentation and subsequent orogenic deformation.

REFERENCES

Bailey, R., J. W. Shkehan, R. Drier, M. Webster, in press, Olistostromes of the Avalonian terrane of southeastern New England, in Melanges and olistostromes of the U.S. Appalachians: GSA Special Paper 228.

Clauer, N., 1976, Geochimie isotopique du strontium des milieux sedimentaire. Application a la geochronologie de la couverture de craton ouest-africain: These Sci., Univ. Strasbourg, 227 p.

LeBlanc, M., and J. Lancelot, 1980, Interpretation geodynamique du domaine pan-africain (Precambrian terminal) de l'Anti-Atlas (Maroc) a partir de donnes geologique et geochronologique: Canadian Journal of Earth Sciences, v. 17, p. 142–155.

Skehan, J. W., D. P. Murray, A. R. Palmer, A. T. Smith, and E. S. Belt, 1978, Significance of fossiliferous Middle Cambrian rocks of Rhode Island to the history of the Avalonian microcontinent: Geology, v. 6, n. 11, p. 694–698.

Pique, A., and A. Michard, in press, Moroccan Hercynides: a synposis. The Paleozoic sedimentary and tectonic evolution at the northern margin of west Africa: American Journal of Science.

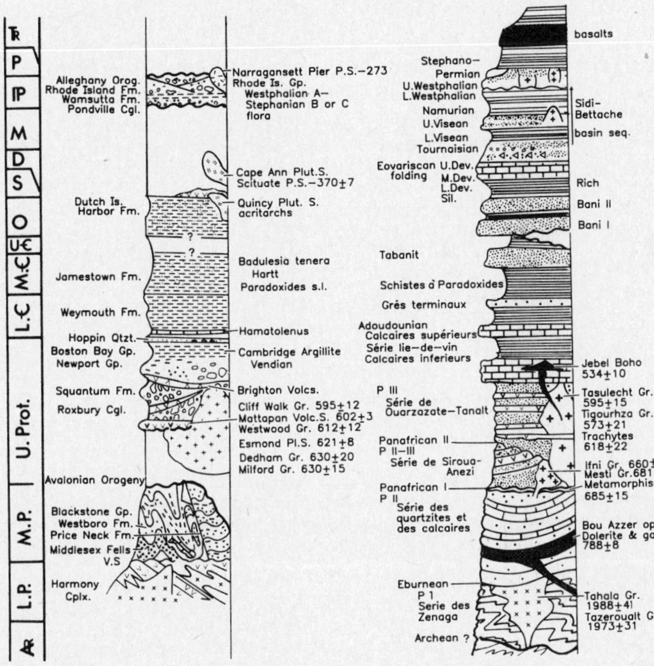

SKINNER, H. C. W., Yale University, New Haven, CT

Crystal Chemistry of FE and MN Biominerals

Biological sequestration of Fe and/or Mn as oxide/hydroxide complexes and the resultant formation of mineral species may be used to indicate environmental parameters (Murray, 1979). A review focusing on the parallelism of the crystal chemistry of these two systems presents an opportunity to examine biological mechanisms that may contribute to the accumulation, precipitation, growth and diagenesis of minerals at low temperatures and pressures.

REFERENCE

Murray, J. W., 1979, Iron Oxides, *in* Roger Burns, ed., Marine minerals: Washington, D.C., Mineralogical Society of America Short Course Notes #6, p. 47–98.

SKLASH, M., University of Windsor, Windsor, Canada

Concepts of Runoff Generation after Two Decades of Natural Isotope Tracer Studies: Implications for Acidification of Streams

In catchments subjected to acid precipitation, the flow path(s) and residence time(s) of the incident precipitation in the catchment play critical roles in governing the degree of both long-term and episodic acidification of streams (Marmorek et al, 1986). Since the soil-water contact time is shorter for "rapid" runoff flow paths than it is for slower subsurface Darcian flow, it follows that catchments conducive to runoff events dominated by "rapid" flow paths afford the acid precipitation less opportunity by runoff events dominated by "slow" flow paths. As a result, streams dominated by "rapid" flow paths should be characterized by greater pH depressions than those dominated by "slow" flow paths.

During the past two decades, natural isotopic tracers (oxygen-18, deuterium, and tritium) have been used in more than 30 studies to assess the flow path(s) and residence time(s) of water in catchments. The results from these studies indicate that in most cases, storm and snowmelt runoff events in humid, temperate areas are dominated by water that has reached the stream via "slow" subsurface flow paths. Unfortunately, most of these studies have not been concerned with the acidification problem.

The main objective of this lecture is to summarize the state of the art of natural isotope tracer studies as it relates to the stream acidification problem by addressing six important questions. How does the natural isotope tracer technique work? What have two decades of natural isotope tracer studies taught us about flow paths and residence times in catchments? What are the general implications of these findings on stream acidification? Is there a physical mechanism that can reconcile the apparently contradictory concepts of rapid stream response to rainfall and snowmelt, and of "slow" subsurface dominance of runoff events? What about macropore flow? How do macropores affect residence times? If "slow" flow dominates, what happens to the acid pulse from spring snowmelt?

REFERENCE

Maramorek, D. R., K. W. Thornton, J. P. Baker, D. P. Bernard, and B. Reuber, 1986, Acidic episodes in surface waters: the state of the science: Final report for the U.S. Environmental Protection Agency, Environmental, Research Laboratory, Corvallis, Oregon, 232 p.

SKUODIS, V., Chicago, IL

New Paleoexogenetic Phenomenon in Baltic States Previously Unknown in Classical Physical Geology

The author of this report conducted engineering geology studies in the Daugava River Basin (Latvia) for hydroelectric power construction projects. Large dislocation structures in Upper Devonian Fran stage deposits were interpreted as resulting from exogenetic processes. These dislocation structures were previously unknown in the geological literature, and have no known analogues in other geologic regimes.

The Fran stage in the Baltic States (Latvia, Lithuania and Estonia) is composed of dolomites with clay interlayers and is underlain by a thick sequence of weakly cemented silts, named the Old Red formation. Within these deposits, which are overlain by Quaternary materials, mapping of borings and excavations revealed spherical collapse structures. The diameter of these features is about 70 m, as is their depth. The upper portions of the collapse structures are filled with dislocated younger terrigenic deposits, which are not found outside the collapse structures. The lower portions of the collapse structures are filled with broken and mixed dislocated rocks which are found in place at much higher horizons outside of the collapse structure. The amplitude of these dislocations reaches 40 m. The surface of the in-filled collapse structures coincides with the general sub-Quaternary erosional surface.

The appearance and the development of the collapse structures is associated with the development of the old pre-Quaternary hydrographic system. Fragments of this system, such as buried river valleys, are found in many areas within the Baltic States. The bottoms of these buried valleys reach 200 m lower than present sea-level in some places. These valleys are bound by upper and middle Devonian Old Red formation deposits.

At the time of the existence of this paleohydrographic system, conditions favored corrosion of Devonian carbonates, undermining (suffusion), and eventually subterranean erosion of the terrigenic deposits. As a result, the winnowing of bottom fines led to undermining and the collapse of the overlying layers. This process was aided by tectonic stresses.

Ten stages in the development of these peleoexogenetic collapse structure have been identified. Corrosive processes dominated in the earlier stages, while later undermining (suffusion), and later still underground erosion processes dominated.

It has been calculated that the density of the collapse structures may reach 20 km^2. They are up to 2 km or greater distance from the buried river valleys. The minimum distance between neighboring collapse structures reaches 100 m.

These same collapse structures have been observed in other areas in Latvia, including the Baltic shore along the Gulf of Riga. They are found in northern Lithuania as well.

Paleogeographic studies indicate that these structures are of Miocene age. At that time, the Baltic States area, as part of the periphery of the East European platform, was 300–400 m higher than its present elevation relative to sea level. This conclusion is dictated by the paleogeographic studies described here by the author. The end of collapse structure development, and the beginning of their preservation began at the end of the Miocene and the beginning of the Pliocene.

The study of these phenomena allows definition of the periods of karst feature development in the Baltic States area from the upper Devonian to the Tertiary inclusively. These studies provide greater detail of the paleogeographic development of the Baltic States as well. They provide the basis in dynamic geology to identify a new exogenetic type, in addition to the known karst and undermining (suffusion) features. The author suggests the name, "karst-suffusional-erosional phenomena."

SLACK, J. F., and W. C. SHANKS III, U.S. Geological Survey, Reston, VA

Geologic and Isotopic Characteristics of Modern and Ancient Besshi-Type Massive Sulfide Deposits

Besshi-type massive sulfide deposits form a class of stratiform ores hosted by clastic metasedimentary rocks (metapelite, metagraywacke) with variable amounts of associated thoeliitic metabasalt (greenstone, amphibolite). For the purpose of this report, Besshi-type deposits are classified on a purely descriptive basis without genetic or tectonic implications. Examples range in age from Proterozoic to Mesozoic and include Besshi, Hitachi, and Shimokawa, Japan; Matchless and Otjihase, Namibia; Kilembe, Uganda; Sulitjelma, Joma, and Killingdal, Norway; Rouez, France; Gairloch, Scotland; Goldstream, Anyox, and Windy Craggy, British Columbia; and Ducktown, Gossan Lead, Ore Knob, and Elizabeth in the U.S. Appalachians. Some of these deposits are very large, such as Windy Craggy (>200 million tons), Ducktown (>180 million tons), Rouez (100 million tons), and Gossan Lead (>50 million tons), and may extend for several kilometers along strike.

The deposits typically consist of abundant pyrrhotite and/or pyrite, variable amounts of chalcopyrite, generally minor sphalerite, and rare galena and arsenopyrite. Many of the sulfide bodies contain only minor amounts of base or precious metals, and as a result have been mined principally for their content of iron sulfides (e.g., for sulfuric acid manufacture). However, some of the larger deposits have significant copper grades (3–4 wt% Cu), and therefore represent important copper resources; Windy Craggy, for example, has 85 million tons of 3.0% Cu in one zone. Minor amounts of zinc are present in many of the ores, typically in the range of 0.5–1.5 wt% Zn, and significant gold and silver values are known locally; high silver (5–10 oz/ton) may occur in deposits lacking galena. Concentrations of Co, Pb, Mo, As, and Se are also present locally, and some of these elements (e.g., Co) may be economically recoverable as well.

Besshi-type massive sulfide deposits are characterized by wall rocks that commonly include several unusual lithologies. The unusual lithologies contrast sharply with the typical country rocks of the massive sulfide bodies (i.e., pelitic schist, metagraywacke, greenstone and/or amphibolite), and may consist largely or wholly of quartz, magnetite, spessartine garnet, sodic plagioclase, tourmaline, chlorite, sericite, biotite, and/or carbonate; graphite concentrations occur locally. Examples include stratabound (and in many cases stratiform) beds or lenses of quartzite or chert, magnetite iron formation, coticule (fine-grained quartz–spessartine rock), albitite, tourmalinite, chloritite, mica-rich schist, or marble. These unusual lithologies are believed to have formed by subsea floor alteration of clastic sediments and/or as exhalative chemical precipitates, and serve as valuable exploration guides. A survey of many Besshi-type deposits suggests that, in greenschist- and lower amphibolite-facies terranes, chlorite- and sericite-rich alteration zones, albitite, and magnetite iron formation occur especially close to the sulfide bodies.

Besshi-type massive sulfides also have a broad range of sulfur isotope values. Sulfur isotope values ($\delta^{34}S$ of sulfide minerals are typically from 0 to about +12 ‰, and reflect derivation of sulfur from the following sources: (1) basaltic sulfide with $\delta^{34}S$ values near 0 ‰, (2) ^{34}S-depleted sedimentary bacterogenic sulfide, and (3) ^{34}S-enriched sulfide derived by reduction of seawater sulfate during reaction of hydrothermal fluid with organic matter in the sediment column. The relative contribution of sulfur from igneous and sedimentary sources and the nature of sedimentary lithologies in the footwall to the massive sulfide ores provide major controls on the sulfur isotopic composition of the sulfide minerals in the deposits.

Lead isotope values of galenas in the deposits show a range from relatively unradiogenic ratios ($^{206}Pb/^{204}Pb$ = 17.5; $^{207}Pb/^{204}Pb$ = 15.4; $^{208}Pb/^{204}Pb$ = 37.4) to relatively radiogenic ratios ($^{206}Pb/^{204}Pb$ = 20.0; $^{207}Pb/^{204}Pb$ = 15.7; $^{208}Pb/^{204}Pb$ = 38.8). The spread of lead isotope data indicates that lead in the ores was derived both from basaltic and sedimentary sources. Some deposits have components of distinctly unradiogenic lead despite being largely hosted by clastic metasedimentary rocks with only minor associated metabasalt (greenstone, amphibolite). The volumes of exposed metabasaltic rocks seem inadequate to account for the amount of unradiogenic lead (and ^{34}S-depleted sulfide) in these ores, thus suggesting that some of the unradiogenic lead was derived from basaltic sources at depth. In contrast to these unradiogenic leads, the relatively radiogenic values imply extensive interaction of the hydrothermal fluids with clastic sediments, in agreement with limited carbon and boron isotope data for the deposits. An unresolved problem is the lack of appreciable lead in such ores, despite evidence for fluid-sediment interaction during subsea floor mineralization.

In the majority of Besshi-type deposits, well-defined footwall feeder zones have not been recognized. Although this may be a function of deformation and metamorphism in some areas, the apparent lack of alteration pipes in many deposits suggests complex mineralizing processes. The absence of feeder zones, together with the blanket or sheet-like morphology of these deposits, may result from any of the following processes: (1) primary deposition in an aerially extensive brine pool which accumulates in a seafloor depression away from vent site(s), (2) physical transport of sulfide precipitates in a hydrothermal plume away from the vent site(s), or (3) diagenetic replacement of permeable sediments in the shallow subsurface below the sediment-seawater interface.

The paleotectonic setting of ancient Besshi-type deposits is poorly understood. Evaluation of such settings is limited by the post-depositional deformation and metamorphism that have commonly affected most of the massive sulfides and their country rocks. Regional geologic information, however, provides some insights, suggesting that Besshi-type deposits probably formed in several different tectonic environments including (1) sediment covered spreading ridges both in continental-margin and back-arc basin settings (e.g., Windy Craggy, BC; Sulitjelma, Norway); (2) within intracontinental rifts (e.g., Ducktown, Tennessee; Gossan Lead, Virginia); and (3) along rifted continental margins (e.g., Elizabeth, Vermont). There appears to be a continuum in the geologic setting of ancient Besshi-type deposits, from largely or wholly clastic metasedimentary sequences (e.g., Ducktown, Rouez), to metasedimentary sequences with minor associated felsic metavolcanic rocks (e.g., Anyox), and to sequences with abundant mafic metavolcanic rocks (e.g., Windy Craggy, Sulitjelma). Deposits which contain minor proportions of felsic metavolcanic rocks suggest proximity to island arcs, whereas deposits within greenstone/amphibolite sequences containing only minor clastic metasediments suggest transitions to ophiolite settings.

Possible modern analogs of Besshi-type deposits have recently been discovered in the northeastern Pacific Ocean in Escanaba Trough (Gorda Ridge), Middle Valley (Juan de Fuca Ridge), and Guaymas Basin (Gulf of California). All of these deposits occur in fine-grained turbidites overlying an oceanic spreading ridge and are distributed along ridge-parallel faults or peripheral faults associated with near-surface basaltic laccoliths or sills. They consist mainly of Fe-sulfides (pyrrhotite, marcasite, pyrite), with subordinate Cu-sulfide (chalcopyrite, isocubanite), Zn-sulfide (sphalerite, wurtzite), galena, calcite, barite, talc, silica, and hydrocarbons. Subsurface alteration in these deposits includes chlorite-rich and chlorite-albite-epidote assemblages.

The actively forming deposits in the Red Sea also may represent a modern analog of some Besshi-type deposits, particularly for those stratiform, sheet-like ore bodies which lack a distinct footwall feeder zone. The deposits of the Atlantis II Deep are similar in their extensive and sheet-like form (14 km × 6 km × 10 m), large size (250 million tons), association with thoeliitic basalt, and abundance of Fe-Zn-Cu with only very minor Pb. However, the Red Sea deposits are different in having dominantly hemipelagic sedimentary host rocks instead of clastic marine turbidites, and in having associated evaporites in the depositional environment.

SLADEN, C. P., British Petroleum Co., London, United Kingdom

Lake Systems: Key Elements during Search for Hydrocarbons

During the search for hydrocarbons in a basin believed to contain a lake system it is necessary to characterize many parameters including: (1) structural history of the basin; (2) favored location of the lake site(s); (3) likely size of the lake site; (4) likely depth of the lake; (5) length of time the lake may have existed; (6) whether or not the basin was a terminal lake site; (7) the shape of the lake; (8) climate and vegetation; (9) the size and topography of the catchment; (10) composition of bedrocks within the catchment; and (11) lake chemistry.

Given an understanding of these controls and their interplay, it becomes possible to predict the infill of a lake basin and subsequently the play types, play fairways, and potential for commercial hydrocarbons.

Four major play fairways develop in lakes formed in rifts and flexural basins:

1. Axial fairways principally involving lacustrine turbidite and lake delta reservoirs.

2. Footwall and thrust belt fairways principally involving lacustrine fan reservoirs.

3. Hangingwall and peripheral bulge fairways principally involving shoreline, shallow lake and lake margin delta reservoirs.

4. Transfer zone and lateral ramp fairways principally involving lacustrine delta reservoirs.

The advantages and disadvantages of each of these fairways will be illustrated, together with the nature of reservoirs/seals (trap configuration) and the producing capabilities of fields in these fairways.

SLOSSON, J. E., Slosson and Associates, Van Nuys, CA, and J. W. WILLIAMS, San Jose State University, San Jose, CA

Current and Future Difficulties in Practice of Engineering Geology

The next five years to ten years are critical to the survival of engineering geology. If the geologic community does not accept the challenge to improve its overall performance record, the contributions that the geologist can make will not be wanted or valued by society.

Many universities have dropped field techniques, field geology, and geomorphology. The number of schools that offer engineering geology and hydrogeology is small. Some universities are producing excellent computer geologists who do not have the ability to identify, collect, or evaluate the data they place into the computer. Ninety to ninety-five percent of the recent graduates in geology have not had the fundamental academic courses to prepare them to work in the fields of hydrogeology, engineering geology, or hazardous waste management.

The reasons for these trends are many and complex. Field camp operations are expensive; enrollments are declining; sophisticated research laboratories are perceived to have greater prestige; newly hired faculty from laboratory-oriented Ph.D.-granting organizations tend to perpetuate the trend.

Few professors are teaching the basic concepts of scientific analysis. Practicing hydrogeologists and engineering geologists have no manuals available for most of the problems that they encounter. They must rely on their educational background. Why should a correct analysis and solution be anticipated if the fundamentals have not been taught and supervision is lacking?

Research engineering geologists are making great strides in the technical aspects of the science. Computers can process large volumes of numerical data using complex mathematical models to help in solving the complex problems of engineering geologists. However, an analysis of this technology and hardware in comparison with the results of the current practice of engineering geology is distressing. Final solutions to investigations appear to rely increasingly on less than realistic and sophisticated but often unrelated geological models that are often inappropriate for the problem. There are larger numbers of omissions, oversights, misinterpretations, mistakes, failures to record, and lack of geological understanding.

If better scientific theories and tools exist, why are the results of investigations not improving but, in fact, appear to be regressing? Even with advanced technology and theories, people who are unable to use that knowledge effectively will not produce quality results. The current inability to produce quality studies is the result of many related factors including: (1) lack of adequate preparation, (2) inexperience in the application of investigation and analytical techniques, (3) lack of motivation or incentive to do the best work, and (4) lack of competent and timely supervision.

Shortages of well-educated and experienced individuals in engineering and applied geology have been documented. These shortages of qualified personnel exist in the consulting firms and the public agencies. There is an inadequate supply of middle management personnel who are adequately trained to supervise entry-level geologists. Thus, most of the recently hired are not receiving the training or supervision necessary during the first two or three critical learning years.

The academic preparation of many field investigators must be evaluated. Generally, the principal of a company does not collect the basic information to determine the final solutions to a problem. The individual collecting the data is often the more junior of the firm's geological employees. These data are reviewed by more senior people, who can only review the data they receive. They seldom have the opportunity to comment upon what has not been noted or reported by the original observer. If the original observer did not appreciate or understand what was seen, critical information would not be recorded.

Decision-making is frequently controlled by judgements rather than prudent scientific concepts and good engineering practices. The practitioner who attempts to please the client and/or politicians thus may filter, deliberately or subconciously, the scientific data bank to assure conclusions that are compatible with the "expected" results.

The concept that "the level of professional standards of work will drop to the lowest level allowed by government" is a characterization of a serious problem. Many clients request that the work effort of the consultant not exceed the minimum required by government. Thus, the professionals are coerced to reduce the work effort. This practice creates a ready market for the less competent consultants—those limited in knowledge, training, and professional skills.

The outlook is not encouraging. The decline in enrollments in the nation's geology departments indicate that fewer students are preparing themselves to enter the profession. Employment trends show that demand for well-educated and well-trained individuals will triple or quadruple during the next decade.

Indications are that the situation is not going to change. Investigations will continue to be hastily done by unprepared and inexperienced individuals and reviewed by those who are not prepared to conduct the reviews or who are so badly overlooked that the quality of their reviews suffers. If the geologic community fails to accept its responsibility, the engineering profession will usurp the work, and the critical role that geology should play will be neglected. It is imperative that the profession reassess and address the educational background and performance record of engineering geologists.

SMALE, J., University of South Carolina, Columbia, SC, M. RICHARDSON, University of Rhode Island, Narragansett, RI, D. Q. COFFIELD, ARCO Int. Oils and Gas Co., Los Angeles, CA, and R. C. THUNELL, University of South Carolina, Columbia, SC

Tectonostratigraphic Evolution of the Morgan Accommodation Zone, Gulf of Suez Rift Basin, Egypt

The structure of the Gulf of Suez rift is characterized by a mosaic of tilted, fault-bound blocks confined within elevated rift shoulders. This tilt-block mosaic can be broadly divided into three structural provinces each delineated by tilt-blocks with similar dip directions. These dip provinces reverse polarity along strike, from southwest dips in the northern rift, to northeast in the central rift, and back to southwest dips in the southern rift. Each of these provinces are separated by accommodation zones, Gulf-orthogonal structures that accommodate the differential rotation and displacement between adjacent blocks. The fundamental role the Morgan Accommodation Zone (MAZ), the southernmost zone, has played in the structural evolution of the southern Gulf of Suez is reflected not only in the structural geometry of the adjoining subbasins, but also in the evolution of early Miocene sedimentation patterns (as observed in outcrop and in well cuttings) and subsidence histories of the adjacent regions.

The initiation of extension and sedimentation in the Gulf of Suez rift appears to have been synchronous in the latest Oligocene. A basaltic intrusion dated at 24.7 Ma near Gebel Gharamul, which is overlain by redbeds lithologically similar to other Oligocene redbeds found in isolated basins of the proto-Gulf of Suez rift, provides evidence for the activation of the MAZ at this time. During the initial rifting phase, the MAZ is characterized by a low relief setting, with fault scarps developing along the mainrift trend, oriented approximately north-northwest–south-southeast. The earliest synrift sediments are continentally derived clastics deposited by alluvial fan systems and record sedimentation associated with a late Oligocene eustatic sea-level lowering rather than an early doming phase along the central rift axis. Rift depocenters also appear to have formed in the early stages of rift evolution as evidenced by thick sequences of red and grey shales that record sedimentation in lacustrine or marginal marine settings in the Gemsa half-graben. Structures at Gebel Gharamul, Gebel Zeit, Morgan, Tor, and Gebel Araba are all truncated by the early Miocene unconformity on which the first marine synrift sediments were deposited. These sediments belong to the Nukhul Formation of Aquitanian age and show that extension, subsidence, and development of the MAZ were well underway by this time. Deposition occurred along a ramped margin suggesting again that topography generally remained subdued.

The overlying Rudeis Formation appears to be divided into two distinct members, often separated by an unconformity. The lower Rudeis Formation was deposited under deeper marine conditions coinciding with a period of accelerated subsidence in the Gulf of Suez. Planktonic foraminifera in the marl successions at Gebel Zeit constrain the age of the intra-Rudeis unconformity to approximately the N7–N8 planktonic foraminiferal zone boundary (16.5 Ma). During this event, increased block faulting and uplift of the rift flanks occurred. The reactivation of structures during the mid-Rudeis event (16.5 Ma) did not occur throughout the MAZ. The Esh el Mellaha block on the southwest margin of the MAZ was tilted during the initial extension in the rift, then had a stable rift-margin carbonate platform develop on its crest form the late Burdigalian through the Langhian indicating it did not undergo rotation during platform development. In contrast, during the mid-Rudeis event, the adjacent Gebel Zeit block was rotated, as was the Tor Nazzazzat high, and subsidence in the intervening axial trough was accelerated. The locking of the Esh el Mellha block on the trailing edge of the northeast-dipping basal detachment while the other blocks on the same detachment were still active may have been the result of isostatic uplift of the basal detachment. Gebel Gharamul, on the opposite side of the MAZ from the Esh el Mellaha, was also reactivated during the mid-Rudeis event as was the Morgan high, and the Ramadan and Gebel Araba blocks. Reactivation of the MAZ during this tectonic episode also led to a marked change in depositional styles. This event, which coincides with a global sea-level lowering, enhanced erosion of the rotated blocks and uplifted margins (i.e., Gebel Zeit, Red Sea Hills). Eroded material from these uplifted margins was transported by paleochannels that followed the inherent fault-block topography and deposited as active fan-delta systems in the adjacent paleodepocenters such as the Morgan trough and the Gemsa and El Qaa subbasins. This increased supply of clastics may account for the decrease in total carbonate (as measured in well samples) across the intra-Rudeis boundary. In other blocks (i.e., Gebel Gharamul, Gebel Gharbi, and Tor Nazzazzat) continued rotation is evidenced by truncation and onlap of the succeeding upper Rudeis and Kareem Formation carbonates. This represented the last major phase of movement within the MAZ. Following the mid-Rudeis event extension and subsidence were generally confined to the axial trough with no significant differential displacement between tilt-block domains. The close of the Gharandal Group sedimentation cycle may have coincided with the emergence of localized sills that eventually isolated the Gulf of the Mediterranean Sea and initiated the formation of the evaporite deposits of the Ras Malaab Group.

SMIRNOFF, L. S., Smirnoff Minerals Corporation, Parker, CO

Binary Multiple Harmonics of Sedimentary and Orogenic Cyclicity, Phanerozoic

There is a major concept of geotectonic cyclicity—the concept of global orogenic phases originated by Stille. The synchronicity of these phases has been endlessly discussed since the concept was introduced. Is there any system in the succession of the orogenic phases and related sedimentologic cycles of the different orders?

Strakhov and Sloss were the first who recognized some global regularity and hierarchy of cratonic cyclicity. Khain (1973) synthesized all data for geosynclinal sedimentation, subdivided by orogenies and silicic magmatism. This pattern of fully developed geosyncline cannot be considered universal. According to Stille and Khain, the successive geosynclinal cycles (Grenville, Baikalian, Caledonian) themselves can be considered as the stages of geosynclinal evolution during Post-Huronian time. The oldest Proterozoic cycles are the cycles of the prevailing initial stages; the youngest ones are the fully developed cycles. A similar concept of evolution has been proposed by Rundkvist (1968) for the cyclical system of magmatism.

The simple binary multiple harmonics (BMH) system of sedimentary cycles with orogenic cycles as their boundaries can put all empirical data in good agreement. For example, the BMH system for Phanerozoic of the North American Craton can be shown on the basis of Stille orogenies, Sloss (1963, 1964) sequences, and Wheeler (1963) pairs of subsequences in Table 1.

The following global data for each order of cyclicity can be confirmed:

2^{n+1} Two major sea-level cycles (Hallam, 1977; Vail et al, 1977); pattern of O^{18} distribution (O'Neil, 1979) and geomagnetic field change (McElhinny, 1978).

2^{n+2}Four cycles based on C_{org} distribution (Ronov, 1958); S34 distribution (Kaplan, 1975); change of atmospheric O (Budko et al, 1979); prime maximum of taxa becoming extinct.

2^{n+3}Four Mz-Cz sequences in West Siberia (Nalivkin et al, 1969); zones of unstable geomagnetic polarity (Kravchinsky, 1979).

2^{n+4}Pairs of subsequences for each sequence of Mz-Cz in West Siberia (Nalivkin et al, 1969).

The main confirmation of the BMH system comes form the statistical distribution of unconformities and radiometric dates for the Phanerozoic. We have been able to illustrate this system in detail for the Cenozoic and Pleistocene cycles. If so, instead of an empirical "pile" of cycles of different orders, we are offering a system with much greater capability for application. This system can be theoretically justified by accepting a specifically pulsating (expanding-contracting) earth with a nonlinear geologic time scale. Some aspects of this theory have been tested and are in good agreement with all data. The proposed system has significant potential because it provides a new basis for global stratigraphy including the Precambrian, which lacks conventional stratigraphic data. The nonlinear nature of geologic time, which has good cosmological justification provides a new approach to stratigraphy. It should be noted that Bode's law of planet distances to the sun is a sequence to 2^n. The exception from this law is explained in terms of the BMH system.

Table 1.

Orders of Cyclicity and their age (m. y.)				
2^n	2^{n+1}	2^{n+2}	2^{n+3}	2^{n+4}
PHANEROZOIC	MESO-CENOZOIC	Alpian (Tejas)		LATE ALPIAN (L. S. 0.01-0.014)
			Late	
			Early	
		Kimmerian (Zuni)		LARAMIAN (50 - 70)
			Late	Austrian (95 - 115)
				Kimmerian (130 - 150)
			Early	Early Kimmerian (180 - 200)
	PALEOZOIC	Hercynian		LATE HERCYNIAN/PFALZIAN (200 - 245)
			Late (Absaroka)	Uralian (272 -290)
				Sudetian (310 - 328)
			Early (kasakaskia)	Bretonian (340 - 355)
		Caledonian		LATE CALEDONIAN/ERIAN (385 - 410)
			Late (Tippecanoe)	Taconian (425 - 440)
				Sardinian/Salairian (480 - 515)
			Early (Sauk)	Missing (L. S.)
				LATE BAIKALIAN/LIPALIAN (555 - 590)

SMIRNOFF, L. S., Smirnoff Minerals Corporation, Parker, CO

Binary Multiple Harmonics of Sedimentary and Orogenomagmatic Cyclicity, Precambrian

There are numerous stratigraphic subdivisions of the Precambrian. We have synthesized one more, which represents a binary multiple harmonic (BMH) system of cyclicity. It includes the Phanerozoic and is based on widely recognized data by Ronov, Salop, Semenenko, Sloss, Vail, and others. This system can be illustrated in great detail for the Paleozoic, Cenozoic, and Pleistocene and could be extended into the Precambrian.

Table 1 shows the proposed system, which is a hierarchy of binary cycles. It is based on the following assumptions for each order of cyclicity:

2^2 Protogean and Neogean after Stille. The major Hudsonian event is an upper boundary of the transitional taphrogenic zone with granites rapakivi about 1.65×10^9 y between mainly geosynclinal and cratonic Precambrian (Salop, 1977, III, IV). Note a similar significance of the taphrogenes with the final silicic magmatism for Paleozoic geosynclines (Kharkevich and Moskaleva, 1969) and the Post-Hudsonian maximum of the continental shift about $1.5–1.6 \times 10^9$ y (after Irving and McGlynn, 1981). The subdivision of Precambrian into two Precambrian mega units has been discussed recently (Shurkin et al, 1984; Voitkevich et al, 1986).

2^3 The cycles based on lithology and stromatolites (Semikhatov, 1974).

2^4 The cycles of Late Archaean–Late Ryphean after Semenenko (1970). Phanerozoic after Vail et al, (1977, Figure 1, sea-level pattern); O'Neil (1979, O^{18} distribution); McElhinny (1978, geomagnetic field pattern); and others.

2^5 The cycles of Ryphean after Semikhatov (1974, Figure 14); Late Arahaean, Early Aphebian, and Ryphean after Semenenko (1970). The Phanerozoic cycles based on four major tectonic epochs, and also after Ronov (1958, C_{org} distribution), and Kaplan (1975 S^{34} distribution). Two inferred events, E. Ar_1/E. Ar_2 and L. Aph_1/L. Aph_2, are the only addition to complete the system.

Three different alternative models (M) of the cycle duration's (D) change with time have been extrapolated from Phanerozoic in the past. Each of them is based on a different concept of the earth's evolution. This leads to discussion of the earth's contraction-expansion, the nonlinear scale of geologic time, and the change of gravity with time. The latter could possibly be investigated by geological means (Smirnoff et al, 1969; 1975).

The proposed system is in agreement with the broad range of the empirical data. It should be mentioned that Bode's law of planet distances to the sun is also a sequence 2^n. The exception from this law is explained in terms of the BMH system.

Orders of cyclicity 2¹?	2²	2³	2⁴	2⁵	Events	Pre-Cambrain "relict" ages, Salop (1972); Phanerozoic ages from Table 1,2	Pre-Cambrian ages by International Commette, (Kolchanov 1979).	Pre-Cambrian pole movement (Irving, McGlynn, 1981).	M-1 $D2^4/D2^5$	M-1 Ages	M-2 Ages	M-3 Ages
					?							
Planetian	Noegean	Phanerozoic (Phz)	Meso-Cenozoic	Future+ Alpian					D_{MzCz}/D_{Apl} ≈3.46			
					Laramian	0.062-0.067				0.065		
				Kimmerian								
					Late Hercynian	0.225-0.245				0.225	0.225	0.240
			Paleozoic	Hercynian								
					Late Caledonian	0.39-0.41			1.97	0.400		
				Caledonian								
					BAIKALIAN	0.555-0.590	0.6±0.05	0.57-0.60		0.570	0.570	0.585
		Riphean (Rph)	Late	L. Rph₂								
					Katangan	0.65-0.68	----------	Gap	3.60	0.703		
				L. Rph₁								
					Grenville	1.00-1.10	0.9±0.2	Grenville Loop 1.050		1.050	1.050	1.065
			Early	E. Rph₂								
					Elsonian	1.30-1.40	1.3±0.1	Logan Loop ?1.2-1.3	2.10	1.357		
				E. Rph₁				Gap				
					HUDSONIAN	1.60-1.70	1.6±0.1	1.65-1.70		1.695	1.680	1.710
	Protogean	Aphebian (Aph)	Late	L. Aph₂				Coronation Loop				
					Inferred (L.S.)	----------	----------	1.75- switch in direction	3.70	1.780		
				L. Aph₁								
					Karelian	1.90-2.00	1.9±0.2	2.0		2.010	2.040	2.040
			Early	E. Aph₂								
					Unnamed	----------	2.2±0.1	?	2.15	2.268		
				E. Aph₁								
					KENORAN	2.6-2.8	2.6±0.2	2.6		2.565	2.670	2.685
		Archean (Ar)	Late	L. Ar₂								
					Kolskian	----------	3.1±0.2	?	3.80	2.782		
				L. Ar₁								
					Saamian	~3.5	>3.5			3.390	3.600	3.705
			Early	E. Ar₂								
					Inferred (L.S.)	----------	----------		2.20	3.915		
				E. Ar₁								
		IX			Beginning of the Earth ~4.5-5.0					4.545	4.860	5.160

Table 1.

SMIT, C. A., C. ROERING, and O. D. VAN REENEN, Rand Afrikaans University, Johannesburg, South Africa

Major Shear Deformation in Southern Marginal Zone of Archaean Limpopo Granulite Belt, South Africa

The high-grade Archaean Limpopo belt (LB) is situated between the Kaapvaal and Zimbabwe Cratons. It consists of a Central Zone (CZ) that is separated from the Southern (SMZ) and Northern Marginal Zones (NMZ) by major east–northeast-trending strike-slip ductile shear zones. The SMZ, an area of predominantly granulite grade gneissic rocks, is characterized by complex and widespread allochthonous deformation and by the presence of major crustal shear zones. These shear zones occur within the SMZ and also bound this zone from the lower grade granite–greenstone terrane of the Kaapvaal craton. The tectonic evolution of the SMZ can be explained by two distinct deformation cycles: an early fold event (T_1), which occurred under granulite conditions, and a younger shear event (T_2), which also occurred under granulite conditions. The resulting interference has led to the development of a complex fold pattern on both outcrop and regional scales (Figure 1).

The transition from the Kaapvaal Craton to the SMZ is characterized by an increase in the density and conductivity of the crust, and the geophysical data suggest that high-grade rocks in the SMZ are thrusted southward onto lower-grade rocks in the vicinity of the Hout River shear zone (Figure 1). The transition is also characterized by the destruction of the original internal stratigraphy of greenstone belts resulting in tectonically distended remnants in the SMZ that are intimately mixed with granitoids (Figure 1).

Important features associated with the shear zones are: (1) All the shear zones are steeply northward-dipping structures with dip-slip sense of movement. The southward vergence is suggested by different types of strain markers. (2) They enclose large crustal wedges that developed at intervals of up to 5–25 km across strike and that extend for distances up to 100 km (Figure 1). (3) The central portions of these crustal wedges display remnants of the older T_1 fold pattern whereas the marginal zones are characterized by transposition parallel to the younger shear direction. (4) Structural elements such as foliation, lineations, and fold axes were rotated into the shear direction during T_2 to produce straightening zones. (5) The shear deformation (T_2) can be linked to a period of regional retrograde (decompression) metamorphism that occurred under granulite conditions. (6) The Hout River shear zone is the major crustal break that bounds the SMZ and therefore defines the contact between the LB and the adjacent lower grade granite–greenstone terrane. It separates virtually undeformed and unmetamorphosed rocks on the Kaapvaal Craton from intensely deformed granulites in the SMZ. The rocks in this zone were buried to depths of at least 35 km. (7) Several other geological phenomena, such as the distribution of granite plutons, copper-gold mineralization, and the development of a regional retrograde orthoamphibole isograd, are closely associated with some of these shear zones. (8) The age of the shear zones is well constrained by the intrusive relationship of 2670 Ma granite plutons that are related to the shear movement.

The high-grade granite–greenstone lithologies of the SMZ and the adjacent low-grade granite–greenstones formed initially at about 3500 Ma and underwent a regional metamorphic and structural change during the Limpopo Orogeny at about 2700 Ma. Furthermore, it is apparent that the entire high-grade terrane of the LB (± 180,000 km^2), including the SMZ, CZ, and NMZ, was depressed to a depth of about 30 km during the Limpopo Orogeny and was subsequently uplifted and the present erosional level of this entire crustal segment represents an isobaric surface. The T2 shear zones in the SMZ clearly participated in the uplift of the high-grade terrane relative to the adjacent granite–greenstone terranes. This movement may have occurred along certain preexisting thrust faults that are related to a family of southward-verging faults in the granite–greenstone terrane south of the granulite area. The influx of H_2O-bearing CO_2-rich hydrating fluids that established the retrograde orthoamphibole isograd and the zone of retrogression on the granulite/granite–greenstone interface of the SMZ was probably caused by overthrusting of hot over cooler crust during uplift. The structural and metamorphic (clockwise P–T time loop) relationships suggest that deep burial of the LB is related to the north to south collision of the Zimbabwe and Kaapvaal cratons before 2700 Ma. The tectonically thickened crust was initially in isostatic equilibrium, but erosion and tectonic erosion at higher levels gave rise to disequilibrium and the development of magmas in the lower crust, which resulted in uplift and the exposure of the high-grade rocks (Figure 2).

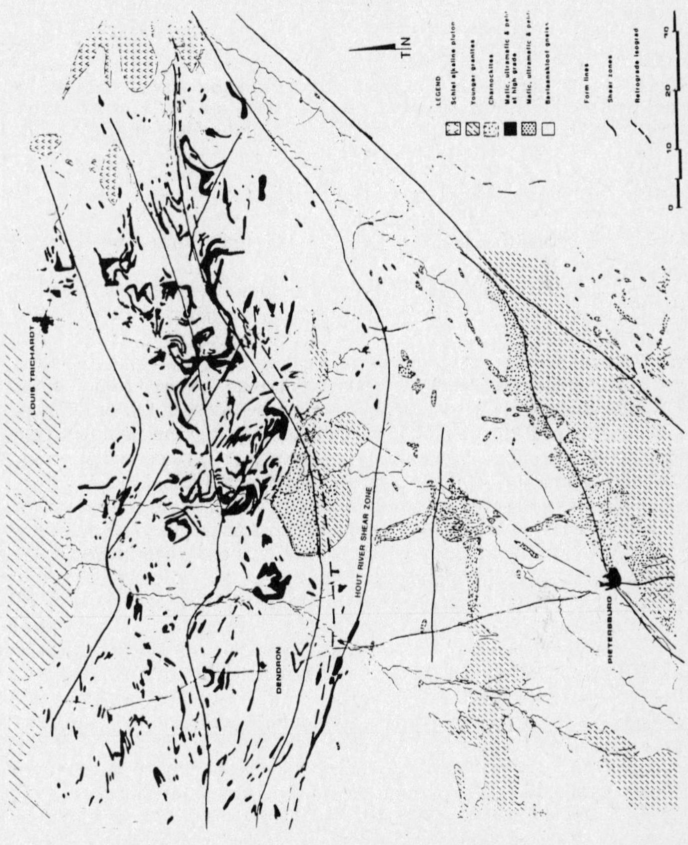

Figure 1—Structural map of the size of the Limpopo Belt.

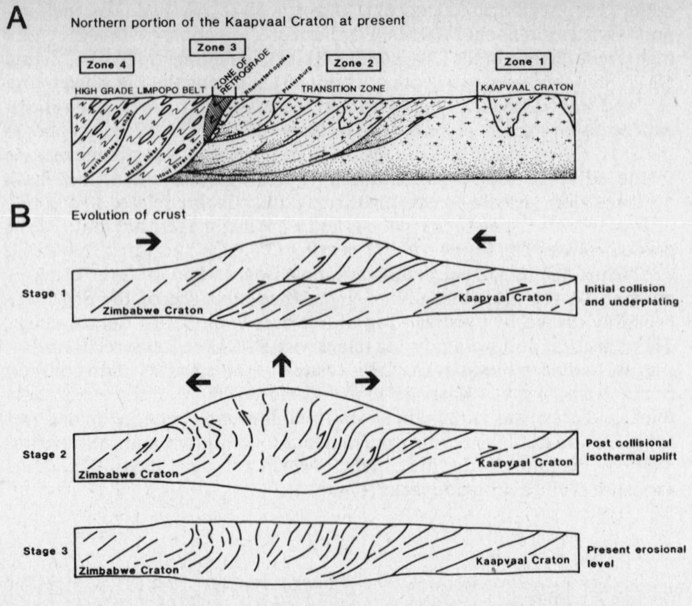

Figure 2—(A) Cartoon section through the northern portion of the Kaapvaal Craton illustrating the relationship between the different structural zones. (B) The evolution of the crust during the Limpopo Orogeny.

SMIT, J., Free University, Amsterdam, Netherlands

High-Resolution Stratigraphy at Cretaceous-Tertiary Boundary

High-resolution studies of high-quality sedimentary sequences across the Cretaceous/Tertiary (K–T) boundary are necessary to test existing extinction models. To separate, for instance, impact triggered extinction from extinction induced by volcanism, we need a geological record with a resolution of 10^3, preferably 10^2 years, rather than the usual resolution of 10^4–10^5 years, although one may question whether such resolution may ever be obtained. In combination the three biostratigraphically complete Spanish K–T sections Agost (AG), Caravaca (CAR) (Smit and Romein, 1985) and Zumaya (ZUM) may approach the required resolution. Comparison of those sections may weed out successfully local sedimentary and diagenetic effects. We report here results of biostratigraphic, trace element (INAA), and stable isotope analyses of these sections. Magnetozonations are known (from ZUM in the Paleocene) and sedimentation rates are estimated from published ages of magnetic reversals (Berggien et al, 1985). The fluctuations in carbonate supply due to extinction of carbonate-producing plankton organisms at the K–T boundary obviously influenced these rates. Therefore, we have kept the noncarbonate flux of detritus to the seafloor constant in the K–T interval. We then estimated the timing and duration of several events at the K–T boundary on the width of measured peaks and shifts in the plotted curves.

Biostratigraphy: Patterns of foraminiferal diversity and production are shown in Figure 1. These patterns do not show any significant changes or trends below the iridium anomaly. Coincident with the Ir anomaly, most species disappear suddenly, and diversity is reduced to a few relicts of Globigerinelloides and Globotruncanella. Guembelitria is relatively abundant. The first Paleocene species appear in AG and CAR at the top of the 6.5, resp. 7 cm thick boundary clay, about 5000 years after deposition of the iridium anomaly.

Trace elements: Trace elements measured by neutron activation analysis show a few distinct patterns in all three sections, shown in Figure 2. A group of largely lithophile elements, (Na, K, Al, Sc, Hf, Th) correlates with clay content. A group of siderophile elements (Ir, Fe, Ni, Cr) correlates normally with clay content, but shows a large anomaly in a 2 mm thick (in AG pure smectite) layer, considered as fallout layer of a large impact event because it also contains microtektite-like spherules and shocked quartz grains. Peak Ir values are remarkably similar in all three sections (26.6 ng/g CAR; 24.6 ng/g AG; and 26.3 ng/g ZUM). Ir shows a tailing in the overlying 1 cm of the boundary clay, interpreted as detrital imput from eroded areas. Rare earth elements (REE) show a pattern almost identical to the first group, with the notable exception of the 2 mm fallout layer, where concentrations are anomalously low (CAR and AG), with low La/Yb (9 AG, 6.5 CAR, all background > 15). As a mix of continental target rocks with meteorite cannot produce this signal, a contribution of low-REE target rock, presumably ocean floor or mantle, is indicated. Some chalcophile elements (Co, As, Sb, Zn, Se, U) show also large anomalies in the fallout layer, but concentrations vary considerably between sections (e.g., As; 0.4–1.4% ZUM; 350 ppm AG; 450 ppm CAR). We interpret these anomalies as a secondary precipitation from percolating groundwater on nondisturbed K–T fallout layers. Wherever the fallout layer is disturbed and broken up by bioturbation (e.g., DSDP site 577), no such anomalous chalcophile element enrichments are found.

Stable isotopes: The $\delta^{18}O$ and $\delta^{13}C$ profiles show strong negative shifts of over 2% directly above the fallout layer in AG and CAR. Even more impressive, the profiles of AG and CAR are almost indistinguishable, which supports the conclusion that these shifts are environmental, not diagenetic effects. The $\delta^{13}C$ shift is generally (Zachos and Arthur, 1986) explained as a consequence of decreased primary productivity. The $\delta^{18}O$ shift is here interpreted as a temperature signal, reflecting a $>10\,°C$ rise in temperature of ocean surface waters, shortly after the impact event. On the width of the peak, the temperature crisis is estimated to last about 500–1000 years whereafter temperatures return slowly, over several thousand years, to normal values. It is probably no coincidence that new Paleocene species come in and carbonate production returns to normal values only when $\delta^{18}O$ values are the same again as in the Cretaceous.

These data suggest the following scenario. A number of impacts, both on land and the ocean, has led to mass-mortality and large-scale destruction of primary (photosynthetic) production within a few months or so. The impacts, or severe drought directly following impact, may have led to widespread wildfires (Walbach et al, 1985), further adding CO_2 to the atmosphere. Simply the severe reduction of photosynthetic plants for ? > 500 years may have prevented CO_2 recycling, keeping the atmosphere in a greenhouse state, leading to high atmosphere and ocean surface temperatures. It is possible that the continued greenhouse has been responsible for the final extinction and the selectibity of the extinctions.

REFERENCES

Berggren, W. A., D. V. Kent, J. J. Flynn, and J. A. Van Couvering, 1985, GSA Bulletin, v. 96, p. 1407.
Smit, J., and Romein, 1985, EPSL, v. 74, p. 155.
Wolbach, W. S., R. S. Lewis, and E. Anders, 1985, Science, v. 230, p. 167.
Zachos, J. C., and M. A. Arthur, 1986, Paleoceanography, v. 1, p. 5.

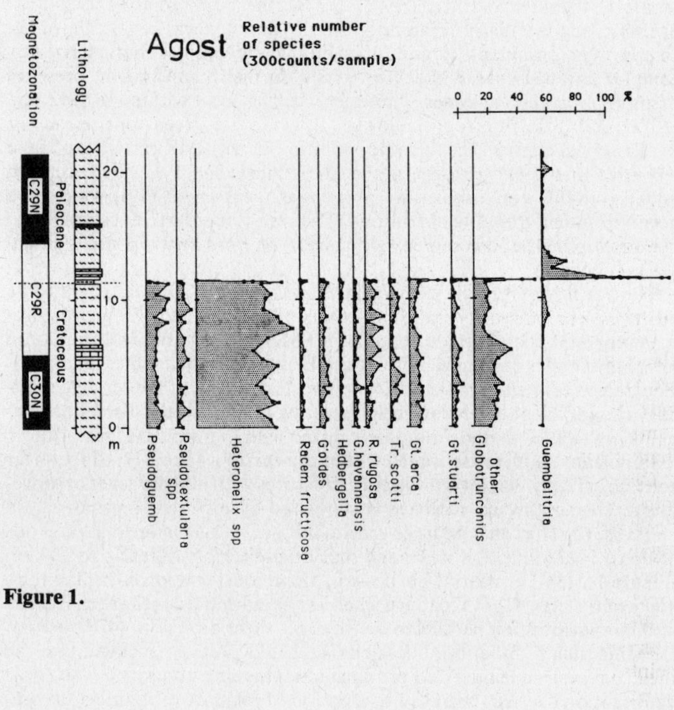

Figure 1.

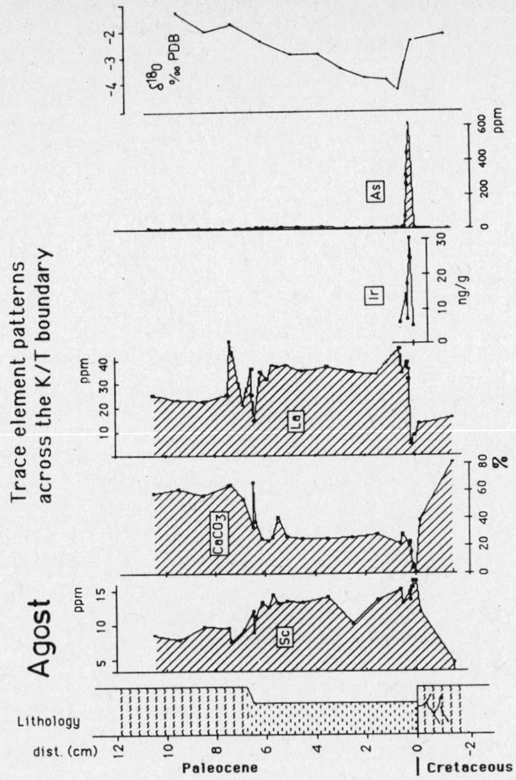

Figure 2.

SMITH, A. G., and W. B. HARLAND, Department of Earth Sciences, Cambridge, United Kingdom, R. L. ARMSTRONG, University of British Columbia, Vancouver, Canada, L. E. CRAIG, Cambridge Arctic Shelf Program, Cambridge, United Kingdom, and D. G. SMITH, BP Research Centre, Sunbury-on-Thames, United Kingdom

Reevaluated Phanerozoic Time Scale

Isotopic dates: Nearly 700 published isotopic dates have been used in the construction of a new Phanerozoic time scale. The dates are roughly equally distributed between Cenozoic, Mesozoic, and Paleozoic eras. Dates include fission-track results, glauconites, high-temperature minerals, and the optional illites for the Cambrian–Precambrian boundary.

Glauconites and nonglauconites: Comparison of glauconite with nonglauconite boundary ages shows that the glauconites give mean ages that are on average about 1 Ma younger than the mean nonglauconite ages for most of post-mid-Cretaceous time. The differences of older periods are consistently greater than analytical uncertainties. Thus, glauconites older than 110 Ma are treated as minimum dates in chronogram assessment.

Chronograms and calibration: Phanerozoic time has been subdivided into 127 "stages," and the age of all stage boundaries has been estimated by "chronograms." Chronograms are simply a graphical method of displaying the relative consistency of a set of isotopic dates for each boundary age. They now take into account not only the chronometric error for each isotopic date but also an estimate of the chronostratic errors.

Well-defined stage boundary ages have been selected as tie points. The intervening ages have been estimated by linear interpolation between tie points using the relative numbers of chrons in each stage to define relative durations. The end result is a time scale defined by the better chronograms and linear chron interpolation between them.

Cambrian–Precambrian boundary: The major uncertainty is still the age of the Cambrian–Precambrian boundary. Four values can be suggested: (1) 575 Ma, the chronogram age, excluding all illites and treating glauconites as minimum dates; (2) 584 Ma, the chronogram age that includes whole rock Rb–Sr data on illites in shales from China; (3) 550 Ma, based on a regression line through three younger Cambrian calibration points and chron extrapolation to the Cambrian–Precambrian boundary; and (4) 570 Ma, based on interpolating between Early Cambrian and Vendian dates.

Magnetic reversal data: Ocean-floor magnetic profiles have been used to refine the Cenozoic, Cretaceous, and later Jurassic timescales. A modified and revised South Atlantic scale, showing the relative spacings of ocean-floor anomalies, was calibrated in Ma. The calibration data are dated magnetic chron boundaries and chronogram ages that can be directly related to the priority reversal time scale.

Visually, the Cenozoic data can be fitted by three line segments. The first is obtained by linear regression of 0–3.4 Ma data and the second by fitting data in the 9–40 Ma range. Because there is a serious discrepancy in the time interval 45–60 Ma between the glauconite dominated chronogram data, the dated magnetic chron C21N(y), dates from Koko seamount and DSDP 516F, none of these data have been used to define the second or third line segments. The data for the calibration of the third line segment from about C23N(y) to C34(y) are quite inadequate, partly because they are scarce and partly because at least three changes in spreading rate may occur on the South Atlantic profile. Thus, the third line segment is arbitrarily fixed as passing through the Tertiary–Cretaceous boundary at 65 Ma and the Campanian–Santonian boundary at 83 Ma. The line segments intersect at 12 Ma and 51 Ma.

Early Cretaceous and Late Jurassic stage boundaries have been precisely related to the M anomaly sequence. A revised nomenclature is suggested for CMO to CM6. A regression line has been drawn through chronogram ages and relative anomaly spacings as given by the Hawaiian M lineations. The calibration ages are similar at the older end of their scale, but up to 5 Ma older than the DNAG time scale and 1982 time scale of Harland et al at the younger end, e.g., the Aptian–Barremian boundary is placed at 125 Ma, rather than 120 Ma. The Cretaceous–Jurassic boundary is placed at about 145 Ma.

Conclusions: In spite of its entirely independent construction and refinement, the new time scale is similar to the Harland et al 1982 time scale, which was based on less than half the current database. Few boundary ages have changed as much as 1%. Change has been in the form of a dramatic reduction in the tie point chronogram errors, which are now about 2% of their age. We think this is the current uncertainty in the calibration of the Phanerozoic time scale.

SMITH, A. Y., and M. TAUCHID, International Atomic Energy Agency, Vienna, Austria

Radioelement Geochemical Maps of the World

As a consequence of the Chernobyl accident in 1986 the International Atomic Energy Agency initiated a program on the Use of Regional Gamma Ray Data to Define th Natural extent and quality of existing data, the Agency sent out a questionnaire to its 113 member states asking for information on details on the type and specifications of past radiometric surveys carried in the world. The index map presented here was compiled from dta obtained from the questionnaire and from other publicly available information. Some of these data are not only suitable for conversion to natural background radiation maps but also represent a potential source for the preparation of the radioelement (U, Th, and K) geochemical maps of the world. In order that interested parties will follow standardized procedures and specifications in the preparations of these maps, the Agency is in the final stage in issuing a Technical Report on the requirements and standards to be maintained in making use of such data. Particular attention has been paid to the problems of back calibration on uncalibrated surveys and leveling between separate surveys.

As part of its ongoing program to improve the quality and use of radiometric and geochemical data for the radioelements the Agency has prepared a Technical Report on the Construction and Use of Calibration Facilities for Gamma Ray Field Equipment. It also has available for distribution, at nominal cost, Certified Reference Materials RGU-1, RGTh-1 and RGK-1, for radiometric analysis of geological methods. Reference Materials for geochemical uranium analysis of tropical soil (IAEA-312) and stream sediments (IAEA-313 and IAEA-314) are undergoing intercomparison analyses, and will be available for general distribution early next year.

SMITH, B. A., Bechtel Civil, Inc., Oak Ridge, TN

Tectonic Origin of Upper Cretaceous Unconformity in East-Central Mexico

A major unconformity overlies the mid-Cretaceous Valles Platform of east-central Mexico. The complex pattern of sedimentation above these shallow-water, carbonate rocks of the Valles Platform cannot be explained by eustatic sea-level changes. The best explanation for the unconformity is uplift and deformation of the platform as a result of tectonic forces in western Mexico.

The Valles Platform was a Bahama-like carbonate platform that was undergoing maximum growth during the Albian and early Cenomanian. Rudist reefs grew along the edges of the platform, and low-lying islands developed behind the reefs. The center of the platform was a wide, shallow lagoon. Topographic relief over the top of the platform was very low; probably no more than 40 m. The platform extended at least 450 km north-south, and had a maximum width of 200 km. The platform rose approximately 1000 m above the surrounding basin and had margins that sloped into the basin in excess of 40°.

The El Abra Formation makes up most of the Valles Platform. The El Abra was deposited in shallow-water environments ranging from rudist reefs to lagoonal environments. Flanking the Valles Platform are the platform-margin deposits of the Tamabra Formation. This formation consists of debris flows of material from the top of the platform, interfingering downslope with basinal sediments. The Upper Tamaulipas Formation is the basinal equivalent of the El Abra and Tamabra Formations and is a thinly to thickly bedded, dark gray to black limestone containing numerous planktonic foraminifers.

Unconformably overlying the Albian and early Cenomanian rocks of the Valles Platform are Upper Cretaceous rocks of varying lithologies and ages. Figure 1 is a generalized stratigraphic chart of east-central Mexico showing the mid-Cretaceous formations that make up the Valles Platform and the principal formations that overlie the platform. There have been no sediments of middle or late Cenomanian age found overlying the El Abra Formation. This suggests that the entire platform was subaerially exposed at this time. In the late Turonian much of the central and western parts of the platform were covered by the sea with both shallow-water (Cuautla Limestone) and deep-water (Agua Nueva Formation) sedimentation taking place. The eastern margin of the platform was exposed subaerially at this time. As more of the platform was drowned in the Coniacian and Santonian, deep-water limestones and shales of the San Felipe Formation were being deposited over both the Agua Nueva and El Abra Formations. By the late Campanian, the entire platform was submerged and the Méndez Shale was deposited over the platform and in the surrounding basin.

The unconformity over the platform represents a hiatus of 7–24 m.y. in duration. The pattern of shallow- and deep-water sedimentation above the unconformity shows that the top of the platform had considerably more relief than it did in the early Cenomanian. Such a hiatus and sedimentation pattern could not have been caused by eustatic sea-level changes.

The sequence of events suggested by these sediments is that the platform was uplifted and deformed during the middle and late Cenomanian. This deformation probably consisted of folding and faulting of the platform. The lowest parts of the platform were flooded as the platform started to subside during the Turonain. Both shallow- and deep-water limestones were deposited on the flooded parts of the platform while other part were still subaerially exposed. Subsidence and flooding continued until the entire platform was covered by the seas and mostly deep-water shales were being deposited over the platform in the late Campanian.

The cause of the deformation of the Valles Platform is most likely due to tectonic activity in western Mexico where subduction along the western margin of the North American plate was speeding up. Engebretson et al (1985) determined that the convergence rate between the North American plate and the Farallon plate to the west increased significantly in the early Cenomanian. There was an intense episode of magmatism in western Mexico starting at about the same time that resulted in abundant batholitic and volcanic rocks (Henry, 1975). The increased rate of subduction and the injection of magmas into the crust brought about compression of the crust in eastern Mexico that caused uplift and deformation of the Valles Platform at the beginning of the Late Cretaceous. These events led to a major unconformity over the mid-Cretaceous sediments of the Valle Platform.

REFERENCES

Engebretson, D. C., A. Cox, and R. G. Gordon, 1985, Relative motions between oceanic and continental plates in the Pacific Basin: GSA Special Paper 206, 59 p.

Henry, C. D., 1975, Geology and geochronology of the batholitic-complex, Sinaloa Mexico: Ph.D. Dissertation, University of Texas-Austin, 158 p.

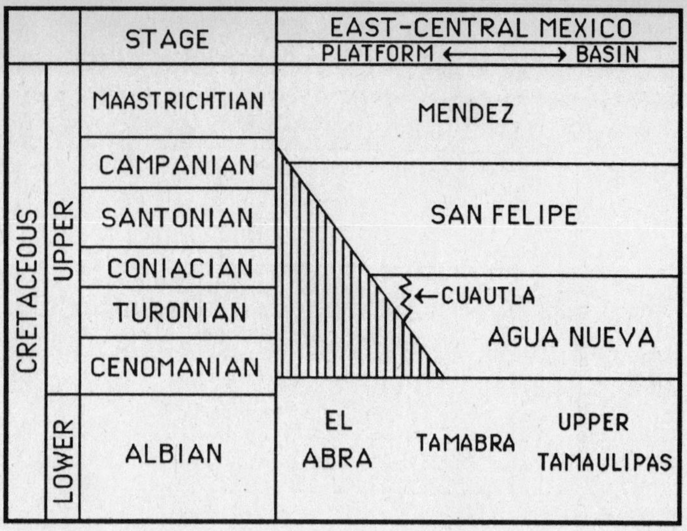

Figure 1.

SMITH, D. G. W., H. OMOUMI, and D. P. LEIBOVITZ, University of Alberta, Edmonton, Canada

MinIdent Database: Some Recent Developments

The MinIdent database and the software for mineral identification (Smith and Leibovitz 1984, 1986; Smith, 1986) have been successfully ported to a PC from the Amdahl mainframe on which they were developed. The "compiled" database (used in mineral identification) plus necessary management programs can be accommodated on a 30 Mbyte hard disc. Developments presently being undertaken will further reduce these storage requirements.

Since the publication by Smith and Leibovitz (1986), many new data have been added and information now exists for some 4200 mineral varieties, species, series, groups, etc. Literature and data-entry errors are being identified using tests for self-consistency, and progressively eliminated.

In particular, that part of the database dealing with unnamed minerals has been greatly expanded and reorganized by year of first description. Data for unnamed minerals are presently scattered throughout 100 years of earth science literature and range from vague descriptions of hand-specimen properties to complete modern analyses. The lists compiled by Hey (1962, 1963) were made the starting point, and then a wide range of journal and other literature sources used to obtain additional information and to bring the list up to date. Only those minerals for which numerical data are available have been included. Presently, nearly 600 unnamed minerals appearing in the literature have been included. Once complete, this subset of the database will provide a unique resource and will allow users attempting to identify unknowns to compare their data with those for all previously described minerals and not only with those species that have received names. The use of the mineral identification software permits a numerical estimate to be obtained of the similarity between an unknown and previously described species. The immediate availability of a compilation of literature data for the latter provides a convenient indication of what other parameters might be obtained for a more unambiguous identification. The possibility also exists of adding a further category of data, for phases that have been obtained as the products of experimental work but that have not so far been found occurring naturally.

Another addition to MinIdent is a substantial list of discredited mineral names and synonyms. Entries are also included for minerals that although of dubious authenticity have not been unequivocally discredited and therefore remain in the database. At present there are about 1500 entries in the list, each of which includes a brief explanatory comment, the synonym (where applicable), and source reference(s). The scheme for naming rare earth-bearing minerals that was recently approved by the IMA has been fully implemented and, as far as possible, data associated with each rare earth variant of that species appropriately assigned. However, the paucity of complete and reliable information on the concentrations of individual rare earths continues to pose a problem. The recent IMA decision to return to the original spelling of many non-English names has been implemented, although present software constraints preclude the inclusion of diacritical marks in such names.

Classification of minerals has been possible since the inception of MinIdent. This facility nas now been expanded so that the following divisions can be recognized where appropriate: variety, subspecies, series, subgroup, group, supergroup, family, class, and type. The top level "type" (e.g., silicates, oxides, sulfides, etc.) has been chosen to avoid ambiguity or overlap with other levels of classification previously used in the literature. Much progress has been achieved particularly with respect to rock-forming and more common minerals. For example, the full IMA amphibole classification scheme has been implemented (Goble and Smith, 1988), and that for pyroxenes is presently being undertaken. However, much remains to be done and progress is hindered by the absence of a generally accepted and definitive classification scheme for minerals.

Other changes include the updating of the JCPDS PDF number and the entry or reentry pf the d-values for the five most intense lines from original literature sources. The algorithm that was specially developed fro MinIdent to identify minerals on a very limited number of d-values has proved to work very well on pure phases. It is not intended for use with mixtures of phases.

Finally, a subset facility has been implemented that allows any group of minerals of interest to be selected from the database, and subsequently only these to be considered in the match/identify procedures. Such subsets are entirely flexible and could include categories such as "silicates," "meteorite minerals," or sets of minerals for instructional purposes. The use of subsets greatly reduces computational time for identification. It could prove extremely useful for the automation of mineral identification in combination with the analytical and image analysis capabilities of modern microbeam instruments.

REFERENCES

Goble, R. J., and D. G. W. Smith, 1988, MinIdent: an application to the identification and classification of amphiboles: Mineral Petrol., v. 38, p. 213–227.

Hey, M. H., 1962, Chemical index of minerals: London British Museum of Natural History, 728 p.

——— 1963, Appendix to chemical index of minerals, London British Museum of Natural History, 135 p.

Smith, D. G. W., 1986, Automation of mineral identification from electron microprobe analyses, *in* A. D. Romig and W. F. Chambers, eds., Microbeam Analysis–1986: San Francisco, San Francisco Press.

——— and D. P. Leibovitz, 1986, MinIdent: a data base for minerals and a computer program for their identification.

——— ——— 1984, A computer based system for identification of minerals on the basis of composition and other properties: 27th International Geological Congress, Moscow, 1984 Abstracts, v. 5, p. 169.

SMITH, G. G., Geological Survey of Canada, Calgary, Canada, and R. M. BUSTIN, University of British Columbia, Vancouver, Canada

Coal Deposits in Rocky Mountain Front Ranges and Foothills of Southern Canadian Cordillera

Canada is considered to have the fifth largest inventory of coal resources in the world. More than one-third of Canada's coal resources that are currently favorable for continuing exploration and possible development are located in the southern Canadian Rocky Mountains. Eleven major coal mines in this region account for approximately 50% of Canada's total coal production, including 97% of the metallurgical coal and a significant amount of the thermal coal. Most of the metallurgical coal and some of the thermal coal production is designated for export to Pacific Rim countries and to a lesser extent, other international markets.

Within the southern Canadian Rocky Mountains major coal deposits occur in three geologically and physiographically distinct areas—Front Ranges, Inner Foothills, and Outer Foothills. The Front Ranges are characterized by major thrust sheets with up to tens of kilometers of displacement and coeval folds involving mainly Paleozoic carbonates and some Mesozoic clastic rocks. Within the Front Ranges major deposits of high to low volatile bituminous metallurgical and rare semianthracite coal occur in Late Jurassic–Early Cretaceous strata of the Mist Mountain Formation.

The Rocky Mountain Foothills comprise mainly deformed Mesozoic and Cenozoic clastic rocks between the Front Ranges and the eastern edge of the deformed belt. The Inner Foothills, immediately east of the Front Ranges, comprise a high relief area dominated by Early Cretaceous coal-bearing sandstones, siltstones and shales. In the Inner Foothills of northeastern British Columbia and west-central Alberta, significant resources of medium and low volatile bituminous metallurgical coal occur in the Early Cretaceous Gething and Gates formations. East of the Inner Foothills the topography is more subdued and underlain by more recessive weathering Upper Cretaceous and Tertiary coal-bearing strata. In these Outer Foothills of western Alberta significant resources of high volatile bituminous thermal coal occur in the Belly River (Campanian) and Coalspur (Maestrichtian–Paleocene) formations.

The major coal deposits of the Front Ranges and Inner Foothills accumulated within coastal plains of the Fernie and Moosebar–Clearwater seas, respectively. Shorelines prograded east and northeast into a foreland basin that developed in advance of, and in response to, the Columbian Orogeny. In the Outer Foothills, coal deposits of the Belly River Formation (Campanian) originated within easterly prograding coastal plains during the withdrawal of the Pakowski Sea. The major coal deposits of the Coalspur Formation accumulated in alluvial environments of the foreland basin during the Laramide Orogeny.

In the Late Jurassic, major coal (peat) deposits accumulated in the southernmost part of the Canadian Cordillera (Mist Mountain Formation). During the Early Cretaceous (Aptian), the site of major coal deposition shifted north to northeastern British Columbia and west-central Alberta (Gething and Gates formations). In the Late Cretaceous (Campanian) the major sites of coal accumulation extended from south of the Canada/United states border to about 50°N latitude (Belly River Formation). The major deposits of coal in the Outer Foothills that originated during the Paleocene occur north of 52°N latitude (Coalspur Formation).

Coal measures of the Canadian Rocky Mountains were variably folded and faulted during Laramide orogenesis. Structural deformation has significantly affected mineability and quality of the coals. Many of the coal beds are sheared, structurally thickened or thinned, and repeated by thrust faults or truncated by normal faults. Shearing of the coal beds and adjacent hanging wall and footwall strata has locally led to pronounced oxidation of the coal and has resulted in weak coal pillars and roof rock in underground mines. Faulting and folding have also resulted in many coal deposits being partitioned into discrete structural domains of varying style and complexity, each of which can present a unique challenge in designing optimum coal extraction methods.

The variation in coal rank in the southern Canadian Rocky Mountains closely reflects, for the most part, the pre-orogenic (pre-Laramide) depth of burial. In some areas, however, there is evidence for syn- and post-orogenic coalification resulting from the emplacement of thick overthrust sheets and/or deposition of thick Tertiary molasse, most of which has been subsequently eroded. With few exceptions, coals of the southern Canadian Rocky Mountains are characterized by low sulfur content (< 1%) and ash values that vary between 10% and 30%. Although the maceral content is variable, the Late Jurassic and Early Cretaceous coals are known for their high percentage of semifusinite. All of the coal-bearing stratigraphic successions include multiple, thick seams of coal and all existing mining operations are designed to extract and, where necessary, blend multiple seams. Medium and low volatile bituminous coals of the Mist Mountain Formation generally yield firm, coherent coke. Weakly coking and non-coking high volatile bituminous and semianthracite coals, and oxidized medium and low volatile bituminous coals also occur in notable quantities in some areas and are used as thermal coals. Medium and low volatile bituminous coals of the Gething and Gates formations commonly have good coking properties except where oxidized, in which case they are used as thermal coals. Coals of the Belly River and Coalspur formations range in rank mainly between high volatile B and C bituminous and are used exclusively as thermal coals.

SMITH, L., University of British Columbia, Vancouver, Canada, F. SCHWARTZ, University of Alberta, Edmonton, Canada, and C. MASE, University of Waterloo, Waterloo, Canada

Large-Scale Computational Requirements for Modeling Solute Transport in Fractured Rocks

Performance assessments for the siting and design of hazardous waste disposal facilities require that hydrogeologic models be used to estimate rates at which contaminants may migrate from a waste cell to the accessible biosphere. In that areas underlain by rocks with low permeability are preferentially targeted as disposal sites, it is probable that fractures will play an important, if not dominant, role in providing pathways for contaminant migration. Modeling of solute transport in a fractured rock mass presents a complex computational problem that, except for highly idealized representations, must be tackled using methods of numerical simulation. Two classes of models are being developed; the continuum approach treats a fractured rock mass as if it were equivalent to a porous medium, while discrete network models are based upon an explicit representation of the fractures. Because of data limitations facing every site characterization program, predictive simulation is likely to be approached from a stochastic, rather than deterministic, viewpoint. Monte Carlo simulation provides a flexible technique to assess uncertainties in model prediction that arise from parameter uncertainty.

Discrete network simulation is based on a probabilistic representation of the network geometry. Multiple realizations of a discrete network are generated, given estimates of the moments of the probability distributions describing fracture dimensions, orientation, density, and aperture. Values of hydraulic head within a network can be calculated using either finite difference or finite element methods. Currently, it is feasible to model steady-state fluid flow in three-dimensional networks of rough-walled fractures that contain from 10^5 to 10^6 nodes. Solute transfer through the network of fractures can be modeled using particle tracking techniques. Methods have recently appeared o take account of diffusion of mass between fractures and the intact rock. However, at this level of detail, only single realization behavior can be modeled. Thus, while useful for concept evaluation, discrete network models have as yet only limited application as predictive tools. Problems involving numerical simulation of transient fluid flow through a discrete fracture network, or incorporation of coupling to geochemical and mechanical processes, lie at the boundary of what today is computationally possible in a practical sense.

Although discrete models may be appropriate for near-field simulation, simulations at the site-scale must be carried out using continuum concepts. The greatest difficulty here lies in formulating the anisotropic dispersion tensor to take account of the interplay between spreading patterns, network geometry, and the orientation of the mean hydraulic gradient. A statistical approach, based on using a discrete subdomain model, has been developed in two dimensions. Particle motion observed in a discrete subdomain is used as the basis for mimicking transport in a larger-scale continuum model where fractures are no longer included as discrete features. Tests are underway to evaluate whether it is computationally feasible to extend this concept to three dimensions.

SMITH, L., University of British Columbia, Vancouver, Canada, and C. FORSTER, Utah State University, Logan, Utah

Interaction Between Fault Zones, Fluid Flow, and Heat Transfer at Basin Scale

Deep-seated faulting is a characteristic feature of mountainous terrain. However, relationships between fluid flow within through-going fault zones and region-scale flow systems have received comparatively little quantitative study. In mountainous terrain, groundwater circulation to depths of at least 3 or 4 km, over horizontal length scales of 5 to 20 km, appears typical. On a geologic time scale, it is likely that a complex dependence exists between hydrologic, thermal, geochemical, and mechanical processes in determining the extent to which major faults influence groundwater flow at the basin scale. In this paper, we focus on the link between the hydrologic and thermal regimes. Site-specific systems are reviewed to summarize observational constraints on the hydraulic behavior of fault zones imposed by the geologic environment, the character of faulting, topographic relief, and where possible, the magnitude of hydraulic parameters. Fault zones have a dual character, in some cases acting as conduits for flow, while in other cases acting as barriers to flow. A key issue concerns the degree to which local-scale structures along and within the fault zone are important in influencing patterns of flow at the basin scale. Model studies are carried out to examine the factors controlling discharge temperatures of thermal springs. Three factors are dominant: (1) the bulk permeability of the mountain massif, which determines the extent to which the background geothermal heat flux is captured by the groundwater flow system and delivered to a permeable fault zone, (2) the fluid mass carried by the flow system, which determines the possible temperature increase for a given geothermal heat flux at depth, and (3) heat transfer mechanisms during fluid ascent along the fault. Our simulations provide support for the growing opinion that local-scale processes are best examined in the context of the regional hydrogeologic setting. Transient simulations show how fluid flow and heat transfer within the fault zone evolves in response to both internal changes in fault-zone permeability, and to external influences of changing hydrologic and thermal conditions at the basin scale.

SMITH, M. A., Texaco E & P Technology Division, Houston, TX

Lacustrine Oil Shales—Their Controlling Processes and Distribution through Geologic Time

Oil shales, rocks that yield substantial amounts of hydrocarbon material during destructive distillation, have been deposited across the globe in a wide variety of lacustrine and marine settings for more than 1 b.y. Coals that form in marginal lacustrine and paludal environments also yield hydrocarbons during pyrolysis, but can be distinguished form oil shales by their lower ash content. Oil shales do not yield appreciable amounts of oil when extracted with solvents, but there is an excellent correlation between the S_2 peak area from Rock-Eval pyrolysis and their Fischer assay oil yield. These finely laminated rocks originated as some of the most kerogen-rich clastic sedimentary sequences in the world, and are therefore of great interest and importance as petroleum source rocks.

Available data on the major Precambrian and Phanerozoic lacustrine deposits in sever hundred basins were reviewed in a company project, part of which may be contributed to IGCP-219 (Comparative Lacustrine Sedimentology in Space and Time). Some large modern lakes and a few ancient ones formed in a fluviatile or coastal setting, but almost all long-lived lacustrine basins have a tectonic origin. Geologic origins—rift valleys and areas of cratonic uplift or downwarp—and the distribution of lakes with significant associated oil shales are shown here on paleogeographic reconstructions for several time slices. Paleolatitude, paleoclimate, and tectonic setting all play a role in the distribution of lakes and their organic-rich sedimentary facies, but the occurrences of lacustrine oil shales suggest that major lakes are most commonly associated with rift, intracratonic, or intermontane basins of various sizes. In general, lacustrine strata are more common and better preserved in younger, Cenozoic basins but many of these lakes were shallow and holomictic. Thus, lacustrine oil shales are more frequently associated with older deposits and are found nearly twice as often in Paleozoic sequences as in Cenozoic sediments.

Precambrian lacustrine sedimentary sequences may be difficult to recognize and need to have escaped erosion for a long period of the earth's history, so it is not surprising that only about one percent of the lacustrine deposits reported in the literature formed during the pre-Phanerozoic. However, bitumen and oil seeps from several Proterozoic lacustrine shales in Australia and North America demonstrate that these rocks still have a sufficient pyrolytic yield to be classified as oil shales. Precambrian and early Paleozoic lacustrine petroleum source rocks contain mostly type-I algal kerogen, but the development of vascular plants during the Silurian provided additional types of organic matter with the result that younger oil shales can also produce gas and waxy crude oils. During the Paleozoic, Australia has the most complete record of lacustrine deposition of any continent and oil shales become most abundant in Gondwana in the Permian Period. The largest lacustrine oil-shale accumulations of late Paleozoic age are the Irati Shale in South America, Albert Formation and Deer Lake sequences in eastern Canada, oil shales in central East Greenland, deposits in the huge Tarim and Zhungeer interior basins of China, Karroo graben fill in East Afric, and Orcadian and Midland basin oil shales of Great Britain. Many of these oil shales are associated with coal seams, and well-defined sedimentary cycles are displayed in their lacustrine record.

Lacustrine oil shales continued to accumulate along various continental borders in rift basins related to the breakup of Pangaea during the Mesozoic Era. Lare, long-lived lakes were present at this time in southeastern Africa and western Australia, as well as in many Newark Supergroup basins of eastern North America and the conjugate African margin. During the Early Cretaceous, synrift lacustrine deposits were widely associated with the rifting episode between South America and Africa. Cenozoic lacustrine oil shales, however, occur in a variety of tectonic settings including rifts, continental sag basins, and compressional belts with significant uplift. Thick late Mesozoic and Cenozoic lacustrine sequences with extensive oil shales, such as the Lower Cretaceous Qingshankou Formation in the Songliao Basin and the Eocene Green River Formation, occur throughout China, southeast Asia, and the western U.S.

The distribution of nonmarine oil shales though geologic time reflects deposition in both shallow and deep lakes in all types of continental sedimentary basins. These deposits are lacustrine petroleum source rocks with remaining hydrocarbon generation potential, and there are many productive analogs around the world of various geologic ages from this depositional environment. As understanding of geoloci controls on the deposition and preservation of high-quality source and reservoir facies improves, petroleum exploration in lacustrine basins will also become more effective.

SMITH, P. E., and R. M. FARQUHAR, University of Toronto, Toronto, Canada

$^{238}U/^{206}Pb$ Dating of Devonian Corals

Direct radiometric age determination of Phanerozoic carbonates has been a long-standing problem in geochronology. The Rb-Sr and K-Ar dating schemes, commonly used to constrain the chronology of young sediments, have so far proved to be unsuccessful in dating these rocks because of poor enrichment in radiogenic ^{87}Sr and ^{40}Ar and also because of analytical difficulties. Recently, Moorbath et al, (1987) have attempted to date carbonates using the Pb-Pb method. This technique may prove to be useful where there are very large enrichments in radiogenic Pb; however, the precision of this method is severely limited by the near extinction of the parent ^{235}U isotope. In this paper, we report the first isochron age of limestones using the $^{238}U/^{206}Pb$ method.

We have studied the classic middle Devonian (Givetian) sediments of the Thedford-Arkona region in southwestern Ontario, Canada. The area consists of a 20 m section of fossiliferous shale and limestone of the Hamilton formation exposed by the Ausable river.

Our preliminary study is focused on samples from a coral zone in the Hungry Hollow member. An analysis of whole rock fossiliferous limestone gives a value of 3.1 ppm for Pb. The Pb concentrations in rugose corals (*Heliophyllum sp.* and *Cystiphylloides sp.*) are much lower, ranging from 280 to 310 ppb, in agreement with values for extant species (Shen and Boyle, 1988). Concentrations in optically clear skeletal calcite are very low

(27-77 ppb). Uranium concentrations in the corals, 50–100 ppb, are also lower than the value of 350 ppb given by the whole rock.

We have also included in our sampling an analysis of pyrite infillings in a brachiopod (*Mucrospirifer sp.*) This allows us to evaluate the relationship of seawater Pb incorporated by the corals to Pb associated with later diagenesis of the rocks. The Pb content of this pyrite, 60 ppm, is typical of magmatic sulfides.

The isotopic data are presented in a plot of $^{206}Pb/^{204}Pb$ versus $^{238}U/^{204}Pb$ in Figure 1. Data points for the two corals and the whole rock are colinear within experimental error. The pyrite analysis from the brachiopod falls significantly below this line, indicating a different age and/or source. The isochron age of 374 ± 12 Ma (1σ) compares very favorably with the boundaries of 385 to 375 Ma inferred for Giventian strata (Odin et al, 1982).

The initial Pb ratios derived from the isochron plot and those similarly calculated for the ^{235}U-^{207}Pb system allow a characterization of the isotopic composition of Pb in Devonian seawater to be made with reference to Pb in the earth's global reservoirs. Figure 2 shows that the initial ratios are relatively radiogenic compared to those ratios predicted by an average earth growth curve (e.g., Stacey and Kramers, 1975) at 374 Ma. This suggests a multistage evolution culminating in a relatively recent enrichment in μ_1 ($= {}^{238}U/{}^{204}Pb_{source}$).

In summary, the apparent closed system evolution of Pb in the specimens and rocks of these carbonates allow the direct determination of their age. This has important implications for the calibration of the Phanerozoic time scale and biological evolution. Finally, since the abundance and residence time of Pb in seawater is very different from that of Sr, the initial Pb isotope ratios of carbonates can provide a valuable complement to the studies of initial $^{87}Sr/^{86}Sr$ in these rocks.

REFERENCES

Moorbath, S., P. N. Taylor, J. L. Orpen, P. Treloar, and J. F. Wilson, 1987, First direct radiometric dating of Archean stromalitic limestone: Nature, v. 326, p. 865–867.

Odin, G. S., D. Curry, N. H. Gale, and W. J. Kennedy, 1982, The Phanerozoic timescale in 1981, *in* G. S. Odin, ed., Numerical dating in stratigraphy, v. 2: New York, John Wiley, p. 957–960.

Shen, G. L., and E. A. Boyle, 1988, Determination of lead, cadmium and other trace metals in annually banded corals: Chemical Geology, v. 67, p. 47–62.

Stacey, J. S., and J. D. Kramers, 1975, Approximation of terrestrial lead isotope evolution by a two-stage model: Earth and Planetary Science Letters, v. 38, p. 67–87.

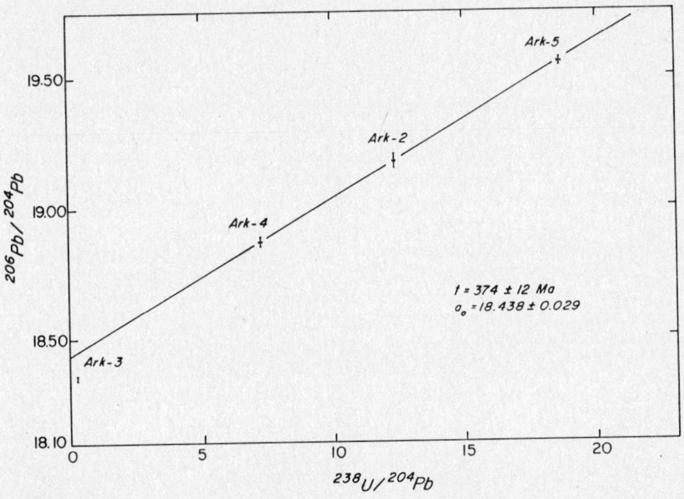

Figure 1.

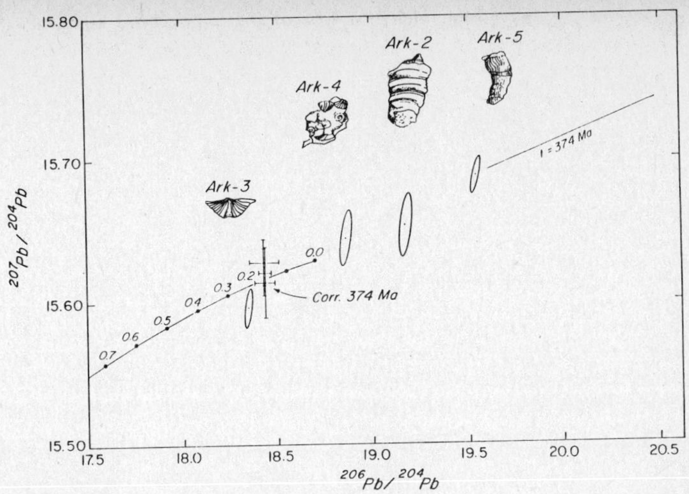

Figure 2.

SMOOT, J. P., U.S. Geological Survey, Reston, VA, and P. W. LETOURNEAU, Wesleyan University, Middletown CT

Wave-Formed Shoreline Deposits on Alluvial Fans: Examples from Modern Pyramid and Walker Lakes, Nevada, and Early Mesozoic Newark Supergroup, Eastern U.S.A.

Wave-formed shorelines produced by climatically-induced lake rises are nearly ubiquitous features on the alluvial fans of modern block-faulted basins in the western United States. Wave-formed shoreline deposits are rarely reported from alluvial fan conglomerates in the rock record, despite the common evidence of climatically-induces lake level changes in many ancient basins. Wave-formed shoreline deposits may be destroyed by flooding and deflation during low lake stands, but the paucity of ancient examples may be due to lack of recognition. The sedimentary features of Modern and Quaternary shoreline deposits of Pyramid and Walker Lakes in Nevada, are comparable to wave-reworked alluvial fan conglomerates in the Hartford, Newark, and Danville-Dan River basins. Wave-formed shoreline deposits may be destroyed by flooding and deflation during low lake stands, but the paucity of ancient examples may be due to lack of recognition. The sedimentary features of Modern and Quaternary shoreline deposits of Pyramid and Walker Lakes in Nevada, are comparable to wave-reworked alluvial fan conglomerates in the Hartford, Newark, and Danville-Dan River basins of the early Mesozoic Newark supergroup in the eastern United States. The preservation of the Mesozoic deposits may be of economic interest because they are potential reservoirs for petroleum within the otherwise impermeable alluvial fan deposits.

The Modern and Quaternary wave-formed deposits associated with alluvial fans are divided into gravel-bar deposits, platform deposits, poorly-ordered deposits, and sandbar and ripple-flat deposits. Each type of these deposits is fairly well sorted, porous and permeable, and commonly areally extensive.

Gravel-bar deposits consist of rare features nearly 2 m high comprised of cobbles and pebbles to more common decimeter-thick ridges comprised of small pebbles and sand. The basic structure consists of steep, shoreward-dipping cross-bed sets overlain and truncated by more gently-dipping lakeward sets. The steep foresets of the large bars are commonly convex-upward and tend to flare toward the base, whereas the foresets in smaller bars are more tabular. Layering is defined by alternations of different grain sizes and grain shapes. Large coarse-grained bars have formed high on alluvial fans where waves to strike the steep fan slope without loss of energy by shoaling. Smaller, finer-grained bars have formed on gentler slopes by lower energy waves.

Platform deposits are wedges of wave-sorted sand and gravel with flat tops and steep lakeward-dipping foresets. Meter-high platform foresets were observed at Pyramid lake extending from a wave-cut terrace a few meters above the present lake level. The foresets consist of graded sequences of moderately-sorted pebble gravel interbedded with sand. The topsets, when present, erosionally overlie the foresets and consist of small gravel-bar deposits or poorly-ordered deposits. Sand and gravel are apparently rounded and sorted on the flat surface by normal wave action, then swept lakeward by storms forming avalanche sets. The platform builds lakeward and parallel to shore via longshore drift.

Poorly-ordered deposits are a common type of wave-formed shoreline deposit consisting of randomly-scattered boulders, cobbles, and/or pebbles surrounded by wave-sorted granules and sand. The coarse clasts were initially introduced by floods or debris flows, then wave action winnowed and sorted the fine-grained material, leaving the coarse material as lags. The large clasts disrupt the wave patterns resulting in localized areas of different grain sizes or local pockets of sand with oscillatory ripples. In cross-section this appears as massive to poorly-bedded conglomerate with the matrix sand forming well-sorted beds or pods, defined by variations in grain size and shape. Commonly the interstitial spaces of a well-sorted granule or coarse sand layer will be filled with a well-sorted fine sand, due to infiltration. The poorly-ordered wave deposits apparently reflect short periods of wave reworking.

Sandbar and ripple-flat deposits consist of decimeter-thick linear sandbars separated by broad flats with oscillatory ripples that occur on relatively flat lake margins. Linear sandbars in water as much as a few meters deep have tabular foresets that dip both shoreward and lakeward, whereas bars near shorelines have lakeward-dipping low-angle planar lamination that overlie shoreward-dipping tabular sets. Oscillatory ripples produce characteristic wavy beds with predominately shoreward-dipping, low-angle, sinusoidal cross-lamination. Near the shoreline, amalgamated ripple cross-lamination is intercalated with planar lamination, whereas lakeward, rippled sand layers are separated by mud flaser and thin beds.

Each type of wave-formed deposit, except the platform deposits, has been recognized in the Newark supergroup. The vertical sequences of sedimentary features in the ancient examples are similar to those in cores from Walker and Pyramid Lakes. Regressive sequences in the Recent and Ancient examples commonly consist of laminated muddy sediments grading upward into thin-bedded mud and oscillatory rippled sand beds. These are overlain by sands with flaser bedding, then coarse sand to pebble conglomerate with sandbar, gravel-bar, or poorly-ordered wave deposit features. In the ancient examples cobble conglomerates with characteristics of the poorly-ordered wave-formed deposits always underlie subaerial alluvial fan deposits. Transgressive sequences are commonly thinner and are dominated by sandbar and ripple-flat deposits.

The common juxtaposition of porous and permeable, wave-sorted conglomerate and thermally-mature, organic matter rich lake laminites surrounded by impermeable alluvial fan deposits is ideal for petroleum entrapment. The wave-sorted, potential reservoir rocks would be thin (1–3 m thick), but extensive deposits that are elongate parallel to basin margins. In all of the Newark supergroup examples the primary porosity is filled with diagenetic and late-stage cements, however, where early hydrocarbon migration into porous wave-formed deposits occurred, attractive targets for exploration may remain.

SNEE, L. W., and K. LUND, U.S. Geological Survey, Denver, Co, J. F. SUTTER and M. J. KUNK, U.S. Geological Survey, Reston, VA, Z. WONG, China National Non-Ferrous Metals Corporation, Kumming, China, and W. C. KELLY, University of Michigan, Ann Arbor, MI

High-Precision $^{40}Ar/^{39}Ar$ Thermochronology of Mineral Deposits

$^{40}Ar/^{39}Ar$ age-spectrum dating, which is a high-precision variant of the conventional K-Ar method of geochronology, provides mineral dates that have analytical uncertainties of about 0.25% of the age and is a powerful tool for solving some long-standing problems in economic geology, including the age, duration, number of episodes, and temperature of mineralization. This method offers the temporal resolution necessary for dating episodes of mineralization and alteration in hydrothermal systems. This technique also provides information about the distribution of argon in the analyzed mineral or "whole rock," and whether the material formed before, during, or after an alteration. Strongly altered wallrock minerals may yield meaningful geochronologic data. Because this high-precision technique can be applied to altered and unaltered materials, this actual duration of mineralization can be addressed. $^{40}Ar/^{39}Ar$ age-spectrum dates are unique because the time at which various minerals close to diffusion of argon can be related to temperature, thus providing a "thermochronology." These time-temperature data are fundamentally significant and provide new opportunities for understanding the details of ore-forming processes. Three mineralized areas where this technique has provided valuable data are Panasqueira, Protugal, the Idaho batholith, northwestern United States, and the Gejiu batholith, southern China.

In a study of the Hercynian-aged (about 300 Ma) Panasqueira tin-tungsten deposit, a high degree of temporal resolution was achieved (Snee et al, 1988). Individual muscovite samples differing in age by only 2.2 m.y. (0.7% of the age of mineralization) are temporarily resolved. Differences in age as small as 0.9 m.y. (0.3%) are resolved by using multiple samples from separate events. The duration of different resolvable thermal activity for the tin- and tungsten-ore-forming stages was at least 4.2 ± 0.5 m.y. Within that time span, the age and the duration of the first substages of the oxide-silicate stage, the main sulfide stage, and the second substage of the oxide-silicate stage (contemporaneous with greisen formation and silica cap alteration), in that order, are resolvable. Minor argon loss from all dated samples occurred during later reheating, probably during the pyrrhotite alteration stage. Studies of cross-cutting relations of muscovite-bearing veins and fluid-inclusion homogenization temperatures of associated quartz (Kelly and Rye, 1979) allow an empirical calculation of the argon retention temperature of muscovite. This retention temperature and age-spectra of muscovites from a thermally complex area, such as Panasqueira, can be combined to quantify the temperatures and the durations of thermal events associated with mineralization. This study formed the basis for a successful test of laser argon techniques as applied to mineralization research.

In several mining districts in the central Idaho batholith, a complex series of hydrothermal events occurred during the Cretaceous. High-precision $^{40}Ar/^{39}Ar$ age-spectrum dates that range from 78 to 70 Ma have been determined fro muscovite and adularia from sulfide-bearing quartz veins that contain associated gold, tungsten, antimony, mercury, lead, and silver. Even though Eocene hydrothermal activity previously has been considered important in this area, $^{40}Ar/^{39}Ar$ age-spectra of muscovite and microcline, which were collected near Eocene dikes, stocks, and volcanic centers, have Cretaceous ages modified by partial loss of ^{40}Ar in post-Cretaceous time. Thus, after the Cretaceous, except in areas near intense Eocene thermal activity, the central Idaho batholith remained below about 150° (argon retention temperature for maximum microcline). The ages of minerals in the quartz vein deposits are similar to emplacement and cooling ages of Cretaceous plutons. The $^{40}Ar/^{39}Ar$ age-spectra provide constraints for the cooling history of plutons and quartz veins and thus yield evidence for depth of formation and genesis of the mineral deposits.

The Late Cretaceous Gejiu batholith in the southern Yunnan province of China is composed of numerous mafic and felsic intrusive rocks, is well-known for its large granite-hosted tin deposits, and has associated copper, lead, zinc, tungsten, beryllium, and sulfur. The batholith is divided into eastern and western parts by the postmineralization, north-trending Gejiu normal fault. The compositions of plutons and types of mineral deposits are markedly different on opposite sides of this fault. Plutons in the portion of the batholith west of the fault, grouped by age from oldest to youngest, include (1) gabbro, diorite, and monzonite, (2) grandiorite, granite porphyry, and granite, and (3) syenite and nepheline syenite. Lead and zinc in galena and sphalerite are disseminated in many western plutons; tin in cassiterite is disseminated in granite porphyry. East of the Geiju fault, minor exposures of a large subsurface granite body crop out within Triassic sedimentary rocks. Core drilling indicates that these outcrops represent parts of a continuous, 20 km long granite pluton of relatively uniform composition but variable texture. Major deposits of tin are associated with numerous cupolas of this pluton and occur in skarn, greisen, and open-space veins. Preliminary $^{40}Ar/^{39}Ar$ age-spectra constrain the age and the duration of mineralization and intrusive activity, showing that emplacement of granites and of associated tin mineralization occurred over at least 2 m.y. and that time-space relations among the various tin-mineralized granites can be resolved. Relative age relations indicate that emplacement of nepheline syenite and formation of associated disseminated galena and sphalerite west of the Gejiu fault took place near the end of, or subsequent to, tin mineralization. Additionally, because lead-bearing mineral deposits, which are genetically related to plutons older than the tin granites, are also present west of the Geiju fault, formation of lead deposits either was a multicycle phenomenon or occurred over a longer time interval than did tin mineralization.

These three examples outline new and powerful applications of $^{40}Ar/^{39}Ar$ thermochronology in the field of ore deposit research. Specific objectives that have been met are (1) determination of the "age" of the deposits, (2) measurement of the durations of mineralization, (3) assessment of the episodic nature of the activity, (4) definition of spatial differences in age of mineralizations, and (5) estimates of mineralization temperatures.

REFERENCES

Kelly, W. C., and R. O. Rye, 1979, Economic Geology, v. 74, 1721–1822.

Snee, L. W., J. F. Sutter, and W. C. Kelly, 1988, Economic Geology, v. 83, p. 335–354.

SNODGRASS, G. B., and J. NEIHEISEL, U.S. Environmental Protection Agency, Washington, DC

Volume Reduction and Chemical Extraction Of Radioactive Contamination in Soils at EPA National Priority Hazardous Waste Sites

The United States Environmental Protection Agency (EPA) has national responsibility to ensure the cleanup of the country's worst hazardous waste sites. The process of site cleanup starts with discovery and proceeds through characterization, risk assessment, and, ultimately, to the selection of a cost-effective remedial action that is protective of public health. Although the EPA has enforced site cleanup through the National Contingency Plan (NCP) for over 10 years, the passage of the 1986 Superfund Amendments and Reauthorization Act (SARA) and a revision of the NCP placed new and stringent requirements on hazardous waste site cleanup, emphasizing treatment over traditional remediation approaches.

The traditional approach has been excavation and disposal in a licensed hazardous waste landfill or permanent storage facility. The new requirements mandate the use of treatment technologies where possible. Additionally, the law states a preference for remedial technologies that would reduce the volume, toxicity, and mobility of hazardous wastes or pollutants.

In October 1986, EPA's Office of Radiation Programs began evaluating the possibility of using physical volume reduction and chemical extraction to remediate radioactive wastes at the Montclair and Glen Ridge Superfund sites in New Jersey. The sites are contaminated with radium from radium industry waste deposition in the 1920's. After preliminary study proved encouraging, the EPA Office of Radiation Programs initiated the Volume Reduction/Chemical Extraction (VORCE) project. Although the project focuses on the New Jersey radium sites, the methods will apply to a broad range of radioactive sites. The project is broadly divided into three major technical elements: soil characterization, treatment studies, and technology implementation.

The characterization of the soil is used to determine the most effective methods of treatment. Characterization involves physical, mineralogical, chemical measurement, and radioassay. A grain-size distribution curve is constructed to characterize the physical texture of the soil by sieving and sedimentation techniques, and representative fractions are obtained from the gravel, sand, silt, and clay-sized fractions for mineralogical, radioassay, and chemical analysis. Special focus is on the size fractions with the highest radioactivity (usually the fines). Special techniques are used for identification of the radioactive minerals, precipitates, or radionuclides adsorbed on mineral surfaces. The radioactive contaminants within the sand-sized materials are predominantly heavy minerals (greater than 2.9 specific gravity) easily separated by simple "sink-float" techniques. Silt- and clay-sized fractions, of less than 15 micron size, are best separated by linear density gradient separations using heavy liquids and centrifugation techniques. Complete mineral identification is easily facilitated by megascopic and petrographic microscopy of the gravel, sand, and upper silt-sized materials. A combination of X-ray diffraction, scanning electron microscopy, and microprobe analysis facilitates identification of the fines. Size fractions virtually free of contamination are identified for simple physical separation and chemical, mineral, and radioactivity data of the "hot" fractions are used to plan the potential chemical treatment studies. Thus, the soil characterization phase of the Montclair/Glen Ridge, New Jersey, Superfund sites has provided invaluable data for the treatment studies.

The treatment studies in the VORCE project include physical separation by both wet and dry screening and sieving methods, washing of soils with water or dilute solutions, chemical extraction, filtration studies, and ion exchange. The water washing studies have the objective of providing a process or unit process that can cost effectively meet standards and be protective of public health. We have found that the most successful washing procedure is wet screening and separation at the lower gravel sizes followed by rigorous washing of the larger fraction in water. Preliminary studies on contaminated soil at 182 pCi/g, result in the generation of a quantity of soil approximately one-third the weight of the starting material, with an activity below 15 pCi/g, a level that approaches a goal of 5 pCi/g, the public health standard.

Future studies will include water washing with more vigorous agitation, extraction of soils with dilute acids, mixtures of acids and inorganic salts, and solutions of complexing agents. Studies to separate dissolved radionuclides will follow the extraction studies. All studies have been planned with priority given to those laboratory processes that are most likely to result in large-scale remediation processes that are inexpensive and safe. Future studies will focus first on smaller particle fractions. These smaller fractions are higher in activity than fractions remediated by soil washing and require chemical methods of removing radioactivity from mineral or chemical "sites" on the particles.

The last project element is "Technology Implementation" and runs concurrently with other project studies. The objective is to produce a design and equipment that will emulate the dynamic functions that have been effective in the laboratory. Where laboratory procedures have been successful in redistribution of contamination by sieving, we are developing methods and identifying equipment to effect the redistribution on the large scale. Similarly, the functions of soil agitation, filtration of effluent, and ion exchange are being scaled up.

It is hoped that the Volume Reduction/Chemical Extraction (VORCE) project at the Montclair/Glen Ridge sites will reduce 40% of the soil, by weight, to below 5 pCi/g. We hope to reduce another 30% to below 15 pCi/g. These levels correspond with standards to be applied to the remedial action.

Estimated costs of the soil washing in an actual remediation are $62.00 per cubic yard. We are hoping to design a secondary treatment process that will operate at $120.00 per cubic yard. This simple and inexpensive treatment scheme will hopefully result in a remedial action that is feasible, cost effective, and protective of public health.

SNYDER, D. B., and the BIRPS Group, Bullard Laboratories, Cambridge, United Kingdom

Mantle Reflectors Around the United Kingdom

Reflections from two-way travel times associated with mantle depths, although uncommon, have now been recognized on migrated BIRPS data from most parts of the continental United Kingdom. Some reflections have reflection coefficients of about 0.1 and can be traced to depths of 80 km. Two of these prominent reflections, located 300 km apart in areas north of Scotland and north of Ireland, respectively, have been imaged in loose (5–20 km spacing) grids of seismic profiles. These grids provide information on the three-dimensional structure of the reflectors and confirm their geologic interpretation—the reflections are not processing artifacts nor noise. The reflectors form complex, plunging anticlinal structures over 90 km in breadth. Other BIRPS profiles from the North Sea and western English Channel provide examples of mantle reflectors confined to the uppermost few kilometers of the mantle. Mantle reflectors around the United Kingdom appear spatially related to major crustal structures such as North Sea rifts, the Great Glen fault, or the Outer Isles Fault. The age of the reflectors remains unresolved and cannot be directly correlated to the age of nearby crustal deformation. Reflectors can be characterized by association with crustal structures as either compressional or extensional in nature, reactivated many times in some cases, but a dominant mode of origin remains unclear. Reflection attributes require a compositional change across a geologic contact or within a shear zone; this change may involve mineralization fluids, magma, or shear deformation of the rock. Co-interpretation of reflector geometries with gravity field anomalies and surface mapping of both rock type and structures constrains possible origins for the mantle reflections but not, as yet, definitively.

SNYDER, D., University of California, Berkeley, CA

On Theory of Relation between Crystal Nucleation Rates and Crystal Deposition Rates in Magma Chambers

A nagging problem in igneous petrology is to unravel the processes responsible for the evolution of magmas from only the final preserved products (the rocks). This study outlines a theory that relates the crystal nucleation rate in a magma chamber to the consequent deposition rate by crystal settling. A unique inverse is derived that permits calculation of nucleation rates from the resulting deposition rate. The model is immediately useful for evaluating a proposed mechanism for the origin of rhythmic layering in layered intrusions. Calculations on the nucleation rate of augite as a function of time in the Kiglapait layered intrusion yield results consistent with kinetic calculations.

We begin by describing a turbulently convecting magma chamber as a linear transducer that sends a nucleation rate function, $g(t)$, to a deposition rate function, $h(t)$. It can easily be shown from the definition of the Dirac delta functional that

$$g(t) = \int_{-\infty}^{\infty} \delta(t - t')g(t')dt'.$$

From this we can conclude that by knowing the response of the magma chamber to a Dirac delta-like nucleation rate function, we can predict the response of that system to *any* nucleation rate function. If we let the response of the chamber to a Dirac delta functional be $k(t)$, it can easily be shown that

$$h(t) = \int_{-\infty}^{\infty} k(t - t')g(t')dt' = \int_{-\infty}^{\infty} k(t')g(t - t')dt'.$$

To derive an expression for $k(t)$, we assume that the rate of crystal depletion in the chamber is proportional to the number of crystals in suspension (Martin and Nokes, 1988), which is given by $N(t)$. If $g(t) = N_0\delta(t)$, then the crystal deposition rate is

$$h(t) = N_0\alpha\exp(-\alpha t)$$

where α is the inverse of the crystal retention time. This parameter can be estimated from Stokes law (Martin and Nokes, 1988; Marsh, 1988). Hence,

$$k(t) = \alpha e^{-\alpha t}.$$

In general, then, the deposition rate, $h(t)$, is related to the nucleation rate, $g(t)$, by convolution:

$$h(t) = \alpha \int_0^{\infty} e^{-\alpha t'} g(t - t') dt'.$$

This integral equation is easily inverted using Fourier transform methods. However, an analytic inverse exists that avoids the truncation effects which are inherent in Fourier transform methods. By suitable operator arguments,

$$g(t) = h(t) + \frac{1}{\alpha} \frac{\partial h(t)}{\partial t}$$

so long as $\alpha \neq \alpha(t)$.

It remains to relate the deposition rate to the modes of the rocks preserved in the cumulate pile. If the concentration of phase i at time t is given by $x_i(t)$, then for n phases,

$$x_i(t) = \frac{h_i(t)}{\sum_{i=1}^{n} h_i(t)}.$$

Also, of $d(t)$ is the depth of the crystal pile as a function of time and we allow each particle to have thickness ϕ_i when lying on the floor, then

$$d(t) = \sum_{i=1}^{n} \phi_i \int_0^t h_i(t') dt'.$$

So long as the deposition rate never drops to zero for any t, the above equations can be combined to calculate the mode of any phase in the cumulate pile as a function of depth from knowledge of the nucleation rates of the phases.

Since nucleation can only occur at finite undercoolings, recent numerical models have suggested that nucleation may occur in pulses of finite duration. If we take these pulses to be approximately Gaussian, then from the above theory we can predict that the consequent deposition rate is of the form

$$h(t) = \frac{\alpha\sqrt{\pi}}{2} \left[1 + \mathrm{erf} \left[\frac{2t - 2\theta - \alpha}{2} \right] \right] \exp \left[-\alpha t + \alpha\theta + \frac{\alpha^2}{4} \right]$$

where $\Theta > 0$ is the time of maximum nucleation rate. Use of this relation for the simulation of the formation of rhythmic layering by oscillatory nucleation has indicated that the mechanism of oscillatory nucleation alone is unlikely to account for the genesis of these layers. The theory of convolution predicts a decline in the mode of the first phase (and in general the densest phase) to nucleate as function of depth which is much more rapid than that observed in nature.

From the mode of augite in the upper zone of the Kiglapait intrusion, we have extracted an estimation of the deposition rate of augite in the intrusion as a function of time. This value was used to calculate a relative nucleation rate as a function of time for augite. The result is a pulse with roughly Gaussian shape. Hence, the calculated form of the nucleation rate function is in striking agreement with that predicted from independent kinetic arguments. This success encourages the further development and use of convolution theory in the plutonic setting.

REFERENCES

Marsh, B. D., 1988, Crystal size distribution (CSD) in rocks and the kinetics and dynamics of crystallization I. Theory: Contributions to Mineralogy and Petrology, v. 99, p. 277–291.

Martin, D., and R. Nokes, 1988, Crystal settling in a vigorously convecting magma chamber: Nature, v. 332, p. 534–536.

SNYDER, S. W., North Carolina State University, Raleigh, NC, A. C. HINE, University of South Florida, St. Petersburg, FL, and W. SNYDER and S. R. RIGGS, East Carolina University, Greenville, NC

Mid-Cenozoic Glacioeustacy: Detailed Record of High-Frequency Sea-Level Changes from North Carolina Continental Margin

Interpretations from a dense network (20,000 km) of very high-resolution seismic reflection profiles collected from the North Carolina estuaries, shelf, and upper slope have defined over 30 regional unconformities. These physical surfaces of discontinuity were used to construct a relative chronostratigraphic framework by stacking the unconformities and intervening sequences vertically according to superposition. Correlations to well/auger/drillhole data were used to establish initial age relationships. These correlations were confirmed by penetrating stratigraphic targets along the seismic lines via 140 continental shelf vibracores. Eighteen Miocene and eight Oligocene sequences have been identified.

A comprehensive biostratigraphic study (planktic foraminifera, nannofossils, radiolarian, and diatoms; Scott W. Snyder, 1988) of the shelf vibracores has rendered relatively tight constraints on the timing of Miocene depositional and erosional events. Biochronologic constraints on the Oligocene sequences are not as well defined. Herein, absolute values of time are relative only to the calibrated biogeochronologic time scheme of Haq et al(1987).

Four biostratigraphic gaps divide the composite Miocene record into three distinct stratigraphic sections.

1. The mid-Burdigalian section consists of six distinct shelf sequences; deposition started near the N-5/N-6 boundary of Blow (1969)(ca. 18.7 Ma) and continued until NN-3/NN-4 boundary of Martini (1971) (ca. 17.6 Ma).

2. The Langhian section consists of several outliers and four prograding shelf-margin sequences; deposition started near the NN-4/NN-5 nannofossil boundary (ca. 16 Ma) and lasted until the boundary between N-9/N-10 (ca. 14.8 Ma); the four shelf margin sequences are confined to zone III of Abbott's (1978) Atlantic Margin Diatom Zonation scheme (AMDZ), which is approximately equivalent to the N-9 global biochron (ca. 0.5 Ma in duration).

3. The Serravillian section consists of eight depositional sequences, although the upper three sequences were found to be nearly barren of index microfossils; deposition began near the N-11/N-12 biochron boundary (ca. 13.2 Ma); the upper age limit for the lower five Serravallian sequences is placed just below the N-12/N13 boundary (ca. 12.1 Ma) based on the identification of diatom species from Atlantic Margin Diatom Zone VI; the upper limit of all eight sequences is not defined.

A large hiatus separates the Miocene and Oligocene deposits. It spans the NN-1 through NN-2 biochrons (ca. 6 Ma. in duration). Available biostratigraphic data also indicate that the Oligocene sequences can be divided into at least two sections separated by hiatuses. (1) The Chattian section consists of five shelf sequences containing nannoplankton assigned to zone NP-25 and planktic foraminifera assigned to the P-22 through N-4 biochrons (latest Oligocene). (2) The Rupeilan section consists of three prograding shelf-margin sequences deposited in associated with the ancestral Gulf Trough, an Oligocene Seaway connecting the Gulf of Mexico with the North Atlantic Ocean. Fletcher (1986) and Jones (1983) found the Rupeilan sequences to contain planktic foraminifera assigned to biochron P-19/P-20 (early Oligocene).

All of the mid-Cenozoic sequences are bound by cross-shelf unconformities. They are therefore interpreted to be the products of discrete sea-level fluctuations. Many of the unconformities are marked by paleofluvial channels that indicate subaerial exposure and render some constraints on the minimum elevation of lowstand events. Benthic foraminiferal assemblages give a semiquantitative description of water depth at the time of deposition; they have been used to help quantify the highstand sea-level positions. Collectively, these data suggest that the amplitudes of the sea-level cycles ranged between ± 50 and ± 200 m. The duration of the Miocene sea-level fluctuations is estimated by dividing the number of sequences by the time limits defined in the biostratigraphic data set. This calculation renders the periodicity for Miocene sea-level oscillations on the order of 10^5 years (4th-order). We conclude that this high frequency of high-amplitude sea-level fluctuations could only be driven by glacioeustacy associated with climatic changes forced by the orbital-perturbations described by Milankovitch (1941). Better resolution in time is required before the periods of the Oligocene sea-level cycles can be quantified.

While most of the mid-Cenozoic sequences are believed to be the products of glacioeustacy, the origin of all five sections and their intervening hiatuses remains more speculative. Each section spans over 1 Ma in duration is separated by hiatuses lasting up to 6 Ma. They may be the products of lower-frequency sea-level oscillations; however, correlations with the 3rd-order cycles of Haq et al (1987) is poor. Many of the hiatuses may have been generated or modified by significant erosional episodes associated with extended lowstand events or Gulf Stream scour. These erosional events could have removed the stratal products of prior sea-level events. Numerous scars of the western boundary current have been identified in the seismic sections. Their role in exhuming and removing records of older sea-level events from the North Carolina margin needs to be quantified by extending the seismic network farther seaward across the slope and Northern Blake Plateau.

SNYDER, S. W., North Carolina State University, Raleigh, NC, A. C. HINE, University of South Florida, St. Petersburg, FL, and S. R. RIGGS, East Carolina University, Greenville, NC

Subsidence Versus Eustatic Flooding Rates: Their Relative Roles in Controlling Preservation of Coastal and Shelf Sequences

Recent stratal models offered by Exxon Production Research (EPR) depict the depositional products of a generic sea-level cycle as a thick aggradational wedge consisting of coastal, shelf, and shelf margin lithosomes or "system tracks" (Haq et al, 1987). In contrast, stratigraphic studies from the North Carolina coast and continental shelf show that the depositional records of Quaternary glacioeustatic sea-level cycles are generally confined to estuarine-filled paleofluvial channels.

Interpretations of high-resolution seismic profiles and vibracores, supplemented by auger data and ^{14}C and amino acid racemization dates from the shelf-barrier-lagoon complex off Bogue Banks, North Carolina (USA), have demonstrated that stratigraphic incision by the shoreface profile has exhumed and removed almost the entire sedimentary record of the Holocene Transgression. Only Tertiary and early Pleistocene sequences crop out on the continental shelf. The erosional mode of this transgression is presently captured by the modern shoreface that is actively exhuming a stacked sequence of barrier-backbarrier coastal lithosomes. These include (1) a basal Pleistocene sequence cut by (2) fluvial channels filled with early Holocene estuarine sediments, which are overlain by (3) a Holocene lagoonal section heavily dissected by a series of tidal inlet channels, and finally capped by (4) a relatively thick section of late Holocene regressive shoreface sands. Lagoonal and inner shelf seismic profiles demonstrate that none of the Holocene barrier-backbarrier lithosomes underlying Bogue Banks will be preserved because they all lie above the depth of stratigraphic incision by the shoreface. Only a few early Holocene paleofluvial channels, which have cut stratigraphically lower into the Pleistocene and Tertiary sequences, have the potential to be preserved as the shoreface continues to migrate landward with sea-level rise.

Analyses of the composite seismic network form the adjacent continental shelf demonstrates that paleofluvial channel-fill lithosomes of Quaternary age are quite abundant. Their surface outcrops cover approximately one-third of the seafloor of the shelf embayment of Onslow Bay. Profiles from inland rivers and estuaries depict the composite Pleistocene section as a mosaic of fluvial and estuarine sediments dissecting a few isolated highstand shorelines. Amino acid racemization dates from in situ mollusca demonstrate that these limestones represent the stratigraphic products of at least eight different Pleistocene sea-level events.

Poor seismic coverage and the lack of drill/vibracore data have limited our analyses of Quaternary records along the outer shelf and upper slope. Existing data depict a series of prograding shelf-margin sequences consisting of oblique clinoforms downlapping on to Tertiary outcrops within the Northern Blake Plateau. A few channels oriented parallel to the shelf margin truncate these progradational wedges. The channels were probably cut by currents associated with the Gulf Stream.

The depositional record of Quaternary glacioeustatic sea-level oscillations along the North Carolina continental margin can be summarized as consisting of a few paralic highstand shorelines separated from a series of prograding shelf margin wedges by paleofluvial channel-fill lithosomes buried in Tertiary continental shelf outcrops. The most complete record of the glacioeustatic sea-level fluctuations is probably preserved in the paleofluvial channels, because successful erosional transgressions have consumed many former highstand shoreline deposits, and Gulf Stream scour has repeatedly attacked the shelf-margin records. The difference between this record and the thick aggradational sequence proposed in the stratal model of Haq et al (1987) is related to absolute shelf flooding rates.

Using the sea-level curve from Bermuda (Neumann, 1971) as a record of Holocene eustacy, maximum flooding rates are calculated as exceeding 800 cm/Ka. Stable passive continental margins overlying continental crust or old, thermally contracted, rift-stage crust is subsiding at rates less than 1 cm/Ka. This is the case for the Carolina Platform. The effective maximum flooding rate is therefore estimated at 801 cm/Ka for the North Carolina margin. At that rate, the shoreline is capable of crossing the continental shelf dozens of times before the margin subsides a few meters. Stratigraphic incision by the shoreface during each cross-shelf transit reworks and removes all coastal and shelf lithosomes, leaving only a few channels cut deeply into the older shelf outcrops.

In contrast, the maximum flooding rate of any Cenozoic 3rd-order sea-level cycle delineated in the curve of Haq et al (1987) was calculated to be 13 cm/Ka (cycle TB 2.3 at 100 m/800 Ka). Subsidence along a stable continental platform margin such as North Carolina (1 cm/Ka) would lower the shelf approximately 10 m during this 3rd-order sea-level event. Thus, some of the coastal and shelf lithosomes would be capable of subsiding below the depth of stratigraphic incision by the shoreface. Along younger continental margins (overlying hot rift-stage crust), or areas of severe isostatic loading (Mississippi delta and fan), subsidence rates range between 50 and 1000 cm/Ka. This significantly increases the potential for preservation of all the system tracks, regardless of the eustatic flooding rate. Thus, the rates of eustatic flooding and regional subsidence can be used to help determine the mode of any transgressive sea-level event: erosional (mostly channels preserved) versus depositional (preservation of a vertical succession of transgressive and highstand system tracks).

SOBA, D., I.R.G.M., Yaounde, Cameroon, and L. LATOUCHE, Museum Natural d'Histoire Naturelle, Paris, France

Metalamprophyres of Lom Series (East Cameroon)

During the 1/50,000 geological mapping in the area of the Lom valley, we recognized a group of dykes of metalamprophyres. These dykes outcrop en echelon in the form of veins in steps about 1 m wide by 100 m long.

Mineralogical, chemical, and geochemical data enable us to link them to minettes (Rock, 1984; Jahn, 1979) on one hand and on the other hand to different syenites known in South Cameroon. Petrographic and structural study as well as the already obtained geochronological data (Lasserre and Soba, 1976) enabled us to estimate the emplacement of these metalamprophyres during the upper Proterozoic.

During the Pan-African, a first phase of deformation characterized by a subhorizontal schistosity and a metamorphism of biotite and staurolite followed an episode of tightening and uprightness of structures coinciding with the development of chloritoids in the zones where the deformation is most intense.

The Pan-African orogenesis ends in this region by a dextral sliding along the Lom fault that prolongs the great Sanaga fault toward the northeast. These metalamprophyre dikes are the last manifestations of a syenitic magmatism that characterizes the Pan-African orogenesis in Cameroon.

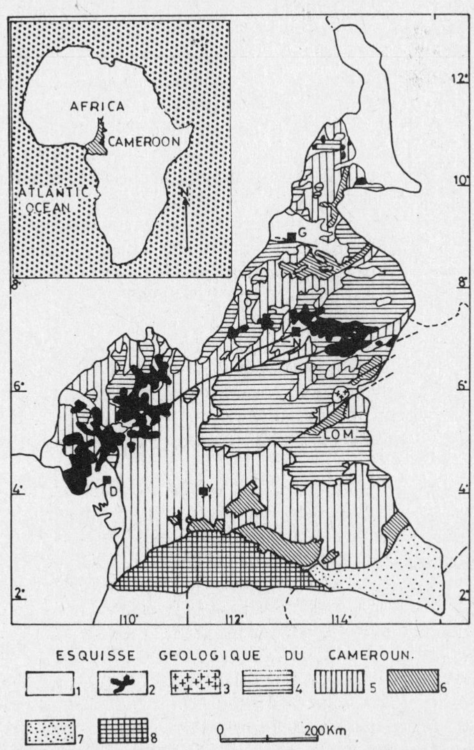

Figure 1—Esquisse Geologique du Cameroun. 1: Sedimentary; 2: Volcanism; 3: Intrusifgranites; 4: Syntectonic granites; 5: LOM; POLI; Mbalmayo–Mbengbis series; 6: Basement Complex; 7: Dja series; 8: Northern end of Congo craton.

SOBCZAK, L. W., J. F. HALPENNY, and D. M. HENDERSON, Geological Survey of Canada, Ottawa, Canada

Gravity Interpretation along Seismic Refraction Lines Surveyed Near Canadian Ice Island, 1985

The structure of the Queen Elizabeth shelf and slope northwest of Axel Heiberg and Ellesmere islands has been analyzed along three profiles using geophysical and geological information obtained largely near the Canadian ice island during 1985; results of a gravity investigation are reported here. Control for sedimentary and crustal thickness along parts of these profiles was obtained from seismic refraction lines (Asudeh et al, in press). Lithologies, densities, and interval velocities were derived from drillholes on the Queen Elizabeth Islands. Surface geology was taken from maps compiled by Thorsteinsson and Trettin (1971). Magnetic anomalies helped delineate areas affected by volcanism. The gravity analysis utilized these data in deriving gravity-density structural sections along these profiles. A novel approach employed in the analysis was the use of enhanced residual isostatic anomalies (ERIA) to examine regional structures. These anomalies provide a representation of the gravity field essentially free from gravity signals related to continent–ocean boundaries, water–rock interfaces, topography/bathymetry, associated isostatic roots/antiroots, and thermal and regional effects (Sobczak and Halpenny, in press).

Negative ERIA generally characterize the western third of the area (340 × 500 km), apart from a large high (maximum +80 mGal) that straddles the shelf-slope break, and positive ERIA the eastern two-thirds. The western negative anomalies delineate a thick (15 km), unmetamorphosed, composite sedimentary basin that incorporates the Franklinian Basin at the base, Sverdrup Basin in the middle, and Canadian Arctic Margin Basin at the top. Evaporites at the base of the Franklinian Basin are usually located at the centers of the local gravity lows and on gravity gradients such as that on the landward, updip, flanks of the thickened Canadian Arctic Margin Basin. The intermediate gravity high (+80 mGal) over the composite sedimentary basin has little structural significance as it merely represents a lens of uncompensated sediments. This extensive ERIA low, which is north of the Sverdrup Rim, crosses the rim southwards along Peary Valley and Prince Gustaf Adolf Sea southwest of Ellef Ringnes Island where it merges with an extensive low over the western part of the Sverdrup Basin (Sobczak and Halpenny, in press), which is also very thick (>15 km), unmetamorphosed, composite basin (Sobczak et al, 1986).

A narrow, northeast-trending offshoot of this low extends from the seaward end of a seismic refraction line perpendicular to the coastline from Axel Heiberg Island to Phillips Trough north of Ellesmere Island. The Canadian Arctic Margin Basin with a seismically determined thickness of 4.5 km correlates with this gravity low. A shallow, high-velocity (5.8–6.2 km/s) structure, about 5 km thick, was also interpreted in this area, which Asudeh et al (in press) explained in terms of an uplifted basement horst of the Franklinan Basin.

However, the gravity-density analysis does not support this concept, because a relative gravity high, which would be expected to correlate with the horst, is not observed. Instead, we propose the Sverdrup and Franklinian basins are continuous through this high velocity feature, which is interpreted as a sequence of thermally metamorphosed sediments heavily intruded with evaporites. This would account for the higher seismic velocities and the normal densities noted in this analysis.

The ERIA highs that cover most of the eastern two-thirds of the study area are of variable magnitudes and can be explained by various degrees of metamorphism and intrusion of mafic igneous rocks within the sediments of the Sverdrup and Franklinian basins. In the drillholes where mafic intrusions are present, metamorphism has apparently increased the density of sediments by 0.17 Md/m^3 for Sverdrup Basin sediments and 0.09 Mg/m^3 for Franklinian Basin sediments. In the region of the seismic profiles, metamorphism usually accounts for a density increase of about 0.02–0.04 Mg/m^3, which is much lower than observed in areas where mafics are common in the stratigraphic column. The greatest concentration of these mafics appears to be nearshore or onshore.

The total thickness of the sediments varies from about 15 km along the outer shelf and slope and gradually thins to about 3 km at the coast where the overlying Sverdrup Basin sediments either are very thin or are nonexistent and the underlying Franklinian Basin sediments are exposed. This thinning of the sediments probably represents the shoreward side of the northeastward extension of the Sverdrup Rim as shown by Sobczak et al (1986, their Figure 1). The lower portion of this combined basin consisting of the Sverdrup and Franklinian basins has been, in some places, thermally metamorphosed and intruded by mafic rocks that have a positive ERIA gravity expression. The unmetamorphosed parts of this combined basin over the continental margin have a negative ERIA signature. Negative ERIA continue southward straddling the Sverdrup Rim and overlying the western Sverdrup Basin within the Queen Elizabeth Islands where significant deposits of hydrocarbons, largely in the form of natural gas, have been found. Similar hydrocarbon prospects are proposed for regions of negative ERIA offshore.

Seaward the crust thins from 40 km at the coast to 15 km below the Canada Basin. A gravity model provides an estimate of maximum thickness of sediments below the Canada Basin; here no independent seismic information is available. These sediment thicknesses are similar to those near the seaward side of the seismic line perpendicular to the coast. The model assumes a limiting case of sediments resting directly on mantle and hence the thicknesses are probably overestimated.

REFERENCES

Asudeh, I., D. A. Forsyth, R. Stephenson, A. Embry, H. R. Jackson, and D. White, in press, Crustal structure of the Canadian polar margin: Part 1: Canadian Journal of Earth Sciences.

Sobczak, L. W., and J. F. Halpenny, in press, Gravity anomaly maps of the Arctic (Free-air, Bouguer, Total Isostatic, Residual Isostatic and Enhanced Residual Isostatic): Tectonophysics, Special Issue on Arctic Geoscience.

U. Mayr, and J. F. Sweeney, 1986, Crustal section across the Polar continent-ocean transition in Canada from Somerset Island to the Canada Basin: Canadian Journal of Earth Sciences, v. 23, n. 4, p. 608–621.

Thorsteinsson, R., and H. P. Trettin, 1971, Geology Maps 1305A, 1309A, 1310A and 1311A, District of Franklin: Geological Survey of Canada.

SOBOLEV, A. V., and A. A. GURENKO, Vernadsky Institute of Geochemistry, Moscow, USSR, and N. V. SOBOLEV, Institute of Geologu and Geophysics, Novosibirsk, USSR

Petrology of Ultrabasic Ultrapotassic Magmas of Continental Lithosphere: Results of Inclusions Studies in Olivine from Lamproites, Kimberlites, and Kamafugites

The problem was to determine composition and crystallization condition (T, P, fO_2) of early liquidus assemblages, composition of fluid, and melt of K-rich ultrabasic magmas of the continental lithosphere (lamproites, kamafugites, kimberlites). On the basis of the information, an estimation of possible conditions of genesis and composition of a source of these magmas were made.

Objects: Lamproites have been studied on an example of the Kimberely Craton, West Australia (olivine lamproites of the pipes Ellendale 11.9; leucite lamproites of Mt. Cedric). Kimberlites from the Udachnaya-East pipe, Yakutia, have been studied. The kamafucites investigated were taken from the West African rift zone (ugandite and mafurite were sampled from lava flows of the craters Kazimiro and Kjambogo, respectively, West Uganda). All the rocks investigated are olivine–porphyric and typical of the most magnesian and less altered varieties of corresponding types.

Results: The following results have been obtained:

1. Olivine compositions, which represent the early stage of crystallization of the magmas investigated, have been determined. Lamproites and kimberlites are characterized by second-generation olivine with Fo contents up to 93.5–90.0 mol% and 94.0–88.0 mol%, respectively. Ugandites and mafurites are characterized by first-generation olivine with Fo contents of 91.0–86.0 mol%, respectively.

2. Inclusions of magnesian low-alumina orthopyroxene (Opx) enriched in Ti (Table 2) and inclusions of Ti-bearing magnesian chrome spinel (CrSp) have been found in all investigated specimens in olivine of the above composition (Table 1). The composition of these CrSp in lamproites and kimberlites is distinctly different from those of CrSp from heavy mineral concentrates from the same pipes. Both CrSp and Opx inclusions in olivines of all samples differ from those in ordinary mantle xenolites by their high Ti content (Sokolev et al, 1975; Hervig).

Besides inclusions of clinopyroxene, Ti-phlogopite, ilmenite, and rutile have been found in the kimberlite olivine, and in the ugandite olivine inclusions of clinopyroxene have been observed.

Primary melt inclusions of lamproites, ugandite, and mafurite are present in the olivine.

3. The estimation of early crystallization temperatures has been done on the basis of two-pyroxene thermometer (Lindsley, 1983), olivine–spinel thermometer (Fabrier, 1979), ilmenite–magnetite thermometer (Spencer and Lindsley, 1981) as well as by the method of melt inclusion homogenization (Roedder, 1984). The following estimates have been obtained for most magnesian olivines: 1200–1060 °C for olivine and leucite lamproites, 1300–1100 °C for kimberlites, 1200–1150 °C for ugandite and mafurite. A subsequent melt fractionation occurred at temperatures up to 1000–950 °C for lamproites and ugandites and 850 °C for mafurites.

4. Primary fluid inclusions have been found in magnesian olivines of all rocks investigated, which evidence the saturation of melts by a fluid component at an early stage of crystallization. According to Raman spectroscopy data, the fluid composition corresponds to CO_2, with trace of CO and N_2 up to 2–3 mol%. The H_2O content in the fluid is less than 1–2 mol% for ugandites and mafurites, and less than 10 mol% for lamproites and kim-

berlites. The minimum fluid pressures at the crystallization stage is estimated to be of 4–5 Kbar for all rocks investigated.

5. The estimation of O_2 fugasity of magma crystallization obtained from the assemblage olivine–orthopyroxene–spinel (Sokolev et al, 1988) corresponds to the FMQ-buffer for lamproites and kimberlites and to NNO-buffer for ugandite and mafurite.

6. The melt compositions are determined at the stage of the second-generation olivine crystallization of lamproites, ugandite, and mafurite. The melts are characterized by extremely high content of alkalies, especially of K, high contents of Ti, P, Ba, Sr, and F, and low contents of Al and Ca (Table 3).

Conclusions:

1. The studied K-rich melts are characterized by lowest temperatures of crystallization among mantle originated magmas and can be due to melting of lithosphere without considerable overheating in relation to the geotherm of stable regions and thus without active mantle diapirism.

2. The most probable cause of the origin of K-rich ultrabasic melts is metasomatism of predepleted harzburgitic material of the lithosphere realized by fluid (melt) of high CO_2/H_2O ratio.

3. Differences in composition of the studied ultrabasic K-rich magmas are due to a different extent of the source predepletion and different conditions of melting. The least depleted and most high temperature source is characteristic for kimberlite, the most depleted and least high temperature source for lamproites with kamafugites somewhere between. The extent of incompatible elements enrichment as well as their isotopy is controlled by composition and origin of metasomatic component.

REFERENCES

Fabries, J., 1979, Contributions to Mineralogy and Petrology, v. 69, p. 329-336.

Hervig

Lindsley, D. H., 1983, v. 68, p. 477-493.

Roedder, E., 1984, Fluid inclusions: Reviews in Mineralogy, v. 12, 644 p.

Sobolev, A. V., et al, 1988.

Sobolev, N. V., et al, 1975, Geologiya i Geophisica, v. 11.

Spencer, R. J., and D. H. Lindsley, 1981, American Mineralogist, p. 1189–1201.

Table 1—Representative compositions of CrSp of lamproite (1), kimberlite (2), ugandite (3), and mafurite (4).

NN	1	2	3	4
SAMPLE	EL11	UNV-4	11529	11497
SIO2	.26	0	.07	.10
TIO2	2.40	4.68	5.68	5.35
AL2O3	4.36	7.76	6.30	5.46
CR2O3	58.53	45.32	41.64	46.52
V2O3	.06	0	.12	.05
FE2O3	6.29	10.97	13.02	10.65
FEO	14.19	18.66	21.27	16.56
MNO	.16	0	.24	.26
MGO	13.52	12.10	10.47	13.11
NIO	.19	0	.25	.33
TOTAL	99.96	99.49	99.06	98.39

Table 2—Representative compositions of Opx of lamproite (1), kimberlite (2), ugandite (3), and mafurite (4).

NN	1	2	3	4
SAMPLE	EL11	UNV-4	11529	11497
SIO2	56.61	56.34	56.30	56.96
TIO2	.11	.23	.50	.29
AL2O3	.62	1.30	.98	1.35
CR2O3	.15	.54	.41	.27
FEO	6.23	6.54	6.61	5.60
MGO	34.60	32.84	33.54	34.05
CAO	.90	1.20	2.04	1.74
NA2O	.21	.24	.24	.21
TOTAL	99.43	99.23	100.62	100.47

Table 3—Representative compositions of homogenized melt inclusions in olivines of lamproite (1,2), ugandite (3), and mafurite (4).

NN	1	2	3	4
SAMPLE	9AC5291	E-11/1	11529	11497
SIO2	44.59	44.70	41.83	40.76
TIO2	5.89	6.20	5.69	5.15
AL2O3	5.54	4.50	10.45	10.03
FEO	6.81	8.90	7.06	6.15
MGO	9.00	8.00	6.26	6.62
CAO	5.77	4.21	6.17	6.57
NA2O	2.64	2.41	8.27	8.01
K2O	10.64	10.50	9.60	14.69
P2O5	1.97	2.13	1.68	1.20
BAO	3.46	3.25	.88	1.15
SRO	.18	.43	.51	.36
ZRO2	.41	.41	.03	.09
F	1.00	1.43	.28	.39
TOTAL	97.90	97.07	98.71	101.17
FO	92.20	91.30	88.10	91.60
T hom	1035	950	1035	1030

SOBOLEV, A. V., and L. V. DMITRIEV, Vernadsky Institute of Geochemistry, Moscow, USSR

Primary Melts of Tholeiites of Oceanic Rifts (TOR): Evidence from Studies of Primitive Glasses and Melt Inclusions in Minerals

The knowledge of the composition and PT conditions of formation of primary melts (PM) of TOR is a key to understanding the origin of global geochemical reservoir within mantle that produce the majority of Phanerozoic basaltic magmas. Thus, the problem is in the center of modern petrological discussion. Two main opposite points of view are known from literature : (1) PM TOR are basaltic in composition and originated at P = 8–10 Kbars and T less than 1300 °C (e.g., Presnall et al, 1979); (2) PM TOR are picritic or even komatiitic and form at P = 15–30 Kbars and T more than 1400 °C (e.g., Gren et al, 1979). However, recent investigations (e.g., Dmitriev et al, 1985) show that the unique PM for all TOR is an oversimplification. Dmitriev et al, (this volume) show that there are at least four separate TOR types that form different fractionalization trends and have independent PMs: TOR 1, TOR 2, TOR K, and TOR Na. Moreover, Duncan and Green (1980) have proposed and Danushevsky et al (1987) have confirmed the existence of an additional rare extradepleted type of TOR, TOR 3.

In the present paper, we make one more attempt to estimate the compositions and PT conditions of formation of PMs for major types of TOR—TOR 1 and TOR 2—as well as for the unique TOR 3 and Iclandic tholeiites (IT). To do this, we use detailed mineralogical studies, investigation of melt inclusions in minerals, and numerical modeling of crystallization of primitive glasses.

Criteria of primarity: Any PM should be in equilibrium with its mantle source (restite). The most reliable signs of such equilibria are as follows:

1. The primary liquidus assemblage (PLA) of PM should contain in every case an Fo-rich olivine. Plagioclase is expected as a member of PLA for PM formed by melting of plagioclase lherzolite (Presnall et al, 1979). Orthopyroxene will appear in PLA if crystallization takes place at the same P as melting.

2. The olivine of PLA should be equally or more Fo-rich than the olivine of restite.

3. Spinel of PLA should be comparable in Cr/Cr + Al ratio with spinel of restite and follow "olivine–spinel mantle array" (Arai, 1987).

4. The composition of PM should be in equilibrium with olivine–orthopyroxene assemblage at PT of formation.

Mineralogy of mantle source and PLA of TOR:

Oceanic periodotites—It is generally accepted that the oceanic periodotites associated with TOR represent restites after their formation. In spite of lack of data (fewer than 100 analyses) the stable range of olivine composition is observed as: $Fo_{89.5-91.5}$ (our data and summary of Bonatti and Michael, in press). Moreover, the most magnesian olivines $Fo_{90.5-91.5}$ are found in the regions occupied by TOR 1 magmas (43 °N, Oceanographer FZ, Atlantic) and least magnesian $Fo_{89.5-90.5}$ in the regions occupied by TOR 2 magmas (Gibbs FZ, DSDP 395, Vema FZ, Romanche FZ, Atlantic) (Bonatti and Michael, in press; Dmitriev et al, this volume).

TOR 1 and IT—The most magneisian olivine was found in olivine-phyric picrites: up to $Fo_{91.7}$ for TOR 1 from Famous area and DSDP 332B, north Atlantic (from 210 analyses) and up to $Fo_{92.2}$ for IT from Teystarykir area, Iceland (from 180 analyses; Gurenko et al, 1988). These olivines are associated only with chromium spinel.

TOR 2—The most magnesian olivine, $Fo_{90.5}$, was found from 250 analyses at Vema FZ and DSDP 396 (central Atlantic) in assemblage with plagioclase and spinel.

TOR 3—This type of TOR was found as a melt inclusion in olivine Fo_{90} in assemblage with orthopyroxene (Mg# 0.907) and spinel at 10 °N, Mid-Atlantic Ridge (Danushevsky et al, 1987).

The compositions of all observed early ol-sp assemblages fit well with "ol-sp mantle array" and are close to the most depleted associated periodotites (Figure 1).

Melt inclusions in the most magnesian olivines: Melt inclusions were investigated by homogenization method (Roedder, 1984) in optical high-temperature cell. Compositions were determined by electron microprobe (Table 1).

TOR 1 and IT—The highest crystallization temperatures are 1330–1320 °C for IT and 1300–1290 °C for TOR 1. The compositions of homogenized melt inclusions are picritic with low Ti and Na.

TOR 2—The highest crystallization temperature is 1240 °C; the composition of melt corresponds to high-alumina magnesian tholeiite relatively enriched in Na and Ti.

TOR 3—The highest temperature estimation is 1220 °C; the composition of melt corresponds to magnesian quartz tholeiite depleted in Ti and Na.

Primitive glasses: From available database of over 6000 analyses of TOR glasses were selected those with Mg/Mg + Fe ratios higher than 0.67: 10 analyses of TOR 1 type, 10 analyses of TOR 2 type and 1 analysis of IT.

TOR 1 and IT—For the selected glasses, the PM estimations were made by back calculation of olivine fractional crystallization up to equilibrium with Fo_{92} with constant $Fe^{3+}/Fe^{3+} = Fe^{2+} = 0.07$ using model of Ford et al, (1983). The calculated PM are picritic (MgO = 13.5–15.6 wt%) and have relatively low TiO_2 (0.4–0.8 wt%) and Na_2O (1.4–2 wt%).

TOR 2—The PM were estimated using back calculation of olivine-plagioclase cotectic (Dmitriev et al, 1985) up to equilibrium with olivine $Fo_{90.5}$ (calculated using Ford et al, 1983, and Drake, 1976) and $Fe^{3+}/Fe^{3+}+Fe^{2+} = 0.07$). The resulting PM are high-alumina tholeiites (Al_2O_3 = 18–19 wt%) enriched in Na_2O (2.2–2.7 wt%) and TiO_2 (0.7–1 wt%).

Origin of TOR:

Composition—Obtained experimental and calculated PMs are similar and correspond to picrite for TOR 1 and IT and to high-alumina tholeiite for TOR 2 (Table 1).

PT conditions—The PT conditions of their equilibria with olivine-orthopyroxene assemblage were determined using experimentally calibrated Ol-Pl-Qz and Ol-Di-Qz projections of basaltic tetrahedron (summary in Falloon and Green, in press) and average liquidus slope of 5°/Kbar.

1. TOR 1 and IT: Estimated ranges of P and T are 10–20 Kbars and 1370–1470 °C, which are similar to estimations of Green et al (1979).

2. TOR 2: The estimated PT conditions of origin are P = 8–10 Kbars, T = 1250–1300 °C, which are almost the same as was proposed by Presnall et al (1979) and are close to low-temperature cusp on lherzolite solidus at the boundary between spinel and plagioclase facies (Pressnal et al, 1979). According to Dmitriev et al (this issue), TOR 2 is the most common type of TOR, so these conditions of melting can be proposed as typical for oceanic mantle in general.

3. TOR 3: The estimated P and T are 5 Kbars and 1250 °C. Low concentrations of Ti, Na, and high Si suggest its origin in the second stage melting process (Duncan and Green, 1980).

Additionally fluid regime of TOR crystallization (on the base of fluid inclusion studies) and geodynamic consequences of obtained results will be discussed.

REFERENCES

Arai, S., 1987, Neues Ib Mineral Mh, v. 8, p. 347–354.

Bonatti, E., and P. J. Michael, in press.

Danushevsky, L. V., A. V. Sobolev, and L. V. Dmitriev, 1987, Dokladi Acad. Nauk USSR, v. 292, p. 1449–1452.

Dmitriev, L. V., et al, 1985, Initial Reports DSDP, v. 82, p. 509–518. this volume.

Drake, M. J., 1976, Geochimica et Cosmochimica Acta, v. 40, p. 457–465.

Duncan, R. A., and D. H. Green, 1980, Geology, v. 8, p. 22–26.

Falloon, T. J., and D. H. Green, in press, Journal of Petrology.

Ford, C. E., et al, 1983, Journal of Petrology, v. 24, p. 256–265.

Green, D. H., et al, 1979, *in* M. W. McElhinney, ed., The earth: its origin structure and evolution: London, Academic Press, p. 265–299.

Gurenko, A. A., et al, 1988, Dokladi Acad. Nauk USSR, v. 301, p. 179–184.

Pressnall, D. C., et al, 1979, Journal of Petrology, v. 20, p. 3–35.

Roedder, E., 1984, Fluid inclusions.

Table 1.—The composition of melt inclusions in most magnesian olivines (1–4) and representative calculated primary melts (5–7).

NN	1	2	3	4	5	6	7
TYPE	TOL INCL	TOR 1	TOR 2	TOR 3	PM TI	PM TOR 1	PM TOR 2
SIO2	48.80	49.00	50.14	52.80	48.40	48.93	49.60
TIO2	.50	.60	.87	.30	.50	.56	.90
AL2O3	14.10	15.00	17.94	15.80	14.10	15.65	18.60
FEO	7.81	7.60	7.11	7.00	8.21	7.68	7.20
MGO	14.00	13.00	9.60	9.60	14.50	13.87	9.40
CAO	13.20	12.30	12.57	13.00	12.70	11.76	12.00
NA2O	1.40	1.60	2.34	1.40	1.40	1.48	2.30
K2O	.02	.06	.06	.03	.02	.01	.04
TOTAL	99.83	99.16	100.63	99.93	99.83	99.94	100.04
Mg#	.762	.753	.707	.710	.759	.763	.699
Host	Fo 92.0	Fo 91.5	Fo 90.1	Fo 90.0	Fo 92.2	Fo 92.0	Fo 90.5
T °C	1325	1290	1240	1220	1336	1324	1230

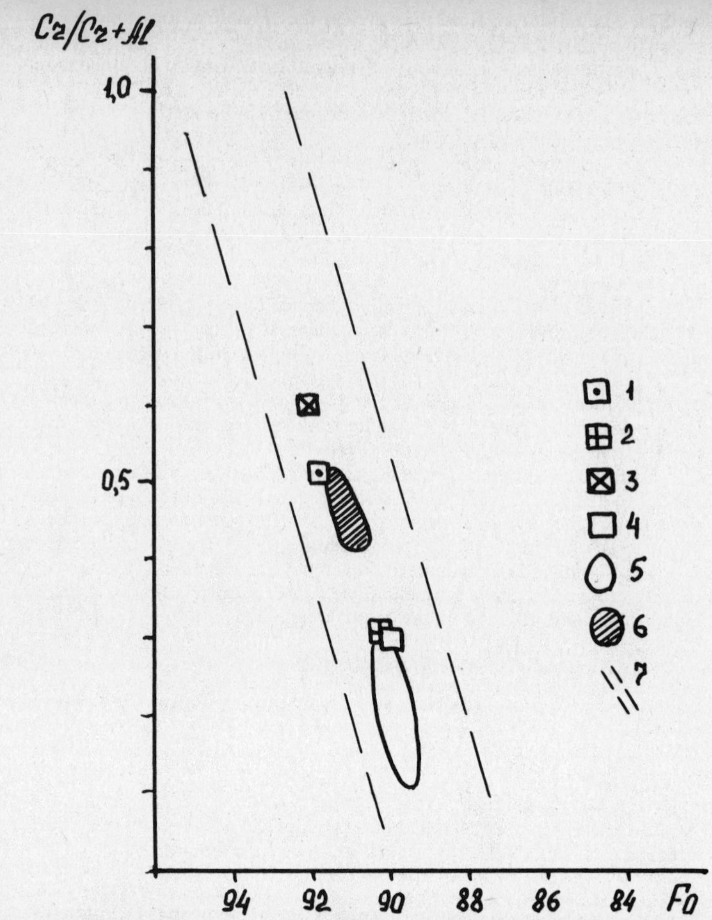

Figure 1—Compositions of coexisting O1 and Sp in PLA of TOR and associated periodotites.
1-TOR 1; 2-TOR 2; 3-IT; 4-TOR 3; 5-periodotites from Gibbs FZ, DSDP 395, Vema FZ, Romanche FZ, Atlantic) (Bonatti and Michael, in press, Dmitriev et al, this volume) 6–43 °N, Oceanographer FZ, Atlantic; 7 O1-Sp mantle array (Arai, 1987).

SOBOLEV, N. V., and V. S. SHATSKY, Siberian Division of Academy of Science of the USSR, Novosibirsk, USSR

Diamondiferous Assemblages in Metamorphic Rocks of Earth's Crust

Inclusions of accessory microdiamonds have been found in garnet, pyroxene, and zircon grains from some metamorphic rocks of eclogite-bearing metamorphic terrain of the Kokchetav massif, Northern Kazakhstan, USSR (Sokolev and Schatsky, 1987, 1988). The mineralogy of eclogites from this metamorphic complex have been recently reviewed (Sobolev et al, 1986).

Metamorphic rocks containing microdiamonds are represented by garnet-biotite gneisses and schists, garnet-pyroxene, and garnet-pyroxene-carbonate rocks. These rocks occur as small lenticular, banded or vein-like bodies within plagioclase gneisses of Zerendinsk rock series. Eclogite lenses and boudines are also typical in the area, but no diamonds have been found in eclogites.

The average diamond size is 12.5 ± − 5.8 microns (n = 501). The majority of crystals are of cube-octrahedral and cuboid morphology. Euhedral graphite inclusions often occur in garnets along with a diamond. Diamond crystals were found also in intergrowths together with mica, rutile, and zircon grains within garnets.

Biotite gneisses and schists contain unusual garnets with high Ca-component contents (from 20 to 40 mol%). Their Fe/Fe + Mg ratios are much lower than in common garnets from gneisses of that type (from 50 to 70 mol%). Biotite, phengite, potassic feldspar, pyroxene, rutile, sphene, and kyanite are found along with diamonds as inclusions in garnets and studied by electron microprobe.

Biotite contains up to 4.5 wt% F, phengite up to 3.5 formula unit Si, sphene up to 11.7 wt% Al_2O_3, and pyroxene up to 14 mol% jadeite.

Garnet-pyroxene rocks contain garnet with Ca-component contents from 40 to 46 mol% and Fe/Fe + Mg ratios 52–61 mol%. Diopside hedenbergite pyroxene contains orthoclase lamallae. Phengite (3.56 formula units Si) and pyroxene (up to .64 wt% K_2O) inclusions are found in garnets.

Garnet-pyroxene-carbonate rocks contain highly calcic garnet up to 50 mol% of Ca-component along with very low Fe/Fe+Mg ratio: 20–22 mol%. Phlogopite, rutile (.33 wt% Al_2O_3), pyroxene (1.15 wt% K_2O), sphene, and calcite inclusions are studied in the garnets.

Potassium-bearing pyroxene inclusions in garnets (.23 wt% K_2O) are found in eclogites of the same area.

The mineralogical indicators, especially K_2O admixture and orthoclase exsolution lamallae in a clinopyroxene (Sobolev, 1977), testify to the ultra-high pressure conditions at metamorphism of rocks containing accessory diamonds and eclogites without diamonds.

Equilibrium temperature estimations using garnet–pyroxene geothermometer (Ellis and Green, 1979) at 40 Kbars resulted in 910–1020 °C for garnet–pyroxene rocks, 780–920 °C for biotite gneisses and 850–1000 °C for eclogites. The results of this study allow along with results of study of coesite-bearing rocks (Schreyer et al, 1987) extend the pressure at metamorphism of crustal rocks not only to 30 Kbars but at least up to 40 Kbars. In attempts to look for the phenomena of a metamorphism at ultra-high pressure conditions, it is important to check the highest temperature crustal eclogites for a presence of potassium in pyroxene. Special attention should be paid to possible occurrences of accessory diamonds in metamorphic rocks of southern part of the East European platform where cubic diamonds have been found in Tertiary alluvials and in North Australia where cube microdiamond populations have been fixed in an east-west belt 1800 km long (Smith, 1985).

REFERENCES

Ellis, D. J., and D. H. Green, 1979, Contributions to Mineralogy and Petrology, v. 71, p. 13–22.

Schreyer, W., H. J. Massonne, and C. Chopin, 1987, *in* B. O. Mysen, ed., The Geochemical Society Special Publication 1, p. 155–163.

Smith, C. B., 1985, *in* Native elements in meteorites and continental lithosphere: Abstracts of papers, Part IV, Novosibirsk, p. 31–36.

Sobolev, N. V., 1977, Deep-seated inclusions in kimberlites and the problem of the composition of the upper mantle: AGU, 279 p.

——— and V. S. Shatsky, 1987, Geologiya i Geofizika, n. 7, p. 77–80 (in Russian).

——— ——— 1988, *in* Composition and processes of deep-seated zones of continental lithosphere: Abstracts of papers of International Symposium, Novosibirsk, p. 182–183.

——— N. L. Dobretsov, A. B. Bakirov, and V. S. Shatsky, 1986, *in* B. W. Evans and E. H. Brown, eds. Blueschists and eclogites: GSA Memoir 164, p. 349–363.

SOCCI, A.D., Ohio University, Athens, OH

Tectonically Predisposed Biogeochemical and Astronomical Controls on Glaciation: Speculative Review of the Last 700 M.Y. of Earth History

Plate movements have likely exerted a first-order influence on the onset and demise of glacials through time. During the breakup and dispersal of supercontinents, the creation of new rifts increases volcanism, ridge volume increases, and old oceanic crust is replaced by hotter, more thermally-expanded ocean crust, younging the ocean basins of the world while forcing sea level to rise. As sea level rises, blanketing and protecting sialic crust from weathering, and as volcanism increases, injecting more CO_2 into the atmosphere, the climate warms, and glaciers recede. During continental assembly and stasis, rifts become sutured, oceanic crust ages and thermally contracts, continental crust thermally expands, and sea level falls. As continents converge and sea level falls, atmospheric pCO_2 drops due to decreased volcanism, and accelerated weathering of increasingly greater area of exposed continents, causing climatic cooling, but not necessarily enough to trigger glaciations, insofar as the short-term rates of glaciation and deglaciation cannot be accounted for by the much longer term rates of plate tectonic processes.

Plate tectonic processes may have predisposed the planet toward glacials and interglacials, on a long-term basis and at long-term rates, insofar as glacials have occurred at times when eustatic sea level was low, while non-glacials have occurred when sea level was high. However, coincident with glacials during the past 700 Ma, are major accumulations of phosphate, biotic innovations, and negative excursions in the organic carbon record suggestive of rapid, biotically-controlled fluxes of carbon between the atmosphere and sediments, resulting in the regulation of climate by organism for at least the last 2.5 Ga, as Lovelock and others have suggested.

In addition, in light of the fact that solar luminosity has increased, the earth should have become increasingly hotter throughout earth history. However, climatic indicators suggest, paradoxically, that global temperatures have remained constrained to within narrow limits, and thus it would seem a requirement that carbon burial has increased throughout geologic time. Furthermore, by most accounts, phosphorous (P) has been, and remains to be, biolimiting, and since P can only be introduced into the oceans and the sedimentary record though weathering of crust, and since rates of weathering appear to be largely controlled by plate tectonic processes which operate at long-term rates, the amount of P carried into the oceans throughout geologic time appears to have been constrained to within relatively narrow limits at any one time. Hence, the biomass, and the volume and rate (long-term) at which this biomass (organic carbon) was buried, would also appear to have been tectonically controlled, by virtue of the dependency of life, and therefore biomass, on the availability of P, thereby fixing the rate of carbon burial to within narrow limits, making it seemingly impossible to account for the increase in carbon burial over time required to have kept the temperature of the planet from having steadily increased due to increased solar luminosity.

However, Worsely and others have presented a plausible explanation of the solar luminosity–carbon–climate dilemma by noting the coincidence of biotic innovations, glacials, and apparent step-function increases in the Corg/P ratio, over time, in the geologic record. It has been suggested that successive biotic innovations in earth history have resulted in step-function improvements in the biotic recycling of P, rapid, step-function increases in carbon burial through time, a relatively constant burial rate of P, buffering of climate over time, in light of increasing solar luminosity, and periodic, rapid decreases in atmospheric CO_2 producing increased and more rapid cooling of climate, and rapidly-induced, biotically-controlled glacials.

Insofar as biotic innovations have occurred when sea level was low, plate tectonic processes appear to have set the stage for glacials which appear to have been ultimately, rapidly triggered by organisms. However, once initiated, Milankovitch-type astronomical perturbations seem to have exerted a detectable, second-order, short-term control on glacials, bringing about a sequence of short-lived glacials and interglacials during long-term glacial events. Although plate tectonic processes are decoupled from Milankovitch-type orbital perturbations, such processes appear to have influenced those times in earth history when orbital perturbations would or would not produce a discernible affect on the onset and demise of glacials.

When supercontinents have rifted, raising sea level, the buffering effect of sea water blanketing more continental areas appears to have overridden Milankovitch-type controls and hence, we observe no glacials and no record of orbital periodicities. When continents have assembled, lowering sea level, the buffering effect of sea water was mitigated due to confinement to a much reduced area, setting the stage for tectonically-induced, biogeochemically- and astronomically-controlled glacials.

Given the nature of suggested controls on glacials in earth history, the exact timing, magnitude, and duration of glacials appear to have been determined, principally, by the degree to which each control was in phase with, timewise, every other control, thus serving to either amplify or dampen glacials, at any one time (e.g., the mid-to-late Ordovician glacial). Consequently, continental position appears to have been far less important in controlling the onset and demise of glacials in the geologic record, and hence low- and/or mid-latitude glacials seem to have been plausible events in earth history.

SOHN, I. G., U.S. Geological Survey, Washington, DC

Video on Ostracodes: Living, Breeding, and Fossil

The living freshwater ostracode species *Darwinula stevensoni* (Brady and Robertson, 1870) was video recorded in September 1987 to study the movements and the functions of the appendages inside and outside the carapace of mature specimens. The vibratory plates of the mandibles, maxillae, and first thoracopods move rhythmically in the closed and the open carapace. The second and third thoracopods can be extruded from the gaping carapace; their movements are not synchronized.

Darwinula is an excellent subject for video observation because it has a thin, translucent, smooth shell, does not swim, and retains the eggs and the first two instars in the posterior part of the carapace. The species was recorded in March 1988 to observe the brooding of egs and young within the carapace. The movements of the appendages within the carapace of the unborn instars could be seen. In June, the feeding habits of the adults, as well as the released instars, were recorded. This genus has been recorded as far back as the Late Mississippian.

The feeding and swimming behavior of several growth stages of cypridopsid were also recorded. The ontogeny, shell, and appendage morphology of fossil and living ostracodes are illustrated in the video by using scanning-electron and light micrographs. Preserved fossil appendages are shown.

SOKOLOV, B. S., and M. A. FEDONKIN, Academy of Sciences of the USSR, Moscow, USSR

Terminal Studies of Precambrian: Problems and Strategy of Studies

Progress in studies of the Late Precambrian reached over the last 20–30 years make it urgent to substantiate the stratigraphic system, which lies directly below the Cambrian. At present, there are several models of the Terminal Precambrian System (TPS), known as the Sinian, Vendian, Ediacarian, etc. Local peculiarities of stratotypes and the history of their studies impose certain traits on concepts of the TPS, developed by different geological schools. However, a generalized idea of the TPS already exists and the geological community is psychologically prepared to consider a possibility of adding one more system to the international stratigraphic scale. This is seen from the number of publications on the stratigraphy and paleontology of the TPS as well as from the fact that recently a working group on the TPS has been organized under the aegis of the International Stratigraphic Commission.

In the course of studies of the TPS some problems have appeared; they can be divided into trivial and specific. Trivial problems are those that are encountered by workers in younger stratigraphic systems, i.e., problems of the stratigraphic potential of different groups of fossils, facila dependence of fauna and flora, and also a problem of the lower boundary of the TPS. A similar problem is not yet solved for the Cambrian. Specific problems of the TPS, which to a large extent determine the strategy of its studies, are as follows:

The animal world of the Late Precambrian is characterized by low diversity compared with older epochs and also by weak evolutionary changes within the terminal Precambrian period, if we can speak about them at all. A soft-bodied fauna now numbering about 100 species characterizes only part of the TPS and does not allow us to use it for more detailed biostratigraphic operations. Experience gained in studies of the Vendian in the USSR (the Russian Platform, the Urals, and Siberia) is indicative of a considerable gap in the Paleozoological record of soft-bodied animals in the second half of the Vendian (the Kotlin Horizon). The diversity of trace fossils of invertebrates is considerably decreased at the same interval. The appearance of numerous and diverse trace fossils of Paleozoic type at the Rovno Horizon is not accompanied, however, by the same increase in the number of body fossils of soft-bodied animals. A change in the phonomic situation through a complicated system of interacting biotic and abiotic factors broke the adequate paleozoological record of soft-bodied fossils and good preservation of these fossils became an exception rather than a rule. This "taphonomic overturn" took place during the Late Vendian and the Early Cambrian. This event makes it difficult to understand the evolution of the animal world at the Precambrian–Cambrian boundary and to establish phylogenetic relations between the Precambrian and Paleozoic organisms since comparing the animal world of the Precambrian and the Cambrian we deal with two kinds of fossils, e.g., imprints (moulds) of soft tissues and remains of skeletons.

Analyzing fossil assemblages from the most representative localities of the TPS, we can indicate at least three different associations; (1) the Ediacarian fauna, which has the widest geographic distribution; (2) the Avalonian, and (3) the Namibian ones. But we cannot definitely say whether it si a phenomenon of provincialism, evolution, paleoecology or taphonomy. This problem is partially connected with another one characteristic of the TPS and the Precambrian as a whole, i.e., the lack of an instrument for precise and detailed correlation on a global scale. It seems, however, that the development of nonbiological stratigraphic methods (climate, event and magnetostratigraphy, geochemical signals, biomarkers, etc.) is a temporary difficulty.

The most important problem for the TPS is how to establish globally correlated parts of the system in a rank of divisions, stages, and more detailed biostratigraphic subdivisions. It should be noted, however, that when establishing the stratigraphic volume of the TPS, preference was given to a historical-geological method and paleontological and biostratigraphic methods had a subordinate character. However, we should not give up the biostratigraphic approach to the problem. This is indicated by experience gained during investigation of the Vendian from the Russian Platform, where complex studies of the distribution of various fossils in the section (soft-bodied Metazoa, trace fossils, vendotaenian algae, acritarchs, sabelliditids, etc.) even now allow us to distinguish three biostratigraphic units (Redkino, Kotlin, and Rovno Horizons) that are promising for global correlation. A more detailed biostratigraphic division of the Vendian should be carried out within taphonomic and paleoecological (ecosystem) models.

Attempts to substantiate stratigraphic systems in the Precambrian inevitably lead to the application of "Phanerozoic methods" down the stratigraphic scale. This is natural because if there is a bridge between methods used in studies of the Proterozoic and Phanerozoic we can understand the dynamics of many processes. But the value and the importance of separate methods in solving stratigraphic tasks can be quite different in the Phanerozoic and the Precambrian. Now it is clear that only the complex utilization of mutually controlling methods can give the most suitable detail for stratigraphic models. Hence a conclusion can be inevitably drawn that progress in the stratigraphy of the Late Precambrian looks relatively expensive and labor consuming. To make it optimal, it is necessary on the basis of international cooperation to develop a set of research methods that must be applied for studying the most complete sections of the Late Precambrian. The possibility of implementing complex studies of all necessary parameters will probably be one of the major requirements for the stratotype of the TPS or its parts.

Speaking about the problems of the stratigraphy of the TPS, we mean only marine deposits, since we do not yet have a key to the stratigraphy of continental units of the Late Precambrian. But even when studying marine deposits that are widespread on all old platforms and also on their edges and in geosynclines, we face difficulties of detailed correlation of remote carbonate and terrigenous facies that have different paleontological characteristics owing to paleobiogeographic and taphonomic circumstances.

SOLER, P., ORSTROM, Paris, France

Andean Alkaline Magmatism in Central Peru

Cretaceous and Tertiary magmatism in the Andes of Central Peru is mainly of subduction-related calcalkaline type. However, minor tholeiitic volcanism has been previously identified in the lower part of the mid-Albian–Cenomanian Casma group of the coastal area, and we have recognized four episodes of alkaline magmatism along a W-E transect around 11°30′S.

The first two are preorogenic and consist of basaltic lava flows interbedded in lower Cretaceous sedimentary formations in the eastern part of the High Plateaus. The first volcanic intercalation, upper Aptian in age, predates the transgression of the carbonated Chulec formation. It has been identified in the same stratigraphic position at numerous localities between 10°30′ and 13°00′S. The second one, Albian in age, in interbedded in the carbonated Pariatam–bo formation and has been recognized only in a limited area near Cerro de Pasco.

Both volcanic intercalations are partly spilitized porphyritic basalts with a trachytic texture and plagioclase, olivine, and clinopyroxene phenocrysts. Transition-elements contents indicate a nearly primitive composition with minor evolution through differentiation. Petrography, clinopyroxene compositions (very high Ti and Al contents, relatively high Ca and Na contents, low Mg, Fe, Mn, and Cr contents) and geochemistry ($TiO_2 \approx 2.30\%$, $P_2O_5 \approx 0.45\%$, high alkalies contents, high highly hygromagmaphile (i.e., LREE, Th, U, Ta, Zr, Hf) elements contents, high Zr/Y, Th/Ta ≈ 1, Ta/Tb ≈ 3.5, $La_N/Yb_N \approx 10$) show those basalts to be typical intracontinental alkali basalts. Their source region is inferred to be the undepleted subcontinental margin.

In the coastal region of central Peru, some 130 km west of the studied basalts, the onset of the subduction-related medium-K calcalkaline Copara volcanism appears to be synchronous with the first alkali volcanic episode while part of the tholeiitic and calcalkaline volcanism of the Casma group would be contemporaneous with the second episode. Despite this relative vicinity of the contemporaneous active subduction, the studied basalts do not show any imprint of the subduction process (e.g., no Ta depletion). This geochemical feature, together with plate-dynamics considerations, suggests that at that time the Benioff–Wadati zone was dipping steeply. The importance of such an observation for paleovolcanic studies must be underlined.

These alkali basalts emphasize the existence of at least two episodes of extensional tectonics at the end of the lower Cretaceous. The precise location of these basalts appears to be controlled by deep-seated fractures that actually played an important part in the paleogeographic evolution of the central Andes from Permian times to the present.

The first and the most important of these extensional events is tightly correlated with the onset of the development of the ensialic Copara–Casma marginal basin and appears to be contemporaneous with the initial rifting episode of the South Atlantic Ocean in the Equatorial zone.

Subsequent episodes are synorogenic and consist of alkali plutonic rocks contemporaneous with the two major subduction-related Cenozoic calcalkaline magmatic pulses.

The third episode, with a mid-Eocene minimum K-Ar age, is located along the physiographic crest of the Western Cordillera only 45 km east of the contemporaneous calcalkaline intrusions. It has been recognized only in the Raura–Oyon area. It consists of remnants of syenite completely included in late Miocene calcalkaline granodioritic plutons. Mineralogy includes mainly plagioclase, K-feldspar, quartz, and arfvedsonite. When compared with the contemporaneous calcalkaline plutonic rocks, the studied syenite shows low MgO, FeO*, U, Th, Zr, Sr, HREE, and transition elements contents, high K_2O (with $K_2O/Na_2O \approx 3$), Rb, Ba, and Cs contents, and nearly similar Na_2O, Ta (with $Th/Ta \approx 3$) and LREE contents.

The fourth episode, with early and mid-Miocene K-Ar ages, is restricted to the amazonian slope of the Eastern Cordillera, 70 km east of the easternmost contemporaneous calcalkaline intrusions. It has been recognized in outcrops in a small area near Oxapampa, but boulders of similar lithology were previously described in various part of the amazonian slope of the Eastern Cordillera of central Peru, so that a regional extension of this episode may be inferred. In the Oxapampa area, it consists of small ($<2\ km^2$), broadly circular stocks emplaced in upper-Cretaceous and Tertiary continental sedimentary layers. Two main types were identified: equigranular nephelinic syenites and porphyric alkali syenites, with a complex mineralogy (primary albitic plagioclase, K-feldspar, aegyrine, ± biotite, ± nepheline, sphene, melanite, zircon, fluorite; secondary calcite, analcime, mesolite, gonnardite, etc). When compared with the contemporaneous calcalkaline plutonic rocks, these stocks show very low TiO_2, CaO, MgO, FeO*, Cs, and transition elements contents, high Al_2O_3, Na_2O, K_2O (with $1.5 > K_2O/Na_2O > 1$), Rb, U, and Zr contents, and nearly similar Th and Ta contents but with Th/Ta ratios as low as 3. Sr and Ba are depleted in the nephelinic syenites but enriched in the porphyric alkali syenites. A slight LREE enrichment appears, but on the whole fractionated REE patterns are similar to those of the calcalkaline rocks.

Both episodes of orogenic alkali rocks show a clear imprint of the subduction process (e.g., low Ti content, $Th/Ta \approx 3–10$). For these rocks, a small extent of partial melting of a subduction-modified aesthenospheric source followed by fractional crystallization may be inferred. This rear-arc alkali magmatism is not a systematic feature of the Cenozoic Andean arc in central Peru. It rather appears to be contemporaneous with the episodes of very high convergence rate between the oceanic Nazca (or Farallon) plate and the continental South-American Plate.

SOLLER, D. R., and J. P. OWENS, U.S. Geological Survey, Reston, VA

Comparison of Soil Formation and Depth of Weathering as Relative Age Indicators, Central Atlantic Coastal Plain, USA

Textural, chemical, and mineralogic study of soils has led to the practice of applying various soil parameters as relative age indicators. From the limited data available, it appears that these soil parameters are not very definitive for differentiating units on the Coastal Plain, perhaps because the soil profile lies only within the uppermost part of a unit's vertical weathering profile. In our studies, we have examined the entire weathering profile for evidence of weathering-induced changes in both sand- and clay-sized mineralogy and have employed two techniques for relative age dating.

One technique depends on accumulation of stable secondary minerals as clay-sized particles in the upper part of the weathering profile, especially in the soil. In the Carolina Coastal Plain, however, weathered sediments commonly do not contain gibbsite, which is a key mineral indicative of deep, protracted weathering, especially in well-drained settings. Deeply weathered profiles that have mature clay-sized mineral suites occur in places and contrast with immature clay mineralogy in profiles developed elsewhere in the same unit; in the Carolinas, therefore, this method does not provide reliable age estimates. In contrast, weathered Coastal Plain sediments to the north, in New Jersey, contain abundant gibbsite. This is not to be expected, because weathering should be more intense under the warmer climate of the Carolinas. This apparent incongruity can be explained, as evidence suggests that many weathering profiles in the Carolinas may have been truncated to some degree. Widespread occurrence of Carolina bays and dunes, as well as the mineralogy of the dunes, indicates that profile truncation may have been accomplished by wind erosion.

A second, more effective tool for relative age dating is the depth of weathering, as recorded by progressive loss of detrital, sand-sized minerals upward in the weathering profile. The depth at which alteration or depletion of detrital minerals occurs is related to the length of time that the deposit has been exposed to weathering processes. Of deposits having similar lithologic profiles and detrital mineralogy, older ones will exhibit deeper weathering. The fact that the depth of weathering as indicated by sand mineralogy is generally greater than the depth indicated by clay mineralogy suggests that sand mineralogy offers a more sensitive measure of weathering. In addition, basing the method on weathering of liable minerals rather than accumulation of secondary minerals yields a more reliable age indicator in areas where profile truncation is suspected.

In southeastern North Carolina and northeastern South Carolina, early Pliocene to Holocene marine, strandline, and fluvial deposits of transgressive-regressive cycles lie at the surface. These deposits have been dated by isotopic and biostratigraphic methods and mapped as formations. The mineralogy of these formations, and especially the weathering profiles, has been studied in detail in order to provide a correlation and relative dating tool to assist in stratigraphic and geologic mapping problems. To minimize the influence of lithologic and topographic variations on the weathering profiles, and hence on the dating technique, we dealt with the sandy marine and barrier deposits, avoiding backbarrier deposits and poorly drained settings. The abundance of feldspar and liable heavy minerals such as hornblende and epidote was found to decrease in older units. In Pliocene sediments, these minerals are nearly absent throughout the weathering profiles. Pleistocene units are richer in these liable minerals, although their abundance does decrease upward in the weathering profiles; i units, liable minerals tend to persist higher in weathering profile than in older units. Paralleling the weathering of liable sand-sized minerals, but occurring at shallower depths, the clay-sized mineral suite also tends to be more mature upward in the profile.

In the Cape Fear River valley, North Carolina, there are five fluvial terraces of different ages, correlative with marine and strandline facies discussed above. Although weathering profiles are not evident, a lower proportion of epidote and hornblende to other heavy minerals does occur beneath successively older terraces. Evidence suggests that this trend is not a function of change in sediment supply over time but rather of weathering. Drainage conditions in the valley are inferred to have caused intrastratal solution of liable minerals throughout the fluvial column. Although relative ages of fluvial deposits cannot be determined by depth of weathering, the relative abundance of liable minerals throughout a vertical profile can serve the same purpose: The older terraces have undergone intrastratal solution for a longer time than the younger terraces and are more depleted in epidote and hornblende. Feldspar abundance shows a similar, albeit less well-defined trend. Clay mineralogy is fairly uniform and cannot be used to differentiate these terraces.

SOMAN, G. R., and R. P. DESHPANDE, Poona University, Pune, India

Prograding Lacustrine–Fluvial Lameta (Upper Cretaceous) Sediments of Central India

In parts of the Nagpur district of Maharashtra, India, rocks belonging to the Lameta Formation occur as detached outcrops unconformably overlying the Precambrian granite, metasediments, sandstones, and siltstones of the Gondwana Supergroup; and in turn are overlained by Deccan Trap basalts. The Lameta Formation is here represented by cherty arenaceous micritic limestones, pink and white nodular marls, and yellowish-brown sandstones, which are equitable with Lower Limestone Member, Mottled Nodular Member, and Upper Sandstone Member, respectively, of the type area at Lameta Ghat near Japalpur, India. The origin of these sediments is controversial; according to some workers these are deposited on marine shelf, others believed them to be of freshwater origin, while few workers consider these to represent transitional sediments deposited in estuarine and coastal complexes.

The heavy mineral analysis of these limestones, marls, and sandstones under study exhibit a great similarity in the species and in their relative abundance. These studies suggest short distance of transport and related nearness of source area, as they contain large amounts of relatively unstable mineral constituents such as staurolite, kyanite, garnet, epidote, and zoisite (together over 50%), and associated low ZTR index of 43%.

These limestones are mainly arenaceous micritic and contain appreciable amount (18%) of detrital sand size quartz, which at the top becomes pebbly micritic limestones, and contain pebbles of vein quartz, quartzite and jasper; and hence exhibit a coarsening-upward sequence. Associated with these are local limeclast breccia, pelmicrite with fenestral fabric, chara-produced pseudobrecia, and in the basal portion these limestones exhibit seasonal warves, consisting of alternate detrital quartz-rich and micritic calcite-rich layers; and hence are here inferred to be deposited in offshore and shoreface regions of intracratonic freshwater lacustrine basin located on earlier rifted Pranhita–Godawari graben. The overlying nodular marl consisting of micritic calcite, illite, fine quartz, and containing abundant worm burrows and frequent calcareous nodules is inferred to be deposited in the upper portions of this freshwater lake, i.e., on mud flats (Figure 1). The sandstones under study exhibit fining-upward sequence. At the base, these contain sandstone, siltstone boulders of Gondwana sediments, and at places become conglomeratic, which exhibit horizontal (parallel) bedding. These are followed upward by medium-grained sandstones containing vein quartz and jasper pebbles, which in turn passes upward into ferruginous horizontal (parallel) laminated siltstones and shales. These clastic rocks are texturally immature, contain large amount (30–50%) of matrix; and also exhibit related upward increase in matrix percentage, decrease in rock fragments content, decrease in mean grain size values, and increase in standard deviation values, i.e., deterioration in sorting; and hence are considered to be fluvial deposits. The bouldery sandstones and parallel bedded conglomerates are inferred to be fluvial channel sediments, with overlying sandstones to be formed by migrating point bars, and topmost horizontal laminated ferruginous siltstones and shales to be formed in levee, or in seasonal standing water bodies, such as meander cutoff channels. Therefore, it is here concluded that these limestones, marls, and sandstones of the Lameta Formation represent a prograding lacustrine-fluvial sequence (Figure 2).

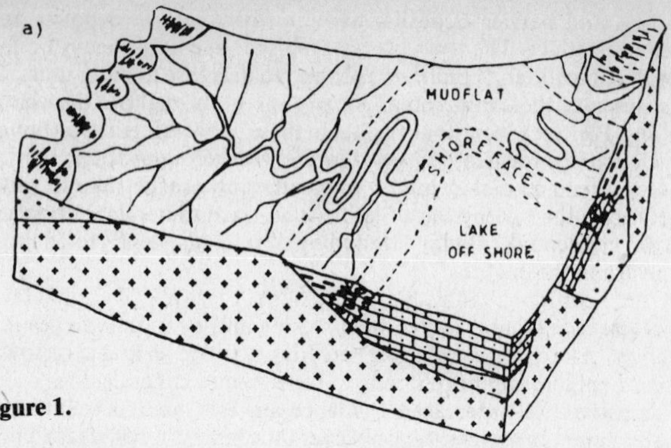

Figure 1.

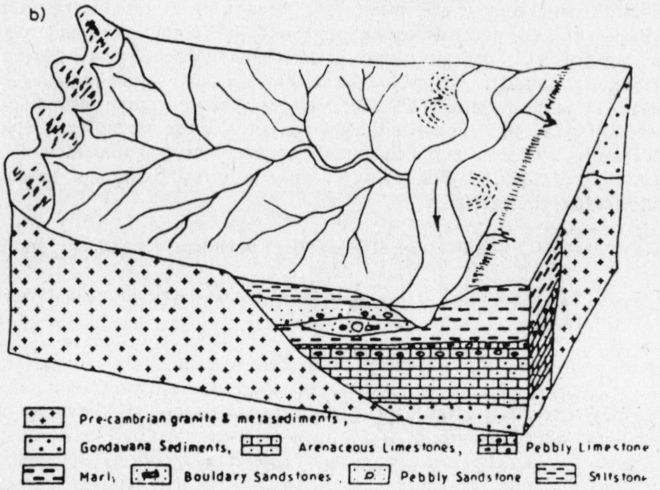

Figure 2.

SOMAN, K., Center for Earth Science Studies, Trivandrum, India, and W. R. VAN SCHMUS, University of Kansas, Lawrence, KS

Khondalite-Charnockite Assemblages of South Kerala, India—Evidence for Proterozoic Crustal Evolution

The khondalite-charnockite assemblages of south Kerala, south of the Achankovil shear zone, were hitherto considered to be Archean in age. Recent investigations, however, bring forth evidence for extensive Proterozoic geologic activity.

The Precambrian crystalline basement complex of south Kerala constitutes the southern part of the Western Ghats continental margin belt and comprises products of granulite-facies metamorphism, belonging mostly to the khondalite suite of rocks encompassing garnet-sillimanite gneiss ± graphite (khondalite *sensu-stricto*) and garnet-biotite gneiss. Occasional bands of charnockite are seen within the gneissic terrain. Both of these units are often enveloped in garnet-quartz-feldspar neosomes. Arrested development of charnockite is also seen within the migmatitic gneisses. Augen gneiss outcrops north of Trivandrum, while isolated bands of cordierite gneiss are seen within the Achankovil shear zone. These are intruded by early Paleozoic pegmatites (Soman et al, 1982) and Cenozoic dolerate dykes (Furnes et al, 1983).

Rocks of the area present field and textural evidences of a major episode of granulite-facies metamorphism. However, arrested development of charnockite in an atectonic environment indicates the existence of granulite-facies P-T-XCO_2 conditions at a later period also. Field evidence suggests that this is later than the migmatization event. Age determinations of zircon from the arrested charnockite in Ponmudi indicate its formation ca. 540 Ma, while the minimum age of the detrital zircons and their parent rocks is suggested to be the upper intercept age of ca. 1930 Ma (Srikantappa et al, 1985). Zircon dating of a charnockite enclave near Nedamnnur, east of Trivandrum by the U-Pb method yielded an age of 2930 ± 50 Ma, while the adjacent garnet-biotite gneiss formed on retrogression of the charnockite gave an age of 2838 ± 40 Ma (Odom, 1982).

Though khondalites were considered to be the oldest rocks among the Precambrians of south Kerala (Crawford, 1969), analytical results of zircon fractions from garnet-sillimanite-gneiss ± graphite and garnet-biotite gneiss of suspected sedimentary parentage define a discordia line that intercepts the concordia at 2190 ± 130 and 2144 ± 130, and 400 ± 130, and 564 ± 177 Ma respectively, suggesting early Proterozoic age of the detrital zircons or their source rocks and a major metamorphic event at ca. 500 Ma.

P-T estimates of 750 ± 50 °C and 5 ± 1 kb for most of the lithological units of the area are consistent with the other reported values for these rocks from south Kerala, while the slightly higher pressure value of 7 kb for the charnockite enclave, though within the accuracy limits of pressure estimates, may be suggestive of the latter's formation at deeper crustal levels. Geochemical features also differentiate the two types of charnockites in south Kerala.

From the available geochronological data, it is, however, very difficult to arrive at the time of the major granulite facies metamorphism. It is evident from field and geochronologic data that the major granulite facies metamorphism is earlier than the 540 Ma event, but later than the ca. 2200 Ma. This conclusion is plausible in the light of the recent SHRIMP data on zircons from the geologically comparable terrain of Sri Lanka (Kroner et al, 1987).

Geologic, geochemical and geochronologic data on south Kerala rocks suggest the following sequence of events:

1. Foundering of an intracratonic basin in south Kerala with Achankovil shear zone as the northeastern boundary (in the present day configuration) during early Proterozoic times and subsequent deposition of sediments. Judging from geochemical affinities, clays, greywacke, polymictic and arkosic sandstones, as well as marl and carbonates constituted the sedimentary sequence.
2. Progressive burial metamorphism resulted in the formation of granulite facies rocks. Judging from the available U-Pb zircon data alone, it is difficult to fix the chronology of this event.
3. Later migmatization resulted in retrogression of the granulites, especially the charnockite, formation of cordierite gneiss, and emplacement of granite plutons and associated later pegmatites. Mineralogical and geochemical features of the granite were controlled by the nature of the source rocks in the area. K-Ar amphibole dating of one such granite yielded an age of 550 ± 25 Ma.
4. Arrested charnockite formation, as evidenced from field relations, is later than the migmatization event and is suggested to be brought about by CO_2-rich fluids accumulated during the migmatization event. This also explains the limited extent of the arrested charnockite formation sites. The source of CO_2, based on geological and carbon isotope ratios of graphite within the khondalite, is suggested to be from an internal buffering system. The available geochronological data suggest that the time of formation of the arrested charnockite is ca. 540 Ma.
5. Charnockite enclaves, geochemically different from the arrested charnockite of sedimentary parentage, are suggested to be large crustal blocks brought up during the migmatization event.
6. U-Pb concordant age of ca. 520 Ma on monazite from garnet-biotite gneiss formed on retrogression of charnockite indicates the time of a major retrograde metamorphism in southwestern India with parallels in Sri Lanka and Madagascar. Metallogeny of south Kerala is mostly controlled by these late Proterozoic thermo-tectonic events.

Predominance of middle to late Proterozoic geologic events appears to have significantly influenced the crustal evolution of southwestern India. Study of this and similar belts within the western continental margin of southwestern India is crucial for the fullest appraisal of the geologic evolution of the region.

References

Crawford, A. R., 1969, Journal of the Geological Society of India, v. 10 (2), p. 117–166.
Furnes, et al, 1983, N. Jb. Miner. Abh., v. 146 (1), p. 82–100.
Odom, A. L., 1982, Final report, U.N. Case No. 81–10084 (Unpublished).
Kroner, et al, 1987, Journal of Geology, v. 95, p. 775–791.
Soman, et al, 1982, Journal of the Geological Society of India, v. 23 (9), p. 458–462.
Srikantappa, et al, 1985, Journal of the Geological Society of India, v. 26 (12), p. 849–872.

SOMAYAJULU, B. L. K., Physical Research Laboratory, Ahmedabad, India, and D. LAL, Scripps Institution of Oceanography, La Jolla, CA

Nutrient Cycling in Oceans Using Cosmogenic ^{32}Si, ^{32}P, and ^{33}P

^{32}Si (half-life = ~140 years), ^{32}P (14.3 days), and ^{33}P (25.3 days) are beta e[illegible]ers produced continuously in the atmosphere due to the spallations [illegible]ar by cosmic ray bombardment. Through their production rates and consequently their concentration levels in the hydrosphere are low, the development of an in situ adsorption technique using ferric hydroxide-dispersed–acrilan fibers has made it possible to concentrate these isotopes from several tons of seawater. Whereas, ^{32}Si measurements in vertical profiles from world oceans have been made, those of ^{32}P and ^{33}P in the Pacific have just begun.

Silicon and phosporous are nutrient elements and their distributions in the ocean along with their radioactive counterparts will provide useful information on the nutrient cycling. Because of their short half-lives, ^{32}P and ^{33}P can be used to gain insight into nutrient dynamics in the euphotic zone only. ^{32}Si extends this information to cover almost the whole ocean.

A brief account of the information so far obtained using these radioisotopes will be presented.

SONDER, L., Dartmouth College, Hanover, NH, and P. ENGLAND, Oxford University, Oxford, United Kingdom

Continuum Calculations of Large-Scale Continental Extension: Effects of Temperature-Dependent Rheology

The effects of a temperature-dependent rheology on large-scale continental extension are investigated using a thin viscous sheet model. In the thin viscous sheet approximation, the continental lithosphere is treated as a thin, isostatically compensated layer of viscous fluid overlying an inviscid halfspace. Vertical gradients of horizontal velocity in the layer are neglected. A vertically averaged rheology is used that is consistent with laboratory experiments on power-law creep of olivine and that depends exponentially on temperature. Solution of the equations of motion is by a finite element method, and temperatures are determined using a Crank-Nicholson finite difference formulation. Results of the calculations depend principally on two parameters: the Peclet number, which describes relative rates of advection and diffusion of heat, and a dimensionless activation energy, which controls the temperature dependence of the rheology.

The calculations show that at short times following the beginning of extension, deformation occurs with negligible change in temperature, so that only small changes in lithospheric strength occur due to attenuation of the lithosphere. However, after a certain critical time interval, thermal diffusion lowers temperatures in the lithosphere, strongly increasing lithospheric strength and slowing the rate of extension. This critical time depends principally on the Peclet number and is short compared with the thermal time constant of the lithosphere. The strength changes the locus of high extensional strain rates to shift with time from regions of high strain to regions of low strain. Results of the calculations are compared with observations from the Aegean, where maximum extensional strains are found in the south, near Crete, but maximum present-day strain rates are largest about 300 km further north. The comparisons support the hypothesis that the observed separation of the loci of maximum strain and maximum present-day strain rates in the Aegean may be the consequence of changes in lithospheric strength arising from the temperature-dependent mechanical properties of lithospheric materials.

SONG Shuhe, and HAN Fa, Chinese Academy of Geological Sciences, Beijing, China

Time and Space Distribution of Major Types of Metallic Ore Deposits of China

In China, many types of mineral deposits of Fe, Cu, Ni, Pb, Zn, Mn, Al, W, Sn, and Mo have been found. Regional geological studies show that these metallic ore deposits have epochal and spatial distribution.

In a broad sense, the ore deposits, ore districts, and ore belts are separately restricted to definite geological regions that consist of the following geotectonic units: the Sino-Korea, Alashan, Tarium, and Sichuan-Kangdian (Precambrian) oldlands. The Precambrian geotectonic units form the main tectonic framework of China. Along the boundaries of these units there occur the following Phanerozoic tectonic units: Tianshan-Neimenggu-Daxinganling, Kunlun-Qilian-Qinling, southeast China Paleozoic geosynclinal systems, and the southwest China Mesozoic-Tertiary geosynclinal system. This geotectonic classification is after T. K. Huang with simplification and some modification by the authors.

The most ancient ore deposits, as we know, occurred in the east part of the Sino-Korea Archean oldland. They are the Qianxi and Anshan siliceous banded magnetite ore deposits which were hosted by submarine volcanics of the volcano-sedimentary type.

More metallic ore deposits have been discovered in the well-developed Proterozoic rock series than in the Archean. Early Proterozoic iron deposits are also volcanogenic in origin, but they are minor in extent and contain siderite and hematite as the ore minerals. The main characteristic of early Proterozoic mineralization is the appearance of special types of copper deposits. For example, the spilite-keratophyre-hosted Dahongshan iron-copper deposit occurred in the geosyncline of Kangdian unit and the Baijiazuizi copper-nickel deposit in a fault zone in the Sino-Korea unit. An ore-bearing ultrabasic rock body is considered to intrude the early Proterozoic eugeosynclinal strata soon after their folding. Other spilite-keratophyre-hosted base metal deposits that are well developed in early Proterozoic rocks are the Tongkuangyu copper deposit and the Xiqiu copper-zinc deposit in the Sino-Korea unit.

In the middle Proterozoic, geosynclinal deposition of thick argillaceous, siliceous, and limy rocks, with comparatively minor submarine volcanics, accumulated within a new geological environment. In this setting, the rare-earth metal-accompanied Baiyunebo hematite-magnetite deposit, the Langshan sediment-hosted lead-zinc ore belt, and the Jingtieshan siderite-hematite deposit were formed. The appearance of large lead-zinc ore deposits and the first formation of an industrial manganese sedimentary ore deposit further show us that the middle Proterozoic was a significant period of new kinds of metallic ore genesis. The late Proterozoic, on the other hand, was a weak period for the formation of associated with basic-ultrabasic rocks, occurred.

After the formation of the Chinese Protoplatform, in Phanerozoic time, the new geological environment generated many new types of ore deposits, even while deposits similar to older types continued to form in remnants of older geological environment. The spilite-quartz-keratophyre-hosted Cambro-Ordovician massive sulfide copper-lead-zinc deposits of the Qilian eugeosyncline and Devonian sediment-hosted lead-zinc deposits and siderite deposits of the northern part of the Qinling geosyncline and the Nanling region may be cited as examples of the older types of deposits. The most prominent of the new deposits, on the other hand, are Carboniferous sedimentary aluminum deposits of the diaspore type and the large tungsten, tin, and molybdenum deposits of Jurassic-Cretaceous age that are structure-, igneous rock- and strata-controlled. In China, the earliest tungsten, tin, and molybdenum mineralization occurred in middle or early Proterozoic time. The tungsten and tin mineralizations apparently are stratabound. But the more valuable of these deposits formed mainly in Mesozoic time. During the evolution of ore-forming processes for these metals, however, the large Dachang stratabound tin-polymetallic deposit and the Fenglin stratabound tungsten-hematite deposit formed earlier, in Paleozoic aulacogens. As in other countries of the world, the Mesozoic and Tertiary were the main periods of polymetallic ore formation. In the Mesozoic, the Dexing porphyry copper deposit and the Gacun volcanogenic lead-zinc deposit formed, and in the Tertiary, the Yulong porphyry copper deposit, and the sediment-hosted Jinding lead-zinc deposit formed.

In addition to aluminum, manganese, and some iron deposits of sedimentary origin are crust-derived, whereas copper, zinc, nickel, and other ferrous metals are considered to be mantle-derived, but the derivation of lead, molybdenum, tungsten, and tin deposits is rather complicated lead, especially, which largely first appeared in the middle Proterozoic, is worthy of discussion. The only reasonable explanation for lead is that it is only concentrated in ore deposits in environments of well-developed crust. Since in the Phanerozoic submarine volcanogenic ore deposits of well-developed crust. Since in the Phanerozoic submarine volcanogenic ore deposits, however, lead also is intimately associated with copper and zinc, it may be inferred that lead is also a mantle-derived metal. As molybdenum, tin, and tungsten mainly form economic ore deposits in complicated geological regions of later Phanerozoic age, they most probably are a combined type of mantle- and crust-derived metals.

SONNENFELD, P., University of Windsor, Ontario, Canada

Greenhouse Effect and Evaporite Genesis

An increase in the air temperature disproportionately increases the water carrying capacity of the atmosphere, reduces the relative humidity, and raises the rates of evaporation. As 70.9% of the earth surface is covered by oceans, an increase in rates of evaporation pumps significantly more water vapor into the atmosphere. This entraps energy as latent heat to be released upon condensation. Half the globe situated in higher latitudes daily loses more energy through radiation to outer space than it receives from the sun. Moisture laden winds and warm ocean currents are the transfer agents of energy from low to high latitudes to make up the shortfall.

Decreases in energy intake result in air masses both swirling out of the equatorial region and descending in the subtropics in a more sluggish and wider circle, the diameter of which defines the subtropical dry belt of descending and adiabatically heating up air. A one-degree drop in temperature in the equatorial belt reduces the absolute humidity more than a similar drop in cooler temperatures of extratropical air and thus reduces the probability and intensity of rains in higher latitudes.

In the Wisconsin ice age, coolness was followed by dryness, a warming trend was followed by a pluvial period. Each of the glacial stages seems to have been a drier period, judging by coeval loess accumulations. In the postglacial period, likewise, several drought periods in the subtropics correlate with cooler temperatures in temperate and polar regions, and we periods with warming trends in the temperate zone.

The carbon dioxide content of the atmosphere has varied throughout the Phanerozoic. Maxima correlate with lush plant growth on land in Pennsylvanian coal forests and with the Late Cretaceous rapid diversification of angiosperms. Both are periods marked by a dominance of humid climates and by the occurrence of hard pans on shallow coastal shelves of tropical seas. The hard pans are the product of redissolution of limestone platforms due to an increased content of bicarbonate and carbonic acid ions in ocean waters produced by a raised partial pressure of carbon dioxide in the air.

A summer without rains produces on land a parched earth that reflects most of its incoming radiation back into the atmosphere, heating it up, lowering its relative humidity, and thus reducing further the chances of rain; the groundwater table drops and only plants equipped to withstand a lengthy dry period can survive. At sea there is no such effect; evaporation coupled with reduced groundwater and surface runoff would only lead to a slight lowering of sea level. In addition, though, cooling temperatures also affect negatively the thermal expansivity of ocean waters and thus lower the sea level further. The Permian, Triassic, and Neogene are periods marked by extensive dry areas, lowered sea levels over entrance sills to marginal bays, large-scale evaporite deposition, a retrenchment of rain-loving vegetation, and a cold climate dominating polar regions.

Upper Cretaceous basic volcanic activity and related carbon dioxide emanation coincided with the appearance of dense angiosperm forests as far north as Ellesmere Island (judging by the preserved lignites), early Neogene cooling of the Antarctic areas coincided with vast evaporite deposits in the Persian Gulf-Red Sea-Mediterranean area.

Wet periods thus correlate with warm periods in the earth history, while dry periods and high rates of evaporite precipitation coincided with or followed immediately upon cool periods and were accompanied by a lowering of sea level and a reduced carbon dioxide content of the atmosphere. Dry periods were prevented or at least alleviated by a greenhouse effect during a raised carbon dioxide content of the atmosphere.

Evidence from the geological record thus suggests that a greenhouse effect induced by an increased partial pressure of carbon dioxide would elevate not only air temperatures, but would also produce higher rates of rainfall outside the equatorial belt, greater recharge of the groundwater table, more lush vegetation, and a contraction of the areal extent of deserts by tightening the whirl and speeding up the movement of subtropical anticyclones. A slight variation in the energy output of the sun could more than adequately counteract or enhance this effect.

SONYUSHKIN, V.E., and O.M. ROSEN, Academy of Sciences of the USSR, Moscow, USSR

Microstructure of Quartz and Flow Stress in Polymetamorphic Rocks of Anabar Shield: Arhaean Folding Flow and Proterozoic Interplate Sliding

The Anabar shield is the north window of crystalline basement of the Siberian platform and composed of two main metamorphic complexes. The Archaean Anabar granulite complex of 2.7 Ga age contains hypersthene-plagioclase gneiss and metabasites of volcanic origin and in some places garnet gneiss (metagraywacke) and metacarbonate rocks. These rocks are crumpled to narrow linear sometimes keel- shaped folds 1–5 km in width and 10–20 km in length of northwest-north strike and deep east-northeast incline, which form a homogenous folding assemblage indicating the regional pressing from east-northeast to the west-northwest.

The Proterozoic Lamujka complex of 1.9 Ga age occupies about one-fifth of the shield area being generated by transformation of granulites along fault-belts of 10–30 km width. At the first stage blastomylonite of amphibolite facies was formed; at the second, migmatite with small plate-like granite bodies; and the third, mylonite of greenschist facies. Data of petrology and fluid inclusions show the metamorphic parameters for the Anabar complex to be P = 7.5–11 Kbars, T = 800–1000 °C; for the Lamujka complex, P = 6–7.5 Kbars, T = 750–850 °C; and for the last tectonites these parameters drop to P = 0.7 Kbars, T = 200–300 °C (Rosen and Sonyushkin, 1987).

Studies of microstructures of deformed quartz have been performed using transmission electron microscopy of ion-thinned preparations. Optical and scanning electron microscopies and selective etching has been also used.

The subgrained structure of quartz, connected with dislocation boundaries, appears to be typical for high-temperature rocks of studied granulite and amphibolite facies. Dislocation climb described shows deformation to realize by high-temperature creep. In greenschist facies quartz shows irregular distribution of dislocations and poorly developed subgrain structure, which allows us to assume the initial stage of recovery process. The deformation lamellies studied seems to be a consequence of dislocation sliding in basal planes. Quantitative measurements, plotted on Figure 1, show an increase of density of free dislocations (ρ) and a decrease of sizes of dislocation subgrains (d) according to decrease temperature from granulite to greenschist facies. In this way also the range of measured values ρ and d increases, being connected with decreasing of recovery process.

The stress values are described as j = 0.18 Kbars according to subgrain sizes and σ = 0.20 Kbars according to dislocation densities for the Archaean Anabar granulite complex. In the Proterozoic Lamujka complex amphibolite facies is characterized by σ = 0.27 Kbars (after subgrains) and σ = 0.4 Kbars (after dislocation densities) and greenschist facies; accordingly σ = 0.45 Kbars and σ = 0.76 Kbars (Fig. 2a,b).

On the whole, the Archaean homogenous folding assemblage seems to be formed through folding stress flow under stress pressure about 0.2 Kbars. This value is in a good correlation with shear stress in front of oceanic plate (Forsyth and Ueda, 1975). Inside the Proterozoic fault belts tectonites were formed under stress pressure as 0.60 Kbars, but at high-temperature stage of migmatite and granite generation the stress pressure measured seems to decrease up to 0.35 Kbars. Such stress values were calculated for continental collision (Molnar et al., 1983).

REFERENCES

Forsyth, D., and S. Ueda, 1975, On the relative importance of the driving forces of plate motion: Geophysical Journal of Royal Astronomical Society, v. 43, p. 163–200.

Molnar, P., W.-P. Chen, and E. Padovani, 1983, Calculated temperatures in overthrust terrains and possible combinations of heart sources responsible for the Greater Himalaya: Journal of Geophysical Research, v. 88, n. B8, p. 6415–6429.

Rosen, O.M., and V.E. Sonyushkin, 1987, Fluid inclusions and deformation of quartz in polymetamorphic rocks of Anabar shield: Zapiski Miner. Soc. of the USSR, v. 116, n. 5, p. 602–609 (in Russian).

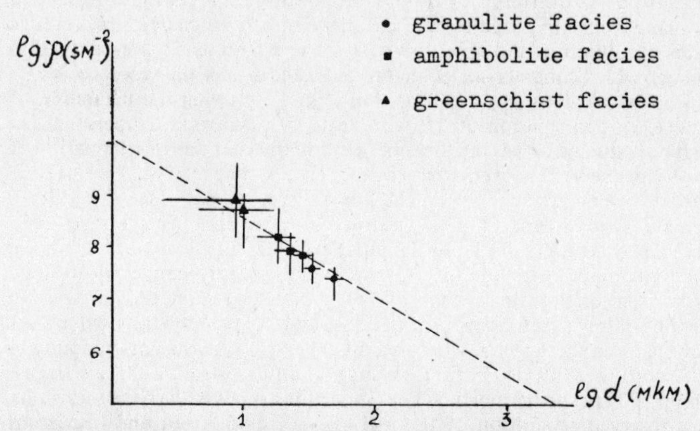

Figure 1—Relationship between subgrain sizes and dislocation densities in quartz from polymetamorphic rocks of Anabar shield.

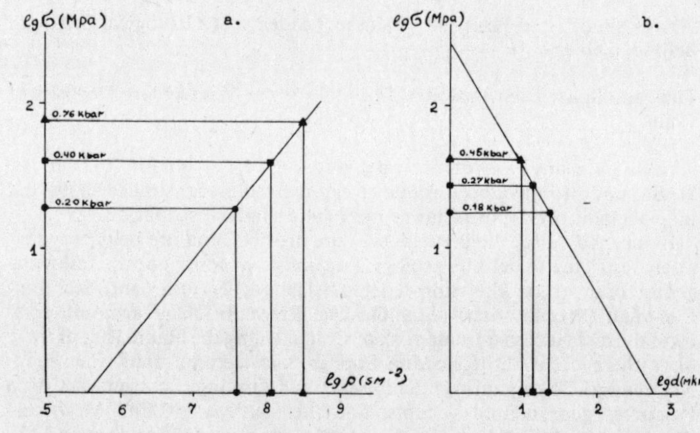

Figure 2—The stress values according to dislocation densities (a) and subgrain sizes (b).

SOPENA, A., and A. RAMOS, Universidad Complutense, Madrid, Spain

Fluvial Architecture Evolution in Permian–Triassic: Cordillera Ibérica, Central Spain

The Buntsandstein facies in the Cordillera Ibérica (Central Spain) consist of fluvial conglomerates and sandstones, late Permian–Early Triassic in age (Ramos and Sopeña, 1983; Ramos et al, 1986). These sediments are related to a graben basin bounded by large northwest–southeast-trending tardihercynian faults. Numerous well-exposed outcrops allow a detailed study of the fluvial architecture, the main "elements" (Miall, 1985, 1988), and the paleodrainage evolution (Figure 1).

Massive sheets of conglomerates interpreted as longitudinal bars and channel-fills are the main elements in the fluvial architecture for the Lower Conglomerates.

Fifth-order surfaces (Miall, 1985, 1988) can be followed laterally across hundreds of meters and are quite often related to large mudclasts.

Channel-fill is preferentially preserved in the Lower Conglomerate (Figure 1.1). Otherwise longitudinal bars form the bulk of the Upper Conglomerate (Figure 1.2). So, they result in two different fluvial architectures.

The basin evolved into a sandy alluvial system where large linguoid or transverse bars are the main elements (Figure 1.6). Those bars are piled up and/or eroded or modified as a result of change in flood stage (Figure 1.5).

The basin evolved in the Upper Buntsandstein into a high-sinuosity system with a fluvial architecture formed by sheets and fine overbank deposits as a result of lateral accretion point-bars.

An attempt has been made to correlate the above described elements to the paleohydrology (Figure 1). So, the proportion of channel fills relative to lateral accretion deposits, grain size of channel variance and distribution patterns, paleocurrent variance, and quantitative reconstruction of channel bar and channel fill have been used to make some estimations about paleochannel patterns and hydraulics (Bridge, 1985).

The system began in a montane area (Galloway and Hobday, 1983) with flashy channel flow and shallow water table. Broad continuous sheets of conglomerates are the main isolith. Mainly vertical aggradation produced a series of vertical stacked bodies (Figure 1.1, 1.2, 1.3, 1.4).

The system evolved into a sandy riverine plain with a more continuous channel flow and shallow to intermediate water table (Figure 1.6). Vertical and lateral stacked bodies formed a complex system of overlapping sheets and ribbons (Friend, 1983).

The final evolution resulted in a mixed load system in a riverine plain, close to the coastal plain with mainly continuous channel-flow and occasional flash flow. Multistory channel fills subordinate to surrounding overbank deposits produced a complex, typically "beaded" belt.

REFERENCES

Bridge, J.S., 1985, Paleochannel patterns inferred from alluvial deposits: a critical evaluation: Journal of Sedimentary Petrology, v. 55, n. 4, p. 579–589.

Friend, P.F., 1983, Towards the field classification of alluvial architecture or sequence, *in* J.D. Collinson and J. Lewin, eds., Modern and ancient fluvial systems: International Association Sediment. Special Publication, v. 6, p. 345–354.

Galloway, W.E., and D.K. Hobday, 1983, Terrigenous clastic depositional systems. Applications to petroleum, coal and uranium exploration: Springer-Verlag, 423 p.

Miall, A.D., 1985, Architectural-element analysis: a new method of facies analysis applied to fluvial deposits: Earth-Science Review, v. 22, p. 261–308.

——— 1988, Architectural-elements and bounding surfaces in fluvial deposits: anatomy of the Kayenta Formation (Lower Jurassic), Southwestern Colorado: Sedimentary Geology, v. 55, p. 233–262.

Ramos, A. and A. Sopeña, 1983, Gravel bars in low-sinuosity streams (Permian and Triassic, central Spain), *in* J.D. Collinson and J. Lewin, eds., Modern and ancient fluvial systems: International Association Sediment. Special Publication, v. 6, p. 301–312.

——— ——— and A. Perez-Arlucea, 1986, Evolution of Buntsandstein fluvial sedimentation in the northwest Iberian Ranges (Central Spain): Journal of Sedimentary Petrology, v. 56, p. 862–875.

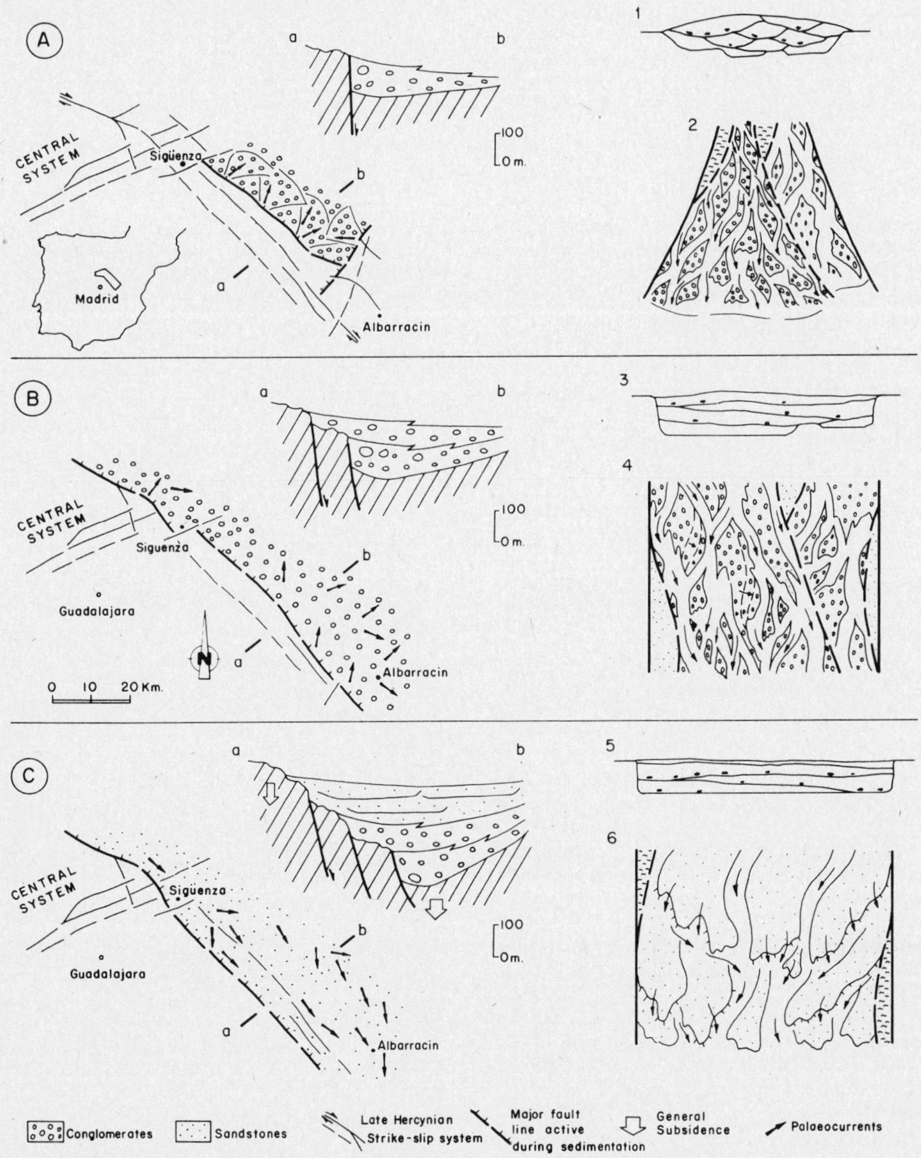

Figure 1—Paleogeographic model, basin-margin settings and alluvial architecture for the Hoz del Gallo Conglomerate (A, lower cycle; B, upper cycle) and the Rillo de Gallo Sandstone (C).

SORRISO-VALVO, M., CNR-IRPI, Roges di Rende, Italy

Influence of Tectonics on Spatial Distribution of Mass Movement: Calabrian Case-Studies

In order to study on a regional scale the relationships between mass movement and tectonic features, an investigation has been undertaken in northwestern Calabria. The study area extends over more than 500 km^2 and is divided into two subzones (Figure 1a, b).

Quaternary tectonic uplift rates are close to 1 mm/y throughout the study zones, with small differences from place to place.

The two subzones are characterized by different tectonic settings: In the north, the north–south-trending normal fault set prevails; in the south, northeast-southwest and west-northwest–east-southeast fault sets are dominant (Figure 1b).

Some 307 major events and more than 50 areas of undifferentiated mass movement have taken place in the two subzones. Phenomena have been classified according to the classification of Varnes (1978). Main detected phenomena, main neotectonic faults, and relevant geomorphology are shown in Figure 1.

In the northern part of study area (Figure 1a), the surface affected by mass-movement is 22%; there phyllites crop out over 20% of territory, and earth material (clay, marl, marly clay, etc.) over 24%. In the southern subzone (Figure 1b), the dominant rock type is ductile phyllite (78.5% of surface). This has eased a great development of rock-flow phenomena that cause the mass movement to affect 47.4% of this territory. Mass movement frequency is then dependent upon rock type. However, in the study zone mass movement is more developed in proximity of faults and outside of remnants of the paleosurface, abrasion platform, and marine terraces. In addition, prevalent direction of mass movement (rose diagram Ra in Figure 1) is perpendicular to the trend of dominant fault set in subzone a, while this directional relationship appears no longer straightforward as far as subzone b is concerned (rose diagram Rb). In subzone a, an analysis has been conducted on directional relationships between mass movement direction and trend of mesostructural discontinuities. One can roughly express a directional convergence index (IP) in terms of the mean of cosines of divergence angles and of the mean square residuals (RIP). There is full directional convergence when IP = 1 and RIP = 0. In the study zone, it results that:

a – IP = .42 and RIP = .46 for stratification and schistosity;
b – IP = .10 and RIP = .58 for neotectonic faults and joints.

The low value of IP and the fairly high value of RIP for mass movement and faults directions confirm the visual display in Figure 1 for subzone a. This is easy to understand because major slopes are structural slopes due to the degradation of fault scarps, so that most mass movement occurs perpendicularly to the main set direction (north-south). The situation is not so clear in subzone b. However, visual inspection permits us to determine that the direction of movement of major mass movement phenomena are consistently perpendicular to major faults. In this subzone, one can also see how the morphology of slopes controls the development of rock flow phenomena: only north-facing slopes, which can be considered as very degraded fault scarps, could develop and extend in length and height beyond the threshold values (Sorriso-Valvo, 1984) for rock-flow to develop.

In conclusion, in the study area fault distribution controls the distribution and direction of mass movement. Terrane quality and geomorphological conditions are also important.

REFERENCES

Sorriso-Valvo, M., 1984, Deep-seated gravitational slope deformations in Calabria: Proceed. Coll. Mouvem. de Terrains, Série Docum. B.R.G.M., n. 83, p. 81–90.

Varnes, D.J., 1978, Slope movement types and processes, *in* R. Schuster and J. Krizek, eds., Landslides, analysis and control: Transp. Board, Nat. Acad. of Sciences, Sp. Rep. 176.

Table 1—Areal incidence (in %) of mass movement in the study zone according to main terranes and movement types.

subzone	a	c	both
type of material			
rock, brittle	18.2	8.4	14.3
rock, ductile	34.8	76.7	51.5
earth	38.7	13.5	28.6
debris	8.3	1.4	5.6
all, average	22.0	47.4	37.6
type of movement			
slide	24.0	17.0	21.2
flow	32.0	59.5	43.2
slide/flow	44.0	23.5	35.6

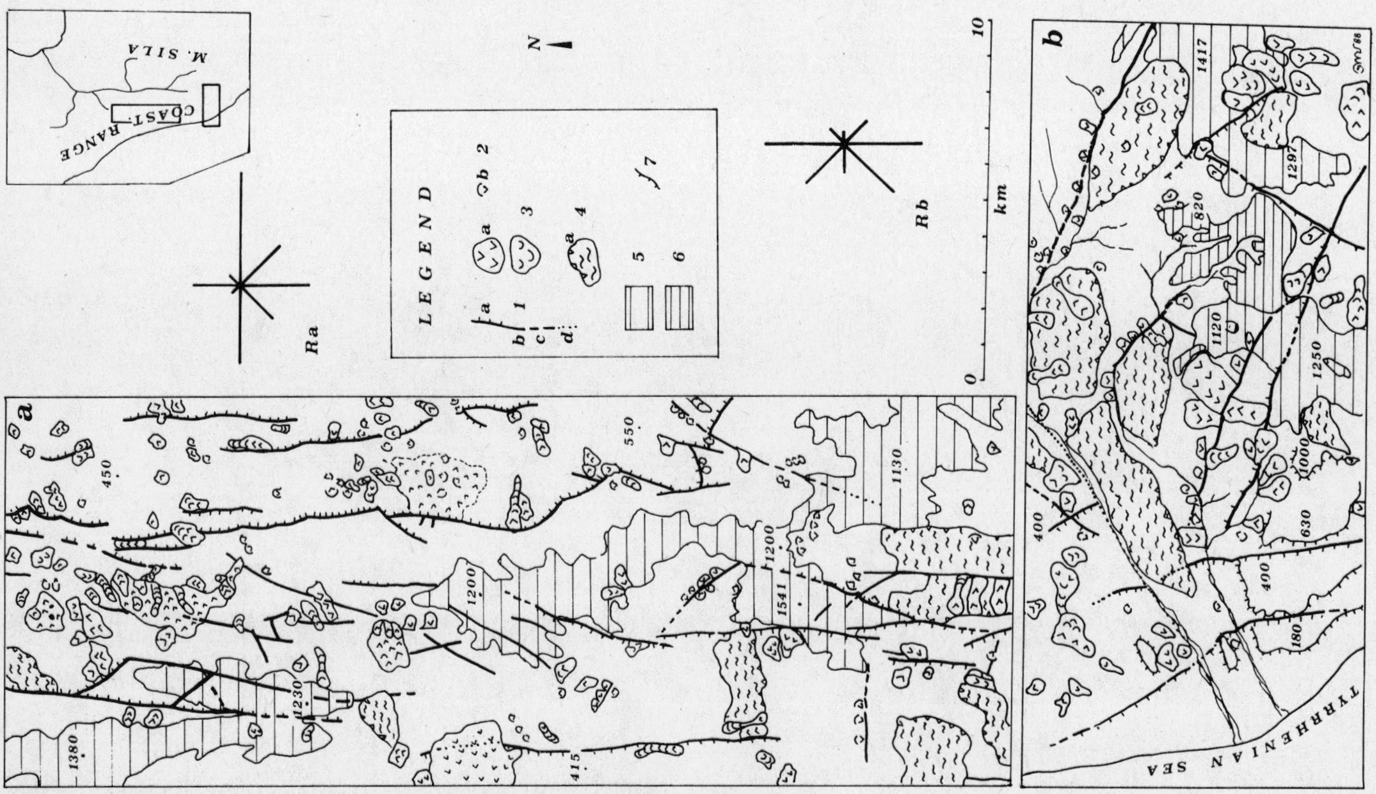

Figure 1—Main mass movement phenomena in the study area (a and b) and rose diagrams (Ra and Rb for subzones a and b., respectively) of direction of landslides (number of cases).
Legend symbols: 1, fault (a, normal fault; b, unknown type; c, faint; d, inferred); 2, slide (a, mappable; b, unmappable); 3, earth flow; 4, rock flow; 5, pre-Holocene paleosurface; 6, Lower Pleistocene abrasion platform.

SORRISO-VALVO, M., CNR-IRPI, Cosenza, Italy

Physical and Economic Impact of Mass Movement in Southern Italy

Mass movement represents one of the main causes of the delay in the development of economy and welfare in Southern Italy. The geographical distribution of mass movement is strongly dependent upon the outcropping of landslide-prone terranes. Clayey tectonic melange, pelite-rich flysch, and phyllite are the weakest rocks. On clayey melange, landsliding may affect more than 90% of the territory. The Apulia region and south Sicily are little affected by mass-movement because carbonatic rocks dominate there. However, sinking due to cave collapse may occur.

The climate of most part of Southern Italy is characterized by extreme meteorological events that are concentrated in the winter season. Alternation of wet and dry seasons also enhances landscape decay.

It is difficult to gather complete information about actual social and economic cost of landsliding in the territory of interest; however, Calabria is fairly representative for most of Southern Italy, and it is possible to give information on my personal experience.

Major extreme meteorological events occurred in 1951 and 1973. They totaled at least 100 victims, mostly because of flooding. There have been fewer than 20 victims of landsliding. In the last 20 years, at least 32 more people died because of landsliding. In 1951, three villages were evacuated, and 15 were isolated for 20 days; in 1973, 77 of the 253 villages and towns of the Reggio Calabria and Catanzaro provinces (less than two-thirds of Calabria) suffered heavy damages. The (underestimated damage was about $750 million (U.S.) in present currency. From 1950 to 1973, there have been 25 cases of landsliding along the 988 km of railroads in Calabria.

The Regional Commission for Geological Hazards has recently (1988) made an estimate of the cost for corrective measures, not including maintenance, for damage caused by mass movement to villages, roads, railways, pipelines, etc. The total is about $3 billion (U.S.). Still underestimated!

SOSSON, M., CNRS-INSU, Valbonne, France, and T. CALMUS, UNISON, Hermosillo, Mexico

Early Late Cretaceous Overthrusting in Northern Sonora Responsible for Large-Scale Displacement of Nevadian Belt

In northern Sonora, three main compressional tectonic events have been described (Rangin, 1982). New structural data collected in an area located between Magdalena de Lino and Cucurpe (La Lamina) attest for large-scale displacement of the Upper Jurassic belt over the North American margin, which has remained undeformed during the Nevadian tectonic event.

Near La Lamina, two main tectonic units are separated by metamorphic bodies sheared and sliced along low-angle thrust.

1. The upper plate is characterized by well-exposed Mesozoic series. Upper Jurassic sequences are composed of detrital and volcanic rock that contains some limestone lenses. The age was determined by Rangin (1982) on the basis of Oxfordian fossils. This monotonous sequence is very deformed by tight folds locally associated with schistosity and shear zones. A south vergence folding associated with reverse faulting has overprinted the prior tectonic event. The lower Cretaceous rests with an *angular unconformity* over the Upper Jurassic sequence. This unconformity indicates that a Nevadian event is responsible for the main deformation of Upper Jurassic rocks. The Lower Cretaceous series constitute extension of the Bisbee Group (Ransome, 1904) which overlays *conformably* the Upper Jurassic volcanic sequence (Canelo Hills) (Kluth, 1982). It consists of a basal conglomerate, red shales, sandstones, and limestone ranging in age from Aptian to Albian.

2. The lower plate is composed by the same Lower Cretaceous series that is unconformably overlain to the north by Upper Cretaceous continental sequences (Rangin, 1982). They are gently affected by east–west-striking folds testifying to the existence of the second deformation also described in the upper plate.

The tectonic contact between both units is marked by slices of sheared metamorphic rocks. These rocks are composed of gabbros, amphibolites, gneisses, calcschistes, and foliated granites. These slices are restricted to the shear zone that also contains some unmetamorphosed volcanics. The whole shear zone materialized the main thrust plane that has carried northeastward the upper plate. The age of the thrusting is post-Albian, and it pre-dates the Late Cretaceous continental sequences deposit that overlie unconformably the Bisbee Group of the lower plate near the contact. Consequently, this main thrusting can be reasonably assigned to the early Late Cretaceous event as described further south in Sahuaripa area (Central Sonora) by Pubellier (1987). The metamorphism pre-dates this overthrusting, which also carried some slices of unmetamorphosed volcanic rocks.

This structural pattern is the result of three main stages of deformation. The first corresponds to the folding of the Upper Jurassic detrital sequences. The metamorphism could be related to this event. In another locality, Morales (1984) has dated a Jurassic metamorphic event (169 ± 0.8 m.y.). The second stage is characterized by a large-scale overthrusting that has carried northeastward the metamorphic rocks, and the Upper Jurassic detrital sequence unknown in the lower plate. The third stage is probably Laramian in age. It is related to a gentle reverse faulting and folding with a south vergency.

These new structural data argue for a main stage of deformation during the Nevadian Orogeny in a western province probably distant from the southern North American margin (Arizona) where the Nevadian Orogeny is unknown. This western province was reactivated after the deposition of the Bisbee Group during the early Late Cretaceous Orogeny and the Laramian event responsible for the structural organization of Sonora and southeastern Arizona.

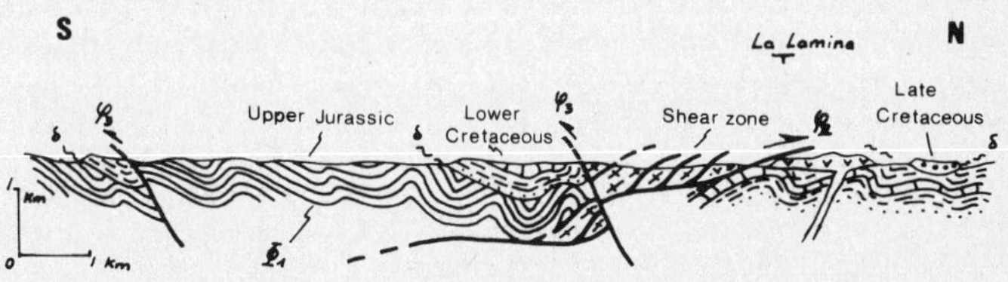

Figure 1.

SOTIN, C., Laboratoire de Géologie Dynamique Interne, Orsay, France

Venus Tectonism: Is Western Aphrodite Terra Equivalent to Terrestrial Spreading Centers?

A basic question concerning Venus is whether plate tectonics can explain the large-scale tectonic features. Despite its similarity to the Earth, the existence of plate tectonics on Venus is very controversial. Anderson (1981) showed that subduction cannot occur because the lithosphere would be always less dense than the underlying mantle. Phillips and Malin (1981) argued that there are no ocean-rise-like features. Kaula and Phillips (1981) analyzed the rises on Venus and reached the conclusion that if spreading exists on Venus, only 15% of its internal energy is delivered to the surface by such a mechanism. Solomon and Head (1982) concluded that there are no arguments whether internal heat is transferred by hot spot volcanism or plate tectonics. All these studies are based on either general concepts or overall analysis and provide no information on why and how the Venus highlands have formed. Recently, Head and Crumpler (1987) have conducted a detailed analysis of Aphrodite Terra (a highland 14,000 km long). On the basis of long wavelength symmetry of the topography and the presence of cross-strike discontinuities, they conclude that Aphrodite Terra is analogous to a terrestrial spreading center. To test this hypothesis, we have analyzed profiles of the Pioneer Venus altimetry and line of sight acceleration over Western Aphrodite Terra. These profiles are then compared with those predicted by models of a spreading center under Venus conditions.

A simple model (Reid and Jackson, 1981) describes a spreading center as the mantle flow driven by the motion of diverging plates. We have developed a model that takes into account the crust and the buoyant flow driven by mantle differentiation due to melt extraction. Two main differences affect models of spreading centers for Venus. First the surface temperature for Venus is 450 °C higher than for the Earth. Parameterized convection therefore predicts a mantle temperature about 125 °C higher. At a spreading center, such a high temperature should produce much more melt. Assuming that the melt segregates and extracts as it does for the Earth, a 10 km thicker crust would be produced at similar spreading rates. Second, there is no water on Venus and therefore no hydrothermal cooling. Combined with a higher surface temperature, the absence of hydrothermal cooling would reduce the thickness of a brittle crustal layer. However, it should not affect the crustal thickness. The absence of oceans and different surface and mantle temperatures affect the expression of isostatic thermal topography.

To test the hypothesis that Aphrodite Terra is a spreading center (Head and Crumpler, 1987), we analyze a domain in Ovda regio where no cross-strike discontinuities have been determined. Since the altitude of the Pioneer Venus Orbiter was between 300 km and 600 km, the maximum value of the line of sight acceleration should coincide with a hypothetical ridge axis. Between longitudes 82E and 90E, the points fit rather well a linear axis along which the values of the LOS acceleration varies between 22.9 and 26.1 mgals. Using the PVO topographic data extending to a distance of 2000 km on each side of the axis, we calculate mean values of both the topography and LOS acceleration in forty 100 km wide regions parallel to the axis. This analysis leads to the following conclusions:

1. A very high degree of symmetry exists for both the topography and the LOS acceleration

2. Three topographic regions are determined. Region 1 is a central trough 300 km wide. Region 2, between 150 and 900 km on each side of the axis, is the only domain in which the topography is fit reasonably well by a thermal boundary layer topography. Region 3, at distances greater than 900 km, is flat. The limit between regions 2 and 3 is characterized by a rapid change in elevation (about 2 km).

To simultaneously interpret the LOS acceleration and the topography, we adapt the simple model of isostatic topography due to plate spreading combined with crustal thickness variations to explain the rapid change in topography between regions 2 and 3. Best model is obtained for a mantle temperature of 1600 °C and a spreading rate of 0.6 cm/y. In this model, the crustal thickness equals 30 km in regions 1 and 2, and 15 km in region 3. From the comparison with the North Atlantic ridge, two main differences exist: the wide trough and the rapid difference in elevation between region 2 and region 3. For the Earth, it seems that the lower the spreading rate, the larger the central trough. Furthermore, rifting of a terrestrial oceanic lithosphere creates rapid changes in topography (Mammerickx and Sandwell, 1986).

A model of spreading center can explain both the observed LOS acceleration and topography over Western Aphrodite Terra. Such a model implies a very low spreading rate (0.6 cm/y), a mantle temperature equal to 1600 °C, and a 30 km thick crust. The same kind of analysis for other parts of Aphrodite Terra should allow us to accept or to reject the idea that Aphrodite Terra is equivalent to terrestrial spreading centers.

REFERENCES

Anderson, D.L., 1981, Geophysical Research Letters, v. 8, p. 309–311.
Head, J.W., and L.S. Crumpler, 1987, Science, v. 238, p. 1380–1385.
Kaula, W.M., and R.J. Phillips, 1981, Geophysical Research Letters, v. 8, p. 1187–1190.
Mammerickx, J., and D. Sandwell, 1986, Journal of Geophysical Research, v. 91, p. 1975–1988.
Phillips, R.J., and M.C. Malin, 1983, *in* Hunten et al, eds., Venus, Univ. Arizona Press, p. 159–214.
Reid, I., and H.R. Jackson, 1981, Marine Geophysical Research, v. 5, p. 165–172.
Solomon, S.C., and J.W. Head, 1981, Journal of Geophysical Research, v. 87, p. 9236–9246.

SPAKMAN, W., University of Utrecht, Utrecht, the Netherlands

Delay Time Tomography of the Upper Mantle

Delay time tomography is an inversion method for investigating the three-dimensional seismic velocity structure of the earth. The data—delay times—are derived from travel times of seismic waves; in most cases P waves. Using travel time data from many earthquake-station combinations offers the possibility to estimate the amplitudes of the three-dimensional seismic velocity field by inversion of the travel time data. An important step in this procedure is the quantification of the way in which seismic waves illuminate the earth's interior. It is assumed that the earth's velocity heterogeneity can be described as a small deviation (few percent) from a radial (one-dimensional) model of seismic velocity. It is advantageous to use the radial model as a reference in tomography. This allows conversion of the nonlinear relation between travel time and seismic velocity into a linear relation between delay times (the difference between the travel time and the model prediction) and lateral velocity heterogeneity. Furthermore, the illumination of the earth by seismic waves can be computed from the reference model.

The number of data and unknowns in delay time tomography can be very large. In large-scale problems, typically $O(10<6)$ data are inverted for estimates of $O(10^4–10^5)$ unknowns. The unknowns represent the three-dimensional velocity heterogeneity but also account for the effects of earthquake mislocation and possible station bias. Inversion problems with that many unknowns can only be solved iteratively. Inversion results will be shown that are obtained with two different types of iterative least squares methods; an algebraic reconstruction technique called SIRT, and a conjugate gradient method called LSQR.

We will focus on applications to the upper mantle and partly the lower mantle (down to 1400 km) in three large areas: (1) Europe and the Mediterranean, (2) Central and South American, and (3) the northwest Pacific subduction zone region.

The velocity heterogeneity in the crust and lithosphere in the regions studied often exhibits a close correlation with larger-scale geological structures at the surface. All studied regions are characterized by the existence of deep subduction of oceanic lithosphere. An example of a tomographic image of a subduction zone is given in Figure 1. It displays a cross-section of the collision between the African (left) and Eurasian plates in the Eastern Mediterranean. The upper panel is a small location map for orientation (dashed, Hellenic trench system; grey, sea areas; dots, shallow events; circles, events deeper than 50 km; black triangles, Pliocene–Quaternary volcanic arc). The lower panel displays the P-wave velocity heterogeneity. The contouring is made in percentages of the ambient Jeffreys-Bullen upper mantle velocity (see legend; lower velocities correspond to negative values). Black symbols indicate the projection of hypocenters with $M > 4$ that are located within 100 km from the plane.

The power of large-scale delay time tomography is to reveal the three-dimensional structure of the earth. However, the results suffer from many uncertainties related to the data, method, and the assumptions involved, which make it very difficult to interpret the results obtained. For instance, the linearity between delay times and velocity heterogeneity is not always warranted, the usefulness of delay time data for tomographic mapping is at least questionable, and it is difficult to obtain reliable and interpretable accuracy and resolution estimates.

REFERENCES

Spakman, W., 1988, Upper mantle delay time tomography with an application to the collision zone of the Eurasian, African and Arabian plates: Ph.D. Thesis, Geologica Ultraiectina, v. 53, University of Utrecht, 200 p.
——— and G. Nolet, 1988, Imaging algorithms, accuracy and resolution in delay time tomography, *in* N.J. Vlaar, N. Nolet, M.J.R. Wortel, and S.A.P.L. Cloetingh, eds., Mathematical geophysics: a survey of recent developments in seismology and geodynamics: Dordrecht, Reidel, p. 155–188.
——— M.J.R. Wortel, and N.J. Vlaar, 1988, The Hellenic subduction zone: a tomographic image and its geodynamic implications: Geophysical Research Letters, v. 15, p. 60–63.

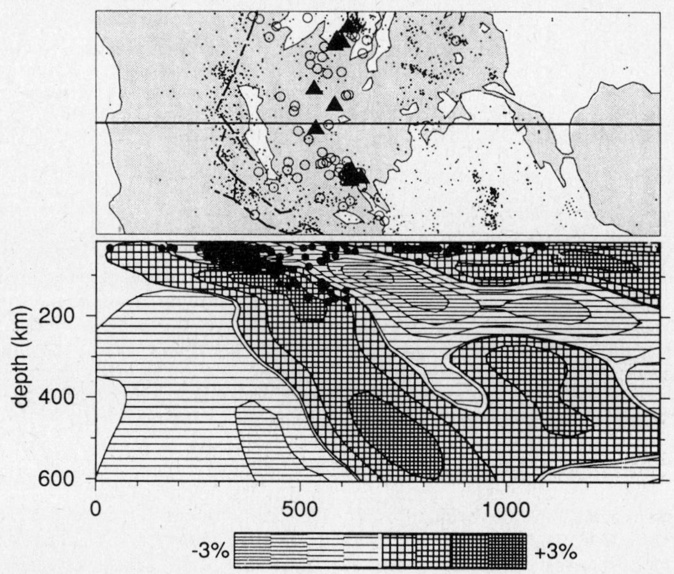

Figure 1.

SPAKMAN, W., University of Utrecht, Utrecht, the Netherlands, S. STEIN, Northwestern University, Evanston, IL, and R. VAN DER HILST, and R. WORTEL, University of Utrecht, Utrecht, the Netherlands

Resolution Experiments for Northeastern Pacific Subduction Zone Tomography

We investigate the resolution of body wave tomography for the subduction zones in the northeastern Pacific by inventing synthetic data. For thermal models of the Kurile, Japan, Izu-Bonin, Mariana and Ryukyu slabs, we generate three-dimensional synthetic velocity anomalies that are then projected onto a cell model for the uppermost 1400 km of the mantle. For regional events (1964–1987), we use the locations and observing stations at epicentral distances of up to 90° reported by the ISC and NEIC, and compute the ~1.4 million Jeffreys-Bullen ray paths through the synthetic models. After adding Gaussian noise ($\sigma = 1$ s) to the corresponding synthetic P wave delay times, we invert the data and compare the resulting velocity structure to the initial synthetic models. We carry out these experiments for slabs restricted to the upper mantle, and for slabs penetrating into the lower mantle, for a variety of slab geometries and temperature (velocity) models.

The results indicate that the Kuriles and Japan slab models are fairly well imaged. In general, the dip of the slab structures can be identified, but, in cases, the depth extent is difficult to resolve. A variety of resolution artifacts, generally of low amplitude, are observed. In some cases slabs appear to broaden, shallow in dip, or "finger."

In the results of inversion of real data, slabs are not, in general, as easily identified. Comparison with the synthetic experiments suggest that resolution artifacts noted in the synthetic case may sometimes be present for the real data. The discrepancies between the real and synthetic cases may result from several effects. The slab models are relatively simple and may differ in important aspects from the real situation. Moreover, the synthetic data do not incorporate some complexities of real delay times such as event location and reference model dependencies.

Figure 1 displays a cross-section through the Kuriles subduction zone. The top panel is a small map for reference. Dots indicate earthquakes. The cross-section is taken along the straight line. The thick line in the lower panel represents the outline of the synthetic slab model. The contouring displays the tomographic image of the slab. All anomalies outside the slab contour are due to lack of spatial resolution and data errors. An important artifact is the deep penetration of the slab image into the lower mantle.

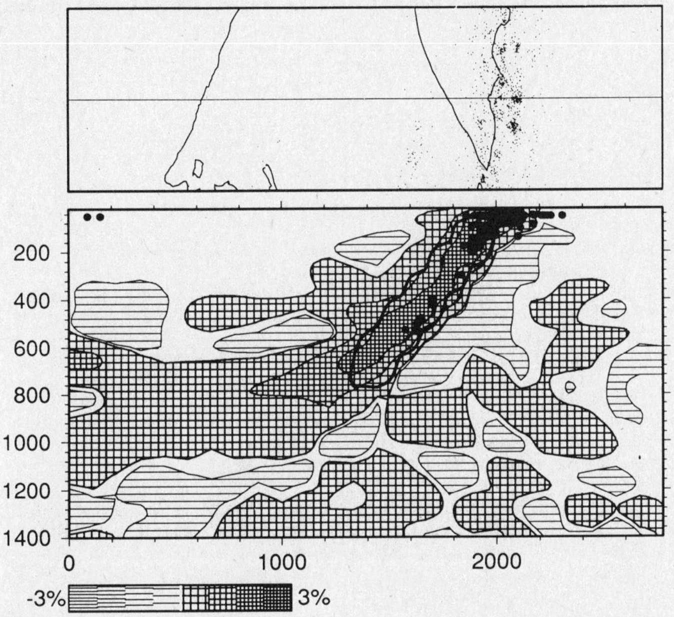

Figure 1.

SPAMER, E. E., Academy of Natural Sciences, Philadelphia, PA

Late Proterozoic Index Fossil *Chuaria Circularis* Walcott, 1899: Understanding an Enigma

Chuaria circularis Walcott, 1899 (Figure 1) is a monospecific genus of planktonic colonia algae. It is generally macroscopic in size, usually occurring as carbonaceous compressions on parting planes, most often in shales. Specimens have been found essentially worldwide in rocks of Late Riphean age (Late Proterozoic, 1000–700 Ma); they are reported from the USA, Canada, Greenland, Sweden, Finland, Norway, Svalbard, Spain, the USSR, Iran, Pakistan, Indian, China, and Argentina, with questionable occurrences in Australia and western Afric. The occurrences of *Chuaria* predate both the glacial event of Varangian time and the appearance of Ediacara-type metazoan assemblages.

Chuaria may have been noted in the literature as early as 1876, but its first positive report was by Holm (1885). Wiman (1894) gave the first figures of this fossil, but it was not given a name until Walcott (1899) erected the genus and species. The morphometric description of the genus *Chuaria* Walcott approaches or overlaps the criteria used to describe some larger forms of microscopic acritarchs, making its identification difficult. Chemical analysis is a prerequisite for positive identification of *Chuaria*, making its utility as a field-identifiable fossil questionable. Its use as an index fossil is likewise of uncertain value; but its macroscopic forms are, in the field, sometimes the only available naked-eye indicators of age. *Chuaria* is thus a useful time-stratigraphic tool in reconnaissance field work.

The genus *Chuaria* has always been problematical to paleontologists. Its enigmatic appearance in late Proterozoic rocks has, over more than a hundred years, led to its identification as one of about 19 different genera, variously as foraminifera, medusoids, brachiopods, gastropod opercula, ostracods, the eggs of trilobites, acritarchs, and algae; it has also been described as inorganic in nature, and noted even in igneous rocks(!). Three taxonomic families, six genera, and 13 species were erected between 1899 and 1985 just to accommodate different morphometric forms of *C. circularis*. The name-bearing taxonomic family in which *chuaria has been placed is Chuariidae Wenz, 1938, erected with question as a monospecific family of Precambrian gastropod.

Current understanding of *Chuaria* places it as a monospecific genus of algae within the Family Chuariaceae Wenz, 1938 (emend. Duan, 1982) together with the monospecific genus *Tawuia* Hofmann *in* Hofmann and Aitken, 1979. *Tawuia*, described as a ribbon-like carbonaceous film with rounded ends, is taxonomically very closely related to *Chuaria*; they often occur in association (Sun, 1987). The *Chuaria–Tawuia* assemblage seems to be a more reliable biostratigraphic index for Late Riphean time than *Chuaria* alone, but they are not always found together. *Tawuia* is not known to occur with *Chuaria* at the latter's type locality in the Awatubi Member of the Kwagunt Formation (Chuar group) in the Grand Canyon of the Colorado River, Arizona.

Ford and Breed (1972) reported to the 24th International Geological Congress that *Chuaria* was monospecific in nature and that many species and genera were taxonomically synonymous with it. They elaborated on this pronouncement in later publications, and they and other workers established that the size range of *Chuaria* overlaps onto the larger end of the size range of acritarchs that otherwise are morphologically very similar to this genus. Among the genera synonymous to *Chuaria*, the genus about which much has been written is *Fermoria* "Chapman" Sahni, 1936, reported from India and Iran. "*Fermoria*," like *Chuaria*, was long recognized by workers in India as an enigmatic fossil; it was erected and often cited as a genus of brachiopod.

Despite the work begun by Ford and Breed, which continues to this day by others, new species continue to be erected within *Chuaria*. Except for emended diagnoses of *C. circularis* itself, all descriptions of the synonymous and newly erected species were based solely on morphometric terms. Nonetheless, all other species erected within *Chuaria* have been placed in synonymy with *C. circularis* because they fail to demonstrate that they are not just variants of that species. Variation could be due to true variation in the living alga or, more probably, to the effects of taphonomic processes, possibly even to effects of laboratory preparation of samples.

Once *Chuaria* was determined to be synonymous with many other taxa in Late Riphean strata around the world, Ford and Breed (1972) noted that *Chuaria*, being a megascopic fossil, might be valuable as a late Proterozoic index fossil visible to the naked eye. However, refined definitions of the genus, requiring laboratory techniques for determinations of identity, and a confusing morphometric overlap with larger forms of arcritarchs, make *Chuaria*-like markings less-reliable field tools in stratigraphic dating. Coincident occurrences of *Tawuia* with *Chuaria*, though, make the field use of the latter genus more sound. In strata which are largely devoid of megascopic fossils, and in which also occur a variety of misleading pseudofossils, any occurrence of *Chuaria*, with or without *Tawuia*, is a welcome aid in biostratigraphic correlation. The identification of *Chuaria* as an alga also presents a more conservative view of late Proterozoic biosystems; the presence at that time of "brachiopods," "gastropods," and such other Paleozoic-like forms is admittedly unsettling to conventional interpretations of evolution.

REFERENCES

Duan, C., 1982, Late Precambrian algal megafossils *Chuaria* and *Tawuia* in some areas of eastern China, Alcheringa, v. 6, n. 1, p. 57–68.

Ford, T. D., and W. J., Breed, 1972, The problematical Precambrian fossil *Chuaria*, 24th Intl. Geol. Congr., Sect. 1, Precambrian Geology, p. 11–18.

Holm, G., 1885, Om vettern och visingsöformationen, Bihang Kongl. Svenska Vetens,-Akad. Handl., v. 11, n. 7, p. 1–49.

Sun, W., 1987, Palaeontology and biostratigraphy of late Precambrian macroscopic colonial algae: *Chuaria* Walcott and *Tawuia* Hofmann, Palaeontographica, Abstract B, v. 203, p. 109–134.

Walcott, C. D., 1899, Pre-cambrian fossiliferous formations, GSA Bulletin, v. 10, 199–244.

Wiman, C., 1894, Paleontologische notizen, Univ. Upsala, Bull. Geol. Inst., v. 2, p. 109–117.

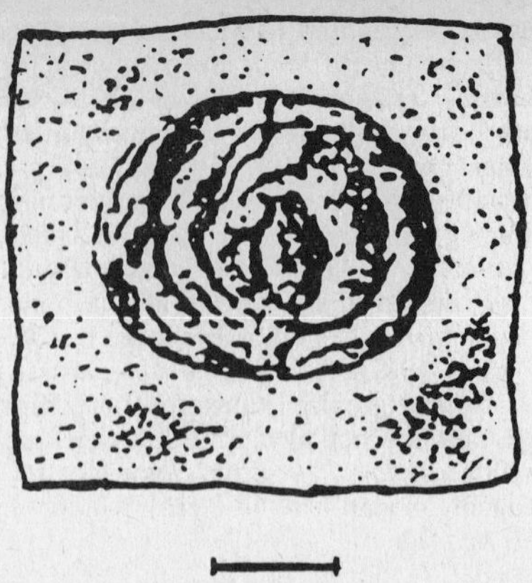

Figure 1—Lectotype of *Chuaria circularis* Walcott, USNM (after Walcott, 1899, pl. 27, fig. 12; lectotype selected by Ford and Breed, 1972). Awatubi Member of Kwagunt Formation (Chuar Group), from Nankoweap Butte, Grand Canyon, AR. Scale bar is 1 mm. Wrinkled discoidal appearance is the result of compaction of originally spherical form.

SPECZIK, S., University of Warsaw, Warsaw, Poland, and W. PUTTMANN, Geochemie und Lagerstatten des Erdöis und der Kohle, Aachen, Federal Republic of Germany

Oxidation of Organic Matter and Its Influence on Kupferschiefer Mineralization of Southwestern Poland

The role of organic matter in ore formation has been discussed in several papers (i.e., Dunham, 1961; Giordano, 1985; Eugster, 1985). Various primary and secondary mechanisms that involved the contribution of organic matter in metal accumulation or precipitation were suggested. The present study provides evidence for the contribution of organic matter and its secondary oxidation to metal precipitation in Permian Kupferschiefer formation of southwestern Poland (central Europe).

The Kupferschiefer formation in central Europe is an organic-rich sediment (~ 5% organic carbon) formed in a euxinic and sappropelitic environment, which in some places is mineralized. The most common component of organic macerals is variously bituminous and oxidized vitrinite. Liptinite (alginite) forms notable accumulations in places that reveal a high soluble organic material content, while allochthonous inertinite dominates in nearshore environments. The amount of soluble organic material in the examined Kupferschiefer samples varies from 58 to approximately 6400 ppm. The above composition of organic matter implied methods that were used to reevaluate organic factors of the mineralization processes. The extracted bitumens were analyzed as total extracts by gas chromatography (GC) and gas chromatography/gas spectrometry (GC/MS). Vitrinite and liptinite studies (vitrinite reflectance and spectral fluorometric measurements) allowed us to measure the degree of maturation of solid organic matter.

An oxidation effect of soluble organic material has been distinguished in certain areas of the Kupferschiefer horizon that reveal ore-grade mineralization (Lubin-Sieroszowice and Konrad deposits). The specific composition of oxidation products (biphenyl, dibenzofurane, dibenzotiophene, and phenanthrene) is quite different from the previously known products of meteoritic weathering, microbial, and diagenetic degradation of organic matter. The relative amount of aromatic compounds is changed by oxidation reactions. Oxygen is incorporated in an aromatic frame system, and carbonyl compounds become important constituents of the bitumen. The ratio of phenanthrene v. the sum of methylphenanthrenes (Ph/ΣMePh) provides a measure for the intensity of oxidation.

The degree of oxidation and copper and silver content changes from the bottom to the top of the elaborated horizon. The most oxidized horizon is at the base of the Kupferschiefer and (or) near the contact with the Rote Fäule facies. With increasing distance from the Rote Fäule, intensity of oxidized compounds decreases relative to a series of saturated hydrocarbons. The correlation between the degree of oxidation (Ph/ΣMePh) to copper and silver content is inversely proportional. Copper and silver content is very low at the bottom, then rapidly increases, and later gradually decreases toward the top of the Kupferschiefer. The composition and distribution pattern of oxidation marking compounds in lower part of Kupferschiefer, relative to its upper part suggest that oxidation of the Kupferschiefer horizon occurred most likely during late diagenesis or early catagenesis.

The maturity of organic matter in the Kupferschiefer horizon of southwestern Poland was found to be principally low. More than 80% of vitrinite measurements fall in the narrow range between 0.5 to 0.9 $\%R_{oil}$. There is a general correlation between the collection depth of samples and vitrinite reflectance, except for samples that either have a high content of organic matter or reveal pronounced oxidation of soluble organic matter. Detailed results of vitrinite reflectance studies, which have indicated the pattern of the paleogeothermal field in Kupferschiefer horizon, were discussed earlier (Speczik and Püttmann, 1988).

Spectral fluorescence measurements reveal an increase in fluorescence intensities with increasing oxidation. Additionally, the green shift of fluorescence maxima is related to oxidation effects. It appears that epigenetic oxidation of vitrinite and associated liptonite caused not only aliphatic groups adsorbed on the surface of vitrinite particles, but affected the entire vitrinite structure.

The above results may to some extent confirm the genetic model presented recently for the Kupferschiefer deposits in Poland (Speczik et al, 1986; Jowett, 1986) with Permian Kupferschiefer acting as a geochemical trap that accumulated metals from ascending, oxidizing solutions. The hydrogen-rich Kupferschiefer provides sufficiently reducing equivalents to enable redox reactions when sedimentary organic matter gets in contact with oxygen- and metal-rich fluids. Low-temperature solutions ascended through the conduits that correspond to epigenetic Rote Fäule facies. The movement of solutions was limited to major tectonic structures associated with the rift zones (the border between Saxothuringian and Rhenohercynian zones of the Variscan externides).

References

Dunham, K. C., 1961, Black shale, oil and sulphide ore: Adv. Science, p. 248–299.

Eugster, H. P., 1985, Oil shales, evaporites and ore deposits: Geochimica Cosmochimica Acta, v. 49, p. 619–635.

Giordano , T. H., 1985, A preliminary evaluation of organic ligands and metal-organic complexing in Mississippi Valley-Type ore solutions: Economic Geology, v. 80, p. 96–106.

Jowett, E. C., 1986, Genesis of Kupferschiefer Cu-Ag deposits by convective flow of Rotliegende brines during Triassic rifting: Economic Geology, v. 81, p. 1823–1837.

Speczik, S., C. Skowronek, G. Friedrich, R. Diedel, G. Schumacher and F. P. Schmidt, 1986, The environment of generation of some base metal Zechstein occurrences in Central Europe: Acta Geol. Polon., v. 36, p. 1–35.

Speczik, S., and W. Püttmann, 1988, Origin of Kupferschiefer mineralization as suggested by coal petrology and organic geochemical studies: Acta Geol. Polon, v. 37, p. 167–187.

SPEED, R.C., Northwestern University, Evanston, IL

Cenozoic Tectonic Evolution of Southeastern Caribbean

The Mesozoic passive margin of northern South America in Venezuela east to longitude 58 °W has been overridden obliquely by Caribbean lithosphere during Cenozoic time. The collision began in the northern Andes early in the Cenozoic and has migrated eastward, suturing progressively after a critical distance of continent was subducted. The collision is still active from the Paria Peninsula to the east.

The main effect of the collision on the continent has been the development of (1) a thin-skinned foreland thrust belt ahead of overriding Caribbean lithosphere, (2) a foreland basin that formed ahead of the thrust belt and is now partly imbricated in it, and (3) east–west-striking strike-slip faults with right slip in suture zones. The strike-slip faults take up tangential slip that can no longer be accommodated by plate convergence.

The collision brought extensive accretionary fore-arc terranes above the continental margin, including the metasedimentary belts of the Cordillera de la Costa, Paria, Araya, and northern Trinidad. These fore-arc terranes have in turn been overridden southward by crystalline and arc terranes of the Caribbean lithosphere due to major imbrication of the margin of the overriding plate. As much as 40 km of unroofing occurred in the overriding plate.

The northern edge of continental crust of northern South America may extend as much as 200 km or more north of the present coastline below the overthrust Caribbean terrane.

The actively overriding lithosphere is fronted by the southern Lesser Antilles Arc (SLAA) system. The pre-mid-Miocene magmatic arc platform of this system contains rocks that record magmatic arc activity from Cretaceous to Mid-Miocene times. This arc complex was probably continuous west to Orchilla and Aruba and was an intra-Atlantic feature, not part of a Pacific-derived Caribbean plate. Neogene motion between the relatively east-moving Caribbean plate and the South American plate may be concentrated along the northern margin of or even farther north than the Aruba–Orchilla–SLA arc system. The pre-Neogene tectonic evolution of the southeastern Caribbean was mainly governed by the oblique collision between this arc system and the passive margin of continental South America.

Effects of the collision on the arc are uplift and north-south contraction in proportion to duration of collision: maximum near Margarita and decreasing eastward. The fore-arc basin has both narrowed and subsided at an increasing rate during collision. Fore-arc basin flanks have apparently thrust out onto adjacent arc terranes, southward onto the accretionary prism and back thrusting onto the arc platform.

The fore-arc of SLAA lies above continent from Tobago west and oceanic lithosphere to the east. The accretionary prism at Barbados probably belongs to an inner zone of Paleogene accretion with north to northwest directed motion that continues southwest to collided, onland equivalents in Trinidad and Araya. The more oceanward zone of the Barbados prism is probably of Neogene age of accretion and caused by westerly motion that is related to subduction of Atlantic lithosphere. The onset of present volcanism on the Lesser Antilles may record an eastward jump in the Miocene of the leading edge of the Caribbean plate from the Aves Ridge to the present subduction trace.

SPEYER, S.E., University of Arizona, Tucson, AZ

Background and Episodic Processes in Fossil Assemblage Taphonomy

Comparative taphonomy of fossil assemblages differentiates taphonomic facies (or taphofacies) which, in concert, reflect complex mosaics of intergrading environmental conditions. These facies, however, represent the superimposed effects of background and episodic sedimentary circumstances. Time-accrued effects contrast with those that result from instantaneous sedimentary events, but both inflict a specific and predictable signature on the resultant fossil deposit. Thus, fossil assemblages portray the superimposed effects of processes operating at two diametrically opposed temporal levels; *background* processes span relatively long periods of time and are interrupted by *episodic* processes that are effectively instantaneous.

Background taphonomic phenomena are intrinsically destructive and account for much of the taphonomic bias incurred during assemblage formation. The taphonomic impact of mechanical processes (disarticulation, sorting, reorientation, fragmentation, and abrasion) are, in part, regulated by the rate of sediment deposition. Sedimentation rate directly affects the duration of surface exposure and the intensity of deposit bioturbation (Figure 1). High sediment input conserves shelly remains from ambient taphonomic processes, but also favors deep, intrastratal bioturbation that results in shell scavenging and enhances fluid flow through the sediment. Conversely, low sediment input and "sediment starvation," although exposing shelly remains to persistent mechanical processing, also limit scavenging to surficial remains and provide a buffer against carbonate dissolution.

Sedimentary events occur at a temporal level considered insignificant in the context of geologic time. Nevertheless, such phenomena are critical to the accumulation of strata, particularly in settings otherwise typified by nondeposition or omission. Sedimentary events interrrupt background conditions and deposit a spectrum of taphofacies across a shore to basin transect of environments; amalgamated shoreface deposits of shells (above normal wave base) grade downslope into medial tempestites (between normal and storm wave bases) and eventually, into smothered bottom assemblages (below storm wave base).

Taphonomic processes acting over short, but catastrophic, intervals produce an array of features that, at least in part, mimic background taphonomic effects (Figure 2). Events, however, superimpose high-energy processes on assemblages that fundamentally are the result of background conditions. Taphofacies, therefore, reflect composite phenomena. Depending upon the environmental setting, sedimentary events may compound or destroy background sedimentary fabrics; e.g., increase the proportion of broken shells in a setting characterized by shell breakage or preserve articulated remains where none normally occur, respectively. Likewise, the taphonomic signature of event processes is also modified, compounded, or mediated by background sedimentary conditions (e.g., condensation, bioturbation, current amalgamation).

Hence, the paleontologic record is a composite of background and episodic effects; both phenomenological categories respond to ambient environmental parameters in specific and predictable ways and both are subject to the effects of the other. The result is a composite taphonomic signature which, in itself, provides important clues regarding shell bed formation and environmental proximity. Each component of the signature is particular to a specific environmental setting, but each represents the influence of different depositional conditions and temporal circumstances. To fully understand fossil assemblage preservation, our models must incorporate this inherent duality.

The interaction between background and episodic sedimentary processes is nowhere better illustrated than in early diagenesis (Figure 3). The requisite conditions for diagenesis depend largely upon the background geochemical setting, but require the spatial isolation afforded by rapid, episodic burial for initiation. Background sediment chemistry is regulated by sedimentation rate that directly affects bioturbation and skeletal concentration. Without rapid burial, however, the ambient geochemical conditions would not be isolated nor driven toward diagenesis. Similarly, without the requisite background circumstances, the necessary chemical species would not be present in significant quantities to effect syngenetic mineralization. Clearly, the diagenetic phenomena evident in many strata are the result of an integration of background and episodic circumstances.

Six general taphofacies are recognized and distinguished on the basis of taphonomic properties. These facies are based on representative fossil assemblages that recur throughout the Middle Devonian Hamilton Group (New York State). These taphofacies illustrate the duality of taphonomic phenomena and argue that strictly uniformitarian or event-biased approaches to fossil assemblage analysis, in themselves, generate inadequate taphonomic models unless modified to incorporate the imprint of reciprocal phenomena.

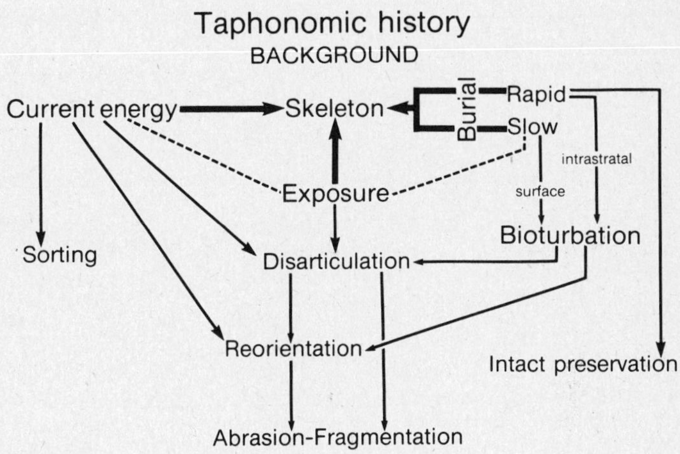

Figure 1.

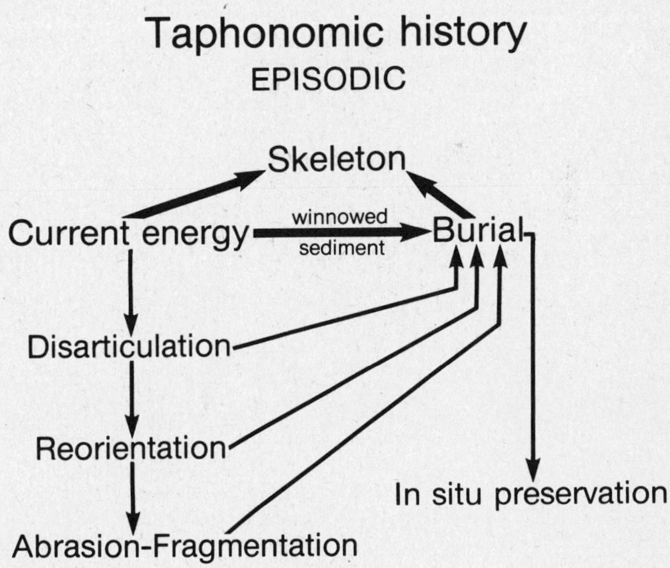

Figure 2.

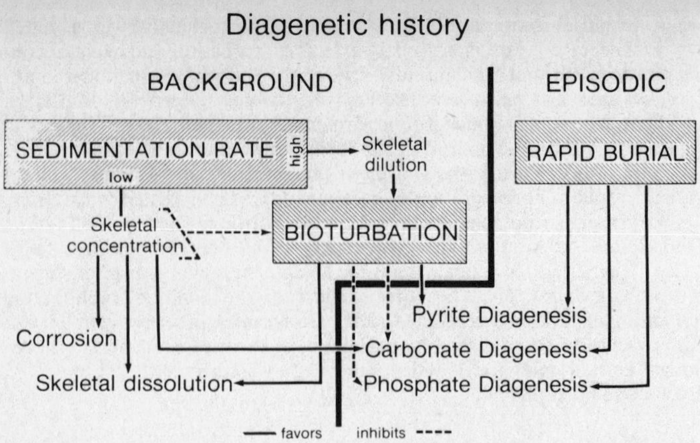

Figure 3

SPIERS, C.J., and C.J. PEACH, University of Utrecht, Utrecht, the Netherlands

Development of Dilatancy and Permeability in Rocks During Creep: Experiments on Rocksalt as Rock Analog

In recent years, it has become widely accepted that there is an intimate association between fluid movement through the deeper crust and localized deformation plus metamorphism. Cataclastic deformation occurring in shear zones under conditions of high fluid pressure is clearly capable of enhancing permeability and facilitating "channelized" fluid flow (Brodie and Rutter, 1985). However, it is not clear to what extent the permeability of rocks can be enhanced during deformation predominated by intracrystalline dislocation mechanisms. Here we report deformation experiments performed on rocksalt. These were designed to investigate (1) the range of conditions under which dilatancy and permeability develop during deformation in the dislocation glide/creep regime, (2) the relationship between induced dilatancy and permeability, and (3) the influence of dilatancy on rheological behavior and deformation mechanisms. The work was performed to generate basic data for use in research on radioactive waste disposal in salt formations. Viewing salt in analog terms, however, the results provide interesting general insight into the likely behavior of other rocks during deformation by dislocation mechanisms.

The experimental program was carried out using cylindrical samples (70 mm diameter by 160 mm in length) of "Speisesalz" rocksalt taken from the Asse Mine, West Germany. These were deformed dry (i.e., without added brine) in a dilatometric triaxial testing machine fitted with an argon gas, transient pulse permeametry system (for description see Spiers et al, 1986; Peach et al, 1987). The tests were carried out in compression at temperatures (T) in the range of 20–200 °C, at confining pressures (P) in the range 0.3–50.0 MPa, and at strain rates ($\dot{\epsilon}$) of 10^{-7} to $10^{-4}s^{-1}$. Testing was performed in constant strain rate and strain rate stepping modes, measuring differential stress (σ, volume change ($\Delta V/V$), and permeability ($\varkappa$) as a function of axial strain. A limited number of experiments were also performed on artificially prepared rocksalt under uniaxial conditions at temperatures up to 400 °C.

The rheological behavior exhibited by the samples was found to be broadly consistent with previous data (Heard, 1972), showing work hardening flow at high differential stresses (σ) and power law creep at lower stresses. In addition, the dilatancy data revealed a transition from nondilatant flow ($\Delta V/V < 0.05\%$) under low σ, high P conditions, to dilatant flow at high differential stresses and low confining pressures (see Figure 1). The volumetric behavior typically observed passing through this transition under conditions of increasing pressure (P) is shown in Figure 2. Note that the maximum dilatation measured in our tests was only $\sim 1.5\%$. However, even these very small amounts of dilatancy were found to lead to dramatic increases in permeability, typically from $\sim 10^{-20}m^2$ in the undeformed or nondilatant condition to 10^{-16} after $\sim 10\%$ deformation in the dilatant field (see Figure 3). Returning to the rheological behavior of the samples, a number of detailed effects were observed. At temperatures below 100 °C, the flow strength was found to be weakly dependent on confining pressure, increasing by 5–10% in the range P = 2 to P = 50 MPa. At higher temperatures, samples appeared to weaken with increasing pressure, the transition from dilatant to nondilatant behavior being accompanied by a transition from work hardening behavior to steady state flow at stresses some 5–10% lower.

Microstructural analysis verified that all samples deformed by dislocation mechanisms, with substantial climb activity (subgrain development) at temperatures > 150–200 °C. The dilatant behavior was found to be associated with widespread inter- and transgranular microcracking, while microcracks were absent or healed in nondilated samples. The "weakening" behavior observed in material deformed at T > 100 °C in the nondilatant field was found to be associated with the onset of fluid-assisted migrational recrystallization (Urai et al, 1986). This process was apparently inhibited by grain boundary microcracking in the dilatant field.

The results reported above have shown that the onset of minor dilatancy (0.1–1.5%) in halite rocks, during macroscopically ductile deformation in the dislocation glide and creep fields, can lead to increases in permeability by up to five orders of magnitude. This implies extensive connectivity of the microcracks developed in the dilatant field. It seems likely that similar effects might occur in other rock materials undergoing deformation by dislocation mechanisms, particularly when the effective pressure is low. On the basis of grain scale strain compatibility considerations, such effects would be most likely to occur in materials with few independent slip systems or in rocks containing dispersed rigid phases. We infer that creep-induced dilatancy might very well provide a mechanism by which high pressure fluid can gain progressive access to otherwise impermeable rock at depth. In the case of salt-based radwaste repositories, numerical modeling indicates that creep-induced dilatancy will occur only in gallery walls and is extremely unlikely to affect containment of the waste (see Spiers et al, 1986).

REFERENCES

Brodie, K.H., and E.H. Rutter, 1985, *in* Advances in Physical Geochemistry, v. 4: Berlin, Springer, p. 138–178.
Heard, H.C., 1972, Geophysical Monograph 16, p. 191–209.
Peach, C.J., C.J. Spiers, A.J. Tankink, and H.J. Zwart, 1987, Vol. EUR 10926 EN, Office for Official Publ. of the Euro. Communities, Luxembourg, p. 91.
Spiers, C.J., J.L. Urai, G.S. Lister, J.N. Boland, and H.J. Zwart, 1986, Vol. EUR 10399 EN, Office for Official Publ. of the Euro. Communities, Luxembourg, p. 131.
Urai, J.L., C.J. Spiers, H.J. Zwart, and G.S. Lister, 1986, Nature, v. 324, p. 554–557.

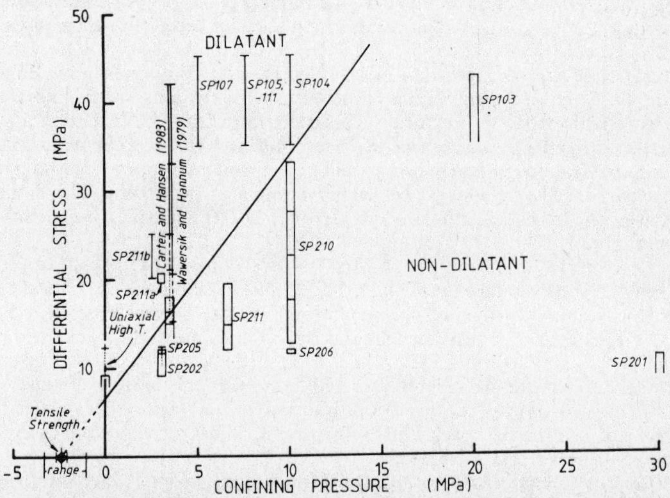

Figure 1—Diagram showing dilatant and nondilatant fields for rocksalt.

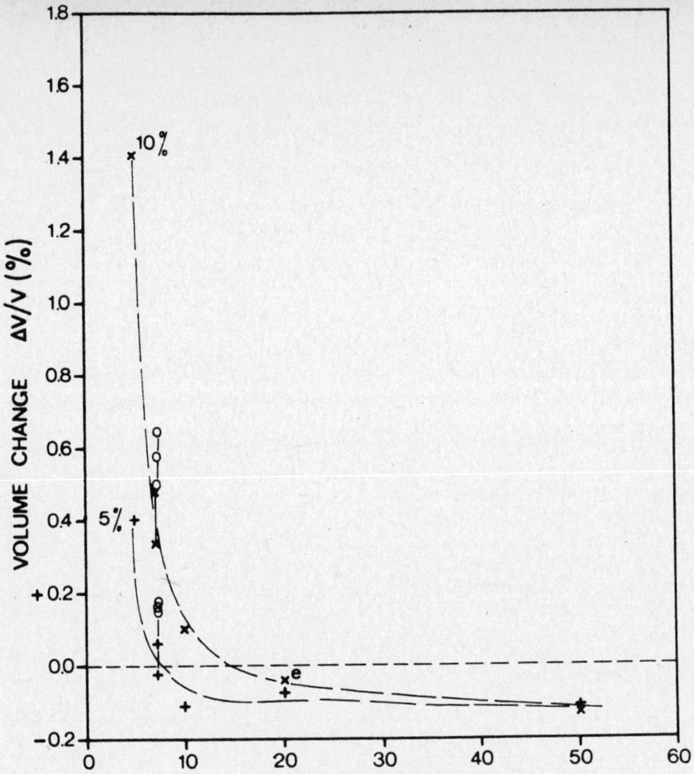

Figure 2—Dependence of sample volume change, measured at 5% and 10% strain, on confining pressure for Speisesalz rocksalt deformed at T = 20 °C, $\dot{\epsilon} \cong 10^{-5}s^{-1}$.

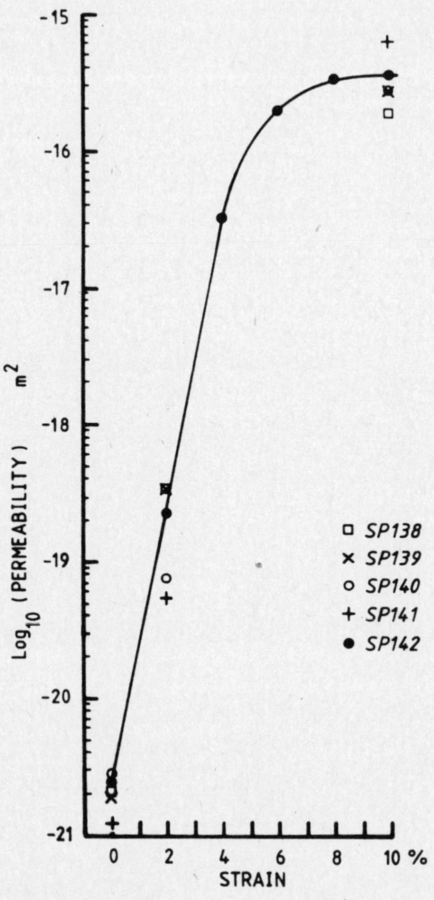

Figure 3—Permeability data for Asse Speisesalz deformed at P = 5 MPa, T = 20 °C, $\dot{\epsilon} \cong 10^{-5}s^{-1}$.

SPIRIDONOV, H. B., Space Research Institute; Sophia, Bulgaria

Ring Structures and Ore Resources

Based on space remote sensing information, dome structures have been defined on the territory of southern Bulgaria. These are mainly metamorphogenic and magmatic geological bodies reactivated during the Tertiary. During their multistage development, various metallogenic formations were intruded along the concentric and radial faults. These formations are important natural resources.

SPRY, P. G., and K. I. MC GOWAN, Iowa State University, Ames, IA

Origin of Archean Lode Gold Mineralization at Atlantic City-South Pass, Wyoming: Fluid Inclusion and Stable Isotope Study

Considerable controversy exists in the literature concerning the source of hydrothermal fluids and ore components involved in the formation of Archean lode gold deposits. Geological, fluid inclusion, and stable isotope studies of Canadian and Australian Archean lode gold deposits have indicated several possible sources of fluid (e.g., Groves and Phillips, 1987; Burrows et al, 1986): (1) a metamorphic source; (2) a magmatic source; (3) a mantle source; and (4) a recirculated or evolved seawater source. Relatively little data are available from Archean lode gold deposits in the United States. To rectify this, a fluid inclusion and stable isotope study was undertaken on gold-quartz veins at Atlantic City-South Pass, Wyoming.

The Atlantic City–South Pass gold deposits occur in a sequence of Archean metasedimentary and metavolcanic rocks (approximately 2.8 Ga) at the southern end of the Wind River Mountains, Wyoming. Ultramafic rocks of the Diamond Springs Formation comprise the basal stratigraphic unit and are overlain by iron formation, quartzites and pelitic schists of the Goldman Meadows Formation. Thin iron formation units also occur in the overlying Roundtop Mountain Formation, which in turn is overlain by metagreywackes, cherts, and graphitic schist of the Miners Delight Formation. The district has been metamorphosed to the upper greenschist-lower amphibolite grade.

Approximately 335,000 oz of gold have been obtained from quartz veins primarily localized in shear zones along contacts of metamphibolites with metagreywacke in the Miners Delight Formation. In many deposits, a particularly close association exists between gold mineralization, metacherts, graphitic schists and tremolite schists. Native gold occurs in quartz–arsenophyrite veins; pyrrhotite, pyrite, calcite, siderite, chalcopyrite and minor amounts of scheelite, tourmaline, and galena are also associated. Surrounding most gold-quartz veins is a narrow (1″–1′) alteration halo composed dominantly of sericite, chlorite, and K-feldspar.

Sulfur isotope analyses obtained on arsenopyrite, pyrite and pyrrhotite range from 0.1 to 3.6 per mil and are consistent with data from most other Archean lode gold deposits. Carbon and oxygen isotope ratios on carbonates associated with gold at Atlantic City-South Pass are unusual (d^{18}O = 5.5 to 23.6 per mil; d^{13}C = -1.5 to -11.0 per mil) in that values of d^{13}C are among the lightest yet recorded for Archean gold deposits. In this respect, it is significant to note that values of d^{13}C from graphitic schists intimately associated with gold veins range from -19.1 to -28.5 per mil. An intriguing part of the stable isotope study is the oxygen and hydrogen isotope data (d^{18}O = -10.7 to 3.5 per mil and dD = -125 to -55 per mil) obtained from fluid inclusions in lode quartz. These data suggest either an evolved seawater or connate water brine or alternately mixing of meteoric water with a magmatic and/or metamorphic component. Compositions of the fluid inclusions, however, make the latter possibility seem less likely.

Four types of primary and pseudosecondary fluid inclusions have been observed: (1) three- and four-phase H_2O-CO_2-NaCl-$CaCl_2$ inclusions; (2) two-phase CO_2-CH_4 inclusions; and (3) one-phase CH_4 or CO_2 gaseous inclusions. Two types of secondary inclusions were also observed: (1) two-phase liquid–vapor aqueous inclusions; and (2) three-phase solid–liquid–vapor aqueous inclusions in which the solid phase is NaCl. Homogenization temperatures for primary inclusions range from 220° to 420 °C whereas those for secondary inclusions range from 52° to 301 °C. Due to teh coexistence of H_2)- and CO_2-rich inclusions and CO_2-H_2O-bearing inclusions with variable CO_2:H_2O ratios, phase separation probably occurred. A salinity of 4 to 10 eq. wt% NaCl was obtained on primary inclusions using clathrate melting temperatures. Secondary inclusions are more saline than primary and pseudosecondary inclusions, but each type has a significant $CaCl_2$ component (initial melt temperatures of ice exhibits a range from -55° to -44 °C). The high $CaCl_2$ content is rare for an Archean gold lode deposit.

The most plausible model for the formation of gold mineralization at Atlantic City–South Pass is for fluids to have been derived by lateral secretion from greywackes in the Miners Delight Formation during compaction. Oxygen and hydrogen isotopes are typical of those associated with a basinal brine. Interaction of the ore-forming fluids with graphitic schists is indicated by the light values of d13C in gold-associated carbonates and by the presence of CH_4 in many fluid inclusions. The high CaCl2 content of fluid inclusions may have formed by the release of Ca to the fluid during conversion of plagioclase in greywacke to K-feldspar. The source of the gold is equivocal, however, it is significant to note that iron formation and graphitic schist contain up to 120 to 49 ppb Au, respectively.

This material is based upon work in part supported by the Bureau of Mines, U.S. Department of the Interior, under Grants G1154119, G1164119, G1174119, as part of the allotment program of the Iowa State Mining and Mineral Resources Research Institute.

REFERENCES

Burrows, D. R., P. C. Wood, and E. T. C. Spooner, 1986, Carbon isotopic evidence for a magmatic origin for Archaean gold-quartz vein ore deposits, Nature, v. 321, p. 851–854.

Groves, D. I., and G. N. Phillips, 1987, The genesis and tectonic control on Archaean gold deposits of the western Australian shield—a metamorphic replacement model, Ore Geology Reviews, v. 2, p. 287–322.

SQUILLACE, P. J., U.S. Geological Survey, Iowa City, IA, and E. M. THURMAN, U. S. Geological Survey, Lawrence, KS

Herbicide Transport in Cedar River, Iowa—Importance of Hydrology and Geochemistry in Nonpoint-Source Contamination

Concentrations of alachlor, atrazine, cyanazide, metolachlor, metribuzin, and dissolved and suspended organic carbon were monitored at six surface-water sites from May 1984 through November 1985 in large subbasins (777 to 15,540 sq km) within the Cedar River drainage basin of Iowa and Minnesota. A computer model separated the Cedar River hydrograph into approximate contributions from ground water and overland flow. The contribution of herbicides from shallow groundwater, including field tile drains and possibly bank storage in the alluvium was normally less than 1 ug/L (micrograms per liter) during the application season. During the remaining part of the year only atrazine was consistently greater than the detection limit (0.1 ug/L) at concentrations of about 0.2 ug/L. Based on the estimated quantity of groundwater contributed to streamflow and the concentrations of atrazine detected during base flow, a conservative estimate of atrazine in the Cedar River transported by groundwater is calculated to be 6 percent of the mass of annual transport. In contrast, overland-flow contributions were 94% of the mass with atrazine concentrations exceeding 16 ug/L during high flow in the spring. During 1984, 70% of the annual transport occurred in June. Virtually all of the herbicides (99%) were in the dissolved phase on the basis of chemical analysis and theoretical considerations of sorption onto suspended sediments. Maximum concentrations of dissolved herbicides were 21 ug/L for alachlor, 16 ug/L for atrazine, 8.7 ug/L for cyanazine, 11 ug/L for metolachlor, and 3 ug/L for metribuzin. Mass-balance calculations for atrazine indicate that 1.5 to 4.0 percent of the herbicides applied during 1 y ultimately were exported from the basin. A significant correlation ($r = 0.86$) was determined between the log of the herbicide concentration and the log of discharge divided by area of drainage for May, June, and July in 1984 and in 1985. Thus, transport of herbicides is critically associated with time of application and the seasonal runoff of dissolved herbicides associated primarily with overland flow.

Additional research has begun in the Cedar River basin that involves 3 months of weekly sampling for nitrate–nitrogen and herbicides at seven surface-water sites and two tile drain sites. Four well nests of three wells each were installed in alluvium in an unfarmed area to determine whether herbicides in the river move into the alluvial aquifers during high river stage and then are released during base flow. The well nests are located from 10 to 80 m from the river and the three wells at each nest range in depth from 3 to 10 m. During base flow, the geographic contribution of herbicides and nitrate to the river will be delineated by water-quality sampling and the determination of discharge at 50 sites throughout the basin.

SQUYRES, S. W., Cornell University, Ithaca, NY

Climatic History of Mars

Mars shows evidence in its geologic record for both long-term and short-term climate change. Evidence for long-term climatic change is provided by small and very ancient water-carved channels that might have required a warmer climate for their formation. Evidence for short-term episodic climatic change is provided by finely layered sedimentary deposits at the martian poles.

The ancient highlands of Mars contain widespread, highly degraded quasi-dendritic valley systems that probably formed by sapping of subsurface water (rather than precipitation). These features may provide evidence for a formerly warmer climate on Mars. Because liquid water could have persisted and flowed for considerable periods under an ice cover, very warm conditions are not required to maintain flow of liquid in the valleys. However, the conditions required to initiate flow may be warmer than those that exist on Mars today. Approximate ages determined from impact crater densities indicate that the epoch in which most of the valley systems were formed persisted for a relatively brief period early in martian history. Other younger and larger fluvial features, the outflow channels, probably could have formed under the present climatic conditions. The geologic evidence suggests, then, that an early warmer climate may have existed for a brief period, but that it was replaced soon after the planet's formation by a cooler one like that at the present. The most likely explanation for a formerly warmer climate, if it existed, is a formerly more dense carbon dioxide atmosphere and its resultant greenhouse warming. An important climatological problem on Mars concerns the fate of this atmospheric carbon dioxide. In the ancient cratered terrain there are many local closed topographic depressions at the confluence of a number of valley systems. These may hvae contained temporary lakes of liquid water, and served as sedimentation basins for the material eroded from the valley systems. Such basins are one possible repository for the carbon dioxide lost from the early atmosphere, in the form of carbonates rocks whose precipitation took place in the liquid water that resided there.

Evidence for more recent episodic climate change is provided by the layered deposits at the martian poles. These deposits apparently are formed of dust cemented by ice. They form when suspended dust that is transported poleward by the winter atmospheric condensation flow is deposited on the surface and becomes intermixed with ice that resides there. The deposits exhibit fine layering that records cyclic fluctuations in martian climate, probably driven by large periodic variations in Mars' obliquity and orbital elements. The dominant periodicities have period of 0.1–1 my. The obliquity and orbital element variations produce periodic changes in high-latitude insolation that substantially alter the capability of the regolith there to contain adsorbed carbon dioxide. The changing adsorptivity may lead to cyclic changes in atmospheric pressure of an order of magnitude, with corresponding variations in temperature, dust-carrying capacity of the atmosphere, and polar deposition rate.

Direct evidence for climatic change on Mars may be preserved in accessible surface materials like ancient water-lain sediments and polar layered deposits. Both sorts of material could be sampled as part of a Mars Rover/Sample Return mission.

SRECEC, I., and E. WOERMANN, Institut für Kristallographie, RWTH, Aachen, Aachen, Federal Republic of Germany

Redetermination of Fayalite Equilibria QFI and QFM by EMF Measurements

Numerous investigations have been performed on fO_2/T relations of the silica–magnetite–fayalite (QFM) and the silica–iron–fayalite (QFI) buffer assemblages. Generally, the literature data show an acceptable agreement.

A reinvestigation of the QFI equilibrium by EMF measurements was performed, employing yttria-stabilized zirconia as solid-state electrolyte and the iron-wüstite buffer pair as reference electrode, relying on the general observation that an EMF signal as a rule is instantaneous and reproducible. The voltage was recorded after thermal equilibrium of the cell was attained and an apparently steady signal was obtained. The temperature was then changed by a certain increment. The results were in satisfactory agreement with former results from our own laboratory (Figure 1) and with literature data.

Interrupting the process for an extended period of time (2–24 hr) at a certain temperature level, however, resulted in a sluggish drift of the EMF signal. Furthermore, it turned out that the direction and magnitude of the drift were reproducible. The arrows in Figure 1 indicate the direction of the drift observed in an EMF experiment performed on a QFI-assemblage that had been equilibrated before at 1273 K for 3 days. Keeping the temperature constant until a steady EMF signal was obtained resulted in a "steady-state" curve in the E/T diagram (Figure 2), the slope of which deviates considerably from the corresponding "fast" curve in Figure 1.

Starting again from assemblages equilibrated for several days at given temperatures between 1173 and 1323 K, several series of cooling and heating cycles were performed, changing the cooling and heating rates from 5.0 to 0.1 $K.min^{-1}$. From Figure 3 it can be concluded that only cooling rates on the order of < 0.1 $K.min^{-1}$ lead to results approximately comparable to the "steady state" data in Figure 2.

A voltage drift due to electronic conduction in the zirconia electrolyte has been observed in the temperature range > 1340 K. Our experiments were thus strictly confined to a temperature range below this limit. Comparable runs wtih an IW | YSZ | fO_2 (air) cell under similar conditions maintained steady voltage readings for many hours or even days. It is thus concluded that the sluggish drift in EMF in the IW | YSZ | QFI cell is a property of the silica–iron–fayalite assemblage. A phase transformation in silica (in these experiments in the form of cristobalite) has not been observed. Metallic iron is considered to be an essentially stoichiometric phase. Fayalite, on the other hand, is a phase characterized by temperature dependent structural defects (Nakamura and Schmalzried, 1983; Simons, 1986, and this symposium).

It is assumed that the "steady-state" voltage of our EMF readings corresponds to the equilibrium state of the QFI assemblage. Any change in temperature results in a corresponding change in the equilibrium defect concentrations of the fayalite, mainly in its O/(Si + Fe) ratio. The adjustment of the fayalite stoichiometry appears to be a sluggish process with a temperature dependent rate.

The data from Figure 1 and Figure 2 are reproduced in a recalculated from in a $\log fO_2/(1/T)$ diagram in Figure 4.

Another series of experiments was conducted on IW | YSZ | QFM cells in order to investigate the properties of the QFM assemblage with emphasis on its time-dependent adjustment of the EMF. A comparable shift in voltage has been recorded. A "steady state," however, is rather difficult to define, due to a considerably greater sluggishness of this system, which is explained by the fact that here not only the O/Si + Fe) ratio but also the Si/(Si + Fe) ratio require an adjustment (Simons, 1986).

Time-dependent equilibrium adjustments, demonstrated here for the QFI and QFM assemblages, amy also be expected for all other systems with temperature-dependent defect concentrations and also for systems with phases characterized by temperature dependent order-disorder processes.

REFERENCES

Nakamura, A., and H. Schmalzried, 1983, On the nonstoichiometry and point defects of olivine: Phys. Chem. Min., v. 10, p. 27–37.

Simons, B., 1986, Temperatur- und Druckabhhangigkeit der Fehlstellenkonzentration der olivine und Magnesiowüstite: Habilitationsschrift Universität Kiel, 175 p.

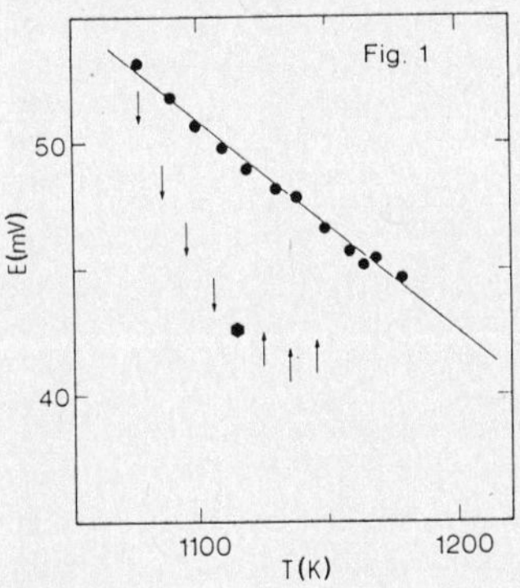

Figure 1.

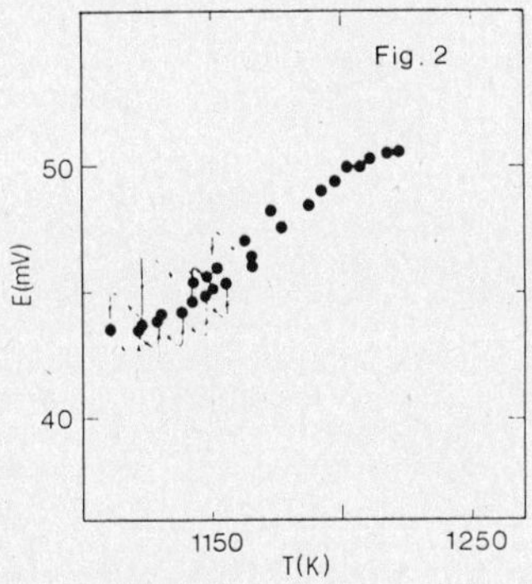

Figure 2.

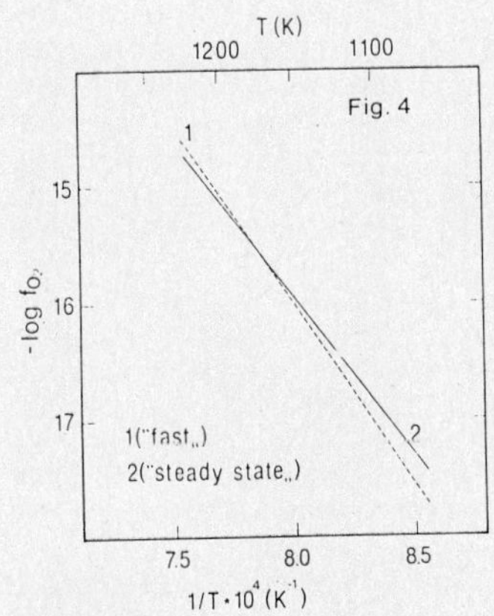

Figure 3.

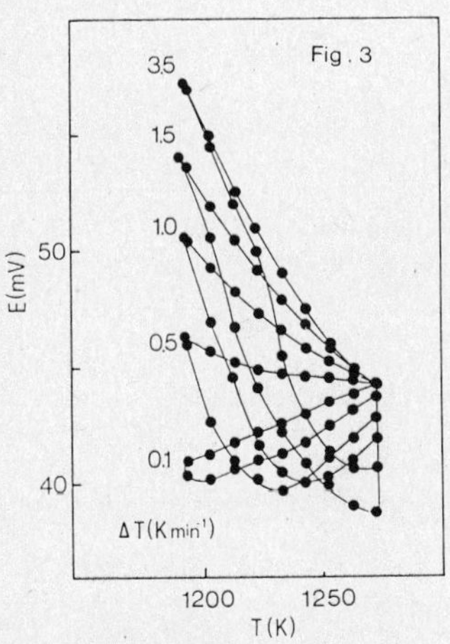

Figure 4.

SRINIVASAN, R., National Geophysical Research Institute, Hyderabad, India

Platform, Shelf and Mobile Zone Associations from the 3.3 Ga to 2.6 Ga Old Archaean Dharwar Sequence of South India

Archaean Dharwar Supracrustal rocks 3.3 Ga to 2.6 Ga old in south India document geological processes that operated in the time that lapsed between the older (>3.4 Ga) and younger (2.7 to 2.6 Ga) greenstone belt cycles in some other parts of the world.

By about 3.3 Ga, a tonalitic "continental" crust had widely developed in the Dharwar province. This sialic crust induced stability, albeit temporarily, into the sedimentary tectonic environment early in the Archaean.

Poorly differentiated psammopelites and dirty quartzites accumulated in the interior part of the tonalitic crustal nucleii, where chromite and titaniferous magnetite-bearing layered igneous complexes are emplaced in them. Braided fluvial plains on the peneplaned surfaces of these platforms were sites of accumulation of mature to supermature quartz pebble conglomerates and quartzites, some of which carry detrital pyrite, uraninite, and gold. The clean platformal quartzites are interbedded with subaerial tholeiites, low silica anesites, lahars, and rhyolitic volcaniclastic rocks, indicating intermittent rifting of the platform. Seaward, quartzite-volcanic rock association makes way for the banded iron formations, which have interlayers of volcanic tuffs. They show good strike persistence. The primary iron formations appear to have been carbonate rich, but their metamorphic derivatives as seen now are a rich repository of magnetite iron ore.

Shallow marine shelf sedimentation is represented by typical quartz arenite-pelite-carbonate-banded manganese and iron formation assemblage. They accumulated in intertidal and subtidal environments. The carbonate members, among other forms, carry morphologically diverse columnar stromatolites. Manganese is concentrated in carbonate rocks and in pelites close to the contact with carbonate rocks. Iron formations show very strong strike persistence and are associated with mafic sills and tuffs. The metapelites of shelf assemblage at places carry a rich concentration of graphite; the δ^{13}C ratio of the carbon in them varying from −21‰ to −28‰ PDB indicates the organic nature of the carbon.

The platform and shelf sequences are succeeded by trough association of 2.7–2.6 Ga mafic to felsic submarine volcanic rocks, polymictic conglomerates, greywackes, and iron formations. Volcanic rocks are of arc type, the polymictic conglomerates and greywackes turbidites of the canyon-fan environment and the iron formations typically impersistant, together indicating termination of stable conditions during trough sedimentation. A mixed continental and arc provenance is indicated by greywackes, and negative europium anomalies in some of them point to the separation of granodioritic upper crust in their provenance. The iron formations are nonoxide type, and some of the sulfidic iron formations carrying interlayers of graphitized carbon (δ^{13}C = −35‰ PDB in one of them) are found to be auriferous.

Stratigraphic relations in the Dharwar sequence show that sedimentation of platformal, shelf, and mobile basin associations occurred in that order. In plate tectonic parlance, the entire sequence can be visualized as having accumulated in a gradually evolving back-arc basin. It is of interest to note that the rock associations and succession of sedimentary tectonic settings, as represented in the Dharwar supracrustal sequence, are similar to those observed in some Proterozoic terranes.

SRIVASTAVA, H.N., Meteorological Department, New Delhi, India

Source Mechanism of Recent Indian Earthquakes and Meizoseismal Areas

During the last 10 years, earthquakes that caused significant damage to property in the Indian region have occurred in northeast India (Manipur-Burma border, 6 August 1988, M = 7.0; Cachar, 31 December 1984, M_L = 5.4) as well as northwest India (Dharamsala, 14 June 1978, M_L = 5.4) Himachal Pradesh (Dharamsala, 26 April 1986, M_L = 5.7) and the India-Nepal border 29 July 1980. A few earthquakes of magnitudes between 4.5 and 5.0 also caused damage to property in Peninsular India, notably Hyderabad (1983), Bangalore (1984), Bhatsa (Khardi 1983), Valsad (1985), Dhenkanal (1985), and Idukki (1988). Of these, the seismic activity around Bhatsa and Valsad regions is still continuing at microearthquake level.

Study of source mechanisms of earthquakes from P wave first motion data in relation to the orientation of the meizoseismal areas in different tectonic regimes has brought to light the following interesting results:

1. In the region of great Kangra earthquake of 1905 (M = 8.5), the meizoseismal area was oriented along one of the northerly nodal planes for the earthquake of 14 June 1978 (M = 5.4) close to the tear faults (Srivastava et al, 1987) while in that for the recent earthquake of 26 April 1986, the meizoseismal area developed parallel to the northwesterly oriented main boundary faults along which the nodal planes were also aligned. This was also supported by the triggering of the strong motion instruments for the

later earthquake deployed under the Indo-U.S. Array program. The nature of faulting also differed from strike-slip (thrust component) of 1978 shock to normal for 1986 earthquake (Figure 1).

2. In the India–Nepal border region, for the earthquake of 1980 meizoseismal area and the isoseismals were oriented along the transverse lineaments while the nodal planes showed their orientation along the northwesterly main central thrust.

3. In northeast India, while the source mechanism of the recent earthquake of 6 August 1988 near the Manipur–Burma border region is being studied, the isoseismal pattern and one of the nodal planes were oriented along the north-northeasterly direction for the Cachar earthquake of 1984.

4. In Peninsular India, the north-northeasterly oriented isoseismal patterns for the earthquakes near Valsad (1986) and Dhenkanal (1986) were associated with the tear faults while the seismic activity for the Bhatsa earthquake was attributed to northwesterly and north- northeasterly lineaments. One of the nodal planes supported its association with northwesterly lineaments.

In view of the complexities of the development of meizoseismal areas near the plate boundary of the Indian and Eurasian plate, these observations need be considered while modeling earthquake source processes in the Himalayan region.

REFERENCES

Srivastava, H.N., R.K. Dube and H. Raj, 1987, Spatio temporal variations of seismicity preceding two earthquakes in Kangra region: Tectonophysics, v. 138, p. 69.

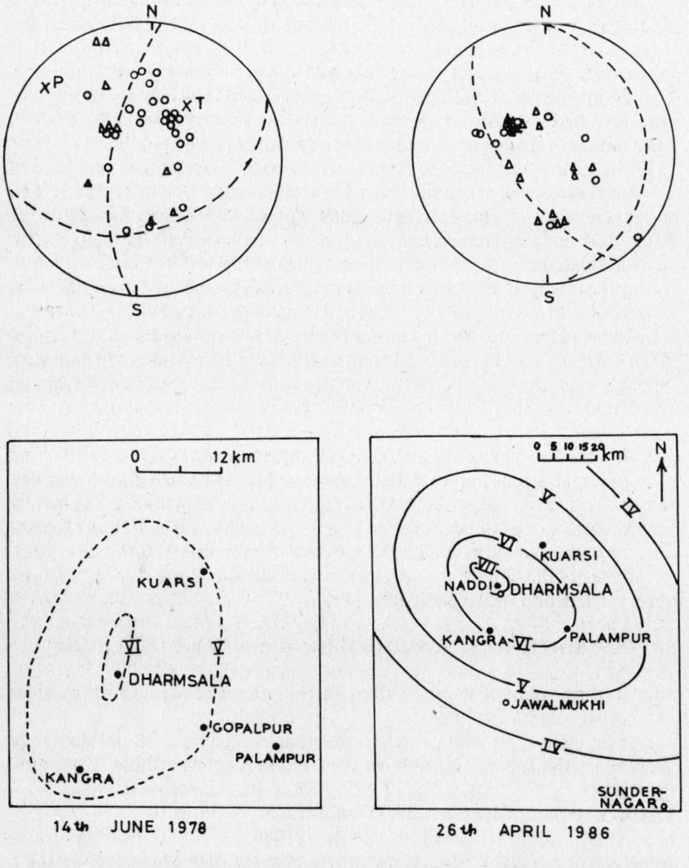

Figure 1—Source mechanism and isoseismals of earthquakes of 1978 and 1986 in Himachal Pradesh (near Dharamsala). The triangles and circles denote the dilatation and compressions, respectively.

SRIVASTAVA, H.N., India Meteorological Department, New Delhi, India

Precursory Seismic Gaps in Himachal Pradesh and Northeast India

Identification of seismic gaps in seismically active regions is a promising method that has led to successful prediction of some earthquakes. In the Himachal Pradesh region, a seismic gap bounded by latitudes 32–33 °N, longitudes 76–77 °E has been identified by Srivastava and Chaudhury (1978) (based on the data during the years 1965–1970 and 1971–1974) through a close network of seismological stations equipped with sensitive Hagiwara electromagnetic seismographs. Located close to the main boundary fault and lying between the great Kangra (1985) and Kinnaur (1905) earthquakes, the region has been inferred to be affected by an earthquake of magnitude 7.5 in the near future. This gap continued to be well marked on the latest microseismicity maps for the period 1981–1984 and 1984–1987. Since there is no significant increase in seismic activity during the last 20 years inside this gap, the region still remains quiescent and calls for seismic tomography to distinguish it from the neighboring seismically active zones where the great Kangra (1905, M = 8.5) and Kinnaur (1975, M = 7.5) earthquakes occurred. In northeastern India, the microseismicity map for the years 1981–1984 based on the seismological stations run by India Meteorological Department and the National Geophysical Research Institute/Regional Research Laboratory, Jorhat, under the Council of Scientific and Industrial Research (Dube et al, 1986) has enabled us to identify five new seismic gaps as follows:1. lat. 25.2–26.2 °N, long 92–93 °E
2. lat. 25.5–26.5 °N, long 93–94 °E
3. lat. 26.5–27.5 °N, long 93–94 °E
4. lat. 26.5–27.5 °N, long 94–95 °E
5. lat. 27.5–28.8 °N, long 94–95 °E
Keeping in view the size of these gaps, the maximum magnitude of the earthquakes may be estimated as 7–7.5 only. Also in addition to Shillong plateau, Kopali lineaments, Manipur Burma thrusts, and the main boundary faults that are known to be seismically active, another lineament along 94 °E in Subansiri region has been identified where there is a marked concentration of seismic activity. The above gaps/zones thus delineated bring to light new hazardous zones in northeast India in addition to those identified so far.

REFERENCES

Dube, R.K., R.S. Dattatrayam, M. Singh, and H.N. Srivastava, 1986, Seismicity of northeast India with reference to the Cachar earthquake of December 1984: Proceedings of International Symposium on Neotectonics in South Asia, Dehradu, p. 379.

Srivastava, H.N., and H.M. Chaudhury, 1979, Precursory seismic observations in Himachal Pradesh and Shillong plateau: Mausam, v. 30, p. 289.

SRIVASTAVA, L.S., University of Roorkee, Roorkee, India

Hill Slope Stability along Tehri Dam and Koteshwar Dam Reservoir Rims

Stability of hill slopes forming the reservoir rims of Tehri Dam and Koteshwar Dam across Bhagirathi river, from the present river bed level to the hill top, has been investigated. The Tehri Dam project is under construction, and investigations for Koteshwar Dam project located 22 km down stream are in progress. The hill slopes are predominantly composed of phyllites and slates and in some sections consist of quartzites, siltstone, and dolomites. The region shows rugged topography. The rock mass is generally covered with thin cover of surficial material consisting of terrace deposits, alluvial fans and cones in the lower reaches, and rock exposures along slopes at higher elevations. Terraces have been developed at various locations for cultivation in areas with fractured and weathered rock mass and surficial deposits and soil cover overlying in situ rocks. Terrace deposits along the bank of the Bhagirathi river and its tributaries, which occur below 1000 m a.m.s.l.

The in situ rock mass shows well-developed joints, cleavage, and foliations. The spacing and condition of such discontinuity surfaces and their dip and strike in relation to hill slope were mapped. Analysis of the data shows that major rock units that could separate and move out resulting in block gliding or wedge failure of the rock mass are not feasible. However, separation of rock wedge along steep road cuttings limited to a few meters in dimension could be identified. The predominant mode of failure of hill slopes is noted to be along failure surface guided by randomly oriented joint surfaces and well-developed cleavages. Parametric studies to identify failure surfaces controlled by such discontinuities with minimum factor of safety show relatively flat circular failure surfaces.

Stability analysis of 55 slope sections around Tehri Dam reservoir rim and 65 sections along Koteshwar Dam reservoir rim has been carried out, in which the stabilizing effect of water load that would be created after the filling of the reservoir has been neglected, and thus provides a conservative estimate of stability under static conditions. Factors of safety computed for shear strength parameter of rock mass, evaluated from rock mass characteristics, results of laboratory testing, and back analysis of existing slopes, indicate stable conditions in the prevailing normal static loads and groundwater conditions.

Tehri and Koteshwar dam project sites lie in seismic Himalayan terrain and displacement analysis of the various slope sections was carried out for design earthquake. The results of the analysis indicate that in the Tehri Dam reservoir rim hillslopes of height 750–1050 m from the present river bed level, with slope angles of 31 ° and 29 °, respectively, with lowest computed factor of safety 1.3 under static conditions, will undergo displacement of 8.7 cm on exposure to maximum credible earthquake with effective peak acceleration of 0.25 g (Srivastava and Jain, 1986). The displacement analysis of hill slopes forming Koteshwar Dam show yield acceleration larger than design earthquake with peak ground particle acceleration 0.12 g. Results of displacement analysis indicate that the rock mass will not undergo insignificant movement on exposure to the maximum credible earthquake in the region.

Surficial materials overlying rock strata in general are noted to be stable as the average slope angle above full reservoir level is less than 10 ° along the Tehri Dam reservoir rim and less than 20 ° in the Koteshwar Dam reservoir rim, as the angle of coefficient of friction of these unconsolidated sediments is about 25 °. However, local failure of these rocks is noted to occur with slope angles greater than 25 °. Such local slopes have been formed due to erosion, excavation for road cuttings, and development of terraces for cultivation. Such parts of alluvial fan/cones in surficial cover with slope angle of more than 25 ° above full reservoir level on failure could laterally spread to a maximum distance of three to four times their height to lie at their angle of repose. Such slopes will gradually attain stability and will not pose any hazard in terms of generating water waves in the reservoir.

Analysis of hill slopes do not indicate large slope failure along the reservoir rim in the main Bhagirathi Valley and side valleys along its tributaries, which could result in sudden collapse of material or debris flow into the reservoir.

REFERENCES

Srivastava, L.S., and S.K. Jain, 1986, Earthquake displacement analysis of hillsides forming rim of Tehri Dam Reservoir: Proceedings of 8th Symposium on Earthquake Engineering, University of Roorkee, Roorkee, India, December 29–31, 1986, v. 1, p. 567–574.

SRIVASTAVA, L.S., University of Roorkee, Roorkee, India

Uncertainties in Evaluation of Seismic Hazard at Engineering Project Sites

Earthquake hazard at a project site expresses the probability of occurrence of strong ground motion within a given time interval (service life) and other accompanying events such as surface faulting, tectonic deformation, ground failure, inundation, and fire, which may cause damage to vulnerable weak, poorly designed, and unsatisfactory and faulty constructions and or injury and loss of life. The hazard is often defined by the seismic source parameters of the maximum credible earthquake (MCE) or the amplitude of strong ground motion attenuated from the source to the specific project site. For engineering applications, the hazard is expressed in terms of parameters that can be utilized to work out response spectra to evaluate seismic coefficients (spectral acceleration) for design of structures of different natural periods of vibrations and damping, or time history of ground motion for dynamic analysis.

Data on earthquake occurrence in many parts of the world, especially in underdeveloped and developing countries, is not available with desired precision. Exact location of events before A.D. 1900 are not known, and available information pertain to places affected by the major events and their effects on structures existing during those times. With the gradual deployment of seismic instrumentation, the seismic sources have been identified. However, location errors were appreciably large in areas where only a few seismological laboratories existed. More accurate locations of events are available since A.D. 1960 and local seismic station networks provide more accurate information. Seismic monitoring of major project sites (e.g., major dams, nuclear power stations, etc.) is therefore undertaken along with feasibility investigation depending on available resources that provide data on contemporary seismicity. Extrapolation of such information consisting mostly of microearthquakes over a relatively short time duration, to evaluate arrival process of MCE and work out the associations with known faults and tectonic features is questioned, as the mechanism and rate of strain accumulation resulting in major earthquakes is not fully understood.

The pattern of occurrence of major earthquakes all over the world has helped in identifying the belts where tectonic deformations and ruptures are taking place in the present geotectonic regime, e.g., lithospheric plate boundaries on the global scale, or mobile crustal block boundaries on a regional scale. Identification and demarcation of the seismic sources and spatial and temporal variation of earthquake activity in such sources around a specific project site are required. This involves preparation of maps on local scale (e.g., 1:250,000) demarcating faults and other tectonic features in the close environs of the project site and interpretation of available information on earthquake occurrence and fault plane solutions to distinguish capable faults or active zones as seismic sources and sort the data to determine frequency magnitude distribution of events in such sources. The objective of the study is to define the seismic sources and assign their maximum seismic potential, and a deterministic approach is generally adopted to prescribe the parameters of MCE at the seismic source for evaluation of seismic hazard at the project site. The methodology consists of establishing the seismic source, characterizing its maximum seismic potential and an appropriate attenuation law keeping in view the geological characteristics of the region, and estimate the intensity of ground motion attenuated from the source to the project site. The uncertainties due to inadequate data and lack of knowledge of operative geotectonic processes in the region are neglected. Often faults with inferred associations with deep-seated tectonic features or plate boundaries do not show evidences of movements during historical times or association with major earthquake occurrence. Inferences in such situations are made from data on earthquake occurrence as probably seismic activity along known faults and tectonic features remain unresolved. Location errors and lack of knowledge about existence of tectonic lineaments associated with seismic phenomena in the area in such a synthesis often indicate floating events unrelated to known faults and fractures in the earth's crust and upper mantle.

In the absence of instrumental data on earthquakes, estimates of MCE are made from maximum intensity observed and recorded during past earthquakes. Strong ground motion data recorded at various parts of the world close (say, within 5–10 km) to the epicenter during earthquakes that have caused damage show comparable peak ground particle acceleration, with increase in duration of strong ground motion with magnitude. Thus, seismic hazard close to the seismic source is appreciably large. Ground failures and significant amplification of ground motion due to local conditions prevailing in the area have been observed. Thus, identification and delineation of seismic sources in the immediate vicinity, as well as larger distances from the project site, becomes a crucial issue for evaluation of seismic hazard. In the absence of precise data on earthquake occurrence, the location of seismic sources based on geological surveys and subsurface explorations is an exceedingly difficult engineering geology problem. The problem is of significance for countries where geoseismological investigations have not been carried out and earthquake hazards have to be estimated from available information. Most of the regulatory agencies prescribe or put greater emphasis on deterministic approach for assessment of earthquake hazards, as it is considered that available information does not permit probabilistic assessment. It is also believed that the approach provides a conservative evaluation of the hazard, which may not be true. In the present state of knowledge in many situations, it is not possible to identify and demarcate seismic sources and predict where earthquakes will occur and how large they will be. Efforts are therefore made to describe the hazard in terms of probabilities by analyzing the uncertainties with clearly stated assumptions and expressing the hazard in terms of likelihood of occurrence of specified ground motion or associated event (e.g., ground breaks, tsunamis, etc.) and its exceedance probability or average return period with respect to service life of the structures. In general, the earthquake occurrence in such analysis is considered a stationary process and provides a relative earthquake hazard estimate, even though the absolute values obtained may be imprecise because of assumptions in the analysis. An engineering judgment keeping in view the uncertainties in the results of deterministic and probabilistic analysis has to be made in prescribing the seismic hazards at engineering project sites.

SRIVASTAVA, L.S., and S. BASU, University of Roorkee, Roorkee, India

Seismic Zoning of India

India can be divided into three major physiographical, geological, and tectonic regions with varying seismic potential. The northern and northeastern mountainous terrains forming part of Alpide–Himalayan belt show high seismic potential, whereas Indi-Gangetic alluvial plains along Himalayan fore-deep and sedimentary basins along marginal depressions, and the Peninsular shield region, show lower seismic potential. Various attempts have been made to schemitize earthquake hazards and prepare seismic zoning maps of India. These maps, showing three to four seismic zones indicating regions with frequent, occasional, or negligible earthquake occurrence, or areas liable to severe, moderate, or slight damage, served limited purpose.

In order to prescribe quantitative values of the seismic hazard in terms of design (response) acceleration to provide adequate earthquake resistance in structures, the Indian standard recommendations for earthquake-resistant design of structures formulated in 1962 (IS: 1893–1962) incorporated a seismic zoning map (SZM). The 1962 SZM delineated Seismic Zones I, II, III, IV, V, and VI corresponding to modified Mercalli intensity (MMI) V, VI, VII, VIII, IX, and X isoseismals, respectively, which were demarcated by enveloping curves from plots of isoseismals of earthquakes with Richter magnitude equal to and greater than 5 for which records were available, with minor modifications for taking into account geological structures and tectonic belts considered as seismogenic features. The region with MMI "less than V" was designated as Seismic Zone 0. Two revisions of the 1962 SZM were made in 1966 and 1970 (Srivastava, 1969, 1974) as more data on earthquake occurrence became available and to give greater recognition to geologic features (faults) and tectonic set-up for demarcation of seismotectonic units of different seismic hazard potential. The 1970 SZM divided the country in five seismic zones as effects below MMI V were considered insignificant in terms of design to call for a separation and areas of MMI "IX and above" show comparable effective peak ground particle acceleration, though with longer duration with increase in magnitude. Modification of 1970 SZM, which continues to be used in the present Indian Standard Criteria for earthquake-resistant design of structures (IS: 1893–1984, Fourth Revision), have been suggested by Basu and Srivastava (1980) and Srivastava and Basu (1982). The 1970 SZM shows zones with many islands. The pattern of earthquake occurrence does not agree with the delineated zones, and the ground motion corresponding to prescribed design (response) accelerations for various zones does not show consistent exceedance probability for expected service life (say, 100 years) of structure or recurrence interval of major earthquakes.

A synthesis of seismic history and operative geological and tectonic processes is therefore required in demarcation of seismic zones, so that for practical purposes the potential for earthquake occurrence and consequent earthquake design requirements may be taken to be approximately uniform within each seismic zone. The predicted design ground motion in seismotectonic provinces with different seismic potential to a large extent depends on the magnitude of the maximum credible earthquake, which in continental regions and along plate boundaries vary from Richter's (local) magnitude 6½ to 7½, and the effective peak ground acceleration estimate is governed by the distance of the seismic source in deterministic approach and the arrival process of the earthquakes in probabilistic analysis. As location of seismic source is more or less a subjective judgment, a restricted probabilistic approach is often adopted in evaluating the relevant earthquake parameters for zoning. Such a methodology provides average seismic hazard level irrespective of the cumulative effect of earthquake occurrence. Seismic hazard evaluation was carried out to account for varied frequency of earthquake occurrence and MCE magnitude potential in different regions. Seismic hazard in terms of effective peak ground acceleration (and corresponding maximum spectral acceleration) were estimated for 100 years' service life, and smoothened contours of unified hazard were drawn. These unified contours do not support delineation of five zones with uniform ratio of seismic hazard among various zones. Four seismic zones have been delineated expressing gross areal hazard, and weighted averages of design (response) acceleration to be specified for each zone have been evaluated.

REFERENCES

Basu, S., and L.S. Srivastava, 1980, Seismic design zoning map of India: Proceedings of Symposium on Earthquake Disaster Mitigation, University of Roorkee, 4–6 March 1980, Sarita Prakashan, Meerut, India, v. 1, p. 99–109.

Indian Standard Recommendations for Earthquake Resistant Design of Structures, IS, 1893–1962, Indian Standards Institution, New Delhi, 1962.

Indian Standard Criteria for Earthquake Resistant Design of Structures, IS, 1893–1984, Fourth Revision, Indian Standards Institution (Bureau of Indian Standards), New Delhi, 1984.

Srivastava, L.S., 1969, A note on the seismic zoning map of India: Bulletin of Indian Society of Earthquake Technology, v. 6, n. 4, p. 185–194.

———, 1974, Seismic Zoning Map of India: Earthquake Engineering—Jai Krishna Sixtieth Birth Anniversary Commemoration Volume, Sarita Prakashan, Meerut, India, Chapter 2, p. 49–65.

———, and S. Basu, 1982, Seismic zoning of India: Proceedings of 4th International Congress on Engineering Geology, IAEG, New Delhi, v. VIII, p. 119–128.

SRIVASTAVA, R.M., FSS International, Vancouver, Canada

Why Use Variograms?

This paper examines spatial continuity measures in the light of the work of several authors who have recently questioned the traditional reliance on the variogram. By drawing together their important practical results and theoretical insights, a case is made for a more flexible approach to the description, modeling, and use of spatial continuity.

For most applications, if clearly interpretable variograms cannot be produced from the available sample data, there are alternatives that may produce more meaningful results and whose use in estimation procedures is not precluded by theory. There are some applications, however, particularly those that involve indicator variables, in which the use of the variogram is not only theoretically preferable, but also practically advantageous.

A complete case study, from initial spatial continuity analysis to final estimates, is performed on a real data set for which hindsight validation is possible. The results show that nontraditional measures of spatial continuity can produce better results than the variogram, not only in terms of the estimates themselves, but also in terms of the assessment of their reliability.

SRIVASTAVA, S.P., and W.R. ROEST, Geological Survey of Canada, Bedford Institute of Oceanography, Dartmouth, Canada

Evolution of North Atlantic: Case of Jumping Plate Boundaries and Microplates

The evolution of the North Atlantic is described using the seafloor spreading magnetic anomalies and the fracture zones from the North Atlantic Ocean, the Labrador Sea, the Norwegian–Greenland Sea, and the Eurasian Basin. A series of poles of rotation determined from these data show that during the entire evolution of the North Atlantic, several of the plate boundaries have shifted from region to region, giving rise to several microplates. The idea that the Lomonosov Ridge has moved as part of the North American plate is evaluated, and it is shown that serious problems arise if other possibilities are considered. It is shown that Iberia has moved with Africa since the beginning of seafloor spreading until anomaly 10 time when it started to move with Eurasia. During this period, the plate boundary between Eurasia and Iberia moved successively to different positions giving rise to a complex magnetic pattern in this part of the North Atlantic. Detailed magnetic measurements carried out off the Grand Banks of Canada have been used to decipher such shifts in this plate boundary. The Porcupine plate, a portion of the Eurasian plate north of the Bay of Biscay and east of anomaly 21, including the shelves of British Isles, got deformed during Eocene. This deformation is observed in the records of the sedimentary basins underlying this region and in the oceanic floor.

The evolution of the North Atlantic is shown to be the result of propagation of spreading from south to north, starting between Iberia and Grand Banks in Early Cretaceous. Following this separation of Iberia from the Grand Banks and from Eurasia, spreading first started in the Newfoundland Basin and the Bay of Biscay and then propagated north, separating Porcupine Bank from Orphan Knoll and Greenland from North America. Greenland continued to move independently of Eurasia and North America until anomaly 13 time when spreading ceased in the Labrador Sea. Since then Greenland has been moving with North America. Spreading commenced in the Norwegian–Greenland Seas and the Eurasian Basin at anomaly 24 time and is continuing. Prior to active seafloor spreading between Greenland and Eurasia, separation was accommodated by crustal extension in the Rockall Trough and in the regions to the north, like the More and Voring Basins. The Rockall Plateau moved as part of the Greenland plate during this time and as an independent plate from anomaly 24 to 21 time. Since then, it has been moving as part of the Eurasian plate.

The evolution of the Norwegian–Greenland Seas has been complex. Several jumps of the ridge axis in these regions left an extinct ridge in the Norwegian Sea and detached the Jan Mayen Island, a continental fragment, from east Greenland. The spreading between the North American and Eurasian plates, from the Azores–Gibralter Fracture Zone to the Siberian Platform, has been continuing since anomaly 7 time.

STACKEBRANDT, W., and H. J. FRANZKE, Central Institute for Physics of the Earth, Potsdam, German Democratic Republic

Alpidic Reactivitation of Consolidated Lithosphere in Central Europe

The Central European foreland tectonics is mainly influenced by Mesozoic–Cenozoic stress fields that depend on structural changes in the Alpine and Carpathian tectogens. A phase of crustal thinning starts in Upper Triassic time and lasts until the Middle Cretaceous producing flexures, normal faults, and graben structures in the sphere of older (Hercynian) blocks margins. Later compressive movements since Upper Cretaceous have overprinted these formerly extensive activated faults in partly compressive types.

The northwest–southeast-trending fault zones were influenced by movements of the Bohemian Massif. The compressional stress fields developing from temporal alternative collisions in Eastern Alps and Western Carpathians gave rise to changes in the senses of motions of northwest–southeast and north-northeast–south-southwest trending fault zones in the northern foreland. A few basin structures are genetically connected with the Alpidic reactivation of the Central Europe fracture zones.

STALIN, B. A., CEPE, Guayaquil, Ecuador

Landslides: Extent and Economic Significance in Ecuador

Landslides are prevalent on the western slopes of the Andes where most of Ecuador's 10 million people live. A landslide near Chunchi in 1983 blocked the Pan American Highway, buried vehicles, and killed at least 150 people. An even larger landslide in the same area in 1985 blocked the highway and a railroad and killed four people. In 1987, two large landslides near Cochancay buried five buses and killed dozens of people. The same year, landslides at the entrance to the Nambija gold mine killed 80 people.

The most disastrous landslides in Ecuador's history occurred on 5 March 1987, when earthquake-triggered landslides near Reventador killed more than a thousand people and severed 33 km of oil pipelines. The damage to the pipeline delayed oil exports for almost one year, reducing government income by nearly 70%.

Landslide fatalities in Ecuador have been increasing, from 10 deaths in the 1950s to 290 deaths in the 1980s, excluding the earthquake-triggered landslides in the Reventador area. Damage has been mainly to highways, railroads, vehicles, homes, irrigation channels, cattle, and cultivated fields.

STALLARD, R. F., U.S. Geological Survey, Lakewood, CO

Weathering Processes and Chemical Composition of Tropical Rivers from Diverse Tectonic Settings

Samples from rivers in the Amazon and Orinoco basins, the Isthmus of Panama, and Taiwan have been analyzed for the major dissolved components in water and for the major-element composition and petrology of solid phases. These data indicate that there is a good correlation between the composition and yield composition of river-borne materials, tectonic setting, bedrock geology, and the nature of landforms that the rivers drain. This correlation is used to delineate essential features of weathering and erosion in the tropics.

Rivers that drain six general categories of morphostructural regions have been examined. These categories are: (1) island-arc type mountain belts (Isthmus of Panama); (2) continental mountain belts (the Andes and Taiwan); (3) minor uplifts in foreland basins adjacent to continental mountain belts; (4) elevated crystalline shields (much of the Brazilian and Guayana Shields); (5) alluvial plains in sedimentary basins (the Amazone Trough and the Andean Foreland Basin); and (6) peneplaned shields (southwestern Guayana Shield). Denudation rates, erosional styles, and the composition of erosion products vary greatly among these regions.

Landscape development reflects the combined effects of tectonic and erosional processes acting over extended periods of time. The concept of "erosional regime" describes the nature of mass-movement processes on slopes and is used to provide a link between the composition of erosion products and the processes that control landscape development. Erosion is transport limited when the rate of supply of material by weathering exceeds the capacity of transport processes to remove the material; whereas erosion is weathering limited when the capacity of the transport process exceeds the rate at which material is generated by weathering. Weathering-limited erosion typically occurs where slopes are steep. Denudation is rapid, and eroded solids often contain much partially weathered material. In regions undergoing transport-limited erosion, denudation is slow; slopes are slight and convexo-concave; and thick soils dominated by cation-poor mineral phases develop.

In tectonically active regions (1, 2, and 3, above), the composition of river-borne materials indicates erosion is weathering limited. Denudation rates, estimated from yields, are rapid (50–5000 m/m.y.). The most rapid denudation occurs where unstable lithologies such as shales, carbonates, and evaporites are eroding. In rivers that drain siliceous sediments and crystalline felsic-to-intermediate igneous and metamorphic rocks, silica concentrations range from 100 to 200 μMol/L. Several important minerals (including potassium feldspars, micas, and 2:1 clays) that commonly occur in the solid load of these rivers contain tightly bound Mg and K. A similar abundance of Na-and-Ca-bearing minerals is lacking; thus, when compared to bedrock proportions, K and Mg are enriched relative to Na and Ca in solid erosion products. Likewise, Na^+ and Ca^{2+} are enriched in solution relative to K^+ and Mg^{2+}. The compositions of secondary mineral suites in the solid load of these rivers correlate with the total cation concentration (after correction for cyclic salt inputs and halite weathering) of the river water. Where the total cation concentrations is less than approximately 500 μEq/L, kaolinite is the most important secondary mineral; at higher total cation concentrations, 2:1 clays are important. The formation of 2:1 clays with their characteristically high Si/Al ratios seems to limit silica concentrations in these rivers. In contrast, silica concentrations in rivers that drain mafic rocks and glassy igneous rocks exceed 800 μmol/L.

In regions of elevated crystalline shield (4), denudation rates (10–50 m/Ma) are less than those of tectonically active areas. The topography of these regions is quite spectacular with steep slopes of bare rock or thin, sparsely vegetated soils. Altitudes reach 3000 m. Bedrock is typically massive and composed of minerals that are resistant to weathering such as quartz, potassium and sodium feldspars, and micas. River silica concentrations range from 40 to 200 μmol/L, and silica is often a dominant component. Major soluble cations (Na^+, K^+, Mg^{2+}, Ca^{2+}) are in bedrock proportions. Dissolved Al and Fe concentrations are low. Soils and suspended sediments in rivers contain kaolinite, quartz, and lesser quantities of illite, vermiculite, and Al/Fe-sesquioxides. Compared to sediments eroded from mountain belts, those derived from elevated shield terrains are compositionally more mature. This maturity is because a greater degree of chemical weathering seems to be needed to erode the lithologies typical of the shield.

Vast tracts of alluvial plains in sedimentary basins (5) and peneplaned shields (6) are classified as undergoing transport-limited erosion. Denudation rates estimated for the shield are less (<20 m/Ma) than those estimated for the sedimentary terrains (<50 m/Ma). In rivers that drain areas of especially slow denudation (<10 m/Ma), silica is the dominant dissolved inorganic component even though its concentration is low (less than 40 μmol/L). Concentrations of major soluble cation concentrations are in near-bedrock proportions. Most of the anion charge is from organic acids, and pH values are low (<5.5). Al/Fe-sesquioxides, kaolinite, and quartz are unstable, and soils in many areas are podzolic. High concentrations of dissolved Al and Fe (>1 μmol/L) are common. The concentration of suspended solids, which consist mostly of organic matter and traces of kaolinite, is exceedingly low. Where denudation is more rapid (>20 m/Ma), the compositions of dissolved and solid phases in rivers are similar to those in rivers that drain elevated shields.

STAMPFLI, G., Institut de Géologie, UNIL-BFSH 2, Lausanne, Switzerland, A. BAUD, Musée de Géologie, UNIL-BFSH 2, Lausanne, Switzerland, M. GAETANI, Instituto di Geologia dell'Università, Milan, Italy, and J. MARCOUX, Université Paris VII, Paris, France

Paleotethyan Margins in Space and Time

Passive margin evolution can be defined using well-known models (Red Sea, Atlantic margins), and although a direct comparison between present-day margins and paleozoic margins might not always be possible, one can expect similar behavior in space and time.

A minimum rifting period of 20 m.y. should characterize any future margin; it will be accompanied by important subsidence and block tilting within the rift, and an important uplift and erosion of the rift shoulders. In applying the simple shear model for rifting, one has to define an upper and a lower plate in which deformation due to extension will be different.

The lower plate margins will generally be broader and consist of a series of tilted blocks. The basins defined by these blocks might be relatively starved (at least in the offshore direction), and condensed pelagic sequences overlying prerift sequences will characterize the offshore part of that type of margin. Close to the break-away zone, thick clastic sequences will record the rift shoulder uplift.

The upper plate margin will be narrower and consist of a major flexure characterized by faults generally dipping landward. Volcanism, if present, will be a characteristic of the rift shoulder area of the upper plate. The former rift area of that upper plate will consist of few basins dominated by clastics (if the prerift sequence is made of a thick carbonate pile, the clastics can be replaced by carbonate breccia and marls). Mafic intrusions can also be found within that plate and can even occupy a large place between the continental crust and the well-layered oceanic crust.

Transgression of the rift shoulders can be used as a good marker of margin maturity and would certainly not occur before 60–100 m.y. after initial break-up. The rim basins (landward from the shoulders) will record the shoulder uplift and the following thermal subsidence accompanied eventually by a major onlapping sequence.

At least three stages of margin formation can be recognized within the Tethyan realm; an example for each stage is presented here:

1. *Lower Paleozoic, passive margin of upper plate type in the Alborz in northern Iran.* It was initiated in upper Ordovician time and was involved in the early Cimmerian orogeny in upper Triassic, lower Jurassic times. Transgression of the rift shoulder is done in Devonian time; from that period until the Triassic collision a carbonate platform develops.

2. *Upper Paleozoic, northern margin of India of upper plate type.* Issued from an upper Carboniferous rifting phase, the margins of that type will be involved in the Himalayan Alpine orogeny. Transgression of the rift shoulder is done in upper Permian and followed by the development of a carbonate platform. (See Baud et al, this volume.)

3. *Lower Mesozoic, northern Apulian margin in the southern Alps, possibly changing from lower plate type in the west to upper plate type in the east.* That margin will be involved in the Alpine orogeny. Transgression of

the rift shoulder happened in mid- to upper Jurassic; a carbonate platform regime was present since the rifting period (upper Triassic to lower Jurassic).

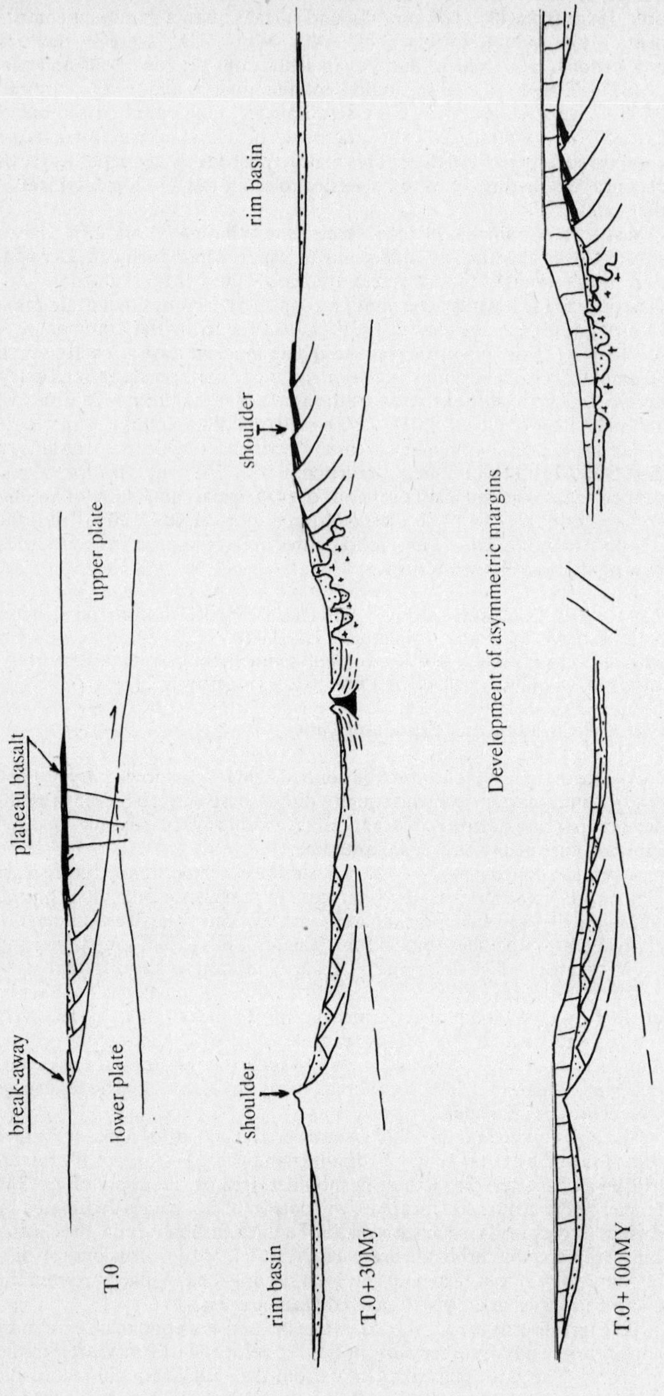

Figure 1—From rifting to passive margin.

STANLEY, D. J., Smithsonian Institution, Washington, DC

Evolution of Nile and Other Mediterranean Deltas during Holocene and Role of Eustatic Sea-Level Oscillations

Deposition of deltas along margins of the Mediterranean Sea resulted from the interplay of climatic effects, transport processes, and tectonics. Fluctuating eustatic sea level is one of the dominant controls. This conclusion is based on the finding that the beginning of deposition of Mediterranean deltas during the Holocene was amazingly consistent, although their geologic and geographic settings were diverse. The database available for the modern deltas is uneven, with more information available for the Nile Delta as a result of petrologic study of numerous (>100) drilled borings, radiocarbon dating of 210 core samples, systematic mapping of lithofacies, and close-grid correlation of core sections. Land-to-sea core traverses show that the Nile Delta began to form between 8000 and 7000 y B.P. Published studies indicate that the Po, Rhône, and Ebro deltas also began to form about 7000 y B.P. Moreover, geometry of the lithofacies sequences detailed in this study (basal transgressive sands overlain by marine, coastal, deltaic, and fluvial deposits) are generally comparable in the four deltas.

The surface configurations of the four deltas are distinct, largely as a function of different tectonic settings, provenance and dispersal patterns, and physical oceanographic conditions affecting the four coastal sectors. Of note, however, is the similar internal "architecture/anatomy" of the four depocenters, i.e., comparable facies sequence distributions in time and space and thicknesses. A likely explanation for the near coincidence of timing of delta progradation in four regions is the slowing of eustatic sea-level rise in the mid-Holocene (~7500–7000 y B.P.) as determined in the Mediterranean and other continental margins. This deceleration occurred during a more humid period characterized by increased river flow and accelerated transport of bed load and suspensates to the coasts. This coupling resulted in the initiation of delta-lobe migration on recently submerged inner shelves. Until recently, the four deltas prograded rapidly as sea level continued to rise slowly. Renewed acceleration of eustatic sea-level rise predicted for the next century and continuation of man-induced changes in drainage basins, along river channels, and in delta plains is likely to modify the late Holocene depocenters. Thus, in years ahead, we will have a unique opportunity to actually measure these eustatic sea-level changes and quantify their effects on lithofacies distributions in Mediterranean deltas.

STANTON, R. W., U. S. Geological Survey, Reston, VA, T. A. MOORE, University of Kentucky, Lexington, KY, and P. D. WARWICK, and S. S. CROWLEY, U. S. Geological Survey, Reston, VA

Facies Development in Selected Coal Beds of Powder River Basin, USA

The petrographic composition of five thick (>6 m) coal beds and the sedimentology of associated strata indicate that three major types of swamps developed in the Powder River Basin: (1) areally widespread, flat-convex domed peat with distinct facies development, which formed in a swamp on a stable platform underlain by abandoned meanderbelt deposits—(Felix coal bed); (2) thick, but local bodies of raised peat that accumulated in swamps in a rapidly subsiding, anastomosed fluvial system of fine-grained clastic sediment—(Wyodak-Anderson coal bed); and (3) local flood plain peats formed in swamps associated with crevasse splays and flood basin lakes. These swamps were subject to detrital sediment interruptions and probably developed only the initial stages of doming—(Anderson, Anderson-Dietz and Smith coal beds).

A total of 325 interval samples of coal from 22 channel and core localities were petrographically analyzed. Analyses consisted of point counts of macerals on etched polished pellet surfaces under white-reflected light- and blue-light fluorescence. Maceral varieties were grouped largely from their statistical correlations; these groupings are inferred to indicate similar peat formation conditions that affected their concentration. These genetic groups are: GI—maceral varieties indicating a high degree aof preservation of plant tissues; GII—maceral varieties indicating a high degree of degradation of plant tissues through microbial means (GIIA) or other means such as physical breakdown (GIIB); and GIII—maceral varieties indicating some degree of carbonization or oxidation.

Sample compositions were divided into four major classes by cluster analysis; compositions of these four classes are shown in Figure 1. One or more of these sample classes were interpreted to comprise a facies of the bed. Figure 1 also contains plots of the interpreted facies on ternary diagrams with end member sample class compositions indicated.

The Felix coal bed, which is over 6 m thick in some places, represents widespread peat formation on an abandoned meanderbelt. The Felix coal bed contains two discrete facies composed of sample classes 1 and 2, and class 4. The peat began as woody accumulations, but as peat accumulation continued, doming caused by ombrogeny (primarily rainfed water source) developed. The doming created drier surficial conditions as evidenced by the enrichment of Group III macerals.

The Wyodak coal bed consists of 12–15 km^2 bodies that commonly reach over 30 m in thickness. In contrast to the Felix coal bed, facies of the Wyodak coal bed are composed of mixtures of sample classes 1, 2, 3, and 4 (Fig. 1). Initial formation of the Wyodak swamp deposits began as accumulation of GI-rich facies. Laterally, more degraded peat facies formed in the interior parts of the swamp under wetter conditions. These conditions, however, were followed by somewhat more woody peat formation prior to shallow lake detrital sedimentation that deposited a clastic facies. A laterally extensive peat facies rich in GI macerals then developed on the lacustrine clastic facies. Swamp cessation was caused by lake development that deposited fine-grained sediments that overlie the bed. Coal facies in the upper part of the bed consist of highly degraded plant material (GIIA). In general, sample locations that represent the margins of the bed contain greater amounts of sample class 1 (high in GI material) than the sample classes that represent the more interior part of the deposit. The parting apparently had minor effect on the vegetational accumulation except for a general increase in the amount of GI-rich facies. Because the parting was relatively thin, we infer that the disturbance to peat formation was minor.

The Anderson and Smith coal beds are up to 6 m thick in places, whereas the Anderson-Dietz is up to 25 m thick. However, facies of the Anderson, Smith, and Anderson-Dietz contain sample classes of similar composition, primarily 1 and 2. The numerous and thick partings that split these coal beds indicate sedimentologically active environments, consequently, peat formation was terminated before substantial doming could develop. The Anderson and Anderson-Dietz facies change from the initial GI-rich facies suggesting robust, woody plant growth to the more degraded GII-rich facies, possibly suggesting some dome development and plant community change. In contrast, peat formation in the Smith bed, which also began as a woody accumulation was followed by lacustrine-type sedimentation upon which woody peat formation was reestablished.

Coal bed isopachs of these beds show indentations in isopach contours that may indicate drainage outward from the thicker part of the peat, thus suggesting doming. Detritus from fluvial systems was probably not a primary contributor to the mineral fraction of the raised peat, except possibly at the outermost margins of the swamp. A limited detrital influx accounts for the low-ash yields (8 wt%) of samples from the coal beds.

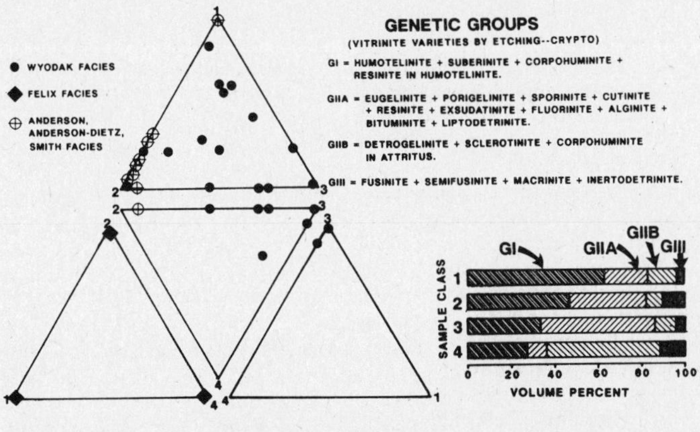

Figure 1—Compositions of facies plotted by sample classes 1, 2, 3, or 4.

STARITSYNA, G. N., Y. I. TOMANOVSKAYA, and S. M. TABUNOV, Ministry of Geology of the USSR, Moscow, USSR

Evolution of Basalt Magmatism in Riftogenic Structures of Pacific Ocean

Magmatic rocks of the world ocean attract attention of investigators as an eventual endogenic source for various metals, which makes the study of oceanic crustal magmatic rocks in terms of their metal content an extremely urgent problem. This problem cannot be solved without step-by-step understanding of the geology of the ocean bed and structural-formational subdivision of the basalt basement top in particular, as well as the study of basalt magmatic history of rift structures, which are active ore- and magma-generating zones in the world ocean.

The structural-formational analysis of the Pacific talasscraton magmatic rocks provided subdivision of rift basalts, earlier regarded as homogenous. From among them, a number of magmatic formations with various metallogenic specializations were distinguished. Rift structural features have revealed their dependence on age, which affected geological and geophysical parameters (earth crustal thickness variations, linear band magnetic anomalies, seismic activity, and heat flow magnitude).

These parameters provided recognition of the three groups of Pacific rift structures varying in age: (1) Pliocene–Quaternary (East Pacific and South Pacific Rises, Galapagos Rift); (2) Miocene (South Galapagos and Chilian Rises), and (3) Cretaceous (Phoenix, Cooper, North-Hass, Japanese, Hawaiian).

Based on petrochemical parameters, Cretaceous rift structural basalts were distinctly separated from those of the Miocene and Quaternary. Despite common petrochemical features typical of basalts from midoceanic ridges (low-K aluminiferous basalts), they revealed significant differences. A peculiar petrochemical feature of Cretaceous rift structural basalts is high Fe with increased Mg, indicating their promising ore potential. Pliocene–Quaternary basalts occurring along EPR margins differ from those of the Cretaceous by low Mg, Fe-, and Ti-rich extreme members of differential rock series, which provides microelemental redistribution. Basalts of this series are Ni- and Cr-poor, while ferrobasalts are enriched in Co. In contrast to flanks, the axial EPR zone shows highly developed basalts of magnesial-aluminoferous series, characterized by increased Ni and sometimes Zn in its low-K variety.

Petrochemical differences in basalts, confirmed by spatial and temporal separation, suggest different environments of the emplacement of rift structures during various time epochs. This provides recognition of several magnetic formations with different metallogenic rock-units from rift structures of the Pacific talasscraton. Cretaceous structures show well-developed komatiit-tholeitic formation, promising magmatic Cu-Ni-mineralization. In rift structures of Pliocene–Quaternary age, a major magmatic formation of low-K aluminiferous basalts–ferrobasalts, subdivided into two subformations (volcanites from flanks and axial zone), corresponding to two evolutional stages of these rift structures, was distinguished. Volcanites from flanks and axial zone are independent evolutional seria, which reflect physical and chemical environmental variations of magma generation due to changed rifting tectonic regimes.

Basalts of the western Pacific rift structures were formed under dispersed spreading from primitive hypebasic-basic magma generated at the upper mantle level under rapid magma rise and crystallization liquational foliation of silicate melt locally occurring. This is supported by komatiit rocks spreading in the section, spinifex structures in these rocks, and by magnesiochromites. In the eastern Pacific rift magmatism began in the Miocene and has not yet ended. Basalts of the Pliocene–Pleistocene stage were formed under linear spreading from basic magma experiencing deep fractionation, from large magmatic chambers generated at shallower depths than during the Cretaceous. Present basalts of the second stage were formed under the mantle diapir extension and vertical transport from hypebasic-basic magma under extremely rapid rise and instant crystallization of liquates of the melanocratic and leucocratic series.

Intermediate chambers situated in various depths are primarily responsible for second-stage basalt formation. These chambers are a thermal source for hydrothermal sulfide-forming processes.

Regional scales of tectonomagmatic activity occurred during the Holocene in the axial zone of rift structures, and its relationships with polymetallic sulfides permits this geological event to be called a Cenozoic mineragenic epoch.

Research in the establishment of oceanic basalt types provides recognition and outlines of major petrochemical provinces of different mineralization in the world ocean. Therefore, two areas have been recognized within the confines of the Pacific talasscraton, i.e., the areas of Mesozoic and Cenozoic rift magmatism in the western and eastern part; they differ not only in evolution but also in metal content.

STARKEL, L., Polish Academy of Sciences, Kraków, Poland

Global Paleohydrology: New Project

Continental hydrology examines the part of the water circulation system comprising the atmosphere, oceans, and land. The increasing deficit of water on land stimulates us to study the changes in global hydrology during various time intervals, related both to natural factors and to man's activity. The reconstruction of water circulation parameters covers the areal pattern of precipitation regime, the run-off and evaporation as well as the surfical and groundwater retention. Information on those parameters is drawn from the vegetation changes (especially from the shift of ecotones), lake level and groundwater level fluctuation, various fluvial activities, advances and retreats of glacials, eolian activity, and archeological and historic records.

When forecasting hydrologic changes, it is necessary to reconstruct the rhythmic hydrologic changes of various duration. The 20 Ka cycle with the shift of perihelion explains the sequence of increased and decreased humidity in various climatic zones from low to high latitudes. On the rhythmic fluctuations of a ~2–2.5 Ka duration and shorter ones there are superimposed the phases of increased deforestation, overgrazing, and cultivation, which cause a rise in the run-off and sediment load. This overlapping leads to the increase of flood frequency during more humid phases and to desertification during drier ones.

Knowledge of the overlapping mechanism of natural factors and of human impact should help to predict the trends of hydrologic changes and teach us how to control and decrease the amplitudes of the run-off. The global research program should include projects on the reconstruction of hydrologic parameters in space and time, as well as the hydrologic simulation models closely correlated with paleoclimatic ones and including the interaction of various spheres.

The working group on Global Paleohydrology has been organized after INQUA Congress within the Holocene Commission of INQUA. This group prepared in 1988 a proposal of the new IGCP project. Its objective will be to establish the nature and causes of hydrological changes over the late Quaternary, emphasizing a time resolution of 10^2–10^3 years and employing an international database. Research will be undertaken by analyzing river and lake basins in terms of changing components of the water balance including precipitation, run-off and surface and groundwaters.

STATTEGGAR, K., Karl-Franzens-Universität, Graz, Austria

Mixing, Outliers, and End Members; Geochemical Compositions of Detrital Garnets from Jurassic Sandstones of North Sea

Detrital heavy minerals are a suitable tool in lithostratigraphy and provenance studies. They reflect source rocks and their variations in time and space from the sedimentary record. Heavy-mineral distributions in sediments are mixing proportions of heavy-mineral distributions in source rocks, often disturbed by transport and diagenetic alteration. Statistically, this raises the questions of mixing, outliers, and end members.

Geochemical compositions obey the constant-sum constraint that becomes of special importance if the compositions consist of few variables. In that case, outliers do influence severely the whole data structure, e.g., in calculating a similarity matrix. The problem can in part be overcome by appropriate data transformation, e.g., additive-logratio, to get out of the simplex sample space. The same problems arise in the modeling of end members to evaluate the mixing proportions of a geochemical data set. An expansion of multivariate Q-mode analysis for that purpose are the EXTENDED CABFAC and QMODEL algorithms of Klovan and Miesch (1976) and Full et al (1981). End-member models that are calculated from Q-mode factor analysis may be seriously affected by extreme sample-compositions so that extreme samples may become end members. Therefore, sample points may be placed outside the calculated mixing polytope that then describes the mixing proportions only insufficiently.

Fuzzy c-means clustering (Bezdek et al, 1984) provides a robust method for grouping and evaluating the mixing proportions of geochemical data. It relies on the collective properties of the data and estimates the similarities of each data point to each group instead of the exclusive classification performed by hierarchical clustering methods. Outliers have only minor influence in the location and composition of group-centers. Fuzzy partitions expressed by membership functions can be used to detect multivariate outliers: Extreme samples have low memberships with all groups and great distances to the cluster centers. Extremity of a sample depends also on the number of groups considered in a data set.

In a case study, 1550 detrital garnet grains of 31 samples from Jurassic sandstones of the northern North Sea were analyzed by electron microprobe. Four geochemical variables were determined for each grain, which sum up to 100%: almandine, pyrope, grossular, and spessartine contents.

Univariate robust statistics show that spessartine has many high value outliers whereas the other variables have only few extreme values. End-member modeling with the CABFAC and QMODEL algorithms resulted in extreme end-member compositions for varimax and oblique solutions. Most samples are situated near the edge of a chosen end-member polytope, and some have negative mixing proportions. That means the QMODEL-based end-member analysis is not very effective in describing the bulk of the data. Data classification by fuzzy c-means clustering with fuzzy partitions for two, three, and four groups using the Euclidian norm and weighting exponents from 1.5 to 2.0 gives more suitable results. In a polytope with the cluster centers as end points, one can find better grouping of the data points; the cluster centers are well aligned to the main distribution of data. Multivariate outliers can be identified easily by their great distances to the cluster centers. They are situated in the central region of the polytope.

Different end-member models calculated for the garnet data are shown in some triangular graphs. The advantage of fuzzy clustering lies in a proper group definition and separation as well as in a clear identification of outliers. This facilitates lithostratigraphic and paleogeographic interpretations of the investigated data set.

REFERENCES

Bezdek, J. C., R. Ehrlich, and W. Full, 1984, FCM: the fuzzy c-means clustering algorithm: Computer & Geosciences, v. 10, p. 191–203.

Full, W. E., R. Ehrlich, and J. E. Klovan, 1981, EXTENDED QMODEL—Objective definition of external end members in the analysis of mixtures: Journal of Mathematical Geology, v. 13, p. 331–334.

Klovan, J. E., and A. Miesch, 1976, EXTENDED CABFAC and QMODEL computer programs for Q-mode factor analysis of compositional data: Computers & Geosciences, v. 1, p. 161–178.

Morton, A. C., 1985, A new approach to provenance studies: electron microprobe analysis of detrital garnets from Middle Jurassic sandstones of the northern North Sea: Sedimentology, v. 32, p. 553–566.

STATTEGGER, K., Karl-Franzens-Universität, Graz, Austria

Heavy Mineral Events: Valuable Tool in Tectonostratigraphy

Heavy mineral analysis is a proven method to decipher provenance of clastic sediments provided that diagenetic alteration is limited. A new approach is the modeling of lithological end members from heavy-mineral data for the geodynamic interpretation of clastic sediments (Stattegger, 1987): Heavy-mineral distributions are mixtures of different minerals depending on the regional distribution of source rocks, transport, and diagenetic alteration. The different mixing proportions are referred to lithologically meaningful end members, which are used to decipher provenance of clastics in different plate-tectonic settings. Different minerals or mineral types (e.g., geochemical varieties) can be used for that purpose.

During an orogenic stage, recurrent heavy-mineral events with sudden changes in the spectrum may occur rather than a single change, reflecting stepwise collisional and succeeding tensional activity in the hinterland in connection with the uplift and erosion of metamorphic core complexes. This phenomenon is demonstrated in two case studies where garnet events overwhelm repeatedly the whole heavy-mineral suite.

1. Neogene molasse-sediments of the Styrian basin (Austria): The mainly clastic rocks of the Styrian basin, situated on the eastern margin of the Alps, reach a maximum thickness of 3500 m. The stratigraphic sequence (Early Miocene–Late Pliocene) records repeated changes between garnet- and epidote-rich heavy-mineral spectra that point at the erosion of the metamorphic Austroalpine hinterland. A concept of heavy-mineral cycles was developed (Stattegger, in press): Such a cycle starts with a garnet-event, continues with a garnet–epidote mixing stage, and ends with the predominance of epidote. Up to four cycles can be distinguished in the Styrian basin. A geodynamic explanation of the cycles is proposed: Garnet events are the sedimentary feedback of Neogene transtensional tectonics after the Alpidic collisional orogeny connected to fragmentation and subsidence of crustal blocks. High relief-energy erodes "fresh" crystalline rocks. More stable tectonic conditions, lower relief and transport energy, and intense weathering are responsible for the enrichment of epidote and the reduction of garnet, respectively.

2. Late Paleozoic orogenic sediments of the Betic Cordillera (Southern Spain): The Silurian to Late Carboniferous sediments of the Malaguide unit represent continuous deep-water environments dominated by flysch deposits. Repeated changes between acid plutonic/recycled cratonic and metamorphic sedimentary input are evidenced by heavy mineral spectra and psephitic clast modes (Herbig and Stattegger, in preparation). Oscillations and amplitudes, especially of metamorphic detrital influx, become more frequent and larger, respectively, near the Variscan unconformity. Permo-Triassic redbeds are clearly postorogenic sediments according to lithofacies and heavy mineral content.

The joint development of lithofacies, clast modes, and especially heavy mineral spectra demonstrates the transition from a passive to an active continental margin during the Paleozoic: The Silurian–Devonian sediments of the Malaguide realm developed during a divergent cratonic stage north of the Betico-Rifean block, which is a detached sliver of the African plate. The Viséan–Upper Carboniferous formations record the collisional orogenic stage between the Betico-Rifean block and the southern margin of the Southern Europe plate. The collision processes induced strong uplift and erosion of successively shallowing carbonate shelf realms and the connected crystalline, mostly metamorphic hinterland.

Garnet events may be investigated further by electron microprobe analysis of individual garnet grains to facilitate recognition of different source rocks and source areas along the stratigraphic record. This can improve essentially any geodynamic interpretation of heavy-mineral data.

REFERENCES

Herbig, H. G., and K. Stattegger, in preparation, Late Paleozoic heavy-mineral and clast modes from the Betic Cordillera (Southern Spain).

Stattegger, K., 1987, Heavy minerals and provenance of sands: modeling of lithological end members from river sands of Northern Austria and from sandstones of the Austroalpine Gosau Formation (Late Cretaceous): Journal of Sedimentology Petrology, v. 57, p. 301–310.

——— in press, Schwermineralstratigraphie im Steirischen Becken: Mitt. Geol. Inst. Innsbruck.

STEIN, R., Giessen University, Giessen, Federal Republic of Germany

Changes in Paleoenvironments in Atlantic Ocean During Cretaceous Times: Results from Studies on Cretaceous Black Shales and on Late Cenozoic Marine Organic-Carbon-Rich Sediments

Large parts of the Cretaceous Atlantic Ocean are characterized by the widespread deposition of organic-carbon-rich sediments ("black shales"). Maximum black shale deposition occurred in the Aptian–Albian and Cenomanian–Turonian. In order to explain the formation of these black shales, (1) changes in the amount and type of organic matter in relation to changes in paleogeography, paleoceanic circulation, and paleoclimate have been investigated in detail. (2) The data derived from these investigations are compared with new results from investigations of Late Cenozoic organic-carbon-rich ODP-sediments. The study is mainly based on (1) data derived from elemental analysis (carbon, hydrogen, nitrogen) of bulk sediments, carbonate-free residues, and kerogen concentrates, (2) Rock-Eval pyrolysis data, (3) kerogen microscopy data, (4) organic carbon/sedimentation rate relationships, and (5) estimates of paleoproductivity.

Because both the amount and type of the organic matter accumulated in the Atlantic Ocean during Cretaceous times changed markedly with location and age, different mechanisms must have caused the accumulation of these organic-carbon-rich sediments:

1. Terrigenous organic matter supply was the dominant factor for black shale formation during Hauterivian and Albian times in the Northwest Atlantic.
2. Increased oceanic productivity was important for the black shale deposition in the Northeast Atlantic during Albian times.
3. The widespread preservation of high amounts of marine organic carbon in the whole North Atlantic during Cenomanian–Turonian times has probably been caused by oceanwide anoxic deep-water conditions, rather than increased supply of marine organic matter due to higher oceanic productivity.
4. Locally, a rapid burial of organic matter by turbidites/slumps may explain the preservation of high amounts of organic carbon in the western and eastern basins of the Atlantic ocean.

These different depositional environments and mechanisms causing the accumulation of organic-carbon-rich sediments have also been recorded and characterized in Late Cenozoic sediments recovered during more recent Ocean Drilling Program (ODP) Legs:

1. The organic-carbon-rich sediments deposited in Baffin Bay (ODP-Site 645) during Plio-Pleistocene times have been caused by high supply rates of terrigenous organic matter.
2. In the upwelling areas off northwest Africa (ODP-Site 658) and off Peru (ODP-Site 679), a distinctly increased productivity of marine organic matter has resulted in the deposition of organic-carbon-rich sediments.
3. In the Mediterranean and in the Black Sea, increased preservation of (marine) organic matter due to anoxic deep-water environments may explain organic-carbon-rich sediments deposited during Plio-Pleistocene times (ODP-Leg 107, DSDP-Leg 42).
4. Several events of high organic carbon deposition have been recorded in turbidite and slump sequences off northwest Africa (e.g., ODP-Site 657).

STEINER, M., University of Wyoming, Laramie, WY, and J. OGG, Purdue University, West Lafayette, IN

Triassic Magnetic Polarity Time Scale

The magnetic field during the Triassic Period alternated frequently between normal and reversed polarity states. A global compilation of the paleomagnetic stratigraphy of the most well dated rock sequences of this time is produced here. The backdrop for the polarity time scale, the Triassic radiometric time scale, has been a subject of debate, and therefore that shown here may need to be changed in the future. The time scale presented by Haq et al (Science, 1987) was used. It is the most recent and possibly the most data inclusive. Although the basis of this scale was not available for inspection, the scale is quite similar to the recent four previous scales (Armstrong, Webb, Odin, Harland).

After many tens of millions of years of constant reversed polarity, two major normal polarity intervals, and perhaps several minor ones, occurred in the latest Permian. The earliest Triassic is of normal polarity and the Early Triassic alternates fairly evenly between normal and reversed states, displaying 10 major polarity intervals. A small number of shorter intervals are also suggested by the excellent data coverage: Four stratigraphic sequences (two marine and two nonmarine), each spanning most of the Early Triassic, were used for this compilation.

The Middle and early Late Triassic are poorly known. Fewer rock sequences are available. Our unpublished and some very old published paleomagnetic data form the basis of this section, which is shown in half width to indicate its much greater uncertainty as compared to the rest of the column. From the information available, rapidly alternating polarity is suggested.

The Late Triassic is better studied than the Middle, but not as extensively as the Early. However, two major terrestrial rock sequences and one smaller marine one all indicate a sequence of largely reversed polarity with many shorter normals followed by a sequence of largely normal polarity with some short reversed intervals. The earliest Jurassic is also poorly known, but the few data indicate an increase in reversal frequency in the Hettangian–Sinemurian (an N-R-N-R sequence) followed by a further increase to very rapidly alternating polarity in the Pliensbachian.

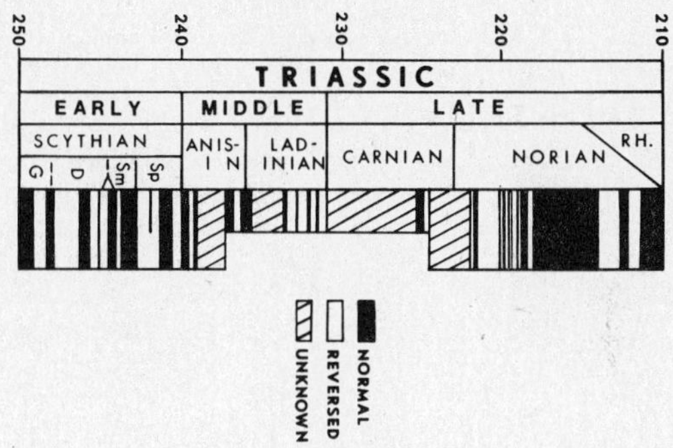

STEINHAUSER, P., University of Vienna, Vienna, Austria

Geodynamic Interpretation of Gravity Statistics and Recent Crustal Movements in Austria

Information upon the geodynamics of the Eastern Alps and its surroundings can be deduced from gravity anomalies. For the balance of crustal masses, the degree of isostatic compensation is decisive, while the total amount of anomalous masses can be derived from the Bouguer gravity field pattern. The isostatic gravity has been calculated based on an Airy-Heiskanen model with local compensation. The model parameters are chosen to a sea-level crustal thickness of 30 km, the density 2.67 g/cm^3 for the topography above sea level, and a density contrast of 0.4 g/cm^3 between the lower crustal and the upper mantle material. Ranging from +58 mgal to −31 mgal the isostatic anomaly pattern shows comparatively small cells of positive and negative anomalies but no relation to the general structures of the Alps. Although the correlation between topographic heights and isostatic anomalies is insignificantly small, for a sea-level crust a positive mean isostatic anomaly of +15 mgals is obtained. This can be interpreted as undercompensation of the alpine topography. Similar results are obtained for the Bouguer anomalies. The statistical relation between mean topographic heights and mean Bouguer anomaly values should correspond in average to the Bouguer plate correction term in case of mass balance. But in reality the statistical analysis proves that the mean Bouguer anomalies are too positive for the existing topography.

Finally, by applying Gauss' theorem to the Bouguer gravity data the total mass anomaly of the causative body can be determined, without any reference either to the shape or the density of the anomalous body being involved. For the Eastern Alps, a mass defect of $3.13.10^{17}$ kg is obtained. On the other hand, the total mass of the mountain topography above sea level amounts to $3.44.10^{17}$ kg, giving a net surplus of 10% of topographic loads.

Thus, all gravimetric evaluations indicate unanimously that on average the Eastern Alps are built up higher than the balance of masses requests. From the isostatic point of view, subsidence of the Alps has to be expected. In contrast to these results, recent crustal movements from repeated precise levelings over several decades indicate uplift rates of 1–2 mm/y for the Eastern Alps relative to the Bohemian Massif, which is considered to be the most stable geological unit within Austria. The uplift data show a significant correlation to Bouguer anomalies, but an insignificant one to isostatic anomalies.

In addition, the correlation coefficients own considerable differences for various regions.

Table 1—Correlation of recent crustal movements with gravity anomalies.

Area	Number of observations	Correlation Bouger	Coefficient Isostatic
Bohemian Massif	8	0.71	0.59
Alpine Foreland and Vienna Basin	18	0.12	0.14
North Central Alps	20	0.80	0.10
Southern Alps	25	0.85	0.56

Therefore, it has to be concluded that the Alpine uplift is not guided by the mechanism of isostasy but by geodynamic forces due to plate tectonic movements between Europe and Africa. This is also supported by the observation that equal Bouguer anomaly values are related with higher uplift rates in the south than in the north, which is consistent with a thrust from the south. In this context, it is interesting to note that the Alps tower up to greater heights in the south than in the north: For equal Bouguer anomaly values, the mean topographic heights are about 600 m greater south of the main crest of the Alps than north of it.

STEINTHORSSON, S., University of Iceland, Reykjavik, Iceland

Mode and Rate of Volcanism in Zones of Crustal Accretion, as Exemplified by Iceland

The volcanic history of Iceland is fairly accurately known from written accounts for the last 300 years, and less completely from a combination of written accounts and tephrochronology since about A.D. 900. During the last 300 year period, some 20 eruptions have taken place per century, and the tephrochronological record indicates similar rates of eruption throughout the 1100 years of settlement in the country. The volcanic productivity during that time is estimated about 4.3 km^3 per century, or 43,000 km^3/Ma. At this rate, the entire volume of Iceland, whose continuous geological history spans 14.5 Ma, could have been produced in one-tenth of that time, indicating powerful "sinks" as well as "source," the Iceland mantle plume. The "sinks" include, in addition to very rapid erosion, the recirculation of rock in the volcanic zones.

Seismological studies show the plane-like boundary between the European and American crustal plates to underlie the rift zones of Iceland which, however, possess tectonics of their own, characterized by an array of elongated volcanic systems. Such systems are represented in the Tertiary formation by dike swarms traversing silicic centers, emphasizing continuity in terms of processes and, by inference, rates throughout the geological history of Iceland. Crustal spreading, although probably continuous between distant points out in the stable plates, occurs in the rift zones sporadically and suddenly at long intervals in events of volcanism and crustal extension; an example in point is the Krafla events of 1975–1984.

Three different facies may be recognized in Icelandic volcanism. The genetic relationship between silicic central volcanoes and the associated volcanic fissures is now recognized, with the concept of volcanic systems. Such systems occur uniformly throughout all geological formations in Iceland. Lava shields form the third facies, which in the Holocene has been largely restricted to the period of isostatic rebound at the end of the Pleistocene glaciation. Several interglacial lava shields are known as well. The question remains, however, whether these structures are to be equated with the lava sequences in the Tertiary formation known as "shield basalt series."

STEMPER, B. A., and J. A. WESTGATE, University of Toronto, Scarborough, Canada T. L. PÉWÉ, Arizona State University, Tempe, AZ

Tephrochronology and Paleomagnetism of Loess in Fairbanks Area of Interior Alaska

Quaternary loess discontinuously blankets the unglaciated interior of Alaska. This loess, with its abundant plant and animal remains and many paleosols, preserves a detailed record of paleoenvironmental change. In addition, numerous distal tephra beds are interspersed throughout its thickness and provide valuable geochronological control.

The loess is up to 57 m thick in the Fairbanks area, where it has been divided into several lithostratigraphic units. The Fairbanks Loess occurs on upper hill slopes and hilltops, and, where traced laterally to lower slopes and valley bottoms, is separated into the Gold Hill Loess (GHL), the organic-rich Goldstream Formation, and the Engineer Loess (see Figure 1). Locally, GHL is separated from the Goldstream Formation by the Eva Formation, an interglacial forest bed.

Most of the tephra beds are very thin, crystal-poor, fine-grained, and contaminated with detrital grains—attributes that make them very difficult to date. However, the grain-discrete isothermal plateau (ITP) fission-track technique has recently been applied successfully to the hydrated glass shards of distal tephra beds. This method gives an age that is corrected for any partial track fading that may have taken place in the glass.

At present, three tephra beds in the GHL have been dated by the ITP f-t method: Old Crow tephra, located near the top of GHL; the Ester Ash Bed, located 3 m above the base of GHL; and the SP tephra, located 25 m above the base of GHL (see Figure 1). Old Crow tephra and the Ester Ash Bed are dacitic units with abundant bubble-wall shards, both being derived from vents in the eastern Aleutians. In contrast, the SP tephra is very pumiceous with few blocky glass shards but with abundant crystals. Its source vent is most likely in the Wrangell Mountains.

Old Crow tephra has a weighted ITP glass f-t age of 149 ± 13 Ka, based on three determinations. The Ester Ash Bed has a weighted ITP glass f-t age of 841 ± 93 Ka, based on four determinations, and the SP tephra has an ITP age of 872 ± 86 Ka.

The magnetic polarity of the loess adjacent to the tephra beds was determined in an attempt to provide additional chronological control. The loess has a normal polarity at the level of Old Crow tephra, but it is reversed at the levels of the Ester Ash Bed and SP tephra. Thus, the magnetic polarity data are in agreement with the ITP glass f-t ages, and the polarity change recorded in the loess 3 m above the SP tephra (see Figure 1) likely represents the Brunhes–Matuyama Boundary.

The tephrochronological and paleomagnetic data demonstrate that deposition of GHL began at the very latest in early Pleistocene time and terminated during the last interglacial, because the Old Crow tephra occurs directly under the Eva forest bed. Numerous radiocarbon age determinations indicate that Engineer Loess is of Holocene age; it follows that the Goldstream Formation must be of Wisconsin age. Additional stratigraphic results from this work are: (1) major unconformities exist in the loess record, and (2) the loess preserved in the Fairbanks area of interior Alaska represents most of Quaternary time—an antiquity not previously recognized.

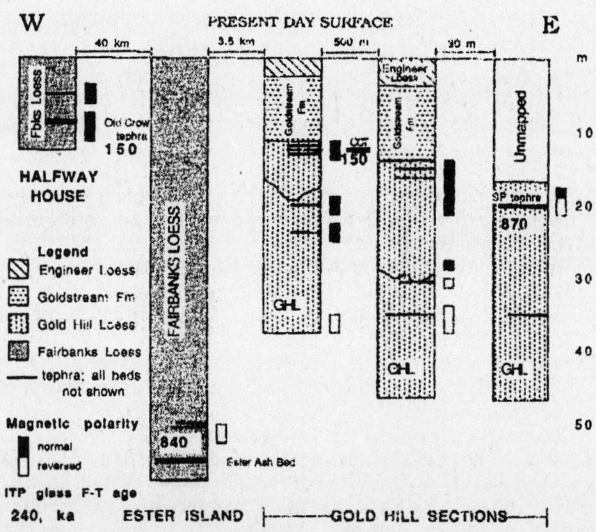

STEMPROK, M., Geological Survey, Prague, Czechoslovakia

The Behavior of Tin, Tungsten and Molybdenum in Felsic Magmas

Recent studies e.g. Mysen (1987) have shown the importance of structure of alumosilicate melts on partitioning of elements. The paper attempts to test the position of Sn, W and Mo in felsic magmas to explain the association of tin, tungsten and molybdenum ore deposits with some granites and rhyolites.

Metallogenetically specialized granites belong to highly evolved granitoids of peraluminous composition characterized by relative excess of silica and the deficiency of some major elements oxides such as CaO, MgO, FeO and Fe_2O_3. Such chemistry is even more distinct in albitic varieties of these granites (apogranites) which are characterized by increased Li_2O (over 0.2 wt%), Rb_2O (0.1%) and very low CaO, MgO and FeO (commonly below 0.5%).

The metls of this composition have fully polymerized three-dimensional network structure (Dingwell 1988) consisting of TO_2 units (AlO_2^- and SiO_2^o) and the ratio of nonbridging oxygens to cations in tetrahedral coordination is less than 0.2 Cationic depolymerizers are Li, Na, K and Rb or Al in excess of the amount required to charge the balance of alkali aluminate tetrahedra. Fluorine and hydroxyl ions belong to anionic depolymerizers.

Two explanations appear possible to ellucidate the dependence of highly charged cations such as Sn^{4+}, W^{6+}, Mo^{6+}, Nb^{5+} and Ta^{5+} on the development of felsic magmas: 1) the accumulation of these elements occurred in consequence of progressive polymerization of less polymerized magmas, 2) the metals were concentrated or added by the action of depolymerizing agents such as fluorine and hydroxyl ions.

Structural position of tin, tungsten and molybdenum in silicate melts may be derived from so far non-systematic studies or binary alkali silica systems or granitic melts with oxides of these metals or melts in equilibrium with hydrothermal solutions containing Sn, W and Mo.

Tin: Data on Mössbauer spectra of binary alkali silica systems with tin dioxide indicate octahedral coordination of Sn^{4+} in the form of SnO_6 structural units. Divalent tin is in tetrahedral coordination and it can probably replace Si^{4+} in silica tetrahedra. In alumosilicate melts of granitic composition tetravalent tin was found to prevail over divalent tin [Sitek et al, 1981) and variable Na_2O did not show changes of Mössbauer spectra. Thus the existence of Sn-O-Si bonds in the melts was postulated.

Tin dioxide (cassiterite) is the first crystallizing phase in binary alkali oxide silica and in dry granite systems with tin dioxide added (Stemprok, Voldán 1975, 19789). In subsolidus regions of ternary alkali oxide-silica-tin dioxide systems several ternary MeO-SnO_2-SiO_2 compounds are reported in the literature. Dry granite melts dissolve about 4 wt% of SnO_2 at 1300 °C and the data extrapolated to 975 °C gave 0.7 wt% SnO_2 dissolved. This value is close to the values of SnO_2 solubility in granitic melts in equilibrium with tin-containing hydrothermal solutions.

Tungsten and molybdenum: Both the elements behave similarly in alkali silicate binary melts. Their oxides are soluble in simple binary Na_2O-SiO_2 melts (Stemprok, Voldán 1974). Alkali silicate and granitic melts can be in stable coexistence with alkali, tungstates or molybdates as immiscible liquids. In contrast to tin, there are no solid liquidus tungsten or molybdenum phases in binary alkali oxide silica or granite melts with oxides of these elements. In granite melts the solubility of WO_3 and MoO_3 is lower than that of SnO_2 (Stemprok, Voldán 1982, 1983). The amount of WO_3 dissolved in granitic melts near the temperature of granite solidus (0.6–0.9 wt%) is close to the values found by Manning and Henderson (1984) in equilibrium with tungsten containing hydrothermal solutions.

The studies of tungsten coordination in tungsten–silica–sodium oxide glasses by optical and electron paramagnetic resonance spectroscopy as reported in the literature indicate the presence of WO_6 units surrounded by silicon and sodium cations. Similar MoO_6 units are postulated to explain the structure of molybdenum oxide glasses with alkali oxides and phosphates (Rawson 1967).

Interpretation and geological application: Radius-ratio conditions for Sn^{4+}, Mo^{6+} and W^{6+} cations (Levin, Block 1957) in silicate melts require octahedral coordination with oxygen. The structural position of these highly charged cations in felsic magmas can be explained by bonds with six bridging and nonbridging oxygens. While bridging oxygens are shared with oxygens of SiO_2 units, nonbridging oxygens are balanced by alkali cations. The close dependence of Sn, W and Mo on SiO_2 units in felsic magmas can explain why these elements are concentrated in granitoid complexes with high silica granites or rhyolites which contain alkali cations as main network modifiers. The release of Sn, W and Mo cations from coordination of SiO_2 units occurred probably by increased alkalies in residual melts of higher evolved granites as it is testified by common alkalization accompanying the origin of tin, tungsten and molybdenum deposits of magmatic affiliation.

REFERENCES

Dingwell, D. B., 1988, The structure and properties of fluorine-rich magmas: a review of experimental studies, R. P. Taylor and D. F. Strong, eds., Recent Advances in the Geology of Granite-Related Mineral Deposits: Spec. Volume 39, The Canadian Institute of Mining and Metal., Montreal.

Levin, E. M., S. Block, 1957, Structural interpretation of immiscibility in oxide systems; II, Coordination Principles Applied to Immiscibility, J. Amer. Ceram. Soc., v. 40, p. 113–117.

Manning, D. A. C., P. Henderson, 1984, The behavior of tungsten in granitic melt-vapor systems; Contrib. Mineral. Petrol. v. 86, p. 286–293.

Mysen, B. O., 1987, Magmatic silicate melts: Relations between bulk composition, structure and properties, B. O. Mysen, ed., Magmatic Processes: Physico-chemical Principles, The Geochemical Society, Special Publication No. 1, p. 375–399.

Rawson, H., 1967, Inorganic glass-forming systems, London and New York, Academic Press.

Sitek, J., Stemprok M., J. Voldán, O. Cicáková, 1981, Mössbauer spectra of granitic glasses containing tin, Silikaáty v. 25, p. 243–249.

Stemprok, M., J. Voldán, 1974, Homogeneous glasses in the systems Na_2O-SiO_2-WO_3 and Na_2O-SiO_2-MoO_3. Silikáty v. 18, p. 19–30.

——— ——— 1975, Solubility in SnO_2 in sodium silicate melts. Silikáty v. 19, p. 211–211.

——— ——— 1979, Solubility of tin dioxide in granitic melts enriched in sodium oxide, Výzkumné práce ÚÚG, n. 22, p. 5–27.

——— ——— 1982, Solubility of tungstic oxide in granite melts, Vest. ÚÚG, n. 57, p. 329–340.

——— ——— 1983, Solubility of molybdenum oxide in granite melts, Vest. ÚÚG, n. 58, p. 79–89.

STEPHAN, J. F., et al, Université de Nice-Sophia Antipolis, Valbonne, France

From Subduction to Collision off Southern and Eastern Taiwan: Results of R/V *Jean-Charcot* "POP 2" Cruise

The island of Taiwan is the result of the active collision between the Luzon arc (which corresponds to the Philippine Sea Plate western edge) and the Chinese continental margin (which belongs to the Eurasia Plate). This collision zone is located in the northern prolongation of the Manila trench subduction zone, along which the South China Sea oceanic basin is subducted to the east-southeast beneath the Luzon arc. Transition from active subduction to active collision is progressive in a northward direction. On the contrary, to the east of Taiwan the contact between the collision zone and the Ryukyu active margin is sharp. The latter is related to the northwestward subduction of the west Philippine Sea oceanic basin beneath the Eurasian margin.

The structural study of the two key areas, South and East Taiwan, was the main goal of the R/V *Jean-Charcot* "POP 2" cruise conducted in October–November 1984. Seabeam mapping together with high-resolution single-channel seismic reflection profilings, gravimetry, and magnetic survey have been completed at the northern end of the Manila subduction zone between 20°15′ and 21°45′ and at the western termination of the Ryukyu active margin between 23° and 24°30′N and 121°45′ and 122°45′E. The main results are as follows.

Off southern Taiwan: South of 21°15′, the Manila active margin consists from west to east of: the Manila accretionary prism; the North Luzon fore-arc trough; the Luzon arc (active on the Batan island, at about 20°30′N). The prism itself is composed of: a western wedge undergoing deformations that consist of westward folding and thrusting; a central zone, tectonically inactive (= paleoprism), which corresponds to the so-called Hengchun submarine ridge; an eastern narrow belt where active deformations are related to the back-thrusting of the prism towards the fore-arc trough.

At about 21°15′, the fore-arc trough (a subduction marker) disappears; it is relayed to the east by a narrower and a shallower trough lying on the Luzon volcanic arc. This last trough is progressively folded and faulted northward and finally emerges to form the Taiwan Coast Range.

The active prism continues, with minor bends and relays, up to the southwest of Taiwan, at about 22°N. The Manila deformation front is not directly connected to the Taiwan frontal thrust located 50–70 km to the east, a right-lateral strike-slip offset is suggested between both fronts.

The inactive prism, corresponding to the Hengchun ridge, passes via the Hengchun Peninsula to the Taiwan Central Range, the most internal and the highest unit of the collision build up; as for the Manila prism, this unit was first submitted to westward folding and thrusting prior to being involved in eastward back-thrusting.

The structural interpretation seems to point to the presence of a thinned continental crust between 20°30′ and 21°30′, south of the continental slope.

The whole data suggest that the subduction/collision transition lies at about 21°30′.

Off eastern Taiwan: East of Taiwan and from south to north, the Ryukyu active margin consists of: the Ryukyu trench; the Yaeyama accretionary prism; the Nanao fore-arc trough; the Ryukyu arc, including the Ryukyu archipelago and, offset to the north, the submarine volcanic arc; the Okinawa back-arc basin at the early stages of oceanization.

The accretionary prism and the fore-arc trough, both trending east-west at about 123°50′, slightly bend to the west-northwest near Taiwan and then take a sharp north-south direction at about 122°E immediately east of the island; the accretionary prism is there in contact with the eastern flank of the northern Luzon arc.

The Ryukyu deformation front and the Nanao fore-arc trough both disappear against the northern end of the N20E trending Taiwan Longitudinal Valley Fault; the connection between the two systems lies near the Taiwan east coast north-northeast of Hualien and south of Suao, between 24°15′ and 24°30′.

The west-northwest–east-southeast trend of the Ryukyu accretionary prism near Taiwan cannot be explained by the northwest-southeast convergence between the Eurasia and Philippine Sea plates. The local kinematic appears to be strongly controlled by the Taiwan Longitudinal Valley Fault in this geometrically complex area; other factors could interfere, such as the bending of the southern Ryukyu arc in relation with the opening of the Okinawa Basin.

STEPHAN, J. F., et al, Université de Nice-Sophia Antipolis, Valbonne, France

Paleogeographic and Geodynamic Evolution of Caribbean–Tethys Realm for Past 200 Ma Illustrated with 13 Printed Maps

In the frame of the French Tethys Program the Caribbean Tethys Group has elaborated 13 paleomaps at 1:20,000,000 in order to describe the paleogeographic evolution and the tectonic history of the Caribbean region for the last 200 Ma. The following key periods have been selected: Triassic-Lias boundary for the fit, Early Callovian (Blake Spur Anom.), Upper Oxfordian (m.25), Neocomian (M.10), Aptian (M.0), Santonian–Campanian boundary (A.34), Cretaceous–Tertiary boundary (A.30), Paleocene (A.25), Early Eocene (A.21), Late Eocene (A.13), Early Miocene (A.6), Late Miocene (A.5), and Present.

Relative position of South America versus North America for each step has been obtained using the poles of rotation proposed by various authors (Klitgord and Schouten, 1987; Olivet et al, 1984 and unpublished, for Africa versus South American; Cande et al, 1986; Pindell et al, in press, for Africa versus South America.). The selection was made on the basis of some key criteria. For example, the loose fit of Olivet et al (1984) between Africa and North America has been preferred to tighter fits; the same choice was made in favor of Pindell et al (in press) fit between South America and Northern Africa.

In the proposed reconstructions, great importance has been attached to all geological data, among others those collected by the members of the Caribbean Tethys group in almost every part of the region. Working that way enabled the group to date precisely regional geodynamic events and to determine major steps in the evolution of the Caribbean.

Hereafter we give as an example the Late Cretaceous to Present period in which five main intervals have been recognized by the group:

1. The Late Cretaceous interval, marked by the first collision between the Antillean paleo-arc—located at the front of the proto-Caribbean plate moving north-northeastwards—and the continental margins of the Caribbean Tethys, i.e., southern Florida and eastern Yucatan to the north, western Colombia to the south.
2. The Maestrichtian–Paleocene interval during which a deep kinematic and tectonic reorganization takes place in the northern proto-Caribbean plate; it corresponds to the opening of the Yucatan oceanic basin and to intense lift-lateral strike-slip faulting through the (present-day) Greater Antilles.
3. The Eocene interval during which the paleo-Caribbean plate starts moving eastwards; this motion is accompanied, to the south, by the oblique collision of the Antillean paleo-arc with the South American Tethyan margin, to the north by left-lateral transpressive deformations (from southeastern Cuba to Puerto Rico and Virgin Islands) in relation with the birth of the Cayman Trough.
4. The Oligocene–Early Miocene interval that follows a period of intense tectonism is characterized by contrast by the development of subsident elongated basins disconformably overlying the previous deformed belts; furthermore an 8 Ma interruption in the magmatic activity is observed along the Lesser Antilles arc at that time; eastward motion of the Caribbean plate seems to have stopped or considerably slowed down during that interval.
5. The Middle Miocene up to Present interval marked by a strong reactivation of movements and tectonism almost everywhere in the Caribbean; northern and southern plate boundary zones are characterized by intense transpressive deformations still active today.

REFERENCES

Klitgord, K. D., and H. Schouten, 1987, Plate kinematics of the central Atlantic *in* B. E. Tucholke and P. R. Vogt, eds., The geology of North America: the western Atlantic Region: GSA DNAG Series, v. 1, p. 351–378.

Olivet, J. L., J. Bonnin, P. Beuzart, and J. M. Auzende, 1984, Cinématique de l'Atlantique Nord et Central: Rapports Scientifiques et techniques n. 54, IFREMER, 2 vol.

Pindell, J. L., S. C. Cande, W. C. Pitman III, D. B. Rowley, J. F. Dewey, J. Labrecque, and W. Haxby, in press, A plate-kinematic framework for models of Caribbean evolution *in* Sager and Scotese, eds., Geodynamic Symposium, Tectonophysics.

STEPHAN, J. F., P. MALETERRE, N. PINET, J. C. RINGENBACH, and J. CHOROWICZ, Université de Nice-Sophia Antipolis, Nice, France

Geometry and Chronology of Transpressive Deformations on Northern Termination of Philippine Fault in Luzon Central Cordillera (Philippines)

The Philippine Fault is one of the main tectonic features of the Philippines archipelago. It is a left-lateral strike-slip fault, more than 1300 km long and extending from Mindanao in the south to northern Luzon in the north. At its northern extremity, the fault splits in various branches forming a horse tail pattern. This tectonized and uplifted zone corresponds to the Central Cordillera, a 300 km long, 50–80 km wide, and 1500–3000 m high range.

A multidisciplinary study including biostratigraphy, petrology, $^{40}K/^{40}Ar$ dating, and tectonics has been conducted on Central Cordillera, especially on the axis and the western flank of the range.

The Central Cordillera axis appears to be a multistage volcano-plutonic arc composed of: (1) a pre–Upper Eocene ophiolitic basement (probably Cretaceous in some places); (2) an Upper Eocene to Upper Miocene volcanic and volcanoclastic sequence intruded by Oligocene to upper Miocene granodiorites; volcanic activity stopped at the end of Lower Miocene, beginning in Middle Miocene when more than 100 m of shallow-water limestones were deposited. (3) Uppermost Miocene and Plio-Pleistocene volcanics and volcanoclastic sediments, locally preserved, disconformably overlying the oldest pile.

The western flank of the Cordillera is a subsident basin (the so-called "locos trough"), filled up with to 5–7 km of Oligocene to Pleistocene volcanoclastic sediments, fed by the volcanoplutonic axis.

The tectonic framework of the Central Cordillera is controlled by different branches of the Philippine fault: the Dalton, Abra river, and Tebbo shear zones in the range axis; Tuba, Pugo, and the coastal strike-slip faults and thrusts on the western flank. These faults delimit tectonic strips trending northwest-southwest to north-south in which east-west shortening is generally small. This is not the case for the coastal strip where folding and thrusting has been intense.

Three main episodes of deformation are recorded in the central Cordillera: Middle to Upper Miocene, Lower Pliocene, and Pleistocene to Recent. The first deformations are limited to the present-day axis of the range and related to the motion of the Dalton and Abra river left-lateral shear-zone; this episode is marked by a strong uplift and is followed by the emplacement of intramountainous basins located along the transcurrent faults in releasing bend portion (for example, the Cervantes basin on the Abra river fault). The most recent deformations are associated with the present-day motion of the Philippine fault. The Quaternary tectonic activity is concentrated in the westernmost part of the Central Cordillera and along the coast of the Lingayen Gulf and the South China Sea. To the south, in the Damortis and San Fernando area the recent thrust is responsible for the Pleistocene to Holocene sediment tilting. Northwards, the continuation of this coastal thrust (the active Vigan–Aggao wrench fault) is characterized by a left-lateral component and is marked out by a positive flower structure.

The present-day tectonic regime shows some important variations along the Central Cordillera: (1) *From the east to the west* with a domain controlled by strike-slip faulting and a domain marked by thrust or wrench fault associated with major folds; and (2) *from the south to the north* with a decreasing of the convergent component. This variation is associated with the change in the orientation of the Philippine fault from N140E in southern Luzon to N10E in northern Luzon.

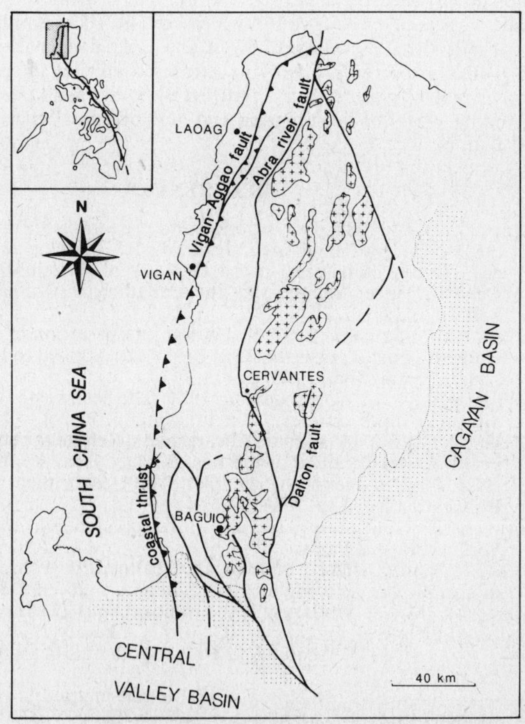

Figure 1.

STEPHANSSON, O., and C. LJUNGGREN, Luleå University of Technology, Luleå, Sweden

Compilation of Stress Measurements in Europe

Stress data from overcoring measurements, hydraulic fracturing, and hydraulic tests on preexisting fractures compiled from the majority of the European countries have been compiled for the World Stress Map Project.

This contribution contains a general description of the methods of rock stress measurements applied to the measurements included in the European database. The data collected have been put in four groups according to a quality ranking system. The outcome of the ranking is discussed, and the data from each group will be presented as stress plots of the direction of maximum horizontal stress.

STEPHENS, M. B., Geological Survey of Sweden, Uppsala, Sweden

Terranes and Their Accretionary History in Caledonides of Scandinavia and Svalbard: A Review

The Caledonides in Scandinavia, Svalbard, and East Greenland form a northerly directed protrusion away from the main assembly of Paleozoic orogens in the circum-Atlantic area. A key factor that separates these mountain belts from those in the main assembly is the completion of orogenic activity during the Devonian. This paper focuses attention on the character and accretionary history of the terraces in the contrasting Caledonide segments of Scandinavia and Svalbard.

On Svalbard, platformal Carboniferous and younger rocks overlie folded late Silurian to mid-Devonian Old Red Sandstone (ORS) strata, and older generally more deformed rocks. The pre-ORS remnants outcrop in three separate areas in northeastern, northwestern, and southwestern Svalbard. The northern areas are separated by a major high-angle fault and an ORS graben, and the western areas by a major fault. Significant late Ordovician to Silurian tectonothermal activity has been envisaged for the pre-ORS rocks. Recent work has emphasized the stratigraphic dissimilarity between these older remnants and suggested that they represent at least three separate terranes.

Medium and high-grade metamorphic complexes of uncertain age comprise the structurally deeper parts of the successions in the northeastern and southwestern terranes. Overlying late Proterozoic and early Paleozoic strata include Vendian tillites and a Cambrian to early Ordovician carbonate-dominated sequence containing faunas with North American affinities. These terranes are distinguished, for example, by the occurrence of volcanites and, locally, eclogites/blueschists unconformably overlain by mid-Ordovician to Silurian rocks, solely in the southwestern segment. The northwestern terrane is composed entirely of medium- and high-grade metamorphic rocks, locally with eclogites, of uncertain age.

The contrasts between these terranes have been explained, somewhat controversially, in terms of large-scale, sinistral strike-slip displacement along their bounding faults and during the late Devonian. Such a tectonic model relies heavily on the stratigraphic similarity of the sequence in the northeastern terrane with that in central East Greenland. Arguments for a pre–late Devonian palinspastic restoration of the southwestern terrane to the Innuitian mountain belt have also been presented. It is apparent, however, that the postulated, final strike-slip assembly of these terranes masks at least one earlier Caledonian accretionary event as witnessed by the blueschists and eclogites in the southwestern terrane.

In marked contrast to Svalbard, the structure of the Scandinavian Caledonides is dominated by thrust-emplaced tectonic units. Recent work has emphasized the possible importance of late extensional tectonics during the Devonian. Viewed in terms of the terrane concept, the Scandinavian Caledonides can be divided into three principal components: the tectonically shortened Baltoscandian margin of the early Paleozoic continent Baltica, terranes strongly suspected but as yet not unequivocally demonstrated to have formed the outer part of this margin, and, finally, exotic terranes that started their history outboard of the continent Baltica. Attention is concentrated here on the outboard terranes. These occur in the structurally higher tectonostratigraphic units and are partly isolated from each other due to the deep erosion down through the mountain belt. Terrane identification concerns pre–late Ordovician signatures, and three principal groups have been recognized: terranes containing vestiges of oceanic sequences, segments composed of metamorphic complexes of less-certain affinity, and, finally, a super-terrane in the highest thrust sheet composed predominantly of exotic continental lithosphere.

Critical signatures in the first group of terranes include ophiolites, bimodal metaigneous suites of low-K character displaying both subduction- and rift-related components, early Ordovician faunas showing both North American and mixed North American-Baltoscandian affinities, and younger clastic rocks derived from distinctive source regions. U-Pb (zircon) data from ophiolite complexes and the bimodal metaigneous suites suggest a narrow age interval for these rocks in the early Ordovician. Linkage of several of these terranes to arc-basin systems marginal with respect to Iapetus has been proposed; systems outboard of Laurentia and Baltica are envisaged.

Common features of the group of terranes containing generally higher-grade metamorphic complexes are their dominance by clastic metasedimentary and subordinate mafic metavolcanic rocks, and the evidence for an early Caledonian tectonothermal evolution. U-Pb (zircon) and Rb-Sr (whole-rock) data in one such volcanosedimentary sequence demonstrate the presence of Ordovician elements and restrict the early Caledonian tectonothermal activity to early or mid-Ordovician. The super-terrane in the highest thrust sheet is composite in character with oceanic and predominantly continental components sealed prior to intrusion of syntectonic granitoids. The continental component is exotic with respect to Baltica.

Recognition of early Caledonian accretionary processes is masked in most of the Scandinavian Calidonides by tectonic disruption during later Silurian–Devonian continent–continent collision. Nevertheless, a complex polyphase accretionary history is apparent. Initial late Cambrian to early Ordovician collision of the Baltoscandian margin with an outboard arc complex preceded uplift and erosion during the Ordovician, and subsequent tectonic shortening of this margin during the Silurian–Devonian. Separate high-pressure metamorphic events, related to subduction of continental lithosphere, are linked to these respective collisions. Evidence for early Caledonian accretionary processes involving the outboard terranes includes recognition of composite terranes, remnants of an early or mid-Ordovician tectonothermal evolution within certain fragments, and eventual sharing of clastic sequences and/or igneous signatures. A key dispute remains between two distinctive accretionary models. Model 1 suggests sealing of the outboard terranes to Baltica in connection with the early Caledonian tectonothermal evolution along its Baltoscandian margin. Model 2, favored here, argues for accretion of many outboard terranes initially to Laurentia, a subsequent extensional regime and an ultimate collision with Baltica during the Silurian–Devonian.

STEPHENS, R., and W. NOVAK, U.S. Armny Corps of Engineers, Jacksonville, FL

Foundation Investigation for Portugues Dam

The Portugues dam is the first thin arch dam designed by the U. S. Army Corps of Engineers and is under the jurisdiction of the Jacksonville District. The dam will be located near Ponce, Puerto Rico, in a seismically active region.

A detailed investigation of the foundation conditions at the Portugues damsite was undertaken to determine the characteristics of the rock mass with respect to geologic structure, hydraulic conductivity, abutment stability, and deformation modulus. The methods used to obtain the required data included drilling vertical and inclined core borings, logging the core holes with a video camera, water pressure testing of the core holes, borehole jack tests, excavation of exploratory adits in each abutment, surface and underground geologic mapping, in situ uniaxial jacking tests, and laboratory testing of soil and rock materials.

The foundation rock is Cretaceous volcanic sediments metamorphosed by an Eocene diorite stock. Metamorphism destroyed bedding and the original clastic nature of the rock, creating a massive, homogeneous unit. The major structural features at the site are joints, shear zones, and dikes. All these features appear related to the intrusion as opposed to regional tectonic activity. Rock quality varied from fresh with joint spacing of 2–3 f to highly weathered shear zones with clay gouge.

Due to the three-dimensional nature of an arch dam and its applied loads, knowledge of the orientations of discontinuities within the foundation rock mass is vital to an assessment of the stability of the abutments. Orientation of the joints, shears, and dikes was determined by mapping in the exploratory adits and logging of boreholes. Over 5000 strike and dip measurements were recorded. These data were analyzed by computer to identify significant orientations. Each abutment was assumed to be a separate domain. Similar values of strike and dip of features were found within each abutment from the two sources of data. Two joint sets were identified on the right abutment and three joint sets on the left abutment. Six shear zones were located in the left abutment and one shear zone on the right.

The hydraulic conductivity of the foundation is important as it relates to seepage, grouting, and uplift on the structure and wedges in the foundation. The Portugues River generally has a low flow, so minimizing foundation seepage is important to filling the reservoir. A grout curtain as well as a drainage system will be necessary to the project due to the upstream-downstream orientation of joints and shears. Permeability of the foundation was determined rationally from water pressure tests in boreholes along the axis. This permeability was used in a two-dimensional finite element seepage program to estimate seepage and uplift. Uplift forces on the wedge planes are used in the stability analyses.

The stability analyses of the abutments were carried out in two steps: first, identification of wedges that were kinematically able to fail by sliding along one or two planes out of the abutment, and second, calculation of the factor of safety for sliding of these kinematically able wedges. No rotational failures were considered due to the confinement from the dam. Additional information required for the analyses included shear strength of discontinuity planes (joints and shears), unit weight of rock, uplift water pressures on the wedge faces, applied loads from the dam and reservoir, and earthquake accelerations for the dynamic cases. Five load cases were analyzed, three static and two dynamic. The factor of safety criteria were chosen to be consistent with the structural design factors of safety for the corresponding load cases.

The dam is relatively thin and subject to deformations under load and the stiffness of the foundation affects the magnitude of deformation and stress in the concrete. To conduct the structural analysis, the magnitude and variability of the foundation stiffness must be known. The relevant stiffness in this case is the deformation modulus or the composite stiffness of the rock mass including its defects. Plate bearing tests (uniaxial jacking tests) were carried out in exploratory adits to determine the deformation modulus in rock representative of the majority of the foundation conditions. Eight tests were conducted in four locations at the site. In order to extrapolate the results to the entire site, it was decided to employ a rock mass classification scheme to correlate the rock conditions at the eight test sits with the value of in situ deformation modulus. Thus, a predictive model could be established for determining the value of modulus throughout the damsite. The rock mass classification scheme chosen was Bieniawski's Geomechanics classification system for rock mass rating (RMR). The literature describes a correlation between deformation modulus and RMR based on previous world-wide experience; the results at the Portugues site agreed well with the correlation. Thus, by logging the rock cores to establish the RMR, the value of in situ deformation modulus throughout the site could be predicted. As a confirmation of the predictive model, the Goodman borehole jack was used in several boreholes to measure deformation modulus. A map of deformation modulus was prepared and used by the structural engineers in developing the finite element model to analyze the stresses in the dam.

The results of the investigations of the proposed dam site indicate a satisfactory foundation for an arch dam. No defects exist that could not be mitigated by usual geologic and engineering practice in dam foundation preparation.

STEPHENSON, J., and A. C. SCOTT, University of London, Egham, United Kingdom

Insect-Related Paleophytopathology

In 1923, E. W. Berry described the study of plant pathology as *"not only symptoms and causes of the maladies which threaten the lives of plants, but* **also** *those that result in abnormalities of structure, form or appearance which are either directly injurious or even merely unsightly."* He outlined a number of fossil specimens that show pathological conditions and attempted to suggest their possible causes. Many were attributed to insects. Here we attempt to update Berry's review by assessing different examples of paleophytopathology that may be due to insect attack. We also describe new material from Carboniferous, Cretaceous, and Lower Tertiary localities.

Insect-related paleophytopathology expresses itself in four major forms. These include the actual evidence of insect feeding on living plant tissues and the related wounding reactions shown by the plant. We will study wood borings both internally and on the surface of the plant tissues, leaf mining, and gall production.

Insect Feeding/Paleophytophagy: All parts of a plant may be subject to insect attack. Spores appear to be the most primitive food supply being utilized as early as the Upper Carboniferous. Feeding on the vegetative structures such as leaves is also known from late Pennsylvanian times (Scott and Taylor, 1983). Cretaceous and Tertiary localities have yielded many angiosperm specimens; some are so well preserved that the responsible insect Order has also been suggested.

When living plant tissues are damaged, a wound reaction is set up. This involves the production of a callus around the infected/wound area. The surrounding cells undergo hypertrophy and produce toxins to deter further feeding. The resultant thickened ridge of cells are often darker in color and may be fossilized, enabling us to discriminate between damage acquired while the plant tissues were alive and after death.

Many insects also act as vectors or carriers of plant diseases and fungi. The presence of these pathogens in the fossil material may often be due to a direct infection by insects feeding on the plant. The appearance of wound areas indicate infections acquired during the life of the plant.

Wood Borings: These fall into two categories. First there are internal borings where the insect burrows have been found deep inside the woody tissues. These have been described from very early material including the presence of coprolites within Marattiaceous fern stems (*Psaronious magnificus*) from the Upper Pennsylvanian (Rothwell and Scott, 1983). However, it is impossible to ascertain whether the plant was alive or dead when used by the insect.

The second type of wood borings involve those produced on the surface of the plant material, either in the bark or on the outer surface of the xylem tissue. These are produced primarily by the family Curculionidae in the Coleoptera. One example that dates to the Wealden deposits of southern England may be attributed to the subfamily Scolytinae, which is significant as a possible indicator of the presence of early insect disease carriers (such as in Dutch Elm disease).

Leaf Mines: Although most common in angiosperm material from Cretaceous and Tertiary deposits, the oldest example described is in *Callipteris conferta* from the Permian (Potonié, 1893). Of the four insect Orders that show this habit—Coleoptera, Diptera, Hymenoptera, and Lepidoptera—only the Coleoptera were present at this time. The larvae of leaf mining insects feed on the tissues between the upper and lower epidermal layers leaving either a long sinuous trail or a large open blotch-like area, both of which may be fossilized. These trails are characteristic of the insect producer and may be used for identification.

Galls: These are a physiological response by the plant to the deposition of insect eggs either on the surface or in the epidermal tissues. The surrounding cells either undergo hypertrophy or rapid multiplication and smother the insect egg in an attempt to seal off the infected area. This ball of tissue is often induced by the insect as a means of protection from the external environment and for food. A great variety of galls have been described; they may be classified and used to identify both the host and the insect infection.

There are other pieces of indirect evidence that may suggest the involvement of insects in paleophytopathology. These include the presence of fungal spores and bacteria in insect coprolites and gut contents, but they must be viewed with great caution. Further study of fossil material will add to our understanding of plant–insect interactions, but it can be seen that the close association between insects and the spread/cause of plant diseases is a very ancient one.

REFERENCES

Berry, E. W., 1923, Pathological conditions among fossil plants, *in* R. L. Moodie, Paleopathology: an introduction to the study of ancient evidences of disease, p. 99–114.

Potonié, H., 1893, Die flora des Rothliegenden von Thüringen: Abd. Preuss. Geol. Landesanstalt., v 9, p. 29, fig. 1 & 2 taf. 1.

Rothwell, G. W., and A. C. Scott, 1983, Coprolites within Marattiaceous fern stems (*Psaronius magnificus*) from the Upper Pennsylvanian of the Appalachian Basin, U.S.A.: Palaeogeography, Palaeoclimatology, Palaeoecology, v. 41, p. 227–232.

Scott, A. C., and T. N. Taylor, 1983, Plant/animal interactions during the Upper Carboniferous: Botanical Review, v. 49, p. 259–307.

STERN, R. J., and W. I. MANTON, University of Texas at Dallas, Richardson, TX and A. KRÖNER, T. REISCHMANN, and I. HUSSEIN, Johannes-Gutenberg University, Mainz, West Germany

Rb–Sr and U–Pb Geochronologic Constraints on Late Precambrian Crustal Evolution in Northeast Sudan

Formation of the continental crust now exposed in the Red Sea Hills of the Sudan, between the Nile and the Red Sea, took place during the Pan African Event (sensu lato) between 900 and 500 Ma ago, with >95% of the crust being formed by 650 Ma ago. We have been studying this episode in crust-formation using Rb–Sr whole-rock (hereafter denoted 'Rb–Sr') and U–Pb conventional ('cZ') and single-crystal ('sZ') techniques. We have taken as our working model the interpretation, based on studies of the Arabian Shield, that Red Sea Hills crust formed by the initiation, maturation, and amalgamation of arc terranes. These terranes must have been accreted to an older Proterozoic or Archean N. African craton. Evidence for the location of the eastern margin of the pre-Pan African craton can be found in several places along the Nile, where rocks giving very old, but massively reset Rb–Sr ages, high initial $^{87}Sr/^{86}Sr$, mid- to early-Proterozoic Nd model ages, or containing old detrital zircons may be documented.

The crust of the Red Sea Hills gives Pan-African ages with low initial $^{87}Sr/^{86}Sr$ (0.704) and high initial Epsilon-Nd (> +6). The isotopic data agree with geochemical data indicating that this crust began as ensimatic arc material. At least five terranes may be identified in the Red Sea Hills (Fig. 1). The Gerf Terrane comprises the SE Desert of Egypt between the Hafafit Culmination (not shown) and the Allaqi Suture; the tract east of the northernmost Hamisana Shear Zone and the Onib-Sol Hamed Suture is probably also part of the Gerf Terrane. Ages as old as 765 Ma (Rb-Sr) have been determined for rocks of the Gerf Terrane, although the bulk of the crust appears to have formed at 740–680 Ma (Rb-Sr, cZ, sZ). The ophiolites of the Gerf region formed at 741 + /-21 Ma (sZ), and those farther west along the Allaqi Suture were emplaced by 732 + /-15 Ma (sZ), considerably younger than crystallization ages of 808 + /-14 Ma (sZ) for rocks of the Onib ophiolite. The Gabagaba and Gebeit Terranes may be a single terrane that was disrupted by the formation of the Hamisana Shear Zone, about 610–560 Ma. Rocks of the Gabgaba Terrane are very poorly known; a large granodiorite pluton in its NE portion is 690 + /-3 Ma (cZ). This body has an initial $^{87}Sr/^{86}Sr$ of 0.7027, indicating its ensimatic affinities. The oldest rocks known from the Gebeit Terrane are the 832 + /-26 Ma andesitic metavolcanics (Sm/Nd isochron). Slightly younger are metamorphosed rhyodacites of the Kashebib Gneiss (778 + /-30 Ma (Rb–Sr)). A very important phase of synorogenic granodiorite plutonism occurred in the Gabgaba and Gebeit Terranes between 720 and 660 Ma (Rb-Sr).

The age of ophiolites in the Nakasib Suture is presently unknown, but this is an important structure that separates generally younger rocks of the Gebeit Terrane from generally older rocks of the Haya Terrane. The oldest intrusive bodies are in the Haya Terrane. The Erkowit granite gives ages of 852 + /-3 Ma (cZ) and 856 + /-5 Ma (sZ), while gneiss and dikes from Khor Ashat, just south of Erkowit, gives ages of 870 + /-5 Ma (sZ) and 860 + /-55 Ma (Rb-Sr), respectively. Bimodal volcanics around Tokar are 854 + /-9Ma (sZ); these commonly have reset Rb/Sr whole rock ages that record late- to post-orogenic igneous activity at 720 to 650 Ma (Rb-Sr, sZ). We do not yet have any geochronologic data with which to evaluate whether or not the Baraka Suture (?) represents an important terrane boundary.

In summary, our data support models calling for the formation of the crust of NE Sudan by the generation and maturation of intra-oceanic arcs into ensimatic terranes, consistent with models for terrane accretion in the Arabian Shield. Our data support the inference that terranes began to develop earlier in the SE, with a progression of primitive arc suites, post-amalgation calc-alkaline plutonism, and post-orogenic alkaline magmatism younging to the NW. This data further indicates that these terranes were amalgamated together and against the flank of older crust now preserved in North Afric west of the Nile. The exposed basement of NE Afric and W Arabia thus appears to represent one of the earliest unequivocal examples of continental crust formation by processes of plate tectonics and terrane accretion similar to those presently in operation.

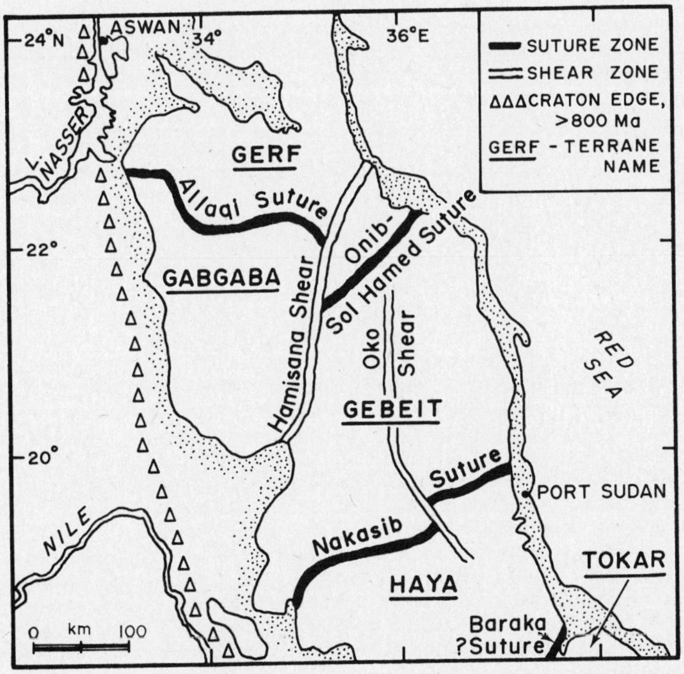

Figure 1—Generalized terrane map for northeast Sudan and southeast Egypt. The location of the pre-800 Ma craton edge is very approximate.

STERNBERG, R. S., Franklin and Marshall College, Lancaster, PA

Use of Laptop Computer and Spreadsheet for Geophysical Surveys

The recent development of the microcomputer has made powerful computing readily available on the geoscientist's desktop for a moderate price. In the last few years, IBM-compatible laptop models have also proliferated. Although largely marketed toward the traveling writer or business executive, these computers are also valuable to the traveling scientist, including the field geologist or geophysicist. A typical laptop might have 640K RAM, two 720K 3 1/2" disk drives, and be able to run on rechargeable batteries for about 4 hours. Hard disks are also available. These computers are small enough to fit in a briefcase, weigh 5–7 kg, and can be bought in the United States for under $1500.

Perhaps the single most useful type of software in geophysical surveying applications is the spreadsheet (see, e.g., Arganbright, 1985; Laudin, 1986; Dory, 1988). The electronic spreadsheet is modeled after the accountant's ledger. It allows for entries, either numeric or text, to be made in individual cells. Numeric cell entries can be algebraically manipulated by entering equations into the cells, which can include a number of intrinsic functions (e.g., trigonometric). The algebraic equations can include absolute references to constants or relative references to variables. Elementary database capabilities and reasonably good graphics are incorporated into many spreadsheet packages. The most popular spreadsheet for the IBM-PC and compatibles is Lotus 1-2-3, although other excellent spreadsheets are also available. These typically can be run from a single 720K disk, require 256K of RAM, and may be purchased for under $200.

As an example of a spreadsheet application, one might wish to reduce gravity data for calculation of free-air or Bouguer anomalies. Values for the universal constant G and base station measurements, times, and meter calibration information for those readings can be set into cells at the top of the spreadsheet. Columns of cells can contain the input data: station numbers, meter readings, north-south distances relative to the base station, elevations, and densities to be used for the free-air and Bouguer corrections. Adjacent columns can contain the equations for the conversion of the readings from dial units to milligals, the drift correction, the latitude correction, the free-air correction and anomaly, and the Bouguer correction and anomaly. The constants are selected from the appropriate cells as absolute references, and the input data for each of the readings are referred to through the use of relative references. After an equation (e.g., for the free-air anomaly) has been entered once for the first reading, it can be copied with a spreadsheet copy command to the equivalent cells for all other readings without manually reentering the equation. When an equation is copied, references to relative cell addresses are also modified accordingly; thus, the anomaly equation in a given row will always utilize the input data for the same row (measurement). The constants can easily be changed at any time to represent a new problem (e.g., new base station), and the dependent spreadsheet values can be automatically updated.

Problems such as these are generally easier to set up with spreadsheets than with programming languages such as BASIC or FORTRAN. A student can learn the rudiments of a spreadsheet package within a couple of hours. Equations can be easily modified, and extra columns and rows can be added to the spreadsheet as necessary. Generation of graphs is easily accomplished once the spreadsheet is set up.

Other spreadsheets that have proven useful in the field and in the lab include: plotting gravity anomaly profiles; reduction of magnetic data (drift corrections) and plotting of profiles; reduction of electrical resistivity data (calculation of apparent resistivities) and plotting of sounding and profiling results; modeling of gravity and magnetic anomalies; plotting of seismic traces.

The laptop computer is also useful in the field for data acquisition. Several modern instruments (e.g., signal enhancement seismographs, proton precession magnetometers, electromagnetic conductivity meters, electronic distance meters) are capable of storing hundreds of digital measurements internally, which can later be transferred to the computer through a standard serial RS232 port, available on most laptops. The data acquisition programs themselves are generally simple programs written in BASIC, which is bundled with many laptops, or can be purchased separately for under $100. These capabilities for data acquisition, reduction, and modeling allow for the possibility of determining whether additional data are necessary while one is still in the field.

A laptop computer will be available during the poster session to demonstrate these applications. Copies of the programs and spreadsheet templates are available from the author.

REFERENCES

Arganbright, D. E., 1985, Mathematical applications of electronic spreadsheets: McGraw-Hill Book Company, 165 p.

Dory, R. A., 1988, Spreadsheets for physics: Computers in Physics, May/June, p. 70–74.

Laudon, R. C., 1986, Using spreadsheet software for gradebooks: Journal of Geological Education, v. 34, n. 2, p. 106–107.

STERNER, S. M., D. L. HALL, and R. J. BODNAR, Virginia Polytechnic Institute and State University, Blacksburg, VA

Fluid Immiscibility in Geologic Environments: Interpretation of Fluid Inclusion Data Using Available Fluid Phase Equilibria

There is increasing evidence, both natural and experimental, that immiscibility exists in salt-water and carbon dioxide–salt-water fluid systems over a substantial portion of geologically relevant P-T space. Fluid immiscibility has been proposed in a number of metamorphic terranes based on the occurrence of fluid inclusions that display a bimodal distribution of compositions, presumably resulting from the presence of two distinct fluids coexisting at equilibrium at the time of trapping. Formation conditions inferred for these inclusions have then been used to help constrain the P-T evolution of the rocks. The compositions of individual inclusions cited as evidence for fluid immiscibility, however, are often incompatible with known phase relations of the appropriate systems and thus are inconsistent with the assertion of simultaneous trapping. This situation is particularly apparent in studies of natural systems where fluid immiscibility has been postulated based on data from H_2O-CO_2-NaCl bearing fluid inclusions. Using presently available data on the pressure–volume–temperature–composition (PVTX) relations in this system, we will outline those criteria that must be met in order to invoke the simultaneous trapping of coexisting equilibrium liquid and vapor phases in fluid inclusions at elevated temperatures and pressures.

The extent of the two-phase region in a one-component system is restricted to a univariant curve in P-T space along which liquid and vapor may coexist at equilibrium. In other words, the simultaneous trapping of fluid inclusions containing both phases at equilibrium can only occur over a very restricted set of P-T conditions. With the addition of a second component, the two-phase region becomes a divariant field in P-T space. In a binary system such as NaCl-H_2O or CO_2-H_2O at fixed pressure and temperature, the compositions of the liquid and vapor phases coexisting at equilibrium are uniquely specified. Furthermore, any fluid at this pressure and temperature having a bulk composition between these two values is unstable and will exsolve into distinct liquid and vapor phases, whereas bulk fluid compositions lying outside this range remain stable one-phase fluids. Of particular interest is that in P-X or T-X projections of binary systems such as NaCl-H_2O or CO_2-H_2O the *mixing lines* are the *tie lines*. Once pressure and temperatures are fixed, the compositions of the coexisting liquid and vapor phases are given by the appropriate tie line on a P-X or T-X projection. This is not necessarily true for ternary and higher order systems in general, and it is almost never true for the system H_2O-CO_2-NaCl.

In studies of fluid inclusions from the H_2O-CO_2-NaCl system, the presence of inclusions containing large amounts of CO_2 together with those containing small amounts of CO_2 but having the same relative salinity [NaCl/NaCl + H_2O)] of the liquid will be many times greater than that of the vapor.

Inclusions formed at a fixed pressure and temperature in the presence of immiscible H_2O-CO_2-NaCl fluids will display a range in both composition and density depending on the liquid to vapor ratio in the inclusion at the time of trapping. Those that have trapped only the liquid phase will homogenize to the liquid upon reheating to their formation temperature. Those that trap only the vapor should also homogenize (to the vapor) at this temperature. In the latter inclusions, however, due to the wetting characteristics of the liquid phase that forms a thin, optically invisible rim at high temperatures, an "apparent homogenization temperature" can be observed several hundred degrees below true homogenization. In the system H_2O-CO_2-NaCl this effect is important in measuring both CO_2 liquid/vapor homogenization to the vapor phase and total inclusion homogenization to the vapor. Thus, failure of these two end-member inclusion types to exhibit simultaneous homogenization does not rule out the possibility of boiling. Inclusions that trap mixtures of the coexisting liquid and vapor phases will contain fluids having bulk compositions and densities intermediate between that of either the liquid or the vapor. Because they did not trap a homogeneous, one-phase fluid, their true homogenization temperatures will necessarily be higher than their formation temperature, but again this may not be observable in the vapor-rich ones.

STEWART, A. J., Bureau of Mineral Resources, Canberra, Australia

Extensional Orogeny at Mt. Isa, Australia

The Mt. Isa Inlier in northwestern Queensland (Blake, 1987, Bureau of Mineral Resources Australian Bulletin 225) is a region of some 50,000 km^2 of Proterozoic crystalline rocks surrounded by younger sediments. It comprises a basement and three overlying cover sequences. The basement was folded and metamorphosed before about 1875 Ma. Sequences 1 and 2 underwent extensional faulting at about 1740 Ma, and this is thought to have initiated half-graben and basins in which sequence 3 was deposited. Compressive deformation began at about 1610 Ma, and was accompanied by low- to medium-grade regional metamorphism. Granite batholiths were emplaced between 1860 and 1500 Ma.

The basement consists of metamorphosed sediments, volcanics, and felsic and mafic intrusions. The overlying sequences comprise shallow-water quartz and feldspathic sandstones and calcareous rock, and largely subaerial felsic and mafic volcanics, all occurring at different stratigraphic levels and metamorphic grades. An unconformity separates basement and sequence 1, and also sequences 2 and 3, but generally no marked angular discordance separates sequences 1 and 2, except in a central tectonic ridge comprising these sequences and basement.

Regional post-basement shortening occurred after deposition of sequence 3, and involved three main episodes (Winsor, 1983, Tectonophysics, v. 92, p. 195). The first (D_1) at about 1610 Ma (Page and Bell, 1986, Journal of Geology, v. 94, p. 365) produced mainly east–west folds and low-angle thrusts; several different transport directions have been proposed. The second deformation (D_2; about 1544 Ma) produced north–south upright folds, commonly with gently plunging axes and steep axial plane foliation, and also reverse faults. The third deformation (D_3; about 1510 Ma) produced local folds and crenulations near Mt. Isa. Major strike-slip faulting occurred at about 1500 Ma; the major faults are dextral, strike northeast, and have local folds, cleavage, and transpressional structures associated.

Low-angle extensional faulting was first documented by Passchier (1986), Geology, v. 14, p. 1008), who described imbricate fault slabs overprinted by and therefore earlier than the regional shortening. The fault imbricates are sandstone lenses up to 200 m thick and 1 km long in sharp contact with metadolerite, felsic porphyry, calc-silicate rock, or sandstone of different composition or bedding orientation. Bedding in the lenses is at 40–60° to their margins, and ramps are present in many, where bedding orientation changes from oblique to subparallel to the margin. Parts of the stratigraphic sequence are incomplete or missing. The evidence indicates a system of normal faults which cut originally horizontal bedding at a large angle, and then underwent extension. Passchier and Williams (in press, Geology Magazine) describe domino-shaped normal imbricate fault blocks; displacement between the blocks indicates about 75% extension. The blocks are surrounded by metadolerite which has been deformed and metamorphosed during, and therefore predates, D_1. Movement of the hanging wall and its imbricate slabs was southward. The extension is dated by synkinematic granite which was emplaced into the extending crust at a deeper level than the fault imbricates, and give U-Pb zircon dates around 1740 Ma (Page et al., 1984, Proc. 27th International Geological Congress 5, p. 25).

The deeper-level extension has been studied in the Wonga Belt 50 km east of Mt. Isa township by Holcombe et al., (1987, Geol. Soc. Aust. Abstr. 19, p. 35) and Pearson et al., (1987, Geol. Soc. Aust. Abstr. 19, p. 37). Here, metasediments, metavolcanics, and gneissic granitoids form an intensely deformed north–south belt about 100 km long and 5–10 km wide. The belt is characterized by a strong subvertical gneissic foliation which was originally subhorizontal, but was rotated during D_2, reaching a largely vertical attitude except where it wraps around hinges of D_2 folds. The foliation has two lineations: a relict subhorizontal pre-D_2 lineation (combined extension and intersection), and a subvertical D_2 extension lineation. Shear sense indicators in the pre-D_2 foliation suggest that the upper plate moved north, but are not wholly unambiguous. The pre-D_2 movement was accompanied by synkinematic intrusion of 1740 Ma-old granite. The belt is interpreted as an originally gently dipping extensional mylonite zone which straddled a mid-crustal detachment surface, and was subsequently folded.

About 10 km west of the Wonga Belt, three outliers of sequence 2 are bounded by pre- or early-D_2 normal or vertical faults on their eastern sides, and are conformable with underlying beds on their western sides (Steart, in press, Aust. Journ. Earth Sci.). Two other outliers are in largely bedding-parallel, pre-D_1 fault contact with the underlying beds. Stratigraphic omissions along the eastern sides of the outliers place younger rocks on older, indicating extension in a southwesterly direction. Extension was followed by D_1 and D_2 shortening, which steepened the early normal faults.

The 1740 Ma date for the extensional faulting predates deposition at about 1670 Ma (Page 1981, Econ. Geol. v. 76, p. 648) of sequence 3, a time-lag sufficient for cooling and sagging to take place after extension. The faulting could have initiated basins and half-graben in which fanglomerate succeeded by basemetal-bearing saline lake sediments of the Mt. Isa Group (sequence 3) accumulated (Neudert, 1986, 12th Int. Sed. Cong. Abstract 229(and also provided channels for metal-bearing brine to flow into the basins during and/or after deposition (Neudert 1986, 12th Int. Sed. Congr. Abstr. 228) and form the ores.

STEWART, B. W., University of California, Los Angeles, CA, and D. J. DEPAOLO, University of California, Berkeley, CA

Mafic Magma Chamber Geometries and Processes: Nd and Sr Isotope Evidence and Constraints

As part of an ongoing study of processes occurring in magma chambers during emplacement, cooling, and crystallization, Nd and Sr isotopes have been measured in representative suites of samples from several layered mafic intrusions now exposed in continental crust. Comparison of the isotope patterns found in the Kiglapait intrusion, Labrador (DePaolo, 1985), the Skaergaard intrusion, East Greenland (Stewart and DePaolo, submitted to Contributions to Mineralogy and Petrology) and the Muskox intrusion, Canada (work in progress) is useful in identifying processes that are common to most mafic magmatism, and in isolating factors responsible for the wide variations in composition and style of basaltic volcanism and intrusion.

In the three intrusions studied, both the $^{87}Sr/^{86}Sr$ and the $^{143}Nd/^{144}Nd$ ratios are sensitive to assimilation of felsic wallrock, and the different mineral/liquid partitioning of Nd and Sr gives additional information about recharge and assimilation. Because the Sr and Nd isotopic ratios measured in a cumulate layer are identical to those of the magma from which the layer crystallized, the isotopic composition of the liquid can be tracked by measuring ratios from successive stratigraphic levels. In Figure 1, Nd isotope ratios (shown in ϵ notation and corrected for the age of each intrusion) are plotted against F, the mass of liquid remaining in the magma chamber relative to the final intrusion mass. The Sr isotopic ratios display complimentary trends but are not shown.

In general, assimilation of felsic wallrock causes a lowering of ϵ_{Nd}. In cases where a chilled margin sample was measured (plotted at F = 1), the margin has a lower ϵ_{Nd} than that of cumulates from the interior of the chamber. This is thought to be due to physical mixing between wallrock xenoliths and basaltic magma near the chamber wall. In both Skaergaard and Kiglapait, a trend of decreasing ϵ_{Nd} in the earliest crystallizing cumulates (shown by arrows) is followed by stabilization or even a slight increase of ϵ_{Nd} in the middle and upper stratigraphic levels. There are as yet insufficient data from Muskox to corroborate this trend, although the upper part of its stratigraphic sequence appears to behave in a fashion similar to that of the other two intrusions. The initial rapid trend toward contamination is probably a result of vigorous convection in the magma chamber in the early stages of crystallization, before extensive crystal accumulation insulates the interior magma from the margins. In addition, all three magma chambers may have started out with liquid volumes smaller than their final volume, with the difference made up by later replenishment. In such a case, the isotopic compositions of early liquids are affected by assimilation to a greater degree than those of later liquids. This effect is augmented by the accumulation of Nd in residual liquids. The stabilization or increase in ϵ_{Nd} (and decrease in $^{87}Sr/^{86}Sr$) shown in the upper levels of these intrusions probably results from an interplay of fresh magma replenishment and decreasing assimilation rates.

The Muskox intrusion is the only one of the three intrusions for which sufficient exposure exists to allow characterization of the chamber roof. Irvine (1970) documented the existence of a granophyric layer at the roof of Muskox consisting of partially molten wallrock that accumulated at the top of the magma chamber. A single analysis from this granophyric layer is shown in Figure 1 ($\epsilon_{Nd} = -11$). The boundary between the upper Muskox mafic cumulates and the granophyric layer is remarkably sharp, suggesting that mixing between mafic Muskox magma and molten wallrock was not extensive. Similar zones of fused wallrock have been surmised for Skaergaard and Kiglapait (see refs. above), although no exposures are preserved. Each displaced a significant amount of felsic wallrock that was not incorporated into the magma, and the isotopic patterns are consistent with such a geometry. As demonstrated by the Muskox data, the existence of such zones does not necessitate large amounts of assimilation of molten wallrock into the magma chamber. Further isotopic measurements from the upper levels of the Muskox intrusion will allow quantitative analysis of the form and extent of mixing between liquid layers, which may provide an in situ view of the plumbing systems underlying bimodal volcanic associations.

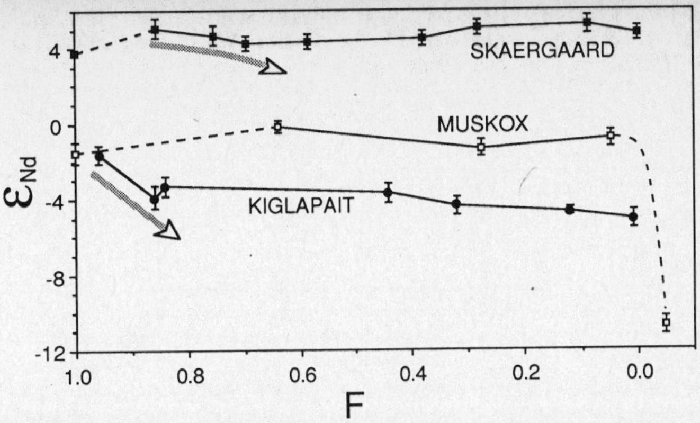

REFERENCES

DePaolo, D. J., 1985, Journal of Petrology, v. 26, p. 925–951.
Irvine, T. N., 1970, Geological Society of South Africa Special Publication 1, p. 441–476.

STEWART, D. B., J. D. UNGER, B. E. WRIGHT, J. D. PHILIPS, and L. M. LIBERTY, U. S. Geological Survey, Reston, VA, C. P. SPENCER and P. CHARLESWORTH, Geological Survey of Canada, Ottawa, Ontario, D. R. HUTCHINSON and K. D. KLITGORD, U.S. Geological Survey, Woods Hole, MA, J. H. LUETGERT, U. S. Geological Survey, Menlo Park, CA, and W. A. BOTHNER, University of New Hampshire, Durham, NH

Digitization of Quebec–Maine–Gulf of Main Global Geosciences Transect

Digitization of a Global Geosciences Transect requires that regional data sets be assembled on a common base from geologic, geographic, geophysical, and other data that originally may have had diverse bases and projections. This process also requires that geologic units on different geologic maps be combined into new groupings of formations, based on inferred tectonic settings, by use of standard color assignments. To meet these requirements as geological and geophysical data sets grow larger and more complex, it is highly desirable to develop generic digital data formats that can be supported widely by geological organizations and by software vendors. Transect data can be then widely disseminated for use within the geological community. Furthermore, the rapid development of capabilities of networked computers to process three-dimensional data and to display in color new combinations of geographic, geologic, and geophysical data sets and their derivatives make it possible for us to obtain images of the earth's crust with unprecedented detail to MOHO depths of 35–45 km along the Quebec-Maine-Gulf of Maine transect. The new computer technologies also permit rapid editing of maps, compact archiving of data and derivative images on digital optical discs (CD-ROM), and machine production of color separates for publications.

The Quebec-Main-Gulf of Maine Transect is completely digitized, and all of the images, sections, and maps have been produced with current and developmental geographic information systems technology. An interactive computer display for this transect is planned to be part of the scientific exhibition by the U. S. Geological Survey. This transect, NA-1 of the Global Geosciences Transect Project (CC-7), extends 880 km from the stable North American craton near Quebec City, Quebec, across the complex composite Paleozoic Appalachian orogen, which was split by continental rifting in the Mesozoic, and reaches the present Atlantic Ocean basin south of Georges Bank.

While technology development has been an important component of our transect project, the principal goals have always been to describe the earth's crust as completely as possible and to deduce the processes that have formed and reformed the crust. Extensive seismic reflection and refraction data, as well as the high-quality gravity, magnetic, geologic, and geochemical data for the region, show much new detail about the several major orogenic episodes that formed the Appalachian orogen during the interval from Late Proterozoic to Late Paleozoic time.

Structures readily imaged in the upper 10 km of the crust include the shapes of batholiths (with indications of their overall composition), folds to depths of 4 km in pyrrhotite-bearing carbonaceous metapelite, cover-basement relationships in the Connecticut Valley-Gaspe synclinorium in Quebec and the Merrimack synclinorium in Central Maine, and major thrust and strike-slip faults in the region.

Details of mid-crustal features can also be seen. One such feature, for example, is the Taconic allochthon in which at least 24 km of crust, containing in its upper part the Precambrian Chain Lakes Massif and its cover sequence of Cambrian and Ordovician ophiolite and melange, was obducted onto the edge of thinned Grenville continental crust that remained after Iapetus Ocean was formed. The boundaries between the tectonostratigraphic terranes in the Northern Appalachians seem to penetrate the entire crust. The Moho at the base of the crust can be readily contoured on the basis of coincident results from seismic reflection and refraction profiles. Taken together these data sets greatly facilitate description and modeling of the processes that created the Appalachian orogen.

STEWART, D. B., J. D. UNGER, B. E. WRIGHT, J. D. PHILIPS, and L. M. LIBERTY, U. S. Geological Survey, Reston, VA, C. P. SPENCER and P. CHARLESWORTH, Geological Survey of Canada, Ottawa, Ontario, D. R. HUTCHINSON and K. D. KLITGORD, U. S. Geological Survey, Woods Hole, MA, J. H. LUETGERT, U. S. Geological Survey, Menlo Park, CA, and W. A. BOTHNER, University of New Hampshire, Durham, NH

Geologic Processes Shown by Digitized Quebec-Maine-Gulf of Maine Global Geosciences Transect

The Quebec-Maine-Gulf of Main Transect is completely digitized, and all of the images, sections, and maps in our poster were produced with current and developmental geographic information systems technology. This transect, NA-1 of the Global Geosciences Transect Project (CC-7), extends 880 km from the stable North American craton near Quebec City, Quebec crosses, the complex composite Paleozoic Appalachian orogen, which was split by continental rifting in the Mesozoic, and extends to the present Atlantic Ocean basin south of Georges Bank.

While technology development has been an important component of our transect project, the principal goals always have been to describe the earth's crust as completely as possible, and to deduce the processes that have formed and reformed the crust. Extensive seismic reflection and refraction data, as well as high-quality gravity, magnetic, geologic, and geochemical data for the region, clearly show that several major orogenic episodes formed the Appalachian orogen from Late Proterozoic to Late Paleozoic time. One of the most prominent episodes was the closing of the Iapetus Ocean during the Taconian orogeny, which culminated in the Late Middle Ordovician with the obduction of at least 24 km of primarily continental crust onto the edge of the thinned Grenville continental crust that remained after Iapetus was formed. This allochthon contained, in its upper part, the Precambrian Chain Lakes Massif and its cover sequence of Cambrian and Ordovician ophiolite and melange. Many details about the base of the allochthon, its internal structures, and subsequent deformation during postorogenic thermal rebound and later orogenies can be seen with our extensive data sets.

In the present regional configuration, the continental block attached to North America during the Taconian orogeny commonly is recognized as a central microcontinental terrane; other continental crustal blocks were later attached to the southeast. The southeastern extent of the central continental block is ambiguous within our data. However, on the basis of the configurations of seismic reflectors, Ordovician Rb/Sr ages of kyanite-grade metamorphic pre-Acadian basement rocks, and the variation of lead isotopic compositions of feldspars in Devonian plutons, we favor the hypothesis that a southeast-vergent, thick pile of nappes formed along the southeastern edge of the block nearly simultaneously with obduction on the northwestern edge, followed by removal of the abutting mass responsible for nappe formation by large strike-slip faults.

By Early Silurian time, a composite continental tectonostratigraphic terrane had been juxtaposed with the central continental block along its southeastern edge along faults that appear to penetrate steeply through the entire crust. This composite terrane includes Cambrian and Ordovician bimodal volcanics of continental rift origin and Cambrian and Ordovician platformal carbonates, quartzites, and sulfidic shales. Exceptionally large amount of turbiditic sediments (1,400–2,000 km^3/km along strike) were deposited in the Central Maine synclinorium during the Silurian and Early Devonian. Much of this sediment could have come from thermal rebound of the central block, as Ordovician kyanite-grade metamorphic rocks were exposed to erosion, but the source of the rest is uncertain. Not much sediment could have been derived from the composite terrane attached just southeast of the central block because present depths of erosion only expose rocks of lowest greenschists metamorphism, probably originally buried less than 6 km. Sedimentation ceased when the Devonian Acadian orogeny greatly shortened the region, with accompanying metamorphism and extensive granitic plutonism. Late in the Devonian, Meguma, another continental terrane, was obducted from the east and southeast along southeast-dipping faults, but with tens of kilometers of strike slip movement. During the Carboniferous and Permian, wrench faulting formed locally thick basins, particularly in the south-central part of the transect. The record of these numerous Paleozoic events is preserved in the brittle upper half of the crust and can be best revealed by mapped geology and potential field data, interpreted together with shallow refraction and reflection seismic data.

The principal process responsible for most of the deep crustal structures present today, and for reactivation of many of the shallower structures as well, was southeastward Mesozoic crustal extension along a deep crustal simple shear. The results of this process seem to be well preserved, and are strongly evident in our transect. Notable is the extension that reverses the throw on inclined Paleozoic thrust faults. Upper Triassic and Lower Jurassic sediments are found in many half-grabenlike basins, and the throw and extent of the basins increase toward the southeast. The footwall faults of some of the Mesozoic basins in the northwestern Gulf of Main terminate against subhorizontal reflections near the Moho. The extended crust is ~35 km thick for hundreds of kilometers and remains unchanged even across several major boundaries between the different Paleozoic tectonostratigraphic terranes. Seismic reflection and refraction results both show that large listric faults break the Maine crust as much as 220 km or more northwest of heretofore recognized Mesozoic faults. Two of these fault zones appear to become listric into laminated crust at two locations where the Moho rises rapidly—4 to 5 km over 15 km. This suggests that the necking of continental crust prior to actual breakup in Early Jurassic time occurred progressively, while also migrating easterly across a 700 km wide zone. The intrusion of many mafic dikes, some of regional extent, accompanied extension. As the amount of extension increased toward the southwest, failure by pure shear with much associated extrusion of basalt occurred. Offshore seismic reflection profiles show that the subsequent opening of the Atlantic was accompanied by thermal subsidence of previously extended crust and prograde deposition of sediments of the coastal plain across the shattered remnants of the upper crust.

STEWART, D. J., KSEPL, Rijswijk, the Netherlands

Regional Seismic Stratigraphy of Tertiary Postrift Succession in North Sea Basin

The Tertiary succession of the North Sea is a classic example of a cratonic failed rift. It has been extensively studied owing to its economic importance as a petroleum province. This paper reports the results of recent studies, aimed at synthesizing the regional stratigraphic history of the Tertiary succession, using newly available regional seismic data from the North Sea.

Extensional faulting of the North Sea Basin began in the Triassic, but by the end of the Cretaceous fault activity had virtually ceased and earlier formed half-graben depressions were almost completely infilled. The top of the Upper Cretaceous, thus, formed a relatively uniform surface on which the postrift Tertiary succession was laid down in a roughly saucer-shaped basin.

The architecture of the postrift depositional succession is essentially similar to that of a passive continental margin with the development of shelf/slope/basin depositional prisms, with geometric configurations controlled by relative sea-level changes. The essential difference is that of basin infill dynamics, with multidirectional sediment input (lateral and axial).

In the Early Tertiary, uplift of the Shetland Platform/Scottish landmass caused large amounts of clastics to enter the basin from the northwest, where a high-energy shelf supplied sediment to a prograding slope. By-pass of the slope resulted in the widespread deposition of turbidites on the basin floor, which today form important oil reservoirs (i.e., the large Forties, Montrose, Frigg, and Balder Fields). The thickness of the Lower Tertiary decreases toward the east and southeast from some 500–600 m in the shelf area to less than 100 m at the basinal extremes. On seismic reflection records, the Lower Tertiary exhibits three major seismic facies belts:

1. An eastward-prograding sedimentary wedge characterized by clinoforms (slope), overlain by flat-lying "topsets" (succession of shelf, coastal plain, and deltaic sediments) containing high-amplitude reflectors caused by the presence of lignites. The slope clinoforms display either a low-amplitude, rather transparent seismic facies, or a characteristically chaotic facies, showing features interpreted as slumps and channels.
2. A basinal area characterized by discontinuous or mounded seismic facies (turbidites).
3. A basinal area (distal fringe of the paleocene basin) characterized by a low-amplitude transparent seismic facies that reflects quiet basinal deposition (hemipelagics and thin-bedded turbidites).

The Paleocene and Eocene shelf edges ran parallel to the old graben boundary, their basinward limit being controlled by the main graben boundary fault. Average water depths on the Shetland Platform were around 200 m increasing to over 600 m in the deepest part of the former graben area.

During the Late Paleocene–Early Eocene, a relative rise in sea level caused sediment supply to the basinal area to be greatly reduced, leading to deposition of black organic shales with abundant graded tuffs (forming a widespread seismic marker) over the entire basin floor. Acceleration of this relative sea-level rise during the lower Eocene resulted in transgression and was followed by a new phase of progradation on top of the former Lower Eocene delta plain.

The approximate base of the Oligocene is marked by the onlap and termination of reflections onto the Eocene. On the western flank of the graben, the shelf was formed by vague low-angle clinoforms, and the shelf break was situated only slightly basinward of the Late Eocene shelf break. The southern part of the Shetland platform shows the preservation only of thin sandy top sets; however, further to the north the presence of clinoforms indicates the development of deeper water conditions. Where the slope has prograded into the old graben areas, it is characterized by chaotic and "broken" seismic reflection packages interpreted to represent faulted, slumped, and deformed sediments. During the Middle Oligocene, clinoforms are observed for the first time on the eastern side of the basin (south Norwegian offshore and north Danish offshore).

The Miocene–Pliocene is characterized by a major reversal in sediment provenance. Slope clinoforms occur on both the east and west side of the North Sea basin. However, the main sediment supply was from the southeast characterized by a prominant seismic clinoform sequence that can be traced from the Danish and Dutch offshore areas northwards, where it axially infills the Central/Northern North Sea trough. Thus, a wide shelf was built out in a northwesterly direction from the Danish Bight. The clinoform exhibits numerous narrow, deep cleft-like features thought to be channels or canyons on the lower slope, which open-out toward the north into small, base of slope turbidite mounds.

Numerous onlapped truncation surfaces within successive Tertiary clinoform developments evidence short-term fluctuations of relative sea level.

The early Pleistocene is represented by a thin clinoform reflection package indicating a northward progradation of deltaic sequences along the axis of the then largely infilled central graben trough. In overall terms, these two packages appear to reflect the presence of a lower marine phase of deposition and an upper glacial dominated phase of deposition.

STEWART, I. S., and P. L. HANCOCK, University of Bristol, Bristol, United Kingdom

Architecture and Degradational Characteristics of Neotectonic Normal Fault Zones Cutting Carbonate Bedrocks in Aegean Region

Much of our current knowledge of fault scarp degradation is derived from studies of piedmont scarps (small, single-event fault scarps offsetting unconsolidated materials [Petersen, 1985]) and fault-generated range fronts in the Basin and Range province, western USA (Wallace, 1978). Furthermore, many of the phenomena and processes regarded as characteristic of fault scarps and their degradation, such as triangular facets and retreat and recline, have been described from this region. In both the Greek and Turkish sectors of the Aegean extensional province, however, many neotectonic normal fault zones cutting Mesozoic carbonates display a pattern of degradation contrasting with that documented from the Basin and Range province. Fault scarps in the Aegean region that separate bedrock carbonates in both footwall and hangingwall, or footwall carbonates from hangingwall Quaternary sediments owe their appearance less to sculpturing by geomorphic processes than to fault zone architecture, generated during faulting and mineralization.

Footwalls of both synthetic and antithetic normal faults are characterized by an ordered arrangement of breccia types (Hancock and Barka, 1987; Steward and Hancock, in press) (Figure 1). Compact breccia sheets, products of attrition during fault movement, comprise indurated microbreccia zones immediately underlying slip planes. Incohesive breccia belts, interpreted as fault-precursor breccias formed ahead of upwards-propagating fault tip-lines (Vita-Finzi and King, 1985), are broad zones of disaggregated and shattered rock, developed within meters of a slip plane. In addition, macro- and mesoscale phenomena indicate that within fault zones there is hangingwall-directed migration of slip planes. This tendency, here called intrafault-zone hangingwall collapse, is an important factor responsible for the stepped appearance of fault scarps. Episodes of scarp retreat punctuated by slip events are less important in the evolution of bedrock normal fault scarps (Figure 2). The denudation of alternating zone-parallel breccia sheets and belts subparallel to normal fault scarps also contributes to their stepped morphology.

The rate and style of degradation is a function of fault zone architecture. Degradation is relatively rapid through incohesive breccia belts but relatively slow through compact breccia sheets which form an armoured carapace on many fault scarps. Tectonic breaches of the carapace, such as pluck holes and comb fractures (Figure 1), are utilized by geomorphic processes such as slab failure and rockfall. With removal of the compact-breccia carapace, fault scarp morphology is determined by the fracture pattern within incohesive breccia belts. Irregular breaching of the compact breccia sheets results in the cavitated appearance displayed by many Aegean region fault scarps.

REFERENCES

Hancock, P. L., and A. A. Barka, 1987, Kinematic indicators on the active normal faults in western Turkey: Journal of Structural Geology, v. 9, p. 419–427.

Petersen, J. R., 1985, Equilibrium tendency in piedmont scarp denudation, Wasatch Front, Utah, *in* M. Morisawa and J.T . Hack, eds., Tectonic Geomorphology: Boston, Allen & Unwin, p. 203–233.

Stewart, I. S. and Hancock, P. L., in press, Normal fault zone evolution and fault scarp degradation in the Aegean region: Basin Research.

Wallace, R. C., 1978, Geometry and rates of change of fault-generated range fronts, north-central Nevada: Journal of Research of the U.S. Geological Survey, v. 6, p. 637–650.

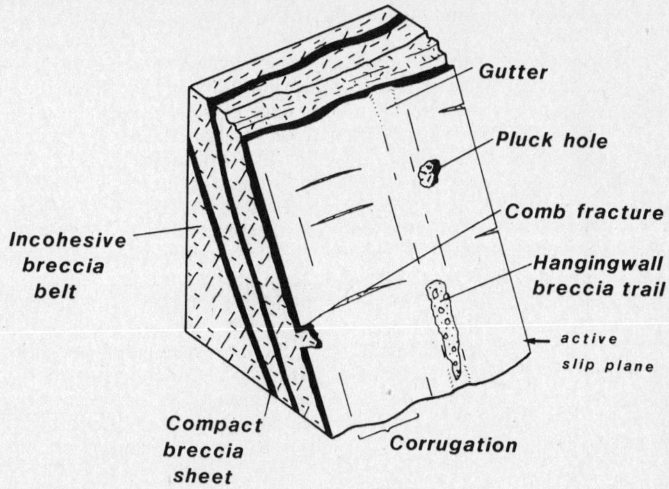

Figure 1—-Slip-plane phenomena influencing fault scarp degradation processes.

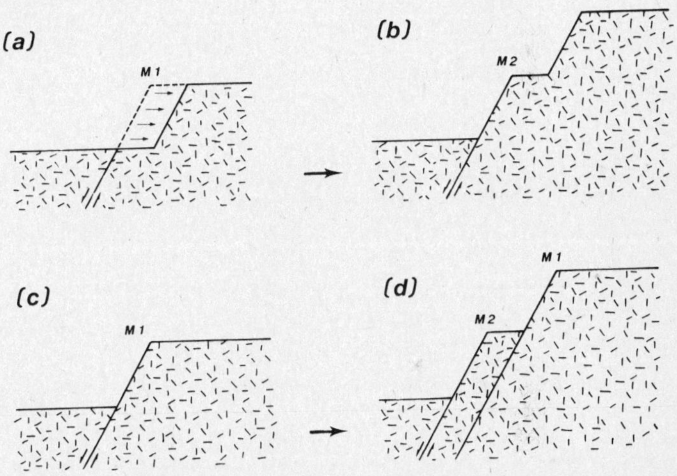

Figure 2—(a) and (b) Development of a stepped fault scarp profile as a result of two increments of motion (M1 and M2) on the same slip plane punctuated by an episode of scarp retreat (arrows). (c) and (d) Development of a stepped fault scarp profile as a consequence of intrafault-zone hangingwall collapse (M1 to M2).

STILLINGS, L, M. MACHESKY, and S. BRANTLEY, Pennsylvania State University, University Park, PA

Potentiometric Acid–Base Titrations of Feldspar–Water Suspensions: Implications for Dissolution Rates

The complexity of feldspar chemistry and feldspar-water reactions has led to a continuing controversy concerning the best model to describe feldspar dissolution. Recently, the surface complexation approach to feldspar dissolution has gained increasing popularity (Stumm et al., 1987; Blum and Lasag, 1988). This model suggests that protonation and deprotonation of surface hydroxyl groups may provide a mechanistic interpretation of the hydrogen ion dependence of the dissolution rate of silicates. We present results of attempts to quantify the protonation–deprotonation behavior of surface hydroxyl groups for several feldspars including adularia, albite, and bytownite.

Samples were crushed and ground to average particle sizes of 5–15 micron as determined by laser light scattering. After grinding, the feldspar was ultrasonically dispersed in acetone and store dry until analysis. Twenty-five ll of a 25 g/l suspension was then washed three times with distilled-deionized water by centrifugation-decantation to remove trances of acetone and to 'precondition' the surface. The volume was then adjusted to 25 ml and salt added to bring the ionic strength up to the range 0.001 to 0.1 m

(NaCl, tetramethylammonium chloride, $CaCl_2$, or KCl). Titrations proceeded with a five minute equilibration period between successive acid or base additions. Separate titrations were also completed with large volume samples to enable us to withdraw, filter, and acidify 10 ml aliquots periodically. These aliquots were then analyzed for dissolved Si, Al, and network cation in order to determine H^+ consumption due to mineral dissolution and ion exchange. An example of several typical titration curves is shown in Figure 1.

These data illustrate the K^+ ion exchange selectivity exhibited by this potassium feldspar. We infer from the data that Na^+ and the $N(CH_3)_4{}^+$ cation are too large to fit into the cation exchange sites optimally, thus a larger surface exchange is exhibited by the K^+ cation. These results are remarkably similar to those obtained for SiO_2 titrations (Tadros and Kyklema, 1968) and suggest that the feldspar-water surface may be chemically similar to the hydrated SiO_2 surface. Presence of an altered Si-rich gel-type feldspar surface layer has been recently proposed by other workers (Petit et al., 1987; Westrich et al., 1988).

We are presently working to interpret these data by modeling the partition of the observed net H^+ consumption in each titration as follows: $H^+_{consumption} = H^+_{surfacehydroxylgroups} + H^+_{ionexchange} + H^+_{solutionreactions}$. By measuring $H^+_{consumption}$ over a range of pH and ionic strength we can solve a system of simultaneous equiations in several unknowns (surface protonation equilibrium constant, ion exchange constant, concentration of surface species including protonated, unprotonated, and ion-exchanged species). The final results are our best estimates for the cation exchange constraints and protonation constants as a function of pH, ionic strength, salt species, and feldspar stoichiometry. Knowledge of such equilibrium constants is essential to the interpretation and prediction of dissolution behavior of feldspar as a function of pH.

REFERENCES

Blum, A., and A. Lasaga, 1988, Nature, v. 331, p. 431–433.

Schott, J., and J. C. Petit, 1987, *in* Aquatic Surface Chemistry, W. Stumm, ed., New York: Wiley-Interscience, chapter 11.

Stumm, W., B. Wehrli, and E. Wieland, 1987, Croat. Chem. Acta, v. 60, p. 429–456.

Tadros, T. F. and J. Lyklema, 1968, J. Electroanal. Chem. Interfac. Electrochem., v. 17, p. 267–275.

Westrich, H. R., W. H. Casey, and G. W. Arnold, 1988, Abstracts of V. W. Goldschmidt Conference, Baltimore MD, p. 82.

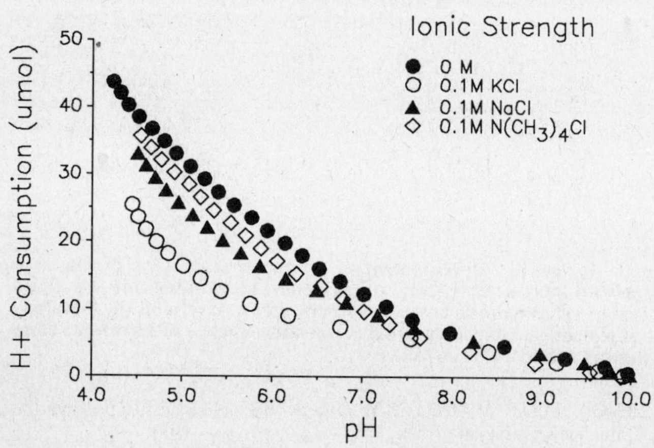

Figure 1—Typical titration curves for adularia-water suspensions with different ionic strength adjustors. See text for further details.

STIX, J., and M. P. GORTON, University of Toronto, Toronto, Canada

Chemical Evolution of Silicic Magmas between Two Caldera-Forming Bandelier Eruptions, Jemez Mountains, USA: Implications for Melt Structure and Development of Compositional Zonation

The Cerro Toledo Rhyolite is a group of small-volume high silica rhyolite domes and fallout and surge tephras that were erupted from Toledo caldera between the Lower Bandelier Tuff (LBT) at 1.45 Ma and Upper Bandelier Tuff (UBT) at 1.12 Ma. Because of the relatively small volumes of the deposits and the chemical continuity with the large-volume Bandelier tephras, the Cerro Toledo Rhyolite appears to have tapped only the top of the Bandelier magma chambers(s). The Cerro Toledo Rhyolite thus provides insight into the temporal evolution of the uppermost liquids over 0.33 Ma and records the restoration of compositional zonation after the formation of Toledo caldera at 1.45 Ma (Stix et al, 1988). This contrasts with ignimbrite deposits that may only provide a geochemical section through a magma chamber at an instant in time.

Incompatible elements such as Cs, Rb, Y, Nb, U, and HREE increase in concentration upsection from the base to the top of the Cerro Toledo Rhyolite tephra sequence. No one trace element is truly incompatible; Cs comes the closest. The other trace elements are fractionated with respect to Cs and are clearly less compatible. This contrasts with similar levels of incompatibility for Cs, Rb, U, Th, Nb, Ta, Yb, and Lu reported for the LBT (Balsley et al, 1987).

We believe that crystal fractionation was the main process by which compositional gradients were restored within the Bandelier magma chamber during Cerro Toledo Rhyolite time. Declining K/Cs, Zr/Nb, and La/Yb ratios upsection through the Cerro Toledo Rhyolite indicate crystallization and removal of feldspar, zircon, and an LREE-rich mineral (allanite and/or chevkinite), respectively.

We believe that diffusive processes did not play a major role in the development of the compositional gradients within the Cerro Toledo Rhyolite because (1) cations of widely different size and charge (e.g., Rb and Nb) are not fractionated with respect to the other, and (2) experimental Soret compositional gradients (Ponader and Mahood, 1984; Lesher, 1986) are opposite to those observed in ignimbrite sequences such as the Bandelier and Bishop Tuffs.

At one locality, a 2 m thick sequence of Cerro Toledo Rhyolite fallout tephras directly overlies the LBT ignimbrite. This basal sequence becomes less evolved upsection (lower incompatible element concentrations) and forms a geochemical continuum with the uppermost LBT pumices when plotted upon an La/Lu vs. Lu diagram. We interpret this trend as progressively deeper tapping of the LBT magma at the closing stages of the LBT eruption, not mixing with more mafic magma. The Cerro Toledo Rhyolite tephras that overlie this basal sequence have a geochemical signature distinct from that of the basal tephras (e.g., lower La/Lu at similar Lu contents). We interpret this difference as a new input of silicic magma after the partial evacuation of the chamber by the LBT eruption.

Because of good chemostratigraphic control, we are able to study the effects of bulk composition, temperature, and volatile content of the Cerro Toledo Rhyolite magmas upon zircon solubility in the melt. The Zr/Cs ratio declines continuously upsection through the Cerro Toledo Rhyolite tephra sequence, indicating crystallization and removal of zircon as well as Zr saturation in the melt. However, the decline of Zr/Cs also slows noticeably upsection, and the Hf/Cs ratio becomes nearly constant. Furthermore, Zr concentrations in the melt rise from 178 ppm in the uppermost Cerro Toledo Rhyolite tephras to 320 ppm in the UBT plinian tephras that immediately overlie the Cerro Toledo Rhyolite. These observations indicate (1) less zircon crystallized and (2) zircon solubility in the magma increased at this point. Although more mafic bulk compositions and increased temperatures can increase zircon saturation levels (Watson and Harrison, 1983), bulk compositions do not change significantly from the Cerro Toledo Rhyolite to the UBT plinian tephras, and it is likely that the top of the Bandelier magma chamber continued to cool during this time. On the other hand, F and Cl contents of matrix glasses also increase dramatically at this transition (F 1070 ppm to 2000 ppm; Cl 1670 ppm to 2760 ppm). If changes in the halogen contents are indicative of changes in the overall volatile distributions at the top of the magma chamber, volatiles may have controlled Zr saturation levels and zircon solubility in the Cerro Toledo Rhyolite and Bandelier magmas by modifying the polymerization state of the aluminosilicate melt.

REFERENCES

Balsley, S. D., J. A. Wolff, D. C. Kuentz, P. R. Kyle, and J. N. Gardner, 1987, The inter-relationship of the Upper and Lower Bandelier Tuffs: evidence from incompatible trace elements: Eos, Transactions of AGU, v. 68, p. 1512.

Lesher, C. E., 1986, Effects of silicate liquid composition on mineral-liquid partitioning from Soret diffusion studies: Journal of Geophysical Research, v. 91, p. 6123–6141.

Ponader, C,. W., and G. A. Mahood, 1984, Soret diffusion in simple peralkaline silicate liquids near magmatic temperatures and pressures: GSA Abstracts with Programs, v. 16, p. 625.

Stix, J., F. Goff, M. P. Gorton, G. Heiken, and S. R. Garcia, 1988, Restoration of compositional zonation in the Bandelier silicic magma chamber between two caldera-forming eruptions: geochemistry and origin of the Cerro Toledo Rhyolite, Jemez Mountains, New Mexico: Journal of Geophysical Research, v. 93, p. 6129–6147.

Watson, E. B., and T. M. Harrison, 1983, Zircon saturation revisited: temperature and composition effects in a variety of crustal magma types: Earth and Planetary Science Letters, v. 64, p. 295-304.

STIXRUDE, L., and M. S. T. BUKOWINSKI, University of California at Berkeley, Berkeley, CA

Compression Mechanisms in Liquid SiO_2

Chemical differentiation of the terrestrial planets is largely controlled by the physical and chemical properties of silicate melts. It is now appreciated that pressure can dramatically change the structure of these melts, and that their density can exceed that of the solids with which they are in equilibrium. If so, processes of differentiation can differ significantly among the planets, and elemental fractionation may be strongly dependent on pressure.

Experimental methods can only detect the spherically averaged structure of amorphous materials, although indirect information changes in bonded coordination can in principle be detected with Raman, infrared, and NMR spectroscopy. A link between spectroscopy and local structure, as well as direct information on liquid structure, can be obtained from Monte Carlo and Molecular Dynamics simulations. Simulations based on ionic potential models have contributed much to our qualitative understanding of liquid structure and transport properties. These models do not reproduce the thermodynamic properties of silicate liquids, however. Realistic simulations of pressure-induced structural changes must take into account the strongly covalent bonding that controls the SiO_2 tetrahedra and the links among them.

We have formulated a parametric covalent potential model that includes three body forces. The model was designed to explicitly incorporate known qualitative features of Si-O bonding. Structural information was not used to constrain the model parameters to avoid biasing simulated liquid structures. Nevertheless, the model closely reproduced the observed correlations between Si-O bond length and Si-O-Si angle, and it gives accurate structures and equations of state for solid phases of silica.

We have carried out Monte Carlo simulations of liquid SiO_2, with the above potential, along several isotherms. Since equilibration required several million configurations, the simulations were performed on a supercomputer. The zero pressure density and bulk modulus at 2000 K are in very good agreement with data, as is the radial distribution function. The nearest neighbor coordination number is 3.9, which is consistent with the experimental value of 3.8 ± 0.3. The coordination increases gradually with pressure, asymptotically approaching 4. The simulations suggest that coordination numbers greater than 4 cannot be achieved without changing the bonded coordination of Si.

With increasing temperature, the volume of the liquid increases, giving an average thermal expansivity of 3×10^{-5} between 2000 K and 4000 K. Also, the bulk modulus decreases with increasing temperature in accord with the behavior of solids. At zero pressure, the model predicts a liquid to vapor transition at 5000 ± 1000 K in agreement with the measured boiling point of silica.

We find two distinct compression mechanisms in the simulated liquid. At 2000 K, the number of 3-coordinated Si atoms decreases from 11% at zero pressure to 8% at 1 GPa and 4% at 20 GPa. The average tetrahedral ring size increases slightly with pressure from 5.6 at 0 GPa to 5.8 at 10 GPa before decreasing to 5.6 at 20 GPa. Thus, an increase in average Si-O coordination number is the dominant compression mechanism at low pressure, while ring distortion and a decrease in average ring size are dominant at higher pressures (20–50 GPa).

STIXRUDE, L., and M. S. T. BUKOWINSKI, University of California at Berkeley, Berkeley, CA

Covalent Bonding in Liquid and Crystalline Phases of SiO_2

The chemical and thermal state of the earth through time is controlled largely by the properties of solid and liquid silicates that make up 80% of the planet by volume. Thus, an understanding of the nature of silicates, over the wide range of pressure, temperature, and composition is central to geophysics. The structure and thus all thermodynamic and transport properties vary widely over the mantle's pressure regime. The structural changes, due to compression and phase changes, ultimately depend on the properties of the SiO bond, the most common chemical bond in the earth. Thus, an understanding of covalent bonding in general and the Si-O bond in particular is required.

In an effort to better understand the physics of Si-O bonding and its relationship to structure, we have developed a simple covalent potential model of bonding in SiO_2. The model is intentionally restricted to describe bonding in tetrahedrally coordinated silicates, since no one model can be expected to describe the properties of both tetrahedral and the more ionic octahedral bonding that occurs at higher pressures (>10 GPa in SiO_2). The important feature of the model that determines its success is the explicit distinction between strong intratetrahedral and weak intertetrahedral bonding. The model incorporates the known qualitative features of tetrahedral Si-O bonding in the form of strong two- and three-body intratetrahedral covalent forces, and weak two- and three-body covalent forces between tetrahedra. Constraints include vibrational properties of small silicate molecules determined from molecular orbital calculations and the equation of state of quartz. No macroscopic structural data was used as a constraint to avoid biasing the model to one particular phase.

The model gives excellent results for the equations of state of α-quartz and coesite. It also reproduces the structure and the change in structure with pressure of these two phases. The model has also been applied to α-cristobalite and β-quartz, and it is consistent with the limited data available for these two phases. Phase transition pressures between the various phases, when extrapolated to room temperature, are also reproduced by the model.

Results for liquid SiO_2 from Monte Carlo simulations are consistent with zero pressure measurements of volume, bulk modulus, and radial structure. At higher pressures, the strong intratetrahedral bonding of the model retains nearly ideal tetrahedral coordination in the liquid up to at least 50 GPa. In contrast, the weak intertetrahedral interactions allow the liquid to freely rearrange its intermediate range structure, causing it to become denser than the symetrically restricted crystalline structures at pressures between 5 and 50 GPa.

STOCK, J. M., Harvard University, Cambridge, MA

Cenozoic Plate Motions in Southwest Pacific Region

Marine magnetic anomalies and bathymetry in the southwest Pacific Ocean, Southeast Indian Ocean, and Tasman Sea yield a history of relative motion of Antarctica, Australia, the Lord Howe Rise, the Campbell Plateau (Pacific Plate), and the Bellingshausen plate since late Cretaceous time. These data and their uncertainties have been used to calculate poles and angles of reconstructions for selected times since the late Cretaceous and to identify the major remaining deficiencies in our knowledge of the tectonic history of this region.

Rifting between the Campbell Plateau and West Antarctica, between East Antarctica and Australia, and between Australia and the Lord Howe Rise began in Late Cretaceous time. Between the times of anomalies 34 and 24 (80–55 Ma), a spreading center separated the Campbell Plateau from West Antarctica. The northeastern end of this ridge was a triple junction between the Antarctica, Pacific, and Bellingshausen plates, after at least the time of anomaly 31 (72 Ma). The Bellingshausen–Campbell Plateau (Pacific) boundary was also a spreading center; the exact geometry of the Bellingshausen–Antarctica boundary is uncertain, although some limits can be placed on its position offshore. Within the uncertainties in the reconstructions, the Campbell Plateau–Antarctica–Australia–Lord Howe Rise plate circuit closes between the times of anomalies 31 and 25, indicating that relative motion either between the Lord Howe Rise and the Campbell plateau, or between East and West Antarctica, was less than the few hundred kilometers of uncertainty in the reconstructions. Thus, the Antarctica–Australia ridge, the spreading center in the Tasman Sea, and the Pacific–Antarctica ridge south of the Campbell Plateau formed three limbs of a triple junction.

At the time of anomaly 24 (55 Ma), spreading ceased in the Tasman Sea, and the Lord Howe Rise plate became attached to the Australia plate. Spreading between the Campbell plateau and Antarctica slowed but apparently did not change orientation. Despite large uncertainties in Pacific–Antarctica motion during the interval between anomalies 24 and 18 (55–42 Ma), the Australia–Antarctica ridge and the Pacific–Antarctica ridge do not form a closed plate circuit. Minor displacement between the Lord Howe Rise and the Pacific plate, or between East and West Antarctica, is needed to complete the circuit.

Sometime between the times of anomalies 24 and 13, Pacific–Antarctica spreading changed direction, becoming parallel to Pacific–Bellingshausen spreading. This is inferred to indicate the cessation of Antarctica–Bellingshausen motion (Stock and Molnar, 1987). However, neither the time nor the manner of this tectonic transition are known in detail from existing reconnaissance data. The parallelism of anomalies 24 through 18 south of the Campbell plateau suggests that this change in direction took place no earlier than the time of anomaly 18. Mayes et al (in press) suggest on the basis of Geosat lineations that it may have occurred as early as the time of anomaly 21 (49 Ma).

Post-anomaly 18 (42 Ma) spreading on the Australia–Antarctica ridge and on the Pacific–Antarctica ridge (between the Campbell plateau and Antarctica) occurred about two different poles, requiring the presence of at least one other boundary to complete the plate circuit. Austalia–Pacific spreading rates and directions inferred from anomalies 18 to 13 on the Australia plate in the South Tasman Sea are within the uncertainties of values needed to close the plate circuit for these times, suggesting that the Australia–Antarctica, Pacific–Antarctica, and Australia–Pacific boundaries formed three limbs of a triple junction at least between the times of anomalies 18 and 13, without significant additional motion elsewhere in the plate circuit, such as between East and West Antarctica.

These reconstructions of the Australia–Antarctica–Pacific plate circuit predict 810 ± 110 km of Pacific–Australia right-lateral strike-slip motion in New Zealand since 36 Ma, and 460 ± 50 km since 20 Ma. This amount of post-20 Ma motion is consistent with geologic data for the amount and timing of right-lateral motion on the Miocene Alpine–Wairau fault system. The amount of relative displacement indicated for the interval between anomalies 13 and 6 (36 and 20 Ma) might partly predate the Miocene inception of the Alpine fault system (Kamp, 1986) but could be coeval with earlier oroclinal bending. The few hundred kilometers of Tertiary displacement between East and West Antarctica (Fitzgerald et al, 1987) is not well constrained in amount or direction and represents an additional uncertainty in the reconstructions.

REFERENCES

Fitzgerald, P. G., M. Sandiford, P. J. Barrett, and A. J. W. Gleadow, 1987, Earth and Planetary Science Letters, v. 81, p. 68–78.

Kamp, P. J. J., 1986, GSA Bulletin, v. 97, n. 3, p. 329–344.

Mayes, C. L., D. T. Sandwell, and L. A. Lawver, in press, Tectonic history and new isochron chart of the South Pacific: Journal of Geophysical Research.

Stock, J. M., and P. Molnar, 1987, Nature, v. 325, p. 495–499.

STOCK, J. M., Harvard University, Cambridge, MA

Rigid Plate Reconstructions of New Zealand Region Since 68 Ma

Reconstructions of the Campbell Plateau (Pacific plate)–Antarctica–Australia–Lord Howe Rise plate circuit (Stock and Molnar, 1982, 1987) have been used to reconstruct the relative positions of the Pacific and Australia plates in the New Zealand region for 10 Ma, 20 Ma, 36 Ma, 42 Ma, 59 Ma, and 68 Ma. Uncertainties in these reconstructions of the Pacific–Australia/Lord Howe Rise plate boundary were calculated from the uncertainties in the data contributing to the individual reconstructions. If internal deformation between East and West Antarctica is ignored, the amount and direction of displacement of the Campbell Plateau relative to the Lord Howe Rise/Challenger Plateau, calculated at positions along the present Alpine Fault, is found to be: 380 ± 65 km along an azimuth of 236° ± 10° since 10 Ma; 460 ± 50 km along an azimuth of 230° ± 8° since 20 Ma, and 810 ± 110 km along an azimuth of 232° ± 5° since 36 Ma. Combination of adjacent reconstructions indicates relative motion of 80–120 ± 80 km, at an azimuth varying rapidly along the boundary, in the interval 20–10 Ma; and 350 ± 100 km at an azimuth varying from 220° to 250° ± 10°, in the interval 36–20 Ma. Between 42 and 36 Ma, the motion of the Campbell Plateau with respect to the Lord Howe Rise is oblique separation, varying from 180 ± 80 km to 140 ± 90 km along the plate boundary, in a direction 187 ± 25° to 190 ± 40°. Between 59 and 42 Ma, the reconstructions suggest that motion of the Campbell Plateau relative to the Lord Howe rise was 230 ± 170 km at an orientation 75° ± 50°. The reconstructions for 68 and 59 Ma overlap within their uncertainties, indicating that the Lord Howe Rise and the Campbell Plateau may not have moved relative to one another during this interval.

Agreement between these predictions and geologic observations from the New Zealand plate boundary varies and is complicated by the nonrigid behavior of much of the region adjacent to the present plate boundary (e.g., Walcott, 1978). Agreement is fairly good for the interval from 20 Ma to the present, since at least 570 km of post-Oligocene right-lateral strike-slip motion has occurred across the Pacific–Australia plate boundary in New Zealand (Walcott, 1978). Agreement is also good for early Tertiary time, in that no deformation in either Antarctica or New Zealand is required prior to 59 Ma. The oblique extension predicted by the plate reconstructions for the New Zealand plate boundary in the interval between 59 and 42 Ma, and slight right-lateral oblique extension between 42 and 36 Ma, is consistent with the development of Late Eocene–Oligocene extensional basins along the plate boundary (Kamp, 1986). However, for the interval from 36 to 20 Ma, the agreement between plate tectonic predictions and geologic data is less clear and deserves closer examination.

Some relative motion occurred between East and West Antarctica during Cenozoic time, although the total amount and direction of this motion are not well known. Fitzgerald et al (1987) estimated that a maximum of 200 km of extension occurred across the Transantarctic Rift system in Cenozoic time. If this motion occurred at a constant rate since the Paleocene, then it has very little effect on the above reconstructions of the New Zealand plate boundary. If it occurred relatively rapidly, then it could affect any one of the intervals studied. The maximum possible effect was estimated by including various reasonable reconstructions of the Transantarctic Rift in the plate circuit, as displacement between East and West Antarctica, during the interval between 36 and 20 Ma. The resulting Pacific-Australia plate circuit predicts 200 ± 100 km of right-lateral strike-slip motion, accompanied by up to an equal amount of compression of extension, between the Campbell Plateau and the Lord Howe Rise during this interval. The amount and direction of this motion vary greatly according to the position of the pole of opening chosen for the Transantarctic Rift system.

REFERENCES

Fitzgerald, P. G., M. Sandiford, P. J. Barrett, and A. J. W. Gleadow, 1987, Earth and Planetary Science Letters, v. 81, p. 68–78.

Kamp, P. J. J., 1986, GSA Bulletin, v. 97, n. 3, p. 329–344.

Stock, J., and P. Molnar, 1982, Journal of Geophysical Research, v. 87, p. 4697–4717.

——— ——— 1987, Nature, v. 325, p. 495–499.

Walcott, R. I., 1978, Geophysical Journal of the Royal Astronomical Society, v. 52, p. 137–164.

STONE, D. B., University of Alaska, Fairbanks, AK

Paleomagnetism and Paleogeography of Arctic Ocean Region

For times prior to the development of the Eurasian Basin, which was formed through an Arctic extension of the North Atlantic spreading center in Cenozoic time, the paleogeographic and tectonic development of the Arctic Ocean is very poorly understood. In almost all global paleogeographic reconstructions of the continents, the Arctic Ocean remains as a largely unfilled space. This is true regardless of whether the present coastlines, present continental margins, or some inferred ancient continental margin is used to depict the paleogeography. If these global paleogeographies are correct, then either the Arctic Ocean space was filled or partially filled with continental material that has since fragmented and dispersed, or the Arctic Ocean must have existed as an oceanic basin throughout much of Phanerozoic time. Many tectonic histories have been proposed, ranging from the concept of a permanent ocean basin (Oceanus Borealis, after Dietz and Holden, 1970) to a now dispersed Arctic Continent (Arctida, after Zonenshain, in preparation). Most of these models are so dramatically different from one another that they should be testable using paleomagnetic techniques.

As part of an attempt to unravel the tectonic development of the Arctic, apparent polar wander (APW) paths have been constructed for key locations for a selection of proposed paleogeographies. These APW paths have been compared with a compilation of all the available paleomagnetic data from relevant areas; however, much of the data show evidence of a remagnetization event or events and are thus difficult to interpret. The best constrained data come from a single locality in Arctic Alaska (Halgedahl and Jarrard, 1987) and indicate an Early Cretaceous or later counterclockwise rotation. Other, less well-constrained paleomagnetic data from Arctic Alaska indicate small mid-Cretaceous or later clockwise rotations and northward translations (Stone, in press), thus there is still considerable ambiguity in the role that Arctic Alaska played in the development of the Arctic Ocean region. Paleomagnetic data from elsewhere in the Arctic are also somewhat ambiguous, but have been used to support both the rotation model and the concept of an ancient Arctic continent, Arctida (Zonenshain, in preparation). All of the paleomagnetic data discussed above are from rocks of late Mesozoic age, thus are only indirectly related to earlier paleogeographies. A comparison of the compiled paleomagnetic data and the predicted APW paths points out the need for reliable paleomagnetic data from rocks of Permian through Late Triassic age, the time of greatest divergence between the various models.

In summary, the data that pass paleomagnetic stability tests tend to support the counterclockwise rotation of Arctic Alaska (and Chukotka?) away from the Canadian Arctic shelf edge, thus opening the Canada Basin, followed by a smaller clockwise rebound of the same landmass, perhaps in response to the accreting terranes of interior and southern Alaska. Paleomagnetic data from localities outside of Alaska show directions that are similar to the adjacent cratons, are difficult to interpret because of possible remagnetization effects, or are not numerous enough to provide adequate stability tests.

REFERENCES

Dietz, R. S., and J. C. Holden, 1970, Reconstruction of Pangea: breakup and dispersion of continents, Permian to present: Journal of Geophysical Research, v. 75, p. 4939–4956.

Halgedahl, S. L., and R. D. Jarrard, 1987, Paleomagnetism of the Kuparuk River Formation from oriented drill-core: evidence for rotation of the Arctic Alaska plate, *in* I. L. Tailleur and P. Weimer, eds., Alaska North Slope geology, v. 2, p. 581–617.

Stone, D. B., in press, Paleogeography and rotations in Arctic Alaska: an unresolved problem, *in* Laj and Kissel, eds., Paleomagnetic rotation and continental deformation, Reidel.

Zonenshain, L. P., and L. M. Natapov, in preparation, Tectonic history of Arctic region from Ordovician to Cretaceous.

STOREY, B. C., et al, British Antarctic Survey, Cambridge, United Kingdom

West Antarctic Mosaic: Tectonic Evolution

West Antarctica is a mosaic of five crustal blocks with high relief, separated from the Precambrian craton of East Antarctica by the Transantarctic Mountains. The crustal blocks are bounded by deep subglacial troughs and basins. In particular, there are two large subsided basinal areas, the Ross Sea and Weddell Sea embayments.

There is still debate about the extent to which the crustal blocks have moved. This has considerable bearing on whether West Antarctica is a collage of accreted exotic terranes or formed of local blocks that were displaced during break-up. There are geological distinctions between all of the blocks, but orogenic events, mainly characteristic of active margin tectonism, can be correlated with neighboring parts of the Gondwanaland supercontinent. For this reason, at least some of the blocks are considered to have been microplates that were displaced during the break-up of the supercontinent and later reassembled to form the present-day West Antarctic mosaic.

The Haag Nunataks block (HN) contains the oldest rocks, 1000 Ma granitic orthogneiss and amphibolite, the only known Precambrian rocks in West Antarctica. These are correlatives of Precambrian domains in Africa and East Antarctica.

The Ellsworth–Whitmore mountains crustal block (EWM) is part of a complex Paleozoic fold belt involving a heterolithic suite of Cambrian silicic volcanic rocks, shelf limestones and locally derived clastic rocks, and a younger platformal cover sequence; the whole succession was deformed during the Permo-Triassic Gondwanide folding. This is part of a larger orogenic system that includes the Cape fold belt in southern Africa, the Pensacola Mountains within the Transantarctic Mountains, and the Sierra de la Ventana fold belt in South America. The sequence in the Pensacola Mountains is similar to the EWM succession, but structural relationships are well-exposed. Here the Cambrian rocks were progressively deformed during a Cambro-Ordovician event and are unconformably overlain by a folded continental fluviatile succession. In contrast, rocks of similar age in the central part of the Transantarctic Mountains are flat-lying. Although the cause of the deformation within the Ellsworth Mountains is uncertain, it is likely to be related to plate margin processes, and the lack of deformation within the central Transantarctic Mountains suggests that the locus of subduction may have migrated Pacificwards on the Marie Byrd Land–New Zealand margin due to the docking of a large terrane or due to accretionary processes.

Evidence for plate margin processes is preserved within the Antarctic Peninsula (AP), Thurston Island (TI), and Marie Byrd Land (MBL) crustal blocks. In AP lower Paleozoic, Triassic, and Jurassic to Tertiary magmatic belts exist; TI and MBL both have Carboniferous and Jurassic–Cretaceous magmatic rocks. An extensive Tertiary alkali basalt province is common to all three of these Pacific margin blocks.

There are enough geological similarities between the crustal blocks to suggest that West Antarctica is not made up of exotic terranes. The HN Precambrian basement has a distinctive aeromagnetic signature, and a geophysical survey shows it underlies, at least in part, the EWM sedimentary rocks and the Mesozoic sediments of part of the Weddell Sea Embayment. Magmatic belts are common to AP, TI, and MBL, and both AP and TI are linked by a distinctive Pacific Margin anomaly belt. Permian cover sequences occur in AP, EWM, and the Transantarctic Mountains. Moreover, structural trends and paleomagnetic data indicate 30° rotation of the EWM, TI, and AP blocks between 120–100 Ma. A widespread suite of Middle Jurassic within-plate mafic and granitic magmatism records an initial break-up phase, although the amount of movement and the rotation of the crustal blocks associated with it are uncertain.

West Antarctica has consequently mainly developed by plate margin processes during the Phanerozoic along the margins of the supercontinent Gondwanaland and contains part of a Paleozoic fold belt and Precambrian orthogneiss. Although subduction may have continued during parts of the break-up proto-Pacific margin dispersed as a series of microplates and reamalgamated to form the West Antarctic mosaic. The present-day troughs and basins that separate and dissect the crustal blocks may represent a further Cenozoic extensional phase forming complex boundaries to the crustal blocks.

STOREY, B. C., P. J. BUTTERWORTH, J. A. CRAME, S. M. HARRISON, P. J. HOWLETT, D. I. M. MACDONALD, M. P. MASLANYJ, P. A. R. NELL, B. A. PIERCY, J. W. THOMSON, M. R. A. THOMSON, T. H. TRANTER, and H. E. WEVER, British Antarctic Survey, Cambridge, England

Antarctic Geoscience Transect: AP2

The Antarctic peninsula was the locus of magmatic arc activity associated with the eastward subduction of proto-Pacific oceanic lithosphere from late Palaeozoic until Tertiary times. Prior to the Middle Jurassic it formed part of the margin of the supercontinent Gondwanaland and then separated during break-up to form the present-day peninsula. The peninsula consists mainly of calc-alkaline igneous rocks, an accretionary complex and large fore- and back-arc basin sequences which formed by the interaction of compressional, extensional and strike-slip tectonism.

This transect, which is 400 km long, is one of a series across the Antarctic Peninsula illustrating the evolution of the plate margin processes that dominated during its formation. It is compiled from many years of geological field data, an aeromagnetic survey and limited gravity data, mainly from the work of the British Antarctic Survey.

The main components of the transect from west to east are as accretionary complex contains a docked ocean island, slices of ocean floor and deformed trench-fill turbidites. The deformation, as illustrated on the transect, is dominated by large-scale melange zones, trenchward-directed thrusting, back-thrusts, and normal and strike-slip faulting. Perched unconformably on the accretionary complex is a 4 km thick fore-arc basin sequence. The basin-fill records continuous sedimentation from Kimmeridgian until Albian times. Sediments were mainly derived from the adjacent volcanic arc and form a shallowing-upward megacycle from submarine fan through shallow marine to delta plain facies.

The fore-arc region is separated from the Mesozoic magmatic arc by a major topographic trench, George VI Sound, which structural and geophysical data suggest may be a Cenozoic transtensional feature.

The magmatic domain preserves evidence for a long history of Cordilleran-type subduction-related activity containing Ordovician-Silurian granitic orthogneiss, and widespread Jurassic and Cretaceous plutonic and volcanic rocks. Lower Jurassic plutons were deformed and metamorphosed during emplacement and Upper Jurassic shear zones show dextral strike-slip displacement.

On the eastern side of the peninsula Middle-Upper Jurassic turbidites and mafic and felsic volcanic rocks are part of a deformed back-arc basin system. Magmatic activity migrated eastwards during the mid-Cretaceous intruding the back-arc system and westwards during the Tertiary intruding the fore-arc accretionary complex.

Except for the accreted ocean island the magnetic signature is quiet over the accretionary complex. There is a large linear anomaly over the western part of the magmatic domain imaging a deep-seated gabbroic body and a number of distinctive circular high amplitude anomalies in the back-arc region. Gravity data are used to model the thickness of the crust which shows some thinning beneath George VI sound.

STOREY, M., University of Leicester, Leicestor, United Kingdom, H. WHITECHURCH, Institut de Geologie, Cedex, France, J. H. SEVIGNY, University of Calgary, Alberta, Canada, V. J. M. SALTERS, MIT, Cambridge, MA, and LEG 120 SHIPBOARD SCIENTIFIC PARTY.

Basement Drilling of Kerguelen Plateau by Ocean Drilling Program (Leg 120): Report on Shipboard Results and Work in Progress

The Cretaceous Kerguelen Plateau in the southern Indian Ocean is the world's largest submarine plateau, stretching approximately 2300 km southeastwards between 46°S and 64°S (Fig. 1). The origin of the plateau is a subject of controversy, and a number of theories have been proposed to account for its formation, viz: (1) It is continental remnant of Gondwana following the breakup of India and Antarctica. (2) It is the product of hotspot-type volcanism. (3) It is thermally or tectonically uplifted oceanic crust.

One of the aims of ODP Legs 119 and 120 was to resolve these conflicting theories by obtaining in situ samples of basement material. Leg 119 cored basaltic basement at one site (738) in the southern part of the plateau. On Leg 120 we obtained basement samples at three sites from the central part of the Kerguelen Plateau (Fig. 1). At Site 747 the basement was overlain by lower Santonian sediments, at 749 by Early Eocene nanofossil ooze and at 750 by a Cenomanian or older ferruginous calcareous claystone rich in wood fragments. In addition, a massive alkali basalt flow interbedded with upper Cretaceous sediments was encountered at Site 748 some 150 to 200 m above the predicted basement.

The Kerguelen Plateau basement samples consisted of slightly- to highly-altered, aphyric to olivine-clinopyroxene-plagioclase-phyric tholeiitic basalts occurring as lavas and breccias. The presence of vesicle, reddened flow tops and the lack of pillow or hyloclastite textures indicates that some basalts were erupted in sub-aerial conditions. The secondary mineral assemblages correspond to the low-temperature (sites 747 and 750) and high-temperature (site 749) zones of the zeolite facies.

Incompatible element abundances, for comparable MgO contents, show systematic inter-site variations, Site 747 basalts having the highest concentrations and Site 750 basalts the lowest (Fig. 2). However, basement from all three sites have similar Zr/Nb ratios (Fig. 3), indicating that the inter-site variations in incompatible element abundances may simply reflect differences in the degrees of partial melting of compositionally similar (at least in terms of Zr and Nb) sources. Kerguelen Plateau basalt basement samples have intermediate Zr/Nb ratios relative to normal Indian Ocean mid-ocean ridge basalt (MORB) and within plate "hotspot" islands but are similar to other oceanic flood basalt provinces such as the Nauru Basin in the Pacific. However, preliminary data shows that in La/Ta-Th/Ta space some samples from the Kerguelen Plateau do not fall between the fields occupied by normal MORB and within-plate ocean island basalts, but instead vector strongly towards values for continental crust.

Plate reconstructions are consistent with the Kerguelen Plateau being a much older manifestation of the major mantle plume which now underlies Kerguelen and Heard Islands. Further, these models indicate that the plume may have been more or less coincident with the spreading axis separating India and Antarctica during the breakup of Gondwana. Excessive on-axis volcanism due to plume interaction with MORB asthenosphere could explain the voluminous nature of the Kerguelen Plateau and perhaps its transitional compositional character in terms of rations such as Zr/Nb. However, the high La/Ta and Th/Ta ratios observed in some Kerguelen Plateau samples preclude simple plume-MORB asthenosphere mixing models. The additional component which appears to be required may have been the adjacent enriched sub-continental Gondwana lithosphere. This model will be reviewed with the benefit of new trace element and isotopic data (work in progress).

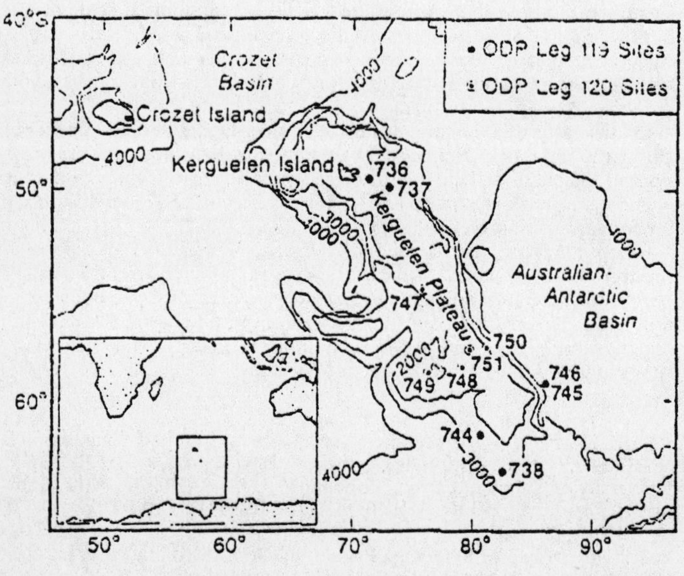

Figure 1—Bathymetric chart of the Kerguelen Plateau showing Leg 119 and Leg 120 drilling locations.

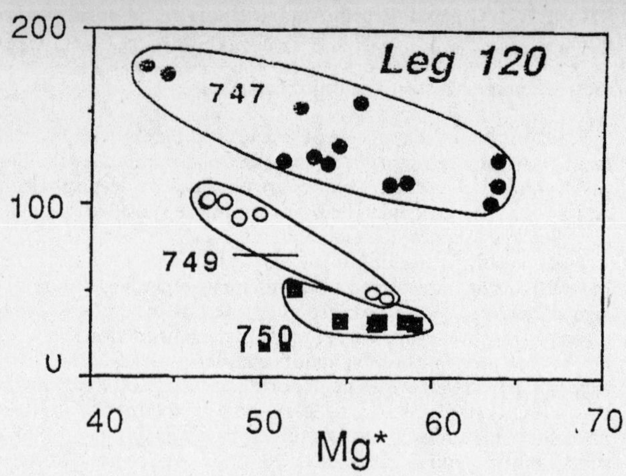

Figure 2.

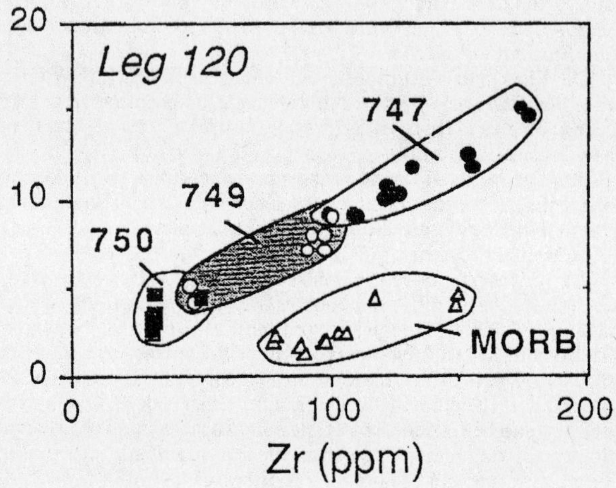

Figure 3

STOW, D. A. V., University of Nottingham, Nottingham, United Kingdom

Role of Geosciences in Development

The role of geosciences in the development of nations is unquestionable, but out understanding of that role has changed and expanded dramatically in recent years. Coal fired the industrial revolution early in the 18th century, metallic and industrial mineral resources then provided the framework for rapid development in the North. Oil and gas provided a major boost early in this century and has moved us into an age of plastics. Many would hold that the nuclear age has already dawned. That the earth's natural resources have provided for such great economic advances in some parts of the world is not debated. The attendant environmental risks and unequal distribution of wealth are, however, cause for global concern.

As the 120 or so nation states that make up the south (or LDCs) move through similar stages of development as witnessed by the industrialized countries of the north, there is an ever increasing demand for energy and mineral resources, and hence for geoscientists to find and explete them. However, real development means improving the quality of life for all a nation's people and involves far more than simple economic development based on exploitation of traditional natural resources. It means providing adequate supplies of food and clean drinking water, achieving good basic health standards, ensuring adequate shelter and a safe environment in which to live, and providing good education and useful employment for all.

The geosciences have an essential role to play in all of these aspects of development. Hydrogeologists are concerned with the search for groundwater for drinking and irrigation, agrogeologists with the search for rock and mineral products that can directly improve soil quality. Industrial geologists must explore for a development a whole range of resources appropriate for building and construction in widely differing conditions. Economic geologists concerned with energy and mineral deposits must consider equally the use of small-scale and alternative resources. Engineering and other geologists must work with environmentalists and land planners toward the prediction and mitigation of natural as well as human-made hazards.

These are the fields in which geoscientists are needed for development just as much as in their traditional roles. These are the main area proposed by AGID past-president P. G. Cooray as "The Way Ahead" for geoscience. These were the topics discussed at length last year at the Geosciences in Development conference in Nottingham, England. We must ask the questions again now. Is the education and training provision adequate for the new generation of geoscientists required to tackle these issues? Are we communicating these urgent needs satisfactorily to policy planners and decisions makers? Geology is a global science - is our perspective sufficiently global?

STOW, S. H., and T. O. EARLY, Oak Ridge National Laboratory, Oak Ridge, TN

Hydrologic and Geologic Studies on Oak Ridge Reservation and their Relationship to Radioactive Waste Management

The Oak Ridge Reservation (ORR) is a large (150 km^2) tract of land in east Tennessee, owned and operated by the U. S. Department of Energy. The ORR is a complex geologic and hydrologic terrain where a wide variety of radioactive and hazardous wastes have been generated and disposed of over the last 45 years. The wastes result from research and production operations at three facilities: (2) the Y-12 plant, built for separation of fissionable isotopes of uranium, and now used for nuclear weapons component manufacturing and nuclear materials processing; (2) the Oak Ridge Gaseous Diffusion Plant (ORGDP) where uranium isotope enrichment was done by gaseous diffusions; and (3) the Oak Ridge National Laboratory (ORNL), where basic research is conducted on energy-related materials and processes. Consequently, the nature of the activities at the plants has resulted in the generation of large amounts of radioactive, hazardous, and mixed wastes.

The wastes have been disposed of by many methods on the ORR. The bulk of the wastes are low-level radioactive material contaminated with Sr-90, Cs-137, and H-3; over 20 million m^3 of low-level waste containing over 6000,000 curies of activity have been disposed of, principally by shallow land burial in trenches approximately 5 m deep. Often the groundwater table is above the bottom of the trench. In addition, depleted uranium was buried in shallow trenches, and liquid radioactive wastes containing transuranic elements and fission products (approximately 2.5 million curies) were disposed of in shallow seepage pits and trenches. In recent years, this waste has been mixed with cement and injected as a slurry into shale at a depth of approximately 300 m. Shallow land burial and neutralization ponds also were used for disposal of hazardous wastes. In recent years, efforts have focused on treatment and above ground storage, rather than disposal underground.

Waste management operations and site characterization activities are under the authority of many regulations. Current emphasis on hydrogeologic studies is principally in response to the U. S. Environmental Protection Agency; the primary regulation is the Resource Conservation and Recovery Act, which was designed to apply to hazardous materials, but which also applies to radioactive wastes. Characterization activities are conducted to develop an understanding of contaminant migration, principally in groundwater, and to permit design of site remediation methods. Remedial actions will be undertaken after site characterization.

The hydrogeologic nature of the ORR is extremely complex. Strata are Lower Paleozoic carbonates and clastics that have been severely folded and faulted. There are at least three major thrust sheets, each bounded by large-displacement thrust faults, and numerous smaller-scale structures. Groundwater flow is controlled principally by the fracture systems: in some thick carbonates Karst features and rapid flow through solution cavities occur. Many hydrogeologic questions are being addresses regarding flow directions and rates and potential contaminant migration.

At ORNL efforts are made to acquire groundwater data from a variety of waste sites; this has involved over 1000 groundwater wells, investigation of steambed gravels with sorbed contaminants from discharge areas, and analysis of streamflow to infer the magnitude of discharge. Well construction has been phased, with piezometers used to identify general groundwater characteristics, followed by installation of high quality groundwater characteristics, followed by installation of high quality monitoring wells. A close correlation occurs among lithology and structure of the underlying rocks, groundwater flow, and associated contaminant migration. A particular challenge involves characterization of the hydrogeology and contaminant extent associated with the radioactive waste injection well, referred to as "hydrofracture." Contaminated groundwater extends at least 330 m in two directions from the well' fortunately it appears to be largely confined to the injection horizon.

At the Y-12 Plant hydrologic monitoring and contaminant migration investigations suggest that, in addition to lithology, structural anisotropy in both weathered and unaltered bedrock strongly influence groundwater flow. Recently, these observations have led to investigation of the potential structural controls on groundwater flow at several different scales. For example, preliminary data from detailed fracture mapping of cores have led to identification of a consistent set of bedding-normal fractures. Furthermore, mapping of structural lineaments from SLAR photos reveals a pattern with a strong resemblance to the core data. Hydrologic testing in different aquifers shows that anisotropy of hydraulic conductivity values is correlative with the dominant fracture sets identified in core and on photos. These results will be factored into hydrologic models to predict contaminant migration pathways and rates, and to facilitate remediation measures.

At the ORGDP, the bedrock geology differs from that of the other facilities as it is more fractured and deformed because ORGDP is in very close proximity to a regional thrust fault zone. In addition to standard characterization techniques, an extensive program of stream sediment sampling has been undertaken. Screening for heavy metals, organics, and radioactive materials has indicated both on-site and off-site sources of contaminants.

In summary, an active characterization program is under way on the ORR to acquire hydrogeologic information to allow regulatory and environmental issues to be addressed. The complexity of the geology and the diversity of waste materials that have been disposed of over decades make this an unusually challenging undertaking.

STOWE, C. W., University of Cape Town, Cape Town, South Africa

Tectonic Settings of Chromite Deposits

Structures, textures and compositions of the podiform chromite orebodies hosted in ophiolitic terranes of the orogenic belts, are contrasted with those of stratiform ores hosted in layered complexes of the stable cratons. The fundamental distinction is in the differing degrees of tectonic instability of their original sites of deposition, in either oceanic, or continental crust.

Large layered complexes like the Bushveld or Great Dyke were emplaced in extensional tectonic settings within stable continental cratons, controlled by pre-existing crustal structures. Compositions, thicknesses and lateral facies changes of the chromite layers are related to syndepositional deformation and crustal thicknesses. The Great Dyke was emplaced in a graben structure and the Bushveld Complex near an aulacogenic trough, possibly part of an early Proterozoic triple junction.

In the ophiolites, podiform orebodies occur within 1000 m of the crustal cumulate/mantle harzburgite contact. Orebodies in the lowermost crustal cumulate layers are disrupted lenticular sheets and display cumulate textures. Orebodies occurring in the underlying mantle tectonites are irregularly shaped, intensely deformed and display resorbtion textures. Both types display structures and textures relating to hot-mode deformation during crystallization at ocean-spreading centers and also cold-mode structural fabrics formed during their subsequent accretion at convergent plate boundaries. Textures and compositions of orebodies in the underlying tectonized mantle harzburgites are different from the crustal cumulates, and differ also from the high-Al chromites originating deeper in the mantle near palaeotransforms.

Chromite compositions provide constraints for tectonic settings and are illustrated for a wide variety of chromium ore fields. Mg/(Mg + Fe) ratios are essentially temperature related and stratiform ores display a wide range, but in contrast to mantle podiform ores their range of Cr/(Cr + Al) ratios are restricted. The Cr/(Cr + Al) ratios are in part pressure dependent and indicate a wide range of depths of crystallization in the podiform chromites. In the continental stratiform complexes chromite crystallization temperatures were around 1200–900 °C and Cr/Fe ratios are lower (2.6–1.4) than in the podiform ores (4.2–2.4) of the hotter oceanic crust–mantle environment. Chromite re-equilibration during metamorphism results in increased trivalent iron.

STOWE, C. W., University of Cape Town, Cape Town, South Africa

Chromite Deposits through Time

Chromite deposits are of all ages, from the 3.5 Ga old, podiform-like deposits in the Selukwe greenstone belt, Zimbabwe, to the Miocene ophiolite–hosted deposits of New Caledonia. They provide constraints for evolution of both the continental and oceanic crustal settings and for crust-mantle interactions. Podiform orebodies are hosted in obducted ophiolite complexes and their occurrence implies both ocean-spreading and plate convergent processes. The oldest known complete ophiolite sequences with podiform ores are of late-Proterozoic age, indicating that plate-tectonics was operating by at least 0.7 Ga. There is a Proterozoic hiatus in formation of podiform chromite deposits and of ophiolites, possibly related to differences in the Proterozoic ocean plates.

In contrast, stratiform ore fields require a stable continental setting and most were formed between 2.5 and 2.0 Ga ago, following cratonization of the Archaean continental nucleii. Layered complexes continued to form, but were defficient in chromite, possibly a function of secularly decreasing continental thermal gradients. Heat-producing isotopes concentrated in continental crust at the end of the Archaean, would inhibit early fractionation of chromite near the base of the crust. If this is a valid assumption, and crustal thicknesses were equal to present, the estimated average local continental geothermal gradient during emplacement of the Great Dyke and Bushveld, would have been as high as 80°/km. The intensely disrupted anothosite-chromite layered complex of west Greenland may be an early Archaean analog of the stratiform ore fields, emplaced prior to continental cratonization.

Early Archaean podiform orebodies in the Selukwe greenstones occur in an ultramafic complex, emplaced in an oceanic-type high-Mg basalt/chert crustal sequence and part of a large nappe structure. Chromite compositions provide estimates of PT conditions and Selukwe compositions are compared with other deposit types. Though closely resembling an ophiolite complex, cumulate textures and geochemical trends indicate the host serpentinite was a layered complex. It was emplaced very close to surface, sheeted-dyke complexes and gabbroic cumulates are absent, magnesium is high (32–40% MgO) compared to recent ophiolitic harzburgites and it was extensively carbonatized. The early Archaean ocean crust was therefore probably very different from the present and thermal plume processes may have been more important.

STRAND, K., University of Oulu, Oulu, Finland

Fluvial-Shallow Marine Transitions in Sedimentation of Early Proterozoic Kainuu Schist Belt in Kurkikylä-Siikavaara Area, Northern Finland

Within the central part of the Fennoscandian Shield the Kainuu Schist Belt is an early Proterozoic quartzite-pelite dominated belt about 220 km long. It runs in a north-south direction and is pressed in a complex way between two late Archaean basement blocks by Svecokarelidic orogenic activity (e.g., Laajoki, 1986). In its northeastern part, in the Kurkikylä-Siikavaara area, the belt comprises the Kurkikylä, Kainuu and East Puolanka Groups.

The Kurkikylä Group (400–700 m) represents the earliest succession deposited nonconformably on the late Archaeean basement, consisting of the deposits of an intracratonic rift basin which remained above sea level (Figure 1.1). It is divided into two interfingering formations of immature conglomerates, diamictites and sandstones of alluvial fan and braided river origin, one terminating in the fine-grained rocks of a lacustrine basin associated with some subaqueous mass flows suggesting a cool to cold climatic regime. Active rifting is indicated by one formation with superimposed subaerial lava flows. The following peneplanation was dominated by intense chemical weathering apparently of global extent.

The Kainuu Group (600–1200 m) is separated by a preserved palaeosol from the Kurkikylä Group and consists of two formations of quartz pebble conglomerates and mature sandstones deposited on a braided alluvial plain of a developing continental margin in its proximal, stable cratonal setting (Figure 1.2). At a narrow ocean basin stage, sedimentation was controlled by reactivation of earlier rift faults.

The East Puolanka Group (2300–2600 m) deposited unconformably on the Kainuu Group consists of six, mostly quartzitic formations exhibiting records of three transitions from fluvial to shallow marine environment and one formation consisting of subaerial lavas. The group forms a large opening wedge of sediments on the marginal parts of a platformal shelf of a divergent continental margin which, in these early stages, was still favored by sporadic volcanic activity. Conglomerates, immature sandstones and fine-grained rocks in the lowermost levels were deposited in the distal parts of an alluvial plain suggesting that the climate was dry and semiaridic. These are overlain by rippled and horizontally laminated sandstones of foreshore and upper shoreface environments which, by succeeding regresion development, become a part of a larger barrier beach complex (Figure 1.3). The prograding of a braided alluvial plain environment with shallow channels and wide flood plains is recorded by a succession up to 1100 m thick of quartz arenitic sandstones (Figure 1.4). Detrital influx was after that so much diminished that shoreline processes dominated. The mudstones represent tidal flat deposits in partial back-barrier regions and a continuing transgression development brings bipolar low-angle cross-bedded mature sandstones of foreshore and deeper water over these (Figure 1.5). The uppermost formation, which presents an erosional surface as an upper contact, consists of coarse immature sandstones deposited in a clastic shoreline environment and the successions preserved are most probably caused by the reworking of an older pre-existing alluvial plain margin (Figure 1.6). Evidence from other parts of the Kainuu Schist Belt indicate that this stage was followed by distal shelf and deeper water deposition.

The sedimentary rocks of the study area attest to a significant variety of depositional patterns for the early Proterozoic basins and prolonged development from the continental rift stage to the low relief platform of a divergent continental margin.

REFERENCES

Laajoki, K., 1986, The Precambrian supracrustal rocks of Finland and their tectono-exogenic evolution: Precambr. Res. v. 33, p. 67–85.

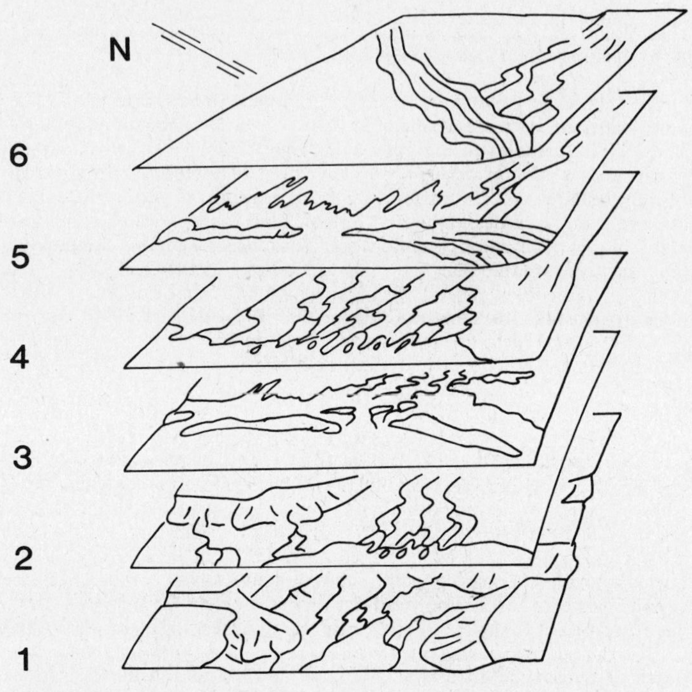

Figure 1—Filling pattern of the sedimentary basins of the study area.

STRIBRNY, B., and H. URBAN, Institut für Geochemie, Petrologie und Lagerstättenkunde, Frankfurt, West Germany

Classification of Sedimentary Rocks in Black Shale Series Based on Their Normative Mineral Compositions

Normative mineral compositions deducted from bulk chemical analyses, e.g., the CIPW-norm, are often used for classifying igneous rocks. Comparable methods for sedimentary rocks in black shale series are lacking. The advantages of such a classification are obvious especially for sediments with grain sizes smaller than 0.02 mm.

Worldwide black shale series are composed of various kinds of fine-grained clastic, chemo-clastic and chemical sediments. The extremely small grain size of these rocks prevents qualitative and quantitative determinations of the rock forming minerals under the microscope. X-ray diffraction studies help to identify the major mineral phases and provide informations on their approximate quantitative relations. Finally, chemical determinations of the major and some minor elements by XRF, AAS, and coulometry combined with optical and X-ray diffractometrical data make the calculation of the normative mineral composition possible. For this, the weight percentages of different elements or oxidic molecules are converted to the corresponding mineral in relation to their atomic or molecular weights:

1. P_2O_5 is alloted for APATITE.
2. TiO_2 is alloted for RUTILE and ILMENITE in relation to their abundance.
3. Na_2O is alloted for ALBITE.
4. C_{org} remains at C_{org}.
5. C_{carb} is allotted for CALCITE, DOLOMITE and SIDERITE in relation to their abundance determined optically or by X-ray diffraction.
6. MgO minus the MgO of dolomite is allotted for MgAl-CHLORITE.
7. K_2O is allotted for ILLITE + MUSCOVITE in relation to their abundance determined by X-ray diffraction.
8. Cu, Ni, Pb, Zn are allotted for SULFIDES.

9. Ba is allotted for BARYTE.

10. S minus the S of sulfides and baryte is allotted for PYRITE.

11. Fe minus the Fe of ilmenite, siderite and sulfides, is allotted for FeAl-CHLORITE.

12. SiO_2 minus the SiO_2 of albite, MgAl-chlorite, FeAl-chlorite, illite and muscovite is allotted for QUARTZ.

13. The percentages of the normative minerals are added up. This sum should be equivalent to the sum of the bulk chemical analyses.

14. For classification a double-triangle is proposed (Figure 1). Quartz (Q), phyllosilicates (P) and carbonates plus the remaining minerals (C) form the corners of the double-triangle. The normative mineral compositions (Q + P + C = 100) of the sedimentary rocks with a content of C_{org} greater than 1% are plotted in the upper part of the diagram. The lower part of the double-triangle is used for sediments with a content of C_{org} smaller than 1%.

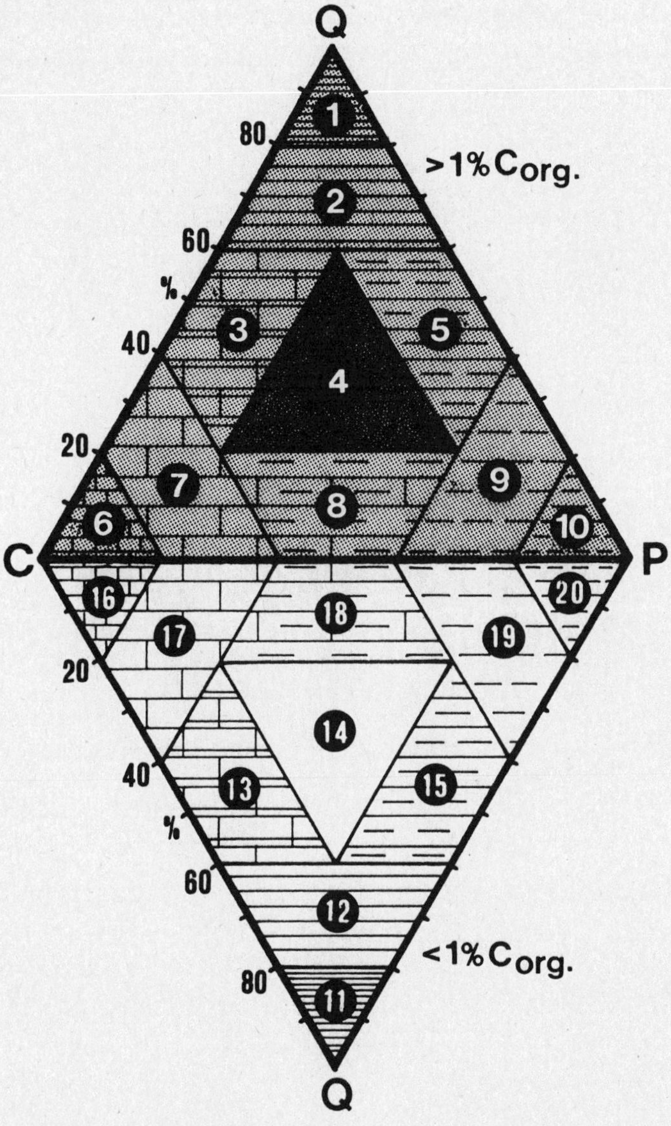

Figure 1—Classification and nomenclature of fine-grained sedimentary rocks associated in black shales series.

STRNAD, J. G., Uranerz Exploration and Mining Limited, Saskatoon, Saskatchewan, Canada, HÖHNDORF, A., and CARL, C., Bundesanstalt für Geowissenschaften, Hannover, West Germany

Athabasca Basin Uranium Province (Canada): Basement Metallotects in Regional Tectonic Context

Introduction: Unconformity-type uranium deposits of the Athabasca Basin are dominated by large and exceptionally high grade orebodies. Since these are in part hosted by the basal portion of the flat-lying unmetamorphosed mid-Proterozoic Athabasca Group sandstone and because the earliest detectable emplacement of uranium occurred 1.4 Ga ago, 300 Ma later the deposition of the sandstone, the host itself is occasionally considered as the source of the uranium. Other models prefer progenitors in the subjacent basement namely in the form of the early Proterozoic black shales which became steeply folded and metamorphosed (graphitized) during the Hudsonian orogeny about 1.9–1.8 Ga. Although the intersection of the sub-sandstone unconformity with the subjacent graphitic basement hosts the majority of orebodies this abundant structural combination becomes statistically only rarely productive. Therefore, there is an obvious need for another control to be added. Currently, notable progress has been achieved in outlining regional specifics in both the early Proterozoic and the Archean basement through still only sketchily within a plate tectonics context. We suggest, that the productivity of the upper structural niveaus (usually covered by prevailing attention) is considerably enhanced only where the Archean basement attains particular development (Strnad, 1981).

Archean: With the exception of a few ensimatic terranes, the predominant part of the Hudsonian-reworked western Churchill (structural) Province (CHP) is underlain by the Archean basement. This is interpreted either as an extension of the surrounding Superior and Slave (Archean) provinces (SSP) or likely as separate domain(s) bounded by ancient collisional or otherwise tectonized boundaries. Current isotopic and other data suggest that both the compositional inhomogeneity and the age of the Archean crust in CHP may not substantially differ from SSP. It is the Hudsonian reworking, known as the principal characterization of the CHP, which is also responsible for its metallogenical distinction from those provinces (SSP) which were left without being overprinted by Proterozoic orogenies. In essentially sialic terranes of the CHP only disseminated progenitors were commonly considered in the Archean. In addition, we postulate that the original Archean in the CHP may have included localized concentrations of massive sulphides, Au-bearing greenstone belts and other types of mineralization (including U). These are indeed (and perhaps necessarily) almost missing in the present Archean of CHP (except for its unmetamorphosed inliers). Such Archean sources could have enhanced the fertility of the superjacent early Proterozoic via a variety of processes. While the contribution of a progressed Archean source of metals to the unconformity-type of uranium deposits remains largely speculative, we think, we can more easily demonstrate the structural link of the deposits with the heterogeneity of the Archean basement. Here, at least at the Key Lake deposit, some Hudsonian granitoids point to the Mesoarchean, 3.0 Ga-old crustal sources (Höhndorf and Carl, 1988). Such sources, by applying updated standards of the Archean geochronology in Canada (Okulitch, 1988), may be potentially separated from Kenoran granitoids by one, possibly by two, unconformities or by other structural boundaries. The Neoarchean (Kenoran) 2.6 Ga-old granitoids from the vicinity of the Key Lake deposit suggest essentially mantle-derived material emplaced into an ancient (Mesoarchean?) deep-seated tectonic zone. Kenoran granites at Key Lake contain 7 ppm U and 27 ppm Th. By temporal position related to the Archean orogenies, such Kenoran granites are extremely young and perhaps affine to the fertile post-orogenic Rapakivi (Haapala, 1977; Huhma, 1988) and late Variscan granites.

Early Proterozoic: Within this (Aphebian) ear of 750 Ma duration, Wollaston Group (WG) represents the earliest widespread sequence of Proterozoic supracrustals including graphitic metamorphics which control the roots of uranium deposits. We assume that WG is relatively young, Mesoaphebian perhaps, and thus resting on the basement which may be consolidated not only during the Archean (as traditionally accepted) but in some regions also by the time of the Blezardian orogeny (+2.1 Ga). Recently, in the Key Lake area, some granitoids have produced Blezardian age correlative with others in the region which are, in addition, fertile either in U or in Au. At Key Lake, Blezardian granites (BG) are assumed to form part of the EW-extending basement beneath the WG basal greywacke and peribasal graphitic metamorphics. Some of these BG yielded a substantial mantle-derived component (Höhndorf and Carl, 1988), another new signature for this particular orogeny in the region. The geometry of Archean and Blezardian compositional and structural boundaries have undoubtedly played a considerable role in adjusting the pattern of the Hudsonian structural overprint. Variously arranged mylonites (2.3 and 2.1–1.9 Ga-old) were followed in upper crustal domain by shearing at 1.76 Ga, both styles mostly in the form of sizeable stratabound thrusts.

Mid-Proterozoic: After the deposition of the Athabasca Group (1.7 Ga) the early phases of the Neohelikian reactivation (1.4–1.3 Ga) have culminated all previous metallogenic processes. This important anorogenic event, synchronous with "intracontinental metallogenic pulse of endogenic activity" (Sokolov in Kratz, 1984), likely represents an intracratonic (and peri-equatorial?) response to the accretional tectonics surrounding the Canadian Shield. Though impressive as the most fertile single episode in the formation of uranium deposits, the location of productive sites is suggested as conditioned by the cumulative inheritance of metallotects which evolved since the Archean.

SUBRAMANYAN, V., M. K. DEVARAJAN, and V. SREEDHARA, Indian Institute of Technology, Powai, Bombay, India

Morphotectonic Evolution of Cardamon Hills, South India

The Cardamon Hills, together with the Anaimalai Hills in the northwest and the Palni Hills in the northeast, constitute a prominent physiographic unit with a triangular outline in the heart of the South Indian high grade terrane. The east-west trending Palghat–Trichy gap constitutes the northern boundary of the Anaimalai–Palni–Cardamon Hills. In the south, the Cardamon Hills taper towards the northwest-southeast trending Achankoil shear zone, a major Proterozoic shear belt.

Geologically, the area consists of the charnockitic suite of rocks and unclassified Peninsular gneisses with a few occurrences of post-kinematic granitoids. A taphrogenic affiliation has been suggested by workers for the origin of the Munnar granite, an elongated body of alkali granite (Nair et al, 1983).

A very conspicuous negative gravity anomaly, culminating in a Bouger anomaly of – 120 m gal is observed in this area, suggesting the occurrence of a thick crust in this region.

Most of the high peaks in the Peninsular India are located on the Anaimudi and Palni hills, the highest being the Anaimudi peak (2695 m) in the Anaimalai Hills. The Cardamon Hills represent an uplifted peneplain currently undergoing erosion.

Figure 1 shows the fracture pattern derived from the visual interpretation of the satellite images (Landsat 5 TM FCC), digitally processes Landsat data and aerial photographs on the scale of 1:60,000. The pattern shows two sets of lineaments: one trending northeast–southwest and the other, northwest–southeast, intersecting at the Cardamon Hills. Two models are suggested for the tectonic evolution of the Anaimalai–Palni–Cardamon Hills based on the interpretation of the fracture patterns.

In the first model it is assumed that the fracture pattern is the resultant of north-south compression between two continental blocks juxtaposed along the present Palghat–Trichy line. Such continent-continent collisions were envisaged in the tectonic models for the Archean suggested by Kroner (1985) and for the South Indian craton by Drury (1983). The thick crust inferred from the geophysical studies and mentioned above could have resulted from the continent-collision. Stress ellipsoid with the maximum bilateral compressional stress from the north and the south explain the two major sets of fractures with northeast–southwest and northwest–southeast trends as conjugate shear faults. A few fractures might well be second order features developed due to local stress reorientations resulting from body forces developed by movement along the major faults. The angle between the two complementary first order shear faults is about 30°. It may be noted that in the vicinity of the Plaghat–Trichy gap, a regional antiform with an east-west fold axis has been reported by Nair and Nair (1975) from an area near Ottapalam. During the present investigations, the occurrence of another regional antiform plunging towards east with an west-east axial trace is suggested in the Annanthampi–Kalyanathandu area in the southern Cardamon Hills. These folds with east-west axial traces constitute the bounding primary folds in an area of north-south compression (Moody and Hill, 1956). A group of en echelon shear fractures are observed in the area northwest of Idukki. This might have developed due to a significant change in the sense of movement along the northeast–southwest trending fractures. The formation of such shear faults by transtensional movement is reported by Pratsch (1982) during his analysis of the tectonics of Paleozoic continental margins of West Germany.

A plot showing the epicentres of the recent earthquakes (from 1856–1985) in the South Indian high grade terrane reveals their marked alignment along the northeast–southwest trending lineaments (Figure 2). The northwest–southeast trending lineaments, such as the Kerala lineament of Katz (1978) seem to be more or less stabilized. The only exception is along the West Coast fault, which is geologically very young in age.

In the second model, a triple junction is envisaged at the vicinity of Idukki where the northwest–southeast trending Anaimalai arm intersects the northeast–southwest trending Palni Hills and the north-south trending Cardamon Hills. In this model the northwest–southeast trending Kerala lineament is considered as the failed arm. Displacement continued, perhaps for hundreds of km, along the northeast–southwest transform faults. Earthquakes, mild to moderate continue to occur along these faults. Alkali granites such as the Munnar granite are intimately associated with the prerift tectonics.

REFERENCES

Drury, S. A., N. B. W. Harris, R. W. Holt, G. J. Reeves-Smith, and R. T. Wightman, 1983, Precambrian Tectonics and Crustal Evolution in South India. Jour. Geol., v. 92, p. 3–20.

Katz, M. B., 1978, Tectonic evolution of the Archaean granulite facies belt of Sri Lanka, South India. Jour. Geol. Soc. India, v. 19, n. 5, p. 185–205.

Kroner, A., 1985, Evolution of the Archean continental const., Ann. Rev. Earth Planet. Sci., 1985, v. 13, p. 49–74.

Moody, J. D., and M. J. Hill, 1956, Wrench-fault tectonics Bull. Geol. Soc. Amer., v. 67, p. 1207–1246.

Nair, N. G. K., M. Santosh, and P. K. Thampi, 1983, Geochemistry and petrogenesis of the alkali granite of Munnar, Kerala (India) and its bearing on rift-tectonics, Nenes Jahrbuch Miner. Abh., v. 148, n. 2, p. 223–232.

——— and E. V. Nair, 1975, Superposed folding of Precambrian rocks around Ottapalam, Palghat District, Kerala Geol. Surv. India, Spec. Pub. n. 5, p. 15–18.

Pratsch, J. C., 1982, Wedge-tectonics along continental margins, *in* J. S. Watkins and C. L. Drake, eds., Studies in Continental Margin Geology, AAPG Memoir 34.

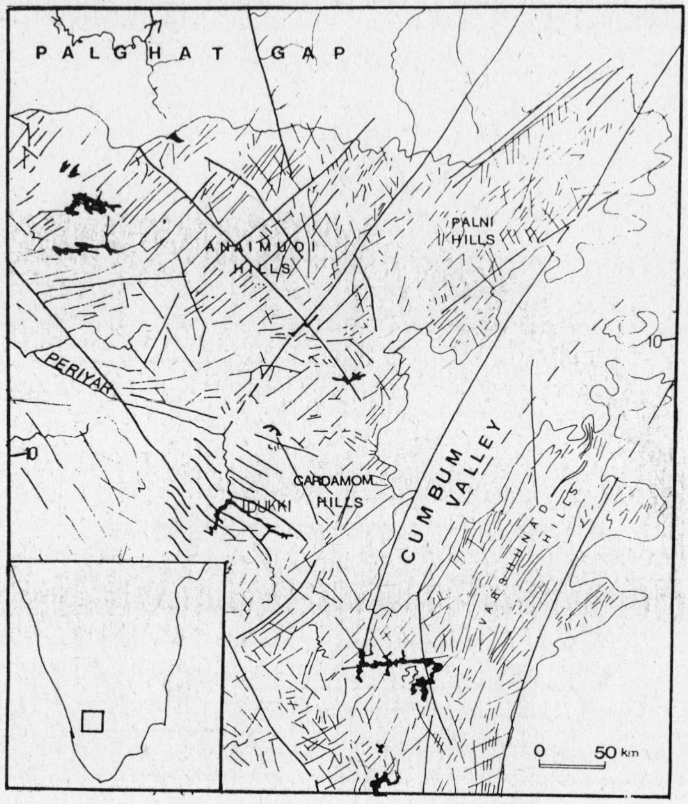

Figure 1—Map showing the fracture pattern of the study area.

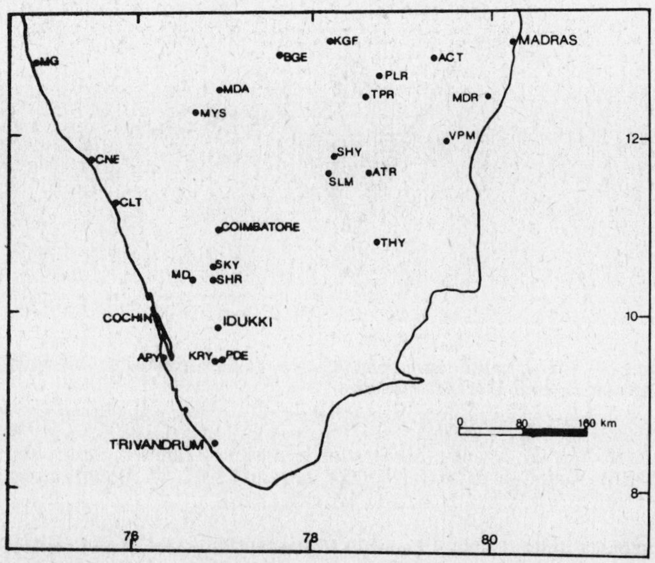

Figure 2—Map showing the epicentres of historic earthquakes in the South Indian High Grade Terrane.

SUBRAMANYAN, V., and N. M. GANTI, Indian Institute of Technology, Powai, Bombay, India

Spectral and Spatial Signatures of Selected Bauxite Deposits of India

The spectral and spatial signatures of selected bauxite deposits were studied on the Landsat image (fale color composites); adequate field work was carried out to collect ground truth. In the southern part of the Indian peninsula, lateritic bauxitic cappings are confined to high altitudes in the Shevaroy and Kollimalai hills. Here, the Eastern Ghats are represented by a group of hills, including the Shevaroys, separated by the east-west running Attur valley. Charnockite is exposed in the Shevaroy and the other hills. From an altitude of about 400 m at the foot of the Shevaroy hills, the slope rises steeply up to 1000 m and then becomes gentle towards the center of the hills where the topography is gently undulating above an altitude of 1550 m. There are six lateritic bauxite cappings between 1560 m and 1630 m with a northeast–southwest trend. An analysis of the Landsat paper print on the scale of 1:250,000 reveals a number of lineaments with a dominant north-northeast–south-southwest trend. The Shevaroy plateau shows up on the imagery as a dark grey patch, highly segmented by two sets of fractures trending north-northeast–south-southwest and north-south. The occurrence of the cappings over a particular altitude on flat-topped hills, their close proximity and the level of accordance of the cappings suggest that they are remnants of an extensive bauxite blanket.

The Kolli hills, lying 30 km to the south-southeast of the Shevaroy hills, form a distinct, oval-shaped feature, rising from the surrounding terrane. An 1200 m, a plateau with rounded hills and "U" shaped valleys and meandering streams, is well defined. Nineteen lateritic bauxite cappings of varied areal extents and altitudes ranging between 1270 m and 1344 m are confined to the southern part of this plateau. Thise cappings occur in the form of small mounds and domes as well as flat to gently sloping hills; these are much smaller than those of the Shevaroy hills. On the Landsat image, the Kolli hills appear as a linear north-south band showing a dark to grey tone along the western edge and medium to pale grey tone over the other places. The dominant lineaments trend north-northeast–south-southwest with one or two trending north-south. The frequency of the lineaments is less than that of the Shevaroys. The southern half of the plateau is almost bereft of the lineaments and all the bauxite cappings are reported only from the southern half. Thus, there appears to be an inverse relation between the frequency of the lineaments and the frequency of the bauxite cappings.

At Amarkamtak in central India is one of the largest, laterite-capped dissected plateaus forming the Maikala range of hills. Bauxite occurs as irregular lenses within the lateritic cappings over the Deccan basalts. An important feature of all these plateaus is the absence of a tree cover on their tops. A study of the false color composite on the scale of 1:250,000 indicates no major lineament but the regional lineament pattern over the entire area suggests north-south as the dominant trend. The whole area has an orange red tone indicating abundant vegetation. One interesting feature is the subtle change in the tone of the plateau tops in comparison to the side-slopes. The tops have a light red tone in contrast to the dark red tone over the slopes. The bauxite areas have a light red tone.

Bordering the States of Madhya Pradesh and Bihar in central and eastern India, tabular bauxite deposits forming the upper part of the lateritic blanket occur in the hill feature locally called the Phutka Pahar. The lineament map prepared from the false color composites on the scale of 1:250,000 brings out three dominant lineament trends: north-northeast–south-southwest, east-west and west-northwest–east-southeast. The northern part of the Phutka Pahar is less dissected and has retained most of the bauxite cappings; these show as grey patches in an area of red tone. This color contrast can be attributed to the absence of vegetation over the cappings and also the spectral signature of the laterites.

The bauxite deposits of Andhra Pradesh in south-central India fall broadly into three groups: the Gurtedu, the Chintapalle and the Anantagiri. In the first group, the bauxite plateaus occur at a higher elevation and are characterized by poly-directional slopes. They occur as a thick blanket over the parent rock with the intermediate zones of altered rock. The bauxite profile has developed from the parent rock (khondalite) that has a linear northeast–southwest trend and southeasterly dips. The deposit is characterized by a large linear escarpment for over 300 m on its northern face with an average height of 20 m. It is important to note that the bauxite deposit has grassy vegetation, slopes that are moderate to steep and a thick cover of trees. A study of the Landsat image indicates that the plateau tops have a very light grey tone compared to the side-slopes. The scarp sections have a very dark tone. The major lineament trend is north-northeast–south-southwest.

In the second group, the area between Sappala and Gudem forms a continuous chain of forested hills. The bauxite-bearing hills in general are of lower altitude than the non-lateritized hills. The deposits of Sappala are moderately dissected having meadows with temporary swamps. The Gudem deposit occurs on small irregular plateaus.

The area comprising the bauxitic duricrust is generally devoid of thick vegetation whereas the slopes support thick vegetation. The sharp contrast in vegetation is a rough index of the areal spread in any of the East Coast Bauxite cappings.

These bauxite deposits show well-developed scarps and slopes, often have irregular boundaries and their surface morphology shows considerable variations.

The Chintapalle area is rugged and well-dissected by streams. The regional lineament has a northeast–southwest trend. The bauxite cappings do not have a common trend but follow the slopes. Most of the cappings have a grey tone and this seems to be due to the presence of the grass species, Saccharum spontaneum occurring over a majority of the laterite caps.

SUENO, S., and I. NAKAI, University of Tsukuba, Tsukuba, Japan, and K. OHSUMI, Laboratory for High Energy Physics, Tsukuba, Japan, and H. MORIKAWA, Toyko Institute of Technology, Yokohama, Japan

EXAFSs Measurement Under High Pressure Using Diamond Anvil Cell

The EXAFS spectra measurement is a useful method to determine the local structure around specific elements in crystals regardless they are crystalline or amorphous. On the other hand, the diamond anvil high pressure technique is a good in situ observation method of materials under high pressure conditions using x-ray.

A combination of the EXAFS method, the diamond anvil technique and the synchrotron radiation beam is expected to provide useful informations concerning the behavior of element in geologically important minerals under high pressure conditions. However, there are two conceivable problems for adopting the diamond anvil to generate high pressure which are diffraction and absorption by the diamonds. On the former effect, the adjustment of the orientation of anvil by tilting the goniometer arc of the stage is good for eliminating the effect of disturbing peaks when the number of diffraction is small. The latter effect is systematically examined by measuring EXAFS spectra of several elements with atomic number from 24 (=Cr) to 32(=Ge). We have so far measured the EXAFS spectra for Cr, Mn, Fe, Co, Ni and Zn K-edges. The intensity of transmitted beam decreased in accordance with the lower atomic number.

We have designed a sample stage for high pressure EXAFS measurements using a miniature diamond anvil cell (27 mm Φ) mounted on a motor driven goniometer. The cell position can be adjusted by a remote controller so that a fine x-ray beam passes through a specimen hole (0.3 mm Φ) of a metal gasket of the diamond cell without touching the gasket. The EXAFS experiments were performed at beam line in the Photon Factory (PF), Tsukuba, Japan with the synchrotron radiation at an energy of 2.5 GeV.

Synchrotron radiation was monochromated by Si (311) channel-cut crystal to provide an x-ray beam with the wavelength around K-absorption edge of Zn, Ni and Co atoms and by Si (111) double crystals for Fe, Mn and Cr atoms. At wavelengths longer than the K-edge of Mn (lower energy absorption edge), the spectra were seriously affected by the absorption effect of a pair of normal 1/8 carats diamond anvils (3.4 mm in total thickness) and at wavelengths shorter than that of the Zn K-edge by Bragg reflections from diamonds disturbed the spectra. The spectra of Cr K-edges and Mn K-edge were also measured using diamond anvils with 1.8 mm thick but this trial was not so successful because of the lower attainable pressure. Cu K-edge spectrum was seriously disturbed by the absorption spikes. In such a case, we may use XANES instead of EXAFS because the energy region of XANES is so narrow (ca. 100ev) that the spikes can be easily avoided by tilting the goniometer arc.

We have carried out EXAFS analysis of Ni-dmg (bis (diimenthyl glyoximato) nickel (II)) under the high pressures of 8p to 5.6 GPa and cubanite ($CuFe_2S_3$ under 9.4 GPa.

Platy Ni-dmg molecule is connected to one dimensionally by the Ni-Ni bounding to form a crystal. The absorption spectra were measured at 1 atm., 3.4 GPa and 5.6 GPa. The peak at around 1.54 Å corresponds to intra-molecular $4 \times (Ni \times N)$ bonds which shorten slightly under the effect of high pressure. Although Ni-Ni inter-molecular distance was expected to be clearly shorter, no peak corresponding to the Ni-Ni interaction is discernible in the Fourier transform.

In orthorhombi cubanite, the structure changes to the NiAs type hexagonal form at pressure above 6 GPa. The absorption spectra about the K-edge of Fe atom were obtained at 1 atm. and 9.4 GPa. Their Fourier transforms showed a significant shortening of the Fe-S bonds. A detailed analysis of the data is currently proceeding.

The present study revealed that EXAFS measurements under high pressure using the diamond anvil cell can be performed on materials containing atoms with atomic numbers from 25 (=Mn) to 30 (=Zn). In future studies, the diamond with minimum allowable thickness combined with strong synchrotron radiation such as might be obtained from a wiggler line or accumulation ring of KEK, may make possible the measurement of the absorption spectra of good quality from geologically important minerals containing Fe, Mn, Cr, V and Ti atoms.

SUGUIO, K., Instituto de Geociências-USP, São Paulo, Brazil, and L. MARTIN, Observatório Nacional (ORSTOM/CNPq), Rio de Janeiro, Brazil

Looking for Indirect Geologic Evidence of Holocene Paleo "El Niño" Along Brazilian Coast

The normal conditions of atmospheric circulation on South America are characterized, during autumn and winter, by transit from the south to the north of air masses, through the middle and high troposphere, and near the earth surface, by corresponding frontal systems. The latter can reach to 10°S, being accompanied by winds and waves from the south, followed by an intense rainfall. During "El Niño" years, the polar advections are blocked by a strong and persistent subtropical jetty current, which crosses the South American continent in a diagonal position, from Southern Peru to Southern Brazil. As a consequence of this blockage, the frontal zones will rest for long periods upon southern and southeastern Brazil, with an occurrence of a concentrated rainfall in these areas and an accentuated dryness in northeastern Brazil (Kousky et al, 1984). Then, the wind regime and consequently the wave climate are modified. A strong "El Niño" phenomenon along the Peru and Equador coasts occurred in 1983, which submitted southern Brazil to a catastrophic inundation while the dryness in the northeastern area was still more severe (Martin et al, 1988). On the other hand, the waves from the southern sectors didn't reach the Salvador area (State of Bahia) during the austral winter, introducing an anomalous situation in longshore transportation of sediments.

Studying the geometry of the beach ridges in Rio Doce (State of Espírito Santo) mouth coastal plains, it was possible to reconstruct the direction of littoral transportation of sediments and then to determine the quadrant in which the dominant waves attained the coastline during the last 5000 years. So it was possible to emphasize some inversions in the directions of littoral transportation of sediments and then to estimate the orientation of the dominant waves (Martin et al, 1984). In normal situations, the transportation occurs from the south to the north, as a consequence of the dominant role played by the waves from the southern sector. On the contrary, during blockage of cold fronts, the waves from the southern sector are absent and the waves from the northeastern sector become active. Similarly, important changes of microclimate in the Cabo Frio region (State of Rio de Janeiro), as a consequence of variations in intensity of local upwelling, can be recorded within carbonate deposits of a small hypersaline lagoon in the area; hence we have seen that the upwelling intensity is modified by the regime of winds. Finally, research in progress will certainly allow us to emphasize changes in rainfall intensities in southern Brazil, mostly by study of the diatoms contained in small lakes of coastal plains.

This evidence suggests that the occurrence of "El Niño," if not monthly or annually, would have a duration of many tens or even hundreds of years.

REFERENCES

Kousky, V. E., M. T. Kagano, and I. F. A. Cavalcanti, 1984, A review of southern oscillation: oceanic-atmospheric circulation changes and related rainfall anomalies: Tellus, v. 36A, p. 490–504.

Martin, L., J. M. Flexor, V. E. Kousky, and I. F. A. Cavalcanti, 1984, Inversion du sens du transport littoral enregistrées dans les cordons littoraux de la plaine côtière du Rio Doce (Brésil): possible liaison avec des modifications de la circulation atmosphérique: C. R. Ac. Sc. Paris, v. 298, Série II, p. 25–27.

———, ———, and J. L. Vincent, 1988, Influence du phénomène océanique Pacifique "El Niño" sur l' "upwelling" et le climat de la région du Cabo Frio sur la côte brésilienne de l'Etat de Rio de Janeiro: C. R. Ac. Sc. Paris, v. 447, Série II.

SUKAMTO, R., Geological Research and Development Center, Bandung, Indonesia

Halmahera, Typical Cainozoic Island Arc in Eastern Indonesia

Halmahera and neighboring islands in eastern Indonesia represent a typical island arc configuration which has come into existence as a result of plate junction in the western Pacific region. Something characteristic is the presence of a double arc system, volcanic in the west and non-volcanic in the east. Its peculiar K-shape resembles very much that of the larger island of Sulawesi to the west. Active volcanism in its western arc is closely related to the active arc-arc collision which is presently developing in the Molucca Sea to west of Halmahera.

The history of Halmahera can be figured out by using as a background the magmatic activities and sedimentological processes inherent to this area. The stratigraphy of the western arc points to the existence of three cycles of magmatism: first, in the Late oligocene–Early Miocene, then in the Plio-Pleistocene, and later during the Holocene. Sedimentation in the eastern arc has commenced since Early Cretaceous.

SUMBATOV, R. A., B. N. KHAKHAEV, and V. N. GRAMOLIN, Objedinenije "Nedra," Moscow, USSR

Drilling Rigs and Technical Facilities for Ultradeep Borehole Drilling

The drilling experience of the Kola Peninsula and other superdeep wells has clarified requirements of the drilling rigs for superdeep drilling in crystalline and sedimentary deposits.

Soviet drilling rigs "4Z," "3D," "Uralmash 15000/400," "Uralmash 15000/500," and "Uralmash 6500," and Romanian rigs "F-400-DHU," and "F-500-DHU" are used for superdeep drilling in the USSR. This paper gives a description of the "Uralmash I5000/500" rig, meant for drilling 12–15 km deep wells. The same rig will be employed for drilling the Ural 15 km deep well, the Krivoy Rog 12 km deep well, and currently the Kola 15 km deep well and the Tyumen 8 km deep well.

The drilling rig consists of 6 units: substructure, derrick with drawworks, mud pumps, power supply unit, compressed air, and circulation systems. The drilling rig is distinguished by a fully mechanized round trip procedure, by the possibility of monitoring previously set bit loads, and by smooth gear change during round trips within 0.5–34.0 mps limits.

The drilling rig incorporates modern scientific and technological achievements in the field of world rig-building industry. There are given technical facilities and technological specifications of drilling, running down and cementing of casings, which have enabled us to reach a depth of more than 12 km in the Kola Peninsula well and which at present are providing the completion of planned wells down to a depth of 14–15 km.

The paper pays particular attention to the use of the string composed of light-alloy drill pipes, special turbodrills, and coring facilities while drilling superdeep wells. We describe an assembly technology for light-alloy pipes with steel tool joints and the functional specifications of the drill column during turbine drilling in crystalline rock.

Requirements are specified for drilling equipment (turbodrills, coring devices, bits) during failure control operations and while tackling some problems at sinking the wells deeper than 10 km in crystalline rock.

In superdeep wells, because of the great string length, it is difficult to imagine real downhole situations by data from measuring devices either run down on a cable or installed on the rotor. That is why one should have some facilities for monitoring the drilling process and transmitting borehole information. The main data sources on a turbodrill downhole operation are turbotachotransducers, which transmit this information through a hydraulic channel.

Specifications are also given for turbotachotransducers by "Obiedinenie Nedra," rotary table torquemeters, and an inclinometer. Permissible string loads are specified for round trips and torques while the string is rotating in a real elliptical borehole.

Difficulties of monitoring the borehole trajectory are the most serious problems in superdeep drilling, especially when the bit diameter is 215 mm or less.

It is essentially difficult to operate in a cavity-rich borehole when the effectiveness of reamers and centralizers is decreasing and there is no way to drill straight down. Approximate drill columns are drawn up to keep present angles of deflection within 5–10°.

Currently, a casing cannot be run down to the depth of 10–12 km and more as a single section. The paper draws up requirements for the tripping of such casings by separate parts and for their cementing.

The drilling experience at 12 km depth enables us to make recommendations on special devices for geophysical surveying of the borehole and for transmitting borehole information.

Well drilling to a depth of 15 km and more is a task of no less complexity than space exploration and requires technological breakthroughs in metallurgy, electronics, chemistry, and other industrial branches. However, studying the earth's interior processes by way of superdeep wells and geophysical research will enable us to draw up certain regularities on formation of mineral deposits and to find out the most effective means of their extraction. And it is the "Globus" program, worked out by Soviet scientists, that suits this purpose.

SUN Shu, LI Jiliang, and LIN Jinlu, Institute of Geology, Academia Sinica, Beijing, China

East Asia Archipelago in Paleotethys Domain

In the Chinese territory, the multitudinous orogenic belts make up a markable geological characteristic. In these belts, the orogenic belts formed by collision in Late Permian and Mid–Late Triassic are superior in numbers. The mountains formed by the orogenics include Wuliangshan, Ailaoshan, Karakoram, Qinling–Tongbo–Dabieshan, Helanshan, etc.

Wuliangshan trends on the east of the Lancang River. It is composed mainly of Triassic–Jurassic magmatic arc, so we called it the Lincang Arc Terrane. In this area, the Baoshan continental block occurred as A-type subduction eastwards under the Lincang Arc and Tongchang Street–Yakou ophiolitic melange zone upthrust onto the metamorphic flysch series on the foreland of Baoshan block. The granitic pluton of Lincang Arc upthrust westward went over the ophiolitic zone emplaced onto the foreland flysch. Meanwhile, the nappes of the Lincang volcanics upthrust onto the melange zone. Based on the fact that the marine Lower Triassic occurs as thrust sheets in foreland fold-thrust belt, it can be inferred that the collision took place at the end of Early Triassic in Middle Triassic.

In Ailaoshan belt, the collision between the Yangtze plate and the Lincang Arc led to complex deformation. There are a number of ductile faults between the Ailaoshan and Yuanjiang faults. In the ductile faults, mylonites with strongly plastic flow was developed. The stretch lineaments in the mylonites orient approximately north-south and horizontal, suggesting the high proportion of the horizontal component. In this belt, the occurrence of garnet-pyroxenite and peridotite blocks in melange indicates deep-seated metamorphism and the features of root zone. The Yangtze plate on the Ailaoshan ductile decollement upthrust westwards onto the Shuanggou ophiolitic melange and then both upthrust onto the foreland belt of the Lincang Arc. According to the geochronological data, the backarc basin opened in Permian and the evidence from the foreland belt suggests that the backarc collapse and collision occurred at the end of Triassic or the beginning of Jurassic.

According to Yin Jixiang (1988, work report), Karakoram is an orogenic belt formed by the collision between South Xinjiang Permian arc and North Tibet continental block in Triassic. The details of this belt will depend on the conclusions from the Karakoram Investigation Team.

Along Burhanbuda–Qinling–Tongbo–Dabie Mountain belt, there are two ophiolitic melange zones parallel to each other and an earliest Triassic granite belt in between. The granite belt trends from the south margin of Qaidam basin eastwards across Lixian and Huixian (Gansu), Fuping and Zhashui (Shaanxi), Xixia and Tongbo (Henan) to Qianshan (Anhui). The granite terrains are possibly autochthonic or parautochthonic. The ophiolitic melange on the north of granite belt extends from Dachaidan pass through the south margin of the Qinghai Lake, Wushan, and Tianshui of Gansu, Taibai and Shangnan of Shaanxi, Tongo of Henan to Fuziling and Mozitan of Anhui. Perhaps this opiolite belt represents a residual oceanic crust of a series of back-arc basins. The ophiolitic melange on the south of the granite belt extends from the south flank of Burhanbudashan, Jishishan and passes through Tanchang and Bikou of Gansu, Liuyang and Mianyang of Shaanxi, Wudangshan and Yingshan of Hubei to Yueyang of Anhui. This ophiolite belt represents possibly the residual oceanic crust of the Paleotethys ocean between the North China and Yangtze plates. The back-arc and fore-arc collisions occurred in Triassic (Wang et al, 1987).

From the Langshan Range of Innermongolia to Helanshan of Ningxia, there is a north–south-trending foreland fold-thrust belt composed mainly of Paleozoic strata, which separates the basement of the Qilian orogenic belt from the North China plate. Along the line from Langshan to Shizuishan, a few ophiolite nappes distribute intermittently. In the foreland fold-thrust belt, there are Permian and Triassic strata as well as metamorphosed Carboniferous marine sediments and volcanics. In foreland molasse basin, the Upper Triassic Yanchang Formation also underwent fold and thrust. The metamorphic basement of Qilian mountain upthrust toward the east-southeast. It can be inferred that the Paleotethys seaway in this area was also closed in Triassic.

Based on the analysis mentioned above, before the collision between the various continental blocks and arc terrains and closing of the Paleotethys seaways in the period from the end of Permian to Triassic, there were Baoshan block, North Tibet block, Qilian block, Lincang Arc, Yangtze block, Burhanbuda–Qinling–Dabie arc series, North China block, Huanan block, Yiyang oceanic arc, South Anhui arc, and a lot of oceanic islands and sea mountains in the East Asia Paleotethys oceanic basin, constituting an archipelago regime. Such a paleogeographic pattern is similar to the present geography of the Indonesia area.

The East Asia Archipelago occurred in a special geological historic period. Its closing and the collisions characterized by multiblock and multisuture at about the same time completed the combination of the main body of the Chinese continent. This type of orogeny provides an important example to kinetic and dynamic analysis in collision tectonics. Further detailed study of the subject has a undoubted significance in the development of tectonics.

SUN Tian-xi, Environmental Monitoring Center, Suzhou, China

On Exploiting Drinkable Groundwater Resource under Suzhou

In view of increasing grave pollution of surface water and splendid prospects for exploiting groundwater in Suzhou, the present paper puts forward a technical scheme for the purpose of exploiting drinkable groundwater resource under Suzhou. This is: drinking water for civil use should be drawn from quality groundwater, whereas the existing waterworks could be completely devoted to the uses of industry. As long as 80,000 T of groundwater a day will be exploited, the drinking needs of all residents of Suzhou City could be satisfied.

The bases of the plan are as follows:

1. Examinations confirmed that 179 trace organic compounds have been examined in Suzhou's running water. Of these, the carcinogenic substances (and their concentrations) are benzene (21.1 μg/L), which is 210 times that of USA; cyclohexane (35.80 μg/L);-heptane (5.90 μg/L); tri-toluene (2.89 μg/L); n-hexane (6.70 μg/L); and propyl cyclohexane (4.89 μg/L). Therefore, reinvestigating Suzhou's drinking policy is a task that brooks no delay.
2. Organic pollution in Suzhou's surface water is becoming more serious every day because of industrial development. The ranges of qualified sources of drinking water will be increasingly narrowed. People will have to consider groundwater.
3. According to the standard of drinking water, the quality of the bearing groundwater under Suzhou is fine and occupies fifth place in China (Table 1). In addition, it hasn't worsened through at least the past decade (Tables 2, 3).
4. The allowable exploiting reserves of Suzhou's bearing water, which will not cause continued earth subsidence, is about 30 million tons a year (about 80,000 T a day). It is enough to satisfy the drinking needs of the 700,000 residents of Suzhou City.
5. The yield of running water will be 540,000 tons a day, whereas the water for Suzhou's industry will be about 500,000 tons per day. It is possible that the existing waterworks will be completely turned to Suzhou's industry.

In brief, the main point of this paper is "quality groundwater for civil use and worse surface water to industry."

SUN Yi-Ying, Institute of Geology, Academia Sinica, Beijing, China, XU Dao-Yi, Institute of Geology, State Seismological Bureau, China, CHAI Zhi-Fang, Institute of High Energy Physics, Academia Sinica, China, and YAN Zheng, Institute of Geology, State Seismological Bureau, China

Latest Results of Studies on Boundary Event Markers in China

Iridium anomaly at the Permian–Triassic boundary in China: Since 1981 a number of Chinese and western scientists have made INAA and RNAA of Ir concentrations at the P-Tr boundary of several sections in China. The results obtained from Asaro (1982, 1985, 1986), Clark et al (1986), Zhou Lei (1986, 1987), and Mao Xue-Ying (visiting Los Alamos Lab invited by C. Orth, 1986) show low Ir detection while the results obtained by Sun et al (1984, Xu et al (1985), Chai et al (1986), and Zhou Shu-Qing (1987, 1988) show Ir anomalies from 0.6–2.0 ppb (see Table 1). In order to cross-check Chai's results (1986), Zhou Shu-Qing of Institute of Uranium Geology, Ministry of Energy Resources, has used RNAA of preradiation repeated concentration method. Her results show that Ir anomaly concentrates at the top of the boundary black mudstone layer of 8 cm in thickness (AG 92 layer in Sun et al, 1984) in Changxing section. Pong Ping of Institute of High Energy Physics, Academia Sinica, has studied the chemical speciation of anomalous Ir in Cretaceous–Tertiary boundary samples from Stevns Lint, Denmark, and Montana, USA, in her master degree thesis and found that most of the Ir are present in insoluble residue, but little in sulfide, carbonate, or complex phases. Her results disfavor the volcanic origin of Ir. Study of mineral speciation of Ir in the boundary clay is underway. Thus, the reason for the deviation of Ir results of P-Tr boundary from various labs is probably due to Ir inhomogeneity, sampling, and analytical procedure since Ir concentration of P-Tr boundary is less than that at the C-T boundary; so, a large amount of sample and repeated concentration of Ir before long radiation would be favorable. P-Tr boundary clay samples collected from Serlung, Tibet, China, Chhidru, and Nammal sections, Pakistan, and from Auronzo, San Antonio, Italy, are being studied by RNAA in Chinese labs among which Ir anomaly nearby the P-Tr boundary in Nammal section has been reported by Pakistani counterparts.

New data of carbon isotope excursions of Permian–Triassic boundary and Frasnian–Famennian boundary in China: ^{13}C excursion shifting from positive to negative are found at or nearby the Permian–Triassic boundary in sections of Changxing, Zhejiang (of a range of 7‰); Guangyuan, Sichuan (of a range of 5‰); Liangshan, Shaanxi (of a range of 5.4‰ from +1.9‰ at top Permian to −3.5‰ at Lower Triassic); Zunyi, Guizhou (of a range of 4‰ from +1‰ to −3‰); Chhidru, Salt Range, Pakistan (from −1.0‰ to +1.5‰), and also at Frasnian–Famennian section at Luoxiu, Guangxi (of a range of 7‰).

Table 1—Comparison of Ir abundance of Permian-Triassic boundary of Meishan and Shangsi sections by various authors.

Section	Sample	Ir(ppb)	Analysed by	Runs	Labs	Reference
Meishan	AG92	5	INAA	2	A	Sun Yi-Yin et al.,(1984)
China	AG92	0.60±0	RNAA	10	A	Chai Zhi-Fang et al.,(1986)
		0.046	INAA		C	Asaro et a.,(1986)
	B4	0.034	RNAA	1	D	Clark et al.,(1986)
	MBS-3	0.0414	RNAA	1	D	Mao Xue-Ying et al.,(1986)
		0.050	RNAA	1	E	Zhou Lei et al.,(1986)
	CG672f	2.1	RNAA	2	F	Zhou Shu-Qing et al.,(1987,1988)
Shangsi	AG253	2.0	INAA	2	A	Xu Dao-Yi et al.,(1985)
China	AG786	2.7	RNAA	2	B	Jin Li-Yun et al.,(1987)
	AG253	0.054	INAA	-	C	Asaro et al.,(1986)
	AG787	0.027	RNAA	1	D	Mao Xue-Ying et al.,(1986)
Auronzo						
Italy		0.035	RNAA	1	G	Nazarow et al.,(1986)
Nammal						
Pakistan	NB	0.7	RNAA	-	H	B. Haq et al.,(1986)

*** A, Institute of High Energy Physics, Academia Sinica; B, Institute of 401, Atomic Energy Academy of China; C, Lawrence Berkeley Lab, USA; D, Los Alamos Lab, USA; E, Institute of Geophysics and Geology, University of California; F. Institute of Uranium Geology, Ministry of Energy Resources; G, Vernadsky Institute, USSR; H, Nuclear Activation Service of Hamilton, Canada.**
Ir anomaly of 3-5 ppb is confirmed at the Precambrian-Cambrian boundary in Kunyang section, Yunnan, by Jin Li-Yun of Institute of 401, Ministry of Energy Resources in 1987.

SUN, S. S. Bureau of Mineral Resources, Canberra, Australia

Chemical and Isotopic Compositions of Archaean Komatiites and Siliceous High Magnesian Basalts: Constraints on Early History of Earth and Formation of Continental Crust

Study of Archaean komatiites (AK) and siliceous high magnesian basalts could provide constraints on models of earth forming processes, thermal condition and the postulated existence of magma ocean in the Archaean mantle, petrogenesis of the early Archaean crust and mineralization of platinum group elements (PGE) associated with Archaean mafic-ultramafic complexes.

AK are commonly depleted in light rare earth elements (LREE) and have positive initial ϵNd (+2 for 3.5 Ga and +4 for 2.7 Ga) suggesting derivation from longterm LREE depleted mantle sources (Jahn, et al., 1987). There is little evidence for the existence of extensive felsic crust before 3.8 Ga, requiring alternative mechanisms to crust formation for such depletion (Galer and Goldstein, 1988). Very low Ti, Zr, and Y abundances and Ti-Cu-Pd ratios similar to fertile spinal lherzolite xenoliths ($\sim$primitive mantle) suggest that AK formed by single-stage large degree of mantle melting (Sun, 1987). High temperature (≥ 2000 °C) sources for AK could originate from a thermal boundary layer at $\sim$670 km phase transformation (layered mantle convection, preferred) or from the core-mantle boundary. The latter model would imply whole mantle depletion and convection.

Two chemical types of AK have been established: (1) Al-normal type with Al-Ti close to chondritic ratio and flat heavy REE patterns; (2) Al-depleted type with Al-Ti about half of chondrites and relative depletion of heavy REE. Sample of Al-normal type have near chondritic rations (/pmn 10$?) for refractory lithophile elements, RLE, (Ti, Zr, Y, Sc, heavy REE, Hf) after minor corrections to compensate for the effect of previous melt extraction which causes light REE depletion in AK. A major exception is Al-Ca which is 20% lower that chondrites. The difference between the two AK types could result from different mantle sources formed during the early history of the earth ($\sim$4.5 Ga) through extensive crystal fractionation from magma ocean or magma generation processes which involve fractionation of garnet. The latter explanation is supported by very similar initial Nd and Hf data for the two AK types (Gruau, 1988).

Estimates of primitive mantle composition (PM) after core formation, derived from studies of AK, ultramafic mantle xenoliths and the composition of the continental crust, are consistent with a model of heterogeneous earth accretion from planetesimals, core formation during accretion and final addition to the mantle of about 1% chondritic material following core formation to account for abundances of PGE. Compared to C1 chondrites, earth forming materials had slightly higher abundances of RLE relative to Mg ($\sim$1.2X) but were poorer in volatile elements. Depletion by volatilization may also explain Si depletion in PM (Ringwood, 1979). AK have fairly constant Ti-Pd ($\sim 3 \times 10^5$), same as observed in fertile ultramafic mantle xenoliths suggesting that PGe addition after core formation has been evenly distributed by convective mixing.

Absolute and relative abundances of RLE and PGE in primitive mantle place strong constraints on layered mantle model, which involves an early magma ocean from which majorite and Mg-perovskite settle into the lower mantle (Ohtani, 1984) and/or olivine floats to the upper mantle (Walker and Agee, 1988). According to the element distribution data of Kato et. al., (1988a,b), 10% majorite or Mg-perovskite separation will significantly fractionate Al, Ca, Ti, Zr, Hf, REE, Y, and Sc but has only minor effect on Si/Mg. A primitive mantle composition with near-chondritic ratios among RLE and Al-Ca about 10% lower than C1 allows small amounts ($\leq 5\%$) of fractionation of majorite and Mg-perovskite from a magma ocean (if existed), but such a mechanism does not explain the difference in Si-Mg between PM and C1 (Kato, et. al., 1988a, b). On the other hand, olivine addition might increase Ir, and Os and significantly decrease Pd-Ir and Re-Os, but not necessarily Co-Ni in the upper mantle. To explain the difference of Si-Mg between PM and C1 by olivine addition, 25% olivine is required resulting in 35% depletion of RLE relative to Mg, in contradiction to the PM estimate.

Thermal consideration, radiometric ages of xenocryst zircons, geochemical and isotope modelling, indicate that LREE-enriched Archaean siliceous high magnesian basalts (SHMB) and related layered complexes were derived from komatiite by crustal contamination. SHMB from Kambalda (Yilgarn Block) and the Negri Volcanics (West Pilbara), in western Australia show a wide range of MgO but constant initial ϵNd within each suite (-2 and -3, respectively). These data support a model involving crustal assimilation at depth, followed by later crystal fractionation at shallow depths in layered flows/sills (Sun et. a., in press).

Chemical and isotopic data of SHMB combined with ages of xenocryst zircons can impose strong constraints on chemical characteristics, age of the felsic basement rocks and their modes of formation. Thus SHMB and related layered complexes could be useful as tracers of unexposed early Archaean sialic rocks. In the case of the eastern Goldfield of Yilgarn Block between Norseman and Kalgoorlie, the inferred early Archaean ($\sim$3.5 Ga) basement rocks may consist of granodiorites and/or tonalites enriched in light REE (La/Am $\sim$6.4 as compared to 1.6 for chondrites) with initial ϵNd about - 5 at 2.7 Ga.

SHMB and layered complexes derived from AK have Ti-Pd, Pd-S, Pd-Cu and Pd-Au similar to primitive mantle ($\sim 3 \times 10^5$, $\sim 2 \times 10^{-5}$, 1.5×10^{-4} and $\sim$4 respectively). Thus, PGE mineralization in Archaean layered complexes is not due to PGE enrichment in the mantle source, but arises by sulfur undersaturation of the parent magma followed by magma chamber mixing and crustal contamination, which result in sulfide saturation and precipitation of PGE.

REFERENCES

Galer, S. J. G. and S. L. Goldstein, 1988, Chemical Geology, v. 70, p. 143.
Gruau, G., 1988, Personal Communication.
Jahn, et. a., 1987, Precamb. Res. v. 34, p. 311-346.
Kato, et. al., 1988a, b, Earth and Planetary Science Letter 89, p. 123-145; v. 90, p. 65-68.
Ohtani, E., 1984, Earth and Planetary Science Letter 67, p. 261-272.
Ringwood, A. E., 1979, Origin of the earth and moon, Springer-Verlag.
Sun, S. S., 1987, Journal of Volcan. Geotherm. Res. 32, p. 67-82.
———, et. al., in press, *in* Boninite and Related Rocks, Allen and Unwin, eds.
Walker, D., and C. B. Agee, 1988, EOS 69, 486 p.

SUNOUCHI, H., T. YUKI, and Y. TSUJI, Japan National Oil Corporation, Chiba, Japan

Two-Dimensional Simulation Model of Carbonate Accumulation Process Controlled by Water Depth

A carbonate sedimentation model, in which sediment types and accumulation rates are a function of water depth, was developed and tested for the Pleistocene carbonate sediments in Irabu Island, southwest of Okinawa.

The model is basically one-dimensional and a result is represented as a color columnar section. But two- and three-dimensional results are possible by arranging the columns.

Assumptions:

1. Accumulation rate is expressed as an exponential function of water depth: accumulation rate $dz/dt = c1 * \exp(-c2 * s - z)$, where z is elevation of sea bottom, $c1$ and $c2$ are positive constants, and x is elevation of sea level. The term of $s - z$) in the equation is therefore the water depth. The accumulation rate decreased exponentially with increasing water depth. At near the sea level, an accumulation amount for a unit time is set to be equal to the water depth. The reason the exponential function was chosen is based on an assumption that carbonate accumulation rate is proportional

to light intensity in water, which intensity decreases exponentially with the water depth.

An analytical solution for the elevation of sea bottom is able to be obtained as a function of time by integrating the equation.

2. Distribution of sedimentary facies is dependent on water depth: For example, off Miyako Islands, southwestern Japan, reef facies is distributed in water depth of 0–60 m, rhodolith and large foraminifera facies is 60–200 m, and pelagic foraminifera facies is deeper than 200 m. These distributions are clarified by a geological survey off Miyako Islands, Okinawa (Tsuji et al, 1988). User of the simulation program can also arbitrarily define the distribution of facies for each depth range.

3. Submarine erosion and reworking do not occur: Accumulation rate is expressed as a resultant rate that includes submarine erosion and reworking.

4. Subsurface erosion is a function of exposure time: Amount of erosion is proportional to exposure time.

5. Amount of compaction is an exponential function of burial depth. Each sedimentary facies has a different response to overburden.

6. Histories of sea-level change and basement movement are known.

Algorithms: Water depth in each time step is calculated from thickness of deposited sediments and data sets of sea-level change and basement movement. Then, sedimentary facies corresponding to the water depth deposits in thickness that is calculated with the equation.

Experiments and results: A number of two-dimensional experiments were performed with conditions of various sea-level changing patterns and with no basement movement. Results were close to real geologic sequences in literatures (Figure 1).

With an experiment for the Pleistocene carbonate sediments in Irabu Islands, predictivity of the model to unknown area was tested: Assumptions were a constant uplifting of the basement from a certain depth. Looking at only one well in six wells in the Irabu Island, sea-level changes were created with trials and errors until a similar lithologic succession to the real succession of the well is obtained. Successions in another well's area were created by the simulation with the sea-level change previously determined. The results were close to the real successions expected from the wells. Prediction of sedimentary distribution in geologic history is thereby possible where hydrodynamic conditions are not essential such as in reef slope margins.

The similarities of the simulated results and the real world shows that the model is appropriate as the first approximation of carbonate accumulation process.

Limitations of the model: There are many limitations to apply the model to real world. Major limitations are that the model does not simulate a forming process of lagoonal sediment and processes of submarine erosion and reworking. These limitations are attributed to lack of the hydrodynamic conditions in the model.

REFERENCE

Tsuji, Y., et al, 1988, AAPG Bull., v. 72, p. 254–255.

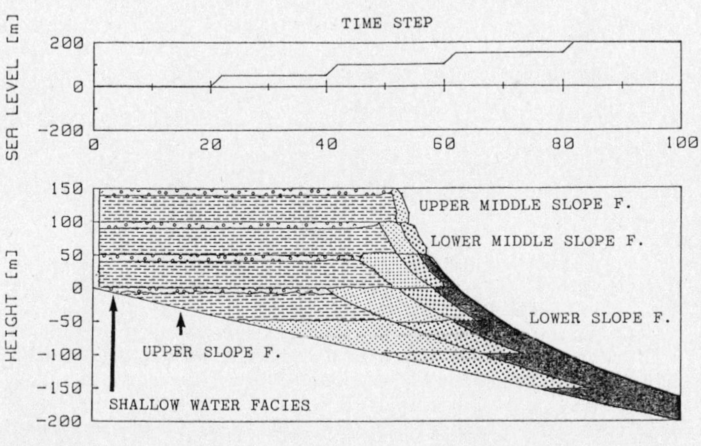

Figure 1—A result of the simulation. The upper diagram shows history of sea-level change in time steps of 100. The lower slope shows the resultant profile.

SUO Shutian, China University of Geosciences, Wuhan, China

Tectonic Boundary Between Proterozoic and Paleozoic Terrane in the Northern Part of the Xixia and Neixiang, Henan Province, China

Structural mapping and special structural analysis allow the recognition of a previously unidentified tectonic boundary between two distinct metamorphic terranes in the eastern Qinling mountains. The boundary crops out in the northern part of the Xixia and Neixiang, Henan Province, China, in a WNW-trending zone roughly 200 km long and 2-5 km wide that narrows to the northwest. The Proterozoic Qinling Group gneisses and marbles are separated from the Paleozoic low-metamorphic volcanic rocks and turbidite sequence by the boundary. The boundary which can be called the Junmahe-Matiwan Fault Belt is major ductile thrust zone. It is marked by mylonitic rocks and straight gneisses. Thick marble units in this zone invariably have the form of tectonic melange, full of rotated and flattened blocks or fragments of various types of granites, silicalites, dolomites, etc. in a mylonitic matrix. The mylonitic foliation within this zone dips gently to the south. It is folded and truncated by younger ductile or brittle shear zones. The kinematic evidence afforded by the major ductile shear zone leads to a model for development of the eastern Qinling mountains through a process of stacking of crustal segments, with translation northward along low angle ductile shear zones originating at considerable depth. The boundary is actually a basal ductile deformation zone, along which the reworked Proterozoic metamorphic belt has been thrusted to the north on the Paleozoic metamorphic belt before emplacement of the Mesozoic granitoids. Thus these new data are clearly favorable to the intracontinental shearing model.

SURDAM, R. C. and D. B. MAC GOWAN, University of Wyoming, Laramie, WY

Predictive Models for Sandstone Diagenesis during Progressive Burial

At temperatures less than 80 °C the diagenetic reactions producing carbonate cements in sandstones can be explained nicely by the model proposed by Curtis and Coleman (1986). Briefly, the distribution of early carbonate cements is controlled by dissolved sulphate concentration and is a function of the processes which affect sulphate concentration (i.e., depositional water composition, microbial sulphate reduction and water mixing). In order to use this model in a predictive sense a knowledge of the original depositional environment hydrology and hydrochemistry is necessary.

Predictive models for sandstone diagenesis in the 80–130 °C thermal interval can be developed based on carboxylic acid/Co_2 distribution and ratios (Surdam and Crossey, 1985). The model assumes that over this thermal interval the alkalinity in the reservoir facies is dominated by carboxylic acids, and a significant portion of Co_2 present is the product of decarboxylation of the acids (assuming there has been no significant mixing of water bodies). Furthermore, it is assumed that the stability of carbonates is a function of the carboxylic acid/CO_2 ratio, and the stability of framework grains is a function of the distribution and concentration of carboxylic acids.

In a specific burial history of a source/reservoir rock system, the essential data to produce this type of model are: (1) time-temperature profiles from stratigraphic intervals of interest, (2) kinetic parameters for the generation and decarboxylation of carboxylic acids (derived from hydrous pyrolysis experiments), (3) oil field water chemistry, and (4) a variety of petrographic and geologic data. The application of this type of predictive diagenetic model will be illustrated using Lower Frontier sandstones from the Bighorn Basin, Wyoming.

At temperatures greater than 130 °C diagenetic reactions controlling the distribution of cements, and the stability of framework grains in sandstones generally can be explained by thermolcatalytic sulphate reduction (Siebert, 1985; Machel, 1987). The determinative aspects of this process are the type of organics present in the system, the sulphase/organic ration, and the presence or absence of iron. In addition to this information if a time-temperature profile and kinetics for the redox reaction of interest are available, the process and resultant mineral reactions can be modeled. The application of this type of predictive diagenetic model will be illustrated using Norphlet sandstones from Alabama.

At this time, the most difficult diagenetic reactions to predict during progressive burial are those resulting in quartz cementation. By constraining the empirical model of Leder and Park (1986) with a knowledge of the distribution of geological factors, such as but not limited to chlorite rims, early migration of liquid hydrocarbons, and overpressuring, it is possible to estimate silica cementation events during progressive burial.

Using the techniques described above it is possible to model from the surface to depths exceeding 20,000 ft, the cementation patterns and framework grain stabilities in sandstones in sand/shale systems.

SURKOV, V. S., M. P. CRISHIN, A. E. KONTOROVICH, A. I. LARICHEV, V. I. LOTYSHEV, and N. V. MELNIKOV, NPO SIBGEO, Novosibirsk, USSR, A. A. TROFIMUK, Siberian Branch of the Academy of Sciences of the USSR, Novosibirsk, USSR, and A. N. ZOLOTOV, USSR Ministry of Geology, Moscow, USSR

Riphean Sedimentary Basins of East Siberian Province and Their Petroleum Potential

Hydrocarbon fields of high commercial importance have been discovered for the first time in world exploration practice in the Riphean of the East Siberian Province.

The Riphean deposits have been developed in a rather complicated geodynamic setting. Split of an extensive Eurasian protoplatform into major lithospheric plates in the earliest Riphean as well as initiation of the Ural-Mongolo-Okhotskii mobile belt structures along zones of spreading have resulted in unstable (transitional type) depositional conditions within the East Siberian craton. Differentiated movements of the earth's crust different-order block elements appeared to be a dominant factor in development of sedimentary cover structures. Herewith in cratonic marginal parts formation of the latter was associated with phasic development of adjacent geosynclinal systems. Peculiar tectonic development of the region is reflected in isolation of sedimentary basins according to formational composition of the Riphean deposits and character of their distribution.

From the location in regional pattern of the East Siberian Province, style, and evolution history, the Riphean sedimentary basins are divided into two groups: "marginal cratonic" and "intracratonic" (Figure 1). The first group includes basins of "pericratonic troughs," "fore-deeps," and "marginal transversal aulacogens." They are characterized by a definite structural stage division according to both formational characteristics and bedding of rocks with different lithology. Within sections, there is a complex combination of formations such as black-shale, flyschoid, volcanogene-terrigenous, carbonate and others.

"Ingressive" position in the platform regional pattern is typical of sedimentary basins in marginal transversal aulacogens. The section is characterized by a complex combination of volcanogene and volcanogene-terrigenous (mainly in the lower part), carbonate-terrigenous and carbonate rocks.

The second incorporates intracratonic sedimentary basins such as "aulacogenic" and cataplatform ones. Main features of aulacogen-type sedimentary basin structure are due to their rift origin and style of subsequent development. Early stages are marked by alternation of volcanogene-terrigenous (rudaceous) and volcaogene-carbonate rocks. There is rhythmic interbedding of terrigenous and terrigenous-carbonate rocks up-section. Several cycles of volcanic activization with prevailing alkaline-potassic, alkaline-basic, and ultrabasic orientation are fixed.

Depositional environments in sedimentary basins of cataplatform depressions have been controlled to a large extent by structural-dynamic setting of final stages in formation of adjacent pericratonic troughs and intracratonic aulacogens. The section is represented mostly by terrigenous-carbonate and carbonate-terrigenous units, i.e., products of rewashing of Early Riphean complexes and ancient basement.

Distribution of particulate organic matter and its catagenetic transformation indicate rather high petroleum-generating potential for the Riphean. Regional development of high-quality seals (Upper Vendian argillaceous evaporite-carbonate units, Lower Cambrian salts) favors high preservation of hydrocarbon accumulations.

Three groups of hydrocarbon (HC) reservoirs are differentiated based on structural-tectonic, lithologic-formational, and geochemical features.

1. Reservoirs of the Riphean deep horizons. Fractured and cavernous-fractured reservoirs are predicted here.

2. Reservoirs in rock disintegration zones of hypsometric uplifts of crystalline basement and in zones of karstification and fracturing in the upper Riphean. In such reservoirs, pools have been determined within the Kamovskii arch of the Baikit Anteclise (the Yurubcheno-Takhomskaya petroleum-bearing zone) and the Beryozovskaya depression of the Predpatomskii regional trough.

3. Reservoirs genetically associated with HC cross-flow due to vertical and lateral migration from the Riphean through Vendian and Cambrian.

There HC pools have been found within the Baikit, Nepa-Botuoba Anteclises, and Katanga Saddle.

All stated above suggests significant petroleum prospects for the Riphean sedimentary basins in the East Siberian Province. HC exploration successes allow more optimistic evaluation of hypothetical petroleum resources in the earth sedimentary cover as a whole.

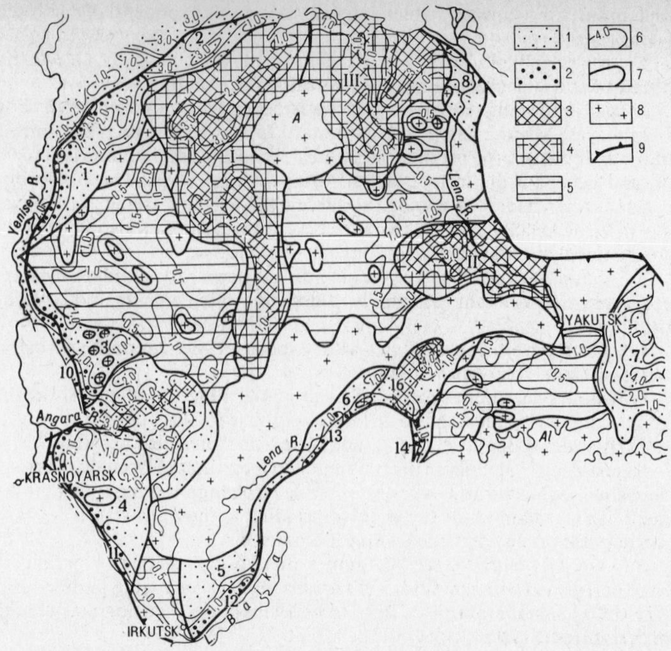

Figure 1—Riphean sedimentary basins of the Siberian Platform. Marginal cratonic sedimentary basins of (1–3): 1, pericratonic troughs (after E. V. Pavlovskii): Turukhan-Khantaiskii (1), Pre-Taimyr (2), Velminskii (3), Chuno-Biryusa (4), Pre-Baikal (5), Nyuisko-Dzherbinskii (6), Uchuro-Maiskii (7), Chekurovskii (8); 2, fore-deeps: Turukhan-Igarka (9), Pre-Yenisei (10), Pre-Sayany (11), Pre-Baikal (12), Ust-Vitim (13), Beryozovskii (14); 3, marginal transversal aulacogenes: Irkineevo-Vanavarskii (15), Urinskii (16). Intracratonic sedimentary basins (4–5): 4, aulacogenic (cross hatching shows the lower structural stage): Kotui-Vovukanskii (1), Vilyui (II), Khastakhsko-Udzhinskii (III); 5, of cataplatform lows and depressions; 6, the Riphean isopachs; 7, basin boundaries; 8, localities lacking Riphean complexes—A, the Anabar crystalline Massif; A1, the Aldan Shield; 9, platform marginal sutures.

SURO-PEREZ, V., HYLSA, S. A., Monterrey, Mexico

Principal Components and Indicator Kriging Approach

An alternative way to multiple indicator kriging is proposed to approximate the full coindicator kriging system applying principal component analysis to the original indicator variables. This transformation is studied in detail for the bigaussian model where the cross-correlations between the principal components are not significant in some cases and in other cases are identical to zero. This result permits inference of the conditional cdf by kriging the principal components and applying a back transformation. A comparison of its performance on an acutal data set (Berea Sandstone) with results derived from multiple indicator kriging is presented.

SUSO, J. M., and M. R. LLAMAS, Complutense University, Madrid, Spain

Impact of Groundwater Extraction on Wetlands: Doñana National Park (Spain) Case

This communication presents some problems caused by groundwater pumpage in the National Park of Doñana (NPD), one of the most important wetlands of Southern Europe.

Until fairly recently, almost all governments offered incentives to convert wetlands into arable land. In some regions, 80–90% of their wetlands have been "reclaimed" in less than one century. In recent decades, however, the particular ecological value of wetlands has been known and many governments have begun plans to protect wetlands. These plans usually show that there is rarely adequate knowledge of the hydrology of wetlands, mainly in relation to the groundwater role. In aria dn semiarid countries, wetlands are more scarce and play an even more important role than in temperate or cold zones. At the same time, these countries usually make special demands of groundwater for agricultural uses. A better knowledge of the hydrogeology of wetlands seem crucial.

The NPD (510 km^2) is situated on an extensive (3000 km^2) geomorphological unit formed by Plio-Quaternary estuarine deposits of the Guadalquivir river. It contains three main ecosystems: (1) the steady eolian sands; (2) the moving dunes; and (3) the marshes. The mobile and steady sands constitute a phreatic aquifer (1900 km^2) and are peripheral. The marshlands, in the center, consist mainly of clay sediments. Beneath these clays there is an extensive confined aquifer in contact with the phreatic aquifer on its periphery. The ecotone contact line between sand and marsh ecosystems constitutes a discharge area of the phreatic aquifer. The marshlands are under 0.3 m of water in winter and totally dry in summer. Average annual rainfall is 0.6 m; average annual temperature, 18 °C. The permanent moisture of the ecotone makes it the most productive and fertile zone in the PND. During the seventies, the Spanish government designed its most ambitious irrigation project (240 km^2) with groundwater (150 million m^3/y) in the proximity of what was subsequently to be the NPD. The warnings from scientists and environmental groups concerned over the ecological impact of this project soon reached the government. The project has currently been reduced to a maximum of 150 km^2. It is supposed to be completed in 1995. Currently, almost 100 km^2 are already irrigated with a pumpage of 60 million m^3/y. Since 1970, several digital flow models of the aquifer have been run. Nevertheless, only on Rodriguez and Llamas (1986) called attention to the possible ecological impact of the increasing decline of the water table at the ecotone. Llamas (1988) did a preliminary comparison between the future problems at the PND and those of the National Park of Las Tablas de Daimiel. This wetland, located in the Spanish plateau, has been destroyed because of the water table drawdown caused by irrigation pumpage. A detailed study of the more critical zone of the ecotone has been made by Suso (1988). Figure 1 shows the observed and computed situation of the watertable at a representative profile. Figure 2 shows the temporal variation of water level in two wells.

In conclusion, it seems that where the demand for fresh groundwater is important, the conservation of wetlands may be difficult, unless good hydrogeological studies are performed and its results are properly transferred to the decision-makers and to the general public.

References

Llamas, M. R., 1988, Conflicts between wetland conservation and ground water exploitation: two case histories in Spain; Environ. Geol. Water Sci., v. 11, n. 3, p. 241–251.

Rodriguez Arevalo, J., and Llamas, M. R., 1986, Ground water development and water table variation in Do–ana National Park (Spain): Memoirs Intern. Assoc. Hydrogeologists, v. 19, p. 203–210.

Suso, J. M., 1988, Estudio hidrogeológico de la zona de El Rocio (Huelva): Complutense University, Madrid (Spain), Master's thesis.

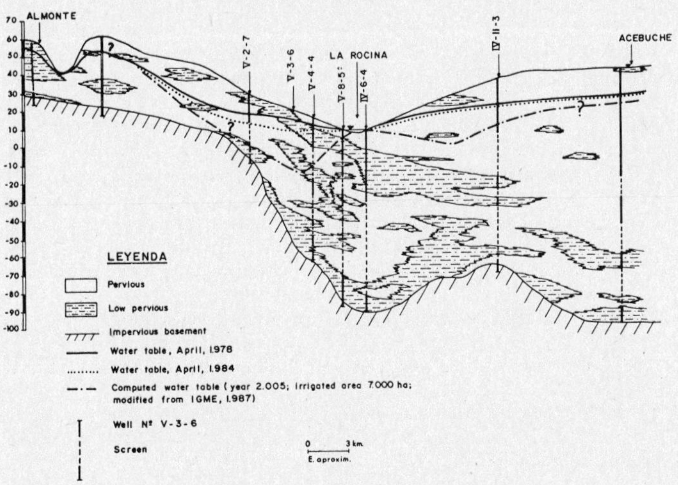

Figure 1—Hydrogeological profile (after Suso, 1988).

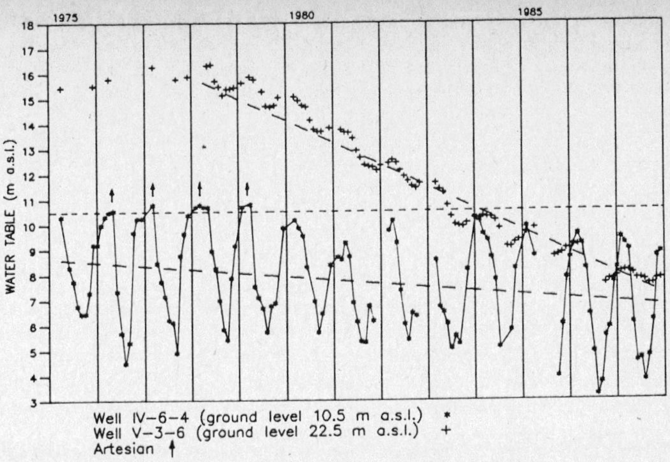

Figure 2—Water level variation in two wells of Figure 1.

SUSSE, P., University of Gottingen, Gottingen, Federal Republic of Germany

MINABS: Mineral Data and Reference File Based on Mineralogical Abstracts

MINABS is a mineral data and reference file based on the Mineralogical Abstracts of the last 40 years. It grew out of practical need during the last 20 years of research work on new minerals and mineral crystal structures. It was not designed to be a file with a full set of geometrical, chemical, and physical data, but rather to contain just enough information to identify a mineral and to give literature reference for further details.

A mineral entry consists generally of the following parameters: mineral name, chemism, crystal system, lattice constants, type of publication, and literature reference. The chemical information is stored qualitatively in form of anions and cations. The lattice constants are rounded off to a tenth of an Ångström and to full degrees. Cations are represented by their regular atomic symbols, anions and the crystal system by three-character code words. The literature is given as volume number and abstract number of the Mineralogical Abstracts plus a three-character code word indicating the type of publication referred to. Examples are given in Table 1.

A special search program is provided for the convenience of the user. A typical search procedure is described in the following example. The user is interested in crystal structural work on sulfates containing antimony. The input search parameters are therefore: SFA for sulfate, SB for antimony, and STR for crystal structure publications. The result, as it appears on the monitor, is given in Table 2. Each mineral entry is printed in one line, containing, from left to right, the mineral name, the main anion, up to four cations, a second anion if present, the six lattice constants, the crystal system, the type of publication, and the literature reference as volume number and abstract number of the Mineralogical Abstracts.

Besides being kept up to date, MINABS is being gradually extended backwards and revised. Currently, it contains about 10,000 mineral entries from the period of 1947–1988. The entire file is packed in the core storage of the computer in order to achieve high-speed performance. Thus, one search process takes only a fraction of a second. The core storage requirement for MINDAT is now 340 Kbyte. It works on all compatible computers with DOS 2.1 or later.

SUTER, J. R., and S. PENLAND, Louisiana Geological Survey, Baton Rouge, LA, and R. BOYD, Dalhousie University, Halifax, Canada

Application of Sequence Stratigraphic Concepts in Interpretation of Modern Deltaic Sedimentary Environments

The interpretation of ancient sediments has traditionally relied on comparison with modern depositional environments and the application of facies models and Walther's law (e.g., Walker, 1984). There have been few attempts to integrate or compare the results obtained from modern depositional environments with those from the seismic-sequence stratigraphic approach (Haq et al, 1987; Vail, 1987). This is largely due to a perceived difference in scale between the two approaches and the long time span apparently required for sequences to develop. In this paper, we compare the modern Mississippi River delta, a well-established and classic shallow-marine depositional model (Frazier, 1967; Penland et al, in press) with the depositional sequences predicted from the Exxon Production Research (EPR) approach (Posamentier and Vail, in press), concentrating on transgressive and highstand systems tracts. This comparison serves both to clarify the current Mississippi delta model and to test the EPR method in modern sedimentary environments. Since both the EPR seismic-sequence stratigraphy approach (e.g., Vail et al, 1977; Haq et al, 1987) and Mississippi delta sedimentation (e.g., Frazier, 1967; Penland et al, in press; Fisk and McFarlin, 1955) are extensively described elsewhere, only a brief summary of each will be presented here.

Common perceptions of the Mississippi River deltaic sequence emphasize the most recent deep water Modern delta (e.g., Coleman, 1980). Comparisons with EPR sequence stratigraphy models correctly identify the Modern delta as a parasequence at the culmination of the highstand systems tract. Sedimentation in the Holocene Mississippi delta began once canyon infilling diminished around 9 Ka and has continued throughout the subsequent transgressive and highstand systems tracts. The full Mississippi delta sequence includes the early Holocene deltas now on the continental shelf, the subaerial deltas of the middle to late Holocene at the end of the Mississippi alluvial valley, and the currently active deep-water Modern delta located at the shelf break.

Previous determination of sequences and systems tracts has been limited to outcrop (Haq et al, 1987) and subsurface studies (e.g., Vail et al, 1984). Testing the sequence stratigraphy approach in modern sedimentary environments enables comparison at much higher stratigraphic resolution and produces a higher degree of confidence in the result. The sequence-seismic stratigraphy methodology of EPR (Vail et al, 1977; Van Wagoner et al, 1987) functions well when applied to Quaternary Mississippi delta sediments. The Type 1 sequence being generated in the Mississippi delta is underlain by a Type 1 unconformity and contains stacked parasequence sets in lowstand, transgressive, and highstand systems tracts. Modern sedimentary facies "geometry" within the systems tracts also conforms well to general EPR model predictions (Posamentier and Vail, in press). The conformity of the Mississippi delta sequence to EPR models is not a trivial correlation. Although short-lived, the Mississippi delta sequence to EPR models is not a trivial correlation. Although short-lived, the Mississippi delta sequence contains a large volume of sediments and would be identifiable on exploration seismic data. Lowstand systems tract deposits in the youngest Mississippi Fan lobe are over 400 m thick (Bouma et al, 1985) and extend downslope for more than 500 km. Sediments deposited in the Mississippi Canyon during the latest Wisconsinan sea-level lowstand are more than 600 m thick (Coleman et al, 1983). Delta complexes in the transgressive and highstand systems tracts average 10–50 m thick (Frazier, 1974; Penland et al, in press) and extend 200 km downdip and more than 400 km along strike. Holocene sediment beneath the Modern delta is more than 175 m thick at the present shelf edge (Fisk and McFarlin, 1955). A large percentage of ancient clastic sequences can be expected to form in tectonic settings similar to the Quaternary in the northern Gulf of Mexico, where passive margin depocenters are infilled by major rivers (Burke, 1977).

Confirmation of EPR methodology for sequence analysis by comparison with modern sedimentary sequences raises serious problems for global cycle chart construction. If ancient cycles are formed by a glacial-eustatic mechanism, as proposed by EPR (Vail et al, 1977; Haq et al, 1987), then some cycles must resemble the Mississippi delta and other cycles of the Late Quaternary in the northern Gulf of Mexico (Beard et al, 1982; Berryhill, 1986). Since the recurrence period of Late Quaternary cycles is less than 100 Ka (Hays et al, 1976), biostratigraphic resolution is incapable of globally correlating such short-lived cycles. Under glacial conditions, thick sequences with high deposition rates could accumulate one order of magnitude faster than they could be resolved. If ancient cycles are not formed by a glacial-eustatic mechanism, then there is no other known process capable of forming rapid, globally synchronous changes in sea level. Without any process available for global cycle formation, it may prove more appropriate to interpret changes in relative sea level as resulting mainly from tectonic processes operating within individual or adjacent basins (Hubbard, 1988).

REFERENCES

Beard, J. H., J. B. Sangree, and L. A. Smith, 1982, Quaternary chronology, paleoclimate, depositional sequences and eustatic cycles: AAPG Bulletin, v. 66, p. 158–169.

Berryhill, H. L., Jr., 1986, Sea level lowstand features of the shelf margin off southwest Louisiana, *in* H. L. Berryhill, Jr., ed., Late Quaternary facies and structure, northern Gulf of Mexico: AAPG Studies in Geology 23, p. 225–240.

Bouma, A. H., J. M. Coleman, and A. H. Meyer, 1986, Introduction, objectives and principal results of Deep Sea Drilling Project Leg 96, *in* A. H. Bouma, J. M. Coleman, and A. H. Meyer, et al, Initial Reports DSDP 96: Washington, U. S. Government Printing Office, p. 533–540.

Burke, K., 1977, Aulacogens and continental breakup: Annual Reviews of Earth and Planetary Science, v. 5, p. 371–396.

Coleman, J. M., 1980, Deltaic sand bodies: AAPG Education Course Note Series 15, 171 p.

——— D. B. Prior, and J. F. Lindsay, 1983, Deltaic influences on shelf edge instability processes, *in* D. J. Stanley, and G. T. Moore, eds., The shelf break, critical interface on continental margins: SEPM Special Publication 33, p. 121–137.

Fisk, H. N., and E. McFarlan, Jr., 1955, Late Quaternary deltaic deposits of the Mississippi River—local sedimentation and basin tectonics: GSA Special Paper 62, p. 279–302.

Fraizer, D. E., 1967, Recent deltaic deposits of the Mississippi River—their development and chronology: Transactions of the Gulf Coast Association of Geologic Societies, v. 17, p. 287–315.

Haq, B. U., J. Hardenbol, and P. R. Vail, 1987, The chronology of fluctuating sea level since the Triassic: Science, v. 235, p. 1156–1167.

Hays, J. D., J. Imbrie, and N. J. Shackleton, 1976, Variations in the earth's orbit: pacemaker of the ice ages: Science, v. 194, p. 1121–1132.

Hubbard, R. J., 1988, Age and significance of sequence boundaries on Jurassic and early Cretaceous rifted continental margins: AAPG Bulletin, v. 72, p. 49–72.

Penland, S., R. Boyd, and J. R. Suter, in press, The transgressive depositional systems of the Mississippi delta plain: a model for barrier shoreline and shelf sand development: Journal of Sedimentary Petrology.

Posamentier, H. W., and P. R. Vail, in press, Eustatic controls on clastic deposition II—sequence and systems tract models, *in* C. K. Wilgus, et al, eds., SEPM Special Publication 42.

Vail, P. R., 1987, Seismic stratigraphic interpretation procedure, *in* A. W. Bally, ed., Studies in geology 27: AAPG, p. 1-10.

——— J. Hardenbol, and R. G. Todd, 1984, Jurassic unconformities, chronostratigraphy and sea level changes from seismic stratigraphy and biostratigraphy, *in* J. S. Schlee, ed., Interregional unconformities and hydrocarbon exploration: AAPG Memoir 36, p. 129–144.

——— R. M. Mitchum, Jr., R. G. Todd, J. M. Widmier, S. Thompson III, J. B. Sangree, J. N. Bubb, and W. G. Hatlelid, 1977, Seismic stratigraphy and global changes of sea level, *in* C. E. Payton, ed., Seismic stratigraphy—applications to hydrocarbon exploration: AAPG Memoir 26, p. 49–212.

Van Wagoner, J. C., R. M. Mitchum, Jr., H. W. Posamentier, and P. R. Vail, 1987, Key definitions of sequence stratigraphy, *in* A. W. Bally, ed., Studies in geology 27: AAPG, p. 11–14.

Walker, R. G., 1984, Shelf and shallow marine sands, *in* R. G. Walker, ed., Facies models (second edition): Geoscience Canada, Reprint Series, p. 141–170.

SUTER, M., and G. SUÁREZ, Universidad Nacional Autónoma de México, México

State of Stress in Mexico and Western Central America

We present new and compiled geologic and geophysical stress measurements that constrain the state of stress in the uppermost 30 km of the lithosphere in Mexico and western Central America.

Northwestern Mexico including the Gulf of California, the Baja California Peninsula, and the part of mainland Mexico that borders the Gulf of California is characterized by active normal and right-lateral strike-slip faulting. The displacement is concentrated at the Gulf of California plate boundary that is being deformed by right-lateral transform faults connecting short spreading centers. The slip vector azimuth from focal-plane solutions of strike-slip events (centroid depth < 20 km) in the Gulf of California is $131 \pm 8°$ (Goff et al, 1987) and is in agreement with geodetic measurements of the same parameter in the central part of the gulf (134°, Kasser et al, 1987) and the strike of the Gulf of California plate boundary ($130 \pm 5°$). Independent stress measurements include the inversion of fault striation data from the Baja California Peninsula (S(h) azimuth: $77 \pm 29°$, Angeleir et al, 1981) and new breakout data from 5 wells (depth < 4.2 km) in Baja California and the Gulf of California (S(h) azimuth: $73 \pm 24°$). These stress data suggest that the plate boundary is oriented at an angle of 55° to the least horizontal far-field stress.

In northeastern Mexico, available stress measurements are limited to focal mechanisms of microseismicity along the Pitaycachi fault in Sonora (Natali and Sbar, 1982) and borehole elongations (depth < 4.9 km) (Suter, 1987), which have in general a west-east to northwest-southeast trend. The neotectonic framework suggests that this low-seismicity intraplate area is under extension with $S(v) > S(N) > S(E)$ and forms the southeastern continuation into Mexico of the southern Basin-and-Range–Rio Grande Rift stress province.

For central Mexico, we define the Trans-Mexican Volcanic Belt stress province, which is characterized by seismically active west–east-trending normal faults. Focal-plane solutions (Suárez and Ponce, 1986), the alignment of Quaternary cinder cones, and striation analysis indicate a stress field with $S(v) > S(E) > S(N)$.

Our data for southern Mexico and western Central America allow us presently to delimit two stress provinces: (1) The S(h) azimuth from six near-surface measurements in the coastal plain of the Gulf of Mexico in southern Veracruz and western Tabasco States based on volcanic alignments and borehole elongations is $19 \pm 17°$ and subperpendicular to the continental margin. Active normal faulting along the continental slope suggest a stress field with $S(v) > S(NNE) > S(ESE)$. Thrust fault focal mechanisms from the same area at a depth of about 20 km, on the other hand, indicate a stress field with $S(NNE) > S(ESE) > S(v)$, which implies vertical polarity changes of the principal stresses. (2) Borehole elongations from 21 wells (depth < 6.4 km) in the seismically active Chiapas–Petén fold belt and in the adjacent foreland indicate a west-northwest–east-southeast to northwest–southeast-trending S(h) azimuth of 118 ± 15° and are parallel to the fold axes of the belt, which suggests a stress field with S(NE) > S(SE) > S(v). A subset of 11 wells closest to the left-lateral Motagua-Polochic transform system has an S(h) azimuth of 122 ± 10°, resulting in an angle of approximately 60° with this plate boundary.

The existence of these regions of relatively uniform S(h) directions suggests that the stress patterns have mainly a far-field origin and are caused by the plate interactions at the Pacific margin of Mexico. An exception seems to be indicated by the upper crust data-cluster from the Gulf coastal plain of southeastern Mexico, which is probably gravity-induced.

REFERENCES

Angelier, J., B. Colletta, L. Chorowicz, L. Ortlieb, and C. Rangin, 1981, Fault tectonics of the Baja California Peninsula and the opening of the Sea of Cortez, Mexico: Journal of Structural Geology, v. 3, p. 347–357.

Goff, J. A., E. A. Bergman, and S. C. Solomon, 1987, Earthquake source mechanisms and transform fault tectonics in the Gulf of California: Journal of Geophysical Research, v. 92, p. 10485–10510.

Kasser, M., J.-C Ruegg, P. Lesage, L. Ortlieb, J. Pagarete, N. Duch, J. Guerrero, and J. Roldán, 1987, Geodetic measurements of plate motions across the Central Gulf of California, 1982–1986: Geophysical Research Letters, v. 74, p. 5–8.

Natali, S. G., and M. L. Sbar, 1982, Seismicity in the epicentral region of the 1887 northeastern Sonoran earthquake, Mexico: Seismological Society of America Bulletin, v. 72, p. 181–196.

Suárez, G., and L. Ponce, 1986, Intraplate seismicity and crustal deformation in central Mexico: EOS, American Geophysical Union Transactions, v. 67, p. 1114 (abstract).

Suter, M., 1987, Orientational data on the state of stress in northeastern Mexico as inferred from stress-induced borehole elongations: Journal of Geophysical Research, v. 92, p. 2617–2626; correction in v. 93, p. 8085.

SUTTON, S. J., University of Texas at Austin, Austin, TX and S. D. RITGER and J. B. MAYNARD, University of Cincinnati, Cincinnati, OH

Metamorphism in Witwatersrand Quartzites

Quartzites of the Archean Witwatersrand Basin have been subjected to lower greenschist facies metamorphism, evidence of which is preserved in the matrix minerals associated with the predominantly quartz framework. Electron microprobe analyses of individual mineral phases and backscattered electron imaging (BEI) of a suite of core samples from the Klerksdorp area have been used to characterize the matrix mineralogy and textural relationships within the quartzites. Metamorphic assemblages and mineral compositions vary vertically within this thick sequence and appear to be dependent on the bulk composition of the premetamorphic sediments. Variations in bulk composition, and hence, mineralogy, are particularly apparent within stratigraphic units, especially in the Central Rand Group and upper part of the West Rand Group, where a pattern of gradual A1 enrichment upward is typical of most formations.

Individual stratigraphic units within the Central Rand Group and upper part of the West Rand Group are characterized by an upward increase in the abundance of Al-rich matrix minerals. The change in bulk A1 content appears to be gradational within stratigraphic units, but more abrupt at their boundaries. The Al-rich matrix assemblages typically comprise chloritoid, pyrophyllite, and white mica. Most also contain some chlorite, and some contain Al-phosphate. White mica and chlorite are common throughout most of the section, but are likely to have somewhat higher Al contents in the Al-rich assemblages. In addition, FE content in chlorite increases, whereas the Mg content decreases, upward within most formations. Feldspar-bearing assemblages occur towards the base of some formations, particularly in the West Rand Group. They typically comprise albite, chlorite, and white mica /pmn sphene and carbonate. Much of the feldspar appears detrital, and, while most is pure albite, some is as calcic as Ab_{85}.

BEI reveals the presence of two generations of white mica in nearly all samples examined. The early mica has been distinguished from the late mica on the basis of incomplete replacement textures and, in some cases, evidence that grains are detrital, e.g. rounding or deformation not reflected in surrounding grains. Rounded mica grains believed to be detrital are about the same size as the surrounding quartz grains. Late mica is typically fine-grained or occurs as an incomplete replacement of larger grains of early mica. Early mica, relative to late mica, is generally Al-poor (typically Fe) + MgO ≈ 8 compared to 1 wt%). In addition, early mica is much more variable in composition than late mica, further suggesting that early mica is detrital.

Late mica in feldspar-bearing samples, compared to late mica in feldspar-free samples, has higher average Na (≈1.5 times), Fe (≈2.5 times), and Mg (≈3 times) contents, and lower average Al content (35 compared to 37 wt% Al_2O_3). The association of Na-rich late micas with Na-rich detrital feldspar suggests that the composition of these micas depends on bulk composition, not Na-metasomatism. Early mica is somewhat more common in feldspar-bearing than in feldspar-free samples, but, in contrast to late mica, its composition appears unrelated to the presence of feldspar, and hence to bulk composition.

Chlorite in some samples also shows evidence of two generations, an early Mg-rich chlorite and a late Fe-rich chlorite. Unlike mica, however, the early chlorite is preserved only where it is rimmed, and presumably protected from equilibration, by late chlorite. Some early chlorites contain several weight percent K_2O, suggesting they may have been detrital phlogopite replaced by chlorite, possibly before deposition. Late chlorites typically show compositional trends within formations, with Fe decreasing and Mg increasing with depth. In chloritoid-bearing samples, late chlorite appears to be in textural and compositional equilibrium with chloritoid. Further, late chlorite in chloritoid-bearing samples, compared to late chlorite in shloritoid-free samples, is higher in average Fe content (33 compared to 27 wt% FeO) and lower in average Mg content) 8 compared to 12 wt# MgO), suggesting that the composition of late chlorite may be dependent on bulk composition. Fe-Mg partitioning coefficients have been calculated for chlorite-chloritoid pairs, and are generally consistent within a given sample, but show a degree of variability between samples that suggests a dependency on bulk composition, as well as on metamorphic conditions.

Taken as a whole, the typical changes in mineral assemblages within stratigraphic units and the compositional data on individual phases, as well as textural distinctions between generations of phyllosilicates, suggest three conclusions that can be drawn about the Witwatersrand quartzites: (1) Prior to metamorphism, a typical stratigraphic unit had well developed vertical trends in composition of the whole rock or the Fe-rich silicates, with Al and Fe increasing and Mg decreasing upward. (2) The entire sequence has been subjected to lower greenschist facies metamorphism, with the assemblages that developed reflecting the bulk composition. (3) Hydrothermal alteration has not made a major contribution to the development of the matrix mineral assemblages.

SUTTON, S. R., M. L. RIVERS, and J. V. SMITH, The University of Chicago, Chicago, IL, and Brookhaven National Laboratory, Upton, NY

Synchrotron X-Ray Fluorescence Micropobe Analysis: II. Applications to Meteorites and Lunar Rocks

Introduction: Detailed compositional data are crucial in understanding the origins and histories of extraterrestrial materials. The high spatial resolution and elemental sensitivity of the synchrotron X-ray fluorescence (SXRF) technique is ideal for such studies (Sutton, et al, 1987a). These methods have been applied to interplanetary dust particles, meteorites, and lunar samples in some exploratory experiments. The initial results from this research are summarized below.

Stratospheric interplanetary dust particles: Micrometeorites (≈ 10 μm) collected by high-flying aircraft have been extensively studied as important unaltered samples of interplanetary dust. The first-order conclusion is that interplanetary dust particles (IDPs) are similar chemically to CI or CM carbonaceous meteorites. However, SXRF analyses have shown general enrichments in volatile elements, particularly bromine (Sutton and Flynn, 1988). The volatile enrichments attest to the low-temperature histories of these particles and also suggest that meteor vaporization may have important atmospheric chemistry effects. Coupled photochemical reactions between halogenated molecules have been implicated in the scavenging of ozone, for example.

Iron meteorite elemental partitioning: Copper and Ni partitioning measurements on metal (Fe-Ni alloy) and troilite (FeS) from nine iron meteorites suggest that subsolidus reequilibration has been significant in these objects (Sutton ct al, 1987b). Further work is in progress to determine the importance of this effect in establishing observed trace element groupings. Subsolidus exchange between troilite and metal should result in strong composition gradients at metal/troilite interfaces analogous to those observed at taenite/kamacite interfaces. A one-dimensional scan across a metal/troilite interface in Four Corners (I) (measurements made with the APS undulator prototype at CHESS) shows ten fold enrichments in Ga and Ge at the grain boundary in a band comparable to the beam size of 10 μm.

Lunar rocks: SXRF analyses of plagioclase from three lunar rocks representing extremes of known lunar rock types, ferroan anorthosite, microgabbro and a glass-rich norite, show intriguing variations in Fe/Mn ratios (Delaney et al, 1988). Fe/Mn are near 30, 40, and 70 for the anorthosite, gabbro, and norite, respectively. Since the bulk Fe/Mn for all three rocks is about 70, these results indicate indicate that Fe/Mn partitioning varies radically. One interpretation is that pristine rocks have low Fe/Mn and shock melting results in the incorporation of a mafic component that increases Fe/Mn. Systematic measurements on a variety of carefully selected lunar samples are being pursued to clarify this partitioning behavior.

Other studies: Other research includes studies of chemical fractionation trends in deep-sea spheres (Sutton et al, 1988) and in situ trace element microdistribution measurements on type B calcium-aluminum inclusions (CAI) from the Allende carbonaceous chondrite (Woolum et al, 1988). Several upgrades currently in progress will greatly improve elemental sensitivity and spatial resolution—a focusing mirror on the incident beam to increase the flux at the sample, a monochromator to allow tunability and selective elemental excitation, a laser-drilled pinhole to produce smaller spots, and a crystal spectrometer for improved energy resolution.

Research supported by NASA Grant No. NAG 9-106, U. S. Department of Energy, Division of Chemical Sciences, Office of Basic Energy Sciences, Contract No. DE-ACO2-76CH00016; NSF Grant No. EAR-8618346.

REFERENCES

Delaney, J. S., S. R. Sutton, and J. V. Smith, 1988, Trace element content of feldspar from achondrites and lunar breccias (abstract): Lunar and Planetary Science XIX, p. 267–268.

Sutton, S. R., and G. J. Flynn, 1988, Stratospheric particles: synchrotron X-ray fluorescence determination of trace element contents: Proceedings of the 18th Lunar and Planetary Science Conference, p. 607–614.

——— M. L. Rivers, and J. V. Smith, 1987a, Applications of synchrotron X-ray fluorescence to extraterrestrial materials: Nucl. Instrum. Methods, v. B24/25, p. 405–409.

——— J. S. Delaney, J. V. Smith, and M. Prinz, 1987b, copper and nickel partitioning in iron meteorites: Geochimica et Cosmochimica Acta, v. 51, p. 2653–2662.

——— G. Herzog, and R. Hewins, 1988, Chemical fractionation trends in deep sea spheres (abstract): 51st Annual Meeting of the Meteoritical Society, Fayetteville, AR.

Woolum, D. S., M. L. Johnson, D. S. Burnett, and S. R. Sutton, 1988, Refractory lithophile partitioning in type B CAI materials (abstract): Lunar and Planetary Science XIX, p. 1294–1295.

SUYAMA, R., O. OHTAKA, T. YAMANAKA, S. KUME, Osaka University, Toyonaka, Japan, and E. ITO, University, Misasa, Japan

Structure of High-Pressure Polymorph of Baddeleyite (ZrO_2

Baddeleyite occurs in a syenite and in a carbonatite rock. Its chemical composition is ZrO_2, with monoclinic symmetry. This monoclinic phase converts to two types of orthorhombic, Ortho I and II, by compression. When the applied pressure is approximately 3 > P > 13 GPa, Ortho I phase appears, and Ortho II becomes stable when 13 GPA < P. Ortho II phase is quenchable and is found isostructural with cotunnite ($PbCl_2$).

In this study, fine powder of ZrO_2 was treated at 1000 °C and 20 GPa for 1 h by a 6–8 anvil type high-pressure device. After being quenched to room temperature in air, the sample was submitted to X-ray diffraction and the intensities were collected in the range 20° < 2θ < 130°.

Assuming the space group of Pnma and the lattice parameters a_O = 0.55807, b_O = 0.33312, c_O = 0.64744 nm, the Rietveld analysis was undertaken. The results are shown in Table 1.

As seen in the magnitudes of R factors, the analysis was considerably satisfactory and this phase was confirmed to be cotunnite type. The result was used to summarize all the data in the form of compressibilities, illustrated in Figure 1, by assuming linear changes of volumes with pressure. It is seen that the values decrease in order of monoclinic, Ortho I, and II, which is the order of stabilities under pressure. This is one example illustrating that denser phase is less compressive in polymorphs.

REFERENCES

McCullough, J. D., and K. N. Trueblood, 1959, Acta Crystallographica, v. 12, p. 507.

Liu, L., 1980, Journal of Physics and Chemistry of Solids, v. 41, p. 331.

Ming, L. C., and M. H. Manghnani, 1985, Solid state physics under pressure: Recent advance with anvil devices, Tokyo KTK Scientific Publishers, p. 135–140.

Kudoh, Y., H. Takeda, and H. Arashi, 1986, Physics and Chemistry of Minerals, v. 13, p. 233.

Block, S., J.A.H. DaJornada, and G. J. Piermarini, Journal of the American Ceramics Society, v. 68, p. 497.

Table 1. Rietveld Analysis of Ortho II (Cotunnite Type) Phase of ZrO_2.

Space gruop : Pnma

Atom	Position	x	y	z	B /nm^2
Zr	4c	0.2463(12)	1/4	0.8893(4)	0.0170(7)
O_1	4c	0.3684(38)	1/4	0.5753(30)	0.0225(51)
O_2	4c	0.4844(40)	1/4	0.1576(26)	0.0149(48)

R_{wp} = 13.79 R_p = 11.02 R_B = 5.35 %

a_o = 0.55807(1) b_o = 0.33312(1) c_o = 0.64744(2) nm

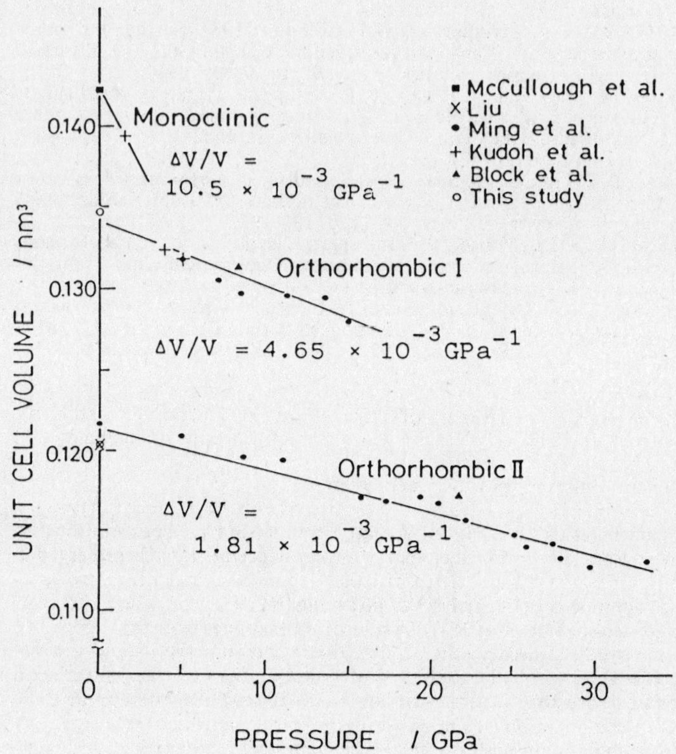

Figure 1—Compressibilities of high-pressure polymorphs of ZrO_2.

SUZUKI-KAMATA, K., U. S. Geological Survey, Vancouver, WA, and U. TADAHIDE, Kobe University, Kobe, Japan

Eruptive History of Ata Pyroclastic Flow, Southwestern Japan

The Ata pyroclastic flow deposit is one of the Quaternary large-scale pyroclastic flow deposits associated with a formation of funnel-shaped caldera. The eruption started with a small-scale plinian eruption followed by a moderate-scale plinian eruption and terminated with phreatomagmetic explosion. Shift of eruption vent at the beginning of climactic eruption was clearly shown by lithic component analysis. Component analysis of lithics is not always useful in locating the vent.

The eruption began with dry plinian eruption leaving alternation of ash and fallout pumice. Mode of eruption converted to phreatomagmatic explosion, which produced thick, fine, ash-fall deposits bearing pisolite. The distribution of this fine ash-fall deposit is limited to the southeastern proximal area. Small-scale and intermittent activity to generate small-scale pyroclastic flow followed. The ejecta of this period is developed only southeastward from the caldera, filling pre-flow valley. Thickness of each flow unit is from a few tens of centimeters to a few meters. Groove cast structure is observed at some of the flow unit boundaries. Surge deposits with a wavelength of several meters are locally observed. Final and climactic phase of the activity was an eruption of normal large-scale pyroclastic flow. The deposits are distributed all around the caldera. A pyroclastic flow plateau has formed at the proxymal to medial areas. The pyroclastic flow deposit of this phase contains a densely welded zone. The maximum thickness of entire pyroclastic flow deposit is about 100 m. Age of the eruption was dated as 80–85 k.y. B.P. The deposit extends at least 100 km away from the center of the caldera. The magma-equivalent volume of the ejecta totals more than 50 km^3.

Preferred orientation of elongated crystals and lithic fragments within matrix of welded deposit are obtained. Orientations radially away from the source were obtained only for samples collected from the surface of the pyroclastic flow plateau. Samples collected along the floor of narrow valleys have orientations parallel to the axis of the valley. Samples collected from valley walls tend to be parallel to the slope of the valley. These results suggest the flow direction was initially radial from the source and later followed local valley and ridge alignments.

Depositional ramp structures, which are asymmetrical distributions within valleys, were first recognized in the Ata pyroclastic flow deposit. The preferred orientation of the depositional ramps suggests that the pyroclastic flow selectively emplaces at the slope of a topographical barrier facing the source of the flow. This is because the velocity decreases when the flow climbs up the slope. The direction of the ramps shows that the flow of the pyroclastic material was radial away from the source.

Ground layer has developed at the western and northern proximal to medial area. Ground layer is absent at the eastern to southeastern proximal area. The predominant lithic component at the proximal area is andesite. Lithics of sedimentary formations are replacing those of andesite at the medial area. Sedimentary formations are predominant basement formation at this area. The ground layer is absent at the proximal area and appears at the medial area in the east of the caldera. Thus, it is clear that lithic components do not always give us information concerning the geology of source vent.

The source of the Ata pyroclastic flow has been controversial. A previously proposed caldera at the southernmost edge of Kyushu is topographically clear. But evidence obtained from flow lineation analysis, lithic component analysis, and depositional ramp structures do not fit this caldera. Another caldera structure has been found at the southern part of Kagoshima Bay by means of sparker subbottom profile analysis. This caldera fits with the results of flow lineation, lithic components, and depositional ramps. Gravity anomaly data also support this site as a funnel-shaped caldera.

Two models for the proximal phase of the eruption are suggested from the evidence of the ground layer. One model suggests that a bay or a lake existed east of the vent at the time of the proximal phase of the eruption, and most of the lithic components could not cross the water. Another model indicates that a new eruptive vent was formed at the west to northwest and that the major part of lithic components derived from the vent were selectively supplied toward the west to northwest.

Climactic welded facies are composed of a single cooling unit. Degree of welding is advanced. Round vesicles are observed within flattened juvenile fragments and also in the glassy matrix near the source. The vesicles within juvenile fragments are thought to be formed by latent degassing after the emplacement. Interstitial gas within the matrix was also trapped to form tiny vesicles. Squeeze-out structures of juvenile lens are found at the surface of the master joint. Secondary flowage structures such as elongated juvenile fragments are also found. These evidences suggest that welding compaction in the Ata pyroclastic flow deposit started soon after emplacement.

SWAMY, A. S. R., and P. SRINIVASA RAO, Andhra University, Visakhapatnam, India

Evolution of Nizampatnam Bay, East Coast of India

Nizampatnam Bay and its adjacent coastal plains, extending over an area of 3450 km^2, are covered by fluvial sediments of the Krishna River. The bay, occupying an area of 1825 km^2, is crecentric in shape and is enveloped between lat. 15°25′–15°55′N and long. 80°07′–80°50′E. Understanding the present and Holocene sedimentary processes in Krishna delta and the bathymetric analysis of the bay have helped to establish the progradation of the delta that resulted in barrier-building and lagoon-filling activity. The average rate of progradation of the Krishna delta is estimated to be 3.34 $m.y.^{-1}$.

Geomorphological and geological investigations in the area reveal the existence of five paleostrand lines at 14, 11, 6, and 3 km onshore and at 27 km in the offshore at a depth of 18 m. ^{14}C dating of the sediments containing the shell fragments, peat, and other paleontological evidence suggests the ages of the above ridges as 6500, 4700, 3650, 2450, and 8300 ± 120 y B.P. The occurrence of the major sea-level transgression at a height of 33 m during middle Holocene period (8500–7000 y B.P.) corresponds to the existence of the present offshore barrier (8300 ± 120 y B.P.) at a distance of 27 km. This in turn resulted in the formation of a shallow lagoon extending 40 km inland. The middle Holocene transgression resulted in the submersion of the entire lagoon. The geomorphic pattern of the paleo river channel distribution in the Krishna delta indicates phase by phase development of the fluvial system. This resulted in a rapid growth of a barrier system in the marine lagoon in northeast-southwest direction toward the bay. The ages and the location of the shore ridges further indicate that the regression might have occurred in four phases.

The borehole lithofacies have proved that the thick muddy sediments occurring at a depth of 5–6 m in the delta area were deposited under lagoonal or estuarine conditions. The absence and or sparse occurrence of microfossils and the presence of laminated (balatino) gypsum crystals in the lower part of the muddy sediments indicate a restricted lagoon environment. A rhythmic sedimentation of gypsum and carbonates (mainly dolomitic nodules) normally occur in a quiet lagoonal basin with a periodic inflow of freshwater.

The results of the flume studies made from a sequential Landsat imagery, textural analysis, major and minor constituent elements, and microfaunal investigations have been utilized in demarcating the depositional and nondepositional regions in the bay. The mean size progressively decreased from 1.76 to 3.21 ϕ. The sorting varied from 1.99 to 1.09 and the skewness ranges from 0.21 to 0.32. The bay sediments are leptokurtic (1.70–1.09) in nature. The major geological formations traversed by the Krishna river are Deccan traps and Dharwars. Montmorillonite is a predominant clay mineral derived mainly from the Deccan traps, the content of which ranges from 78 to 99% in the upper, middle, and lower parts of the Krishna River. The bay sediments also contain considerable quantities (70–90%) of montmorillonite.

Organic matter is higher in the bay sediments relative to the river sediments due to the increase of fine materials. Calcium carbonate is mainly associated with biogenic and relict sediments. Phosphorus is high in river sediments due to the formation of ferric and calcium phosphates. Calcium is found occurring mostly as calcium carbonate, calcium phosphate, and fixed in the structural lattice of smectite. Manganese content might have been formed due to authegenisis in the bay sediments. Potassium is carried away into the bay from weathered khondalitic rocks through the adjacent canals. Sodium is found fixed in the smectite lattice in saline water conditions. The trace element concentrations are low in bay sediments. The positive correlation between trace elements and smectite confirm their structural presence in smectite and as absorbed surfaces in clay minerals.

Ammonia beccarii, Asterorotalia trispinosa, Amphistegina lessonii, and *Quinqueiloculina* and *Sprioloculina* spp. are the common foraminiferal assemblages in the bay sediments. *Ammonia* and *Asterorotalia* species occupy the central part of the depositional regions of the bay. *Quinqueiloculina, Spiroloculina,* and *Amphistegina* spp. are observed as earthy colored, dull-lustered, highly abraded and large-sized tests. They are confined only to nondepositional sandy regions and are reworked in nature, supporting the lowered sea-level conditions.

The analysis of bathymetric charts (1891, 1976) with the present acquired data (1987) and the aerial distribution of sediment types in the bay reveal that the sediments are mainly accumulated in the central portion of the bay, which are redistributed from the Krishna River mouth by the nearshore currents. The nearshore currents (0.5 kmph) of the northeast direction indicated the debouching of Krishna River sediments into the bay year round. The diverging currents might transport terrestrial sediments advected into the sea during southwest monsoon toward the Nizampatnam Bay while the weak currents in the bay allow the suspended sediments to settle and shift the bathymetric contours toward the sea. Currently, the bay area is under rapid siltation forecasting its ultimate disappearance in the course of time.

DISTRIBUTION OF GEOCHEMICAL ELEMENTS IN KRISHNA RIVER AND NIZAMPATNAM BAY SEDIMENTS

Major elements (%):

	OM	$CaCo_3$	P	Fe	Mn	Ti	Ca	Mg	Na	K
Krishna river	0.92	8.10	0.36	7.40	0.80	0.48	1.23	1.58	1.63	0.77
Nizampatnam bay	1.75	8.50	0.70	5.32	0.62	0.62	2.11	2.33	2.44	1.79

Minor elements (ppm):

	Cu	Co	Ni	Pb	V	Cr
Krishna river	131	41	93	50	65	95
Nizampatnam bay	17	20	28	17	63	94

SWAMY, A. S. R., G. KRISHNA RAO, and M. PRAKASA RAO, Andhra University, Visakhapatnam, India

Sediment Characteristics of Continental Shelf, East Coast of India

he continental shelf along the east coast of India, spreading into an area of around 1×10^5 km^2, is characteristically dominated by the presence of terrigenous sediments transported by the rivers Ganges, Brahmaputra, Mahanadi, Godavari, Krishna, Penner, and Cauvery. The total catchment area of all these rivers is 2.1×10^6 km^2. The annual sediment discharge into the Bay of Bengal by these rivers is reckoned to be 2.6×10^{15} g, which is a significant fraction of the world's total sediment discharge into the seas, estimated at 18.3×10^5 g. The shelf width varies from 2.5 to 150 km south of Madras, while it reaches 210 km near the Gangetic delta. However, the average width and slope of the shelf are around 50 km and 21 ′, respectively. The river mouth valleys were formed due to lowered sea levels during Pleistocene that do not seem to extend beyond shelf and slope regions. The bathymetric data do not suggest any direct connection between these valleys and the main turbidity channels of the Bengal Fan.

The major sediment load of the peninsular rivers is produced during southwest monsoon and is carried in a northeasterly direction to swatch of no ground by the long-shore currents. Most of the sea surface sediments are muddy sands. However, fine-grained sediments occur near the river mouths and the seaward side of the shelf break. These sediments represent an admixture of several river phases, recent river run-off, reworked paleosediments, and biogenic material. The inner shelf (0–50 m), middle shelf (50–100 m), and outer shelf (100–200 m) are characterized by sandy mud, mud, and muddy sands, respectively. Modern sedimentation is concentrated near the river mouths extending to depths of 90–100 m with a general increase in fineness toward the deep sea. However, the occurrence of coarse sand (coarser than the present beach sands) at 18–24 m water depths in Nizampatnam Bay and the oolites in the outer shelf area are interpreted to be the relicts from the earlier lowered sea levels of Holocene and Pleistocene periods, respectively.

Heavy minerals off the east coast of India between Visakhapatnam and Madras include opaque (mostly magnetite) garnet, sillimanite, zircon, monazite, epidote, apatite, hornblende, hypersthene, rutile, tourmalene, and glauconite. The heavy mineral assemblage indicates that the sediments were mainly derived from khondalites (metamorphic rocks) and charnockitic and granitic gneisses (igneous rocks) occurring in the provenances. of the Godavari, Krishna, and Penner rivers. The variations in the heavy mineral assemblage appear to be related to variations in source areas, specific gravity of the minerals, underwater currents, and shelf characteristics. Smectite, illite, kaolinite, and chlorite are the clay mineral groups that normally occur in the east coast shelf sediments. Smectite (70%), which is derived from Deccan Trap weathered rocks, is the predominant mineral in the vicinity of river deltas and is observed to be widely spread over the drainage basins of Godavari and Krishna Rivers. The differential settling, flocculation, and source areas have resulted in the lateral variations of clay minerals along and across the shelf. Diagenesis does not seem to have affected these clay minerals.

The organic matter in the nearshore sediments is supposed to have been derived primarily from plankton of overlying waters and secondarily from the plant materials of swamps and interior drainage basins of Godavari, Krishna, and Penner rivers. The organic content (1.7%) is much lower than the world's average (2.5%) along the east coast of India. The distribution of major elements (Fe, Mn, Na, K, Ca, Mg, Ti, and P) in the shelf sediments are interpreted in terms of their source, the rate of deposition of continental material, physicochemical conditions of the depositional environment, the nature of clay minerals, and associated chemical constituents. Low concentrations of Fe, Mn, and Ti in the deeper parts of the shelf are attributed to the decreasing influence of the source and variations in physicochemical conditions. Calcium occurs mostly as calcium carbonate, calcium phosphate, and in a structural form of smectite. Magnesium is generally associated with clay minerals and more significantly in smectite and chlorite. Phosphorus is formed in the form of ferric phosphate and calcium phosphate. The covariance of Cu with Fe and Mn in the shelf indicates scavenging of Cu by hydrous manganese and ferric oxides. The decreasing contents of Fe and Mn and the low pH and Eh conditions cumulatively account for the dimunition of Cu content in the deeper parts of the shelf. The positive correlation of smectite with Cr, Ni, and Co indicates the structural bonding with smectite and as adsorbed surfaces on clay minerals. Correlation of Li and Pb with Fe and Mg suggests that smectite favors a better fixation of trace elements in the sediment.

Detailed sedimentological and geochemical studies revealed zonal sedimentary distribution pattern: occurrence of coarse sand in the outer shelf, presence of heavy mineral placer deposits in the inner shelf, and lime deposits in the shelf between off Godavari and Mahanadi rivers.

SWART, P. K., and R. N. GINSBURG, University of Miami, Miami Beach, FL, D. M. AISSAOUI, University of Miami, Miami Beach, FL and Université de Paris-sud, Orsay, France, B. H. PURSER, Université de Paris-sud, Orsay, France, V. C. VAHRENKAMP, University of Miami, Miami Beach, FL and KSEPL, Rijswijk, the Netherlands, and D. BUIGUES, Commissariat à l'Energie Atomique, Bruyères le Chatel, France

Common Origin for Atlantic and Pacific Tertiary Dolomites?

Shallow-subsurface dolomites replacing late Cenozoic marine carbonate sediments are a common diagenetic feature in the Atlantic and Pacific oceans. These occurrences of dolomites have many similar petrographic and geochemical properties, suggesting perhaps a common mode of occurrence. Here we compare certain geochemical data from two large dolomitized carbonate bodies, Mururoa Atoll in the Pacific and Little Bahama Bank (LBB) in the Atlantic. The distribution, petrography and geochemistry of dolomite in LBB [Williams (1985); Vahrenkamp (1988)] are based on five cores which penetrate a maximum of 100 m. The dolomite and limestone found in these cores are however, only the uppermost carbonates of a sequence which extends for over 5000 m. In contrast, at Mururoa Atoll there are between 180 and 450 m of dolomites and limestones overlying a volcanic conglomerate (Aissaoui et al., 1986). Despite these differences in proximity to basement, certain geochemical (mol% Mg CO_3 and $\delta^{18}O$) and petrographic parameters are remarkably similar between the two occurrences (Fig. 1). In both locations dolomitization is associated with evidence of dissolution indicating contact with waters undersaturated with respect to calcium carbonate. In addition, within comparable dolomite textures, the cores nearest to the bank or atoll margin, GBI (LBB) and Francoise (Mururoa) contain the most stoichiometric dolomites (44 to 49 mol% $MgCO_3$), whereas the interior cores are calcian rich (36 to 43 mol% Mg CO_3). Such variations in mol% $MgCO_3$ are correlated with changes in the $\delta^{18}O$, the interior dolomites being isotopically lighter than the margin ones. Two of us (PKS and VCV) have previously argued that in the case of LBB, such trends in $\delta^{18}O$ were an artefact linked to varying stoichiometry of the dolomite. By correcting all the data to a constant stoichiometric coefficient, it was suggested that the dolomites in LBB really contained very similar oxygen isotopic ratios. Using this approach the previous range of over 2 o/oo was reduced to 1 o/oo yielding a mean of +3.9 o/oo. The constancy of the $\delta^{18}O$ over such a large dolomite body argues for a single homogeneous source of dolomitizing fluids, a criteria unlikely to met by mechanisms such as a zone of mixing between isotopically light fresh ground water and heavier seawater. Differences in the mol% $MgCO_3$ of the dolomite relative to the bank/atoll margin may be a result of proximity to the source of dolomitizing fluid not yet depleted in Mg, or relate to difference in the original sediment type.

It was previously suggested that the dolomites at Mururoa were formed in the mixing-zone, while in LBB dolomites were produced from normal seawater driven through the platform by the demand of the mixing-zone. However, the similarity in association between the oxygen isotopic values and Mg composition of LBB and Mururoa, however, leads us to believe that in fact similar fluids and processes may be responsible for the dolomitization in both locations.

REFERENCES

Aissaoui, D. M., D. Buigues, and B. H. Purser, 1986, Model of reef diagenesis: Mururoa Atoll, French Polynesia, *in* Reef Diagenesis, J. H. Schroeder and B. H. Purser, eds., Springer-Verlag.

Williams, S. C., 1985, Stratigraphy, facies evolution and diagenesis of late Cenozoic limestones and dolomites, LBB, Bahamas University of Miami, 462 p.

V. C. Vahrenkamp, 1988, Constraints on the formation of platform dolomites: A geochemical study of Late Tertiary dolomite from Little Bahama Bank, Bahamas, University of Miami, 434 p.

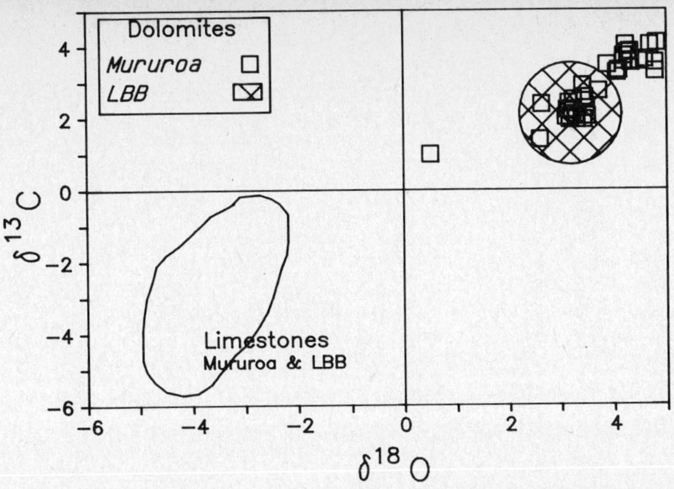

Figure 1—Th δ13C and δ18O of dolomites from Mururoa and LBB. Limestones overlying the dolomites fall in the same field and are plotted for comparison.

SWEENEY, J. F., Geological Survey of Canada, Vancouver, BC, Y. KRISTOFFERSEN, University of Bergen, Bergen, Norway, J. R. WEBER, and L. W. SOBCZAK, Geological Survey of Canada, Ontario, Canada, and A. GRANTZ, U. S. Geological Survey, Menlo Park, CA

Arctic Ocean Crustal Profile

The Arctic transect crosses the north polar seafloor between Norway and Alaska. It follows available seismic refraction profiles across the continental margins, deep basins, and submarine ridges of the Arctic Ocean. Constraints on crustal structure and age along the profile are provided by geophysical measurements, shallow seabed samples and, near shore, by extrapolation of mapped on-shore relations.

Seismic refraction measurements indicate substantial crustal thinning of 25 km or more across the continent-ocean transitions in both polar North America and Europe. The crust of the Eurasia Basin is considerably thinner, at 3 to 7 km, than that below either the Makarov Basin (9 to 10 km) or the Canada Basin (10 to 12 km). The Lomonosov Ridge has a 27 km deep crustal root and a crustal P-wave velocity structure similar to that of the outer Barents Shelf from which the ridge is thought to have been derived. The Alpha Ridge crust is up to 40 km thick and it has a P-wave velocity structure similar to that below Iceland.

Sediment cover is generally thinner than one km over the submarine ridges and generally thicker than two km in the deep basins. Up to 10 km of sediments lies below the continental rise of polar North America.

Seismicity along the Nansen-Gakkel Ridge and symmetric marine magnetic anomalies over the adjacent abyssal plains indicate that the Eurasia Basin has been expanding at slow rates since the early Paleogene.

Shallow seabed fossils, marine heat flow measurements and structural/stratigraphic evidence of continental breakup along the North American polar margin suggest that the Amerasia Basin and the Alpha Ridge formed during the Cretaceous Period between about 120 and 80 Ma.

Gravity measurements indicate that the aseismic submarine ridges are in isostatic balance and that some of the magnetic anomaly stripes in the Makarov Basin are associated with basement relief. High gravity and low magnetic anomalies over the polar margin of Canada reflect the presence of very thick sediments below the outer shelf and slope. Intraplate seismicity along this margin and off northeast Alaska indicate that this part of the Arctic is presently under compression.

SWEET, G., University of Winnepeg, Winnepeg, Canada

Subcutaneous Drainage and Cave Development in Northern Interlake, Manitoba, Canada

The interlake area of Manitoba is made up of several thousand square kilometers of carbonate bedrock. In the south this is overlain by varying thicknesses of glacial till, in the north the glaciers scraped the surface clean and the bedrock has been exposed to aerial weathering for about 8000 y. Dolomites and limestones of the Silurian and Devonian form the surface, with carbonates from the Ordovician underlying the whole. These beds form the eastern lip of the Williston Basin, a large synclinal system extending into the northwestern states.

Recently deglaciated (9000 BP), these carbonates represent a tremendous potential for Karst development. There are thousands of hectares of pavement and sinkholes poljes and caves litter the area. Over forty caves have been discovered in the last three years. They are small, ranging from 10 to 160 m in length and up to 10 m high (Fig. 1). They are found within 25 m of the surface, with vertical entrances and are generally independent of each other, even when they are close together.

There is some controversy with respect to the processes of formation and the age of these cavities. It has been suggested that they are the remains of a much more extensive system, the upper sections of which have been erased, primarily by glaciation. A second theory is that the caves are of relatively recent origin and were formed interglacially, during a warmer wetter period, than the present, or perhaps immediately post glacially, when there was a great deal of very aggressive water available.

Presently the mean annual precipitation of the region is less than 600 mm, much of which accumulates as snow between November and April. Spring melt is a very important source of water for solution and cavities that are totally dry for most of the year fill with water for a period of three weeks to two months at that time. During heavy summer storms water streams down the walls of several caves, although they are dry at other times. These two sources of water may play an important part in the modern development of the caves. Until recently the caves were thought to be relic features with little or no modern solution. This view has been changed on the basis of the finding in the 1988 field season. Secondary precipitation in several of the caves adds a farther dimension tȯ the complex development which appears to be continuing, despite low annual rainfall and wide temperature fluctuations.

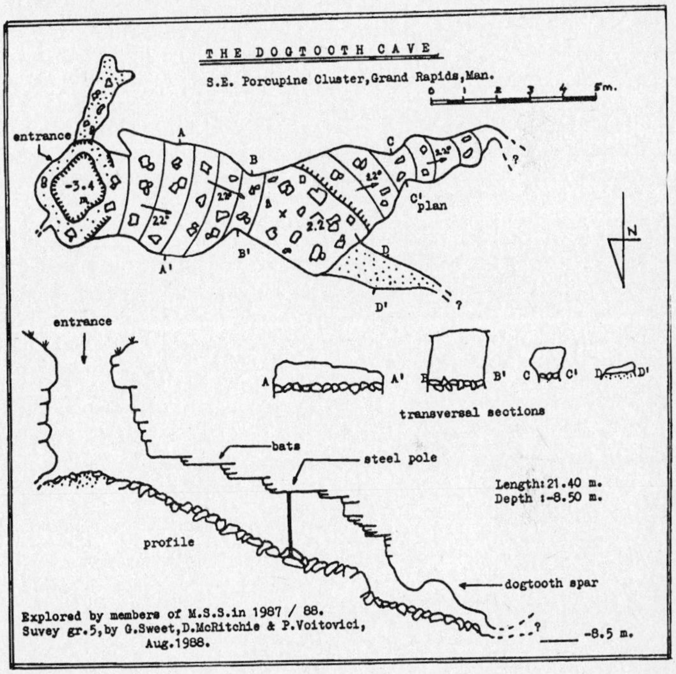

Figure 1

SWEET, M. L., and G. KOCUREK, University of Texas at Austin, Austin, TX

Empirical Model of Eolian Dune Lee-Face Flow Dynamics

Most eolian cross-strata record depositional processes that occurred on the lee of dunes. Refined interpretations of the paleoflow conditions that produced these cross-strata require an understanding of the aerodynamic interactions between the dune and primary winds. Unfortunately, these interactions are not well understood.

An empirical model is presented, based on four components: (1) a quantitative record of near-surface wind speed and direction at several locations on and leeward of the dune; (2) a record of wind speed and direction from the surface up to 1 dune height; (3) the above data gathered from a range of dune sizes and shapes in winds of varying speed and direction; and (4) a consideration of the effects of atmospheric thermal stability.

Atmospheric thermal conditions can be classified as neutral (momentus is transferred vertically by turbulent mixing), unstable (momentus is transferred vertically by convection), and stable (atmosphere is thermally stratified and flow is essentially laminar). Stable conditions were not considered because they require wind speeds below the threshold velocity for sand.

Experimental data were gathered on a 60 m tall, complex-crescentic dune (i.e., a large crescentic dune with smaller superimposed crescentic and star dunes) in the Algodones dunes of California; 10–13 m tall simple crescentic dunes in the White Sands dune field, New Mexico; 1–3 m tall reversing-crescentic and bachanoid dunes; and 0.1–1.0 m tall bachanoid protodunes on Padre Island, Texas. In addition, a 0.5 m tall "model" dune was constructed.

In the Algodones dunes, an array of recording anemometers and vanes (instrument height 1.5 m) were used to generate a record of wind speed and direction at various locations across the lee face and the adjacent interdune flat. A reference station was maintained near the crest of the dune. Measurements taken leeward of the dune were compared to data gathered simultaneously near the crest to determine the occurrence and nature of dune-modified flow. In addition to data from near-surface sensors, data from an instrument-carrying balloon were used to determine vertical profiles of wind speed, direction, and temperature from 10–80 m above the adjacent interdune flat. Meteorologic data were gathered several times over the course of a year to determine the effects of changes in atmospheric thermal stability, wind speed, and direction.

On the dunes of White Sands and Padre Island, and on the model dune, small anemometers (instrument height 0.1–0.6 m) and an array of flags for flow visualization were used.

Flow leeward of the dune can be characterized as attached (deflected or not deflected) or separated (Figure 1). Dune shape, as measured by aspect ratio (dune height/lee half length), and the incident angle between the dune brinkline and the primary wind direction are the principal factors controlling lee-face airflow. Atmospheric thermal stability exerts a secondary control on lee-face airflow. Transverse flow (incidence angle 80–90°) separates leeward of dunes with aspect ratios from 0.2–0.7 in unstable and neutral conditions. Leeward of dunes with aspect ratios from 0.1–0.2, transverse flow is attached and deflected in neutral conditions and separated in unstable conditions. Flow leeward of dunes with aspect ratios less than 0.1 is attached and not deflected in all stability conditions. When flow is oblique to the dune (incidence angle 10–80°) lee-face flow is attached and deflected for dunes with aspect ratios from 0.7–0.1 and all stability conditions, while on dunes with aspect ratios less than 0.1 flow is attached and not deflected. For flow longitudinal (incidence angle 0–10°) to the dune, lee-face flow is attached and not deflected for all aspect ratios and stability conditions.

Where flow on the lee face is attached and deflected, it is directed obliquely across the lee face regardless of the primary flow direction. This zone of attached deflected flow extends, at most, 2 dune heights into the adjacent interdune area. The zone of deflected flow is restricted to a layer near the dune surface.

As first proposed by Tsoar (1983), flow speed on the lee face is a cosine function of the incidence angle. Maximum flow speed (80% of crestal speed) occurs with a 0° incidence angle while minimum flow speed (30% of crestal speed) occurs with a 90° incidence angle. Vertically, the velocity profile is divided into two distinct components: one below dune height and the other above dune height. This modification of the velocity profile occurred across the entire width of the interdune area.

Flow separation is characterized by a narrow zone of weak (<10% of crest speed) return flow. This zone of return flow extends at most 1–2 dune heights into the adjacent interdune area.

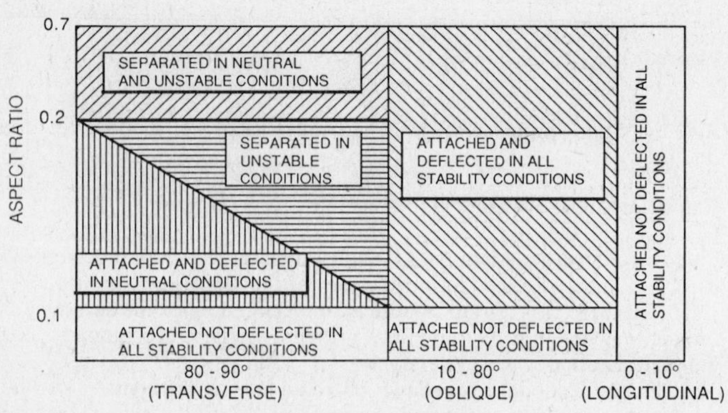

Figure 1—The effects of incidence angle, aspect ratio, and atmospheric stability on lee-face airflow patterns.

SWINT-IKI, T. R., and D. Z. PIPER, U.S. Geological Survey, Menlo Park, CA

Seafloor Coverage by Manganese Nodules in Pacific Basin

Ths distribution of deep-ocean manganese nodules within much of the Pacific Ocean was ascertained from bottom photographs and sediment cores. In the northeast quadrant of the Pacific Ocean, three areas have nodule coverage that commonly exceeds 50% of the seafloor. One area lies between the Clarion and Clipperton Fracture Zones, the second is in the Central Pacific Basin, and the third is in the region of the Musicians Seamounts. These three areas were characterized by noncalcareous surface sediments and low sedimentation rates during the Neogene.

Four generally nonoverlapping groups of these nodules can be distinguished by the following chemical criteria: (1) greater than 1.8% Ni plus Cu, (2) 1–1.8% Ni plus Cu, (3) greater than 1% Co, and (4) greater than 35% Mn. No simple relation is apparent between the distribution of the four groups and the amount of coverage, although nodules from the area between the Clarion and Clipperton Fracture Zones consistently contain greater than 1.8% Ni plus Cu.

Nodule distributions and compositions in the other three quadrants of the Pacific are not as well documented. However, nodule coverage at the central part of the South Pacific and the area around Antarctica at approximately 50°S is high, possibly as extensive as the areas in the northeast Pacific. The nodules in both areas apparently have fairly low Ni and Cu concentrations.

SWISHER III, C. C., G. H. CURTIS, and R. E. DRAKE, Berkeley Geochronology Center, Berkeley, CA

Single Crystal $^{40}Ar/^{39}Ar$ Dating and Its Application to Calibration of North American Land Mammal "Ages"

The correlation of North American Land Mammal "Ages" (NALMAs) to standard European stage/ages depends upon a detailed integration of mammalian biostratigraphies, magnetostratigraphies, and accurate isotopic ages. The biochronologic basis for the North American Land Mammal "Ages" has recently been reviewed and magnetostratigraphies of important representative fossil-bearing sections covering much of NALMA time are currently available (Woodburne, 1988). Unfortunately, many of the isotopic dates used to calibrate these fossil-bearing sections and their accompanying magnetostratigraphies still stem from the pioneering work of Evernden et al (1964). Most of these published ages are based on single argon analyses without error estimates (±). Furthermore, the accuracy of these ages has not been reviewed despite the technological advances in geochronology.

Recent advances in mass spectroscopy now allow the measurement of smaller amounts of argon gas than was possible in 1964, such that we are now able to measure the argon and obtain a date from a single volcanic mineral grain. Single crystal dating at the Berkeley Geochronology Center has been shown to yield high precision results. Standard errors of the mean on replicate analyses are typically <0.2% for minerals as young as Pliocene, as low as 0.04% at 20 Ma, and as low as 0.1% for sample at the Cretaceous–Tertiary boundary. This high precision is obtained because of the high purity of selected crystals, the low total system blank (M/e 36 < 1 × 10^{-18} moles, M/e 40 < 1 × 10^{-16} moles), the small volume of evolved gases resulting in rapid purification and low air Ar correction, and computer-timed, reproducible analytical conditions. Geologic accuracy is constrained by errors inherited from current international potassium-argon standards but will improve with the development of new homogeneous single-crystal $^{40}Ar/^{39}Ar$ standards.

Using this new technology, we have recently initiated a program to refine the calibration of North American Land Mammal "Ages," the associated magnetostratigraphy, and as a result of collaborative work, direct age calibrations with European standard stage/ages. In North America, we are currently focusing on the late Cretaceous–early Paleocene (Puercan NALMA), the Eocene–Oligocene boundary (Chadronian NALMA), and middle–late Miocene (Hemingfordian through Hemphillian NALMA).

SYKES, M. L., University of Texas at Arlington, Arlington, TX

Ascent of Granitic Magmas: Constraints from Thermodynamic and Phase Equilibrium Model

Model calculations for the ascent of granitic magma through the crust permit evaluation of common assumptions made in physical models of magma transport and constraint of conditions and mechanisms of magma transport. Combination of thermodynamic properties and phase relationships of the system albite-H_2O provides an analog system for the fluid-absent, closed system ascent of "granitic" (qz. or by-normative calc-alkaline) melt plus crystals through continental crust. Pressure-temperature trajectories are based on the dependence of albite liquidus temperature on pressure and H_2O activity for a constant, pressure dependent, crystallization or melting rate (CR or MR), and a given H_2O content. Based on the observed transition from liquid- to solid-like behavior at about 60% crystals, the maximum amount of crystals is assumed to be that value. Trajectory slopes dP/dT (Figure 1) range from $-15°$ to 35°/Kbar (-40 to 9.5°/km) for 1 to 8 wt% H_2O, but are generally positive. For a given CR or MR, trajectory slopes are steeper and start at lower temperature for highest volatile contents. Physical properties for the model magma system calculated along each trajectory using the P, T, and component data allow calculation of physical magma properties (volume, viscosity, enthalpy, etc.) and application to fluid dynamic and heat transfer models of ascent.

The magnitude of change in magma viscosity during ascent depends both on the proportion of crystals to melt *and* on the H_2O content of the melt. Because a closed system has been assumed, crystallization of anhydrous melt component results in an increase in the proportion of H_2O in the melt, tending to reduce magma viscosity. Thus, magma viscosity for intermediate (4 wt%) to high (8 wt%) water contents remains nearly constant during ascent even for high CRs (6–10%/Kbar) despite temperature drops of more than 200°. At low H-2O contents (<1%), magma viscosity increases up to four orders of magnitude over 10 Kbars of ascent with high CR. Similarly, magma density in such closed systems remains approximately constant.

In most physical models of magma transport, volume of the magma body is assumed to be constant. However, due to the increase in the H_2O content of the melt resulting from closed system crystallization of magma, magma volumes are constant only at high crystallization rates (>6%/Kbar). For model magma systems with a constant proportion of crystals to melt (CR = 0%/Kbar), magma volume increases by a total of 4–6% during ascent through the crust, with the larger value for higher H_2O content and low initial crystal (restite) content. Total volume increases of 9–14% are attained by systems undergoing fusion, at MR = 6%/Kbar. These volume changes can be expressed as strains, assuming a spherical magma body increasing equidimensionally in radius; total strain expressed as t/t_o varies from about 0.015 to 0.02 for CR = 0%/Kbar and from 0.03 to 0.04 for MR = 6%/Kbar, for ascent through the crust. At characteristic diapiric ascent velocities (10^{-5} to 10^{-4} cm/s), such strains would result in fracture of the surrounding wall rock.

If magma volume is assumed to be constant in the model system, the amount of overpressuring that would ensue during ascent is calculated by determining the pressure necessary to maintain zero volume change. For CR <2%/Kbar, overpressures exceed 200 bars (approximate tensile strength of wall rocks) in the first kilometer or so of ascent. For crystallization at 5–6%/Kbar, this limit is reached at midcrustal levels. For greater amounts of crystallization with zero volume change, no overpressuring occurs. Deformation or fracture of the wall rocks around the magma body must ensue when the tensile strength is exceeded.

These volume data suggest that closed system ascent may not always be a viable means of magma transport; volume-reducing processes involving loss of crystals and/or melt must operate. Possibilities include loss of crystallized material attached to the side walls during ascent, or loss of an H_2O-rich boundary layer through the top and sides of the magma body during ascent. The latter process could facilitate the formation of microfractures in roof rocks and enhance ascent. Alternatively, the failure of wall rock around a buoyantly ascending and expanding magma body could initiate transport by fracture propagation through the remainder of the crust.

Energetics of magma transport can also be evaluated using model ascent trajectory data. Adiabatic ascent trajectories have steep but positive slopes dP/dT (Figure 1) and require fusion of preexisting crystals (restite) when present. Application of energetics data to previous heat transfer models of magma ascent (Figure 1) suggest that failure to consider the effects of H_2O in closed system evolution of magma bodies can yield results that are unrealistic in terms of phase equilibria.

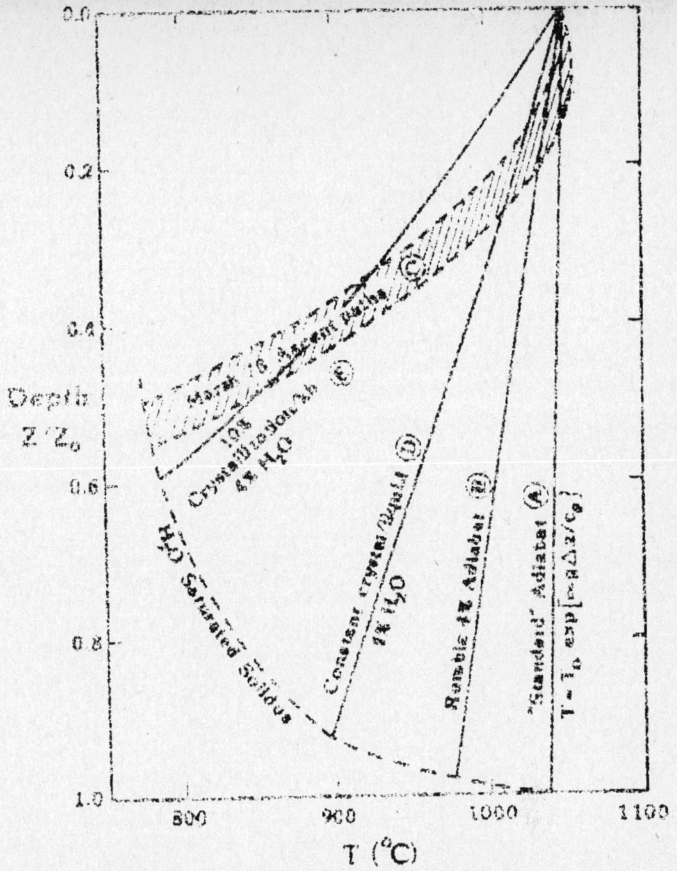

Figure 1—Comparison of ascent trajectories calculated using the present model and other models for magma pressure-temperature paths. Dashed line is the H_2O saturated solidus. A, standard one-phase adiabat; B, multiphase adiabat from Rumble (1976) for 4 wt% H_2O in the system; C, general shape of trajectories calculated after Marsh (1978) for diapiric ascent, assuming CR = 0%/Kbar, diapir radius = 104 to 106 cm, velocity = 10^{-4} to 10^{-8} cm/s; D, ascent trajectory, this model, with constant crystal/melt (CR = 0%/Kbar); E, ascent trajectory, this model, CR = 10%/Kbar.

SYLVESTER, A. G., University of California, Santa Barbara, CA

Triggered Vertical Strain across Active Strike–Slip Faults in Southern California

Minor vertical displacements and tilts have been observed geodetically in the days to weeks following recent minor to moderate size earthquakes along various fault strands of the San Andreas fault system in southern California. Some of these triggered displacements were observed on adjacent faults up to 45 km aways from the associated earthquake epicenter. To date, no anomalous pre-seismic tilt or strain changes have been observed; however, the geodetic data do not represent continuously monitoring. The co-seismic and post-seismic fault movements are determined from repeated first order, second class surveys of precise leveling arrays established at 50 sites across various faults in southern California. Between 1978 and 1988 some of the sites have been resurveyed as many as 30 times. The leveling arrays vary in length from 300 to 7000 m and contain as many as 70 permanent bench marks. Resurveys are done at least annually to determine time, place, and magnitude of long-term, nearfield vertical strain, as well as after major regional earthquakes to determine if the strain field around the faults has been perturbed. Leveling precision of each survey is typically a few parts in ten million.

Three northeast-trending line arrays transect the southern San Andreas fault near the Salton Sea on segments of that fault where triggered horizontal slip was observed after the 1968 Borrego Mountain (M 6.5) and 1979 Imperial Valley (M 6.4) earthquakes. No surficial slip, horizontal or vertical, was visually observed in these arrays within 30 hours after the 24 November 1987 Superstition Hills earthquake (M6.6) located at the southern end of the adjacent San Jacinto fault zone, although a small amount of triggered dextral creep (2 mm or less) was recorded by a creepmeter on the southern San Andreas fault within three minutes of the earthquake epicenter (McGill, et. al., 1988). Releveling of the 2200 m long line array (65 bench marks) over Durmid Hill (45 km northeast of the epicenter), which straddles the San Andreas fault near the creepmeter that exhibited the triggered slip, showed that the hill rose, or that the Imperial Valley subsided, about 1 mm between 3 September and 10 December 1987. Between 10 December 1987 and 22 March 1988 an additional millimeter of vertical displacement was measured in the northeastern half of the array. Even though the observed displacements were very small, this pattern of horizontal and vertical strain is consistent with the long-term pattern of strain accumulation which is expressed in the youthful physiography of Durmid Hill itself, as well as with that expected from the complex interaction of northwest-striking right-slip faults (the San Andreas fault) and a northeast-striking left-slip fault (the Elmore Ranch fault). Fifteen kilometers farther northwest along the San Andreas fault, an 870 m long line array (23 bench marks) across the fault at North Shore (65 km north–northwest of the epicenter) tilted southwestward 2 microradians two weeks after the Superstition Hills earthquake, then recovered 1 microradian northeastward some time in the following 16 weeks. Eight of the 45 bench marks in the 1500 m long line array across the San Andreas fault in the Mecca Hills (90 km north–northwest of the epicenter) rose about 1.5 mm between September and December 1987, but returned to their September 1987 level by March 1988. Resurveys in July 1987 and July 1988 show no changes among the 23 bench marks in the 720 m long line array across the San Andreas fault at Miracle Hill, 115 km northwest of the epicenter.

In 1982, no height changes occurred along a 375 m long, W-shaped leveling array across the right-slip San Jacinto fault near Anza either before, during, or after a M 4.8 earthquake on 15 June. The earthquake was located about 3 km south of the array, beneath the community of Anza itself, at a focal depth of 12 km. The array, consisting of 19 permanent bench marks, was leveled seven times between January 1981 and May 1982. The earthquake was sensibly felt by the surveyors during the eighth survey when they were standing on the San Jacinto fault itself. The ninth releveling was done three days after the earthquake. The relative heights of bench marks remain unchanged among all nine surveys, showing that no measurable surficial movement had occurred as a result of the nearby earthquake, nor has any been detected in the results of 21 more resurveys to August 1988.

Two north–south-trending line arrays, 790 m and 1900 m long, cross the left-oblique, reverse-slip Sierra Madre fault zone at the south edge of the central Transverse Ranges and were in the northern half of the south edge of the central Transverse Ranges and were in the northern half of the meisoseismal area of the 1 October 1987 Whittier Narrows earthquake (M 5.9). Distances from the earthquake epicenter for each leveling array were 13 km north–northeast and 16 km north–northwest, respectively. Both arrays were releveled several times between 1984 and 1987, and then again nine months after the earthquake. Neither surficial slip nor geodetic offset or north-south tilt were observed at either location.

Similarly, no height changes were detected in a 375 m long line array across the Furnace Creek fault in Fish Lake Valley, eastern California, eight days after a magnitude 5.3 earthquake that occurred in the valley on 24 September 1984, 50 km north of the array, or 45 days after the 21 July 1986 Chalfant Valley earthquake (M 6.2), 50 km west–northwest of the array.

The 10 June 1988 earthquake (M 5.3) near the north Garlock fault caused no perceptible height changes in an 800 m long, north–south line array across the Pleito thrust, 13 km west–southwest of the epicenter, or in a 450 m long, S-shaped array across the San Andreas fault, 20 km southwest of the epicenter. However, six days after a M 4.8 earthquake on 4 November 1981, 10 km northwest of the Pleito thrust leveling array, all bench marks on the south, or hanging wall of the fault rose an average of two millimeters relative to those on the footwall. Six months laters, the array returned to its pre-earthquake datum.

These leveling data thus record both the absence of any significant triggered slip effects as well as what appears to be dynamically-triggered strain transients following a nearby earthquake on an adjacent fault. These strain transient represent a permanent deformation and are typically observed on fault segments which have previously exhibited creep, and which are evidently pre-stressed themselves (largely as a result of accumulated creep), or because the faults have variable shear strength along their lengths. Faults in broad zones of shear strain and that are near failure will slip small amounts either because of dynamic shaking induced by the main earthquake (Allen, et al., 1972; Fuis, 1982; Sieh, 1982), or because regional static stress changes associated with the mainshock dislocation are concentrated on those faults which exhibit sympathetic triggered slip by the process of creep amplification (Wesson, 1988). Along other faults, such those that do not evince creep, strain transients appear to pass through the fault with the result that any minor vertical strain is this study are surficially unaffected by minor earthquakes, either because they are not primed for slip with accumulated elastic strain, or as has been observed in underground nuclear tests (McKeown and Dickey, 1969), the faults are simply too far beyond the zone of effective radiated seismic energy to be measurably perturbed.

REFERENCES

Allen, C. R., M. Wyss, J. N. Brune, A. Grantz, and R. E. Wallace, 1972, Displacements on the Imperial, Superstition Hills and the San Andreas faults triggered by the Borrego Mountain earthquake, U. S. Geological Survey Professional Paper 787, p. 87–104.

Fuis, G. S., 1982, Displacement on the Superstition Hills fault triggered by the earthquake, *in* The Imperial Valley, California, Earthquake of October 15, 1978: U. S. Geological Survey Professional Paper 1254, p. 145–154.

McGill, S. F., C. R. Allen, D. Johnson, and K. E. Sieh, 1988, Slip on the Superstition Hills fault and on nearby faults associated with the November 23 and 24, 1987, earthquakes, southern California, Seismological Research Letters 59, n. 1, p. 49.

McKeown, F. A., and D. D. Dickey, 1969, Fault displacements and motion related to nuclear explosions, Seismological Society of America Bulletin 59, n. 6, p. 2253–2269.

Sieh, K. E., 1982, Slip along the San Andreas fault associated with the earthquake, *in* The Imperial Valley, California, Earthquake of October 15, 1979: U. S. Geological Survey Professional Paper 1254, p. 155–160.

Wesson, R. L., 1988, Dynamics of fault creep, Journal of Geophysical Research 93, (B8), p. 8929–8951.

SYVITSKI, J. P. M., Geological Survey of Canada, Dartmouth, Canada

Basin Sedimentation and Growth of Prograding Deltas

A theoretical model is developed for the infilling of a basin by the progradation of a delta. The model simulates four mechanisms for the transfer of sediment from the land to the sea: (1) bedload dumping along the delta front; (2) hemipelagic sedimentation of particles carried seaward by the river plume; (3) proximal slope bypassing by turbidity currents; and (4) the combined effects of both short-term (wave and tidal action) and long-term (creep and small slides) downslope diffusion of the accreting sediment mass. The model requires a priori knowledge or estimates of: (1) fluvial discharge rates of water, suspended sediment, and bedload; (2) the size distribution of the fluvial suspended sediment input and the basin removal rate constants for the various size fractions; (3) the river mouth channel geometry; (4) the plume dynamics; (5) the quantity of sediment bypassing the proximal slopes and the run-out distance for these gravity flows; and (6) initial basin geometry and the present seafloor bathymetry.

Particle scavenging model: SPM enters the basin from the river at an initial concentration, C_o, and subsequently undergoes both settling and advection down the basin. SPM removal from the basin is considered solely by sedimentation at a rate, $Z(x)$ in units of mass per area per time, and for a plug flow:

$$Z(x = \lambda Q_s (u_o b_o)^{-1e-(\lambda/u_o)x}$$

where λ is a first-order removal rate-constant, in units of $time^{-1}$, Q_s is the suspended load, u_o is the longitudinal plume velocity, b_o is the width of the river mouth, and x is distance along the plume. This simple model predicts that the sedimentation rate under a river plume will decrease exponentially with increasing distance out from the river mouth.

To model the velocity distribution within highly stratified marine basins, a buoyancy-dominated, free, two-dimensional jet is considered. The longitudinal and lateral components of the surface water velocity are residual or tidally averaged values. The velocity distribution within three dynamic zones of the river plume are considered, i.e., zone 1 where the plume is not yet established (nearest the river mouth) and where the center of the plume continues to behave as a plug flow, followed by zone 2 where the established flow decreases as the plume spreads, and zone 3 where plume spreading is affected and constrained by the basin boundaries. The analytical solution for predicting the sedimentation rate along the axis of such a plume may take the form of:

$$Z(x,y) = Z_o e^{-(\lambda/u_o)x} \text{ for } x < 5.2b_o$$

$$Z(x,y) = Z_o \exp[-\lambda(1.76b_o/u_o + 0.29x^{1.5}/u_o b_o^{0.5})] \text{ for } 5.2b_o < x \leq x_b$$

$$Z(x,y) = Z_o \exp[-\lambda(1.76b_o/u_o - 0.15x^{1.5}/u_o b_o^{0.5})] \text{ for } x > x_b$$

As the SPM load is composed of a mixture of particle sizes, each particle size of class s is characterized by a unique removal constant, λ_s. Thus the total sedimentation rate, $Z_T(x)$, is equal to the summation of the individual sedimentation rates $Z_s(x)$ for each size class:

$$Z_T(x) = \Sigma \lambda_s l_{so} e^{-\lambda_s t} = \Sigma Z_{so} e^{-\lambda_s t}$$

Model predictions favorably compared with sediment flux data from four basins (from Alaska, Norway, and British Columbia) that differ widely in their sediment concentrations and discharge conditions. All measured flux data were similarly derived from anchored sediment traps.

The accumulation of sediment deposited from hemipelagic sedimentation, Z, at any given position may be given as

$$Z = \Sigma\rho_{bs}^{-1}\int Z_s(x)dt$$

where ρ_{bs} is the appropriate bulk density of size fraction s. Sedimentation rates are integrated to the lateral boundaries of the plume.

Delta front bedload dumping: Only the fluvial bedload, Q_b', that enters the marine environment is considered in the model. We also assume that a portion of the bedload, α_b, may be removed by turbidity flows and that the remainder, α_d, is dumped close to the river mouth. Bedload is distributed evenly over the width of the active delta front, W, and out to some distance, L. The dumping function of bedload, D, in units of thickness per unit time is given by

$$D = \alpha_d Q_b'(W_d L_{\rho b})^{-1}$$

where α_d is the fraction of the bedload that remains within the delta front environment and ρ_b is the bulk density of the deposited sediment (i.e., $r_b = n^{-1}\Sigma\rho_{bs}$ where n is the number of size fractions considered). The longitudinal cross-section of this deposit is rhomboidally shaped.

Bypassing transport and deposition: Some transport processes operate by removing sediment from the nearshore, whereby a significant portion of the prodelta environment is bypassed before the sediment is deposited. Highly fluid and initially turbulent gravity flows are the prime mechanism. Such flows are generated from failure along the distal portion of mouth bars as a result of rapid progradation over steep foreset beds. Submarine channels normally serve as conduits for these sandy gravity flows.

In the model, the amount of material deposited by turbidity currents at a distance, *x*, from the river mouth is a function of the slope near the river mouth and the slope of the basin floor. If the foreset slope is gentle (less than some critical value, θ, we assume that no delta front failure occurs and channelized flows are not generated (i.e., bedload is deposited wholly by dumping near the river mouth). After a foreslope failure, material is deposited where the prodelta slope falls below some critical angle,θ, and therefore at some distance x_θ, out from the delta front. θ is dependent on changes in the flow characteristics, including the friction forces acting on the flow, flow viscosity and velocity, and the type of particles transported by the flow. The run-out distance, x_R, depends primarily on the rate of flow deceleration.

Accumulation of bypass material, B, in units of thickness per time, is assumed to decrease linearly as a function of x in such a way that:

$$\int_{X_\theta}^{X_R} B(x)dx = B(X_R - x_\theta) = \alpha_p Q_b' p_b^{-1} W_f^{-1} \qquad x_\theta \le x < X_r$$

where α_p is the fraction of the bedload entering the delta front environment that fails and moves downslope along the conduits of the submarine channel system and W_f is the average width of the basin floor. Values of ρ_{bs}, x_R, and x_θ are averages for the possible variations in the gravity flows.

Down-slope diffusion: This action includes all processes, other than channelized sediment flows, that move previously deposited material away from the proximal prodelta slopes into deeper water. Diffusion may be influenced by both short-term (e.g., tidal action) and long-term (e.g., creep) processes. Diffusion is most important on steep basin slopes which, at least initially, are a consequence of the original basement geometry. Diffusive processes operate at a rate proportional to the bathymetric slope, such that

$$\partial h/\partial t = K\partial^2 h/\partial x^2.$$

Where K is the coefficient of diffusion and $K = -S[\partial h/\partial x]^{-1}$, S is the rate of sediment transport, and $\partial h/\partial t = -\partial S/\partial x$ is used to conserve mass. Thus, the effect of downslope diffusion decreases with decreasing concavatiy of the slope.

In the model, diffusion operates independent of the primary methods of sediment deposition. A delta profile can therefore be adjusted even in the absence of sediment input so as to simulate a marine transgression without evoking a relative sea-level adjustment. This method can also be used to simulate the action of a short-lived mass transfer of sediment, i.e., earthquake-generated submarine slope failure, by adjusting K with time.

Numerical delta model: Sediment accumulation is modeled with a parabolic partial differential equation,

$$\partial h/\partial t = K\partial^2 h/\partial x^2 + D(x,t) + B(x,t,dh/dz) + Z(x,t)$$

where K is the coefficient of the sediment diffusion effect and the remaining three terms represent contributions to the seafloor from bedload dumping, bypassing, and hemipelagic sedimentation, respectively. The corresponding finite difference equation is solved numerically using an explicit method.

Model simulations are most sensitive to changes in: (1) the rate of transport of accumulated sediment by diffusive processes, (2) the fraction of bedload moved into the deep basin by turbidity currents, (3) the runout distance of turbidity currents, (4) changes in the bathymetry that affect the frequency of turbidity currents, (5) the distribution of sediment between suspended load and bedload, (6) grain size fractionation within the suspended load, (7) bathymetry of the river mouth, and (8) the critical slope upon which turbidity currents begin to deposit their sediment load.

The final model consists of: (1) a high-resolution particle-scavenging model capable of predicting, on time scales less than a year, hemipelagic sedimentation rates beneath a river plume; and (2) a lower resolution portion of the model that predicts the accumulation of sediment by bedload dumping at the river mouth, deposition by turbidity currents, and mass transfer processes such as creep and slides. The resultant model allows for the prediction of changing lithologies with time and space and at a predictive resolution of tens to hundreds of years.

The model has been used to simulate infilling processes for Knight and Bute inlets, two graben-like basins along the coast of British Columbia. Model simulations of delta progradation over the last 10,000 years favorably predicted the modern seafloor bathymetry and sediment accumulation rates to within a factor of two of measured (Pb-210, Cs-137, and sediment trap) and in an environment with three orders of magnitude variation in accumulation rates. This supports the value of a unified sedimentation-accumulation model accounting for downslope diffusion acting on slopes affected by inputs of bedload and suspended load.

SZULC, J., Jagellonian University, Cracow, Poland

Features of Seismically Controlled Sedimentary Basin: Example from Muschelkalk, Southern Poland

The Lower Muschelkalk (Pensonian) sediments of the Upper Silesia Upland comprise greyish carbonate muds, light crinoid-oncoidal packstones, coarse-grained bioclastic grainstones, and spongean and corallian build-ups.

Extensive synsedimentary deformations—wavy and crumpled structures, creeping, flow and folding structures, rolled-up deformations, and convolutions—are particularly significant for the described sediments. It seems that origin of deformations was influenced by seismic events that triggered deforming movement of the unconsolidated and semiconsolidated sediments.

The following features support the supposition:

1. Syndepositional faults.
2. Initial crumples and folds linked to the fault planes.
3. Neptunic dikes.
4. Large slides and slumps.
5. Liquefied firmgrounds—in situ homogenized mud containing dispersed, early cemented wall of Balanglossites burrows.
6. Bipartite coquinas beds, consisting of the lower member containing convex-down stacking terebratula shells (direct effect of shock) followed by layered, convex-up stacking detrital member (effect of the involved tsunami).
7. Abrupt lithological transistions both from bed to bed and set to set (e.g., muddy firmground → coarse-grained megabed, or dark, euxinic limy mud → crinoid-oncoidal pale deposits).

Other features of seismites (Seilacher, 1984) common in siliciclastics and caused mainly by sediment dewatering are less distinctive in carbonates because of their specific early lithification.

Because of the broad lateral extension (up to 200 km), the deformations have long been used as particular, lithostratigraphical, regional correlation markers.

Earthquakes resulting in the above-described sedimentary features may be connected with the initial extension of earth's crust in the Western and Central Paleo-Europe (Ziegler, 1982).

REFERENCES

Seilacher, A., 1984, Sedimentary structures tentatively attributed to seismic events: Marine Geology, v. 55, p. 1–12.

Ziegler, P. A., 1982, Triassic rifts and facies patterns in Western and Central Europe: Geol. Runcsch., v. 71/3, p. 747–772.

T

TABESH, E., Syracuse University, Syracuse, NY

DRILL-PAT: an Expert System to Design an Optimum Drilling Pattern

Introduction: By the beginning of the twenty-first century, the mineral exploration industry will be faced with significant economic hardship. Most of the deposits within the first 100 m of the earth's crust have been identified; the detection of deeper deposits will necessitate more sophisticated, and clearly, more expensive exploration techniques. Unfortunately even the best deep probing techniques are associated with high probability error. These circumstances will require a new exploration philosophy that is based on minimizing costs and reducing risk, which can eventually be achieved by maximizing a desired goal within financial constraints.

Statement of problem: Mineral investigation generally consists of geological, geophysical, and/or geochemical surveys followed by a drilling program to evaluate deposits. The drilling is mandatory in order to accurately evaluate the deposit, regardless of what mineral deposits are being investigated. The critical decision, whether or not to mine, must be made based on drilling results. If the deposit is a commercial orebody, the design for the mine must incorporate this information. The results acquired from drilling mark the division between a geological investigation and mining activities. Though drilling is expensive and time consuming, one cannot over emphasize the importance of accurate drilling data to a mineral investigation.

A number of attempts have been made to optimize sampling patterns; however, the available techniques are not suitable to the drilling stage of prospect evaluation, nor do they take practical factors into account. The purpose of this project is to introduce an optimization strategy that can be practically implemented at the drilling stage while taking into account empirical factors associated with mining.

Theory: A knowledge-based expert system is designed to provide economic and financial expertise, as well as control the rest of the system. For structured decision making, linear programming techniques are used, and for unstructured decision making fuzzy set techniques are applied. Fuzzy set method has been borrowed from the field of artificial intelligence and applied to map data. A classical (crisp) set is defined as a collection of elements or objects $x \epsilon X$ that can be finite and countable. Each element can either belong or not belong to a set. A fuzzy set does not have certain boundaries and each element can belong to a set to a certain degree. One can define each point on a map as a descriptive variable to some extent. For example, if a point on a map is not located on a granite body, the degree of membership for this point belonging to a granite is 0. The fuzzy set concept helps to quantify almost all the descriptive variables; the max-min method is used to combine the stack of membership degree matrices and results in a single matrix.

Expert system: An expert system has been designed to advise the exploration geologist as to the economic forecast and the cutoff grade for a particular material. The economic forecast is based on price fluctuations during the last 10 years while the cutoff grade recommendation is dependent on economic factors and the physical properties of the ore body such as the approximate size, grade, and the type of deposits. The next task is to outline the anomalies in the geophysical and geochemical data set and use them as input data for the implementation of the sampling algorithm. The output will be a preliminary drilling pattern.

Structured decision making: The structured decision involves processing the preliminary drilling pattern using a clear, well-defined optimization algorithm. The input data for this technique is the preliminary drilling pattern, and the output is the optimized pattern with assigned priority scores.

Unstructured decision making: The unstructured decision making process assumes an incomplete database in order to be able to consider descriptive variables. Descriptive variables include geologic parameters, topographic maps, and road maps, for example. Each of these variables are evaluated in the form of grids whereby each grid cell is weighted according to favorability to drilling. The size of the grid node depends on the variation of features, environmental factors, geologic favorability related to specific deposits, and other controlling factors.

Given several maps showing grid nodes and weighted values assigned to each node, the next step is to combine the variables using the fuzzy set technique and the max-min concept. The result is a single matrix of numbers called the *suitability factor matrix*.

Finally, the suitability factor matrix is combined with the preliminary drilling pattern simply by multiplying the priority scores and suitability factors. The result will be a new matrix of values that is contoured to show optimal drilling locations.

Comparison of results: The map determined by the structured decision method is compared with the map generated using the unstructured decision technique.

System performance will be tested using geological, geochemical, and geophysical data from the Big Hill deposit in Pembroke, Maine. This deposit is located at the faulted limb of a synclinally folded breccia horizon. It is a poly-metal sulfide mineralization showing anomalies of copper, lead, silver, and zinc on geochemical analysis of soil maps, as well as spontaneous potential data showing zones of strong oxidation that could indicate areas of silver enrichment. Equivalent lead-zinc but lower silver grades continue in all directions beyond the main mineralization zone.

TAGHIZADEH, N., Khiaban Khalili, (Amir Ahmadi Zamany); Tehran, Iran

Manganese Mineralization in Iran: Geneses and Relations with Mn Mineralizations in Surrounding Countries; Southeast USSR, Turkey, Southern Europe, Afghanistan, Pakistan, India, East and North Africa

Several Mn indications and deposits in Iran have been known and produced during the last 50 years. Also, many have been found recently.

The Mn mineralizations have been recognized in mesozoic, tertiary, and quaternary sediments and in volcanics of mesozoic tertiary, and prequaternary ages. The intrusives and subvolcanics of tertiary age contain some veins and indications of manganese minerals. Many Mn deposits and indications have been found in radiolarites of the colored melange sequence (ultrabasic containers) which were deposited during the Paleozoic but tectonically deformed during the Precretaceous.

All of the above mentioned mineralizations formed in different types of formations: stratiform, sedimentary, and hydrothermal.

The paper will describe the above mentioned mineralizations and their geneses.

It will also describe the relations with different Mn mineralizations in the surrounding countries such as Anatoly peninsula, Balkan peninsula, Italy, Iberia peninsula, Asian part of USSR, Afghanistan, Pakistan, India, east and north Africa.

Different models will be described for different geneses of Mn deposits in this large part of the world.

TAGUCHI, K., Teikoku Oil Co., Ltd., Tokyo, Japan

Formation and Expulsion of Immature Oils in Carbonate and Siliceous Sediments

Recently, the number of reports on the occurrences of "immature oils or condensates" from various sedimentary basins in the world is increasing rapidly, though much controversy still persists concerning the secondary alteration of oils due to the migration or biodegradation. Only a few studies, however, have touched on the possible formation mechanism of immature oils. In this paper, the formation and expulsion of immature oils are discussed in relation to the formation of protokerogen and the time of hydrocarbon expulsion controlled by a variation of sedimentary matrices and organic components associated with the matrices in the possible source sediments. The common sedimentary rocks can be divided in terms of mineralogical composition into three major extremes, shales, carbonates and silica sediments, which are regarded as potential petroleum source rocks. In this paper, important roles of the two extremes, carbonates and silica sediments exclusive of sandstone for the formation of immature oils are discussed on comparing with clayey sediments (shales).

It is recognized generally that carbonates are effective source rocks at lower Total Organic Carbon (TOC) contents compared with shales on the basis of higher hydrocarbon/TOC ratio in carbonates. And the lower TOC and higher hydrocarbon/TOC ratio in carbonates are attributed mostly by many researchers to a result of the greater preferential distribition of non-hydrocarbon organic matter in highly oxygenated sediments. In connection with this, however, the present author rather emphasizes that carbonates, both Recent and ancient, usually contain I or II type kerogen consisting mostly of amorphous one and relatively small amounts of kerogen compared with clayey sediments. Interestingly, laboratory simulations in melanoidin formation show that the rates of melanoidin formation is much higher in the presence of clay mineral (montmorillonite), while the $CaCO_3$ association reactions for the same experiments show extremely low rates of melanoidin formation. In addition, Gehman's data (Gehman, 1962) show that the average absolute amounts of distribution of hydrocarbons of ancient limestones and shales are nearly identical showing the increase of those contents from the Recent to the ancient in both sediments. In the Recent sediments, however, the average absolute amounts of hydrocarbons in clayey sediments is significantly less than that of carbonates. Namely, the average absolute amounts of hydrocarbons in clayey sediments shows significant increase from Recent to ancient ones, while that in carbonate sediments show only slight increase. This may be interpreted that carbonate sediments contain relatively much higher hydrocarbon contents than clayey sediments even at the initial depositional stage and the increase of hydrocarbons with lithification—increasing of burial depth in carbonates is not significant in comparison with clayey sediments. Namely, this suggests that generation of hydrocarbons from carbonates does not always follow a pattern of hydrocarbon generation model as seen in shales. If we consider the above data and the earlier consolidation of carbonate-evaporite rocks, carbonate sediments seem to be the apparent source of immature oils.

Recent researches on siliceous source rocks also suggest that oils from siliceous sediments are generated at a lower temperature than conventionally thought possible. Here, generation and expulsion mechanisms of immature oils is given through a case study of siliceous rocks of Miocene age in Japan. In the area of the Takanosu, northeastern Japan, several beds of hard, highly siliceous opal-A-bearing rocks (porcelanites) with 0.2–0.7 m thick are interbedded, contrasting sharply with massive, soft opal-A-bearing rocks (diatomites). Namely, opal-A-bearing and opal-CT-bearing rocks which belong to the different diagenetic stage essentially are found co-existing in the same pressure/temperature regime. Porcelanite rocks showed relatively lower bitumen ratio, S_1 value and Production Index, and at the same time relatively higher Hydrogen Index and CArbon Preference Index (CPI) than those values of diatomites or diatomaceous mudrocks. The mechanism of the high concentration process of the release of dissolved silica, as a monomeric silicic acid and as some kinds of organic-silica complexes, such as catechol-like materials from siliceous organisms (diatoms), to the interstitial water near the uppermost layers of the sediments. As compaction and temperature increase with increasing burial depth, the pH in the interstitial water of the porcelanite bed decreases rapidly, owing to the generation of carbonic acid from the protokerogen predominant in porcelanite beds and from related organic acids. At the same time, this promotes the gelatinization and crystallization of amorphous silica and the ejection of aluminosilicate components and of the organic matter present in silica mineral as contaminated matter from porcelanite beds. Thus, some hydrocarbons have been expelled from porcelanite beds at a lower temperature than conventionally thought possible. The observation that organic maturity tends to be inversely related to the silica phase transformation and that porcelanites are associated with extremely poor content or clay minerals sustain strongly the above possibility. As a result, it is suggested that if a thick formation consisting of a vast number of porcelanite-like layers, such as the Onnagawa formation or the Monterey formation in the U.S.A., is developed covering a wide area, those siliceous formations will expel considerable amounts of oil hydrocarbons at a low maturation level.

Conclusively, it is suggested that most of immature oils from carbonates or siliceous sediments will be generated directly from organic matter such as lipids or proteins, and/or from protokerogen at a lower level of organic maturity.

TAIRA, A., University of Tokyo, Tokyo, Japan

Geoscience Transect of Southwest Japan and Izu-Bonin Arcs

Compilation of geological and geophysical data on 1) the Southwest Japan across the Nankai Trough, Shikoku to the Sea of Japan and 2) the Izu-Bonin arc at about 30 °N latitude was conducted to delineate a geoscience transect which is a part of the GGT program in Japan. The objective of the Transect is to show a comparison between two different arc-trench systems. The Southwest Japan arc is a mature arc which shows a long history of sediment accretion and magmatic activity since the Jurassic time. The Izu-Bonin arc is a younger oceanic island arc system which has been formed since the Eocene time.

The geologic structure of the Southwest (SW) Japan is characterized a general young trend of accretionary prism from the Sea of Japan side toward the Pacific side, from the Permian accretionary prism to the Pleistocene accretionary prism at the Nankai Trough landward slope. These accretionary prisms are piled-up primary by a series of napped, thus the SW Japan has been dominated by compressional tectonics. The basement of the SW Japan arc is truncated by a major strike-slip fault, the Median Tectonic Line of Japan and also intruded widely by Cretaceous–Miocene granitic intrusives. The thickness of the continental crust is about 35 km and we interpret that the lower crust of the SW Japan arc is composed of granulites. The distribution of crustal seismicity indicates a concentration of seismic activity along the boundary between the lower and upper crustal layers, which may suggest a detachment between two layers.

The cross section of the Izu-Bonin arc is quite different from the SW Japan arc. The surface geotectonic units of the Izu-Bonin arc are composed of four domains: 1) lower trench slope with serpentite diapirs, 2) forearc massif composed of the Eocene arc basement (boninite and related rocks), 3) volcanic front and sedimentary basins related to moat flexure, 4) bacakarc depressions, and 5) backarc volcanic complex. The forearc region is dominated by extensional tectonics. The subduction of the hydrous oceanic plate serpentinize the forearc basement causing a chain of diapiric mounds at the surface. The forearc massif is a rifted portion of older volcanic arc of the Eogene age which probably is a conjugate part of the Kyushu-Palau ridge. The successive volcanic arc and surrounding volcaniclastic sedimentary basins. These sedimentary basins are developed primarily due toby a gravity induced flexure (moat) due to volcanic addition. The backarc depression is a narrow rift basin which generally shows an asymmetric graben characteristics. The crust of the Izu-Bonin arc is about 15–20 km, much thinner than the SW Japan. We interpret the lower crust is composed of a gabbroic layer. This geoscience transect revealed the basic geological and geophysical difference of two contrasting arc types.

TAKASU, A., Shimane University, Matsue, Japan, and R. D. DALLMEYER, University of Georgia, Athens, GA

$^{40}Ar/^{39}Ar$ Mineral Ages and Implications for Tectonothermal Evolution of Sambagawa Metamorphic Terrane, Southwest Japan

Three contrasting pre-Miocene lithotectonic units are exposed within the Japanese archipelago. Northwestern elements include the Pre-Jurassic Hida and Sangun terranes, whereas the southeastern units are represented by a Cretaceous–Paleogene subduction complex (Shimanto belt). Intervening pre-Miocene elements include the Mino–Tamba, Chichibu, Sambagawa, and Rhoke belts. The Mino–Tamba and Chichibu belts are composed of a mixed sedimentary facies where Carboniferous–Jurassic cherts, limestones, and greenstones are intercalated with Jurassic terrigenous sedimentary successions. The high P/T Sambagawa and low P/T Ryoke belts are presently separated by a major strike-slip fault (Median Tectonic Line: MTL) of uncertain total offset. The contrasting belts may, in part, be metamorphic equivalents of the Tamba and Chichibu belts.

The Sambagawa metamorphic belt is an intermediate high-pressure terrane mainly composed of pelitic, psammitic, and basic schists with subordinate siliceous schists. Four distinct metamorphic zones have been outlined on the basis of mineral assemblages developed in pelitic, psammitic, and basic schists with subordinate siliceous schists. Four distinct metamorphic zones have been outlined on the basis of mineral assemblages developed in pelitic schists. In increasing grade these are chlorite, garnet, albite–biotite, and oligoclase–biotite. The metamorphism has been classically considered to have developed in association with Jurassic–early Cretaceous subduction, although detailed chronological calibration has been lacking. A recent description of Jurassic radiolarians from metamafic tuffs in the chlorite zone provides a lower age constraint. Blocks of Sambagawa-type metamorphic rocks occur within several younger conglomeratic units, including the Onogawa (Santonin) and Kuma (Eocene) Groups. These provide an upper limit for the timing of the high P/T metamorphic event.

Hornblende concentrates prepared from amphibolite within the oligoclase–biotite zone record well-defined $^{40}Ar/^{39}Ar$ plateau dates ranging between ca. 90 and 95 Ma. Hornblende concentrates from the albite-biotite zone record slightly older dates ranging between ca. 100 and 105 Ma. The $^{40}Ar/^{39}Ar$ plateau ages are considered to date postmetamorphic cooling through those temperatures required for intracrystalline retention of argon (~500 ± 25 °C). Because the maximum metamorphic temperatures attained within the albite–biotite zone (~450–600 °C) were not significantly greater than the argon retention temperatures, it is likely that the ca. 100–105 Ma hornblende plateau ages closely date the Sambagawa metamorphic culmination. They suggest an Early Cretaceous event and are therefore consistent with the available geological controls. Slightly younger cooling ages are likely recorded in the oligoclase–biotite zone because of the initially higher metamorphic temperatures.

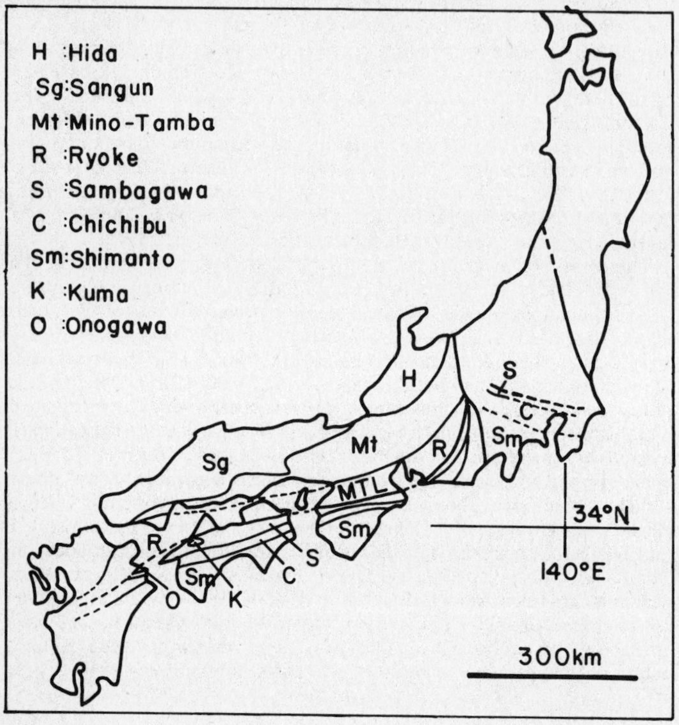

Figure 1.

TAKEDA, H., and J. SAITO, University of Tokyo, Tokyo, Japan

Processes Responsible for Evolution of Materials in Ureilite Parent Body

Achondrites provide us with information responsible for materials evolution in primitive planets from primitive materials. The HED (howardites, eucrites, diogenites) achondrites experienced extensive melting and crystallization (Ikeda and Takeda, 1985), and the primitive achondrites are least modified with respect to their chemistry (Prinz et al, 1983). Ureilites are coarse-grained ultramafic rocks and appear to be differentiated (Goodrich et al, 1987). However, their oxygen isotope anomaly (Clayton and Mayeda, 1988), high noble gas contents, and anticorrelation of MnO versus FeO suggest that ureilites still restore detectable remnants of primitive materials related to some kinds of carbonaceous chondrites. The oxygen isotope trend, including newly discovered Antarctic ureilites (Takeda, 1987), shows a mixing trend similar to dark inclusions in carbonaceous chondrites of type CV3 (Clayton and Mayeda, 1988). They suggested on the basis of the oxygen isotope anomaly that ureilites can be subdivided into three subgroups, which are also related to mg numbers = 100 X Mg/(Mg + Fe) variation. They also mentioned that within each oxygen subgroup, the oxygen isotope data are on the possible mass fractionation and are compatible with differentiation from a common reservoir. In order to find mineralogical implication of this subgrouping, we investigated pyroxene chemical variations of all available Antarctic ureilites.

Because only olivines and pyroxenes are crystallizing in the ureilite formation event, we cannot apply a diagram normally employed when plagioclase is coprecipitating such as An mol% versus mg number. The absence of chromite and plagioclase crystallization in ureilites enable us to use Na, Ca, and Cr contents in their pyroxene structure, as an indicator of possible fractionation. The pyroxene quadrilateral with Mg-Fe-Ca end members of ureilitic pyroxenes revealed Ca variation with nearly constant mg numbers for the Fe–Ca-rich subgroup.

The most Mg-rich subgroup including Y74659, Y791538, and LEW85440 does not show much variation in the Ca content, but modal abundances of pigeonite, orthopyroxene, and augite are consistent with variation with temperature and Ca contents. The intermediate subgroup shows mg number variation between the low-Ca portion of the two other subgroups.

There are abundant augites in the MET78008 and Y74130 ureilites recognized by us and their Na-Cr pyroxene components are high (Na_2O 0.8 Cr_2O_3 1.8 wt%). Minor augites are also present in ALH82106/130 and LEW85440 of the Mg-rich group, but their components of the Na-Cr pyroxene do not exceed those of augites (Na_2O 0.2 wt%) produced in the shock partially melted pigeonites in Y790981.

LEW85328 and Y790981 belong to the Ca-Fe group, but they contain augites only as rim materials and small inclusions in pigeonites produced by shock partial melting. Y790981 preserves glass veins produced by the final shock event, but LEW85328 includes disturbed areas, which are possibly annealed shock partial melt. Interstitial metal grains contain Ni and P, which can be used as an indicator of fractionation. The LEW85328 pyroxene is one of the most calcic pigeonites. The Ca-Fe group is characterized by remnants of shock melting.

If we assume the mixing or removal of residual liquids produced possibly during crystal growth from shock partial melts (Takeda, 1987), we can explain variation of pyroxene trends on the basis of Na_2O verus mg number (in wt%) diagram (Figure 1). This liquid could also be intercumulus liquid if one favors the cumulate model for the ureilite origin (Goodrich et al, 1987). the Ca–Fe-rich subgroup may contain more residual components, but the Mg-rich group has higher degree of refractory mafic minerals with a few residual components.

If the oxygen isotope anomaly is inherited from the Mg-rich mafic silicate fragments in carbonaceous chondrites, the oxygen isotope anomaly may be more likely preserved in the Mg-rich subgroup. The Ca-Fe subgroup may have more chances of homogenization of oxygen isotope anomaly during the mixing of the residual liquids at high temperature. By such process it may be easy to produce ultramafic rocks from primitive materials.

We thank National Institute of Polar Research and U.S. Antarctic Meteorite Working Group for the samples and the Mitsubishi Foundation for scientific grant.

REFERENCES

Goodrich, C. A., J. H. Jones, and J. L. Berkley, 1987, Geochimica et Cosmochimica Acta, v. 51, p. 2255–2273.

Ikeda, Y., and H. Takeda, 1985, Proceedings of 15th Lunar and Planetary Science Conference in Journal of Geophysical Research, v. 90, Suppl. p C649–C663.

Prinz, M., C. E. Nehru, J. S. Delaney, and M. Weisberg, 1983, Lunar and Planetary Science XIV, p. 616–617.

Takeda, H., 1987, Earth and Planetary Science Letters, v. 81, p. 358–370.

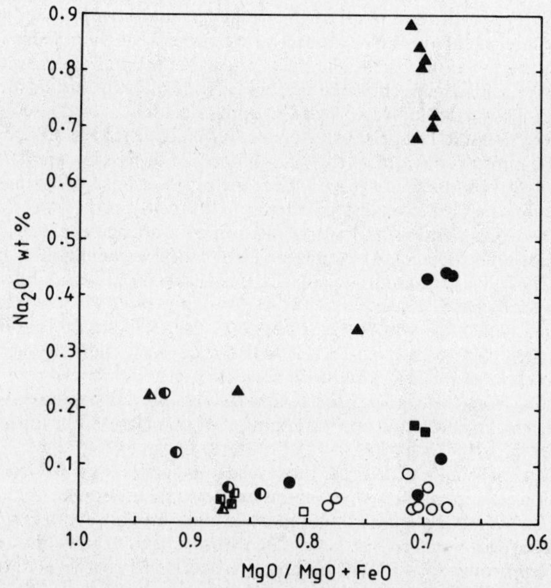

Figure 1—Na_2O (wt%) versus MgO/(MgO + FeO) (wt%) in the ureilitic pyroxenes for three subgroups of oxygen isotopes. Solid symbols, Ca-rich; half-filled, Mg-rich; open, intermediate group. Triangles, augites; circles, pigeonite; and squares, orthopyroxenes.

TALUKDAR, S. O. GALLANGO, and A. RUGGERIO, INTEVEP, Caracas, Venezuela

Integrated Geochemical Study and Timing of Oil Generation and Migration in Maturin Subbasin, Eastern Venezuela

The Maturin Subbasin which occupies a large part of the Eastern Venezuelan Basin, has known oil accumulations along two separate east-northeast-trending belts, one to the north (Maturin area) and the other to the north (Cerro Negro area, Orinoco Oil Belt). The northern oil belt, located on a fold-thrust zone, is under active exploration and has a number of important oil and gas discoveries in recent years. The present work attempts to show how an integrated geochemical study has helped in evaluating the hydrocarbon potential of the northern oil belt.

The Maturin Subbasin can be classified as a "polyhistory" type basin that includes Lower Cretaceous to Middle Eocene sediments deposited in a passive continental margin setting and Upper Oligocene to Pliocene sediments deposited in a foreland basin setting. The pre-Upper Miocene sediments of the northern part of the subbasin were deformed during the Miocene (12–6 m.y.) into an east-northeast-trending fold thrust belt that is largely exposed and eroded in the Interior Mountain Front. The thrusting episodes appear to be progressively younger southward and were responsible for the southward migration of the foreland basin axis with time.

Regional geochemical evaluation of the possible Cretaceous and Tertiary source rocks and oil-oil and oil-source rock correlation studies have identified the fine grained, calcareous shales and argillaceous limestones of the Upper Cretaceous, Querecual and San Antonio Formations of the Guayuta Group as the source rocks for the oils in the subbasin. These source rocks are highly oil prone and contain mainly amorphous marine, type II, organic matter. Regionally significant variations in source rock characteristics have been noted within the rocks of each formation and the group as a whole. In the Interior Mountain Front to the north, the rocks show a TOC average of 2.5% with more than 85% amorphous marine organic matter, a Hydrogen Index of 700 mgHc/g TOC (immature stage), an atomic H/C ratio of 1.3 (immature stage) and an oil generating capacity of 166×10^6 bbls/km^3 rock. Towards the south this organic facies is intercalated with a slightly different facies showing a TOC average of 1.8% with less than 60% amorphous marine organic matter and more than 30% woody material, a Hydrogen Index of 400 mgHc/g TOC (immature stage), an atomic H/C ratio of 1.1 (immature stage) and an oil generating capacity of 68×10^6 bbls/km^3 rock. Both source rock facies apparently disappear in the south of Maturin.

Geochemical studies indicate that the oils found in different reservoirs (Cretaceous, Eogene, Oligocene, Miocene and Pliocene beds) of the Maturin Area (northern belt) as well as from Miocene reservoirs of the Cerro Negro area (Orinoco Oil Belt) are all of mature, marine type, genetically related to the Upper Cretaceous, Querecual and San Antonio source rocks. The oils of the Maturin area include light to medium gravity as well as heavy and extraheavy oils. The heavy oils are formed by biodegradation in shallow reservoirs. The extraheavy oils found in relatively deep Oligo-Miocene reservoirs appear to have formed by the in-situ alteration of normal oil or natural gas. In addition, some normal oils of the Maturin area show evidence of mixing with the highly altered residue of a previously accumulated oil, while a few heavy oils from the area show two phases of biodegradation. The oils from the Cerro Negro area are heavy and extraheavy, and are formed by biodegradation.

Based on thermal maturity data and maturity modeling of the source rock interval, the timing of oil and gas generation in the entire subbasin has been outlined. The areas that describe the hydrocarbon generation through time include 1) the internally deformed overthrust slab of the Frontal Thrust (actual Interior Mountain Front) and 2) zones beneath principal thrusts (Maturin area). It is concluded that the source rocks of the Interior Mountain Front were in active oil and gas generation during the Lower and Middle Miocene, prior to the first phase of thrusting (12 m.y.). Long distance (150–300 km), southward migration of mature oil from this generating area during the Middle Miocene and subsequent biodegradation in reservoirs gave rise to the heavy and extraheavy oil accumulations in the Cerro Negro area. On the other hand, new oil and gas generating areas were developed beneath the principal thrusts of the Maturin area during 12 m.y. to Present. The actual structural configuration remained practically unchanged since 5 m.y. which permitted the accumulation of a large quantity of oil in different structural traps of the Maturin area. The existence of a single oil-source rock interval, active generating conditions beneath the principal thrusts during 12 m.y. to Present, and the structural configuration characterized by repetition of beds, indicate the possibility of having medium to light oils in any potential reservoir of relatively deep structural traps.

The present study is an example showing how an integrated geochemical study with a sound geological framework can be useful in exploration even when the detailed geology is still to be worked out.

TALWANI, M., J. DIEBOLD, M. CALKIN, and L. Z. HU, Houston Area Research Center, The Woodlands, TX

Exploration Seismic Data Processing at HARC

HARC is a research consortium of four universities in the state of Texas. The members include Rice university, the University of Texas at Austin, the University of Houston, and Texas A & M. One of HARC's principal research tools is its NEC SX-2 1.3 gigaflop supercomputer. The Geotechnology Research Institute at HARC uses the SX-2 for the development and application of high-speed seismic processing methods. These include standard procedures, with optimized versions of the well-known DISCO and GEOVECTEUR processing packages, as well as more specialized methodologies, including various forms of migration.

The great speed of the SX-2 reduces the cost and time required for complex processing jobs (DISCO, for example, runs more than 100 times faster on the SX-2 than it does on an array processor-equipped VAX) and also brings the goal of interactive data processing closer to reality. Data display and interactive functions are performed on a network of SUN workstations, whose server is connected directly to the SX-2 via a hyperchannel. The SUN server also serves as a node to the national computer networks, providing access from anywhere in the world. Graphics-based interactive tasks such as migration velocity analysis and parameter selection can thus be performed at remote sites, while the computationally intensive processing functions are performed at HARC, where the database also resides.

TALWANI, M., Rice University, Houston, TX

International Geological–Geophysical Atlas of Atlantic Ocean

The international geological–geophysical atlas of the Atlantic Ocean has been compiles under the auspices of the Intergovernmental Oceanographic Commission (IOC) of UNESCO, from the latest available data.

It complements the Geological/Geophysical Atlas of the Indian Ocean, published in 1975, to bring together the scientific results of the IOC's International Indian Ocean Expedition (IIOE), 1959–1965. An International Geological/Geophysical Atlas of the Pacific Ocean is due for publication in 1991.

The main idea behind the compilation of these new atlases was to bring together the large amount of data which had already been gathered by research scientists but had still to be published or had only been published on a small scale, often with a fair amount of interpretation. The extremely rapid development of the technology used for marine scientific research in the years since the Second World War had given rise to a whole avalanche of new data on the structure of the ocean floor.

Careful study of factual data and their critical analysis and synthesis are only possible if they become accessible in an objective form that is convenient for purposes of analysis. Precisely such a form is provided when primary data are published with a minimum of distortion, on a large scale and with wide use of graphics; this involves in particular the use of color in atlases such as the Geological/Geophysical Atlas of the Indian Ocean.

This main idea complements the need to illustrate the new technical equipment and methods used in research, providing examples of their application.

The history of the study of the oceans and the development of human conceptions of the relief of the ocean floor is illustrated by fragments of old maps. A special section contains information on a selection of the main research vessels from which the work on advancing the study of the oceans has been carried out since the science of oceanography was in its infancy. This is followed by a section on information concerning new research methods into the geology and geophysics of the ocean floor, and also information on the methods used for compiling some of the maps contained in the atlases.

The main selection of maps on scales of 1:10 million (at 45° latitude) Mercator Projection, and 1:30 million Transverse (Pseudoazimuthal) Projection contains materials on the relief of the ocean floor, the magnetic total intensity anomalies, gravity anomalies, sea surface topography, geothermal data, seismicity, the deep structure of the earth's crust, the thickness of sediment cover, types of sediments, on magmatic and metamorphic rocks of the earth's core and on the results of deepsea drilling. This selection is supplemented by more detailed maps illustrating certain areas that are particularly interesting from the geological point of view, or which have been subjected to particularly detained study on a larger scale.

The team of contributors included specialists from many countries, well known for their work on the study of the oceans. As a rule there was a Curator coordinating the work of individual contributors in charge of each group working within a particular field. The Central Editorial Board set itself the goal of ensuring that all contributors presented their data in as much detail and in as objective a way as possible.

The contributors' original materials were collected and approved by the Central Editorial Board and then handed over to the cartographic staff where, under the supervision of Dr. D. I. Zhiv, one of the members of the Central Editorial Board, the original maps were prepared for final publication. The originals, and later black and white and color proof stages of preparation through which the materials passed prior to publication, were presented to the Central Editorial Board to be checked and edited.

Preparation of the final authors' originals was completed by the end of 1987 and as a result the atlas will reflect the level of knowledge relating to the geology and geophysics of the Atlantic Ocean attained by the period 1984–1987. All maps submitted earlier have, wherever possible, been supplemented with data that had become available by 1987.

The Soviet Geophysical Committee of the USSR Academy of Sciences, the USSR Ministry of Geology and the Main Administration of Geodesy and Cartography under the Council of Ministers of the USSR are responsible respectively for managing, financing and final publishing of the atlas. A set of selected maps from the Atlas can be viewed in the exhibit area.

TALWANI, P., University of South Carolina, Columbia, SC

Characteristic Features of Intraplate Earthquakes and Models Proposed to Explain Them

Although much less frequent than their plate margin counterparts, intraplate earthquakes have a large potential for destruction. Their cause and nature are not well understood, especially since in most cases they are not associated with any surface faulting. Hence, potential seismic source zones have to be identified by indirect means.

The tectonic cause of intraplate earthquakes has remained engimatic. As newer data became available, several common features were apparent for intraplate earthquakes occurring in a wide variety of geologic terranes. An evaluation of various aspects of intraplate earthquakes suggests that they occur by reactivation of a variety of preexisting features in response to the current stress field. In this review, I first list the various methods of identifying these features and then describe the characteristic elements of intraplate earthquakes.

Reactivation occurs by one or more of three ways. These include a localized build-up of stress on the potentially seismogenic feature due to the ambient stress field, the superposition of a triggering stress, and the reduction of strength of the feature(s) by mechanical and/or chemical means. The various models that have been suggested to explain intraplate earthquakes contain one or more of these elements.

Models in the first category include stress amplification near plutons, the initiation and termination of seismicity at fault bends, the concentration of stress at intersections of seismogenic features, and stress build-up on faults due to localized strain in the mid-lower crust. In the second category are triggering stresses due to glacial rebound and due to lateral variations in the density. In the third category, which is an extension of the mechanism of reservoir-induced seismicity to subcrustal depths, failure occurs by the reduction of strength of rocks by mechanical or chemical means. This is the hydroseismicity model. Of all the models, I prefer the intersection model.

The intersection model: Seismicity occurs near the intersection of, and by the reactivation of, preexisting zones of weakness. The intersections are the foci of anomalous stress build-up in response to the ambient stress field due to plate tectonic forces. This anomalous stress build-up is relieved by strike-slip motion on a suitably oriented fault and, due to kinematic adjustment, is followed by vertical or horizontal movement on the intersecting fault. The intersection model has several elements that complement other models proposed to explain intraplate earthquakes.

TALWANI, P., University of South Carolina, Columbia, SC

Induced Seismicity at Monticello Reservoir, South Carolina: Case Study

The small (0.5 km^3) and shallow (h 50 m maximum depth) Monticello Reservoir that serves as the source of the cooling water for a plant in central South Carolina was impounded between December 1977 and March 1978. The background seismicity was monitored for about 3 years preceding the impoundment and has been continuously monitored since. To understand the nature of seismicity, a variety of multidisciplinary studies were carried out. These include monitoring the seismicity on a dense network of seismic stations; water levels in observation wells; and geological and potential field mapping of the reservoir and its vicinity. In situ studies were carried out in two deep (h 1 km) boreholes drilled into seismically active areas around the reservoir. The results of the studies are as follows:

After impoundment, the seismicity rapidly increased from a preimpoundment level of about one event (M c 0.0) per week to over 100 events per day in early 1978. Following this peak, in the ensuing decade, the seismicity has gradually decayed to slightly greater than the preimpoundment levels. The seismicity has been shallow (<5 km), of low magnitude ($M<3.0$), and limited to discrete zones beneath and in the immediate vicinity of the reservoir. To obtain accurate hypocentral locations, a velocity model was developed from velocity logs of the deep boreholes and from calibration blasts.

Borehole televiewer analysis of fracture patterns revealed major differences in orientation and density of fractures with depth and over a horizontal distance of less than 5 km. The orientation of poles to fault planes obtained from composite focal mechanism solutions of Monticello reservoir earthquakes are in agreement with those of fractures observed in the deep boreholes. Based on this and other lines of evidence, it was concluded that earthquakes occurred on preexisting fractures. These focal mechanisms and in situ stress data imply predominantly thrust faulting with northeast-trending maximum compressive stress.

Integration of geological, gravity, magnetic radioactivity, and seismicity data suggests that the induced seismicity is associated with fractures on the flanks of shallow intrusive plutons. Results of hydrogeological investigations at the reservoir site indicate that water flow is controlled by discrete fracture systems. Permeabilities measured in wells ranged between about 0.01 and 1.0 millidarcy.

Various data were integrated to develop a model to explain the observed seismicity. Earthquakes in the area occur on preexisting fractures that are distributed nonuniformly. In situ stress data suggest the presence of large deviatoric stresses at shallow depths. The presence of clays in cores from the deep boreholes argue for relatively low frictional coefficients. Time lags between impoundment and onset of seismicity argue for the diffusion of pore pressure to hypocentral locations.

Together, these observations suggest the following scenario: Following impoundment, pore pressures were diffused along preexisting fractures (more abundant on the flanks of the plutons) to hypocentral depths. At these locations there was a decrease in the effective stress values, and on suitably oriented faults, many of which were filled with clays (thus having low coefficient of friction), failure occurred in accordance with the Mohr-Coulomb criterion.

Based on the heterogeneity observed in reservoir rocks, source parameters of a few events recorded digitally and other statistical data, the largest event that could be induced was estimated to be about magnitude 4.0.

TAN Tjong Kie (CHEN Zong-ji), XIONG Shao Bai, and OU Yang Ting, Institute of Geophysics, Academia Sinica, Beijing, China

Geotectonics and geothermics of Fujian Continental Margin

This paper reports on a part of the widescale investigation of the interrelations between temperature field, crustal structure, and geotectonics in order to study the origin of geothermal heat, its distributional pattern, and the feasibility for economic exploration. The geodynamic evolution with intense magma outpour during volcanic activity in the Mesozoic (>135 Ma) and the formation of the granitic basement 75 Ma ago has been studied. Further, the volcanic eruptions of about 2–1 Ma ago in the Longhai and Mingxi regions are proofs that the deeply cutting large faults were excellent conduits for magma outpour. The heat reservoir formed in that period may still contribute to the present heat anomaly. The hot water is of meteoric origin, which is infiltrated through the fissures in the basement. Land upheavals with a maximum rate of 2 m/s in the region northwest of the Zhangzhou basin are also proofs of present geodynamic activity.

The northeast-trending coastal zone is characterized by a negative Bouguer anomaly with a negative northwest gradient. The Fuzhou and Zhangzhou basins, where geothermal wells are most abundant with maximum temperatures of 107 °C and 121.6 °C, respectively, are indicated by a northwest-trending zone of positive Bouguer anomaly superimposed on a strong negative background.

Two main families of deeply cutting large faults are the northwest-striking extensional shear faults and the northeast-running compressional shear faults that obviously govern the pattern of geothermal wells distribution. The preferential distribution along the northwest faults are indications that they act as good conduits for the outpour of hot fluids. Further, the wells in the southern part of Fujian by far outnumber those in the northern part. Two seismic profiles cross-sectioning the Zhangzhou basin, the first running via Anxi in a nearly northeast direction and the second in south-north direction show that strong Pn wave energy absorption in the upper mantle in the southern part is a very strong indication of a viscous upper mantle under high temperature. In this region the Vp velocity at the roof of the upper mantle is 7.9–8.0 km/s, whereas in the northern part Vp = 8.1 km/s and Pn wave energy absorption is much lower.

These findings may give an explanation for the fact that wells are more abundant in the southern than in the northern part of Fujian. In fact, a large number of geothermal wells have been found in the continental margin of southern Fujian to neighboring Guangdong province. So it is concluded that the high-temperature upper mantle as a deeply seated heat source supplemented by residual heat sources, in combination with extensional shear fractures running into the upper mantle, form the basic condition for fluid and heat transport under the present geodynamic activity.

TANAKA, H., Yamagata University, Yamagata, Japan

Corlandtite and Its Associated Rocks from Abukuma Mountains, Northeast Japan

The Abukuma Mountains are famous as the type terrane for the metamorphism of low-pressure/high-temperature type (Miyashiro, 1961). A large amount of granitic rocks and minor amounts of gabbroic and ultramafic rocks are distributed in the mountains in close relation with the metamorphic rocks. Most K-Ar ages of the plutonic rocks are 85–110 Ma (middle Cretaceous). The Cretaceous plutonic rocks were formed by an intense igneous activity that occurred throughout Japan in Cretaceous–Paleogene time. Ueda and Miyashiro (1974) pointed out that the igneous activity was closely related to the subduction of the proto-Pacific plate and the proto-Pacific ridge under the Asiatic continent at the time.

The gabbroic rocks in the Abukuma Mountains are commonly abundant in amphibole and plagioclase and are sometimes accompanied by a minor amount of olivine and pyroxene. Small bodies of cortlandtitic rocks, intimately associated with the gabbroic rocks, are found at some places. Cortlandtite is a peridotite that contains large amphibole poikilitically enclosing olivine, and the name is from the Cortlandt complex, New York (Willaims, 1886). Cortlandtite and its associated rocks from two places, the southern Abukuma and the central Abukuma, are described and their origins are discussed.

In the southern Abukuma mountains a cortlandtitic pluton of small size (600 m long and 400 m wide) occurs, consisting of various rock types. The pluton is intruded into high-grade metamorphic rocks that consist mainly of mica gneiss, mica schist, and amphibolite of the amphibolite facies. Two kinds of fractional crystallization formed the diversity of the pluton. Corlandtite, quartz hornblende gabbro, monzonite, and syenite are products of a clinopyroxene-dominated fractionation of a picrite-basaltic magma. Whereas an amphibole-dominated fractionation played an important role in the formation of hornblende gabbro, quartz diorite and granite form a fairly MgO-poor basaltic magma. The various rock types commonly occur in intrusive relationships. The two rock groups are probably not comagmatic. However, the two parental magmas have a common character of alkali basalt affinity and might have originated from a common parental material enriched in K and Rb and depleted in Ni. The corlandtite had crystallized under conditions of 7–9 Kbars $> P_{total} > 5 \pm 2$ Kbars, $P_{H\frac{2}{3}20} \geq 2$–3 Kbars, and had equilibrated at temperatures from 1000 to less than 750 °C. When the almost congealed cortlandtite had intruded into the present position, situated at about 5 ± 2 Kbars, the latest liquid of the clinopyroxene-dominated fractionation was segregated to form the syenite and monzonite.

In the central Abukuma mountains, a circular cortlandtitic pluton of about 1.5 km in diameter occurs as a roof pendant in biotite granite. The pluton is mainly composed of cortlandtite accompanied by a minor amount of hornblende metagabbro and hornblende gabbro. The associated gabbros are right in Al_2O_3 (18–19 wt%) and relatively poor in SiO_2 (40–47 wt%), and a high-alumina basalt magma (Kuno, 1966) can be assumed as their parental magma. The cortlandtite has large glittering crystals of amphibole up to 15 mm in size, which contain olivine (Fo_{72-77}), orthopyroxene (En_{78-81}), and diopsidic pyroxene in various amounts. It is very rich in MgO, Cr, and Ni compared with the associated gabbros. Numerical calculation using distribution coefficients given by Roeder and Emslie (1970) indicate that the olivine and orthopyroxene in the cortlandtite are in equilibrium with a gabbroic melt of $Mg/(Mg + Fe^{2+})$ ratio of about 0.50. This $Mg/(Mg + Fe^{2+})$ ratio of gabbroic melt is within that ratio of the associated gabbros and neighboring gabbroic rocks in the central Abukuma mountains, ranging from 0.74 to 0.49. These findings suggest that the cortlandtite represents cumulative facies from the magma of high-alumina basalt composition.

TANG, Chunfa, Heilongjiang Geoexploration Corporation, CNNC, Harbin, China

On Characteristics of Diwa-Type Gold Deposits in Northeast China

Gold deposits formed during diwa stage are defined as diwa-type gold deposits, which are controlled by various tectonisms, such as tecto-deformation, tectno-magmatic activity and tectno-metamorphism. The gold deposits are not only products of diwa stage, but formed during inheriting various stages of geological evolution history in pre-diwa stage, then remigrated, accumulated, concentrated and located after reformation of diwa movement. Resources of diwa-type gold deposits in northeastern China are distributed extensively, the deposits are usually existed at the edge of deep-seated fracture and the contiguous position of Mesozoic diwa faulted basin. The metallogenic epoch is mainly associated with Yanshanian magmatic activity, in which the tectono-magmatic activity in Late Jurassic—Cretaceous was the height of gold mineralization, and the metallization is generally characterized of "polygenetic compound-type." The author divided diwa-type gold deposits into three classifications in accord with geology, tectonics, magmatic activity and metallization of gold deposits in northeast China:

1) *Porphyry gold deposit associated with volcano—subvolcano (magma) activities:* This type of deposit generally occurred in igneous plug, stock, dyke and sill, most of the deposits are network-vein or irregular diffusive disseminated ore bodies. Gold-bearing zone is closely related to albite rock, albite quartz porphyry and quartz porphyry and so on. This is because that the porphyries or their nearby rocks are easy to break and also easy to be infiltrated by gold or other minerals, supplying ore bodies with existing places. For the gneiss of various porphyries in gold-bearing zone, both are formed from magma and from metasomatism or diffusion in solutions. Under the frequent tectono-magmatic activities, ore-bearing porphyries intruded into gold-rich old strata (ore source bed), and made them reactivate, remigrate and relocate with gold in ore-bearing porphyries, finally concentrated to industrial ore body in the certain position, which type of deposit being a massive scale and shallow bury. Profitable compositions are mainly gold, silver, the grade is lower, but easy to mine and dress. The main prospecting guide is to the edge of diwa basin in which volcanic (magmatic) activity has been intense since Mesozoic and the bottom wall (20–40 km) of deep-seated fracture, it is the best target to find out large scale porphyry-type gold deposit.

2) *Broken altered rock(zone)-type gold deposit associated with tectono-magmatism(fracture):* The type of deposit usually existed in structure broken and shear zones. This is because that during forming great fracture and shear zone in diwa stage, carbon dioxide, water and sulfur are up to high concentration, at the same time, chemical equilibrium of various minerals was intensely changed, chloritization, carbonization and pyritization in a large number of volcanic rocks were taken place, leading to separate out SiO_2, K, Ca, Fe, Au, Ag as well as other metal elements contained in altered rocks. These separated compositions and elements, with the elements which entered into shear zone from wall rocks by the diffusion, migrated laterally and vertically into the joint position of shear zone and other structural position (contoured zone or crack) "two level" low-pressure expand zone. In these places, quadric reaction caused by low-pressure and low chemical potential resulted in precipitations of quartz, carbonate and nature gold, forming common gold—quartz—sulfide deposits which are large size, formed wide ore field of gold, ore bodies occurred in groups, high grade and more associated compositions. The prospecting guide is mainly to the region in which regional magmatic evolution and metamorphism were intense, structure movement was frequent and there was certain ore source bed since pre-geosyncline.

3) *Quartz polyvein-type gold deposit associated with metamorphic thermal:* Migmatization generally occurred in gold-bearing quartz lenticle and silicified zone of crack, fracture and shear zone. Most of gold in deposit existed in pyrite, only a few are in magnetite and other various minerals, the gold is distributed along the edge of rocks' grains and various kinds of structural break faces (i.e., microcracks), when suffering regional metamorphism or igneous intrusion, a series of complex reflects would be happened, that is once gold is free from trammel state of rocks, it could be migrated by material transportation and diffusion (or combined with them), under the high temperature and high pressure, gold and associated gangue elements were dissolved (activated) from bury rocks, at the mean time, in the case of gas drive pressure, gold and other elements in solution were along fracture system to go up, and in the expand zone of temperature and pressure being markedly dropped, quartz, gold and sulfide were deposited; or by atomic, ionic and molecular diffusions, various kinds of compositions (SiO_2, CO_2, H_2O, S, As, Cu, Ag, Au, etc.) are in gas state or in ionic and molecular states, migrated along rocks' pores or grain borders of minerals and other break faces, finally deposited to form gold deposit under suitable condition (temperature, pressure and medium), the type of deposit is large scale, more associated profitable compositions, but it is complicated in gneiss, and deeper bury. The prospecting guide is approximately similar with the second one above.

TANG, Tianfu, Nanjing Institute of Geology and Palaeontology, Nanjing, China

Carbonate Deposits and Depositional Environments During Early Tertiary in East China

In East China, including Shandong, Jiangsu and Guangdong provinces, Early Tertiary strata are well-developed. They consist chiefly of lacustrine terrigenous deposits, containing red and dark argillaceous, dark calcareous shale, red siltstones and sandstones, carbonate rocks and evaporites, generally up to more than 2000 m in total thickness. The carbonate rocks are mainly composed of various types of limestones such as algal limestones, polychaetic boundstones, stromatolitic boundstones, oncolitic limestones, pellopackstones, biograinstones, algal ooidograinstones, micrite root mold limestones and caliche, about 10–200 m or more in total thickness. In various kinds of the carbonate rocks mentioned above, worm disturbing and burrowing traces can be usually observed, especially in biolimestones. The above mentioned characters show that the organic action was most important for the origin of carbonate rocks accumulated under a very shallow water environment and sometimes exposed above water level due to the climatic changes.

Recently, the sedimentary environments of these carbonate rocks are argued, because of the appearance of both continental fossils (ostracods, charophyte and gastropods) and marie fossils (foraminifers-Milliolids, Discorbidae, Nonionidae and Spirillinidae, worm tube-typed calcareous tube like Serpula, algae and stromatolite) in them.

Based on the studies of palaeontology, sedimentary petrology, mineralogy and chemistry, it is considered that in Early Tertiary carbonate rocks were formed in the lakes with the influence of the sea-transgressions. The carbonate rocks, especially algal limestones, polychaetic boundstones and roof mold limestones associated with dark argillaceous rocks are suitable for reservoiring oil and gases.

REFERENCES

Tang Tianfu et al., 1979, Science Press, Beijing, China: Late Tertiary biolimestones and their stratigraphic significance in Eastern Coast, China, (in Chinese), Proceed. of Symp. on Mesozoic–Cenozoic Red strata in S. China, p. 182–189.

———, Xue Yaosong, et al., 1980, The characteristics of carbonate rocks and their sedimentary environment of the Lower Tertiary Buxin Group in Shanshui, Guangdong, Acta Geologica Sinica, v. 54, n. 4, p. 249–270.

———, ———, and Yu Congliu, 1984, Carbonate rocks of the Second Formation of the Tertiary Funing Group in Jinhu Area, N. Jiangsu, Bulletin of Nanjing Inst. Geol. & Paleont., Acad. Sinica, v. 7, p. 165–187.

Shuai Defu et al., 1988, Tertiary lacustrine deposits. Shandong Province, Guidebook excursion B-3, IAS, International Symposium on sedimentology related to mineral deposits, Beijing, China, 1988.

TANIMOTO, T., California Institute of Technology, Pasadena, CA

How Much Do We Know about Global Scale Earth Structure?

Knowledge of the interior structure of the earth has advanced significantly in the last five or six years due mainly to the high-quality digital seismic data. Various studies, using body and surface waves and free oscillations of the earth, have clarified three-dimensional structures at different depths. Any structure study, however, is limited in resolution from the wavelength of waves that are used to probe the structure. In this aspect, models obtained by body waves, for example, represent different scale of averaging from those obtained by surface waves or free oscillations and should contain more details in the results. However, it is very common to find that various results are interpreted as though they give point estimate of seismic velocity in the three-dimensional earth. In this paper, we review how each type of data gives us information of earth structure and how it should be interpreted in terms of its resolution and error.

One of the most robust features that emerged in the three-dimensional structure study is the near-surface velocity variations that correlate with surface geology. Fast shear wave velocities are found in old continents such as shields in Canada, Eurasia, Africa, Antarctica, and western Australia and also in old ocean such as western Pacific. Resolution analysis shows that they represent averaging over a thousand kilometers and give only the very large wavelength features in the earth structure. Shear wave velocity variations in the results are about $\pm 4\%$ with estimated errors(1σ) of $\pm 1\%$. We can thus say that we have just begun to see the three-dimensional structure but the pictures we obtain are not quite focused yet.

The above results were obtained for the global scale structure that is difficult to get because earthquakes occur in restricted regions and also seismometers are not evenly spaced. Note that we do not yet have many seismometers in the oceanic area. If we restrict the region for study, however, we could get a better coverage that should lead to better resolution in the velocity maps. One such region is the Pacific ocean. There are not many seismographs in the Pacific ocean yet, but there are many earthquakes and seismometers in the surrounding regions. Because of these, we can get better results for the Pacific ocean than the global scale structure. The figure shows the resolution for the Kuril region from such a study. In general, we can now obtain resolution kernels with about 1000 km width in the Pacific ocean, which is to say that we can obtain the structure averaged over 1000 km. These results are for phase velocity of Rayleigh waves with the velocities ranging $\pm 3\%$ and the errors ranging $\pm 0.5\%$.

Below the depth of 1000 km, surface waves do not give us much information. Structures in the lower mantle and the core have been obtained from body waves and free oscillations. Among the body wave studies, studies based on ISC (International Seismological Center) catalog are hard to judge because we cannot see the raw data and it is thus impossible to obtain resolution and error in a rigorous sense. We will take long period waveform study of body waves and also free oscillation study to discuss the limit of resolution we are currently obtaining. In general, body wave study gives us a few hundred kilometer averaging and free oscillation study gives us a few thousand kilometer averaging. Obviously, free oscillation provides very low resolution, but we can view it as a long wavelength constraint on the structure in the future high-resolution study. Because of its large averaging nature, free oscillation gives us high-quality results in terms of both precision and accuracy.

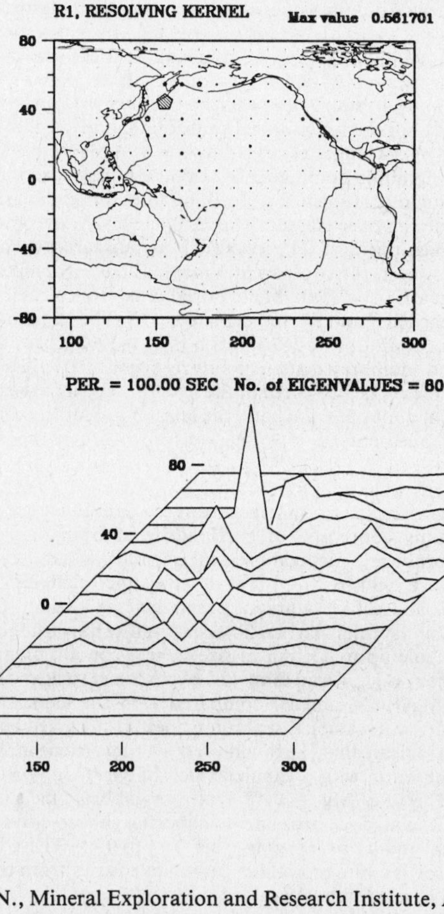

Figure 1.

TARHAN, N., Mineral Exploration and Research Institute, Ankara, Turkey

Genesis, Age and Nomenclature of the Granitic Rocks

In the field, the high grade metamorphic rocks are gradually transitional to the granitic rocks by a migmatitic zone. Within the granitic rocks, the relict crystal aggregates and remnants of the metamorphic rocks (1) which are transitional to them are similar to the xenoliths, (2) but actually they are not.

The mineral paragenesis observed in the almandine-amphibolite facies (Table 1), are stable minerals at the high T/P conditions. These minerals stable at higher pressure lose their stability with respect to the amount of decreasing pressure. They either dissolve in situ, or transitionally turn into stable mineral paragenesis under these conditions (Table 1). As the amount of the pressure decreases, both garnet, disten, sillimanite and andalusite, cordierite gradually turn into other minerals. By the way, the amount of the felsic mineral increase. Due to decreasing pressure, either from transformation of minerals by the reactions in solid state or mineral dissolution formed by sudden decreasing pressure, formed the recrystallization of the minerals stable at low-medium P/T conditions in the idiomorphic pattern. Because, in this stage the directional pressures are disappeared. During decreasing the P/T conditions, the recrystallization of the dissolved minerals result in growing larger crystal grains. As a result of these event cycles, the genesis of the holocrystalline magmatic textured granitic rocks come to existence.

According to the petrographic data, as indicated in Table 1, a number of minerals in the green schists facies at high P/T conditions are assembled and turned to make fewer but stable basic and ultramafic minerals. As a result of the decreasing pressure, these minerals are again resolved to the mineral paragenesis from which they are formed (by reversible reaction). During decreasing pressure, the basic and ultramafic minerals are reduced due to gradual transformation of them to the acidic and felsic minerals. Metamorphic rocks are gradually transformed to the granitic rocks by recrystallization involving reactions in solid state, in situ without passing through any magmatic stage. At the same time, the granitic rocks are also gradually transitional with the ophiolitic rocks in the field. As explained above, the ophiolitic rocks could form the granitic rocks by indicating the transformation-cycles of the mineral paragenesis appeared indifferent P/T conditions. The granitic rocks formed no ore mineralization, recrystallization, intrusions and contact metamorphism in the surrounding formations.

The existence of the granite rocks, carbonatite and feldspathoids together is an unsolved problem in literature and still under question. Granitic rocks indicate a heterogeneous structure. In this structure, the feldspathoic alkali rocks (unsaturated insilica) and granitic rocks (saturated in silica) are alternated, gradually transitional, wedge and lens-shaped with each other. These features insist that, the granitic rocks were not passing through any magmatic stage. Because the rocks have undergone any magmatic stage loose their heterogeneous structure due to formation of a homogeneous magma. The differentiation of a homogeneous magma produces granitoids which are homogeneous structure and saturated insilica (Tarhan, 1986). This type of the magma forms at high T/P conditions. The feldspathoids (e.g. leucite) are not stable at higher pressures. In the literature, it is claimed that, the feldspathoic minerals are formed either in the intrusions of the granites which passing through a magmatic stage or in a calcareous syntexis by magma (Bowen, 1928; Larsen, 1942; Huang, 1962; Streckeisen, 1979). Feldspathoids are unsaturated in silica. However, the feldspathoic rocks and quartz-feldspar rich granitic rocks are as one within the other and gradually transitional into each other in the field. Hence, the formation of them together is not possible from an acidic or basic magma. The mineral paragenesis form the granitic rocks (Table 1) reflect low-medium P/T regional metamorphism conditions.

Consequently, granitic rocks and related migmatites are derived by granitization (3) of the already present rocks (oceanic, continental and island-arc originated) in situ, solid state reactions.

Age of the retrogressive regional metamorphism which formed granitic rocks, assign the age of them.

The granitic rocks which the granitization is not perfect should be named according to the either primary recognized rock types (field and petrographic features) or to the metamorphic equivalent of them. The name of the primary recognized rock must be written after "granitic" term (e.g. granitic gabbo, granitic limestone, etc.) or should be used by writing granitic term first and name of the metamorphic rocks later (e.g. granitic amphibolite, granitic gneisse and granitic schist). Because the granitic rocks indicate the features of the metamorphics rather than magmatics, the rocks in which granitzation is perfect, the nomenclature have to be done according to their mineralogical compositions as in the granitoids (Streckeisen, 1967) (e.g. lmonzonite, foid syenite).

References

Bowen, N. L., 1928, The evolution of the igneous rocks: Princeton N.J. University Press, 325 p.

Huang, W. T., 1962, Petrology: McGraw-Hill Book Co., Inc., 480 p. New York.

Larsen, E. S., 1942, Alkalie rocks of Iran Hill, Gunnison Country: U.S. Geological Survey, Professional Paper, 197A, 1-64.

Streckeisen, A., 1967, Classification and nomenclature of igneous rock: N. Jb. Mineral., Abh., 107, 144-240.

Streckeisen, A., 1979, Classification and nomenclature of volcanic rock, lamprophgnes, carbonatites and melilitic rocks: Geology, 7, 331-355.

Tarhan, N., 1986, Evolution and origin of granitoid magmas related to the closure of Neo-Tethys in Eastern Taurus, Turkey.

Tipes of metamorphism	PROGRESSIVE METAMORPHISM (Barrowian type)			REGRESSIVE METAMORPHISM (Abukuma type)	
T/P	Increasing			Decreasing	
Rock types	METAMORPHITE		MIGMATITE	GRANITIC	
Metamorphic facies	Greenschist	Almandine-amphibolite	Almandine-amphibolite Kordierite-amphibolite	Kordierite-amphibolite	Greenschist
Mineral paragenesis	Sericite+Calcite +Chlorite+Epidote +Albite+Quartz +Actinolite ∓Plagioclase ∓Hornblende ∓Opaque mineral	Plagioclase+Hornblende +Sillimanite+Garnet ∓Pyroxene∓Kyanite +Mica∓Opaque mineral	Plagioclase+Zoned Plagioclase+Hornblende ∓Pyroxene∓Sillimanite +Andalusite∓Garnet +Cordierite+Orthose +Quartz+Mica∓Biotite Pyroxene∓Resolved opaque mineral	Andalusite+Cordierite Quartz+Orthose +Residual Plagioclase ∓Zoned Plagioclase ∓Zoned Pyroxene ∓Residual Pyroxene +Blue-green hornblende ∓Residual pyroxene having biotite ∓Feldspathoid ∓Resolved opaque mineral	Quartz+Orthose+Albite +Microcline+Epidote +Calcite+Sericite +Chlorite+Muscovite +Talc+Serpentine +Actinolite+biotite ∓Feldspathoid+Tremolite +Phlogopite +Browen-green hornblend∓Residual opaque mineral
Texture and structure	Metamorphic		Metamorphic and Magmatic	Magmatic	

Table 1. Mineral paragenesis and rock types produced by progressive and retrogressive metamorphisms following each other.

TARUNA, M. F., Padua, Italy

Alimentary Chains from Triassic to Recent

The alimentary chains from the Triassic to the Recent are represented in six paleogeographic maps.

The top of the feeding sequence varies with the geologic time: It is given by amphibians, reptiles, and then mammals (man included).

The geographic distribution and the stratigraphic position of fossil plants and animals are the base of the reconstruction.

The original drawings are in color.

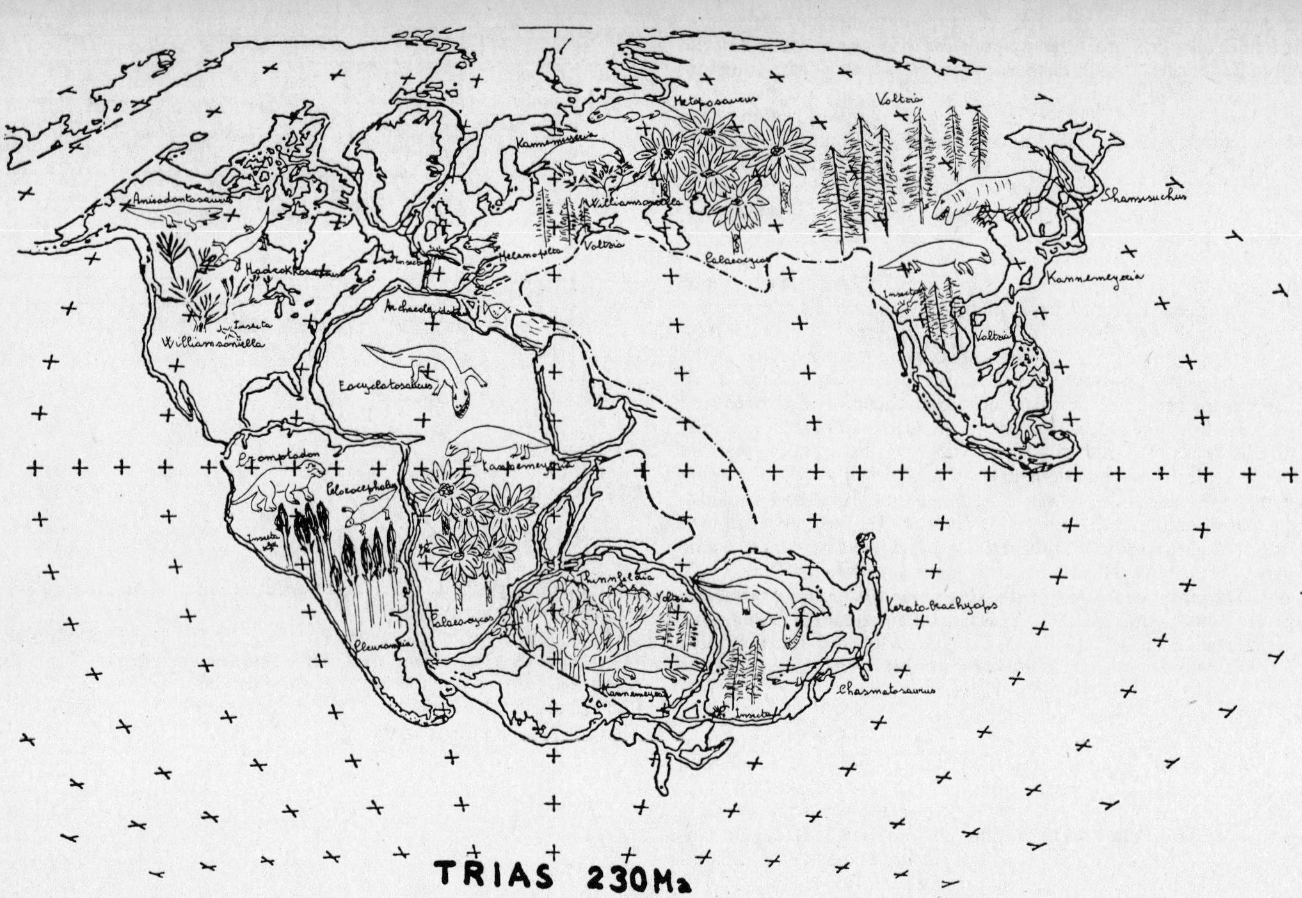

Figure 1.

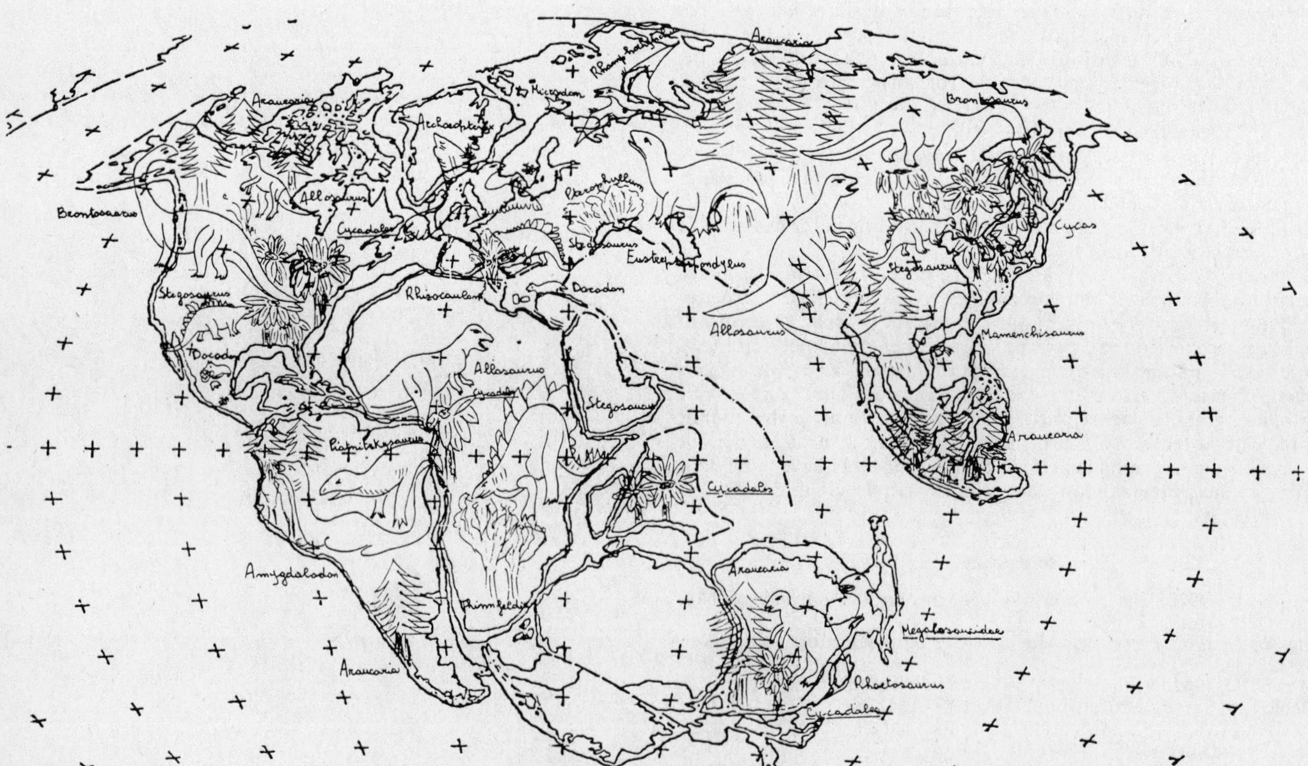

Figure 2.

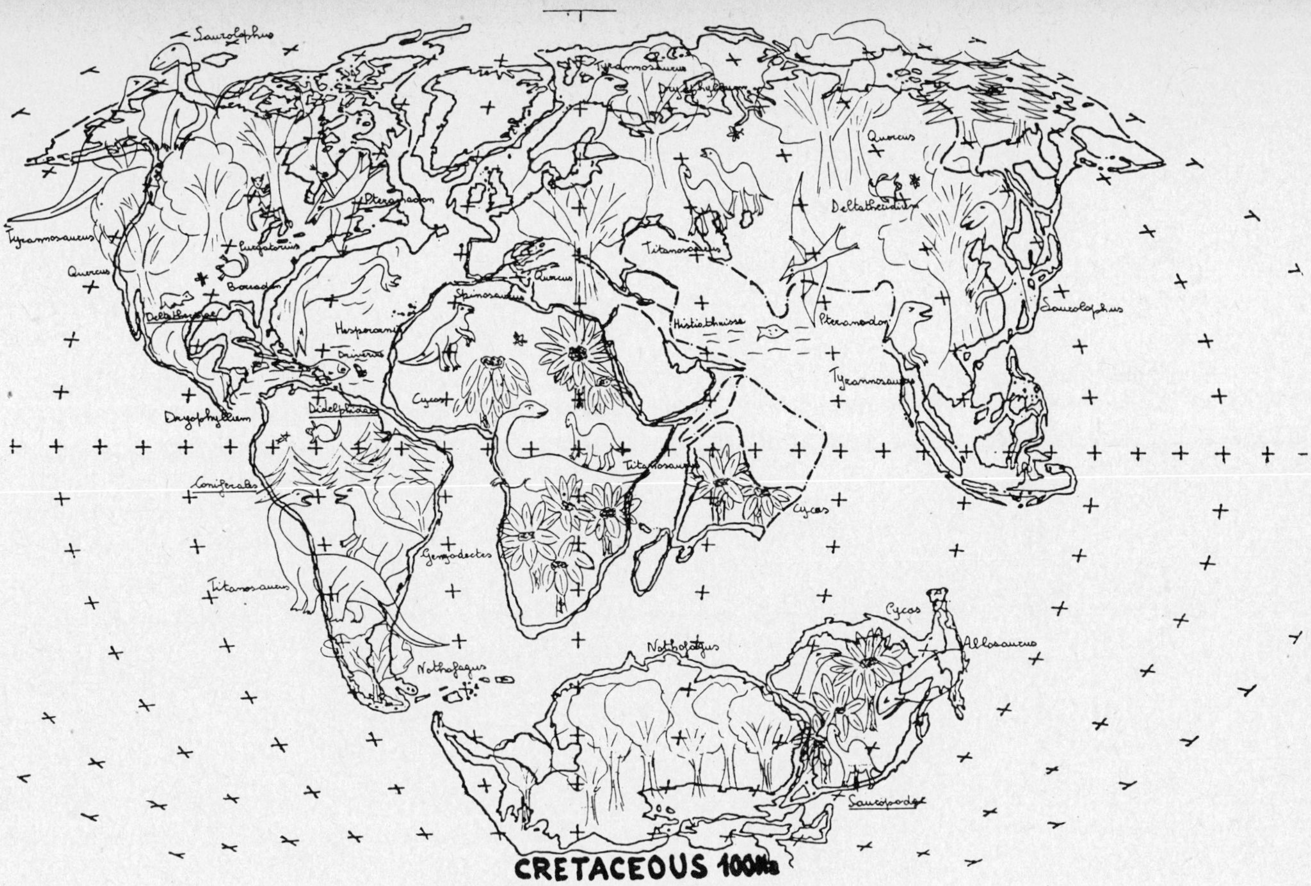

Figure 3

Figure 4.

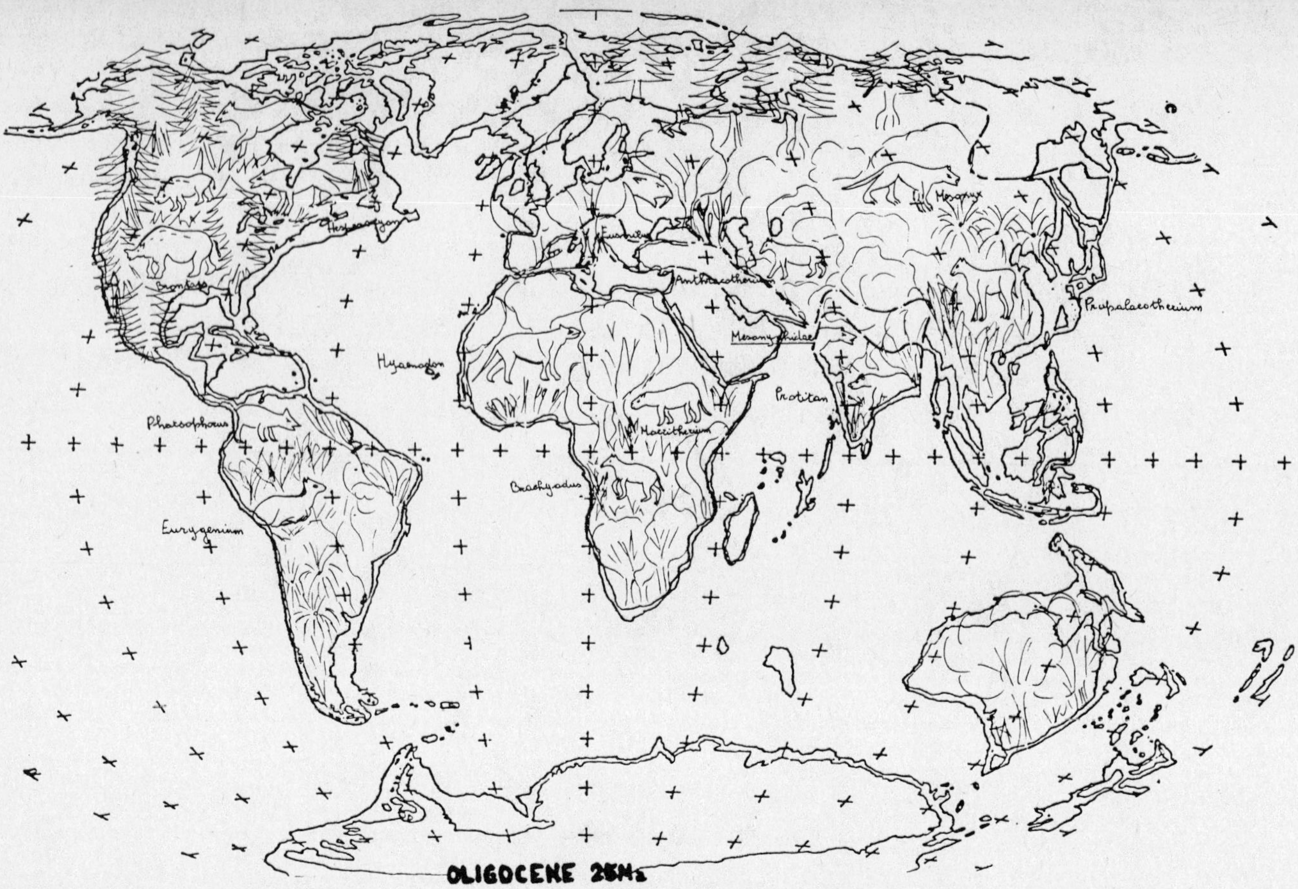

Figure 5.

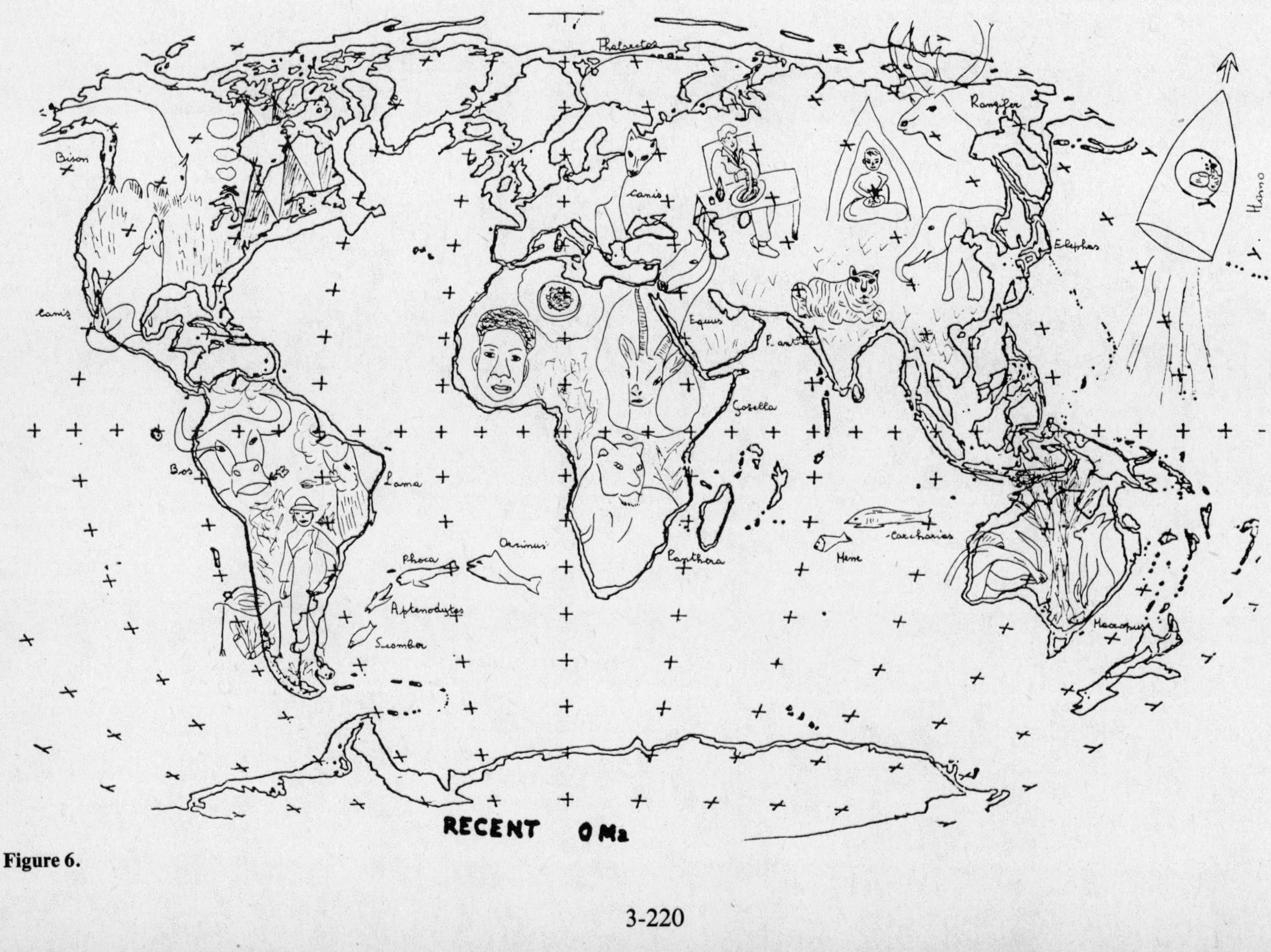

Figure 6.

TASSINARI, C. C. G., W. TEIXERIA, and O. S. JUNIOR, University of São Paulo, Brazil

Archean Early Proterozoic Crustal Evolution of Amazonian Craton, Brazil

This review deals with the interpretation of the main geochronological results obtained during the activities of the IGCP Project 204 (1983–1987), with emphasis to the study of the Archean and Early Proterozoic continental crust of the Amazonian Craton, Brazil.

The Amazonian Craton (Fig.) located in the north of South America, is a large stable unit which acted as a stable foreland in Late Proterozoic times. In a few words, this unit is composed by an Archean cratonic domain (Central Amazonian Province) which is surrounded by mobile belts of Early- to Mid-Proterozoic ages: Maroni-Itacaiúnas (2.25–1.9 Ga); Rio Negro-Juruena (1.75–1.65 Ga); Rondonian (1.55–1.3 Ga) and Sunsás (1.25–0.95 Ga).

The present geochronological knowledge of the investigated area is mainly based on Rb-Sr whole rock isochrons and K-Ar mineral determinations, although Pb-Pb whole rock isochrons, U-Pb Concórdia zircons ages and Sm-Nd data are also available for some areas. Concerning the Archean and Early Proterozoic ages, these are found within the Amazonian and Maroni-Itacaiúnas (MI) provinces, in the north-eastern portion of the Craton.

The Central Amazonian (C.A.) province, which is surrounded to the north an northeast by the Maroni-Itacaiúnas province, comprises granodioritic-throndjemitic granite-greenstone assemblages in association with high-grade metamorphic terranes. The Archaen province is partially overlain by unfolded and unmetamorphosed volcanic-sedimentary sequences (1.9–1.5 Ga old), as well as intruded by cratogenic-type granites (1.9–1.3 Ga old) and mafic dykes.

The available geochronological data on metemorphic terranes from the C.A. province, are concentrated in the Carajás ridge (Fig.). According to them, three rock-forming events are identified for the Late Archean time (3.1–2.5 Ga). The most representative results are in the 2750–2650 Ma time interval, and the younger (2.5 Ga) seems to be related to partial melting process of pre-existent continental rocks as suggested by high values of the Sr initial ratio (0,7072).

So, the primitive continental crust in the Carajás area was probably formed about 3.1 Ga ago, but the main tectonic-magmatic event occurred around 2.65 Ga ago, just after the supracrustal assemblages that possibly developed along the period 2.8–2.7 Ga.

The Maroni-Itacaiúnas (M.I.) province is characterized by a large exposure of metavolcanic and metasedimentary sequences, deformed and metamorphosed at greenschist to amphiobolite faces, as well as by granulites and gneissic mesozonal terranes.

The geochronological pattern of this province is very complete, including U-Pb, Rb-Sr, Sm-Nd and Pb-Pb radiometric data. In general, these ages are very concordant along the province, so demonstrating that its evolution took place during a major event of the Early Proterozoic (Transamazonian Orogeny; 2.2–1.9 Ga). Indeed, Rb-Sr, Pb-Pb and Sm-Nd isochrons on granulitic and gneissic rocks indicated ages in close agreement, around 2.0 Ga, with coherent $^{87}Sr/^{86}Sr$ i.r. (0,7018–0,7024); Pb μ_1 values (8,09) and $\epsilon_{Nd} = +2{,}1$. In special, the Nd model age (TDM) suggests that the primitive crust was separated from the upper mantle about 2.25 Ga ago. Therefore, according to the geological control, the coherent structural trends, and the agreement of radiometric ages, a model based on a mantle-derived magmatic arc can be considered for a significant part of the M.I. province.

In turn, remnants of some old basement (Archean) rocks within the Maroni-Itacaiúnas belt have been identified. They generally consist of high grade polimetamorphic rocks, such as the Imataca Complex in Venezuela and the Amapá region in Brazil, where radiometric ages (around 2.9 Ga) were found. In the southernmost portion of this province, below the Amazonas phanerozoic sediments, Rb-Sr ages obtained on mesozonal rocks yielded results around 2.9 Ga, but with high values for the Sr initial ratios (0,710). These data are suggestive of reworking of pre-existing continental crust during the Transamazonian Orogeny, so indicating that part of the M.I. province was also of ensialic character.

In the same way, other nuclei of old basement rocks occur within the Amazonian Craton, such as in Bolivia (Lomas Manèches granulites) and in Brazil (Jauru-Mato Grosso area). In both areas, Early Proterozoic and Archean ages were found, although not well established analytically. Such nuclei include many kinds of migmatites, granlites and greenstone belt sequences, and probably represent an old fragment added to the Amazonian Craton during its Mid-Proterozoic evolution (around 1.5 Ga).

In conclusion, based on the integrated isotopic, geophysical evidences and geological informations it is possible to make an estimative of the Archean and Early Proterozoic continental crustal growth for the Amazonian Craton. In our view, considering the whole primitive cratonic area, about 45% of the crust was differentiated during the Archean time, followed in the Early Proterozoic by more 15% of continental crust accretion. However, the Early Proterozoic evolution was also responsible by the reworking of significant part of the Archean domains. Later on, during the Middle Proterozoic evolution of the Amazonian craton, the estimated growth of the crust was approximately 40%.

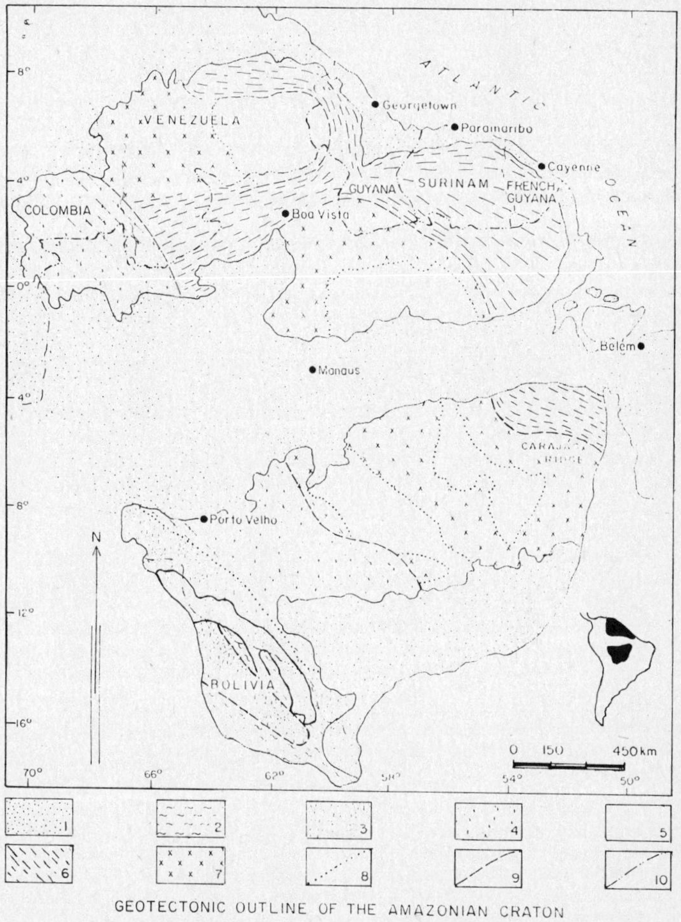

Figure 1—Geotectonic outline of the Amazonian Craton. Key: 1. Phanerozoic sediments; 2. Sunsás mobile belt (1.25–0.95 Ga); 3. Rondian mobile belt (1.55–1.3 Ga); 4. Rio Negro-Juruena mobile belt (1.75–1.6 Ga); 5. Proterozoic platform covers (1.9–1.5 Ga); 6. Maroni-Itacaiúnas mobile belt (2.25–1.9 Ga); 7. Central Amazonian province (> 2.3 Ga); 8. Transition zone between belts; 9. Approximate contacts between belts; 10. Country limits.

TAYLOR, A. E., and A. S. JUDGE, Geological Survey of Canada, Ottawa, Canada

Terrestrial Heat Flow, Canadian Arctic Archipelago: Thermal Conditions in Arctic Sedimentary Basin and Surrounding Platform Region

Precise ground temperatures to depths ranging between 500 and 1100 m have been measured over the past two decades at 40 suspended petroleum exploration wells in the Canadian Arctic Archipelago, 75 °N–81 °N. While the data provided immediate and timely information on the variation in permafrost thicknesses across the region (e.g., Judge et al, 1981), the data set has also been used to study the terrestrial heat flow and the underlying thermal regime.

Permafrost dynamics and the effects of recent shoreline emergence due to uplift complicate the thermal analyses at many of these onshore wells. The effect on the measured temperature profile of permafrost aggradation is frequently so pronounced that traditional analyses are unable unambiguously to separate the transient disturbance from the background, or steady state regime. However, in 1985, an offshore well in the central Sverdrup Basin, where such transient environment effects are minimal, was instrumented to 800 m depth with an automatic well temperature acquisition system. Detailed analysis of the latter data (Taylor et al, in press) has established with some confidence the terrestrial heat flow at this well: 39–52 mWm^{-2}.

The above range embodies the residual and unavoidable uncertainty in thermal properties of the sandstone and shale formations penetrated by the well. Detailed analysis (Taylor et al, in press) of the precise temperature data from this offshore well and the laboratory conductivity measurements on well cuttings provides strong evidence for the thermal anisotropy of the shale formation penetrated by the well: thermal conductivity parallel to bedding may be up to 2.6 times that perpendicular to bedding. Hence, the range in heat flow values has been calculated using the measured thermal conductivities of the sandstone formation (argued as a lower limit, 39 mWm^{-2}) and, for comparison, using median values of conductivity quoted in the literature (giving 52 mWm^{-2}, considered an upper limit unlikely to be exceeded). The range is supported independently by an analysis of bottom-hole and drill-stem test temperature data from deeper in the well (Taylor et al, in press).

The analysis resulting from this unique offshore instrumentation has been crucial in constraining similar analyses at the onshore wells. At half of these wells where transient effects are relatively small, it has been possible to calculate the terrestrial heat flow; values lie generally in the range 50–70 mWm^{-2}, both in the Mesozoic Sverdrup Basin and the surrounding Arctic Platform. At the remaining wells, temperature profiles show large effects of recent shoreline emergence, and the heat flows established above have provided the crucial boundary condition necessary to model their recent geomorphic history.

The upper value of the range of heat flow, 70 mWm^{-2}, is similar to the world average heat flow for continental areas that experienced Mesozoic to Tertiary orogenies, while the lower limit and the heat flow at the offshore well are more typical of shields. It appears that little heat remains in this region from orogenic events that terminated largely in the Tertiary. There is some evidence, comparing the heat flow calculated within and below the permafrost, that the deep permafrost observed in this region is currently aggrading.

These results complement other studies on the geothermal regime undertaken in the region. A somewhat lower range of heat flow, 40–60 mWm^{-2}, is reported for the same study area from an analysis on less precise, but much deeper industry bottom-hole temperature data (Jones et al, in press). Terrestrial heat flow values for the Canada Basin, Arctic Ocean to the northwest of the Archipelago are 50–57 mWm^{-2}) (Lachenbruch and Marshall, 1969), while higher values (mean 73 mWm^{-2}) have been reported along the northwestern continental margin (Langseth et al, 1988), somewhat north of the present study area.

REFERENCES

Jones, F. W., J. A. Majorowicz, and A. F. Embry, in press, Preliminary estimates of the deep heat flow along a regional profile across the Sverdrup Basin, Canadian Arctic Islands: Geophysics.

Judge, A. S., A. E. Taylor, M. M. Burgess, and V. S. Allen, 1981, Canadian geothermal data collection—northern wells 1978–80: Geothermal Series #12, Earth Physics Br., EMR Canada, 190 p.

Lachenbruch, A. H., and B. V. Marshall, 1969, Heat flow in the Arctic: Arctic, v. 22, p. 300–311.

Langseth, M. G., A. H. Lachenbruch, and B. V. Marshall, 1988, Constraint on the evolution of the Amerasia Basin from seafloor heat flow: Abstract, Geological Association of Canada Annual Meeting, St. John's, NF, v. 13, p. A71.

Taylor, A. E., A. S. Judge, and V. S. Allen, in press, The automatic well temperature measuring system installed at Cape Alison C-47, Offshore well, Arctic Islands of Canada. 2. Data retrieval and analysis of the thermal regime: Journal of Canadian Petroleum Technology.

TAYLOR, E. L., and T. N. TAYLOR, Ohio State University, Columbus, OH

Silicified Triassic Peat from Antarctica: Geologic and Biologic Implications

Silicified plants occur within a peat-like deposit in a col north of Fremouw Peak in the central Transantarctic Mountains. Although the site was first collected in 1969–1970, the significance of the flora and the excellent level of preservation have only recently been realized. The peat occurs near the top of the Fremouw Formation in rocks that are believed to be Middle Triassic based on the palynoflora within the peat. Plant remains occur as permineralized blocks of peat that were apparently rafted into a sand stream channel. The presence of a paleosol at the base of the peat blocks suggests that they may have been part of a forested island that was undercut during a flood. Permineralization apparently took place rapidly and was enhanced by the dissolution of siliceous volcanic detritus that is abundant in both formations.

The flora in the peat is remarkable for its exquisite preservation of cellular detail, comparable in many ways to Carboniferous coal ball floras. This particular deposit is important, not only for the quality of preservation, but also because of the age of the deposit. There was a substantial floral change between the Triassic and the Permian in the southern hemisphere. Following the relative monotony of the Permian *Glossopteris* flora, the Triassic in Gondwana saw the rise of abundant diverse seed fern groups, as well as a number of vascular cryptogams and conifers. This period in geologic time is a significant one for plant evolution since many of the groups present during the Triassic have been implicated at one time or another as ancestors to the angiosperms. The discovery of a permineralized Triassic flora allows for the examination of many more characters that may have evolutionary significance than is possible with compression or impression floras. This is true despite the fact that permineralized peat deposits have been described as evolutionarily "conservative" deposits. These structurally preserved plants provide us with information about basic features of particular plant groups that would not otherwise be available or that cannot be obtained utilizing two-dimensionally preserved preservational types. The flora is highly diverse and dominated by members of the corystospermalean seed ferns, including *Dicroidium* fronds and reproductive structures that are comparable to *Pteruchus*. Also present are members of the Cycadales, filicalean and marattialean ferms, fungi, sphenophytes, and probable confiers. Reproductive structures are also well represented, including the *Pteruchus*-like pollen organ, dispersed seeds, and seeds contained within a cupule structure.

Like other examples of permineralized plant deposits, the Fremouw Peak flora appears to be a case of fortuitous preservation. Also like many deposits of terrestrial plants (as opposed to stromatolitic deposits), the excellent and rapid preservation of the plants can probably be ascribed to the proximity of volcanic detritus that resulted in a high concentration of silica within the groundwater. Although the peat in the col at Fremouw Peak is limited in this extent, additional silicified peat is present at the base of Mt. Falla, approximately 7 km away, probably also within the Fremouw Formation. In addition, a similar deposit of Permian age occurs 60 km southeast of Femouw Peak at Skaar Ridge.

TAYLOR, L. A., and C. R. NEAL, University of Tennessee, Knoxville, TN, and R. N. CLAYTON and T. K. MAYEDA, University of Chicago, Chicago, IL

Eclogite Petrogenesis by Ancient Subduction: Recognition of Crustal Signatures with Mantle

The origin of eclogite xenoliths in kimberlites and alkali basalts is at present the subject of much controversy. There are three contrasting petrogeneses proposed for these "mantle-derived" eclogites: (1) as high-pressure igneous cumulates (garnet pyroxenites) within the upper mantle (O'Hara and Yoder, 1967; MacGregor and Carter, 1970; Smyth and Caporuscio, 1984); (2) as metamorphic products of a subducted oceanic crustal protolith (Hemstaedt and Doig, 1975; Jagoutz et al., 1984; MacGregor and Manton, 1986; Shervais et al., 1988); (3) as relicts of the earth's primary differentiation shortly after accretion (e.g., Anderson, 1981; McCulloch, 1987). Basically, it is the differences in the progenitors (i.e., crust versus mantle) that will impart distinguishing chemical characteristics upon these rocks.

We have determined the mineral chemistry from 28 eclogite samples from the Bellsbank kimberlite (DeBryun and Martin Mine). All eclogites have experienced metasomatism, witnessed by interstitial phlogopite, amphibole, K-feldspar, and celsian. Nine of the freshest samples were chosen for "ultrapure" mineral separation (handpicking; $HC1/HNO_3/Hf$ acid leaches; K spikes) and whole rock (INAA and XRF) analysis. Isotope (Sr, Nd, O) and INA analyses were also performed on these separates. These data have allowed a three-fold classification of eclogites to be constructed (Table 1).

Smyth and Caporuscio (1984) have performed an extensive mineralogical investigation of the pyroxenes in eclogites from the Bobbejaan kimberlite (a different fissure of the Bellsbank kimberlite). Figure 1 shows the compositions of the Bobbejaan pyroxenes, along with 28 from our eclogites. Although three groups can be discerned, this correlation could be attributed to fractionation (Smyth and Caporuscio, 1984). However, the isotope and REE data, and whole-rock compositions, are inconsistent with such a petrogenesis.

Isotopes: Group B eclogites have ϵ Nd values of $\cong +200$, which are inconsistent with the high $^{87}Sr/^{86}Sr$ values of $\cong 0.710$ or the low $\delta^{18}O$ values of 2.8–4.1 ‰. Group C eclogites have similar $\delta^{18}O$ values (3.1–4.9‰). Group A eclogites exhibit ϵ Nd values of $\cong -15$, $^{87}Sr/^{86}/Sr$ of $\cong 0.704$, and $\delta^{18}O$ of 5.0–5.8‰, consistent with a mantle origin. Isotope ratios from Groups B and C are similar to *oceanic crust that has experienced high-temperature hydrothermal alteration* (Muehlenbachs and Clayton, 1972).

REE chemistry: The whole-rock REE analyses demonstrate the metasomatic contribution to the LREE (Fig. 2). Modal analyses were performed, paying particular attention to the alteration associated with the primary phases (e.g., garnet, cpx), and reconstructed whole-rock REE patterns were calculated. The quantitative differences between the measured and calculated REE patterns represent the part contributed by the metasomatism/alteration. The reconstructed whole-rock REE patterns show that three eclogite groups can be identified (Table 1 and Fig. 2). Note that *Group C eclogites display a distinct positive Eu anomaly.*

Whole-Rock Data: K and Ba abundances of the eclogites are directly related to the host kimberlite, which plots at the end of a positive correlation between these two elements. Elements not attributable to metasomatism (i.e., Zr, Nb, and Ta that have no correlation with K or Ba) demonstrate that three groups rather than a continuum are present.

Synopsis: The positive Eu anomaly, which is superimposed on the garnet, pyroxene, and whole-rock REE patterns of the Group C samples, is characteristic of plagioclase accumulations. The major element whole-rock chemistry is similar to a high-alumina basalt or gabbro and the $\delta^{18}O$ values are consistent with high-temperature hydrothermal alteration of a plagioclase-rich oceanic crustal component (e.g., anorthositic gabbro).

Group B eclogites also display features that are consistent with an oceanic origin. During subduction, the hydrous, spilitized, basaltic rocks would have undergone dehydration and possible partial melting. This would affect the resulting eclogite in two ways. First, the slightly LREE-depleted REE pattern of a MORB would become LREE depleted during dehydration. This depleted signature must have been isolated for at least 2 Ga in order to generate observed $^{143}Nd/^{144}Nd$ ratios. A recent influx of LREEs has decoupled the Nd isotopes [depleted $^{143}Nd/^{144}Nd$ ratio ($\cong$ 0.520) and an enriched Sm/Nd ratio ($\cong$0.2): Shervais et al., 1988]. Second, Na would be lost to some extent during the dehydration process, as reflected by less jadeite-rich clinopyroxenes in these eclogites. The Group B Sm-Nd data give model ages (relative to Bulk Earth) of 2.3–2.5 Ga. These represent the age of the final depletion experienced by these rocks. The highly radiogenic Sr in Group B eclogites could result from either derivation from Rb-enriched protolith or isotopic exchange with a radiogenic Sr reservoir, or both.

Kimberlite infiltration has affected both the LIL elements and isotope ratios of the entrained eclogites. The Bellsbank kimberlite is a Group II micaceous type. On the basis of whole-rock isotope data, no groups of eclogites can be distinguished as they all fall within the range of Group II kimberlites (Smith, 1983). Whole-rock Sm-Nd model ages range from 150 to 200 m.y., which is within error of the age of Cretaceous kimberlite magmatism in the Kaapvaal Craton. It is the mineral separates that yield kimberlite-free signatures.

Our eclogite investigation tends to support an oceanic crustal origin for Group B and C eclogites (cf., Shervais et al., 1988), but Group A eclogites are probably a high-pressure mantle cumulate (e.g., high MG's and $^{18}O \cong$ mantle values). The importance of the interpretation for subducted oceanic protoliths cannot be overemphasized, since it suggests: (1) the preservation of possible Archean oceanic crust beneath the Kaapvaal craton; (2) late Archean–early Proterozoic plate tectonics similar to those of the present day; and (3) the possible presence of a stratified mantle in this region due to underplating of the Kaapvaal craton by Archean oceanic crust.

REFERENCES

Anderson, D. L., 1981, Hotspots, basalts, and the evolution of the mantle: Science, v. 213, p. 82–89.

Helmstaedt, H., and R. Doig, 1975, Eclogite nodules from kimberlite pipes of the Colorado Plateau—samples of Franciscan-type oceanic lithosphere: Phys. Chem. Earth, v. 9, p. 95–111.

Jagoutz, E., J. B. Dawson, S. Hoernes, B. Spettel, and H. Wanke, 1984, Anorthositic oceanic crust in the Archean Earth: Lunar and Planetary Science XV, p. 395–396.

McCulloch, M. T., 1987, Sm-Nd systematics in eclogite and garnet perikitite nodules from kimberlites: implications for the early differentiation of the Earth: Proceedings of 4th Kimberlite Conference.

MacGregor, I. D., and J. L. Carter, 1970, The chemistry of clinopyroxene and garnets of eclogite and peridotite xenoliths from the Roberts Victor mine, South Africa: Physics Earth Planet. Inter. v. 3, p. 391–397.

———, and W. I. Manton, 1986, The Roberts Victor eclogites: ancient oceanic crust: Journal of Geophysical Research, v. 91, p. 14063–14079.

Muehlenbachs, K., and R. N. Clayton, 1972, Oxygen isotope studies of fresh and weathered basalts: Canadian Journal of Earth Science, v. 9, p. 172–184.

O'Hara, M. J., and H. S. Yoder, Jr., 1967, Formation and fractionation of basic magmas at high pressures: Sccotish Journal of Geology, v. 3, p. 67–117.

Shervais, J. W., L. A. Taylor, G. Lugmair, R. Clayton, T. Mayeda, and R. Korotev, 1988, Archean oceanic crust and the evolution of subcontinental mantle: eclogites from southern Africa: GSA Bulletin, v. 100, p. 411–423.

Smith, C. B., 1983, Pb, Sr, and Nd isotopic evidence for sources of southern African Cretaceous kimberlites: Nature, v. 304, p. 51–54.

Smyth, J. R., and F. A. Caporuscio, 1984, Petrology of a suite of eclogite inclusions from the Bobbejaan kimberlite: II. Primary phase compositions and origin, *in* J. Kornprobst, ed., Kimberlites II: The mantle and crust-mantle relationships: Elsevier, p. 121–132.

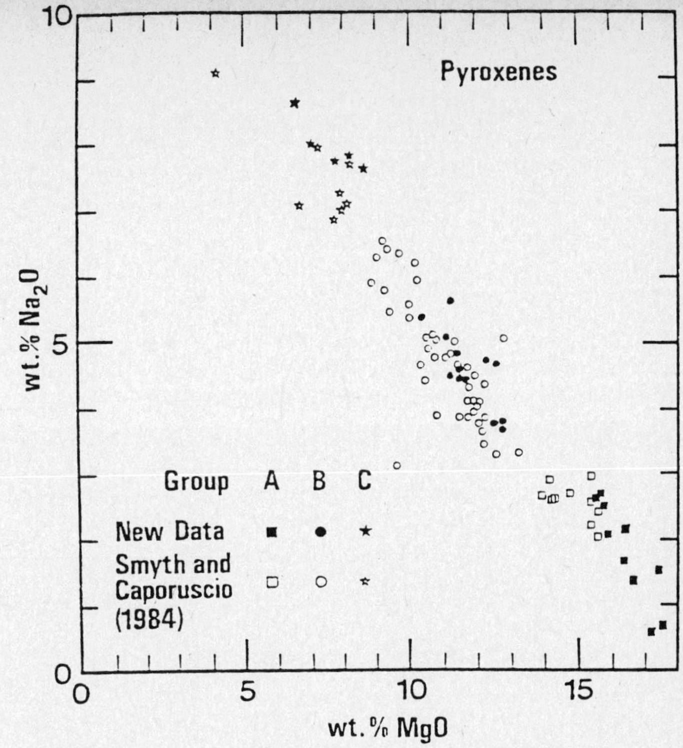

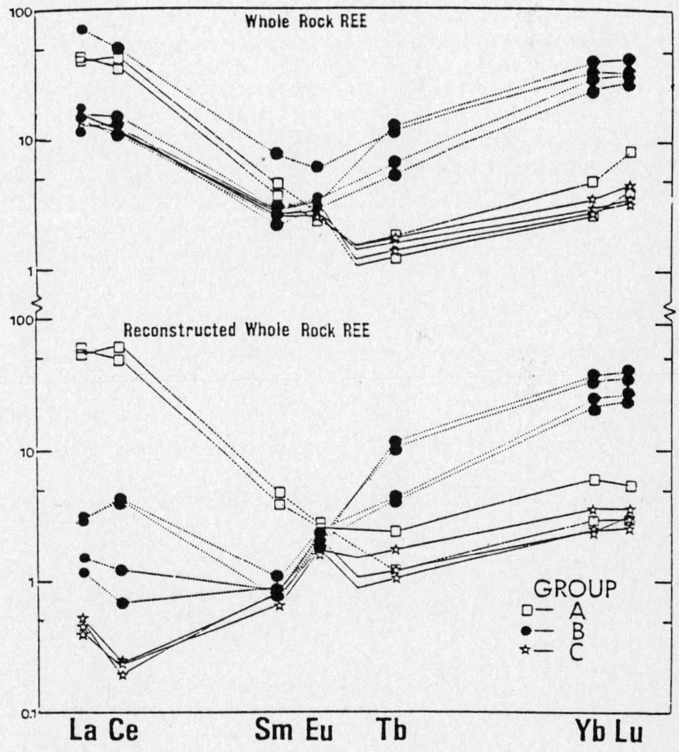

Table 1: Chemical characteristics of Bellsbank Eclogites.

	Group A	Group B	Group C
Clinopyroxene			
Na_2O	0.6-2.8	3.7-5.7	7.6-8.7
Al_2O_3	0.9-3.5	5.5-8.6	14.8-17.0
Cr_2O_3	0.3-1.3	<0.1	<0.1
MG#	90-96	86-89	90-93
Orthopyroxene			
MG#	89-93	----	----
Garnet	Mg-rich	Fe-rich	Ca-rich
Cr_2O_3	1.0-2.2	<0.1	<0.1
MG#	78-83	57-59	67-73
Whole Rock MG#'s	86-91	65-69	75-79
REE's	LREE-enriched HREE-depleted	LREE-depleted HREE-enriched	LREE-depleted HREE-depleted
Eu Anomaly	No	No	YES
Isotopes			
$\delta^{18}O$	5.0-5.8	2.8-4.1	3.1-4.9
ϵ Nd	-10 to -19	+120 to +235	no data yet
$^{87}Sr/^{86}Sr$	0.7042	0.7094-0.7100	no data yet

TAYLOR, S. R., Australian National University, Canberra, Australia

Geochemical and Cosmochemical Significance of Trace Elements and Isotopes in Meteorites

Meteorites present us with tantalizing evidence about the early solar system, our current understanding being aptly summarized by the comment made in 1492 about the fall of the Ensisheim (Alsace) meteorite: "Many know much about this stone, everyone knows something, but no one knows quite enough."

The composition of the C1 carbonaceous chondrites, even though they have been subjected to aqueous alteration, is close to that of the solar photospheric abundances for the nongaseous elements. These data give us our best estimates for the composition of the primitive solar nebula. The oldest reliably dated objects are the refractory calcium–aluminum inclusions (CAI) at 4559 ± 4 m.y., with an initial $^{87}Sr/^{86}Sr$ ratio (ALL) of 0.69877 ± 2. They also preserve isotopic anomalies, carrying a memory of presolar events. The CAI record multiple episodes of evaporation and condensation.

The ubiquitous chondrules are slightly younger and were formed by melting of preexisting dust in the nebula: Volcanic and impact origins for chondrules can be ruled out. Neither did they condense from a hot vaporized nebula; cooling times are too fast to allow hot nebula, which is ruled out also by the preservation not only of local isotopic anomalies in CAI, but also because of the ubiquitous oxygen isotopic variations: All would be homogenized in a hot nebula.

The chondrules are also depleted in siderophile and chalcophile elements, but not in volative elements such as Na. They are individually variable in elemental and oxygen isotopic composition, but their mean compositions for all meteorite groups are relatively uniform: The matrix accounts for the intragroup variations. The total range in ages from 1-Xe data appears to be about 5 m.y. They preserve rare evidence of CAI among their precursors, so that the CAI predate chondrule formation. Chondrules melted and cooled quickly (^{1}h).

The common H, L, LL, and E chondrites all have similar whole-rock ages, the best estimate being 4555 ± 4 m.y. (Rb decay constant 1.402 × 10^{-11}/y) and an initial $^{87}Sr/^{86}Sr$ ratio of 0.69885 ± 10. Sm-Nd ages for phosphates from chondrites yield 4550 ± 45 m.y. (the $^{143}Nd/^{144}Nd$ initial ratio for Juvinas = 0.506705.). The Pb-Pb and Rb-Sr ages give the time of depletion of volatile Pb and Rb relative to refractory U and Rb. All meteorite data (CAI, H, L, LL, E chondrites) give effectively the same age for this event, indicating that the volatile element depletion occurred close to T_o (4560 m.y.). Mineral isochons yield higher $^{87}Sr/^{86}Sr$ initial ratios and Rb-Sr ages consistent with the establishment of thermal equilibrium (mineral isotopic closure) for periods up to 200 m.y. compared to unequilibrated chondrites such as Tieschitz (H3). These periods and the degree of metamorphism (Types 4–6) do not correlate with cooling rates or 1-Xe closure ages; the whole concept of chondrite "metamorphism" remains to be evaluated.

The terrestrial planets (Venus, Earth, and Mars) exhibit low K/U ratios; the meteorite data establish that this volatile-refractory fractionation thus occurred effectively at T° and was not due to later separation in planetesimals or under planetary conditions. Probably the event is due to early intense solar activity as the sun settles onto the main sequence and clears the inner nebula of volatiles and water out to a "snow line" at about 5 A.U.

Igneous processing in asteroidal bodies began within a few million years of T_o. Angra dos Reis (ADOR) is an igneous meteorite with an age of 4551 ± 2 m.y. and an initial $^{87}Sr/^{86}Sr$ ratio of 0.69883 ± 2. Basaltic achondrites have an average crystallization age of 4539 ± 4 m.y., notionally only 20 m.y. younger than the oldest material dated by the Allende CAI. The basaltic achondrites are true basalts similar to lunar basalts. Clearly, however, the achondrite data reveal that planetary type differentiation was occurring on a small but extensive scale in asteroids very close to T°. Iron meteorites, perhpas the ultimate in differentiated objects, come from at least 60 separate bodies.

Meteorites are derived from the asteroid belt, which is zoned, with differentiated objects occurring in the inner belt, while primitive carbonaceous compositions dominate in the outer belt. The presently observed zones have been broadened by collisions from initial widths of perhaps 0.1 A.U. The asteroid belt represents a planetary accretion zone arrested at an early stage of development most likely by early growth of Jupiter.

Is the present population of meteorites representative of the building blocks of the terrestrial planets? There are many differences in K/U ratios, oxygen isotopes, noble gases, density, and other characteristics that rule out any of the known meteorite classes as having appropriate chemical or isotopic compositions to make them suitable source material for the inner planets. The chief significance of meteorites is in providing unique age information on early solar system processes and compositional information about the asteroid belt and the early solar nebula.

REFERENCE

Kerridge, J. F., and M. S. Matthews, 1988, Meteorites and the early solar system: Arizona University Press.

TAZHIBAEVA, P. T., A. P. SLUSAREV, and I. B. SAMATOV, Institute of Geological Sciences of Kasakh Academy of Sciences, Alma-Ata, USSR

Kinetics of Mineral Formation and Useful Minerals in Weathering Crusts

Mineral composition of weathering crusts zone and mineral complexes changing during crust formation were examined. In studying the transformational processes of rocks and crust development, we used thermodynamic and X-ray powder diffraction analysis of serpentines, chlorites, smectities, and kaolinites to determine a variation of composition, structure, and crystal perfection. While distinguishing separate stages of development and verifying margin zones, the results of structural and thermochemical mineral peculiarities such as coefficient of crystal perfection, cell parameters, polytype composition, and energetic parameters of thermal decay of weathering products were taken into account.

Analysis of thermochemical peculiarities of hypergine formations allowed us to determine the mechanism of crystallostructural alterations and stages of crust development. Sequence and zonation of mineral phase formation across the profile of weathering was verified. It is shown that the position of the margin zone is controlled by the intensity of evacuation of rock-forming and migration of ore components.

The regularities of distribution and accumulation of useful minerals were determined based on our research of stage transformation of mineral complexes. Concentration of nickel, cobalt, magnesium, aluminum, and iron related to the stages of weathering and structural state of mineral was established.

TEDESCO, L. P., and H. R. WANLESS University of Miami, Miami, FL

Biogenic Generation, Transformation and Destruction of Sedimentary Facies

Excavating burrowers are an extremely important influence on sedimentary sequences throughout the Phanerozoic. Yet analysis of influenced sedimentary sequences using traditional stratigraphic, sedimentologic, and paleoecological principles will produce misleading conclusions and paleoenvironmental interpretations. The following discussion examines the role of the burrowing mud shrimp, *Callianassa,* in modifying modern sedimentary sequences. The principles, however, are broadly applicable.

Callianassa and other deep, excavating burrowers rework shallow-marine sequences to depths in excess of 150 cm. Burrow excavation processes of *Callianassa,* coupled with subsequent catastrophic infilling of open burrow networks results in (1) obliteration of precursor sequences, (2) subsurface generation of false facies and stratification, and (3) production of preserved depositional fabrics and faunal assemblages that are unrelated to either the original or the surficial depositional conditions and fabrics. Initial application of these results indicates these integrated processes have been far more widespread and influential in molding modern and ancient sedimentary sequences than is currently recognized.

Excavation and textural segregation: The deep excavator *Callianassa* mines extensive subsurface networks of thumb-sized chambers. During excavation, it expels finer grained material (< 1-2mm) to the surface but concentrates encountered coarser material (> 1-mm) in subsurface side chambers. Through repetitive excavation, *Callianassa* may completely remove, and thus obliterate, all evidence of the previous sedimentary sequence except for the coarse skeletal fauna.

Infilling as depositional subsurface events: Open burrow networks can be catastrophically infilled by hurricanes. These burrow fills (tubular tempestites) consist of a slurry of surficial coarse material in addition to any coarse material packed in the chambers during the excavation process. Importantly, burrow fills contain fauna from both the sequence being excavated as well as the surficial facies. Tubular tempestites, resulting from one moderate hurricane, infilled all open burrow networks and accounted for 10% of the upper 75 cm of sequence. Where burrow fills are coarse, reexcavation should avoid fills and focus on finer grained host sediment. In areas of gradual sediment accumulation (< 1 m/100 years), the rate of hurricane passage is sufficiently high that tubular tempestites dominate the formation of a sedimentary sequence. It is the fabric, texture, and fauna of these burrow fills that define the preserved lithology. After repetitive excavation of host sediments and infilling with surficial sediment, individual tubular tempestities are not easily distinguished. This is especially true in muddy sediments where burrow networks are not lined. The resultant lithology may be easily misinterpreted as a homogenized marine sequence with a transported fauna, whereas, in fact, it represents a false sequence possessing (1) fauna from a now phantom sequence and (2) textures related to excavation and infilling of burrow networks.

Subsurface facies generation: Repetitive excavation and infilling processes exert an overwhelming influence on the character of preserved depositional sequence. This gradual subsurface excavation and replacement process ultimately results in the loss of the original sequence and the generation of new, apparently unrelated lithologies. Importantly, these biogenically induced lithologies are created in the subsurface and have little direct relationship to surficial depositional sedimentary characteristics.

Precursor facies destruction and modification: Examples of false and phantom facies come from shallow-marine carbonate environments of south Florida and the Bahamas.

Carbonate mudbanks—Many of the carbonate mudbanks in Florida Bay and Biscayne Bay, south Florida, are formed as stacks of layered mudstone units 20–100 cm thick. Between mudstone events or after the bank has shoaled to near sea level, seagrasses recolonize and a burrowing benthic community repopulates the substrate. Repetitive excavating burrowing and storm infilling of these burrows with coarser sediments gradually converts the layered mudstones to a bioturbated skeletal wackestone or even packstone.

Transgressed coastal deposits—Carbonate tidal flat deposits, peritidal peats, and other coastal sedimentary sequences, when transgressed by marine environments containing excavating burrowers, are partially to completely obliterated by repetitive burrow excavation and infilling. This process has eliminated direct evidence of basal transgressive facies from Holocene sequences in south Florida and the Bahamas.

Transgressed coastal levees—Storm-generated supertidal coastal levees defined the coastline of Florida Bay through the late Holocene. With rising sea level, individual levees become transgressed by marine environments. Repetitive excavation and infilling processes gradually convert the peritidal sequence contained in these linear sediment bodies into a marine deposit. This serves as an example of how sediment body geometry can be inherited although the internal facies are now transformed.

Faunal segregation: Others have recognized that the packing of coarse material at depth by burrowers during the excavation process, combined with the repetitive ejection of fine particles to the surface, exerts an important influence on the preservation potential of carbonate grains. The combination of continuous fine sediment recycling to the surface and loss by transport and biocorrosion can lead to selective preservation potential of carbonate grains. The combination of continuous fine sediment recycling to the surface and loss by transport and biocorrosion can lead to selective preservation of coarser macrofauna and a bias in the fossil record that can result in incomplete and skewed faunal assemblages and erroneous environmental reconstructions.

Faunal mixing: In the process of facies transformation, although finer faunal components are eliminated, coarser components are concentrated toward the base of the excavated zone. In addition, fauna from the present surficial environment are injected toward the base of the excavated zone by both storm infilling as well as excavation processes. The result, in a transformed facies sequence, will be a mixed faunal assemblage with coarse fauna concentrated near the base. Such burrow generated skeletal concentrates may be easily mistaken for basal transgressive lags or misinterpreted as depositional faunal assemblages.

TEHRANI, K. K., University of Tehran, Tehran, Iran

Summary of Cretaceous Stratigraphy in Central Iran

The study area is located in the middle part of the Central Iran Geological Province. Central Iran Geological Province is an intensely folded and faulted region situated between two great mountain ranges of Alpian Age. The Alborz range (northern branch with an approximate east-west trend) and the Zagros range (southern branch with northwest-southeast trend) constitute the region of the study. These two mountain ranges meet in northwest Iran in the Azerbaijan province.

The great and major transgression of the sea in Central Iran during Cretaceous time occurred from Barremian to Aptian stage. This transgressive series usually begins with conglomerates, sandstone, dolomitic sandstone, and dolomites containing iron, and, finally, grades into limestones and shales containing *Orbitolina*.

Upper Cretaceous sediments are characterized by major variations in thickness and facies change: In Nakhlak the Cenomanian contains *Ovalveolina ovum* (d'Orb.), *Cuneolina pavonia* Henson, and *Dicyclina schlumbergeri* Munier & Chalmas, and in Chupanan and Khur areas is characterized by detrital limestones and *Holaster revestensis* Lambert, *Epiaster restriclus* Gauthier, *Hemiaster desvauxi* Coquand, *H. consobrinus* Peron & Gauthier, *H. proclivis* Peron & Gauthier, and *Rhynchostreon columbum* (LMK). This stage is present in regions of south of Esfahan with glauconitic sandstones and numerous ammonites: *Mantelliceras mantelli* (Sow.), *Acanthoceras rhotomagens* (Brongn.), *Cayoceras qentoni* (Brongn. et Defrance), etc. (Seyed-Emami, 1971).

A stratigraphic gap of Turonian and Coniacian age is present to the east of the Naein-Baft fault, which trends northwest-southeast (Fig. 1). However, to the west of this fault in regions of south Esfahan are pelagic carbonaceous facies of Turonian and Coniacian age, characterized by Inoceramus limestone with faunas such as *Pithonella ovalis* Kaufmann, *Calcishaerula innominata* Bonet, *Stomiosphaera sphaerica* (Kauf.), *Praeglobotruncana gibba* Klaus, *P. stephani* (Gandolfi), *Globotruncana helvetica* Bolli, *G. imbricata* Mornod in Turonian, and *G. lapparenti* Brotzen, *G. lapp. coronata* Bolli, and *G. marginata* Reuss, in Coniacian time.

To the east of the Naein–Baft fault, sediments of Santonian age containing *Dicyclina schlumbergeri, Cuneolona pavonia, Pseudolituonella reichli Marie, [Valvulammina picardi Henson, Nummofallotia apula Luperto-Sinni, Minouxia lobata Gendrot, Dictyopsella sp., Vidalina sp., and "Coskinolina" sp. are present. In regions of Nakhlad and Khur, Santonian sediments form an erosional unconformity that is on top of Cenomanian age sediments. But these sediments in Jandaq are covering the metamorphic rocks of Precambrian age.*

Rocks of Santonian age in Esfahan region consist marls and echinoid limestones and contain Echinocorys vulgaris and Micraster coranguinum. These rocks are overlaid by Campanian age sediments consisting of Rudist limestone, but there are no rocks of Maestrichian age. Campanian rocks to the east of the Naein–Baft fault (Nakhlad–Chupanan and Hafttoman) are also characterized by Rudist limestones (Vaccinites veredenburggi Kuhn and Siderolites vidali Douville, Cuneolina pavonia, Dictyopsella tenuissima Reuss, Dicyclina schlumberger, Goupillaudina debourlei Marie, Orbitoides sp., G. elevata, G. arca, G. calcarata Cushman).

Maestrichtian rocks of Jandaq–Khur and Kerman areas show outcrops in these regions and have fossils of Orbitoides media (d'Arch.), O. apiculata (Schlum.), Leidorbitoides minor (Schlum), L. socialis (Leymerie), Simplorbites gensacicus (Leymerie), Clyperobis mamillata (Schlum.), Siderolites calcitrapoides (LMK), Omphlaocyclus macroporus (LMK), Nummofallotia cretacea (Schlum.), Goupillaudian debourlei Marie, Hellenocylina beotica Reichel, Planorbulina cretae (Marsson), and Sulcoperculina sp.

REFERENCES

Teherani, K. H., 1977, Etude stratigraphique du Creétacé supérieur et du Paléocène de l'Iran central: Thèse Etat, Univ. Paris 6.

———, 1978a, Role des failles majeures dans la paléogeographie du Crétacé sup, et du Paléocène de la partie médiane de l'Iran central: Rev. Geogr. Physi. & Geol. Dyn., v. 20, fasc. 2.

———, 1978b, Le Maestrichtien et la paléocène dans la partie médiane de l'Iran central: Rev, Micropal. v. 21, n. 1.

———, 1980, Mis en evidence d'une lacune stratigraphique dans le Crétacé sup. du Mont de Nakhlak (Iran central): Rev. Géol. Médit., Tome 7, n. 3.

Stocklin, J., 1968, Structural history and tectonic of Iran: AAPG Bulletin, v. 52, n. 7.

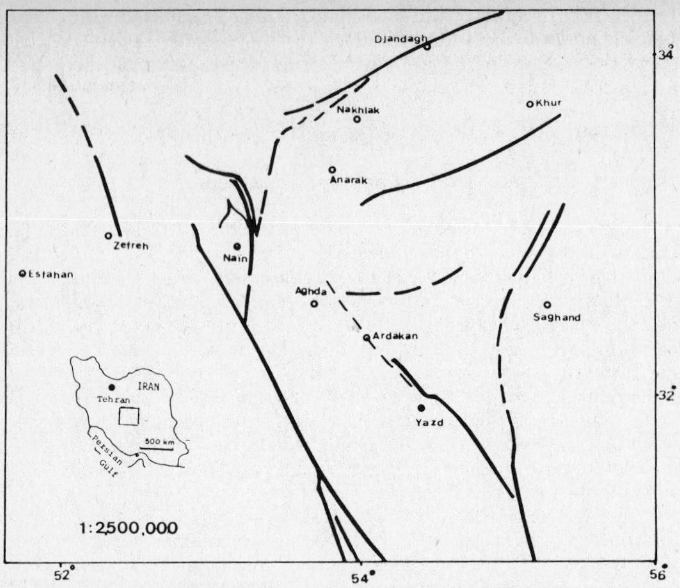

Figure 1.

TEICHMULLER, Dr. M., Geologisches Landesamt Nordrhein-Westfalen, Federal Republic of Germany

Genesis of Coal From the Viewpoint of Coal Petrology

The genesis of the microscopic constituents of coal (macerals) and of maceral associations (microlithotypes) representing various coal is discussed in light of recent advances that have been made in the study of coals and modern peats. Peat microscopy especially promoted the understanding of humification and biological gelification, both of which are decisive for the genesis of huminites/vitrinites in brown coals and hard coals respectively. Microbial activity during peatification has been testified by chemists, microbiologists and petrologists, and recently also by electron-microscopical observations in coals. It leads to the incorporation of lipid substances into the humic matter and, later, influences the chemical, physical and technological properties of vitrinites. The geochemical gelification transforms the huminites of brown coals into the vitrinites of bituminous coals and is probably caused by the onset of bitumen generation in coal. In this process, the hydrogel brown coal is transformed to the bitumogel bituminous coal. "Coal as a source rock for oil" has become an important research subject for coal petrologists (and oil exploration!). The formation of more or less liquid bitumen and its later cracking to gaseous hydrocarbons leads to "coalification jumps" of the macerals, especially liptinites and vitrinites and to the generation of secondary macerals like exsudatinite (bitumen) and micrinite (dead carbon). Bituminization in coal has been revealed by the use of fluorescence microscopy which also showed the presence of new liptinite macerals (bituminite, fluorinite and exsudatinite) and suggests a higher proportion of algal material in coals that was thought earlier. The genesis of various inertinites will probably become an increasingly important subject of research, especially with regard to "rank-inertinites" which seem to attain their high inertinitic reflectance initially during coalification.

Besides the origin of macerals, the origin of various coal facies in terms of maceral associations is considered in relation to the paleogeographic depositional milieu, the peat forming vegetation, paleo climate, water and nutrient supply, acidity, marine and calcareous influence and fire, with examples from Euramerican Carboniferous coals, Gondwana coals and Tertiary brown coals. Vegetations, water and oxygen supply (EH-conditions) are most important for the coal facies. Vegetation has been reconstructed with the help of paleobotany, including palynology, cuticular analysis and wood anatomy (in brown coals) and with the study of petrified peats preserved in coal balls. The microscopical investigation of botanically recognizable plant remains isolated from coals or accompanying rocks has been especially informative. However, the results of palynological and cuticular analyses should be evaluated with caution considering their varying resistance and, in the case of palynomorphs, their possible extrapalustral origin. Microscopical studies of modern peats derived from different types of vegetation growing in different hydrological environments have been very helpful in interpreting coal facies, especially in brown coals. The much disputed origin of brown coal lithotypes is considered in detail. For hard coals there seems to be agreement that coal layers rich in vitrite and liptinite-poor clarite were deposited in wet forest swamps, that microlayered liptinite-rich clarites are subaquatic humic muds and that inertinite-rich durites represent oxidized peats. The origin of liptinite-rich durites is controversial (either subaquatic muds or oxidized peats, the latter perhaps from raised bogs). A recent trend is the assumption that many if not all mineable coal seams have been deposited in raised bogs under ombrogeneous, oligotrophic conditions. Although raised bogs with a dense arborescent vegetation are known from Borneo it seems not probable that most seams (which were deposited in foredeeps or steadily sinking graben zones) are of ombrogenous origin although a succession from relatively wet and eutrophic to relatively dry and oligotrophic conditions may be true for occasional seams. But, as in modern peats, striking changes of seam facies are commonly caused by hydrological events, for example, inundations as a consequence of crevasses or of fires that burned the peat. The origin of the inertinite-rich Gondwana coals is discussed in relation to their specific climatic conditions and a comparison with sub-arctic peats is recommended. Coals influenced by marine and calcareous conditions are characterized by high contents in sulfur and pyrite, as well as hydrogen-rich vitrinites with a relatively low reflectance and high fluorescence, all due to a strong bacterial activity in peats of low acidity.

Proposals for further work are made in the conclusions.

TEIXERIA, W., and M. C. H. de FIGUEIREDO, University of São Paulo, São Paulo, Brazil

Outline on Early Proterozoic Crustal Evolution in São Francisco Craton, Brazil

The proposed review deals with the Early Proterozoic (E.P.) crustal evolution of the São Francisco Craton, which is a Late Proterozoic tectonically stable unit in the eastern part of the South American continent. Within the Craton, the oldest known rocks are Archean and they characterize granite-greenstone assemblages associated with medium- to high-grade metamorphic terranes.

Likewise, the Early Proterozoic domains are well represented by supracrustals rocks of greenschist to amphiobolite facies, many times included as greenstone assemblages (e.g., Itapicuru, Contendas-Mirante). In the southern portin of the Craton, they contain very important banded iron formations (e.g., Minas Supergroup), while in the northern part, conglomerates from the Jacobina Group contain syngenetic gold, and the Rio Capim metabasites include a highly concentrated sulfide level. Moreover, such E.P. domains are characterized by granulites and extensive granitization. The geologic framework of the Early Proterozoic is completed by intrusive granites (e.g., Campo Formoso, Porto Mendes(, alkaline rocks (Itiuba and Matola) and mafic dyke swarms (e.g., Uauá, Salvador, Pará de Minas, Bom Sucesso areas).

The regional structures within the E.P. provinces display conspicuous linear trends. In addition, the existence of some Archean "cratonic" nuclei bounded by the E.P. belts suggests that a certain duality in the tectonic processes occurred during the Early Proterozoic evolution. Such a tectonic style is also typical within these strongly developed linear belts, like the sigmoidal features observed for some supracrustals associated with Archean mantle gneissic domes.

According to the U-Pb, Sm-Nd, Pb-Pb and Rb-Sr radiometric data, the E.P. tectonic-magmatic events are concentrated in the 2.4–1.9 Ga time interval. Such events are defining two geochronological provinces surrounding an Archean stable domain. Within both provinces the basement rocks are exposed in two main areas (in the north and south of the Craton), partially covered by Mid and Late Proterozoic sequences.

The most important events are concentrated in the Transamazonian cycle (2.1-1.9 Ga), along which intensive reworking of continental crust have occurred, on mesozonal and catazonal conditions. This is supported by the relatively high $^{87}Sr/^{86}Sr$ initial ratio range (0.705–0.720) and μ_1 values (8.6–8.5) from the basement rocks. However, the Sr and Pb isotopic evidences combined with geochemical data also support, at that time, the existence of mantle-derived accretionary processes, particularly in the case of the Itabuna granulite belt. Besides, in terms of the Transamazonian tectonic context, the ages for the late tectonic alkaline intrusions (2.1 Ga) and of the post-tectonic granites (1.9–1.8 Ga) were followed by regional cooling of the basement rocks (1.9–1.7 Ga), so indicating events of final cratonization of the region.

Preliminary geochemical data are available for the majority of the above mentioned E.P. sequences. According to this, in the northern part of the Craton, for example, the Jacobina supracrustals may be interpreted as the upper pelitic unit of the adjacent gneissified and migmatized rocks. These paragneisses have granitic composition and REE patterns resembling those of S-type granites, with large negative Eu anomalies. In the same way, in the southern part of the Craton, E.P. granulites (Acaiaca area) are also considered as the higher grade metamorphic equivalent of the metasedimentary rocks. In this part of the Craton, the exposed kinzigites may be interpreted as an anatetic residue from pelitic metasediments, as suggested by the chemical and petrographic studies.

Again in the northern part of the Crato, the low-Fe-Ti basal tholeiitic rocks units from the R. Itapicuru Group show LREE, Eu and HREE depletion similar to those of Morb, while its felsic to intermediate volcanics have a REE fractioned pattern with a small to pronunciated Eu anomalies. The latter pattern is comparable to those of modern cal-alkaline and high K-andesites and dacites. In addition, REE major and trace elements for the felsic to intermediate volcanics support a local derivation for the upper clastic sequence.

In the same geographic portion, the Contendas–Mirante volcano-sedimentary sequence indicated two geochemically distinct volcanic series: a tholeiitic and a transitional calc-alkaline. The correlation of ultramafic and anorthositic cumulates with the volcanic series also suggests two distinct mantle sources (of different composition) with varying conditions of partial melting and fractional crystallization.

Lastly, in the northeast part of the Craton, the Itabuna belt can be geochemically separated into four sequences: tholeiitic sequence; calco-alkaline sequence; shoshonitic basic rocks; and Fe-Ti rich basic rocks. The REE pattern for the calc-alkaline granulites is similar to Archean andesites or high-Al tonalites, while the tholeiitic and Fe-Ti-rich basic rocks resemble Archean tholeiites of the TH_2 type.

From the available data, the following model for the E.P. geodynamic evolution are being proposed:

The E.P. evolution probably started with crustal rifting of the Archean domains along the period 2.4–2.0 Ga with development of elongated intra-cratonic basins, simultaneously with the injection of mafic dyke swarms. During this period, important tectonic-magmatic events also occurred leading to reworking of continental crust and accretion from mantle sources. The most important of such events was the Transamazonian cycle along which about 30% of new crustal material was accreted from the mantle sources.

Depending on the intensity and combination of the dynamic processes, the E.P. intraplate tectonics are distinctive along the Craton: in its southernmost part, a large basin opening is supported by the vast stable shelf deposits (including BIFs) of the Minas unit. In the same way, a relatively large basin is suggested for the Contendas-Mirante sequence (in the northern part), but with its closing stage being probably associated to an island-arc system. In turn, the Jacobina unit may be considered as formed in an aborted rift, although the observed tectonic differences among the mentioned units can be also due to later E.P. crustal shortening which should be strongest in the cast of the Jacobina belt.

The complementary tectonic scenario is illustrated by the R. Itapicuru greenstone and the Itabuna belt, which are considered as a back-arc basin and a continental margin magmatic arc, respectively. Moreover, the E.P. shortening-type processes eventually allowed granite magma production due to subduction, as preliminary described.

Finally, the modeled tectonic evolution is consistent with the documented metamorphic complexity of the E.P. provinces. According to geothermobarometric data, temperatures and pressures were variable along the medium- to high-grade terranes (T = 730–900 °C and Pt = 4/5kb up to 10kb). Likely, this situation is suggested for the Rio Itapicuru supracrustals (T = 400–600 °C and P_T < 2 Kb), as well as for some BIFs in which the variability of the regional metamorphism is also supported by Oxygen isotope studies.

TELLERIA, J. L., Universidad Mayor de San Andres, La Paz. Bolivia

Bolivian Global Geosciences Transects Program

The Bolivian Global Geosciences Transects Program (CC-7, ICL) involved three transects executed in the area of Bolivia (SA_5, SA_1 and SA_2 from south to north). These corridors are connected with the corresponding transects of the surrounding countries (Argentina, Paraguay, Brazil, Peru and Chile).

The transects represent compilations (in map and cross-sections, 1;1,000,000 scale) of existing geological and geophysical data along selected corridors crossing structures crucial for a better understanding of the nature and evolution of the lithosphere.

Each transect cover an area 100 km wide with a longitude variable—according to the different drafts (700, 1200 and 1400 km).

These three transects cut from west to east the following major geological domains: 1) Occidental Andes, a volcanic arc; 2) Altiplano a T-K sedimentary basin; 3) Eastern Andes and Subandean Ranges, a wide Paleozoic–Mesozoic sedimentary basin, folded and faulted; 4) Chaco Plains and "Llanos", a T-Q Sedimentary basin; and, 5) Brazilian Shield, a Precambrian cratonic terrane.

The strip maps shows the direct relation between the major geological terranes with the gravimetric anomalies. The interpretative cross sections shows the main geotectonics features and the Moho depths.

The Moho depths calculated thicken toward the central Altiplano reaching 70 km and Bouguer Anomaly about to ~400 miligals.

All the results are presented in a typical GGT display (one m^2 for each transect).

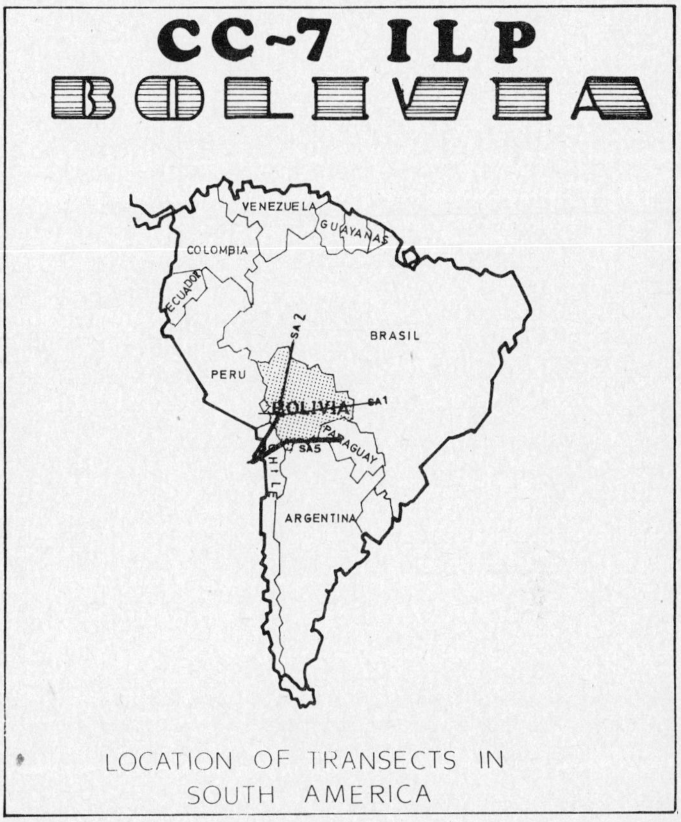

Figure 1.

TERA, T., and J. MORRIS, Carnegie Institution, Washington, DC, and A. A. TSVETKOV, Institute of Geology and Geochemistry of the Academy of Sciences of the USSR, Moscow, USSR

^{10}Be-^{9}Be and Geochemistry of Kruile-Kamchatka Arc

^{10}Be, a radioactive isotope ($t_{1/2} \approx 1.5 \times 10^6$ y) produced in the atmosphere, is transported to the earth's surface in rain and snow and concentrated in soils and sediments. ^{10}Be has never been found in pristine basalts from midocean ridges, oceanic islands, or continental rifts. In contrast, concentrations of 0.1 to 24 × 10^6 atoms/g have been measured in numerous pristine lavas from subduction related volcanoes (Brown et al., 1982; Tera et al., 1986). This clear-cut distinction, involving hundreds of samples, is a strong indicator of the incorporation of subducted sediments into magmas generated at subduction zones, an interpretation strengthened by combined measurements of ^{10}Be and the stable isotope ^{9}Be. Recent measurements of constant $^{10}Be/^{9}Be$ in mineral separates from historic volcanic rocks (Monaghan et al., 1985, 1988; Morris and Tera, 1987, in preparation) provide clear evidence that ^{10}Be was incorporated in the magmas prior to crystallization, and rule out sample alteration and cosmic ray bombardment as mechanisms for ^{10}Be incorporation in historic lavas.

The $^{10}Be/^{9}Be$ ratio can also be combined with other geochemical data to identify and characterize the end-members and geochemical processes responsible for the incorporation of ^{10}Be in island arc lavas (Morris and Tera, 1987, in preparation; Morris et al., in press). Thanks to ongoing American–Soviet cooperation, we are currently extending this endeavor to the Kurile–Kamchatka volcanic arc and provide here a progress report on this collaboration.

^{10}Be has been measured in 23 historic (1857–1983) lavas from the Kurile and Kamchatka volcanic arcs. The eight Kamchatka lavas are from six volcanoes; five are basaltic and three are andesitic. Fifteen analyses from the Kuriles represent nine volcanoes; eight are basaltic lavas and seven are andesitic. Results are presented in Figure 1. Kamchatka volcanoes have low

^{10}Be in the narrow range of 0.5–1.2 × 10^6 a/g. In contrast, those of the Kuriles show higher concentrations in a wider range, from 2.2 to 7.9 × 10^6 a/g; values above 4.2 × 10^6 a/g are seen only in adesitic rocks. The pattern of low values in Kamchatka and high and variable values in the Kurile segment is also seen in the ^{10}Be/^{9}Be ratios shown in Figure 1. The ratio is not changed by partial melting or crystal fractionation and can be used to "see through" these processes and to look specifically at the effects of sediment incorporation in the arc source region. Both the ^{10}Be concentrations and the ^{10}Be/^{9}Be ratios argue *against* significant incorporation of young sediments in Kamchatka and *for* such incorporation in the Kurile arc.

Because of its short half life, ^{10}Be in oceanic sediments decays during subduction from the trench to the zone of magma generation. Subduction times in the Kuriles and Kamchatka are 2.1 and 2.4 Ma, respectively; a difference that would reduce ^{10}Be in the Kamchatka lavas by ~15% relative to the Kuriles. Clearly other factors, sedimentological or geophysical or geochemical, contribute to the observed disparity. Possible explanations are:1. The ^{10}Be budget of the sedimentary pile is controlled by the amount of young, ^{10}Be-rich, pelagic sediment. The lower ^{10}Be values for Kamchatka could be due to dilution of the pelagic sediment with relatively ^{10}Be-poor continental sediment, as seen in DSDP core 192.2. Reflecting its short half-life, most of the ^{10}Be is found in the uppermost 10–100 m of the sediment column. Sediment accretion off Kamchatka and sediment subduction beneath the Kurile islands would explain the pattern.3. Lavas from Kamchatka may assimilate more old (> 10 Ma) ^{10}Be-poor continental crust.4. More sediment is incorporated in lavas generated in the Kuriles than in Kamchatka.

In summary, the Kamchatka arc is characterized by low ^{10}Be concentrations and ^{10}BE/^{9}BE ratios while the Kurile arc shows higher and variable concentrations and ratios. Continuing ^{9}BE measurements will show either a step change between Kamchatka and the Kuriles or a gradual change along the arc. The possible explanations given above will be further tested through analyses of near-trench sediments and through geochemical modeling of the Be isotope system in conjunction with major and trace elements and Sr, Nd, and Pb isotope compositions.

REFERENCES

Brown, L., et al., 1982, Nature, v. 229, p. 718–720.
Monaghan, M. C., et al., 1985, Eos, v. 66, p. 1110.
———, 1988, Earth and Planetary Science Letters, v. 89, p. 288-298.
Morris, J., and F. Tera, 1987, Eos, v. 68, p. 1517.
——— ———, in preparation, Geochimica et Cosmochimica Acta.
——— ———, and W. P. Leeman, in press, Eos.
Tera, F., et al., Geochimica et Cosmochimica Acta, v. 50, p. 535–550.

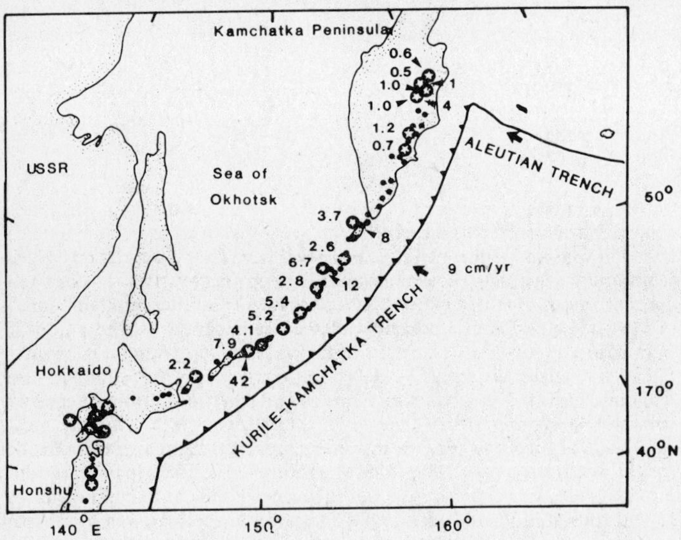

TESSENSOHN, F., Bundesanstalt für Geowissenschaften und Rohstoffe, Hannover, West Germany

Northern Ross Sea Transect, Antarctica

A status report on a fairly advanced Antarctic transect is presented based on the following material: geological strip map, cross section, and aeromagnetic map for the terrestrial part; and depth-to-basement map, seismic profile, magnetic map and gravity map for the marine part. A crustal interpretative section is not yet completed.

The northern Ross Sea transect crosses the upfaulted basement block of the Transantarctic Mountains and the rifted Ross Sea embayment, both major Cenozoic tectonic features. In the basement block of the Transantarctic Mountains, much older structural elements of the early Paleozoic Ross Foldbelt are exposed by the young uplift.

Onshore, the traverse begins on the ice of the Polar Plateau, where dome-shaped magnetic anomalies have been interpreted as indicating the presence of Precambrian rocks of the East Antarctic shield. There are no rock outcrops here. In the Transantarctic Mountains, two terranes of the Paleozoic Ross Foldbelt are traversed: High-grade metamorphic rocks and Paleozoic granites of the Wilson Terrane are thrust over Cambo-Ordovician sediments of the Bowers Terrane. The terrane boundary is a several-km-wide. southwest-dipping zone of ductile deformation. Originally flatlying Mesozoic cover rocks in the Transantarctic Mountains now dip away from the Ross Sea depression and thus form direct evidence for young uplift and tilting of this tectonic block.

Offshore, the traverse crosses two basement highs and an intervening sedimentary basin. The highs seem to be similar in structure to the adjacent Transantarctic Mountains. Thus, a province of very strong magnetic anomalies extends from /the Mountaineer Range of the Transantarctic Mountains offshore over shallow basement areas of the Coulman High. These anomalies are interpreted as plug-like mafic intrusive bodies related to Cenozoic alkaline volcanism of the Hallett province. This province marks the faulted transition between the Transantarctic Mountains and the Ross Sea. Halfgrabens of the highs in the Ross Sea are filled with up to 2 km of strata and covered by glacial sediments. The eastern end of the transect crosses a major basement ridge that extends north-south across the Ross Sea (Central High) and forms Iselin Bank at the outer continental margin. This ridge is underlain by Paleozoic marble and gneiss (DSDP site 270). Oceanic basement is reached at the foot of the continental slope northeast of Iselin Bank.

TEUTSCH, G., Universität Stuttgart, Stuttgart, West Germany

Delineation of Zones of Protection using Numerical Models and Field Techniques at Regional and Local Scale

The delineation of zones of groundwater protection around domestic water supplies requires an appropriate combination of field and numerical investigation techniques at regional and at local scale.

At local scale the "Schutzzone II" is defined as the area bounded by the 50 days groundwater travel time contour, the concept being to protect the wells mainly against biodegradable contaminants. Within this zone only a very limited use of the land is accepted. Agricultural activities are allowed only with severe constraints on the use of organic fertilizers.

In many cases, however, the groundwater quality is threatened by pollutants like chlorinated hydrocarbons which are hardly biodegradable. Here the 50 days travel time concept is insufficient. An accurate determination of the regional groundwater flow regime is required in order to identify adequate zones of protection respectively areas of high pollution potential (e.g., industrial sites).

In this paper the results of the first two years of an ongoing model and field investigation groundwater project are presented.

The work refers to a new waterworks which is planned to be located in the Rhine valley near the city of Karlsruhe. The aquifer consists of about 34 m of sand and gravel with some interbedded silt layers. Due to several non-hydrogeological constrains (e.g., available open land, distance to the city, distance to the Rhine floodplains, etc.) the location of the future pump wells was a priori fixed (Fig. 1).

At first, based on the available field data a horizontal, two-dimensional finite element model was developed to simulate the regional groundwater flow in the whole catchment area prior to the anticipated pumping. A fine mesh with a 100 m spacing was used in an area somewhat greater than the expected 50 days travel time contour in order to facilitate subsequent small scale investigations (Fig. 2). The regional model was carefully calibrated on a monthly base over a period of 5 years. It was used to describe as accurately as possible the flow conditions prior to the installation of the planned waterworks. Furthermore, it was used to calculate the expected regional drawdowns assuming different pumping scenarios.

Using sensitivity analysis it was then shown that the remaining uncertainty in respect to the available aquifer parameter data was acceptable at the regional scale but not at the local scale. Based on the available hydrogeological information a reliable delineation of the 50 days travel time contour could not be achieved.

Therefore, further local scale investigations were proposed including the drilling of 11 additional monitoring wells, the analysis of core samples and the recording of geophysical borehole logs. A 100 l/s pump- and tracertest running for 50 days was designed in order to obtain reliable aquifer parameters at the anticipated waterworks location. The tracer flowing towards the pump well coming from different directions and from different depths is monitored by 24 tracer sampling tubes placed in the gravel pack.

For the numerical simulation of the pump and tracer tests a local three dimensional groundwater flow and transport model had to be developed. This model has been coupled to the regional flow model (Fig. 3). In the local model the flow is calculated using a finite-difference approximation whereas the transport is calculated with the random-walk method.

Results of the two-dimensional regional and three-dimensional local modelling are presented.

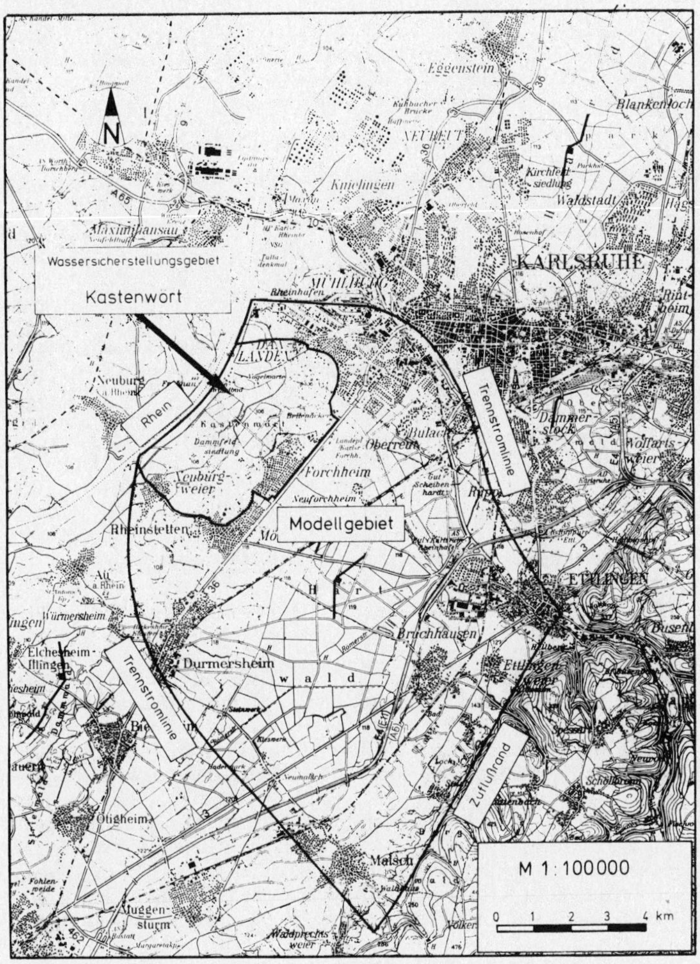

Figure 1—Model study area.

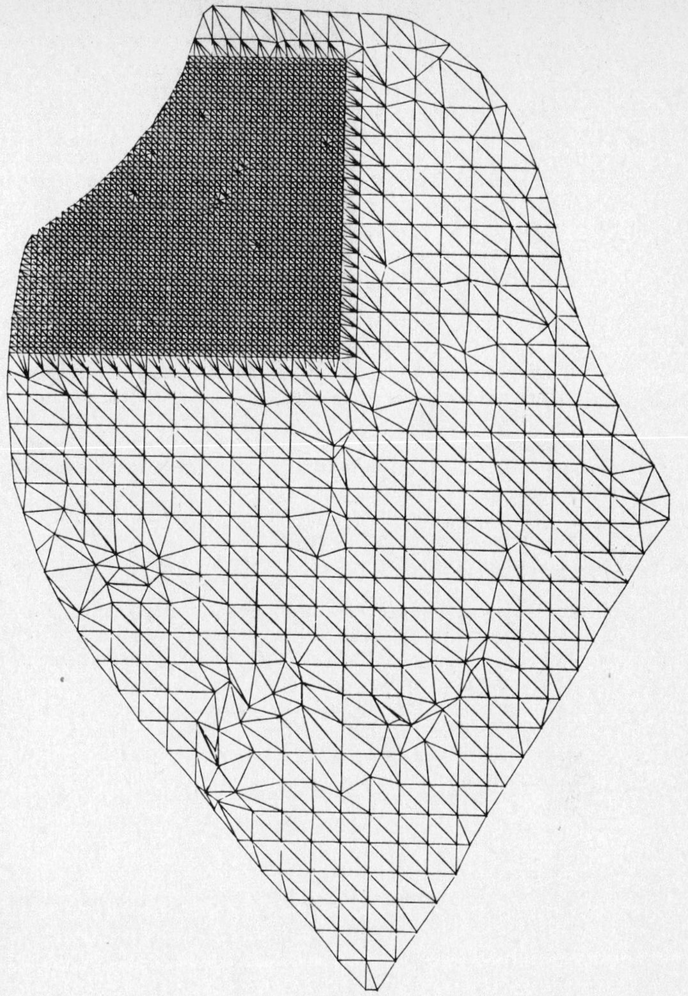

Figure 2—Regional two-dimensional finite element flow model.

TEXTORIS, D. A., University of North Carolina, Chapel Hill, NC, E. I. ROBBINS, U. S. Geological Survey, Reston, VA, and P. J. W. GORE, Emory University, Atlanta, GA

Origin of Organic-Rich Strata in an Upper Triassic Rift Basin Lake, Eastern USA

Three connected large half-graben basins developed in central North Carolina during the Karnian, and are filled with strata defined as the Chatham Group of the Newark Supergroup. The northern-most Durham basin contains up to 2300 m of predominantly fluvial conglomerates, sandrocks and mudstones which show a southern axial flow. The southern-most Wadesboro basin contains up to 1300 to 2700 m of fanglomerates and fluvial sandstones and mudrocks. Each of these basins is connected to the middle Sanford basin by restricted narrow zones of thinner strata caused by transfer faults which formed hinged highs. We estimate that at least half of the original strata thickness remains in each of the three basins.

The Sanford basin contains three formations: Pekin Fm, consisting mainly of fluvial conglomerates, sandstones and mudrocks deposited during a warm and humid climate and during active rift faulting; Cumnock Fm, consisting mainly of coals, lake shales and other fine basinal siliciclastics deposited during a warm and humid to drying climate and during tectonic stability; and the Sanford Fm, consisting of fluvial sandstones, conglomerates and mudrocks deposited during an arid climate and renewed tectonic activity. Overall, the basin contains up to 2000 m of strata, 300 m of which constitute the lacustrine Cumnock Fm.

The Sanford basin was tectonically active during the middle and late Karnian, for perhaps 6 my, at latitude 10 °N. The basin originated in a warm and humid climatic zone and ended in an arid zone. We ascribe this large scale change to northward drift of the North American plate from one climatic zone to another. Local climatic variations were due to periodic tectonic activity along the major Jonesboro system of listric faults on the east side of the basin causing periodic blockage of rain-bearing easterly winds.

The Cumnock Fm consists of coals representing swamps (forested wetlands), carbonaceous shales representing marshes (emergent and shrub wetlands), black shales, representing evolving lake stages (oxygenated, suboxic, and anoxic waters), and other fine basin-fill siliciclastics. Authigenic dahllite, siderite, ammonium illite, pyrite and chaosite are common in the mudrocks.

The Cumnock lake was fed primarily by southward flowing axial rivers from the northern Durham basin, and by major rivers from the less-faulted to non-faulted river-delta-beach roll over system to the west and northwest of the Sanford basin. The lake reached dimensions of at least 20 x 60 km, and was usually hydrologically open for much of its 1 to 2 my duration. At various stages, the lake evolved from oxygenated to suboxic to anoxic sulfidic to anoxic-nonsulfidic under methanic conditions.

The coals, near the base of the Cumnock Fm, are dominated by wood cells and gelified wood. TOC ranges from 12 to 80. Sulfur ranges from 0.4 to 7.1. Palynomorphs are dominated by spores of seed ferms and ferns. This swamp facies is particularly well-developed in the western and northwestern part of the Sanford basin.

Carbonaceous shales, representing marsh facies, are dominated by wood cells, also, but horsetail and lycopod remains dominate the palynomorphs. TOC ranges from 4 to 34.

Black shales, representing the various stages of lake facies, are dominated by sheet-like algal kerogen that may be the remains of cyanobacteria. Algal balls and zooplakton fecal pellets are abundant. TOC ranges from 1 to 13.

Data regarding burial history all indicate temperatures of 95 to 105 °C putting Cumnock lake strata well within the petroleum window: TAI ranges from 3 to 4, R_0 ranges from 0.8 to 1.5, petroleum shows in basinal deep wells, methane is common, and high volatile A bituminous coal is present. This information, plus the high TOC values and the thickness and extent of the organic-rich strata, indicate the presence of an excellent petroleum and natural gas source rock.

Possible reservoir sandstones exist in the underlying fluvial channels of the Pekin Fm and the overlying Sanford Fm, the north and south parts of the Sanford basin which may contain beach and channel sandstones of axial drainage deltas both entering and exiting the open lake system, beach and related sandstones in the subsurface in the eastern part of the basin, and interbedded lake and delta siltstones and sandstones within the Cumnock. However, poor to moderate sorting, presence of matrix, compaction, and diagenetic cements and recrystallization have eliminated primary porosity and permeability in the sandstones. Secondary porosity, as fractures, especially associated with diabase intrusives or faulting, may serve as local conduits for hydrocarbon migration and trapping in all rocks. Numerous syn- and post-tectonic intrabasinal faults blocks provide favorable structures.

Probably the best reservoir beds, those associated with tectonic quiescence and more mature sorting and extent, were formed as roll over sediments of channel, delta and beach origin to the west of the Cumnock lake facies outcrop; however, they have been eroded.

THABET, K. M., and S. A. KHATTAB, Mosul University, Mosul, Iraq

Correlation of Geotechnical Properties of Limestone in North Iraq

The most widely used rock for building and foundation purposes in North Iraq is limestone.

Outcrops of many types of limestone are available in this region. Previous works have centered mainly on the geological aspects and geotechnical properties of these limestone.

In this study seven types of limestones of differing geological ages were classified, and correlations between their geotechnical properties (based on previous investigations) were carried out. The study included the following relations based on statistical treatment: 1) bulk density with porosity; 2) longitudinal sonic velocity with shear wave velocity; 3) longitudinal and shear wave velocities with density, porosity, and water absorption; 4) compressive strength with tensile strength, porosity, bulk density, longitudinal and shearing sonic velocities as well as static and dynamic Elastic Modulus.

A model was arrived at. This model can be used to predict the strength characteristics of the limestone rock from values of tests that can be carried out more rapidly and economically.

THAKUR, V. C., Wadia Institute of Himalayan Geology, Dehra Dun, India

Sedimentation-Tectonics of the Tethyan Basin of Northwestern Himalaya

In Kashmir, Ladakh, and Himachal sectors of northwestern Himalaya, the Tethyan basin sequence is exposed in three belts. From southwest to northeast, these belts are Kashmir–Chamba, Zanskar–Spiti–Tso Morari, and Karakoram; of these the Karakoram belt lies north of the Indus Suture, whereas the rest of the Tethyan belts are located south of the suture. The stratigraphic divisions and their correlation are given in Table 1.

In Kashmir, Zanskar, and Spiti, the basal part consists of low-grade metasedimentaries together with 500 Ma dated granite intrusive bodies. An epeirogenic break above the Cambrian characterized by Ordovician–Silurian succession missing in several Lower Paleozoic sections can be related to the 500 Ma Pan-African orogeny. The Panjal volcanics of Lower Permian is a marker stratigraphic unit in Kashmir and Zanskar but absent in Spiti and Kumaun. Geochemistry of these volcanics indicate rift-type tectonic setting. Gondwana plants were reported form the Gangamopteris Bed of Middle Permian in Kashmir and from the Po formation of Middle Carboniferous in Spiti. A Hercynian gap, indicated by the presence of Conglomeratic Bed in the Upper Carboniferous of Kashmir and Spiti and the absence of Carboniferous succession in Kinnaur and Kumaun has been reported. In Tso Morari area, which lies adjacent to the north of Spiti basin, the Taglang La Formation of Upper Permian–Lower Triassic directly overlies the Puga Formation, which contains 500 Ma dated granite bodies. To the south of the Zanskar basin, metamorphosed Panjal volcanics together with crystalline limestone of Permian–Triassic overlies the Suru Crystallines of late Pre-Cambrian. In the Tandi area of Lahaul, condensed succession of Permian, Triassic, and Jurassic directly overlies the late Pre-Cambrian Haimanta Formation. These occurrences are interpreted as representing deposition over the uplifted blocks that were developed north and south of the main basin as a result of rifting. The Mesozoics represent major carbonate accumulation, characterized by high rate of sediment fill, complex subsidence history, and varied depositional environment. Early Triassic to Early Jurassic sedimentation in Kashmir exhibits continental shelf, continental margin, shallow-marine, and tidal flat complex. Deep-sea followed by continental margin, continental shelf, and tidal deposits are recognized in Zanskar; whereas in Spiti, in addition to these depositional environments, oolitic sand bar and reefal build-ups are reported. The Early–Middle Jurassic sediments were laid down in shallow-marine and continental margin in Zanskar, and shallow-marine, continental shelf, outer shelf, and subtidal and intertidal sediments were deposited in Spiti. Late Jurassic is dominated by mudstone facies representing deepening of the basin. Early Cretaceous represents outer-shelf in Zanskar and continental shelf to slope facies in Spiti, and Late Cretaceous shows deep-sea environment in both the regions.

The Karakoram Tethyan basin lies north of the Indus Suture. The Pangong Tso group consisting of schists, gneisses, marble, and metabasics together with intrusive granites constitute the base of the Karakoram sequence. It is overlain by black slates, arenites, argillites, and basic volcanics of Upper Carboniferous and Permian. This Upper Paleozoic succession is succeeded upward by shale and volcanosedimentary which in turn is followed by megolodon bearing Triassic carbonate of Morgo Formation. The Jurassic is represented by plant fossils bearing fluvial deposits and limestone and black shales. Qazilangar conglomerate overlying the limestone and black shale is of Cretaceous age. The post-Triassic sedimentation history of the Karakoram basin appears to be different from that of the Tethyan basin of Zanskar and Spiti.

Period	KASHMIR	ZANSKAR	SPITI	KARAKORAM
Lower Palaeogene		Chulung La Spanboth Kangi La		Kumdan
Cretaceous		Chikkim Giumal	Chikkim Giumal	Qazi Langar
Jurassic	Lower	Spiti Shale Kioto Limestone	Spiti Shale Kioto Limestone	Burtse Morgo
Triassic	Upper Middle Lower	Zangla	Nimaloksa Hanse Tambakurkur	Aqtash
Permian	Zewan Gangomopetris Panjal Volcanics	Kuling Panjal Volcanics	Kuling	Chongtash
Carbonife-rous	Agglomerate Slate Fenestella Shale Syringothyris Limestone	Ganmachidam PO Lipak	Ganmachidam PO Lipak	Sasser Brangsa
Devonian	Muth Naubug Beds	Muth	Muth	
Silurian	Harpatnar Beds		Pindolomite	
Ordovician	Gauran Beds	Thaple	Shian Quart-zite	
Cambrian	Upper Middle Lower	Kurigiakh Karsha Phe	Haimanta	
PreCambrian	Sulkhala	Suru (Zanskar)		Pangong Tso metamorphics

THAKUR, V., Wadia Institute of Himalayan Geology, Dehradun, India

Structural framework and tectonic evolution of Kashmir and Chamba nappes, Zanskar Crystallines and Lesser Himalayan formations in Kashmir, Himalaya

The Kashmir synclinorium comprising of late Precambrian to Jurassic sequence of Tethyan facies constitute the Kashmir nappe which is overthrust to the south along the Panjal Thrust overlying the Autochthon zone (Wadia, 1928; Thakur and Gupta, 1983). The rocks of Chamba nappe, showing some lateral facies variation, represent southeast extension of the Kashmir sequence. An axial culmination (Thatri Dome) located along the Chenab Valley in Doda district exposes the Salkhala group overlying the medium- to high-grade Zanskar Crystallines. The Salkhala group, forming the basal part of the Kashmir and Chamba sequences, consists of phyllite, quartzite, quartzitic phyllite with occasional limestone and basic lenses together with 500 Ma dated deformed granite bodies of Dalhousie, Kundkaplas, Ramban and Kangan.

The Salkhala group extends from Kashmir and Chamba eastward to Beas and Satluj valleys where it joins with the Chail group; and the Panjal Thrust, which demarcates the base of the Salkhala group, joins with the Chail Thrust. Earlier workers (Gansser, 1964) defined this Chail–Panjal Thrust as the MCT. The Salkhala group of the Mandi–Pandoh area extends towards northwest enclosing the closure of the Paleozoic succession of the Chamba basin and joins with Haimanta Formation of Keylong area in Lahaul. Similarly, the Salakhala group of the northern margin of the Kashmir nappe extends into the Phe Formation of the Zanshar Valley. These tectonostratigraphic relations indicate that the Salkhala group, the Chail group and the Phe Formation have been designated with different names, but they all belong to the same lithostratigraphic and tectonic unit.

The Zanskar Crystallines, occupying the Great Himalayan range, is located between the Zanskar and Chenab valleys. They represent northwest extension of the Central Crystallines of Kimaun Himalaya but showing different tectonic setting. In Keylong region of Lahaul an axial depression exposes the Haimanta group rocks; and the Central Crystallines of Kumaun Himaaya and the Zanskar Crystallines plunge northwest and southeast respectively underneath the Haimanta group. In Kumaun Himalaya the MCT brings the Central Crystallines of the Higher Himalaya to override the Lesser Himalayan formations, whereas in the Kashmir Himalaya this tectonic framework and the Kumaun type MCT are not observed.

In my tectonic interpretation the Panjal Thrust and the Chail Thrust are the same, and this thrust can be designated as the MCT-I. The Jutogh Thrust in Simla Hills separates the low grade Chail group (Salkhalas) from the medium- to high-grade Jutogh group. The Jutogh Thrust separating the underlying Salkhala group can be extended from Satluj valley northwestward to Kulu valley and further west to Lahaul where near Sissu village it joins with the Tethyan Thrust which separates the high grade Central Crystallines from the overlying low grade Haimanta group. The Jutogh Thrust is designated as the MCT-II.

eBetween Satluj and Beas valleys, the Central Crystallines of Higher Himalaya zone is underlain by the Salkhala group and overlain by the Haimanta group (=Salkhalas). The Jutogh Thrust (MCT-II) separating the Central Crystallines from the underlying Salkhala group constitute the floor thrust, and the Tethyan thrust separating the former from the overlying Haimanta defines the roof thrust; both these thrusts meet at Sissu village near Keylong in Lahaul. This tectonic relation indicates that the Central Crystallies form a duplex on a crustal scale. A similar type of duplex is interpreted for the Zanskar Crystalline slab. On its northern margin the Zanskar Crystallines is overlain along the Tethyan Thrust by the Phe Formation which is equivalent to the Haimanta Formation. On its southern margin, The Zanskar Crystallines slab is seen to override the greenschist facies rocks of the Kishtwar group and underlie the Salkhala group rocks in Chenab valley along thrust contacts; the thrusts in both the cases will be equivalent of the Jutogh Thrust (MCT-II)The occurrence of duplex structure on crustal scale of the Zanskar Crystallies and the Central Crystallines point to contradict the classical concept of root zone in the crystallines of Higher Himalaya.

Srikantia and Bhargava (1974) showed that the "Autochthonous" zone of Wadia represents northwest extension of the Shali structural belt. Based on recent work (Jangpangi et al., 1986) and my own observations, it is now fairly well established that the "Autochthonous zone" is largely made of Lesser Himalayan formations. These formations include Sunder Nagar Formation, Mandi Darla volcanics, Shali Formation and Simla Slates. These Lesser Himalaya formations extend in a narrow belt from Simla Hills to Kashmir. They occur as imbricate slices located between the MBT and the Panjal Thrust.

REFERENCES

Gansser, A., 1974, Geology of the Himalayas, Interscience Publisher, John Wiley and Sons, London, 1-289.

Wadia, D. N., 1928, The Geology of the Poonch state (Kashmir) and adjacent portions of the Panjab: Mem. Geol. Surv., India., v. 51, n. 2, p. 185–370.

Srikantia, S. V., and O. N. Bhargava, 1974, The Salkhala and the Jutogh relationship in the Kashmir and Himachal Himalaya—a reappraisal: Himalayan Geology, v. 4, p 396–413.

Thakur, V. C., and V. J. Gupta, 1983, Regional stratigraphy, paleontology and structure of Kashmir and Ladakh Himalaya. In contributions to Himalayan Geology: Stratigraphy and Structures of Kashmir and Ladakh Himalaya, V. J. Gupta, ed., Hindustan Publishing Corporation, Delhi, v. 2, p. j1–32.

Jangpangi, B. S., G. Kumar, D. R. Rathore, and S. Dutta, 1986, Geology of the "Autochthonous Folded Belt", Jammu and Kashmir Himalaya with special references to Panjal Thrust,; Jour. Pal. Soc. India, v. 31, p. 36–51.

THIEDE, J., GEOMAR Research Center for Marine Geosciences, Kiel, Federal Republic of Germany

Eastern Arctic Sediment Histories: New Chapter in Polar Research

The evolution of the depositional environment of the eastern Arctic deep-sea basins has remained an unknown chapter of Northern Hemisphere paleo-oceanography. After Nansen sampled the sediment surface of the newly discovered Arctic abyss, no sediment cores had been recovered from this area prior to the late 70s when the first short sediment sections became available during the FRAM-I ice island expedition. Since then rapid progress has been achieved because: (1) numerous expeditions have been organized for coring the region; (2) the stratigraphic uncertainties plaguing central Arctic sediment records have been resolved; and (3) it could be shown that eastern Arctic sedimentation rates are orders of magnitude faster than in the central Arctic, allowing resolution of its paleo-oceanographic history in considerable detail and correlation with North Atlantic and Norwegian–Greenland Sea records.

Long and high-quality sediment cores along a transect from the poleward flank of the eastern Arctic Nansen–Gakkel Ridge across the Yermak Plateau and through Fram Strait into the Norwegian–Greenland Sea consist of dominantly terrigenous, but fossiliferous clays, silts, and sands with intercalated horizons rich in relatively coarse ice-rafted clasts whose lithologies can be traced to very specific source regions along the Eurasian continental margins. By means of bio-, litho, O-, C-, Th, and Be-isotope and magnetic stratigraphies, it could be shown that these cores document the past 300,000 years, but that the sedimentation rates were highly variable, exceeding several centimeters to decimeters per thousand years, depending upon their depositional site. Scattered occurrences of ice-rafted coarse clasts throughout the cores document the continuously cold surface waters and the presence of an ice-cover over the past few hundred thousand years in this part of the Arctic. The dominance of the fine-grained sandy, silty, and clayey material that resembles material found included into the modern ice pack of the Transpolar Drift and the scarcity of coarser ice-rafted components for most of the time documented by these cores suggests an ice setting, in principle similar to the modern one, for long spans. It was interrupted repeatedly by two different types of events that resulted in intercalations of quite different lithologies.

The first ones produced distinct horizons with relatively frequent coarse ice-rafted dropstones floating in a fine-grained clastic sediment matrix. These horizons can be correlated across the morphologically elevated Nansen–Gakkel Ridge into the Amundsen deep-sea basin and coincided with rapid changes of the O-isotope ratios in planktonic foraminiferal shell material suggesting that these horizons represent bursts of iceberg production which, contrary to today, were able to reach the entire Arctic Ocean, to melt there and to release their load of coarse terrigenous materials. The other sediment type consists of variably thick horizons of mud clasts that may be related to the production of sediment pellets on the ice pack as can be observed today in the Transpolar Drift, which carries thick and old pack ice from Siberian shelf seas to Fram Strait.

Despite indications of a continuous ice-cover over the eastern Arctic for the past 300,000 years, the repeated occurrences of calcareous and organic-walled pelagic microfossils and of planktonic as well as benthic foraminifers (virtually no siliceous materials have been preserved) as well as thc thorough oxygenation of the basin sediments document the stratification and the relatively rapid turnover of the Arctic Ocean water masses. Planktonic foraminiferal O-isotope ratios are much lighter in the Arctic than in the Norwegian–Greenland Sea during interglacial times because of the repeated development of a brackish surface water layer whereas they coincided with those from the subpolar deep-sea basins during glacial settings. Calcareous nannofossils and planktonic foraminifers have been found both in glacial and interglacial sediments. Their occurrences are very problematic because of the poor understanding of their present ecology in the eastern Arctic. Floral and faunal distributions are today related to the boundary between the Atlantic subsurface water masses (reaching the Arctic Ocean through Fram Strait after they have passed the Svalbard continental margin as the Westspitsbergen Current and after they have disappeared under the ice pack) and central Arctic water masses. We can only speculate how these water masses might have been structured during earlier parts of the Quaternary.

THIGALE, S. S., University of Poona, Pune, India

Processes Conducive for Slope Failures along Western Ghats of Maharashtra, India

The objective of this paper is to analyze the episode of slope failures that occurred along the foothills of the western flank of Western Ghats of India (between latitudes 73°30′ and 73° 50′E; longitudes 17°10′ and 17°35′N) within a short span of 14 hours (from 4 P.M. of 28 June and 6 A.M. of 29 June 1983). Geological mapping was carried out, rainfall and seismological data were collected, and structured interviews were conducted for obtaining desired information during 1983–1986.

In the study area, which is otherwise not susceptible for slope failures, over 350 events of soil creeping and slumping and 15 major events of landsliding were recorded. It has been observed that most of the events of soil creeping and slumping occurred respectively along gentle to moderate (4–10°) and moderate to steep (8–30° slopes. The landslides, on the other hand, took place along steep to very steep (25–45°) slopes. Further, the majority of slope failures have taken place between two adjacent lower order streams backed by fractures, removing the earth material from intervening spur portions.

A generalized geological succession encountered in the study area is as follows (Fig. 1): (1) hard but jointed laterite; (2) moderately hard, jointed, and porous laterite; (3) lithomarge; (4) weathered basalt; and (5) fresh, unaltered basalt.

The study area has been experiencing earthquakes since 1964. The major earthquake of magnitude greater than 7 on the Richter Scale was recorded in December 1967. Figure 2a shows the plots of epicenters of earthquakes recorded in the area. It is interesting to note that the affected area overlaps with the area where maximum epicenters are clustered (Fig. 2b).

The analysis of rainfall data reveals that the affected area started receiving monsoon showers on 21 June followed by intensive rains of increasing order (from 50 mm to over 350 mm) between 26 June and 29 June, 1983. It is noteworthy that average rainfall received by the unaffected area had more or less similar intensity during the period of episode (Fig. 3).

Structured interviews of over 125 people who witnessed the events suggest that slope failures took place within 2–5 min when huge masses of regolith and country rocks collapsed, with a loud rumbling noise, converting the scenic slopes into ugly scars and bottomlands into dumps of debris. The majority of landslides occurred along fractures developed during the major earthquake of December 1967. A number of springs had developed at the base of these fractures during the monsoon following the earthquake. A few of them turned out to be perennial while the rest were seasonal. The reports in the local newspapers, published from time to time after the 1967 earthquake, also supported the findings of the interviews.

From the foregoing description it is suggested that the geological setting, mountainous topography, and earthquake-proneness of the affected area were the factors conducive for initiating the process of slope failures as percolation of rainwater, through jointed lateritic crust up to impervious lithomargic layer, can take place. Heavy rains during the episode ensured percolation sufficient to overcome the stability of slopes. In all probability, repeated tremors of varying intensity were responsible for widening of inherent joints of lateritic crust. A slow creeping of regolith and rocks due to percolation of rainwater every year further enhanced the widening process. The resulting fractures received a large amount of water during heavy rains of 26 June and 29 June, saturating the pervious lateritic crust overlying impervious lithomarge. This weakened the earth material by displacing air and increasing the pore-water pressure along slide surfaces as a consequence of which slopes were brought into inequilibrium condition and huge masses of regolith and rocks suddenly collapsed.

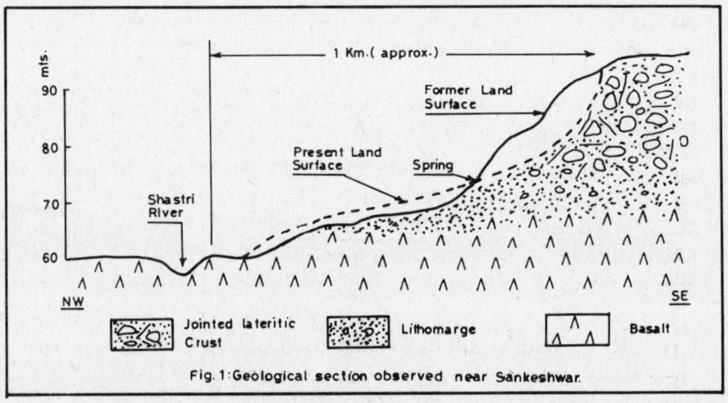

Fig. 1: Geological section observed near Sankeshwar.

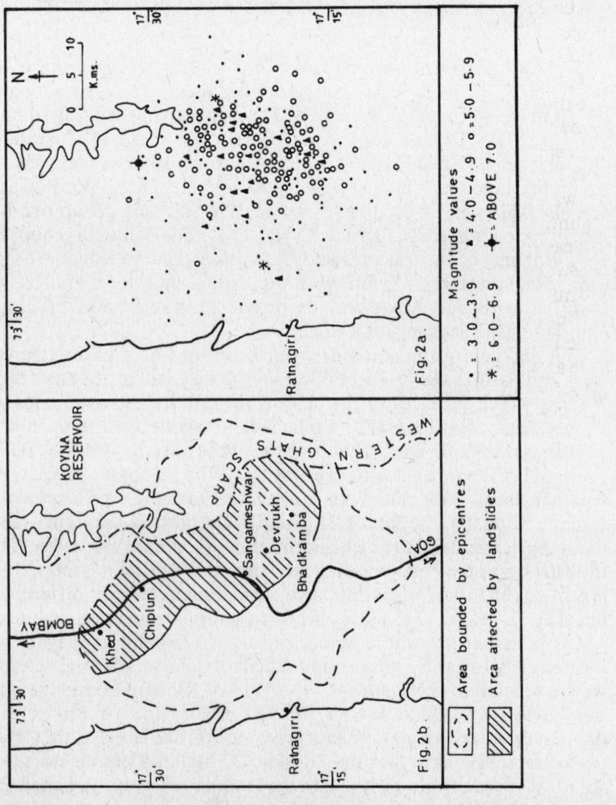

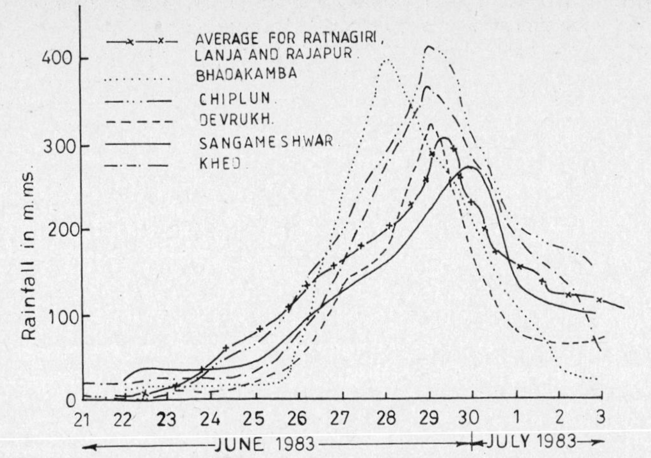

Fig. 3

THIGALE, S. S., and P. S. PARASWAR, University of Poon, Pune, India

Impact of Borewells on Hydrogeological Regime and Agricultural Development: Case Study of Ahmednagar District, Maharashtra State, India

Ahmednagar, a famous sugarcane producing district, is covered by horizontally disposed basaltic flows of Cretaceous–Eocene age. These flows are hydrogeologically regarded as hard rocks. The occurrence of groundwater in them is erratic mainly due to unpredictable occurrence of weathering and jointing patterns, lack of significant amount of primary porosity, and presence of redbeds of impervious nature. Repeated drought conditions during the last two decades and an increase in the density of dug wells have caused a considerable depletion of the shallow water table in the district. As a result, the dug wells, the main source for tapping groundwater for agricultural purpose, have failed to supply adequate quantities of water required for cultivation of sugarcane. The need for tapping deeper aquifers by drilling has, therefore, been increased considerably. It reached a peak after the drought years of 1983–1984 and 1984–1985, when the Ahmednagar District Central Co-operative Bank, The New India Assurance Company, and sugarcane factories made joint efforts to undertake an extensive drilling program.

About 535 borewells drilled under this program were studied. The data suggests that the depth of borewells varies between 50 m and 70 m, and the yield between 0 and 19,000 L/hr. The data also reveal that zeolithic basalt and/or jointed basalt, encountered at about 30 m and 55 m depth, constitute the deeper aquifers. It has been further observed that only 37% of the borewells are successful per the criteria of 4500 L/hr yield laid down by the insurance company. Further, 40% of borewells yield up to 2000 L/hr and 23% borewells yield between 2000 and 4500 L/hr water. In certain areas, the water table at shallow depth has completely disappeared because the groundwater from this zone has been contributed to the deeper depths. In a few cases, successful borewells have failed to yield water because the new borewells have been drilled without any consideration of safe distance.

Thus, adverse hydrogeological setting, inadequate rains, and haphazard drilling have been responsible for lowering of the shallow groundwater table in the district. High requirements of water for cultivation of sugarcane and lack of appropriate geotechnical approach are the factors responsible for failure of the drilling program. It has been suggested that due consideration for safe distance between two adjacent borewells and also between borewells and dug wells is necessary. A change in crop pattern from high to low water requirement could help in utilizing the 23% borewells yielding 2000–4500 L/hr water, which are otherwise declared as failed.

THIRY, M., Ecole des Mines de Paris, Fontainebleau, France, and MILNES, A., C.S.I.R.O. Division of Soils, Glen Osmond, Australia

Development and Evolution of Pedogenetic and Groundwater Silcretes of Central Australia

The thin Cainozoic sequence of continental origin in northern South Australia displays many forms of secondary silicification. In the regions between Lake Torrens and Lake Eyre, silcretes with pedogenic structures occur as cappings on plateau and mesa surfaces. Silica cements, with respect to the primary sedimentary structures, are dispersed deeper in the profiles.

Section description: A type section can be described at Stuart Creek Opal Field (northern edge of Lake Torrens).

1. At the base, grey Cretaceous shales are capped by an irregular opaline silica layer in which sedimentary structures are well preserved and with cracking pattern due to dehydration.

2. The Tertiary sequence overlies the Cretaceous sediments with an irregular erosional boundary. It begins with a massive lens of silicified sand, varicolored, lustrous in appearance with zones of porous sandstone.

3. Beige to ochre-colored sands, with abundant rhizoliths structures, overlay the silicified lens. The sands contain silicified nodules of irregular shape and size and a horizon of discontinuous lenses of hematite-mottled quartzite.

4. Variable indurated, red and purple hematitic sands form a distinctive horizon. These are commonly disrupted by vertical percolation structures or pipes that penetrate the section from the land-surface and usually exhibit features of significant collapses.

5. A well-structured silcrete caps the profile. The basal interval is composed of centimetric silica-impregnated white sand nodules, and larger silicified sand masses, in an unconsolidated white sand matrix. Toward the top, the nodules aggregate and form clear vertical columns 0.3 m in diameter and 0.5 m high. The cores and top of the columns display numerous illuviation features that record percolation of solutions and successive depositional phases.

Mineral sequence: At the base of the section, silicification commences with the pseudomorphic replacement of clay minerals by opal, continues with the development of void linings of microlaminated opal, and then fibrous silica in residual cavities.

In the middle part of the section, clay minerals are replaced by microcrystalline quartz, while the silicification of clay-free sand is achieved by overgrowth of the detrital quartz grains. In both cases, the residual voids are filled with chalcedony and automorphic quartz.

In the upper part of the section, the silicification develops predominantly microcrystalline quartz in the matrix and in the titania-rich illuviation laminae at the base of voids.

Thus, a mineralogical profile develops with essentially opal-CT at the base of the section, while the upper parts are composed almost exclusively of quartz.

Water regimes and chemistry: The uppermost columnar silcrete displays numerous features relating to near-surface water infiltrations. The contrast between deposition and dissolution of silica records successive periods of deposition and leaching. The presence of microquartz reveals solutions without very high silica content, but with impurity elements that affect the size of the precipitated quartz crystals.

In the deepest horizons, the preserved sedimentary structures, as well as silica deposits around pores and the horizontally distributed lenses, have to be related to groundwater silicification. The silica deposits, mainly in the form of opal, indicate solutions with high silica content.

In the middle part of the section, structures are similar to those in the deeper part, but opal has recrystallized into quartz. Silicification started at groundwater level and proceeded with dissolutions and recrystallizations in a more dilute environment, probably above the water table.

Mechanisms of silicification: These different silicification processes occurred in the same landscape and are due to different mechanisms. At the surface, in the pedological silcretes, the solutions take silica during their infiltration and are concentrated by evaporation during dry periods. At depth a general acidification of the environment occurs, resulting in destruction of the clay minerals and providing silica phases with high solubility. Clay minerals are epigenetized and opal precipitated in the groundwater environment. During the plateau dissection, the water table drops and the former silicified level rises above the water table, in a less silica concentrated percolation zone, and the former opal cement recrystallizes into quartz. At a new water table level, another silicification horizon can develop.

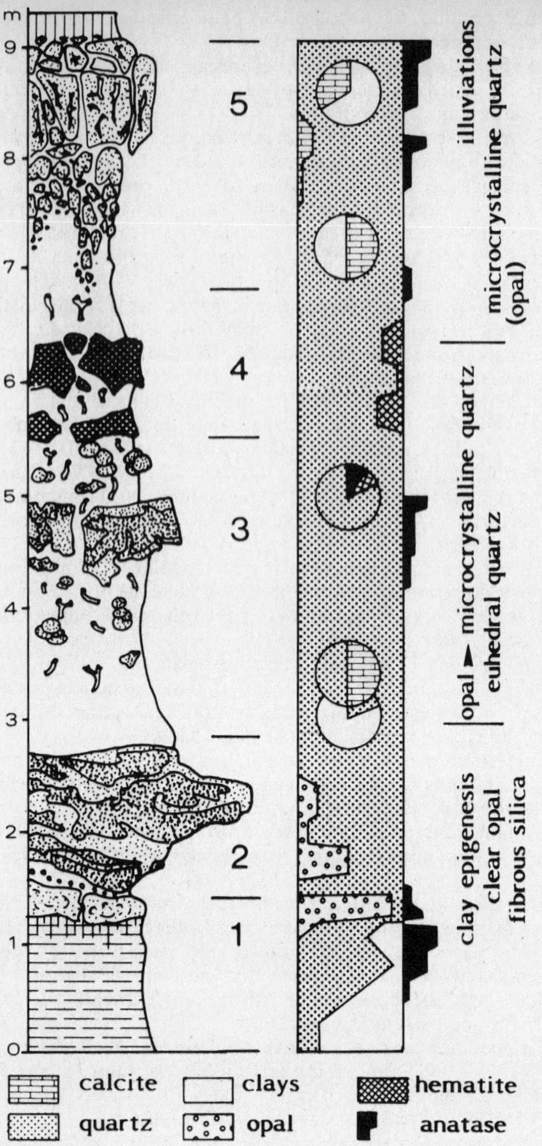

THOMAS, E., Wesleyan University, Middletown, CT

Late Cretaceous through Miocene High–Latitude Deep–Sea Environments

Recent drilling by the Ocean Drilling Program in the southern oceans has provided material for evaluation of the climatic development of Antarctica, a region crucial to our understanding of long-term climatic change. The earth's climate changed considerably during the Cenozoic, from warm and equable to the present-day climate with polar ice caps and strong oceanic latitudinal and vertical temperature gradients. Cooling probably proceeded step-wise, with the largest decrease in temperature at high latitudes. Polar cooling affects the formation of deep waters in the oceans: in the absence of large ice caps deep water may have formed at high latitudes by cooling, or at low latitudes by strong evaporation. Sources of deep waters in the oceans may thus have changed during the Cenozoic, leading to changes in deep-water characteristics. Deep-sea benthic foraminiferal faunas reflect such changes, and here I will discuss faunas from Ocean Drilling Program Sites 689 and 690 (Maud Rise, Antarctica).

At these sites (Site 689: 64°31.01'S, 03°05.99'E, depth 2080m; Site 690: 65°09.63'S, 01°12.30'E, depth 2914 m) we recovered Pleistocene through lowermost Maestrichtian siliceous, mixed siliceous-calcareous, and calcareous biogenic oozes. Ages were determined from the calcareous nannofossil zonation by Pospichal and Wise in Barker et al., in press, with the timescale of Berggren et al., 1985.

Changes in benthic foraminiferal faunal composition at the two sites were coeval (Fig. 1), and occurred in the Maestrichtian (/pmn69Ma), at the early/late Paleocene boundary (/pmn61.6Ma), in the latest Paleocene (/pmn58 Ma), in the middle early Eocene (/pmn55.0Ma), in the middle Eocene (/pmn45.5Ma), in the earliest Oligocene (/pmn36.5Ma) and in the early middle Miocene (/pmn14.5Ma). Assemblages were long-lived in the Paleocene and before, and in the late middle Eocene and after. The long-lived assemblages are dominated by trochospiral and planispiral species ("spiral species"), with fluctuations in relative abundance of species of a triserial-biserial group ("buliminids"). Short-lived assemblages occurred in the early and early middle Eocene, and are dominated by "buliminid" species. The "spiral species" probably libed as epifauna, and indicate environments with well-oxygenated water and fairly low levels of organic carbon. The "buliminid" species, probably infauna, may indicate lower levels of oxygen or high levels of organic carbon. Therefore faunal fluctuations within the long-lived Maestrichtian through Paleocene assemblages as well as in the late middle Eocene through Miocene assemblages may reflect fluctuating primary productivity.

There was no major benthic foraminiferal extinction at the Cretaceous/Tertiary boundary: there were fluctuations in relative abundance and less than 25% of the species disappeared. These minor effects probably resulted from the collapse of surface productivity, while deep waters were not strongly disturbed. In the latest Paleocene, however, large extinctions in deep-sea benthic faunas occurred globally. On Maud Rise the extinction occurred in less than 20,000 y: many common species disappeared and diversity dropped by 50%. This extinction was probably not caused by changes in surface-water productivity: the faunas had survived the great mass-extinction and collapse of surface productivity at the Cretaceous/Tertiary boundary.

Faunas living after the late Paleocene extinction were not diverse and dominated by "buliminids", suggesting that deep waters were poor in dissolved oxygen in the early and early middle Eocene (assemblages 4 and 5, Fig. 1). A composite oxygen isotopic record (Fig. 1) shows that at that time deep waters of the oceans were warmer than at any other time in the Cenozoic. This warming might have been caused by a change in sources of deep waters, which probably formed at high latitudes during the Late Cretaceous-Paleocene, resulting in well-oxygenated deep waters at Maud Rise. During the latest Paleocene deep-water production may have switched to lower latitudes, causing extinction of the deep-water faunas as a result of decreased availability of oxygen in the warmer waters. The change in deep-water sources might have been caused by late Paleocene changes in plate arrangement such as the collision of India and Asia, or opening of the North Atlantic. Increased plate tectonic activity could also have influenced deep water formation indirectly by increased production of CO_2 at mid oceanic ridges, thus increasing global temperatures.

Deep-water formation shifted back to the high latitudes, possibly in the middle Eocene, during a long-term decrease in temperature and changes in the deep-sea benthic assemblages (Fig. 1). An overall decrease in diversity from the middle Eocene on reflects the continuing cooling of the deep waters and the decrease in availability of $CaCo_3$, as shown in decrease in size and wall-thickness of the benthic faunas. A strong increase in relative abundance of the species Nuttallides umbonifera in the latest Eocene indicates increased corrosivity of the waters to $CaCO_3$. Thus deep-sea benthic faunal assemblages suggest that corrosive (very cold) deep water formed from the latest Eocene, which suggests that ice caps started to form at that time.

Benthic foraminiferal faunas thus indicate that Cenozoic changes in the formation of deep waters did not proceed gradually from Cretaceous circulation to the present-day patterns of formation of deep water, but constituted a complex process.

REFERENCES

Barker, P. F., J. P, Kennett, et al., eds., in press, Proc. ODP 113, pts. A and B.

Berggen, W. A., D. V. Kent, J. J. Flynn, and J. A. Van Couvering, 1985, Cenozoic Geochronology, GSA Bulletin 96, p. 1407–1418.

Miller, K. G., R. G. Fairbanks, and G. S. Mountain, 1987a, Tertiary isotope synthesis, sea level history, and continental margin erosion, Paleooceanography, v. 2, p. 1–20.

analyses indicate calorific values ranging from 6000 to 8000 Btu on an as-received basis.

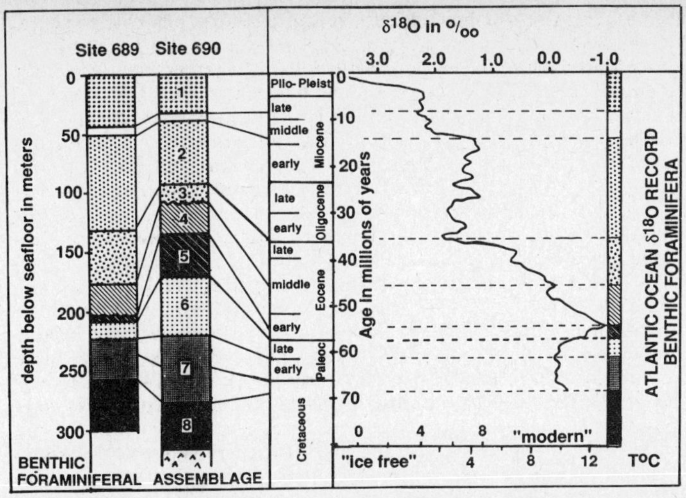

Figure 1—Benthic foraminiferal faunal assemblages at Sites 689 and 690 plotted vs. sub-bottom depth, and correlated with the time scale of Berggren et al., 1985, and the composite oxygen isotopic record of benthic foraminifera from the Atlantic Ocean (Miller et al., 1987a).

THOMAS, R. E., U. S. Geological Survey, Reston, VA and M. R. KHAN and S. A. KHAN, Geological Survey of Pakistan

Coal Exploration in Sind Province, Pakistan

Twenty-five coal exploration drill holes were completed between April 1986 and February 1988 in a 1400 km^2 area located approximately 80 km north of Karachi in southern Sind Province, Pakistan. The drilling was part of a Geological Survey of Pakistan, U. S. Geological Survey, and U. S. Agency for International Development Coal Resource Exploration and Assessment Program (COALREAP). This effort identified 11 major coal zones within the Paleocene Bara Formation. The upper 300 m of the Bara Formation consist predominantly of sandstone, claystone, and siltstone and contain most of the coal. The Bara Formation is overlain by approximately 120 m of Paleocene to Pleistocene limestone, sandstone, siltstone, and shale. Recent erosion has removed most of the Eocene-Pleistocene sediments in the study area. The majority of the coal resources in the Bara Formation are contained in the Sonda and the Inayatabad coal zones. The Sonda coal zone (average depth 200 m) contains the thickest coal beds. One coal bed, for example, penetrated in drill hole UAS-4, ranges from 6.3 m thick in the Sonda area to 2.4 m thick in drill hole UAK-15 east of the Indus River and apparently is restricted to a 50 km^2 area. The Inayatabad coal zone (approximately 60 m above the Sonda coal zone) contains coal beds that range from 0.35 to 1.50 m thick within a 272 km^2 area. These two coal zones contain most of the coal in the study area. Other coal zones contain coal beds that are extremely variable in thickness and are not as well defined as those in the Sonda and Inayatabad coal zones.

Thick (average 240 m) overburden on the Sonda and Inayatabad coal zones makes surface mining unlikely. Drill-hole data indicate that underground mining will be hampered by thick, unconsolidated sandstone beds that overlie many of the coal beds in the Bara Formation. Sections measured in the northern part of the study area (near Ganjo Takkar) have provided data on the stratigraphy and have demonstrated that the Indus River alluvial silt and sand are extremely thick, sometimes as much as 110 m. Collapse of this sand near the surface also presents a major roof problem in underground mining.

Coal beds are difficult to correlate because of abrupt pinching out and swelling. They are lenticular and grade laterally to carbonaceous shale and rooted siltstone and shale within 2 to 5 km. The coal beds commonly grade downward to carbonaceous shale and (or) darkgray to black claystone. Partings of carbonaceous shale and claystone are extremely difficult to distinguish from fresh coal, which is usually dark gray to brownish black and cleated in core. Seat rock can be claystone, sandstone, or siltstone or any combination of these. Carbonaceous shale commonly overlies the coal beds, but sandstone can be present also. Lithologies between coal beds within a zone generally consist of sandstone and (or) siltstone and a few thin claystone beds. As the spacing between drill holes increased, the relation of intertonguing sandstone, shale, siltstone, and coal became more difficult to define. A developmental program in which drill holes are closely spaced is needed to ascertain the stratigraphic relation in the coal-bearing sequence.

Analysis of 90 coal samples indicates a rank of subituminous C with average contents of 13.7 percent ash and 3.6 percent sulfur. These coal analyses indicate calorific values ranging from 6000 to 8000 Btu on an as-received basis.

THOMPSON, K. F. M., Texas A & M University, College Station, TX

Comparison of Petroleum Migration Phenomena in Clastic and Carbonate Sequences in the U. S. Gulf Coast Region

Over three hundred and fifty oil and gas-condensates from reservoirs in the deltaic Tertiary section of the offshore U. S. Gulf of Mexico, and eighty petroleums from reservoirs in Mesozoic carbonate-evaporite sections adjacent to the basin margins, have been analyzed. Striking difference in molecular weight distributions distinguish the two suites. The majority of petroleums in Tertiary reservoirs are depleted in gasoline-range hydrocarbons, while virtually all petroleums in Mesozoic carbonate reservoirs are rich in gasoline and light ends, preserving a pristine, exponentially decreasing gas-chromatographic envelope.

The compositional contrast between suites is principally attributable to fundamental differences in their modes of migration. Two models are suggested. In carbonate-evaporite sequences reservoir integrity appears to be extremely high, so that fault pathways for migration are few. Migration is intraformational and fractionation due to fugacity contrasts rarely takes place. Petroleums in deeply buried reservoirs undergo progressive cracking without loss of gas-condensate by dismigration.

In the clastic Tertiary section, dismigration is the rule, with evident effect in approximately 60% of the analysed fluids. Dismigration appears to involve the repeated segregation of a gas phase from gas-saturated accumulations of oil, and loss of the gaseous phase by migration, principally along faults. The term evaporative fractionation was suggested (Thompson, 1987) to describe the principal features of the modification petroleum light ends by such purely physical means.

The majority of Tertiary-reservoired Gulf Coast petroleums are depleted in light ends. Progressive losses exhibited by the contents of various reservoirs in South Pass Block 61 field were accompanied by proportional changes in the composition of the remaining gasoline fraction. Concentrations of aromatic and naphthenic hydrocarbons having six to eight carbon atoms increased (relative to normal-alkanes of matching carbon numbers) in proportion to the extent of gas and condensate loss. Evaporative fractionation phenomena were reproduced experimentally by repeatedly contacting an oil sample in a sealed cylinder with methane at elevated temperature and pressure. It is suggested that evaporative fractionation is driven by migrating natural gas generated at advanced maturity levels.

Vertical migration is principally responsible for the emplacement of Gulf Coast Tertiary petroleums as evidenced by the distribution of genetic characteristics. Over 350 petroleums have been fingerprinted employing carbon and sulfur isotopic ratios, as well as sulfur, nickel and vanadium contents. One of the recognized genetic groups comprises comparatively tightly clustered data: $\delta^{13}C$ = -26.5 /pmn 0.4o/oo, $\delta^{34}S$ = -5.6 /pmn 2.4o/oo, and V/9V + Vi) = 0.27 /pmn 0.10, suggesting generation from a specific, unique source formation. Close isotopic comparison with Mesozoic oils from the Campeche area of Mexico suggest that a substantial proportion of the Tertiary-reservoired Gulf Coast oils could be of Mesozoic origin. They are, however, broadly distributed in reservoirs of Miocene, Pliocene, and Pleistocene age, due to vertical migration.

In contrast, oils having a known Jurassic source occur in the Oxfordian Smackover Limestone of Alabama and have been emplaced by intraformational migration. The oils are readily distinguishable on the basis of isotopic characteristics: $\delta^{13}C$ = -23.4 /pmn 1.13o/oo, $\delta i^{34}S$ = +2.4 /pmn 7.3o/oo. Their source region is suggested to lie within the Mississippi salt basin, adjacent to the west. The near absence of vertical migration is indicated by the limited occurrence of Jurassic oils in Cretaceous reservoirs in Alabama. Of 68 producing Mesozoic fields, 61 are Jurassic.

There is evidence that intraformational (or pincipally bedding-parallel) migration may repeatedly follow identical pathways at different times in the catagenetic evolution of the source region. A cluster of fields in close geographic and stratigraphic proximity in Escambia County, Alabama, exhibits an association of mature oils and supermature gas-condensates. Maturity differences are confirmed by carbon and sulfur isotopic data. Significantly, the gas-condensates are not depleted in light ends. Dismigration and evaporative fractionation are absent. Relative to naphthenes, light aromatics and paraffins are substantially elevated, due in this instance to advanced maturity.

Petroleum maturity was investigated employing two systems, one based on gasoline range compounds, the other on three-ring aromatic hydrocarbons. The first system employed isoheptane values (I) (Thompson, 1979) and heptane ratios (F - n- heptane/methylcyclohexane), while the second employed methyphenanthrene indices, MPI1 (Radke, 1982) and MP 13 (Garrigues, 1985). Paired petroleum data were compared with the appropriate "generation curve," or "concordia," based on rock extract compositions representing series of well samples of increasing, measured, maturities. Petroleum compositions which differ substantially from the generation curve are construed as altered. Equivalent, calculated, vitrinite reflectances were determined for concordant petroleum compositions.

In Tertiary reservoirs, over 95% of oil and gas-condensate accumulations are altered by evaporative fractionation and/or biodegradation. Both processes result in diminished values of F and I, relative to the generation curve, leading to spurious indications of immaturity. In oils, paired values of MPI1 and MPI3 are distributed close to concordia. A sub-set of gas-condensates is distinguished by abnormally low values of MPI1, suggestive of low maturity, accompanied by gasoline compositions indicative of extremely high maturity, accompanied by gasoline compositions indicative of extremely high maturity (mean equivalent reflectance 1.4%). A process of solvent extraction of immature phenanthrenes from reservoir sandstones, that is, migration contamination, is postulated in these instances.

Data representing MPI1 and MPI3 in Gulf Coast carbonate-reservoired oils and gas-condensates define a narrow band that progressively departs from the relevant concordia, exhibiting increasingly anomalously high values of MPI1. Similarly, in certain cases, values of F and I trend away from concordia, involving elevated values of both parameters. Loss of light aromatics suggest interaction with water in numerous instances. Published aqueous solubility and experimental data suggest that observed departures from Concordia in both systems, in the cases of certain carbonate-reservoired petroleums, are attributable to aqueous fractionation during migration.

THOMPSON, R., University of Edinburgh, Edinburgh, United Kingdom

Quantitative Core Correlation

The problem of matching geological and geophysical sequences has been investigated using the method of sequence slotting. An optimal slotting is found by minimizing the combined path length of the sequences based on pairwise dissimilarities. Dynamic programming procedures allow a very efficient solution to the slotting problem.

The sequence slotting approach has recently been modified and extended in order to: (1) allow neighborhood and precedence constraints, e.g., the occurrence of tephra layers of marker beds; (2) include block size constraints, e.g., to limit relative accumulation rates; (3) perform multicore studies, e.g., on transects or on grids; and (4) assess the quality of match through confidence limits and similarity/correlation statistics.

The sequence slotting method has been applied to paleomagnetic, mineral magnetic, chemical (carbonate and heavy metal), isotopic, radiolaria, diatom, and pollen data. Magnetic susceptibility has been found particularly useful for within-basin correlations.

Multivariate data can be analyzed as easily as univariate data, thus allowing paleomagnetic data to be analyzed on the sphere or combinations of data (e.g., chemical and diatoms) on equivalent horizons to be matched simultaneously.

As in many quantitative methods in the earth sciences, the question of scaling (both through data transformation and relative weightings) is of paramount practical importance. The inelegant but robust approach of simplex optimization has been found to be a useful method of assessing linear data scaling transformations, while equal weightings have been given to either the unit normalized observations or their principal components.

Various approaches have been tried in order to assess the quality of match of sequences. These include slotting statistics, simulation experiments, sensitivity tests and partioning procedures. The most useful measure of the goodness of fit of two sequences has been found to be provided by an interpolated correlation coefficient. A correlation value of 1.0 indicates a perfect match.

Radiolarian data from ocean cores RC14-106 and V20-123; calcium carbonate concentrations from sequences DSDP-157 and RC12-66, and O^{18} data from cores P6304-9 and P6408-9 have similar interpolated slotted correlation coefficients of 0.8.

THOMSON, M. R. A., J. A. CRAME, and P. DOYLE, British Antarctic Survey, Cambridge, United Kingdom

Mesozoic Paleogeography of Antarctica: Seaways and Oceanic Connections

With good reason it can be claimed that the Mesozoic was a truly critical period in earth history. It was a time that witnessed the breakup of the supercontinents of Laurasia and Gondwana, and a fundamental rearrangement of the areas of continent and ocean on the surface of the globe. Seaways opened to provide new migration routes for some marine organisms and, eventually, deep-water barriers to the distribution of shelf faunas. Once-contiguous tracts of land became separated by new oceans, whereas others were to move closer together, providing new challenges for the land biotas.

Antarctica formed the center of the southern supercontinent of Gondwana. East Antarctica, a huge cratonic area the size of the USA, was probably mostly land through Mesozoic time and initially connected directly with the southern margins of Africa, India, and Australia. However, the possibility that some of its present subice depressions were even then low enough to have received incursions from Mesozoic seas should be borne in mind. By contrast, West Antarctica had a more complex paleogeographic setting as a volcanic archipelago that formed part of the proto-Pacific margin and acted as a connection between South America and New Zealand, subject to major changes in the distribution of land and sea.

Paleogeographic reconstructions can be made on the basis of seafloor and paleomagnetic data and by the mapping of facies as displayed in rock outcrops on- and offshore. However, the test of their validity rests largely on whether they satisfy the observed distribution of fossil biotas and the paleoenvironmental conditions the fossils and their enclosing rocks indicate. We maintain, however, that, except in rare instances confined essentially to the Cenozoic, chronostratigraphy is not well enough constrained to determine the dispersal directions of fossil organisms. Care must also be taken not to underestimate facies control when comparing the distribution of fossil biotas. Ideally direct comparisons can only be made when dealing with analogous facies, otherwise true differences and similarities may be easily distorted.

In Triassic time Gondwana was entire. The presence of *Glossopteris* floras and a *Lystrosaurus* vertebrate fauna in East Antarctica points to the free dispersal of land animal and plant groups throughout much of the supercontinent. Two poorly diverse marine faunas, both from the northern part of the Antarctic Peninsula, contain bivalves with circum-Gondwana and circum-Pacific affinities. Early Jurassic marine faunas are present in only one restricted area of Alexander Island but include an ammonite of cosmopolitan affinities. However, all of these marine faunas occur in accretionary complexes and could be exotic. The Middle Jurassic faunal record for Antarctica is again sparse. In the Northern Hemisphere, it was a time of major faunal differentiation between the Boreal and Tethyan Realms, but in the Southern Hemisphere marine faunas seem to have been generally more cosmopolitan and of largely Tethyan affinities.

The first splits in Gondwana, sufficient to allow incursions of the sea, probably appeared in Late Jurassic times with the initiation of a seaway along the Eastern coast of Africa—The "trans-Gondwana seaway." At first this connection is likely to have taken the form of a series of barred anoxic basins, and it is difficult to assess precisely when a free marine connection was established between Tethys and the Antarctic Peninsula region on the Pacific margin. Affinities between the marine faunas of the Peninsula, Madagascar, the Himalayan region, Indonesia, and New Zealand can be explained just as well by circum-Gondwana migration as by movement between a shorter, but probably restricted seaway across Gondwana. However, some faunal groups, especially among the belemnites, appear to have been endemic to the trans-Gondwana seaway, identical species being restricted to Madagascar, Antarctica, and Patagonia.

The initiation of opening in the South Atlantic began in earliest Cretaceous times and progressed northwards until a connection with the North Atlantic was established in Aptian–Albian times. During its earlier history, the Rio Grande–Walvis Ridge and the Falkland Plateau acted as shallow water barriers across the developing ocean. Australia was still essentially part of Antarctica, and much of its central regions was flooded by an epicontinental sea. There are extensive deposits of richly fossiliferous rocks of this age in the Antarctic Peninsula region that permit a good assessment of the faunal affinities and migrational pathways. Clear examples of cosmopolitan forms contrast with biotas of distinctly austral aspect that mirror those found in the boreal region. At this time, or possibly somewhat earlier, the Pacific margin of Antarctica fragmented into a number of discrete crustal blocks that underwent a degree of movement and rotation relative to each other. The region thus became an archipelago of islands, shelves, and intervening basins. Nevertheless, any developing barriers were clearly not substantial enough to prevent the dispersal of the marsupial mammals and plants, such as the southern beech, *Nothofagus*.

In the Late Cretaceous, first New Zealand and then Australia broke away from Antarctica. The progressive increase in the size of the new oceanic areas during the Late Cretaceous, together with the dispersal of the continental fragments, led to increasing endemism and, in Antarctica, the formation of the so-called Weddellian province. There is evidence from Australia for the development of small ice caps during the Cretaceous but none as yet from Antarctica, where glaciation seems to have been an essentially Cenozoic event.

THOREZ, J., Liège State University, Liège, Belgium

Paleosol Setting, Diversity, and Evolution in Deltaic Complex of Condroz Sandstones (Upper Famennian) in Belgium

The Upper Famennian (Upper Devonian) Condroz Sandstones in Belgium display a whole spectrum of laterally and vertically interfingering depositional environments as a result of the progradation of a deltaic complex. Among these environments occur: (meandering) fluviatile, alluvio-lagoonal, evaporitic and tidal lagoonal facies, inshore and offshore barrier complexes, tide-, wave-, and storm-influenced siliciclastic and mixed carbonate-siliciclastic deposits, tempestites, fluxoturbidites, and outer shelf muds. All these environments have diachronically developed during the progradation that took place on the Condroz Shelf, south of the London–Brabant Massif. Moreover, they were influenced by the basin paleomorphology and affected by local differential subsidence, sea-level changes, synsedimentary tectonics (block tilting), and paleoclimatic oscillations. During the deposition of the Condroz Sandstones, the prevailing climate was generally arid to subarid, but short wetter periods occurred that favored the fluviatile input and, more important, the development of a variety of paleosols. The latter appear in the fluviatile, evaporitic lagoonal and in the supratidal environments. These paleosols comprise red inceptisols, vertisols, and dolocretes usually developed at the top of fining-upward sequences. Some of the dolocretes are typically polycyclic and were former vertisols. In the evaporitic lagoonal environment, some of the "primary" dolomitic layers that cap the sedimentary rhythms are often calicified.

Besides field characterization, these various paleosols have been investigated by micromorphology—a technique usually referred to in the study of the recent soils. The micromorphological analysis was systematically completed by XRD and geochemical analyses, their data confronted to the characteristics of the sediments not affected by pedogenesis. Despite the influence of diagenesis, the clay mineral contents of the inceptisols, vertisols, and colocretes are still different from those of the unaffected sediments. Macroscopic and microscopic paleopedological features are well recognizable and interpretable: pedotubules, skew-planes, desiccation cracks, concentric dolomitic nodules and glaebules, pseudogley patches, calcitic crystallaria, plant root residues, dolomitic clasts, color mottling, k-fabric, etc. The stratigraphic and paleogeographic setting and diversity of the Upper Famennian paleosols further document the paleogeographic and paleoclimatic reconstruction of the Condroz Sandstones in Belgium.

THUNE, T. G. O., University of Stockholm, Stockholm, Sweden

Plant Communities of Younger Dryas and Their Evolution by Influence of Friesland and Piottino Phases East of Mt. Billingen, Sweden, Found in Studies of Macrofossils Compared with Observed Results of Pollen Analysis

With the open mine at Ranstad, an industrial innovation in the early 1960s, a new place of reference of fossil plants was discovered. Alfred G. Nathorst about a hundred years ago used fossil plants as the basis of the documentation when he showed the existence of the glacial tundra in southernmost Sweden and on the North European Continent.

The tremendous increase of fossil plants at Ranstad dated at the end of Younger Dryas at 10,110 ± 170 B.P. (St 3433) confirm the contemporaneous rise of *Pinus* pollen in the warm Friesland phase (Behre, 1967).

At the places of research, Ranstad and the town of Hjo, edaphic conditions with lime in the soil have favored habitats of calciphilous plants. These two locations east of low uplands have provided for snow accumulations, generating show-bed habitats of chianophilous plants in addition to oceanic climate at the end of Younger Dryas (Berglund, 1966, p. 101, 108).

As a method of research, I have applied the cover degree of plants in the field layer of recent snow-bed communities in the Swedish mountains (Gjaerevoll, 1956) to the frequency of different macrofossils resulting in ancient plant communities.

On one hand, *Salix polaris* have been found predominant in litter assemblages accompanied by the less frequent *Polygonum viviparum* supplemented with *Silene aucaulis* and *Taraxacum*. On the whole, the common character is of a *Salix polaris* heath (Gjaerevoll, 1956, p. 259).

On the other hand, *Saxifraga oppositifolia* as litter is just as frequent as *Salix polaris*. Besides, it is predominant below snow-bed sheets in different meadow communities with companions of species preferring oozing water. Less frequently, *Oxyria digyna* is found supplemented with *Saxifraga aizoides* (Gjaerevoll, 1956, p. 302) and *Armeria sibirica*. The moss genus of Polytrichum is supplemented in the heath and the meadow communities.

These kinds of stable plant associations are climax communities where snow forms the vegetational environment.

As to the Dryas flora, the older idea of windswept habitats on drumlins is being realized in the fossil record as Dryas, *Juncus trifidus* (wind-eroded straw-basis), *Loiseleuria procumbens, Saxifraga groenlandica, Salix reticulata*, and *S. herbacea* where snow masses.

In the stratigraphy at Ranstad, I found an Allerod flora on macrofossils of *Artemisia, Gramineae, Betula*, and the unique species of *Potentilla egedii*. Disregarding the last one Digerfeld (1979, p. 62) found through pollen analysis the same assemblage west of Mt. Billingen. According to my research at Ranstad, all of these became extinct early in Younger Dryas. The first two above together with *Cyperaceae* reappear in dominant frequency at the end of Younger Dryasa and concerning *Betula* in the Preboreal as the pollen analysis at Ranstad records (Hedin, 1985, SGU, p. 9). Judging from these facts, the reappearance of the Allerod plants at the end of Younger Dryas emanate from a process of resedimentation and does not agree with the basis of a real tundra flora of the Younger Dryas.

The oceanic climate above gave rise to heavy snowfall and increased erosion, which is the reason litter and detritus of the snow-bed communities accumulated in depressions of old drumlins dated at 9980 ± 175 B.P. (St 4592) and 9935 ± 180 B.P. (St 4593). After flourishing for 130–175 years (= ^{14}C years) the arctic alpine flora was extinct in 45 years at a new glaciation initiated during the cold Piottina phase (Zoller et al., 1966).

At a glacioisostatic transgression in the basin of Lake Vättern in the Preboreal, the arctic alpine plants still alive were washed down on the proximal sedimentary-plain at Hjo, at the country-seat of Å sen, dated at 9200 ± 255 B.P. (U-4330). Until further research is done, this survival here for about 800 years will remain a mystery.

The afforestation at Ranstad began with *Pinus* established at 8500 ± 160 B.P. (St 4590), which means 1400 years after the extinction of the Younger Dryas flora. The accepted idea that the heliophytic arctic alpine plants "succumbed in the shade of the postglacial forest" (Iversen, 1954) is not valid in this part of Sweden.

The continental times of the Friesland and Piottino phases have their upper boundaries dated at 10,250 and 10,100 B.P. Compared with ^{14}C dates east of Mt. Billingen they are delayed about 150 and 100 years, respectively. The permanent glacial state at Mt. Billingen for about 700 years ending at a disintegration of the glacial-ice dated (collected by SGU) at 9335 ± 120 B.P. (St 6402) and 9290 ± 110 B.P. (St 6400) delays the lower boundary of the Piottino phase here from 9700 B.P. to about 9300 B.P. This event is stratigraphically indicated as an inundation that has had consequences in the view of resedimentation at the afforestation by *Betula* in the Preboreal.

REFERENCES

Behre, K. E., 1967, The late glacial and early postglacial history of vegetation and climate in northwestern Germany: Rev. Palaeobotany and Palynology, v. 4, p. 149–169.

Berglund, B. E., 1966, Late-Quaternary vegetation in Eastern Blekinge, south-eastern Sweden: Opera Botanica, v. 12, n. 1, p. 1–180.

Digerfeldt, G., 1979; The highest shore-line on Hunneberg, southern Sweden: Geological Society Sweden (GFF) Stockholm, v. 101, n. 1, p. 44–64.

Gjaerevoll, O., 1956, The plant communities of the Scandinavian alpine snow-beds: Det Kongl. Vidensk. Selskabs Skr., n. 1, p. 1–406.

Hedin, K., 1985, Biostratigrafisk dokumentation av lagerföljder från Ranstad: Geological Survey Sweden (SGU), p. 1–11.

Iversen, I., 1954, The late-glacial flora of Denmark and its relation to climate and soil: Geological Survey Denmark, Ser. 2, n. 80, p. 87–118.

Zoller, H., C. Schindler, and H. Röthlisberger, 1966, Postglaciale Gletscherstände und Klimaschwankungen im Gotthard-massiv und Vorderrheingebiet: Verhandlungen der Naturforschenden Gesellschaft in Basel, v. 77, n. 2, p. 98–137.

THUROW, J., Universitaet Tuebingen, Tuebingen, Federal Republic of Germany

Late Mesozoic Supra-Regional Paleo-Oceanographic Events: Sedimentary, Geochemical, and Faunistic Evidence from Atlantic to Central Tethyan Realm

The Mesozoic pelagic realm is a favorable environment for the preservation and detection of paleo-oceanographic events. These events contribute important information about the sensitivity of the ocean/atmosphere system to outside influence. Such short-termed events are well studied now for the Cretaceous, while the resolution for the Jurassic is less precise. We anticipate that ODP Leg 123 will provide valuable data to improve this lack of understanding of Jurassic paleo-oceanography.

Several major irregularities in the geological record, even disrupting major sedimentary cycles, on various scales of intensity and distribtution have been detected in the Late Mesozoic. All events have in common: (1) onset or end of "event-characteristic" lithologies (e.g., black shales, radiolarites), (2) abrupt faunal changes or distinct extinction/evolution patterns, and (3) sudden changes in sedimentary composition and geochemical patterns (detrital input). Furthermore, isotopic excursions and trace metal anomalies are recorded from several Cretaceous events.

We studied Mesozoic sequences in various types of settings and within the scope of a multidisciplinary study combining sedimentological/geochemical and paleontological/stratigraphical data. (1) DSDP/ODP Sites in the North Atlantic, (2) onshore sections along bathymetric transects in the Western Mediterranean as well as in the Himalayas, and (3) ODP Sites in the former Central Tethys. The most distinct and widespread global events are: (1) the K-T boundary event (65 m.y., not discussed in detail); (2) the LCE (Lower Campanian Event) (84 m.y.); (3) the Cenomanian–Turonian Boundary Event (CTBE) (91 m.y.); and (4) the "Selli Event" (Lowermost Aptian) (117 m.y.).

Not very well correlated yet are events in the Valanginian, in the Kimmeridgian–Tithonian, and in the Callovian–Oxfordian.

Sedimentary, geochemical, and faunal changes at these horizons are related to paleo-oceanographic changes such as the distribution of water masses, position of the CCD, and seawater chemistry (especially the degree of oxygenation). Initiating factors are various, and often no single process is responsible for the event. Such processes may include: (1) changes in the current pattern and/or ocean topography, (2) major climatic changes, (3) extraterrestrial impact, (4) volcanic (tectonic) processes that led to: CO_2/SiO_2-exhalations, trace metal enrichments, and (5) transgression/regressional-cycles (variations in terrigenous-detrital input, variations in availability of nutrients, decline or increase of life habitats). We will discuss these events in the light of: (1) global vs. regional/local factors, (2) litho- and biofacies of event-intervals, (3) diachrony or synchrony of these events on a global scale, (4) high fertile/low fertile oceans/stagnant oceans, (5) tectonics and crustal evolution (intensity of spreading rates), and (6) possible cyclicity in the event record, and if so, can it be traced into the paleozoic? (evidence for a 26 m.y. periodicity of the "extinction/evolution" rates in the Phanerozoic, as it was postulated by several authors in recent years, is given by the 65 m.y./91 m.y./117 m.y. of the K-T, CTBE, and Selli Events). Other distinct events, however, do not match this pattern.

THUROW, J., Universitaet Tuebingen, Tuebingen, Federal Republic of Germany, M. GIBLING, Wollongong University, Australia, Lost Ocean Expedition, and Shipboard Scientific Party Leg 123

Hydrocarbon Potential, Organic Matter Diagenesis, Sedimentology, and Paleoenvironment of Upper Jurassic Dark Shales, Northern Himalayas and Argo Abyssal Plain

The Upper Jurassic was a time favorable for the deposition of "black shale"-type sediments in shallow environments as known from circum-North Atlantic basins, North Sea, and Himalayan Tethys regions. Locally, these shales have excellent hydrocarbon source potential.

The site of the Spiti-Shales in the Thakkola region of North Central Nepal provides the opportunity to study a long-term (Oxfordian–Tithonian) stable, shallow, and oxygen depleted environment. Horizons with calcareous benthic communities show that the environment was not always anoxic. Organic geochemical analyses and sedimentological studies of sediments obtained from the Spiti Shales (Oxfordian–Valanginian) were done in order to understand the hydrocarbon potential, organic matter diagenesis, sedimentology, and paleoenvironment of these sequences. The depositional environment, driven by tectonoeustatic and climatic events, changed from an open shelf (approximately 250 m) with low amount of detrital input and characterized by rich macrofossil communities and siderite concretions to an extremely shallow, partly continental environment with intercalations of quartzose clastic channel-fills, silty shales, rare lumachelle layers and coal seams ("Wealden"-type facies). The thickness of this sequence is at least 250 m. Paleocurrents in the "Wealden"-type sandstones indicate a transport toward the north, hence it is likely that the continental margin faced northward, bordering the Tethys ocean. The organic matter changes from more marine (Jurassic) to purely terrestrial material in the Cretaceous, highest enrichment found in the Tithonian "Saligram" sequence. The latter might be also the source for the eternal flame in the Hindu Temple of Mukhtinat.

Kerogen and vitrinite reflectance point to a mature or slightly overmature hydrocarbon potential, thus allowing a rough estimation of the tectonic history of the area. Our carbon and sulfur isotope studies will confirm the paleoenvironmental interpretation.

Analysis of coeval strata, deposited in the deep marine environment off the northern Indian shelf (contiguous with the present-day Argo Abyssal Plain) demonstrates the changing hydrocarbon potential from shallow- to deep-marine settings. It reflects, to some extent, the more advanced organic matter maturation of the onshore material due to Himalayan tectonics and allows us to trace the transport of the organic matter.

THURSTON, P. C., Ontario Geological Survey, Toronto, Canada,

Platform Sequences of the Archean Superior Province: Evidence for a Central Sialic Nucleus

Archean sediments have been considered to represent the alluvial-fluvial and the abyssal environments with few representatives of an extensive shelf or platformal environment such as quartz arenite-carbonate sequences. This has led to speculation about a unique environment of deposition in the Archean without the normal proportion of platformal environments (Ojakangas 1985). Widespread crustal stability was considered a mainly Proterozoic and Phanerozoic phenomenon.

In the Superior Province of the Canadian Shield, Wood et al., (1986) described quartz arenite-bearing platform sequences in four greenstone belts of the Wabigoon and Sachigo Subprovinces. They are similar in stratigraphy, primary structures, paleoenvironment and age (approaching 3 GA). The sequences are older than adjacent mafic to felsic volcanic cycles. The generalized stratigrpahy consists of lower quartz arenites and associated granitoid-rich conglomerates /pmn stromatolitic carbonates, succeeded by iron formation and mafic/ultramafic volcanics. Subsequent work has shown that quartz arenite-bearing sequences occur in every major greenstone belt of the Schigo subprovince, representing scattered occurrences over an are of 60,000 km^2 (Thurston et al., 1987). Similar sequences occur in the Wabigoon Subprovince over an area of 35,000 km^2 (Thurston et al, 1987) and scattered occurrences are found in the La Grande Riviere Subprovince (Roscoe and Donaldson, 1988) and the Slave Province (Ashton, 1982; Covello et al., 1988).

The quartz arenites are supermature sediments with coarse (~ 5 mm) to fine highly rounded quartz and common clast rich beds with weathered felsic volcanics and chrome-rich mica clasts. A granitoid provenance for the quartz arenites is based upon the petrography and heavy mineral suite of the arenites and abundant granitoid clasts in the conglomerates (Thurston et al., 1987; de Kemp, 1987; Wilks and Nisbet, 1988). Primary structures in the quartz arenites include sandstone interbeds in conglomerate units, trough cross beds, herringbone cross beds, and hummocky cross beds (deKemp, 1987; Cortis, 1988; Thurstone et al., 1987) all indicative of a shallow water origin. Stromatolite-bearing sedimentary carbonates and siliciclastics suggest deposition in the photic zone. Iron formations are associated with karst topography on underlying carbonates, subaerial volcanics, and they include parallel lamination and slump structures suggesting deposition in shallow water. Primary structures in the komatiitic and tholeiitic volcanic upper units include accretionary lapilli at Steeprock Lake (Wilks and Nisbet, 1988) and delicately crossbedded, thin-bedded komatiitic tuffs and sediments at North Spirit Lake (Wood et al., 1986) suggestive of proximal, shallow water deposition.

The quartz arenite-carbonate lithological association, the persistence of the shallow-water environment, and the geographic extent of the occurrences suggests the quartz arenite-bearing sequences represent a widespread shallow water platform analogous to modern platformal sequences. Limited facies analysis suggests the dominance of proximal platformal sequences but some limited submarine fan and abyssal environments.

The sequence in some instances lie unconformably upon granitoid units (Wilks and Nisbet, 1988) or volcanic sequences (Piroshco and Shields, 1985). Where relations are clear, platform sequences are overlain by mafic to felsic cyclical volcanics. The stratigraphic data, provenance, assemblage of primary structures and facies analysis suggests these quartz arenite-bearing sequences may represent a widespread supracratonic platform over much of the Sachigo Subprovince.

Volcanism associated with these sequences is generally minor volumes of subaerial felsic ash-flows and greater volumes of tholeiitic or komatiitic subaqueous flows or shallow water pyroclastic units. Geochemically, the mafic/ultramafic magmas are unevolved liquids with unfractionated REE patterns which fractionate to evolved basalts with no related andesites or siliceous derivatives. This geochemical pattern, paleoenvironmental data, the stratigraphic position and volume relations suggest the volcanism is related to generally late rifting after development of the quartz arenite-bearing sequences.

The platform sequences stratigraphically underlie or yield older absolute ages than mafic to felsic volcanic cycles which typify much of the Superior Province (Thurston and Chivers, 1988). The age of the second deformation, related to subhorizontal transport along deformation, related to subhorizontal transport along subprovince boundaries, youngs more or less systematically southward from the Sachigo Subprovince, the region with most extensive platform sequences (Stott, et al., 1987). From this relationship and the granitoid provencance of the platform sequences, it is suggested that the Sachigo subprovince is an ancient sialic nucleus around which the subprovinces to the south accreted.

REFERENCES

Ashton, K. E., 1982, Geological Survey Canada Paper 82-1A, p. 151–157.
Cortis, A. L., 1988, Ontario Geological Survey Misc. Paper 141.
Covello, L., S. M. Roscoe, J. A. Donaldson, D. Roach and W. K. Fyson, 1988, Geological Survey of Canada Paper 88-1C, p. 223–232.
De Kemp, E., 1987, unpub. M. Sc. thesis Carleton U, 93 p.
Ojakangas, R. W., 1985, Geological Association of Canada Spec. Paper 28, p. 23–47.
Piroshco, D., and H. Shields, 1985, Ontario Geological Survey Misc. Paper 126, p. 277–286.
Roscoe, S. M., and J. A. Donaldson, 1988, Geological Survey of Canada Paper 88-1C, p. 117–121.
Thurston, P. C., A. L. Cortis, and K. M. Chivers, 1987, Ontario Geological Survey Misc. Paper 137, p. 4–24.
Wilks, M. E., and E. G. Nisbet, 1988, Canadian Journal of Earth Science, v. 25, p. 370–391.
Wood, J. P. C. Thurston, F. Corfu, and D. W. Davis, 1986, GAC Abstracts, v. 11, p. 146.

THURSTON, P. C., and G. M. STOTT, Ontario Geological Survey, Toronto, Canada

Secular Progression and Episodicity in Superior Province

New data have demonstrated that the Superior Province, comprised of fault-bounded subprovinces of contrasting lithologies, is also a more heterogenous framework comprising widespread older sialic crust separated by either voluminous volcanics extruded over a short time span (e.g. the external zone of the Atitibi Subprovince over 50 Ma) or belts of fault-bounded panels with volcanics of a range of ages (e.g. Uchi Subprovince over a 300 Ma age range).

The Superior Province contains a cyclic pattern of volcanic lithostratigraphic associations, volcanic ages, sedimentary basin development, the evolution of individual granitoid complexes and the sequence of deformation events. An exception to this cyclicity is the restriction of shallow-water supracratonic platform sequences to units 2.85 GA old. We note the widespread distribution of platform sequences comprised of: basal quartzite / pmn stromatolitic carbonate, iron formation and an upper komatiitic unit. These sequences are found in most greenstone belts of the older Sachigo Subprovince and older parts of the Wabigoon Subprovince. The restriction of these platform sequences to older parts of the Superior Province may reflect a long term secular trend commencing with the platformal conditions of an Archean supercontinent.

A second lithostratigraphic association of mafic volcanic sequences represents early oceanic volcanism. These sequences comprise voluminous tholeiitic to komatiitic flows with abyssal muds, sequences analogous to submarine lava plains and central volcanic complexes.

There is a consistent secular progression to the next lithostratigrpahic association-mafic to felsic volcanic cycles (2.8 Ga) that typify Superior Province greenstone belts with a basaltic base overlain by lesser dacite to rhyolite. Deep water shield volcanoes shoal upward from basaltic to bimodal volcanism succeeded by volumninous andesite-rhyolite volcanism in the Abitibi and Uchi Subprovinces. The facies variation, volcanic style, geochemistry and petrography suggest comparison to modern magmatic arc sequences.

A fourth lithostratigraphic association consists of "Timiskaming"-type fault-bounded basins with fluvial conglomerates and sandstones, and calcalkalic to alkalic volcanics. These sequences are 2700–2685 Ma old. The facies variation over the length and breadth of the fault-bounded basins suggest comparison to modern pull-apart basins.

Precise U-Pb zircon dating of felsic volcanics across the Superior Province shows a clustering of ages: >2.9 Ga, 2800–2850 Ma, and 2775 to 2700 Ma. The clustering of ages is accompanied by variation in the age span of volcanism in various subprovinces. In the Abitibi and western Wabigoon Subprovinces volcanism is voluminous over a short time span (2775 to 2685 Ma) yet in the Uchi Subprovince there is a rangc from ~3.0 Ga to ~2.7 Ga volcanism, younging southward. We now recognize major discontinuities forming boundaries of repeated sequences or out-of-sequence packages of supracrustal strata in the Uchi Subprovince.

The large granitoid batholiths in granite-greenstone subprovinces are complex yet possess similar patterns-older foliated genissic tonalite forming domes within the core of the complexes and younger plutons as central stoping intrusions piercing the domes or as crescent-shaped sheet intrusions invading along the greenstone-batholith contacts. The crescentic plutons typically mark the termination of magmatic evolution in the granitoid complex: This pattern of domes and crescents is preserved in the 3 Ga Marmion Lake granitoid complex of Wabigoon Subprovince, in the 2.86 Ga North Caribou Lake complex of Sachigo Subprovince and in the 2.7 Ga Lake St. Joseph granitoid complex of Uchi Subprovince.

Sedimentary subprovinces are largely metamorphosed sandstones with rare conglomerates, mafic volcanics and mafic sediments cut by calcalkaline and peraluminous batholiths. Zircon studies sediments from Quetico and Pontiac Subprovinces suggest a provenance involving contributions from 3.0–2.7 Ga sial, volcanic and granitoid debris ~2.7 Ga, and slightly younger material suggesting sedimentation ceased after adjacent greenstone belt volcanism. Using Quetico Subprovince as a model, Percival and Williams (1988) described north-younging north-verging thrust panels of sediments and volcanics against the older Wabigoon granite-greenstone subprovince to the north. With increasing distance southward away from the greenstone, the thrust panels reflect alluvial through submarine fan to abyssal environments. Kinematic criteria indicate a consistent south-down sens of movement. The lateral facies changes, wacke-rich sediments, consistent younging direction, fold pattern, and kinematics of the faults suggest origin as a northward converging accretionary prism. The English River and Pontiac Subprovinces may be similar, but intense deformation precludes direct comparison.

Posttectonic plutons are clustered in age and type. In the Uchi subprovince they young southward and parallel the age asymmetry of the volcanism. Some suites of plutons bear a spatial relation to the faulted subprovince and to "Timiskaming"-type volcanics.

Finally, elements of a secular progression occur in regional deformation patterns. A cyclic pattern of orogeny involving volcanism, plutonism and deformation is repeated in different parts of Superior Province. The assembly of volcanic suites of diverse ages to form a greenstone belt culminated in broadly penetrative deformation including thrusts that produced tectonic repetition and out-of-sequence stacking of tectonic slices. This is followed by folding with moderately plunging axes and lineations and development of shear/fault zones. Strike-slip or moderately-plunging slip along shear zones terminates the main deformation in greenstones apart from pluton-induced strain aureoles. These moderately plunging structures are preserved in broader greenstone belts (e.g., Abitibi-Wawa Subprovince). Smaller belts are more prone to imposition of contact strain aureoles upon adjacent supracrustal rocks by rising granitic diapirs.

A more regional stage of transpressive deformation is centered on the metasedimentary subprovinces and affects adjacent granite-greenstone subprovinces mainly near their margins. This deformation occurs over the length of subprovinces involving thrusting along subprovince boundaries and within volcanic units forming tectonic slices and upright folding with moderately plunging fold axes. Folds in the metasedimentary belts are shallower in plunge. The culminating stage in the transpression event produced dextral strike-slip faults along subprovince boundaries and similar, less numerous, sinistral, northeast-striking faults.

A secular progression exists in the Superior Province beginning with platformal sequences in the older cratonic blocks, followed by mafic plain volcanism, are volcanism and alkalic-calcalkalic volcanism marking the cratonization and termination of volcanism, Plutonism, regional and local deformation complement the progression. Such a progression invites comparison with the sequence from the platformal environment of a Sachigo supercontinent, through breakup of the supercontinent, formation of magmatic arcs culminating in cratonization with alkalic volcanism and major transpression. This deformation developed between subprovinces and compares with oblique convergence between island arcs and continents. The deformation progressed southwards from the Uchi/English River Subprovince boundary where it culminated at ~2700 Ma to the Quetico/Wawa Subprovince boundary at ~2686 Ma. Late strike-slip faults across subprovinces and along subprovince boundaries mark the last crustal movement in the final cratonization of the Superior Province. The entire cycle from platform sequences on a supercontinent through breakup to final cratonization took about 320 Ma, comparable to the 440 Ma estimated for a post-Archean mega-cycle (Nance et al., 1986).

REFERENCES

Nance, R. D., T. R. Worsley, and J. B. Moody, 1986, Geology, v. 14, p. 514–518.
Percival, J. M., and H. R. Williams, 1988, Geology: accepted.

THY, P., NASA, Johnson Space Center, Houston, TX, and Y. DILEK, University of California, Davis, CA

Troodos Ophiolite, Cyprus

The Troodos ophiolite is one of the best-preserved eastern Mediterranean ophiolites that represent remnants of the Neotethyan ocean crust. The ophiolite is composed of a central plutonic complex, overlain successively by a sheeted dike complex, extrusive sequences, and pelagic sediments. The plutonic complex is exposed in a central uplifted core as tectonized harzburgites and serpentinites overlain by ultramafic and gabbroic cumulates.

The International Crustal Research Drilling Group's (ICRDG) Cyprus Crustal Study Project was initiated in 1980 with the major objective of reinvestigating and drilling through the ophiolite.

The presence of axial grabens and normal faults parallel to the grabens within the sheeted dike and extrusive complexes are reminiscent of oceanic spreading centers. Complex intrusive and tectonic contact relationships between the sheeted dike complex and the underlying complex indicate multiple and episodic intrusion of magma and spatial variation in volcanic and tectonic activities. Coupled with the existence of the Arakapas transform fault to the south, these features suggest an origin at an intersection between a spreading ridge and transform fault.

One of ICRDG'S drill holes sampled the lower sheeted dike complex, upper gabbros, and ultramafic cumulates. A major intrusive contact is indicated in the drill core, separating lower coarse-grained gabbroic and websteritic cumulates from upper medium-grained gabbroic cumulates. The upper cumulates contain plagioclase ($An_9$0–86), clinopyroxene (high-Ti), and orthopyroxene as cumulus phases throughout the sequence and Fe-Ti oxides appear locally. The lower cumulate sequence contains clinopyroxene (low-Ti) and orthopyroxene as cumulus phases throughout, olivine appears in the ultramafic rocks, and plagioclase (An_{95}) only in the gabbros. The upper sequence shows complex cryptic variation with three reversals to more primitive mineral compositions (one reversal for each 200 m of cumulates). The lower cumulate sequence shows only one reversal (one reversal for each 1000 m). The reversals are gradual over intervals of 50–100 m and are interpreted as reflecting magma chamber replenishment events. The upper gabbro sequence reflects a higher replenishment rate, relatively to the lower mafic and ultramafic sequence, and, therefore, may be related to spreading center magmatism. However, there is in none of the Troodos cumulate sequences indications for the development of steady-state magma chambers.

The Troodos ophiolite can be viewed as composed of two distinct magmatic suites. An early, relatively high-TiO_2 basaltic suite constitutes the upper gabbros, the sheeted dike complex, and the andesitic to rhyodactic lower pillow lavas. A late, low-TiO_2 basaltic andesite suite constitutes the lower ultramafic to gabbroic cumulates and the upper pillow lavas. Parental magmas to the early suite originated as partial melts from hydrous, relatively undepleted, peridotite diapirs rising in a mantle wedge in response to dehydration of a subducting crust. The late basaltic andesite suite is highly depleted in incompatible elements and shows similarities with second-stage basalts from oceanic spreading centers. The Troodos basaltic andesites and associated cumulates may thus have formed in response to continued adiabatic upwelling as second-stage melts from the hydrous, water-undersaturated peridotite diapirs from which the earlier spreading-related magmas were extracted. The basaltic andesites formed in a relatively slow- or nonspreading tectonic environment and dominantly crystallized in magma chambers showing multiple intrusive relationships and underplating the Troodos ophiolite. The extrusion of the basaltic andesites of the upper pillow lavas occurred simultaneously with still active evolving volcanic centers belonging to the earlier magmatic suite. An alternative possibility would be that the basaltic andesites separated from renewed diapirs rising as a consequence of reinitiation of subduction and liberation of fluids from the subducting slab into the overlying depleted peridotite. The existence of LREE and LIL enriched lavas among the upper pillow lavas support this possibility. The chemically distinct groups found in the upper pillow lavas are consistent with the multiple intrusive character of the lower plutonic complex.

The reconciliation of this apparently conflicting geological and geochemical evidence implies a subduction zone setting in with both magmatic and tectonic spreading. Regional reconstructions suggest that the Mesozoic Neotethys formed as a marginal basin both to the Afro-Arabian continent and the Paleotethyan ocean over a south-dipping subduction zone. This postulated subduction zone consumed the Paleozoic Paleotethys to the north. The model implies that the Troodos ophiolite and some of the eastern Mediterranean ophiolites formed along east–west-trending spreading centers separated by north–south-trending transform faults within this marginal basin. It is proposed that the subduction-related magmas of the Troodos ophiolite formed in relation to these spreading centers. A rapid change in relative plate motion between Eurasia and Afro-Arabia, following the South Atlantic opening, and northward convergence of Afro-Arabia created a regional compressive regime that terminated the spreading activity along the ridge segments. Continued compression resulted in plate boundary reorganizations within the Neotethyan ocean and initiated north-dipping subduction zone(s) beneath the Troodos ophiolite and other ophiolites in the region. The well-documented anticlockwise rotation of the Troodos ophiolite is interpreted to have occurred during the initiation and evolution of a northward oblique subduction beneath Cyprus. Rapid uplift of the Troodos ophiolite since the Pliocene may have resulted from an attempted subduction of the Eratosthenes seamount beneath Cyprus.

TIAINEN, M., E. KOISTINEN, and I. SUPPALA, Geological Survey of Finland, Espoo, Finland

Prediction of Ore Potential Areas in Southwestern Finland—Application of Statistical and Image Processing Techniques

A project with the purpose of predicting Cu-Zn, Ni- and Au-potential areas in southwestern Finland is being executed by the Exploration Department, Geological Survey of Finland in the year 1987–1991. The size of the study area is 240 km × 320 km. The work is carried out in three major stages: 1) collecting and pre-processing of data; 2) analysis of data; 3) integrating of data.

This presentation shows examples of methods applied and results in a subarea of 160 km × 120 km size. The processed geodata comprise four major data groups: 1) bedrock geological data, 2) geophysical aeromagnetic and gravimetric data, 3) till geochemical data and 4) photogeological data. All data are processed by computerized statistical methods (factor analysis, clustering and similarity analysis).

Collecting and preprocessing of data: Geological data and lineaments from geological and photogeological maps are digitized using the Geographic Information System FINGIS developed by the National Board of Survey of Finland. Other digitized data available are:

geophysical aeromagnetic data from high altitude survey,

gravity data from measurements of the Geodetic Institute of Finland, with an average distance between points of 5 km, and

geochemical data based on till samples collected 1 sample/2 km × 2 km.

All data are interpolated in a 1 km × 1 km grid.

Analysis of data is carried out by visual and statistical methods and derived variables are calculated from the primary data. A structural interpretation map is derived by visual evaluation of geological and geophysical maps. Various derivatives are calculated from aeromagnetic and gravity data. For geochemical data the correlations between the element contents and with other geodata is studied. All variable derived from various geodata are compared to the maps with known mineralizations and bedrock geology.

Integration of data is done using factor analysis, image analysis and deposit modelling. Factor analysis is used to integrate geochemical data. Image analysis is used to integrate various variables for visual inspection. Deposit modelling is used to determine the regional mineral potential of the study area processing a great number of variables by the SIMANA program (Similarity analysis). Deposit models used are the known Ni-Cu, Cu-Zn-Pb and Au mineralizations of the area. The variables are calculated from geological, geophysical, geochemical and photogeological data.

TICKSHAEV, V. V., M. T. ABDULVALIEV, V. G. OSIPOV, N. P. SMILEVEZ, and V. A. GLECHIKOV, USSR Ministry of Geology, Saratov, USSR

Oil and Gas Prospecting Based on Common Seismic and Electromagnetic Technology

The combined geophysical method giving the improved data on seismic and geoelectrical formation conditions has been developed for searching of unanticlinal traps for oil and gas, zones with better reservoir properties and oil- and gas-bearing rocks. This method makes it possible to compare the electric image with seismic relative to detailed pattern and resolution. The seismic reflection survey and electromagnetic near-field transient sounding (NTS) are most in agreement with the field propagation and prospecting procedures.

The wave equation describes a seismic field propagation, and the equation with diffusion and save components describes an electric field. The electric current takes effect on induced polarization that resulted in additional alternating currents. The induced polarization gives a rise in apparent value of dielectric permeability and accordingly in wave component in electromagnetic equation. Under certain conditions, it is possible to measure this component and as a result to use the multiple coverage system and summation methods of seismic survey during electromagnetic survey. The basic components of the common technology are as follows: impulse and code-pulse excitation of seismic and electromagnetic fields, signal sounding in near zone of source, use of multiple coverage systems, space-time cumulation and filtration of signals, multichannel recording, and so on.

The seismic and electric patterns appear in common format on the display. The time conversion of electromagnetic data into "seismic time" with the special program that reveals the relationship between elastic and electromagnetic properties of a section is used for a comparison of vertical sections.

The application of this method for the Pricaspian depression exploration has allowed us to detect the sand lenses associated with the terrigenous rocks in carbonate bed, the reef bodies, better carbonate and terrigenous reservoir areas, and the local salt bodies. Often none of the above-mentioned methods can help in locating the above geological bodies.

TIDBALL, R. R., U. S. Geological Survey, Denver, CO

Digital Geochemical Mapping

Geochemical mapping techniques in the U. S. Geological Survey have evolved along pathways that parallel the development of computer technology. Early computer-generated maps were merely automated extensions of simple hand-drafted point plots of single elements either with or without hand-drawn contour lines. On some maps more than one element was accommodated by devising a complex symbol that incorporated element identities as well as the quantity of each element.

Computerized contouring replaces operator bias with an objective and consistent procedure that (1) transforms the map data into a digital grid by estimating a series of additional points to partially fill the blank area between the measured points, and (2) plots contours among the gridded values according to a specific algorithm. The validity of the resulting map depends on the effectiveness of the sample—that is, whether the sample was collected at a scale that allows the largest amount of between-sample variance. Assuming sample effectiveness, validity also depends on the continuity of the variable being mapped, the density of sampling, and the method of estimating the grid.

Sample effectiveness can be judged by a two-stage iterative sampling process: (1) the distribution of variance as a function of scale and the sampling requirements are estimated by a hierarchical analysis of variance, and (2) a main sampling design based on the estimates are applied to the entire area and population of interest in the form of a simple random, stratified random, or uniform grid design. As experience was gained in sampling soils in various regions, the preliminary stage was bypassed, and the sampling density sufficient to avoid excessive error either from oversampling or from undersampling was estimated on basis of that experience. After the samples were collected, however, there was still the requirement to know whether those samples represented the population over the scale of the study, so we have combined the two stages into one by including within the main sample design a limited analysis of variance set.

This set estimates the spatial variance between and within localities and the analytical variance. For investigation of soils on a regional scale (for example, an area of several thousand square kilometers) a sampling interval of 5 to 10 kilometers is used in a stratified random design.

The method of choice for gridding is universal kriging (Tidball, Grundy, and Sawatzky, 1986), which yields a minimum-variance, unbiased estimate of the grid values. Outliers in the data are initially removed by the method of Krige and Magri (1982).

Interpretation of geochemical maps is enhanced by the following refinements:(1) Plotting a natural multivariate association of elements on one map rather than plotting a series of single-element maps. The multivariate association as derived by factor analysis is expressed as a single index value (factor score). Thus one map may portray a common controlling process.(2) Expanding from a point-plot type map to a continuous surface that brings out gradients and more subtle details.(3) Displaying the surface as a color-raster pattern rather than as contour lines. The color-raster pattern is a specialized variation of contours on which the grid values shown as a color pattern replace contours. Such displays have a significant visual impact.(4) Digital maps are also readily integrated into geographic information systems (GIS) with other types of geologic data. Relationships among these data are readily determined and should improve the prospect of deriving meaningful interpretation about overall geologic control of element distribution.

REFERENCES

Krige, D. G. and E. J. Magri, 1982, Studies of the effects of outliers and data transformations on variogram estimates for a base metal and a gold ore body: Mathematical Geology, v. 14, no. 6, p. 557–564.

Tidball, R. R., W. D. Grundy, and D. L. Sawatzky, 1986, Kriging techniques applied to element distribution in soils of the San Joaquin valley, California: Proceedings of HAZTECH International Conference, Denver, Colorado, August 11–15, 1986, p. 992–1009.

TIETZ, G. F., University of Hamburg, Hamburg, Federal Republic of Germany

Chemical Weathering Processes in Quartz Grains: Etching Versus Recrystallization

Since Krinsley and Doornkamp published their atlas of quartz sand surface textures in 1973, the SEM-supported study of quartz grains has become a widely accepted method in sedimentary petrography. By the rapid improvements in both, scanning electron microscopy and adaptation of photography techniques, today it is much easier to recognize surface features on grains more in detail than in those old days. One aspect of surface studies is the search for etch or recrystallization marks to characterize the weathering environment whether it was aggressive or not. For a proper interpretation of the weathering history of grains (soils, sediments), it is very important to decide (surface) pits or holes are representing etching only or show already the beginning of recrystallization. This decision became complicated by the rise of a serious controversy: different shapes of pits and scars on chemically attacked grains were reported etch pits by one group of geologists while another group called them recrystallization marks. The result of this controversy: many articles already published are reporting confusing misinterpretations. In fact, from a crystallographic point of view the different steps for growing a crystal are exactly the same as for dissolving a crystal, but this is restricted to an ideal crystal only, grown and subsequently dissolved in a closed system which is similar in composition. But these ideal conditions are hardly met in nature. Quartz, like other rock forming minerals, is derived originally from a crystalline source rock, formed in an environment completely different from any environment controlled by chemical weathering processes at or near the earth's surface. Comprehensive studies on quartz grains from various saprolites revealed accurate morphological and chemical criteria for dissolution and recrystallization:

A: Morphology:

Two areas on/in quartz grains are preferably designated for the formation of etch pits/holes: (1) Generally, etching starts at dislocation sites within the quartz lattice structure; (2) Additional etching may develop or proceed wherever some net planes of the lattice structure become attacked/broken, either chemically (i.e. continuation of (1) or mechanically (i.e., abrasion, "rounding")). The more net planes become affected, the better is the chance for etch pits development.

Due to inhomogeneities/changes in weathering solutions, etching is *not restricted to dissolution along a distinct net plane,* but is attacking several net planes at the same time resulting in *IRREGULARLY SHAPED PITS AND HOLES,* that may, later on, penetrate the entire grain/crystal. The same is with recrystallization, it starts at that point that needs lowest energy to restore/heal a net plane. Any change in solution concentration, i.e. increase in oversaturation with respect to silica, is answered by the reconstruction of those net planes/face edges energetically in the most suitable positions. This energetic balance is responsible for the development of *EVEN PLANES* that may proceed to form crystal faces/edges. Irregularly shaped etch holes are getting restored to even plated channels.

A completely different behavior is found at those places where the quartz crystal is lodging other (smaller) minerals. Most susceptible for dissolution attacks during weathering are aluminosilicates like feldspar and mica. In general, these mineral inclusions were already dissolved before quartz etching started. More or less curved scars in quartz grains indicate former mica flakes, often engraved by continuous etching (=incomplete/disturbed net planes of quartz). Rhombic holes are a characteristic proof for the presence of former microcline. In most acid rocks microcline was formed prior to quartz, therefore no net planes of quartz "host" in the neighborhood of microcline were hampered during formation. The smooth but even planes of these rhombic holes never show any trace of etching. During recrystallization of the quartz host these rhombic holes of former microcline indicating monoclinic symmetry are going to change to triangular holes representing the usual trigonal symmetry of quartz. The general restoration of chemically attacked quartz grains will end up finally with either well shaped quartz crystals or polygonal (sub-) grains.

B: CHEMISTRY:

Dissolution and recrystallization of quartz during weathering process is promoted by the absence of alumina-ions, whereas the presence of alumina inhibits quartz dissolution and recrystallization: with respect to silica undersaturated solutions will attack aluminosilicates (feldspar, mica, clay) only. Oversaturation will create ± amorphous Al-Si-crusts or induce clay mineral formation (i.e. kaolinite).

TIGHE, S. A., P. J. FOX, R. S. DETRICK, and R. C. TYCE, University of Rhode Island, Narragansett, RI

Synthesis of SeaBeam Bathymetric Data from East Pacific Rise between Orozco Transform (16 °N) to Easter Microplate (24 °S)

In collaboration with colleagues from L-DGO, and with the sponsorship of JOI/USSAC, we have carried out a synthesis of geological and geophysical data from the East Pacific Rise (EPR) between 16 °N and 24 °S. The East Pacific Rise is part of the midocean ridge system, forming the accretionary plate boundary between the Pacific plate to the west, and the Cocos and Nazca plates to the east. The study area is located between the Orozco Transform (16 °N) and the Easter Microplate (24 °S) and contains the fastest known spreading rates (84–162 mm/y whole rate) along the ridge system. Included in the synthesis are all available SeaBeam bathymetry, SeaMarc I side-scan sonar, magnetics, gravity, multichannel seismics, and petrologic data. Here we present the compiled SeaBeam bathymetric database from 12 U. S. cruises and 3 French cruises, comprising over 300 days of shiptime. SeaBeam is a high-resolution, wide angle, multibeam echo sounder that can image a swath of seafloor as wide as approximately two-thirds of the water depth with a resolution of 1 m. The navigation-registered data from all cruises have been combined into a single digital database by interpolating onto a 200 m gridded surface using a weighted, biharmonic cubic spline-fitting technique. The data was then machine contoured and plotted. Using the SeaBeam results of the synthesis, variations in the architecture of the ridge crest along strike and across strike can be examined from within a single digital data set, rather than between several "postage stamp survey" data sets.

The general morphology of the EPR has been defined by earlier surveys: In plan view, the rise axis is segmented by over 30 overlapping spreading centers and 9 transform faults; in profile the axis undulates up and down hundreds of meters over distances of 30–200 km, the deep areas occurring at axial discontinuities. The synthesis database provides additional insight into detailed topographic variations along the ridge axis, as well as morphotectonic patterns developed off-axis, providing constraints on how the template of accretion behaves in time. The along-strike continuity of off-axis, narrow, steep-sided ridges that are in close parallelism with the strike of the present ridge axis is remarkable. This fabric extends uninterrupted for tens of kilometers, except in the vicinity of overlapping spreading centers where V-shaped (in plan view) bands of disturbed and discontinuous terrain are recognized. Scattered along the flanks are seamounts that overprint the ridge-flank structural grain. The synthesis database has also given us a better picture of more poorly defined (and therefore probably less active) sections of the ridge crest. We present the results of the first-order quantitative variations in the along-strike linearity of the ridge axis and the cross-axis evolution of features using statistical analysis of the digital database.

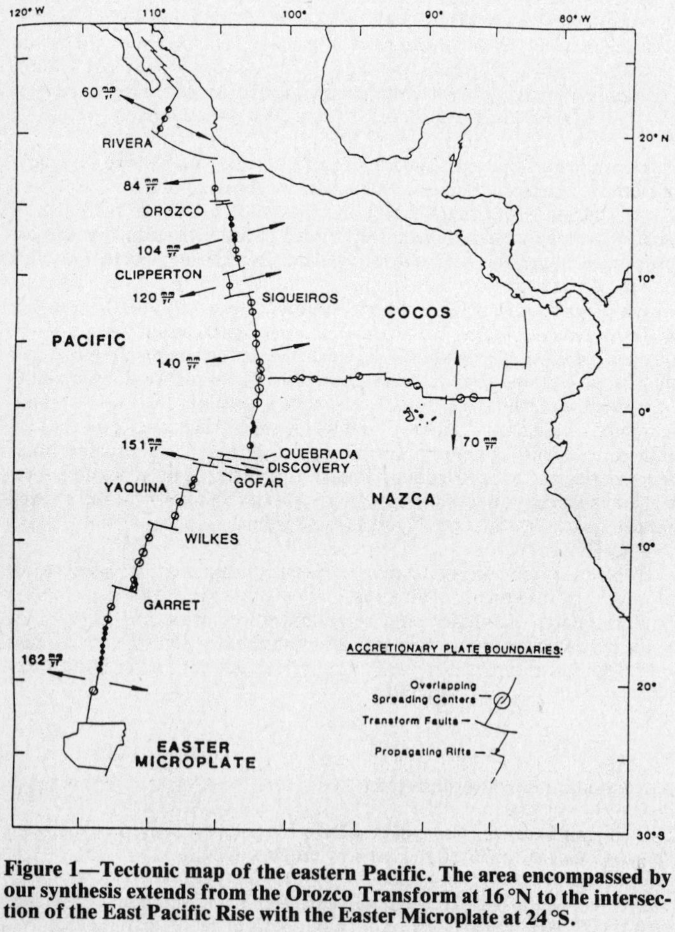

Figure 1—Tectonic map of the eastern Pacific. The area encompassed by our synthesis extends from the Orozco Transform at 16 °N to the intersection of the East Pacific Rise with the Easter Microplate at 24 °S.

TIJANI, A., A. BELKASMI, and L. LASRI, Bureau de Recherches et de Participations Minières, Rabat, Morocco

Precious Metals Prospecting Tied to Precambrian Volcano: Sedimentary Formation of Moroccan Anti-Atlas

During the last years, the B.R.P.M. applies with success the combination of the indirect methods especially the stream-sediment geochemistry and the lithogeochemistry accompanied by the mercurometry for the search of the silver-bearing mineralizations linked to Precambrian volcano-sedimentary formations in the Moroccan Anti-Atlas.

The Anti-Atlas chain correspond to vast anticlinorium, where the Precambrian socle appears in many areas from the Atlantic in the West until Jbel Ougnat in the East.

In strategic prospecting, we use the streamsediment geochemistry. The sampling density of 1 to 5 samples per square km has been used. Locally, the number of samples has been increased until 15 samples per square km. The test about the granulometric fraction which shows the optimal contrast between the background content and the threshold of anomalies is that of 0.10 mm.

The lithogeochemistry and the mercurometry are used in tactical scale according to profile, whose the sampling density depends on the results of strategic prospecting.

The different exploration phases are ended to locate several deposits and anomalies; some of them was the subject of detail mining works in Imiter–Zgounder–Sidi Flah districts. These sectors belong (in general) to the same lithostratigraphic formations of middle Precambrian (PII, PII–III). The structures of top wall of PII which we interest here, in Imiter for example had constructed with alternation of metapelites with pyrite layers, black metapelites with organic matter (graphitic), greenish levels decimetric of tuff, and metric lava. The average chemical analysis of these black shales gives the following content:

* SiO_2	Al_2O_3	FeO	Fe_2O_3	Na_2O	K_2O	CaO	MgO	TiO_2	MnO	P_2O_5	P.F	Cu
56.19	14.38	4.04	1.94	2.87	3.99	2.69	3.72	0.53	0.14	0.18	5.77	147.33

Pb	Zn	Ag	Cd	V	Ni	Cr	Zr	Ba	CO	As	B	Bi
338.83	253.83	4.72	< 10	92.5	107	218.33	88.33	112.66	30	225	38.33	<10

Sn	Sb	Ta	Nb	MO	Au	Hg
<10	16.66	<10	<10	16.5	0.11	114.79

A sedimentary and volcano-sedimentary series of PII presents stratiform polymetallic-mineralization (Ag, Hg, Ti, Fe, Zn, Pb, Cu...) Vargas and Samama (1983). These metals are located either dessimination in rocks or in stratiform layers with a weak thickness.

These microlayers correspond sometimes to the association like rutile-organic matter; rutile–quartz–native silver. Many samples shows the pyrite mineralization, sometimes associated to marcassite, pyrrothite on grains shaped conformable with bedding.

Geochemical study of trace elements in carry levels confirmed the existence of precious metal with silver content greatly higher than clarke (until 20 ppm counter 0.08 ppm).

The Imiter mineralizations represents the case of successive and superposed reconcentration of initial polymetallic stock. These reconcentration has been carried out during the last two majors mineralizer phases, structuro-metamorphic mineralization by lateral secretion on the one hand; the hydrothermal mineralization associated to granitic intrusion on the other hand. At last the supergenes process when they affect these rich precious minerals to lead towards to the reconcentration of the high content economically workable.

The Imiter mine is composed with three mineralization morphotypes (BELKASMI and al. 1987).

1. The principal deposit results from different process of concentration to name above.

2. At East of this principal deposit, the conceal mineralization (Hill B_3) are recently decovered, and they are like veins stockwork in schistous on stratiform socle, crackling in detretic level and tuffaceous under cover of terminal Precambrian.

3. In the South, the principals massive are known in veins structure enclosed in sandstone like metapelite series.

The structural and lithostratigraphic guides constituted the determinative metallotect to search the silver mineralization–like Imiter type.

TING, F. T. C., West Virginia University, Morgantown, WV

Facies in Lower Kittaning Coal Bed, Appalachian Basin (USA)

Facies changes in the form of maceral and palynomorphic composition and textures are observed, in thin sections and polished sections, both vertically and horizontally in the Lower Kittanning coal bed (Allegheny Formation, Middle Pennsylvania, Westphalian D) over a large area in western Pennsylvania, U.S.A. The lower half of the coal bed is characterized by abundant well-defined telocollinite and thin-walled lycopod spores which are interbedded with thin layers of mixtures of clearly outlined inertodetrinite and vitrodetrinite. The uppermost layer, in places where the coal is though to be influenced by marine conditions, is characterized by a densospore-rich desmocollinite with abundant inertinite cluster (quasigelifusinite). The transitional layers between the lower half and the uppermost layer are dominated by fine-grained vitrinite with disseminated spores of distinctive shape and size (*Punctatisporites* sp. and *Schopfites* sp.). In many places, a clay- or mineral-rich parting with abundant alginite, fragmented megaspores, and clay filled microspores are also observed in the transitional layers. Any layer composed of unique plant assemblage (a plant community representing a specific environment in time and space) is termed a phytostrome. The Unique association of macerals is termed a lithotype. Five genetic lithotypes are recognized: kittosite, herbosite, klastosite, arborosite, and allosite, representing (1) brackish, (2) transitional, (3) limnic, (4) freshwater arboreous, and (5) reworked peat, respectively. The accumulation of the last phase of the Lower Kittanning coal appears to have been influenced by a transgressive sea.

TJONG KIE, Tan, Xiong SHAO BAI, Ou YANG TING, Institute of Geophysics, Beijing, China

Geotectonics and Geothermics of Fujian Continental Margin

This paper reports on a part of the widescale investigation of the interrelations between temperature field, crustal structure, and geotectonics in order to study the origin of geotherma heat, its distributional pattern and the feasibility for economic exploration. The geodynamic evolution with intense magma outpour during volcanic activity in the Mesozoic (> 135 Ma) and the formation of the granitic basement 75 Ma ago has been studied. Further, the volcanic eruptions of about 2–1 Ma ago in the Longhai, Mingxi regions are proofs that the deeply cutting large faults were excellent conducts for magma outpour. The heat reservoir formed in that period may still contribute to the present heat anomaly. The hot water is of meteoric origin which is infiltrated through the fissures in the basement. Land upheavals with a maximum rate of 2m/s in the region northwest of the Zhangzhou basin are also proofs of present geodynamic activity.

The northeast running coastal zone is characterized by a negative Bouguer anomaly with a negative northwest gradient. The Fuzhou and Zhangzhou basins, where geothermal wells are most abundant with maximum temperatures of 107 °C and 121.6 °C respectively, are indicated by a northwest running zone of positive Bouguer anomaly superimposed on a strong negative background.

Two main families of deeply cutting large faults are the northwest striking extensional shear faults and the northeast running compressional shear faults which obviously govern the pattern of geothermal wells distribution. The preferential distribution along the northwest faults are indications that they act as good conducts for the outpour of hot fluids. Further the wells in the Southern part of Fujian by far outnumber those in the Northern part. Two seismic profiles cross-seecting the Zhangzhou basin, the first running via Anxi in nearly northeast direction and the second in south-north direction show that: strong Pn wave energy absorption in the upper mantle in the Southern part is a very strong indication of a viscous upper mantle under high temperature. In this region the V_p velocity at the roof of the upper mantle is 7.9–8.0 km/2, whereas in the Northern part $V_p = 8.1$ km/s and Pn wave energy absorption is much lower.

This findings may give an explanation for the fact that wells are more abundant in the Southern than in the Northern part of Fujian. In fact a large number of geothermal wells have been found in the continental margin of Southern Fujian to neighboring Guangdong province. So it is concluded that the high temperature upper mantle as a deeply seated heat source supplemented by residual heat sources, in combination with extensional shear fractures running into the upper mantle, form the basic condition for fluid and heat transport under the present geodynamic activity.

TOMADIN, L., Università di Urbino, Urbino, Italy, R. LENAZ, Istituto di Geologia Marina del CNR, Bologna, Italy, and R. SARTORI, Dipartimento di Scienze Geologiche, Bologna, Italy

Clay Minerals as Natural Tracers in Sediments, Water Column, and Lower Atmosphere of Tyrrhenian Basin (Mediterranean)

During the last 20 years, extensive studies of clay minerals in the Mediterranean sediments allowed us to define the main mineralogical provinces, to recognize the role of fluvial supply, and to focus on general aspects of clay sedimentation in the marine environment (Chamley, 1971; Tomadin, 1981). In the case of the Tyrrhenian Sea, clay minerals were studied not only in the seafloor sediments, but also in the suspensates, both in the water column and close to the seawater–air interface.

Clay minerals in present-day sediments of the Tyrrhenian Sea are mostly illite (45–60%), with lesser kaolinite (15–20%), chlorite (10–20%), and smectite (10–15%) (Tomadin, 1974, 1981).

Their patterns of distribution depend on the source areas, on the dispersal and transport paths, and on seafloor morphology. The main results regarded the occurrence of various source areas: Illite and chlorite mainly derive from the crystalline rocks of Sardinia and Calabria, whereas smectite largely derives from the perityrrhenian volcanic edifices and kaolinite does not appear to have an obvious source in the nearby land outcrops. Its source area should be located south of the basin. Different transport and dispersal processes were also recognized, including transport by turbidity/density currents as well as mixing by marine currents; eolian transport was hypothesized as an important factor for particles coming from distant sources.

Clay minerals suspended in the water column were subsequently studied. Total concentrations of tyrrhenian suspensates, collected at 100 m water depth, range from 0.14 to 0.52 mg/L (Sartori, 1977). The inorganic fraction is only a small part of such figures and includes abundant clay minerals, quartz, feldspars, and occasional carbonates. Distribution patterns and trends of various minerals (e.g., the feldspars/quartz ratio) appear quite close to those observed in the sediments, with the exception of abundant kaolinite in the waters east of Sardinia, due probably to strong northwest winds blowing during the sampling period.

Dispersion of inorganic suspensates appear controlled by the circum-Tyrrhenian fluvial input, by the cyclonic gyre of currents, and by the eolian input from prevailing winds (Sartori, 1977; Sartori and Tomadin, 1980).

More recently (EOLO Project of CNR, 1980–1983), studies of suspensates from lower atmosphere allowed us to directly investigate the role of eolian transport on mineral dispersal and deposition. African and European sources of wind-blown dusts have been defined on the base of mineral content (Tomadin et al., 1984).

Taking into account the meteorological conditions occurring during sampling, a "background" indirect transport of dusts has been pointed out.

The importance of this contribution for sedimentation, in comparison with the more rare episodes of direct transport (dust storms), has also been recognized (Tomadin and Lenaz, in press). While indirect transport acts by continuously carrying dusts of variable composition, direct transport is ephemeral, is volumetrically more signifiant (Prodi and Fea, 1979), and supplies dusts with composition reflecting the source areas from where wind episodes arise (Tomadin et al., 1984).

By comparing the mineral patterns obtained at the different sampling levels (seafloor, water column, seawater–air interface) the following considerations can be drawn on the dynamics of sedimentation in the Tyrrhenian Basin.

1. The transport and deposition of sediments by turbidity current is presently very efficient in the basin; illite and chlorite are the main tracers of these processes, which appear to act mostly along the west, north, and east continental slopes with eventual deposition if the bathyal plain.

2. An eolian supply from African sources is also volumetrically important; kaolinite is the main tracer of this process, as demonstrated by its dispersion pattern both in sediments and in suspensates (water column and lower atmosphere). Eolian deposition from southern sources is evident across the southern margin of the basin, which is offshore Sicily down to the bathyal plain, as well as in all areas free of turbiditic deposition such as the seamount tops. In areas where turbiditic deposition prevails, such as off eastern Sardinia, the wind-blown supply from the northwest, registered by high amounts of kaolinite in suspensates, is not recorded in sediments because it is masked by high amounts of illite and chlorite.

REFERENCES

Chamley, H., 1971, Recherches sur la sédimentation argileuse en Méditerranée: Science Géol., Mém., v. 35, 209 p.

Prodi, F., and G. Fea, 1979, A case of transport and deposition of Saharan dust over Italian peninsula and southern Europe: Journal of Geophysical Research, v. 84, p. 6951–6960.

Sartori, R., 1977, Fattori che influenzano la distribuzione del materiale sospeso nel Mare Tirreno: dati di due crociere pilota: Giorn. Geol., v. 42, n. 1, p. 1–12.

———, and L. Tomadin, 1980, X-ray mineralogy of inorganic suspended matter from the Tyrrenian Sea: 4th Meeting of European Clay Group, Freising 1980, Abstract, p. 126–127.

Tomadin, L., 1974, Les minéraux argileux dans les sediments actuels de la Mer Tyrrhénienne: Bull. Groupe Franç. Argiles, v. 26, p. 219–228.

———, 1981, Provenance and dispersal of clay minerals in recent sediments of the Central Mediterranean Sea, *in* F. C. Wezel, ed., Sedimentary basins of the Mediterranean margins: Bologna, Tecnoprint, p. 313–324.

———, and R. Lenaz, in press, Eolian dust over the Mediterranean and their contribution to the present sedimentation: Contribution to NATO Advanced Research Workshop: Paleoclimatology and meteorology, modern and past patterns of global atmospheric transport, November 15–19, 1987, Oracle, Arizona.

———, ———, V. Landuzzi, A. Mazzucotelli, and R. Vannucci, 1984, Wind-blown dusts over the Central Mediterranean: Oceanological Acta, v. 7, n. 1, p. 13–23.

TOMEK, C., Geofyzika Brno, Brno, Czechoslovakia

Subsequent Closure of Tethyan Basins as Recorded in West Carpathians

Comprehensive geological and geophysical work in the West Carpathians has resulted in significant progress in the understanding of the evolution of this particular mountain belt. The history of Carpathian tectonic studies goes back to Dionys Stur, Eduard Suess, Victor Uhlig, and Maurice Lugeon in the last century. More than a century of intensive research has therefore been done in the West Carpathians. Modern structural studies and more than 900 km of deep seismic profiles (mostly 16 twt) have revealed detailed information concerning the subsequent evolution of the West Carpathians.

The Carpathians today represent an Oligocene–Miocene subduction orogen with developed Krosno (Moldavian) accretionary subduction complex. Subsequent collision was weak and did not much change the shape of subduction orogen. In the West Carpathians, the lower plate is represented by Brunnia and its passive margin a Late Precambrian terrane mostly untouched by the Variscan collision that occurred farther westwards.

There is significant rheologic difference in behavior of the Variscan Crust and Late Proterozoic Brunnia crust (Podolian Platform), which was noticed by Eduard Suess. Significant lower crustal layering has been observed beneath Brunnia and its passive margin. It was strongly bent together with the upper crust during the Miocene subduction as observed on deep seismic lines 2T and 8HR.

It is suggested in this contribution that the arcuate shape of the orogen is due to the arcuate shape of the passive margin of the Krosno Sea. This sea originated in the Lower Cretaceous, most probably as a back-arc basin. In my opinion, the shape and area of the Krosno Sea, which was transformed by subduction processes and the Pannonian extension to the Carpathian arc and the Pannonian Basin, may be compared to the Black Sea before the Oligocene.

The Magura flysch and the Pieniny Klippen belt form the northernmost terrane of the upper plate in the West Carpathians. The Penninic metamorphosed flysch of the East and Central Alps forms a parallel to the Paleocene and Eocene nonmetamorphosed flysch of the West Carpathians. I therefore regard the Magura flysch as an accretionary complex of the North Penninic trough whose closure preceded the Tertiary collision in the Alps. Magura flysch is in places more than 10 km thick and forms typical accretionary subduction imbrications on all seismic lines.

The Pieniny Klippen belt is a major puzzle; 4–20 km wide and 800 km long, it is closely associated with the Magura flysch tectonically and genetically. It is composed of rocks deposited in shallow and deep sedimentary environments as well as of accretionary complexes of Mesozoic and Cenozoic age in an incredible tectonic melange. At least one subduction accretionary wedge of Late Cretaceous age is present in the Pieniny belt. The Pieniny Klippen belt was obviously subjected to intensive strike-slip deformation in the course of the Oligocene Mesoalpine collision. The sinistral strike-slip fault along which the Pieniny Klippen formed served as the tectonic escape of the Inner West Carpathians from the Tertiary Alpine collision into the zone of the Krosno Sea subduction. In some places (e.g., in Orava area), transpression with significant backthrusting along the Pieniny Klippen belt occurred.

It has been known for more than 50 years, from the times of Alois Matejka and Dimitrij Andrusov, that the Inner West Carpathians are Late Cretaceous orogen with significant nappe development. They are formed by the Tatricum, Veporicum, and Gemericum units. Tatricum and northern Veporicum are closely associated genetically through similar tectonic development in the course of Hercynian orogeny. Southern Veporicum and Gemericum are different. Deep seismic reflection line 2T shows significant whole crustal suture more than 25 km thick beneath Veporicum dipping 30°S. This suture where all Inner West Carpathian cover and basement nappes root is most likely the result of Late Cretaceous (Turonian and later) collision between the Tatric and Veporic terranes in the north and Gemeric and other southern terranes in the south. This collision closing the South Penninic ocean was present also in the Alps. Their features, however, had been destroyed there and presented only in the West Carpathians due to their earlier escape from the Cenozoic collisional setting.

Summarizing, at least four ocean basin closures of the Tethyan realm are geologically recorded in the West Carpathians. The northernmost basin—Krosno Sea—was closed after Oligocene–Miocene subduction. This basin was formed in the Lower Cretaceous as a back-arc basin of similar extent like the present Black Sea. Its position was in the whole Carpatho-Pannonian realm. Positions of other seas with oceanic crust are, of course, unknown. The Magura flysch basin we connect with the North Penninic basin in the Alps. Ocean basin record in the Pieniny Klippen belt reminds us of small ocean basins between Africa and Europe during the last 150 m.y. Gigantic continental collision in the Veporicum followed after subduction of unknown (most probably South Penninic) ocean.

TOMEK, C., Geofyzika Brno, Brno, Czechoslovakia

Eastern Boundary of European Variscides: Large-Scale Oblique Continental Collision Zone

Modern structural geologic studies, deep seismic reflection profiling, and scientific borehole Potstát-1—4.2 km deep—suggest that the eastern boundary of European Variscides most probably forms a large-scale continental collision zone with significant amount of dextral transcurrent movement. This zone is localized in present geography between Polish Lowland southeast of Wroclaw and the Danube river in Lower Austria. Most of it forms today the eastern boundary of the Bohemian Massif in Moravia (Czechoslovakia); a minor part is beneath the Carpathian fore-deep and overthrust sheets.

From east to west, the following geologic units are present:

1. Autochthonous Brunnia basement of Late Precambrian age covered by Devonian and Dinantian platform sediments that passes diachronously from west to east into diastrophic Culm sedimentation.
2. Allochthonous Brunnia basement with sedimentary cover. Culm sediments are in external position completely detached and thrust as thin-skinned overthrust sheet over autochthon. More internally, however, Brunnia crystalline rocks form the imbrications within thrust sheet 8–10 km thick today according to keep seismics. Significant amount of dextral shearing (Moravian shear zone) is present in this zone.
3. Moravicum: Zone of considerable dextral transpressive shearing (Moravian shear zone). Extremely deformed Moravian rocks consist again of lower plate Brunnia basement; the Culm flysch cover is not, however, present there. Moravicum crops out in windows and half-windows beneath the overthrust sheet of metaophiolites and sometimes directly beneath the upper Moldanubian plate.
4. Metaophiolite zone consisting mostly of amphibolites, metagabbros, tonalites, and metasediments is, like Moravicum, strongly sheared and deformed. Metabasalts belong geochemically to the MORB types and their new dating by Sm-Nd method resulted in Devonian original ages. This zone is most likely a remnant of ocean basin between the upper Moldanubian plate and lower Brunnia plate.
5. Moldanubicum: Strongly tectonized terrane but not much dextrally sheared. Katazonal high-grade metamorphic rocks are present here with numerous granulite massifs, amphibolites, serpentinites, and eclogites. Mode of emplacement of lower crustal and upper mantle rocks to the surface is unknown. Some part of them were already present as pebbles in Upper Visean Culm conglomerates.

Near the Czechoslovakian–Polish frontier in the Jeseníky area, more than 70 km long overthrust sheet of Culm and Upper Carbonaceous (Ostrava beds) is emplaced over Brunnia autochthon. There is significant imbricating of Culm slices observed on seismic reflection lines and no reflectors in the autochthon. This thrusting was confirmed by scientific borehole Potstát-1 drilled to the final depth of 4.2 km. Beneath the 2.8 km of Culm allochthon, slices of paraautochthonous platform Devonian and Dinantian sediments are present. Allochthonous Culm sediments have significantly higher vitrinite reflectance coefficient than underlying paraautochthonous slices. Potstát-1 borehole enables, moreover, detailed study of the nature of reflections. Major reflections are caused there by cataclastic and brecciated rocks within the thrust fault zones. Master decollement fault zone of the whole Variscan overthrust sheet is about 100 m thick.

More internally in tectonic sense, but farther to the south, deep seismic reflection profiling on the 8HR line nearby Brno revealed imbrications and stacking of Brunnia granitoid slices dipping some 40° west-northwest in thickness of 8–10 km. Similarly dipping, but more pronounced, are imbrications of Moravicum. Reflections from Moravian rocks continue to a depth of 15 km and probably more. The thrust zone between upper plate and Brunnia allochthon is extremely narrow compared with the Variscides in Western Europe.

Geologic, geophysical, and borehole data confirm long-term Lower Carboniferous continental collision in the eastern boundary of the European Variscides. Subsequent stacking and piggy-back thrusting of basement Brunnia slices over Brunnia autochthon is proved geologically and seismically. Thrusting finished before Namurian R time. Emplacement of the high-grade metamorphic rocks of the Moldanubian upper plate took place before Late Visean. Large-scale normal faulting was probably present during this emplacement; its proof is, however, not certain.

TONG, L. Department of Geology, University of Science and Technology of Beijing, Beijing, China

Chemical Composition of Continental Crust in China

China's continental crust means the crust which is districted by the boundaries of Chinese territory, about 9.6 million km^2 in land area, and its lower limit is the Moho. So that it is corresponding to the upper part of the lithosphere in China.

According to the model of China's continental crust, its average thickness is 47 km, its volume is 451 million cubic km, and its total mass is 12,437 Gg ($1Gg = 1 \times 10^{14}T$). The average thickness of its sedimentary layer is about 5 km, and of its Sial and Sima layers are 26 and 16 km, respectively. The mass percentage of its sedimentary layer is 9.21%, and the mass ratio of its Sial/Sima layers is 1.55.

About 2,252 chemical data of rocks occurred in China are used to calculate the average chemical composition of China's continental crust. The calculated result is tabulated in Table 1.

As compared with the average chemical composition of the global continental crust, the chemical characteristics of China's continental crust has the lower contents of silica and alumina, and the higher content of potassium oxide.

In China's continental crust, the chemical characteristics of its sedimentary sphere are of the extremely high content of calcium oxide, and the extremely low content of sodium oxide. Its upper part (including sedimentary layer and Sial layer) has higher contents of silica, alumina and potassium oxide, and the lower contents of ferrous oxide, magnesium oxide and calcium oxide.

Table 1. The average chemical composition of China's continental crust (Wt%)

	China's Sedimentary sphere	China's Upper continental crust	China's continental crust	Global cont. crust (Li et al. 1966)
SiO_2	57.61	64.84	57.77	60.74
TiO_2	0.44	0.42	1.10	0.86
Al_2O_3	9.61	14.88	14.22	15.35
Fe_2O_3	2.37	1.78	2.90	2.80
FeO	2.13	1.95	3.92	3.51
MnO	0.03	0.06	0.10	0.13
MgO	2.27	1.30	3.57	3.41
CaO	10.43	4.55	6.04	5.61
Na_2O	0.30	3.08	3.19	3.30
K_2O	2.00	3.09	2.82	2.36
H_2O^+	2.96	1.39	1.91	1.22
P_2O_5	0.16	0.17	0.44	0.27
CO_2	9.69	2.49	2.02	0.44
Total	100.00	100.00	100.00	100.00

TORRENS, H. S., Dept. of Geology, University, Keele, United Kingdom

Transmission of British Ideas to America on Uses of Fossils in Stratigraphic Analysis 1800 to 1840

The transmission of technical ideas and inventions across the Atlantic in this period has been examined, as have British attempts to stem the flow of machines and technicians. The flow of scientific ideas has been less investigated. This is despite the fact that many had practical significance; particularly the identification of rocks aided by fossils, which had revolutionized the search for raw materials in England from 1800. This paper discusses the trans Atlantic flow of ideas in this field and how it was, or might have been, achieved through avenues which did not involve publication. The focus on British ideas here does not imply that these were any earlier developed than in other European countries, or that no American ideas flowed back to Britain.

The first problem faced by the historian is the sheer scale of emigration to the New World, after an unsettling period of revolution. Joseph Priestley (1733–1804), who emigrated in 1794, is the tip of an iceberg, 100,000 persons left Ulster alone for the U. S. between 1783 and 1812.

The second problem is the actual period over which ideas, that fossils could help in the identification of rocks, were actually transmitted from Britain. That they could help was here first properly elucidated by William Smith (1769–1839), late in 1795 while a canal engineer. But he did not publish his own results until 1816. Some historians, used to a modern notion of the role of publication, has dismissed the part personal contacts could have played in spreading his discoveries before 1816. There is however much evidence that Smith started to disseminate his ideas widely, from a base in the most popular space in Europe, by word of mouth and personal inspection of his fossil collection arranged in stratigraphic order. Other individuals were here also busy disseminating his ideas from 1799, as were corporate bodies like the second Bath Philosophical Society founded in 1798, or the Bath and West of England Agricultural Society from 1800; that also opened its Chemical laboratory in 1805, at which Smith's results were again demonstrated.

This paper discusses how the discoveries of Smith and others were disseminated from Britain through a series of migrants. These include R. C. Taylor (1789–1851), who had been Smith's pupil in 1811 and a major supplier of fossils and associated stratigraphic data to the Sowerby's project on Mineral Conchology from 1813. He emigrated to the U. S. in 1831. He there published a long series of papers and, perhaps more significantly, took his personal fossil collection from Britain. This was offered for sale from November 1831 and had been purchased for the Academy of Natural Sciences in Philadelphia by 1846.

An earlier migrant to the U. S., under less happy circumstances, was John Finch (1791–1854); Priestley's grandson following a family tradition. Finch however did not go to escape political persecution but to escape creditors! Finch had published a number of papers on English geology in the period 1818 to 1821 and was a member of the Birmingham Philosophical Society. In 1822 he was imprisoned for debt but soon made his escape to the U. S. in November 1822. He is not yet known to have had any direct contact with the Smith school but, by 1822, Smith's results were so widely accepted into English geology that Finch was simply following current practice. In Jly 1823 he read his first paper in America, in which the use of fossils to help with the identification of rocks was advocated. Finch returned to England by 1833 and now took his American fossils with him, which were sold to the British Museum in 1834.

In 1815 a more significant migrant had crossed the Atlantic, the Moravian schoolmaster Rev. Henry Steinhauer (1782–1818), who from 1812 until his departure from England was based in Bath and closely involved with the circle of Smith's friends there. Steinhauer read his only American paper, on Fossil plants, to the American Philosophical Society in May 1817, to which he was elected in October. He died just after it was published, in July 1818. His fine stratigraphically arranged collection of English fossils had also gone with him to America and again passed to the Academy of Natural Sciences in Philadelphia.

English ideas could also be transmitted to Americans visiting England. Smith himself had contacts with several American, from a subscriber to his 1815 map to visitors like William Maclure in 1815. Later the Anglo-American G. W. Featherstonhaugh was personally instructed by Smith on a visit in 1827. The attempt to get Smith to the United States in person in 1819 was not successful, but it brought him into contact with yet more Americans. Benjamin Silliman had earlier met Smith's disciple Rev. Joseph Townsend at one of mineralogist Charles Greville's dinner parties in London in June 1805.

Documenting such links is quite possible but demonstrating their significance for the spread of ideas is more problematic. The presence of stratigraphically arranged collections of actual English fossils in Philadelphia may well have been the most significant. The only way by which these questions can be properly answered is a joint Anglo-American project, which can hopefully explore other European ideas and put them in their proper contexts.

TORRES, A., Universidad Nacional Mayor de San Marcos, Lima, Peru

Peruvian Andenes as an Alternative of Land Space Planning in Hillside

Peru, geographically located in the Central Andes presents peculiar characteristics: very variable topography from wide plains and gentle hills to steep canyons. Stratigraphic relations with complicated variations in sedimentary, volcanic-sedimentary and volcanic rocks either horizontally as vertically in sequences particularly related to environments of continental deposition. Peculiar tectonic is a matter of genetic and evolutives controversies referred to its development areal and regional complexity, from the climatic point of view related to seasonal periods, mainly dry and wet, with their respective transitions.

In relation to the above mentioned, either geographically as geologically the morphodynamics phenomena are conditioned favorably, being of different kind, composition, morphology, structure, cause and effect; thus there phenomena are extremely violent in occurrence; due to obvious topographic differences, reaching in average to 30° of inclination, making easy the development of variated morphodynamic phenomena in close relation to the strong variations of the pluviometric regimens.

Since the Pre-Inca period and mainly during the Inca civilizations, the Andean settlers had known and understood the different natural phenomena that affected them. Evident testimonies of the mentioned are observed in many places of the Peruvian territory (Lima, Ayacucho, Apurimac, Arequipa, Cusco, etc.) where it is possible to observed the Peruvian models based on sophisticated artisinal technics, working properly.

Administration of the Inca State was closely related to control and coherent handling of the multiplicity of ecologics floors developed in the Andean regions, as for in the agricultural societies is organized the relations of production necessarily interrelated with the organization of the laboral time, this principle had great importance in the andean civilizations and was sustained in the calendar organization of the agricultural labors.

In the other hand research about the theoretical roots of such models are still in process, but considering the splendid results on the Sierra of our territory, related to recovering of land for agriculture and morphodynamic phenomena control, the Peruvian andenes are considered as an alternative for planning the land use on hillsides, because the high level of efficiency and development of this artisianal model.

Nowadays, the lack of land space planning is associated to numerous cases of natural disasters, involving loss of human lives and loss of materials as well as economic resources. This have originated a peculiar interest about the knowledge of the Andenes, which will be possible only based on multidisciplinary effort, related intrinsically with the proper analysis in relation to Peruvian technics.

Particularly, the Andenes are a sort of structure as grader or steps, which are situated one based on the other to a variable distance, its width is predominantly short on the slopes; its length is direct relation of the volume of unconsolidated material, the height changes according to the thickness of the unconsolidated material. Therefore the dimensions are in relation to the slope and volume of unconsolidated material, missing a general rule referred to construction and dimension, which obviously is related to nature, composition, morphology and structure of the material over which they are built.

For the construction of the Andenes, in first place is located the basement, wherever is sustained the construction of a contention wall, which change of size according to the slope; immediately the wall is built with stones more or less smoothed with plain surfaces putting the big blocks down and front, the little blocks are placed between the wall and the natural slope, then is refilled with uncultivated land in the base and agricultural land or soil in the upper part, forming a plain zone with light inclination against the hillside, this procedure begins in the upper parts advancing downward on the hillslopes and concludes building channels of lateral derivation (vertical, diagonal and central following the direction of flowing); is important to precise that the fine grained material (clay or sand) between the blocks must be damp, in the construction.

From the morphodynamic point of view the Andenes favours the use and control of water; helps in the control of the erosion, due to superficial drainage, regulation of the water percolation due to the great porosity and permeability of the coarse material of the base and contention wall in the Andenes; also attenuate the frost, helping the use of hillsides; permitting to realize adequate planning either for locating of villages and works of engineering foundations. Moreover, increases the insulation (favoring the photosynthesis) and the temperature of the soil, facilitating the efficient programming of the work; helps in the retention of the humidity of the soil, increasing the oxygenation of the soil, thus the farms productivities increases with oxygenation of the soil, thus the farms productivities increases with a low cost concerning in their executions; its estimated cost of U.S. $1,800 a 2,000 each hectare (2.47 acres) in the rehabilitation of ruined Andenes; the advantages are based on the short lapse of execution of the works, thus it has a multiplicator effect that no have increases greatly the economic cost and motive on autoadministration.

TOSELLI, A. J., Universidad Nacional du Tucumán, Tucumán, Argentina, and H. FOURNIER, CRICYT, Mendoza, Argentina

Transect VI

Transect VI cuts the central-north zone of Argentina, from its boundary with Chile (27°15' S latitude; 69°00' W longitude), along 1669 km, to its boundary with Brazil (25°40' S latitude and 53°50' W longitude).

From W to E, the section intersects the transitional area between the Frontal Cordillere and the Puna which is constituted by igneous rocks and neo-paleozoic or even older. In this area, cenozoic strato-volcanoes are characteristic, and it is also the only region with evident Andean magmatism. To the east there is the Famatina range system and the Western Pampean Ranges. This region has a lithology of metamorphites and granites which belong to the early Paleozoic and also outcrops of continental sedimentary rocks, that belong to the late Paleozoic and the Tertiary. The mountains are characterized by a relatively stable neo-tectonic regime, which is controlled by vertical movements and by tilted range blocks as a reaction to subhorizontal compressive regimen, which produce reverse faults of high angle, which are supposed to become listric in depth. The dominant orientation of the fracture from the Argentina-Chile border to the Sierra de Velasco (approx. 67° W longitude) is SSW-NNE; while to the east the dominant one is from north-south to NNW-SSE. It is interpreted that the former is related to the ancient Pre-Cenozoic structures, and the latter would be Cenozoic. Once the Pampean Ranges is finished, a plain corresponding to the Chaco-Paranense Flat-slab begins. There, it is only possible to recognize Quaternary outcrops that go up to the Parana River. In this region it is possible to appreciate fluvial and marine sediments from the Tertiary. There are also volcanic rocks, with continental sandstone belonging to Parena Basin largely developed in Brazil.

The geophysical data, on the Transect VI, shows:

Seismology. The seismological observations collected, include data about hypocentres between 20° and 31°, S latitude and between 62° and 74° W longitude.

The penetration angles with which subducted Nazca plate continues beneath the continental margin make it appears as a contortion or scissors fracture. This contortion becomes more neat when passing from the area between 20° and 25° S latitude, with an inclination of 23°, to the area between 25° and 31° S latitude with an inclination of 17°.

Magnetometry. The systematization of magnetic data, corresponds to ISODINAS Z and H, with important variations in Z at the extremes of Transect VI of about 2000 gammas in decrease from west to east.

The most remarkable variations of horizontal component H occur in accordance with basaltic rocks of the Parana Basin.

Gravimetry. Along the section 360 data point were collected from Gravimetry Stations.

Maps on gravimetrical anomalies Free-Air and Bouguer show, along the axis of Transect VI a stability of low values between 56° and 63° W longitude with an important increase to the west up to AgB = –250 mgal at 66°W and AgB = –350 mgal at 68.5 west, taking the values 2.67 g/cm^3 densities.

TOSELLI, A. J., and F. G. ACEÑOLAZA, Instituto Superior de Correlación Geológica, Tucumán, Argentina, and J. SAAVEDRA, Instituto de Recursos Naturales y Agrobiologica(CSIC), Salamanca, Spain

Ordovician Magmatic Arc of Sierra de Famatina, Argentina

The Sierra de Famatina, Figure 1, is formed by low-grade metamorphic rocks—Negro Peinado Formation—intruded by Ordovician–Silurian granites (De Alba, 1979) with thick sequences at west and northwest zones of volcanic materials very altered of a large petrographic compositional, with dominant pyroclastics types, of dacitic-andesitic to rhyolitic composition, alternating with subordinate intercalations of basaltic flows (Toselli et al., in press). Sediments suggest continental proximity and the fossils show shallow waters of a low Ordovician age (Aceñolaza and Toselli, 1984). The system is partially covered by post-Devonian sediments (Turner, 1979).

The granitoids essentially display compositions from gabbroic to granodioritic terms, with amphibole and/or biotite in variable amounts. Muscovite is limited or does not appear and aplites and pegmatites are presented only in few places. These granites and gabbroic rocks have gradations of mixing and mingling types. According to the chemical data, they reflect calcalkaline features typical of mature volcanic arc (Brown, 1982).

The strong hydrothermal alteration of the volcanic rocks allows only the roll of immobile elements as characteristic indication of genesis from a geochemical point of view. The diagram TiO_2-Zr suggests that it is a mainly magmatic sequence, with a fractionation of plagioclase, apatite, olivine, and pyroxenes, continued by amphyboles (Fig. 2).

The low contents of Y and Nb discount a within-plate or oceanic ridge nature (Pearce, personal communication, in Twist and Harmer, 1987), and the domain of moderately acid or basic terms, Figure 3, suggests maturity of one calcalkaline in the limit to alkaline magma.

Consequently, all data point to a volcanic arc magmatism, of active continental margin, of Ordovician age, manifest in plutonic as well as volcanic terms.

REFERENCES

Aceñolaza, F. G., and A. J. Toselli, 1984, Lower Ordovician volcanism in North-West Argentina, *in* D. L. Bruton, ed., Aspects of the Ordovician system: Paleontological Contributions from the University of Oslo, n. 295. Universitetsforlaget, p. 203–209.

Brown, G. C., 1982, Calc-alkaline intrusive rocks: their diversity, evolution, and relation to volcanic arcs, *in* R. S. Thorpe, ed., Andesites: orogenic andesites and related rocks: John Wiley & Sons, p. 437–461.

De Alba, E., 1979, Sistema del Famatina, *in* Geologia Regional Argentina, Segundo Simposio, v. I, p. 349–395. Academia Nacional de Ciencias Córdoba.

Floyd, P. A., and J. A. Winchester, 1975, Magma type and tectonic setting discrimination using immobile elements: Earth and Planetary Science Letters, v. 27, p. 211–218.

Pearce, J. A., and J. R. Cann, 1973, Tectonic setting of basic volcanic rocks determined using trace element analyses: Earth and Planetary Science Letters, v. 19, p. 290–300.

Pearce, J. A., and M. J. Norry, 1979, Petrogenetic implications of Ti, Zr, Y and Nb variation in volcanic rocks: Contributions to Mineralogy and Petrology, v. 69, p. 33–47.

Toselli, A. J., J. Saavedra, E. Pellitero, and J. N. Rossi de Toselli, in press, Geoquímica y petrogénesis del vulcanismo ordovícico de la Formación Las Planchadas, Sistema de Famatina, Argentina.

Turner, J. C., and V. Mendez, 1979, Puna, *in* Segundo Simposio Geol. Reg. Argentina, v. I, p. 13–56.

Twist, J. A., and R. E. Harmer, 1987, Geochemistry of contrasting siliceous magmatic suites in the Bushvels complex: genetic aspects and implications for tectonic discrimination diagrams: Journal of Volcanology and Geothermal Research, v. 32, p. 83–98.

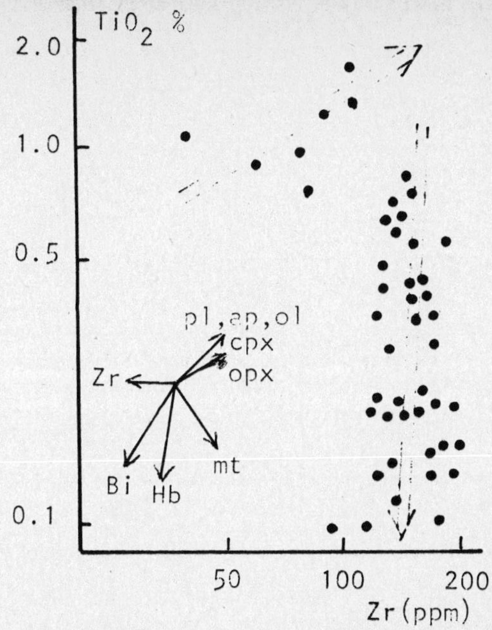

Figure 2—TiO_2-Zr diagram of Pearch and Norry (1979) for the volcanic rocks of Sierra de Famatina. The direction vector shows the fractionation trends. Zr, zircon; pl, plagioclase; ap, apatite; ol, olivine; cpx, clinopyroxene; opx, orthopyroxene; mt, magnetite; Hb, hornblende; Bi, biotite.

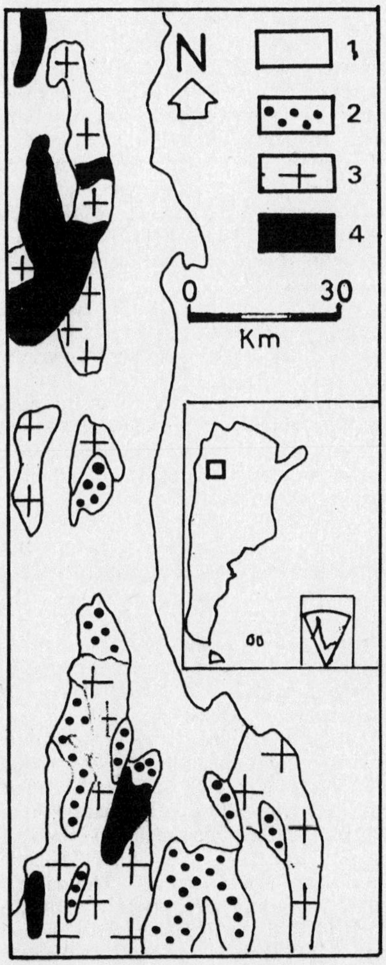

Figure 1—Geological map of Sierra de Famatina. 1, Post-Devonian cover; 2, Negro Peinado Formation (metamorphites); 3, granitoids; and 4, Ordovician volcanic rocks (Las Planchadas Formation).

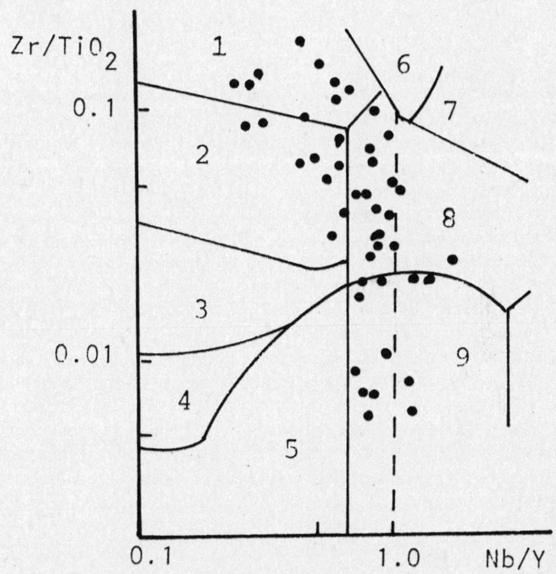

Figure 3—Diagram of Winchester and Floyd (1977). The limit in broken line of the alkaline and subalkaline fields is from Pearch and Cann (1973) and Floyd and Winchester (1975).

TOURET, J. L. R., Free University, Amsterdam, the Netherlands

Free CO_2 and Fluid-Absent Domains in Lower Crust: Model Based on Fluid Inclusion Studies

Many lower crustal granulites contain early fluid inclusions that fall basically into two categories: high-density CO_2 and NaCl-rich brines. Other components, such as N_2 and CH_4, seem also to occur in some high pressure granulites (Calabria, Italy, Massif Central, France), but their role needs to be further substantiated. Many inclusions have been trapped or reequilibrated during post-peak metamorphic conditions, but estimating that all inclusions are late is at odds with some fundamental observations:

1. Some fluids, notably brines, are related to specific lithologies (metasediments, acid volcanics) and they have been inherited from a supracrustal premetamorphic stage (Touret, 1985). This does not mean that they witness the existence of a permanent free fluid phase, but they indicate that some inclusions, at least, were not completely obliterated during high-grade metamorphic recrystallization.

2. Some granulites, notably from India, Southern Norway, Brazil, etc., contain many more CO_2 inclusions than any other comparable sample from another metamorphic grade. Some of these inclusions, which occur not only in quartz, but also in garnet, pyroxenes, feldspars, etc., are demonstrably trapped under peak metamorphic conditions (Touret and Hansteen, 1988). Their occurrence indicates a local influx of CO_2 during peak metamorphic conditions and the existence, at this emplacement, of a free synmetamorphic CO_2-rich fluid phase. A number of arguments plead in favor of a deep-seated, mantle-derived origin for the CO_2 and its magmatic transportation in the lower crust by CO_2-saturated synmetamorphic intrusives (Frost and Frost, 1987). At the magmatic source, the CO_2 was either already gaseous and dissolved in the magma or possibly combined in the form of Mg/Fe-rich carbonates or eventually carbides. It must be noted that at the temperatures which prevailed during the CO_2 liberation in the lower crust (about 1000 °C), the oxygen fugacity imposed by mineral assemblages is compatible with the existence of graphite. Local CO_2 streaming, which remains limited to the immediate vicinity of the synmetamorphic intrusions, will therefore not lead to any graphite precipitation. A rapid decrease of fo_2 during cooling may eventually lead to the reduction of some CO_2 into graphite, as actually seen in some inclusions.

Many CO_2 inclusions are trapped or reequilibrated after peak metamorphic conditions. The evolution of their density, compared to P-T conditions recorded in neighboring mineral assemblages, constrain possible postmetamorphic P-T trajectories between the lower crust and the surface. Granulites originating below 30 km (high-pressure granulites) are characterized by a constant, regular decrease of both P and T ("pseudo-isochoric P-T path"). For the other types, notably for the low-pressure granulites (depth of formation; about 15 km), a period of isobaric cooling (between roughly 900 °C and 500 °C) is then followed by a rapid uplift at more or less constant temperature. This complicated type of P-T path has been documented in various regions from different ages, and it seems to characterize many Archean and Proterozoic granulites.

These observations support a model in which vapor-absent and CO_2-dominated domains may coexist in time and space. Refinements of this model require: (1) a better understanding of fluid transfer between lower crust and upper mantle in continental environments; (2) more information about the role and origin of nitrogen; and (3) some knowledge of the timing of fluid pulses during and after peak metamorphic conditions.

REFERENCES

Frost, B. R., and C. D. Frost, 1987, CO_2 melts and granulite metamorphism: Nature, v. 327, p. 503–506.

Touret, J. L. R., 1985, Fluid regime in Southern Norway: the record of fluid inclusions, *in* A. C. Tobi and J. L. R. Touret, eds., The deep Proterozoic crust in the North Atlantic provinces: NATO ASI C158, Reidel Pub., p. 517–549.

———, and T. H. Hansteen, 1988, Geothermometry and fluid inclusions in a rock from the Dodda Betta charnockite complex, Southwest India: Rend. Soc. Ital. Min. Petr., v. 43, p. 65–82.

TRAYNER, P. M. and T. J. CHAPMAN, Geco Exploration Services, Surrey, United Kingdom

Structural and Stratigraphic Controls on Hydrocarbon Prospectivity West of Berlin

Unlike the prolific hydrocarbon province of the North Sea the continental shelf west of the UK mainland has so far failed to live up to early expectations. Just two producing fields are now being exploited and exploration activity has reduced drastically from the heady days of the early 1980's.

Detailed stratigraphic and structural studies have now explained the reasons for many of the early failures and identified many encouraging plays and prospects for the future. This paper reviews the major basins from south to north based on the interpretation of a seismic database of more than 27,000 km.

In the south in the Western Approaches Basin the lack of good source rocks is a major weakness. However, some of the thickest and deepest Jurassic has yet to be tested and furthermore the possibility of a deep Stephanian coal basin remains to be investigated.

In the Celtic Sea a thick Jurassic sequence is at peak maturity in the basin center. Early exploration targets were the prominant Tertiary inversion structures, but evidence now shows that unless secondary migration has taken place, hydrocarbon generation in the early Cretaceous predated the formation of Tertiary traps. Future targets must concentrate on deeper Jurassic fault blocks especially at the basin margins. In the adjacent Cardigan Bay Basin the Jurassic is even deeper, but a complex early Tertiary inversion period has removed the entire Cretaceous sequence. Prospectivity is enhanced by the recognition of a Carboniferous coaly facies.

North of the established Morecambe Bay field a series of half grabens, controlled by deep reactivated basement faults, locally bury the Jurassic to levels of maturity. Although undoubtedly still frontier acreage these basins contain source units in the Jurassic and evidence suggests they are underlain by Carboniferous in a coaly facies. Traps are associated with the major listric faults such as the Minches Fault, and although these have been reactivated from Proterozoic through to Tertiary times, potential still remains.

TREAGUE, J. E., R. A. D. PATTRICK, and S. CURTIS, Department of Geology, University, Manchester, United Kingdom

Mineralization, Igneous Activity and Faulting in Dalradian of Scotland

The Dalradian supergroup of Scotland and Ireland is a c.20 km thick sequence of metasediments with minor volcanics. The sediments, which range in age from Late Proterozoic to Cambrian, were deposited in a northeast–southwest ensialic basin on Archaen basement. The sedimentary environment progressed from deltaic and stable shelf through a development of basins with local volcanicity, faulting and stratabound mineralization, to deep water turbidite basins and increased volcanicity.

These rocks initially suffered polyphase deformation, associated with nappe formation and greenschist to amphibolite metamorphism, in the early Caledonian Orogeny (Grampian, 520–490 Ma). Subsequent deformation, latterly related to uplift, continued to c. 430 Ma. The closure of the Iapetus Ocean to the southeast from late Silurian to early Devonian, was accompanied by late Caledonian intrusives and northeast to north–northeast-trending strike-slip faults, which affected both the Laurentian margin as well as the Lower Paleozoic rocks of Iapetus. The main period of intrusive activity was at 420–395 Ma, closely followed by faulting, whose major movements finished before the mid-Devonian.

One of these faults, the northeast-trending Highland Boundary Fault (HBF), is the southeast boundary to the Dalradian terrane. This fault is now considered to have had a long history of sinistral strike-slip movement from the Ordovician to the early Devonian and is probably the master fault to the smaller north–northeast-trending fractures that affect the Dalradian and older rocks to the north (Figure 1). These faults include the Great Glen Fault (GGF), for which various estimates of pre-mid Devonian strike-slip displacement, up to 200 km, have been made. Within the Dalradian of Scotland there are five major fractures, and many smaller, ones, that appear to splay from the HBF in the manner of 1st order kiedel shears. The major fractures certainly have a sinistral component of displacement which cumulatively amounts to over 20 km. However, they also can be shown to have a vertical component of displacement, which might be the consequence of a separate movement phase.

Most of the major faults have associated mineralization. Two phases of economically significant vein mineralization are associated with one of these fractures, the Tyndrum Fault (T on Figure 1) and related minor fractures. The older phase is Au-bearing pyrite-quartz veins, which are hosted by lowest Dalradian quartzites. fluid inclusions reveal mineralizing temperatures $T_{(m)}$ in the range 300–330 °C, salinites of about 7.0 wt% NaCl and CO_2 concentrations of about 7.0 mol%. This mineralization has similar characteristics to low-grade porphyry-style mineralization associated with late Caledonian (420–400 Ma) granitic intrusions in the region (Figure 1). Gravity surveys suggest the presence of such a granite beneath the veins at Tydrum. The Tyndrum fault also ouis and late Caledonian pluton but is intruded by a flesite dyke of similar composition. It is apparent that intrusive activity, faulting and mineralization were broadly contemporaneous.

The later phase of mineralization is manifest by Pb + Zn veins, occupying both the major fault plane and sub-parallel fractures in quartzitic metasediments, and these are cut by small uraniferous veins. Dating of K-feldspar suggests a Lower Carboniferous age (about 360 Ma) for the Pb + Zn veins and $^{87}SR/^{86}Sr$ data indicate that the U-bearing veins are a late stage of this episode. Strontium isotopes suggest the presence of granitic rocks beneath the veins. Fluid inclusions indicate that the mineralizing temperature was in the range 140–240 °C and salinities about 20 wt% NaCl.

The fractures in the Dalradian metamorphic terrane record the evolution from an epithermal Au mineralization related to late orogenic magmatism to low temperature base-metal mineralization formed in a post-organic extensional regime.

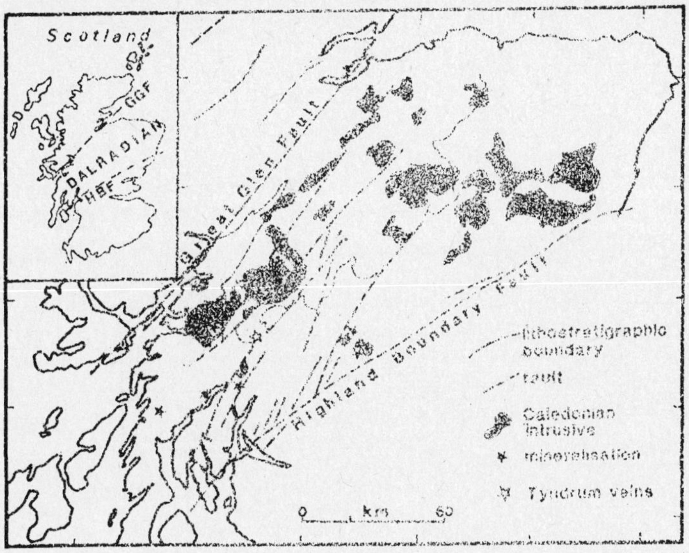

Figure 1.

TREAGUS, S. H. Manchester University, Manchester, United Kingdom

Strain Modeling in Contrasting Lithologies

Introduction: At the scale at which most geological structures are observed, differences in lithology appear to control deformation behavior. Traditional models of folding and boudinage are based on the mechanics of layer-parallel compression/extension of layers of contrasting competence (measured by ratios of elastic or viscous moduli). Materially different layers in equal shortening or extension must, by definition, be in unequal stress; the stiffer the layer, the greater the principal layer-parallel stress. However, for the more general case of layers under non-layer-parallel principal stress, the principal stresses must not only be unequal in value from layer to layer, but change orientation. This may be termed stress refraction.

Stress refraction: Stress refraction can be modelled theoretically for obliquely stressed planar Newtonian viscous layers with contrasting viscosity, or by correspondence, in Hooksan elastic layers. The model is based on two equalities at interfaces between different layers: (1) equal layer-parallel longitudinal strain rates and (2) equal layer-parallel shear stresses. The corollary of 2 is that the shear strain rates across layering must be unequal in different layers, and their ratios will be the inverse viscosity ratios.

Strain refraction: The refraction of principal stresses in viscous layers signifies an equivalent refraction in principal strain rates. The model can thus be applied to an incremental strain history across contrasting layers, but will it allow integration to finite strain refraction?

The two equalities for stress refraction must also apply throughout finite strain development in viscous layers. First, finite strain ellipsoids must refract, such that their layer interface strain ellipses are identical for all layers. Second, for Newtonian-viscous layers it can be shown that the finite shear strain ratio across an interface is equivalent to the viscosity ratio. This rule allows finite strain refraction to be modeled and quantified.

Finite strain refraction can be described in a manner comparable to many analyses of heterogeneous strain. The two rules of continuity that underlie the model can be regarded as two components of the finite strain pattern: (1) a homogeneous layer-parallel/perpendicular strain that is equal in all layers, and (2) a system of layer-parallel simple shear zones, of layer width, with simple shear strain inversely proportional to viscosity.

Examples of finite strain refraction patterns in two and three dimensions for variably viscous layers illustrate some points that may be of geological significance.

1. Strain will only be homogeneous across contrasting layers, where layering is a principal plane of strain.
2. If principal strain axes are oblique to layering in a particular layer, they will refract and change value across a viscosity contrast.
3. Stiffest layers will be least strained, with strain axes closest to layer-parallel/perpendicular.
4. Weakest layers will be most highly strained, with strain axes oblique to layering. Their strong component of layer-parallel shear strain could lead them to be interpreted as shear zones.

If the viscous model is applicable to ductile straining of layered rocks, the four points above lead to the conclusion that strain data from particular lithologies should not be used to indicate regional (tectonic) strain (likewise for stress measurements). So can strain data from one lithology be used in conjunction with other, structural data to indicate regionally significant strain?

Cleavage refraction for strain modelling: It has long been recognized that cleavage refraction could represent strain refraction. As yet, this is far from proved for foliations in all kinds of lithology, and for a range of finite strain values. Nevertheless, cleavage is probably the most ubiquitous single indicator of strain available to geologists. Cleavage refraction may thus provide the best data at hand for measuring strain variation in contrasting lithologies.

Strain refraction modeling, combined with the assumption that refracting cleavage approximately represents orientations of principal (XY) planes of finite strain in different lithologies, leads to the following conclusions:

1. No cleavage refraction indicates no change in strain.
2. For a set of planar layers with cleavage refraction, if the strain ellipsoid is known for one layer, it can be deduced for the others.
3. The "mean" cleavage for a set of layers provides a method of determining their "bulk" strain.
4. Estimates of bulk strain across folded terrains would need to consider subregions such as fold limbs, derive mean strains for these, and integrate them to the large scale.

TRENDALL, A. F., J. G. BLOCKLEY, M. F. MIDDLETON and J. S. MYERS, Geological Survey of Western Australia, Perth, Western Australia

Geological Evolution of Western Australia and Its Control of Mineralization

In early 1989, a new monograph describing the geology and mineral resources of Western Australia was published by the Geological Survey of Western Australia, as Memoir 3. It was accompanied by a revised geological map of the State at a scale of 1:2,500,000, two maps (one devoted to gold) summarizing its mineralization, and a comparative stratigraphic chart of the Phanerozoic sedimentary basins.

The publication of this volume provided an opportunity to review previous tectonic nomenclature and to establish the relationship between mineralization and tectonic development. The continental crust of this region is divided into three major tectonic components: cratons, orogens, and basins, (Figure 1). Cratons are those parts of the crust which have remained stable since about 2.4 Ga. Orogens consist of metamorphic and magmatic rocks that represent the eroded roots or orogenic belts younger than 2.4 Ga; they consist partly of reworked older craton as well as newly generated crust, including sedimentary basins. Basins are depositaries of sedimentary rocks that are not intimately associated with orogens.

There are two cratons (Figure 1) with very different geological histories. The Pilbara Craton consists of granite-greenstone terrane that mostly formed between 3.6 and 3.0 Ga, and overlying 2.8–2.6 Ga volcanic rocks (Fortescue Group) and 2.6–2.4 Ga banded iron formation (Hamersley Group). In contrast, the Yilgarn Craton consists of 3.0–2.6 Ga granite-greenstone terrane and a relatively minor portion of 3.8–3.0 Ga gneiss terranes. The gneiss terranes are largely derived from granite and minor components of layered basic intrusions and predominantly silicic sedimentary rocks.

The Yilgarn Craton is surrounded by the Capricorn, Albany-Fraser and Pinjarra Orogens (Figure 1) that formed at about 1.8 Ga. The Capricorn Orogen marks the collision and fusion of the Pilbara and Yilgarn Cratons, and subsequent relatively minor tectonic activity in this belt led to the formation of the Bangemall and Savory Basins. The Albany-Fraser and Pinjarra Orogens were also sites of intense tectonic and magmatic activity at about 1.3–1.1 Ga, and further major activity occurred in the Pinjarra Orogen at about 750–650 Ma.

The northeast part of Western Australia contains a number of Precambrian orogens and basins whose tectonic evolution is poorly known. Orogenic activity occurred in the King Leopold, Halls Creek, Arunta and Paterson orogens (Figure 1) at about 1.8 Ga, and again at about 750–650 Ma. The latter activity in the Paterson orogen led to the fusion of this belt with the combined Pilbara and Yilgarn Cratons and associated orogenic belts in the southwest of Western Australia. The Kimberley, Birrindudu and Amadeus Basins formed at about 1.8, 1.6 and 0.75 Ga respectively.

During the early and middle Paleozoic, the Canning Basin formed as an intracratonic sag, which was connected to the foreland Amadeus Basin, probably by a narrow seaway. Also at this time, initiation of the Bonaparte Basin occurred as an intracratonic sag. During the middle Paleozoic, the Perth and Carnarvon Basins formed initially as interior fracture basins, which developed into intracratonic sag basins. In the late Paleozoic, the Canning, Bonaparte and Perth Basins developed further as interior fracture basins.

Rifting and continental breakup of Australia occurred in the Mesozoic, during which time minor intracratonic sag subsidence occurred in the Canning Basin. Continental breakup occurred initially on the northwestern margin of Western Australia in the Middle Jurassic, and opening gradually proceeded in an anticlockwise direction until the Late Cretaceous. Large marginal sag basins (Bonaparte, Browse, Offshore Canning, Carnarvon, Perth Basins and southern margin) developed on the Western Australian continental margin from the Jurassic to Holocene.

The distribution of mineralization is indicated in Table 1. Stratabound nickel, vanadium and copper-zinc deposits formed contemporaneously with the greenstone belts of the Yilgarn and Pilbara Cratons, and widespread gold mineralization accompanied metamorphism and granite intrusion, the latter also producing mineralized pegmatites. Epigenetic gold, uranium and base-metal deposits also formed during periods of tectonic activity in the orogens. Supergene enrichment of banded iron-formations within the cratons produced iron-ore deposits. Diatremes, comprising lamproite, kimberlite and carbonatite carry diamond phosphate. Extensive lateritization of Precambrian rocks during the Tertiary produced bauxite deposits, while uranium and limonitic iron ore formed in Tertiary drainage channels. The Phanerozoic sedimentary basins were depositaries for coal and mineral sands while diagenesis produced accumulations of oil and gas, in addition to Mississippi Valley-style lead-zinc deposits.

MINERAL	ESTIMATED VALUE PER UNIT (Aus$ Million)*				
	CRATONS	OROGENS	BASINS	DIATREMES	REGOLITH
Base Metals					
Copper	3,500	600	**		
Lead	50	**	530		
Zinc	3,750	**	4.250		
Bauxite					55,000
Coal			40,000		
Diamond				3,900	600
Gold	17,000	2,000	**		3,000
Iron Ore	300,000	**	330		85,000
Mineral Sands					
Garnet			270		
Ilmenite			5,000		
Monazite			290		
Rutile			2,200		
Zircon			1,800		
Nickel	25,000	430			11,500
Pegmatite Metals					
Lithium	2,800	**			
Tantalum	1,400	**			**
Tin	50	**			**
Petroleum					
Condensate			12,000		
Gas			80,000		
Oil			4,000		
Phosphate			**	1,200	
Uranium		1,720	1,020	**	1,710
Vanadium	2,600				100

Notes:- * Value based on 1987 production returns, where possible, or recent metal prices

** Minor mineralization present

Table 1—Demonstrated resources of principal western Australian minerals by major geological units.

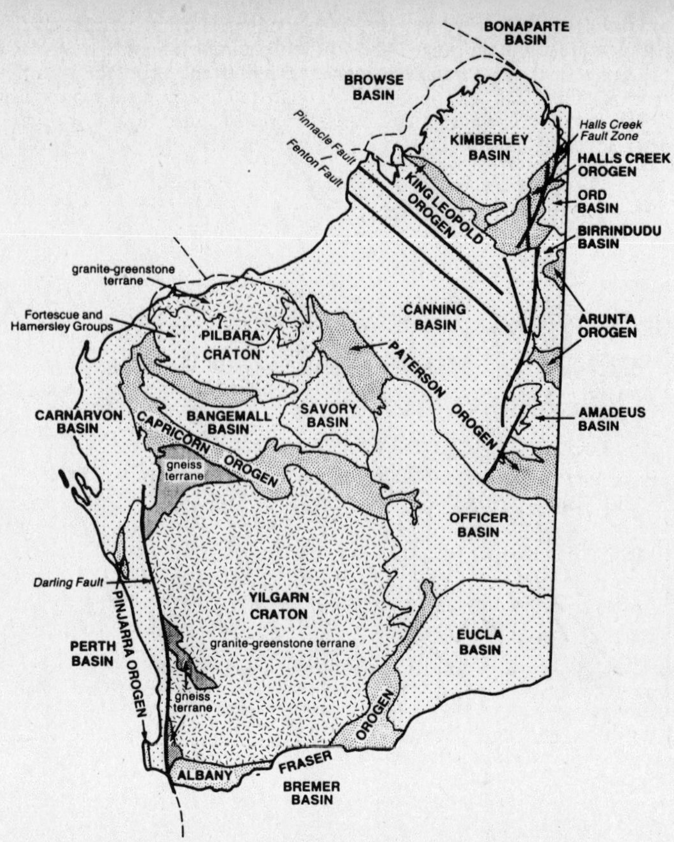

Figure 1—Map of Western Australia showing main tectonic units.

TRETTIN, H. P., and A. V. OKULITCH, Geological Survey of Canada, Calgary, Canada

Tectonic Framework, Canadian Arctic Islands and North Greenland

Major elements of the Canadian Arctic Islands and North Greenland are the Canadian Shield, Arctic Platform, Franklinian and Caledonian mobile belts, Sverdrup and Wandel Sea basins, and the Arctic Continental Terrace Wedge. The mid-Paleozoic Ellesmerian Orogeny and the early Tertiary Eurekan Orogeny together formed the Innuitian Orogen that merges with the Caldedonian Orogen off northeastern Greenland. The Carboniferous to lower Tertiary Wandel Sea Basin straddles the junction.

The Canadian Shield borders and extends under the Arctic Platform on the south and east, appearing in salients and inliers such as the Boothia Uplift, Minto Arch, Melville Horst, and the Wellington and Duke of York highs. It comprises metamorphic rocks of Late Archean–Early Proterozoic age, and unconformably overlying Proterozoic strata. Rifting at about 1.2 Ga produced dikes, sills, flows, and coarse clastics along much of the craton bordering the Arctic Islands and North Greenland.

The Arctic Platform includes relatively undeformed sedimentary rocks of Cambrian to Devonian and Jurassic to Paleogene ages that unconformably overlie the Shield.

The Franklinian Mobile Belt, defined as the active site of sedimentation, volcanism, and tectonism bordering the Shield and Platform, probably originated during Late Proterozoic rifting and emplacement of the Franklin dike swarm and flows of the Minto Arch (700–750 Ma). Lower Paleozoic strata were deposited on an unstable shelf and in an adjacent deep-water basin. Pearya, a Middle Proterozoic to Late Silurian composite exotic terrane presently adjacent to the mobile belt, is related to the Caledonian Mobile Belt by common ages of basement (1.0–1.1 Ga) and orogenies (Middle Ordovician). Accretion of Pearya, by presumably sinistral transpression, likely occurred in the Late Silurian. Northerly structures in the Canadian Shield responded to east-west compression in the Siluro-Devonian, forming the Boothia and Inglefield uplifts, and related tectonism formed the Northern Heiberg Fold Belt. Clastic wedges were shed from these uplifts. A thick Middle–Upper Devonian clastic wedge, derived from Greenland and the northern Innuitian region, prograded southwestward across the basin and adjacent shelf. Northwest-southeast and north-south compression terminated the Franklinian Mobile Belt in the Devono-Carboniferous (Ellesmerian Orogeny), and in northern Ellesmere and Axel Heiberg islands was accompanied by regional metamorphism to amphibolite grade and minor granitic plutonism.

The Sverdrup Basin was created by Carboniferous to Permian rifting followed by thermal subsidence superimposed on the basin and shelf of the Franklinian Mobile Belt. It contains up to 15 km of clastics, and lesser carbonates and evaporites, with local volcanics of Carboniferous, Permian, and Cretaceous ages. The northwestern margin of the basin (Sverdrup Rim) appears to have been affected by Jura- Cretaceous rifting and spreading in the western Arctic Ocean. Diapirism of Carboniferous evaporites occurred throughout the Mesozoic and during the Paleogene Eurekan Orogeny.

Paleogene northwest-southeast compression (Eurekan Orogeny) terminated the Sverdrup Basin and affected the northeastern Innuitian region. Older compressive structures were reactivated while extensive normal faulting occurred in southeastern parts of the region. Both compression and extension are interpreted as intraplate deformation caused by seafloor spreading in Labrador Sea and Baffin Bay.

The northwestern margin of the Arctic Islands is unconformably overlain by Neogene to Recent clastics of the Arctic Continental Terrace Wedge deposited along a rifted margin during continued opening of the western Arctic Ocean.

REFERENCES

Okulitch, A. V. (compiler), 1988, Geology of the Candaian Arctic Archipelago and North Greenland; scale 1:2 million: Geological Survey of Canada Open File 1850.

Trettin, H. P., ed., in press, Innuitian Orogen and Arctic Platform: Canada and Greenland: Geological Survey of Canada, The Geology of Canada, n. 3 (*also* Geological Society of America, The Geology of North American, v. E).

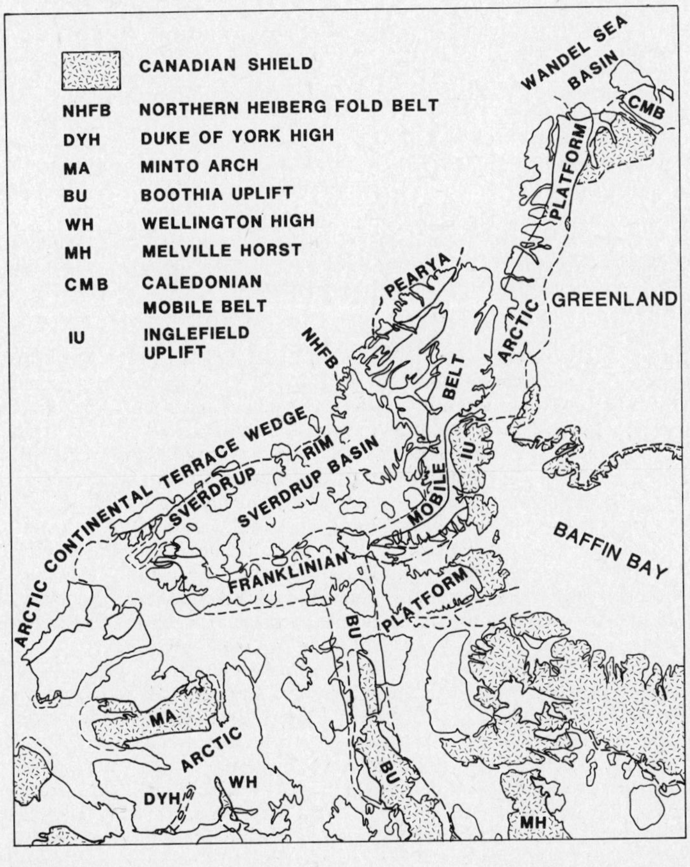

Figure 1.

TRIFENOV, V. G., Geological Institute of the USSR Academy of Sciences, Moscow, USSR

Geological Hazards in System of Geological Processes

Geological hazards are result of combination of different geological processes and properties of rocks, and they are interrelated within an intricate system. There is a group of natural disasters, which belong directly to the competence of geology. These are earthquakes, volcano eruptions, effects of fluid and gas activity, landslides and avalanches of many varieties, some permafrost-related phenomena, karstic and suffosional collapsing, inundations of underground arrangements by ground waters, erosion and sedimentation. Another group of disasters do not themselves belong to geology but which can be caused by geological reasons. For instance, a tectonic upheaving or sinking of a territory will alter the surficial drainage, the hydrogeology and the hydrochemistry of the groundwaters, which in its turn will bring about changes in the vegetation and the soil, uplifts developing erosion and formation of mobile sands, and depressions getting salinization swamped. There also exists a feedback. Alterations in vegetation influence the surficial drainage and the hydrogeology. Human life and activity transform the environment, directly or indirectly influencing geological processes. The most apparent are the geological consequences of exploiting mineral deposits and of building structures and communications. It has been established that the surface is sinking in the sites of heavy consumption of ground water, and on a larger scale, under all big cities.

Geological estimation of risk of strong crustal earthquakes is founded on combination of data about geological structure of the area and properties of the rocks, as well asneotectonic and recent tectonic activity. It requires using all of remote sensing data and ground geological, geomorphological, geophysical and geodetic observations.

TRIFONOV, V. G., Geological Institute of the USSR Academy of Sciences, Moscow, USSR

Neotectonics and Evolution of Tectonic Processes

Possibility for extrapolation of the regularities of neotectonics to the geological past is limited by changes of geological processes and properties of the lithosphere during the earth's evolution. There are some neotectonic characteristics that can be extrapolated to all the Phanerozoic with small corrections. These are: rates of relative lateral movements of the lithospheric masses masses; structural patterns in the continents; irregularity and even impulsivity of the tectonic motion. Other tectonic characteristics could change monotonously or cyclically. The monotonous changes depended on increase of volume and silification of the earth's crust and on growing of volume of fluid products of the mantle differentiation. The thin and ductile Archaean lithosphere had lower ability for creation of mountains and brittle destruction than recent one. Cyclic changes were expressed in the Phanerozoic by alternation of geocratic and talassocratic epochs. The neotectonic epoch belongs to the first group. Talassocratic epochs differ from geocratic ones by gentle topography, smaller mountains, transgressions in the continents, warmer and indifferentiated climate, and, perhaps, by the more ductile and tectonically layered earth's crust. These peculiarities of talassocratic epochs were correlated with larger magmatism reflected more intensive penetration of products of the mantle differentiation to the lithosphere.

TRIFONOV, V. G., V. I. MAKAROV, V. S. PONOMAREV, Geological Institute of the USSR Academy of Sciences, Moscow, USSR

Regularities of Neotectonics

Analysis of modern (Neogene and Quaternary) deformations of the land surface, geophysical fields, seismicity, distribution of modern volcanism, fluid and gas activity proved tectonic layering of the lithosphere, i.e. differences of structural patterns, geodynamic situation and intensity of neotectonic movements in different lithosphere layers. The parts are differentiated in the continental lithosphere: upper crust with brittle deformation, less solid lower crust with plastic and quasi-plastic deformation and more solid mantle part. The tectonic layering is more complicated in orogenic belts with the thickened crust.

Tectonic system is the aggregate of the natural processes that are connected to each other in some volume of the rocks and produce directly or indirectly tectonic motion and formation of the structures. There are global tectonic system and local ones. Main energetic source for the global system is progressing differentiation of the mantle rocks. The system is characterized by uplifting of the hot products of the deep-seated differentiation to the upper layers of the earth. These products increase the oceanic lithosphere and inisiate its motion towards the continents. In active continental margins the oceanic lithosphere is not only subducted to the mantle, but largely worked into and increase the continental lithosphere. As a result, the last is compressed and thickened. Deep-seated erosion of the thickened lithosphere compensated its increasing in the oceans. Tectonic processes of the global system are the outer factors to initiate local tectonic systems. These are: volcano-tectonics, gravitational tectonics, arching of areas of granitization and sinking down of the land surface because of eclogitization of the lower crust, etc. The local system of the isostatic compensation of the exogeneous removing of the rock material is interesting from the viewpoint of the important role played by solar energy in the triggering of tectonic motion. Global distribution of the phases and some episodes of the tectonic activization support a conclusion about global connection of the tectonic processes.

TRINH Long, Viet Nam

Geodynamic Characteristic of Whiteschist and Crossite Bearing Greenschist in the Northern Edge of the Kontum Massif

The whiteschist is considered as a high pressure metamorphic rock (Schreyer 1973, 1977). Also, the high crossite content of Ca-amphibole in greenschist is regarded as a guide to high pressure of metamorphism (Brown, 1977). The coexistence of whiteschist and high crossite content bearing greenschist in the northern edge of the Kontum massif of Vietnam is a proof of high pressure metamorphism caused by a special geodynamic regime.

For the first time the whiteschist has been described by Schreyer (1973) as a guide to a high-pressure metamorphic rock. Brown (1977) points out that the crossite content of Ca-amphibole in greenschist is also a guide to pressure of metamorphism. Generally, as pressure increases the Na-amphibole (crossite) component increases too. In Vietnam the whiteschist has been found out for the first time by Phan Truong Thi and Trinh Long (1986). The rocks are closely associated with greenschist that contains a high Na-amphibole (crossite) component and with metaophiolite formations forming a special metamorphic belt occurred in the northern edge of the Kontum massif.

The main purpose of this paper is to give some preliminary results of a comprehensive study of above mentioned metamorphic belt in order to examine its geodynamic regime.

Geologic evidence: The Kontum massif is the eastern part of the Indonesian geoblock which occupies a large area of Vietnam, Laos, Thailand and Kampuchee. In the western edge of the geoblock there occurs a series of geologic formations characterizing an active continental margin such as calc-alkaline igneous rocks, metabasalt-metaandezite-metaliparite formations.

Within the Vietnam territory, the Indonesian geoblock occurs as an uplift named Kontum massif. This is situated in the central Vietnam occupying an area of about 30,000 sq km. The southern border of the massif is in contact with Triassic and younger rock and the northern border – with late Paleozoic folded zone (Truong Son zone).

The Kontum massif is composed predominantly of: Archean metamafic, metapelite (Kanak complex) and igneous rocks related to them, Early Proterozoic metavolcanic, metapelite (Ngoc Linh complex), Late Proterozoic metavolcanic, metapelite, metacarbonate (Kham Duc Formation) and in the northern edge it consists of Early Paleozoic metaophiolite, metavolcanic and metapelite (A Vuong Formation). In addition calc-alkaline igneous rocks and granitogneiss originated from granitzation (all Paleozoic in age) are widespread in the terrane of A Vuong Formation. Within this, the metavolcanic formations present a full serie composing of metabasalt-metaandezite-metadacite-metaliparite which compare in part with standard tholeiitic types and predominantly with calc-alkalic rocks.

The A Vuong Formation is overlain by the rocks of Kham Duc Formation, the contact between them is an evident unconformity. On surface of the Kham Duc formations there occurs an old weathering zone that underwent metamorphism together with those of the A Vuong Formation under the same P-T condition in Early Paleozoic (greenschist facies). The age of this old weathering zone is 487 m.y. Generally the A Vuong formations underwent regional metamorphism belonging to two stages: embryonal metamorphism and inversion one. The embryonal stage takes place during the sinking of basin materials at respective depth. This is metamorphism of low temperature and medium pressure depending on types of basin. Among the products of embryonic stage, the most important are metaophiolite formations associated with metabasalt-metaandezite-metadacite-metaliparite such as serpentinized peridotite, epidote-albite schist, actinolite-chlorite schist, thick layers of siliceous sediment. Especially, there occur thick layers of whiteschist intercalated with greenschist. All above mentioned rocks form a special metamorphic belt about 35 km long (east to west), 5 km wide and occupies an area of about 170 sq km (Figure 1).

The inversion stage of metamorphism is corresponding with uplifting of basin formations or with formation stage of island arc in active belt of continental margin. This is characterized by metamorphism developing in increasing of temperature but in descending of pressure. The products of inversion stage in the studying area are: greenschist, micaschist, crystalline limestone. The inversion stage is closely relative to granitization process with increase in K and decrease in Na. This is followed by regional hydrothermal metamorphism with considerable intensity.

The A Vuong formations are covered by Devonian sedimentary rocks unmetamorphosed.

Metamorphic characteristic: The essential characteristic of rocks from the metamorphic belt are very fine grains, chlorite, actinolite porphyroblats (up to 1.5 = 1 mm) and often fabric relics. Crystallization of the constituents was unoriented or weekly oriented.

Characteristic mineral assemblages in the belt are:

1. Albite-actinolite-pumpellyite-chlorite-quartz.
2. Albite-quartz-actinolite-prehnite-pumpellyite.
3. Albite-epidote (zoizite)-actinolite.
4. Albite-zoizite-actinolite-chlorite-epidote-quartz-magnetite.
5. Quartz-kyanite-talc-pyrophylite-stilpnomelan.
6. Quartz-sericite-chlorite.

The chemical composition of several representation rocks corresponding to this belt is presented on Table 1.

According to the data obtained, the rocks of this metamorphic belt represent originally a series of geologic formations characterizing a basin that developed on transitional crust.

They are metsophiolite, metabasalt, metaandezite-metadacite-metaliparite and metapelite formations.

Among the above mentioned typical mineral assemblages of the belt, it is noted that the assemblages 4 and 5 are the most important. The first one characterizes the greenschist with a high content of crossite in its amphiboles and the second one, the whiteschist.

Using the Brown method of calculation of crossite content (Brown 1977) in order to estimate the pressure condition of metamorphism, it is seen that the crossite content (NeM_4) of Ca-amphiboles in greenschist of the A Vuong strata is about 1.3-1.5 (Table 2) corresponding to a pressure of 7-8 Kb.

The whiteschist associated with the greenschist is appreciated to be also formed at 7-8 Kb.

This is obtained directly from the mineral assemblages that contain kyanite, talc, quartz. According to Schreyer (1974) and Korikovski's (1979) experimental, field and petrographic data, these mineral assemblages are formed at least at 7-8 Kb.

The thermodynamic analyses of the greenschist and whiteschist within the territory confirm the high condition of pressure under which metamorphism took place.

Considered in terms of the above mentioned characteristics of metamorphic belt in the Northern edge of the Kontum Central massif, it is seen that these rocks are products of metamorphism that took place during the growth of the Kontum Continent. The Kontum massif, a fragment of Gondway, began its growth northwards in Archean. Analyzing petrologic assemblages during Archean, proterozoic and paleozoic time suggests a history of crustal development by growth of the Kontum massif. At least four stages of this are postulated in the Kontum massif. The Precambrian and Paleozoic continental growth is thought to have resulted in development of a mobile continental mass characterized by a thicker and thicker crust which are similar to that of orogenic belts at continental margins, including island arc systems. The continental growth northwards is marked by four stages of crust evolution characterized by the Kannak (in Archean), the Ngoc Linh (in Early Proterozoic), the Kham Duc (in Late Proterozoic) complexes and the Z Vuong (in Early Paleozoic) Formation. In its turn, each stage is characterized by different grade of metamorphism (in decreasing). These are granulite (in Archean), pyroxen gneiss (in Early Proterozoic), amphibolite (in Late Proterozoic), and greenschist (in Early Paleozoic) facies. However, Paleozoic continental growth northwards is the last stage in crustal evolution within the Kontum massif and characterized by volcanic and sedimentary formations of a marginal basin including island arc system developed on the Kontum continental margin. These formations underwent (in embryonal stage) high pressure metamorphism in Early Paleozoic. This is thought to be caused by subduction of the A Vuong basin formations in northern edge of the Kontum massif. The materials were sunk down to 20-24 km deep, in consequence the whiteschist and the greenschist with high content of crossite in its amphibole were formed. However, the pressure was not high enough to form glaucophane (blueschist). Thus, the whiteschist and crossite-bearing greenschist occurred in the terrane are considered as a transition to low pressure part of the blueschist facies.

In the inversion stage of metamorphism, the basin materials were uplifted and metamorphosed under condition of high temperature and low pressure. This stage is followed by granitization with notable intensity. The products of metamorphism and granitization of inversion stage are widespread in the northern part of the studying area playing an important role in forming new continental crust, expanding active continental boundaries, marking the end of folded active belt. The presence of unmetamorphic formations overlying those above said rocks confirms the consolidation of continental crust in the studying area.

Conclusion: Coexistence of whiteschist and greenschist with high content of crossite in its amphibole is a special type of metamorphism being, however, a guide to high pressure of metamorphism. A part of geological formations (whiteschist and crossite-bearing greenschist) occurred in the Kontum massif northern edge are products of embryonal metamorphism under high pressure condition. The embryonal metamorphism of the Kontum massif northern edge took place during the subduction of materials of the A Vuong basin belonging to the last stage of the growth northwards of the Kontum continent in Early Paleozoic.

References

Brown, E. H., 1977, The crossite content of Ca-Amphibole as a guide to pressure of metamorphism: J. Petrol., v. 18, p. 1, p. 53-72, London, Oxford.

Phan, Truong Thi, Trinh Long and others, 1986, Metamorphism formation and facies series map of S. R. Vietnam at the 1:1,000,000 scale: Report of the First Conference on Geology of Indochina, V. I., 1986, Ho Chi Minh City, p. 191-207.

Trinh Long, Phan Truong Thi, 1986, Finding Kyanite-talc facies in Lower Paleozoic formations in Quang Nam-Da Nang area: J. Geol. 173-1986-2, p. 28-30, Hanoi (in Vietnamese).

Trinh Long, 1986, Regional metamorphism in the Northern edge of the Kontum massif during the Early Paleozoic: J. Geol. 174-175, 1986-3-4, p. 24-26, Hanoi (in Vietnamese).

Schreyer, W., 1973, Whiteschist: A high pressure rock and its geologic significance: J. Geol. 81, p. 735-739, Chicago.

Schreyer, W., 1977, Whiteschist: Their composition and pressure, temperature regime based on experimental, field and petrographic evidence: Tectonophysics, v. 43, p. 127-144, Amsterdam.

Table 1—Chemical analyses of typical rocks from the A Vuong strats.

Oxides	1 (TI- 1)	2 (3123/1)	3 (TI-18)	4 (TI 19/3)
SiO_2	60.34	61.74	65.04	49.23
TiO_2	0.41	0.57	0.58	2.16
Al_2O_3	17.38	19.60	16.21	15.01
Fe_2O_3	2.19	1.18	2.42	3.17
FeO	4.02	2.69	2.01	9.56
MnO	0.24	0.30	0.06	0.25
MgO	3.18	4.86	1.47	6.95
CaO	6.89	5.42	3.14	10.19
Na_2O	3.76	3.62	3.98	2.87
K_2O	1.64	0.51	1.24	1.01
H_2O	0.56	1.49	1.51	-
P_2O_5	-	-	0.18	-
	100.61	101.98	97.64	100.04

1. Actinolite - crossite schist.
2. Talc - kyanite schist.
3. Quartz - Chlorite schist.
4. Actinolite - Chlorite schist.

Table 2—Analysis of Ca-amphibole in greenschist of the A Vuong strata.

	SiO_2	Al_2O_3	FeO	MgO	CaO	Na_2O	K_2O		Al^{IV}	NaM_4	P Kb
TL 1/2	50.65	8.78	16.37	10.66	3.74	5.09	0.10	95.39	0.73	1.43	7.5
TL 4	51.95	7.18	16.92	10.75	3.49	6.90	-	96.60	0.83	1.47	7.5
TL 6	47.63	10.07	17.12	8.24	4.51	5.70	3.26	97.13	0.85	1.28	7.0

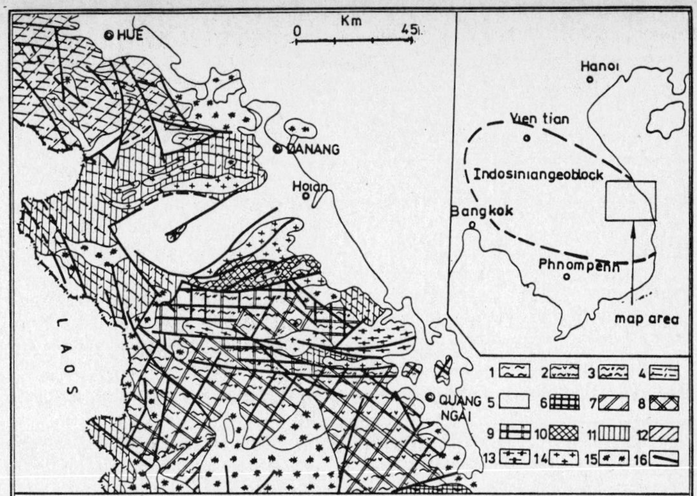

Fig 1. Metamorphic facies scheme of the northern edge of the Kontum massif

1. Archaean rocks of the Kannak complex
2. Early Proterozoic rocks of the NgocLinh complex
3. Late Proterozoic rocks of the Kham Duc complex
4. Early Paleozoic rocks of the AVuong - Long Dai strata
5. Unmetamorphic rocks
6. Granulite facies
7. Garnet - Cordierite gneiss facies
8. Biotite - Sillimanite gneiss facies
9. Mica gneiss and epidote amphibolite facies
10. Talc - Kyanite facies
11. Greenschist facies
12. Prehnite - Pumpellyite facies
13. Proterozoic igneous rocks
14. Paleozoic igneous rocks
15. Mezozoic igneous rocks
16. Tectonic fault

Figure 1—Metamorphic facies scheme of the northern edge of the Kontum massif.

TRIPLEHORN, D., University of Alaska, Fairbanks, Al

Applications of Tonsteins to Coal Geology: Some Examples from Western United States

Tonsteins are thin, distinctive, widespread clay bands or partings that are associated with coal beds. They are now generally accepted as the product of alteration of volcanic ash, mostly as distal air-falls from violent eruptions of siliceous material. In tonsteins, glass predominates, but is usually accompanied by phenocrysts of quartz, zircon, magnetite/ilmenite, feldspar, and other minerals. The glass and some of the phenocrysts were altered to kaolinite, less commonly to smectite.

Tonstein studies can yield geologic information of great variety and significance. However, many tonstein occurrences are overlooked in the field. Most important, this means that the potential value of tonstein studies goes largely unrecognized and undeveloped. Following are some examples of untested ideas as well as some applications that merit further development. Where available, examples are given, mainly from western North America.

Tonsetines as Marker Beds: Assuming that tonsteins represent air-fall volcanic material deposited within a short time (days to months), each one is of the same age everywhere and distal occurrences may be laterally extensive. Recognition of their widespread character has lead to their empirical use in coal-bed correlation for half a century (Williamson, 1970).

Their use as marker beds has much more potential. Tonsteins have been observed to enter the top of a coal, pass diagonally through it, and out the bottom. This allows recognition of changes in the geographic position of coal formation through time. Similarly, multiple tonsteins in a given coal delineate it into exact time slices regardless of changes in thickness of petrographic composition (Ryer et al, 1980).

There is some chance that tonsteins may serve as marker beds on an intercontinental or even global scale. Eruptions of magnitudes vastly greater than any in historic times have been documented in the geologic record.

Extrapolating from modern examples, one can speculate on the likely record of such larger events, although the interpretation will be complicated by plate tectonics and changes in wind patterns as well the improbability of simultaneous coal-forming environments (to preserve tonsteins) in widely separated locations. Efforts are currently underway to correlate tonsteins between continental Europe, Britain, and the eastern United States.

Tonsteins as Absolute (Numerical) Time Horizons: Recent technological improvements in radiometric dating provide great opportunities. Precision is available to less than a my in the Carboniferous, step-wise heating permits evaluation of sample alteration and laser microprobe techniques can produce age measurements from single small grains. The latter vastly reduces the sample size required so that even cores and cuttings from bore holes are potentially datable. The small sample size also permits distinction of detrital contaminants or other grains not related to the age of the volcanic event.

Coals, with their well-preserved phenocryst-bearing tonsteins could become the most datable part of the sedimentary record, along with bentonites in marine shales. Furthermore, increased availability of precisely dated coal beds should lead to precise radiometric ages for palynological zones, thereby increasing the usefulness of palynology for time determination.

Tonsteins in Interpretation of Paleodepositional Environments: Beneath any given ash is the land surface that existed at the time of the eruption. The nature of the contact between the resulting tonsteins and subjacent coal should reflect the character of that surface, including topographic irregularities, type of vegetation, as well as the presence of open water. Lateral variations in these should record the distribution of sub-environments; that is, local paleogeography. Evidence of early compactional effects may indicate the water content and physical properties of the substrate.

In a similar fashion, the re-establishment of vegetation may be recorded in the nature of the contact and composition of the coal immediately above a tonstein. Lateral changes may reflect differences in the time or kinds of plants involved in resumption of organic accumulation. On a larger scale, differences in plants might reflect climatic changes related to world-scale eruptions.

Tonsteins in Coalification, Diagenesis, and Coal Quality: Alteration of the volcanic material involves substantial loss of chemical components in solution—mainly Si, Na, Mg, Ca, and Fe—as well as a host of minor and trace elements. Where do they go? Presumably these elements pass from the ash parting through the adjacent peat and are totally removed from the system. However, to my knowledge, there are no studies of the possible effects tonsteins have on the composition of overlying or underlying coal.

In some cases, it is clear that elements have been added to tonsteins but without clear indication of the source, timing, or pathways involved. For example, some tonsteins are impregnated with or partly replaced by siderite; where unaltered feldspars are preserved, this suggests early diagenesis (Triplehorn et al, 1984). Aluminum phosphate minerals of the crandallite group are becoming more frequently recognized and sometimes almost totally replace kaolinite. The source, timing, and pathway for the phosphorous is currently being investigated, as well as those for Sr, Ba, rare earths and other elements (depending on the mineral variety present).

The inorganic part of mineral matter in coal is generally interpreted as detrital sediment washed into the coal-forming environment, or as later chemical precipitates introduced by ground or diagenetic water. Wind transport, including but not restricted to volcanic ash, should also be considered. Such material may be thoroughly mixed with the organic matrix so that it is not recognizable as a discrete mineral-rich layer and certainly not recognizable as a tonstein in core or outcrop.

Potential for Advances in Coal Geology: Tonsteins in coals have the potential to provide better time control than the fossils in marine rocks, to permit detailed analysis of paleoenvironments of deposition, and to increase understanding of the composition of coals and the coalification process.

REFERENCES

Kunk, M. V. and Sutter, 1984, $^{40}Ar/^{39}Ar$ Age Spectrum dating of biolite from Middle Ordovician Bentonites, Eastern North America: *in* D. L. Brunton, ed., Aspects of the Ordovician System, Palaeontological Contributions from the University of Oslo, n. 295, Universitets-Portaget, p. 11–22.

Ryer, T. A., R. E. Phillips, B. F. Bohor, and R. M. Pollastro, 1980, Use of altered volcanic ash falls in stratigraphic studies of coal-bearing sequences—an example from the Upper Cretaceous Ferron Sandstone Member of the Mancos Shales in Central Utah: Geological Society of America Bulletin, v. 91, pt. 1, p. 579–586.

Triplehorn, D. M., D. L. Turner, and C. W. Naeser, 1984, Radiometric age of the Chickaloon Formation of south central Alaska; Location of the Paleocene–Eocene boundary: GSA Bulletin, v. 95, p. 740–742.

Williamson, I. A., 1970, Tonsteins—Their nature, origins, and uses. Mining Magazine, v. 122, pt. 1, n. 2, p. 120–125; v. 122, pt. 2, n. 3, p. 203–210.

TRIVEDY, A. N., Geological Survey of India, Patna, India

Tertiary Vertebrates of Tripura and Their Significance in Biostratigraphy

Several fossiliferous horizons have been located by the author in the late Tertiaries of Tripura in the course of his field work in the area. Of them, two intraformational conglomerate horizons west of Teliamura (22°47':91°34':79M/9) within upper Bokabil Formation of Baramura range are promising. The fossils recovered from them are listed below:

Mammalia
— *Gomphotherium anqustidens*
— *Dorcatherium* sp.
Reptilia
— *Crocodilus palustris*
— *Gharialis quangeticus*
— *Gharialis* sp.
Pisces
— *Isurus* sp.
— *Isurus cf pagoda*
— *Galiolamna (Prinodon) gangeticus*
— *Sliluroid indet*
Lamellibranchiata
— *Venus (clementia) papyracea Grey*
— *Cytheria (Callista) pseudoumbonella,* sp. nov.
Gastropoda
— *Turritella pinfolde*

A sequence of shale, silt, and massive ferruginous sandstone of Surma and Tipam Groups constitute the main rock types in Tripura. Bhuban, a mainly argillaceous facies, occupies the basal part of Surma followed by a sequence of shale, silt, and sandstone of Bokabil Formation. Tipam is mainly an arenaceous facies. There is no sharp boundary between any of the divisions and subdivisions; the lower one can be seen to pass gradually into the next higher one without any break. There are lateral variations as a matter of rule. It is not possible to correlate the litho units and time plane and any attempt in this regard is fraught with suspicion of time plane intersecting the lithological plane. Fossil evidences are of great help in correlation of such units.

Among the fossils recorded, the mammalian forms that were reported for the first time from the area by the author have a direct bearing on the stratigraphy. These include molars (M^3-M_2) of *Gomphotherium* and lower molar (M_3) of *Dorcatherium.* The gomphotherian teeth have been recovered from the base of upper Bokabil whereas the dorcatherian tooth and other bones are from the top of the formation.

The superior molar of *Gomphotherium* has a nearly rectangular crown with slight tapering on the posterior side. Longitudinally, it is divisible into two symmetrical halves by a median line. The valley between the two ridges is wide and deep and except for blocking by accessory turbercule on the median line, it is open and straight. The hind talon is prominent, broad, and single coned. With these characteristics, this specimen is not in conformity with any of the Indian forms. On the other hand, these are strikingly similar to the characteristic of *Gomphotherium angustidens* Cuvier reported from Simmore and Sansan beds of France and as such it has been assigned as a variety of Cuvier's *angustidens.*

In the mutational history of *Gomphotherium* (which is strictly orthogenetic and rectigradational evolution of cones, conelets, conules, ridge-crest, and trefoil), the specimen from Tripura is considered more progressive than *G. Cooperi* Osborn reported from Bugti beds (Burdigalian) and retrograde than *G. palaeindicus* (lydekker) Osborn reported from Kamlial formation (Helvetian). *Gomphotherium angustidens* Cuvier has also been placed in this range. This specimen suggests upper Burdigalian to lower Helvetian age for the basal part of upper Bokabil.

The dorcatherian tooth depicts general characters very similar to *Dorcatherium minus*, but it is 40% smaller in size and also less hypsodont. On the other hand, it closely conforms with the characters of *Dorcatherium* sp. Colbert in having strong development of mesostyle, slight development of cingulam, and in shape and size of the tooth. The measurements of the M_3 of the three specimens are given below for comparison.

The specimen from Tripura undoubtedly belongs to *Dorcatherium* sp. described by Colbert from northwest India.

Dorcatherium sp. has been reported from Chingi Formation of Lower Siwaliks (Tortonian) of northwestern India. The present specimen is the only reported occurrence from northeastern India. *Gomphotherium angustidens* indicates Burdigalian–Helvetian age for the basal part of the upper Bokabil. Since *Dorgatherium* sp. has been recovered from a horizon higher than *Gomphotherium*, the age of upper Bokabil may range up to lower Tortonia.

The reptilian, piscine, and invertebrate fossils from the area are wide ranging. However, the assemblage is similar to those reported from Bokabils of Garo Hills. Mukherjee (1939), on comparison of Garo Hills fauna with Koma and Pyalo fauna of Burma, has assigned Aquitanian-Burdigalian age to this formation.

Gomphotherium has been reported from Europe, India, and Africa. In Europe, it has a wide distribution in France, Switzerland, Hungary, Spain, etc. From North America a form in the name of *Gomphotherium obscurs* has been reported from Mio-Pliocene. The last lower molar of this form has relatively small tubercules and nearly straight lateral border indicating that while farthest removed in space, it is likewise farthest removed in structure from Indian race. The Indian and European forms on the other hand have striking similarity.

Dorcatherium has been reported from Europe, Africa, and India. The Indian forms are distinct from the African ones but appear almost cogeneric with Miocene form of Europe.

The close affinity between Indian and European forms and occurrence of allied form in higher horizons in North America indicate that the original home for these mammalian fauna might have been Europe and Western Asia from where they migrated east and west.

TROESTER, J. W., U.S. Geological Survey, San Juan, Puerto Rico and W. B. WHITE, Pennsylvania State University, University Park, PA

Carbonate Chemistry and Hydrogeology of Rio Camuy Drainage Basin, Puerto Rico: Humid Tropical Karst

The carbonate chemistry of the Rio Camuy, a humid tropical karst river, was studied at several sites along its length and on its tributaries every month from January 1983 to January 1984. The calcite saturation index (SI_C), total hardness, and the partial pressure of carbon dioxide (PCO_2) were calculated with WATEQF and compared in two ways, (1) by examining the change with distance downstream and (2) by comparing the values at each station throughout the year. A mixing model describing the hydrology and carbonate chemistry of the river was constructed from discharge data and the concentration of conservative ions (K, Na, Si, and Sr) in the river and its tributaries.

The water in the Rio Camuy entering the northern karst belt of Puerto Rico is undersaturated (average SI_C = -0.60), low in hardness (average hardness = 71 mg/L as $CaCO_3$), and has a PCO_2 only slightly above atmospheric (average PCO_2 = 0.11 volume percent). The hardness increases systematically downstream, with the largest increase occurring shortly after the river enters the karst. The chemical mixing model indicates that in this section an average of 97% of the increase in calcium and 95% of the increase in bicarbonate is from the addition of calcium- and bicarbonate-rich, autogenic water from ground-water tributaries. Very little of the additional calcium and bicarbonate is from dissolution of the limestone by the river.

The water in the tributaries contains more dissolved carbon dioxide than the Rio Camuy; the highest PCO_2 measured was 0.95%. Because the river is flowing in an open cave system, the excess carbon dioxide, added by the tributaries, degasses and causes the calcite saturation index to become positive within a relatively short distance. When the river flows through a flooded cave passage, carbon dioxide cannot degas, the PCO_2 reaches its maximum (average PCO_2 at the resurgence = 0.35%), and the saturation index approaches equilibrium.

The total hardness and the SI_C of the Rio Camuy and its tributaries show some seasonal variation in that they are somewhat lower during months of higher rainfall. However, no seasonal variation is apparent in the PCO_2 in either the Rio Camuy or its tributaries.

On two occasions the river was samples while it flowed through both its main passage and a flood bypass. On the first occurrence the mixing model was used to calculate a discharge of 1.00 m^3/s for the main passage and 0.23 m^3/s for the flood bypass. The second time, the calculated discharges were: 0.48 m^3/s for the main passage and 1.19 m^3/s for the flood bypass. The difference between the two calculated discharges for the flooded main passage of the Rio Camuy may be due to blockage by flood debris and sediment.

TROIAN, F. L., M. A. F. HANSEN, and H. C. FENSTERSEIFER, University "Vale do Rio dos Sinos"-UNISINOS, Rio Grande do Sul, Brazil

Geological Contribution to Harmony Point in Nelson Island South Shetland, Antarctic

In this paper, we evaluate and describe the association of volcanic and volcanoclastic rocks as to their petrographical, structural, and geochemical characteristics, which represent part of the geological research work developed during the subproject "Geological Mapping of Nelson Island" (scale 1:15,000) Antarctic. The fieldwork has been done in the austral summers of 1985–1986 and 1986–1987, respectively, by researchers of University of Vale do Rio dos Sinos-UNISINOS and sponsored by the Brazilian Antarctic Program-PROANTAR. The aim is to show continuity of the geological mapping work in detail to contribute to a better understanding of the information about the sequences of the polimitical breccias that are interbedded with lava flows, as well as intrusions of Harmony Point. The structural details should also contribute to a better visualization of the spacial distribution of intrusive features, such as dikes and plugs, which appear in the southwestern sector of Nelson Island as different acid rocks with higher contents of SiO_2 that may be correlated with the ones appearing at the Tue Point. This presents mainly green andesite and basaltic andesite lava flows, intensely metasomatized and sulfetized. There are also intrusions of andesites, microleucogabres, and rocks of more acidity (dacites). Various of these intrusions seem to be contemporaneous to the same volcanic manifestations that are also associated to polimitical volcanic breccias with carbonized trunks.

TROMPETTE, ROLAND, Université d'Aix-Marseille III, Cedex, France

Ensialic Evolution of the Centrifugal Aracuai West Congo (Zaire) Pan Africa Belt—Role of Middle Proterozoic Fracturation

The Araçuai (Brazil) and West-Congo (Gabon, Congo, Zaire and Angola) Pan African Fold Belts represent two halfs of a single huge orogen cut by the opening of the South Atlantic Ocean. This orogen shows a centrifugal vergence: its African part being thrusted onto the Congo on Zaire Craton and its Brazilian part on the southern edge of the Saõ Francisco Craton.

Five main lithostratigraphic units are outcropping in the Araçuai-West Congo Belt. The Saõ-Francisco or Bambui (s.l.) or West Congo Supergroup, the age of which is bracketed between 1000 and 600 Ma. It starts with a diamictite horizon, which is proved to be of glacial origin in Brazil, and continues with alternating fine detrital sediments and carbonates with stromatolites similar to those of the Upper Riphean of U.S.S.R. The underlying Espinhaço (Brazil) or Mayombe Supergroup, the age of which is bracketed between 1750 (?) and 1000 Ma, is detrital with, locally, an acidic volcanic lower part. Pan African reworked Precambrian Basement forms the internal unit of the West Congo Belt in Africa and parts of the external units of the Araçuai Belt, East of Belo Horizonte. Varied gneiss and kinzigites widely outcropping in Brazil are probably time-equivalent to the Sao-Francisco supergroup. Finally, a voluminous suite of Pan African granitoids is geographically associated to the gneiss and kinzigite facies.

The Bambui-West Congo Supergroup, subhorizontal on the edges of the two cratons, constitutes most of the metasediments of the external units on both sides of the orogen and probably the internal gneissic and kinzigitic unit in Brazil. The Espinhaço or Mayombe Supergroup fills up two long and narrow Middle Proterozoic aulacogens, which broadly delimit the extension of the belt, or is widely transgressive on the adjacent craton. The Espinhaço and Mayombe aulocogens show no indications of Middle Proterozoic tectonogenesis: in Brazil, the external units of the Araçuai Fold Belt are affected by a first phase of ductil deformation generating a prominent stretching lineation. They are detached from the basement and carried as a whole to the West, without folding. In the West Congo Belt, the tectonic regime is different: deformation and metamorphism are linked to a first phase of upright to inclined folding which become recumbent to the West. Most of the Congo Belt can be interpreted as a parautochtonous Province folded in front of internal Basement nappes. The geometry of the Araçuai-West congo "geosyncline" is dependent on the location of Middle Proterozoic fractures. In Brazil, the origin and signification of the Paramirin Belt is disputed. It seems to represent a Pan African arm or diverticule develops by reaction of a Middle Proterozoic aulacogen.

The geodynamic evolution of the Araçuai-West Congo Orogen is entirely ensialic. In Brazil a huge thermal swelling is superimposed to the classical barrowian metamorphism. It is responsible for the "mise en place" of numerous S-type granitoids, partly derived from melting of the gneissic and kinzigitic unit. The shortening, moderate in the external units of the belt where it can be evaluated, is probably accommodated in the basement by a type-subductions.

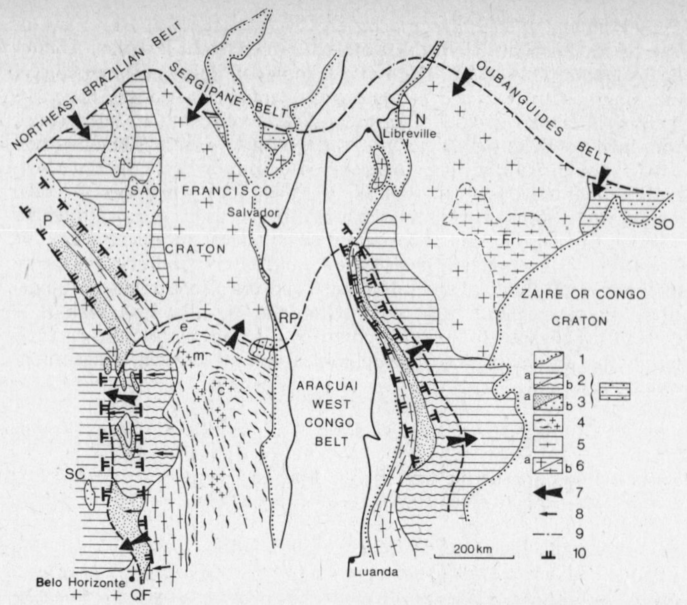

Figure 1—Schematic geological map of the Pan African Araçuai-West Congo Fold Belt.
1—Phanerozoic, mainly Meso-Cenozoic, covers; 2: Sao-Francisco or Bambui (s.l.) or West Congo Supergroup (1000-600 Ma) (a) cratonic and subhorizontal or (b) folded by the Pan African tectogenesis; 3: Espinhaco or Mayombe Supergroup (1750 (?)-1000 Ma) filling up narrow aulacogens (a) or widely transgressive on the basement (b); 4: Varied gneiss and kinzigites probably equivalent to the Sao Francisco Supergroup and cut by Pan African granitoids; 5: Precambrian Basement reworked in the Pan-African Orogen; 6: Precambrian Basement strongly deformed (a) or platformal (b); 7: tectonic vergence; 8: prominent stretching lineation; 9: Isograds in the Aracuai Fold Belt; 10: limits of the Middle Proterozoic aulacogens. Fr : Francevillian; N : Noya; P: Paramirin Pan-African arm superimposed on a Middle Proterozoic aulacogen; QF: "Quadrilatero ferrifero;" RP: Rio Pardo sequence; SC: Serra do Cabral; SO: Sembe-Ouesso Sequence. e., m., c.: epizonal, mesozonal and catazonal subdivisions of the Aracuai Fold Belt.

TROTSUK, V. Y., Academy of Sciences of the USSR, Moscow, USSR

Geodynamic Control in Formation of Black Shales

The computerized correlation of the chronostratigraphic distribution of organic carbon (C_{org}) in geodynamically different elements of the oceanic stratisphere with the variability of sedimentation conditions throughout the history of the present-day oceans has revealed that the time intervals associated with the accumulation of deposits rich in C_{org} coincide with early manifestations of main stages of the intensive spreading and volcanic activity. Accumulation of sequences rich in C_{org} is most typical of low-velocity sedimentation occurring either in early stages of the continents' rifting or under the conditions of back-arc spreading or else in the zone of juncture of midoceanic ridges (or their ancient branches) with the continents (or blocks–microcontinents). Occurring on a global scale in the oceanic (as well as in the continental) stratisphere are C_{org}-rich deposits pertaining to the following time intervals: Middle Cretaceous (80–112 m.y. ago), Pliocene–Quaternary (0–5.3 m.y.), Middle Eocene (41–45 m.y.), and Early Miocene (17–25 m.y.). This order agrees well with the degree of occurrence of deposits high in C_{org} in the stratosphere. Chemical composition of the organic matter (OM) dispersed in the deposits of these time intervals attests, as a rule, to their high petroleum-generating properties.

The author's researches in the course of surveys carried out in geologically different areas of seas and oceans suggest that the contribution of autotrophic chemosynthesis processes into the elevated fossilization of OM in a water basin is appreciable. The OM fossilization increases stepwise due to the activation of fluid flows from the sedimentary cover entrails up to the near-bottom environment, with this process evolving on a greater scale in the near-continent sedimentation zones on the young, newly formed or destructing ("hot") crust. The conditions thus combined cause an intensive release of energy and chemical elements into the near-bottom environment of water areas, which first stimulates chemosynthesis and then provides a trophic overloading in the functioning of bottom organism communities. This brings about short-period changes in the biomass and hence outbursts of OM fossilization in the deposits.

The discussed data on the conditions of oil-source deposit formation can be employed in the forecast of hydrocarbon potential as regards the entrails of firm land and water areas with inadequate data coverage.

TRURNIT, T. P., Burgwedel, Federal Republic of Germany

Precollisional Origin of Cordilleran-, Andean-, and Island Arc-Type Margins and Postcollisional Formation of Tectonostratigraphic Terranes and Mediterranean-, Caribbean-, and Basin and Range-Type Extensional Provinces

Presumably on account of the eccentric rotation of the spinning earth around the gravitational center of the earth–moon (–sun) system (principle of hypocycloid gearing), the Pacific basin (the shape of the Pacific; not the Pacific crust), the oceanic stages included in the Wilson Cycle (Rift/Red Sea state, Atlantic state, Pacific state, Collisional/Himalayas state), and most probably also the lower mantle are constantly displaced eastward relative to the earth's crust or the convecting upper-mantle crust system (one full turn around the globe each 200 to 250 m.y.), thus causing the Global Tectonic Megacycles (Oceanic/Wilson Cycle, Cycle of Continental Margins, Cycle of the Oceanic Crust, Orogenic Cycle, Cycle of the Collisional Mountain Belt, etc.).

The continents migrate westward around the Pacific basin both in the north and south (i.e., the Pacific basin is displaced eastward between the continents of the Northern and Southern Hemispheres). They collide at the equator to the west of the Pacific, continuously adding segments to a collisional mountain belt that becomes older toward the west (sequential collision; zip fastener principle) and since the Permian has lapped some 1 1/3 times around the cratonic nucleus of Laurasia in the form of a spiral (Alpine mountain belt of Southern Eurasia, Laramide–Kimmerian mountain belt of south and central Asia, Urals). The ophiolites in the suture zones of all collisional mountain belts or orogens should mark the trail of the eastward migrating Pacific basin (Fig. 1).

Following half an east drift lapping of the earth's crust by the Pacific basin (some 100 to 125 m.y.), the Pacific appears or "rises" again in the west. In the extensional Mediterranean and Caribbean settings, the continents of the Southern Hemisphere that had previously collided at the equator to the west of the Pacific against the continents of the Northern Hemisphere disengage again from the northern continents (postcollisional rift propagation toward the East; zip fastener principle), leaving their former northern margins in the form of tectonostratigraphic terranes attached to the northern continents. These later migrate westward around the Pacific basin in the north accompanied by basin and range-type extensional provinces (extensional unroofing and isostatic rise of metamorphic core complexes) (Fig. 1).

The earth's crust is subdivided into a Pacific area and a continental or Pangaea area with Intra-Pangaea Oceans (Atlantic, Red Sea, Indian Ocean, etc.). The Pangaea area in turn is subdivided into a North Pangaea area and a South Pangaea area with the North and South Pangaea continents (Laurasia, Gondwana) broadly distributed over the Northern and Southern Hemispheres. The earth's history appears to be subdivided into alternating North Pangaea growth/South Pangaea break-up eras (Permian to present-day Uralian–Indosinian–Kimmerian–Laramide–Alpine Cycle; Late Proterozoic Panafrican–Brasiliano Cycle; etc.) and South Pangaea growth/North Pangaea break-up eras (Late Proterozoic and Early to Middle Paleozoic Baikalian–Cadomian–Assyntic–Caledonian Cycle; Middle to Late Proterozoic Kibaran–Grenvillian–Uruacuano Cycle; etc.). Lateral growth of a Pangaea and of the continents as well as the cyclical repetition of orogenic events for a certain continental margin are effected by the joining of the younger lap of a collisional mountain belt spiral to the older, from the equator toward a pole (Fig. 2).

In the hemisphere of a Pangaea growing (since the Permian the Northern Hemisphere), the continents are subjected to pendular movements (alternating clockwise and counterclockwise rotations with movements between high and low latitudes). They always face either the equator or the Pacific with the same margin. Otherwise a collisional mountain belt spiral would not form. The remaining two passive margins alternate between an Arctic- and a North Atlantic-type of setting. In consequence, the Cordilleran-type margin of the northeast Pacific is the forerunner of the northwest Pacific island arc-type margin, and both types are one-sided, embryonic states of the two-sided collisional mountain belt forming at the equator to the west of the Pacific. since the Jurassic–Cretaceous, the Pacific margins of the Northern Hemisphere are remobilized segments from the older lap of the North Pangaea collisional mountain belt spiral (Figs. 1, 2).

In the hemisphere of the Pangaea break-up (since the Permian the Southern Hemisphere) a continent rotates through some 120° while passing through the Antarctica setting (clockwise in the Southern Hemisphere during a South Pangaea break-up era—Permian to present; counterclockwise in the Northern Hemisphere during a North Pangaea break-up era—Late Proterozoic and Early to Middle Paleozoic) and breaks up into several

India-, Australia-, and Antarctica-sized fragments. The one-sided Andean-type margin of the southeast Pacific (previously evolved from a passive West Africa-type) develops therefore (without passing through the southwest Pacific island arc-type setting) into a one-sided New Guinea-type and into the equatorward facing thrust zone of the two-sided collisional mountain belt forming at the equator to the west of the Pacific. On the other hand, the southwest Pacific island arc-type margin has evolved from a passive North Africa-type that might still carry fragments from the collisional mountain belt (Atlas, parts of the northern and Caribbean Andes, West Antarctica, New Zealand), the main parts of which migrate around the Pacific basin in the north in the form of tectonostratigraphic terranes (Fig. 1).

Through reverse rotations of the continents from the hemisphere of the Pangaea break-up (since the Permian the Southern Hemisphere) to their Early Mesozoic and Paleozoic orientations one arrives at the correct reassembly of Gondwana, South Pangaea, and the South Pangaea collisional mountain belt spiral (Figs. 1, 2).

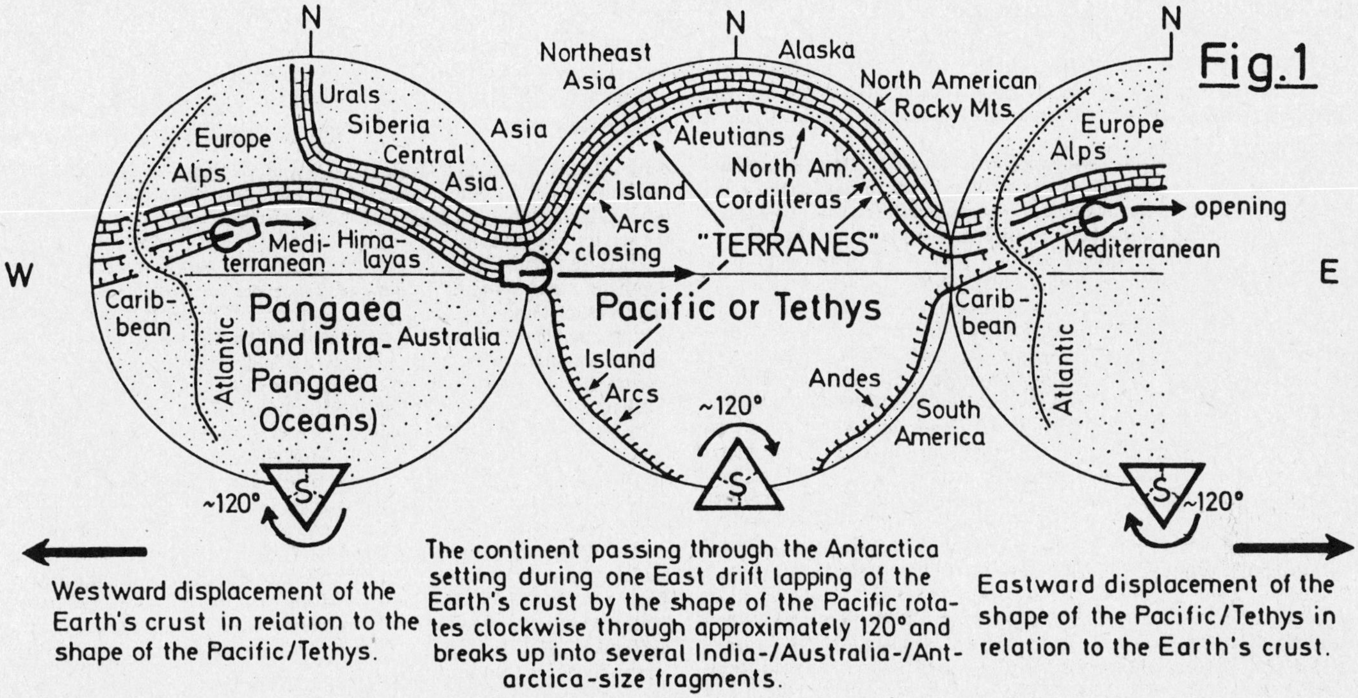

Fig.1

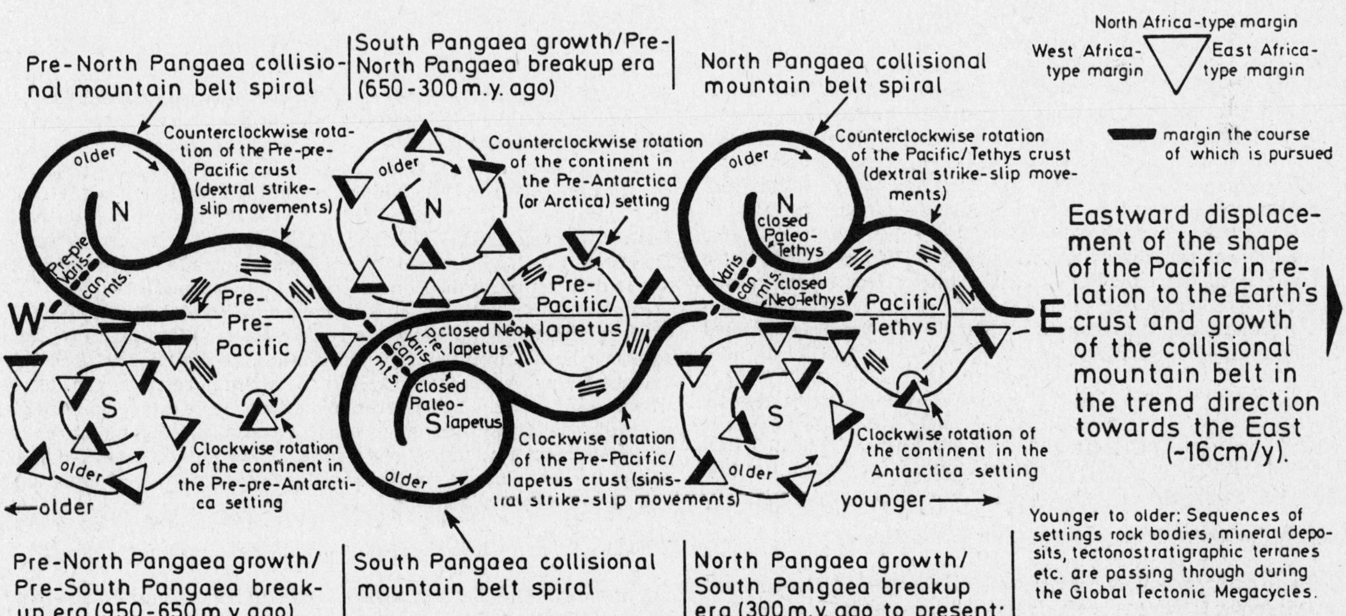

Fig.2

TRURNIT, T. P., Burgwedel, Federal Republic of Germany

Fit between West and East Gondwana and Reconstruction of Pangaea (Gondwana, Laurentia) in Southern Hemisphere

Due to the westward-directed off-center rotation of the spinning earth around the gravitational center of the earth–moon (–sun) system, the lower mantle, the Pacific basin (the shape of the Pacific; not the Pacific crust), and the oceanic stages of the Wilson Cycle (Rift/Red Sea state, Atlantic state, Pacific state, Collisional/Himalayas state) are displaced eastward relative to the earth's crust (convecting upper mantle-crust system; once around the globe in 200 to 250 m.y.; principle of hypocycloid gearing), thus causing the Global Tectonic Megacycles (Oceanic/Wilson Cycle, Cycle of Continental Margins, Orogenic Cycle, Cycle of the Collisional Mountain Belt, etc.). The continents migrate westward around the Pacific basin both in the north and south (i.e., the Pacific basin is displaced eastward between the continents of the Northern and Southern Hemispheres). They collide at the equator to the west of the Pacific, continuously adding segments to a collisional mountain belt that becomes older toward the west (sequential collision; zip fastener principle) and since the Permian has lapped some 1 1/3 times around the cratonic nucleus of Laurasia in the form of a spiral (Alpine mountain belt of southern Eurasia; Laramide–Kimmerian mountain belt of western North America and northeast Asia; Indosinian mountain belt of south and central Asia; Urals). The ophiolites in the suture zones of all collisional mountain belts or orogens should mark the track of the eastward migrating Pacific basin (Fig. 1).

Following half an east drift lapping of the earth's crust by the Pacific basin (100 to 125 m.y.), the Pacific appears or "rises" again in the west. In the extensional Mediterranean and Caribbean settings the continents of the southern hemisphere that had previously collided at the equator to the west of the Pacific disengage again from the northern continents (rift propagation toward the east; zip fastener principle), leaving their former northern margins in the form of tectonostratigraphic terranes attached to the northern continents. These later migrate westward around the Pacific basin in the north accompanied by basin and range-type provinces (extensional unroofing and isostatic rise of metamorphic core complexes) (Fig. 1).

The earth's crust is subdivided into a Pacific area and a continental or Pangaea area with the Intra-Pangaea Oceans (Atlantic, Red Sea, Indian Ocean, etc.). The Pangaea area in turn is subdivided into a North Pangaea area and a South Pangaea area with the North and South Pangaea continents (Laurasia, Gondwana) broadly distributed over the Northern and Southern Hemispheres. The earth's history appears to be subdivided into alternating North Pangaea growth/South Pangaea break-up eras (Permian to present-day Uralian–Indosinian–Kimmerian–Laramide–Alpine Cycle; Late Proterozoic Panafrican–Brasiliano Cycle; etc.) and South Pangaea growth/North Pangaea break-up eras (Late Proterozoic and Early to Middle Paleozoic Baikalian–Cadomian–Assyntic–Caledonian Cycle; Middle to Late Proterozoic Kibaran–Grenvillian–Uruacuano Cycle; etc.). In the hemisphere of the Pangaea break-up the continents rotate through some 120° while passing through the Antarctica setting (clockwise in the Southern Hemisphere during South Pangaea break-up eras; counterclockwise in the Northern Hemisphere during North Pangaea break-up eras) and break up into several India-, Australia- and Antarctica-sized fragments. (Figs. 1–3).

Through reverse rotations of the continents from the hemisphere of the Pangaea break-up (since the Permian the Southern Hemisphere) to their Early Mesozoic and Paleozoic orientations one arrives at the correct reassembly of Gondwana, South Pangaea, and the South Pangaea collisional mountain belt spiral (Figs. 2, 3).

TSAI Chao-Ming, J. H. C. WANG, Radwaste Administration/Atomic Energy Council 65, Taiwan, Republic of China, KUO-LIANG Soong, Institute of Nuclear Energy Research, Atomic Energy Council, KUO-LIANG Pan, Energy & Mining Research/Service Organization Institute Technology Institute

Geological Survey for Potential Repository Sites in Taiwan

Because of the very limited energy resources in Taiwan, the development of nuclear energy has always been one of the top priority in the energy policy of Republic of China. As in other industrialized countries, the rear end of nuclear fuel cycle—high level waste repository—is an important task in Taiwan nuclear industry. Taiwan is situated at the eastern margin of continental Asia bordering the Pacific Ocean. It hosts one of the most recently formed mountain chains of the Earth with rapid uplift of the land and conspicuous earthquake activity. The oceanic Philippine Sea plate moves towards north and northwest and overrides the continental Asian plate. The resulting mechanism causes the past volcanism and possible future geological disturbance in Taiwan. The general geology of the region from west to east can be divided into the following units:

1. An essential passive, continental platform, submerged under the Taiwan Straight, characterized by a block structure and rifting with a relatively thin series of cover rocks.

2. An intermediate Foreland, consisting of the Western Coastal Plain and the Western Foothills, characterized by westward trending deformation, both folding and faulting.

3. The Central Highlands, composed of metamorphic cover rocks as well as basement rocks. All rocks show strong deformation, and the present position of the basement rocks indicates rapide uplift and erosion during the past six million years.

4. The East Coastal Range, taken to the leading edge of the Philippine Sea plate overriding the buried eastern edge of the continental Asian plate in the west.

The active deformation of the bedrocks seems to calm down about 130 thousand years ago, as evidenced by the near horizontal, undeformed younger sediments of the Coastal Plain. However, rapid land uplift and frequent strong earthquakes still indicate the continuous geological unrest in Taiwan. From the consistent geological arrangement, it shows that the peak tectonic activity has, for the last 30 million years, been located in the eastern Taiwan, and the seismic intensity has decreased towards the West. All available evidences suggest that this situation will persist throughout the service of a repository.

Potential host rocks for a safe repository may be found at appropriate depths in many parts of Taiwan, ranging from rocks of crystalline basement to shales and mudstone in the cover series. Therefore, the availability of host rock is not seen to be a critical issue. However, the host rocks can not be evaluated only on their technical and physio-chemical properties but also on their geological context and local socio-economic factors. Based on mechanical properties, the rocks in Taiwan can be classified as hard rocks and soft rocks. Man-made openings in hard rocks may usually be expected to persist for long times, thus providing safe accessibility and even retrievability. Although the hydraulic conductivity in hard rocks normally is very low, the solute transport in rock fractures is complicated and difficult to model. Proper description and evaluation of fracturing is a major task of site investigation and important for safety and performance assessments. Hard rocks occur mainly in the Eastern Coastal Range and in the Basement Range of the Central Mountains. All information indicate that Western Platform appears to offer considerable potential for a safe repository in hard rock, and the Penghu Islands seem to make its basement rocks reasonbly accessible. Outside the Western Platform other hard rocks may deserve some consideration. Andersites and decites in Northern Taiwan mark dormant volcanism. However, their associated high subsurface temperatures exclude a deep repository in this area.

Regarding the soft rocks, they usually have low hydraulic conductivity, coupled with higher transport porosities and stronger sorption than most hard rocks. Additionally waterbearing fractures in these rocks generally are rare, notably at depths. Plastic rocks and clays will tend to close around emplaced waste packages, and require special attention to stability aspects during construction of a repository. The thick mudstone is found in the Western Foothills and the shale is widespread over the cover rocks of the Foreland. Concerning the future denudation, information on the present erosion processes in the Foothills and on the Plain, and on the uplift of the Foothills relative to the Coastal Plain, is of vital interest. Other questions relate to the mechanical properties of the soft rocks may call for special attention to the techniques of emplacement, operation and restoration, and to the response of these host rocks to future tectonic activity.

Minimum future denudation and maximum predictability may be tentatively suggested for the Taiwan Strait on the Western Platform for potential repository area. Cover rocks can also provide extra geological barriers for valuable safety. Very deep emplacement in mudstone represents an alternative disposal option to against erosion problem. More detailed planning of the geological investigation must await the results of ongoing work. The availability and accessibility of potential candidate sites, and alternative uses of the land must also be considered in the selection of candidate areas and sites for investigation.

TSOAR, H., Ben-Gurion University of the Negev, Beer Sheva, Israel

Importance of Flow Separation in Processes of Sand Transport and Deposition on Dunes

Wind magnitude is known to abate on the lee side of a dune, a process followed by deposition of sand. This basic principle rules the fundamental dynamic mechanism of barchan and transverse sand dunes that have erosion and movement of sand across their gentle windward slope and deposition on their lee slope.

Lee side wind abatement and deposition derive from divergence of wind streamlines or from separation of flow. There are two causes for separation of flow. When the body is blunt and rounded like a flat semicylinder, separation ensues under the increasing pressure existing on the lee side, which prevents the flow from penetrating there. This mode of separation, which occurs downwind of the crest, is rare on dunes and may occur on zibars, vegetated, and other rounded or subdued dunes. On sharp-edged bodies with a precipitous topography of surface inclination, the air flow is not capable of achieving an infinitely large velocity at sharp brink or crest and becoming separated there. This type of separation does occur on all dunes with sharp crests (such as seifs) and brink-line over slip-facies (such as on barchan and transverse dunes).

When separation (mostly of the second type) occurs, there is, on the lee slope, both deposition of sand and flow diversion that conduces longitudinal sand movement parallel to the crestline. The ratio between deposition and longitudinal transport depends on the angle of attack of the wind. On barchan and transverse dunes, where the angel of attack is 90° or near to it, there is mostly deposition and less longitudinal sand transport. On seif dunes there are segments where, at an angle of attack of up to 50°, longitudinal movement is greater than deposition whereas, on other segments where the angle is higher, deposition predominates.

Simulation studies in the wind tunnel of several fixed-bed models allow for studying this process under controlled conditions of angles of attack and of model shapes, with visualization of flow. Results indicate that the effect of the wind's angle of attack has two contradictory sequels. As the angle approaches 90°, the profile line of the model parallel to the airflow is sharpest and, therefore, the rate of increase of wind velocity above the crest is higher. Hence, the separation is more pronounced and the reattachment velocity is increased. On the other hand, the smaller the angle of attack, the less sharp the profile line parallel to the airflow, and the lower the velocity increase above the crest; however, the flow component parallel to the crestline is higher, and this brings about an increase of the flow velocity on the lee slope parallel to the crestline. Accordingly, the critical angle of attack is between 30° and 40°.

TSU-MING Han, Cleveland-Cliffs Iron Company, Ishpeming, Michigan, USA

Origin and Evolution of Magnetite in the Upper and Lower Cherty Members of the Biwabik Iron Formation, Mesabi District, Lake Superior Region, USA

the Biwabik iron formation in the Mesabi District of the Lake Superior region extends east-northeast for about 192 kilometers west of Grand Rapids to Birch Lake, east of Babbitt, Minnesota. Stratigraphically, it consists of four members: Lower Cherty, Lower Slaty, Upper Cherty, and Upper Slaty in ascending order. Magnetite is one of the principal constituents of the cherty members and is essentially the only oremineral now mined and beneficiated in the District.

The eastern end of the iron formation has been subjected to a progressive contact metamorphism by the intrusion of the Duluth Gabbro Complex. French (1968) established four metamorphic zones within the iron formation along the strike toward the Complex: unaltered (Zone 1), transitional (Zone 2), moderately metamorphosed (Zone 3), and highly metamorphosed (Zone 4). This paper deals with the origin of magnetite and its changes in the four metamorphic zones within the cherty members.

In addition to the petrographic study, a substantial number of magnetite-rich specimens from the cherty members throughout the District were polished and microscopically examined before and after induced oxidation. The oxidation was induced by heating the polished sections in a stagnant air furnace at about 300°C for approximately 72 hours. The oxidation was rather selective and occurred along either the octahedral planes or the grain boundaries of preexisting inclusions within the magnetite or both, depending on the degree of metamorphism. The study revealed the following:

1. In Zones 1 and 2, the magnetite is chiefly associated with quartz, hematite, siderite, ankerite and layered iron silicates. It occurs as octahedra, sawtoothed blades, wedges and rhombs present either in scattered grains or coalesced in granules, laminae or irregular clusters. Traces of hematite are often present in some of the magnetite; these are actually remnants of preexisting hematite crystals revealed by induced oxidation. The magnetite oxidizes along the grain boundary with the preexisting hematite inclusions. The hematite generated by induced oxidation divides the existing magnetite into the inner pseudomorph and the outer overgrown rim. The outlines of the pseudomorph are platy, bladed, hexagonal, rhombohedral, wedge-shaped, or colloform and exhibit vuggy, microgeodic or botryoidal microstructures. The edges of the outer overgrown rim are either sawtoothed or coalesced octahedra, depending on the magnitude of the overgrowth. The size, orientation, and morphologist of the magnetite granules, laminae, irregular clusters and single grains are usually controlled by those of the preexisting hematite nuclei.

2. The magnetite in Zone 3 mainly coexists with quartz and amphiboles, and the mode of occurrence is essentially the same as that in Zones 1 and 2. However, the induced oxidation takes place along its octahedral planes rather than along the preexisting hematite boundaries within the magnetite. Nevertheless, the outlines of the preexisting hematite are still evident in some magnetite grains.

3. In Zone 4, the magnetite is only in octahedral form and is mainly associated with quartz, amphiboles, pyroxenes, and is locally with faylite. It is both coarse- and fine-grained. The former occurs as single, subangular grains or as coalesced octahedra and the latter as random, irregular clusters. Both oxidize along the octahedral planes. Under induced oxidation, the fine-grained variety is relatively inert compared to the magnetite in the first 3 zones but is more susceptible to oxidation than the coarse-grained variety. The magnetite granules that are typically present in Zones 1, 2 and 3 are now extinct.

In summary, based on the microstructural relationship of the existing magnetite to its preexisting hematite inclusions, hematite was apparently one of the chief constituents of the cherty members of the Biwabik iron formation before the development of magnetite. The magnetite was formed through centripetal diffusion of Fe_{++} to hematite nuclei which probably occurred under late diagenetic or low-grade metamorphic conditions. The mineralogy of the cherty members was significantly altered by the intrusion of the Duluth Gabbro Complex. This intrusion not only caused the successive transformation of the structures of the silicate minerals; the recrystallization of quartz; the extinction of siderite, ankerite and existing hematite; but also led to the following changes of the coexisting magnetite:

a. A significant increase in grain size and thermal stability.

b. Identity loss of sedimentary structures typically present in Zones 1, 2, and 3-granules, laminae, etc.

c. The extinction of bladed, rhombohedral, and wedge-shaped morphologies often observed in Zones 1, 2 and 3.

d. The progressive obliteration of the outlines of preexisting hematite inclusions that are revealed after induced oxidation.

These changes resulted from grain growth and recrystallization that are particularly pronounced as the moderately metamorphosed zone grade into the highly metamorphosed zone.

REFERENCES

French, B. M., 1968, Progressive contact metamorphism of the Biwabik iron formation, Mesabi Range, Minnesota. The University of Minnesota Press, Bulletin 45.

TSURUTA, H., Yokahama Research Institute for Environmental Science, Yokohama, Japan

Acid Precipitation Study in Japan

Recent study on acid precipitation in Japan will be presented. Japan is located at the edge of the Asian Continent and of the North Pacific Ocean, and accordingly its atmospheric condition is affected by polar continental air masses in winter and by maritime tropical air masses in summer. Coastal areas facing toward the Pacific Ocean may have much rain between June and September, whereas coastal areas facing to the Sea of Japan have much snow in winter due to the strong northwesterlies or westerlies. The contribution of sea salt particles is not negligible to the constituents of precipitation in Japan accordingly.

The annual average of the molar ratio of nitrogen oxides to sulfur dioxide in anthropogenic emissions is 1.5 in Japan as a whole, while it is ~3 in such areas of heavy industrialization and urbanization as Tokyo and Osaka. This ratio in Japan is higher than in eastern North America, Europe and China.

Acid precipitation study has started all over Japan since 1983. The bulk sampling of precipitation was made by collectors with filtration and snow collectors at urban and rural sites in seven areas, with a sampling period of one week during five years (wet-only sampling was also made at seven urban sites). Precipitation weighted annual average pH at 14 sites was in the range 4.4-5.3 between 1984-1985. The pH value obtained by the bulk collector was higher by 0.2 at most than that by wet-only sampler, and hence the minimum pH value in Japan seems to be slightly higher than in eastern North America and Europe. The annual deposition of excess sulfate (non-marine sulfate) was in the range 2-4 g/m^2 and that of nitrate was 1-3 g/m^2 at urban sites except for the areas facing to the Sea of Japan. The molar ratio of nitrate to excess sulfate in precipitation was below 1.0 except for the Tokyo Metropolitan area.

A field study on mesoscale precipitation chemistry was carried out at 24 sampling sites in and around the Kanto district where the Tokyo Metropolitan area is located, by using the same types of bulk collectors as those previously mentioned for one year between September 1984 and August 1985. The highest value of ~1.5 for the molar ratio of nitrate to excess sulfate in precipitation was found just around the Tokyo Metropolitan area. The annual deposition of excess sulfate in the area facing to the Sea of Japan reached ~6 g/m^2 primarily due to much snow in winter, but the concentration of excess sulfate was lower than that in the Kanto district. The seasonal variation of water soluble constituents in precipitation revealed that the increase of the molar ratio of nitrate to excess sulfate corresponded to the decrease of pH from spring to summer in the Kanto district, while the decrease of the ratio corresponded to the decrease of pH from fall to winter in the coastal area facing to the Sea of Japan.

The effects of acid precipitation in soils and lakes have been also studied. Japanese soils were classified into six levels according to the relative susceptibility to acidification or to their relative capacity for buffering to the acidification, and a map for assessing the susceptibility of the Japanese soils to acid precipitation was prepared. The pH value of surface soils at 12 sites was in the range 4.3-6.8 and no year-to-year trend was found between 1983-1985.

TSUSUE, A., Kumamoto University, Kumamoto, Japan, and H. SHIMAZAKI, University of Tokyo, Tokyo, Japan

Genesis and Geochronology of Granitoids and Associated Ore Deposits in South Korea

Four geological provinces have been recognized in South Korea on the basis of tectonics and geology (Kim, 1975). The provinces have a northeast-southwest trend across the Korean peninsula, known as the Sinian direction, and from north to south, are referred to as the Kyeonggi massif, Ogcheon zone, Ryeongnam massif, and Kyeongsang basin. In the Kyeonggi and Ryeongnam massifs Precambrian systems and granite gneisses are well exposed. Jurassic granitoids occur widely along the Sinian direction. The Ogcheon zone lies between the Kyeonggi and Ryeongnam massifs, and runs along the Sinian direction. The Ogcheon group of unknown age, Paleozoic and Mesozoic sedimentary formations, and Jurassic and Late Cretaceous granitoids are found in the Ogcheon zone. Nonmarine sedimentary formations and associated volcanic rocks of the Cretaceous age are found in the Kyeongsang basin. Late Cretaceous and Paleogene granitoids intrude the basin as stocks and small masses (Kim, 1975).

Radiometric age data for the granitoids in South Korea are bimodal with an older mode of 180-140 Ma and a younger mode of 120-40 Ma, although older plutons such as, Puncheon and Hongjesa granite complexes are known in the northeastern part of the Ryeongnam massif (Kim, 1971; Geological Survey of Korea, 1972; Lee and Ueda, 1977; Lee, 1980; Lee, 1987).

The initial 87Sr/86Sr ratios of Cretaceous granitoids in the Kyeongsang basin are 0.704-0.707; the same ratio of the Cretaceous quartz porphyry in the Ogcheon zone is 0.716. Although many data for the initial 87Sr/86Sr ratios of the Jurassic granitoids and Precambrian gneisses are available, reliable data for the initial ratios of the Jurassic granitoids range from 0.710 to 0.718. Fullagar and Park (1975) suggested an age of 1,790 Ma and initial 87Sr/86Sr ratio of 0.720 for the Precambrian gneiss in the Kyeonggi massif near Seoul.

To clarify the mode of differentiation of granitoids of various ages, a neutron activation analysis has been made to measure the rare earth element (REE) abundance of the granitoids. The REE patterns of the Jurassic granitoids in South Korea are quite different from the patterns of the Jurassic granitoids in South Korea. Namely, the [Eu]/[Eu*] ratio range from 1.1 to 0.62, the (La/Lu)cn ratio changes from 20 to 80, and the total REE content varies from 100 to 180 ppm for Jurassic patterns, while the same ratios and the total REE content range from 1.1 to 0.04, from 2.4 to 38, and from 100 to 240, respectively for the Cretaceous patterns. We used the Rayleigh fractionation model as a crystallization model of the granitic melts. Since the model calculations fit the REE data for the Cretaceous granitoids very well, it is suggested that the Cretaceous granitic rocks in South Korean were mainly differentiated by fractional crystallization. The REE patterns for the Jurassic granitoids suggest the presence of garnet or zircon as residues of partial melting of old metamorphic rocks (Tsusue et al, 1987).

Although it is not possible to specify the intrusive depths of Mesozoic granitoids in South Korea, there are several facts indicating the Cretaceous granitoids in the Kyeongsang basin are high level intrusions compared with the Jurassic granitoids in South Korea. A high level intrusion is one of the potential characteristics of granitic magmas which generate hydrothermal ore deposits of base metal, such as copper, lead, and zinc (Urabe, 1984). The data for intrusive depths of Jurassic and Cretaceous granitoids are consistent with the fluid inclusion data for the Jurassic and Cretaceous gold deposits in South Korean (So and Shelton, 1987a, b).

Radiometric age data for base, heavy, and precious metal deposits clearly revealed the major Zn-Pb, W-Mo, and Au deposits in South Korea were formed in the Late Cretaceous age, with some exceptions of Precambrian and Jurassic tungsten and gold deposits (Fletcher and Rundle, 1977; Farrer et al, 1978; Yun and Silberman, 1979, Sato et al, 1981; Shibata et al, 1983; So et al, 1983, Park et al, 1985, 1986; Shimazaki et al, 1986, 1987; So and Shelton, 1987a).

There are at least three epochs, Precambrian, Jurassic, and Cretaceous for gold and tungsten mineralization in South Korea. Precambrian gold mineralization is represented by the Geumjeong alaskite-quartz vein deposit and Precambrian tungsten mineralization by the Ssangjeon tungsten deposit, both in the Taebaegsan area. Jurassic Au-Ag mineralization is found in the Cheonan area, and Jurassic Mo and W-Mo mineralization is found in Geumseong (Mo), Jangsu (Mo), and Garisan (W-Mo) deposits.

The tungsten and molybdenum mineralization occurred in the early Jurassic age, while Cretaceous W-Mo deposits have shown the radiometric ages of about 75 ± 10 Ma. Among them, radiometric ages of typical W-Mo deposits in South Korea, such as the Sangdong, Daehwa, and Cheongyang, are about 85 Ma, ages of Zn-Pb dominant deposits in the Taebaegsan area, such as the Shinyemi and No. 2 Yeonhwa are about 75 Ma, and ages of W-Mo deposits in the Kyeongsang basin are about 65 Ma.

Although the reason is not obvious at present, it has been shown that W-Mo mineralization occurred prior to Au-Ag mineralization in the Jurassic age, whereas vice versa in the Cretaceous age.

TU Kan, M. F. J. FLOWER, and Ming ZHANG, University of Illinois, Chicago, IL, Guanghong XIE and Junwen WANG, Institute of Geochemistry, Academia Sinica, Guiyang, China, and R. W. CARLSON, Carnegie Institution of Washington, Washington, DC

Lead Isotopic Data for Cenozoic Basalts from Eastern China: Evidence for Cratonic and Circum-Cratonic Mantle Domains

Cenozoic basalt magmatism in eastern China constitutes a distal part of the Circum-Pacific volcanic belt, lying behind the southwest Japan, Izu-Bonin, Marianas, and Philippine volcanic arc systems. These basalts penetrate the Archean–Proterozoic Sini-Korean craton, Proterozoic–Paleozoic Yangtze craton, and later circum- cratonic crust to north and south. New and published Pb isotopic data for the basalts plotted in the Pb-Pb diagram (Table 1, Fig. 1) define an array lying subparallel to the NHRL (Hart, 1984) straddling the Geochron. Variation of $^{207}Pb/^{204}Pb$, $^{208}Pb/^{204}Pb$ is coupled with that of $^{206}Pb/^{204}Pb$, and an overall increase in Pb isotope ratios is observed from north to south China. As a first- order approach, we have classified the Pb isotopic data into one cratonic and two circum-cratonic groups, on the basis of geographical location, tectonic setting, and basement age.

The cratonic group is defined by three linear arrays of data from Hebei, Shanxi Province (HS), and Shandong Province (SD) (Sino-Korean craton), and Anhui, Jiangsu Province (AJ) (Yangtze craton) corresponding to secondary isochron ages of 2.2, 2.6, and 2.1 Ga, respectively. This age range is consistent with that of craton formation. Most HS data plot on the intersection of the Geochron and NHRL, indicating a depleted mantle source. If linear trends of SD and AJ are explained as mixing lines, low $^{206}Pb/^{204}Pb$ values correspond to a presumed subcontinental lithosphere component EM-1, while high values may be attributed to a sedimentary component EM-2. Thus, heterogeneous subcratonic source materials are believed to comprise depleted mantle (DM), and two enriched types, (EM-1), and (to a lesser degree) EM-2 (cf. Zindler and Hart, 1986).

In northeastern China (NE), the Sino-Korean craton is extended and transitional to a late Paleozoic accretionary fold belt. Pb isotope data for northern circum-cratonic basalts cluster near the Geochron. Enriched and variable $^{207}Pb/^{204}Pb$ contrast strikingly with low $^{206}Pb/^{204}Pb$ ratios and may reflect recycling of continental material via subduction of the Pacific plate. Pb isotopic ratios for high-K shoshonitic lavas from Wudalianzhi, Heilongjiang Province, are markedly less radiogenic and resemble those of high-K lavas from the western USA, and appear to have tapped a low- mu source.

South of the Sino-Korean and Yangzte cratons, continental crust is younger and thinner than that to the north and was partially disaggregated during formation of the South China Basin. Basalt Pb isotopic compositions define a distinctive southern cratonic group and plot well to the right of the Geochron. More significantly, $^{207}Pb/^{204}Pb$ variation is large and scattered, lying consistently above the NHRO and showing high-angle, even negative slopes on the Pb-Pb plot. Plotted in $\Delta 7/4$ versus $\Delta D8/4$, these data occupy the Dupal quadrant and confirm existence of a Dupal-type mantle domain. This domain appears to incorporate CFB and OIB intraplate magmas of the South China Basin (e.g., Hainan, SCB seamounts, Paracels, Penghu Islands) and arc eruptives at its eastern margin (Taiwan, southwestern Luzon). Combined isotope and magmaphile element systematics for the SCB and Hainan Island magmas, although decoupled, confirm regional heterogeneity, and suggest at least three source components are involved: DM, OIB, and a $^{207}Pb/^{204}Pb$-, $^{208}Pb/^{204}Pb$-rich Dupal type. The Dupal signature must reflect an ancient component in view of the rapid decay of ^{235}U in the early earth. Due to the absence of HFSE depletion in (at least) magmas from South China and the SCB, the Dupal signature cannot be solely ascribed to recent subduction-derived sediment, but may reflect asthenosphere-generated delamination or thermal erosion of subcontinental lithosphere.

REFERENCES

Hart, S. R., 1984, A large scale isotope anomaly in the Southern Hemisphere mantle: Nature, v. 309, p. 753–757.

Peng, Z. C., R. E. Zartman, K. Futa, and D. G. Chen, 1986, Pb, Sr and Nd isotopic systematics and chemical characteristics of Cenozoic basalts, Eastern China: Chemical Geology, v. 59, p. 3–33.

Sun, S.-S., 1980, Lead isotopic study of young volcanic rocks from mid-ocean ridges, ocean islands and island arcs: Philosophical Transactions of Royal Society of London, v. 297, p. 409–445.

Zindler, A., and S. R. Hart, 1986, Chemical geodynamics: Annual Review of Earth and Planetary Science, v. 14, p. 493–571.

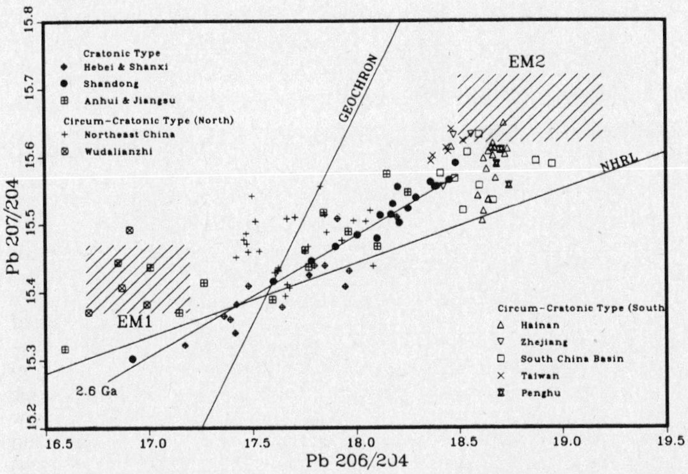

Figure 1—Plots of $^{207}Pb/^{204}Pb$ versus $^{206}Pb/^{204}Pb$ data for Cenozoic basalts from southern, eastern, and northeastern China, representing cratonic and northern and southern circum-cratonic mantle domains. Data sources: Sun (1980), Peng et al. (1986), Xie et al. (in press), and Tu et al. (unpubl.). Geochron, NHRL, EM-1, and EM- 2 from Zindler and Hart (1986). The 2.6 Ga secondary isochron for Shandong samples is shown for reference.

area	time	rocks	sample	206Pb/204Pb	207Pb/204Pb	208Pb/204Pb	slope	R	age	data
Hebei & Shanxi	Q-N	TH, ALB	14	17.175-18.199	15.322-15.510	37.112-38.216	0.130	0.74	2.2	1,4
Shandong	Q-N	ALB	18	16.921-18.454	15.302-15.565	37.284-38.538	0.178	0.98	2.6	2
Anhui & Jiangsu	Q-N	TH, ALB	12	16.594-18.252	15.317-15.574	37.352-38.469	0.124	0.86	2.1	2
Northeast China	Q-N	TH, ALB, BAS, TRA	27	17.422-18.082	15.394-15.556	37.554-38.290	0.042	0.19		1,2
Wudalianzhi	Q	PHO, SHO	5	16.712-16.992	15.371-15.492	36.515-36.998	0.135	0.28		1,2
Hainan	Q	TH, ALB, BAS	16	18.464-18.738	15.505-15.649	38.636-38.988	0.659	0.68		3
South China Basin	Q-N	TH, ALB, NE, TRB	10	18.411-18.954	15.520-15.632	38.402-38.991	0.05	0.24		3
Zhejiang	N	TH	3	18.425-18.562	15.555-15.632	38.507-38.911	0.50	0.79		2
Taiwan	Q	TH	6	18.366-18.523	15.593-15.640	38.636-38.791	0.175	0.65	2.6	5
Penghu	Q	TH, ALB	4	18.684-18.745	15.557-15.610	38.821-38.875	-0.47	0.61		5

ALB=alkali basalt, BAS=basanite, NE=nephelinite, PHO=phonotephrite, SHO=shoshonite, TH=tholeiite, TRA=trachyte, TRB=trachybasalt, R=regression factor, age=secondary isochron age (Ga)Data sources: 1 Xie et al. in press, 2. Peng et al. (1986), 3 . Tu et al. in prep. 4. Carlson (unpub.). 5. Sun (1980).

TUEZOV, I. K., Institute of Tectonics and Geophysics, Academy of Sciences of the USSR, Khabarovsk, USSR

Asthenosphere and Geodynamics of the Asia-Pacific Active Margin

The thesis is suggested that it is expedient to separately specify geothermal, geoelectrical, velocity (wave guide) and density asthenospheres while analyzing the data on the upper mantle structure in the Asia-Pacific Active Margin and the adjacent regions of Asia and the Pacific Ocean. It facilitates the problem of study of the asthenosphere, because the discrepancy of the data, obtained by different methods, should not be regarded as erroneous but as reflecting the real pattern, in case they are reliable enough.

The above asthenospheres form two pairs, geothermal and geoelectrical, and velocity and density ones in the southern part of the Okhotsk Sea. Obviously, this division testifies to the proximity of the physical nature of the asthenospheres, included in each pair.

A sharp lateral and vertical variation of the composing matter, also the sufficient variation of the angles of dipping, thickness and depth are characteristic of each of the above asthenospheres. Thus, the velocity values vary from 7.2 to 7.9 km/s, densities ranging from 3.4-3.08 to 3.25 g/cm^3. The complex density variations within the upper mantle make the situations, when the density inversions within the upper boundary of the asthenospheres yield their general distribution, i.e. the increase with depth, while the asthenosphere is distinguished by the decreased density gradient with depth within the limits of the former.

The presence of the highly conducting zones under Lake Baikal, the Tatar Strait and Sakhalin, and of the low velocity zone under the central part of the Okhotsk Sea allows us to suggest the presence of the highly melted zones with a large amount of magma melt respectively.

The depth to the upper boundary of the asthenosphere on the route Lake Baikal—the Pacific Ocean varies from 40-50 to 200 and more km, the thickness ranging from 35-40 to 160 km within the central part of the Okhotsk Sea. The angle of dipping varies from zero to 30°. The same sufficient depth variations to the asthenosphere and the thickness variations of the former are noted on the latitude of the Philippine Sea and China. The depth to the density asthenosphere reaches 30-50 km, the thickness ranging from 40 to 200 km. The structure of the geoelectrical asthenosphere is well studied in the northeast of China. The depth to this asthenosphere ranges from 40-50 km to 160 km here. The thickness is valued about 70 km. The depth to the velocity asthenosphere in Tibet reaches 100 km, and the thickness is valued about 70 km. The decrease of the velocities in the asthenosphere is accompanied by the decrease of Q parameter from 250 km in the lower crust to 70-100 in the upper parts of the asthenosphere.

The above stated testifies to the fact that the thickness and the depth to the asthenosphere vary rather sharply from region to region. The asthenosphere does not always occur subhorizontally and represent the continuous formation either. It may have skip zones along the strike, as for example, in the west of the Okhotsk Sea and west of the Baikal. Simultaneously, it may be characterized by the sufficient angles of dipping and form the ascending uplifts, reaching the Earth's crust by their tops.

The asthenospheric lenses with the thickness of 200-400 km are distinguished under the Baikal rift zone and the Asia-Pacific junction zone. The asthenosphere is absent or is rather weakly manifested in the Far East. It is known that the first two regions are notable for the high rates of Late-Cenozoic movements, while the latter, vice versa, for their weak manifestation. It allows us to suggest, that these facts are directly correlated, i.e., the thick asthenosphere, occurring not very deeply in the Baikal rift zone and the Asia-Pacific junction zone, is conditioned by Late Cenozoic activity, while the absence or the weak manifestation of the asthenosphere in the Far East in determined by low rates of Late Cenozoic movements here. It corroborates the conclusion, that probably, the upper boundary of the asthenosphere migrates upward and downward along the section due to the character of tectonic movements during either epoch. Its thickness is also varied. The other conclusion, which follows from the data, is that the asthenosphere is not continuously developed, and is mostly widespread in the tectonically active regions during the recent epoch, while in relatively stable ones it is sufficiently weakly manifested.

Numerical modeling allows us to suggest that the most active formation of the uplift of the geothermal asthenosphere in the Okhotsk Sea is related to the time span of 40-25 m.y. ago. During this time span its upper boundary ascended from 100 km depth to that of about 20 km in a number of sections of the Okhotsk Sea. The velocity of ascending reached 0.6-1.0 cm/year. Then the velocity of formation of the asthenospheric uplift decreased to 0.04 cm/year i.e., 25 times reduced. It is similar to the available geological and geophysical data on stabilization of tectonic movements in the Okhotsk Sea during the Oligocene-Miocene epoch and the origin of the steeply occurring Cenozoic assemblages of the sedimentary cover, beginning from that time period.

The so called "blind zone" is traced in the Kuril-Kamchatka seismic focal zone at depths of 200-350 km where the earthquake hypocenters are absent or nearly absent. It appears that its presence is due to the influence of the asthenosphere, resulting in the decrease of viscosity of the subducting oceanic plate. In respect with this, the probability of occurrence of the earthquakes decreases sufficiently. The greater depth of the "blind zone" than that of the asthenosphere in the adjacent continental and oceanic blocks is obviously explained by the downward movement of the lithospheric plate.

TUFAR, W., Fachbereich Geowissenschaften der Philipps-Universität Marburg, Marburg/Lahn, Federal Republic of Germany, P. MÖLLER, Hahn-Meitner-Institut Berlin GmbH, Berlin, Federal Republic of Germany, and W. PÖHL, H. RIEDLER, and A. OLSACHER, Radex Austria AG für feuerfeste Erzeugnisse, Radenthein, Austria

Comprehensive Evaluation of Radenthein (Carinthia, Austria) Sparry Magnesite Type Deposit: Genesis, Geological Setting, Polymetamorphic Overprinting, and Chemical Composition

Over many decades, the Radenthein sparry magnesite deposit, one of the largest of its kind in the world, was regarded as a typical epigenetic-metasomatic sparry magnesite deposit of Alpine age.

Recent research shows that this genetic interpretation cannot be supported.

The Radenthein deposit represents the type locality of a magnesite deposit in a crystalline complex. It occurs in a thick garnet–mica–schist series (Radenthein mica schist series) of the Middle Austro Alpine Altkristallin. This originally represented a thick volcanogenic-sedimentary (geosynclinal) series of Paleozoic age in which the magnesite deposit can be traced back to an evaporitic basin and its magnesite layer clearly marks a change of shorter duration in the sedimentation of the original deposition area within this original sedimentary-volcanogenic (geosynclinal) series.

The Radenthein mica schist series and its embedded magnesite deposit were subjected to a pre-Alpine medium-grade progressive regional metamorphism and an Alpine diaphthoresis, thus revealing a striking metamorphic history. The polymetamorphic overprinting is responsible for the typical characteristics of the present Radenthein mica schist series as well as its magnesite deposit and its typical sparry magnesite or magnesitic marble. The polyphase overprinting is illustrated by different types or generations of garnet, kyanite, plagioclase, magnesite, and dolomite.

The textural heterogeneous magnesitic marble and associated dolomitic marble in places even display static recrystallization of previously tectonically strained magnesitic material overprinted by dynamic recrystallization. Furthermore, old magnesite porphyroblasts exhibiting strong postcrystalline deformation are overgrown ("oversprouted") by idioblastic magnesite. Around these idioblasts the typical features of mineral growth in pressure shadows are present. Coexisting magnesite and quartz suggest high CO_2 fugacity during metamorphism and recrystallization.

Specific contents of trace elements, such as the rare-earth (positive Eu-anomaly), carbon and oxygen (showing slight negative values in their isotopic composition relative to PDB, with a higher degree of scattering of the oxygen values) in the magnesite, are consistent with its complex history, which is also reflected by homogenization temperatures (with several different maxima) of fluid inclusions within the magnesite and their trapped gases (CO_2, CH_4, N_2).

TUISKU, P., University of Oulu, Oulu, Finland

Geothermometry, Geobarometry and Their Bearing on the Evolution of the Late Archaean Kuhmo-Suomussalmi Greenstone Belt, Eastern Finland

Metamorphic pressures and temperatures were calculated for the supracrustals of the last Archaean Kuhmo-Suomussalmi Greenstone Belt by several garnet-biotite Fe-Mg exchange geothermometers and two sphalerite geobarometers.

The greenstone belt has undergone two Archaean metamorphic episodes M1 and M2. During M1, the edges of the greenstone belt synclinorium being formed at this time were heated up to 660 °C. This caused partial melting and generation of migmatites at the marginal zones while the center of the synclinorium did not reach temperatures of garnet grade. The M1 phase was followed, at least at the marginal zones of the belt, by retrogressive stage, after which the entire belt was heated up to 550 °C.

The pressure estimates (mean 6.3 Kbar) are independent of the fold structures which indicate that the sphalerites were re-equilibrated during M2, after the formation of the main synclinorium.

The metamorphic evolution of the belt can be explained by a thermal relaxation model of the isograds, after the formation of the major synclinorium. The belt was probably kept at isostatic equilibrium by movements along marginal shear zones so that M2 phase with evenly distributed temperatures was possible.

TUISKU, P., and K. LAAJOKI, University of Oulu, Oulu, Finland

Metamorphic and Structural Evolution of the Mn-Bearing Metapelites of the Early Proterozoic Puolankajärvi Formation

The early Proterozoic Kainuu schist belt forms a major north-south trending synclinorium between two Archaean blocks at central Finland, contains both autochtonous and allochtonous units and is cut by faults and shear zones. The allochtonous, amphibolite-facies Puolankajärvi Formation (PjF) lies on the western side of the KSB. The formation has undergone five major deformation phases (D_1-D_5) during which the maximum principal stress (O1) rotated clockwise from southwest-northeast (D_1) to northeast-southwest or east-northeast–west-southwest (D_5) and the medium principal stress (O2) changed from horizontal to vertical between D_2 and D_3. The interval between D_2 and D_3 denotes transition from thrust even to strike slip tectonics.

The PjF consist of turbiditic psammites and pellites and tempestitic semipelites. The pelitic lithologies include quartz—two mica—plagioclase schists with variable amounts of garnet, staurolite, andalusite and biotite porphyroblasts as well as sillimanite and cordierite segregations. The relic structures in the porphyroblasts indicate the following mineral growth-deformation (MG-D) evolution in the PjF: (1) Successive crystallization of garnet (syn-D_1), porphyroblastic biotite (inter-$D_{3/4}$) and staurolite (inter-$D_{3/4}$) during pre-D_4 progressive stage. (2) Syn-D_4–inter-$D_{4/5}$ crystallization of kyanite, sillimanite (fibrolite), porphyroblastic tourmaline, magnetite, rutile, cordierite and muscovite-biotite-plagioclase pseudomorphs after staurolite. (3) Syn-post-D_5 generation of andalusite, ilmenohematite and sheet silicates after staurolite and cordierite.

Paragenetic relations between kyanite, sillimanite, andalusite, chlorite, cordierite, biotite, garnet and staurolite and their compositions are used to construct a partial petrogenetic grid for metapelites with significant Mn content (MnO = 0.1-0.6%) by installing a six phase invariant of earlier existing grid for Mn-free AMF-phases.

Tectonic interpretation to the PjF and the western part of the KSB is based on the structural observations and the constructed P-T-path. Thrusting (D_1) and subsequent folding of the thrust sheets vertical (D_2) caused a considerable thickening of the crust. Increased heat production and structural control created a thermal antiform west of the KSB, with partial melting of the supracrustals and production of granitoids. Wrench fault deformation began to work in the thickened eroding crust and late granites and pegmatites intruded into the associated shear zones which also served as channels for fluids derived from the granitoids.

TURCOTTE, D. L. Cornell University, Ithaca, NY

On Role of Magma Fracture in Magma Transport

Magma is initially generated on grain boundaries by pressure release melting or another mechanism. This magma can be buoyantly transported upwards along grain boundaries; a porous flow model with matrix collapse is generally applicable to this problem. However, this mechanism cannot result in magma transport through the cold lithosphere for which the dominant mechanism is almost certainly magma fracture. We have obtained a series of solutions to this problem including the flow in the fracture (laminar and turbulent), the elastic deformation of the country rock, and the propagation of the fracture. Similarity solutions for pressure-driven fractures have been obtained for the full problem. These solutions show that the fracture resistance of the country rock can generally be neglected compared with the viscous resistance of the magma. We have also obtained a time-dependent solution for buoyancy-driven crack propagation in which both elastic distortion and fracture resistance are neglected. The crack shape satisfies the nonlinear kinematic wave equation. For laminar flow, the cavity initially deforms but does not propagate upwards; at a critical time the solution becomes multivalued indicating the presence of an upwardly propagating shock wave. The tail of the crack closes but does not move upwards. Inclusion of elastic deformation and fracture initiation determines the structure of the shock wave, i.e., the shape of the crack tip, but does not influence the propagation of the crack and magma. Solutions for turbulent flow have also been obtained. Applications of this solution have been obtained for kimberlitic eruptions, magma ascent beneath Hawaii, and the generation of the sheeted dike complex in normal oceanic crust. As a particular example of a kimberlitic eruption, we obtain propagation and flow velocities from a depth of 200 km of about 40 m/s in a 1 m wide dike. These are consistent with inferred eruption velocities and field observations of dike widths. For Hawaii, magma transport from a depth of 50 km requires dike widths of 10–100 cm and velocities of 1–10 m/s in order to prevent solidification during ascent. Consistent results are also obtained for ascent through the oceanic crust at midocean ridges. An unresolved question is the transition from flow on grain boundaries at the solidus and fracture propagation through the cold lithosphere. One hypothesis is that melt ponds at the base of the lithosphere and then ascends in response to deviatoric horizontal tensional stresses. A second hypothesis is that the magma on grain boundaries generates microfractures that coalesce to form magma-driven fractures.

TUSAK, B., Ina-projekt, Zagreb, Yugoslavia

Hydrogeologic System of the Eocene Flysch in Water-Deficient Regions, i.e., the Islands of the North Adriatic Area

When studying regional hydrogeologic relations in karst, flysch is considered a barrier, as a rule. Regional importance of the barrier depends on the structure of the layers. By such a scheme real hydrogeologic characteristics of flysch and flysch-like layers are usually neglected. Even in Yugoslavia where karst appears as locus tipicus in global relations, flysch and flysch-like layers are studied in how much they slow down and limit groundwater movement in carbonate layer.

According to the author's experience hydrogeologic structures appear even in flysch, and they accumulate groundwaters convenient for local water supply. This refers to the North Adriatic tourist region, i.e., to the major North Adriatic islands in particular, where local water supplies are of major economic significance.

Otherwise water would have to be supplied from distant areas. According to the results of the investigations, the hydrogeological characteristics of the area can be concluded as follows:

Aquifer appears in the synclinal structure of rough clastic sediments with macrofossils.

Under the influence of radial tectonics rocks break in the form of faults and accompanying fissure systems, creating secondary porosity.

The position of the aquifer in the sedimentation sequence of flysch layers, depends on the basin dynamics and conditions that prevailed.

Groundwater storage is regenerated through precipitations.

Groundwater drainage is accomplished by natural runoff and by wells.

Well capacities range from 10 1/s to more than 20 1/s.

Depending on the regime and quantity of precipitation and rate of pumping, within the period of ten years, the ratio between minimum and maximum underground water level ranges approximately 50 meters.

After reaching the maximum draw down (twenty five meters under the sea level) in 1987, the process of recovery was also supported by means of artificial recharge from carstic spring.

This research has already been applied in explorations carried out in several major tourist centers of the region (i.e. islands Rab and Krk). The successful exploitation and increasing capacities confirm the anticipations based on the above research.

TUSAK, B., Ina-Projekt, Zagreb, Yugoslavia

Water-Bearing Horizon in Upper Cretaceous Limestone

The specific characteristic of the water supply needs of the islands of the Northern Adriatic is its unevenness; relatively big quantities are concentrated in the summer tourist season, which falls within the time of minimum water levels. The karst area is in question built of Cretaceous-eocene carbonate rocks and Eocene flysch. There prevail the concentrated streams, i.e. an intensive draining is performed immediately after rainfalls. In such conditions catchment of underground waters in dry periods is not possible on the locations, where tectonic relations cause concentrated streams.

The explorations I have carried out recently have been aimed at the water-bearing horizons, in which there exist areas for collecting significant quantities of underground waters, provided that hypsometric and structural relations enable a retention of water even in dry periods.

Several specific examples have been ascertained:

The island of Krk, the Baska Valley is a morphologic depression, which slopes from 200 m to the sea level. It is built of Eocene flysch and Cretaceous and Eocene carbonates have been discovered in the valley flanks (sides). The central part of the valley has been explored. Three wells are located from 60 to 90 m above sea level. The flysch deposits have been drilled in and the upper Cretaceous limestones have been drilled on at the height of 40 to 130 m above sea level. All of them have drilled more than 10 m limestone. After the perforation of the water-bearing horizon the piezometric pressure reached even 17 m above sea level.

By a pumping test of the well (the well head 90 m above sea level, and the water bearing horizon almost 40 m under sea level) an efficiency Q = 30 l/s has been gained at the end of the dry period. The testing lasted five days (together with conquering 9 days), and the maximum drawdown on the piezometer 2 m far from the well amounted to 3.30 m (the elevation 13.11 m above sea level).

The recovery was measured for 2170 m and the difference between the initial level amounted to 28 cm. When analyzing this result it should be emphasized that this is a smaller difference from the natural trend of drainage in the dry season for the period observed. By its chemical composition this is a calcium-carbonate water with the concentration of Cl^- to 125 mg/L without a trend of increase during pumping. A small mineralization about 400 mg/L and the total hardness about 11 U °d H make this water very convenient for drinking. The refreshment of the underground water reserves is very well expressed after rainfalls in a short period.

It is a relevant fact that exploration drilling to a depth of 199 m on this locality reached tectonically undamaged rocks, particularly towards the contact with the upper Cretaceous aquifer.

By testing Eocene limestones in the depth of 186 m of the borehole, low permeability was established. With further drilling the overburden of the upper Cretaceous aquifer had been bored through. Piezometric pressure rose to above the sea level. By "air-lift" testing a yield of Q = 4.6 l/s was obtained from the exploration well, 0 = 86 mm. It was a first-rate indication that the quantity was not reduced, not even after a seven-day-long pumping. (e.g. on the basis fo the exploration well with the yield Q = 3.21 l/s, obtained by "air-lift," the location of the well yielding Q = 30 l/s was selected). Chemical composition of water is the same as the composition of the well water.

The Rab island, locality Barbat is of Cretaceous Eocene Carbonate rocks and Flysch Eocene layers. Of the five exploration wells, two of them reaching the upper Cretaceous carstified limestones yielded good results. Permeability was so obvious that the draft quantity on the exploration well was limited by the freshwater-seawater interface. Examples of other localities on the islands Krik Rab and Cres speak in favor of this concept. However, no exploration drilling has been performed here so far. Nevertheless, these examples are worth considering.

The island Krk, lake Njivice. The edge of the lake, in the contact zone with the younger, less permeable chrono-stratigraphic members is full of springs, and due to the upper-Cretaceous aquifer there are springs in the lake bottom, and that only along the tectonic contact with the Flysch layers.

The island Cres is of the Cretaceous Eocene Carbonate rocks. The exploration aim was the Vrana carst lake, of very specific characteristics. On this occasion the same functioning of the upper-Cretaceous aquifer was observed. Since this element is structurally placed rather unfavorably for major groundwater accumulation, abundant submarine springs are formed on its wings. proportional to the precipitation quantity. Aquifers have been established in the upper Cretaceous limestones on all the explored localities, independent of tectonic fissures which might have caused ground-water flows to concentrate.

It is obvious that carstification of the upper Cretaceous limestone has led to regionally significant secondary permeability of layers; therefore the presence of tectonic fissures need not mean there is groundwater there. Permanent saturation in these layers is due to the structural set-up and hypsometric relation to the sea-level, which causes retention of the groundwater drainage.

TWICHELL, D. C., U.S. Geological Survey, Woods Hole, MA, L. M. PARSON and N. H. KENYON, Institute of Oceanographic Sciences, Wormley, United Kingdom, B. A. McGREGOR, U.S. Geological Survey, Reston, VA, and G. ROTHWELL, Institute of Oceanographic Sciences, Wormley, United Kingdom

Depositional and Erosional Processes in Deep Gulf of Mexico

GLORIA long-range sidescan sonar imagery that covers the entire northern Gulf of Mexico extending from the southern tip of Florida to the U.S.-Mexican border has provided a regional perspective of this small ocean basin that has greatly enhanced our understanding of the sedimentary processes that were active during the latest Pleistocene and Holocene. The data, collected by the U.S. Geological Survey in 1985, show that the study area can be separated into three provinces: the salt province of the northwestern Gulf, the Mississippi Fan province of the central Gulf, and the carbonate province of the eastern Gulf. The GLORIA imagery indicates that mass-wasting has been an important process in all three provinces.

The complex morphology of the northwestern Gulf is the result of deformation by salt tectonics. The Sigsbee Escarpment marks the seaward edge of the salt province, which is intruding basinward into late Pleistocene fan deposits. GLORIA sonographs reveal aprons of debris that cover areas as large as 3–10 km^{+2} that have been shed from the escarpment. Mobility of salt is inferred to be a major contributor to sediment instability in this area. Within the salt province, sediment failures vary in size from small localized deposits that are associated with the flanks of individual salt domes to a major slide that covers approximately 2250 km^{+2}. This large failure originated at a shelf-edge delta, and its 110 km path was strongly controlled by the positions of the salt domes.

The Mississippi Fan is a large, passive-margin fan that has been fed most recently by a single channel that originates in the Mississippi Canyon. The sonographs of the fan reveal three primary backscatter patterns: (1) A series of narrow, linear, high-backscatter stripes that radiate southwestward from the main channel. These stripes appear to be associated with overbank deposition derived from the main channel. (2) A series of high-backscatter swirls that cover a large area on the middle and upper fan. These are the surficial expression of a large slide deposit that originated in the Mississippi Canyon area. (3) A series of elongate lobes with high-backscatter surfaces. These lobes, which cover the largest area, radiate either across the lower fan from the end of the channel or across the middle fan from a sharp bend in the channel. A study of one representative lobe shows that it consists of a series of smaller elongate high- backscatter regions (5–40 km in length) located at the ends of previously unrecognized small channels (<5 m deep). These small channels extend as far as 120 km from the main channel, and the high- backscatter regions at their ends are interpreted to be sites of deposition. Small channelized turbidity currents or debris flows appear to be the dominant transport process constructing these lobes.

The Florida Escarpment, a steep carbonate cliff, marks the western edge of the Florida–Bahama platform. It is the boundary between the carbonate province of the eastern Gulf and the fan province of the central Gulf. GLORIA imagery and bathymetric data show that the Escarpment is incised by two types of canyons, suggesting that different erosional processes take place along its length. North of 27 °N, canyons are narrow and straight, are spaced 3–8 km apart, and are fed by dendritic tributaries. These canyons extend from near the top of the escarpment to its base; many have small fans of debris extending from their mouths. The small fans interfinger with the lobes that are associated with the Mississippi Fan. South of 27 °N, large, flat-floored, steep-sided box canyons deeply incise the escarpment and the West Florida Slope above it. These canyons are spaced 10–25 km apart and have nearly vertical headwalls; they incise the escarpment as much as 18 km. Talus deposits are found at the foot of the headwall cliffs of most of the box canyons, indicating that cliff retreat is ongoing. Slumps, common on the West Florida Slope, mostly feed into the box canyons. The morphology of the northern-type canyons that have fans at their bases suggests formation by downslope transport. The box canyons in the southern section have very small or no fans, and they are similar in appearance to canyons on land whose origin has been attributed to head wall erosion by groundwater sapping. Evidence of groundwater discharge has been found along the base of the escarpment in localized areas south of 27 °N; this evidence suggests that sapping could have contributed to erosion of the box canyons and to the overall erosion of this part of the escarpment.

The GLORIA imagery reveal that mass-wasting is widespread in the Gulf of Mexico. Deposits from slope failures vary in size from a few square kilometers to several thousand square kilometers. Styles of failures include rock falls along the Florida Escarpment, slumps and slides in the salt province, and, on the Mississippi Fan, slides as well as debris flows and turbidites originating from slope failures on the upper part of the fan. The triggering mechanisms include tectonic deformation associated with salt intrusion, groundwater sapping, and failure of shelf-edge deposits possibly associated with sea-level fluctuations.

U

UCHMAN, A., Jagiellonian University, Cracow, Poland

"Shallow-water" Trace Fossils in Paleogene Flysch, Carpathian Mountains, Poland

In Paleogene flysch (southern part of the Magura Nappe), Polish Outer Carpathians, there appear the so called "shallow-water" (sensu Seilacher, 1967) trace fossils. They are represented by Rhizocorallium ichnosp., Ophiomorpha nodos Lund., Ophiomorpha ichnosp., Thalassinoids ichnosp., Pelecypodichnus ichnosp., which occurring together with "deep-water" forms such as: Paleodictyon, Urohelminthoida, Spiroraphe and others.

According to Crimes et al, 1981, "shallow-water" trace fossils may be present in proximal deepwater flysch. However the mentioned forms were found (Ophiomorpha in high density) in proximal flysch as well as in rather dystal thin-and medium-bedded one. It leads to conclusion that the "shallow-water" also can occur in deep-water flysch deposits, irrespective to the flysch subfacies, hence the determining of bathymetry based on trace fossils must be very careful.

REFERENCES

Crimes, et al, 1981, Eclog. Geol. Helv., 74: 953-995.
Seilacher, A., 1967, Marine Geol., 5/6: 413-426.

UDINTSEV, G. B., Geological Institute, USSR Academy of Sciences

Equatorial Segment of the Mid-Atlantic Ridge: A Structural Barrier Between the North and South Atlantic?

The Equatorial Segment of the Mid-Atlantic Ridge, limited in the north by 4 degrees North Fracture Zone (proposed new name: Nikolai Strakhov Fracture Zone) and in the south by São Paulo Fracture Zone, was studied in detail during the 7th cruise of R/V Akademik Nikolai Strakhov (1 April–4 August 1988). The expedition was organized by the Geological Institute of the Academy of Sciences of the USSR. Scientists from several Soviet Institutions, and also from Brazil, the UK and the USA, participated in the expedition. The studies carried out included multibeam echo-sounding, continuous (single channel) seismic profiling, magnetic observations, and collection of samples of bottom sediments and igneous rocks by core sampling and dredging. The work was carried out along a system of latitudinal tracks of length 600-800 miles, with spacing of 5 miles. The majority of the tracks (14 in no.) were continued from Sierra Leone Rise in the east to Guyana Basin in the west, but six additional short latitudinal tracks were run in the central part of the area. Eight longitudinal cross-tracks were also run to tie in with the latitudinal tracks. Detailed surveys were made in the area of the offset of 4 degrees North Fracture Zone along the system of longitudinal tracks, with spacing of 3 miles, and in the eastern part of the offset of São Paulo Fracture Zone, also along longitudinal tracks, with spacing of 5 miles.

A number of differences were observed between the features in the Equatorial Segment on the one hand and in the North and South Atlantic Ridges on the other. Whereas the width of the clearly riftogenous parts of both these ridges is almost 2,000 miles, in the Equatorial Segment the width of spreading of the riftogenous part is no greater than 60-80 miles. A typical riftogenous morphostructure and a regular linear structure of the anomalous magnetic field exist only in this part of the Segment; here there is no sedimentary cover and the surface consists of a typical riftogenous association of igneous rocks. The name "São Pedro e São Paulo ridge" has been proposed for this part of the Segment.

On both sides of the riftogenous axial ridge of the Equatorial Segment are very wide and almost level flank plateaux; these are characterized by a mosaic structure of the basement and a similar mosaic structure of the anomalous magnetic field. A typical feature of the basement structure is horst blocks separated by flat, low and local grabens. Both plateaux are covered by regular and thickish sedimentary cover. Its maximal thickness reaches 800-1,200 m but it narrows down to 100-300 m and also to complete interruption on the steep fold scarps of the horse blocks.

The thickness of the sedimentary cover increases very quickly when the boundary of the riftogenous part of the Segment is crossed into the area of the flank plateaux, but after this there is no tendency for the thickness to increase further with distance from the axis of the Segment. The eastern flank plateau is directed connected with the Sierra Leone Rise and the latter is practically a part of this plateau. The western flank plateau continues into Guyana Basin and almost without interruption reaches Ceara Rise.

The lower part of the basement of the flank plateaux and axial riftogenous part of the Equatorial Segment of the Mid-Atlantic Ridge is composed of an hyperbasit-gabbro-amphibolite complex. The hyperbasits of this complex are analogous to alpinotype hyperbasits of continental rifts. The uppermost part of the basement is composed of pretertiary associations of ferrous basalts of either a trap or sill complex. the upper suite of the riftogenous part, overlapping the basement, is formed by young, typical, riftogenous basalt complexes of tholeiite magnesial pillow lavas. On the flank plateaux in the upper suite, rocks of alkali rich trahy-basalt complex and intrusive hyperbasit-gabbro-amphibolite complex are found locally, with evidence of high metasomatos and milonitization.

The flank plateaux are most likely of nonspreading origin. Their sedimentary cover is mainly biogenous but includes terrigenous components related to erosion of local outcrops of the basement. There is evidence of shallow water sedimentation in the recent geological past when the Equatorial Segment could have been an intercontinental structural barrier separating the North and South Atlantic. Later this barrier would have been cut through by propagating rifts which united the rift system of the North Atlantic ridge with that of the South Atlantic.

UGOLINI, F. C., R. DAHGREN, W. NUHN, and J. ZACHARA, University of Washington, Seattle, WA

Examples of Weathering in Recent and Holocene Tephra Deposits of Pacific Northwest

Mineralogical evolution of tephra deposits in the Cascade Range of Washington, USA is influence by time, vegetation, and natural acid deposition. We have observed these relationships in the central Cascades and on Mt. St. Helens where repeated depositions of tephra have occurred throughout the Holocene and as recent as 1980. In the central Cascades, specifically at Findley Lake, the tephritic soils examined are Spodosole mostly developed from Holocene airfall tephra with the inclusion of rock fragments derived from the underlying andesitic till or bedrock.

These Spodosols have formed in the Abies amabilis zone, (Ugolini et al, 1977) about 1100 m of elevation in a cool to cold climate.

The weathering regimen at these high elevation ecosystems is imposed by both the harsh climate and the biota. Directly or indirectly the biota is the main source of the major proton donors: —organic acids and CO_2—that actuate the weathering processes. Because of the low rate of decomposition (Edmonds, 1984) humic substances tend to accumulate at the surface of the soil while water soluble organic acids such as fulvic acid are readily leached from the humus layer (Dawson et al, 1978). Fulvic acid is the major proton donor in the upper mineral horizon (E) where minerals are dissolved and the metal cations complexed (Dawson et al, 1978). Because of the strong acidity in the E horizon, (pH = 3.5–4.0) carbonic acid dissociation is suppressed. In this intense weathering environment smectite is the dominant phyllosilicate while opaline silica is the dominant noncrystalline phase, allophane/imogolite were not detected.

In the Bs1 and Bs2 horizons, where the pH averages around 5.0, carbonic acid is the major proton donor rather than fulvic acid, the latter being effectively arrested at the surface of the Bs. Aluminum hydroxy smectite is the dominant phyllosilicate, while allophane/imogolite constitute the major noncrystalline component (Zachara 1979; Dahlgren, 1987).

The most probable origin of smectite is the alteration of primary chlorite, present in the underlying andesitic clasts, in a weathering environment dominated by organic acids such as fulvic acid (Schnitzer and Kodama, 1977). Always in the E horizon opaline silica is favored by the presence of fulvic acid that complexes the aluminum and leaves an excess of silica that eventually precipitated as amorphous material (Wada, 1977; Shoji and Saigusa, 1978; Yoshinaga, 1986). Aluminum in solution in this horizon is between 80 to 100% organically complexed (Dahlgren, 1987; Dahlgren and Ugolini, 1988).

Formation of Al-hydroxy interlayered smectite becomes possible in the Bs horizons where the pH is above 5.0 and where the low level of fulvic or other organic acids does not compete for the polymerization of Al. Imogolite/allophane in the carbonic acid compartment (Bs, BC, and C horizons) is proposed to be forming *in situ* via carbonic acid weathering. Carbonic acid formation and dissociation is favored by the high pCO_2 and the pH above 5. Weathering by this proton donor, in the pH range of 5.5 to 6.0, leaves an Al and Fe rich residue while some of the Si and cations are leached. Aluminum in the soil solution in this lower compartment is almost 50% in inorganic form (Dahlgren and Ugolini, 1988). In the Bs and C horizons interlayer $Al(OH0_3$ regulates Al activities while imogolite buffers H_4SiO_4 activities near $10^{-4.1}$ (Dahlgren and Ugolini, 1988). Monomeric silica may then control the formation of imogolite.

Iron liberated by carbonic acid weathering is less soluble than aluminum and tends to precipitate as a separate cryptocrystalline phase such as ferrihydrate (Schwertmann 1985; Schwertmann, 1988 personal communication). Ferrihydrate may eventually crystallize into goethite (Schwertmann, 1985).

The weathering regimen occurring in the Holocene tephra of the forested central Cascades, is now contrasted with the regimen existing on the young barren pyroclastic deposits of Mt. St. Helens. The soils analyzed were excavated on a pyroclastic flow emplaced on 18 May, 1980. Samples were secured in 1981, 1983, and 1986.

Samples collected in 1981 show, in addition to a triotahedral smectite, probably saponite, reported by Pevear et al (1982), a low-charged vermiculite not previously reported. In some samples the saponite and low-charged vermiculite appear in equal amounts. We interpret both phyllosilicates as detrital minerals originated in the old cone of Mt. St. Helens from hydrothermal alteration and incorporated as lithic fragments into the pyroclastic flow (LaManna and Ugolini, 1987).

Samples collected in 1983 show that the upper 1.5 cm of the soil is depleted of trioctahedral vermiculite. The depletion of this mineral from the 1983 surface is interpreted as due to weathering caused by the natural acid rainfall (pH = 4.0–4.6) (LaManna and Ugolini, 1987).

The most dramatic mineralogical changes were observed in the samples collected in 1984 and 1986. These samples were obtained both from depressions and associte mounds that constitute the microtopography of the surface of the pyroclastic flow. Depressions show more morphological development, more acid conditions and more advanced weathering than level or mounded sites. In the depressions weathering is progressed to the extent that smectite has been transformed into a poorly crystalline kaolinite. Colonization of the barren pyroclastic surfaces by lupin plants had a considerable impact on the weathering. The lupin plants neutralize the acid precipitation and maintain a soil with higher pH than the barren sites. Consequently smectitic minerals are still present. Direction of mineralogical evolution of the 1980 Mt. St. Helens pyroclastic deposits has been dictated by the acid deposition, the colonization of lupine and the microtopography (Nuhn, 1987).

In conclusion weathering environments existing in the forested high elevations of the central Cascades and the barren pyroclastic flow surfaces of Mt. St. Helens show that mineralogical evolution is a function of vegetation, acid deposition and time. Mineral stability in the central Cascades is controlled by the chemical compartments determined by the organic and carbonic acids. On Mt. St. Helens acid depositions from the volcano and the youthfulness of the surfaces dictates the mineralogical transformations. Lupine plants mitigate the acidification of the surfaces.

REFERENCES

Dahlgren, R. A., 1987, Aluminum biogeochemistry in a subalpine abies amabilis ecosystem, Ph.D. dissertation, University microfilm n. 88-02212, University of Washington, Seattle.

——— and F. C. Ugolini, 1988, Aluminum fractionation of soil solutions from unperturbed and tephra treated spodosols, Cascade Range, Washing, U. S. Soil Sci. Soc. Am. J. (in press).

——— ——— Formation and stability of imogolite in a tephritic spodosol, Cascade Range, Washing, (Submitted Geochem. Cosmochim. Acta).

Dawson, H. J., F. C. Ugolini, B. F. Hrutfiord, and J. Zachara, 1978, Role of soluble organics in the soil processes of a podzol, central Cascades, Washington, Soil Sci., v. 126, p. 290–296.

Edmond, R., 1984, Long-term decomposition and nutrient dynamics in Pacific silver fir needles in western Washington, Can. J. For. Res., v. 14, p. 395–400.

LaManna, J. M. and F. C. Ugolini, 1987, Trioctahedral vermiculite in a 1980 pyroclastic flow Mt. St. Helens, Washington, Soil Sci., v. 143, p. 161–167.

Nuhn, W., 1987, Soil genesis on the 1980 pyroclastic flows of Mt. St. Helens, M.S. thesis, University of Washington.

Pevear, D. R., D. P. Detheir, and D. Frank, 1982, Clay minerals in the 1980 deposits from Mt. St. Helens, Clays Clay Miner., v. 30, p. 241–252.

Schnitzer, M., and H. Kodama, 1977, Reactions of minerals with soil humic substances, *in* Minerals in Soil Environments, J. B. Dixon and S. B. Weed, eds., Chap. 21, p. 741–770., Soil Sci. Soc. Am., Madison, WI.

Schwertmann, U., 1985, The effect of pedogenic environments on iron oxide minerals, Advances in Soil Sci., v. 1, p. 171–200.

Ugolini, F. C., R. Minden, H. J. Dawson, and J. Zachara, 1977, An example of soil processes in the Abies amabilis zone of central Cascades, Washington, Soil Sci., v. 124, p. 291–302.

Wada, K., 1977, Allophane and imogolite, *in* Minerals in Soil Environments, J. B. Dixon and S. B. Weed, eds., Chap. 16, p. 603–658, Soil Sci. Soc. Am., Madison, WI

Yoshinaga, N., 1986, Mineralogical characteristic II, Clay minerals, *in* Ando soils in Japan, K. Wada, ed., p. 41–56, Kyushu University Press.

Zachara, J., 1979, Clay genesis and alteration in two tephritic subalpine podzols of the central Cascades, Washington, M.S. thesis, University of Washington.

UI, H., Toyama University, Toyama City, Japan

Radon Discharge from Atotsugawa Fault and Hida Triangle Zone: Possibility of Earthquake Prediction

The radon alpha track etch method has been applied to earthquake prediction since the 1970s (King, 1980). I started the earthquake prediction program in 1980 and improved "radon pot" placed directly on a shear zone of the Atotsugawa fault, which runs northeast-southwest and extends about 70 km in the Hida region. Two pieces of alpha detector film inside the "radon pot" were exposed for 1-3 weeks. The exposed detectors were etched by ordinary chemical treatment and then the number of tracks was counted under the microscope. In 1983-1984, tracks were calculated by manual counting and in 1983-1986 by a particle analyzer (Figure 1).

The monitoring stations are Sakagami, Kamura, Jogashima, and Amo from east to west; the stations are about 4 km apart and 13.1 km long from end to end. Since three observations are taken at Amo, six observations are given (Figure 1). The radon emanation tends to increase in summer and decrease in winter. This tendency is a long-term variation over the course of a few months. On the other hand, it tends to increase or decrease in rainfall. This is a short-term variation over the course of 1-2 weeks. Cumulative rainfall greater than 50 mm/week are shown by solid bars in Figure 1.

The western Nagano Prefecture earthquake (WNPeq) occurred on September 14, 1984 (M 6.8, and epicentral distance about 65 km), and aftershocks are shown by arrows on Figure 1. The radon emanation increased before and decreased after the WNPeq as counted by both methods. The variation of radon emanation must be checked by the following criteria: (1) changes in radon emanation have the same tendency in the same tectonic framework of the monitoring sites, (2) intrinsic variation especially weekly rainfall of each site can be excluded.

Seismic activity along the Atotsugawa fault is present now (Mikumo et al, 1987). Also, the nearly north-south-trending Hida mountains, including volcanoes such as Yakedake, are seismically active. On the other hand, the northwest-southeast-trending active Atera fault has no remarkable seismic activity. These three tectonic lines (about 70 km long each) make up the Hida Triangle Zone, which seems to be a tectonic block in central Japan. The Bouguer anomaly is low in the Hida Triangle Zone and values change parallel to the tectonic lines (Kono et al, 1982). Radon emanation changes monitored on the Atotsugawa fault, Hida mountains, and the Atera fault could be good indicators for future earthquakes occurring in the Hida Triangle Zone.

REFERENCES

King, Chi-Yu, 1980, Episodic radon changes in subsurface soil gas along active faults and possible relation to earthquakes: Journal of Geophysical Research, v. 85, p. 3065-3078.

Kono, Y., T. Hibi, M. Kubo, O. Michigami, K. Shibuya, K. Suzuki, and N. Furuse, 1982, Gravity anomaly over the northern part of the central Japan (1): Sci. Rept., Kanazawa Univ., v. 27, p. 72-83.

Mikumo, T., H. Wada, and M. Koizumi, 1988, Seismotectonics of the Hida region, central Honshu, Japan: Tectonophysics, v. 147, p. 95-119.

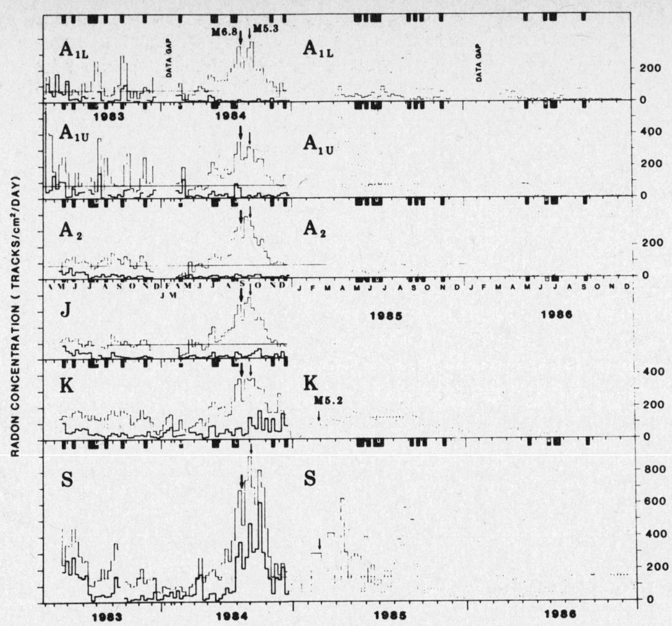

ULMER, G. C., Temple University, Philadelphia, PA

Redox Equilibria: Progress in Laboratory Techniques

Just over 200 years ago, Cavendish and Lavoisier studied hydrogen and oxygen, respectively, as newly discovered and purified elements. The death of the phlogiston theory came with the first quantitative redox experiment on the oxidation of mercury by Lavoisier. The quantification of oxidation-reduction equilibria in geology did, however, come only as recently as the later 1950s with the Eugster thrust into hydrothermal solid-state redox buffers in the research done at Johns Hopkins and the simultaneous Muan and Osborn application of gaseous buffers at one atmosphere pressure to phase equilibria research at Penn State. Prior to the 1950s, the main concern with geologic oxidation state was via ferrous-ferric data from wet chemistry, an art perfected at that time for difficultly soluble silicates and oxides by Ingamells. Borrowing from metallury and physical chemistry, Sato in 1964 pioneered the use of ZrO_2 cells for the direct determination of fO_2, and initiated the so-called "intrinsic oxygen fugacity" (IOF) measurement technique. In the 1960s geoscientists such as Shaw introduced diffusion membranes as a technique to control and/or measure redox at ever higher total pressures.

In this IGC F-22 Session, highlights of some of the newest technique studies will be presented either by contributed papers or by this speaker. Topics will include (1) ZrO_2 cell studies in which Mössbauer spectroscopy was used to compare IOF interpretations with actual ferrous/total iron measurements for mantle ilmenite nodules; (2) ZrO_2 cell studies for pH measurements at hydrothermal conditions; (3) comparative testing on spinel and glass standards; (4) Zeebeck potential measurements compared to IOF data for the establishment of fO_2 values; (5) the use of gas chromatography for the direct determination of evolved hydrogen (DHEM) for establishing redox state in high-pressure gold-lined autoclaves; (6) results of comparative tests of several redox techniques applied to the same samples; and (7) the discovery that some redox buffers are still dependable at temperatures as low as 200–300 °C.

Now, 30 years after the Eugster–Muan–Osborn impetus, progress has been spurred in geological research techniques on redox equilibria! New hydrothermal buffering techniques for chlorine, hydrogen, and even oxygen have evolved; the modified Redlich-Kwong equation of state has been used at least to some limited degree to extrapolate CHO buffers to higher pressures; practitioners of wet chemistry have continued, particularly on oceanic rocks, to gain redox insights; and several labs are using ZrO_2 cells either for fO_2 monitors or for IOF and pH measurements even including some work at higher pressures. Additional permeability membranes to sense or control redox species such as hydrogen have been developed from both alloys and polymers for use at elevated pressures. New buffers for acidity and oxidation and sulfidation are numerous.

As shown above, complimentary techniques have been "borrowed" and include Mössbauer spectroscopy, gas chromatography, thermogravimetric analysis, and Zeebeck potential measurements. The integration of these "borrowed" techniques into redox studies has strengthened the redox state characterization.

REFERENCES

Ulmer, G. C., ed., 1971, Research techniques for high pressure and high temperature: New York, Springer-Verlag, 367 p.

——— and H. L. Barnes, eds., 1987, Hydrothermal experimental techniques: New York, Wiley Interscience, 523 p.

——— D. E. Grandstaff, D. Weiss, M. A. Moats, T. J. Buntin, D. P. Gold, C. J. Hatton, A. Kadik, R. A. Koseluk, and M. Rosenhauer, 1987, The mantle redox state: an unfinished story: GSA Special Paper 215, p. 5–23.

Virgo, D., R. W. Luth, M. A. Moats, and G. C. Ulmer, 1988, Constraints on the oxidation state of the mantle: an electrochemical and $^{+}57$Fe Mössbauer study of mantle-derived ilmenites: Geochimica et Cosmochimica Acta, v. 52, p. 1781–1794.

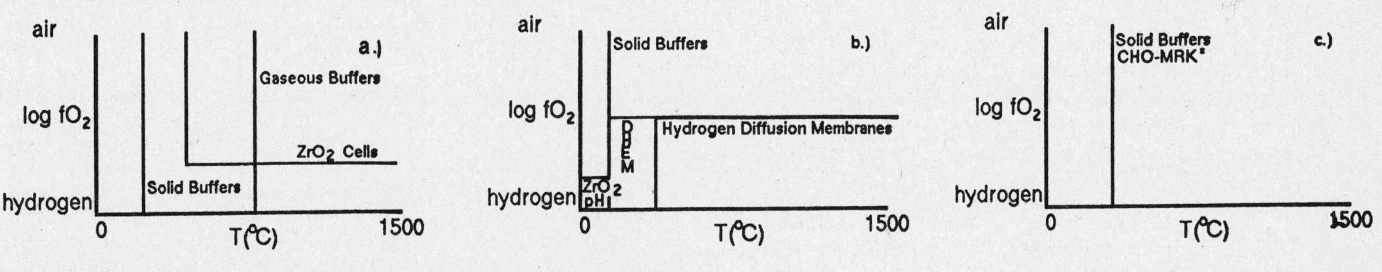

Figure 1—The polybaric slices in the diagrams below, in T-fO_2 space, summarize the P-T ranges over which these redox laboratory techniques have been utilized/calibrated: (2) 1 bar; (b) 0.2–5 Kbars; and (c) > 5 Kbars.

UNDERWOOD, M. B., and E. G. HATHON, University of Missouri, Columbia, MO, and M. T. MUCK, University of California, Santa Cruz, CA

Dispersal of Volcaniclastic Sediment in Aleutian Fore-Arc–Trench System, North Pacific Ocean

The Aleutian arc-trench is ideally suited for the study of volcaniclastic sediment dispersal because there are systematic changes in the composition of sediment sources from east to west. The easternmost (continental) segment of the active volcanic arc is constructed above a polymictic basement complex composed primarily of accreted graywacke–argillite belts, older plutonic-volcanic assemblages, and their metamorphosed equivalents. Basement sequences on the Alaska Peninsula (transitional arc) are mostly sedimentary, and the intraoceanic arc segment (Aleutian Islands) rests above oceanic crust (mostly volcanic flow rocks and volcaniclastic strata). Holocene sediments transported away from the arc system include materials from Quaternary volcanic eruptions and detritus eroded from underlying basement strata.

Petrographic analyses of deep-water sands, together with provenance linkage to the appropriate arc segment, can be used to reconstruct the general dispersal pathways for turbidity currents. Turbidite sands on the trench floor include the products of both axial (west-directed) and transverse (south-directed) flow. However, most of the sands cored from the thick sediment wedge within the central trench segment are enriched in volcanic glass shards, volcanic rock fragments, and fresh plagioclase, suggesting a dominance by transverse influx from the intraoceanic arc. This mode of trench sedimentation is somewhat paradoxical because existing bathymetric and sonar data show no large, through-going erosional conduits (submarine canyons) connecting the arc platform directly to the trench.

Fore-arc bypassing must occur, therefore, via unconfined turbidity currents rather than channelized flows. There are many bathymetric obstructions on the accretionary trench slope (e.g., thrust ridges, anticlinal ridges, and diapirs), but the relief is not enough to prevent turbidity currents from flowing over (or around) the obstructions.

Additional evidence favoring upslope sediment transport comes from the seaward slope of the Aleutian Trench, where seismic-reflection and sonar data show the presence of small basins formed through extensional faulting of the downgoing oceanic slab. Based upon echo character, many of these grabens appear to be filled with turbidites. More important is the fact that thin turbidite sands actually cored from the seaward slope show the same east-to-west variations in detrital mineralogy as do deposits on the insular slope, thereby suggesting a common provenance for common longitudinal corridors. Some of the coring stations on the seaward slope are elevated 1-2 km above the trench floor. Evidently, the largest arc-derived turbidity currents accelerate down the insular slope, traverse the trench floor, and still maintain enough kinetic energy to transfer small volumes of sand well up onto the opposite slope before back-sloshing onto the trench floor. Upslope transfer of sand is accomplished through a combination of large flow thickness and a true vertical shift in the center of gravity of the flow.

To test the validity of the above interpretation, we examined the behavior of saline density flows as they encounter a reversal in gradient, using a small (190 × 50 × 20 cm) plexiglass flume. We also used existing mathematical models pertaining to the hydrodynamic evolution of turbidity currents to estimate the boundary conditions required for upslope flow. Flow thickness appears to be the most important variable, and some bypassing of bathymetric obstructions should occur if the flow thickness is equal to 55–60% of the barrier height (assuming relatively small frictional coefficients). Upslope movement to positions 2 km or more above a trench obviously requires the generation of very thick turbidity currents.

Clay minerals in muds cored from the Aleutian arc-trench system were analyzed using X-ray diffraction and scanning electron microscopy coupled with energy dispersive X-ray. These data show that the dispersal patterns for fine-grained sediment differ significantly from those associated with turbidite sands. Clay minerals eroded from continental sources bordering the Gulf of Alaska are dominated by illite and chlorite. As expected, percentages of mixed-layer clay minerals (derived from volcanic rocks and ash of the active Aleutian Arc) increase from east to west. However, contents of mixed-layer clay minerals (mostly smectite) in the fore-arc are significantly lower than smectite percentages for comparable muds deposited in the back-arc. Moreover, smectite contents on the insular slope are actually lower than values on the seaward slope, even though the insular slope is closer to the inferred source. We believe that the transverse outflux of volcanogenic mud is diluted within the near-bottom nepheloid layer, largely because of mixing with muds carried by contour-parallel bottom currents. Bottom circulation on the insular slope moves from east to west, whereas currents affecting the trench floor and seaward slope move in an easterly direction.

Collectively, our studies show that regional-scale dispersal patterns for volcaniclastic sediment can be difficult to predict in deep-marine environments; moreover, many different physical processes must be taken into account when interpreting the total sediment budget.

UNLU, T., Mineral Research and Exploration, Ankara, Turkey

A New Approach on the Genetic Model of Divrigi Region Iron Ore Deposits, Central Anatolia, Turkey

Within the Inner Tauride suture zone, defined by Robertson and Dixon, 1984, based on a model by Sengör and Yilmaz, 1981 the iron ore deposits of Civrigi region are located. This zone, situated between Kirsehir Block and the Eastern Tauride Block, is composed of an ophiolitic complex, which is mainly represented by serpentinites. The complex is in contact with a composite silisic pluton of quartz-syenitic to dioritic composition, and a formation of limestones (Zeck and Unlü, 1987a, Zeck and Unl; 1987b, Zeck and Unlü, 1988). In the previous studies, these iron deposits have been described as contact metasomatic calc-skarn (Bottke, 1981) and typical contact metasomatic deposits (Petrascheck and Pohl, 1982).

According to the ore microscopic studies, the magnetite of massive magnetite ore contains many silicate inclusions. Further phase is the differentiation of magnetites and silicates, and the enrichment in magnetite content. This phase, established by multivarial geostatistical analysis, is geochemically characterized by Si versus Cr and Fe expressions.

Disseminated primary magnetite occurs together with chromite in the serpentinized ultrabasic rocks (host rock). Fe and Cr differentiation in the secondary magnetite in the host rocks, shows a negative correlation among these elements. Serpentinization effect becomes clear with the geostatistical expression of Fe versus Mg, Cr and Ni.

Hydrothermal phases are generally characterized by various element associations of mobile elements, and are verified by rare earth element geochemistry, and Rb/Sr isotopic studies. These phases in the same time are the cause of alterations in the rocks (for example in the host rock). Sulfide mineralizations, which consist of mobilized Fe, Co, Ni and Cu elements of mafic and ultramafic rocks are worth attention. Meanwhile, Fe, Cu, Ni, and Co bearing primary sulfides are also observed in the ultramafic rocks.

Age of the granitic rocks in the mineralized area has been determined as 110 ± 5 Ma by Rb/Sr isotopic studies. The magma was intruded subsequently to the obduction of the ophiolite complex which is thus suggested to have taken place more than 110 ± 5 Ma ago (Zeck and Unlü, 1987a, Zeck and Unlü, 1987b, Zeck and Unlü, 1988). Serpentinization is earlier than the magmatic intrusion and probably related with the tectonic emplacement.

On the contrary of Klemm's (1960) genetic model of pneumatolitic-metasomatic skarn type iron mineralization, related with granitic rocks, this new model implicates that the Divrigi region iron ore deposits may be interpreted as a formation by the combination of serpentinization of ultrabasic and basic rocks, and the further alterations of various hydrothermal phases (Unlü and Stendal, submitted, Unlü and Stendal, in preparation).

Thus, as a result of interpretations of various laboratory methods, mentioned above, together with field observations, it will be postulated that the formation conditions of ore deposits can be clearly understood.

UNRUG, R., Wright State University, Dayton, OH

Early Proterozoic Eburnean Continental Crust of South–Central Africa

Two large areas of Early Proterozoic crust formed by magmatism of the Eburnean tectono-thermal event are present in South-Central Africa: The Bangweulu terrane of NE Zambia centered on 10°S, 30°E, and the west-central region of Angola centered on 13°S, 15°E, with "regional granites" and associated rocks. In both areas low initial Sr isotope ratios suggest new continental crust formation.

The overall similarity of the evolution of the Bangweulu terrane and the Eburnean terrane of west-central Angola is remarkable. Both regions include gneisses and migmatites of Eburnean age formed of older protoliths and narrow belts of low grade metasediments deformed and metamorphosed in the Eburnean tectonothermal event. The latter occur as rafts and synclinal keels among rhyolitic to andesitic volcanics and related syntectonic granitoids, which are followed by post-tectonic porphyroblastic granites (Mansa and Mambwe granites in the Banweulu terrane, Quibala granites in Angola). Unmetamorphosed sedimentary covers (Mporokoso Group in the Bangweulu terrane, Oendelongo group in Angola) overlie the igneous rocks and metasediments.

The Banweulu terrane is separated from the Tanganyika shield by the Ubendian mobile belt, formed in the Eburnean tectonothermal event and repeatedly reactivated as a shear zone since. Middle and Late Proterozic mobile belts developed around the Bangweulu terrane: the Middle Proterozoic Kibaride foldbelt in the NW, the coeval Irumide foldbelt in the SE, the Late Proterozoic Lufilian foldbelt in the southwest. During Middle and Late Proterozoic tectonic events the Bangweulu terrane was subject to minor shearing only.

The evolution of the west-central Angola Eburnian terrane was similar: it remained undeformed since its formation in the Eburnean tectonothermal event. The Eburnean terrane of west central Angola forms a rigid block between the Mayombian and West Congo, Kaoko and Damaran foldbelts that rework to a large degree older continental crust. The SE margin of the Eburnean terrane of west-central Angola between 11°S, 18°E, and 15°C, 14°E shows a pronounced southwest–northeast foliation trend (Andula-Matala region). Here, Archaean protolith gneisses intruded by Eburnean granitoids may form another mobile belt.

Proterozoic mobile belts of south–central Africa seem to follow repeatedly reactivated zones of weakness that originated in the Eburnean tectonogenetic cycle or earlier.

UPADHYAYA, R., V. AGRAWAL, and M. K. PANDIT, University of Rajasthan, Udaipur, India

Utility of Clustering Technique in Classification and Geochemical Stratigraphy of Volcanic Suites from Western India

Modern methods of geochemical analysis can generate large amounts of data quickly; their interpretation requires multivariate statistical techniques. Two major problems of igneous petrology, namely classification and geochemical stratigraphy, can be tackled efficiently by the use of multivariate statistical method of cluster analysis based on Euclidean distance between any two rock analyses as a similarity measure and agglomerative algorithm as the clustering techniques.

In fine-grained and glassy rocks where modal percentage cannot be determined accurately, their partitioning into conventional classes utilizing a few oxides at a time (e.g., TAS diagram) becomes problematical especially for samples falling on or near empirical boundaries of different classes. For a cogenetic suite of volcanics from Tavidar, Jalore district, Rajasthan (India), a solution was sought in clustering. Samples collected from an area of approximately 70 km^{+2} between latitudes 24°46'10''N and 24°45'10''N and longitudes 72°5'40''E and 72°10'29''E form a cogenetic suite belonging to the "Malani group" of upper Proterozoic age. The main rock types are basic, intermediate, and acid volcanics, many of which are indistinguishable in the field. Clustering of 79 sample analyses leads to a dendrogram having four major clusters, A, B, C, and D, with a threshold distance of 2. Conventional techniques lead to their identification as rhyolites (A), trachytes (B), andesites (C), and basalts (D). Rhyolite cluster (A) is further split into A_1, A_2 of potassi-sodic, and A_3, A_4 of sodipotassic rhyolites. The trachytic rocks are split into three subclusters of quartz trachyte (B_1), trachytes (B_2), and the low silica trachytes (C_2) indicating considerable chemical variations in these rocks. Basaltic rocks are split up into hawaiites (D_1) and mugearites (D_2) with high mutual minima possibly due to variable degree of weathering and phenocryst contents. Partitioning of igneous rocks of cogenetic suite using conventional nongenetic geological techniques is thus seen to coincide with their classification based on cluster analysis showing the complimentary in the present case.

The second case is concerned with overcoming the natural drawbacks of monotonous flat terrain, absence of continuously exposed vertical section, and isolated nature of outcrops from the northern extremity of Deccan trap sequence covering the Malwa plateau in building the flow stratigraphy through conventional methods. Cluster and discriminant analyses together can overcome them to a large extent.

The plagioclase-pyroxene basalts with rare phenocrysts of olivine exposed between 340 and 550 m MSL near Pratapgarh (74°40'E:24°05'N) overlie the Precambrian pre-Aravalli and Aravalli supergroup rocks. A major north–south-running fault passing in the north and a northeast-southwest-running fault to the south of Pratapgarh demarcate the study area into four sectors, A, B, C, and D. Clustering of major analyses based on 96 tholeiites and 2 alkali olivine basalts of this cogenetic population shows 12 major clusters consisting of four or more samples. Mahalnobis "d^{+2}" distances using the four variables SiO_2, MgO/FeO, CaO, and (Na_{2O} + K_{2O}) between various clusters and their subsequent tests for significance as "F" tests indicate the presence of only nine subgroups that are chemically distinct from each other. Two of them are quartz tholeiites with low CaO (<7%) (LCT), three each are quartz tholeiites (QT) and olivine tholeiites (OT), and one is samples that are transitional between QT and OT (<2% normative quartz or olivine).

A log according to the height of the traps as they occur in different sectors builds up the chemical stratigraphy of different flows showing a minimum of nine flows. Other workers have also reported presence of nine flow units among the Deccan traps in the Malwa plateau region including the present area and adjoining Banswara district, respectively, on the basis of detailed field investigations.

UPADHYAYA, R., M. W. CHAUDHARI and P. KATARIA, University of Rajasthan, Udaipur. Raj., India

Discrimination Between Basement (Archaean) and Cover (Proterozoic) Rocks Using Chemistry of Orthoamphibolites of Central Rajasthan

The Precambrians of Rajasthan constitute the Banded Gneissic Complex (BGC) of Archaean age comprising metaigneous rocks with subordinate metasedimentaries; Aravalli supergroup (< > 1900-2500 Ma) which is a thick metasedimentary unit with base marked by quartzite and metavolcanic rocks and Delhi Supergroup (< > 1400-1900) which is a metasedimentary unit with intermittent volcanics. Basic rocks form an important lithology in BGC and Delhis whereas they are restricted to noly basal Aravallis. Extensive tectonization involving basement and cover rocks has made basement-cover boundary (BCB) diffuse in most of the region, necessitating chemical discrimination between Archaean and Proterozoic rocks. Since Archaean geochemistry is distinct from that of other periods this approach holds exciting possibilities. This problem has been tackled here by developing diagrams and statistical functions from chemical data of orthoamphibolites of known ages. Subsequent extension of this study to other analyzed Archaean and Proterozoic orthoamphibolites leads to geochemical identification of BCB in Rajasthan.

Fifteen analyses of BGC orthoamphibolites and ten each of Aravalli and Delhis published by others were selected for statistical study by using the following screen: (1) analyses falling in the region of unaltered rocks in the CaO/Al_2O_3–$MgO/10$-$SiO_2/100$ triangular plot of Davis et al quoted by Schweitzer and Kronter (1985) and (2) discriminant function for identification of orthoamphibolites of Chaudhari et al 1987. The available orthoamphibolites of this region plot in subalkali (tholeiite) field in alkali-silica diagram. Even after making allowances for standard deviations it can be seen that SiO_2 and FeO are lower and TiO_2, Al_2O_3 and P_2O_5 are higher in basal Aravallis than in either BGC or Delhis.

The averages for BGC orthoamphibolites from various parts of Rajasthan are comparable and show low Mg numbers indicative of their fractionated nature. However, Mg numbers and S.I. values of orthoamphibolites show a rough increase in moving from south to north in BGC of southcentral Raj. These subtle variations may indicate cognate relationship between these amphibolites and/or different levels of exposure. The latter inference is also corroborated by concommittent increasing grade of metamorphism from middle amphibolite to granulite facies from south to north. Additionally these chemical relationships could also be related to partial melting which is so common in the Precambrian i.e. some of the amphibolites may represent residues of different degrees of melting as has been proved in various Archaean TTG terrains all over the world. This warrants a thorough study of the region (work along these lines is in progress).

On the basis of low mobility of TiO_2 and P_2O_5 during low grade metamorphism and low mobility of MnO during alteration as well as its systematic variation seen in rocks of present study, TiO_2–$P_2O_5 \times 10 -$ MnO $\times$ 10 triangular plot of Mullen 91983) has been used here for recognition of BCB in Rajasthan. A line separates Archaean from Proterozoic amphibolites of the present study in this diagram. The separation is quite good with a little overlap.

To elaborate upon this distinction recourse was taken to statistical methods. Eigenvalues and eigenvectors were extracted from variance-covariance matrix. The canonical variables so defined are mutually uncorrelated and hence ideal for plotting and discrimination. A line separates Archaean and Proterozoic amphibolites in the plot of canonical variables 2 and 3. The overlap is now negligible.

Conclusions: The present study attempts discrimination between Archaean and Proterozoic orthoamphibolites on the basis of their chemistry and its extension to recognition of BCB in Rajasthan. Amphibolites showing igneous parentage on the basis of field, chemical and statistical criteria were utilized. In TiO_2, $P_2O_5 \times 10$, MnO $\times$ 10 diagram a line separates Archaean from Proterozoic amphibolites with a little overlap which gets narrowed down in F_2 vs. F_3 plot utilizing all the major oxides. Figures developed for discrimination of Archaean and Proterozoic orthoamphibolites from Rajasthan seem to have wider applicability. The work along these lines is under way.

REFERENCES

Chaudhari, M. W., R. Upadhyaya, O. P. Goel, 1987, Discriminant function analysis of amphibolites: Ind. J. Earth Sci., v. 14, p. 142-151.

Mullen, E. D., 1983, $MnO/TiO_2/P_2O_5$: a major element discriminant for basaltic rocks of oceanic environments and its application for petrogenesis: Earth Planet. Sci. Lett., v. 62, p. 53-62.

Schweitzer, J., and A. Kroner, 1985, Geochemistry and petrogenesis of early Proterozoic intracratonic volcanic rocks of the Ventersdorp Supergroup, South Africa: Chem. Geol. v. 51, p. 265-288.

UPSON, C., Stellar Computer, Inc., Milpitas, CA

Role of Visualization in Geosciences

As numerical simulations are increasingly used as diagnostic tools by scientists and engineers, the emphasis is shifting from questions regarding numerical computability to those related to simulation comprehension. Up to now, the computational scientist's main concerns have been related to his ability to mathematically formulate the laws of nature and the translation of those forumulae into numerical approximations whose behavior is stable for the parameters of interest. As the numerical techniques used in this process become more and more common, and scientists in ever increasing numbers begin to extend their studies to transient, three-dimensional domains, they realize that a major impediment to their research lies in their ability to comprehend the results of their simulations. As the simulation size increases (due to larger and faster supercomputers and near supercomputers) and as the simulated complexity increases (due to improved numerical methods) the researcher soon finds himself unable to digest the tremendous volume of data produced.

The visualization of the simulation results can greatly reduce the "comprehension" burden on the scientist and thus minimize the amount of time required to understand the results of the numerical simulation. Three-dimensional computer graphics techniques such as volume rendering can rapidly give an overall view to a large, complicated three-dimensional domain, allowing the scientist to isolate regions that require more in-depth, quantitative analysis. The ability to view a portion of the computational domain in a variety of different ways frequently generates insight into either the fundamental mechanisms governing the simulation or at least stimulates a stream of more detailed questions regarding the simulation methodology.

This talk will formalize the major steps performed by the computational scientist while simulating a physical process. The usage of visualization techniques in each of these steps to speed up the comprehension process will also be presented for the analysis of simulation results.

The process of analyzing the simulation results can be viewed as either a postprocessing function, one that occurs after the simulation has completed, or as one that occurs concurrently with the simulation. As the volume of data produced from three-dimensional simulations grows, it becomes infeasible to store the results on disk for later postprocessing. This makes the possibility of interactive visual co-processing more attractive to the scientist. The goal of this technique is the ability to interactively guide the computation as it progresses, just as a pilot steers a flight simulator. The techniques needed for co-processing will be described and demonstrated during the presentation.

URAI, J. L., Instituut voor Aardwetenschappen, Utrecht, the Netherlands, and P. F. WILLIAMS, University of New Brunswick, Frederickton, Canada

On Formation and Significance of Fibrous Veins

Fibrous veins are commonly found in low-grade metamorphic terrains. They can be both syn-and antitaxial in origin, depending on material contrast between wall rock and vein material. Repeated crack-seal events are important in the formation of the fibrous morphologies; quite often the fibers are markedly curved but appear internally strain-free. This has led to the hypothesis that the growing fibers track the opening history of the vein (Durney and Ramsay, 1973). If this is the case, important information on the progressive deformation history of the rock may be obtained, and fibrous veins are often used as a tool by structural geologists. Unfortunately, there is often little attention paid to check the validity of the tracking hypothesis.

A simple test is to check if fibers indeed connect suitable markers such as layering across the vein. Recent papers have given examples of fibers not connecting markers, casting doubt on the general validity of the tracking hypothesis.

For this study, a number of slate samples containing antitaxial calcite veins were studied in detail in the optical microscope, using transmitted light and cathodeluminescence.

The veins studied apparently belong to one generation. Most of them contain well-developed fibers that are more or less undeformed-looking optically, with minor twinning and undulose extinction. Some are more massive, containing only a few large grains. However, they all show very similar cathodeluminescence zoning.

The fibers are often curved in shape and are commonly nonperpendicular to the vein wall.

Abundant marker layers allow checking if fibers connect markers across the vein. Although this is the case in most veins, a number of them were found with a large mismatch, implying that these could not have tracked the opening direction.

Alternative processes to produce curved fibers are crystal plastic deformation, which will in general produce significantly less lattice curvature than is suggested by the shape of the fibers; and recrystallization after a strong plastic deformation event, producing strain-free grains that inherit the shape of the old distorted grains. In the case studied, the second alternative is unlikely, because recrystallization is likely to have erased the cathodeluminescence zonation. The constant orientation of fibers in strongly curved veins suggests that plastic deformation was absent too, but no proof of this can be given as yet. TEM work is in progress to study dislocation configurations across areas of sharp changes in fiber curvature. In a number of cases, the veins are fractured and slightly pulled apart, with the fracture at high angle to the fiber axis. In these locations, the fibers change sharply, connecting markers across the secondary fracture, showing that fibers can develop a curvature when growing at a new fracture with a new opening direction.

The data are at least consistent with the interpretation that in most cases curved fibers indeed grew parallel to the opening direction. However, there are clear exceptions, with fibers growing in a direction closer to the normal to the vein wall.

A model explaining both types of behavior uses the following features and assumptions. The interface where growth occurs is nonplanar, with a certain roughness that is essential in controlling fiber morphology. Assuming the simple rooftop morphology shown in Figure 1, and that all crystal faces grow with the same velocity, an opening history as shown can be seen to introduce a curvature in fiber shape. Simulations using this model show that variations in the shape of the interface and growth velocities can produce quite natural-looking fiber morphologies. The model also predicts that fibers will grow straight and normal to the vein wall if the interface is smooth or only gently irregular.

To check the validity of the model under controlled conditions, a series of experiments are in progress in which polycrystalline sodium nitrate is grown in an opening crack, with growth rate being approximately equal to the opening rate. Opening direction of the vein can be varied, together with parameters such as opening rate and pressure gradient in the vein.

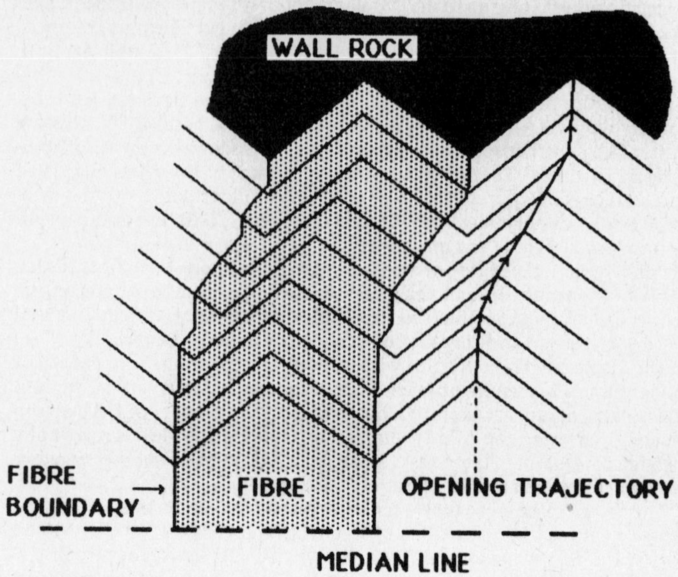

Figure 1.

URBANSKI, W. S., A. L. DEVERNOE, and M. S. WALKER, Intermagnetics General Corporation, Guilderland, NY

Application of a Magnetohydrostatic Centrifuge to Concentration of Galena for Lead Isotope Dating

The concentration of galena (SG = 7.58) for lead isotope dating is traditionally accomplished in three steps using halogenated heavy liquids for gravimetric concentration, an isodynamic magnetic separator for magnetic separation and hand-picking to obtain highest purity. This paper describes results obtained with an alternate means of concentration using a magnetohydrostatic (MHS) centrifuge operating with a non-toxic, water-based ferromagnetic fluid. A separation procedure is described that should be applicable to most sulfide ores when it is desirable to create a clean galena concentrate.

The MHS centrifuge consists of a rotating duct surrounded by a cylindrical magnet. Minerals to be separated are slurried with the ferromagnetic fluid and fed downward through the duct. An outward magnetic attraction of the fluid drives mineral particles radially inward against an outward centrifugal force of rotation. Separation is achieved by balancing these forces for a specific gravity between those of the minerals to be separated. Separation is possible because all minerals exhibit a unique range of specific gravity and magnetic susceptibility properties. The MHS centrifuge uses a combination of these physical properties to provide a specific gravity equivalent separation.

Clean galena concentrates are achieved in two passes through the MHS centrifuge. The crushed and appropriately sized samples were first separated at a specific gravity split point of 4.00. This pass removed the predominantly silica rich gangue from the sulfide mineralization. A second pass run on the sulfide mineralization at a specific gravity split point of 6.50 isolated the galena as the heavy product. Subsequent analysis performed on the galena as the heavy product. Subsequent analysis performed on the galena concentrate showed a grade of 99.93% galena with a recovery of 90.08%. The bulk sample was graded at 68.73% galena.

As the gerromagnetic fluid is water based and non-toxic, each mineral separate is easily washed with water. By using this method the researcher may eliminate exposure to halogenated heavy liquids, reiterative passes through an isodynamic magnetic separator and hand picking through a large residual sample.

URRUTIA FUCUGAUCHI, J., Universidad Nacional Autonoma de Mexico, Mexico

Chaotic Dynamics and Origin and Evolution of Pull-Apart Basins

Strike-slip motion along systems of lateral faults with discontinuities (en echelon or sharp bends) results in extension and subsidence at releasing bends or in uplift and deformation at restraining bends. The resulting structures are known as pull-apart basins and pressure ridges, respectively (or simply pull-apart structures for short). Pull-apart structures are common at transform plate boundaries and in intraplate tectonic settings such as the San Andres–Gulf of California right-lateral transform system, the Polochic–Motagua left-lateral transform system, or along central Mexico (Trans-Mexican volcanic belt). Despite the widely varying tectonic settings and scales, pull-apart structures are characterized by a remarkably constant geometry, with generally rhomb-like shapes and a half-length to width ratio of about 1.6. A simple mechanism to account for the origin and evolution of pull-apart structures is proposed and discussed; it involves block rotation as the main factor in the origin and evolution. The relationships examined allow us to briefly discuss similar relations involved in widely different fields such as those of spiral growth patterns of plants and animals, those used in design of paints and buildings, and those arising from geometric progressions (Fibonacci series). The relationship found for the half-length to width ratio of pull-apart structures (the so-called golden section of arts and architecture) also arises in certain chaotic systems where randomness presents an underlying geometric form. Simple deterministic systems that despite containing few elements may still generate random behavior (i.e., chaos), yet they suggest causal relationships.

UTADA, M., University of Tokyo, Tokyo, Japan

Sequence of Zeolitization in Green Tuff Region, Miocene Back-Arc Basin of Japanese Arc-Trench System

In marine and freshwater environments, some types of zeolitization are mainly controlled by temperatures. Therefore, the geothermal gradient can be roughly estimated by manner of zoning of zeolite and related mineral zones.

The Green Tuff Region in Japan had been a typical back-arc basin through Neogene period. It is characteristic that volcanic activities were violent and produced a large amount of volcanic rocks and volcaniclastic sediments which attain to about a half of the whole sediments. Briefly speaking on the history of this basin, the Green Tuff basin was formed in the early Miocene as a back-arc basin of new arc-trench system. In early Miocene, a part of the basin was probably situated on land and the remainder was shallow marine. A transgression in middle to late Miocene extended over the whole basin, and thick sediments were piled up. Before a retrogression, the basin was divided into several terrains that were distributed subparallel to the trend of the Japanese Islands. A specific terrain uplifted on land since latest Miocene, while others remained in shallow marine until the Quaternary.

A remarkable zeolitization of the sediments, especially vitric materials, occurred near the periphery of submarine volcanoes in a rift zone along the probably volcanic front in middle Miocene. The hydrothermal activities related to felsic volcanisms produced Kuroko-type ore deposits on the seafloor and formed clay and zeolite zones in the surrounding sediments. The typical zoning of alteration zones is arranged from the center to the periphery as follows: K-feldspar→sericite–chlorite→mixed layer clay→smectite→analcime→mordenite zones.

A regional zeolitization by diagenesis progressed in the whole basin through the Neogene period. The burial depth of sediments attained a maximum in late Miocene to Pliocene except in a special terrain. The diagenetic zeolitization is typically found in felsic to intermediate volcaniclastic sediments, and the following five zones are recognized in descending order from the shallower to the deeper: fresh glass→clinoptilolite → mordenite → analcime–heulandite → laumontite → albite zones. Vertical zonings can be grouped into four types, according to combination of zones (Utada, 1970). The upper and lower fronts of zeolite zones were at different burial depths in each type, but they were probably formed at nearly the same temperatures, which were measured in drillholes of several oil fields (Huzioka and Yoshikawa, 1968; Iijima and Utada, 1971; Aoyagi and Kazama, 1980; Sasaki et al., 1982). On the other hand, the combination of zones may be partly caused by chemical differentiation controlled by temperatures. Therefore, the classified four types of zoning were probably differentiated by areal variation of geothermal gradient in the Neogene period. It is noteworthy that each type is distributed in a terrain subparallel to the trend of the Japanese Islands. The terrain having the highest geothermal gradient was along the probable volcanic front at the time.

In progress of diagenesis, a large number of igneous masses intruded and made metamorphic aureoles in the surrounding sediments. Seki et al. (1969) first clarified a contact metamorphism containing amphibolite, actinolite greenschist, and zeolite zones around a huge intrusive complex at Tanzawa Mountain. However, common zoning surrounding small intrusive masses is arranged from the contact to the periphery as follows: propylite → laumontite → analcime → mordenite → clinoptilolite zones (Utada, 1973). This zoning is rather similar to that of diagenesis, but both are different in morphology and dimension. The zonings can be grouped into three types according to combination of zones, but each type has no clear character in distribution.

A number of calderas have been formed in the terrain uplifted along the volcanic front since late Miocene. A characteristic zeolitization occurred in caldera-filling sediments. The zeolite zones are arranged from the top to the bottom as follows: fresh glass → clinoptilolite → mordenite → laumontite → wairakite zones. This zoning is also similar to that of diagenesis, but wairakite appears as the highest-grade zone of this zeolitization instead of albite zone of diagenesis. The morphology and dimension of alteration zones is rather similar to those of alteration around intrusive masses mentioned above. This zeolitization probably occurred in very shallow places by thermal effects of uplifting program.

The zeolitization in active geothermal areas is the last one in the Green Tuff Region that had wholly uplifted on land in Quaternary period. Active geothermal areas are concentrated in several volcanic zones, especially in a terrain along the volcanic front of the Japanese Arc-Trench System at present. Calcium zeolites such as wairakite, laumontite, yugawaralite, stilbite, heulandite, chabazite, and others are characteristically found from active geothermal areas, though zeolitization is rather local.

Thus, various kinds of zeolitization have occurred in the Green Tuff Region through the Neogene period. The sequence indicates a change of geothermal conditions in each terrain and the distribution seems to be closely related with the tectonics of the Japanese Arc-Trench System.

REFERENCES

Aoyagi, K., and T. Kazama, 1980, Sedimentology, v. 27, p. 179–188.

Huzioka, K., and T. Yoshikawa, 1969, Journal of Japanese Association of Petroleum Technology, v. 34, p. 145–154.

Iijima, A., 1978, *in* Sand and Mumpton, eds., Natural zeolites, occurrence, properties and use: Oxford, Pergamon Press, p. 175–195.

——— and M. Utada, 1971, Advances in Chemistry Series, v. 101, p. 342–349.

Sasaki, A., N. Fujioka, and K. Huzioka, 1982, Journal of Japanese Association of Petroleum Technology, v. 47, p. 367–373.

Seki, Y., Y. Oki, T. Matsuda, K. Mikami, and K. Okumura, 1969, J. Jap. Assoc. Min. Petr. Econ. Geol., v. 61, p. 1–24 and 50–75.

Utada, M., 1970, Sci. Paper Coll. Gen. Educ., Univ. Tokyo, v. 20, 191–262.

——— 1973, Sci. Paper Coll. Gen. Educ., Univ. Tokyo, v. 23, p. 167–216.

V

VAI, G. B., Università di Bologna, Bologna, Italy

Pre-Jurassic Strike-Slip Rift Pulses in Circum-Mediterranean Western Tethyan Realm

Five main rifting cycles supported by different types of field evidence (including magmatism, sedimentation, and tectonics) have been recognized in the circum-Mediterranean area (western Tethys realm) before the onset of the Jurassic Tethys proper. They occur roughly in late Cambrian–early Ordovician, early Silurian, late Devonia-Dinantian, early-mid-Permian and mid-Triassic times.

Each of these rifting cycles starts developing from intracontinental, mainly shallow-marine conditions evolving usually to basinal environment of quite different bathymetry (Paleozoic CCD has been proved to be much less deep than post-Liassic one).

Minor suboceanic seaways cutting across and fragmenting the Gondwana–South European plate are expected to have intermittently developed following the early Paleozoic rift cycles.

The long mid-Paleozoic (Devono-Dinantian) rifting cycle might possibly have resulted in a wider system of suboceanic seaways connecting a Paleotethys Ocean to the east of Gondwana with the Rheic Ocean to the west of it. The birth of Paleotethys or its westward spreading would have been directly linked with the gradual death of the Devono-Mississippian Rheis Ocean. A complex, large-scale, dextral transform system would have connected the two respectively expanding and contracting oceans. Fragmentation and mutual displacement of microblocks in the connection area (the circum-Mediterranean area and Southern Europe) would have resulted from strike-slip, oblique rift pulses with coeval, local contractional and extensional processes.

A major change of the Gondwana rotational vector would have produced the mid-late Carboniferous transpressional strike-slip Hercynian orogen. This is a good explanation of the peculiar thermal regime of the Hercynian diastrophism all over Europe.

The previous scenario does not support the hypothesis of a very wide (more than 3000 km) Paleozoic ocean separating, inside or outside the Alps, Northern Europe from Southern Europe or Gondwanaland. This idea is mainly based on poor paleomagnetic data and is heavily contradicted by paleobiogeographical and geological data. Furthermore, it discards the dramatic shortening of the Hercynian orogen all over Europe.

The late Paleozoic to early Mesozoic history of the circum-Mediterranean area appears to repeat during early-mid-Permian and mid-Triassic the cyclic recurrence of the early Paleozoic oblique rifts, culminating in the early Jurassic strike-slip rift and subsequent narrow opening of the Tethys proper.

To avoid confusion in terminology, it is suggested that we use prefixes referring to age, bathymetry-physiography, and geotectonic pertinence for qualifying the general term Tethys as appropriate (e.g., shallow Cretaceous T., basinal Devonian intracontinental T., oceanic Jurassic T., etc.).

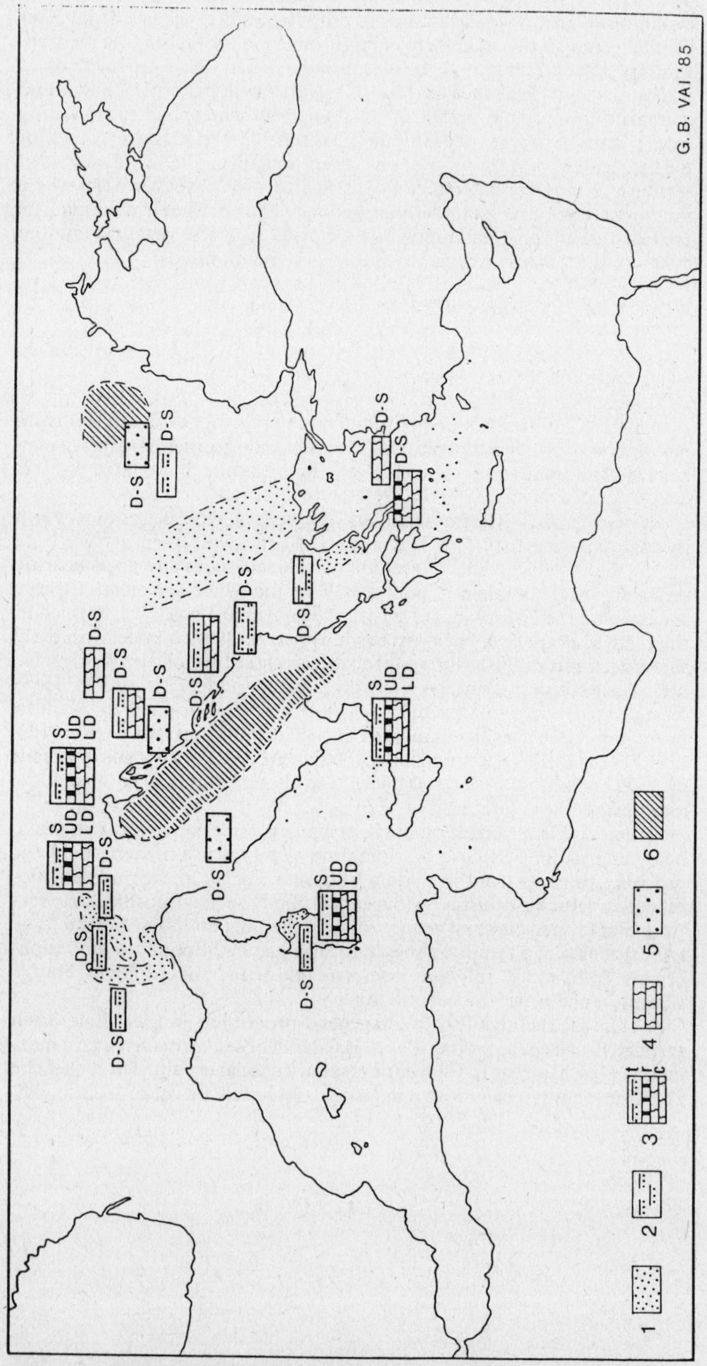

Figure 1—Simplified distribution of Dinantian to Serpukhovian tectofacies. 1, emergent erosional internide areas; 2, persistent flysch basins; 3, exposed transition from pre-flysch basinal carbonates (c) to pre-flysch carbonate basinal radiolarites (r) to flysch-stage turbidites (t); 4, persistent pre-flysch carbonate basins; 5, persistent fore-deep basins; 6, emergent foreland (and/or nonsedimentation) areas. LD, Lower Dinantian; UD, Upper Dinantian; S, Serpukhovian.

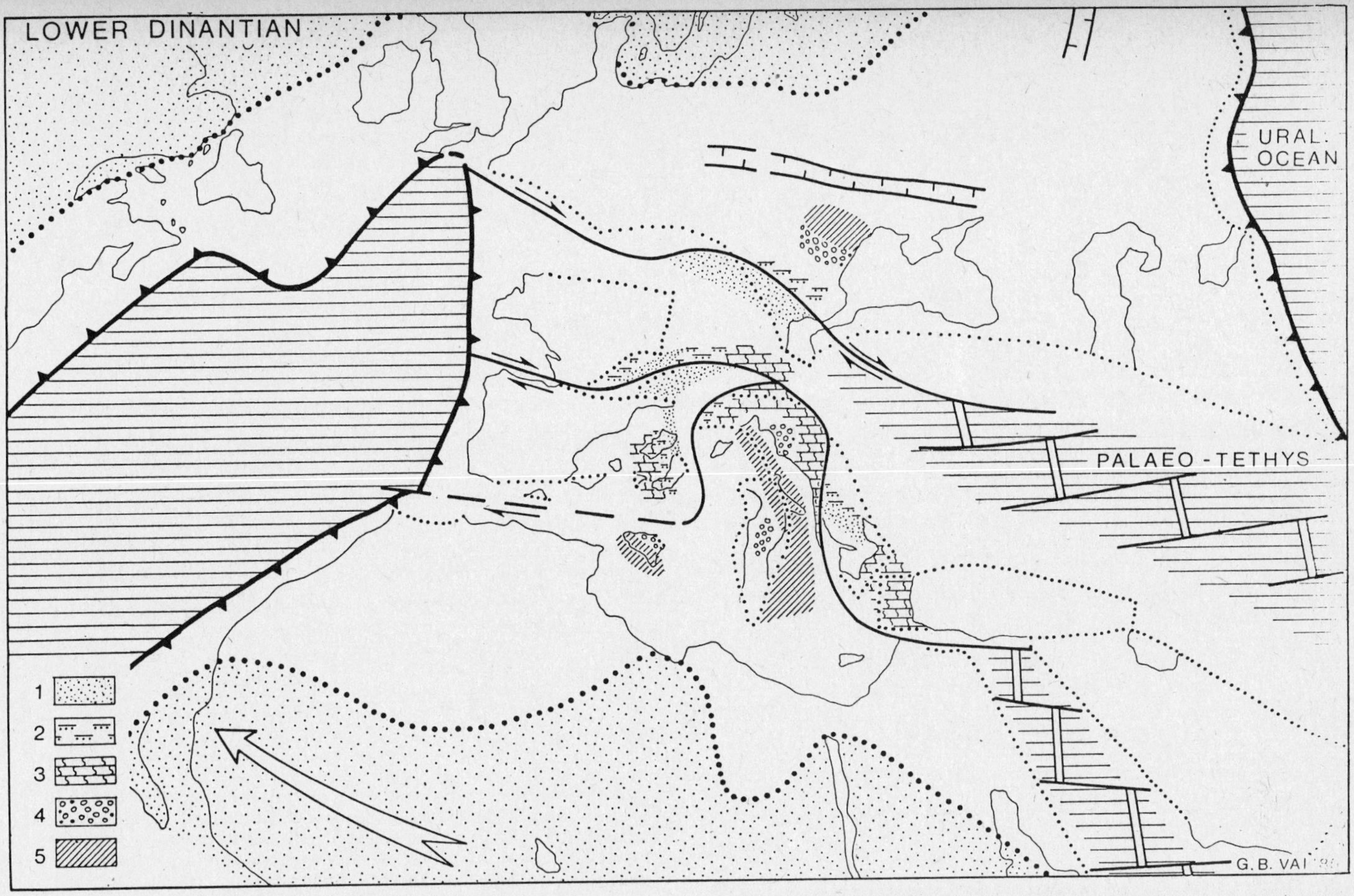

Figure 2—Inferred paleogeography and geodynamic setting of the circum-Mediterranean areas during the Lower Dinantian. 1, major land areas with coastline and minor emergent areas; 2, flysch basins; 3, pelagic carbonate basins; 4, fore-deep basins; 5, foreland areas.

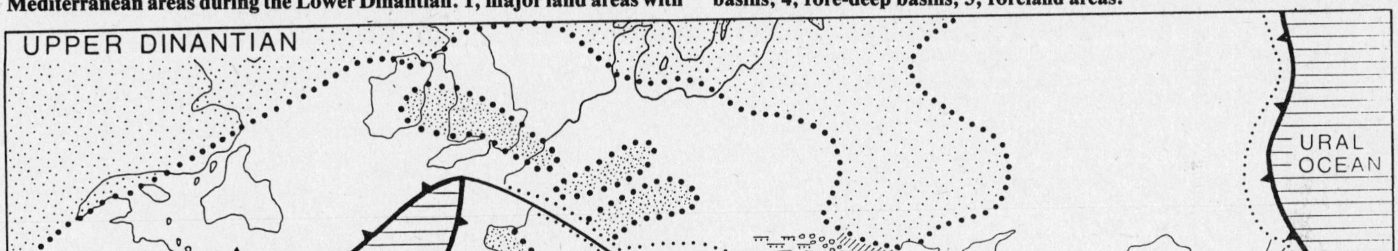

Figure 3—Inferred paleogeography and geodynamic setting of the circum-Mediterranean areas during the Upper Dinantian. 1, major land areas with coastline and minor emergent areas; 2, flysch basins; 3, radiolarite basins; 4, pelagic carbonate basins; 5, fore-deep basins; 6, foreland areas.

VAI, G. B., Università di Bologna, Bologna, Italy

Lower Crustal Sections in Italy: Exhuming Ages, Basement Relation, and Structural Style

Lower crust wedges and slices are known in the western Alps (Southalpine Ivrea–Verbano Zone and Austroalpine II Diorito–Kinzigitica Zone), in northern Calabria (Diorito–Kinzigitica or Polia–Copanello Unit of Catena Costiera and Sila nappe system) and in northeastern Sardinia (partly retrogressed granulite slices within the amphibolite facies metamorphics of the Axial Zone). The two first areas show evidence of strong Hercynian deformation and crustal consolidation followed by weak (Southalpine Ivrea Zone) to severe (Austroalpine and Calabride Units) involvement in the Alpine remobilization. The third area was preserved from Alpine orogeny, thus providing a standard case for comparison.

The exhuming age appears to be definitely Hercynian in Sardinia, almost likely late Hercynian for the Ivrea Zone, uncertain (Hercynian or/and eo- to meso-Alpine) for the II Diorito–Kinzigitica Zone, and possibly eo-Alpine for Calabria. One might therefore suggest Sardinia as a test site for checking the different interpretations of the Ivrea–Verbano (plus II Diorito–Kinzigitica and Calabria) thermal history (thermal overprint by Permo-Mesozoic rifts, Alpine diastrophic rejuvenation, slow cooling) some of which contrast markedly with basic geological data.

Common geotectonic features characterizing the lower crustal sections examined include (1) stretching, rifting, rotational block faulting, and formation of thinned, warm, partly buoyant continental lithosphere quite close to spreading axes, followed rapidly by (2) B-type subduction and subsequently by A-type subduction with opposite vergence. The inversion tectonics combined with the converging subductions provide the ideal condition for squeezing lower continental crust or even mantle wedges at the surface. However, some additional features can be recognized. Obduction of lower continental crust is not ubiquitous along this setting but appears to concentrate at or close to the intersection with transform faults, fracture zones, and oblique rift belts. Lithospheric inversion tectonics and oblique convergence appear as the most effective system by which lower crustal slabs can rise at the surface, owing to the incremental squeezing effect provided by transpression.

A Neogene obduction of the Ivrea Zone was suggested in recent years. It is based on the assumption that the Pogallo Line, instead of a late Hercynian thrust, was first a Liassic low-angle extensional fault. This assumption would require no Hercynian tectogenesis in the Southern Alps, which was, however, very severe indeed, involving shallow and deep crustal levels. Furthermore, it would require new difficult explanations for the Mesozoic to late Paleozoic north–south-trending facies belts in an area of east–west-trending Alpine structural axes. The north–south trend is in fact interpreted as an ancestral feature inherited from the Hercynian structure. In this view, an extensional Liassic reactivation of the late Hercynian Pogallo thrust would be welcome but of poor help for the obduction of the Ivrea Zone. With convenient geometry, however, such extensional fault would be relevant for the subsequent blind thrust of the buried geophysical "Ivrea body."

The recurrence in time of completely comparable settings during the Hercynian and Alpine geotectonic cycles in the area has to be remarked. However, of unique importance is the recurrence of obductions in almost the same place (Ivrea–Verbano and II Diorito–Kinzigitica). This implies a similar geological evolution at different times in the circum-Mediterranean area. [This paper was presented at the Verbania Workshop on "The nature of the lower continental crust," 23–27 May 1988 and will be published in a special issue of Tectonophysics.]

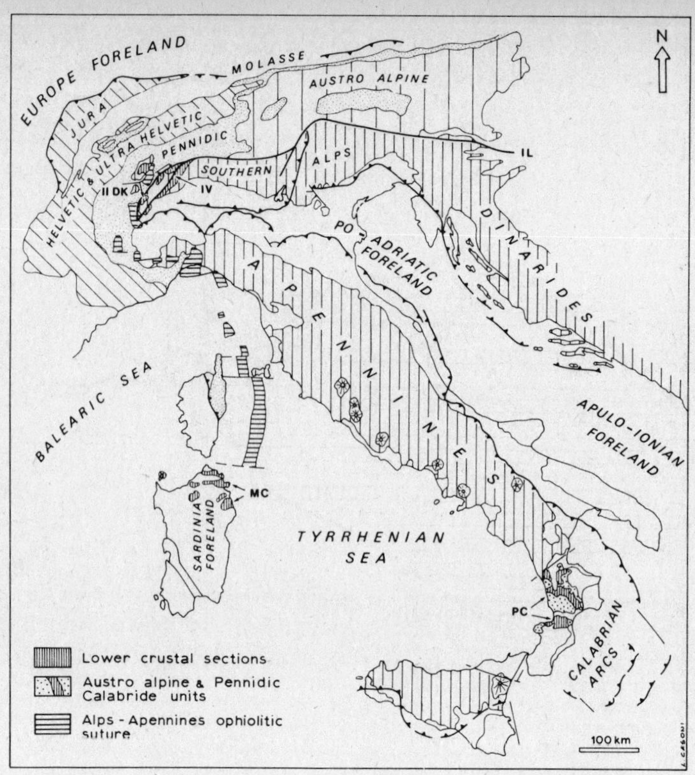

Figure 1—Schematic structural map of Italy with location of the lower crustal sections dealt with in this paper. MC, migmatite complex of northeast Sardinia; II DK, 2nd Diorito–Kinzigitica Zone (Austroalpine, western Alps); PC, Polia–Copanello Unit of northern Calabria; IV, Ivrea–Verbano Zone (Southern Alps); IL, Insubric Lineament.

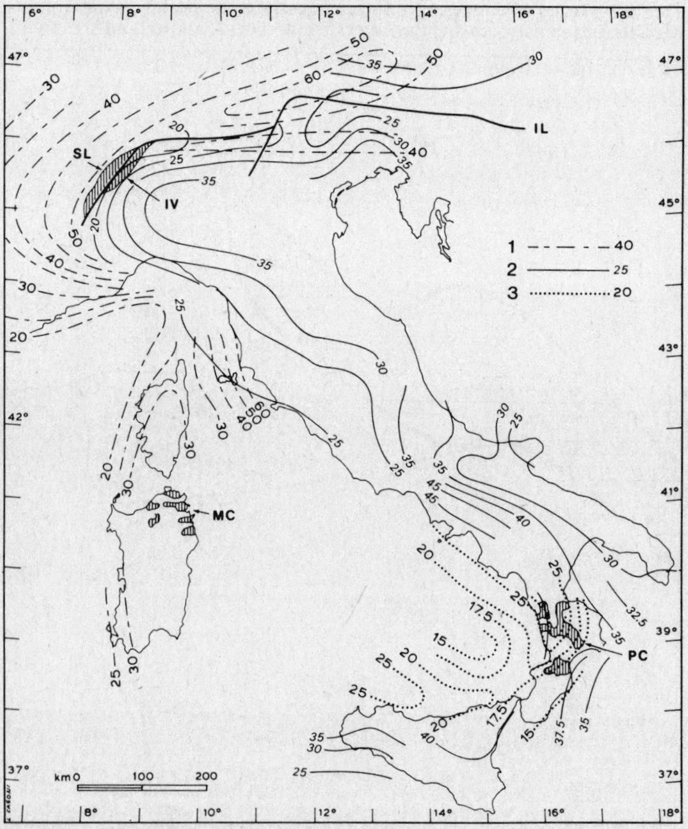

Figure 2—Extent and overlap of different Moho types compared with location of the Italian lower crustal sections discussed. SL, Sesia–Lanzo Zone; 1, European; 2, African; 3, Tyrrhenian Moho. Moho depths in km. (See Fig. 1 for other symbols.)

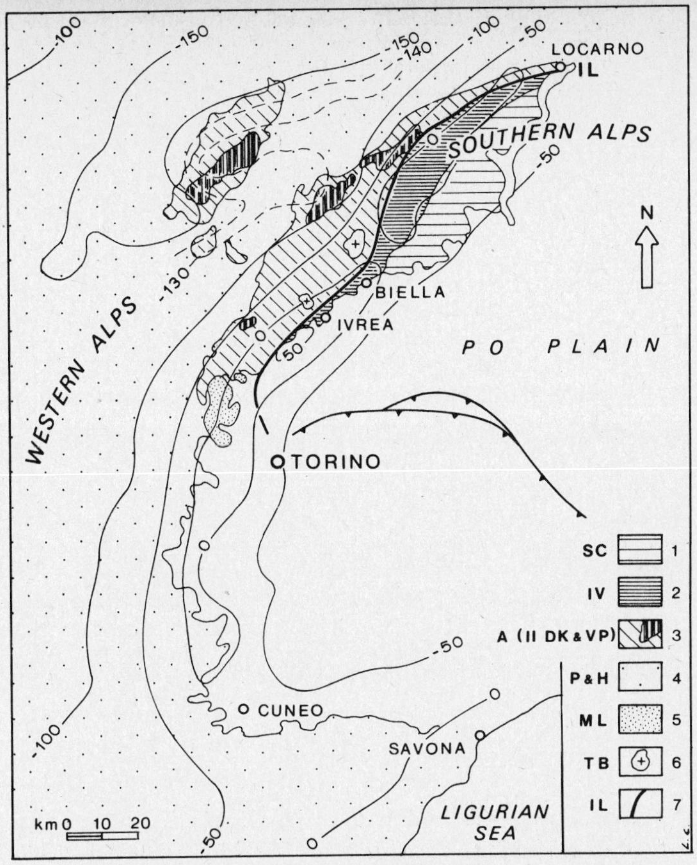

Figure 3—Bouguer anomaly map of the western Alpine Arc compared with distribution of Ivrea and II Diorito-Kinzigitica lower crustal sections. A, Austroalpine Units; VP, Valpelline "series"; ML, Lanzo Massif; TB, Traversella and Biella plutons (other symbols as in Fig. 1). Notice the low gravity relief produced by the flat-lying Valpelline outcrop.

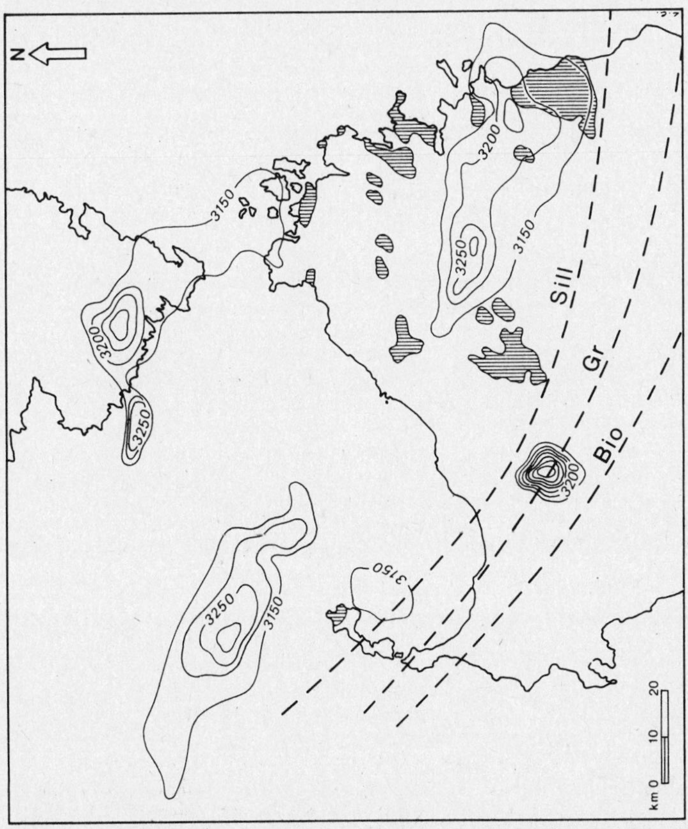

Figure 4—Main positive magnetic anomalies of northeast Sardinia compared with the migmatite complex outcrops (including partly retrogressed granulite-facies rocks). Contour interval: 50nT.

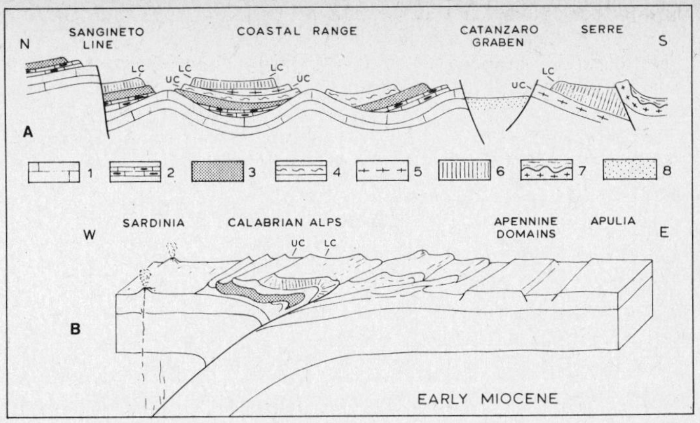

Figure 5—A: Present geometric relationships among the Calabrian Alpine units in the Coastal Range and the Serre. 1, Apennine units; 2 Frido Unit; 3,ophiolitic units; 4, Bagni Unit; 5, Castagna Unit; 6, Polia–Copanello Diorito-Kinzigitic Unit; 7, Stilo Unit; 8, Neogene to Quaternary Graben fill. B: Palinspastic cross-section from Sardinia to Apulia showing lower crust (hatching) already obducted upon upper crust and ophiolites (dark) in eo-Alpine times, and being involved in the later Apennine thrusting with opposite vergence (modified after Scandone, 1982).

VAIL, P. R., M. ULIANA, R. M. MITCHUM, Jr., and J. HARDENBOL, Rice University, Houston, TX

Global Correlation of Uppermost Jurassic and Lowermost Cretaceous Sequences and Systems Tracts

Do sea-level fluctuations at scales of 1–5 m.y. correlate globally? Sea-level changes within this frequency range are postulated by Haq et al. (1987) and Vail et al. (1987) to cause the third-order cyclicity that produces the discontinuity bound depositional sequences and systems tracts described in these articles. This paper discusses the correlation of the uppermost Jurassic and lowermost Cretaceous sequences and systems tracts between three widely separated areas: the Subalpine Basin of France, the Neuquen Basin of Argentina, and the Exmouth Basin of northwestern Australia.

For each area of study, the regionally distributed major transgressive and regressive facies cycles (MTRFC) and the globally distributed sequences and systems tracts are defined. Backstripping analysis is used to show the relation of tectonic subsidence to the MTRFC cycles. Geohistory analysis and correlation between the study areas are used to show the relation of sequences and systems tracts to global eustasy. In this manner, the long-term, high-magnitude, slowly changing tectonic signal can be separated from the short-term, low-magnitude, rapidly changing eustatic signal.

Sequence stratigraphy provides a new way to subdivide, correlate, and map sedimentary rocks observed in outcrop, core, well, and seismic data. A predictable succession of physical stratigraphic units called sequences, systems tracts, and parasequences are defined on the basis of internal stratal "geometries" and evolve in response to changes in shelfal accommodation. The sequence and systems tract boundaries are believed to be globally synchronous. Consequently, a global chronostratigraphic correlation scheme can be established from physical stratigraphic analysis that is independent of formation boundaries and rock type; biostratigraphy is essential to accurately date the physically defined sequence cycles. Since the sequences, systems tracts, and parasequences are defined within a chronostratigraphic framework, they are composed of genetically related depositional systems and lithofacies tracts. Each interval can be interpreted in terms of the depositional environments, lithofacies, and paleogeography that existed at specific time periods in the past. Depositional systems produce time-transgressive lithofacies units that are controlled by local episodic depositional processes. They may transgress across sequence stratigraphic interval boundaries.

VAITHIYANATHAN, P., Jawaharlal Nehru University, New Delhi, India

Basin Geology-Sediment Yield Relationships in the Peninsular Rivers of India

The major peninsular Indian rivers Godavari, Krishna and Cauvery together cover over 60% of the tropical Indian drainage system and 25% of the Indian drainage system. They show extreme variations in their sediment load. For example, Godaveri river has an annual sediment load of 170 million tons and ranks as the ninth largest sediment transporting river in the world. On the other hand neighboring rivers Krishna and Cauvery yield an extremely low annual sediment load of 4 and 1.5 million tons respectively.

Thus, the controlling factor in the sediment yield in the peninsular Indian rivers seems to be not the size of the basin but the basin geology.

To understand the extent of basin geological influence on the sediment transport of the peninsular Indian rivers the distribution of rock types in the drainage basin and their relative contribution to the annual sediment load are estimated. In Godavari basin granites and hard rocks (pre-Cambrian or older), Deccan traps (Tertiary), and sedimentary rocks (pre-Cambrian or younger) occupy 39, 48 and 13 percent of the basin area respectively and contribute nearly 16.51 and 13 percent of the total annual sediment load respectively. Thus the actual contribution from granites and hard rocks to the annual sediment load is significantly low compared to their areal distribution whereas the deccan traps and sedimentary rocks in the drainage basin contribute higher sediment load because of their high degree of erodability. Over 80% of the drainage area of Krishna and Cauvery are made up of granites and hard rocks as a consequence of which the sediment yield from these basins is very low.

The estimated erosion rates suggest that there are two major zones of erosion in the peninsular Indian region. The erosion rate is higher in the Central Indian region (>500 t/km^2/yr) whereas it is low in the South Indian part (<20 t/km^2/yr). Major part of Central India is made of deccan traps and Gondwana sedimentary formations while the southern region is made up of the shield area hard rocks thus suggesting that local geology controls the sediment transport in the peninsular rivers of India.

Over 95% of the sediment load in Godavari and Krishna and over 90% of the sediment load in Cauvery river are transported during the monsoon period. The maximum recorded sediment load during the non-monsoon season at the Godavari river mouth is 6.7% of the annual sediment load.

Sediment deposition rates at the river mouth are estimated by the excess Pb^{210} technique and are compared with the sediment erosion rates. The negative relationship between them indicates that when the sediment load carried by the river is very large as in the case of Godavari bulk of the sediments escape deposition in the river basin and enter the ocean. On the other hand in rivers carrying lower sediment load such as Krishna and Cauvery significant quantity of sediment load is deposited at the river mouth itself. Subsequent delivery of sediment to the Bay of Bengal would indicate that in the deltaic region of all these rivers localized erosion would cause a small supply of sediment to the adjacent marine environment.

VALENTINE, P. C., and J. A. COMMEAU, U.S. Geological Survey, Woods Hole, MA

Rutile in Gulf of Maine Quaternary Sediment—Accumulation of Fine-Grained, Authigenic TiO_2 from Sandstone and Shale Source Rocks

The Gulf of Maine, an embayment of the New England margin of the northeastern United States, is floored by shallow, glacially scoured basins that are partly filled with Pleistocene and Holocene silt and clay. These sediments contain 0.7 to 0.8 weight percent TiO_2, chiefly in the form of fine-grained rutile and anatase. Collectively, 11 basins in the gulf are estimated to contain 563×10^6 metric tons (t) of TiO_2 (to a depth of 10 m) in the U. S. Exclusive Economic Zone and 169×10^6 t in Canada, on the basis of analyses of surface sediment and of piston cores. The U. S. annually consumes approximately 1×10^6 t of TiO_2 (as rutile, leucoxene, ilmenite) of which 75 percent is imported. The inferred amount of TiO_2 in the basins is large, and the unconsolidated deposits lie in shallow water (300 m) and are relatively near the coast. No attempt has been made yet to beneficiate the fine-grained ore.

The Bay of Fundy and surrounding regions of new Brunswick and Nova Scotia are underlain by Carboniferous and Triassic basins that contain extensive exposures of red and gray clastic sedimentary rocks. These relatively soft rocks have been a major source of the fine-grained sediment in the Gulf of Main basins during and since the Wisconsinan Glaciation. Silt- and clay-size rutile crystals are present in the Carboniferous and Triassic rocks, in intertidal mud of the Bay of Fundy, and in the Gulf of Maine water column and bottom sediment. We suggest that the rutile formed authigenically in the clastic sedimentary rocks which underlie and surround the Bay of Fundy after the titanium-rich detrital minerals (ilmenite, ilmeno-magnetite) contained in them had dissolved. Following glacial and postglacial erosion and transport, the fine-grained rutile was deposited from suspension in the gulf basins. Additional sources of fine-grained, authigenic TiO_2 during the Quaternary probably include: sedimentary rocks of inferred Triassic age that underlie the Gulf of Maine; Ordovician, Silurian, and Devonian sandstones and shales in northern Maine and northwestern New Brunswick (that also may be the source of rutile found the fine-grained Presumpscot Formation of southern Maine); and reworked, coarse glacial debris, from all the foregoing sources, that is present in the Gulf of Maine, in New England, and in the Bay of Fundy region.

Rutile and its polymorphs are stable minerals, and their stability ensures that fine-grained, authigenic TiO_2 crystals will be passed through multiple sedimentary cycles involving both coarse and fine clastic rocks and that many of them eventually will be concentrated in fine-grained deposits. Gulf of Maine mud is comparable in TiO_2 content to typical shale. The Cody Shale (Upper Cretaceous) of Wyoming contains 0.60 weight percent TiO_2 as well as small rutile crystals that resemble those found in Gulf of Maine sediment, in the Presumpscot Formation, and in the Bay Fundy Carboniferous and Triassic clastics. A determination of whether the fine-grained TiO_2 in shales occurs primarily as detrital titanium minerals from igneous and metamorphic sources or as recycled, authigenic titanium oxides from sedimentary sources may assist in interpreting shale provenance and environment of deposition.

The geologic processes and settings outlined here provide a useful model for predicting and prospecting for unconsolidated deposits of fine-grained rutile. Primary features of a Gulf of Maine model would include: (1) coarse-grained sedimentary source rocks (especially arkose and graywacke) that originally contained an appreciable amount of detrital iron-titanium oxides and/or titaniferous silicates; (2) past intrastratal solution of host minerals and the formation of authigenic titanium oxides; (3) fine-grained sedimentary source rocks (shale) that contain recycled, authigenic titanium oxides from sandstone sources; (4) past chemical weathering of igneous and metamorphic source rocks, which contained iron-titanium oxides and titaniferous silicates, and the formation of authigenic titanium oxides in soils; (5) a sedimentary environment that includes vigorous erosion of source rocks, geographic segregation of bedload and suspended sediment, and a depositional area that receives fine-grained sediment predominantly from the source areas outline above.

VALETTE-SILVER, N. J., University of Maryland, College Park, MD, M. J. PAVICH, U. S. Geological Survey, Reston, VA, and F. TERA, Carnegie Institution, Washington, DC

Salton Sea Scientific Drilling Program: Story from ^{10}Be-^9Be Results

The Salton Sea Geothermal system (Southern California, USA) is located in the Salton Trough, a rift valley filled with Colorado River sediments that represents the on-land extension of the East Pacific Rise. Under the Salton Sea Scientific Drilling Program (SSSDP), the California State 2-14 well was drilled between September 1985 and April 1986. Cores were collected between 475 m and 3250 m and flow tests at 1850 and 3250 m were performed. We studied fluids (in situ and flowline, filtered and unfiltered, diluted and concentrated samples), rocks, cuttings, and ore deposits for their ^{10}Be and ^9Be concentrations.

^{10}Be is a cosmogenic radioactive isotope ($T_{1/2} = 1.5 \times 10^6$ y) formed in the upper atmosphere and is delivered to the earth's surface mostly by rain (flux ≈ 0.04 atoms/cm^2.s. For ^{10}Be, the chemical separation used is based on the isotope dilution technique, and the measurements were performed on more than 110 samples (fluids and solids) by acceleration mass spectrometry (Tandem Van de Graaff machine, University of Pennsylvania). ^9Be was analyzed by ICP on about 35 samples after chemical preconcentration, using ^7Be ($TB_{III1/2}$ = 54 days) to measure the yield.

In the solid samples, ^{10}Be concentration ranges from 9.44 to 259×10^6 atoms/g and ^9Be from 0.36 ppm to 2.52 ppm. The ^{10}Be concentrations in the geothermal waters ranges from $<1 \times 10^3$ to 4.9×10^6 atoms/g and ^9Be from 0.7 to 16.6 ppb. Compared to other natural waters, these values are at least two orders of magnitude higher for both isotopes. The ^{10}Be inventory measured in the core is three orders of magnitude higher than the steady-state inventory expected from rain deposition alone indicating that most ^{10}Be is inherited and that the pile of sediments hosting the geothermal field are young, less than a few million years old. In the core $^{10}Be/^9Be$ ratios exhibit a strong decrease at the depth of both aquifers, whereas a reverse trend exists for the fluids. The decrease from surface to bottom of ^{10}Be and ^9Be distribution coefficients (K_ds, 3333 to 48 for ^{10}Be and 727 to 393 for ^9Be), associated with the large-scale variations observed in the $^{10}Be/^9Be$ ratio and in ^{10}Be concentration, suggests enrichment of the fluids by suspended ultrafine material and/or leaching of both isotopes from the solid phase. This observation is in accord with the physicochemical parameters of the environment, low pH (<5.5), high salinity ($>2.5\%$), and high organic content, characteristics that enhance the solubility of Be.

In a graph of ^{10}Be K_d versus ^9Be K_d (Fig. 1), some of the samples plot on a line for which the ratio of K_ds is one. In theory, such a mixing line could arise from simple mechanical mixing of a homogeneous, ^{10}Be-rich sediment (such as the average core sediment containing $\approx 120 \times 10^6$ atoms/g ^{10}Be and 1.55 ppm ^9Be) with a Be-poor fluid (such as pure water). If this was the case, then all samples would have a ^{10}Be K_d/^9Be K_d ratio of one, irrespective of the range of the measured values for the individual isotopes. Deviation off the line and off the wedge enclosed by the two dashed-lines representing an uncertainty of $\pm 20\%$ would result from sampling of a solid phase not representative of the average sediment (such as drilling mud) or from the preferential leaching of one of the isotopes. The tick marks on the right side of the figure indicate the amount of suspended sediment required to slide a data point along the mixing line, for example from a K_d of 2000 (intermediate value among those observed) to the position of the tick mark. About 0.1% suspensoid in the fluid drops the K_d to 1000, half its original value. This figure helps us to check which samples have a chemistry that could arise only from contamination by fine suspensoid; these are the points plotting in the wedge. The samples plotting under the field, close to the ^{10}Be axis, have a fluid that is ^{9}Be-enriched or ^{10}Be-depleted perhaps through adsorption on surfaces. The samples plotting close to the ^{9}Be axis represents ^{10}Be-enriched fluids due to selective leaching of ^{10}Be out of the young sediments; this observation is consistent with the exogenic origin of ^{10}Be and its occupation primarily of adsorption and exchange sites. Apparent ^{10}Be enrichments could also arise from incorporation of ^{10}Be-rich fine clays. However, the fact that the levels of both ^{9}Be and ^{10}Be are orders of magnitude higher than the average value obtained for surface waters strongly suggests the involvement of leaching.

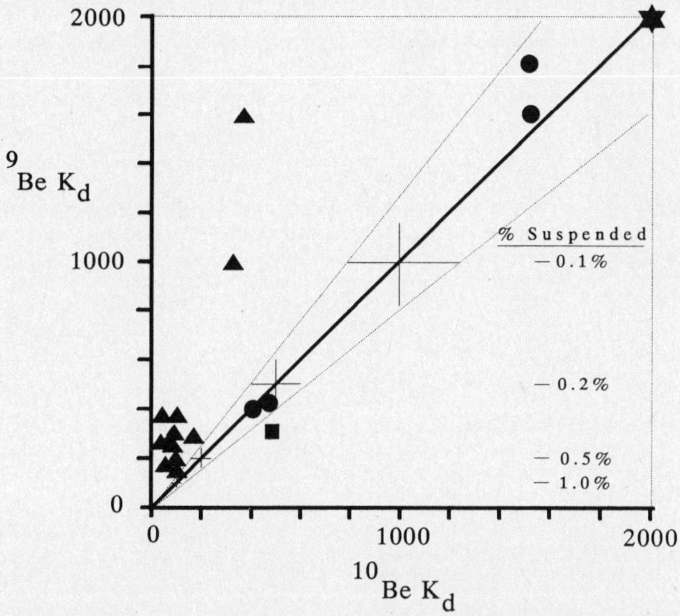

Figure 1—^{10}Be K_d vs. ^{9}Be K_d for samples collected in the Salton Sea Geothermal System. Solid line is the theoretical mixing line of suspended sediment and pure water. Crosses represent calculated K_ds using various percentages of average suspended sediment with pure water assuming ±20% uncertainty in average sediment composition. Circles, triangles, and squares are K_ds calculated from collected fluids. The star with K_ds >2000 represents a reasonable value in the Salton Sea system.

VALETTE-SILVER, N. J., University of Maryland, College Park, MD, F. TERA, Carnegie Institution, Washington, DC, and M. J. PAVICH, U. S. Geological Survey, Reston, VA ^{10}Be Behavior in Natural Systems

^{10}Be Behavior in Natural Systems

^{10}Be is a cosmogenic radioactive isotope ($T_{1/2}$ = 1.5 × 10^6 y) formed in the upper atmosphere and delivered to the earth's surface by precipitation with a flux of ≈0.04 atoms/cm^2.s. The isotope has been used in a few cases for dating young systems (< × 10^6 y) and, because of its exogenous origin, as a tracer of young sediment participation in geological processes (Tera et al., 1986). For the past few years, we have investigated the isotope's distribution in surface and near-surface systems. These include hydrothermal vents from the East Pacific Ridges, geothermal systems (USA, Mexico, Italy), rivers and groundwaters from the eastern United States, from which waters, sediments, rocks, and ore samples were analyzed. ^{10}Be concentration was determined by the isotope dilution technique using a Tandem Van de Graaff Accelerator as a mass spectrometer (Middleton, 1977; Klein et al., 1982).

In parallel, we have also undertaken an experimental investigation in order to determine the parameters responsible for the observed patterns of distribution for natural systems. These studies were greatly facilitated through use of ^{7}Be, a γ-emitter of $T_{1/2}$ = 54 days. Results from the experimental work show that ^{7}Be is rapidly adsorbed on sediments above pH = 5.3 (distribution coefficient, K_d, ≈ 10^4 at pH = 5.5) and that the equilibrium condition is nearly attained in about 1 hour. Iron was found to be a powerful scavenger of Be, more so than Al, at pH ≥5.5. At a concentration of <20 ppm, the readily hydrolyzed Fe was found to remove 75% in 1 hour and over 90% in 2 days of the radioactivity from the aqueous phase at pH = 5.5. As expected, the scavenging power of Fe, at equilibrium, is independent of the chemical form in which it is added. However, at pH >5.5, the initial rate of tracer removal is much faster in case of $Fe(OH)_3$ addition relative to salt. In contrast, when Al is added to river waters, it was incompletely precipitated even at pH >5.5, and as much as 40% of Al may remain soluble, apparently through complexation with organic matter. The effect of salinity on Be adsorbability is pronounced mostly at concentration >0.1 mol. At a given pH, the higher the salt concentration, the lower the K_d of Be, due in part to ion competition for exchange sites. In addition, we have observed that temperature, carbonates, and sulfates do not affect K_d in any pronounced manner. Reverse experiments of desorption from the solid have shown Be to be leached out of the solid phases at pH <5.3 and at high salinities.

Of all the parameters investigated, the pH appears to be the most important. Under alkaline conditions, Be was shown to be removed from water even by fine-grained quartz. A corollary of this is the solid-liquid distribution of ^{10}Be in the Potomac river (pH ≈6) and the Virginia Piedmont groundwater (pH ≈6.3), where the concentration in the sediments and waters are 10^9 and 10^3 atoms/g, respectively, corresponding to K_d ≈ 10^6. Contrasting examples that further substantiate the overriding role of pH are the hydrothermal systems: Cerro Prieto (Mexico), pH = 5.2; Salton Sea (USA), pH ≈5; and East Pacific Rise (13° and 21°N), pH <4, where K_d values are 5 × 10^3, <10^2, and <10^2, respectively. In a New Jersey site (Pine Barrens), a low pH of 3.5 is combined with very high organic content. As a result, ^{10}Be remains largely in the solution in correspondence to K_d <10. The role of organic matter is currently under investigation.

In the East Pacific Ridges hydrothermal deposits, the high ^{10}Be concentrations (>10^6 atoms/g) found in sulfides overlying thick sediment piles (Gorda Ridge, Guaymas Basin) compared to the low ^{10}Be concentrations (<10^6 atoms/g) observed in sulfides covering basalts can be interpreted mostly by leaching of the isotope out of the young sediment and its subsequent deposition with the heavy metal sulfides (Valette-Siler et al., 1987). Finally, our study of the Salton Sea hydrothermal system revealed a disparity in behavior between ^{10}Be and ^{9}Be. The hot, saline, and acidic fluids enrich the two isotopes in the fluids, presumably through leaching of the sedimentary pile. However, ^{10}Be is more effectively leached than ^{9}Be, a condition ascribed to the fact that ^{10}Be primarily occupies adsorption and exchange sites.

Although the mobility of ^{10}Be limits its potential as an accurate dating tool, it may enhance its value as a tracer for the participation of young sediments, or their derivatives, in geological processes.

REFERENCES

Klein, J., R. Middleton, H. Tang, 1982, Nucl. Instrum. Meth. Phys. Res., v. 193, n. 3, p. 601–616.

Middleton, R., 1977, Nucl. Instrum. Meth. Phys. Res., v. 144, p. 373–399.

Tera, F., L. Brown, J. Morris, S. Sacks, J. Klein, and R. Middleton, 1986, Geochimica et Cosmochimica Acta, v. 50, p. 535–550.

Valette-Silver, N. J., F. Tera, J. Klein, and R. Middleton, 1987, Journal of Geophysical Research, v. 92, p. 11364–11372.

VAN DEN BOGAARD, P., H. U. SCHMINCKE, Ruhr-Universität Bochum, Bochum, Federal Republic of Germany, C. M. HALL, and D. YORK, University of Toronto, Toronto, Canada

Tephra Deposits in Loess and Paleosol Sequences: Chronostratographic Markers of Terrestrial Climatic Conditions in Central Europe 158,000 and 214,000 Y. B.P.

Marine oxygen isotope studies provide most useful continuous records of the Pleistocene changes in $^{16}O/^{18}O$ fractionation, global ice volumes, and "apparent climate." Paleosols in terrestrial loess sequences, however, are direct evidence of interglacial climatic conditions that favored weathering, plant growth, and soil formation. The precise dating and correlation of climatic changes is a fundamental problem in both environments, which can be overcome by integrated tephrostratigraphic studies and single grain $^{40}Ar/^{39}Ar$ laser dating. Here, we present data on the age, composition, and stratigraphic and paleoclimatic context of the late Pleistocene Glees and Hüttenberg tephras, which were erupted by Wehr volcano in the East Eifel volcanic field (EEVF, FRG). Our results shed new light on the emplacement process of apparent excess ^{40}Ar in volcanic rocks, contribute to the controversial problem of the beginning of the second-last interglacial (oxygen isotope stage 7), and enable the correlation of marine and terrestrial stratigraphic records in the 150,000 to 250,000 y age ranges.

Glees tephra was deposited on loess *overlying* periglacially folded mafic tephra layers (Dachsbusch scoria cone) at its type locality. The tephra consists of base surge and minor fallout deposits with highly evolved, strongly silica-undersaturated phonolite pumice lapilli. Single grain $^{40}Ar/^{39}Ar$ laser analyses of 10 anorthoclase phenocrysts (0.5–1.5 mm in diameter) from essential lapilli yield argon isotope compositions that are colinear in isotope correlation diagrams, and a best-fit isochron age of 159,000 ± 15,000 y.B.P. with an initial $^{40}Ar/^{36}Ar$ ratio of 299.6 ± 15.6 (all errors at 1 sigma level). With an initial atmospheric $^{40}Ar/^{36}Ar$ ratio of 295.5, 8 out of 10 analyses yield an error-weighted average age of 158,000 ± 4000 y. B.P. (MSWD = 1.48) for the eruption and deposition of Glees tephra. Atmospheric ^{40}Ar contents typically range from 40% to 80%.

Hüttenberg tephra deposits *underlie* the Dachsbusch scoria cone deposits in the rim of Wehr crater but are also exposed at 7 other sites over 400 km^2 in the Middle Rhine and Eifel area. The tephra deposits are covered by humus-rich paleosol layers (A-horizons of paleosol profiles), but underlain by loess and pedogenic Bt-horizons wherever the base is exposed (6 sites). Hollows of tree trunks and branches are preserved exclusively in the overlying soil deposits, and leaf imprints document a plant association with present-day climatic needs.

The tephra comprises base surge, ash flow, and fallout deposits. Steep major and trace element gradients in composite tephra sections (18–85 ppm Th; 400–1200 ppm Zr; 0.04–1.16 wt% MgO) reflect the systematic evacuation of a strongly compositionally zoned magma chamber: The eruption started with phenocryst-poor magmas (<5 vol%) with highly evolved, nepheline-normative phonolitic magma compositions, and terminated with phenocryst-rich magmas (≥30 vol%) with mafic but quartz-normative trachytic compositions. All members of the Hüttenberg tephra, however, can be unambiguously distinguished from Glees tephra (and all other tephras in the EEVF) by their xenolith suite, phenocryst assemblage, and the chemical composition of its essential lapilli. The entire Hüttenberg eruption lasted a few days or weeks at most.

K-feldspar phenocrysts from late-erupted, phenocryst-rich Hüttenberg pumice lapilli show highly consistent argon isotope compositions and apparent ages. Fourteen analyses on samples from the near-vent type section yield a best-fit isochron age of 213,000 ± 4000 y B.P. and an initial $^{40}Ar^{36}Ar$ ratio of 292.7 ± 2.2 (MSWD = 1.00). With an initial $^{40}Ar^{36}Ar$ ratio of 295.5, 29 out of 33 phenocryst analyses from 4 sites yield an error-weighted average age of 214,000 ± 3000 y B.P. (MSWD = 1.26). Atmospheric ^{40}Ar contents typically range from 30% to 70%.

Analyses of K-feldspar crystals from evolved pumice lapilli confirm the results obtained from the mafic members of the Hüttenberg tephra, but also show the early erupted, highly differentiated Hüttenberg lapilli to be severely contaminated with older xenocrysts: >60% of the 105 crystals analysed yielded apparent ages well in excess of 213,000 y. These xenocrysts were picked up by the Hüttenberg magma prior to its eruption and are probably derived from older Pleistocene to late Tertiary intrusive rock bodies that are common in the crust beneath the entire East Eifel volcanic field.

The stratigraphic position and $^{40}Ar^{39}Ar$ laser age show that the Glees tephra of Wehr volcano was erupted and deposited during a loess-forming, glacial climatic episode 158,000 y B.P., which corresponds to stage 6 in marine oxygen isotope stratigraphies. The Hüttenberg tephra, in contrast, is sandwiched by biogenic-sedimentary A-horizons and pedogenic Bt-horizons of mature paleosol profiles. It shows primary pyroclastic bedding characteristics, a sharp contact to underlying loess deposits but diffusive contacts to overlying soil strata at all sites, indicating that it was erupted and deposited at the very beginning of an interglacial climatic period 214,000 y B.P. to 275,000 y B.P. The age and stratigraphic position of the Hüttenberg tephra show that interglacial climatic conditions following the third last major glaciation were only established at 214,000 ± 3000 y B.P. in terrestrial Central Europe. This age also coincides with the most prominent positive $\delta^{18}O$ excursion in orbitally tuned oxygen isotope curves of the 150,000 to 300,000 y age range.

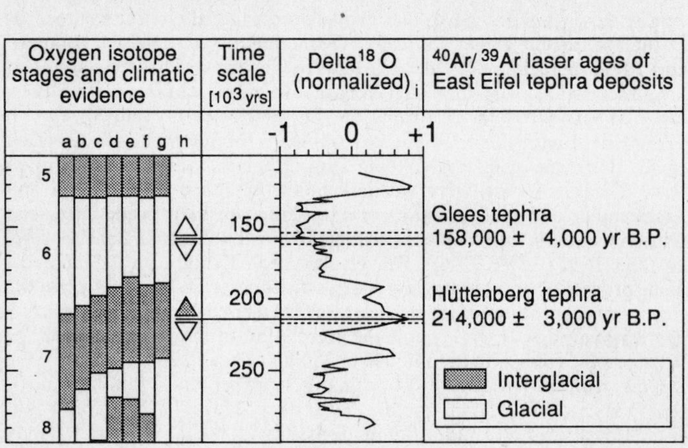

Figure 1—Terrestrial paleoclimatic evidence (h) and $^{40}Ar/^{39}Ar$ laser ages of East Eifel tephra deposits versus marine oxygen isotope stages from Herterich and Sarnthein (1984) (a,e), Shackleton and Matthews (1977) (b), Shackleton and Opdyke (1973) (c), Kominz et al (1979) (d), Morley and Hays (1981) (f), and Imbrie et al. (1984) (g). Orbitally tuned oxygen isotope curve (i) is from Martinson et al. (1987). Dashed lines indicate 1 sigma errors of tephra ages.

VAN DEN KERKHOF, A. M., and J. L. R. TOURET, Free University, Amsterdam, the Netherlands

Systematic Classification for Gaseous Inclusions in CO_2-CH_4-N_2 System

Water-free inclusions containing CO_2, CH_4, and N_2 and their mixtures occur in a number of geological environments, notably in lower crustal and upper mantle rocks. Only pure CO_2 (maximum homogenization temperature 31 °C) may be biphased showing L(iquid) and V(apor) at room temperature. All other compositions appear to be gaseous as they contain a single supercritical fluid (F). Therefore, they cannot be distinguished from open, empty cavities. But upon cooling, they show complicated phase transitions to liquid (L) and/or solid (S) in the temperature range from +31 to −180 °C. At low temperatures, inclusions containing CO_2-CH_4-N_2 fluids present a number of unusual features:

1. Possible transition S → F (liquid *or* vapor) upon warming ("sublimation"), without the occurrence of a three-phase equilibrium (S + L + V).

2. Successive homogenization in the same inclusion at intermediate (+31/−60 °C: L + V → F) then at low temperatures (below −70 °C: S + L + V → S + F).

3. Existence of metastable homogenization.

4. For binary mixtures (CO_2-N_2 and CO_2-CH_4), small variations in the fluid melting temperature and large decrease in the homogenization temperature compared to pure CO_2. Therefore, low-density mixtures can be easily confused with high density CO_2. In practice, the composition cannot be derived from phase transition temperatures, and it needs to be measured directly, notably by Laser excited micro Raman analysis.

Based on theoretical considerations, the study of many natural samples and experimental data, a unified classification system is proposed that can cover in principle any composition and density in the CO_2-CH_4-N_2 system (van den Kerkhof, 1988).

1. Two major types, H (for homogenization) and S (for sublimation) are determined from the final phase transition observed at rising temperature: L + V → F and S + V or S + L → F, respectively (where F denotes L or V).

2. Each of the H or S inclusions may show a succession of phase transitions, defining 9 subtypes (H1 to H5 and S1 to S4) according to Table 1.

Type H3 and H1 inclusions are characteristic for CO_2-rich and CO_2-poor fluids, respectively. H2-type inclusions are unique for CH_4-rich fluids; H4-type inclusions are unique for N_2-rich compositions. S2-type fluid inclusions are common for higher N_2-contents and lower molar volumes, but they may occur for both CO_2-N_2 and CO_2-CH_4 fluids for molar vol-

The final multi-phase assemblage upon warming = L+V (H-type)

	S+L	S+L+V	S+L or S+V	S+L+V	L+V	L or V
H1					... $T_h^{L/V}$	...
H2				... T_m	... $T_h^{L/V}$	...
H3			... T_i^{V}	... T_m	... $T_h^{L/V}$	...
H4		... T_{hs}	... $T_i^{V/L}$	... T_m	... $T_h^{L(V)}$	...
H5	... T_i	... T_{hs}	... T_i^{L}	(...) T_m	... T_h^{L}	...

The final multi-phase assemblage upon warming = S+L or S+V (S-type)

	S+L+V	S+L	S+L+V	S+L or S+V	L or V
S1				... T_s	...
S2			... $T_{hs}^{L/V}$	... T_s	...
S3		... T_i^{L}	... T_{hs}^{L}	... T_s	...
S4	... T_{hs}^{L}	... T_i^{L}	... T_{hs}^{L}	... T_s	...

metastable (S-type)

	L+V	L or V
S1 to S4	... $T_{hm}^{L/V}$	...

Type H3 and H1 inclusions are characteristic for CO_2-rich and CO_2-poor fluids respectively. H2-type inclusions are unique for CH_4-rich fluids;

umes below 65 cm^3/mole. H5-type inclusions (all of CO_2-CH_4 composition) are extremely rare and can be seen as a special type of H4; S3 and S4-type inclusions represent CO_2-N_2 fluids of very low molar volume (about 40 cm^3/mole). S1-type inclusions are mostly of very high molar volume, but sometimes extremely low molar volume.

Inclusions belonging to each subtype have been actually observed in natural samples. For binary mixtures (CO_2-CH_4 and CO_2-N_2), it is now possible to calculate the molar volume of the inclusions, knowing the temperature of the phase transition *and* the composition.

REFERENCE

Kerkhof, A. M. van den, 1988, The system CO_2-CH_4-N_2 in fluid inclusions: theoretical modelling and geological applications: Phd. Thesis, Free University, Amsterdam, 206 p.

VAN DER MERWE, S. W., University of the Orange Free State, Bloemfontein, South Africa, M. J. SLABBERT and A. L. D. AGENBACHT, Geological Survey, Upington, South Africa

Aspects of the Enigmatic Relationship Between Thrusting and Apparent Refolding in the Namaqua Mobile Belt, South Africa

The Namaqua Mobile Belt, in continental perspective, forms part of the 1000–1200 Ma Kibaran system of mobile belts and separates Archaean cratonic rocks in the east from Euburnian (2000 Ma) provinces towards the west (Blignault eastwards directed ductile thrusting which imparted a penetrative foliation and extension lineation on the metamorphites. A set of open folds with subvertical axial planes and fold axes parallel to the extension lineations, is superimposed on the thrust fabric and suggests shortening normal to the transport direction. This relationship is discussed with reference to the Hartbees River Thrust Zone (Figure 1).

Movement along the Hartbees River Thrust Zone occurred during the later stages of the Namaqua tectonogenesis. Lower grade metamorphic conditions prevailed; hornblende is replaced by chlorite. The thrust plane dips toward the north and the extension lineations plunge north-east. The thrust plane appears to be folded by north-east trending open folds with subvertical axial planes. Geometric analyses have shown that the axes of these structures are co-axial with the extension lineations. Local changes in the orientation of the lineations are met by similar changes in the orientation of the fold axes, even on outcrop scale, and the lineations are never folded. This relationship and the fact that these folds are tightest near the main locus of movement, dying out towards the top of the hanging wall, suggest that the folds and the thrusting are process related. The folded attitude of the thrust plane is therefore regarded as only a reflection of the original thrust geometry which consists of an arrangement of lateral, oblique and frontal ramps, in such a way that fold structures are formed. Shortening normal to the transport direction is locally present when the hanging wall is wedged between two lateral ramps, but the total strike length of the thrust, however is unchange.

Open folds in the rest of the Namaqua Mobile Belt probably formed in a similar way as those described above. The existence of a component of contraction normal to the transport direction, traditionally inferred for the Namaqua Mobile Belt, therefore questioned.

REFERENCES

Blignault, H. J., G. Van Aswegen, S. W. Van der Merwe, and W. P. Colliston, 1983, The Namaqualand geotraverse and environs: part of the Proterozoic Namaqua mobile belt. In: Botha, B.J.V., Ed. Namaqualand metamorphic Complex, Spec. Publ. Geol. Soc. S. Afr, v. 10, p. 1–30.

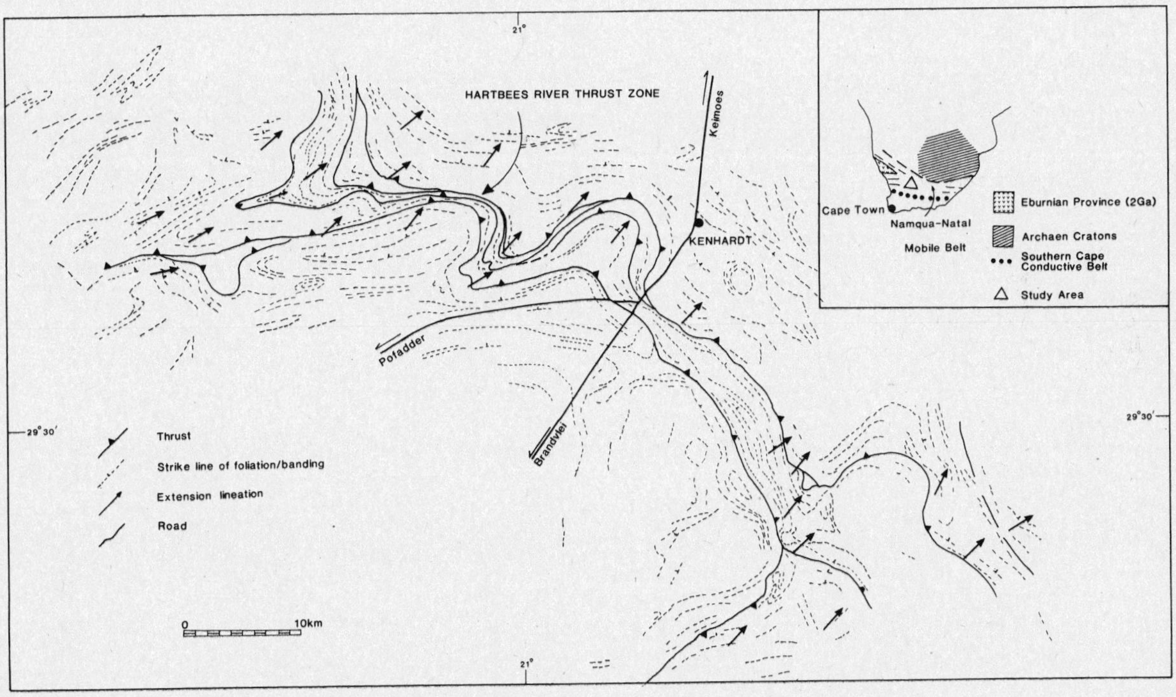

Figure 1—Structural map of the Hartbees River Thrust Zone, South Africa.

VAN DER VOO, R., and Wu FANG, University of Michigan, Ann Arbor, MI, and Qizhong LIANG, Yunnan Institute of Geological Sciences, Kunming, China

Magnetostratigraphy of Precambrian–Cambrian Boundary Section at Meishucun, China

A sequence of platform carbonates and phosphorites at Meishucun, near Kunming in southeast China, contains the Precambrian–Cambrian boundary and has been proposed as a global stratotype. Thermal demagnetization of 159 samples indicates that a majority of the rocks in the section have been strongly overprinted in recent geomagnetic fields. However, 57 samples revealed stable ancient remanent magnetizations with shallow antipodal directions to the north and south (mean declination/inclination 4.2°/+7.1°, α_{95} = 6.6°). The results show a zonation of at least eight reversals and yield a paleopole that is different from any paleopoles thus far obtained for Phanerozoic rocks of the South China block (paleopole at 68.8°N, 270.7°E). Further sampling of parallel sections is necessary, however, to confirm these reconnaissance results. A comparison of our preliminary magnetostratigraphy (see Fig. 1) with records obtained from Siberia, Australia, and the western United States shows that all sections are characterized by frequent reversals, but that detailed correlations are not yet possible.

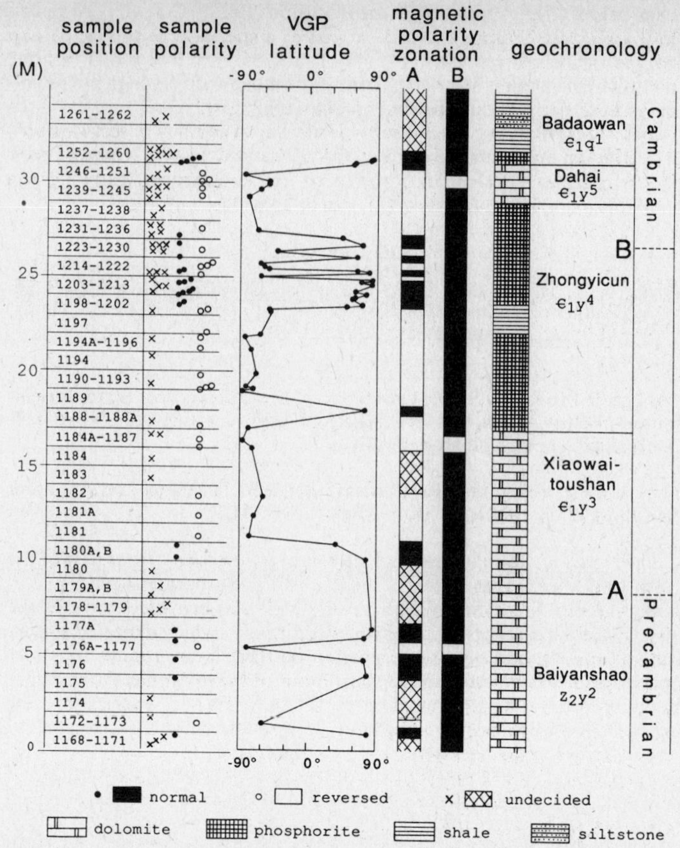

Figure 1—Magnetic polarity zonation derived from the characteristic directions of the lower Meishucun section. Formation names and positions A and B for the Precambrian–Cambrian boundary are from Luo Huilin et al. (1984) and the IGCP Working Group 29 recommendations, respectively. Stratigraphic position (in meters) is indicated on the left, followed by a column with our sample numbers; the sample polarity column contains symbols for remagnetized present-day field directions (crosses, not used), and characteristic normal and reversed directions (dots and open circles); the Virtual Geomagnetic Pole (VGP) latitude is plotted according to the method of Lowrie et al. (1980). The Column A magnetic polarity zonation is from this study, whereas column B is based on the very preliminary results of Liang et al. (1984).

REFERENCES

Liang, Q. Z., and many others, 1984, Report on the regional geological investigation in the Kunming District: Unpublished paper by the Yunnan Geological Bureau.

Lowrie, W., W. Alvarez, I. Premoli Silva, and S. Monechi, 1980, Lower Cretaceous magnetic stratigraphy in Umbrian pelagic carbonate rocks: Geophysical Journal of the Royal Astronomical Society, v. 60, p. 263–281.

Luo Huilin, and many others, 1984, Sinian-Cambrian boundary stratotype section at Meishucun, Jinning, Yunnan, China: Yunnan, China, People's Publishing House, 22 plates, 154 p.

VAN DER VOO, R., University of Michigan, Ann Arbor, Michigan

Paleomagnetic Constraints on Continental Motions During Late Precambrian

Much of the available geological evidence suggests that plate tectonic processes were operative during the Proterozoic, whereas paleomagnetic data have been inferred to support a relatively constant supercontinental assembly of all the major continental blacks (e.g., "Palaeomagnetism and the Continental Crust", by Piper, J. D. A., 1987, Open University Press, p. 434). The orogenic belts found within the supercontinental assembly are thought by Piper to have no associated ocean opening and closing and the orogenies would therefore be ensialic, in sharp conflict with the geological evidence. This paradox may result from the typical level of resolution of the paleomagnetic data, if continental divergence and the opening of oceans never exceeded values of about 15°. Such a compromise solution would involve small oceans and eventual closure of these oceans would result in ensimatic orogenies. On the other hand, it is equally possible that the available paleomagnetic data are simply too "permissive", i.e., that their imprecise dating or lack of structural control simply allow them to be fitted onto a common supercontinental apparent polar wander path without violation of their temporal and spatial constraints (age and error limits). The latter suggestion clearly calls for more reliable, well-dated paleopole determinations based on sufficient structural control, and it is surprising to note how few of the existing Precambrian paleopoles satisfy such reliability criteria.

For Late Precambrian times, the results from Africa also appear to support a relatively permanent assembly of the cratonic elements, as argued by Kroner, McWilliams and colleagues. However, recent paleomagnetic comparisons between east and west Gondwana, and between Laurentia and Baltica, have been used to suggest that after about 1000 Ma, relative motions may have occurred. After the beginning of the Paleozoic, the paleomagnetic data clearly support large relative motions in a plate-tectonic framework. In this talk a discussion of modern reliability criteria will be followed by the presentation of case histories illustrating the above statements.

VAN HOUTEN, F. B., Princeton University, Princeton, NJ

Global Distribution of Paleozoic Oolitic Ironstones

Paleozoic oolitic ironstones have been found on the paleocontinents of Laurentia, Baltica, Kazakhstan, Siberia, South China, Southeast Asia, Australia, Africa, and South America, as well as on peri-Gondwanide microplates. To some extent this distribution may be an artifact of geological field work. Nevertheless, in spite of much exploration the records of Laurentia and Australia remain sparce whereas that of less extensively explored North Africa is prolific. This contrast suggests that the present available data are reasonably representative.

Most of the known Paleozoic ironstones, on both large and small cratonic blocks, were deposited along trailing margins, and less commonly in interior basins, locally broken into swells and troughs by syndepositional faulting. A smaller number accumulated on convergent margins during either an episode of extension or an orogenic lull. Within these different tectonic frameworks ironstones were associated most commonly with widespread second-order transgressions, and consistently developed during hiatuses in the normal sequence of facies.

The initial Paleozoic high stand of sea level, in Middle and Late Cambrian time, produced a very meager record of oolitic ironstones. These were essentially limited to minor examples in southern Laurentia, while glauconitic deposits were common on many continents.

During the most extensive Paleozoic transgression, in Ordovician time, accumulation of ironstones reached its peak in both geographic extent and thickness of deposits. Significantly, most were in high southern latitudes. Some of the numerous Ordovician sites on North Africa and peri-Gondwanide blocks continued to be favorable places during Silurian and Devonian times. On South China and most of the peri-Gondwanide blocks, there is no Silurian ironstone between Ordovician and Devonian ones. The latter accumulated after the peri-Gondwanide blocks had drifted into low latitudes and docked in Middle Europe during the late Caledonian orogeny. In contrast to repeated development of Ordovician, Silurian, and Devonian ironstones, no Ordovician ones were deposited on the southeastern Russian Platform or on South America, where ironstones developed later in the Era. Moreover, Paleozoic ironstones never accumulated again after Ordovician time on low-latitude northeast and southwest Baltica, Kazakhstan, Siberia, Southeast Asia, Australia (except in late Permian), and southern Laurentia (except a minor mid-Devonian one in Arizona).

Development of ironstones declined severely during latest Ordovician–earliest Silurian Southern Hemisphere glaciation. Accumulation of ironstones was renewed only modestly in the Silurian Period as southern-latitude cratons drifted equatorward. Ones that developed in new tectonic frameworks include a major deposit in the long foreland basin of eastern Laurentia and a few minor examples on South America, on West Africa, and on peri-Gondwanide northeast Italy and Sardinia. Those in Guinea, north-central Brazil, and eastern Laurentia were succeeded by minor Devonian ironstones, whereas the ones in northwest and southeast Argentina, in Mauritania, and in Sardinia and Italy had no Devonian successors.

Widespread accumulation of ironstones was renewed in Middle and Late Devonian time in low and middle latitudes. In addition to development on many of the Ordovician and Silurian sites, major Devonian ironstones were deposited in several new settings, notably on the southeastern Russian Platform and South China, after the Caledonian orogeny. Minor ones accumulated on southwest Laurentia and the Meguma block. The significantly different paleogeogrqaphic regimes of Ordovician and Devonian examples point to different combinations of factors controlling their development.

Accumulation of Paleozoic ironstones ended as Pangaean cratons converged in late Paleozoic time. A few minor Middle and early Late Carboniferous ones in central England were associated with marine members of coal measures. In Middle Permian time, shortly after waning of Gondwana glaciation, a single well-developed ironstone accumulated in a marine embayment in north-central Australia. After the demise of Paleozoic ironstones, none accumulated until Late Triassic time. These few developed along the margin of Tethys (Czechoslovakia) and the Arctic Sea (Sverdrup Basin), where ironstones accumulated more abundantly in Early Jurassic time. (An IGCP Project 277 contribution.)

VAN SCHMUS, W. R., and M. E. BICKFORD, University of Kansas, Lawrence, KS, and J. J. PAPIKE, and C. K. SHEARER, South Dakota School of Mines and Technology, Rapid City, SD

Project UPPER CRUST: Use of Holes of Opportunity and Regional Geophysics to Define Proterozoic Orogenic Provinces in Central United States

The Precambrian crust of the central United States is covered by 0.2–15 km of Phanerozoic strata, requiring extensive use of geophysical data and subsurface samples from drillholes to define the subsurface geology and to correlate it with adjacent exposed terranes. Project UPPER CRUST is a DOSECC/NSF-funded program designed to retrieve samples from "holes of opportunity" drilled to Precambrian basement in the central United States. Typical holes of opportunity include oil, gas, and mineral exploration drillholes plus occasional engineering or stratigraphic test wells. Geochronologic, petrologic, and geochemical studies are carried out on the samples obtained to define the nature and origin of the rock units found at depth, to correlate them with exposed terranes using geochronologic and geophysical data, and to interpret the results in terms of continental lithospheric evolution for North America.

We have also obtained samples from a few localities by means of shallow scientific drilling. Continuous wire-line drilling is very effective to depths of 1–2 km and can return samples at modest costs. By determining our drilling sites beforehand on the basis of regional geophysics and current subsurface interpretations, we can obtain samples that will test specific hypotheses. Such drillholes can also be used for limited ancillary studies such as heat-flow, in situ stress measurements, and permeability studies with relatively low costs.

The core of our analytical program is U-Pb geochronology on zircon samples separated from samples of core and drill cuttings. Modern analytical techniques allow us to analyze samples of less than 0.1 mg, so that it is possible to obtain precise and accurate geochronologic data from the very small sample sizes often returned from the subsurface. Sm-Nd model ages for many samples can also be used to extend our interpretations by determining whether specific rock units represent primary orogenic crust, anatectically remelted crust, or tectonically reworked crust. The isotopic data are used in conjunction with the geochemical and petrologic data to determine the petrogenetic nature of the rocks encountered and their lithotectonic affinity.

A major limitation to working on subsurface samples is a lack of detailed field relationships. However, through use of the appropriate analytical techniques, the information obtained can be extrapolated far beyond the specific drillhole, including extrapolation to deep crustal levels. For example, one result of our research shows that there are many anatectic plutons in the subsurface of the continental interior that can be readily recognized through their aeromagnetic signature. Therefore, these plutons can be used as targets for scientific drilling with a high level of confidence regarding the nature of the rock encountered at depth. Such plutons can be used as probes to the lower crust, for the isotopic and petrologic character of the plutons must reflect the nature of their lower crustal source rocks. Hence, data obtained from the plutons can be inverted to deduce the age and composition of the lower crustal host terrane for the plutons.

An example is the Quimby magnetic anomaly in northwestern Iowa. Geophysical data and sparse subsurface samples suggest that the southern boundary of the Archean Superior Craton cuts across the northwest corner of Iowa and is bounded on the south by early Proterozoic orogenic units of possible Penokean age. Several small diameter aeromagnetic anomalies south of the boundary have been interpreted as middle Proterozoic anorogenic, magnetite-bearing granites. A 315 m deep scientific drillhole was drilled into a 1200 gamma aeromagnetic anomaly near Quimby, Iowa, to test these interpretations. Nearly 100 m of undeformed megacrystic, magnetite-bearing granite core was obtained. Chemical analyses show enrichment of incompatible trace elements, consistent with derivation by melting of preexisting crust. The U-Pb zircon age is 1433 ± 4 Ma, showing that this unit is part of an older anorogenic granite-rhyolite suite in the United States. Sm-Nd data yield a depleted-mantle separation age of 1860 Ma, which confirms that the source for the pluton's magma was Early Proterozoic crust, as hypothesized.

Many major uncertainties remain in our tectonic reconstructions. We suggest that a comprehensive program of shallow scientific drillholes in the continental interior of the United States can yield substantial improvement in our understanding of Proterozoic lithospheric evolution for this region at reasonable costs.

VAN SCHMUS, W. R., and M. E. BICKFORD, University of Kansas, Lawrence, Kansas

Early Proterozoic Crustal Evolution in the Central United States

Over the past 15 years substantial improvement has been made in our understanding of the chronology and genesis of Early Proterozoic orogenic provinces and lithologic terranes in North America. Much of this has been possible as a result of precise U-Pb age measurements on zircons separated from plutonic and volcanic rocks from these regions, in conjunction with Sm-Nd data on crustal formation ages and petrologic data on the genesis of igneous crustal units. However, one of the major problems in trying to decipher the Precambrian history of the continental interior of the United States is that most of the Precambrian crust is covered by a veneer of Phanerozoic sedimentary rocks 0.5 to 10 km thick.

In order to sample this basement, it is necessary to obtain samples from drill holes that penetrate the Phanerozoic cover. We have focused on dating zircons separated from basement samples; such ages generally are insensitive to weathering and lower grades of metamorphism, and modern techniques allow use of very small samples, so that we can obtain data from many drill sites. For larger, fresh samples the primary age information can be supplemented by other isotopic studies (such as Sm-Nd) and petrologic studies. These data are, in turn, complemented by a larger suite of general petrographic data from hundreds of drill sites throughout the region. Through extensive use of geophysical maps and the data from subsurface samples it is possible to define the major crustal provinces in the central United States, to correlate them with adjacent exposed terranes, and to use that information for defining processes of crustal genesis in this part of North America.

Early Proterozoic crust of the central United States formed adjacent to the Canadian Shield, which mainly consists of Archean cratons that were welded together with and locally bordered by Early Proterozoic orogenic complexes 1830 to 1900 Ma old (including the Trans-Hudson Orogen of Canada and the north-central United States and the Penokean Orogen of the Lake Superior region). Formation of the stable shield was essentially completed by 1800 Ma, and subsequent development occurred to the south and east. Most of the primary Precambrian crust south of the 1.8 Ga Canadian Shield consists of late Early Proterozoic orogenic terranes 1800 to 1600 Ma old and shows decreasing ages southward. This has been referred to as the Central Plains Orogen and two major subprovinces can be recognized. The older of these is 1700 to 1800 Ma old. It is correlative with the Early Proterozoic terrane of Colorado and southern Wyoming, and it probably extends to the Yavapai terrane of the southwest United States. This subprovince is bounded on the south by 1600 to 1700 Ma orogenic crust that has a similar southwest-northeast trend. The younger terrane is known from northern Kansas and southern Missouri. It is probably correlative with the Mazatzal terrane of southern Arizona and New Mexico; the eastward limit of this subprovince is unknown. The total width of the Central Plains Orogen is about 1000 km, and Sm-Nd data indicate that it consists of juvenile crust which was derived directly from the mantle with little contamination by older continental material. It is not possible to define detailed domains within the Central Plains Province, but comparison with exposed equivalents in the southwest United States and with younger orogenic belts suggests that southward growth of North America during the Early Proterozoic was probably accomplished by accretion of small arc complexes or exotic terranes, and there was no single large orogenic belt extending fully along the continent.

The Early Proterozoic primary crust underwent extensive partial melting during two distinct episodes in the Middle Proterozoic. The older episode occurred 1450 to 1500 Ma ago and affected most, if not all, of the 1600 to 1800 Ma old terrane. A similar, but more restricted, episode occurred about 100 Ma later, at 1340 to 1400 Ma, and affected the south-central United States. Nd isotopic signatures show that most of the felsic magmas were derived from melting older, Early Proterozoic crust, but recent results indicate that some of the 1450 to 1500 Ma old "anorogenic" rocks in the southeastern part of the orogen may have been derived from juvenile crust of the sam age.

Since 1300 Ma ago Proterozoic continental growth was been primarily to the east and southeast and is represented mainly by units within the Grenville Province to the east and the similar Llano Province of southern Texas. Major intracontinental rifting occurred about 1100 Ma ago, and this Midcontinent Rift System may be related to the tectonic regime that produced the Grenville Province.

VAN SICLEN, D. C., Bellaire, TX

Early Mesozoic Rifting West of Gulf of Mexico as It Opened from Fully Closed Condition

As North America (NA) moved away from *future* South American (SA) and Africa, then rigidly joined in the virtually immobile megacontinent of Gondwana, the rifting that was opening the formerly closed Gulf of Mexico (GOM) continued out to NA's West Coast. The accompany figure shows the inferred positions of the four largest microcontinental blocks beneath the lands now located west to south of the GOM, and of northern SA's present Guyana shield, at the beginning of Mesozoic time. Both northwestern microcontinents belonged to pre-Paleozoic NA, while the others then formed an undivided northwestern extension of the Guyana shield, which had collided with southern NA to create the Late Paleozoic Ouachita–Marathon fold/thrust belt. These blocks subsequently arrived "near" their present positions by remaining with SA when NA pulled away, then being transferred to NA by a series of spreading-center jumps that opened new rifts progressively farther into primitive SA. This occurred because the GOM spreading was *passive*, and the jumps involved significant *cessations* of spreading, during which the recently formed rifts' crust became stronger. These cessations are evident in the *active* Central North Atlantic Ocean (CNAO), where the first three spreading centers became extinct as each one migrated too far from the supporting mantle plumes alongside Africa, which would then start a new center back there, until a self-sustaining one finally developed on the fourth try (see Van Siclen, this volume). Pacific plate interactions subsequently moved the blocks to their present positions.

Rifting began all across southern NA in the Late Triassic, when the eastern Pacific plates *briefly* drove it clockwise around the bulge of northwestern Africa, until mantle plumes in the CNAO started the normal spreading. This earliest rifting created highlands along the entire International border, indicated by conglomerates and voluminous redbeds from southern sources that are almost continuous from the Ouachita–Marathon foreland in western Texas into southeastern Nevada, overlain farther north by spectacular Early Jurassic aeolian sandstones. Southward around the Arizona–California line Cenozoic uplift and erosion have exposed, beneath the Cordilleran overthrusts, autochthonous Precambrian basement overlain by metasediments of the cratonic western Grand Canyon succession. These are faulted against the west–northwest-trending McCoy basin, which holds over 24,000 f of Early Jurassic–Cretaceous nonmarine clastics that are unusual so near the Pacific in having a strictly cratonic provenance. Precambrian basement crops out intermittently for at least 90 m along the north side of the basin, but nowhere in or south of it (Harding and Coney, 1985), giving this the look of a new edge to the NA craton. The McCoy basin passes east-southwestward into the Bisbee basin, which features a substantial section of Early Cretaceous carbonate and other GOM-like strata. This basin continues onto the Pedregosa (basin) block, which extends east to the Rio Grande. Its Paleozoic strata resemble those of Texas's Permian Basin, of which this looks like a detached former extension. The Pedregosa block experienced only the *first* rifting event (equivalent to opening of the Interior basin in the northern Gulf Coasst region) before the first spreading-center jump returned it to the NA plate. Intense Laramide thrusting has since driven it back against the craton.

The West Mexico block (WM) remained with SA through the *second* rifting and spreading event, during which receding NA left behind a broad strait that connected the GOM with the Pacific. In this strait somewhat south of the McCoy basin, *marine* sediments from similar cratonic sources were deposited during the Middle Jurassic (Haxel and Tosdal, 1986). Also, the extensive Callovian salt in northeastern Mexico accumulated near its eastern end, probably continuing into the deep GOM salt. The *third* rifting event then left WM with NA as the salt-free strip of GOM seafloor formed between it and Yucatan. The *final* rifting transferred Yucatan, with Northern Central America (NCA) attached, to NA and inaugurated spreading in the Caribbean Sea beyond.

The Pacific plates subsequently pushed NCA and Yucatan northward while the Cuban volcanic arc was leading a tongue of Pacific lithosphere across the westmost Caribbean. When the plate interactions became easterly, WM was driven across the salt-free strip back against Yucatan, and the developing North Caribbean transform system began to slip NCA past Yucatan, thereby transferring NCA to the SA–Caribbean plate(s).

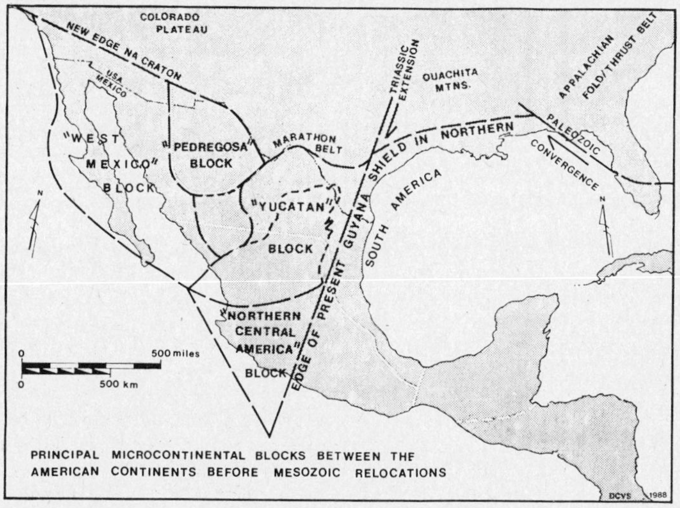

Figure 1.

VAN SICLEN, D. C., Bellaire, TX

Opening of Closed Gulf of Mexico and Early Central North Atlantic Ocean

When the Gulf of Mexico (GOM) began to open, *future* Africa and South America (SA) were rigidly joined, with northern SA's Guyana Shield fitting snugly between convergent North American (NA) wrench faults, which localized the southern Interior graben system shown on Figure 1. The Paleozoic *eastern* wrench had torn the edge of Gondwana (southeastern Pangaea) so that Africa could drive the Appalachian thrusts over southeastern NA while SA passed by on the south to create the Ouachita–Marathon fold/thrust belt farther west. The western wrench transferred the rifting of the Late Triassic Interior basins (Fig. 2) along the south side of the strictly *Ouachita* thrusts south-southwest to the Rio Grande embayment area. That's where nuclear Yucatan (not shown) was located, off the northwest corner of SA's Guyana shield, inside the arc of Marathon thrusts it had created.

NA's very earliest movement, clockwise translation around the bulge of northwest Africa, rifted the southern side of the *entire* Ouachita–Marathon belt. Extension of about 50 mi (80 km) formed the Interior basins of the northern Gulf Coast; that action then jumped south to start opening the main GOM basin, thereby transferring the Interior basins to NA, while Yucatan stayed with SA. In the Central North Atlantic Ocean (CNAO), this rotation opened the Carolina trough and Blake Plateau basin south of Cape Hatteras, by southern NA's swinging away from this recessive sector of Africa.

Figure 2 illustrates conditions when the CNAO's Blake Spur magnetic anomaly lineament (BSMA) formed. The BSMA and East Coast magnetic anomaly (ECMA) lineaments are magnetic maxima, accompanied by gentle seafloor ridges. They arise from dike complexes that record the early development of a new spreading center above the parent mantle plumes alongside Africa (which remained *stationary* relative to them) to replace an expired predecessor that had migrated beyond their support. Thus, these lineaments disclose discrete stages in the CNAO's (and GOM's) opening during this unstriped Jurassic magnetic quiet interval. The East Coast magnetic anomaly (ECMA) marks the eastern limit of the Carolina trough, and north of Cape Hatteras, of NA continental lithosphere. Its southern section split down the middle as spreading resumed there, leaving its eastern "half" with Africa's Senegal basin! The second resumption of spreading broke the newly formed BSMA into southern NA and northern African segments. In the GOM, this resumption generated the median strip of salt-free seafloor along a spreading center that passed between Yucatan and southern Mexico, thereby transferring the latter back to NA.

A third CNAO spreading-center jump to near San Salvador Island about 200 mi (320 km) southeast of the BSMA is interpreted, in the absence of extensive magnetic mapping, from the bathymetry of the unique spur that projects northward from the Bahama Escarpment, accompanied by a slight change in its direction and by right-lateral offset of magnetic trends. Recognition of the nonmagnetic effects is based on the more complete BSMA data, where the Bahama Escarpment (represented northwest of the Northeast Providence Channel by the southern slope of the Great Abaco Canyon) and a parallel, regionally persistent linear steepening of the magnetic gradient, are offset right-laterally about 60 mi (100 km) across the BSMA. The offset demonstrates that during the cessation of plume-powered spreading this developing dike complex functioned as a transform that accommodated a persistent external force directed northeastward against NA. The ECMA had functioned similarly during the initial translation of NA around northwest Africa, the driving force in each case being northward-moving eastern Pacific plates.

The CNAO's southwestern transform boundary recorded these offsets, hence the jumps, because Gondwana remained virtually stationary relative to the mantle plumes, and the one inside the "Florida" reentrant between future SA and Africa supported a *fixed, uni*-directional spreading center that produced relatively buoyant lithosphere. This provided the higher-standing substrate with a distinct northeast edge on which carbonate organisms built the Florida–Bahama Platform. The third, and final CNAO spreading-center jump to opposite San Salvador was accompanied by termination of spreading in the Gulf, which transferred Yucatan to NA near its present position as spreading began in the Caribbean Sea beyond.

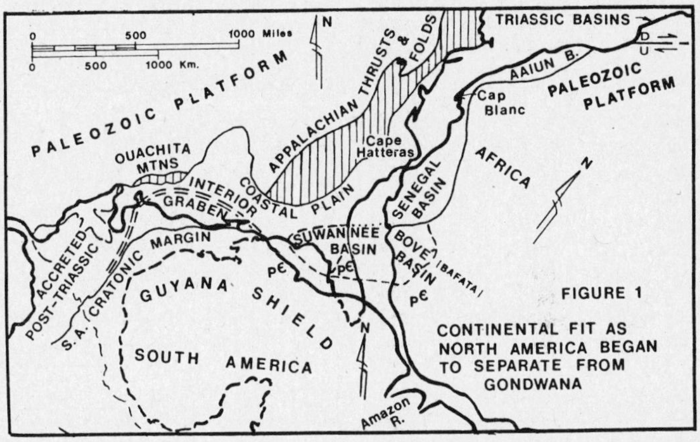

Figure 1.

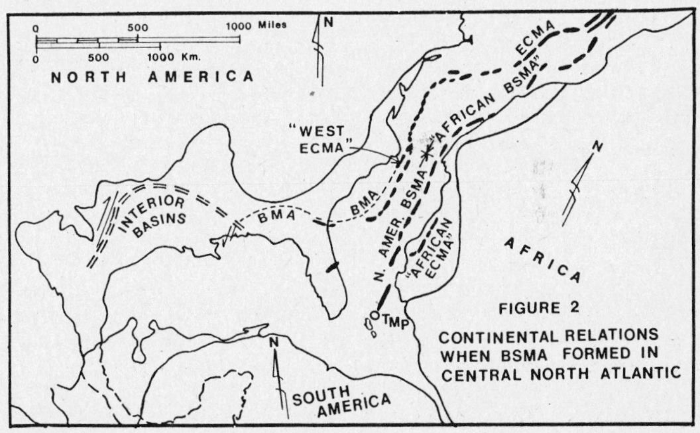

Figure 2.

VAN TASSELL, J., Eastern Oregon State College, La Grande, Oregon

Upper Devonian (Frasnian) Catskill Delta Milankovitch Cycles, Central Appalachians, U.S.A.

Recent field studies and examination of previously-published sections of Upper Devonian sequences in central Pennsylvania (Walker, 1971, Geol. Soc. America Bull.; 1972, Pa. Geol. Survey Gen. Geology Rept. g–62) and New York (Walker and Sutton, 1967, J. Sediment. Petrology; Woodrow and Nugent, 1963, NY Geol. Assoc.; Rickard, 1975, NY State Mus. Geol. Surv. Map and Chart Ser. No. 4) have confirmed previous suggestions by Van Tassell (1987, Geol. Soc. America Bull.) that transgressive-regressive cycles with periods ranging from approximately 0.1–2 m.y. are superimposed on the overall pattern of tectonically-triggered filling of the Appalachian Basin during the Late Devonian. Individual cycles with periods of approximately 0.1 m.y. evident in Walker and Sutton's (1967) ABC index data for turbidites in the Rock Stream Member of the Sonyea Group in New York match similar variations in the pattern of ABC index values illustrated by Walker (1971) in the Brallier Formation at Shamokin Dam, Pennsylvania and Woodmont, Maryland. Similar comparisons of cycle patterns suggest that individual cycles in the Scherr Formation section at Briery Gap Run, West Virginia may be the time-equivalents of the Catskill Formation in central Pennsylvania and the West Falls Group of New York. These correlations suggest that major black shale events in the New York sequences may be mirrored in the Pennsylvania and West Virginia areas by cycles of slope and shallow-shelf sediments. Similar cycles are also present in the overlying Foreknobs Formation of West Virginia.

The match between the overall patterns of Frasnian deposition in the New York, Pennsylvania, Maryland and West Virstriking. this overall depositional pattern is strongly cross-associated with the long-term pattern of the minimum values of successive Quaternary 0.1 m.y. eccentricity cycles calculated by Berger (1976, Astron. and Astrophys.), which includes cycles with periods of approximately 0.4, 1.3, and 2.0 m.y. Thus, the evidence indicates a hierarchy of Milankovitch cycles which can be traced over a large portion of the Upper Devonian (Frasnian) Catskill Delta paleomargin that goes clear up to the level of the mesothems of Ramsbottom (1975, J. Geol. Soc. London), the third-order cycles of Vail and others (1977, AAPG Mem. 26) and the fourth-order T-R units of Busch and Rollins (1984, Geology). Johnson and others (1985, Geol. Soc. America Bull.) have matched some of the cycles in the New York sequences with similar variations in other Devonian sequences throughout the world, suggesting the possibility that long-term orbitally-triggered cycles similar to those in the Catskill Delta sequence may occur in Devonian sequences throughout the world. Further study is needed to document this possibility.

VAN VESSEM, E. J., and T. L. GAN, Chevron UK Limited, London, England

Ninian Field, United Kingdom

The Ninian Field lies in the northern North Sea on the continental shelf between the UK and Norway, at a location centered on geographical coordinates 60°50'N and 1°30'E. Water depths range from 440 to 490 ft. The field, straddling blocks 3/3 and 3/8, is held by two separate groups of license holders and is developed under a unitization agreement with Chevron UK as operator.

The Field was found in the early seventies as North Sea exploration moved progressively northwards into deeper water areas from the initial southern North Sea gas basin discoveries.

The location of the Field is on the western side of the Viking Graben, the northern part of a north-south trending rift system that runs through the North Sea.

The Ninian structure is westward-tilted fault block that trapped migrating hydrocarbons generated from Upper Jurassic source rocks buried more deeply in the Graben.

The main reservoir rocks are the marine-deltaic sandstones of the Middle Jurassic Brent Group.

Estimated original in place reserves for the Ninian Field are 2995 million barrels, of which about 35-40% is recoverable. The oil is a paraffinic-naphthenic type, with a sulfur content of 0.39 weight percent. It is undersaturated, having a gas/oil ratio (GOR) of 301 standard cubic ft per stocktank barrel. Viscosity if 0.91 centipoise at saturation pressure of 1330 pounds per square inch.

The Field, with a productive area of 22,000 acres, is developed with three platforms: one gravity-type concrete structure and two steel-piled jacket structures, installed in 1978 and 1979. Because the natural drive of the field is negligible, the entire production will be achieved through water flooding. All three platforms have fully integrated production, injection, drilling and accommodation facilities. Available drilling slots total 109.

The Field has produced over 811 million barrels of oil in the ten years since production first started in December 1978. Peak oil production of 315,000 barrels per day was reached in the third quarter of 1982. Production in the early part of 1989 averaged 130,000 barrels per day. Both oil and associated gas liquids (NGL) are transported via pipelines to the Shetland Islands (Sullom Voe). Lean gas, surplus to fuel gas requirements were sent to the Scottish mainland (St. Fergus). This was discontinued in September, 1988 when the surplus no longer existed.

VAN WAGONER, J. C., and R. M. MITCHUM, Jr., Exxon Production Research Co., Houston, TX

High-Frequency Sequences and Their Stacking Patterns

A higher order of sequence development is widely observed in siliciclastic strata than is reflected by the Exxon Global Cycle Chart. These higher-frequency, or fourth-order sequences, form in 100,000 to 400,000 years, have all of the stratal attributes of sequences including constituent parasequences and systems tracts, and play a dominant role in controlling reservoir, source, and sealing rock distribution. These additional sequences are recognized from well logs, cores, and outcrops in Paleocene and Eocene strata in south Texas, Cretaceous strata in the western United States, Miocene strata in south-central Louisiana, and Pliocene and Pleistocene strata in offshore Louisiana. Fourth-order sequences have also been identified in Pennsylvanian strata from south-central New Mexico, Oklahoma, and in Valanginian-aged (Cretaceous) sediments of the Exmouth Plateau, offshore western Australia.

In the examples studied, a consistent hierarchy of stratigraphy is observed. Parasequences (Van Wagoner, 1985) stack to form fourth-order sequences, and groups of fourth-order sequences are deposited between major third-order boundaries forming third-order units. The stacking patterns of the fourth-order sequences suggest that they occur in lowstand, transgressive, and highstand sequence sets of the third-order units. These third-order units are deposited during the 1–3 m.y. cycles of the Exxon Global Cycle Chart, but because they are internally composed of sequences, we believe that they should not be defined as sequences in the strict sense of Mitchum et al. (1977) and Van Wagoner et al. (1987). The lowstand fourth-order sequence set within the third-order unit consists of two or more fourth-order sequences stacked in a progradational or aggradational pattern so that successively younger sequences step farther basinward in the former, or stack vertically with little or no basinward shift in the latter. In general, sequences in this sequence set extend farther into the basin than do those of other sequence sets and commonly have well-developed submarine-fan deposits in basins with a shelf-slope break. In basins with a ramp margin, sequences in the lowstand sequence set do not develop submarine fans. The transgressive sequence set consists of two or more fourth-order sequences stacked in a retrogradational or backstepping pattern so that successively younger sequences develop farther landward. Sequences in this sequence set extend farthest landward and may have the best-developed marine source rocks. The highstand sequence set consists of two or more fourth-order sequences stacked in a progradational pattern.

Within thick siliciclastic depocenters, third-order boundaries are marked by greater basinward shifts in facies, by larger, more widespread incised valleys, and by more extensive onlap than are fourth-order sequence boundaries, although not always by significantly deeper erosion. Seismic sections commonly exhibit third-order sequence boundaries with reflection onlap and truncation, as well as downlap surfaces. Fourth-order boundaries are commonly poorly defined on seismic lines because of the lack of seismic resolution. However, fourth-order sequences are clearly defined on well-log cross-sections, in outcrops, and cores where vertical stratal resolution is good.

As the total sedimentary section becomes thinner away from thick siliciclastic depocenters, third-order units become more condensed as fourth-order sequence definition is lost due to downlap, stratal thinning, and facies change to finer grained strata. Consequently, fourth-order sequences away from the depocenter are commonly represented by thin mudstone units containing one or more distal parasequencs that are often equivalent to more proximal parasequences within the depocenter. Commonly, only third-order boundaries are recognized in these thinner sections; fourth-order sequence boundaries may be distinct marker beds but usually do not exhibit criteria characteristic of unconformities.

Recognition of third-order boundaries is critical for providing a regional framework for correlating and mapping, and for tying the stratigraphy from depocenters to time-equivalent areas of low-sediment supply. Recognition of fourth-order sequences is critical for developing accurate reconstructions of sea-level change through time, and for developing a detailed picture of the reservoir, source, and seal distribution within a stratigraphic unit at the play and prospect scale.

REFERENCES

Mitchum, R. M., P. R. Vail, and S. Thompson, III, 1977, Seismic stratigraphy and global changes of sea level, Part 2: The depositional sequences as a basic unit for stratigraphic analysis, *in* C. E. Payton, ed., Seismic stratigraphy applications to hydrocarbon exploration: AAPG Memoir 26, p. 53–62.

Van Wagoner, J. C., 1985, Reservoir facies distribution as controlled by sea-level change, abstract: SEPM Mid-Year Meeting, Golden, Colorado, August 11–14, p. 91–92.

——— R. M. Mitchum, and H. W. Posamentier, 1987, Seismic stratigraphy interpretation using sequence stratigraphy, Part 2: Key definitions of sequence stratigraphy, *in* A. W. Bally, ed., Atlas of seismic stratigraphy: AAPG Studies in Geology, n. 27, v. 1, p. 11–14.

VARELA, M., Instituto Valenciano de Investigaciones Agrarias, Madrid, Spain

Transport and Evolution of Nitrates Through the Unsaturated Zone in Areas of Intensive Irrigation—Application to Coastal Plains in Spain

The Planas de Castellón and Valencia are two coastal plains bordering the Mediterranean Sea in an intensively developed agricultural area of southeastern Spain.

La Plana de Castellón (LPdC), presents a triangular shape and occupies an extension of 430 km2, surrounded by hills which grade into mountains. Topographic elevations range from mean sea level (msl) to 175 m above msl at Sichar reservoir.

About 95% of LPdC is under intensive irrigation, with orange trees covering a big part of the area. Consumption of water varies from 5000 to 17,000 m^3/ha/yr, depending upon crop type. Ground water abstraction totals 170 $\times$ 10^6 m3/yr. Excessive pumping has caused seawater intrusion and abandonment of numerous wells.

Another source of groundwater contamination is related to the overuse of fertilizers. Irrigation with groundwater shows an increasing development in this century, accompanied by the use of synthetic fertilizers, frequently applied in a very inefficient way. The irrigation return flows percolate through the detritic levels of Plioquaternary aquifer and carry the excess of nitrogen in nitrate form.

Surface water regulation through reservoirs has achieved maximum use of available supplies. Agricultural development is nearing an upper limit, utilizing all available land. The consequence is an extended process of cumulative pollution: nitrate levels in the aquifer exceed 50 mg/L all over the area, with 250 mg/L in some wells located in the southern portion.

The situation in La Plana de Valencia (LPdV), located 30 km south of LPdC is very similar to the one described above. The irrigated land occupies a high proportion of the area, with citrus and vegetables as the predominant crops. Part of the water used in agriculture is pumped from plioquaternary, detritic, heterogeneous aquifer which outcrop areas cover the whole coastal plain. Nitrate content in groundwater frequently surpasses 150 mg/L and has been increasing for the last fifteen years.

Various modelling efforts have been carried out in order to understand the dynamics of the nitrate pollution process in both regions. Although the flow models developed were partially successful in determining the characteristics of the aquifers a few uncertainties remain with respect to the parameters controlling the solute transport, especially in the unsaturated zone.

In order to estimate nitrate flux below the root zone, the (NO3-)/(Cl-) ratio has been used in two adjacent plots in the vicinity of Valencia. In recent years, alternated crops in these pilot zones included onions, potatoes and artichokes. Two fertilizer application and irrigation schemes have been compared in the two plots: in one case, conventional nitrogen fertilization was performed; in the other field, half the amount of nitrogen was used, but twice as frequently.

Vacuum samplers with ceramic cups were installed in the fields, at 1.8-2.0 m depth, distributed in four different rows, to obtain soil solution at various points. The evolution of the (NO3-)/(Cl-) ratio with time was used as an index of nitrate leaching below the root zone. Two assumptions were made in the analysis: steady state conditions for soil chloride and same chloride inputs and removal by crops on both treatments. Under this assumptions, nitrate leaching is proportional to (NO3-)/(Cl-).

Decrease of yield in the crops has not occurred decreasing the input rate of mineral fertilizer. Not significant differences in nitrate leaching between the two management treatments have been observed during a three year period, probably due to the high spatial and time variability of both (NO3-) and (Cl-).

In January, 1988, sixteen suction samplers have been added to the observation network, distributed along the four previous rows. The new samplers have been installed at depths of 1.7, 3.50, 7.00 and 12.0 m, in order to obtain information on the evolution of the nitrate plume in the unsaturated zone, which presents an average thickness of 12.30 m in the pilot fields. Concentrations varying from 800 and 200 mg/L have been measured, showing a clear reduction with depth but also a high spatial variability. Two observation wells have also been constructed, with the purpose of determining the effect of dilution on the contaminant front in the upper part of the saturated zone. The nitrate concentration shows a sharp decrease when the leachate reaches the piezometric surface.

Although the silty-argillaceous soil column in the pilot plots is rather homogeneous, complex mechanisms involving solute exchange can explain the vertical and lateral differences observed in the nitrate contents of the pore-water samples. Sampling is being done on a monthly basis and will continue, thus providing us with a longer time series of observations which might improve the understanding of the parameters controlling the changes in the nitrate front. The use of a tracer is being envisaged as a means of determining the residence time of the leachate in the unsaturated zone.

VARENTSOV, I. M., V. A. DRITS, A. I. GORSHKOV, A. V. SIVTSOV, and B. A. SAKHAROV, USSR Academy of Sciences, Moscow, USSR

Geochemistry of Major and Trace Elements of Oxyhydroxide Crusts from Krylov Seamount, Atlantic Ocean

Seamounts are the elevated elements of the oceanic lithosphere, their quantities and sizes increasing with increasing substrate age (Batiza, 1982; Smith, 1983). It is evident that generation of such notable amounts of oceanic lithosphere in the form of seamounts could influence the geochemistry of sedimentation and mineralization in some areas of the ocean.

The Krylov Seamount, chosen as the model site, is a volcanic edifice of the central type, which was formed in association with the effusion of subalkalic and alkalic weakly differentiated basalts along the strike of the sublatitudinal faults in the region of Cape Verde plate (17°32.8′–17°33.1′N; 29°59.7′–30°07.4′W; water depth 1270–4700 m). The seamount rocks especially the hyalocastites, are strongly hydrothermally altered into ferrous palygorskite.

Ferromanganese mineralization occurs as encrustation 1–10 cm thick, and as patches of pervasively impregnated volcanic rocks, and was studied by a combination of techniques: optical and electron scanning and transmission microscopies, X-ray diffraction, electron microdiffraction, microprobe X-ray energy dispersion analysis. Chemical analyses were carried out by means of plasma and emission spectroscopy and traditional wet chemical techniques.

The Mn–Fe oxyhydroxide crusts have a laminated structure. The lower layers at the contract with altered substrate are composed dominately of Fe-vernadite and geothite, with mixed-layered asbolane-buserite and traces of todorokite. The same intervals have relatively high Mn-Ti and Ba-Ti values and minimum Ti concentrations.

In the upper layers, the major components are Fe-vernadite, Mn-feroxyhyte, and subordinate amounts of birnessite, mixed-layered asbolane-buserite, and goethite. These layers have higher Ti contents, as compared to the basal layers but minimum Ba concentrations. The Mn-Fe rations vary over a small range: from 0.64 to 0.89.

The changes in the mineral and chemical compositions reflect a general trend both in the change of the hydrothermal and hydrogenetic inputs and in diagenetic transformations. The hydrothermal influence was greatest during the early stages of formation of Mn–Fe crusts when the Krylov Seamount was situated near the axial zone of the Mid-Atlantic Ridge. As this site of the Cape Verde plate moved east, the role of hydrothermal processes decreased and the part of hydrogenetic precipitation increased, thereby influencing the outer crust layers the most.

The postdepositional alterations are associated with the separation of ferruginous minerals in the crusts. The upper layers contain Mn-feroxyhyte as the main Fe-phase with admixtures of goethite, whereas in the lower layers, Mn-feroxyhyte is absent and foethite becomes one of the dominant components. In the light of experimental data, these relationships are interpreted as evidence of oxidative transformation of feroxyhute into goethite.

Thus, we generalize the data on the distribution of Mn–Fe minerals in the fine layers of the crusts and provide the following succession of diagenetic transformations on the basis of the structural features of the oxyhydroxides (Chukrov, Drits, Gorshkov, 1987): (1) products of direct precipitation from hydrothermal and/or hydrogenetic media (todorokite, Fe-vernadite, Mn-feroxyhyte); (2) products formed after the matrix of initial minerals and from their partial dissolution and redeposition (buserite-I,(?); (3) products of solid phase transformations (goethite, birnessite, mixed-layered asbolane-buserite).

The general scheme of Mn–Fe oxyhydroxide crust formation can be represented as a combination of hydrolytic mechanism with antocatalytic oxidation and precipitation of Mn and Fe oxyhydroxides when hydrothermal fluids mix in variable proportions with seawater. In the early stages of accretion, this interaction is accompanied by sorption and possibly by microbiological phenomena. In the course of formation of the upper crust layers, the dominant mechanism of accumulation of Mn–Fe, and associated elements was by interactions between the oxidized crust substrate and the bottom seawater.

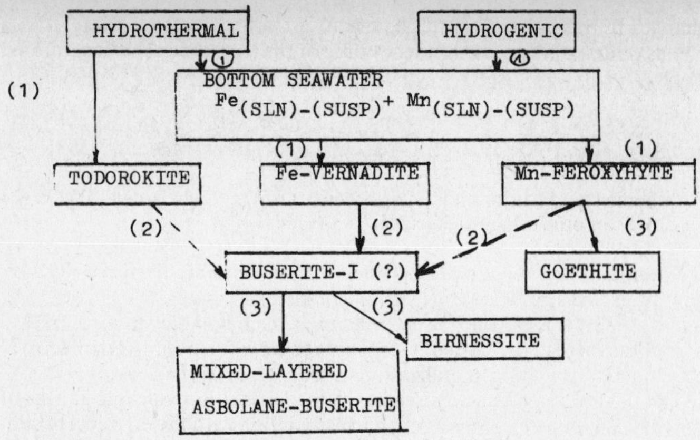

VARENTSOV, I. M., V. A. DRITS, and A. I. GORSHKOV, USSR Academy of Sciences, Moscow, USSR

Geochemistry of Rare Earth Elements in Processes of Mn–Fe Crust Formation (Krylov Seamount, Atlantic)

Krylov Seamount, chosen as the model site, is a volcanic edifice of the central type and is located on the Cape Verde plate. Mn–Fe oxyhydroxide crusts 2–10 cm thick are found on the upper to middle parts of the seamount. They are characterized by fine laminations. The upper 5–7 cm are composed mostly of Fe-vernadite and Mn-ferohydryte with subordinate amounts of birnessite, mixed-layered asbolane buserite, and goethite. The Mn-feroxyhyte content decreases towards the base of the crust. The basal cryst layers do not contain Mn-feroxyhyte, and goethite becomes the major Fe oxyhydroxide phase; Fe-vernadite is the dominant Mn-oxyhydroxide. On the whole, the crusts have relatively uniform Mn–Fe rations (0.64–0.89; seldom 1.13).

The studied crusts show a decrease of ΣREE (ppm) from the surface layers to the basal layers: 2093.2→1453.4. There is a direct correlation between the REE content and (Mn + Fe))correlation coefficient, γ = 0.74), which reflects the dominance of hydrogenetic sorption mechanism in REE accumulation by the Mn and Fe oxyhydroxides.

There is a reduction in the light REE fraction and corresponding increase of the heavy REE fraction from the basal layers to the surface of the crusts. Moreover, there is a good correlation (r = 0.8) between Mn–Fe and La–Sm ratios, which means that the major carrier of LREE is Mn oxyhydroxides. HREE are associated with the Fe oxyhydroxides as there is a good correlation between Fe and Yb/Sm values (r = 0.98). The observed differences in the behavior of light and heavy REE are interpreted from the physicochemical features of the surface of Mn-Fe oxyhydroxides, from the REE modes of occurrence in seawater, and from the relative amount of hydrothermal components in the hydrothermal fluid-seawater mixture. The LREE are found in seawater mostly in the cation form and are preferably adsorbed by Mn oxyhydroxides with a low pH points of zero charge (pH PZC 3), whereas HREE are present in seawater chiefly as complexes and are adsorbed mostly by Fe oxyhydroxides whose pH is 7.5. The decrease of relative LREE contents from the base to the surface of the crust indirectly implies a relative reduction of components of hydrothermal origin at later stages of accumulation of the Mn–Fe oxyhydroxides.

Ce is characterized by decreasing concentrations and decreasing Ce/Ce*, Ce/La, Ce/Yb, and Ce/Sm from the basal layers of the crust to the surface. There is a distinct correlation between Fe and Ce/Sm values, which shows an affinity of Ce to HREE. On the whole, when hydrogenetic precipitation dominates (Ce/Ce* 4.15–2.33). The Ce distribution is largely controlled by the subsidence of Krylov Seamount in the course of its movement with the Cape Verde plate from the axial zone of the Mid-Atlantic Ridge towards the African continent. It has been established (DeBaar et al., 1983; 1985) that shallow-water horizons of the Atlantic are considerably rich in REE, in particular, in Ce as compared to abyssal waters. This distribution is reflected in the REE of the Mn–Fe oxyhydroxide crusts.

There is a decrease in the Eu/Sm ratio from the lower layers of the crust (0.30–0.32) to the surface (0.26). Hydrothermal Mn deposits of the Galapagos Rift have Eu/Sm = 0.31 (Migdisov et al., 1983) and in seawater it is 0.26. The composition of hydrothermal fluids in the active volcanic zones of the ocean is 1000 times richer in REE than seawater with Eu/Sm of about 2.6 (Micharf et al., 1983; Corliss et al., 1978; Klinkhammer et al, 1983). These distributions of REE indicate that the role of components of hydrothermal origin in the early stages of Mn-Fe oxyhydroxide cruet formation can be important.

Thus, the mineral composition and distribution of REE in the Mn–Fe oxyhydroxide encrustations from Krylov seamount in Eastern Atlantic indicate that at the early stages of formation, when the seamount was situated close to the axial zone of the Mid-Atlantic Ridge and when the sublatitudinal (sub-transform) faults were active, the role of hydrothermal contributions was conspicuous. At later stages, as the Cape Verde plate moved to the east, the importance of hydrogenetic processes increased.

VARTANYAN, G. S., and L. V. BOREVSKY, Ministry of Geology of the USSR, Moscow, USSR

Modern Problems of Regional Hydrogeodynamics

Presently, regularities in formation, movement, and localization of groundwater in a zone of active water-exchange are studied in reasonable detail, whereas knowledge about mechanisms of groundwater movement in the deep hydrogeosphere continue to be schematic and contradictory. This results from a shortage of factual material and reliability of data on processes that take place in this zone. Analysis of the available information about fluid hydrodynamics in deep horizons of lithosphere shows that a uniform interpretation of all diverse thermodynamic phenomena (being found during studies of the subsurface) in terms of classic theory is failing.

The most complete mathematical model of mass and heat transfer was suggested by Ogilvi (1970). Unfortunately, practical application of this model is presently impossible due to the great number of phenomenological parameters involved, which cannot be determined by existing methods.

The discovery of a natural phenomenon of global transient pulsating changes in hydrogeosphere, so-called hydrogeodeformation (HGD) field of the earth (Vartanyan and Kulikov, 1982), made a contribution to understanding of mechanisms and basic motive forces that govern formation of hydrogeodynamic and hydrogeochemical situations in deep zones of the earth's crust. HGD-field manifests itself in global distribution of variations of barometric and physical-chemical parameters of a fluid-bearing lithosphere medium in time and space and reflects regional time-dependent changes in stress situation and the related changes in physical properties (filtrational properties, in particular) of rock masses. A transformation curve of HGD-field can be considered as a characteristic of intensivity and direction of the above changes.

Short-term processes of microdeformations, simultaneously covering huge territories, cause an effect of "wandering sieve" when at some instances of extension period distributed seepages of confined water can occur through usually impervious rock masses (e.g., clays). During compression, groundwater discharge stops (or reduces), increasing at the same time in the neighboring areas where at this moment extension states prevail.

It is obvious that regional extension results in universally occurring hydrogeological anomalies during preseismic periods, which embrace large areas and degrade some time after seismic stress relaxation.

In direct connection with the above processes are hydrodynamic evolutions associated with neotectonics.

Up to now it was considered, in spite of relations theoretically established between movements of a rock skeleton and geofiltration processes, that tectonic motions cannot affect the current piezometric surface. This opinion was based on a supposition that linear relations between discharges and gradients within the whole range of their changes lead to relaxation of head gradients, occurring practically parallel with these changes.

A comparison of piezometric head gradients and amplitudes of neotectonic motions for a number of regions of the USSR (Borevsky and Borevsky, 1983) enables us to draw the conclusion that relaxation of head gradients, caused by the above-mentioned processes, proceeds much slower than in theoretical calculations and in some cases is not identified by the current methods at all. In deep stratified zones of the lithosphere, this can result in formation of abnormally high and abnormally low pressures. Removal of pressures has, as a rule, an impulsive nature and represents one of the mechanisms of forming HGD-field of the earth.

It follows from the above that at low head gradients an evaluation of filtration rates using Darcy's law may lead to substantial errors and false conclusions concerning water-exchange velocity and even existence or absence of groundwater movement. This should be taken into account, too, when solving regional filtration problems using mathematical modeling.

Among the most poorly studied problems of regional hydrodynamics is the problem of a lower boundary of free groundwater distribution in the earth's crust and the related problem of global vertical water-exchange.

The results of deep and super-deep drilling have shown that free water occurs within the whole profile studied up to a depth of over 11 km and according to indirect data—even to 20 km. Here its occurrence is not of a local nature, but it forms rather extensive water-bearing zones of a considerable thickness (to 4–5 km).

Solution to the problem on formation of regional water-saturated rock masses at great depths under high temperatures and pressures lies in the discovered phenomenon of forming horizons and zones of hydrogenic deconsolidation of rocks in the course of their progressive transformations (Borevsky et al., 1984). The cause of formation of these zones is a change in state and properties of water due to rock dehydration at certain thermobaric levels. This phenomenon influences the origin of physical stratification of the earth's crust, the formation of crustal pore waveguides, ore-bearing hydrothermal fields, scaly thrusting displacements, and so on.

Due to the conditions of fluid formation, hydrostatic pressures in the above-mentioned zones exceed lithostatic ones, and pore formation is attributed to volume decrease of solid phase, microhydrofracturing, and dissolution. Geological events and modeling results give an evidence of pulsating water-discharge from these zones, which causes in overlying horizons a rise of fluid pressure (up to super-lithostatic one). This pressure is transmitted upwards and serves as one of the effective mechanisms of HGD-field functioning in the earth.

From the above statements it follows that in the near future a theory must be developed for fluid dynamics taking into account reconstructions in macro- and microstructure of huge rock masses and interrelations between groundwater state and water-bearing medium.

VASLET, D., Bureau de Recherches Géologiques et Minières, Orléans la Source, France

The Paleozoic of Saudi Arabia and Correlations with Neighboring Regions

A new lithostratigraphic succession (Figure 1) is proposed for the Paleozoic of central Arabia, based on detailed geologic mapping of Central Saudi Arabia by the Directorate General of Mineral Resources.

A widespread peneplaned erosion surface characterizes the major unconformity (Pan-African orogeny) between the Phanerozoic cover rocks and the Proterozoic basement. However, a newly discovered (possibly early Paleozoic) unit of basement-derived sedimentary rocks is preserved in places under this peneplaned surface and above the basement.

Early Paleozoic sedimentary rocks were deposited in paleoenvironments which evolved from continental to shallow marine. Sedimentation was entirely siliciclastic, and it is assumed that the first major alluvial-to-deltaic Early Paleozoic deposits over the peneplaned surface were not derived from the Arabian basement but originated from an area farther south. Three brief marine transgressions took place during Late Llanvirnian, Llandeilian and Middle- to Late-Caradocian times.

It is demonstrated for the first time that Late Ordovician glacial and periglacial continental to subaquatic deposits are ubiquitous in Early Paleozoic outcrops of Arabia. A precise link is now established with the ice cap which existed on the Gondwana paleocontinent in the Late Ordovician. Several phases of advance and retreat of the ice, each marked by a pronounced erosional unconformity, are recorded in central and northwest Arabia.

After the melting of the ice cap, marine sedimentation began in the Early Silurian (Middle Llandoverian), but there is no evidence of Late Silurian deposition.

During the Late Paleozoic, the Devonian to Early Permian deposits are now regarded as a single megacycle, evolving in paleo-environment from continental to marine and reverting to continental. Early Permian glacial deposits occur only in the south of the peninsula.

Finally, pre-Late Permian crustal movements in the Arabian craton caused major erosion of all older rocks. The erosion surface cuts obliquely into the Proterozoic basement, and is marked by paleolaterite. Deposition resumed in the form of shallow-water marine sediments dating from the Late Permian.

The regional biostratigraphy is modified, on the basis of new paleontologic studies. In particular, the surface-graptolite zonation is now more accurate, and its relationships with subsurface microflora are shown for the first time in the Arabian peninsula.

Updated paleo-facies maps were integrated with lithologic comparisons to establish correlations between central Arabia, other regions of the Arabian peninsula, North and East Africa, Turkey, Iran, and India. These Correlations demonstrate the close relationships of the Paleozoic geodynamic evolution of Saudi Arabia to the broader paleogeographic context of the Gondwana paleocontinent and its margins.

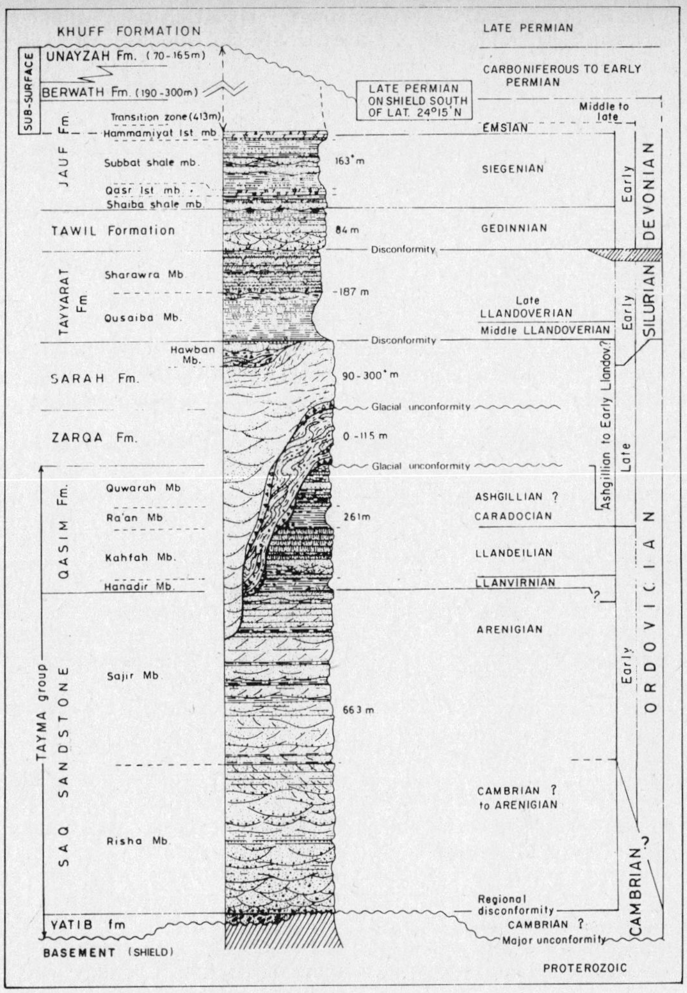

Figure 1—The pre-late Permian Paleozoic succession in Central Saudi Arabia.

VASUDEVAN, V., Centre for Earth Science Studies, Trivandrum, India, and A. SEETARAMASWAMY, Andhra University, Waltair, India

Shallow-Marine Sediments of Palk Bay, East Coast of India

The textural and mineralogical studies of the sediments of Palk Bay, a sheltered sea between India and Srilanka, have been carried out. The mean grain size decreases toward the east from medium sand to coarse silt. The sand fraction is quantitatively minor (<6%) throughout the bay. The major concentration is located in the northwest portion; however, the central and eastern part of the bay has a mud zone comprising dominantly silt and minor amounts of clay. The finer fraction composes montmorillonite, kaolinite, and illite in the order of abundance. Montmorillonite has a negative correlation with illite probably due to absorption of potassium in the marine environment. The montmorillonite increases toward the deep bay from 56% to 64% contrasting to its low concentration (14%; Siggiquie, 1967) farther east in the Bay of Bengal. The predominance of montmorillonite in the Palk bay reveals a distinct source from that of the sediments of the Bay of Bengal. The high smectite content of Palk bay is comparable to that observed in the sediments of Godavari (Sambasiva Rao, 1979), Krishna (Swamy et al., 1973), and Cauvery rivers (Seralathan and Seetaramaswamy, 1982) in the north. The similarity of the clay mineral composition in the light of the surface circulation patterns off the east coast of India implies that all three of these peninsular rivers are the main source for the sediments of Palk bay.

REFERENCES

Sambasiva Rao, M., 1979, Ph. D. Thesis, Andhra University, 209 p.
Seralathan, P., and A. Seetaramaswamy, 1982, I. J. M. S., v. 11, p. 167–169.
Siddiquie, H. N., 1967, Marine Geology, v. 5, p. 249–291.
Swamy, A. S. R., N. V. N. D. Prasada Rao, and M. P. Rao, 1973, Journal of Marine Biology Association of India, v. 15, n. 2, p. 625–631.

VEBLEN, D. R., and E. S. ILTON, Johns Hopkins University, Baltimore, MD, and C. M. SCHLINGER, University of Utah, Salt Lake City, UT

Submicroscopic Inclusions in Rock-Forming Minerals and Glasses: TEM Observations and Geological Implications

Numerous observations with transmission electron microscopy (TEM) and X-ray analytical electron microscopy (AEM) show that populations of submicroscopic inclusions are a common feature of many rock-forming minerals and glasses. Although they cannot be observed and identified with more conventional techniques, such inclusions can have important implications for geochemistry (e.g., for the distribution of trace and minor elements) and for the geophysical properties of rocks and glasses (e.g., they can strongly control magnetic properties).

Microinclusions typically form as a result of phase separation or reactions, including oxidation. For example, in the reaction of biotite to chlorite, relatively insoluble reaction products may be kinetically precluded from diffusing to grain boundaries (e.g., Veblen and Ferry, 1983; Eggleton and Banfield, 1985). Instead, microinclusions of minerals such as iron oxides and titanite can precipitate with sizes in the 1–100 nm range. Morphologically similar crystals of native copper are observed to form during hydrothermal alteration of biotites and chlorites associated with porphyry copper deposits (Ilton and Veblen, 1988). The resulting anomalously high Cu contents in the sheet silicates previously have been interpreted as resulting from solid solution and equilibrium partitioning of Cu. This incorrect interpretation results from the inability to resolve the copper inclusions with standard analytical and petrographic techniques and provides a clear example of how microinclusions can drastically affect the concentrations and apparent partitioning behavior of trace/minor elements.

Large populations of microinclusions also have been observed in lower crustal pyroxenes. For example, augites from pyroxene granulites from Lofoten and Vesteraalen, Norway, contain numerous inclusions in the 10–100 nm range (Schlinger and Veblen in preparation). These include apatite, monazite, and zirkelite, as well as a host of silicate and oxide minerals. The former contain high concentrations of rare-earth elements (e.g., La, Ce, Nd), in addition to other "exotic" elements such as Th and Zr. The fact that appreciable REE are concentrated in inclusions, rather than being in solid solution, is problematical for trace-element modeling of processes such as partial melting of lower crustal and upper mantle rocks. If such inclusions are rehomogenized prior to melting, then a treatment of REE partitioning using measured equilibrium melt/solid distribution coefficients for pyroxenes and other silicates may be warranted. However, if the inclusions are not dissolved before melting, then such a treatment is incorrect.

Just as microinclusions such as monazite may partially control the behavior of trace elements, oxide inclusions in the submicrometer size range can control the magnetic properties of both crystalline and amorphous phases. In the granulites noted above, cubic and rhombohedral oxide inclusions with large shape anisotropy dominate the magnetic behavior of the pyroxenes (Schlinger and Veblen, in preparation). Similarly, populations of oxide microcrystals that form in volcanic glasses by precipitation or oxidation can determine the magnetic behavior (Schlinger and Smith, 1986; Schlinger et al., 1986, 1988a, 1988b). Glasses from rapidly cooled tuffs can possess large populations of cubic oxide in the 2–10 nm size range and are superparamagnetic. With slower cooling, strong stable remanence may develop, as oxide crystals coarsen into the single-magnetic-domain size range and in many cases develop strong shape anistropy. Correlations between such microinclusion populations and magnetic properties require the combination of TEM, AEM, electron diffraction, and magnetic measurements on the same materials.

REFERENCES

Eggleton, R. A., and J. F. Banfield, 1985, American Mineralogist, v. 70, p. 902.

Ilton, E. S., and D. R. Veblen, 1988, Nature, v. 334, p. 516.

Schlinger, C. M., and R. M. Smith, 1986, Geophysical Research Letters, v. 13, p. 729.

———, and D. R. Veblen, in preparation, Journal of Geophysical Research.

———, R. M. Smith, and D. R. Veblen, 1986, Geology, v. 14, p. 959.

———, J. G. Rosenbaum and D. R. Veblen, 1988a, Geology, v. 16, p. 556.

———, D. Griscom, G. C. Papaefthymiou, and D. R. Veblen, 1988b, Journal of Geophysical Research, v. 93, p. 9137.

Veblen, D. R. and J. M. Ferry, 1983, American Mineralogist, v. 68, p. 1160.

VEEGER, A., S. DAVIS, and A. LONG, University of Arizona, Tucson, Arizona, and E. CUSTODIO, Universidad Politecnica de Cataluna, Barcelona, Spain

Ground-Water Chemistry Related to Sources of Recharge on the Island of La Palma, Canary Islands, Spain

La Palma is the northwestern most of the seven main islands of the Canary Archipelago. The island has an area of 728 km^2 and was formed by a series of volcanic events dating from the Miocene to the present. The northern half of the island is dominated by older Miocene and pliocene basalts, with Pleistocene and recent volcanics occurring almost exclusively in the southern half of the island. A deep canyon, the Caldera de Taburiente, cuts through the island from the center of the northern half to the western coast.

Rainfall on the island varies from less than 250 mm/yr along the southwestern coast to greater than 1000 mm/yr on the northeastern flank of the island at an elevation of about 1500 m. Variations in precipitation are due primarily to topography, wind direction, and the elevation of the tropical inversion layer. An undetermined amount of fog condensation with resultant fog drip adds to the water available for recharge in the higher forested slopes of the north-central portion of the island.

Agriculture is primarily confined to the lower slopes along the coast. In the north, an extensive broad-leafed forest covers the intermediate elevations up to about 1200 m. Above this, a pine forest covers all the higher slopes except those left barren by recent volcanic activity.

Except for minor amounts of unconsolidated alluvium, colluvium, and marine-terrace sediments, the geologic materials exposed on the island are volcanic and low-grade metavolcanic rocks. The basal complex, the oldest unit exposed on the island and thought to be pre-Miocene to Miocene, crops out in the lower part of the Caldera de Taburiente. The basal complex includes volcanic breccia, pillow basalts, pyroclastics, and intrusive igneous rocks, which are intersected by a dense network of basaltic dikes that may comprise more than 60% of the rock mass. The basal complex forms an almost impermeable base for the island.

Pliocene basalts, in the form of hundreds of thin lava flows, overlie the basal complex and crop out only in the northern half of the island. These permeable flows form the most exploited aquifer on the island.

Qyaterbart volcanics, including cinder cones and lava flows, are found almost exclusively in the southern half of the island. Although porous and permeable, these units are generally above the water table and are not a major source of water.

Chemical characteristics of the ground water reflect the diverse origins of the water. One of the most chemically distinct is return irrigation water draining areas planted in bananas. Agricultural chemicals together with evaporative concentration of original dissolved solids give rise to water with 10 to 372 mg/1 NO_3, 35 to 640 mg/1 SO_4, and total dissolved solids commonly in excess of 1000 mg/1. Ground water originating from recharge at elevations above areas of cultivation and in zones of higher precipitation, represents the other chemical extreme. total dissolved solids content of this water is commonly less than 200 mg/1 and dominant constituents are dissolved silica, sodium, and bicarbonate. Ground water originating as recharge in drier portions of the island has higher total dissolved solids, but can be distinguished from return irrigation water by having lower amounts of sulfate and nitrate than the return irrigation water.

A drill hole and a spring in th lower portion of the Caldera de Taburiente produce water with about 1000 mg/1 SO_4, and saturated with respect to calcite. Although geochemically interesting and distinct, this water is probably not quantitatively important in the water budget of the island.

Water samples collected in July and August 1988 are being analyzed for chemical and isotopic composition. Sampling was completed in a broad band from northeast to southwest, across the Caldera de Taburiente, in order to cover a wide range of elevations and precipitation patterns. Samples were collected from wells, springs and water collection galleries (Figure 1).

Preliminary results of chloride and bromide analyses show that the mass ratio of these two anions varies across the island (Figure 1). The chloride to bromide ratio in seawater is about 290:1; values from ground-water samples range from about 100:1 to 760:1. The highest ratios were found in samples from springs in areas where annual precipitation is moderate to low, and evaporation rates are high. The lowest ratios were found in samples from wells in highly cultivated areas. The mechanism for this apparent fractionation is not known. Laboratory experiments are being conducted in an attempt to identify the fractionation mechanism.

Complete results of the chemical and isotopic analyses will be presented. With these data we expect to further define the recharge systems on La Palma. Particular attention will be given to identifying the importance of fog drip as a source of ground water recharge.

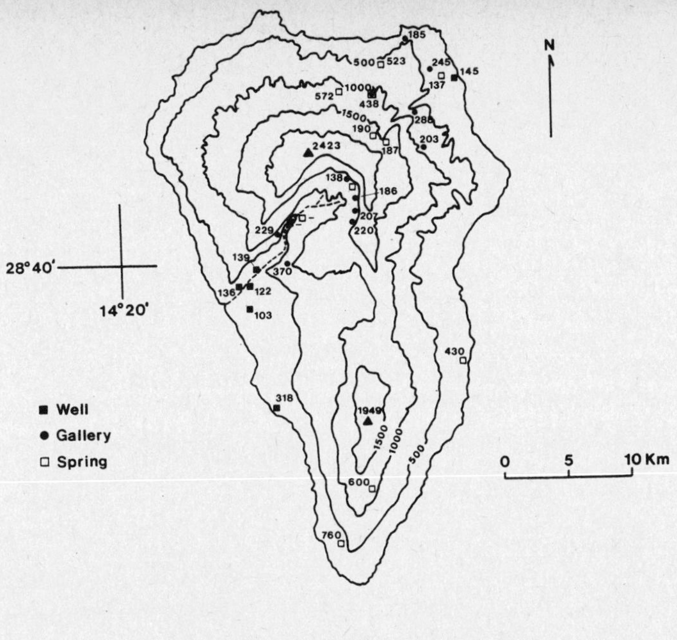

Figure 1. Cl-Br ratios in ground water, La Palma, Canary Islands.

VEIZER, J., University of Ottawa, Ottawa, Canada

Sedimentation and Mineralization Through Geologic Time: Recycling Perspective

The tenents of plate tectonics demand that tectonic evolution follows the sequence mantle → oceanic crust → volcanic arc → orogeny → craton (including its non-sedimentary basement). As the material moves along this chain, it undergoes progressive fractionation and refinement. For example, the melt generated in the mantle forms the oceanic crust. The latter is, in turn, mostly resorbed back into the mantle, but a subordinate portion feeds the successor tectonic domain, the island arc. This pattern continues downstream along the mantle → craton succession. As a consequence, each tectonic domain is involved in a continuous generation/destruction ("birth/death") cycle, with the rate of recycling decreasing downstream due to the necessity to process ever diminishing quantities of material per unit time. The above systematics is amenable to quantitative treatment, which is analogous to that of the populations dynamics in living systems. The "birth/death" cycle for a given population — in this case tectonic domain — imposes an internal age structure on its constituent units (rocks). The proportion of progressively older units diminishes, usually exponentially, and the rate of this decay is inversely proportional to the rate of recycling. For steady-state systems, the calculated theoretical half-lives (or 50% probabilities of destruction) for the rocks of global tectonic domains are the following: active margin basins ~ 27 Ma, oceanic intraplate basins ~ 51 Ma, oceanic crust ~ 59 Ma, passive margin basins ~ 75 Ma, immature orogenic belts ~ 78 Ma, mature orogenic belts (roots) ~ 355 Ma, platforms ~ 361 Ma. For continental basement, the half-life depends on the relative ease of resetting of ages established by different radiometric techniques and therefore varies between ~ 673 and ~ 1728 Ma limits. Consequently, the probability of preserving tectonic domains with half-lives of ≤ 100 Ma (active margin basins → orogenic belts) into the Precambrian is remote. If so, the Precambrian terranes should encompass mostly rocks of the metamorphic continental basement, roots of former mountains (mature orogenic belts) and platformal sedimentary sequences. This is not to say that one cannot recognize vestiges of, for example, a former oceanic crust in the assemblage of rocks of a basement or orogenic root. However, these are only reprocessed remnants of former oceanic crust, now incorporated into, and being part of, a downstream tectonic domain.

The age structure for the global and continental sedimentary mass represents a polymodal distribution, with age slope shallower with increasing antiquity. If representative, such pattern could result from progressively older sediments being recycled at ever slower rates. In my view, the wasting of sediments proceeds mostly en masse. The principal agent is physical weathering and transport, but controlled by the rate of uplift and hence tectonic setting. With increasing geologic age, the continental (or global) sedimentary mass is dominated by sediments associated with progressively more stable tectonic domains; each domain having a characteristic package of sediment lithologies. It is these complex interplays of packages which have been responsible for the apparent time dependent lithological evolution of sediments summarized in the work of Ronov and associates and modelled as differential recycling by Garrels and Mackenzie. The existing inventories for sedimentary basins are as yet insufficient to model such tectonically controlled secular evolution of the sedimentary system.

In contrast, utilizable inventories are available for mineral deposits. In this concept, the half-life (τ_{50}) of generation/destruction — or recycling — of petroleum during the Phanerozoic has been ~ 87 Ma. coal and gas attained their "modern" steady-state only in the Late Paleozoic, following the colonization of land by vascular plants. Subsequently, they have been recycled with half-lives of ~ 265 and ~ 120 Ma, respectively. The present-day age distribution of reserves of metallic ores suggests that the system attained a steady-state at ~ 1.75 ± 0.25 Ga ago, coincident with the growth of continents to their near present-day cumulative size. The subsequent history has been mostly that of perpetual recycling, the rates varying with the genetic categories of ores rather than specific metals. The genetic categories, in order of their ascending recycling rates, are the following: (1) magmatogene, associated with mafic-ultramafic rocks (τ_{50} ~ 1700 Ma), (2) metamorphic associations (~ 1350 Ma), (3) volcano-sedimentary (~ 800 Ma), (4) chemical-sedimentary (~ 414 ± 194 Ma), (5) hydrothermal (~ 95 ± 41 Ma), (6) detrital-sedimentary (~ 50 ± 20 Ma), and (7) weathering crusts (~ 3 ± 2 Ma). This progression reflects the diminishing tectonic and geological stability of the host environments. The simulated overall metallogenic evolution of the Earth, deconvoluted for recycling and combined with geological considerations, indicate five partially overlapping development stages. These are: (1) the greenstone belt stage, dominant in the Archean and petering out ~ 1.8 Ga ago; (2) the cratonization stage, with the peak in the Late Archean and Early Proterozoic; (3) the rifting stage, at ~ 1.8 ± 0.3 Ga ago; (4) the stable craton phase, ~ 1.7–0.9 (or 0.6) Ga ago; and (5) the Phanerozoic stage of continental dispersal.

REFERENCES

Veizer J., 1988, The evolving exogenic cycle. In: Chemical Cycles in the Evolution of the Earth, Wiley, p. 175–220.

Veizer, J., P. Laznicka and S. L. Jansen, 1989, Mineralization through geologic time: recycling perspective, Am. J. Sci., in press (RMGarrels volume.

VEKSLER, I. V., Vernadsky Institute, Moskow, USSR

Conditions for Crystallization and Concentration of Perovskite-Type Minerals in Alkaline Magmas

Perovskite-type titano-niobates are multicomponent solid solutions of four main end-members: perovskite $CaTiO_3$, lueshite $NaNbO_3$, loparite $Na_{0,5}TR_{0,5}TiO_3$ and tausonite $SrTiO_3$. Minerals of this group occur in the intrusions of ultramafic-alkaline rocks, agpaitic nepheline syenites, in carbonatites, kimberlites, albitites and greusens. Usually they are no more than accessory minerals, but in olivinite and pyroxenites of ultramafic-alkaline intrusions and in agpaitic nepheline syenites they may form big accumulations. In the ultramafic rocks chemical composition of the perovskite-type solid solutions is close to pure $CaTiO_3$; in agpaitic nepheline syenites their main component is loparite. Rocks enriched in perovskite-type titano-niobates are of economic value as titanium and rare element ore.

For the lack of the physico-chemical information about the behavior of perovskite minerals in silicate systems at high temperatures there have been no definite view of their genesis (this is especially true for loparite in agpaitic nepheline syenites). That is why we have carried out an experimental study of several phase diagrams with perovskite and loparite important from petrological point of view.

With the help of simplified model systems (nepheline - sphene, nepheline - albite - sphene, nepheline - diopside - sphene) we have studied the perovskite-sphene equilibrium, which is believed to be an important indicator of the SiO_2-activity in magmas (Carmichael, et al., 1970). We have found out a reaction of sphene with nepheline melt leading to perovskite crystallization:

$$2CaTiSiO_5 + NaAlSiO_4 = 2CaTiO_3 + NaAlSi_3O_8$$

Albite component produced in this reaction accumulates in the residual melt and enters the nepheline solid solutions. Perovskite crystallizes as the first liquidus phase from a broad range of intermediate mixtures of nepheline and sphene in $NaAlSiO_4$ - $CaTiSiO_5$ system and in more complex systems containing the nepheline - sphene pseudo binary section.

In the system nepheline - diopside - sphene (Fig. 1) titanate-titanosilicate equilibrium is much complicated by significant isomorphic entrance of titanium into the clinopyroxene solid solutions. Investigations of the chemical composition of the equilibrium phases in this system by microprobe method made it possible to calculate the constants of the reactions within the liquidus-solidus temperature interval.

We have carried out experimental modelling of the loparite solid solutions' magmatic crystallization. At first nepheline - loparite phase diagram was investigated, then we studied loparite's crystallization from the multicomponent agpaitic melt, which composition corresponds to lujavrite. Quench experiments with lujavrite melt were carried out in dry and water-saturated conditions with natural loparite as a starting material for charges.

It was determined that loparite is stable within all temperature interval studied (1480–750 °C). Due to cerium partial oxidation by air atmosphere small amounts of CeO_2 crystallized together with loparite. In products of quench experiments under water pressure 1 Kbar rare britolite crystals were detected in the association with loparite.

Special efforts were aimed on quantitative determination of loparite's saturation concentrations and lopartie's solubility in alkaline melts. As a simple measure of loparite's solubility we used summary concentration of TiO_2, Nb_2O_5, TR_2O_3 of the melt in equilibrium with loparite solid solutions.

It was determined that the composition of the agpaitic melt has little effect on loparite's solubility, that is why it is possible to put all the data for nepheline and lujavrite melts on a single curve of temperature dependance (Fig. 2). Water content has no observable effect on loparite precipitation either.

Reaching of laparite's saturation by magmas in the course of their differentiation is the main and obvious condition for crystallization of this mineral. Experimental data on loparite's solubility provide a realistic quantitative criterium for prediction of the loparite occurrence in agpaitic rocks (loparite's formation is possible only in agpaitic environment). For example, calculating the gross summary concentrations $TiO_2 + Nb_2O_5 + TR_2O_3$ in Khibina, Lovozero and Ilimaussaq intrusions on the base of the data about trace element's abundances in them (Gerasimovsky et al., 1968) and comparing resulting values (1.13, 1.46, and 0.85 wt%) with the experimental data presented in Figure 2 one can conclude that loparite's early magmatic crystallization have been quite possible in the Lovozero intrusion and unreal in the other two. This conclusion is in agreement with relative loparite's abundances in this greatest alkaline intrusions (loparite is common in Lovozero and absent in Ilimaussaq).

Experimentally determined chemical evolution of the loparite solid solutions with falling temperature is parallel to loparite's cryptic layering in the section of the Lovozero intrusion. It is a strong evidence in favor of the Lovozero loparite's magmatic genesis.

Processes of rhythmic layering in large magma chambers, though still not well understood, may be considered to be an effective and universal mechanism of mineral separation. Early magmatic crystallization of the perovskite minerals assists their accumulation and concentration in the course of magma solidification.

REFERENCES

Carmichael, J., J. Nichollis, and A. Smith, 1970, Silica activity in igneous rocks, Amer. Min., v. 55, n. 1–2, p. 246–264.

Gerasimovsky, et al., 1968, The geochemistry of the Lovozero alkaline massif, Part II, Canberra: A.N.U. Press, 370 p.

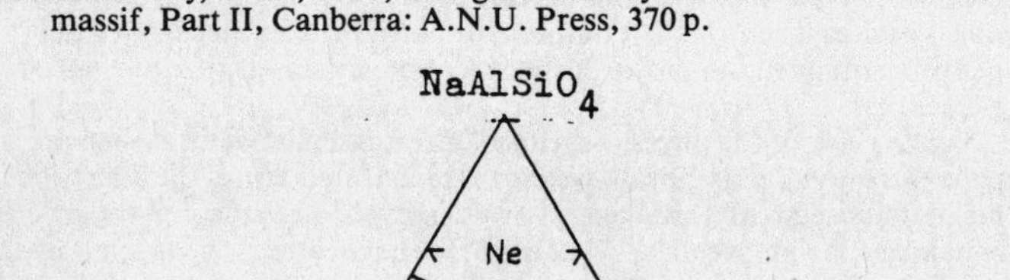

Figure 1—Phase diagram nepheline - diopside - sphene. Ne- nepheline; Ol-olivine; Mel-melilite; Di-diopside; Pst-perovskite; Sph-sphene.

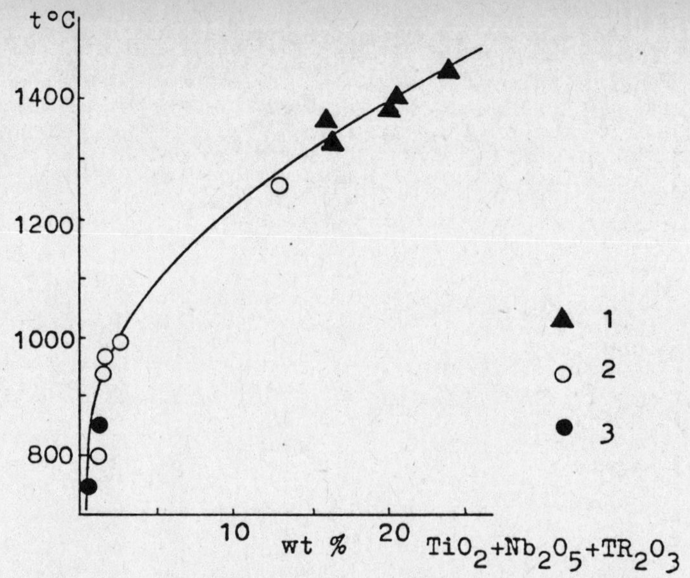

Figure 2—Loparite's solubility in alkaline melts. 1, nepheline melt; 2, dry lujavrite melt; 3, water saturated lujavrite melt.

VELBEL, M. A., and A. B. TAYLOR, Michigan State University, East Lansing, MI

Geochemical Mass Balances and Weathering Rates in Forested Watersheds of Southern Blue Ridge. II. Effects of Botanical Uptake Terms

Geochemical mass balance (input-output budgeting) is a common means of investigating elemental mobility in natural hydrogeochemical systems. Geochemical mass balance methods are particularly common in small-watershed studies of cation budgets, nutrient cycling, and landscape susceptibility to acid deposition. In these systems, geochemical mass balance is used to quantify uptake of elements into, or release of elements from, the small watershed. Because the ultimate source of exported elements is the weathering of primary rock-forming minerals, geochemical mass balance studies are often used to estimate rates of primary-mineral weathering and soil formation.

Many researchers employ the "balance sheet" approach, a system of simultaneous linear equations with constant coefficients. This system of linear equations represents the steady-state behavior of the modeled system. Most workers also assume that the steady-state assumption for the entire system means that botanical factors can be ignored; in other words, that there is no elemental transfer between biomass and inorganic compartments of the system. However, a recent mathematical derivation of the mass balance equations (Velbel, 1986) shows that this second assumption is not mathematically included in the first. Individual compartments within a steady-state system constantly change; for example, even in a steady-state watershed, the primary-mineral compartment is constantly shrinking as minerals are consumed by weathering reactions. Nevertheless, the input-output behavior of the overall system remains at steady-state; the shrinking mineral reservoir supplies the cations that maintain the steady-state efflux. The assumption that biomass is constant in a steady-state system is mathematically equivalent to stating that the size of the mineral compartment remains constant, which is clearly not the case. Therefore, the widely invoked assumption that biomass is at steady-state is mathematically a second, explicit assumption, not a built-in mathematical consequence of the assumption of overall steady-state.

The "overall-system-steady-state" assumption is the foundation of the mathematical formulation actually used; the "steady-state-biomass" assumption is an additional and separate assumption, which may or may not apply to any individual system, and which is independent of the "overall" steady-state assumption. However, the "hidden" assumption of steady-state biomass is inadvertantly applied by many workers, with the possible result that their mass balances are erroneously missing an additional source or sink term for certain elements. For example, if the steady-state formulation is mistakenly applied to a system in which the botanical compartment is actively taking up nutrients, the measured net efflux is not the only output term; biomass is also an output from the weathering compartment. In such systems, cation budgets and weathering rates based only on net aqueous fluxes will be underestimates, because there is an internal sink that is not accounted for.

To quantify the implications of the separate mathematical assumptions, weathering rates for several forested watersheds in the southern Blue Ridge mountains were calculated twice, using different sets of assumptions. In the first set of calculations, only the overall system was assumed to be at steady-state, and mathematical coefficients were introduced into the equations to allow elemental uptake by biomass. In the second set of calculations, *both* steady-state for the overall system *and* steady-state biomass were assumed; in this set of calculations, biomass did not exchange elements with the rest of the system. The second set of calculations, which is more mathematically restrictive and less likely to represent the behavior of natural botanical systems, is directly analogous to most geochemical mass balance studies undertaken to date. The same (measured) elemental fluxes, and the same stoichiometric coefficients for mineral weathering reactions, were used in both sets of calculations.

Rates of plagioclase weathering were not significantly affected by changing the assumptions, because the weathering rate of plagioclase is determined almost exclusively by the mass-balance equation for Na. Na is not a major plant nutrient, so the botanical uptake coefficient is very small, and there is little difference between including or excluding biomass from the Na mass-balance equation. Rates of garnet weathering were 101–138% higher in the calculations that included botanical uptake than in the commonly used method which ignores biomass. Garnet weathering rates are calculated from the mass balances of Ca and Mg, both of which are important plant nutrients. Biotite mica weathering rates were 118–300% higher in the calculations that included botanical uptake than in the commonly used method which ignores biomass. The biotite weathering rate is calculated primarily from the mass balance equation for K, a major plant nutrient.

Our calculations show that misunderstanding the mathematical constraints that are built into the most widely used geochemical mass balance equations causes errors (underestimates) of up to 300% in the calculated rates of mineral weathering and soil formation. This problem is most pronounced for minerals that contain major nutrient elements and for rates calculated from mass balances of those nutrient elements.

REFERENCES

Velbel, M. A. , 1986, *in* S. M. Colman and D. P. Dethier, eds., Rates of chemical weathering of rocks and minerals: Academic Press, p. 439–451.

VELBEL, M. A., and N. L. ROMERO, Michigan State University, East Lansing, MI

Amphibolite Weathering and Cation Budgets in Southern Blue Ridge and Colorado Front Range, USA

Hydrolysis reactions during chemical weathering of primary rock-forming silicate minerals consume hydrogen ions. The same silicate-mineral weathering reactions are also the ultimate source of base cations to the soil exchange complex. Consequently, silicate-mineral weathering reactions are fundamental processes in the reaction of acid precipitation with susceptible landscapes. The role of common silicates (especially feldspars) in supplying cations is widely studied; often, however, the weathering of these silicates does not explain observed cation budgets, and researchers are forced to invoke post hoc "unobserved" minerals or phenomena to explain the observed budgets. Excess Ca in watershed effluxes is a prime example—to explain high Ca fluxes from granitic or gneissic rocks, previous researchers have reluctantly invoked dissolution of unobserved calcite, which is postulated to occur as "hidden" veins; as carbonate dust introduced by eolian processes; or as anthropogenic (agricultural) lime. Our investigation of two "high-Ca" watersheds, one in the southern Appalachians and one in the Colorado Front Range, shows that amphibolites occur in both "high-Ca" watersheds, and that weathering of amphibolite minerals can directly explain observed calcium budgets without invoking unobserved calcite sources. In both cases, mineral compositions from actual or closely related parent rocks were used in the calculations.

The amphibolite in the southern Appalachian study area is from the Carrol Knob Complex in watershed 3 (WS3) of the Coweeta Hydrologic Laboratory (U. S. Department of Agriculture, Forest Service), located near Franklin, North Carolina, USA. These rocks consist of hornblende and plagioclase, with minor quartz, epidote, and other accessories. Bedrock in the surrounding area consists of high-rank (amphibolite facies) metasedimentary schists and gneisses. All crystalline rocks of the study area are extensively weathered to saprolite.

Of 25 watersheds in the Coweeta Hydrologic Laboratory, streamwater from WS3 has the highest calcium concentration. Previous workers attributed at least part of this anomaly to applications of agricultural lime in the 1940s. New geologic maps of the study area reveal that WS3 is the only experimental watershed underlain by mafic rocks. It is therefore possible that the Ca anomaly in streamwater results from the unusual composition of the parent rock in WS3 and not from historical agricultural liming.

To test the hypothesis that weathering of mafic-rock minerals could account for the observed elemental concentrations in streamwaters of WS3, we constructed a simple geochemical balance sheet for WS3, following the method of Garrels and Mackenzie (1967). Published analyses of hornblende (ferroan pargasite) and plagioclase (bytownite, An_{80}) from the nearby petrologically similar Laurel Creek Complex were used in the mass balance calculations.

Weathering of hornblende (6.1 μmol/L of water), biotite 13.3 μmol/L), and calcic plagioclase (157 μmol/L) can account quantitatively for all alkali and alkaline earth cations in streamwater. The calcic plagioclase of the Carrol Knob Complex weathers 3–10 times more than the sodic (oligoclase-andesine) plagioclases of the surrounding gneisses (Velbel, 1985), consistent with laboratory kinetic studies and empirical generalizations (e.g., the Goldich weathering series). The high Ca content of streams draining Coweeta WS3 is due to weathering of mafic, Ca-rich parent material, not to prolonged dissolution of agricultural lime. The hornblende weathering rate (in moles per liter of throughgoing solution) is only 3.9% of the plagioclase weathering rate. However, if conservation of aluminum is assumed for weathering products of plagioclase, hornblende weathering consumes 17.3% of the total acidity consumed by all silicate-mineral hydrolysis reactions. The magnitude of acid-neutralization by hornblende weathering is comparable to that determined in other mass-balance studies.

The second study area, the Loch Vale watershed in Rocky Mountain National Park (Colorado, USA), is underlain predominantly by granitic gneisses and schists, but scattered amphibolites are also reported. Geochemical mass-balance calculations for this watershed indicate the following (areally averaged) weathering rates for primary rock-forming silicates (in mole/ha/y): from the granites and gneisses; biotite, 23; and An_{35} plagioclase, 63: from the amphibolites; actinolite, 22; and An_{65} plagioclase, 94. Although the amphibolite is not areally extensive, its ferromagnesian silicates and more calcic plagioclase are so much more reactive in the weathering environment (compared to the more abundant but less-reactive granitic/gneissic minerals) that the amphibolite minerals make a disproportionately large contribution to cation budgets.

Calcic/mafic compositions and consequent high reactivities cause minerals in mafic rocks to supply cations in abundances and proportions that are uncharacteristic of the surrounding more silicic rock types. Even small amounts of amphibolite can supply significant proportions of total cation budgets. Cation budgets of some "high-Ca" streams can be explained by weathering of known amphibolites, without resorting to "unobserved" calcite or other hypothetical sources.

REFERENCES

Garrels, R. M., and F. T. Mackenzie, 1967, *in* W. Stumm, ed., Equilibrium concepts in natural water systems: American Chemical Society Advances in Chemistry Series, v. 67, p. 222–242.

Velbel, M. A., 1985, American Journal of Science, v. 285, p. 904–930.

VELICHKO, A. A., Academy of Sciences of the USSR, Moscow, USSR

Loess Formation of East Europe: Stratigraphy, Paleogeography, Origin

The East European loess region includes vast areas of the middle and southern parts of the Russian Plain and northern Caucasus foothills. Loesses of the western Russian Plain (Dnieper and Dnester drainage basins and coasts of the Black and Asov Seas) are most typical, with thickness up to 40 m and predominance of silt fraction. Eastward (Don drainage basin, right bank of the Volga River) loess decreases in thickness and its composition is less homogenous. East of the Volga there are only separate patches of loess. Exceedingly thick loess had been recorded north of the Caucasus near the Black Sea and Caspian basins, which can be attributed to tectonics within the Caucasian fore-deep and to a special environment of sedimentation.

Studies of loess stratigraphy and chronology based on physical, paleontological, lithological, and geochemical methods revealed a complete sequence of horizons within the loess-soil formation that embraces the whole Quaternary period; 14 loess horizons and 15 fossil soils have been identified so far. The loess-soil formation merges into till horizons northward and is associated with certain marine terraces in the south, which permits us to correlate the three continental areas, i.e., glaciated, periglacial (extraglacial), and maritime areas. Loess horizons cannot always be matched with a corresponding till of an ice sheet. Such a connection can be traced for several tills of the Don (Lower Pleistocene), Dnieper (Middle Pleistocene), and Valdai (Upper Pleistocene age), while fossil soils correspond to marine transgressions. A general stratigraphic scheme has been worked out by the author with the assistance of T. D. Morozova, A. K. Markova, V. P. Nechayev, V. P. Udartsev, and A. I. Tsatskin (Laboratory of Paleogeography, Institute of Geography, USSR Academy of Sciences) (see Figs. 1, 2). At present we try to correlate the scheme with chronostratigraphy of oceanic sediments; the correlation has been established only within Late Pleistocene series; older formations require further investigations and probably joint efforts of specialists of loess and marine sediments.

Paleogeographic analysis permits us to distinguish six climatic macrocycles within the Pleistocene and at least two or three within the Eopleistocene. Warm epochs of the Eopleistocene are marked with red soils of subtropical character; soils of Early and Middle Pleistocene are transitional from subtropical to warm temperate; Late Pleistocene soils vary from temperate to subarctic types. Loesses correspond to cold epochs and arid environments. In general, the aridity increases toward the Late Pleistocene. Since the Middle Pleistocene, the loess-soil formation shows cryogenic structures that indicate permafrost development. The permafrost area reached its maximum expansion during the Late Pleistocene.

Investigations into the loess genesis (including the quartz grain morphometry and surface texture studies) indicate eolian processes to have been the main agent of sedimentation, a postsedimentary redistribution of the material by noveodeluvial processes being slightly pronounced. Some attributes of loess were acquired under the influence of vegetation and cryogenic processes.

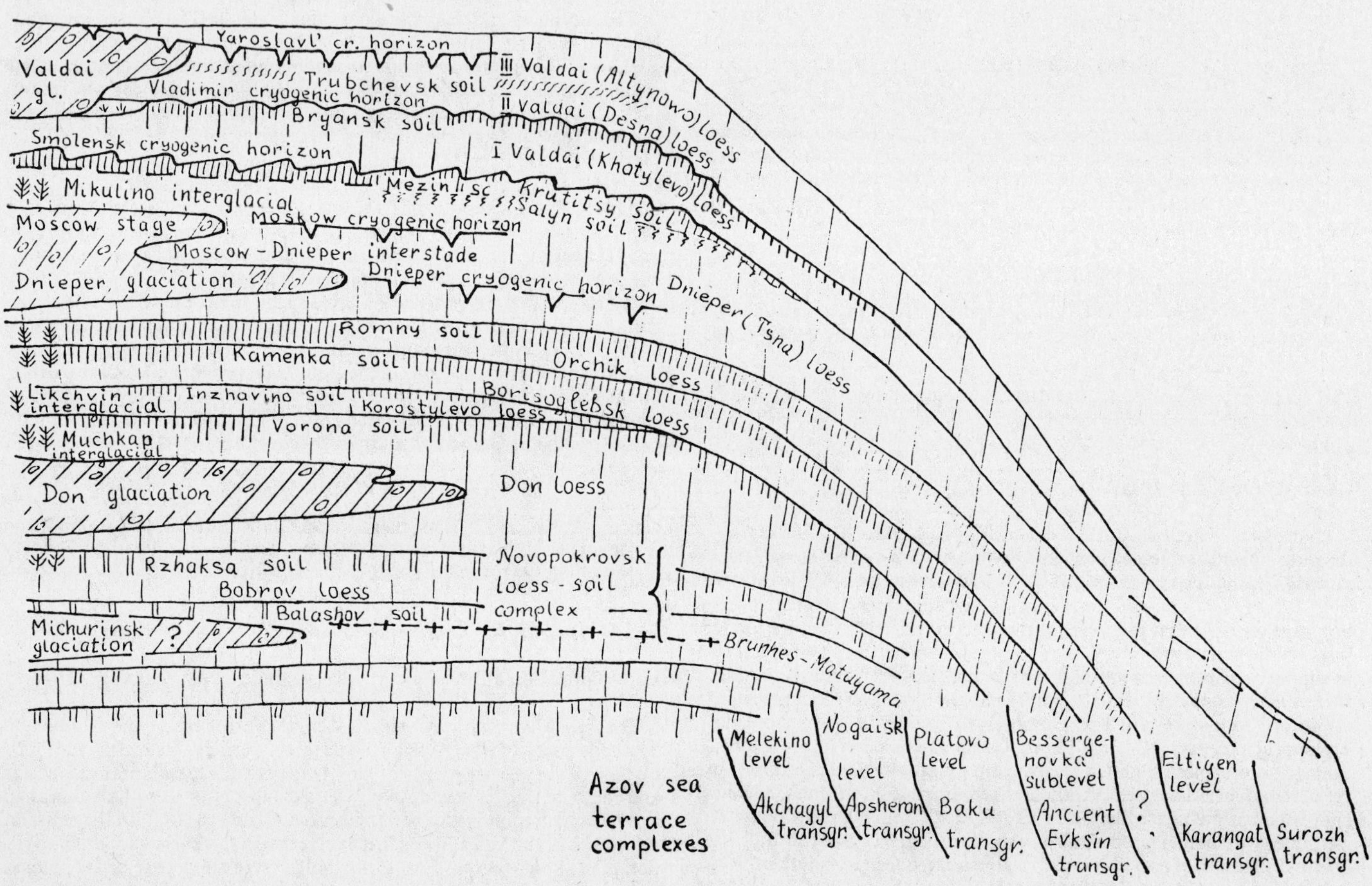

Figure 1—East European loess formation and its facial relations with glacial and marine deposits (Velichko, 1984).

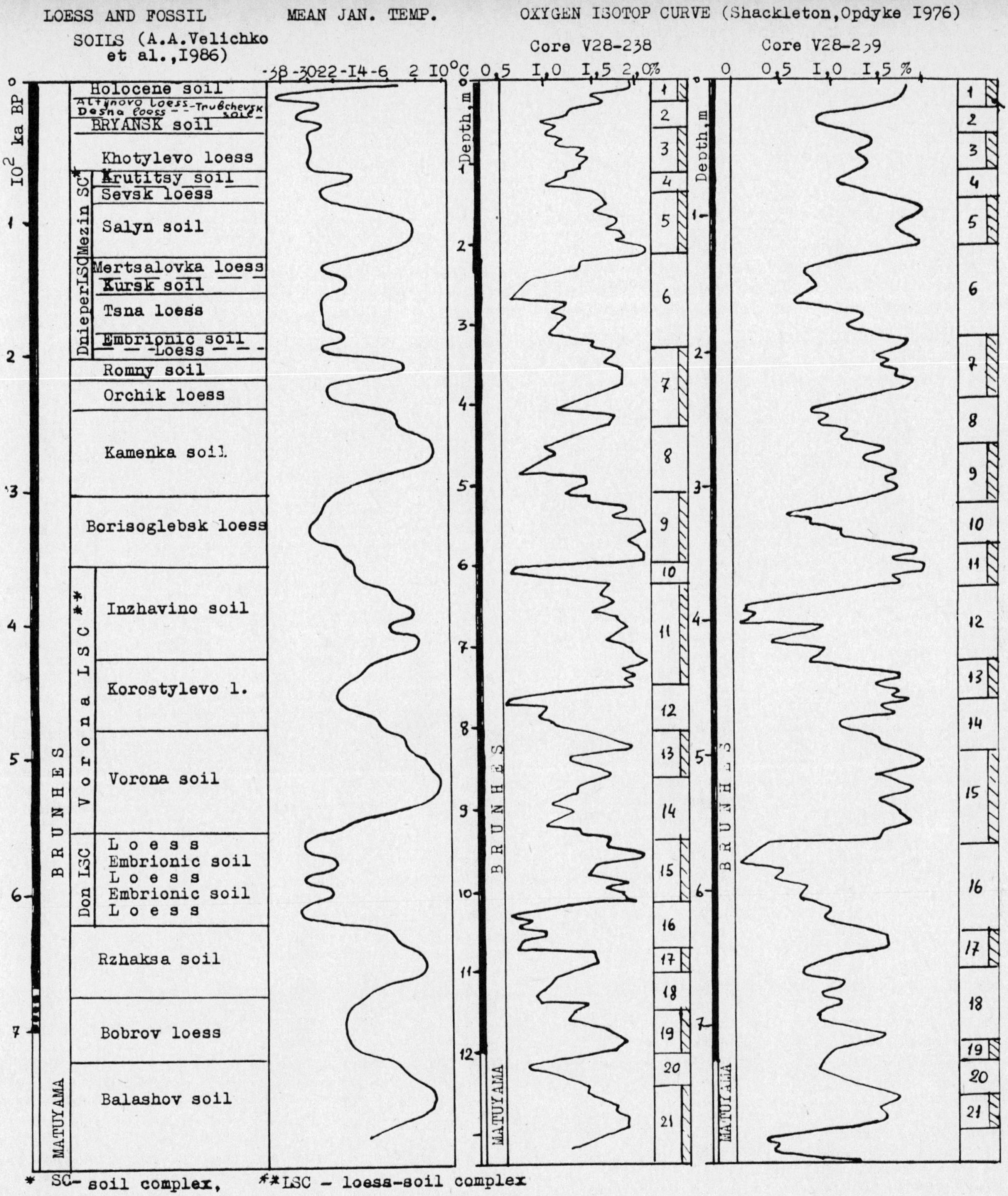

Figure 2—East European loess formation and its correlation with oceanic sediments.

VERDON, A. J., JR., American Geological Institute, Alexandria, Virginia

A Kindergarten Through Grade 12 Earth Science Education Framework for the 21st Century

Science education in the United States has received significant attention during the 1980s. To fulfill its mission to coordinate improvements in earth science education during this time of renewed interest in precollege education, the American Geological Institute initiated conferences, workshops, projects and programs through its National Center for Earth Science Education. These activities have been supported by the National Science Foundation and other research agencies, i.e., the United States Geological Survey, National Aeronautics and Space Administration, and the National Oceanic and Atmospheric Administration. AGI has also sought cooperation with its 19 member societies and other professional earth science and science education societies. The plan to redefine the nature of instruction in the earth sciences at the precollege level includes continual interaction between the research community, school teachers and other science educators with the goal to publish a new framework (set of guidelines) for earth science education from Kindergarten through Grade 12 for all students.

VERGNIOLLE, S., and C. JAUPART, Université Paris 7 et Institut de Physique du Globe, Paris, France

Degassing in Viscous Liquid at Constriction: New Model for Basaltic Eruptions

Basaltic volcanoes erupt in several different regimes that have not been explained. At Kilauea volcano (Hawaii), eruption can take the form of "fire fountaining," where lava clots are propelled to great heights in the atmosphere in a powerful gas jet. Another possible regime at Kilauea and other basaltic volcanoes is the effusive one, where vesicular lava oozes out of the vent to form a lava flow. At Stromboli, eruptive activity often takes the form of intermittent explosions due to the bursting of large gas pockets that are almost as wide as the volcanic conduit. Hawaiian eruptions are particularly important because they show cyclic alternation between fire fountaining and effusive activity. During an episode of fire fountaining, the eruptive characteristics remain steady during a time that exceeds the rise-time from chamber to vent, implying that conditions at the conduit entrance in the chamber are also steady. The same conclusion holds for effusive activity. Both regimes are therefore in steady-state and hence correspond to different conditions in the reservoir that feeds the conduit. Cyclic activity thus reflects cyclic changes in the magma chamber.

To propose a framework that allows an interpretation of all these different volcanic manifestations, we present a study of the degassing magma in a reservoir that empties into a small conduit. Laboratory experiments are made in a cylindrical tank topped by a thin vertical tube. The tank is filled with silicone oils and glycerol solutions to investigate a range of liquid viscosity and surface tension. Gas bubbles are generated at the tank bottom with known characteristics (bubble diameter and total gas flux). The bubbles rise through the tank and accumulate in a foam layer at the roof. Depending on the behavior of this foam layer, three different regimes are distinguished: (1) steady horizontal flow of a foam made of bubbles from the tank interior, (2) alternating regimes of foam build-up and collapse leading to the violent eruption of a single and large gas pocket, and (3) persistence of a foam layer made of coalesced bubbles leading to intermittent slug flow in the conduit. When the gas flux is below a critical value, regimes 2 and 3 are not observed, and only regime 1 occurs. leading to steady bubbly flow in the conduit.

The three regimes have natural counterparts in basaltic volcanoes. Effusive activity corresponds to regime 1, Hawaiian cyclic behavior of alternating episodes of fire fountaining and effusive outflow to regime 2, and Strombolian intermittent explosive activity to regime 3. An important feature of Hawaiian eruptions is that they start by exhibiting cyclic behavior, and, in a second phase, proceed in the effusive regime only without fire fountaining. This is also explained by our findings: As the eruption wanes, the degassing rate falls below the critical value needed to achieve foam collapse, and hence regime 2 is no longer possible. This implies that regime 1, i.e., effusive activity, is the only possible one.

A simple theory is proposed to explain regimes 1 and 2. The bubbles in the foam deform under the action of buoyancy forces. Using a geometrical model for the packing configuration, the evolution of the bubble shape is calculated as function of the foam thickness. We find that the foam is no longer stable if its thickness exceeds the critical value $h_c = 2\sigma/\epsilon\rho_1 gR$, where σ is the coefficient of surface tenison, ρ_1 the liquid density, g the acceleration of gravity, R the original bubble radius, and ϵ the gas volume fraction in the foam. The foam thickness is determined by a balance between the input of bubbles from below and the output into the conduit and scales with:

$$\left(\frac{\mu m.Q}{\approx^{2}.\rho_1.g}\right)^{1/4}$$

where μ_m is the foam viscosity and Q the imposed gas flux. All else being equal, the higher the gas flux, the larger the foam thickness. A necessary and sufficient condition for foam collapse is that the foam reaches the critical thickness h_c. To achieve foam collapse, the gas flux must therefore exceed a critical value that depends on viscosity, surface tension, and bubble diameter. Experimental determinations of the critical gas flux and of the time between two events of foam collapse are in agreement with this simple theory. The quantitative model can be applied to well-known eruptions and allows constraints on the size of the magma chamber roof, the bubble-size, and the gas content of magma at depth.

VERHOEVEN, J. T. A., and M. B. SCHMITZ, University of Utrecht, Utrecht, the Netherlands

Nitrogen and Phosphorus Mineralization in Peat Mires in Relation to Vegetation and Water Chemistry

Peat mires differ in vegetation and hydrology and are commonly classified into "ombrotrophic" bogs with precipitation as the only water supply and "minerotrophic" fens with inputs of rain, groundwater, and surface water. Peat mosses (*Sphagnum* spp.) form a main vegetation component in bogs, whereas grasses and sedges dominate in fens. Strong differences in hydrochemistry exist, with much higher values of pH, EC, [Ca], [HCO_3], and most other macro-ions in fens (Moore and Bellamy, 1974).

Although bogs are usually considered as nutrient-poor ecosystems and fens as (moderately) nutrient-rich, there have not been many studies of nitrogen and phosphorus dynamics in mires to support this view. Waughman (1980) studied soil inorganic N and P in a range of mires in Germany and found no differences among bogs and fens for N, and higher values in bogs for P.

We studied N and P mineralization in 10 peat mires in the Netherlands: 4 "rich" fens, 4 bogs, and two "poor fens" intermediate between fens and bogs. Mineralization was measured at 10 and at 25 cm below the mire surface by an in situ incubation technique (Verhoeven and Arts, 1987). A relative measure of decomposition rate was obtained by measuring tensile strength loss of cotton strips placed at all sampling sites.

Two-way ANOVAs for the data on soil inorganic N and on N mineralization showed significant effects for the factors of site depth; for soil inorganic N the interaction factor is also significant. A stepwise multiple regression showed that the variance in soil inorganic N is to some extent explained by the factors of depth and presence of *Sphagnum*. Variance in N mineralization is explained by the factors of depth, total soil P, and soil pH. There are no significant differences between bogs and fens in soil inorganic N or N mineralization. In all mires studied, N mineralization is higher at 10 than at 25 cm below the surface.

A two-way ANOVA for the data on soil inorganic P showed a significant effect for the factor site only. For P mineralization, there are significant effects for site and depth. The interaction factors are not significant. Stepwise multiple regression showed that the variance in the data for soil inorganic P is explained by soil pH, presence of *Sphagnum*, total soil P, and total soil C. The variance in P mineralization is explained by soil pH and depth. Inorganic soil P and P mineralization are significantly higher in bogs and in "poor" fens than in "rich" fens. P mineralization is higher at 10 than at 25 cm below the surface.

The ANOVA for the cotton tensile strength loss data showed a significant effect for the factor site only. The variance in the data is primarily explained by the soil pH. Cotton decomposition is much faster in the "rich" fens than in the bogs, whereas "poor" fens show intermediate values.

The results on N and P mineralization clearly indicate that fens have by no means lower N and P availability than bogs. For P rather the reverse is true, whereas there are no significant differences for N. The view that bogs are nutrient-poor and fens nutrient-rich is not supported by our data on the most important plant nutrients N and P.

The results of the cotton strip experiment show that decomposition rates are distinctly lower in bogs than in fens, presumably as a result of the low pH in the former. This means that N and P mineralization rates are directly related to decomposition rates. The fact that a much slower decomposition in bogs results in a similar or even higher mineralization of N and P indicates that dead *Sphagnum* material loses inorganic N and P much faster than dead grass and sedge material, probably as a result of strongly different chemical compositions of these materials. The low P availability in fens is probably caused by the adsorption of phosphates to Fe, Al, and Ca ions in the fen water, which is strongly influenced by groundwater inputs rich in these three cations.

REFERENCES

Moore, P. D., and D. J. Bellamy, 1974, Peatlands: London, Elek Science, 221 p.

Verhoeven, J. T. A., and H. H. M. Arts, 1987, Nutrient dynamics in small mesotrophic fens surrounded by cultivated land. II. N and P accumulation in plant biomass in relation to the release of inorganic N and P in the peat soil: Oecologia (Berl.) v. 72, p. 557–561.

Waughman, G. J., 1980, Chemical aspects of the ecology of some South German peatlands: Journal of Ecology, v. 68, p. 1025–1046.

VERMA, P. N., Geological Survey of India, Calcutta, India

Slope Failures around Bichu Khola, Sikkim, India: Its Impact and Remedies

Environmental degradation caused due to slope failures, subsidences, debris avalanches, etc., are common and occur frequently in Eastern Himalayas where rainfall is heavy and intense. In turn, such failures result in mass wastages and their movements inflict heavy damages to various civil structures coming their way and endanger many others. The failures that occur in remote localities generally go unrecorded or even unnoticed and others that interfere with developmental activities are required to be thoroughly studied and suitably treated for their control and stabilization.

Bichu *Khola* (= stream) catchment, a part of Rora*Chhu* (= river) valley in Sikkim Himalaya, presents a classic example of fast deterioration of its morphological character over a short span of a decade during which a major slide generated all along the stream course to a height of about 1000 m caused by headward erosion and lateral slope failures reaching to the crestline of the ridge. The slope failures along Bichu *Khola* not only affect the smooth functioning of a medium hydroelectric project located close by but have also dislocated the line of communication across it. As the problem aggravated over the years, the water conductor channel of the hydroelectric project was relocated many times across the stream. Ultimately, a single-span 110 m long three-hinged arch bridge was constructed to carry the water conductor pipe across this stream. The recent slope failures, which occurred about 5 years ago close to the bridge abutment, endangered the safety of the bridge built during 1982–1983 to carry the water conductor pipe. The rock slide that occurred on 3 July 1983, on the right abutment immediately downstream of the bridge, is a very small fraction of the major Bichu *Khola* slide, but it proved to be of serious concern because of its proximity to the bridge. It was caused due to percolation of rainwater into the gneissic rockmass of Pre-Cambrian, traversed by a number of adversely oriented sets of cross joint that constituted the right abutment. Containment and control of this slide and related subsidences were the prime concern that called for immediate study of the problem and undertaking of urgent measures for stabilization and protection of the failed bridge abutment.

Stabilization measures commonly utilize the concepts of removal, consolidation, drainage, and restraint. Removal concepts include detachment of unstable overhanging rock mass both manually and by controlled blasting techniques. Consolidation methods involve grouting of jointed rock mass. Drainage methods involve dewatering using vertical and subhorizontal drainage holes and gravel filled surface drains. Restraint techniques involve the use of buttressing concepts such as buttresses, concrete buttresses, and rock bolts. Common protection methods include river training works, slope dressing by making berms and benches, vegetation turfing, shot-creting, and wire-meshing.

Most of the stabilization and corrective measures have been designed and executed on both the banks of the Bichu *Khola* around the water conductor bridge. A new technique to provide a 30 m deep grout curtain on the uphill slope behind the crown of the Bichu *Khola* slide of 3 July 1983, extending both upstream and downstream from the bridge alignment on the right abutment, has been attempted. The main objective of this grout curtain is to arrest the flow of subsurface water through the joint planes present in the abutment rock causing instability to the hill slope. These measures, however, posed the problem of build-up of water pressure in the rock mass behind the grout curtain, which is likely to cause a major slope failure of the abutment including the grout curtain itself. To overcome this problem, a number of vertical pressure release holes have been provided behind the curtain and, in addition, a number of subhorizontal drainage holes have been provided at three different levels below the crown so as to facilitate the guided draining of the accumulated water behind the curtain to avoid failure of the abutment.

The paper describes the nature of the fast deterioration of the slopes of the Bichu *Khola* catchment and its impact on the Lower Lagyap Hydroelectric Project, and identifies the causes of the slope failures. The various slope stabilization and protective measures adopted and their efficacy are discussed in detail.

VERWOERD, W. J., University of Stellenbosch, Stellenbosch, South Africa

Genetic Types of Ore Deposits Associated with Carbonatites

The economic geology of carbonatites is receiving renewed attention as a result of (1) an upsurge in demand for some of the rare earth elements (REE), especially europium, samarium, neodymium, and yttrium; (2) realization that in certain metals not previously sought there may also be sufficiently enriched in carbonatites, e.g., Ti, Mo, Ta, U; (3) economic advantages of multielement ore deposits where fluctuations in demand and market price in single commodities can be compensated provided that metallurgical difficulties are overcome; (4) strategic importance of phosphate resources leading to trebling of production of apatite from carbonatite complexes in ten years (reaching 8.5 Mt in 1985); and (5) increasing awareness that carbonatites are not rare.

Typical carbonatite bodies can be regarded as the roots of carbonatite volcanoes but others form dikes, sills, irregular intrusions, tuffs, and lava flows. Associated ore deposits show a surprising variety: They run almost the full gamut of the generally recognized genetic types excluding the biogenic, pneumatolytic, and placer categories. More than one type is often present at a single locality (Table 1).

Like granite pegmatites, carbonatites are enriched in large-ion-lithophile (LIL) elements, but each rock type has its own characteristic suite, with only a few elements in common. Carbonatites are characterized by Sr, Ba, Ce, La, Nb > Ta, Zr, Th, U, P, and F, and pegmatites by B, Be, Li, Cs, Sn, Ta > Nb, Y, W. Th, U, and P.

The best-known carbonatite-related ore deposits are the pyrochlore (Nb) mines of Araxá (Brazil), Oka (Canada), and Fen (Norway); the copper, vermiculite, and apatite mines of Phalaborwa (South Africa); the Mountain Pass rare earth deposit in California, USA; the fluorite mine of Amba Dongar, India; and the vanadium mine of Potash Sulphur Springs, Arkansas, USA. These and others have been reviewed by Deans (1966, 1981), Verwoerd (1986), and Mariano (in press). The purpose of the following paragraphs is to focus on some recent developments and a few lesser-known occurrences.

Phalaborwa (South Africa): During 1987, production was maintained at approximately 3 Mt phosphate concentrate, 128,692 t Cu, 17 t precious metals (Ag, Au, Pt, Pd), 176 t U_3O_8, 7134 t ZrO_2, 446 t Ni sulfate, 150,638 t sulfuric acid, 70,931 t magnetite concentrate, and 185,838 t vermiculite concentrate. Phalaborwa is also in a favorable position to become an important REE producer. The apatite concentrate contains 0.5% REE as oxides and is unusually rich in europium (1.2% of total REE). The recoverable reserve is estimated at 2.16 Mt, the third largest in the world after China and the USA.

Powderhorn (Colorado, USA) is currently under investigation as a titanium mine. Perovskite occurs in biotite pyroxenite and the indicated reserve is 500 Mt assaying 11.5% TiO_2. Associated carbonatite of Big and Little Iron Hill contains several potential pyrochlore ore bodies.

Similar projects to exploit anatase (alteration product of perovskite) from deeply weathered carbonatite complexes in Brazil have reached an advanced stage.

Huanglongpu (Shaanxi, China): This is an Mo-Pb deposit "of industrial scale" consisting of lenticular quartz–carbonatite veins up to 500 m long and 20 m wide, emplaced in Proterozoic metaspilites 206 m.y. ago. Ore minerals are pyrite, galena, and molybdenite with calcite, quartz, microcline, barytocelestine, and minor amounts of plumbodavidite, monazite, bastnaesite, etc. Carbonatitic affinities are indicated by fluid inclusion, rare earth, and Pb, S, Sr, C, and O isotope studies (Huang et al., 1985).

Kangankunde (Malawi): Following unsuccessful attempts to develop a strontiantite mine, work by BRGM of France is now underway to open up an REE deposit grading 5.58% monazite in ankeritic carbonatite.

Khanneshin (Afghanistan) is a dissected early Quaternary carbonatite volcano with concentrations of bastnaesite, strontianite, magnetite, apatite, pyrochlore, galena, and secondary uranium minerals (Alkhazov et al., 1977).

Baiyun-Ebo (Inner Mongolia, China): This is a huge stratabound magnetite–hematite deposit intimately associated with the world's largest resource of REE. Bastnaesite and monazite occur in a typical carbonatitic assemblage of pyrochlore, columbite, barite, apatite, fluorite, phlogopite, and riebeckite. Proven reserves are given as 36 Mt REO, but this may include other localities in China. According to Bai and Yuan (1986), the ore bodies are lenticular and genetically related to K-rich alkaline volcanism. They are interpreted as submarine volcanosedimentary carbonatite deposits of exhalative or effusive origin.

Olympic Dam (Roxby Downs, South Australia): Originally described as a disseminated sulfide ore deposit hosted by sedimentary breccias, this important new Cu-Au-Ag-U mine that came into production in mid-1988 may perhaps be related to carbonatite. Drilling and underground development has led to the idea that the ore body, rich in hematite, is located within a very large volcanic breccia pipe with alkaline affinities. The ore contains up to 1% REE as bastnaesite and florencite. The immense size (in excess of 2000 Mt grading 1.6% Cu) makes it difficult to relate this deposit to ordinary mineralized carbonaitites.

Morro do Seis Lagos (Amazonas, Brazil) is now reputed to be the world's largest niobium deposit with 2800 Mt of ore grading 2.81% Nb_22O_5, mostly occurring in alteration products of pyrochlore. Araxá with 460 Mt at 3.0% Nb_2O_5 will probably remain the major producer for many years to come, but many other pyrochlore-bearing carbonatites are known. A new discovery of 36 Mt grading 1.5% Nb_2O_5 plus unassessed quantities of REE, Cd, and Ti was announced in 1987 by Gabon.

REFERENCES

Alkhazov, V., Z. Atakishiyev, and N. Azimi, 1977, International Geology Review, v. 20, p. 281–285.

Bai, G., and Z. Yuan, 1986, Bulletin Inst. Mineral Deposits, Chinese Academy of Geological Science, v. 2, p. 126–136. (English abstract).

Deans, T., 1966, *in* O. F. Tuttle and J. Gittins, eds., Carbonatites: Wiley, p. 385–413.

———, 1981, Proceedings of 1st International Symposium on Carbonatites, Brazil, p. 123–133.

Huang, D., Y. Wang, and F. Nie, 1985, Acta Geologica Sinica, v. 59, p. 241–258 (English abstract).

Mariano, A. N., in press, *in* K. Bell, ed., Carbonatites, Allen and Unwin.

Verwoerd, W. J., 1986, *in* C. Anhaeusser and S. Maske, eds., Mineral deposits of southern Africa: Geological Society South Africa, p. 2173–2191.

Classification			Minerals	Uses	Examples
Hypogene	Magmatic	Major phase crystallization	1. Calcite	cement	Sukulu, Tororo
			2. Bastnaesite	REE	MouNtain Pass
			3. Apatite	phosphate	Jacupiranga
			4. Magnetite	Fe ore	Phalaborwa
		Dissemination	1. Pyrochlore	Nb	Oka, Fen, Sokli
			2. Apatite	phosphate	Phalaborwa, Tapira, Patrocinio, Catalão I, Otjisazu.
			3. Baddeleyite	Zr	Phalaborwa
			4. Uranothorianite	U	Phalaborwa
		Segregation	1. Pyrochlore	Nb	Araxa, Sarfartôq
			2. Perovskite	Ti	Powderhorn
			3. Apatite	phosphate	Chishanya, Ondurakorume, Tundulu
			4. Phlogopite	mica	Phalaborwa
	Hydrothermal	Cavity filling and Replacement	1. Chalcopyrite etc.	Cu (Ni, Ag, Au, Pt, Pd)	Phalaborwa
			2. Molybdenite, galena	Mo, Pb	Huanglongpu
			3. Rutile, brookite	Ti	Magnet Cove
			4. Pyrochlore (Th)	Nb	Sokli
			5. Strontianite	Sr	Kangankunde
			6. Monazite	REE	Kangankunde, Wigu, Nkombwa
			7. Apatite	phosphate	Dorowa, Glenover
			8. Fluorite	CaF_2	Amba Dongar, Okorusu, Tchivira
			9. Sodalite	gemstone	Swartbooisdrif
			10. Hematite (Th)	Fe ore	Fen, Kalkfeld
	?Volcanogenic	?Exhalation	1. Magnetite, fluorite bastnaesite, quartz monazite etc.	Fe, REE, Nb	Baiyun-Ebo
			2. Hematite, bornite, chalcopyrite, fluorite, bastnaesite etc.	Cu, U Au, Ag	Olympic Dam
Supergene	Residual	Mechanical concentration	1. Anatase	Ti	Tapira, Catalão I, Patrocinio
			2. Apatite	phosphate	Araxá, Cargill, Mt.Weld, Sukulu, Sokli
			3. Pyrochlore	Nb	Araxá, Mrima, Lueshe
	Chemically enriched	Secondary crystallization	1. Apatite, francolite	phosphate	Bukusu, Mt.Weld, Sokli, Salitre
			2. Monazite, florencite, churchite cerianite	REE	Araxá, Mrima, Mt.Weld, Morro do Seis Lagos, Sandkopsdrif
			3. Montroseite etc	V	Potash Sulphur Springs
			4. Carnotite ? etc.	U	Khanneshin
			5. Pyrochlore (altered)	Nb	Morro do Seis Lagos, Lueshe, Mt.Weld
			6. Strontianite	Sr	Kangankunde
			7. Vermiculite	mica	Phalaborwa, Mud Tank
			8. Trona etc.	soda	Lake Natron, Lake Magadi, Zoutpan

VIDAL, C. E., Perubar S.A., Lima, Peru, D. C. NOBLE, Mackay School of Mines, Reno, NV, E. H. McKEE, U.S. Geological Survey, Menlo Park, CA, J. E. BENAVIDES, Lima, Peru, and H. C. de los RIOS, Mauricio Hochschild y Cia. Ltda. S.A., Lima, Peru

Hydrothermally Altered and Mineralized Late Pliocene-Quaternary Central Volcanoes in the Andes of Southern Peru

Large glacially eroded central volcanoes of late Pliocene and Quaternary age are found in southern Peru both along and behind the late Neogene volcanic front. The stratovolcanoes Ccarhuaraso and Palla-Palla are situated on the southwestern part of the high plateaus. Together with Nevado Portugueza 140 km NW of Ccarhuaraso, they define an extinct northern extension of the active frontal arc, which came into existence in late Pliocene time. The Nevado Tetón-Nevado Huayunca-Cerros Chancoaña central volcanic complex, within which the Tetón and Millo centers of alteration and mineralization are situated, is located in the Huanzo Cordillera in the central part of the high plateaus of southern Peru approximately 100 km behind the frontal arc. In this back-arc region calcalkalic intermediate lavas are associated with widely distributed flows of basalt and basaltic andesite of late Pliocene to Recent age.

The central volcanoes, about 25 to 35 km in diameter, are composed largely of gently outwardly dipping flows of intermediate lava, mainly pyroxene andesite. Domes and short flows of dacite are present in the central parts of the volcanoes. Reconnaissance suggests that Ccarhuaraso was erupted within a large collapse caldera that formed about 2 Ma ago (E. Farrar and D. C. Noble, unpub. perlim. K-Ar dates) concurrent with the eruption of a rhyolitic ash-flow sheet exposed around the volcano. Reconstruction of Ccarhuaraso and Palla-Palla indicates that glaciation has removed about 500 m from these structures. The Nevada Tetón-Nevado Huayunca-Cerros Chancoaña complex has been highly modified by glacial erosion.

K-Ar ages of 1.5 ± 0.1 and 2.6 ± 0.1 Ma have been obtained on separates of phenocrystic biotite from unaltered dacite lavas from Ccarhuaraso and Palla-Palla, respectively. Hypogene alunite from the core of Ccarhuaraso has been dated at 1.2 ± 0.15 Ma. The dacite dated at Palla-Palla predates hypogene advanced argillic alteration. Separates of hypogene alunite from Tetón and Millo have yielded ages of 2.3 ± 0.1 and 2.8 ± 0.11 Ma, respectively, and biotite from unaltered dacite overlying hydrothermally altered rock at Millo has been dated at 0.45 ± 0.2 Ma. The unaltered dacite lavas dated postdate late Pliocene-early Quaternary glacial erosion at Ccarhuaraso, Palla-Palla and Millo.

The highly altered central part of Ccarhuaraso has been explored for copper deposits of vein and porphyry type and, more recently, precious-metal deposits. Formation of early dikes and bodies of quartz-tourmaline-pyrite altered intrusive breccia and tuffisite was followed by one or more episodes of intense hypogene advanced argillic alteration. Fractures, which are generally oriented radial to the center of the volcano, are surrounded by intense quartz + alunite + pyrite selvages that grade outward into zones characterized by kaolinite-group minerals and prophylitically-altered rock. Secondary fluid inclusions in quartz phenocrysts within quartz + alunite + pyrite rocks give T_h of 330°–380°C, with T_t of about 330°C, and variable salinities of from about 7 to 18 equiv. wt. % NaCl. These inclusions represent the sulfur- and chlorine-rich and highly oxidized magmatic fluids, apparently mixed to greater or lesser extents with meteoric waters, that formed the hypogene quartz + alunite ± pyrite assemblage. Later fractures have localized zones of hydrothermal brecciation and intense silification. In these silicified zones alunite has been destroyed, and enargite, pyrite, barite, and native sulfur deposited within the resulting vuggy silica rock. Gold and silver is present largely as micron-sized argentite and electrum that occur both free and as inclusions in enargite. Au/Ag ratios are typically low, ranging from 0.01 to 0.05. Primary dispersion patterns as evaluated in diamond drill cores show progressive increase from kaolinite to alunite to vuggy silica zones; mean values of this progression are 150-450a-3600 ppm Cu, 250-150-1700 ppm As, 25-35-120 ppm Ag, and 0.14-0.18-1.0 ppm Au. Concentrations of Cu, Ag, and Au in talus has proved useful in delineating blind vein extensions. Ore shoots have been found in the Rasuhuilca vein; targets in other parts of the central zone have not been evaluated.

Central and satellite vent areas with dacitic domes, hydrothermal alteration, and silver-bearing veins are present in Runtos, Cieloccacca, Ccorihuiri, and Pirhua areas of Palla-Palla. Large zones of barren quartz-alunite altered rocks are exposed at Runtos, and pebble dikes and silicified vein structures are present between Runtos and the center of the Palla-Palla complex. At Pirhua, on the southwest flank of Palla-Palla, narrow northwest-trending veins are mined for silver and base metals. Vein mineralogy consists of pyrite, galena, sphalerite and chalcopyrite accompanied by quartz, rhodonite and adularia. Silver values are mostly carried by minute inclusions of tetrahedrite and ruby silver in galena.

The Tetón and Millo areas are characterized by well-developed hypogene advanced argillic alteration. Although well-defined veins and quartz-alunite ledges possess clear Au, Ag, As, and Sb signatures, and finely disseminated native sulfur is present over large areas at Millo, no potentially economic deposits have been identified.

VIDIC, N., and F. LOBNIK, University of Ljubljana, Ljubljana, Yuogslavia, and M. J. PAVICH, U. S. Geological Survey, Reston, VA

Soil Stratigraphy of Sava Terraces, Slovenia, Yugoslavia

This report presents chronostratigraphic and soil characterization data obtained from the Sava River terraces. These terraces are among the most significant geomorphic elements of the Sava River valley and the Slovenian forealps, and their gravels are correlative with the glacial moraines mapped by Sifrer in the Dobrave region of western Slovenia. Moraine and terrace ages are related to and can be correlated with periods of Alpine glaciation calculated by Milankovitch. Between periods of fluvioglacial aggradation and gravel deposition, the Sava River eroded to new base levels, probably in response to continuing uplift of the forealps. From a study of terrace soil profiles and soil geochemistry, we can postulate the climatic and geomorphic record of this tectonically active area.

Nine soil profiles were characterized on four terraces of different ages. Using Milankovitch model ages, the terraces are of Gunz (~575 Ka), Mindel (~425 Ka), Riss (~100 Ka), and Wurm (~30 Ka) age. A comparison of soils on different terraces shows that the thickness of the weathering profile, the thickness of the argillic Bt horizon, the degree of base leaching and acidification, the amount and the thickness of clay coatings, and the amount of concretions increase in soils with increasing age. The ratio of cation exchange capacity, (CEC) to percent clay shows a decreasing trend in the soils of the older terraces. These changes are consistent with soil characteristic changes in chronosequences of other temperate-humid areas.

Analyses of textural composition of soils suggest that the fine particles in the terrace soils are derived from in situ weathering processes and are not the result of colluvial or aeolian deposition. The initial product of the weathering of predominantly limestone and dolomite gravel deposits is fine silt, which alters to clay within the solum. Results of X-ray diffraction analyses of clay fractions of soil samples from the terraces show a trend of increasing 14.7-A clay and decreasing 10.2-A clay from the base of the profiles. The transitional nature of the change from 10.2-A to 14.7-A through a mixed-layer intermediate phase in interpreted as further evidence for pedogenic rather than aeolian origin of the clay.

Analyses of the cosmogenic ^{10}Be content of the soil profiles show that ^{10}Be inventories increase with increasing soil development and stratigraphic age of the terraces. The preliminary ^{10}Be analyses of selected soil profiles indicate minimum ages of 36 Ka and 410 Ka for two of the terraces, consistent with two of Milankovitch's calculated ages. A 100 Ka terrace has not yet been identified from the ^{10}Be analyses. The present results indicate a major interglacial period between about 410 Ka (Mindel) and 36 Ka (older Wurm). Such a long interglacial interval provided a period of weathering sufficient to alter the older fluvioglacial deposits and to modify the landscape of the Sava River valley.

Continued sampling and analysis over the course of the project and statistical analyses of the soils data will provide better chronostratigraphy and improve our ability to distinguish soils and terraces of different ages.

VIELE, G. W., University of Missouri, Columbia, MO, and W. A. THOMAS, University of Alabama, Tuscaloosa, AL

Tectonic History of Ouachita Orogenic Belt

Lying mostly in the subsurface, the late Paleozoic Ouachita orogenic belt flanks the southern margin of the North American craton. The orogenic belt, exposed in the Ouachita Mountains of Arkansas and Oklahoma and in the Marathon region of western Texas, closely mimics the trace of a deeply buried, early Paleozoic continent–ocean boundary, defined by gravity and seismic reflection-refraction surveys.

The tectonic history of the Ouachitas starts with late Precambrian–Cambrian rifting of the southeastern and southern margins of the North American craton. An ocean opened with northeast-trending rifts offset by northwest-trending transform faults. Shallow-water carbonates were deposited on the southern margin of North America, but equivalent age preorogenic sediments of the Ouachita orogenic belt were deposited in two large embayments off the southern margin of North America. Outcrops in the Ouachita Mountains and the Marathon region record Late Cambrian–Ordovician deposition of shale, sandstone, and micrite on the North American continental slope and rise. A few debris flows contain clasts of carbonate rocks and granitic rocks derived from the continent. The ocean widened and deepened, during a span of 100 m.y., and middle Paleozoic siliceous shale, chert, and manganese-bearing novaculite were deposited at abyssal depths. These preorogenic Ouachita sediments were deposited on oceanic crust of the North American plate, adjacent to but not on the North American craton.

During the Carboniferous, south to southeasterly subduction of the North American plate led to continent–arc collision—the Ouachita orogeny. A remnant ocean closed diachronously from east to west. First evidence of the closing comes from the subsurface Balck Warrior foreland basin of Mississippi: There, Carboniferous, shallow-water sandstones, prograding northeastward over North American continental crust, record a multipart provenance containing a volcanic arc and an uplifted subduction complex. To the west, in the still-open remnant ocean, as much as 16km of Carboniferous sandstone and shale in the Ouachitas and 4.5km of sandstone and shale in the Marathons were deposited. This synorogenic deep-water flysch formed a complex of submarine fans, generally prograding westward, although fans and olistostromes entered the ocean from point sources along both margins. By middle Pennsylvanian time (Middle Atokan), the sediments filled the ocean and lapped onto the North American continent. Farther to the southwest, in the subsurface, east Texas segment of the Ouachita orogenic belt, the ocean was closed by Atokan time, and a shallow-water clastic wedge was prograding westward from the uplifted orogen.

While the deep-water sediments were being deposited, subduction of the oceanic crust and off-scraping of its sedimentary cover formed an offshore accretionary prism composed of preorogenic rocks that were stacked in fold and thrust nappes. Carboniferous flysch that entered the subduction zone was also incorporated into the accretionary prism. Some flysch was underplated; much was mixed into belts of melange. South of the accretionary prism, flysch was deposited in a fore-arc basin flanked on the south by a volcanic arc, now in the subsurface. In the Ouachita Mountains, preorogenic strata in the accretionary prism form a lower deck of tightly folded rocks, and synorogenic strata of the fore-arc basin form an upper deck of broadly folded rocks. During the middle Pennsylvanian, the complexly deformed accretionary prism and part of the fore-arc basin were obducted onto the southward-subducting North American crust, which broke down under the load by syndepositional high-angle faulting. In some places, such as the Devils River uplift of western Texas, strike-slip faulting accompanied obduction of the Ouachita rocks. The widest part of the Ouachita allochthon extends about 150km over the North American craton. Salients and recesses of the allochthon copy of the embayments and promontories of the southern margin of early Paleozoic North America. Zones of faulting and folding, some parallel to Ouachita fold and fault trends and some at high angle to Ouachita structures, extend far inland from the Ouachita tectonic front, defined as the most inboard thrust sheet that carries rocks deposited off the craton.

South of the Ouachita Mountains, in the subsurface, upper Pennsylvanian and Permian strata, deposited in successor basins, unconformably overlie broadly folded Ouachita rocks. In the frontal Ouachita Mountains and in the Marathon region, structures of the orogenic belt extend under unconformable Permian strata.

On the seaward side of the Ouachita orogenic belt, a collage of terranes, including Yucatan, filled the present-day area of the Gulf of Mexico. Triassic rifting signaled the start of the opening of the Gulf of Mexico, and the subsequent subsidence and burial of much of the Ouachita orogenic belt.

VIGNERESSE, J. L., Université de Nantes, Nantes Cedex, France

Shape and Volume of Granitic Intrusions Deduced from Geophysical Data

Introduction: Seismic velocities contrasts between granitic facies are usually very poor and the granitic bodies are transparent to seismic waves (Matthews 1986). The paucity of magnetic minerals excludes the magnetic methods except the delineation of the surrounding metamorphic aureoles or facies mapping using the magnetic susceptibility variations. Electric methods are often perturbed by water-filled fracture zones, especially within granites. These methods help in determining the internal post-magmetic structures of the massifs, but can hardly solve the mode of settlement problems. Gravity surveying is the only appropriate method to determine the three-dimensional shape of the granitic intrusions at depth. To obtain enough resolution, data should be measured on a small mesh (about 1 km) covering the whole pluton. We only considered 3D gravity data interpretations. We sorted between anorogenic and syntectonic granites.

Anorogenic granites: Ring complexes or alkaline granites are usually related to a large thermal perturbation and have a deep seated origin. They are geographically concentrated and form series of intrusions of circular shape. In general, the magnetic survey is not directly representative of the granitic body due to the mixing of acid and basic magmas. The variety of facies results in scattered density values, leading to oversimplified gravity models such as a vertical cylinder. The root may extend down to 12 km (West Africa or Corsica). Alkaline granites from the Oslo Graven (Ramberg 1976) present similar shapes. From a statistical study over ring complexes (Bonin 1982), the plutons present regular features, with an average diameter of 14 km. Their average volume is then about 5000 to 6000 km^3. Alkaline volcanism and plutonism display a spatial periodicity with wavelength varying from 60 to 100 km, depending on the underlying lithosphere thickness. This confirms the deep seated origin of the alkaline intrusives.

Syntectonic granites: More evolved granites are often associated with a complex tectonical activity. During their diapiric ascent, the massifs suffer the regional deformational field and registrate the local strain pattern (Brun 1981). The final shape results from both the effects of the intrusion and regional deformation.

Within a weak deformational field, the plutons present a subcircular shape with a metamorphic aureole in the sedimentary cover. The general shape of the later massif is subcircular with a bulk symmetry axis. The roots are located under the more differentiated material and extend down to 6 to 8 km.

When the granitic massifs are syntectonic to a deformational field, they registrate the regional deformation and act as an active marker. Their bulk shape corresponds to a vertical cone with an elliptical cross-section, the lateral extent of which superates the depth. In the French Armorican Massif, the vertical extension of the Hercynian massifs ranges between 3 to 5 km, while the elliptic horizontal cross-section is about 9 to 12 km in radius (Vigneresse 1983). Those values correspond to those determined on the granites in Maine and New Hampshire (Hodge et al 1982) or on the granites of UK (Lake District: Lee 1986, Donegal granite: Hutton 1928). Their depth is smaller than their lateral extent, resulting in a flattened shape. The plutons are spatially organized with an average wavelength of 60 km (Vigneresse and Brun 1983).

Their final shape depends on the deformation field during the upwelling. In case of transcurrent deformation (the shear plane is vertical), the orientation of the major and minor axis of the pluton with respect to the shear plane controls its final shape. A flattened massif, the minor axis of which is vertical, is totally deformed by a transcurrent shear zone. Its minor axis remains vertical, but the horizontal cross-section of the massif changes from subcircular to an elliptical shape (Brun 1981). This is observed in the granodioritic massif of Plouaret (Guillet et al 1985), or in most leucogranitic plutons of the Armorican Massif (Vigneresse 1983). Conversely, if the major axis of the massif is vertical, it results either in an intense shearing along the walls or an offset of the massif in two parts. This is observed in the Main Donegal granite (Young 1974, Hutton 1982) or in the Mortagne granite, western france (Guineberteau et al 1987).

If emplaced during thrust faulting, the pluton is involved in the thrust and is completely laminated within the deformation, as observed on the southern leucogranites in the Armorican massif (Pont l'Abbé, Guérande). Those massifs are very flat (maximal depth shallower than 2 km) and present horizontal magmatic structures. This is also observed in the Saint Sylvestre massif (Central Massif, France), the average depth of which averages 2 km over several hundred of km^2 (Audrain et al 1988).

Other types of plutons: Other batholiths have extension over thousands of km^2, as the South Mountain Batholith in Nova Scotia, Canada or the Cornwall Batholith in southern England. From seismic data, the mean depth of the Cornwall Batholith averages 10 km and may reach 20 km from gravity data (Tombs 1977). The South Mountain Batholith also presents a mean depth of about 10 km, locally increasing to 20 km (O'Reilly 1976). Both massifs are characterized by the settlement of small (about 30,000 km^3 massifs, very closed in space (about 35 km between them) and on time (some 10 Ma). Their origin is not easily understood since it involves a very large amount of heat to explain the melting of the crustal material on such a scale. They are both postorogenic.

REFERENCES

Audrain, J., Vigneresse, J. L., Cuney, M. and Friedrich, M., 1988, C. R. Acad. Sc. Paris (in press).

Bonin, B., 1982, Les granites des complexes annulaires: Bur. Rech. Géol. Min. Orléans, Manuals et méthodes, v. 4, p. 184.

Brun, J. P., 1981, Unpublished Thèse d'Etat, Univ. Rennes, p. 212.

Guillet, P., Bouchez, J. L. and Vigneresse, J. L., 1985, Bull. Soc. Géol. France, v. 8, p. 503-513.

Guineberteau, B., Bouchez, J. L. and Vigneresse, J. L., 1987, Geol. Soc. Am. Bull., v. 99, p. 763-770.

Hodge, D. S., Abbey, D. A., Harbin, M. A., Patterson, J. L., Ring, M. J., and Sweeney, J. F., ;1982, Am. J. Sci., v. 282, p. 1289-1324.

Hutton, D. W., 1982, J. Geol. Soc. London, v. 139, p. 615-631.

Lee, M. K., 1986, J. Geol. Soc. London, v. 143, p. 425-435.

Matthews, D. M., 1986, Annales Geophysicae, v. 5B, p. 353-356.

O'Reilly, C. T., 1975, Sc. Thesis, Dalhousie Univ., Halifax, 84 pp.

Ramberg, I. B., 1976, Norsge Geol. Undersokelse 325, Oslo, 194 pp.

Tombs, J. M. C., 1977, Cornwall. Miner. Rec. Program Res. Int. Geol. Sc. 11, London, 16 p.

Vigneresse, J. L., 1983, Bull. Soc. Géol. Min. Bretagne, Rennes, 15C, p. 1-15.

Vigneresse, J. L., and Brun, J. P., 1983, Bull. Soc. Géol. France, v. 25, p. 357-366.

Young, D. G. G., 1974, The Donegal granite a gravity analysis, Proc. Royal Irish Acad., v. 74B, p. 63-73.

VILADKAR, S. G., St. Xavier's College, Bombay, India, and D. CHANDRASEKHARAM, Indian Institute of Technology, Bombay, India

Geochemistry of Composite Dike of Mandaleswar–Deccan Volcanic Province: Constitution of Crust Below Deccan Volcanics

The Narmada lineament is one of the prominent east–northeast-trending features in the Deccan volcanic province. This lineament hosts a composite dike having a similar trend and a width of about 80 m near Mandalswar (lat: 22°15′; long: 75°37′). One half of the dike is acidic while the remaining part is basic, and the contact between these two members is very sharp. Petrographic examination reveals only thin cracks filling acid veins within the basic part at the contact. Grain-size variation is exhibited by both the rocks. The acid unit contains quartz, k-feldspar, albite, and biotite set in a fine-grained matrix. The volume percent of the matrix is minimum at the contact with the phenocrysts showing larger size while away from the contact the texture is microporphyritic. On the other hand, the basic unit at the contact shows chilled characteristic features with microphenocrysts of plagioclase, and pyroxene set in a glassy matrix; away from the contact the basic member shows typical medium-grained basaltic texture with abundant plagioclase laths and clinopyroxene grains. The ranges of major and trace elements concentration of these rocks are shown below:

Acid			Basic		
SiO_2	67.97–76.60	Rb:131–286	SiO_2	47.67–50.18	Rb:9–28
TiO_2	0.06–0.79	Sr:143–441	TiO_2	2.26–3.27	Sr:183–237
Ai_2O_3	12.30–14.51	Ba:128–1274	Al_2O^3	13.31–14.37	Ba:98–188
Fe_2O_3	0.54–2.67	Y:18–42	Fe_2O_3	4.12–7.35	Y:33–45
FeO	0.68–2.00	La:24–167	FeO	7.27–10.74	La:27–34
MgO	0.14–1.35	Zr:151–309	MgO	4.44–6.95	Zr:129–173
CaO	0.58–2.41	K/Rb:166–302	CaO	10.00–10.67	K/Rb:200–378
Na_2O	3.03–3.49	Rb/Sr: 0.38–2.00	Na_2O	2.1–2.56	Rb/Sr: 0.04–0.15
K_2O	4.68–6.60	Sr/Ba: 0.20–1.87	K_2O	0.41–0.69	Sr/Ba: 0.98–2.07

In the TAS diagram of Cox et al (1984), the acid members fall in the dacite–rhyolite fields whereas the basic part of the dike falls in basalt field. The geochemical parameters do not support a simple differentiation process for the origin of the dacite and rhyolite.

In order to see whether the acid rocks have originated by mixing of basaltic liquid with crustal rocks, we mixed average Deccan tholeiite rock chemical values with the average continental crust in different proportion (5, 10, 20, 30, and 50%) following the procedure suggested by Allegre and Minster (1978) to produce high silica, low MgO rocks similar to dacite and rhyolite. The trace element values do not support such an assumption. We feel that the dacite and rhyolite represent crustal melts intruded into the Narmada linear feature during the course of the basaltic dike activity.

The chondrite normalized Ba, Rb, K, Sr, Zr, and Y values and Rb/Sr, K/Rb, and Sr/Ba ratios of the dacite and rhyolite match very well with certain older crustal rocks from the Indian and Canadian shields and early Precambrian rocks of Minnesota, USA. Among the older crustal rocks, the above values of quartz monzonites (Minnesota) and adamellites (Kolar and Chitoor, India) match very closely with the values obtained for dacite and rhyolite. These values are shown below:

	Ba	Rb	K	La	Sr	Zr	Y	Rb/Sr	K/Rb	Sr/Ba
	Rock/chrondrite values							Ratios		
Dacite	332–335	374–483	330–351	292–530	23–40	53–55	9–12	0.38–0.51	249–302	0.2–0.34
Rhyolite	34–180	660–837	320–451	111–171	13–23	23–31	7–17	1–2	166–182	0.26–1.87
Qtz. monzonites (USA)[1]	118–212	434–691	261–354		10–126			0.32–1.58	134–542	0.25–0.36
Adamellite (Kolar)[2]	227	511	295		41			0.39	198	0.52
Adamellite (Chittoor)[2]	110	837	323		11	32		2.32	162	0.30

[1]Arth and Hanson (1975).
[2]Saha (1979).

In the absence of any assimilation-fractional crystallization processes, the acid members of Mandaleswar do indicate the presence of quartz monzonites/adamellites in the crustal rocks below the Deccan volcanic province of the area. REE analyses on these rocks are being carried out, and we hope the data will throw more light on the nature of the crust below the Deccan province.

REFERENCES

Arth, J. G., and G. N. Hanson, 1975, Geochemica et Cosmochimica Acta, v. 39, p. 325–362.
Allegre, C. J., and J. F. Minster, 1978, Trace elements in igneous petrology: New York, Elsevier Scientific Pub., 272 p.
Cox, K.G., J.B. Bell and R.J. Pankhurst, 1984, The interpretation of igneous rocks: London, George Allen and Unwin, 450 p.
Saha, A.K., 1979, Geological Society of India, v. 20, p. 375–392.

VILAS, J. F., and A. E. RAPALINI, Universidad de Buenos Aires, Buenos Aires, Argentina

Tectonic Significance of Late Paleozoic Paleomagnetic Data from Argentine Andean Chain

Paleomagnetic data of Late Paleozoic rocks exposed in the Precordillera and Cordillera Frontal at about 32°S (Argentine Andean Chain) are analyzed regarding their tectonic significance.

The paleomagnetic pole (PP) positions for the clastic sedimentary rocks of the Majaditas Fm. (Upper Carboniferous) and the Yalguaraz Fm. (Upper Carboniferous), and for the volcanic rocks of the Portequelo del Cenizo Fm. (Lower Permian, 275 ± 10 Ma) are not coincident with the South American apparent polar wander path (APWP). This is interpreted as caused by large clockwise rotations of the respective sampling areas around vertical axes (Majaditas Fm: 88.7° ± 7.5°). The PP positions for younger rocks exposed at neighboring localities, such as the volcanics of the Tambillos Fm. (Upper Permian?) and some ignimbrites of the Choiyoi Group (Lower Triassic, 228 ± 10 Ma), as well as the preliminary paleomagnetic pole for the volcanic rocks of the Horcajo Fm. (Lower Triassic?) exposed at the same sampling area of the Yalguaraz and Portezuelo del Cenizo Fms., are coincident with the South American APWP, showing no evidence of rotation. This suggests that the rotations occurred prior to the Early Triassic and probably, according to the age assigned to the Tambillos Fm., prior to the Late Permian.

The Carboniferous rocks were deposited in back-arc basins while the Permian and Triassic volcanics correspond to a magmatic arc located on the Pacific continental edge of Gondwana. In this geotectonic setting, important dextral movements along transcurrent faults parallel to the continental margin—caused by a northward oblique subduction of the Proto-Pacific Oceanic Plate—could have easily produced the rotations observed. This interpretation agrees with the Late Paleozoic stress trajectories found by Frutos (1988) in the Chilean Andes, north-northeast–south-southwest trend of which was interpreted by this author as being caused by an oblique subduction of the Proto-Pacific Oceanic Plate under South America. The larger values of rotation observed in the Carboniferous rocks suggest that this tectonic mechanism was probably active at least since the Late Carboniferous. The nonrotated post-Early Permian PP's indicate an important change in the tectonic evolution of this section of the Southern Andes, which probably occurred in the late Early Permian or during the Late Permian (the Appalachian tectonic event?).

C7 and C8 also show lower paleolatitudes than those corresponding to the cratonic poles. This may suggest important southward displacements of these rocks parallel to the continental margin during the Late Carboniferous–Early Permian. These displacements seem to be unlikely because they must have caused rotations of opposite sense to that observed. Thus, the paleolatitude anomalies of C7 and C8 could be explained arguing that they correspond to the poorly defined fast Late Carboniferous–Early Permian section of the South American APWP (Valencio and Vilas, 1985).

P14 is the PP of the Permo-Triassic s.l. volcanic rocks of the Choiyoi Fm. exposed at 39°S in the Argentine Andean Chain. Its position indicates a large clockwise rotation of the sampling area (78.8° ± 9.0°). A similar tectonic process could have caused this rotation although other mechanisms have also been proposed (Rapalini et al, in press).

An alternative interpretation of the anomalous pole positions, which assumes that these rocks are part of an allochthonous terrane that accreted the western margin of South America prior to the Early Triassic, is against all known geological evidence. thus, it must be considered as merely speculative.

REFERENCES

Frutos, J., 1988, Evolución de las trayectorias de esfuerzo en la Tectónica de los Andes Meridionales: V Cong. Geol. Chileno, v. 1, p. 261–281.
Rapalini, A. E., J. F. Vilas, M. L. Bobbio, and D. A. Valencio, in press, Geodynamic interpretations from paleomagnetic data of Late Paleozoic rocks in the Southern Andes: American Geophysical Union, Monograph Series.
Valencio, D. A., and J. F. Vilas, 1985, Evidence of a microplate in the Southern Andes?: Journal of Geodynamics, v. 2, p. 183–192.

Table 1.

	Age	Sampling area	Symbol	P.P.	A95
Majaditas Fm.	Cu(?)	31.8°S 69.4°W(1)	C8	200°E 42°S	8°
Yalguaraz Fm.	Cu	32.1°S 69.5°W(2)	C7	195°E 56°S	7°
P.del Cenizo Fm.	Pl	32.1°S 69.5°W(2)	P13	212°E 30°S	7°
Choiyoi Fm.	P-Tr	39.1°S 70.5°W(5)	P14	232°E 21°S	8°
Tambillos Fm.	Pu(?)	32.3°S 69.4°W(3)	P15	320°E 79°S	6°
Horcajo Fm.	Trl(?)	32.1°S 69.5°W(2)	P16	263°E 57°S	14°
Choiyoi Group	Trl	31.5°S 69.5°W(4)	P12	290°E 67°S	10°

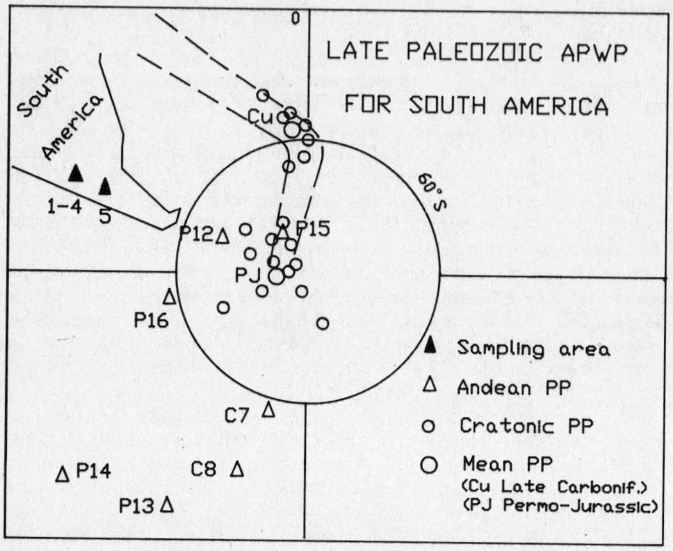

VILLASUSO, M., PACHECO, J., GAMBOA, J., CABRERA, A., SANCHEZ, I., GONZALEZ, R., FRIAS, J., CORONADO, V., GOMEZ, A., VAZQUEZ, J., CAMARA, A., PEREZ, I., Universidad Autonoma de Yucatan, Yucatan, Mexico

Hurricane Gilbert's Effects on the Karstic Aquifer North-Northwest of Yucatan, Mexico: Hydrochemical and Geohydrological Historical Data

This abstract has been written during an emergency monitoring campaign due to the invasion of Hurricane Gilbert. Data have not been completed at this stage. The cyclone event gave the opportunity to study the influence of the storm on the groundwater properties. Two hundred fifty sites have been monitored with history on head and water quality records.

North-northwest region of the Yucatan Peninsula has been studied by the authors since 1984. Water quality and piezometric records which show the variations present in the aquifer because of the dry and rainy seasons, have been collected.

The hydrogeology of this region, in general, presents a karstic aquifer containing a fresh water lens, 40 mts. thick, overlaying a salt water lens with a thickness which varies, in terms of the bottom of the aquifer, between 50 and 80 mts.

Groundwater movement is through fractures and dissolution cavities, from higher precipitation zones to the coast. The heads clearly define a flow direction to the north and northwest which coincides with the trend of the main fractures system of the geological formations in the studied region. Analysis of heads in micro-zones show variations of ±20 cms., because of the horizontal strata through which thew water passes, given that not all the wells tapped the same depth.

Heads have varied in terms of recharge between 0.8 and 1.0 mts., during one hydrological cycle. Water quality presents an increase in its nonmetallic ions as high as 40% in rainy seasons, being the nitrate of special importance because it increases in 100% because of the organic matter which comes from the soil, reaching the squifer. Hurricane Gilbert invaded the Yucatan Peninsula with an average speed wind and precipitation of 200 kph and 14 cm respectively in all the study area. The water precipitated went directly to the aquifer by infiltration. This is because the soil had already been saturated because of the rainy season. Records before the event show 780 mm of precipitation from June up to mid-September which made the water levels increase 76 cm in average. The day after Hurricane Gilbert passed, heads had increments ranging from 0.64 mts up to 1.1 mts in the region.

This extraordinary event generated tides of up to 7 mts high, which invaded approximately 5 km inland. This phenomena modified significantly the aquifer dynamics 25 km from the shoreline in the study area. Heads in this influenced zone suffered additional increments of 50 cm with respect to other sites located south of this 25 km belt. It has been observed, at standard conditions of recharge, that in both zones water levels have increased in similar ratios of some centimeters, maintaining a flow natural gradient to the sea.

The aquifer hydrochemistry has been studied for four years. Reactions of the elements vary due to the circulation of groundwater in the rock matrix. Calcium and magnesium ions, in general, get their values increased during wet seasons (Cabrera et al, 1984). Their average values are 96 ± 13.6 mg/L and 35 ± 11.9 mg/L respectively in a four year observation period. Chloride ion also increases during wet season, having average values of 135 ± 55.3 mg/L. On the other hand, bicarbonate ion is depleted recording average values of 384 ± 61.3 mg/L in a four year observation period.

The partial data collected at this stage in the field which is being processed so far show no variation with respect to the recorded average values. In general, the cationic behavior (Ca^{2+} and Mg^{2+}) of the waters sampled has been constant with no brusque variation, differing from bicarbonate and chloride which had significant increments with respect to the previously observed average. Nitrates stayed with their normal observed variation.

Nitrate can be used as a tracer in groundwater. The nitrate content in groundwaters surrounding a wastewater injection well is almost zero. It increases in concentration following the flow direction (Villasuso et al, unpublished; Villasuso, 1988).

Taking the aquifer of the area under study as a closed system and considering that Hurricane Gilbert distributed the precipitation uniformly, it is possible to attempt to trace the groundwater using nitrate ion in order to determine its main flow direction. On the other hand, NO_3^-/Cl^- ratios can be used as parameters to identify chloride origin and confirm Back and Lesser's theory (Back and Lesser, 1981) over the origin of groundwaters in the Yucatan Peninsula.

REFERENCES

Back, W., and Lesser, J. M., 1981, Chemical constraints of groundwater management in the Yucatan Peninsula, Mexico: Journal of Hydrology, v. 51, p. 119-130.

Cabrera, A., Pacheco, J., Coronado, V., Gomez, A., 1984, Evolucion de la Calidad del Agua al Norte de Merida, Yucatan: Memorias del IV Congreso de Ingenieria Sanitaria y Ambiental, v. II, p. 441-414.

Villasuso, M., 1988, El Suministro de Agua Potable y la Disposicion de Efluentes en el Acuifero Karstico de Merida, Yucatan: Simposio sobre Proteccion de Aguas Subterraneas, WHO/CEPIS, San Juan, Puerto Rico.

Villasuso, M., Sanchez, I., Mendez, R., Coronado, V., Deep well injection of waste in the Yucatan Aquifer, Mexico (unpublished).

VILLENEUVE, M., Université d'Aix-Marseille III, Marseille-St.-Jérome, France

Southern West Africa Geotraverse

Within the context of the International GGT project (profile number AF 31) a proposed geological traverse is supposed to be carried out in the southern part of the West African craton. The profile oriented north-northwest–south-southeast extends from the Atlantic border of the Craton from Guinea-Bissau up till the Volta Basin in Ghana. Thus, from the west to the east, it cuts across: the Meso-cenozoic Senegalese basin, the Paleozoic basin of Guinea, the Pan-African fold belt of the Bassarides/Rokelides, the Proterozoic basin of Madina-Kouta, a large part of the Birrimian (approximately 2,000 Ma) of the dorsal of Ivory Coast (Leo dorsal) and the proterozoic and paleozoic Volta Basin which constitutes the foreland of the Pan-African fold belt of the Dahomeyides.

The main interest of this geological traverse is to show the interrelationships between the principal geological units which form the southern part of the West African Craton. These units are: the Lower Proterozoic substratum actually being studied (principally) for its richness in minerals and including its Upper Proterozoic cover, the Pan-African fold belts attached to the Craton on its two borders, the Lower Paleozoic sedimentary basins partly deformed tectonically by the Hercynian Orogenesis during the time of the collision of Africa and North America and finally, the Meso-Cenozoic basin which was installed in the West African margin during the opening of the Atlantic Ocean. Available geological data should be able to lead to a generalized synthetic view of the folded Birrimian troughs installed on an Archean Basmenet. There are no general geological sections of these troughs up till now. The localization and identification of the limit between the Archean Basement of Sierra Leone and the Birrimian formations east of Guinea would be one of the points of interest of this study. A second point of interest would be the study of the Pan-African accretion at the end of Proterozoic on the margin of the West African Craton. Certain microblocs of the Pan-African fold belt can actually be found on the American margin. The deformation of the northern part of the Paleozoic Guinca basin represents therefore, the most external limit of the Hercynian fold belt of the Appalachian mountains and the Mauritanides.

The only main geophysical work carried out in the area is the gravimetric coverage which is about to come to an end very soon. It is intended to lead to a better understanding of the structure of the continental crust of the West African craton as well as models for the Birrimian troughs and the Pan-African fold belt. The structuration of this crust by megafaults related to the successive orogenesis into blocks or structural units with various densities represents the final major point of interest of this West African geotraverse.

VILLENEUVE, M., Faculté des Sciences de St-Jérôme, Marseille, Francé

Cross-Section in Southern Part of Mauritanide Orogen

This southern cross-section through the Mauritanide belt in eastern Senegal crosses at the northwest Koulountou belt and at the southeast Bassaris belt, both separated by the sedimentary Youkounkoun basin.

The Koulountou belt is mainly formed by volcanisedimentary rocks and by a calcalkaline volcanic complex (Niokolo-Koba complex). The Bassaris belt is mainly made up of basic volcanisedimentary rocks with tholeiitic affinites (MORB-type volcanism). All these formations are overlapped in angular unconformity by folded clay deposits and by sandstones and conglomerates filling up the Youkounkoun basin.

Geodynamically speaking, the Bassaris rocks correspond to intracontinental rift deposits (Red Sea type, and the Koulountou rocks to magmatic materials intruded in an active margin. This belt is interpreted as resulting from the closing of an intracontinental rift, owing to the collision between the West African craton to the east and a western block, overthrust ca. 650 Ma (Dallmeyer and Villeneuve, 1987) on the West African craton.

The schematic cross-section takes into account field data and gravimetric interpretations (Ponsard, 1984; Villeneuve et al, 1984) and magnetotelluric data (Ritz and Robineau, in press). It shows that the Bassaris units overthrust to the east the West African craton and the Koulountou units themselves overthrust the Bassaris units. The suture between the West African craton in the east and the western block in the west is supposed to extend westward below the Niokolo-Koba units, where gravimetric data suggest the existence of a heavy mass at depth (probably granulites).

The Bassaris units have been strongly deformed and metamorphosed (greenschist facies). The Koulountou units are less deformed and less metamorphosed, but they show many thrusts. These two sets of units constitute the Pan african 1 cycle (800–600 Ma). They have been remobilized during the Pan african 2 cycle (650–660 Ma) and during the Hercynian orogeny.

The Pan african 2 is responsible for thrusting and folding in the Mali group (Cambrian ?) and the Hercynian orogeny of the mild deformations of the northern part of the Youkounkoun group and probably as well of some thrusting in the volcaniclastics of the Koulountou (270 Ma ages on mylonites to the east of Simenti after Dallmeyer and Villeneuve, 1987).

REFERENCES

Dallmeyer, R. D., and M. Villeneuve, 1987, $^{40}Ar/^{39}Ar$ mineral age record of a polyphase tectonothermal evolution in the southern Mauritanide orogen: GSA Bulletin, v. 98, p. 602–611.

Ponsard, J. F., 1984, La marge du craton ouest-africain du Sénégal à la Sierra Leone: interprétation géophysique de la chalcine panafricaine et des bassins due Protérozoîque à l'Actuel: Thesis Univ. Aix-Marselle III, Fr., 198 p.

Ritz, M, and B. Robineau, in press, Tectonic interpretation of electrical structures beneath the west-african craton edge in Eastern Senegal.

Villeneuve, M, 1984, Etude géologique de la bordure SW du craton ouest-africain: Thesis Doct. Etat, Univ. Aix-Marseille III, Fr., 552p.

———, A. Lesquer, J. F. Ponsard, and J. Roussel, 1984, Géologie et géophysique due Sud des Mauritanides et du Nord des Rokélides. Les témoins d'une suture panafricaine en bordure occidentale du craton ouest-africain: Klerkx and Michot, ed., African geology, Tervuren, Belg., p. 57–65.

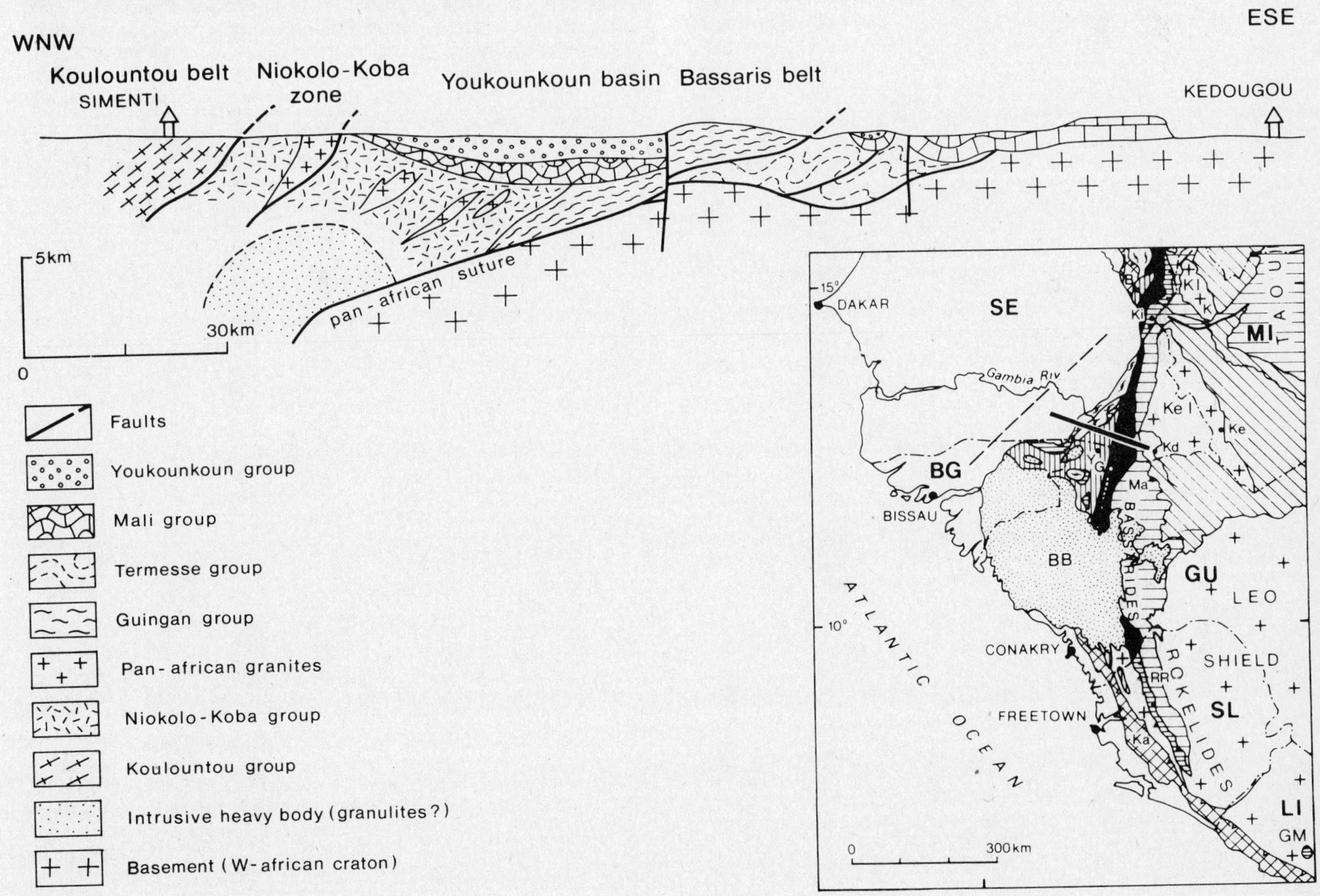

Figure 1.

VILLEY, M., F. BÈCHENNEC, M. BEURRIER, J. LE MÈTOUR, D. RABU, BRGM, Orléans, France

Geology of a Tethyan Mountain Belt: The Oman Mountains

The Oman Mountains are a composite orogenic belt that was built during two major orogenic cycles, the Eo-Alpine cycle, that terminated with the overthrusting of the Hawasina Nappes and the obduction of a part of the Tethyan oceanic crust onto the Arabian Platform, and the Alpine cycle, during which the present mountain range was uplifted.

The Eo-Alpine cycle began in the Late Permian with the transgression of the Fusulinid Sea, and was characterized by crustal extension that gave rise to the Hamrat Duru Basin and whose effects were felt as far as the northeastern edge of the platform. Renewed extension during the Triassic led to the opening of the composite Hawasina Basin with its constituent morphostructures (Umar Basin, Misfah Horst, Aridh Trough and including the Hamrat Duru Basin). The Jurassic was a period of quiescence. In the late Tithonian-Berriasian, a further phase of extensional tectonism provoked the foundering of the northeastern edge of the Arabian Platform and the westward retreat of the continental slope, with deposition of majolica facies sediments on the former platform and subsidence of most of the Hawasina Basin below the CCD. During the early Cretaceous the submerged areas were covered by a regressive mega-sequence that can be recognized both on the platform and in the Hawasina Basin. In the Late Cretaceous the Eo-Alpine Oman mountain belt was built on the northeast edge of the Arabian Platform with the closure of the Samail basin (which had opened during the Albian-Cenomanian in the older, Triassic-Jurassic, neo-Tethyan crust close to the continental edge) and the thrusting of the ophiolitic and sedimentary allochthonous units onto the foreland. Several extensional and compressive phases have been distinguished:

In the autochthon, the first major tectonic phase, which occurred during the late Turonian-Coniacian, was accompanied by regional HP/LT metamorphism and resulted in the development of northeast (oceanward) verging structures in the Saih Hatat region, with partial northeastward subduction of the Arabian Plate beneath the neo-Tethyan oceanic lithosphere. In the second phase, during the Campanian, mylonitic schistosity and southwest (continentward) verging intra-autochthonous thrust slices were formed. This phase was directly related to overthrusting of the Hawasina Nappes and the obduction of the Samail Nappe onto the Arabian Platform.

In the Hawasina Nappes obduction occurred in three main stages during the Cenomanian:

(1) Eo-obductive stage - compression, with delamination of the Hawasina from its basement to form an obduction prism characterized by an abundance of southwest-verging fold and shear structures.

(2) Obductive phase (s.s.) - essentially gravitational, with large displacements and overthrusting of the Hawasina Nappes onto the Oman platform, such that the most external of these thin nappes were overthrust by the most internal ones. Reconstruction of the pre-tectonic disposition enables estimation of the original extension of the main palaeogeographic domains and the displacement of the principal tectonic units.

(3) Late-obductive stage - this is only in evidence locally, with north-northeast verging retrofolding, in places with an associated schistosity.

The ophiolitic complex, after its opening during the Albian-Cenomanian, underwent two further geodynamic phases. The edge of the newly formed oceanic crust was first subducted, the neo-Tethyan lithosphere attached to the Arabian Plate passing beneath the marginal basins. This movement was then blocked, leading to the inception of obduction.

The Alpine orogenic cycle, between the Maastrichtian and the Burdigalian, gave their present architecture to the Oman Mountains.

VIRDI, N. S., Wadia Institute of Himalayan Geology, Dehra Dun, India

Cretaceous Marginal Basins of Indus–Kohistan Collision Zone: Development and Evolution

The Indus–Tsangpo Suture Zone (ITSZ) in the Himalaya represents the crustal expression of a Western Pacific type subduction zone encompassing petrotectonic assemblages representing different segments of the zone. In spite of severity of postsubduction tectonic activity and high rates of uplift and erosion, the ITSZ still preserves fairly complete sequences that were formed from north to south in a back-arc or marginal basin, the magmatic arc, the fore-arc basin, and a trench. Sufficient geological information is available now on different segments of the suture zone in Kohistan, Ladakh, and Tibet and the evolutionary models thus proposed are well constrained. Some difference of opinion still prevails about the type of subduction zone at the southern margin of Tibet and the timing of the closure of the back-arc in relation to cessation of subduction activity along the ITSZ.

Two types of models have been proposed for the ITSZ. The first type proposes the southern margin to be akin to an Andean margin where oceanic crust of the Tethys was subducted under a continental arc without the development of a back-arc basin. The second type models, more widely accepted now, envisage the margin to be similar to a western Pacific type with a back-arc basin and fore-arc basins on the northern and southern sides, respectively, of a magmatic arc.

The trench and the back-arc sequences were metamorphosed under high-pressure and low-pressure conditions, respectively, to give the paired metamorphic belts now described from Kohistan, Ladakh, and Tibet.

There is, however, some difference of opinion regarding the timing of the closure of a back-arc along the northern or the Shyok Suture vis-ž-vis the closure of the main subduction along the ITSZ. While the author favors the first closure of the back-arc, some others favor the closure along the Shyok or the northern suture, after the closure of the ITSz.

The paper discusses the development and evolution of the back-arc basin in the Kohistan and Ladakh sectors of the ITSZ in relation to northward movement and consumption of the oceanic crust of the Tethys along a Benioff zone at the southern margin of Tibet. The study shows that during the build-up of the island arc, the back-arc basin was formed between the Tibetan plate and the magmatic arc. With continued activity related to subduction along the ITSZ, the back-arc closed along the northern or the Shyok Suture during early Cretaceous. The subduction activity continued and finally the Neo-tethys closed along the ITSZ during Early Eocene (50 m.y.).

VIRDI, N. S., Wadia Institute of Himalayan Geology, Dehra Dun, India

Late Precambrian–Early Paleozoic Paleotectonic Evolution of Southern Margin of Proto-Tethys: Indian Scene

The Lesser and Higher Himalayan sedimentaries and metamorphics constitute the northward extension of similar rocks exposed in the northern half of the peninsular shield (Virdi, 1988). The Middle Proterozoic (1600 Ma onward) was a period of stable platform sedimentation represented by the Shali–Deoban and its equivalents in the Lesser Himalaya and the Lower Vindhyan (Semri Group) in the Peninsular region. The carbonate sedimentation continued until around 900 m.y. and was followed by dominantly clastic facies (with occasional carbonates) represented by the Simla–Jaunsar sequence and their equivalents in the Lesse Himalayas and the Upper Vindhyans (Kaimur and Rewah groups) in the Peninsula. To the west of the Aravalli massif, however, we have magmatic and metamorphic activity dated around 870–860 Ma (Erinpura, Kirana group) and 750 Ma (Malani Group).

After 750 Ma, the pericratonic margin of the Indian plate was differentiated into a number of basins delimited by major basement highs, faults, and megalineaments. While the Aravalli itself acted as a barrier between the Trans-Aravalli and the main Vindhyan basin, one of its splays, the Chandigarh–Bilaspur high, divided the Krol basin (sensu stricto) from the western basin in the Punjab reentrant and J & K parautochthonous belt (Figure 1). Though the overall sedimentation during Vendian and partly Upper Riphean (700–570 m.y.) was under shallow tidal flat environment, the sedimentation in the basins to the west of the Aravalli massif and in the Lesser Himalaya was characterized by evaporitic horizons. Thick horizons of salt and gypsum occur in the Salt Range, Lesser Himalaya in Jammu & Kashmir and Himachal Pradesh, and the Trans-Aravalli Vindhyan basin, while in the main Vindhyan basin, limited dessication is observed in the uppermost Vindhyans—the Bhander Group. The evaporite-forming conditions were favored during 650–525 m.h., when the Gondwana margin in India, Arabia, and Persia lay in the tropical and subtropical latitudes (Virdi, in press).

In contrast to the evaporitic basins, the Krol basin was nonevaporitic, and continuous sedimentation from Blainis up to Tal (650–500 Ma?) is observed. In the initial phases, the Blaini, Infra-Krol, and Krol are more or less uniformly distributed for about 400 km, however, the Tal sediment (Cambrian onward) are restricted to the central half only.

After the Cambrian deposition, most of the region was uplifted and underwent erosion. Thus, the Ordovician to Carboniferous sequence is missing and Permian rocks occur over older rocks in Salt Range, J & K, Garhwal, the Trans-Aravalli Vindhyan basins, and the main Vindhyan basin. The period 700–500 Ma is also important in that it coincides with the Pan-African Cycle and the Lesser and Higher Himalaya is characterized by numerous anorogenic granitic plutons (600–500 Ma) and metamorphic activity.

The paper deals with the paleotectonic evolution of the pericratonic margin of the Indian plate during the Middle Proterozoic to Early Paleozoic period.

REFERENCES

Virdi, N. S., 1988, Pre Tertiary geotectonic events in the Himalaya: Z. Geol. Wiss, v. 16, p. 573–587.

——— in press, Vendian-Lower Cambrian dessication along the southern margin of the Proto-Tethys: Journal of Geological Society of Indian.

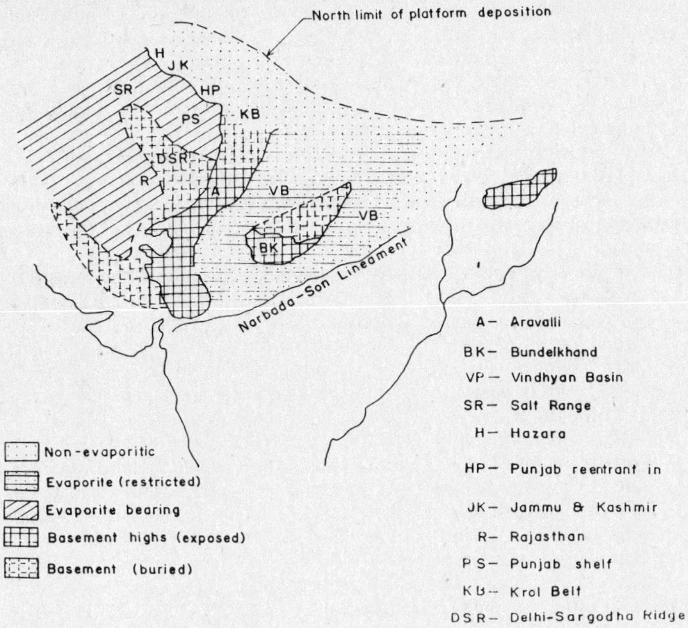

Figure 1.

VISSER, J. N. J., Department of Geology, Bloemfontein, South Africa

The Distinction Between Ancient Mountain and Shelf Glaciations

Ancient glacial deposits preserved in structural basins are commonly being seen as evidence for continental glaciations. However, where the tectonic setting of the stratigraphic sequence is also considered in the interpretation past glaciation models are biased towards a marine setting which is entirely due to the poor preservation potential of glacial sediments deposited in a continental interior. The above contradictory views are not conducive for the analysis of earth's history and it is therefore essential that distinction should be made between the different phases of a continental ice sheet in space and time.

A glaciation occurs either when ice accumulates over the continental interior (or highlands), forms small caps over mountainous regions, or expands as large sheets or tongues onto the shelf. The preservation of glacial or glacially related sediments from each of these morphological regions ;is tectonically controlled as shown in the accompanying table. In basin analysis the distinction between deposits left by a mountain and a shelf glaciation is easier and also of greater importance in the reconstruction of the geological history than trying to prove the presence of an extensive ice cover over the continental interior. To achieve this it is necessary to define a set of parameters reflecting the tectonic setting, erosional and depositional processes, sediment source and the characteristics of the environment.

The tectonic setting of the depositional basin can be inferred from the stratigraphic sequence in which the presumed glacial rocks occur. Such a simplistic approach, however, may not be always indicative of either a shelf or a mountain glaciation. The lithology of the diamictite sequence itself may be diagnostic of a shelf or a mountain glaciation as depicted in the accompanying figure. The typical glacial sequence deposited during a shelf glaciation shows a predominance of massive diamictite which formed by lodgement, melt-out and rain-out processes. Sequences deposited in valleys or along slopes during mountain glaciations show a much more varied lithology with stratified diamictites and mudrocks being abundantly present. Sedimentation occurred predominantly by rain-out of glacial debris, subaqueous outwash and sediment gravity flow.

As a result of the heterogeneous lithology of sequences deposited during a mountain glaciation lateral and vertical facies relationships are complex. Vertical facies changes occur at a much higher frequency rate, whereas lateral correlation between facies is almost impossible unless marker beds are present. The clast composition may also assist in distinguishing between the types of glaciation as clasts in diamictons deposited near mountainous regions show a higher frequency of local derivation than those in shelf diamictons. In the latter deposits clasts are generally smaller and mostly distantly derived. Fossils in valley sequences consist of plant remains, transported and in situ invertebrate marine species, and traces. Ancient glacial shelf deposits are commonly barren and only marine microfossils were reported. Bedrock erosion features may also be an indication of tye type of glaciation. During a mountain glaciation bedrock is commonly eroded, sculptured and striated, whereas the advance of an ice sheet over the shelf results in glaciotectonised bedrock and the formation of boulder and soft-sediment pavements with lodged clasts.

The results from such a study are essential in advanced basin analysis. In palaeoclimatological reconstructions problematic "glacial" deposits which have no correlatives in the stratigraphic record, may be attributed to mountain glaciation. Land surface elevation is a major controlling factor in the formation of mountain ice caps and these deposits probably represent resedimented glacially derived debris in even low latitudes. Expansion of an ice sheet onto the continental shelves is indicative of true polar conditions and only thos occurrences should be considered as diagnostic of a fluctuating global climate.

In palaeogeographic reconstructions the role of mountain glaciations in providing sediment to a depository is totally underestimated as thick sequences were deposited by tidewater glaciers having their source in mountainous regions. Ancient mountain glaciations may also be an indicator of rifting with uplift of the rift shoulders or subduction with alpine-type island-arc systems during sedimentation. The recognition of mountain glaciations may thus prove to be a valuable tool in the understanding of the tectonic development of a region.

	GLACIATION	DEPOSITION
HIGHLAND	ICE SHEET OVER CONTINENTAL INTERIOR	GLACIAL SEDIMENTS ONLY PRESERVED IN FAULT-CONTROLLED BASINS
MOUNTAIN	SMALL ICE CAPS OVER MOUNTAINOUS REGIONS IN A CONTINENTAL OR MARINE SETTING	GLACIALLY RELATED SEDIMENTS PRESERVED IN RIFTS GLACIALLY RELATED SEDIMENTS PRESERVED ON SLOPE OR IN ADJACENT TRENCH
SHELF	CONTINENTAL ICE SHEET EXPANDS ONTO THE CONTINENTAL SHELVES (MARINE ICE SHEET)	GLACIAL SEDIMENTS PRESERVED IN BASINS ON THE PLATFORM SPILL-OVER OF GLACIALLY RELATED SEDIMENTS ONTO SLOPE AND INTO TRENCH

VISSER, W., Petroleum Development Oman, Muscat, Sultanate of Oman

Thermal History of Precambrian Sedimentary Basin of Oman

The thermal history of the interior sedimentary basin south of the Oman Mountains has been studied in an attempt to model the maturation history of the Precambrian source rocks that have charged most of Oman's known oil accumulations.

The sedimentary basin measures 800 by 300 km and reaches a maximum of 8 km depth in its southwest–northeast-trending axis. Its stratigraphy spans the interval from the Vendian to the Late Tertiary. The eastern margin of the basin is very pronounced, with Precambrian sediments and basement outcropping in places. In the west the originally separate, older Oman basin grades into the Paleozoic and younger Rub Al Khali basin of Saudi Arabia.

The geological and thermal development of the interior of Oman is dominated by its position along two major paleorift axes that developed into the Indian and Thetys oceans after the break-up of Gondwana in the Late Paleozoic. The main erosional events and heat flow pulses are related to this event.

The simplified stratigraphic record shows two sedimentary sequences separated by 150 Ma gap during the Silurian–Carboniferous. The lower sequence comprises the Vendian–Lower Cambrian Huqf Group and the Cabro-Silurian Haima Group sediments. An important wrench-related compressional tectonic event in the Late Vendian–Early Cambrian caused large-scale (4–5 km) uplift in the southwest of Oman (Ghundun–Khasfah high), separating the South Oman Salt Basin from basins that may have existed to the west.

Thermal doming during the Late Silurian–Carboniferous, heralding the creation of the Indian ocean, caused uplift and erosion along a southwest-northeast trend: The erosion increases from less than 500 m at the Saudi border to about 3000 m along the eastern coast line and possibly 5000 m over the future (Triassic) break-up axis some 50–100 km offshore present-day Oman.

The uplift and erosion associated with the thermal doming along a west-northwest–south-southeast trend in the north, which preceded the creation of the Tethys ocean, has only mildly affected the present-day interior of Oman due to the remoteness of the spreading center.

The upper sedimentary sequence comprises the Permian Haushi and Akhdar Group sediments, which are separated from the almost continuous Mesozoic and Tertiary by the Base Jurassic unconformity. This unconformity is mainly caused by the final break-up of Gondwana that created the Indian ocean in the Late Triassic–Early Jurassic. It had a less profound effect on the thermal/subsidence development of interior Oman than the thermal doming phase. Only the eastern flank of the basin was affected, where less than 1000 m of uplift occurred. The rest of the basin only had a period of nondeposition or slight erosion, which in part may be due to the generally low eustatic sea-level stand during that period.

The large-scale thrusting and wrenching during the Late Cretaceous that emplaced the Semail (Oman Mountains) and Masirah Ophiolites have not significantly affected the heat flow in interior Oman. Also the Tertiary wrenching, which in the Oligocene was associated with volcanics in the northeastern part of the country, has not had a significant effect on the regional heat flow in interior Oman.

A present-day surface heat flow map was constructed based on about 80 deep oil exploration wells with reliable temperature data. From east to west, the surface heat flow generally increases from 40 to 60 mW/m^2; however, over the deepest part of the basin the heat flow is as low as 33 mW/m^2 in places. Gravity and magnetic data suggest that basement is shallower in the east than in the west of the country. However, seismic data suggest that a very thick (8 km) package of (thrusted?) sediments may be present in the subsurface below the Huqf outcrop area.

For the maturity modeling heat flow through time curves were constructed based on general assumptions on decay in heat flow, on which heat pulses were superimposed related to the main tectonic events. Surface heat flow during the Cabro-Silurian Haima deposition varied over the basin from 70 to 85 mW/m^2. At the end of the main uplifting period, surface heat flow reached 90 mW/m^2. At the end of the main uplifting period, surface heat flow reached 90 mW/m^2 along the eastern flank of the Oman Salt basin, while in the west it did not exceed 75 mW/m^2. The final stages of the creation of the proto-Indian ocean caused onshore only a heat flow anomaly in the eastern part of the country, because the spreading axis was some 500–100 km offshore the present-day coastline.

Calculated maturity histories show that oil expulsion was pre-Mesozoic over large parts of Oman. This is mainly caused by the missing Paleozoic overburden and the decaying heatflow. Therefore, complex (re)migration histories are required to explain the existing oil accumulations.

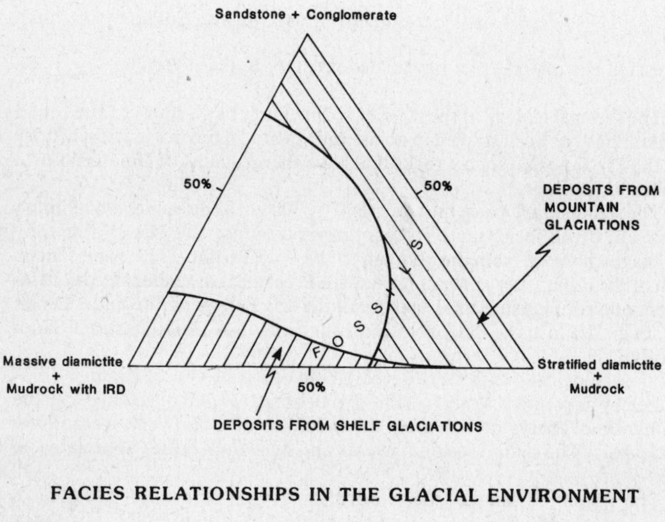

FACIES RELATIONSHIPS IN THE GLACIAL ENVIRONMENT

Figure 1.

VISTELIUS, B., O. GRAUNOV, and V. PAVLOV, Academy of Sciences of the USSR, Leningrad, USSR

Main Stages of Zonal Massifs Formation: Mathematical Models of Commingling, Self-Cleaning, and Preeutectic Stabilization

Solution of the problem of the origin of the zonal massifs composed by basic and ultrabasic rocks requires careful examination of numerous details. These zonal massifs are usually big, have no good outcrops, surround different country rocks that they cut, and in many cases are not bored. It seems that it is reasonable to begin the solution of the mentioned problem on some small body readily available for examination and surrounded by the same country rocks.

We studied massif Gul'shad (Near Balkhash Lake, Kazakhstan). This massif is composed of two bodies. The smaller body is located at the southwestern end of the massif and is joined with the northeastern body by a groove about 5 m thick that cuts monzonites. A whole mass of the southwestern body is immersed in homogeneous monzonites as is clear from the map of the body and its cross-sections. Their age is about the same (325 m.y.), as was determined by Levchenkov using U-Pb. The contact zone of this body with monzonites is composed of pyroxene gabbro. Olivine pyroxenites encircle the whole body. The nucleus of the body is composed of gabbro with olivine of forsterite type and very basin plagioclase (An_{85}–An_{95}). Separate grains of common hornblende and biotites are dessiminated irregularly throughout the whole body. There are also about 5–10% titanomagnetite and apatite.

Modal analyses of olivine gabbro and pyroxenites show that they corresponded to crystallization of a system fo-di-an where olivine corresponds to fo, pyroxene to di, and plagioclase to an. Investigation of relationships between gabbro and monzonites indicates that there are rocks of intermediate composition between them with $c = \alpha a + (1 - d)\beta$ where c is the investigated rock, a is a monzonite, β is some gabbro, and α a coefficient more or less constant for the main oxides.

The comparison of the geological observations and modal and chemical compositions of rocks shows that the intrusion was generated the following way.

Basic magma from the mantle was intruded into the basement of the crust composed of monzonite melt. All this mass with basic magma inside and monzonitic magma outside had migrated in the upper part of the crust. The main processes between basic and monzoinite magmas were realized as follows. At the beginning, basic magma and monzonitic magmas were mixed by diffusion with alternating coefficient of diffusion in Fickian equation. This process will be referred to as comingling. After some cooling, migrations of clusters of olivine and pyroxene, generated along the cotectic line, gave accumulation around intrusion of shell of pyroxenites with about 10% olivine. It is a process of self-cleaning. At last the residue of the magma at the end of the self-cleaning process remains near to eutectic in the system fo-di-an. This residue is subjected to relatively short period tide oscillations. The final magma stabilization is carried out under such conditions. These conditions should be expressed by some model of preeutectic stabilization. Now let us look at these models.

The model of comingling for mutual diffusion is based on Fikian equation.

$$\frac{\partial c_1}{\partial t} = \frac{\partial}{\partial x} (D(t) + a(c,\eta) \frac{\partial c_1}{\partial x}$$

Cauchy problem is the following for this equation on $\infty, < x < \infty, t \geq 0$ there is the solution $c_1(x, t)$ for the equation that satisfies condition

$$c_1(x,0) = \left[\begin{matrix} a_2 & x<0 \\ a_1 & x\geq 0 \end{matrix}\right.$$

This solution we seek as a function $\eta = x/(2\sqrt{\int_0^t D(\tau)d\tau})$. If $D = D(t)a(c_1,\eta)$ we obtain solution of integral nonlinear equation. This equation can be slowed by iteration. If $D = D(t)a(\eta)$ (if D is independent of c_1 we obtain solution as follows

$$\frac{\int_{\infty}^{\infty} \exp \left(- \int_0^{\eta} \frac{2sds}{a(s)}\right) \frac{d\eta}{a(\eta)}}{\int_{\infty}^{\infty} \exp \left(- \int_0^{\eta} \frac{2sds}{a(s)}\right) \frac{d\eta}{a(\eta)}}$$

In particular if $a = 1$

$$c_1(\eta) = \frac{1}{2} [1 + \phi(\frac{\chi}{2\sqrt{\int_0^t D(\tau)dt}})]$$

where

$$\phi(t) = \frac{2}{\sqrt{\pi}} \int_{0}^{t} e^{-\mu^2} d\mu$$

(Laplacian function).

If the melts before the beginning of diffusion can be described by the bectors of components $\vec{b_1}$ and $\vec{b_2}$ the further composition should be as follows

$$\vec{b_1}(x,t) = \vec{b_1}c_1(x,t) + \vec{b_2}[1 - c_1(x,t)].$$

Geological data supporting these models are given in Vistelius' report for 27 IGC (1984, Moscow).

The model of self-cleaning is solved as the boundary value problem for the inverse equation of diffusion. Since this problem is very complicated, it was developed for the case at plane circle that corresponds to the contours of observed pyroxenite section of the investigated intrusion (its southwestern body). Here we will give the precise state of the problem and will omit mathematical side of solution that requires a lot of room. The model supposes that the process of self-cleaning sums up the diffusional random motion in some area with the first exit on the termination of the area. The distribution of the migrating particles is uniform. The particle that reaches the termination of the are remains in fixed position at this termination. For this case, the probability of finding the migrating particle on the termination of the circle is expresses as some series that can be found by computation. The general outline of the density of probability in this case is like the outline of the thickness of the pyroxenite zone (if taking into consideration a possibility of some deformation of the contour by tectonic transformation).

The model of stabilization, before eutectic temperature is reached, supposes that the melt before crystallization is composed of clusters of chain silicates (pyroxenes), carcase silicates (especially $Al_2Si_2O_8^{-2}$), and ions of Ca, Mg, and Fe^{+2}. A whole mass of the indicated composition with volatile components is subjected over a long time, since cooling is very slow process, to an oscillating process generated by tide oscillations of the crust. If intrusion is located in a convenient position concerning direction of thes oscillations, they can generate stratification before crystallization. It is well-known stratified gabbro. If position of magma is less convenient concerning direction of oscillation, magma will not be stratified but oscillation will generate some random stationary process that determines preliminary arrangement between plagioclases and pyroxenes. This process should be of the type of autoregression. Counting of the content of plagioclase is a very slow process, but plagioclases are enriched by stroncium, when content of Sr in pyroxenes is negligibly small. Thus the random process should be reflected in the distribution of stroncium. Technology of high precision quantitative stroncium determination is easily available. We cut two pieces of core sample (boreholes 1 and 46) into sequences of specimens through 1 cm, determined content of SR, and counted correlation functions. These functions are like the functions of autoregressive process. That coincides with our model, but at this moment the number of specimens is small (in one specimen about 45 and in another about 90), and the obtained results only suggest further examinations, which we hope to continue.

VITENKO, V. A., Institute of Geology, Lvov, USSR, B. P. KABYSHEV, Institute of Geology, Chernigov, USSR and P. F. SHPAK, Institute of Geology, Kiev, USSR

Connection between Retrotiferous Structures and Their Formation Time

For two main productive complexes in the Dnieper–Donets Depression (DDD) were established different regularities as to connection between trap age and their productivity. In Lower Carboniferous complexes, oil and gas deposits tend to old (Paleozoic) traps: later ones concentrate up to 96% of total hydrocarbon reserves, while combine traps (Paleozoic–Mesozoic) contain 4% and recent ones (Mesozoic) are empty. Contrary regularity is observed in Lower Permian–Upper Carboniferous complex where a great number of oil and gas pools (50%) in the western part of the DDD and major volume of its hydrocarbon reserves (82%) occur in recent traps, while in the eastern part 48% of hydrocarbon reserves were discovered in recent traps and another 48% are contained in traps of combined structures.

Both mentioned regularities of connection between traps age and their productivity are inherit to the productive complexes of various genetic types. Deposits in oil-and-gas-producing complexes, primarily as to forming conditions, are in close connection with paleotectonics of zones and traps. The biggest hydrocarbon reserves usually refer to old and combined structures. Paleotectonic factors may be evidently treated as an independent on contemporary tectonics medium for evaluating oil and gas potential of regional zones or local traps and practical criterion for prospecting oil-fields as well. Such conditions are characteristic for oil and gas deposits in DDD Devonian and Lower Carboniferous deposits, as well as for numerous oil pools or early formed traps in other regions (Volga–Ural, Southern Mangyshlak, Precaspian, Pripyat, Western Siberia, etc.).

Another regularity is established in productive complexes containing no proper sources of hydrocarbons. Here oil and gas pools were formed during vertical oil and gas leakage from reforming pools in deeper sited syngenetically productive complexes. In such a case, paleotectonics lost its controlling function on oil-and-gas potential. Moreover, recent traps had even advantage in hydrocarbon accumulation against old ones, which growth during neotectonical phases was too slow in order to grant favorable conditions for trapping large amounts of hydrocarbons in the processes of the primary pool reformation. This was the case for oil and gas pools in Lower Permian–Upper Carboniferous productive complex of the DDD, as well as for gas pools in Albian reservoir at Karpinsky shaft, in many Upper Cretaceous traps of Middle Asia, Cenomanian layer of northern part of WEstern Siberia, etc.

For this reason, oil and gas accumulation in recent traps was obviously caused by special conditions during prolonged processes of primary pool formation and reformation. Therefore, the latter regularity cannot be used as a basis for rejecting connection between age of deposits and traps and, moreover, as ground for accepting rather late (up to Neogene) time of deposit formation. The mentioned connection simply becomes more complex, like other phenomena being subjected to more detailed study.

VLADIMIROW, V. D., Sofia University, Bulgaria, Europe

Geodynamical and Petrophysical Conditions of Mobilization and Remobilization of Ore Substance in Chelopech Cupry-Ferrous Pyritic Deposit, Bulgaria, Southeast Europe

The Chelopech cupry-ferrous ore deposit is situated in Alps-Himalayas belt 60 km east of Sofia, capital of Bulgaria, Balkan Peninsula, southeast Europe. It is tied to an area consisting of Precambrian metamorphites, thrustfolded upon older Paleozoic metamorphites, as well as of Upper Cretaceous sedimentary, subvolcanic and effusive andesite-dacite and andesite rocks. The ore bodies are with irregular, lenticular or sometimes tube-like form. The borders are gradual and the mineralization contains 48 minerals. The main ore minerals are pyrite, tennantite, halcopyrite, luzonite, enargite, bornite and a bit of galenite and sphalerite. The ore bodies are localized in monotonous volcanic rocks without obvious relationship with any traditional geological structures.

In this paper is described the influence of petrophysical and geodynamical conditions of ore bodies localization on the basis of data, accepted by investigation of 213 samples from the whole volume of the Chelopech ore deposit. About effective pore space, quantity of big ($>10^{-2}$ mm), middle (10^{-2} to 10^{-4} mm), little ($<10^{-4}$ mm) pores, constant of saturation, density, the information is obtained by free water-saturation and by measuring velocity of longitudinal and transversal ultrasonic waves to compute different elastic modules and ratios, Debye temperature and complex petrophysical coefficient. For every sample is determined the mean dimension of grains in groundmass, hardness (Brinell, HB) and contents of 20 elements.

Computer data processing include statistical methods and cluster analysis, once only for petrophysical data, second time only for geochemical data and finally together for both types of data. Thus in different cases are received 7 or 8 groups of rocks united in two complexes. The first of them includes rocks with good filtration properties and high metal contents. In this complex the hydrotherms flowed intensively and there were not conditions for deposition of ore components. If they were in it, they were extracted and carried away. The second complex includes rocks with bad filtration capabilities, high values for elastic-density properties and high contents of ore and other attendant components. At the time of ore formation the rocks of this complex were barriers-concentrators for ore substance. They built blocks with wedge-shaped form. The ore bodies are located in their peripheral areas at places of their wedging.

There are two stages of formation of structures and ore bodies of deposit. The first, sin-volcanic was reconstructed by ultrasonic structural petrophysical analysis and second, post-volcanic, by tectonophysical analysis. The sin-volcanic stage includes two substages × early and late. At the early substage compressive forces of paleotectonic field of tension were oriented in a northwest direction and the stretch forces in a northeast direction. The main part of the ore was deposited then. At the late substage an inversion of the paleotectonic regime occurred. Compressive forces acted in a northeast direction and stretch forces in a northwest direction. At that time the typical process was remobilization of ore components accompanying the formation of little ore bodies and final formation of main ore bodies, with insignificant addition of a new ore substance. Keeping in mind the volcanic nature of the ore field the blocks with vertical or steep orientation of compressive forces were outlined. Those blocks were areas where the hydotherms flowed and ore bodies are localized around them. At the time of post-volcanic stage was reconstructed five paleotectonic regimes. The compressive forces were oriented in horizontal direction and stretch forces in a vertical direction.

VLAHOVIC, M., Civil Engineering Faculty, Belgrad, Yugoslavia

Evaluation of Engineering-Geological Parameters in the Selection of the Most Favorable Trace of Road

The engineering-geological conditions in the construction of roads are considered to be one of the basic factors of the selection of the most favorable variant among the concurrent ones. For this purpose engineering-geological maps with general data have been used. These maps, however, cannot meet the needs of designers in many cases, and for the following reasons: (a) presented data are not classified for the purpose of designing and constructing roads, and (b) the synthesis of the overall engineering-geological effects has not been performed.

That is why the engineer-designer cannot gain the complete insight into the features of the soil. In order to improve these maps V. Jovanović et al. (1974 and 1977) proposed a work methodology for a new type of map, so-called synthetized engineering-geological maps. This abstract deals with some supplementary data of the mentioned methodology by which the author is trying to contribute to the completeness of the soil classification.

The proposed work methodology comprises several successive stages, such as:

1. The evaluation of individual effects of engineering-geological parameters is carried out according to the following criterion: (a) stability of soil – 35%, (b) relief of soil – %, (c) compressibility of soil – 25%, (d) hydrogeological features of soil – 5% and (e) necessary conditions of earthworks – 5%.

The suitability of rock-masses for building in roads is not implied by this criterion. This datum can only be perceived through the degree of soil compressibility. Significantly more relevant consideration of the quality of rock-masses could be gained by the introduction of CBR California Bearing Ration). Data about compressibility cannot be eliminated because by CBR data and correlational equation Youngs modulus of elasticity can be defined.

2. Classification of engineering-geological parameters has been carried out within five categories by which conditions of construction are determined. The most favorable quality is denoted by Category 1 and the most unfavorable one by Category 5.

Classification of stability of soil:

(1) Highly stable soil cannot be thrown off its natural balance even under the influence of the most intensive natural processes.

(2) Stable soil can be slightly thrown off its natural balance only under the influence of some intensive changes of natural processes.

(3) Soil becomes stable under the influence of minor changes of natural processes.

(4) Moderately stable soil is the soil in which only infrequent engineering-geological processes (landslides, liquifaction, etc.) of minor dimensions occur.

(5) Highly unstable is the soil in which numerous active engineering-geological processes of major dimensions are present.

Classification of soil relief is made accordind to the following criterion: 1 – slope milder than 1:10; 2 – slopes from 1:10 to 1:5; 3 – slopes from 1:5 to 1:2, 4 – slopes from 1:2 to 1:1 and 5 – slope over 1:1.

The criterion of suitability of rock-masses for building in roads is the following: 1 – CBR >20%; 2 – CBR = 10-20%; 3 – CBR = 5-10%; 4 – CBR = 3-5%; and 5 – CBR <3%.

Classification according to hydrogeological features is based on the maximal position of the underground waters level (H) as follows: 1 – H > 10m; 2 – H = 5-10m; 3 – H = 3-5m; 4 – H = 1-3m; and 5 – H = 0-1m.

Necessary conditions of earthworks are classified according to the following sequence: 1 – mining is not necessary; 2 – here and there mining is necessary; 3 – mining in weak rocks; 4 – mining in hard rocks; 5 – mining in very hard rocks.

The overall indicator of the quality of soil is obtained by summation of multiplication of individual effects and the number denoting the quality (1 to 5).

The presented method is considered very practical when more concurrent variants of road are correlated in reference to various geological characteristics. For this purpose the sections with different engineering-geological features should be determined along one variant. For each section an overall indicator of quality according to the explained method should be defined. By summation of all indicators for all the sections the overall estimation of the variant is made. By division of the overall estimation (of the total number of points) with the total length of variant the total coefficient of suitability (Kp) is obtained.

By the results of one of engineering-geological analysis of 12 concurrent variants of trace of the road between two cities the following data have been obtained:

—variant with the shortest distance between the cities 6,750 m; coefficient of suitability was Kp = 1.20;

—variant with the longest distance between the cities 8,760 m; coefficient of suitability was Kp = 1.106;

—most favorable variant was with the length of 8,560 m and coefficient of suitability Kp = 1.05.

When applying this methodology it is recommended to draw a network on the map of investigated territory. The quality of each one quadrant on the network should be defined according to the mentioned procedure. Afterwards the total indicators of quadrants on the trace of road are summed up. The data obtained by this procedure can be adequately used for computer elaboration.

VOCHTEN, R., State University of Antwerpen, Antwerpen, Belgium

Transformation of Schoepite into Becquerelite, Billietite, and Wölsendorfite

Schoepite [$UO_2(OH)_2 \times H_2O$] obtained by the action of H_2O, CO_2, and O_2 on uranites [U_3O_8] in a continuous reaction system has been discussed (1988). The obtained rhombic crystals with cell parameters a, b, c: 14.33; 16.79; and 14.73 Å are represented in Figure 1. The crystalforms show the faces (100), (101), and (011).

Until now synthetic becquerelite [$CaO \times 6UO_3 \times 11H_2$)], billietite [$BaO \times 6UO_3 \times 11H_2O$], and Wölsendorfite [$PbO \times 2UO_3 \times 2H_2O$] were only obtained from pure chemical (Protas, 1956).

In our experiments, schoepite is directly transformed into becquerelite, billietite, and wölsendorfite by the reaction of the proper ions with schoepite at 100 °C over 2 weeks.

Generally the transformation can be represented as:

$$nUO_2(OH)_2 \times H_2O + M^{2+} \rightarrow MO \times nUO_3 \times 3nH_2O$$

The obtained species were identified by chemical analysis and X-ray diffraction (Guinnier-Hägg, $CuK_\alpha 1$) as becquerelite, billietite, and wölsendorfite.

The cell parameters correspond with those of Christ and Clark (1960). The results are summarized in Table 1.

The natural relationship between U_3O_8, schoepite, and the three uranates will be discussed and is summarized below.

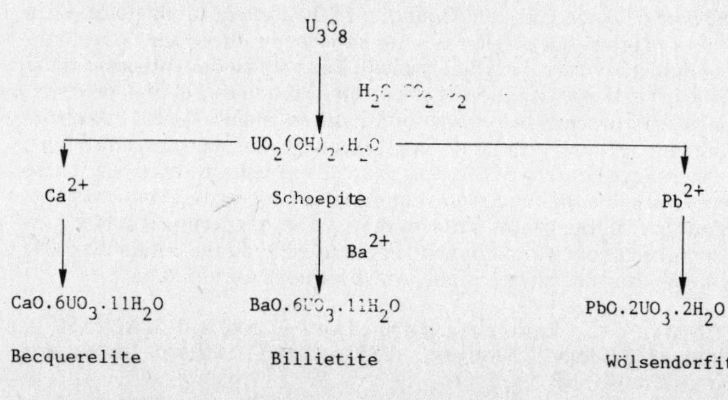

The variation of the solubility of schoepite, becquerelite, billietite, and wölsendorfite as function of pH is represented in Figure 2.

It is clear that the solubility of schoepite and becquerelite is nearly the same while billietite and wölsendorfite are characterized by a much lower solubility.

If we compare the solubility as function of pH with the zeta potential (Figure 3) (microelectropheric determination), we may conclude that the ζ potential is always negative (hydroxylation effects) for pH values higher than 4 and that the solubility increases for pH values which correspond with positive ζ potentials.

In the presence of CO_2 (meteoric water), the solubility increases considerably. This effect is illustrated for schoepite in Figure 4. The increase of solubility can be explained by the formation of uranyldicarbonate (UDC) and uranyltricarbonate (UtC) complexes, generally represented as:

$$UO_2(OH)_2 \cdot H_2O \longrightarrow UO_2(OH)_2 + H_2O$$

$$UO_2(OH)_2 \rightleftharpoons UO_2^{2+} + 2OH^-$$

$$CO_2 + H_2O \rightleftharpoons H_2CO_3 \rightleftharpoons HCO_3^- + H^+$$

$$HCO_3^- \rightleftharpoons CO_3^{2-} + H^+$$

$$UO_2^{2+} + 2CO_3^{2-} \rightleftharpoons \underset{UDC}{[UO_2(CO_3)_2]^{2-}}$$

$$[UO_2(CO_3)_2]^{2-} + CO_3^{2-} \rightleftharpoons \underset{UTC}{[UO_2(CO_3)_3]^{4-}}$$

REFERENCES

Christ, C. L., and J. R. Clark, 1960, American Mineralogist, v. 45, p. 1026–1061.

Protas, J., 1956, Bull. Soc. Fr. Minér. Crist., v. 79, p. 350–351.

Vochten, R., E. De Grave, and H. Lauwers, in preparation, American Mineralogist.

Table 1—Chemical and Crystallographic Characteristic of becquerelite, billietite, and wölsendorfite.

	BECQUERELITE	BILLIETITE	WÖLSENDORFITE
THEOR. FORMULA	$CaO.6UO_3.11\ H_2O$	$BaO.6UO_3.11\ H_2O$	$PbO.2UO_3.2\ H_2O$
EXPERIMENT.FORM.	$(CaO)_{0.98}(UO_3)_{6.006}.11H_2O$	$(BaO)(UO_3)_{6.200}.10.5H_2O$	$(PbO)_{0.99}(UO_3)_{2.003}.2.06\ H_2O$
DENSITY CALC.	5.10	5.27	6.80
DENSITY MEAS.	5.064	5.244	6.716
A (Å)	13.76	7.143	11.90
B (Å)	12.38	12.08	13.98
C (Å)	14.96	15.10	6.85

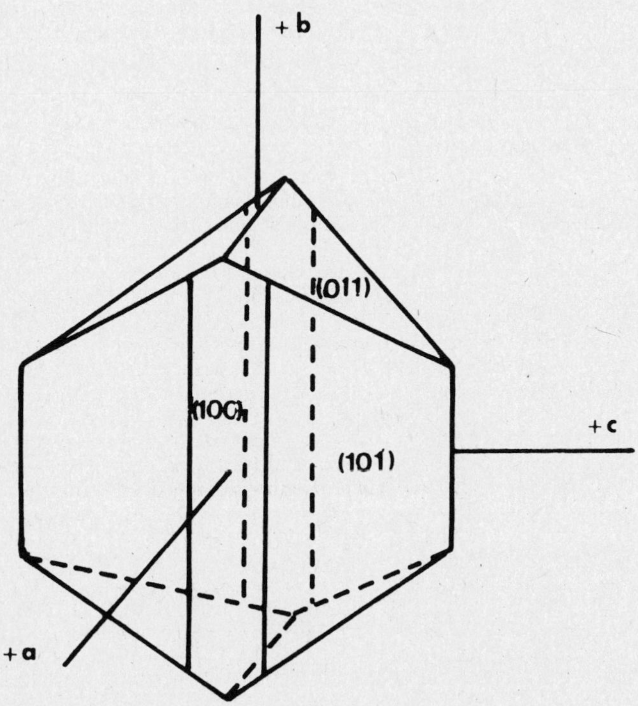

Figure 1—Crystal of synthetic schoepite with faces (100), (101), and (011).

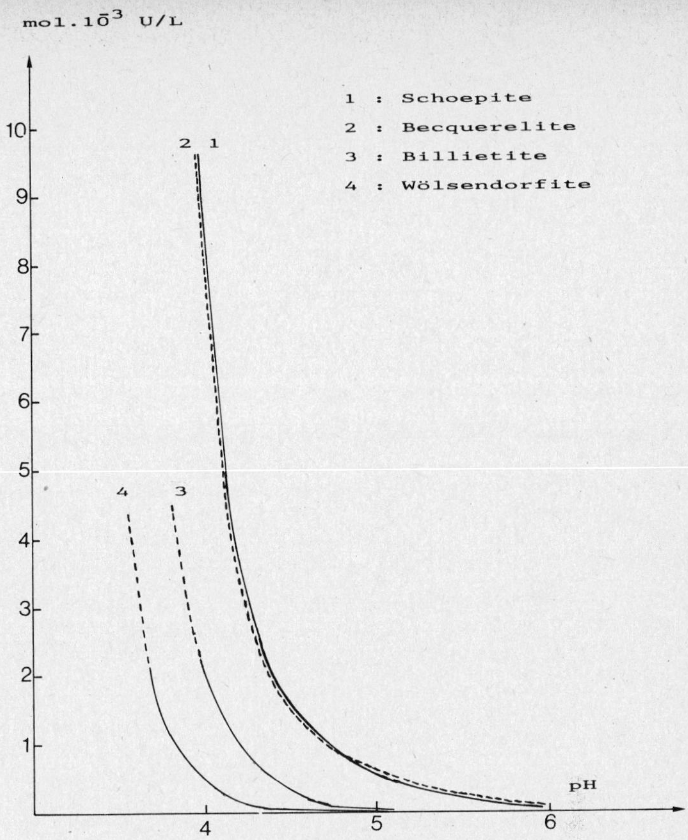

Figure 2—Variation of the solubility of schoepite, becquerelite, billietite, and wölsendorfite as function of the pH at 25 °C.

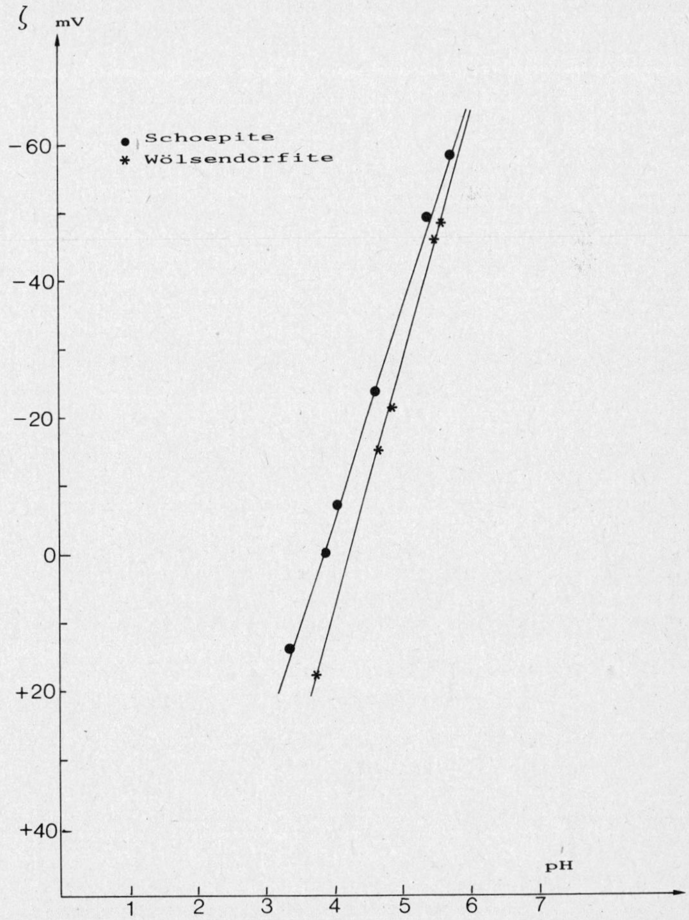

Figure 3—Variation of the zeta-potential of schoepite and wölsendorfite as function of the pH at 25 °C.

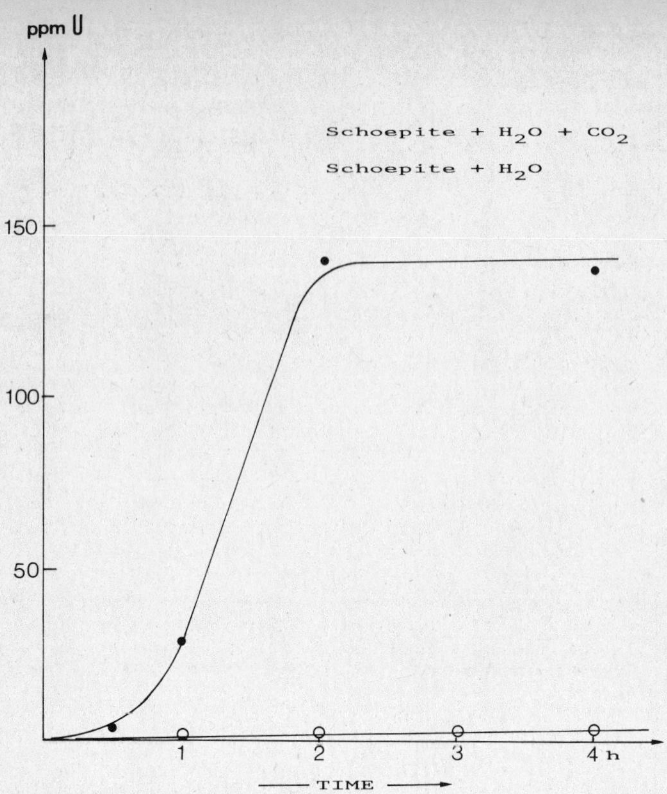

Figure 4—Variation of the solubility of schoepite in the presence of H_2O, with and without CO_2 at 25 °C.

VOGEL, K., and R. SLINGERLAND, Pennsylvania State University, University Park, PA, J. S. BRIDGE, State University of New York, Binghamton, NY, A. VAN NIEKER,, Chamber of Mines of South Africa, Johannesburg, South Africa, and P. FLEMINGS, Cornell University, Ithaca, NY

Origin of Downstream Fining in Alluvial Sediments: Numerical Investigation

Dynamic stratigraphic modeling of fluvial deposits requires understanding the origin of variations in bed slope and bed sediment composition in alluvial streams. Toward this end we have constructed a numerical model simulating the differential bedload and suspended load transport of heterogeneous size-density mixtures under quasi-unsteady, nonuniform flows in single channels. Bedload transport rates are calculated using a modified Bagnold function; suspended load rates are computed by integration of a Crank-Nicholson formulation of the mass-balance equation for suspended sediment in two dimensions. Grain hiding in the channel bed is modeled using the concept of an active layer and the Komar entrainment function. The model uses a shear stress distribution to simulate stress variability due to turbulence, unlike other models that use a time-averaged bed shear stress. The model was calibrated using data from the flume armoring experiments of Little and Mayer (1976), among others. Simulation of the evolution of the bed slope and sediment composition of a 5 km long river and alluvial fan, of channel width 2.5 m and constant downstream clear water discharge 1.7 m^3s^{-1}, produced the following results: (1) After 6 months of channel forming discharge, an initially linear longitudinal profile evolves to a slightly convex-upward profile as expected, and an initially uniform bed caliber evolves to coarser textures upstream that decrease exponentially downstream. The convexity of the profile depends upon the magnitudes of time elapsed, discharge, stream width, and initial bed material. The clast size decrease is an exponential function of distance ($d/d_0 = e^{-0.11*x}$) and arises as the slope and sediment transport rates adjust to upstream armoring.

VOGELSANG, E., and M. SARNTHEIN, Geologisch-Paläontologisches Institut und Museum der Christian-Albrechts Universität, Kiel, Federal Republic of Germany

Temperature Transect across Norwegian–Greenland Sea for Last 350,000 Years

Plantonic $^{18}O/^{16}O$-isotopic profiles were measured on *N. pachyderma sinistral* of numerous long sediment cores, which form an east-west transect across the Norwegian–Greenland Sea.

The isotopic data allo us to estimate water temperatures of the sea surface, more accurate at 100–150 m water depth, for the last 350,000 years. The glacial to interglacial isotopic shift amounts to 2.5‰. According to Duplessy et al (1987), from present to isotopic Stage 6, at the most 1.05‰ is caused by the variable ice volume effect (= volume of continental ice). The ice volume effect of the earlier interglacial Stages 7 and 9 is extrapolated by means of global records assuming constant deep-water temperatures. The remainder of the isotopic signal is divided into the temperature effect and the salinity effect; the latter can be neglected for water depths of 100–150 m. After deduction of the ice volume effect, the isotopic record can be converted into a temperature record by means of an equation from Shackleton (1974).

The results of out temperature transect are: (1) Stage 1 (Holocene) is by far the warmest period in the measured interval; only Substage 5e is almost as warm; (2) the remaining ~80% of the measured intervals are much colder, especially Stage 8 (250,000 y P.P.), early Stage 6, and late Stage 5 (80,000 y B.P.); (3) in all warm stages there is a distinct east-west gradient, indicating the advection of Atlantic waters into the Norwegian Sea.

REFERENCES

Duplessy, J. C., L. Labeyrie, and P. L. Blanc, 1987, Norwegian Sea deep water variations over the last climatic cycle: paleo-oceanographical implications, *in* Long and short term variability of climate.

Shackleton, N. J., 1974, Attainment of isotopic equilibrium between ocean water and the benthonic foraminifer genus Uvigerina: isotopic changes in the ocean during the last glacial: Colloque International du Centre National de la Recherche Scientifique, n. 219, p. 203–210.

VOKURKA, K., Geological Survey of Prague, Prague, Czechoslovakia, and B. KOBER, University of Heidelberg, Heidelberg, Federal Republic of Germany

Heterogeneity of Earth's Mantle below Bohemian Massif

Equilibrium P-T conditions ranging from 1 to 2.5 GPa and from 900° to 1050 °C were estimated for 18 sp peridotite nodules from Bohemian Cenozoic alkali basaltic rocks. The pressures indicate the peridotite to have existed in the upper mantle at depths of 30–80 km for geologically significant periods of time.

$^{87}SR/^{86}Sr$ ratios of Tertiary alkali basaltic rocks vary between 0.7031 and 0.7035 and that of a Silurian basalt is 0.7031. This range is consistent with the derivation of possible magma from a common lower mantle source.

Peridotite nodules seem to represent upper mantle fragments brought to the surface by a genetically unrelated basaltic magma ascending from the depth.

The varying degrees of nodule depletion in LIL reveal heterogeneities with respect to incompatible elements existing in the upper mantle of the Bohemian area.

$^{87}Sr/^{86}Sr$ ratios of three Cr-diopsides from sp lherzolite xenoliths exhibit a range from 0.7032 to 0.7036. Their $^{87}Rb/^{86}Sr$ and $^{87}SR/^{86}Sr$ values are decoupled. Thus, they might reflect a late metasomatic event in the upper mantle.

Slightly increased $^{87}Sr/^{86}Sr$ ratios of Tertiary trachytes compared to the basaltic data ranging between 0.7037 and 0.7040 exclude a substantial involvement of crustal material in generation of their parental magmas.

VOLPE, A. M., and J. D. MACDOUGALL, University of California, San Diego California

Trace Element and Nd Isotope Geochemistry of Mafic Rocks in Western India: Implications for Rifting During the Mid-Late Proterozoic

Regionally metamorphosed rocks in the Upper Proterozoic Delhi Supergroup form the main part of the Aravalli mountain range, which strikes northeast across the state of Rajasthan in western India. Ultramafic-mafic rocks (amphibolites) exposed along the western flank of the Aravalli range have been termed the Phulad Ophiolite by the Geological Survey of India (Gupta et al., 1981). The occurrence of ophiolitic rocks in this foldbelt is intriguing in light of the debate about whether the same type of plate tectonic processes operating today or alternate tectonic models are applicable during the Proterozoic. This is controversial in part because of the lack of preservation of ophiolites in foldbelts older than 1.0 Ga. During fieldwork we found little evidence for complete ophiolite sequences (GSA Penrose 1972). Individual units are present, but highly dismembered as a result of intense deformation and shearing. Evidence for pillowed meta-basalts at Phulad is equivocal, and a sheeted dike complex has yet to be found.

Amphibolites in the Delhi Supergroup were collected at localities along the western flank of the axial section of the Aravalli range. the majority of samples are from Jetgarh (southwest of Ajmer), the Bheru ki Baryia river section near Barr, Phulad, and Ranakpur-Sadri. These rocks are composed of amphibolite grade mineral assemblages, principally magnesio-tshermakitic hornblend and andesine plagioclase. The amphibolites are interfolded with meta-sedimentary rocks (bio-hblde calcareous schists and gneisses) of the Ajabgarh Series, and are pervasively intruded by a variety of syn to post-tectonic granitic rocks. Alkali granites and granodiorites in the axial section of the foldbelt have whole rock Rb-Sr ages between 900–800 Ma, while mineral isochrons for the same rocks indicate more recent disturbance at 600–500 Ma (Choudhary et al., 1984).

The amphibolites have tholeiitic compositions; hyperstene normative mineralogy, subalkaline minor and trace element abundances, and FeO enrichment relative to alkalies with progressive fractionation. Phulad-Barr-Jetgarh (PBJ) meta-basalts have minor and trace element abundances which are indistinguishable from mid-ocean ridge basalts (MORB). The data are strong evidence that the meta-basalts formed by partial melting of lherzolitic mantle and the magmagenesis did not involve emplacement through continental crust or an unusual fractionation history (e.g. continental flood basalts, arc basalts). In contrast, Ranakpur meta-basalts have high Al_2O_3 and low TiO_2 for similar MgO (Mg# = 54–62), high LIL element (Rb, Ba, K, Sr), and relatively low HFS (Ti, Zr, Nb) and rare earth element (Sm, Nd) abundances. Uniformly low normalized Zr abundances relative to Nd-Sm (Zr/Zr*1.0) suggest derivation from a HFS element depleted source. Meta-basalts and diorites at Ranakpur (SiO_2 = 51.6–53.8%) exhibit a low-K differentiation trend different than the alkali enrichment observed in calc-alkaline rocks.

Ranakpur diorites define a Sm-Nd whole rock isochron corresponding to an age of 1008 ± 78 Ma, with an initial Nd isotopic composition (ϵ_{Nd}–0.6) that is similar to bulk earth, or undifferentiated mantle. The age is consistent with field evidence that the diorites are older than the intrusive granodiorites (850 ± 50 Ma; Choudhary). Sm-Nd mineral isochrons for two of these rocks (quartz diorite and diorite) and a meta-gabbro correspond to younger ages of 795 ± 45 Ma, 837 ± 44 Ma, and 838 ± 36 Ma respectively. These younger ages represent the time of metamorphic re-equilibration due to widespread tectono-thermal events in this region during the Late Proterozoic. Given the spatial and chemical affinity between the diorites and meta-basalts at Ranakpur, the simplest explanation is that the rock suites are cogenetic and broadly (± 100 Ma) contemporaneous. However, major unresolved questions are the ages of the ophiolitic and meta-sedimentary rocks in the Delhi Supergroup.

Regional metamorphism and a combination of initial source heterogeneity and restricted isotopic compositions limits the extent to which geological ages for the meta-basalts can be determined by Rb-Sr and Sm-Nd radiometric dating. Model Sm-Nd ages relative to differentiation from a chondritic reservoir range from 1.3 to 1.7 Ga. Assuming an emplacement age of 1.0–1.1 Ga, the meta-basalts and meta-gabbros have variable initial Nd isotopic compositions: $\epsilon_{Nd}T = -3.6$ to $+5.5$. Most of the samples have positive initial ϵ_{Nd} and Sm/Nd ratios greater than chondritic. Those with lower Sm/Nd and negative initial $_{Nd}$ are mostly from Ranakpur.

High initial ϵ_{Nd} (+5.5) is unequivocal evidence that the PBJ meta-basalts were derived by partial melting of long-term LREE depleted mantle. Those samples with lower initial ϵ_{Nd} suggests incorporation of an enriched component, possibly older Proterozoic or Archean crustal material. This precludes the possibility that these meta-basalts are fragments of mid-ocean ridge lithosphere. One explanation is that these rocks represent an episode of spreading and basaltic magmatism in a back-arc tectonic setting situated along a continental margin. Another possibility, which could be tested by detailed field and U-Pb zircon studies, is that they formed during intracratonic rifting. In this case the ophiolitic rocks in Western India may be evidence for the breakup and subsequent closing of a proto-continent during an episode that was a precursor to widespread tectono-thermal activity in the Late Proterozoic.

VON DER BORCH, C. C., and P. DI BONA, Flinders University, Bedford Park, Australia, and N. CHRISTIE-BLICK, Lamont-Doherty Geological Observatory of Columbia University, Palisades, NY

Implications of Kilometer-Scale Base-Level Changes, Late Proterozoic Adelaide Geosyncline, South Australia

A series of kilometer-deep erosional paleocanyons forms a spectacular element associated with Late Proterozoic sediments of the Adelaide Geosyncline, South Australia. The canyons occur over an area spanning hundreds of kilometers. They have eroded down from a specific stratigraphic level within the Wonoka Formation of the Wilpena Group, approximately 2–4 km below the Proterozoic–Cambrian boundary, and are infilled by slumped mudstones, carbonate breccias, and calcareous sandstones and mudstones of the lower Wonoka Formation.

The canyons formerly were described to have been submarine in origin, eroded and filled in an inferred basin-slope setting by subaqueous processes (von der Borch et al, 1982, 1985). They were considered to represent ancient submarine canyons, possibly analogous to many situated on present-day continental slopes. Infilling sediments were described variously as slumped wall-rocks, channelized turbidite sandstones, and suspension-settled mudstones, with the sandstones representing the deposits of an onlapping channelized submarine fan-head.

Subsequent detailed sedimentological studies have cast considerable doubt on a submarine origin for the canyons and suggest that they were eroded subaerially (Eickhoff et al, 1986, in press; von der Borch et al, in press). Evidence in favor of a subaerial origin includes the observation that many of the sandstone units within the canyon-fill are characterized by sedimentary structures reflecting wave-induced oscillatory processes, particularly combined-flow ripples, which suggest deposition above fairweather wave-base. Hummocky cross-stratification is also locally present. In addition a thin (~1 m) diffusely laminated micritic to microsparry carbonate unit, which veneers upper portions of canyon walls, has a stable isotope signature that suggests its nonmarine origin. Locally, the veneer exhibits large-scale (~1–3 m) tepee-like structures, and microscopically some samples of the veneer have a clotted fabric suggesting an algal influence during its formation. The carbonate veneer, on existing evidence, is interpreted to have developed as a travertine, from groundwater discharge on canyon walls. The commonly occurring carbonate breccia units, interbedded with canyon-filling sediments, were largely derived from the veneer.

The currently favored model, in which the canyon-filling sediments were deposited mostly above fairweather wave-base, has important implications for syndepositional processes that were active during Wonoka Formation sedimentation. The most likely scenario, in fact, requires what appears to have been geologically rapid base-level changes of the order of 1 km. Relative sea level first would have had to fall at least 1 km over a major portion of the depositional basin represented by the presently exposed Adelaide Geosyncline. Subaerial, or combined submarine and subaerial erosion, could then have produced the canyons, by a combination of headward slumping and possible fluvial action. A subsequent relative sea-level rise of the order of 1 km next would have been required, to explain the shallow-water nature of onlapping canyon-filling sediments.

Base-level changes of this magnitude far exceed the extent of normal eustatic excursions. Ice-volume changes, for example, produce sea-level variations that have a maximum of about 200 m. Spreading-ridge volume changes, operating over a considerably longer time-scale, can account for a maximum of only 200–300 m of eustasy. Thus, the only remaining mechanisms available to explain the Wonoka Formation paleocanyons are (1) tectonic uplift within the basin and subsequent subsidence, or (2) a Messinian-style evaporitic draw-down event, confined to a restricted portion of the Late Proterozoic rift.

The tectonic uplift/subsidence model requires a region within a basin, measuring at least 300 × 200 km, to have been uplifted and then to have subsided a minimum of 1 km, during a relatively small (~300 m) depositional interval. However, existing precanyon and postcanyon paleocurrent data do not show obvious perturbations immediately before and after such an event; rather, the regional paleoslope maintained its northerly trend, at least in the northern portion of the basin. Moreover, a second canyon-forming event, of a similar nature to the first, occurred after the original canyons were completely filled, requiring a second tectonic uplift/subsidence event. Considered overall, it is difficult to accept a double tectonic event of this type.

The Messinian-style evaporitic draw-down model, on existing evidence, more favorably explains existing data. It also could account for the sudden appearance of calcareous sediments at the onset of Wonoka Formation deposition (evaporative concentration of basin waters), and for the apparently rapid base-level fluctuations suggested by the stratigraphic evidence. If this model is the correct one, then the paleocanyons represent the eroded portion of a rapidly exposed basin-margin ramp.

REFERENCES

Eickhoff, K. H., C. C. von der Borch, and A. E. Grady, 1986, Incised meandering canyon in the late Proterozoic Wilpena Group; problems of timing and thalweg migration: International Sedimentological Congress, Canberra, Abstracts 92.

——— ——— ———, in press, Proterozoic canyons of the Flinders Ranges, South Australia: submarine canyons or drowned river valleys? Sedimentary Geology.

von der Borch, C. C., R. Smith, and A. E. Grady, 1982, Late Proterozoic submarine canyons of Adelaide Geosyncline, South Australia, AAPG Bulletin, v. 66, p. 332–347.

——— A. E. Grady, R. Aldam, D. Miller, R. Neumann, A. Rovira, and K. Eickhoff, 1985, A large-scale meandering submarine canyon: outcrop example from the Late Proterozoic Adelaide Geosyncline, South Australia: Sedimentology, v. 32, p. 507–518.

——— ——— K. H. Eickhoff, P. Di Bona, and N. Christie-Blick, in press, Late Proterozoic Patsy Springs Canyon, Adelaide Geosyncline: submarine or subaerial origin? Sedimentology.

VON STACKELBERG, U., and H. BEIERSDORF, Bundensanstalt für Geowissenschaften und Rohstoffe, Hannover, Federal Republic of Germany

Hydrothermal Mineralization in Lau and North Fiji Basins

Hydrothermal Mn-rich and Fe-rich crusts as well as sulfide-impregnated volcanic rocks were recovered by dredging or observed with a photo sledge at various locations in the Lau and North Fiji Basins. Such hydrothermal deposits accompanied by fields of dead clams are widely distributed mainly at an active spreading ridge in the southern Lau Basin, the Valu Fa Ridge. Here, mostly low- to medium-temperature precipitates occur at the seafloor surface while higher-temperature mineralization is found in xenoliths, Therefore, an extended sulfide stockwork mineralization is assumed. Still higher formation temperatures must be assumed for massive sulfides (sphalerite, pyrite, chalcopyrite, galena) found at one location on the Valu Fa Ridge. While the spreading ridge of the northern Lau Basin is composed of massive MOR basalts, scoriaceous andesited or basaltic andesites prevail at the Valu Fa Ridge. Mineralization mainly occurs at segments of the Valu Fa Ridge characterized by tectonic destruction.

The extended occurrence of low-temperature surface mineralization and of high-temperature stockwork mineralization may be explained by the high permeability of the andesites, which may lower the hydrothermal front.

VORDER BRUEGGE, R. W., and J. W. HEAD, Brown University, Providence, RI

Tectonic Evolution of Fortuna Tessera, Venus: Evidence for Multiple Stages from Mapping

The Fortuna Tessera region of Ishtar Terra, Venus, displays evidence of extensive tectonic deformation (Barsukov et al, 1986; Basilevsky et al, 1986; Vorder Bruegge and Head, 1988a, 1988b). This high plateau (2–6 km above the mean planetary radius) exhibits a transition in morphology along with a corresponding change in topography from west to east (Vorder Bruegge and Head, 1988a, 1988b). The change in morphology extends east from the Maxwell Montes mountain range (6–12 km above m.p.r.) and is reflected in the unit map of Figure 1.

Previously, we have proposed a model of east-west convergence and crustal thickening to account for the large-scale morphology and topography of western Fortuna Tessera, including Maxwell Montes and the adjacent Arcuate-Ridged Tessera (Figure 1) (Vorder Breugge and Head, 1988a, 1988b). In this paper, we suggest that there have been two additional directions of tectonic deformation within central and western Fortuna Tessera, based on observed patterns of features within and around the Chevron Tessera and Normal Tessera units.

Observations: Maxwell Montes—Maxwell Montes represents the highest topography on Venus, with a height of 6–12 km. The topography is asymmetric with the western face representing the steepest slope on the planet (up to 2.4°) (Sharpton and Head, 1985), while the eastern slope is more gentle (<1.5°). The morphology of Maxwell reflects this topographic asymmetry in that the western slope is characterized by long (up to 200 km), linear, parallel ridges spaced 5–10 km apart and striking north-northwest, while broader ridges with a spacing of 10–20 km occur on the summit and eastern slope. Some of these features are represented in the schematic diagram of Figure 2.

Arcuate-Ridged Tessera—This unit mainly occurs on a broad plateau immediately east of Maxwell Montes at a height between 5 and 6 km, and consists of short- to intermediate-length parallel ridges (50–150 km) with a spacing of 15–20 km (Figure 2). The strike of these ridges broadly parallels the strike of ridges in Maxwell Montes, and the pattern of ridges closely corresponds to the topographic contours of this unit in that both the ridges and the contours tend to broadly wrap around Maxwell Montes, in a roughly north-south orientation.

Chevron Tessera—The topography of this unit ranges from 5 km to below 2 km. In the southern part of this unit, this decrease occurs rapidly from west to east (a slope of 0.5° over 400 km). In the north, however, topography decreases toward the northeast, with a slope of less than 0.3° over 600 km. A variety of morphologies are present within the Chevron Tessera, and this variation corresponds to changes in topography. The western Chevron Tessera, between 4 and 5 km, resembles typical tessera terrain (Basilevsky et al, 1986) in that it consists of very short (<50 km) ridges and valleys intersecting at a variety of angles. Below 4 km, the small-scale morphology changes to a pattern of short- to intermediate (50–100 km)-length, subparallel ridges and valleys which have a characteristic spacing of approximately 12 km and strike northwest. Some of this pattern is represented in Figure 2.

The small-scale morphology of the Chevron Tessera is often disrupted by linear troughs that have up to 1 km of relief and are between 50 and 100 km wide and several hundred km long (Figure 2). These troughs have two dominant orientations, striking both northwest and northeast, and it is the intersection of these troughs that produces the chevron-like features. In addition to the large troughs, there are also three large swales in the southern Chevron Tessera that have a spacing of 75 km, up to 0.5 km of relief, and strike northwest (Figure 2). These swales have a small-scale morphology of subparallel ridges and troughs with a spacing of approximately 15 km. Finally, the northern edge of the Chevron Tessera is characterized by several large, northeast-facing scarps that strike northwest for up to 400 km and have a characteristic spacing of approximately 150 km. The combination of the small-scale morphology, the large scarps and swales, and the interesting nature of the northwest- and northeast-striking linear troughs gives the Chevron Tessera a highly disrupted, or buckled, appearance.

Normal Tessera—While the southern and northern Chevron Tessera has topography that slopes to the east and northeast, respectively, the Normal Tessera, instead, exhibits topography which can be characterized as a broad (1000 km wide) swale striking northeast, that is from 2 to 4 km in altitude. No large-scale structures disrupt the small-scale morphology in this unit, which is primarily a typical tessera area with ridges and valleys intersecting one another at a variety of angles. However, while the typical tessera-like area within the Chevron Tessera exhibited no consistent pattern, the Normal Tessera does contain traces of some continuity of structure. In particular, the southwest Normal Tessera contains some subparallel ridges and troughs striking northeast, up to 100 km in length, and with spacings of approximately 12 km. Adjacent to, and in some places coincident with, these northeast-striking ridges and valleys, there is another set of ridges and valleys that strike northwest. These northwest-striking ridges appear terminated by a northeast-striking ridge and valley pair (Figure 2). Another set of northwest-striking ridges is present in the northern Normal Tessera with a spacing of approximately 13–14 km. This region, however, does not exhibit any northeast-striking features.

Plumose Riges and *Ridge Belt*—Two other units occur to the south and north of Fortuna Tessera (Figure 1). The Plumose Ridges are narrow ridges with a spacing of 6–8 km and are individually continuous for up to 100 km (Figure 2). These subparallel ridges occur within a valley that starts in the northwest at an altitude of over 5 km and fans out to the south and east until they abruptly terminate near the 3 km altitude contour. As these ridges fan out of the southeast, those along the northern edge parallel the northwest-striking large-scale swales and parallel ridges and valleys in the southern Chevron Tessera. The Ridge Belt, however, consists of arcuate-to-sinuous, subparallel ridges and valleys that strike northeast, are up to 200 km in length and exhibit a spacing of approximately 10 km. This unit occurs along the northwest edge of the Fortuna Tessera plateau, paralleling a scarp with over 2 km of relief.

Interpretations: (1) The variation in style, scale, and orientation of deformational features within Fortuna Tessera, as shown in Figure 2, lead us to suggest that it is unlikely that a single tectonic event could be responsible for all of the observed deformation. (2) Instead, the presence of three dominant trends in the observed small- and large-scale features (north-northwest, northeast, and northwest) indicates that there have been three preferred orientations for tectonic activity. (3) While a complete sequence of events is yet to be established for this region, we can define at least one relative age relationship based on the structural stratigraphy. The northwest-striking pattern of ridges and valleys within the Chevron Tessera appears truncated by the large northeast-striking troughs. Similarly, the northwest-striking ridges and valleys in the Normal Tessera are sometimes truncated by northeast-striking ridges and valleys (Figure 2). These relationships indicate that the northwest-striking ridges and valleys formed before the northeast-striking troughs and ridge-valley pairs. (4) The nature of the ridges and valleys and their correlation to the large-scale topography within Maxwell Montes, the Arcuate-Ridged Tessera, and the Ridge Belt led us to suggest that these features were the result of large-scale lateral convergence of crustal material (Vorder Breugger and Head, 1988a, 1988b). The similarity of the small-scale structure within the Chevron Tessera and the Normal Tessera to that within the Maxwell Montes, Arcuate-Ridged Tessera, and Ridge Belt units leads us to suggest that these features were also the result of horizontal compression and the lateral convergence of crustal materials. Further evidence of compressional deformation is the presence of the northwest-striking scarps to the north of the Chevron Tessera. These features are morphologically identical to scarps north of Freyja Montes in western Ishtar Terra, which have been interpreted as evidence of large-scale convergence and crustal imbrication (Head, 1988).

Conclusions: The correlation of the morphological patterns and large-scale topography within Fortuna Tessera is consistent with a model of crustal convergence and related compressional deformation. The multiple orientations of features (north, northwest, and northeast) suggest that several stages of compressional deformation may have taken place with different orientations. The morphology and topography of Maxwell Montes and the Arcuate-Ridged Tessera indicates a stage of east-west compression and crustal convergence, while the patterns within the Chevron Tessera and Normal Tessera indicate at least two stages of compressional deformation—northeast-and northwest-directed events.

REFERENCES

Barsukov, V. L., et al, 1986, JGR, v. 91, p. D378.
Basilevsky, A. T., et al, 1986, JGR, v. 91, p. D401.
Head, J. W., 1988, LPSC XIX, p. 467.
Sharpton, V. I., and J. W. Head, 1985, JGR, v. 90, p. 3733.
Vorder Bruegge, R. W., and J. W. Head, 1988a, LPSC XIX, p. 1218.
——— ——— 1988b, LPSC XIX, p. 1220.

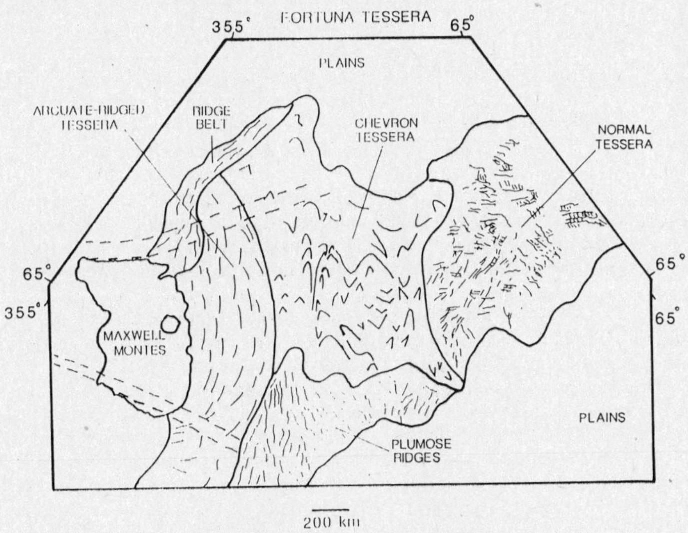

Figure 1—Geomorphic-tectonic map of Fortuna Tessera.

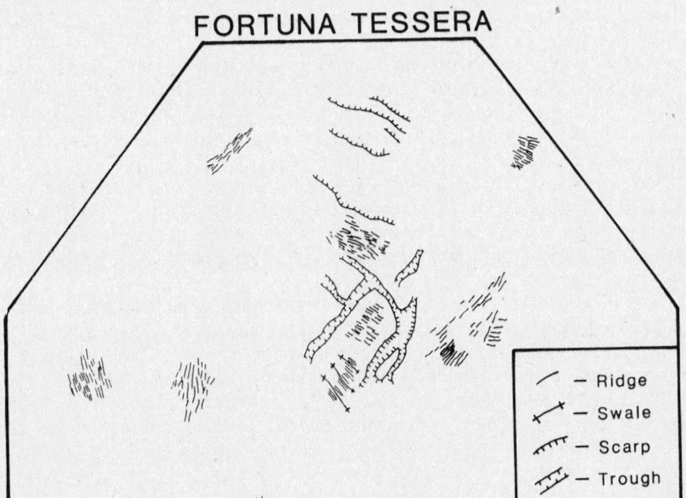

Figure 2—Sketch map of Fortuna Tessera.

VOROBYOV, Y. K., Institute of Geology of Ore Deposits, Mineralogy, Petrography and Geochemistry, Moscow, USSR

Origin of syngenetic Inclusions in Minerals

An experimental study of the crystallization process in multicomponent media (solutions, melts and crystalline matrixes) has shown that the growth of crystals is always accompanied with the separation of media components. Besides, a local heterogenization of the medium almost always takes place at the growth front due to a higher concentration of impurities near the crystal and the appearance of impurity phases. It has been discovered that the trapping of inclusions is an indispensable stage in the growth of crystals in impure natural media.

Syngenetic inclusions along the growth zones are formed from the most impure part of the crystallization halo and are always either impurity phases (solid, liquid or gaseous) or an exhausted and impure medium. Mineralothermometry is particularly interested in the frequent phenomenon of local heterogenization and the accompanying exsolution of a gas, as the released bubbles are trapped in the crystal at a random ratio within the liquid. This is the reason why primary syngenetic inclusions cannot serve as a basis for geothermometric measurements. As different crystals (and even different faces of a crystal) have differing distribution coefficients, the composition of syngenetic inclusions in paragenetic crystals of different minerals will also be different.

Syngenetic inclusions may be close in composition to the average composition of the crystal-forming medium, whereas their homogenization points may coincide with the real crystallization temperature only when the process of inclusion filling is separated from the growth process. This separation may occur during the filling of growth cracks or the intumescence of crystalline material on the etched surface of the seed.

VRBA, J., Stavebni Geologie, Praha, Czechoslovakia

Ecological, Economical and Social Aspects of Groundwater Protection

Groundwater protection should be based on the fact that prevention of pollution is always less expensive than aquifer reclamation which is a costly, long-term and technologically demanding process with unpredictable results. The consequences of groundwater pollution and depletion observed during the last few decades in various countries are producing undesirable economic, social and ecological impacts on their populations and on national or regional development.

Groundwater protection should not be perceived as an isolated action but rather a long-term program controlled and supported by governments and their institutions.

Groundwater protection management should be focused on overall preventive protection of groundwater resources over the whole national territory, and on comprehensive protection around groundwater public supplies. The conflicts and restrictive measures produced by protection of groundwater resources must be identified and analyzed from the economic and social points of view. The task of decision-makers is to choose a suitable strategy in defining priorities and preferences, with respect to national or regional development and to find balance between environmental protection policy and economic incentives. The monitoring activity significantly supports the groundwater protection management. The benefits derived from the financial means expended on monitoring should always be compared with the value of the information obtained. Statistical methods help to optimize the relations between monitoring costs and information losses. However, it is still difficult to express the value of groundwater monitoring data in financial terms and economic criteria.

Legislative measures based on Water and Environment Acts are essential for groundwater protection policy and strategy. Special attention should be devoted to the regulations concerning the licenses and permits for groundwater quality conservation. Regulations for groundwater quality protection should determine first of all the responsibility of potential polluters, rules of their penalization, and related prohibition or permission systems. Within the anti-pollution programs, various advantages could be established, such as tax benefits or low-interest loans, endowment, bonuses, etc. In the case of polluted water supplies, it is not the population which is to be financially afflicted. The financial responsibility for securing alternative supplies is borne by the polluter and the government regulations, particularly by well developed charging system.

Control should be extended particularly over the observance of legislative measures and regulations, and over the protection management process and monitoring activity. A groundwater protection program is demanding in terms of technology, human energy and financial funds, and therefore control over the whole program is desirable. A functioning control system completes the process of groundwater protection; it should guarantee that measures for pollution prevention are implemented and observed. In such a case, the benefits in the economic, social and ecologic spheres will be attained.

Research of groundwater protection is primarily focused on the impact of farming activities and related nitrate diffuse pollution on the hydrogeological system. The movement of nitrogen in the crop-soil-water system controlled by natural cyclic processes, anthropogenic influences and random events is under study in Czechoslovakia.

The nitrogen and carbon balance is essential for gaining insight into the physical, chemical and biological processes that take place in the unsaturated zone. These processes strongly control the amount of nitrogen leached into the saturated zone.

Perturbation of the organic carbon and nitrogen balance in soil, with consequences for the groundwater quality, occurs especially when the traditional crop rotation is replaced by monocultures. Over a period of four years at a same intensity of fertilization, the nitrate content increase in the experimental plots' groundwater under monoculture (barley) was nearly twice as high as for plots with traditional crop rotation (potatoes, wheat, barley, oat and clover).

The nitrogen balance equation indicates that *for a long time*, excessive amounts of nitrogen have been supplied to the soil through fertilizer; this is reflected in the instability of the soil organic matter and in considerable losses of nitrogen into the groundwater system.

However, research has also revealed the significance of *short-term* changes for the total nitrogen balance due to the immobilization and denitrification processes in a year's cycle. Monitoring of the soil-groundwater quality, with a high frequency of sampling in the early spring period, is therefore especially emphasized for European climatic conditions.

During the past thirty years, nitrate content in groundwater in areas where arable land is farmed has doubled, as have cereal yields wile the amount of fertilizer applied has grown nearly eightfold. The aquifer's rapid response to natural and anthropogenic impacts is positive in the sense that if agricultural activities and groundwater protection are mutually optimized and under control there will exist realistic possibilities for a rapid restoration of the hydrogeological system's quality. On the other hand, the hydrogeological system's fast response also has negative consequences because it signals easy vulnerability of the groundwater and the possibility of its quality's rapid deterioration.

Short-term cyclic changes in nitrate content have been identified in the aquifer's upper parts, depending mainly on the climatic conditions. A fast response of shallow aquifer's system to extreme climatic situations is particularly emphasized. The long-term changes and the increasing trend in nitrate concentrations in groundwater reflect anthropogenic, especially farming influences.

Comprehensive management of the agro-groundwater system based on scientific information sound is strongly recommended in regions with conflicting interests of the agricultural and water sectors, i.e. in regions with fertile soils under intensive farming activities having simultaneously significant groundwater resources.

VRUBLEVSKY, A. A., Far East Branch of the Academy of Sciences of the USSR, Khabarovsk, USSR

Faults of Fold Systems of Soviet Far East

The faults determine the cover-fold structure of Mesozoic and Cenozoic fold systems of the Far East and their break into en echelon blocks.

In the modern structure of the fold systems, boundary, intrastructural, and through faults and their systems are distinguished. To the first category belong the faults that at their full length or most of their length coincide with the boundaries of the fold systems; for example, the Arsenievsky, and Alchansky faults and systems of the East-Bureya faults of the Skikhote–Alin fold system. To intrastructural faults belong a number of known faults, in particular the Central Sikhote–Alin, Central Sakhalin, Kamuikotan, and others. Through faults are represented on the surface most often by transregional zones of heightened jointing.

In the fold systems, together with structural-material series, are ascertained dislocational vertical and lateral series where heterogeneous rupture dislocations play an important role. Incidentally, the sequence of these faults of this or that genetic variety in these series may be superposed and even repeat depending on the general direction of the region's tectonic evolution. The full vertical and morphogenetic series of faults of the fold systems of the Far East South includes: (1) faults of the foundation stage of the fold systems; (2) consedimentation nappes accompanied with mixtites; (3) scaly-thrust dislocations; (4) strike-slip fault late-fold dislocations; (5) postfold systems of riftogenic superimposed basins, normal fault-tension zones, and strike-slip faults of marginal-continental volcanic belts, and others. They are most vividly apparent and fully investigated in the Sikhote–Alin fold system. Numerous observations were also made of the Sakhalin–Hokkaido system.

The Jurassic–Cretaceous consedimentation covers and scaly dislocations of the South and East Sikhote–Alin and East Sakhalin are well correlated by the time and character of their manifestation with the close assemblages of the Japanese and Philippine Islands.

The Sikhote–Alin Late Cretaceous marginal-continental strike-slip fault system, the Cenozoic strike-slip faults of Sakhalin, and the Amur–Ussuri system of Cenozoic riftogenic basins are related to particular types of deposits of mineral and energy raw materials.

Manifestation of incipient magmatism with specific metallogeny is related to the faults of the initial stages of the fold systems development. Though the consedimentation nappes did not directly serve as canals for ore-bearing fluids, in many respects they determined spacial distribution of olistostrome complexes in which polymetallic and tungsten mineralization is often localized. The scaly-fold dislocations formed a complex system of screens for the ore-bearing fluids that resulted in concentration of minerals, in particular, gold and gold-rare metal ores in the underthrust parts and in the curves of the folds complicated by upthrow faults. The underthrust type of oil- and gas-bearing deposits broadly manifested in similar conditions in the other parts of the Pacific framing is not excluded in the sedimentary basins of the Far East South. Strike-slip fault dislocations are ore-controlling structures for many syngenetic deposits of the volcanic belts and fold systems at the final stage of their foundation. It is these dislocations that in most cases determine the en echelon structure of the ore fields, location of the ore knots, columns, etc.

VUKASINOVIC, S., and A. ANTONOVIĆ, Geoinstitut, Belgrade, Yugoslavia

The Fundamental Magmatogenic-Minerogenic Structures in Yugoslavia

Central Part of Balkan Peninsula, occupied by eastern Yugoslavia, is traversed by numerous deep megastructures. This surface of 60,000 sq km exhibits one of the most intensive concentrations of lead-zinc (Trepca, Zletovo, Sase, etc.) and copper (Bor, Majdanpek) mineralizations in the world, including antimony (Zajaca) and molybdenum (Mackatica), besides chromium (Radusa), nickel and magnesium (Goles), directly related to ultrabasic rocks.

Most of the studied deep fractures in the described surface are typical segments of the very long megastructures, extending up to 10,000 km, accompanied by various magmatic rock varieties, where the two structure types are distinguished. The first one is related to Internal Dinarides (Nos. 1, 2, 3 and 5 in the map), continuing along over Greece, Turkey and further to the east. The second type is restricted to the Carpathian-Balkanids (Nos. 6 and 7), passing into Bulgaria, northern Turkey and Iran. Other structures are of limited (Nos. 9a and b) or very short distances (Nos. 8 and 10). All of the mentioned structures are characterized by different position, extension, morphology, magmatism, mineralization, age and others.

The deep channels of magmatic rocks, recognized by magnetometery, are indicated as hot spots (basic and acidic ones), knotting that surface west of the longitude 18 °E has not been investigated.

Basic and ultrabasic rocks (mostly diabases and serpentinized ultramafites) connected with the mantle riching deep channels, are related to the Dinaric structures (1, 2, 3, 5), accompanied by a specific sedimentary-magmatic rock unit, named diabase-Chert Formation (corresponding to Ankara Mélange in Turkey and Coloured Mélange in Iran) and copper mineralizations, being of greater importance out of Yugoslavia. The age of magmatism is Jurassic in Yugoslavia or Upper Cretaceous-Eocene in Turkey and Iran.

Carpathian-Balkan structure 7 is accompanied by the Upper Cretaceous (Yugoslavia, Bulgaria, northern Turkey) up to Eocene (Iran) volcanics and subvolcanics, with the large and locally very rich Cu-ore deposits. Very limited volcanic bodies, Paleogene in age, occur along the practically non mineralized structure 6.

Miocene volcanics and rich Pb-Zn ore deposits (among the largest in the Eurasian continent) occur mostly in the domain of structures 2, 3 and 5 and partly east of structure 10. The structure 5 is accompanied by basites in its central part and by intermediate volcanics and granitoids to the south, but it is identified in the north by geophysical methods and drilling.

The structure 9a, marked by small granitic intrusions (Prosara, Motajica, Cer, Bukulja) and some other buried hot spots in between, bears the U, Th and Sn mineralizations of lesser importance, having the most probable continuation as the structure 9b. This is one of the oldest structure, possibly Early Paleozoic in age, reactivated in Miocene. The north-south structures, penetrated by Tertiary intermediate volcanites and subvolcanites bear the Sb (structure 8) or Mo (structure 10) mineralizations. The structure 4, buried under the sedimentary cover, is restricted to the Panonian basin.

The Dinaric structures coincide with the suture or collision zone of megacontinents, but the curved part of these, including the Carpathian structures 6 and 7, might surround a huge circular (probably Pre-Tertiary impact) structure. The Pb-Zn mineralizations, occurring mostly in swarms, are located likely in the southwest external part of this (hypothetic) ring structure.

REFERENCES

Antonovic, A., 1972, Geol. Map of Iran 1:100,000, sheet 6851—Kamrud.

Antonovic, A., Vukasinovic, S., 27th Intern. Geol. Congr., Moscow, 1984, section C-12.

Metallogenic map of Carpathian-Balkan Area 1:1,000,000, CME, Standing Commission of Geology, Sofia, 1978.

Vukasinovic, S., Serbian Geol. Soc. Proceedings for 1974, Belgrade, p. 21-34.

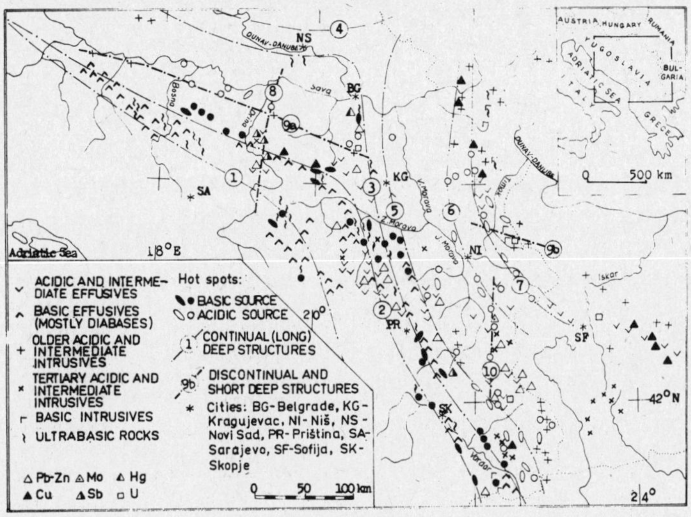

Figure 1

VUOLLO, J., and T. PIIRAINEN, University of Oulu, Oulu, Finland

Primary Minerals in the Serpentinites of the Outokumpu Complex, North Karelia, Finland: Contributions to the Origin

The serpentinites of the Outokumpu complex located in the early Proterozoic schist belt of North Karelia in Eastern Finland are investigated petrographically, mineralogically and chemically, with an emphasis on relict minerals, especially chrome spinel (Figure 1), and also olivine and orthopyroxene. The relict minerals and textures provide evidence for an ophiolite of the Alpine type in the complex, and for the opinion that the serpentinites have originated from residual mantle peridotites, being serpentinized to crysotile rocks, contain relicts of olivine (Fo_{91-93}), orthopyroxene (Mg/(Mg + Fe) = 92-94) and anhedral, euhedral or interstitial Cr-rich chrome spinel (Cr/(Cr + Al) 0.75-0.98 and Mg/(Mg + Fe) 0.1-0.4), while the cumulates have been serpentinized completely to antigorite rocks, being characterized by euhedral Al-rich chrome spinel (Cr/(Cr + Al) 0.65 and Mg/(Mg + Fe) 0.6).

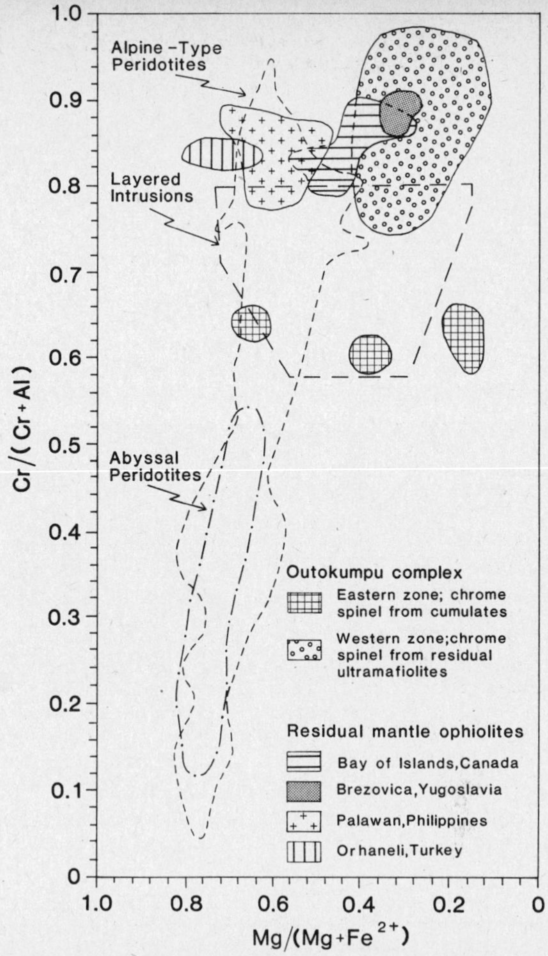

Figure 1—Composition fields of chrome spinel from the Outokumpu complex and some selected fields from residual mantle ophiolites (Malpas et al., 1975, Obradovic, 1986, Rammlmair, 1986, Tankut, 1980). Fields of layered intrusions chrome spinels from Irvine (1967) and alpine-type peridotites, abyssal peridotites from Dick and Bullen, 1984).

REFERENCES

Dick, H. J. B., and T. Bullen, 1984, Chromian spinel as a petrogenetic indicator in abyssal and alpine-type peridotites and spatially associated lavas: Contrib. Mineral. Petrol., v. 86, p. 54-76.

Irvine, T. N., 1967, Chromian spinel as a petrogenetic indicator, part 2, Petrologic applications: Can. J. Earth Sci. v. 4, p. 71-103.

Malpas, J., and D. F. Strong, 1975, A comparison of chrome spinels in ophiolites and mantle diapirs of Newfoundland: Geochim Cosmochim. Acta 39, 1045-1060.

Obradvic, L. J., 1986, Short review of the chemistry of chromites from Brezovica, Yugoslavia, *in* Karamata, S., ed., Chromites. Unesco's IGCP-197 Project. Metallogeny of ofiolites. Theosphrastus publ. Athens., p. 91-106.

Rammlmair, D., 1986, Chromite in the Philippines: Its relationship to the tectonic setting of the host ophiolites; examples from Zambales and Palawan. *In* Karamata, S., ed., Chromites. Unesco's IGCP-197 Project. Metallogeny of ofiolites. Theophrastus publ. Athens. p. 285-310.

Tankut, A., 1980, The Orhaneli massif, Turkey. *in* Panayiotau, A., ed., Ophiolites. Proceedings International Ophiolite Symposium, Cyprus, 1979, Cyprus Geol. Surv. Dep., p. 702-713.

Vuollo, J., and Piirainen, T., 1988, Mineralogical evidence for an ophiolite from the Outokumpu serpentinites in Northern Karelia, Finland (in press).

W

WADA, H., T. TOMITA, K. MATSUURA, and K. IUCHI, Shizuoka University, Shizuoka, Japan

Graphitization of Carbonaceous Matter During Metamorphism

We studied the graphitization process during metamorphism as an indicator of metamorphic grade. The graphitization can be described by its mineralogical structure and chemical and isotopic compositions. The degree of graphitization depends mainly on temperature, pressure, strain rate, duration of a metamorphism, and rock type, which includes carbonaceous matter. We compared the trends in graphitization for some different types of metamorphic terrains (contact aureole, high-temperature and high-pressure types) and in different bulk compositions (calcareous and pelitic rocks) using X-ray, DTA, and chemical and isotopic data.

1. Basal spacing d(002) decreases with increasing metamorphic grade and is mainly controlled by temperature. Decreases in basal spacing of the carbonaceous matter in limestone is accelerated at temperatures higher than about 300 °C. Carbonaceous matter in pelitic rocks shows a more gradual decrease of d(002) spacing.

2. We can demonstrate that the graphitization process involves progressive construction process of three-dimensional crystal structure. The sequence of appearance of d spacing correlates with metamorphic grade.

3. Carbon isotopic ratios of graphite in limestone drastically change from − 28 per mil to ^{13}C-enriched values that indicate equilibration with calcite. The beginning of this enrichment of ^{13}C in the carbonaceous matter corresponds to the stage of acceleration in graphitization mentioned above.

4. Carbonaceous matter in pelitic rocks is sometimes a mixture of poorly crystallized organic material and detritus of well-crystallized graphite. Differential thermal analysis is a good tool for the distinction of detritus graphitic material.

WAGNER, J. F., Karlsruhe University, Karlsruhe, West Germany

Transfer and Retention of Heavy Metals in Clayey Landfill Subsoils

A large number of heavy metal dusts and sludges are dumped on natural clayey host rocks or artificial clay liners. Therefore the migration process of these pollutants was analyzed by field and laboratory investigations in order to predict the migration velocity resp. the danger potential for groundwater contamination.

The transfer of heavy metals in porous media in composed of the transport with the carrier medium water (advection and hydrodynamic dispersion) and of the transport due to a concentration gradient (molecular diffusion). In "reactive" porous media like clays, both transport mechanisms are superposed by retention processes like precipitation and ion exchange resp. adsorption.

The description of the exact transport phenomena may be determined in three different ways:(1) By mathematical prediction models. An often applied model is the differential advection–dispersion equations, describing one dimensional transport of a reactive solute in a discrete porous medium under steady waterflow condition:

$$dC/dt = Dd^2C/dx^2 - vdC/dx - \rho\, dS/v\, \delta\tau$$

where C is the concentration of the solute in the percolating solution, D the dispersion coefficient, v the average pore-water velocity, ρ the dry bulk density, S the amount adsorbed, Θ the volumetric water content, x the distance and t the time. The rock-solution interaction is determined in batch equalibration tests and expressed by a distribution coefficient (k_d).(2) By direct measurements of breakthrough data from column experiments. Column experiments were performed with (percolation test) and without a hydraulic gradient (diffusion test). In the diffusion test an apparent diffusion coefficient $D^* - D_oT/R$ (D_o is the diffusion coefficient in free solution, T the tortuosity factor and R the retardation factor), which includes a sorption constant, is measured.(3) By field measurements below a waste disposal site or at a "natural analogue" (e.g. ore body).

The three methods have been applied to describe the transfer of different heavy metals (Cd, Cr, Cu, Pb and Zn) in different types of clays and mudstones. Mathematically predicted breakthrough curves using a distribution coefficient k_d were compared to measured breakthrough curves of column tests and heavy metal profiles below waste sites. The results indicated great differences between the three methods, because the use of distribution coefficients in mathematical prediction assumes that the rock-solution interaction is at equilibrium and that is reversible, which is generally not given. The differences between field situations and column experiments are due to simplifications in the laboratory tests and the high extrapolation in time and space.

A back calculation from field data leads to an effective k_d, which shows how to ameliorate the performance of batch and column experiments including further solution and especially rock constants like rock pH and eH, microfavric, mineralogy, organic compounds, zeta potential, etc.

WAGNER, J.-F., Karlsruhe University, Karlsruhe, West Germany

Transfer and Retention of Heavy Metals in Clayey Landfill Subsoils

A large number of heavy metal dusts and sludges are dumped on naturaly clayey host rocks or artificial clay liners. Therefore the migration process of these pollutants was analyzed by field and laboratory investigations in order to predict the migration velocity resp. the danger potential for groundwater contamination.

The transfer of heavy metals in porous media is composed of the transport with the carrier medium water (advection and hydrodynamic dispersion) and of the transport due to a concentration gradient (molecular diffusion). In "reactive" porous media like clays, both transport mechanisms are superposed by retention processes like precipitation and ion exchange resp. adsorption.

The description of the exact transport phenomena may be determined in three ways:

(a) By mathematical prediction models. An often applied model is differential advection-dispersion equation, describing one dimensional transport of a reactive solute in a discrete porous medium under steady water flow condition:

$$dC/dt = Dd2C/dx2 - vdC/dx - \varrho\, dS/\theta dt$$

where C is the concentration of the solute in the percolating solution, D is the dispersion coefficient, v the average pore-water velocity, ϱ the dry bulk density, S the amount adsorbed, θ the volumetric water content, x the distance and t the time. The rock-solution interaction is determined in batch equilibration tests and expressed by a distribution coefficient (k_3).

(b) By direct measurements of breakthrough data from column experiments. Column experiments were performed with (percolation test) and without a hydraulic gradient (diffusion test). In the diffusion test an apparent diffusion coefficient $D^* = D_oT/R$ (D_o is the diffusion coefficient in free solution, T the tortuosity factor and R the retardation factor), which includes a sorption constant, is measured.

(c) By field measurements below a waste disposal site or at a "natural analogue" (e.g. ore body).

The three methods have been applied to describe the transfer of different heavy metals (Cd, Cr, Cu, Pb and Zn) in different types of clays and mudstones. Mathematically predicted breakthrough curves using a distribution coefficient k_d were compared to measured breakthrough curves of column tests and heavy metal profiles below waste sites. The results indicated great differences between the three methods, because the use of distribution coefficients in mathematical prediction assumes that the rock-solution interaction is at equilibrium and that it is reversible, which is generally not given. The differences between field situations and column experiments are due to simplifications in the laboratory tests and the high extrapolation in time and space.

A back calculation from field data leads to an effective k_d, which shows how to ameliorate the performance of batch and column experiments including further solution and especially rock constants like rock pH and eH, microfabric, mineralogy, organic compounds, zeta potential, etc.

WALKER, A. T., E. I. Dupont de Nemours Co., Wilmington, DE, and C. B. SCLAR, Lehigh University, Bethlehem, PA

Subsolidus Redox Reactions Recorded in Fe-Ti Oxides of Preston Gabbro, Southeastern Connecticut

The Preston gabbro of southeastern Connecticut is an early Paleozoic differentiated mafic pluton enclosed in high-rank gneisses from which cataclastic gneisses and mylonites were locally developed. The latter constitute the Honey Hill and Lake Char faults at the southern and eastern contacts of the gabbro, respectively. The Preston gabbro is a plagioclase-rich (high Alp_2O_3) differentiated pluton that includes olivine and pyroxene gabbro of cumulate origin and late stage differentiates that are expressed as (Fe-Ti-P)-rich gabbro, quartz gabbro, and trondhjemite dikes (Sclar, 1958). A strong iron-enrichment trend occurs between the olivine and pyroxene gabbros and the quartz and (Fe-Ti-P)-rich gabbros. The concentrations of the rare-earth elements (REE) also increase in this sequence, but the slopes of the chondrite-normalized heavy REE patterns remain parallel. These trends are attributed to crystallization of early olivine–plagioclase cumulates that depleted the magma in Al^{+3}, Sr^{+3}, and Ni^{+2}, and enriched the residual liquid in Fe^{+2}, Ti^{+4}, and REE (Walker and Sclar, 1983).

The late-stage igneous rocks (quartz gabbro and Fe-Ti-P gabbro) are rich in iron (12–16 wt% total Fe expressed as FeO) and contain abundant Fe-Ti oxides (>10 vol%). The magmatic oxide assemblage of these rocks consists of granular magnetite and of hemoilmenite (original composition $Il_{80}Hem_{20}$–$Il_{70}Hem_{30}$) that exsolved into oriented ilmenite–hematite intergrowths in which the topactic plane is (0001). Early exsolution platelets of Hem_{ss} were reduced to magnetite, which occurs as oriented lamellae surrounded by hematite-drained ilmenite. Also present in the ilmenite are irregular granular aggregates of rutile enclosed in magnetite. This rutile–magnetite assemblage was produced by oxidation of the host ilmenite. The magnetite lamellae are apparently contemporaneous with and in places physically continuous with the magnetite of the retile–magnetite assemblage. Locally, the magnetite of the rutile–magnetite assemblage was partly oxidized to hematite along grain boundaries. The subsolidus cooling history deduced from these relationships is as follows: (1) magmatic crystallization of magnetite and homogeneous hemoilmenite; (2) subsolidus exsolution of hemoilmenite to Il_{ss} and Hem_{ss}; (3) subsolidus reduction of the oriented hematite exsolution lamellae in the ilmenite to oriented magnetite lamellae according to the reaction of $3Fe_2O_3 \rightarrow 2Fe_3O_4 + 1/2\ O_2$; (4) oxidation of ilmenite to magnetite plus rutile according to the reaction $3FeTiO_3 + 1/2\ O_2 \rightarrow Fe_3\ O_4 + 3\ Ti^{O}2$; and (5) Incomplete oxidation of magnetite of the rutile–magnetite assemblage to hematite.

Any reconstruction of the redox evolution of these oxides must be constrained by the hematite–agnetite buffer (reaction 1 of Figure 1), and the redox reaction ilmenite → magnetite + rutile (reaction 3 of Figure 1). Subsolidus redox conditions may well have been controlled by the composition of the associated hydrothermal fluid that was responsible for the extensive development of hydrothermal hornblende at the expense of the magmatic pyroxenes in these rocks. However, the region between reaction 1 and reaction 3 in Figure 1 is too reducing to be governed by the breakdown of pure water (reaction 2 of Figure 1). this suggests that a reducing species was present in the hydrothermal fluid. It is important to note that pyrite and chalcopyrite are commonly associated with the Fe-Ti oxides in the Preston gabbro and that the effect of sulfur on FO_2 can be very large. The FO_2-temperature relationship of a hydrothermal fluid with a molecular ration of $H_2O/S_2 = 100$ (curve 4 in Figure 1) is such that it will result in the reduction of hematite to magnetite. Thus, the change with decreasing temperature from relatively oxidizing to relatively reducing conditions may have been controlled by chemical variations in the hydrothermal fluid, specifically a decrease in the ration of H_2O to S_2.

REFERENCES

Ellis, A. J., 1957, Chemical equilibrium in magmatic gases: American Journal of Science, v. 255, p. 416–431.

Robie, R. A., B. S. Hemingway, and J. R. Fisher, 1978, Thermodynamic properties of minerals and related substances at 298.15°K and 1 bar (10^5 pascals) pressure and at higher temperatures: U. S. Geological Survey Bulletin, v. 1452, 456p.

Walker, A. T., and C. B. Sclar, 1983, Petrochemistry of the Preston gabbro of southeastern Connecticut: tectonic implications: GSA Abstracts with Programs, v. 15, p. 118.

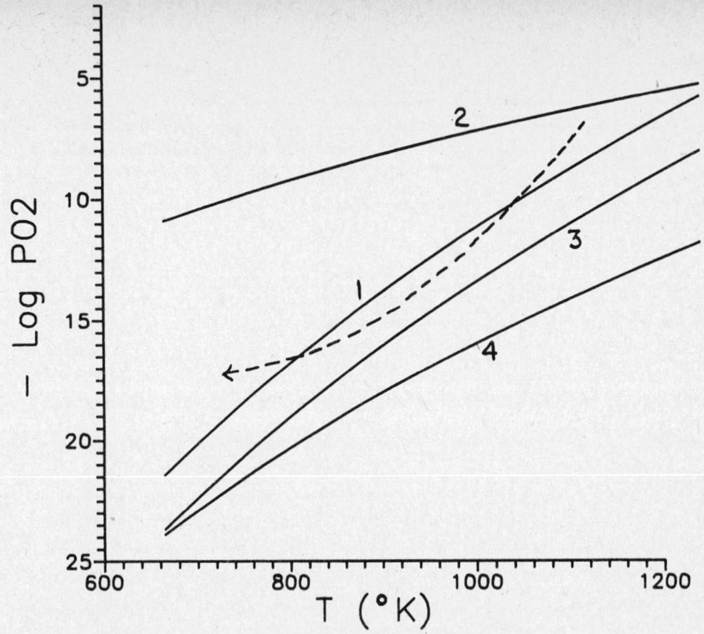

Figure 1—Negative log PO_2 versus temperature for several redox reactions: 1, $3Fe_2O_3 \rightarrow 2\ Fe_3O_4 + 1/2\ O_2$ (calculated using data from Robie et al, 1978); 2, $H_2O\ 2H_2 + O_2$; 3, $3FeTiO_3 + 1/2\ O_2 \rightarrow Fe_3O_4 + 3TiO_2$ (calculated using data from Robie et al, 1978); and 4, $H_2O/S_2 = 100$ (molecular) (Ellis, 1957).

WALKER, P. H., CSIRO Division of Soils, Canberra, Australia

Soil Formation on Transported Parent Materials

Most soils develop on transported parent materials; few develop in place on bedrock. Transported parent materials have a variety of origins and generally relate to landscape transport processes of the Quaternary. Some of these transport processes represent distinct geological episodes (e.g., glaciation) after which prolonged pedogenesis occurred. Other transport processes are more or less continuous (e.g., hillslope movement, aeolian accession) so that pedogenesis must be considered in the context of sediment fluxes. In many landscapes both kinds of transport process have contributed to soil parent materials. Data are presented here from two locations in southeastern Australia and one location in midwestern USA to illustrate these principles.

Flights of late Quaternary alluvial terraces occur along all major rivers in the uplands of southeastern Australia. Close to rivers, alluvial surfaces are often flooded and profiles of alluvial sediment are little modified by pedogenesis. At successively higher levels, alluvial sediments are older and surfaces are less frequently flooded; profiles have successively greater soil organization. Regionally, the stages in profile development from lowest to highest surface have been named: stratic–cumulic–low contrast solum–high contrast solum–extended subsolum. In formal soil terminology, these stages are approximately equivalent to Entisols–Mollisols–Alfisols–Ultic Alfisols–Ultisols.

Radiocarbon dating of sediments beneath the soils of these sequences has allowed limiting ages to be assigned to the first four stages. The stratic stage is contemporary; the cumulic stage is generally less than 2000 y B.P.; the low contrast solum stage is generally less than 10,000 y B.P.; the high contrast solum stage is generally less than 30,000 y B.P.; the extended subsolum stage is generally beyond the limit of radiocarbon dating. The soil stages have been interpreted as a regional chronosequence. On this basis, a soil Bt horizon can develop within 30,000 years in alluvium. In terms of the introduction of this paper, however, the stages represent contrasting conditions of alluvial sediment flux. For example, stratic profiles represent sediment flux rates > rates of pedogenesis; cumulic profiles represent sediment flux rates = rates of biotic pedogenesis; profiles with a high contrast solum represent sediment flux rates < rates of pedogenesis. Within a valley cross-section, the status of alluvial parent materials ranges from that of more or less continuous input on the lowest surfaces to that of a geologic episode on the highest surfaces.

In central Iowa, Cary till deposits of the last major glaciation have been subjected to varying rates and conditions of water erosion, deposition, and pedogenesis for the last 11,000 years. The history of development of the Clarion-Webster soil catena of these landscapes has been reconstructed from the stratigraphy of enclosed depressions. Between 8000 and 3000

B.P., rates of water erosion on hillslopes were high and very little remains of the previous soil cover. From 3000 B.P. to the present, rates of hillslope erosion were much slower, favoring the development of soils (Mollisols) under a prairie grassland environment. During the last 3000 years, soils of the Clarion series (Hapludolls) developed on upper hillslopes essentially in the till deposit; on lower slopes and in adjacent depressions soils of the Webster series (Aquolls) developed in hillslope detritus that accumulated during the previous phase of erosion. In different parts of the catena, soils developed in different parent materials and in response to rates of net removal or deposition. The resulting soil pattern is determined largely by the properties and distribution of hillslope sediments.

At Sutton, in the tablelands of southeastern Australia, a Udic Haplustalf on a hillslope summit was selected for study because the presence of undisturbed quartz veins and dikes indicated soil development in place on granodiorite. Initial field and laboratory data were consistent with development from the underlying bedrock. Detailed particle-size distributions, however, contained a conspicuous coarse silt mode (31–62 μm) that decreased down the profile. ^{18}O values for coarse–sand-sized quartz in the soil were the same as quartz in the bedrock (9.6‰); through the profile quartz of the coarse silt fraction had a different ^{18}O value (11.7‰) from quartz in the coarse sand fraction and the rock. The foreign coarse silt component is thought to be from continental dustfall during the late Quaternary. It has penetrated more than 0.5 m into the soil profile. Clay and fine silt would also have been deposited and must have profoundly affected pedogenesis. Rough estimates indicate that as much as 30% of the A horizon of this soil has an aeolian source.

Interpretations of pedogenesis depend to a considerable extent on the identification of parent material. The three examples given above show that the definition and identification of soil parent materials are difficult tasks, particularly where the parent material is added as the soil develops and is not necessarily locally derived. Resolution of these problems can be achieved, however, through a multidisciplinary approach to field and laboratory research.

WALKER, R. J., S. B. SHIREY, and R. W. CARLSON, CIW-DTM, Washington, DC, and J. D. FASSETT, NIST, Gaithersburg, MD

Re-Os Isotopes and Chemical Evolution of Mantle

The Re-Os isotope system has numerous applications in the study of the earth's mantle. Re and Os, both of which are siderophile and chalcophile, have very different geochemical characteristics compared with the lithophile elements that comprise the other long-lived radiogenic isotope systems. Hence, the Re-Os isotope system, when used together with the other systems, can provide complementary but also unique insights into the chemical evolution of the mantle.

Os isotope measurements of osmiridiums of presumed ages by Allegre and Luck (1980) provided some of the first evidence that the earth's mantle evolved with an essentially chondritic Re/Os. Thus, it is likely that the initial ^{187}Os/^{186}Os of the earth was approximately 0.8 and is now approximately 1.04–1.08. Recent measurements reported for modern mafic and ultramafic rocks (Martin and Turekian, 1987; Pegram et al, 1988) confirm that at least some portions of the mantle have developed with a chondritic Re/Os.

The similarity in the Os isotope evolution of the mantle compared with that in chondrites is somewhat surprising. The abundances of Re and Os in the mantle (approximately 0.2 and 3 ppb, respectively) are approximately a factor of 300 lower than abundances reported for carbonaceous chondrites (40–70 and 500–1200 ppb, respectively). This presumably resulted from the incorporation of most of the earth's Re and Os into the core during its formation. The retention of a chondritic Re/Os ratio following core formation is less easily explained. It seems unlikely that the partitioning characteristics of Re and Os between the silicate earth and the metallic material that formed the core would be identical. Alternate explanations include the possibility that certain phases that contain high Re and Os abundances and a chondritic Re/Os were retained in the mantle following core formation and "buffer" the abundances at their present level. Another possible explanation is that most of the Re and Os contained within the mantle was added by the accretion of extraterrestrial material onto the earth following core formation. A definitive resolution of this puzzle could provide new insights into the processes of core formation and the rates of both the accretion that followed core formation and the subsequent homogenization of the silicate earth. However, much additional information regarding the Os isotope evolution of the mantle through time, the sites for Re and Os in the mantle, and silicate-metal partitioning characteristics for Re and Os must be obtained before this problem can be resolved.

Some of this work is underway. Ultramafic rocks of known age provide some of the most useful information about the Os isotope evolution of the mantle. These rocks provide a sufficient range of Re/Os for precise isochron work. Also, unlike for the other radiogenic isotope systems, Os isotope compositions of ultramafic rocks are unlikely to have been modified by contamination because of the two orders of magnitude higher abundances of Os in these rocks relative to most possible crustal contaminants. For example, the initial Os isotope composition determined for a series of komatiite flows from the late-Archean Abitibe belt of Ontario, Canada (Walter et al, 1988), is consistent with their derivation from a portion of the mantle evolving with a chondritic Re/Os. Compared with ultramafic rocks, most mafic rocks typically have very high Re/Os and much lower Os abundances. Hence, except for young rocks, mafic rocks provide less precise information about the isotopic compositions of their mantle sources.

In contrast to the "chondritic" source indicated by the komatiite data, significant deviations from chrondritic evolution have been noted in certain peridotite xenoliths (Carlson et al, 1988). Re is incompatible during mantle melting while Os is compatible to highly compatible, so the extraction of melt from the mantle severely retards ^{187}Os growth in the residue. Thus, some of the observed isotopic deviations from chondritic growth can probably be attributed to ancient melt depletion events. The magnitude of such deviations can provide information about the timing of the generation of crust from the mantle. For example, some xenoliths from the subcontinental lithosphere underlying the Kaapvaal craton of southern Africa have experienced long-term Re depletion resulting in as much as 13% depletion of ^{187}Os relative to rocks evolving with a chondritic Re/Os. The intersection of the Re-Os growth trajectory of a depleted xenolith with the chondritic growth trajectory provides chronologic information regarding the depletion event and, by inference, crust formation. Re-depletion ages for xenoliths are analogous to Nd T_{DM} ages, which are based on the isotopic compositions of crustal rocks. The results from some of the Kaapvaal xenoliths indicate melt extraction from the mantle a minimum of 2.8 b.y. ago. These Re depletion events presumably correspond with the formation of the Kaapvaal craton.

The growing Re/Os database is beginning to reveal the exciting possibilities for the applications of the system to the study of the mantle. Expected improvements in instrumentation should soon lead to the accumulation of more precise Re-Os data that will likely significantly advance our understanding of the mantle.

REFERENCES

Allegre, C. J., and J. M. Luck, 1980, EPSL, v. 48, p. 148.
Carlson, R. W., et al, 1988, Transactions of AGU, v. 69, p. 502.
Martin, C. E., and K. K. Turekian, 1987, GSA Abstracts with Programs, v. 19, p. 761.
Pegram, W. J., et al, 1988, Transactions of AGU, p. 69, p. 510.
Walker, R. J., et al, 1988, EPSL, v. 87, p. 1.

WALLACE, P. J., and I. S. E. CARMICHAEL, University of California, Berkeley, CA

Mineralogy and Chemistry of Late Pliocene, Lamprophyric Lavas and Associated Leucitites from Los Volcanes Region, Western Mexican Volcanic Belt

During the last three million years (Ma), small volumes of alkaline and calcalkaline magma have been erupting in the Western Mexican Volcanic Belt in a region that is much closer to the Middle America Trench than the main line of andesite stratovolcanoes. Volcanism near the town of Los Volcanes occurred between 2.9 and 1.8 Ma, and subsequently shifted 30 km to the northwest, where it is now focused near the town of Mascota. In the Los Volcanes region, a variety of volcanic rock types are present. Alkalic varieties include minettes (biotite–lamprophyres), leucitites, vogestites (hornblende–lamprophyres), absarokites, and high-K basaltic andesites. These rocks occur in association with calcalkaline spessartites (hornblende–lamprophyres) of intermediate silica content. All of these types, with the exception of the leucitites and vogesites, are also very abundant in the Mascota region. The minettes of Los Volcanes differ from those of the Mascota region in having lower MgO and compatible element concentrations, in their lack of olivine phenocrysts, and in the occasional presence of leucite microphenocrysts. Both of these minette types occur as porphyritic (<25 vol% phenocrysts), rapidly sequenced volcanic rocks, making them ideal for studying the compositions of minette magmas. The importance of this cannot be overemphasized as minettes typically occur as more fully crystalline dikes, making the identification of magma compositions and the details of their petrogenesis obscure.

In the Los Volcanes region, flows of absarokite and minette are far more abundant than any of the other rock types. Both of these are Ne normative and contain phenocrysts of augite in a groundmass composed predominantly of alkali feldspar. In addition to augite, the minettes contain phenocrysts of phlogopite and commonly apatite, while the absarokites bear olivine with inclusions of spinel. A single flow of felsic minette contains olivine with spinel inclusions, along with the other phases found in the minettes. Leucitites are much less abundant, contain little or no modal feldspar, and have phenocrysts of augite along with either phlogopite or olivine. Relative to the minettes, the leucitites are higher in MgO and K_2O, and lower in SiO_2. Vogesites are characterized by phenocrysts of hornblende and augite and, in some cases, contain olivine. The major difference between them and the minettes is the much higher K_2O/Na_2O ratio of the latter.

Concentrations of major and selected trace elements for representative samples of the various types presented below. Compatible trace element abundances are generally similar in all of the minettes, and their Ni and Cr contents are much lower than those found in most of the minettes from other parts of western Mexico and from around the world (Luhr et al, in press; Rock, 1984). Extreme enrichment in many of the incompatible trace elements (Ba, Sr), and mild enrichment in some of the Ti-group elements (Ti, Zr) is found in all of these volcanic rocks. However, Rb is not as enriched as the other large ion litholphile elements (LILE), resulting in high K/Rb and low Rb/Sr. These lavas are strongly enriched in the light rare-earth elements (REE) relative to the heavy REE, with (La/Yb)N ratios ranging from 60 in the minettes to 65 in the olivine leucitite. In addition, enrichment in the light REE is in parallel with that of the LILE resulting in relatively unfractionated (Ba/La)N ratios.

The Mg/Mg + Fe ratio and comparable element concentrations in the minettes are lower than the values that characterize primary magmas derived by melting of mantle peridotite. Results of least squares mass balance models demonstrate that with the exception of a single felsic sample, the minettes are unlikely to be formed by crystal fractionation from a parent magma similar in composition to the more mafic, olivine-bearing minettes which occur in western Mexico. This conclusion is also supported by the high-ressure phase equilibria experiments on mafic minettes conducted by Esperanca and Holloway (1987). These observations suggest that the minette magmas from the Los Volcanes region may have been generated in an enriched source region similar to that of the mafic minettes, but in which olivine was a relatively minor phase, perhaps due to either metasomatism (Luhr et al, in press), or hybridization involving silicic magmas (Wyllie and Sekine, 1982) in the mantle wedge above the subducting Rivera Plate. Trace element abundances and ratios in the other lava types indicate that similar enrichment processes have occurred in their source regions. The high concentrations of MgO and compatible trace elements, and the presence of olivine and spinel in the absarokites and the olivine leucitite indicate that olivine was a major phase during melting, while the lower K_2O/Na_2O ratios and the occurrence of hornblende phenocrysts in the vogesites suggest a greater role for amphibole, relative to phlogopite, in their source region. These petrological and geochemical features provide direct evidence of the extreme heterogeneity that may exist in magma source regions at convergent margins.

REFERENCES

Esperanca, S., and J. R. Holloway, 1987, On the origin of some mica-lamprophyres: experimental evidence from a mafic minette: Contributions to Mineralogy and Petrology, v. 95, p. 207–216.

Luhr, J. F., J. F. Allan, I. S. E Carmichael, S. A. Nelson, and T. Hasenaka, in press, Primitive calc-alkaline and alkaline rock types from the Western Mexican Volcanic Belt: Journal of Geophysical Research.

Rock, N. M. S., 1984, Nature and origin of calk-alkaline lamprophyres: minettes, vogesites, kersantites and spessartites: Transactions of the Royal Society of Edinburgh: Earth Sciences, v. 74, p. 193–227.

Wyllie, P. J., and T. Sekine, 1982, The formation of mantle phlogopite in subduction zone hybridization: Contributions of Mineralogy and Petrology, v. 79, p. 375–380.

	minette	felsic minette	phlogopite leucitite	olivine leucitite	vogesite	absarokite	basaltic andesite
SiO_2	47.4	55.6	42.7	44.0	50.1	50.0	52.6
TiO_2	2.4	1.3	3.6	2.6	1.6	1.4	1.8
MgO	6.2	5.4	8.1	10.0	7.3	11.3	6.6
Na_2O	2.9	4.0	2.8	1.6	4.8	3.6	3.8
K_2O	4.9	4.5	7.0	6.1	3.0	2.2	2.3
P_2O_5	1.8	0.9	0.9	2.0	1.1	1.1	0.6
Ni	117	185	160	320	160	315	85
Cr	128	140	205	400	190	440	–
Rb	60	50	80	170	25	35	35
Sr	3220	2480	2775	4375	2380	2930	1310
Ba	2595	2015	3600	3330	1440	1995	1045
Zr	395	305	295	510	230	220	250
La	120	75	110	145	70	120	40
K / Rb	680	730	710	300	1045	520	545
Rb / Sr	0.02	0.02	0.03	0.04	0.01	0.01	0.03
$(Ba/La)_N$	1.0	1.3	1.6	1.1	1.0	0.8	1.2
Sr / Zr	8.2	8.1	9.3	8.6	10.4	13.3	5.2

major elements in weight %; trace elements in ppm; minette is the average of nine (9) samples

Table 1—Chemical characteristics of volcanic rocks from the Los Volcanes region.

WALLISER, O. H., Inst. Geol. Palaeont., Göttingen, Federal Republic of Germany

Use of Global Changes for Stratigraphical Subdivision

For the discussion of the question if stratigraphic boundaries indicate global changes, we need beforehand some definitions: *Lithostratigraphy* describes and subdivides sequences of rocks on account of changes in lithology and/or facies. In this case, fossils are taken only as indicative lithological and/or facial components. *Ecostratigraphy* is based on the evolution of ecosystems. Those characterize corresponding facies. Hereby facies is defined as the entirety of abiotic and biogenetic components. *Biostratigraphy* describes and subdivides sequences of rocks on account of the evolution of organisms. Out of the biostratigraphy a *biochronology* can be abstracted, which is independent of the rock sequence. *Event stratigraphy* uses distinct changes in lithology, sedimentology, facies, fossil content, physical and/or chemical conditions to describe and to subdivide rock sequences (a linguistically better term would be *katastratigraphy*). *Chronostratigraphy*, whatever this may be, is taken as the fundamental principle for the decisions of the International Commission on Stratigraphy (ICS). *Commission boundaries* are those boundaries that have been accepted by the ICS since 1972. They are defined by a certain point within the sequence of a boundary-stratotype. *Natural boundaries* are what I call those boundaries that are recognizable on account of a distinct change, e.g., in lithology or facies or fossil content. Insofar they are event stratigraphic boundaries. In addition to these definitions I wish to emphasize that here I am dealing with boundaries only in the Phanerozoic, excluding the Quaternary.

Most of the lately established commission boundaries do not coincide with natural boundaries for the following reasons: If a boundary is caused by a gobal event that is combined with an extinction event, new radiations, mostly comprising good index fossils, occur only after a certain time interval. This interval is less indicative: The former index fossils have disappeared, but the new forms have not yet evolved. The understandable request for good index fossils, i.e., taxa that are easily recognizable, part of an evolutionary line, short living, and widely distributed, lead in most cases to the definition of commission boundaries at a time level that coincides with the first radiation after a global event.

Regarding the above-mentioned connections, we should in this discussion neglect the commission boundaries, but rather concentrate on natural boundaries. The stratigraphic units of high order (systems, series, stages) were established in the last century. The scientists of this time recognized that there are main differences between subsequent units. These differences concern in most cases the fossil content, e.g., a "Devonian faunal aspect" versus a "Carboniferous faunal aspect." In many cases, these faunal changes are connected with a change in lithology and/or facies. Insofar, ancient stratigraphic subdivision was based in some respect on a combination of bio-, eco-, and lithostratigraphy. Nevertheless, the main aspect was a biostratigraphical one.

In that pioneer time of stratigraphy, the exact position of changes within the sequences was not always exactly fixed. In numerous cases there was also the assumption that the changes were only of local or regional importance. All the more we have to admire the achievement of our scientific ancestors: former and recent investigations proved that most of the boundaries between those traditional high-rank units are caused by global events. This will be demonstrated at the Phanerozoic system boundaries and at the boundaries between series and stages within and around the Devonian system.

In order to complete the discussion, it should be mentioned that on the one hand numerous boundaries of minor order are also caused by global events, and that on the other hand some boundaries of higher order are not connected with major changes.

Conclusion: In the Phanerozoic, most boundaries of higher order are event boundaries, i.e., they are indicative of global changes. Unfortunately, in most cases the Commission boundaries do not coincide with these natural boundaries. This demonstrates that global events play a decisive role in the stratigraphical development as well as in the evolution of organisms and ecosystems, but not so much in the decisions of stratigraphical commissions.

WALLISER, O. H., Inst. Geo. Palaeont., Göttingen, Federal Republic of Germany

Global Biological Events: Complex Story

Global biological events show a great variety of causes, patterns, extent, and effects. Several categories can be recognized:

1. *Innovation events:* Biological innovations, such as, e.g., the creation of bisexuality, the ability to mineralize skeletal elements, or the first coiling of the Cephalopods, may lead to major changes in the world of biota. Insofar, biological innovations played an enormous role especially during the

early evolution in the Precambrian and early Phanerozoic. But also minor innovations with respect to construction (Bauplan) or physiology may cause remarkable changes in the concerned communities.

2. *Radiation events:* In most cases, the radiation events occur either after an innovation event or after an extinction event. In both cases, there is a certain time interval between the causing event and the subsequent radiation. In cases of postinnovation radiations (R_{pi}), the new acquisition becomes stabilized by further evolution. The following radiation reflects the now reached ability of "play" with the new acquisition and to intrude into adjacent biotopes. In cases of postextinction radiations (R_{pe}), the radiation mirrors the occupation of niches that occurred due to the extinction event. During the mentioned interval the selectional stress is more or less strongly reduced. This enables the taxon by relatively rapid evolution to enlarge its potency to occupy the niches.

3. *Spreading event:* In comparison with the other mentioned events, this one is of minor importance. Nevertheless, it is remarkable that a sudden spreading out or a burst of a blooming of a certain taxon can be globally observed in numerous cases within the Phanerozoic. Those spreading events might be caused by environmental changes either of anorganic origin or by the occurrence of a disturbance of the biological balance.

4. *Extinction events:* Of course, the extinction of a taxon is an event and always global. If this extinction occurs in the "normal" course of evolution (within times of nomismogenesis), it can be taken as "background extinction." If within a higher taxon, e.g., a class, the rate of background extinctions is higher than the rate of originations, then the diversity of the taxon decreases and may lead to the final extinction of the entire group. We are still far from understanding the mechanisms of these background extinctions; therefore, future investigations should also concentrate on this problem.

Here we are dealing with those extinction events that happen in times of major global changes. The latter are short termed in comparison with the intermediate times of relatively stable conditions. The extinction event may occur in a sudden, often catastrophic manner, or it may be stepwise, even if it happens within a very short time. Extinction events may range from the extinction of only a few taxa within only one biotope up to affecting entire ecosystems.

With respect to causes, we also may distinguish several categories:
1. Extra-telluric = cosmic causes, such as (1a) changes caused by the revolution of the solar system within the galaxy, and (1b) the impact of cosmic bodies.

2. Telluric = earth-born causes, such as (2a) biological causes (mostly biological innovations) and (2b) abiotic = geological causes. The latter can be manifold, e.g., sea-level changes, climatic fluctuations, changes of chemical conditions of seawater or atmosphere, changes or disruptions in the oceanographic patterns or parameters, extraordinary activities in tectonics and/or volcanism, etc.

All these extra-telluric and telluric causes may trigger by themselves or in any combination those processes that finally lead to global bioevents, might they be sudden or stepwise, short termed or long termed.

The discovery of the iridium event at the Cretaceous–Tertiary boundary and the subsequent hypotheses about Nemesis and periodicity stimulated extraordinarily the investigations on global bioevents. Many of these investigations were carried out in order to verify the general validity of the impact hypotheses. But all these activities, especially those within the IGCP-Project 216, led to the following recognition: Impacts may play a role in some cases, but in most cases, global bioevents show a very complex story with respect to causes as well as to processes, patterns, and effects.

WALTER, M. R., Bureau of Mineral Resources, Geology and Geophysics, Canberra, Australia

Proterozoic Petroleum in Australia

A major period of plate rifting commenced about 1 b.y. ago and led to the formation of narrow depositional basins comparable with those of the Cretaceous proto-Atlantic and Tethyan regions. Within the Cretaceous basins, organic-rich sediments were deposited as a result of limited aeration of deep waters. These are the source rocks of some of the giant petroleum fields of the Middle East and elsewhere. The climatic, tectonic, and paleogeographic conditions that obtained in the later Proterozoic and Cambrian in Australia and elsewhere seem to have been comparable with those of the Cretaceous. These conditions would have been reinforced by high atmospheric concentrations of CO_2 coupled with low concentrations of O_2 that would have stimulated productivity and decreased oxidative degradation of organic matter. Furthermore, during the Proterozoic, there was no significant bioturbation, due to the lack of burrowing metazoans. Some small horizontal burrows occur in the very latest Proterozoic, but deep vertical burrows indicative of thorough bioturbation are first found in earliest Cambrian rocks. As a result, sediments would not have been irrigated with oxygenated water even in situations where the overlying water body was oxygenated. Conditions would seem to have been ideal for the formation of unusually rich petroleum source rocks. At the same time, potential reservoirs were formed as extensive well-sorted sand bodies, ooid grainstones and abundant stromatolite reefs, bioherms and biostromes. Evaporites occur at several stratigraphic levels and form potential cap sequences. Examples of such sequences occur in the Adelaide Geosyncline and the Amadeus, Georgina, and Officer Basins of Australia; they are similar to sequences from which petroleum is being produced in China, Siberia, and the Sultanate of Oman.

As elsewhere in the Late Proterozoic, carbonates of this age in Australia are enriched in ^{13}C, suggesting a high rate of burial of organic carbon enriched in ^{12}C. Potential source rocks occur as a carbonate-evaporite sequence and black shales in the Amadeus Basin and elsewhere in Australia are prospective for significant accumulations of hydrocarbons.

Perhaps most remarkable has been the discovery of rich source rocks 1.7–1.5 Ga old in the McArthur Basin of northern Australia. One occurrence is accompanied by live oil. Despite the age of this sequence, in large areas of the basin the source rocks are immature to mature. Both marine and lacustrine facies are present. Organic carbon contents range up to 7% with organic-rich intervals up to 200 m thick. An active exploration program is underway in this basin.

Biomarker studies have indicated that the Proterozoic kerogens and oils have unique characteristics that allow their differentiation from younger examples, and which give new insights into the biology of Proterozoic times.

WALTER, M. R., Bureau of Mineral Resources, Geology and Geophysics, Canberra, Australia

Major Features in Record of Proterozoic Stromatolites

Already by Early Proterozoic times carbonate platforms had developed that show the range of reef and carbonate clastic facies once thought typical of the Phanerozoic. Throughout the Proterozoic, stromatolites formed reefs of many types, from huge barriers to patch and pinnacle reefs. Just as metazoans and calcareous algae were the sources of carbonate sediments during the Phanerozoic, stromatolites fulfilled that role in earlier times. Coherent benthic microbial communities covered vast tracts in both marine and lacustrine environments, forming a wide range of stromatolite morphologies each characteristic of its own paleoenvironment.

Against this background we can examine the temporal distribution of stromatolite types for any regularities that might encode information on environmental or biological evolution. The Proterozoic record does reveal some interesting temporal patterns, the most outstanding of which are:

1. An abundance of ministromatolites with a radial fibrous fabric, characteristic of peritidal environments, in the Early and Middle Proterozoic, with a marked decline thereafter.

2. An abundance of *Conophyton* and related stromatolites, characteristic of quiet subtidal environments, in the Early and Middle Proterozoic, with a marked decline thereafter.

3. The decline in abundance and diversity of all stromatolites about 700 Ma ago.

4. The absence or near absence of thrombolites, which became abundant during the Early Cambrian.

Significant clues to the interpretation of these four trends have come from studies of extant microbial mats. The ministromatolites can be compared to calcified mats of the cyanobacteria *Rivularia* and *Phormidium,* in which elongate calcite crystals parallel the radially arranged filaments. Grotzinger has proposed that an abundance of such stromatolites which record in situ precipitation of calcium carbonate is consistent with other evidence, suggesting that Early and Middle Proterozoic seawater was greatly oversaturated with respect to calcium carbonate and perhaps also had a high $[HCO^-{}_3]$ to $[Ca^{2+}]$ ratio.

The other three events have been related by various authors to the ecological effects on microbial mats of the earliest metazoan grazers and burrowers. This interpretation is still controversial, and a number of alternatives have been postulated. No one disputes that grazing and burrowing metazoans disrupt the fabric and restrict the distribution of extant microbial mats, but unequivocal evidence of these same processes in ancient mats is not found until the Cambrian. Burrows are abundant in thrombolites of Cambrian and younger times. A recent study by Farmer and his colleagues in Baja, California, has examined the effects on the fabric of extant microbial mats of the meiofauna that inhabits them. Microscopic and mesoscopic animals such as nematodes and tiny annelids are abundant in many stromatolitic mats, including those in Shark Bay, Western Australia. A comparable meiofauna probably inhabited Proterozoic stromatolites, and a search for evidence for this has begun, so far without success.

So in dealing with the decline in the abundance and diversity of stromatolites that began about 1 Ga ago we can still only make reasonable inferences. Several lines of evidence support the hypothesis that the decline can be attributed to the earliest effects of metazoans on benthic microbial populations. The most convincing alternative hypothesis attributes the decline to the effects of the addition of algae to the cyanobacterial communities, suggesting that this would have produced less coherent mats, or that macroscopic algae would have out-competed the microorganisms for space and nutrients. This seems unlikely as our evidence for the existence of algae now extends back to at least 1.7 Ga ago, long before the changes in stromatolite-building microbial populations.

The discovery by Southgate that some of the stromatolites and most of the described microfossils of the Late Proterozoic Bitter Springs Formation are lacustrine, not marine as was long assumed, indicates the danger of interpreting the Proterozoic fossil record without the support of a sedimentological framework. That framework is still very limited, particularly in the Late Proterozoic. As a result, it would be foolish to be too emphatic in insisting on any particular interpretation of regularities in the record of stromatolites. But the regularities mentioned here, and many others, exist and indicate that the stromatolite record will continue to be a rich source of information about the history of benthic microbial communities.

WALTHER, C., and E. R. FLÜH, University of Kiel, Kiel, Federal Republic of Germany, and U. LUOSTO, University of Helsinki, Helsinki, Finland

Combined Interpretation of P- and S-Wave Data along POLAR Profile in Northern Finland

In 1985, a 440 long refraction profile called the POLAR Profile was observed in northern Finland, aiming to resolve the crustal structure in an area of Archaean to Early Proterozoic lithosphere formation. From 9 shotpoints, high-quality P- and S-wave record sections were recorded at 2 km station spacing.

From the P-wave seismograms, a complex upper crust with high velocity bodies such as the Karasjok Kittilae Greenstone Belt and the Lapland Granulite Belt have been inferred (Luosto et al, in press). In their interpretation, the crustal thickness varies between 40 and 47 km, and a pronounced boundary between the middle and lower crust is seen.

In general, the S-wave data show interesting and striking similarities to the P-wave data, but in a number of details differences are also encountered. To allow for a detailed and reliable phase correlation of the S-wave seismograms, which should not be biased by the P-wave seismograms, so called "energy controlled gain" (ECG) sections were produced. These are obtained by adding all amplitudes in a specified time and space window (the energy trace), and then weighting the original trace with the energy trace to obtain the ECG sections. As three-component recording was made, this had to be done for every component. On several sections, S- to P-converted phases could be identified, whereas P to S conversions are not easily recognized, since they are normally hidden in the P-wave coda. These phases are used to determine more reliably some of the velocity boundaries for P- and S-waves.

No clear evidence for S-wave splitting is encountered that could be indicative of anisotropy.

From the P- and S-wave velocity field the Poissons ratio can be determined. In the upper crust, it is highly variable but generally low. In contrast, for the lower crust and upper mantle the Poissons ratio seems to be higher. The middle to lower crust boundary, which is seen as a prominent reflection in the P-wave data, produces only a very weak S-wave reflection.

The combined use of P- and S-wave data allows for further discrimination between the various geological and tectonic interpretations of the Lapland Granulite Belt and its deeper structure, as discussed in detail by v. Knorring and Lund (in press).

REFERENCES

Luosto, U., E. R. Flüh, and C. E. Lund, in press, The crustal structure along the POLAR Profile from seismic refraction investigations: Tectonophysics.

v. Knorring, M., and C. E. Lund, in press, The Earth's crust along the POLAR Profile. An integrated geophysical geological study in the Archaean part of the Baltic Shield: Tectonophysics.

WALTZ, J. P., Colorado State University, Fort Collins, Colorado, and R. E. BOYLE, Grid Company, Fort Collins, Colorado

Cost-Effective Enhancement of Ground Water Production in Crystalline Rock Aquifers

The problem of developing adequate ground water supplies from crystalline rock aquifers is widespread. Over 25 percent of the land surface area of the earth is directly underlain by plutonic igneous and metamorphic rocks. In many of these areas the only reliable source of water is from the crystalline rock aquifers. In spite of using state-of-the-art exploration techniques, 15 to 50 percent of drilled wells fail to produce adequate supplies of ground water. The reason for these failures often is linked to near-misses of permeable fracture zones by the borehole. Often the best water-bearing fractures are nearly vertical. Since most well bores are also oriented vertically, near misses of good fractures may be the rule rather than the exception. A second reason for failures is that many major fracture zones have associated chemical alteration zones where porosity may be relatively high but permeability is very low. the low permeability results from collapse of fractures due to incompetence of the clay-rich altered rock mass.

Significant enhancement of low-yield wells has been achieved through use of the FLEXIFRAC procedure of hydraulic fracturing and well development. This procedure involves controlled injection of water and proppant into packer-isolated well bores with injection volumes and rates adjusted in accordance with aquifer responses to the fracturing procedures. In additional to hydraulic fracturing, the equipment is used to stimulate the well with other mechanical and chemical treatments, e.g., high-velocity jetting and acidizing. The FLEXIFRAC equipment uses injection pumps which provide discharge rates up to 170 gpm and pressures up to 3000 psi. Inflatible packers are lowered into the wellbore by means of a continuous flexible hose assembly. A control panel permits monitoring and control of down-hole pressures, injection rates, proppant emplacement and packer security. Data acquisition is automated and provides for data storage on diskettes and graphical output of key data in the field. Aquifer Response Curves are prepared as part of the FLEXIFRAC procedure. These curves provide insight into the hydrogeologic condition of the fractured rock aquifer.

To date more than 600 water wells in the U.S., Europe, and Africa have been treated with the FLEXIFRAC technique. Over 90 percent of these wells have shown a significant increase in yield as a result of the treatment. The high probability of success and the low cost of the procedure generate a high benefit/cost ratio. Evaluation of cost-effectiveness shows that cost per unit of water produced by use of the FLEXIFRAC procedure is usually less than on half that of conventional exploration and drilling procedures.

WAN Minghao, JIN Xingchun, Tongji University, Shanghai, and WANG Shengli, National Bureau of Oceanography, Beijing, China

Primary Study of Gravity Sections from Shanghai to Caroline Islands

During November 1983 to January 1984, two gravity survey sections from Shanghai to Caroline Islands were made by the Chinese scientific survey ship "Xiang Yanghong No. 10." The section I, 4010 km long from northwest to southeast, went through the East China Sea, the Ryukyu Arc-Trench System, the Philippine Sea, the Mariana Trench (deepest one in the world) and into the East Mariana Basin; the section II, 5371 km long, went through the East China Sea, the Ryukyu Arc-Trench System, the Philippine Sea, the Yap Trench and into the East Carolina Basin. The purpose of survey lines setting is to cross over major geological tectonic units. The accuracy of automatic rudder is $\pm 1°$, positioning accuracy about ± 0.3 nautical miles and sounding accuracy about 1%. The gravity data are got automatically and synchronously with two oceanic gravimeters made in China. The free-air anomalies, Bouguer anomalies and two dimensional isostatic anomalies have been got on the two sections. The Moho depths of various submarine units are acquired after spectral analysis of separated parts in two sections (see Table 1).

From see in Section I (Figure), Bouguer anomalies rise from East China Sea shelf to Ryukyu Trench and Philippine Sea. It shows the feature of transition from continental crust to oceanic crust. The bouguer anomaly is up to 400 mgal in the West Philippine Basin, which is one of the highest values in the world. A double-V type of negative gravity anomaly zone is found respectively at the trench and fore-arc areas along the Ryukyu Arc-Trench System. At the negative anomaly zone of the trench, Bouguer anomalies are not low, and the negative free-air anomaly disappears after the Bouguer correction, which shows it resulted generally from the landform (greater water depth). At the anomaly zone of fore-arc, the Bouguer anomalies show clearly low values, which may be relative to greater thickness of crust or huge sedimentary beds. 10860 m sounding data and the great negative free-space anomalies −300 mgal have been got at the Mariana Trench. The sea floor of Pacific plate which will be subducted soon was older (Mesozoic); denser the plate, steeper the subduction zone and much deeper the trench.

Therefore, there are much greater free-air negative anomalies along the trench.

The free-air anomalies and isostatic anomalies in the West Philippine Basin are the lowest and most stable, showing an isostatic state of crust. The lower anomalies also appeared in the Parece Vella Basin. These properties support the idea that these two Basins are inactive marginal basins. Of them, the West Philippine Basin has lower amplitude and lesser change in anomalies because it stopped spreading earlier (Early Tertiary Period) than the Parece Vela Basin, where the squeezing stress by the recent collision at Yap Trench might pass into Parece Vela Basin and make it have a little higher positive isostatic and free-air anomalies. The obvious free-air and isostatic anomalies are in the Okinawa and Mariana Troughs, which are actively spreading marginal basins with higher thermal flows, suggesting a unisostatic state of crust. Presumably it is uplifting of rising hot mantle material that makes extra mass in two Troughs. The Mariana Trough has higher isostatic anomalies because of its more actively spreading, and its higher Bouguer anomalies implied the thinner crust and nature of clastic oceanic crust. The relatively lower Bouguer anomalies in the Okinawa Trough result from its thicker crust.

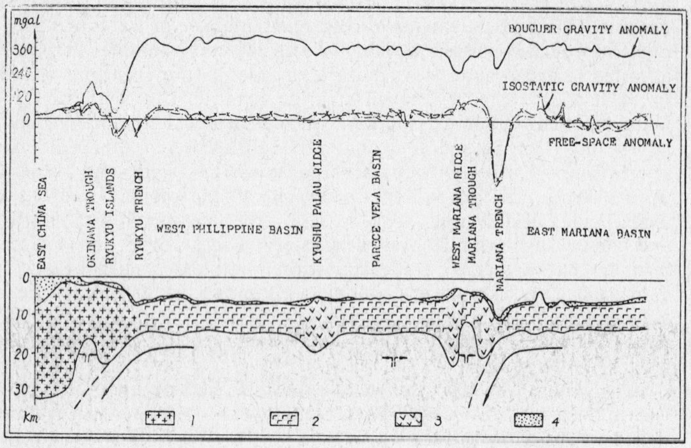

Figure 1—Comprehensive gravity section I from Shanghai to Caroline Islands.
1. Continental crust or subcontinental crust
2. Oceanic crust
3. Transitional crust
4. Depositional formation

WANG Binhai, and QIAN Kai, Geological Scientific Institute of ShengLi Oil Field, China

Tectonic and Sedimentary Evolution and Accumulation of Oil and Gas in Meso-Cenozoic Basins, Northern Shandong, China

For more than 20 years, the Meso-Cenozoic basins have been the main areas for oil and gas exploration and development in northern Shangdong province. These basins are also the main productive-area of Bohai Bay petroliferous province.

In these basins, there are not only the common properties of continental rift basin, but also some characters of continental margin basin. Its tectonic and sedimentary evolution can be approximately divided into three stages. Stage I (from Jurassic to Cretaceous) is the initial stage. During this stage, the crust arched, extended, and fractured, the volcano exploded, and the initial basins which are characterized by containing formation of coarse clastic rock-andesite were formed. Stage II (palaeogene) is the principal stage of the formation of basin. During this stage, the crust extended, faulted, and depressed strong, associating with explosion of basic basalt, and the hemi-graben type or graben type basins were formed. During the early period of Stage II, the sedimentation was characterized by filling evaporite sediments, with a major cycle consisted of evaporite and carbonate rock. There was probably some influence resulted from a transient transgression in some part of these basins during the depositing of carbonate rock. During the middle period of Stage II, the sedimentation was characterized by depositing thick dark source rocks which were widespread and contained various type of turbidite body. During the late period of Stage II, the sedimentation was characterized by depositing sandstone body of fluvial-delta facies which were widespread and resulted from rapid lateral progradation. Stage III (Neogene) is the filling and ending stage. During this stage, the Bohai Bay region subsided as a whole and filled up with sediments. The sedimentation in this stage is characterized by depositing sandstone and conglomerate of alluvial facies, especially fluvial facies.

As a result of the mentioned characters of evolution, there are three structural layers which contain oil and gas accumulation in these basins. In the lower structural layers (pre-Tertiary), the major oil field are of predominantly basement rock oil pool. In the middle structural layer (Palaeogene), the major oil field are of predominantly syngenetic anticline oil pool, with some stratigraphic-lithologic pool. In the upper structural layer (Neogene), the major oil field are of predominantly dropping anticline oil pool. The geological reserves of oil in the middle and upper structural layer are estimated to be more than 75% of the total geological reserves.

WANG Chengshan, YU Guangmin, ZHANG Shaonan, Chengdu College of Geology, Chengdu, China

Sedimentary Basin Evolution of Mesozoic along Yalung Zangbo River, Tibet

It is confirmed that there are two ocean-floor sediment associations deposited in the Yalung Zangbo Ocean Basin, which are different not only in age but also in tectonic setting, distributed in both the southern and northern of the Yalung Zangbo Ophiolite Suite. One of them, the radiolarites-layered basalts association on the south side, which ranges from late Jurassic to early Cretaceous in age, probably was produced during the rapid-spreading stage of Tethyan Tibet like a pattern of Pacific Rise type. The ophiolitic clastic rocks-pillow lavas association of Mid-Cretaceous age, which is determined by *Orbitolina* fossils found recently, explored on the north side of the ophiolite suite, and this sedimentary association was produced in a plate tectonic setting of slow-spreading condition of Atlantic Ridge type. The Yalung Zangbo ophiolite suite is only a remnant of the ocean crust in this stage because the north sedimentary association covers the Yalung Zangbu ophiolite suite with sedimentary contact.

During the late Cretaceous, Yalung Zangbo ocean plate provides the first "tectonic emplacement"; i.e., the Yalung Zangbo ophiolites were trapped between a trough to the south and Gandese Mountain arc to the north and turned into the basement of the southern Gandese fore-arc basin. Consequently, a sequence from the ophiolite suite and the sedimentary association on the north side to Xigaze Group turbidites deposited in the fore-arc basin presents more or less a complete succession and seems to accumulate in a unified sedimentary basin. As a matter of fact, it may be the sequence of Dickinson's (1979) residual fore-arc basin. Therefore, it is necessary to discuss that the conclusion of the Yalung Zangbo ocean basin in the Mesozoic was a small ocean basin or Yalung Zangbo tectonic zone is not a zone of plate suture.

WANG Dezi, CHENG Kerong and ZHOU Jincheng, Nanjing University, China

Mesozoic subvolcanic granitoids in Southeastern China and Their Metallogenesis

Mesozoic (upper Jurassic and lower Cretaceous) volcanic rocks, most of which is intermediate acidic-acidic, are widely distributed in Southeastern China. There exist various kinds of subvolcanic granitoids associated with the volcanic rocks. The following rock types are referred to as the so-called subvolcanic granitoids: They belong to granitoids in chemical composition and are closely related to intermediate-acidic to acidic volcanic rocks in petrogenesis; they emplace as high-level intrusive body near surface in a certain place of volcanic apparatus and are controlled by volcanic structures; they usually are the products of late stage of a volcanic eruptive cycle.

In recent years, some outstanding volcanic apparatuses in which there are various kinds of subvolcanic granitoids were studied in detail by us. According to the difference of their occurrences, six types of subvolcanic granitoids may be divided as following: (1) the subvolcanic granitoids occurred as central stock within caldera, (2) the subvolcanic granitoids in the shape of volcanic neck, (3) the subvolcanic granitoids injected along ring and radial fractures, (4) the subvolcanic granitoids emplaced in explosive pipe and parasitic crater, (5) the subvolcanic granitoids distributed along the border of fault-depression basin, (6) the subvolcanic granitoids intruded into extrusive dome and the covering strata of volcanic rocks or occurred as the subterraneous part of extrusive dome.

These subvolcanic granitoids possess following charateristic features: (1) they have close relations with volcanism of intermediate-acidic and acidic magma in space and time. They are usually formed at the late stage of eruptive cycle. Both volcanic rocks and their corresponding subvolcanic granitoids often possess similar evolutional trend, (3) the occurrences of most subvolcanic granitoids are controlled by certain volcanic structures, (3) the general texture of volcanic rocks are often present in the subvolcanic granitoids, (4) the oxidation coefficient ($Fe_2O_3/FeO + Fe_2O_3$) of subvolcanic granitoids is between those of the hypabyssal and volcanic rocks, (5) the temperature of formation of these subvolcanic granitoids is lower than that of corresponding volcanic rocks, (6) the K-feldspars in subvolcanic granitoids are usually of low sanidine or high orthoclase. For the comagmatic volcanic intrusive complex occurred in a volcanic apparatus, the K-feldspars in the subvolcanic granitoids possess intermediate structural state, (7) the REE distribution patterns of subvolcanic granitoids are generally similar to those of associated volcanic rocks. It indicates that the volcanic-intrusive complex in a certain volcanic apparatus is often comagmatic, (8) the initial ratios of ($^{87}Sr/^{86}Sr$) of most subvolcanic granitoids are usually less than 0.710. Their values of O^{18} are almost from + 6.9 to + 10.0.

The subvolcanic granitoids have close genetic relations with the formation of many ore deposits, such as those of Cu, Pb, Zn, Mo, Au, Ag, U and some W, Sn. These ore deposits are located either within the subvolcanic granitoids to form the porphyry deposits, or around the contact zone to form the veinlet-disseminated and vein deposits. The subvolcanic granitoids intruded along volcanic neck, explosive pipe, parasitic crater, ring and radial fractures are fine carriers of mineralized materials. These volcanic structures are also the conduits of postmagmatic hydrothermal solutions and the accumulated place of ore-forming materials. Finally, authors discuss the ore-forming specialization of each kind of subvolcanic granitoids from petrology, petrochemistry and isotopic geochemistry of subvolcanic granitoids.

WANG Dongpo, Changchun College of Geology, P.R. China, and P. K. LINK, Idaho State University

Depositional Setting of Sinran Strata, Dalian Peninsula, Liaoning Province, China

Sinian and lower Cambrian strata on the Dalian Peninsula, Liaoning Province, China (Figure 1) are up to 9000 meters thick and consist of alluvial fan and fluvial strata at the base, overlain by carbonate platform and bank and siliceous shelf facies strata (Figure 2). The marine rocks of the middle and upper parts of the section bear evidence of syn-sedimentary folding at several stratigraphic horizons. Unlike some other Late Proterozoic strata worldwide, the Sinian of Liaoning does not record volcanism or fault-related sedimentation which would suggest continental rifting, nor does the section contain evidence of Late Proterozoic glaciation.

The strata were deformed in the Mesozoic Yanshan orogeny. They now crop out in northward overturned folds, and in places have prominent spaced cleavage. In Xiangshuisi, 20 kilometers northeast of Dalian, the Sinian strata (Nanfen Formation) is juxtaposed in a ductile fault zone against Archean? gneiss and granite with relations that suggest a metamorphic core complex.

The lowest exposed strata are terrigenous detrital rocks of the Yongling Group. These thick and coarse-grained strata are interpreted to be alluvial fan deposits at the base, grading to fluvial and deltaic rocks at the top. The basal Sangshu Formation (310 m) lies unconformably on Archean gneiss, and contains conglomerate and arkoses. The overlying Duozishan Formation (1515 m) contains rhythmic interbeds of conglomerate and arkosic sandstone. The thick Miaoshan Formation (2645 m) contains fluvial and deltaic sandstones interbedded with conglomerates.

All the overlying Sinian strata are interpreted to be marine. Clastic strata of the Diaouytai Formation (112 m) and Nanfen Formation (561 m) contain intertidal to offshore facies, with evidence of synsedimentary folding at several stratigraphic horizons. Because these rocks are not of slope facies, earthquakes may have caused this deformation before lithofaction and while the rocks were still ductile. Medium bedded quartz sandstone of the overlying Qiaotou Formation (565 m) contains both soft sediment folds and load structures.

The middle of the Sinian section (Changlingzi Formation, 1500 m, Nanguangling formation, 150 m, and Ganjingzi Formation, 398 m) is shallow water cryptalgal-laminated carbonate containing abundant 1 to 3 meter thick shallowing upward cycles, teepee structures, dolomite, and chert. The Yinchenzi Formation (874 m), an open subtidal platform deposit, contains small stromatolites and mudrocks with abundant evidence of slumping and syn-sedimentary folding.

The upper part of the Sinian section is stromatolitic limestone and mudstone of shallow marine and mudflat facies. Columnar subtital stromatolites of the Shisanlitai Formation (155 m) are a favorite ornamental stone in eastern China. The overlying Chuijiatun Formation (80 m) is an offshore mud unit that coarsens upward to slump folded deltaic sandstones of the Xingmincun Formation (240 m). The upper part of the formation is interti-

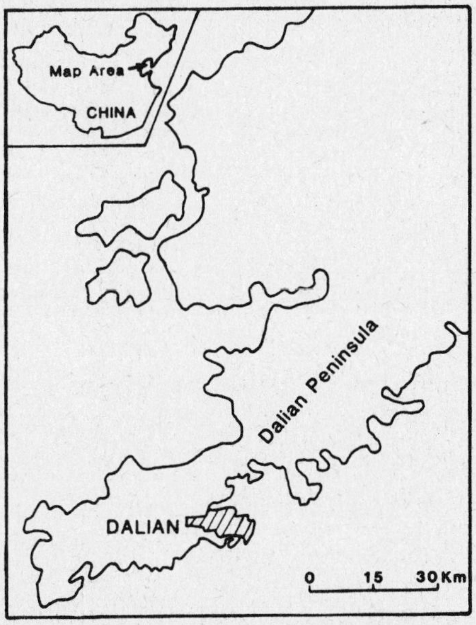

Figure 1—Map showing location of Dalian Peninsula in northeastern China.

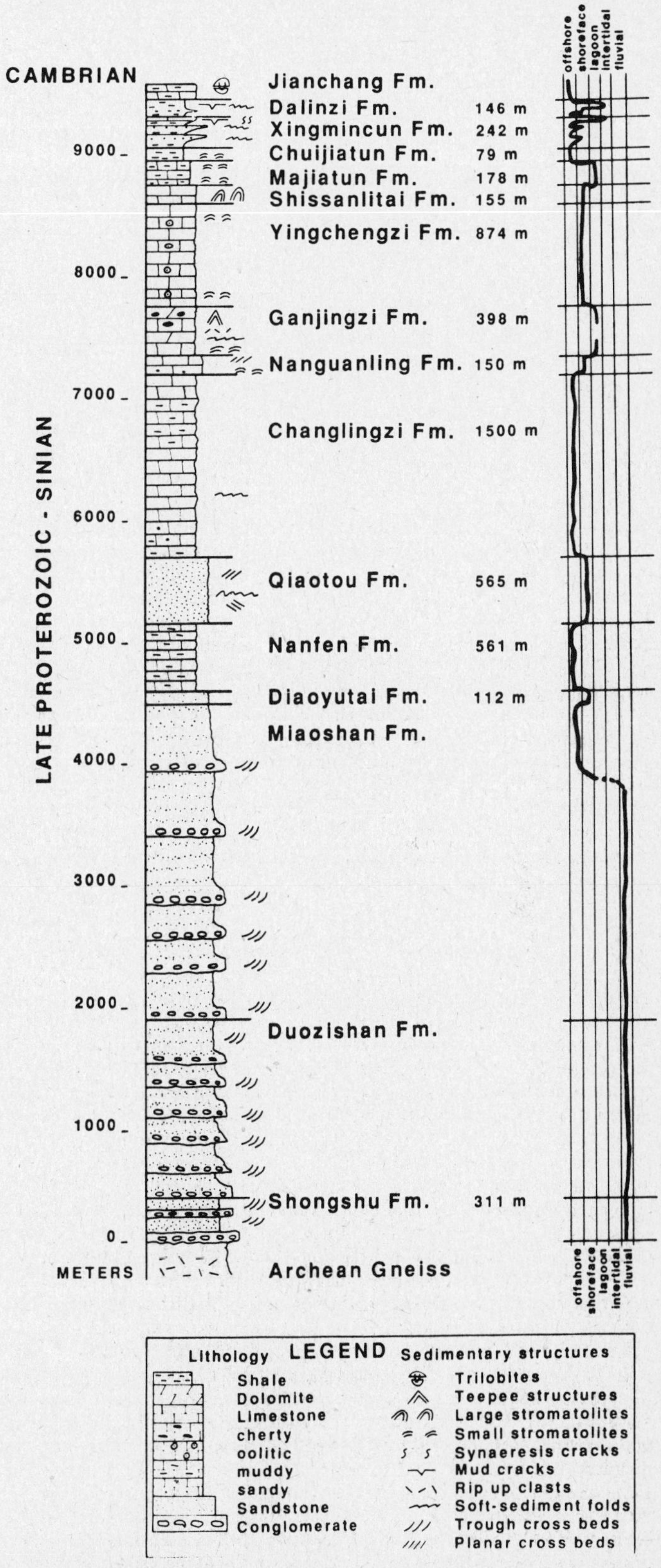

Figure 2—Stratigraphy of Sinian Rocks, Dalian Peninsula, Liaoning Province, People's Republic of China.

dal with synaeresis cracks. A disconformity may exist at the base of the overlying lower Dalinzi Formation (146 m), which contains the Ediacarian fauna and is an offshore or slope deposit, containing soft-sediment folds and a problematic pebble layer. The formation shallows upwards, and the upper part contains intertidal mudflat deposits with abundant mudcracks and raindrop imprints. These rocks are overlain by gray trilobite-bearing limestone of the Cambrian Jianchang Formation.

WANG Erkang, LIU Cong, and SHAO Yuying, Nanjing, University, Nanjing, China

Tectonic Transformation of Hercynian–Indosinian Paleocontinental Margin in Southeast China

Extension and collision are the fundamental dynamic evolutionary mechanism of paleocontinental margin during the Hercynian–Indosinian stage in Southeast China.

Extension in Hercynian Stage: Along the paleocontinent margin of Southeast China there is a large marginal sea basin: Southwest Fujian–East Guangdong Hercynian fault depression. The basement of the depression is early Paleozoic metamorphic flysch formation. As a result of the development of synsedimentary faults, the depression is divided into one graben and two horsts. According to the field observation on the arrangement of gravels and inclined bedding of Devonian to early Carboniferous clastic sediments, it is believed that there exists a mature paleocontinent surviving as erosional source both in the inner side (continental side) and in the outer side (oceanic side) of the depression. Taking into account the appearance of turbidites and intermediate-basic volcanic rock series, it is obvious that the extension is most intense at the middle Carboniferous period. The REE pattern, the contents of FeO, MgO, SiO_2, and bimodal volcanic rocks show that they are products of intense extension, and that the thickness of crust is thinned down to 15 km (Wang Erkang et al, 1986). During the Permian period, intense extension still exists in some areas. The authors discovered a rock record of a great paleoearthquake that took place in the Permian period corresponding to magnitude 8 on the Richter scale, which is the earliest earthquake ever discovered in the world (Wang Erkang et al, 1987). The extension results in both regional bedding skarn by regional geothermal anomaly and economic iron, sulfur, and polymetallic ores (Makeng type ore deposits, one of the famous ore types in China; Wang Erkang and Liu Cong, 1987). It is concluded that there exists an even earlier continental massif in the paleocontinental margin of Southeast China that rifted in Hercynian stage, and the Southwest Fujian–East Guangdong fault depression is exactly a rifting zone. This rifting zone coincides with typical backarc basin in the light of structure features, sedimentation, and magmatism, showing that Hercynian plate subduction exists on the outer side of the depression. It should be pointed out that the tectonic framework of the Hercynian stage in the continental margin of Southeast China is not a simple trench-arc-basin model, but a special plate subduction-erosion mechanism. The continental margin is not accretion; on the contrary, the erosion velocity of paleocontinental margin is faster than that of the depletion of oceanic crust. This is why there is no Hercynian accreted continent, nor paleoisland arc in Southeast China.

Collision in Indosinian stage: The authors discovered an overthrust block group in north part of the depression (Wang Erkang et al, 1985). Further study indicates that the Mesozoic overthrust-nappes are widespread in the whole depression. There are decollements between basement and overlying strata, and intralayer gliding within the overlying strata. They form a large-scale thin-skinned fold-thrust system. It shows a northeast strike linear complex foldbelt, all of the anticlinorium and synclinorium folds are unsymmetrical, and the axis planes are inclined to northwest. From northwest to southeast of the depression the structural features become more complicated and the scale of nappes becomes larger as well. It shows that the basement subducted in the direction of the continent and the overlying layers obducted in the direction of the ocean (Wang Erkang and Liu Cong, 1986). The geophysical data indicate that there exists a crust thickened zone beneath the thin-skinned thrust zone, which coincides with a granitoid zone of 150–100 Ma. We conclude that it is a Mesozoic A-type subduction zone, which forms a new type granitoid series-deep fault granitoid in the area (Wang Erkang, et al, 1984). They are different from the syntaxis granitoids of B-type subduction, and also different from the transformation type series granitoids. Tungsten and tin mineralized granitoids in Shichen of Jianxi Province, Xingloukeng of Fujian Province, and Lianhuashan of Guangdong Province are typical examples.

Conclusion: (1) The extension in Hercynian stage results in formation of a rift zone of paleocontinental margin, accompanied by submarine volcanic hydrothermal-sedimentary ore deposits. (2) The collision in Indosinian stage exactly coincides with the rift zone of Hercynian stage. It forms an early Mesozoic A-type subduction zone and results in formation of specific granitoids associated with specific mineralization.

REFERENCES

Wang Erkang, et al, 1984, Journal of Nanjing University, n. 2, p. 345–368.
——— 1985, KEXUE TONGBAO, v. 30, n. 9, p. 1207–1211.
——— 1986, Proceedings of Symposium of Geology of Continental margins, p. 132–133.
——— 1987, KEXUE TONGBAO, v. 32, n. 17, p. 1200–1204.

WANG Fangguo, Chendu Geological Survey, Peoples Republic of China

Primary Discussion on the Genesis Classification for Alpine-type Chromite

Occurrences of all the refractory chromite (high-aluminum chromite) and part of the metallurgical chromite (high-chromium chromite) in all parts of the world relate to the Alpine-type ultramafic rocks. In this paper, subjects about the Alpine-type chromite are discussed, such as the rock sequence for partial melting of the upper mantle, two kinds of genesis mechanisms about the ophiolite, drop-downward process of chromite body, high-aluminum chromite or high-chromium chromite, as well as the occurrence position of chromite deposits. A new classification is suggested: I, the chromite deposit from partial melting of upper mantle; II, the chromite deposit from crystallization — differentiation in the pocket of magma; III, the chromite deposit from the drop-downward process.

Many geologists considering the Iherzolite as an original rock type of mantle. With the increasing of the degree of partial melting, Iherzolite is turned into harzburgite, and then dunite; a rock sequence against the typical evolution of magma. It is clear that chromium come from the mantle and the concentration of the chromium is related to the disappearance of pyroxene throughout the whole process of the partial melting of the mantle.

A. Nicolas et al (1981) suggested two kinds of genesis mechanisms of the ophiolite, i.e., horizontal expansion and the diapiric expansion of the asthenosphere. The author is much interested in the differences of the shape, scale, state and the compositions of the chromite in the ophiolite of different genesis mechanisms. The high-aluminum chromite occurs in the ophiolite from horizontal expansion of asthenosphere, and lens-like high-chromium chromite occurs in the ophiolite from diapiric expansion of the asthenosphere.

S. D. John et al (1975) made a conjecture based on experiment, chromite body with a thickness of 50 m can drop downward a distance of 600 m in the first m.y. at the temperature of 1250 °C. The data show that there is a possibility for drop-downward of the chromite body formed in the pocket of magma. This is why that chromite body occurs in the upper mantle beneath the cumulative rock rather than occurs at the bottom of the cumulates in the pocket of magma.

The genesis classification for Alpine–type chromite

chromite classes	partial melting of upper mantle	crystallization differentiation	drop–downward
genesis mechanism	diapiric expansion of the asthenosphere	horizontal expansion of the asthenosphere	process of drop–downward
position of chromite	top of upper mantle	bottom of the pocket of magma	top of upper mantle
shape of the ore body	pod	sublayer	sublayer or pod
scale of the ore body	small, many (length < 100m)	big, fewer (length 100–200m)	medium
composition of the ore	high–chromium chromite	high–aluminium chromite	high–Cr&high Al chromite
ore structure	massive knotty	disseminated (band disseminated)	massive
deformation	strong, ductite deformation	weak, brittle deformation	polygenetic deformation
relationship with parent rocks	discordant	concordant	subconcordant concordant
example	Luo Busha, China Acoje, Philippines	Ilarshan, China Honggu, China Oman	Coto, Philippines Camaguey, Cuba Worinous, Greece

Chromite can be divided into two kinds by the difference of the content of aluminum. The study of many chromite deposits shows that the high-aluminum chromite occurs in a position near or beneath rock Moho, while high-chromium chromite occurs further down from rock Moho. In other words, there is a difference in the sequence that the high-aluminum chromite lies above the high-chromium chromite if both of them occur in company in an area which is worth notice for the exploration geologist.

It seems possible to conclude a pattern about the position of the occurrence of Alpine-type chromite which occurs in the top of ultramafics with metamorphic fabrics, and at the bottom of the ultramafics. In simple words, one top and one bottom near the rock Moho. Chromite occurrence does relate to the assemblage of harzburgite + schlieren of dunite.

A new program on the genesis classification for Alpine-type chromite has been proposed. Such factors as the genesis mechanism for rocks, genesis of deposits, shape and scale of the ore body, position of chromite body, texture and composition of chromite, deformation and the relationship with parent rocks, etc., have been considered as important by the author.

WANG Hongzhen, WANG Ziqiang, ZHANG Linghua, China University of Geosciences, Beijing, China

Proterozoic Tectono-Sedimentary Development of China

The Proterozoic used here covers the time span of 2500-600 Ma, including the Sinian (740-600), although the latter may better be attributed to the Palaeozoic. The boundaries within the Proterozoic are taken at 1800 and 1000 Ma, mainly based on widespread unconformities as criteria for regional chronostratigraphic units. Three continental cratons may be discerned in China, Sino-Korea with well developed Archaean, and Yangtze and Tarim with Archaean and Lower Proterozoic usually inseparable. Recent work on the Sino-Korean Platform has shown that the Wutai Supergroup is mainly Late Archaean and represents greenstone belts coalescing and surrounding the high grade regions to form greater nuclei. The Lower Proterozoic has undergone regional metamorphism and intensive deformation and is a part of basement, while the Middle and Upper Proterozoic represent the first non-metamorphic cover strata.

The Lower Proterozoic comprises two tectono-sedimentary types, the rift-geosyncline type and the semistable marine basin type. The former, represented by the Hutuo and correlated strata, comprises variously trending "proto-aulacogens" in Shanxi, developed in the mobile region between the Ordos nucleus in the west and the Jilu-Hehuai nucleus in the east. Mg-rich carbonates, reddish clastics, dark argillites and basic, sometimes bimodal pattern volcanics and the unconformably overlying molasse are the common rocks, which were strongly folded and metamorphosed during the Luliangian orogeny (1850-1800 Ma). Well sorted clastics and carbonates with few volcanics of marine basin type are found in Henan (Songshan Group), Anhui (Fengyang Group) and Liaoning (upper part of Liaohe Group). They are mildly folded but also metamorphosed.

The middle and Upper Proterozoic contain genuine aulacogen sequences of both continental interior and marginal types. The former is exemplified by the Changchengian and lower Jixianian terrestrial clastics, normal to deep water carbonate flysch and K-rich volcanics in Yahsnan region and the Changchengian clastics and basic effusives in Luliangshan, while the latter is represented by the Xiongер volcanic sequence to the north of Qinling and the Zhartai clastics and carbonates along the northern margin of the platform. The transgressive platform cover type consists of the widespread Upper Changchengian and the Qingbaikouan in the north and the open shelf sequence from littoral to outer shelf facies in the south. The genuine marginal sequence including continental slope deposits may be found in the Kuanping Group volcani-sedimentaries in northern Qinling and the Bayan Obo Group of Nei Mongol. The Sinian is represented by the Jiaoliao-Xuhuai marine basin deposits and the glacigene deposits along the southern and western border of the platform.

On the Yangtze Platform, the ancient nucleus formed of Archaean and Lower Proterozoic is found in Upper Yangtze and the Middle and Upper Proterozoic consist of accretional parts of newly formed sialic crust. The Lower Proterozoic island arc sequence is represented by the Tahongshan Group of central Yannan, resting on the Archaean Ailoashan gneiss. To the east is the Middle and Upper Proterozoic back-arc basin sequence of Kunyang Group. Arc-basin sequences of Middle (Sibao Group and correlatives) and Upper (Bauxi Group and correlatives) Proterozoic age with ophiolite suites, calc-alkaline volcanics and fine clastic turbidites are known all along the southern marbin of the Jiangnan belt. Sometimes both back-arc and fore-arc basins may be designated. Similar arc-basin strata denoting active continental margins are known in southern Shaanxi, which may be continuous through western Sichuan to central Yunnan, all along the western margin of the platform. Thick Middle and Upper Proterozoic sequence with occasional volcanics comparable with those of North China occur in Shennongjia, Hubei. The Sinian System forms genuine cover strata on the Yangtze, although the Lower Series consists often of aulacogen (Suxiong volcanics) and molasse (partly Chengjiang clastics) deposits. The northern and southern margins of the platform were converted into a passive nature as shown by the tranquil and deeper water deposits. An ill-defined Proto-Cathaysia, independent of the Yangtze, formed of medium-grade metamorphic basement supplying detritus to the northwest, is recently designated in Jejiang and northern Fujian.

In Tarim, Archaean rocks are found on the northern slope of Aldyn (Milan Group), and probably under the southern Tarim basin. The medium-grade metamorphosed Aksu, Muzart and Xingxingjia groups along the northern border of Tarim, sometimes containing spilites and blue schist, may well denote an active margin of Early Proterozoic age. The Middle and Upper Proterozoic clastics and carbonates over 7000 m thick in Quruktagh and Central Tianshan are moderately deformed and show a northward deepening in facies. The Sinian in Quruktagh, ca 5000 m thick, contains diamictites and effusives and represents a marginal type aulacogen.

WANG Hongzhen, CHEN Jianqing, and ZHANG Linghua, China University of Geosciences, Beijing, China

Ordovician and Silurian Rugose Coral Biogeography and World Reconstruction of Paleocontinents

Ordovician and Silurian rugose coral faunas in various parts of China and of the world are discussed. Two epochs of Early Paleozoic rugose coral diversification, Caradocian-Ashgillian and Late Llandoverian-Early Wenlockian, are chosen for analysis and calculation for Otzuka coefficients and sometimes for Q-type cluster analysis at genus and family levels, according to an overall revised classification of the rugosa. World reconstruction of paleocontinents are based on paleomagnetic, paleoclimatic, paleotectonic, as well as paleobiogeographic data. The maps were constructed through an autograph program on IBM PC-AT microcomputers.

A three-rank category of biogeographic classification is used. The realms are deemed to be mainly controlled by changes of world frame of continents and oceans. The provinces owe their origin chiefly to geographic isolation rather than climatic differentiation. The subprovinces may result from tectonosedimentologic and therefore overall ecologic conditions.

In Caradocian and Ashgillian times, the continents were mainly concentrated in the equatorial and low-latitude regions. North Europe was then far away from North America, and South Europe may have been near North Africa in high southern latitudes. East Gondwana and West Gondwana were probably not yet entirely united. The Caradocian and Ashgillian rugose corals contain 15 familites and 52 genera, the genera distribution being 15% Cystiphylida, 54% Streptelasmatida, and 27% Columnarida. Three biogeographic realms may be discerned. The West Prototethyan realm includes three provinces: the Siberian, the North American, and the Kazakhstan–Hinganian. Caradocian faunal assemblages comprise the widespread Lambelasma–Primitophyllum fauna, while the Ashgillian assemblages include the Siberian Favistina–Cyathophylloides fauna and the North American–Kazakhstanian Dalmanophyllum–Kenophyllum fauna. This realm is also characterized by the Monorakid–Remopleuridid trilobite fauna of Whittington.

The East Prototethyan realm also includes three provinces, the Southeast Asian (Yantgzean), the Australian, and the North European. Among the Caradocian assemblages are the Southeast Asian Yohophyllum–Ningnanophyllum fauna, the Australian Hillophyllum fauna and the European Streptelasma–Calostylis fauna. The Ashgillian assemblages comprise the Borelasma–Grewingkia fauna of the Southeast Asian and the European provinces and the Bowanophyllum–Rhabdelasma fauna of the Australian province. The Pliomerid–Calymenid trilobite fauna of Whittington is characteristic of the Southeast Asian and the Australian provinces, but is also known in western South America. The nautiloid Sinoceras–Eomichelinoceras fauna and the brachiopod Hirnantia–Eostropheodonta fauna are known in the Southeast Asian province. The latter is also present in the European province.

The third realm, the Gondwana, is devoid of corals and is characterized by the Trinucleid–Homalonotid fauna of Whittington.

In Early Silurian, as a result of the northward floating of Siberia, the eastward shifting of Kazakhstania and the mutual approach of North Europe and North America, the boundary between East Prototethyan and West Prototethyan realms was shifted to the east of North Europe. A new realm, the Australo-Pacific realm, may be recognized, probably including Antarctica and India. The Malvinokaffric or Gondwana realm become better defined.

In the Late Llandoverian–Early Wenlockian epoch, the rugose corals amount to 28 families and 130 genera, the genera being distributed as 24% Cystiphyllida, 30% Streptelasmatida, and 44% Columnarida, with only two genera of Zaphrentoidida. The West Prototethyan realm contains three provinces, the Siberia–Mongolian, the North European, and the American. The Llandoverian and Late Landoverian–Early Wenlockian coral assemblages include the Siberian Tungussoplyllum–Nipponopnyllum fauna and the Spinolasma–Protopilophyllum fauna, the European Goniophyllum–Onychophyllum fauna, and the American Ryderophyllum–Craterophyllum fauna. Early Wenlockian assemblages consist of the Siberian Cyathactis–Altaja fauna and European Holmophyllum–Xyphelasma fauna. The Brachiopod Tuvaella–Taunuspirifer fauna is characteristic of the Siberia–Mongolian province.

The East Protothethyan realm consists of two provinces, the Southeast Asian (Yangtzean) and the Kazakhstan–Sino-Korean, which bear a common coral assemblage, the Ceriaster–Amplexoides fauna. The Llandoverian assemblages include the Yangtzean Kodonophyllum–Maikottia fauna, the Kazakhstanian Gissarophyllum–Kungejophyllum fauna, and the Crassilasma–Tungussophyllum fauna known in western Sino-Korea. Among Wenlockian assemblages are the Yangtzean Gyallophylloides–Idiophyllum fauna and the Kazakhstanian Paliplyllum–Junggarophyllum fauna. The nautiloid Szechuanoceras fauna is characteristic of the Southeast Asian province.

East Australian province is the only known province in the Australo-Pacific realm, which controls the Angullophyllum–Spongophyllum fauna. The Malvinokaffric realm is characterized by thebrachiopod Clarkeia fauna.

WANG Kuiren, ZHOU Youqin, University of Science and Technology, Hefei, Anhui, People's Republic of China

Geochemistry Study on a W-Ore Deposit in Xinjian, China

One W-ore deposit is situated in Bole state of Xinjian province, China. Therc is biotitc-monzonitic granitc which formed in Hercynian nearby the deposit. The petrochemical characteristics are high-silica, rich-alkali, poor-Fe, Mg, Ca, Ti, P, etc. The deposit belongs to quartz vein type in external zone. The type of mineralization is W-Cu mineralized. Mineral associations are (1) tourmaline-muscovite-wolframite, (2) arsenopyrite-wolframite-cassiterite, (3) chalcopyrite-chalcocite-wolframite.

The geochemistry characteristics of deposit and granite is as follows:

(1) Metallogenetic epoch: the age for biotite in granite is 300.8 Ma (K-Ar), the age for full rock, is 281 Ma (Rb-Sr) which refers to the time of rock formed. The age for muscovite in the deposit is 259.6 Ma (K-Ar) which refers to metallogenetic-epoch.

(2) REE geochemical characteristics: REE pattern of the granite is that of the typical granite derived from crust, e.g. Eu shows strong depletion. It can be sure that the origin of matter which formed granite is from upper crust.

(3) Trace elements geochemical characteristics: the contents of some elements in granite have the value more than the average value of that in rock of crust, e.g. W, Sn, Cu, Pb, Bi, Cr, etc. W content for quartz in W-bearing vein is always higher than that for quartz in granite. It is very important to search for ore body.

(4) Sr isotope: $^{87}Sr/^{86}Sr$ initial ratio is 0.7065, which is between S-type and I-type granite.

(5) H, O isotope composition: $\delta^{18}O$ value (+12.89‰) shows that the granite were formed by deep-melting from crustal sediment. δ D value is −72‰ for quartz in granite, −85‰ for quartz in ore veins.

(6) S isotope composition: $\delta^{34}S$ values are +0.59 — +0.84‰ and +1.24 — +3.24‰, which have the property of sulfur from magma.

(7) Metallized fluid: the main component of gas-phase is CO, H_2, CH_4. The components from liquid-phase are K^+, Na^+, Ca_+, SO_4^{2-}, F^-, Cl^-, etc.

(8) Physical-chemistry conditions of deposit formation: the inclusion homogenization temperature (326–346 °C) for quartz and oxygen-isotope temperature (390–380 °C) for quartz-wolframite show that this deposit is high-temperature hydrothermal deposit. It also can be obtained from the MnO/FeO ratio and the Nb, Ta contents of wolframite that the mineralization was formed under the conditions of low-alkali and oxygenation.

To sum up, it can be sure that this W-ore deposit is a middle scale type, which has very important industrial value and interests to geologists and geochemists. It also can be sure that much more information would be obtained if much more research work would be done.

WANG Kuiren, ZHOU Youqin, University of Science and Technology, Hefei, Anhui, People's Republic of China

Geochemistry Study on a W-Ore Deposit in Xinjian, China

One W-ore deposit is situated in Bole state of Xinjian province, China. There is biotite-monzonitic granite which formed in Hercynian nearby the deposit. The petrochemical characteristics are high-silica, rich-alkali, poor-Fe, Mg, Ca, Ti, P, etc. The deposit belongs to quartz vein type in external zone. The type of mineralization is W-Cu mineralized. Mineral associations are (1) tourmaline-muscovite-wolframite, (2) arsenopyrite-wolframite-cassiterite, (3) chalcopyrite-chalcocite-wolframite.

The geochemistry characteristics of deposit and granite is as follows:

(1) Metallogenetic epoch: the age for biotite in granite is 300.8 Ma (K-Ar), the age for full rock, is 281 Ma (Rb-Sr) which refers to the time of rock formed. The age for muscovite in the deposit is 259.6 Ma (K-Ar) which refers to metallogenetic-epoch.

(2) REE geochemical characteristics: REE pattern of the granite is that of the typical granite derived from crust, e.g. Eu shows strong depletion. It can be sure that the origin of matter which formed granite is from upper crust.

(3) Trace elements geochemical characteristics: the contents of some elements in granite have the value more than the average value of that in rock of crust, e.g. W, Sn, Cu, Pb, Bi, Cr, etc. W content for quartz in W-bearing vein is always higher than that for quartz in granite. It is very important to search for ore body.

(4) Sr isotope: $^{87}Sr/^{86}Sr$ initial ratio is 0.7065, which is between S-type and I-type granite.

(5) H, O isotope composition: $\delta^{18}O$ value (+12.89‰) shows that the granite were formed by deep-melting from crustal sediment. δ D value is −72‰ for quartz in granite, −85‰ for quartz in ore veins.

(6) S isotope composition: $\delta^{34}S$ values are + 0.59 — + 0.84‰ and + 1.24 — + 3.24‰, which have the property of sulfur from magma.

(7) Metallized fluid: the main component of gas-phase is CO, H_2, CH_4. The components from liquid-phase are K^+, Na^+, Ca^+, SO_4^{2-}, F^-, Cl^-, etc.

(8) Physical-chemistry conditions of deposit formation: the inclusion homogenization temperature (326–346 °C) for quartz and oxygen-isotope temperature (390–380 °C) for quartz-wolframite show that this deposit is high-temperature hydrothermal deposit. It also can be obtained from the MnO/FeO ratio and the Nb, Ta contents of wolframite that the mineralization was formed under the conditions of low-alkali and oxygenation.

To sum up, it can be sure that this W-ore deposit is a middle scale type, which has very important industrial value and interests to geologists and geochemists. It also can be sure that much more information would be obtained if much more research work would be done.

WANG Peijun, Geological Institute for Chemical Minerals, Beijing, China

Boron Deposits in Relation to Plate Tectonics

In the present paper, plate tectonics is meaningfully applied to illustrate the formation, development and evolution the ore-making materials of boron deposits, and the mechanism by which active plate boundaries cause the mineralizations. All boron deposits, despite the genesis varied, can be included under a model of plate-controlled mineralization, a concept that is suitable to some extent to explaining the formation of the deposits. It follows that studies of age-different plate boundaries will have important implications for prospecting the ore.

Source of ore-forming materials: Though boron deposits can be distinguished in genetic terms into endogenetic, exogenetic or polygenetic, they are products from the same ore-forming materials under different ore-generating conditions. These source materials are dominated by hot brines or boron-bearing fluids fluxing from calcalkalic magma and incorporating sea-and meteoric-waters during late migration.

Time and space distribution of the ores: In geological past, subduction of oceanic plate tended to be accompanied by calc-alkalic magmatism at medium depths and the filtering of boron-containing fluids. At the same time, upward migration and compositional fraction of these ore solutions, finally leading to concentration of boron, were considerably exacerbated by the volcanic, volcanic-sedimentary, metamorphic and granitizational events manifested by any subduction zones. Therefore, boron deposits must have a connection with the tectonic activity along active boundaries, and the problem will get solution of where they deposited.

Mineralization epochs: Careful compilation of the existing geological data has given the differentiation boron mineralizations into five epochs. These include (1) Proterozoic, (2) Permian, (3) Early to Middle Paleozoic, (4) Tertiary and (5) Quaternary.

Ore classification based on plate tectonics: 1-Convergent plate boundary environments; 1.1-Alpine Himalayas type. The boron deposits of this type are principally distributed along the Alpine-Himalayas orogenic belt where there exist a number of ore occurrences involved in extensive calcalklic magma activity, such as volcanic-exhalative, volcanic-sedimentary, Quaternary saline lake and lesser skarn. Of these the most important is volcanic-sedimentary deposits which cannot be formed in broad terms unless high development of tectonic faults along which the hot mineralizing brines rise after volcanic eruption, relatively restricted basins for trapping the rising brines, and arid climate with low annual average temperature, all of these premises existed prior to mineralization. In addition, the ore-forming epochs are confined to Tertiary time.

1.2 Andes type. Cordilleran-Andex orogenic belt runs along the western margin of American plate, characteristic of narrower shelf, steeper continental slope and more pronounced relief as compared with other active plate boundaries. Here calcalkalic magma regime has also developed a large amount of Tertiary volcanic-sedimentary deposits and minor Quaternary saline lake sedimentary ones.

1.3 Island arc type. Generally in the tectonic setting of island are, volcanic activity frequently takes place sometimes in parallel with upflowing of hot hydrothermal fluids from the volcanics and their plutonic equivalents, finally to the surface as hot boron-bearing springs. Such represents to be the case from Kamchatshiy P-ov, through Kuril'skiye O'van and Japan, to Taiwan along which zone there is possibility that the hot springs may be utilized as a potential boron resources.

1.4 Anatectic granite type. Plate convergency tends to cause granitic or S-type complexes along the boundary. If these granitic terranes happen to encounter country carbonatite during their emplacement probably some boron deposits will generate in relation to the Calcareous or magnesian skarnization.

2 Divergent plate boundary environments; 2.1 Proterozoic rift-type. Actualistic example of such type is represented by rift boron deposits scattering throughout the rift zone in the eastern Liaoning province, China. This huge extensional structure stretches 400 km from Yingkou in the west to Tonghua in the east, widening 45 km in the south-north, in which lie the ore-hosting metamorphic rocks, stratigraphically categorical to the Jingeryu formation of the Liaohe group and isotopically aged 2,000 m.y., or Early-Middle Proterozoic times. The whole succession of the formation consists, in increasing order, of at first bedded migmatite, then iochroite-biotite granulite intercalated with serpentized dolomitic marble, at the end leptite. All these rocks are the metamorphic products from basalt, andesite, dacite, rhyolite and sandwiched argillaceous-arenaceous-ferrugunous carbonate sediments. The accommodation of the boron-bearing series or the Jingeryu formation within a granite-corded arch means that it was once extensively tectonized, as indicated by its appearance of closed fold. The commercial types of the ores are bimodal, szaibelyite-dominated and ludwigite-dominated.

2.2 Paleozoic rift type. Paleozoic (particularly Permian) rift-related boron deposits tend to be identical in time and space with other salt deposits, so their host rocks are lithologically the same, essentially composed of sandstone, mudstone and evaporite. The enrichment of B element took place in response to desalination at the late stage of sulfate precipitation.

WANG Shan Shu, Offshore Oil Exploration and Development Research Center, Heibei Province, Peoples Republic of China

Formation and Distribution of Oil and Gas Fields Under Chinese Continental Shelf

Author summarizes in this paper characterization of oil and gas fields under Chinese continental shelf, discusses their formation and distribution, and points out prospective offshore areas for future exploration. Author's main points are presented as follows:

1. Offshore oil and gas fields discovered in China are predominantly anticlinal, and structural-stratigraphic, with multiple pay zones among which Miocene, Oligocene, Eocene Series and Paleozoic carbonates are important. Those important ones except Eocene are distinguished by high production.

2. Consisted of petroleum pools of different types, oil and gas fields discovered are defined as composite oil and gas fields, and divided further into following eight kinds:

(a) composite field associated with draping structure
(b) composite field associated with growth anticline
(c) composite field associated with compressional anticline
(d) structural-stratigraphic composite field
(e) structural-lithologic composite field
(f) composite field associated with fault block
(g) composite field associated with organic reef
(h) composite field associated with buried basement hill

3. Geologic settings particularly petroleum geologic characteristics dictate the formation and distribution of oil and gas fields at offshore China, and are outlined as following five aspects:

(a) Three tectonic-sedimentary layer systems control the hydrocarbon distribution

(b) Each type of sedimentary basin is of its own oil and gas pool sequence and particular kinds of the composite fields. Author established the pool sequences and composite field models for cratonic, epicontinental and back-arc basins.

(c) It is emphasized that source sag is the basic unit which controls distribution of oil and gas fields. Moreover, author summarizes significant characteristics of rich source sag and distribution of the fields in it.

(d) Each sedimentary system encompasses its particular kind of play. This paper presents a summary of seven kinds of play as well.

(e) Upon analysis of several regional seals occurred under the continental shelf of China, it is believed that regional seal is a critical fact which affects formation of oil and gas field groups.

4. It is indicated that geologic settings exist for development of large-middle scale oil and gas fields at offshore China. Subsequently, author points out 10 offshore blocks and specifically 21 structural zones which are relatively realistic to hunt for new fields.

WANG Shiguia, Bureau of Geology and Mineral Resources, Henan Province, China

Evaluation Indexes of Irrigation Water Quality and Soil Salinization in Eastern Henan

It is inconsistent with actualities that the quality of groundwater in Eastern Henan is evaluated by the indexes of other countries. The water of some wells evaluated as suitable for irrigation has made the crops necrotic, while the water as unsuitable for irrigation makes bumper harvest. According only to the statistics of both Lankao and Yucheng counties, there are 51,667 ha put under "unsuitable for well-water irrigation" of which 39,733 ha represented suitable for well-water irrigation.

To counter this, surveying irrigations of 600 wells and analyzing 3,230 water samples and 6,562 soil samples, we have proposed a new evaluation index of irrigation water quality and an evaluation index of soil salinization.

Difference between the old and the new indexes is that in the old the harm of water to crops is divided only on the basis of harmful ion content, whereas in the new the qualities of water and soil are evaluated from both harmful ion content and chemical types of water and soil.

Sodium salt is dominant harm in saline areas and soil salinization areas. There are three combinations of it: combination of sodium with bicarbonate as alkaline harm, sodium with sulfate and chlorion as saline harm, and sodium with bicarbonate, sulfate and chlorion as saline and alkaline harms dually. Added the combination of calcium and magnesium with bicarbonate which is unharmful there are four combinations (Figure 1). This classification is compatible with quantitative division by Q cluster analysis and the classification by direct perception. Thus, natural water and soil may be divided into four types (Figure 2).

The implication and calculation of new indexes are shown in Table 1 and the evaluation indexes of the irrigation water quality and the soil salinization are represented. The harmfulness of single harm type is inconsistent with dual harm type, so the indexes of water or soil are listed into two tables.

By the new indexes we have resolved some problems of old indexes. Let us compare the new index of water quality to the index of US. Example, in the water of a well located at Zhangge Village of Xiayi County and of another well located at Dacuizhuang Village of Yucheng County TDS is 1.400 and 1.499 g/L, HCO_2 20.67 and 9.9, SO_4 3.6 and 7.98, Cl- 1.65 and 8.84, Na⅔ 18.72 and 10.55, Mg⅔ 4.6 and 8.65 and Ca⅔⅔ 2.6 and 7.58 meg/L respectively. Both well-waters are divided into II type by the index of US; but by the new index, the former belong to heavily saline-alkaline water, whereas the latter is saline water (salinity 10.55 meq/L and alkalinity 0), it is in unharmful range.

图1 新分类化学类型组合模式

Figure 1—The combination pattern of chemical types of the new classification.

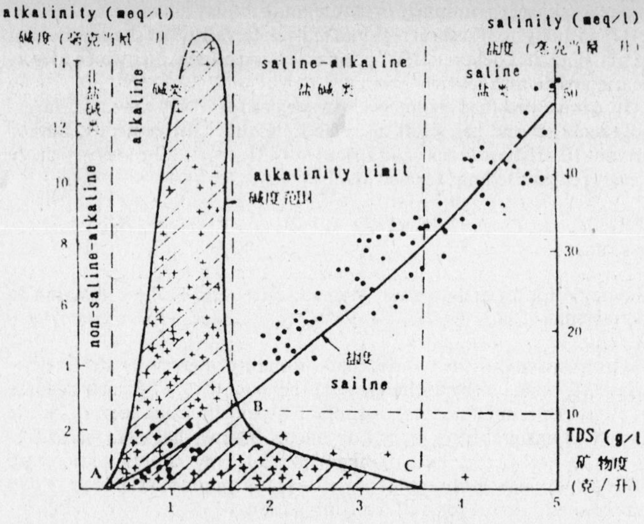

图2 天然水盐类碱类存在的一般规律

Figure 2—Common occurrence regularity of saline and alkaline waters in nature.

WANG Shou Lun, Tianjin Geological Academy of MMI, Beijing, China

The Iron-Bearing Formation and Its Tectonic Environment in the Northern Region of the North China Platform

The North China Platform is an ancient massif consolidated at the end of the Early Precambrian. In the northern vast region of this platform the Archean iron-bearing formation are developed mainly at the following five areas: Northern Liaoning, Western Liaoning, Eastern Hebei, Northern Shanxi provinces and Miyun district. The iron-bearing formations occurred within the metavolcanic-sedimentary rock series of Late Archean greenstone-granite terrain may be divided into three rock associations: (1) Granulite-pyroxene plagioclase amphibolite-gneiss including Anshan Group in the Fushun area of Northern Liaoning, Jianping Group in Western Liaoning, Qianxi Group in the Shuichang area of Eastern heibei and Miyun Group if Miyun district; (2) plagioclase amphibolite-biotite leptite-schist, including Anshan Group in the Waitoushan-Beitai area of Central Liaoning, Qianxi Group in the Zunhua area of Eastern Hebei and Wutaishan Group in the Northern Shanxi; (3) Albite amphibolite-fine grained biotite leptite-phyllite, including Anshan Group in the Yingtaoyuan-Gongchangling area of Central Liaoning, Dantazi Group of Eastern Hebei, Miyun Group in the Fengjiayu area of Miyun district and Jianshan Group of Northern Shanxi.

The essential evolution mode of these three rock associations is as follows: the Late Archean sea basins occurred on the basement of Pre-Early Archean, and the mantle materials of deep sea upwarping during the early stage, involving the eruption of basic-intermediate-acidic volcanic rocks, almost at the same time or shortly later, depression occurred at the right margin, appearing as the eruption of tholeite and komatite of deep origin, then the sea basins were uplifted with local metamorphism of granulite and amphibolite facies, and early deformation accompanied by diapir emplacement of tonalite and intrusion of the hypersthene granite occurred at the hot spot of granulite facies. During the late stage, shallow sea basin formed at the marginal active belt outside the depression at the margin of deep sea, involving the eruption of calcalkali volcanic rocks and tholeiite of island-arc type, which is driven from the deep melting of the early greenstone, then masses of terrigenous sediments were deposited in the shallow basins forming epidote amphibolite facies in the central part and greenschist facies in the edge part consisted of a multicolor metamorphic belt, while these shallow basins were uplifted. At last late deformation, k-granite intrusion and migmatization superimposed on them. This evolution process reflects the trend of the decreasing of geothermal gradient and the increasing of continental crust accretion from the deep sea to the margin sea.

The evolution of above mentioned is well expressed by Anshan Group of Northern Liaoning. The region is divided into following parts: deep sea deposition area of Fushun in the North part; depression area of the margin of deep sea at Waitoushan-Beitai in the Central part, and the shallow deposition area of continental margin at Yingtaoyuan-Gongchangling in the south part.

The BIF (iron-bearing formation) deposited in the above-mentioned environment are quite different in their features. The BIF in north and central parts is proximal volcanic type, mostly of middle and small size and occasionally large size deposits, nevertheless, deposits in south part are of distal volcanic type and of concentration region for large even huge size. In the north part the ore bodies occurred within metaintermediate-basic volcanic rocks as polylayers, with individual layer thickness 10-20 m, and short stretching; the ore type is of diopside (hypersthene) magnetite quartzite, with low oxidation coefficient f = 0.59. In the central part, ore bodies occurred at the upper boundary of meta basic volcanic rocks, where meta fragmental rocks and marble lenses with less thicknesses occurred, polylayers, with individual thickness 20-40 m, and larger stretching; the ore types is amphibolite (actinolite, tremolite) magnetite quartzite with higher oxidation coefficient f = 0.65. In the south part ore bodies occurred at the upper boundary of meta basic volcanic rocks or upper metasedimentary rocks, polylayers or single layer with individual thickness tens to 150-200 m, stretching largely; ore types are amphibolite (cummingtonite) magnetic quartzite and chlorite magnetite-hematite quartzite with higher oxidation coefficient f = 0.77.

From the view of whole region, the BIF in North China Platform is controlled by the cycles of volcanic sedimentation, and occurred in a certain position stratigraphically. The favorable metallogenetic conditions are: high differentiation of volcanic rocks and distinct sedimentary boundary, which indicate that the sedimentation of BIF mainly occurred during the intermittent period of the eruption of basic volcanic rocks, when the sedimentary environment was stable. At the same time, the mixture process of terrestrial materials made it possible for sea water to be weak-alkalic condition. Later, the folding structure thickened the ore bodies in the synclines and the local dynamic metamorphism generated brittle shearing zone, where the regression metamorphism brought large size of rich iron deposits into formation.

The BIF-bearing metamorphic rock sequence developed in linear-form folding over the gneiss basement of Pre-Earth Archean brach fold system. The biotite plagioclase gneiss in Teijiashan forming a dome structure, which reflects the "structural pattern of deep layer crust" providing zircon and sphene dating 3,300–3,400 Ma; while the overlaying BIF metamorphic rock series has the age of 2,800 Ma for early tonalite, and 2,500 Ma for late K-migmatization.

WANG Sijing, ZHAO Zuoquan, DAI Hua, Institute of Geology, Beijing, China

System Evaluation of Engineering Geological Environment in Urban Planning

Engineering geological environment can be considered as a dynamic system of interaction between engineering construction and geological environment. System engineering would be helpful in prediction of the engineering geological hazards and evaluation of the quality of engineering geological environment. Briefly speaking, the main task of system evaluation is to determine the constitution, objectives, trends and coordinatabilities of the systems.

Rational utilization of the engineering geological environment (to ensure safety and economy of the engineering projects) is the objective of system evaluation in urban planning. Trend analyses of engineering geological environment systems are used to predict the state of the whole systems and subsystems according to the distribution curves of system properties. Based on the analysis of the boundary characteristics, internal structure and substantial composition of rock-soil system. The stability level, quality and classification can be evaluated. The coordinatability level between subsystems is to be determined in the coordinatability analyses, which may be considered as the core of the system evaluation. The coordinatability can be characterized by parameter Cco (coordinatability coefficient) quantitatively. The higher the value of Cco is, the more easily the objective be realized, or engineering project will keep safe in the optimal economic conditions. Using the comprehensive evaluation, cluster analysis and other mathematical methods, the engineering geological environment then can be evaluated in three steps: evaluation of single factor, determination of power coefficient and calculation of Cco value.

$$C_{CO} = \sum_{i} PiVi$$

here

Pi—power coefficient of factori;

Vi—standardization coefficient of factor i;

Based on coordinatability analysis mentioned above, a classification and zonation of the engineering geological environment could be made.

A systematic study of the engineering geological environment in Dukou city–a newly developing metallurgical industry area with a population of 400 thousand has been conducted by the authors. First, a series of engineering geological maps were prepared, including bedrock geological map, map of rock slope stability zonation, geological hazards prediction map, microseismic zoning map (Wang Sijing, Yu Suolong, 1987, 1). During the course of rapid urbanization of Dukou city, the geological hazards due to the interaction between engineering construction and geological environment (such as landslide, mud-debris flow, foundation swelling and cracking), not only express the ordinary problems in mountain areas, but also represent the particularity to some extent in Dukou area (Wang Sijing, Yu Suolong, 1987, 2).

The system analysis of engineering geological environment in Dukou shows that the objective of engineering geological environment system is restricted by land elevation, gradient, surface smoothness, strength and stability coefficient of rock-soil system. They are the main parameters which may affect the objective, including proper foundation bearing capacity, less geological hazards, smooth terrain, a good transportation condition and a convenient water supply. According to the results of detailed study in 30 typical zones, we calculated the power coefficient of various factors using the distribution curves. Then a comprehensive evaluation of the coordinatability was made for 1933 subzones of the whole city (15 km × 30 km). The evaluation shows that, the coordinatability coefficient Cco = 0.32 - 0.97, its curve is of normal distribution intimately, and the curve peak value Cco = 0.68. The engineering geological environment in Dukou area could be divided into 7 classes of zones: A, B, C, D, E, F, and G, which are expressed in the map of engineering geological environment system evaluation. The zonation is suited to the distribution of existing buildings and 90 percent of buildings in Dukou city are located in zones A, B and C with a high coordinatability; and there is approximately 35 percent area in the zones A, B and C, which could be used in top priority for further urban development. The system evaluation map of engineering geological environment is the comprehensive expression of a series of engineering geological maps, and could provide a scientific basis for the urban planning in Dukou area.

REFERENCES

Wang Sijing, Yu Suolong, Engineering geological evaluation for urban construction in Dukou city, Proc., Int. Symp. on Engrg. Geol. Envir. in Mountainous Areas, 1987, Beijing, p. 3-16.

Wang Sijing, Yu Suolong, On the interaction between the geological environment and engineering construction in urbanization. Geological Society of Hong Kong, Bulletin, v. 3, October 1987, p. 141-152.

WANG Zhitian, Beijing Research Institute of Mineral Resources and Geology of CNNC, Beijing, China

REE Geochemical Character of Porphyry Copper Molybdenum Multimetal Metallogenic Series and its Application of Distinguishing Ore Bearing Quality of Porphyries in Manzhouli-Xinbaerhuyouqi Area of Inner Mongolia

Manzhouli-Xinbaerhuyouqi area is situated to the northwest side of the northeast-trending Eerguna-Hulun deep seated fracture, belonging to Eerguna fold system of early Caledonian. Eerguna-Hulun deep seated fractures revived intensely and took thrust-compressing movement as dominant with multicycle of volcanic-magma action in Yenshanian. Northwest-trending secondary tenso-shear fractures are quite developed, there are Hanigou fracture in the north and Muhaer fracture in the south.

The northeast-trending fracture system controls the distribution of porphyry series metallogenic belt, and the northwest-trending fractures control the formation of ore cluster. The intersection of these fractures is often the volcanic-magma active center and ore-controlling structures are directly related to volcanic apparatus. Major metallogenic epoch is late Yenshanian. Intrusion occurs mainly as subvolcanic stocks. Main rocks are monzogranite porphyry, granite porphyry, quartz porphyry, rhyoporphyry, and granite. Main alterations are quartz potassium feldspar alteration (Q-Kf), quarta-sericite alteration (Q-Ser), illite-hydromuscovite alteration (I-H), kaolinization, propylitzation and ferromanganocarbonatization. The acidity and differentiation index (DI = 77-90) of porphyries are higher. DI has a negative correlation with δEu and positive correlation with SiO_2. In this area, Wunugetushan (or Wushan for short) is a large porphyry Cu-Mo deposit ascertained. In recent years a number of new mineralization stocks have been discovered in succession. The types of mineral deposit are porphyry, subvolcanic hydrothermal and skarn type. The major commodities are Cu, Mo, Ag, Pb, Zn, etc. They all can be considered as a porphyry metallogenic series with genetic relationship. REE patterns can be divided into 3 types. Type (1), ΣREE lower and Eu-depletion notable. ΣREE = 86-157.2 ppm, δEu = 0.37-0.66, ΣCe/ΣY = 2.8-3.4. It includes all altered rocks and unaltered biotite granite (wall rock) of Wushan, Changling No. 2 and Huanghuacaigou altered bodies. Type (2), ΣREE higher and Eu-depletion remarkable. ΣREE = 220.4-336.8 ppm, δEu = 0.46-0.56, ΣCe/ΣY = 3.8-4.3. It comprises Jiawula and Halasheng bodies, their alterations are outside zones. Type (3), right deviation smooth curve without Eu-depletion, ΣREE = 80.15-150.34 ppm, δEu = 0.85-1.08, ΣCe/ΣY = 4.8-8.9. It consists of unaltered monzogranite porphyry (parent rock) of Wushan, Changling No. 1, Toudaojing No. 1 and No.2 altered bodies:

Wushan complex, monzogranite porphyry (138 Ma), ΣREE = 80.15 ppm, δEu = 1.08, ΣCe/ΣY = 4.8, Eu/Sm = 0.35, Rb/Sr = 0.22, is derived by partial melting from lower-crust which provides Cu, S, Ag, Pb, and Zn. Biotite granite (187 Ma), ΣREE = 149.6-157.2 ppm, δEu = 0.61-0.63, ΣCe/ΣY = 2.9-3.3, Eu/Sm = 0.19-0.22, Rb/Sr = 1.4-2, is remelting product from lower part of upper-crust which provides mainly Mo. Wushan deposit is middle-deep denudation. Alterations can be divided into 3 zones, Q-Kf, Q-Ser, and I-H from the center outwards. Changes in REE patterns during alteration process reflects the evolution of metallogenic fluid. REE occurs respectively as chloride, F^- complex, F^- and CO^{-}_3 complexes in Q-Kf, Q-Ser and I-H 3 zones. The alterations are overlapping of two phases. Early phase: Q-Kf (magmatic water dominates), I-H (meteoric water joins in), late phase: Q-Ser (with more meteoric water), I-H (meteoric water dominates). After Q-Kf ΣREE of porphyry increases (HREE increases more), ΣCe/ΣY decrease and Eu-depletion occurs. I-H in wall rock causes REE to be leached strongly (MRE is leached more), accompanied activation and transport of Mo. Inner I-H zone is overprinted by Q-Ser, leading to redisposition of previously leached REE and reactivation and reenrichment of Cu, Mo, Ag, Pb and Zn. This process has a great significance to mineralization. Late I-H continues to extend outward.

According to changes of REE pattern in every zone of Wushan, 5 distinguishing criteria of ore bearing quality of porphyries are suggested. For the same kind of alteration, the more obvious the Eu-depletion (δEu ≤ 0.65), the better the mineralization. Otherwise the mineralization is poor. Other REE parameters, such as ΣCe/ΣY, LRE, MRE, HRE% can also indicate ore bearing property of bodies. ΣREE and alteration zoning can be used to infer the depth of erosion and burning of ore-body. We divide surrounding bodies into 3 kinds according to metallogenic prospect. (1) optimum prospect sites: Jiawula, Halashang and Changling No. 2 bodies; (2) potential prospect sites: Huanghuacaigou and Toudaojing No. 2 bodies; (3) unpromising sites: Changling No. 1 and Toudaojing No. 1 bodies. Now, the veined ore bodies of Pb, Zn, Ag and commercial Cu mineralization in porphyry in Jaiwula have been discovered by drilling. It is a large subvolcanic hydrothermal type Cu, Pb, Zn, Ag deposit. Our prediction has been verified.

WANG Zhitian, and QIN Kezhang, Institute of Mineral Resources and Geology, Beijing, China

Types, Metallogenetic Environment and Features of Temporal and Spatial Distribution of Copper Deposits in China

Based upon the comprehensive study of 123 major copper deposits of China, the types, metallogenetic environment and features of temporal and spatial distribution and evolution of copper deposits in China are outlined.

(1) China extends across the Pal-Asian, Marginal Pacific and Tethys-Himalayan tectonic zone. The solidification time of continental crust of China is later. The Sino-Korean Platform solidified 1700 Ma ago and the Yangtze Platform solidified 700 Ma ago. The Chinese continent was grown up by several cratons with developing of geological history, which led up to the temporal and spatial distribution of igneous activity, sedimentation and its related types' copper deposits, moving toward the Chinese plate margin.

(2) Types of copper deposits are comparatively complete. Porphyry type is the most important, marine facies massive sulfide type (includes volcanic type and clastic rock-carbonate type), skarn type and marine sedimentary metamorphic type are secondary. Mafic-ultramafic rock Cu, Ni(Co) type and terrestrial facies sand-shale type are third. The massif sulfide type in transitional beds between the marine clastic rock and carbonite is classified definitely in this article.

(3) The carbonite strata is widespread in China, which led up to the larger proportion of copper deposits occurred in the carbonite strata than other districts in the world. The multicycle movements in Chinese geological history resulted in well developed reworked and metamorphic copper deposits after syngenetic sediment, such as massive sulfide type in transitional beds between the marine clastic rock and carbonatite, and sedimentary metamorphic type.

(4) There are 7 big metallogenetic epochs of copper deposits. The Mesozoic is the major, the Middle (Late) Proterozoic and Cenozoic are secondary, the Early Proterozoic, Late Paleozoic, Early Paleozoic and Archeozoic are third. That is, the metallogenetic era of copper deposits have multiepoch but the younger era (Yenshian) is the major.

(5) The metallization shows simple in the early geological history, and shows varied in the late geological history in China. The basaltic magma, marine deposit process and its related types copper deposits are predominant in the early geological history, and the granitic magma, terrestrial

deposit process and its related types copper deposits are predominant in the late geological history.

(6) There are 13 major copper metallogenic concentration provinces in China, in which the Yangtze depression zone, Sanjian fold system, Jiangnan axis and Xikang-Yunnan axis metallogenic concentration provinces are possessed of larger reserve proportion. The Inner Mogalia axis, South China fold system, Zhongtiao upwarping are secondary. Then, Inner Mongolia-Daxinganling fold system, Qilian fold system, Eerguna fold system, Ganzi fold system, East Yunnan depression and East Liaoning upwarping are in order of precedence. There are more copper metallogenic concentration provinces and the southwestern and eastern are taken as the dominant.

(7) The crust activity in China is intenser and frequenter, which is few in other countries. The metallogenic inheritance, types coexistence and multi-genetic complex metallogeny, which were led by multicyclic evolution, show clearly.

WANG, Chi-yuen, University of California, Berkeley, CA, and Yaolin SHI, Academia Sinica, Beijing, China

Tectonic Processes at Convergent Margins: Insight from Modeling

Much impetus to recent quantitative modeling of the tectonic processes of convergent margins has come from advances in geologic and geophysical observations, which have both provided the data needed for constraining models and exposed the need to use quantitative modeling as a tool for interpreting measurements and observations. Quantitative modeling is also useful in testing the accuracy and completeness of our understanding of the tectonic processes. Advances in modeling technique, both in simulating the kinematics of the accretionary processes such as thrust-bound folding, duplexing, and underplating, and in incorporating real sediment and rock properties in the model, have also contributed to the effectiveness of this approach.

The lower slope of the fore-arcs in sediment-rich convergent margins is the site of rapid dewatering of deformed sediments. Pore fluids are expelled by progressive compaction in response to the increased load that is produced by the tectonic stresses that build an accretionary complex and migrate to the seafloor along various paths. The results of modeling of the sediment-rich Barbados accretionary prism show that tectonic overburden may be the most important mechanism for generating excess pore pressures below the decollement; the direct effect of horizontal compression may be of secondary importance, and the effect due to temperature increase with subduction is of little significance. Modeling of the Washington–Oregon margin shows that pore pressures may approach lithostatic values below the decollement. Maximum expulsion of pore fluids is predicted under the marginal ridge where porous sediments under the ramp of the frontal thrust experience the greatest increase in overburden. Temperatures at this fluid source may be significantly below that required for producing thermogenic methane, consistent with the existing observations. Fluid flow in this region is controlled by the permeability ratio between the permeable fault zones, which cut the accreted sediments, and the surrounding sediments. On the other hand, high pore pressures below the decollement are not significantly affected by these faults. Focused venting of fluids, such as that observed by recent ALVIN dives at the Oregon margin, is predicted if faults have permeability significantly higher than that of the surrounding sediments. Fluid venting via faults was also supposed to be the cause of methane concentration observed in the master decollement under the Barbados accretionary prism and the high heat flow near the Japan trench. The high pore pressures created below the sediment-rich accretionary prisms can cushion the friction between the convergent plates. Heat flow at such margins generally reflects the effect of active thickening of sediments, which causes a landward decrease of heat flow across the lower slope. Across the sediment-starved convergent, on the other hand, pore pressure is low and frictional heating between the subducting and the overriding plates can be high to dominate the thermal evolution of the convergent margin. Across the Peru margin, for example, heat flow is high over a broad region, from near the trench to the upper slope and the shelf.

With increasing temperatures, either due to increasing depth of burial or due to frictional heating on faults, dehydration of hydrous minerals of various kinds can occur. Significant amount of bond water is released during these phase transitions. Thus, in sediment-dominated convergent margins rich in diatomaceous and clay components, dehydration processes can create fluid sources and influence fluid and heat transport during sediment burial and subduction. Because sediment compaction becomes a less effective mechanism in generating fluids with diminishing compressibility, mineral dehydration may become more important as a fluid source at increasing depth. The maximum depth above which this mechanism remains effective depends on mineral composition, temperature, pore pressure, and thus on the thickness and the lithology of the subducted sediments, the age of the subducting plate, and the subduction rate and angle.

The deep thermal states and the dynamics of the subduction margin of the North America Pacific margin are examined for the conditions by which blueschists are formed and exhumed to the surface. Models are constructed on the basis of the LITHOPROBE seismic results for controlling subsurface structures, regional tectonic data for controlling subduction rate, and heat flow measurements across the Pacific Northwest for constraining the models. Temperatures in the underplated rocks in the accretionary prism depend on the age of the subducting plate, the radioactive content of the accreted sediments and the continental crust, subduction rate, and the subduction angle. It is found that at subduction rates of several centimeters per year, the age of the subducting plate plays the dominant role in determining the thermal structure of the prism. Subduction of plate older than 20 Ma is likely to produce P-T conditions in the accretionary prism suitable for the formation of blueschists. Uplift and denudation rates and the radioactive content in the prism determines whether the blueschists, once formed, can be preserved. Concurrent uplift and erosion during underplating are most favorable for blueschist preservation. Tectonic settings of oceanic–oceanic and oceanic–continental convergence are more favorable than that of continental–continental convergence. Ridge subduction may heat up the accretionary prism and destroy the blueschists formed at depth greater than 20 km. Variations in these factors around the world may determine the characteristics in the distribution of blueschists in space and time.

At depths of 100–150 km, partial melting of the subducted oceanic crust, sediments, and the overlying mantle leads to magma formation and island-arc volcanism. The lithosphere vertically above is weakened by the rising temperature associated with these processes and can become the favored location in initiating back-arc extension.

WANG, JIANHUA, M. A. GARDINER, D. E. GRANDSTAFF, and GENE C. ULMER, Department of Geology, Temple University, Philadelphia, PA

Differentiation in Mesostasis and Effect on Hydrothermal Redox Buffering Capacity of Basalt

A number of studies have investigated the nature of redox control achieved during water-rock reactions (e.g., Giggenbach, 1980; D'Amore and Gianelli, 1984; Arnorsson and Gunlaugsson, 1985). These studies have shown that alteration minerals control the redox state of the rock-water system, buffering it to log f_{O_2} values between the magnetite-hematite (MH) and fayalite-magnetite-quartz (FMQ) buffers. However, few studies have investigated factors controlling buffer capacity in water-rock systems; that is, the ability of rock to remove oxygen and other oxidants from reacting fluids. In the present study mass balance calculations, autoclave reinjection experiments, and SEM investigation of hydrothermally etched polished basalt sections have all been combined to evaluate the redox buffering capacity and reactivity of a tholeiitic basalt from the Cohassett Flow (Grande Ronds Formation) of the Columbia Plateau relative to synthetic Grande Ronde groundwater in the temperature interval of 200-300 °C and at pressures up to 30 MPa.

The buffer capacity and behavior of a rock is governed by the content of oxidants in the fluid, by the ferrous iron content of the rock, and by the distribution of Fe2 + in and reactivity of individual phases. Rock samples and polished thin sections of Columbia Plateau Basalt were documented (whole rock chemical analyses, modal analyses, reflected light microscopy, pixel mapping [Friel et al., 1986], and SEM). These analyses revealed that the mesostasis within individual interstitial pools was differentiated and composed of several distinct phases including: two types of glass, one potassium-rich and the other sodium and calcium-rich; Ca-Fe-P-bearing microglobules (generally less than 1 μm diameter), and fine dendritic magnetite. This differentiation may have resulted from both liquid immiscibility and spinoidal decomposition. Iron distribution within the basalt was shown by electron microprobe and by SEM to be (1) within the various ferromagnesian silicates, (2) as iron within glass, (3) indendritic iron and titanium oxide opaques within the glass, and (4) as discrete iron-rich (up to 38 wt% FeO) microglobules of calcium silicate and calcium phosphate within the glass. About half of iron in the Mesostasis is contained in the microglobules.

Well-documented polished basalt thin sections of Columbia Plaeau basalt were subjected to brief hydrothermal etching (up to 3.3 hr with a peak temperature of 167=C at a flow-rate of 72 mL/hr) in synthetic groundwater in flow-through systems. Other basalt samples were reacted in reinjection hydrothermal experiments (300=C, 30 MPa) having durations greater than 7600 hours. Glass is the most rapidly dissolving phase in basalt hydrothermal experiments (at 300=C in low-salinity simulated groundwater the order of dissolution rate is: K-rich glass > sodic glass > olivine > > magnetite > clinopyroxene > plagioclase). Glass was strongly etched in experiments, magnetite dendrites were slightly enlarged by oxidation, and many of the microglobules were released; however, most silicate minerals were virtually unaltered. It appears that dissolution of mesostasis, and perhaps the microglobules as well, must control redox capacity during short-term experiments and in glass-bearing rocks in geothermal fileds. Magnetite oxidation may also contribute ro the redox capacity. Theoretical considerations of $Fe^{2+}/\Sigma Fe$ in basalt and its phases lead to the prediction that, at the nominal dissolved oxygen content of groundwater, in closed-system conditions water/rock mass ratios in excess of 190 would be needed to exhaust the buffering capacity of mesostasis in a Columbia Plateau basalt, i.e., to react and oxidize all ferrous iron during dissolution of glass.

To test this calculation, Dickson-type autoclave experiments were done on Columbia Plateau basalt with Grande Ronde synthetic groundwater. To reach water/rock mass ratios as high as desired suggested that either flow-through or reinjection experiments (fresh aliquots of Grande Ronde groundwater) should be tried. Concentration-time data for soluble species and time dependencies of redox were measured. After reinjection, most solution parameters quickly (hrs-days) return to near preinjection values. However, silica concentrations stabilized near quartz saturation values rather than the cristobalite saturation value found in closed-system experiments. Short-term relationships between cations appear to be controled by ion-exchange between solutions and clay minerals. The redox state was measured both by the evolved hydrogen (DHEM) technique of Kishima and Sakai (1984) and by measured $SO^{2-}{}_4/H_2S$. Each reinjection stresses anew the redox buffering capacity of the basalt remaining within the autoclave system, and the redox recovery could be followed over a period of days as the basalt buffering capacity reestablished redox control at f_{O_2} values slightly more reducing than (MH). The effective water/rock ratio was raised to about 140 without exhausting the buffer capacity of the basalt. Higher water/rock ratios could not be obtained due to the flexure-failure of the gold liner during subsequent reinjections.

REFERENCES

Arnorsson, S., and E. Gunlaugsson, 1985, New gas geothermometers for geothermal exploration-calibration and application: Geoghimica et Cosmochimica Acta, v. 49, p. 567-577.

D'Amore, F., and G. Gianelli, 1984, Mineral assemblages and oxygen and sulfur fugacites in natural water-rock interaction processes: Geochimica et Cosmochimica Acta, V. 48, p. 847-857.

Friel, J., G. H. Kacandes, D. E. Grandstaff, and G. C. Ulmer, 1986, Image analysis of an Icelandic basalt (abstract): GSA Abstracts with Programs.

Kishima, N., and H. Sakai, 1984, A simple gas analytical technique for the Dickson-type hydrothermal apparatus and its application to the calibration of MH, NNO and FMQ oxygen buffers: Geochemistry Journal, v. 18. p. 19-29.

WANG, Y., Department of Geology, National Taiwan University, Taipei, Taiwan

Tectonic Features of Arc–Continent Collision Around Taiwan

Subduction of the South China Sea plate at the southeast China continental margin has produced an arc–continent collision system between Taiwan and the western edge of the Philippine Sea plate since about 10 Ma. This collision has resulted in dramatic structural and sedimentary changes around Taiwan. Before the collision, the Taiwan region enjoyed relatively calm cyclic sedimentation from Paleocene to Miocene on the passive margin of the eastern Asiatic continent that had been built up by the subduction and collision of the Kula plate in late Mesozoic time.

Toward the end of Miocene, oblique indentation of the Taiwan–Luzon arc representing the western edge of the Philippine Sea plate from southeast into Taiwan, eastern margin of Asia, caused not only the rapid uplifting of the Taiwan Central Range but also formation and bending of the thrust-fold belt of the western foothills. The uplift rate of the Central Range has been accelerated from about 6 mm/y up to about 10 mm/y for the past 1.1 Ma. The folding and thrusting have been especially severe in the weak Tertiary clastic wedge of the foothills belt to the west of the Central Range, having northwestern vergence toward the scarcely influenced foreland represented by the Taiwan Strait belt underlain by the hardened Yenshanides. The wave-front of this thin-skinned tectonics has evidently moved progressively from east to west resulting in about 200 m high abrupt cliffs skirting the foreland belt, that is, the coastal plain of western Taiwan. It has been dated that the thrusting and folding in the eastern portion of the foothills belt were completed between 4.0 and 2.0 Ma, in the central portion between 0.5 and 0.1 Ma, and are still active in the western margin of the belt.

The rapid uplift of the Central Range due to constant northwest push of the Philippine Sea plate along the north–south-running transform in eastern Taiwan caused on one hand downflexing of the foreland basins in western Taiwan and forming of en echelon ridges and troughs on the colliding volcanic arc, that is, the Coastal Range of eastern Taiwan, on the other.

These basins and troughs have been rapidly filled up by collision sediments at a surprising rate of 1.5 mm to 2.5 mm per year as strikingly contrasted with that of about 0.15 mm per year for the Miocene sedimentation of passive continental margin type that is pervasively observed in the western foothills belt.

Presence of klippes, fission-tract dating, and sedimentation of the collision sediments strongly suggest obduction of oceanic crust on the underthrusting plate that lies west of the suture which is represented by the longitudinal valley in eastern Taiwan. On the metamorphic Tananao belt east of the suture, several flat-lying tectonic blocks have been spotted on the polymetamorphsoed terrain. They are composed of sheared ophiolitic sequence with glaucophane schist and slate on the top. The glaucophane was dated 4–6 Ma, evidently younger than the underlying schist, which is 80–90 Ma in age. Fission-track dating of clastic zircon from the Paleogene, Miocene, and the metamorphic Tananao belts shows partial and complete resetting of the zircon clock. Completeness of the resetting was observed in the terrain immediately west of the suture, and as we go further west, the resetting varies from partial to nonexisting. The reset ages range from 10 Ma to 2 Ma, evidently older than the formation ages of these underlying sedimentary and metamorphic rocks. Further, petrography of the late Miocene–Pleistocene collision sediments in southern Taiwan and the Coastal Range shows vividly that huge and small fragments of ophilitic association were contained in the conglomerates and sandstones of turbidite sequences. The above evidence all points to a conclusion that at the outset of the collision, almost half of the Taiwan Central Range was covered by oceanic-crust materials obducted from east to west.

Thus, the late Miocene-Pliocene collision of the Coastal Range with Taiwan brought about tectonosedimentary phenomena including obduction and accretion of an oceanic volcanic arc, swift uplift of the Central Range, and rapid filling in the suture zone and the collapsed volcanic arc.

It is geologically worthy of note that Taiwan has experienced subduction, passive margin rifting and sedimentation, and arc–continent collision in such a relatively short geological time as from late Mesozoic to the present.

WANG, Yifeng and E. MERINO, Indiana University, Bloomington, IN

Textural Self-Organization of Agate: Dynamic Modeling of Crystallization

Agates are rounded bodies ($\leq$ a few dm across) made up of concentric layers of microcrystalline quartz fibers and commonly found in the vesicles of acid and intermediate lavas. Many agates display three self-organizational textural patterns at once (Merino, 1984): concentric layers with inward-increasing spacing; layers 1,3,5, etc., consist of *twisted* (and length-fast) quartz fibers, whereas layers 2,4,6, etc., consist of *non*-twisted fibers (Frondel, 1978); and chevron extinction. The genesis of these autonomous patterns poses fascinating problems with crystallographic, geochemical, and dynamic aspects. We propose below an isothermal reaction-transport feedback that accounts for spontaneous, self-organizational layering and fiber twisting in agate.

The very occurrence of these associated patterns precludes an origin by successive precipitation of silica layers out of groundwater in an empty space (as this would imply that, time and again, the three patterns come about by chance), and implies self-organization within an initially uniform lump of silica gel formed in advance (compare with Sunagawa and Ohta, 1976). Layered chalcedony has in fact been grown experimentally from silica gels (Oehler, 1976) Also, there is independent reason to believe that length-fast quartz fibers form from silica polymers (Folk and Pittman, 1971).

Mechanism of agate crystallization: Amorphous, silica-rich lumps containing also other components become differentiated (perhaps by unmixing) during the cooling of acid or intermediate lava. Each lump is assumed to start to crystallize into quartz from the outside inwards. We assume that Mg is a catalyst for this crystallization (Kastner, 1977; Williams and Crerar, 1985). As quartz crystallizes, it sweeps the catalyst (and also other minor elements) out of the amorphous silica. The increase in catalyst concentration at the reaction front increases the quartz growth rate, which in turn pushes the catalyst concentration even higher. Increased quartz growth also depresses silica concentration at the front. Eventually the process becomes diffusion-limited; the catalyst diffuses into the gel, and the silica diffuses to the front. When the crystallization rate decreases enough, the process again becomes limited by surface reaction and resumes accumulating catalyst at the front; the cycle repeats itself. At the peaks in concentration of minor elements, which coincide with those of catalyst, the crystallizing quartz is forced to take up, in substitution for Si, a significant amount of a minor element such as Al or Mg, which, being of larger ionic size than Si, causes the quartz fibers to grow twisted (Keith and Padden, 1961).

Modeling: The process above can be modeled by coupled mass conservation equations for silica and catalyst applied to the (moving) reaction zone, or front—the thin region between advancing quartz and retreating gel. The equation link diffusion, advection, and crystallization of quartz. The crystallization rate is assumed to grow with silica concentration and with a power of the catalyst concentration.

These equations are scaled and linearized. A stability analysis yields the conditions for temporal oscillatory behavior of the concentrations of silica and catalyst in the front (Hlavacek, 1969). Numerical simulation of the full nonlinear equations confirms the oscillatory nature of the phenomenon (Figure 1).

Conclusions: At least two repetitive textural patterns typical of agate can be accounted for by a reaction-transport feedback involving silica and a catalyst of quartz crystallization. Internal evidence from the dynamic model proposed above, plus published experimental and petrographic evidence, together imply that agate results from the self-organizational crystallization of gel lumps previously exsolved from lavas. The alternation of layers made up of twisted fibers and layers made up of nontwisted fibers implies that the concentration of some minor elements in agate layers should vary oscillatorilly along a radius (Frondel, 1985).

Y. Wang thanks the Overseas Exchange Program between Indiana and Nanjing Universities for a fellowship.

REFERENCES

Folk, R. L., and J. S. Pittman, 1971, Journal of Sedimentary Petrology, v. 41, p. 1045–1058.
Frondel, C., 1978, American Mineralogist, v. 63, p. 17–27.
——— 1985, American Mineralogist, v. 70, p. 975–979.
Hlavacek, V., et al, 1969, Journal of Catalysis, v. 15, p. 17–30.
Kastner, M., et al, 1977, Geochimica et Cosmochimica Acta, v. 41, p. 1041–1059.
Keith, H. D., and F. J. Padden, 1961, Journal of Polymer Science, v. 51, n. 156, p. S4–S9.
Merino, E., 1986, *in* G. Nicholls and F. Baras, eds., Chemical instabilities: NATO Adv. Sci. Series C., v. 120, p. 305–328, Dordrecht, Reidel.
Oehler, J. H., 1976, GSA Bulletin, v. 87, p. 1143–1152.
Sunagawa, I., and E. Ohta, 1976, Sci. Rep. Tohuku Univ., v. 13, p. 131–146.
Williams, L. A., and D. A. Crerar, 1985, Journal of Sedimentary Petrology, v. 55, p. 312–321.

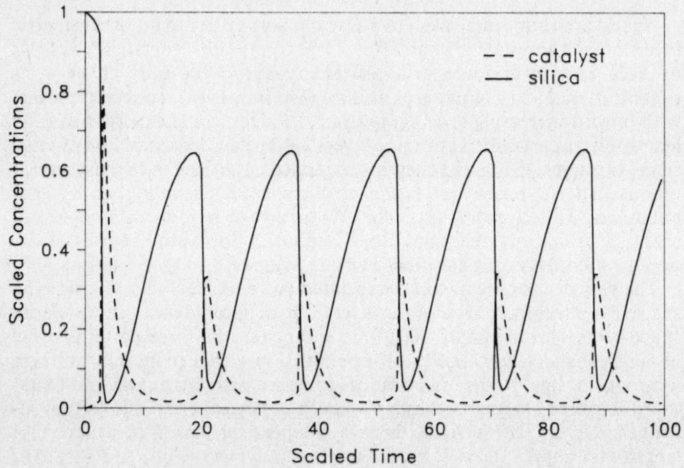

Figure 1—Predicted temporal variation of silica and catalyst concentrations at the agate crystallization front, obtained by numerical solution of coupled mass conservation equations. The curve for catalyst concentration also applies to other minor elements in the gel.

WANKE, H., Max-Planck-Institut für Chemie, Mainz, Federal Republic of Germany

Chemistry and Accretion History of Earth

Today there is an almost absolute agreement about the composition of the upper mantle. With the assumption that the upper mantle composition holds also for the whole mantle, i.e., a homogeneous mantle—the composition of the primitive mantle can be simply obtained by adding the respective contribution of the elements present in the crust. The normalized abundances of the primitive mantle are plotted in Figure 1, taken from estimates by Wänke (1988). However, there are no significant differences in the abundance values published by other authors.

The following trends are to be seen from Figure 1. All lithophile refractory elements have abundances of about 1.3; Mg is somewhat less enriched, possibly reflecting fractionation according to volatility. Alternatively a Si deficiency could be compensated by a higher Si abundance in the lower mantle or by some Si in the core in metallic form.

For depletion of the elements V (abundance a = 0.73), Cr (a = 0.56) and Mn (a = 0.28) volatility may be responsible especially in the case of Mn. Depletion due to volatility may be responsible especially in the case of Mn. Depletion due to volatility is less likely for Cr and V, although CO-meteorites show a similar abundance trend (Newsom and Drake, 1987). In CO-meteorites the abundance of V exceeds, however, that of Si. Removal of these elements into the core either in reduced form or also in form of oxides is a viable alternative. Iron and moderately siderophile elements (Ga, Cu, Co, Ni, W) have abundances between 0.1–0.2. Especially the abundance of Ni is incompatible with all models, requiring chemical equilibrium of the mantle with a pure FeNi core. Hence, either a core-mantle equilibrium was not achieved—inhomogeneous accretion (Wänke and Dreibus, 1988)—or the respective partition coefficients have been considerably lowered, for example by the presence of significant amounts of oxygen in the core material (Ringwood, 1979). Moderately volatile elements Na, K, Rb, Cs, and F and even the under solar nebula condition highly volatile element in show similar abundances of about 0.1–0.3, i.e. the abundances of these elements fall within the abundances of the moderately siderophile elements. The abundances of the elements Na to F can best be explained by the two component model (Ringwood, 1979; Wänke, 1981).

Both the highly siderophile and highly volatile elements are strongly depleted in the earth's mantle. Volatility resp. siderophility is the obvious explanation. However the highly siderophiles are considerably more abundant as one would expect. The low abundances of P and As are still miraculous.

Inhomogeneous accretion of the earth from two chemically different components can successfully describe more of the observed abundances. In this model (Wänke, 1981), component A is highly reduced an all elements with equal or high volatility than Na are depleted, while component B is oxidized and is thought to contain all elements in Cl-abundances. It is proposed that accretion started with the reduced component A, containing all Fe in metallic form. It is assumed that segregation of FeNi, i.e. core formation to be contemporaneously with accretion. The silicate mantle of the growing earth will be almost free of FeO and siderophile elements during this stage. After the earth has reached about 60% of its present mass more and more oxidized material (component B) is thought to be added. Small amounts of metal still present but in decreasing amounts as accretion proceeds is thought to be responsible for complete extraction of the highly siderophiles (Ir, Au, etc.) into the core, while the moderately siderophile elements would be considerably less affected.

The conveyance of H_2O and Fe^{3+} from component B is thought to be the reason that metallic iron finally became unstable towards the very end of accretion and the highly siderophile elements contained in the last 0.4% of the mass added to the earth remained in the mantle. Variants and alternatives of this model have been described by Jones and Drake (1986).

REFERENCES

Jones, J. H. and M. J. Drake, 1986, Nature, v. 322, p. 221.
Newsom H. E. and M. J. Drake, 1987, Lunar Planet. Sci., v. XVIII, p. 716.
Ringwood, A. E., 1979, Origin of the Earth and Moon, Berlin: Springer Verlag.
Wänke, H., and G. Dreibus, 1988, Phil. Trans. R. Soc. Lond., v. A325, p. 545.
——— 1981, Phil. Trans. R. Soc. Lond., v. A303, p. 287.

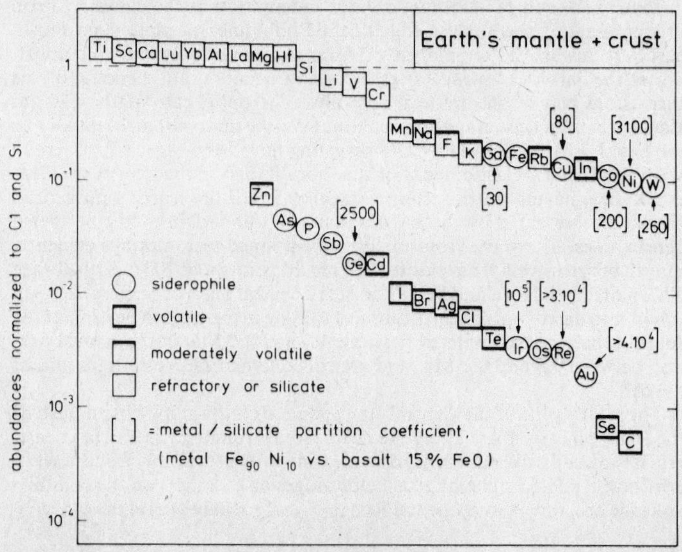

WANKE, H., and G. DREIBUS, Max-Planck-Institut für Chemie, Mainz, Federal Republic of Germany

Chemical Evolution of Mars

From the Viking X-ray fluorescence experiment (Clark et al, 1982) we know that FeO is rather high on both landing sites at which in fact an almost identical regolith composition was found. These high FeO concentrations at the Martian surface correspond in all probability to an FeO-rich mantle. The low SiO_2 content and the low abundances of K and other trace elements in the Martian soils point to a mafic crust with a considerable smaller degree of fractionation compared to the terrestrial crust.

Another indication of a low degree of fractionation of the Martian crust is the low concentration of the radiogenic ^{40}Ar in the Martian atmosphere. Normalized to the mass of the planet, the Martian atmosphere contains 48 $\times$ 10^{-8} cm^3/g ^{40}Ar compared to the terrestrial value of 612 $\times$ $10^{-8}cm^3g$ (Anders and Owen, 1977). It is the general consensus that on Earth about 50% of the ^{40}Ar produced during the age of the Earth found its way into the atmosphere. As also about 50% of the Earth's inventory of K resides in the crust, this means that the material of the crust has been almost fully degassed while only a small fraction of the ^{40}Ar produced in the mantle has found its way to the surface. Even if we assume that the bulk concentration of K in Mars is identical to the K of the Earth (as discussed below K in bulk Mars probably exceeds the terrestrial value), we obtain for Mars a degassing efficiency of about 3% only. It is highly unlikely that this difference is mainly due to a difference in the degassing efficiency of the crustal material. It seems that the low ^{40}Ar concentration in the Martian atmosphere is related to a low overall concentration of K in the Martian crust, i.e., to a low degree of fractionation (Dreibus and Wänke, 1987).

In recent years, evidence has been collected that points toward Mars being the parent body of SNC-meteorites, and, hence, these meteorites have become a valuable source of information about the chemistry of Mars. Using element correlations observed in SNC-meteorites and general cosmochemical constraints, Wänke and Dreibus (1988) have estimated the bulk composition of Mars. The data of their most recent estimate are given in Table 1 and plotted in Figure 1 together with data of the Earth's mantle. Again using the SNC-meteorites, a water content of 36 ppm for bulk Mars is obtained, which leads under the unrealistic assumption of a 100% release to an ocean of 130 m, covering the whole planet (Dreibus and Wänke, 1987).

The composition of the Earth's mantle can successfully be described by a two-component model. Component A is highly reduced and volatile-free; component B is oxidized, containing all elements in C1-abundances including volatile elements. The same two components can be used as building blocks for Mars, if one assumes that contrary to the inhomogeneous accretion of the Earth, Mars accreted almost homogeneously (Wänke and Dreibus, 1988). In the case of the Martian mantle, the striking depletion of all elements with chalcophile character indicate that on Mars a chemical equilibrium between component A and B was achieved that led to the formation of significant amounts of FeS which on segregation extracted the elements according to their sulfide-silicate partition coefficients rather than their metal-silicate partition coefficients. While for the Earth a mixing ratio A:B = 85:15 was derived, the Mars ratio of 60:40 indicates considerably higher concentrations of moderately volatile elements such as Na, K, and sulfur on Mars as compared to the Earth. A homogeneous accretion of Mars could also explain the obvious low abundance of water and primordial rare gases (Dreibus and Wänke, 1987).

REFERENCES

Anders, E. and T. Owen, 1977, Science, v. 198, p. 453.
Clark, B., et al, 1982, Journal of Geophysical Research, v. 87, n. B12, p. 10059.
Dreibus, G. and H. Wänke, 1987, Icarus, v. 71, p. 225.
Wänke, H., and G. Dreibus, 1988, Philosophical Transactions of the Royal Society of London, v. A325, p. 545.

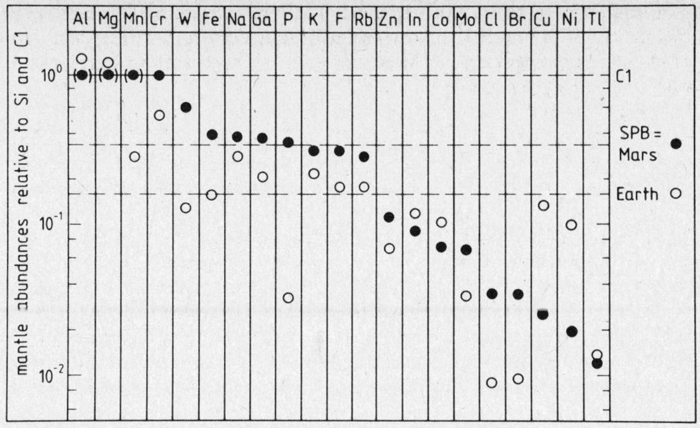

Figure 1

Mantle + crust			
MgO %	30.2	Co ppm	68
Al_2O_3	3.02	Ni	400
SiO_2	44.4	Cu	5.5
CaO	2.45	Zn	62
TiO_2	0.14	Ga	6.6
FeO	17.9	Mo ppb	118
Na_2O	0.50	In	14
P_2O_5	0.16	Tl	3.6
Cr_2O_3	0.76	W	105
MnO	0.46	Th	56
K ppm	305	U	16
Rb	1.06	**Core**	
Cs	0.07	Fe %	77.8
F	32	Ni	7.6
Cl	38	Co	0.36
Br ppb	145	S	14.24
I	32	O	-
		core mass %	21.7

Table 1—Bulk composition of Mars.

WANLESS, H. R., L. P. TEDESCO, J. J. DRAVIS, and J. D. EMERSON, University of Miami, Miami, FL

Alternate Models of Bahamian Platform Carbonate Sedimentation: Caicos Platform, British West Indies

Caicos Platform is a moderate-sized carbonate platform (100 km by 70 km) located in the southeastern Bahamian Archipelago. In contrast to the carbonate platforms of the northern Bahamas and south Florida (rainy and subjected to gentle prevailing easterlies), Caicos Platform is situated in a subarid climate and is subjected to brisk prevailing easterlies. Because of these differences, marine and peritidal carbonate sedimentation on the interior of Caicos Platform is significantly different from that documented to the north. Further, Caicos Platform represents a sedimentary province that is in an intermediate climatic setting between the wet northern Bahamas and the arid Arabian Gulf. Specifically, the marine platform interior (1) is dominantly grainstone, (2) is shallow and organized into broad banks, (3) contains an abundance of oolitic grains throughout most of the environments of the platform interior, and (4) has a variety of types of ooid grains reflecting a variety of agitation settings. The peritidal salina and tidal flat environments (1) are sites of surficial evaporite sedimentation and (2) contain significant deposits of Holocene reflux dolomite.

Holocene physiography: The present physiography of Caicos Platform is defined by five positive bathymetric/topographic features. (1–2) A barrier reef and Pleistocene–Holocene island trend separate the northern margin from the platform interior. These islands form a nearly continuous barrier to tidal flow, wave energy, and circulation and effectively isolate the northern skeletal grainstone/boundstone rim from the nonskeletal sands of the platform interior. An extensive tidal flat occurs on the southern (platform interior) margin of this island trend. On the northeastern and northwestern margins, emergent islands form a discontinuous barrier and focus sites of onbank and offbank circulation and sediment transport through breaks in the reef and island chain. (3) A broad subtidal platform-margin levee of grainstone with only scattered reefs defines the unprotected southern edge of Caicos Platform. (4) An elongate midplatform oolitic-grapestone grainstone shoal extends east-west across the north-central portion of the platform interior. (5) An elongate, cross-platform oolitic grainstone shoal extends westward across the south-central portion, isolating the southernmost region.

These five positive topographic features define four sedimentary troughs on the platform: (a) a northern back-reef lagoon between the barrier reef and the emergent islands; (b) a very shallow protected platform-interior lagoon between the southern margin of the tidal flats and the midplatform oolitic-grapestone shoal; (c) a central platform trough defined by the midplatform oolitic-grapestone shoal on the north and the oolitic shoal and the subtidal levee rim on the southern margin; and (d) the deeper, storm-swept southernmost portion of the platform that does not have sufficient protection for accumulation of Holocene sediment.

Holocene sediments: Coralgal framestones, rudstones, and skeletal and oolitic grainstones dominate the windward Caicos platform margins. On the northern margins, reef-derived sands and rubble are swept across the backreef lagoon toward the shore and contribute to lagoonal infilling and strandplain growth. Where the northern backreef lagoon is very wide, a molluscan-foraminiferal grainstone dominates interior lagoonal and adjacent strandplain sediment. Oolitic sands dominate the northern and eastern platform margin in areas of strong wave agitation that are not provided with reefal sands. On the southern and western margins, transport of oolitic and grapestone grainstones from the platform interior provides an important contribution to the platform-margin sediment. Heavily biocorroded skeletal and grapestone grainstones dominate the nondepositional eastern portion of the central-platform trough.

Peloidal grainstones to low-mud packstones dominate the protected platform-interior lagoon and the adjacent tidal flats. Significant amounts of carbonate mud are accumulating only in the most protected environments adjacent to the tidal flats.

Concentrically laminated oolitic grainstones dominate the cross-platform oolitic shoal and the nearshore and beach sediments of east-facing (windward) coastlines in the platform interior where either waves or tidal currents persistently agitate the bottom. Fine-grained oolitic grainstones also form the shore sands on (a) the beaches on exposed north- and east-facing platform margins and (b) the beaches on interior south-facing island shores.

The remaining bulk of platform interior deposits are a mixed nonskeletal grainstone of grapestone, "crinkly-laminated" ooids, superficial ooids, and peloids. This sediment dominates the midplatform shoal (depth 0–3 m), the central platform trough (3–5 m), the flanks of cross-platform oolitic shoal (2–5 m), and the subtidal levee rim (2–5 m) on the southern margins. These environments contain active to relict ripples and low sand waves that indicate frequent, but intermittent, bottom agitation. Coated grains are as common in the deeper zones as they are in the shallower areas.

The widespread distribution of ooids on Caicos Platform includes settings not normally associated with oolitic sand generation and provides alternative models for environments of oolitic sand production and deposition.

A more thorough documentation of carbonate sedimentation and diagenesis on Caicos Platform is provided in the IGC Fieldguide for trip T374.

WARDLE, R. J., and C. F. GOWER, Department of Mines, Newfoundland Canada

The Eastern Canadian Shield Global Geoscience Transect

The transect is designed to provide a synoptic view of the crustal structure of the eastern Canadian Shield in Labrador and Quebec. This region, which includes portions of two Archean cratons and four Proterozoic orogens, represents a crustal history of 3.2 Ga (3.8 to 0.6 Ga). It is in many ways a microcosm of the Canadian Shield and offers opportunities to establish the style and mechanism of Precambrian cratonic growth. This growth includes the initial coalescence of continental nuclei to form the Laurentian-Greenland Shield (3.8 to 1.8 Ga) and subsequent southerly accretion in the period 1.8 to 0.6 Ga. An important component of the transect will be an offshore seismic reflection profile which will provide critical constraints for modelling of three-dimensional crustal structure, especially across Precambrian orogen boundaries, and will also yield information on the evolution of the modern continental margin.

The transect is divided into 6 legs as shown in Figure 1.

Recent work has shown that much of the crust in the eastern part of the orogen formed in the period 1.7 to 1.6 Ga in association with the newly-named Labradorian Orogeny. This was followed by rift-related plutonism in the period 1.45 to 1.1 Ga and finally by the Grenvillian Orogeny ca 1.0 Ga. The principal effect of this event was to assemble the crust into its present configuration of major thrust-bounded terranes that were, in general, transported northward onto the Archean-Lower Proterozoic foreland. An exception concerns a major transcurrent shear system, associated with east-dipping thrusts, located on an oroclinal flexure in the easternmost part of the orogen.

The Grenville Orogen has long been proposed as the deeply-eroded root of a collisional orogen, which recent studies have tended to support. Although various features have been proposed as sutures, the original nature of most structural junctions remains uncertain and can probably only be elucidated by seismic profiling. Such profiling will assist in explaining the cause of a huge positive gravity anomaly in southeast Labrador. This feature, which is one of the most pronounced anomalies in the Canadian Shield, must be of lower crust or mantle origin and is perhaps (amongst other hypotheses) related to an upthrust wedge of upper mantle.

The western Makkovik Orogen consists of Archean crust in part remobilized and locally overlain by, or intercalated with, Proterozoic cover rocks. Major uncertainty exists concerning the eastward extent of Archean crust. Eastern parts of the orogen which consist of felsic volcanic and metamorphic terranes potentially represent accreted crust. The supracrustal record appears to indicate rifting of Archean crust in the period 2.4 to 2.0 Ga, resulting in formation of MORB-type basalts, followed by closure and production of extensive granitoid plutonic intrusions and associated volcanic cover sequences in the interval 1.86 to 1.80 Ga. These volcano-plutonic rocks may represent the remnants of either arc-related or syncollisional magmatic belts.

The Nain Craton forms the Western part of the North Atlantic Craton and is exposed in cross-section along the Labrador coast. The southern Nain Craton (Leg. 3) is composed largely of amphibolite-facies gneiss of 3.0 to 2.5 Ga age with intercalated 'greenstone type' volcanic belts. The northern Nain Craton (Leg. 5) is composed of amphibolite-to granulite-facies gneiss (3.8 to 2.5 Ga) and preserves extensive areas of early Archean (pre 3.0 Ga) crust.

Problems associated with the evolution of the Nain Craton concern (i) the mechanism by which the relatively low-grade 'greenstone belts' were incorporated into the structure of the craton, (ii) the means by which the lower crustal rocks of the northern Nain Craton were brought to the surface, which inturn poses questions regarding the nature of the present lower crust.

The northern and southern components of the Nain Craton are separated by the Nain Anothosite Complex (Leg. 4), and Middle Proterozoic suite of anorogenic anothhosite and granite typical of those intruded along the southern margin of the Laurentian Shield in the interval 1.45 – 1.1 Ga. The coastal cross-section through the Nain Complex offers an unrivalled opportunity to study the three-dimensional architecture of such intrusions.

The final leg of the transect crosses the entire southeastern arm of the Trans-Hudson Orogen along the south shore of Ungava Bay. The western margin of the orogen comprises supracrustal rocks of the Labrador

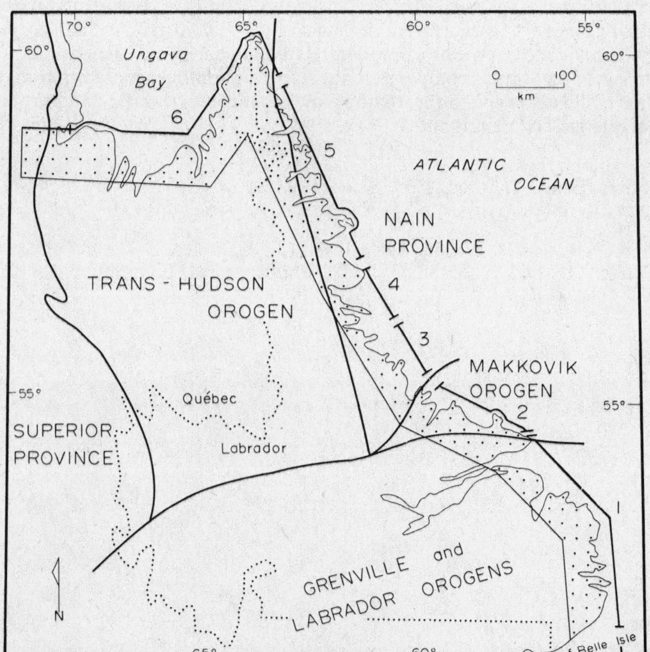

Trough. These rest unconformable upon Archean gneiss of the Superior Craton and have been deformed into a west-verging fold and thrust belt. The eastern margin is dominated by reworked Archean gneiss and minor Lower Proterozoic supracrustal rocks deformed within a major sinistral shear system. The central part of the orogen is the most poorly known part of the entire transect but appears to consist largely of reworked Archean gneiss. The orogen apparently evolved in a environment of oblique convergence, resulting in transpressional deformation.

In summary the transect will contribute to a synthesis of the three-dimensional structure of a critical region of the Canadian Shield and offer new insights into the varying styles and mechanisms by which shields evolve.

WARREN, P. H., University of California, Los Angeles, California

Growth of the Continental Crust: A Mechanistic Model

The net growth through time of the total mass and volume of the continental crust (CC) is one of the great unresolved problems of geology. In this discussion, net growth should not be confused with sporadic growth of individual continental masses (e.g., by accretion of island arcs), a process that does not necessarily require net growth. Geochemists tend to emphasize differences between CC and oceanic crust (OC), but from a planetary-mantle perspective CC and OC are remarkably similar in composition and density. The distinction between CC and OC is canonically defined in terms of elevation. However, the Earth's bimodal elevation pattern clearly reflects a genetic distinction. The OC is that portion of the crust which participates directly in mantle convection, whereas the CC is the remaining, relatively stable portion of the crust. The Earth's advective cooling mechanism requires massive subduction of lithosphere, which due to the inherent buoyancy of crust tends to require a low crust/total lithosphere thickness ratio. Given relatively constant lithosphere thickness (except near thermally buoyant mid-ocean ridges), the Earth responds by heaping most of its crust into a few thick continents, allowing subduction to freely operate over a large area of relatively thin crust. This crustal thickness bimodality engenders the bimodal elevation pattern.

Prevailing wisdom tends to hold that the rate of CC growth has been slow and ongoing; typical models show CC volume roughly proportional to 1/h, where h is the total heat generation by the Earth's K, Th and U. However, such models are generally vague as to the mechanism that supposedly holds back CC growth, even after 4.5 Gy. The strong T-dependence of mantle rheology probably ensured that despite extremely high initial T the mantle took only a few My to cool within a few hundred degrees of its modern T. Gradual compositional evolution of the CC from a plausible initial composition (e.g., basaltic, anorthositic) might have slightly lowered the mean CC density, but this effect is of second-order importance in comparison to the general increase in crust/total lithosphere thickness ratio (for any given crustal thickness) engendered by temporal cooling of the mantle. Intense cratering, occurring coevally with early intense crust/mantle recycling, probably reconstituted virtually all of the CC extant prior to ~ 3.9 Ga, but CC was probably not recycled into the mantle any faster than it was regenerated.

Lunar samples, meteorites, and photogeologic data for Mars, Mercury, and the Moon, provide strong circumstantial evidence that in primordial times the outer few hundred km of all terrestrial planets were largely molten. Factors such as gross mass and core/silicate ratio probably tended to make intense primordial heating even more inevitable for the Earth than for most other terrestrial planets. Detailed outcomes of primordial fractionation processes must have been strongly dependent on the pressure regimes within the planetary interiors. In general, the far greater dT/dP of silicate melting curves vs. silicate adiabats probably tended to foster a layered magmasphere, with a partially molten or paracumulate layer, constrained by dT/dP relationships to a P range of roughly 3 SPa, surmounted by a fully molten "magma ocean" layer, also limited in thickness by efficient rapidly-convective cooling.

In the Earth, P-stabilized garnet and A1-rich pyroxene were key factors limiting ultimate yield of crustal feldspar. Constraints on high-P melt-solid equilibria are provided by recent experimental petrology studies. For example, Takahashi's (1986) results for composition KLB-1 (a peridotite that strongly resembles the estimated bulk composition of the silicate portion of the Earth) indicate that as P increases form 1 to ~ 14 GPa, the compositions of partial melts of KLB-1 tend to approach the bulk composition of KLB-1 itself' i.e., the A1 that is required for production of crustal feldspar is not appreciably enriched in partial melts at high P. Implications for models of crustal growth have been examined with quantitative models of high-P fractional crystallization; including, in some models, periodic replenishment of the magma ocean from below (Warren, 1988). Initial magma ocean compositions were assumed to be similar to the estimated bulk mantle composition ($\sim$KLB-1), or else (for relatively "shallow" models) partial melts of that composition (Takahasi, 1986). Model results indicate that a plausibly deep ($\sim$300–400 km) magma ocean would produce a crust comparable in both composition and thickness to that of the modern Earth. Mildly worrisome discrepancies involve CaO (for which high-P pyroxene compositions are poorly constrained) and SiO_2. The crust probably evolved to its present SiO_2-rich composition through differentiation catalyzed by H_2O, which presumably was not as abundant in the magma ocean as in the crust soon afterward. Nominal model results for mean crustal thickness are typically ~ 1.2–$1.3 \times$ the modern global mean, but for this parameter the model is designed to give conservatively high results, so the discrepancy is probably not significant. After the chaotic first few My of solar system history, I suggest that planetary size and composition, far more than time, govern crustal volume.

REFERENCES

Takahashi, E., 1986, Melting of dry peridotite KLB-1 up to 14 GPa: implications on the origin of the peridotitic upper mantle, J. Geophys. Res. 91, p. 9367–9382.

Warren, P. H., 1988, Growth of the continental crust: A planetary-mantle perspective, Tectonophysics, in press.

WARTEL, S., Royal Belgian Institute of Natural Sciences, Brussels, Belgium

A New Classification of Inorganic Mud Sediments

During the last 20 years many excellent scientific contributions have been published dealing with cohesive, very fine marine or estuarine sediments, generally called "mud" or "fluid mud." Also, considerable progress has been made in the understanding of the sedimentological setting of these sediments. However, it is surprising to ascertain that in contradistinction to the soundness of the scientific analysis of the "mud," a clear description and classification of "mud"-sediments is still lacking. Not only are many different names used (mud, slingmud, hedoro, fluff), but also the given descriptions (stationary suspension, mobile suspension, particles finer than 2, 50, or 63 microns transported in suspension) does not say very much about the sediment itself. This incomplete description is not only confusing, but also hinders a comparison, based on published data, between different environments.

This contribution aims at clarifying this situation and attempts to arrive at an acceptable description and classification based on analyses of Schelde and North Sea muds and on data taken from the literature. The name "mud" is maintained as the general name. However, to arrive at a clearer description, major and minor modifiers are added. The major modifier relates to the clay content, which in the case of mud is more relevant than the mean grain size. The first minor modifier gives the ratio between the silt fractions and the clay content. This parameter can be correlated with the degree of flocculation of the sediment when still in suspension and provides information on the hydrodynamic conditions prevailing during deposition. As far as the Schelde estuary is concerned, the ratio has also an environmental meaning and can thus be of interest for the interpretation of older clay deposits. The second minor modifier deals with the nature of the bedding and some internal structures, such as clay fabric and gas bubbles. Gas bubbles can be observed in radiographs and SEM analyses. From observations in North Sea muds it has been assumed that expanding and migrating gas bubbles induce changes in the properties of the clay owing to changes in hydrostatic pressure.

Some different types of recent and Pleistocene muds are discussed and an example of a paleogeographic interpretation of lower Pleistocene muds evolving from intertidal to fresh water deposits is shown.

WATANABE, T., Hokkaido University, Sapporo, Japan, R. H. GRAPES, Victoria University of Wellington, New Zealand, Y. KAWACHI, Otago University, New Zealand, T. ITAYA, Okayama University of Science, Okayama, Japan, and I. M. TURNBULL, Geological Survey, Dunedin, New Zealand

Formation of Biotite Accompanying a Later Fracture System in the Haast Schists, New Zealand

Prograde chlorite, biotite, and garnet zones in quartzo-feldspathic schists of the Haast schists, New Zealand, have been well documented (e.g., Suggate, Grindley, et al, 1972). However, the biotite zone is not textually and mineralogically gradational from the chlorite zone.

The low-grade boundary of the biotite zone in the Haast schists obliquely cuts across the textural chlorite IV subzone, so that the biotite zone appears on the higher grade side of chlorite subzone III in the northern Haast schists area, e.g., Maruia River area, whereas it grades into chlorite subzone IV in the southern Haast schists area, e.g., the Haast River area. In the Maruia River area, biotite first appears along fracture planes that obliquely crosscut the Alpine (S_2) schistosity planes. In the Dunstan Range area of western Otago, biotite is also found within fracture planes that are subparallel to the axial plane cleavages of later folds. In the Franz Joseph–Fox glaciers area of the central part of the southern Alps (central westland), some biotite also occurs along fracture planes that are oblique to the main "Alpine" schistosity. All these occurrences of biotite indicate that it has been produced during a late stage of deformation. K-Ar age determination of biotite suggests that the deformation may be as old as 70 Ma.

The formation of biotite is commonly considered to be a product of consumption of the phengitic (or celadonitic) molecule of white mica. For example, the Al_2O_3 contents in white mica from the Franz Joseph–Fox glacier and the Haast River areas increase from the higher grade part of the chlorite zone (chlorite subzone IV) to the lower grade biotite zone as shown in Figure 1. However, in the Dunstan Range area, biotite coexists with phengitic mica similar in composition to that in the lower grade chlorite zone (chlorite subzone II–III). This latter relationship demonstrates that biotite grade metamorphism has overprinted lower grade chlorite zone rocks during a later phase of metamorphism.

It is well known that K-Ar "ages" of the metamorphic rocks become younger toward the Alpine Fault with increasing metamorphic grade (Adams and Gabites, 1985). The important point is that the "ages" become drastically younger from around the area of the biotite zone (100 Ma ↔ 70 Ma) and may support the textural and mineralogical data for later production of biotite mentioned above.

We also note that 70 Ma roughly corresponds to the opening of the Tasman Sea, and the southwestern extension of the Alpine Fault is located to the margin of north–south-trending magnetic lineation of the Tasman Sea. This leads us to speculate that the formation of the fracture system containing biotite in the Haast schist may have been produced as a consequence of rifting and opening to form the Tasman Sea and that it was accompanied by a change in the geothermal regime leading the formation of the later generation of biotite in the Haast schists.

The biotite zone in the Otago region discussed by Brown (1971) should be reexamined in detail in this regard. Two stages of generation of biotite are suggested.

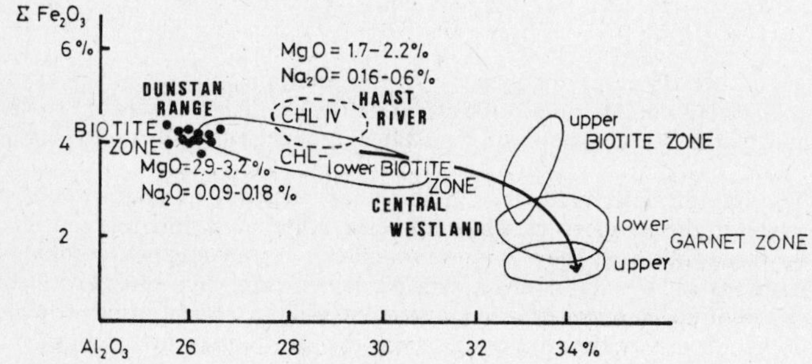

Figure 1—Graphic plot of chemical compositions of white mica coexisting with biotite in the Haast schists.

WATANABE, T., Hokkaido University, Sapporo, Japan E. C. LEITCH, University of Technology, Sydney, Australia, M. IWASAKI, Tokushima University, Tokushima, Japan, H. ISHIGA, Shimane University, Matsue, Japan, T. ITAYA, Okayama University of Science, Okayama, Japan, and Y. KAWACHI, University of Otago, Dunedin, New Zealand

Timing of Accretion in Southern New England Fold Belt, Eastern Australia

Radiolarian biostratigraphy (Ishiga et al, 1987, 1988; Aitchison, 1988; Blake and Murchey, 1988), K-Ar age determinations of metamorphic rocks (Watanabe et al, 1988), a reassessment of Rb-Sr date for S-type granites (Iizumi and Honma, 1987), and stratigraphic relationships (Allan and Leitch, in press), indicate that a major disturbance occurred in the southern part of the New England Fold Belt in the Late Carboniferous. This disturbance marked the final stage in the development of the accretionary prism associated with Devonian–Carboniferous plate convergence. Structures within the accretionary rocks indicate the affects of both compression and strike-slip movements, the latter manifested by steeply plunging fold axes and the fabric of serpentinite masses (Offler and Williams, 1987). Recently discovered sheath folds suggest simple shear was important during deformation. The overall affect of deformation was to thicken the prism, and S-type granites were produced within a short time span.

Microfossils indicate that Pre-Devonian rocks occur in parts of the accretionary prism (Macdonald Block). These are interpreted as exotic blocks embedded within melange that contains elements as young as Early Carboniferous. Serpentinite melange with rare inclusions of high-pressure metamorphic rocks is of uncertain origin but had been emplaced within the crust by the end of the Carboniferous.

Early Permian rifting marked the onset of a new tectonic regime in the New England Fold Belt. Sediments were deposited unconformably on top of the accretionary rocks. Later compressive and strike-slip deformation was closely followed by the rise of many I-type granitic bodies.

WATANABE, Y., Geological Survey of Japan, Sapporo, Japan

Regional Tectonic Control on the Vein-Type Mineralization in Japan

The major tectonic setting of the Japanese Islands since the Middle Miocene is represented by subduction of the Pacific and the Philippine Sea plates under the Eurasian plate. The Pacific plate has also subducted under the Philippine Sea plate at the Izu-Ogasawara (Bonin) trench, and the Izu block on the Philippine Sea plate collided with the Japanese Islands. These subductions have caused the volcanism related to the formation of numerous epithermal base and precious metal deposits in the Japanese Islands.

Distribution density and production of more than 2000 known Japanese ore veins are approximately proportional to the intensity of related volcanic activity in each region of Japan. The majority of the veins are distributed in Hokkaido, northeastern Honshu, Izu Peninsula, and southern Kyushu, where contemporaneous intensive volcanism is observed. The production of metals also shows a regional variation. For example, gold and silver are concentrated in Hokkaido, Izu Peninsula, and southern Kyushu.

The rosette diagram of vein directions (Figure 1) exhibits another characteristic of this region. East–west to east-northeast–west-southwest veins are predominant in Hokkaido and southern Kyushu, northeast-southwest ones in northeastern Honshu and northwest-southeast ones in Izu Peninsula. The analysis of vein patterns of individual deposits indicates that these veins of predominant directions are regarded as shear fractures of dextral strike-slip type. The stress fields deduced from these shear fractures indicate east-west to west-northwest–east-southeast compression both in Hokkaido and southern Kyushu, and north-northwest–south-southeast compression in Izu Peninsula, which observations are compatible with the movement of the subducting Pacific and Philippine Sea plates and colliding Izu block. This implies that the formation of these veins has been strongly influenced by the regional stress caused by the subduction and collision.

According to the convergence mode among the Pacific, Philippine Sea, and Eurasian plates, the Japanese Islands are roughly divided into (1) oblique subduction area (Hokkaido and Kyushu), (2) normal subduction

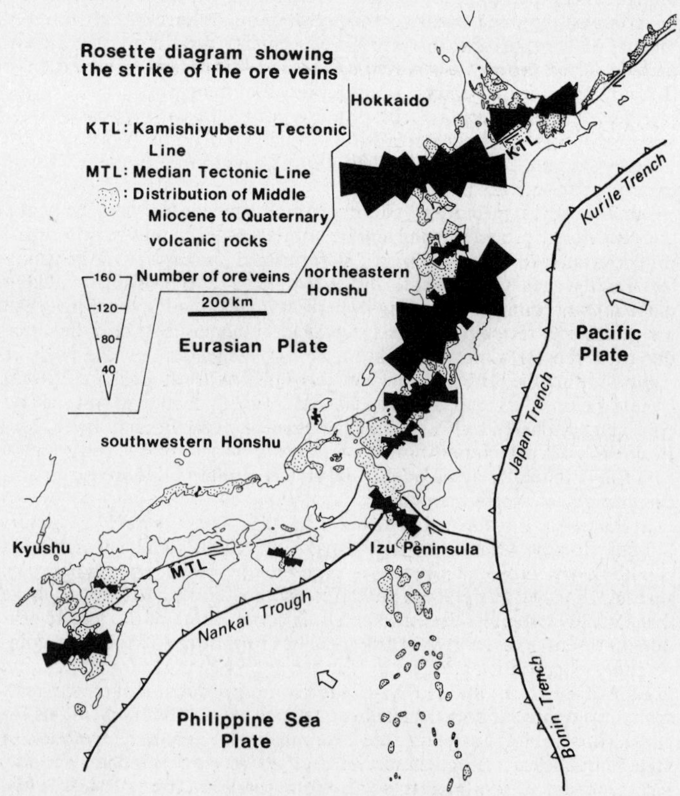

Figure 1—Rosette diagram showing the strike of the ore veins.

area (northeastern Honshu), (3) transform fault area (southwestern Honshu), and (4) collision area (Izu Peninsula). Shear fractures may be more easily formed in the oblique subduction and transform fault areas than in the normal subduction area. Actually, dextral strike-slip faults through Kurile Arc and Hokkaido (Kamishiyubetsu Tectonic Line), and through southwestern Honshu (Median Tectonic Line) run parallel to the Kurile trench and the Nankai trough, respectively. In contrast to this, volcanism has been more active in the normal subduction area (northeastern Honshu) than in the transform fault area (southwestern Honshu).

Two important factors for the formation of ore veins are active volcanism, which supplies ore fluid, and well-developed strike-slip type shear fractures. Circumstances that satisfy these two antagonistic conditions appear to be found at the transition zone between the oblique subduction (or transform fault) and the normal subduction. This is evidenced by the existence of large vein-type Au-Ag deposits predominantly in Hokkaido, Izu Peninsula, and southern Kyushu, which are all situated at the transition zones. Consequently, arc-arc junctions are considered to be the most favorable tectonic setting for the formation of the vein-type deposits.

WATERS, D. J., University of Cambridge, Cambridge, U.K.

Melt Extraction from Granulite Facies Migmatites

Migmatitic rocks in granulite facies terrains commonly consist of coarse anhydrous patches, stringers or vein networks (mobilizates) in a finer-grained, foliated matrix. We interpret these features as the result of incongruent dehydration melting reactions (Waters, 1988). The mobilizates do not represent the composition of segregated melt, but rather represent varying proportions of (i) melt, (ii) solid products of incongruent melting and (iii) excess phases not consumed during the reaction. In metapelites and biotite gneisses, where the near-solidus melt composition is granitic and its water content can be estimated as a function of T, the melting reactions can be balanced and relative volumes of product phases predicted.

We have compared the calculated products with the observed distribution and abundance of phases in granulite facies migmatites from Namaqualand, South Africa. At the amphibolite-granulite transition (c.750 °C), the modal composition of mobilizates indicates that 10–15 vol% melt was produced, but was not extracted from its site of generation. At higher grade ($\geq$800 °C) some mobilizate patches are dominated by solid products (e.g. garnet, K-feldspar) implying the loss of 10–20 vol% melt. The existence within a small outcrop area of an apparently complete gradation from garnet-or pyroxene-rich patches to elucogranitic stringers and sheets suggests the the extracted liquid has not migrated far.

A second set of predictions follows from an estimation of the physical properties of water-undersaturated silicic melts. The dihedral angle for melt in a quartzofeldspathic aggregate is probably less than, though close to, 60°. Therefore, if textural equilibrium is approached, the melt phase should be interconnected. The efficiency of melt extraction will be critically dependent of the viscosity of the melt (McKenzie, 1985; Wickham, 1987). Calculated values for the water-undersaturated case are several orders of magnitude greater than those for water-saturated melts, and fall in the range 10^6 - 10^8 Pa.s. These values correspond to a length scale for compaction on the order of centimetres, which is similar to that of the mobilizates, but imply a long time scale of 10 - 1000 Ma. In the absence of deformation, short-range controls such as the minimization of surface energy may determine the distribution of the melt phase. If this is the case, the matrix regions are predicted to contain a few percent melt in an equilibrium distribution along grain boundaries, while the mobilizates contain pooled excess melt.

Many, but not all, of these predictions are supported by the observations on natural rocks. The solid products of incongruent melting are concentrated in the mobilizates, and are not distributed through the rock. Commonly, evidence suggests that mobilizates develop along compositional or structural inhomogeneities in the rock. For example, at the granulite transition, small mobilizates appear to have nucleated on pre-existing matrix garnets. These features indicate that melt is generated within, and is perhaps confined to, the mobilizate volumes. The matrix regions of rocks are texturally similar to their amphibolite facies equivalents, and give no indication that the melt has penetrated the grain boundary network outside the mobilizates. Possibly, textural equilibrium is not achieved on the time scale of the melting event, or else the true dihedral angle is greater than 60°. Elsewhere, some mobilizate follow earlier fabric elements, or are aligned in the hinge zones of decimetre-scale crenulations and in minor shear zones. The structural control of mobilizates, particularly those with more leucocratic compositions, suggests that melt extraction is assisted by deformation or tensional fracturing. In summary, field evidence suggests that some mobilizates represent sites of melt generation, while others represent structurally controlled zones where melt has locally accumulated.

Amphibolite facies migmatites, in contrast, commonly show leucosomes whose compositions are similar to minimum melts, and are markedly different to the matrix (palaeosome) composition. A melanosome is commonly also present. These features imply a much more efficient separation of solids from melt at low melt fractions, consistent with the lower viscosity predicted for melts nearly saturated in H_2O.

We conclude that water content has a critical effect on the viscosity (and/or possibly the dihedral angle) of siliceous melt, so that small melt fractions are even less likely to be extracted from partially melted granulite facies rocks at about 800 °C than from amphibolite facies migmatites. Large, mobil magma bodies will result only when the melt fraction becomes large enough to cause disaggregation and convective instability of the whole rock mass. A higher melt fraction may result from higher temperature, a higher initial proportion of hydrous minerals, or an external source of H_2O. In the case of Namaqualand, the region of magma generation lay at hotter, deeper levels in the crust. The magmas are represented in part by late-tectonic, approximately synmetamorphic granitoids, which intrude the migmatites, and which probably assisted in the transfer of heat to the exposed levels.

REFERENCES

McKenzie, D. p., 1985, Earth Planet Sci Lett 74: p. 81–91.
Waters, D. J., 1988, J. Metamorphic Geol. 6: p. 387–404.
Wickham, S. M., 1987, J. Geol Soc London 144: p. 281–297.

WATERS, D., University of Cambridge, Cambridge, United Kingdom

Hercynite-Quartz Granulites: Phase Relations, and Implications for Crustal Processes

Assemblages in high-grade metapelitic rocks are valuable indicators of metamorphic conditions within the granulite facies. The assemblage Zn-poor spinel + quartz, recorded from a handful of terrains world-wide, is stable in a wedge-shaped P-T field opening upgrade from about 770 °C, 3 Kbars. The stability field of the rare K-Mg-aluminosilicate mineral osumilite overlaps that of spinel + quartz. These high-grade index minerals can potentially be used to map grade variation in granulite terrains.

The granulite facies region of the Namaqualand Metamorphic Complex, South Africa, contains a zone $\sim$10,000 km^2 in the area in which the diagnostic assemblage in metapelites is hercynite-rich spinel + quartz (+ Crd + Kfs ± Grt ± Sil ± Pl). A typical spinel composition is $Hc_70Sp_25Mag_5$, with negligible Zn and Cr. Thermobarometry on these and associated rocks suggests P-T conditions of >800 °C at 5 Kbars, with little variation across the zone. The AFM facies type can be determined from cofacial assemblages in varied metapelitic compositions. Hercynite + quartz developed in Fe^{2+} and Al-rich rocks. Rarer, highly oxidized metapelites contain osumilite (+ Crd ± Opx) with Hem and Mag. Under these conditions, osumilite is restricted to high Mg/(Mg + Fe^{2+}) bulk compositions by the stability of the join Crd-Opx-Kfs. Many rocks of intermediate Mg/(Mg + Fe^{2+}) contain Grt + Crd + Kfs + Qtz, and lack the diagnostic assemblages of the higher grade zone. Pressures were nowhere high enough to stabilize sapphirine + quartz.

The mineral association hercynite + quartz developed during prograde metamorphism by replacement of large cordierite grains. During initial cooling, osumilite broke down to largely anhydrous Crd-Kfs-Qtz-Opx symplectite, and narrow coronas of Crd, Grt, Sil, or Pl formed around hercynite by diffusion-controlled retrograde reaction with quartz. Subsequently, it appears that (1) the peak spinel phase first unmixed to form large grains of hercynite + Fe oxides, and (2) at a later stage hercynite exsolved 4–5% fine, dendritic magnetite.

Reaction processes and thermobarometry indicate that the rocks followed a flattened anticlockwise P-T loop, which is consistent with a thermotectonic environment dominated by magmatic accretion at both deep and shallow crustal levels. Few other tectonic settings are capable of generating the high T at relatively low P required for the regional development of hercynite–quartz assemblages.

WATERS, J. A., West Georgia College, Carrollton, Georgia

Evolutionary History of the Blastoidea (Echinodermata)

Blastoids are an extinct class of blastozoan echinoderms that ranged from the Late Ordovician (Caradocian) to the Late Permian (Kazanian). The 100 blastoid genera belonging to two orders Fissiculata and Spiraculata, are known from every continent except Antarctica. Throughout most of their evolutionary history, blastoids were a minor component of echinoderm communities which were principally dominated by crinoids. Blastoids tended to be ecologically opportunistic, are often found on the leading edges of transgressive sequences, and seemed to benefit from crises in crinoil communities. Diversity changes in blastoids throughout the Paleozoic tended to parallel trends in the camerate crinoids, although the timing of extinction events did not always coincide within those of other elements of the benthic community.

Current evidence suggests that blastoids originated in North America which remained a major center of evolution throughout the Lower and Middle Paleozoic. The oldest blastoid genus, Macurdablastus, is found in the Caradocian of Tennessee. Siluran genera are mostly endemic to North America but Polydeltoideus is also know from Czechoslovakia. In contrast, Devonian blastoids are well known from a variety of localities worldwide, many strata yield diverse, relatively abundant blastoids with several genera co-existing for the first time. Although maximum blastoid diversity remained in North America and Europe, additional Devonian blastoids are known from Bolivia, and South Africa. For the first time blastoids migrated into North America, with many genera originating in Spain and South America. The major Late Devonian extinction event affecting the blastoids occurred at the Givetian-Frasnian boundary rather than the Frasnian-Fammenian boundary. Recent data suggest a modest increase in blastoid diversity in the Fammenian which continued into the Carboniferous.

Blastoid diversity reached a maximum in the Lower Carboniferous. Most of the dominant families originated in either North America or Europe and later migrated eastward into the Tethyan seaway in the Late Paleozoic. During the Tourniasian-Lower Visean, monobathrid camerate crinoid communities reached a zenith on the broad carbonate platforms such as the Burlington Limestone. Blastoids were an integral part of this community and 17 genera (mostly globose spiraculates) are known from the Burlington. A major extinction event radically altered the composition of blastoid faunas in the Lower Visean, although blastoids continued to diversity throughout the Visean. Visean blastoids are best known from North America and Europe, but are also found in Africa, the Middle East, Australia, China and Japan continuing their migration around the Tethyan seaway in the Late Paleozoic. A second center of origin developed in the eastern Tethys at this time. In many Visean communities, blastoids were the dominant echinoderm. Blastoids suffered a sudden decline in diversity and abundance in the Upper Carboniferous at a time when advanced inadunate crinoids were diversifying rapidly.

The most diverse Upper Carboniferous-Permian faunas are found in southeast Asia and Australia. North American faunas are poorly known, although recent work has yielded a small poorly preserved fauna of paedomorphic fissiculates. Fifteen blastoid genera are known from Timor. Fissiculates are the most diverse (13 genera) but Deltoblastus (a spiraculate) dominates in terms of abundance and is known from tens of thousands of specimens. Although the Timor fauna is very well known from museum collections, its palaeoecology is very poorly understood.

WATKEYS, M. K., University of Cape Town, Rondebosch, South Africa

Archaean Limpopo Belt and Proterozoic Namaqua Belt: Comparison Between Granulite Terrains in Southern Africa

The Kaapvall Craton of southern Africa is flanked by the Archaean Limpopo Belt and Proterozoic Namaqua Belt (Figure 1). Although of different ages, these two "mobile belts" are of comparable size and have some superficial similarities. Each belt records an initial crust-forming event followed by deposition of a dominantly "miogeosynclinal" supracrustal sequence prior to reworking in a crust-deforming event. Furthermore the structural history of many regions in both belts may be simplified into early recumbent folding superimposed by upright folding which was deformed by late buckling and ductile shearing. Pb isotopic studies reveal higher U values than on the Kaapvaal Craton while stable isotope investigations suggest partially closed systems during metamorphism. Consequently these two belts afford the opportunity to compare and contrast processes operating during the Archaean and the Proterozoic, with particular emphasis on the mechanism causing granulite metamorphism.

The Limpopo Belt is situated between the Kaapvaal and Rhodesian cratons. It may be subdivided into Northern and Southern Marginal Zones separated by a Central Zone. The Marginal Zones represent granite-greenstone material at high metamorphic grade which has been thrust onto the adjacent lower grade cratons. In contrast, the Central Zone is dominated by a mainly metasedimentary succession at high metamorphic grade overlying a ca 3.2 Ga orthogneissic basement. It has been thrust onto the Northern Marginal Zone and is separated from the Southern Marginal Zone by a major transcurrent shear.

The Central Zone records a continuous PTt loop commencing with a high pressure granulite event followed by rapid depressurisation to low pressure granulite conditions and then rehydration to amphibolite facies. The uplift was due to the northward thrusting of the Central Zone and caused adiabatic melting which gave rise to ca 2.7 Ga charno-enderbitic syntectonic intrusions. A similar PTt loop has been recognised recently in the Southern Marginal Zone where the high grade rocks have been thrust southwards over the Kaapvaal Craton.

The Namapua Belt is part of a network of early to mid-Proterozoic terrains that wrap around the Archaean nuclei of southern Africa. It has been subdivided into a western Bushmanland Subprovince and an eastern Gordonia Subprovince. In the north of the Bushmanland Subprovince, the Vioolsdrif Terrain consists of ca 2.0–1.8 Ga calc-alkaline volcanics and plutonics at low metamorphic grade. The Groothoek Thrust separates this terrain from the rest of the Bushmanland Subprovince which is characterised by the presence of a thin ca 1.7 Ga metasedimentary sequence that hosts the Aggeneys-type massive sulphide deposits. This succession is underlain by ca 2.0 Ga grey orthogneiss and is intruded by a variety of ca 1.2–1.0 Ga granitoid batholiths. Structural trends are dominantly east-west while metamorphism is medium to high grade.

The north-west trending Gordonia Subprovince is separated from the Bushmanland Subprovince by the Hartebees Thrust Belt and from the Kaapvaal Craton by a series of major thrusts and shears. Adjacent to the craton is the Upington Terrain, consisting of a ca 2.0 Ga quartzite-pelite succession. A wedge of ca 1.35 Ga greenstones delineates its north-eastern boundary while the southwestern boundary is represented by a line of calc-alkaline volcano-plutonic amphibolite complexes with a similar age. To the west lies the Kakamas Terrain which contains high grade metasediments intruded by ca 1.2–1.0 Ga syn-to late tectonic granites and charnockites, including juvenile crustal material.

Although metamorphism across the Namaqua Belt varies from low to high grade, pressure was constant at about 5 kb. Steep paleogeothermal gradients are due to magmatic heat transfer in the crust caused by the intrusion of ca 1.2–1.0 Ga granitoid bodies. This contrasts to the Limpopo Belt where syntectonic intrusions are rare and pressure, rather than temperature, was the varying factor.

Overall the Limpopo Belt represents a "passive mobile belt", being a continental margin which was subducted, deformed and uplifted without much crustal reworking. The Namaqua Belt, however, is an "active mobile belt", involving the amalgamation of various terrains, extensive crustal reworking and addition of juvenial material. The reason for this difference may be due to the Limpopo Belt being underlain by a colder, thicker lithospheric root than the Namaqua Belt.

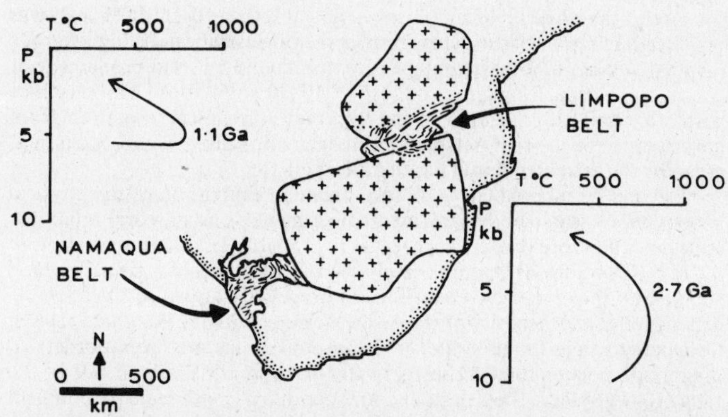

Figure 1—Simplified geological map of southern Africa showing the localities of the Limpopo Belt and Namaqua Belt, with their PTt loops, as well as the Archaean cratons.

WATNEY, W. L., Jan-Chung WONG, AND J. FRENCH, Kansas Geological Survey, University of Kansas, Lawrence, KS

Computer Simulation of Coastal Marine Sedimentary Sequences Influenced by Glacial-Eustatic Sea-Level Fluctuations

A process-response stratigraphic model has been developed to simulate marine sedimentation of a coastal carbonate-dominated system. Sea-level fluctuation, subsidence, sedimentation rate, and elevation of depositional surface, coupled with special functions that affect rate of sedimentation, are input either via keyboard or digitization. Output consists of time versus elevations of sea level, original sediment surface elevation, and subsidence curve.

The program is written in Turbo Pascal for IBM compatibles. Graphic routines are written without special utilities in order to facilitate portability. Operation is menu driven to facilitate ease of use. The program operates using monochrome, EGA, or VGA graphics and can output results to a dot matrix printer of HP/pen plotter.

General objectives of the simulations are: (1) to improve understanding of the relative contributions of potential causal mechanisms on sedimentation by providing means to test scenarios of process-response and develop constraints for potential causes; (2) to identify the types of geologic data that need to be collected to more precisely address causal mechanisms; (3) to eventually be able to predict sedimentary succession based on limited information about depositional conditions; (4) construct time-stratigraphic section via an inversion of the process-response model used to reconstruct sequence of depositional events.

The focus of the modeling has been to describe the effects of glacioeustatic sea-level fluctuations on marine carbonate-dominated sedimentation. It has been proposed by workers for many years that glacial eustatic sea-level fluctuations were a major cause of Late Paleozoic cyclical sedimentation, and recent information continues to support this hypothesis. Yet, the magnitudes of associated sea-level change, and its interaction with other potentially significant and perhaps overwhelming causal mechanisms are poorly understood.

A potentially excellent analog of Late Paleozoic glaciation is the Pleistocene, which is much better understood and documented in terms of the effect of ice advance and retreat on sea-level changes. The effect of the earth's orbital parameters on insolation, and the interaction of unique factors that control the geometry and extent of ice advance would necessarily produce differences between these two periods of glaciation. Appropriate adjustments to parameters can be afforded through modeling.

Modeling was initially directed to characterizing Pleistocene sedimentary sequences. Cyclical Late Carboniferous Missourian (Stephanian) strata of the Midcontinent, USA, are being studied concurrently. Depositional sequences are being delineated and described in detail from three juxtaposed depositional settings: carbonate platform, sediment-starved lower shelf, and siliciclastic-dominated shelf, with the later locality markedly affected by proximal orogenic activity. Geometries of temporally equivalent depositional sequences are established through correlation of bounding surfaces and condensed sections. Results suggest widely differing accommodation potential for sedimentation because of inferred variations in both elevation of depositional surface and subsidence rate. The influence of eustatic change on individual depositional sequences is markedly different across the area; various scenarios have been illustrated through modeling.

The simulation illustrated in Figure 1 is for coastal marine carbonate-dominated sedimentation in the Pleistocene. Special features include sea level, which is approximated by a digitized curve of O–18 analyses of planktonic foraminfera from deep-sea cores from the Caribbean, and the incorporation of a retardation factor applied to sediment accumulation that varies as a function of the rate of sea-level rise. The function is complex and is related to sediment supply, type of sediment, and depositional slope.

Modeling for Pleistocene sedimentation as illustrated in Figure 1 includes the following results, which are compared with actual coastal marine sedimentary sequences. (1) The magnitude of sea-level change during the Pleistocene is large (± 110 m) and rates were rapid, especially during rise. The pattern of sea-level fluctuation is unique from cycle to cycle but has an average periodicity of 100,000 years. (2) The depositional sequences that are generated in the model are bounded by subaerially exposed surfaces (also referred to as the hiatal surfaces or unconformities of sequence stratigraphy). These are also observed as the bounding surfaces in the Missourian depositional sequences. Sediment bypassing over much of the shelf and local channeling on the shelf are likely processes that would accompany subaerial exposure. (3) Initial sediment accumulation is low during rapid sea-level rise, leading to reduced sediment accumulation. Rapidly rising sea level and eventually deeper water (>50 m) result in very slow accumulation rates (0.001 m/Ka) and produce a condensed (sediment-starved) interval. The condensed section may represent 30% of the total time reflected by the entire cycle. (4) A considerable amount of sedimentation occurs during slowly falling sea level when accommodation is available and both clastic wedges and carbonate banks can develop and prograde or aggrade.

Similar observations are seen in the Missourian rocks. Interactive simulation modeling of Pleistocene and ancient depositional sequences provides a more rigorous, quantitative means to establish analogous relationships. The method has many potential applications, including prediction of petroleum and coal resources.

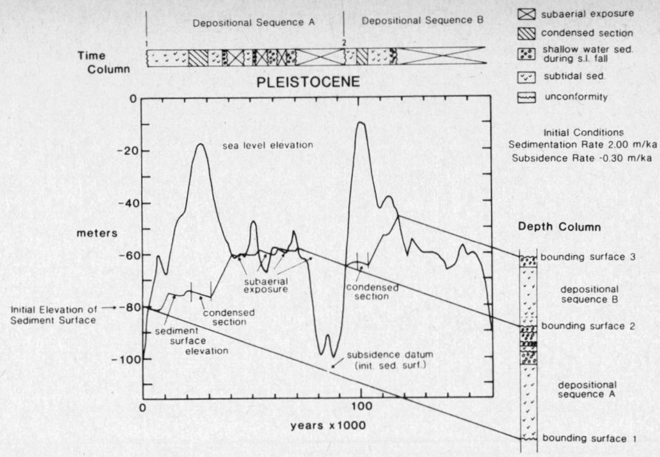

Figure 1

WATSO, D. C., University of Illinois at Urbana-Champaign, Urbana, IL, Present address: Gulf Coast Division, Shell Western E & P, Inc., Houston, TX, and G. deV. KLEIN, University of Illinois at Urbana-Champaign, Urbana, IL

Tectonic Origin Of Cambro-Ordovician Platform Cycels, Upper Middle West, USA

Cambro-Ordovician platform stratigraphy of the Wisconsin dome is organized into five depositional cycles consisting of a basal unconformity overlain by quartzose sandstone which is, in turn, overlain by glauconitic sandstone, and capped by limestone or dolostone. Unconformities at the base of each cycle are found to be coeval with known times of rapid tectonic subsidence in the Illinois Basin at 510, 500, 495, and 460 Ma (Figure 1). Subsidence on the edge of the Wisconsin Dome was slow and gradual, but was greater than estimated concomitant sea-level rise (Figure 1). Times of rapid subsidence in the Illinois basin are interpreted to represent resurgent events of mechanical, fault-controlled subsidence, which was superimposed on an overall trend of thermal subsidence following basin formation by mechanical extension. Associated lateral heat flow with each repeated mechanical subsidence event caused concomitant minor uplift of the Wisconsin dome above sea level and development of the erosional unconformities observed at the base of each Cambro-Ordovician stratigraphic cycle. Because sea level rose at a gradual global rate during deposition of these Cambro-Ordovician cycles, periodic uplift of the Wisconsin dome controlled localized regression and initiation of subsequent cycles. Each time uplift of the Wisconsin dome occurred, the sedimentary sequence underwent erosion and local unconformities developed. Global eustatic rise resulted in onlap onto the Wisconsin dome, causing deposition of reworked clastic sedimentary rocks and initiation of a new sedimentary cycle. Exceptions to this model include initial Late Cambrian transgression onto the midcontinent, and a global drop in sea level at 485 Ma.

These finding pose important implications for cycle stratigraphy, sea-level history, and basin subsidence. In this particular case, local tectonic processes, rather than global eustasy, controlled cycle stratigraphy. The subsidence curve (Figure 1) for the depocenter of the Illinois basin demonstrates that a significant component of mechanical basin subsidence after initial thermal subsidence began. Although this basin formed by original mechanical subsidence in response to extension, mechanical fault-controlled subsidence processes temporarily overlapped thermal basin subsidence. Consequently, the tectonic subsidence curve shows sharp positive deflections for a true thermal curve (Figure 1). We interpret our findings to mean that such overlapping mechanical subsidence processes during thermal subsidence may be common to other continental interior basins and may, in fact, characterize such basins when compared to other domains that evolved from a mechanical and extensional to a thermal subsidence regime.

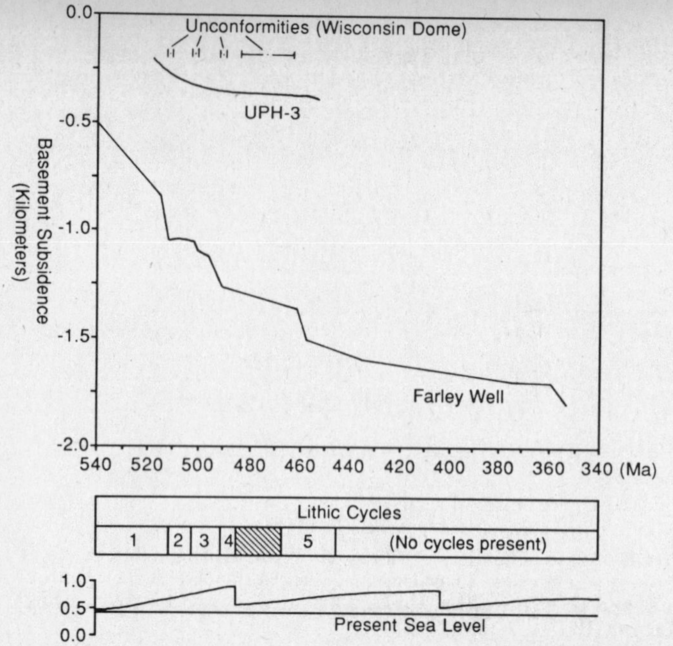

Figure 1—Correlation diagram showing tectonic subsidence curves for Hole UPH-3 on the edge of the Wisconsin dome in northern Illinois, and Farley well (lower curve) in the depocenter of the Illinois basin. Also shown are distribution of Cambrio-Ordovician cycles, Cambro-Ordovician unconformities on Wisconsin dome, and Cambro-Ordovician second-order sea-level rise. Unconformities plotted with both mean age (dot) and duration of each unconformity with possible errors shown as horizontal bars. Shaded interval represents pre-Tippecanoe unconformity in southern Illinois.

WATSON, E. B., and D. LAPORTE, Rensselaer Polytechnic Institute, Troy, New York

Textural Aspects of Partially-Molten Granitoids: An Overview of Experimental Rationale, Results and Possible Implications for Migmatite Development

Almost by definition, natural migmatites and felsic intrusives do not cool sufficiently fast to preserve crystal/melt textural relationships on the scale of individual grains. Experimental studies at lower crustal conditions can thus provide unique constraints on small-scale textural aspects of crustal rocks undergoing partial fusion or solidification. Questions that come to mind in this regard include: To what extent are individual grains wet by partial melt during anatexis? Is the melt phase continuous and likely to be homogeneous? Is there any tendency for the melt to coalesce and segregate from the residue?

A parameter of primary importance in answering these questions is the dihedral angle, Θ, formed between contacting surfaces of two crystals bounding a melt pocket. The value of Θ is determined by the relevant interfacial energies:

$$\Theta = (2)\text{arcos}(\gamma_{ss} 2\gamma_{sl})$$

where γ refers to the interfacial energy per unit area of the subscripted interface, either solid-solid (ss) or solid-liquid (sl). (Ideally, one would like to know γ values for all natural crystals and melts; however, first-principles computation or even experimental measurement of γ for any range of conditions or materials is a formidable task, because γ is likely to depend not only upon the identity of phases but also their major- and trace-elements composition, temperature, pressure and crystal orientation. Moreover, very accurate knowledge of interfacial energies is needed for even a rough estimate ($\pm 20°$ of Θ. The only practical approach is to measure Θ directly, as was originally done for the basalt/peridotite system by Waff and Bulau (1979)).

The first-order significance of Θ is that it determines whether a partial melt 1) wets all grain surfaces; 2) forms an interconnected network of trigonal prism-like tubules along grain edges; or 3) forms isolated pockets at grain corners. For small melt fractions ($f \leq 0.01$) the necessary conditions for each of these three possibilities are $\Theta = 0°$, $0° < \Theta \leq 60°$, and $\Theta > 60°$, respectively. Von Bargen and Waff (1986) have shown that the 60° cutoff for melt interconnectivity increases for $f \geq 0.01$, reaching -65° at $f \sim 0.02$ and $\sim 80°$at $f \sim 0.05$.

The first "wetting" measurements involving granitic melts, made by Jurewicz and Watson (1985), revealed that Θ varies from as high as -60° for a quartz/quartz boundary to $\sim 44°$ for a feldspar/feldspar boundary at 1000° and 10 kbar. This result implies that even very small amounts of melt will form interconnected grain-edge channels in leucocratic rocks undergoing partial fusion (It also means that, unless the rock is disaggregated by shear, melt will generally be excluded from the grain surfaces). The authors also proposed the concept of an "equilibrium melt fraction" (f_{eq}), by which is meant a specific melt proportion that represents a minimum in bulk-system interfacial energy relative to higher or lower melt abundances. It was suggested that if the actual melt fraction were to exceed f_{eq}, the system would respond by forming microsegregations or small melt pockets surrounded by several grains. Although the principle of an equilibrium melt fraction was confirmed in a simple, 2-dimensional model involving circular grains, the actual f_{eq} values computed for various Θ's probably have no quantitative significance. Nevertheless, the process of microsegregation appears to have been documented in the Jurewicz-Watson experiments, in which irregular melt poos up to 50 μm across were observed to form during both crystallization and partial melting (typical crystal size ~ 10 μm in the partial melting experiments and 5 μm in the crystallization experiments).

Here we report the first measurements of Θ for a hydrous, felsic melt in contact with a mafic mineral, specifically biotite: $\Theta \sim 40°$ at 800° and 10 kbar. Combined with the value of 33 °reported by Vincenzi et al., 1988, for a melt of intermediate composition contacting amphibole, this value suggests that mafic minerals in general have relatively low crystal/felsic melt interfacial energies (i.e., low Θ's) — lower, in fact, than values for feldspar and quartz. This relationship raises the interesting possibility that any early-formed, granitic melt produced in the leucocratic part of heterogeneous source region will quickly penetrate nearby areas dominated by mafic minerals, thereby establishing a pathway for rapid diffusional equilibration between leucosome and melaonsome.

In addition to its role in melt distribution, Θ may also be important in determining rheological properties of systems experiencing anatexis. Of particular interest in this regard is the contiguity (C) of the system, defined as the fraction of crystal surface area that is shared with other crystals (German, 1985). the value of C depends mainly upon f and Θ, and secondarily upon grain size distribution. For $\Theta \sim 50°$, ;the contiguity may be sufficiently high even at $f \sim 0.5$ to sustain a rigid, continuous skeleton of interconnecting grains. This skeleton is likely to impart high yield strength to a partially-molten region, and possibly high bulk viscosity as well, given strain rates low enough to avoid disruption of solid-solid contacts.

REFERENCES

Waff, H. S., and J. R. Bulau, 1979, Equilibrium fluid distribution in an ultramafic partial melt under hydrostatic stress conditions, J. Geophys. Res. 84, p. 6109–6114.

Von Bargen, N. and H. S. Waff, 1986 Permeabilities, interfacial areas and curvatures of partially-molten systems: results of numerical computations of equilibrium microstructures, J. Geophys. Res. 91, p. 9261–9276.

Jurewica, S. R., and E. B. Watson, 1985, The distribution of partial melt in a granitic system: The application of liquid-phase sintering theory, Geochim. Cosmochim. Acta 49, p. 1109–1121.

Vincenze, E. P., and E. B. Watson, 1988 Crystal-melt wetting characteristics in partially-molten amphibolite, EOS, Trans. Am. Geophys. Union 69, p. 482.

German, R. M., 1985, The contiguity of liquid-phase sintered microstructures, Metall. Trans. 16, p. 1247–1252.

WATSON, E., and J. M. BRENAN, Rensselaer Polytechnic Institute, Troy, NY

Small-Scale Distribution of C-O-H Fluids in Deep-Seated Rocks and Its Implications for Diffusional Mass Transport

Recent experimental studies in our lab have provided fundamental information concerning the geometry of C-O-H fluid distribution on the scale of individual grains in high-grade rocks (900–1200 °C, 5–20 Kbars; see Watson and Brenan, 1987). Two types of distribution are well documented: (1) the fluid may exist as isolated pores at grain corners, or (2) it may form an interconnected network of trigonal prism-like tubules along grain edges. No case has yet been observed in which C-O-H fluid wets all grain boundaries. For fluid volume fractions less than 1-2%, the crossover in behavior from type 1 to type 2 occurs when the dihedral angle (Θ, formed between surfaces of two contacting crystals bounding of a fluid-filled cavity) assumes a value of 60° or lower. The value of Θ for any given mineral/fluid assemblage is determined by the relative magnitudes of the interfacial energies involved, according to the well-known "wetting" relation

$$\gamma_s s / 2\gamma_s f = \cos(\Theta/2)$$

where γ refers to the energy per unit area of the subscripted interface, either solid-solid (ss) or solid-fluid (sf). For the variety of silicates and fluid compositions investigated thus far, Θ ranges between ~ 40 and $\sim 90°$, the lowest values pertaining to aqueous saline fluids in quartzite and the highest values to CO_2 in quartzite, dunite, and "feldsparite." Qualitative values for CO and CH_4 in dunite are also high ($\geq 90°$). Accordingly, on the basis of data available at the time of this writing, it can be concluded that the rocks most likely to have an interconnected fluid phase are quartz-rich, mid- to lower-crustal rocks containing saline H_2O.

The effects of changes in pressure and temperature upon Θ have been investigated in detail only for San Carlos olivine ($\sim Fo_90$; see Brenan and Watson, 1988). In this case, there exists a clear negative dependence of Θ upon temperature (6–7 degrees/°C) for both H_2O and CO_2. With increasing temperature at 10 Kbars, the crossover from $\sim > 60°$ to $\Theta > 60°$ occurs at ~ 1100°C. As pressure increases from 5 to 20 Kbars at 1200°C, Θ_{H_2O} decreases from ~ 65 to $\sim 45°$, with most of the change occurring over the 5–10 Kbars interval. (Note that the apparent dependence of wetting behavior on pressure may actually reflect changing fluid composition, inasmuch as the solubilities of silicates are significantly pressure dependent.) The measured P-T effects on Θ_{H_2O} for dunite raise the possibility that an interconnected fluid exists in the mantle over a limited range in depth.

For all types of rocks, the distinction between interconnected ($\Theta \leq 60°$) and noninterconnected ($\Theta > 60°$) fluid is of paramount importance when considering chemical transport: Isolated, grain-corner pores ($\Theta > 60°$) provide no path for fluid flow, nor do they contribute to diffusional transport on a scale greater than their own dimensions. Three-dimensionally interconnected grain edge channels ($\Theta \leq 60°$), on the other hand, may constitute avenues not only for Darcy flow but also for diffusion of ions dissolved in a static fluid. We have investigated the latter possibility specifically for diffusion of Fe in synthetic dunites containing CO_2-H_2O fluids.

The experimental technique involves sealing coarsely ground (~ 50 μm) San Carlos olivine ($\sim Fo_90$) plus fluid-producing compound (H_2O, $Ag_2C_2O_4$ or oxalic acid dihydrate, in quantities chosen to produce 2–3 vol% fluid at run conditions) in a platinum container, which is then subjected to 10 Kbars (1 GPa) and 900–1200°C for 1–6 days in a piston-cylinder apparatus. As shown by our previous studies, the granular olivine + fluid assumes at least a quasi-equilibrium microstructure within minutes at these conditions.

As most experimental petrologists know, Pt is an excellent sink for Fe, and post experiment electron microprobe analysis shows that all fluid-bearing dunites lose Fe from the region contacting the Pt container. The extent of Fe loss is markedly dependent upon the identity of the fluid, however, being far greater for pure H_2O than for pure CO_2. Detailed energy-dispersive microprobe analysis employing a $\sim 10 \times 500$ μm beam "spot" is used to characterize Fe diffusion gradients in the dunite perpendicular to the container wall, and from these we can extract bulk diffusivities for the fluid-bearing samples. In dunite containing 2–3 vol% pure H_2O, $D_{Fe} \sim 10^{-8}$ cm^2/s at 1200°C, with an activation energy (E_a) of ~ 50 kcal/mol. This latter bulk diffusivity value is similar to that for lattice diffusion of divalent cations in olivine (see Jurewicz and Watson, 1988, and references therein), suggesting that neither the CO_2 fluid nor the olivine grain boundaries contribute significantly to diffusive transport of Fe at these conditions. This conclusion is not unreasonable in view of the isolated nature of CO_2-filled pores. The much higher bulk diffusivity characteristic of H_2O-bearing dunite is readily attributed to the complete interconnectivity of a fluid of this particular composition. Considering the small volume fraction of fluid and its probably modest capacity to dissolve the Fe component at the experimental conditions, the measured bulk-rock diffusion coefficient implies that diffusion in supercritical H_2O is extremely fast.

REFERENCES

Brenan, J. M., and E. B. Watson, 1988, Wetting characteristics of CO_2-H_2O fluids coexisting with olivine as a function of P and T: EOS, Transactions of American Geophysical Union, v. 69, p. 482.

Jurewicz, A. J. G., and E. B. Watson, 1988, Cations in olivine, Part 2: Diffusion in olivine xenocrysts, with applications to petrology and mineral physics: Contributions to Mineralogy and Petrology, v. 99, p. 186–201.

E. B. Watson, and J. M. Brennan, 1987, Fluids in the lithosphere, 1. Experimentally-determined wetting characteristics of CO_2-H_2O fluids and their implications for fluid transport, host-rock physical properties, and fluid inclusion formation: Earth and Planetary Science Letters, v. 85, p. 497–515.

WATTS, A. B., and COAKLEY, B., Geological Observatory of Columbia University, Palisades, New York

Tectonics and Sedimentation: Feedback Processes in Basin Modelling

A number of examples exist in the geological record of the control of tectonics on sedimentation. Backstripping is an attempt to isolate tectonics by correcting stratigraphy for the effects of sedimentary processes. On a basin-wide scale, backstripping fails because it is based on a point-wise isostatic scheme. Previous forward models for basin stratigraphy are based on the assumption of a fixed paleobathymetry through time. Because sedimentation is a dynamic process, a description of the sedimentary environment is required that includes the amount of sediment, angle of repose, grade and their feedback effects with subsidence and uplift. A physically based model provides the best means to both minimize the input parameters and incorporate the physics of the problem. One such model that describes how slopes (or grade) evolve is be diffusion. This paper examines the interaction of sedimentation and tectonism in both foreland and passive margin basins and examples of each will be discussed.

WAYCHUNAS, G., Stanford University, Stanford, CA, W. A. DOLLASE, University of California, Los Angeles, CA, and C. R. ROSS, Universität Bayreuth, Bayreuth, Federal Republic of Germany

Short Range Order (SRO) Determination in Mineralogical Solid Solutions via X-Ray Absorption Spectroscopy

Considerable information about the crystal chemistry, phase equilibria, reaction kinetics, and crystallization history of mineral solid solution can be obtained, in principle, from a complete description of the short range order (SRO) among substituting cation species. For example, a system that exhibits clustering behavior (positive SRO parameters) will have a solvus T-X topology and can show sings of incipient phase separation well before actual nucleation and growth of the equilibrium phases. In this case, not uncommon at low temperatures or in systems with slow kinetics, SRO measurements of the clustering give direct evidence of the presence of the solvus. Characterization of SRO in natural samples of differing compositions formed at various temperatures, or subjected to laboratory annealing, could enable more detailed reconstruction of the shape of the solvus. Measurement of SRO in natural systems can also be compared to experimental, theoretical, or modeling results to indicate the duration of natural anneal or the anneal time-temperature profile.

Ordering systems (negative SRO parameters) can be studied analogously via SRO measurements. Phase transformation boundaries can be mapped out by detection of incipient ordering. Laboratory annealing of natural samples, which alters the SRO, can yield information on the kinetics of the ordering process and give insight into its atomistic mechanism.

A common type of mineral solid solutions exhibiting SRO behavior is those with charge coupled substitutions (CCS). In CCS a set of *n* ions are replaced by *n* differing valence ions or vacancies with the constraint that total valence be conserved. The substitutions may involve identical structural sites, or couple two or more distinct structural sites. For example, in rutile the CCS $2Ta^{5+} + Fe^{2+} = 3\ Ti^{4+}$ occurs, involving only one type of site. In pyroxenes, the M1-M2 substitution of Al-Na for Mg-Ca involves two separate sites. Al-Ca for Si-Na substitutions are also possible, where tetrahedral and octahedral sites are involved. Systems with CCS can demonstrate both clustering and ordering behavior and can have complex phase relations, such as in the omphacites. They also bring up many questions basic to mineral crystal chemistry, such as: Is CCS maintained locally or only statistically over long ranges? What factors determine this range? How does a structure adjust to CCS, and can the mechanism of adjustment give insight into the solid solution limits or shape of solvi and phase boundaries? How are ordered vacancies compensated structurally? SRO measurements can answer these questions.

For most mineral systems, the quantitative measurement of SRO is difficult or impossible. SRO is classically determined via measurement of the diffuse X-ray scattering associated with atom substitutions of varying atomic number. Both the differences in Z and bond length create diffuse intensity, and these effects must be separately handled in order to accurately extract the SRO. More important, the intensity of the diffuse scattering depends linearly on $(Z_1\text{-}Z_2)^2$ so that the atomic number difference must be large (typically 30 or more) for practical experimentation. This restricts application to mineral solutions. Even worse, if the SRO species are dilute, as for small quantitites of acmite dissolved in hedenbergite, the diffuse intensity is correspondingly weakened.

Non-X-ray methods of SRO measurement are generally qualitative as well as limited in applicability. Mossbauer studies yield information on the clustering and ordering of ^{57}Fe, but few studies yield unambiguous results. NMR measurements can sense the identity of next nearest and more distant neighbors about a probe ion, but are also limited in application. Other types of spectroscopy offer only indirect measurements of SRO. An exception is synchrotron-based EXAFS spectroscopy, which allows direct extraction of pair correlations about literally any probe ion, thus yielding the SRO information explicitly.

We have been pursuing a set of studies utilizing EXAFS analysis along with atomistic computer modeling (distance least squares or DLA) and statistical thermodynamic calculations (cluster variation method or CVM) in order to gain insight into CCS and other SRO problems in mineral solid solutions.

Our work began with studies of the model systems MgO-FeO and MgO-$LiFeO_2$ (Waychunas et al, 1986). The former demonstrates the resolution of EXAFS SRO analysis in an ideal system. Clustering of less than one next nearest neighbor atom out of the 12 in a full second shell is detectable, and the system behaves randomly at the synthesis temperature of 1140 °C. Atomistic modeling of the solution is consistent with such behavior and also demonstrates how the solution adjusts to cation size disparity. The MgO-$LiFeO_2$ CCS system is much more complicated. Both modeling and EXAFS analysis reveal a tendency of Fe^{3+} to be surrounded by excess Li^+, even at significant MgO compositions. This ordering results from a strong tendency to maintain local charge balance, forcing localization of the CCS.

CCS in systems with incomplete solid solution are commonplace. An example is the replacement of Ge^{4+} for Fe^{3+} in hematite and goethite. The Ge^{4+} is charge compensated by Fe^{2+} replacing Fe^{3+}, or by deprotonation of a hydroxyl group, respectively. Ge EXAFS reveals the substitutional site unambiguously, and the Ge second neighbor distance indicates that the CCS occurs adjacent to the Ge (Bernstein and Waychunas, 1987). Another example is the substitution of Ti^{4+} for lower valence species in silicates. While, as yet, we do not have confirming evidence for localized CCS in Ti-bearing silicates, the EXAFS and XANES information obtained to date is consistent with closely coupled charge compensating ions (Waychunas, 1987).

Our most recent work has focused on the Di-Hd, Ac-Di, and Hd-Jd pyroxene solutions, which have been studied with synthetics produced at 800–900 °C. Most of our results are consistent with mainly random site occupations in these solutions. The Di-Hd system appears to show some clustering behavior at low Hd component. We intend to collect additional data on these pyroxene systems using natural single crystals and polarized EXAFS methods. Such techniques will allow us to separate the SRO effects as a function of direction, and hence unravel SRO between M1-M1 and Mi-M2 sites.

We are also beginning studies of clustering behavior in TiO_2-SnO_2 solutions, ordering behavior in substituted rutile and perovskite solutions, ordering and clustering behavior in several carbonate systems, and disordering kinetics in materials with ordered vacancies. Besides developing a powerful and unique SRO structure probe, our EXAFS and modeling work also improves our understanding of basic crystal chemical tendencies and paves the way for predictions on new or experimentally inaccessible systems.

REFERENCES

Bernstein, L. R., and G. A. Waychunas, 1987, Geochimica et Cosmochimica Acta, v. 51, p. 623–630.

Waychunas, G. A., 1987, American Mineralogist, v. 72, p. 89–101.

———— W. A. Dollase, and C. R. Ross, 1986, Journal de Physique Collonque, v. C8, n. 47, p. 845–848.

WEBB, F., JR., Appalachian State University, Boone, North Carolina, and SHUHE, H., Northeast University of Technology, Shenyang, China

Basin Analysis Based on Slump-Folded Proterozoic Turbidites in Northeastern People's Republic of China

The ~850 Ma Nanfen Formation exposed at its type area in Liaoning Province, People's Republic of China, consists of from 300 to 900 m of shales, marls, breccias, carbonate-clast diamictites, and thinly laminated lime mudstones. The lower half of the formation is composed mostly of laminated lime mudstones and marl, with lesser amounts of thinly interbedded grayish-red calcareous shale, and minor amounts of nodular chert. Carbonate content decreases in the upper half of the formation, whereas the quantity and grain size of the fine-grained terrigenous content increase. Although beds in the highest and lowest portions of the formation are predominantly light greenish-gray and grayish olive-green, respectively, the middle third of the sequence is grayish red.

Exposures of the formation are numerous, but it is difficult to measure a complete, unbroken stratigraphic section at any one site, as the area is cut by many high-angle normal faults. The formation is only gently folded and has not been metamorphosed.

In at least five localities, there are individual napp-like slump folds that are traceable parallel to bedding for distances of up to 70 m and involve up to 5 m-thick stratigraphic intervals. Although the slump-folded sections are most common in the middle, calcareous, shale-dominated portion of the formation, the basal unit of lime mudstones also contains prominent folds. The cores and lower limbs of all of the slump folds contain diamictites and breccias composed mostly of lime mudstone and marl fragments. Although imbricated pale gray lime mudstone clasts in grayish-red shale matrix are the most common, at a few localities the breccias consist of pale greenish-gray chert clasts in pale gray lime mudstone matrix. At all localities, the limestone and chert clasts commonly show soft-sediment folding. Individual soft-sediment faults that parallel bedding are traceable downslope where they merge into isoclinal recumbent floods characterized by internal brecciation and chaotic soft-sediment folding. Splay faults that cut across bedding and bound smaller folds diverge upward from the dominant bedding plane faults. Analyses of geometry and direction of transport of slump folds, fabric of associated breccias, and displacement sense for soft-sediment faults at eight localities indicate that the Nanfen paleoslope dipped toward the west.

The association and abundance of extremely thin laminae (commonly >40/cm), the predominance of terrigenous and carbonate grain sizes ranging from 3 to 10 μm, and dominance of fining-upward beds indicate deposition of hemipelagic sediment below storm wave base. Both the overlying Qiaotou and underlying Diaoyutai formations contain coarse-grained quartz arenites characterized by cross-bedded ripple and dune bedforms deposited above fair-weather storm base. In addition, some of the Diaoyutai beds contain glauconite. These features indicate a depositional environment quite similar to modern shallow-water marine settings.

The formational contact between the bottom of the Nanfen Formation and the top of the underlying Diaoyutai Formation is gradational although the gradation occurs within a stratigraphic thickness of less than 10 m. Initiation of Nanfen deposition occurred in response to a relatively abrupt deepening of the water accompanied by accumulation of laminated lime muds and minor amounts of clay. As the water deepened in response to fault-induced subsidence of the basin, the volume of carbonate sediment decreased, while increasing amounts of clay sediment were deposited as hemipelagics and distal turbidites. Submarine slumps, mud flows, and grain flows were initiated by failure of unconsolidated fine-grained sediments on the oversteepened west-facing submarine slope. Slope instability probably resulted from the relatively thick buildup of laminated clay and carbonates, but continued faulting along the basin margins may have created earthquakes that triggered submarine transport of sediment. Near the end of deposition of the Nanfen Formation, clay, silt, and very fine-grained sands were deposited typically as Bouma T_{cde} sequences that are characteristic of the D lithofacies of Mutti-Ricci Luchi (1972). Euxinic deposits, consisting of several meters of black shale, occur at the top of the Nanfen Formation and mark the end of subsidence of the sea floor and/or uplift of the slope. Deposition of the overlying Qiaotou Formation began when shallow-water marine conditions were reestablished as sea level dropped and sands from a rising source area prograded into the basin. Black shale pebble rip-ups and dark shale partings with polygonal cracks in the basal Qiaotou Formation demonstrate that it disconformable overlies the Nanfen Formation.

Our interpretation is that deposition of the Nanfen Formation took place in a relatively short-lived, marine basin that was probably produced by normal faulting of the shelf. Faulting occurred in an extensional tectonic setting on the continental margin of the North China plate. Our study of the Nanfen Formation has not yet extended into the western part of the basin.

WEBB, J., and T. BERNECKER, La Trobe University, Bundoora, Australia, B. G. FORDHAM, Queensland Department of Mines, Brisbane, Australia

Early Silurian to Early Devonian Limestones and Cherts of Chillagoe Formation, Northeast Australia

The northernmost sedimentary province of the Tasman Orogenic Belt in eastern Australia is the Hodgkinson Basin; the Chillagoe Formation outcrops for 300 km along the western margin of this basin. The formation consists predominantly of limestones, which outcrop prominently as bluffs up to 100 m high, as well as minor cherts, clastic sediments, and basalt. To the east the Chillagoe Formation is faulted against the Hodgkinson Formation, a very thick sequence of siliciclastic turbidites with occasional limestone clasts.

Reconnaissance sampling for conodonts has revealed that the limestones range in age from late Llandovery to early Emsian, and this time span appears to represent a continuous period of carbonate accumulation. No conodont zones are absent, and there are no obvious breaks in the sequence, except for one in the Early Devonian. Thus, the Chillagoe limestones represent the longest sustained period of carbonate sedimentation known from Australia, and this must reflect the tectonic stability of the area over at least 30 m.y. in the mid-Paleozoic. The conodonts from the limestone clasts in the Hodgkinson Formation are Early to Late Devonian in age, i.e., generally younger than the Chillagoe Formation.

The conodont biostratigraphic studies also showed that the Chillagoe Formation is cut by numerous faults running parallel to bedding; these have caused extensive repetition of the sequence. The strata are all more or less vertical and consistently young westward, so the outcrop thickness of the formation (about 4 km) was previously regarded as the true thickness of the unit. Recognition of the faulting has reduced this estimate to less than 1 km, despite the long time range involved.

The extensive faulting may also have been responsible for the very widespread brecciation in the limestones. The cavities in between the breccia fragments are sometimes filled with subhorizontal laminated sediment, indicating that the brecciation postdated deposition of the limestones, which have uniformly subvertical bedding. The laminated infill and the surrounding breccia have been contact metamorphosed by Permo-Carboniferous intrusives in the area, so the brecciation event must have occurred in the Late Devonian or Carboniferous. This period of time corresponds to the major thrusting event that forced the limestone beds into their present subvertical position and split the sequence along the numerous bedding plane faults.

The environment of deposition of the Chillagoe Formation has been the subject of some dispute. Recent studies have interpreted the formation as a shallow-water carbonate shelf that collapsed into deeper water; the limestone bluffs were believed to be allochthonous blocks of shallow-water limestone surrounded by deep-water sediments, mainly thin-bedded limestone and chert. However, detailed mapping has shown that continuous shallow-water marker horizons can be traced along strike for over 4 km, so that the limestone bluffs are portions of continuous beds and are not isolated blocks with the stratigraphy. The whole sequence displays abundant evidence for shallow-water sedimentation, e.g., algae are very common. Thus, the Chillagoe Formation in the area mapped in detail is an intact shallow-water limestone; there is still the possibility that allochthonous blocks are present elsewhere.

The limestones can be divided into several facies: peloidal grainstone (often bioclastic), bioclastic wackestone/packstone (the dominant fossils include megalodont bivalves, brachiopods, corals, and stromatoporoids), crinoidal grainstone (often with large stem fragments and calices), boundstone (tabulate coral bindstone and stromatoporoid framestone), thin-bedded mudstone, and marl (containing sponges and bryozoans). The marl facies is relatively rare. In addition, stromatactis occurs at several localities, indicating the presence of mud mounds.

Apart from a thin coral bindstone growing directly on a basalt flow, there is no evidence for the presence of reef or slope deposits, and the overall environment of deposition is best interpreted as a shallow-water gradually sloping shelf or ramp. Scattered mud mounds occurred in a fairly quiet, restricted environment containing a mosaic of different biological communities, often dominated by particular faunal groups, e.g., *Amphipora,* megalodont bivalves, solitary rugose corals, pentamerid bivalves. Periodic inputs of higher energy (?storms) are indicated by the bioclastic lags throughout the sequence.

Intertidal carbonates are presented by very uncommon occurrences of cyanobacterial mats; the rarity of this facies may merely reflect its lack of exposure. Deep-water clay-rich carbonates are also virtually absent, perhaps for the same reason.

At least two separate chert horizons are present in the Chillagoe Formation. The lower one is thin-bedded and radiolarian rich and may represent siliceous turbidites. The upper chert is finely laminated and consists predominantly of sponge spicules. Whether it represents a deep-water sediment or accumulated in restricted areas on the shelf has yet to be ascertained.

At the top of the limestone sequence is a prominent unit of clastics, consisting of conglomerates and sandstones, often showing graded bedding. These sediments are composed of clasts of limestone (ranging in age from Early Silurian to Early Devonian), together with chert fragments and grains of quartz, feldspar, and muscovite, probably derived from the high-grade metamorphics to the west. Broad channels up to 250 m deep have been eroded into the underlying limestone and filled with these sediments, which are interbedded with coarse crinoidal grainstones of Early Devonian age.

This upper unit reflects the final phase of deposition on the shelf, with nearby uplift causing erosion of some areas of limestone. Subsequently, the whole limestone sequence may have been downfaulted and buried by the siliciclastic turbidites of the Hodgkinson Formation. Deposition in the basin ceased altogether in the Late Devonian–Early Carboniferous, when the major thrusting episode occurred.

WEBB, J., La Trobe University, Bundoora, Australia, and B. L. FINLAYSON, Melbourne University, Parkville, Australia

Use of Karst Geomorphology to Infer Uplift Histories: Example from Buchan, Southeastern Australia

The densely cavernous karst barre at Buchan in eastern Victoria, southeastern Australia, contains considerable evidence relating to the post-Eocene history of the landscape in terms of preserved surface features (particularly river channel gravels and terraces), as well as levels of cave development and datable speleothems. These can be used to reconstruct the geomorphological evolution of the area in considerable detail.

In the Eocene, a low-gradient river valley 3–4 km wide, with at least one major tributary, ran more or less north-south through the area. This valley had formed since the initial uplift of the southeastern highlands in the Late Cretaceous, and its width and low gradient reflect the post-uplift stability of the area. Since the late Eocene no rivers in the region have been able to erode valleys of similar size.

Where this ancestral river and its tributary flowed over limestone bedrock, phreatic cave systems developed along vertical joints below the water table. Areas of relatively pure limestone became densely cavernous as a result.

Between 38 and 42 Ma ago (late Eocene), basalt lavas from at least two separate eruption points flowed down the valleys of the ancestral river and its major tributary. The remnants of both the flows and the river channel gravels allowed the above reconstruction of the late Eocene drainage system.

Twin lateral streams developed either side of the basalt flows filling the valleys, effectively doubling the number of rivers in the area. Uplift concomitant with the eruptions resulted in extensive stream diversions and valley incision, draining the preexisting phreatic caves. Part of the ancestral river valley was abandoned, probably as a result of a small domal uplift in the south of the area, and is largely intact today.

Following this phase of downcutting, which resulted in a fall in river level of around 100 m, was a period of stillstand, and a moderately broad river valley developed (1–2 km wide). A second series of phreatic cave systems formed along vertical joints in the limestone beneath the river and its tributaries. These caves are very similar in morphology to those formed under the late Eocene rivers, in that they are typically joint-controlled mazes of narrow passages with high ceilings. Roof and wall pendants, large wall scallops, and occasional spongework all attest to a nothephreatic mode of origin. The period of stillstand that formed these caves may correspond to the maximum onlap of the sea in the Gippsland Basin to the south during the Middle Miocene.

More recently, there has been a further period of downcutting (approximately 80 m at Buchan), which drained the phreatic caves. Another phase of stillstand followed, accompanied by the development of river terraces and the formation of horizontal epiphreatic cave systems, largely along the strike of the limestone. These caves are frequently elliptical in cross-section, with flat roofs, and formed at the water table. The systems are often quite extensive, being up to several kilometers long, and all have developed beneath the valleys of tributaries flowing into the main river in the area (these valleys are now dry except after very heavy rain).

The formation of epiphreatic caves at this time in the geomorphological evolution of the area appears to be the result of a combination of favorable circumstances: relatively high-gradient surface drainage systems developed on limestone areas containing abundant preexisting (phreatic) cave systems. Development of an epiphreatic cave beneath a surface stream will allow the entire surface flow to be captured and diverted underground once the water table falls below the level of the surface stream.

Both phreatic and epiphreatic caves at Buchan contain abundant speleothems, and several from the epiphreatic caves have been dated by the uranium series method (those in the phreatic caves are beyond the range of this technique). These dates indicate that the epiphreatic caves were drained by a final minor episode of downcutting at least 99,000 years ago (and perhaps over 350,000 years ago).

The detailed picture of landscape evolution presented herein enables a number of general conclusions to be drawn. First, cave development in the area dates back at least 40 m.y., and it is possible that other karst landscapes in Australia are even older. Second, uplift in the southeastern highlands of Australia since the Eocene has been episodic and often localized rather than gradual and continuous, suggesting that it may be difficult to correlate phases of uplift across the highlands as a whole. In addition, there appears to be good evidence that volcanism can be associated with uplift.

WEBB, P., Ohio State University, Columbus, OH, P. J. BARRETT, Victoria University of Wellington, Wellington, New Zealand, K. BIRKENMAJER, Geological Institue, Krakow, Poland, A. K. COOPER, U. S. Geological Survey, Menlo Park, CA, A. ELVERHOI, Norwegian Polar Research Institute, Oslo, Norway, and D. FUTTERER, Alfred Wegener Institute, Bremerhaven, Federal Republic of Germany

Cenozoic Sedimentary Basins of Antarctica

Although Cenozoic investigations were initiated in Antarctica at the turn of the century, comprehensive research efforts commenced less than 30 years ago. Cenozoic outcrops of sedimentary rocks are scarce since most of the continent is enveloped by the east and west Antarctic ice sheets. Cenozoic outcrops are present in the James Ross Basin area (Seymour Island, Cockburn Island) and the South Shetland Islands (King George Island) on the Atlantic and Pacific Ocean sides of the Antarctic Peninsula, respectively. Limited outcrops are present in the Transantarctic Mountains, west of the Ross Sea region. Most Cenozoic sedimentary basin data are derived from the more than 40 stratigraphic drillholes completed in rift basins close to the continent, in glaciated valley fjord basins within major mountain systems, and at numerous abyssal sites around Antarctica.

The distribution and tectonic settings of the major Cenozoic basins have been established at a reconnaissance level, using a combination of geological and geophysical data. Lateral and vertical crustal displacements controlled the location of basin margins, connections between basins migration of sedimentary depocenters, terrestrial paleotopography and sediment sources, ice nucleation centers and reservoirs, transgressive and regressive sedimentary patterns, terrestrial and marine biotic migration routes, and marine connections to the Southern Ocean.

Early Cenozoic basin histories were strongly influenced by structural fabrics inherited from Jurassic and Cretaceous break-up and dispersal of Gondwana. Crustal displacements continued throughout the early and late Cenozoic. The most obvious crustal events having an important influence on Cenozoic basin development include: separation of Australia, New Zealand, and Antarctica; possible extension within the Wilkes Basin, extension and rifting in the Ross Sea; transcurrent motion and extension in West Antarctica, and up to 5 km of uplift in the Transantarctic Mountains. Cenozoic basins have, therefore, been characterized in terms of major tectonic settings such as fore-arc and back-arc basins, intracratonic basins, rift basins, epicratonic basins or embayments, and failed rifts or aulocogens. To date, basin studies have emphasized relatively shallow-water marine shelfal environments close to the continent and peri-Antarctic abyssal environments. The Victoria Land basin, in the western Ross Sea, is probably the most thoroughly understood at this time. Shelf sediments are strongly differentiated into packets of contrasting clastic sediments, often separated by major and widespread erosional contacts. Lithofacies are dominated by breccias, conglomerates, diamictites, sandstones, siltstones, and biogenic claystones. Multiple sedimentary cycles, representing marine transgressions and regressions, are apparent in many shelf successions and result from regional tectonics (uplift/subsidence), sea-level fluctuation (probably related to terrestral glacial/interglacial events), or some combination of the two. Glaciation in the marine basins of the Ross Sea dates from the Early Oligocene. Abyssal sediments include claystones, diatomaceous and calcareous nannoplanktonic oozes, cherts, and chalks. Polar abyssal basins first show evidence for glaciation in the Neogene. A composite section comprising Paleocene–Eocene successions of the inner Weddell Embayment and Oliogocene–Recent successions of the Ross Sea basins provide the most complete stratigraphic coverage for the Cenozoic of Antarctica at this time. It is estimated that less than 50% of all Cenozoic time is represented by the sedimentary record of these two relativity well studied areas. More complete temporal representation is anticipated when data become available from drillhole successions in deeper parts of the Weddell and Ross Sea basins.

At different times during the Cenozoic, the Antarctic continent was enveloped by alpine ice, regional ice caps, and pan-Antarctic ice sheets. During repeated regional and continent-wide ice recession phases, marine waters penetrated deep into and even across parts of the Antarctic continent. The effects of transcontinental marine circulation have not as yet been integrated into hemispheric and global climate and ocean circulation, sea-level fluctuation, and sedimentary basin cyclicity-episodicity trends and models.

WEBER, C., Bureau de Recherches Géologiques et Minières, Orléans, France

Borehole on Paris Basin Magnetic Anomaly: Geophysical Methods Put to the Test

Although the Sancerre-Couy borehole has not solved the riddle of the origin of the Paris Basin magnetic anomaly, it has at least led to the discovery of a steeply dipping, mafic metamorphic series that probably represents the trace of a large-scale Variscan thrust-suture.

It has also been a laboratory for the study of the layered crust by direct comparisons between the cores recovered from the hole (continuous oriented coring to 1800 m and 10% coring from there to the bottom of the hole) and the various geophysical measurements made in the hole and on surface. From this point of view, the interpretation of the magnetic and seismic data appear to be representative of the problems encountered in the study of basements.

The large magnetic anomaly has been known since 1885. Airborne surveys carried out in the sixties showed that the anomaly approached closest to the surface in the southern part of the basin, where most two-dimensional interpretations place the top of the magnetic body at a depth of 5–7 km. New techniques of three-dimensional modeling (linear programming, inverse stochastics) (Bayer et al, 1987) gave a shallower origin, between 3 and 4 km, for an average magnetization of 1–2 A/m, assuming the base of the magnetic body to be limited at about 8 km by a strong seismic discontinuity. The Sancerre-Couy borehole was sited on the basis of this hypothesis. It has provided important data that now place precise constraints on modeling: The magnetic body is at a depth greater than 3500 m, average magnetization of the metamorphic basement is of the order of 0.2 A/m, the vertical gradient of the magnetic field has been measured between 0 and 3500 m, and it is now known that identification of seismic reflectors is ambiguous.

The new three-dimensional model drawn up on the basis of the information yielded by the borehole is appreciably different from the previous model (Bayer et al, 1988). It shows a first magnetic level between 4 and 5 km deep, with magnetization between 0.5 and 1 A/m, and a second, more intensely magnetized level from 7 km depth. We are now approaching the limits of capability of inversion of magnetic field data.

Interpretation of the seismic reflection data acquired during drilling of the borehole has also raised problems, due to the fact that the foliation in the metamorphic rocks intersected is dipping 80°–85° to the southeast. In particular, a first interpretation of vertical seismic profiles (VSP) and offset profiles (OSP) by conventional methods gave horizontal reflectors that it was impossible to correlate with those detected by the surface seismic profile passing through the borehole.

A new method of processing the data was then used (O'Rourke, 1988). A migration was applied to the VSP-OSP along the velocity field given by the sonic log, using a nonhorizontal model, without, however, "forcing" the data in order to reveal steep dips.

The surface seismic section was also processed by integrating the VSP data so as to assist the selection of velocities and to provide a deconvolution operator. A migration in cascade, with a second-order algorithm, in finite difference at 45°, facilitated demonstration of the highest dips.

The results of the reprocessing are in good agreement with the borehole data and bring out new information enabling better assessment of the structural setting of the body at the source of the Paris Basin magnetic anomaly. The main reflectors visible on the VSP correspond well with those of the seismogram compiled by calculation from the log of velocity in the well. The OSP with eastward offset shows steep dips. In contrast, that with northward offset, close to the strike of the foliation, shows no important dip.

Finally, the reprocessing makes the interpretation of the surface seismic profile credible. The horizontal continuity of a certain number of reflectors nevertheless poses problems. The overall picture of the steeply dipping series in the Sancerre-Couy borehole is that of a fault zone dipping at about 80° to the east. Two sets of deep reflectors, one at about 5 km and the other at about 7 km, could be interpreted as due to major thrust.

The experience acquired in the interpretation of seismic reflection data around the Sancerre-Couy borehole show the inadequacy of standard processing in a metamorphic basement context. However, specific processing, guided by the data themselves and not constraining the results by an a priori fixed model, give a more satisfactory approach to the geological reality.

REFERENCES

Bayer, R., A. Guillen, and C. Weber, 1987, The recognition of magnetic anomaly of the Paris Basin (area of Sancerre) by deep drilling: gravity and magnetic data inversion: CR. Acad. Sci. Fr, t. 305, II, p. 203–208.

———, and D. Rousset, 1988, Réactualisation de l'interprétation de l'AMBP sur la région de Sancerre-Couy: Doc. BRGM 138, France, p. 167–178.

O'Rourke, T. J., 1988, Reprocessing of borehole and surface seismic well Couy 1b-line Cher 34: Doc. BRGM 138, France, p. 245–258.

WEBER, K., Institute of Geology and Dynamics of the Lithosphere, Göttingen, Federal Republic of Germany

History of Accretionary Processes in Variscan Orogen of Central and Western Europe

The Variscan basement of central and western Europe is composed of a number of composite terranes that are arranged from north to south in three large zones: the Rhenohercynian, the Saxothuringian, and the Moldanubian. The boundaries between them are late Hercynian (Upper Carboniferous) major collision zones, which are characterized by extensive northwest-facing crustal imbrication and overthrusting of more than 100 km.

The Rhenohercynian basin developed during the Devonian from a stretched passive continental margin, composed of basement rocks of Cadomian and Caledonian age. Subduction took place during the Devonian and the lowermost Carboniferous. Collision began during the Visean.

The Saxothuringian is more heterogeneous in structure and composition than the Rhenohercynian. Its eastern part is represented by the Lugicum, its central part by the Mid-German Crystalline Rise, and its western part by the Amorican Massif. During the Devonian, an accretionary wedge developed along the northern boundaries of this terrane formed by the Lizzart of southwest England and the Northern Phyllite Zone of Central Europe. Its southern boundary in central Europe represents a passive continental margin.

The Moldanubian comprises large parts of the Bohemian Massif, the Black Forest, the Vosges Mountains and the Massif Central. It is composed of different terranes that are particularly well outlined in case of the Bohemicum and the Brno–Vistulicum of the Bohemian Massif.

The Variscan metamorphism is pressure dominated during the subduction stage (ca. 450 Ma to 360 Ma) and of the HT/LP type during the climax of the collisional stage (ca. 320 Ma). The Variscan crustal structure of central Europe evolved from south to north during multiphase deformations in which two main events can be distinguished. The older events are characterized by northwest-directed tectonic transports in the Rhenohercynian, Saxothuringian, and the northern Moldanubian and southeast-to east-directed transports in the southern part of the Moldanubian. This is also revealed by the seismic crustal structure. Backfolding and backthrusting is widespread and seems to be restricted to middle and upper crustal levels.

The younger main event developed during the advanced collisional stage particularly in the Moldanubian and Saxothuringian. It produced indentation structures and wrench fault systems accompanied by a reorientation of local and regional strain fields around several terranes.

Variscan crustal convergence was probably much stronger than previously assumed. Initial oceanic subduction zones have been completely overridden by continental crust. Thereby, the complexly deformed cryptic sutures seen at the present outcrop level of the crystalline basement have lost their contacts with the previous subduction zones. During collision, terranes previously separated by oceanic crust were thrust over one another, so that suture zones widely separated before collision are now closely neighboring or overridden by other terranes. The latter is presumably the case, e.g., at the northwestern margin of the Bohemian Massif, where the Bohemian terrane overthrust the Saxothuringian–Moldanubian suture zone.

WEEKS, E., U. S. Geological Survey, Lakewood, CO

Convective Air Flow through Fractured Rock in Hilly Terrain

Convective air flow due to temperature and humidity-induced density effects has been observed in open boreholes and in open vertical fractures at sites of hilly terrain in Idaho, Nevada, and Arizona. This convective air flow is explained as follows. In winter, the column of air extending from a hillside fractured-rock outcrop to the height of the hilltop is colder, drier, and hence more dense than the air column extending from the outcrop through the fractured rock and open conduit to the hillcrest. This density contrast induces air flow into the outcrop and out the conduit. During summer, the density drive is reversed, with hot air entering the conduit, cooling, and draining from the hillside outcrop. In addition to density-induced circulation, flow occurs in conduits due to changes in atmospheric pressure. This pressure-change induced circulation occurs because the gas pressure in the fractured rock mass responds more slowly to atmospheric pressure change than does the gas column in the conduit. Hence, if atmospheric pressure is declining, a pressure gradient is induced from the rock to the conduit that causes gas exhaust from the conduit, and conversely, increasing atmospheric pressure induces air flow into the conduit.

Convective air flow observed in two open boreholes (wells UZ6 and UZ6S) located near the crest of Yucca Mountain, Nevada, is of particular interest because the site has been proposed for a high-level radioactive waste repository. Topographically affected air circulation occurring in the absence of boreholes could reduce deep percolation into the proposed repository by providing a mechanism for the net discharge of water vapor. Alternatively, the potential for gaseous radionuclide release from the proposed repository might be increased, as convective air flow represents a potentially more rapid transport mechanism than gaseous diffusion alone. Consequently, air circulation in the wells is being studied to determine its effect on undisturbed unsaturated zone conditions and to obtain data needed to estimate natural density-driven rock gas circulation.

Yucca Mountain forms a north–south-trending ridge of tuffaceous rock with a steep western scarp rising about 250 m above the Solitario Canyon floor. At the well sites, the mountain is capped by a 125 m thick unit of fractured, highly permeable welded tuff that dips about 6° to the east. The east flank of the mountain is a highly dissected dip slope merging into an alluvial flat. A 30 m thick unit of nonwelded and bedded tuff of relatively small permeability separates the caprock from an underlying 300 m thick sequence of highly fractured permeable welded tuffs.

Well UZ6S is open from a depth of 1 m to 130 m and penetrates the caprock and some nonwelded tuff. Well UZ6 is open from a depth of 100 m to 525 m and penetrates both welded tuff units outcropping at Yucca Mountain. Air-flow, temperature, relative-humidity, and air-chemistry data are being collected in the wells. Simultaneous records of atmospheric pressure, temperature, and relative humidity are being obtained nearby. These data indicate that net gas transport occurs from the hillside outcrops to the wells. Quantitative estimates of air flow are available only from well UZ6S; net annual discharge of rock gas is about $10^4 m^3$.

The net discharge in well UZ6S arises in part because the discharging rock gas is nearly saturated with water vapor, and hence its virtual temperature (the temperature at which the wet air would have the same density as dry air) is about 2.5 °C greater, on an annual basis, than that of the atmosphere. In addition, spectral analysis indicates that temperature-induced atmospheric pressure fluctuations reinforce the temperature-induced density effects to enhance net air exhaust from the wells.

Convective air flow in well UZ6S results in the net annual discharge of about 10^4kg of water as vapor and about 600kg of CO_2. These large net discharges of water and CO_2 have undoubtedly affected moisture distribution, CO_2 distribution, and rock-moisture and rock-gas isotope chemistry in the vicinity of the wellbore, and it may be many years after the hole is instrumented and stemmed before measurements will represent undisturbed conditions.

Topographically affected air flow may be significant at other sites. Open fractures that discharge sufficient moisture to maintain moss growth around their outcrop perimeter during winter have been observed on the flank of Devil's Hole Ridge and on the crest of Point of Rocks Ridge, two ridges of carbonate rock located about 50 km southeast of Yucca Mountain. The natural circulation at these sites may be large enough to substantially affect the rock-moisture balance in the ridges. Topographically affected circulation may enhance radon discharge near hillcrests. Radon concentrations in house basements located near hilltops tend to be greater than those in houses located in valleys. Topographically affected rock-gas circulation may also prove significant in other situations as its occurrence becomes more widely recognized.

WEERASOORIYA, S. V. R., K. B. P. N. JINADASA, and C. B. DISSANAYAKE, Institute of Fundamental Studies, Kandy, Sri Lanka

Decontamination of Fluoride from Community Water Supplies

Many water supply schemes, particularly in developing countries such as Sri Lanka, where groundwater forms the major source of drinking water, have excess levels of fluoride normally detrimental to dental health. A simple and rapid method of defluoridating this water is required. It was observed that serpentinite forms a suitable starting material in the defluoridation process. Serpentine as a group name denotes a class of minerals having a layer structure similar to that of kaolinite but with a di-octahedral Al layer. The major varieties of serpentine include chrysolite, lizardite, and antigorite. When serpentine particles occur in such proportions as to make up a major portion of the entire rock mass, it is termed serpentinite. However, the identification of individual serpentine group minerals in such rock is often difficult.

Previous studies have shown that the efficiency of serpentinite in removing excess fluoride from fluoride-rich water is limited. Serpentinite tends to be deactivated with repeated use.

In this paper, a chemical treatment is described for the enhancement of fluoride uptake by serpentinite in an aqueous medium. The powdered raw material is acidulated with concentrated Hcl to a stoichiometric ratio for generating protonated serpentinite (SH). This product is hydrolized with a dilute base/distilled water, to yield SO^+H_2. In principle, there are two different ways for fluorides to be adsorbed by serpentinite. One is chemisorption through an ion exchange at the solid surface, and the other is by forming inorganic precipitates.

Investigations in this laboratory indicate that the uptake of fluoride by chemically treated serpentinite does not depend on temperature. Therefore, any inorganic precipitation of fluoride in the present study is not a deciding factor in the equilibration between the liquid and solid phase. Ion exchange reactions are fast, and as such will not be affected by temperature. Therefore, it is suggested that the uptake of fluoride by chemically treated serpentinite is an adsorption process involving anion exchange. A surface complex formation model is suggested to describe the absorption of fluoride by the solid substrate.

According to this model, reaction between fluoride and serpentinite is thought to occur via the formation of solid fluoride complexes (SF). The reaction can take place in two independent steps that are determined by the pH of the solution. One involves the surface hydroxyl groups with the release of OH^-, and the other involves the water molecules on the solid SO^+H_2 without the release of OH^- or H^+.

Therefore, the exchange sites of chemically treated serpentinite show amphoretic behavior for the variation of solution pH. The sites saturated with fluoride ions can be reactivated with the treatment of NaOH, followed by HCI.

WEI Jiayong, Regional Geological Survey Team of Guizhou, Bagongli, Guiyanag, China

Sauk Sequence in South China

The Sauk sequence is extensively distributed on the Yangtze platform in South China, and particularly well developed and displayed in upper Yangtze region. It is divided by a remarkable unconformity (or disconformity) from the underlying Presinian (lower Late Proterozoic) rocks. The basal part is marked by extensive glaciation deposits and a set of interglacial manganese-containing rocks accumulated during the time span of ca. 760–700 Ma. Upward, these were followed by phosphatic rocks and platform-wide shallow-water dolomites deposited during the time span of ca. 700–580 Ma. These strata yield abundant microphytolites, algae, acritarchs, and small shelly fossils and trace fossils and provide important manganese ores and most of the phosphorite reserves in China. And the dolomites are locally reservoirs for gas and hosts for mercury, lead, and zinc ores, or intercalated with halites. The middle part is characterized by siliciclastic rocks and platform-wide carbonates containing abundant trilobites, deposited largely during the Cambrian Period. The carbonates are locally hosts for mercury, lead, and zinc ores, or locally interlayered with gypsum and anhydrite. The upper part consists of carbonates (including bioclastic limestones, widespread biostromes, and buildups) and clastic rocks. The later are repeated in shale–sandshale–thick-bedded sandstone order demonstrating a series of shoaling-upward cycles at the top part. These strata form a conspicuous regressive succession. Its upper limit coincides with the top of graptolite Biozone Glyptograptus austrodentatus corresponding to the Arenigian–Llanvirnian boundary.

The sequence passes upward into the overlying Llanvirnian–lower Llandovery limestones-black graptolitic shales without obvious stratigraphic discontinuity. The limestones generally contain chamosite and glauconite, and the black graptolitic shales are distributed world-wide. These suggest that the rocks were accumulated in the oxygen-deficient environments caused by a major transgression. So the upper boundary of the sequence itself corresponds to a major turning point of global changes in sea level.

WEI Siyu, Institute of Geophysics, Academia Sinicu, Beijing, China, LIU Shaocheng, and AO Guanghea, Geological Bureau of Fujian Province, M G M R, TENG Hongyuan, University of Science and Technology of China, Heife

The Geothermal Activities, Distribution of Heat Flows, and Comprehensive Geophysical Study in the Fujian Continental Margin of China

Fujian Province is located in the southeast coastal area of China. It has strong seismic activities and faces the Taiwan channel, which is affected by Pacific, Philippines, and Eurasian plates. Hydrothermal activity is frequent in the area. Fujian Province is one of the three large geothermal fields of China. We'll discuss four aspects of this matter.

The geological tectonic background of geothermal activities region and regions for measuring heat flows: The regions for measuring heat flows; Longyan, Zhangzhou, Dashen, Datian, and Fuzhou, are all on the southeast of Caledonian fold belt in northwest Fujian, the central segment of Mesozoic igneous rock belt of Zhejiang, Fujian, and Guangdong Provinces, and on the south ridge tectonic belt, which is oriented east-west. Geologically, they are within the area delineated by three major great faults named Changle-Nan'ao, Fu'an-Minqing-Nanjing, and Zhenghe-Dapu, which belong to Neocathaysian tectonic system. These regions belong to the Pucheng-Liancheng mantle depression area and anomalous area of Moho depth of southeast Fujian. These two areas are deep tectonic units. Both shallow and deep structures are very complex. The Himalayan orogeny in this area appeared as extensive faulting different uplifting of blocks and reactivity of heritage faults.

The characteristics of hydrothermal activities and effect of shallow hole temperatures: Under the influence of maximum major stress directed northwest-southeast, the faults with northeast strikes are in the state of extension. This formed good channels for the movement of underground hot water, and also controlled the distribution of hot springs. Thus, the main belt of hot springs was formed; so far 192 hot water sites have been found. The temperature of the water is 40–60 °C, and the average temperature of hot water for whole province is 51 °C. The highest temperature is 89 °C. The highest temperatures of hot water (the depth of hole is usually 300–500 m) are 120 °C and 105 °C for Zhangzhou and Fujian, respectively. Eighty-five percent of the hot springs are within granite areas, and distribution is concentrated in coastal areas. The hot water comes mainly from faults. The flowing water quantity of the whole province is 1325 liter/s.

The temperature measurements of shallow flows for Zhangzhou, Fuzhou, and Xiamen show that isothermal lines can delineate the regions and directions of underground hot water zones.

The measurement of heat flows and characteristics of distribution: The measurements of temperature gradients and thermal conductivities were conducted at 12 drill holes located in five regions mentioned above. The average values of heat flows for these areas are higher than continents and the distribution of heat flows is not uniform at different areas. The temperature gradients and heat flows are high at Zhangzhou and its vicinity, a little lower at Fuzhou, and lower yet at Longyan, Datian, and Dashen. The heat flows have high values on the east side of Zhenghe-Dapu, the modern tectonic active area, and have low values on the west side. This indicates that the Changle-Nan'ao great fault plays an important role in controlling the geothermal activities of Fujian. More important is the action of the Pacific plate. The geothermal activities of this region are due to the movement of hot materials in the deep part of the earth, cycling and compensating of fissure water.

The utilization of geothermal energy along the coastal area of Fujian and its evaluation: The hydrothermal activities along coastal area of Fujian Province occur widely, especially southeast of Fujian. The hot water resources with low and median temperatures are abundant. The key problem is to find hot water with high temperature and body of heat sources. However, the prospecting of shallow and deep layers is not carried out widely. Much prospecting work needs to be done. At present the preliminary results are as follows:

1. The hot water with low temperature of Fujian can be exploited and utilized widely. It is very useful to agriculture, medicine, breeding, and daily human activities. However, the comprehensive utilization and water balance must be taken into consideration.

2. The median and higher temperature waters are important energy sources for median factories and mining enterprises. This should be investigated carefully. The exploiting and consuming of these hot waters should be carried out region by region.

3. Careful investigations need to be done to detect whether hot waters with high temperatures and heat sources exist. Attention should be paid to mantle convection patterns in this area.

4. The comprehensive studies of geothermal activities, tectonic movement seismicity, and the ore-forming and distribution rules suggest the possibility of looking for deposits of rare and precious metals and increasing present ore reserves. Of course, the combined considerations of shallow and deep exploration will increase this possibility.

WEIDNER, D., State University of New York, Stony Brook NY

Elasticity of Mantle Minerals and Chemistry of Earth's Mantle

The comparison of seismic data with elasticity data recorded at room pressure and temperature provides constraints on the chemical and physical properties of the minerals of the mantle. We have measured the single-crystal elastic properties of virtually all magnesium silicate mineral phases that are expected to occur in the pressure-temperature regime of the earth's mantle using Brillouin spectroscopy. This information, together with recent accurate phase equilibria data, is integrated to test mineralogical modes of the mantle.

The four most diagnostic seismic characteristics include the velocities at the base of the upper mantle, the velocity jumps at the 400 km discontinuity, the velocity gradients in the transition zone, and the velocities at the top of the lower mantle. The velocities at 400 km depth allow a wide range of mineralogical models that encompass both eclogite and pyrolite.

The 400 km velocity discontinuity can be due either to chemical layering or to a phase transformation. In the latter case, the olivine to modified spinel transformation is the main candidate. The magnitude of the discontinuity will constrain the amount of olivine that is present once the pressure and temperature derivatives of the elastic properties of the high-pressure phase are known. Reasonable assumptions for the values of these variables allow 30–70 vol% olivine to be present at this depth.

The steep gradient of seismic velocities in the transition zone must be due to either a gradual phase transformation throughout this zone or to a broad variation of chemical composition. Pyroxene to garnet and garnet to spinel plus stishovite provide phase transformations that could generate this steep gradient under appropriate yet uniform chemical environments. The recent phase equilibria data are examined to determine the constraints that the transition zone places on the aluminum, calcium, and sodium abundance in this region.

Chemical variations can cause the large gradient in the transition zone. However, the primary signature of chemical variations would be through the balance between the abundance of spinel and garnet. Since garnet has the slower acoustic velocities, the steep gradient can be explained by an increase in spinel with a coincident decrease in garnet. This will obtain if there is a decrease in the Si/Mg ratio as a function of depth in this region.

The high shear modulus we have obtained for $MgSiO_3$ perovskite has very significant implications for the physical properties of the material of the lower mantle. Considering large uncertainties in Fe/Mg and Si/Mg, as well as a large uncertainty of the pressure and temperature derivatives for the shear modulus, the magnitude of the observed shear modulus cannot be rationalized with that observed for the lower mantle. We conclude that the shear modulus of the lower mantle perovskite is softened owing to the influence of soft-mode dominated phase transformations, which are quite common in other materials with the perovskite structure. Such a ferroelastic transformation is consistent with the observations that perovskite spontaneously twins with exposure to moderate light levels, to temperatures of only 100 °C, and to lightly nonhydrostatic stresses. The apparent temperature derivative of the shear modulus is expected to be quite large while the temperature derivative of the bulk modulus may be unaffected. This model also explains the anomalously large lateral variation of shear velocity compared to longitudinal velocity implied by seismic tomography.

Anomalous behavior can also be expected in thermal expansion and specific heat. As such ferroelastic transformations are often associated with a domain type structure with very mobile domain walls, the diffusion could be enhanced and rheology may become Newtonian and the viscosity could be anomalously low. Owing to this complex behavior, it is not now possible to conclude the chemical composition of the lower mantle.

WEIGEL, W., V. GEBHARDT, R. HERBER, A. DEHGHAMI (Hamburg), H. MILLER, W. JOKAT (Bremerhaven), E. R. FLäschen (Kiel) (Reporter: E. R. Flüh, University of Kiel, Kiel, West Germany

Deep Structure of Continental Margin of East Greenland between 70° and 72°N from Seismic Refraction Investigations

Along the eastern margin of Greenland between Iceland and the Jan Mayen Fracture Zone, oceanic lithosphere of Early Oligocene Age is found. This lithosphere is formed at the Kolbeinsey Ridge, which separated the continental Jan Mayen block from Greenland at around anomaly 6 times after spreading along the Aegir Ridge and the Iceland Plateau had stopped.

Thus, along a 400 km long east-west profile from the Kolbeinsey Ridge into Scoresby Sund a sequence from the youngest oceanic lithosphere to the paleozoic–Precambrian basement in East Greenland is traversed.

To test the various hypotheses for the deep crustal structure and the ocean–continent transition, combined onshore/offshore seismic refraction measurements were carried out in August 1988. Using the icebreaking R/V *Polarstern* an east to west striking profile from inside Scoresby Sund to the Kolbeinsey Ridge was observed. On the southern coast of Scoresby Sund 7 landstations with small apperture arrays (10-20 seismometers) were operated, and up to 10 ocean-bottom seismometers were deployed from the coast toward Kolbeinsey Ridge. Signals were generated by a 32 ltr. BOLT airgun at 300 m intervals; in addition explosive charges (50 kg) were detonated at approximately 5 km intervals east of the shelf.

Further on, two north-south–trending profiles on the Liverpool Land Shelf using OBS's and another land/sea profile from Kong Oskar Fjord toward the southeast across the shelf were collected.

Throughout the survey area, the ambient noise level was extremely low and airgun signal can be identified on single channel recordings to distances of at least 170 km. Using the array configurations of the landstations, stacking techniques will be applied to further improvement of the signal-to-noise ratio.

All profiles were supplemented by low fold seismic reflection measurements using a 24-channel streamer, and the gravity field was continuously monitored.

First results of this investigation will be presented.

WEILAND, T., Georgia Southwestern College, Americus, GA

Early Island-Arc Volcanism of Northeastern Puerto Rico: Partial Mantle Melts Modified by Multilevel Crystal Fractionation

Puerto Rico, the easternmost island of the Greater Antilles, is part of an inactive oceanic island arc. The island contains one of the better-exposed and more complete island-arc series found in the Caribbean. The core of the island consists of a nearly continuous, Lower Cretaceous to Eocene, volcanic sequence that has been subsequently uplifted and partially eroded. The oldest rocks of this volcanic-plutonic core belong to the Lower Cretaceous pre-Robles formations. These rocks were deposited during a distinctive and extensive episode in the early magmatic history of the island arc. The pre-Robles formations crop out in both central and northeastern Puerto Rico; however, this study was restricted to the northeastern area where stratigraphic relationships are better understood.

The Lower Cretaceous volcanic rocks of northeastern Puerto Rico have been subjected to low-grade (prehnite–pumpellyite to lowermost greenschist facies) burial metamorphism; therefore, the following petrologic and tectonic interpretations are based primarily on whole-rock concentrations of immobile elements, mineralogy, and geochemistry of relict phases, and isotopic compositions of selected samples. These deposits are subdivided into three separate formations: (in ascending order) Daguao Formation, Figuera Lava, and Fajardo Formation.

Volcanic breccias, lava flows, and tuffs are the predominant rock types of the Daguao Formation. They are generally more mafic than rocks from the other pre-Robles formations, as shown by the low SiO_2 contents (45–56 wt% SiO_2), the high modal concentrations of clinopyroxene (avg. = 9.2 modal%), the calcic nature of the unaltered plagioclase (An 75 to 90), and the common occurrence of olivine pseudomorphs. Variations in the modal mineralogies and whole-rock compositions of the Daguao samples indicate that several units within the formation are petrographically distinct, with separate crystallization histories. Clinopyroxene-rich, plagioclase-depleted basaltic flows and tuffs appear to have crystallized phenocrysts at subcrustal levels where plagioclase was unstable. By contrast, the plagioclase-rich, highly porphyritic, intrusive, and tuff breccias underwent extensive shallow-level crystallization. Much of the geochemical variation found in the Daguao samples appears related to both the accumulation and the fractionation of crystals at different levels within the crust and upper mantle.

The Figuera Lava is a thick sequence of lava flows with some locally interbedded tuffs, volcanic breccias, and sediments. Petrographic and geochemical data indicate that the flows grade upward from basal basaltic andesites and andesites to pyroxene-depleted dacites and rhyolites. These flows are generally much finer-grained and more silicic (53–72 wt% SiO_2) than other pre-Robles volcanic rocks of the northeastern block. The silicic flows were apparently derived from mafic or intermediate magmas by fractional crystallization of plagioclase, augite, and iron oxides as shown by: (1) geochemical trends defined by whole-rock and mineral data; (2) temporal relationships of samples; and (3) good fit of computer-derived fractionation models.

The predominance of volcaniclastic rocks and the scarcity of flows in the Fajardo Formation distinguish these deposits from the other Lower Cretaceous formations of Puerto Rico. Most of the basal Fajardo volcaniclastic units are highly siliceous, as shown by the apparent replacement of glass by feldspar and quartz, the presence of quartz phenocrysts and perlitic clasts, and the general scarcity of mafic minerals. These units appear to represent a continuation of the silica-enrichment trend defined by the underlying Figuera samples. By contrast, the interbedded flows and overlying volcaniclastic rocks mark a change to more intermediate-composition, pyroxene-rich volcanism. These rocks commonly contain abundant orthopyroxene pseudomorphs, which distinguish them from other pre-Robles volcanic rocks of the northeastern block. Whole-rock and mineral compositions from these lava flows generally plot in the area of overlap between the silicic Figuera rocks and the mafic Daguao samples. They appear to have been derived from a similar magmatic source but to have undergone a somewhat different crystallization history than the adjacent pre-Robles volcanic rocks. Major-element computer models show that the Fajardo flows could have been derived from mafic magmas, similar in composition to the Daguao basalts, by fractional crystallization of plagioclase, clinopyroxene, orthopyroxene, and iron oxides.

Based on the abundance of normative hypersthene, absence of normative nepheline, absence of groundmass olivine, absence of alkali-rich modal phases, low whole-rock alkali contents, and the compositions of relict clinopyroxenes, the pre-Robles volcanic rocks of the northeastern block are subalkaline tholeiites. The low $^{87}Sr/^{86}Sr$ ratios (avg. = .7040) of these rocks, the Cr-rich diopside cores of the basaltic clinopyroxene grains, the major- and trace-element composition of the samples, and the similarity between the pre-Robles basalts and experimentally derived melts indicate that the pre-Robles magmas were probably derived from the partial melting of mantle peridotite. Subsequent crystal fractionation at various levels within the crust and upper mantle appears responsible for much of the geochemical and mineralogic variation found in the pre-Robles samples.

WEIMER, P., University of Texas, Austin, Texas

Sequence Stratigraphy and Depositional History of the Mississippi Fan, Gulf of Mexico

The Mississippi Fan is a large, mud-dominated submarine fan over 4 km thick that was deposited in the deep Gulf of Mexico during the late Pliocene and Pleistocene. Analysis of 19,000 km of multifold seismic data across the fan defined 17 seismic sequences, each characterized by a series of channel, levee and associated overbank deposits, and other mass transport deposits. At the base of nine sequences are a series of seismic facies consisting of mounded, hummocky, chaotic and subparallel reflections, which constitute 10–20% of the sediments in the sequence. These facies are externally-mounded in cross section and occur in two general regions of the fan. In the upper and middle fan, they occur below channels and are elongated in shape, mimicking the channel's distribution. In the middle fan to lower fan, they have a fan-shaped distribution, increasing in width downfan. These facies are interpreted to have formed as disorganized slides, debris flows and turbidites, and are informally called "mass transport complexes."

Overlying this basal interval and characteristic of all sequences are well-developed channel-levee systems, which constitute 80–90% of the fan's sediments' Channels consist of high amplitude, subparallel reflections. Levee sediments have subparallel reflections that have moderate to high amplitudes at the base changing upward to low amplitude. The vertical change in amplitude may reflect a decrease in the grain size of the levee sediments. Overbank sediments consist of interbedded subparallel to hummocky and mounded reflections suggesting both turbidites from the channel as well as slides and debris flows derived both locally and from the slope.

Plio-Pleistocene eustatic cycles are interpreted to be the major factor controlling the timing and style of sedimentation in the fan. "Mass transport complexes" are interpreted to have formed during a lowering of sea level, and reflect sediments derived from retrogressive slumping during the formation of submarine canyons along in the upper slope and outer shelf. Channel-levee systems were deposited when sea level was near its lowest position and sediment derived from deltas were transported into the deep basin via submarine canyons. During highstands in sea level, a thin layer of hemi-pelagic sediment was deposited on the fan surface.

WEIMER, R. J., Colorado School of Mines, Golden Colorado

Valley Fill Deposits of Depositional Sequences: Stratigraphic Trap Frontiers in the Pennsylvanian and Cretaceous, Central U.S.A.

A valley fill is deposited during a rising sea level in an incised drainage that was cut during a relative sea-level lowstand. Where thickest and most complete, the valley fill commonly has a vertical zonation from lowermost fresh water deposits to brackish and then marine deposits. The zonation results from deposition during a rising sea level (transgression) by aggradational processes within the valley. This vertical facies order is contrary to the normal facies arrangement in typical regressive deposits.

Valley fill sequences, related to late Quaternary sea-level changes, are described for the Mississippi, Brazos and Sabine Rivers of the Gulf coast, and the Delaware River for the Atlantic coast.

Valley fill sequences, related to late Quaternary sea-level changes, are described for the Mississippi, Brazos and Sabine Rivers of the Gulf coast, and the Delaware River for the Atlantic coast.

Valley fill deposits are bounded by surfaces of erosion. They overlie a lowstand surface of erosion (LSE) associated with the valley incisement phase and a lowering of base level (this surface is a major sequence boundary in sequence stratigraphy analysis). They may be truncated at the top by a transgressive surface of erosion (TSE) associated with shoreface erosion related to the transgressing marine shoreline. Marine shale, commonly an organic rich condensed section (source rock), overlies the valley fill. Minor erosion surfaces which are found within the valley fill, associated with depositional processes, are called diastems.

Valley fill deposits in the Pennsylvanian and Cretaceous are up to 120 ft thick, are less than a mile to as much as tens of miles in width, and may be hundreds of miles in length. The nature of the fill varies in percentage of sandstone, siltstone and shale. the most important petroleum reservoir in the valley fill is the lenticular porous and permeable fluvial channel sandstones, normally in the lower portion. Of less importance are the estuary, tidal flat and bayhead delta (estuarine) sandstones because they are thinner (10 ft) and tighter. Typical stratigraphic and unconformity traps in valley fills are discussed for the Rocky Mountain and Mid-continent regions. Recent exploration successes in Wyoming and southeast Colorado indicate that exploration targets are fields with reserves in excess of 30,000,000 bbls of oil equivalents.

Some valley fills can be recognized on seismic, but the future exploration frontier is to integrate improved seismic and geologic models, to improve predictability.

WEIMER, R., Colorado School of Mines, Golden, CA

Sequence Stratigraphy, Lower Cretaceous, Western Interior Basin, USA

The Lower Cretaceous of the Western Interior Basin consists of intertonguing marine and nonmarine detrital strata. Distribution of facies is related to transgressions and regressions of the shoreline, tectonics, and sea-level changes. The stratigraphic section over the cratonic eastern portion of the basin varies from 100 to 200 m in thickness, whereas in the western portion, near the tectonically active thrust belt, the section is more than 1000 m thick. Because the strata are Albian and Late Aptian(?) in age, a large hiatus exists between the Jurassic and Lower Cretaceous strata.

Sequence stratigraphy is the study of genetically related strata that are bounded by unconformities. An unconformity is defined as a sedimentary structure of regional occurrence in which two groups of rocks are separated by an erosional surface; the erosion may be by subaerial or submarine processes.

Three sequences can be observed in the Lower Cretaceous of the Western Interior Basin. Sequence 1 is the nonmarine Lytle (or Lakota and Fuson) Formation. Sequence 2 includes the coastal plain deposits of the Plainview Sandstone (Fall River), the overlying marine Skull Creek (Thermopolis) Shale, and the overlying delta and marine sandstone genetically related to the Skull Creek, but which for mapping purposes is placed in the Muddy Formation. Sequence 3 includes incised valley fill or coastal onlap deposits of the Muddy Formation, and the overlying marine Mowry Shale, Graneros Shale, Greenhorn Limestone, and that portion of the Carlile Shale underlying the erosional surface at base of the Codell (upper Frontier) Sandstone. All strata above the Mowry are designated as Upper Cretaceous.

Two types of major erosional surfaces, related to unconformities, are observed either within or at the formation contacts of the Muddy. Each is associated with major changes of sea level, but the magnitude of erosion may be influenced by local tectonic events. One type, a sequence boundary, is called a lowstand surface of erosion (LSE) related to a lowering of base level that causes subaerial exposure and incisement of drainages into older deposits. The second type is a transgressive surface of erosion (TSE) related to shoreline and shoreface (marine) erosion that is caused by a rising sea level and associated water deepening. The unconformity associated with the TSE occurs within a depositional sequence. Minor erosional surfaces associated with depositional processes within environments of deposition are called diastems (e.g., scour at the base of a channel).

The erosional surfaces and facies related to the sea-level and other changes have been mapped regionally for the 3 sequences of the Lower Cretaceous. Sequence 2 is used to illustrate the application of sequence stratigraphic terminology to outcrop and subsurface data. The marine Skull Creek Shale and genetically related underlying and overlying sandstones are highstand deposits (HSD). The lower and middle Skull Creek Shale are dominantly transgressive deposits (TD) related to deeper water anoxic conditions. The upper Skull Creek Shale and the overlying Fort Collins Sandstone Member of the Muddy are dominantly regressive deposits (RD) related to water shallowing and progradation. A drop in sea level occurred that subaerially exposed a large portion of the depositional basin and a lowstand surface of erosion formed the sequence boundary. Paleosols and/or chemical changes caused early diagenesis below the LSE. A paleovalley system was incised to as much as 50 m with valleys varying in width from 1 to 30 km. The first strata above the sequence boundary are valley fill deposits of sequence 3 (Horsetooth Member of the Muddy Formation) that are related to sea-level rise and transgression. The incised valley fill (IVF) may be zoned—more freshwater into the lower part and brackish to marine in the upper part. The zoning reflects aggradational fill associated with coastal onlap on the LSE. Valley fill deposits are thickest in structural low areas that formed during deposition because of recurrent movement on basement fault blocks.

The two types of unconformities observed in the Lower Cretaceous are related to the cyclic interplay of emergence and submergence of the basin. Emergence, causing a lowstand surface of erosion (LSE), resulted from a lowered sea level with tectonic influence on all or portions of the basin. Submergence resulted from either sea-level rise, subsidence (tectonic), rates of sedimentation, or a combination of these components affecting the depositional system. The transgressive surface of erosion (TSE) and overlying anoxic condensed section (CS) are related to water deepening in the marine basin.

Sequence stratigraphy incorporates all of the above factors to improve the facies and geologic models for interpreting the Lower Cretaceous. However, the ultimate causes of the water depth changes within the Cretaceous Basin remain uncertain.

WEISSERT, H., Geological Institute, ETH-Z, Zürich, Switzerland

Early Cretaceous Paleo-oceanography: Tethyan Perspective

A north-south transect through the Early Cretaceous alpine Tethys documents the paleo-oceanographic evolution of different paleobathymetric and paleogeographic environments. A carbonate platform along the northern continental margin documents changes in shallow water masses while the pelagic sediments of the central Tethys and its southern continental margin inform about intermediate and deep-water evolution. The pelagic sediments contain additional information on changes in the oceanic carbon reservoir (C-isotopes) and on fluctuations in accumulation rates of authochthonous, oceanic particles and of allochthonous, detrital components. Two significant paleo-oceanographic events are recorded in these alpine Tethyan sediments:

1. During the Valanginian, white or grey pelagic limestone (Calpionellid Limestone) was replaced by alternating black siliceous shales and dark pelagic limestones (Palombini Formation) in the central Tethys. A synchronous change from thick-bedded white nannofossil limestone to thin-bedded dark nannofossil limestone is observed in sections from the southern continental margin (Maiolica Formation). On the carbonate platform of the northern Tethyan margin, shallow-water carbonates (Betliskalk) are replaced by limestones rich in detrital quartz (Pygurus-Schicten) and by a glauconite-phosphorite hardground marking the top of the Valanginian (Gemsmättli-Schicht).

2. The Aptian of the central Tethys is characterized by change from siliceous marls and limestones to black siliceous marls (Lavagna shales). Nannofossil limestones and organic carbon enriched marls (Maiolica) accumulated along the southern continental margin are replaced by siliceous foraminifera marls with periodic intercalations of organic-carbon rich marls or shales (Scaglia). This transition is dated as mid-Aptian. A carbonate platform (Schrattenkalk) of the northern Tethyan margin was destroyed during the early Aptian and replaced by glauconitic-phosphoritic sandstones and shales (Gault).

The changes in the Valanginian and Aptian Tethyan facies seem synchronous with (1) a rise in global sealevel (Haq et al, 1987), (2) a positive γ^{13}C-excursion by 1‰ during the Valanginian (Cotillon and Rio, 1984) and by 2‰ during the Aptian (Weissert, in press) and (3) a rise in the global CCD (Thierstein, 1979).

The C-isotope record is an indicator of a changing oceanic carbon reservoir and monitors the rate of environmental change leading to the observed facies pattern. The transition time between the two modes of sedimentation lasted up to several hundred thousand years. Possible explanations of these low-frequency paleo-oceanographic changes include processes limited to the oceanic reservoir and processes related to changes in the global geochemical cycles:

1. Changes in *basin-shelf calcite fractionation* leading to CCD changes and additional changes in calcite production rate affecting the position of the CCD.

2. Changes in the *ocean-sediment flux* of organic matter and associated materials (nutrients, silica).

3. Changes in *material transfer from continents to oceans* balancing the accelerated flux of material from continents to oceans.

Simple model calculations indicate that the described changes in pelagic and in shallow-water facies are not only the result of sea-level controlled basin-shelf carbonate fractionation. Times of elevated CCD coincide with periods of accelerated production of calcite and of associated accelerated transfer of organic matter and associated nutrients and silica from oceans to sediments. Accelerated organic matter and nutrient burial was favored by (1) increased productivity, (2) low oxygen levels in deep water, and (3) high bulk sedimentation rates. Considerations of material balance lead to the postulate that times of accelerated burial of organic matter, nutrients, and silica were related to times of accelerated material transfer from continents to oceans.

Tracers that record changing rates of weathering and drainage on continents include quartz and clay accumulation rates and the clay mineralogy. Times of elevated CCD and positive carbon-isotope excursions coincide with periods of kaolinite enrichment and smectite depletion in clay mineral assemblages (Persoz, 1982; Deconinck et al, 1985).

The Tethyan data indicate that low frequency paleoenvironmental changes leading to the accelerated geochemical cycling triggered paleo-oceanographic changes recorded in pelagic and carbonate platform facies. Intensified hydrological cycling due to elevated atmospheric CO_2 may explain the observed changes.

REFERENCES

Cotillon, P., and M. Rio, 1984, Cyclic sedimentation in the Cretaceous of DSDP Site 535 and 540 (Gulf of Mexico), 534 (Central Atlantic) and the Vocontian Basin (France): In. Rep. DSDP, v. 77, p. 339–376.

Deconinck, J. F., B. Beaudoin, H. Chamley, P. Joseph, and J. F. Raoult, 1985, Contrôles tectonique, eustatique et climatique de la sedimentation argileuse du domaine subalpin français au Malm-Cretacé: Revue de Géologie Dynamique et de Géographie Physique, v. 26, p. 311–320.

Haq, B. U., J. Hardenbol, P. R. Vail, 1987, Chronology of fluctuating sealevels since the Triassic: Science, v. 235, p. 1156–1167.

Persoz, F., 1982, Inventaire minéralogique, diagenèse des argiles et minera-lostratigraphie des séries jurassiques et crétacées inférieurs du Plateau Suisse et de la bordure sud-est du Jura entre les lacs d'Annecey et de Constance: Beiträge zur Geol. Karte der CH, N.F., v. 155, 52 p.

Thierstein, H. R., 1979, Paleoceanographic implications of organic carbon and carbonate distribution in Mesozoic deep sea sediments, *in* M. Talwani, W. Hay, and W. B. F. Ryan, eds., Deep drilling results in the Atlantic Ocean: continental margins and paleoenvironment: Maurice Ewing Series 3, AGU, p. 249–274.

Weissert, H., in press, C-isotope stratigraphy, a monitor of paleoenvironmental change: a case study from the Early Cretaceous: Surveys in Geophysics.

WENG Shijie, Ministry of Geology and Mineral Resources, Beijing, China, CHEN Hushen, East China Bureau of Petroleum Geology, MGMR, China, and ZHOU Xueqing, and CUI Zichen, The Sixth Geophysical Prospecting Party, MGMR, China

Seismic Reflection Studies for Tectonic Framework of Yangtze Platform

The Yangtze platform is one of the petroliferous regions in China. A set of deep seismic profiles have been applied to study the deep crustal structures. Several unexpected findings have resulted.

Tengchen–Lujiang fault was formerly considered to be a long-activated, left-lateral displacement wrench fault. It consists of four parallel faults. Two seismic lines cut across the fault 150 km apart. Surface mapping suggests the eastern boundary fault is also the divide between two stratigraphic regions; some writers also consider the western boundary fault as a dividing line between stratigraphic regions. Reflection lines clearly show the western fault is the deepest, nearly vertical, displacing all events including the Moho. Its configuration is similar to that of a strike-slip fault in petroleum exploration and confirms the left lateral movement. However, the eastern fault is expressed as thrust on the profile, which denotes that it was originally thrust between stratigraphic terranes. The present-day figure is altered by late-stage lateral movements of shallow nature. A two-stage model should be established. On both lines, the lower crust of the east block, i.e., Yangtze platform, reached the western boundary. Related events are inclined toward northwest, or in other words underthrusted below the North China craton.

One of the profiles runs over a granitic batholith, the Suzhou granite. In the profile, the batholith is transparent and has a flat base around 10 to 11 km deep, which is in accordance with calculations from gravimetric data. Below the batholith, a sequence of densely spaced, thin reflectors with medium amplitude occurred and extended down to the Moho. By comparison with the results of BIRPS groups, these are the resitite of partial melting; these findings agree with the petrological evidence of its crustal remelting genesis.

The study region has a lot of Meso-Cenozoic sedimentary depressions. In gravimetric survey, they commonly show negative anomalies. The inversion of gravimetric data suggests the Moho surface is the mirror image of the depression basement. Depressions are regarded as the superposition of early stage rifting and late stage subsidence. Seismic lines run across twelve depressions, except the Kaoyu depression, which is an elongation of the large North Jiangsu–Southern Yellow Sea basin, the mirror image of Moho and the basement doesn't exist. They may be put into various categories, some are the sags of the hanging block for thrusts, some are the pull-apart depressions on the eastern side of strike-slip fault, or the stress released subsidence after the thrust front, or domino faulting induced by stretching, or simple half-grabens. Meanwhile, the Moho rises below the anticlinorium and is concave below the depression. These findings lead us to the conclusion that these depressions are only the results of upper crustal extension, not lithosphere extension. Luzhi depression is very interesting. It is situated before the front of a low-angle thrust, the base of Cenozoic and all the reflectors below it show the same style of curvature, suggesting that compressional effects should be taken in consideration.

Much information about the nature of the lowercrust is obtained in the survey. Four types of lower crust may be distinguished that are equivalent to four tectonostratigraphic terranes. North China craton has Archean rocks; lower crust reflectors are nearly continuous and strong, with some boudinage-like reflectors intercalated among them, similar to the results of Lithoprobe. Yangtze platform is consolidated in middle Proterozoic. Its lower crust has long reflectors, but they are thin with medium amplitude and suddenly disappear below the Moho, similar to the layered Proterozoic terrane shown by COCORP. Jiangnan terrane's basement is the youngest. The interpretation of seismic data denotes that the metagrade is also the lowest. The lower crust is wholly reflective along all the lines. This may suggest that the accretion of the four terranes wasn't accompanied by the subduction of oceanic crust, or the oceanic crust had been totally consumed or altered in late stage. In other words, these terranes are broken pieces of a megacontinent; a wide ocean basin didn't exist between them. On the western part of each terrane, the lower crust reflectors are commonly inclined. After migration and spacial calibration, the inclinations become steeper. These phenomena need more research; perhaps they are the remnants of terrane collision. The driving force from the east pushed blocks westward, the upper crust thrusted upward to form low-angle thrust sheets, while the lower crust could only underthrust. The inclined events are clear and dense in the lower crust; some of them may extend below the Moho, but most of them disappeared.

WENJI Bai, Chinese Academy of Geological Sciences, Beijing, China, and Jingsui YANG, Dalhousie University, Halifax, Canada

Can Mountains Be Formed by Spreading of Basins in Continents?

Apparently, this question consists of two aspects, i.e., whether continental basins are formed by spreading, as ocean basins are formed, and what is the genetic relationship between basins within continents and the surrounding mountains?

As research in basins has developed, a stretching process has been proposed in recent years by many authors for the origin of some basins within continents, e.g., the Aegean Sea Basin (McKenzie, 1978a), the North Sea Basin (Sclater and Christie, 1980), the Great Basin (Proffett, 1977), and the North China Basin (Liu et al, 1980), etc. Most previous, traditional concepts, in which basins were regarded as rigid massifs, inactive tectonic units, graben or down-faulted depressions, have now been challenged. Here we propose the views of basin spreading and basin-mountain system in continental areas, including those basins well within cratons. Although the evidence for spreading for the origin of continental basin is now only available for some basins within the continental parts of plates, we identify the characteristics of active, or formerly active spreading basins found in continents as follows:

1. Normal fault spreading and listric faults are commonly developed in basins and can be shown to be associated with crustal thinning.
2. During the stretching phase of basin formation the values of heat flow are higher in the basin than those in the neighboring unstretched crust.
3. The center of a basin has thinner crust, equivalent to a mantle uplift.
4. During the active phases of basin formation, earthquakes mostly showing normal fault type first motions occur in basins, e.g., in the Aegean Sea (McKenzie, 1978a) and the North China Basin. At an advanced stage of extension, basic magma may extrude at the surface along extensional fissures, e.g., in the Bo Hai Basin of northeast China, etc.

These features also can be observed within some intercratonic basins, as well as those active spreading basins known on the continental margins, e.g., the Tarim Basin (Tian et al, 1985), the Juggar Basin (Tan, 1985), and the Ordos Basin (Zhang and Wu, 1985). A striking feature is that these basins are surrounded by high mountains.

It appears that there can be two possible relationships between these basins and surrounding mountains. One is that a basin opens as a passive response to differential movement within the mountain belt. The other is that a basin is an active spreading system that causes the differential movement including uplift of the mountain belt. At present, it seems that most people only consider the former and neglect the latter.

For the latter mechanism to be operative, it is implied that if a basin is spreading actively, it is increasing in lateral dimensions. Thus, space is an essential prerequisite for basin spreading. So, we identify a further requirement, i.e., the space for basin spreading should be gained by the compression to the surrounding area. As a result of this process, the strata around the basin will be shortened, folded, and uplifted into mountains. From this point of view, basin spreading would be the active and mountain (perhaps including plateaus) elevation would be a passive responsive process. This is in contrast to the current idea of basin formation being a passive response to other tectonic activity (McKenzie, 1978b). We call a process of this type of basin spreading and basin-mountain collision a basin-mountain system. Some possible indicators of basin-mountain collision of this type are outlined below.

1. The strata that display around the margins of basins would be folded and curved and uplifted into mountain-arcs as the result of the spreading of the basin.
2. The marginal strata in the basin underthrusts beneath the adjacent mountains, or, relatively, the mountains overthrust onto the basin.
3. The rocks distributed along the margin of basin, due to the compression process from the basin, would metamorphose or melt to give rise to granitic or other remelted magmatic lithologies.
4. Earthquakes occur concurrently both in the marginal area of the basin (with thrust and probably also strike-slip fault motions) and in the center of the basin (with normal and probably strike-slip fault motions).
5. If a mountain adjacent to a spreading basin is uplifted in a quite different period from the larger scale plate orogeny in the area, it is probably related with basin-mountain collision, e.g., the South Tethys Sea closed at about 50-54 Ma; however, the Himalaya began a period of major uplift only since 20-23 Ma.
6. Plate orogeny zones are usually distributed along plate margins, while mountain-building resulting from basin-spreading may occur in the interior of continents, even within cratons, which are far away from plate convergent zones.

The undulating form of the upper mantle is a key factor that may be used to identify the present or former tectonic regime. The tectonic movement in the crust in an uplifted area of mantle will point to stretching, while depression of the mantle is a characteristic of the shortening process. Mantle uplifting indicates that the overlying strata are thinned and spreading basins are formed, while mantle depression implies that the overlying crust is thickened and mountains are formed. The inference of mantle convection that explains the formation of ocean basins should be extended to the model of the formation of continental basins, and the difference is just one of relative dimensions and rates of formation.

WENNER, D. B., University of Georgia, Athens, GA

Postcrystallization Low-Temperature Hydrothermal Alteration in Proterozoic Basement of Midcontinent of North America

the common occurrence of disturbed stable and radiogenic isotopic systems in rocks from the Proterozoic silicic plutonic and volcanic basement of the Midcontinent of North America may reflect a low-temperature hydrothermal alteration that occurred long after crystallization and cooling of the magmas. This type of alteration is not commonly recognized since primary igneous textures, mineralogy, and major and trace element chemistry are largely unaffected.

Two styles of hydrothermal alteration appear to exist. The first is documented in drill cores from the Illinois Deep Hole Project (IDHP). Here, the upper part of the granitic basement lying beneath ~335 m of Paleozoic sedimentary units (see Coates et al, 1983, and references therein). The manifestations of this alteration are principally: (1) $^{18}O/^{16}O$ quartz-feldspar disequilibrium fractionations, largely reflecting isotopic exchange of feldspar, extending about 70 m below the unconformity; (2) whole-rock Rb/Sr model ages (1350–1450 m.y.) that are younger than the age of ~1485 m.y.) of crystallization (as evidenced from U/Pb zircon ages) throughout a 1 km core; (3) variably younger whole rock U/Pb ages; and (4) secondary minerals such as calcite, fluorite, and chlorite. Alteration temperatures were probably $<350\,°C$ since the primary $^{18}O/^{16}O$ of quartz is preserved.

A second style of low-temperature hydrothermal alteration affected most Proterozoic rocks exposed in the St. Francois Mountains of Missouri. Here, systematic quartz-feldspar $^{18}O/^{16}O$ fractionations exist, with Δ-values ranging from near normal (for igneous rocks) of +1 to strongly disequilibrium values of < -1.5. These fractionations correlate with the original volcanic-plutonic stratigraphy (Wenner and Taylor, 1976). This isotopic disturbance, which principally affects the feldspars, may have been caused by low-temperature hydrothermal fluids that penetrated to depths of at least 2–3 km. Unlike at the IDHP site, this style of alteration is almost certainly Precambrian as evidence by observations that: (1) some Precambrian mafic dikes intruding the felsic igneous rocks are unaltered; (2) disequilibrium quartz-feldspar fractionations mimic the stratigraphy of the Proterozoic plutonic-volcanic complex and not the overlying Paleozoic sedimentary units; (3) the degree of discordancy in Rb/Sr whole-rock model ages correlates with Sr content—the low Sr rocks are 200–300 m.y. younger than the age of crystallization (Bickford et al, 1981; Cullers et al, 1981), hydrothermal processes may account for partial loss of Rb and K and anomalous Ce/La in some rocks. Other evidence for hydrothermal alteration includes: (1) the occurrence of calcite, fluorite, and secondary chlorite; (2) brick-red coloration of the alkali feldspars that correlates with $^{18}O/^{16}O$ quartz-feldspar fractionations; and (3) micron-sized cavities in alkali feldspar that may reflect a dissolution process. Concentrations of submicron-sized iron oxide inclusions adjacent to grain boundaries and fractures in alkali feldspars imply grain boundary fluid migration. Like the unconformity-controlled hydrothermal alteration at the IDHP site, this also probably occurs at $<350\,°C$ since the primary $^{18}O/^{16}O$ of coarse-grained quartz is preserved.

Possible causes for this postcrystallization alteration in the St. Francois Mountains include: (1) an IDHP type process in which brines, perhaps originating from now missing Precambrian sedimentary units, migrated into the underlying basement; or (2) circulation of hydrothermal fluids, long after the cessation of magmatic crystallization and cooling of the granite–rhylite terrain, induced either by intrusion of mafic dikes or by radiogenic heating by nearby high heat-producing granites. This latter hypothesis is tenable since numerous high U and Th granites occur in the region (Kisvarsanyi, 1982).

Much of the Proterozoic granite–rhyolite basement underlying the Mid-continent of North America may have been affected by one or both of these styles of hydrothermal alteration. Many samples have whole-rock Rb/Sr ages that are quite discordant from their U/Pb zircon ages (e.g., Thomas et al, 1984). Brick-red alkali feldspars are quite common in these rocks. Evidence for this same kind of low-temperature alteration process can also be found in felsic igneous rocks of layered complexes such as the Bushveld, Muskox, and Duluth Gabbro (Taylor, 1969), among high heat producing granites (e.g., Moller et al, 1985), adjacent to major detachment zones (e.g., Kerrich and Rehrig, 1987), and in basement rocks beneath closed hydrographic basins (Chapin, 1987). This implies that low-temperature hydrothermal alteration of "fresh" felsic igneous rocks is probably more common than currently recognized.

WENRICH, K. J., H. B. SUTPHIN, B. S. VAN GOSEN, U.S. Geological Survey, Denver, Colorado, and G. H. BILLINGSLEY, U.S. Geological Survey, Flagstaff, Arizona

Exploration for Music Mountain Type Quartz-vain Gold Deposits on the Hualapai Indian Reservation, Arizona

The 4000-km² (1550-mi²) Hualapai Indian Reservation lies along the southwestern margin of the Colorado Plateau in northwestern Arizona. The reservation is bounded on the north by the Colorado River, and on the west by the Grand Wash Cliffs. Because of the reservation status of this land, little or no gold exploration has occurred there during the past 50 years. Gold mines in the Music Mountain district are located at the base of the Grand Wash cliffs, adjacent to part of the Hualapai Indian Reservation. In 1987 the U.S. Geological Survey began an exploration program aimed at searching for gold-bearing quartz veins in the reservation, similar to those in the Music Mountain mines.

Hualapai Valley, which lies along the base of the Grand Wash Cliffs, contains several other gold districts in addition to Music Mountain, most notably the Gold Basin and Lost Basin districts farther to the north. Mineralization of these districts was probably controlled by the Grand Wash fault, along which the Colorado Plateau was raised over 10,000 ft (as estimated by Ivo Lucchitta in 1966) to form the Grand Wash Cliffs. All three districts contain nearly vertical quartz-gold veins, which are associated with diabase and granite porphyry dikes intruded into Precambrian granite, schist and gneiss. The veins range in width from several centimeters to one or two meters. Some of the veins follow the margins of the diabase dikes, while others occur adjacent to shear zones parallel to, but offset from, the dikes. The quartz-vein gold deposits are strongly controlled by the local structures: (1) The Music Mountain district contains parallel quartz veins and diabase dikes that are oriented n50°W, parallel to the Grand Wash fault, (2) Deposits in the Gold Basin and Lost Basin districts are concentrated in a 15-km-wide belt that trends N15°E to N20°E.

Recorded production from the Gold Basin and Lost Basin districts between 1901 and 1942 indicates that 382,952 grams (13,508 ounces) of gold and 194,369 grams (6,857 ounces) of silver were mined, along with some byproduct copper and lead. Production records for the Music Mountain district are unavailable.

The Music Mountain district was discovered in 1880, and according to F.C. Schrader (1909, USGS Bull 397), "the ore contained 7 to 8 ounces of gold to the ton" (210–240 ppm). Analyses of samples collected during this study show that gold ranges from a trace to 7.2 ppm (0.24 oz/ton) and silver to as much as 15 ppm (0.5 oz/ton), with corresponding minor enrichment of As, Cu, Pb, and Zn. The following table shows geochemical data for three samples from the Music Mountain mine:

	Sample #1	Sample #2	Sample #3
Ag	15 ppm	5 ppm	10 ppm
As	330	50	80
Au	7.2	0.1	0.65
Co	50	15	10
Cr	30	50	15
Cu	200	70	150
F	600	100	200
Hg	0.14	0.06	0.34
Mo	7	<5	5
Ni	15	70	15
Pb	700	200	1000
Sb	<2	<2	<2
V	20	70	50
Zn	230	1400	1200

The gold is present as native gold, some occurring near contact of the quartz vein with the country rock (hornblend-biotitie grandiorite); no gold tellurides have been observed. Feldspars in both diabase dikes and Precambrian granodiorite have been altered to sericite, and pyroxenes within the diabase have been altered to chlorite. Emplacement of quartz was accompanied by sulfide mineralization that included pyrite, sphalerite, galena, and chalcopyrite. Pyrite is the most abundant, chalcopyrite the least. Fractures have been filled by calcite and pyrite indicating that the sulfide mineralization postdated the intrusion of the diabase and a later fracturing event. Calcite mineralization was distinctly later than the sulfides as it fills in quartz-lined vugs, and the quartz appears to have formed contemporaneously with the sulfides.

Precambrian rock exposed in the Grand Canyon and its tributaries throughout the Hualapai Indian Reservation offers good potential for quartz-gold vein occurrences similar to those in the Music Mountain district. The Hurricane fault bisects the Hualapai Reservation, and the area where this fault and the Precambrian rocks are exposed offers good potential for the location of gold-bearing quartz veins. Like the Grand Wash fault, the Hurricane fault has been reactivated repeatedly since the Precambrian, and represents considerable vertical displacement (as much as 730 m (2400 ft)), and extends for a great distance (in the case of the Hurricane fault for over 480 km (300 mi). Diabase dikes, oriented parallel to the Hurricane fault, are exposed around the mouth of Diamond Creek at the Colorado River.

A stream-sediment survey was completed for all drainages in the Precambrian rock on the Hualapai Reservation. One panned concentrate and one stream sediment, sieved to less than 80 mesh, were collected from each of 269 sample sites. Six of these sites showed anomalous gold concentrations, with three of them located in the lower part of the Diamond Creek drainage about 6.5 km (4 mi) from its confluence with the Colorado River. One sample, containing 528 ppm gold, was taken from a plunge pool at the base of a 2.5-meter (8-ft)-high waterfall in Precambrian schist. The following are concentrations of selected elements in this sample

Ag-- 8 ppm	As--58 ppm	Au--528 ppm
Bi--<0.2	Cu-- 8	Cd-- 0.46
Ga--<0.5	Hg--<0.09	Mo-- 6
Pb--21	Pd--<0.5	Sb-- 1.4
Se--<0.9	Te--<0.5	Tl-- <0.5
	Zn--42	

Additional detailed stream-sediment and lithogeochemical sampling is planned in the Diamond Creek area. It appears that areas of the Precambrian cut by the major northeast-and northwest-trending basement faults and shear zones of northwestern Arizona hold moderate to high potential for quartz-vein gold deposits.

WERNER, B. T., Naval Weapons Center, China Lake, CA

Computer Model for Evolution of Sand Surface Under Eolian Saltation

A definitive theory for the formation of sand ripples in eolian saltation is lacking. A computer simulation algorithm for the purpose of studying, at the level of individual grains, the evolution of a sand surface in the wind from a nearly flat bed to a rippled surface is presented.

Recent research has confirmed Bagnold's notion that grains on a sand bed are mobilized (into *reptation*) by the low-angle impacts of high-energy saltating grains. The rearrangement of surface grains in reptation leads to the formation of ripples. Previous studies have suggested that the ejection of surface grains can be divided roughly into two classes: *target reptation,* where the grain struck by the incident grain and those grains surrounding it are ejected from the surface, with a wide distribution of velocities and a slight bias toward the downstream direction, and *brink reptation,* where a grain on the upwind edge of a downward step in the surface topography is ejected with significant forward momentum as a result of an impact upwind.

To simulate ripple formation, circular grains are placed on a two-dimensional hexagonal lattice with the requirement that each grain rests on two grains underneath it. The lattice is periodic in the horizontal direction, and it is stretched from a close-packed configuration along the horizontal so that the angle of repose is 30°. Grains are propelled at the bed one at a time with constant incident angle and from random positions. Surface grains at or near the impact point are dislodged and transported according to an algorithm that simulates target reptation, brink reptation, or a combination of the two. The ejected surface grains are moved prescribed horizontal distances, where they are "dropped" and fall downhill into the closest resting spot on the lattice. The incident grain is presumed to rebound, and it is removed from the calculation. In this system, time is measured by the number of impacts that have occurred.

In the simplest case, where the grain hit by the incident particle is moved a constant distance downwind, surface features that resemble natural ripples in cross-section and wavelength form (see Figure 1). The mean ripple wavelength grows with time through mergers between adjacent ripples. These mergers are promoted by the stochastic nature of the transport in reptation. As the wavelength grows, and the relative fluctuations in ripple velocity decrease, mergers become less frequent. The production of ripples remarkedly occurs for almost any algorithm that transports grains "downwind," excepting pure brink transport, which levels a roughened surface.

Sedimentation and some sorting processes in eolian saltation may be amendable to study using the algorithm outlined here.

The author is deeply indebted to P. K. Haff for his many contributions to the research. Supported in part by the United States Department of Defense: Office of Naval Technology, Defense Nuclear Agency and Naval Weapons Center.

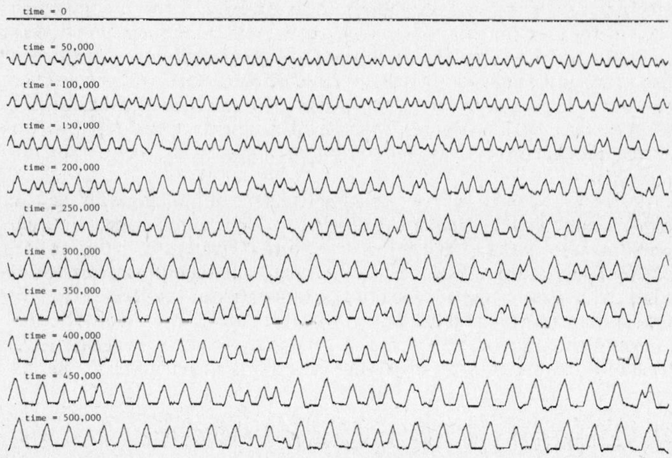

Figure 1—Surface profile development with time on a lattice of horizontal length 2000 full lattice spacings. A target reptation algorithm is used, whereby the surface grain struck by the incident grain is placed five full lattice spacings downwind. The incident angle is 10°. The surface profile are exaggerated in the vertical by the factor of 8.

WEST, D. B., GECO Geophysical Company, Inc., Houston, Texas

Salt Morphology and Implications for the History of The Gulf of Mexico Basin

Variations in the original extent and thickness of the Jurassic Louann Salt combined with variations in the history of sedimentary loading in different parts of the Gulf of Mexico Basin have resulted in a wide range of salt forms. Deep penetration seismic data clearly show a base of salt well above the level of original Jurassic deposition in many areas. Displaced or "unrooted" salt may very well be the rule instead of the exception.

Examples of interpreted deep penetration seismic sections show relatively minor salt on the shelf in the westernmost Gulf, isolated (and potentially unrooted) salt diapirs on the shelf in the central Gulf, extensive unrooted salt sheets on the slope of both the western and central Gulf, and obviously Jurassic rooted salt structures in the eastern Gulf. Development of these various forms was dependent on the original thickness and extent of salt, as well as the history of sedimentary loading. Investigation of sediment patterns and resulting morphology of salt structures can be used to reconstruct the geologic history of individual features, and by extension, the entire basin.

WEST, R. R., Kansas State University, Manhattan, KS, and H. B. ROLLINS, University of Pittsburgh, Pittsburgh, PA

"Anatomy" of Condensed Sequence: Example from Coastal Georgia

Our studies at St. Catherines Island, Georgia, revealed an exhumed (475–1300 y B.P.) marsh surface along the seaward side of the island. This exhumed surface, acting as a hard/firm ground, has been colonized by intertidal nestling, *Brachidontes,* and boring, *Barnea, Petricola,* and *Crytopleura,* bivalves and barnacles. This palimpsesting, or heterochronous paleocommunity replacement, has been superimposed on the relict marsh fauna of in situ *Spartina, Crassostrea, Mercenaria, Geukensia,* and *Tagelus.* In addition, some of the marsh taxa, *Mercenaria* and *Geukensia,* have recolonized this same surface, recording a second palimpsesting, or heterochronous paleocommunity replacement, event. Thus, recorded on what would appear as a single bedding plane in the rock record are three different events from three different times. Would we be able to recognize and separate these events if we encountered such a situation in the rock record? We think so.

While all the taxa are preserved in situ, there are ways to separate the three events. The relict marsh bivalves are all represented by individuals that exhibit the full size range for that particular species, that is, normal adults are a conspicuous part of the assemblage. The processes of exhumation by tidal forces and wave action have truncated in situ clusters of the four marsh bivalves noted above, and indeed, one can see different stages of this truncation. Also, relict marsh epifaunal bivalves lack encrusters, like *Balanus,* probably because the marsh environment was too stressful (low salinity and high turbidity). The shell morphology (thin valves with sharp, toothed ridges along the anterior) of the infaunal intertidal bivalves (*Petricola, Barnea,* and *Crytopleura*) indicate a firm to hard substrate and the borings of dead individuals of these taxa are filled with clean sand (not common in the marsh). Thus, the first palimpsesting event is recognized. *Mercenaria* and *Geukensia* are the numbers of the second palimpsesting event and are commonly found burrowing near, or attached to, the dense root masses of relict *Spartina* clumps. These two bivalves are easily recognized as being in an atypical habitat. Individuals of both taxa are small, the shell morphology of *Mercenaria* reflects a coarse (sand) substrate rather than the common marsh substrate of mud, and specimens of *Geukensia* are heavily encrusted by *Balanus* because conditions of the intertidal zone are ideal for this sessile organism. Thus, the third event is recognized. Using shell form, taphonomy, and other subtle features one can separate the several events of a condensed sequence preserved on a single bedding plane. Studies of bedding planes in the rock record should be aware that palimpsesting occurs and that it can be recognized.

We acknowledge, with thanks, grants from the St. Catherines Island Research Program of the American Museum of Natural History, supported by the Edward John Noble Foundation.

WESTERMANN, G. E. G., McMaster University, Hamilton, Canada

Limitations Placed on Jurassic Seaway Reconstructions by Ammonite Biogeography

Jurassic ammonoids have for a long time been treated biogeographically with implications to past continent positions. Phylloceratids, Lytoceratids, and homeomorphic Ammonitids ("leiostracans") inhabited oceans, mostly continental slopes, of most or all latitudes and, presumably, water temperatures; the large (mostly Cretaceous) heteromorphs were pelagic drifters in epicontinental seas as well as oceans, whereas the majority of Ammonitids (the "trachyostracans") were consistently neritic. The duration of the epiplanktonic post-hatching "pseudolarva," however, remains unknown, possibly varying from a few days in typical "trachyostracans," to several weeks in some "leiostracans." Thus, only the former had their dispersal restricted to epicontinental seas; they tended to be more restricted paleogeographically, often delineating bioprovinces. But species (and genus) distributions are much more complex and endemism (versus pandemism or cosmopolitanism) can in no way be predicted from ammonite shell morphology or structure. Like Nautilus, most ammonites could presumably swim actively as much as several kilometers per day and thus had the potential for rapid dispersal.

Another, poorly known variable in paleobiogeography is postmortem shell drift. Probably, shells of deep-water species tended to sink after death, but once rare shells had reached the surface they could have drifted for some time. Shallow-water shells presumably drifted more easily. As a rule, "leiostracans" drifted maximally a few tens of kilometers onto the shelf, whereas "trachyostracans" could perhaps drift as much as several hundred kilometers.

Biogeographic evidence for seaways is by homeotaxis across its length and/or at both ends. Epicontinental seaways should act as a filter against deep-water taxa, but their pandemism usually prevents recognition of filter usage. As a general rule, the degree of endemism should vary with the constriction of seaways and, if the latter are epicontinental, with the global eustatic cycle.

Northern Pacific gate: The Arctic ocean was generally wide open to the northern Pacific at least since the Simemurian, with still ill-defined "gates." The Early Jurassic Arctic Ocean connected via the epicontinental northern North Sea (Greenland–Norway Seaway) with western Europe. Earliest Jurassic ammonites, however, were extremely pandemic, probably because of unexplained exceptional eurytopy. From the late Toarcian to the Early Bajocian, however, the Greenland–Norway Seaway was closed, and the Arctic Ocean together with the northernmost Pacific contained only a low-diversity, Boreal endemic fauna (Bering Province). In the Late Bajocian, the Greenland–Norway Seaway opened again and a new circum-Arctic Boreal fauna spread widely, from the northern North Sea, through the Arctic Ocean, far into the northern Pacific.

The Arctic was occupied by a true ocean, at least temporarily, as evident from locally abundant "leiostracans." A second seaway, across eastern Europe, soon improved the trans-Arctic marine connections that lasted throughout most of the Late Jurassic.

Eastern Pacific gate: The faunistic connection of the eastern Pacific with the western Tethys (and North Atlantic) via the epicontinental Hispanic Corridor began in the Pliensbachian, perhaps already in the Sinemurian when rifting is clearly evident in the Huayacocotla Graben of eastern Mexico. The Hispanic Corridor acted as a decreasingly restrictive two-way dispersal route for "trachyostracans," until the end of the Early Bajocian. Sudden mid-Bajocian closure caused isolation that lasted for two stages and resulted in almost total endemism of eastern Pacific ammonites (East Pacific Realm). The described change in the degree of endemism is correlated with the global eustatic sea-level changes. Middle American ammonite faunas of Mid-Jurassic age occur only in the Mixteca Terrane of southern Mexico, located east (or the eastern part) of the Guerrero Plate. Its mid-Bajocian assemblage combines North-Cordilleran (USA), Andean and West-Tethyan (Eurafrican) taxa; but the late Bathonian–early Callovian assemblages are only Andean and West-Tethyan. These biogeographic data are compatible with current plate-tectonic models in which, during the early rifting stages, the Guerrero Block remained in proximity to the northwestern part of South America, while eastern Mexico, together with the American Craton, moved northward. The first overwhelming immigration from western Tethys occurred in the later Early Callovian and, perhaps for the first time, also across the Gulf of Mexico. This event coincides with the first appearance of Mid-Atlantic deep-sea floor, i.e., the beginning of true ocean. Subsequent Jurassic ammonite assemblages of the central and southern East Pacific resembled those of Tethys, attesting to good oceanic connection. Moderate provincialism, however, prevailed in the eastern Pacific, with the Oxfordian ammonites of Cuba somewhat linked to the Andes, and the Tithonian ammonites of the Southern Andes giving the first evidence of marine connections to Central Tethys via the Mozambique Channel.

WETZEL, K., Academy of Sciences of the German Democratic Republic, Leipzig, German Democratic Republic

Geochemical Evolution of Basic and Intermediate Volcanic Rocks of the German Democratic Republic

Basic and intermediate igneous rocks of the German Democratic Republic comprise early Phanerozoic, Proterozoic, and Tertiary volcanic rocks having ascended from deep regions of the crust or from the upper mantle, respectively, on tectonic inhomogeneities of the crust. Comprehensive geochemical and geochronological investigations (see References, this paper, and Gerstenberger, et al, 1987) have revealed that at least the majority of these rocks have formed by crystallization of mantle-derived magmas without or with small crustal contamination, the degree of crustal contamination increasing with decreasing age.

We have investigated the secular trends of minor element concentrations at given SiO_2 mass percentages as an indication of the evolution of the crust/mantle system in the region of GDR. It turned out that such a trend does exist and that the following regularities are valid:

1. The concentrations of the incompatible elements (elements with bulk solid/liquid distribution coefficients <1: K, Sr, Ba, P, Be, Li, Rb, Pb, Nb, Zr) in basic volcanic rocks of a given SiO_2 content increase in the course of the earth's history.

2. The bulk concentrations of the compatible elements (elements with bulk solid liquid distribution coefficients >1: Mg, Ca, Cr, Ni, Co, Mn) in basic volcanic rocks of a given SiO_2 content decrease in the course of the earth's history.

3. Chemical elements with distribution coefficients that are larger or smaller than unity in dependence on the SiO_2 content of the rock or magma, respectively (Ti, Fe, Na V), behave like compatible elements or like incompatible elements in dependence on the SiO_2 content of the volcanic rocks the secular trends of which are investigated.

4. The succession of the chemical elements with respect to those secular trends corresponds approximately with their bulk solid liquid distribution coefficients.

5. The concentrations of the most compatible elements in the youngest basic and intermediate rocks (Tertiary nephelinites, phonolites, and andesites) are smaller by a factor of about ten than their concentrations in the oldest basic and intermediate rocks (Phanerozoic harzburgites, garnet lherzolites, garnet websterites, garnet pyroxenites, garnetites, pyroxene granulites, and eclogites of the Saxon Granulite Mountains) at the same SiO_2 content.

6. The concentrations of the most incompatible elements in the youngest basic and intermediate rocks are higher by a factor of about thirty than their concentrations in the oldest basic and intermediate rocks at the same SiO_2 content.

The investigated rocks comprise basic and intermediate volcanic rocks of the Saxon Granulite Mountains and its surroundings, the age of which is supposed to be higher than 1000 Ma (pre-dalslandic metabasites), rocks of the ophiolite complex with ages clustering around 750 Ma, metabasites of the late pre-Cambrian era with ages of about 650 Ma, Cambrian metabasites with ages of about 540 Ma, Rhenohercynian and Saxothuringian variscic initialites with ages of the order of 400 to 370 Ma, subsequent volcanic rocks clustering around an age of 285 Ma, and Tertiary basaltic rocks with ages of the order of 25 Ma.

The regularities are discussed in terms of the geochemical evolution of the mantle beneath the territory of the GDR, in terms of the crustal evolution in this region (in view of a possible contamination of the mantle magmas by crustal material) and in terms of the evolution of the crust/mantle system (oceanic–geosynclinal–cratonic evolution) during the earth's history.

The following hypotheses leads to a full understanding of the regularities of minor element behavior during the earth's history:

Continuous thickening of the crust gives rise to shifting mantle magma generation toward greater depths. Thus pressure and temperature at which partial melting of mantle material proceeds are enhanced in the course of crustal evolution. Since the diminishing effect of pressure surpasses the enhancing effect of temperature on the degree of partial melting under such conditions, the degree of partial melting decreases during the earth's history. In view of a continuous supply of "fresh" material by mantle convection, the secular trends of minor element concentrations can thus be qualitatively attributed to a secular trend of the degree of partial melting in the mantle. Model calculations, however, reveal that closed system partial melting does not yield quantitative agreement with the experimental data. (The observed temporal changes of minor element concentrations are larger than those calculated from a simple partial melting model.) But a modified model of mantle anatexis taking into account a continuous extension of the melting regime (Wetzel, 1988) leads to a not only qualitative, but also satisfactorily quantitative agreement with the experimentally found extent of the temporal changes in minor element concentrations of basic and intermediate volcanic rocks in the territory of the GDR. It is an additional argument in favor of this extending system partial melting model that fast heat transfer provides both spatial expansion of the melting regime and fast magma generation as a precondition of the magma's becoming capable of penetrating the crust.

REFERENCES

Collective of authors, 1986, Vergleichende Geochemie basischer Magmatite auf dem Gebiet der DDR, part I and II: Freiberger Forschungshefte C 406 und C 407, Leipzig, VEB Deutscher Verlag für Grundstoffindustrie.

Gerstenberger, H., et al, 1987, Isotope geochronological, isotope and elemental geochemical investigations of the Sächsisches Granulitgebirge: A synopsis of the results: ZfI-Mitteilungen, v. 133, p. 353–368.

Wetzel, K., 1988, Natural element and isotope separations by phase conversions—synopsis of element and isotope separation by crystallization of magmas and partial melting of silicic rocks: Z. Geol. Wiss. Berlin, v. 16, p. 185–201.

WETZEL, K., M. REMER, and K. HIRSCH, Academy of Sciences of the German Democratic Republic, Leipzig, German Democratic Republic

Effects of Combined Fractional Partial Melting and Crystallization on Minor Element Concentrations in Igneous Rocks

The trace element patterns of igneous rocks have so far been conceptualized in terms of a series of models with more or less idealized stages in the generation and evolution of magmas (DePaolo, 1981; Allégre and Minster, 1978; Shaw, 1970; Wetzel, 1988). But all these models of magmatic processes take either the influence of partial melting as the process generating the magma or the effect of crystallization as the main process governing the evolution of the magma into account.

It is improbable, however, that this condition is fulfilled for the majority of magmatic processes to a sufficient degree. If the regions of generation of the magma on one hand and of its evolution on the other hand are of comparable size and if the distance between these two regions is small in comparison to their depths, timing of partial melting and fractional crystallization will be of the same order of magnitude. Furthermore, there will be no reason to assume that there can generally be established a continuous disjunction between the two regions.

We present a geochemical model of magmatic processes accounting both for generation of a magma by partial melting and crystallization of this magma after its ascension to a magma chamber. Both partial melting and crystallization are assumed to follow Rayleigh fractionation conditions, the magma chamber being continuously refilled by the products of partial melting.

The scheme of this combined fractional partial melting and crystallization (FPMC) model is shown in Figure 1.

The upper circle in this figure represents a magma body generating a continuous flow W of crystallization products under Rayleigh conditions. The concentration of a given minor element in this flow is denoted by $c(t)$; the respective concentration in the magma chamber is denoted by c^1. The lower circle in Figure 1 symbolizes a partial melting region generating a continuous flow ^{M}W of magma that refills the magma chamber. The concentration in the rocks before anatexis is denoted by c_o, the concentration in the anatectic melt by $^{M}c^1$.

The mathematical treatment of the model yields

$$\frac{c(t)}{c_o} = F^{-1}\left[F^{\alpha} + \frac{\varrho}{1-\varrho}\int_0^{1-F}\left(\frac{1-t}{F}\right)^{-\alpha}(1-kt)^{-\beta}\,dt\right]$$

where

F = the ratio of the residual amount to the initial amount of the magma in the magma chamber

ϱ = the degree of recharge of the magma chamber ($^{M}W/W$)

α = the quotient of the bulk solid/liquid partition coefficient D divided by (1-ϱ)

β = the quotient (D-1) divided by D

t = the time

k = a coefficient characterizing the timing of melting and solidification relative to each other ($k << 1$ for magmatic processes distinguished by a long period of magma generation in comparison to magma evolution and $k \approx 1$ for timing of melting and solidification to be of the same order of magnitude).

Numerical evaluation of this equation reveals the following geochemical features of magmatic rocks formed by coupled fractional melting and crystallization.

1. Despite the counteracting effects of partial melting and crystallization on the chemical composition of igneous rocks, the differentiating power of FPMC is not under all circumstances less, but under certain circumstances higher than that of single partial melting or single crystallization, respectively. This is particularly true if the degree ϱ of recharge of the magma chamber is small.

2. Due to the counteracting effects of partial melting and crystallization, the chemical composition of FPMC products shows a characteristic inversion in the course of propagating solidification of the magma body in a wide region of the parameters ϱ, k, and D, in the first stages preferentially following laws of fractional crystallization and in the latest stages progressively tending toward law of fractional partial melting. This is equivalent to a sequence mafic → sialic → mafic in a series of successively formed magmatic rocks with respect to minor elements as well as with respect to major elements.

Indeed, a series of such sequences can be found in the earth's crust, the generation of which might be attributed to combined fractional partial melting and crystallization. Thus, the global secular trend mafic → sialic in continental magmatism is followed by a widespread regressive tendency toward mafic igneous rocks during late stages of continental evolution.

Evolution of magmatism before, during, and after orogenic periods of the development of geosynclines is probably a second example of FPMC acting in the earth's crust. More or less all orogenic periods are distinguished by a development from mafic to felsic magmatism during the pre- and synorogenic stage and a subsequent inversion toward mafic or intermediate igneous rocks.

REFERENCES

Allégre, and J. F. Minster, 1978, Quantitative models of trace element behavior in magmatic processes: Earth and Planetary Science Letters, v. 38, p. 1.

DePaolo, D. J., 1981, Trace element and isotopic effects of combined wallrock assimilation and fractional crystallization: Earth and Planetary Science Letters, v. 53, p. 189–202.

Shaw, D. M., Trace element fractionating during anatexis: Geochimica et Cosmochimica Acta, v. 40, p. 73.

Wetzel, K., 1988, Natural element and isotope separations by phase conversion—synopsis of element and isotope separation by crystallization of magmas and partial melting of silicic rocks: Z. Geol. Wiss. (Berlin), v. 16, p. 181–197.

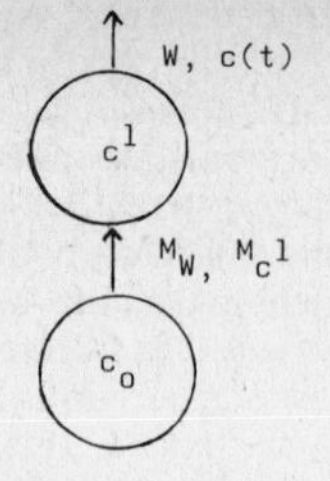

Figure 1—Combined fractional partial melting and crystallization (FPMC) model. (For explanation see text.)

WETZEL, K., and H. K. BOTHE, Academy of Sciences of the German Democratic Republic, Leipzig, German Democratic Republic, and M. BOTHE, Academy of Sciences of the German Democratic Republic, Dresden, German Democratic Republic

On Origin of Granitic Rocks of Meissen Monzonite Massif as Derived from Their Trace Element Patterns

The Meissen Monzonite Massif covers an area of at least 60 km length and up to 20 km width. Striking northwest-southeast, it follows the valley of the Elbe river, in a southwesterly direction adjoining the Saxon Ore Mountains (Erzgebirge) and in the northeast adjacent to the Lusatia Granodiorite Massif. The massif consists of several concentric dome-shaped intrusions. From margin to the center the following phases are to be found:

pyroxene monzodiorite and other monzodiorites	PMD/MD
hornblende monzonite, augitic	HMA
hornblende monzonite	HM
monzonite	M
biotite monzogranite	BMG
biotite monzogranite, porphyric	BMGP
monogranite	MG

This order corresponds to the age sequence of the several partial intrusions. All rock types except the monzogranite are subalkaline. The outcropping areas of the predominate types HM (including HMA), M, and BMG are 265, 50, and 220 km^2, respectively (Pfeiffer, 1964).

According to geological and petrological investigations the Meissen Monzonite Massif is assumed to have formed by fractional crystallization of a deep magma chamber episodically ejaculating magma into the top side. The aim of the present paper is to verify or refute, respectively, this hypothesis by geochemical investigations on the basis of the relation

$$\lg {}^1c = m \lg {}^2c + q \tag{1}$$

valid for rock series formed by crystallization of a uniform magma reservoir (Wetzel and Haberlandt, 1987), where ^1c and ^2c are the concentrations of pairs of elements 1 and 2 belonging to the alkaline, alkaline earth, rare earth, or first transition group, and m the function

$$m = ({}^1D - 1)/({}^2D - 1) \tag{2}$$

of the solid/liquid partition coefficients ^1D and ^2D of these elements calculated using a semiempirical approach of continental magmatic evolution (Wetzel, 1986).

If the different magmatic intrusions of the Meissen Monzonite Massif would stem from the same magma, relation 1 should be valid for several element pairs with correlation coefficients r close to unity ($0.7 < |r| < 1$). Using results of chemical analysis for major elements given in | 1 | and our own results for trace elements gained by X-ray fluorescence and neutron activation analysis, we indeed obtained correlation coefficients close to unity and good agreement between m values from equations 1 and 2 for a large number of element pairs in more than 300 samples taken from different sites. This gives evidence that the granitic rocks of the Meissen Monzonite Massif formed by fractional crystallization of a uniform magma, but does not allow us to discriminate between in situ crystallization and crystallization of a magma body situated beneath the emplacement area. In order to clarify this question, an in situ crystallization model is used assuming that the amounts of crystallization products are proportional to the outcropping areas of the appropriate types of rocks and that the totality of every type of rocks is in thermodynamic equilibrium with a magma whose chemical composition corresponds to the average concentrations of the magmas at the beginning and at the end of the solidification of the appropriate type of rock. Such a batch crystallization model permits the evaluation of solid/liquid partition coefficients, which are compared with those derived from the mentioned semiempirical approach. These investigations reveal severe discrepancies between partition coefficients calculated the two ways. Moreover, the partition coefficients for the successive crystallization steps (HM–M–BMG) for most of the investigated chemical elements differ from each other by far more than can be expected regarding the similarity of major element concentrations of the rocks under consideration.

These two arguments exclude in site fractional crystallization and give evidence of crystallization of a deep magma reservoir episodically delivering magma toward emplacement area of the granitic rocks of the Meissen Monzonite Massif.

Conclusions:

1. The granitic rocks of the Meissen Monzonite Massif, including their northern and southern offsets, have formed by fractional crystallization of uniform magma.

2. In situ crystallization can be excluded in favor of crystallization of a deep magma chamber delivering magma toward emplacement area of the granitic rocks.

3. Investigation of lg ^{1}c to lg ^{2}c correlations according to equations 1 and 2 combined with batch crystallization models and model calculations of solid/liquid partition coefficients | 3 | can be a good tool for investigating the history of igneous rocks.

REFERENCES

Pfeiffer, L., 1964, Beiträge zur Petrologie des Meissener Massivs: Freiberger Forschungshefte C 179.

Wetzel, K., 1986, Ein halbempirischer Ansatz zur Berechnung der Verteilungskoeffizienten einer Reihe von Kationen zwischen natürlichen Alumosilikaten und ihren Schmelzen: Chem. Erde, v. 45, p. 251–266.

———, and R. Haberlandt, 1987, Fractional crystallization models for calculating distribution coefficients of chemical elements between magmas and their crystallization products: Isotopenpraxis, v.3, p. 95–103.

WEZEL, F. C., Istituto di Geologia dell'Universita, Urbino, Italy

Helicyclic Tectonics

Geotectonic phenomena involve a series of events that repeat themselves in both space and time with a cyclical rhythm that is more or less regular and has various spatiotemperoal dimensions. Temporal cycles give rise in space to morphotectonic forms that express the different zonal stages of geotectonic and magmatic evolution reached by the various crustal units. This zonal structure has created a series of parallel tectonic belts that show an areal succession in which each outer belt represents a younger tectonic stage of development than that of the next contiguous belt on its internal side. Within each geotectonic system, the tectonic zones are progressively younger the nearer they are to the external convex side of the system.

The crustal deformation that occurs within a cycle is created by three principal geotectonic systems, namely the rift , island-arc, and Alpine-type fold-mountain systems. The zones of the *rift system,* from the youngest to the oldest, are as follows: (1) the continental rift, (2) the platform basin, and (3) the passive margin. These involve zonal stages that show the initial phases of the cycle (krikogenetic and early geosynclinal development), in which the continental crust becomes torn and fragmented into slivers by shearing, which also involves huge outpourings of basalt lavas. The post-rift regional (geosynclinal) subsidence is considered to be caused by the activation of compressional tectonics. The main contractional phases occurred during the Upper Jurassic, Mid-Cretaceous, and Upper Cretaceous. They correspond to the "peak" transgressions.

The tectonic zones of the *island-arc system* are: (1) the trench, (2) the fore-arc, and (3) the marginal back-arc. This is a shear-dominated system of deformation that is characterized in its most highly evolved (Mariana-type) phase by deep-reaching lithospheric root. In all the back-arc zonal stages anticlockwise block rotations (cyclonic-type vortices) are found within the region of geosynclinal subsidence. The back-arc zonal stage represents the eugeosynclinal phase (with associated ophiolites) of the geotectonic cycle.

The geosynclinal weak mobile strip is compressed by a tectogenesis that gives rise to the fold belts. Some of these major powerful phases of horizontal compression are actually of planetary extent, such as a the Late Eocene and Mid-Miocene compressional paroxysms. The fold-and-thrust belts become morphologically high mountains in the wake of a morphogenetic phase of the rapid vertical uplift of broad areas, which is preceded by acid plutonism.

In the *Alpine-type fold-mountain system* different zones of development can be distinguished: (1) the foreland, (2) the externides, (3) the internides, and (4) the hinterland. The presence of extremely high pressure minerals in the internides requires an extraordinary vertical component of movement, which is more than 100 km according to some authors. The remarkable change from horizontal compressional movements to substantial vertical uplifts indicates a transition from tangential contraction (of the order of 4–8 cm/A) to radial expansion (of the order of 0.1–0.5 cm/A). The compressional hemicycle is characterized by major transgression and submergence phases, while the expansional hemicycle shows regressional phases of emergence.

All phases of the tectonic cycle correspond to precise depositional sequences. The krikogenetic stage is characterized by continental clastics that fill graben-like depressions, and by trap basalts. The geosynclinal stage presents evaporites and black shales, siliceous cherts, limestones, and a thick sequence of clastics (pelagic and turbidite deposits). In contrast, the morphogenetic stage is characterized by molasse-type clastics in the foredeep trough. Moreover, a correlation exists with periodical changes in earth's magnetism, volcanic activity, climate, and life history. The coincidence of geological rhythms could hardly occur without worldwide fundamental changes in the mantle.

Thus, all geological phenomena and events are intimately connected with each other so that they form an inseparable whole characterized by dynamic equilibrium. Earth contracts and expands in endless cycles. The phases of each cycle are no more than a transient part of the continuous process of change and transformation taking place in the earth's crust. Such a global conception is not easily compatible with simple conventional geodynamic mechanisms. In fact, the above described regularities in spatial and chronological structure manifest even a certain unidirectional pattern that is not allowed for in current geodynamic models, which instead favor irregularity and chance as prime factors in the evolution of tectonic phenomena.

The link between the geotectonic cycle and phases of north-south contraction is believed to be caused by the inception of a period of acceleration in the earth's rotation, which culminates at the spin maximum in a phase of orogenic disturbance. A fundamental change in the stress field of the crust then occurs as a result of deceleration in the earth's rotation rate, during which there is expansion that reaches its maximum during the spin minimum.

Geotectonic cycle

(Diagrammatic)

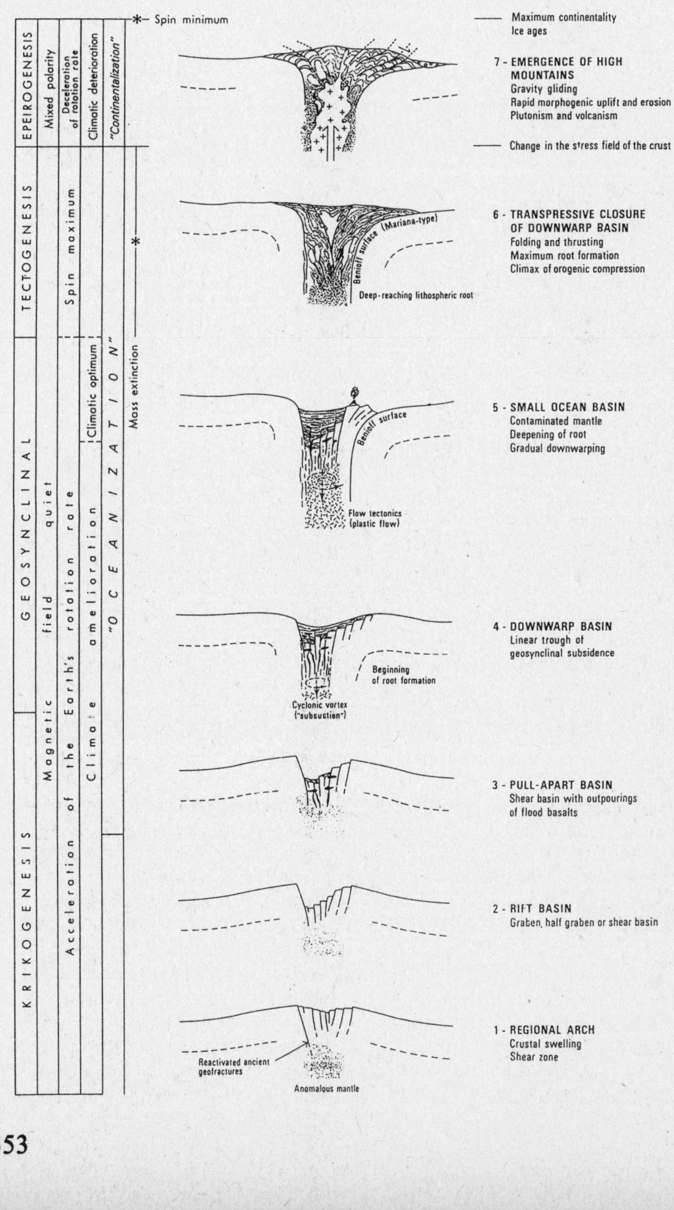

REFERENCES

Wezel, F. C., 1986, The Pacific island arcs: produced by post-orogenic vertical tectonics?, *in* F. C. Wezel, ed., The origin of arcs: Developments in Geotectonics, v. 21, Amsterdam, Elsevier, p. 529–567.

——— 1988, Earth structural patterns and rhythmic tectonism, *in* F. C. Wezel, ed., The origin and evolution of arcs: Tectonophysics, v. 146, p. 1–45.

——— in press, Influence of Earth dynamics on innovations in human thought: Modern Geology, v. 12.

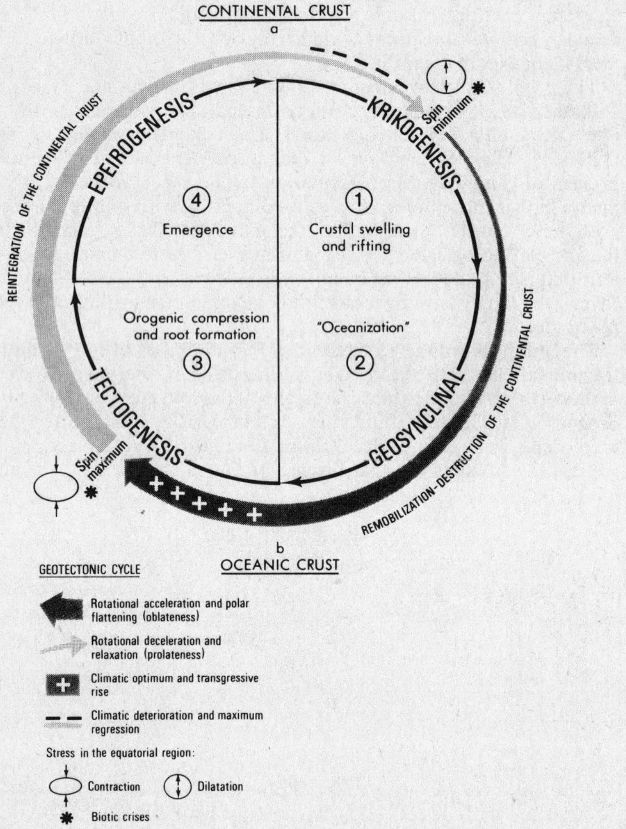

WEZEL, F.-C., Istituto di Geologia dell'Universitá, Urbino, Italy

The Indoysian: A Young Fold Belt 3000 km South of the Himalayas

On the basis of already published seismic reflection profiles and deep-sea drilling results (DSDP and ODP), the existence on the equatorial belt of the Indian Ocean, from the south of Sri Lanka to Sumatra for 2200 km, of a young south-verging fold belt, which we propose calling "Indoysian fold belt," can be shown. Its structural style (asymmetric concentric anticlines, displace by high-angle reverse faults) and the age of folding (post-Messinian) are features characteristic of the compressive fronts of the external zones of mountain folds.

It is postulated that this Jura-type fold belt represents the youngest distal part of the compressive Himalayan front, situated between a deep Molasse-type trough (Bengal Basin) and a submerged, possibly pre-Mesozoic, rigid foreland (present Central Indian and Wharton basins), cut by very deep graven-like troughs. These tectonic elements of the eastern region of the Indian Ocean are considered characteristic of the compressional deformation of the Himalayan foreland. This distal deformation, occurring 3000 km south of the Himalayas, would imply the presence in the foreland of deep detachment horizons (delamination) either inside or at the base of the crust or below the crust, and besides a mechanical coupling between the orogen and the foreland, necessary for the transmission of compressional stresses over so long a distance. These tectonic domains lie on a crust, which on the basis of DSDP coreholes, is postulated to be composed of eroded continental crust, which underwent phases of regional subsidence and progressive basinward (southward) transgression. It is a noteworthy fact that a parallelism exists between linear magnetic anomalies, gravity lineations, high heat-flow zones and the tectonic trends of the Indoysian fold belt. The geological evidence is not unfavorable to the idea that the oceanic flood basalts (Layer 2) might represent the covering of a foundered continental basement, viz., the ancient "Lemurian" foreland basement.

Seen in this perspective, the central and eastern Indian Ocean could be considered not as separated domain but rather as the *most distal* frontal part (foreland deformation) of the immensely huge bivergent lithospheric funnel of the Himalayan orogenic belt.

REFERENCES

Bowles, F. A., W. F. Ruddiman and W. H. Jahn, 1978, Acoustic stratigraphy, structure, and depositional history of the Nicobar Fan, eastern Indian Ocean: Marine Geology v. 26, p. 269–288.

Moore, D. G., J. R. Curray, R. W. Raitt, and F. J. Emmel, 1974, Stratigraphic-seismic section correlations and implications to Bengal Fan history, *in* C. C. von der Borch et al, Initial Report Deep Sea Drilling Project: v. 22, U.S. Govt. Printing Office, Washington D.C., p. 403–412.

Weissel, J. K., R. N. Anderson and C. A. Geller, 1980, Deformation of the Indo-Australian plate: Nature, v. 287, p. 284–291.

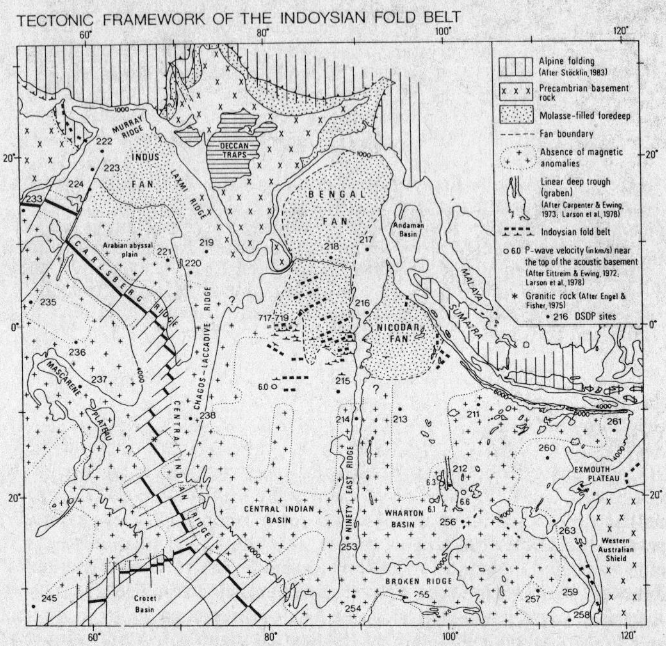

Figure 1—The *Indoysian fold belt* and the surrounding principal tectonic elements and major structures. The anticline axes and reverse faults are taken from Weissel et al (1980) and Bowles et al (1978).

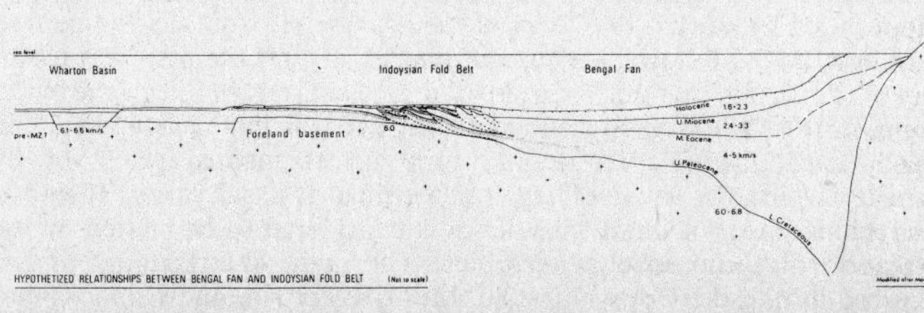

Figure 2—Diagrammatic structural cross-section through the *Indoysian fold belt* to illustrate in a simplified way the supposed relationships between the different tectonic elements (modified after Moore, Curray, et al, 1974, DSDP v. 22). Note the inferred progression through time of the compressive deformation of the fold belt, from the internal (northern) toward the external (southern) zones.

WHITE, R. R., T. J. ALCOCK and R. A. NELSON, Amoco Production Company, Denver, Colorado, USA

Anschutz Ranch East Field

The Anschutz Ranch East Field of northeastern Utah and southwestern Wyoming, USA, is a giant gas and retrograde condensate field with an estimated 800 million barrels of oil equivalent in place. This field, consisting of two anticlinal closures, is located in the Utah-Wyoming Thrust Belt and is the largest of 17 fields in this producing trend. Hydrocarbon production is from the Jurassic Nugget Sandstone, an extensive eolian deposit correlative with the Aztec and Navajo sandstones of the Colorado Plateau Province, western United States. The west and east lobes of the field occur in a slightly echelon pattern forming a leading edge fold pair in the hanging wall of the Absaroka Thrust sheet. The west lobe, the largest of the two structures, is approximately 2,000 ft higher than the east lobe and has an independent gas water contact. Discovered in 1979, this field is now fully developed on an 80 acre spacing after nearly 10 years of development drilling. The retrograde gas condensate produced from the field has its dew point only 230 psi below the original reservoir pressure. The continuing development challenge is to produce the gas and gas condensate with a minimum loss of the liquid hydrocarbons. A strict pressure maintenance program injecting a mix of nitrogen and residual field gas has been utilized. To date, 15% of the estimated hydrocarbons in-place for the west lobe have been produced and replaced with injection gas. The east lobe, a much smaller structure, is entering into the pressure depletion phase of production.

WHITEHEAD, J., Woods Hole Oceanographic Institution, Woods Hole, MA

Magma Waves in Diapiric and Compaction Dynamics

The roots of magma sources lie at great depths where melt first begins to form. As the melt collects, it must ultimately rise upward by either of two distinct mechanisms: by bulk diapirism, where an entire partially melted mass buoyantly rises or by porous flow (either around grains or through vein cracks, etc.). This talk will focus principally on the first mechanism and will end with some discussion of certain similarities between the first and second.

In early stages, a layer of material with enhanced melt will tend to develop a lower density and also a lower bulk viscosity than the surrounding regions. Such a layer is prone to develop fluid dynamic instabilities. Both observations of laboratory experiments and theory show that a relatively thin and horizontal planar layer of unstable material progresses through the following stages (Whitehead, 1986): The first stage is a Rayleigh-Taylor instability, in which disturbances of one specific wavelength grow most rapidly. If e is the ratio of the viscosity of the ambient material to the viscosity of the melt-enhanced layer, fastest growth is usually for wavelength of order $\epsilon^{1/3}$, though in some cases $\epsilon^{1/6}$. In the second stage, distortion of the interface is large, and it is found experimentally that the fluid moves out of the thick layer as cylindrical columns surrounded by relatively broad regions of descending material. In the third stage, fully matured structures are formed with development of a rim-syncline and pronounced overhang. Eventually, the original material will ascend as a spherical pocket of fluid trailed by a conduit. The presence of rigid horizontal surfaces below the layer does not change this result. If the original body has a form closer to a horizontal cylindrical tube, the wavelength can be considerably shorter—of the order of the radius of the tube. For a continuous melt source, the trailing conduit transports upwelling material through steady Poiseuille flow in which the stress of the low viscosity fluid balances the buoyancy force.

Theory and laboratory experiment show that the walls of these conduits support solitary waves. Among other interesting properties, these waves behave like solitons (solitary waves that are conserved upon collision with any other waves). In laboratory experiments (Helfrich and Whitehead, in preparation), wave characteristics (shape, amplitude, and speed) are very nearly conserved after collision with another wave. There was a slight tendency for the large wave to increase in amplitude and the small wave to decrease. The phase speed of both interacting waves decreased by $<4\%$. Measurement of phase shifts compared well with direct numerical calculations using the full conduit equations. KdV theory, formally valid for small amplitudes, gave poor results. It is also shown theoretically that these waves have closed streamlines in a frame moving with the wave. Hence, they directly transport unmixed material upward. Calculations (Whitehead and Helfrich, in press) indicate that material in these parcels will be far less contaminated by diffusion from the surroundings than material in ordinary pipe flow. Thus, such waves might convey material from deep in the earth to the brittle region (order of tens of kilometers) with little mixing.

In magma chambers (Huppert et al, 1986), conduits of one type of magma rising through another type of magma are also a possibility. When Reynolds number $[(g'Q3)^{1/5}]/\nu$ is large for either fluid (g' is buoyancy difference between the two magmas, Q is volume flux, and ν is kinematic viscosity), flow in the conduit is "turbulent." The "turbulence" is accompanied by severe axisymmetric or meander type distortion of the conduit walls, and there can be extensive mixing between the two magmas. When flow is nonturbulent, and especially when there are solitary waves, there may be almost no mixing between the two magmas.

The flow and nonlinear waves in conduits are analogous to flow and waves in the compaction problem (Scott et al, 1986) and share similar governing equations. It does not appear, however, that compaction waves have closed streamlines.

REFERENCES

Helfrich, K. R., and J. A. Whitehead, in preparation, Solitary waves on conduits of buoyant low viscosity fluid.

Huppert, H. E., R. S. J. Sparks, J. A. Whitehead, and M. A. Hallworth, 1986, Replenishment of magma chambers by light inputs: Journal of Geophysical Research, v. 91, p. 6113–6122.

Scott, D. R., D. J. Stevenson, and J. A. Whitehead, 1986, Observations of solitary waves in a viscously deformable pipe: Nature, v. 319, p. 759–761.

Whitehead, J. A., 1986, Buoyancy-driven instabilities of low-viscosity zones as models of magma-rich zones: Journal of Geophysical Research, v. 91, p. 9303–9314.

——— and K. R. Helfrich, in press, Wave transport of deep mantle material: Nature.

WHITTAKER, A., and T. C. PHARAOH, British Geological Survey, Nottingham, United Kingdom

Proposed Program of Scientific Deep Drilling for United Kingdom

In 1987, the UK Natural Environment Research Council established a Working Party to look into the feasibility of a scientific deep drilling program. The UK geoscientific community was asked to provide suggestions on possible sites and drilling-related activities that could be carried out as part of the project.

Although the surface geology of the United Kingdom is known reasonably well, detailed knowledge of the deeper geology rarely extends below a few hundred meters, particularly away from the major sedimentary basins. However, the BIRPS offshore data network of 11,000 km of high-quality seismic reflection data has imaged the deep, internal structure of the earth's crust beneath Britain better than anywhere else in the world. A British scientific drilling program would complement and build upon the BIRPS work, testing geotectonic and other models of major structure interpreted from surface-acquired data.

The recommended depth of drilling is in the range 3 to 6 km, within which there is wide scope for addressing many scientific and economic objectives in the United Kingdom. It is considered that the scientific return from a program that began with a number of carefully selected deep holes is potentially greater than that which would arise from a concentration of resources on a single superdeep borehole. The full program would include regional geophysical surveys, local and site-specific investigations, shallow pilot boreholes, the main deep drilling operations, downhole investigations and the complementary, essentially laboratory-based research. Each borehole will achieve many objectives. In addition to providing unique scientific information on sequence and structure, they will yield data on in situ physical and chemical conditions in the British crust. Many of the proposals fell within three thematic project areas: Tertiary Igneous Centers, Iapetus Suture Zone, and Variscide Belt.

The Tertiary Igneous Centers of Western Scotland and Northern Ireland characteristically show granitic and basic/ultrabasic rocks in close spatial and temporal association. Despite the dominance of granite at surface outcrop, all centers have large (up to 50 mgal) positive Bouguer gravity anomalies indicating the presence of underlying bodies of dense basic/ultrabasic rocks. Problems to be addressed include evaluation of the classic cauldron-subsidence models of emplacement, evaluation of recently published models on physical and chemical evolution of magma chambers, refinement of geophysical models, and the nature and distribution of hydrothermal alteration. In addition, the results would complement those from drilling projects in the shallow levels of active volcanic provinces in other parts of the world.

The Iapetus Suture Zone has been the subject of geological debate for over a hundred years. It is one of the most important features of the Caledonian mountain chain stretching in a continuous belt from Greenland and Scandinavia through Britain and Iceland to the eastern seaboard of North America. The structural complexity of the suture zone itself involves subduction imbrication, accretion, thrusting, and probably transcurrent displacements on subvertical discontinuities. The so-called Iapetus Suture is not exposed in the United Kingdom but has been imaged at depth on offshore BIRPS seismic reflection profiles. The proposed drilling program will concentrate initially on proving the nature of different, poorly exposed terrane elements rather than penetrating the discontinuities between them. Major problems to be addressed include definition and characterization of terranes in terms of their early Paleozoic cover sedimentary rocks and underlying basement, nature of terrane amalgamation processes (specifically the timing and kinematics of displacements and internal deformation and metamorphic patterns), and evolution of postamalgamation basins.

The Variscide Orogenic Belt and equivalents extend from central Europe through Britain to North America. Basement structure established by the Variscan Orogeny has played a significant role in the development of the younger Mesozoic sedimentary basins, many of which are prospective for oil and gas. Gently dipping, high-amplitude, deep crustal seismic reflectors are observed on deep seismic profiles in the Variscide belt of southern Britain and are usually interpreted as thrusts; reactivation of such structures is thought to have strongly influenced Mesozoic sedimentary basin development and inversion. The proposed drilling program would investigate the nature of the reflectors, locate the frontal detachment of the Variscide fold belt, investigate the structural and metamorphic history of the nappe stack, and examine the role of Variscide basement structures in controlling subsequent basin history.

Wide-ranging investigation of the in situ physical and chemical conditions of the British continental upper crust are proposed by the Working Party and recognized to be of prime importance. Among the geophysical work proposed, in addition to extensive logging, are special VSP experiments, determination of in site stress, paleomagnetic studies, heat flow, and other geothermal research. Geochemical experiments will incorporate testing and sampling of fluids to assess their flow, driving forces, sources, residence time, mixing processes, and their possible influences on the interpretation of surface-acquired geophysical data.

WICHROWSKA, M., Geological Survey of Poland, Warsaw, Poland

Lithology and Petrology of the Upper Precambrian Sediments in the Polish Platform Area

In Poland, the sedimentary cover of the Precambrian Platform forms a narrow belt along the Polish-Soviet border in the Podlasie and Lublin area. On the crystalline basement or on the Ryphean series beneath the Lower Cambrian rests the Vendian formation composed of two complexes: the upper (Valdai complex) built of arcoses (the Siemiatycze series) and claystones and mudstones (the Lublin series) and the lower (the Volhynia complex) effusive rocks (the Slawatycze series).

The unsorted or conglomeratic arcosic sandstones of the Slawatycze series lie upon volcanogenic rocks washed and weathered at the top. Locally, they grade into sandy conglomerates or better sorted and finer clay-sandy sediments. They represent continental sediments of alluvial fan type or coastal sediments.

The Lublin series shows clear evidence of shallow and unsteady sea. It is represented by alternating shales, mudstones, and fine-grained sandstones. Algae scales of the Vendotaenides group have been noted on the lamination surfaces in shales. This series grades into Lower Cambrian sandy sediments with glauconite as well as mudstone and claystone intercalations. The Cambrian-Precambrian sediments have been the subject of a comprehensive and thorough analysis by a team of Polish and Soviet geologists and paleontologists and an extensive report has been published in Russian and in English (Paleogeography and lithology of the Vendian and Cambrian of the Western East-European Platform, ed. of the English version, A. Yu. Rozanov and K. Lydka, 1987.)

Extensive petrographic studies of the Cambrian-Precambrian formations based on the quantitative analysis of the clastic sediments, including the grain size distribution and mineral composition, revealed considerable varieties in the apparently fairly uniform complexes. The biostratigraphic and sedimentological studies were coupled with a mineralogical and geochemical analysis of clay minerals from the claystone-mudstone complex of the Lublin series and earlier Lower Cambrian members, which permitted us to define the paleofacial conditions in the depositional environment.

The characteristics of the Polish Upper Vendian sediments are based entirely on the deep drilling record. The annexes contain two stratotype profiles and sketches showing the location of the boreholes. Petrographic and mineralogical results for the sandstones and claystones are presented in the form of diagrams and maps, while the variability of mineral composition and basic structural features are shown in photographs.

WICKHAM, S. M., Department of the Geophysical Sciences, Chicago, Illinois

Melt Transfer at Low Melt-Fraction and its Role in the Geochemical Modification of the Crust

Recent advances in our understanding of compaction, and other fluid dynamical processes associated with the extraction of liquid from partially molten rock by intergranular flow, have been successfully applied to the extraction of basaltic liquid from mantle rocks and to the petrogenesis of ocean-ridge basalts. As yet, there have been few applications of compaction theory to crustal melting environments involving the extraction of dominantly silicic liquids. Due to their high viscosity, these liquids are much more difficult to extract at low melt fraction. The situation is also complicated by the enormous variation in silicic melt viscosities (at least six orders of magnitude) and by the difficulty of accurately estimating melt viscosity in a given geological setting. However, notwithstanding these difficulties, compaction theory explains very nicely the centimetre to metre scale segregations that are commonly observed in migmatite terranes. A key question that follows is, can these small-scale segregations move away from their source regions and be collected to form large (kilometre scale) volumes of silicic liquid, or ascend to higher levels in the crust, stripping the partially molten region they came from of important chemical components in the process? This question is fundamental to the geochemical evolution of the continental crust, and remains unresolved. Although removal of low melt fracations is widely alluded to, (as a mechanism for dehydrating the crust to form granulites for example) it has yet to be convincingly demonstrated to have operated over large volumes of the crust.

Low melt-fraction silicic liquids (20% melting) derived from typical continental crustal source materials such as pelites, are often close to minimum melts in composition and contain few mafic minerals. They also tend to be depleted in various trace elements that are concentrated in relatively refractory phases such as biotite or amphibole (e.g. transition metals). These liquids will crystallize to form quartz-K feldspar-plagioclase (± muscovite) rocks, commonly termed leucogranites. Although leucogranites are common constituents of some high grade metamorphic terranes, (e.g. the Hercynian of Western Europe), in total they represent a volumetrically small fraction of the granitic rocks that occur worldwide. If we exclude those leucogranites that have not formed by liquid extraction at low melt fraction, the proportion becomes even smaller. Furthermore, these rocks often occur as relatively small bodies (less than a kilometre and often only tens of metres in size,) and never take on batholithic proportions. The exception that proves this rule is the Manaslu leucogranite (the best studied of the Himalayan leucogranites) which although being several kilometres thick, is texturally and isotopically very heterogeneous and therefore thought to be a composite body, made up of many hundreds of metre-scale dykes and sills.

The combined evidence suggests that although silicic liquids can be extracted on a small scale at low melt-fraction, large scale removal of this liquid over huge volumes of the crust does not occur. Even if segregation in the source is accompanied and assisted by other physical processes such as extensional fracturing or deformation, the rate limiting step is ultimately always going to be the compaction process, which for silicic melts will always be slow. thus, both the sluggishness of grain-scale segregation mechanisms by two phase flow, and the relative scarcity of minimum melt magma bodies in orogenic belts, both indicate that melt transfer at geochemical effects, is of relatively minor significance. This process should not be appealed to as a general model for dehydrating the deeper parts of the crust.

WICKHAM, S. M., University of Chicago, Chicago, Illinois

Problems in the Use of Granites as Indicators of Tectonic Setting

Although several geologists have previously sought to correlate the chemical and isotopic composition of granitic rocks with their tectonic setting, there is no a priori reason why such a link should exist. A sisicic magma will form if a pre-existing volume of rock of suitable composition (i.e. silica saturated) is subjected to the appropriate P-T-XH2O conditions at which large scale (>50%) melting may occur. This simple physical criterion has no direct logical connection with tectonic setting. Thus, very contrasting settings can give rise to a wide variety of different magma types and magma volumes, depending in large part on the local composition of the crust.

For example, pelitic material lying within the deeper parts of a thick (~12 km) supracrustal sedimentary sequence will melt to produce large quantities of silicic magma if heated to temperatures to 800 °C or more (e.g. if thick (~100 m) sills of mafic magma are emplaced at mid-crustal levels immediately below and adjacent to it.) If the same sedimentary sequence was composed mostly of quartzites and carbonates, but was subjected to exactly the same thermal regime, virtually no magma would be produced. Hence, vastly different quantities of magma may form with a wide range of different compositions, depending respectively on how refractory the crust was and what was the average composition of the region being melted. These simple considerations highlight one of several important reasons why granites are relatively insensitive indicators of tectonic setting.

This problem is well illustrated by comparing the tectonic and metamorphic setting of granitic rocks from the Hercynian of Western Europe and the Basin and Range of the western United States. In the Pyrenees, the fortuitous preservation of supracrustal sediments, volcanics and granites that have all formed at approximately the same time (during the Late Carboniferous Hercynian orogeny) allows us to place powerful constraints on tectonic setting. New geochronological results indicate that the metamorphism and melting that gave rise to the granites was probably not the product of continental collision and regional compressive deformation (as suggested by other models) and most likely took place within a rift zone environment. The very fact that there is an ambiguity of possible permitted interpretations, even in this well preserved and extensively studied terrane, clearly illustrates the weakness of using the granites as sole indicators of tectonic setting.

In the Basin and Range, several thermal events, resulting from both compressional and extensional tectonics during the Mesozoic and Tertiary have led to the generation of a wide variety of crustally-derived silicic magma types. The chemical and isotopic composition of many of these magmas is strikingly different to those in Hercynian Europe, despite there being strong similarities in structural style, tectonic history and thermal structure in some cases. This further emphasizes the important role of pre-existing crustal composition, rather than tectonic setting, in controlling granite geochemistry. Although granites are ubiquitous in orogenic belts, they should not be used as definitive evidence of tectonic setting. This is much better constrained by other lines of evidence (e.g. structural, stratigraphic and paleomagnetic studies.)

WICKHAM, S., University of Chicago, Chicago, IL, and A. L. GRUNDER, Oregon State University, Corvallis, OR

Evolution of Silicic Magmas from Source to Eruption Site: Examples from Northeastern Nevada

Tertiary crustal extension in northeastern Nevada was accompanied by voluminous silicic volcanism and by high-grade metamorphism and the generation and intrusion of various plutonic magma types. The volcanics are well dated by stratigraphic and radiometric techniques (e.g., the 40–32 Ma volcanic suite related to upper plate extension of the flanks of the Snake Range, east-central Nevada). The age range of the dominantly granitoid plutons in this area is not so well constrained, but although Mesozoic plutons occur throughout the region, well-dated Tertiary granitoids are also found (e.g., the Harrison Pass pluton in the Ruby Mountains dated at 37 Ma and the widespread suite of biotite monzogranite dykes and sills in the East Humboldt Range, dated at 29 Ma).

The extreme differential uplift affords a unique opportunity to study the petrology of synextensional volcanic rocks and the environment in which they were erupted, and to compare this supracrustal record with processes occurring in the deeper level melting environment within which the same magmas were generated. The volcanic record provides quenched samples of an evolving magma system and gives us a series of "geochemical snapshots" of a midcrustal melting environment at successive moments in time. The plutonic rocks provide us with a time-integrated record of crustal anatexis and allows us to evaluate the source materials and the mechanisms by which these magmas have been mobilized.

The volcanic rocks define two major groups, each representing about half of the estimated >9000 km^3 of magma erupted throughout this region. The older suite (40–36 Ma) ranges in composition from basaltic andesite to high-silica rhyolite, but this variation is not continuous; compositions between 66.5 and 72.0 wt% silica are absent. This compositional gap is filled by the dacites and rhyolites of the younger suite, (36–32 Ma), which comprise areally extensive dacite lavas and domes and three voluminous ash flow tuffs, zoned from rhyolite to dacite or rhyodacite to dacite.

Notwithstanding the fact that relatively few reliable age determinations have been made on the plutonic rocks of the Ruby–East Humboldt Range, and that the proportion of Tertiary granitoids in several of the high-grade terranes of this area is still relatively uncertain, it is clear that large volumes of crustally derived silicic magma formed at this time during crustal extension, and that some of this material crystallized at depth rather than being erupted at the surface. As with the volcanics, these granitoids can be crudely divided into two groups on the basis of size, composition, and level of occurrence. Within the high-grade parts of the Ruby–East Humboldt Range, enormous numbers of small bodies occur (typically on a scale of meters to tens of meters) covering a wide range of compositions. The principal lithologies are leucogranites (including 2-mica granites), biotite monzogranite (dated at 29 Ma), hornblende diorite, and rare amphibolite. This compositional heterogeneity in part mimics the heterogeneity of the early volcanic suite (although it is yet to be established if there is also a compositional gap). The other plutonic group includes much larger granitoid plutons such as the kilometer-scale Harrison Pass pluton, which intrudes at much higher structural levels than the suite of smaller bodies and comprises several distinct compositional variants (monzogranite, granodiorite, and granite). This type of magma body may be analogous to the younger volcanic suite.

Although more detailed comparisons await the collection of more compositional and isotopic data, we have the opportunity to use this terrane to evaluate directly the physical processes that govern the evolution of silicic magmas from the midcrustal regions where they were formed, through segregation into discrete magma bodies, to high-level magma chambers and supracrustal lavas. This will also allow us to compare metamorphism, melting, and magma transport within this extensional terrane, with magmatism in other orogenic belts where the tectonic setting is less well constrained.

WILBER, R., SEA, Woods Hole, MA, J. D. MILLIMAN, WHOI, Woods Hole, MA, and R. B. HALLEY, USGS, Woods Hole, MA

Late Quaternary Stratigraphy of Western Great Bahama Bank: Sea-Level Highstands Recorded in Progradational Slope Sequences Along Leeward "Megabank" Margin

The western margin of Great Bahama Bank between 26°N–23°N comprises a 325 km segment of the largest single carbonate platform in the western tropical Atlantic. Extensive high-resolution seismic profiling, detailed submersible reconnaissance, and sediment cores have shown that this "megabank" margin has undergone extensive lateral growth during Late Quaternary sea-level highstands.

Present-day sediment distributions show that large quantities of Holocene bank-top sediment are being deposited in two major depocenters located at the shallow bank edge and along the adjacent slope. The shallow bank areas are open (nonreef) systems characterized by a continuous sand wedge extending from the platform margin (at ~50 m water depth) eastward toward the bank interior (Figure 1). This sediment wedge is up to 25 m thick near the slope-break and thins to less than 5 m toward the bank interior.

Seaward of this depo-center, a relatively steep bypass slope occurs between water depths of 50 m and 160 m. Below 160 m, most of the slope is characterized by a thick accretionary wedge of fine sand and mud that is up to 10 km wide. Slope sediments in this wedge are strikingly monomineralic and monomorphologic, being comprised of $>80\%$ aragonite needles (primarily 1–4 microns in size) of a bank-top origin. In the southernmost part of the study area, the slope accretionary wedge achieves a maximum thickness of approximately 80 m in the relatively current-free Santaren Channel. Rates of slope deposition in this area may be as great as 10–15 m/Ka.

Holocene slope muds are not present in those areas where the Florida Current "sweeps" north along the embayed margin of the bank between 25°10'N and 25°40'N. Here the slope is composed of submarine lithified hardgrounds covered with a thin mobile layer of coarse sands and gravels. In these areas, the finer fractions have been winnowed and transported northward along the bank to "downstream" sites of deposition.

For the entire bank margin, first-order calculations indicate that the amount of fine-grained sediments "exported" to the slope may be volumetrically equal to the amount of sediment presently found on the bank-top itself.

In those slope areas with a thick Holocene sediment build-up, the underlying strata consist of a series of progradational sequences that are similar in thickness and lateral extent to the uppermost Holocene unit (Figure 1). Up to eight major sequences are present on seismic lines obtained from the Santaren Channel between 23°20'N and 25°00'N. These units are laterally continuous along strike throughout this interval exhibiting little variability in thickness or seismic signature. Sequence geometry is altered locally by post-depositional erosion or slumping. In current-swept areas, these underlying sequences are thinned, compressed, and seismically opaque.

By analogy to the Holocene, we interpret these units to have been deposited during repetitive sea-level highstands (bank-flooding events) in the northern Bahamas during Pleistocene time. These deposits thus represent one of the most highly expanded records of Late Quaternary sea-level fluctuations known from pure carbonate deposits in the western North Atlantic. The timing of deposition of these units is approximately a half-wavelength out of phase (with regard to sea-level curves) when compared to similar slope deposits found in terrigenoclastic systems.

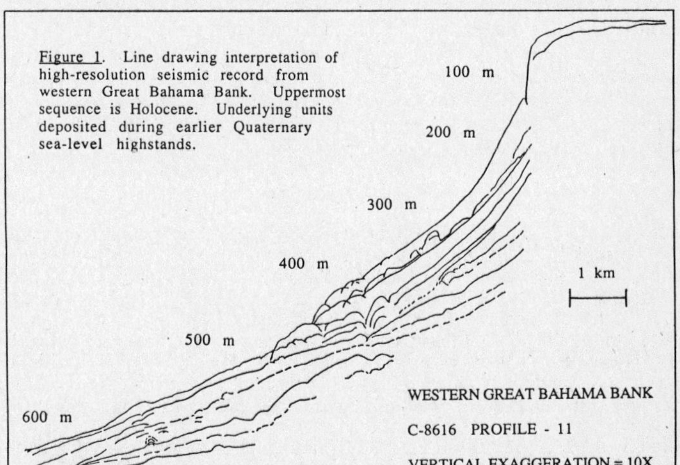

Figure 1. Line drawing interpretation of high-resolution seismic record from western Great Bahama Bank. Uppermost sequence is Holocene. Underlying units deposited during earlier Quaternary sea-level highstands.

The pattern of Holocene deposits found along western Great Bahama Bank demonstrates the mechanisms of lateral (leeward) growth of large carbonate platforms during sea-level highstands as well as the influence of strong bank-parallel currents in limiting such growth. The pre-Holocene progradational slope sequences illustrate the importance of leeward growth processes in the overall geological evolution of such "megabanks."

WILDE, P., M. S. QUINBY-HUNT, and W. B. N. BERRY, University of California, Berkeley, CA

Maintenance of Anoxic Oceanic Conditions by Thalassocratic Ridge Circulation and Coincidence of Formation of Pangaea with Oceanic Ventilation in Paleozoic

The geothermal activity on midocean ridge-rise systems has a profound effect on oceanic chemistry (Edmond et al, 1979; Von Damm et al, 1985) due to incorporation and modification of oceanic waters into the hydrothermal circulation at the ridge crest. Well-oxygenated deep waters enter the circulation system at the flanks of modern active ridges, react with ridge rocks, and are converted to hot anoxic fluids. These fluids exit the system at the ridge crest, where they react with normal ocean water. As a result of the ventilation of the modern ocean, such anoxic fluids are rapidly cooled and oxidized so that the thermal and anoxic effects are localized. However, the thermal and anoxic effects would be more extensive during warmer climates when the background oceanic waters would be less ventilated and potentially warmer. Maximum effect would be during thalassocratic times with active and extensive ridges. At such times, the ridge crest would be generating volumetrically more anoxic waters against a smaller redox gradient in the background ocean. Secondarily, more active ridges would swell the seafloor and produce high stands of sea level (Mackenzie and Piggot, 1981).

During the thalassocratic interval in early and middle Paleozoic, with anoxic waters in the pycnocline as a result of residual anoxicity (Berry and Wilde, 1978; Wilde, 1987) and relatively low atmospheric oxygen (Budyko et al, 1985), anoxicity could be maintained below the wind-mixed surface layer through ridge circulation independent of the oxygen demand of organic material and even against increased ventilation due to the production of biogenic oxygen and periodic cold climates. Anoxic water would be emitted at the ridge crest as a line source in the pycnocline and would spread laterally away from the ridge defining the anoxic maxima. This midwater anoxia would prevent or diminish ventilation and nitrification from below by more oxygen-rich deep waters. The ridge-controlled anoxic maxima would decrease with distance from the source at the ridge crest due to vertical advection of oxygen from the surface mixed layer above and deep, more oxygen-rich layer below the pycnocline. However, anoxia would be enhanced near shore and along belts of upwelling and higher organic productivity resulting in anoxia extending from the pycnocline into the mixed layer below the photic zone. At the same time, the high stands of sea level produced as a result of extensive ridge area also would result in transgression of anoxic water at the top of the pycnocline onto the continents. During such transgressions, bottom waters on the outer shelf would be anoxic, providing an environment for extensive accumulations of black shales.

The cessation of many active ridge segments during the incorporation of continental blocks into Pangaea, with the reduction of anoxic water production at the ridges coupled with the increase in atmospheric oxygen due to the development of land plants, would have enhanced ventilation of the pycnocline and vastly reduced the midwater anoxic layer as seen by a shift in the Ce anomaly (Wright et al, 1986) by the end of the Devonian. The colder climates during the Permo-Carboniferous glaciation would have ensured ventilation of the pycnocline and eliminated the potential for an anoxic layer until conditions favorable to anoxia returned in the anoxic events and episodes of black shale deposition of the Jurassic and Cretaceous (Schlanger and Jenkyns, 1976; Wilde and Berry, 1982; Jenkyns, 1988).

REFERENCES

Berry, W. B. N., and P. Wilde, 1978, Progressive ventilation of the Oceans—an explanation for the distribution of the Lower Paleozoic Black Shales: American Journal of Science, v. 278, p. 257–275.

Budyko, M. I. A. B. Ronov, and A. L. Yanshin, 1985, History of the earth's atmosphere: Berlin, Springer-Verlag, 139 p.

Edmond, J. M., C. E. Measures, R. E. McDuff, L. H. Chan, R. Collier, B. Grant, L. I. Gordon, and J. Corliss, 1979, Ridge crest hydrothermal activity and the balances of the major and minor elements in the ocean: the Galapagos data: Earth and Planetary Sciences Letters, v. 46, p. 1–18.

Jenkyns, H. C., 1988, The early Toarchian (Jurassic) Anoxic Event: stratigraphic, sedimentary, and geochemical evidence: American Journal of Science, v. 288, p. 101–151.

Mackenzie, F. T., and J. G. Pigott, 1981, Tectonics control of Phanerozoic sedimentary rock cycling: Journal of the Geological Society of London, v. 138, p. 183–196.

Schlanger, S. O., and H. C. Jenkyns, 1976, Cretaceous oceanic anoxic events: causes and consequences: Geologie und Mijnbouw, v. 55, p. 179–184.

Von Damm, K. L., J. M. Edmond, B. Grant, C. L. Measures, B. Walden, and R. F. Weiss, 1985, Chemistry of submarine hydrothermal solutions at 21 °N, East Pacific Rise: Geochimica et Cosmochimica Acta, v. 49, p. 2197–2220.

Wilde, P., 1987, Model of progressive ventilation of the Late Precambrian-Early paleozoic Ocean: American Journal of Science, v. 287, p. 442–459.

——— and W. B. N. Berry, 1982, Progressive ventilation of the Oceans—potential for return to anoxic conditions in the Post-Paleozoic, *in* S. O. Schlanger and M. B. Cita, eds., Nature and Origin of Cretaceous organic carbon-rich facies: New York, Academic Press, p. 209–224.

Wright, J., H. Schrader, and W. T. Holser, 1986, Paleoredox variations in ancient oceans recorded by rare earth elements in fossil apatite: Geochimica et Cosmochimica Acta, v. 51, p. 631–644.

WILDI, W., and P. FAVRE, Département de Géologie et Paléontologie, Geneva, Switzerland

History of Seaway Linking Western Tethys to Opening Atlantic Ocean

The history of Mesozoic seaways linking the western Tethys to the opening Atlantic is determined by the large amplitude (some 600 km), slightly divergent, counterclockwise strike-slip movement of Africa with respect to Europe between the Late Triassic and Early Cretaceous (Olivet et al, 1987).

Late Triassic (Figure 1): Open marine and restricted platform facies in the eastern Alpine and Adriatic realms are associated with subsiding salt basins both in areas of future opening sutures of the Tethys and intracratonic basins (Germanic and Sahara Basin).

Early Liassic: Moderate crustal stretching starts to affect the Alpine area of the European margin and the Atlas trough within the North African continental crust. Furthermore, a northern and a southern seaway separates the small Alboran plate from Africa, Central Europe, and the Adriatic block (Wildi, 1983).

Middle and Late Liassic: Open marine faunas in the Rif Basin indicate the opening of an Atlantic seaway between Northern Europe and the Portugese and North African Atlantic basins. Shortly afterwards, the Tethyan ammonite faunas penetrate through the seaways forming between Africa and Iberia as the marginal basins of the Rif and the External Betics.

Middle and Late Jurassic: The External Betic, Rif, Tell, and Alboran continental margins subside rapidly, whereas the clastic-filled Atlas Basin is affected by strike-slip movements. Ophiolites in the Penninic and Ligurian realms of the Alps indicate the formation of oceanic crust in the central part of the Tethys (Dercourt et al, 1985).

However, these rocks do not extend into the westernmost Tethys, where only few submarine volcanics are known between Africa and Iberia.

Early Cretaceous: Huge amounts of siliclastics are derived from the Arican, Iberian, and Alboran continental blocks and deposited in the Tethyan basins as large deep-sea fans. The main siliciclastic cycles may be correlated with periods of low global sea-level stand.

Late Cretaceous: The counterclockwise rotation of the Iberian plate associated with the opening of the Gulf of Biscay makes place for the

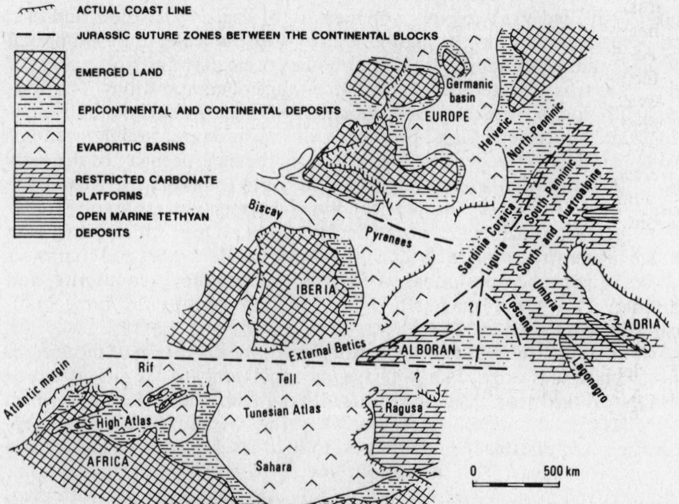

Figure 1—Late Triassic paleogeography and sediment facies distribution between Africa and Europe.

Alboran microcontinent between Iberia and Africa. This is the beginning of the collision of the Alpine orogenesis.

REFERENCES

Dercourt, J., L. P. Zonenshain, L. E. Ricou, et al, 1985, Présentation de 9 cartes paléogéographiques au 1:20'000'000 s'étendant de l'Atlantique au Pamir pour la période du Lias à l'Actuel: Bull. Soc. Géol. France, v. 1/5, p. 637–652.

Olivet, J. L., J. Goslin, P. Beuzart, et al, 1987, The breakup and dispersion of Pangea, map: Elf Aquitaine and Ifremer, Pau.

Wildi, W., 1983, La chaîne tello-rifaine (Algérie, Maroc, Tunisie): structure, stratigraphie et évolution du Trias au Miocène): Rev. Géol. Dyn. Géogr. Phys. v. 24/3, p. 201–297.

WILDI, W., B. LOUP, and E. AMATO, Department of Geology and Paleontology, Geneva, Switzerland, and H. FUNK, Geological Institute, ETH-Zentrum, Zurich, Switzerland

Mesozoic Subsidence Pattern of European Marginal Shelfs and Basins of Tethys

During the opening of the marine basins of the Alpine Tethys between Africa and Europe, the marginal shelfs and intracratonic basins subsided, from the Permian (Paris Basin; Brunet and Le Pichon, 1982) or the Triassic (eastern Helvetic realm; Funk, 1985) to the Cretaceous or Early Paleogene. Subsequent inversion and uplift was due to the Alpine orogenic events during the collision of Africa and Europe. The following geological units derived from the former Mesozoic marginal shelf areas (Figure 1a,b): (1) the Helvetic units, including the Variscan basements and the sedimentary covers of the External Massifs, and the allochthonous Helvetic nappes that have been thrust from south to north over the top of the External Massifs: (2) the Swiss Plateau, where the Mesozoic sediments of the European borderland are buried below thick deposits of the Oligo-Miocene Molasse; (3) the Folded Jura mountain range, where Mesozoic and some Cenozoic sediments have been folded and thrust to the north by the Alpine compression; and (4) the Tabular Jura, where the Mesozoic sediments are almost undeformed on top of the Variscan basement.

After palinspastic restoration of this Mesozoic platform, the subsidence history of the area has been reconstructed in order to obtain a better understanding of the relationship between the timing and the subsidence mechanisms of the European marginal shelfs with respect to the basins of the Tethys. In the computation of subsidence (Figure 1c), the following data from about 50 stratigraphic sections have been introduced: stepwise compacted sediment thickness, age in million years, sedimentation depth, and long-term eustatic sea-level fluctuations (Haq et al, 1987).

During Middle and Late Triassic times (240–210 m.y.), the Jura and the northern part of the Swiss Plateau subsided relatively rapidly, a trend that decreased in the Early Jurassic. The Helvetic realm remained stable. A limited acceleration of basement subsidence started at the beginning of the Middle Jurassic at about 180 m.y., and a little earlier in the adjacent Northpenninic Tethyan area. Up to the end of the Middle Jurassic, the northeast-southwest-striking Jura basin was clearly lined out. Subsidence rates were higher in the west than in the east (Figure 1). A similar east-west trend may also be seen in the Helvetic area. From the Triassic to the Middle Jurassic there is little evidence for pronounced crustal extension. Thus, other mechanisms have to be tested to explain the subsidence.

At the beginning of the Late Jurassic (150 m.y.), strong subsidence affected all realms. Extension structures are known from the Helvetic area. Between 130 and 120 m.y. the subsidence slowed down definitively, with the exception of the southernmost Helvetic realm, where a late Early as well as Late Cretaceous subsidence phase is known. The extension structures in the Helvetic realm indicate a linkage between the Late Jurassic and Cretaceous evolution of the Helvetic margin and the deformation of the Tethyan basin.

REFERENCES

Brunet, M. F., and X. Le Pichon, 1982, Subsidence of the Paris Basin: Journal of Geophysical Research, v. 87, n. B10, p. 8547–8560.

Funk, H., 1985, Mesozoische Subsidenzgeschichte im Helvetischen Schelf der Ostschweiz: Eclogae Geol. Helv. v. 78/2, p. 249–272.

Haq, B. U., J. Hardenbol, and P. Vail, 1987, Chronology of fluctuating sea levels since the Triassic: Science, v. 235, p. 1156–1166.

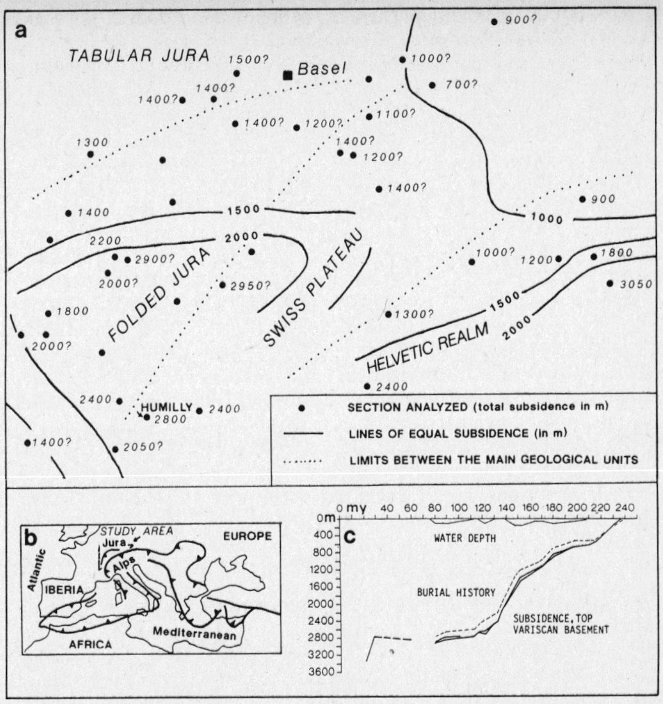

Figure 1—Subsidence history of the European marginal shelfs and basins of the Tethys between 240 m.y. and 100 m.y. (a) Palinspastic map of the European marginal realms; evaluation of the total subsidence of the top of the Variscan basement between 240 and 100 m.y. (? = extrapolation of local subsidence history diagrams). (b) Location of the study area. (c) Subsidence history diagram for the Humilly well near Geneva.

WILDING, M.C., and B. HARTE, University of Edinburgh, Edinburgh, Scotland, and J. W. HARRIS, University of Strathclyde, Glasgow, Scotland

Evidence of Asthenospheric Source for Diamonds from Brazil

About 50% of the eclogitic paragenesis garnets included in diamonds from an alluvial diamond mine at Sao Luiz, Brazil, show high silica contents, which suggests that they originated at depths greater than 300 km. Two major groups of inclusions are recognized according to whether the garnets have high or low silica contents, and these groups are subdivided on the basis of their clinopyroxene characteristics.

Group I: This group includes all garnets with normal silica content and two subgroups are recognized.

IA—Homogenous garnets of composition 0.38 Mg/R^{2+}. By comparison with eclogitic paragenesis garnets from other localities, they are relatively rich in calcium, titanium, and sodium (12.8–13.7 wt% CaO, 1.1–1.9 wt% TiO_2, 0.2–0.4 wt% Na_2O).

IB—Garnets of more variable composition (0.38–0.42 Mg/R^{2+}) are intergrown with clinopyroxenes of $Mg/(Mg + Fe)$ 0.56 to 0.76 and significant jadeite content (up to 10 mol%). The grain boundaries between the mineral species are highly curved and cuspate, unlike normal equilibrium solid-solid boundaries or crystallographically controlled boundaries.

Group II: This group is characterized by high silica garnet and is subdivided into two subgroups.

IIA—These inclusions consist of a single phase, which when calculated on a normal garnet atomic basis of 12 oxygens, shows silicon atomic proportions of 3.1 to 3.4. This chemistry represents a solid solution between garnet and pyroxene with silicon atoms occupying aluminum sites in a garnet structure, and the compositions show that there is up to 40 mol% of pyroxene relative to the garnet. High sodium contents (0.6–1.25 wt% Na_2O) correlate with high silica while calcium and aluminum are depleted relative to Group I. This suggests the coupled substitution $NaSiCa^{-1}Al^{-1}$. The Mg/R^{2+} ratios of these high silica garnets are similar to those in Group I and show a range of 0.36 to 0.42.

IIB—As shown in Figure 1, these inclusions are composite with high silica garnets (garnet–pyroxene solid solution) in the cores usually surrounded by separate pyroxene phases that are distinguished from IB pyroxenes by having up to 45 mol% jadeite. In the margins of some of these inclusions there is evidence of a breakdown of high-jadeite to low-jadeite pyroxene through a network of an unidentified alumino silicate (composition 66 wt% SiO_2, 28 wt% Al_2O_3). Occasionally, surrounding the garnet–pyroxene solid solution cores are garnets of normal silica content that have

Mg/R^{2+} of 0.36 to 0.42, and these are probably a product of the breakdown of the garnet-pyroxene solid solution.

Phase boundaries are usually cuspate. A few inclusions show complicated lamellae of clinopyroxene in garnets of normal silica content, suggesting exsolution; but generally the volume of clinopyroxene appears to be too great to have simply exsolved from a high silica garnet composition.

Paragenesis: The Group I diamond inclusions, although distinct from other localities, do not show any unusual pressure-related characteristics. For Group II garnets, however, a higher than normal pressure of derivation is required because pyroxene only becomes soluble in garnet at pressures in excess of 100 Kbars. Garnet-pyroxene solid solutions are stable at depths from 400 to 900 km and coexist with a jadeitic pyroxene before transition into a single garnet structured phase (Ringwood, 1967, 1982). The Group II inclusions from Sao Luiz diamonds, therefore, show clear evidence of derivation from depths where the mantle is considered to be asthenospheric. Transition of garnet-pyroxene solid solution to garnet of normal silica content and of high-jadeite pyroxene to low-jadeite pyroxene appears to have occurred following incorporation into the diamond. The fact that discreet groups of normal and high silica garnet occur and that Group II inclusions contain both high and low silica garnets suggests that the diamonds and their inclusions were formed and resided at two different mantle depths prior to transportation to the surface.

REFERENCES

Ringwood, A. E., 1967, Earth and Planetary Science Letters, v. 2, p. 255–263.

——— 1982, Journal of Geology, v. 90, p. 611–643.

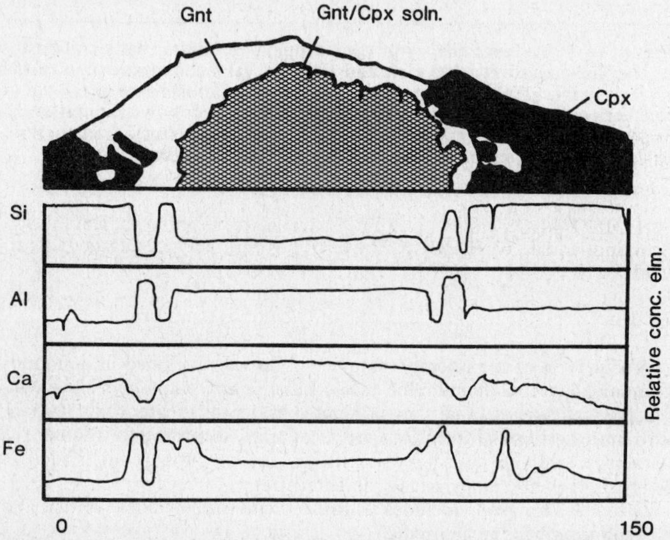

Figure 1—Schematic plan view of half a Group IIB inclusion (upper part) showing the relative concentration of elements (lower part). The central core of garnet-pyroxene solid solution shows a gradual variation in iron and calcium suggesting some diffusion. Surrounding the core is garnet of normal silica content, which probably represents the breakdown of the garnet-pyroxene solid solution. A homogeneous jadeitic clinopyroxene forms a discontinuous rim surrounding the inclusion.

WILKERSON, G., Bureau of Land Management, Bakersfield, California, USA

Cyclic Production of Monterey Shale

The Monterey Formation is the source for much of the hydrocarbon in oil and gas fields of Miocene and younger age in Southern California. It is also a reservoir formation for some of these oil fields. Oil or gas shows are encountered in the Monterey practically everywhere it is tested. Many Monterey fractured shale reservoirs are tabular, the boundaries of which parallel the axial plane of folds. These reservoirs are fragile and easily damaged. Many marginal quality fractured reservoirs have been and continue to be destroyed during drilling. What is done to these reservoirs in the first few minutes of drilling affects their behavior and producibility for several years. The permeability of many reservoirs is too low to sustain economic production rates using traditional techniques for more than a few months. Artificial fracturing and acidization improves the production rates of marginal quality Monterey fracture reservoirs, but only for a short period of time. Because of the plastic behavior of shale at depth, it is possible to produce some Monterey fractured reservoirs cyclically. Average production rates of three to ten bbls/day for several years can be expected from marginal quality Monterey fractured reservoirs which have traditionally been abandoned. The cyclic production of Monterey fractured reservoirs will significantly increase the recoverable hydrocarbon from this formation. Drilling practices and production procedures must protect the fragile and easily damaged Monterey reservoirs. These practices are different from those used to explore for hydrocarbon accumulations in the turbidite or beach sand facies.

WILLIAMS, G. E., University of Adelaide, Adelaide, Australia

Precambrian Tidal Rhythmites and the Earth's Paleorotation

Since John W. Wells published "Coral growth and geochronometry" in 1963, the study of growth rhythms in fossils has provided the only means with the potential to directly determine the Earth's paleorotation and the Moon's orbital periods in the geological past. Information on the dynamics of the earth Moon system for Precambrian time in general has not been forthcoming, however, because of the absence of shelly fossils and the doubtful reliability of geochronometric data so far obtained from Precambrian stromatolite growth patterns.

Sedimentary rhythmites of laminated siltstone and fine sandstone from late Proterozoic (~650–Ma) glaciogenic formations in South Australia (Elatine Formation, Reynella Siltstone, Chambers Bluff Tillite) therefore are of particular importance as they are best regarded as tidal deposits recording a full spectrum of paleotidal cycles ranging from semediurnal to the lunar nodal cycle (Williams, 1988a,b). Although the rhythmites have no known modern counterparts, they are interpretable as the deposits of distal ebb-tidal jets (see Ozsoy, 1986) built below wave base in a distal ebb-tidal delta to marine shelf setting. Cyclic changes in the thickness of semidiurnal and diurnal laminae, and of conspicuous fortnightly "bundles" of such laminae, are taken to reflect variations in the amount of suspended material jetted offshore through cyclic variations in the velocity and range of paleo-ebb tides.

Time-series obtained from drill core of the ~650-Ma-old Elatina Formation contain strong periodicities that are evident visually or revealed by Fourier spectral analysis. These data allow determination of the Earth's paleorotation and the past dynamics of the Earth-Moon system with an accuracy previously unobtainable for the Precambrian. The late Proterozoic (~650 Ma) year contained 13.1 (±0.5) lunar months and 400 (±20) days, and the late Proterozoic lunar month 30.5 (±1.5) days. The periods of the lunar apsides (perigee) and lunar nodal cycles were then 9.7 and 19.5 (±0.5) years respectively. the data indicate an Earth-Moon distance ~650 Ma ago no less than ~95% of the present distance, which implies that the Moon was not in close proximity to the Earth at any time during the Proterozoic.

Paleotidal and paleorotational data for late Proterozoic time ~650 Ma ago as determined from the rhythmites of the Elatina Formation and Reynella Siltstone, and those calculated for that time by Lambeck (1978) based on paleontological data, are compared with their modern equivalents in Table 1. The value of 13.1 (±0.5) lunar months per late Proterozoic year derived from the Elatina series is consistent with most paleontological data (Scrutton, 1978). Importantly, the rhythmite data imply, by comparison with values derived from paleontological data (Table 1), an average equivalent phase lag (the angle between the Earth-Moon line and the Earth's tidal bulge, derived from the response of the solid and ocean tides) near 3° since late Proterozoic time rather than the present value of 6°. An average phase lag angle smaller than the present value may be a principal cause of the apparent reduced tidal friction in the past.

Overall, the late Proterozoic rhythmites in South Australia provide a unique set of data that represents the first firm evidence for Precambrian paleotidal values and paleorotation. Study of other sedimentary rhythmites, particularly those of early and middle Proterozoic age, may further advance our understanding of the Earth's paleorotation and the history of the Earth-Moon system. For example, very thin microbands in certain early Proterozoic banded iron-formations, although usually regarded as annual increments, may well be dirunal or semidiurnal (see Walter, 1972). The maximum count of 28 microbands per rhythmite cycle in the 2.5-Ga-old Weeli Wolli Formation of the Hamersley Basin, Western Australia (Trendall, 1973), therefore may suggest that these cycles are fortnightly or monthly bundles of microbands. Through study of such ancient rhythmites, sedimentology promises to extend geochronometry deep into the Precambrian.

REFERENCES

Lambeck, K., 1978, the Earth's palaeorotation, In: Brosche, P. and Südermann, J. (eds.), Tidal Friction and the Earth's Rotation, Springer-Verlag, Berlin, p. 145–153.

Ozsoy, E., 1986. Ebb-tidal jets: a model of suspended sediment and mass transport at tidal inlets, Estuarine, Coastal and Shelf Science, p. 22, 45–62.

Scrutton, C. T., 1978, Periodic growth features in fossil organisms and the length of the day and month, In: Brosche, P. and Südermann, J. (eds.), Tidal Friction and the Earth's Rotation, Springer-Verlag, Berlin, p. 154–196.

Precambrian Hamersley Group, Western Australia, Econ. Geol., 68, p. 1089–1097.

Walter, M. R., 1972, A hot spring analog for the depositional environment of Precambrian iron formations of the Lake Superior region. Econ. Geo., 67, p. 965–980.

Wells, J. W., 1963. Coral growth and geochronometry, Nature, 197, p. 948–950.

Williams, G. E., 1988a, Late Precambrian tidal rhythmites in South Australia and the history of the Earth's rotation. Jour. Geological Society of London, in press.

Williams, G. E., 1988b, Precambrian sedimentary cycles and the Earth's paleorotation, Eos, in press.

Table 1. Late Proterozoic (~ 650 Ma) and modern tidal and rotational periods

Parameter	Late Proterozoic			Modern
	Lambeck (1978)[1]		Rhythmite data	
	(a)	(b)		
Days in lunar month	30.7	30.2	30.5 (± 1.5)	29.53
Lunar months in year	14.4	13.2	13.1 (± 0.5)	12.4
Days in year	~440	~400	400 (± 20)	365
Lunar apsides cycle (years)			9.7	8.85
Lunar nodal cycle (years)			19.5 (± 0.5)	18.61

[1]The figures from Lambeck (1978) are derived from Phanerozoic paleontological data. they are based on average equivalent phase lag angles of (a) 6° (the present value) and (b)3°, and assume that tidal friction is the only phenomenon responsible for secular changes in the earth's rotation and the Moon's revolution.

WILLIAMS, H., Memorial University of Newfoundland, St. John's, Canada

Temporal and Spatial Subdivisions of Rocks of Canadian Appalachian Region

Rocks of the Canadian Appalachian region can be subdivided into four broad temporal categories: late Precambrian to early Paleozoic, middle Paleozoic, late Paleozoic, and Mesozoic. Rocks of all temporal categories are lithologically distinct, and in most places they are separated by unconformities.

Rocks of the late Precambrian–early Paleozoic temporal category display sharp contrasts in lithology, stratigraphy, and thickness across the orogen, and these contrasts define spatial *zones*, which from west to east are the Humber, Dunnage, Gander, Avalon, and Meguma. The divisions also exhibit structural, faunal, geophysical, plutonic, and metallogenic contrasts, thus enhancing their distinctiveness. Because zones are based on the oldest rocks of the orogen, they are fundamental to understanding its protoliths and controls of subsequent development. Obviously, the zone boundaries are defined best where rocks of appropriate age are exposed over broad areas. This is the case in Newfoundland, but each zone can be traced by limited exposure across the entire Canadian Appalachian region.

Middle Paleozoic rocks also exhibit lithic, stratigraphic, and thickness variations across the orogen, and as in the case of early Paleozoic zones, they define middle Paleozoic *belts*. The belts are clearest among the extensive middle Paleozoic rocks exposed in Quebec, New Brunswick, and Nova Scotia. From west to east, the belts are Gaspe, Fredericton, Mascarene, Arisaig, and Annapolis. The middle Paleozoic record in Newfoundland is fragmentary, and there is no belt for belt equivalence with the mainland. Different names are used, which from west to east are Clam Bank, Springdale, Cape Ray, Badger, Botwood, and Belleoram.

The contrasts among rocks of middle Paleozoic belts are less severe, compared to those of early Paleozoic zones. Lithologic and stratigraphic contrasts are most important, but eastern mainland belts also have faunal, plutonic, and metallogenic distinctiveness. In areas affected by Ordovician deformation, middle Paleozoic belts cross early Paleozoic zones. Thus, in the Gaspe–New Brunswick area, the Gaspe Belt crosses the Humber, Dunnage, and Gander zones, indicating that these early Paleozoic zones were already in proximity during the middle Paleozoic. Similarly, the Newfoundland Springdale Belt straddles the Humber–Dunnage boundary. In areas unaffected by Ordovician deformation, early and middle Paleozoic rocks are conformable and middle Paleozoic belts coincide with early Paleozoic zones. Thus, the Annapolis Belt and Meguma Zone are one and the same, and the Badger Belt lies within the Newfoundland Dunnage Zone.

Late Paleozoic rocks of the Canadian Appalachian region are mainly terrestrial cover sequences that unconformably overlie deformed middle Paleozoic and older rocks. The rocks are everywhere essentially the same, dominantly redbeds with local volcanic units toward the bases of the sections. Their thicknesses define a number of *basins*, mainly located in New Brunswick and Nova Scotia. The distribution of these basins has no relationship to the geometry of belts or zones.

Mesozoic rocks, mainly Triassic, are redbeds and basalts that unconformably overlie deformed late Paleozoic and older rocks. These are restricted onland to the Cape Split *Trough* in the Bay of Fundy area. Coeval dykes are more widespread.

The temporal rock divisions are separated by structural events, although these are not everywhere coeval and controls may be different from one area to another. Taconic Orogeny (Early and Middle Ordovician) affected the Humber Zone and parts of the Dunnage and Gander zones. It is absent in the Avalon and Meguma zones and parts of the northern Newfoundland Dunnage Zone. Acadian Orogeny affected the entire orogen, except for parts of the Newfoundland Avalon Zone. It is generally of Devonian age, but there is growing evidence for Silurian orogenesis in the eastern Dunnage and Gander zones. Alleghanian Orogeny of Carboniferous–Permian age affected a narrow area along the Avalon–Meguma boundary in Nova Scotia and another along the Humber–Dunnage boundary in Newfoundland.

In a general way, variations in structural style, plutonism, metamorphism, and other features correspond to lithic and stratigraphic contrasts exhibited among rocks of the four temporal divisions, thus adding to their importance. This is expectable, as each temporal division represents a different depositional environment and tectonic setting: entirely marine early Paleozoic zones, mixed marine and terrestrial middle Paleozoic belts, mainly terrestrial late Paleozoic basins, and an entirely terrestrial Mesozoic trough.

Zones are discrete entities. The Humber represent that part of the orogen that is linked stratigraphically to the North American craton, or the Appalachian miogeocline. Others represent accreted parts of the orogen or suspect terranes.

Belts may represent a number of tectonic elements. Some trend acutely across the miogeocline and already accreted terranes. Others are sited along zone boundaries, either above earlier structural junctions or located between early Paleozoic terranes. Still others form internal parts of accreted terranes unaffected by early Paleozoic orogeny.

Late Paleozoic rocks represent a uniform terrestrial cover across the whole orogen with local anomalously thick sections controlled by rifts and uplifted blocks.

Mesozoic rocks occur in a single rift coincident with a thick Carboniferous basin and the earlier Avalon–Meguma zone boundary.

WILLIAMS, J., San Jose State University, San Jose, CA, and J. N. FALLS, Upp Geotechnology, Inc., San Jose, CA

Extrapolation of Subsurface Geology Data as Basis for Designing Select Regional Geological (Liquefaction Potential) Hazard Maps

Geologic hazard maps designate areas in which particular geologic hazards are believed to exist and are popular as aids in land use planning. Governmental agencies use these maps as published evidence to require predevelopment geological investigations. Most mpas appropriately contain the statement, "This map is not a substitute for a site evaluation and should be used only for regional planning purposes." The question of how much extrapolation of geological data should be made, however, must be addressed. During the development of a number of liquefaction potential hazard maps for portions of the Central Coast Ranges of California, extrapolations were made with too little information about the region's subsurface geology.

In the Santa Clara Valley, immediately south of San Francisco Bay in central coastal California, liquefaction has occurred, notably during the 1906 (8.3 magnitude) San Francisco Earthquake and the 1984 (6.2 magnitude) Morgan Hill Earthquake. Five independent regional studies to identify areas of potential liquefaction were completed between 1974 and 1988. Comparisons of the maps document significant disagreements as to areas designated as having a high potential for liquefaction. Early maps showed as much as 80% of the City of San Jose portion of the valley floor as being in the "high or very suspect" category. Recent studies assign less than 25% of the area to the high risk category, and large areas are unassigned as to hazard level. These great variations in assigned hazard level from presumably credible studies create confusion, unneeded and expensive supplementary geotechnical investigations, and the potential for undetected hazardous areas.

Early studies were based on simple assumptions about the subsurface geology such as regional continuity and extensive vertical extent. Additionally, the presence or absence of liquefiable materials was defined by elementary techniques, and crude estimates were made of the position of the water table (treating it as a static surface). The 1988 study by Falls utilized expected worst-case shear stresses generated by a postulated maximum credible earthquake of RM 8.3 on the northern San Andreas fault, calculated overburden stresses, corrected for sampling techniques, accounted for fluctuations in the level of the water table, and incorporated a more realistic interpretation of the complex subsurface geology. The use of these parameters reduced the area earlier rated as having high liquefaction potential by more than 50%.

Because critical subsurface data are obtained from borings, which, by their nature are data sources of very restricted geographical extent, the critical question is: "How large an area should be considered to have characteristics based on data from a single boring site?" The answer is very dependent on the nature of the geology in the vicinity of the boring site. A variety of extrapolation techniques can be used. The simplest of these involves uniformly expanding the regions of influence of individual boring sites until the entire map area is covered. The geology of the Santa Clara Valley makes this an inappropriate approach.

The Santa Clara Valley is a tectonically active, colluvium, alluvium, and bay mud filled valley characterized by rapid vertical and lateral changes in materials. Sedimentary deposition in this valley by alluvial fans, meandering streams, and fluctuating sea-level stands has produced an extremely complex prism of sedimentary materials. The three-dimensional understanding of this highly variable prism is poor, although some regional interpretations have been made based on geophysical studies and correlations of water well logs. These interpretations are not at a scale appropriate for extrapolation of data regarding materials that are susceptible to liquefaction. Potentially liquefiable materials are restricted generally to localized depositional environments such as certain portions of stream channels or dune areas. As a result, the lateral and vertical extent of these materials is often very limited.

Simple extrapolation (expansion of an area of assumed geologic similarity) around a borehole that has penetrated a potentially liquefiable deposit may result in the designation of a hazardous area far larger than is appropriate. Overestimation of hazardous areas incur needless expenses for unnecessary investigations or foundation designs. The opposite of this situation also poses a threat. If an inappropriate extrapolation is made around a borehole that encountered nonliquefiable materials, the presence and distribution of potential hazards may be underestimated; it is possible that liquefiable materials may be adjacent to the borehole, but not actually penetrated.

The efforts to categorize terrain with respect to geologic hazards frequently have not included an appropriate consideration of the subsurface geology of the area. In highly complex geological environments, such as the Santa Clara Valley, geologically reasonable hazard maps that are based on subsurface information should be "spot maps." This means that at a geographically restricted point (the borehole), there exists data which indicate the presence or absence of a specific hazard. Until the three-dimensional subsurface geology is better known, it is very misleading to do otherwise. Geologically uncontrolled extrapolations create a false sense of security or risk with respect to the specific geologic hazard.

WILLIAMS, P. J., University of London Goldsmith's College, London, U.K.

Gold Deposit at Calumet, Quebec (Grenville Province): An Example of the Problem of Metamorphic Versus Metamorphosed Ore

The distinction between ores which have the same metamorphic history as their host rocks and those which formed during metamorphism is of primary importance to ground selection for mineral exploration because the distribution of metamorphic ores may not be constrained by stratigraphy. Recent developments in the understanding of the metallogeny of low grade terranes (e.g. Mt. Isa, Cour d'Alene, Archean lode Au deposits) have shown the importance of regional metamorphic processes and we (particularly those of us in the West where syngenetic models of ore genesis have become so popular) should be critically examining the criteria that are used to establish the metamorphic histories of ores in higher grade terranes which typically undergo heterogeneous retrograde metamorphism under conditions which may be favorable for ore generation.

The disseminated Au deposit on Calumet Island, some 90km N of Ottawa, is the second recent economically-significant Au discovery in the Grenville Province of the Canadian Shield. It exhibits many geological similarities to the earlier discovery at Mountauben near Quebec which has been quite extensively studied and is generally regarded as a premetamorphic element of the Late Proterozoic Grenville Supergroup. At both Montauban and Calumet, disseminated Au mineralization occurs close to but spatially separated from relatively massive Zn-Pb-Ag deposits. The host rocks in both cases underwent peak metamorphism in at least the upper amphibolite facies and both mine sequences are characterized by the presence of peraluminous rocks whose distinctive compositions are consequences of hydrothermal metasomatism.

The dominant lithologies in the Calumet sequence are biotite-sillimanite gneisses and basic amphibolites. These rocks exhibit sequentially-developed structures and mineral assemblages related to history of retrograde metamorphism at the end of the Grenville event from peak (preserved) conditions of 600–700 °C and <4kb. The structural sequence was punctuated by multiple granitoid intrusive events.

Trace element geochemistry indicates that the gneisses and amphibolites are derived respectively from evolved calc-alkaline igneous rocks (and/or their sedimentary derivatives) and basalts all of which were generated in a volcanic arc environment. The peraluminous assemblages in the ore sequence which include the immediate hosts of the most important Au mineralization consist of profoundly varied mixtures of quartz, anthophyllite-gedrite, biotite-phlogopite, sillimanite, gahnite-spinel, garnet (variable almandine-pyropespessartine solid solutions), cummingtonite, hornblende, plagioclase, magnetite and sulfides. These were formed by the alteration of both of the main lithologies or their protoliths but most of them are derived from the gneisses by addition of Mg, Zn, Pb, Cu, S (also Fe and Mn in some cases) and depletion of Ca (and Na in most cases). Zones of localized strong Na-enrichment are also present in the form of albite gneisses with up to 8 wt% Na_2O. The base metal-enriched character of the Au zone is unusual for a disseminated Au deposit and only small and inconsistent enrichments have so far been found of the more typical suite of As, Sb, Bi, Mo, and W. the auriferous zone contrasts strongly with the Mn-enriched calc-magnesian (calcity, dolomite, diopside, tremolite, phlogopite, spinel) rocks which enclose the Zn-Pb-Ag mineralization and which are also probably alteration products of the gneisses.

At Calumet there are other forms of alteration which are characterized by high temperature assemblages whose development transgressed the transition from penetrative tectonic fabric-formation/ductile strain to heterogeneous semi-ductile and brittle strain in the host rocks,-i.e. demonstrably synmetamorphic alteration. This is best shown by the presence in the basicamphibolites of K + Rb-enriched, Ca-depleted biotite ± garnet rocks (Potassic alteration) and Ca + CO_2-enriched diopside-plagioclase-epidote-scapolite-calcite rocks (calcic alteration) which occur as both strongly foliated bodies and as strongly discordetn veins and irregular replacements. Both these types of alteration are commonly associated with late metamorphic silicification and minor granite intrusions. Disseminated magnetite-hemantite-ilmenite mineralization lying structurally beneath the auriferous zone is also related to quartz veining and other minor forms of alteration including feldspathization and tourmalinization are locally associated with granites.

All of these features are overprinted by low temperature sericite-chlorite-carbonate-quartz-sulfide alteration of variable intensity which is controlled by closely-spaced microfractures. The distinctive pyrite-pyrrhotite-chalcopyrite-rich sulfide assemblage of the auriferous rocks is paragenetically part of this late low temperature association. These sulfides occur in microfracture arrays which cross brittle high temperature minerals such as garnet and amphibole or are localized in other textural settings such as silicate grain boundaries which are likely to have bee loci of late metamorphic permeability.

In summary therefore, a late metamorphic-epigenetic origin seems possible for the Au mineralization at Calumet in view of the petrographic evidence and of the association with a package of late metamorphic alteration which is chemically similar to that which characterizes tectonic vein and replacement deposits elsewhere. the Zn-Pb-Ag mineralization which in places is characterized by annealed high temperature metamorphic textures may predate the Au and it is possible that this is a polyphase deposit in which early or premetamorphic massive sulfides created a chemical and/or mechanical locus for late orogenic strain, fluid flow, alteration, and Au concentration. This raises the possibility that similar Au deposits could be independent of stratigraphy in other parts of the Grenville Province.

WILLIAMS, P., University of New Brunswick, Fredericton, Canada, and G. P. PRICE, C.S.I.R.O., Mt. Waverley, Australia

Origin of Kinkbands and Shearband Cleavage in Shear Zones: Experimental Study

An artificial KCl/mica schist has been deformed in simple shear with its initial schistosity (S_1) inclined to the imposed simple shear plane, at angles varied in 10° intervals, between 0° and 170°. Angular shear strain in most experiments was 30° and in two-stage experiments was 50°. All experiments were conducted at room temperature, confining pressures between 10 and 80 MPa, and a strain rate of 4.8×10^{-4}/s.

Deformation of the specimens is achieved largely by slip. There is slip on S_1, and in addition two new slip surfaces may develop. They take the form of kinkbands and a locally penetrative shear-band cleavage. Both structures occur singly and as conjugate pairs and form in specimens in which the orientation of S_1 coincides, respectively, with the shortening and extensional fields of the instantaneous strain ellipsoid. We discuss the various factors involved in the development of these structures.

Our experiments indicate that kinkbands or shear-band cleavage forms in shear zones in which S_1 is inclined to the margin of the zone. Which structure forms and whether it occurs as a single set of structures or as a conjugate pair depends on the initial orientation of S_1. In situations where S_1 is parallel to the margins of the zone, neither structure develops if the deformation is a simple shear. The presence of kinkbands or shear-band cleavage in such zones, where S_1 is parallel to the margins, is an indication that the zones were transtensional or transpressional, respectively.

We conclude that shear-band cleavage is not only an excellent indicator of shear sense but also a qualitative indicator of the ratio of S_1-normal shortening to S_1-parallel shear. Conjugate shear-bands indicate that S_1-normal shortening was relatively strong and single shear-band sets indicate that S_1-parallel shear was relatively strong.

WILLIAMS, Q., University of California, Santa Cruz, CA

Bound Volatiles in Minerals at Lower Mantle Conditions: Results for Hydrous Magnesium Silicate and Calcium Carbonate

In order to characterize the structural state of volatiles in minerals at high pressures, we have measured the infrared spectra of a natural sample of chondrodite ($2(Mg,Fe)_2SiO_4 \cdot Mg(OH)_2$), and of the calcite and aragonite polymorphs of $CaCO_3$, each to pressures in excess of 40 GPa at 300K, and following quenching from high temperatures at pressure. In chondrodite, a possible hydrous mantle phase, the effect of pressure on the three most prominent hydroxyl stretching vibrations is variable: Pressure shifts of these modes (with ambient pressure frequencies between 3380 and 3700 cm^{-1}) vary between $-2.7\ cm^{-1}$/GPa and $1.8\ cm^{-1}$1/GPa. At 300 K, no evidence for structural changes over the pressure range to 44 GPa is evident within this mineral. Following laser-heating of samples to 1600 K, however, two new hydroxl stretching bands appear at pressures above 20 GPa and 31 GPa, respectively. The first of these modes appears at the comparatively low frequency of 3180 cm^{-1}, indicating that it is associated with relatively strong hydrogen bonding. Above 31 GPa, a second new band appears at approximately 3400 cm^{-1}. From the width and position of each of these bands, it is inferred that the hydroxyl is present in a crystalline environment. Notably, none of the characteristic vibrational bands of water are observed in the samples quenched from high temperatures: All hydrogen remains bound as hydroxyl groups. Thus, within a magnesium silicate assemblage at lower mantle pressures and temperatures, hydrogen is retained as hydroxyl groups, and the overall effect of pressure appears to be to increase the degree of hydrogen bonding.

Spectra of aragonite and the high-pressure phase calcite (III) under pressure demonstrate that the CO_3^{2-} group is stable at 300 K to at least 37 GPa. However, changes in band positions strongly indicate that the carbonate group becomes progressively more distorted in both materials with increasing pressure. The separations between different bands generated by the asymmetric stretch of the carbonate group increase dramatically at high pressures in both calcite and aragonite. This splitting, which for an ideal trigonal carbonate group is zero, represents an indicator of the degree of distortion of the carbonate group. It increases from 20 cm^{-1} at 4 GPa to 170 cm^{-1} at 31 GPa in calcite, and from 22 cm^{-1} at zero pressure to 87 cm^{-1} at 38 GPa in aragonite. This increase in distortion of the CO_3^{2-} unit may be a precursor to an increase in the coordination of carbon with respect to oxygen at still higher pressures. Heating of calcite (III) to approximately 2000 K at pressures of 40 GPa does not produce decarbonation, but rather a conversion to aragonite. Thus, in both the case of chondrodite and the carbonates, it appears that lower mantle pressures increase the tendency of volatile components to be stably bound into mineral structures. It is possible, therefore, that significant amounts of both hydrogen and carbon could be retained in stable crystalline phases within the lower mantle.

WILLIAMS, R. T., and W. M. DUNNE, University of Tennessee, Knoxville, Tennessee and L. GLOVER III, Virginia Polytechnic Institute, Blacksburg, Virginia

Central Appalachian Geoscience Transect

The central Appalachian transect of the global Geoscience Transects Program extends from the cratonic interior of North America, southeastward across the Appalachian orogenic belt. The transect is situated between the northwestern structural front of the Grenville Province and the Atlantic coast. The character of the continental crust along the transect results from two Wilson cycles during Proterozoic and Phanerozoic time, comprised of Grenvillian orogeny, Eocambrian rifting, late Paleozoic orogeny, and Mesozoic rifting. The purpose of the transect is to address issues related to the geologic development of the eastern margin of North America during Paleozoic time. These issues include: (1) structural development of the middle and lower crust during rifting and orogeny; (2) sources of major potential field anomalies, including the New York-Alabama magnetic lineament (King and Zietz, 1978); (3) displacement termination of the thrust system in the Appalachian foreland; and (4) thrust displacement transfer from the crystalline terranes into the foreland.

The Plateau, Valley and Ridge, and Great Valley provinces in the northwestern portion of the transect dominantly expose Paleozoic stat. Thickness of the sedimentary section ranges from 1 km on the Findlay Arch near the northwestern end of the transect, to 8 km at the Eocambrian rifts of the Appalachian Plateau. As shown by seismic reflection profiles and a few basement wells, the strata of the Plateau are relatively underformed, but the underlying Precambrian crystalline basement is tectonically shortened (Beardsly and Cable, 1983). The Valley and Ridge, and the Great Valley, contain an Alleghanian-age blind thrust system where the tectonic shortening is about 50 percent.

The Blue Ridge separates the Great Valley, and the Valley and Ridge, to the northwest from the Piedmont to the southeast. It is a narrow anticlinorium that extends about 900 km along strike. The Blue Ridge within the transect has a central core of largely Grenvillian-age granitoid rocks. The core is flanked on both sides by younger Proterozoic metasedimentary and metavolcanic rocks. Recent interpretations of seismic reflection data suggest that the Blue Ridge is entirely allochthonous, underlain by platform sedimentary rocks of low metamorphic grade.

Igneous rocks and lower crustal metamorphic facies dominate in the Piedmont, and continue beneath the Atlantic Coastal Plain at the southeastern end of the transect. The I-64 seismic reflection profile shows an antiformal structure beneath the Piedmont, extending to about 20 km depth (Harris and others, 1982; Pratt and others, 1988). the antiform results from imbricate thrusting of crystalline tectonic slices, and some thrust faults apparently were reactivated several times during the Paleozoic. The Piedmont contains accreted terranes, and is cut by subvertical fault zones due to late Alleghanian transpression.

Mesozoic extension associated with the opening of the Atlantic resulted in the formation of Triassic sedimentary basins, crustal thinning, and igneous activity. Seismic reflection, seismic refraction (James and other, 1968), and gravity data all indicate that depth to Moho decreases from about 40 km beneath the northwestern edge of the Piedmont, to about 33 km beneath the Piedmont/Coastal Plain boundary.

REFERENCES

Beardsly, R. W., and M. S. Cable, 1983, Overview of the evolution of the Appalachian Basin, Northeastern Geology, v. 5, p. 137–145.

Harris, L. D., W. DeWitt, K. C. Bayer, 1982, Interpretive seismic profile along interstate I-64 from Valley and Ridge to the Coastal Plain in central Virginia, United States Geological Survey Oil and Gas Investigation Chart OC-123.

James, D. E., T. J. Smith, and J. S. Steinhart, 1968, Crustal structure of the Middle Atlantic states, Journal of Geophysical Research, v. 73, p. 1983–2008.

King, E. R., and I. Zietz, 1978, The New York-Alabama lineament, geophysical evidence for a major crustal break in the basement beneath Appalachian Basin, Geology, v. 6, p. 318–321.

Pratt, T. L., C. Coruh, J. K. Costain, and L. Glover, 1988, A geophysical study of the earth's crust in central Virginia and its implications for Appalachian crustal structure, Journal of Geophysical Research, v. 93, p. 6649–6667.

WILLIAMS, S., U.S. Geological Survey, Reston, VA, A. H. SALLENGER, JR., U.S. Geological Survey, St. Petersburg, FL, and S. PENLAND, Louisiana Geological Survey, Baton Rouge, LA

Coastal Erosion, Barrier Islands, and Nearshore Geologic Framework: Overview of U.S. Geological Survey Study in Louisiana

With nearly three-quarters of the United States' population living within an hour's drive of an ocean or Great Lakes coast, coastal erosion and wetland loss are serious problems with long-term economic and social consequences. Developed coastal areas face billions of dollars in property damage and potential loss of life as a result of long-term erosion and storm impacts, and valuable coastal wetland habitats are being altered at rapid rates. All 30 states bordering an ocean or one of the Great Lakes are undergoing some erosion; of these, 26 states presently experience an overall long-term net erosion of their shorelines. Furthermore, in view of the National Academy of Sciences' recent forecast of increasing rates of sea-level rise in the near future, erosion and coastal land loss are likely to accelerate in the future.

Louisiana, due to natural causes and man-made activities, has the highest rates of coastal erosion and wetland loss of any region in the United States. Louisiana contains 41% of the nation's wetlands, supporting a $1 billion a year commercial fishery industry. These wetlands are fronted on the Gulf of Mexico by low-relief sandy coastal barriers. In the Mississippi River delta plain, rates of wetland loss exceed 130 km^2 per year, and Louisiana's barrier islands are eroding in places an average of 20 m/y.

All of the coastal barriers are decreasing in area as they migrate landward. For example, since 1880, Louisiana's barriers have decreased in total area by nearly 50%. If this rate of land loss continues, many of the barrier islands will completely erode, which, in turn, will accelerate the deterioration and destruction of valuable wetlands. For example, at this rate of erosion the life-expectancy of the Isles Dernieres west of the Delta is about 50 years; while that of the Chandeleur Islands, east of the Delta, is approximately 225 years.

In 1986, the U. S. Geological Survey in cooperation with the Louisiana Geological Survey began a planned 5 year study focused on the geologic processes that cause barrier island erosion. Many processes contributing to the erosion of barrier islands are poorly understood and are not quantifiable with any degree of confidence. These processes must be better understood in order to predict future shoreline change and thus allow better management of the coastal resources in Louisiana, as well as in other coastal regions.

Long-term erosion of Louisiana's barrier islands is due primarily to sea-level rise relative to the land and to diminishing volumes of sand transported to the coast. The primary objectives of this joint study are to better quantify the processes related to sea-level rise and sand supply and to present the scientific results in a form that can be applied to practical problems such as identification of coastal hazards, prediction of future coastal erosion, and methods to mitigate the effects of erosion. The study consists of three main parts:

1. Investigations of the geological framework of the coastal plain, within which such barriers have originated and migrated landward during the past several thousand years, are done using integrated analyses of data from high-resolution seismic reflection and sidescan sonar techniques and 12 m long vibracores. These detailed analyses and interpretations of the shallow subbottom stratigraphy and structure provide a broad regional understanding of the development of the delta plain and barrier island coast during the Holocene epoch and an understanding of the physical processes that cause the barriers to erode and migrate landward. The information also allows us to locate and assess hard-mineral nonenergy resources present in continental shelf deposits. For example, sand of suitable grain-size and sorting characteristics is valuable as fill for barrier-island nourishment projects. These studies show that large quantities of sand are often available in marine sand bodies in the form of shoals and subbottom channels.

2. The second part is development of an improved quantitative understanding of the various physical processes responsible for erosion. We are focusing on only a few of the many processes, including relative sea-level rise, storm overwash, onshore-offshore sediment transport, and gradients in longshore sediment transport, because these seem most responsible for the erosion. A series of field experiments and modeling efforts are being undertaken. For example, sediment overwash during storms is being measured on the Isles Dernieres, west of the Mississippi River Delta.

3. The third component of the study involves assembling the technical research results in the form of maps, reports, and workshops that can be transferred and more readily used by coastal zone planners and engineers. Applications of the study results include developing better techniques for determining the frequency at which artificially nourished beaches should be renourished, and predicting future shoreline erosion so that coastal planners can establish set-backs and zoning regulations to ensure that new construction is properly located a safe distance landward from the eroding shoreline.

The results from the studies conducted thus far in the Isles Dernieres region demonstrate that the Louisiana coastal barriers undergo evolutionary changes, marked by distant stages that occur over periods of several thousand years. The barriers appear to owe their origins to frequent delta-switching by the Mississippi River, followed by severe erosion due to rapid subsidence and sediment winnowing by storm waves and currents. With continued rapid rise in relative sea level, as much as 1 cm/y due to compactional subsidence and eustatic rise, and dissipation of sand by inlet formation and storm overwash, the barriers undergo long-term transgressive submergence. Rates of landward migration average several kilometers per century, combined with large reductions in total island area.

From a dense array of seismic reflection and coring results, the relatively thin Isles Dernieres barrier sands are shown to be overriding back-barrier, sandy, organic muds that overlie a basal delta-plain ravinement surface. With continued island migration and loss of littoral sand, the barriers will eventually submerge and convert to an inner shelf marine sand body. Continued reworking of the shoal by storm waves and infauna will result in landward migration and destruction of barrier sedimentary structures. These processes will continue until the shoal is so spread out and low in relief that it is no longer affected by waves and tidal currents.

WILLIAMSON, A., U.S. Geological Survey, Austin, TX

Possible Cause and Substantial Effects of Brine Distribution in Gulf Coast Aquifer Systems, South-Central United States

Three-dimensional simulations of groundwater flow in the Gulf Coast aquifer systems have lead to the following preliminary conclusions about the interaction of flow in the brine and freshwater parts of the systems:

1. The brine part of the aquifer systems is not in equilibrium.

2. There is potential for density-driven groundwater flow to transport solutes long distances from salt sources, such as domes onshore and on the Continental Shelf and massifs on the Continental Slope.

3. Large-scale groundwater pumpage of freshwater has markedly changed flow patterns in the brine part and induced flow updip toward pumping areas; however, it will take decades or centuries for the brine to reach the pumping areas because of the groundwater velocities, which are relatively slow in proportion to the travel distance.

4. The variations in density not only have a substantial effect on the brine part of the system, but they also affect the freshwater part.

The aquifer systems simulated underlie 750,000 km^2 onshore and on the Continental Shelf, and consist of sediments of Tertiary and younger age. A 10 layer, finite-difference, variable-density model, with blocks 16 km on a side, was used to simulate conditions before development and through 1985. The base of the simulated system was the shallower of the top of the Midway confining unit or the top of the geopressured zone. Groundwater density was estimated from spontaneous potential and other log data from about 20,000 sand beds penetrated in 989 wells and was assumed to be constant during the less than 100 year simulation period.

The flow system in the brine part of the aquifer systems is not in equilibrium because of the long time needed to respond to changes, such as sea-level change, additional sedimentation, and salt dissolution. A long time is needed to respond to changes because of the slow groundwater velocity and the long flow paths necessary for water to move to a discharge point to accommodate the changes. Predevelopment simulations indicate that there is a complex three-dimensional flow pattern of brine and that average velocities of about a meter/year occur in some blocks containing brine. Although density estimates vary substantially from one sand bed to the next in an aquifer, and from one well log to the next, the regional trend is quite certain due to the number of logs used. Similar large-scale flow patterns from a simulation where the density data was smoothed demonstrate that the complicated flow pattern is not a function of the local density variation.

There are many hypotheses about the source of salt causing the widespread distribution of very saline water or brine that is present downdip in all sediments of the Gulf Coast aquifer systems. J. S. Hanor (Kilometre-scale thermohaline overturn of pore waters in the Louisiana Gulf Coast, Nature, 11 June, 1987, p. 501–503) has presented "geochemical and physical evidence for the existence of density inversions in Gulf Coast pore fluids sufficient to drive large-scale convective fluid flow at rates of several feet per year." The inversions are caused, in part, by the dissolution of salt diapirs and the formation of dense brines at shallow depths. Our simulations support this as a possible mechanism causing the widespread distribution of very saline water or brine in the Gulf Coast aquifer systems, even though salt domes occur only in parts of the area. The massive salt formations that blanket most of the Continental Slope have received little consideration as a source of salt because they occur far offshore and downdip. Our simulations demonstrate the substantial effect that density has on the direction and magnitude of groundwater flow. Because the Continental Slope contains a large source of dissolvable salt, and because temperatures probably are abnormally high because of the salt, the Continental Slope is a possible source for the very saline water and brine that have moved long distances coastward and updip.

Large-scale groundwater pumpage in the freshwater part of the aquifer systems has markedly changed flow patterns in the brine part and induced flow updip toward pumping areas. However, it will take decades or centuries for the brine to reach these pumping areas because of the relatively slow groundwater velocities when compared to the long flow paths. For example, freshwater pumping onshore in and around Houston has changed flow directions more than 45° or flow rates more than 50% in the brine part of the upper Pleistocene and lower Pliocene deposits throughout an area of about 15,500 km^2 that extends as far as about 30 km offshore. Block-averaged Darcy velocities in the brine are as fast as 3 m/y. Assuming an effective porosity of 20%, this would mean average particle velocities of about 15 m/y or about 100 years to move 1 to 2 km. These estimates could be in error by an order of magnitude or more.

Variations in density not only have a substantial effect on the brine part of the system, but also affect the freshwater part. In a simulation where the brine part of the aquifers was assumed to be fresh, the effect was as much as an 0.3 m difference in hydraulic head at a point 320 km upgradient from the freshwater–saltwater interface.

WILSON, A., University of Natal, Pietermaritzburg, South Africa, and A. J. NALDRETT, University of Toronto, Toronto, Canada

Fluid Dynamic Controls on Formation of Cyclic Unit 1 of Great Dyke, Zimbabwe

Cyclic Unit 1 of the Great Dyke represents the uppermost part of the Ultramafic Sequence and is of major importance because it is host to several sulfide layers bearing platinum group elements. In this paper, we describe the stratigraphy and petrology of this cyclic unit and the setting for the development of the sulfide zones in the Great Dyke discussed in detail in a companion paper (Naldrett and Wilson, this volume).

The layered succession of the Great Dyke consists of a lower Ultramafic Sequence (2000 m) overlain by gabbroic rocks of the Mafic Sequence (1000 m). The Ultramafic Sequence is broadly divided into the lower Dunite Succession and upper Bronzitite Succession, the latter comprising cyclic units with a basal chromitite overlain by dunite or harzburgite and passing upward through olivine bronzitite into bronzitite. Cyclic Unit 1 (Figure 1) represents the critical transition from olivine and bronzite cumulates through websterites to the overlying mafic rocks. Two sulfide zones (S2 and S3) are contained entirely within brozitite of Subunit 1b and the upper sulfide zone (S1) occurs immediately below the websterite layer in Subunit 1a. Unlike many other layered intrusions (Naldrett et al, 1987), sulfide mineralization in the Great Dyke occurs prior to the formation of plagioclase cumulates.

Cyclic Unit 1 formed as a result of a series of pulses of hot primitive magma emplaced into cooler and more evolved resident magma in the chamber. Several pulses were of sufficient volume and turbulence to drive the resulting hybrid mixture into the primary field of chromite to give the thick chromitites at the base of Subunits 1c and 1d. Subunit 1b was initiated by a smaller pulse of magma that failed to produce chromite but caused olivine to replace orthopyroxene as the dominate cumulus phase. With continued fractionation, olivine once again gave way to cumulus orthopyroxene with bronzitite forming the dominant lithology of the cyclic unit. Subunit 1a formed from the last magma emplacement event of the Ultramafic Sequence, and this caused olivine to appear only temporarily as a narrow olivine bronzitite. The emplacement of magma probably took place as fountains (Campbell et al, 1983; Campbell and Turner, in press), and the hybrid liquid resulting from entrainment of the more evolved resident magma by the turbulence in the fountains would have broken down to a series of double diffusive layers giving rise to a compositionally stratified magma column. Successive pulses of new magma would have sustained the double diffusive convection system.

Immediately above the olivine bronzitite of Subunit 1a, the cumulus orthopyroxene shows a greater iron enrichment than at any previous stage in the Ultramafic Sequence. Augite also appears for the first time as major phase in the websterites to be followed within the next 40 m by cumulus plagioclase of the overlying mafic rocks. This sequence cannot be simply the result of the in situ fractional cyrstallization, and we consider that mixing took place of the denser and relatively primitive magma undergoing crystallization of pyroxene with a cooler, more evolved liquid originating from higher up in the magma chamber. Plagioclase crystallization may have commenced from these upper liquid layers causing a marked increase in density of the release liquid. The plagioclase would have tended to have remained buoyant while at the same time the highly evolved liquid separated and became incorporated in the lower layers accentuating the differentiation process taking place by in situ crystallization at the base of the magma chamber. This process led to the formation of the upper S1 sulfide zone to be followed soon afterward by the websterites and the rapid transition to the mafic rocks. Eventually, the entire liquid system would have become involved in a major density overturn, and the deep scour-and-fill structures observed in the websterites suggest erosive action of a crystal (plagioclase) laden liquid as the entire magma body was destabilized and the stratification set up by earlier emplacement events was destroyed.

The stratigraphy and lithologies of Cyclic Unit 1 are similar in all parts of the Darwendale Subchamber, and for the Great Dyke in general, but significant lateral variations are present. All subunits and individual lithologies become highly compressed toward the margins of the subchamber and the pyroxenes become markedly more iron rich. This facies variation is a consequence of the thermal regime imposed on the Great Dyke magma chamber by its flared or trumpet-shaped cross-section. In the marginal facies, heat was lost both by convection in the liquid and through the floor by conduction by virtue of the lowest liquid layers being juxtaposed and in close proximity to the floor over wide areas. Nearer the axis the effect of conduction became progressively less with the net result that the liquid layers in the double diffusive system would have been thinner toward the roof and margin of the chamber and therefore the liquid reservoir at the zone of crystallization would have been much reduced compared to the axial facies.

REFERENCES

Campbell, I. H., and J. S. Turner, in press, Fountains in magma chambers: Journal of Petrology.

——— A. J. Naldrett, and S. J. Barnes, 1983, A model for the origin of the platinum-rich sulfide deposits in the Bushveld and Stillwater Complexes: Journal of Petrology, v. 24, p. 133–165.

Naldrett, A. J., G. Cameron, G. von Gruenewaldt, and M. R. Sharpe, 1987, The formation of stratiform PGE deposits in layered intrusions, *in* I. Parsons, ed., Origins of igneous layering: NATO ASI Series, v. 196, p. 313–398.

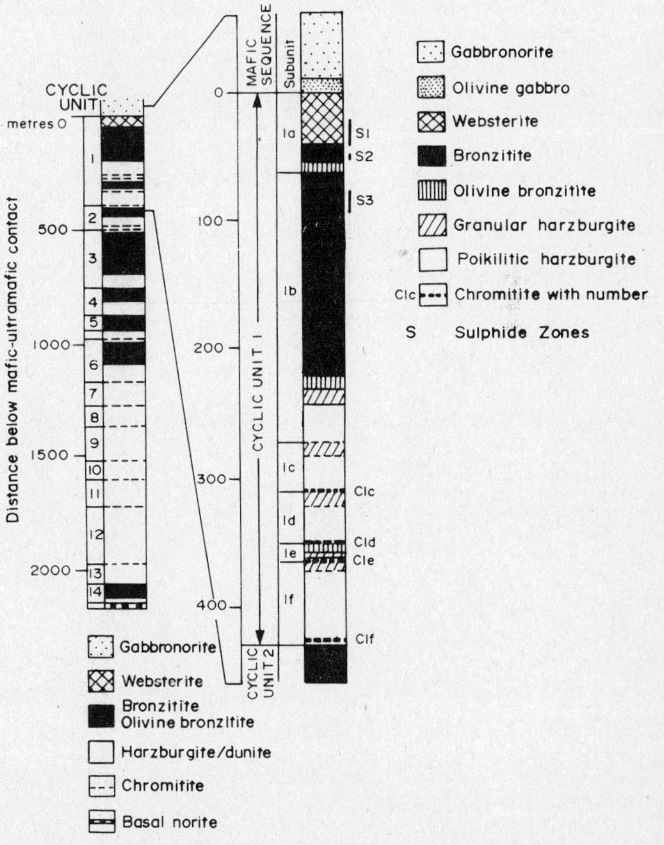

Figure 1—Cyclic units in the Ultramafic Sequence of the Great Dyke together with subdivision and lithologies of Cyclic Unit 1.

WILSON, B. H., University of Oklahoma, Norman, Oklahoma, and T. IMBRIGOTTA, U.S. Geological Survey, Trenton, New Jersey

Behavior of Chlorinated Hydrocarbons and Alkylbenzenes in Anoxic Aquifer Material from the Picatinny Arsenal, New Jersey

The most common ground water contaminants are the chlorinated solvents used as degreasers and the soluble constituents of petroleum products. Of these, trichloroethylene (TCE), 1,1,1-trichlorethane (TCA), tetrachloroethylene (PCE), benzene, toluene, and the xylenes are most commonly found. TCE and TCA are not biologically degraded in oxygenated ground waters and persist in those environments. In anoxic subsurface materials, they may undergo a reductive dechlorination (chlorine is replaced with a hydrogen) to form new compounds that are more mobil than the parent compounds and, in the case of vinyl chloride (VC), are more carcinogenic.

The ground water from Picatinny Arsenal in north-central New Jersey has been contaminated by two partially overlapping plumes of discharged wastewater from a metal-plating wastewater treatment system and spilled diesel fuel. The plumes are found in an unconfined glacial-stratified-drift aquifer approximately 11 m thick underlain by a confining unit. The material was highly reducing, with redox potentials of −375 to μg/L in the plume exposed to diesel fuel; no methane was detected in the plume exposed only to TCE.

TCE is the predominant contaminant from the wastewater treatment system with PCE, TCA, cis-1,2-dichloroethylene (cis-DCE), and VC also present. TCE and its daughter products of reductive dechlorination, cis-DCE and VC, were observed in material from both plumes of contamination. No information on the composition of the ground water from the diesel fuel spill has been provided, but the presence of significant concentrations of methane indicates methanogenic processes may be occurring.

Laboratory studies were conducted on aquifer materials influenced by both plumes of contamination to confirm field evidence of the biotransformation of TCE and TCA. Because of the potential for codisposal of the chlorinated solvents plus the soluble constituents of petroleum, the fate of TCE and TCA was observed botyh with and without the presence of benzene, toluene, chlorobenzene, and o-xylene to determine any potential effects of the additional compounds on reductive dechlorination.

Microcosms were constructed using aquifer material exposed only to the chlorinated solvents and exposed to the chlorinated solvents plus the diesel fuel. The material from each site was divided into half. One-half received an aqueous solution containing TCE and TCA; the other half received an aqueous solution containing TCE, TCA, benzene, toluene, chlorobenzene, and o-sylene. Approximately 40 g of aquifer material (wet weight) were added to fill 50-mL serum bottles. Each microcosms was dosed with the addition of 1.0 to 1.5 mL of the dosing solution, then immediately capped with a Teflon-coated butyl-rubber septum and a crimp cap seal. The controls were autoclaved overnight at 121 °C, returned to the glovebox, and dosed in the same manner as the living samples. The microcosms were incubated inverted in the dart at 15 °C under a nitrogen atmosphere.

Biotransformation of all compounds was observed in materials from both sites that received TCE, TCA, benzene, toluene, chlorobenzene, and o-xylene. Biological removals of one-third to one-half were noted for all compounds with the exception of toluene, which had dramatic removals of up to 98 percent. The observed biotransformations did not vary greatly for TCE with or without the presence of benzene, toluene, o-xylene, or chlorobenzene. This study confirms the biotransformation of TCE in the anoxic aquifer material from Picatinny Arsenal. The aquifer material studied has been shown to be biologically active against the compounds found as contaminants in that same aquifer. With the exception of toluene, the biological removals observed after 4 weeks of incubation do not appear subtantial; however, they may be environmentally significant when coupled to the long residence times that typically prevail in aquifers.

WILSON, J. L., and W. R. LINDERFELT, New Mexico Institute of Mining and Technology, Socorro, New Mexico

Influence of Streams, Barriers, Groundwater Divides, and Other Hydrogeologic Features on Well Capture Zones

Well water quality depends on the relative amounts of water drawn from different sources, and the particular capture zones within these sources. A pollutant spilled within an aquifer capture zone will eventually enter the well. A spill outside the capture zone may eventually discharge elsewhere, perhaps to a nearby stream of lake; if the spill is upstream of the well, some pollution may still enter the well by induced infiltration from the stream. From a technical or scientific point of view the well head protection area is defined by the capture zone.

We are developing generic mathematical models that help define how capture zones, sources of well-water, and pollutant travel times are influenced by a variety of factors: aquifer geometry and properties, well locations and depts, well pumping rates and schedules, recharge sources and temporal distribution (ambient aquifer flow, local vertical recharge, streams, etc.), macro-dispersion, and uncertainty in all of the above. Our approach is to determine the velocity field for a given conceptual model, including its critical points (eg., stagnation points). We use these points and particle tracking to define the boundaries of the capture zones. Front tracking is used to determine travel times within the capture zone. Both analytical and numerical mathematical tools are employed in these two and three dimensional analyses.

Our results show that capture zones may extend far out along groundwater divides and aquifer barrier boundaries, grabbing pollutants from sources that might otherwise be inaccessible. Wells near streams may induce infiltration of stream water. It the stream is contaminated the well water quality will be impaired. The amount of induced infiltration depends on the pumping rate, the distance between the well and stream, the stream penetration and gradient, nearby aquifer boundaries, and the ambient aquifer flow rate and direction. The well capture zone can extend under a partially penetrating stream to capture pollution from sources on the other side of the stream. Parameter and model uncertainty, parameter spatial variability, and time varying conditions make the delineation of "exact" capture boundaries impossible. We examine the concept of "fuzzy" capture boundaries through the application of Monte Carlo techniques.

REFERENCES

Schafer-Perini, A., H. L. Kenrick, J. C. Lee, J. M. Newsom, and S. C. Leppert, Present work is being conducted under USGS Grant #1423620 and New Mexico State University, Water Resources Research Institute Grant #1345682.

WILSON, J. T., and D. H. Kampbell, U.S. Environmental Protection Agency, S. VANDEGRIFT, NSI Technology Services, Inc., Ada Oklahoma, C. H. WARD, Rice University, Houston, Texas, and J. ARMSTRONG, The Traverse Group, Ann Arbor, Michigan

Preferential Removal of Alkylbenzenes (BTX) From Fuel Hydrocarbons During Bioremediation of a Spill From an Underground Storage Tank

In 1969 at least 10,000 gallons of aviation gasoline was spilled from an underground storage tank at the U.S. Coast Guard Ari Station at Traverse City, Michigan. The fuel contaminated the water-table aquifer, resulting in a large plume that moved off the Air Station and contaminated more than 40 private drinking-water wells. In 1988 the U.S. Coast Guard and the U.S. EPA installed a pilot scale study of bioremediation in the area of the original spill. Cores were acquired from the source area to determine the vertical and horizontal extent contamination, the concentration of total hydrocarbons in the contaminated interval, and concentrations of individual alkylbenzenes. The alkylbenzenes are the object of the regulatory concern, and the bioremediation will be finished when their concentration is brought to a level specified by the Michigan Department of Natural Resources.

Injection wells were installed to perfuse the contaminated area with mineral nutrients, and oxygen or hydrogen peroxide. The water contained 380 mg/liter ammonium chloride, 190 mg/liter disodium phosphate, and 190 mg/liter potassium phosphate. The temperature was 11-12 °C, and the pH near neutrality. The flow rate of injected water in the aquifer was 10 feet per day. Injection began the first week of March, 1988. The injected water contained 40 mg/liter of oxygen during the first 12 weeks of operation, then the oxygen was replaced with 50 mg/liter hydrogen peroxide for one week, 110 mg/liter hydrogen peroxide for a second week, and finally 250 mg/liter for the subsequent eight weeks.

The concentration of alkylbenzenes in monitoring wells declined dramatically (Table 1).

The alkylbenzenes were selectively removed. There was no significant reduction in the total quantity of fuel hydrocarbons in the spill area (Table 2).

Toluene concentrations in the core material were much lower after bioremediation, indicating that toluene was being selectively removed. Because toluene is the most important component of the alkylbenzene complex in the spill, the time and cost to bioremediate the spill should be much less than originally anticipated.

Table 1. Changes in concentrations of alkylbenzenes in monitoring well during bioremediation of an aquifer contaminated with aviation gasoline.

Well and Sample Number	Benzene	Toluene	Ethylbenzene	Xylenes
	µg/liter			
1-1-M31-8083, Background conditions in an unweathered part of the spill area.				
April, 1988	54	12,600	33	850
BD-31-3, thirty one feet down gradient from the injection wells in the most contaminated depth interval.				
March 7, 1988	0.0	4.1	4.3	38
April 6, 1988	0.0	2.9	2.4	39
May 5, 1988	0.0	7.3	0.6	3.6
June 6, 1988	28 (?)	0.0	0.0	3.2
June 28, 1988	0.0	0.0	0.4	1.2

Table 2. Changes in concentrations of alkylbenzenes and total fuel hydrocarbon in core material during bioremediation of an aquifer contaminated with aviation gasoline.

Date Core Number	Oil and Grease	Fuel Hydro-Carbon	Benzene	Toluene	Ethyl-benzene	Xylenes
	mg/kg wet sample					
Background conditions in an unweathered part of the spill area.						
June, 1988						
50R6		12,150	1.0	107	57	218
50R7		5,220	1.0	170	24	100
Preliminary sampling used to design the bioremediation project near monitoring well BD-31-3, August 1987.						
50A3	4,310	5,590	0.6	235	33	121
50I14	4,130	6,500	0.3	544	12	48
50D18	1,130	2,500	0.7	112	11	39
Sampled after four months of perfusion with mineral nutrients and oxygen, June, 1988.						
50T3		3,330	1.4	1	7.3	23
50W3		4,800	1.5	1	13	41

WILSON, J., and E. M. Habekost, Aarhus University, Aarhus C, Denmark

Raft-Like Metabasic Inclusions in Fongen–Hyllingen Layered Complex, Norway

The Hyllingen Series (HS) forms the southern part of the Caledonian, synorogenic Fongen–Hyllingen (F-H) layered mafic intrusion, situated southeast of Trondheim, Norway. This HS has a maximum thickness of about 4200 m, consisting dominantly of layered gabbroic and dioritic rocks that grade upward to quartz-bearing ferrosyenites which vein country rock amphibolites that form the roof. Highly evolved rocks at the base become more primitive upward through about 100 m. This basal reversal (Stage I) is followed by layered diorites with only minor cryptic variation except at the top where they become more evolved (Stage II). This is followed by a gradual regression (Stage III), reaching the most primitive compositions in the HS. Stage IV covers a wide compositional range reaching extremely evolved syenitic rocks at the roof. Modal layering strikes directly toward the southern wall where layered diorites interfinger with metabasalt and metapelite. Modal and cryptic layering are discordant, with increasingly evolved compositions approaching the southern wall. These discordant relations developed as a result of the in situ crystallization of compositionally zoned magma along an inclined floor, and the regression of Stage III reflects crystallization during elevation of the zoned magma in response to the influx of dense, primitive magma along the floor.

Large, raft-like inclusions form an important part of the HS. Most of these are of metabasic hornfels, although metapelitic hornfels inclusions are prevalent in the southeast, consistent with the lithology of the adjacent country rock. The metabasic inclusions are dominantly fine-grained, equigranular rocks of basaltic composition that are sometimes cut by plagioclase–phyric metabasic dykes which have been contact metamorphosed together with the metabasalt. The lithology of these inclusions matches that of the country rocks around most of the HS.

These raft-like bodies are generally lens-shaped with their long axes concordant to modal layering in the host rocks. They are particularly abundant in Stage II in the HS where they form about 22% of the area studied. Here they are often branching and give the impression that they form a three-dimensional network. These inclusions vary greatly in size; they can be followed for up to about 1500 m along the strike and vary in thickness from a few meters up to about 150 m. The bottom and top contacts of the metabasic rafts are strikingly concordant with modal layering in the host gabbroic rocks. Thin rafts taper out laterally, and when they are less than a few tens of centimeters thick they often have igneous-looking textures and are difficult to distinguish from the host layered gabbroic rocks.

Relict ophitic textures and phenocrysts are sometimes preserved in the inclusions. The contact metamorphism produced high-temperature pyroxene hornfelses and olivine-pyroxene hornfelses, grading down to hornblende hornfelses. There is evidence for local partial melting in some of the high-temperature metabasic inclusions. The country rocks have usually been affected by later amphibolite facies regional metamorphism that has only locally influenced the HS and its inclusions.

Mafic silicates in the metabasic inclusions have compositions roughly equivalent to only the more primitive HS layered rocks. Plagioclases, however, cover a wide compositional range, reaching An90 in a relict plagioclase phenocryst compared with the most calcic composition of An75 in the HS. The whole-rock chemical composition of the metabasic inclusions confirms that they are equivalent to the local country rocks of similar lithology.

The Hyllingen magma chamber developed as a southerly expanding, thin wedge. The HS represent only the uppermost ~40% of the F-H intrusion, and by the time the HS started to develop, compositional zoning of the magma was well established. The roof of the wedge-shaped chamber became elevated, and the leading edge propagated southward as a result of the elevation of compositionally stratified magma up the inclined floor in response to the influx of dense, primitive magma in the northern part of the chamber. The basal reversal of Stage I was produced on the inclined floor by crystallization of increasingly primitive magma during expansion of the wedge-shaped chamber.

The raft-like inclusions developed at the roof of the HS magma chamber. The highly evolved magma at the top of the chamber, instead of consistently pushing up the roof, must periodically have penetrated along cracks that became filled with magma to form veins which spread out laterally in sill-like bodies. The roof zone therefore consisted of a network of veins and sills penetrating an interconnected framework of metabasic hornfels. Crystallization occurred predominantly on the inclined floor of the wedge-shaped magma chamber and not at the roof; this was partly due to the low liquidus temperature of the evolved, hydrous, felsic magma at the roof, and the relatively high temperature of the metabasic roof rocks owing to the heat flux through the roof of the extensive magma chamber crystallizing under synorogenic conditions at a depth of 16–20 km. Impact structures below some relatively small metabasic inclusions with identical lithology to the rafts imply that the roof rocks had a density above that of the entire zoned HS magma. The large raft-like inclusions are therefore considered to be in situ bodies that remained attached to the roof and became enveloped by the advancing crystallization front.

WILSON, K. M., University of Colorado, Boulder, CO

Coevolution of Marginal Seas and Terranes Since Triassic

Speculation on the close relationship between marginal sea evolution and terrane accretion has been the subject of several recent papers by terrane researchers (Ben-Avraham and Uyeda, 1983; Silver and Smith, 1983; Dercourt et al, 1985; Burke, 1988; Wilson et al, in press). These workers have noted similarities between the geologic setting of such regions as the marginal Indo-Pacific and the presumed analogous paleogeographic setting of such regions as the Mesozoic North American Cordillera. Recent global tectonic reconstructions developed by the present author confirm the role played by marginal seas in the development and evolution of suspect terranes over the last 250 m.y.

A number of paleomarginal seas comprising separate microplates are hypothesized to have played significant parts in post-Paleozoic terrane evolution, including "Paleochina," "Paleojapan," "Paleoburma," "Cordilleria," "Yukonia," "Mexicania," "Paleococos," "Paleohotsk," "Paleocelebes," "Hunsteinia," "Paleoscotia," "Brooksia," "Paleocaribbea," "Talkeetnia," "Beringia," "Paleoalpidea," and "Shyokia." "Paleochina" was actually an entire system of related paleomarginal seas that bordered the Asian margin during the Triassic–Jurassic assembly of China. "Paleojapan" was the marginal sea separating the Hida–Abean and the Akiyoshi–Mino–Tanba terranes; the latter collided in the Late Triassic to close this basin. "Paleoburma" was a large back-arc basin lying between the Western Burma Block and the Western S.E. Asian Block during the Triassic and Early Jurassic.

"Cordilleria" was a very large marginal sea lying between the Stikinia terrane and North America and was consumed by subduction during Late Triassic–Middle Jurassic time. "Yukonia" was a narrow sea formed west of the Yukon–Tanana and Nixon Fork terranes after the collapse of "Cordilleria" in the Late Jurassic. "Mexicania" was a small paleomarginal sea lying along the west coast of Mexico in the Jurassic; it interacted with the Cortez, Baja, and other terranes during its evolution. "Paleococos" was a large sea formed by the back-stopping of Alexander terrane against Central America in the Jurassic.

"Paleohotsk" was formed in the Late Jurassic and evolved through the Cretaceous in response to entrapment of the Okhotsk Arch plateau and subsequent development of the Kamchatkan terranes. "Paleocebes" is postulated to have formed as a result of the initiation of a subduction zone in front of the Cebu–Bohol arc terrane during the Early Cretaceous. "Hunsteinia" formed shortly afterward when Banggai–Sula and Buton were back-stopped against the "Paleocebes" subduction zone; the Hunstein–Jimi–Bismarck terrane then formed as a result of subduction beneath the new "Hunsteinia" plate.

The "Paleoscotia" marginal sea formed in the opening gap between South America and the Antarctic Peninsula terrane during the Early Cretaceous breakup of Gondwana; the South Georgia terrane was strongly affected by subduction along one margin of this sea. "Brooksia" formed in front of the rotating Arctic Alaska block in the Early Cretaceous and was consumed by subduction just prior to Arctic Alaska's collision with the mainland. "Paleocaribbea" was formed by the entrapment of an oceanic plateau against Central America, which led to the development of the Greater Antilles arc terrane in the Early Cretaceous. "Talkeetnia" formed in the Late Jurassic when subduction jumped outboard of Terrane II after amalgamation of its parts (Wrangellia, Peninsular, Alexander) along the Central American margin. "Beringia" developed following entrapment of two oceanic plateaux against a Late Cretaceous–Early Tertiary southern Alaskan subduction zone; the Aleution arc terrane originated along the outboard edge of the newly formed sea.

"Paleoalpidea" was actually an entire system formed during the late Mesozoic. The largest of these seas eventually became the land-locked Black Sea and Caspian Sea basins following the collision of Afro-Arabia with Eurasia. A number of small terranes (e.g., Pontides, Alborz) were involved in the tectonic evolution of these seas. "Shyokia" was a large marginal sea lying behind the Ladakh–Dras arc terrane during the latest Cretaceous. The arc was involved in the first stage of India's collision with Eurasia; "Shyokia" was then consumed by subduction prior to the main collision.

Current evidence supports the notion that marginal seas have been a common feature of plate tectonic evolution. In the Mesozoic and Cenozoic, marginal seas formed in several ways, including plateau entrapment, conversion of transforms into subduction zones, and back-stopping of continental terranes against subduction zones. One theme seems to join these basin-forming events together: Terranes have coevolved with most of these basins in such a way that no discussion of marginal seas is complete without consideration of the role of terranes in their development and evolution.

REFERENCES

Ben-Avraham, Z., and S. Uyeda, 1983, Entrapment origin of marginal seas, *in* T. W. C. Hilde and S. Uyeda, eds., Geodynamics of the Western Pacific—Indonesian Region: American Geophysical Union Geodynamics Series, v. 11, p. 91–104.

Burke, K., 1988, Tectonic evolution of the Caribbean: Annual Review of Earth and Planetary Science, v. 16, p. 1–40.

Dercourt, J., et al, 1985, Presentation de 9 cartes paleogeographiques au 1/20.000.000. s'etendant de l'Atlantique ar Pamir pour la periode du Lias a l'Actuel: Bull. Soc. Geol. France, v. 8, t. 1, p. 637–652.

Silver, E. A., and R. B. Smith, 1983, Comparison of terrane accretion in modern Southeast Asia and the Mesozoic North American Cordillera: Geology, v. 11, p. 198–202.

Wilson, K. M., M. J. Rosol, and W. W. Hay, in press, Global Mesozoic reconstructions using revised continental data and terrane histories: a progress report in J. W. Hillhouse ed., Deep structure and past kinematics of accreted terranes: American Geophysical Union Monograph Series.

WILSON, K., C. N. WOLD, C. A. SHAW, and W. W. HAY, University of Colorado, Boulder, CO

Model for Tectonic Evolution of New World

At least 70 tectonostratigraphic terranes have been identified in the North and Central American Cordillera, the Caribbean Sea, and the South American Andes. A likely debate has developed on the tectonic evolution of these terranes and their origins in space and time (e.g., Howell et al, 1985; Debiche et al, 1987; Wilson et al, in press). A portion of this debate results from the difficulties encountered by modelers in their attempts to define the tectonic evolution of these terranes using paleomagnetic and paleobiogeographic data. Regional terrane workers have presented conflicting views on the history of various terranes, based on the type of data they employ and the relative quality of that data.

Some of the conflicts in interpretation result in tectonic "traffic jams," wherein some terranes must undergo intermittent reversals in their directions of motion, and others must pass them going in the opposite direction, all on the same plate. It is clear that some of the data in these cases must be unreliable, and it is essential that we resolve these conflicts using some kind of data reliability ranking system. We have developed such a system and applied it to the tectonic interpretation of the 70 suspect terranes in the New World, for the time interval from Triassic to Recent. The system of reliability weighting is based on examination and interpretation of a global database. The system consists of the following hierarchy of relative data reliability:

1. The most reliable and useful data are "departure" (rifting) and "arrival" (collision) times.
2. Also crucial are episodes of strike-slip motion and geologic evidence that allow tectonic environment (e.g., subduction) to be inferred.
3. High-quality paleomagnetic data, as defined by a detailed list of positive reliability indicators (Wilson et al, 1988).
4. Vertebrate paleobiogeography involving well-preserved large tetrapod forms and/or primary freshwater forms.
5. Good paleomagnetic data, i.e., where most reliability criteria are met but doubt exists about some of them.
6. Invertebrate paleobiogeography utilizing well-preserved benthic marine faunas considered to have had low dispersal capacities.
7. Paleomagnetic data of intermediate quality, if use of these data does not produce conflicts.
8. Floral paleobiogeography where definition of realms is clear and dispersal capacity is low.
9. Invertebrate paleobiogeography utilizing nektonic forms.
10. Poor-quality paleomagnetic data, if no other data are available.

Central to this analysis is the critical evaluation of the relative reliability of all paleomagnetic data. This is true because the only well-dated position evidence available for many terranes is that provided by paleomagnetic data. In the case of North America, these data are relatively abundant, but data quality varies across a wide spectrum. In our ranking system for paleomagnetic data, 22 reliability factors are used in evaluating each locality, and a rank ranging from "category 1" (excellent) to "category 5" (very poor) is assigned. This system has been applied to more than 140 separate studies in the Americas; unfortunately, some 45% of these studies were ranked below "category 3" and were therefore rejected for use in the interpretation of terrane evolution presented here.

The remaining paleomagnetic and marine biogeographic data (there are apparently no useful vertebrate data and limited floral data from these terranes), together with geologic data constraining "arrival" and "departure" times, have been used to develop a new model for the tectonic evolution of the New World. This model will be presented in the form of computer-plotted maps displaying regional paleogeography for continents and terranes, as well as all inferred plate boundaries, at 10 m.y. intervals from Triassic to Recent.

Many of the terranes are shown to have traveled great distances latitudinally, but few have demonstrably traveled large distances longitudinally. Thus, the importance of oblique convergence and strike-slip transport of terranes along the Pacific margins of the Americas is underscored by this study. This may reinforce suggestions put forward by Silver and Smith (1983) and Sarewitz and Karig (1986) regarding the appropriate analogs to be used in reconstructing terrane histories. In any case, the model presented here is internally consistent and has resolved all major data conflicts. Although many of the solutions provided by the model are nonunique, they are testable and should serve to help focus future research.

REFERENCES

Debiche, M. J., A. Cox, and D. C. Engebretson, 1987, The motion of allochthonous terranes across the North Pacific Basin: GSA Special Paper 207, p. 1–49.

Howell, D. G., D. L. Jones, and E. R. Schermer, 1985, Tectonostratigraphic terranes of the circum-Pacific region: *in* D. G. Howell, ed., Houston, Circum-Pacific Council Energy Min. Res., p. 1–28.

Sarewitz, D., and D. E. Karig, 1986, Processes of allochthonous terrane evolution, Mindoro Island, Philippines: Tectonics, v. 5, p. 525–552.

Silver, E. A., and R. B. Smith, 1983, Comparison of terrane accretion in modern Southeast Asia and the Mesozoic North American Cordillera: Geology, v. 11, p. 198–202.

Wilson, K. M., R. T. Bakker, and W. W. Hay, 1988, Paleobiogeography, paleomagnetism and the tectonic history of Southern Asia: Nature, v. 336.

——— M. J. Rosol, and W. W. Hay, in press, Global Mesozoic reconstructions using revised continental data and terrane histories: a progress report, *in* J. W. Hillhouse, ed., Deep structure and past kinematics of accreted terranes: American Geophysical Union Monograph Series.

WILSON, W. E., U.S. Geological Survey, Denver Federal Ctr., Denver, Colorado, and J. L. YOUNKER, Science Applications Int'l Corp., Las Vegas, Nevada

Role of Hydrogeologists in Evaluating Yucca Mountain, Nevada, as a Repository Site for High-Level Radioactive Waste

Plans are being developed to evaluate the hydrogeology of the Yucca Mountain, Nevada, region to help assess the suitability of Yucca Mountain for a mined geologic repository for high-level radioactive waste. Yucca Mountain is the only potential repository site currently (1988) under consideration by the U.S. Department of Energy. Among all sites previously considered, Yucca Mountain is the only one in which the waste would be placed in the unsaturated zone. The thick unsaturated zone and probable small moisture flux are expected to be effective natural barriers to radionuclide transport. In order for construction of any repository to be licensed by the Nuclear Regulatory Commission, the Department of Energy needs to demonstrate that the repository design and performance are expected to conform to regulations that have been established to protect the health and safety of the public during the next 10,000 years. Thus, scientists are being expected not only to demonstrate a comprehensive understanding of a complex hydrogeologic system, but also to predict the functioning of this system for at least 10,000 years, taking into account the effects of climatic changes and tectonic processes and events.

Some understanding exists of the general hydrogeologic framework of the Yucca Mountain region. Yucca Mountain is in the southern Great Basin, a region of active extensional tectonics. Numerous faults that have had movement along them during Quaternary time have been mapped in the area. Yucca Mountain consists of a series of east-dipping fault blocks, each bounded by steep west-dipping normal faults. The repository site is near the center of a closed ground-water flow subbasin of about 8,000 square kilometers. In this subbasin, saturated-zone flow occurs in alluvium, volcanic tuff, and carbonate rocks; generally, flow is from north to south. Recharge results from infiltration of rainfall, snowmelt, and streamflow, principally in the northern part of the subbasin, where land-surface altitudes are higher and precipitation is greater than at Yucca Mountain. Discharge is primarily by evapotranspiration in the southern part of the subbasin, and perhaps by spring flow. The unsaturated zone beneath Yucca Mountain is 500 to 750 meters thick and consists mostly of alternating layers of welded and nonwelded turr. Precipitation averages about 150 millimeters per year, and unsaturated-zone flux through the potential repository horizon is estimated to be about 0.5 millimeter per year. Flow probably is predominantly in the matrix of the tuff units, but flow also may occur in faults and fractures under local or intermittent conditions.

Within the context of this general framework, hydrogeologists need to develop a comprehensive understanding of the present-day and future hydrologic properties, processes, and conditions that are significant to wast isolation. The effects of future climatic change on hydrologic conditions will be evaluated by numerical modeling, by interpreting paleohydrologic evidence, and by studying sites that are modern analogs of Quaternary hydrologic conditions. The effects of tectonic events will be evaluated by examining evidence of past tectonically induced hydrologic conditions, by observing modern transient effects, and by utilizing couplelenges for hydrogeologists will be to: (1) identify and charactìcant for wast isolation, (2) recognize the degree of uncertainty associated with their analyses, (3) convince peers of the reasonableness and validity of their results, and (4) convey to decision makers judgments about the degree of conaddition, these judgments will be used to help evaluate the extenet to which costly engineered barriers nees to be considered, versus the extent to which the effectiveness of natural barriers need to be relied on for waste isolation.

WINCHESTER, J. A., Department of Geology, University of Keele, Keele, United Kingdom, and M. D. MAX, Naval Research Laboratory, Washington DC

Characterization of Terranes Through Geochemistry: Examples From Late Precambrian and Lower Paleozoic rocks in West of Ireland

Metavolcanic and metasedimentary rocks within late Proterozoic and Cambrian Dalradian successions in the NW Mayo and western Ox Mountains inliers of western Ireland have proven difficult to subdivide into tectonostratigraphic successions. Despite detailed stratigraphic and structural analyses, there are often no good correlations in detail across important tectonic discontinuities because of often undistinctive stratigraphies and similar tectonothermal histories. Although the problems of lithostratigraphic correlation within the Dalradian supergroup in Ireland have been largely resolved in Donegal and Scotland to the east, the NW Mayo and western Ox Mountains inliers and small inliers in the Clew Bay region have been shunted along shear zones, slides and thrusts, and stratigraphic passages are much interrupted and often telescoped.

The easily identifiable lower part of the Dalradian stratigraphic succession is not widely seen in the NW Mayo and other inliers in the far west of Ireland, except for parts of Connemara where there are important structural breaks whose significance remains to be fully assessed in a terrane context elsewhere; the more undistinctive upper part of the Dalradian stratigraphy occurs broadly. In addition, more than one metavolcanic horizon occurs within the upper part of the Dalradian succession; in the field these are not distinguishable from each other.

Detailed geochemistry has allowed us to characterize various of the metasedimentary successions of previously equivocal stratigraphic position so that correlation can be made. Detailed geochemical characterization thus has been used as the final arbiter between conflicting stratigraphic correlations. Geochemically characterizing the metavolcanic suites not only allows for correlation between structural blocks, but also allows for identification of the tectonic position of the original volcanism. Strong contrasts between the tectonic positions of metavolcanic successions that are now geographically in close proximity have allowed us to reassess the significance of certain tectonic discontinuities, whose importance was not previously indicated from detailed structural and metamorphic analyses. At least one tectonic slide in what was regarded as a single tectonostratigraphic succession is now regarded by us as a probable terrane boundary because of the implied telescoping of once widely separated successions across it.

The technique of geochemical characterization of metasedimentary and metavolcanic beds and successions can be broadly applied in metamorphic rocks so as to facilitate identification of way up, local and regional correlation, and original tectonic position. Geochemistry provides a powerful tool for dealing with terrane for geochemical terrane analysis. Although this present work deals with individual metamorphic inliers and correlation within the northwest of Ireland only, as geochemical characterization on the terrane scale becomes more widely available, correlation throughout the Appalachian–Caledonian orogen would provide valuable insights into the construction of the Lower Paleozoic mobile belt.

WINSTON, D., University of Montana, Missoula, Montana

Middle Proterozoic Alluvial Apron, Playa and Lacuine Facies: Ravalli Group, Belt Supergroup, U.S.A.

Although the Middle Proterozoic Belt and Purcell supergroups of Montana, Idaho, Washington, Alberta and British Columbia are widely regarded as the initial deposits of the Cordilleran marine miogeocline (Price, 1964; Harrison and others, 1974; Mechan, 1981; Reed, 1988), the Belt has alternately been interpreted as an intracratonic deposit, bordered on the west by a continental terrane (Winston, 1986 a,b: Frost and Winston, 1987). Recent correlations in the Ravalli Group across the central part of the Belt basin provide more detailed fluvial, sand flat, playa flat, and lacustrine interpretations, largely based on recognition of recurrent sedimentary facies, called sediment types.

During deposition of the Ravalli Group, sediments derived from both west and east were transported toward an enclosed basin axis that lies in the eastern part of the Belt outcrop. Most Ravalli Group rocks were deposited by floods that drained a large western continent and flowed eastward down alluvial aprons crossed by braided channels in their upper and medial parts. Medium-to fine-grained sand was deposited from the lower flow regime in broad channels floored by large-scale bedforms forming the crossbedded sand sediment type. Where channels broadened and shallowed down slope, flow shifted to the upper regime and accretionary crossbeds outline channel and bar edges (also crossbedded sand sediment type). On the lower parts of the alluvial aprons, channels blended into sheetflood surfaces where tabular beds of fine-grained flat-laminated sand capped by mud drape accumulated (flat-laminated sand sediment type) Floods crossed sandflats that stretched lakeward from the alluvial apron beds 3 to 10 cm thick that grade from flat-laminated and climbing-ripple crosslaminated fine sand up into desiccated mud (even couple sediment type). Floods crossing exposed playa flats deposited the mudcracked even couplet sediment type composed of graded layer, 0.3 to 3 cm thick, of: 1) basal fine sand and silt deposited from traction load as floods crossed exposed playa flats, 2) capping mud layers deposited from suspension as the lakes expanded across the playa flats in response to the floods. Suspended silt and clay carried beyond the limit of fine sand transport settled as thin mud layers on playa flats, forming the mudcracked mud sediment type. Continual evaporation contracted the lakes, exposing and drying the playa flats. Flood waters entering perennial lakes formed suspension clouds, which deposited graded uncracked even couplets on the lake bottoms. Silt and clay interlaminations at a millimeter scale accumulated on perennial lake bottoms remote from sediment influx and produced the microlamina sediment type, which closely resembles Phaqnerozoic lacustrine varves and oil shale. Silt and clay deposited in shallow water was commonly reworked by waves into ripple lenses capped by mud and comprise the lenticular couplet sedimart type.

Sediment from the east came from a coarse- to medium-grained mature quartz sand blanket that veneered the Archean Wyoming province east of the Belt basin. Floods transported sand in the lower flow regime across the playa flats of the easter part of the basin, depositing it in tabular beds of coarse-to medium-grained, nearly white, crossbedded sand sand, which is interstratified with red silt and clay derived from the west.

Graded beds, at a range of scales that reflect episodic flood deposition, are a hallmark of Belt sedimentology. Absence of barrier spits and tidal channels in thick sequences of alternately flooded and exposed flats of the mudcracked even couplet and lenticular couplet sediment types militates against a marine tideflat interpretation. Instead, the inferred submergence of the flats in response to flooding supports an enclosed basin interpretation. Similarity of the microlamina sediment type to Phannerozoic lacustrine varves and oil shale further supports a lacustrine setting.

REFERENCES

Frost, C. D., and D. Winston, 1987, ND isotope systematics of coarse-and fine-grained sediments Examples from the Middle Proterozoic Belt-Purcell Supergroup: Journal of Geology, v. 97, p. 309–327.

Harrison, J. E., A. G. Griggs, and J. D. Wells, 1974, Tectonic features of the Precambrian Belt basin and their influence on post-Belt structures: U.S. Geological Survey Professional Paper 866, 15 p.

McMechan, M. E., 1981, The Middle Proterozoic Purcell Supergroup in the southwestern Purcell Mountains, British Columbia and the initiation of the Cordilleran miogeocline, southern Canada and adjacent United States: Bulletin of Canadian Petroleum Geologists, v. 29, p. 583–621.

Price, R. A., 1964, The Precambrian Purcell System in the Rocky Mountains of southern Alberta and British Columbia: Canadian Petroleum Geologists Bull., v. 12, p. 399–426.

Reed, J. C., Jr., 1988, Precambrian Geology of the U.S.A: Episodes, v. 10, p. 243–247.

Winston, D., 1986a, Sedimentology of the Ravalli Group, middle Belt carbonate and Missoula Group, Middle Proterozoic Belt Supergroup, Montana, Idaho and Washington: in S. Roberts, ed., Belt Supergroup: A Guide to Proterozoic Rocks of Western Montana and Adjacent Areas: Montana Bureau of Mines and Geology, Special Publication 94, p. 85–124.

Winston, D., 1986b, Sedimetation and tectonics of the Middle Proterozoic Belt basin, and their influence on Phanerozoic compression and extension in western Montana and northern Idaho: in Paleotectonics and Sedimentation in the Rocky Mountain Region, United States, J. A. Peterson (ed.) AAPG Memoir 41, p. 87–118.

WINTERER, E., Scripps Institution of Oceanography, La Jolla, CA

Tethyan Pacific

During Mesozoic times, the Pacific Ocean occupied almost an entire hemisphere, plus a great wedge of ocean that lay between eastern Gondwana and Laurasia and extended westward from the Pacific into the gradually widening Tethys sea in the Mediterranean, Atlantic, and Gulf of Mexico. As the Atlantic widened, the Pacific narrowed apace, and the Pacific Plate expanded at the expense of bordering plates. During the Late Jurassic and Cretaceous, connections between the Atlantic and Pacific, via passageways between small continental plates in Central America, gradually widened, permitting a globe-circling, mainly west-flowing current system in low latitudes.

Except for tectonic slices preserved along the adjacent American and Asian continents, all pre-Jurassic Pacific oceanic sediments have been subducted. Middle and Late Jurassic oceanic crust is inferred from magnetic data, but no certainly dated Jurassic sediment samples have yet been raised on cores, in part because most Jurassic oceanic crust is buried beneath thick piles of mid-Cretaceous intraplate volcanics.

Cretaceous sediments deposited in low (Tethyan) latitudes are known from some 50 different drill sites in the central and western Pacific, which taken together with dredge hauls provide a sampling of paleoenvironments ranging from shallow-water (seamount) platforms through moderate-depth plateaus to normal oceanic crustal depths. A zigzag northwestward motion of the Pacific Plate of about 30° since Early Cretaceous time resulted in subduction of most Cretaceous sediments deposited in the northern hemisphere, and we have few samples of shallow-water Cretaceous sediments from close to the equator.

Pacific Cretaceous pelagic sediments are typically nannofossil limestone above the calcite compensation depth and brown, zeolitic clay and radiolarian mudstone below. Chert is common as nodules in both limey and clayey sediments, and siliceous sediments appear to have been deposited over a wide band of latitudes in the Cretaceous, suggesting a more diffuse distribution of fertile waters than in modern oceans. Mid-Cretaceous black shale is present on several plateaus, at paleodepths of 500–1500 m, but contemporaneous sediments in adjacent basins are oxidized, indicating that Pacific anoxia resulted from intensification and/or expansion of the oxygen minimum, and not from basin-wide stagnation.

In the Lower Cretaceous, poor preservation and lack of distinctive Tethyan zonal fossils have prevented any but crude (stage-level) age assignments. Nannoconids are common, but no calpionellids are yet reported from the Pacific; these animals may have preferred the more hemipelagic environments of the narrow seas of the Mediterranean and Atlantic Tethys to thc ultra-oceanic Pacific. Upper Cretaceous calcareous pelagic sediments, with better preserved planktonic and benthic foraminifers and nannofossils, can generally be correlated confidently at the substage or zone level.

Cretaceous vulcanism was on a stupendous scale: in the western equatorial Pacific, a region of about 10^7 km^2—the size of Europe—is covered almost entirely by midplate tholeiitic flood basalts flows and sills, probably emplaced mainly during the Aptian. The thickness locally may be several kilometers and the volume dwarfs famous continental plateau basalt piles. Emplacement of these huge masses in a relatively short time may have had important consequences not only for global sea level, but also for the chemistry of the oceans and atmosphere. Early Cretaceous volcanism also built both oceanic plateaus (Ontong Java, Shatsky, Hess, Magellan, and Manihiki) and chains and clusters of seamounts, especially of Aptian–Albian and of Campanian age. Repeated episodes of vulcanism along seamount chains was common.

A striking link to Tethyan facies in the other hemisphere is the presence of shallow-water platform and reefal sediments on the summits of many Pacific Cretaceous seamounts. More than 100 Early Cretaceous seamounts built to heights above sea level, and doubtless many more existed on crust now subducted along the margins of the Pacific. If the close spacing of preserved Cretaceous seamounts was typical of the entire Cretaceous Pacific, then there was no significant barrier to transoceanic shallow-water faunal dispersal along the island stepping stones. As the seamounts subsided, shallow-water rudist/coral platforms gradually covered them, eventually forming atolls on some seamounts. The platforms are the Pacific expression of the "Urgonian" platforms of the Tethys. Many—perhaps most—of the Pacific platforms drowned in mid-Cretaceous times and are guyots today, with summits typically at about 1500 m depth. New volcanic chains rose later in the Cretaceous, especially during the Campanian, and these were in turn capped by reefs that flourished through the Maestrichtian, but then probably drowned. Faunal similarities with other parts of Tethys, especially with the Caribbean, are strong for the Late Cretaceous; indeed, it may be that Caribbean crust is partly a fragment of a plate from the Pacific, trapped by Central American tectonics. Cenozoic atolls crown some Pacific Cretaceous edifices, possibly because renewed volcanism raised the drowned foundations into shallow waters.

WINTLE, A., University of London, Egham, United Kingdom

Thermoluminescence Dating of Loess

The thermoluminescence (TL) properties of minerals such as quartz and feldspars enable sedimentary grains to be dated directly. Loess is particularly well suited to the application of the TL method because exposure to sunlight during transport prior to deposition removes the earlier TL signal. The TL signal then grows with time as the grains are exposed to radiation from the decay of naturally occurring radioactive elements within the sediment. Laboratory procedures involve the measurement of the natural TL signal and the TL sensitivity of the sample to laboratory irradiation. Together these are used to calculate the Equivalent Dose (ED) used in the age equation

$$\text{TL age (years)} = (\text{ED [grays]})/(\text{dose rate [grays/year]})$$

The dose rate is obtained by measurement of the radioactive content of the sediment allowing for appropriate dosimetric consideration. Thick source alpha counting is used to determine the uranium and thorium contents and thus the alpha and beta dose rates, the ^{40}K contribution is calculated from the bulk potassium content obtained by atomic absorption spectrometry, and the gamma dose rate is obtained in the field using a portable gamma spectrometer.

In the last three years, the author has applied the method to loess deposits in northern central Europe (Wintle, 1987; Juvigné and Wintle, 1988), Eastern Europe (Wintle and Packman, 1988), and Alaska.

The end of loess deposition in northern Europe occurred about 12,500 years, on the basis of consistent TL and radiocarbon age estimates. About 3 m of loess was deposited in about one and a half thousand years after a short warming. This warming resulted in a tongued horizon being found in a number of sections in Belgium. Until now the age of this feature was thought to be about 22,000 years based on two radiocarbon dates on humus, but 3 TL dates gave an age of about 15,000 years. Before this warming loess was deposited from about 19,000 to 15,500 years. A date of 25,000 years for the underlying peniglacial loess is in agreement with radiocarbon evidence for the onset of loess deposition in the second part of the Weichsel in northwestern and central Europe.

In Hungary, the thick loess sections have been divided into older and younger loess, with the boundary occurring at the Mende-Base (MB) Soil complex, which was thought to date from the last interglacial. TL ages for the humic horizons at the top of the upper loess were in agreement with the radiocarbon dates indicating loess deposition around 16,000 to 18,000 years and around 23,000 years, in agreement with north-central Europe. However, dates of around 76,000 years were found for the uppermost chernozem (MF), in disagreement with published radiocarbon dates. This is likely to be an underestimate of the age because of the particular TL method employed in the laboratory. This also implies that the much lower MB soil is much older than the last interglacial.

WISE, S. W., et al, Florida State University, Tallahassee, Florida

Geologic and Tectonic Evolution of the Kerguelen Plateau as Revealed by ODP leg 120 Drilling

Between February and May, 1988, five ODP Leg 120 drill sites were emplaced in the central portion of the submergent Kerguelen Plateau in the Antarctic sector of the Indian Ocean (between 54° and 59° S Latitude), including 3 sites in the Raggatt Basin. The primary objective was to obtain as complete a record as possible of the origin and subsequent development of this remote, 2500 km long feature.

Basement rocks were drilled at 3 sites, and all were transitional tholeiites (T-MORB) in trace element characteristics. No continental crust was drilled. Erupted prior to the Cenomanian, the basalts are overlain by nonmarine sediments at Site 750 in the eastern Raggett Basin, where kaolinitic clays contain colorful soft-pebble conglomerates and large cm-size wood fragments representing a relatively warm, humid, well-forested depositional environment. Intense subaerial weathering over a considerable area would have been necessary to produce the kaolinitic clays, and the conglomerates probably represent fluvial conditions.

At Site 748 in the western Raggatt Basin, the Cretaceous basalt flows are overlain by ocean island basalts (OIB), the first evidence of hot spot activity in this region. Subsidence of this site below sea level during the turonian is indicated by cobbles and boulders of OIB basalt set in a matrix of fine glauconitic siliciclastics with fragments of thick-walled molluscs. Slow subsidence thereafter produced a thick (200 m) sequence of Turonian to Coniacian glauconitic silts with up to 0.5% organic matter (mostly terrestrial), denoting a restricted marine environment which lacked the calcareous benthic and planktonic microfauna and flora that flourished elsewhere at this time on the plateau. This sequence is overlain apparently disconformably by an equally thick, shallow water Campanian to Maestrichtian bioclastic limestone, again highly glauconitic (up to 58% of the sediment). Intermittently silicified, this unit contains mostly fragmented bryozoans, crinoid columnals, inocerimid bivalves, echinoid spines, siliceous sponge sipcules, radiolarians, coralline "red" algae, benthic foraminifers, common bone debris and some teeth attributed to swimming reptiles. Cool climates prevailed as the Maestrichtian planktonic foraminiferal and nannofossils assemblages generally have a distinctive austral aspect, similar to those fro the Falkland Plateau and Maud Rise in the high latitude South Atlantic.

Although the western portion of the Raggatt Basin remained structurally high throughout the Cretaceous, the eastern margin at Site 750 began to subside earlier, as documented by the change to open marine sediments by Cenomanian times and the presence of reworked Cenomanian planktonic foraminifers and benthics washed down from inner shelf depths. By late Campanian time, the site was at upper slope depths with no indication of the shallow-water, glauconitic facies developed at Site 748 to the west.

Foundering of the remaining portions of the central plateau resulted from a major rifting event that apparently affected the entire plateau and opened the extensive 77 °E Graben, among others. The first indications of tectonic activity are noted at Site 747, north of the Raggatt Basin in lower Maestrichtian chalks which contain silt to sand size clastics eroded from volcanic basement. These are overlain by massive debris flows consisting of clay, sand and cobble size clastics and breccias derived from nearby uplifted basement. Wireline logs suggest that these debris flows span up to 40 m of section, and micropaleontologic evidence shows that they spill over into lowermost Danian sediments. The climax of this event around the time of the Cretaceous/Tertiary boundary is coincident with or followed immediately by the precipitous subsidence of the plateau close to its present day water depts.

Except for the upper Paleocene through middle Eocene section, where interbedded cherts impeded recovery of biosiliceous and calcareous oozes, and excellent pelagic record was obtained for interpreting the Cenozoic paleoceanographic history of this high latitude region. Particularly significant is the high percentage of carbonate in the Oligocene through lower Pliocene sediments at sites as far as 900 km south of the present day Polar Front, because these will allow high resolution shore-based stable isotope, magneto-and biostratigraphic correlations of the sections. The first ice-rafted sediments from the Antarctic continent reached Site 748 (58°26.45 S) during the early Oligocene and consist of predominantly zed angular quartz grains, but with an assortment of other minerals exotic to the area. This further signifies the intensity of Antarctic glaciation during the early Oligocene. Ice-rafted debris, however, did not reappear at these sites until the late Miocene. Another unusual finding was a very young (early Pliocene age) porcellanite bed encountered only 9 m below sea floor at Site 748, an exam by ODP in Antarctic waters, the sedimentary facies became predominantly biosiliceous by Pliocene times, and the record is truncated by hiatuses, in this case a reflection of deteriorating climate and the increasing strength of the Circumpolar Current, which became a dominant control over sedimentation during the Plio/Pleistocene.

WISE, S., Florida State University, Tallahassee, FL, and Ocean Drilling Project Leg 113 and 120 Shipboard Parties and Shore-Based Collaborators

Mesozoic Paleogeography of Seaways and Ocean Connections in Southern Ocean and Antarctica (Dispersal of Southern Gondwanaland)

Geophysical reconstructions of Gondwanaland for the early Mesozoic bring the Mozambique Ridge, the eastern Falkland Plateau, and the Dronning Maud Land margin of East Antarctica into close juxtaposition (Lawver and Scotese, 1987; Lawver et al, in press). Jurassic and/or Lower Cretaceous sediments have been cored on each of these features by the DSDP/ODP Legs 25, 36, 71, and 113, and thereby provide sedimentary and paleontologic records that can be compared with those from land outcrops to trace the early breakup history and subsequent evolution of this portion of Gondwanaland. Site 330 on the Falkland Plateau records the Middle to Late Jurassic invasion of shallow-marine (shelf depth) waters with the development of calcareous nannofossil assemblages of a high latitude aspect by Oxfordian–Kimmeridgian times. Belemnite data suggest migration of Tethyan elements into this Southwest Gondwana Sea by a circum-Gondwana route during the Kimmeridgian–lower Tithonian, after which a more direct route opened across Ethiopia/Madagascar during the late Tithonian. This hypothesis, which could be tested by searching for nannofossil assemblages of a Falkland Plateau affinity in the Madagascar region, is in accord with geophysical reconstructions of Lawver et al (in press) for 140 Ma. They show East Gondwanaland separating from West Gondwanaland (South America, Africa) along a sheared transform margin, which is today seen as the Explora Escarpment off Dronning Maud Land and the Mozambique Escarpment off Africa.

Kimmeridgian sediments of the Falkland Plateau are black mudstones that denote restricted environments and are characterized by total organic contents that average 3%. Restricted conditions also prevailed within the narrow, incipient deep-marine basins of the Early Cretaceous despite the initiation of seafloor spreading in the Weddell Sea and the initial rifting of the South Atlantic Basin by 130 Ma. The best documented drill record for this time slice is at ODP Site 692 off East Antarctic, where dark organic-rich Valanginian mudstones were cored near the base of a 1000 m thick basinal sequence deposited in about 500 m of water. Total organic contents range up to 8%, the highest at any of the DSDP/ODP sites mentioned. These are predominantly marine-derived (Type II) hydrocarbons. Blockages of the developing seaways by major plateaus (Falkland, Mozamabique) and land masses (Madagascar) are sufficient to maintain restricted conditions over the Falkland Plateau and the East Antarctic Margin until late Aptian–early Albian times despite the fact that rifting between Indian and Antarctic had begun by MO time (118 Ma). Blockage of that developing seaway to the east was probably achieved by the massive outpouring of the transitional (T-MORB) basalts that underlie the present-day Kerguelen Plateau.

Ventilation of the Atlantic sector basins was finally achieved by the introduction of deep ocean circulation as the eastern tip of the Falkland Plateau cleared South Africa and eustatic sea levels began to rise. This is evidenced by the deposition of progressively thicker carbonate-rich sediments interbedded with the Aptian black mudstones at Falkland Plateau Site 511. Nevertheless, nannofossil assemblages there contain distinctive provincial elements, reminiscent of those of equivalent age within the boreal North Sea region. This underscores the geographical restriction of the high-latitude basins of both hemispheres before the major sea-level rise during the Aptial–Albian and late Cretaceous stages plus continued tectonic fragmentation opened these regions up for wholesale invasion by warmer water Tethyan elements from the north. The overturn of nutrient-rich waters over the East Antarctic margin at the close of black shale depositon during the early Albian is evinced by the accumulation of carbonate-free olive green diatomite, which contains spectacular assemblages of exceptionally well preserved diatoms and silicoflagellates, some of the oldest known from the geologic record.

With the subsequent dispersal of the continents during the Late Cretaceous, the Southern Ocean plateaus and rises plus the marginal basins of West Australia provide a much larger database by which to trace paleo-oceanographic development. This larger database plus the free communication between major water masses displays latitudinally zoned planktonic microfaunal and microfloral assemblages, the distribution of which are controlled by paleotemperatures and water mass characteristics, a pattern not as easily discerned in the Lower Cretaceous sequences.

REFERENCES

Lawver, A. A., and C. R. Scotese, 1987, A revised reconstruction of Gondwanaland, *in* G. D. McKenzie, ed., Gondwana Six: structure, tectonics, and geophysics: AGU Geophysical Monograph 40, p. 17–24.

——— J. Y. Royer, D. T. Sandwell, and C. R. Scotese, in press, Evolution of Antarctic continental margins, *in* M. R. A. Thompson, et al, eds., Proceedings of the 5th International Symposium on Antarctic Earth Sciences: Cambridge, Cambridge University Press.

WITHERSPOON, P., and H. F. MORRISON, University of California, Berkeley, and Lawrence Berkeley Laboratory, Berkeley, CA, and J. C. S. LONG, E. L. MAJER, and L. R. MYER, Lawrence Berkeley Laboratory, Berkeley, CA

State of the Art of Fractured Rock Characterization

Fractured rocks have been extremely difficult to characterize in providing the groundwater hydrologist with the kind of details that he must have if he is to develop reliable models of such systems. This paper presents a new approach to this problem that consists of a combining of several investigative methods from the field of earth sciences. The basic idea is to combine the results of geological and geophysical field measurements with the hydraulic signature of the system and generate a three-dimensional flow model. The final model must, of course, contain the essential features of the flow system and yet be constrained so that it is consistent with all sources of information. An important element of this approach is that of using seismic methods to locate fracture networks in rock masses. To provide a basic understanding of the phenomena when a seismic wave passes through fractured rocks, laboratory investigations have been carried out on single fractures. A comprehensive suite of measurements have shown that a single fracture in crystalline rock will attenuate P- and S-waves even though the fracture thickness is much smaller than the propagating wavelength. These experimental observations were found to be in both qualitative and quantitative agreement with the displacement discontinuity model of elastic wave propagation across a nonwelded interface. For the past several years, we have been carrying out field experiments at various fractured rock sites to determine the fundamental nature of the propagation of seismic waves in fractured media. The experiments have utilized high frequency (1000 to 10,000 Hz) signals in cross-hole configurations at scales of several tens of meters. A second effort has been to develop wave propagation models that account for fracture properties. This has resulted in the development of codes to model anisotropy due to fracture systems in two- and three-dimensional models. The results indicate that both P- and S-wave results can be used to map the location of fractures. In addition, fractures that are open and conductive are much more visible to seismic waves than nonconductive fractures.

Another geophysical method that we are also exploring is the use of electrical conductivity. Low frequency electromagnetic methods can be used in subsurface configurations to map conductivity distribution with far greater confidence than can be obtained with conventional surface-based methods. The basic equipment needed to begin field experiments is available, and we have made a first step in developing direct imaging methods for the conductivity distribution. Used in conjunction with seismic methods, the electromagnetic approach provides an added window on properties that are of direct application to the problem of characterizing fractured rocks.

In working with fractured rocks, the problem becomes that of generating a model of the fractured system that incorporates the geophysical type of data with other field measurements into a unified model of the fractured rock. The approach that we are now investigating involves three steps. The first step is to start with available geologic data and make conditional simulations of those parameters that should have direct (or indirect) relationship with the flow behavior of the fractured rock. We do this by analyzing the statistical structure of each source of data to build a probability model of each parameter and then construct a simulation of the parameter in three-dimensional space.

The second step is to utilize each of the parameter maps in constructing a three-dimensional map of the flow properties of the system. For example, a tomagram of seismic slowness developed in the first step can presumably be related to the probability of observing a certain permeability at any given location. From the laboratory results, the slowness is a measure of interference in wave propagation and if fractures are the cause, the rock should be more permeable where there is more evidence of seismic slowness. A three-dimensional map of randomly chosen values of permeability can be developed such that the zones of higher permeability correlate with zones where maximum slowness is evident in the seismic simulation. In a similar manner, the results from the other parameter maps can be used to further condition the three-dimensional distribution of permeability. The result is a probabilistic joint inversion of the several simulations developed in the first step into a permeability distribution. The map obtained represents a summary of the "soft" information and is only a nonunique estimate of the permeability distribution in three dimensions.

The third step is to use this permeability map as the initial conceptual model in solving the hydrologic inverse problem. Field results from various flow testing are used in this step. The idea is to modify the three-dimensional model of permeability until a best match with field data is obtained while still retaining the essential features of the result developed from the "soft" data described above. One way of doing this is to represent the system as a fractal process where the fractal nature of the system is a measure of the degree on interconnection between the relatively more permeable regions of the fractured rock. We have found that such systems have distinct hydraulic signatures in flow tests of hypothetical systems.

Another way is to use the annealing process to find the best fit between the three-dimensional permeability distribution and the observed field data. The annealing process is a procedure for making arbitrary changes in the infrastructure of the permeability model and then comparing the flow behavior of the new model with that of the fractured system. As successive changes in the model are made, the agreement may or may not improve. The annealing process is designed to make changes in an appropriate manner to avoid converging on an answer that does not represent the global system.

We believe the procedure described above can provide a much more comprehensive method of characterizing fractured rocks than is currently available.

WITT, G., D. OPPER, S. KRAMM-GLADE, and H. A. SECK, Universität Köln, Köln, West Germany

Reheating and Secondary Enrichment of Amphibole-Bearing Spinel Peridotite: Implications for Mantle Metasomatism Beneath Eifel

Amphibole-bearing spinel peridotites with mosaic and tabular recrystallized textures are widespread among mantle xenoliths from the Eifel. Temperatures derived from orthopyroxene geothermometry plot into a remarkably narrow interval of 900 ± 20°C except for a larger portion of xenoliths affected by a young reheating. The reheating is evident from element zonation in orthopyroxenes with Ca, Al, and Cr increasing from core to rim, which is due to the temperature-dependent reactions of orthopyroxene with clinopyroxene and spinel, respectively. From a detailed study of the concentration gradients of these elements in orthopyroxenes from more than 10 xenoliths and a statistical evaluation of the data, we infer that the diffusion of Ca in orthopyroxene is faster by a factor of 4 than that of Al and Cr whose diffusion constants are nearly identical. Concentration profiles of Ca, Al, Cr, and Ti in orthopyroxene from four xenoliths are shown in Figure 1. Differences in element profiles of the same xenolith result from the differences in the pertinent diffusion rates. Variations of the profiles for the same element in the xenoliths reflect the duration of heating. In orthopyroxene 81 (Figure 1a) all elements slightly increase toward the rim. The plateaus reflect concentrations typical for amphibole-bearing xenoliths not subjected to heating. In the case shown in Figure 1d, Ca is nearly homogenously distributed at a high concentration level appropriate to elevated temperature. Al and Cr, however, still reveal marked concentration minima in the center of the grains. Temperatures pertinent to the orthopyroxene rims as calculated from the solubilities of Ca, Al, and Cr in enstatite (Witt and Seck, 1988) are between 1000° (Figure 1a) and 1050°C (Figure 1c,d). Thus, xenoliths 81 and 766 (Figure 1a,b) reflect a shorter duration of reheating and lower temperatures compared with xenoliths 62 and 815 (Figure 1c d).

In spite of reheating, TiO_2 in orthopyroxenes 81, 766, and 815 remains at the low level typical for orthopyroxene in the amphibole-bearing peridotites from the Eifel. This is reasonable because of the lack of a Ti-rich phase the orthopyroxene could exchange with. The extremely strong increase of TiO_2 at the rim of orthopyroxene 62 (Figure 1c) cannot be a temperature effect but must result from an enrichment that according to experimental evidence is likely to be due to an exchange with a melt (Schneider and Eggler, 1986).

The correlation between reheating and the Ti-enrichment is also evident from the TiO_2 contents of the amphiboles. Amphiboles of the recrystallized peridotites from the Eifel are typically TiO_2-poor with concentrations ranging from 0.15 to 0.5% that are strongly correlated to Cr/(Cr + Al) ratios of spinel. This indicates that the content of TiO_2 is controlled by liquid/crystal equilibria. All xenoliths with amphiboles in which TiO_2 exceeds the level of 0.5% are reheated and lack the correlation between TiO_2 and the Cr/(Cr + Al) ratio. However, some of the reheated xenoliths show no TiO_2 enrichment.

The causal interrelation between reheating of the recrystallized xenoliths and the TiO_2 enrichment is apparent in composite nodules of amphibole-bearing spinel periodotite and TiO_2-rich micaceous hornblendites. The chemical interaction between a hornblendite vein and peridotite is shown in Figure 2. At a distance greater than 1 cm from the vein, the amphiboles of the peridotite are typically TiO_2-poor and chemically homogeneous. Closer to the vein, the amphiboles are zoned with TiO_2 increasing toward the rim of the grains. The TiO_2 content of the center of the grains also increases as a function of the distance from the vein. In amphiboles closest to the vein TiO_2 has nearly homogenized at a high concentration level of more than 2%. According to orthopyroxene analyses, the peridotite has been reheated to a temperature close to 1000°C. Two conclusions may be drawn from this: First, the hornblendite vein intruded in a peridotite in which a TiO_2-poor amphibole already existed. That is, the interaction of the peridotite with the vein caused the TiO_2 enrichment but not the formation of the amphibole. Second, the TiO_2 enrichment is related to a heating event. This does not necessarily mean that the vein was the sole heat source. An alternative view is that a portion of the upper mantle beneath the West Eifel was already heated at the time of vein formation.

We prefer the following hypothesis to account for the interrelation between reheating and TiO_2 enrichment recorded in a great number of the recrystallized amphibole-bearing peridotites. The TiO_2-poor amphiboles of these peridotites were formed due to a metasomatic event predating the reheating and the concomitant TiO_2 enrichment. The reheated samples reflect a thermal aureole around magma body. Xenoliths such as 815 (Figure 1d) are closer to the heat source than nodules such as 81 (Figure 1a). Within this thermal aureole, samples adjacent to veins that intrude into the aureole were subjected to chemical interaction with basic melt flowing through the veins. Because of the lack of reliable diffusion coefficients for the elements studied in orthopyroxene, the duration of the heating can be roughly estimated at best. According to the data available, a diffusion coefficient of 10^{-16}s cm_{-2} at 1050°C seems reasonable. If this is about correct, a heating period of about 150,000 a is calculated for xenolith 62.

REFERENCES

Witt, G., and H. A. Seck, 1988, EOS, v. 69, p. 503.

Schneider, M. E., and D. H. Eggler, 1986, Geochimica et Cosmochimica Acta, v. 50, p. 711–724.

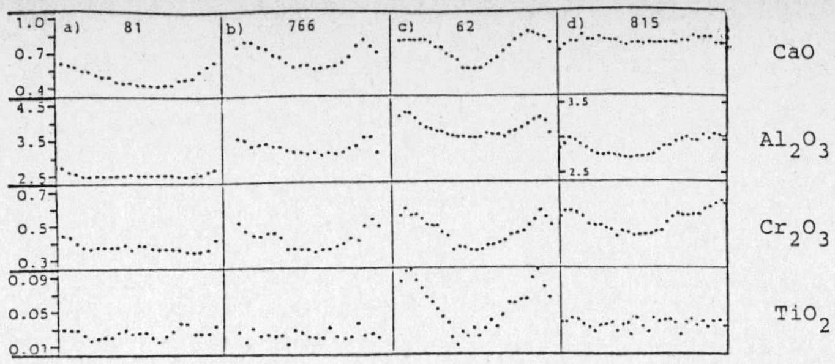

Figure 1.

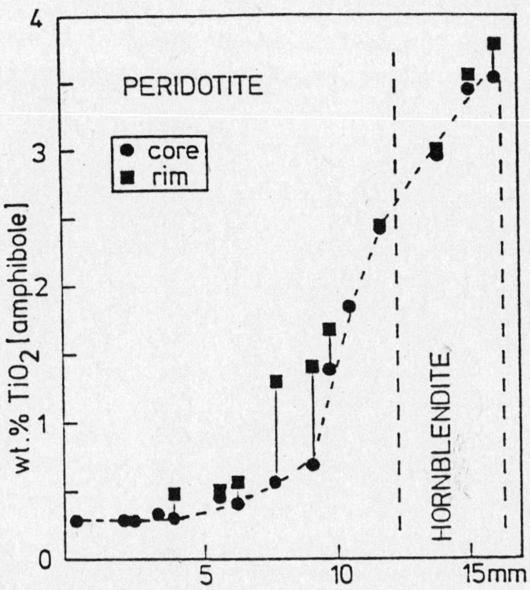

Figure 2.

WNUK, C., U.S. Geological Survey, Reston, Virginia

Diversity of Middle Pennsylvanian Clastic Swamp Communities

Prevailing reconstructions of Pennsylvanian plant communities suggest that there was little diversity in the types of plant associations that colonized lowland clastic swamp environments. to test this hypothesis, the species composition of in situ underclay floras was determined at 84 localities in the Western Middle Anthracite Field, Pennsylvania. (Evidence supporting the in situ nature of underclay floras has been presented by Wnuk and Pfefferkorn, 1987.) The 84 fossil assemblages occurred on underclay bedding plane surfaces in the basal part of the Middle Pennsylvanian Llewellyn Formation between the Buck Mountain (No. 5) coal bed and the Bottom Split of the Mammoth (No. 8) coal bed. The analyses of species composition involved detailed quadrant counts and rapid-survey line transects covering 18,000 m of outcrop surface. The results of this study indicate that the stated hypothesis is incorrect. A significant diversity was observed among the types of plant communities that colonized clastic swamp environments.

Of the 84 localities investigated, 20 contained mono-specific assemblages of Stigmaria ficoides roots. It is presumed that the edaphic conditions at these 20 sites did not favor the preservation of typical underclay trunk floras. Floras at the remaining 64 localities can be divided into 4 broadly defined community types (listed below), and further subdivided into 23 compositional variants.

encounterred.1. Lycopod dominated associations: Twelve lycopod dominated fossil assemblages were encountered. These assemblages were divided into six community subtypes. The lycopod-dominated associations included monospecific communities composed of Sigillaria or Lepidodendron, communities containing a combination of Sigillaria and Lepidodendron, and communities containing a small percentage of Calamites among the lycopods. In many instances, the lycopod stems were decorticated and therefore unidentifiable. Additional studies of these communities will undoubtedly reveal many additional lycopod dominated subtypes.

2. Pteridosperm dominated associations: Ten pteridosperm dominated plant fossil assemblages were encountered. These were divided into two community subtypes: pteridosperm communities and communities containing a small percentage of Calamites stems. Since pteridosperm stems cannot be identified to genus or species, the diversity among community

subtypes is difficult to assess and probably will always be underestimated. Analyses of the variation in petiole architecture suggest that a variety of genera occur in these communities and that there is variation in the genera present at any given site.

3. Associations containing mixtures of pteridosperms and lycopods. Thirty nine pteridosperm-lycopod plant fossil assemblages were encountered. These were divided into 12 community subtypes. Subtype diversity is underestimated significantly because the pteriodsperms cannot be identified precisely and because the decorticated lycopods are difficult to identify. A variety of lycopods occur in this community type, including Lepidodendron (L. aculeatum, L. rimosum, and L. bretonense), Bothrodendron punctatum, Lepidophloios, and Sigillaria. Some subtypes also contain a small percentage of Calamites.

4. Associations containing tree ferns. These community types are not common. Only three have been identified, each its own subtype. Tree fern communities typically include lycopods and may also contain pteridosperms and Calamites.

The line transect studies have shown that, over the length of a single bedding plane, the species composition of plant fossils can vary from one community type or subtype to another. Results based on analyses of floral composition of underclays containing more than one trusite may belong to significantly different community types. Investigations are currently underway to determine which environmental and edaphic factors are involved in promoting the development of this observed community heterogeneity.

REFERENCES

Wnuk, C., and H. W. Pfefferkorn, 1987; A Pennsylvanian-age terrestrial storm deposit: using plant fossils to characterize the history and process of sediment accumulation. Journal of Sedimentary Petrology 57, p. 212–221.

WOERMANN, E., Institut fur Kristallographie, RWTH Aachen, Aachen, Federal Republic of Germany, and M. ROSENHAUER, Mineralogisch-Petrologisches Institut der Universität, Göttingen, Federal Republic of Germany

Oxygen Fugacity in Earth's Mantle

It is now universally accepted that oxygen fugacity plays an important role in the control of petrogenetic processes in the earth's mantle. Thus, it is not surprising to see the great number of attempts to evaluate this parameter qualitatively and quantitatively.

Mantle material for investigations is provided only by volcanic processes, more directly in the form of basaltic or kimberlitic xenoliths, but also by indicator minerals, fluid inclusions, volcanic magmas, and gasses.

From all information collected at the present, it can be stated beyond all doubt that the oxygen fugacity in the earth's mantle covers a wide spectrum—from relatively oxidized, in the order of the OFM-reference level, to reduced, in the range of the IW-level or below. Apparently two outstanding groups of higher and of lower oxygen fugacity levels can be distinguished, but much more information is required to prove or disprove a "dichotomy" of the mantle (Ulmer, 1987).

Peridotite xenoliths from alkali basalts are the easiest accessible original mantle material. Thus, it is not surprising that experimental data from basaltic nodules are numerically strongly dominating—even to such a degree that here and there a bias seems to exist projecting results from the basalt source regions to the earth's mantle as a whole.

In order to determine the oxygen potential of a sample, essentially two independent methods have been employed: (1) intrinsic oxygen fugacity measurements (IOF), developed by Sato and by Ulmer in a great number of contributions since 1971 (compare also this symposium), and (2) thermodynamic calculations from oxide-oxide equilibria (Haggerty and Tompkins, 1983) or from oxide silicate equilibria (Eggler, 1983; O'Neill, 1986; Ender and Wuermann, this symposium).

In broad outline, the data on peridotite nodules in basalts from all over the world confirm an approximately constant level of oxygen fugacities on the order of "QFM." A closer inspection, however, reveals local differences, such as slightly higher or lower absolute fO_2 values, a closer clustering or a wider scattering of all data points from one and the same locality.

There must be a reason for the accumulation of fO_2 data around "QFM." Some process with a sufficiently potent buffering capacity has to be responsible for the control of the mantle fugacity in the regions tapped by basalts.

Any redox reaction has to contain one element in different oxidation states.

In a typical mantle environment, the major elements magnesium, calcium, aluminium, the alkali elements, and silicon occur in one stable valence state only. The same statement essentially also holds for chromium and titanium.

Typical elements, occurring in different oxidation states, are iron, carbon and hydrogen. Only reactions involving elements from this latter group are considered as possible mantle buffering processes.

In basalt derived xenolites, iron is present both in the divalent and in the trivalent state. Under defined temperature and pressure conditions the basic redox reaction:

$$6\,FeSi_3 + 2\,FE_2O_4 = 6\,Fe_2SiO_4 + O$$

is defined in the system Fe-Si-O, but it is multivariant in complex peridotite compositions. The Fe^{2+}/Fe^{3+} ratio of the mantle assemblage is variable and reacts sensitively on changes of fO_{-2}. For this reason the Fe^{2+}/Fe^{3+} ratio is not capable of controlling fO_2, but it may instead be employed as a sensitive indicator for the fO_2 installed.

Carbon is present in several mantle phases and species as elementary carbon (diamond or graphite), in carbonate phases (dolomite or magnesite), or as fluid species (CO_2, CH_4). Only one possible mantle reaction involving carbon as a component:

$$MgSiO_3 + MgCo_3 = Mg_2SiO_4 + C + O_2\,(EMOG)$$

is precisely defined under given temperature and pressure conditions. Eggler et al (1980) thus proposed that EMOG actively controls the oxygen fugacity of the mantle. Unfortunately, however, coexisting forsterite + enstatite + a carbonate + elementary carbon have never been verified for mantle nodules. This paragenesis can safely be excluded for normal source regions of basalts and probably also for deeper levels. It is thus improbable that carbon-carbonate reactions control the oxygen fugacity of the upper mantle.

A fluid phase in the C-III D system does not possess an own buffering capacity in a graphite-free assemblage. Even in assemblages containing elementary carbon, the oxygen fugacity of the system is under relevant P-T conditions sensitively dependent on the specific composition of the fluid phase.

Since neither the Fe^2/Fe^3 ratio nor the reactions with elementary carbon is capable of controlling the oxygen fugacity of the earth's mantle, the possibility should be discussed whether it could have been adjusted by the dissociation reaction of H_2O:

$$2\,H_2O = 2\,H_2 + O_2.$$

Several facts seem to be in favor of this model:

1. It is now a generally accepted fact that the earth's mantle contains fluids that are—in the stratigraphic levels under discussion—composed of CO_2 and H_2O, probably with high H_2O concentrations.

2. The existence of fluids in the mantle requires equilibrium conditions with the solid peridotite phases, including equilibrium fO_2.

3. The oxygen fugacity of the above reaction corresponds closely to "QFM" under the relevant temperature and pressure conditions.

In order to confirm or to refute this model, it will be necessary to pay attention to the properties of the xenoliths analyzed for fO_2 calculations: whether they are depleted or undepleted, whether they show metasomatic alterations or whether they contain hydroxyl-bearing phases.

WOHLETZ, K. H., and G. A. VALENTINE, Los Alamos National Laboratory, Los Alamos, New Mexico

Computer Simulations of Volcanic Eruptions

Models of volcanic eruptions, based upon numerical solutions of fundamental differential equations of hydrodynamics, give precise results applicable to analysis of field data and prediction of hazards. We have adapted a computer cose that solves the full set of two-phase, compressible Navier-Stokes equations in two-dimensions to compute pyroclastic flows with initial and boundary conditions appropriate for explosive eruptions. The governing equations consist of eight coupled, nonlinear, partial differential equations with sixteen dependent variables that express conservation of mass, momentum, and energy for a solid and a gaseous phase. Eight algebraic relationships for internal energy, ideal equation of state, and interphase momentum and heat transfer close the set of differential equations. The numerical solution technique fully retains the nonlinear terms of the equations and is stable for flow fields containing a wide range of velocities from subsonic to supersonic; these features are necessary for modeling the complex physical system of a volcanic eruption. When a solution time-step of 0.02 s is combined with a computational domain of 5182 cells, each representing two-dimensional space of 100 x 100 m, nearly 2.5 hours of Cray-1 computer time is required to model 200 s of eruption time (note that this machine vectorizes arrays and executes about 10^8 floating point operations per second). Each model run produces more than 200,000 pages of printed data and over 400 contour and vector plots of flow parameters. Figure 1 shows the computational domain.

Initial computer models of the first moments of large explosive eruptions showed similarities to shock-tube behavior, with propagation of a bow shock into the atmosphere and a rarefaction wave down the volcanic conduit. The propagation of the rarefaction followed characteristic speeds predicted for multiphase fluids of less than several hundred meters per second. Fluid flow in the conduit accelerates across the rarefaction wave as water expands from supercritical fluid to vapor, such that flow is essentially incompressible in portions of the conduit not decompressed by the rarefaction wave. The chamber decompression requires several passages of the rarefaction wave and its reflections. During this time, which may last from several tens of seconds to several minutes, flow from the conduit is highly unsteady and contributes to blast phenomena.

A second phase of modeling computes the steady phase of eruptive behavior. Figure 2 shows results for one of over 50 models calculated. These results show collapse of an eruption column to form a pyroclastic flow that moves over the substrate away from the vent. In analysis of various eruption models, we find that column collapse can be predicted by values of four nondimensional parameters. The ratio of exit pressure to atmospheric pressure is important in determining the column structure. Vent overpressure causes underexpansion of the erupted column so that if behaves as a supersonic jet with a shape maintained by shock waves that reflect off atmosphere/eruption column interfaces and converge upward in the column to form a Mach disk. Pressure-balanced columns are most likely to experience gravitational instability and collapse with subsequent formation of pyroclastic flows. The other three parameters are: a thermogravitational parameter that is a ratio of thermodynamic driving forces to those of buoyancy; the Richardson number, a ratio of inertial to buoyancy forces, which embodies all variables considered by previous analytical models; and the Rouse number the measures the ratio of clast settling velocity to upward flow velocity. Overall, in contrast to previous models, our results do not support the assumption that eruption column behavior is largely controlled by air entrainment.

Because these numerical simulations model the intrinsic nonlinearities involved in multiphase fluid flow and are generalized, we have found that solutions can vary by over an order of magnitude from those predicted by linearized, analytical methods. Sets of eruptive conditions that control eruption column behavior aer those determined by movement of magma from the chamber to the vent. New solutions for conduit flow, using the full capabilities of numerical hydrodynamics, are needed to more accurately constrain the range of volcanic eruption behavior on Earth and other planetary bodies.

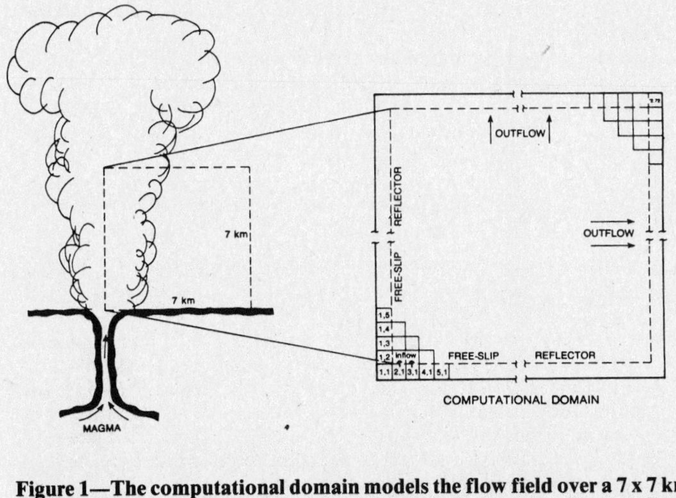

Figure 1—The computational domain models the flow field over a 7 x 7 km area adjacent to the volcano, and it is organized into 72^2 cells that form toroids in three dimensions for the cylindrical coordinate system used for calculations. A symmetry axis dissects the vent and modeled eruption plume, and the outer rows and columns of cells are solely for calculation of boundary conditions.

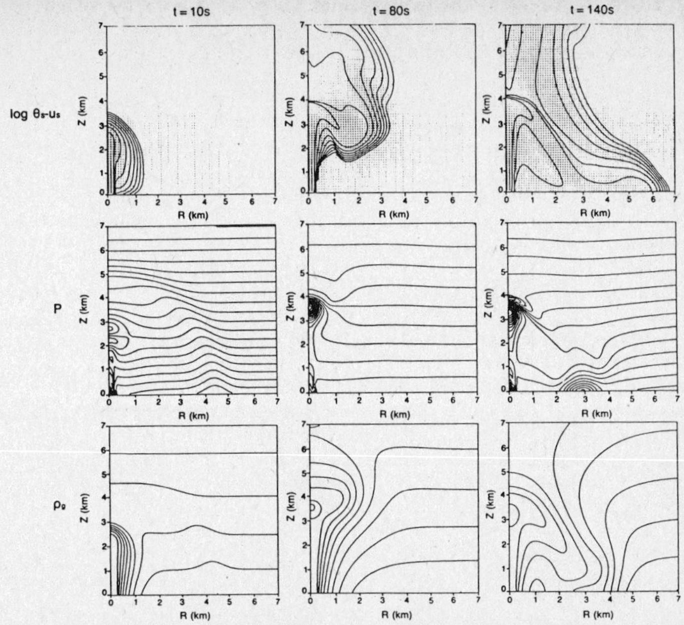

Figure 2—Contour plots of solid-volume fraction (θ_s) overlaying solid particle velocity vectors (u_s), pressure (p), and gas density (ρ_g) show development of a collapsing eruption column at three times. The innermost θ_s contour is 10^{-3} and the maximum u_s is 300 m/s, occurring at the exit plane.

WOHNLICH, S., University of Karlsruhe, Karlsruhe, West Germany

Preferential Flow Paths in Soils of Low Permeability

Soils of low permeability, e.g., clay and silt, are frequently used as natural and technical barriers against the transport of pollutants from hazardous waste disposals. Most of the pollutants are transported by the liquid phase. Therefore, the mechanisms that dominate the waterflow through these soils are vital for the risk assessment of barrier systems.

Due to the overall small flux density of compacted clay and silt up to recent times, little effort was made to investigate the mechanisms of water flow. Laboratory tests of saturated hydraulic conductivity are still the most common measurement used to estimate the quality of barrier materials. These tests, like any small scale tests, presume homogeneous and isotropic conditions of the material.

The comparison of field and laboratory tests have often revealed significant discrepancies. These discrepancies indicate that there are major inhomogeneities that occur in small as well as in large-scale experiments. In contrary to this knowledge, the discussion up to now focuses on the applicability of Darcy's equation and the existence of a stagnation gradient, but not on the effective pathways that are used.

Investigations in a field test area near Frankfurt am Main, West Germany, were done in order to detect preferential flow paths in a clay layer (hydraulic conductivity $K_f = 1 \times 10^{-9}$ m/s) with a thickness of 50 cm. Tracer experiments were done with flourescent dyes. The content of solutants and the natural isotope ^{18}O ($\delta^{18}O$) were observed in the seepage water over a 2 year period. The experiments showed that water flow through clay soils consists of a mixture of several flow paths. Three major pathways could be determined:

1. Flow along the matrix pores with travel times of more than 2 years for the 50 cm profile (microsystem).
2. Flow along structural discontinuities of the soil with travel times of 5 to 6 months (medium system).
3. Flow along cracks and biotic channels where travel time of a few days was observed (macrosystem).

The ratio of flow that contributes to the total flow varies mainly with the hydraulic gradient. This means that with very low gradients ($I = < 1.1$), the microsystem becomes essentially stagnant and only the medium system is involved in the total flow. With rising gradients, the proportion of microporewater increases nonlinearly. Within the range of the tested gradients ($I = 1.2$–3.4) the major part of the flow takes place on preferential flow paths, which are bound to the medium system.

This leads to the conclusion that the travel times through five-grained sediments have to be considered as much faster than currently thought. From the point of view of the storage of hazardous wastes, it may be said that very low hydraulic conductivities are no protection against a fast breakthrough of contaminants.

WOJCIK, A., Sztygarka-School of Mines, Dabrowa, Gornicza, Poland

Sedimentology of the Andelska Hora Beds in the Sudetes, Poland

The Opawskie Mountains are built up by Devonian and Carboniferous-Culm deposits. The Devonian and Culm series were folded during the Variscan orogeny. In general, the folding proceeded from west to east. The Andelska Hora Beds cover an extensive area in the eastern part of Opawskie Mountains, between the Vrbno Beds in the west and the Horni Benesov Beds in the east. Beds are mostly affected by slight epimetamorphism, which decreases in intensity toward the east. These rocks, with a thickness of 2000 m, represent typical flysch deposits.

The study area consists of fine-laminated clayey-sandy slates, which are metamorphosed conglomerates, greywackes. It is a thick complex of flysch sediments characterized by rhythmic alternation of clayey and sandstones, greywackes. The Andelska Hora Beds display the most characteristic features of shore-zone deposition. During the sedimentation of the Andelska Hora Beds, turbidity currents must have played an important role. They are probably responsible for the presence of graded bedding, and also partly responsible for the lenticular bedding. The latter type may occur as a result of the mutual superposition of turbidity currents differing in velocity and in content of the sand fraction.

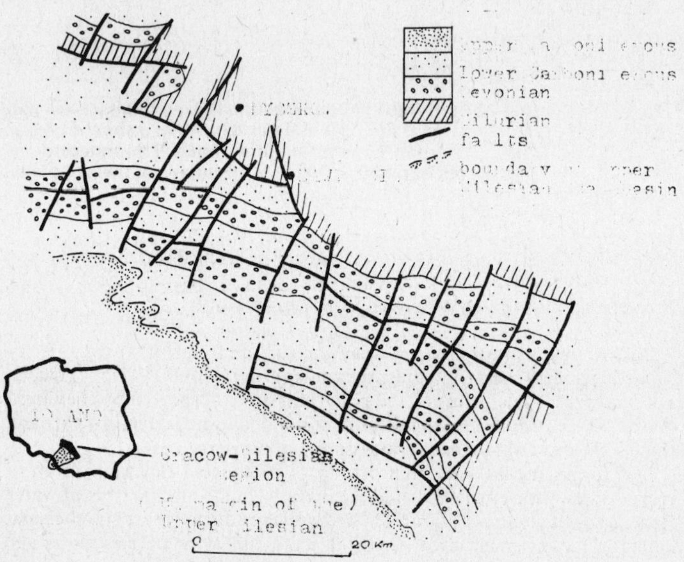

Figure 1—Geological map of the Paleozoic Basement of the northeast margin of the Upper Silesian coal basin.

WOJCIK, A., Sztygarka-School of Mines, Dabrowa Gornicza, Poland

Litho-Facies of the Metasedimentary Formations in the Central Part of the Cracow-Silesia Region, Poland

In the area of Myszkow-Zawiercie (central part of the Cracow-Silesia region) in the northeastern border zone of the Upper Silesian Coal Basin (Upper Carboniferous), the occurrence of polymetallic Cu-Mo mineralization associated with a porphyritic copper deposit has been stated in boreholes.

The analysis of these data made it possible to carry out litho-stratigraphic correlation of the Silurian. The obtained characteristics of lithology and petrography of individual links of the Silurian are as follows:

—the Lower Silurian complex (Llandoverian = Lower Lludlovian): formed by dark gray, sometimes greenish-gray chlorite-sericite and chlorite-biotite metaaleurites and metapelites with lidyte intercalations. Its thickness amounts to approximately 800 m.

—the Middle Silurian complex (Middle-Upper Lludlovian): formed by flysch deposits represented by clay and siltstone, gray-greenish or red-brownish shales, intercalated by graywackes and conglomerates, some thousand meters thick.

—the Upper Silurian complex is composed of reddish-brown sandstones, mudstones, and locally conglomerates.

The Caledonian structural stage, consisting of Lower Paleozoic metamorphic rocks, shows a folded structure, which is characteristic for geosynclinal areas; it is discordantly covered by Triassic rocks, ca. 180 m thick.

The oldest paleontologically dated rocks are Ordovician metaleuritic, conglomeratic, and carbonate deposits and Upper Silurian metaaleurite and metapelite series.

All these rocks are slightly metamorphosed under conditions of the greenschist facies and show symptoms of phyllitization, boudinage, ptygmatic structures, brecciation, and jointing of several generations. Mineralizations are often connected with these joints.

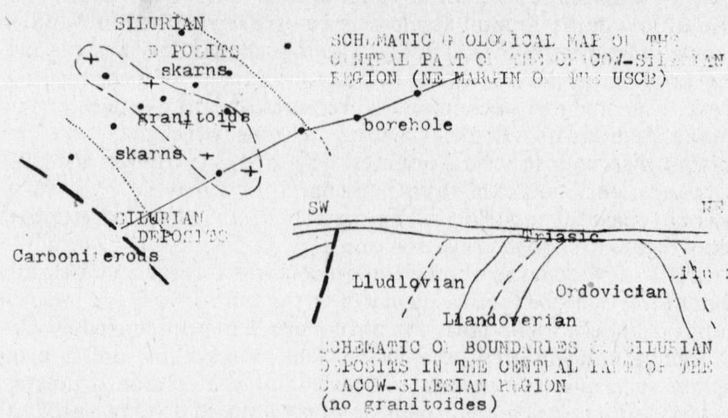

Figure 1—Schematic geological map of the central part of the Cracow-Silesian Region (northeast margin of the USCB).

WOKOMA, W. D., and AKPOKODJE, E. G., University of Port Harcourt, Port Harcourt, Rivers State, Nigeria

Use of a Database Information System in Gully Erosion Studies

This paper demonstrates the use of a database information system in studies of an erosion-prone area of Imo State in Nigeria. The study area was divided into 400 land units of 0.25km^2. This is a 10 km by 10 km area in the Amucha-Okwudo area, near Orlu in Imo State. Data on rainfall, geology, geomorphology, land slope, soil types, hydrology, vegetation, and human activities (agriculture and civil engineering construction) were used for the study.

The system is structured to compute simple correlation coefficients of data segments as well as to allow for other analysis in order to isolate the main risk factors of gully erosion in the area. The major thrust of the study is to organize four basic data files: human activities—construction, farming, residential, fauna, etc; geomorphology—rock types, land slope, rainfall, and hydrology; soil—depths of soil, soil types; and locational information—coordinates, etc. Efficient organization of these data files will help to define zones of increasing risk of gully erosion.

A fully developed information system of this nature can be used to evaluate gully susceptibility. Moreover, if this is correctly applied, it might constitute a warning service to the public, thereby minimizing the adverse effects of this natural phenomenon. The system could be used for studies focusing on controls and prevention of gully erosion also.

WOKOMA, W., and L. A. U., KAKULU, University of Port Harcourt, Nigeria

Development of Multipurpose Road Network Information System for Semirural Environment

Most developing countries suffer from a great dearth of consistent collection of information in nearly every aspect of human and environmental activities. The idea of structuring and properly organizing relational spatial information for efficient manipulation and handling is rather far-fetched in Nigeria in particular. This study aims at developing a road network information system using one local government administrative area of Rivers State, Nigeria, for demonstration purposes.

The aim of the research is to develop a computer-aided road information system that addresses questions of linkage, accessibility, route-finding, and orientation as well as future road network development, to be used by different categories of people. A fully developed system beyond the demonstration level will meet the following objectives:

1. Search for and provide all neighborhood and relational spatial information for any identifiable segment (edge) of any type of road in the study area.

2. Compute distances, shortest paths, and circuits from any given node to another point as well as compute accessibility indices for given nodes.

3. Suggest best routes and direction to different locations based on the type of road needed by the user.

System data is held in four basic files, containing information about links (road segments)—their lengths and connections; roads—different types and qualities and the links they contain; nodes or junctions; and settlements and their characteristics. Relational operations are performed on these files by a set of dedicated routines to provide answers to a defined set of queries bordering on transportation demand and allocation, accessibility routes, and distances.

Four categories of roads that are common to all places and regions in Nigeria—concrete, paved roads, tarred roads, motorable unearth roads, and bicycle-footpaths—are considered. Further classification is done according to the parties responsible for building and maintenance: national, state, local government administrative area, and individual community functionaries. The system therefore encompasses all types of roads in all urban and rural areas of the study region.

Anticipated users of the system include building investors who may need much relational spatial information on a given site before starting any investment. The system provides well-structured data on road networks for the study region for transportation modeling and road network analysis. The system is a great asset to farmers and housing investors, tourists and travelers, as well as government planning personnel. Once developed, the system could be subscribed to by any agency, and continuous updating will be undertaken by the operators.

WOLD, C., Institut fur allgemeine und augewandte Geophysik der Universitat Munchen, Munchen, Federal Republic of Germany

Digitizing the Geological Atlas of World

The Geological Atlas of the World (published by UNESCO and the IUGS, and also available through the American Association of Petroleum Geologists) is the most widely available, most complete set of geologic maps of all regions of the globe. Hence, it must inevitably become the standard set of geologic maps for global geological studies. However, it is obvious that in the preparation of the atlas little thought was given to the idea that parts or all of the atlas might need to be digitized for other studies.

In the course of preparing a database for a new generation of global plate tectonic maps with terranes, it has been necessary to solve the problems of digitization of the maps of the Geological Atlas of the World. Many of the maps do not state the projection used and in any case do not give the projection centers, standard parallels or earth radius used in their construction. To manipulate information from the atlas, the coordinate points of a geologic entity, such as a terrane, must be transformed into coordinates of latitude and longitude (spherical). The problem is that the maps must first be digitized in the x-y coordinates (or some scaled version of Cartesian coordinates) of the digitizing table. This step is entirely independent of the map projection. To transform Cartesian (x-y) coordinates into coordinates of latitude and longitude, the geometric equations used for the projection must be applied to the data. To construct a projection, coordinates of latitude and longitude must be transformed into Cartesian coordinates; this is termed the *forward transformation*. Calculation of coordinates of latitude and longitude from Cartesian coordinates is termed a *reverse transformation*.

Antarctica and the Arctic region are each presented on a polar stereographic projection, for which the transformations are straightforward. North, Central, and South America are presented in four sheets on the Bipolar Conic Conformal Projection developed by O. M. Miller and William A. Briesemeister in 1941. This projection was tailored specifically for the presentation of North and South America; it shows the two continents in their present-day orientation with minimal distortion. Both transformations for this projection have been published by Snyder (1982). The remaining continental land masses (e.g., Africa, Europe, Asia, Indonesia, and the Australian region) are presented in 11 sheets on another specifically tailored but very complex projection called the Miller Oblated Stereographic Projection. This projection was initially intended for a conformal map of Africa and Europe and was later extended to include Asia and Australia, eventually being completed by O. M. Miller in 1955. The Miller Oblated Stereographic Projection consists of three conformal zones centered in central Africa, central Asia, and northeastern Australia, and four nonconformal "fill-in" zones between the conformal zones. The "fill-in" zones are adjusted so that lines of the graticule are smooth between the eleven sheets of the map. The forward transformation for this projection has been published (Miller, 1955; Sprinsky and Snyder, 1986). The reverse transformation has not been published, and due to the nonlinearity of the forward transformation equations, it is unlikely that an exact solution to the reverse transformation will ever be possible. However, following a suggestion from J. R. Southam, I have developed the functional equivalent of a reverse transformation, which is illustrated in the following.

To transform the digitized Cartesian coordinates of an area into spherical coordinates, two grids or matrices covering that area are created. The first grid contains spherical coordinates spaced 0.05° apart; this is the resolution of the maps published on this projection in the Atlas. A forward transformation is applied to the spherical coordinate grid creating a second grid of Cartesian coordinates. The elements of the second grid are the Cartesian coordinate counterparts of the elements of the first grid. The Cartesian coordinates for the area are then found in the Cartesian coordinate grid using a binary searching technique. The latitude and longitude coordinates for the area are elements from the spherical coordinate grid that correspond to the elements selected from the Cartesian coordinate grid.

REFERENCES

Miller, O. M., 1955, Specifications for a projection system for mapping continuously Africa, Europe, Asia and Australia on a scale of 1:5,000,000: Am. Geogr. Soc. Rpt. to the Army Map Service, contract number DA 49-018-ENG-996, including corrections by Mr. Miller, June 1955, and supplement, Jan. 1956.

Sprinsky, W. H., and J. P. Snyder, 1986, The Miller Oblated Stereographic Projection for Africa, Europe, Asia and Australia: American Cartographer, v. 13, n. 3, p. 253-261.

Snyder, J. P., 1982, Map projections used by the U. S. Geological Survey: U. S. Geological Survey Bulletin, v. 1532, 313 p.

WOLF, C. W., Geological-Paleontological Institute and Museum of Christian-Albrechts University, Kiel, Federal Republic of Germany, and G. BOHRMANN, Alfred-Wegner-Institute for Polar and Marine Research, Bremerhaven, Federal Republic of Germany, and J. THIEDE, GEOMAR Research Center for Marine Geosciences at the Christian-Albrechts-University, Federal Republic of Germany

Coarse Fraction Composition of ODP-Leg 104 Drill Sites: Implications to Paleoceanographic Changes

Sediments of ODP-Leg 104 were drilled at three sites in the eastern Norwegian Sea. Site 643 is located on the lower slope of the outer Vøring Plateau, and Site 644 is located on the inner Vøring Platea and represents the landward end of this northwest-to-southeast–tending transect.

One of the principal objectives in drilling this transect was to understand paleoclimatic and paleoceanographic evolution of the Norwegian Sea during the Cenozoic. Because of its high latitude location adjacent to the cold Arctic Ocean, the Norwegian-Greenland Sea is one of the most important regimes globally for modern deep-water renewal. This study is concentrated on the documentation of compositional variations of coarse fraction particles ($>63\ \mu m$).

The results of coarse fraction analyses on largely undisturbed sedimentary sequences permit a first lithostratigraphic comparison and correlation between these three sites. We distinguished four particle assemblage types and used changes in these particle assemblages to characterize the most significant shifts in coarse particle deposition, which are believed to be linked to paleoceanographic changes.

Type 1 consists predominantly of variable amounts of planktonic and benthic calcareous foraminifers. Admixtures of minor components are terrigenous (quartz, feldspar, rock fragments, and mica) and biogenic siliceous particles (sponge spicules, radiolarians, and diatoms). Bulk sediment properties of this assemblage normally display intermediate to high carbonate contents (<10 to 50%) and low to intermediate organic carbon contents (0.1 to 0.5%). Type 1 particle assemblage dominates the Plio/Pleistocene interglacial sediments and foraminiferal-nannofossil oozes deposited during Miocene and Pliocene times. This type reflects well-oxygenated surface water and newly formed deep water.

Type 2 consists of variable amounts of radiolarians, sponge spicules and a minor amount of diatoms and biogenic calcareous particles. Terrigenous components are almost absent or occur as minor traces. Low bulk carbonate contents are carbonate-free deposits and intermediate to high bulk organic carbon contents (0.8 to 5.0%) are observed. During Miocene and Pliocene, this type reflects an increased fertility of surface water and sluggish deep-water renewal.

Type 3 consists of variable amounts of quartz, feldspar, rock fragments, mica, and heavy minerals. Minor amounts of biogenic calcareous and siliceous particles are found. In this assemblage type, bulk carbonate contents are low (0 to 5%) and organic carbon contents are intermediate (0.4 to 1.3%). Terrigeneous particle assemblages are restricted to the glacial/interglacial cycles during Plio/Pleistocene. This type is mainly composed of ice-rafted debris and occurs mainly at the transitions from glacial to interglacial and from interglacial to glacial environments.

Type 4 consists exclusively of ash-fragment particles dominated by vitric unaltered glass shards and strongly altered tephra components. The altered tephra is comprised of diagenetic aggregates, composed of authigenic clay minerals (smectites and glauconites) with impregnations of pyrite. Coarse fraction components during Eocene times are mainly composed of these tephra particles.

WOLF, T. C. W., Christian-Albrechts University, Olshausenstrasse, Federal Republic of Germany, R. D. MÜLLER, University of Texas, Austin, TX and J. THIEDE, Christian-Albrechts University, Wischhofstrasse, Federal Republic of Germany

Diagenesis and Burial History of Sandstone from Central Tertiary Marginal Strike-Slip Basin of Spitsbergen

The Central Tertiary Basin of Spitsbergen originated as a marginal strike-slip basin in the Early Tertiary adjacent to the plate boundary between Greenland and Eurasia. A shallow marine/fluvial clastic sequence was deposited in the basin during Paleogene, unconformably overlying Aptian/Albian shales of the Carolinefjellet Formation. The first phase of basin development (Paleocene to earliest Eocene) was characterized by transtension/strike-slip motion along the fault zone, initiating asymmetric subsidence of the basin floor. The first transgression resulted in the deposition of a succession of delta plain/shoreface sandstones and offshore shales (Firkanten Formation). This succession is overlain by an outer shelf mud complex (Basilika Formation) and lower shoreface sandstones (Grumantbyen Member). The overlying Hollendardalen Member, a deltaic sandstone wedge, is restricted to the northwestern part of the basin. A contemporaneous transgression gave rise to the deposition of prodelta shales (Gilsonryggen Formation) in all other parts of the basin. During the second phase of basin development (early Eocene to early Middle Eocene) the tectonic regime changed to transpression, resulting in the formation of an eastward prograding fold and thrust belt along the western margin of Spitsbergen. During this phase, the final basin regression commenced, resulting in deposition of an eastward prograding delta front/barrier coast system (Battfjellet Formation and Aspelintoppen Formation). Both Formations consist of interbedded sand- and siltstones (Steel et al., 1985). In the early Oligocene rifting began between Greenland and Svalbard due to a change in seafloor spreading direction between Greenland and Eurasia (Talwani and Eldholm, 1977).

Paleogene sandstones from the Central Tertiary Basin of Spitsbergen are characterized by strong consolidtion, low porosities, and high seismic velocities Bjós1.6prlykke et al., 1979, Elverhói and Grónlie, 1979). No comprehensive study, however, has been published to explain the diagenetic processes leading to the unusually high ° of lithification. Our main objective was to examine the mineral composition and diagenetic alteration of the sandstones of the Grumantbyen and Hollendardalen Members, and the Battfjellet- and Aspelintoppen-Formation. In addition we estimated the maximum burial temperature and burial depth of the Tertiary succession.

The unusually low porosities are due to extensive quartz cementations as revealed by cathodolumninescence methods without evidence for significant pressure solution. Other late diagenetic mineral reactions include the formation of patchy carbonate cement, leaching of kaolinite, and precipitation of chlorite postdating the onset of silica cementation.

We estimate a maximum burial temperature of 120 to 140°C for the upper part of the succession (Battfjellet- and Aspelintoppen formation) based on $\delta^{18}O$-values from calcite cement and sparse occurrence of partly disintegrated kaolinite. A proposed burial depth of 2000 m for the base of the uppermost formation (Aspelintoppen Fm.) in the Central Basin resulting in a geothermal gradient of 6 to 7°C/100m for the time of maximum heating.

These data suggest that the onset of oblique rifting between Greenland and Svalbard after Early Oligocene times induced a rising geothermal gradient in the Tertiary basin, which resembles the present geothermal gradient in the Rhine Graben (West Germany).

REFERENCES

Bjórlykke, K., A. Elverhói, and A. O. Malm, 1979, Diagenesis in Mesozoic sandstones from Spitsbergen and the North Sea—A comparison, Geologische Rundschau, v. 68, n. 3, p. 1152–1171.

Elverhói, A., and G. Grónlie, 1981, Diagenetic and sedimentologic explanation for high seismic velocity and low porosity in Mesozoic-Tertiary sediments, Svalbard region, AAPG Bulletin, v. 65, n. 1, p. 145–153.

Steel, R., J. Gjelberg, W. Helland-Hansen, K. Kleinspehn, A. Nóttvedt, and M. Rye-larsen, 1985, The Tertiary Strike-slip basin and orogenic belt of Spitsbergen, *in* Strike-Slip Deformation Basin Formation and Sedimentation, SEPM Special Publication 37, p. 339–359.

Talwani, M. and O. Eldholm, 1977, Evolution of the Norwegian-Greenland Sea, GSA Bulletin, v. 88, p. 969–999.

WOLF, T., Geological-Paleontological Institute and Museum of Christian-Albrechts University, Kiel, Federal Republic of Germany, R. D. MÜLLER, University of Texas, Austin, TX, and J. THIEDE, GEOMAR, Research Center for Marine Geosciences at the Christian-Albrechts University, Kiel, Federal Republic of Germany

Diagenesis and Burial History of Sandstones from Central Tertiary Marginal Strike-Slip Basin of Spitsbergen

The Central Tertiary Basin of Spitsbergen originated as a marginal strike-slip basin in the Early Tertiary adjacent to the plate boundary between Greenland and Eurasia. A shallow-marine/fluvial clastic sequence was deposited in the basin during Paleogene, unconformably overlying Aptian–Albian shales of the Carolinefjellet Formation. The first phase of basin development (Paleocene to earliest Eocene) was characterized by transtension/strike-slip motion along the fault zone, initiating asymmetric subsidence of the basin floor. The first transgression resulted in the deposition of a succession of delta plain/shoreface sandstones and offshore shales (Firkanten Formation). This succession is overlain by an outer shelf mud complex (Basilika Formation) and lower shoreface sandstones (Grumantbyen Member). The overlying Hollendardalen Member, a deltaic sandstone wedge, is restricted to the northwestern part of the basin. A contemporaneous transgression gave rise to the deposition of prodelta shales (Gilsonryggen Formation) in all other parts of the basin. During the second phase of basin development (early Eocene to early Middle Eocene), the tectonic regime changed to transpression, resulting in the formation of an eastward prograding fold and thrust belt along the western margin of Spitsbergen. During this phase, the final basin regression commenced, resulting in deposition of an eastward-prograding delta front/barrier coast system (Battfjellet Formation and Aspelintoppen Formation). Both formations consist of interbedded sand- and siltstones (Steel et al, 1985). In the early Oligocene, rifting began between Greenland and Svalbard due to a change in seafloor-spreading direction between Greenland and Eurasia (Talwani and Eldholm, 1977).

Paleogene sandstones from the Central Tertiary Basin of Spitsbergen are characterized by strong consolidation, low porosities, and high seismic velocities (Bjørlykke et al, 1979; Elverhøi and Grønlie, 1979). No comprehensive study, however, has been published to explain the diagenetic processes leading to the unusually high degree of lithification. Our main objective was to examine the mineral composition and diagenetic alteration of the sandstones of the Grumantbyen and Hollendardalen Members, and the Battfjellet and Aspenlintoppen Formation. In addition, we estimated the maximum burial temperature and burial depth of the Tertiary succession.

The unusually low porosities are due to extensive quartz cementations as revealed by cathodoluminescence methods without evidence for significant pressure solution. Other late diagenetic mineral reactions include the formation of patchy carbonate cement, leaching of kaolinite, and precipitation of chlorite postdating the onset of silica cementation.

We estimate a maximum burial temperature of 120° to 140°C for the upper part of the succession (Battfjellet and Aspelintoppen Formation) based on $\delta^{18}O$ values from calcite cement and sparse occurrence of partly disintegrated kaolinite. A proposed burial depth of 2000 m for the base of the uppermost formation (Aspelintoppen Fm.) in the Central Basin resulting in a geothermal gradient of 6° to 7°C/100 m for the time of maximum heating.

These data suggest that the onset of oblique rifting between Greenland and Svalbard after Early Oligocene times induced a rising geothermal gradient in the Tertiary basin, which resembles the present geothermal gradient in the Rhine Graben (West Germany).

REFERENCES

Bjørlykke, K. A. Elverhø and A. O. Malm, 1979, Diagenesis in Mesozoic sandstones from Spitsbergen and the North Sea—a comparison: Geologische Rundschau, v. 68(3), p. 1152–1171.

Elverhøi, A., and Grønlie, G., 1981, Diagenetic and sedimentologic explanation for high seismic velocity and low porosity in Mesozoic-Tertiary sediments, Svalbard region: AAPG Bulletin, v. 65(1), p. 145–153.

Steel, R., J. Gjelberg, W. Helland-Hansen, K. Kleinspehn, A. Nøttvedt, and M. Rye-larsen, 1985, The Tertiary Strike-Slip Basin and orogenic Belt of Spitsbergen, *in* Strike-slip deformation basin formation and sedimentation: SEPM Special Publication 37, p. 339–359.

Talwani, M., and O. Eldholm, 1977, Evolution of the Norwegian-Greenland Sea: GSA Bulletin, v. 88, p. 969–999.

WOLFF, F. Chr., Geological Survey of Norway, Trondheim, Norway, O. SELINUS, Geological Survey of Sweden, Uppsala, Sweden, and H. TRANSKANEN, Geological Survey of Finland, Espoo, Finland

Geology for Environmental Planning in Northern Scandinavia—A New Challenge for Producing Small-Scale Maps for Land-Use Planning

The inter-Nordic geological correlation program, "The Nordkaiott Project," carried out by the geological surveys of Finland, Norway, and Sweden from 1979 to 1986 produced a series of 14 geoscientific maps covering the northern part of the three countries (Figure 1) at a scale of 1:1,000,000 concluding with an overall mineral assessment map. These maps will be exhibited in the exhibition hall during the 28th International Geological Congress.

The succeeding "Mid-Norden Project" (Figure 1) started in 1987 and includes a new subproject for environmental geology. The major goal of this subproject will be to produce environmental geological maps of the Mid-Norden area. This includes the production of large-scale geological maps for land-use planning in certain test areas, but also the production of an overall small scale (1:1,000,000) map of the entire project area.

The production of a geoscientific map for land-use planning at such a small scale is not easy and represents, therefore, a new challenge to the subproject's participants.

The first step, to establish a suitable legend, has been completed (Figure 2) and will be discussed.

While waiting for the other subprojects to complete their maps and data sets, from which the information in the small-scale environmental geological map will be derived, a map of the Nordkalott area, using these already available maps, has been compiled to test the legend. Problems encountered during this work will be discussed, and the demand for geological information in land-use planning in Scandinavia will be presented.

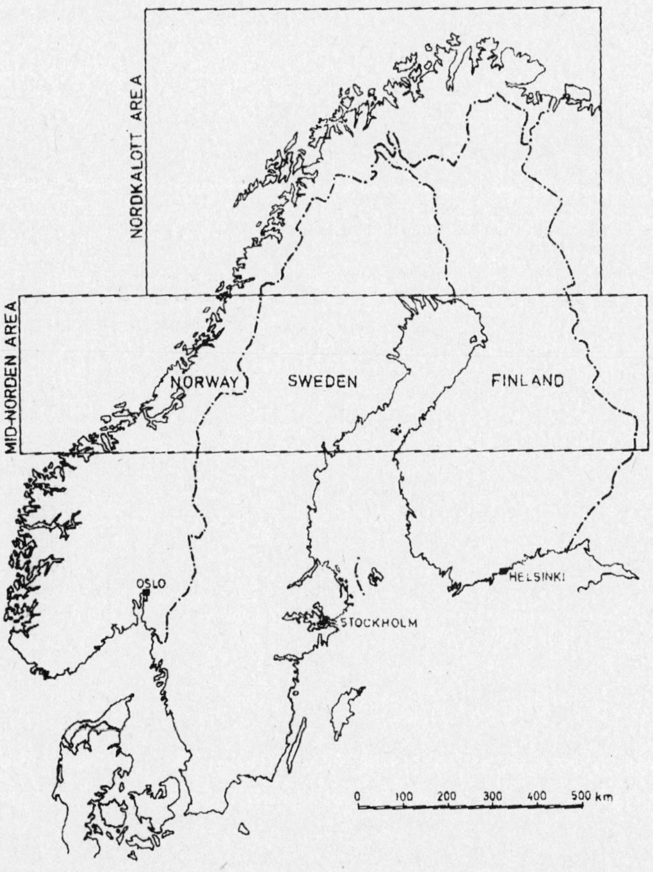

Figure 1—Index map of the Nordkalott Project area and the Mid-Norden Project area.

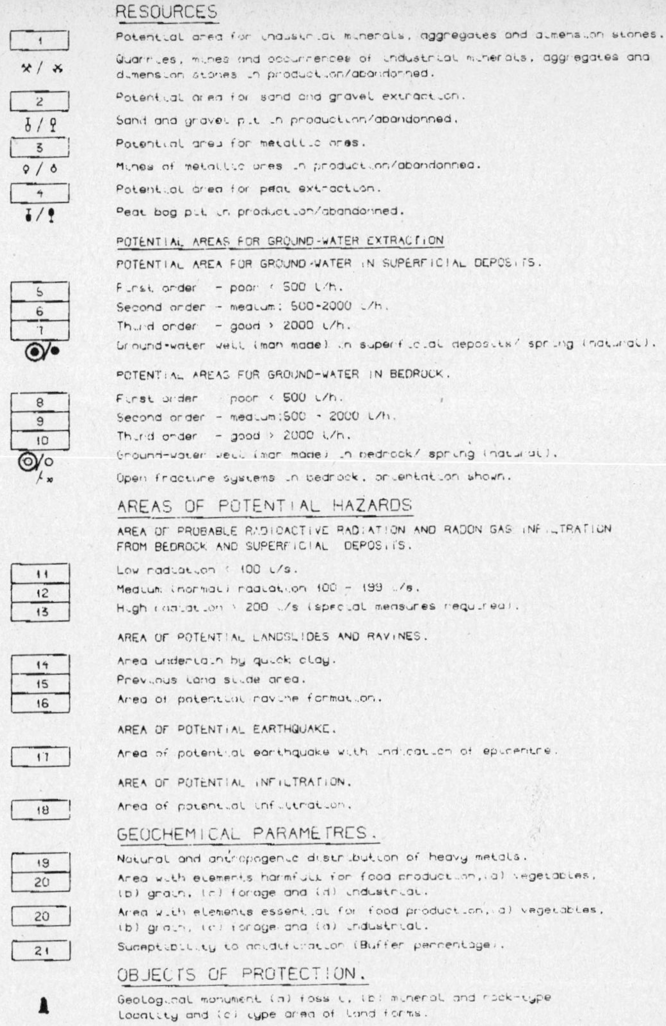

Figure 2—Legend to a map of environmental geology in the scale 1:1,000,000 for northern Scandinavia.

WONG, I. G., Woodward-Clyde Consultants, Oakland, CA

Recent Seismologic Observations of Colorado Plateau and Their Relationship to Its Thermal and Uplift History

The Colorado Plateau, a major crustal block and tectonic province in the western United States, has been tectonically stable during the Phanerozoic but has also been epeirogenically uplifted approximately 2 km in the Cenozoic. Possible mechanisms for plateau uplift fall into three classifications: thermal expansion, magmatic crustal thickening, and phase changes (Morgan and Swanberg, 1985). Seismological studies in the past 10 years, aided principally by the installation and operation of seismographic networks in and around the plateau, have provided new data on its lithospheric and thermal structure and state of stress. These observations place new constraints on possible uplift mechanisms.

Based on this seismic monitoring, earthquakes in the interior of the plateau can be characterized as being of small to moderate magnitude ($M < 6$), widely distributed, and, contrary to earlier views, of a low to moderate rate of occurrence (Wong and Humphrey, in preparation). These characteristics are consistent with a highly faulted Precambrian basement and crust and a moderate level of differential tectonic stresses. In contrast, a high rate of occurrence and large earthquakes ($M > 6$) characterize the western and southeastern margins of the plateau, the Intermountain Seismic belt, and the Rio Grande rift, respectively.

Heat flow measurements for the Colorado Plateau suggest the existence of a low heat flow thermal interior (average 60 mW/m^2) surrounded by a 100 km wide periphery having substantially higher heat flow (average 80–90 mW/m^2). This transition in heat flow has probably resulted from lateral warming and weakening, in a rheological sense, of the sides of the plateau lithosphere (Bodell and Chapman, 1982). Such weakening eventually leads to crustal failure and is consistent with the greater seismicity observed along the plateau's margins.

Earthquakes in the plateau typically occur in the upper crust to a depth of 15 km; however, many earthquakes appear to originate deeper than 20 km (Wong and Humphrey, in preparation). These events appear to occur throughout the lower crust, with a few events in the uppermost mantle at depths as great as 58 km. Several investigators have suggested that the maximum crustal temperature allowable for brittle failure and hence earthquakes is 250°–450°C (which corresponds to approximate depths of 15–20 km in most intraplate/intracontinental regions) and 600°–800°C in the upper mantle (Chen and Molnar, 1983). Based upon the observed low heat flow in the plateau interior and average continental geotherms, temperatures appear to range from 300° to 600°C in the lower crust (depths of 20–40 km) and from 600° to 800°C in the uppermost mantle(?) (depths of 40–60 km). (Deep reflections on COCORP data suggest a Moho at 50 km beneath the Colorado Plateau, substantially deeper than the previously suggested 40–43 km depth based on refraction data [Hauser, 1988]). Temperatures less than 450°C in the lower crust and these upper mantle temperatures agree well with those suggested for earthquake occurrence. However, if temperatures in the seismogenic lowermost crust exceed 500°C, then the Moho beneath the Colorado Plateau may be transitional in terms of composition and strength rather than a sharp boundary. A transitional Moho is consistent with COCORP data that show an absence of prominent, laterally extensive reflections normally indicative of a sharp velocity boundary (Hauser, 1988).

The relatively low upper mantle temperatures of 600°–800°C conflict with high temperatures (1000°–1200°C at 50km) suggested by high electrical conductivities (Pedersen and Hermance, 1981). Yet it was also previously thought that low P_n velocities of approximately 7.8 km/s were consistent with these high temperatures. Recent studies of seismic travel times, however, suggest that P_n velocities are relatively high, approximately 8.1 km/s, which is consistent with low temperatures (Beghoul and Barazangi, 1988).

The state of stress within the plateau based on earthquake focal mechanisms appears to be characterized by northeast-southwest extension, in contrast to an earlier belief that the plateau was being subjected to east-west compression (Wong and Humphrey, in preparation). Morgan and Swanberg (1985) have suggested that prior to 10 m y ago, when the plateau first rose above the Basin and Range, this stress field would have been reversed (i.e., extension), especially on the plateau's margins. Thus, reversal of the plateau's stress field may be nearly complete, with the possible exception of an area in the plateau's northern interior, the eastern Wasatch Plateau and Book Cliffs.

Based on recent seismologic observations, the Colorado Plateau appears to be a relatively cold and coherent block, at least to upper mantle depths. Due to heating and uplift of the plateau's lithosphere during the Cenozoic, the stability of the cool plateau may be decreasing and it may be breaking up, especially along its margins (Morgan and Swanberg, 1985). The state of stress in the plateau suggests that in addition to possible uplift-generated extension, it is being subjected to the same tectonic extension influencing much of the western Cordillera, including the adjacent Basin and Range. The difference in the extension direction compared to the surrounding regions further suggests that the plateau's response to these tectonic forces is different—not surprising, given its unique lithospheric characteristics and uplift history.

REFERENCES

Beghoul, M. N., and M. Barazangi, 1988, Relatively high P_n velocity beneath the Colorado Plateau constrains uplift models: EOS Transactions, American Geophysical Union, v. 69, p. 496.

Bodell, J. M., and D. S. Chapman, 1982, Heat flow in the north-central Colorado Plateau: Journal of Geophysical Research, v. 87, p. 2869–2884.

Chen, W., and P. Molnar, 1983, Focal depths of intracontinental and intraplate earthquakes and its implications for the thermal and mechanical properties of the lithosphere: Journal of Geophysical Research, v. 88, p. 4183–4214.

Hauser, E. C., 1988, Reflection Moho at 50 km (16 s) beneath the Colorado Plateau on COCORP deep reflection data: EOS Transactions, American Geophysical Union, v. 69, p. 496.

Morgan, P., and C. A. Swanberg, 1985, On the Cenozoic uplift and tectonic stability of the Colorado Plateau: Journal of Geodynamics, v. 3, p. 39–63.

Pedersen, J., and J. F. Hermance, 1981, Deep electrical structure of the Colorado Plateau as determined from magnetotelluric measurements: Journal of Geophysical Research, v. 86, p. 1849–1857.

Wong, I.G., and J. R. Humphrey, in preparation, Contemporary seismicity, faulting, and the state of stress in the Colorado Plateau.

WOO, K., Kangweon National University, Chuncheon, Korea, K. H. PAIK, Korea University, Seoul, Korea, and Y. A. PARK, Seoul National University, Seoul, Korea

Middle Miocene Paleo-Oceanographic Implications of Pohang Area Based on Isotopic-Chemical-Textural Investigation of Marine Bivalves, Southeastern Korea

The Middle Miocene marine bivalves such as *Patinopecten, Vicaryella, Vicarya,* and *Anadara* sp. in Pohang area of the southeastern part of Korea have been investigated to delineate paleo-oceanographic histories using coordinated textural-isotopic-chemical data.

Except for the calcitized portions of the originally aragonitic shells, both the aragonitic and calcitic bivalves still retain the original mineralogy, microstructure, and chemical composition. Aragonitic bivalves show a crossed lamellar microstructure and *Patinopecten* sp. shows a foliated microstructure. High Sr content of aragonitic shells (2000–4000 ppm) and calcitic shells (1000–1500 ppm) strongly suggests that some skeletons have not undergone the diagenetic alteration. Mg, K, Na, and Zn compositions of the shells are within the reasonable range of the modern counterparts, whereas Fe and Mn compositions are slightly higher.

Carbon isotopic values are within the range of marine carbonate carbon [−1.3 to +1.7‰ (PDB)]. Oxygen isotopic composition of *Patinopecten* sp. range from −1.7 to −2.2‰, which suggests that the Middle Miocene seawater in this region was 1-2°C warmer than the present-day shallow seawater. However, aragonitic bivalves didn't appear to precipitate their shells in isotopic equilibrium with the ambient seawater due to two reasons: (1) Oxygen isotopic values of aragonitic bivalves are significantly depleted (−2.8 to −4.2‰), which indicates the abnormally high temperature of 29–34°C for the shallow seawater of the temperate region, and (2) aragonitic bivalves show more negative oxygen isotopic values than the coexisting calcitic ones.

WOO, Ming-ko, McMaster University, Hamilton, Canada, and T. C. WINTER, U.S. Geological Survey, Denver, CO

Hydrologic Characteristics of Northern Wetlands

Vast wetlands occupy northern North America, and this paper relates their hydrologic characteristics to the physical processes operating in these wetlands. The type and magnitude of various hydrologic processes show distinct seasonality.

In the arctic and the subarctic, winter freezing of the wetlands is accompanied by ice formation on the rivers, lakes, and ponds. Small creeks cease to flow, and even large rivers (basin area over 5,000 km^2) have mean discharge falling below 10 m^3/s. Spring arrives with intense snowmelt, and most of this meltwater cannot infiltrate the frozen soil, but floods the wetland extensively. The snow and ice in the river channels also break up, often producing the maximum flow of the year. At this time, wetland storage is completely saturated and there is little retardation to surface flow. As melt terminates, the ground thaws and vegetation growth quickens. The abundance of water near the surface and the increased transpiration result in large evaporation losses, and consequently the water table falls. Many ponds shrink and most puddles dry up, until rainfall causes the depressions to be filled and then overflow. Overland flow is the dominant mechanism that conveys water from the wetland to its rivers so that when overland flow diminishes after the cessation of rainfall, streamflow is also reduced.

In the cool temperate latitudes, snowmelt is also important in filling up the depressions in the wetlands and recharging the groundwater reservoir. There, groundwater plays a relatively more important role in terms of wetland runoff, and the intensity of evaporation is increased because of a longer thawed season with higher energy receipt than in the more northern zones.

WOOD, J., and A. Hashimoto, Harvard-Smithsonian Center for Astrophysics, Cambridge, MA

Effects of Redox State on Stable Mineral Assemblages in Solar Nebula

Chondritic meteorites are aggregations of structural components (*chondrules, CAI's, matrix grains*) that were formed out of earlier materials by high-temperature processing while they were dispersed in the solar nebula. The mineral assemblages that would be thermodynamically stable under these conditions have been calculated by Grossman (1972); his "condensation sequence" (CS) of minerals has been the basis for most attempts to understand the processing history of chondrite components.

The petrographic nature of chondrules, the most abundant component in chondrites, does not correspond to an equilibrium state of affairs anywhere in Grossman's CS. Chondrules are clearly igneous objects, yet the

CS predicts their minerals would vaporize at a lower temperature than their melting range. Many chondrules contain ferrous olivine (up to Fa_{50}) phenocrysts, yet the CS predicts that significant concentrations of fayalite are stable only at < 500 K, far beneath the temperature ranges of melting or diffusive transfer. These anomalies are customarily explained as due to failure of equilibrium: melting of precursor material rich in ferrous oxide produced the chondrules, and they were molten for too short a time to vaporize away or have their Fe^{2+} reduced.

In his calculations, Grossman assumed the system to contain the chemical elements in their cosmic abundances. We have recently recalculated the CS for systems with various fractionated compositions. (Fractionated system compositions would have been produced, e.g., whenever gravitational settling of involatile dust particles concentrated dust near the nebula midplane.) Results are reviewed in this talk. Among other things, we find that in systems where dust/gas has been enhanced to 1000 × the cosmic value or more, the assemblage observed in many chondrules (ferrous olivine and pyroxene, metallic Fe, silicate melt phase) would be stable in the temperature range ~ 1400–1800 K.

REFERENCES

Grossman, L., 1972, Condensation in the primitive solar nebula: Geochimica et Cosmochimica Acta, v. 36, p. 597–619.

WOODHOUSE, J., and A. M. DZIEWONSKI, Harvard University, Cambridge, MA

Seismic Tomography by Inversion of Waveforms and Spectra

Seismological studies of the earth's three-dimensional structure, carried out over the last few years, have employed several different kinds of seismic data, spanning three orders of magnitude in frequency (1 Hz to 1 mHz), which yield different but overlapping information on interior structure. It is possible, for instance, to model lower mantle P-velocity using travel time data, lower mantle S-velocity using long period, to multiply reflected body waves and to provide constraints on both P- and S-velocity using various modes of free oscillation. This ability to investigate the same region of the earth using different data and different methods is very valuable in that it provides a check on the various modeling techniques and also allows the comparison of heterogeneity in different structural parameters (v_p and v_s).

Some of the major outlines of the earth's large-scale three-dimensional structure are becoming clear, and the process of interpretation is underway. In the upper mantle, the continental shields are characterized by deep (~300 km) high-velocity anomalies, and the strongest low velocity features are associated with the fast-spreading oceanic ridges (Woodhouse and Dziewonski, 1984). While the latter can be understood in terms of the upwelling of hot material as the plates move apart, the deep extensions of the continents present some difficulty since it may be expected that vigorous convection would rapidly destroy them. Probably they are compositionally different, cooler and of higher viscosity than the suboceanic mantle (Jordan, 1978). By translating the observed velocity anomalies into density anomalies and performing a dynamical calculation of fluid flow in a mantle of assumed viscosity structure, these models have been used to infer the expected fluid motion at the surface (Forte and Peltier, 1987). Comparison with the (poloidal component of the) observed motion of the plates shows good agreement, and thus it is possible to begin to understand the internal forces that drive plate motion. A limiting factor, however, is the unknown relationship between density and velocity variations beneath the continents; conventional assumptions lead to difficulty in explaining surface topography and the geoid.

While the debate on the issue of the trajectory in the mantle of material subducted at convergent plate boundaries will doubtless continue for some time to come; the most natural inference from both global (Dziewonski, 1984; Woodhouse and Dziewonski, 1986) and regional (e.g., Creager and Jordan, 1986; Grand, 1987) seismological studies is that cold subducted material penetrates the boundary, although a significant increase in viscosity may act to reduce the velocity and to increase the cross-sectional area of the descending flow (Richards and Hager, 1984). Models of the lower mantle show that a very long wavelength pattern of high velocities around the Pacific persists throughout the lower mantle (Dziewonski, 1984; Woodhouse and Dziewonski, 1986; Morelli and Dziewonski, 1986) and, moreover, its morphology bears the closest resemblence to what may be expected due to subduction in the upper part of the lower mantle. This feature, which fills a major fraction of the earth's volume, is undoubtedly of fundamental significance in understanding the earth's long-term dynamics; an important question is whether this feature *explains* or whether it is *explained by* the fact that subduction zones have tended to be in fixed locations for long periods of earth's history. That is: Is the tomographically observed anomaly to be attributed to material subducted in the same locations for long periods of time, or are the locations of convergent plate boundaries governed by a long-lived thermal anomaly in the lower mantle? Just outside the core the pattern of high and low velocities is such that approximately 85% of "hot spots" at the surface are above regions of lower than average velocity in the lowermost mantle, lending support to the hypothesis that hot spots are the surface manifestation of plumes rooted in the deepest part of the mantle. Instantaneous fluid dynamical calculations have had major success in explaining the very long wavelength geoid perturbations as the result of density anomalies, proportional to the seismic velocity anomalies in the lower mantle and acting as internal loads in a viscous mantle and have led to the conclusion that the viscosity of the lower mantle is substantially higher than that of the upper mantle (Richards and Hager, 1984).

Methods of inversion based upon the modeling of complete waveforms or spectra—as opposed to those employing travel time data—will be reviewed; these may be broadly termed *waveform modeling* techniques. Updated shear velocity models of the upper and lower mantle based upon a large data set consisting of mantle waves (f < 1/135 Hz; duration ~ 4 hr) and body waves (f < 1/45 Hz; duration such as to include all arrivals up to the first surface waves) from 220 events will be presented.

REFERENCES

Creager, K. C., and T. H. Jordan, 1986, Slab penetration into the lower mantle beneath the Mariana and other Island Arcs of the Northwest Pacific: Journal of Geophysical Research, v. 91, v. 3573–3589.
Dziewonski, A. M., 1984, Mapping the lower mantle: determination of lateral heterogeneity in P velocity up to degree and order 6: Journal of Geophysical Research, v. 89, p. 5929–5952.
Forte, A. M., and W. R. Peltier, 1987, Plate tectonics and aspherical Earth structure: the importance of poloidal-toroidal coupling: Journal of Geophysical Research, v. 92, p. 3645–3679.
Grand, S. P., 1987, Tomographic inversions for shear structure beneath the North American plate: Journal of Geophysical Research, v. 92, p. 14065–14090.
Jordan, T. H., 1978, Composition and development of the continental tectosphere: Nature, v. 274, p. 544–548.
Morelli, A., and A. M. Dziewonski, 1986, 3D structure of the Earth's core inferred from travel time residuals: EOS, Transactions American Geophysical Union, v. 67, p. 311.
Richards, M. A. and B. H. Hager, 1984, Geoid anomalies in a dynamic Earth: Journal of Geophysical Research, v. 89, p. 5987–6002.
Woodhouse, J. H., and A. M. Dziewonski, 1984, Mapping the upper mantle: three dimensional modelling of Earth structure by inversion of seismic waveforms: Journal of Geophysical Research, v. 89, p. 5953–5986.
——— ——— 1986, Three dimensional mantle models based on mantle wave and long period body wave data: EOS, Transactions of American Geophysical Union, v. 67, p. 307.

WOODWARD, R., and G. MASTERS, University of California, San Diego, La Jolla, CA

Global Upper Mantle Structure from Long Period Differential Travel-Times

We have made over ten thousand measurements *PP–P* and *SS–S* differential travel-times from long period GDSN data. We have examined all events with $m_b \geq 5.5$ which occurred during the years 1976 to 1986, making measurements from those events that contained sufficient signal. We have obtained reasonable global coverage, with very good coverage of regionals such as the Eurasian continent and the Southwest Pacific.

We use a cross-correlation technique to measure differential travel-times, using vertical component seismograms for the *PP–P* measurements and transverse components for the *SS–S*. For the purpose of studying global upper mantle structure, differential travel-times have several important advantages over using direct travel times: reduced sensitivity to structure in the vicinity of the source and receiver; reduced sensitivity to errors in the location of the source; and the ability to obtain measurements in regions of the world where there are no sources or receivers. By using long period data (dominant period approximately 25 s), sensitivity to source and receiver structure is further reduced.

Our measured differential travel-times show residuals, relative to PREM, of −5 to 5 seconds for *PP–P* and −12 to 12 seconds for *SS–S*. Because of the geometry of the ray paths, we would expect that structure in the vicinity of the reflected phase's bounce point is the dominant contributor to these residuals. However, we must consider the contribution to our measured differential travel-times from lower mantle structure (as the direct and reflected phases bottom at different depth) as well as source and receiver region structure.

As an experiment, we have used the lower mantle *P* velocity model LO2.56 of Dziewonski to calculate the *PP–P* residuals that would result from lower mantle structure alone. We find that these residuals average approximately ± 0.5 s. Experiments using comparisons of subsets of our measurements to examine the effects of source and receiver region structure show average contributions of approximately ± 0.5 and ± 1.0 s to our *PP–P* and *SS–S* differential travel-times, respectively. Therefore, our original hypothesis, that the observed residuals are primarily due to structure in the vicinity of the reflected phase's bounce point, seems reasonable.

By mapping our observed residuals to the location of the reflected phase's bounce point, we can compare the large-scale patterns present in our measurements to those predicted by the tectonic regionalization GTR1 of Jordan. We find a general qualitative agreement, with strong agreement for those regions that one would expect to yield significant residuals, such as young oceans (spreading centers) and Precambrian shields. There is also a clear differentiation between the residuals for young and old oceanic areas.

We have also compared our measured residuals to those predicted by global three-dimensional earth models. For example, we compare our *SS–S* residuals to those predicted by the upper mantle shear velocity model M84A of Woodhouse and Dziewonski. While M84A only provides a small variance reduction for our raw data, the overall pattern of our residuals is quite similar to the pattern of synthetic residuals predicted by M84A. One of the chief differences between these patterns is that the measured residuals reveal variations in structure over shorter length scales than M84A.

Finally, we can perform the inverse problem to obtain models of upper mantle structure directly from the data, for those regions of the globe where we have adequate sampling. We are currently working on this aspect of the problem, giving careful attention to methods of parameterizing the model and dealing with the effects of variations in the density of measurements.

WORRALL, D. M., Shell Offshore Inc., New Orleans, LA, and S. SNELSON, Shell Oil Company, Houston, TX

Evolution of Northern Gulf of Mexico, with Emphasis on Cenozoic Growth Faulting and Role of Salt

The Gulf of Mexico was initiated in the late Middle to early Jurassic as a result of crustal attenuation and seafloor spreading associated with the breakup of the supercontinent, Pangaea. Although authors differ on the original position and translation paths of South America, Yucatan, and other crustal elements, there is general agreement that the Gulf of Mexico and the Central Atlantic developed at the time North America separated from South America and Africa. The Basement terrain affected by this rifting event was thc latc Paleozoic Appalachian–Mauritanide megasuture, an orogenic belt generally believed to have resulted from the closing of a proto-Atlantic Ocean.

Following the development of this basement terrain and prior to the development of the Cenozoic Gulf Basin, five main episodes are recorded in the northern Gulf Basin: (1) Late Triassic–Early Jurassic terrestrial synrift deposition (Eagle Mills Formation) during crustal attenuation; (2) Late Middle Jurassic evaporite (Louann) and aeolian sand deposition (Norphlet), and subsidence over a major, regional unconformity; (3) Late Jurassic–earliest Cretaceous carbonates (e.g., Smackover) and updip evaporite deposition (Buckner) followed by progradation of terrigenous clastics (Cotton Valley and Hosston); (4) two major shelf margin reef cycles during the Early (e.g., Sligo) and Middle Cretaceous (E.g., Stuart City); and last (5) widespread Late Cretaceous drowning of reefs and associated extensive sedimentation of chalks, marls, and shales, following some broad upwarping of the shelf region, local intrusives (e.g., Monroe Uplift and Jackson Dome), and the development of a post-"mid"-Cretaceous unconformity. In the western Gulf region of northeastern Mexico, the early Tertiary was a time of compressional folding and local volcanism associated with the Laramide orogeny.

The major Cenozoic growth fault systems of offshore Texas and Louisiana are of two main styles: Type I, short, arcuate fault systems associated with minibasins and salt domes, mostly typically found in offshore Louisiana; and Type II, laterally extensive, linear, listric fault systems, most common in coastal Texas. Both styles are the result of updip extension and related downdip, gravity-driven, lateral spreading of strata and salt.

The short, arcuate fault systems (Type I), typically occur when rapid, local sedimentary loading has displaced thick salt. Downdip spreading of this thick salt and its clastic cover resulted in the Sigsbee salt nappe offshore Louisiana, earlier hypothesized by others. Seismic data now show that this nappe has overthrust abyssal strata a distance of at least 60 miles (90 km). The basal nappe surface dips north and is last seen truncating lower Tertiary strata beneath the middle slope, where extensive allochthonous salt can exceed 25,000 ft (7.6 km) in thickness.

Palinspastic reconstructions of major, linear Type II fault systems offshore Texas suggest that at least some of the extension in these systems is also compensated by downdip deformation of salt. Mostly Paleogene salt-cored (?) folds (Perdido foldbelt) at the base of the lower Texas slope are too old and exhibit far too little horizontal shortening to compensate for the tens of kilometers of Neogene extension under the Texas shelf. Downdip compensation for these Neogene faults occurred in the salt-rich middle and lower slope. It is likely that Paleogene fault systems farther updip also were compensated by downdip mobile salt, probably in conjunction with stratal shortening by Perdido folds.

The contrast in the nature of faulting between the two types of fault systems may well be a function of depositional style. Short, arcuate fault systems associated with minibasins and trapped salt, (Type I), are more likely the result of irregular point loads provided by shifting deltas. Long, linear, and listric fault systems, (Type II), are generally associated with strike-oriented barrier island/interdeltaic deposition.

WORTEL, R., University of Utrecht, Utrecht, The Netherlands, S. CLOETINGH, Free University, Amsterdam, The Netherlands, and T. HERKINK, University of Utrecht, Utrecht, The Netherlands

Dynamics and Stress Field of the Pacific Plate: Results from Quantitative Modeling

The aim of this study is to increase our understanding of the dynamics and the intraplate stress field of oceanic lithosphere. The Pacific plate provides an excellent opportunity to study the dynamics of oceanic lithosphere. First, it is almost entirely oceanic in nature. Second, the subduction zones bordering the plate exhibit a large variety ranging from zones consuming relatively young lithosphere (Alaska, Aleutians) to zones in which the oldest oceanic lithosphere is being subducted (Marianas). This enables us to explore in detail the role of subduction zones in plate dynamics. As in our earlier modeling of dynamics and stress field of other lithospheric plates (e.g., Wortel and Cloetingh, 1981, 1985; Cloetingh and Wortel, 1985, 1986) we model the ridge push force as a pressure gradient integrated over all contributing parts of the lithosphere. The slab pull forces are modeled as being dependent on the age of the subducting lithosphere. We parameterize the resistive forces and determine the unknown parameters by requiring that the total torque of all forces acting on the plate must vanish.

An important trade-off appears to exist between the resistive forces acting in the convergence zones and the shear at the base of the plate. The relationships between such resistive forces are studied in detail, with special attention to the differences between resistance in deep subduction zones and that in zones with maximum focal depth in the range of 200–300 km.

We increase the resolution of the analysis by a simultaneous study of thc dynamics of the Nazca plate, which is also entirely oceanic in nature. The advantage is that the age distribution of the Nazca plate and the ratio of the length of convergent plate boundary to area of the plate differ significantly from those of the Pacific plate.

The stress modeling is done by finite element analysis, in which we employ linear strain elements. The data compiled in the course of the World Stress Map project, including oceanic intraplate earthquakes (see e.g., Berman, 1986) are used to test the resulting stress field, both in orientations and magnitude. A particular aim of our modeling is to provide a theoretical framework for the analysis of stress measurements that will hopefully be provided by future Legs of the Ocean Drilling Program in the Pacific Ocean (see recommendations in: Report of the Second Conference on Scientific Ocean Drilling "COSOD II," Strasbourg, France, 6–8 July 1987).

REFERENCES

Bergman, E. A., 1986, Intraplate earthquakes and the state of stress in oceanic lithosphere, Tectonophysics, v. 132, p. 1–35.
Cloetingh, S., and R. Wortel, 1985, Regional Stress field of the Indian plate: Geophysical Research Letters, v. 12, p. 77–80.
Cloetingh, S., and Ro Wortel, 1986, Stress in the Indo-Australian plate: Tectonophysics, v. 132, p. 49–67.
Wortel, R., and S. Cloetingh, 1981, On the origin of the Cocos-Nazca spreading centre: Geology, v. 9, p. 425–430.
Wortel, M. J. R., and S. A. P. L. Cloetingh, 1985, Accretion and lateral variations in tectonic structure along the Peru-Chile Trench: Tectonophysics, v. 112, p. 443–462.

WORTEL, R., and R. GOVERS, University of Utrecht, Utrecht, The Netherlands, S. CLOETINGH, Free University, Amsterdam, The Netherlands, and C. STEIN, University of Illinois, Chicago, IL

Stress Level in the Indian Plate: New Constraints from the Depth Distribution of Oceanic Intraplate Seismicity

An important motivation to study intraplate stress fields is to improve our understanding of the dynamics of plate motion. With this goal in mind Richardson et al (1979) performed numerical calculations of the stress field in all major plates. Wortel and Cloetingh (e.g., 1981, 1985) performed similar numerical studies, with some important new aspects in their force modeling procedure. In contrast with the earlier models, the ridge push was not represented as a line force but as an integrated pressure gradient distributed over all contributing parts of the lithosphere. The slab pull was incorporated as an age-dependent force giving rise to significant lateral variations in the forces representing the subduction process. Whereas, the force modeling for the Nazca plate yielded stress values in agreement with Richardson et al's (1979) odels, a later application of the same modeling to the Indo-Australian plate (Cloetingh and Wortel, 1985, 1986) resulted in a significantly higher stress level than in Richardson et al's (1979) work (typically: kilobars versus a few hundred bars).

In view of the relevance of such discrepancies for future numerical modeling, in particular, and stress analyses, in general, we address this issue by investigating the information that the unusually high seismic activity in the Central Indian Ocean (see e.g. Wiens et al, 1985) provides concerning the level of intraplate stress. Intraplate earthquakes have hitherto been used in the context of stress modeling only insofar as the orientations of principal stress directions are concerned. See, for the Indian plate, Richardson et al (1979), Cloetingh and Wortel (1986), and also Bergman and Solomon (1985). Here we focus on the *depths* of intraplate earthquakes because they contain, via the depth-dependent strength of the lithosphere, information on the magnitudes of the principal stresses. The Central Indian Ocean is particularly suitable for a comparison between stress levels resulting from model calculations and those inferred from seismicity because it is the only area where the combination of calculated stress models and high seismicity with well-determined focal depths is available.

As a key element in our analysis, we employ a depth-dependent rheological model encompassing Byerlee's law and the flow laws derived from olivine (Goetze and Evans, 1979). Through the temperature-dependence of the flow law, the dependence of the lithosphere's rheology on age is implicitly taken into account. Because the strength value is taken to be the upper limit of the differential stress that can be supported elastically, the basic premise is that earthquakes do not occur at a certain depth if the stress is below the strength value. Hence, seismic activity indicates that the strength is reached. This enables us to derive a minimum value for the differential stress from the depth of an earthquake's hypocenter in combination with the depth-dependent strength distribution. Uncertainties in the age of the lithosphere and in strain rate (both affecting the temperature-dependent ductile branch of the strength envelope), and in focal depth result in an uncertainty in the inferred value for the differential stress. It is noted that, in the present context, the most pertinent parameter of the stress distribution is not the maximum value, but rather the integral of stress over the depth interval comprising the lithosphere.

If we incorporate all sources of uncertainties, the depth distribution of oceanic intraplate earthquakes does not provide very tight constraints on the magnitude of the intraplate stress field. Nevertheless, our calculated stress level is fully consistent with the range of stress levels required to account for the observed seismicity. If, however, we take into account the calculated orientations of principal stress directions (which are in good agreement with focal mechanism data), much closer constraints can be derived. It appears that the observed seismic activity is not only consistent with, but, in several areas, *requires* the high level of stress Cloetingh and Wortel (1985, 1986) found for the Central Indian Ocean part of the Indian plate.

Stress levels of an order of magnitude lower, as advocated by Richardson et al (1979, see also Richardson, 1987), are not sufficient to explain the observed seismicity. Thus, seismicity data point in the same direction—of high intraplate stress—as recent studies of a completely different nature, such as those by McAdoo and Sandwell (1985) and Zuber (1987) on the stresses associated with the folding of the oceanic lithosphere in the Central Indian Ocean.

REFERENCES

Bergman, E. A., and S. C. Solomon, 1985, Earthquake source mechanisms from body wave inversion and intraplate tectonics in the northern Indian Ocean: Phys. Earth Planet. Inter., v. 40, p. 1-23, 1985.

Cloetingh, S., and R. Wortel, 1985, Regional stress field of the Indian plate: Geophysical Research Letters, v. 12, p. 77-80.

Cloetingh, S., and R. Wortel, 1986, Stress in the Indo-Australian plate: Tectonophysics, v. 132, p. 49-67.

Goetze, C., and R. Wortel, 1986, Stress in the Indo-Australian plate: Tectonophysics, v. 132, p. 49-67.

Mcadoo, D. C., and D. T. Sandwell, 1985, Folding of oceanic lithosphere, Journal Geophysical Research, v. 90, p. 8563-8569.

Richardson, R. M., 1987, Modeling the tectonics of the Indo-Australian plate: EOS, v. 68, p. 1466.

Richardson, r. M., S. C. Solomon, and N. H. Sleep, 1979, Tectonic stress in the plates: Rev. Geophys. Space Phys., v. 17, p. 981-1019.

Wiens, D. A., C. DeMets, et al, 1985, A diffuse plate boundary model for Indian Ocean tectonics: Geophysical Research Letters, v. 12, p. 429-432.

Wortel, R., and S. Cloetingh, 1981, On the origin of the Cocos-Nazca spreading center, Geology, v. 9, p. 425-430.

Wortel, M. J. R. and S. A. P. L. Cloetingh, 1985, Accretion and lateral variations in tectonic structure along the Peru-Chile Trench: Tectonophysics, v. 112, p. 443-462.

Zuber, M. T., 1987, Compression of oceanic lithosphere: An analysis of intraplate deformation in the Central Indian Basin, Journal of Geophysical Research, v. 92, p. 4817-4825.

WRIGHT, J., Battelle, Pacific Northwest Laboratories, Richland, WA

REE in Fossil Apatite Record Phanerozoic OAE

Rare earth element (REE) contents of marine biogenic apatite of conodont elements, ichthyoliths, inarticulate brachiopod shells, and other phosphatic fossil material show secular variations in samples from Paleozoic and Mesozoic stratigraphic horizons. The variations probably result from changes in the oxidation-reduction regime that was prevalent in these ocean basins at the time of deposition. Fossil apatite samples are from normal, shallow-water carbonate sequences that were deposited under oxidizing conditions. However, at certain time intervals oceanic anoxic events (OAE) occurred. The signature of anoxic conditions that dominated other portions of the water column was transferred to oxic surface waters and recorded in the REE patterns of biogenic apatite samples deposited under normal, shallow seawater conditions (J. Wright, 1985, Ph. D. Dissertation, Univ. Oregon, Eugene, OR). These data support earlier models which predicted that oceanic anoxic conditions had been widespread in Cambrian through Devonian and again in Permian–Triassic seas (E. T. Degens and P. Stoffers, 1976, Nature, v. 263, p. 22–27; W. B. N. Berry and P. Wilde, 1978, American Journal of Science, v. 278, p. 257–275). Modern analog studies of fish debris deposited under varying redox conditions are the control for interpretation of the fossil sample data (Wright, 1985).

The rare earth element group exhibits similar chemical behavior unless a mechanism is present in the system to produce fractionation of one or more of the REE. Because cerium has two possible oxidation states in aqueous solution, Ce^{3+} and Ce^{4+}, it can be fractionated relative to the other REE. In normal oxidizing seawater, Ce is depleted relative to lanthanum and neodymium (H. Elderfield and M. J. Greaves 1982, Nature, v. 296, p. 214–219; H. J. W. De Baar, et al, 1985, Geochimica et Cosmochimica Acta, v. 49, p. 1943–1959). The removal of Ce from seawater under oxidizing conditions is dominated by co-precipitation with metallic oxides. This process governs the fractionation of Ce from the other REE producing the negative Ce anomaly of modern seawater. However, below an undetermined low pO_2 threshold there is no fractionation of Ce by the metallic oxide co-precipitation mechanism. Therefore, under dysaerobic anoxic conditions, Ce is at the same concentration level or enriched relative to the other REE (J. Wright, et al, 1987, Geochimica et Cosmochimica Acta, v. 51, p. 631–644). The variation of Ce relative to measured neighboring REE, lanthanum and neodymium, can be stated mathematically by: $Ce_{anom} = \text{Log}\,[3Ce_n/(2La_n + Nd_n)]$, where n signifies shale-normalized concentrations (after Elderfield and Greaves, 1982). When Ce is depleted under oxidizing conditions, Ce_{anom} values are negative, and when Ce is enriched under low oxygen to anoxic conditions, Ce_{anom} values are positive.

The graph in Figure 1 shows the distribution of Ce_{anom} values for seawater of three different redox signatures and for fish debris from three distinct redox environments. In the shaded area of the graph, anoxic conditions are dominant. The Atlantic seawater bar shows the range of normal values from open ocean seawater profiles, exclusive of localized anomalous phenomena (Elderfield and Greaves, 1982; De Baar et al, 1983, 1985). The Pacific seawater bar shows the range of values of normal seawater from a profile taken in the eastern Pacific basin (De Baar et al, 1985). The bar representing a profile of Black Sea water shows that even in surface, oxidizing waters of the Black Sea, Ce_{anom} values are positive. This indicates that in a seawater basin where deeper waters are anoxic, the cumulative buildup of cerium concentrations at depth is transferred across a concentration gradient to surface waters and reflected in an overall signature of anoxia throughout the basin (Wright et al, 1987).

Fish debris from red clay sequences on the Pacific abyssal plain (M. Bernat, 1975, Cahiers ORSTOM Ser. Geol., v. 7, p. 65–83) show a range of values within the range of Pacific seawater. These samples carry an oxic signature. Fish debris samples from the Peru Margin overlap and are greater than the most positive values of Atlantic seawater. Fish debris from Guaymas Basin in the Gulf of California have even more positive values encompassing the entire range of Black Sea water values. These regions have dysaerobic and anoxic water conditions, respectively, which are recorded in the biogenic apatite samples deposited in each location (a detailed discussion of locations and samples is given in Wright et al, 1987). Modern seawater studies indicate that biogenic apatite acquires and retains the signature of the prevailing redox conditions of the seawater environment. The signature can subsequently be read in the REE pattern and the Ce_{anom} value of the fossil apatite material.

The next figure shows the secular variation of Ce_{anom} in Upper Cambrian through Upper Jurassic oceans. This curve is based on Ce_{anom} values of fossil apatite samples from continental margin carbonates with a widespread geographic distribution. Values are calculated for groups of samples at a single time interval, and the curve was drawn through the average value for the time interval showing a two sigma error bar associated with each value. (Full details of analyses and locations are given in Wright, 1985, 1987).

The secular variation of Ce_{anom} indicates that Lower Paleozoic oceans were dominated by anoxic conditions. There is a shift in the Devonian to generally oxidizing conditions that persist throughout Upper Paleozoic seas until Late Permian to Early Triassic times when the paleoredox environment may have returned to low oxygen to anoxic conditions. Oxidizing conditions became dominant again in Upper Triassic and Jurassic seas. The bimodal distribution in the Upper Jurassic samples from Solnhofen Basin reflects shifts from normal oxidizing conditions (samples from clay layers) to catastrophic deposition of thick coccolith layers during episodic red tides when anoxic conditions were dominant (P. H. De Buisonje, 1972, Koninkl. Ned. Akad. Wetenschap., Proc. Ser. B, v. 75, p. 152–177).

Studies of Upper Devonian sections across the Frasnian–Famennian boundary are being conducted. These examine in detail some of the paleochemical changes that occurred in the Late Devonian when the world ocean shifted from the anoxic environment that had dominated it for hundreds of millions of years, to perhaps the first recorded occurrence of generally oxidizing conditions that are similar to conditions in the modern ocean.

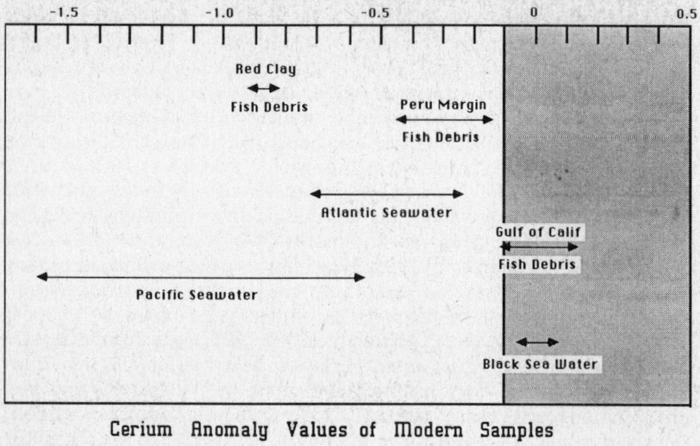

Figure 1

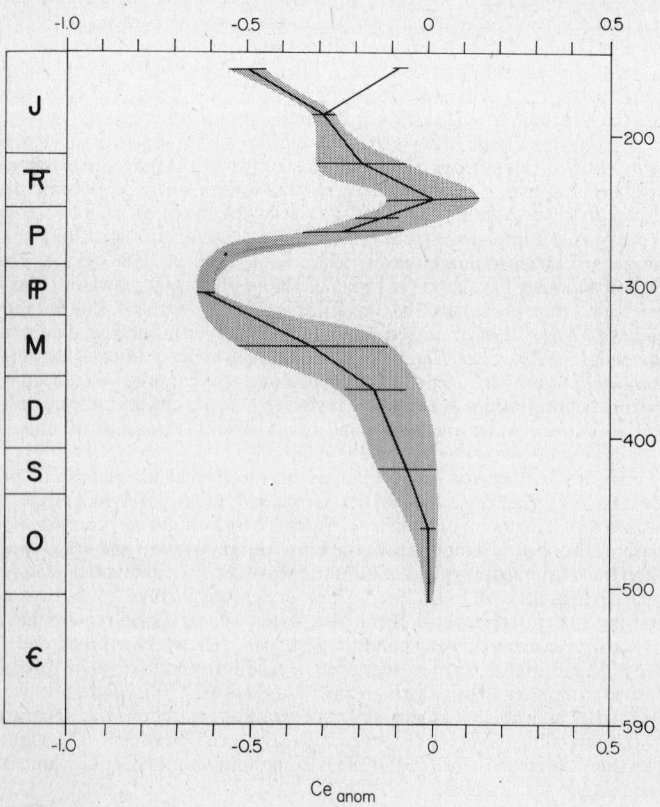

Figure 2

WRIGHT, T., U.S. Geological Survey, Hawaii National Park, HI

Hot-Spot Volcanism in Space and Time: Hawaiian Example

The Hawaiian–Emperor chain of islands, atolls, and seamounts, extending across 5900 km in the north-central Pacific Ocean, is the best-known and longest lived example of hot-spot volcanism. The chain extends from Meiji seamount, close to the subducting margin of the Pacific plate, to Loihi seamount, the active submarine volcano southeast of the Island of Hawaii. The source of magmas has remained very close to 19°N, the present latitude of the Island of Hawaii, even through a change in motion of the Pacific plate about 43 m.y. ago. The rate of propagation of the chain has been variable, averaging 8.6 cm/y over the 70 m.y. lifespan of the chain. During the last 5 m.y., the time in which the Hawaiian Islands were formed, the rate of lava production has been accelerating; the Island of Hawaii is estimated to be growing 50% faster than the average rate of volcanism for the Kauai–Hawaii part of the chain.

The islands and seamounts of the entire chain have a similar and identifiably Hawaiian chemistry, defined by major oxides, trace-elements, and isotopic ratios. The volcanoes of the Hawaiian Islands are, in addition, arranged along two subparallel trends, the "Loa" and "Kea" loci. Volcanoes along a given locus line each have a unique chemical signature within the overall "Hawaiian" chemistry and also have certain chemical similarities, such as ranges of Pb isotope ratios, that distinguish them from volcanoes on the other locus line. Volcanoes on different locus lines may be simultaneously active. Thus the hot spot appears to be segmented with a degree of chemical heterogeneity.

1. An early alkalic stage, represented by Loihi seamount, that presumably represents the initial melting near the leading edge of the hot spot.

2. A tholeiitic shield-building stage that creates the bulk of the volcanic edifice and represents magma production directly over the hot spot. Magma production rates of 0.1 km^3/y, estimated for currently active Kilauea, indicate that the shields were probably built within about 500,000 years.

3. A post-shield, dominantly alkalic stage represents the breaking of the connection with the hot spot. Magma reservoirs are trapped within the uppermost mantle where they cool and differentiate to form a series of lavas ranging from alkali olivine basalt to trachyte.

4. Following a hiatus lasting from 0.2–2.0 m.y., renewal of magmatic activity occurs, represented by eruption of highly alkalic lava ranging from basanite through nepheline- and melilite-bearing lavas. These are presumably *not* derived from the hot spot, now located far down the chain, but reflect instead the process of crust and mantle adjustment to the volcanic load imposed during the three earlier stages of volcanism

Current activity at the hot spot comprises eruption on both locus lines at four volcanoes: Kilauea, Mauna Loa, Hualalai, and Loihi seamount. Haleakala, from which an eruption took place about 1790, is already in the rejuvenated stage and, therefore, lies off the hot spot. Of the four active hotspot volcanoes, Hualalai stands out by virtue of its bimodal post-shield association of alkali olivine basalt and trachyte, the only Hawaiian volcano to exhibit such an association.

Kilauea and Mauna Loa have erupted only tholeiitic lava over the exposed subaerial and the dredged submarine surfaces of their shields. Mauna Loa may have reached its maximum height and be close to the end of its tholeiitic stage. Evidence for this is its massive size, lower rate of historic eruption compared to Kilauea, and the occurrence of slightly fractionated and olivine-poor lavas erupted at its summit. Kilauea is intensely active, currently in continuous eruption, with a magma supply rate that appears to be increasing during the 200 years of observation.

Early workers in Hawaii were concerned with the relationship between Kilauea and Mauna Loa, tidal influences on their respective cycles of activity, and possible sympathetic response of one volcano to the activity of the other. It now appears that the two volcanoes are truly independent; lunar, solar, and earth-tidal effects may be important in triggering activity when the system is erupting episodically.

There remains a problem regarding the geometry of the source region. Deep, long-period seismicity defines an area located midway between Loihi, Mauna Loa, and Kilauea, in which magma is assumed to be collected for eruption at each of the volcanoes. Yet, as noted above, each volcano retains a chemical identity, defined by major oxides, trace-elements, and isotopes, that precludes a simple common source region. The data appear to require a premelting mantle heterogeneity, at least over short periods of time, that allows discrimination of the source regions for these active volcanoes.

The Hawaiian Island story is not duplicated in other areas of the world where hot spots have been identified. The principal reason for this is most likely related to tectonic setting. The Hawaiian hot spot is located beneath the middle of the Pacific Ocean plate, far from any continental masses, plate boundaries, or subduction zones. Other hot spots are not so favorably situated and have shorter lifetimes with a more complex array of volcanic products.

WU Gongjian, ZIAO Xuchang, LI Tingtong et al, Chinese Academy of Geological Sciences, Academica Sinica, Chinese University of Geosciences, Changchun College of Geology, Aeromagnetic Survey of the Ministry of Geology and Mineral Resources, Beijing, China

Integrated Study of GGT from Yartong to Golmud

The 1200 km long Zinghai Tibetan transect, extending from Yartong in Tibet to Golmud in Qinghai province, is supported by the China National Science Foundation and the Ministry of Geology and Mineral Resources. The project is carried out by the Lithospheric Centre affiliated to CAGS in cooperation with geoscientists from eight other research institutions and universities. The major achievements obtained can be roughly summarized as follows: (1) Six terranes subdivided along the transect line on the basis of integrated studies of various geological, geophysical, and geochemical data available; (2) A long geomagnetic profile completed along with this transect; (3) A high-precision gravity profile accomplished from Ando to Golmud along with the transect; (4) Heat flow measured on several sites in the northern part of the GGT; (5) MT sounding made on 30 sites of observation along the transect line; and (6) about 2000 paleomagnetic samples from six terranes collected and analyzed on the basis of the data available, a draft GGT map at 1:1,000,000 scale from Yartong to Golmud has been compiled, with its third dimension extending down to the upper mantle.

The terranes covered by the GGT from south to north are (1) Himalayan terrane, (2) Gangdise terrane, (3) Qiangtang terrane, (4) Bayan Kar terrane, (5) South Kunlun terrane, and (6) North Kunlun terrane. All the terranes mentioned are separated by faults that can be recognized from field geology in combination with gravity, magnetic, and seismic observations. Evidence indicates that they welded together through tectonic events occurring successively in the history of evolution of the Qinghai-Tibetan plateau. Moreover, paleomagnetic and biostratigraphic studies also show that some terranes did travel long distances in terms of horizontal displacement. A space-time diagram was constructed to show clearly the differences of each terrane in stratigraphy, tectonic structure, magmatism, metamorphism, and biozone, which are used for the characterization of the terranes.

Subdivision of these terranes is well supported by geophysical measurements. According to paleomagnetic data, these terranes were inferred to have migrated from middle or lower latitude areas of southern hemisphere. The northern Tsaidam basin was located at about 20° N in the late Carboniferous period. During the late Triassic period, South and North Kulun had drifted to 22.6° N. Qiangtang terrane might have come together with the South and North Kulun terranes during the period of Jurassic. Gangdis terrane is estimated to have merged with Qiangtang terrane during the late Jurassic to Cretaceous. Himalayan terrane and Gangdis terrane merged during the late Tertiary, and the whole continental began to form, which indicated progressive crustal accretion of Eurasian continental. The collision of Indian plate and the Eurasian plate led to thickening and strongly uplifting of Qinghai-Tibetan plateau. Structures and boundaries of terranes are identified to a certain degree by using magnetic anomalies after their direction filtering and continuation and Bouguer anomalies after their correlation with surface elevation and high-order differential derivation. Moho dislocations are given by seismic technique through solution of inverse problems.

2-D Seis81 forward program and in part 3-D Chander forward program are used to model crustal structures. As a result, the crust shows clear later and vertical inhomogeneities, and it can thus be divided into four or five layers, with one or two low-velocity layers existing within it. The study shows that Moho occurs as a gradient zone or transition zone in the north and as an interface in the south. The upper and lower limits of the magmatic interface were derived from short-wave and long-wave length magnetic anomalies, respectively. More information on distribution, mode of occurrence, and emplacement of magmetic rock bodies was obtained from gravity and magnetic measurements. Furthermore, integrated interpretations of gravity and seismicity, gravity and magnetics, and MT and geothermics were also done for the transect, and finally an integrated multiparameter model was constructed by using an interactive constraint system.

As it is suggested, the terranes disposed from north to south in the Qinghai-Tibet are considered to be originally separated by a series of small oceanic basins caused by crustal extension and consequent subsidence in various times. The closing of these basins occurred subsequently owing to seafloor subduction caused by tectonic compression. These events now can be recognized through records of some active continental margins and arc-trench systems, as well as ophiolites, glaucophane metamorphic rocks, and migmatic zones. In addition, sediments and igneous rocks are found, as considered to be formed in environments related to trench, volcanic arc, and/or back-arc basins.

The existence of various terranes implies the history of disintegration of an ancient supercontinent, presumably during the period from Caledonian to Yanshanian, their successive drifting from south to north, and final accretion together to form the present Qinghai-Tibetan plateau, which was amalgamated with the modern Eurasian plate. Apparently, intensive squeezing, collision, and intraterrane thrusting were common during accretion of these terranes, as seen in the Himalayan, north Tibetan, and Kunlunian districts. As a result, the Tibetan plateau has been strongly uplifted, shortened, and thickened ever since, and has been complicated by deformation, metamorphism, and granitic intrusions.

WU, S. P., Z. W. ZHOU, and Z. Q. GUO, University of Science and Technology of China, Hefei, China

Electric Phenomenon before Earthquakes and Production of Free Charge Carriers during Deformation of Rock

We have recently observed electron emission during the fracture of granite. It seemed plausible that deformation might also create free charge carriers on the surface of rock. In this paper, we present the evidence of the production of deformation-induced free charge carriers by measuring the current between the ground and electrode adhering to the sample subjected to uniaxial compression. The rock samples used in the experiments were of rectangle granite from Huaiyuan, Anhui, China. Their dimensions were $50 \times 50 \times 90\ mm^3$. All samples were polished, cleared, dried, and coated with silver on the side as an electrode having area of $30 \times 30\ mm^2$.

In the experiments, an electrode is connected through a micrometer with the ground to measure the current during compression of the sample. The current increases as the pressure on the sample is increased. When the pressure reached 30 Mpa and maintained this value, the current measured in micrometers was about 9×10^{-9} A and did not change.

These experimental results indicate that the positive free charge carriers did take place on the entire surface of the sample. From the current measured in our experiments, the average charge density was estimated to be about $20 \times 10^{-9} c/m^2$. It is important that a piezoelectric mechanism based on the quantum effect of compressed atoms has been proposed. Free-carrier generation may serve as a probe of lower-energy electronic processes induced by deformation before fracture and related to the atmosphere electric phenomenon before earthquakes.

WUCHENG, Sun, Ma BAOLIN, Song SONGYAN, Hu HONGXIANG, Song ZHANLONLG, and Sun JIE, State Seismological Bureau, Honan Province, China

Chifeng—Suizhou Geoscience Transect

This transect runs from Chifeng, Hebei Province to Suizhou, Hubei Province (118°51 'E, 41°51 'N to 113°23 'E, 32°09 'N). It is 1220 km long and crosses north-southwards the first order tectonic units in the east of the continental of China, i.e., the North China craton and the suture zone of its alternation with Yangzi platform. The North China platform is composed of three secondary tectonic units, i.e., southern part of North China basin, North China basin and earth's axis of Inner Mongolia Autonomous region. The suture zone consists of three units, i.e., Hercynian foldbelt, Dabieshan massic block and Caledonian folding.

The basements of the North China craton were formed during Luliang–Zhongyue orogenic movement in lower Proterozoic era. The crystalline basements were composed of the Pre-Cambrian metamorphic rocks of various periods. The covering sedimentary strata of Palaeozoic were developed but the depositions from upper Ordovician to lower Carboniferous are absent. The crystalline basements and covering strata form the North China Platform. Laterally undergoing two rift and downcast cycles, which are relative but different, the North China Platform disconnected and a series of continental dislocation basins was formed, accompanied by intermediate and basic volcanic activity. During the post-Mesozoic Yanshan orogeny these basins were cutoff, accompanied by the intrusion of a great of intermediate and acidic magma.

The suture zone is consisted of the complex folded chains with approximately east-west strike, i.e., Suixian–Guangji high pressure metamorphic belt, Dabieshan high pressure complex and north Huaiyang high pressure metamorphic belt. During early Devonian north Huaiyang geosyncline thrusted anti-polarizely and accreted together with Dabieshan massive block on southern edge of the North China platform. From late Palaeozoic to early Mesozoic Suizhou–Guangji geosyncline thrusted polarizely and accreted on the southern side of caledonian folding belt locating on southern edge of the North China Platform. Finally the Yangzi Platform coalesced with the North China Platform and a series of collisional structures formed in coalescing process. A large-scale thrust slip tectonic belts developed on both sides of Dabieshan massive block and a large slip dislocating belt developed on the surface.

Both the extent and the accuracy of deep geophysical research on the northern half section of the transect are higher but on the southern half section are lower.

The deep geophysical data show that the northern half section of this transect has a boundary line near Guan county. In Guan and its northern area the strike of gravity anomaly is northeastward or near east-westward. There exists a sharp gravity gradient belt in Xinglong region. The contours of gravitational anomaly are sparse and smooth southwards and their orientation is north-northeast or north-south. The gradient values are one of the lowest parts in the North China Platform.

In Guan and its northern region the gradients of magnetic anomaly are dense, showing locally positive anomaly. They are characterized by mainly east-west strike and accompanying northeast direction. Toward south there is a silent negative magnetic anomaly field with direction of north-northeast or north-south. Around Guan area the top of higher conductivity layer downdips northwards and southwards respectively and the depth to the top is about 15 km. The results from the electromagnetic sounding indicate that the lithospheric subface shows the same state as that depicted above. The depth to the lithospheric top is 60 km and the maximum depth dipping towards south and north is 90 km or more. The geothermal heat flow is higher toward north but lower toward south.

The results from artificial seismic sounding show that the crustal structure near Guan may be divided into two parts. The southern part is simple tectonically and composed of some positive velocity gradient layers with various gradient values. The velocity increases monotonously with the depth. To the depth of 13 km there exists a positive gradient layer with mean velocity of 6.0 km/s, less 0.2 km/s than that in the north part. At depth of about 24 km appears unobvious velocity leap, corresponding to the Conrad discontinuity. The north part consists of several velocity gradients with alternation of positive and negative gradients. Above the depth of 13 km there is a positive velocity gradient. The layer between 14–23 km is a lower velocity layer of 6.0 km/s. The layer at depth of about 24 km is a successive velocity discontinuity and the velocity contrast is 0.15 km/s (from 6.55–6.70 km/s), corresponding to Conrad discontinuity. At depth of 33 km there is another lower velocity layer of 6.6 km/s. The Moho discontinuity is more continuous along whole transect and its depth increases gradually to 38 km at the south to Xingtai area, 32 km from Xingtai to Shijiazhuang, 34–35 km at the north to Shijiazhuang and 32.5 km northwards from Xinglong.

The earthquake activity along this transect is same as that observed on other profiles in North China, that is to say the earthquake activity is correlated to the wedging out zones of the lower velocity layer and to the wings of wave-shaped tectonics of the Moho discontinuity.

The features of the south half section of this transect and the geological interpretation are shown in the figures.

WUCHENG, Sun, JIE, Xu, Zheng SIWEI, Zhang XIANKANG, Zhu ZHIPING, and Yang ZUEN, State Seismological Bureau, Honan Province, China

Shanghai–Zhengzhou–Alashan Geoscience Transect, North China

This geoscience transect runs from Fengxian near to Shanghai to Alashan in Inner Mongolia Autonomous Region. It is about 1700 km long and goes through the east of the continent of China. The transect is perpendicular to the strike of main tectonic lines. Three great tectonic units, i.e., the Yangzi platform, the North China craton and the Alashan geobody (or the Alashan geoblock), are transversed by this transect. The extent of geological and geophysical study along this transect is high. Since the eighties the deep geophysical works have been carried out in these area, especially some deep seismic sounding profiles have been conducted.

The basements of the Alashan geobody and the Yangzi platform were formed during the Ningjin orogenic movement in middle proterozoic subera. The basements of the North China craton were formed during Luliang-Zhongyue orogenic movement in lower proterozoic subera. In middle and upper proterozoic subera the aulacogen was formed around their edges and in their interiors. From Palaeozoic era to triassic period the covering strata were typically developed, but lower ordovician and lower carboniferous sediments are practically absent. During Hercynian–Yangzi orogeny these two tectonic units merged into a unit continent. Then, underground two relating and different cycles of rift and downcast during mesozoic and cenozoic eras, this unit continent was failed and disjointed, and a series of continental dislocation basins was formed, accompanied with the intermediate and basic volcanic activity. During the post-mesozoic Yanshang orogeny these basins were cut off, accompanied with the intrusion of intermediate and acidic magman, and the cenozoic basins were developed to a certain degree.

The Yangzi platform and North China craton were developed to the minor tectonical units, such as Xiayangzi platform and fold belts, Subei fault depression, Luxi uplift, Yuhuai uplift, Yuhuai fault depression, Shanxi fault uplift, Eerdousi platform depression and Eerdousi marginal fold belts.

From Shanghai, the Bougure gravity anomaly reduces gradually towards the north by the rate of 0.13 mgl/km. The western half-section of the transect is a negative value region, and the eastern half-section is a positive one. The boundary of these two parts is located at kaifen region. This fact indicates the inhomogeneous variation of the crust density and crust thickness. The horizontal gradient for the Bougure gravity anomaly is stronger in the east-section, whereas it is weaker in the west-section with the exception of some specific area. That indicates the lateral inhomogeneity of the density structure at the top of the upper crust. This kind of obvious inhomogeneity indicates strong tectonic movement in the upper crust. In most areas of the west-section the tectonic movement on the top of the upper crust is not very strong.

The features of calculated deep gravity anomaly indicate that generally tectonic fluctuation of the crust is stronger in the eastern part of the transect than in the western part where only in some areas the tectonic fluctuation of the crust is obvious.

The calculated magnetic anomalies are circle-shaped. The magnetic anomalies, which are related to the shallow tectonic features, consist of a series of circle-shaped anomalies. Their trend is coincident with the geotectonic strike. The shallow and deep magnetic anomalies are related with each other. They are associated with major gravity gradient belts and tectonic anomaly belts. The anomalies, in the middle segment of the transect, from Xuiqian to Qingian, are relatively stronger, especially from Taikang to Lingfen. There are found stronger positive magnetic anomaly zones in this segment (Xuqian, Taikang, Lingfen and Qingbian). They reflect the fracturing process in the deep crust.

The results of deep seismic sounding with perfect observation system reveal the structure and tectonics features of the deep crust. Along the transect the crust consists of three layers, i.e., upper, middle and lower crust. The depth of the Moho is about 30–35 km in the southeast to Jingcheng is Shanxi plateau. The wave-shaped tectonics of the Moho with the length of 5 km is formed within the range of decades kilometers. The Moho has a depth of 40 km in most segments in the northwestern area, and it recheves about 45–47 km in some segments. The wave-shaped tectonics of the Moho is characterized by large amplitude and low frequency, and its range is over 200 km.

There are low velocity layers with thin thickness in the crust in some segments along the transect. However, there are some large scale low velocity layers with large thicknesses and high-conductivity layer in the crust from Jingcheng to Lingfen and from Yanchuan to Alashan.

The earthquakes, $M = 6 \sim 8$, which occurred in the west of the transect, are all located at the flanks of the Moho uplifts and zones with large scale low velocity layers. The magnitude of the earthquakes occurred on the eastern part of the transect is not big. They are also located at the flanks of the Moho uplifts.

The deep geophysical data show that there are two huge anomaly zones of crustal structure and tectonics along this transect, i.e., Yingchuan–Alashan zone and Jingcheng–Lingfen zone. The Yangzi fractured zone does not exist. In addition, the boundary of the first-order tectonic units between the North China and Yangzi platforms is located where the Xiangshuei river fault crosses over the Tanlu fault. The deep geophysical data indicate that the Tanlu fault is characterized by the heave, but not by the obviously vertical displacement.

The results of comprehensive geology interpretation are shown in figures.

WURZER, F., C. REIMANN, T. LEITNER, H. KÜRZL, and K. KÖCK, Joanneum Research Association, Leoben, Austria

SGP: Computer Package for Graphical Statistical Analysis of Geochemical Data

A computer package for the graphical statistical analysis of geochemical data using the most modern methods of exploratory data analysis (EDA) including resistant and robust statistics has been developed. It is especially suited for the graphical analysis of regionalized geochemical data and includes the following features:

- command language driven user interface including the use of command files for batch and interactive mode;
- interactive window technique for free graphical layout;
- enhanced data management and transformation facilities including:
  - –handling of missing values
  - –case selection
  - –definition of new variables using logical and mathematical operators and functions

–transformation methods: ladder of powers, box-cox, logit, lot10, Nlog, standardization, range linear transformations, and ranking

–possibility to save updated data sets anytime

- metafile concept for device independent preparation of hardcopies;
- interactive text facility using up to 20 different hard- and software fonts
- flexible automated mapping with a wide range of layout facilities including symbol and color selection, different types of legends, handling of background drawings;
- interactive sample identification and grouping facilities in most statistical graphics and geochemical maps;
- the standard version can handle up to 4000 samples with 60 variables each.

The following graphical and tabular outputs and facilities are incorporated in the package:

- Histogram
- Density Trace
- One dimensional Scattergram
- Boxplot (including boxplot comparisons of different data sets)
- Any combination of above in one graphic
- CDF-Plot
- XY-Plot with interactive sample identification by sample number and/or cursor and encoded variables to color, symbols, and sizes
- Scattergram with interactive grouping function
- Ternary Diagram with "zoom" function for enlargement of selected subtriangles, interactive sample identification, and encoded variables
- Draftsman's Display: draws a "graphical" correlation matrix with encoded variables for colors and symbols
- Streamplot: draws profiles for selected variables along interactively chosen sampling lines (three different symbols and/or colors)
- Levels: calculates class limits for mapping according to boxplot, percentiles, or freely chosen values
- Legends: 10 different legend types available
- Geodisp: geographical display (mapping) of sample points with interactive grouping function
- Ustatistics: calculates a table of univariate statistics including robust mean and spread and outliers identification
- Correlation: calculates and displays a correlation matrix of selected variables (optionally in a robust manner)
- Covariance: calculates and displays a covariance matrix of selected variables (optionally in a robust manner).

WYLLIE, P., California Institute of Technology, Pasadena, CA

Experimental Boundaries for Origin and Evolution of Carbonatites

Carbonatites may occur alone or in complex plutonic or subvolcanic associations with rare volcanic occurrences. Despite considerable variety, there is a persistent association of (1) carbonatite dominated by calcite or dolomite, with many accessory minerals including several of economic value, (2) nepheline-normative alkalic igneous rocks, including ultramafic types, and (3) fenites, country rocks subjected to alkali metasomatism caused by emanations from carbonatite rather than silicate magmas. Experimental phase equilibrium studies involving representatives of the carbonatites, silicates, and metasomatizing solutions define boundaries for petrogenetic interpretations. The natrocarbonatite lavas of Oldoinyo Lengai extend the range of relevant materials to sodium carbonate.

Among the processes proposed for the origin of carbonatites are: (1) Several deny the existence of carbonatite magmas. (2) Carbonatite magmas are primary, possibly very rich in alkalis. (3) Carbonatite magmas are residual melts derived by fractional crystallization of a parent carbonated alkali peridotite magma. (4) Carbonatite magmas are immiscible liquids separated from kimberlitic, melilitic, or nephelinitic parent magma.

Information from experimental phase equilibrium studies elucidates the following problems: (1) the chemical and physical nature of carbonate-precipitating melts; (2) the conditions for the formation of carbonatite melts from peridotite at mantle pressures, or of CO_2-rich parent silicate melts that may yield carbonatite derivatives; (3) the locations of liquidus paths between high-temperature silicate melts with dissolved volatile components and low-temperature synthetic carbonatite magmas; (4) the compositional range of liquid immiscibility between silicate and carbonate melts; (5) the compositions of vapors evolved during crystallization of synthetic carbonatite magmas; and (6) the paths of crystallization causing differentiation of carbonatite magmas, and the precipitation of accessory minerals such as phosphates, niobates, fluorides, sulfides and sulfates, bastnaesite, and other rare earth element minerals.

Experiments have demonstrated the following: (1) Carbonatite magmas can occur. Melts containing moderate amounts of water or alkalis can precipitate calcite or dolomite down to temperatures of about 650 °C, low enough to match those estimated for the emplacement of natural carbonatites. Large carbonatite bodies are surely cumulates. (2a) There is a ledge on the solidus for peridotite-CO_2-H_2O at about 26 Kbars (corresponding to 75 km depth) where dolomite becomes stable instead of CO_2, and where the solidus temperature drops through several hundred degrees; at greater depths the near-solidus liquid composition corresponds to that of a dolomitic carbonatite magma, with Ca:Mg > 1, with about 10% dissolved silicates, and enriched in alkali carbonates to the extent that alkalis are available in the peridotite. (2b) At lower pressures, $(CO_2 + H_2O)$ vapors react with peridotite at temperatures below the peridotite-H_2O solidus to generate traces of Ca-Na carbonatite melt, immiscible with higher-temperature silicate melt. (3) At shallow crustal pressures, it appears that carbonatite melts are separated by liquidus thermal barriers from many high-temperature silicate melts. However, fractional crystallization of nepheline-normative silicate melts containing CO_2 and H_2O does yield a complete series of silicate-carbonate and carbonatite mineral assemblages analogous to those occurring in some igneous complexes (e.g., Oka). At pressures above 26 Kbars, silicate melts derived from peridotite-CO_2-H_2O fractionate to dolomitic carbonatite melts. (4) Nepheline-normative melts with sufficiently high $(Na+K)/(Ca+Mg+Fe)$ can crystallize and yield immiscible sodic carbonatite melts through crustal and mantle pressures; crystallization of these melts precipitates calcite while Na is stored in residual melt (or partitioned into coexisting vapor), and the residual melt could correspond to natrocarbonatite. (5) Crystallization of synthetic carbonatite melts can be accompanied by the evolution of aqueous, Na-enriched vapors, analogous to fenitizing solutions. (6) The solubilities of most silicate components and accessory minerals in low-temperature synthetic carbonatite magmas are less than about 5% by weight, but the solubility of $La(OH)_3$ in such liquids ranges from 20% to 40%, increasing with increased CO_2/H_2O in the system.

The experimental results are consistent with the following scheme. Mantle carbon, derived at least partly through subduction, is involved in the generation of silicate melts in the asthenosphere, with the melts then processed in the lower lithosphere, and with parental nephelinitic and melitic melts forming magma chambers beneath major rifts at about 75 km depth. The nephelinitic magmas may evolve to a stage where immiscible Na-Ca carbonatite magmas are formed, allowing both to be erupted through the same conduit system. Alternatively, the parent silicate magmas may yield carbonatites only at shallow levels in the crust, either by immiscibility or by fractional crystallization, depending on factors such as Na/Ca and perhaps H_2O content. Experimental data are consistent with the idea that primary dolomitic carbonatite magmas (with alkalis) could be erupted from depths greater than 75 km, presumably in diatremes similar to those of kimberlites, but the high ratio of silicate:carbonatite in most igneous complexes supports the view that most carbonatites are derivative rather than primary.

X

XENOPHONTOS, C., Ministry of Agriculture and Natural Resources, Nicosia, Cyprus

Troodos Ophiolite—An Ancient Analogue of a Special Type of Marginal Basin

Diversity in ophiolite genesis is now well established and most ophiolites are thought to represent marginal basin or oceanic island-arc crust. Oceanic and post-emplacement metamorphism and tectonism may either mask or obliterate those characteristics normally invoked for deciphering the environment of ophiolite genesis. Most ophiolites exhibit variable lava stratigraphies, geochemical and petrologic characteristics and internal structures. Undoubtedly such diversities reflect variations between and within marginal basins due perhaps to selective preservation of parts of marginal basins or in rare cause to preservation of young island-arc crust.

The Troodos ophiolite of Cyprus is generally considered to represent the earliest stages of marginal basin or island arc development. A well-developed ($\approx$1.5 km thick) sheeted dyke complex, at least 120 km across strike, provides indisputable evidence for spreading. Diffuse, slow spreading, along a migrating ridge defines a number of ridge—parallel asymmetric grabens whose axes are flanked by normal listric faults and whose geometry is considerably more complex in the vicinity of ridge-transform intersections. Supra- or inter-lava volcanogenic sediments are completely absent which suggests that the ophiolite is not related in time and space to any subaerial arc volcanoes.

The Troodos ophiolite lava pile is devoid of any MORB or Marginal Basin basalts and differs markedly from these in being uniformly more depleted. Basaltic andesites, andesites and dacites are topped by high-SiO_2, high-MgO basalts with "boninitic" affinities. All are characterized by the crystallization sequence olivine-clinopyroxene-orthopyroxene-plagioclase.

The volcanics are enriched in LFS relative to REE and HFS elements and have negative Ta-Nb anomalies, characteristics that typify lavas of the early stage of intraoceanic island arc development. Unlike similar Back Arc Basin (BAB) basalts they do not occur together with N-type or E-type MORB; isotope evidence confirms their island arc affinity and points to a derivation from a depleted and modified mantle source. An origin through varying or repeated hydrous partial melting of this source enriched by the addition of an LFS-rich fluid phase and possibly minor amounts ($\approx$1%) of pelagic sediment is suggested.

In the absence of a well-developed island arc and considering the geological and geochemical evidence, an origin of the Troodos ophiolite above a north-dipping subduction zone in the Upper Cretaceous is more plausible than its generation in a back-arc basin. Trenchward migration brought the newly formed crust into juxtaposition with the trench by the removal of pre-existing oceanic crust through tectonic erosion.

Three important features are proposed for the Troodos Marginal Basin: (1) the northern part floored by trapped pre-Upper Cretaceous oceanic crust, (2) young Upper Cretaceous oceanic island-arc crust that extended as far south at the trench and (3) the trench, now marked by an arcuate belt of pre-Upper Cretaceous accretionary melange, that forms the southern boundary of the ophiolite and extends into northwest Syria.

Subduction and ophiolite generation were arrested prior to arc-maturation in the Maastrichtian by the collision of the Arabian plate, and other microcontinental fragments, with the trench. Subsequent complete marginal basin closure was confined to the north of Cyprus where basin crust remnants are to be found in southern Turkey. Thus the Troodos Marginal Basin is possibly unique in that there are no present day identical analogues.

XIANG, W., China University of Geoscience, Wuhan, China

Sedimentary Model of Red-Bed Under the Foundation of the Gezhouba Project on the Yangtze River and Their Engineering Geological Significance

The weak intercalations under the foundation of the Gezhouba Project on the Yangtze River are so well known that experts and scholars from every organization of this field in China have made extensive and in-depth studies. But there have been few studies on the depositional environment of red-beds, or the depositional law of the weak intercalations.

On the basis of detailed analyses of a large amount of data and information acquired from field and laboratory, the author believes that the red-beds under the foundation are the strata of alluvial facies developed from a set of fluvial environments. Weak intercalations are formed in subenvironments including flood basin, floodplain, natural levees, channel and point bar, and abandoned channels. The deposit law is described thus: depositional environment determines the distribution of intercalations, and the spatial occurrence is controlled by the environment. The change of environment affects the distribution of intercalations. The intercalations distribute generally on the top of the semirhuthmite (from bottom to top and from the coarse to the fine). The clay mineral ingredients and chemical ingredients of the weak intercalations are consistent on the whole, but the grain sizes are variant. Therefore, even the same intercalations exhibit different behaviors, in terms of engineering geology.

This article also discusses the respective evaluations of intercalations formed as a result of various environmental conditions, and points out that adequate study should be given at the early stage of engineering survey, in order to guide engineering geological investigations.

XIANG, WEI, China University of Geoscience, Wuhan, China

Engineering Geological Significance of Deposition Environment of the Weak Intercalations under the Dam Foundation

The Gezhouba Project on the Yangtze River, whose rock foundation is composed of Eogene–Cretaceous red clasolite of continental facies with all together 69 weak intercalations of clayrocks distributing in it. On the basis of detailed analyses of large amount of data and information acquired from field and laboratory, the author beleives that the red–beds under the foundation are the strata of alluvial facies developed from a set of fluvial environment. Weak intercalations are formed in sub–environments including flood basin, flood plain, natural levees, channel and point bar, and abandoned channels

The weak intercalations of the Gezhouba dam foundation formed in different sub–environments of river environment, and so with then respective characteristics of spatial distribution the weak intercalation of different environments have their own specific positions in the evaluation of engineering geology. The intercalations originated from the flood basin environment have the worst engineering geological conditions, those are very thick, well continuity and stable in distribution, and perform generally as the structural faces controlling the stability of engineering rock masses. Clay grains are much more in such intercalation than in the others, ranging from 40 to 60%. Therefore intercalations are highly hydrophilic, low in shearing strength and deformable. At the meantime, because they are thick layers of clayey rocks, they could be developed completely into shearing zones when they are subject to the change of tectonics and further be argillized on a large scale.

The weak intercalations of the flood plain environment can be divided into three types. The aggradated thin layers of clayey rocks on top of siltstone are spread constantly and as structural faces control the stability of the engineering rock masses. Due to the fine drains and generally low shearing strength, plus the fact that they are connected above and below with some relatively hard siltstone, they are easily damaged by shearing when Tectonics changes so as to become argillized faces with continuity. The intercalations deposited in the low land of the flood plain, generally thick and fine, but not constant in distribution, are not easy to allow uniform sheared zones under the change of tectonics. That can be easily treated by excavation during construction. The siltstone intercalations deposited in the flood plain, which is thick with good continuity and constant in distribution, can become comfortably sheared zones between layers; but because their grains are coarse and hard to be argillized, the engineering geological property is better than those described above. The intercalations formed in the natural levee environment consists mainly of interbedding of clayey stone and siltstone and fine sandstone, with rather flat faces. The clayey stone, composed of fine grains and high content of clays, plus water conductive bodies above and below them, are easy to be argillized on faces, indicating a low shearing strength. Because the natural levee environment could be disturbed by water flow in flood season, the tectonics of scour and fill in is developed, so that the continuity of such intercalations is affected, that concerning the stability of the rock masses at some sections. The intercalations of this category are subject to splitting along the faces. Therefore, those who are close to the foundation surface should be removed, while the ones buried in depth should be bolted and grouted. The weak intercalations deposited in the channel bar and point bar environment consist mainly of boulder clay layers (psephitic clayey stone intercalation) and strip carbonaceous intercalations. Generally the boulder clay layers distribute unconstantly, and mostly on the scouring faces uneven (low) at cross-section. There is no uniform intercalation face, with the head and tail over-lapped. Such intercalations varied according to their density of boulder clay and thickness, impact the sliding stability of individual sections but they can easily result in large and uneven settlement. The distribution of carbonaeous strip intercalations is not constant, affecting only individual sections in terms of stability. Such intercalations can be remarkably treated by bolting and grouting. If the boulder clay intercalations are very thick, they should be removed by excavation. The intercalations formed in the abandoned channel environment which are worst in distribution and overlapped each other upper and on a horizon, is easy to be split during construction deterisrating the behavior. Such intercalations are subject to uneven settlement thus requiring only normal bolting and grouting.

The evolution of deposition environment affects directly the distribution of intercalations. If the environment changed into flood plain and basin after the intercalations had been deposited, they were good in continuity and constant in distribution, with the maximum impact on the project. While it changed into river channel as well as channel and point bar, the clayey stone intercalations suffered a damage of their continuity due to bottom scouring and side erosion, such a case relatively beneficial to the stability of the project against sliding on one hand and subject to uneven settlement. Meanwhile, because the deposits after erosion are composed of permeable rockmasses including sandstone and conglomerate rocks, the ground water within the intercalations moves horizontally to promote argillization of them. Actual excavation shows also that when the sides of many intercalations had been scoured away, the remainders were all argillized.

XIANG, Wei, China University of Geoscience, Wuhan, China

Engineering Geological Significance of Deposition Environment of Weak Intercalations under Dam Foundation

The Gezhouba Dam Project on the Yangtze River has encountered rock foundation composed of Eogene–Cretaceous red clasolite of continental facies with altogether 69 weak intercalations of clayrocks distributed in it. On the basis of detailed analyses of large amounts of data and information acquired from field and laboratory, the author believes that the redbeds under the foundation are the strata of alluvial facies developed from a fluvial environment. Weak intercalations are formed in subenvironments including flood basin, floodplain, natural levees, channel and point bar, and abandoned channels.

The weak intercalations of the Gezhouba dam foundation formed in different subenvironments of river environment, and so with the respective characteristics of spatial distribution the weak intercalations of different environments have their own specific positions in the evaluation of engineering geology. The intercalations originating from the flood basin environment have the worst engineering geological conditions. They are very thick, continuous, and stable in distribution, and perform generally as the structural faces controlling the stability of engineering rock masses. Clay grains occur much more frequently in such intercalation than in the others, ranging from 40 to 60%. Therefore, the intercalations are highly hydrophilic, low in shearing strength, and deformable. Because they are thick layers of clayey rocks, they could be developed completely into shearing zones when they are subject to the change of tectonics and further be argillized on a large scale.

The weak intercalations of the floodplain environment can be divided into three types. The aggradated thin layers of clayey rocks on top of siltstone are spread constantly and as structural faces control the stability of the engineering rock masses. Due to the fine grain and generally low shearing strength, plus the fact that they are connected above and below with some relatively hard siltstone, they are easily damaged by shearing when tectonics changes so as to become argillized faces with continuity. The intercalations deposited in the low land of the floodplain, generally thick and fine but not constant in distribution, do not easily allow uniform sheared zones under the change of tectonics. That can be easily treated by excavation during construction. The siltstone intercalations deposited in the floodplain, which are thick with good continuity and constant distribution, can become comfortably sheared zones between layers; but because their grains are coarse and hard to argillize, their engineering geological property is better than those described above. The intercalations formed in the natural levee environment consist mainly of interbedding of clayey stone and siltstone and fine sandstone, with rather flat faces. The clayey stone, composed of fine grains and high content of clays, plus water-conductive bodies above and below them, are easily argillized on faces, indicating a low shearing strength. Because the natural levee environment could be disturbed by water flow in flood season, the tectonics of scour and fill is developed, so that the continuity of such intercalations is affected, which impacts on the stability of the rock masses at some sections. The intercalations of this category are subject to splitting along the faces. Therefore, those close to the foundation surface should be removed, while the ones buried at depth should be bolted and grouted. The weak intercalations deposited in the channel bar and point bar environment consist mainly of boulder clay layers (psephitic clayey stone intercalation) and strip carbonaceous intercalations. Generally the boulder clay layers distribute discontinuously and mostly on the scouring faces uneven (low) at cross-section. There is no uniform intercalation face, with the head and tail overlapped. Such intercalations, varied according to their density of boulder clay and thickness, impact the sliding stability of individual sections, but they can easily result in large and uneven settlement. The distribution of carbonaceous strip intercalations is not constant, affecting only individual sections in terms of stability. Such intercalations can be treated by bolting and grouting. If the boulder clay intercalations are very thick, they should be removed by excavation. The intercalations formed in the abandoned channel environment that are worst in distribution and overlap each other at upper levels and on a horizon are easy to split during construction. Such intercalations are subject to uneven settlement, thus requiring only normal bolting and grouting.

The evolution of deposition environment affects directly the distribution of intercalations. If the environment changed into floodplain and basin after the intercalations had been deposited, they were good in continuity and constant in distribution, with the maximum impact on the project. While it changed into river channel as well as channel and point bar, the clayey stone intercalations became discontinuous due to bottom scouring and side erosion. Such a case is relatively beneficial to the stability of the project because it prevents sliding on one hand and uneven settlement on the other. Meanwhile, because the deposits after erosion are composed of permeable rock masses including sandstone and conglomerate rocks, the groundwater within the intercalations moves horizontally to promote argillization. Actual excavation shows also that when the sides of many intercalations had been scoured away, the remainders were all argillized.

XIAO, Xuchang, LI Tingdong, LI Guance, GAO Yuanlin, and XU Zhiqin, Chinese Acadmeny of Geological Sciences, Beijing, China

Tectonic Evolution of Zinghai–Xizang (Tibet) Plateau

Tectonic units of the Zinghai–Tibet Plateau and neighboring regions: Five plates—Sino-Korean (I), Tarim–Qaidam (II), Yangtze–Bayan Har (III) 1–2), Qiang Tang–Gangdise (IV), and India–Himalayan (V)—were respectively sutured by four major suture zones: (1) Qilian (Chilian)–Qinling, (2) Eastern Kunlum–Southern Qinling, (3) Southern Karakorum–Dengqen–Lancangjiang, and (4) Yarlung Zangbo (Tsangpo) suture zone. The first two suture zones, resulting from the consumption of the paleo-ocean basin in the north, are named Qin–Qi–Kun ocean. The last two, representing the Mesozoic ocean basins in the south, are respectively named north or early Tethys ocean basin and south or late Tethys ocean basin, having been subducted and consumed respectively in the Early–Middle Mesozoic and Early Tertiary.

The tectonic evolution of the plateau is actually the history of the opening, and of the spreading to the consumption of these ocean basins that consequently formed the Qin–Qi–Kun Paleo-tectonic domain and the Tethys–Himalayan domain. The southern Karakorum–Southern Hohxilshan–Lancanjiang suture zone (3) might constitute the boundary between the Eurasia and the Gondwanaland during the Late Paleozoic.

Tectonics and geodynamics of Yarlung Zangbo (Tsangpo) suture zone and its adjacent area: Two different types of ancient continental margins—In the Mesozoic, the Yarlung Zangbo suture zone resulting from the consumption of the south Tethys ocean basin and its adjacent area occupied the space between two different continental margins: the southern passive continental margin and the northern active continental margin. The former is constituted, from south to north, by the platform sediments ranging in age from the Paleozoic to the Early Tertiary and the slope-rise sediments of Mesozoic. The latter consists of the trench, fore-arc basin, and the volcanomagmatic arc.

The marker for the Yarlung Zangbo suture zone are the ophiolite zone, the Melange zone, and the high-pressure, low-temperature metamorphic zone.

Microstructures of upper mantle rocks in Yarlung Zangbo ophiolite and their geodynamic significance—The microstructural study of the Yarlung Zangbo ophiolite permits tentatively paleotectonic reconstruction; that is, this ophiolite may have originated in a small oceanic basin in which three possible evolution stages of the oceanic crust-upper mantle thrusts might occur due to the oceanic convergence.

Information provided by microstructures has been applied also to the research of the ductile deformation and the geodynamical analysis of the Himalayan Chain.

Causes of the crustal doubling and uplifting: Since the Hipparion, the subtropical flora such as Eucalyptus, Ficus, Cassia and Rhamnus of Early-Middle Tertiary have been found both in the region north and the region south of the Yarlung Zangbo (Tsangpo) suture zone, implying that the elevation of the plateau is not more than about 2000–2500 m during that time. So the crustal uplifting might be after the period of the suturing of the plateau. In the Late Tertiary and Early Quaternary, the post-suturing doubling and shortening of the plateau were caused by both the crustal thrusting and the strike-slip fault.

The specific deep structure of the plateau is also related to the doubling and uplifting. Recent geophysical data reveal that the deep structure of the Qinhai-Tibet plateau may be divided into three structural layers. The first structural layer to be considered as the upper crust is between 10 and 30 km. A low-velocity, low-resistivity layer occurring below the upper crust might act as a decoupling level along which the listric thrust causing the crustal thickening converged. The second structural layer to be considered as the lower crust is between the low-velocity, low-resistivity layer and the depth of 50–60 km. The third structural layer is between the depth of 50–60 km and that of about 70–80 km, which is characterized by anomalous mantle or the crust-mantle transitional layer due to the upwelling of mantle materials (harzburgite?) and their interaction with crust.

Recent paleomagnetic data of Eocene sediments and volcanics from Gangdise microplate, the present latitude of which is 29°30′N, give a lower paleolatitude of 10°N, and that from the Himalayan microplate, the present latitude of which is 28°N, also give a lower paleolatitude of about 5°N. So the Gangdise microplate probably moved northward about 2100 km, and the Himalayan microplate about 2400 km. Hence about 300 km was consumed by crustal shortening and thickening. So we may infer, from above paleomagnetic data, that the plateau crust was thickening, shortening, and consuming during the Eocene to probably Pleistocene when these microplates moved northward.

But the drifting velocity of the lithosphere of the plateau may have become very slow after the Pleistocene. Since then the isostatic compensation causing the plateau uplift has been instead of the compression caused by the crustal shortening. So the plateau crust has been uplifted about several kilometers by the isostatic adjustment and become the highest region in the world.

XIE Douke, and DUO Kunyi, Nanjing Institute of Geology and Mineral Resources, Nanjing, China

The Geological Character of the Liner Metamorphic Belt of the Continental Margin of Eurasia

The growth fault trend north-northeast developed along the coast of Fujian to Guangdong provinces, occurred on the continental margin of coast of southeast China. The Changle-Nanao metamorphic belt was produced during the process of the fault growing. This belt fall can be divided into three metamorphic lithofacies subunits from coast to intercontinent according to paragenic association of metamorphic minerals. I belt belongs to greenschist facies, which occurred on the side of the continent. M belt was low-amphibolite facies, and C belt was high-amphibolite facies.

-: P-t-c information and trend obtained from silicate rock-forming mineral during the process of metamorphism.

1. Plagioclase: The composition of plagioclase is different in the three facies. For the C belt, SiO_2 = 59–66.34 wt%, which represents silica unsaturated. For the M belt, SiO_2 = 67.41–70.72 wt%, which displays silica saturated.

Feldspar texture stage: An and Si/AI of plagioclase are different in the three lithofacies from An = 16–19, Si/AI = 3.0–3.5 in the M belt to An = 28–76, Si/AI = 2.1–2.9 in the C belt, but for the I belt An = 25–29, which shows that the metamorphic temperature decreases along with the increase of the order degree.

2. Geological temperature and pressure estimated by mica. Mica. In the C belt, geochemical features of muscovite are characterized by Si^{+4} = 3.23–3.3, Si/AI values of tetrahedron coordinate are over 3/1, the content of AI is about 2.6–3.27, and the values of Mg in octahedron range from 0.104 to 0.226. In the M belt, the content of Si^{+4} is more than 3.0, which shows that it belongs to phengite, and Na/(Na + K) = 0.08. Biotite. Studying the results shows that the biotite of this belt all belongs to picrobiotite. The geochemical features display that s ^{18}O = 0.56%, $\Delta\ ^{18}O$ = 4.18, $\delta\ ^{18}O_{H2O}$ = 3.13–3.11%. The values of KD, Ti increase along with the increase of the values of Mg/Fe^{2+}.

3. Garnet. It is pyralspite system in the C belt (mainly manganalmandine-iron-manganesian garnet), and is calc-garnet system in the M belt. The host rock of greenlandite within metabasite of C belt is characterized by that Fe^{2+}/Mg ratio, and content of MgO are higher, but content of CaO is lower, the values of $K_{D(Fe\text{-}Mg)}$ of G^t-Bi was strongly influenced by the change of the content of Ti in pelite, and the average values of Fe^{2+}/Mg of garnet in C belt was 12.7, which shows that the content of Mn Fe^{2+} are higher but the content of Mg-Ca is lower. The geochemical features in M belt are of that values of Ca content range from 2.733 to 2.375, content of Fe is more than 1.419 and extreme components are And = 71%, Gro = 20%.

4. Amphibole. Amphibole occurred both in the C belt and the M belt and all belong to calc-amphibole subunit. The geochemical feature of amphibole is characterized by $\delta\ ^{18}O$ = 4.84%, Δhb-Q = 3.08, $\delta\ ^{18}O_{H20}$ = 7.23–7.22%. In C belt, $\delta\ ^{18}O$ = −0.95%, Δhb-Q = 7.97, $\delta\ ^{18}O_{H2O}$ = 4.39% in M belt. Recrystallization temperature, obtained from hb-Q are 693 °C in the C belt, 438 °C in the M belt.

REE character: LREE is positively correlated with HREE in the C belt. The values of δ Eu range from 0.7 to 1.2. In the M belt LREE is linearly correlated with HREE. The values of δ Eu range from 0.2 to 0.6 in the I belt, and are rich than HREE, δ Eu = 0.4–0.7.

O and Nd isotopes geochemical: Since different metamorphic grade: for C belt, $\delta\ ^{18}O$ (whole rocks) = 2.65–6.49%. $\delta\ ^{18}O$ (single mineral) = 0.985–7.915%, $\delta\delta\ ^{18}O_{H2O}$ = 3.13–7.23, the temperature measured by O-isotope paired minerals is 510–792 °C, for the M belt, $\delta\delta\ ^{18}O$ (whole rocks) = 2.84–6.41%, $\delta\delta\ ^{18}O$ (single mineral) = −4.96–7.02%, $\delta\delta\ ^{18}O_{H2O}$ = 1.27–5.72, and the temperature was 3.44–593 °C, for the I belt, $\delta\delta\ ^{18}O$ (whole rock) = 4.92–5.26%, $\delta\delta\ ^{18}O$ (single mineral) = 5.26–11.95%, and temperature was 170–593 °C, which suggested that the values of $\delta\delta\ ^{18}O$ and $\Delta\delta\ ^{18}O$ decrease along with the increase of temperature, but the values of $\delta\delta\ ^{18}O_{H2O}$ increase along with increase of the temperature.

Geologic setting: Changle-nanao high temperature thermal dynamic metamorphic belt developed along the paleo-continental margin, southeast China, and was probably formed during the period of 35–75 Ma, since the influence of magma intrusion during the process of the fault growing.

XIE Douke, GUO Kunyi, and SHANG Yuqiang, Nanjing Institute of Geology and Mineral Resources, Nanjing, China

Dynamical Features of Lithosphere, Eastern China and Western Pacific: Collision, Breakup, and Growing

The east part of China, located on the west Pacific ocean, has undergone two collision events: the Gondwana continent collision with the Eurasia continent plate, i.e., the Yangtze slab subducted from south to north at 1200 Ma age, and Cathaysian slab obducted on the southern Yangtze plate during the period of 500 Ma.

Unity continent plate was formed in eastern China since 350 Ma. After this tectonic event, it disintegrated beginning in east China. New tectonomagmatic terrane was formed accompanied by a large-scale eruption of the volcanic complex within the break-up belt.

A suite of accretionary prism of ocean-type ophiolite, which occurred on the active continental margin of the south part of the lower Yangtze plate, comprised the first basement since Sinian. Different types of volcanic complex, owing to different basement structural environment, occurred in the Yangtze plate: "Trench type" and "marginal sea type" are characterized by spilite-buschtaunite association; "volcano-arc type" is of basalt-andesite association; "continental margin type" consisted of basalt-andesite-liparite association, and turbidite, jasper rock, and oceanic limestone covered on above volcanic complex.

Break-up of the continental plate during the period of 190–100 Ma, along with the boundary of eastern part of China, took place in the great break-up belt trending north-northeast to south-north. Three kinds of volcanic rock area, accompanied by uplifting of asthenosphere and thermal-pulsating, occurred within the eastern part of the plate during the process of deepening and break-up: (1) lower Yangtze type, which consisted of latite system of silica-saturation to weakly saturated silica, erupted along the east–west-trending basement downwarping belt that formed during the period of Jinning cycle; (2) Nanling type, which is characterized by dacite system of unsaturated to supersaturated silica, occurred in southward growing continental margin and east–west-trending magma-tectonic belt that formed during the period of Caledonia cycle; (3) break-up growth type, which displays some characteristics by liparite system of strongly saturated silica, is newly magma-terrane developed in the continental break-up belt of eastern continental margin. This type of volcanic rock area is different from the lower Yangtze and Nanling types in geochemical composition and forms during the period of 125 to 100 Ma.

The west Pacific Islands, such as Sahalin, Japanese Islands, and Taiwan Islands, were formed by the break-up of continental fragments.

The break-up growth belt of west Pacific Ocean was formed during the period of 75–5 Ma and migrated gradually from the Eurasian continental plate to eastern part of the continental margin to Itoiigawa–Shizuoka tectonic line trending nearly south-north. Sahalin and Northeast Japan–Izu–Bonin comprise a tectonomagmatic terrane.

A large scale of eruption of alkali-basalt and magmatism occurred at the same time, but displayed cross-wise unhomogeneity in the mantle.

The information obtained from geophysical field show features as follows: (1) The whole area from the Chinese continent to Japanese Islands and Taiwan Islands is the continental crust; (2) seismic information suggests that the crustal structure of the east part of China, the west Pacific continental plate and some microplates is different; (3) forming mechanism of break-up belt, east part of China and west Pacific ocean, is controlled by hot-spot and mantle-convection; (4) the break-up belt is characterized by gravitational-high anomaly, low-velocity zone on the top upper mantle "break-up cushion" that occurred in high-hot flow area; (5) the compositions of trace elements and isotopic growth of tectonomagmatic terranes is different from the one of MOB, and magma derived from the break-up belt is of the features of the submantle; (6) the features displayed by magma-differentiation fusion also suggest that upper mantle, geochemical-boundary zone, and hot-boundary zone are uncontinuous in crosswise.

We conclude that the tectonic model of the west Pacific is characterized by the system of continental margin-forward break-up belt-trench, but not by system of island-arc–trench–back-arc basins.

XIE Xuejing, Institute of Geophysical and Geochemical Exploration, Hebei, China

International Geochemical Mapping—Problem of Unification and Diversification During Implementation

International geochemical mapping (IGM) is the inevitable outcome in the historical development of geochemical exploration from a very local scale to regional and then to a national scale. But there are essential differences of IGM with its early stages of development because of its international or global nature. Many problems confronting IGM were not encountered or taken into consideration by most of the world's national geochemical mapping projects. The aims and emphasis, the environments, the traditional methods of approach, the facility and capability of different nations or regions in carrying out geochemical mapping are all different. So, diversification in IGM practices has to be allowed. But in order to make international correlation of the regional data, the element abundance, the large-scale anomalous features, the regional, provincial, and global trends of element distributions, we must have unification in the procedures and approaches of different nations and regions. Such unification is needed to make the data globally comparable. This report proposes the solution of these contradictions.

Sampling method: Wherever possible, stream sediment will be used as the unified sampling medium with minor modification about weight, size fraction, and sample collection methods. But there are a variety of environments where stream sediment survey is not feasible, and other geochemical sampling methods have to be used. According to China's experience, a map showing physiographic-climate zones for the optimization of geochemical method and sampling medium should be compiled at an earlier stage. This map will serve as a guide for the global layout of orientation studies or pilot surveys and also for the systematic search of literature.

Sampling density: Sampling densities have to be diversified according to the dimensions of the area covered in a project, the time and money the interested nations or organizations could afford, the nature of the environment surveyed, and sampling medium used. But an allowable range of sampling densities has to be set in each physiographic-climate-socioeconomic zone for the sake of reference.

Rapid coverage of the earth's land surface: Rapid coverage of the earth's land surface by super-low density sampling proposed by Dr. Bjorklund has been under discussion. The sampling problem of what kind of medium and what order of stream could be representative of a very large area has to be studied. After the solution of this problem, demonstration surveys of this kind could be carried out in the interested nations in order to convince the rest of the world to do the same kind of job. An orientation survey to study this problem has been initiated in China.

Element analyzed: According to our experience, the list of elements analyzed has to be unified in IGM. After a study of the elements analyzed in existing national geochemical mapping projects, elements of first priority and second priority could be chosen for analysis by the nations or organizations concerned, according to their interest and capability.

Detection limit: Many of the elements that are in the lowest end of crustal abundance sequence are key elements in mineral exploration and environmental study. In China, we have made great efforts to bring the detection limits of all elements down below their crustal abundance. This effort has already increased the strategical significance of the RGNR project and changed the outlook for mineral exploration in China. For a global project like IGM, it is necessary to lower the detection limits of elements below their crustal abundance in order to obtain far more important information from the samples. A proposed list of detection limits for the elements analyzed is given and discussed in the report.

Digestion method: Total digestion methods are proposed to be used in IGM in order to obtain the most basic data first. Sample banks have to be established. Samples could be reanalyzed quickly using the partial extraction techniques, if necessary after the problem targets are sorted out for more detailed or specific study. Some partial extraction elements, most important in environmental study are propsoed for inclusion in the list of elements of second priority. The aging problem as it affects stored samples has to be studied.

Analytical methods: Analytical methods used in IGM might be diversified because the facilities and capabilities of different laboratories are different. But the detection limits, precision, and accuracy must be unified. The critical problem at present is that the detection limits of many key elements could not be reached by most existing national mapping projects using the low-cost and high-turnover instrumental analysis capable of determining 30 to 40 elements simultaneously. In China, this critical problem has been solved by the unique combination of these factors: plenty of experienced analysts, highly sophisticated analytical instruments, and a comparatively low wage. About two dozen competent laboratories are working for the RGNR project. Several examples of the multielement, multimethod analytical systems used by these laboratories are cited here, along with their costs and productivities. So, we think Chinese laboratories could make contributions in helping to produce qualified demonstration products outside China, especially in the first five years project of IGM, when new technology and instrumentation that could achieve the abovementioned requirements have not yet been worked out.

Data monitoring system and standard samples: To maintain the consistency and comparability of IGM results over the whole world is very difficult but not impossible. According to China's experience, the choice of analytical methods with small bias by individual laboratories can be fulfilled by participating in the collaborative analysis activity of Chinese standard samples GSD 1-12. And consistent results have been sustained by using a strict data quality monitoring system during mass production of analytical data. The data quality monitoring system used in China is described with some details.

Data processing and map generation: Methods of data processing and map generation are multifarious in different geochemical mapping projects. It seems to be possible to classify the geochemical maps into two categories: the basic geochemical maps, the method of compilation and the format of which must be unified, and the interpretation maps, which could be highly diversified.

XIE Xuejing, Institute of Geophysical and Geochemical Exploration, Hebei, China

A Decade of China's Regional Geochemistry—National Reconnaissance Project

Progress of China's RGNR Project during the past ten years is shown by maps and statistical tables and diagrams. A variety of multielement geochemical maps is demonstrated, including maps of 1:200,000 quadrangles in different environments, maps of several quadrangles combined, and maps of a whole province, showing important Au, Ag, Sn W REE, base metals, etc., trends and promising targets. Several case histories are also presented, demonstrating the role of regional geochemical data in the discovery of new economic deposits, especially gold.

XING Yusheng, LIU Guizhi, YIN Chingyu, YUE Zhao, and GAO Linzhi, Chinese Academy of Geological Sciences, Beijing, China

Evolution of the Terminal Precambrian Biotas and its Geological Significance

The evolution of the Micropalaeophyta (microscopic algae): In the early Terminal Precambrian, i.e., the present concept of early Sinian (800–700 Ma), the main characteristics of the Micropalaoephyta were similar to that of the Qingbaikou Period (1000–800 Ma), with some representatives of Sphaeromorphida as the main elements of the both periods. Although some new genera, such as Megasacculina, Bavlinella, Teophipolia, and Monotrematosphaeridium, appeared during the Early Terminal Precambrian.

In the late Terminal Precambrian, i.e., the late Sinian (700–600 Ma), the Micropalaeophyta changed obviously by the various forms. The newly appeared taxa included Archaeodiscina, Scaphyta, Polyedrixium, Tylosphaeridium, Cymatiosphaera, Fuchunshania, Pseudodiacrodium, Hubeisphaera, Baltisphaeridium, Comasphaeridium, Tianzhushania, Obruchevella, Nostocomorpha, of more than 20 genera. What is the most important is the appearance of the large-sized Balticphaeridium, Comasphaeridium, and Tianzhushania, and the wide distribution of small-sized Micrhystridium in Group Acanthomorphida. This change was very obvious, which shows that in the late Terminal Precambrian, i.e., in Doushantuocunian and Dengyingxiaian stages of Sinian, the Micropalaeophyta had already begun to show some of early Cambrian character. In the early Cambrian, the taxa from Group Acanthomorphida increased greatly and have a wide distribution in Europe, Asia, and America. The basic character of the early Paleozoic micropalaeophyta was founded in this time.

The changes of the macroscopic algae: The Term "macroscopic algae" used in this paper refers to the large-sized (can be seen by the naked eye) multicellular algae belonging to Metaphyta. In China, the earliest reliable macroscopic algae appeared in Gaoyuzhuang formation, Changcheng system, which had a isotopic age of 1500–1400 Ma. They are straight, "S" shaped, or coiled ribbonlike algae, under the name of Sangahuania by Dr. Du Rulin (1986). These fossils are very similar to Crypania from the Belt Supergroup in the USA (M. R. Walter, 1976), and both genera have about the same age. The earliest Chuaria in the world appeared in the strata as early as 1100 Ma. In China, its early representatives were reported by Dr. Xing Yusheng (1982) from the Hongshuizhuang formation, Jixian system in the Jixian county, which has an age of approximately 1200 Ma, a little earlier or about the same age as in the other parts of the world. Chuaria flourished in China in the Qingbaikou Period (1000–800 Ma), in which macroscopic algae Longfenshania and Shouhsienia were found coexisting with Chuaria. The character of the early Terminal Precambrian, i.e., the early Sinian, macroscopic algae flora is similar to that of the Qingbaikou Period. Both are dominated by Chuaria, Shousienia, and Tawuia, with rare specimens of Vendotaenides occurring in a few localities. In late Terminal Precambrian, i.e., the Doushantuocunian and Dengyingxiaian (700–600 Ma) of the Sinian, the algae flora changed drastically. Besides the continuing flourishing of Chuaria, Shouheienia, and Tawuia, the Vendotaenides represented by Vendotaenia and Tyrasotaenia appeared in abundance and had a worldwide distribution. The Epiphytalean flora from Lantian formation (Doushantuocunan stage of Sinian) in southern Anhui province reported by Dr. Xing Yusheng et al. in 1985, is according to the present data, the earliest appearance in the earth history of a large-sized lagal flora with obvious branching and brushlike algal prothallium. This flora includes Epiphyton cf. semirotundum Korde, E. cf. durum Korde, Lantania complexus Xing, Bi et Wang, L. nematoformis Xing, Bi et Wang; Cyathophyton simplicis Xing, Bi et Wang; C. formosus Xing, Bi et Wang; Semisphaerophyton gracile Xing, Bi et Wang; Gordonophyton sp.; Xiuningia lantianensis Xing, Bi et Wang; Paraxiunungia densa Xing, Bi et Wang, etc. The branching algae Entromorphites has also been found in the same period. According to the present study, the brush-formed Epiphytaleans should belong to the Pyllum Rhdophyta, while Entromorphites may belong to the Pyllum Chlorophyta. In the early Cambrian, both the fossil taxa and geographical distribution of Epiphytaleans increased markedly.

The emergence and evolution of the metazoans: In China, the earliest soft-bodied metazoan in earth's history is a worm. Up to the present, the Precambrian worm-bearing horizons in China include Changlingzi formation (723 Ma) in southern Lianing province, Jiuliqiao (younger than 750 Ma), Shijia and Jinshanzhai formations in northern Anhui province. This fauna consists of Sabellidites, Paleolina, and Huaiyuanella from families of Sabelliditidae and Huaiyuanellidae.

In south China, Precambrian worms have been found only in the upper Sinian, mainly in the Dengyingxiaian stage, with the fossil localities confined to the Yangze platform. The worm fossils include Sinotubulites, Micronemaites, and genera of the Family Sabelliditidae.

The appearance of the medusoid fauna is another characteristic of the Late Terminal Precambrian (Late Sinian) faunas. Up to the present, three horizons which contain the medusoid fossils have been found in China. The lower horizon is the bottom of the Nanguanling formation (Upper Sinian) in southern Liaoning province, which is stratigraphically a little younger than the Changlingzi formation which contains the worm fossils. The middle horizon is the lower part of Xingmincun formation in southern Liaoning province. The lower horizon is dominated by the species of Cyclomedusa, while in the middle horizon some new taxa are added to the preexisting Cyclomedusa. The age of the two horizons is 700–600 Ma, so they undoubtedly belong to the Terminal Precambrian. The third (upper) horizon is the Qiongzhusian in eastern Yunnan province, which, of course, belongs to the Lower Cambrian.

Apart from the metazoan fossils mentioned above, some spongespicules have been found from Zhangqu formation in northern Anhui province, Doushantuo and Denying formations in East Yangtze Gorges. And a sea pen fossil Charnia has been also reported from Dengying formation in East Yangtze Gorges.

The geological significance of the biotic evolution: On basis of the facts mentioned above, two big changes in biotic evolution that happened during the Terminal Precambrian (800–600 Ma) can be found. The first one happened some 700 Ma ago, which greatly altered the organic world. Before this change, the organic world had only bacteria and the algae that had rather simple forms and structures, such as Chuaria, Tawuia, and Shouxiania in macroscopic algae and some taxa of Sphaeromorphida in microscopic algae (Micro-paleophyta), represented mainly by Baltisphaeridium, appeared in the vegetable kingdom: in animals, worms and medusoids in the soft-bodied metazoans appeared and began to flourish and the spicule bearing might have appeared. This is a change of global significance. Now, a great many soft-bodied metazoan fossils have been found from the strata of 700–600 Ma ago in many places of the world including Australia, the Soviet Union, Britain, and southwest Africa. These fossils are what we called "Ediacaran Fauna." The world change in plants during this time is similar to what happened in China. So this drastic change in the organic world can be considered biostratigraphically as the beginning of a new geological system. In China, this biological event can be used as beginning of Sinian System: that is, ca. 700 Ma, i.e., the base of Doushantuocunian, should be chosen as the lower boundary of Sinian, instead of the former Lower Sinian boundary. From the biostratigraphic point of view, the new lower boundary of Sinian can also be taken as the lower boundary of the Paleozoic.

The second big change (evolutionary diversification leap) happened during 600–570 Ma, which was marked by the mass appearance of small shelly fossils of diverse groups. This change is now internationally considered as the Precambrian-Cambrian boundary. In China, it is the Sinian-Cambrian Boundary.

XU Huifang, and LUO Gufeng, Nanjing University, Nanjing, China

First Discovery of Superlattice Orthoclase with Large Period and Study of Its Genesis

The authors have found that there are a series of sharp satellite spots around every diffraction spot in SAD of orthoclase, through study with transmission electron microscope (TEM), high-resolution transmission electron microscope (HRTEM), and mineralogical methods for orthoclase ($Or_{65}Ab_{24}An_{11}$) from augite=monzonite (subvolcanic rock) in Wulian, Shangdong, North China. There are more than four satellite spots around each main diffraction spot. All the satellite spots are paralleled to b* direction. This is the first discovery of superlattice orthoclase with large period. The parameters of monoclinic unit cell of the orthoclase are: $A_0 = 8.568$, $b_0 = 12.980$, $c_0 = 7.191$ (Å), $\beta_0 = 116.01°$. Its triclinic degree is: $\Delta_{131} = 0.04$. The extended cell parameters of the superlattice are: $a = a_0$, $b = 8d_{010} \approx 8b_0$, $c = c_0$, $\beta = \beta_0$, and its possible space group is Pm. The discovery of the superlattice orthoclase verifies the correctness of the ordering process of Al^{IV} in tetrahedral site and the existing of orthoclase with intermediate state structure.

The results obtained from bright field image (BF), dark field image (DF), and its selected area diffraction (SAD) with a series of satellite spots prove that there is a long period structure in the structure of the orthoclase along b-axis. The period is about 104 Å ($= 8b_0$). The triclinic domains in the relationship of albite twin law form one-dimensional commensurate modulated structure. There is albite-pericline complex twin in some parts of the orthoclase, also, which constitutes two-dimensional commensurate modulated structure. The superlattice dislocation that may appear in intermediate plagioclase can also be seen in the superlattice orthoclase also.

The results obtained from HRTEM observation show that triclinic domains in the relationship of albite twin law by step of four triclinic domains constitute one-dimensional modulated structure with period of $8b_0$ along b-axis. Even though every unit cell in the crystal displays triclinic symmetry, the superlattice orthoclase shows monoclinic symmetry in average structure.

The superlattice orthoclase is a kind of modulated structure that is formed through the ordering of Al_{IV} in site T, and also is an intermediate state structure formed in the process of phase transformation. The orthoclase that is crystallized under the condition of high temperature possesses characteristics of An-rich in chemical composition and high symmetry (C2/m) in structure. Under special geological condition, two orientation domains will be formed as Al^{IV} in site T_2 segregating into site T_1o and site T_1m, simultaneously. If two of these orientation domains develop simultaneously in proximity to one another, either an albite twin pair or a pericline twin pair will result. It shows albite twin pair mainly in the orthoclase. The superlattice is formed by the regular stacking of two orientation domains along b-axis through the ordering of Al^{IV} in site T. Satellite spots of the orthoclase disappear when the orthoclase is heated for 20 hr at a temperature of 700 °C.

The crystallization temperature of the orthoclase is about 720 °C, which is obtained from Ti-Fe oxides mineral thermometry. The authors consider that the superlattice orthoclase with large period is probably a typomorphic mineral that can indicate subvolcanic genesis. The period and the direction of modulation may vary with changing geological conditions (e.g., temperature, pressure, super-cooling and alkalinity of rock, etc.), which the orthoclase is crystallized in. Similar phenomena have been found in the orthoclases from subvolcanic rocks in Zhejiang and Jiangsu, East China.

XU Renchao, LI Yuan, and ZHOU Chengqun, Bureau of Geology and Mineral Resources, Shaanxi Province, China

Fuzzy Prediction on Regional Slope Stability in Southern Shaanxi Mountainous Areas

Introduction: A model of fuzzy synthetic judgment and analysis is established to predict regional slope stability in the south of Shaanxi, a province in northwest China, with an area of about 85,000 km^2. Geographically, it is a part of Qinling and Bashan mountains with an elevation of 1500–3700 m. The tectonic section is the eastern part of the so-called Qinling–Kunlun latitudinal system, with greatly developed folds and fractures. The precipitation in the district averages 600–1200 mm per year, gradually increasing from the north to the south.

From a system analysis point of view, the system elements, i.e., the influence factors of slopes, may be divided into group factors and state factors. The relative frequency of the slope movement affected by each state factor has been obtained by statistics according to the investigating information.

Algorithm: We assume that X is a fuzzy set of the factor, $x_1, x_2, \ldots$ and x_n are its elements, and that Y is a fuzzy set of the estimation, $y_1, y_2, \ldots$ and y_n are its elements. Then a mathematic model can be established on the principle of fuzzy synthetic judgement:

$$B = (B_D{}'B_s)' = [(A_D R_D)'(A_s D_s)]' \tag{1}$$

Where B is a fuzzy set of the synthetic estimation of estimation of the dynamic factors; B_s is a fuzzy set of estimation of the static factors; R_d is a fuzzy relation matrix of the dynamic; R_s is a relation matrix of the static; A_d is a is a fuzzy subset of weight coefficients of the dynamic; A_s is a fuzzy subset of weight coefficients of the static; and is the composite operation notation of fuzzy matrixes. Then, the parameters of regional slope stability are discussed from the Relative Slave Degrees, weighting coefficients, and so on.

Results and validating: Mapping—The area is divided into about 5000 mesh units at intervals of 3' longitude and 2' latitude, and each unit is about 17 km^2. The static state factors in every mesh are drawn by statistics in the corresponding mesh on the geologic and topographic maps (scale 1:100,000 or 500,000), one by one analogy. As for the dynamic factors, they can be also divided into some "states" in terms of the theory of Grey System (Deng Julong, 1982) and those "states" may be described by Slave Degree as well.

To input all x_1, y_1, and the other parameters into DPS (data processing system) in the computer together, a series of prediction maps can be printed out by means of controlling the different combinations of the dynamic and static state factors. Because of space limitations, the concrete operational examples are omitted herein, besides, some topical expression is only summarized in this paper.

Resulting and validating—The results represent the large-scale evolutionary trends of southern Shaanxi regional slopes. Those combinations of the dynamic often show themselves here. Each predicted subregion's area on the maps is discussed, according to the following reasons that (1) 71.9% of the existing landslides and collapses fall into the predicted unstable subregion; (2) here occur in groups the four catastrophes since 1981, namely, the landslides and debris flows; most of them happen in the predicted unstable or substable subregion; and (3) those maps tally with the subjective estimation of the researcher's by and large; generally speaking, the predicted results are reliable.

XU Yan Qiang, Bureau of Geology and Mineral Resources of Heilongjiang Province

The Basic Characteristics and Economic Significance of the Graphite Ore Belt in the Eastern Part of Heilongjiang Province

The graphite ore belt in the eastern part of the Heilongjiang Province is a crystalline one of great economic value. The probed graphite reserve is much more than 100 megaton in the whole belt. Up to now, we have found three large-scale deposits, each of which has a reserve over 25 megaton.

The belt, distributing in SN direction, is situated on the Jiamusi rise in the eastern end of the Inner Mongolian Great Xing An Fold System. The Heilongjiang Group and the Mashan Group in Early Proterozoic era distribute widely on the rise. Its protolith age is 2251–2539 Ma.

Mashan Group has at least experienced regional metamorphism twice, with the metamorphic age being 1861 Ma and 954 Ma respectively. Graphite ore beds lie in the graphitic metamorphic rock beds (graphite schists, graphite-quartz schists, graphite gneisses, graphite leptynites and so on) of Ximashan Formation and Liumao Formation in Mashan Group. The average content of fixed carbon is about 7%, with a small amount having the content of 10%. The ore contains higher V, and a certain amount of U.

The factors that control the ore-forming of crystalline graphite in the region are as follows:

The sources of carbonaceous materials and sedimentary environment. It has been proved, by means of A. Simonen's diagram and aluminous coefficient, that protolith of Ximashan Formation and Liumao Formation occurred in the graphite ore body belong to fine-grained clastic and carbonaceous sedimentary rocks. Its C-V-U assemblage and sedimentary formations are similar to the black shale that widely spreads in the Lower Cambrian and V and U associations in south China. The organic carbon contents in the shale amount to 5–10%, with some more than 20%.

The values of δ13C ‰ (P DB) in the graphite of the region range from -16.8 to -26.5, which is similar to those of present algae and plankton in Qingdao bay where δ13C ‰ is - 15.6--27.7.

Therefore, it is inferred that organic carbon in the region might come from the procaryotes propagated widely in the Early Proterozoic era. Shallow-sea facies and organic terrigenous clastic carbonate formation have formed under the quite dynamic condition in semi-confined environment. Their paleogeographic environment is considered to be the type of fresh lagoon.

Two stages regional metamorphism has made organic terrigenous-carbonate formation become amphibolite facies and a part of it become granulite facies in the progressive metamorphism at last (500–550 °C, 2.5–3.9 kb). In the process, the organic carbon has turned into crystalline graphite. The second regional metamorphism is stronger in the southern part than in the northern part. As a result, the ore-forming belt has larger lake graphic content in the southern part than that in the northern part.

Graphite deposit is reformed by migmatization and granitification. Migmatiziation has acted on graphite deposit beneficially and harmfully. But it's more harmful than beneficial. For example, remelting process has made graphite in the ore body diluted, and has even made the ore body disappear. Feldspar and quartz components of lower fusion point occur as vein and inject along their schistosity and gneissic schistosity or fissure. This has made gangue composition in the ore and horse-stone in the ore body increase.

The graphite ore belt in the eastern part of Heilongjiang Province is the largest in China at present. Most of those deposits are close to railways, with some large-scale coal fields such as Hegang, Shuangyashan, Qitaihe and Jixi coal fields located nearby, which provides people with better energy resources and convenient transportation conditions to use and exploit graphite resources. Exploiting the graphite resources and making full use of them will no doubt benefit us economically and financially.

XU Yanqiang, Bureau of Geology And Mineral Resources of Heilongjiang Province, China

The Basic Characteristics and Economic Significance of the Graphite Ore Zone in Eastern Heilongjiang Province

The graphite ore zone in eastern Heilongjiang Province is a crystallogenic one with great economic value. The proved graphite reserve is more than 100 megaton in the zone. Three tremendous-scale deposits have been found so far, which is more than 25 megatons.

The ore, ranging from south to north, is situated on the Jiamusi rise in the east of the Inner Mongol-Great Khingan Variscan fold system. The Heilongjiang and Mashan groups of the Early Proterozoic Era are widely distributed over the rise, diagenetic age of which is 2251–2539 Ma.

The Mashan group is the only metamorphic terrane bearing graphite deposits. It composed of leptynites, schists, gneisses, dolomitic marbles, and superior-type iron formation. the Mashan group, with a total thickness of 6300–7800 m, has been divided into the Ximashan, Liumao, and Shizishan formations from the lower to upper. It has undergone at least two periods of regional metamorphis,. their ages are 1861 Ma and 954 Ma, respectively. The metamorphic facies is mainly amphibolite facies, while granulite are found in the southern area. The rock has been widely affected by migmatation, and formed migmatites, wide-ranging migmatitic granites, and granodiorites. The migmatization is divided into two periods, too; the age of the later period is 935–946 Ma.

Graphite orebeds exist in the middle part of Ximashan formation and the lower part of Liumao formation. In space an important ore-forming bed is the Ximashan formation in the south, and the Liumao formation in the north. The orebodies exist in graphitic metamorphic rocks such as graphite schists, graphite-quartz schiste, graphite gneisses, graphite leptynites, and so on, and in a few of graphite marbles and graphite migmatitic granites, and become bedded and lensoid in shape. Because of the effect of late tectonism and migmatization, it appears obvious as widening, pinching, and converging. the average content of fixed carbon is approximately 7%, and locally more than 10% in the ore area. In brief, the tenor of the ores and content ratio of macroplaty graphite in the southern ore zone are higher than the northern and higher V and some U intergrowth with graphite.

The metallogenic factors are as follows:

The source of carbon: By A. Simonen's diagram and coefficient of aluminum, the protolyte assemblage of Ximashan and Liumao formation containing graphite belong to fine-grained clastic and carbonatite sedimentary rocks, and pelitic rocks containing C-V-U assemblages, which are similar to the black shale distributed in southern China at Cambrian System. The organic carbon content is 5–10%, and locally over 20% in this shale, containing a large amount of algae.

Chemical analyses of graphite show that 2.38% organic carbon of the total carbon remained. The values of $\delta^{13}C$% (PDB) of graphite in the area range from -16.8 to -26.5, and are similar to those present algae and plankton in Qingdao bay where $\delta^{13}C$% is -15.6 to -27.7.

Therefore, at Early Proterozoic Era we speculate that the organic carbon is from the procaryotes propagated widely in the area. The organic terrigeneous cast-carbon formation of neritic facies had formed under quiet conditions of dynamic in semiconfining environment. The paleogeographic environment is the fresh lagoon-type.

Metamorphism: There are two regional metamorphisms in the area, in the first, the organic terrigeneous clast-carbonite formation had finally reached amphibolite facies (500–550 °C, 2.5–3.9 Kb) and locally granulite facies in progressive metamorphism, while, the organic carbon had been become the first close graphite (schist facies), and the later crystal graphite (amphibolite facies); in the second, regional metamorphism is stronger in the southern part than the northern, resulting in higher macroplaty crystal graphite in the southern part than the northern.

Migmatization: The migmatization has acted on graphite deposit both beneficially and harmfully. It resulted in largely reformed deposits, progressive reform gave rise to increased ore grades and graphite size and formed a new ore type, while retrogressive reform gave rise to decreased ore grades, broken orebodies, and increased horse-stock.

The graphite ore zone is the largest in China, and the deposits are close to railways, and near to or beside some large-scale coal fields: Heygang, Shuanyashan, Qitiahe, and Jixi.

The graphite has wide use in traditional industries and high-level technical industries, and the need of graphite is increasing in international markets. Full development and use of graphite resources of the ore zone may bring great economic and social benefits.

XU Zhigin, and TANG Yaoqin, Institute of Geology, CAGS, China, and LU Yilun, Geological Bureau of Shanxi Province, China

Formation of Qinling Mountains as Backbone of China: Convergence, Collision, and Intracontinental Subduction

The Qinling Mountains, including Qinling Mountain, DaBei Mountain, and Jiaonan Mountain, the backbone of China, are an east–west-elongated zone, over 1500 km in length, which formed by convergence, collision, and intracontinental subduction between the North China Plate (NCP) and the South China Plate (SCP).

The Qinling Mountains are complex orogenic belts composed of a northern narrow Paleozoic orogenic belt with the age of 450–330 Ma and a southern broader Indosinian belt (210–230 Ma). The tectonic feature of the Qinling Mountains is dominantly characterized by large nappe, large strike-slip, and large detachment.

Nappe tectonics of the Northern Paleozoic belt: The belt has an old metamorphic basement (2500 Ma) and covers from the Middle–Upper Proterozoic to the lower Paleozoic, characterized by fold and large nappe tectonics. The fold structures are simple in the north with Jura-type open folds; southward they become more tight and overturned with a strong slatey cleavage, which were involved in the pennine-type fold nappe as Alpes. In this belt there exist three main ductile thrusts, which are manifested by gentle flow cleavages, A-type folds, nearly north-south down-dip stretching lineation, and mylonitization. The finite strain of deformed oolites, rotational strain, and fabric analysis indicates a simple shear mechanism and a southward shearing direction.

The early Paleozoic ophiolites as oceanic remnants are presented in the southern margin of the belt. A large ophiolitic massif (Shannan) consists mainly of Cr-rich metaperidotites, metagabbro, and metavolcanic rocks as an allochthonous slab that has been thrust on the lower Paleozoic Qinling Microplate. It is estimated that the high-temperature (1300 °C) creep strain rate 6.14 × 10^{14}/s and paleostress deviation 138 Pa in the upper mantle, and the subsequent superimposed low-temperature strain (800 °C), are both related to southward shearing.

Indosinian folds and the double-level detachment structures in the Southern Indosinian belt: The strongly folded sedimentary strata of upper Sinian–Triassic constituted a huge slab about 10- km in thickness, which is related to detachments at the deep-level crust.

The deformation structures are manifested by listric structure in the covers and shear strain in the basement.

The transition from the vertical fan shape in the upper part to the horizontal cleavage in the lower throughout the folded sedimentary slab is progressive, and the multilevel detachment structures and the imbricate structure formed by flant faults and ramp faults are developed in the covers.

The stable stretching lineation (N0 °E–N30 °E), A-type folds, shear, and mylonite are developed within the range of 1500 km from west to east in the epimetamorphic basement. The middle-high metamorphic zone has been found along the main detachment plane, where glaucophane-schists and jadeite-bearing glaucophane schist facies exist.

Deeper detachment structure between the epimetamorphic basement (Pt_3) and the sypometamorphic basement (Pt_1–2) deeper detachment, which is characterized by the stretching (150 °–160 °), A-type folds, the large-scale gentle mylonite zone with the high-temperature (650 °) fabric of quartz, and the high-pressure metamorphic zone composed of glaucophane–chrotoite have been found between both basements in the DaBei Mountain. The magnificent Indosinian imbricate nappe tectonics are developed under the detachment structure.

Formation, evolution of the Qinling Mountains and its place dynamics: The Northern Paleozoic belt—The belt originated in Caledonian–Variscan time.

Caledonian: During the early Paleozoic, a narrow oceanic basin existed most likely at the southern part of NCP. The intraoceanic obduction, as the first convergence, conducted under the high temperature creep strain in the upper mantle. Subsequently, the ophiolites were obducted southward onto the Qinling Microplate and superimposed by low-temperature strain.

The presence of pebbles of various crystalline rocks coming originally from the north in the Middle–Upper Devonian molasse fore-deep between the two belts suggests that the age of the closure of Paleozoic ocean and plate collision occurred before the Middle–Upper Devonian.

Variscan: The intracontinental deformation initiated (1) the large-scale southward thrusting with metamorphism (350–320 Ma); (2) the thrusting migrated northward (less than 350–320 Ma); (3) the formation of radial structures under the north-south contraction; (4) the occurrence of abundant early Variscan granites; and (5) large-scale sinistral movement of the SCP relative to the NCP since the beginning of Carboniferous (319 Ma).

The Southern Indosinian belt—A typical listric structure in the covers (Sinian–Trias) and the shearing strain in the metamorphic basement indicate that the large-scale ductile detachment developed between both covers and basement. Behind the detachment, the deformation accretion prism may reach 20 km thickness, which may have caused formation of high-pressure glaucophane. It is estimated that the crust was shortened with a total displacement of 100–150 km.

The multi-level detachment and accretion prism structure of the crust-scale at the northern margin of SCP under the NCP was perhaps caused by the opening of the southern Indosinian ocean.

XU Zhinghuai, YU Yanxiang, and WANG Suyun, State Seismological Bureau, Beijing, China

Joint Action of Plate Collision and Basal Drag: An Interpretation of the Stress Field of China's Continent

Runcorn (1967) proposed a mantle convection theory from which the drag forces exerted on the base of the lithosphere can be derived. Simplifying the lithosphere as a homogeneous elastic spherical shell, Fu and Huang (1983) calculated the stress field in the shell generated by the basal drag of Runcorn's type and suggested that the drag forces may be a major source for the lithospheric stress field. In Asia, they declared an agreement between the calculated stress and the average stress field indicated by focal mechanisms.

Recently, we have obtained the directional pattern of the principal stress axes on China's continent from both composite and individual focal mechanism analysis. Using this new observational data as a constraint, we have tested whether the drag forces of Runcorn's type can serve as the major source of the stress field of China's continent. The result shows that in western China the agreement between the computed and observed stress directions is rather good, while in other regions, a discrepancy is obvious, especially in the southern part of China's continent.

Simplifying the part of the plate corresponding to China and its surrounding regions as an elastic plate with two layers, we have used a three-dimensional finite element method to simulate the intraplate stress coming from both the lateral actions of adjacent plates and the basal drag of Runcorn's type. An attempt is made to seek a boundary loading model that can give the best fit between the calculated and observed stress orientations. The grid includes 588 nodes and 334 elements. 8-nodes isoparameter elements are used. The dead weight of the lithosphere is not considered. On the northern boundary, displacements in the north-south direction are constrained, while in the east-west direction, they are free. On the western boundary, displacements in the east-west direction are constrained, while in the north-south direction, they are free. On all other lateral boundaries forces are loaded. At the base of plate, vertical displacements are not allowed, and the basal drag forces of Runcorn's type are loaded. The plate surface is free.

Adjusting the direction and relative magnitude of the lateral boundary forces, and also the relative magnitude of the basal drag, a large number of loading models are tested. Applying the superposition principle, we made a large number of tests on boundary forces with limited elementary 3-D finite element computations. Each elementary computation was made by applying a unit lateral force in either X or Y direction on one boundary section, where X and Y are horizontal coordinates. The best fitting model indicates that, under the constraint of the newest stress axis observation, a boundary-loading model with the primary lateral lace action and minor basal drag may be a possible solution to the problem of forces generating the crustal stress field of China's continent. In the best fitting model, the force acting on the India-Eurasian plate boundary doubles that on the others. On the whole, the drag force of Runcorn's type cannot be the major source of the stress field, but its influence on the stress field in western part of China's continent can not be excluded yet.

REFERENCES

Fu, R. S., and P. H. Huang, 1983, Phys. Earth Planet. Inter., v. 31, p. 269-276.

Runcorn, S. K., 1967, Geophys. J. R. Astr. Soc., v. 14, p. 375-384.

XU ZUYE, HuBei Bureau of Geology and Mineral Resources, Wuhan, China

On Modern City Geology

At present, people are more concerned with the changes of the city geological environment with the development of the city construction. There are some geological calamities arising because the human and engineering action have been remaking the geological environment, such as the pollution of water resources, landslide, collapse and etc. The major tasks of the city geology are to research the property, texture, structure, stability, reformation, changes, feedback effect of the geological body localized in the city area. By means of these jobs, the reasonable adviser and programmer will be provided for the plan and distribution of the city as well as the development and coordination of the city function. Therefore, the city geology not only researches the plan and construction range of the city, but also shows the developing prospects of the city.

Since a city is a multiple network system, the research range of the city geology will be extensive. The content includes the basic geology of city, environmental geology, mineral resources, energy sources, water resources and pollution and city administration. Therefrom, an integrated system of function will be formed, that is, city geological system. The characteristics of this system are regional and synthetical, dynamical and protractional as well as functional and practical.

The paper still discusses the research methods and means of the city geology. The author considers the principal characteristics of them to be the high division and synthesis of the multiple subjects, being involved in various subjects, such as geology, geography, resources, soil, environment and tourism. It is consisted of a city system of the geological research and inspection in order to guide and serve the development of the city construction.

To coordinate and unitedly manage the research of city geology better, the author would like to suggest to establish a "Land Department of resources research of the city" in our country.

XU, Xin, Nanjing University, Nanjing, China

A Preliminary Study of the Paleo-Climate and Paleo-Environment in Eastern China since 18,000 Years B.P.

The paleoclimatic and paleoenvironmental changes in eastern China since 18,000 years B.P. are characteristic of the cycles, which are clearly expressed in the paleontology, stratigraphy, sea level, dissected depth, depositional thickness, etc. Based on the paleontology, Quaternary deposits, and chronology in eastern China, the following points are made in this paper:

1. The Dali (Wurm II or Wisconsin II) Glacial Period from 18,000 to 15,000 years B.P.

In the early part of the period the climate was wet and cold. The region was covered by the coniferous forest. However, in the late part of the period the climate with the annual mean temperature of 9–13 °C, which is lower than that of the present, was dry and cold, and the arid steppe landscape was dominated.

The lowest sea level (– 156 m) appeared in the 14,780 ± 700 years ago and the lowest shell ridge in East China Sea was deposited. During that time, the greater part of the continental shelf was exposed and formed the lowest sea-level plain. As a result, the debouch shifted to the sea, and the strong headward erosion of the rivers occurred, alluvial deposits were deposited, and the peat horizon of lake-bog facies was formed on the East-Yellow Sea seabed plain. However, late Pleistocene Mammuthus-Coelodont fauna moved southward to 34 °N in eastern China as the fauna's fossils were found in the Bohai sea seabed. All of these show that during the period the climate was very cold and the environment was very adverse.

2. The warm subinterglacial period from 15,000 to 13,500 years B.P.

The spore-pollen analysis shows that the coniferous and broad-leaf mixed forest contained a few subtropical evergreen broad-leaf trees in the Lower Yangtze River Valley, which means the climate was warmer and wet, sea level rose and slowly the deposition in the small valleys along the Yangtze River occurred. In the Northern China Plain, the deciduous broad-leaf forest and herbaceous hydrophytes increased the dry herbaceous organisms decreased. However, the sea level was lower than that of the present and there were still some parts of the lower sea plain.

3. The Late Glacial Period from 13,500 to 10,500 years B.P.

During this period the climate changed frequently. The spore-pollen analysis shows that in the period there were two warmer periods with the mixed coniferous and broad-leaf forest of the temperate and warm-temperate zones from 13,200 to 12,600 B.P. and from 11,600 to 10,800 years B.P., as well as three colder periods with the temperate steppe and the forest-steppes from 13,700 to 13,200, from 12,600 to 11,600 and from 10,800 to 10,200 years B.P.

These reflect the two climatic and environmental changes.

Five wave crests of Fir appeared in 13,100 ±, 12,700 ±, 12,560 ±, 12,400 ±, and 11,850 ± years B.P. and the broad-leaved forests or the forest-steppes in the four warmer climatic periods between them were found in the Beijing area.

(3) The new period of higher temperature from 6000 to 4000 years B.P.

The spore-pollen analysis shows that the coniferous forest or the deciduous broad-leaf forest-steppe, predominated from 6,000 to 5,500 years B.P. in the Lower Yangtze Valley. The temperature was 2–3 °C lower than that of the present.

In Beijing, the fir and spruce appeared by 56,000 ± years B.P., and the temperature was 4–6 °C lower than that of the present. It may correspond to the second new glacial period.

In the Lower Yangtze Valley from 5400 to 4300 years B.P., the temperature was very warm, and the pollen of the subtropical evergreen broad-leaf trees has been found. The subtropical evergreen broad-leaf forest distributed extensively from the Chandong Peninsula to southern China. Northern China was covered by the deciduous broad-leaf forest. At that time, the annual mean temperature was 2–5 °C higher than that of the present.

A widespread distribution of the hydrophytes shows that there was an extensive water area, the precipitation was abundant and the peat was very well developed.

A widespread distribution of the marine layer marked that there were the transgressions which caused the sea level to rise to 1–2 m higher than that of the present. The discovery of the foraminifers, mussel-shrimps, and diatoms fossils indicated that the seawater was warmer.

(4) The period with the dropping temperature and the descent of sea level since 4000 years B.P.

In eastern China since 4000 years B.P., the climate changed to become cooler and drier. The evergreen broad-leaf trees decreased and the deciduous broad-leaf trees and herbs increased. During the period the Yangtze River delta expanded gradually to the East China Sea, the area of marine deposits on the continental margin shrank, the shell or sand dikes on the coastal zone were moving to the sea, the land surface was dissected again, and greyish-brown secondary loess was deposited in the valleys. All of these show that the climate was becoming cool and the environment was becoming more severe.

However, the fluctuations in climate have continued since 4000 years B.P., and it can be divided into following cold and warm periods:

The cool period from 4000 to 3800 years B.P.;

The warmer period from 3800 to 3000 years B.P.;

The cold period that may correspond to the third new glacial period from 3000 to 2700 years B.P.;

The warm period from 2700 to 2000 years B.P.;

The cold period from 2000 to 1500 years B.P.;

The warm period from 1500 to 1000 years B.P.;

The cold period from 1000 to 800 years B.P.;

The warm period from 800 to 700 years B.P.;

The Little Glacial Period (the fourth new glacial period) from 600 to 100 years B.P.; and

The warmer period from 100 years B.P. to the present.

Y

Y-F. ZHENG, Geochemisches Institut der Universität Göttingen, Göttingen, Federal Republic of Germany, Z-H. Zhang, and W-Z. Shen, Nanjing University, Nanjing, China

Genesis of Reddening in the Dabu Uranium Deposit and Its Relationship to the Uranium Mineralization

The Dabu uranium deposit is located in the contact zone of two granitic bodies inside the Taoshan granitic batholith, south China. The uranium mineralization is determined to have a hydrothermal origin. Periore reddening (including hematitization and geothitization) is observed to be widespread in the hydrothermal uranium mineralization. It is superimposed on alkali feldsparization zone. Calcite-pitchblende and fluorite-pitchblende veins are generally distributed in the reddening halos, but dereddening occurs surrounding the veins. The reddening usually shows an antithetical correlation with pyritization. Investigation of the paragenetic association of alteration mineral assemblages reveals that the reddening is formed before deposition of the uranium mineralization, and the visible pitchblende vein is copecipitated with the pyrite. Physiochemical conditions of the preore alterations are estimated to be of high oxygen fugacity (about 10^{-24} to 10^{-34} atm). The alkali feldspars associated with the reddening have low $\delta^{18}O$ values (+2.9 to +5.6‰), indicating a poor-^{18}O, meteoric alteration solution (having $\delta^{18}O_{H2O}$ values of about -1.7 to +2.0‰ at alteration temperature of 300-400 °C). Thus, the pre-ore solution dominant in the meteoric water component could have high oxidative potential, resulting in the reddening associated with the alkali feldsparization.

In order to decipher the relationship between the reddening and the uranium mineralization in the Dabu deposit, a thermodynamic calculation is performed at temperatures between 150 and 300 °C using combined thermodynamic data of Naumov (1971) and Robie (1978). The following reactions are taken into account:

The reactions (1) and (2) do not proceed in the right direction owing to the positive Gibbs free energy (i.e., $\Delta G_r > 0$), implying that the hematitization cannot result redox between UO_{2+}^{2} and Fe^{2+}. The reactions (3) and (4) also have positive Gibbs free energy, showing that even if H2S exists in a solution containing Fe^{2+}, the uranyl complexes cannot be reduced to form pitchblende. Therefore, the formation of the hydrothermal uranium mineralization needs a given redox environment. The redox properties of uraniferous solution are too strong or too weak to be favorable for reduction and precipitation of uranium. The reactions (5) and (6) possess the negative Gibbs free energy (i.e., $\Delta G_r > 0$). This means that when Fe2O3 and H2S coexist with the uranyl complexes, the pitchblende can then be formed. These are consistent with the geological observation. It implies that the hematited granite is favorable for the hydrothermal uranium mineralization. In other words, when a solution containing both the uranyl complexes and H2S encounters the hematited rock in hydrothermal conduits, the hematite is then reduced to pyrite, concomitant with the reduction of uranium through combined oxidation of UO_{2+}^{2} and H2S. If a solution contains only H2S without FeO3, the H2S would not be oxidized by the uranyl complexes before the solution is contacted with the hematited rock, because reaction (7) and (8) have positive Gibbs free energy.

The conclusions are thus drawn that the reddening in the Dabu uranium deposit results from oxidation of the pre-ore solution to the ambient granites. But the reddening rocks provide a suitable redox condition, whereby the uranium in the solution can be reduced to deposit the pitchblende.

Y-F. ZHENG, Geochemisches Institut, Göttingen, Federal Republic of Germany

Geneses and Evolution of the Taoshan Granitic Batholith in South China: Geochemical Evidences

The Taoshan granitic batholith occurs within a 1100 km2 area and is spatially related to hydrothermal W, Be, U, and fluorite mineralizations. Tectonically, it is situated in the northwestern marginal belt of the paleo-South China-Southeast Asian plate. The development of the batholith can be described in two petrogenic series.

Series I contains major complexes that can be assigned to three petrologic associations: a gneissic migmagranite, a prophyroid biotite granite, and a two-mica granite. These granites are relatively enriched in silicon, higher in potassium to sodium ratio, oversaturated in aluminum, and show a higher total REE content and LREE/HREE ratio. Whole-rock samples are determined to have the higher $\delta^{18}O$ values (10.5 to 13.2‰) and initial 87Sr/86Sr ratios (0.7197 to 0.7233). Calculation of the U-Pb three-stage model for zircons indicates that their protolith is probably the Proterozoic sediments (about 1320 Ma).

Series II is dominated by two coeval granitic associations: biotite granite and adamellite. They contain hornblende. Biotite occurs in the form of isometric flaky. Petrochemically, they are intermediate-acid and calcalkaline, and relatively richer in calcium and magnesium. Whole-rock has the lower $\delta^{18}O$ values (7.8 to 8.2‰) and initial 87Sr/86Sr ratio (0.7124). Mass balance calculations for Sr, Pb isotopes, and rare-earth elements put constraints on their origin, which is related to a contamination/assimilation of mantle-derived magma with upper crustal material. Computation of the crust-mantle lead mixing model shows the identical mantle-crust differentiate age (3198-3232 Ma) and characteristic ‰mix value (9.10 to 9.25).

The series I granites have S-type features and originate from anatexis of the upper crustal material, while the series II rocks have I-type features and may be formed by syntexis of the mantle-derived magma with the upper crust. The coexistence of such two genetic types of granites in the same region is observed to be related to a deep-seated fault situated in the northwestern marginal belt of the paleo-South China-Southeast Asian plate. The Mesozoic subduction of the West Pacific plate beneath the Eurasian continental margin is probably an important process to remobilize materials of the previous plate. In conjunction with this process, the regional deep-seated fault is produced, which leads to remelting of the crustal materials and mixing of the mantle-derived magma with the crustal component.

The following preliminary model is postulated to account for the geologic evolution of the Taoshan region. The crust-mantle differentiate and ancient continental crust formation occurred in the Archaean (about 3200 Ma ago). The flysch formation of the Sibao Group was formed during 1100-1400 Ma by sedimentary cycle. Then it was deeply buried underground owing to subduction of the paleoplate in the Sibao age, constituting the protolith of the series I granites. During Late Proterozoic and Cambrian, a geosynclinal environment was formed in this region to deposit a suite of huge thick sandy, muddy, and silty sedimentary unit. Toward the end of Cambrian time, folding-uplifting and regional metamorphism took place under the influence of the Yunan Movement. Subduction of the paleoplate in the Late Caledonian converted the previous sediments into metamorphic rocks, migmatite, and migmagranite. During Hercynian and Indidsinian cycles, this region uplifted. In Early Yenshanian, the subduction of the West Pacific plate beneath the Eurasian plate caused the generation of north-northeast and northeast deep fractures. During this time, partial fusion of the basement occurred to provide the source of magma for numerous syntectonic intrusions, such as the series I granites. With development of the tectono-magmatic activity, magmatic process reached its peak in this region, accompanied by an increase in heat flow, crustal thinning and generation of the deep-seated fault, so that the mantle-derived magma could ascent and assimilate the crustal materials to form the series

II granites. To Late Yenshanian time, the rate of subduction of the plate decreased, and the continental crust was in the state of relative relaxation. So tensile tectonics were produced to initiate the development of fault basins and sedimentation of redbeds, which constitute covering strata of the batholith and the metasedimentary unit. Between Late Mesozoic and Early Cenozoic, the fault tectonics in this region were reactivated to produce new fractures of different patterns within the batholith and its edges, accompanied by the formation of intermediate-basic volcanic rocks and the hydrothermal mineralizations.

YAGHUBPUR, A., Tarbiat Moalom University, Tehran, Iran

Zagros Crush Zone: Proposed Hypothesis for Action of Early Miocene Rifting in Early Carboniferous Time

The region containing the Gulf of Aden, Red Sea, and the East African Rifts has been so far the type example of an early stage in the development of continental drift. Apparently, the assumption is that the true oceanic basement in the Red Sea ranges in age from 30 to 40 m.y., and the basaltic flow that has been available by coring is much later in age. It is also suggested that the opening of the Red Sea started sometime in Early Miocene time.

The Zagros Thrust Zone, located in western Iran, which trends northwest-southeast for about 1300 km from the Iraq border to the Hormoz Region in south-central Iran, is an arcuate belt that is supposed to be the result of a collision between the Arabian Shield and central Iran. This structural feature is presumably located in the place of an earlier rift that supposedly began opening sometime in Early Carboniferous time.

The geological, geophysical, tectonic, and geographical maps of the world, southwest Asia, and northeast Africa show a subparallelism between the Zagros Crush Zone and the Red Sea that indicates the subparallelism between the Red Sea and the Rift which was located in the place of present Zagros thrust fault. This phenomenon may be merely a coincidence, but it could also be interpreted as a similarity in the origin of these two features. In other words, the causative mechanism that produced the opening of the Red Sea during the Early Miocene was similar to that responsible for opening the Zagros rift in the Early Carboniferous time.

YAGHUBPUR, A., Tarbiat Moslem University, Tehran, Iran

Possible Oil Shale Potentials in Iran

Oil shale is a relatively new potential source of vast quantities of liquid fuels. These sedimentary rocks contain various amounts of solid organic materials that can be decomposed by heat into a liquid resembling petroleum gases and a carbonaceous residue.

Two localities in Iran have so far been recognized as potential oil shale resources. Structurally these shales are: (1) a member of middle Cambrian age sediments, and (2) a member of Oligo-Miocene age sediments. Although there are several outcrops of these two members in different parts of Iran, significant indications of bituminous materials associated with shale are observed in two localities: (1) part of Mila Formation (Cambrian age) in eastern Alberz, in northeast Iran (Khosh-Yeilagh region), approximately 10 to 20 m thick, and (2) C_2 member of Qom Formation (Oligocene-Miocene age) in central and north-central Iran, approximately 15 to 25 m thick. The Cambrian member could be traced for several kilometers, and the C_2 member in just one location covers a large area more than 50 kilometers long. These two places are located in more or less remote areas, so there would be few environmental problems if they were to be exploited in future.

Preliminary examinations have shown that these two shales are associated with some 25 to 30 gallons of oil per ton. The rocks of Cambrian age are mostly composed of dolomite and clay minerals, and the C_2 member of Qom Formation is a shale containing mostly clay minerals and carbonates.

YAGI, T., and Y. TSUCHIDA, University of Tokyo, Tokyo, Japan

High-Pressure Phase Transformations in Silica

Phase transformations in silica are very important in considering the stable mineral species in the earth. So far rutile-type polymorphs of silica, stishovite, are known to be the most dense form of silica. Although further transformations into either α-PbO_2 type or Fe_2N-type structures are reported either by shock compression or static compression, no stability fields are confirmed in these two polymorphs. Based on the systematics of the phase transformations in AX_2 type compounds, it is generally considered that rutile-type structure may ultimately transform into fluorite structure. Recent theoretical study, however, clarified that fluorite-type SiO_2 has much higher energy than that of stishovite, and pyrite-like structure is the candidate for the post-stishovite transformation (Park et al, in preparation).

In the present paper, an experimental study on the pressure-induced phase transformations in silica is reported which was performed up to 110 GPa using diamond anvil cell and laser heating technique. Quartz, glass, and cristobalite were used as starting materials. In order to study the stable phases, equations of state, and quench products, X-ray diffraction studies were performed three times for each sample. First measurement was made after the sample was compressed to the desired pressure at room temperature. Second measurement was made after the sample was heated under pressure by a YAG laser. These two high-pressure in situ observations were performed using film technique, and exposure time for each film was about 3 days. The last observation was made on quench products recovered from high pressures.

Diffraction lines observed after the heating are summarized in Figure 1. It is clear from this figure that up to 110 GPa, no large change of the diffraction pattern from stishovite into other phase was observed. Above 60 GPa, however, discrepancy between observed and calculated d-values become larger and larger when tetragonal rutile-type structure was fitted to the observed diffraction pattern. An orthorhombic $CaCl_2$-type structure, which is formed by a slight distortion of rutile structure, can explain the observed diffraction patterns successfully, but further study is required to clarify the real nature of this transformation.

All the quench products were well explained by stishovite throughout the entire pressure range. No difference of the experimental results was found between two starting materials, quartz and glass. It was also found that when quartz or glass was subjected to a pressure above about 60 GPa, stishovite was formed even without any heating of the sample. It was reported that when quartz or coesite was compressed at room temperature, it transforms into amorphous solids at 25–35 GPa (Hemley et al, 1988). Present results indicate that amorphous solids thus formed are gradually crystallized into stishovite under very high pressures. When cristobalite was used as a starting material, transformations to unknown structures were observed at about 10 and 30 GPa at room temperature. Room temperature transformations in silica, therefore, depend strongly on the crystal structure of the starting materials.

REFERENCES

Hemley, R. J., A. P. Jephcoat, H. K. Mao, L. C. Ming, and M. H. Manghnani, 1988, Nature, v. 334, p. 52.
Park, K. T., K. Terakura, and Y. Matsui, in preparation.

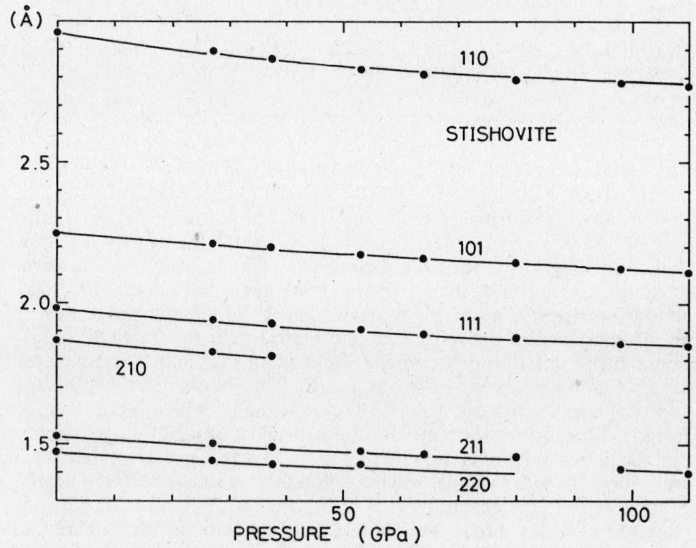

Figure 1—D-values of the X-ray diffraction patterns of silica observed after laser heating under pressure.

YAIR, A., Hebrew University of Jerusalem, Jerusalem, Israel

Climatic Changes and Environment in Northern Negev Desert

The relationship between annual rainfall amount and the physical and biological properties of the environment in the desert fringe attracted the attention of workers in many disciplines. Climatologists use aridity indices to express this relationship. These indices tend to imply that the acuteness or aridity in the desert fringe is inversely correlated to annual precipitation; annual rainfall amount being considered as the predominant factor in water availability. The same trend is assumed by pedologists and botanists. They assume better leaching of salts from the soil profile and increased biomass production with increasing rainfall. When soil erosion processes are considered, a positive relationship between annual rainfall and denudation rate is assumed by many authors up to 300 m of annual rainfall.

All these assumptions encounter serious difficulties when applied to the northern Negev desert area where a converse relationship, between annual rainfall and environmental variables, seems to fit the field conditions better. A detailed paleogeographical-ecological study of the evolution of the northern Negev (where annual rainfall now ranges from 70-250 mm) covering the last glacial period to the present points to a direct relationship between annual rainfall and aridity in terms of soil-forming processes, composition of the vegetation, and human activity. Parallel to this, erosion rates decrease from 70 mm to 250 mm annual rainfall. The explanation proposed for such a situation is that water availability and runoff energy in the study are strongly controlled by the efficient transformation of rainfall into runoff; annual rain amount being only a secondary factor. As rain events in the study area are scarce and low to very low in their amounts, while evaporation losses are very high, only the concentration of water from a large into a small area can allow, within the latter area, a deep infiltration of water beyond that allowed by the direct rainfall. The deeply infiltrated water is an efficient agent in the leaching of salts from the soil profile and in the development of a water reservoir, available for plant growth, at a depth below that influenced by evaporation from the topsoil.

The process of water concentration described above can be achieved in those areas where the efficiency of transformation of rainfall into runoff is spatially nonuniform due to nonuniform surface properties. In that respect, the study area is characterized by two basic surface units: rocky areas composed of dense limestone, where the soil cover is patchy and shallow, and soil-covered areas, where the soil is contiguous and thick. Detailed hydrological studies conducted on both surface units indicate that the soil-covered areas are characterized by a high infiltration rate and therefore by runoff and erosion processes of low frequency and magnitude. At the same time as rainstorms are infrequent and low in their amounts the depth of rainwater infiltration is limited. The infiltrated water is quickly lost by evaporation leading to the development of a dry and saline environment. On the other hand, rock areas, which have a low infiltration capacity, respond quickly to rainfall and are characterized by a high frequency and magnitude of erosion events. The runoff generated over the rocky areas is quickly absorbed by soil strips and soil pockets within the rocky areas and downslope of it, leading to the local development of well-leached soils and to the formation of a water reservoir at depth, available for plant growth. As extensive rocky areas prevail in the southern climatologically dry part of the region and soil-covered areas in the wet north, a negative correlation between annual rainfall and environmental aridity is obtained.

Of great interest is the fact that the extensive soil cover in the north is not derived from local weathering processes. It is represented by a thick eolian loess unit that penetrated into the area during the last glacial period when wetter climatic conditions prevailed in the area. However, due to loess penetration the former rocky surfaces were gradually covered with a soil mantle leading to a decrease in the extent of bare rocky surfaces. The drastic change in the surface properties led to a decrease in runoff and thus to a decrease in water availability expressed by the development of saline soils and limited biological and human activity. The decrease in runoff reduced the frequency and magnitude of erosion events. As loess penetration continued, the former drainage network was almost completely buried under the thick loess cover, and saline lakes, where gypsum was deposited, developed.

In summary, data obtained show that two important factors controlling the rate of geomorphic processes changed during the last glacial period: first the amount of rainfall, and second, the extent of soil cover; both of which increased. The analysis of the influence of each of these factors on landscape evolution clearly indicates that the increase in soil cover resulted in an increased aridity and in a general obliteration of the preexisting drainage network, predominating over the possible positive effect of rainfall increase on environmental variables and on erosion rates.

YAMADA, E., Geological Survey of Japan, Ibaraki, Japan

Postcaldera Volcanotectonism in Onikobe Caldera, Japan, Compared with That of Some Other Calderas

In assessing the hydrothermal resources and the mineral resources associated with calderas, it is important to clarify the nature of postcaldera volcanotectonism, which is also important to clarify geological structures of caldera regions.

The Onikobe caldera in northern Honshu, Japan, is a silicic caldera about 10 km across, collapsed about 1 km and at about 1.7 to 2.7 m.y. ago (Yamada, 1988). Its postcaldera volcanotectonism is very peculiar.

In the northeastern part of the caldera, a rectangular block about 2 × 3 km across of the precaldera formations was uplifted about 1 km after the caldera collapse. As a result, the caldera-fill has been domed surrounding the block. The doming was accompanied by debris flows, slumping, slide faults, etc. In the final stage of the doming, debris avalanches were also generated. The uplift took place probably from about 1.0 m.y. to 0.1 m.y. ago, at first slowly and in the final stage very rapidly. This uplift was probably caused by subsurface intrusion of silicic viscous magma, but no surface volcanism took place in association with the intrusion.

Meanwhile in the southeastern part of the caldera, distention of the caldera floor took place. As a result, the caldera fill has been cut by extensional joints, clastic dikes, and highly inclined minor normal faults with a throw commonly less than 1 m. The distention probably occurred due to intrusion of silicic magma, part of which has been extruded to the ground surface and has formed a rhyodacite lava dome. The directions of those fractures were primarily controlled by the fracture systems that preexisted in the basement and the local stress field generated by the intrusion of the magma but not by the so-called regional tectonic stress field at the time. The fracture system formed also from 1.0 m.y. to 0.1 m.y. ago.

In Valles type calderas, general doming of the caldera floor with associated fracturing has been reported as the common type of postcaldera tectonism. In some calderas, quite different postcaldera tectonism has been reported; for example, in the Platoro caldera the core of the collapsed block was uplifted to form an unbroken homoclinally tilted block, and in the Hakone caldera postcaldera volcanism was associated with the dike swarm formation. In the Onikobe caldera, the postcaldera volcanotectonism was the uplift of part of the foundation of the caldera and the distention in another part of it. Thus, the nature of postcaldera tectonism varies greatly in each caldera depending probably on the preexisting upper crustal structures and the nature of associated magmas.

REFERENCE

Yamada, E., 1988, Geologic development of the Onikobe caldera, Northeast Japan, with special reference to its hydrothermal system: Report of Geological Survey of Japan, n. 268, p. 61-190.

YAMAGISHI, H., Geological Survey of Hokkaido, Sapporo, Japan

Submarine Lavas/Dikes and Associated Volcaniclastic Rocks in Green Turf Belt, North Japan

The North Japan Green Tuff Belt, corresponding to island-arc orogenic zone, includes abundant volcanic rocks of mafic to felsic through intermediate compositions. In particular, the Oshima Peninsula and Shimokita Peninsula are dominated by Neogene submarine lavas/dikes and volcaniclastic rocks. These volcanic rocks can be considered primary structures and facies because of their young age and lack of alteration.

The submarine lavas are subdivided into pillow lobes, lava lobes, sheet flows (massive lavas), and pseudo-pillow lavas (Table 1). The volcaniclastic rocks are classified into hyaloclastites, pyroclastic rocks, and epiclastic volcanic rocks. Particularly the hyaloclastites are predominantly distributed.

The pillow lobes are cylindrical or tubular and mostly andesitc to basaltic. Their surface structures are recognized as ropy wrinkles, spreading cracks, tensional cracks, and contraction cracks. The former three are formed during the growth of the pillow lobes, whereas the latter two are made after emplacement. The ropy wrinkles are similar to those on a subaerial pahoehoe lava. The corrugations are formed by internal convection or scratching. The spreading cracks are formed across or along the elongation of the pillow lobes, and play the most important role in their growth. The tensional cracks are produced by expansion of the pillow lobes due to further supply of the interior magma after emplacement of the pillow lobes. The contraction cracks are produced by gradual cooling of the pillow lobes.

The internal structure of the pillow lobe is characterized by joints, crusts, vesicles, and hollows. The joints correspond to the contraction cracks and are radial, irregular, and tortoise-shell in shape. The crusts are produced by shear jointing during the growth of the pillow lobes. The vesicles are ellipsoidal to spherical. The inner vesicles are larger than the outer ones. Pipe vesicles develop in places along the border zone of the pillow lobes. Hollows can be observed in the core of some of the pillow lobes. Microscopic features of the andesitic pillow lobes are of intersertal or intergranular texture in the inner part, and quench crystal overgrowth on the phenocrysts or microphenocrysts in the outer part.

The lava lobes are characteristic of andesitic to rhyolitic subaqueous lavas. They are ellipsoidal a few meters across and are composed of crystalline cores and glassy rims. The cores are characterized by flow layers conformable to the outlines and radial rude joints. In places, the cores have vesicles and hollows. The glassy rim is considerably thick and grades outward into hyaloclasites through in situ breccia. Small lava lobes develop from the glassy rim and are scattered around the lava lobes. The lava lobes are formed by expansion of sheet flows due to further supply of the internal magma.

Sheet flows are tabular-shaped bodies with columnar joints. They are a few tens of centimeters to a few meters thick. They have glassy margins on the top, bottom, and toe. The margins usually grade outward into hyaloclastites through in situ breccia.

The pseudo-pillow lavas are one of the andesitic to rhyolitic subaqueous forms. They consist of polyhedral blocks closely fitted with each other. They are tens of centimeters to a few meters across. They are formed by developing of rude quench jointing in massive lavas or dikes. Most of the polyhedral blocks have been, however, spalled off from the main massive lavas or dikes and then transported by avalanche or debris flow of hyaloclastite.

The hyaloclastites are divided into (A) and (B) types. The (A) type is basaltic to andesitic monolithologic breccia and usually includes pillows (ellipsoids with glassy margin). These pillows are formed by disintegration of pillow lobes or apophysis-like feeder dykes. Particularly, those spalled off from the apophysis-like feeder dykes have concentric jointing; they are called concentric pillows. The (B) type is andesitic to rhyolitic monolithologic breccia. It is a product formed by brittle fracturing of the pseudo-pillow lava, massive lavas, and massive feeder dyke, caused by quenching with water.

Massive hyaloclastites are accumulated around the lavas and feeder dikes and grade laterally into stratified hyaloclastities that are foreset-beds showing a primary dipping of 20°-30°, in places overlain by topset-beds. Each foreset-bed displays reverse or reverse-to-normal grading, suggesting that it was formed by debris flow.

The Neogene subaqueous lavas/dikes and hyaloclastites in Green Tuff Belt in North Japan are highly vesicular and associated in places with submarine pyroclastic rocks, such as scoriaceous agglomerates and pyroclastic pillow breccias, suggesting that they are the products of shallow submarine volcanism.

Table 1—Morphological features of pillow lobe.

	pillow lobe	sheet flow (massive lava)	pseudo-pillow	lava lobe
Rock type	basalt, px andesite	basalt, px andesite, rhyolite	px andesite, ho andesite	px andesite, rhyolite
Size	a few tens centimeters to a few meters across	a few tens centimeters to a few meters thick	a few tens centimeters to a few meters across	a few to ten meters across
Shape in three dimension	cylinder, tube	tabular	polyhedron, ellipsoid	ellipsoid
Feature of surface	ropy wrinkle corrugation, spreading crack	in-situ breccia	even, tortoise-shell joints	in-situ breccia
Chilled margin	thin glass, tiny joint	thin glass	tiny joints	thick glass
Inner joint	radial-columnar, tortoise-shell joint	columnar joint	radial-columnar, tortoise-shell joint	radial-columnar joint
Flow layer	conformable to outer surface	conformable to outer rim	oblique to outer surface	conformable to outer rim
Amount of vesicles	abundant	common	rare or common	common
Arrange-ment of vesicles	elongated, normal or in parallel	conformable to outer rim	oblique to outer surface	conformable to outer rim
Shape of vesicles	pahoehoe type	pahoehoe and aa type	aa type	pahoehoe and aa type
Hollows	common	rare	rare	common
Associated hyalo-clastite	(A) type	(A) and (B) type	(B) type	(A) and (B) type

px: pyroxene, ho: hornblende

YAMANAKA, T., Osaka University, Osaka, Japan

Structure Refinement and Kinetic Study under High Pressures and Temperatures by Energy Dispersive Powder Diffraction Spectra Using Synchrotron Radiation

An attractive X-ray light source of synchrotron radiation (SR) has been utilized for diffraction studies applied to mineral physics and other broad geoscience fields.

X-rays emitted from the bending magnets or insertion devices (e.g., undulater, wiggler) offers the following outstanding properties for diffraction studies in comparison with those from an ordinary bremstrung X-ray source: (1) great brightness, (2) highly collimated beam, and (3) smooth polychromatic (white) X-ray.

Structure refinement at high pressure: The SR source is composed of a continuous energy that enables us to utilize a desired wavelength by monochromator. The SR white X-ray is also advantageous for energy dispersive (ED) diffraction studies, in which an optical alignment of the assembly can remain fixed under various controlled conditions around the sample, as many ED experiments under high pressure have been reported.

A new X-ray powder diffractometer was designed for multiple uses of SR source (2.5 Gev 300 mA) at Photon factory, Tsukuba (Yamanaka et al, 1983). A diamond anvil pressure cell was set on the spinner with 1/20 rad/sec, which was attached on the goniometer. The spinner was simultaneously oscillated within 5° around the ω axis, which was prepared for the randomization of the sample orientation and for reducing an effect of the grain growth. A platinum ring heater around the diamond anvil and another heater are installed in the pressure apparatus. The later was prepared for a subheater because the pressure cell is built of a good heat conductor.

Integrated intensity of diffraction peaks—A profile decomposition method of the ED spectra can separately assign integrated intensities to the superposed diffraction peaks. The energy positions E_H of the ED diffraction peaks H_i can be determined from the energy conversion equation. Integrated intensity I_H of the reflection H can be derived by structure factor F_H, when the SSD is set at 2θ:

$$I_H = I_o(E_H) \cdot p \cdot [(1+\cos^2 2\theta)/\sin^2 2\theta \cdot \cos\theta]$$
$$I\, F_H I^2 \cdot \exp(-2B \cdot E_H{}^2 \sin^2\theta / h^2 c^2) A(E_H)$$

Diffraction intensity $V_O\,(E_i)$ at each energy is derived from the integrated intensity I_H in accordance with the following formula:

$$Y_0(E_i) = B(E_i) + \Sigma Y(2\Delta E_{ih}) I_H.$$

where $B(E_i)$ is a background, and $Y(2\Delta E_{iH})$ is a profile function.

A new FORTRAN 77 program EDSYSTEM introduced the pseudo-Voight profile function to ensure higher accuracy. The function is expressed by the sum of Gaussian function and Lorentzian function. The program takes into account many intensity-correction parameters of source intensity distribution, background, escape efficiency, asymmetry peak, absorption in terms of energy. The structure parameter were optimized by the least-squares method using EDSYSTEM, on the basis of the diffraction intensity $YO(E_i)$ of each energies observed in 4096 channels of multichannel analyzer.

Structure of GeO_2 at high pressures—ED spectra of GeO_2 polymorphs, trigonal α-quartz type structure ($P3_221$ z = 3) and tetragonal rutile type ($P4_2/mm$ z = 2), were measured under high pressures and temperatures using a diamond anvil pressure cell with a platinum ring heater.

Two sets diffraction intensities y(Ei)(i = 1 ~ 4096 channel) measured at 13.0° and 20° in 2θ were treated with the above-mentioned corrections.

From the least-squares pattern fittings at several pressures and temperatures, variations of Ge-O bond lengths in both structures were clarified as a function of pressure. An example of the fitting is shown in Figure 1.

Kinetics of polymorphic transition: The SR having a great number of photons gives a brightness 10^4 times stronger than the common X-ray source and provides a high resolution of profiles due to a large S/N and low background. A small amount of sample and a small region of domain and exsolution can thus be detected and analyzed. The strong source intensity is capable of shortening the measuring period, which is suitable for the kinetic studies of transition, decomposition, oxidation, dehydration, recrystallization, and melt.

An in situ observation of the polymorphic transition of GeO_2 under 30 Kbars at several temperatures has been performed using ED spectra observed for every 15 min. The transition curve was composed of three stages; (1) a considerably long inductance period, (2) an initiation period of the rapid growth of the transformed phase, and (3) the conversion tail of the slow growth. The transition rate from α-quartz type to rutile type structure was calculated with the solid-solid reaction equation. The kinetics of the transition of GeO_2 is confirmed by the transition-curve fitting to several models of Jander, Avrami-Erofeev, and Prout-Tompkins equation. The transition mechanism was estimated from the fact that the Avrami model was best fitted to the curve (Figure 2). The growth rate rather than nucleation rate resulted in the rate determinative of the transition. An apparent activation energy of the transition was also calculated by the Arrhenius equation.

Compressibility and thermal expansion coefficient: Compared with the para-focusing beam from the ordinary X-ray tube, the SR is an extremely collimated parallel beam (about 0.2 mm rad), and the diffraction profile brings a very narrow full width at half maximum. The sharp profile provides a high angular resolution and ensures a precise measurement of cell constant of the sample. An accurate analysis of mixture, solid solution, or exsolution can be expected.

SR white X-ray emitted from wiggler has a practically wide energy range (0.35 ~ 70 KeV) in comparison with the SR from bending magnet, which is a great advantage for ED diffraction studies through diamond anvil pressure cell. Lattice constants observed under variable pressures elucidates a compressibility and a bulk modulus.

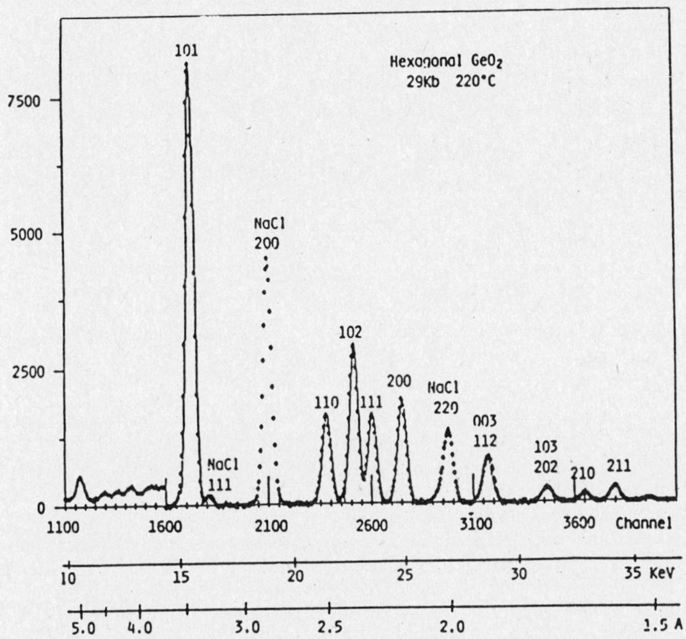

Figure 1—The observed and calculated spectra are represented by dots and solid line, respectively.

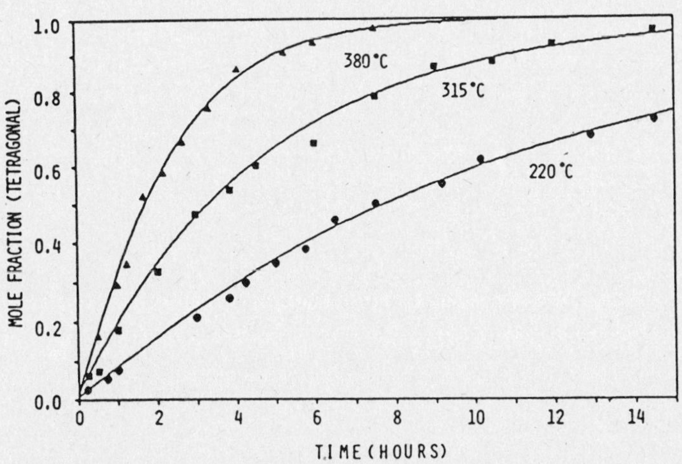

Figure 2—Fitting of the Avrami-Erofeev model to the ED spectra at 30 Kbars.

YANG Qi, PAN Zhigui, LI Zhen, WANG Zhenping, China University of Geosciences, Beijing, China

Accumulation and Evolution of the Late Paleozoic Coal and Coal-Baring Formation in North China

The Late Carboniferous Taiyuan Formation and Early Permian Shanxi Formation are widely developed in north China. In the southern belt of north China, the Shihezi Group of late Early Permian and early Late Permian bears coal also. The topographic feature in north China then was higher in the north, the sea was limited to the south, so rivers ran mainly southward. The lithofacies zones of the Paleozoic sediments trend latitudinally. During the Late Carboniferous on the south part of the northern belt, there was a coastal area cut by rivers—an environment favorable for sustained development of peat bog. The middle belt of north China was often transgressed, so that environments such as bays, delta plains, and tidal flats developed. In the south part of the middle belt were lagoons and barrier bars, and rather thick coal formed. In the southern belt, the Taiyuan Formation, a mixed coastal deposit, consists of epeiric carbonate and clastics; in the east area, deltas, lagoons, barrier bars, and thicker coals developed, while on the west area only thin coal or shed coal were formed on the tidal flats owing to too frequent marine transgression. On the south margin of north China, lagoons, barrier bars, and thicker coal seams occurred under tidal influences.

During the hanxi epoch with the gradual regression of the sea, peat formed near river channels and of the floodplains in the northern belt and the north part of the middle belt and the southern belt peat bog occurred extensively on the tidal flats. As the shoreline had retreated southward successively, the coal accumulation also extended southward. Coals of both the Taiyuan Formation and Shanxi Formation were better developed trending latitudinally along the shoreline. The river-dominated and tidal-influenced shallow-water deltaic environment appeared in the upper section of the Shanxi Formation, and the coal accumulation was weak. Further on, the early transitional deltaic environments of the Shihezi epoch developed in the south belt where numerous coal seams were later formed.

In short, the Late Paleozoic coal accumulation mainly occurred on alluvial plains, deltaic plains, lagoons, and tidal flats. As the shoreline retreated southward during the Late Paleozoic, so thick coal seams of the Taiyuan Formation are now mainly distributed in the northern belt of north China; those of the Shanxi Formation occupy the middle belt and the northern part of the southern belt; those of the Shihezi Group are limited to the southern margin of the southern belt.

Late Paleozoic coals in north China are mostly humulith. The Taiyuan Formation coals were formed in the coastal region, so they contain a high percentage of vitrinite and sulfur. The vitrinite being mostly collinite and partly perhydrous; while coals of the Shanxi Formation were formed in an environment transitional from marine to terrestrial facies. The vitrinite and sulfur contents are less but inertinite more as compared with Taiyuan coals, and the vitrinite is mostly telinite.

The coal ranks in the central and southern belts of north China are much higher than those in the northern belt and the southern margin. Such a feature is the combined result of three stages of coal metamorphic evolution. During the stable stage of the north China platform, the Late Paleozoic coal evolved through geothermal metamorphism only to low rank except in individual subsiding belt. In the early mobile stage of the platform owing to the strong intrusion of the Yanshan Movement, part of the low rank coal was promoted to medium and/or high ranks through telemagnetic metamorphism. The telemagnetically metamorphosed coal is controlled by the Yanshan magmatic intrusions, which intruded discontinuously along three latitudinal and two north-northeast-trending zones. In the late mobile stage, the Zijingguan Fault moving dextrally in the Tertiary removed the latitudinally distributed high rank coal zones east to the fault from several tens of kilometers to about 100 km southward.

The distribution of Late Paleozoic coal of various ranks in north China is predicted on the basis of geological analysis and types of coal metamorphism, and predicted reserves of coking is increased.

YANG Wenjin, WANG Liankui, ZHANG Shaoli, and XU Wenxin, Academia Sinica, Guiyang, Guizhou Province, China

The Geochemistry of Micas from Granites and the Evolution of Two Petrogeno-Mineralization Series in South China

Micas, including biotites and muscovites, crystallize through the petrogeno-mineralization evolution of granite and are widely distributed over the complex of granites in south China. It has been proved that, based on the data of chemical components, crystal chemistry, trace elements, REE, and DTA of approximately 100 micas, the granites of south China can be divided into two petrogeno-mineralization series (Nanling Series and Yangtze Series) by means of the geochemical criteria of micas.

The evident differences can be found between micas from the two series of granites. The typical mineral assemblage concerned with mica is Fe-biotite + muscovite + ilmenite + silicate minerals rich in Al (cordierite and garnet) in the Nanling Series, and Mg-biotite + hornblende + Magnitite + silicate minerals poor in Al (epidote and sphene) in the Yangtze Series. The chemical components of biotites from the Nanling Series granites are rich in Al_2O_3 (>15%), FeO (>20%), Li_2O (>02%), F (0.74%, average value), but relatively poor in Mg, Ti, Ca, Cl; while those from the Yangtze Series granites are rich in MgO (>10%), TiO_2 (>3%), but relatively poor in Al, Fe(+2), Li, and F. The parameter M (M = Mg/ $Mg+Fe^{2+}$ +Mn) of biotite can be regarded as a reliable criterion to distinguish the genesis of granite in Mg-Fe^{2+} +Mn-Al^{vI} +Fe^{3+} +Ti^{vI} triangle, the biotite from the Nanling Series plotted in Fe-biotite and siderophyllite area and their M<0.45; while that from the Yangtze Series plotted in Mg-biotite area and their M<0.45. In F/Cl-OH/O illustration, the biotites from the Nanling Series reflect the magma relatively higher in fH_2O and lower in fo_2, but the opposite is true of the Yangtze Series. This observation coincided with the calculation results based on Fe^2/Fe^{2+} +Mg of biotite, fH_2o 500–1200 bars, and fo_2 10^{-13}-10^{-16} in the Nanling Series, while f_{H2O} 100–600 bars and fo_2 10^{-11}-10^{-14} were observed in the Yangtze Series. The DTA curve of biotite is very effective in distinguishing the genesis of granite, the heat absorption valley of the curve appears in lower temperatures (about 1150 °C) and very weak in Nanling Series, but in higher temperatures (about 1200 °C) and very strong in Yangtze Series. It is mainly controlled by the contents of Mg and Fe(+2) and bond energy of Mg–O and Fe(+2)–O in biotites.

The micas from Nanling Series granites are evidently rich in Nb(204 ppm, Average value), Ta(33), Sn(68), Li(1272), but poor in Co(38), Ni(78), while those from the Yangtze Series are rich in Co(50), Ni(98), but poor in Nb(49), Ta(4.4), Sn(10), Li(739). These trace elements can be regarded as indicators for petrogeno-mineralization evolution of the two series granites. Tin, Nb, Li in micas increased and Co, Ni decreased with the petrogeno-mineralization evolution in two series, respectively. Tungsten variation in micas is similar to Sn in the Nanling Series, but irregular variation occurs in the Yangtze Series. The contents of Nb(Ta), Sn(W), Li in micas can be regarded as direct indicators of mineralization themselves in the Nanling Series, and the contents of Co, Ni in biotites as indirect indicators of Fe, Cu, Au, Mo(W), Pb, Zn, Nb, REE mineralization in the Yangtze Series. The corresponding relationship between petrogenesis and mineralization for the two series can be evidently reflected in the Li-Mg-Fe^{2+} +Mn trianglc of mica, respectively.

REE data about 20 micas from south China granites are shown in this paper. In general, the REE-distributed model of micas is similar to that of granites, Eu<0.4 in the Nanling Series and Eu and LREE/HREE smaller with the evolution of the Yangtze Series. The content of REE in biotite is 180 pm (average value) in the Nanling, more than that (53 ppm) in the Yangtze Series, and both of them are controlled by REE contents in rocks. The REE contents in micas decreased with the REE contents in rocks. The REE contents in biotites decreased with the petrogenesis evolution in the Nanling Series, but increased in the Yangtze Series.

The biotites from K-feldspar granites and other related granites in Min-Zhe coast have been specially discussed in order to indicate the genesis of these granites. It has been shown that the biotites all are Mg-biotite, M>0.45, and rich in Mg, Ti, Co, Ni, Cr, but poor in Fe(2+), Al, Sn, Nb, Ta, Li, Rb. The contents of Sn, Nb, Li, F increased and the contents of Ti, Co, Ni, Cr decreased regularly in biotites with the petrogenesis evolution. Combined with data from petrochemistry, geochemistry, field geology, chronology, Sr initial ratio(<0.7100) and ^{18}O(>10%). It has been proved that these granites may be mainly derived from upper mantle, and K-feldspar granites are different from A-type granite in genesis.

Spatially, the Fe-biotites are mainly distributed over the Nanling area, while the Mg-biotites are mainly in the Ninwu area, Min-Zhe coast area, and rift zones inside the Nanling area in south China. The Fe-biotite distributions indicate the lower energy environment of Nanling Series granites, and the Mg-biote distributions reflect higher energy environment of Yangtze Series granites.

YANG Yu Chung, Ministry of Metallurgical Industry, Tianjin, China

Two Kinds of Special Habited Manganites and Their Origin

Pyrolusite and nsutite with special habits have been found respectively in the Ge Chuan Po slide manganese deposit, Lian Chen county, Fu Jian province, and Hou Zi Po sedimentary manganese deposit. He Qin county, Yun Nan province during our investigation of manganese geology.

Macrocrystal pyrolusite—polinaite and its significance in Ge Chuan Po manganese deposit: Pyrolusite usually occurs as fibrous, granular, and fine-grained aggregates, however, it occurs extremely rarely as monocrystal pyrolusite present as macroidomorphic acicular-prisms. During a survey in 1984, idiomorphic pyrolusite was found in this area. It has a length of up to 4 cm width 0.5 cm, and is closely intergrown with manganese hydroxide, hematite, and ferrohydrite, etc, and occasionally has a girdle-like, nodular texture. They occur in a slide manganese ore deposit of Quaternary age, which occupies nearly 100 m^2. The top and bottom margins of the deposit are siliceous. The idiomorphic crystals are brittle with pale steel color, the growth lamanae can be seen faintly, but no metasomatism can be found. Two endothermic peaks on the DAT diagram were caused by the phase conage from pyrolusite to B-manganosite and from B-manganosite to hausmannite. X-ray reflections include: 3.115(100), 2.405(50), the cell constants are: a_o = 3.38168, c_o = 2.86441. The mineral's MnO_2 content determined by EPM is 96.15%. The reflective indices (R)% are: 437 nm (34.64), 480 nm (35.61), 546 nm (35.82), 591 nm (36.03), 654 nm (34.95). Microhardness Hv = 4 12–835 kg/mm^2 (5.1–6.6 Moh's).

Origin and significance of the pyrolusite-poliante: The mineral crystallized directly after formation of colloform MnO_2. $2H_2O$ which came from manganese-bearing strata and was deposited on the surface and then formed after gradual upward growth.

This mineral reflects late stage oxidation and indicates that the climate during that time was warm and humid and tectonics was relatively stable.

It also indicates that manganese was abundant at depth or adjacent to this area.

The origin and significance of nsutite in sedimentary manganese deposit at Hou Zi Po: Both upper and lower margins of the deposit are sedimentary rocks with basalts nearby. Associated minerals include: cryptomelane, manganite, birnessite, and hausmannite. Under the microscope, the mineral has isometric texture and locally girdle-like colloid texture in which a large amount of fine-grained hausmanite with strong anisotropy prismatic habit and pale color are involved. The MnO_2 content of the nsutite analyzed on EPM was 99.37%. The nsutite and hausmannite are of crystals with special microsometric texture (like algae in form). Therefore, the formation of the nsutite might have experienced four stages of evolution:

(1) Accumulation of blue algae.

(2) The algae was replaced by original rhodochrosite.

(3) The rhodochrosite was decomposed into CO_2 and Mn and hausmannite formed through thermometamorphism.

(4) Hausmannite was changed into nsutite by weathering and oxidation, but relict hausmannite can still be observed.

The origin and significance:

(1) The nsutite in this area might be regarded as evidence that the original sedimentary manganese deposit experienced oxidization, thus we can look for original rhodochrosite deposits along the strata.

(2) The rhodochrosite was metamorphosed and the hausmannite was formed as a result of the basalt eruption, so we can conclude that basalt was later than the sedimentary manganese deposits.

(3) The appearance of the nsutite indicates that there was an interruption of sedimentation in the area during the metallization process and was subjected to sedimentation, metamorphism, and weathering stages.

(4) The mineralogy can serve to divide the types of deposits.

YANG, Bo, and ZHOU Jizuo, Guizhou Academy of Sciences, Guiyang, China

Collapse of Karst Overburden: Causes and Forecast

Collapse of karst overburden is a dynamic geological phenomenon that may be caused by natural forces or human activities. Disastrous collapses are often caused by human activities. For example, draining in the mines, digging tunnels, overpumping underground karst water to effect a massive fall of water level, rapid rise of underground karst river level, frequent and violent shaking, etc., can all cause collapse on a large scale.

Collapse of karst overburden occurs under three basic conditions. First, the bedrock under the covering strata has karst development and a system of karst homes with upward openings. Second, the overburden is of loose deposit or accumulation as soil, sand, or gravel. Third, underground water level changes frequently, and it changes around the interface of bedrock and overburden.

Collapse occurs mostly when the natural balance of rock, soil, water, and air in the karst area is broken. That is, collapse is closely related to rock, soil, water, and air. It is controlled by physical properties of rock and soil, their combination with water and air, and properties of their interaction. Thickness and dynamic properties of the overburden, frequency and range of water level changes, and degree of karst development, are all factors affecting collapse. Karst collapse occurs in various forms. The mechanism of its genesis, development, and formation is rather complicated. In different regional environments, the causes are often varied. Even in the same area, there are often different types of causes of collapse, only because of different geological structures. The generally accepted causes are as follows: (1) Negative pressure corrosion. When natural forces or human activities cause underground water to subside rapidly on a large scale, a "vacuum" is formed in the upper part of the karst hole, which has a strong negative pressure to denude the loose overburden. Consequently, the denuded substance is quickly carried away by the underground water. At a certain degree, sudden collapse occurs. (2) Mechanical corrosion. With the fall of underground water level, water slope becomes more steep and water flows more quickly to wear away the loose overburden. Sometimes seepage of surface water accelerates this process. As more material is carried away, the hole becomes larger. Collapse occurs when the arch of the hole cannot support the overburden. (3) Chemical corrosion. This is similar to mechanical corrosion, only the material is work away by water dissolution. (4) Shaking. Shaking waves generated by earthquake or some human activities transfer in different wave forms and at different speeds in water, rock, and the overburden, breaking the stable geological structure and thus causing collapse. (5) Air explosion. This mainly happens in the area where the overburden is fairly airtight and where in the bedrock karst tubes exist. When underground water rises rapidly and the air in the hole (in the interface of overburden and bedrock) is pressed to a degree beyond the downward stress of the overburden, air explosion occurs to cause collapse.

Karst collapse is disastrous to human beings, so it is important to study the forecast of possible collapses. Accurate forecast remains an unsolved problem. While the quantitative assessment is unfeasible because there are numerous factors to cause collapse, the fuzzy comprehensive assessment is adopted. A fuzzy comprehensive forecast model has been set up by employing the idea of fuzzy group. In the process of comprehensive forecast, the normal subordinate function is the fuzzy numeralization of raw data, as in function $\phi(x)$.

$$\phi(X) = \begin{cases} \exp\left\{-\left(\frac{x-ai}{bi}\right)^2\right\}, & X \in I \\ I, & X \bar{\in} I \end{cases}$$

1. Factor set $u = [u_1, u_2, \ldots u_n]$

The main factors are: u_1 = quantity of pumped water; u_2 = range of changing water levels; u_3 = frequency of water level changes; u_4 = thickness of overburden; u_5 = dynamic properties of rock and soil; u_6 = degree of karst growth; u_7 = homogeneous degree of karst growth; u_8 = degree of geological structure growth; u_9 = stress added by human activities.

2. Assessment set $V = [V_1, V_2, \ldots V_n]$

Assessment of stability: V_1 = stable; V_2 = fairly stable: V_3 = average; V_4 = unstable; V_5 = extremely unstable.

3. Single factor forecast matrix R

For factor i, $R = [R_1, R_2, \ldots R_n]^r$

4. Weight determination $A = [a_1, a_2, \ldots a_n]$

5. Comprehensive forecast model. The comprehensive forecast of a certain unit results from single factor forecast matrix R and weight distribution A.

$$B = A \cdot R$$

This model has been applied to the forecast of collapse in Shuicheng karst basin caused by irrational over-pumping of underground water. The result indicates that serious collapse would happen in the area where the overburden is 0.5-6m thick, underground water level falls to 5-20m deep water level decrease ration I > 3%, karst grows strongly in the bedrock, and structural joint cracks are concentrated. Collapse would not occur in the area where the overburden is more than 25 m thick, underground water level decrease ration is smaller than 1%, karst is not quite developed, and joint cracks are not developed. Between these are moderate collapse areas. The result of repeated calculation in a 2.5 km^2 collapsed area agrees with the fact. Now the same method is adopted in the forecast of collapse in a 147 km^2 area of the whole Shuicheng karst basin, and the result is presented in a drawing, awaiting verification.

YANG, C. S., and S. D. NIO, SedCon International Geoservices BV, HE Zeist, the Netherlands, and J. B. SOUTHARD, Massachusetts Institute of Technology, Cambridge, MA

Mechanics of Giant Subaqueous Sand Bar in Eocene Roda Sandstone, Southern Pyrenees, Spain

The Eocene Roda Sandstone in the southern Pyrenean foreland basin of Spain is a good example of an ancient tide-dominated shoreline-attached system. Its sedimentation pattern was controlled by an active clastic supply from the north and a more or less continuous basin subsidence. Sea level was rising due to this basin subsidence and superimposed on that a regional eustatic sea-level rise. Biochronostratigraphic data indicate that the first coarse clastic influx reached the Roda area during the Maximum Emergence at about 35 m.y. The major sandstone units of the Roda Sandstone were built up as the subsequent Transgressive Systems Tract. The overlying Plateau Limestone represents the Maximum Flooding at about 52.5 m.y.

Giant transverse sandbars are well exposed in the Estuary Mouth Facies Zone of the Roda Sandstone. Figure 1 illustrates a 25 m high transverse sandbar. It exhibits a coarsening-up sequence with a giant cross-bedded structure. The bottom-sets consist of strongly bioturbated, fine to medium sand. Foraminifera (Nummulites), bivalves, and pyrites are present. Individual bottomset beds are less than 10 cm thick. The foresets consist of coarse sand. Individual foreset beds are ~15-30 cm thick, but become thinner (~10 cm) at the lower part of the set. All foresets dip toward the southwest, i.e., the inner shelf.

The giant cross-beds were formed by avalanching of coarse sand along the slip-face. Some small megaripple cross-beddings with an orientation opposite to that of the foresets can be observed within toesets (Figure 2). These megaripple cross-beds are ~20 cm thick. They represent backflow structure, suggesting that a large separation vortex developed at the lee side of the bar. The structure of the separated flow behind the bar can be described by the distance L downstream from the crest to the reattachment point, which depends upon only the Reynolds number Re of the flow, specified as $Re = \rho UH/\mu$ where U is the flow velocity over bar crest, H is the sandbar height from trough to crest, ρ fluid density, and μ fluid viscosity. For a flow of water with density of 1 g/cm^3, viscosity of about 10^{-2} poise, and velocity of 0.5 m/s over a sandbar 10 m high.

$Re = (1\ g/cm^3)(50\ cm/s)(10^3\ cm)/(10^{-2}\ poise) = 5 \times 10^6$. For a wind with air density of $10^{-3}\ g/cm^3$, air viscosity of 2×10^{-4} poise, and velocity of 8 m/s over a subaerial sand wave of the same 10 m height.

$Re = (10^{-3}\ g/cm^3)(8 \times 10^2\ cm/s)(10^3\ cm)/(2 \times 10^{-4}\ poise) = 4 \times 10^6$. Note that Re is almost exactly the same in the two cases; the lower velocity in the subaqueous case is just about offset by the lower value of the ratio μ/ρ. So if there is pronounced flow separation and reattachment a distance L downstream of the crest in the subaerial case, the same must be true in the subaqueous case.

A distinct feature of the giant sandbar is the variation of the foreset dip angle. A gradual transition of the foreset dip from low angle (13°-14°) to high angle (23°-30°) can be observed within a distance of 20-30 m (Figure 3). This transition reflects the change of the flow pattern (i.e., the distance L), which was probably caused by variations of water and sediment discharge from the fluvial system that fed the estuary. High fluvial discharges and abundant coarse sediment supply during wet seasons favor the accretion of sediments on slip-face, resulting in the aggradation and rapid progradation of the transverse sand bar, producing high-angle cross-bedding. While during stages of low fluvial discharge and low sediment supply (e.g., dry seasons), wave and tidal actions were dominant. Consequently, the bar front was reworked and the bar degraded.

Another characteristic feature of the giant sandbar is the presence of erosional reactivation surfaces (Figure 1). They reflect the interruption of the bar progradation by storm events, which could erode the bar front. A storm could also change the morphology of the whole bar and channel system. The resultant change in flow pattern would cause some variation in the direction of the bar progradation as shown.

The top of this transverse sand bar was truncated (Figure 1). The truncated upper surfaces show strong bioturbations and abundant marine fauna, indicating periods of nondeposition and erosion. The frequent appearances of such surfaces reveal that active bar progradation was a rather short-period event.

The transverse sand bar developed in a wide, relatively straight, high-gradient estuarine channel, which was over 30 m deep and several kilometers wide. This north-northeast–south-southwest-oriented estuary was fed by a braided river system from the north-northeast. The preservation potential of the transverse sandbar would depend on basin subsidence. The stratigraphic record is characterized by a vertical stacking of truncated sandbar units.

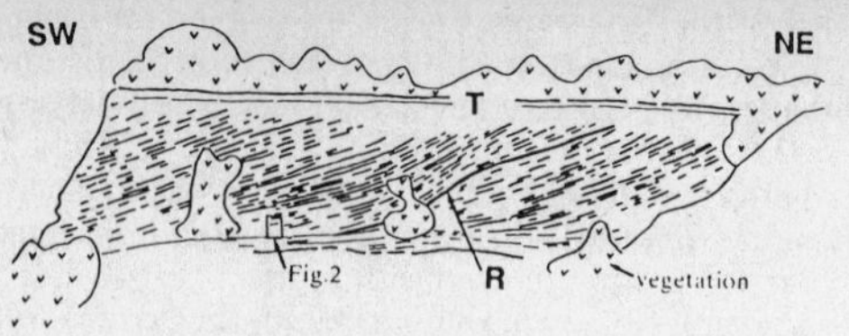

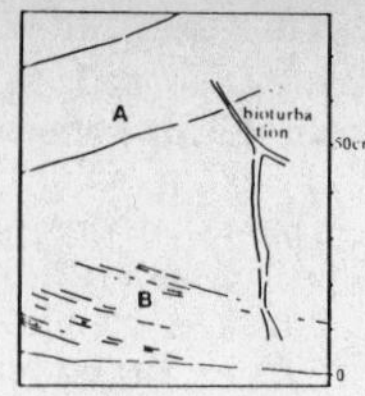

Figure 1—Giant cross-bedded sandstone body of the transverse sandbar. R, erosional reactivation surface; t, truncated upper surface.

Figure 2—Toesets of the giant sandbar. Note small megaripple cross-beddings (B) with an orientation opposite to that of the foresets (A). For location of Figure 2, see Figure 1.

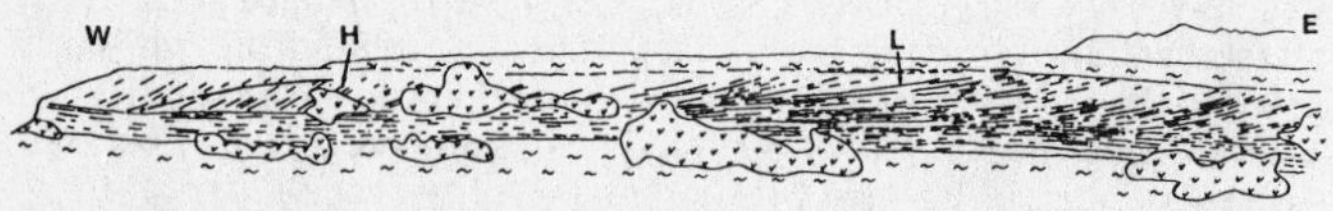

Figure 3—Giant cross-beddings of a sandbar. A gradual transition of the foreset dip from low angle (L) to high angle (H) can be observed within a distance of 20-30 m.

YANG, Hua, Aero-Geophysical and Remote-Sensing Center, MGMR, Beijing, China

Earth Sciences Interpretation of Aeromagnetic Maps in Qinghai-Xizang Plateau and Discussion of Earth Sciences Features of Crust and Structural Evolution in This Plateau

An integrated earth science interpretation of 1/1,000,000 aeromagnetic map in eastern Qinghai-Xizang Plateau and adjacent area is presented. This area, located at 22°-38°N latitude, 88°-104°E longitude, covers 220 km^2. This plateau is in the hinterland of Asia, in southwest China. It is known as the third pole of the earth. Its crustal earth science features, structural evolution history, and construction mechanism are popular topics for earth scientists around the world. Mineral resources in this area are very important for future Chinese economic development.

Interpretation of aeromagnetic maps involves transformations of magnetic field, computation of depth, thickness, and size of magnetic bodies and crusts, comparison between magnetic parameters and another geological and geophysical data, and determination and discussion of earth science models of the plateau.

Aeromagnetic maps of the plateau demonstrate clearly an inlaid pattern consisting of different background of magnetic field and characteristic anomalies, surrounded in the margin or cut apart inside this map by magnetic lineaments. This map reflects some crustal blocks that are different from one another in the structure and in the evolution history in the plateau and deep fractures filled with strong magnetic volcanic rocks between blocks. In view of the fact that this area is between Eurasian and Gondwanian landmasses the situation mentioned above reveals complicated contact between two landmasses.

Integrated interpretation shows that modern magnetic fields in the plateau mainly indicate structural pattern formed in the process of the Cenozoic crustal evolution. Before the Cenozoic, this area went through the following evolution periods: During growth period of continental crust (from the Proterozoic to the early Paleozoic), the Eurasian and the Gondwanian continental crust grew oppositely and finally combined along Tanggula-Sanjiang structural zone of the plateau; during stable-development period of thin continental crust (the early and the middle Paleozoic), there were platform-type shallow-sea deposits in this area; during fission and evolution period of continental crust (from the late Paleozoic to the early Cenozoic), the continental crust first pulled apart, thinned due to etching, subsided from north to south, then closed, grew, napped and superimposed, and upwarped; during upwarping due to heat expansion of the continental crust (the late Cenozoic, for the last 200 m.y.) crust-mantle mixed layers developed in the lower continental crust held and warmed this crust; as a result, this crust as a whole expanded and upwarped, and counterrushed and nappled outward.

In the process of the continental-crust evolution paleogeographic location of the plateau drifted northward from high-cold area in Southern hemisphere during the early and the middle Paleozoic; crossed paleoequator from the end of the Paleozoic to the late Mesozoic; converged south to the middle-latitude area in Northern hemisphere from the late Mesozoic to the early Cenozoic. The drift and structural evolution of the continental blocks was related to paleolatitude and thermal convection of mantle under lithosphere.

Magnetic field and many kinds of earth science data show that the structures in the plateau are very special and have a deep and thick root. Asthenosphere is 100-140 km thick, 80-160 km deep; upper mantle (lower lithosphere) is 40-90 km thick, 50-70 km deep, with the shape of concave lens. The top of the crust (upper lithosphere) is 45 m higher than that of adjacent area; the bottom is 35 km deeper than that of adjacent area and is characterized by lenticular structure and layering. A crust-mantle mixed layer developed beneath the crust, which is 5-25 km thick (the middle is thicker than its margin). The top of the layer is 50 km deep and has a small relief. The middle of the crust is a low-velocity and low-resistivity layer, 5-8 km thick. This layer divides the crust into two different parts in shape and structure. The depth of the Curie surface from the computation of magnetic field is the same as that of this layer. This layer with a temperature of 500°C benefiting the formation of magma and mineralized heat fluid is a decollement surface in the crust benefiting structural development. Many deep fractures detached and disappeared along this surface. The upper crust faults clearly and there develop nappe and magma-intrusion structures, is 25-35 km thick, and the middle is thinner than its side. Satellite magnetic and gravity maps show that the structures in the plateau are characterized by a united negative anomaly which indicates that the lithosphere in the plateau consists of weak-magnetic and low-density rocks. As mentioned above, these rocks are formed due to structural activities such as compression, napping and superimposition, melting and growing, as well as temperature-rising and expansion. Temperature-rising and expansion not only lowers bulk density of lithosphere but also eliminates magnetism of the bottom of original magnetic crust and this bottom changes into non-magnetic one, thus lowering general magnetism.

At present, the Qinghai-Xizang plateau is still in the process of heat expansion and upwarping. The evidence is as follows: Magnetic observation reveals that the absolute magnitude of magnetic field in the plateau has been attenuating. The annual attenuation is 15-25 nT, and it is stronger at the side than in the middle; the gravity map shows that there are isostatic anomalies in the margin of the plateau and that this area needs isostatic readjustment; earthquakes are frequent and of the punching-type in the margin and of the sliding-type in the middle, release stress, dredge the channel of heat passage, accelerate the upwarping and napping in the margin of the plateau and spreading and high-temperature heat flow and extensive thermal spring, which predicts that there are buried-magma room and high-temperature mineralized-heat-water accumulation structures in the upper crust of the middle plateau.

In conclusion, this paper proposes a new mode of structural evolution of continental crust in the plateau. This mode is as follows: (1) growing period-stable development of platform shallow sea of Pan-continent; (2) fission period-secondary oceanic-basin development and continental-block sinking; (3) contraction period-island-arc basin and continental-margin mountain development; and (4) heating and upwarping period-expansion and upwarping of continental crust, formation of plateau plate. Evolution of continental crust took place in the process of drift and was related to paleolatitude, crustal-stress field, and heat convection of mantle.

This mode deepens the understanding of structural features of crust of the plateau, demonstrates the prospect of rich geothermal resources, noble metal and nonferrous metal mineral resources, and points out new direction of oil and gas exploration in the napple zone of the margin of the plateau.

YANG, JICHENG, Ministry of Water Resources, Beijing, China

Soil Erosion Mapping from Space in China

Landsat image characteristics-synoptic view, multispectral data collection, ability to collect temporal data can reveal soil erosion conditions and investigations into control methods for soil erosion. These data are better than the statistics gathered in the early fifties. There are three characteristics: Landsat data are quasi-quantitative analysis. Seven kinds numeric code-soil erosion degree, potential dangerous degree, soil erosion type, topograph, soil kinds, landform, and vegetation will stored in a database for monitoring data replacement, and output at any scale on any topic maps. Three kinds of soil erosion type—water erosion, wind erosion, and frozen and melting erosion were classified in the country.

China is a country with a vast territory. There are a lot of big rivers, lakes, and high mountains. The whole country has an area of 9.6 million km^2. Of the total territory, plateaus and mountainous areas cover 59%, hilly areas 10%, plains 12%, and basins 19%. Soil erosion control is a problem. We have often faced, and we urgently need to use space technology for soil erosion investigation and soil erosion control management. Landsat (MSS and TM) imagery characteristics-synoptic view, multispectral data collection, ability to collect temporal data can help us evaluate soil erosion conditions and investigate control methods for soil erosion.

According to statistics made in the early fifties, appreciable water erosion areas occurred over an area of some 1.5 million km^2 in the country, besides 1.2 million km^2 of wind erosion areas. Water erosion areas are located mainly in the loess region in the northwest, red loam rolling and mountainous regions in south China, earth and rocky mountainous regions in north China.

YANG, Jiduan, QU Lifan, ZHOU Huiqin, CHENG Zhengwu, SHOU Tongahun, HOU Jingpeng, LI Peixian, SUN Suying, Chinese Academy of Geological Sciences, Beijing, China

Nonmarine Triassic Stratigraphic Units and Their Boundaries in China

The Triassic System is extremely well developed in China. The Lower and Middle Triassic are dominated by continental sediments, but with minor alternation of marine and continental deposits in the areas north of the Kunlun-Qinling-Daibieshan Mountains. In southern China, the marine Triassic is dominant, but with either alternation of marine and continental deposits or continental sediments in some parts of the area. The Upper Triassic is comprised of continental sediments in northern China, and is dominated by alternating marine and continental deposits, with minor continental sediments, in southern China.

In the past decade, great progress has been made in studies of the characteristics of fossil assemblages, stratigraphic-time division, and determination of boundaries, as well as in studies of time sequence and distribution, of nonmarine strata.

This paper deals mainly with the sequence of strata, the characteristics of fossil assemblages, and the determination and correlation of boundaries of the nonmarine Triassic stratigraphic units (see Tables 1 and 2). Using both biotic and nonbiotic changes, we have made a revision of boundaries of the units.

Table 1—Simplified correlation chart of Triassic deposits in China.

SERIES \ REGION	NORTHERN CHINA		SOUTHERN CHINA		
	Shaanxi-Gansu-Ningxin Basin	Junggar Basin	Sichuan Prov.	Guizhou Prov.	Hunan-Jiangxi Prov.
LOWER JURASSIC	Fuxian Fm.	Badaowan Fm.	Baitianba Fm.	Zhiluijing Fm.	Zaosan Fm.
UPPER TRIASSIC	Yanchang Fm.	Haojiagou Fm.	Xujiahe Fm.	Ergiao Fm.	Anyuan Group: Sanqiutian Fm.
		Huangshanjie Fm.	Xiaotangzi Fm.	Huobachong Fm.	Anyuan Group: Sanjiachang Fm.
			Maantang Fm.	Banan Fm.	Anyuan Group: Zijiachong Fm.
MIDDLE TRIASSIC	Tongchuan Fm.	Karamay Fm.	Tianjingshan Fm.	Falang Fm.	Yangjia Group
	Ermaying Fm.		Leikoupo Fm.	Guanling Fm.	
LOWER TRIASSIC	Heshanggou Fm.	Shaofonggou Fm.	Jialingjiang Fm.	Yongningzhen Fm.	Daye Group
	Liujiagou Fm.	Jiucaiyuan Fm.	Feixianguan Fm.	Feixianguan Fm.	
UPPER PERMIAN	Shiqianfeng Fm.	Guodikeng Fm.	Zhangxing Fm.	Dalong Fm.	Dalong Fm.

Table 2—Characteristic taxa of fossil assemblages of the Triassic.

SERIES \ REGION	NORTHERN CHINA	SOUTHERN CHINA
LOWER JURASSIC	Cyathidites-Dictyophyllidites-Cerebropollenites; Horstisporites-Minerisporites; Baiera, Sphenobaiera, Czekanowskia, Phoenicopsis	Classopollis-Cyathidites- Klukisporites; Marattiopsis-Otozamites
UPPER TRIASSIC	Dictyophyllidites-Apiculatisporis-Alisporites- Minutosacous; Trileites-Hughesisporites; Danaeopsis-Bernoullia (T_2^2-T_3); Shaanxiconcha (Utschamilla)-Sibireconcha	Corollisporites-Micrhystridium, Kyrtomisporis-Ricciisporites; Ptilozamites-Lepidopteris; Shuniscus sp.
MIDDLE TRIASSIC	Punctatisporites-Caytonipollenites-Colpectopollis; Tuberculatisporites-Dijkstraisporites Shaanxiconcha-Unio; Luthevichinlla, Shensinella; Parakannemeyeria spp.	Verrucosisporites-Corisaccites; Annalepis zeilleri- Neocalamites meriani
LOWER TRIASSIC	Limatulasporites-Lundbladispora-Taeniaesporites; Trileites-Maexisporites-Pusulosporites inflatus; Pleuromeia-Voltzia (T_1-T_2^1); Darwinula spp. Lystrosaurus spp.	Leiotriletes-Pteruchipollenites; Neuropteridium-Voltzia;
UPPER PERMIAN	Limatulasporites-Lueckisporites; Triangulatisporites spp.; Callipteris-Iniopteris-Comia; Panxiania-Darwinula; Palaeanodonta-Palaeomutela; dicynodonts	Yunnanospora radiata- Gardenasporites; Triangulatisporites spp.; Gligantopteris-Labatannularia; Panxiania-Darwinula; Palaeonodenta

YANG, Weiran, China University of Geoscience (Wuhan), Wuhan, China

Tectonic Evolution of Circum-Pacific Continental Margins in South China

In South China, the coastal area is the modern continental margin of the Circum-Pacific; other areas are paleocontinental margins. Consequently, it is an ideal base for studying tectonic evolution of continental margins.

The continental margins in South China border on the Pacific in the east, and are sandwiched between Yangtze Continent and South China Sea-Indosinia Continent, both facing each other. Thus they display the complex nature of continental margin area.

The tectonic evolution of continental margins in South China can be divided into four stages.

Pre-Sinian period: The continental margin was of island-arc type in double or multiarrays. At least two zones can be determined. The subduction zone in the north of the area was situated on the line of Guiyanf-Anhua-Hangzhou. A number of islands in groups were formed to the north of the line due to Sibao movement, back-arc basin lying to the north of the islands. Another subduction zone lay to the south of Sibao Group and Danzhou Group in North Guangxi province. An inter-arc basin was located between both subduction zones. Furthermore, the Mayuan Group and Dikou Group in Fujian province as well as Chencai Group in Zhejiang province might be the islands scattered in a broad oceanic crust area and consolidated earlier, i.e., a microcontinent.

The southern edge of Hainan island and the South China Sea are belonged to another paleocontinent and its continental margin.

Caledonian stage: This was the period that the continent and ocean were obviously differentiated, and the development of paleocontinent was varied and perfect. Two types of continental margin could be recognized. One was due to the development of fault "activization" of Jinning basement. There were siliceous, carbonaceous, and abundant organic sediments representing detention of flow within shelf margin basin. The structures were represented by "belts" and "blocks." The "belts were composed of northeast- and northwest-trending faults and surrounded the "blocks." Domes and brachyanticlines were developed in "blocks." Magmatism was not very strong. Volcanism was largely of terrestrial-eruptive facies and alkali content was rather high, representing a comparatively stable tectonic setting. Intrusive rocks were essentially crust-derived reformed granite or fault-anatexis origin, compared with typical S-type granite, the features of high -Fe, Mg, Ca, Ti, low-K, low differential index, high solidified index, rather low A/CNK, and low K_2O/Na_2O are extremely similar to Caledonian type granite determined by W. S. Pitcher (1983). The region is defined as Jiangnan reformed continental crust area ascribed to the outer zone of the paleocontinental margin. It is characterized by the development of spilite-karatophyre suite and flysch deposits, polyphase and complicated deformation, intense regional metamorphism, widespread migmatization and granitization, the volcanism of island arc and active continental margin character, and granite of crust-derived reformed type, corresponding to Hercynian type granite determined by Pitcher (1983).

Hercynian-Indosinian stage: The continental margin moved to the East China Sea, Taiwan, and other places. Late Paleozoic flysch deposits in Nanxi area, Fuding country, Fujian province, represented the western boundary of the continental margin. Tailuge metamorphic belt of high-temperature and low-pressure type in Taiwan might be a paleoarc zone, and Yili belt of low-temperature and high-pressure type was the position of paleotrough and subduction zone. The oceanic crust area was to the east of the subduction zone. The northern continental margin of South China Sea Continent became increasingly narrow. The South China Sea Continent and the Yangtze Continent converged with each other in the Hercynian stage. In addition, the Caledonian movement was not obvious in Qinzhou-Yulin area, Guangxi province. Dongwu movement brought about strong folds followed by the intrusion of granite. It is called remnant marine trough, and its nature is similar to the continental margin.

Yanshanian-Himalayan stage: The continental margin is still in the coastal area of the southeast of China, but there are different understandings of the nature and evolutionary history of the continental margin. We regard that it was the passive continental margin in J_1-J_2, and then was transferred into the continental margin of Andes type in J_3-K_1, and the continental margin of the island arc type occurred in K_2-E, constituting the basic features of modern continental margin.

The evolution of Circum-Pacific continental margins in South China reflects the dialectical relation of mutual transformation between the continental crust and the oceanic crust, as well as the features of stages of the continental marginal evolution. There are the characteristic minerals on the continental margin of South China at the different stages, and the existence of the continental margin and the changes in its nature exert different influences on the continental internal structures and ore deposits.

YANG, Zun-Yi, China University of Geosciences, Beijing, China

Permo-Triassic Events of East Tethys Region

Permo-Triassic sequences in the East Tethys region have been carefully studied for over one decade, and they are correlated and discussed here. Both biotic and nonbiotic events are recognizable from the P-Tr boundary. Event (1) The most evident and well-known event is the biotic alternation near or at the P-Tr boundary where one notices the mass extinction of Paleozoic life and the appearance of Mesozoic organisms. Event (2) Changes of sea level, i.e., transgression and regression at the P-Tr boundary as evidenced by the change in lithofacies. Event (3) Geochemical changes at the P-Tr boundary as exemplified by the Meishan section in Changxing, Zhejiang Prov. and Shanghai section near Guangyuan, Sichuan Prov., China and by other sections in Himalayas, Salt Range, etc. Event (4) Abrupt changes of stable C13 and O18 isotopes at the P-Tr boundary. Event (5) Volcanism at or near the boundary as shown by volcanic ashes (with volcanic shards, microspherules). Event (6) Magnetic reversals near the P-Tr boundary sections at Shangsi and Yanjing, Sichuan Prov., China. Event (7) Presence of Ir anomalies in Shangshi, Meishan sections.

Among the above events, 1 to 4 have been proved to be true for sections in China, India, and Pakistan, thus explaining their possible universality and helping to determine more exactly the position of the P-Tr boundary. Consequently, in the light of event stratigraphy, other important boundaries (both interperiodic and intraperiodic) may be studied in the same way.

YANG, Zuosheng, Shandong College of Oceanography, Qingdao, China

Mineralogical Assemblages and Chemical Characteristics of Clays from Sediments of Huanghe, Changjiang, and Zhujiang Rivers and Their Relationship to Climate Environment in Their Sediment Source Areas

Clay fraction of more than 150 sediment samples taken from the lower streams and delta areas of the three largest Chinese rivers, the Huanghe, Changjiang, and Zhujiang Rivers, were analyzed mineralogically and chemically. The results show significant differences in mineralogical and chemical characteristics among the clays of the three rivers. Three types of clays with different mineral and chemical compositions were distinguished based on the result of analyses: the Huanghe type, Changjiang type, and Zhujiang type. Their mineral and chemical characteristics are the same as the clays formed under three different climates (semidry and chill, warm and wet, hot and rainy) types corresponding to those of the main sediment source areas of the Huanghe, Changjiang, and Zhujiang Rivers, respectively.

The three largest Chinese rivers are placed from north to south in sequence. The relief of China is high in the west and low in the east, making the three rivers flow eastward. The above mentioned unique geographical location of the three rivers makes their main sediment source areas locate under different climate types and parallel more or less to latitude. Thus, the clays, being products of different sediment source areas, have corresponding different mineralogical and chemical characteristics to distinguish them among themselves and in the sediments.

YASVIN, L., Ministry of Geology of the USSR, Moscow, USSR

Methodological Bases of Underground Water Resources and Regime Control

Groundwater is one of the main components of geological environment, closely connected both with the other components and elements of nature as a whole. Accordingly, any changes in groundwater regime and resources, caused by natural factors and human activities, bring about changes of hydrogeological conditions of some territories as well as considerably affect nature in general. It follows that groundwater resources and regime need to be controlled. Control or management is suggested to be a system of measures guaranteeing such changes of groundwater regime and resources that are permissible for functioning of certain branches of industry and conservation of other components of the environment at a given level.

The theory of rational management of groundwater resources and regime is at the initial stage of development. It should include:

1. Revealing the regularities of natural and disturbed groundwater regime formation, groundwater resources and safe yield, and also peculiarities of the influence of hydrogeological condition changes on the other components of the environment (vegetation, soil, surface water, etc.).
2. Developing and improving methods of groundwater regime prediction in nature, when water management measures are carried out in the course of groundwater development and also methods of prediction of natural condition changes with a change of groundwater resources and regime.
3. Creating the theory of rational management criteria that are specified by the interconnection between hydrogeological condition changes and changes of nature (ecological changes) and possibility to control different changes in the ecosystem on the whole. These criteria should be estimated for certain natural and water management regions considering their economic and social development.
4. Scientific proving and development of the methods of groundwater resources and regime management.

Choice of the methods of groundwater resources and regime management is mainly determined by the factors causing hydrogeological condition changes. In this case, all the factors should be subdivided into three main groups: (1) a change of climate and other natural conditions; (2) human activities that are not connected with groundwater production (hydrotechnical engineering, irrigation with surface water, etc.) and (3) human activities connected with groundwater production (groundwater development for water supply and irrigation, groundwater pumping out when developing mineral deposits, oil and gas, drainage of meliorated land, and protection of territories from water logging). Groundwater resources and regime management is of particular concern when groundwater is used for water supply and irrigation. In this case, control of the resources is closely connected with the problem of hydrogeological proving of groundwater rational use. Solving the problem requires us to adhere to policy when using groundwater, to regulate groundwater storage study and development, to plan groundwater development considering its interconnection with surface water and other components of natural environment, and also to carry out measures to control groundwater development and protection from contamination and depletion.

The main methods of groundwater resources and regime management include:

1. Limiting of groundwater development by the amount of groundwater safe yield revealed, estimated considering given nature conservation criteria. Accordingly, groundwater production should follow the estimation of groundwater safe yield, made using the results of exploration on the basis of physical-mathematical models of groundwater flow to water extraction plants in different natural and water management conditions.
2. Carrying out prophylactic measures to protect groundwater from contamination and groundwater safe yield from depletion.
3. Control of groundwater production and groundwater development area state and corrections of the operational regime. The latter, in turn, can be carried out using different methods-regulation of groundwater production, maintaining groundwater at a given level, artificial water recharge, constructing double water extraction plants for simultaneous production of potable and mineralized groundwater (for instance, when developing potable water lenses), hydrobarriers, water-proof barriers when exploiting mineral deposits, etc.

Corrections of groundwater development regime according to the data observed should be carried out on the basis of permanent functioning models.

YATES, M. V., University of California, Riverside, California

Wellhead Protection: Assessing the Potential for Microbial Contamination of Drinking Water

Ground water traditionally has been considered safe for consumption without treatment because the soil acts as a natural filter to remove contaminants. As a result, private wells in the United States generally do not receive treatment, nor do a large number of public water supply systems. In recent years, numerous reports of ground-water contamination have led people to question the quality of their drinking water. Although most of the concern has focused on contamination by toxic organic chemicals, the majority of the waterborne disease outbreaks in the United States are caused by microorganisms. The class of microorganisms responsible for a large number of these outbreaks is the enteric virus group. Viruses are extremely small, approximately 20 to 200 mn in diameter, thus they are not readily removed by filtration through the soil. A major source of viruses in the subsurface is septic tank effluent; there are over 22 million septic tanks in the United States which dispose of approximately 4 trillion liters of waste to the subsurface every year. Indeed, septic tanks are the most frequently reported causes of contamination in ground-water disease outbreaks associated with the consumption of untreated ground water. In the United States, the placement of septic tanks is generally by county- or state-wide regulations with little consideration given to the local hydrogeologic, climatic, and land use conditions. This study was undertaken to examine the variation in septic tank setback distances (the distance from a septic tank to a drinking water well) that would be calculated using local ground-water flow characteristics and considering the potential for virus contamination of the water. Ground-water samples were collected from 71 continuously pumping municipal drinking water supply wells in Tucson, Arizona. Viruses were added to the samples, and the rate at which they were inactivated was determined over a 30-day period. Using virus inactivation rates, transmissivities, and gradient values for each well, and average values for the aquifer thickness and porosity, setback distances were calculated for each well. The setback distances were calculated such that there would be 10^7 viruses inactivated in the time required for the ground water to travel from the septic tank to a drinking water well. The setback distances ranged from less than 15 m to over 300 m in a 200 km^2 area. After determining that the setback distances were spatially correlated by calculating and validating a semivariogram, geostatistical methods were used to further analyze the data. Using a kriging program, a contour map was constructed which shows the estimated septic tank setback distances for the entire area. The analyses were then extended to calculate the conditional probabilities associated with the estimated setback distances using another geostatistical technique, disjunctive kriging. The results are presented in two different ways: 1) given a setback distance (e.g. stipulated by a regulation), the spatial distribution of the probabilities that the number of viruses in the ground water will be within acceptable limits were calculated and 2) given a desired probability level (e.g. 90%) the setback distances that would be required to achieve that level of confidence that the water will be free of virus contamination were calculated. The results of these analyses showed that, to be 90% confident that virus numbers would be within acceptable limits, in many areas the setback distances would have to be severalfold higher than stipulated by current regulations. The methods described have potential for use by local government officials for land use planning purposes.

YE Lianjun, and CHEN Qiying, Academia Sinica, Beijing, China

Deductions of the Metallogeny of Chinese Sedimentary Mineral Deposits

Chinese sedimentary mineral deposits possess characteristics that are worthy of consideration regarding metallogeny. They disclose two kernel points of basic importance: (1) sedimentary mineral deposits are not just simple hydrogenic chemical reactions, but are, rather, products of the pore water imbibition during early diagenesis on terrestrial clastics and ore-bearing organic remains, and later concentrated through sedimentary recycling or physical emplacement; and (2) genesis of sedimentary mineral deposits is conducted by composite sedimentary processes and accomplished under polyepisodic stages. They are really the result of a rather complex procedure, being constrained by a combination of both geochemical and biochemical processes, sedimentary, biological, and climatic environments, and paleogeological background, including structural and eustatic conditions, as well as any possible astronomical influences. The basic cell of sedimentary mineral deposits is the sedimentary basin. Different kinds of sedimentary basins possess different sedimentary systems, different sedimentary associations, different genetic types and associations of sedimentary ores. Migration and differential evolution of sedimentary basins complicate the distribution of mineral deposits.

Sedimentary mineral deposits occur largely in the lower or basal part, and properly the frontal belt of marine transgressive series where the rate of transgression is comparatively slower, and the frequency of sea-level oscillation is more intense. Ores of the Inner Continental Shelf are used to be well-sorted, intraclastic in texture, and thick bedded in structure. They were products of mechanical enrichment above the wave base. Stratabound discontinuities or omission surfaces are numerous. All the rocks possess a high degree of constituent maturity. Ore-bearing series of the Outer Continental Shelf are mostly blackish in color, rich in organic material. Sedimentary differentiation is rather weak and generally finely laminated silty, cherty, or calcareous turbiditic shales, with calc-debris flow, occasionally. Constituent maturities are all low. They were deposits of deep water with deficiency of oxygen. Primary orebeds are not stable and well defined from country rocks. Industrial ores are mostly veinlets stratabound in nature, being predominantly formed by relevant processes of deep-burying diagenesis.

In eastern China, the Late Proterozoic and Paleozoic continental shelves seem to be rather gentle and very wide, being at times approximately 1000 km wide. Sometimes, they were dissected by growth longitudinal deep-down-saggings so that ore deposits of both shallow and deep water are alternated.

The primary concentration of ore-forming elements such as Fe, Mn, P, etc. seems to have been through the absorption of microalgae and microbes as disclosed under SEM recently in many of the Chinese ores. This means that the Upwelling Theory, as set up by Sheldon, might well be applied to many kinds of sedimentary mineral deposits other then phosphorites.

YEATS, R. S., Oregon State University, Corvallis, OR

Contemporary Tectonics of Himalayan Frontal Fault System

The northern edge of the Ganga and Jhelum plains is commonly marked by a thrust fault bringing Sisalik and older strata over floodplain deposits. At the foot of the Salt Range, the thrust fault follows Eocambrian salt, does not involve basement rocks, and has undergone 12-23 km displacement at 7–14 mm/y (Baker et al, 1988). High-angle block faults in basement produce an irregular surface to which moving thrust plate must conform, resulting in ramping at the north flank of the Salt Range and transpression at the Kalabagh tear fault. The more interior north-over-south Main Boundary Thrust (MBT) stopped moving by early Pleistocene time in Pakistan (Yeats and Hussain, in press); in India and Nepal, the MBT has been reactivated as south-side-up back thrusts or strike-slip faults (Nakata, in press). The active-fault pattern of the upper plate of the Himalayan frontal fault system is dominated by (1) northeast- and northwest-trending strike-slip faults that tend to narrow the thrust wedge normal to strike and elongate the wedge parallel to strike, and by (2) back thrusts that tend to elevate and to thicken the thrust wedge, particularly within the zone between the MBT and the Himalayan Front thrust (Subhimalaya). The Himalayan Front thrust is discontinuous, suggesting that only part of the Himalayan-Indian convergence is taken up at the frontal thrust. An important part may be reflected by flexural-slip folding underlain by blind thrusts that never reach the surface. Blind thrusts related to the Himalayan Front thrust may have elevated folds in the Subhimalaya, including those of the dun valleys of Indai and Nepal (Dang Dun, Dehra Dun, Pinjore Dun, etc.) and the Pabbi Hills of Pakistan. Geodetic monitoring of these folds may provide evidence of fold growth that could be the beginning of a program to evaluate earthquake hazards or buried thrust faults. A second component of this program should be confirmation of Indian-Himalayan (as opposed to Indian-Eurasian) convergence rates using space-based geodesy.

I thank my colleagues in the Himalayan frontal fault study group of IGCP 206 (K. S. Valdiya of India, A. Farah and M. A. Mirza of Pakistan, M. R. Pandey of Nepal, and T. Nakata of Japan) for their contributions to the fault atlas and to my understanding of modern tectonics of the Himalaya.

REFERENCES

Baker, D. M., R. J. Lillie, R. S. Yeats, G. D. Johnson, M. Yousuf, and A. S. H. Zamin, 1988, Development of the Himalayan frontal thrust zone: Salt Range, Pakistan: Geology, v. 16, p. 3-7.

Nakata, T., in press, Active faults of the Himalaya of India and Nepal, in L. Malinconico and R. J. Lillie, eds., Tectonics of the Western Himalaya: GSA Special Paper.

Yeats, R. S., and A. Hussain, in press, Zone of late Quaternary deformation in the southern Peshawar basin, Pakistan, in L. Malinconico and R. J. Lillie, eds., Tectonics of the Western Himalaya: GSA Special Paper.

YEGORKIN, A. V., B. V. ERMAKOV, V. A. ERHOV, G. N. KRASNOPEVTSEV, V. N. SEMOV, Yu. K. SCHUKIN, Ministry of Geology of the USSR, Moscow, USSR

A Deep Structure of the USSR Territory on the Basis of Regional Geophysical Researches

Principally new information has been obtained concerning the location of deep subvertical, subhorizontal and inclined boundaries within the Earth's crust and upper mantle by means of carrying out the complex program of the USSR territory studies by system of the first and second class geotraverses together with deep and superdeep drilling data. The model concepts of medium velocity anomalies have been improved, and medium absorption properties, degree of sedintegration, and types of relationship between subsurface and deep structures have been also investigated. Those complex studies permitted us to state the wide range of the Earth's crust model concepts from the horizontally layered, block layered, to tessellated and scaly ones.

The three-layer model has been chosen as a base of the modeling done. Interval velocities within the range of 5.5-6.3 km/s characterize the upper layer; the intermediate and the lower ones are characterized by interval velocities of 6.4-6.6 km/s and 6.7-7.4 km/s, respectively. The Moho boundary is marked by velocities of 7.8-8.4 km/s.

Consolidated crust description from layer to layer is based on the thickness-velocity change schemes within each layer and also on some statistical data of the analysis of the seismic parameter distribution of each layer.

The following features are specific for the upper layer: main depths are about 0-6 km; Vp does not depend on depth of the top of consolidated crust; the main values of Vp are 5.6-6.2 km/s with peaks of 5.6, 5.8, and 6.0 km/s; the values of Vp on the surface and near it (depth marks are equal to zero or close to it) vary widely from 5.0 to 6.7-6.8 km/s with a peak on 6.0 km/s; the thickness of the upper layer changes from 5 to 20 km with peaks on 7, 10, and 15 km.

The intermediate layer has thicknesses from 4-5 to 25 km with peaks on 5-15 km; the values of Vp vary from 6.2-6.3 to 6.7-6.8, with peaks on 6.4-6.5 km/s; there is no reliable dependence between Vp and the depth or the thickness, however one can note that the more the thickness of the layer, the less its velocity; the crustal waveguide occurs mainly within the blocks of comparatively low (6.3-6.4 km/s) Vp values within the intermediate layer, and thus the waveguides seem to increase the velocity fall within the intermediate part of the crust.

The top depth of the lower layer of the Earth's crust varies widely from 10 to 40 km with peaks on 18-30 km; the layer thickness changes within limits of 4-30 km, mainly 12-21 km; velocity values amount to 6.5-7.4 km/s with peaks on 6.7, 6.8, and 7.0 km/s; Vp increases slightly with depth, especially from 25-27 km and deeper. Thus the crust property changes take place on different levels and within the whole number of crustal layers. It is natural to suppose the construction of certain types to be caused by the arrangement of specific features of each layer.

The consolidated crust structure specific features are the waveguides' presence and that of high-velocity layers (Vp 7.2 km/s). The former occur on different deep levels, the latter are located mainly within the lower part of the crust. Both play as important a role in crustal types formation as in its geodynamical regime.

Joint interpretation of different type wave seismograms Vp and Vs (refracted and reflected)—and converted ones provide the detailed division of the crust and mantle especially by Vp-Vh relation value, which is strongly connected with total thickness of consolidated crust. The Vp/Vs coefficient decreases with crustal thickness growing for young plates and folded areas, and increases for ancient plates.

According to the relationship between Vp/Vs and SiO2 contained in rocks, the fact was revealed that there is no decase of SiO2 contained with depth growth. Continental crust consists mainly of felsic and near-felsic rocks and the so-called "basalt layer" is practically absent.

The main results of upper mantle researches are:

- velocity changes have been studied to depths of 750 km;
- Vp changes from 7.8 to 8.8 km/s within the depths of 40-400 km;
- Vp sufficiently increases from 8.8 to 9.4 km/s and from 10.2 to 11.0 km/s at the depths 410-415 and 670-700 km, respectively;
- in eastern and western Siberian cross-sections, there are sometimes four low-velocity layers (70-80, 90-110, 140-160, and 205-275 km). The velocity drop at waveguide boundary amounts to 0.35-0.45 km/s;
- velocity horizontal heterogeneities within the upper mavite of the USSR territory occur as wide as the first hundreds to one thousand km;
- there is no regionally spread asthenospheric layer.

In tectonically active areas, the asthenospheric layer is characterized by a thick low-velocity layer and high electrical conductivity within the depths of 100-250 km composed of partly melted mantle material. As for ancient platforms, the asthenospheric layers may be assumed at great depths to be composed of solid-crystallic material.

YILMAZ, A., General Directore of Mineral Research and Exploration, Ankara, Turkey

Tectonic Zones of Caucasus and Their Continuations in Northeastern Part of Turkey: A Correlation

The study area covers the Caucasus and the northeastern part of Turkey. Tectonic zones of the Caucasus and their continuations in northeastern Turkey, and also the relationships, lateral and vertical variations, similarities and differences of both districts, will be presented.

Ketin's (1966) classification has been adopted for the tectonic zones of Turkey, Adamia's (1974) and International Geological Congress XXVII Session Guidebook's (1984) classification for the tectonic zones of the Caucasus.

On the basis of the main geologic characteristics, the rock units of the Caucasus are divided into tectonic zones. Each zone has pre-Liassic, Liassic, and post-Liassic units reflecting different geotectonic environment.

The Caucasus was bordered by the Scythian platform to the north and by the Anatolide-Tauride and Iranian platforms to the south. The northern part of the Caucasus is called the Great Caucasus, the median part is the Transcaucasus, and the southern part is the Lesser Caucasus. The tectonic zones of the Great Caucasus lying from north to south are presented below: the Laba-Malka Zone (the Bechasinian Subzone and the Forerange Subzone), the Main Range Zone, and the Southern Slope Zone. In the Transcaucasus, the Gagra-Djava Zone and the Drizula massif and its coves are situated to the north, the Somcheti-Kafan (Karabakh) Zone to the south and the Adjara-Trialetian and Talysh Zones, which are the continuation of each other, are situated between the previous zones. In the Lesser Caucasus, the ophiolitic belt (the Sevan-Akeran Ophiolitic Zone to the north, the Vedi Ophiolitic Zone to the south) and the Miskhan-Zangezur Zone and the Araks Zone of the northern part of the Iranian platform have been distinguished from each other.

The Caucasus tectonic zones are mainly delineated from each other by the overthrust planes dipping 70-80° to the north. The Oligocene-Recent molasse showing enormous lateral and vertical facial changes overlies the rocks of the Caucasus tectonic zones conformably or unconformably in places.

The results, presented below, can be obtained by the correlation of the Caucasus tectonic zones and tectonic zones of northeastern Turkey:

1. The tectonic zones of the Great Caucasus and northern part of the Transcaucasus cannot by followed in northeastern Turkey.

2. The Adjara-Trialetian Zone continues along the Black Sea shores. The rock units are similar in both districts of the zone.

3. The Somcheti-Kafan Zone, which is the south part of the Transcaucus, corresponds to the Pontial Zone. The pre-Liassic units of both regions are made up of a complex whose units are composed of gneiss, schist, amphibolite, marble, granites cutting them, and ophiolitic slices with tectonic contacts. Carboniferous-Permian volcanosediments of the shallow environment overlies granite-schist complex unconformably as well.

In the Jurassic-Cretaceous and Tertiary sequences, alkaline and calcalkaline volcanics are alternated with each other along both sides of the zone. In addition, the volcanics show lateral and vertical transitions. In the same sequences, volcanics are dominant in the Somcheti-Kafan Zone, but sediments are dominant in the Pontian Zone. Nevertheless, in the Eastern Pontides, volcanics are dominant in the north, but sediments are dominant in the south in the Jurassic-Cretaceous sequence, whereas the sediments are dominant in the north, but volcanics are dominant in the south in the Tertiary sequence. On the other hand, open folds are observed in the north, but tight folds and overthrusts are observed in the south in the Jurassic-Cretaceous sequence. There are clear unconformities in the Mesozoic-Tertiary sequence of the Eastern Pontides, but there are only local unconformities in the Mesozoic-Tertiary sequence of the Somcheti-Kafan zone.

4. The Lesser Caucasus Ophiolitic Belt corresponds to the North Anatolian Ophiolitic Belt. There may be Paleozoic, Jurassic-Lower Cretaceous, and Upper Cretaceous ophiolites in both regions of the belt. It is accepted that Jurassic-Lower Cretaceous ophiolites occurred in the midoceanic ridges, but Upper Cretaceous ophiolites occurred depending on the ensimatic arcs.

The age of ohiolitic complex in the Lesser Caucasus is of pre-Upper Coniacian, but in the south of the Eastern Pontides is of pre-Upper Campanian. However, in the south of the belt mainly around Erzurum Hinis area, the ophiolitic complex is unconformably overlain by Maestrichtian clastic rocks. That is, the units overly the ophiolitic complex unconformably and become younger from north to south in eastern Anatolia.

The ophiolites are reworked considerably in both regions of the belt in the Late Eocene.

5. In general, the Iranian platform of the Lesser Caucasus corresponds to the Anatolide-Tauride platform. The pre-Liassic units of the Miskhan-Zangezur Zone may be equivalents to the Central and East Anatolian massifs. Jurassic-Cretaceous units are represented by clastic rocks in the south of the Lesser Caucasus and by carbonates on the Taurus platform.

If the characteristics presented above are reviewed again, it can be postulated that tectonic zones of the southern Caucasus and northeastern Turkey may be interpreted as the products of different levels of the same basin as well as different independent basins.

REFERENCES

Adamia, S. A., 1974, 1/1.000.000 ölçekli Kafkasyanin tektonik haritasi açciklamasi, Tbilisi, 23 p. (Translation by Hediye Özkok, MTA archive, Ankara.)

International Geological Congress, 1984, XXVII Session, Moscow, Georgian Soviet Socialist Republic, Excursions Guidebook, publishing house "Khelovneba," Tbilisi, 224 p.

Ketin, I., 1966, Tektonic units of Anatolia (Asia Minor): MTA Bulletin, v. 66, p. 23-34.

YILMAZ, A., General Directorate of Mineral Research and Exploration, Ankara, Turkey

Origin and Setting of Two Melange Belts in Turkey

Tow melange belts in Turkey are important. One of them is situated along the southern part of the Pontian Zone, pre-Liassic in age. The other is to the south of the first belt, Upper Cretaceous in age, which is known as Anatolian Ophiolitic Belt.

The western part of the first belt, known as the Karakaya formation, has been interpreted as deposits of rifting during Triassic (Bingöl et al, 1975). The eastern part of it was considered as a subduction melange (Tekeli, 1981). A different interpretation is that of Sengör and Yilmaz (1981), who regard the Karakaya basin as a marginal sea that opened above the Paleotethyan subduction zone.

The eastern part of the first belt, the Tokat area and its eastern continuations, is renewed again. Three different rock units can be differentiated in this belt. These are, from north to south, greenschists, melange, and ophiolites in turn. The northern part is composed of metavolcano sediments in greenschist facies, which were intruded by small granitic rocks. The age of some intrusions of these magmatic rocks is probably pre-Liassic. This part may represent an island arc lithology. The second part is made up os a metavolcano sedimentary sequence with ostoliths in different origin and has undergone greenschist metamorphism. In this part, neritic limestone intercalations are dominant rock units in places. The ages of these parts are changing from Permian to Triassic without any break. And the third part is ophiolites that are made up of serpentinite, periodotite, and diabase dikes that are cutting others. Metavolcano sediments that are directly beneath ophiolites turned into the blueschists. Although the data are insufficient, it is postulated that the units of this belt may come together by the obduction mechanism after forming in back-arc and related ensimatic arc systems during Permian-Triassic times, rather than a subduction event.

The second belt, which is situated along Izmir, Ankara, between Tokat and Sivas, Erzincan, and Erzurum, is accepted in common as a melange prism that formed during Upper Cretaceous.

The eastern part of the second belt, between Tokat-Sivas, Erzincan, and Erzurum, areas are reviewed again as well. There are two parts of this melange too. The pre-Upper Campanian part consists of juxtaposed tectonic slices of rocks with different origin. There is no matrix in this melange part, and constituents of it are mainly made up of cherts, volcanics, limestones, ophiolites, and metamorphics that derived from Tokat and Central natolian Massifs. The contacts between constituents are tectonic. The thrusts delineating the tectonic slices dip northward at 30°-70°. It is possible to interpret this part as a subduction melange. The Upper Campanian-Eocene part is made up of magmatic, metamorphic, sedimentary, and volcanic blocks in a matrix of sandstone, siltstone, and volcaniclastic rocks. Products of submarine slumps are common in some levels of this unit. The second part overlies forst part unconformably and may represent fore-arc deposits as pointed by Koçcyigit (1987). Some of melanges argued by Tekeli (1980) and Özkaya (1982) may be reworked materials of the first parts. In that connection, the Central and Eastern Anatolian massifs may be tectonic windows (Yilmaz et al, 1986) beneath obducted ophiolites plus subduction melange.

REFERENCES

Bingöl, E., B. Akyürek, and B. Korkmazer, 1975, Biga yarimadasinin jeolojisi ve Karakaya Formasyonu'nun bazi özellikleri: Cumhuriyetin 50: Yili Yerbilimleri Kongresi, Tebligler, MTA-Ankara, p. 70-77.

Koçcyigit, A., S. Özkan, and B. Rojay, 1987, Examples from the forearc basin remnants at the active margin of northern Neo-Tethys: emplacement age of the Anatolian Nappe, Turkey: Melih Tokay Geology Symposium '87, Abstracts, Ankara, p. 18-19.

Özkaya, I., 1982, Origin and tectonic setting of some melange units in Turkey: Journal of Geology, v. 90, p. 269-278.

Sengör, A. M. C., and Y. Yilmaz, 1981, Tethyan evolution of Turkey: a plate tectonic approach: Tectonophysics, v. 75, p. 181-241.

Tekeli, O., 1980, Structural evolution of Aladag mountains in Taurus belt: Bulletin of Geological Society of Turkey, v. 23, p. 11-15.

——— 1981, Subduction complex of Pre-Jurassic age, northern Anatolia, Turkey: Geology, v. 9, p. 68-72.

Yilmaz, A., I. Terlemez, and S. Uysal, 1986, Erzurum güyeydogusunda yer alan Hinis, Tekman ve Karayazi arasinin jeolojisi: MTA Rep., 8089 (unpublished).

YILMAZ, A., General Directorate of Mineral Research and Exploration, Ankara, Turkey

Thoughts on Crustal Structure of Eastern Anatolia, Turkey

The study area is located within the framework of the Eastern Mediterranean Alpides between the Eastern Pontian and the Bitlis-Pötürge fragments. The aim of this study is to demonstrate the different approaches on the crustal structure of the area depending on the basis of geological evidence and to present new evidence about the structure of the area.

The units occupying lowermost section in the Eastern Anatolia are overlain by the young formations and therefore crop out in limited exposures. This causes some contradictory approaches in interpreting the lowermost crustal structure of the region. So, there are different approaches on the crustal structure of the Eastern Anatolia. For instance, on one hand Seymen (1984) proposes that the melange obducted upon the metamorphic crystalline basement and underlain by a volcanosedimentary cover sequence of Eocene-Quaternary age. On the other hand, the basement of the Eastern Anatolia is commonly accepted as a melange prism (Arni, 1939; Ketin, 1977; Sengör and Yilmaz, 1981). But if the data presented by Yilmaz et al (1986) and Senel (1987) or observations made on the other ophiolitic exposures in the Eastern Anatolia are reviewed, it can be proposed that the Eastern Anatolia has a napped structure. The metamorphic crystalline basement is the lowermost level and may be a fragment of the Anatolide-Tauride platform. Ophiolites constitute only in a limited extent related to the ophiolite nappe emplacement. These units are unconformably overlain by Maestrichtian-Quaternary volcanosedimentary cover.

According to both areal distribution and boundary relationships of the ophiolitic units, it can be concluded that the ophiolitic melange might have been formed due to the contraction and slicing of the oceanic crust during Cenomanian to Campanian.

The post-tectonic Maestrichtian-Quaternary cover is made up of successive and regular transgressive and regressive sequences. There are frequent lateral and vertical facial changes in the sequence. Maestrichtian-Eocene part of the cover contains reworked materials of ophiolitic rocks in places. The study area became a piece of land as a whole, as being parallel to the development of the last regressive sequence that began to form during Middle?-Upper Miocene.

Eocene deposits are cut by gabbros and overlain by andesites. Oligocene deposits contain only thin andesitic basalt interbeds. Middle and Late Miocene-Quaternary deposits include dacite and andesite in the lower levels, but basaltic pyroclastics and lavas in the upper levels.

As a consequence of the compressional events occurring in the Late and post-Miocene times, parallel northwest-southeast–trending right-lateral and parallel northeast-southwest–trending left-lateral oblique fault zones were developed in addition to the overthrusts and folds trending approximately east-west. During these times, new basins are developed under the control of the oblique fault zones in which the lateral offset is dominant.

As a result, it can be postulated that the recent crust has a napped structure, and crustal shortening and thickening in Eastern Turkey (Saroglu et al, 1980) are the result of compressional oblique system that has begun since Late Cretaceous.

REFERENCES

Arni, P., 1939, Tektonische Grundzüge Ostanatoliens und benachbarter Gebiete: Veröff. Inst. Lagerstöttenfosch, Türkei, Ser. B, 4, 90 p.

Ketin, I, 1977, Van Gölü ile Iran siniri arasindaki bölgede yapilan jeoloji gözlemlerinin sonuçclari hakkinda kisa bir açiklama: Türkiye Jeol. Kur. Bull., v. 20, p. 79-85.

Seymen, I., 1984, 30 Ekim 1983 Horasan-Norman Depreminin Oluçs Mekanizmasi Açcisindan Deprem Fayi ve Bölgenin Kabuk Yapisi Üzerine Bir Yaklaçsim: Kuzey Anadolu I. Ulsal Deprem Simpozyumu, Bildiriler, p. 335-348.

Saroglu, F., Y. Güner, W. S. F. Kidd, and A. M. C. çSengör, 1980, Neotectonics of Eastern Turkey: new evidence for crustal shortening and thickening in a collision zone: EOS Transections, American Geophysical Union, v. 61, n. 17, p. 360.

Senel, M., 1987, 1:100.000 ölçcekli açcinsama nitelikli Türkiye Jeoloji haritalari serisi, Baçskale-H38 paftasi: MTA Publ., Ankara, 12.

Sengör, A. M. C.,, and Y. Yilmaz, 1981, Tethyan Evolution of Turkey: a plate tectonic approach: Tectonophysics, v. 75, p. 181-241.

Yilmaz, A., I. Terlemez, and çS. Uysal, 1986, Erzurum güneydogusun da yer alan Hinis, Tekman ve Karayazi jeolojisi: MTA Rep., 8089 (unpublished).

YILMAZ, Y., I.T.U. Maden Fakultesi, Jeoloji Bolumu, Ayazuga, Istanbul, Turkey

The Tethyan Evolution of Turkey

The tethyan evolution of Turkey began following the Carboniferous assembly of Pangea and comprised two major episodes: the Paleo-Tethyan evolution lasted from Westphalian B (or possibly earlier) time to the final elimination of Paleo-Tethys in the Middle Jurassic. The second episode, the Neo-Tethyan, began in the early Triassic (? possibly late Permian) rifting of the eastern Mediterranean and is still continuing. The products of the first episode of the Tethyan evolution are the Cimmerides, whereas those of the second episode are the Alpides. The Alpide evolution has a paleotectonic period lasting until the collision of Arabia with Eurasia and a neotectonic period followed that.

The Cimmerides in Turkey cover mostly the northern part of the country, the Rhodope-Pontide fragment and the Sakarya Continent with only local ingressions into the Menderes-Taurus block (e.g., Lycian Nappes, Engirek Mountains). They are divided into two major units: a northern "main trunk" and a southern "Karakaya branch." The main trunk represents the collisional orogen that resulted from the closure of the Paleo-Tethys, whereas the southern "Karakaya branch" is a smaller orogen produced by the earliest Jurassic collapse by southern-dipping subduction of a Paleo-Tethyan marginal basin. Both these branches were constructed mainly on a Devonian basement that belonged to Gondwanaland. Only in northeastern Turkey, the Cimmeride Istanbul nappe may have originally (i.e., before the late Triassic) belonged to Eurasia.

The rifting of the Neo-Tethyan oceans, now represented by the Intra-Pontide, Ilgaz-Erzincan, Izmir-Ankara, Sungurlu-Erzincan, Inner Tauride, Antolya, and Assyrian, was likely a back-arc phenomenon and implies tremendous later (mainly Cretaceous-early Cainozoic) strike-slip widening of the Turkish orogen to double a number of Neo-Tethyan sutures back onto themselves. The Assyrian, Antalya, and Inner Tauride oceans began opening in the Triassic and the others during the Liassic (mainly Sinemurian-Hettangian). Opening times are constrained mainly by structural evidence for normal faulting, stratigraphic evidence for rift sedimentation, and subsequent thermal cooling and foundering, and magmatic evidence for rift-related magmatism.

Neo-Tethyan convergence may have begun in Turkey earlier than hitherto believed (pre-Albian versus late Cretaceous), but it still substantially postdated the collisional obliteration of Paleo-Tethys. This causes certain problems of kinematics and may be the result of paucity of data. Widespread convergence began in the ocean (ophiolite sole aureole ages of pre 100 Ma) and rapidly migrated to the northern continental margin in the Rhodope-Pontide fragment, to the southern margin of Sakarya, and possibly to the southern margin of the Kirsehir block. On all of these, extensive arc magmatism characterized the Turonian (in places Albian) to Eocene development.

Vast sheets of mainly supra-subduction zone ophiolites were emplaced onto the Sakarya, Mendered-Taurus, Kirsehir, and the Arabian blocks. Emplacement onto the Kirsehir was clearly pre-Cenomanian, and emplacement onto Arabia Campanian was early Maestrichtian. Emplacement onto the Menderes-Taurus block was accompanied by widespread blueschist metamorphism along its northern margin.

Continental collisions in the Alpide era marched from north to south. The intra-Pontide ocean disappeared at the end of the Maestrichtian. The Izmir-Ankara ocean vanished in the late Paleocene, the Inner Tauride in the middle Eocene, and the Assyrian in the latest Eocene (?) early Oligocene. The Antalya suture terminally closed, probably in the early Cainozoic.

Collisional thickening in western Turkey generated a crust probably >60 km by late Oligocene, which led to extensive, post-collisional Tibet-type magmatism. A similar event in the east followed the mid-Miocene tightening of the Assyrian suture system and created the east Anatolian high plateau and led to coeval onset of westward escape of Turkey. This escape may have triggered the extensional collapse of the overthickened west Anatolian highland leading to the present-day Aegean extension.

Important episodes of sideways motion, in places exceeding 1000 km, have punctuated the convergent evolution of Turkey during its Tethyan evolution. Reconstructions will be presented to show these and to accommodate them in the overall basin evolution in the Tethysides.

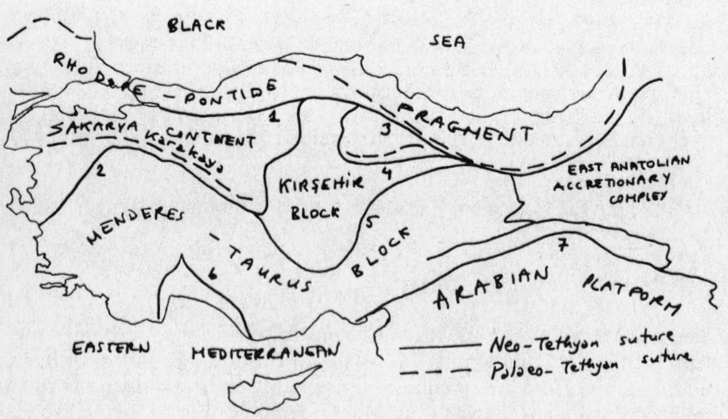

Figure Legend—1. Intra-Pontide Suture; 2. Izmir-Ankara suture; 3. Ilgaz-Erzincan suture; 4. Sungurlu-Erzincan suture; 5. Inner Tauride suture; 6. Antalya suture; 7. Assyrian suture

YIN Hanhui, Academia Sinica, Changshan, Hunan, China

A Huge Pb-Zn Deposit in Western Yunnan Diwa Region

The western Yunnan Diwa region lies in the Hengduan Mountains in southwestern China. Tectonically, it is a first order tectonic region, lying in the transition zone between the Indian and Chinese platforms, at the juncture of some tectonic regions of different nature. Thus, it is a famous active tectonic zone. Before the Sinian, it was a geosynclinal region, but in the Jinning orogeny, it turned into a platform region. During the Mesozoic-Cenozoic, tectonic activity became intense so that the platform was reactivated and converted into a Diwa (geodepression) region. This Diwa region can be divided into two second-order tectonic regions, the Baolan geodome system and Puer Diwa system. The sandstone-conglomerate-hosted Pb-Zn deposit occurs in the Diwa structural layer of the northern part of the Puer Diwa system. This Diwa structural layer is composed of Mesozoic-Cenozoic redbeds consisting mainly of red sandstone, argilite, conglomerate, breccia, and intercalated salt-bearing beds with a total thickness of more than 8000 m. In the Diwa stage, tectonomagmatic activity was intense. The Nujiang, Lancangjiang, and Red River deep fractures cut through the region from north to south. Between the fractures lie the Chongshan and Cangshan regional metamorphic belts, each several hundred kilometers long with a width of several tens of kilometer north-northwest-trending, composed mainly of slate, marble, green schist, gneiss, and migmatite with evident schistose, gneissic, and mylonitic structures. The Yanshanian granitic magma intruded into the metamorphic belts along the fracture zone. But more intense magmatic activity is the emplacement of hypabyssal and superhypabyssal alkaline magma of late Yanshanian-Himalayan age into the Diwa structural layer. Basic and ultrabasic rocks are distributed along deep fracture zones as a string of beads, which demonstrates that deep fractures cut through the crust, and mantle material was involved in tectonomagmatic activity in ge in the region.

The huge Pb-Zn deposit is located at the center of the northern part of the Puer Diwa system. Here different trending fractures intersect and the Paleogene contains abundant ore-forming materials and attains a thickness of more than 4000 m with intercalation of many salt-bearing beds, but without magmatic rocks. Therefore, the site where the deposit is located in an intersection of fractures, a sedimentary center, and a mineralization center in the Paleogene.

This huge Pb-Zn deposit consists of a large bedded orebody concordant with the Paleogene and some small orebodies. The large one occurs in the Yulong Formation of the Paleogene, associated with salt-bearing beds. It is several ten of meters in thickness and more than 1000 m in dip length. The ore could be divided into two kinds, sandstone hosted and breccia hosted. The major ore minerals are sphalerite, galena, pyrite, celestite, berite, gypsum, quartz, and calcite.

The sulfides in the deposit are enriched in S^{32}, with μS^{34} values ranging from -0.4 to -21.0(%). The average value is -17.4(%) for sandstone-hosted ore and -10.9(%) for breccia-hosted ore. Gypsum has μS^{34} values ranging from + 14.5 to sedimentary rocks without obvious relation to magmatism. Thirty-nine samples of galena in the deposit have been analyzed for lead isotopic composition by different research workers, but they all got similar values of lead isotopic compositions. There are average values: $^{206}Pb/^{204}Pb$ is 18.281, with deviation being 0.09; $^{207}Pb/^{204}Pb$ is 15.455, with deviation being 0.09; $^{208}Pb/^{204}Pb$ is 38.203, with deviation being 0.233. Homogenous lead isotopic composition is noticeable. Average % value is 8.75. They are similar to those of tholeiite from midocean ridges, indicating a deep-seated derivation from a homogeneous U-Th-Pb system. Results of the stable isotopic studies on H_2O in primary fluid inclusion of celestite are: μD value is -100.55 (AMOW), $\mu^{18}O$ value is -8.64(SMOW). These values are similar to those of underground hot water.

The mechanism of formation of the rift-type Diwa region and the huge Pb-Zn deposit is shown in the following:

Geological and geophysical data indicate that before the Mesozoic, the Indian and Chinese platforms were separated, but after the Triassic they were sewed up, and the united Indian-Chinese platform was formed. The Red River fracture zone might have been the suture line.

In the early Mesozoic, creep of mantle materials resulted in updoming of the crust, and the central part of the geodome bore tension faults, along which intermediate-acid volcanics erupted.

In the late Triassic and Jurassic, a large Diwa basin of graben type (a first-order graben) occurred in the central part of the dome. Transgression and regression took place frequently, and mixed marine and continental sediments were deposited in the graben.

In the Cretaceous, the crust experienced further uplift and seawater completely withdrew from the region. A second-order structural basin occurred in the large Diwa basin of graben type, and Cretaceous strata with salt-bearing beds were lain down in it under a dry, hot, and closed paleogeographical conditions.

In the Paleogene, a large rift occurred in the central part of the Cretaceous second-order structural basin and Paleogene strata with more than 4000 m in thickness and many salt-bearing beds were lain down as a red molassoid formation. At that time, ore-forming materials of the Pb-Zn deposit came from the deep and ascended along fractures into the rift lake so that source beds and orebeds were formed.

In the Himalayan orogeny, both vertical and horizontal movements were strong, and an unconformity occurred between the Paleogene and Neogene; large nappes made the Upper Triassic, Jurassic, and Cretaceous thrust onto the Paleogene; dynamometamorphic belts got a further development; alkaline, basic, and ultrabasic igneous rocks were intruded. This indicates that tectonically the region was in the maximum-mobility period of the Diwa stage. At that time, ore-forming materials in the Paleogene were reworked and entered into hydrothermal solution. In the late Himalayan orogeny, tectonic activity dropped in intensity. In the residual-mobility period of the Diwa stage, the ore-forming materials in the hydrothermal solution were enriched and precipitated, resulting in the huge Pb-Zn deposit with high-grade ore.

REFERENCES

Chen Guoda, 1987, Methods of metallotectonic researches: Geological Press.

Yin Hanhui, et al, 1984, Sedimentary-reformed Pb-Zn deposits in China: Scientia Sinica, B, v. 8, p. 852-864.

YOCHELSON, E. L., National Museum of Natural History, Washington, D.C.

Geologic Time; As Indicated by the Sedimentary Rocks of North America

August 17, 1893, Charles Doolittle Walcott (see Yochelson, 1967) delivered an address with the title quoted above, retiring as vice-president of section E of the American Association fo the advancement of Science. This talk was published almost simultaneously in the October-November issue of the inaugural volume of the Journal of Geology, and in the American Geologist in October (v. 12). The following year it was published in the AAAS Proceedings (v. 42); reprints are known from all three sources. Also in 1894, it was printed in the Annual Report of the Smithsonian Institution (for 1893) and again in the United States Congressional Record (Walcott, 1894). As a result, by several orders of magnitude, this was the most widely distributed geologic paper in North America during the 19th Century; in the 1890s, the geologic community consisted of about 500 persons. No detailed concordance of the five has been prepared, but all printings appear identical.

Walcott began "Of all subjects of speculative geology few are more attractive or more uncertain in positive results than geologic time." He then went on to cite the views of earlier and current workers on the age of the earth or on selected parts of the geologic column. those quoted form an eclectic list; in order, they are: Hutton, Lyell, Houghton, Croll, Wallace, Winchell, Reede, Dana, McGee, and Upham.

Calculations of geologic time based on the thickness of measured sections and assumptions as to rates of erosion or of deposition were not new. However, Walcott was the first to use this approach with information from western North America. He chose the Cordilleran region, in part because of his familiarity with its geology in Utah, Nevada, and Arizona; published data on stratigraphy in Idaho, Montana, and British America was also included in his discussion and calculations.

Walcott went to lengths to support his methodology. Change in shape of the continent and paleogeography of Paleozoic seas of the Cordilleran region were summarized before he considered sediment source. He distinguished between mechanically- and chemically-formed sediments in the region was its predominately calcareous Paleozoic section. Walcott emphasized the overall uniformity of the rock column throughout this elongate region, and the shallow water sedimentary features of the strata. In 1881-1882, he had measured a section of more than two miles Algonkcks in the Grand Canyon, and used these data for a far-reaching observation. 'THere is absolutely nothing to indicate more rapid denudation and corresponding deposition in this early pre-Cambrian series than we find in the Paleozoic, Mesozoic or Cenozoic formations."

From extrapolations on rates of denudation and of deposition, Walcott derived a minimum of 500,000 years for duration of pre-Upper Cambrian Paleozoic. He contrasted these mechanical sediments with younger Paleozoic ones, presumably deposited more rapidly, and arrived at about 660,000 for that part of the column.

Considerably more attention was given to the carbonates. Following suggestions of Bailey Willis, Walcott estimated the enormous volume of limestone in both cubic miles and mile-feet of thickness. He next discussed what was then known of present-day depositional rates of lime, by larger organisms and by chemical precipitation, drawing mainly on the "Challenger" reports. In contrast to speculations of some authors of the time, who suggested vastly different conditions in the Paleozoic waters, another important generalization he drew was "...judging from the deposition that is going on today in the great oceans, the accumulations of calcareous sediments has gone on in the past as steadily and uninterruptedly as at present, subject to varying conditions of temperature, life, depth of water, etc."

Walcott was greatly hampered in that most information on modern carbonates then came from deep water deposits; the Bahamas Banks and similar areas were essentially unknown and could not enter into his considerations. However, he did emphasize that conditions in the ancient western seas seemed particularly favorable for deposition of limestones. As the final step, he converted thickness/rate of limestone deposition into years.

Walcott estimated 17.500.00 year for Paleozoic time and 45.150,000 years for post-Archean time. He ended, "In conclusion, geologic time is of great but not infinite duration. I believe that it can be measured by tens of millions of years, but not by single millions or hundreds of millions of years." Little more than a decade later, radiometric dating would prove him in error and eclipse this work. However, Walcott provided a longer time span for the geologists than was accorded them by the physicists. Further, his proportioning of time among Paleozoic, Mesozoic, and Cenozoic was a departure from the accepted standard of his day, but is essentially unchanged from current calculations based on radiometric dates.

REFERENCES

Walcott, C. D., 1894, 53rd Congress, 2nd session. Misc. Document 184, part I. (This is a reprint of the Smithsonian Institution Annual Report for 1 p. 470540. (Contains a summary of Walcott's career, including his field work in the west prior to 1892, plus a listing of his publications).

YOGODZINSKI, G. M., S. KAY, and R. W. KAY, Cornell University, Ithaca, NY

Geochemical Evolution of Early Series in Aleutian Arc, Alaska

Tertiary sequences in the western Aleutians are geochemically distinct from those in the eastern and central Aleutians. These data suggest that the basement in the central and eastern Aleutians was built by arc processes, while that in the western Aleutians was principally built by Tertiary back-arc magmatism with only minor arc magmatism. The complex tectonic evolution of the western Aleutians is underscored by the presence of the Bowers and Shirshov ridges and the relatively young marginal Komandorsky and Bowers basins in the back-arc. Models suggesting that western Aleutian basement was formed along a continuous arc with the central and eastern Aleutians at approximately 55 Ma do not account for these features.

The oldest rocks in the central Aleutians, the Finger Bay Volcanics on Adak Island (52 °N, 177 °W), have arc geochemical signatures that are generally similar to those in younger central Aleutian arc rocks. These Finger Bay Volcanic rocks, some of which have minimum ages of 50 Ma ($^{40}Ar/^{39}Ar$ age, Rubenstone, 1984), apparently resulted from subduction that initiated at an old fracture zone about 55 Ma (Scholl et al, in press). Compared to MORB, the Finger Bay volcanics are enriched in light rare earth elements (LREE) (La/Sm = 2.0-2.8), depleted in high field strength elements (HFSE) (La/Ta = 40-60) and have high Th/Ta ratios (4-12) and low ϵNd (7-8) (Rubenstone, 1984). Some of them have relatively high La/Yb and low La/Th ratios, which could reflect the influence of the fracture zone. Tertiary plutonic rocks intrusive into the Finger Bay volcanics show similar incompatible trace element and isotopic signatures while more compatible elements reflect the thickening and maturing of the arc crust. Compared to older Adak magmatic rocks, Adak Plio-Pleistocene volcanic rocks are notably enriched in Th, U, and Ba relative to the REE. These trace element differences are most easily explained by a change in a subducted sedimentary component in the arc magmas related to the build-up of a subduction complex in the central Aleutians in the last 5 Ma (Scholl et al, in press).

Tertiary basement volcanic rocks from the Amchitka Formation and from near the bottom of 2.1 km deep drillhole on Amchitka Island (51 °N, 179 °W) have geochemical characteristics similar to the Finger Bay Volcanics (La/Sm = 2.0-3.3, La/Ta = 25-75, La/Hf = 2.4-5.8, Th/Ta = 2.7-7.1). These data suggest that the Aleutian basement as far west as the intersection of the Bowers Ridge (179 °W) was formed by island-arc processes.

In contrast, Tertiary basement rocks in the western Aleutians on Attu Island (53 °N, 171 °E) are primarily MORB-like basalts (some pillows) and Basaltic andesites associated with minor amounts of intermediate composition volcanic rocks. This dominantly mafic basement is intruded by small bodies of isotopic gabbro (5% of the total exposure). No periodotites or rhythmically layered gabbros have been observed. All of the Attu rocka have been bivariably affected by seawater alteration. Geochemically, the basalts show MORB-like characteristics; i.e., light REE depleted REE patterns (La/Sm 0.9-1.7), low La/Ta (15-20), La/Hf (1.4-2.5), and Th/Ta (0.2-1.0) ratios, and high Nd isotopic ratios (ENd = +10 to +11) (Rubenstone, 1984). The gabbros have whole rock compositions that range from nearly liquid compositions similar to the basalts to cumulates of plagioclase + clinopyroxene + oxide + amphibole with variable amounts of trapped liquid. Incompatible trace elements ration in the gabbro are similar to those in the basalts, but some gabbros have slightly lower ENd values (+8.9 to +9.5) that could reflect open systems behavior during cooling within the crust.

Almost all intermediate composition Attu rocks (55-70% SiO_2) are intrusive. An important sequence of these amphibole-bearing diorites, quartz diorites, and tonalites crops out in the north-central portion of Attu. Whole-rock compositions suggest that these plutonic rocks range from nearly frozen liquids to crystal cumulates with variable amounts of trapped rhyolitic liquid. The most evolved intrusive rocks (70-75% SiO_2) (2-10 km^2) are occasionally surrounded by margins of glassy rhyolite (10-100 m thick). These rhyolites are high in Na2O (up to 7.4%), but can have up to 3.3 wt% K2O and thus are not like the low-K trondjemites commonly observed in ophiolites. These rhyolites, which have ϵNd from 8.8 to 9.0, appear to be the product of extended fractionation (85-95%) of a MORB-like mafic parent.

The Attu rhyolites bear a striking resemblance to quartz diorites dredged from the Mid-Atlantic Ridge (Aldiss, 1981) and to some Icelandic rhyolites. This similarity of the Attu rhyolites to those formed in midocean ridge settings and the MORB-like character of the basalts suggests that the Attu basement formed in an oceanic rift setting, possibly a back-arc basin. Comparison of the geochemistry of the Attu basement with other marginal basin sequences suggests that if the Attu rocks formed in a marginal basin, that basin was well developed. Only in the final stage of volcanic activity on Attu (middle-to-late Miocene) were a small volume of calcalkaline andesites and dacites erupted.

REFERENCES

Aldiss, D. T., 1981, Plagiogranites from the ocean crust and ophiolites: Nature, v. 289, p. 577-579.

Rubenstone, J. L., 1984, Geology and geochemistry of Early Tertiary submarine volcanic rocks of the Aleutian Islands, and their bearing on the development of the Aleutian island arc: PhD. thesis, Cornell University, Ithaca, NY, 350 p.

Scholl, D. W., T. L. Vallier, and A. J. Stevenson, in press, Geologic evolution and petroleum geology of the Aleutian Ridge: Circum-Pacific Council for Energy and Mineral Resources, Earth Science Series.

YONG NING, Tao, China Offshore Oil Exploration and Development Research Center, Hebei Province, China

The Geophysical Features of Southern China Sea

The offshore area of China includes the Bohai Sea, Yellow Sea, Eastern Sea, Gaiwan Strait, Southern Sea, and Baibu Gulf (Tonkin Gulf). Since CNOOC cooperated with foreign companies to explore the offshore oil and gas in 1980, more than 350,000 km of seismic line has been completed in the above area. Meanwhile, approximately 1200 prospects have been discovered and 394 wells drilled and 12 oilfields established. They produce more than 1.0-1.2 million tons of oil each year. In the last eight years, $2.5 billion has been invested by the contractors.

The southern China Sea (SCS) is the biggest one among these, but it is less fully explored which provided with very few data. However, all around the continental shelf (encircling the land) has favorable geological condition for hydrocarbon accumulation. Each country has obtained information.

(1) The region from north Xisha (Paracel IS)-(Zhingsha (Macclesfield) toward the China Coast, which drilled 26 oil-bearing prospects and confirmed 10 billion ton oil reserve.

(2) The sea located in west Luzon exists more than 10 west-east-trending magnetic strips, spreading from north to south and stretching about 700 km. This is a sea-basin expansion region, the sediments is thinner than others. Hence, this region is unfavorable for hydrocarbon accumulation.

(3) Palawan and Reed Bk are two shelf basins, they coincide with all geophysical data. The reserve in this region reaches 0.3 billion ton, and produces 0.4 million tons of oil each year.

(4) Brunie-Sabah and ZhenMu Basins are also coincide completely with gravity, magnetic, seismic information. This region's oil reserve reaches 0.85 billion tons and produces 12 million tons oil each year.

(5) Southwest Block consists of Wan-An Bank Basin and four or five other basins. Sediments 5 km thick, thus it is a favorable region. In addition, Malay Basin is adjacent to this region. There are sediments 9 km thick. It possesses .55 billion ton reserve and produces 13.8 million ton oil each year.

(6) Sarawak Coast: Reserve is 0.55 billion ton. Production is 7.8 million ton each year.

(7) The scope of ocean crust is situated as following: west-Vietnam coast fault; east-Luzon marine; north-Xisha (Paracel) and Zhingsha (Macclesfield); south-Nansha (Spratly Arch); where is the deep-water region, positive gravity anomaly. Moreover, the sediments are thinnest. The feature is shown symmetry like a mirror.

On Bouguer gravity, the islands and reefs are reflected as negative anomalies. On free-air gravity, they are reflected as positive anomalies. For example, Taiwan -90/200 mg; Hainan -30/+45 mg; Dongsha (Pratas Grp) -18/+20 mg; Xisha +2/+40 mg; Zhingsha -80/+60 mg; Nansha -50/+50 mg. Mostly, their magnetic tensions are low, but individuals show a bit high.

Besides, some seabed mountains also have similar reaction on the map.

YONG-CHANG, Xu, and PING Shen, Academia Sinica, Lanzhou, Gansu Province, China

Isotopic Geochemistry of Fossil Fuel in China

Fossil fuel is a general term for energy resources such as coal, petroleum, and natural gas, which were generated by organic matter during a long long period of geologic time. The geological and geochemical studies show that they were usually associated with each other, namely syngenesed or interlaced with each other in space or transmitted on a vertical section in the same basin. The formation of fossil fuels has much similarity and many characteristics. On the basis of studying isotopes of gaseous hydrocarbons, light hydrocarbons, crude oils, and organic matter from coal series, the paper discusses the generality and characteristics of isotopic composition in different fossil fuels and their significance, and also provides useful information for exploration and exploitation of various fossil fuels.

Isotopic geochemical characteristics of organic matter from coal series: The carbon isotopes of coals from various coal-accumulation basins are widely studied in China. (1) Carbon isotopic composition of total components in coals: 200 coal samples were collected from different coal rank in the coal-accumulating basins. The studies show, generally the $\delta^{13}C$ values of whole coals are -21 to -25 ppt, the average value is -24 ppt. In the mass, the thermorevolution process of coals is not obviously correlated with the $\delta^{13}C$ values of whole coals. (2) The carbon isotopic composition of macerals: The different macerals have different type of organic matter from coal series. The carbon isotopic composition of different macerals in the same coal sample differs and can be determined. The studies on carbon isotopic composition of macerals from the Junggar basin in Xingjiang province show the exinite normally has a lighter carbon isotopic composition. (3) The characteristics of carbon isotopic composition from coal extracts: Both family components of choloraform "A" belonging to coal extracts and characteristics of carbon isotopic composition of kerogen are studied by use of eight coal samples whose characters are similar but whose coalification is different. The results show: The fraction of aromatic hydrocarbons has the character enriching ^{13}C. In the stages of flame-coal and fat gas coal the carbon isotopic composition of family fraction of the coal changes more heavily with the intensification of polarity, and it ought to be the main stage mature of oil and gas. The $\delta^{13}C$ of kerogen from eight coal rank is very stable, it becomes -24;1 ppt. (4) The characteristic isotopic composition of thermoranalogue products (gases, liquids, solids, and remaining slag extract from organic matter of coal series.

Characteristics of the carbon and hydrogen isotopic composition of gaseous hydrocarbon in natural gases: The carbon and hydrogen isotopic compositions of methane in more than 160 gas samples from ten basins in China are analyzed. On the basis of the results, they can be divided into six types: bioganic gas, biothermocatalytic gas, gas associated with petroleum, gas associated with condensates, over-maturity gas, and coal-type gas. Isotopic composition of these gases is related to sedimentary environment, the type of organic matter, the stage of thermorevolution and characteristics formed in different gas reservoirs of depositional basins.

Isotopic geochemical characteristics of light hydrocarbon: The light hydrocarbons mainly refer to light oils and condensate associated with natural gases. Researching their geochemistry not only has significance for their formation and evolution, but also provides us with useful scientific information in studying geochemical characteristics of natural gas and its relation with gas source rocks. (1) The petroleum and condensate from identical sources have similar carbon isotopic composition, $\delta^{13}C$ ($\delta^{13}C_{condensate} - \delta^{13}C_{crudeoil}$) less than 0.5 ppt; (2) In the same area, the carbon isotopic composition of condensate generated by organic matter from coal series and of that by source rock series might be distinguished. The former is rich in ^{13}C, the latter in ^{12}C. (3) The carbon isotopic values of condensates with ^{13}C of kerogen from source rocks are similar, the latter is rich in ^{13}C, correspondingly. (4) The decisive factor of carbon isotopic composition of condensate is the isotopic inherit effect of oil- and gas-generated mother matter. (5) Natural gas-petroleum-condensate-kerogen come under a series, and they are rich in ^{13}C in order. In a same basin, the difference between carbon isotopic composition of a series related to coal series and that to oil-generated mother material is quite obvious. Conditions permitting the carbon isotopic contrast of the series might be used to distinguish the difference between coal-type gas and oil-type gas.

Isotopic geochemistry characteristic of petroleum: The studies on continent petroleum of China greatly enrich the understanding of characteristics of carbon isotopic composition of crude oils. The carbon isotopic composition of terrestrial petroleum has the widest change range, its value is from -21 to -32 ppt. The isotopic composition of carbon and hydrogen from crude oils related to typical freshwater, deep-lake facies is lighter; that from crude oils related to typical saltwater lake facies deposits heavier than that from marine facies crude oils. The crude oils related to petroleum and coal series also have heavier isotopic composition, $\delta^{13}C$ is about -24 to -25 ppt, which is similar to carbon isotopic composition of whole coals and of liquid hydrocarbon from products of thermal analogue. Contrasting carbon isotopic of crude oil family and extractant family might make comparison of oil and source better. The thermorevolution, migration, and other action have less effect on fraction of carbon isotopic composition of crude oils.

YORK, D., and P. W. LAYER, University of Toronto, Toronto, Canada, M. Lopez-Martinez, CICESE, Ensenada, Mexico, and A. Kröner, Institut fur Geowissenschaften

Thermal Histories from the Barberton Greenstone Belt, Southern Africa

The Barberton Greenstone Belt and the surrounding granite-gneiss terrane have a thermal and tectonic history spanning over 1500 my from 3600 Ma to 2000 Ma. The remarkable state of preservation of this area of the Kaapvaal Craton has made it one of the most thoroughly studied regions of Archean crust in the world. Recent geochronologic investigations involving U-Pb and $^{40}Ar/^{39}Ar$ dating methods have begun to piece together the complex thermal history of this area. Critical in these investigations is the use of single grain dating techniques of zircon evaporation for $^{207}Pb/^{206}Pb$ dating (Kober, 1986) and $^{40}Ar/^{39}Ar$ laser step-heating of hornblende, mica and feldspar. These new techniques have been combined with more traditional bulk sample methods of U-Pb and $^{40}Ar/^{39}Ar$ dating in our study.

In this paper we focus on four of the principal geologic events in the area and discuss the thermal evolution of the region as recorded by these events. The earliest is the extrusion and subsequent hydrothermal alteration of komatiites and komatiitic basalts of the Onverwacht Group, the basal member of the Barberton Greenstone Belt. $^{40}Ar/^{39}Ar$ step heating experiments on whole rock fragments of knmatiite have yielded plateau ages in the range 3450 Ma to 3490 Ma (Lopez-Martinez et al, 1984). Amphibole separates from the basalt show a well defined $^{40}Ar/^{39}Ar$ plateau at 3439/pmn4 Ma. Since the argon dates here are a measure of the age of metamorphism, their closeness to the Sm-Nd age of 3540/pmn15 Ma (Hamilton et al, 1979) implies that hydrothermal alteration followed soon after emplacement of the lavas.

The second event is the intrusion of the tonalite-trondhjemite plutons which surround the greenstone belt. Examples of this include the Kaap Valley Pluton and the Nelshoogte Pluton. The tonalitic Kaap Valley Pluton has a single grain $^{207}Pb/^{206}Pb$ age of 3227 ± 3 Ma, and single grains of hornblende have a $^{40}Ar/^{39}Ar$ plateau age of 3207 ± 20 Ma, indistinguishable within experimental errors from the zircon age. This agreement implies rapid initial cooling of the body. The trondjhemitic Nelshoogte pluton has a $^{207}Pb/^{206}Pb$ zircon age of 3212 ± 2 Ma. Although no hornblende is found in the Nelshoogte Pluton, hornblende from a crosscutting microgabbro dyke has a single grain $^{40}Ar/^{39}Ar$ plateau age of 3163 ± 13 Ma, which obviously represents a minimum hornblende age for the pluton. Muscovite and biotite separates from the trondhjemite have $^{40}Ar/^{39}Ar$ integrated ages which fall in the range 3080 Ma to 2860 Ma. The spectra from these minerals have plateaux which are slightly upwardly tilted, so that the oldest fractions from the muscovite are about 3110 Ma. These data imply that the Nelshoogte Pluton has experienced a prolonged cooling history.

The third event in the history of this area of the Kaapvaal Craton is the intrusion of a series of granitic plutons, one of which is the Mbabane Pluton in northern Swaziland. These plutons represent the last major pluton emplacement episode in the region. Single grain zircon $^{207}Pb/^{206}Pb$ ages (mean age of 2691 ± 4 Ma) and single grain $^{40}Ar/^{39}Ar$ hornblende ages (mean age of 2693 ± 8 Ma) are in perfect agreement. This coincidence of ages implies that the Mbabane Pluton cooled relatively rapidly after emplacement. Biotites from the pluton show evidence of some low temperature resetting roughly two billion years ago.

This resetting may have been caused by the fourth event, the intrusion of dolerite dykes. Whole rock dating of these dykes, from samples which crosscut the Nelshoogte Pluton, shows $^{40}Ar/^{39}Ar$ ages of 1900-2000 Ma.

The integration of U-Pb and $^{40}Ar/^{39}Ar$ dating methods is proving extremely useful in unravelling the one and a half billion years of complex thermal history of the Barberton Greenstone Belt. The use of recently developed single grain dating methods has greatly increased the rate at which precise, meaningful ages can be obtained. Even at this early stage of the application of this approach, there seems reason to suppose that a detailed understanding of the thermal history of this classic greenstone belt is an attainable objective.

REFERENCES

Hamilton, P. J., N. M. Evensen, R. K. O'Nions, H. S. Smith and A. J. Erlank, 1979, Sm-Nd dating of Onverwacht Group volcanics, southern Africa, Nature, 279, p. 298300.

Kober, B., 1986, Whole-grain evaporation for $^{207}Pb/^{206}Pb$-age-investigations on single zircons using a double-filament thermal ion source, Contrib. Mineral. Petrol. 93, p 482490.

Lopez-Martinez, M., D. York, C. M. Hall, and J. A. Hanes, 1984, Oldest reliable $^{40}Ar/^{39}Ar$ ages for terrestrial rocks: Barberton Mountain komatiites, Nature, 307, p. 352-355.

YOUNG, G. M., University of Western Ontario, Ontario, Canada

Plate Tectonic Setting of Some Proterozoic Glacigenic Successions

Glacigenic rocks are rare in Archean successions. They are common in the Early and Late Proterozoic but no well documented examples are known from the long period from about 2.0 Ga to 1.0 Ga. The purpose of this paper is, using stratigraphic context and basin analysis techniques, to attempt to identify the plate tectonic setting of some glacigenic successions in the Proterozoic and to speculate on the possible relevance of tectonic setting to the genesis of glacial conditions.

Early Proterozoic glacial deposits are best developed in North American as represented in Ontario by the Huronian (Gowganda Formation and less extensive older Huronian formations), in Michigan by the Reany Creak, Fern Creek and Enchantment Lake Formations, in southeastern Wyoming by the Headquarters Formation and two older diamictites of the Snowy Pass Supergroup, in the Northwest Territories by the Padlei Formation of the Hurwitz Group (and possibly diamictites of the underlying Mongomery Group) and by the Chjibougamau Formation in northern Quebec. Early Proterozoic glacigenic rocks have also been reported from Finland, and there are possible occurrences in central India.

In three regions of North America, the Great Lakes, Wyoming and the NWT, preservation is sufficiently good to permit basin analysis and an attempt at plate tectonic interpretation. In these areas, glaciation appears to have taken place during continental rifting and, in particular, following what is interpreted as the rift-drift transition of the plate tectonic cycle.

In the southern hemisphere, well documented Early Proterozoic glacigenic rocks occur in Western Australia (Meteorite Bore Member) and in South Africa (Makganyene Dismictite Formation and Rietfontein Diamictite Member). Within broad limits, these deposits are about the same age as those of North America. In both areas, the glacigenic deposits are locally preserved, occur above a "chemical miocline", and near the base of a thick silicilastic sequence here tentatively interpreted as a foreland basin succession.

Thus, the plate tectonic setting of Early Proterozoic glacigenic rocks in the northern and southern hemispheres appears to be different. In North America they lie above a siliciclastic-dominated succession, possibly related to continental rifting. In Australia and South Africa the glacigenic rocks overlie a mioclinal succession dominated by chemical deposits (dolostones and BIF). The North American, siliciclastic-dominated successions probably reflect a wet climate with abundant fluvial input, and possibly also a tectonically unstable setting. On the other hand the essentially chemical and biochemical mioclinal successions of the Hamersley and Transvaal basins may reflect aridity and a setting characterized by tectonic stability. The Early Proterozoic glacigenic rocks of the southern hemisphere may be associated with the closing stages of ocean basins, development of foreland basins and uplift of adjacent orogenic mountains.

In North America Early Proterozoic glaciation could be genetically linked to the rifting process, with opening up of new interior seaways and uplift of rift shoulders providing the conditions necessary for initiation of glaciation. The ocurrences in the southern hemisphere could likewise be ascribed to mountain building episodes related to ocean closure. On the other hand, if it can be shown that all of these occurrences are about the same age, then some global control, such as an "icehouse" effect, related to changes in atmospheric composition might be the prime cause, and tectonics could then be relegated to a secondary role in the development of an Early Proterozoic ice age (or ice ages) and in the formation and preservation of glacigenic rocks.

Late Proterozoic glacigenic rocks are known from all continents and span a period of at least 400 Ma. Some appear to be platformal deposits but the vast majority are preserved in troughs or basins with clear evidence of contemporaneous tectonic activity in the form of significant facies and thickness changes. In many areas, such as the Cordilleran region of North America, the Adelaide "Geosyncline" of South Australia, and in the Damaran of South Africa Late Proterozoic glacigenic rocks have been interpreted as having formed in a rift environment. This interpretation is reinforced by the occurrence, in intimate association with glacigenic deposits, of iron-formations and manganese-rich sedimentary rocks, interpreted as the products of hydrothermal activity in a rift setting. As in the North

American Early Proterozoic occurrences, there is a spatial and temporal association between rifting and many Late Proterozoic glacigenic successions.

In both the Early and Late Proterozoic there is a common association between glaciation and rifting. Some Early Proterozoic glacial successions of the southern hemisphere may be associated with mountain building during ocean closure. Whether these associations are fortuitous, and there were other more important controls such as changes in atmospheric composition, or whether they played a primary role in initiation and preservation of glacigenic deposits, remains unresolved.

YOUNG, G. M., University of Western Ontario, London, Canada

Evolving Tectonic Styles: Evidence From Some Early Proterozoic Basins of the Canadian Shield

In recent years there has been a growing trend towards plate-tectonic interpretation of Early Proterozoic basins. The "type" succession for the Early Proterozoic is the Huronian Supergroup, on the north shore of Lake Huron in Canada. The Huronian succession (up to 12 km thick) lies unconformably on Archean basement rocks from which it is, in some areas, separated by a paleosol. The lower Huronian consists largely of terrigenous clastic rocks of fluvial origin, associated with volcanics, some turbidites and some chemical sediments (dolostones and limestones). There are also some glacigenic diamictites in the lower Huonian. These rocks, because of their facies and limited distribution, are interpreted as having formed in a rift setting. The upper Huronian, in contrast, is much more widespread and consists largely of shallow marine rocks, interpreted as part of a terrace wedge or mioclinal succession developed on a passive margin. Widespread glacigenic rocks (Gowganda Formation) occur at the base of the mioclinal sequence. The lower portion of the Early Proterozoic successions in the Lake Superior region (Chocolay Group and Mille Lacs Group) are considered to be correlative with the upper (mioclinal) part of the Huronian Supergroup. The unconformably overlying Early Proterozoic rocks, comprise a thin siliciclastic succession, followed by Superior-type iron formation and a coarsening-upward flyschoid sequence. These rocks are considered to be the products of a migrating foredeep in a foreland basin setting, signaling closure of the ocean inferred to have formed during deposition of the mioclinal succession. Thus, the combined successions of the Huronian and the Lake Superior area can be interpreted the near-complete record of an Early Proterozoic Wilson Cycle of ocean opening and closure. Only the molasse is missing. A near-identical succession of rocks is present in the Snowy Pass Supergroup of S.E. Wyoming, about 2,000 km to the S.W. and presumably reflects the same sequence of tectonic events, suggesting the possibility of large scale plate tectonic processes.

The Huronian succession also contains evidence of evolving atmospheric composition: pyritic and uraniferous conglomerates occur only in the lower Huronian; red beds are present in the upper Huronian but are absent from fluvial deposits of the lower Huronian; ferric/ferrous ratios in both sandstones and mudstones of the Huronian succession show a marked increase in the upper Huronian. These lines of evidence support the idea of at least some free oxygen in the atmosphere during deposition of the upper Huronian.

Even more complete records of the Wilson Cycle are present elsewhere in Early Proterozoic supracrustal successions of the Canadian Shield. For example the Hurwitz Group on the west side of Hudson Bay and probable correlatives in the Amer and Montresor Belts to the north may be interpreted in terms of a rift episode (Montegomery Group), followed by development of a miocline (lower Hurwitz Group), succeeded by a flyschoid succession (including Fe-rich rocks, as in the Lake Superior succession) and cross bedded arkosic rocks and conglomerates interpreted as molassic deposits reflecting filling of the foreland basin in which the flysch was deposited. A similar succession is also present in many parts of the Circum-Ungava belt. Where paleocurrent data are available they commonly show a major reversal between the mioclinal succession, derived from the carton, and the flysch/molasse deposits which are either derived from an oceanward direction or show evidence of longitudinal transport.

These few examples serve to illustrate that the supracrustal record of Early Proterozoic times in the Canadian Shield can be interpreted in terms of lithospheric plate interactions comparable, in scale and general style, to those of the Phanerozoic. On the other hand there are unique features of the Early Proterozoic basins, such as the vast accumulations of iron-formations, and probably also the unique pyrite-gold-uranium-bearing conglomerates, is probably related in part to crustal evolution but must also reflect changes in the atmosphere and biosphere.

In Canada, most of the characteristics of these Early Proterozoic basins are in marked contrast to those of the Archean. On a global scale, the recently proposed 2.5 Ga boundary between the Archean and Proterozoic separates sedimentary basins of very similar sedimentary and tectonic style (e.g. Huronian and Witwatersrand) and places dissimilar basins in the same category. In some ways therefore, this time subdivision obfuscates what appear to be real differences among these basins, presumably reflecting significant stages in the evolution of the lithosphere. The fact that these differences developed diachronously, over a period of at least 500 Ma does not make them any less significant, nor does it detract from the dramatic nature of the differences at any one locality.

YOUNG, K. D., B. VOIGHT, and N. ORKAN, Pennsylvania State University, University Park, PA

Thin-Skinned and Thick-Skinned Crustal Rotations along Transform Faults

Macroscopic block rotations resulting from distributed shear have occurred along the Tjornes fracture zone (TFZ) of northern Iceland. Comparative examples from southern Iceland, exposures of ophiolite, and rotational structures found in continental crust suggest that rotations of crustal rocks along transform plate boundaries have the following associations:

1. Transpression or discontinuous master faults within continental crust; rotations are probably accommodated at depth along subhorizontal detachments.

2. Simple shear generated behind migrating transforms connected to paired propagating/failing rift boundaries.

3. Penetratively deforming, broad subvertical shear zones, possibly related to abnormally high heat flow under back-arc or mantle plume conditions.

Block rotations of style 1 have been detected by paleomagnetism and seismicity along the San Andreas system (Luyendyk et al, 1985; Nicholson et al, 1986) and the Dead Sea transform (Ron, 1987). This style of deformation along continental transforms is typically associated with restraining bends subjected to transpression. Mechanisms include rotation of antithetic shears. COCORP seismic reflection profiles and earthquake focal mechanism solutions suggest that near-surface block rotations are accommodated by subjacent subhorizontal detachments. Rotations occur by lateral coupling, and may be regarded as "thin-skinned."

Side-scan sonar surveys of a propagating rift along the Cocos-Nazca Rise exemplify type 2 rotations behind a migrating transforms (Searle and Hey, 1983). Distributed, heterogeneous shear apparently results from a gradual rather than instantaneous acquisition of the full spreading rate along the propagating rift.

Broad zones of shear associated with fossil transforms have produced rotations of type 3 in sheeted dike complexes at Bay of Islands, Newfoundland, and at the Troodos massif in Cyprus. At Bay of Islands, a 90° rotation of sheeted dikes has been produced by penetrative flow (Karson and Dewey, 1978), and shear may be regarded as "thick-skinned." North of the east-west-trending Arakapas fault belt at Troodos, the north-south-trending dikes of the main massif swing progressively westward into an east-west alignment as the fault is approached. Formation of the bend to sheeted dikes by simple shear has been verified by paleomagnetism (Bonhommet et al, 1988).

Along the TFZ in northern Iceland, rotations as much as 90° resulted from right-lateral shear south of the Husavik-Flatey fault system (HFSS), and rotations up to 120° occurred in rocks of the Tjornes peninsula north of the HFFS. Although the lava pile has been deformed by the brittle processes of fracturing, faulting, veining, and cataclasis, the spatial arrangement of dike orientations resembles that of a heterogeneous simple shear zone deformed by penetrative flow. Displacements are thought to have accommodated heterogeneous simple shear and block rotations along second-order antithetic and synthetic shears. Major antithetic faults at 0.5 to 2 km spacing reactivated the structural grain of north-northeast-trending faults and dikes inherited from crustal accretion, and experienced left-lateral slip that effected clockwise rotation of the faults and fault-bounded domains. Synthetic shears of northwest trend and minor right-lateral slip separate domains of incrementally greater clockwise rotation approaching the HFFS. Increases in rotation across domain-bounding, northwest-trending zones are about 2-4° at 0.5 km fault spacings. Regions of large rotation experienced the sequential formation of two or three antithetic fault sets, as earlier faults rotated to orientations of low shear stress.

Clay cake models of wrench faulting illustrate that antithetic shears become dominant when displacement is distributed over a significant width of the substrate (Hoeppener), rather than along the single slip surface used in typical shear box experiments (e.g., Wilcox et al, 1973). A shear zone deformed by penetrative flow is therefore suggested to underlie and be of comparable width to the zone of rotated crust exposed at the surface along the TFZ. Similar "thick-skinned" relationships may hold for Troodos and Bay of Islands.

In light of this model, transform deformation along the sinistral South Iceland Seismic Zone may be in the primitive stages of "thick-skinned" shear zone formation under association 3 and/or 2. Surface ruptures and distributions of damage associated with historic seismicity over the past 1000 years indicate an accommodation of strike slip via dextral antithetic faults with a 3-6 km spacing (Einarsson et al, 1981). Approximately 15° of counterclockwise rotation of the antithetic faults may be inferred from their present orientation with respect to the seismic zone. A possible southward migration of the deformation zone may be linked to southward propagation of an axial rift zone to the east (Einarsson, 1988).

REFERENCES

Bonhommet, N., P. Roperch, and F. Calza, 1988, Geology, v. 16, p. 422.

Einarsson, P., 1988, Geodyn. Symp., Jan. 26-27, Woods Hole Oceanographic Institute, Woods Hole, MA.

S. Bjornsson, G. Foulger, R. Stefansson, and T. Skaftadottir, 1981, Maurice Ewing Ser. 4, AGU, p. 141.

Karson, J., and J. F. Dewey, 1978, GSA Bulletin, v. 89, p. 1037.

Luyendyk, B. P., M. J. Kamerling, R. R. Terres, and J. S. Hornafius, 1985, Journal of Geophysical Research, v. 90, p. 12454.

Nicholson, C., L. Seeber, P. Williams, and L. R. Sykes, 1986, Tectonics, v. 5, p. 629.

Ron, H., 1987, Tectonics, v. 6, p. 653.

Searle, R., and R. Hey, 1983, Journal of Geophysical Research, v. 88, p. 6433.

Wilcox, R. E., T. P. Harding, and D. R. Seely, 1973, AAPG Bulletin, v. 57, p. 74.

YOUNG, R. A., State University of New York, Geneseo, NY

Laramide Paleogeomorphology and Evolution of Western Grand Canyon Region

The Grand Canyon is incised into a late Cretaceous-early Paleocene cuestaform landscape that was blanketed by Paleocene and Eocene fluvial sediments. Development of the modern canyon landscape has had a minimal effect on the broader elements of the Laramide physiography of the western Colorado Plateau. The paleogeomorphology consists of remnants of a system of deeply incised canyons leading from the margins of Laramide uplifts in central Arizona toward basins in south-central Utah (Graf et al, 1987). The relief on these paleocanyons attests to late Cretaceous-early(?) Tertiary uplift in the western Grand Canyon region of at least 1200 m-1500 m above sea level (present elevation 1500 m-2000 m).

The maximum age of the basal Laramide gravels filling the channel remnants is constrained by the occurrence of Cretaceous-Paleocene volcanic clasts with K-Ar whole-rock ages of 80-63 Ma (Young et al, 1987). An additional age constraint is represented by freshwater viviparid gastropods of probable early Eocene age (extinct in middle Eocene time) in lacustrine limestones contained within the arkosic sediments, which buried much of the regional Laramide erosion surface.

The Laramide paleogeomorphology demonstrates both regional and localized control of the evolving landscape by the rock structure. The basic landscape developed on northeast-dipping Paleozoic rocks by scarp recession of resistant sandstones and limestones separated by less resistant shales. This regional, consequent drainage system with its strike-valley tributaries stripped away the Mesozoic and upper Paleozoic rocks at cliff recession rates that may have approached 1 km per million years (Young, 1985).

Eventually, the headward reaches of the north- to northeast-flowing streams became deeply incised adjacent to the uplifted source regions, whereas their lower (northern) reaches were adjusted to aggrading lacustrine basins. Much of northern Arizona was then surfaced with Kaibab Limestone with divides capped by the lower, resistant sandstones of the Moenkopi Formation.

Streams crossing this erosion surface were deflected by individual Laramide structures, especially monoclinal folds and upwarps. Paleocanyon remnants converging on the Hurricane fault zone south of the Colorado River exhibit hanging relations with the modern Grand Canyon tributaries. The elevations and fluvial imbrication of these discontinuous, sinuous paleocanyons suggest that their courses were progressively deformed or severed by intermittent structural movements along an older, east-verging monoclinal fold. Fission track studies by Naeser et al (in press) tend to confirm the pulsed nature of the Laramide deformation. The modern Hurricane fault, coincident with the older monoclinal axis, is downthrown to the west, possibly indicating the response of a preexisting, west-dipping basement fault to middle Tertiary extension.

Paleochannel remnants sloping northeastward down to the regional dip toward the monocline are relatively straight, whereas those approaching the structure from the south and east exhibit marked sinuosity. Convergence of four or more separate drainages along this major structure created a number of large paleovalleys close to the modern course of the Colorado River. The younger Grand Canyon and its tributary drainage have incorporated some of these paleocanyon segments and are in the process of capturing other remnants, which are presently being exhumed.

The Laramide sediments and associated erosion surface are disconformably overlain by younger fluvial sediments interbedded with Miocene volcanic rocks, all of which predate the modern Grand Canyon. A significant weathering profile had formed on the arkosic Eocene sediments prior to the extrusion of the 16-19 m.y. old volcanic rocks. This deeply weathered regional surface of post-Laramide (late Eocene?) age has been noted throughout many parts of western North America from Washington State to Mexico and has been called the Telluride surface by other workers. The interval of vigorous Laramide erosion was separated from the modern canyon-forming episode by a period of relative tectonic quiescence (deep weathering), by Basin and Range extension and volcanism, and by Miocene fluvial aggradation.

The Laramide paleogeomorphology and fluvial stratigraphy document the gradual unroofing of the adjacent source terranes, the onset of Laramide volcanism, and the dynamic effects of contemporaneous structural deformation of the drainage. The new geologic evidence demonstrates that most of the uplift on the western Grand Canyon had occurred by late Laramide time. In addition, the field relations demonstrate that pre-Tertiary landforms and erosion surfaces can be well preserved, even in areas subjected to extensive younger erosion and characterized by extreme relief.

It is not possible to evaluate adequately how the burial and reexhumation of the paleotopography might have enhanced the preservation of the Laramide landforms. However, it appears that a significant degree of landform inheritance from early Tertiary events could go undetected in modern landscapes where adequate geomorphic age control is lacking. If the remaining traces of the paleocanyons and associated fluvial sediments had been completely removed from the western Colorado Plateau by Grand Canyon erosion (which has been relatively efficient), it would not have been possible to ascertain that the basic cuestaform landscape of the western plateau had formed prior to Eocene time. Nor would it have been possible to demonstrate that scarp erosion since Paleocene time has been documented by irregular headward erosion (scarp dissection) and relative stagnation rather than uniform scarp recession. Conversely, the present example makes a strong case for the importance of relatively short, dramatic episodes of dynamic landscape development, driven by regional tectonic events such as the Laramide Orogeny, with significant intervening periods of much less dramatic change.

In the present case, it is clear that the interval of Laramide landscape evolution was more significant in the shaping of the regional physiography than the more dramatic, late Tertiary canyon-cutting episode. In fact, the cuestaform Laramide landscape, created by consequent drainage on northeast-dipping rocks, is unlikely to have developed during an episode of regional drainage incision by a southwest-flowing Colorado River.

REFERENCES

Graf, W. L., R. Hereford, J. Laity, and R. A. Young, 1987, Colorado Plateau, in W. L. Graf, ed., Geomorphic systems of North America: GSA Centennial Special Volume 2, p. 259-302.

Naeser, C. W., I. R. Duddy, and D. P. Elston, in press, in D. W. Elston, G. Billingsley, and R. A. Young, eds., IGC Guidebook for Grand Canyon River trips, 28th International Geological Congress, Washington, DC, Ch. 17.

Young, R. A., a tectonic model to distinguish between causes of rapid symmetrical scarp retreat and scarp dissection, in M. Morisawa and J. T. Hack, eds., Tectonic geomorphology: Boston, Allen and Unwin, p. 261-278.

D. W. Elson, and E. H. McKee, 1987, Laramide deformation and sedimentation in northern Arizona: Post-Turonian through Ypresian events coincident with Cordilleral Fold and Thrust Belt chronology (abs.): GSA Abstracts with Programs, v. 19, n. 7, p. 903.

YOUNG, R. A., State University of New York, Geneseo, NY

Late Cretaceous-Paleocene Paleodrainage System of Western Grand Canyon: Evidence for Structural Controls

A paleodrainage system consisting of numerous discontinuous canyon segments with relief of up to 1200 m converges on the southern end of the Hurricane fault zone in the western Grand Canyon region. At least four distinct drainage channels were incised from the Mississippian Redwall Limestone through the lower Paleozoic rocks and into the Precambrian basement. Unlike the modern Grand Canyon, these paleocanyons have smooth, U-shaped profiles and smoothly curved to sinuous trends. Minor tributaries to the major canyon remnants have a more dendritic form than Grand Canyon tributaries and lack the theater-heated valley terminations associated with many of the headwardly eroding side canyons of the modern Colorado River. The basal fluvial gravels within the paleodrainages are Paleocene in age, based on their included volcanic clasts with K-Ar ages between 80 and 63 Ma (Young et al, 1987), and on the occurrence of lower

(?) Eocene gastropods in lacustrine limestones found in higher stratigraphic sections. It is assumed that the period of canyon incision, which necessarily predated the deposition of the basal gravels, was largely confined to late Cretaceous through early Paleocene time. The canyon cutting is presumed to have occurred in response to Laramide uplift, which created a regional surface sloping from the Plateau margin toward the late Cretaceous-Eocene basins of Utah. The resulting regional drainage system eroding the northeast-dipping Paleozoic strata produced the broad cuestaform landscape that still dominates the physiography of Arizona. The western Grand Canyon region was near the edge of the Laramide uplift, and the incised paleocanyons represent the headward reaches of the regional drainage system. The region had to have been raised at least 1200 m above sea level to account for the relief still preserved on the largest channel.

The most unusual aspect of the paleodrainage system is the evidence indicating that the drainage was forming contemporaneously with compressional Laramide monoclinal folds. Conversely, the unique structural style associated with the Laramide Orogeny can be used to reinforce other evidence for the early Tertiary age of the paleodrainage. Overall, the relationships between the paleodrainage and the Laramide structures document episodic disruption of large valleys, followed by a change in the tectonic regime from compression to extension. This contrasting structural history was directly related to the long-term regional drainage reversal (from northeast-flowing to southwest-flowing) between early and late Cenozoic time.

The Hurricane fault zone has a complex deformational history in Peach Springs Canyon, the Colorado River tributary that has extended southward along the structure by headward erosion and has intercepted several paleochannel segments. This late Tertiary, west-dipping normal fault closely coincides with an older monoclinal fold, which has an opposite sense of displacement. Most of the major fault-monocline structures throughout Grand Canyon proper have similar, complex deformational histories. Typically they began as west-dipping, Precambrian basement faults, underwent Laramide compressional reversed motion accompanied by monoclinal folding of the near-surface rocks, and then reverted to normal faults in response to late Tertiary extension.

The paleochannels converging on the Hurricane fault-monocline can be divided into two groups. Those trending approximately downdip and approaching the structure from the west or southwest are relatively straight, whereas those approaching the east-verging monocline from the southeast or east side are very sinuous. This conspicuous difference in the relative sinuosity of the two sets of channels is reminiscent of similar patterns seen in modern channels incised across other folds, such as the Kentucky River system where it approaches the east edge of the Lexington Dome. Another good example is the change in channel from where the Green and Colorado Rivers encounter the Monument Upwarp. Compressed, sinuous channel forms are common along reaches where streams flow "up structure." One explanation for the sinuosity involves the isolation and superimposition of an alluvial upstream reach as it slowly uncovers a bedrock structure. A similar pattern could develop if the deforming structure were gradually to deflect the drainage. The sinuosity might be initiated in younger, more erodable surface rocks preserved on the structurally lower side of the uplift. The recent analysis of fission track data by Naeser et al (in press) demonstrates an extended Laramide deformation history on several Grand Canyon structures and lends support to this suggestion.

Data in support of the extended period of time represented by the evolution of the paleodrainage also are provided by pebble counts along the main Peach Springs Canyon channel remnant. Counts taken vertically through several hundred feet of intact fluvial gravels show a marked upward decrease in Precambrian quartzite clasts accompanied by an increase in granitic and metamorphic clasts. This documents the gradual unroofing of the adjacent Precambrian source terrane. An erosional event of this magnitude must represent, at a minimum, a time interval of hundreds to thousands of years. This reinforces the suggestion that the paleodrainage evolved over an extended period, with ample time for significant structural adjustments. In other stratigraphic sections, the uppermost gravels become dominated by Laramide volcanic clasts, providing further evidence of the passage of a lengthy period of time involving substantial changes within the orogenic belt. In addition, thick lacustrine facies within the fluvial sediments attest to significant drainage disruptions late in the interval of deposition.

A final piece of evidence bearing on the assertion that drainage evolution was accompanied by a lengthy period of structural change concerns the discordant elevations of the different channel segments. Although the straightforward reconstruction of adjacent channel segments appears to create an integrated drainage pattern, the elevations of the several segments are too varied. Channels approaching the Hurricane fault zone from the west are nearly 300 m higher than those on the upthrown (east) side of the fault. Restoration of the offsets on the numerous subsidiary faults does not eliminate these large discrepancies. The alternative would be to have a series of large valleys joined by high waterfalls. It is more likely that the different channel segments represent separate stream courses, which were progressively abandoned during times of structural adjustment as the drainage and structure evolved.

This paleodrainage system provides an excellent example of the complexity of compound landscapes. Very old Tertiary or late Cretaceous landforms can coexist within elements of a much younger landscape. The present situation has involved climatic, structural, and regional geologic changes of significant magnitude and has been preserved in a region of extreme erosion. Even a reasonable understanding of the most basic relationships required a lengthy, multidisciplinary effort and the fortuitous preservation of evidence bearing on the ages of fluvial gravels. The obvious conclusion to be drawn is that geomorphic field relationships often may be more complex than is apparent. Unfortunately, the data required to resolve (or even recognize) such problems are often either lacking or inadequately exposed.

REFERENCES

Naeser, C. W., I. R. Duddy, and D. P. Elston, in press, in D. P. Elston, G. Billingsley, and R. A. Young, eds., IGC Guidebook for Grand Canyon River trips, 28th International Geological Congress, Washington, D.C., Ch. 17.

Young, R. A., D. P. Elston, and E. H. McKee, 1987, Laramide deformation and sedimentation in northern Arizona: Post-Turonian through Ypresian events coincident with Cordilleran Fold and Thrust Belt chronology (abs.): GSA Abstracts with Programs, v. 19, n. 7, p. 903.

YOUNG, R. E., NASA Ames Research Center, Moffett Field, CA

Computer Modeling of Dynamic Meteorology of Plateary Atmospheres

The principal purpose of studying the dynamic meteorology of plateary atmospheres is to understand the general characteristics of atmospheric circulations as a function of varying solar insolation, atmospheric infrared opacity, planetary rotation rate, gravity, atmosphere-surface interactions, internal heat sources, and obliquity. The atmospheres of the planets for which we have reliable information on wind and thermal structures (Venus, Earth, Mars, Jupiter, Saturn) each have atmospheric circulation states that exhibit very different properties, a consequence of significant variations in the quantities mentioned above.

The major characteristics of Venus are an atmosphere 100 times as massive as Earth's, surface temperatures over 700 K, sulfuric acid clouds 50 km above the surface, and atmospheric winds corresponding to a rotation rate 60 times faster than the solid planet. This dynamical regime appears to be relatively steady. In contrast, terrestrial wind and temperature patterns vary substantially with season, and on shorter time scales as well. Water Vapor plays an extremely important role in terrestrial meteorology, and unlike any other planet, atmosphere-ocean interactions are present. Mars possesses the most terrestrial-like dynamical regime, exhibiting many of the sorts of atmospheric features seen on Earth, even though it has only about 0.02 of the terrestrial atmospheric mass. However, Martian climatology also involves unique features, such as the fact that the total Martian atmospheric mass varies by a substantial fraction seasonally as the principal constituent carbon dioxide condenses to and sublimes from the polar seasonal ice caps. Further, global dust storms produced by atmospheric winds sometimes cover the entire planet, affecting global wind and temperature fields. Jupiter's atmospheric wind patterns are characterized by a zone-belt structure, alternating zones of east-west winds on the order of 100 m/s. There is no solid surface for the atmosphere to interact with, but as on Earth, the hydrological cycle of water vapor is thought to significantly influence atmospheric motions. A particularly different aspect of Jovian meteorology is the existence of an internal flux of heat from the planet interior comparable to the received solar flux. The large size of Jupiter combined with the rapid planet rotation are important factors for atmospheric dynamics. Saturn has dynamical considerations similar to those of Jupiter, but atmospheric winds are observed to be considerably stronger than on Jupiter with less of a banded structure.

Modeling of the above planetary atmospheres has progressed to the point that three-dimensional simulations of the general circulations are being attempted. Although much has been learned from simple modeling efforts, full three-dimensional models are required to properly understand many dynamical processes, as well as how physical, radiative, and dynamical processes interact to produce the observed wind and temperature patterns. However, even on the largest and fastest presently available computers, various approximations must by made with regard to the treatment of dynamical and physical processes that have small enough length scales that they cannot practically be treated explicitly, but which nevertheless are important for the general circulation. Additionally, it must be realized that it is only the statistical properties of the solutions that are significant in most three-dimensional simulations. The governing equations have the property that even very small changes in initial conditions lead in a relatively short time to large changes in the solutions at a given time, but similar initial conditions give similar statistical properties, such as time averaged wind, etc.

Examples will be given of particular modeling efforts, pointing out successes as well as deficiencies. Numerical considerations and the factors that most affect them will be discussed. Finally, an attempt will be made to assess the future directions of general circulation models in light of currently predicted computer capabilities.

YOUNG, R. W., University of Wollongong, Wollongong, Australia

Evolution of Tertiary Landscapes in Southeastern Australia

Extensive remnants of ancient landforms that are preserved under Tertiary basalts over some 1200 km of the highlands of Southeastern Australia provide a perspective on long-term denuadation that is at odds with widely accepted models which have provided the basis not only for the interpretation of many landscapes, but also for numerous tectonic histories. Two fundamental tenets of those models are challenged. The first is that, given tectonic stability and adequate runoff, the passage of several millions of years will result in stream profiles declining in gradient and becoming increasingly independent of lithological constraints. The second is that general rates of erosion are such that 10-30 Ma are sufficient for an area of continental extent to be reduced to a surface of low relief.

Detailed studies of the origin and dissection of landforms preserved beneath Eocene to Miocene basalts have demonstrated six fundamental morphological features.

1. Landforms of this region cannot be explained simply in terms of cycles of erosion graded to migrating nickpoints initiated near sea level. Erosion has occurred simultaneously above and below nickpoints, and geological constraints are no less important than regional base levels.

2. Rather than showing evidence of long-term decline of gradients, modern stream profiles are roughly parallel to their Tertiary counterparts.

3. Local adjustment to variable lithology is as evident on modern channels as it is on Tertiary channels.

4. The amount of incition throughout drainage systems is more closely linked to variations in stream power than to any regional base level.

5. Despite the passage of 15-30 Ma the local relief in individual catchments has tended to remain static or increase, rather than decrease.

6. Drainage patterns have been remarkably stable over this period of time and show no evidence of significant capture or derangement.

Despite substantial elevations (300-200 m) and humid climates throughout much of the Tertiary, rates of erosion have been much slower than the model rates cited in most textbooks.

1. The large coastal escarpment south of Sydney has retreated at an average rate of about 170 m/Ma.

2. Sandstone cliffs in upland valleys have retreated at rates ranging from about 10 to 25 m/Ma.

3. Stream incision in upland valleys (Shoalhaven, Lachlan, Murrumbidgee, Hunter, Murray) on varied lithologies has ranged from about 4 to 12 m/Ma.

4. The most prominent erosion has been caused by the headward extension of deep canyons below the upland surfaces, but even there rates have been slow. The 600 m Shoalhaven canyon, for example, has retreated at less than 5 km/Ma and its mean cross-sectional depth gives a rate of only 10 m/Ma of denudation.

So far from being reduced to planation surfaces in 10-20 Ma these valleys have not even passed from one stage of the classical cycle to another. At these rates of erosion, probably more than 100 Ma would be needed to planation, even without isostatic compensation.

The evidence reviewed here supports and extends the reports of very slow rates of change from other cratonic areas, and certainly prompts caution against the uncritical application of traditional models of long-term landform evolution either locally or in intercontinental comparisons. Indeed, rather than searching for a generally applicable model, we should now recognize that there have been fundamental differences between landform evolution in the stable and in the mobile parts of the world.

YOUSEFI, E., Geological Survey of Iran, Tehran, Iran

The Application of a Combined Airborne Magnetic and Radiometric Survey to the Geology and Mineral Exploration in Central Iran

The present paper deals with the results obtained from a combined aeromagnetic and airborne radiometric survey carried out over 43.300 km^2 in the Anarak area of central iran. Because the extremely complex geology of central Iran contains many environments highly favorable for mineral accumulation, this survey was designed to assist in geological mapping and in selection of the promising zones for mineral exploration.

The surveyed area is confined to the central Iranian Highland. Its relief is typical of a mountainous desert and represents rocky massifs and ridges alternating with intermontane valleys covered with rock debris, boulders, and sands.

Geologically, the area is explained in brief as follows. Basement rocks and Paleozoic-Mesozoic platform form the lower structural stage of the area. These include pre-Alpine and early Alpine structures. Such a wide range of the lower structural stage could be attributed to a unity of platform regime of the area. The geophysical properties of the rocks composing these stages are rather uniform, and, therefore, do not permit their detailed differentiation.

The Precambrian rocks are represented by a thick sequence of crystalline micaceous and green schists, marbles, and quartzites, along with metavolcanics and amphibolites. The rocks have undergone regional metamorphism of essentially low grades. Paleozoic and Mesozoic rocks have identical lithological composition, and are represented by massive and layered limestone, dolomites, and conglomerates. The Paleozoic rocks in the east area contain much detrital material. Middle structural complexes of the area are composed mainly of Paleogene, volcanic-sedimentary rocks, most of the effusive eruptions are attributed to Eocene. The rocks composing this structural stage occupy vast areas, mostly in the western portion of the area. Upper structural complexes are assigned to Neogene-Quaternary age; these characterize the late Neogene and postorogenic stages of the area. The sediments are represented by a mollassa-sandstone-conglomerate sequence; these are overlain by gypsum, clays, marls, carbonate rocks, and radiolarities.

A complex of "Coloured Melange" is dominant in the central part of the area of the upper Cretaceous-lower Eocene age. With respect to mineral deposits, the area is one of the most promising regions in Iran, and from the ancient times the deposits of lead, copper, nickel, cobalt, iron, and manganese have been identified. The area was flown employing an AN-2 biplane; and traverses were flown with 2-km intervals perpendicular to the general trends of geological structures; all profiles were flown with terrain clearance of 70-120 m for 90% of covered area and the remaining 250-350 m under conditions of strongly rugged relief.

Data of total-filed magnetic intensity differences (ΔT) a, gamma-ray intensity of rocks and U (Ra), Th, and K were collected along the flight lines.

Some areas were flown more in detail, including three known deposits, to obtain more information, and to get qualitative interpretation of data. In addition, some amount of express ground control verification was performed over some locations to assist in the interpretation of magnetic and radioactive fields. Some samples were collected from the anomalous zones to study the magnetic and radioactive properties of rocks are ores.

Data recorded along the flight lines are interpreted using the information obtained from the physical properties of rocks and ores, field observations, and available information about the geology of the area as follows:

1. Analysis and classification of the magnetic and radioactive field through zoning the area according to: (a) type of anomalies observed (amplitude, degree of differentiation, U (Ra) contents, shape and strike of anomalies, relationship between anomalies of various potential field. (b) results of quantitative estimate of magnetic field variations.

2. Analysis of data available on the rock magnetic properties, including results of statistical processing.

3. Quantitative estimate of parameters of buried bodies.

4. Analysis and estimates of statistical parameters of distribution of U (Ra), Th, and K in various rocks.

5. Calculations of depth to the magnetic bodies were done using the program involving determination of H values of anomalous bodies by five variants of the tangent method introduced by Yu. N. Grachev (two modifications H_1 and H_3), V. K. Pyatnistsky (H_2), Peters (H_5), and A. K. Weigberg (H_4).

Based on results obtained from the combined airborne magnetic and radiometric survey, it has been concluded that the survey has provided lots of new information with respect to the geology and selection of promising zones for mineral explorations and also appears to be very essential for geological mapping and determining the tectonic aspects of the area. As a result of geological analysis of geophysical data, the sketch map of tectonic zoning of the surveyed area at scale of 1:500,000 has been compiled indicating 18 localities that appear to be promising with respect to searching for nonferrous, rare and noble metals, and seven local magnetic anomalies promising for magnetic ore.

Results form magnetic data have singled out four large pre-Alpine basements that are separated by tectonic ophiolitic suture zones, and deep-seated faults. The ophiolite sutures occur at the great depth, and some single tectonic blocks of ophiolite crop out and represent the coloured Melange formation. Each of the above blocks is of a complex structure with its own way of development. The trends of the faults are expressed well in potential fields, and based on their characteristics, such as depth, and the roles they play in the tectonics of the area, are divided into three categories: (1) Great Kavir fault and torkemani, which are deep-seated faults, (2) seven faults having interblock characters, and (3) local faults that play minor roles in the area.

YRIGOYEN, M. R., Trend Argentina S.A., Buenos Aires, Argentina

Petroleum Basins of South America

Oil wealth in South America has long been known by its indigenous population, but Europeans took notice of its existence only during the third voyage of Christopher Columbus to the southern coast of the Caribbean Sea in 1499. What might be regarded as the "first oil concession" in America was the authorization granted by emperor Carlos V in 1532 to the inhabitants of Cubagua for the free use of the oil seep discovered in the island, which was described as "an oil liquor very useful for the treatment of peoples's diseases."

More than four centuries later, oil reserves discovered in South America amount to 42 billion cu m (163 B brl) EUR and 7249 B cu m (256 trillion cubic feet) of natural gas, representing over 9% and 7.5%, respectively, of the world reserves. Present oil production in South America from 33,500 oil wells exceed 620,000 cu m/day (3.9 million brl/day), comparable to the production from western Europe.

These significant volumes of oil and gas are reservoired in 22 productive sedimentary basins whose extension, polihistory, and geologic setting differ considerably. In South America, there are four oil-producing intracratonic basins, including the recently discovered Solimoes or Upper Amazon Basin. No doubt, the most relevant system of hydrocarbon-producing areas is found within the subandean pericratonic-backarc domain and in broad, intermontane depressions of the north Andean Orogenic Belt (e.g., Maracaibo Basin). The marginal Atlantic and Pacific Basins also embody very important exploitable accumulations, the most significant being in the Campos and Talara Basins, in that order. Finally, reference should also be made to the fields located within the intra-arc basins, such as the Upper-Middle Magdalena and Titicaca.

South America on land has the largest crude oil accumulation in the world; the supergiant Orinoco Oil Belt with 187.8 cu m (1181 B brl) of oil-in-place; while the Brazilian Atlantic coast exhibits the largest offshore oil field worldwide; the Marlin Field, with estimated crude oil reserves of 635 million cu m (4 billion brl).

South American hydrocarbons were sourced from several marine and lacustrine organic shale units and subordinate limestones of a wide range of ages. The main source rock intervals are of Mid-Devonian, Late Triassic, Late Jurassic, Cretaceous, and Early/Middle Tertiary ages, but the peak of generation is in most cases restricted to Mid-Tertiary times.

Sandstones of different ages and depositional environments are the dominant type of reservoir facies, followed by limestones, fractures volcanic/intrusive bodies, and even altered crystalline basement rocks. Oil and gas productive reservoir units are recorded from Ordovician/Silurian age up to the Neogene time, the most important being the base and top wedge clastics, deltaic, and turbidite deposits conforming traps of structural and stratigraphic control.

By far, the richest oil country in South America is Venezuela and the most prolific region up to now is the Maracaibo Basin with a cumulative oil production of 5024 M cu m (31.6 B brl) and an Estimated Ultimate Recovery of 12,763 M cu m (80.3 B brl). On the other hand, the largest single gas field in the continent is the giant Loma de la Lata Field, in the Neuquen Basin of Argentina, with an EUR of 396,443 M cu m (14 trillion CFG) plus 122 M cu m (767 million brl) of oil and condensate.

There still remain in South America many large sedimentary basins that deserve systematic exploration. Moreover, large part of them have never been reached by the drill. In other cases, recent deeper pool discoveries in areas with a mature exploration indicate that significant findings may be expected in them, like the recent discoveries in the Eastern Venezuela Basin (e.g., El Furrial, EUR 310 M cu m or 1950 M brl). A similar situation occurs in the offshore. Well after well, the Atlantic margin of Brazil is adding new reserved, although some of these pools are located in deep ocean waters that make exploitation very difficult and expensive and sometimes impossible at the present technological stage. In the South Atlantic, many attractive areas from an exploration viewpoint remain unexplored and undrilled. The same situation applies to some portions of the Caribbean and Pacific offshore. Petroleum possibilities of the Antarctic Continent and the Falkland-Georgia Plateau are even more speculative.

YU Guanming, WANG Chengshan, and ZHANG Shaonan, Chengdu College of Geology, Chengdu, China

New Recognition of Tectonic Nature of Bangong Lake-Dingqing Fault Zone

The Bangong Lake-Dingqing fault zone is thought to be an active-continental margin by most geologists, of the "trench-island arc-back-arc basin" tectonic model. Our studies suggest that it is an interplate transform margin in Jurassic Period. The depositional system of Jurassic Period is of apparent transform basin character: (1) The major sedimentary assemblages are ocean bottom assemblage-turbidite-shallow sea deposit-fan delta deposit assemblages; (2) deposit thicknesses are much different in the area, varying from several hundred to more than 10,000 m; deep-water basin has the greatest thickness; (3) sedimentary facies change quickly along the strike being of cut apart deep water deposit framework; (4) sedimentary facies of each deep-water basin vary greatly from vertical section and contain lots of conglomerates and also have a complete transform basin sequence of "subfan-olistrostrome-fan delta-radiolarian silicalite"; (5) there is no unifying paleocurrent model in each basin; (6) turbidites do not contain volcanic clasts and have paragenesis with oceanic molasse and also are of the character of Califonia-type flysch assemblages; (7) unconformities occurred in limited area; (8) fault blocks of Paleozoic result from transform rising existed between deep-water basins; (9) ophiolites are well developed and are of small and isolated magma chambers and mantle sources at different depths, and so forth.

YU Puzhi, Institute of Oceanography, Academia Sinica, Qingdao, China

Gravity Field and Tectonics of Ryuku T-A-BA System and Adjacent Sea Area

Based on the gravity anomaly centered around Ryuku T-A-Ba system, the author did many calculations, such as Moho calculating, spectrum analysis, smooth filter, analytic extension, horizontal derivation, correlative analysis, and so on. By contrast to seismic profiles, the analyses show that oceanic and continental crusts demarcated by Ryuku T-A-Ba system are very different from each other in gravitational and tectonic characteristics, while in the T-A-Ba system region there are very complicated structures and linearity is very obvious.

The shelf basis of the East China Sea is the depression since late Cenozoic Era to be based on different structural basement. It has undergone development from rift valley to folded zone to depression. The Okinawa Trough is an active back-arc basin up to now. It began to be formed in late Miocene and was at a fixed stage in late Pliocene. There is an extension axis in the central part of the southern trough filled with the sediment of Pliocene and Pleistocene. Therefore, the extension axis is different from spreading axis of midoceanic ridge. The crustal thickness is 15-17 km in the southern trough. The middle-northern part of the trough is still developing in the rifted-depression phase, while the southern parts in the spreading phase. Ryuku Arc separated from the shelf edge, as the crust of Okinawa Trough was rifted and extended.

From the shelf of the East China Sea to Ryuku Arc, Bouguer anomaly increases from 0 mGal to 80 mGal, while the crustal thickness changes within the limits of 30 ± 3 km, based on this gravity background. Bouguer anomaly zone in Okinawa Trough is a superimposed linear gradient zone.

From Ryuku Arc to Trench, Bouguer anomaly increases at a rate of 20 mGal/10 km, while the crustal thickness decreases at a rate of 0.4/km. The crustal thickness is 12 km in the Trench. Obviously, the crustal transformation is sudden between East China and Phillipine Sea. Seismic profile shows that submarine layer of Phillipine Sea lies under the basement of Ryuku Arc. Geologic evolution of the Asian continental margin has been directly controlled by the processes of subduction of the Phillipine Sea plate.

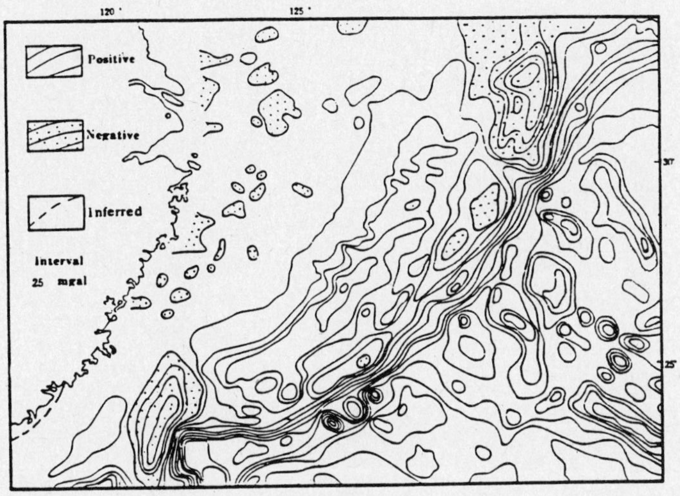

Fig.3 Bouguer gravity field

Figure 1

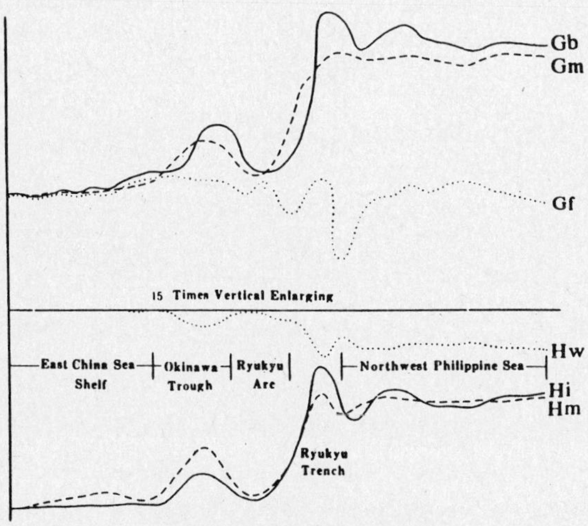

Fig.9 Cross profile of crustal thickness

Hw: Water depth, Hi: Isostatic thickness

Hm: Moho depth, Gf: Free air anomaly

Gb: Bouguer anomaly, Gm: Anomaly caused by the Moho

Figure 2

YU Rongbing, ZHAO Zhen, and YAN Yanjin, Tianjin Institute of Geology and Mineral Resources, Tianjin, China

Oldest Stratigraphic Age in China

The author determined an age of bronzite by K-Ar method in 1975 from pyroxenolite within the Qianxi Group in East Wangzhudian Shaidunying of Qianxi region, East Hebei Province, that obtained an age of 3913 Ma for the stratum, the oldest age in China. Subsequently, a series of older ages had been obtained in the same place: a 3411 Ma for minerals by K-Ar isochronous method (Sun Jiashu, 1975), a 3670 ± 230 Ma for whole-rock isochronous Rb-Sr method (Institute of Geol., A S China, 1978), a 3620 Ma by ^{40}Ar-^{39}Ar (Wang Sonshan, 1985), and a 3515 ± 115 Ma (Jiang Baiming, 1983) and 3500 ± 80 Ma (Huang Xuan, 1986) by Sm-Nd. These show that the oldest stratum in the crust truly distributed from this location.

The authors have studied the age geochemistry and mineralogy from the above mentioned site and used four samples—two bronzites, an amphibole, and whole-rock pyroxenolite—from this Group, and have obtained an age of $3760 \pm ^{131}_{136}$ Ma by isochronous K-Ar method. It is the oldest age of stratum in China thus far. The correlative coefficient of isochron (r = 0.983) shows that the isochronal character is very strong; an angle of 35° between isochron and abscissa demonstrated that the excess of Ar existed in the rock and minerals. The content of excess Ar is $(0.0058 \pm 0.00314)^{10\text{-}6}$ g/g. On that score, it proved that the age as stated above of single mineral (3913 Ma) is older than isochronous K-Ar age. It attributes to upper mantle period, which provides beneficial geological information in studying the earliest form of upper mantle in China.

YU Shuhe, Geology and Minerals Bureau of Qinghai Province, China

The Geological Characteristics and the Mineralizing Controlled Condition of the Asbestos Deposit in Mang'Ai, Qinghai, China

The Mang'ai asbestos deposit is located in the middle section of the ultrabasic rock belt of the A'erjin mountains. Its direct enclosing rocks are gray-green altered basalts and clastic rocks with griotte of the lower Ordovician. The polygenetic-inherited mobilities of the major faults in the direction of east-northeast are strong, while the north-northwest faults and their cracks are the main structure containing ores. The Rb-Sr ages of the basalt and liparite are respectively 4.94 and 3.76 hundred million years.

The original rocks of the rock bodies are harzburgite and augite-periodotite. All of them have been serpentized. Antigorite is the main rock. The inclusions and veins in the early wallrocks and the rodingitization in the periphery of the basic rocks can be generally seen. The asbestos veins are controlled by the tectonic fractures and often have the shapes of branches, thinning out and so on. The main types of ore structure are massive and rudaceous. The asbestos is mainly composed of clinochrysotile ($2Mc_1$) and a small amount of orthocrysotile ($20rc_1$), and has a Povlen texture formed by $2Mc_1$.

The mineralization of asbestos is one of the stages of serpentinization of the ultrabasic rocks from the mantle. The inclusions research shows that the serpentinization is a slow process under the condition of a middle-low temperature and pressure. According to the calculative results of oxygen isotopes using lizardite (or asbestos) with magnetite as the mineral pairs, the serpentinization temperature is 134–137 = C, the asbestos-forming temperature is 145-192 = C. Both processes may be completed under conditions of high oxygen fugacity and neutral-acidic solutions. The water for the serpentinization, with δD value of -60 to -111, δO^{18} value of -1.69 to -5.49, belongs to mixed water. The genetic type of Mang'ai asbestos deposit belongs to the metasomatic-filling type of mesoepithermal and tectonic fractures.

The Mang'ai asbestos deposit is controlled by four conditions restricted each other: The best type of rock body must be harzburgite-periodotite, then the dunnite-augite periodotite type; the serpentinization must be complete and the physiochemical condition must be suitable; the faults and fractures must develop; there must be the ultrabasic rocks of Palaeozoic Era or earlier, which originated from the fold system.

YU Yuanchang, Geological Society of China, Beijing, China

Skarn Copper, Iron, and Gold Deposits in the Southeast Hubei Province, China

The southeast Hubei province is part of the important metallogenic belt of Cu, Fe, Au, S, Pb, and Zn. It is famous for its skarn iron and copper ore as well as associated gold. Tectonically, it is located in the west end of the Lower Yangtze platform fold belt of the Yangtze paraplatform, which is the base consisting of Precambrian system and depressive belt from Sinian period to Triassic period. During the Yangshan Movement period, a large-scale magmatic intrusion and volcanic eruption took place in this area. Many different genuses of skarn deposits (Fe, Cu-Fe-Au, Cu-Au, Cu-Mo, W-Mo) are closely related to the magma intrusion of the Yangshan Movement period. The ore deposits occur in the contact between carbonates and intermediate-acid intrusion. The compound deposits are formed in different structures and various wallrock environments along the contact zone.

Ore deposits are mainly divided into four types:

1. Ore magma-skarn type (Tieshan iron-copper deposit). Tieshan is a large iron deposit associated Cu, Co, Au, and Ag. The mineralization has undergone multiple stages. First mineralization stage is characterized by simple association of iron-bearing minerals. The temperature of magnetite determined by the explosion is 750-770 = C. Besides massive magnetite ore, there is special holey tubular ore, which shows the iron-magma injection process. The second mineralization stage has widespread skarn and a complicated association of minerals. The temperature of magnetite is 320-560 = C. It is evident that this ore deposit appears to be the contact-metasomatic mineralization.

2. The contact-metasomatic skarn type (Tonglushan copper-iron deposit). Tonglushan is a large copper-iron deposit, with multiple elements including Mo, Au, Ag, Co, Ga, Se, Te, and S. The orebodies mainly occur in the contact fractures and shatter zone between hidden peninsula and island chainlike marbles of the Lower Triassic period and quartz syenite-diorite porphyrite of Yangshan Movement period. The skarn exhibits widespread development in the external or internal contact containing anorthosite related to metamorphic rocks of contact metasomatism. And skarn mineralization exist a obicous zoning.

3. The prophyry Cu-Mo skarn deposit (Tongshankuo copper deposit). Tongshankuo deposit is predominantly characterized by copper ore including multiple associated elements (S, Ag, Se, Re, Ge, and W). Mineralization is circumferentially distributed around the porphyry intrusive body from the center of porphyry to the external contact: (1) molybdenite zone with potassic alteration; (2) copper zone with quartz-sericitization; (3) copper zone with skarnitization. The occurrence of orebodies is controlled by the contact structures of circumferential porphyry stock. Mineralization from the center of stock to wallrock follows as: Mo-W, Mo-Cu, Cu, Pb, Zn. This deposit has not only the character of common skarn copper deposit, but also the character of potassic metasomatism and hydrogen metasomatism zoning with porphyry copper.

4. The stratabound skarn type. This type of ore deposit is widespread in all regions and is not developed. Some copper and sulfur orebodies are discovered on discordant plane between Carboniferous period and Silurian period ore interband of Lower Triassic period.

According to stable isotopic data of oxygen, sulfur, deuterium, and carbon, the mineralization liquor mainly comes from mixing magma water. Therefore, we consider that these deposits may be generalized skarn deposits with , and Au metallogenic series, which is related to intermediate-acid intrusions in genesis.

YUAN Daixian, Institute of Karst Geology, Guilin, China

Man's Impact on Karst Hydrological System

A better understanding of the relationship between human activities and karst hydrological system is the key to the rational exploitation and protection of karst groundwater resources.

Many environmental problems in karst areas are related to the deterioration of its hydrological system. The origin of deterioration may be natural, e.g., water table fluctuation brought about by seasonal or long-term climatic change, harmful water quality that comes from the hydrochemical processes between natural water and rocks. However, a lot of them are induced by human activities. The latter could be classified into three categories: human activities on the recharge area of a karst hydrological system; water regulation projects within the system; and pumping water from the phreatic zone.

The conventional approach to protecting groundwater from pollution is to prohibit human activities in the recharge area of the groundwater system and to establish a sanitary zone. However, the situation on China's karst regions makes it usually inevitable for human beings to live and carry out economic activities in recharge areas of karst water. For instance, in the karst areas of Yunnan and Guizhou provinces of Southwest China, land resources are mainly distributed on plateaus of different altitude, so it is usually impossible to forbid people to grow crops, establish manufactories, and build cities on the catchment area of karst hydrological system (Figure 1). The situation is the same in Shanxi province of North China, where the Carboniferous coal measures overlaying the recharge area of major karst springs flowing out from Cambrian or Ordovician carbonate aquifers. The latter are usually important sources for local agricultural, industrial, and domestic use (Figure 2).

The aim of building regulation engineering in a karst hydrological system is to make the karst water resources more useful to the people by means of changing the natural conditions of water flow in the system and readjusting its distribution in space and time, such as to conduct water from underground streams for irrigation; and to build underground reservoirs to store the flood for the use of dry season. There are many successful examples of this kind in the provinces of Guangxi, Guizhou, etc. However, environmental side effects may happen if such engineering is not properly planned and designed. For instance, the upper reach of farmland may be inundated after the underground stream is dammed up.

Using the storage capacity below the water table in a karst hydrological system to pump more water during the dry season and let it be compensated naturally in the following wet season is also a good way for people to regulate karst water resources. However, a series of environmental problems may occur if the total exploitation from the system is in excess of the overall recharge into the system both in rainy season and dry season, such as the declining of regional water table; drying up of wells for water supply; damage from tourism attractions; and even surface collapses. These kinds of problems have taken place in Jinan City, Shandong Province; at Jinci karst spring, Taiyuan, Shanxi province; the Green Lake scenic area, Kunming, Yunnan province; and the section of Tianjiang–Nanjing Railway near Tai'an, Shandong province.

To meet people's long-term demands in many respects, it is necessary on the basis of overall investigation and analysis of karst hydrological system to take into account the exploitation and protection of different kinds of natural resources (mineral deposits, land, energy, forest, water, tourism attractions, etc.) comprehensively, work out a rational environmental plan and design, and implement it seriously.

REFERENCES

Karst and Man: Proceedings of the International Symposium on Human Influence in Karst, 11–14 September 1987, Postojna, Yugoslavia.

Yuan Daoxian, 1983, Problems of environmental protection of karst area: Paper presented to the annual meeting of AAAS, Detroit, Michigan, 15 p.

———, 1986, Some characteristics of groundwater protection in karst area: IAH Memoires, v. XIX, P. 2, p. 135–142, Novinar publishing House.

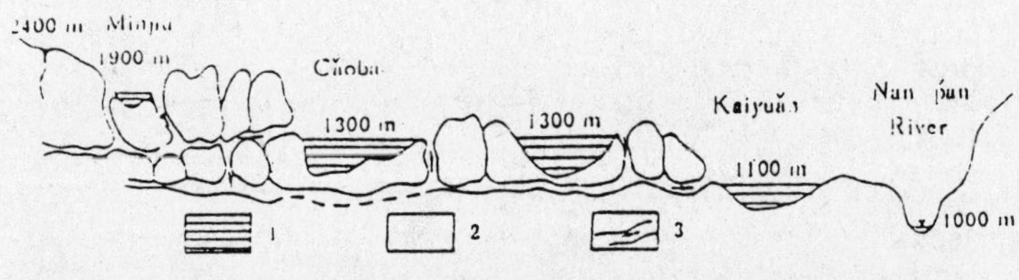

Figure 1—The relation between plateau farmland and karst water drainage system; an example from Caobá-Kaiyuan area, southern Yunnan province. 1, impermeable strata (Quaternary and Tertiary); 2, karstified limestone (Middle Triassic); 3, underground drainage conduit and flow direction.

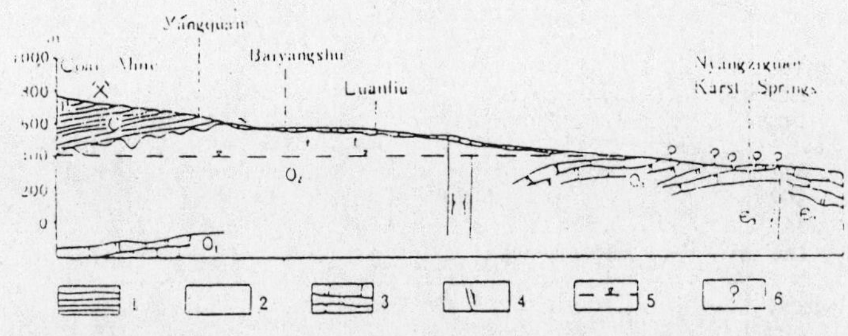

Figure 2—Profile along Tahoe river, Shanxi province, Northern China, showing the relation between coal mining and karst aquifer system. 1, sandstone and coal measure (Carboniferous and Permian); 2, limestone (Middle Ordovician); 3, dolomite (Lower Ordovician and Upper Cambrian); 4, fault; 5, water table of limestone aquifer; 6, karst spring.

YUAN Jun, and H. H. MURRAY, Indiana University, Bloomington, IN

Siderite Concretions in Shales of Pennsylvanian Brazil Formation, Brazil, Indiana

The investigation of carbonate concretions has specific importance in understanding the sedimentary processes of siliciclastic rocks as indicated by Hudson (1978) and Raiswell (1971, 1987). Typical siderite concretions in the Brazil Formation in Indiana were selected to study the relationships of mineralogy, structure, time and space using the diagenetic concretion model from Curtis et al (1986). Laboratory techniques used include X-ray diffraction, SEM, and stable isotope analyses.

The Brazil Formation (Lower Pennsylvanian) is comprised of shale, coal, underclay, and sandstone. Abundant siderite concretions along the bedding-planes were found in the gray to light-gray shales overlying the Upper Block Coal Member of the Brazil Formation; however none were found in the dark-gray to black shales in the same formation. The concretions had various shapes including spheroidal, loaf-like, and cow-horn-like, with a diameter from a few centimeters to several tens of centimeters. A septarian structure was commonly observed inside the concretions, which was the result of the intersection of two sets of cracks, i.e., vertical cracks perpendicular to the bedding-plane and horizontal or curved cracks parallel to the bedding-plane caused by compaction and syneresis. Sometimes residual plant fossils were perfectly preserved in the concretions without any compaction.

X-ray and SEM analyses show that the main bodies of all concretions are comprised mostly of massive, pure, euhedral to semi-euhedral micritic siderite crystals with trace amounts of euhedral quartz and pyrite. The fillings in the septarian cracks of the concretions are kaolinite, calcite, and sphalerite. The kaolinite replaces calcite, sphalerite, and even the earlier-formed kaolinite. A general paragenetic sequence of the authigenic minerals is recognized as follows: [siderite, pyrite]→[early stage kaolinite, sphalerite, calcite]→[late stage kaolinite]. The heavy carbon isotopic values (+3.816 to +4.319) of the siderite indicate that anaerobic bacteria activity generates the source-carbon for $FeCO_3$ during very early diagenesis (Curtis et al, 1986). However, the sulfur isotopic values (+3.30 to +6.67) of the sphalerite in the concretions coincide with that of the sphalerite from the Mississippi Valley-Type Sulfide deposits and the Illinois Basin coals (Heyl et al, 1974).

It is concluded that the growth of siderite concretions in the shales was in a fresh or nearly freshwater environment and took place mainly during very early diagenesis. Evidences include: parallel and curved laminae, preservation of plant fossils, preservation of randomly oriented and well-crystallized kaolinite in cracks, septarian structure, and carbon isotopic values. The siderite body grew first in a colloidal state at a relatively faster rate than the sedimentation rate. A weak or absent sulfate reduction zone beneath the water/sediment interface resulted in the absence of the pyritiferous margin on the concretions (Raiswell, 1971). The sites forming concretions in the sediments were related to the presence of a certain type of organic materials that provided the needed carbon source through microbial processes, The Fe^{2+} derived from the detrital iron oxides reacted with HCO_3^- to form $FeCO_3$ in alkaline and reduced conditions. Such a geochemical environment (reducing, high pH) favored the dissolution of the detrital clay minerals in the surrounding sediments and released aluminum and silica that is proposed to be the major source for the kaolinite precipitation in the concretion cracks. The concretions may stop growing in deeper burial but start compacting and form septarian structures. Later during diagenesis, a regional thermal event in Mid-Continental United States related to the Mississippi Valley-Type deposits (Keller, 1988) may have influenced the formation of authigenic mineral fillings in the cracks of the concretions.

REFERENCES

Curtis, C. D., et al, 1986, Pre water evolution during sediments burial from isotopic and mineral chemistry of calcite, dolomite and siderite concretions: Geochimica et Cosmochimica Acta, v. 50, p. 2321–2334.

Heyl, A. V., et al, 1974, Isotopic evidence for the origin of Mississippi Valley-Type mineral deposits: a review: Economic Geology, v. 69, p. 992–1006.

Hudson, J. D., 1978, Concretions, isotopes, and the diagenetic history of the Oxford Clay (Jurassic) of Central England: Sedimentology, v. 25, p. 339–370.

Raiswell, R., 1971, The growth of Cambrian and Liassic concretions: Sedimentology, v. 17, p. 147–171.

——— 1987, Non-steady state microbiological diagenesis and the origin of concretions and nodular limestones, *in* J. D. Marshall, ed., Diagenesis of sedimentary sequences: Geological Society Special Publication, v. 36, p. 41–54.

YUAN Xuecheng, Ministry of Geology and Mineral Resources, Beijing, China

The Geoscience Transect from Taiwan to Heishui, Sichuan Province

The Geoscience Transect from Taiwan to Heishui, Sichuan Province, is the eastern section of the long transect Taiwan-Altai, which is one of the transects sponsored by the Ministry of Geology and Mineral Resources during the seventh Five-Year Plan (1986-1990).

The long transect Taiwan-Altai begins from the eastern flank of Taiwan, traverses the Zhongqing, Sichuan province, the Dari, Qinghai province, to Fuyun, Xinjiang autonomous region, which cuts through the whole country and has a total length of 4600 km. It might be divided into three sections. The western section starts from the Altai, crosses the eastern part of the Zhungel basin, the Tianshan, the Tulufan-Hami basin, the Beishan, the eastern part of the Tarim massive, to the Altunshan, penetrates almost all of the tectonic unit of the northwestern China. The middle section extends from the Altashan to the Longmenshan, through the northeastern part of the Qinghai-Tibet plateau, which is exceptionally valuable to study the structure of the Qinghai-Tibet plateau. And the eastern section is from Taiwan to Heishui displaying just here. It runs over the Yangzi platform, the south China fold belt, and Cathaysia, extends for about 2200 km.

All of the area which the transect traverses underwent a geological, aeromagnetic, and gravitational survey with a scale not smaller than 1:500,000. Along the transect, explosive seismology, MT, and heat flow was conducted, and in some places geomagnetic differential sounding, also. Various geological and geophysical data were compiled. Besides, a seismic reflection profiling extending 100 km was carried out at the Xuefengshan area.

The Yangzi platform is believed to have double basement. The katametamorphic basement is high density, and intense magnetism distributes in the center of the Sichuan Basin. The upper side is a series of epimetamorphic flysch sediments, it transformed to platform at the end of the Proterozoic period and formed the epimetamorphic basement. In the late Paleozoic period, shallow sea sediment was deposited. At the end of the Silurian perh China fold belt became platform and joined to the Yangzi platform. After the Devonian, both Yangzi platform and the south China fold belt had the same sediment circumstance, depositing carbonate and sometimes clastic rocks and coal series. Since the Triassic period, it entered into active continental margin stage, deposited continental sediments, deformed widely, and extensively activated magmatism.

The existence of the Cathaysia was confirmed by increasing data, but identification of its extent was difficult. The boundary between the south China fold belt and the Cathaysia is perhaps situated along the Taiwan Strait or somewhat east or west.

Both the gravitational or geomagnetic field int the south China area are gentle, heat flow is not very high, and he geomagnetic differential sounding is devoid of anomaly. The crustal velocity structure appears monotonic, increasing with depth. The velocity of the upper crust is about 4.9 km/s, the middle crust is about 6.0 km/s, and the lower crust is about 7.0 km/s. The velocity of the top of the upper mantle is about 8.1 km/s, somewhere (Sichuan Basin) is greater than 8.2 km/s. All of these geophysical characteristics show that the deep structure in the south China area is relatively homogeneous and stable.

The second characteristic of the deep structure of the south China area is that there are decollement faults of great scale. The seismic reflection profiling in Xuefengshan area recorded several gentle dipping reflections under the Banxi epimetamorphic series of the Proterozoic period, revealed there might exist young sediments underneath. All the granites in the south China area give rise to a gravity low, the bottom of the granite is several kilometers to more than 10 km deep by quantitative interpretation. This draws an inference that the bottom of these floating-like granites might be a decollement fault.

The third characteristic of the deep structure of the south China area is the lack of coordination between surface structure and deep structure. At the surface, there are numerous folds and faults, but at depth, there appear arrested anticline and syncline.

In summary, the south China fold belt is a arrested sea between the Yangzi platform and the Cathaysic massif in early Palaeozoic; its sedimentary environment is stable. The extension in the late Paleozoic caused massive rupture of the basement at the western flank of the Sichuan Basin; extensive basalt erupted. In the Mesozoic-Cenozoic period, the brittle surficial crust slipped along a decollement fault of depth about 10 km under the compression from southeast, caused tremendous thrusts and nappes in the upper crust and gentle folding in the lower crust. Therefore, in contrast with the north China, the south China area has few basic rocks of Mesozoic-Cenozoic in the crust, and is a light, cold crust. And hence, the gravity field corresponding to the south China is 40 mgl lower than the north China, notwithstanding its thickness of crust is thinner.

YUI, Tzen-Fu, Academia Sinica, Taipei, Taiwan, and LU, Chia-Yu, LO, Ching-Hua, and LEE, Chihming Wang, National Taiwan University, Taipei, Taiwan

Tectonic Evolution of Tananao Schist Complex of Taiwan

The Tananao Schist Complex is the basement of Taiwan. It has been poly-metamorphosed and deformed. The outcrop area of the complex has been divided into the Tailuko and the Yuli belts. Based on the available data, a speculative model for the tectonic history of this complex is proposed as follows.

1. During late Permian to Triassic, the Farallon plate subducted westward(?) beneath the ancient Asiatic continental margin. This subduction may have produced some calcalkaline continental-margin type igneous

activity along Southeast China. East of the subducting trench, the Changchun seamounts moved westward.

2. During early to middle Jurassic, the Changchun seamounts reached the trench. Due to the effect of buoyancy, the seamounts could not be subducted and therefore collided with the continental margin and finally collapsed. This subduction/collision complex formed the Tailuko belt. The collision might have caused the uplift of the igneous body in Southeast China (160–190 Ma) and might also have choked the trench. The subduction therefore might have retreated oceanward.

3. During late Jurassic to late Cretaceous, the oceanic plate (the Izanagi-Kula plate) continued subducting westward. The subduction complex of this stage constitutes the Yuli belt. This subduction also produced some igneous activity in Southeast China as well as in the Tailuko belt. It should be noted that the Tailuko belt was once the site of intermediate- to high-pressure type metamorphism during the previous subduction/collision event. The original records for that metamorphism therefore might have largely been obliterated by this low-pressure type metamorphism. This late Mesozoic subduction process must have soon ceased since no subduction-related calcalkaline igneous activity younger than 70 Ma was found in the area.

4. During Paleogene to Miocene, the Tananao Schist Complex was covered by a thick sequence of shale, mudstone, and sandstone. During late Miocene, the newly formed South China Sea plate subducted eastward beneath the Phillipine Sea plate, producing the Luzon arc, and possibly also subducted westward beneath the Eurasian continent. Either the aborted westward subduction or the deep burial have caused the initiation of the late Tertiary greenschist facies metamorphism of the Tananao Schist Complex.

5. The South China Sea plate continued to be consumed under the Phillipine Sea plate. During Plio-Pleistocene time, the Luzon arc collided with the Asian continent. The late Tertiary metamorphism continued during this stage and the collision process also caused the uplift of the Tananao Schist Complex. The process is still going on at the present time.

YUMUL, Jr., G. P., University of Tokyo, Tokyo, Japan

Complexities in Upper Mantle–Crust Relationships: Insights from Zambales Ophiolite Complex, Luzon, Phillipines

A detailed petrological study of the Zambales Ophiolite Complex located at the island of Luzon, Phillipines, revealed the complex history it has undergone involving several stages of partial melting, fractional crystallization, and volcanism. This ophiolite complex has been recognized to be made up of at least two upper mantle–crust sequences, the Acoje Black (AB) and the Coto Block (CB), that generally preserve island-arc and MORB characteristics, respectively (e.g., Bacuta, 1978; Hawkins and Evans, 1983; Geary et al, 1986). These two blocks differ in their respective mantle–crust thicknesses, crystallization phases, type of mineralizations, petrofabric directions, and, of course, geochemical signatures (e.g., Nicolas and Violette, 1982). However, in spite of all these differences, the present data indicate that the AB and CB are basically transitional in chemical characteristics with each other and do not require a large-scale tectonic movement to juxtapose them together.

Plagioclase from the mafic cumulates of the two blocks show that the melt responsible for the AB (An90–97) is more calcic compared to the CB (An80–90). Other workers (e.g., Thy, 1987; Browning, 1984) had hinted that the calcic nature of the plagioclase suggests high water pressure that they consider as probably indicative of a subduction event. Plotting the appropriate data on Beard's (1986) Fo# vs. An# diagram shows that the Acoje samples plotted on the arc field compared to the Coto samples, which plotted on the MORB field. This suggests that a subduction event had played a major role during the formation of the AB. The spinel data from the residue to the cumulate sequences are far more revealing in terms of the probable environments of formation and residue-cumulate relationships of the two blocks. The XCr (Cr/Cr + Al) vs. XMG (Mg/Mg + Fe^{2+}) diagram for spinels shows that the Coto residue-cumulate are within the MORB-abyssal periodotite fields, whereas the Acoje ultramafic cumulate straddle the chromium-rich portion of the diagram. However, the Acoje residue spinels plot in the aluminum-rich portion (0.2–0.45 XCr) of the diagram, suggesting that the Acoje residue is not directly parental to the overlying Acoje cumulates. The Coto residue-cumulate, on the other hand, exhibits a normal parent-daughter melting relationship.

Field and laboratory evidence also suggest that there were at least two distinct magma chambers responsible for the formation of the AB and the CB. Contact of the two blocks is transitional since no large-scale tectonic contact had been mapped in the field. Aerial photographs and Landsat imageries do not support the presence of a major tectonic boundary for the two blocks. Unfortunately, the available data are not yet sufficient to quantitatively and qualitatively describe the spatial and temporal relationships between the magma chambers. Nevertheless, Taylor and Karner (1983) had pointed out the possibility that distinct back-arc basin basalt (BABB) and island-arc tholeiite (IAT) magma sources may be in extremely close proximity during the major rifting of island arc.

The geochemistry of the volcanic rocks is consistent with the data derived from the cumulates of the two blocks. The Acoje volcanics are generally IAT while the Coto volcanics plot on the MORB field. The Acoje volcanics are characterized by lower TiO_2, Zr, and Y compared to the Coto volcanics, which suggests that either the sources of the two blocks had different trace element abundances or the source of the Acoje volcanics had undergone a higher degree of partial melting. The second possibility would again suggest that the Acoje residue is not directly parental to the Acoje cumulates/volcanics since on the basis of the spinel data, it seems that the Acoje residue had undergone a lower degree of partial melting compared to the source of the CB. A probable explanation for this is that the source of the Acoje cumulates/volcanics could be located far below the Acoje residue. This source had undergone a higher degree of partial melting compared to the overlying Acoje residue and for that matter, also to the Coto residue. Passageways could have brought the generated magmas above the Acoje residue leading to the formation of the Acoje cumulates and extrusions of the Acoje volcanics.

Combining the mineralogical and geochemical data with field evidences makes it impossible to conclude that the Zambales Ophiolite Complex could have formed in a marginal basin (supra-subduction zone environment) whose rifting is subduction related. Tectonic reconstruction also makes it feasible for this ophiolite complex to have come from the Paleocene–Eocene proto-South China Sea Basin.

Z

ZAHEDI, M., Geological Survey of Iran, Tehran, Iran

Study on the Upper Paleozoic in Zardkuh of Bakhtiari Region (High Zagros Mountains)

A sedimentary sequence composed of shales with intercalations of limestone layers and sandstone beds has been observed during the field work in the north side of the Bazuft River. This river is located in the southern slopes of the Zardkuh Mountain (High Zagros Mountain ranges).

The sediments belonging to Ordovician-Silurian are overlain disconformably with Permian deposits. Devonian and Carboniferous sediments are absent at the studied section. The collected samples from this section contain Fusulinas, Brachiopods, Trilibites, Graptulites, condonants, and other fossils.

These fossil samples are partly studied and will be studied further in the future.

Moreover, the chemical analyses of the rock samples indicate that the Ordovician-Silurian layers contain phosphate minerals, and the transitional part forming the main phosphatic zone seems to be richer in phosphate. The average percentage of the collected samples shows 21.4% of P2O5.

It should be noted that the above-mentioned sediments are extensive in the southern slopes of Zardkah. They have a length of approximately 70 km in the area. The previous authors have not mentioned Silurian deposits up to now, but on the contrary, the Carboniferous sediments that have been mentioned are probably absent.

ZAILIANG, Ji, Memorial University of Newfoundland, St. John's, Newfoundland, C. B. BARNES, Geological Survey of Canada, Ottawa, Ontario, Canada

Preliminary Lower Ordovician Conodont Zonation in the Midcontinent Province

Lower Ordovician conodonts of the Midcontinent Province have received only moderate biostratigraphic study, mainly in the last two decades. Many conodont faunas or zonations have been based on incomplete stratigraphic sequences or limited and poorly preserved material. Over 430 samples, collected from ten sections of a nearly complete, 600 m thick, Lower Ordovician carbonate succession through the St. George Group in western Newfoundland have yielded more than 45,500 excellently preserved conodonts. These faunas permit major revision to the reconstruction of Lower Ordovician multielement taxa, as well as the correlation of the St. George Group with Lower Ordovician sequences elsewhere in the world. These correlations are based on the establishment of a new and systematic Lower Ordovician conodant zonation for the Midcontinent Province from this remarkable succession of undeformed strata and thermally unaltered conodonts (CAI 1).

Based on this study, eight conodant Assemblage Zones can be recognized as following: (1) Teridontus Nakamurai-Semiacontiodus Nogamii Zone associated with *Cordylodus intermedius* Furnish and *C. lindstromi* Druce and Jones, which can be related to previously established informal Midcontinent Fauna B of Ethington and Clark (1971); (2) Loxodus Bransoni-Variabiloconus Bassleri Zone with *Cordylodus angulatus* (Pander), *Clavohamulus densus* (Furnish), *"Acodus" oneotensis* (Furnish) and several other important species which are indicative of the lower Fauna C (middle Tremadoc age); (3) Rossodus Manitouensis-Clavohamulus N. Sp. B. Zone, which is equivalent to middle Fauna C and is of late Tremadoc age; (4) Glyptoconus Floweri Zone has a short stratigraphic range which corresponds to the upper part of Fuana C, and is of the earliest Arenig age; (5) "Scolopodus" Triangularis-Macerodus Dianae Zone which is of early Arenig age, and corresponds to the uppermost of Fauna C and lower most Fauna D; (6) Scolopodus aff. Rex-Acodus Delicatus Zone with extremely abundant *Glyptoconus quadraplicatus* (Branson and Mehl) and *Drepanodus concavus* (Branson and Mehl) that is an important assemblage of within Fauna D, and is of middle Arenig age; (7) Glyptoconus Emarginatus-Oepikodus Communis Zone associated with several important species, such as *Oneotodus coastatus* Ethington and Brand, and with distinctive *Scolopodus carlae* Repetski, *Oneotodus* n. sp. A. and "O" N. sp. B. in the lower part, this Zone is equivalent to the Fauna E and of middle Arenig age; (8) Pteracontiodus Cryptodens Scolopodus N. Sp. A. Zone that is of late Zrenig age, and corresponds to the Fauna E/1.

As indicated in the Figure, the eight conodont Assemblage Zones represent limited stratigraphic intervals in the Lower Ordovician, most can be found in other low latitude cratons (Australia, Siberia and North China), and may be correlated with the zonation for the North Atlantic Province. Because the western Newfoundland sequence was deposited close to the ancient continent margin, it was periodically invaded by conodants from the North Atlantic Province. It is therefore possible to establish critical biostratigraphic tie-points to enable correlations between the two provinces. This preliminary zonation provides a succession of refined Assemblage Zones. The completeness of the sections and the abundant faunas are enabling a phylogenetic zonation to be developed based on evolving lineage of taxa such as *Clavohamulus, Glyptoconus,* finely straited *Scolopodus,* costate *Oneotodus, Loxodus* and *Acodus.*

ZAINUDDIN, S. M., M. P. ALAM, and A. RAHMAN, Aligarh Muslim University, Aligarh, India

Tectonic Set-Up of Dudran Granite and Its Implication for Himalayan Orogeny

The Dudran (named after a nearby village), a constituent of central Himalayan axis, lies in the Rampur-Apharwat section of the Salkhalas in the Pir-Panjal Range and is exposed in the Gulmarg anticlinorium. The area lies ~90 km north-northwest of Srinagar in Baramula District (J & K, India) on the Uri-Srinagar Highway. Of the various lithologies found in the area, arenaceous phyllites are the most abundant rock type; the metamorphites are intruded by Dudran granite of Tertiary age. A study of the tectonomagnetic cycle is being attempted here to formulate the evolutionary and tectonic history of this complex mountain range—the Himalayas.

The granite shows distinct variation both in terms of color and grain size. Compositionally, it ranges from granodiorite to monzogranite. Plagioclase in the major mineral constituent and quartz is next in abundance. K-feldspar is generally low or absent. Xenoliths of phyllite found in the granite indicate a younger age of the granite than the metamorphites.

The concentration of oxides and certain trace elements in granite have been plotted on a series of bariation and discriminant diagrams. It has been observed that with increasing SiO_2, the concentration of Rb, Ba, and Sr increases. The average K/Rb ratio (Avg. 242) is similar to that of the crust (Avg. 240), which may suggest that the granite was formed by melting of the crust. The Sr-Rb relationship and the low Ba/Rb ratio indicate its differentiated nature, which is further corroborated by high K/Ba ratio and low Ba/Rb ratio. The plots of Dudran granite on El-Bouseilly and El-Shokkary's (1975) Rb-Ba-Sr ternary diagram are also concentrated toward the highly differentiated granite zone.

Pearce et al (1984), by employing a trace element discrimination diagram for tectonic interpretation of granitic rocks, have divided them into four main tectonomagmatic groups: ocean-ridge granites, volcanic-arc granites, within-plate granites, and collision granites; collision granites have further been divided into syncollision granites and postcollision granites.

The plots of Dudran granite on the diagram of Pearce et al, (1984) fall in the syncollision granite field (Figure 1, 2).

With respect to the evolution of the Himalayas, there is no common agreement among geologists. Hashimoto et al (1973), Kaila and Narain (1976), Saxena et al (1985), Ahmad (1988), and many others have suggested vertical movement and block faulting to be the cause of Himalayan orogeny. They have cited lithological, paleobiological, and like evidence against the possibility of continent–continent collision along the Indus Suture Zone. Nevertheless, a large number of geologists and geophysicists are plate-tectonic protagonists and have opined that the evolution of the Himalayas is a result of northward drift of Indian plate and its subsequent collision with Eurasian plate, with the Indus Suture Zone being the junction of the two colliding plates.

The study of the geochemical fingerprints and trace element discrimination diagram for Dudran granite suggests that the granite is a collision granite, which is the major product of most collision belts (Pearce et al, 1984).

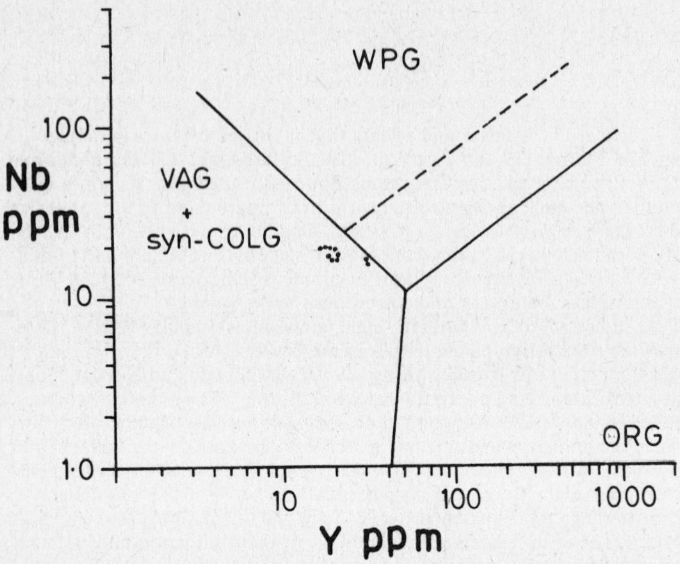

Figure 1—Plots of dudran granite on the Y-Nb discriminant diagram.

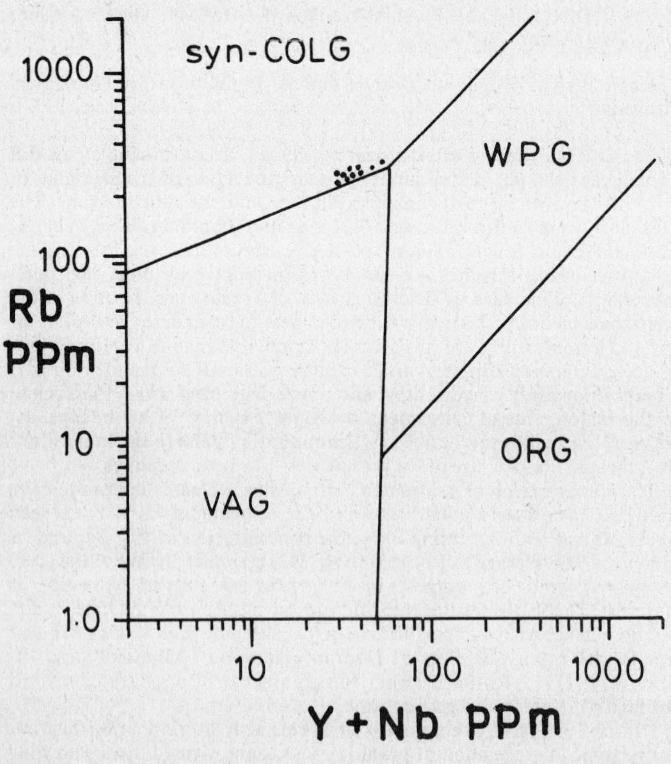

Figure 2—Plots of dudran granite on the Rb − (y + Nb) discriminant diagram.

ZAKHAROV, V. A., Institute of Geology and Geophysics, Novosibirsk, USSR

Cretaceous System Boundaries and Event Stratigraphy Conceptions

The stratigraphic situation on the lower and upper boundaries of the Cretaceous system differs greatly. There are no striking traces of events of biotic or abiotic nature on the lower boundary either in Tethian (Tithonian–Berriasian) or Boreal (Volgian–Boreal Berriasian) realms. Gradual evolution of all groups of organic world is observed, neither geochemical nor mineralogical anomalies are found, and the sedimentary reorganization in different regions is situated on different stratigraphic levels. There are essential disagreements among specialists in determination of the Jurassic–Cretaceous boundary location. No less than six levels are discussed as candidates for the system boundary.

The thin layer of "boundary clays" that has anomalous geochemical and mineralogical characteristics is situated on the upper boundary between Cretaceous (Maestrichtian) and Paleogene (Danian). In the Maestrichtian–Danian interval, the radical reorganization of major groups of organic world and replacement of sedimentary formations took place. Investigators are unanimous about the fact that the Cretaceous–Paleogene boundary should lay between the Maestrichtian and Danian, and they agree to consider the first zone on the planktonic foraminiferas above the "boundary clays" to be the base of Danian. There is uncertainty about fixing the boundary at the base of "boundary clays" layer or above its base.

Many investigators perceive different methodological approaches in defining the lower and upper boundaries of the Cretaceous system: biostratigraphic in relation to lower boundary and "event" in relation to the upper boundary. But it is not so. In both cases the leading role belongs to the biostratigraphic method.

The first major requirement for the system boundaries is their isochroneity, which can be proved by way of zone correlation in both cases: ammonite on the lower boundary, planktonic foraminifera and nannaplankton on the upper one. Isochroneity of the "boundary clays" themselves, as a trace of the global event, which formally has no specific features, can be proved by their position at the base of the paleogenic Fringa–Eugubina zone (basal zone of the Danian stage) and its correlatives all over the world.

The question about the choice of the boundary is solved with the help of two successive procedures of boundary determination and its fixing. Both the lower and the upper boundaries of the Cretaceous system are determined (theoretically) by the trace of only one (single) event of biotic nature by the appearence in the section of the orthostratigraphic group of species, chosen for the development of the zonal scale: ammonites on the lower boundary and planktonic foraminiferas on the upper one. In both cases, the boundary (the base of a zone as a point in a section) is fixed in a base of elementary stratum in the limits of which the species is found, i.e., according to the "trace of event" of abiotic nature.

Questions about priority, convenience, and stage conformity are also taken into consideration but as of secondary importance. The position of the lower boundary of the Cretaceous system at the foot of Berriasian has priority. So far, as in terminal stage of Jurassic and in basal stage of Cretaceous systems none of the competing boundaries have obvious advantages over the historically established, one can consider the aspiration of the majority of specialists to define the Jurassic–Cretaceous boundary near the traditional Tithonian–Berriasian boundary to be motivated (the fate of the Jacoby–Grandis zone is at stake). The situation of the boundary between Cretaceous and Paleogene at the base of Danian reflects the stage conformity in the evolution of the organic world, on the other side the clay layer—the trace of a "global event," discovered at the base of the Danian basal Fringe–Eugubina zone or its correlatives provide good comfort for the diagnosis of the boundary in the Maestrichtian–Danian continuous monotonous sections.

ZAKHAROV, V. A., Institute of Geology and Geophysics, Novosibirsk, USSR

Paleogeographic and Paleoenvironmental Reconstructions of Mesozoic Epicontinental Seas of Soviet Arctic and Ocean Connections of Arctic and Northern Pacific Basins from Paleontological Dates

During the whole Mesozoic era, the territory of recent shelf and nearshore part of the land of the Soviet Arctic from the Kola peninsula in the west to the Chukchi peninsula in the east has been occupied by the seas of epicontinental = platform (the Barents, Timan–Uralian, West Siberian, Enisei–Lena, Lena–Viljui), subplatform, and geosynclinal types. The extent and outlines of the platform basins have been largely affected by transgression and regression, and geosynclinal basins by the processes of orogenensis.

Predominant in the composition of sediments for all types of basins are clastic terrigenous sediments: sands, silts, and clays. Pure carbonaceous rocks are known from Triassic only. Tufogene component is characteristic of the deposits of geosynclinal type. Intensive volcanicity is fixed in Induan epoch of the Triassic period (Kazakov et al, 1982). Essential biogenic sedimentation is noted for Late Cretaceous basins. High bituminous clay shales formed in platform seas during the Olenek epoch of the Triassic period and at boundary between the Jurassic and Cretaceous periods (Bogolepov, 1983).

During the Mesozoic era, the main source areas ablation were the lands at present occupied by the Kola Peninsula, Timan, the Ural mountains, the Siberian Platform, and by the islands in the north of the Barents Sea, on the Novaya Zemlya islands, the Taimyr Peninsula, in the Laptev Sea, and the island arcs in the northeastern USSR (Saks and Ronkina, 1979).

Each period has diverse genetic types of sediments: open basin, shelf, shallow water, littoral-marine, lagoonal, deltaic, alluvial.

In the territory of Siberia during the Mesozoic, the predominant climate was warm (in Early Triassic) and temperate warm (in Jurassic and Cretaceous), more humid in the north and arid in the south. Average annual temperature during the Mesozoic period has fallen from +(20–25) °C in the Early Triassic to +(12–15) °C at the end of the Cretaceous (Golbert, 1987). Thermal stratification existed in the seas. For example, the difference in the temperature of surface and benthonic waters in the Volgian–Berriassian West Siberian basin was no less than 10 °C. Water salinity in Mesozoic seas in the north of Eurasia was unlikely to have deviated greatly from normal marine for that time (Bogolopov, 1983).

Considerable similarity between the basins of the arctic part of Eurasia and Arctic Canada is noted in type of sediments and fauna and flora composition (Balkwill et al, 1983). These were high boreal (arctic) seas with predominance of terrigene sedimentation and peculiar impoverished biota over that of south seas. Fauna evidence suggests great influence of North Pacific waters on the creation of arctic fauna in the Triassic and Early Jurassic; in the Middle Jurassic, this influence grows weaker, and in the Late Jurassic and Cretaceous the influence of the North Atlantic and East European seas predominated.

Within the Mesozoic, Arctic basins always existed free interchange of marine fauna and therefore circumarctic interchange of waters. This phenomena needs to palinospathic reconstructions for its explanation. On the contrary, close similarity of mollusks from Arctic Canada and Arctic Siberia suggests the absence of deep sea between the Chukchi Peninsula and Alaska or at the place of the Cherskiy ridge. Paleontologic evidence offers no reason to assume different position in Mesozoic, as it is at present, of such terranes as the Omolonian, Kolymian, Olyutorian, and Anadyrian. All these terranes have been accreted to the Siberian craton at least in Triassic period. One terrane (Koryak) has probably been situated further south than at present and has likely been accreted at the end of Mesozoic.

REFERENCES

Balkwill, H. R., D. G. Cook, R. L. Detterman, A. F. Embry, E. Hakansson, A. D. Miall, T. P. Poulton, and F. C. Young, 1983, Arctic North America and Northern Greenland, *in* The Phanerozoic geology of the world, II, The Mesozoic B: Netherlands, Elsevier Science Publishers, p. 1–31.

Bogolepov, K. V., ed., 1983, Jurassic Paleogeography of the north of USSR: Novosibirsk, Nauka, 190 p. (in Russian).

Golbert, A. V., 1987, Basis of the regional paleoclimatology: Moscow, Nedra, 222 p. (in Russian).

Kazakov, A. M., A. S. Dagis, and N. I. Kurushyn, 1982, The main features of paleogeography of North Central Siberian Triassic, *in* Geology and presence of oil and gas of the Enisei-Khatanga Basin: Trans. Inst. Geol. Geophys., v. 514, Moscow, Nauka, p. 54–75 (in Russian).

Saks, V. N., and Z. Z. Ronkina, 1979, New data on the development of Siberian topography during the Mesozoic: Geol. Geophys., n. 10, p. 3–16 (in Russian).

ZAMBRE, M. K., D. B. F. Dayanand College of Arts and Science, Solapur, India

Replenishment of Basaltic Aquifers by Percolation Tanks: Case Study of Solapur District

In India, use of large-diameter wells (dug wells) to withdraw groundwater for irrigation and domestic use is a common practice. In drought-prone areas, because of low recharge, these dug wells either yield very poorly or dry up in the summer. As a result, crop failure is frequent. In Solapur district out of total cultivable land of 1.33 million hectares, only about 14% of the land has access to a canal irrigation system. More than 90% of farmers cultivate their land using large-diameter dug wells only. The water potential of these dug wells largely depends upon the rainfall, geomorphological set up, and hydrogeological characteristics of rock units in the area. Under these circumstances, one has to think how the aquifers can be recharged to increase the groundwater potential. During the last decade in Maharashtra State the construction of percolation tanks played a dominant role in replenishment of basaltic aquifers. In Solapur district, a total of 1386 percolation tanks have been constructed on various streams of Bhima, Sina, Man, and Bhogawati river basins. The storage capacity of these tanks ranges between 0.0294 m.M^3 and 0.6938 m.M^3. Expected irrigation land under each tank ranges between 6 and 113 hectares.

Rainfall and physiography: The Solapur district (17°.00'–18°.20'N and 74°.15'–76°30'E) (Figure 1), covering an area of about 15,057 km^2, is geographically divided into 11 tahasils out of which four tahasils (Madha, Mohol, Karmala, and Sangola) are known to be drought prone. The district receives an average rainfall of about 580 mm. Over the greater part of the district, the topography is undulatory with altitudes ranging between 410 and 470 m above mean sea level.

The physiography of the district has given rise to three major characteristic land forms (Figure 3). The hill and ghat section covers 2.52%, foothill zones 23.84%, and plains and plateau area 73.64% (G. S. D. A., 1975).

The Bhima river has a sinuous course and flows in an approximately southeastern direction through the central part of the district, dividing it into northern and southern sectors (Figure 4). The Sina river, which flows along a more or less parallel course, and the Man river with its northeastern course flow into the Bhima river.

Geology and hydrogeological conditions: Preliminary data on hydrogeological conditions of the district have been given by Adyalkar and Mani (1972) and Balakrishna et al. (1974). The Deccan Volcanics of Cretaceous–Eocene age constitute the predominant rock unit of the Solapur district. The volcanics are represented by basaltic flows having almost horizontal disposition, though easterly dipping flows are met in Akkalkot tahasil (Adyalkar and Mani, 1972). The thickness of each flow varies from 2 to 20 m. In Solapur district, 27 flows have been reported so far (G. S. D. A., 1975). The flows are totally devoid of hypabyssal intrusives and intertrappean beds. Throughout the district, there are layers of red-colored soil with calcareous, zeolitic, and quartz-bearing material. These layers vary in thickness from a few centimeters to 6 m.

On the basis of field observations, the basaltic flows have been grouped in three types:

1. Hard and massive, fine-grained flows of grey color.
2. Vesicular and amygdaloidal flows of pink and grey color in which the amygdules may be filled with green, earthy zeolitic material.
3. Redbed formations.

The second type is more susceptible to weathering. At places, the first two types are found to be traversed by close-spaced joints and fractures. The weathered, jointed or fractured and vesicular horizons of the flows are usually considered to be ideal sites for groundwater development (Deshpande, 1975; Radhakrishna Nair et al., 1979). Groundwater in such horizons occurs both under watertable and confined conditions. The porosity in such horizons occurs both under watertable and confined conditions. The porosity in such horizons was found to be very high. However, lateral and vertical distribution of these potential horizons is erratic, and their recognition and delineation are the main problems in prospecting for groundwater.

Redbeds are interstratified layers between two flows that are composed of clayey material of tuffaceous nature. The position and thickness of redbed layers are important factors in hydrogeological investigations. Redbed layers moe than a meter thick and exposed within a depth of 15 m from the ground level in open wells act as an aquiclude, unless associated with zeolitic basalt rocks. To work as a good aquifer, a thin redbed layer should be overlain by highly weathered fractured and jointed basalts whose thickness should be more than 10 m and underlain by hard and massive basalt formation.

Percolation tanks: The percolation tanks are constructed on second- and third-order streams of Deccan volcanics in Solapur district. In all, 1386 tanks were constructed during the last decade by Zilla Parishad Solapur (Figure 2). The average storage capacity of the tank is 0.1387 m.M^3. The designs of the tanks are modified for local requirements as well as for geological and geomorphological set up of the terrain. The salient features of a typical percolation tank constructed on a stream in the village Salgar in Akkalkot tahasil is as follows:

Catchment area	5.00 km^2
Storage area	0.7428 m.M^3
Height of the tank	13.55 m
Top width of the tank	3 m
Length	450 m
Expected irrigation	160 hectares

The seat of the dam is cleaned by removing shrubs, bushes, and grass, then a cut off french is taken throughout the central length of the dam to remove the sand, muroom, and soft rock material, and it is then refilled with black cotton soil, which is derived mainly from decomposition of Deccan Trap flows. It is clayey to loamy in nature; when it becomes wet it comes very sticky and acts as an aquiclude or as an impervious material.

The soil portion is termed the hearting zone and is supported on both sides by hard muroom. These sides are known as casing zones. The pitching is provided with hard rubble on the upstream side of the tank for absorbing water waves and to protect the casing zone from scouring.

Table 1 shows the distribution of these tanks in the district. The maximum number of tanks has been constructed in Sangola tahasil because the area has been affected by drought for many years, and the geomorphology is quite favorable for the construction of the percolation tanks. Solapur North and South tahasils are situated on the planar surface of Sina and Bhima river basins, hence there are fewer tanks.

A number of pumping tests have been carried out on the dug wells that are situated downstream of selected percolation tanks in different tahasils. The results of pumping tests show that there is a lot of variation in the values of specific capacity, transmissivity, and storage. This is because of unpredictable weathering patterns and its lateral extension. The pumping tests determine the direction of groundwater movement and thus aquifer pattern distribution. The program of construction of percolation tanks for replenishment of basaltic aquifer proved to be successful. The author is of the opinion that such a program should be adopted on a large scale in drought- prone areas of other states in the country.

The author would like to thank Dr. N. G. Zunjurwad, Principal, D. B. F. Dayanand College of Arts and Science, Solapur, for his continued support and encouragement in completing this paper. He also thanks the Executive Engineer, Minor Irrigation Division, Zilla Parishad, Solapur, for his valuable help in furnishing the information of Percolation tanks of the district.

REFERENCES

Adyalkar, P. G., and V. V. S., Mani, 1972, An attempt at estimating the transmissibilities of trappean aquifers from specific capacity values: Journal of Hydrology, v. 17, p. 237–241.

Balakrishna, S., B. Ventata Narayana, K. R. Ramanujachari, and B. S. Gogate, 1974, Ground water investigation in Solapur-Wadi-Raichur section of Solapur division: S. C. Rly. N. G. R. I. Tech. Report.

Deshpande, B. G., 1975, Some aspects of hydrogeology of the Deccan Traps: Proc. Indo-German Workshop, N. G. R. I. Hyderabad, p. 105–114.

G. S. D. A., 1975, An appraisal of hydrogeological conditions of Salapur district: Ground Water Survey and Development Agency, Maharashtra State, 59 p.

Radhakrishna Nair, P. V., A. S. Pendharkar, A. S. Navada, S. V. Rav, 1979, Groundwater recharge studies in Maharashtra: Isotope Hydrology, I. A. E. A. Vienna, v. Z, p. 803–826.

Table 1 : Percolation tanks in different tahasils of Solapur district, 1988.

No.	Tahasil	Plan	Complited	Storage capacity in m.M.	Expected irrigation in Hectres
1	Akkalkot	189	114	19.26	4121
2	Karmala	157	132	17.43	3283
3	Barshi	282	156	19.02	3117
4	Madha	169	123	9.72	849
5	Malshiras	143	118	18.85	3096
6	Mangalvedha	176	147	14.50	3304
7	Mohol	198	150	22.26	2724
8	South Solapur	120	71	14.73	1695
9	North Solapur	88	61	8.55	1111
10.	Sangola	306	279	35.60	6376
11	Pandharpur	58	35	5.46	828
	Total	1842	1386	185.38	30504

Table 2 : Effect of Percolation Tanks on Groundwater potential of Solapur district.

Sr.No.	Types of Aquifers	Depth range of Dug wells in mtrs.	Diameter range of Dug wells in mtrs.	S.W.L. range in mtrs. b.g.l.	Specific capacity range C in lp m/m	Transmissivity range T in M^2/day	Storativity range S in fraction
1	Weathered Basalts of various Types	5-15	6-15	3-6.5	475-525	135-148	0.016-0.021
2	Vesicular Zeolitic with Red bed mixed basalt	7-18	8-12	4-9	320-415	85-89	0.037-0.042
3	Fractured and jointed basalt	10-14	7-12	6-10	150-197	51-85	0.048-0.062
4	Hard and Massive Basalt	13-20	8-12	9-17	320-195	41-51	0.052-0.055

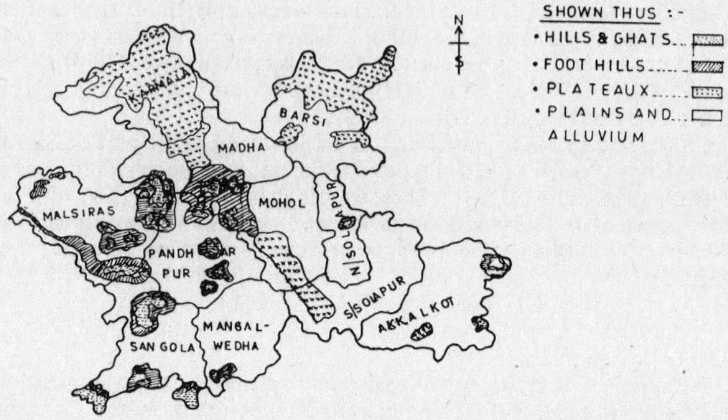

FIG 3. PHYSIOGRAPHIC DIVISIONS OF SOLAPUR DISTRICT.

Scale:-1CM=10KM

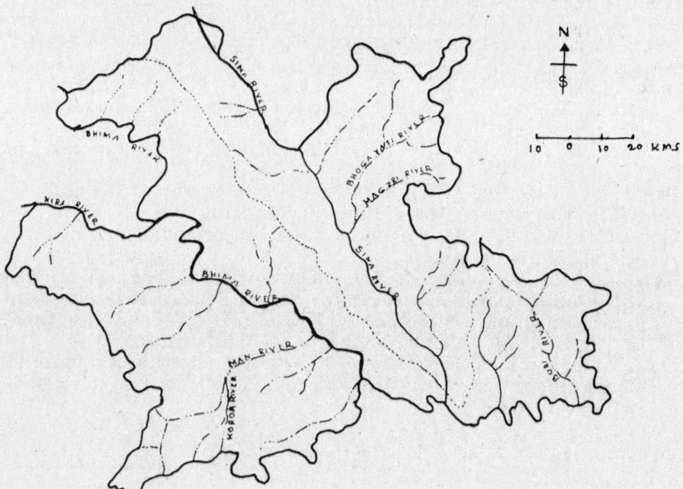

FIG 4-MAJOR RIVER BASINS

ZARAYAN, S., University of Tehran, Tehran, Iran

Petrology of Magmatic Series of Intrusive Complex, Alvand, Iran

The Alvand plutonic massif is situated west of Hamadan, Iran, and has an elliptical shape with approximate diameters of 35 and 20 km and an area of 400 km². This plutonic body has intruded into rocks of pre-Jurassic formations of Sanandaj–Sirjan Geologic Province. These Formations are slightly metamorphosed and have an epi-amphibolite facies character.

The chemical characteristics of this massif complex are acidic and basic in nature. Variation in the rock types range from gabbro on the basic side to granitic on the acidic side. The whole massif assemblage is of post-Jurassic age. Age determination by K-Ar method has been carried out on granitic and gabbro complexes of this massif. The granitic mass has an age of 68–78 m.y., which corresponds to upper Cretaceous. The basic assemblage has an older age of 78.5–89.1 m.y. (Senonian) determined by K-Ar.

The oldest basic (mafic) rocks of this intrusion series are 2.5 km thick at the north of the complex and consist of norites, troctolites, pyroxenolites, diorites, quartz diorite, and some intermediate rocks. The basic rocks form a crescent-shaped outcrop that can be delineated as an aerole in the northern half of Alvand, the maximum thickness of which is about due north.

The granitic assemblage that makes up the most part of the massive is a biotite granite porphyry (perthitic orthoclase 29%, quartz 24%, biotite 23%). In the central areas of the granitic mass there is a very light-colored granite also having an elliptical shape of 2 by 1.5 km, trending in a northwest-southeast direction. This "white" granite has the following composition: perthitic orthoclase 29%, plagioclase 45%, quartz 22%, and average sphene and rutile content of 3.5%. On the basis of rutile and sphene variations two types of this "white" granite have been identified: one with the sphene and rutile content ranging from 3.5% to 10% maximum; the other in which only rutile is present up to amounts of 6%.

The structural trend of Alvand massif has a general direction of northwest-southeast, which corresponds to the general trend of the Zagros Range. The deduced structural trends have been interpreted from the schistosity studies of the country rocks of Sanandaj–Sirjan range, and conform with structural trends of the region and that of the Zagros Range. Other structural features that are present as a consequence of modifications present in this massif are mylonitization and presence of a major fault in the middle part of the massif. The mylonitization of the granitic rocks at the center of the granitic mass is along a north–south-trending fault. The compressive stress field that originated from two different directions of southeast and northwest has caused rock breakage at the fault zone after granite emplacement.

Chemical differences within the two mafic and felsic assemblages present in Alvand are confirmed and represented by an association of calcalkaline composition with a potassic character. This association has been indicated by variation in $Na_2O + K_2O/SiO_2$ index. This alkalinity index varies from 4% in mafic rocks to more than 7% in felsic ones.

Chemical evolution of major elements for this complex has been deduced from Harker's diagram (1976). The CaO, FeO variations are within the normal expected range, but MgO content varies very substantially. The abnormal low values of MgO could be interpreted, in part, by formation of olivine and magnesium clinopyroxene. This could cause further depletion of the MgO content of crystallization of biotite. Garnet represents the contamination produced by inclusions in the slightly metamorphosed country rocks and is present in the peripheral zones of Alvand. Sphene could be attached to hydrothermal activity associated with faulting at the middle parts of the massif. Presence of sphene-rich rocks is interpreted as a result of tectonic activities within the faulted zones of the massif.

Finally, it is concluded that the Alvand plutonic massif is formed by "plutonite" of the orogenic type comparable to the calacalkaline granites of "des Ballons-Vosges" in France. The Alvand massif is aligned with the Zagros suture, and it is about 60 km from that Range. The potassic character of this association appears to be "abnormal" considering other volcanic activities present in Iran, i.e., Avaj and Manjil Volcanisms, due to its age and geologic setting.

ZECHUN, L., Laboratory of Quaternary and Peat Geology, Nanjing Normal University, Nanjing, China

Study Results of Archaeology Geology in Area of Zhoukoudian Site, Beijing, China

Zhoukoudian, a village in front of Beijing West-mountain, about 50 km southwest of Beijing, is renowned for fossils of Peking Man (*HOMO ERECTUS PEKINENSIS*) discovered nearly 60 years ago. Only in the Dragon Bone Hill at the area, less than 1 km², are there over ten limestone caves containing lots of animal fossils, and four of which have excavated out early man's fossils and artifacts, from *HOMO ERECTUS* to *HOMO SAPIENS*.

Those caves were formed before 1 m.y.b.p., and yet have the self-filling processes, which depend on conditions of cave and changes in environment, and largely influence on the living activity of early man. So there are chronologically various sequence of sediment in the caves. Based on the karst evolution, the distribution of ash layers and stone tools found, the filling development of the caves in relation to early man's activity has various stages: the Peking Man's Cave can be divided into 5 stages during 1–0.23 m.y.b.p.; the New Cave with early *HOME SAPIENS*, 3 stages during 0.2–0.07 m.y.b.p.; and the Upper Cave lived by late *HOMO SAPIENS*, 4 stages during 0.3–0.01 m.y.b.p. This research explains why caves formed in same time were occupied by early man for different periods, and relations between cave deposits inside and outside.

According to the sedimentology, climatic stratigraphy and dating by several methods, the sedimentary sequence of cave in Zhoukoudian site, about 4–40 m thick, 17 layers of deposits in Peking Man's Cave and deposits of the 15th Cave, New Cave and Upper Cave, can be correlated with events both in China and farther afield and placed in the global framework of the Quaternary layer climatic change. All the cave deposits from the 14th layer of Peking Man's Cave upwards were sedimented in the Brunthes Normal Epoch, and may be correlated with the great loess sequence (S7–L1) in China and with deep-sea-core oxygen–isotope stages (19–2).

ZEEH, S., B. MAUL, and T. ECHSTÄDT, Geologisches Institut der Universität Freiburg, Freiburg, West Germany, and J. A. McKENZIE, and P. J. KINDLE, Geologisches Institut, ETH Zentrum, Zürich, Switzerland, R. RIEKEN, Institut für Geologie und Dynamik der Lithosphäre der Universität Göttingen, Göttingen, West Germany, and D. K. RICHTER, Institut für Geologie, Ruhruniversität Bochum, Bochum, West Germany

Cement-Stratigraphy of Upper Wetterstein Limestone (Triassic, Austria): Clue to Timing of Lead-Zinc Mineralization

In the Eastern Alps during the Middle Triassic and early Upper Triassic (Cordevolian) time carbonate platforms developed; they were rimmed by marginal reefs. Inner parts of many of these Wetterstein platforms, at quite some distance to the reefs, contain Lofer-type cyclic sediments, caused by sea-level fluctuations (Milankovitch cycles). Sub-, inter-, and supratidal sequences are present, and emersion periods led to weathering, erosion, and meteoric karstification of the sediments (Bechstädt and Döhler-Hirner, 1983). Most of the Pb-Zn ore deposits of the Eastern Alps are situated in this type of facies. According to many authors, these mineralizations are synsedimentary to early diagenetic: Depositional fabrics within some of the ores have been interpreted as indicative of external precipitation on the seafloor (Schneider, 1964; Schulz, 1964), whereas Bechstädt (1975) interpreted these occurrences as infills of meteoric cavities. Dzulybski and Sass-Gustkiewicz (1977) proposed the model of ore-bearing solutions, causing the almost synchronous dissolution of carbonates ("thermal karst") as well as emplacement of the ores.

The diagenetic evolution of the upper Wetterstein limestone answers some of these problems. For this study, the following techniques have been applied up to now: sedimentologic and petrographic methods, cathodoluminescence, geochemistry and isotope geochemistry of the different diagenetic stages, and fluid inclusions.

The diagenetic sequence is as follows: (1) early cements like meniscus, dripstone cement, mamillary crusts, botryoidal aragonite, which are generally non-luminescing; (2) radiaxial fibrous cements with a weak speckled, red luminescence; (3) scalenohedral (dog-tooth) cements with bright yellow luminescence of marginal parts, indicating the change to reduce pore water conditions; (4) epigenetic dolomite crystals and dolomitization; red luminescence; (5) saddle dolomite I; the luminescence of this cement varies from dark over red to orange, often wih an intensive zonation; (6) blocky calcite I with distinct yellow to red luminescing zones; a connected partly dedolomitization of rock and dolomitic cements leads to a bright yellow luminescence; (7) saddle dolomite II shows an orange or red luminescence without any zonation; and (8) blocky calcite II with red to dark luminescence.

Internal sediments occur from very early (during diagenetic stage 1) to late stages of diagenesis (up to stage 6). These internal sediments are often dolomitized and/or fluoritized.

With progressive burial and temperature increase, oxygen isotopes become strongly negative, relative to the PDB standard (blocky calcite I $\delta^{18}O$ −6.92; blocky calcite II $\delta^{18}O$ −5.50; saddle dolomite I $\delta^{18}O$ −5.50; saddle dolomite II $\delta^{18}O$ −9.55). First results of fluid inclusions studies of late diagenetic stages reveal homogenitization temperatures (Th) between 90 °C and up to 250 °C (saddle dolomite II and blocky calcite II). Especially the homogenization temperatures of the saddle dolomites II are distinctly higher than those reported for saddle dolomites by Radke and Mathis (1980).

According to the lead isotope data of Koppel and Schroll (e.g., 1988) the ore lead can be derived from Paleozoic metasediments; lead isotopes from the host rocks as well as from Triassic volcanic rocks strongly differ. Our data show that ore-mineralization started at a later diagenetic stage, after the change to an overall reducing pore-water environment, that is after the precipitation of dog-tooth cements. The so called "bottom ores" (Schulz, 1968) re not external precipitates, but took place within internal cavities (as has been assumed already by Siegl, 1956) long after deposition of the host rocks. Mineralization continued until stage 7 (saddle dolomite II). The Triassic lead-zinc mineralization of the Eastern Alps therefore cannot be considered "syngenetic"; they are normal epigenetic Mississippi-Valley-type lead-zinc deposits.

REFERENCES

Bechstädt, T., 1975, Lead-zinc ores dependent on cyclic sedimentation (Wetterstein limestone of Bleiberg-Kreuth, Carinthia, Austria): Mineralium Deposita, v. 10, p. 234–248.

——— and B. Döhler-Hirner, 1983, Lead-zinc deposits of Bleiberg-Kreuth; AAPG Memoir, v. 33, p. 55–63.

Dzulnski, S., and M. Sass-Gustkiewicz, 1977, Comments on the genesis of the eastern Alpine Zn-Pb deposits: Mineralium Deposita, v. 12, p. 219–233.

Köppel, V., and E. Schroll, 1988, Pb-isotope evidence for the origin of lead in strata-bound Pb-Zn deposits in Triassic carbonates of the Eastern and Southern Alps: Mineralium Deposita, v. 23, p. 96–103.

Radke, B. M., and R. L. Mathis, 1980, On the formation and occurrence of saddle dolomite: Journal of Sedimentary Petrology, v. 50, p. 1149–1168.

Schneider, H. J., 1964, Facies differentiation and controlling factors for the depositional lead-zinc concentration in the Ladinian geosyncline of the Eastern Alps: Developments in Sedimentology, v. 2, p. 29–45.

Schulz, O., 1964, Lead-zinc deposits in the Calcareous Alps as an example of submarine hydrothermal formation of mineral deposits: Developments in Sedimentology, v. 2, p. 47–62.

——— 1968, Die synsedimentäre Mineralparagenese im oberen Wettersteinkalk der Pb-Zn-Lagerstätte Bleiberg-Kreuth (Kärnten): Tschermaks Mineralogisch-Petrographische Mitteilungen, v. 12, p 230–289.

Siegl, W., 1956, Zur Vererzung der Blei-Zink-Lagerstätten von Bleiberg: Berg-und Hüttenmannische Monatshefte, v. 101, p. 108–119.

ZEN, E-AN, U.S. Geological Survey, Reston, Virginia

Pre-University Earth Science Education and the Role of Scientists in Fostering Scientific Literacy

There is much current interest among U.S. earth scientists in improving our pre-university earth-science education. This is a practical concern because many of our high-school students do not go to a university; yet each is a consumer of resources and a member of the society in which his or her future vote could affect actions at local and national levels. In planning the education of our future citizens, we need to ensure that they become knowledgeable (more literate) about the Earth we live on.

We need to clarify what we mean by "scientific literacy" in order to formulate a clear, focused, and achievable set of objectives that would help us design a suitable instructional approach and convince funding sources, the taxpayers, and policy makers of the need. I see three major reasons for improved scientific literacy of our citizens:

1. To demystify science and to provide access to a communal knowledge base, in other words, to help build a common cultural base that binds civilized people together. What are some cardinal concepts of science that ought to be part of such a cultural base? I would include the Laws of Motion; the nature of physical forces and their scales of interaction; the concept of chemical reactions; the meaning of the Law of Conservation of Energy; the meaning of thermodynamic irreversibility; the genetic code and DNA; age of the Earth and of the universe; the concept of evolution; the place of the Earth in the heliocentric solar system; the origin of Earth's resources and the effects of human activities on the Earth; the processes and factors that shape the land; and the concept of geologic time. We should use this knowledge to bridge the gap between science and humanistic studies. For instance, we should know how a person's cultural heritage affects his perception of the need to master and command and how it affects his very concept of the nature of scientific inquiries.

2. To allow a better intellectual understanding of the method of objective reasoning, of the notion that a good scientific hypothesis must be its ultimate falsifiability, of the challenge of reconciling the worldviews of science and of religion, of the responsibilities of scientists to society and the moral implications of scientific endeavours; as well as to allow an appreciation of the richness of the world we live in. Scientists, being human, are not only fallible but temptable; the structure of scientific research organizations often fails in self-supervision. What can a nonspecialist trust of the supposedly good and socially useful results? Should some aspects of scientific work be regulated? If so, how and by whom? These are real and current social issues, and a literate person should be conversant with the significance of the issues, if not with their solutions.

3. To enable students to better cope with the demands of the society and with the need to make intelligent social or individual decisions that will affect the quality of life for them and form their children. This is likely what many people have in mind by "scientific literacy" and is a large component of our social motivation. World-wide population growth accompanied by increased use of resources imposes stresses on our social structure as well as on our natural environment: not just fuels, minerals, soil, and water are becoming depleted; our social fabric and local and national policy options are equally being tested. Our citizens and leaders must understand the Earth system well enough to use its resources with wisdom, and to avoid its hazards with knowledge. This knowledge needs to be imparted at an early stage of education. We need to keep in mind also that the ability to make policy and personal decisions based on a sound geologic understanding is important no matter what the level of technological sophistication. For instance, the need to know how to avoid geological hazards is endemic to living. The timely access to such information, however, becomes increasingly urgent because of population growth and our persistent encroachment onto marginal living spaces and irreplenishable resources. Only the availability of the information is being promised and promoted by technology.

In order to promote greater scientific literacy among our students, we need to address such issues as:

* What are our aims, and how can we attain them? What are the responsibilities of professional scientists in the endeavour?

* How may we effectively cooperate with educators, teachers, and others in the education structure, in other professional disciplines, and in our own professional earth-science communities, so as to give any improvement permanence?

* Should pre-university science education be integrated or discipline-oriented? If the former, how can we accomplish this goal? If the latter, how should the disciplines be sequenced so as to be mutually supportive? Is there a good middle ground?

* Should pre-university education be reinforced and supplemented by adult education? What is an effective strategy?

* What are the concepts, facts, and methodologies that students ought to learn? Which of these are central to all science, and which are peculiar to the earth sciences? In what ways can earth-science help teaching other topics, and vice versa?

Discussion at this Congress is opportune because the problem of scientific literacy is global in nature, if for no other reason than the simple proposition that nature responds to man's activities and abuses without regard to political boundaries.

ZEN, E-AN, U.S. Geological Survey, Reston, Virginia

Pre-University Earth Science Education and the Role of Scientists in Fostering Scientific Literacy

There is much current interest among U.S. earth scientists in improving our pre-university earth-science education. This is a practical concern because many of our high-school students do not go to a university; yet each is a consumer of resources and a member of the society in which his or her future vote could affect actions at local and national levels. In planning the education of our future citizens, we need to ensure that they become knowledgeable (more literate) about the Earth we live on.

We need to clarify what we mean by "scientific literacy" in order to formulate a clear, focused, and achievable set of objectives that would help us design a suitable instructional approach and convince funding sources, the taxpayers, and policy makers of the need. I see three major reasons for improved scientific literacy of our citizens:

1. To demystify science and to provide access to a communal knowledge base, in other words, to help build a common cultural base that binds civilized people together. What are some cardinal concepts of science that ought to be part of such a cultural base? I would include the Laws of Motion; the nature of physical forces and their scales of interaction; the concept of chemical reactions; the meaning of the Law of Conservation of Energy; the meaning of thermodynamic irreversibility; the genetic code and DNA; age of the Earth and of the universe; the concept of evolution; the place of the Earth in the heliocentric solar system; the origin of Earth's resources and the effects of human activities on the Earth; the processes and factors that shape the land; and the concept of geologic time. We should use this knowledge to bridge the gap between science and humanistic studies. For instance, we should know how a person's cultural heritage affects his perception of the need to master and command and how it affects his very concept of the nature of scientific inquiries.

2. To allow a better intellectual understanding of the method of objective reasoning, of the notion that a good scientific hypothesis must be its ultimate falsifiability, of the challenge of reconciling the worldviews of science and of religion, of the responsibilities of scientists to society and the moral

implications of scientific endeavours; as well as to allow an appreciation of the richness of the world we live in. Scientists, being human, are not only fallible but temptable; the structure of scientific research organizations often fails in self-supervision. What can a nonspecialist trust of the supposedly good and socially useful results? Should some aspects of scientific work be regulated? If so, how and by whom? These are real and current social issues, and a literate person should be conversant with the significance of the issues, if not with their solutions.

3. To enable students to better cope with the demands of the society and with the need to make intelligent social or individual decisions that will affect the quality of life for them and form their children. This is likely what many people have in mind by "scientific literacy" and is a large component of our social motivation. World-wide population growth accompanied by increased use of resources imposes stresses on our social structure as well as on our natural environment: not just fuels, minerals, soil, and water are becoming depleted; our social fabric and local and national policy options are equally being tested. Our citizens and leaders must understand the Earth system well enough to use its resources with wisdom, and to avoid its hazards with knowledge. This knowledge needs to be imparted at an early stage of education. We need to keep in mind also that the ability to make policy and personal decisions based on a sound geologic understanding is important no matter what the level of technological sophistication. For instance, the need to know how to avoid geological hazards is endemic to living. The timely access to such information, however, becomes increasingly urgent because of population growth and our persistent encroachment onto marginal living spaces and irreplenishable resources. Only the availability of the information is being promised and promoted by technology.

In order to promote greater scientific literacy among our students, we need to address such issues as:

* What are our aims, and how can we attain them? What are the responsibilities of professional scientists in the endeavour?
* How may we effectively cooperate with educators, teachers, and others in the education structure, in other professional disciplines, and in our own professional earth-science communities, so as to give any improvement permanence?
* Should pre-university science education be integrated or discipline-oriented? If the former, how can we accomplish this goal? If the latter, how should the disciplines be sequenced so as to be mutually supportive? Is there a good middle ground?
* Should pre-university education be reinforced and supplemented by adult education? What is an effective strategy?
* What are the concepts, facts, and methodologies that students ought to learn? Which of these are central to all science, and which are peculiar to the earth sciences? In what ways can earth-science help teaching other topics, and vice versa?

Discussion at this Congress is opportune because the problem of scientific literacy is global in nature, if for no other reason than the simple proposition that nature responds to man's activities and abuses without regard to political boundaries.

ZENG Xuesi, LIU Yingkai, CHEN Xianqun, LI Shushun, RAO Kemin, Geological Society of China, Beijing, China

The Accumulation Conditions and the Exploration Prospects of Natural Gas of Paleozoic Marine Carbonate Rocks in the Northeastern Part of Sichuan Basin

Since the 1970s, there have been a series of discoveries during exploration for natural gas of Paleozoic marine carbonate rocks in Sichuan basin. Up to now, we have had a number of Carboniferous and Permian gas pools and obtained some commercial gas flows or important shows in Ordovician and Silurian.

The northeastern area of Sichuan basin, located on the south flank of Daba and Micang mountains, covering an arca of about 50,000 km^2 and having a thickness range from 3000 to 6000 m of Paleozoic, is a new domain to be exploited. Conditions controlling natural gas accumulation are extremely complicated in this area. Now, a few main factors will be discussed as follows:

High maturation of hydrocarbon and sufficient gas source: Because Paleozoic formation is older and deeply buried, most of hydrocarbons have been overmatured and transformed to dry gas in the basin. In the view of organic abundance, the residual organic carbon content is 0.1-0.6% of the carbonate rock and 0.3-2% of the shale. The total thickness of source rock may range from 1000 to 2000 m, so the gas source is sufficient.

Composite reservoirs associated with pore, cave, and fracture: Because of diagenetic influence, the porosity of reservoir becomes low to 1.6% in the Paleozoic carbonate rocks, which experienced early deposition, deep burial, and complicated geologic history. The dolomitization pores and the solution pores hold an important position. When accompanied by fracture and solution cavity, they can still produce commercial gas flow.

Paleostructure conditions timed to the stage of hydrocarbon maturation: The paleouplift and the paleofault block control the regional distribution of oil and gas. According to the findings of the paleogrotemperature and organic matter maturity, most of the source rocks in Paleozoic matured in Hercynian-Indosinian. The Kaijiang paleouplift formed then plays an important role in early accumulation of hydrocarbons.

Anticline traps in vertical variation and superposition: A great number of results of the seismic exploration show there is a triple structure in the area of Daixan and Xuanhan, that is, above the Middle Triassic, the fold is simple, gentle, and of few fractures; under the Silurian, the fold is also gentle, but between the Middle Triassic and the Silurian, the fold is strong, the fracture is more developed, and many anticline traps are formed. There are evident differences in the pattern and direction of fold and fracture in each deformed layer. These characteristics reflect vertically not only the mechanical properties of rocks, but also the stress processes in individual times and various directions. As a result of this in this area are formed the superimposed-anticline traps, where the fractures are usually developed.

Organic reefs and nonanticline traps: The Permian and Cambrian organic reefs have been found in the outcrop of Daba mountain. Among them, the Upper Permian sponge reef is especially typical. These reefs controlled by paleostructure and sedimentary facies are characterized by zonated distribution. Otherwise, there may be nonstructure traps related to buried hills, unconformities, lithological wedges, and diagenesis.

Importance of seal conditions: The depth of the Paleozoic group is more than several thousand meters. And gypsum or thickish shales are developed in the Paleozoic and Mesozoic groups in this area. Therefore, the seal conditions are generally good. But the local caprock close to the gas pool must be emphasized. In the areas adjacent to the Paleozoic outcrop in Daba and Micang mountains, to the erosional window of high-steep anticline in southeastern part of the area, and to the place where the cover beds are destroyed by the fractures, the gas pools may be subjected to destruction in different degree. Also, the seal conditions and hydrodynamic factors are considered in the course of exploration.

As mentioned above, there are sufficient gas sources, better composite reservoirs, timely paleostructure, many kinds of effective traps (dominated by anticline), and better seal conditions in this area. Now, some Carboniferous and Permian gas pools have been found there, indicating that it is a better prospective area. Of course, there are also some disadvantages, such as bad quality in reservoirs in some areas, complications in structure, rough relief in most of the work area, and bad conditions for transportation. Considering that the favorable accumulation conditions are more than favorable ones, which are only relative and may be overcome, it is necessary, from the view of domestic economics, to accelerate the exploration in this area. Meanwhile, to study many complicated and interesting academic problems in this area may also cause the further development of natural gas geology.

ZENTILLI, M., Dalhousie University, Halifax, Canada, and V. MAKSAEV, SERNAGEOMIN, Santiago, Chile

Genesis of Porphyry Copper Deposits in Andes of Northern Chile in Relation to Tertiary Magmatism and Tectonic Uplift: Fission Track Evidence

The metallogenic evolution of the Andes of Chile is typified by a strong dominance of copper mineralization related to the volcano-plutonic processes on the active continental margin of South America (e.g., Clark and Zentilli, 1972; Sillitoe, 1988). Magmatism and the formation of associated mineral deposits have been virtually continuous in the Andes since the Jurassic (Boric et al., 1985), however the major (world class) ore deposits were formed in relatively short periods of time, during which the capacity for metallic concentration appears to have been optimum due to the concurrence of particular tectonic and paleogeographic conditions (Maksaev et al., 1988). The premier copper deposits of northern Chile (21°–26°S) are porphyry-typc dcposits related genetically to the epizonal emplacement of Late Eocene–Early Oligocene (41–31 Ma) stocks. They are distributed within the domain of a regional north-south, in part strike-slip fault system that roughly coincides with the western limit of a large uplifted block (Domeyko Cordillera), cored by Paleozoic crystalline rocks, which in turn forms the western limit of the high Andes (Puna) at these latitudes. The Paleozoic basement, which according to stratigraphic evidence remained under the sea during the Jurassic and part of the Early Cretaceous, has been uplifted to 4000 to 4500 m a.s.l. The uplift began in the Late Cretaceous, but has been accomplished mainly in Tertiary times (e.g., Mortimer, 1973, 1980) as a consequence of two compressive tectonic phases, the Incaic and Quechua phases (Late Eocene and Miocene in age, respectively; e.g., Maksaev, 1979). Apatite samples from Paleozoic intrusive rocks of the Domeyko Cordillera give fission-track ages ranging from ca. 45 to 35 Ma. These ages show a regular variation with altitude, evidencing an important uplift of this crustal block that took place in the Late Eocene–Early Oligocene with a mean exhumation ratio of 0.21 km Ma^{-1}. Our fission-track data support the notion that the formation of major porphyry copper deposits of this Andean segment was preceded by and partly simultaneous with an important crustal thickening event. Tectonic uplift was also a precursor to the formation of major porphyry copper deposits in southern Peru in the Paleocene, and in Central Chile in the Miocene (Maksaev and Zentilli, 1988), a pattern that suggests a genetic relationship between the formation of large porphyry copper deposits and crustal thickening during compressive tectonic events. We interpret this relationship to reflect the deepening of the zone of magma generation under the active plate edge of the South American continent as a consequence of lithospheric shortening, which permits the tapping of a deep, homogenized, undepleted mantle source for the porphyry magmas as suggested by Zentilli et al. (1988) on the basis of Pb and Sr isotope data.

Apatite fission track ages from the Chuquicamata and El Abra deposits are identical (within error limits) with biotite ^{40}Ar-^{39}Ar ages from the same rocks samples (Maksaev et al., 1988), which confirms the epizonal emplacement of the porphyry stocks. In spite of considerable uplift, the amount of erosion has been relatively meager in this arid region. This fact explains the remarkable preservation of the shallow porphyry copper systems with their supergene enriched blankets. Samples from different Jurassic intrusives in the Coastal Cordillera of northern Chile give Cretaceous apatite fission track ages (Maksaev and Zentilli, in press) suggesting that this block was possibly uplifted significantly earlier than the Oligocene time proposed by Mortimer (1980), and thus acted as a persistent barrier for the drainage toward the Pacific Ocean.

REFERENCES

Boric, R., F. Diaz, and V. Maksaev, 1985, Magmatic events and related metallogenesis in the Antofagasta Region, Northern Chile: Comunicaciones, Departamento de Geologia, Univ. de Chile, v. 35, p. 37–40.

Clark, A. H., and M. Zentilli, 1972, The evolution of a metallogenic province at a consuming plate margin: the Andes between latitudes 26° and 29°S: Canadian Mining Metallurgical Bulletin, v. 65, n. 719, p. 37.

Maksaev, V., 1979, Las fases tectónicas Incaica y Quechua en el Norte Grande de Chile: Actas, II Congreso Geológico Chileno, p. B63–B77.

——— and M. Zentilli, 1988, Marco metalogénico regional de los megadepósitos de tipo pórfido cuprifero del Norte Grande de Chile (w. English abstract): Actas, V Congreso Geológico Chileno, v. 1, p. B181–B212.

——— ——— in press, Tertiary uplift and formation of porphyry copper deposits in the Andes of northern Chile: fission track evidence: Proceedings, VI International Fission Track Dating Workshop, Besancon, France, 5/9 September 1988, and Abstracts, C2-14(O).

——— R. Boric, M. Zentilli, and P. H. Reynolds, 1988a, Significado metalogénico de dataciones K/Ar, ^{40}Ar-^{39}Ar y trazas de fisión de zonas mineralizadas en el Norte Grande de Chile (w. English abstract): Actas, V Congreso Geológico Chileno, v. 1, p. B65–B86.

——— M. Zentilli, and P. H. Reynolds, 1988b, Geocronologia ^{40}Ar-^{39}Ar de depósitos de tipo pórfido cuprifero del Norte Grande de Chile (w. English abstract): Actas, V Congreso Geológico Chileno, v. 1, p. B109–B133.

Mortimer, C., 1973, The Cenozoic history of the southern Atacama desert: Journal of the Geological Society of London, v. 129, p. 505–526.

——— 1980, Drainage evolution of the Atacama desert of northernmost Chile: Revista Geológica de Chile, v. 11, p. 3–28.

Sillitoe, R. H., 1988, Epochs of intrusion-related copper mineralization in the Andes: Journal of South American Earth Sciences, v. 1, p. 89–108.

Zentilli, M., B. R. Doe, C. E. Hedge, O. Alvarez, E. Tidy, and J. A. Daroca, 1988, Isótopos de plomo en yacimientos de tipo pórfido cuprifero comparados con otros depósitos metaliferos en los Andes del norte de Chile y Argentina (w. English abstract): Actas, V Congreso Geológico Chileno, v. 1, p. B331–B369.

ZHAI Guangming, Scientific Research Institute of Petroleum Exploration and Development, Beijing, China, and LIU Hefu, China University of Geosciences, Beijing, China

Classification and Tectonostratigraphic Evolution of Sedimentary Basins in China

Tectonic framework and tectonostratigraphic evolution of China: China comprises a mosaic of continental blocks separated by fold belts of different phases, some of which may be suture zones, because remnants of oceanic crust are found in them. The main suture zones in China are the products of the obliteration of the Central-Asia-Mongolian sea between Siberian and North China blocks, of the Kunlun-Qinling sea between North China and South China blocks, and of the Tethys between South China Block and Indian plate.

With respect to tectonostratigraphic evolution, carbonate deposition was predominant, with some shale, on North China Block in the Cambrian and Ordovician, while along the southern border of the block and in its aulacogen, gravity flow carbonate or clastic turbidite was deposited. In the Carboniferous, tidal deposition prevailed. Continental sedimentation was initiated at the beginning of Permian. Throughout the Paleozoic to middle Triassic, carbonate and shale deposition were predominant on South China Block. To its west, turbidite up to ten thousand meters thick was deposited in Songpan-Ganzi Basin during Triassic time.

Beginning from the late Triassic, China went into a period of continental deposits. The Late Triassic and Jurassic was the time for the main coal-bearing formation, while the Cretaceous was the time for the main oil-bearing formation. Lacustrine sediments and evaporites prevailed in rift basins in Cenozoic. Moreover, the closures of Tethys and the formation of fold-thrust belts in different phases led to the deposition of thick molasse sediments in foredeeps.

Classification and structural styles of sedimentary basins in China: The Indosinian movement was a time of important tectonic change in China. The previous east-west-trending grain was superimposed by the south-north-trending Helanshang-Longmenshan inbricated thrust belt, resulting in the formation of two distinct tectonic domains in eastern and western China and of the various types of sedimentary basins, which are extensional, compressional, and pull-apart basins.

In eastern China, the double effects of the Pacific Plate and the closure of Tethys resulted in the mainly north-northeast-trending extentional basins, including rift basins and marginal sea basins. According to Songliao, Bohai Bay, and Zhujiangkou (Pearl River mouth basin) basins, these extentional basins mainly have the following four features. (1) Listric normal faults develop widely. Tilted blocks and rollover anticlines occur in the basement and in the cover rocks, respectively. (2) Horst-graben structures or half-grabens normally occur in the basins. Listric normal faults usually belong to two stages. The earlier ones are mainly antithetic faults, while the latter ones synthetic faults. (3) The sediments are usually 5000-7000 m thick, the difference between isostatic and cooling subsidence imposed on the basins. (4) Various types of traps, e.g., rollover structures, trapdoor structures, salt domes, or diapir structures, stratigraphic traps, or lithologic traps, superimposed to form composite hydrocarbon accumulation zones.

In western China, the closures of Tethys in different phases formed the east-west-trending compressional basins, including foredeep and relict back-arc basins. Their features are as follows: (2) Fold-thrust belts prevail. Upthrusts and compressive blocks usually occur along the borders of the basins, while anticlinal structures occur in the interior of the basins. The loading of thrust sheets is the main cause of the subsidence of the basins, making the lithosphere flexure and subside. The wedge like sediments deposited in them are usually 10,000 m thick. Sometimes they can be up to 15,000 m thick, e.g., in Tarim basin. (3) Various types of structural traps were regularly in paralle form or en echelon for m within basins, thus forming a large hydrocarbon accumulation zone. For example, the ke-Wu fault zone in Junggar basin consists mainly of concentric folds, trapdoor structures, thrust propagation folds, and footwall traps of thrusts.

Because they lie between eastern and western tectonic domains, the basins in central China are composite ones, including Ordos and Sichuan basins. The compression of Tethys to the west affected them so much that imbricated thrust zones, concentric folds, and gentle domes were formed in order of space from west to east.

Hydrocarbon potential of sedimentary basins in China: The sedimentary basins in China are characterized by their large area, thick deposition, and deep depression. Giant oil and gas fields are controlled by oil-generation sag, they amount to 70% of the total hydrocarbon reserves in China. Superimposition of various types of traps in each basin provides a favorable condition to form a large composite hydrocarbon accumulation zone.

China has established four large hydrocarbon-bearing districts. The sedimentary basins in the eastern district have an area of 600,000 sq km. Oil production from this district accounts for 90% of total production in the country. The sedimentary basins in the western district have an area of more than 1,280,000 sq km. But it has been only poorly explored. The sedimentary basins in the central district have an area of about 2,000,000 sq km. Marine carbonate prevail there. At present, it is the main gas production base of the country. The offshore district is the continental shelf, with an area of 1,300,000 sq km, within water depths less than 200 m. Though it has just entered into a primary exploration stage, dozens of high-production oil or gas wells have been found. These marine and nonmarine sedimentary basins in China provide great potential for future hydrocarbon exploration.

ZHAN BING-SHAN, and JIAN WEN-BIN, Geotechnical Engineering Department of Tongji University, Shanghai, China

Nonlinear Time Series Model of Discharge of Longzi Karstic Spring, Shanxi

Such models as AR(p), ARMA(p,q), and ARLMA (p,d,q), which deal with linear time series, are widely used in hydrogeologic calculations. However, when a series of data is complicated, it is difficult to simulate by the above mentioned models. In 1983, Tong founded the threshold auto-regressive model—self-exciting threshold auto-regressive model (SETAR $[2,k_1,k_2]$): $x_t = a_o^{(j)} + \Sigma^{K_j}_{i=1} a_i^{(j)} x_{t-i} + \epsilon_t^{(j)}$, $x_{t-d} \epsilon R_j$, $J = 1,2$. and the mixed threshold auto-regressive model (TARSO $[2, (m_1, S_1), (m_2, S_2)]$): $x_t = a_o^{(j)} + \Sigma^{M_j}_{i=1} a_i^{(j)} x_{t-i} + \epsilon_t^{(j)} + \Sigma^{S_j}_{i=0} b_i^{(j)} y_{t-i} + t\epsilon_t^{(j)}$, $y_{t-d} \epsilon R_j$, $j = 1,2$—which is a nonlinear time series model. The basic concept of these models is to treat nonlinear systems by linearizing each section.

Background information: The Longzi Spring is situated 13 km southwest of Linfen city. It is a large, famous karstic spring in Shanxi Province, covering an area of 246 km^2 of the spring drainage basin. The Longzi spring drainage basin is a karstic groundwater system with relatively independent recharge, run-off, and drainage. Its average annual discharge is 5.88 m^3/s, and the unsteady coefficient is $R = 1.12–1.41$. In light of the connection degree (Den Julong, 1985) analysis, we know that the spring is alimented by precipitation of many years, and the specific yield of the spring in a given year is related to the precipitation of the 11 years preceding that year. According to the analysis of cross-correlation function, we also know that there is a time lag of 6.3 months between the precipitation and the discharge. Finally, as a result of the analysis of frequency spectrum, we know that the discharge of the spring has a first cycle of 6 months and a second cycle of 8.4 months.

Nonlinear time series model: Self exciting threshold auto-regressive model (SETAR $[2, p_1, p_2]$)—Let (Q_t, t = 1,2,...n) be a time series of the spring discharge, then

$$Q_t = \begin{cases} a_o^{(1)} + \sum_{i=1}^{p_1} a_i^{(1)} Q_{t-i} + \epsilon_t^{(1)}, & Q_{t-d} \le r \\ a_o^{(2)} + \sum_{i=1}^{p_2} a_i^{(2)} Q_{t-i} + \epsilon_t^{(2)}, & Q_{t-d} > r \end{cases}$$

where p_1, p_2 are the step numbers of the first and the second section dividing models, $a_i^{(j)}$ is the coefficient of the model ($i = 0,1,...p_j$, $j = 1,2$), $[\epsilon_t^{(j)}]$ is the white noise with an average equal to zero and its variance being $\sigma^2_t(j)$ ($j = 1,2$), $[\epsilon^{(1)}_t]$ and $[\epsilon^{(2)}_t]$ are independent, r is control value of the threshold, and d is the lag parameter.

Given the maximum steps L and maximum lag parameter D of the two sections of the model, let $n_0(d) = \max[d,L]$, $d = 1,2,.....D$, which is denoted by n_0 to stationary d. The threshold value of r and the lag parameter of d can be determined using AIC. To each stationary r and d, data of $Q,Q,...Q_{no+1}, Q_{no+2}...Q_N$ are divided into two kinds. If $Q_{no+i-d}, \le r$ ($i = 1,2,...N\ n_0$), Q_{no+i} belongs to the first kind; if $Q_{no+i-d}, > r$, Q_{n0+i} belongs to the second kind. Substituting the data of the first and second kinds, respectively, into the first and second section-dividing auto-regressive models, we have coupled equation: $Q^{(1)} = A^{(1)}\Theta^{(1)} + \epsilon^{(1)}$ and $Q^{(2)} = A^{(2)}\Theta^{(2)} + \epsilon^{(2)}$. $a_o^{(j)}$, $a_i^{(j)}$, $a_p^{(j)}$, $a_{p2}^{(j)}$, and $\sigma^2_{et}(j)(j = 1,2)$ can be obtained by solving the above equations. Step numbers p_1, p_2 can be determined using AIC (Akaike's Information Criterion).

Calculating the Longzi spring discharge with SETAR model, respectively taking a month and a year as time unit, very good simulating results are obtained. For example, we took a series of the Longzi spring discharge from January 1955 to December 1970, altogether 192 months. Predetermining that the maximum step number L is 1 to 7 and the maximum lag parameter D is 0 to 3, and making up different calculation plans, we find that when $p_1 = 2$, $p_2 = 1$, d - 2, and r = 6.38 m^3/s, the optimum model of SETAR (2,2,1) is

$$Q_t = \begin{cases} 0.59094 + 0.44278 Q_{t-1} + 0.47245 Q_{t-2}, & Q_{t-2} \le 6.38\,m^3/s \\ -0.11788 + 1.00997 Q_{t-1}, & Q_{t-2} > 6.38\,m^3/s \end{cases}$$

Mixed threshold auto-regressive model (TARSO)$[2, (p_1, q_1), (p_2, q_2)]$—Basically, the model-establishing steps and the parameter discerning of this model are similar to the SETAR model.

$$Q_t = \begin{cases} a_o^{(1)} + \sum_{i=1}^{p_1} a_i^{(1)} Q_{t-i} + \sum_{i=0}^{q_1} b_i^{(1)} P_{t-i} + \epsilon_t^{(1)} & P_{t-d} \le r \\ a_o^{(2)} + \sum_{i=2}^{p_2} a_i^{(2)} Q_{t-i} + \sum_{i=0}^{q_2} b_i^{(2)} P_{t-i} + \epsilon_t^{(2)} & P_{t-d} > r \end{cases}$$

According to the TARSO $[2, (p_1, q_1), (p_2, q_2)]$ model, we took the series of the Longzi spring discharge and precipitation for 192 months. Predetermining that L is 2 to 6 and d is 0 to 2, with a time delay of from 5 to 7 months to make up various calculation designs, then we have the optimum TARSO [2,(2,1),(2,1)] model where $p_1 = 2$, $q_1 = 1$, $p_2 = 2$, $q_2 = 1$, $d = 1$,

$Q_t = [0.2952 + 0.5798Q^+t1 + 0.3618Q^+t2 + 0.0006p^+t5 + 0.0007p^+t6$, $p^+t6 \le 55.44$ mm

$-0.0053 + 0.5989Q^+t1 + 0.3850Q^+t2 + 0.0008p^+t5 + 0.0008p^+t6$, $p^+t6 > 55.44$ mm

Conclusions:

1. Only if a model-establishing series approaches a certain length can the system be characterized.
2. Considering not only the precipitation but also the succession of the spring discharge leads to a more objective variation law of the spring discharge.
3. Through an examination of efficiency of the model, we find that the SETAR (2,2,1) and the TARSO [2, (2,1),(2,1)] models are efficient, taking a month as a time unit which can be used in insertion of the spring discharge. As the long model-establishing series presents the characteristics of the whole system, the model can also be used as a long-term forecasting model.
4. The above mentioned nonlinear time series model can be widely applied to the simulation of the regime of spring discharge and water level of similar karstic areas.

REFERENCES

Den Julong, 1985, Grey control system: Published by the Middle China Engineering Institute.

Tong, H., 1983, Threshold models in nonlinear time series analysis: Lecture Notes in Statistics, n. 21, New York.

ZHANG FUQIN, Nanjing University, Nanjing, China

Origin of Seamounts

After studying the leaking of transform faults, the process of seafloor spreading, and stress-strain conditions of the oceanic lithosphere in the process, the author points out that seamounts can result from three types of mechanisms, which will result in three types of seamounts:

1. The Hawaii–Emperor type: Such seamounts, because of the leaking of transform faults and the spreading of midocean ridges, were caused by the stretching of ridges and heat compression of lithosphere. So the mechanism that makes them appear to be oblique to the transform faults is rotation extended fracture zones. Seamounts of this type include Hawaii–Emperor, Tuamotu-Line, and Gilbert- Marshall in the Pacific ocean and Iceland–Farroe in the Atlantic ocean.
2. The southwest Pacific seamount type: Such seamounts are formed oblique to the Hawaii–Emperor type in a southwest or west- southwest direction in the southwest Pacific Ocean. They were caused by a shear-make-pull-apart mechanism.
3. The tectonic collision of oceanic lithosphere type: Such seamounts commonly occurred in the regions between two midocean ridges; for instance, the 90°E seamount in the Indian ocean.

The author analyzes the data on a few dozen transform faults in the Pacific, Atlantic, and Indian oceans and finds that the transform faults are commonly leaky. The reason for the leaking is anomalous mantles, i.e., the leaking results from extending. The strips of finite width underlain by a special type of crust are generated within them. The author regards the leaky fracture zones as the second spreading centers in addition to the midocean ridges. Now that the transform faults are leaky, the stretching of midocean ridges can't be avoided. The very stretching makes midocean ridges subduct beneath continental lithosphere. A good example of ridge subduction is California Bay. In past work, we neglected the effect of such stretching on the velocity of plate motions, only taking into consideration the spreading of midocean ridges. As a matter of fact, the direction of plate motions is oblique to the trends of transform faults, i.e., the Euler Law can't quantitatively describe the plate motions. Such properties are unexpected in the ideal concept but are compatible with the proposition that transform faults are leaky. If $\vec{V}_e$ is the vector from the ridge spreading and $\vec{V}_t$ stands for the vector from the ridge stretching, the vector of the plate motions can be expressed by the equation:

$$\vec{V} = \vec{V}_e + \vec{V}_t \quad (1)$$

According to equation 1, the author has the equation for the loci of the plate motions. The pattern of the locus is similar to the one of exponential function or of parabola, which exactly conform to the distribution of Hawaii–Emperor island chain. The author regards the origin of the Hawaii–Emperor seamount as the inherent process of seafloor spreading. In light of equation 1, the $dV \neq 0$, nor are the dV_e and the dV_t, so there not only occurs shear zones along the loci of plate motions, but also results in extending perpendicular to the shear zones. The en echelon faults along the Hawaii island chain and other chains in the West Pacific Ocean identify the existence of such shearing. The extension perpendicular to the shear zones along their trends will compensate for the width of new crusts generated within transform faults. The heat compression of oceanic lithosphere will occur in the fracture zones while the plates are moving apart from mid-ocean ridges. So new volcanos will only be produced at the end near the midocean ridges. The mechanism results in the seamounts of Hawaii-Emperor type. The author also points out that "hot spots" can't explain why the Hawaii–Emperor island chain, etc., in the West Pacific Ocean are all oblique to the small circles centered on the pole of relative plate motion. The dynamic and kinematic stability of mantle convection plumes is far to reach.

Because of the shearing along the seamounts of the first type, pull-apart units will occur in the regions among them, and new volcanos will be generated within them. The second type is realized through such shear-result-in-pull-apart model. The above seamounts of two types make net-like patterns in the oceans that can be explained with Riedel model.

As to the seamounts of the third type, they commonly occur in the regions between two midocean ridges. The 90°E seamount in the Indian ocean is a good example of this type of formation. This type suggests that tectonic collision can be formed not only in the continents but also in the oceans.

ZHANG GUOWEI, YU SAIPING, ZHOU DINWU, SUN YONG, GUO ANLIN, and LI TAOHONG Northwestern University, Xian, China

Phanerozoic Tectonics and Intracontinental Orogeny of Qinling Belt

The Qinling is a famous composite continental orogenic belt in China characterized by complex composition due to a prolonged history (Huang Jiqin, 1984; Li Chunyu, 1978). It consists of three major tectonic units: (1) North China craton southern marginal belt, (2) Northern Qinling belt, and (3) Southern Qinling belt. During the course of its history, the Qinling belt has undergone several distinctive stages of evolution characterized by different tectonic regimes (Zhang Guowei et al., 1986). Four such major stages of evolution are recognized, which are marked by, in chronological order, (1) formation of an integrated early Precambrian craton, (2) development of a middle–late Proterozoic Qinling rift system, (3) late Proterozoic to early Mesozoic modern-style plate tectonic evolution, and (4) Mesozoic–Cenozoic intraplate tectonic evolution and intracontinental orogeny.

Phanerozoic plate subduction and collision orogenies: Plate tectonics commenced in the Qinling belt with initiation of an ancient Qinling ocean between the North China and the Yangtse continent toward the end of the late Proterozoic, which was followed successively during the Paleozoic by spreading, subduction, and collision orogeny, and terminated during the Indochinese period of early Mesozoic age.

The Southern Qinling represents a Paleozoic passive margin on the northern side of the Yangtse continent. It is characterized by a middle–late Proterozoic basement of Yangtse type and distinctive late Proterozoic–Paleozoic–Triassic sedimentary cover successions. The cover rock associations, sedimentary structures, and facies variations are consistent with a passive margin sedimentary pattern in which sedimentary facies become progressively thicker from south to north. Despite the occurrence of a marginal fault trough in early Paleozoic times and some adjustments in facies pattern between the early and the late Paleozoic, no marked orogenic disturbances occurred and the passive margin always persisted on the northern border of the Yangtse plate throughout the Paleozoic.

By way of contrast, an active continental margin developed at the same time on the southern border of the North China plate on the other side of the Qinling ocean, which is now represented by the Northern Qinling. Its Paleozoic contains Danfeng and Erlangping ophiolite complexes of dominantly arc-marginal sea type (Coleman, 1984), lacking any typical ridge ophiolites and pelagic sediments but associated with large amounts of terrigenous turbidite rocks. Along the southern border of the Northern Qinling occurs an enormous lineament, known as Shangdan fault zone, which consists of a complex system of faults of ductile and brittle series formed at various structural levels, forming the important boundary between the Northern and the Southern Qinling belt. In the Qinling Group crystalline rocks to the north we also find subduction-related granitic intrusions that occur as aligned masses parallel to the fault zone. These different units are tectonically juxtaposed together in the Northern Qinling, collectively forming the integrated active margin at the southern border of the North China plate.

The great Shangdan fault zone mentioned above marks the major plate suture of the North China with the Yangtse continent. It is not a simple fault zone but an intricate tectonic border terrane with complicated material compositions (Zhang Guowei et al., 1986). In its complex and involving system of brittle faults and ductile thrust and strike-slip shear zones are juxtaposed Danfeng ophiolite blocks, obducted slices of a sedimentary prism and subduction- and collision-related granitic intrusions. Gravity and magnetic anomaly patterns and seismic reflection profiles as well as geological and geochemical studies all suggest that this is a significant dividing boundary in the crust and upper mantle and represents the lithospheric plate suture in the Qinling belt.

To summarize, the Qinling belt is characterized by the following plate tectonic history. Toward the end of the Proterozoic the continent broke up and an ancient Qinling ocean of limited areal extent came into existence. On either side of this then occurred a passive continental margin. As the Qinling oceanic crust rafted on the Yangtse plate was being subducted beneath the North China plate during the Caledonian period, the southern margin of the North China plate began to acquire characteristics of an active margin, in which developed its appropriate volcanosedimentary successions (402 ± 17.4 Ma Sm-Nd, 447 ± 41.5 Ma Rb-Sr, etc.). Despite the arc–continent collision in local places during the late Caledonian to early Hercinian period, the terminal collision of the Yangtse with the North China continent occurred in the Indochinese period (granites and rock matrix of shear zones related to this collision have yielded isotopic dates 211 ± 8 Ma U-Pb, 219 Ma Rb-Sr, etc.) (Li Chunyu et al., 1978; Mattauer et al., 1985; Coward et al., 1986, Hsu et al., 1987).

Mesozoic–Cenozoic Intracontinental Orogeny: After its incorporation into a unified China plate during the Indochinese collisional orogeny, the Qinling belt is still characterized by tectonic instability, as is indicated by continued northward subduction at depth and large-scale shortening in the upper crust as well as tectonic disturbances in response to lateral motion of the Indian and the Pacific plate toward the China plate. The Mesozoic–Cenozoic history in general involves thrusting and interstacking of continental crust, strike-slip and block faulting, and related formation and deformation of pull-apart basins, accompanied by extensive granitic magmatism and associated mineral migration and ore concentration. This intraplate orogeny is of no less importance than the preceding subductional and collisional orogenies, and it has produced large-scale crustal compressed zones, stacked thrusts, and nappe structures that contribute equally significantly to the make-up of the present Qinling belt.

REFERENCES

Coleman, R. G., 1984, The diversity of ophiolites: Geol. en Mijnbouw, v. 63, p. 141–165.

Coward, M. P., and C. Ries Alison, eds., 1986, Collision tectonics: Oxford, Blackwell Sci. Pub.

Hsu, K. J., Sun Shu, and Li Jiliang, 1987, South China—an orogenic belt rather than a craton: Sci. Sinica, v. 10.

Huang, Jiqin, 1984, New researches on the tectonic characteristics of China: Bull. Chin. Acad. Geol. Sci., v. 9, p. 5–18.

Li Chunyu and Liu Yangwen, 1978, Tectonic development of the Qinling and Qilianshan Mountains: A Collection of Geological Papers for International Academic Exchange.

Mattauer, M., P. Matte, J. Malavielle, P. Taponnier, H. Maluski, Xu Zhiqin, Lu Yilun, and Tang Yaoqin, 1985, Tectonics of the Qinling belt: built-up and evolution of eastern Asia: Nature, v. 317, p. 496–500.

Zhang Guowei, Mei Zhichao, Zhou Dinwu, Sun Yong, and Yu Zaiping, 1986, Formation and evolution of the Qinling tectonic belt, *in* Symp. 45th Anniversary Geol. Dep. NW Univ. Shaanxi Sci. & Technol. Pub. House, Xian.

ZHANG HUIFEN, CHEN DAMEI, and JIANG ZECHUN Institute of Geochemistry, Academia Sinica, Guiyang, China

Study on Mineralogy and Mineral Physics of Chinese Pyrophyllite

China is a country famous for its carving in pyrophyllite. This paper is intended to study the mineral composition and thermophysical properties of Chinese pyrophyllites and the diagnostic features by means of spectroscopic experiments. A series of samples collected from main pyrophyllite localities of China were studied by X-ray diffraction, TEM, EPMA, thermophysical measurements, absorption, and laser Raman and NMR spectroscopies. The results of the study are as follows:

1. The main constituent mineral of pyrophyllite ore is pyrophillite, but it contains kaolinite, dickite, and sometimes quartz, diaspore, and alunite as impurity phases. The chemical composition of pyrophyllite and kaolinite minerals is consistent with their theoretical values.

2. Two polymorphs of pyrophyllite have been revealed. The famous Qingtian carving stone consists of pure monoclinic pyrophyllite, and the Shangyu and Emei stones are composed of pure triclinic phyrophyllite, but in some other samples besides the monoclinic pyrophllite we also found the triclinic polymorph. The measured unit cell parameters are: a = 5.163–5.170 Å, b = 8.954–8.968 Å, c = 9.348–9.368 Å, α = 90.99–91.27°, β = 100.15–100.47°, γ = 89.88–89.10° for ticlinic pyrophyllite, and a = 5.175–5.180 Å, b = 8.904–8.919 Å, c = 18.660–18.648 Å, β = 99.86——100.47° for monoclinc pyrophyllite.

3. Thermal expansion coefficient of pyrophyllite measured by laser interference method ranges from 7.71 to 8.6 10^6/°C, and the thermal emission coefficient at 250 °C and 2–2.5μ is 0.41–0.53. The endothermal reaction temperature is 600°–700 °C; no exothermal reaction was found. The thermogravimetric and infrared spectroscopy studies at various temperatures show that pyrophyllite loses its hydroxl at 500°–700 °C and the mineral decomposes and turns into mullite and crystabolite at 1200 °C.

4. The computer decomposition of absorption spectra measured for pyrophyullite, kaolinite, and dickite, respectively, shows that they usually contain an absorption edge and three absorption bands, but the height of the main absorption band at 5620 Å for these three minerals is quite different. The highest band appears in dickite and the lowest in pyrophyllite. So they can be easily distinguished from the absorption spectra. Furthermore, with the increase of iron content in pyrophyllite the wavelength of main absorption band shifts toward shorter one, from 5680 Å to 5632 Å, and the band height increases from 195 to 413. The color of pyrophyllite depends mainly on the position and height of the main absorption band. According to the band position in spectra, we may deduce that the absorption edge is due to the Fe^{3+}-O^{2-} charge-transfer band and it can be interpreted as the low-energy tail part, and the absorption bands we measured are due to the crystal field transition of Fe ions.

5. Pyrophyllite, kaolinite, and dickite can also be distinguished by their Raman spectra. Pyrophyllite gives bands at 110, 146, 200, 270, and 710 cm^{-1}, among which the 270 cm^{-1} band is the strongest, and the 200 and 710 cm^{-1} bands are moderate strong. Kaolinite gives bands at 137, 205, 250, 275, 340, 445, 470, and 755 cm^{-1}, among which the strongest is at 137 cm^{-1} and the moderate strong bands are at 340 and 470 cm^{-1}. The Raman spectra of dickite is quite similar to that of kaolinite, but it has only one strong band at 150 cm^{-1}, and the bands at 343 and 470 cm^{-1} are much weaker than those of kaolinite. When a sample contains another impurity mineral phase, its Raman spectra also show the characteristic bands of both main and impurity minerals.

6. The NMR absorption pattern of pyrophyllite has a trigon shape, but when the sample contains a certain amount of active water, a narrow absorption peak appears at the top of the trigon. With the increase of the active water content, the peak becomes higher and sharper. The line-width of the trigon is about 3–3.5G. The NMR absorption pattern of kaolinite has a shape similar to a stalactite, with a sharp peak at its top. Its line-width is 6–8G, so the area of spectra for kaolinite is much larger than that of pyrophyllite. The overlap of narrow peaks over the wide bands indicates that pyrophyllite and kaolinite contain at least two kinds of water in their structure. One is more active, such as adsorption water, which corresponds to the narrow component of the NMR absorption pattern; another is less active, such as hydroxyl, which corresponds to the wide component of the pattern. It is also revealed that in pyrophyllite there exist two types of hydroxyl of different activities.

ZHANG Huifen, and XIE Xiande, Academia Sinica, Guiyang, Guizhou, China

Mineralogy and Mineral Physics of Rutile

A detailed mineralogical study has been conducted of rutile samples from various types of ore deposits in different localities of China. The occurrences and chemical analyses of representative rutile samples are presented in Table 1. It can be seen from the table that the content of Fe (converted to Fe_2O_3) in natural rutile varies from 0.17 to 2.58.

The existing form of iron in rutile: In reflected light, rutile samples are greyish white. Their reflection spectra are almost the same. Table 1 shows that the most significant difference among these rutiles lies in whether there is exsolution of solid solutions and to what extent the exsolution occurs.

Two types of exsolution mineral are recognized. One is white in reflected light and its reflection spectrum is similar to that of pure hematite. Containing 90.68% Fe_2O_3 and 9.11% TiO_2, it is named titano-hematite. The other is brownish gray and its reflection spectrum is the same as that of ilmenite. Bearing 48.00% FeO and 51.43% TiO_2, it is named ilmenite.

Titano-hematite normally occurs in the form of acicular exsolution lamallae arranged in two parallel groups with an angle range of 90°-140°. Titano-hematite is often associated with ilmenite to form veinlets (1-3μ) in rutile grains. Sometimes, these two types of exsolution mineral are concentrated locally, and sometimes they occur as aggregated in the veinlets.

Typical exsolution of titano-hematite is also observed in exsolved ilmenite grains. It is mostly vermicular in form, and occasionally it is around ilmenite grains to form a bright ring, and then extends as acicular crystals in a certain direction.

EPR measurements of single rutile crystals have shown: (1) No matter whether exsolution occurs or not, all the rutile samples show extremely strong Fe^{3+} spectra with a line width of about nx10 Gauss in general. In addition, the spectra are characteristic of tetragonal crystals, indicating a replacement of Ti^{4+} by Fe^{3+}. (2) In rutile with relatively high Fe_2O_3, a wide spectrum is also observed in addition to the narrow Fe^{3+} peak mentioned above, whose line width $\Delta H > 500$ Gauss and $g \cong 2$, indicating another existing form of Fe^{3+} in rutile, i.e., in the form of inclusions of an Fe-rich independent mineral.

A heating experiment showed that when the temperature rose to 500 °C, acicular exsolution crystals became coarser, and when the temperature continued rising to 600 °C, the exsolution crystals disappeared. The above phenomenon can be well accounted for in terms of phase diagrams of the corresponding components.

The influence of iron on the physical properties of rutile:

(1) The unit cell parameters of rutile tend to increase progressively with increasing Fe^{3+} in rutile, which is isomorphously substituted for Ti^{4+}, showing the correlation a = 4.5914 + 0.054x^2, r = 0.9019; c = 2.9583e$^{0.000919x}$, r = 0.9124.

(2) The magnetic susceptibility of rutile increases rapidly with increasing the iron content. The results of calculation by spin-only formula are in good agreement with experimental values.

(3) The position of the absorption edges shifts in the direction toward long wavelength from 403 to 460 nm with increasing the Fe^{3+} content in rutile.

(4) The microwave dielectrical constant ϵ decreases from 20.80 to 19.82, and the dielectrical loss $tg\delta_\epsilon$ increases from 0.001 to 0.1 when the iron content (Fe_2O_3) in rutile increases from 1.22 to 2.51.

The vibration spectra of rutile:

(1) The unit cell parameters at high temperatures, obtained by means of high temperature X-ray diffraction method, indicate that the relationship between the unit cell parameters and the temperature is linear with correlation

$$a(\text{Å}) = 4.5925 + 4.0853 \times 10^{-4} T\,(°C),\ r = 0.998,$$
$$c(\text{Å}) = 2.9583 + 3.3273 \times 10^{-4} T\,(°C),\ r = 0.999,$$

(2) Nearly all samples have similar Raman spectra. Four characteristic lines were observed and the assignment of each line was made. They are: B_1g, 134-146 cm^{-1}, E_g, 410-450 cm^{-1}, A_{1g}, 614-618 cm^{-1}, B_{2g}, 788-830 cm^{-1}.

Conclusions:

(1) The extent of exsolution depends mainly on the content of Fe in rutile. 0.88% Fe_2O_3 is regarded as the lower limit of Fe content for the occurrence of exsolution in ferrominerals. With increasing Fe content, exsolution will tend to become more and more obvious.

(2) Iron in rutile exists mainly in the form of Fe_{3+}, which is isomorphously substituted for Ti4+, and exsolution mineral, titano-hematite.

(3) The iron content in rutile determines the differences in physical properties of rutile samples.

Table 1. Characteristics of rutile samples

Sample No.	Occurrence	TiO_2	Fe_2O_3	Exsolution
5	Metamorphic rock	98.08	0.17	No
7	Metamorphic rock	97.76	0.21	No
16-2	Intrusive rock	97.19	0.88	No
16-1	Intrusive rock	97.19	0.88	Minor
1	Pegmatite	98.03	0.91	Minor
6	Intrusive rock	97.82	1.00	Obvious
4	Valley placer	97.69	1.78	Obvious
2	Eclogite	97.91	1.91	Extensive
3	Eclogite	97.54	2.51	More ext.
9-1	Placer	96.65	2.58	No

ZHANG Huimin, Tianjin Institute of Geology and Mineral Resources, Tianjin, China, D. P. ELSTON, U.S. Geological Survey, Flagstaff, AZ, and ZHANG Wenzhi, Tianjin Institute of Geology and Mineral Resources, Tianjin, China

Paleomagnetic Poles from Middle and Late Proterozoic Dagushi, Huashan, and Sinian Groups, Jingshan County, Hubei Province, South China

An approximately 640-m-thick Middle and Late Proterozoic sequence (Changhe sequence), about 130 km northwest of Wuhan, was sampled for paleomagnetic analysis. The strata, which include the Middle Proterozoic Dagushi Group and the Late Proterozoic Huashan and Sinan Groups, are located on the southwest flank of a large anticlinal structure; attitudes of sampled strata were highly variable (15° to overturned). More than 200 oriented core samples were collected from redbed units, from a few representative gray units, and from volcanic strata near the base of the sampled section. All sampes were subjected to progressive, stepwise, thermal or alternating-field demagnetization treatment. Large secondary overprints encountered in many of the samples could not be completely removed by the cleaning.

A provisional paleomagnetic direction and pole (pole 1) has been obtained from normal- and reverse-polarity directions from two cores in the uppermost part of the Dagushi Group measured at the 670 °C cleaning step. Calculations for pole 2 (11 of 13 normal-polarity samples) and pole 3 (9 of 12 reverse-polarity samples) are from redbed sections in the middle and upper parts of the Huashan Group. Pole 4 was calculated from 27 of 90 normal-and reverse-polarity samples from heavily overprinted redbed strata collected from across the entire section of the overlying Liantuo Formation of the Sinian Group. The Liantuo Formation is directly overlain by the glaciogenic Nantuo Formation, which was not sampled. Directions used for calculation of poles 2,3 and 4 are reported from the 674 °C cleaning step.

Poles 3 and 4 are in an area where paleomagnetic poles have been reported from lower and middle Paleozoic rocks of the South China Block; that grouping of poles is bounded on the south by a Late Proterozoic pole (pole pЄu; Lin, 1984). A 60° clockwise rotation of the Changhe terrane would move poles 3 and 4 of the Changhe sequence out of the cluster of Paleozoic poles and would bring pole 4 into coincidence with pole pCu. This simple correction suggests that the Changhe terrane was structurally rotated at the time of its folding and faulting. The polar path for the Middle and Late Proterozoic Changhe sequence, both before and after such structural correction, is unlike the polar path obtained from the Middle and Late Proterozoic Jixian section of North China (Elston et al, this volume). The dissimilarity between the Proterozoic polar paths for south and north China thus appears to extend a general dissimilarity already noted for Paleozoic pole positions reported from the North and South China blocks.

REFERENCES

Lin, Jin-lu, 1984, The apparent polar wander paths for the north and south China blocks: PhD Dissertation, University of California, Santa Barbara, California, 248 p.

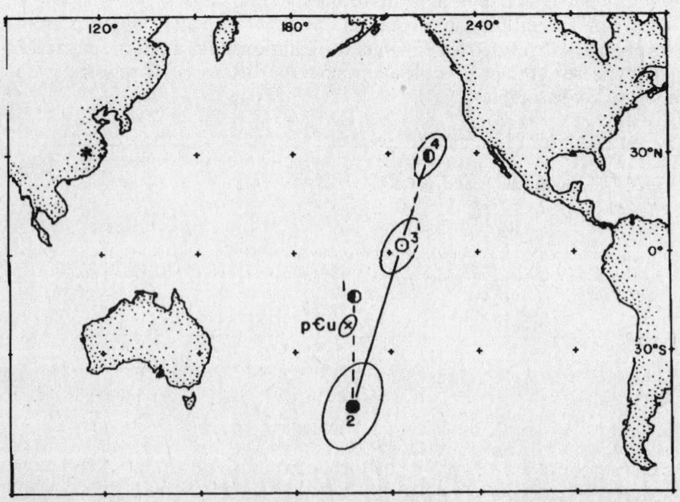

Figure 1—Plot of paleomagnetic poles calculated from Proterozoic redbeds of the Changhe sequence, south China. Poles 2, 3, and 4 enclosed by ovals drawn at the 95% confidence level. Solid symbol—normal polarity; open symbol—reverse polarity; half solid symbol—mixed polarity; X—from Lin (1984); *-sampling area.

ZHANG LIANG, LIU JINGXIU, ZHOU BO, and CHEN ZHENSHENG, Yichang Institute of Geology and Mineral Resource, CAGS, Yichang City, China

Oxygen Isotope Equilibration Systematics Between Quartz and Water: Salinities Effect Problem

Because quartz is the most common and predominant mineral facies in crustal rocks and ore its formation temperature and pressure are the widest. It is also most important in oxygen isotope geochemistry. Oxygen isotope fractionation factor of quartz-wolframite/cassiterite/magnetite is the largest, so it is the most sensitive temperature meter. For this reason, oxygen isotope equilibration studies between quartz and water are an interesting subject for stable isotope geochemists.

Various investigators since 1964, including O'Neil and Clayton (1964), Wenner (1971), Jensen (1971), Clayton et al. (1972), Shiro et al. (1972), Botting (1973), Becker (1976), Matthews (1979), Matsuhisa (1979), Kawabe (1978), and Zhang (1985), have developed a series of differential oxygen isotope fractionation curves based on experimental, theoretical, and experiential examination.

In our experiments, the $_{18}O/_{16}O$ fractionation accompanying the hydrothermal crystallization of quartz from silica gel at 180°–55 °C has been studied in order to assess the influences of (1) rate of quartz formation, (2) the nature of the reaction mechanism, (3) salinities effect, and (4) temperature.

Experimental methods: The oxygen isotope composition of the reagents is given in Table 1. Typical experimental charges consisted of 30 mg solid reactant and 200–400 mg solution (depending on temperature and salinity) sealed in gold or silver capsules. All experiments were performed in stainless steel cold seal bomb, using conventional hydrothermal techniques.

Oxygen was extracted from the silica phases using BrF_5 and a modified version of the method described by Clayton and Maycda (1963), and subsequently converted to CO_2 for isotopic analysis. All CO_2 was analyzed on the Mat-251 mass spectrometer. All analyses are reported relative to SMOW, as defined by repeated analysis of the NBS-28(= 9.6‰). The isotopic compositions of the solutions at the completion of experimental runs were calculated by material balance.

Results and discussion: Table 2 lists the results of some experiments at ρH_2O = 500–100 bars in which to vary from 180° to 550 °C and different reaction times were used, and salinity concentration was varied from 0 to 40 wt% (NaF).

Based on the data from Table 2, the linear equation defined by least-squares regression:

$$10001n\alpha_{Q\text{-}H2O} = 3.48 \times 10^6 T^{-2} - 2.9$$

This equation and curve may be compared with three other calibrations obtained at temperatures less than 500 °C (Clayton et al., 1972; Matsuhisa et al., 1976, 1978; Matthews et al., 1979). They define the line, respectively, as:

$$10^3 \ln\alpha = 3.38 \times 10^6 T^{-2} - 2.90$$

$$10^3 \ln\alpha = 3.31 \times 10^6 T^{-2} - 3.26$$

$$10^3 \ln\alpha = 3.05 \times 10^6 T^{-2} - 2.09$$

For the purpose of comparison, the former calibration (Clayton et al., 1972) has been recalculated for water analysis using a CO_2-H_2O fractionation factor of α(25 °C) = 1.04120 (rather than 1.0407).

We found also that oxygen isotope fractionation factor did not depend fundamentally on salinity. This conclusion is in contrast to oxygen activity effects in solutions at elevated temperatures of the type discussed by Truesdell (1974).

This investigation is part of the subject supported by National Nature Science Foundation (4860149).

ZHANG Ligang, Yichang Institute of Geology and Mineral Resources, CAGS, Hubei, China

Stable Isotopic Compositions of Three Major Types of W-SN Ore Deposits in Southern China

In the past ten years, many detailed stable isotope geochemical studies have been made in wolframite-quartz vein deposits and others, including Xihuashan (Zhang, 1982; Mu et al, 1982; Ding et al, 1985), Yangchuling (Zhang Dachun, 1984), Lianhuashan (Zhang, 1985), Woxi, China (Zhang, 1985); Pasto Bueno (Landis and Rye, 1974), San Cristobal, Peru (Campbell et al, 1984); Panasqueira, Portugol (Kelly and Rye, 1979; Schuiling et al, 1984); Carrock Fells (Sheppard et al, 1976), Tungsten Queen, America

(Casaderall and Rye, 1980); Dae Hwa, Korea (Shelton et al), etc. In the deposits such as Xihuashan, Panasqueir, etc., the explanations are not completely consistent with each other except for Tungsten Queen and Woxi. They can be divided into three, namely, (1) magmatic water, (2) meteoric water, and (3) mixture of magmatic and meteoric water.

The published and unpublished data have been used in this paper to study synthetically the related problems of tungsten ore deposits geneses in southern China. The data include more than 800 oxygen isotope analyses of minerals and rocks, and large number of hydrogen, sulfur, carbon, lead, and strontium isotope data of minerals and fluid inclusions in the 34 W/Sn ore deposits. Based on the greatest amount of the stable isotope data, I propose that the W/Sn ore deposits in China may be subdivided into three extreme types.

Reequilibrium magmatic hydrothermal W/Sn veins: For the W-bearing granitoids, such as Xihuashan, Dajishan, etc., the Pb isotopes of feldspars are the richest radiogenic (Pb-206/204 being 18.5-19.0, 207/204 15.6-15.8, 208/204 38.5-39.2) ore lead are also the richest radiogenic, the initial $^{87}Sr/^{86}Sr$ ratio (0.710 to 0.745) and $\delta_{18}O$ values (+9.6 to +14.5) are also the highest in the more than 120 Mesozoic granitoids in eastern China, so there is little doubt that the W-bearing were originated from the upper-crustal metasedimentary rocks-that is the Proterozoic Erathem-the rocks of lower Banxi or/and Sibao Group exposed in Jiangnan old land.

The $\delta^{18}O$ values of quartz samples of the main ore-forming stage in W ore veins are from +11 to +14%, wolframites have $\delta^{18}O$ ranging from +4 to +7%. Calculated $\delta^{18}O_{H2O}$ values for fluids in equilibrium with quartz range from +4.5 to +8.5%. And not only in southeast China but in southwest China, the δD_{H2O} values of the fluid inclusions in the quartz range from -40 to -70% similarly.

The $\delta^{18}O_{H2O}$ and δD_{H2O} values of primary magmatic water for W-bearing granitoids are from +9 to +13.5 and -60 to -90%, respectively. Owing to the $\delta^{18}O$ and δD values of the feldspathization rocks generally higher and lower than the unaltered rocks, respectively, and calculated W/R ratio is lower, in this case, $\delta^{18}O$ and δD values of evolved magmatic water should be decreased and increased, respectively. Because of the $\delta^{18}O_{H2O}$ and δD_{H2O} values of the main ore-forming stage for 20 tungsten ore veins in southern China are falled on the range of the evolved W-bearing granitoids primarily magmatic water, that is reequilibrium magmatic water, so the tungsten ore veins are related to reequilibrium W-bearing granitic magmatic water.

But in late sulfide or/and carbonate stage, the $\delta^{18}O$ and δD values indicated which is related to Mesozoic meteoric water. As compared with δD value contours of present meteoric water in China, it would seem that the δD values of Mesozoic meteoric water are systematically lower than those of present meteoric water, by about 20%. This fact may be accounted for by the observation that the east coast of Asia in the Mesozoic era was somewhere east of the present coast.

The $\delta^{34}S$ values of sulfides are about 0% and carbonates have $\delta^{13}C$ ranging from -6 to-10% in the veins, suggest that the sulfur and carbon was derived from the leaching of the W-bearing granitoids.

Meteoric hydrothermal W/Sn ore deposits: Mesozoic meteoric hydrothermal W/Sn ore deposits are veinlet disseminated or brecciaed style. The deposits are situated in the contact zone of subvolcanic intrusions, such as Lianhuashan, Xiling, etc. or in fault zones within sedimentary rocks such as Dachongshan, etc. In the former, quartz and wolframite have $\delta^{18}O$ values ranging from +10 to +7 and +2 to -2%, respectively, but in the latter, the $\delta^{18}O$ values are about 17 and -2%, respectively.

Stratiform W/Sn ore deposits: These types of deposits, including Woxi, etc., four ore deposits which the age of mineralization is Precambrian, but at Qingjia is Cambrian and at Fengling is Devonian. The $\delta^{18}O$ values of the 80 quartz and 25 scheelites/cassiterites/wolfrimites in these deposits are about +18 to +21 and +3 to +5%, respectively. In the stratiform W-Sb-Au ore deposits, for example, Woxi mine, W-Sb-Au-bearing siliceous beds with sedimentary rock beds are of similar folds, and silicalite are of recrystalline texture. The δD values of fluid inclusions in quartz are from -60 to -80%, and the sulfides have $\delta^{34}S$ values vary within the stratigraphic sequences in nine ore deposits. The origin of the stratiform W/Sn (Cu, Sb, Au) ore deposits is related to submarine exhalative activity, and sometimes associated with contemporaneous submarine volcanic eruption.

reactant	symbol	^{18}O (‰)
Water*	Y	+33.32
	S	-8.15
	G	-5.96
Silica gel	C	+14.80
	J	+34.55

* Isotopic analyses obtained by the CO_2 equilibration procedure described by Epstein and Mayeda(1953), using a value of $\alpha^{25°C}_{CO_2-H_2O} = 1.04120$ (O'Neil et al. 1975)

Table 2. Fractionation factor of quartz-water in different temperature

Run No.	Time (h)	Reactants			$10^3 \ln\alpha$
		Solid	water	Wt. (salinity)	
180°C					
167	206	C	S	25	13.46
250°C					
186	125	C	S	15	9.40
157	34	J	S	25	9.56
87	90	J	G	25	9.51
109	40	J	Y	25	9.45
122	36	J	S	25	10.82
128	76	C	G	40	9.66
168	54	C	S	40	9.25
169	34	C	S	40	8.86
520°C					
104	48	J	S	5	6.57
96	52	J	Y	25	6.04
110	48	J	Y	25	6.25
144	32	C	G	25	6.92
155	48	J	G	25	6.68
124	60	J	Y	40	6.48
130	48	C	Y	40	6.44
320°C					
19	48	J	G	25	5.95
63	20	J	G	60	6.04
450°C					
196	104	J	G	5	3.61
165	30	C	Y	25	3.60
220°C					
206	104	C	S	5	2.73
84	48	J	G	5	2.91
75	48	J	G	25	2.20
126	24	C	S	40	2.74
129	24	C	G	40	3.30
133	16	J	S	40	2.94
134	24	J	S	40	2.00

ZHANG Ligang, Yichang Institute of Geology and Mineral Resources, CAGS, Hubei, China

Relationship Between the Mineralization and Pb, Sr, O, Nd Isotopic Compositions of Some Mesozoic Granitoids in Eastern China

Triassic to Cretaceous plutonic rocks associated with contemporaneous volcanic rocks are extensively distributed in the variant tectonic element in eastern China, in which including (A_1) the northern China platform with a Yanshan fold belt in the north margin of the platform, (A_2) the northern China platform periphery fold belt, and (A_3) the Neimong-Xinganling fold belt is on the northwest of the platform in the northern China series. In the southern China series, including (B_{1-1}) the Nanjing-Zhengjiang-Suzhou mountain range, (B_{1-2}) the Yantze platform including the Yangtze fold belt is in the northeast part of the Yangtze craton, and (B_2) the Nanling fold belt. In addition to the series, (C) the southeast coastal volcanic rock zone. A unique feature of each tectonic element is the widespread distribution of Mesozoic granitoids, many accompanied by ore deposits to have a unique economic minerals. In A_1 and A_2, the Mesozoic granitoids are mainly considered as the parent rocks of Cu, Mo, Au, (PB, ZN) et c. ore metals, but in

A_3, in recent years, a number of Sn/W/Cu mineralizations were discovered. In $B_{1\text{-}1}$ and $B_{1\text{-}2}$, there are the most kinds of economic minerals, including Fe, Cu, Mo, W, Pb, Zn, U, Be, etc. are prospected, and also in C, but in B_2, that in Nanling area, there is little doubt that the reserves of W/Sn ores are the largest in the world. But, why is it so? During the past eight years, the author has been working intermittently on this problem, approaching it from the angles of the stable isotopic geochemistry.

Based on the data including recently available data in the literature and to be based are stored in Pb, Sr, O, Nd isotope data Data Bank of the isotope geology department of YIGMR, about 500 Pb isotope data of the feldspars and rocks in the 65 Mesozoic granitoids and Sr, O, Nd, isotope data of more than 120 Meoszoic granitoids in eastern China. In addition, in consideration of the ore Pb isotope data for 45 hydrothermal ore deposits, which is related to the granitoids. The author suggests classifying the crustal basement rocks in eastern China into three discrete series. The three series may be further subdivided into several mineralization provinces and related plumbotectonic provinces.

Pb isotopes: For the Pb isotope data, a summary plot of the Pb isotope ratio from which I and others define the modern fields for broad geological environment in eastern China is shown in Figure 1. Every field is drawn around ~95% of all data for the Mesozoic granitoids, but it is showing the points of the feldspars in Figure 1. Obviously, characteristic patterns of Pb isotopic compositions of feldspars and others from the Mesozoic granitoids obtained by using the plumbotectonic model of Zartman and Doe (1981) have defined three discrete series (that is A, B, C) and six provinces, at least, for the crustal basement rocks in eastern China. The Pb isotopic composition within each province reflects the age and geochemical behavior of the source materials, maybe that is basement rocks.

From Figure 1, we can find that in A_1 and A_2, the granitoids are least radiocenic and have the lowest % values, with a remarkably broad range of isotope composition indicating mainly an Archaeozoic lower crustal source. The granitoids are not related to W/Sn mineralization. In B_1, feldspars are variable in isotope composition and μ values, and contain lead derived from early to middle Proterozoic basement rocks with different buried depths, age, and compositions. But in B_2, the Pb isotopes are the most radiogenic and have the highest μ values, so they contain lead derived principally from middle Proterozoic upper crustal metasedimentary rocks from which related to the granitoids are closely associated with W, Sn, Li, Be, Nb, Ta, U, Pb, Zn, REE, etc., mineralization.

Sr, O, and Nd isotopes: Fr the Sr, O, and Nd isotope data, the histograms and plots of the Sr, O, and Nd isotopic compounds in broad geological environments in eastern China are shown in Figure 2-5 (omitted). From the pictures, we found that the richest radiogenic Sr and ^{18}O granitoids are in B_2.

It is well known that because the enrichment of ^{87}Sr, ^{18}O, ^{206}Pb, ^{207}Pb, ^{208}Pb, and ^{143}Nd in granitoids is due to entirely different mechanisms, therefore, based on the data mentioned above, I am sure that the relationships among the isotopes of granitoids, source rocks and their evolution, and related to ore metals in the crustal rocks are very close.

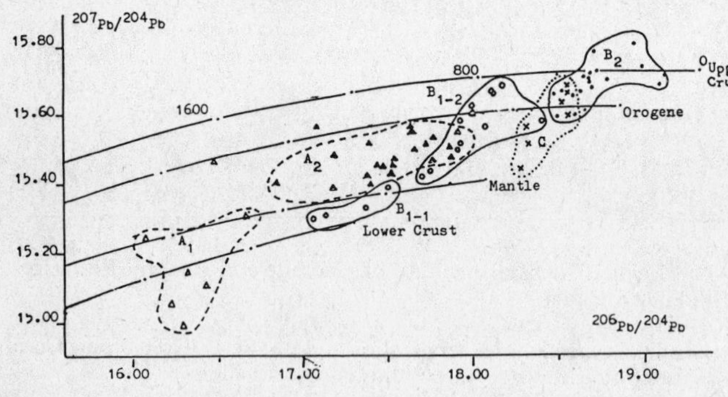

Figure 1—Plot of Pb isotopic compositions for feldspars and whole rocks analyzed in some Mesozoic granitoids in eastern China shown on a $^{207}Pb/^{204}Pb$ vs. $^{206}Pb/^{204}Pb$ diagram with the Pb evolution curves generated by the Zartman and Doe (1981).

ZHANG PENFEI and JIN KUILI, China Institute of Mining and Technology, Beijing, China, and LIU HUANGJIE, China Institute of Mining and Technology, Xuchou, China

Aspects of Mid-Hunan Coal Field, South China

Mid-Hunan coal field with an area of 7000 km^2 is situated in the central part of Hunan Province, South China and is a famous and typical Lower Carboniferous coal field in our country.

Stratigraphy: The Carboniferous of Mid-Hunan naturally falls into two divisions comparable with Mississippian and Pennsylvanian of North America. The Lower Carboniferous admits of a threefold division, of which there exists Visean coal measures; whereas the Upper Carboniferous is a unicomponent system composed of pure limestones; therefore, the whole Carboniferous shows a huge cycle of sedimentation reaching 1500 m in thickness. In this region, the geochemical method was also taken to subdivide the stratigraphy.

Structure: There are anticlines and faults trending, in west-northwest, north-northeast, north-south, and east-west directions. Based upon study of stress fields by different methods, especially the improved Ting's Vitrinite methods, the authors determined five stages of variance of stress field by the Mesozoic Yanshan movement.

Sedimentology: On the basis of the study of lithology, sedimentary structures, body fossils, and trace fossils, the sedimentary aspects of lower Carboniferous are discussed briefly as follows: The environments of the lower part consist of lagoon and local tidal flats; the middle part includes coastal environments such as tidal flats, coastal lakes, and marshes and barrier islands; the upper part mainly comprises lagoon. In the other hand, the authors reconstructed these three stages of paleogeographic maps showing the changes of facies in this area. Morever, the authors used the methods of sedimentary geochemistry, organic geochemistry, and organic petrology and achieved some new ideas of the whole Carboniferous sediment in terms of salinity, temperature, redox potential and their changes of waters; for example, the temperatures of waters for Lower and Upper Carboniferous might be 29 °C and 17 °C, respectively.

Coalification: The coal-bearing strata contain seven seams, but only No. 3 and No. 5 seams with 1–2 m are minable. In this coal field, the dominant coal is anthracite. Because the anthracite varies in rank, we plot coalification maps. By means of geological and geophysical investigation, there are some subsurface igneous bodies. It is clear that the thermal metamorphism overlapped the regional metamorphism for coal. Because of this special type of metamorphism, the coal macerals, especially exinites, are still visible. In order to aid in observation, the authros also used the dyeing method.

ZHANG RENJIE, Yichang Institute of Geology and Mineral Resources, Yichang City, China

Discovery of *Chuaria* and *Tawuia* Flora (Late Precambrian) in Shilu Group, Hainan Island, South China

The Shilu iron mine of Hainan Island is well known in China for its very high grade ore and great reserves. However, in the past no valid fossil evidence was known from the ore bed, so the geological age of the Shilu Group has been disputed for a long time. Hence, the outstanding issue for understanding the geological evolution and ore-prospecting in the Island, as well as in South China, has remained unsettled, although great efforts have been made since 1949. In 1986, fairly abundant Late Precambrian algal megafossils (*Chuaria* and *Tawuia*) were discovered from the Shilu Group by the author. Such a discovery contributes tremendously to the solution of the problem of the age of the latter.

The rocks from which the fossils were collected: The Shilu iron mine is located in Shilu Township, Changjiang County, west Hainan Island, South China. Geologically, it is surrounded by Mesozoic batholith of granodiorite and quartz-diorite. The Shilu Group mostly crops out near Shilu Township and consists of a suite of medium-light metamorphic rocks, such as schist, slate phyllite, carbonaceous slate, and quartzitic dolomite, with a total thickness of 1700–2600 m. The very high grade iron ore deposits, the so-called Shilu-type iron deposits, and the copper and cobalt ore deposits measuring tens to hundreds of meters thick occur in Bed 6 in the upper part of the Shilu Group. This group belongs sedimentologically to shallow-water, embayment, or lagoon sediments of a geosyncline, and was lithologically subdivided into seven megabeds, Beds 1–7. The Shilu Group is overlain unconformably by the Sanlengshan Formation containing the Early Carboniferous biota such as plants, bivalves, and dendroid graptolites. But its underlying rocks are unknown yet, because the base of the Shilu Group was cut out by Mesozoic granodiorite. Hundreds of well-preserved algal megafossils have been found in carbonaceous slate in intercalations of dolomite of the upper part of Bed 6, Shilu Group.

Features of the Chuaria–Tawuia *Biota of the Shilu Group:* The Late Precambrian algal megafossils of Shilu Group consist most of *Chuaria circularis* Walcott and *Tawuia dalensis* Hofmann. The morphology of *Tawuia dalensis* in this group is varied, including I-, J-, C-, and U-shaped forms. Generally speaking, the U-shaped individuals are larger in size, 2.5–5 cm in length, 5–8 mm in width. In some specimens, the marginal zone (1 mm width) and rough folds are visible, respectively. Most U-shaped individuals are preserved as carbonate compressions (about 1 mm thick). C- and J-shaped forms are smaller in size, mostly 2 cm in length, 3 mm in width, and even-surfaced, having no marginal zone or folds. I-shaped individuals are

8-35 mm long and 2.5-6 mm wide. The last three forms, C-, J-, and I-shaped ones, are preserved as impressions. Except for a few overlapping individuals, most *Tawuia dalensis* are preserved as isolated specimens. The species *Chuaria circularis* Walcott is also markedly abundant in Bed 6 of the Shilu Group and occurs with *Tawuia dalensis*. *Chuaria circularis* Walcott of the Shilu Group is circular to oval, from 1-2 mm to 10.5 × 13 mm in diameter. Most individuals of this species are scattered on bedding planes with only a few overlapping each other. Marginal ring-shaped wrinkles are visible on some specimens. These longitudinally oval algal megafossils are large in size, with maximum dimension of 16 × 24 mm.

Geological age of the biota: At least three Late Precambrian biotas of algal megafossils can be recognized in China so far. They are: (1) Longfengshan biota (or Huailai biota)—found in the Changlongshan Formation of the Qingbaikou system in northwestern Hebei. (2) Huainan biota—found in the Liulaobei through Jiuliqiao formations of Souxian County, Anhui Province. (3) *Chuaria circularis* and *Tawuia dalensis* associated closely with a great number of metazoans—found in the Sinian System of Anhui and the Yuhucun Formation of Jinning County, Yunnan Province.

The macrobiota of the Shilu Group in Hainan Island is comparable with those of the Longfengshan, Huainan biotas of China, and Little Dal biota of northwestern Canada, all of which are marked by the presence of the dominate taxa *Chuaria* and *Tawuia*. However, the biota of Bed 6 of the Shilu Group differs from the others in lacking *Longfengshania* and metazoans (fossil worms) and *Vendotaenia*.

The time range of the *Chuaria–Tawuia* assemblage is put at 0.6–1.0 Ga. Thus, Bed 6 of the Shilu Group may be placed in the Upper Proterozoic and is a likely equivalent of the Liulaobei Formation, which is assigned in general to the Qingbaikou System in China. The existance of Upper Precambrian rocks in Hainan Island is confirmed for the first time with fossil evidence.

ZHANG RENJIE, Yichang Institute of Geology and Mineral Resources, Yichang City, China

Devonian Bivalves of China

Devonian bivalves are well-developed, highly diversified, and widely distributed both in platform and basinal areas of China. Up to now, 349 species of marine bivalves, placed in 113 genera and 39 families, have been found from 122 localities in China; most of the localities are in South China. Geologically, Devonian bivalve faunas occur in most depositional regions of China. Guangxi, Hunan, East Yunnan, and the Longmenshan district of northern Sichuan are the areas where Devonian bivalves are most abundant and best known. Stratigraphically, the Lower and Middle Devonian rocks of China have well-developed bivalve faunas. During late Early Emsian and Eifelian times, bivalves flourished in China; however, in the Upper Devonian rocks, bivalves sharply declined. All six subclasses of bivalvia are represented in the Devonian of China. Among them, Palaeotaxodonta accounts for 15.9% of the genera; Cryptodonta 5.4%; Pteriomorphia 38% (Pterineidae 11.5%, Ambonychiidae 8.9%); Palaeoheterodonta 17.7%; Heterodonta 8.8%; and Anomalodesmata 14.2%. These faunas are characterized by: (1) well-developed and highly diversified bivalves, the Pterineidae and Ambonychiidae being especially flourishing taxa; (2) times of high endemism, with the number of endemic genera decreasing sharply from Emsian to Frasnian (from 13, 7, 4 to 1, respectively); a cosmopolitan fauna exists in China in the Upper Devonian; (3) two different depositional types and corresponding bivalve groups, the shallow-water nearshore depositional type (Xiangzhou type) dominated by benthic bivalves, and the deep-water (basinal) type (Nandan type) dominated by epiplanktonic bivalves; (4) this is comparable with bivalve faunas of West Europe and eastern North America; (5) most Devonian species of China are endemic, but some species show affinities with areas adjacent to China and to Europe and eastern North America.

Early–Middle Devonian bivalve paleobiogeography of China: Analysis of the Early–Middle Devonian bivalve faunas shows that the South China Region is a well-defined paleobiogeographic bivalve province, distinguishable from the Junggar–Hinggan Region. Considering the features of other biotic composition (brachiopods, corals, and nonmarine fossils), the Tarim–North China and West Yunnan–Tibet regions are preliminarily defined as informal paleobiogeographic provinces, although only a few bivalves are at present known from these areas. Above mentioned four regions may be regarded as paleobiogeographic provinces of the Devonian Old World Realm. They may have been paleogeographically located in tropic to subtropic belts during the Early–Middle Devonian. This conclusion is drawn from the following evidence: (1) well-developed and highly diversified bivalves, with a number of endemic genera; (2) the presence of such tropical taxa as *Megalodon* (*Megalodon*) of the Hippuritoida and tropical to subtropical Rhenish and Bohemian endemics; (3) also 14 Appalachian genera and 2 western North America genera are found in China, mostly in the South China and Junggar–Hinggan regions.

In comparing the Chinese bivalve fauna, the Appalachian Region is herein treated as a Devonian bevalve region (province) of the Old World Realm, rather than an independent realm. With many endemics, it is clear the South China Region is one of the endemic centers in the world during the Early–Middle Devonian. On the basis of the differences and similarities of the Devonian bivalves, the South China Region may be further subdivided into certain subprovinces when detailed studies have been done.

The communities and life-habits of Devonian bivalves of China: Among the known 113 genera of Devonian bivalves of China, 47 genera (41.6%) are infauna (burrowing); 41 genera (36.3%) are epifauna with byssal-attachment, and 25 genera (22.1%) are semi-infauna, mostly endobyssate. All Devonian Palaeotaxodonta are deposit feeders and burrowers. Most Anomalodesmata, certain Heterodonta, and some Palaeoheterodonta employ the burrowing life-habit. Among them, *Parina,* and *Palaeolucina* may be deep burrowers. The epifauna are mostly byssally attached Pteriomorphia. However, *Kochia* and *Antipleura* may be reclining species, being partly buried in soft substrata. *Guangxiconcha, Pterochaenia,* and *Buchiola* may be epiplanktonic. Most semi-infauna bivalves are endobyssate.

Three marine bivalve communities are recognized in the Xiangzhou type: the platform, nearshore, and shallow-water clastic facies. They are: (1) /*Dysodonta–Leiodysodonta* community—principally a low-energy, tidal flat community of early–middle Gedinnian age, including *Cypricardella* and *Modiomorpha*. Most of them are suspension feeders and free burrowers, with the exception of the endobyssate *Modiomorpha*. Palaeotaxodonts are not known in the community. It is noted that the endobyssate *Leptodesma* increase in number from the tidal flat environment toward the normal marine shallow-water environment. (2) *Newellipecten–Lyriopecten–Sinodora* community of late Early Emsian—a shallow-water shelf community, dominated by the byssally attached suspension feeding epifauna and associated with certain burrowers including the suspension and deposit feeders. (3) *Fasciculiptera–Tanaodon* community—also a shallow-water shelf community of Givetian age, consisting most of epifauna and accompanied by the semi-infauna *Tanaodon* and infauna free burrowers.

The only community of the Nandan sedimentary type living on the outer platform, in a deep water, stable environment (basinal facies) of Zlichovian–Dalejian age, is the *Guangxiconcha–Pterochaenia–Buchiola* community. It is dominated by the small-sized, thin-shelled pseudoplanktonic *Guangxiconcha, Pterochaenia* associated with a few epiplanktonic *Buchiola,* and very few *Actinopteria,* which is byssally attached epifauna.

ZHANG Ronghua, Chinese Academy of Geological Sciences, Beijing, China, and J. POSEY-DOWTY, R. HELLMANN, M. BORCSIK, AND D. CRERAR, Princeton University, Princeton, NJ

Chemical Kinetics of Dissolution of Minerals in Circulating Solution at Elevated Temperatures and Pressure

This paper emphasizes analyses of new experimental results pertaining to chemical kinetics of mineral dissolution in circulation flow system at high temperatures and pressure. The sample bag is made of a metal network and the mineral particles put inside; the bag is then suspended in the vessel. The liquid pump is used to drive water (or NaCl solution) through the sample bag; it reacts to the mineral and drops out from the tubing system. The output solution is sampled and the reaction product analyzed. The flow system can be approximately considered as a CSTR reactor (continuous stirred tank reactor). The flow system is apparently divided into three parts: fluid transfuse network—nearly to PFR flow system (plug flow reactor), the resident time apparently equal to space time but above and under the bag are near to CSTR system, the average resident time t equals space time. According to the above assumption, the dissolution rate $-r = (dc)/(dt)$, c stands for the concentration of the output solution.

The calcite–water reaction kinetics experiments are generated in the steady processes, and temperatures (T °C) range from 25 to 350 °C and pressure (P) = 1600 psi, with constant flow rate. The flow rate (U) can be controlled and changed from 13 to 130 mL/h and sampling of the output solution is at constant T,P,U,SA (surface area).

In the calcite–water system: as T < 200 °C the dissolution rate equation with zero order, and the system comes to the far equilibrium state, the rate of the reaction is mainly controlled by the surface reaction, $-r = k_1$(T)SA. As T ranges from 200 to 350 °C, the rate of the reaction is related to the concentration of products, and limited by the transport rate. The system is to near equilibrium state and an apparent first-order reaction is presented. The dissolution rate can also be slowed down by the flow rate decreasing in the above cases. $-r = k_2(T)SA(K_{sp} - a_{ca2+} \cdot a_{co3^{2-}})^{\alpha}$; α is the kinetic order of the reactions.

In the experiments, an apparent CSTR performance under steady state condition, but a nonsteady state process is still involved in the whole dissolution processes, while the experiment was proceeding for more than 3 days or above 1 week. The accumulation of reaction time (t_t) varying affects the general dissolution rate, and this change is considered as $-(dNi)/(dt_t)$. Then the general dissolution rate can be written as follows:

$$-r = (dc)/(dt) = k(T)SA(K_{sp} - a_{ca2+} \cdot a_{co3^{2-}})^{\alpha} - (dNi)/(dt_t)$$

steady state processes — nonsteady state processes

Where SA means surface area of calcite in the sample bag, k refers to the reaction rate constant at temperature < 200 °C, k_2 reaction constant at T ranging from 200 to 350 °C. K is the function of T °C, K_{sp} is the solubility product of calcite in water at each temperature. Ni, numbers of mole of chemical species i; for instance, Ca^{2+}. Generally, the surface reaction and the transport rate both affect the dissolution rate of calcite in the flow system.

The authors also study the dissolution processes of fluorite and albite in the packed bed reactor by using the axial-dispersion model. It is found that the fluorite dissolution rate in the aqueous solution at low values of pH will go up with respect to flow rate rising; if flow rate is getting rather high, the dissolution rate becomes stable without changing as flow rate continues to drop; when flow rate is very high, the dissolution rate will drop down with respect to the flow rate increase. This process shows that the transport rate affects the dissolution rate as the system nears equilibrium state. And also it means the system changes from near equilibrium state to far from equilibrium state. A transport-controlled process is one in which the overall reaction is limited by the molecular diffusion of solutes toward or away from the mineral surface. The distance from the equilibrium state may be a worthy value to modify the dissolution mechanism in open-flow system.

The following equation can modify the axial-dispersion model, in our tubular reactor. The vessel has 1/4″ inner diameter and z = 210 mm.

$$D_L(d^2c)/(dz^2) - u(dc)/(dz) + r = 0$$

Here, D_L is the axial dispersion coefficient. The boundary condition of the equation for the tubular system can be determined by using input tracer concentration in response to pulse or stop change. Open-flow system experiments are very important ways of revealing the chemical oscillation, self-organization, and respective structure processes in natural systems.

ZHANG Shao Nan, Chengdu College of Geology, Chengdu, China, and V. SCHMIDT, Calgary, Canada

Diagenesis of Tertiary Litharenites, Reindeer, and Kugmallit Sequences, Beaufort/Mackenzie Basin, Canada

Composition, depositional textures, and diagenetic overprint were examined for 130 core samples of deltaic hydrocarbon-bearing sandstones from two well locations of the Beaufort/Mackenzie Tertiary Basin employing petrographic microscopy, SEM, and XRD analysis. The polymict litharenites and associated siltstones and mudstones of the Paleocene Reindeer sequence and of the Oligocene Kugmallit sequence were deposited as thick prograding deltaic wedges. The samples of the onshore well Taglu C-42 were obtained from normal pressured sandstones of the Kugmallit and Reindeer sequences and range in depth from 2583 to 3508 m. At the offshore well Issungnak 20–61 samples of the geopressured Kugmallit sequence range in depth from 2383 to 3322 m subsea. Salinities of formation waters are low basin-wide and average about 10,000 ppm.

Mean grain-size ranges in Reindeer sandstones and conglomerates from 0.08 mm to 15.00 mm and in Kugmallit sandstones from 0.08 mm to 0.40 mm. The grain framework composition of the generally poorly to moderately well sorted sandstones is dominated by quartz, chert, metamorphic, and sedimentary rock fragments and syngenetic clasts of reworked fine-grained deltaic sediments. Subordinate grain types include mainly K-spar, plagioclase, muscovite, ± altered biotite and volcanic rock fragments, organic debris and reworked synsedimentary siderite. Ductile grains comprised 20–60% of the original framework grains.

The varied present compositions of the grain framework is related to depositional environment and grain size as well as to the preferential dissolution and replacement of labile grains and the compaction of highly ductile grains resulting in pseudomatrix.

Eodiagenetic alteration is weak to moderate and includes mainly: (1) cementation and replacement by siderite, chlorite, and minor pyrite and (2) hydration of biotite.

Mesodiagenesis was very extensive both during and after the destruction of primary porosity and mechanical and chemical compaction. The chief diagenetic processes operating during the elimination of primary porosity are: (1) pervasive mechanical and chemical compaction; (2) replacement and cementation by iron-rich calcite, chlorite, kaolinite, illite, and minor quartz; and (3) dissolution of labile silicate minerals and authigenic calcite. Apparently, dissolution/reprecipitation processes created on a large scale secondary porosity at the expense of primary porosity.

The main mesogenetic changes after loss of primary porosity include: (1) large-scale dissolution of carbonate, aluminosilicate, and silica minerals; (2) cementation and replacement by kaolinite, calcite, and quartz, and (3) chemical and subordinate mechanical compaction, reducing secondary porosity. Progressive burial apparently did not affect the stability of K-spar.

In the majority of sandstones, late mesogenetic dissolution significantly enhanced macro- and microporosity. Porosity enhancement is best developed in the geopressured Kugmallit section where it often resulted in diagenetically disaggregated loose sands.

The late porosity enhancement required large-scale subsurface mass transport of the chemical constituents of carbonate, aluminosilicate, and silica minerals by compactional and thermobaric formation waters in advective and convective fluid flow regimes.

ZHANG Shaohua, Geological Institute, National Bureau of Building Materials Industry, Beijing, China

Preliminary Study of the Fibrous Sepiolite Deposit in East Qinling, China

The fibrous sepiolite deposit in the east Qinling district is now the only one that can be mined economically in China. We have, for the first time, systematically studied minerogenetic-geological characteristics, mineralogical features, and physical-chemical properties of this minerogenetic zone extending several hundreds of kilometers, exploration-assessment methods and industrial use of sepiolite.

Minerogenetic-geological characteristics of fibrous sepiolite:

(1) The fibrous sepiolite occurred as veins and reticulated veins in Precambrian dolomite-marble in which the dolomitic matter and siliceous matter have close minerogenetic relations with sepiolite and are the substance source of its mineralization.

(2) The sepiolite deposits are distributed along the paleoplate junctures. The sepiolite is slowly crystallized out from the postmagmatic minerogenetic hydrothermal solution in the confined zone during the plate tectonic activity. The plate multiphase tectonic activity has resulted in the multiphase sepiolite. The sepiolite veins are controlled by the pressure-torsion interlayer fissures or X-type joint fissures.

(3) There are three types of spatial contact relations between the sepiolite-bearing rock system and the granite mass: (a) sepiolite occurred in the outer contact zone; (b) sepiolite occurred in magmatite xenolith; and (c) there is no granite around the sepiolite-bearing rock system. The quality, the vein-bearing rate, the physical conditions of minerogenesis, and the mineralization type of these three kinds of sepiolite were compared and studied in detail.

(4) The sepiolite is closely associated with the hydrothermal alteration minerals such as tremolite, taxoite, talc, calcite, dolomite, and fluorite. The sepiolites in different locations are different in their assemblages.

(5) The study of the composition of pneumato-hydatogenic inclusions, minerogenic temperature-pressure, and isotopes of sepioile indicates that sepiolite is formed under weak alkaline conditions, whereas strong ones are unnecessary. Sepiolite is formed as plasma Si and Mg in wallrock are activated, segregated, and deposited as a result of thermal process. Homogenization temperature measurement indicates that the sepiolite is crystallized out in the minerogenic hydrothermal solution at 70-85 °C.

(6) H and O isotopic studies indicate that the minerogenic hydrothermal solution is related to magmatic activity, and is stacked by metamorphic water and atmospheric water.

(7) The ore types of sepiolite include fibrous, thin paper, and earthblock ones. In the fiber growth form, the longitudinal fiber is dominant, clinofiber secondary, and transversal fiber minor. These three types of fibers could be developed simultaneously in the same ore veins, and this development phenomenon is considered to be due to the different tectonic stresses during crystallization.

(8) The sepiolite is closely related to wallrock alterations such as tremolitization and talcization.

Mineralogical feature and physical-chemical properties: X-ray diffraction analysis, infrared absorption spectroscopy, monomineral chemical analysis, differential thermal analysis and electron microscope study indicate that the fibrous sepiolite in this district is a well-crystallized, high-aquiferous, and high-magnesian sepiolite, and its ordering is relatively good. The strength of extension and specific surface area of the sepiolite fiber is found to be remarkably changed when it is being heated. When the sepiolite is heated to 380-450 °C, its strength of extension is 1.2-1.7 time higher than that at 25 °C, and 1.3-1.9 times higher than that at 100 °C, but when its temperature exceeds 450 °C, its strength of extension will abruptly decrease. The specific surface area of natural fibrous sepiolite decreases as the temperature increases, and when the temperature exceeds 200 °C, it will abruptly decrease. The reason for such changes of strength and of specific surface area is believed to be that when the sepiolite is heated to 380-450 °C, the zeolitic water and half of the coordination water in the crystal body will be removed, so that the crystal body will be bent and coaction of Mg^{2+} in octahedron and O_2 in the basal plane of tetrahedron will take place.

A series of experimental studies of specific resistivity, electrodynamic potential, magnetism, acid resistance, alkali resistance, heat resistance, heat conductivity, filtrability anion exchangeability, specific gravity, and tarnish capacity have been made, and a large quantity of firsthand data were obtained.

Exploration and assessment methods of the ore deposit: Having analyzed the important deposits and compared them with asbestos deposits, we worked out more economic and more reasonable methods of exploration and assessment, and provided some basic data for the assessment of such a deposit, and a basic working process.

Study of industrial use: The experimental study of using the fibrous sepiolite to make nonasbestos friction material has already been successful, and its performance index has met the specifications of the asbestos friction material.

ZHANG Wei, Center for Analysis and Prediction, State Seismological Bureau, Beijing, China

Possibility and Effectiveness of Hydrogeochemical Method Applied to Earthquake Prediction

In this paper the results and experience in earthquake monitoring and prediction by the hydrogeochemical method in the last 20 years in China are outlined, based on a thorough and systematic review and research on the observation techniques, instruments, environments, contaminations, effectiveness of the predictions, the mechanisms of hydrogeochemical precursors, etc. Numerous observation data show that the gas-chemical constituents in the groundwater did vary significantly before and after large earthquakes. Variations in the chemical constituents in the deep and shallow layers of the crust might rapidly propagate to the earth surface through the groundwater. The detection of geochemical precursors to earthquakes have extended the traditional geophysical ways of detecting precursors.

Some progress has been made in earthquake prediction by the hydrochemistry method. But variations in hydrochemical constituents are affected by many factors; the present monitoring system has shortcomings and limitations, and observation and prediction is still at a low level. Existing problems and future research orientation are discussed.

ZHANG Xianqi, and LI Guanghe, Changchun College of Geology, China

Experimental Research on Migration of Pb(II), Cu(II), Hg(II), and As(III) in Soil

In the areas of industry, sewage irrigation, and urban development, the main problem of groundwater pollution is heavy metal and other toxic elements. These elements seriously affect the health of people who live in those areas where groundwater is the main drinking water source. Therefore, while engaged in research on groundwater quality pollution and protection we pay special attention to the pollution mechanism of heavy metal and other toxic elements to groundwater.

Zhengzhou City is the capital of Henan province. With the development of industry and the increase in population, the pollution of heavy metal and other toxic elements such as Pb(II), Cu(II), Hg(II), and As(III) have been the main problems of groundwater pollution. Especially in the sewage irrigation area of suburbs, heavy metals are highly concentrated in the cultivated horizon, and their concentration in groundwater is obviously increasing. Because groundwater is one of the area's important drinking water sources, it is imperative to evaluate the effect of sewage irrigation on groundwater quality. For this purpose, employing a leaching experiment of prepared solutions of Pb(II), Cu(II), Hg(II), and As(III) passing through original soil, we have done a comprehensive study on the migration characteristics of Pb(II), Cu(II), Hg(II), and As(III) in loess aeration zone of west Zhengzhou City and the aeration zone of alluvial sandy clay in the east of Yellow River of the city. The conclusions from experimental research are as follows:

Under the condition of less organic materials and the oxidizing environment of soil, the main mechanism of action that Pb(II), Cu(II), Hg(II), and As(III) are purified in soil is the precipitation of oversaturated heavy metal elements at the state of chemical equilibrium and the heavy metal ions adsorption-displacement between water and soil. The product of dissolution-precipitation and adsorption-disadsorption equilibrium in the water–soil system is the solution that has passed through the aeration zone and reached the aquifer. The content of Pb(II), Cu(II), Hg(II), and As(III) in the irrigation solution is the greatest of these ions in groundwater.

As(III) is the most difficult to adsorb; therefore, it can pass through the aeration zone with its saturated concentration and enter into groundwater. Contrarily, because of being easily adsorbed by soil, concentrations of Pb(II), Cu(II), and Hg(II) in the infiltration solution is very low.

A great deal of PB(II), Cu(II), and Hg(II) is mainly fixed in soil 30–45 cm below the surface, below the accumulation layer. They migrate in forms of $PbOH^{\cdot}$, $CuCO_3^{\cdot}$, $Hg(OH)_2^{\cdot}$, and H_3AsO_3 in the soil, respectively.

Under experimental conditions, Pb(II) and Cu(II) pass through the aeration zone with lower concentration ($n \times 10^{-3}$mg/L) and enter into groundwater. Therefore, if the sewage with Pb(II) and Cu(II) cannot directly reach the aquifer by the abandoned well, sewage river (or canal), sewage irrigation cannot cause the pollution of Pb(II) or Cu(II) in groundwater.

Because As(III) migrates easily in soil, it is not sewage with concentrations of As(II) cannot be used for irrigation. Under the condition of weak alkali, because migration capability of Hg(II) in soil is weak, the migration concentration of Hg(II) is stabilized at around 0.002 mg/L and sewage irrigation cannot cause the pollution of Hg(II) in groundwater. When $[\Sigma Cl^-] > 10^{-3.5}$mol/L in sewage with Hg(II), it is favorable for migration of Hg(II) in soil, and sewage irrigation can cause the pollution of Hg(II) in groundwater.

ZHANG Xiqi, and WANG Mingzhen, Mining Institute of Shandong, China

Series of Palynological Assemblages from Jurassic of North China

The Jurassic spores and pollen in North China may be subdivided into four assemblages: early Jurassic, early middle Jurassic, late middle Jurassic, and late Jurassic.

Cycadopites–Cyathidites *assemblage:* This palynological assemblage of early Jurassic is represented by the Fuxian formation of the Shanxi–Gansu–Ningxia Basin. The assemblage is characterized by the alternative dominance of Gymnospernous pollen and Pteridophytic spores, and the fact that among bisaccate pollen grains prevails the type of incomplete separation of the bladder from the corpus (e.g., *Protoconiferus, Protopicea, Piceites,* etc.). Besides, there exists a certain quantity of *Cycadopites* of *Bennettitales, Cycadales,* and *Ginkgophytes. Chordasporites* and *Taniaesporites* of the old type are found on individual occasions in the assemblage. Rather monotonous are Pteridophytic spores, among which *Cyathidites* and *Deltoidosporites* are more frequent while *Lycopodiumsporites, Verrucosisporties, Dictyophyllidites,* etc., are comparatively rare. The above mentioned assemblage has also been found in the Xiaomeigou formation of the Qaidamu Basin and Daxigou formation of the Minhe Basin of Qinghai, and the Beipiao formation of the Beipiao region of Liaoning.

Cyathidites–Piceaepollennites–Cycadopites *assemblage:* This early middle Jurassic spores and pollen assemblage proceeds from the Yanan formation of the Shanxi–Gansu–Ningxia Basin. Gymnospornous pollen in the assemblage has been distinctly replaced by Pteridophytic spores, among which the contest of *Cyathidites* is the highest, but that of *Lycopdiumsprotes, Neoraistrikia, Osmundscidites Duplexisprites, Klukisporites, Cibtiumspora,* is lower. The Gymnospermous pollen consisting mainly of biasaccate pollen of pinaceae and monocolpate *Cycadopites* is more abundant. Besides, *Inaperturopollenites, Cerebropollenites, Callialasperites, Quadraeculina, Psophosphaera,* etc., are insignificant in quantity, but commonly present in the formation.

Based on the characteristics of the above mentioned assemblage, similar spores and pollen have also been found in the Zhaogou formation of Shiguaizi of Baotou and the Yanan formation of Dongsheng region, Inner Mongolia, the Dameigou formation of the Qaidam Basin, Qinghai, the Datong formation of the Datong Basin, Shanxi, and the Fangzi formation of Shandong.

Cyathidites–Quadraeculina *assemblage:* This late middle Jurassic assemblage is represented by the Zhiluo formation of the Shanxi–Gansu–Ningxia Basin. This assemblage is characterized by the dominance of Gymnospollenites pollen with *Pinuspollenites, Piceaepollenites, and Podocarpidites* as main elements and a higher content of *Quadraeculina.* In addition, *Classopollis* and *Callialasperites* obviously increase in quantity. The Pteridophytic spores mainly contain different general of *Cyatheaceae* and have a certain quantity of *Concavissimisporites* in the assemblage. The above assemblage of the Zhiluo formation has been found in the Zhiluo formation of the Dongsheng region in Inner Mongolia and the Yungang formation of the Datong Basin in Shanxi.

Classopollis–Concavissimisporites *assemblage:* This late Jurassic assemblage has been located in the Anding formation of the Shanxi–Gansu–Ningxia Basin. The assemblage is characterized by the dominance of Gymnspernous pollen grains with the highest content of *Classopollis* reaching 54% and a certain quantity of other *Callialasporites, Podocarpidites, Quadraaeculina,* and *Cerebropollenites.* The Pteridophytic spores that consist mostly of *Cyathidites Concavissimisporites* are in small quantities in the assemblage. A small quantity of *Schizacoisporites* and *Cicatricosisporites* have been obtained from the late stage of late Jurassic. This approximative palynological assemblage has also been spotted in the Hengtang formation of Gansu, the Tuchengzi formation in the west of

Liaoning, the Xiguayuan formation of Zhangsanying of Longhua, Hebei, the Mengyin formation in central and west Shandong, etc.

We have obtained some results in the study of the Jurassic spores and pollen assemblages in North China. Due to the insufficiency of scientific research, the series of the Jurassic spores and pollen assemblages remains a rough generalization and requires further work.

ZHANG Yifei, XU Shanren, WANG Chongwu, Yunnan Bureau of Geology and Mineral Resources, Kunming City, China

The Supergene Enrichment Deposits and Their Effective Exploration Methods of Phosphorite in Lower Cambrian, East Yunnan, China

Abundant practical information from a systematic prospective survey in an area of 31,000 km^2 of east Yunnan and the explorations for several large-scale phosphorous ores around the Dianchi lake of Kunming, have demonstrated since 1980 that the shallow parts of the phosphorite deposits in Lower Cambrian were, in varying degrees, widely weathered because of the influence of internal and external agents, and that the phosphorous-rich ores delineated were almost formed by the supergene enrichment process despite the fact that there was a great disparity in weathering intensity and developed depth. It is worthwhile to point out that many scholars have been using the view of transformed supergene enrichment to demonstrate the primary sedimentary environment, enriched law, and other relative problems of Yunnan phosphorous deposits for tens of years. It seems as if "the facts were covered by some prejudice." Therefore, it is significant to reconsider them both in theory and practice for studying and prospecting the large deposits of supergene enrichment.

The ore types differ greatly from one another including the Deze type phosphorous ore predominated by siliceous content sedimentated in shallow sea, the Nazhang type one predominated by clastic sediments near the paleocontinental margin, and the Kunyang type predominated by carbontite and represented by several large deposits around the Dianchi lake. Although the shallow parts in the first two have undergone weathering in varying degrees, no larger phosphorous-rich deposits have been found in them. The last one consisting of the phosphorite, dolomite, siltstone, and silicalite occurred in the tidal flat environment. Its phosphoric carbonate rocks and ores, being easy to produce the supergene enrichment, are high-quality ores. A few sperficially enriched large deposits explored in Yunnan all belong to this type.

The outcrop line of phosphorous orebeds in Yunnan is 1315 km long, of which the economic ores are 763 km and the supergene enriched ores, approximately 300 km. Although the supergene enrichment is widespread along the outcrops of orebeds, the rich orebodies extend from tens to a few thousand meters in length only with the dip depths ranging from 250 to 350 m (max. 14100 m, min. 10 m more) and the reserves of an individual rich orebody, a few hundred tons to tens of million tons.

The basic factors on the formation of the supergene enriched zone: In general, the phosphorous deposits in Yunnan are divisible into the weathering (oxidation), transitional, and primary zones with their distributions controlled mainly by the water table. The first two are brought together and labeled the supergene enriched zone. The first one is located above the water table, the transitional zone located between the highest and lowest ones, and the last one, below it. Therefore, the properties and the reservoir migration features of the surface filtration and underground circulation water become the essentially external factors in forming the supergene enrichment zone. The alternation of relatively dry and moist climate in Yunnan is more favorable for its formation. The subordinate factors comprise dip angles of the orebeds, topographic relief, slope direction, thickness and lithology of the cover, and extent of the structural development. Relatively flat topography and gentle occurrence are the basic conditions preserving the thick, deep oxidation zone. Gentle consequent slope, especially the topographic slope almost coincident with the occurrences of strata and orebeds, and with a thin covering (only a few to tens of meters) are good environments for the weathering zone to develop.

The internal factors controlling the supergene enrichment are lithology, lithofacies, phosphoric abundance, and the orebed's thickness in phosphoric rock formation. The primary ores in large, mesoscopic deposits ascertained are mostly the calc- or calc (magnesian)-phosphorous ores of 20-25% P_2O_5 and some supergene enriched ores also formed by phosphoric dolomite. The form, occurrence, and changes of the enriched orebeds are interdependent with the primary ones. Wherever there were thick, stable primary orebeds, there would be more chance to form the huge supergene enriched ones. With downward extension of the orebodies, the oxidation intensity and thickness decrease gradually.

The paleo- and recent supergene enriched deposits are similar in genesis, but consolidation of the former is better.

Geologic features of phosphorous orebeds in the weathering zone:

(1) Peculiar minerals, textures, and structures

The peculiar minerals in ores of this zone consist of the brazilianite, epigenetic fluor-apatite, cacoxene, and variscite. The hydromica clay minerals increase, while the dolomite, calcite, crystalline grains decrease distinctly in the weathering zone. The developed degree in it can be reflected by kaolin-clay rock (the sensitive mineral).

The textures resulting from epigenetic leached metasomatism, filling, and special gelation are clearly developed in this zone.

In the weathering zone, decoloration, increase of mineral grain sizes, lattice parameter, and ore porosity are found in collophane. The weight changes for unit volume of ores depend on ore types showing different laws. The Eh and pH values vary with weathering intensity in a regular way.

(2) Variations of chemical components

The increment components in the weathering zone are P_2O_5, F, Al_2O_3 + Fe_2O_3, and effective P_2O_5; the decrement components are MgO and CO_2. SiO_2 and acid-insoluble materials belong to the increment ones. The CaO varies selectively, depending on the relative proportions of P_2O_5 to MgO + CO_2 and mostly showing an increasing trend. In general, with the increase of depths, the useful components decrease, while the harmful ones increase gradually. The progressive rates of increase (decrease) for each component display a certain tendency.

Formation of the supergene enriched phosphorous ore was characterized mostly by chemical action. In this paper, the geochemical specialization of ore components (carbonate, silicate, and phosphate), the leaching, migration, gelation, and centered-deposit features during weathering and relative factors on aggradation and denudation, etc., are discussed in detail.

The qualitative/quantitative indexes discriminating the supergene enrichment ores: In recent years, various experts have attempted to classify the supergene enrichment zones from different sides and taken the chemical components as the discriminant indexes and characteristic value. For instance:

Zhu Longguang uses the CaO/P_2O_5 as the discriminant value in the Chengjiang mine.

The K value was recommended by Wang Dongsheng

$$[K = P/C = P_2O_5 + (CaO + F\text{-}0.84 \times CO_2)/MgO + 1.84 \times CO_2).$$

Xi Chubin provided that one could discriminate whether the ore is the low-magnesium washable weathering one by the characteristic value b in different deposits and ore types, which is determined on the trivariate (silicate, carbonate, and phosphate) phase diagram after the mineral compositions are calculated from the chemical ones. The formula is: b = P + S = P_2O_5/K + A.I, where b = discriminant value, P = phosphate mineral, S = silicate mineral, P_2O_5 = percent content, A.I = content of acid insoluble material, K = analysis value for pure phosphate in ore from various mines.

The authors draw on others' strong points (zoning by chemical components) combining with the physical properties to systematically demonstrate a comprehensive classification plan applied.

A set of methods suitable for exploring the supergene enriched deposits has been grouped in recent years.

Unfolding the prospect survey for phosphorous resources: Systematic specimens are collected along the control sections (2-3 km interval for general phosphoric zones, 0.5-1 km for rich ones) in order to find out the lithology, facies. ore grades, and structures, and thus get the macriscopic understanding, and their scale. On this view, the Malu, Jianshan, Jiangchuan, Chengjiang, Anning, and Deze mines have been predominantly explored, and the auxiliary engineerings have been properly densified on the basis of normal net density (generally, the interval in dip is two times more than that in the strike direction).

The "cleaning" results for supergene enriched ores: The experimental data from tens of "cleaning deargillation" specimens show that the P_2O_5 content in ores of rank I and II can rise 2-6%, meanwhile the R_2O_3 in them drops 1-3%, and the filtering intensity also improves, reaching a better effect. It is expected that this type of ore might become the main resource of domestic composite phosphorous fertilizer with high efficacy in this century.

The reserves of supergene enriched phosphorous ores in Yunna are up to thousands of million tons, of which the rich ores reach 1000 million or more and centralize around the Dianchi area.

ZHANG YiGang, Ministry of Geology and Mineral Resources, Beijing, China

Classification and Correlation of Natural Gas and Source Rock with Their Application to Eastern China

The classification and correlation of natural gases and their source rocks of eastern China are studied in depth. Gases are genetically classified into abiogenic deep-seated gas, thermogenic gas, biogenic gas, shallow-mixed gas, and deep-mixed gas, based on carbon isotopic compositions of individual C1 to C5 hydrocarbons. The thermogenic gases are further classified into marine sapropelic type, nonmarine sapropelic type, and humic type according to their C1 to C9 hydrocarbon distribution.

(1) Abiogenic deep-seated gas is found in the East China Sea and could be related to the activities of the subduction zone. It is characterized by $\delta^{13}C_1 > -20$ o/oo and $\delta^{13}C_1 > \delta^{13}C_2 > \delta^{13}C_3$. (2) Nonmarine sapropelic thermogenic gases of eastern China are mostly of associated gas that occurs in the Lower Tertiary of North China and Lower Liaohe Basin and the Cretaceous of Songliao Basin. It is characterized by $\delta^{13}C_1 = -40$ to -52 o/oo, $\delta^{13}C_2 < -29$ o/oo, much normal alkanes, less isoalkanes, and less cyclohexane homologues (iC5/nC5 <1, cycC7/(n+i)C7 > 0.5, MCH/nC7 <4, where MCH stands for methylcyclohexane). (3) Marine sapropelic thermogenic gases of eastern China are exclusively linked with carbonate sequences occurring in the Sinian and Lower and Middle Triassic of the Sichuan Basin and the Ordovician of the Ordos Basin. They are highly mature to overmature and are characterized by $\delta^{13}C_1 = -28$ to -38 o/oo, $\delta^{13}C_2 < -29$ o/oo, less n-alkanes and less cyclohexane homologues (iC5/nC5 ≥ 1, cycC7/(n+1) C7 < 0.5, MCH/nC7 < 4). (4) Humic thermogenic gas of eastern China occurs in coal measures, mainly nonmarine, i.e., the Permo-Carboniferous of north China an the Ordos Basin, the Upper Triassic of the Sichuan Basin, the Lower Cretaceous of southern Songliao Basin and the Tertiary of the East China Sea. Their maturity ranges widely from low mature to overmature. They are characterized by $\delta^{13}C_1 = -30$ to -40 o/oo, $\delta^{13}C_2 > -29$o/oo, less n-lakanes and much cyclohexane homologues (iC5/nC5 > 1, cycC7/(n+i)C7 >0.5, MCH/nC7 > 4). (5)Biogenic gas is found in the Tertiary of Huanghua Depression and the Quaternary of Changjiang River Delta. Not even its $\delta^{13}C_1$ is very light (<-52 o/oo), but its $\delta^{13}C_2$ is also light (<-30 o/oo). (6) Shallow-mixed gas is formed by mixing biogenic gas with thermogenic gas with its $\delta^{13}C_1$ very light (<-52 o/oo) and its $\delta^{13}C_2$ heavy (> -29 o/oo). It is encountered in drilling wells in north China. Additionally, most seam gas can be assigned to this kind of gas. (7) Deep-mixed gas is formed either by mixing of abiogenic gas and thermogenic gas or by mixing of early thermogenic gas generated before uplift and later thermogenic gas generated after resubsidence. The later case occurs in the Permo-Carboniferous carbonates of the Sichuan Basin. It is characterized by its $\delta^{13}C_1$ close to or even heavier than its $\delta^{13}C_2$, but its $\delta^{13}C_2$ still lighter than its $\delta^{13}C_3$.

Gas and gas could be correlated with each other on the basis of their source material and maturity. Even the source materials of the same type may have different carbon isotopic composition and hydrocarbon distribution. Using $\delta^{13}C_1$ and the difference between $\delta^{13}C_2$ and $\delta^{13}C_1$ combinatively, the gases of the Sichuan Basin representing one of the most complicated cases in the world are clearly divided into six groups, i.e., Jurassic, upper complex of Upper Triassic, Lower complex of Upper Triassic, Middle Triassic, Permo-Carboniferous, and Cambrian-Sinian. The ratios of methylpentane to methylcyclopentane, methylcyclohexane to dimethylcyclopentane plus ethylcyclopentane, etc., could also be effectively used for the correlation purpose. The result of thermocatalytic experiements reveals that the maturity of a gas expressed as the H/C atomic ratio of its source material when it was generated could be determined according to the differences between $\delta^{13}C_1$, $\delta^{13}C_2$, $\delta^{13}C_3$, and $\delta^{13}C_4$. Therefore, these differences could be used for correlation purposes, too.

Direct gas to rock correlation is not an easy task. A technique was developed to analyze the $\delta^{13}C_1$ to $\delta^{13}C_4$ of the residual gas in the rock by acidolysis. Good correlation was made for the Ordovician carbonates of the Ordos Basin, the Permian carbonates of the Sichuan Basin, and the Lower Tertiary clastic rocks of north China. In other cases, the situations are more complicated, and detailed correlation integrated with diagenesis study is needed. For gas/rock correlation, the ratios of hydrocarbons of the same homologue should be used, i.e., the ratios of methylbutane to dimethylpropane, ethylcyclopentane to dimethylcyclopentane, ethylcyclohexane to dimethylcyclohexane, etc.

The classification of source rock is based on the estimation of the amount of gas that it already generated in geological history. Thermocatalytic experiment was advanced in geological history. Thermocatalytic experiment was advanced to become a routine technique to analyze the residual gas productivity of a rock. On the other hand, simulated gas generation experiments were conducted to establish a nomogram for the calculation of historical gas productivity from residual gas productivity (HGP). Accordingly, a source rock could be ranked as excellent (HGP > 6 m3/T) good (> 3 m3/T), fair (> 1 m3/T), poor (> 0.1 m3/T) and nonsource rock (<0.1 m3/T).

ZHANG Zhongying, Nanjing University, Nanjing, China

Earliest Bryophyte-Like Fossil from Late Precambrian of North China

A large number of round, oval, elliptical, pyriform, and fan-shaped carbonaceous films with stalk-like projections at their proximal ends occur in shales of the late Precambrian Changlongshan Formation (900–840 Ma) in the Yanshan Range of Hebei Province, North China. Du (1982) first studied these specimens and established a new genus *Longfengshania* for them. Hofmann (1985) described bundled specimens of *Longfengshania* exhibiting subradia arrangement and superposition from the mid-Proterozoic Little Dal Group in the Mackenzie Mountains, northwest Canada. *Longfengshania* was regarded as an algal fossil (Du, 1982; Du and Tian, 1985, 1986) or possibly an alga of phaeophyte or rhodophyte affinities (Hofman, 1985). No vascular tissues have been found in either Chinese or Canadian specimens.

Over the years I reexamined the rather vaset collections of *Longfengshania* at the Hebei College of Geology gathered by Du Rulin and his colleagues. When comparing the morphology of *Longfengshania* with that of some early fossil bryophytes, it is clear that there is a basic similarity between the genus and the lower Devonian fossil *Sporogonites,* which consists of stalks about 5.0 cm long that terminate in an elongate capsule. Andrews (1960) believes that *Sporogonites* had attained a bryophytic level of evolution, possibly related to the liverworts or representing a distinct and independent group. Another fossil plant that morphologically resembles *Longfengshania* is *Tortilicaulis* from the late Silurian (Downtonian) of South Wales, which consists of elongate, fusiform to oval bodies, thought to represent sporangia, attached to smooth, stout axes. It is interesting that *Tortilicaulis* appears remarkably similar to the modern bryophyte *Pellia.*

It is thus tentatively assumed that *Longfengshania* probably represents the sporophyte of bryophytic affinities consisting of a slender sporangiophore (seta) that bears a terminal sporangium (capsule), and has an occasionally preserved foot at its base, which attached it to the gametophyte. The common carbonaceous mass in Canadian specimens (Hofmann, 1985, pl. 38, fig. 4, text-fig. 5) might be interpreted as representing the remainder of the gametophyte. If this alternative interpretation can be accepted, *Longfengshania* does represent an earliest record of bryophyte-like fossils, possibly related to the liverworts. In other words, the fossil record with probably bryophytic affinities can be traced with confidence back to the late Precambrian. Conceivably, *Longfengshania* evolved from aquatic ancestors, probably from the Chlorophyta (or Phaeophyta?, Rhodophyta?), and represents the transition from an aquatic to a terrestrial habitat. It stands to reason that discovery of *Longfengshania* in the late Precambrian implies that the first appearance of the bryophytes on the earth might be much earlier than was expected before, and also supports the view that the bryophytes evolved from algae.

The systematic description of *Longfengshania* and its type species, *L. stipitata* Du, 1982, is emended here. The emended diagnosis of the genus is as follows:

Carbonaceous compression interpreted as sporophytes consist of three parts (not always fully developed): a round, oval, pyriform, ligulate, or gourd-shaped terminal sporangium (capsule); a slender sporangiophore (seta); and a tuberous or discoid basal foot. Sporophytes attached directly to a common carbonaceous mass (gametophyte?). No conducting tissues observed.

REFERENCES

Andrews, H. N., Jr., 1960, Palaeobotanist, v. 7, p. 85–89.
Du Rulin, 1982, Geological Review, v. 28, p. 1–7.
——— and Tian Lifu, 1985, Precambrian Research, v. 29, p. 5–14.
——— ——— 1986, The macroalgal fossils of the Qingbaikou Period in the Yanshan Range: Hebei Science and Technology Press, p. 1–114.
Hofmann, H. J., 1985, Palaeontology, v. 28, p. 331–354.

ZHANG, Y. G., D. HICKMOTT, J. D. FRANTZ, and T. C. HOERING, Carnegie Institution of Washington, Washington, DC

Use of Zeolites and Silica Gel for Loading Gases in Hydrothermal Experiments

An interesting property of the synthetic zeolite, molecular sieve 5A, is that at low temperatures it can sorb large quantities of noncondensible gases such as N_2, Ar, CH_4, and O_2. These adsorbed gases are released with rising temperature, and, under hydrothermal temperatures and pressures, the zeolites break down to a feldspar and water. Silica gel exhibits similar properties and recrystallizes to quartz under hydrothermal conditions.

Because of the large adsorption capacity of gases by zeolites and silica gel, they can be used to load common gases or gas mixtures in noble-metal capsules used in hydrothermal experiments. Possible experimental uses of this gas-loading technique include: (1) studies of metamorphic phase equilibria over the full range of naturally occurring fluid compositions, (2) diffusion studies of gases into minerals, and (3) studies of supercritical mixed-volatile fluids using synthetic fluid inclusion techniques.

To test the technique, a piece of fractured quartz and small amounts of zeolite and silica gel or just silica gel (approximately 100 mg) were added to a gold capsule. The capsule was connected to a gas pipette containing a known volume of gas. The gas or gas mixture was then condensed into the gold capsule by cooling with liquid nitrogen. After the volatiles were drawn into the capsule, the capsule was crimped, cold-welded, and then electrically welded. The capsule was heated under confining pressure in a cold-seal hydrothermal pressure vessel (500 °C, 2000 bars). After quenching experiments containing zeolite, the phase was found to have reacted with the silica gel to form feldspar, thus releasing water. Synthetic fluid inclusions containing the gas mixtures were found to have formed in the fractured quartz prism. Gas compositions in the capsule were measured using gas chromatography.

The experimental studies of synthetic fluid inclusions, the phase equilibria of gas mixtures, and the equilibria between gas mixtures and mineral assemblages have long been hampered by difficulties in loading known amounts of volatiles in hydrothermal charges. The technique presented above should greatly facilitate further advances in these areas.

ZHAO Pengda, China University of Geosciences, Wuhan, China

Theories and Principles of Mineral Deposits Prediction in Large Scale

Increasing need for mineral resources and increasing difficulties in prospecting have promoted the development of scientific prospecting. Scientific prospecting includes aspects of theoretical prospecting, composite prospecting, stereo prospecting, and quantitative prospecting. The prediction of mineral deposits in large scale according to geological mineralization theories relies on composite information from three-dimensional spatial and quantitative assessment. Prediction of mineral deposits in large scale fully reflects the demands of scientific prospecting.

The theories and principles of mineral deposit prediction in large scale are discussed in this paper as follows:

1. Similarity-analogy theory in prospecting: Similar deposits should develop in similar geological environments; the same quantity of mineral resources should be contained in the same size of predicted areas.

2. Theory of finding differences and anomalies: Geological anomalies may be the essential condition of forming the special type and new-type or scope deposits. Therefore, we must not only make analogies but also find differences and anomalies.

3. Theory of quantitative combination factors controlling the formation of mineral deposits: The quantitative combination of metallogenic factors are the essential and ample conditions of mineralization.

4. Principle of composite prediction: Mineral predictions are carried out by making use of synthetical information; every paragenetic and associated ore is predicted compositely.

5. Principle of identification between the prediction scale and variable scale: The scale of geological, geophysical, and geochemical factors and variables used for mineral prediction should be suited for the prediction scale.

6. Principle of using the distribution laws for mineral values: The distribution laws of geological, geometrical, and economical parameters, i.e., the size of ore deposit, the grade of ore, the total value of deposit, etc., are significant for the prediction of the same kinds of deposits.

7. Principle of quantitative prediction: The locations of minerogenetic resources, the probability of finding deposits, and the prospecting information and the favorable data intervals for prospecting in controlling the formation of mineral deposits are determined.

8. Principle of assessment: The prediction precision and reliability, the permutation of prospecting advantage, and the economy in minerogenetic prospective region should be evaluated.

9. Principle of discovery rate analysis: Analysis of the potential of finding the new deposits and the degree of exploration and investigation in prospective regions.

ZHAOLI, S., Z. ZUOXIN, and W. MIN, China University of Geosciences, Beijing, China

On Acid Rain in China

About 90% of acid rain with pH<5.6 arises in the south area of the Yangtze River, especially in the city of Chongqing, Sichuan Province and Guiyang, Guizhou Province in China. The average pH values of acid rain in the two cities are 4.14 and 4.02 respectively. Both of the two cities' frequencies of acid rain are above 80%. On the contrary, acid rain seldom occurs in the cities of northern China. The annual average pH values of precipitation in Beijing in 1981–1983 were all above 7. There is acid rain in only a few northern cities. Qingdao is a typical example. In 1981–1983 its average annual pH values of precipitation were 6.4, 4.98 and 4.13 respectively and the annual frequencies of acid rain were 16.7%, 98.3% and 87.5%. The ionic concentrations in the acid rain water generally occur as follows: ${SO^{2-}}_4 > Cl^- > {NO^-}_3 >$ FT $Ca^2 > NH^+ > {M^{2+}}_9 > K^+$. The acidity chiefly comes from sulphuric acid. The question is that in some places, especially in the northern cities, although the sulphur dioxide contamination is serious, there is little acid rain. The explanation is this: the formation of acid rain is related not only to SO_2 air pollution, but also to the nature of the atmospheric particles. The situations of Beijing (non-acid rain region of north China) and Guiyang (serious acid rain region of south China) are an example. The average initial pH values of the atmospheric particles in the two cities are 8.2 (Beijing) and 5.72 (Guiyang). The former is obviously bigger than the latter. So the buffering capacity with respect to acid rain in Guiyang should be much lower than that in Beijing. This could be the main reason why the Guiyang region has the most serious acid rain in China. The formation of acid rain is connected closely to the geographical environment. Generally, the soil of the north area is alkaline while the soil of the south area acidic in China. For instance, the pH of Beijing soil equals 7–8 but the pH of Guizhou soil is not higher than 5.5. Two-thirds of the total atmospheric particles come from the dust in the wind in north China. However, it is only one third in south China. This explains why the pH value of the atmospheric particles in the south is lower than that in the north. In the small city of Ji'an in the south, industry is not developed, its concentration of gaseous pollutants is very low. Sulphur dioxide and nitrogen oxide are only 0.013–0.034 mg/m^3 and 0.013–0.031 mg/m^3 respectively. But the frequency of acid rain is as high as 84%. The soil of this city is red clay with a pH range of from 4.5 to 5.5. Owing to the high frequency of the wind, the atmospheric particle concentration is rather high and equals 0.577–0.612 mg/m^3. Therefore the formation of acid rain is evidently related to the atmospheric particles coming from the red clay. The washout of the acid particles by the rain water in the atmosphere is the major cause of acid rain. Acid rain is not caused completely by the industrial air pollution. This typical example shows that the weathered products at the earth's surface have significant influence on the formation of acid rain.

The research project entitled "The Influence of Acid Rain on the Ecological System" has been carried out for the past 3–4 years in the provinces of Sichuan, Gangxi, Zhejiang and Guizhou in south China. Following are the results of this project: (1) Above 65% of the branches and twigs in the tested plants are acutely harmed when the rain's pH value is 2.0–3.–. (2) As the rain water comes through the forest crown cover, most ionic concentrations in the water increase and some kinds of the plants are directly influenced; the chloroplast is injured, the nutrient transmissibility of the branches becomes decreased, the root system is undeveloped and the nutrients in the crown cover are lost. (3) The direct influence on the crops doesn't seem so serious; wheat, soybean and Chinese cabbage are hurt only when the rain's pH<; likewise with rice plants when the pH<2.0. (4) Soils are more directly influenced; with the rain's pH<4.0, the amount of leaching out of ions greatly increases. The concentration of exchangeable aluminum becomes much higher and the exchangeable calcium is leached out. The soils will at last inevitable become acidic. (5) Different soils have different adsorption characteristics to ${SO^{2-}}_4$; the tawny soil and the grey-brown soil obey the Temkin equation; the Quaternary red soil is in keeping with the linear equation; and the yellow soil, the crimson soil and the forest soil conform to the Freundich equation. The SO ${SO^{2-}}_4$ concentration in the leaching out solution is controlled by the dynamic equilibrium of SO ${SO^{2-}}_4$ in the soil and in the rain water. It is the mobility of SO ${SO^{2-}}_4$ that results in the leaching out of a large number of the cations.

The influence of acid rain on ancient buildings including grottoes built in the sandstone, granite and limestone, the influence on the geological environment, and the synthetic influence on the total ecological system are all being studied.

ZHENG Baoshan, and HUANG Ronggui, Institute of Geochemistry, Chinese Academy of Sciences, Beijing, China

Fluorosis in Inhabitants of Southwest China and Environmental Geochemistry

Endemic fluorosis is now well known in many countries and districts of the world. It also harms the health of numerous Chinese people. To date, most of the fluorosis cases reported in various countries are caused by drinking groundwater of high fluorine concentration. However, the fluorosis cases found in inhabitants in Southwest China are caused, upon the authors' investigations, by food and air pollution resulting from burning of coal indoors. The authors have made a systematic study of the environmental features in the areas of this type of fluorosis and on the pathogenetic ways and rules.

Two different viewpoints about fluorine source: There are two different viewpoints regarding the source of the fluorine. One holds that the rocks and soils in the areas of fluorosis in Southwest China are characterized by high fluorine contents, resulting in the concentration of fluorine in grains and causing fluorosis. Another viewpoint claims that fluorine comes from the coal smoke. In order to clarify the problem of fluorine source, the following work has been done.

1. The average fluorine content of 22 fresh wheat samples and 102 fresh corn samples (both sets of samples were collected from the fluorosis areas) has been determined to be 0.38 and 0.41 mg/kg, respectively, showing no

evident difference from the grains from other areas in China.

2. A total of 102 profiles of rock-soil-corn and 22 profiles of rock-soil-wheat were studied with the result that no correlation exists among the fluorine contents of rock, soil, and grains. Even the corn that was grown in phosphorous ore stratum with as high as 30,000 mg/kg fluorine yields a fluorine concentration of only 1 mg/kg.

3. A total of 139 rock samples representing various times and various lithologies were systematically collected from fluorosis areas. The fluorine concentrations in the samples and their corresponding distribution areas have been analyzed and calculated. The results show that the average fluorine content in rocks from fluorosis areas is less than that in the earth's crust. The analysis indicates also that the fluorine content in soils from the areas of interest is higher than the average fluorine content for soils in the world.

4. Simulation experiments were done to explore the origin of high fluorine food. We hung corn that had been collected from one field in the houses of two groups of families. One group burned coal and the other burned wood. After a certain period of time, the corn was taken down for analysis. The results indicate that the fluorine content of the corn hung in the houses of coal-burning families had increased markedly, while the fluorine content in the corn hung in the houses of wood-burning families remained basically the same. The experiments also revealed that the concentration process of fluorine from air in grains and vegetables is not a simple process of mechanical contamination or absorption, but a physiologico-biochemical process similar to respiration of plants. At certain temperatures, the greater the water content in grains and vegetables is, the quicker the absorption of fluorine from air.

Based on the above work, we analyzed the components of the intake of fluorine by the inhabitants in fluorosis areas. For example, in Zhijin County, Guizhou Province, the average intake of fluorine per person per day is 9.7 mg, and 81.6% of it originates from pollution by coal smoke; only 18.4% (1.78 mg) comes from the natural background. In addition, over 90% of the total intake of fluorine comes from grains, vegetables, and peppers that have been contaminated by fluorine contained in coal smoke.

Therefore, it is evident that the fluorosis in inhabitants of Southwest China is caused by two correlated processes: first, the indoor air is polluted by the fluorine contained in coal smoke; second, the polluted air further contaminates grains, vegetables, and other foods that are stored indoors.

Distribution of coal-polluted type fluorosis and related environmental geochemistry: The altitude, quantity of coal burned, method of coal burning, fluorine content in coal, proportion of corn in staple food, living style, and annual income may all affect the distribution of fluorosis areas and the seriousness of the disease. Of course, there exist certain interrelationships among all these factors.

In Southwest China the vertical differentiation of climate is very significant. With increasing altitude the precipitation becomes greater, the temperature becomes lower, and the weather becomes more humid, thus the overcast and rainy period in fall and winter becomes longer. In such a case, the period of heating becomes very long and the quantity of coal burned for heating becomes very great, resulting in a long-term high fluorine concentration in indoor air. At the same time, in such environments the most appropriate grain to grow is corn. However, the harvest of corn is coincident with the overcast and rainy period, so the inhabitants have to dry their corn indoors with coal heat. Because fresh corn has an exuberant metabolism, it strongly absorbs fluorine from the air during drying.

The typical climate in such fluorosis areas is a subtropical warm and wet climate, and the regional soil is yellow soil. Thus, the areas can be assigned to iron- and aluminum-rich wet geochemical environment. The annual accumulated temperature is between 3500 and 4500 °C; the annual precipitation ranges between 1000 and 1500 mm. The aridity is about 0.5–0.6, and the altitude is usually over 1000 m above sea level.

Relation between fluorosis and high fluorine coal: If other factors are equal, the higher the fluorine content in coal, the more serious the fluorosis in an area. The average fluorine content in the coal produced in Southwest China is about 200 mg/kg, higher than the average for coal in the world. According to the authors' results, coal with higher ash proportion also contains higher fluorine. The major source of coal in Southwest China is the Longtan Group of Permian System. The bottom of this coal series contains higher fluorine, which is probably associated with the eruption of Emeishan basalt. The investigation shows that for all fluorosis areas where the handicapped patients total more than 5%, the fluorine content in coal is exclusively higher than 500 mg/kg.

Conclusions: Necessary attention should be paid to finding similar type of fluorosis in other districts in the world. In particular, special attention should be paid to the hazard of fluorine to our environment.

ZHENG Jiandong, Institute of Geology, Beijing, China

A Preliminary Study on Geodynamics of the Tibetian Plateau

The Tibetian plateau, the highest, and youngest plateau in the world, has an average elevation about 4500 m, crustal thickness about 50 km, covers an area about 3.3 million km2, and intensively uplifted during the later Cenozoic Era. It is also the most intensively active seismic region in China, in which 345 earthquakes with $M \geq 6$ have happened from 192 A.D. to 1986. Based on geophysical exploring, Tibet is a relatively soft lithosphere with inhomogeneous features both in horizontal and vertical, and its tremendous crustal thickness and high elevation plateau plays an important role in geodynamic evolution in the Asian continent.

ZHENG Jiandong, State Seismological Bureau, Beijing, China

Altun Tagh Fault

The Altun Tagh fault, a famous left-lateral strike-slip fault in the Asian continent, originates from western Lazhulong and extends eastward to the Jinta basin, the total length of which is about 1600 km. Based on field investigations, the Altun Tagh fault is very active and complex, representing not only strike-slip and compressional features, but illustrating seismic activities as well. The fault's principal strike-slip movement occurred in the early Cenozoic Era during the collision between the Indian and Eurasian plates contemporaneously. After collision, the total displacement of the Altun Tagh fault is about 280 km. The fault is also an active boundary of the Asian continent, separating the Tarim and Qiadam basins, playing an important role in geodynamics of the Tibetan plateau.

ZHOU Huitang, YANG Qi, LI Baofang, LI Zhen, China University of Geosciences, Beijing, China

Depositional Environments and Coal-Bearing Characteristics of Coal Bed of Shanxi Formation in Shanmian Coalfield, Western Henan

Shanmian coalfield, which is located in the western part of Henan Province, is one of the major coalfields in the southern zone of the Late Palaeozoic coal-forming basin of north China. The coal-bearing strata consist of the Taiyuan Formation of the Late Carboniferous, the Shanxi Formation of the early Permian, and the Shihezi Group of the late Early Permian to early Late Permian.

Coal B1, located in the lower part of the Shanxi Formation, is the main minable coal bed in the study area. The deposits underlying coal B1 bed are mainly formed on the shoreline and in the shallow sea. They can be divided, from bottom to top, into four members: Bottom Bauxitic Mudstone Member, Lower Clastic Rock Member, Middle Limestone Member, and Upper Clastic Rock Member. The first three members are part of Taiyuan Formation, the last member belongs to top of the Taiyuan Formation and the lower part of Shanxi Formation. On the lower Clastic Rock Member, the less important coal A3 is developed. The Bottom Bauxite Mudstone Member is believed to be formed in coastal lakes on the strandplain. The Lower Clastic Rock Member is interpreted as deposits of open coast with clastic tidal flats and associated tidal channels and tidal creeks, whereas those of Upper Member are mainly deposited on microtidal coast with extensive clastic tidal channels (medium-to-fine sandstone, fining-upward sequence, sandbody shows branching from south to north land direction in plan view and lenticular shape in trend section) and lagoon flats (grey silt mudstone with abundant pyrite nodules, interbedded bedding, lenticular, and flaser bedding, and bioturbate structure), undeveloped barrier islands (medium-to-fine quartz sandstone, lower angle planar cross-bedding). In the light of microfacies, sedimentary structures, and fossils preserved in the limestone, the Middle Limestone Member is considered as intertidal and low-energy subtidal zone deposits of the epicontinental sea, which constitute a deepening upward sequence.

The middle-upper part of the Shanxi Formation (which overlies the coal B1) is regarded as fluvial-dominated, tide-influenced delta system filling in a shallow bay. It is composed of extensive developed distributary channels (with abundant mica in bed plane, sandbody shows branching and thinning from north to south in plan view and lenticular in shape in trend section, paleocurrent is mainly single direction, some with reverse direction), interdistributary bay-tidal flats (interbedded bedding, small ripple cross-bedding, horizontal and incline biologic hole) and less developed subaqueous delta plan (distributary mouth bar, distal bar, and prodelta). On the whole, the deposits overlying coal B1 show a regressive sequence from the deposits of subaqueous and lower delta plain (middle upper part of the Shanxi Formation) toward those of upper delta plain (Shihezi Group).

The thickness and quality of coal bed are directly affected by the precedent and contemporaneous depositional environments. Coal bed (coal B1)

formed on barrier-lagoon-tidal flats of microtidal coast tends to be thick (0-14.10 m, mean 3.66 m), laterally continuous, and high in sulfur (2.5%-4.0%, mean 3.91%). Conversely, delta plain coals are thin (0-3.22 m, mean 0.52 m), laterally discontinuous, and low in sulfur (0.60-0.73%, mean 0.66%). Coal bed A3 formed on progradational open coast tidal flats is medium in thickness (0-2.22 m, mean 0.68 m) and highest in sulfur (1.74-12.26%, mean 4.66%). As far as a single coal bed, such as coal B1 is concerned, the thickness, laterally extent and quality of coal bed are mainly dependent on sedimentary environments. Where the tidal channels developed, the overlying coals tend to be thin (0-3 m), low in sulfur (0-2.5%), but high in ash (higher than 40%). Where the lagoon-tidal flats developed, the overlying coals tend to be thick (thicker than 3 m), high in sulfur (higher than 2.5%), but low in ash (10-25%). The thickness and lateral extent of coal A3 are mainly affected by the present depositional environments. Where the tidal channels and low-lying land on the tidal flats developed, the overlying coal tends to be thin (0-0.5 m). Conversely, the coal overlying other parts of tidal flats is thick (thicker than 0.5 m). The sulfur of coal A3 is mainly affected by the contemporaneous and postdepositional environments. The coal with thin mud limestone s apparently high in sulfur. The widespread high sulfur and ash of coal A3 result from the deposits of deep-sea water carbonate.

ZHOU X., and CAO Y., Bureau of Geology and Mineral Resources of Tibet Autonomous Region, Lhasa, China

Plate Tectonics and Its Evolution of Xizang (Tibet): Explanation of Plate Tectonic Map of Xizang (Tibet)

The plate tectonic map of Tibet is a regional geotectonic map that shows the tectonic evolution and the crust structure in Tibet from the viewpoint of tectonic lithofacies.

The map has revealed that Tibet Plateau is a huge collision accretion body between north and south continents, being formed by combination of four slabs. In Mesozoic, these four slabs formed two pairs of continental margins (folded zones) one after the other and were respectively confined to Jinshajiang and Yarlun Zangbo suture zones. Jinshajiang suture zone is the obvious trace of Paleotethys, but it does not represent the north boundary of early Gondwanaland. Actually, it is the combination of a series of small oceans developed in the southern margin of Cathaysian composit oldland, probably having the property of a transform fault tectonic zone. Yarlun Zangbo suture zone, where the main domain of Neotethys was, may have a sutural texture with two natures. One is that today's ophiolite zone there is mainly the trace of the secondary spreading small ocean basins developed on the Cretaceous Gangdise continental margin; the other is that the Triassic–Liassic flysch well developed on the southern side of the ophiolite zone may provide information about the development of the eastern Tethyan "wedge-like ocean" coexisting with the "joint continent." The Bangongcuo–Nujiang suture zone, which is located in "Kimmeridgian intermediate continent" (Sengör, 1979), is actually the trace of apron-arc shaped marginal sea that once occurred in the Neotethys area. The east segment of this zone might have partly superimposed the fore-deep caused by Paleotethyan closure, thus leading to comparatively complex structure.

The key problem to be solved in the study of the geotectonic evolution of Tibet Plateau relates to the northern boundary of Gondwanaland. Through analysis of the slab structure for mapping, this boundary has been found to exist in the composite slab of Qiangtang–Sanjiang. The slab, having the character of a composite terrain, is made up of terrain of two types that come, respectively, from the southern margin of Cathaysia and the northern margin of Gondwanaland. In the zone, the trace of Paleotethys that represents the north boundary of Gondwanaland is Lancang River–Middle Malayan suture zone in west Yannan. The main mark of this zone is a huge tectonic-magmatic complex zone of island arc-overthrust composite type, the north-extending part of which just the Taniantaweng island chain zone in Tibet Plateau area. On the north side of the island chain zone existed a Mesozoic depression (Tanglha Depression); it was likely to have the characteristic of the backdeep depression and might have concealed the trace of the early oceanic crust zone. The history of its corresponding area can be traced back, according to the information obtained from the research of the paleoisland arc, to Early Paleozoic or Late Precambrian. This indicates that there once existed a "Proterozoic ocean" between Gondwanaland and Yangtze paleocontinent, and the evolution of Tethys just began with the gradual subduction of this oceanic crust. The subduction zones of both sides show a distribution pattern that reflects a probable sinistral transform convergence movement of Yangtze paleocontinent to Gondwanaland. There were two important turning points in the course of the following plate movement: One was an oblique collision between the southern segment of Cathaysia and the northeastern margin of Gondwanaland in Late Permian; the other, a collision of Indian block with the Asian continent in Eocene–Oligocene. The two collisions have marked two important changes in the framework of the plate movement in this area.

Through study of the plate movement history in Tibet Plateau and its adjacent area, a model has been provided by this map about the mechanism of the Tethyan plate movement. That is, from the viewpoint of geotectonics, Tethys was actually a kind of transform tectonic system that occurred near the equator (middle and low latitudinal region). It was formed by transforming the early circum-oceanic margins by cutting, just as in the process occurring in the present Caribbean sea region. The background of this transform tectonic system has been the asymmetric distribution of oceans and continents in Southern and Northern Hemispheres, and the periodical change of this asymmetry in geological history. Although each continent had its own drifting characteristics, there was a general tendency that they were all restricted by the periodical change of this axial asymmetry, which probably depended on the pattern of the global mantle convection. Obviously, it is involved with an abstruse geodynamic mechanism, which is probably related to the movement of the solar system around the center of the galaxy. As a transform tectonic system derived in the equator area, Tethys performed two tectonic functions: One was to regulate two opposite tendencies—divergence and convergence—displayed respectively in Southern and Northern Hemispheres during the continental drifts; the other, to regulate the difference of rotational inertia caused by the axial asymmetry of the two hemispheres. It is certain that further study of this mechanism will help to develop the hypothesis of plate tectonics and the theory about global tectonic genesis.

ZHOU XINHUA, Academia Sinica, Beijing, China, and Zhu Bingquan, Academia Sinica, Guiyang, Guizhou Province, China

Chemical Mapping of Subcontinental Mantle Beneath East China: Further Evidences for the Multiple Components

The complexity of subcontinental mantle as revealed by the global study has become an attractive topic in mantle geochemistry. In the last decade, enormous reservoirs beneath east China using Sr-Nd-Pb isotope and geochemical systematics. A three-component model proposed by the authors is able to fit most of the available data, however, it leaves unsolved problems. Recently, attention has been drawn to the recognition of various end-members or components in their sources. Two regions, northeast China and the southmost China (Guangdong and Hainan provinces), have been selected for this purpose. Based on more than 100 new data and the summary of the published data, a variety of components have been recognized as follows: LOMU/EM1, Quasi-PUM, DM)A), ASZ, and the PREMA component.

LOMU (Low %)/EMI, represented by the K-rich basalts from the Wudalinchi area, northeast China, is characterized by the enrichment of nonradiogenic Pb (206/204< 17.0, 207/204< 17.5 and 208/204< 37.0) and radiogenic Sr and Nd. It should be emphasized that the recognition of the low % component opens a new approach to solve the long-existing Pb paradox. In addition, geochemica mantle metasomatism in this component, which is further suolite xenolith from the surrounding area.

Quasi-PUM is represented by the Jingpohu basalts, northeast China. Their Sr-se to that to Bulk Earth, or BSE.

DM(A) is represented by the basalts from Guangdong and Hainan provinces. It shows the most depleted signature, with respect to the Sr-Nd-Pb system, in continental basalts in the world.

ASZ (recycled ancient subduction zone) is recognized from both northeast China (Changbaishan area) and south China (Shanshui basin). The Sr-Nd system of ASZ is more or less similar to that of a modern island arc, indicating the contribution of recycled and subducted materials.

PREMA is the major component of the dominant rock type of Cenozoic basaltic rocks in east China, Na-Alk (Na-al basalt). The isotope and geochemical system of this component is basically close to that of some well-documented OIB, such as Hawaii and Iceland.

Based on above results, a general secular trend in the composition of mantle has been observed. There is a transition from enrichment (LOMU/EM and PUM) in the north, to depletion (PREMA and DM) in the south. It can be attributed to the regional heterogeneity, tectonic environment, and mantle evolution history.

ZHOU, L. and Y. WANG, Institute of Geochemistry, Academia Sinica, Guiyang, China, and Y. KONG, Geologic Research and Analysis Center, Zhejiang, China

Minerals and REE Geochemistry of Zhijinshan Alkaline Complex, China

Zhijinshan alkaline complex, 23 km^2 in area and 134.8 Ma in age, is hosted by Triassic sandstones and shales, located in Shanxi anticline and west to Lüliang deep fracture zone. From outer to inner, the complex is composed mainly of monzonite, augite syenite and nepheline syenite. Of these, monzonite is predominant, accompanied by some brecciated trachyte and phonolite. The rock body belongs to a miaskitic nepheline syenite with an agpaitic coefficient of 0.77, a K/Na ratio of 0.91, and a progressively increasing Fe^{3+}/Fe^{2+} ratio from the early to late stages.

Minerals: *Orthoclase*—Monzonite thin sections show that ortohclase always forms the outer part of a euhedral plagioclase phenocryst. In augite syenite, orthoclase may be associated with plagioclase (Pl) or nepheline (Ne). Measured by EPMA, the composition of orthoclase from different rocks is listed as follows:

	Or	Ab	An
Monzonite	0.502-0.549	0.428-0.499	0.022-0.027
Augite syenite (Pl-bearing)	0.602-0.797	0.227-0.386	0.025-0.033
Augite syenite (Ne-bearing)	0.857-0.953	0.001-0.140	0.003-0.046
Nepheline syenite	0.952-0.957	0.029-0.032	0.011-0.019

As can be seen, from early to late, Or increases and An and Ab decrease.

X-ray diffraction data indicated that all orthoclases in intrusive rocks are partially ordered belonging to monoclinic system with $t_1 = 0.70–0.80$, $\Delta Y = 0$, $\Delta Z = 0.4–0.5$.

Plagioclase—Usually with a zonal structure, from inner to outer plagioclase varies in composition from andesine, $Ab_{0.585}An_{0.412}Or_{0.003}$, to oligoclase, $Ab_{0.818}An_{0.168}Or_{0.014}$. Most of them and andesine, commonly with $An_{0.3–0.4}$, $Or_{0.006–0.017}$, $Ab_{0.7}$, and $t_{10} = 0.658–0.735$, $t_{1m} = 0.145–0.199$, $t_1 = 0.083–0.934$, $\Delta Y = 0.513–0.536$, $\Delta Y = 0.513–0.536$, $\Delta Z = 0.606–0.830$.

Nepheline—In nepheline syenite and augite syenite, nepheline normally fills in interstices of euhedral orthoclase or intergrows with orthoclase showing a graphic structure.

Pyroxene—Pyroxene is widely distributed in the rock body. So far, it has been known that diopside and salite, aegrine-bearing diopside, and aegrine-bearing salite as well as aegrine augite and aegrine occur mainly in monzonite, augite syenite, and nepheline syenite, respectively. In augite syenite, some pyroxenes exhibit zonal structure and diopside or salite always constitutes the core and aegrine-bearing ones or aegrine augite forms the outer part. The composition of pyroxenes is shown as follows:

	Ac	Di	Hd
Monzonite	2.07-8.79	55.22-85.26	9.58-35.97
Augite syenite	5.97-30.28	41.12-81.02	7.75-26.08
Nepheline syenite	62.09-71.17	23.34-26.29	3.26-10.80

Pyroxenes always have a small number of Al^{VI}(0–0.06), and with the magmatic evolution, Di, Hd, and Ca, Mg, Fe contents gradually decrease and Ac, Na, Fe^{3+}, and $Fe^{3+}/(Fe^{2+} + Fe^{3+})$ increase.

Hornblende and biotite—Minor amounts of hornblende and biotite have been observed in monzonite and augite syenite. Hornblende belongs to Mg-hastingsite with a higher $Fe^{3+}/(Fe^{2+} + Fe^{3+})$ ratio (0.37) and a lower $Mg/(Mg + FE^{2+})$ (0.56) in monzonite than those in augite syenite (0.14, 0.42). The Mg/Fe ratios of biotite are always <2.

REE geochemistry: Individual REE minerals have not been found in this alkline complex. Studies show that apatite and sphene are REE principal carriers. Determined by ICP, the average REE content of apatite (0.65%) and sphene (0.649%) are very close, and they possess similar concentration ratio ($\Sigma REE_{hostrock}/\Sigma REE_{mineral}$), too: 11–60 and 12–58 respectively. Their REE chondrite-normalized patterns are similar to those of their host rocks and all belong to LREE-enriched type, but with the LREE/HREE, $(La/Yb)_N$, $(La/Sm)_N$, and $(Gd/Yb)_N$ ratios varying in the order of apatite > host rock > sphene. A pronounced negative Eu anomaly of apatite, δ_{Eu} 0.6–0.7, and a positive Ce anomaly of sphene, δ_{Ce} 1.10, have been observed.

Thc Σ REE content of Zijinshan complex is on the average 196.94 ppm with the $(La/Yb)_N$, $(La/Sm)_N$, $(GD/Yb)_N$, and LREE/HREE ratios being 16.1, 4.3, 2.0, and 11.1, respectively. Unexceptionally, all the REE chondrite-normalized patterns of the intrusive or extrusive rocks slope down toward the right side. There is no Eu anomaly in those intrusive rocks, but a weak negative Eu anomaly is observed in extrusive rocks. It appears that the intrusive stage, from early to late, ΣREE content, LREE/HREE, $(La/Yb)_N$, and $(La/Sm)_N$ ratios were fairly close with a little declination, but in the late extrusive stage, they obviously increased.

Based on REE geochemical features, geological environment, and Rb-Sr isotope data, it is believed the rock under study was derived from mantle. A genetic model, partial melting model, has been formulated, using a garnet lherzolite (55% olivine, 25% orthoclase, 5% garnet, and 15% clinopyroxene) as source material, $C_o = 1.8 \times REE_{chondrite}$ content, F = 0.5% and partition coefficient cited from Hanson (1980). the results of calculation coincide well with those measured.

Conclusion: Similarities of REE contents and patterns of different intrusive rocks might imply that they were derived from the same mantle source and have not undergone an intensive differentiation. The magma might move and ascend along Fenhe River deep fractures. A partial melting model can simulate the genesis of the rock body.

The structural state of feldspar suggests that feldspars were crystallized neither under high nor low temperature and have not experienced a rather slow cooling history. Therefore, the alkaline complex might form in a shallower depth. The low Al^{VI} content in pyroxene also shows that pyroxenes were formed under the above mentioned condition, because the lower Al^{VI} is related to a lower crystallizing pressure. Using two feldspars geothermometer, the initial crystallizing temperature of monzonite has been estimated to be about 800 °C. From the variation tendency of $Fe^{3+}/(Fe^{2+} + Fe^{3+})$ ratios of the minerals and rocks, it is inferred that during magma intrusion stage F_{O2} increased gradually.

ZHOU, Xinhua, Academia Sinica, Beijing, China

Isotope and Geochemical Constraints on the Accretion and Tectonic Evolution of SE China

One of the major controversies in geology of China is the formation and tectonic evolution of SE China and its relation to Taiwan. SE China consists of the Proterozoic Yangtze Craton and the so-called SE China Foldbelts. It has been interpreted in various ways, from the classical to thin-skin tectonics. To test the hypotheses proposed by different authors, the isotope and geochemical study on the sediments, granitoids and metamorphic rocks is performed, with the emphasis on the Sm/Nd crustal residence age of sediments and the depleted model age of the granitoids. The following observations and conclusions have been reached:

(1) The bulk of the SE China foldbelts and the Yangtze Craton is essentially mid-early Proterozoic in age with respect to their ultimate sources. The Nd and Sr isotopic evidence indicates that the foldbelts in SE China represent remobilized or recycled ancient continental crust with variable but small contribution from depleted mantle sources. They do not indicate a wholesale growth of continental mass during the Phanerozoic periods of magmatism; and the SE China Block has essentially grown to the present size by about 1 Ga ago.

(2) Granitoids of the SE China Foldbelts are likely to have multiple origins as viewed from their chemical and isotopic characteristics. Secular geochemical variation has been established for inland granitoids of Proterozoic to Mesozoic ages. Their high Sr initial ratios (0.710-0.737) suggest that most granitoids are basically anatectic products of ancient continental crust. This seems to be compatible with crustal remelting via microcontinental collision process or terrane accretion.

(3) The isotopic signature of the Mesozoic granitoids in coastal region (excluding A-type alkaline granites) imply that substantial amount of fresh mantle component has been added to ancient continental material in the generation of these rocks. This means that the Mesozoic magmatisms in the coastal region were resulted from subduction of paleoPacific plate beneath the Asiatic continent.

(4) The subordinate amount of TTG and syntectic series in the continental interior suggest that the influence oceanic plate subduction is relatively limited. This in turn implies that the Paleotethys seaways were small and short-lived, which is not inconsistent with the rare preservation of ophiolites suites in the SE China Block.

(5) The overall geochemical evidences seems to be consistent with a model of serial collisions between microcontinental blocks for the bulk of the SE China foldbelts.

ZHOU, Y., Virginia Division of Mineral Resources, Blacksburg, VA

Evaluating Impact of Mining Underlying Seam on Recoverable Coal Reserves in Overlying Seam

A reliable estimate of mineable coal reserves is essential to long-range planning and for conservation of energy resources. In Virginia, the majority of the coal beds are found in multiple horizons. In the Vansant Quadrangle, located in south-central Buchanon County, there are up to 25 coal beds in a given stratigraphic section. Ideally, to maximize coal recovery the sequence of mining should proceed in a descending order, starting with the uppermost bed, with complete extraction of each coal bed before the next bed is mined. However, in practice the decision traditionally has been based on factors such as ownership, economy, and accessibility, without considerations for the potential effects on subsequent mining operations.

Today, many coal operators are mining above existing or abandoned mines, some of which exhibit subsidence, pressure arching, and large-scale caving. As a result, special ground control problems from such interactions may occur in an upper bed that is subsequently mined. Coal recovery from coal beds that overlie mined-out beds may be significantly reduced, depending upon the magnitude of interaction. Some coal beds lying in close vertical proximity may be completely lost to caving or intense fracturing, whereas others may only be slightly affected. Complete and partial extraction also tends to create different interaction effects, both in terms of interaction mechanism and magnitude. The percent of coal loss is therefore a function of extraction methods, geological structures, and the innerburden features and thickness.

In overmining operations, there is no absolute minimum safe innerburden thickness, because mining can affect coal beds lying any distance above the mined-out bed, as evidenced in surface subsidence effects. From the point of view of ground control practice, there is a minimum innerburden thickness between the two coal beds below which the upper seam will be completely lost. This distance is defined by the caving and fracturing zones induced by mining of the lower seam. Based on average stratigraphic conditions (that is, from soft to medium hard) and coal bed thicknesses (1–2 m) in Virginia coal fields, an average minimum M-Index of 5 (ratio of innerburden thickness to lower bed mining height), or a minimum innerburden thickness of 5–10 m has been established.

Even if a coal bed is lying above this minimum innerburden thickness, complete recovery may not be achievable. The stratigraphic interval the upper coal bed must lie above the mined lower bed to be essentially unaffected is defined as the recoverable innerburden for zero loss and has been determined for partial extraction and complete extraction. For coal beds lying between the minimum innerburden and the recoverable innerburden, there is a corresponding loss of coal due to interaction.

In order to analyze the effects of mining an underlying coal bed on the recoverable reserves in an overlying bed, remnant structures in the lower mined coal bed must be identified. Drillhole data obtained from mining companies are superimposed onto the mine map to obtain geological information. The two coal beds under consideration are then evaluated for stress concentrations, uppcr scam subsidence, and potential damage to the upper seam using various interaction mechanisms. The amount of recoverable coal in the upper coal bed is determined based on minimum safe mining conditions calculated for the upper bed.

ZHU Xingshan, Shandong Institute of Mining and Technology, Taian, China

The Control of East Shandong Land and Tanlu Fracture to the Deposition of Coal Series in West Shandong

Very famous Tanlu Fracture divides Shandong Province in east China into two parts, west and east (Figure 1). In the west, Archean, Sinian system and Paleozoic eratham develop well. In the east, Middle and Lower Proterozoic suberathem develop, but Sinian system and Paleozoic erathem are absent. There has been contention on whether east Shandong land existed and when Tanlu Fracture began to move for a long time. After researching, I consider that Tanlu Fracture began to move in Sinian period at least and controlled the deposition in west Shandong together with east Shandong land since then.

Before the deposition of coal series: Sinian system and Wushan formation of Lower Cambrian series in west Shandong only exist near Tanlu Fracture. Their thickness increases from west to east, isopach and the boundary line of lithofacies are parallel to Tanlu Fracture. There is transgressive overlap in west border. Sandstones bear unstable heavy minerals. The facts above prove that deposition occurred in a channel resulted from the movement of Tanlu Fracture. After the deposition of Wushan formation, the range of deposition expanded, and the situation was similar. The thickness of every formation of Cambrian system increased from west to east. Conglomerate 30 m thick of the Lower Cambrian series was discovered in Fangzi (inside Tanlu Fracture). The terrigenous constituent content of Yeli formation of the Ordovician system is higher in Ju County (inside Tanlu Fracture) than in other parts of west Shandong.

During deposition of coal series:

Characters of sand bodies.

In order to determine the terrigenous province and orientation of paleocurrents during deposition of the coal series, I have studied the sand bodies in coal series. They are netted and trend northwest or NWbN in plane. From southeast to northwest, the paleocurrents branch gradually; thickness, grain size of clast, and unstable constituent content of sandstone decrease, and roundness of clast and weathering degree of feldspar increase. These show that the sediments were provided from east Shandong land and carried from southeast to northwest.

Laws of the development of coal series and seams

North China basin rose after the Ordovician period. Profusulinella Zone of Lower Bengxi formation was deposited in Lingyi (near Tanlu Fracture) in Middel Carboniferous epoch. Until the Fusulina-Fusulinella Zone, over which the Profusulinella Zone was deposited, did the sediments cover all west Shandong. Thickness of every formation increased from west to east. The isopach of total, limestone and clastic rock are parallel to Tanlu Fracture. In main paleocurrents, the thickness of total, limestone and clastic rock are larger and coarse sediments are more. This is because east Shandong land was smooth enough to provide only chemical substance and fine clast, so few coarse sediments developed and the places near Tanlu Fracture and main currents are provided more chemical substance and clast so thickness of total, limestone and clastic rock are larger. The development of coal seams is controlled by the distance from land and the position of main paleocurrents. They are well developed between main paleocurrents and in lower reaches.

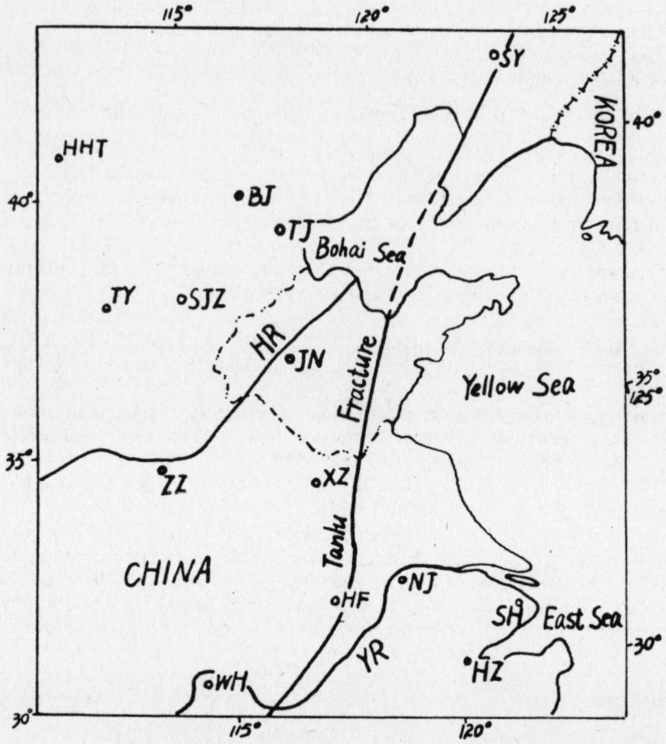

Figure 1—Index map of Shandong Province. SY-Shengyang; HHT-Huhehaote; BJ-Beijing; TY-Taiyuan; SJZ-Shijiazhuang; TJ-Tianjing; JN-Jinan; ZZ-Zhengzhou; XZ-Xuzhou; HF-Hefei; NJ-Nanjing; SH-Shanghai; HZ-Hangzhou; WH-Wuhan; HR-Yellow R.; YR-Yangtze R.

ZHUANG P., and XU C., Chinese University of Geosciences, Beijing, China

Tectonic Evolution and Mechanism of Rifting of North China Plain–Bohai Bay Basin

Since the formation of the Palaeo-Eurasian Continent during the Triassic period, a series of tensional fault basins, which are superimposed on the basement consolidated during different periods, have developed subsequently in the continental margins of East China. The North China Plain–Bohai Bay (NCBB) Basin is typical among this basin series. This paper attempts to discuss and prove the tectonic evolution and the mechanism of rifting of the NCBB Basin, based on field structural geological investigations and geophysical remote sensing data as well as the background of deeper structure, especially the transformation of the tectonic stress fields during the Meso-Cenozoic eras in East China. The conslusions are achieved as follows:

1. The opposite sides of the NCBB Basin are bounded by the Tan-Lu Fault Belt and Taihang Structural Belt whose structures have been complicated by secular activities. Since Mesozoic, the two belts have mainly undergone three stages of stress field evolution and have exerted an important influence on the evolution of the basin. The values of paleoshear stress as estimated from the dislocation densities in deformed quartz crystals in the fault zones are approximately 400–750 bars.

2. The seismic profiles and drilling records of the basin reveal that the basin is posses of three-layered texture vertically, reflecting that the basin has undergone three different stages: (1) the early (J3-K1), interior faulted basin stage, reflected by the upper part of the third seismic wave group; (2) the middle (K2-E), main rift basin stage, reflected by the second wave group; and (3) the later (N), depression basin stage, reflected by the first wave group. During the three stages, the different rock associations and structural deformations have developed respectively, corresponding to the different tectonothermal events and tectonic stress fields, which coincide

with the tectonic evolution of the adjacent areas. In addition, the paleotectonic maps of the basin in these stages reveal the various tectonic frameworks and the sedimentary distributions, and each geological architecture has afforded boundary conditions of the stress field in succeeding stages.

3. The magnetic field reveals the network frame constituted of the basement lineaments trending north-south, east-west, northeast, and northwest, reflecting the characteristics of stable block type of the basin basement. The gravity field and the relief of Moho display and upper mantle uplift in the central part of the basin and possess local variations that reflect the alternating rises and depressions extending approximately northeast. The relation between the relief of Moho and that of the basin basement is a mirror image. Seismic sounding data reveal that the lithospheric architecture may be divided vertically into seven layers, and the buried depths of various layer-boundaries change considerably in amplitude, thus forming an undulating faulted block texture. The values of heat flow of the whole basin and of the Meso-Cenozoic depression are 0.6–1.3 and 1.5–1.9 HFU, respectively, these are higher than the average values of similar regions in the world.

4. By synthesis of the materials mentioned above, we recognize that the formation and development of the basin probalby result from a complex process of crustal extension and thinning and mantle diapirism, and are closely related to the interaction between Eurasian, Kula-Pacific oceanic, and Indian plates as well as the periodic variations of the velocity of the earth's rotation.

After comparing the various current models of basin subsidence mechanisms so far suggested, a model associated with regional extensional stress field accompanied by uplift force from deeper part has been chosen, and the directions of the maximum and minimum principle stress and the values of shear stress from the planar and the vertical stress field in the basin have been simulated by adopting a program of limited element increment for the elastoplastic mechanics. The results support the thesis suggested previously and indicate: (1) The initial stress field in the rifting basin stage is evidently confined to the crustal architecture during the earlier stage and the mechanism of rifting is controlled by the preexisting faults; (2) there is a "higher shear stress layer" in the shallow crust, in which possible weak zones of rock deformation and damages are distributed and (3) normal faulting may result from the stress amplication in upper crust.

ZHUANG Peiren, Lei Qingxia, Yang Wenli, Chinese University of Geosciences, Beijing, China

Research of Deep Faults in Precambrian Continental Block in East China from Remote Sensing and Geoscientific Data

The aim of this paper is to discuss how the deep faults are investigated by use of remote sensing and geoscientific data, and to describe the application of the results for mineral exploration and assessment of regional crustal stability, etc.

1. A number of deep faults, formed in several periods, trending in various directions, and occurring in different layers of the lithosphere, have developed in the Precambrian continental block in east China. These deep faults may be distinguished from those confined only to the cover strata on the remote sensing images, and may be verified by the geological and geophysical data. After interpreting the images in the investigated areas, some synthetic criteria for discriminating the deep faults based on remote sensing, geology, and geophysics have been established.

2. By way of a serial digital image processing to enhance the image display of deep faults in test areas, including Xiao Qinling Precambrian terrain in Shanxi, Jin-yang Mesozoic volcanic basin in Liaoning, northern part of north China-Bohai Bay sedimentary basin, the Tan-Lu fault zone in Shantong, and the Tongling minerogenetic province in Anhui, all situated within the East China Precambrian Continental Black, the optimal processing programs and methods including the spectral/spatial information enhancement and the multitemporal phase information processing have been selected. The controlling factors that influence the efficiency of the digital image processing have been discussed.

3. Two textural analysis methods, i.e., graylevel cooccurrence matrix and Fourier power spectrum, have been adopted and programmed for extraction of more geological information from both MSS and TM images of Tongling area in Anhui. Textural images produced by these two methods present a large amount of lineaments. Many of them, especially the large-scale ones, which are closely related to the deep faults, are obscure on original images. As a quantitative information of remote sensing, textural information has also been used in the statistical prediction of copper deposits. Results in test areas show that the classification accuracy among areas with different mineralizing conditions can be enhanced when textural information is used.

4. By use of the methods of statistical analysis of image lineament, such as the equidensity of length/intersection of lineaments, and the central symmetric degree of azimuth of lineaments, the deep faults along which paleovolcanic edifices and mineralized centers are distributed, have been determined. Results in test areas suggest that there is a correlation between the statistical anomalies and the volcanic edifices/mineralized centers.

5. A specific image processing procedure aiming at synthetic processing of two different kinds of geophysical images has been proposed and used in aeromagnetic and gravity data. The images produced by the method of related matrix analysis from both aeromagnetic and gravity data show a number of large-scale lineaments which are closely related to the deep faults. The textural analysis method has also been applied to both kinds of geophysical data.

6. Through synthesis of the results of previous studies, we have defined the tectonic frameworks in test areas, and we have good reason to believe that deep faults have played an important role in the magmatism, mineralization in some test areas, and have exerted an influence on earthquake activities in others. By comprehensive analysis or integrated image processing and statistical predictive device, several previously unknown promising areas have been predicted with success.

ZIEGLER, W., Forschungsinstitut Senckenberg, Frankfurt, Federal Republic of Germany

Conodont Evolution and Event Stratigraphic Boundaries (Mid-Devonian to Mid-Carboniferous

Platform (pectiniform) conodont elements of pelagic Mid-Devonian to Mid-Carboniferous record at least five evolutionary cycles (Ziegler & Lane, 1987). Each cycle consists of successive low-and high-diversity episodes followed by an extinction event before returning to a low-diversity episode in the succeeding cycle.

Low-diversity episodes are (1) basal-Givetian *varcus* Zone, (2) basal-Famennian Lower *triangularis* Zone, (3) Late Famennian post-Lower *praesulcata* Zone and pre-*duplicata* Tournaisian, and (4) Early Viséan post-*anchoralis-latus* Zone.

High-diversity episodes are (1) Givetian *hermanni-cristatus* Zone to Late Frasnian *linguiformis* Zone, (2) Lower Famennian Middle *triangularis* to Late Famennian early Middle *praesulcata* Zone, and (3) Tournaisian *duplicata* to *anchoralis-latus* Zone.

During the short-term or abrupt extinction events preceding the low-diversity episodes, at least near the Frasnian–Famennian boundary, the Devonian–Carboniferous boundary, in the Early Viséan and Early Namurian, almost all or many of the previous conodont taxa became extinct.

Associated with these extinctions, many species of other organic groups were eliminated resulting in large biomass reductions. Conodont biofacies analyses (Sandberg, et al., 1988) demonstrate that abrupt eustatic rise and fall immediately preceded the mass extinction at the Frasnian–Famennian boundary and most probably also at the Devonian/Carboniferous boundary. The conodont extinction occurred within the fall episodes.

After a short-term innovative phase following the extinction event and low-diversity phase a radiative outburst produced a gradualistic evolution in which many species give rise to the well-known rapid evolutionary development on which the world standard conodont zonation (species of *Polygnathus*, *Palmatolepis*, *Siphonodella*) is based.

It seems, thus, that severe global events have genetically influenced conodont evolution during mass extinction phases, which can be precisely recorded by biostratigraphic and biofacies analysis of conodonts. Resulting boundaries may be easily recognizable both physically and biostratigraphically, and thus be suitable for major stratigraphic boundaries. In the intermediate gradualistic phases, however, pelagic conodonts underwent "quiet" phylogenetic evolution uneffected by physicochemical influences, such as less severe eustatic falls (which might terminate or isolate shallower faunas). The standard conodont zonation preferredly constructed from pelagic conodonts results in phylogenetic zones that can evolve and be distributed world-wide within time intervals so short that it appears synchronous in the geologic record.

Because of the rapid evolution in the gradualistic phases, standard conodont zones produce a finer time scale than almost any other paleozoic organisms. Conodont zones are therefore excellent time markers and correlative tools.

REFERENCES

Sanberg, C. A., W. Ziegler, R. Dressen, and J. Butler, 1988, Late Frasnian mass extinction: conodont event stratigraphy, global changes, and possible causes: Cour. Forsch.-Inst. Senckenberg, v. 102, p. 263–307.

Ziegler, W., and H. R. Lane, 1987, Cycles in conodont evolution from Devonian to mid-Carboniferous, Aldridge, ed., Palaeobiology of conodonts, p. 147–163.

ZIERENBERG, R. A., U.S. Geological Survey, Menlo Park, California, W. C. SHANKS, III, U.S. Geological Survey, Reston Virginia, R. A. KOSKI, J. L. MORTON, W. R. NORMARK, and K. A. KVENVOLDEN, U.S. Geological Survey, Menlo Park, California, and M. L. Holmes, U.S. Geological Survey, Seattle, Washington

Massive Sulfide Deposition and Sediment Alteration in the Escanaba Trough, Gorda Ridge, North East Pacific Ocean

The Escanaba Trough is the slow-spreading (2.3 cm/yr total spreading rate) southern portion of the Gorda Ridge, and is bounded by the Medocino transform on the south and a right-lateral offset of the ridge axis near 41°25′N. latitude, 127°25′W. longitude on the north. The Escanaba Trough has the characteristic morphology of slow-spreading ridges with a 5 to 15 km wide flat-floored axial valley at a depth of 3300 m flanked by steep ridge crests rising 800 to 1200 m above the axial valley floor. The southern two thirds of the Escanaba Trough is covered by hemipelagic and turbiditic sediment that is 300500 m thick over the spreading axis and thins to the north. Several volcanic edifices 35 km in diameter rise up through and locally breach the sediment cover. The volcanic edifices are built by both intrusion and extrusion of basalt at a rate similar to the rate of burial by sediment. Detailed geologic mapping and sampling at two of the volcanic edifices utilizing surface ship and submersible, and reconnaissance work at two other sites, have discovered at least six major massive sulfide deposits are associated with the volcanic edifices, they are all developed in sediment. Hydrothermal vents have been sampled near 41°N. latitude, 127°30′W. longitude; other sulfide occurrences do not appear to have active hydrothermal vents. Clear hydrothermal fluid at temperatures as high as 220°C issues from porous anhydrite-barite chimneys on top of large pyrrhotite-rich massive sulfide mounds. Active vents are colonized by dense clusters of tube worms associated with palm worms, anemones, limpets, gastropods, and bacterial mats. Bivalves are common in the sediment near some sulfide deposits.

the presence of a sediment blanket over the hydrothermal system influences the form of the sulfide deposits by changing the hydrologic regime and by chemical interaction of the hydrothermal fluid with the sediment. The relatively impermeable sediment cover acts as a thermal blanket preventing large-scale convective cooling of the ridge crest. Hydrothermal circulation is restricted to recharge and discharge zones were tectonic and volcanic processes provide cross-stratal permeability. Sand and silt turbidites are presumed to be hydrothermal aquifers because of their higher permeability. The volcanic edifices are both topographic and thermal foci for hydrothermal fluids, resulting in long-lived discharge systems capable of forming large massive sulfide deposits. Sulfide veining in sediment cores and sedimentary breccias having a sulfide matrix are indications of subsurface deposition of sulfides. the extent of subsurface mineralization can not be determined from surface observations, but the large amount of sulfide exposed on the sea floor may represent only a portion of the deposits.

The chemistry and mineralogy of the sulfide deposits indicate that chemical interaction of the hydrothermal fluid with sediment is an important process in forming these deposits. Relative to massive sulfide deposits formed on sediment free ridge crests, the sulfides are enriched in Ag, As, Au, bi, Pb, Sb, and Sn. Pyrrhotite is the dominant sulfide mineral and is accompanied by variable amounts of sphalerite, isocubanite, and galena with minor to trace amounts of arsenopyrite (FeAsS), marcasite, chalcopyrite, loellingite ($FeAs_2$), tennantite ($Cu_{12}(As,Sb)_4S_{13}$), and alabandite (MnS). This reduced, low sulfur fugacity assemblage results from interaction of the hydrothermal fluid with organic matter in the sediment. Some samples also contain asphaltic petroleum formed by hydrolytic cracking of terriginous organic matter in the sediment. The predominant gangue mineral is barite which forms by interaction of barium-rich hydrothermal fluid with sulfate-bearing sea water.

The predominant process which has altered near-surface sediment is magnesium metasomatism, which is locally intense. Compared to unaltered hemipelagic sediment, the altered sediment is enriched in Mg and Fe and is depleted in Na, K, Ca, and Si. Mg content increases from 4 wt. percent in unaltered hemipelagic sediment to 25 wt. percent in intensely altered samples. Unaltered sediment is predominantly quartz with lesser amounts of feldspar, mica, chlorite, illite, and, in the finest grain fraction, poorly crystalized smectite. Sand grains from turbidites veined by sulfide are coated by well-crystalized smectite. Sand grains from turbidites veined by sulfide are coated by well-crystallized autoigenic smectite. More intensely altered sediment contains talc and(or) chlorite. Quartz is completely destroyed in the most altered samples, which are nearly monomineralic Mg-rich chlorite. The undiluted hydrothermal fluid is Mg-free. The intense Mg-metasomatism occurs in shallow secondary circulation cells where magnesium-bearing sea water mixes with high temperature hydrothermal fluid.

ZOBACK, M. D., Stanford University, Stanford, CA, J. H. HEALY, U.S. Geological Survey, Menlo Park, CA, and G. SHAMIR, Stanford University, Stanford, CA

State of Stress along Major Transform Faults: Implications of Cajon Pass Scientific Drilling Project

Some of the most challenging problems in the earth sciences are related to understanding the forces in the lithosphere that cause plate movement and crustal deformation. The Cajon Pass Scientific Drilling Project is aimed at addressing the fundamental question of the magnitude of crustal forces required to cause plate movement and earthquakes along the San Andreas fault. The state of stress along plate boundaries like the San Andreas fault is of global importance to understand the nature of forces driving and resisting plate motion, the nature of crustal deformation along the plate boundary, and the physics of earthquake nucleation and propagation.

The uncertainty about the magnitude of stresses required to cause movement of the San Andreas fault is sometimes referred to as the stress/heat flow paradox. The essence of this paradox is that a variety of indirect data obtained over the past 20 years have lead to completely contradictory inferences about the magnitudes of the stress along transform faults and the physical processes occurring in the fault zone. On one side of this paradox, scores of laboratory studies on the frictional strength of rock suggest that the coefficient of friction of most crustal rocks is relatively high and fairly uniform (~0.6–1.0). In the context of conventional frictional faulting theory, these coefficients of friction suggest relatively high average shear stresses (~80 MPa) in the seismogenic part (upper 15 km) of the crust along the San Andreas fault. Stress measurements made at depth in a number of areas of active faulting around the world appear to confirm the validity of the laboratory-derived coefficients of friction, the applicability of frictional faulting theory, and generally high levels of shear stresses at depth.

A variety of data indicate that shear stresses may be quire low along transform faults like the San Andreas and either that coefficients of friction are quite low (0.1) or that the concept of a frictional constitutive law may not be applicable to the fault zone. Most notable, abundant heat flow measurements in relatively shallow holes (mostly ~300 m deep) along the San Andreas fault indicate a complete absence of frictionally generated heat. These observations have led to the suggestion that there must be relatively little resistance to plate motion, with average stresses along the fault that are quite low (~20 MPa). If this concept is correct it not only challenges the applicability of the laboratory data but, by doing so, challenges the conceptual basis for crustal earthquakes that is now widely accepted. In the mid-1970s, low frictional resistance along transform faults was called upon to explain the orthogonality of ridge-transform intersections. Both of these arguments have been recently reviewed by Lachenbruch and Sass (Geophysical Research Letters, v. 15, p. 981–984, 1988). Zoback and others (Science, v. 238, p. 1105–1111, 1987) and Mount and Suppe (Geology, v. 15, p. 1143–1146, 1987) have suggested that the San Andreas fault system must be quite weak as right-lateral strike-slip motion occurs even though the observed directions of maximum horizontal compression within a broad zone adjacent to the fault system are almost perpendicular to it. Low shear resistance along transform faults helps explain why reverse faults and folds are subparallel to continental transforms in many places as well as why in other places normal faults and basins are subparall to transforms. Similarly, the weak fault concept is useful in explaining how subhorizontal detachment faults near the San Andreas might move. Accompanying the reduction of shear stress parallel to the weak fault, a rotation of the vertical principal stress occurs that generates a component of shear stress on the subhorizontal planes.

O'Neil and Hanks (Journal of Geophysical Research, v. 85, p. 6286–6292, 1980) suggest that the low heat flow values obtained in the shallow holes adjacent to the fault are contaminated by thermal convection although Lachenbruch and Sass (Journal of Geophysical Research, v. 85, p. 6286–6292, 1980) have argued that this is not the case. The Cajon Pass experiment has been designed to help put this issue to rest and measure directly the state of stress and heat flow at seismogenic depth (5 km) adjacent to the San Andreas fault. In addition to in situ stress and heat flow, a suite of hydrologic studies, borehole geophysical measurements, and geologic, geochemical, and petrophysical studies on the core and cuttings are also being conducted. If the inferences of high shear stress at depth are correct, at 5 km the stress magnitudes should be sufficiently high to rule out the ~20 MPa limit implied by the heat flow data. At the same time, the heat flow data between the surface and 5 km will reveal whether near-surface thermal convection is important. Five kilometers is also deep enough to minimize the effects of local erosion and topography.

To date, drilling and downhole measurements have been completed to 3.5 km. Taken together, the spectrum of scientific studies in the Cajon Pass borehole yield important new insight into the relationship between in situ stress, faulting, physical properties of the crust, and circulation of pore fluids and heat transport. Critical data include changes of stress magnitude and orientation and heat flow with depth; geochemical, isotopic, petrophysical, and seismological evidence of major structural discontinuities in the hole; and in situ permeability, pore pressure, and pore fluid geochemistry. The preliminary interpretation of heat flow data shows no evidence of frictionally generated heat, and the in situ stress measurements show a complete absence of right-lateral shear stress on planes parallel to the San Andreas fault. Nor is there evidence of sufficient in situ permeability to allow significant heat transfer through convection. Although the Cajon Pass data are not yet from sufficient depth to be definitive, the preliminary results seem to support the case for low stress faulting.

ZOBACK, M. D., Stanford University, Stanford, CA, and M. L. ZOBACK, U. S. Geological Survey, Menlo Park, CA

State of Stress along San Andreas Fault System: Inferences about Development of Transtensional and Transpressive Deformation along Transform Margins

Utilizing well-constrained earthquake focal mechanisms, stress-induced wellbore breakouts, and in situ stress measurements at depth a detailed stress map in the vicinity of the San Andreas fault has been compiles (Zoback, et al., 1987, Science, v. 238, p. 1105–1111). The pattern of stress, the style of geologica deformation along the San Andreas during late Tertiary time, and the similarity between the state of stress and deformation along the San Andreas and other continental transform faults around the world all lead to similar inferences about the physics of faulting and the nature of tectonic deformation along transform plate margins.

The data on in situ stress along the San Andreas fault come from a variety of sources. The most comprehensive data are in central California where stress-induced wellbore breakouts (Mount and Suppe, 1987, Geology, v. 15, p. 1143–1146) and earthquake focal mechanisms adjacent to the San Andreas system (Eaton and Rymer, 1988, USGS Professional Paper, 1487; Oppenheimer et al., 1988, Journal of Geophysical Research, v. 93, p. 9007–9026) consistently shows that the direction of maximum horizontal compression is nearly perpendicular, it suggests that the San Andreas must be quite weak, as previously suggested on the lack of a frictionally generated heat flow anomaly along the fault (see recent review by Lachenbruch and Sass, 1988, Geophysical Research Letters, v. 14, p. 981–984). As the direction of relative motion between the Pacific and North American plates is nearly parallel to the San Andreas in central California, the origin of the fault normal compression is not immediately obvious. Zoback et al, (1987) argued that the direction of maximum horizontal compression was approximately perpendicular to the San Andreas because it was weak. In other words, in an otherwise strong and highly stressed crust, if the San Andreas was a nearly vertical plane that could resist little shear stress, the horizontal principal stresses *must* be almost perpendicular to the fault to minimize the amount of resolved shear stress. This concept leads to a rather simple model of stress and deformation along a transform plate margins. If the relative plate motion is slightly convergent, the maximum horizontal stress will be nearly perpendicular to the transform and crustal shortening will occur nearly perpendicular to it. Conversely, when the relative plate motion is slightly divergent, the minimum principal stress will be almost perpendicular to the transform and crustal extension will occur nearly perpendicular to it.

The weak-transform model we propose seems quite consistent with the style of deformation along the San Andreas fault in late Tertiary time and provides an alternative to models that require extensional or compressional deformation near transform faults to be restricted to areas between en echelon fault segments. For example, a remarkably consistent pattern of compressional deformation is observed along the length of the San Andreas in central California with numerous folds and thrust faults nearly parallel to the San Andreas. This deformation has been occurring for about the past 4 m.y., the same period during which the relative plate motion between the Pacific and North American plates has been slightly convergent (Cox and Engebretson, 1985, Nature, v. 313, p. 472). Prior to 4 m.y. ago, the relative plate motion was slightly divergent (Engebretson et al., 1986, GSA Special Paper 206) and, consistent with the model we propose, a number of sedimentary basins are observed that strike subparallel to the plate margin and indicate extension perpendicular to the San Andreas fault system. Deposition in these basins ended at the time of the plate motion change and the abrupt change from divergence and fault normal extension to convergence and fault normal compression.

The state of stress and pattern of crustal deformation along other continental transform faults around the world seems to be consistent with the simple model we propose. For example, as shown by Ben-Avraham (Journal of Geophysical Research, v. 90, p. 703–726), geophysical data along the Dead Sea Rift indicates extension perpendicular to the rift zone that is not explainable in terms of pull-apart models. in transpressional environments, both the Alpine fault in New Zealand and Bocono fault in Venezuela show nearly perpendicular compressive stress and crustal shortening, similar to what is observed along much of the San Andreas fault.

ZOBACK, M. L., U.S. Geological Survey, Menlo Park, California

World Stress Map Project: Global Patterns of Stress and Constraints on Intraplate Deformation

The goal of the World Stress Map Project of the International Lithosphere Program is to compile data on the orientation and relative magnitudes of the contemporary in-situ tectonic stress field. The project, a collaborative effort involving over 20 scientists from 16 different countries, was initiated in early 1986 and is scheduled for completion by the end of 1990. At that time the entire digital database will be released through the World Data Center System. As of September 1988 the database contained over 1800 points although the coverage is rather non-uniform.

Four different categories of geological and geophysical data are being used to infer tectonic stress information: earthquake focal mechanisms, wellbore breakouts, in-situ stress (hydraulic fracturing and overcoring) measurements, and young geologic data including fault slip and volcanic alignments. While the geologic and in-situ stress measurement data are generally restricted to the surface or near surface (less than 12 km depth), the earthquake focal mechanisms provide data for depth between about 5-20 km. Wellbore breakout data, utilizing the natural deformation around a wellbore due to stress concentration around the hole, come primarily from petroleum wells commonly 1-4 km deep and in come cases as deep as 5-6 km. The breakout data thus provide a link between the near surface and the very deep stress data.

All data in the database are assigned a quality using a ranking system based both on the accuracy of the determination and the relative reliability of the technique as a tectonic stress indicator. Four qualities are use, A-D, with A being the highest quality. D quality data are included in the database for completeness, however these data are not considered to be reliable stress indicators.

The data confirm the results of earlier studies which indicated that stress orientations inferred from the four different data types are consistent despite the different techniques used and the range in depths sampled. This consistency with depth implies a relatively uniform stress field throughout the upper brittle crust. Regional consistency of both stress orientations and relative magnitudes permits the definition of regional stress provinces.

Global Stress Patterns

Several first order patterns can be observed in the dataset. The most obvious feature is the lateral extent of regions presently experiencing a nearly uniformly-oriented stress field. The best and most well-documented examples include a $\sim 2 \times 10^7$ km^2 region of "midplate" North America (including most of the central and eastern United States, much of Canada, and possibly the western Atlantic Basin) and a $\sim 6 \times 10^6$ km^2 area of western Europe.

Probably the most striking feature of the dataset is a strong correlation between S_{Hmax} orientation and the azimuth of absolute plate velocity observed for many plates: midplate North America, South America, western Europe, and the younger portions of the Pacific plate. However, in all these regions the absolute velocity direction is nearly parallel to the expected ridge push direction with the exception of southern Europe where both S_{Hmax} orientations and absolute velocities are aligned nearly parallel with NNW convergence of the African plate. Similarly, in western South America the absolute plate velocity directions are nearly parallel to the convergence direction between South America and the Nazca and Pacific plates.

In four major regions there is no correlation between S_{Hmax} and absolute velocity: eastern China, western United States, most of the Indian-Australian plate, and in the older portions of the Pacific (where the available data suggest that S_{Hmax} directions are nearly orthogonal to absolute velocity). Data for the African, Antarctic, Nazca, and Cocos plates are too sparse to confirm or contradict this correlation.

Without detailed modeling and knowledge of how stress magnitudes vary across the plates, the sources of the observed regionally uniform stress provinces can not be easily constrained. Thus, while the strong correlation between S_{Hmax} and absolute velocity direction can be interpreted to indicate that asthenospheric drag is a major source of stress, as indicated above, ridge push and forces due to plate collision and subduction can not be ruled out. the most simple view of the strong correlation between S_{Hmax} and absolute velocity is that the net plate boundary forces responsible for moving the plates are also responsible for the stresses measured. In this contest, the regions where S_{Hmax} and absolute plate motion directions are not correlated will provide the best constraints on relative magnitudes of forces acting on the plates.

A comparison between the entire database and the extensional stress subset indicates that most midplate regions are characterized by compressional deformation (strike-slip or reverse faulting). Areas of active extension within the continents occur primarily in anomalously high (generally thermally elevated) regions: western United States Cordillera, East African rift, Tibet and parts of the Andes.

Constraints on Intraplate Seismicity

In order to understand the mechanism(s) responsible for intraplate deformation we are attempting to characterize earthquakes occurring in midplate regions. Three main hypotheses have been suggested to explain intraplate seismicity: 1) deformation is occurring in a relatively uniform regional stress field on fault zones well-oriented (high resolved shear stress) with respect to that stress field; 2) deformation is occurring on fault zones of extremely low shear strength which may be randomly and/or unfavorably oriented with respect to the local stress field; and 3) deformation is primarily the result of localized stress concentrations (e.g. around plutons) which should appear as local anomalies in both stress magnitude and orientation.

Independently-derived, high-quality information on stress orientations (primarily from wellbore breakouts and in-situ stress measurements) are compared with well-constrained focal mechanisms for intraplate events in central and eastern North America, Europe, and China. Two criteria are used to test for favorably-oriented zones of weakness 1) the angle between the inferred S_{Hmax} direction and the fault is between 30–45°, and 2) the observed slip vector is consistent with the S_{Hmax} direction and a strike-slip or thrust faulting stress regime with ϕ values ($\phi = (S_2-S_3)/(S_1-S_3)$) in the allowable range of 0 to 1.

Preliminary results of the above comparison on 20 well-constrained focal mechanisms in the central and eastern United States indicates that most events meet both of the above criteria. This fact, coupled with the observation that deviations from a mean S_{Hmax} direction in this region are small ($\leq 15°$), suggests that the observed intraplate seismicity in this area is primarily occurring on well-oriented, pre-existing faults.

ZODROW, E. L., University College of Cape Breton, Sydney, Canada

Is Compression/Impression Palaeobotany Capable of Contributing Data for Upper Carboniferous Bio- and Palaeogeography

The literature shows that upper Carboniferous macrofloral biostratigraphy is based almost exclusively on the use of sterile compression/impression fossils, and on the morphologic information they are only capable of yielding. In assessing the posed question, it is therefore appropriate to indicate summary arguments that point to problems as consequence of their use as data. A common problem is range bias in the biostratigraphic data through change of a name of a macrofloral name. An example of this is the taxonomy of the marattialean fern frond *Pecopteris cyathea* (Schlotheim 1804) emend. Stur 1883 non Brongniart 1834, i.e., the customary "*P. cythea*" of the *Pecopteris arborescens* Group *sensu* Corsin. Over the past six years I have been doing research studies of this Group (including their types) and believe that I can support the claim of having documented 18 synonyms of *P. Cyathea*. These are:
P. arborescens, P. arborescens var. integripinna, P. candolleana, Cyatheites candolleanus, P. hemiteloides forma longipinnata, P. lepidorachis, C. lepidorrhachis, P. lesquereuxii, P. nyranensis, P. raconensis, P. pectinata, P. permica, C. schlotheimii, P. squamosa, P. hispanica, P. laxenervosa, P. melendezil, and P. robbii.

While 16 of these name changes add "noise" to local biostratigraphy by muddling ranges, potential range bias is introduced with the confusion between *P. cyathea* and *P. arborescens*. *P. arborescens* is truly a Stephaian floral element, whereas *P. cyathea* occurs in the record as low as Wesphalian C (in the Canadian and Czech Ostrava Coalfields, and in Floetz Dickenberg, Ibbenbuehren, Germany).

The species question is another problem. In modern biologic thought, intraspecific variability, polymorphism, or phases in the life cycle are central concepts in species models. The concepts cannot be accommodated with information from sterile compression/impression floras. Moreover, as Carboniferous plant species are extinct, eliminating the possibility of comparative study with extant plants, it is axiomatic that a palaeobotanic species cannot be compared with the "biologic" species, although the gap narrows in younger geologic age. Therefore, the Carboniferous species concept is severely limited biologically as it only provides data on morphology (i.e., the binomial name of the fossil plant may represent nothing more than a convenient identification label). It follows that taxonomic assignments, as well as synonymy lists, of compression/impression floras are of questionable biologic value, and comparative work is idiosyncratic, as mental images are based on a designated type.

In summary, the inescapable conclusion is that macrofloral compression/impression biostratigraphy is suspect of being able to provide data for factual upper Carboniferous bio- and palaeogeographic reconstructions. From the latter it follows that interpretations of such data in respect to distribution and composition of plant communities, degree of developed floral provincialism, and evolutionary changes within and across the provinces require circumspect critical interpretations. The suggestion is made that the biostratigraphic data base for upper Carboniferous homotaxial correlations and reconstructions should be composed of groups of morphologically similar taxa, cutting across the lines of taxonomy. In the absence of such revision, palaeobotanic contributions are at best marginal.

ZODROW, E. L., University College of Cape Breton, Sydney, Canada

Is Compression/Impression Palaeobotany Capable of Contributing Data for Upper Carboniferous Bio- and Palaeogeography

The literature shows that upper Carboniferous macrofloral biostratigraphy is based almost exclusively on the use of sterile compression/impression fossils, and on the morphologic information they are only capable of yielding. In assessing the posed question, it is therefore appropriate to indicate summary arguments that point to problems as consequence of their use as data. A common problem is range bias in the biostratigraphic data through change of a name of a macrofloral name. An example of this is the taxonomy of the marattialean fern frond *Pecopteris cyathea* (Schlotheim 1804) emend. Stur 1883 non Brongniart 1834, i.e., the customary "*P. cythea*" of the *Pecopteris arborescens* Group *sensu* Corsin. Over the past six years I have been doing research studies of this Group (including their types) and believe that I can support the claim of having documented 18 synonyms of *P. Cyathea*. These are:
P. arborescens, P. arborescens var. integripinna, P. candolleana, Cyatheites candolleanus, P. hemiteloides forma longipinnata, P. lepidorachis, C. lepidorrhachis, P. lesquereuxii, P. nyranensis, P. raconensis, P. pectinata, P. permica, C. schlotheimii, P. squamosa, P. hispanica, P. laxenervosa, P. melendezil, and P. robbii.

While 16 of these name changes add "noise" to local biostratigraphy by muddling ranges, potential range bias is introduced with the confusion between *P. cyathea* and *P. arborescens*. *P. arborescens* is truly a Stephaian floral element, whereas *P. cyathea* occurs in the record as low as Wesphalian C (in the Canadian and Czech Ostrava Coalfields, and in Floetz Dickenberg, Ibbenbuehren, Germany).

The species question is another problem. In modern biologic thought, intraspecific variability, polymorphism, or phases in the life cycle are central concepts in species models. The concepts cannot be accommodated with information from sterile compression/impression floras. Moreover, as Carboniferous plant species are extinct, eliminating the possibility of comparative study with extant plants, it is axiomatic that a palaeobotanic species cannot be compared with the "biologic" species, although the gap narrows in younger geologic age. Therefore, the Carboniferous species concept is severely limited biologically as it only provides data on morphology (i.e., the binomial name of the fossil plant may represent nothing more than a convenient identification label). It follows that taxonomic assignments, as well as synonymy lists, of compression/impression floras are of questionable biologic value, and comparative work is idiosyncratic, as mental images are based on a designated type.

In summary, the inescapable conclusion is that macrofloral compression/impression biostratigraphy is suspect of being able to provide data for factual upper Carboniferous bio- and palaeogeographic reconstructions. From the latter it follows that interpretations of such data in respect to distribution and composition of plant communities, degree of developed floral provincialism, and evolutionary changes within and across the provinces require circumspect critical interpretations. The suggestion is made that the biostratigraphic data base for upper Carboniferous homotaxial correlations and reconstructions should be composed of groups of morphologically similar taxa, cutting across the lines of taxonomy. In the absence of such revision, palaeobotanic contributions are at best marginal.

ZOLNAI, G., Pau, France

Intracontinental Mobile Belts and Passive Tectonism

Continents are not homogeneous, monolithic blocks, but should rather be considered as mosaics, divided by belts of relative weakness, which are important parts of their heritage and loci of significant ongoing stress and deformation. These belts are most frequently composed of a succession of segments and branches put end-to-end, such as: arches, lineaments, geofractures, linear uplifts, rifts, and aulacogens. They may also include separate pull-apart basins and isolated uplifts. They are able to carry deformation across the continents, from the active margins to the passive margins, and can trigger various anomalies (structural, paleogeographic, thermal, etc.) along their paths.

Such first-order fracture systems penetrate the continental crust to its full thickness, as demonstrated by upwellings of the Mohorovicic discontinuity and the presence of basic volcanism. They form transcontinental structural webs, parts of which may be rejuvenated by the successive stress regimes and pick up most of the intracontinental deformation, mostly in the wrenching mode (incl. transpression and transtension). They zig-zag or meander between individual cratonic blocks, "nuclei," or miniplates, which may undergo relative displacements (and sometimes slight rotations) following the above megashear belts. The spacing of these suture zones seems to exhibit some regularity, the blocks they separate being 5 to 600 km in diameter.

Deformations within the continental interiors are not triggered by coast-to-coast orogenic compression; they are instead initiated by slight adjustment movements along such shear zones between the more-or-less loosely bound continental blocks. The rearrangement between the neighboring cratonic masses may create, locally, compression. The orogenic compression proper, with crustal shortening, remains limited, in most cases, to a narrow strip along the active margin, directly exposed to the collision. In contrast, blocks along sheared margins or trailing edges may undergo rotation and eventually break away.

Most large intracontinental downwarps, depressions, or depocenters seem to coincide with, and be genetically related to, underlying rifts, aulacogens, and fracture systems: e.g., the Williston, Hudson's Bay, Illinois, the Paris (France), and possibly the Pannonic (Hungary) basins. Their frequent, periodical subsidence is accounted for by preferential stretching of the crust at or near the crossroads of zones of weakness or sutures. One may call these downwarps "fontanels," where wrenching and normal faulting may be present at depth during phases of subsidence at shallower structural levels. Fracturing and above-average heat flow may ensue. Sharper "funnels" appear in orogenic forelands, where foldbelts intercept fault systems or rifts of the underlying shield.

Some of the transcontinental fault swarms and lineaments (e.g., the Texas and Lewis-and-Clark fault systems), radiating out of the continent's interior, do not stop at the leaking edges of the overthrust belts, but may be followed, though in a subdued manner or indirectly, across the orogenic complexes. At the intersection of these fundamental faults and weakness belts, the trace as well as the style of the foldbelts often change (polarity, frequency, and anatomy of the folds and overthrusts). While the leading edge of the overthrust complexes is often bent or shifted, in the seaward continuation of such geofractures one may often observe oceanic ridges.

Folds and fold systems in cratonic interiors, as well as in orogenic forelands and along passive margins, exhibit in many cases draping or dragging, coincident with deep shear zones. None of these necessitates shortening (regional compression) perpendicular to the fold axes. Such fold groups can form "passive orogenic belts" of great extent and relief, far away from any "active" orogen.

Continental blocks may be calved from the passive margins and induce considerable tectonic disturbance in the form of wrenching and transpression related to the rotation of the detached landmasses. Secondary compression, vaulting of elongate slabs, block uplifts, etc. may initiate drape and drag-folding, enhanced by halokinesis and accompanied by basic volcanism (e.g., the Greenland-Innuitian Tertiary foldbelt).

The absence of penetrative metamorphism, of large-scale acidic intrusions and arc-type volcanism, and that of ophiolitic complexes leads to the conclusion that this type of (passive) tectonism definitely stands apart from the true orogenic process, with crustal shortening, subduction and/or collision, typical of the active margins.

All the above phenomena may be easily understood if we admit that continents are pulled rather than pushed during their wanderings. They may lose break-away blocks along their passive margins, which may be reintegrated later, when and if the direction of the motion is changed or inversed. Furthermore, collapse systems, "peripheric" graben, furrows, or troughs may be formed near or along passive margins, through crustal thinning (elongation). These may become intracontinental foldbelts or foreland foldbelts, when the passive margin is exposed to crustal shortening or shearing.

A global crustal motion will create one leading edge and one trailing edge, with various amounts of jostlings in between. If the continental drift is driven, at asthenospheric levels, by convention cells, the driving force must be "self-contained" for each continent and act in a "front-end-drive" manner; as if the leading edges of the convention cells were lined up, to march along the active continental margin. The mantelic undercurrents, acting as "conveyor belts," may drag the various segments of the continental plates at slightly different speeds and azimuths, the result of being some transcontinental shearing. This may occur at any moment of the global motion cycles; although such intraplate movements (incl. rifting, wrenching, pulling-apart) are currently synchronous with the major diastrophisms.

ZOLOTOV, A. N., Ministry of Geology of the USSR, Moscow, USSR, B. A. SOLOVIEV, and O. S. OBRYADCHIKOV, VNIGNI, Moscow, USSR, D. L. FEDOROV, and Yu. S. KONONOV, NV VIIGG, Moscow, USSR, and S. U. UTEGALIEV, RazNIIGRI, Moscow, USSR

Discovery and Exploration of Unique Fluidal Systems in the Subsalt Complex of Pre-Caspian Depression

In recent years, the most promising oil and gas discoveries in the European part of the USSR, characterized by a unique composition of their fluidal systems, have been made in the Pre-Caspian depression. The fields have been discovered in Upper Paleozoic subsalt deposits at depths ranging from 2.5 to 3 km near the flanks of the depression to 8 to 9 km in its central part.

The particular geologic evolution of the pre-Caspian depression during the Upper Paleozoic resulted in an abundance of tectono-sedimentation and sedimentation structures composed of carbonate rocks within its subsalt complex. The formation of these structures was probably related to a progressive development over a long period of time (Middle Devonian to Early Permian) of noncompensated paleodeeps adjacent to paleoshelf areas. The formation of the latter was accompanied by the growth of multistaged organic buildups both in their marginal parts and within the relatively deep-sea zones. As a result, extensive sequences of shallow marine carbonate deposits up to 2000 m thick or more, characterized by a wide or local area of distribution, formed in the subsalt complex of the depression.

Most of the fields discovered here are associated with these structures. Among the discoveries, we should mention Astrakhan gas condensate field, Tengiz oil field, Karachaganak gas condensate-oil field, and Zhanazhol' oil-gas condensate field. The analysis of the geological information obtained has allowed us to establish regularities in the distribution of hydrocarbon accumulations of various phase states all over the area. Oil-gas and oil fields are confined to the east and southeast flanks of the depression, while gas condensate and oil-gas condensate accumulations are located and predicted in the remainder of its territory.

Gas dissolved in the Tengiz oil is represented by 40% methane, up to 17% hydrogen sulfide, 3% carbon dioxide, and 1% nitrogen; the remaining gases are ethane and higher hydrocarbons. At Karachaganak, there is a change in gas composition with depth: the methane concentration in the in-situ mixture decreases from 77% at a depth of 3800 m to 62% at 5000 m, with a respective enrichment in heavier hydrocarbons. Such a differentiation of the gas composition may be attributed to the presence of oil "pillow."

The properties of condensates and oils also vary with depth within each field. Gas-condensate density at Karachaganak shows a progressive increase from 0.7762 g/m^3 to 0.8206 g/m^3, and below 5000 m, the fluid phase density reaches 0.8352–0.8667 g/m^3. At Tengiz, one can observe the vertical differentiation of oil properties: its density increases from 0.8055 g/m^3 to 0.8500 g/m^3, the gasoline fraction yield decreases, and asphaltene content increases from 0.09 to 1.2%. Such changes in the properties of oils, gases, and condensates are not observed in the pools Astrakhan and Zhanazhol' fields characterized by relatively small thickness of pay zones. The pools in the subsalt complex of the depression have unique fluidal systems. The latter contain commensurable quantities of gaseous and fluid hydrocarbons. The fluidal systems of gas condensate and oil-gas condensate fields have an extremely high condensate content (up to 795 g/m^3 at the Karachaganak and more than 220 g/m^3 at the Astrakhan fields), while oil and oil-gas fields are noted for very high gas rations (603 m^3 at Tengiz). The gas-condensate pools exhibit uniquely high contents of acid gas components. Their total amount is 6 to 10% in the northeastern regions of the depression and as much as 50% in the southwest. Of these, hydrogen sulfide accounts for 3–4% in the southeast and about 20–25% in the southwest.

The available information shows that gas condensates are predominant in the hydrocarbon systems of the subsalt deposits. The conditions for gas-condensate generation have been provided by a deep burial of subsalt deposits organic matter into zones of intensive hydrocarbon generation at rapidly elevating temperatures and pressures. From another point of view, the peculiarity of the fluidal systems of the pre-Caspian depression fields can be explained by their formation mechanism, particularly by transformation of oil paleopools into oil-gas condensate and gas-condensate reservoirs. Based on this viewpoint, an extensive formation of "secondary" gas-condensate pools is suggested. Differences in the hydrocarbon composition of the subsalt condensates are explained in this case by genetic differences in original oils.

The developed model explains the structure of the subsalt complex and preferential concentration of oil and gas resources in carbonate reservoirs and helps to predict their fluidal systems, which allows efficient prospecting for new hydrocarbon accumulations. The exploration strategy in the pre-Caspian depression consists in activity concentration in areas of thicker carbonate sequences, with which both identified and predictable oil-gas accumulations are known to be associated. These carbonate sequences contain complicated traps of a great capacity. Studies of their structure and regularities in distribution of sedimentary rocks with different filtration-capacity properties is very important in orienting exploration.

ZONENSHAIN, L. P., Institute of Oceanology, Moscow, USSR, M. I. KUZMIN, Institute of Geochemistry, Irkutsk, USSR, and L. M. NATAPOV, Aerogeologia, Moscow, USSR

Plate REconstructions of USSR Territory

Available paleomagnetic, paleoclimatic, kinematic, and geological data were used to compile tentative plate reconstructions of the USSR territory for Paleozoic, Mesozoic, and Cenozoic times (see figures). As can be seen from the reconstructions, this territory (and Eurasia as a whole) is a composite continent amalgamated during the Phanerozoic from numerous smaller blocks of different nature and provenance that approached one another and eventually collided, accreting a large continental mass.

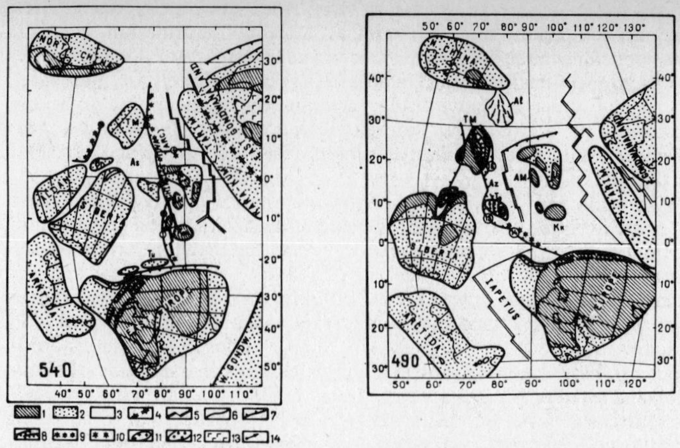

Figures—1, dry land; 2, shallow sea; 3, oceanic basin; 4, continental rift; 5, spreading axis; 6, transform fault; 7, subduction zone; 8, collision zone; 9, subduction related volcanics; 10, hot spot related volcanics; 11, granite batholith; 12, molasse depression; 13, flood basalt; 14, paleomagnetic vector.

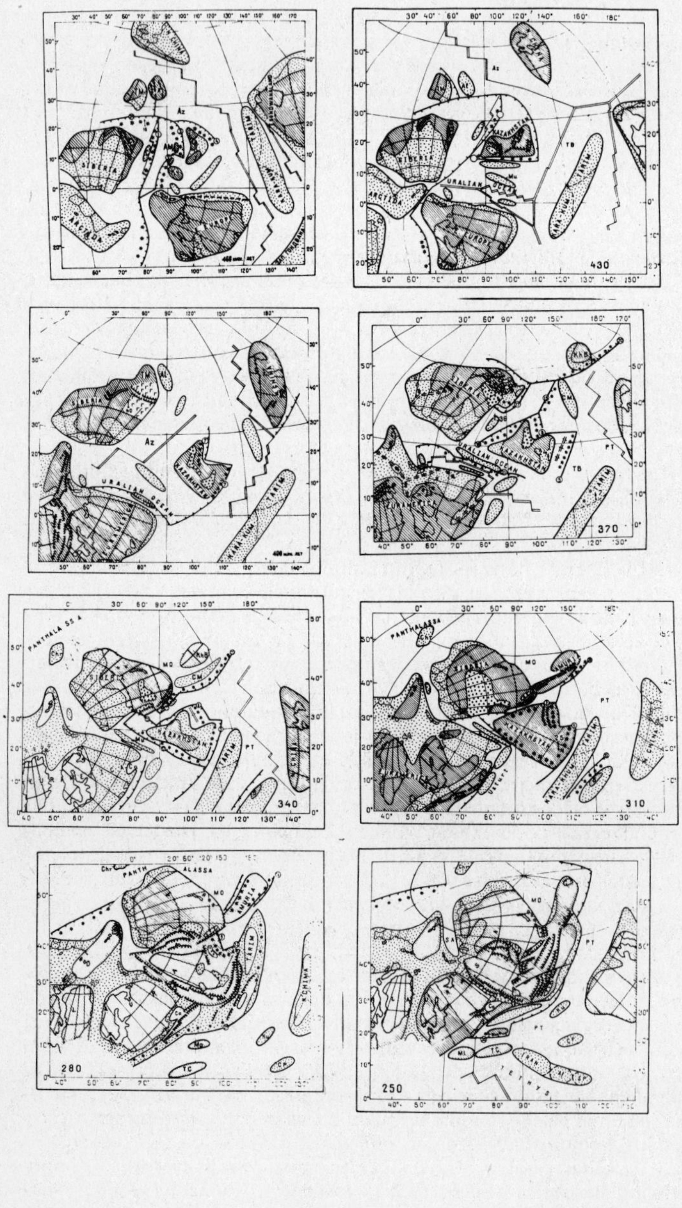

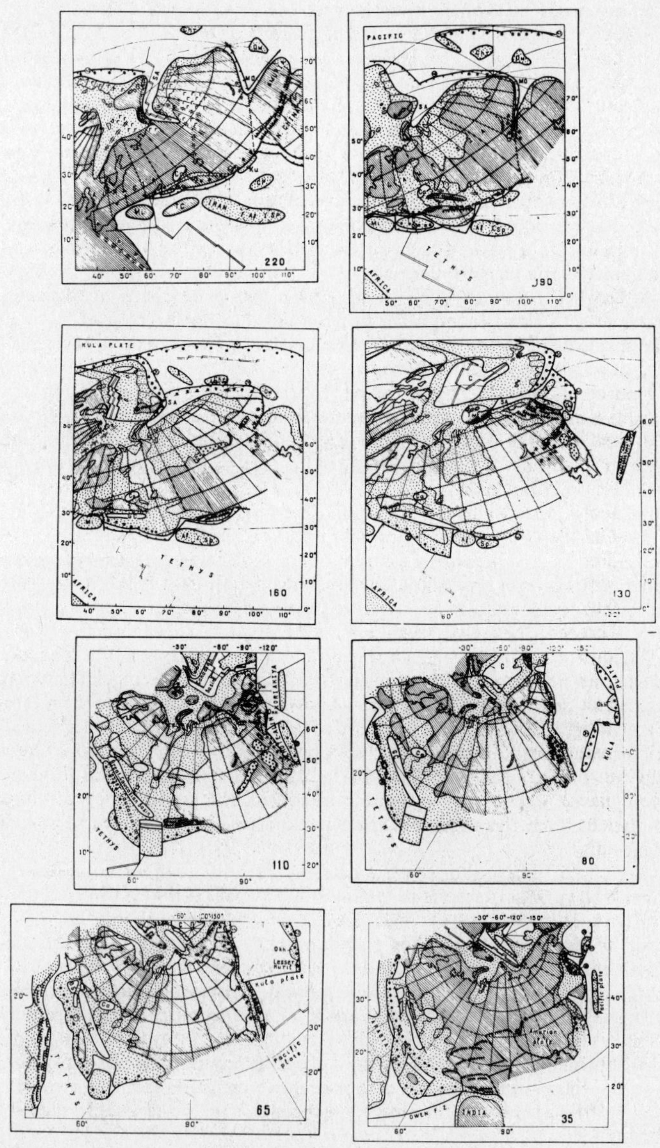

ZONENSHAIN, L. P., Institute of Oceanology, Moscow, USSR, and M. I. KUZMIN, Institute of Geochemistry, Irkutsk, USSR

Continent Accretion: Eurasia as an Example

The Eurasia continent has a mosaic-like inner structural pattern consisting of numerous stable blocks surrounded by a matrix of orogenic belts. It is clear that Eurasia is a composite continent amalgamated from numerous smaller blocks of different nature and provenance that approached one another and eventually collided to form a large continent. Orogenic belts are sutures along which smaller blocks were welded together. Main collisional events occurred in the latest Precambrian and Phanerozoic. The largest stable masses within Eurasia are cratons, or previous continents: East European, Siberian, Indian, North China, Yangtze. They constitute a framework of Eurasia. Besides, many small massifs, or microcontinents, such as Tarim, Kazakhstan, Indo-China, Tuva-Mongolian, Central Mongolia, Khingan-Byreya, Kankay, Okhotsk, Omolong, and others, were included in the growing continent within orogenic belts.

Cratons, like the whole Eurasian continent, have been accreted from several smaller blocks during the Proterozoic. For example, the Baltic and Ukrainan shields of the East European craton include Archean greenstone-granite domains (Murmansk, Kola, Belomorian, Karelian, Dneprovsky, Kirovograd, and other blocks) that are divided by Early Proterozoic thrust-nappe zones. These can be easily interpreted as sutures along which ancient blocks collided and joined together. There is evidence for Middle and Late Proterozoic collisional events as well (correspondingly in the Svekofennian and Sveko-Norvegian zones). The same amalgamation process is restored for the Siberian craton.

The younger history of the Eurasia accretion can be described as follows:

1. Latest Precambrian and Cambrian: All the constituting continents were dispersed and situated mostly in the southern hemisphere. Barentsia collided with East Europe accreting the East European continent, and Aldan shield collided with Angaria forming the Siberian continent.
2. Ordovician and Silurian: East Europe collided with North America to create the Euramerica continent. Siberia was enlarged due to collision with Tuva-Mongolian, Tomsky, and Altay microcontinents. Between the East European and Siberian continents, the Kazakhstan microcontinent was formed in the accretionary way as a result of numerous arc/arc and arc/suspected terrane collisions.
3. Devonian: It was mostly the time of continental dispersion.
4. Late Paleozoic: Almost all continental blocks converged and collided. The East European, Siberian, Kazakhstan, Tarim, and some other continental blocks were joined in a single Laurasia megacontinent. The North China continent was enlarged by attachment of the Central Mongolian block; it possibly collided with the Yangtze continent. They both remained isolated from Laurasia and Gondwanaland, situated within the Paleotethys ocean.
5. Triassic and Jurassic: Siberia (East Laurasia) collided with the China continent and with intervening Khingan-Bureya microcontinent. Also, the Indo-China block joined Laurasia.
6. Cretaceous: The northeast Eurasia margin was accreted at that time when the Okhotsk, Omolon, and Chukotka continental blocks were brought to the margin.
7. Tertiary: The main events were India-Eurasia and Arabia-Eurasia collisions. Simultaneously, many suspected terranes approached from the Pacific and attached to the eastern Eurasian margin.
8. Recent time and the nearest future: Australia is going to collide with the southeast Eurasian margin.

Eurasia is not the only example of a composite continent. Gondwanaland also was accreted from numerous smaller blocks, but this accretion took place in the Late Precambrian. From the accretionary nature of the continent, it can be predicted that deep crustal structure must involve flatly deepening discontinuities as remnants of previous collisional suture zones.

ZUCHIEWICZ, W., Jagiellonian University, Cracow, Poland

Morphotectonic Phenomena in Northern Flysch Carpathians as Compared to Southern Alps

The internal part of the Northern Flysch Carpathians comprises two types of mountain groups of various altitudes and depths of erosional dissection. The southern group (1200–1700 m a.s.l.) reveals features indicative of stronger neotectonic uplift, as compared to the northern one (700–1000 m a.s.l.). Boundaries to morphostructures frequently coincide with the strike of Miocene fault and thrust zones or follow those fractures, which cut across the substratum of emplaced flysch thrust sheets. Of special interest are neotectonic escarpments, ranging from 200 to 400 m in height. The origin of these features seems to be connected with Pliocene and Early Pleistocene times, as suggested by the displacement of Late Neogene planation surfaces. A comparison of relative altitudes of Quaternary terraces within various mountain groups makes it possible to date the main episodes of neotectonic uplift for the Neogene-Quaternary boundary, the Early Pleistocene, the Brunhes-Matuyama boundary, as well as for the Holsteinian. The last episode was of greatest intensity and extent.

A peculiar alternation of elevated and subsided neotectonic structures seen in a north-south-running section across the Northern Carpathians may resemble Quaternary large-radius folding, the wavelength of which could have attained a dozen or so kilometers. These movements might have been associated with the still continuing collision between the East European Platform and the Pannonian Block. One can consider some analogies with the Southern Calcareous Alps where the Late Eocene-Oligocene collision of the Adria microplate with the European Plate was followed by folding and uplift, while the inner part of the Pre-Alpine Zone revealed thrusting and associated uplift of nappe fronts. The outer part, however, showed mostly vertical movements. Frontal parts of nappes, active during Quaternary times, are now being uplifted, whereas their rear parts subside. Small-scale grabens, perpendicular to the fold axes, are also being formed.

Similar processes, although much less intensive, can be seen in the Plio-Quaternary evolution of the Northern Carpathians. Their mobility represents, however, a later stage of development in relation to the Alpine area, and it is characterized by relaxation of stresses accumulated during early and middle Miocene nappe emplacement.

ZUCHIEWICZ, W., Jagiellonian University, Cracow, Poland

Stratigraphy of Early Quaternary Fluvial Deposits in Northern Carpathian Valleys

Main valleys of the Northern Flysch Carpathians were formed in late Neogene times under diversified tectonic conditions. In the West Carpathians, drainage development commenced at the early Serravalian while in the East Carpathians it followed the late Serravalian. Alluvial deposits dated from earlier Neogene stages are now buried under the emplaced flysch nappes. The primary river network was oriented north-south, being deflected in the East Carpathians toward north-northeast and northeast. Principal water courses were being controlled by Miocene fracture zones. Minor changes in the valley orientation have been associated with Plio-Quaternary reactivation of longitudinal elevations and depressions.

Relative altitudes of Early Pleistocene terraces within individual valleys of the Northern Carpathians do not show any remarkable differentiation, as compared to Meso-Pleistocene ones. Differences seen in terrace altitudes along river profiles, however, seem to indicate neotectonic uplift of some elevations and subsidence within intramontane troughs.

Three distinct terrace levels dated back as far as Early Pleistocene are now preserved on valley sides, being separated from each other by rock socles cut during Tiglian A (Ponurzyca Interglacial) and Tiglian C (Waalian, Celestynow Interglacial) stages. The oldest terrace deposits occur at altitudes ranging from 120 m within intramontane depressions to 160 m in river gorges dissecting mountain ranges. Coeval deposits in the Carpathian Foredeep are represented by the Witow Series, Kozienice Fm, and the Polaniec and Osiek Gravel Series, deposited either at the turn of Plio-Pleistocene or in the earliest Quaternary. A younger terrace level, preserved at 90 to 110 m above the present river beds, was most probably formed during the Tiglian B (Eburonian, Otwock) stage. The upper part of the Krasnystaw Fm in the Carpathian Foredeep, as well as gravel series occurring in southern part of the Sandomierz Basin (eastern Carpathian Foredeep), may be regarded as time equivalents. The youngest Early Pleistocene fluvial deposits build a terrace preserved on valley sides at 70 to 90 m, its altitude being the highest in middle part of the Northern Carpathians. These deposits, together with their equivalents from the southernmost part of the Carpathian Foredeep (Sub-Carpathian Furrow), can be assigned to the Menapian (Narevian) stage.

Early Pleistocene fluvial series attain a dozen or so meters in thickness in mountainous valleys whereas in the Carpathian Foredeep their thickness increases rapidly to 80–100 m (Witow Series). Poor preservation of the gravel series, the lack of organic remains, as well as unsuccessful attempts at TL datings force one to apply morphostratigraphic and lithostratigraphic criteria aided, in some cases, by palynological analyses. The sequence of terrace flights, however, together with their relation to slope covers and younger, more precisely dated fluvial deposits (Elsterian, Saalian, Vistulian), enable a tentative stratigraphic subdivision.

ZUCHIEWICZ, W., Jagiellonian University, Cracow, Poland

Geomorphic Methods Applied to Neotectonic Studies: A Northern Carpathian Example

A number of geomorphic techniques have been applied to study the morphotectonic pattern of the northern Carpathians. Apart from classical methods, aiming at the reconstruction of Neogene planation surfaces, altitudinal differentiation of terrace rock socles within individual valleys, mutual relationships between the ridge and valley patterns, and the structure and lithology of solid rocks, a wide range of morphometric methods has also been used. Among these, the construction of base-level, goniobase, and isolong maps require most serious consideration.

Neotectonic implications can also be drawn from careful analyses of geometries of the valley network. The highest bifurcation ratios are usually confined to areas showing the most intensive Quaternary uplift. The same applies to increased drainage density measures and the N_1/N ratios, showing the ratio of first-order valleys (classified according to the Horton–Strahler's hierarchy) to the total number of valleys, although this parameter is more strongly controlled by diversified lithology of solid rocks. Parameters of longitudinal riverbed profiles are also indicative of tectonic tendences. The north Carpathian rivers reveal an eastward-directed increase in order, length, and relief, being accompanied by a decrease in the main riverbed gradient. The highest concavity measures express the river profile maturity, while the position of inflection points, marked on standardized 0–1 riverbed profiles, delimit sectors modeled by erosional and accumulational processes.

Another opportunity for morphotectonic reasoning is provided by the studies on orientation of principal stress axes, inferred from the orientation of the river network, and compared with parameters of the stress field, reconstructed on the basis of mesostructural investigations.

ZWOLINSKI, Z., Adam Mickiewicz University, Poznán, Poland

Geomorphic Adjustment of the Parseta Channel to the Present-Day River Regime

Attention was concentrated on outling the trends in the adjustment of the river channel to the regime of a lowlands meandering river of the temperature climatic zone. The supreme task was to indicate the preconditions, the directions, the rates and the effects of the changes of the Parseta River channel on the grounds of a repeated research into processes, sediments and forms occurring within the natural river channel in a short time-scale. A detailed research was undertaken on three study reaches along the longitudinal profile of the Parseta River. Three channel cross-sections were chosen within each reach differing in location and shape: a straight reach, a bend with a larger curvature radius and a bend with a smaller curvature radius.

The undertaken research task and the methodology applied revealed a different hydraulic, sedimentary and geomorphic character of bends with lower curvature ratios in comparison to bends with higher curvature radios. The indicated differences in functioning of the two types of river bends and the characteristics of the bends with lower curvature ratios substantiate the thesis that it is necessary to revise the existing opinions on development of meander bends, as well as of the processes and phenomena within them and in implications and geomorphic effects of appearing of this type of bends on the bottom of river valleys in actualistic, paleogeographic and comparable studies. A dynamic allometric analysis and adjustment models document a considerable geomorphic convergence of these bends with straight reaches of the river.

An analysis of the documentation material, relationships of at-a-station and downstream hydraulic geometry, dynamic and static allometric orientation of the river channel, values and relations of exponents of discharge cooperation (connections of water and total load discharges) and the models of adjustment of the river channel to the present-day river regime permit to formulate the following heirarchy of geomorphic adjustments of the Parseta channel system:

(a) positional adjustment: the reaches of the river channel at the bends of the higher curvature ratios are better adjusted to receive a broader range of discharges than the straight reaches of the river and the bends with lower curvature ratios; these cross-sections, therefore, are characterized by the highest morphological changeability: the cross sections of bends with lower curvature ratios in turn are characterized by a considerable stability of channel banks and a mobility of channel bottom; finally, a cross section of the straight reaches of the river characterized by a relative morphological stability favor more rapid overflows of flood waters from the channel onto the floodplain:

(b) regional adjustment: isometry of the discharge cooperation is the threshold value for determining the character of the river channel adjustment in the regional aspect; the lower study reach at Bardy is characterized by isometric features of discharge cooperation which evidences about the most natural (normal) reaction of the channel to changes of the river regime; the middle study reach at Debczyno characterized by positive allometry of the water discharge and negative allometry of the total load discharge is adjusted to the existing variability of the river regime; the majority of discharges, therefore, including high water levels and excluding the highest do not exceed the capcaity of the channel cross sections; the upper study reach at Krosino characterized by an inverse pattern of discharge cooperation is featured by the poor adjustment of the channel morphology to receive even slightly raised discharges and there are frequent overflows of flood waters as a result;

(c) system adjustment: antithetic patterns of the downstream hydraulic geometry relationships indicate explicitly to an increasing degree of downstream channel adjustment to the changing range of water and load total discharges; the Parseta River channel is particularly well adjusted in its downstream profile to receive of discharge cooperation at low and intermediate water levels zones; meaningful geomorphic metamorphosis of the river channel at high water levels determines the scale of its lack of adjustment to infrequent high and extremely high discharges; high and bankful discharges, therefore, be considered channel forming.

The presented heirarchy of the Parseta channel geomorphic adjustments analyzed against a broader background of the fluvial system permits to reveal the following characteristics of the Parseta River channel system:

(a) the river channel changes are of an episodic character both in time and space; periods of a longer relative geomorphic stability of the channel are interspersed with periods of short term rapid changes and reaches of a relatively stable morphology are interspersed with reaches characterized by more sizable and faster morphology changes;

(b) a differentiated mechanism of modern channel changes results from the impact of numerous factors and elements controlling the fluvial system, among which of major importance are widespread influence of intrinsic thresholds rather than extrinsic ones, continuous and strong interrelationships between fluvial processes within the river channel and geomorphic processes within the whole drainage basin, marked adaptation of river channel adjustment rate to modes and time of reaction and relaxation in a given geomorphic and hydrometeorological situation;

(c) periodical, although meaningful preponderance of the aggradation trend over continuous, mainly in time and less pronouncedly in space, degradation trend of the river and an insignificant rate of the channel migration determine a short time dynamic equilibrium stage with a not too high amplitude of geomorphic and hydrological disturbances in the channel and in the drainage basin, an amplitude decreasing downstream for the Parseta River channel system.

The following abstracts were received too late to be included with the major alphabetical listing.

A

ABLAEV, ALBERT G., Pacific Oceanological Institute of the Far East Branch, Academy of Sciensis of the USSR, Vladivostok, USSR; Igor V. VASILIEV, All-Union Geological Institute, Ministry of Geology of the USSR, Leningrad, USSR.

Data on Ecological Crises and Cenozoic Florogenesis of the Far East

During the century-long history of studies of Tertiary floras of the Far East enormous material has been gathered, whose study enabled a construction of the framework based on the chronology of the taxa at different levels of hierarchial and phylocoenogenesis. The works of the last years present a clear trend of the association of the distinguished floral assemblages with the established climatic boundaries, regarding the history of floral evolution as an ecological process. A significant example is the so-called beech horizon (with an elevated content of beech pollen, up to 35-40%) of the climatic optimum of the Miocene epoch. The equivalents of this horizon discovered by G. G. Kartashova (1974) for the Yana-Omoloi Interfluve were subsequently reported from many areas of East Asia, including Primorye and the Korea Peninsula, and placed onto a single age level with thermophilic biota of the end of the Early-beginning of the Middle Miocene of the Daijima stage in Japan. It is necessary to have a differentiated approach to an estimate of space-time relations between certain paleobiocoenoses from the viewpoint of ecological-phylocoenogenetic features. By all means, beds with a high content of *Fagaceae* in Primorye (Rettikhovka, Kraskino etc.) and Korea (Engelhardtia, Hamjinian etc.) should be excluded from "coeval" ones: floras have a temperate or moderately warm appearance and do not mark the warming peak in the Miocene.

Our observations of the Cenophytic floras of the Far East, primarily, the Sea of Japan area, allow regarding the evolution as a process of transition of quantitative changes into qualitative ones by means of a long-term transformation: the succession is established from the Late Mesophytic floras and an abrupt change in the composition of floras confined to boundaries of adjacent systems (Cretaceous and Paleogene, Paleogene and Neogene, Neogene and Anthropogene) taken as "turning-points" is rejected. The organized combination of the live matter maintains the succession and retains basic (physiognomic) characters of the organism under a continuously changing habitat. In accordance with such an interpretation, the absence of demarcation lines in flora (and fauna) evolution and natural phenomena over the planet is confirmed, as well as an non-coincident rhythmic pattern of the live and inanimate matter. Biogeocoenotic links are a source of self-motion.

AFSHAR, M. EHTESHAMZADEH, Faculty of Engineering, University of Teheran, Teheran, Iran.

Evolution of East Central Iranian Major Faults During Late Alpine Foldings.

Major deep seated faults have played important role in the structural evolution of the Iranian plateau since late Pre-Cambrian time. Iranian plateau contains a broad zone of compressional deformations between two rigid blocks of EURASIA and ARABIA.

The compressional motions between these two blocks resulted in continuous shortening of the continental crust in approximately NE-SW direction. Consequently an elongation in perpendicular direction (Maximum principal strain) took place.

N-S trending fault systems in east central Iranian Block (Lut-Block) remained unchanged during late alpine orogenic movements. In west of this Block, the same fault system underwent pronounced deformations. Lut Block as a rigid mass has diverged the direction of the movements. This divergence resulted in formation of Wedge Shape structures, bending the fault planes and rearrangement of fold axes. The new displacements of reactivated strike-slip-faults would be right-lateral north of the wedge and left-lateral south of the wedge.

The mentioned mechanism will be discussed in this paper, with regard to application of plate tectonic concepts in Iran.

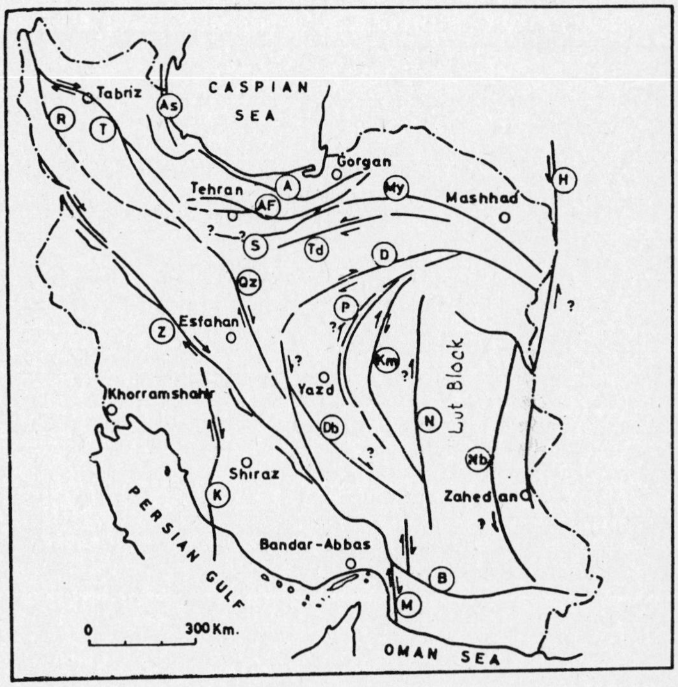

Faults of unknown horizontal displacement

Strike-slip faults, dextral

Strike-slip faluts, sinistral

Letters inside of circles indicate names of faults: A) Alborz; AF) Abyek-Firuzkuh; As) Astara; D) Doruneh; Db) Dehshir-Baft; M) Minab; H) Hari Rud; K) Kazerun; Km) Kalmard; B) Beshagerd; My) Mayamey; N) Nayband; Nb) Nehbandan; P) Posht-e-Badam; Qz) Qom-Zefreh; R) Rezaiyeh; S) Semnan; T) Tabriz; Td) Torud; Z) Zagros; Main faults in Iran.

ALLOWAY, B.V. and R. B. STEWART, Department of Soil Science, Massey University, Palmerston North, New Zealand

Late Quaternary Aeolian Quartz Influx in a Terrestrial Environment, Western North Island, New Zealand

A unique record of late Quaternary climate change is preserved within a terrestrial coverbed sequence of dominantly andesitic provenance in Western North Island, New Zealand. The coverbed sequence comprises loesses, tephras and paleosols (Figure 1) on a marine bench cut by the c. 130 ka B.P. high sea level stand correlated with last Interglacial oxygen isotope stage 5e. Stratigraphic control is provided by the presence of both rhyolitic tephras from central North Island and distinctive andesitic tephras from Egmont volcano. The stratigraphic record indicates the presence of five loess units, identified on their physical properties and field characteristics (L1 - L5, Figure 1), deposited during the last c. 130 ka B.P.

Quartz contents of the coverbed material were determined by quantitative X-ray diffraction (XRD). The presence of quartz in any quantity in these materials is anomalous, in as much as the andesitic parent materials contain negligible quartz. Given that the prevailing wind is westerly and that andesitic coverbeds extend to the present western coast, the only source of aeolian quartz possible is the continental shelf. This would only be exposed during a period of low sea level such as would occur under glacial conditions. Additional quartz in the fine grain sizes (<20μm) would also be contributed from southern hemisphere continental land masses as aerosols, particularly Australia, and would reflect intensified circulation of the southern hemisphere westerly wind belt during glacial times. (Thiede, 1979; Petit *et al.*, 1981).

Steward and Neall (1984) demonstrated that quartz accumulation increased in marine sediments off the east coast of North Island, New Zealand, during the last stadial of the last glacial. Petit *et al.*, (1981) showed that microparticle concentration in an Antarctic ice core from Dome C also increased during the last stadial and De Angelis *et al.* (1987) obtained similar results from an Antarctic core at Vostok using Al concentration as a measure of continental microparticle input. De Angelis *et al.*, (1987) also used Na_m (marine sodium input) as an indicator of marine aerosol generation, and therefore wind strength to show that wind intensity had increased during the last glacial.

The results from the ice core studies also showed that during the last c. 150 ka only three periods of markedly higher microparticle and Na_m concentrations at c. 150 ka B.P., 60 ka B.P. and 20 ka B.P. These were interpreted as the only times where 'full' glacial conditions (i.e. similar to the last stadial) were developed.

In the coverbed sequence examined in this study, spanning the last c. 130 ka B.P., only two periods of elevated quartz accumulation occurred, the oldest corresponding to L3 at c. 70-80 ka B.P. and the youngest to L1 at c. 22-13 ka B.P. We interpret our results as indicating that full glacial conditions were only reached during these time periods, corresponding to oxygen isotope stages 2 and 4 (Imbrie *et al.* 1984), when sea level dropped sufficiently to expose a source of quartzose material on the continental shelf to the west of the North Island. The remaining three loess units were derived from localized sources within the andesitic terrain during periods of possibly cool, but not full glacial, climate.

Loesses L4 and L5 (Figure 1) are correlated with cooler intervals (oxygen isotope stages 5b and 5d respectively) of the last Interglacial.

This study shows that a unique record of paleoclimate changes during the last 130 ka has been preserved in a volcanic terrain. The data is consistent with paleoclimatic changes from both marine and ice core sediment accumulation and can be correlated with the now well established marine sediment oxygen isotope record.

REFERENCES

De Angelis, M., Barkov, N.I. and Petrov, V.N. (1987). *Nature* 325: 318-321.

Imbrie, J., Hays, J.D., Martinson, D.G., McIntyre, A., Mix, A.C., Morely, J.J., Pisias, N.G., Prell, W.L.C. and Shackleton, N.J. (1984) *in* Berger, A., Imbrie, J., Hays, J., Kukla, G. and Saltzman, B., *ed.* Milankovitch and climate — understanding the response to astronomical forcing. Part 1. Holland, Riedel pp 269-305.

Petit, J.R., Briat, M. and Royer, A. (1981). *Nature* 293: 391-394.

Thiede, J. (1979). Geology 7: 259-262.

Stewart, R.B. and Neall, V.E. (1984). *Nature* 311: 47-48.

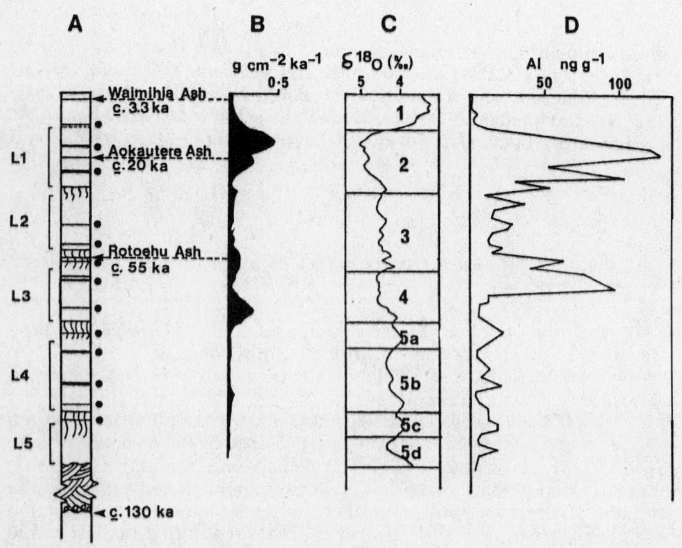

Figure 1.

A. Onaero Section. L1-L5 are loess units, paleosols are shown by wavy lines and dots indicate andesitic tephras from Egmont Volcano. Arrow heads indicate the stratigraphic positions of named rhyolitic tephras from central North Island.

B. Quartz accumulation rates from the Onaero Section.

C. Oxygen isotope curve from Imbrie *et al.*, (1984). Numbers indicate oxygen isotope stages.

D. Aluminium concentration in the Vostok core of De Angelis *et al.*, (1987).

ALFÖLDI, LÁSZLÓ, Research Centre for Water Resources Development — VITUKI, Budapest, Hungary

Hydrogeological Aspect of Toxic and Hazardous Wastes in Hungary

In any study on the geological aspects involved in the safe disposal and control of toxic and hazardous wastes, the formations which control the spreading of polutants and especially the conditions in the vicinity of surface deserve special attention.

The vulnerability of groundwaters to pollution depends basically on the geological environment. Geological events and processes determine the paths of seepage flow, of which the geological structure particular to Hungary provides a broad spectrum of examples.

The loose sediments in the alluvial plains, debris cones, lacustrian and maritime deposits contain a variety of open, unconfined groundwaters and artesian aquifers. Pollutant infiltration from the surface is controlled decisively by the granulometry, stratification and cross bedding of the layers. Simple mass transfer combined with physico-chemical processes influence pollutant spreading in the course of vertical infiltration. The relative values of vertical and horizontal permeability of the aquifers are also important factors. In thick and multi-horizon groundwater reservoirs natural of artifically induced horizontal flow may deflect vertical downseepage, giving thus rise to a variety of spreading patterns. The differences in pressure propagation and material migration, in the mineralogical composition, micro- and macrostructure of the aquifers may influence the spreading process considerably. The impervious or barrier layers in loose sediments are more effective seals than consolidated rocks, where secondary porosity and transmissivity tend to play predominant roles.

The outcrops of fissured or karstified formations in consolidated rocks are most voulnerable, most exposed to pollution, while under a covering layer their behaviour depends on local, particular conditions. Since human measures, such as withdrawals, structures, etc., may modify fundamentally the natural intake areas of, the flow paths and velocity distribution in aquifers, hydrogeological conditions must be viewed in their dynamic variability espectially when considering alluvial, bank filtered systems and generally shallow aquifers, karstified groundwater reservoirs, etc.

The success of any mathematical model formulated to describe and simulate the vulnerability of, and pollutant spreading in particular aquifers depends on how accurately the hydrogeological conditions can be exploted and how true a hydrogeological model can be constructed.

In any problem related to hazardous and toxic wastes the analysis of surface conditions and the shallow zone assumes paramount importance, so that the geological factors must be considered in their interrelations with meteorological, pedological, ecological and other factors. In the vicinity of the surface geochemical and geomicrobiological processes play an especially important role, which is influenced also by the type (continuous, or intermittend) and conditions (saturated or unsaturated) of downseepage.

In view of the fact that toxic and hazardous wastes represent a point-source of pollution, the particular site must be explored thoroughly for any geological analysis. By studying the geological conditions in a region and by outlining the areas round vulnerable and safe, limitations can be imposed on the disposal sites, and conversely, sites can be selected where the disposal of toxic and hazardous wastes is likely to involve the smallest hazard of pollution.

ANDREESCU, ION, N. TICLEANU, Institute of Geology and Geophysics, Bucharest, Romania

Middle-Upper Neogene Coal-Generating Phases in Romania

The Sarmatian-Pleistocene deposits on the Romanian territory are assigned to two paleogeographic areas: the Dacic Basin, external to the Carpathians, and the Pannonian Basin, inwards.

Structurally, these deposits either belong to the cover of both the Moldavian and Moesian platforms, or constitute the upper molasse members within the Carpathian fore-deeps and intramountainous depressions.

Specific sedimentation conditions of the mentioned areas favoured, during certain time intervals, the presence of coal-generating phases (Table 1).

The study of the sequence of those phases leads to the following main conclusions:

1. They succeed always to tectonic events;
2. Within a sedimentation cycle they occur either at the beginning of regressions, or at their end, but never occur during maximum regressional or transgressional stages;
3. Against the regressive tendency, coal-generating phases occur after the freshenning of basin water. There is an interdependence between the diversity and frequency of freshwater rheophid molluscs and the importance of coal-generating phases. During those intervals, dominated by coal-generating phases, the freshwater rheophid molluscs diminish drastically and the stagnophid ones develop (e.g. Anodonta, Hyriopsis, Cristaria, Planorbis etc.). At the same time the vegetal associations consist exclusively of plants adapted to coal-generating freshwater marshes. The main taxa of the phytomass are: Glyptostrobus europaeus, Acer tricuspidatum, Phragmites oeningensis, Typha latissima (A-E in tab. 1); Byttneriophyllum tiliaefolium (B-E); Taxodium dubium, Myrica lignitum (A, ?B); Liquidambar europaeum (A-C); Alnus cecropiaefolium, Betula macrophylla (B,C); Osmunda parschlungiana, Stratiotes tuberculatus (C); Salix macrophylla, Carya ex gr. leiodermis, Osmunda regalis (D); Nelumbium hungaricum (A,D); Salix varians, S. pliocenica, S. fragilis, Stratiotes intermedius (D,E);
4. The coal-generating phases lack from certain Upper Miocene freshwater periods (tab. 1) due to the absence of optimum subsidence conditions, of alluvial plains, or to climate conditions. In the Paratethys area the Upper Sarmatian-Upper Meotian time-span is marked by an arid climate which led to the proliferation of xerophite flora and the expansion of Vallesian and Turolian land mammals;
5. The significant extension of Pliocene coal-generating phases points to favourable deposition environments, i.e. optimum subsidence within vast alluvial plains, and luxuriant development of Pliocene Arcto-Tertiary flora as well.

Since the phytomass accumulation, lithotypes setting, spatial development and coalification degree of coal seams are dependent on many factors (tectono-sedimentary, paleogeographic, paleoclimatic, paleofloral etc.), it is to infer that the coal-generating phases can be regarded as suitable indicators of geological history of a given area.

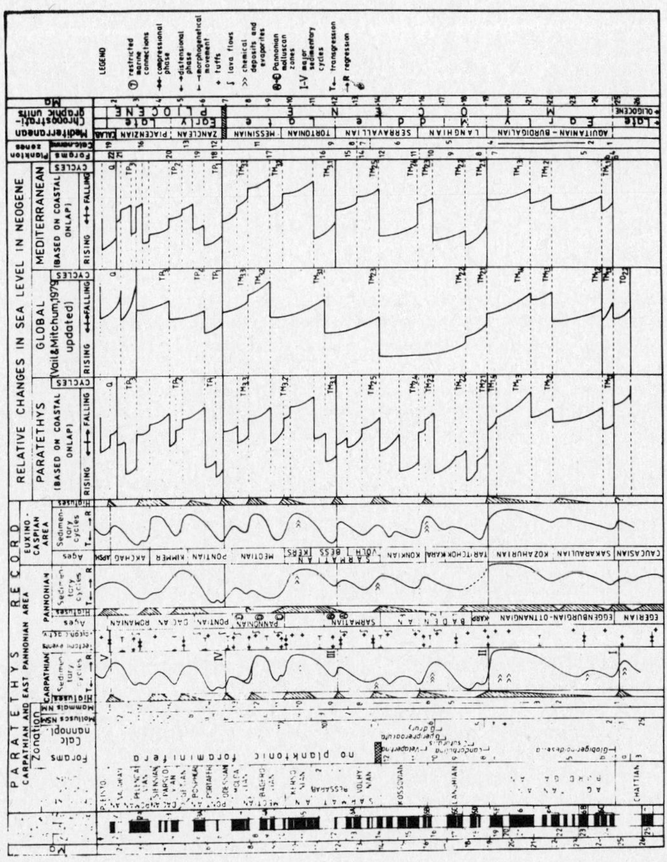

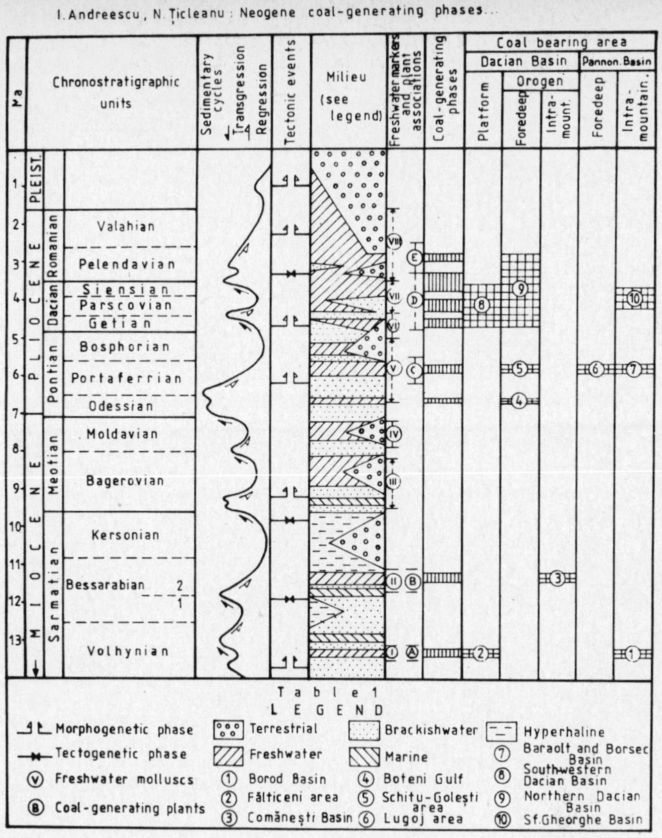

ANDREEV, G. V., The Institute of Geology, the Siberian Department, the USSR Academy of Sciences, Ulan-Ude

The Origin of Salic Alkaline Rocks of South Siberia

The belt of Middle Paleozoic alkaline rocks is traced in the south folded frame of the Siberian platform the north end of the Baikal lake at the distance of 450 km in the north-eastern direction. The intrusives composing the belt are united into the Synnyr complex. Nine intrusives are established. They are controlled by the large regional faults of north-eastern strike. Their sizes vary from 0,5 to 464 sq.km and the forms are isomeric, ellipsoidal and sometimes platy in a plane. They are formed by alkaline, nepheline, pseudoleucite syenites and synnyrites. Alkaline granites, quartz syenites and granosyenites are met as secondary rocks.

A brief description of typical intrusives is given below. The largest Synnyr intrusive has an isometric form in a plane. Its cross sizes vary from 23 to 26 km and its area is 564 sq.km. Concentric zonal structure is distinctly expressed. Alkaline trachytoid syenites (pulaskites and hedrumites) are exposed in the central part of the pluton. They form a stocky body with the cross sizes of 16,0 x 6,5 km. The most part of the intrusive is composed of nepheline and pseudoleucite syenites which form a pipe-like body of inconstant thickness reaching 3,5 km in the western pluton part and 12 km in the eastern part. Pseudoleucite syenites are exposed mainly in the inner zone of ring intrusion bordering on alkaline syenites. The line width of these rocks varies from 750 m to 3,5 km. The places formed by synnyrites are marked out among pseudoleucite syenites. The Kalyumna place, the largest of them, has the sizes of 7,0 x 2,7 km. Vein and dike formations of the intrusive are presented by sviatonossites, shonkinites and alkaline pegmatites. The strong predominance of potassium oxide over sodium oxide is characteristic for all the rocks of the intrusive.

The Goudzhekit intrusive is located at the distance of 30 km to the west from the north end of the Baikal lake, in the zone of a large regional fault. It presents a typical fractured body oriented by the azimuth of 40° and having high, almost vertical dip. The intrusive is 2,5 km long, its thickness is persistent and equals 200 m. It is exposed in the touch of Early Proterozoic granites and granite gneisses with the crystalline schists and gneisses of the same age. The rocks erupted are located in the north-west and metamorphic ones — in the south-east. North-western and south-eastern touches become complicated because of high-dipping tectonic dislocation. The intrusive is composed mainly of biotite nepheline syenites. Small lens-like bodies of biotite-riebeckite mepheline syenites and melteigites are marked out among them. The distinctive feature of the chemism of the rocks is the predominance of sodium oxide over potassium oxide.

The Akit intrusive has an isometric form in a plane. Its area is 4,5 sq.km. Enclosing rocks are presented by Early Paleozoic granites and granosyenites. The intrusive is formed mainly of pyroxene-amphibole syenites. In its central part a shoe-shaped body of fine-grained aegerite-augite granites and granosyenites which are turned with its convex side to the south is exposed among syenites. The rocks of the third phase are presented by nepheline syenites which form dike-like bodies of north-eastern strike. The largest body is exposed in the northern part of the intrusive. It is traced along the strike to 55 m and its thickness reaches 150 m. The sizes of other bodies are less. They are traced along the strike to 70-300 m and their thicknesses vary from 15 to 40 m. The most part of nepheline syenite dykes is located in the exocontact of the intrusive.

The main peculiarities of the petrochemistry, geochemistry and mineralogy of the Synnyr complex rocks are as follows. Leucocratic rocks predominate in the composition of intrusives: syenites, nepheline syenites, synnyrites, granosyenites and sometimes granites. Mesocratic nepheline syenites are exposed at small places only in the Synnyr intrusive. Alkaline gabbroids prsented by the dyke-like bodies of shonkinites and fergusites in the Synnyr intrusive compose less than 0,005% of the whole mass of alkaline rocks. The Synnyr complex rocks are closer to granites than to basic rocks and alkaline gabbroids in uranium and throium content and in throium-uranium ratio. The contents of the most part of trace-elements in the rocks of the Synnyr intrusive, viz. tantalum, boron, vanadium, copper, zinc, lead, tin, rubidium and caesium, are comparable with their clarks in acid rocks (Osokin, 1970). Lithium and beryllium contents occupy an intermediate position among the clarks of these elements in basic and acid rocks. And only zirconium and rare-earth elements are contained in the rocks of the intrusives in quantities near to clerks in basic rocks.

The author and other researchers have studied the peculiarities of macro- and microadmixture distribution in the magnetites from the rocks of peridotite-pyroxenite-gabbroic formation, from the granites of various complexes and from the Synnyr complex alkaline rocks. It has been resulted that the magnetites from the Synnyr complex rocks are identical to the magnetites from the rocks of crustal origin in macro- and microadmixture content and distribution. G. A. Murina and others (1980) have determined Sr^{87}/Sr^{86} ratio. It is equal to 0,715. Smaller figures (9,708) have been got by V. F. Posokhov and V. L. Shalagin. Both figures are characteristic for crustal rocks.

According to the data cited above on petrochemistry, geochemistry, the peculiarities of magnetite composition and the isotopy of the Synnyr complex rocks it is possible to draw a synonymous conclusion of parent magma formation at the expense of a crustal substratum of "granitic" composition. However, it is necessary to note that getting alkaline melt from granitic substratum is possible only under the one passing the stage of metasomatic transformation. The possibility of such a process is dealt with in detail in G. V. Andreev's work (1981).

Its essence is as follows. The intensive entering of water fluids from the mantle is taking place in certain zones of the earth's crust during some epochs. The fluids have potassium, sodium and aluminum as main components in their composition (Korzhinsky, 1952; Vinogradov, 1961, 1964). Granitic substratum is changed into alkaline rock under their influence. Further the delivery of a light heated material to these zones is taking place (Artyushkov, 1979). The influence of this material on metasomatic rock results in the formation of alkaline sialic melts. In some cases a magmatic hearth encroaches a little-changed granitic substratum. As a result, the melts of granitic and syenitic composition appear. The suggested mechanism of alkaline melts' formation satisfactorily explains the 1250° high temperatures of the homogeneity of melt inclusions in the Synnyr complex rocks (Panina, 1978). D. K. Bayley (1970) has suggested the model near to the one under consideration.

AVIAS, J. V., Cerga, Montpellier, France

Geologic Control on the Evolution of Regional Aquifers of the World

Geologic control on the evolution of Karstic Aquifer Systems of Mediterranean type the example of "Source du Lez" K.A.S. (North Montpellieran Karstic Area-France S.E.)

Karstic aquifer systems evolution controls are essentially climatic (and paleoclimatic) geologic and biologic.

After climatic controls ruling water imput in the system, geological controls are the main ones.

They included.

1st - Lithological (and paleosedimentological) controls

2d - Tectonical controls

3d - Glyptogenic and base levels controls

4h - Other geological controls (magmatism, metamorphism, vulcanism. Mediterranean type K.A.S. are located in the middle (Bajociam) and mainly upper (Berriasian to Portlandian) jurassic, and upper cretaceous (urgonian) coral reef and open sea limestones or dolomites.

Finegrained Limestones of great purity of *upper jurassic* age are the main rocks involved in K.A.S. of "Source du Lez" (Les Spring). Evolution of Karstification processes, is geographically linked with space distribution of different facies of the *coral reef and open sea sedimentary body* involved. The maximum karstification processes are linked with the fine grained open sea carbonate rocks.

Tectonic control explains the evolution of the sedimentary body geometry and fracturation through folding and breaking processes which began at the end of Hercynian times, linked with European and African plates drifting interaction between Hercynian times till present.

After the tension relaxes distensions of the end of Hercynian foldings then occured *progressive compression* of South and South West towards North and North East direction. This compression gave rise to "Cevenol strike-slip faulting system and inverse faulting and foldings of E - W. general direction. The paroxystic phase of upper cretaceous to lower oligocene age (Pyrenean foldings) ended with two E-W lines of *overthrust* - South, to north - ("Pli de Montpellier" and "Voils le Fort anticline - Pic St Loup line") which bring up to surface impermeable layers and play a dam role for upper part of the aquifer (700 meters thick intotal) and drainage system.

Then began *oligo-aquitanian distension* major phase with Pyreneo-Provençal chain collapse which gave rise to "Golfe du Lion" new gulf, and a net of major faults of S.S.W. - N.N.W. direction like "Faille de Corconne" major drain of the Lez Spring Karst Aquifer system, and "Faille de Nimes" with vertical displacement til 4 km deep, and filling of tertiary and quaternary deposits actually covering limestone aquifers.

A very slight compression of Alpine age, but not reaching truly the Montpellieran Karstic area, which is still in distendion, with mediterranean "golfe du Lion" collapse.

Combined with the paleosedimentologie controls and, finally, sea level fluctuations and stagnations which give or gave main "base levels" for the Karstic system ("base levels" ruling levels of main chanelisation and cavitation) tectonics explain the powerfulness of K.A.S. of North Montpellieran Karsts and its very efficient underground drainage system. This system is, as a matter of fact constituted of a main drain (Corcome Fault and Lez Fault, feeded by latera drain along the northern overthrust, and communicating with Karstic aquifer natural units ("Mangin systèmes Annexes") on both sides of the main drain.

To add that, isotopic studies are showing that to the alimentation basin of the limestones outcrops must be added a part of the northern background Cevennes Mountains Cristalline massives surface running waters; A part of these surface water going down in the Karstic system, along with erosional boundary between cristalline substratum rocks and carbonate rocks.

B

BACKARDJIEV, SVETOSAR DIMITROV, Higher Institute of Mining and Geology, Sofia, Bulgaria, Georgi Kostadinov ZAHARIEV, Bulgarian Academy of Science, Sofia, Bulgaria

Geological and Structural Causal System and Emperic — Fractial Modeling of Ore Deposit

The ore structure is viewed as attracting for the orebearing streams (geochemical field); limiting the volume of space-disipation, forming conditions for ergodicity concentration into one part of the space is compensated by absence in another part of the space; forming quazi-scale nonvariable fractal nature.

The ore bearing system is dynamic since variations in the concentration of the solutions can be identified.

The basis for this methodology is a computer program, which is utilized to study the gradiets of the concentration and their velocities related to the space coordinates. The System NPLRG was reported by us at the 27 IGC (Moskow, 1984) and this work represents a natural continuation of original research.

An important result is Hausdorff metrics for the complexity of the object obtained by original methodology. By solving the inverse problem it allows us to model deposits with high complexity. If we take into account the special points it is possible to model the space position and partially the morphology of the deposit. In that sense our results are closer to those of Benua Mandelbroth (IBM Research Center) for modeling natural objects. In our case we utilize partial basic points and emperical complexities which increases the validity of the modeling.

We will explain and demonstrate the basic algorithm, test results from real geological objects together with qualitative and quantitative interpretations.

BAKUN-CZUBAROW, NONNA, Institute of Geological Sciences, Polish Academy of Sciences, Warszawa, POLAND

On the Mantle Mineralogy of Terrestrial Planets

The paper deals with the two aspects of the mantle mineralogy: silicon coordination in the interior of terrestrial planets and with the 3 GPa norm assemblages within the planet mantles, that might be to some extent dependent upon the cosmochemical models applied.

In all silicate minerals under crustal conditions silicon is coordinated to four oxygens, but at high-pressure conditions there are known phases with octahedral silicon (Mao et al. 1977, Yagi et al. 1978). Taking into account the complex models of the terrestrial planet interiors and possible boundaries of the high-pressure transformations it is reasonable to conclude that Moon and Mercury are devoid of octahedral silicon, in the Mars lower mantle should appear a small amount of silicon coordinated by six oxygens in garnet solid solutions, whereas in the mantles of Venus and Earth ^{VI}Si dramatically dominates over ^{IV}Si.

In order to model the 3 GPa mantle assemblages and their STP densities for the terrestrial planets the program MANTLE (Smyth, 1981) was applied for two sets of cosmochemical data, i.e. the planet compositions obtained in chondrite model as well as those predicted by modified equilibrium condensation model (BVSP, 1981). For both sets of data comparable results have been obtained. Moreover they are consistent with the experimental results for equilibrium condensation model compositions (BVSP, 1981). In the MANTLE algorithm (Smyth, 1981) the terrestrial planets follow either Si-poorer (wehrlitic) or Si-richer (lherzolitic) path of Si-rich branch of differentiation of the Al and Ca-poor material.

The major determinant of the both the silicon coordination and the 3 GPa mineral assemblages is size of the planets (Fig. 1). Mantles of the small planets Moon and Mercury are most probably devoid of octahedral silicon, whereas in the mantles of large planets Venus and Earth there is a strong predominance of octahedral over tetrahedral silicon, as the magnesium metasilicate in perovskite-type structure is supposed to be the most abundant mineral of their mantles.

The 3 GPa assemblages of small planets (Moon, Mercury and Mars) are of wehrlitic (orthopyroxeneless) nature, whereas in the mantles of large planets at comparable p-T conditions the lherzolitic (olv + opx + cpx + gar) assemblages should be stable. Moreover the content of the main mineral olivine increases with the planet size, while that of clinopyroxene & garnet decreases.

REFERENCES

BVSP (1981) Basaltic Volcanism Study Project (1976-1979), Basaltic Volcanism on the Terrestrial Planets, 1286pp.

Mao H.-K., Yagi T. and Bell P.M. (1977), Carnegie Inst. Washington Yearbook 76, 502-504.

Smyth J.R. (1981), Computers and Geosciences 7, 27-34.

Yagi T., Mao H.-K. and Bell P.M. (1978), Phys. Chem. Minerals 3, 97-110.

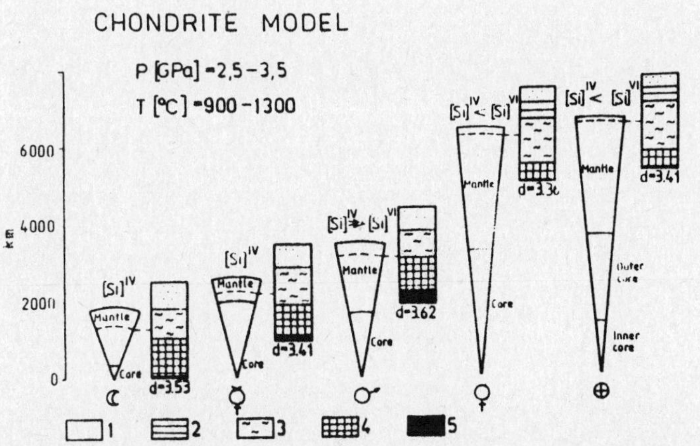

FIGURE 1. Interiors of the terrestrial planets. The sections are given in the following increasing radius order: Moon, Mercury, Mars, Venus and Earth. Apart from the internal layers and 3 GPa isobars, modes and densities of the chondrite cosmochemical model mantle materials under the pressure of 3 GPa are given as well as silicon coordination throughout the mantles is emphasized. 1 — clinopyroxene, 2 — orthopyroxene, 3 — olivine, 4 — garnet, 5 — oxides.

BENITEZ, STALIN, Guayaquil, Ecuador

Landslides: Extent and Economic Significance in Ecuador

Ecuador, one of the South America smallest country is however one of the most hazardous region from the point of the landslide phenomena because of the very high regional slopes associated to the Andean cordillera which crosses the country from north to south.

The present article appears to be the first that targets a synthesis on landslides problems throughout Ecuador. This approach has been done studying the most important events reported by the national communication media in the 1.947-1.987 period. Though this is a very restricted data source, there have been delineated the main concentration zones of landslides events. In order of importance these are: "The Pallatanga-Chunchi Alley" in the west flank of the Andean Cordillera, and the cities of Guayaquil (the more populated of the country) and Quito (the capital of the country). Other most important events zones are located along the roads which traverse the highest slopes of the cordillera.

The landslide that caused more casualties was the Chunchi landslide which killed more than 150 people in the 1983. In that year occurred the highest number of events caused by the strongest rainny season of the century.

There have been reported in this work only the gravity induced phenomena, however, other very big magnitude landslide events tectonically or seismic induced haven't been reported like those of the 5-III-87 earthquake which caused very big landslides in the skirts of volcan Reventador.

BERGMAN[1], J., P. WEIHED[2] & U. BERGSTRÖM[2], [1]Geological Institute, Uppsala University, Uppsala, Sweden, [2]Department of Geology, University of Gothenburg/Chalmers, University of Techology, Gothenburg, Sweden

Genesis and Structural Evolution of Early Proterozoic Gold Lode Deposits in the Skellefte District, Northern Sweden.

The early Proterzoic Skellefte district consists of a volcanic succession with mainly acid, marine pyroclastics, intercalated in their upper parts with basic volcanics and sediments. The supracrustal sequence is interpreted as subduction related in a volcanic arc setting. Several generations of gabbro-granitoid suites intrude the supracrustals. The two main suites are the primitive synkinematic Jörn tonalites-granites, coeval (1.89 Ga) and comagmatic with the volcanics, and late to post kinematic, crustal derived, Revsund granites (1.75 Ga).

Historically the district has produced base- and precious metals from massive sulphide deposits, including the well known Au-rich Boliden deposit. In recent years, however, intense prospecting for gold has revealed several types of epigenetic gold mineralizations. These mineralizations occur both in supracrustal rocks and in large gabbroic to granitoid batholiths. on a large scale the deposits are all associated with, and spatially controlled by tectonic zones. On a deposit scale brittle-ductile tectonic features such as faults, shear zones and quartz veins are ubiquitous. A rough classification of the Au mineralizations can be made with respect to host rock lithologies (note that this is by no means a genetic classification):

A. Mineralizations associated with intrusives

1. Cu-Au mineralizations within shear zones in synkinematic granitoids.
2. As-Au mineralizations spatially associated with small porphyries within sedimentary sequences.
3. Quartz vein related gold in contacts between synkinematic granitoids and supracrustals.
4. Gold in fracture zones within gabbroic intrusives.
5. Minor Cu-Ni-Au mineralizations within large basic dykes.

B. Mineralizations without any known intrusive relationship, within the supracrustals.

1. Quartz vein related As-Au within sediments.
2. Quartz vein related As-Au within volcanics.

C. Epigentic Au in quartz veins overprinting massive sulphide deposits.

The mineralizations within synkinematic granitoids (**A1** and **A3**) are both related to quartz veins with variable amounts of calcite, chlorite, and epidote. In **A1** type deposits electrum gold is situated on edges or joints in pyrite grains, the main ore mineral. Alteration is often sericitization (phyllic) in proximal parts and propylitic alteration in distal parts.

The **A2** mineralization type is characterized by small scale brittle deformation accompanied by quartz veins with arsenopyrite and gold. Alterations involve sodium enrichment (albitization).

The newly discovered fracture zone related monometallic gold occurence in a synkinematic gabbro (**A4**) is very poor in sulphide minerals (<1%). It occurs as a tabular body of very closely spaced thin quartz veins and amphibole filled fractures, probably formed by hydraulic fracturing of the gabbro in an extensional direction of a large shear zone. Gold is not confined to the quartz veins, but occurs also in the gabbro adjacent to the veins. The mineralization is cut by pegmatites belonging to serorogenic granitoids dated to 1.80 Ga.

Mineralizations with a supracrustal setting (**B1** and **B2**) generally consist of very small arsenopyrite rich en echelon quartz vein systems with small alteration zones, mainly silicification and sericitization, in the host rocks. So far, no differences have been noted between the sediment and volcanic hosted mineralizations.

The epigenetic type **C** deposits, overprinting massive sulphides, have traditionally been treated as "bonanza type" gold rich quartz-tourmaline parts of the massive sulphides. Au contents of >100 ppm over several meters are known from these systems. The epigenetic character of this type is, however, evident from the cross cutting relationships between the quart-tourmaline-sulfosalt veins and the massive sulphides.

A late introduction of gold into structurally controlled zones is evident in all mineralization types. The absolute timing of this event is not known, but the age is limited upwards by the cross cutting 1.80 Ga pegmatites previously described, and downwards by the 1.90-1.85 Ga host rocks. In some intrusive related mineralizations late basic dyke swarms, later than or coeval with the mineralizations are correlated with the 1.87 Ga Vargfors magmatic event. This indicates that gold was introduced slightly after the main period of volcanic and magmatic activity into settings controlled by brittle-ductile deformation probably within the uplifted newly formed crust. This is in contrast with the earlier deformation of the massive sulphide deposits which was mainly ductile and formed during the most active period of the orogeny. The structural control and the spatial association of many mineralizations in time and space to basic dykes, suggest that the system was open for mantle and lower crust fluids at that time.

In summary, despite the undoubtable fact that the majority of gold was introduced into the upper crust at Archaean and perhaps recent times, the global early Proterozoic epoch of volcanic and magmatic activity around 2.0-1.7 Ga produced a substantial amount of economic gold lode deposits. Proterozoic gold lode deposits are now found in all parts of the world and several are now being taken into production in the Baltic shield.

BISCHOF, JENS, Kiel University, FRG

Dropstone Distribution in the Norwegian-Greenland Sea — Indications of a Pleistocene Circulation Pattern?

In this study, an investigation of dropstones in deep-sea sediments of the Norwegian-Greenland Sea is presented showing that dropstones can be used as a fascinating instrument in paleocurrent modelling.

Terrigenious particles enclosed in an iceberg are representing the bedrock composition of the region where the iceberg was once generated. While the iceberg drifted in the ocean, these rock fragments dropped to the sea floor by continuous meltout and developped a carpet of specific rocks marking the ancient iceberg trajectories. By the fact that icebergs are driven by surface currents, the dropstone distribution is nothing less than the projection of pleistocene surface current directions in deep sea sediments.

For this purpose, 32 short sediment cores from the Mid-Norwegian shelf to Jan-Mayen have been investigated in order to ascertain the dropstone input rates and the qualitative composition of the ice rafted material in the $\geq 500\mu m$ fraction). The regional frame is confined to an area from 66°N - 72°N and 10°E - 12°W. The uppermost 50 cm of the sedimentary column have been studied, representing up to 30.000 years of continuous sedimentation.

Glacial marine sediments in the Norwegian-Greenland Sea are characterized by highly varying amounts of ice-rafted terrigenious debris. Three regional settings of different quantitative dropstone input can be recognized along the transect:

1) Ice rafted detritus is most abundant in close proximity to the Norwegian shelf break. Dropstones are found throughout the entire sediment column investigated and the sea floor is covered with pebbles.

3) On the Jan-Mayen Fracture Zone and the northern Vöring Plateau, under 15cm of Holocene calcareous clay dropstones occur in comparable abundance to shelf break samples. Thin, almost dropstone-free layers are intercalated in the dropstone-bearing sections.

2) Sediment samples from the western and southern Vöring-Plateau don't have any dropstones in the upper 50 cm.

The dropstones are largely composed of crystalline rocks, reflecting the bedrock geology of Norway and Greenland. These continental areas are consisting of medium - high grade metamorphic rocks of precambrian and caledonian origine and many instrusive bodies composed of granites and granodiorites.

Generally, crystalline rock fragments account for approx. 80% of the total dropstone volume. They consist mainly of light coloured, fine-banded gneisses, variegated quartzites and granites, the latter mostly composed of white or red feldspar. Amphibolites are found frequently, but marbles, phyllites and mica-schists are rare.

In should be mentioned that schistose metasedimentary rocks are covering approx. 50% of the surface of the adjacent Norwegian Caledonides between 64° and 68°N, but the proportion of these rock types among dropstones is almost less than 10%. Percentages like these have been observed only near the Norwegian shelf break. In most distal environments like the Vöring-Plateau, only 1% phyllites and mica-schists are present in the dropstone fraction. This demonstrates how fast Mid-Norwegian rocks were mixed up with foreign material of unknown derivation in late Pleistocene times.

In this study, an attempt has been made to connect some dropstones to their source regions. Throughout the whole investigated area, chalk was found in minor amounts, but always present (even at Jan-Mayen) in samples older than 10.000BP. Chalk occurs only south of 59°30'N latitude in the North-Sea and onshore Denmark, indicating that some icebergs must have been carried from southern sources into the Norwegian-Greenland Sea by a northward moving current.

In contrast, some samples younger than 13.000BP from the northern Vöring Plateau are composed of more than 50% non-metamorphic argillaceous rocks which must have been derived from locations further north. Silt- and sandstone accumulations like that have not been observed elsewhere in any Pleistocene samples. Possible source regions of these rocks are Spitsbergen and the underground of the adjacent Barents-Sea, where vast areas are consisting of sand- and siltstone formations.

A huge ice-shield existed in that region during late Pleistocene times, eroding and incorporating these rocks in the ice. As this ice-shield disintegrated, lots of icebergs were generated at the seaward front that drifted to the south. A strong current, probably out of the Barents-Sea area, must have existed periodically since at least 22.000BP, carrying the siltstone-bearing icebergs to the northwestern Vöring-Plateau. This current interacted with a current flowing northward along the Norwegian Shelf, forming a large counterclockwise gyre centered over the western Vöring Plateau.

Between these two current regimes with active iceberg rafting even at the Pleistocene/Holocene boundary, a third, northward thinning wedge without any dropstone input is located on the western and southern Vöring-Plateau. This zone represents the area where warm North Atlantic watermasses first intruded into the Norwegian-Sea, inhibiting continuing iceberg rafting in this area.

BOURROUILH-LE JAN, FRANÇOISE G., Laboratoire de Géologie-Sédimentologie comparée et appliquée, Université de PAU et des PAYS de l'ADOUR, Avenue de l'Université, France

Oceanic Phosphates: An Alternative Model of Phosphatogenesis

Based on modern sedimentological and geochemical studies of phosphate oceanic ores from Pacific Islands of Nauru, Makatea, Mataïva and Clipperton, a new model of phosphogenesis is presented. As a matter of fact, the geochemical, mineralogical and sedimentological composition of the phosphatic facies show links with the phosphato-bauxitic soils of High Carbonate Islands of the Pacific, enriched in phosphate oolites and pisolites. These soils are known as coming from the pedogenetic evolution of volcanic accumulations (tephra, pumices). The modern meromictic lake of Clipperton atoll, in place of the marine lagoon, seems to be a good sedimentological model to explain the formation of modern phosphate, which deposit from a reworked pleistocene or more ancient soil. The Clipperton meromictic lake is characterized by stratified anoxic waters, rich in H2S, with low pH and low salinity. Phosphatic and double carbonate (Ca, Mn, Mg) are forming today in this exceptionnal meromictic lake, meanwhile in its upper layers, freshwater fauna and flora and occasional trapped marine fauna survive.

Three main phosphogenesis steps can be recognized:

1. accumulation of volcanic products on a carbonate emerging platform
2. pedogenesis of these volcanic accumulations and evolution toward bauxitic soils under a high rainfall climate
3. submersion and reworking of these deposits by anoxic waters of meromictic lakes, forming pure phosphate accumulations.

Dispersed modern Oceanic islands illustrate each of these steps, which can be clearly observed on the Pacific Ocean High Carbonate islands. The geochemistry of soils of Rennell (Solomon Islands), Maré and Lifou (Loyalty Islands) and Tongatabu (Tonga Islands), compared to the geochemistry of phosphatic cover of phosphate islands as Nauru, Makatea and Mataïva, show a similar geochemical signature between soils and phosphates ores and thus a genetical link between the both which pertain to a same genetical evolution.

However, differential geological evolution appears between bauxitic and phosphatic covers and are due to their different geological history, due again to their different position and evolution on the tectonic plates of the Pacific, and to global sea level changes. From their sedimentological and geochemical signatures, some covers have undergone continuous karstification and emersion due to slow tectonic uplift, meanwhile other covers have undergone emersion and submersion due to the combination of slow plate tectonic movements and global sea level fluctuations since Early Miocene.

BOURROUILH, ROBERT J., Département de Géologie-Océanographie, Universite de Bordeaux, Cédex, France

Bay of Biscay-Pyrenees, Gibraltar Arc-Rif Betic and Scotia Arc : Three Responses of Divergent-Convergent and Transform Tectonics During the Atlantic Evolution.

The Atlantic Ocean opened in a reunified but very heterogeneous Pangea which presents three main area :

— to the North, the break-up of Pangea occured in the northern Atlantic in a Precambrian to Variscan multi-sutured contiental crust. After several attempts of divergence, from Late Paleozoic to Triassic, the opening of the North Atlantic occured progressively during the Jurassic and then the Cretaceous. The Atlantic structures developped by divergence became antagonistic with transform to convergent remnant Tethys tectonics, which developped Eastward in the future Western Mediterranean.

— in the Central Atlantic, between the African and the European plates, the existence of dispersed North-saharian and South-european continental blocks represents a permanent crustal heterogeneity and provokes the formation of a very complicated orogenic belt. Its Western part or the Rif-Betic orogen is primarly a linear belt, in which transform tectonics prevails, and then become a crushed domain, between the African plate and the Iberian block.

— to the South, the American and African plates remain joined at least 2 billion years but during Lower Cretaceous, the divergence of Antarctica, African and South America occurs in close relationship with the Pacific subduction and transform tectonics. Combined subduction, break-up and tranform tectonics on both sides of the future Drake Passage allow the progressive formation of the Scotia Arc.

During the Atlantic evolution, Bay of Biscay-Pyrenees, the Gibraltar Arc-Rif Betic orogen and the Scotia Arc will represent three responses of divergent-convergent and transform tectonics.

The Bay of Biscay is due to the divergence of North-America and European plates and to the opening of the North Atlantic Ocean. The European continental mass has undergone deformations in the trans-tensive and trans-pressive modes, which reactivated parts of its inherited structural network, and differencial motions of inherited heterogenous blocks, some of them proceeding from Palaeozoic transform tectonics. Mechanisms are proposed for the formation and deformation of the North-Pyrenean Interplate basin, of the South-Pyrenean Intraplate basin and for their evolution, regarding the formation of the Pyrenees, the African Plate migration and its progressive collision towards Southern Europe (Upper Cretaceous to To-day).

The Gibraltar Arc is the consequence of Tethyan transform tectonics of South European continental blocks moving Westwards, in response to the vanishing of the Tethyan Ocean and of an ultimate phase of crushing. Following a California tectonic model, the Rif-Betic orogen is the result of transform tectonics and then of the collision between Africa, the South-European moving blocks and the Iberian block. The Rif-Betic orogen is primarly a linear orogen which is slightly bowed by the moving African plate in the Gibraltar Arc.

The closure of the Bay of Biscay and the Pyrenean orogeny begin during Upper Cretaceous on the Northern edge of the Iberian block, meanwhile on its Southern edge, in the same time, transform tectonics always prevails in the Rif-Betic orogen. Thus, migration of the African plate is not the whole responsible of the formation of the Pyrenees and of the Rif-Betic orogens which appear in fact as the result of the antagonistic behaviour of Atlantic divergences and Tethyan convergences.

The Scotia Arc develops between the South Atlantic Ocean, the Weddell Sea and the Pacific Ocean in response to the break-up of South America-Antarctica and South Africa. Although the Scotia Arc has a specific genesis, it is an on-the-way forming Arc related to convergent-divergent and transform tectonics. Thus, some of its tectonical characteristics as well as its related basinal and sedimentological evolution can provide good comparizon framework with some aspects of the tectonical and/or basinal and sedimentological evolution of Bay of Biscay-Pyrenees and Gibraltar Arc-Rif Betic orogens. For example, on its northern side, intensive transform tectonics allows the migration of dispersed continental blocks as for example South Georgia. Mechanism of this transform tectonics can be compared with some aspect of tectonic migration of Pyrenean and/or Rif-Betic blocks. For basinal and sedimentological evolution, ODP Leg 114 has given large attention to the opening and development of oceanic passageways between progressively-created divergent marine basins and transform tectonics basins and underlined the importance of the tectonics on the formation of oceanic current circulation, and then on sedimentological evolution of adjacent marine basins. This can also provides good examples for a comparizon with basinal and desimentological evolution of some areas of Pyrenean and/or Rif-Betic domains.

In conclusion, heterogeneity of the continental crust, multi-sutured in the North Atlantic, constituted by dispersed continental blocks between Africa and Europa, more homogeneous in the South Atlantic, a perennial inherited structural network provoke three different responses to the Atlantic divergence. These responses are closely controled by transform tectonics which represents in fact a compensation of the nearboring vanishing Ocean.

REFERENCES

BOURROUILH, R., RICHERT J.P. and ZOLNAI G., 1988. North Pyrenean Interplate Basin : Evolution and Hydrocarbons. AAPG Mediterranean Basins Conference, Nice, and abstract, AAPG Bulletin, v.72/8, p.990.

LAUBSCHER H.P. 1988. The Arcs of the Western Alps and the Northern Apennines : an updated view. Tectonophysics, 146, p. 67-78.

ODP-LEG 114., 1987-1989 : Site 698-704, Subantarctic-South Atlantic, Initial Reports, Proceedings of the Ocean Drilling Program, v. 114, in press. College Station, TX.

TUBIA J.M. and CUEVAS J., 1987. Structures et cinématique liées à la mise en place des péridotites de Ronda (Cordillères Bétiques, Espagne). Geodinamica Acta (Paris) 1, 1, p. 59-69.

BRANCA*, M.E., G. CALDERONI** and V. PETRONE*, *Centro di Studio per la Geochimica e la Geocronologia delle Formazioni Recenti (CNR), c/o University of Rome, Rome, Italy. **Department of Earth Sciences, University of Rome, Rome, Italy.

Geochemical and Palaeoenvironmental Significance of Dating Reversal in Radiocarbon Chronostratigraphy of Lacustrine Sediments: A Case Study from Castiglione Valley (Latium, Italy)

An interdisciplinary study including geochemical, sedimentological, paleontological and palynological investigations was carried out on the uppermost 10 m of a lacustrine sedimentary suite accumulated since Pleistocene through Holocene on the bottom of a former volcanic lake located in an hilly volcanic area close to Rome. The results provided consistent information on the basin history apart from a trouble concerning the chronostratigraphy established by means of radiocarbon dating of the organic matter (OM) dispersed in the sediment. In particular 9 out of 10 samples from a 10 m long sediment core displayed ^{14}C ages that bracketed a 3480-31,300 yr BP time-span and exhibited a regular aging trend with increasing depth. In contrast, though sedimentological, paleontological and palynological data pointed to a continuous, undisturbed sedimentation pattern and ruled out sediment reworking, sample from 1.90-2.00 m in depth yielded a dating reversal, its age being far exceeding those determined for two deeper samples.

By considering that the ^{14}C signature of OM in lascustrine sediments besides being time-dependent reflects the source of the primitive organic carbon input, the dating reversal could be attributable to notable modification of the paleoenvironmental conditions which governed the sedimentation process. In this view the appearance of OM anomalously ^{14}C-aged relative to its stratigraphic position could be explained by a change of the pathways of OM recruitment from the drainage basin, say intensive leaching of aged organic carbon from topsoil in response to deforestation. Such hypothesis, however, was neither supported by palynological data, pointing to a stead, blooming forestation phase, nor by the $\delta^{13}C$ value exhibited by the OM ,-34 permil, which is hardly compatible with that, averaging -24 permil, consistent for aged, diagenetically decayed soil-OM. Furthermore, as it is known that OM undergoes significant ^{12}C loss during early diagenesis, it is reasonable to infer that the primitive organic input was isotopically lighter than its present residue. This turns unlikely that the parental OM originated *via* a one-step isotope fractionation of atmospheric CO_2 ($\delta^{13}C$ = -7 permil) by terrestrial and/or aquatic plants as well as lacustrine plankton.

Though atmospheric CO_2 represented the initial source for the whole organic carbon production in the watershed of Castiglione, dating reversal records an episode of carbon isotope depletion (lighter and older) which likely occurred owing to significant injections of biogenic CO_2, produced *via* diagenetic decay of aged OM, in the lacustrine reservoir. Determination of carbon-number distribution of *n*-alkanes extracted from the OM contained in the sediment suite are of some concern to support the above hypothesis as they revealed that dating reversal is paralleled by a dramatic predominance of those hydrocarbons enriched in plankton (C_{15}-C_{20}) over those typical of higher plants (C_{25}-C_{33}). Such OM feature is in line with the great abundance of algal remains which were detected and brings to infer that dating reversal stemmed because of the increased phytoplankton production which took place in the ^{14}C- and ^{13}C-depleted lacustrine carbon reservoir. Algal isotope fractionation of such inorganic carbon along with possible though small heterotrophic fractionation effects likely were responsible for the resulting low ^{13}C value.

Finally, owing to a striking higher concentration of even-carbon than odd-carbon number C_{15}-C_{32} *n*-alkanes there is strong suggestion that bacterial activity played an important role in the production process.

BROOKINS, DOUGLAS G., Department of Geology, University of New Mexico, Albuquerque, New Mexico, USA

Hazardous and Radioactive Waste Mobility: Predictions from Eh-pH Diagrams

Eh-pH diagrams are useful in addressing problems of mobility of toxic elements from both hazardous wastes and from radioactive wastes. In hazardous wastes, elements such as Pb, Cd, Hg, As, Se, As, Sb, Tl are of concern because they are all toxic, and in some cases carcinogenic, if taken into organisms in large amounts. In radioactive wastes, all the actinides are toxic, as are many of the fission products as well as actinide daughters. Eh-pH diagrams can be used to address the mobilities of Pb, Cd, Hg, As, Se, Sb, Tl, U, Np, Pu, Am, Bi, REE and some others. These are discussed in turn.

The mobility of Pb in waters is strongly dependent upon dissolved sulfur (as SO_4^{2-} or S(-II)) and carbon (as HCO_3^- usually). The Eh-pH diagram indicates limited mobility due to the presence of solids in the diagrams. Pb transport as an organic form, such as from gasoline, results in transport independent of Eh-pH conditions until the water-soil surface is encountered, at which time the mobility is restricted.

The mobility of Cd is dependent on pH conditions under mildly oxidizing conditions, and strongly dependent on presence of S under reducing conditions. Cd is readily mobile under oxidizing, acidic conditions.

Hg owes much of its transport to its highly volatile nature. The Eh-pH diagrams also show that dissolved Cl will facilitate Hg mobility under very acidic and oxidizing conditions. Mobility is high under many oxidizing conditions.

Sb, As, Se and Tl also show considerable mobility under oxidizing conditions, whereas under reducing conditions the presence of S will play a major role in As and Tl mobility.

Uranium and the transuranics, Np, Pu, and Am show limited mobility under many Eh-pH conditions. These elements are quite immobile as M(IV) species, although $M(OH)_5^-$ may encroach on $M(OH)_4$ species in Eh-pH space prior to ageing of $M(OH)_4$ to MO_2 species. Dissolved carbonate facilitates some U(VI) mobility but helps immobilize Am(III) and Pu(III) species. At higher Eh, Pu, Np and Am form important complexes with carbonate in solution.

Bi (^{209}Bi is the stable end product decay daughter of ^{237}Np) is immobilized under much of Eh-pH space, but may be readily soluble at low pH under mildly oxidizing conditions. The REE normally contain only M(III) species, although Ce(IV) and Eu(II) are known. In Eh-pH space, the presence of a field of $M_2(CO_3)_3$ is very important in preventing REE migration.

The World's only nuclear reactor, that at Oklo in Gabon, convincingly shows that Eh-pH diagrams can be used with confidence to predict (and/or confirm) the migration potential of U, Pu, Np, Bi, Cd, Sb, and other species. Studies of uranium mill tailings and uranium deposits allows similar tests of Eh-pH predictions for As, Se, Sb and Pb. Zonation of Hg, Tl, Sb, As in oxidizing metallic ore deposits allows testing of these elements by use of Eh-pH diagrams.

The conclusion reached is that, despite uncertainties introduced due to lack of kinetic information, the Eh-pH diagrams are most useful in predicting the behavior of these elements in the earth's surface environment. This, in turn, means that predictions about the behavior of certain hazardous wastes and most radioactive wastes can be addressed.

BUKHAROV, A.A., Institute of the Earth's Crust, USSR Acad.Sci., Irkutsk, USSR

Volcanic Belts in Zones of Protoactivity in the Crust and Their Metallogenic Properties

The formation of volcano-plutonic belts of sialic profile within crustal structures during the period of protoactivity was important for the evolution of continental crust. At all the continents this period lasts from 1800±100 Ma to 1600±50 Ma. The tectonic regime of this stage, set in after such regimes, refered to as protogeosyncline or protoplatform ones, was also quite peculiar.

For the first time geologically meaningful manifestations of volcanic, mostly continental formations of rhyolite composition within the structures of ancient continental platforms, occured. During this stage the first noticeable potassium leap in the chemical evolution of the crust also took place. Besides, red colour terrigenic formations of molasse type, comagmatic volcanic series and volcano-plutonic associations of sialic profile as well as specified granitic formations of rapakivi granite type appeared for the first time.

Volcano-plutonic belts with typical rhyolite volcanism occuring in such amounts for the first time in the earth's history, were in framing of ancient platforms and in intracratonic zones of all the continents. Thus, the formation of belts was worldwide and almost synchronous.

Then andesite-rhyolite and basalt formations of Kivino, Dabaunt, Atabasca, Coppermine, Cameron, Whitewater and other series were formed in suture sags framing ancient cores of the Canadian shield. Rhyolite volcanic formations of Carpentary systems were formed in narrow trough depressions separating stable massifs from fold terrains of the West-Austalian shield. These formations are very similar to Middle Proterozoic volcanic formations (akitkanides) of the Siberian platform framing and include Scraton, Edith-River, Argilla, Creudon, Hatchescreek and some other groups. Similar trough depressions of the Guiana shield in South America are known to contain volcanic-terrigenous complexes of andesite-liparite formation, which are comparable with Middle Proterozoic complexes of East Siberia and the Baltic shield. These include Paramaka series, Barama and Mazaruka groups, Pastora-Carigano complexes and their analogs. The western edge of Guinea-Liberia shield in Africa is bounded by Rockel graben synclinorium composed of Rockel series of rhyolite-dacite lavas and tuffs. The same stage is represented by subiotnic rock masses and also Smalland and Dala porphyres in Norway.

It was the period of Middle Proterozoic activity when the consolidation of continental crust was accomplished, but the mobility of continents was relatively low at that time. Only when the formation of mature continental crust was finished, the crust began developing according to a classic geosyncline-platform pattern and the plate tectonic mechanism started to work with all force.

Volcano-plutonic belts in the zones of ancient activity form separate metallogenic provinces with ore content determined by conditions of mobilization of ore material in the crust and its deposition in favourable tectonic setting during volcano-plutonic process. Various types of ore occurences correspond to different depth facies of magma sources and different facies of volcanics and volcano-terrigenic formations.

Comparison of metallogenic features of the zones of Precambrian activity with younger volcanics makes it possible to predict discovering within volcanic belts of ancient activity zones hydrothermal deposits of rare and precious metalls, as well as polymetalls including epithermal volcanic deposits, quite unusual for Precambrian.

Epithermal deposits actually lacking in Early Precambrian rocks could not have originated prior to the manifestation of ore-generating geological formations such as basalt-andesite rhyolite volcanic continental formations that were active in the Middle Proterozoic. New theoretical concepts on a possible occurence of epithermal deposits in the regions of Precambrian volcanism make the prospecting of low temperature metallic ores more promising.

CIMBALNIKOVA[1], A., Z. FRAIT[2], O. RODE[3], T. ZEMCIK[4], [1]Institute of Geology and Geotechnics, Czechoslovak Academy of Sciences, Prague, Czechoslovakia; [2]Institute of Physics, Czechoslovak Academy of Sciences, Prague, Czechoslovakia; [3]V.I. Vernadsky Institute of Geochemistry and Analytical Chemistry, USSR Academy of Sciences, Moscow, USSR; [4]Institute of physical Metallurgy, Czechoslovak Academy of Sciences, Brno, Czechoslovakia

FMR and Mössbauer Measurements of Fine-Grained Fractions of LUNA 16, 20 and 24 Regolites

Introduction

One of the important genetic characteristics of the lunar regolith is the absolute iron metal content and its form. The metal distribution in individual fractions contributes to the knowledge of the geologic history of the regolith, estimates the degree of its exogenic reworking, and the conditions of its formation. For that purpose the ferromagnetic resonance (FMR) and the ^{57}Fe Mössbauer spectroscopy (MS) method were used.

The samples of the LUNA 16 (1609) mare regolith, LUNA 20 (2001) highland regolith and the LUNA 24 (24118 and 24160) mare regolith were divided into fractions within the grain size limits of 370-200, 200-90, 90-70, 70-40, 40-10, 10-5, 5-2 and <2 μm. The samples of about 0.5 mg to 1.5 mg were sealed into thin-walled cylindrical polystyrene containers of 1 mm diameter and of 2 mm height; the same specimens were used for both experimental methods (FMR and MS).

Methods

The ferromagnetic resonance: The diameter of iron particles detected by FMR ranges from a few nanometers (in smaller particles the superparamagnetic effect) to tens of nanometers (Housley et al., 1972). Our FMR measurement were performed ad R.T. in the Q-band (around 35.7 GHz, the resonance field ammounts to 1.3 T). At these fields the iron particles with diameter d' 3 nm are magnetically saturated; as the microwave penetration depth in iron ammounts to 50 nm (at 36 GHz, Fraitová, 1983), FMR detects still almost fully the volume of particles with d' ~100 nm.

The Mössbauer spectroscopy: Measurements were made in transmission geometry at room temperature using a ca 10 mCi/Rh high specific activity source. The measuring procedure was simillar Zemcík and Raclavský (1976).

The absolute amount of the fine grained metallic iron Fe^{o}_{FMR} was determined by comparing the FMR line intensities with electron paramagnetic resonance (EPR) absorption of paramagnetic standards.

In the Mössbauer spectra all substantial Fe-minerals in the regolith were detected, i.e. the non-magnetic olivine, pyroxene (M1 and M2 positions), glass and ilmenite contributions plus the magnetic metallic phase. From the area ratio of the Zeeman component and the whole absorption spectrum, the relative metal content Fe^{o}_{MS}/FeO was determined. The values of the effective magnetic induction B_{eff} at the ^{57}Fe nuclei in the magnetic Fe-Ni phase made it possible by using the known B_{eff} dependence on the Ni content to estimate the degree of its nickel alloying.

Results

The values of absolute Fe^{o}_{FMR} content (I_A) obtained from the measurements, given in per cent units of the lunar soil fraction weights are inversely dependent on the mean grain diameter (d) of the fractions. The I_A values for LUNA 16 soils are about 20 times larger than the LUNA 20 data, the quantities obtained for LUNA 24 soil laying between the LUNA 16 and 20 values. The FMR study has shown that the mean Fe^{o}_{FMR} content, reflecting the degree of regolith reworking under exogenic factors (maturity indexes MI), form the sequence LUNA 16 — LUNA 20 — LUNA 24. The highest Fe^{o}_{FMR} content is observed in the "mature" regolith from LUNA 16, the lowest one in the LUNA 20 "submature" regolith.

The MS method enables to determine the relative content of Fe^{o}_{MS} consisting of both the reduced metal from FeO on the lunar surface, the primary lunar metal and that of the meteoritic matter. Thus determined Fe^{o}_{MS} distribution in the grain size fractions exhibits generally the same tendencies as that from FMR: the Fe^{o}_{MS} content decreases going from the fine towards coarse grained fractions. Nevertheless, the MS values of

Fe^{o}_{MS} always exceed the FMR ones, mainly due to the meteoritic metal contribution.

REFERENCES

Fraitová D. (1983): An analytical theory of FMR in bulk metals. II. Penatration depths. — Phys. Stat. Sol., b 120, 659-665.

Houseley R.M., Grant, R.W., Abdel-Gawad M. (1972): Study of excess Fe metal in the lunar fines by magnetic separation, Mössbauer spectroscopy, and microscopic examination. — Geoch. Cosmoch. Acta, Suppl. B, 1065.

Zemcík T., Raclavský K. (1976): Mineralogical analysis of lunar rock particles. — J. Phys., 37, C6-833-836.

COUNTIN CORREA, D.P., J. Martinez SALCEDO, Institute of Geology and Paleontology, Ministry of Basic Industry, Cuba

Map of Non Metallic Ore Deposits and Combustibles of the Republic of Cuba, Scale 1:500 000. Principles of Its — Confection

This map shows to geologists and the related specialists, an actualized information of the non metallic minerals and combustibles of Cuba. The map shows the geographical and structural geology distribution of different kinds of non metallical and combustibles, that have been grouped and classified in order to origin, use and size. Besides are indicated if it is an ore deposit or a mineral manifestation in actual explotation or abandoned.

The principal data for the confection of the map proceed from the generalization and sistematization of the former informations of non metallics and combustibles about these minerals have existed until 1985 in different geological institutes of non metallics and petroleum in the country. With the compiled information was prepared a general catalogue for all raw materials and a structural and lithofacial scheme with the location for all objectives that will be represented in a general map with a zonal legend related to the Geological Map of Cuba, printed in 1985, scale 1:500 000. In such a manner the actual informations of the relation between formational and structural framework of the ore deposits is bonded.

The mineral objects represented on the map were selected according to the principles determined by the map scale. Were considerated the ore deposits and the raw materials accumulations with calculated reserve until C_2 category and to reach stablished volume for small ore deposits, as soon as the acumulations without evaluated geological works, but which have been in extraction several years ago, and makes possibles to consider them as an ore deposit.

The classification used in the national map of mineral resources (non metallic and petroleum) are based on genesis, industrial kind and size.

The map includes nine enlarged areas per some regions with a dense distribution of objectives, to permit a better geostructural entailment of the ore deposits.

Furthermore the map shows a geological profile with the principal stratigraphic and structural characteristics of the oil fields and a relation of the names, petrographical types, dimension and respective geological formations.

In the map are represented information about 50 different raw materials, grouped in 653 objects; 91% of them correspond to non metallic minerals and the 9% to combustibles. In the group of non metallics the 53% correspond to ore deposits and the 47% to manifestations. For the group of combustibles, 34% are oil deposits and the rest only manifestations.

The mineral objectives conforms the information of the map and are plotted by graphic literal symbols, colour and numbers, that means the genetic types of mineral accumulation, its size means the reserves volume according to adopted classification. The symbol corresponds with mineralogical and petrographical type of raw material. The colour is related to industrial use; numbers correspond to names and other specifications. Also are added symbol related to its actual or older explotation; this symbol is represented by two crossed hammers.

Non metallic ores represented in the map are: magmatic and carbonated rocks for use in construction; sands for natural arids; soft limestones for blocks of building materials; recristalized limestones, marbles, antigoritic serpentines and granites as ornamental stones; also there are jadeite, chalcedony, agate, quartz, amethyst, jaspers for jewellry and artesan work; the clay, standstones, limestones, marls, puzzolanic tuffs and gypsum for cement factories; kaolin, bentonite, feldspars and kaolinitic clays for refine ceramic; clays for red ceramic; quartz sands, dolomites and feldspars for glasses; quartz for electronic industry; pure and phosphatic limestones, dolomite, turf and zeolitic rocks for agricultural uses; quartzite, bentonite, garnet, rock salt, talc, high purity gypsum, magnesite, asbestos, ochre, asphaltite, baryte, graphite and cyanite for various purposes. Combustibles are represented by oil and gas.

This map has an actual importance because in Cuba for the first time is published a summary of the principal characteristics of all ore deposits and manifestations of raw material non metallic and combustibles, dispersed in all country and also is very useful for geologists, technologist and managers related with this matter and its utilizations.

CROOK, KEITH A. W. and Robert J. MUSGRAVE, Department of Geology, Australian National University, Canberra, Australia

Aborted Subduction of Spreading Centres as an Explanation for Near-Trench Island Arc Volcanism

Migration of a ridge-trench-trench triple-junction along a trench has been invoked as a cause for episodes of igneous intrusion within accretionary prisms anomalously close to an active trench (Marshak & Karig, 1977). In this model, the spreading centre associated is subducted beneath the accretionary prism partially melting the accretionary prism sediments.

We here summarise a rather different model, based on studies of the Woodlark triple-junction, Solomon Is (Taylor & Exon, 1987; Crook et al., 1988), to explain near-trench calc-alkaline volcanism. The model is applicable to cases where the ridge and trench intersect at $>25^{\circ}$ and the trench makes a high angle with the directions of the plates' convergence.

When a ridge segment impinges on the trench, both spreading and subduction cease at the point of impingement, presumably because of thermo-mechanical constraints. Spreading soon ceases along more remote parts of the ridge segment as well. Windows to the asthenosphere are opened, and magma moves obliquely from beneath the over-riding plate to construct calc-alkalic volcanic edifices on the formerly subducting oceanic plates, close to the convergent boundary. Subsequently the convergent plate boundary steps out; the now-inactive ridge segment and the newly-constructed volcanic edifices, together with their magma sources, are uplifted and incorporated into the over-riding plate.

A triangular region defined by Rendova and Tetepare islands and Kavachi, an active submarine volcano 30km inboard of the trench (Johnson & Tuni, 1987), adjoins the Woodlark triple-junction. This geometry and the >2.5km Quaternary uplift of Tetepare (Hughes et al., 1986) is consistent with transfer of this region to the Pacific plate during the Quaternary. The model is also invoked to explain the Gallego Volcanics of NW Guadalcanal and the puzzling occurrence of deep-sea ooze in the centre of that island discussed by Coleman & Haig (1988).

REFERENCES

Coleman, P. G., Haig, D. W. & Wilkins, H. A. 1988 *Search* 19, 90-91

Crook. K. A. W., Taylor, B., Exon, N. F. & Johnson, R. W. 1988 *Geol. Soc. Aust. Abs.* 21, 104-105

Hughes, G. W., Varol, O. & Dunkley, P. N. 1986 *Roy. Soc. N.Z. Bull.* 24,397-408

Johnson, R. W. & Tuni, D. 1987 *Circum-Pacific Council for Energy & Mineral Resources Earth Science Series (CPCEMR-ESS)* 7, 89-112

Marshak, R. S. & Karig, D. E. 1977 *Geology* 5, 233-236

Taylor, B. & Exon, N. F. (eds) 1987 *CPCEMR-ESS* 7,363p.

CSASZAR, GEZA, Hungarian Geological Survey, Budapest, Hungary

Urgonian Facies in the Circum Mediterranean Area

The paper will summarize the present state of knowledge in the framework of project IGCP 262, dealing with global study on a typical Tethyan Cretaceous formation.

The term Urgon understood as facies reflects the distribution of the formation (Fig. 1.). It shows occurrences of this shallow-marin rudistid facies in the wider Mediterranean (S. Europe, Caucasus, Near and Middle East). The last 15 years produced rich literature on the subject, and based on it and own section-like elaboration I try to separate facies types, and to outline their distribution in space and partially in time. The grouping is based on the following aspects:

- relation to transgression and regression
- age (duration) of the formation
- lithological composition
- dominant fossil content

Based on transgression and regression 3 types may be separated. Regression type is understood as carbonate platform develops from basin facies. In this case the Urgon facies is followed by subaereal or submarine discordance. In case of transgression after the, generally significant, discordance freshwater carbonate or clastic sediments lead to the rudistid facies. In its basement, especially on carbonate surfaces, mineable bauxite lenses are frequent.

The third type is represented by the continuous platform where the Urgonian facies is not of transgression and/or regression origin but of other factors, e.g. as disappearance of fossils. The disappearance of the Urgon facies is usually related to similar factors. The inside partition of the facies is very varied, depending on the eudtatic sea level changes.

According to the duration of the formation the grouping is rather difficult as begin and end are again varied. For the basis of grouping this time we decided for the begin.

According to the first interpretation of the term the first group includes rudistid sediments from the Late Barremian-Early Aptian (Alps, Carpathians and Balkan). The second group includes of earlier Urgon formations (Hauterivian) start. According to the basic principle the Urgonian limestone in Georgia (Caucasus) ending in the Barremian belongs also here. The same is valid for the Estremadura sequence in Portugal which end s only in the Albian.

The third group of the Urgonian formations is formed after the main phase, like those in the Transdanubian Midmountains in Hungary.

As to lithologic composition of the formation the following groups may be separated:

In case of pure limestone type at best some small marly intercalations are found (Villány Mts., Hungary, Helvetic zone, in Eastern Alps, etc.). Limestone with dolomite intercalations are characteristic for the type developing on continuous platforms like the outer Dinarids in Yugoslavia. The position of the Tunesian sequences due to the abundance of terrigenic material is rather in doubt but due to the presence of evaporite may belong also here. The characteristic feature of the terrigenous type is the separation of the typically Urgon facies into two or more rock bodies under effect of the huge amount of terrigenous material. The Pyrenean and Balkanian sequences belong here. The Urgon between volcanic layers is a special type, the so called volcanoclastic Urgon. The best examples are found in the Little Caucasus, in Azerbaidzhan. The redeposited calcarenitic type appears as a very peculiar one, containing alian material, mainly bioclastic calcarenite, transported from the platform into the basin (Tristel Formation in the Alps).

Based on dominant or characteristic fossils first of all Orbitolina-rich, gastropoda and coral bearing types seemed to deserve separation. Besides, ocassionally, some facies may be characterised by dominant rudist geneva or species.

The aforementioned types are characterised in the paper on the basis of real sequences.

Based on ongoing investigations and evaluations it became clear that in the Circum Mediterranian part of the Tethys at the beginning of the Early Cretaceous the shallow-marin sedimentation became common feature in the archipelago and also at the margin of the self. Within it the formation and long lasting preservation of the Urgonian biosedimentary system (Rat et Pascal, 1979) are formed and due to the coincidence of two special conditions. Once, the submersion of platforms and ridges were not overstressing the carbonate precipitation in these huge areas, on the other hand rudists had their peak just in this periode. The role of the increased importance of rudists was helped by the agitated, nutrient-rich warm waters, similar to those required by corals, though of a lower energy level.

On shallow platforms with slightly uneven bottom surface weak agitation is typical. Thus it is understandable why became rudistid platforms so important during this period, instead of coral colonies. During the Early Cretaceous the African and European plates and intermediate were moving facing one another and collided. The horizontal displacements and tiltings produced areas, even during collision and nappe formations, where carbonate sedimentation went continuously on, or where the conditions just has been created for it apart from the fact that because of the great orographic differences on the hights denudation and consequently on the coastal regions and in the basins clastic accumulation started.

The Urgonian facies has been ceased on the Eurasian plate from N to S continuously from the Aptain onwards to the Late Albian, due to new connection to the boreal realm. More to the south carbonate sedimentation went on and the end of the Urgonian sedimentation is caused by the genetical decline of rudists, unable to adapt until the Late Albian time, or maybe due to some climatic change.

Further studies in space and time of the aforementioned and other conditions may help us to elucidate a more realstic evolutionary history of the enventually Early Cretaceous.

During presentation of the paper features of the Urgonian between Circum Mediterranean and Central and North America will be compared.

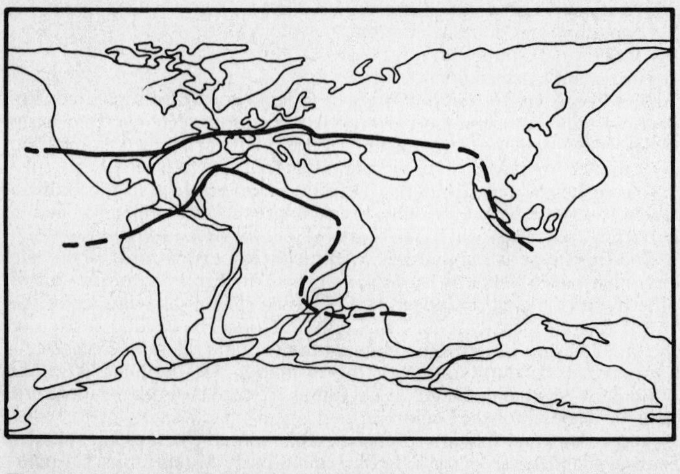

Figure 1 — Location map of the Urgonian after R.Rat et Pascal 1979 — Location of continents in Aptian according to E. I. Barron, 1987., Modified

DETAY, M. and P. POYET

Influence of the Development of the Saprolite Reservoir and of its State of Saturation on the Hydrodynamic Characteristics of Drillings in Crystalline Basement

The role played by the weathered zone of the crystalline basement was sensed by the hydrogeologists but quantifying it remained purely empirical. After the analysis of 1.500 drillings carried out in North Cameroon, Gabon and Central African Republic, we are able to bring out arguments outlining this relation and the role of the saprolite reservoir has been quantified.

We have been able to determine the statistical hydrogeological characteristics of the drilling in the crystalline basement. Thanks to a probabilist approach, the influence of the thickness of the saprolite reservoir on the yield, the specific yield and the probability of finding a minimum yield (infra-yield) has been determined.

Thanks to kriging, we have been able to represent geometrically the relation between the thickness of the saprolite reservoir, the importance of the saturated weathered zone and the yield.

Finally, we propose some probabilistic charts and diagrams to determine the hydrodynamical characteristics of the wells in the crystalline basement aquifer.

DETAY, MICHEL, Maurice CORNACCHIA and Loïc GIORGI

New Data on Central African Hydrogeology

Since 1984, several water supply programmes for village and town have been initiated in the Central African Republic. After having located the study area in its geological, geomorphological and structural context, the major aquifer system will be identified through analysis of data from 570 drillings.

First unpublished data concerning Central African hydrogeology are presented. The statistical hydrodynamic characteristics of the different aquifer units are discussed. The hydrogeological application and the survey methods tested in order to identify the different reservoirs are emphasized. The geoelectrical characteristics of the different horizons are presented and commented. Probability equation are suggested to explain the hydrodynamic behavior of the different aquifers.

Second, the Central African hydrogeological complexes are compared to recognized units in other African countries. The aquifer of the schisto-carbonated unit located in the paleo-crypto-karst of the Bangui region is compared to the schisto-calcareous unit in the Nyanga syncline (Gabon). The hydrogeology in the argillaceous-sandstone series of Carnot is compared to the Cretaceous basin formation in the North Cameroon. The hydrodynamic characteristics of the crystalline basement are identified and considered in the general context of basement complex hydrogeology.

Finally, the aquifer potentialities of the major hydrogeological complexes are estimated and global mapping for water resources planning in Central Africa are offered too.

DIJON, ROBERT E., Interregional Advisor, Water Resources Branch, Natural Resources and Energy Division, Department of Technical Cooperation for Development, United Nations, New York.

Ground-Water Development and Management in Developing Countries: Twenty-Five Years of United Nations Activities

In the course of the last 25 years, the Secretariat of the United Nations in New York City, through its substantive office specializing in water resources, as part of a natural resources unit, has been engaged in about 150 ground-water projects. These have involved ground-water resources exploration, assessment, development, management, planning and policies in some 80 countries around the world.

The international funding for the group of projects listed in an annex to the paper provided, *inter alia,* for the services of 400 experts assigned to projects for one year or longer, an equivalent number of short-term consultants, about 100 drilling rigs, 150 trucks and 400 other vehicles. The projects also provided for the training of more than 1,000 national counterpart professionals and technicians, through fellowships abroad, seminars, workshops and other group training activities, including study tours and attendance at international scientific meetings.

The major objectives of those projects were: strengthening governments or parastatal organizations involved in ground-water activities; assessing ground-water resources and carrying out pilot projects as part of prefeasibility studies; solving emergency water situations due to droughts, earthquakes, or inflow of refugees; transferring technologies; and promoting planning and policy-making through, in particular, the organization of computerized data banks and mathematical modelling activities. Project types were related to conditions prevailing in the countries concerned in terms of: geographical, geological and climatic conditions; needs to be served; technical knowledge of local personnel; and financial resources of the country.

By and large, the United Nations was involved in areas in which groundwater conditions were particularly difficult, mostly in "least-developed countries" (LDCs), with low per capita incomes, and "most seriously affected" (MSA) countries such as those having suffered from war (civil or foreign), natural disasters or detrimental geographic conditions (land-locked or small island countries).

Projects have been established in response to a variety of requirements in urban areas and industrial installations as well as for dispersed populations in rural areas, and in some cases for agricultural and pastoral needs.

They were carried out in practically every region and subregion of the world: the Mediterranean; the Middle East; Central Asia; the Indian subcontinent; sub-Saharan Africa, including the arid Sahelian belt and the humid coastal areas; Central and South America; and island countries in the Caribbean, the Pacific and off the coast of Africa.

With respect to geological conditions, most projects may be grouped in one of four general categories of geological terrain: crystalline basement rocks; volcanics; marine sedimentation, mostly standstone and limestone; and continental sedimentary and alluvial formations.

Government contributions to the projects were mainly in the form of national personnel: professional, subprofessional, ancillary and labour. In some cases, equipment, including drilling rigs and supplies, was also supplied.

External financing of most projects has been provided mostly by the United Nations Development Programme and to a lesser extent by UNICEF, the United Nations Regular Programme funds, the United Nations Capital Development Fund and bilateral assistance funds.

Problems encountered in the execution of projects included: delays in the recruitment of experts and consultants and in acquisition of equipment; national and international conflicts; insufficient counterpart inputs as compared with government commitments.

Project outputs included: water resources inventories, including computerized data banks; selection of well sites; the results and interpretation of geophysical surveys; a substantial number of exploratory tests and exploitation tubewells (500 per year in the early 1980s); establishment of specialized ground-water services; water resources development plans based on an assessment of water resources supply, current consumption and future needs, evaluation of costs and design of development projects, and elaboration of laws and regulations.

These activities have been carried out on an unusually large scale including a considerable geographic dispersion. The technical supervisory team in UN Headquarters never exceeded four professionals (hydrogeologists and a drilling engineer), in charge of the formulation, staffing and monitoring of projects which were visited once a year on average, and finalization of technical and terminal reports.

Project reports, normally printed in 100 copies, are the property of the Government. They may be consulted in UNDP and government offices of the countries concerned and also in the reference collection of the United Nations (DC-1 building, 10th floor).

The impact of UN projects with a ground-water component can be summarized as follows:

- developing national capabilities in ground-water related activities;
- developing a sense of awareness of the value of ground water;
- improving living conditions;
- attracting funds for water resources development;
- lowering costs for well construction and ground-water exploration;
- developing water resources management and planning.

Trends noted in recent project designs are towards: poorer regions; more difficult hydrogeological conditions (crystalline rocks, volcanics); integrated approaches to water resources (joint surveys and development plans for ground water and surface water); increased use of short-term highly specialized consultants in preference to long-term experts; expanded training programmes; and increased use of computer hardware and software for data banks, water resources modelling and water planning/management scenarios.

UN projects with a ground-water component current being implemented numbered about 50 by March 1989 including approximately 15 projects related to rural water supply, 10 projects in water resources, 2 projects for computerized data banks and 1 project for training of water well drillers.

DJEDDI, M. (I.N.H.), N. KETRANE (Division Exploration), K. AMEZIANE (EREM).

Combinaison of Geophysical Methods With Results of Satellite-Photos for the Resolution of Structural Problems in Oil Exploration, in the Region of Bechar Oued Namous, Algeria.

The target of this work is to test the possibilities of combining the geophysical methods (gravity, magnetic and seismic reflection) with the results of satellite-photos, for the search and delineation of positive geological structures that may contain hydrocarbons, in the region of Bechar Oued-Namous, north-west of the Algerian Sahara.

The result of interpreting the residual gravity anomalies, seismic time contour map, structural map (photogralogy), and magnetic basement map, helped in deliniating many positive and negative structures, as well as giving few evidences on the existence of many geological contacts. Also, the combination of the methods determined exactly the tectonics, morphology and structure of the magnetic basement.

In this study an great number of anomalies has been proved and were confirmed by 2 or 3 methods.

The magnetic materials of shallow depth, with reference to the magnetic map, led to distinguishing a favourable zone, which is generally similar to that of TAMAZAIA, the west-side of Oued Namous, and the northwest part of the TIMIMOUN depression.

In conclusion, each method needs the support of the others, and their combination gives very rich information and allowing also the confirmation or not of the anomalies, as well as limitting the structural problems.

DUBROVSKIY, V.A., Institut Physics of Earth, Moscow, USSR

A Possible Mechanism for Tectonic Waves

The evidence of tectonic waves (waves of rather long period) can be explained by the presence of a state of stationary unstable equilibrium in some mechanical systems. Two important geologic and geophysic examples are: (1) the lithospheric-asthenospheric system with a density inversion and (2) a fault in an elastic medium with an inverse dependence of the friction coefficient on relative displacement along the fault. The unstable stationary disturbance has a definite dimension, depending on the density-elastic properties of the mechanical system. In the lithospheric-asthenospheric system the nontrivial (displacement not equal to zero) state of unstable equilibrium takes place when the Archimedean buoyancy force, arising from the inversion of density with depth is compensated by the elastic forces as a result of deformation of the elastic lithosphere. In this case there is a definite wavelength for a periodic equilibrium disturbance. The Archimedean force does not depend on the wavelength of the disturbance but the elastic force is inversely proportional to a wavelength (a rate of deformation) and consequently the two forces compensate each other at some definite critical wavelength Lc of the disturbance. For wavelengths larger than Lc the elactic forces cannot oppose the Archimedean force. Here the instability comes about without interference. For wavelengths less than Lc the elastic forces exceed the Archimedean and therefore make possible an oscillating process in this case. The nearer the system comes to a state of balanced force, i.e. to an unstable equilibrium (when $L<Lc$) the lower the frequency is. Every initial disturbance of wavelength $L<Lc$ generates a traveling tectonic wave. The nature of the tectonic waves is such that their wavelengths are specified by the elastic-density parameters of the lithospheric-asthenospheric system. At the same time the period of these waves can vary over a rather wide range: everything depends upon how close the system is to the critical state — that of unstable equilibrium. This property of tectonic waves makes it possible for us to obtain information concerning the state of the earth's interior and its geophysical-mechanical characteristics: the average density, effective shear modulus, thickness of the lithosphere and the value of the density inversion between lithosphere and asthenosphere.

Similar considerations lead to the conclusion that there can exist a slow wave process along faults in an elastic medium if the dependence of the coefficient of friction on relative displacement along the fault varies inversely. This slow moving or standing wave connects in the same manner as before with the existence of a state of unstable equilibrium. Consequently, a decrease in the frequency of the slow wave means that the movements along fault will soon become unstable. This important conclusion implies that a decrease in the frequency of slow wave motion in regions of earthquake preparation may serve as a reliable short-term forecaster.

DZHEVANSHIR, R.D., L.A. BURYAKOVSKY, G.G. KULIEV, E.A. ALIEV, M.T. ABASOV, Baku, USSR

Mathematical Theory of Linear Folding

The authors have suggested a mathematical model of one general folding mechanism based on a three-dimensional linearized stability theory. The difference of this model is that the reasons resulting in folding loss and edge conditions accompanying this process are taken into account more completely.

While setting a problem the nature of forces stipulating the folding process is not of great importance and sufficient community of a model allows to give these forces and other parameters the definite geological, physical and mechanical, mining and geometrical interpretation of different nature and to make qualitative and quantitative analysis of specific results.

It is necessary to distinguish stability problems for geosynclines and platforms according to their setting and methods of their solution. The boundary conditions bring in essensial difference at surfaces of these objects.

Folding, according to the model suggested, may take place with or without horizontal compressive loads at the lateral surfaces of a layer observed. These loads may appear as a result of weight of overlying layers, spacial density heterogeneity, abnormal pressures of pore fluids, tectonic movements and other "external" and "internal" factors.

The authors have developed a computer program that helps to simulate folding according to numerical information on parameters of deformed layers.

E

EADINGTON,[1] P.J., P.J. HAMILTON[1] and P. GREEN[2], [1]CSIRO Division of Exploration Geoscience, North Ryde, Australia; [2]Queensland Department of Mines, Minerals and Energy Centre, Brisbane, Australia

Fluid Inclusion and Stable Isotope Investigation of Diagenesis and Oil Migration in the Hutton Sandstone, S.W. Queensland

Fluid inclusion temperatures, thermal maturities, stable isotope measurements and chemical kinetic modelling have been combined to investigate the diagenetic and hydrocarbon fluid history of the Jurassic Hutton sandstone in the Eromanga Basin, south west Queensland.

The Eromanga Basin overlies Permian sediments of the Cooper Basin and comprises Jurassic fluvial and lacustrine sediments with coal measures grading into overlying Cretaceous fluvial and shallow marine sediments. Structures were formed in at least three periods, syn - depositional drape folds in the Jurassic, broad folds in the Paleocene and Eocene and minor post - Miocence downwarps. The Hutton sandstone is the lowermost of three producing sands in the study area which are also aquifers for artesian water.

The sequence of diagenetic minerals and their oxygen isotope composition in $\delta^{18}0$ notation relative to SMOW is siderite (11.94‰), followed by kaolin (7.67 to 9.58‰) then quartz overgrowths (13.4 to 14.6‰) and dolomite. The $\delta^{13}C$ PDB composition of siderite is -7.04‰.

Siderite, the earliest formed cement, formed at a temperature of about 55°C from parental groundwater that was little modifed in oxygen isotope composition from surface water at the time ($\delta^{18}0$ -11 to -13‰). If fluid inclusion temperatures are used for mineral-water isotope fractionations then the $^{18}0$ composition of formation waters increased to -6‰ for formation of quartz overgrowths. Kaolinite precipitated at temperatures and oxygen isotope compositions between those of siderite and quartz.

This trend of increasing $^{18}0$ content parallels the normal diagenetic trend for evolving pore fluids in sedimentary basins. Subsequent uplift has changed fluid flow patterns as evidenced by a reversal in the $^{18}0$ content of modern groundwates to about -8‰ at the present day.

Quartz overgrowths have trapped drapes of clays that sometimes fluoresce (adsorbed hydrocarbons ?) and aqueous fluid inclusions at the boundary with the detrital core. Liquid hydrocarbon inclusions, identified by their fluorescence, occur mostly in late fractures in quartz grains, but occur at the boundary of detrital core and overgrowth in two samples. Therefore hydrocarbon migration in part overlapped and in part followed formation of quartz overgrowths.

Fluid inclusion homogenization temperatures and isochores indicate quartz overgrowths were formed at temperatures of 91 to 116°C and dolomite crystallized at about 114°C. These temperatures assume formation of the diagenetic minerals at depths 300 m greater than the current depths. This amount of erosion and the fluid isochores indicate a geothermal gradient about 4.5°C/100 m in the mid Cretaceous, smiliar to the current gradient. Erosion associated with the Eocene-Miocene folding is assumed.

Previously the thermal history of the rocks was ill defined. Current geothermal gradients are high (ca 4.0 to 5.0°C/100 m) and give calculated vitrinite reflectances that are higher than the measured reflectances leading to suggestions of lower paleo-geothermal gradients in the study area (Pitt, 1986).

Fluid isochores impose additional conditions on a suggested lower paleogeothermal gradient and demand higher formation temperatures since on a temperature versus depth plot the line for a lower temperature gradient and the isochores intersect at a greater depth of burial. This results in calculated maturities even higher than if calculated with the current geothermal gradient.

Calculated maturities using the fluid inclusion temperatures (constant geothermal gradient) are significantly greater that vitrinite reflectances and Rockeval Tmax values for formations just above and below the Hutton. Rockeval Tmax values for sediments higher in the sequence, where vitrinite reflectances are not available, suggest maturities similar to the calculated values.

Contributing factors to the discrepancy between calculated and measured maturities could be inadequate definition of the thermal history, the correlation that is used between TTI and vitrinite reflectance (Issler, 1983), or suppression of vitrinite reflectance by hydrogen rich fluids (Price, 1985).

If organic metamorphism is measured in terms of progress of oil generating reactions using kinetic modelling then the current geothermal gradient and fluid inclusion temperatures result in calculated oil yeilds that indicate late Cretaceous to early Tertiary oil migration. This closely follows the formation of quartz overgrowths and is consistent with the petrography of liquid hydrocarbon inclusions. It is also consistent with some structures being only partly filled with oil since structures of Eocene age formed at a late stage of oil generation.

REFERENCES

Issler, D.R., 1983. Calculation of organic maturation levels for offshore eastern Canada – implications for general application of Lopatin's method. Can. J. Earth Sci., V.21, pp. 477-488.

Pitt, G.M., 1986. Geothermal gradients, geothermal histories and the timing of thermal maturation in the Eromanga-Cooper Basins, pp. 323-352 in Gravestock, D.I., Moore, P.S., and Pitt, G.M., Contributions to the Geology and Hydrocarbon Potential of the Eromanga Basin, Special Publication No. 12, Geological Society of Australia Inc.

Price, Leigh C. and Barker, C.E., 1985. Suppression of vitrinite reflectance in amorphous-rich kerogen – a major unrecognised problem. J. Petroleum Geol., V8, pp 59-84.

ELDER, WILLIAM P., U.S. Geological Survey, Menlo Park, CA

Molluscan Extinction and Paleocommunity Patterns in the Cenomanian-Turonian Boundary Interval, Western Interior, United States

High-resolution stratigraphic analysis of 18 sections spanning the Cenomanian-Turonian (C-T) Stage boundary in the western interior of the United States has allowed determination of the magnitude and pattern of molluscan species extinction and community disruption. This study shows that the molluscan extinction across the stage boundary occurs in a series of seven narrow stratigraphic zones defined by multiple last occurrences, separated by intervals displaying little or no extinction. Two of the observed extinction steps (bottom and top of the *Neocardioceras juddii* Biozone) appear to be intercontinentally developed. The additional extinction pulses, as well as origination pulses, correspond to shale-limestone cycles on a nearly one to one basis. The extinction levels correlate with the lower parts of shale intervals, and the origination pulses correlate to limestone beds in the central western interior basin. This lithologic cyclicity probably reflects orbital forcing of climate, and thus, water mass and substrate characteristics in the basin. Therefore, at least some of the extinction episodes may be confined to the western interior region.

The ammonites (85% species level extinction) and inoceramid bivalves (92% species level extinction) were the most affected mollusks in the C-T boundary interval. Both of these groups contain many species that had intercontinental distributions, and that display similar relative timing of extinction in North America and western Europe. The widespread distribution of these taxa indicates that they utilized planktotrophic larval dispersal mechanisms, and suggests that larval disruption played a role in preferentially increasing their extinction and speciation rates near the C-T boundary. The nekto-benthonic ammonites were affected earlier and to a greater degree than the pelagic forms. This pattern implies an upward progression of habitat disruption, probably due to intensification and expansion of the oxygen minimum zone prior to the stage boundary. Each of the two intercontinentally developed extinction steps were preceded by a period (ca. 50 Ka duration) of changing community structure and general biotic deterioration, suggesting a non-catastrophic forcing mechanism.

Molluscan assemblage data indicates onshore to offshore and upsection community changes through the upper Cenomanian *Sciponoceras gracile* and *Neocardioceras juddii*, and lower Turonian *Watinoceras coloradoense* Biozones. Diverse infaunal/epifaunal bivalve assemblages are best developed in the *Sciponoceras gracile* Biozone throughout the western interior basin, especially in the limestone beds and equivalent strata. The most diverse assemblages are found along the clay-rich western margin of the basin. Gastropod and infaunal bivalve abundance and diversity decreases in an offshore and upsection direction as carbonate and organic carbon percentages generally increase. Stenotopic species are progressively replaced by low-oxygen tolerant taxa, and many species that required firm substrates have last occurrences at several levels in the *Sciponoceras gracile* Biozone. These patterns imply pulsed decreases in benthic oxygen levels and increases in sedimentation rates during deposition of that biozone.

The base of the *Neocardioceras juddii* Biozone is marked by an influx of southern ammonites and unusual inoceramid bivalve species. Following this immigration event, there was the near loss of all taxa except low-oxygen tolerant inoceramids in the central western interior basin. Along the basin's western margin, this latest Cenomanian interval is characterized by pulsed increases in organic content associated with zones containing abundant detrital-feeding gastropods and infaunal bivalves adapted for soft-substrate conditions. These data suggest that the stage boundary extinctions were associated with episodic increases in organic detrital resources and decreases in substrate firmness and oxygenation. The disruption of pelagic ammonites in this biozone also suggests that low oxygen conditions extended well into the water column. However, low surface-water salinity also might explain this ammonite disruption.

The C-T boundary is marked by the first occurrence of characteristic Turonian *Mytiloides* bivalve species and a decrease in organic carbon percentages throughout the western interior basin. In addition, there is an abrupt change from infaunal/epifaunal assemblages to depauperate epifaunal-dominated associations along the western margin of the basin at this level. This change is accompanied by an abrupt increase in carbonate content, which, when coupled with continued low benthic oxygen levels, may have been detrimental to molluscan infaunal habitation. Ammonite immigration events and pulsed, progressive recovery through this biozone were associated with the carbon-rich depositional periods (now limestones).

In conclusion, a complex set of fluctuating physical parameters produced repeated taxonomic and community turnovers in the C-T boundary interval. Benthic to water column oxygenation and associated organic resource levels appear to have played the strongest role in limiting taxonomic distributions and causing extinctions, but fluctuations in clastic sedimentation rates were also important, particularly in nearshore regions. Salinity changes also may have affected molluscan distributions, but there is no direct macrofaunal evidence of this being the case.

EL-GABY, S., E. R. PHILOBBOS & S. RIAD, Geol. Dept., Assiut Univ., Assiut, Egypt

Crustal Structure Along Geotransect Gabal Oweinat — Red Sea, Southern Egypt

Archean rocks crop out in the extreme western part along the geotransect Gabal Oweinat-Red Sea followed eastward by Early Proterozoic (1800 Ma) which might extend eastward in the Eastern Desert below Late Proterozoic rocks. The basement is unconformably overlain by Early Paleozoic sediments in the western part which are, in turn, unconformably overlain eastwards by Cretaceous and Tertiary sediments all the way into the Eastern Desert. Synrift Neogene sediments accumulated along the Red Sea coastal area.

The prominent surface fractures and shear zones along the transect display the trends (in degrees) 90, 60, 120, 20, 0 and 150, arranged in decreasing order of abundance. It is believed that these trends are inherited from the Late Proterozoic tectonic trends and were reactivated several times with different magnitudes during the Phanerozoic.

From available geophysical data the crustal thickness was computed and gravity and magnetic anomalies were interpreted in terms of deep and shallow structures and faults. Crustal thickness decreases smoothly from about 34 km in the western part to about 32 km in the Eastern Desert, but suddenly to about 28 km at the Red Sea coast. Different fault sets occur with different order of abundace in the different parts along the transect. The northeasterly trends prevail in the western part while the easterly trends predominate in the eastern part of the Western Desert.

Conformity of geophysical structural data with surface structures indicates that basement tectonics controlled the tectonic set up of the different areas along the transect throughout the Phanerozoic. The stresses that affected the area during the Late Proterozoic acted from the ENE. They acquired other directions during the Phanerozoic and were controlled during its latest phases by then northerly drift of the African continent.

EMSELLEM, YVES, Michel DETAY, Pierre ALLA and Pierre SUZANNE

A Model for Self-Purification of Nitrates in Groundwater Example of the Ansereuilles Aquifer (France)

Nowadays, nitrate pollution is well known in all developed countries. The causes, the general physical and chemical rules of this phenomenon have been identified, but the magnitude of time fornitrate bio-degradation in the soil is not very well-known.

First, the reduction in nitrate content in groundwater by bacteria in an anaerobic environment is studied. This approach is illustrated by a field study of the Ansereuilles aquifer (Géolab-Arlab mathematical model).

Second, the two aquifers being considered in their geological, structural and hydrodynamic context, the model for self-purification of nitrates in groundwater is developed and adjusted. The suggested model takes into consideration the dilution in nitrate poor waters, the assimilation of nitrates through the root system and the ground retention. The role played by the oxygen in nitrates is emphasized and the phenomena observed for each geological horizon are detailed. General schemes for nitrate circulation and retention are also suggested.

Finally, the major information of the model are considered in detail. It appears that time is a major factor in the reduction of nitrates in the soil. It corresponds to the bacteria's ability to absorb more or less oxygen according to their weight of dry matter per year. The model outlines an essential relation between the nature and the thickness of the geological zone and the rate of nitrate self-purification. For a period ranging from six months to one year, the reduction of nitrates in a confined aquifer, with no contact with the oxygen of air easily excess 40% to 50%.

Monitoring the model is carried out step by step from known and measured in the field and through a mathematical model. Finally, the generalization of the model of nitrate self-purification in groundwater is offered together with a methodology.

FED'KIN, V. V., Institute of Experimental Mineralogy USSR Academy Sciences, Moscow district, USSR

Staurolite geothermobarometry: interconsistent evaluations of T, P, f_{H_2O}

More than 100 samples have been analyzed on the microprobe to determine the composition of coexisting minerals in the paragenesis Sta + Gr + Bi + Mu + Al_2SiO_5 + P1 + Q. Using the data available on the stability of staurolite and its mineral assemblages, the experimentally evaluated biotite-garnet geothermometer (Lavrent'eva, Perchuk 1981) and three interconsistent geobarometers: Bi = Gr = Pl - Al_2SiO_5 - Q; Bi - Gr - Mu - Al_2SiO_5 - O and Bi - Gr - Mu - Pl - Q (Aranovich, 1983), a new Sta-Gr mineral geothermometer has been derived (T = ± 22°):

$$T, K = \frac{3496.9+0.0078 \bullet P+1079.278(n_{Fe}-n_{Mg}) - 488 \bullet n_{Zn}}{R\ln K_p(1) + 2.99865 + 1.98\,(n_{Fe}-n_{Mg})} \quad (1)$$

where n_{Fe}, n_{Mg} and n_{Zn} are the coefficients in the staurolite formula calculated to 23.5 oxygen atoms; p is the pressure, bar; R - is the universal gas constant = 1.987 cal/degree.mole; $K_p(1) = Q_{Mg\text{-}Sta} \bullet a_{Alm}/a_{Fe\text{-}Sta} \bullet a_{prp}$, a - is the activity of the corresponding minal per 1 atom of the cation being exchanged. The nonideality of staurolite and garnet solid solutions as well as the effect of Ca into Gr and Zn introduction into Sta have been accounted for. Activities of Sta end-members have been calculated, based on the multiple ionic solid solution model with due regard to the site occupancy:

$$a_{Fe\text{-}Sta} = (X_{Fe}^{IV})^2 \bullet (X_{A1}^{3A,3B}) \bullet (X_{Si}^{IV})^4 + \exp X_{Mg}^2 W_{FeMg} + X_{Zn}^2 + X$$

$$X_{Zn}(W_{FeMg}-976)\,/RT$$

$$a_{Mg\text{-}Sta} = (X_{Mg}^{IV})^2 \bullet (X_{A1}^{3A,3B}) \bullet (X_{Si}^{IV})^4 + \exp X_{Fe}^2 W_{FeMg} + X_{Zn}^2 \bullet 1976 +$$

$$X_{Fe} \bullet X_{Zn}(W_{FeMg}+976)\,/RT,$$

where $W_{FeMg} = -\,2158{,}6 + 3.96 \bullet T$

The activities for Alm, Prp, Mu and Bi iron minal are from Aranovich (1983).

The staurolite-garnet geohydrometer has been derived by correlating the available experimental data on the stability of staurolite and its assemblages in the system FeO - $A1_2O_3$ - SiO_2 - H_2 - 0_2 (Fed'kin, 1975). The equilibrium 2Fe-Sta+3.667Q=1.333Alm+7.66Ky(Sill, And)+H_2O (2) has been calculated to give:

$$f_{H_2O} = -\,\exp \Delta G^{\circ}_{T} + \Delta V_s P_s + RT\ln K_p(2)\,/RT$$

$$\Delta G^{\circ}_{T}(2) = 19770-32.541 \bullet T \text{ (Ky)}$$

$$\Delta G^{\circ}_{T}(2) = 32320-53.953 \bullet T \text{ (Sill)}$$

$$\Delta G^{\circ}_{T}(2) = 27134-49.01 \bullet T \text{ (And)}$$

where $K_p(2) = a_{Alm}^{1.333}/a_{Fe\text{-}Sta}^{2}$

The f_{H_2O} value for 87 parageneses Sta + Bi + Gr + Mu + Al_2Si0_5 + Q has been found to agree well with the Mu + Q stability for the same T and P.

Based on the data obtained for T and f_{H_2O}, five interconsistent geobarometers have been calculated for staurolite-bearing metapelites.

No	Equilibrium	ΔH°_T, cal	ΔS°, cal/degree	ΔV_s cal/bar	T, bar
1.	2Fe−Sta+0.667Mu+3Q= 0.667Ann+0.667Alm+9Sill +H_2O	42007	64.205	0.4624	480
2.	2Fe−Sta+3.833Ann+ 7.5Q= 5.167Alm+3.833Mu+H_2O	−32148	−18.299	−1.811	1165
3.	2Fe−Sta+1.333Mu+ 2.333Q= 1.333Bi+10.333Sill+H_2O	51433	74.154	0.7992	525
4.	2Fe−Sta+2.333Alm+ 3.667Mu= 15Sill+3.667Ann+H_2O	84468	109.090	1.978	573
5.	Alm+Mu=Ann+2Sill+Q	14396	15.249	0.5052	620

REFERENCES

Aranovich, L. Ya., 1983. Biotite-garnet equilibria in metapelites. I Thermodynamics of solid solutions and mineral reactions. Nauka (in Russian), p. 112-127

Fed'kin, V. V., 1975. Staurolite Composition, properties, parageneses, formation conditions, Nauka (in Russian), 272 p.

Lavrent'eva, I. V., and Perchuk, L. L., 1981. Phase correspondence in the system biotite-garnet. Experimental data. Dokl. Akad. Nauk SSSR. 260, No 3. (in Russian), p. 731-734

FINN, V.J., J.C.WILT, K.C. HORSTMAN, and G.S. BARNETT, University of Arizona, Tucson, Arizona, U.S.A.

Morphostructural Analysis of the North American Cordillera as Indicated by Tectonics, Geomorphology, Geophysics, Geology, and Metallogenesis

Morphostructural analysis reveals concentric, circular or oval, or linear dislocations of basement blocks that are hidden by sedimentary or volcanic cover, deformational events, or intrusions. The techique also delineates the evolution of the rocks and events covering and affected by the dislocation blocks in the basement and describes the periodic movements of these blocks as uplifts or depressions. The intersections of these basement dislocation blocks and linear discontinuities are zones that localize emanations (magmas, gases, or mineralizing fluids) from deep regions of the crust or mantle, to produce different ore deposits. The presence, size, and shape of these dislocations or basement blocks can be detected by the use of topography, present elevation or relief, drainage patterns, distribution of geology, paleotectonic patterns, facies patterns, lineaments from satellite images, structures, geophysical anomalies, and distribution of mineralization.

The scales of concentric structures or megadomes range from first order or global (1,500-2,000 km diameter), regional (900-1,200 km diameter), subregional (500-800 km diameter), provincial (300-450 km diameter), to district (80-200 km diameter). The scales of lineaments or zones of faults, which indicate basement belts of linear dislocations, also range from global or first order features to local features. Global lineaments (300-500 km wide, 1,000-10,000 km long, 500 km apart) allow fluids to penetrate the crust and localize instrusions; the lineaments traverse different geological terranes, such as the Mendocino zone in both oceanic and continental terranes. Transregional linear features (100-200 km wide, 1000 km long, 150 km apart) generally are boundaries between regional tectonic features (such as geosynclines or hinge lines). Regional linear features (20-50 km wide, 100 km long, 50-100 km apart) are tectonic elements that are interior to the larger concentric features or megadomes, such as a graben or caldera. Subregional linear features (5-10 km wide, 10-100 km long) include concealed basement faults. Fracture zones (0.1-1 km wide, 1-10 km long) are fault zones at the local scale.

Current research on the Pacific margin reveals specific basement dislocation blocks in the North American Cordillera from Alaska to Guatemala. The evolution of this mountain chain includes geosynclinal systems from late Precambrian through Paleozoic time and easterly dipping subduction zones during Mesozoic and Cenozoic time. The block structures or basement dislocation blocks inherited from Precambrian features have influenced specific elements of the geosynclinal or subduction systems. The global concentric domains and/or lineaments have been reactivated as uplifts or basins throughout geologic time in the same general area.

The morphostructural analysis of the North American Cordillera using topographic maps of regional scales (1:5,000,000) reveals two global scale oval structures joined by two linean structural belts. The global scale oval or linear features from north to south along the Pacific margin of North America (the Cordilleran transregional linear) are the "Alaska" circular domain, the "British Columbia" regional linear belt, the "Colorado" circular domain including much of the western United States, and the "Western Mexico" regional linear belt.

The common features of the giant circular domains or basement dislocation blocks include a concentric zoning of geological, tectonic, and paleotectonic elements which are correlated in space with hypsometric levels (elevation categories). The characteristic features of the regional linear belts are a correlation of the linear hypsometric plan with linear elements of geologic, tectonics, and paleotectonic maps. The global circular domains, such as the "Colorado" domain, are later segregated into smaller blocks of second or third order. The sequence of formation of these smaller blocks and their interrelationship as shown by the morphostructural analysis at a more detailed scale (1:1,000,000) indicates the history of development of these structures.

The results indicate a coincidence of these morphostructural features in space and in time with different types of zoning in geosynclines, magmatic belts, areas of uplift and subsidence, paleotectonic features, distribution of mineralization. This strongly implies an inheritance from older or Precambrian features and continued activity of the basement dislocation blocks.

Morphostructural analysis at more detailed scales reveals disintegration of the first order magnitude structures (global circular domains) into second order and third order features, which allows us to trace a relationship between the larger and smaller scale features in space and time. The development of the Gulf of California, Baja California, and the British Columbia islands are specific structures within the Pacific Belt whose movement through time has been traced through time.

The distribution and zoning of endogenic mineralization and its control by the various elements revealed by morphostructural analysis constitute one of the best practical applications of this technique.

GALETSKY, L. S., A. O. SCHMIDT, V. K. TITOV, USSR

Tectonics and Metallogeny of the East-European Platform on the Basis of the Concept of Geoblock Development and Activation of the Earth's Crust

Since the early 60-es the Soviet researchers Krasny, Tyapkin, Schmidt, Dedeyev and others have studied regularities of the block structure of lithosphere and at present a comprehensive concept of geoblock development and activation of the Earth's crust or geoblock tectonics has been elaborated /I and others/.

The concept covers peculiarities of individual development of geoblocks and specific features of their constituents, i.e. continental and oceanic megablocks of the Earth's crust that proceed through the stages of mobile structures with plutonogenic mechanisms of forming vertical zonality of lithosphere and through the stages of stabilized crustal blocks complicated by tectogenic mechanisms of forming its bedding in combination with vertical and lateral dislocations in overthrust, riftogenic, spreading and other zones.

At present inherited transformation of lithosphere-forming plutonogenic processes in the Earth evolution that prevailed at certain stages (the authors identify blocks with granite and sulphur-gneiss crust of Early Archean relict craton cores as the subtypes of the first type of continental crust and blocks with protogranite, plagiogranite crust of Late Archean cratons as the second type of continental crust), replaced by tectogenic processes (crust of relict riftogenic greenstone belts, riftogenic-orogenic crust of granite-greenstone regions, tectonically bedded crust of overthrust zones) has been well-grounded.

Approximately since the middle of Early Proterozoic period there form blocks of developed granite crust (the third type of continental crust) under the action of plutonogenic tectogenic and exogenic processes.

In the early periods of the Earth's crust evolution and probably in the evolutio of lithosphere as a whole the role if interblock and other sutural zones of intensive exchange of substance and energy (geotectonic units /2/, interblock zones, linear activation zones /3/ increases in the course of time. The role of similar geostructures in forming both lateral zonation of lithosphere and vertical zonation of the crust in separate blocks becomes very important in Early Proterozoic and reaches its culmination by the end of this megachrone.

In the present paper the authors suggest their own definition of geoblock tectonics concept. The following members have been distinguished on the exaple of the Pre-Cambrian East-European Platform (EEP): geoblocks, their constituents presented by megablocks and blocks of the first order as well as interblock and transblock sutural zones named lineamental mobile zones (ILMZ and TLMZ correspondingly). Superimposed activation zones constituting TLMZ, less frequently ILMZ develop on the basis of regmatic systems of intersected faults and play the leading role in formation of metallogenic zonality.

The suggested EEP zonation enables to define more precisely the evolution of geoblocks and megablocks in modern continental crust and, what is especially interesting, to obtain a new and more detailed explanation for a number of peculiarities of EEP metallogeny. Some of these peculiarities emphasize the evolutional history of formation of geoblocks, megablocks and ILMZ, in other typical cases metallogenic zoning is governed by the structures of activation of transblock zones, thus its superimposed character is distinctly pronounced.

Unfolding the ideas of paper /I/ the authors distinguish four geoblocks instead of seven featured by predominant structural and compositional complexes and metallogenic provinces, systems of metallogenic zones specific for each geoblock.

Here belong: the Byelorussian-Baltic geoblock with predominant development of iron-ore, siderophile and chalcophile shows in provinces and zones of megablocks that constitute it; the Karelian-Kola geoblock with distinct chalcophile (nonferrous metals) specialization of the crust and of its activation structures in the presence of lithophylic ore shows as well; the Volga-Kama geoblock with siderophiles and oil content in the basement and plate complexes of aulacogens; the Voronezh-Ukrainian geoblock with unique concentrations of iron ores of different types including the high grade ones, ore shows of non-ferrous metals associated with greenstone belts.

The geobocks in question are divided by large intergeoblock sutural zones of riftogenic nature. Here belong the Volin-Dvina volcano-plutogenic belt striking north-eastwards for over 2000km and the Vorontsovo and Ladoga-Outokumpu zones following one after another north-westwards.

The zones under scope specialize in non-ferrous metals mainly associated with basite-ultrabasite complexes. Metallogenic specialization of megablocks and interblock sutural zones has also been analysed.

A special position is occupied by continuous cross-cutting activation megazones of different, usually interrelated orthogonal trends. They consist of formations of alkaline series and of rare-metal specialization (the Central, Central-Ukrainian, North-Ukrainian, Rossoshany-Pri-Azov and Polkanov's zone).

Performed generalization reveals new perspectives in interpretation of tectonic development and metallogeny in Pre-Cambrian.

REFERENCES

L. I. Krasny: Global geoblock system; Nedra Press, Moscow, 1984 (in Russ.).

G. N. Shcherba: Geotectonic members and ore belts (Kazakhstan); Nauka Press, Alma-Ata, 1970 (in Russ.).

A. O. Schmidt et al.: General regularities of deep structure of shields and their metallogenic significance; Proceedings of the All-Union Geol Inst., new series, 1977, v.240, 126-131 (in Russ.).

L. S. Galetsky, V.A. Kolosovskaya, G. I. Ilkevich et al.; Correlation of structures, geologic and ore formations of the Ukrainian shield, Byelorussian and Voronezh rock masses; Geol. Journ., 1987, v.47, N4, 52-59 (in Russ.).

L. S. Galetsky et al.: Lineaments in Ukrain and their metallogenic significance; in book: Geophysical methods of investigating the Earth's crust fault systems and principles of their application for prognostication of ore deposits; edited by Prof. K. F. Tyapkin; Dniepropetrovsk, 1988 (in Russ.).

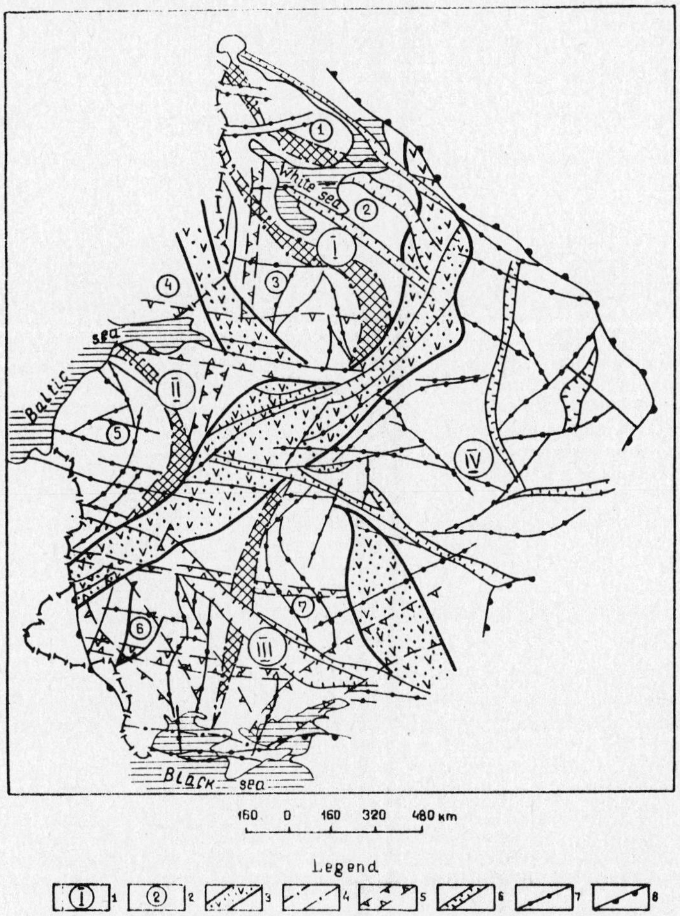

Legend — 1 = Geoblocks: I — Karelian-Kola, II — Byelorussian-Baltic, III — Voronezh-Ukrainian, IV — Volga-Kama; 2 = megablocks: 1 — Kola, 2 — Laplandian-White Sea, 3 — Karelian, 4 — Svekofen, 5 — Byelorussian, 6 — Volin-Podol, 7 — Kursk-Dnieper; 3 = intergeoblock suturual zones (ILMZ); 4 = interblock sutural zones (ILMZ); 5 = transblock activation zones (TLMZ); 6 = aulakogens; 7 = boundaries of East-European platform.

GALETSKY, L. S., Yu. N. YERMAKOV, G. I. ILKEVICH, V. A. KOLOSOVSKAYA, N. D. KONONOV, L. I. MATRUNCHIK, USSR

Metallogenetic Zoning and Evolution of Pre-Cambrian Ore Formation in the South-Western Part of the East-European Platform

A comparative study of the Ukrainian shield, the Byelorussian and Voronezh crystalline rock masses constituting the indivisible Sarmatian shield in Pre-Cambrian has been accomplished. The fragmenting of the shield began in Late Proterozoic and continued the most intensively in Late Palaeozoic. At the present time these crystalline rock masses are divided by the deep Dnieper-Donetsk basin and Pripyat depression of riftogenic type and also by Krestsy and Roslavl troughs. For the first time an attempt has been made to reconstruct the indivisible Pre-Cambrian basement of the region under consideration, to correlate structural and compositional peculiarities of complexes and metallogenetic zones of the presently disconnected regions. Their metallogeny is of common character: besides of the typical Pre-Cambrian deposits of iron, manganese, titanium, nickel, copper, chromite, graphite, pyrophyllite, sillimanite, the mineralization of rare metals, tin, tungsten, molybdenum, apatite, fluorite is widely spread.

The general structural metallogenetic zonation has shown a certain metallogenetic zoning of megablocks, of sutural structures that divide them, of metallogenetic regions and zones, of ore areas and fields. Structures of various ranks characterized different geodynamical environments of formation, by certain features of the Earth's crust structure are first of all notable for metallogenetic specialization and productivity. Thus the Kursk-Dnieper megablock arisen against the protooceanic crust background distinguishes itself by distinct and inherited iron ore specialization. The marginal sutural zones (Krivoy Rog-Kremenchug-Krupetsk) and intraplate troughs (Belgorod-Mikhailov, Alekseyev-Voronetsk) are notable for the utmost productivity. The granite-greenstone belts with copper-nickel mineralization are confined to the same megablock. The oldest charnokite-granulite belts (West-Byelorussian and Podol blocks) are featured by mineralization with low concentration grade (iron-, carbon- and alumina-bearing series). Deep zones of riftogenesis and activation with ultrabasic alkaline formations containing apatite-rare-earth carbonatites and with troughs containing rare-metal pegmatites develop around old enderbite-charnokite cores.

Large sutural zones of folded thrust type (East-Byelorussian-Byelaya Tserkov) containing old ophiolites are characterized by the mineralization of iron, chromite, nickel, by very large graphite deposits. Blocks of mature Earth's crust (micro-continents) are notable for powerful intraplate magmatism (Korosten-Korsun-Novomirgorod plutons and others) with phosphorus-titanium and rare-metal mineralization as well as for intensive regional and local metasomatism producing rare-metal mineralization.

Linear zones of tectonic-metasomatic activation have continous development, they concentrate rare-metal mineralization, fluorite. Intracontinental aulacogens (Dnieper-Donetsk basin) are featured by a different mineragenetic profile: polymetals, mercury, antimony, rock and potash salts, sulphur, hard coal, petroleum, gas. It is interesting to note that longitudinal metallogenetic zoning of aulacogens is defined by cross structures of Pre-Cambrian basement.

The transformation of ore-forming systems has been studied in time: from regional slightly differentiated systems that provide disseminated accumulation of ore material to local, relatively closed, ore-concentrating ones. Thus the verticle extent of ore-forming systems and remoteness of deposits from primary sources of ore material increase. In early periods (Archean) there appear the deposits the ore bodies of which require a slight increase of concentration in comparison with background values (Fe, Mn, Ti, Al). In late periods there originate the deposits of rare metals the concentration of which demands more complicated, prolonged and multistaged ways of material differentiation. The leading genetic types of ore formations also change from volcanogenic-sedimentary, metamorphic, magmatic to metasomatic and hydrothermal-veined. The degree of differentiation of ore formations, the variety of qualitative and quantitative composition of ore minerals, the degree of concentration of ore material increase in the same order. The inherited development of the Earth's crust, its differentiation, formation of granitic and sedimentary series, transformation and inversion of geodynamical environments represent the main factors of evolution of Pre-Cambrian formations and sructures that frame them. As a rule, productive mineralization and the largest deposits appear against the background of contrast events in the Earth's crust development: accretion, maturing, stabilization alternating with destruction and activation processes.

GEYER, GERD, Institut für Paläontologie, der Universität Würzburg, Würzburg.

"Supercycles" and Complex Cyclicity in the Late Precambrian and Cambrian of Morocco

The lithological succession of Late Precambrian through Cambrian in southern Morocco is divided into groups which each reflects a supercycle (Monninger, 1979) that starts at preferably transgressive and ends at regressive depositional conditions. These supercycles are comparable in many aspects with the "grand cycles" of the North American Paleozoic. However, they differ remarkably by the longer duration and by the lithological sequence of generally carbonate through fine-clastics due to the deposition in extreme shallow-water environments. Therefore, boundaries are additionally strongly diachronous.

The Cyclizity

The stratigraphic interval dealed with is part of a column deposited in a slightly expanding epicratonic basin existing from the Late Precambrian up to the Silurian. Several transgressions and regressions provided for a variable deposition of shallow marine to terrigeneous sediments (supercycles). The long lasting **supercycles** do not necessarily include cyclothems, but are composed at least partly of small cyclic lithosequences. They are exactly represented in the Taroudannt (supercycle I) and Tata Groups (supercycle II) and less strictly in the Feijas internes and Tabanite Groups (supercycles III and IV). They usually commence with carbonatic or siliciclastic transgression layers. By means of the rather quickly advancing seas the depositional area moved into a carbonate belt (supercycles I and II) or an equivalently situated shallow siliciclastic shelf offshore (supercycles III and IV), respectively, forming the lower half-cycle. The upper half-cycle reflects increasing regressional conditions caused by protruding shorelines from east to the west. The area therefore shifted into more and more terrigeneously influenced sedimentary environments.

The **cyclothems** are composed of several similar lithosequences. They were controlled by rhythmic bathymetrical fluctuations and/or changes of salinity and terrigeneous influx.

The rock units

The **Taroudannt Group** (representing supercycle I) commences with the Adoudou Formation, which is composed of a clastic to carbonatic Basal Member and a Limestone Member. Cyclothems within the formation were only recognized from the Limestone Member. The lithological change within a cyclothem is usually weak and has more the character of a single rhythm, except the larger scale. Towards the basinal margin cyclothems tend to efface.

The Lie de vin Formation is composed essentially of fine-grained siliciclastic rocks with numerous intercalations of carbonate beds, each representing complex cyclothems. Supercycle I ends with an extremely shallow lagoonal situation (uppermost Lie de vin Fm.).

The overlying **Tata Group** (supercycle II) commences with the Igoudine Formation, essentially consisting of massive carbonate beds. Cyclothems may possess the character of a single rhythm with thick core beds of crystalline limestones and thin interbeds of shale or siltstone. Facial trends are reflected by changing interbed lithology.

The Amouslek Formation is composed of coloured slates with intercalated limestone beds. Genetically, the sequence is a product of varying water depths and shoreline oscillations under shallow marine shelf conditions, causing cyclothems of various lithological changes. Towards the east the content of detrital elements increases with shallowing water depths.

The Lemdad Formation is a variable and rhythmic alternation of marlstone to siltstone sequences and intercalated limestone and dolostone units. Within the formation, there is a long term trend towards regressive conditions, best reflected by the microfacies of the carbonate units within the cyclothems. The cyclic succession is masked by the capricious intercalations of volcanogeneous material.

Within the Issafen Formation, sedimentary cycles are indicated by rhythmic intercalations of limestone complexes into the slate sequence, giving evidence for lagoonal siliciclastic deposition.

The Asrir Formation is a succession of resisting arenitic shales and sandstones with tuffaceous intercalations. In the western Anti-Atlas, the Asrir Formation consists generally of cyclic shale and sandstone sequences. Towards the east, volcanic eruptions are responsible for cyclothems initiated by slaty tuffites. Finally, the influence of a huge delta system protruding from the east gave rise for local sedimentary breaks.

The **Feijas internes Group** (supercycle III) commences with deltaic influence in the sound and a shallow sea with synchronous volcaniclastic deposition in at least the central High Atlas. The lower half-cycle is dominated by shales in the west and argillaceous sandstones in the east. The west of the basin probably reached nearly the same offshore position on the platform as in Late Precambrian to Early Cambrian times, but only sparse carbonate sedimentation is visible, interpreted by a drifting into more temperate latitudes. However, the marine regime advanced towards the east during sedimentation of the initiating carbonatic and bioclastic Bréche à *Micmacca* Member. In the upper half-cycle deposition took place in a basin that gradually silted up: sediments are preferably terrigeneously influenced, mainly siltstones and sandstones with increasing grain-sizes from Central High Atlas towards the east and south.

Within the **Tabanite Group** (supercycle IV) bathymetrical oscillations caused again marine environments in the *Bailiella* Formation. This last supercycle is less clearly defined, due to less clearly recognizable differences in lithology. Moreover, the end of the supercycle is rarely represented as vast areas were eroded in late Upper Cambrian times. However, there is evidence for again weakly transgressive layers within the Jbel Lmgaysmat Formation at the top of the group.

GOODFRIEND, GLENN A. and Mordeckai MAGARITZ, Isotope Department, Weizmann Institute of Science, Rehovot, Israel

$\delta^{18}O$ of Land Snail Shell Carbonate: Paleoclimatic Interpretation and the Quaternary Record in the Negev Desert

The factors affecting $^{18}O/^{16}O$ ratio of the shell carbonate of land snails have been studied by means of 1) survey of geographical variation in $\delta^{18}O$ values in relation to climate and 2) a field study of short-term variations of the $\delta^{18}O$ values of land snail body water in relation to rainfall events. These studies indicate that shell $\delta^{18}O$ values are controlled mainly by the oxygen isotope composition of water vapor in the air at the site, by isotopic exchange with body water and/or by uptake of dew derived from condensation of water vapor. Because the isotopic composition of water vapor reflects the isotopic composition of rain, the oxygen isotope composition of fossil land snail shell carbonate can be used to reconstruct past rainfall ^{18}O variations.

The results of analyses of $\delta^{18}O$ in a series of radiocarbon-dated land snails from the Negev Desert in southern Israel are presented. A detailed $\delta^{18}O$ curve for the Holocene is given.

GOODWIN, ALAN M., Department of Geology, University of Toronto, Toronto, Canada

Precambrian Granulites and the Dynamic Lower Crust

Earth's Precambrian crust (exposed+buried) underlies $106 \times 10^6 km^2$ or 72% of the total continental crust. Precambrian proportions by tectonic age are: Archean (+2.5Ga)=15%; early Proterozoic (to 1.8Ga)=19%; middle Proterozoic (to 1.0Ga)=23%; and late Proterozoic (to 0.57Ga)=43%. Excluding extensive flat to subtabular, non-to little-metamorphosed late Proterozoic platform cover, e.g., Congo Basin, Precambrian terranes of the world contain on average 13% greenschist and lower-, 62% amphibolite-, and 25% granulite-facies metamorphism. This implies the removal of 12-15 km of overlying crust. The granulite terranes in turn require exhumation from depths of 17-31 km. Although clearly representing lower crust, it is questionable if even the most mafic presently exposed granulite represents the lowermost crust. However the lower crust has obviously been dynamically active throughout Precambrian time.

Granulite complexes vary in size from the small but intensely studied and historically important Lewisian Complex, Scotland at $0.5 \times 10^3 km^2$, to the Ashuanipi complex of Labrador, Canada, at $100 \times 10^3 km^2$ one of the largest areas of unretrogressed granulite in the world. Granulites are characterized by: (1) dryness (pyroxenes); (2) depletion in some but not necessorily all of Rb, Th, U, K and heavy REE; (3) CO_2-rich primary fluid inclusions; (4) high metamorphic temperatures (650-950°C) and pressures 5-11 Kbrs, (5) common shallow marine (shelf) -type metasupracrustal associates (marbles, quartzites, metapelites, paraamphibolites); and (6) instrusive charnockites (pyroxene-bearing quartzofeldspathic rocks) with associated anorthosite massifs. Virtually all granulite terranes show evidence of large-scale horizontal tectonics e.g. recumbent folding, usually early in the metamorphic cycle. The excessively high temperature range requires some sort of thermal anamaly over and above the ambient condition of the lower continental crust. The high pressure range suggests that some unique control initiated granulite metamorphism at crustal depths of 17-30 km; the extreme burial of surficial layers implies great crustal thickening attendant on granulite-facies metamorphism. Massif-type anorthosites are conspicuous in middle Proterozoic high-grade terranes of which the largest were emplaced as flat sheets at shallow depths, probably in rifted continental settings. Finally, a specific "drying" mechanism is required to provide for the CO_2- dominated fluid inclusions in contrast to the H_2O-dominated fluid inclusions and abundant anatexis of lower metamorphic grades.

Granulite complexes occur in Archean and Proterozoic terranes of all ages, and even in Cretaceous (Big Sur) and Eocene (Prince Rupert) assemblages. A wide variety of tectonic and developmental proceses has been invoked to explain the return transfer of metasediment-bearing granulites of the lower crust to their present position astride a normal thickness of continental crust. No single mechanism of granulite-facies metamorphism satisfies all the following examples:

The Napier Complex, Enderby Land, Antarctica comprises complexly interleaved orthogneisses and metasediments that exhibit well-developed horizontal layering to form a recumbent gneiss terrane. A protolith model age of 3930 Ma makes this the oldest known area on earth. The main granulite metamorphiam occured 700-900ma later at 3070ma, with intense deformation and massive tectonic and magmatic thickening of the crust. Peak regional metamorphic conditions of 950°C and 7-10 kbs represent crustal depths of 21-30 km. The complex was stable for the next 500 Ma. The presently exposed granulites record a near-isobaric cooling path through 200-250°C over this time, involving maximum uplift of only 6-9 km from the previously attained 21-30 km depth. Shallow cooling paths of this type are possible in crust thickened by accretion of large volumes of magma (30-60% of crust) over a sufficiently long period of time. The near-isobaric cooling path is inconsistent with models of tectonic thickening of crust by continental collision or overthrusting which would have resulted in near-isothermal decompression paths derived from rapid erosion rates. Napier Complex probably remained deeply buried until at least 1000 Ma, when partially exhumed as a result of marginal underplating by the adjoining Rayner Complex.

The Lewisian Complex, Scotland, represents a bimodal association of (1) mafic gneisses characteristic of tholeiitic low-pressure fractionation, with metasedimentary and mafic-ultramafic associates, and (2) tonalitic-trondhjemitic gneisses representing partial melts at moderate depths of a mafic source under hydrous high-pressure conditions, with consequent crustal thickening by underplating of tonalitic magmas. Lewisian granulites rank amongst the most "depleted" in the world, with extremely low U, Th, Rb and K levels. This depletion occurred at 2900 Ma, which is the age of generation of the tonalitic gneisses. The Badcallian granulite-facies metamorphism peaked at 2700 Ma, i.e., 200 Ma after the major crustal generation. P-T conditions are amongst the highest recorded for granulite terranes (+1000°C and 15kb). However there was no major addition of magma at the time of the Badcallian event to supply the thermal energy. As a result the cause of the granulite-facies metamorphism remains unresolved.

The Limpopo Belt of southern Africa is a zone of Archean high-grade metamorphic and igneous rocks situated between the kaapvaal and Zimbabwe Cratons. It consists of a central zone of largely epicontinental rocks separated by major shear zones from marginal zones of granitoid-greenstone lithologies. Central Zone rocks were subjected to high-grade conditions at about 3150 Ma and again, along with the marginal zones, at about 2700 Ma. The belt is essentially symmetrical and lacks evidence for a collisional structural ramp or "crust-on-edge" relation. Instead the favoured model involves crustal thickening to at least 65 km in the Central Zone at 2700 Ma as a direct result of continental collision in the manner of the Himalayas. The thickened crust responded by rapid uplift accompanied by diapirism. The granulite terranes presently exposed were driven upwards from deeper crust along high-angle thrust and reverse faults. The Central Zone was transported to the west, the Northern Marginal Zone to the north and the Southern Marginal Zone to the south. Large volumes of anatectic melt created gravitational instabilities, resulting in widewspread diapirism.

The Grenville Province, Canadian Shield, is 2000 km long and 300-600 km wide. The subsurface extension is equally long to Texas and even Mexico. The 4000 km-long Grenville Front, on the NW margin, is one of the world's great structural discontinuities marking the junction of assorted Archean-Proterozoic terranes on the west with the middle Proterozoic Grenville Orogen. This orogen is notable for its widespread high-grade metamorphism, complex deep level structures, and abundant anorthosites. Granulites are concentrated in a broad zone of allochthonous slices thrust northwestward towards the Front. Rocks of these allochthons, dated at 1700-1350 Ma, are significantly older than the Grenvillian orogeny at 1100-1000 Ma. Boundaries of the slices are marked by ductile tectonite zones. The Grenville Front represents a major decollement developed upon collision of an allochthonous terrane with the North American craton, and NW-directed stacking of microterranes with the upward-ramping of deeply buried granulite-facies rocks to the near surface.

Possible mechanisms of granulite-facies metamorphism in these examples include (1) continental-scale underthrusting, (2) accordion-style thickening, (3) nappe-stacking, and (4) hot-spot development. Although applicable to one or other of the tranulite terranes, no single mechanism explains them all and certain critical features remain unresolved.

REFERENCES

Newton, R.C. 1987. Petrologic aspects of Precambrian granulite facies terranes bearing on their origins. AGU, Geodynamics Ser., *17:* 11-26.

Van Reenen, D.D., Barton, J.M., Roering, C., crustal response to continental collision: The Limpopo belt of southern Africa, Geology, *15:* 11-14.

Park, R.G., and Tarney, J. 1987. The Lewisian complex: a typical Precambrian high-grade terrain? *In* Evolution of the Lewisian and Comparable High Grade Terrains, (eds.) R.G. Park and J. Tarney. Geol. Soc. Spec. Public, *27:* 13-25.

Newton, R.C. and Hansen, E.C., 1983. The origin of Proterozoic and late Archean charnockite-evidence of field relations and experimental petrology, Geol. Soc. America, Memoir *161:* 167-178.

Moore, J.M. Davidson, A. and Baer, A.J. (eds.) 1986. The Grenville Province. Geol. Assoc. Canada, Spec. Paper 31.

GRIGORIAN, SERGEI V., Institute of Geological Sciences, Academy of Science, Armenian SSR, Yerevan, USSR

Geochemical Aspects of Modelling the Ore Deposits

The research of the past few years has established that in modelling the ore deposits most informative results are yielded by geochemical research and by the primary geochemical halos in the first place.

The primary geochemical halos encompass ore deposits of different groups, types and classes regardless of their origin, age or geological conditions of localization. They are the formations genetically integrated with the ore bodies multiply surpassing the ore bodies in size and containing considerable quantities of commercially viable components often in excess of the ore body mass.

The major feature of the primary halos is their geochemical zonality. Studied in detail by the present time is the geochemical zonality of the primary halos of hydrothermal deposits which is most contrastive towards the displacement of ore-carrying solutions and is stipulated by the differentiation in element distribution along the axis of this displacement. For steeply dipping ore bodies this zonality is vertical with a predominant accumulation of one type of element associations in the head sections of the halos and other types in the tail sections.

The practical significance of zonality of the primary geochemical halos is that it enables us to reliably evaluate the level of halo intersection (the erosion section in a particular case) with regard to the position of ore body and to predict not only the occurence of ore formation but also the depth of its possible localization and even scale.

Generalized results of compositional and structural research of the primary geochemical halos of deposits differing in compositions and conditions of formation shows that at the present time the significance of the primary halos cannot be restricted to prospecting alone. Coming to light now is a great geological and genetic significance of the primary halos as well.

In connection with what has been previously mentioned it seems relevant to consider the concept of ore deposit. The experience of studying the primary halos of ore deposits shows the deposit to be not a mere conglomerate of ore bodies but a consistent and inextricable combination with the enclosing primary geochemical halo. The total mass of metals in ore deposits is thus constituted by the masses of halos and ore bodies. The mass within the delineations of the commercial ore bodies, unlike the total mass may be regarded as concentrated.

The ore body and its primary halo is spatially and genetically an integrated whole, the delineation between them being "floating" because it is determined by economic factors.

The ratio between the masses of chemical elements in halos and ore bodies is fluctuating depending on the morphostructural types of the deposit, however in most genetic types of deposits the mass in halos is predominant, often substantially. The "floating" subjective nature of ore body delineations in its turn makes the nature of the deposit subjective as well. By contrast, embracing of the primary halos (marginal sections of the deposit) unlike the ores have an objective character and are not function of the current economic situation (in contrast to the ore body delineations determined by the commercial conditions).

Implementing the proposal to regard the primary geochemical halos as constituent parts of deposits will, it seems, substantially boost the efficiency of deposits on the basis of their comparative study since the objectively defined boundaries of deposits will obviously supply generalized research data obtained on a quantitative basis.

A major component of geochemical modelling is to be the study of the so-called ionometric halos of deposits recently established by E.M. Kopin and the author of this paper. Both experimental and methodological studies performed on deposits differing as to composition and conditions of formation have shown that by using special ion receivers recordings can be made of elevated concentrations of different element-indicators in the periterrestrial atmosphere above the deeply buried (very deeply at times) ore bodies and deposits. Studies were conducted on typically hydrothermal deposits of the non-ferrous (polymetalic) and precious (gold, silver) metals, diamond-carrying kimberlites, copper pyrite deposit of Urup having a complicated genesis and sulphur deposit of Gauzdak. It is to be noted that in each case without exception there were distinct temporally stable anomalies of some heavy metals (according to repeated sampling data) recorded above the buried deposits in the periterrestrial atmosphere.

It should also be noted that the elementary composition of the ionometric anomalities under investigation are truncated (characterized by a smaller number of chemical elements) as compared to ore bodies and their primary halos, which is explained by the available analytic capabilities and the formidable difficulties still confronting the analysis of electrolite samples of multiple elements in ultra-small amounts.

Achieved at the present time is a sufficient sensitivity in determining concentrations in electrolyte of a limited number of elements. Therefore, solving the problems of analyzing the ionometric samples is a major prerequisite for a successful development of this method.

As has already been stated, the primary halos are efficent indicators of the blind ore bodies. In cases where the ore body (the blind one including) is overlapped by younger (in relation to ore formation) deposits (buried ore deposit), ionometric halos can be used as prospecting indicators since in case of buried deposits their primary halos are buried as well, while the ionometric halos punch through the younger deposits overlapping the buried ore bodies.

GUREVICH, A. V.[1], M. S. VERKHOVSKY[2], D. N. TOPOR[2], A. A. PEK[2]
[1]U.S.S.R. Academie of Science, Lebedev Ins. of Physics, Moscow
[2]U.S.S.R. Academie of Science, Ins. of Geology of Ore Deposits, Petrography and Mineralogy, Moscow

The Influence of Highly Permeable Fault on Hydrodynamics of Homogeneous Fluid Flow (2-D Problem)

There are a number of dynamic problems concerned with the transport of fluids of different origin by the faults to the earth's surface.

The solution is presented that makes it possible to determine mass flow rate in the fault zone taking regard its geometrical and hydrologic properties.

Let's consider the following 2-D hydrodynamic problem. The vertical fault with the length equal to L and transverse width equal to H is contained in homogeneous media. Fault zone permeability K_f is much higher than media intrinsic permeability K_r. The steady upwelling flow subjected to Darcy law moves through the region comprising the fault. Taking into account that L>>H (usually L/H>100) we can consider the fault zone to be infinitesimaly thin but possessing finite hydraulic conductivity K_fH. The fault will be assumed to lie on $[-1,1]$ segment of the X axis (vertical) in a system of cartesian coordinates. We determine dimensionless fault fluid conductivity $F(x)=K_f(x)H(x)/(K_rL)$ as known function defined in the interval $[-1,1]$. The fluid pressure P(x,y) satisfies Laplace equation $\triangle P = O$ with boundary condition at the fault

$$\frac{\delta}{\delta x}\left(F(x)\frac{\delta P}{\delta x}\right)+\left[\frac{\delta P}{\delta y}\Big|_{(x,o)}\right] = O, \qquad -1<x<+1,\ y=O \tag{1}$$

where $[\ \ ]$ is the jump of the first pressure derivative. The boundary condition at infinity takes the form

$$P_o(x,y) = x\mathrm{COS}\,(a) + y\mathrm{SIN}\,(a)$$

Where a — is the angle between the direction of the fluid movement and the fault zone axis.

For $F(x)=F_o$=const the problem was investigated by the methods of theory of functions of complex variations /1/. Solutions were obtained for $F_o>>1$ and $F_o<<1$. The corresponding flow direction patterns were given for a=O at Fig. 1a and 1b. The fluid mass flux along the fault (integrated velocity) Q(x) achieves its maximum value a the centre of the fault zone: $Q_{max}=Q(o)$. For $F_o>>1$ a dimensionless Q_{max} value equal to

$$q=Q(o)/(K_rL|\mathrm{grad}\ Po|) = F(o)\ \frac{\delta P}{\delta x}\Big|_{(o,o)} \tag{2}$$

is determined by the expression

$$q = \left(1-\frac{1}{F_o}\left(\frac{6+4G}{\pi}\right)\right)\ \mathrm{CO}^{5}\,(a) \tag{3}$$

Here G = 0,916 — is the Katalan constant. If $F_o<<1$ then

$$q=F_o\mathrm{COS}\,(a) \tag{4}$$

Here the general solution of the problem can be found from a one-dimensional integro — differential equation /2/

$$P(x,o) = P_o(x,o) + \frac{1}{2\pi}\int_{-1}^{1}\frac{\delta}{\delta s}\left(F(s)\frac{\delta P}{\delta s}\right)\ln\frac{1}{|x-s|}ds \tag{5}$$

with boundary condition $F(-1)\ \frac{\delta P}{\delta x}\Big|_{(-1)} = F(+1)\frac{\delta P}{\delta x}\Big|_{(+1)} = O.$

Taking into account the Chebyshev's polynomial propertyes $T_m(x)$, $U_m(x)$ the equation (5) can be expressed in a set of equations

$$\int_{-1}^{1}\frac{F(x)}{\sqrt{1-x^2}}\ \frac{\delta P}{\delta x}N_m(x)\,dx = b_m, \qquad m=1,2,\ldots \tag{6}$$

where $N_m(x) = \frac{1-x^2}{F(x)}\ U_{m-1}(x) - \frac{\sqrt{1-x^2}}{2}\ \frac{d}{dx}\ T_m(x),$

$$b_m = -m\int_{-1}^{1}P_o(x,o)\ \frac{T_m(x)}{\sqrt{1-x^2}}dx=\begin{cases}-\frac{\pi}{2}\mathrm{COS}\,(a), & m=1\\ 0, & m>1\end{cases}$$

We seek the solution to equation (6) in the form of a series

$$F(x)\frac{\delta P}{\delta x} = \sqrt{1-x^2}\ \sum_{m=1}^{\infty} g_mM_m(x) \tag{7}$$

where $\{M_m(x)\}$ is obtained by ortogonalization of $\{N_m(x)\}$ system.

The plot for value $q=F_o\frac{\delta P}{\delta x}\Big|_{(o,o)}$ in case when $F(x)=F_o$=const computed according to (7) is given at Fig. 2. When $F_o>>1$ and $F_o<<1$ the general solution coincides with special solutions (3) and (4).

Conclusion. Highly permeable faults are preferable passageways for discharge of deep fluids. In the general case the fluid flux in the fault zone depends on hydraulic and geometrical characteristics of the fault. The fluid flow rate (integrated velocity) focused by the fault reaches its maximum value in the centre of the fault where it can be determined from solution (7). The maximum flow rate of highly conductive faults when $F_o>>1$ (i.e. $K_f/K_r/>>L/H$) can be found with the help of solution (3). When $F_o<<1$ (i.e. $K_f/K_r<<L/H$) the maximum integrated velocity is defined by (4).

Thus while increasing fault hydraulic conductivity a change of mechanism limiting fluid mass discharge by the fault takes place: for low-conductive zones ($F_o>>1$) this is a value of conductivity (K_fH), while for highly conductive ones it is a vertical length L of the fault.

REFERENCES

A. V. Gurevich, A. L. Krylov, D. N. Topor. Solution of 2-d flow problem of porous media in the vicinity of vertical fault by complex potential method. 1988, Dokl. AN ASSR, v.288, #4.

M. S. Verkhovsky. Use of Chebyshev's polynomials for the steady state fault-controlled hydrodynamics. 1988 (in press).

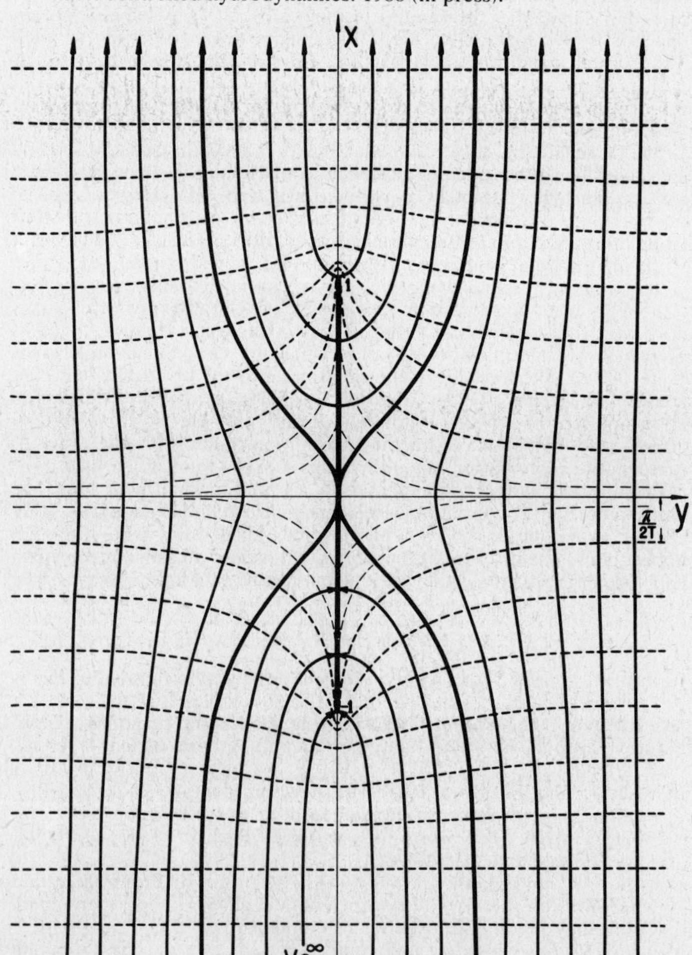

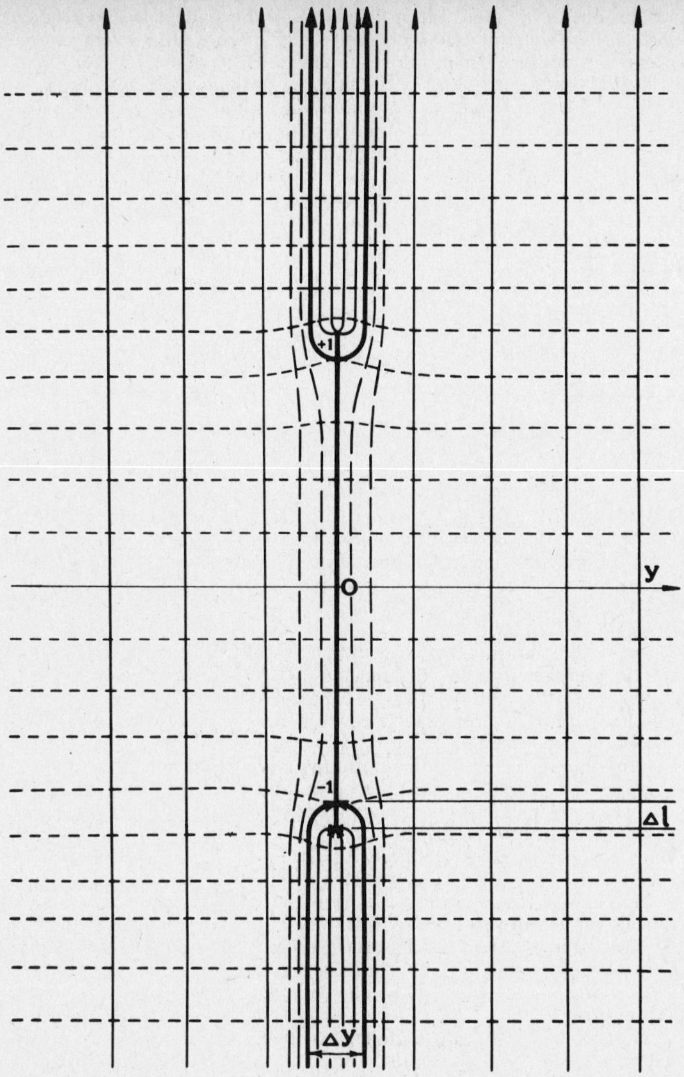

Figure 1 — Directions of upwelling flow controlled by a vertical highly conductive fault (a) and low-conductive fault (b). The fault is a segment of the straight line y=0, −1<x<+1. The undisturbed fluid flow is parallel to the fault. Full curves are the streamlines, dashed lines are hydrodynamic pressures.

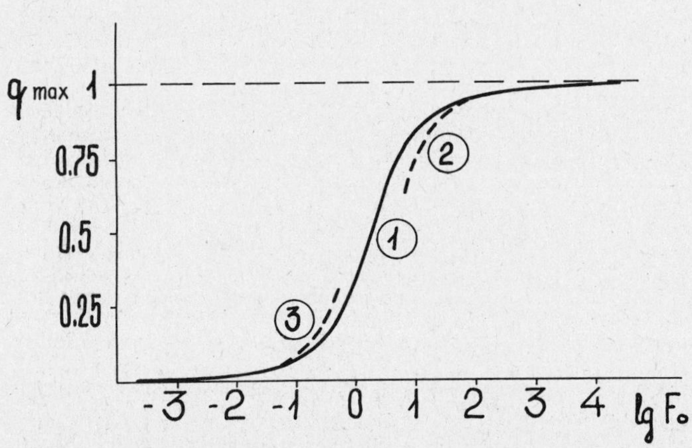

Figure 2 — The relationship between dimensionless maximum integrated fluid velocity q in the centre of the fault (see Fig. 1) and its hydraulic conductivity:
1. according to solution (7),
2. according to solution (3),
3. according to solution (4).

HERNANDEZ, G., R. RODRIGUEZ, A. PEREZ, S. TOUCET, C. SUYI.
Empresa Nacional Geofísica, Habana, Cuba.

Main Characteristics of Geophysical Fields in the Cuban Shelf and its Relation with Oil Exploration

During the last six years (1982-1987) in the cuban shelf and surrounding areas has been done a great amount of geophysical work (about 20 000 km of lines) by different methods.

The objective of these surveys was to study from a regional point of view the geological structure of the area and to look for zones with good possibilities of gas and oil.

The seismic data was collected with 1175-2375 m, 48 channel streamers or baycables. Air or water guns served as the energy source. Shot points were 25-50 m apart in order to obtain a 12-24 fold stack. The gravimeter system consisted of three seagoing gravimeters. In addition to CDP and gravity data, total field magnetics measurements were obtained using a proton magnetometer. Navigation was supplied by means of satellites or by systems like Syledis or Trisponder.

Having in mind that the geological structure in the south of Cuba (the southern sector), which geographically corresponds to the northwestern end of the Caribean sea, is completely different with respect to the north (the northern sector) in the southeastern part of the gulf of Mexico, in Nicholas and Old Bahama channels and Atlantic ocean, the results are presented referring to each sector.

In the southern sector, surveyed mainly in deep waters, a strong reflection, followed all over the area, is considered the frame for distinguish major tectonic elements. In some places this reflection is recorded just below the sea bottom and in others a sequence of coherent reflections appear over it. The seismic lines showed no significant sedimentary thickness except in the gulf of Cauto and Batabano. The gravity and magnetics fields are characterized by intensive anomalies of both signs.

In the western part of the northern sector, the seismic sections show a different character of reflection data which varies from flat-lying reflections on top and bottom of a 2.0-3.0 sec time interval where no information is obtained at all near the coast to strong, planar, continuous band of coherent reflections down to 7 sec in bassinal settings far from the coast.

In its central part in the north-south lines it is not possible to follow any reflections below 0.5 sec probably due to the presence of allocthonous formations of the fold-thrust belt near the coast. At a certain distance from the coast, the band of coherent reflections mentioned before, inferred to represent and accumulation of deep water basinal carbonate deposits no less than 10 km thick is recorded.

In this same part, in eastern Nicholas channel two essentially seismic facies are revealed. One represented by a sequence of coherent and generally undisturbed reflections (already described) and another facies which lack coherent internal reflections. The zone between them, accompanied by a positive free-air gravity anomaly, is characterized by reflections typical from a reef trend and by the presence of a broad anticline structure. These two features, which may have a continuation in shallow waters, have been reported by Ball in 1985 (AAPG, vol. 69, No. 8).

In the eastern part of the northern sector the seismic lines are similar to those of the central part. However, the band of coherent reflections is interrupted abruptly in its north side by a change in the seismic field which is inferred to represent a reef trend or salt domes.

In the northern sector the magnetic anomalies are of long wavelength indicating a great depth to magnetic basement. There are no indications in the potential fields about the extent of the fold-thrust belt into the sea.

Considering the onshore oil fields and the features revealed, the central part of this sector is evaluated as the most important from the point of view of oil and gas exploration.

Authors' main conclusions are that measurements revealed the sedimentary thickness in the northwestern end of the Caribean sea, allowed to know the character of the geophysical fields in the marine part of the fold-thrust belt and its extent northward off the coast, showed the presence of a reef trend and confirmed the existence of a sedimentary section of considerable thickness in the offshore eastern part of Cuba and the presence of reefs or salt domes.

The interpretation of the characteristics of the geophysical fields measured allowed to increase the geological and geophysical knoweldge of the island's shelf and to evaluate its hydrocarbon possibilities from a modern scientific base.

In this paper the authors present the most significant characteristics of the geophysical fields measured during the period when the greatest amount of marine geophysical work has ever been done in Cuba and its importance for oil exploration.

HICKS, B.G., Rogue River National Forest, Medford, Oregon

Geohydrologic Hazard Study of Big Butte Springs Watershed, Implications for Land Management

The Rogue River National Forest located in southwestern Oregon has completed a geohydrologic study of a watershed which is primarily on National Forest land underlain by the following volcanic rock units:

- young High Cascade (YHC) — less than about 100,000 years old.
- old High Cascade (OHC) — about 3 to 8 million years old.
- Western Cascade (WC) — about 15 to 23 million years old.

This study was undertaken jointly with the Medford Water Commission which has supplied water from this basin since 1927 to Medford, Oregon and other local customers. Big Butte Springs (BBS), a potential 36 million gallon per day or about 56 cubic feet per second (cfs) source originating from this 87 square mile basin was developed (in 1927 and 1951) by tapping closely grouped natural springs which issue from YHC and OHC lava flows along Willow Creek, an uppermost tributary of Big Butte Creek.

The study utilized contracts, consultants, universities, Forest Service staff, U.S.G.S., and other Federal and State agencies to complete geophysical studies, core drilling, pump testing, exploration trenching, geologic mapping, special photography, radar imagery, chemical analyses including natural tritium levels, groundwater constituents, potassium-argon rock dating, rock neutron activation analyses, special soil mapping (infiltration characteristics), drill hole hydrographs and hydrologic and tracer studies.

The principal objectives included definition of the following: (1) geohydrologic boundaries of the watershed, (2) primary sources of groundwater, (3) groundwater flow quantities and velocities, (4) groundwater storage vs spring flow, (5) groundwater infiltration and surface water inflow hazard zonation, and (6) relationship of land management of the watershed to potential impacts on the groundwater system (quality and quantity).

The watershed consists of three major drainage basins. Fourbit Creek basin (26944 acres) is composed of approximately 47% highly permeable YHC lava flows and 53% less permeable, OHC lava flows. Mosquito Creek basin (7325 acres) is essentially a closed basin consisting of 65% highly permeable YHC lavas and 35% less permeable OHC lavas. All Mosquito basin streamflow recharges directly into the groundwater system except during rare flood events. The Willow basin (21532 acres) is essentially composed of relatively impervious weathered WC rocks. This basin differs significantly from the other two in its hydrologic properties.

The study revealed that the basin contains from about 50 feet (at the downstream end) to an estimated 4,000 feet (on Mt. McLoughlin) of YHC lava flows, the primary groundwater units. These latest Cascade lava outpourings flowed onto and down a relatively linear valley surface created by an approximate 3 million year period of erosion of OHC lava flows and WC volcanic rocks. The basin is interpreted to contain groundwater flowing through conduit-type zones formed of fractured rock created by rapid quenching of hot lava flowing onto stream and spring outlet water flows.

Two core holes were used to perform tracer tests to determine groundwater flow rates and hydraulic conductivities at two spring outlet areas in YHC units. These tests yielded an average velocity of 100 and 600 feet per day and average hydraulic conductivities of about 700 and 2000 feet per day at these two sites. These values are believed to be higher than the parameters controlling the conduit-type groundwater flow through the basin.

Hydraulic conductivity tests in core holes yielded maximum values of 16 to 36 feet per day in YHC rocks near BBS and about 1.5 feet per day in upstream areas. Maximum values in the OHC rocks ranged from 0.6 to 26 feet per day near BBS and 0.12 to 0.15 feet per day in upstream areas. These values are interpreted as in the low range of the full properties of these rocks.

The pump testing gave results representative of the flow properties in the slower flowing, diffuse-type zones in the groundwater system. Transmissivity values from observation hole data ranged from 1300 to 1700 gallons per day per foot, and storativity values from 8×10^{-4} to 11×10^{-4}.

The average precipitation within the watershed varies by basin from about 40 to 52 inches per year. Runoff has been calculated ranging from a low of 14% of precipitation from the highly permeable YHC terrain to a high of 40% from the weathered WC terrain of the Willow basin. Evapotranspiration percentages of 35% in the less vegetated YHC a high of 50% in the Willow basin (with deeper soils) were calculated and estimated. Groundwater recharge percentages varying from a high of 51% for the permeable YHC terrain, to a low of 10% for the deeply weathered WC were used in the water balance table.

The BBS Watershed receives 295 cfs of precipitation in a normal year. The following estimates of percentage distribution of annual precipitation (and water quantities) were developed and assigned: Runoff: 23% (± 67 cfs); Evapotranspiration: 43% (± 126 cfs); and Groundwater Recharge: 34% (± 102 cfs). This recharge is subdivided as follows:

- groundwater emerging as springs: ± 33% or about 98 cfs;
- groundwater underflow/export (plus equation error): ± 1% or about 4 cfs;

Thus one of the most important results of the study is that 96% (about 98 cfs) of the precipitation recharged to the groundwater system emerges as spring flows into streams downslope near BBS.

Hazard zonation of the basin showing groundwater infiltration and surface water inflow to rock units combined with the soil infiltration data produced maps detailing the relative potential for contamination of, or influence to the groundwater system. These products and the accumulated basic data from this study form the basis for decision-making regarding management of entry to, and impact on the BBS Watershed to insure that this relatively vulnerable basin with fast flowing groundwater is adequately protected.

HILL, DAVID P., Jerry P. EATON, and Lucile M. JONES*, U.S. Geological Survey, Menlo Park, CA; *Pasadena, CA

Seismotectonics of the San Andreas Fault System

The Pacific and North American plates slide past one another in a right-lateral sense along the San Andreas transform boundary at a rate of 48 to 50 mm/y. Approximately 70% of this motion is concentrated along the San Andreas fault system, which extends for over 1100 km from the head of the Gulf of California northwestward through coastal California to the Mendocino triple junction at Cape Mendocino. The remainder is distributed across a broad region on either side of the San Andreas fault system from the continental margin on the west to the Basin and Range province in western Nevada on the east as suggested by the distribution of $M \geq 1.5$ earthquakes in California and Nevada for 1980 through 1986 plotted in Figure 1.

Most of the displacement along the San Andreas fault system accumulates during great, strike-slip eqrthquakes ($M \approx 8$) that recur along specific segments of the fault at intervals ranging from one hundred to several hundred years. Ruptures associated with these great earthquakes typically involve several hundred km of the fault trace with right-lateral offsets of meters extending through the entire thickness of the brittle crust, which averages 10 to 15 km beneath most of the fault system. Sections of the fault that produce great earthquakes apparently remain locked and seismically quiescent between the great events. The 1857 Fort Tejon and 1906 San Francisco earthquakes are the two most recent $M \approx 8$ events along the San Andreas fault system, and corresponding sections of the fault stand out in Figure 1 as nearly devoid of earthquakes down to the $M \approx 1.5$ level during the seven-year period from 1980 through 1986. Although the section of the San Andreas fault south of the 1857 break has not ruptured with a $M \approx 8$ earthquake in historic time, geologic evidence indicates that the currently quiescent Coachella stretch has produced four major ruptures since 1000 AD with the most recent approximately 300 years bp.

In contrast, the central section of the San Andreas fault between the 1906 and 1857 ruptures is characterized by the recurrence of numerous small-to-moderate earthquakes (Figure 1). Slip along this section of the fault is dominated by fault creep with background of persistent microearthquake ($m \leq 3$) activity with occasional moderate ($3 \leq M \leq 6$) earthquakes. Geodetic measurements indicate that the crustal blocks on either side of this section of fault slide by one another at rate of roughly 34 mm/y without accumulating significant shear strain, and it appears that this 200-km-long section of the fault rarely, if ever, produces major ($M \geq 7$) earthquakes. Fault creep and dense lineations of microearthquakes also characterize the branches of the San Andreas fault in central California that extend northward along the east side of San Francisco Bay as well as along the southern-most branches of the fault inthe Salton Trough near the U.S.-Mexico border (Figure 1).

Although the smaller, more frequent earthquakes that dominate the seismicity along the San Andreas fault system between the recurrence of the great ($M \approx 8$) events contribute only marginally to relative plate motion, they provide important clues to the tectonic processes common to earthquakes of all sizes along the transform boundary. The vast majority of these smaller earthquakes located on branches of the San Andreas fault system have strike-slip focal mechanisms with the right-lateral nodal plane parallel with the local strike of the fault. Earthquakes located adjacent to the fault, however, have mechanisms that range from strike-slip to thrust with slip directions typically oriented at high angles ($> 45°$ and $> 60°$, respectively) to the local strike of the San Andreas fault. These "off-fault" earthquakes emphasize that the crustal convergence responsible for uplift of the Transverse Ranges and Coast Ranges continues to be a tectonically important process.

The importance of crustal convergence associated with the left-stepping (restraining) bend in the San Andreas fault through the Transverse Ranges has been recognized for some time. Structural complexities at the northern (Tejon Pass) and southern (Banning) ends of this bend are marked by major left-lateral faults conjugate to the San Andreas trend and dense clusters of earthquakes that extend to depths of 20 km (Figure 1). More broadly, the restraining geometry of the bend results indistributed deformation across much of southern California as reflected in irregular seismicity clusters generally associated with the east-west trend of the Transverse Ranges. The importance of crustal convergence in the Coast Ranges was emphasized by the M=6.7 Coalinga earthquake of May 2,

1983, which involved low-angle reverse slip on a plane sub-parallel with the San Andreas fault. Although slip on this blind thrust did not reach the surface, it increased the amplitude of an overlying anticline, suggesting a genetic relation between folds within the Coast Ranges (the axes of which generally parallel the strike of the San Andreas system) and seismicity adjacent to the San Andreas system.

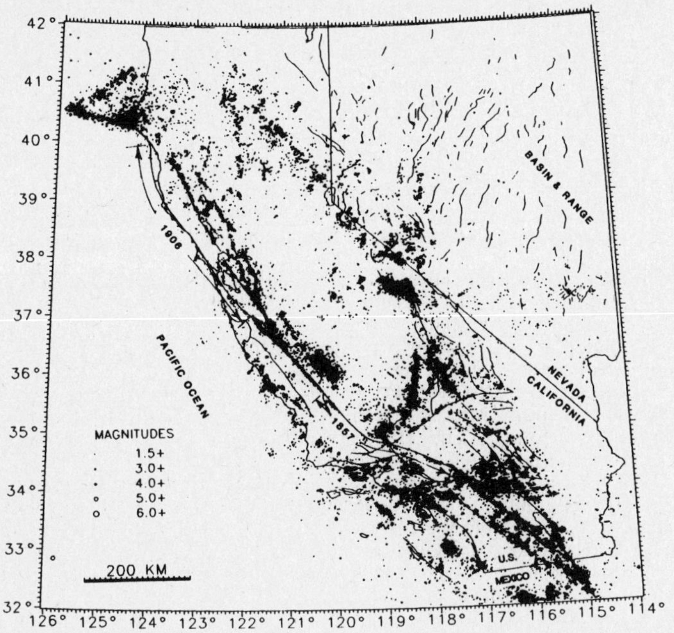

FIGURE 1. Seismicity in California and Nevada for 1980-1986 Surface ruptures for the great (M ≈ 8) Fort Tejon (1857) and San Francisco (1906) earthquakes indicated by arrows.

J

JONES, PETER A., Rogue River National Forest, Medford, Oregon

Geology and Engineering Geology of the Incised Pumice Canyons, Mount Mazama, Oregon, U.S.A.

No event in the long volcanic history of Oregon was more dramatic than the cataclysmic eruption and destruction of Mount Mazama. The focus of this presentation is on the geology and engineering geology of the deep narrow canyons carved into ash flow deposits on the west flanks of the volcano.

About 6,900 years ago the towering stratovolcano violently erupted more than 50 km3 of magma. Wind-born pyroclastic ash from the Plinian phase extended 2,000 km northeast of the summit and blanketed more than one million square kilometers of the continent. Further devastation quickly followed. So rapidly was the feeding magma chamber drained of its contents by the eruption, that support for the upper slopes was withdrawn and the summit began to founder into the void below. During the collapse, the ring-fracture second phase produced highly mobile pyroclastic flows which hurtled down the flanks of the volcano at hurricane rates. The flows scoured the upper slopes, then covered the lower flanks for at least 14 km from the vent, and continued by racing nearly 60 km down glacier and stream carved canyons. The dissected mountainous terrain had instantly been transformed to broad flat ash flow sheet deposits up to 100 meters thick. The pumice was hot enough to weld densely close to the summit, while partially welded ash formed further away. Distal and upper portions of the pumice sheet are non-welded. The resulting caldera is 10 km in diameter and 1,200 meters deep.

The 15 meters of annual snowfall has played a major role following final eruptions by both partially filling the caldera creating Crater Lake and by sculpturing the surrounding topography. Surface and subsurface erosion and mass wasting have carved dramatic steep narrow canyons into the ash flow sheet. They average 175-300 meters across and 75-100 meters deep. The V-shaped canyons have straight slopes ranging from 70% to 120%, averaging 85%. The majority of canyon bottoms have toe slopes directly adjacent to the creek edge. In reaches where deposition of debris and pumice is occuring the canyon bottom may have a meander plane up to 100 meters across. At various locations along the canyon bottoms the pre-eruption topography has been exposed by stream erosion. These sites form resistant 'nick points' in the channel, preventing further downcutting through the pumice at points above them.

The infiltration rates of the pumice mantle is many times greater than the precipitation rates. Yet the slopes in the drainage headwaters and the flat slopes adjacent to the canyon walls show abundant examples of dry gully swales. These are interpreted to have formed by collapse of subsurface erosion tunnels. The ash flow sheet possesses the following characteristics necessary for subsurface pipe formation. They have a significant permeability contrast between the highly porous non-welded surface pumice and the relatively dense low permeability underlying welded to partially-welded ash. There is a sufficient regional gradient of 90 meters/km to provide for drainage. And finally, the pumice material in the upper non-welded layer is clastic enough for particle or aggregate detachment, but angular enough to prevent premature collapse. Dendritic tunnel voids up to two meters diameter and eight meters long have been found. Eventually enough material is removed such that collapse of the tunnel void will result. With subsequent erosion of the piping failure, it may then have the appearance of gully erosion or debris slides.

Debris slides and translational slides are dominant mechanisms in sculpting the steep canyon walls. The loose non-welded pumice mantle overlying the partially welded ash is highly susceptible to failure, particularly where groundwater emerges. Laboratory testing shows the rough textured pumice has a very high angle of internal friction of 42 degrees. This laboratory value and field measured failure plane angle coincide exactly. The material is typically cohesionless. There is nearly a 100% increase in unit weight between dry and saturated conditions. Thus, it is apparent why thinly mantled saturated slopes fail while adjacent vertical slopes maintain a high degree of stability.

Once the slopes have been made barren by slope failures, other forms of erosion and mass wasting interact with one another to keep the slopes raw for long periods of time. The cohesionless soils on the steep slopes are continually subjected to creep and ravel. Frost action during the majority of the year provides fresh material for downslope movement. Winds erode the barren slopes by removing the finer ash particles which form matrix support for larger particles. The wind also abrades the soil by particle impact. The impact of rain drops upon the surface dislodges the pumice particles and deposits them in a net downslope direction.

Vegetation is critical in maintaining the stability of the canyon walls. Where the walls have not been made barren by erosion and mass wasting, dense old growth conifer forests are found. The canopy and understory minimizes effects of creep, ravel, and erosion by wind and rain by holding the particles in place. Computer modeling with infinite slope stability programs show that roots add significant cohesive strength to the mantle; enough to provide the margin between stable and failed slopes. Where the mantle is less than two meters thick, conifer tap roots anchor into the jointed partially welded ash below. As large conifers die and fall on the slopes and canyon bottom they provide barriers behind which the dislodged pumice accumulates.

JORDAN, H., K. FRÖHLICH, D. HEBERT, R. GELLERMANN, N. MEINERT, Bergakedemie Freiberg, I.A.E.A. Vienna, VEB Hydrogeologie Nordhausen, G.D.R.

Groundwater Protect — A New Method Assuming Groundwater Resources Protection

The protection of groundwater resources in respect of the vulnerability of quantity, quality and temperature regime as well plays an important role in all developed countries and the developing nations too. In the G.D.R. special conditions on behalf of a very low natural debit and the very intensive use of water resources ought to an extreme groundwater protection management. The best protection consists in prophylactic procedures, concepted as a method of assumption of groundwater protection. This method uses environmental isotopes, especially tritium and ^{14}C, and geohydrodynamical modelling in connection with a fundamental hydrogeological analysis. Additionally the acceptable concentration of any contaminant is under consideration. A new coefficient is proposed, which summarizes all processes, physical and chemical as well, influencing mixing, fixation or migration of pollutants between input and output.

One of the most important parameters primarily derivable from measurements of environmental isotopes in groundwater is the mean groundwater residence time. Through this "isotopic" parameter, suitable model considerations enable the characterization of groundwater dynamics in such terms as origin, replenishment, flow and mixing of groundwater. By this "isotopic" parameters a quantity vulnerability estimation can be made. But also as far as the assessment of volnerability of groundwater resources to pollution is concerned, the mean residence time becomes an independent hydrogeological parameter. The mean residence time is considered to be a measure of the degree of vulnerability of groundwater resources to pollution. Bomb-tritium, additionally, is an indicator of young groundwater and potential pollution.

The paper discusses this rather new approach for utilizing environmental isotopes in water resources development and protection, especially. Considerations to derive definitions and quantitative expressions and algorithms of application of this concept are presented.

K

KAMENICKÝ, LANDISLAV and Jakub KAMENICKÝ

Precambrian in the Alpine Belt of the West Carpathians

In the Alpine belt of the West Carpathians the structural framework of the crystalline basement was disturbed by young Alpine orogenie affecting even the distribution and structure of the Dalslandianf, Cadomian, Caledonian and Variscan lithostratigraphic units. Now the crystalline basement, in its prevailing extent, is cowered with younger stratigraphic units or tectonic structural units. It is also, of course, polymetamorph.

In compiling a new synthesis of the lithostratigraphic division and history of the West Carpathian crystalline we interpreted only a part as Precambrian in opposition to all at times prevailing opinions. This concept is based on following proves. The recent palynologic study determined the three lithostratigraphic sequences Paleozoic as units. Only a small rest of crystalline belt corresponds to the Precambrian.

The Precambrian was only sporadically documented palaeontologically /B. V. Timofeev, personal communication/. Also geochronological data proving its existence within the range of 1150-505 Ma are scarce. This is why our studies may be based mainly on lithostratigraphical study, the geologic mapping, the tectonic structure and correlation with adjacent regions or with a appropriate model mountains.

We may assign to the Precambrian particularly the areas with orthogneisses, some paragneisses passing into synkinematic migmatites with sporadic intercalation of amphibolite. They can be ascribed to the Jarabá Group, which is divided into Podbrezová Subgroup/including Boca Orthogneiss Formation/and Bystrá Subgroup.

If we wish to prove shortly the history of polystage development of the crystalline we can use the concordia — discordia diagram of the geochronilogical zircons data from magmatic and metamorphic rocks/ Kamenický and Král, 1979/. There also the two preliminary geochronological analysis of orthogneisses from Lubietová zone of Veporides near Banská Bystrica, made in the last year by T. V. Gracheva, are on record there. We can see some discordia i.e. 640-290, 1160-450, 2100-240 Ma and for the orthogneisses exclusive perhaps the better discordia 630-255 Ma. This large scale of data and discordia proves polystage history of the central West Carpathian crystalline.

For a correct idea of the geologic evolution was accepted also a correlation with adjacent regions/L. Kamenický, 1973/and at the same time a comparison with the Appalachian Mountains in USA.

REFERENCES

Kamenický, L. /1973/. "Relation of the central West Carpathians to the adjacent areas", Geol. Zborn. Geol. Carpathica, 24, 2, 303-313

Kamenický, L. and Král, J. /1979/. "Poznámky k problematike vývoja a stavby krystalinika v sirsom priestore bradlového pásma Západných Karpát", in Tektonické profily Západných Karpát, KÚDS, Bratislava, 59-63.

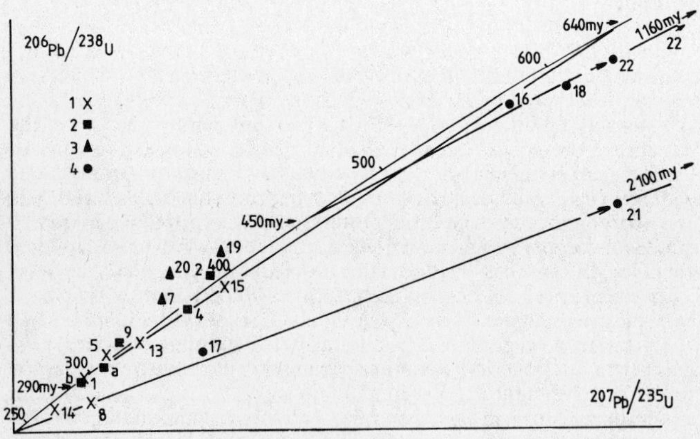

Figure 1 — $^{206}Pb/^{238}U$-$^{207}Pb/^{235}U$ diagram of zircon geochronological data of the West Carpathians: 1 — Granitoids; 2 — Orthogneisses with the two new analysis a and b; 3 — Porfyroids; 4 — Metasediments

KANGXIANG, LIU, Karst Research Section, Guizhou Academy of Sciences

Conceptual Model of Carbonate Aquifer in Shuicheng Basin in Western Guizhou

Groundwater circulates from time to time within entire geological history. Water quality and water amount varies mainly with time and space. There are many very complex hydrogeological phenomena and natures. This is especially true that hydrogeological study can not be well suited to the development only by means of one discipline during the recent years. To determine the hydrogeological conditions accurately, an emphasis will be placed on physical and mathematical models. So it is necessary to simplify and to summarize the hydrogeological conditions in study areas. Hydrogeological conceptual model is just a comprehensive data map with both qualitive and quantitative expression on it as hydrogeological characters were studied interdisciplinarily on a regional scale. Thus, it is a basis for the evaluation and management of groundwater resources.

Shuicheng basin, locating at the divide zone between Shacha river and Beipan river, is a karst faulted basin in the NW-SE direction with a length of 22.5 kilometers and with a width of 9 kilometres on the average. Carbonate rock layers crop out widely in the basin with a area of 120 square kilometres, almost faking up 83.1% of total area of the basin. Various karst forms develop very well. They are primarily classified as two types, i.e, mantled karst and exposed karst. There is distinctive change in elevation from 2582 meters in northwest to 1760 meters in southeast meters above sea level. However, the region is in subtropical climate, with all year-round mean temperature of 12.3 celsius, and with mean rainfall of 1231.1 millimeters. Xiongshui river is a sole surface drainage channel, flowing from the NW towards the SE and discharging off the basin at Kingjuling Hydrological Station, which is also a only outlet of groundwater and surface water of this karst basin. The discharge of this surface stream during dry period is about 0.07 cubic meter/second. There is little doubt that Shuicheng basin is a complete hydrogeological unit.

In fact, this study aims at solving the water supply of Shuicheng Steel Plant and Seeking the unpolluted sources of drinking water for all Lupanshui city. Many investigation on hydrogeological boundary of the basin, genesis of the basin, geomorphology, recharge and runoff and discharge of groundwater, and environmental hydrogeology were made in the last decades to a significant extent. The tracing test to seven underground river were conducted to indicate regional patterns of water flow in this heavily karstified area.

Shuicheng karst basin consists of a sequence of carbonate rock, clastic rock and igneous rock with a multi-layer tecture, which is carboniferous and permian limestone, dolomite, sandstone, shale, mudstone, and basalt. The final configurations of cave passage and the development of underground river are controled by geological structure, lithic character of the bedrocks, topographic conditions, and hydralic gredient. In conclusion, it has been show that Shuicheng basin as a multilayer waterbearing medium made up of small soltion pore, and soltion fissure and karst cave conduit has a storage amount up to 120000000 cubic meters. The thickness of aquifer in the central part of basin is about 60 meters, and the thickness in the upstream area and southern recharge area is about 50 meters. However, extensive karst collapse and severe groundwater polution took place in a few past years due to exploitting unreasonably the groundwater. According to the investigation and calculation, recharge amount of the basin in low discharge was 1.723 cubic meters /second and producing amount of pumping well was 1.542 cubic meters/ second in 1984. Groundwater chemistry at Shuicheng basin is main controled by man-made factor, rainfall, and lithic characters. Because of the influence of industrial and living waste waters, karst groundwater has been poluted to varying extent. Studying result indicates the saturated index of conduit flow ground water is less than the saturated index of diffuse flow groundwater, this is to say that karst waters from conduit system are more aggressive on the average than waters from diffuse flow system. It can be seen that the groundwater chemistry in Shuicheng basin is similar in terms of this study to other karst areas in different climate zones in China and is comparable to the karst zones in the world.

REFERENCES

William B. White, Conceptual models for carbonate aquifers: revisited, Hydrologic Problems In Karst Regions, Western Kentucky University Press, 1978.

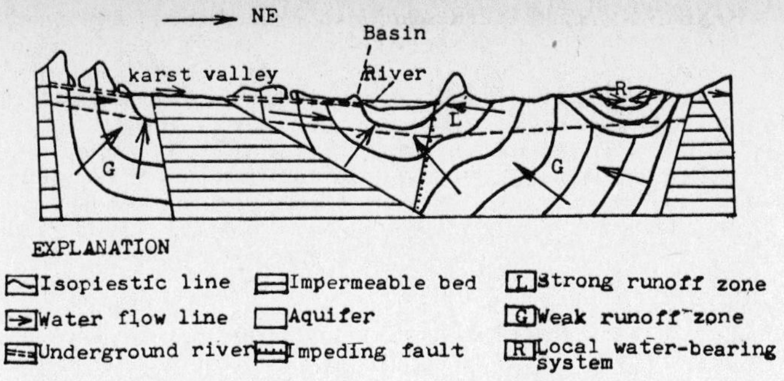

Figure 1 — Flow model of groundwater in Shuicheng basin

KASIPATHI, C., Reddy S. VENKATA, Department of Geology, Andhra University, Visakhapatnam, Andhra Pradesh, India

Genesis of Gibbsite from Granite Gneiss in Parts of Ranchi District, Bihar, India

Significant geological data is not available from the lateritic formations of parts of Ranchi District, Bihar to delineate genetical aspects of bauxite. For the first time, this aspect has been dealt by the present authors based on detailed mineralogical and geochemical studies. Geological mapping is carried out on the two important plateaus of the area, i.e., Pakhar Pat and Mandua Pat which comprises their lithological characters, soil and drainage patterns, colour, compactness and specific gravity of the material. Samples were collected representing the total area of more than 30 sq. kms. along different valley slopes of fresh, semi-weathered and completely weathered rocks, which shows all stages of fresh to alumina-rich bauxite. Different laboratory methods including optical studies, chemical studies using sophisticated instrumentation, DTA and XRD studies were taken up.

The present area of investigation is located between 23°30′ and 23°37′ North Latitude and 84°34′ and 84°40′ East Longitude. The bauxite occurs as flat or nearly flat plateaus at an elevation of about 1,075 m above M.S.L. and nearly 450 to 600 m above the surrounding country. The geological formations of the region consists from Archaeans to Recent constituting phyllites, mica schists, quartzites, granite gneisses, dolerites, grit, basaltic lavas, laterite, conglomerate and alluvium. The Central India is geologically dominated by Deccan basalts, but typically in the study area granite gneisses are the country rocks for the resultant bauxite deposit. It is due to the prominent valley slope of this plateaux, which as a part of Chota Nagpur grainte gneiss. This is an enormous intrusive mass which shows wide variations of granite from massive to highly gneissic type. The vast change in the strike suggest common local plications and crumpling. Shearing and faulting is also found on large scales. The granite gneiss is porphyritic which is exposed in valley terrains due to its resistance to weathering.

Bauxite occurs as segregations, boulders and bedded with varied colours. The average thickness of bauxite capping varies from 11 to 13 m. The weathering profiles of the area show soil, murrum, mixtures of bauxite and murrum, nodular bauxite and laterite, semi-weathered granite gneiss and finally at the bottom fresh rock, granite gneiss are found. The XRD and DTA instrumental data show predominant gibbsite with minor constituents of goethite, halloysite, hematite, kaolinite and montmorillonite etc. The analytical study of bauxite show the average values of SiO_2 34.77%, Al_2O_3 36.9%, Fe_2O_3 9%, TiO_2 7.1%, Na_2O 0.18%, K_2O 0.2%, CaO 0.24%, MgO 0.13%, MnO 0.11% and LOI 14.62%. Geochemically on an average the ppm values of Ga 3.34, Co 0.35, Sr 0.48, Ni 0.33, Cu 0.18 and Cr 0.50.

The detailed petrological and ore mineralogical study reveals five stages of the process of bauxitisation. First stage represents fresh rock (granite gneiss), where no alteration is observed. In the second stage, alteration starts along the mineral cracks, cleavages/boundaries of the feldspars, where the plagioclase is the first mineral which show effects of alteration and followed by microcline. In the third stage, feldspars are completely altered and in the fourth stage most of the primary minerals are altered except quartz, which indicates the initial stages of replacement. Gibbsite and geothite are observed to be pseudomorphous after feldspar and biotite. In this stage redistribution of the secondary minerals is also noticed. In this stage, a few relicts of quartz are found and the primary textures are obliterated. Comb-like, warm-like, lensoidal, thread-like and cavity filling textures of gibbsite crystals are found common. Two generations of gibbsite and goethite are observed, where the first generation being the secondary products from feldspars and biotite; and the second generation being found as replacements and cavity fillings. The flow structures, cavity fillings and selective concentrations of gibbsite and goethite indicate a resolution of the hydroxide minerals, their migration and precipitation. The triangular diagrams of weathering (Goldich, 1938 and Reiche, 1943) of the area indicate that there is a gradual loss of silica content with the progressive weathering. A comparison of chemical constituents from fresh to weathered product of bauxitisation show the tendency that the depletion of major ions of silica and corresponding enrichment of alumina in all the profiles. Iron shows both depletion and concentration. Alkalies and alkaline earth elements show a gradual loss in some profiles and abrupt loss in others. There is an increase in hydroxyl content with progressive weathering, which is a good correlation with alumina than iron content.

Environmental factors such as rock type, topography and climate might have played an important role in the formation of bauxite. Plagioclase feldspar appears to be more favourable for gibbsite formation than potash feldspar where it reflects in the weathering of granite gneisses at higher altitudes. This is attributed to alkaline pH and low silica content in the altering mineral face. High rainfall helped by undulating topography at higher altitudes, facilitated intensive drainage conditions for weathering. Low temperature and high humidity providing continuous moisture in the weathering front, provide active chemical reaction throughout. These conditions favoured the formation of gibbsite at higher altitudes. The bauxite of the present study is comparable to the other terrains of Tamilnadu, Andhra Pradesh, Orissa, Kerala, Maharashtra, Gujarat and Karnataka of India; S.W. Australia and S. America of the other continents which are classified as lateritic residual deposits.

KAVANAGH*, R., H. BOYD**, P. COFFEY*, and M. WELDON**, *Bord na Mona, Newbridge, Co. Kildare, Ireland, **Wavin Ireland, Balbriggan, Co. Dublin, Ireland

Protection of Ground and Surface Waters from Septic Tank Effluents

The pollution of ground and surface waters arising from septic tank run-off has become a serious problem in many countries and is causing serious concern to planning and health officials at local and central Government levels. Few cost-effective solutions are available for treating run-off for groups of houses (up to 45-50) or for single family homes.

This paper describes the research, development and commercial aspects of a new biofibrous peat-based system which achieves very high reductions in pollutant levels. The product ensures protection of ground and surface waters, and reduces the requirements for perculation significantly. A modular option for single family homes as well as a custom-designed system for groups of houses are described. The technical aspects of the systems including active material selection and optimization, loading regimes, sizing parameters, residence times and distribution systems are outlined. The results from a number of installations are provided.

KIM, JONG HWAN, Korea Institute of Energy and Resources, Seoul, Korea

Middle Palaeozic Isotopic Ages of the Ogcheon Group in Korea and Their Significance

Precambrian basement of the Korean Peninsula has been separated diagonally, from northeast to southwest by a tectonic element called the Ogcheon Belt. Thick miogeosynclinal sediments were deposited in long geosynclinal sinks termed the Ogcheon Geosynclinal Zone. The geosynclinal zone is further divided into a northeastern part occupied by lower Palaeozoic rocks of the Joseon Super Group, a central part by rocks of the Ogcheon Group of controversial age and a southeastern part by migmatites.

The Ogcheon sequence is made up of thick terrigenous clastics which are associated with shallow intrusives of basic and ulrabasic rocks. It distributes 120Km in its NE extension with 25-30Km width. The Ogcheon sediments have experienced low to medium grade regional metamorphism and foldings of more than two ages. Metamorphic grade of the Ogcheon Group increases progessively northwestward from greenschist to amphibolite facies toward a high heat flow zone which lay parallel to the Ogcheon Basin axis that became a locus of huge granitic intrusion of Mesozoic age.

In the Goesan-Deukpyong area, the Ogcheon Formations with 18Km width of outcrop perpendicular to strike are flanked by Mesozoic granites on either sides, the Jurassic Goesan granites (160 Ma) to the northwest and Cretaceous Sokrisan granites (80 Ma) in the southeast. Deukpyong uraniferous black slate which locates at 13Km southeast and 5 Km northwest from both granite boundaries, respectively.

Deukpyong black slate core samples containing high graphitic carbon and uranium yield 392± 13.4 Ma (IR : 0.7207± 0.0004) by Rb-Sr whole rock methods, which indicates diagenetic or early metamorphic age of the Ogcheon marine black shale. A systematic variation of K-Ar dates of biotites and fine muscovite separated from psammites of Goesan-Deukpyong area ranges 318.95 ± 4.71 Ma to 94.59 ± 1.34 Ma as a function of distance from the contacts with the Jurassic and Gretaceous granites. The K-Ar dates of the metamorphic minerals increase gradually away from the contact to the distance of 8Km where they reach a maximum value of 318Ma.

With these rock age dates and other observations, the Ogcheon Group is designated to be a Silurian sequence above the lower Palaeozoic Joseon

Supergroup. The Great Hiatus, a long believed geologic break characterized by neither sedimentation, erosion nor deformation and metamorphism in the Middle Palaeozoic times in geology of Korea, is known to be the time of deposition of the Ogcheon Group and succeeding metamorphism and deformation in its lower part. These are the first recorded events of the Caledonian Orogeny in Korea.

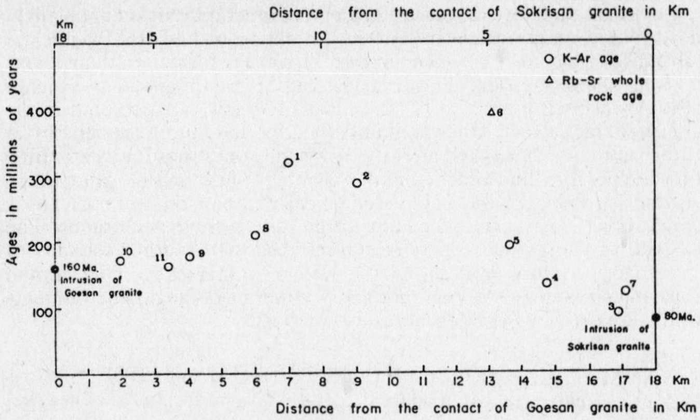

Variation of Rb-Sr whole rock isochron age of uraniferous black shale and K-Ar ages of fine biotites of metapsammites of the Ogcheon Formation in the contact metamorphic aureoles of the Mesozoic granites. The data reflect varying losses of radiogenic ^{40}Ar as a function of distance from the granite contacts, while Rb-Sr whole rock age remains unaffected by the intrusions which indicates either diagenetic or metamorphic age of the rock.

KHOLIEF, MAHMOUD M., Petroleum Research Institute, Nasr City, 7th Region, Cairo, Egypt

Sedimentological Characteristics of some Subsurface Jurassic Sediments at North Western Desert, Egypt.

Three formations are under investigation in the present study, namely : Wadi El-Natrun, Khatatba, and Massajid, which span nearly all the Jurassic time from Liassic to Oxfordian and even the early Kimmeridgian.

Wadi El-Natrun Formation is composed mainly of bio-micritic limestone with some siltstone and shale intercalations at the type locality of Wadi El-Natrun well. There is an increase in the thickness of this formation in the north-east direction of the Western Desert. It represents for the most part of its occurrence an inner shelf environment. This formation may correspond to the badly defined clastic series of the so-called Eghei Formation which extends in the western and south-western parts of the Western Desert.

Khatatba Formation is confined between two calcareous facies of Wadi El-Natrun Formation at the bottom and Massajid Formation at the top on vast areas of its Extension. It is composed mainly of quartz arenite and quartz wacke sandstone with some shale, anhydrite and dolomite associations, indicating deposition in a continental facies. However, the continental influences decrease in the north-east direction with a gradual replacement of a shallow marine facies. The depocenters of both Wadi El-Natrun and Khatatba Formations are nearly of the same configuration.

Massajid Formation represents the marine transgression of Upper Jurassic age from Callovian to Lower Kimmeridgian on the north Western Desert. It varies in composition between biomicrite and microcrystalline limestone to entirely dolomitized limestone. It is often cherty and argillaceous with a depositional facies of shallow to open marine environment.

Care samples from all these formations were petrographically studied and some of them were examined under the scanning electron microscope. A detailed description for the composition of the rock samples is given. Also, the diagenetic features of sediments and the effect on porosity development are carefully considered. The importance of these formations has been increased after the discovery of commercial oil in some layers of Khatatba Formation more recently.

KOLTIPIN, A. V. and V. P. KHVOROSTOV, Earth Study Museum of Moscow University, Production Association Kamchatgeologia

Late Mesozoic-Cenozoic Island Arc Complexes of the Extreme North-East of the USSR

Field geological investigations carried out by us in different regions of Koryak upland, Kamchatka, Chukotka, as well as paleogeodynamic analysis of published data on this region have shown much greater development of island arc complexes of the Late Mesozoic and Cenozoic ages than it was believed earlier. These complexes in total account for about 30% of the area of Kamchatka-Chukotka region. In another 30% of the area they are concealed under the cover of younger deposits where they are traced from geophysical data. The island arc complexes are subdivided in terms of age into Late Jurassic-Neocomian, Late Cretaceous-Danian and Eocene-Oligocene.

Upper Jurassic-Neocimian complexes are predominantly developed in the northern and eastern parts of Koryak-Chukotka region. Their outcrops are found also on the peninsulas of Knoni, Pyagin, Taigonos and in West Kamchatka. We combine them in Koni-Murgal, Tnurer-Zolotogorsk (its eastern continuation is Koyukuk belt of Alaska), West-Kamchatka, Vanetat-Ust Belsk, Khatir-Koryak and Mainitsk-Pekulney volcanic belts (see Fig. 1 in our abstract "A new scheme of Tectonic Zoning of the extreme North-East of the USSR" in the present collection).

In the structure of Koni-Murgal and Tanurer-Zolotogorsk belts take part moderately deformed calc-alkali, more seldom high-alumina tholeiite volcanites of medial, acid and basic composition associated with diorites, granodiorites, granites and volcanoclastic terrigenous rocks of shallow-sea and continental genesis. They compose large blocks of lithosphere limited by faults and ophiolite formations, which correspond to Late Jurassic-Neocomian volcanic uplifts.

The paleogeographic reconstructions carried out show that Tanurer-Zolotogorsk belt marks a large (200 km wide) volcanic island with a Pre-Cambrian-Paleozoic metamorphic foundation. This island was a plain periodically flooded by sea, with volcanic massives elevating over it. The extended positive magnetic anomalies extending along the northern border of the belt together with the outcrops of ultrabasites and tufo-terrogenous formations of pre-arc troughs suggest an underthrust from the north simultaneously pointing to the fact that the islands under reconstruction was separated from the margin of Chukotka (then belonging to North America) by a wide strip of the oceanic lithosphere. At the same time, the recently found outcrops of ultrabasites towards the south of the belt do not make it possible to exclude a possibility of underthrust, in some period of history, of Tanurer-Zolotogorsk island arc from south.

Koni-Murgal volcanic belt includes two quite different, in terms of structure, segments — Koni-Taigonos and Murgal separated by Mametch-Shambovey shear-fault. Koni-Taigonos segment is characterized by the presence of a Pre-Upper Jurassic foundation which is represented by Upper Triassic-Middle Jurassic island arc and also, on the peninsula of Taigonos, Pre-Cambrian and Paleozoic metamorphic complexes. Murgal segment corresponds to the island arc formed on the oceanic crust. The Late Jurassic-Neocomian Koni-Murgal island arc system morphologically resembled the contemporary Kuril-Kamchatka system only deployed in the opposite direction. The marked lateral zonality (from west to east: volcanic uplift-accretion cline and pre-arc trough) testifies to the underthrust from the cast, i.e. from the Pacific paleocean.

Vanetat-Ust-Belsk, Khatir-Koryak and Mainits-Pekulney volcanic belts developed in the north of the Koryak highland, as well as almost completely buried under the water of the Okhotsk sea and Cenozoic sediments of West Kamchatka West-Kamchatka belt are of quite different structure. The volcanic uplifts of corresponding island arcs are overlapped over a considerable area by shallow terrigenous sediments of the Cretaceous time. They crop out only in the Pekulney ridge, the mouth of the river of B. Kuybivey, the cape of Omgon in West Kamchatka, in a number of places of the southern setting of Anadir depression, and are also revealed by three wells in East-Ozero uplift of the foundation of Anadir depression. They are composed of moderately deformed calc-alkali and high-alumina tholeiite volcanites of basic, medial and acid composition associated with gabbro, diorites, plagiogranites and volcanoclastic terrigenous rocks of shallow sea genesis.

Much better preserved are accretion prisms of the island arcs in question which constitute Vayezh, Ust-Belsk, Raritkin, Mainits and Khatir uplifts. They are composed by more than 70% by tufo-terrogenous-siliceous deposits of pre-arc troughs and, also, include numerous tectonic blocks and lumps of Late Jurassic-Neocomian calc-alkali volcanos, Paleozoic (Permian, possibly, black coal), Triassic and Late Jurassic oceanic tholeiite basalts, pelagic limestones, siliceous, siliceous-argillaceous sediments, ultrabasic rocks, gabro, amphibolites, lawsonite-glaucophane shales and exotic formations of different ages and genesis.

Analysis of particular sections of volcanogenic-sedimentary formations of the belts shows that the island arcs corresponding to them are dismembered volcanic structures formed on the oceanic crust (of Kuril and Aleut arcs type). They were accompanied with extended pre-arc troughs and powerful accretion prisms brought to the sea surface in a frontal non-volcanic island arc (like the islands of Mentawai and Barbados). We consider them as fragments of once a whole island arc system (other fragments of this system crop out in the basin of the river Mulchatna, in the Alaskan ridge and some other places in Alaska), connected with the underthrust of the Kula plate.

The island arcs mentioned above seem to have existed, for a long time (up to their attachment to the margin of Eurasia) as aseismic submarine uplifts which submerged below the sea level after the thrust under them was over.

The Upper Cretaceous-Danian island arc complexes form considerable areas in East Kamchatka, in the southern part of the Koryak highland

and on the island of Karagin. They are also found in the north-eastern part of the Koryak highland along the northern border of Khatir upland and on Bering peninsula, as well as in Western and Central Kamchatka. We divide these formations into East-Kamchatka-Ivtigin, East-Olyutorsk (the submarine ridges of Shirshov and Bauer of the Bering sea appear to be its continuation), Iruney (extending farther to the Okhotsk sea shelf up to the island of Sakhalin), and Khotir-Bering (extending along the edge of the Bering sea shelf up to the river of Kuskokwim in Alaska) volcanic belts.

They are composed of moderately deformed high-alumina tholeiite and calc-alkali volcanites of basic and media composition associated with gabbro, diorites and volcanoclastic terrigenous rocks of shallow sea genesis. They form large blocks of the lithosphere limited by faults and ophiolitic formations which correspond to Late Cretaceous volcanic uplifts, formed on the oceanic crust.

The paleogeographic reconstructions performed show definite differences in the morphology of the distinguished island arcs due to the length of their evolution.

The "youngest" Iruney (K_2st-cp) and Khatir-Bering (K_2m-P_Id) island arcs are submarine uplifts with separate island-volcanoes (of Marian arc type). East-Kamchatka-Ivtiginsk and East-Olyutor island arcs (K_2st-P_Id) developing for a long time turned into more dismembered volcanic structures (like New Hebrides arc) to the end of the Cretaceous. the marked lateral zonality (the consequtive change of rocks of tholeiite, calc-alkali and shoshonite series, and for Khatir-Bering belt volcanic uplifts by the accretion and pre-arc trough) testify to the occurrence of an underthrust: under Khatir-Bering arc from the south from the Pacific paleoocean, and under East-Kamchatka-Ivtigin and East-Olyutor from the west and north, i.e from the marginal sea. It is difficult to determine the direction of the underthrust under Iruney island arc because of the fragmented character of the outcrops.

Paleogenic island arc complexes are also widespread in the southern part of the Koryak highland, in East Kamchatka and on the island of Karaginski. They form East-kamchatka-Piligin volcanic belt overlapping East-Kamchatka-Ivtiginsk island arc structure. Its structure does not differ basically from that of Murgal sector of Koni-Murgal belt. The island are marked by it was a sharply dismembered volcanic structure like Kuril arc separated from the margin of Eurasia by a narrow shallow sea basin. Subduction under it took place from the east, from the Pacific Ocean, which proves the existence of a buried trench of Paleogenic age at the continental base of Eurasia.

KONCZ, I., Hungarian Hydrocarbon Institute, and A. SZALAY, Oil Exploration Company.

Quantity, Type and Maturation of Organic Matter in Upper Pliocene Sediments of Pannonian Basin

The sub-basins in the part of Pannonian Basin in Hungary are filled by Pliocene sediments of several thousand meters thickness. The Upper pliocene sediments of 1400-1600 meters thickness have been deposited in delta plain of fluvial-lacustrine environment during deltaic sedimentation, consisting of interbedded sands, sandstones, siltstones and — as regards source rocks — shales, coaly shales, coals.

Geochemical profiles of three deep boreholes will be presented on the basis of cuttings. The thermal maturation is low (vitrinite reflectance o,35-o,55%) in the Upper Pliocene sediments, but the free hydrocarbon content is relatively high, which is due to high organic carbon content and more favorable kerogen type. The Upper Pliocene sediments may act as good source rocks for gas, gascondensate, oil accumulations in Hungary.

KONO, MASARU, Department of Applied Physics, Tokyo Institute of Technology, Tokyo, Japan and Working Group for Japanese GGT

Transects across Trench-Arc-Backarc Systems: Japanese Case

We have prepared five routes for the Global Geoscience Transect program (Figure 1). In the Japanese transects, the emphasis is laid on the traverse of the open sea — island arc — backarc basin — continent system, which will provide a unique opportunity to study the subduction-accretion process associated with the plate boundaries between the Pacific, North American, Eurasian, and Philippine Sea plates.

Four of the transects crosses Honshu island while the last one is a long traverse of the Philippine Sea and East China Sea.

(1) Hokkaido Transect (Coordinator: Gaku Kimura)

This transect crosses the Kuril trench, Hokkaido island, and the northern end of the Japan Sea. Hokkaido is thought to have been produced by the collision of two island arcs from east and west. The collisional structure is most well developed near Hidaka mountains, which will be the focus of this transect.

(2) Northeast Japan Transect (Coordinator: Kensaku Tamaki)

This route passes through the Tohoku (Northeastern) district of Honshu island at about 40°N latitude, and extends to Shatsky rise in the Pacific basin to the east and Japan Basin in the Japan Sea to the west. This will supply a typical section of the well-developed island arcs.

(3) Central Japan Transect (Coordinator: Nobuaki Niitsuma)

This goes through the complex Fossa Magna area which is the site of the collision between the Philippine Sea plate (Izu, Tanzawa, etc.) and Honshu island. The oceanward extension passes through the TTT triple point of Pacific, Philippine Sea, and Eurasia (North America?) plates.

(4) Southwest Japan Transect (Coordinator: Asahiko Taira)

It starts from Shikoku Basin in the Philippine Sea, crosses Nankai Trough, Shikoku island, Seto Inland-Sea, Chugoku district of Honshu island, and extends to the western part of the Japan Sea.

(5) Philippine Sea Geotraverse (Coordinator: Seiya Uyeda)

This will be produced by an international cooperation between Japan, U.S.S.R., and China. The transect lies on the 18°N latitude in the Philippine Sea, changes direction to the northwest, crosses Ryukyu Trough and East China Sea, and finally connects with the SE-NW transect of China.

The progress made in the preparation of these transects will be reviewed, and some interesting examples of the sections will be shown.

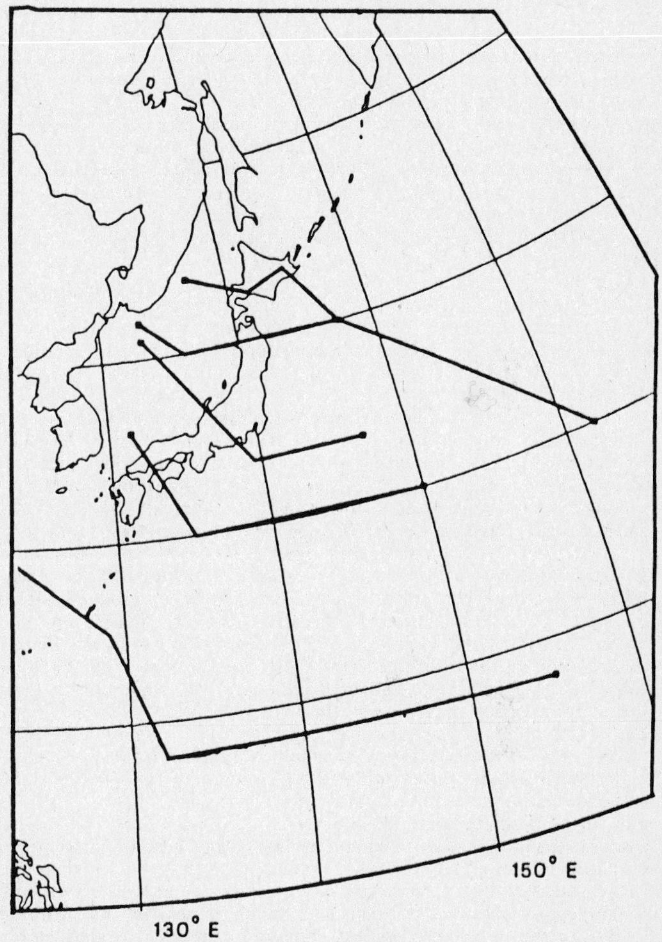

Figure 1 — The Japanese GGT transect routes.

KOVALEVSKY, V. S., D. Sc., Water Problems Institute, USSR Academy of Sciences, Moscow, USSR

Management of Ground-Water Regime and Resources

Ground water, due to its known advantages, is being increasingly used for public water supply and irrigation. In the USSR, about 80% of cities and towns are supplied with ground water. The ground-water withdrawal rates increase (up to 10 cu m per second and more in some localities). Cases of deep ground-water level recession and depletion of ground-water resources are frequent and this motivates the importance of the management of ground-water resources. The need for the management of ground-water resources is generated not only by the quest for preventing ground-water resources depletion, but also by the quest for enhancing the efficiency (rationality) of ground-water development even where ground-water resources are on the brink of depletion. The problem of the management of the ground-water regime and resources often emerges in connection with the need for eliminating the negative after-effects in the environment produced by the human impact on hydrogeological conditions.

The management ways are various and frequently based on: (a) control of the ground-water withdrawl regime and ground-water level recession as well as on determination of the optimal development regime: (b) artificial recharge of ground water: (c) conjunctive use of surface and and ground water; (d) on consideration of the laws of generation of ground-water resources during a year and long-term periods; and (e) on construction of barriers for preventing the intrusion of sub-standard water to water intakes.

In keeping with the above, the conditions for the management of ground-water resources are: (a) knowledge of intra-annual and long-term laws of the formation of the ground-water regime and resources under natural and disturbed conditions: (b) availability of comprehensive information on the geological environment (distribution of aquifers, their boundary conditions, and water-conducting parameters); (c) availability of the necessary initial information on the regime of ground-water withdrawal, levels, and quantity as well as on river discharges, i.e. ground and surface water monitoring data; and (d) general data on the total water balance of a drainage area and the perspective rise in water demands for various purposes within the drainage area.

The prerequisities for the possibility of the management of ground-water resources are various. They often rely on the consideration of the irregularity of ground-water recharge during a year and long-term periods, on the statistical analysis and finding of the duration of series of dry and wet periods, on the substantiation of the possibility of the decrease in aquifer storage in dry periods and the guarantee of ground-water recovery in subsequent wet periods. Such studies in the USSR showed the possibility of an increase in the yield of ground-water reservoirs under development by a minimum of 20-25% that demonstrated the efficiency of the exploration and development of these ground-water reservoirs. In the case of conjunctive use of surface and ground water, the problem involves either the computation of covering the water deficit only in dry periods by speeding up ground-water development with the recovery of decreased supplies in a subsequent wet period or the consideration of the changing use of surface water, induced to water-supply wells, with a hydralic connection of surface and ground water that varies depending on water availability.

The theoretical principles of the management of ground-water resources are based on ground-water regime predictions as well as the consideration of these predictions when assessing the safe ground-water yield or when substantiating the optimal ground-water reservoir development regime to be recommended. Considering the complexity of calendar predictions of the infiltration recharge of ground water for distant future, these predicted estimates may be largely based on the probabilistic and statistical analysis of long-term series of observations of the ground-water regime and river discharges; and baseflow, or ground-water discharge to rivers, is separated from the river discharge hydrograph.

The probabilistic duration of dry and wet period series may be found by determining the concealed periodicity in time series, by the Monte-Carlo method using the statistical parameters of series found for the observation period, based on statistical computation of the series number and duration, as well as by an empirical way using generalized regional characteristics of the regime of ground water or ground-water discharge, expressed in relative characteristics (e.g., in percent of probability). Such estimates are given in the report.

In these predictions, of importance is the consideration of long-term trends in the ground-water recharge regime caused both by anthropogenic climatic changes and by various regional human impacts on ground water (irrigation, drainage, urbanization, etc). These trends in ground-water recharge may amount to $\pm$ 0.03-0.4 l/S per sq km per year. As the author's studies of the effect of predicted climatic changes on ground-water resources showed, in the European area of the USSR at the turn of the century we may expect sharply different trends in ground-water discharge from an appreciable increase in recharge in the northern area to minimal recharge values in the southern area.

The prediction of the ground-water regime under disturbed conditions should be based on a combination of deterministic hydrodynamic models and stochastic predictions of ground-water recharge for the future. In addition to the solution of individual problems of the management of ground-water resources of any local water resources area, problems of the effect of the water resources area on the total water balance of the drainage basin as a whole as well as on the geological and ecological environment are to be solved.

The main theses of the report are illustrated by examples from the USSR experience.

KOZERSKI, BOHDAN, Technical University of Gdańsk, Gdańsk, Poland

The Transformation of Groundwater Règime in the Gdańsk Region

Introduction

The Gdańsk region forms an area of differentiated hydrogeological conditions. There has been found water in Triasic, Jurassic, Cretaceous, Tertiary, and Quaternary formations. A particular role in the hydrogeology of that area rests upon the sea which is a fundamental drainage base of the whole groundwater circulation system. In the Gdańsk region we are also dealing with an unusually intensive groundwater consumption amounting to 400 000 m^3/day. The groundwater resources are so abundant that up to now they satisfy the demand for Gdańsk, Gdynia, and the adjacent suburban areas whose population reaches approximately one million. The progressing urbanization and the development of industry create a threat to the groundwater quantity and quality and change conditions its occurrence.

Main factors transforming these conditions are:

- a considerable lowering of the piezometric surface
- variation in the natural direction of groundwater flow
- sea water intrusions into the coastal aquifer
- groundwater pollution by industrial and municipal wastes.

The changes in hydrogeological environment concern all utilized aquifers.

General Hydrogeological Conditions

The analyzed region is supplied with groundwater of Cretaceous, Tertiary and Quaternary aquifers. In lower Mesozoic series there occur iodine and bromine rich brines/Kozerski, 1984/. The Cretaceous aquifer is represented by qlauconitic fine-grained sands of thickness reaching 100 m and transmissivity amounting to 400 m^2/d. The sands lie upon clay sediments and are covered with limestones and marls. Such an arrangement of Cretaceous strata makes it possible to form an artesian basin of an area to approximately 4000 km^2. The water of the Tertiary aquifer, connected with lignite formation sands, is of minor importance on account of its limited resources.

The Quaternary aquifers appear in the region of the Kashubian moraine plateau, the Vistula Delta, and Reda ice marginal valley, the seaside terrace, and the sea barr. Regardless of the origin of the water-bearing series, the conditions of the groundwater occurrence are similar in the above mentioned geomorphological units. There appears free water table or semiconfined conditions. The water-bearing fluvioglacial sands of transmissivity up to 1000 m^2/d can be found at depths down to 40 m on the average.

Adopting Toth's /1963/ scheme of groundwater flow the aquifers listed here can be regarded as regional groundwater flow system. The system constitutes the Gdańsk hydrogeological unit the boundaries of which are determined by the Cretaceous artesian basin /Sadurski, 1988/.

Changes in the Groundwater Regime

Changes in the groundwater regime in the Gdańsk region result from its high extraction. That extraction has led to the development of a wide, regional depression cone. In this way natural hydrodynamic conditions have been changed. These changes concern quantities and direction of groundwater flow. These factors are the main cause of changes in the chemistry of groundwater, so that its composition is anomalous in comparison with hydrochemical background. The observed anomalies count as antropogenetic, hydrodynamically induced. In the Gdańsk region they are manifested through groundwater salinization, iron and fluorine compound increase.

Groundwater salinization

The greatest salinity is observed in the case of intakes from the Western part of the Vistula River delta. The majority of these intakes are located along the Motlawa River and the Dead Vistula River. These rivers reveal salinity of to 3000 ppm.

Salinity front spread to the south and enclosed one of the large municipal intakes. In 1971 the salinity was about 300 ppm, in 1978 it was 1000 ppm and nowadays it sometimes reaches 2000 ppm.

Iron content increase

Iron content increase has been observed in the intakes where organic sediments appear in the stratum overlaying the given aquifer.

Such is the situation in the Vistula River delta and in Reda Ice marginal Valley. The development of depression cone and fluctuation of groundwater level have led to iron increase of about 1 ppm at the beginning of exploitation to 5-6 ppm and occasionally 10 ppm after a few years of exploitation.

These processes are stimulated by such factors as oxygen, carbon dioxide, carbonic acid and humus acids in young aluvial soil.

Fluorine content increase

Anomalous contents of fluorine ion in Quaternary groundwater, appearing locally in the Gdańsk region, are caused by ascension of Cretaceous-aquifer water /Kozerski B., et all, 1987/. Cretaceous aquifer water is of HCO_3-Na character and contains about 4 ppm fluorine ion in some parts of the aquifer. Ascension to the main Quaternary aquifer from Cretaceous aquifer takes place. In these areas Quaternary aquifer water reveals anomalous chemical composition, similar to the composition of Cretaceous aquifer water. Concentration of fluorine ion can reach 3.5 ppm against the natural background of typical Quaternary water, which is less than 0.5 ppm.

Conclusion

Intensive extraction of Quaternary aquifer water in the Gdańsk region has led to changes in conditions of its occurance. Recently, however, changes in its quality have been observed, apart from evident changes in its dynamics. Those anomalies are mainly of antropogenetic character.

REFERENCES

Kozerski, B. 1984, Groundwater resources of Gdańsk region, Proc. IAH Conf., Montreal, 116-125.

Kozerski, B., Macioszczyk A., Pazdro Z., Sadurski A., Fluor w wodach podziemnych w rejonie Gdańska. Ann Societ.Geolog. Polonia, vol. 57.

Sadurski, A. 1986/, Hydrogeological evolution of Upper Cretaceous artesian basin in Gdańsk region, Ann.Soc.Geol.Po., Cracov.

Tóth, J. 1963. A theoretical analysis of groundwater flow in small drainage basins, J.Geoph. Res., 68, 4795-4812.

KOZHURIN, A. I., Geological Institute Academy of Sciences, Moscow, USSR

Late Quaternary strike-slip fault in Kamchatka.

The present structure of Kamchatka has mainly resulted from Middle and Late Quaternary tectonic movements. Its base is formed by a system of linear asymmetric graben-synclines and horst-anticlines manifested in the relief by depressions and uplifts (fig.). Most Middle and Late Quaternary volcanic manifestations are associated with these depressions (Erlich, 1968).

Whether there occur horizontal movements along the Kamchatka structures is a controversial problem as regards the Middle and Late Quaternary stage of its evolution. Such movements may possibly occur along the fracture zone trending mainly between the basin of Central Kamchatka depression and the Eastern Range uplift (fig.). Only in the north the zone crosses the uplift with a gentle curve and reaches the eastern coast of the peninsula.

Studies of displacements manifested in relief have shown that the latest movements along faults were recognized in the Early Holocene through out the most part of the zone whereas its northern end (the eastern slope of the Kumroch Range, north of the Kamchatka river, Kamchatka peninsula) accommodated them in the Middle Holocene. Dextral displacements are confidently registered as well as evident vertical (normal fault) ones. The amplitude of dextral strike-slip faults ranges from 1.7-4 m to 30-35 m. It is determined by the relative age of the displaced relief element and the rate of their movement. Maximum values are registered along the marginal canal displacements occurring during the second stage of Late Pleistocene glaciation. Maximum amplitudes of dextral slip increase up to 70-75 m in the zonal area active in the Middle Holocene as well. They always surpass the vertical displacements' amplitude. The value of the shift/normal fault ratio varies in different faults within the zone and grows regularly in the following series: NNE faults located at the scarp's foot between the depression and the ranges' uplift (1-2), parallel faults within the uplift (up to 4), NE (6-9) and ENE (12-14) faults separated from the depression (the Kamchatka peninsula region). Meanwhile, dextral slip velocity in zonal areas of different trend remains relatively constant — some 1 cm/y. Hence, only vertical movements magnitude and rate is being changed. As far as their maximum is observed in faults stretching at the scarp's foot between the depression and the ranges' uplift, the fault movements are most possibly associated with the depression evolution.

Structural association analysis leads us to conclude that horizontal movements along the structures of Kamchatka are caused by the Earth's crust sublatitudinal compression. This is evidenced both by directions of horizontal movements and the system of smaller NE dextral slips and CE sinistral slips existing in the depressions and uplifts. En eshelon relative disposition of a single ranges that enter as parts into the Eastern range's uplift as well as existence of the Nachikin zone of transversal (northwestern) dislocations forming on the eastern wing in the southern part of the zone may be treated as the result and manifestation of the above movements.

However, dislocations along the faults within the zone are evidently unaccountable. These movements reflecting the formation and evolution of horst-anticlines and graben-synclines (uplifts and depressions) may assumably have another source, i.e. be unrelated with the shift ones. The direct field studies support the assumption on the dynamic and kinematic independence of normal faults and horizontal movements within the same fracture zone.

Consequently, we may possibly deal with two synchronous, directly unrelated processes that existed in the tectonic evolution of Kamchatka in the Middle and Late Quaternary: formation of the NNE system of linear uplifts and depressions on the one hand and the right-sign horizontal movements along their trend on the other hand.

The available data allow to suggest that horizontal movements and associated structures evolve only in the upper-crust 20-25 km-thick layer in Kamchatka. In deeper lithospheric horizons, tectonic processes possibly occur mostly due to eastward transportation of matter, lateral in relation to the low-density upper mantle lens. The direction of the lens matter' dislocation corresponds to the trend of sublatitudinal deep (upper mantle-lower crust) shifts (fig.). The effect of the described deep masses movement on the Kamchatka upper crust layer may account for the emergence of asymmetric depressions separated by uplifts (Neotectonics and recent, 1988).

Synchronous development of the two processes may indicate tectonically layered lithosphere in Kamchatka.

The described fracture zone is similar by its parameters to other largest strike-slip faults within the Pacific fringing (Kozhurin, Trifonov, 1982).

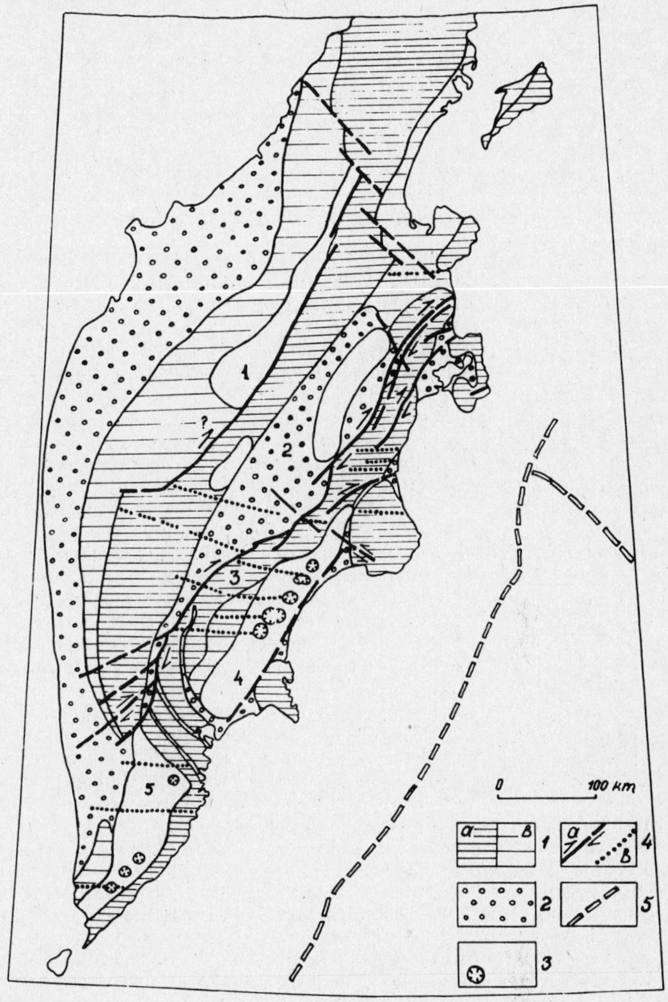

KUBOVICS, I., Cs. SZABÓ, K. SÓLYMOS, and Zs. MOLNÁR

Petrology and Geochemistry of Metasomatized Ultramafic Xenoliths from Cretaceous Lamprophyres and Neogene Alkaline Basalts /Hungary/

Cretaceous lamprophyres and Pliocene-Pleistocene alkaline basalts of the Carpathian-Pannonian region contain varied ultramafic xenoliths of upper mantle origin. The lamprophyric-carbonatitic-picritic assiciation, concentrated mostly in Triassic carbonates in NE Transdanubia contains xenoliths of extremely variable mineral composition. The small number of lherzolite nodules with sheared texture bear Mg- and Cr-rich but Al- and Ti-poor phlogopites, orth- and clinopyroxenes. Opaque minerals are Cr-spinels. The calculated PT conditions indicate 120-150 km depth of origin for most xenoliths, and approximate a continental geotherm like those of kimberlites.

Young basalts, alkaline basalts and basic pyro clasts are widely scattered in the Carpathian-Pannonian region. Some of the latter ones are especially rich in ultramafic xenoliths, mostly in peridotites/lherzolites, wehrlites/ and in mafic megacrysts. There are some amphibole — or phlogopite — bearing lherzolites among the peridotites. Their texture do not display deformation traces, but ortho- and clinopyroxenes and spinels are characterized by high Al contents. Calculated PT conditions indicate not more than 80 km depth for the alkaline basalt melt.

Mineralogical composition and geochemistry /main element and REE composition/ of these xenoliths indicate mantle metasomatism. Its contribution dominated the formation of host rock magma of lamprophyres, while was less important to that of alkaline basalts.

KULKARNI, K.G. and M.A. GHARE, Department of Geology and Palaeontology M.A.C.S. Research Institute, Pune, India

Repichnical Traces from the Jurassics of Kutch, Western India

In the present communication nine repichnial traces are described among which *Annerepichnites* is a new ichnogenus. The occurrence of *Aulichnites* and *Cochlichnus* in association with *Phycodes, Diplocraterion, Teichichnus,* etc., has been used to delineate the Precambrian-Cambrian boundary in known sequences. Their presence in Jurassic and younger rocks suggests that traces fossils can act only as local stratigraphic markers.

Repichnial traces, in Jurassic of Kutch, are associated with either pascichnial or fodinichnial traces. This association points to shallow but quiet, poorly agitated waters which the creators of repichnial traces preferred.

Minor oscillations in water level are detected by the characteristic replacement of *Curvolithus* by *Gyrochorte* in a vertical succession.

L

LAGO, M.; M. B. MARTINEZ: H. MILLAN; J. J. NAVARRO and A. POCOVI, Dept. of Earth Sciences, University of Zaragoza, Zaragoza

The Upper Triasic Tholeiitic Magmatism in the Pyrenean Area (Northern Spain – Southwestern France): Petrology, Geochemistry and Emplacement Conditions

A tholeiitic magmatism intrudes at the end of Triasic time (Keuper-Infralias) in the Cantabro-Pyrenean zone (Fig. 1). Its analogies in nature and age with the NE America and W Africa ones allow to relate them with the rifting preceding Atlantic opening which extends through the Bay of Biscay.

Magmatic materials mainly appear like hypovolcanic bodies (lopoliths and great sills) which set up in the saliferous Keuper sediments, producing a weak contact metamorphism (400-500°C). The bigger magmatic bodies are located at the W of the Cantabro-Pyrenean domain, decreasing in size and frecuenzy to the E. Radiometric age (K/Ar) obtained by several authors is compatible with the observed intrusive-countryrock relationship.

Tholeiitic bodies have a remarkable petrological and geochemical uniformity all over the pyrenean zone. At outcrop scale, a petrological zonality, shown by mineralogical composition and texture, from chilled margins to the internalparts of each body is observed. Pegmatoid-diferentiated bands, fortuitously located, are ocasionally found in the internal zones of magmatic bodies.

Primary mineralogical composition is make up of olivine (Fo_{80-54}), plagioclase (An_{85-50}) and clinopyroxene (Ca-augite and pigeonite). There is litle ortopyroxene and it can only be found in olivine-riched zones. Compositional variations in clinopyroxenes clearly express the evolution of this tholeiitic magma.

Good examples of well preserved contact surfaces between sediment and intrusive rocks have been found in several outcrops of the south-pyrenean domain. Some fluidal structures (lava-toes, flow vrinkles...) observed on these contacts inform about viscosity of envolving sediment and magmatic bodies during lava outpouring.

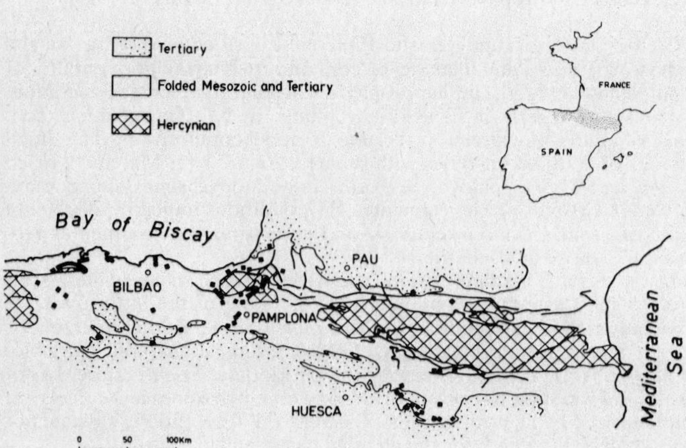

LAING, W. P., Geology Department, James Cook University of North Queensland, Australia

A System of Structural Analysis in Drillcore, and How to Use it to Find Ore Deposits

Explorationists are only too aware of the lack of a system of analysis of geological structures in drillcore. In response to this need the writer has formulated a *comprehensive and unified system of structural analysis in drillcore.* The system is grounded in fundamental geometry and is practical in application.

The central problem posed by drillcore is this: "How do I obtain unique, useful information from a length of drillcore which has rotated by an unknown amount during its withdrawal from the drillhole, and whose only known orientation is its axis?". The answer begins with an important division of structural information into three types of parameters: *scalar* parameters, *non-polar vector* parameters, and *polar vector* parameters. This threefold division has the following features:

(a) all aspects of the structural world can be expressed in terms of these parameters; eg. planes and lines are non-polar vectors, angles between planes/lines and the strength of structures are scalar parameters, and stratigraphic younging vergence and shear direction are polar vectors;

(b) each parameter is capable of analysis in its own way in drillcore; *three different styles of geometric analysis* result;

(c) in particular, structures which can be expressed in scalar terms are easily measured in drillcore, and hence we should be alert for this form of expression of structure.

The system is built around a *fundamental geometric reference frame,* which is directly analogous to the reference frame used on the earth's surface in normal structural analysis. Within this frame any structure is uniquely specified by two of three fundamental angles: α, δ and ϵ (see Figure), which are directly analogous to strike, dip and pitch respectively in the surface reference frame. The relative orientation of any other structure in the same piece of drillcore can be determined by measurement of these angles.

Methods are then available to *uniquely* fix the orientation of all structures in the drillcore. These methods are of two types:

(1) Relating all the structures to the orientation of one structure which is known.

(2) Measuring one structure at a minimum of three locations with different drillhole orientations.

Method (1) can be implemented via *stereographic procedures* or by using a *core orienting frame.* Stereographic procedures have been developed to cover all possible combinations of planes and lines in drillcore. A core orienting frame has been developed which dramatically improves the structural output from a core logging programme. Method (2), a powerful new technique based on the principle of simultaneous equations, is known as the *multiple data point method,* and it permits the general determination of planes and lines without any external orientational input.

In order to uniquely fix the orientation of a structure by one or other of these methods, a certain minimum amount of information is required. Understanding of the foregoing methods enables the explorationist to decide what information he/she needs, and most importantly which drillcore will yield the information. This recognition of *structurally fertile and infertile drillcore* significantly increases structural logging efficiency by minimizing the time spent on infertile core. Informed decisions can also be made on

(a) *how we are going to treat the data* (for example, a stereographic net may be unnecessary):

(b) whether we require *surface information;*

(c) whether the structural geometry in a particular section of core is what we may term a *special case,* because these simplify structural interpretation.

The notion that drillcore has varying structural fertility leads to the concept of *structural drill targeting.* In any specific structural situation some drillholes inherently yield more structural information than others, depending on the orientation of the drillhole. Thus, when targeting a drillhole, we need to consider two separate targets:

(a) the *spatial target,* that is, the co-ordinates of any body of rock we wish to intersect;

(b) the *structural target,* that is, the type and amount of structural information we wish to obtain.

There are a number of paths to any given spatial target, and careful selection of the most fertile path will greatly enhance the structural interpretation. Guidelines to structural targeting have been developed.

Using structural analysis in drillcore to finding ore deposits comes from understanding how the geometry of specific deposit styles provides vectors toward one. Several examples follow.

(1) Many vein deposits are characterized by ore shoots parallel to the intersection line of two vein sets. Solution: determine the orientation of the intersection line.

(2) Victorian Bendigo-style gold deposits occupy saddle reefs in antiformal crests. Solution: log vergence and/or fold interlimb angle. Ex-

tensions of existing reefs lie along the crestlines. Solution: determine the orientation of the fold axis.

(3) Massive sulfide deposits in deformed terranes characteristically localize high strain in their adjacent wallrocks, and fold axes in the wallrocks rotate more than normally toward the principal stretching direction. Proximity to a massive sulfide deposit is thus characterized by an anomalously small angle between the fold axes and the stretching lineation. Solution: this angle, a scalar parameter, is easily measured in drillcore.

The system of structural analysis in drillcore has the following valuable characteristics:

(a) its geometric reference frame is intuitively straightforward;

(b) it does not require oriented drillcore;

(c) all structures can be routinely determined in totally blind deposits;

(d) core logging effectiveness is optimized by minimizing the time taken to extract the structural data and maximizing the amount of on-the-spot interpretation.

(e) the system has a logical sequential structure amenable to computerization;

(f) the system is capable of routine use by non-specialist explorationists.

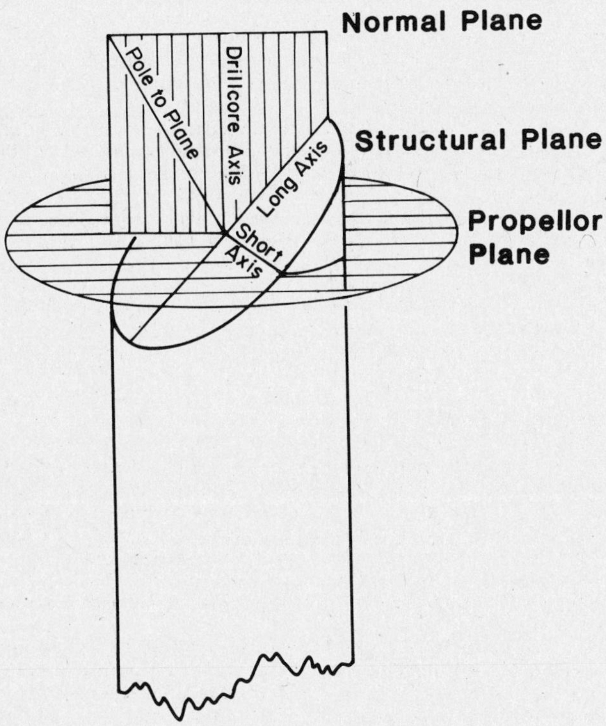

LAP, NGUYEN KIM, Institute of Geophysics, National Centre of Scientific Research of Vietnam Nghia do - Tu liem - Hanoi, Vietnam

Seismogeodynamic of Indochina Peninsula

Indochina peninsula occupies the central part of the structural junction of planetary tectonic elements i.e. three lithospheric plates - the Eurasian, the Indoaustrian and the Pasific. These features partly determine the seismic activity of Indochina peninsula. Data related to more than 1100 earthquakes which occured in Indochina peninsula from 1900 to 1987 were collected. In Indochina peninsula, almost all earthquakes hypocentres were located along the main faults and in the Earth's crust. This region has experienced several destructive earthquakes with magnitude m = 6.0 - 7.0 during the period 1900-1987. The seismicity of Indochina peninsula was studied qualitatively by showing the seismic activity, maximum expected earthquakes, seismic risk, the stress field, the connection between the seismicity and geophysical fields and some characteristics of the strong earthquakes. The motion type of fault is determined according to the motion signs of seismic wave P and fault orientation. The slip rate along fault zones are calculated from the sum of moments of earthquakes. The seismotectonic deformation in the main fault zones was studied by showing the velocity of seismotectonic deformation.

Long-term prediction of strong earthquakes is studied according to the results of seismogeodynamic study of Indochina peninsula.

LAYNE, A. W., U.S. Department of Energy, H. J. SIRIWARDANE, West Virginia University, and C. W. BYRER, U.S. Department of Energy

Assessment of Gas Production Potential From the Strata Above Longwall Mining

Pipeline quality gas production potential from the highly fractured methane-bearing strata above longwall coal mine panels may be many times that of the gas-in-place estimate for the mineable coalbed. The "strata-gas" is released from relatively shallow gas-bearing coals, organic shales, and sandstones. The permeability in these strata is increased significantly due to the fractured zone generated during longwall mining activity, and hence, strata gas could be produced at levels many times that of an average conventional gas well. This paper assesses the potential production and influence of well spacing for drainage of gas from the strata above coal mine panels.

Published methane production from longwall mining vent wells indicates the magnitude of gas production potential. One mining operation in the southern Appalachian Basin, recently produced 4 billion cubic feet of pipeline quality methane in just 2 years. Many vent wells associated with longwall mining initially produced 1 to 2 million cubic feet of gas per day. The gas source is mostly from broken gas-bearing strata in the overburden above the mined out coal.

Coalbed methane reserves for the Appalachian Basin are 129 trillion cubic feet (ref). However, this estimate does not include the gas available from the overlying interbedded shales and sandstones which contribute to vent well production. In this paper, strata-gas reserves were quantified with the predicted failure zone extent, formation gas contents, and geologic cross sections present throughout the Appalachian Basin. Here, a finite element method was used to determine the size and shape of the affected stratigraphic areas and gas contents were derived. Additionally, a simulation analysis was performed to compute reserves for a mine in the southern Appalachian Basin. This was completed to confirm results from the previous study with actual gas production data. Again, the finite element method was used to determine the size and shape of the affected stratigraphic area. (ref. #2)

A two-phase reservoir model was used to compute increased production rates from methane wells. Production data from actual wells above a longwall panel in the Appalachian Basin were history matched to characterize flow mechanisms and reservoir properties, such as permeability. These properties were then evaluated in parametric studies to determine the influence of well location on gas production. Results indicate that cumulative gas production can be increased by placing wells farther than 1,000 feet apart. The model appears to make reasonable predictions of the size of the multi-reservoir area in complex geologic conditions. The study considered the influence of an advancing mine face on gas production. Thus, the methodology presented in this paper is applicable to the optimum design of gas wells in highly fractured reservoirs and prediction of long-term gas production.

Additional results show that significant gas production potential can be expected in the Appalachian Basin from longwall mining operations. The Central Appalachian Basin has the highest gas production potential per longwall panel. Estimates from the three regions of the Appalachian Basin show potential gas estimates per panel from 700 Bcf in the Northern Appalachian Basin to 1.3 Bcf in the Central Appalachian Basin. Bcf = billion cubic feet of gas. (ref. #3)

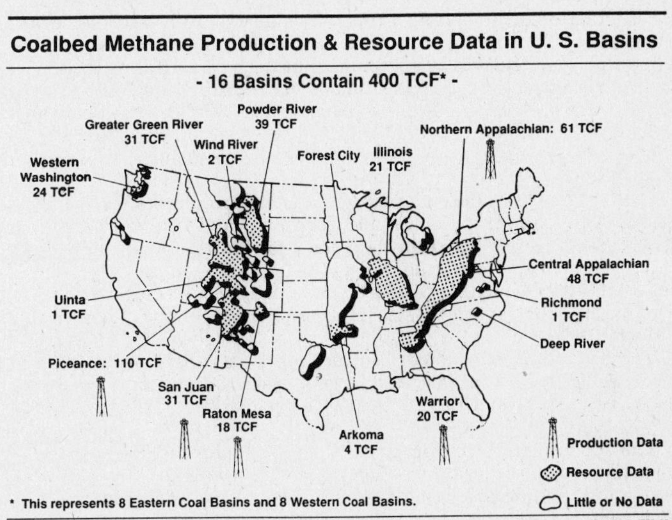

Methane Potential from Gob Wells in Mined Out Area is Large

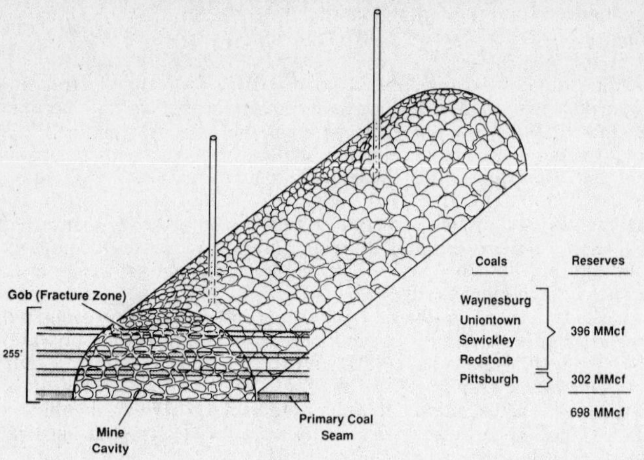

Region 2: Central Appalachian Basin

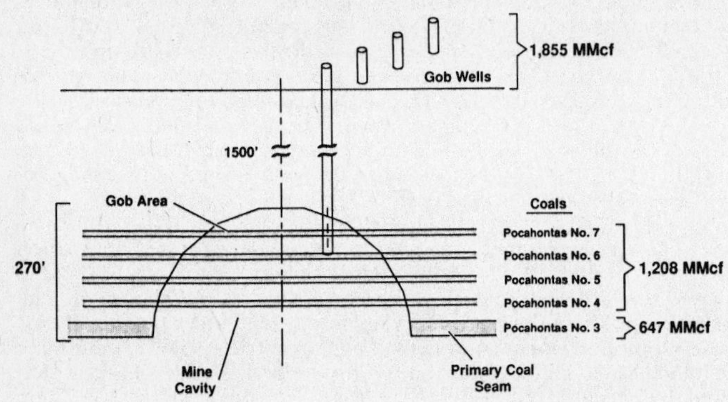

LECA, XAVIER, B.R.G.M. Geological Mission with D.G.M.R., Jeddah, Saudi Arabia.

The Discovery of Concealed Massive Sulfide Deposits in South Portugal Review of Methodology and Main Features of the Neves-Corvo Deposit

(Translated from Chronique de la Recherche Minière nos. 479 and 481 BRGM Ed. [1985])

The Neves-Coro polymetallic sulfide deposit, discovered in May 1977, provides a good example of successful methodology.

The first prospecting phase lasted from 1969 to 1973. After standard prospecting based largely on gravimetry, which was given precedence over a rapid geologic survey, a core hole was drilled on the Neves anomaly. The result was negative, which at the time seemed surprising in view of the promising geologic, geophysical and geochemical context.

The second phase spanned the years 1973 to 1977 and can be divided into two phases: between 1973 and 1975, data were accumulated from the various prospecting methods used in the Portuguese province of Baixo-Alentejo.

From 1975 to 1977, the data collected, covering geophysics, geochemistry and geology (lithostratigraphy, paleogeography and structure), were applied to the Neves area, held in abeyance since the negative drilling result.

This comparison led to the drilling of a second core hole close to the site of the first, which, in May 1977, revealed 50 m of massive sulfides. Three other bodies — Corvo, Zambujal and Graça — were discovered in 1977 and 1978.

These discoveries resulted from the use of a variety of prospecting methods. It can be said that without geophysics, Neves-Corvo would not have been discovered but neither would geophysics alone have led to its discovery.

The deposit consists of four massive sulfide bodies and stockwork-type disseminated mineralization. Four types of ore have been recognized, with a total sulfide potential of 250 Mt. The deposit is characterized by high copper grades (average 7 percent), the presence of tin in relatively high amounts (1 kg/t of sulfides), low silver content, and absence of gold and tellurium, in addition to normal zinc and lead grades.

LOGATCHEV[1], N., P. LIPMANN[2], Ch. CHAPIN[3], Yu. ZORIN[1], P. MORGAN[4], S. RASSKASOV[1]. [1]Institute of the Earth's Crust, USSR Acad. Sci., Irkutsk, USSR. [2]U.S. Geological Survey, Federal Center, Denver, Colorado, USA. [3]New Mexico Bureau of Mines and Mineral Resources, Socorro, New Mexico, USA. [4]Northern Arizona State University, Flagstaff, Arizona, USA

Baikal — Reio Grande Rifts: Some Points of Similarity and Difference

The available geological and geophysical data on Baikal and Rio Grande rifts were discussed during two Soviet-American field workshops. Comparative study of these two continental rifts is supposed to be a subject of joint Soviet-American project "Baidal - Rio Grande Rifts". However, some preliminary characteristic of points of similarity and difference between these regions is possible already now.

As for points of similarity, they are: the position of chains of rift depressions within wide mountainous uplifts, almost the same orders of length, width and depth (sedimentary thickness) of major basins (except for a very deep Baikal depression itself), local crustal thinning beneath rift depressions and regional lithospheric thinning beneath wide mountainous uplifts. The fields of volcanics do not have close spatial connection with basins in both rifts.

The majority of rift basins are halfgrabens: normal faults with significant displacements are attributed to the one side of the basin. However, in the Baikal zone, the asymmetry of basins is uniform: step normal faults are at north-western and northern sides of basins. In Rio Grande rift, the style of asymmetry changes along the strike. Such changes occur at transverse accommodation zones. The nature of these zones is not always clear. Along some of them strike-slip displacements are discovered. The prerift and synrift volcanic fields are often attributed to the accommodation zones.

The two rift zones differ mainly in their prerift history, in the character and volume of magmatic rocks, in heat flow values and in recent seismicity.

Baikal rift began developing 30-35 Ma ago in the region, where last intensive tectonomagmatic event has occurred in the Middle Palaeozoic. In the Jurassic and Cretaceous, a part of this region was subjected to a certain reactivation with formation of relatively shallow coal-bearing basins which underwent unsignificant compressional deformation with thrusting. This reactivation was not accompanied by significant magmatism. Jurassic magmatism was widespread in the Trans-Baikal area, adjoining the rift zone from south-east.

Two rifting stages are distinguished. The first stage lasted from the Oligocene to the Early Pliocene, and the second one continued from the Middle Pliocene to the present. Some coarsening of clasic material is characteristic of the sediments related to the second stage. This indicates the rise of relief contrasts. Nevertheless, in the second stage, the position of depressions remained practically the same.

Both rifting stages in the Baikal rift zone are associated with relatively uniform magmatism. Volcanic eruptions are mainly represented by alkaline olivine basalts. Basanites and olivine tholeiites are of certain importance. In north-eastern part of the zone trachites are present. The total volume of Late Cenozoic magmatic rocks in the Baikal rift zone is not so significant as in Rio Grande rift.

The Baikal rift zone possesses relatively high seismicity: about 3,000 earthquakes with magnitude more than 2 are recorded annually. The Muya earthquake with magnitude 7.6 had occurred in 1957. In the Rio Grande rift zone, first signs of extension have manifested also about 30 Ma ago. This extension was preceded by significant Laramide compressional deformations accompanied by intermediate siliceous calc-alkaline magmatism. In the first rifting stage (till 20 Ma ago), similar magmatism occurred with minor basaltic eruptions. The basins of early stage are practically not reflected in present topography.

After tectonomagmatic lull, occurring from 20 to 13 Ma ago, the second rifting stage began, involving formation of sedimentary basins which are reflected in recent topography. In the second stage, bimodal magmatism took place, that is typical of continental rifts. Volcanics are represented mainly by alkaline basalts, but in central parts of some depressions tholeiites are present. Silicic volcanics are minor. About 3 Ma ago rifting activity began decreasing.

Thus, the stages of development of Rio Grande rift appear to be of some different geological sense than that of Baikal one. Judging by the character of magmatism and by tectonic setting, only second stage of development of Rio Grande rift has the features of typical continental rifting. The first stage has some signs of phenomena which can occur above the zone of shallow subduction.

Higher heat flow values and lower seismicity are characteristic of the Rio Grande rift zone as compared to the Baikal one.

It can be assumed that difference of geological and geophysical features of the two considered rifts is caused by their different prerift development. There is no large time interval between rifting and preceding tectonomagmatic event in Rio Grande rift. This have conditioned the more intensive heating of the mantle and the crust, that is reflected in the specifics of magmatism and in highheat flow values. More heated crust is less brittle, that can lead to decrease of seismicity, since crustal deformations are mostly realized by creep.

Our inference about causes of difference between the two rifts (as well as the list of points of this difference) is preliminary. Further joint study of Baikal and Rio Grande rifts should be objected to a closer definition of their similarity and difference, and must favour our better understanding of the nature of continental rifting.

LÓPEZ-ESCOBAR, L. and M. A. PARADA, Departamento de Geologia y Geofisica, Universidad de Chile, Santiago-Chile

Geochemical Reconnaissance of the Upper Cenozoic Igneous Rocks in the Andes of Southern Chile Between 41°-42°S: Petrogenetic Implications

An extensive Miocene epizonal plutonic belt outcrops in the Southern Andes south of 39°S (Munizaga et al., 1985). Between 41°-42°S, this belt includes three units: Cayutué, Reloncaví and Lago Chapo. Cayutué unit consists of hornblende + clinopyroxene + orthopyroxene bearing gabbros and diorites. Reloncaví unit is made of biotite + hornblende + two pyroxe nes tonalites and Lago Chapo unit consists of biotite ± muscovite leucogranites. This latter unit is restricted to small plutons that do not exceed 5% of the belt. Quaternary strato volcanoes, such Osorno and Calbuco, and many small cinder ± lava flow cones distributed along the Liquiñe-Ofquie fault zone (LOFZ; Thiele et al., 1986), overly the plutonic belt.

Little rocks developed in inmature arcs, Cayutué and Reloncaví plutonics exhibit low to normal K_2O contents and high alkali-lime index (Parada et al., 1987); Lago Chapo leucogranitoids are distinctly peraluminous. Reloncaví tonalites and Cayutué diorites have similar LREE abundances and incompatible element ratios, but the Reloncaví plutonics are depleted in MREE and HREE. This suggests that Reloncaví magmas could be genetically linked with Cayutué magmas by amphibole controlled crystal fractionation. However, it is difficult to explain by this model the similarity in Nb, Y and Zr existing between the Reloncaví tonalites and Cayutué diorites. Reloncaví magmas are also consistent with a generation by partial melting of a garnet-bearing source (granulite or eclogite).

Two Lago Chapo leucogranitoids were analyzed. Both have subparallel REE patterns, similar to those of the Reloncaví tonalites. Compared to the SiO_2 — poor sample, the SiO_2 — rich one is enriched in REE (La = 58.8 vs 42.1 x chondrites; Yb = 7.2 vs 2.9 x chondrites), exhibits a significant negative Eu anomaly (Eu/Eu* = 0.36) and is strongly depleted in K, Rb, Cs, and Ba. Probably, the peraluminous character of these leucogranitoids resulted from an alkali leaching process.

The less evolved vocanics from Osorno and Calbuco stratovolcanoes and Pichihuinco, Cayutué and La Vigueria cones have variable La contents (10-30 x chondrites), uniform Yb abundances (7-10 x chondrites), and normal Eu content. Negative Eu anomalies coupled with high REE abundances are observed in some evolved rocks from the Osorno volcano. Likewise, some porphyritic samples of this center have significant positive Eu anomalies coupled with a low REE contents, which is consistent with a plagioclase assimilation process. Variable LREE abundances can be explained by variable degrees of partial melting of a compositional similar source.

Compared to the volcanics, the Cayuté, Reloncaví and Lago Chapo plutonics are significantly depleted in HREE. According to our petrogenetic modelling, the source for volcanics is similar to that of the Cayutué diorites. However, garnet is less important, as a residual phase, in the source for volcanics, suggesting that were generated at lower pressure. We postulate that the LOFZ have played an important role in controlling the magma generation pressure of the volcanic centers located along this fault zone.

ACKNOLWEDGEMENTS

Nuclear irradiations of volcanics were performed at the MIT research reactor. Financial support was provided by NSF grants EAR 8115850, 8319666 and 8520644 to Prof. Frey (MIT); DTI (Universidad de Chile) grants E-1703 to LLE and E-2457 to MAP and FONDECYT (Chile) grant 83/01261.

REFERENCES

Munizaga, F., Hervé, F., Brook, M., Pankhurst, R., Snelling, N., Drake, R., 1985. Geochronology of the granitoids of the Lake Region, Chile (39°-42°S Lat.). Communicaciones, N°35, p. 167-170.

Parada, M. A., Godoy, E., Hervé, F., Thiele, R. 1987. Miocene calcalkaline plutonism in the Chilean Southern Andes (41°00'-41°45'S). Extended Abstract. International Symposium on Granites and Associated Mineralization (ISGAM), Salvador, Brasil. Geociencias, V. 17, p. 450-455.

Thiele, R., Hervé, F., Parada, M. A., Godoy E. 1986. La megafalla Liquiñe-Ofqui en el Fiordo Reloncaví (41°30'), Chile. Comunicaciones N°37, p. 31-47.

M

MAKARIKHIN, V.V., Institute of Geology, Petrozavodsk, USSR

Lower Precambrian Stromatolite Assemblages in Karelia

Nowadays one can hardly doubt that fossil evidence for ancient life, essentially algal in nature, can be preserved and is fairly often found in large quantities in rocks of both pre- and post- 2 Ga age. Lower Precambrian organic remains fall into four large groups:

1) phytogenic structures comprising stromatolites, stromatolite-like units and microphytoliths (oncolites and catagraphites);

2) organic-walled microfossils or acritanchs i.e. various small organic shells isolated, as a rule, by maceration;

3) cellular remains which occur as microscopic units stadied in transparent thin sections;

4) problematics represented by systematically obscure biogenic units including traces and relict organic matter.

Of the above phytogenic groups algal units have so far been most properly studied.

The Institute of Geology in Petrozavods has conducted systematic paleontological investigations since the late 1950's. Lower Precambrian phytoliths have been reported from about 60 localities. Lower Precambrian deposits of Karelia are represented by two complexes: the Lopian complex of Upper Archean age and the Karelian complex of Lower Proterozoic age (Fig.). Their upper boundaries are dated at 2600-100 Ma and 1650-50 Ma, respectively. The Lopian complex is composed of medium and acid volcanic tuffs, schists, iron formation, carbonaceous and carbonate rock, metabasalts, komatiites and metaandesites. The complex is 2000-3000 m thick. It is here that the first problematic algae Careliella and microphytofossils were encountered.

The Karelian complex is subdivided into three subcomplexes: lower, middle and upper separated by large stratigraphic inconformities at the boundaries with an absolute age of 2450-50 Ma and 2100-50 Ma. Each subcomplex includes two superhorizons, has its own name derived according to a conventional ethnographic principle: Sumian, Sariolian, Jatulian, Ludicovian, Kalevian and Vepsian. Each superhorizons is characterized by a unique phytolith biocoenosis in which the leading role is played by stromatolites. The maximum among of stramatolites was reported from the carbonate rocks of the Jatulian superhorizon. The above stromatolite associations provided a basis for subdividing the lithologica uniform Jatulian carbonate sequence into 5 stratons ranking as bed which contain the following phytolites (from bottom upwards); Nucle-phyton, Sundosia, Omachtenia kintsienis, Butinella, Calevia ruokanensis.

The distribution of organic remains in the oldest complexes of Karelia is fairly uneven. However, most of the aforementioned superhorizons contain specific biocoenoses of their own. This allows to use paleontological information in distinguishing and tracing these deposits. The use of organic remains for discerning smaller units is mainly restricted to the Jatulian superhorizon.

The unique character of the Precambrian should be considered in discussing the importance of the above groups for the stratigraphy of the Lower Proterozoic or the role of biostratigraphy at this stage. The stratigraphic units of the Precambrian are not adequate to those of the Phanerozoic although their taxa are similar in nomenclature. The Precambrian is remote chronologically, and the time boundaries of its units are by one order higher than of the Phanerozoic. On the other hand, the animated nature was still at a very lower level. The bulk of organisms were so primitive that their evolution can only be a matter of speculation. One should also in mind the repeated effects of later processes, widespread metamorphism and other factors which markedly disturb the probable pattern of paleocoenonses.

Therefore, the paleontological method used in stratigraphic studies should be combined with all other possible methods for age determination, and biostratigraphic construction must be based on all available data on organic remains. It would be premature in this connection to distinguish "leading" complexes and "leading" species. The names of beds derived from those of individual fossils should be regarded as tentative.

Available paleontological information indicates that the evolutionary level of the Early Precambrian world was high enough for accumulation of a considerable amount of biogenic matter and for providing a basis for big qualitative changes in the subsequent evolution of the biosphere.

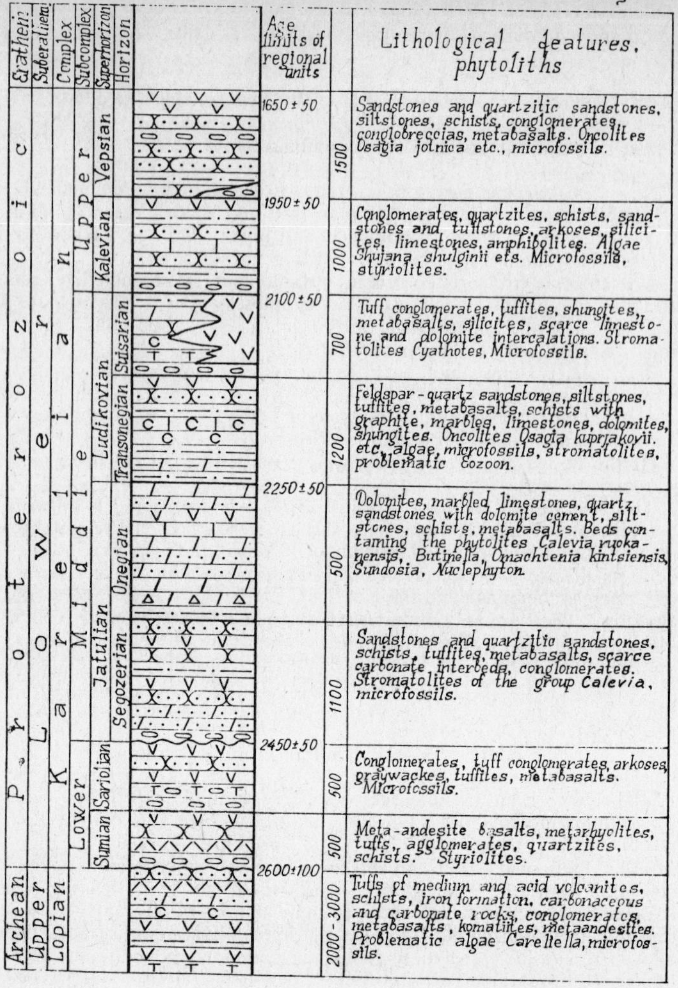

MCCALPIN, James, Department of Geology, Utah State University, Logan, UT

Quaternary Geology and Neotectonics of the Molesworth Section of the Awatere Fault, South Island, New Zealand

The Molesworth section is a distinct 55 km-long reach of the Awatere Fault in the Marlborough district, New Zealand, which strikes 10° to 15° more easterly than adjacent segments. The Awatere Fault is one of the four major active dextral-reverse faults in the Marlborough-Wellington shear zone, which comprises the plate boundary north of the Alpine Fault. The research described herein had 2 goals: 1) to determine if this fault section has acted independently of adjacent ones during Holocene surface-rupturing earthquakes (i.e., is a fault segment), and 2) to confirm that this segment ruptured during the 1848 Marlborough earthquake (maximum intensity IX).

Quaternary deposits were mapped at 1:25,000 in a 5 km-wide strip along the Awatere Fault between Maling Pass (42° 13′ S., 172° 41′30″ E.) and Molesworth Station (42° 03′30″ S., 173° 18′ E.). Quaternary deposits are dominated by glacial tills and outwash gravels of at least three glaciations, correlated with the early Otiran and late Otiran (Kumara 2_2 and 3) glacial stages of Suggate (1965). For most of its length the fault offsets either alluvial fans or outwash plains of Kumara 2_2 age (approximately 22 ka), the aggradational outwash surface of Kumara 3 age (approximately 14 ka), or Holocene stream terraces incised into the aggradational surface. The ages of faulted deposits are constrained by 50 radiocarbon samples (dates pending) and 50 weathering rind stations calibrated against the rind thickness/age curve of Whitehouse and others (1986).

Although the Holocene offsets on the Molesworth segment have been predominantly dextral, both normal and reverse components occur related to changes in fault strike. Vertical slip components increase up to 30m as the strike of the fault trace deviates from the overall fault strike of N 70° E. Fault scarp profiles were measured at 135 locations where tectonic relief had been formed. Profiles across dextral-normal scarps were generally symmetrical, similar to normal fault scarps worldwide, while profiles across dextral-reverse faults were asymmetrical, a result of successive overriding of the colluvial wedge sequence by the hanging wall.

At 35 locations the fault offsets multiple geomorphic surface and has left geomorphic evidence of recurrent Holocene slip. The best sequences occur where multiple Holocene terraces are dextrally displaced. At the Saxton River, five Holocene terraces risers inset below the Kuara 2_2 outwash plain have been dextrally offset a minimum of 7m, 12m, 35m, 62m, and 64m respectively. Elsewhere, dextral offsets of stream channels tend to be a minimum of 6-8m, and larger offsets are some multiple thereof. Evidence from the Saxton River implies that 8 to 9 paleoearthquakes of 7-8m dextral slip each have occurred since about 22 ka, yielding an average recurrence interval of 2.4-2.9 ka.

Four trenches were excavated across fault scarps at the Acheron River (5 km E. of the Saxton River) to confirm the number of paleoevents. Two trenches across scarps 5.1m and 8.0m high yielded multiple colluvial wedge stratigraphy. In the deepest trench dextral-reverse faulting had thrust fractured graywacke over six colluvial units (Fig. 1) each one inferred to represent a faulting event postdating 22 ka. Five thermoluminescence dates from this trench, and three radiocarbon dates from a nearby natural exposure are now pending, and promise to constrain the latest 6 paleoevents on this segment.

REFERENCES

Suggate, R. P. 1965, Late Pleistocene geology of the northern part of the south Island, New Zealand. N.Z. Geol. Survey Bull. 77, 91 p.

Whitehouse, I. E., McSaveny, M. J., Knuepfer, P.L.K. and Chinn, T.J.H., 1986, Growth of weathering rinds on Torlesses Sandstone, Southern Alps, New Zealand in Colman, S.M. and Denthier, D.P. (eds.), Rates of Chemical weathering of rocks and minerals: Academic Press, New York, p. 419-437.

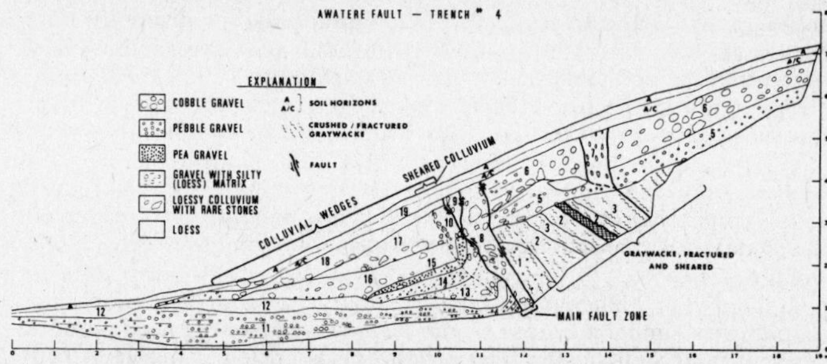

Figure 1 – Log of trench across the Awatere Fault 0.5 km W of the Acheron River. Units 13-18 are interpreted as scarp-derived colluvium deposited after individual surface-faulting events.

MELEZHIK, V.A., Geological Institute, Kola Branch of the USSR Academy of Sciences, Apatity, USSR

Geochemical Conditions of the Early Precambrian (The Baltic Shield)

1. In geochemical investigations of the Precambrian exogenesis (2.8 - 1.6 Ga) a comprehensive analysis of sedimentary, volcanogenic-sedimentary, biogenic-sedimentary diagenetic systems has been realized as well as examination of the evolution of carbonaceous, carbonic, oxide-ferruginous and sulphate sedimentary ones. Determinations of some parameters is the basin-atmosphere system have been made.

2. The main results of a physical-chemical and hydrochemical study of the basins over the range 2.8-1.6 Ga based on some works (Landergren, Carvajal, 1974; Melezhik, 1987) demonstrate that the bulk of waters (2.8-2.2 and 2.0-1.8 Ga ago) was characterized by the low salinity (1-10‰), reducing parameters with respect to $Fe^{2+} \rightleftarrows Fe^{+}$, low sulphate concentration ($SO_4^{2-} < 0.3$‰) and high acidity. Minimal salinity (0.5 - 0.1‰) is typical of the Sariolian (Guronsky) basins (2.3 - 2.2 Ga) for which the features of glacial origin have been defined. Essential acidity and reducing character of the waters are exhibited in a high migration activity of Fe and Mn simultaneously in the exogenic conditions and extremely limited carbonate accumulation.

The basins dated 2.2 - 2.0 Ga proved to be anomalous according to the hydrochemical and physical-chemical parameters. They were characterized by the high alkalinity and salinity (16-60‰), in a number of cases by the high sulphate concentration, essential and irregular oxidation power. A "slash" of the total water mineralization is established within the entire Baltic Shield and coincides with the "peak" of dolomite accumulation, formation of sedimentary sulphates and magnesites. For the first time the displayed oxidation power of the aqueous media finds its manifestation in the abundance of the basin red rocks, appearance of the higher syngenetic oxides of Fe and Mn, change of the single-stage sedimentary rock formation to the double-stage one and in the wide-spread occurrence of diagenetic concretion formations.

3. The cardinal change in physical-chemical and hydrochemical conditions at the turn of 2.2 Ga ago is caused by the endogenic evolution of the shield and biosphere. The oxidation power of the waters in a basin

is due to the wide-spread occurrence of stromatolith-building algae and their photosynthesis activity. The high mineralization is provided by the substance inflow over endogenic channels.

4. The peculiarities of composition of waters and sedimentation are attributed to the specific interaction of sedimentary, endogenic and biogenic processes. This manifests itself in the irregular occurrence and close time and space connection of the oxidation conditions, areals of the life activity for stromatolith-building algae and endogenic relief zones. The distribution of oxidation conditions in the basins is irregular and zonal. The positional connection and symmetry in the change of oxidizing properties and alkalinity in the aqueous medium, realized in the geochemical parameters of sedimentary rocks with respect to the areals of stromatolith-building algae occurrence, enables us to consider the last ones as a zone of microbiological photosynthesis formation of O_2. An analysis of geochemical profiles (Example, Fig. 1) shows that the photosynthetical O_2 was entirely spent on the polyvalent metal oxidation within the local zone of cianobacterial community life activity and its inflow in the atmosphere - meteoric waters - continental waters system was hindered until the 1.6 Ga boundary. The additional source of a fixed oxygen were exhalation products and alkaline magma lavas providing the endogenic hematite ($\delta^{18}O$ - 2.3‰SMOW) different from the "hydrogenous" one ($\delta^{18}O$=+3.8‰) in the oxygen isotopic composition.

5. The reconized changes in composition and physical-chemical parameters of waters in the Precambrian basins of the Baltic Shield are correlated with geochemical (Budyko, Ronov, Yanshin, 1985) and isotopic (Kempbell, Krouz, 1985) studies in the framework of two models: one suggesting the existence in the history of the secondary Earth atmosphere of a transitional stage (from 2.2 till 1.6 Ga) with unstable and very changable (up to full disappearance) resources of O_2 in the hydrosphere and partly in the atmosphere, and the model of "Oxygen oases" in hydrosphere accounting for the different in space and time manifestations of the oxidizing aqueous media up to the 1.6 Ga boundary.

6. The essence of interaction of sedimentary, biogenic and endogenic systems in the Early Proterozoic red rock formation and a retrospective analysis of the conditions for generating Paleozoic red rocks enables us to infer that the former are: i) the result of the water oxidizing effect due to a photosynthesis activity of the stromatolith-building algae; ii) the effect of high life activity processes; iii) the characteristic of the aqueous media oxidizing conditions rather than of the climate regimes. This implies that the Phaneorozoic and Precambrian sediments, similar from the geochemical point of view, may denote quite different cause-and-effect relations. This conclusion is of great importance and has a more general methodological meaning in the geochemical research of Precambrian basins.

7. In the geochemical evolution of Precambrian sedimentary rock formation of the Baltic Shield three main boundaries are distinguished over the range 2.8 - 1.6 Ga: 2.2, 2.0 and 1.8 Ga which manifest themselves in the change of the saline composition, oxidizing and alkaline properties of the basin waters (2.2, 2.0, 1.8 Ga), regular cyclic change in limestone and dolomite accumulations (2.2, 2.0 Ga), inversions of the utilization forms with atmospheric CO_2, alteration of the life forms (2.0 Ga), realization of the two-stage sedimentation and wide-spread occurrence of the diagenetic concretion formations (2.2 Ga). The boundary of 2.2 Ga is of global significance.

REFERENCES

Budyko M.I., Ronov A.B., Yanshin A.L. The history of the atmosphere. L.: Gidrometeoizdat, 1985, 208p.

Melezhik V.A. The composition of waters in Precambrian basins from the geochemical data. Izv. Akad. Nauk SSSR, 1987, N 9, p.100-111.

Hattory K., Kempbell F.A., Krouz H.R. Abundance of sulphur isotopes in sedimentary rocks in connection with the evolution of Precambrian atmosphere. - Geohimiya, 1985, N 6, p.834-849.

Landergren S., Carvajal M.C. Contribution to the geochemistry of boron. 111. The relationship between boron concentration in marine clay sediments, the salinity of the depositional environments expressed as an absorption isoterm. - Ark.mineral. geol. 1974, v.5, N 11, p.11-22.

Ojakangas R.W. Evidence for the Early Proterozoic glaciation; the dropstone unit - diamictite association. - Bull. Geol. Surv. Finl. 1985, N 3, p.51-72.

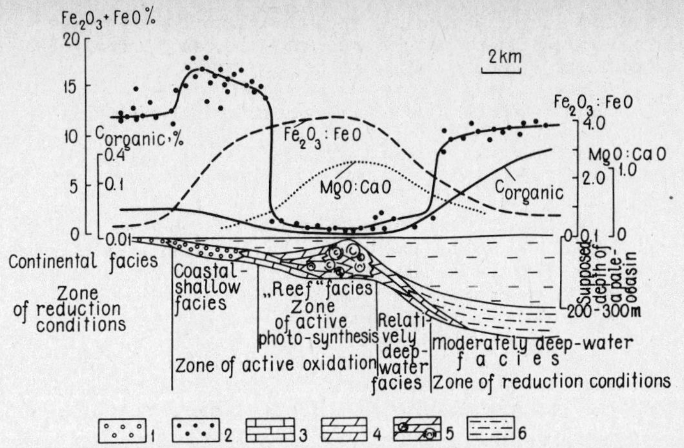

Figure 1 — An example of the facies-geochemical profile in the basin with a local zone of the life activity of stromatolith-building algae. The age — 2.8 Ga (The Pechenga -Varzuga Belt, Il'mozero formation).

1 - gritstones, sandstone; 2 - red aleurosandstones; 3 - red chemogenic limestones; 4 - chemogenic dolomites; 5 - partially silicified stromatolithic and oncolite dolomites; 6 - gray aleuropelites; Fe_2O_3 : FeO and MgO :CaO - molecular ratio. C_{org} content in % mass. The distribution of elements is based on the analysis of 202 mean samples.

MELIORIS, LADISLAY, Comenius University, Czechoslovakia

Conditions of the Rise and Hydrogeochemical Stability of Mineral Waters of the Levice Mineral Water Springs Line in the West Carpathians

On the inner side of the West Carpathians arc significant resources of mineral waters can be found with the specific use of the so-called "Levice Mineral Water Springs Line". The Pre-tertiary basement of the territory of this line is created by the Permian rock and carbonates of the Triassic. The geological structure of the Neogene system was significantly affected by the neovolcanism. In Sarmatian and Badenian its centres were situated in the near surroundings of the mineral water springs line. The nappe and fold structures of the Pre-Tertiary basement were created by paleoalpine movements. The structure of the whole region was completed by fault tectonics which was especially active in the Neogene system. It resulted in significant block structures (the Turovsko-Levice Horst and others) which are of decisive importance in the localization of mineral waters.

According to the salt composition of the so-called "Levice Mineral Water Springs Line", higher contents of all main ions (Na^+, K^+, Ca^{2+}, $C1^-$, SO_4^{2-}, HCO_3^-) are typical. The waters contain a large amount of carbon dioxide (CO_2> 1 mg. l^{-1}, max. 2842 mg . l^{-1}), and the content of H_2S < 10 mg . l^{-1}. The values of $K^{N_2}_{O_2}$ > 10 in most waters indicate a reducing environment. The values of K^{He}_{Ar} indicate that exchange can be limited or intensive. Trace elements are represented especially by J, Br, F, Li, Sr, HBO_2. The isotopic composition of mineral waters was examined. Maximum temperature of the waters is 33°C. The components of Ca-Mg-HCO_3, Na-HCO_3 and $NaSO_4$ are the result of water interaction with volcanic and sedimentary rocks. The interaction is affected by thermodynamic conditions of circulation and by the amount of CO_2. The occurence of Na-C1 and partially also the occurence of Na-HCO_3, as well as higher contents of Br and B are affected by marinogenic mineralization. In the final phase the resultant chemical composition and the mineralization of waters are determined by the waters mixing processes.

Genetically, but also from the point of view of balneotherapy the waters of the "Dudince type" are distinguished here. They had the greatest stability of the chemical composition since 1893. Only the content of H_2S fluctuates. The ratio of HCO_3/C1 is 5.0 up to 6.0. The temperature of the water is about 30°C. Quite different are the waters of the "Slatina type". The temperature of the waters is lower (12-13°C), the high content of CO_2 and H_2S has only a rare occurence. The ratio of HCO_3/C1 is 3,0 up to 4.0. The waters are of a highly fluctuating chemical composition. Destabilization of the hydrogeochemical condition occurs as a consequence of unsuitable exploitation. The waters of the "Santovka type" are cold and they have a high content of CO_2 and the content of H_2S max. 4 mg. l^{-1}. The balance hydrogeochemical evaluation has shown that in the water springs structure deep mineral waters (containing CO_2) are mixed with waters of subsurface origin in the ratio of 1:0.8 up to 1:3.

Mineral waters from this springs line represent cold and thermal waters of the complex chemical composition and an unstable hydrogeoochemical condition. The common origin, the verified hydrological and hydraulic bond make heavy demands on the optimum exploitation of individual resources according to their rank of importance, as well as on the permanent preservation of their specific parameters — hydrogeochemical stability in dependence on time.

MÉNDEZ, AMARILIS SALINAS and Armando Simón MÉNDEZ

A Geochemical Evolution Model of Jurassic Stratabound Deposits in Pinar Del Rio Province, Western Cuba

The Santa Lucia-Castellanos and Nieves-Matahambre ore fields are located in the Northwestern part of Pinar del Rio province, Cuba, in the Guaniguanico structuro-facial zone. The most common rocks are quartz-feldspatic sandstones, siltstones, shales and minor limestones, belonging to the San Cayetano Formation (Ji-2).

Quartz-feldspatic sandstones are the enclosing rocks of the footwall cross-cutting cupriferous mineralization (Matahambre, La Esperanza, Castellanos, Nieves), whereas carbonaceous shales comprise the stratiform pyrite-sphalerite-galenite ore (Santa Lucia, Castellanos).

At Nieves and Matahambre the pyrite-polymetallic ores are located near the contact between the carbonaceous shales and the quartz-feldspatic underlying sandstones.

The following alteration processes have been described in the Santa Lucia-Castellanos ore fields:

— Carbonaceous shales are recrystalized into microlepidoblastic aggregates of sericite, chlorite and hydromica, characteristic for the low grade regional metamorphism.

— Secondary carbonates are present, together with the sulfide mineralization.

— Baritization and dolomitization.

These alteration processes are reflected by weak geochemical anomalies, with similar composition to the ore bodies.

Indicator elements for the stratiform pyrite-polymetallic mineralization are Ba, Ag, As, Sb, Pb, Zn, Cd, Te, Ge and Cu (all determinations by the semi-quantitative spectral emission method). In the ore outcrops zones, high concentrations of Ba, Ag, As and Sb are revealed.

The most representative profile through the oxidation zone in the ore field can be studied at the Castellanos deposit. It has an extremely complex composition. Intense weathering of carbonates and sphalerite and oxidation of Fe and Pb sulfides led to formation of limonite-barite-quartz aggregates, with Pb ochres. The secondary enrichment zone is absent.

According to the mineralogical composition, oxidated ores are represented by a mixture of Fe oxides and hydroxides, quartz and barite, forming kidney and earthy aggregates and including argilaceous minerals, gold, silver and Pb carbonates and sulfates. The Zn is almost wholly lixiviated.

The gold is finelly and extremely deseminated within hidrogoethit caolinite-micaceus rocks, which in turn develop around the hematite-goethite collomorphic formations.

At Matahambre-Nieves ore field can be obserbed alteration processes of hidrothermal and metamorphic nature. The former are represented by intense quartizification and carbonatizacion, whereas the latter are represented by recrystallization of the siliceus-argilaceus matrix to sericite and chlorite.

Beside these two kinds of mineralizacion we have mentioned, a third kind, cross-ctting pyrrhotine-pyrite, has been revealed, which is characterized by the following elements: As, Sb, Ag, Pb, Cu, Zn, Co, Bi and Mo.

The chalcopyrite mineralization is represented by the following indicator elements: Cu, Ag, Co and Ga. In deeper zones, pyrrhothine and cobaltite have been reported.

The pyrrhotinic ores have the following composition: Co, Bi, Ag and Mo.

Between the polymetallic and cupriferous ores it is usually present a stock-werk with the following distinctive features:

— weak contents of Ag, Pb, Sn and Ga.

— lack of Cd, As, Sb and Ba.

— significative contents of Cu, Co, Ga, Ni, Bi, Mo and Pb.

The chemical composition of the different kinds of mineralization, together with a clear and well known zonality of geochemical anomalies indicate a probable deposition from common solutions, whith properties gradually changing in space and time.

Brecciated textures, substitutions, corrosions, and cementation of some minerals by others are evidences of repeated affluence of mineralized solutions.

MESHREF, Dr. Wafik M. and Mohamed Salah Abu El KARAMAT

New Exploration Concepts for Oil in the Gulf of Suez

Many oil fields were discovered in the Gulf of Suez rift, some of which produce from Miocene, pre-Miocene and others from both Miocene and pre-Miocene rocks. The exploration history in the Gulf of Suez exceeds 100 years, accordingly, cumulative information and experience is growing through time. Some major factors that play a very important role in oil exploration are discussed below. Among these factors are:

1. *Cross Gulf Trend Faults:* Major known oil fields in the Gulf terminate abruptly from north and south at major cross elements. The exact location, orientation and magnitude of throw along these cross elements are of vital importance in exploration.

2. *Attitude of Fault Blocks:* The Gulf of Seuz rift is divided into three main tectonic provinces that are separated by two accommodation zones. The northern and southern provinces are characterized by regional SW dip, while the centeral province is characterized by regional NE dip. The accommodation zones are characterized by much gentler angles of dip. It is a well known fact that major oil accumulations occur at the up-dip part of the tilted fautled block. So it is very important to know the amount and direction of dip of the explored block in question. In this regard it is worth mentioning that local reversals of dip within the regional dip provinces exist which necessitates special attention in the exploration process.

3. *The Sealing Rocks:* Most workers in the G.O.S. consider the South Gharib Formation is the ultimate seal. This is true as far as the central and southern tectonic provinces. Meanwhile, the L. Rudeis Formation could be regarded as an important seal rock specially in the northern tectonic province.

However, the distribution of the South Gharib Formation and its relation to the hydrocarbon accumulations is critical but seems to be understood.

4. *Migration:* Usually it is accepted that oil migrates up-dip into favourable reservoirs, however, close examination of some oil traps suggest that migration takes place along fault planes, specially the G.O.S. region is an area of tension. Moreover, it is feasible that strike migration may be taking place in the G.O.S. specially around the accommodation zones.

5. *Reservoir:* The distribution of the Miocene and pre-Miocene sands are now mapped with great accuracy that permits exporation for pinch-outs.

6. *New Technology:* This factor will be always there, having a great impact on oil explorationin any basin. This is because the fact that the advancement of new technology permits re-acquisition and/or reprocessing of geophysical data which may lead to improvement of signatures from deeper prospects in the G.O.S. Moreover, re-interpretation and new fresh approaches may lead to new prospects or leads.

MEZHELOVSKY, N. V., L. M. NATAPOV, and L. P. ZONENSHAIN, 1. Ministry of Geology of the USSR 2. Objedinenije "Aerogeologia" 3. Institute of Oceanology, USSR

Yakutsk-Kamchatka Traverse: Performance and Planned Research

In the USSR geological and geophysical studies began along the Yakution-Kamchatka Geotraverse (YAKAG) across the Siberian platform margin, Verkhoyansk-Kolyma folded belt, Okhotsk-Chukotka volcanic belt, northern part of the Sea of Okhotsk near the Tinro Trough, Koryak-Kamchatka folded belt and Kuril-Kamchatka trench. The traverse will help to understand the lithosphere structure in the North-East of the USSR, to determine a role of accretionary tectonics and to interprete a structure of main suture, volcanic and seismic zones.

The Geotraverse begins near Yakutsk city, runs eastwards to Khandyga-town on the Aldan river, then runs along the Magadan-Khandyga highway. It crosses Koni Peninsula, Penzhina Bay and northern Kamchatka going across Karaginsky Peninsula and the Kuril-Kamchatka subduction zone up to Obruchev Rise in the Pacific ocean near the DSDP Site 192. New geological gravity and magnetic fields maps are compiled along the traverse in 1:1 000 000 scale.

An additional geotraverse is branching from the YAKAG near Indigirka River towards Chukotka and Wrangler Island. This is so-called Yakutian-Chukotka Geotraverse (YACHUG).

The position of the YAKAG is chosen along the Magadan-Khandaga highway where multichannel seismic reflection studies with CDP are proposed to be done. Along with the American TALI Geotransect in Alaska the YAKAG allows to solve many problems of the structure and evolution of the continental and transitional lithosphere type in the north-western Circum Pacific.

The part of the north-east of the USSR (the Siberian platform passive margin) along the YAKAG was mainly deformed in the Middle Cretaceous time and was accreted by the Omalon and Okhotsk microcontinents and by very complicated Koni-Murgal terrane consisting of amalgamated Late Paleozoic and Mesozoic island arc complexes. A collisional suture zone separating the Okhotsk massif from Siberia was mapped. It is represented by thrust belt involving greenschist metamorphics. A further accretion of the Siberian platform margin took place in the Late Cretaceous time when the Andean type Okhotsk-Chukotka volcanic belt was originated along the newly shaped continental margin above the subduction zone. Volcanic polarity allowed to calculate a burried trench position nearly 160 km to the east of the volcanic belt and to estimate the subduction zone dipping to be equal to 35°.

A small area which lacking the granitic crust is proposed to be in the northern part of the Sea of Okhotsk. Formation of the area is believed to be related to the Late Cretaceous rifting and a separation of the fore-arc terrace from the Okhotsk-Chukotka volcanic belt. Kamchatka consists of the Late Cretaceous island arc and back arc complexes which are superimposed by the Oligocene-Neogene island are related to the Kuril Kamchatka subduction zone. The YAKAG crosses the Kronotsky terran in the eastern-most Kamchatka. It is a part of a system of terranes extending from the Small Kuril islands to Karaginsky Island which collided with Kuril-Kamchatka arc in the Miocene time.

The YACHUG crosses the collisional Chersky suture zone along which the Siberian platform passive margin and the Late Jurassic island arc collided in the beginning of the Cretaceous time. Further, the YACHUG goes through: the Late Jurassic fore-arc basin, Pre-Kolymian and Omolon microcontinents, Late Jurassic Oloy island arc which incorporates numerous Paleozoic and Early Mesozoic terranes, the South Anni suture zone with Late Jurassic ophiolites, and the Chukotka microcontinent. The YACHUG ends near the Wrangel-Herald-Brooks suture zone marking the Laramide collision. Two geotraverses can be connected with the TALI Geotransect through the Arctic Basin. Along the YACHUG, the structure and evolution of suture zones and related thrust fronts are fairly well studied. Volcanic polarity within the Late Jurassic Island arc near the Siberian platform margin allowed to reconstruct the former subduction zone position being 125-130 km eastward from the Chersky Ridge. This zone is dipping 45° westwards. The YACHUG crosses also young grabens and the present seismic belt of the Moma rift zone along the Eurasia-North America plate boundary. Information along the YACHUG would significantly increase if multi-channel seismic reflection study could be performed. But this work is very difficult to fulfil because of hard geographical conditions.

Main collisional sutures and other large structural units are well pronounced along both traverses on the gravity and magnetic field maps.

MIMOUNI, O. and S. LARABI – IST/USTHB Dar-El-Beida – BP 9 Algiers, Algeria

Study of the Ain-Ouessara Plain Groundwater Flow Pattern for Water Management (Algeria)

Ain-Ouessara is a vast plain located in the high plains of Algiers region with an area of approximately three thousands kilometers square. Due to the built of a new city (nearly 200.000 inhabitants) and its economic development, a thourough study of the aquifers and a drawing of water management, is required. A geological study, lithostratigraphy, tectonic and geophysics, of the plain, located in the preatlasic unit between Saharian domain characterized by box folds and Tellian domain containing autochton, para-autochton and nappe slides, was the main part of this work. Hydrochemical analysis resulted with isocon maps and a general scheme of the evolution in concentration of the different chemical elements. An increasing pollution problem du to waste water was also noticed. The hydrogeological study completed the knowledge of the different aquifers, their importance and therefore their futur management were emphasized. This complete study will be a tool for water management in that region.

MOHANTI, MANMOHAN, Abhay K. SINGH, and Subhas C. SRIVASTAVA, Department of Geology, Utkal University, Vani Vihar, Bhubaneswar, India

Microfacies and Sedimentation of Larger Foraminifera-Rich and Associated Carbonates (Palaeogene) in the Northwestern Kutch Basin, India

Carbonate microfacies and sedimentation pattern of Harudi and Fulra Limestone Formation of Middle Eocene and Maniyara Fort Formation of Oligocene age comprising a thickness of ca.100m of an outcrop belt of ca.45Km between Lakhpat and Rodasar (Fig. 1) in the pericratonic rift basin[4] of northwestern Kutch have been investigated. Attention has been given to larger foraminiferal associations in interpreting depositional environments.[1,2,3,5]

Microfacies of the Harudi Formation are argillaceous pelecypod - rich rudstones, *Nummulites* - rich floatstones and carbonate mudstones associated with gypsum-rich and carbonaceous layers and greenish shales. The succeeding Fulra, principally of biogenous rudstones and floatstones abound in larger Foraminifera like *Discocyclina, Nummulites, Assilina* and *Alveolina* with coralline algal fragments and rhodoliths becoming remarkable upwardly. Microfacies of the Maniyara Fort Formation with basal part having conspicuous admixture of "glauconites"/quartz sand, are composed of rudstones - floatstones dominated by *Nummlites, Lepidocyclina,* and *Spiroclypeus.* Coralline algae (some rhodolith-type) and dasyclad algal fragments also occur. Mudstones - floatstones with coral growths (locally, dolomisized muddy, mound-like coral-buildings), dolomites, argillaceous and sandy beds occur as well.

Sedimentary structures principally include cross-bedding (at places "large-scale"), channelling, scour-fills, lenticular and wavy bedding, imbricated bioclasts, desiccation cracks, burrows, borings, trail marks, intraformational carbonate breccias and pebbles (some show borings), karstic-solution features, laminated calcite (calcrete-like) and ferruginous crusts and pockets and layers of leached, empty formaniferal cavities.

Primary pore spaces are largely cemented by blocky, sparry calcite and/or filled by internal sediment. Neomorphism has varyingly affected micrite. Secondary porosity is conspicuous as solution channels, vugs and fossil moulds. Argillaceous matter may coat micrite as thin films affecting permeability. Cementation and dissolution processes possibly occurred under phreatic to meteoric-vadose environments.

Palaeogene sedimentation started with marine transgression through the "lows" of structure-controlled post-Mesozoic land surface cover by Deccan Trap volcanics (Late Cretaceous-Early Palaeocene) west of a synchronous "Median High". Close to the undulating margin of Mesozoic-Deccan Trap country, Palaeogene microfacies were deposited in a lagoon to sheltered open embayment-like marine setting (Fig. 2). A quiet, to some extent restricted lagoonal sedimentation of Harudi gradually developed to a more open empayment-like environment during Fulra where sheet-like carbonate banks and bars of larger Foraminifera existed under moderately low to espisodically high-entery conditions in shallow, subtidal environment. Open to relatively restricted quiet environments possibly of lagoonal to back-bank type were associated. Fulra sedimentation terminated under upwardly shoaling "regressive" conditions when sedimentation possibly exceeded subsidence. After a sharp break in sedimentation (paraconformable), transgression occurred and with influx of "glauconites" and quartz sand Oligocene sedimentation commenced in shoal-lagoonal environments. Dolomitisation occurred probably in peritidal environment. Corals thrived in a shelf-lagoonal setting and have also aided formation of mound-like, muddy, buildups which locally developed probably on a "palaeohigh" (Fig. 3). Mound-like carbonate buildups developed under favourable hydrodynamic and biologic influence where branched corals, coralline algae and perhaps other benthic microbial community (cynobacteria etc.) and/or seagrass played a role in trapping, binding and stabilising mud which subsequently underwent early diagenetic dolomitisation. Associations of larger Foraminifera had thrived in shallow subtidal conditions with subtle fluctuations in depth and other palaeoecological factors. Two transgressive phases separated by a regressive phase of discontinuity in sedimentation and quartz sand influx is apparent in the Oligocene. Siliciclastic sedimentation in the Miocene terminated Palaeogene carbonate sedimentation. Palaeogene facies associations may be valuable in offshore exploration for hydrocarbons.

REFERENCES

(1) Adams, G.C., 1965, The Foraminifera and Stratigraphy of the Melinau Limestone, Sarawak, and its importance in Tertiary correlation. Quart. Jour. Geol. Soc. Lond., V. 121, P.283-338.

(2) Aigner, T., 1983, Facies and origin of Nummulitic buildups: an example from the Giza Pyramids Plateau (Middle Eocene, Egypt). Neues Jahrb. Geol. Paläont. Abh, V.166, P.347-368.

(3) __________, 1985, Biofabrics as dynamic indicators in Nummulite Accumulations. Jour. Sed. Pet., V.55, P.0131-0134.

(4) Biswas, S.K. and Deshpande, S.V., 1983, Geology and Hydrocarbon Prospects of Kutch, Saurashtra and Narmada Basins. Petroleum Asia Journal (November), P.111-126.

(5) Hallock, P. and Glenn, E.C., 1986, Larger Foraminifera: A tool for Palaeoenvironmental Analysis of Cenozoic Carbonate Depositional Facies. Palaios, V.1, P.55-64.

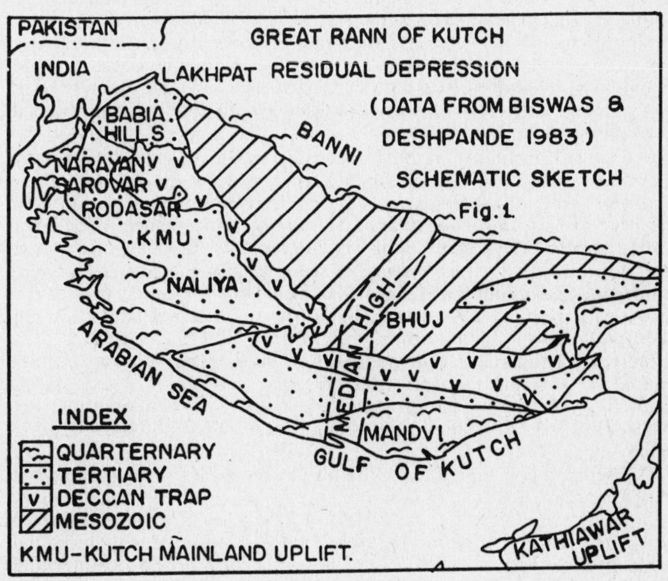

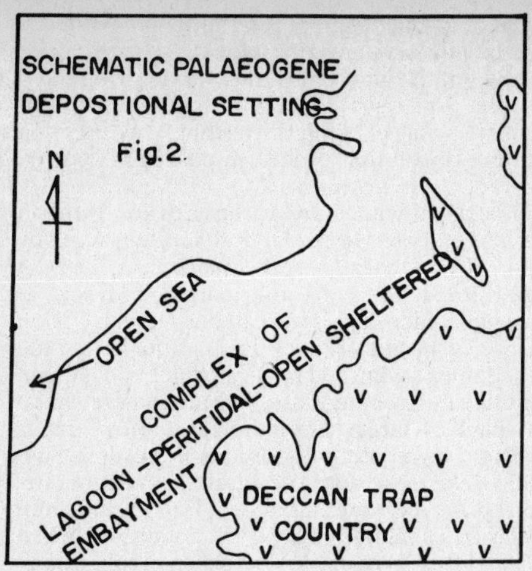

Fig.2.

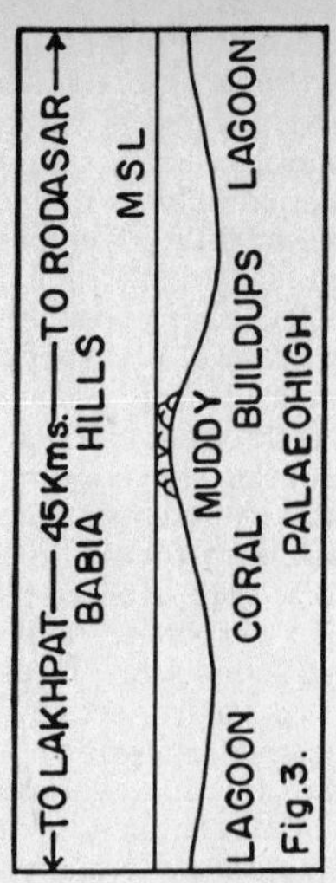

Fig.3.

MOHANTI, MANMOHAN, Department of Geology, Utkal University, Vani Vihar, Bhubaneswar (Orissa), India.

Microbial Carbonates from Neogene-Holocene Interval, Orissa, East Coast of India.

Carbonates (microbialites)[2] possibly formed by benthic microbial community (BMC) occur in Neogene (Baripada Beds) and Holocene deposits (Chilka Lake margin)[3] in Orissa, East Coast of India (Fig. 1).

A sequence of greenish-bluish, darkish and yellowish clay and shaly beds (in some horizons with carbonaceous concentrations), darkish mudstones, thin sand layers, sands admixed and alternating with clays, yellowish-brown sandy carbonates and yellowish argillaceous carbonates constituting "BARIPADA BEDS", a Neogene stratigraphic unit are exposed along the Burhabalang River banks and occur in subsurface as well in Mayurbhanj District, Orissa State, east coast of India.

Fossil fauna chiefly include molluscs (oysters locally forming dense colonies) and locally rich occurrences of shark teeth and other fish remains which may be associated with mammalian and reptilian teeth, some bones, balanids, bryozoans, corals, echinoids, serpulids, crabs, foraminifers and ostracods. Sedimentary structures include low-angle and wedge-shaped cross-bedding in sands, flaser-linsen and lenticular bedding, burrows, borings (lithophagid boring prominent), desiccation cracks, laminated argillaceous carbonates and local pyrite stringers.

Near the village Mahulia, ca. 5 km south from the Baripada Town, yellowish argillaceous carbonates and yellowish-brown sandy carbonates occur along river bank underlain and overlain by clay beds and measuring 1m in thickness. Yellowish argillaceous, dense limestones of flattish to ovoidal masses occur as several discrete bodies measuring 10-50cm across and ca. 6 cm in thickness in yellowish-brown sandy limestones that appear irregularly protruberating. The dense carbonates exhibit faint irregular laminations, thin stringers of quartz sand, desiccation cracks, lithophagid borings and other conspicuous tube-like borings of 1-6cm diameter penetrating nearly perpendicular to the surface and filled largely with quartz sand. In the sandy carbonates, oyster and some other molluscan remains, shark teeth, bones, lithophagid borings and bored pebbles occur. Oysters locally form in-situ mounds. The surface of the carbonates with borings and oyster growths marks a surface of discontinuity that probably formed a hard-ground.

The dense carbonates in thin sections appear to be greyish to cloudy micritic, showing very fine undulating laminations of darker and lighter material. Laminations may show slight arching and may also be broken and discontinuous probably due to desiccation and/or burrowing. Dark clotted and "pelleted" patches also occur. Angular to subangular quartz sand occur in small patches, lenses or stringers and may also form undulating bands associated with fecal pellets. Some microspar may be present. Cross-sections of roundish thin to slightly dense, thick-walled borings are filled with sparry calcite. Some borings have fillings, recrystallised to pseudospar. The sandy carbonates may have 30-40% coarse to very coarse, angular to subangular quartz sand. Irregular patches of blocky sparry calcite may be burrow fillings. Pelecypod fragments are neomorphosed. Other bioclastic fragments also occur. Microsparry patches occur in muddy matrix.

The sedimentary-faunal associations and the facies setting indicate a coastal-marginal environmental complex close to an estuary with bordering tidal swamps and intertidal bay. The discrete bodies of weakly laminated dense carbonates (stromatolitic)[4] were probably formed due to microbial sedimentation[1] where microbial mats growing in tidal pools in an intertidal environment, were binding and trapping detrital sediments. Quartz sand influx occurred probably at a slow rate. The microbial-complex was probably lithified early as indicated by borings.

In Rambha region,[3] on the southwestern margin of Chilka Lake, loosely distributed carbonate nodules and relatively more continuous, irregularly protruberating sheet-like dense carbonates of ca. 2-3m thickness are found closely associated with argillaceous siliciclastics with some quartz sand and very fine to fine gravels at a depth of about 3m in the subsurface. The deposits occur between the high-water line of Chilka Lake and foothills of the Pre-Cambrian Eastern Ghats. Petrographic characters have been reported.[3]

The carbonates probably represent microbial lithostromes that formed during the Holocene in very shallow waters of the lake margin. Calcification around microbial community was biologically associated in some way. Microbial growth and associated carbonate precipitation were probably controlled by seasonal changes in supply of siliciclastics and water chemistry. Salinity might have fluctuated seasonally.

Pelecypod and gastropod shell accumulations occurring nearby in sandy-argillaceous deposits in subsurface, in a vast marginal lacustrine-nearshore tract without any carbonate nodules/sheets may indicate that benthic grazing pressure was probably an important factor in inhibiting microbial growth.

The Neogene and Holocene occurrences of microbial carbonates show dominance of trapping-binding and biologically influenced carbonate precipitation process respectively.

REFERENCES

(1) Awramik, S. and Margulis, L., 1974, Definition of Stromatolite : Stromatolite Newsletter (unpublished), V. 2, P.5.
(2) Burne, R.V. and Moore, L.S., 1987, Microbialities : Organo-Sedimentary Deposits of Benthic Microbial Communities. Palaios, V. 2, P.241-254.
(3) Mohanti, M., 1988, Carbonates of possible microbial origin on southwestern margin of Chilka Lake (Rambha Region, Orissa), East Coast of India. Benthic Microbes and Reefs, Discussion Sessions, 6th Int. Coral Reef Symp., Townsville, Australia, Abstracts, P.5.
(4) Riding, R., 1988, Classification of Microbial Carbonates. Benthic Microbes and Reefs, Discussion Sessions, 6th Int. Coral Reef Symp., Townsville, Australia, Abstracts, P.5.

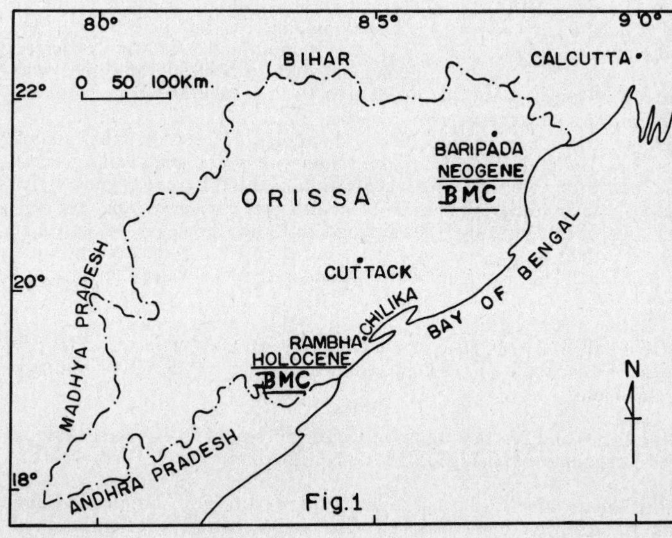

Fig.1

MONJARET, M. C., H. BELLON, GIS Océanologie et géodynamique, Université de Bretagne Occidentale, France; MAILLET, P., ORSTOM, IFREMER, France

Petrology and K-Ar Geochronology of the New Hebrides Back-Arc Troughs. Relationships with the Arc Volcanism

The SEAPSO (Leg 2) cruise of the R/V Jean-Charcot was devoted to a geophysical survey and a geological sampling (dredges located in figure) of the New-Hebrides back-arc troughs. A petrological and geochemical study together with a detailed K-Ar geochronological analysis of the dredged samples allow to compare the nature and the timing of the volcanic activity in the back-arc troughs and on the island arc.

Geochemical affinities of the dredged lavas — Basic lavas (i.e. basalts and basaltic andesites) can be classified into 6 groups, according to their contents in K_2O, Rb, Ba, Sr; Mgo, Cr, Ni; TiO_2.

Type a/Mid Ocean Ridge Basalts (MORB), with low K_2O (0.09-0.25%), Rb (2-3 ppm), Ba (10-70 ppm) and Sr (150-180 ppm); relatively high TiO_2 (1.25-1.55%) MgO (7.25-9%), Cr (230-300 ppm) and Ni (130-190 ppm).

Type b/Island Arc Tholeiites (IAT), with medium K_2O (0.5-0.9%), Rb (6-12 ppm), Ba (65-300 ppm), Sr (200-500 ppm), TiO_2 (<0.85%), MgO (<7%), Cr (<100 ppm) and Ni (<40 ppm).

Type c/IAT which are slightly enriched in TiO_2 (0.85-1.25%).

Type d/Ankaramitic IAT, i.e. enriched in MgO (7-11%), Cr (200-350 ppm), Ni (100-275 ppm).

Type e/Calc-alkali basalts, enriched in K_2O (0.95-1.9%), Rb (19-30 ppm), Ba (100-650 ppm) and Sr (250-550 ppm).

Type f/ Basalts with geochemical characters intermediate between MORB and IAT. 3 groups of acid lavas (acid andesites and dacites) have been dredged in the New-Hebrides back-arc troughs:

Type g/Low-K acid lavas (K_2O<0.95%) with high TiO_2 (0.75-1.80%), which resemble primitive island arc (PIA) tholeiitic volcanics.

Type h/K-dacites (K_2O: 2.95-3.70%; TiO_2: 0.55-0.65%).

Type i/High-K dacites (K_2O: 4.6-6.25%).

Geochronological and geochemical evolution of the back-arc troughs volcanism — 3 specific groups of volcanics occur in the back-arc troughs: MORB-like basalts (a), intermediate basalts (f) and low-K acid lavas (g). All other groups correspond to orogenic lavas, either tholeiitic [groups (b), (c), (d)] or calc-alkaline [(e), (h), (i)], which are found in the troughs as well as on the islands. However, lavas from the troughs are preferentially enriched in TiO_2, and/or Mg, Cr, Ni, and the proportion of tholeiitic versus calc-alkaline lavas is greater in the troughs than on the islands. MORB-like lavas are limited to the northern troughs (Vanikoro and Hazel Holme areas), and show quite old K-Ar ages (13-12 Ma and 6-5 Ma). They presumably correspond to remnants of the North Fiji Basin crust. Basalts with chemical characteristics intermediate between MORB and IAT only occur in the Vanikoro area between 2.6 and 1.25 Ma. They are followed by a low-K acidic volcanism, resembling PIA tholeiites, between 1.75 and 9.3 Ma, which may mark the onset of the island arc volcanic line in this area. K- and high-K dacites are found in the central and southern areas (Hazel Holme, Vate and Erromango troughs, Vate and Tanna islands) and particularly in the Vate (southern) zone (island and trough). Their K-Ar ages range between 3.5 and 2.25 Ma. Younger volcanics usually are less-K enriched. However, high-K dacites have been dated at 1.5 Ma on the Vate island, and at 0.5 Ma in the Vate trough.

This geochemical and geochronological analysis leads to the following evolutionary scheme: the formation of the northern New-Hebrides back-arc troughs began at least at 2.75 Ma in Vot Tande area and at 2.6 Ma in Vanikoro area. The opening of the southern troughs is earlier, i.e. older than 4 Ma in Erromango trough and perhaps a little more recent in Vate trough. This fact agrees with the greater structural maturity of the Coriolis troughs where a multistage opening is probable. Most of the volcanics dredged in the New-Hebrides back-arc troughs present chemical similarities with the islands volcanic rocks. K-Ar ages of 6.5-6 Ma in the southern troughs, and of 4.8 Ma in the northern ones, obtained an orogenic lavas, correspond to the first phases of the arc building. A second — and quite general — volcanic phase began around 4 Ma. The Vanikoro trough (the northernmost trough), however, corresponds to a unique feature of the New-Hebrides back-arc area, i.e. a possible initiation of a volcanic arc building phase.

Finally, the great variety of K-Ar ages and geochemical affinities of the volcanics replaced and compared within their geochronological framework dredged in the New-Hebrides back-arc troughs reflects one of their most striking features, i.e. their volcano-tectonic unstability, still effective today.

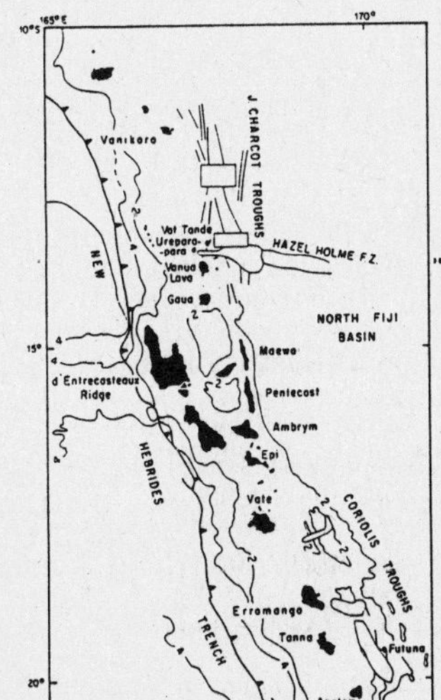

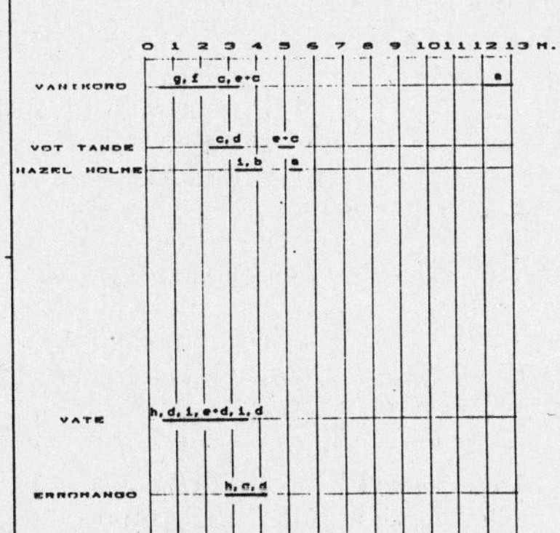

MOSE, DOUGLAS and George MUSHRUSH, Center of Basic and Applied Science, George Mason University, Fairfax, VA

Winter to Winter Variations of Indoor Radon in Virginia and Maryland

Indoor radon concentrations using three-month alpha-track radon monitors in a study of 1500 homes show that in Virginia and Maryland, indoor radon normally is at its greatest concentration in the winter and lowest in the summer. However, the winter of 1987-88 was warmer and dryer than normal, and the median radon was 25% lower than the median radon for the prior winter. The winter-to-winter decrease more closely simulates winter-to-summer variations, and can be demonstrated using data subsets based on geological units, home construction and home heating system.

Variations of this magnitude between winter seasons, which have been assumed to be the time of greatest indoor radon concentration, casts concern on on-going studies designed to evaluate regional radon potential. Several problems develop: a. If indoor radon measurements are used to develop radon potential maps for adjacent states, the radon potential for in fact similar areas may appear quite different if the radon measurements are made in different years. The difference may be as great as would develop if the measurements were made in different seasons. b. If soil parameters currently being used to establish radon potential maps are as variable as indoor radon, radon potential maps based on these parameters will contain inherent inaccuracies related to weather changes between seasons or between years. c. Estimation of annual radiation dose related to radon and its progeny may vary considerably from year to year, particularly for the winter when homes are most often sealed against the weather.

Studies currently in progress indicate that radon in drinking water may show similar variations in concentration, for homes using private water wells. However, comparisons between water radon and cancer show a higher cancer rate than comparisons between indoor (i.e., room atmosphere) radon and cancer. Emphasis may change from studies of indoor radon to studies of radon in water, since even in urban areas such as are found in Virginia and Maryland, more than 10% of the homes use private wells.

MURTY, K. S., University Dept. of Geology, Law College, Compound, Nagpur India

Natural and Man-Induced Geological Hazards: India

Natural hazards that India faces every year include earthquakes, landslides, floods, cyclones and droughts. Of these, earthquakes and landslides can also be man-induced, while he is also indirectly responsible for floods. At least forty earthquakes have hit India on a large scale, causing loss of life and damage to property. In mountain regions such as the Himalaya and the Nilgiri, the problem of hillside instability has become very serious. The Darjiling Himalaya, for example, recorded more than 20,000 landslides in one day! More than 300 floods occurred in India between 1963-1988. Some 250 cyclones have hit the east coast of India between 1891-1982 and 46 hit its west coast, killing millions of people and cattle. The country has been in the grip of severe drought for the last 4 years, eased to some extent this year due to a good rainfall.

Earthquakes and landslides have also been caused by human agency. Dams on rivers have been built for storage of water in their reservoirs, but filling in these reservoirs has induced earthquakes, the Koyna case being the most significant one in the last 20 years. Indiscriminate deforestation in the mountainous regions has caused loosening of the soil cover which, in turn, resulted in mass movements, and landslides in the Himalayan region are legion. India has hardly 21% of its area under forests, against 33% as stipulated in the official forest policy. Hazard-zoning maps have yet to be prepared so that preventive steps could be taken well in advance. A Natural Disaster Management Institute has been proposed to be set up. Its purpose would be to identify the hazard prone belts and deparcate on maps of the areas and belts prone to floods, cyclones and other natural hazards. A seismic map of the country has already been prepared, after taking into account the previous maps and subsequent seismic events, particularly in peninsular India which was for long considered to be aseismic, but not now after the Koyna, Bhadrachalam and other earthquakes.

NAZAROV, V.I., I.E. STUKALOVA, V.P. ALEKSEEV, Geological Institute, USSR Acad. Sci., Moscow, USSR

Lithology and Coal-Bearing Potential of Mesozoic Deposits in the South Yakutian Coal Basin

The South Yakutian coal basin is a series of isolated depressions in southern Siberian platform filled with Jurassic-Cretaceous terrigenous coal-bearing sediments. Coal-bearing stratum more than 3500m thick is made up by sandstones, gravelstones, conglomerates, aleurolites and argillites.

Sedimentary environments have been reconstructed, a cyclic structure of coal-bearing sequences analyzed, material-petrographic composition of the deposits and source areas of terrigenous material have been shown as well as zoning of catagenetic alternations of mineral and organic matter.

Genetic types and facies have been recognized of proluvial, alluvial and lacustrine-swamp sediments.

Peat bogs formed in mobile, relatively stable and stable peat swamps of river valleys and sapropelic lakes that are largely water-filled and stagnant. The valley and lacustrine-swamp types of peat accumulation predominate. Vast peat bogs that had transformed into very thick beds (up to 25 m thick) are related with the alluvial-valley type of peat accumulation.

Studies of the deposits' composition allowed to isolate 4 terrigenous-mineral associations. The first one is represented by feldspar-quartz sandstones and arcoses proper with few quartzite, granite and rarely dolomite fragments. This is a typical Lower Jurassic association.

The second association features greywacke arcoses, feldspar-quartz and quartz-feldspar greywackes with high contents of felsitic and quartzy porphyrys, granites, gneisses and rare trachytes. It is characteristic for greater part of the coal-bearing stratum dated by the Lower to Upper Jurassic.

The third association is typical of Upper Jurassic and Lower Cretaceous deposits in the Aldan-Chulmanian region of South Yakutian coal basin. It shows proper arcoses, greywackes with high content of quartzy and felsitic porphyrys, quartzites and granites.

The fourth association is mostly encountered in Lower Cretaceous deposits in the Tokinian region of South Yakutian coal basin. It comprises feldspathic greywackes and greywacke arcoses with high content of rhyolites, felsitic porphyrys, trachites and their tuffs.

To estimate catagenetic modifications of organic matter, a vitrinite reflectance parameter (R^O, %) was used, measured in oil immersion.

In the Aldan-Chulmanian region of the South Yakutian coal basin R^O of organic matter (OM) varies from 1.44% to 2.15%. In the Tokinian region it equals 0.67-1.15% and in the Usmunian region R^O is from 0.65% to 1.00%.

Terrigenous rocks are altered up to the deep catagenesis and initial metagenesis stages.

A subzone of lomontite development comprising three typomorphic autigenic mineral associations has been recognized within the deep zone of catagenesis.

The first association is mostly developed in Lower Cretaceous sediments in the Tokinian region. It includes the corrensite-like clay minerals, lomontite, chlorite and mixed-layer phase of the smectitie-mica composition with swelling packages content up to 10-15%.

The second association is represented by rectorite-like clay mineral, lomontite, chlorite and the mixed-layer phase of smectite-mica composition with swelling packages content up to 10-15%. This association is widespread in Upper Jurassic sediments in the Tokinian region as well as in Middle and Upper Jurassic strata in the Aldano-Chulmanian region.

The third association is represented by chlorite, lomontite and the mixed-layer phase of smectite-mica composition with swelling packages content up to 10-15%. It is developed in Upper Jurassic-Lower-Cretaceous sequences of the Aldano-Chulmanian region as well as in Middle and Upper Jurassic strata in the Usmunian region.

The subzone of lomontite development displays autigenic mineral associations related both with the terrigenous-mineral ones and with the genetic types of sediments.

Lomontite forms mainly in central parts of pores or substitutes other minerals such as plagioclase, hornblende, biotite and pyroxene. Among other authigenous minerals typical of lomontite subzone of deep-seated catagenesis are epidote, sphen, and quartz.

Vitrinite reflectance in specimens from lomontite subzone sequences varies from 0.65% to 2.00%. Taking into account that R^O% depends mostly upon palaeotemperature in rock-forming basin, we may conclude that formation of lomontite occurred at 100-175°C.

The initial metagenesis zone affects Lower Jurassic deposits whose main clay minerals are caolinite, dickite and mica of the IMd polytype modification. The organic matter R^O equals 2.15%.

The analysis of succession of deposits genetic types in a sequence and variations of terrigenous-mineral associations allow to reconstruct the palaeogeographic evolutionary features of the South Yakutian coal basin. Stage-by-state palaeogeographic reconstructions presented as a series of palaeogeographic schemes permit to reproduce the spatial and temporal evolution of basic palaeolandscape sedimentary and peat accumulation environments, to lay the basis for evaluation of coal-bearing prospects and the directions for coal exploration.

In Lower Jurassic South Yakutian coal basin sedimentation occurred in piedmont valleys associated with debris cone of prolluvial deposits. In Middle Jurassic period the sedimentation area had considerably extended within the vast alluvial plain environments replaced by lacustrine-swamp lowland by the close of the Jurassic. The period was associated with accumulation of numerous peat bogs.

In the Late Jurassic a large alluvial plain has formed, trending along the South Yakutian foredeep and passing into the pre-Verkhoyansk foredeep. The river had great meanders; the palaeovalley was being swamped. Isolated areas accommodated thick valley-type peat bog accumulations.

The alluvial-valley piedmont landscape had formed by the Early Cretaceous. Sedimentary area had reduced. Swamping had been confined to only the lower palaeovalley areas in the Tokinian region, where thick peat bogs were being accumulated.

Younger deposits are absent from the regions but the deposition evidently lasted for a certain period.

NEDJAI, R. and A. ISSAADI, IST/USTHB Labo d'Hydrogéologie. Alger, Algerie

Hydrogeological Study of the North Algerian Center Thermalism

The Algerain thermalism is caracterized by a high concentration of hot springs in the north, generally related to great fractures affecting deeply the sedimentary formations and sometime the basement. This fractures are often injected by triassic evaporites. There respective directions are NE-SW, NW-SE or E-W and they limit the great geostructural units. Two differents facies can be recognized:

• A high percentage of the springs, stakings out the great accidents have chlorured or sulphated facies and they are often hot (hyper or mesothermals)

• A second group related to the calcareous massifs or basement schow bicarbonate calcic or sodic facies. These springs are sometime carbogazeous.

The necessary warming up energy come principally from great geothermic reservoirs (Bibans) and secondary can be related to the geothermic gradient, exothermic reactions or/and seismic activity.

The reservoirs are carbonated or sandstone-liked.

So, the Algerian thermalism supports great analigies with that of the franco-italian alps, particulary for its relation with the overthrusting front limiting geostructural units.

NEUMANN, A. CONRAD, Hans A. PAERL, Brad M. BEBOUT, Linda R. McNEESE (Univ. North Carolina, Marine Sciences Program), Charlotte, NC, and Robert COMMEAU (USGS, Woods Hole, MA).

Modern Stromatolites and Associated Mats: San Salvador, Bahamas.

Stromatolitic growths of Mg-calcite are forming subtidally in several hypersaline lagoons on San Salvador, Bahamas. A study in Storrs Lake reveals that they range from locally thickened laminated crusts to club-shaped heads over a meter in diameter and up to 1 meter tall in water depths up to 2m deep. The surface of the sediment and the heads is everywhere draped by a dark green, cohesive, microbial mat. Heads in water depth less than 1m are larger, have a deeply pitted top, and protrude 10cm or more above the sediment surface, while those in deeper water protrude only a cm or more above the mat level and have a smooth, crenulate surface. The exterior of the heads is densely cemented while the interiors often crumble into loose stacks of mm-thick laminae and thick intervals of thrombolitic, non-layered growth. The top exterior few cm of all sampled heads exhibit thrombolitic structure while walls, bases and inner portions can be finely laminated. The enclosing muddy sediment is composed of carbonate, gypsum and organic matter. Measured salinities range from 60 to over 100 parts per thousand and may vary throughout the seasons. The ubiquitous microbial mat is made up of filamentous and unicellar cyanobacteria as well as anaerobic photosynthetic bacteria. Gradients of oxygen were measured by microelectrodes to be over 5 times atmospheric concentration during daylight in the upper 5 mm below which anoxic conditions abruptly set in. Corresponding pH in the high oxygen zone went as high as 10 and fell to close to 7 in the anoxic zone beneath. SEM photos reveal freshly precipitated rosettes of Mg-calcite on the small hard nodes that form along the mat/head contact. Elsewhere, the surfaces evidence episodes of dissolution indicating a complex and rapid alteration of deposition and dissolution across the calcified heads. It appears at present that these calcified heads are both thrombolitic and stromatolitic, and that they accrete by chemical precipitation at the mat/head contact. Foreign material incorporated into the stromatolites indicates some accretion by trapping but it appears to be a secondary process.

NIINI, HEIKKI, Helsinki University of Technology, Laboratory of Engineering Geology and Geophysics, Helsinki, Finland, and Eero SLUNGA, Helsinki University of Technology, Laboratory of Soil and Mechanics and Foundation Engineering, Helsinki, Finland

Landslides in Finland

This paper reviews the occurrence of landslides in Finland. It describes the geotechnical conditions, the causes, and the consequences of both natural landslides and those induced by human activities.

Large landslides are infrequent in Finland. Natural landslides are most frequent in the southwestern lowlands, where they occur along the valleys of rivers which cut into Littorina clay deposits. A typical feature of these clays is their high humus content that leads to cracking in drying and lowers the tension strength, diminishing slope stability.

Landslides have occurred commonly in connection with canal and other construction, for example, at the Saimaa Canal connecting Lappeenranta in SE Finland and Vyborg in the USSR and at the Kimola Canal east of Lahti. Slides of varying extent have occurred at half a dozen open pit mines, in some places causing material damage. Minor landslides are common in connection with any kind of massive construction on weak ground, especially in railway, road, and waste hill constructions. Also other human activities such as the vibrations of traffic, machinery, and blasting, the removal of earth materials and snow, the draining of waterlogged areas, and certain agricultural practices contribute to landsliding.

The surface of rupture in most of the Finnish landslides is composed of weak clay (shear strength often 10...20 kPa), although it may also be the surface of the bedrock.

The greatest number of landslides in Finland occurs in May — caused almost exclusively by the melting of the snow cover — and in autumn — due to high precipitation.

The economic effect of a landslide may be considerable, depending on the nature of the work in question. In recent years, the loss of human life in the natural landslides has been rare. Finnish legislation includes regulations concering ground stability, but too little interest so far is taken in creating and following any consistent national landslide policy.

NYGAARD, ERIK, Geological Survey of Denmark, Copenhagen, Denmark

Sequential Ichnology, a Supplementary Approach to Sedimentologic Interpretation of Chalk, Cretaceous-Tertiary, Denmark.

Close inspection of apparently homogegenous chalk reveals its dominating bedded nature. This may be obvious as where hardgrounds separate beds, or it may consist of diffuse rhythms Mortimore and Pomrol (1987). The beds indicate depositional pulses with intervening periods with reduced rate of or cessation of deposition. Also variations in the subordinate admixture of fine siliciclastic material may discriminate beds.

The endobenthic trace fossil producing fauna always pentrated the sediment within a close distance from the sediment/water interface. Therefore the traces reflect characteristics of the sediment itself, as well as the processes on the sediment/water interface.

In chalk sequences, the pauses in sedimentation seem to be equally influentive as deposition on the trace fossil assemblage. Thus the means to explore the concealed time intervals is to study the traces of the animals, which were active during these periods, e.g. Ekdale and Bromley (1984).

Presently three relations between the sediment, the sediment/water interface and the burrowing organisms, as reflected in characteristic tracefossils, may be established:

1. *Systems of open burrows*

These open permanently on the sediment surface. The responsible animals were dependent on a cohesive sediment and a stable position of its surface. This implies depositional hiati, very low sedimentation rates, or even slight erosion. It is suggested that these are the most, important requirements for the dominant *Thalassinoides* domichnia in the chalk.

Within the limited facies differences found in European chalk the lithology of the sediment appears to be less important for the formation of these traces.

The relative duration of the period in which the sediment surface conditions were favourable for the dwellers of *Thalassinoides* is suggested to be reflected in the diametre of its single branches. These diametres are very constant within the boxwork relating to a specific surface. Diametre differences occur in between boxwork horizons, which based on other criteria seen to be associated with bed surfaces that have been stable for deviating time spans.

2. *Endobenthic foodichnia*

These appear to have been produced with less dependance on the depositional conditions on the sediment surface as long as it was within fair reach. For the activity and presence of the casual animals the lithology, presence of nutrients, and oxygen content of the pore water seem to be decisive. Despite the relative independance of the processes on the contemporaneous sediment surface, the density and association of these trace fossils may still reflect responses to depositional mode. *Zoophycos* with up to 27 windings in sequence have been observed and are taken to indicate that the sediment accumulation has been slow. The uppermost windings may have dimensions less than 1 x 10 mm in cross-section. The dimensions increase successively downwards to 5 x $\geq$110 mm. The conclusion in these cases is that the burrowing animal grew while the trace was produced, and that the animal was not forced to move laterally. Where Zoophycos is recognized with only few windings the animals are assumed to have been disturbed by depositional events causing them to relocate themselves.

The common foodichnion *Chondrites* maintain a constant diametre within all branches of a "cluster". Assuming that the animal producing *Chondrites* grew bigger during its life time (as do most organisms), the constant diametre implies that the animal shifted position regularly through its life. This is supported by the presence in the same levels of clusters having differing constants of branch diametres. As mobility therefore in this case seems to be a common behaviour, the responsible animal is assumed to have been more tolerant to e.g. rapid and episodic deposition than the producer of *Zoophycos.* This tolerance is supported by other sedimentological observations, such as the relation to anoxia in the intertitial water, Bromley and Ekdale (1984).

3. *Pressumed fugichnia*

These consist of vertical shafts that are known sometimes to be very densely spaced within individual beds, whereas in other beds trace fossils with vertical orientation are practically absent. The density of these vertical shafts is such that they are frequently obtained also in drilled cores despite their crosssection of less than a square cm. In key examples these beds contain primary lamination in their lower part and hardground with associated *Thalassinoides* along their upper limit, while the median part is devoid of other trace fossils. This context supports the interpretation of the vertical shafts as fugichnia from animals, which responded to rapid deposition.

Most conspicuous among the fugichnia are *Bathichus* with single or paired vertical shafts. Traditionally this trace fossil is interpreted to reflect the behaviour of an animal, which survived several episodes of rapid deposition. This is indicated by a.o. the continuity of the diagenetic flint paramoudras around the shafts. These are frequently observed (albeit slightly offset) to continue through several beds each of which are m-thick. Considering the great density of vertical shafts it is suggested that the diagenetic effects picked a track along a closely positioned set of genetically unconnected vertical shafts. Therefore the individual animals need not to have survived more than one depositional episode.

An unbranched vertical trace fossil with diametre ca. 1 mm is called *"Trichichnus".* The dimensions of this trace compare to the dimensions of the internal central trace in the core of the *Bathichnus* shafts, and these may stem from the same precursor.

The interplay of the above characteristic trace fossils with one another and with the mixed *Planolites* facies, and adding sedimentary structures of physical origin; is the basis for an integrated scheeme of sedimentologic interpretation.

REFERENCES

Bromley, R.G. and Ekdale, A.A., 1984: *Chondrites:* A Trace Fossil Indicator of Anoxia in Sediments. Science, Repr. Ser. Vol. 224, pp. 872-874.

Ekdale, A.A. & Bromley R.G., 1984: Cretaceous Chalk Ichnofacies in northern Europe. Geobios Mem. Special no. 8, pp. 201-204.

Mortimore, R.N. & Pomrol, B., 1987: Correlation of the Upper Cretaceous White Chalk (Turonian to Campanian) in the Anglo — Paris Basin. Proc. Geol. Ass. 98 (2), pp. 97-143.

O

OLSON, EARL P., USDA Forest Service, Intermountain Region, Ogden, Utah

Thunderbolt Mountain Slope Failure (Wyoming-Utah)

Location:

Thunderbolt Mountain slope failure is the largest slope failure in the transition between the Green River Basin and the Central Rocky Mountains. The slope failure is J shaped in plan view. The largest portion of the slope failure is found in Uinta County, Wyoming, and the smallest portion in Summit County, Utah. The Wyoming portion is found in T. 12 N., T. 13 N., and R. 115 W., and R. 116 W., and is the largest slope failure in Wyoming.

The late Pleistocene failure mobilized the slopes of a strip of pediment land that dips north toward the Green River Basin. This strip of pediment land is like an elongate butte, but not detached. The pediment portion of the slope failure is about 6.6 miles (10.6 kilometers) long and about .9 miles (1.4 kilometers) wide and forms a tableland. The failure mobilized three side slopes off the tableland. The highest elevation is the southern end at about 9,700 feet (2949 meters). The north end is the lowest elevation at 9,320 feet (2840 meters).

Format:

The stem of the plan view is opposite Blacks Fork on the west and is labeled the west limb. The lip of the plan view is opposite Smiths Fork on the east and is labeled the east limb. Both Blacks Fork and Smiths Fork were glaciated. Of especial interest at the north end is the landform called Buck Fever Ridge. This prominent form established the name of the northern 7½-minute topographic quadrangle. It is recognized as a head ridge, a new term for slope failure anatomy. It was formed by opposed cycloidal extension which left a symmetrical knife-edge ridge about 1,600 feet (488 meters) long.

The format is cycloidal extension away from the tableland pediment. There is no uniformity in the extension segments either in width or length. However, from a third to a half the distance downslope, including the terminus is plastic flow format (earthflow of U.S. authors).

The original mapping was completed on colored air photographs. The mapping was transferred to orthophoto quadrangle maps by machine and visual methods by Geometronics group, Division of Engineering. The mapping was then transferred to a topographic master map of the same scale as the orthophoto map. High altitude black and white photographs were also used. All planimetric measurements used in this report were taken from the topographic base maps. Measurements on high altitude photographs differed from topographic mapping only in the third decimal. Volumetric measurements were made without benefit of drilling.

Dimensions:

Total overall dimensions:

9116.7 acres = 3688.6 hectares

14.24 square miles

1.43 billion cubic yards = 1.09 billion cubic meters

These dimensions establish that this cyclopean (a cyclopean failure is anything larger than one billion cubic yards = 760 million cubic meters) failure is equivalent to or exceeds all cyclopean slope failures in the Western U.S. not mobilized by volcanic processes.

The Wyoming portion is 79 percent of total failure. The Utah portion is 21 percent of the total failure.

Utah dimensions:

1910.75 acres = 910.2 hectares

about 3 square miles

300 million yards = 229 million cubic meters

There are more than ten slope failures in Utah larger than this portion.

Geology:

The drainageways off the Uinta Mountains are crudely aligned in a north-south direction. Their depth varies from about 500 feet (152 meters) near the periphery of the mountain mass to about 1,700 feet deep (518 meters) near the crest of the mountains. The Uinta Mountains are the largest east-west mountain range in the contiguous United States.

The Bishop conglomerate and its eroded equivalents caps the pediment tablelands on the north side of the Uinta range (see Bradley in references). The Wyoming portion is underlain by the Bridger formation. The formation is made up of marlstone, tuffaceous mudstones, and some sandstone. The formation disaggregates with moisture and commonly forms badlands slopes. The exposed rock units are all Tertiary in age. The Utah portion is buried by glacial debris in many areas.

History:

There is no evidence of ice erosion on the pediment tableland. It is assumed that late Pleistocene snow fields provided the excessive moisture to mobilize the slopes below the tableland surface. Other slope failures have been mapped in the immediate vicinity. Tableland slope failures can be seen north of the mapped area.

The glacial ice both east and west of the slope failure may have kept the local environment cool in late Pleistocene time.

There are eight smaller slope failures touching the cyclopean mass. Four of these are active failures in the contemporaneous environment (moisture per year 20 to 24 inches = 508 millimeters to 609 millimeters). With the eight adjacent slope failures, the late Pleistocene volume exceeded 1.5 billion cubic yards = 1.15 billion cubic meters and covered over 15 square miles - 3,885 hectares.

REFERENCES

Bradley, W. H. 1964. Geology of Green River Formation and Associated Eocene Rocks in Southwestern Wyoming and Adjacent of Colorado and Utah, U.S. Geological Survey, Professional Paper, 496-A, pp A1-A86, three geologic maps.

Olson, Earl P. February 1988. 21 Hazard Analysis Maps, North Slope Uinta Mountains. USDA Forest Service, (available at Supervisor's Office, Wasatch National Forest, 8230 Federal Building, 125 South State Street, Salt Lake City, Utah 84138).

U.S. Geological Survey. 1971. Topographic Sheets, Bridger Lake Quadrangle, Buck Fever Ridge, 1/24000, 7½-minute series.

ORDOÑEZ, S. (1), A. G. MEJIAS (2) and J. MENDUIÑA (3)
(1) Dpto de Petrología y Geoquímica. Fac. C. Geológicas.Univ. Complutense. Madrid. Spain. (2) REPSOL. (3) Instituto Geológico y Minero de España.

Stratigraphic Position of Karstic Bauxite Deposits Northeastern Spain

Bauxite deposits in northeastern part of Spain may be classified as Mediterranean Type Bauxites (S.H. PATTERSON et al, 1986). Some recent investigations accomplished by the I.G.M.E. (Geological and Mining Institute of Spain), allow us to stablish the stratigraphic position of spanish bauxite deposits and bauxite clay deposits.

These deposits are located as follow: 1) North of Lerida, *Southpyrennees Zone,* 2) in the coastal area of Catalonia, between Barcelona and Tarragona, *Catalanides Zone,* 3) in the Ebro area between Tarragona and Teruel, *Maestrazgo Zone.*

The bauxite deposits of Southpyrennees Zone are outcropping along 70 Km at a narrow band with an orientation mainly SW-NE. The morphological type of deposits varies from small lenticular bauxite bodies to stratiform deposits, in both cases filling a paleokarstic surface developped on middle jurasic dolostones. The bauxite ores are allways covered by senonian sandstones. The mineralogical composition is mainly boehmite, kaolinite, hematites, rutile or anatase, and lithofacies are commonly of "Bean" type.

The bauxitic deposits of Catalanides Zone are located probably in two stratigraphic horizons. The bedrocks of the early stratigraphic horizon corresponding to dolostones of rhaetian age and the cover terrigenous sediments of senonian age. The bedrocks of later stratigraphic horizon are laminated dolostones of upper senonian age and are covered by foramminiferal limestones of eocene age. Indeed, it has been located some reworked bauxite deposits. All these bauxites show some textural and mineralogical aspects to those observed in bauxites from southpyrennees zone.

Stratigraphical position of the bauxites and bauxite clay deposits in Maestrazgo zone are more complex, because they involve a NS stratigraphic zonation, so there or more stratigraphic horizons may be stablished in South Maestrazgo Zone. The earliest horizon is on a karstic surface developped on upper jurassic dolostone, and is fossilized by sediments of barremian age. The upper bauxitic clay horizons (two or three) are located within barremian sediments, and their bedrock is allways a karstified continental charophytas limestone. Their mineralogical composition consists mainly of kaolinite and hematite, and their lithofacies belong to bean type. With regard to North Maestrazgo Zone, it is possible to stablish two stratigraphic horizons of bauxite accumulation. The earliest one is the most important. It belongs to the small lenticular type and it fills a developped karst surface on middle-upper jurasic dolostones and limestones, and it is covered by detrital deposits of upper albien age. Post folding and overthrusting events reactived karstic processes leading to later transformations of bauxitic clays in a high quality-gibbsitic bauxite. Fire clays and lignite bearing deposits of lower and middle albien age may be related with these karstic bauxites.

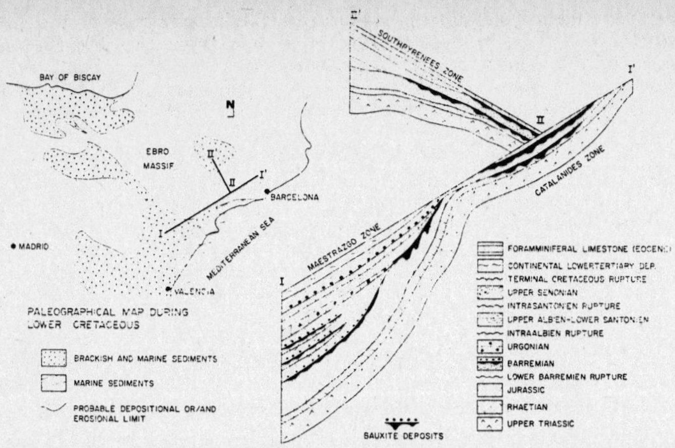

ORDOÑEZ, S. (1), M. BUSTILLO (1), R. FORT (2), and J. MENDUIÑA (3), (1) Dpto.Petrología y Geoquímica.Fac.Geología.Univ.Madrid.Spain. (2) Inst.Geología Económica.C.S.I.C.Madrid.Spain. (3) Inst.Geológico y Minero.Madrid. Spain.

Trace Element Geochemistry of Karstic Bauxites from Northeast Spain

Karstic bauxites from Northeast Spain are located in three different zones: Pyrenees, Catalánides, and Maestrazgo. All of them are karstic, Meditterranean sub-type, and similar to those situated at Central Europe. Pyrenees bauxites refill a dolomitic karst of Middle Jurassic age covered by Santonian sandstones. Catalánides bauxites refill also a dolomitic karst but from Rhaetian age. Reworked bauxites are very common in this zone. Finally, Maestrazgo bauxites are located in Upper Jurassic dolomites covered by terrigenous sediments of Upper Cretaceous age. A stratigraphic synthesis of these bauxites is presented at the Congress by ORDOÑEZ, MEJIAS, and MENDUIÑA.

Analytical data of major elements and mineralogical composition by X-Ray Diffraction are used to obtain a quantitative estimation of normative composition. The most important Al-mineral is the boehmite, being only present in some cases minor quantities of gibbsite. Clay minerals are kaolinite with various degress of crystalinity and very minor quantities of halloysite related to gibbsite. Other minerals in these bauxites are hematites, rutile, anatase, quartz, and calcite.

Trace element analysis (Zr,Ga,Cu,Ni,Mn,Cr,V,Y,Sr,Zn) were performed by using X-Ray Fluorescence technique with BX-N international bauxite standard. The results for the three different zones studies are presented in table 1.

Mathematical studies on trace elements reveal a close relations among different elements: Zn-Y-Mn-Ni, Zr-Ga-Ni-Sr, and V-Cr. These results are very similar to those obtained by Schorin and Puchelt (1987) in lateritic bauxites. According to Ozlu's model (1983), Northeast Spain bauxites are derived from a source rock of alkaline character. Factor analysis of trace elements shows that the parent rock for the three zones were the same. This conclusion is agree with palaoegeographic scheme of the region in Mesozoic age, which shows the three bauxite zones surrounding a same hinterland, the so-called Ebro Massif.

REFERENCES

OZLU, N. (1983). Trace-element content of "Karst Bauxites" and their parent rocks in the Mediterranean Zone. Mineral. Deposita, 18: 469-476.

SCHORIN, H. & PUCHELT, H. (1987). Geochemistry of a ferruginous bauxite profile from southeast Venezuela. Chemical Geol., 64: 127-142.

	I	II	III
	$\bar{X}$ -	$\bar{X}$ -	$\bar{X}$ -
Zr (ppm)	594-302	655-242	455-221
Ga (ppm)	81-43	99-36	65-32
Cu (ppm)	16-7	10-5	22-7
Ni (ppm)	176-87	120-38	163-121
Mn (ppm)	21-14	19-4	40-50
Cr (ppm)	316-58	311-57	215-37
V (ppm)	329-64	327-60	234-50
Y (ppm)	72-45	54-20	68-43
Sr (ppm)	173-75	632-280	377-183
Zn (ppm)	24-28	17-10	62-49
N	33	9	8

I = Pyrenees
II = Catalánides
III = Maestrazgo

N = number of samples

Table 1.

OUABADI, A. and B. SEMROUD, Institut des Sciences de la tierre, Alger, Algeria

Anatectic Peraluminous Granites from the Cap Bougaroun Area, NW Constantine (North Algeria)

In Cap Bougaroun (COLLO) located a hundred KM NW of Constantine, outcrop an intrusive granitic pluton of Miocene age (16 to 22 M.A.) that belongs to Maghrebide internal zone. This magmatic complex (granites and micrgranites) is limited to the South by the Kabile Oligo-Miocene, Cretaceous flyschs, the Numidian and the burdigalian marls; to the East by a crystallophillian serie. Beside a granulometric variation, these rocks are caracterized by biotite, cordierite, garnet, spinel and ilmenite association. Analysis of different mineral phases show that:

- Biotites have high iron and titan level and Fe/Fe+Mg ratio ranging from 0.42 to 0.63.
- Cordierites are Mg rich and present a Fe/Fe+Mg ratio ranging from 0.47 to 0.50.

Chemical composition to different encountered phases are comparable to peraluminous granites phases. Major element geochemistry shows a thout trend wi iron enrichement typical of calc-alkaline series. These rocks low in ferromagnesian are caracterized by normative corundum >2% and A/CKN >1.1. Magma crystallization occur at approximately 700 C, 23K bars and low oxygen fugacity.

Petrographic, mineralogic and geochemical data show that Cap Bougaroun granitoids derive from aluminous metasediment rocks and might belong to "S" type granites (Chapell and White, 1974). All these points allow to distinguish these granitoids from other tertiary plutonic rocks of Northern Algeria and are in favor of an anatectic origin.

P

PERINÇEK, Dogan, *Orhan DURAN, Dogan ŞEMŞIR, Ismet SEZGIN, Turkish Petroleum Corporation, Ankara-TURKEY

Stratigraphy and Sedimentology of the Midyat and Silvan Groups and Their Possible Hydrocarbon Potentials, Southeast Turkey

This study consists of Stratigraphic and sedimentological studies of the Midyat and Silvan Groups ranging from the Late Paleocene to the Early Miocene in Southeast Turkey. The area studied covers around 120.000 square kilometer.

The contact relationship of all the units, their typical localities, stratigraphic positions and extentions, facies and microfacies features, sedimentary depositional environments and modelling, diagenetic and reservoir characteristics have been examined and defined.

Effective transgression took place during the Middle Eocene and marine waters occupied the complete area resulting in the disposition of carbonates ranging from tidal flat to slope/basinal environments.

The transgression was followed by the regression which occurred during the last stages of the Late Eocene leaving some remarkable deposits of shallower carbonate facies. Due to the regression, the carbonates deposited in the north of the study area became exposed, effective from the last stage of the Late Eocene and marine waters gradually withdrew from the north to the south.

The results obtained from this study have revealed that the Midyat Group was terminated as a regressive sequence upon which the Silvan Group was lain transgressively and that the two seas depositing the Midyat and Silvan Groups are thought to be two different seas.

The stratigraphic and sedimentological results and the depositional history in the region have been compared with the tectonic events and have been found comformable.

The carbonates of the Midyat Group have fair to good reservoir capacity whereas the carbonates of the Silvan Group's capacity are poor to fair. Fracture contribution is, however, expected to have developed beneath the thrust belts in the northern portion of the study area which might have improved the reservoir quality. The Ashmari Limestone in the Middle East is equivalent to the carbonates of the Silvan Group and has had a very important hydrocarbon potential.

PICARD, CHRISTIAN and Donald M. FRANCIS, Mineral Exploration Research Institute, Montréal, Qué, Canada. Department of Geological Sciences, McGill University, Montreal, Qué, (Canada)

Rare Earth and Trace Element Behaviour During Successive Stages of Rifting of the Southern Ungava Trough, New Quebec

The southern part of the early Proterozoic Ungava trough (1960 to 1922 Ma) in northern Québec, resulted from the progressive opening of a rift in the Archean Superior craton by: 1) the creation of a continental fault trough into which shallow-water sediments accumulated forming the southern part of the Povungnituk Group; 2) the formation of an ensialic prot-rift into which continental tholeiitic basalts (MgO<10%, Ti02=1.2-3.6%) of the northern part of the Povungnituk Group were emplaced; 3) the opening of an oceanic rift accompanied by the emplacement of successive sequences of komatiitic olivine-phyric basalts (Mg0=19-11%, Ti02<0.9%), the latter evolved to tholeiitic pyroxene-phyric basalts (Mg0-12.5-7%, Ti02=0.8-1.1%) in the southern part of the Chukotat Group; and 4) the formation of an oceanic crust characterized by N MORB plagioclase-phyric basalts (Mg0<8%, Ti02=1.5-2.8%) in the northern part of the Chukotat Group.

Rare earth and trace element distributions indicate that the Povungnituk plagioclase-phyric basalts were formed by partial batch melting (F=10-20%) of an enriched spinel Ihezolite mantle source, followed by crystal fractionation of plagioclase, pyroxene, olivine and ilmeno-magnetite in magma chambers. The most primitive liquids of the successive Chukotat olivine-phyric and pyroxene-phyric basalt sequences were formed by successive melting processes (F=20-35%) from a increasingly LRR depleted spinel Iherzolite mantle source. In each of the southernmost sequences, the rare earth and trace element behaviour of the olivine-phyric basalts reveal that the successive lavas were produced by progressive melting from the same mantle source.

During the ascent of basaltic liquids, olivine cristallied in feeder systems, producing periodotitic intrusions and olivine-phyric basalts. Subsequently, olivine and pyroxene fractionation in the last liquids produced pyroxene-phyric basalts in the upper part of the sequences. Finally, plagioclase phyric basalts were formed by a new phase of partial melting ($F \geq 15\%$) from the depleted mantle source, and by crystal fractionation of plagioclase, pyroxene ± olivine in new magma chambers.

Thus, successive basaltic sequences evolving from continental to oceanic type rifting of the southern Ungava trough resulted from successive melting processes of a Iherzolite mantle source, which became more and more LREE depleted with time.

POCOVI, A.[1]; M. LAGO[1]; J. BASTIDA[2]; D. ZACHMANN[3] & R. VAQUER[4].
[1]Department of Earth Sciences. University of Zaragoza. Spain. [2] Department of Geology. University of Valencia. Spain. [3]Institute of Geology and Paleontology. Technical University of Braunschweig. F.R. Germany. [4]Department of Petrology and Geochemistry. University of Barcelona. Spain.

The Alkaline Magmatism in the Triasic-Liassic Boundary of the Iberian Chain (Spain): Emplacement Features, Petrological and Geochemical Characters. Rate of Spilitization.

Alkaline magmatism of Upper Trias - Lower Lias age outcrops in different zones of Iberian Chain and Nort Range in Mallorca (fig. 1).

As a whole, some explosive (cinerites, bombs, etc.) and lavic builds are found in Tarrgona (fig. 1B) and Mallorca (fig. 1C) zones, but they have not been observed in the North-eastern Iberian Chain (fig. 1A). Furthermore, dikes and sills, with different degrees of spilitization - alteration processes appear in the Tarragona and Mallorca zones too.

In the North-eastern Iberian Chain (fig. 1A) the alkaline manifestations are only constituted by very alterarted - spilitizated hipovolcanic sills.

Peridotite xenoliths (olivine, pyroxene and spinel) are usually enclosed in these lavic bodies, and granitoid enclaves can be observed in aglomeratic type outcrops.

From a petrological point of view, these alkaline builds are olivinic, nefeline absent, basalts, with a low differentiation degree. They present a wide textural variety from porphidic microlitic (more or less vesiculated) to doleritic texture with some fluidal features. There is a petrographic zonation from border zones, olivine enriched, to central areas, pyroxene and plagioclase enriched over olivine proportions. Locally, pegmatoid bodies, mainly constituted by feldspar, plagioclase and Fe - Ti oxides (Ti-magnetite and ilmenite), with low proportion of pyroxene and without olivine.

Primary paragenesis are formed by olivine (Fo_{76-68}), zoned Ti rich augite (diopside - salite type) with $En_{41.5}$ $Fs_{10.3}$ $Wo_{48.2}$ to En_{36} Fs_{18} Wo_{46} compositions and Ti enriched during differentiation. Chemical composition of plagioclases varies from An_{73} - An_{62}.

Development of secundary processes (spilitization) leads to formation of albite, chlorite (clinochlore), carbonate, etc.

Geochemical analysis (ICP, XRF) appear to indicate the alkaline feature of these rocks and the Ti/V - SiO_2 relation prove the observed petrological differentiation. Spilitization processes originate important variations in the proportions of compatible and incompatible elements as, for example, Li (52 - 462 ppm). With regard to mobility, it must be pointed out: a) high to medium values for Rb, Sr, Ba, and Cu; b) low values for Sc, V, Co, Y, and LREE; c) very low values for Nb, Zr, Th, Cr, Ni, and HREE).

These basaltic bodies are located above Keuper materials and under Hettagien sediments. The existence of a conglomerate constituted by spilitised basalte fragments and other sedimentary rocks seem to confirm the proposed age.

The important lateral extension and small thickness of lava bodies indicate the existence of lava flows enclosed by unconsolidated wall sediments. This last feature can explain the low degree of the observed contact metamorphism.

Similar chemical composition and emplacement age of these lava bodies appear to indicate the existence of only one alkaline province associated to intracontinental rifting in the Tethys border, from French Alps to the studied zone.

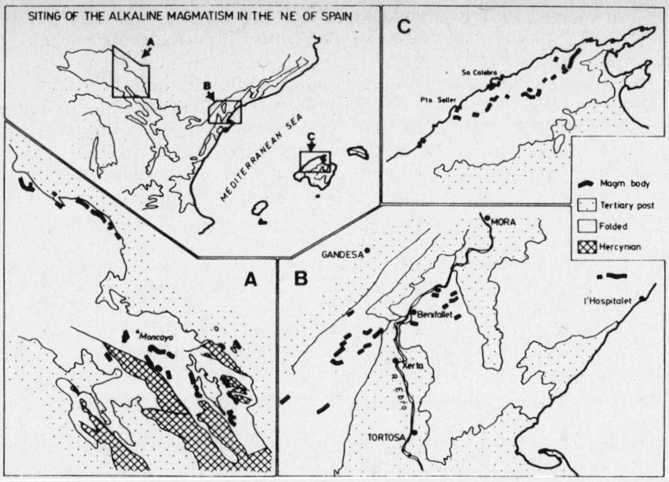

POYET, PATRICE* and Michel DETAY**,
*ILOG S.A., Gentilly Cédex — France.
*Université de Nice, Lab. de Géologie et Géochimie, Nice Cédex — France
**GEOLAB, Valbonne Cédex — France
**Institut Méditerranéen des Géosciences, Draguignan — France

Hydrolab: A New Generation of Compact Expert Systems for Village Water Supply

During the last twenty years, African countries have been improving the water-supplies for their rural populations. The search for water has become crucial, and in 1988 more than 80 percent of the rural population in the developing countries are still without drinkable water. In the past twenty years hydrogeologists have carried out numerous studies which have led to the definition of well-location techniques for village water-supply programmes.

The primary importance of a drinkable water-supply, the existence of a large amount of statistical data that could be applied to the location criteria, the repetitive nature of the steps to be carried out for a hydrogeological search and the African technicians' training needs, all led us to develop a new kind of compact expert system. HYDROLAB embodies the major techniques used by experts in the field and draws on a considerable wealth of practical knowledge gained from numerous drilling campaigns that we carried out in Africa, particularly in North Cameroun.

The software engineering techniques we used and developped enabled us to create a very compact expert system operational on microcomputers and useful to geologists who are faced with the real problems of developing countries.

As far as the user is concerned, the system acts at the dialog level as a diagnostic-like expert system, asking questions, analysing user answers, and building plans in order to schedule the appropriate set of actions and to explore constrained parts of a large search tree. Interaction with the system is achieved through a natural language module, and we developed a wide set of self-explanation functionalities in order to use this system in computer aided learning, mainly for African technicians, in the water-supply domain.

A control panel is associated with the inference engine, in order to display the internal activity of the system at a high symbolic level. The user is informed in this way of the current goal and subgoal pursued by the system and of the branching factor of the search tree. The panel also shows the current position and the depth in this tree, the current rule fired from the conflict set and the number of solutions encountered so far during the run.

The aim of the system is to build a model of the user's problem and to try to match it initially, using a fuzzy logic, with a set of typical known well patterns that have emerged from well-drilling campaigns in Africa and are stored by HYDROLAB in a data-base-like structure.

When this reasoning step fails, the system builds a new plan and schedules it in order, first, to get more data related to information that is lacking, which could help to reach a solution. It then tries to rematch the currently-described situation with real solutions and finally with generic ones.

For all cases, the system is able to make a diagnosis about the user's situation, to give advice in order to increase the user's average success, and to evaluate the hydrodynamic characteristics of the future well, according to the experience based on the statistical processing of 1080 bore holes in Africa (Detay, 1987), (Bernardi *et al.*, 1988).

REFERENCES

Albert, P., 1985, Prolog et les objets: Proc. Fifth international workshop on Expert Systems and their Applications, Avignon, 1985, France, p. 332-350.

Arrivet, L. 1987, Introduction du flou dans un système expert: Mémoire d'Ingénieur de l'Ecole Supérieure d'Informatique d'Electronique et d'Automatique, Paris France, 44 p.

Bernardi, A., Detay, M, 1988, Corrélations entre les paramètres géoélectriques et les caractéristiques hydrodynamiques des forages en zone de socle, to be published in Revue de BRGM, 15 p.

Borland International, 1986, Turbo Prolog Owner's Handbook: Borland eds., 221 p.

Clayton, B., 1985, ART™ reference manual: Inference Corporation eds., Los Angeles, California, Vol 1, 138 p., Vol 2, 158 p., Vol 3, 102 p.

Colmerauer, A., 1977, Programmation en Logique du premier ordre: Actes des journées La compréhension, I.R.I.A., France, p. 124-132.

Colmerauer, A., 1983, Prolog in Ten Figures: Proc. of Eighth International Joint Conference on Artificial Intelligence, p. 487-499.

Clocksin, W.F., and Mellish, C.S., 1984, Programming in PROLOG: Springer-Verlag, Berlin, 304 p.

Detay, M. Goulnik, Y., Casanova, R., Ballestracci, R., Emsellem, Y., 1986a, HYDROLAB an Expert System for groundwater Exploration in Africa: in Bull. IWRA — International Conference on Water Resources Needs and Planning in drought prone areas — Khartoum Soudan, 6 p.

Detay, M., Goulnik, Y., Casanova, R, Ballestracci, R., 1986b, HYDROLAB un Système Expert pour la recherche des eaux souterraines en Afrique: Actes du deuxième colloque Int. Eau Gestion des données et aide à la Décision, Montpellier — France, p. 94-99.

Detay, M., 1987, Identification analytique et probabiliste des paramètres numériques et non-numériques et modélisation de la connaissance en hydrogéologie sub-sahélienne: Thèse de Doctorat d'Etat Sciences, Université de Nice — France, 456 p.

Forgy, C. L., 1982, A fast Algorithm for the Many Pattern — Many Object Pattern Match problem: A.I. Journal n° 19, 1982, p. 17-37.

Haren, P., Neveu, B., Corby, O., Montalban, M. 1985, MEPAR: Un moteur d'inférences pour la conception en ingenierie: 5 ème Congrès Reconnaissance des Formes et Intelligence Artificelle, Grenoble 27, 29 Novembre 1985 — France, p. 1273-1280.

Haren, P., Neveu, B., Giacometti, J. P., Montalban, M. Corby, O., 1985, SMECI: Cooperating Expert System for Civil Engineering Design: SIGART Newsletter, (April 1985), n° 92, p. 67-69.

Hayes-Roth, B., 1984, BB1: An architecture for blackboard systems that control, explain and learn about their own behavior: Heuristic Programming Project, Report n° HPP-84-16, Stanford University, 22 p.

Konolidge, K., 1979, An inference net compiler for the PROSPECTOR rule based consultation system: Proc. of Sixth International Joint Conference on Artificial Intelligent, p. 487-489.

Nii, P., 1986, Blackboard systems: the blackboard model of problem solving and the evolution of blackboard architectures. A.I. magazine, July 1986, p. 38-53.

Nii, P., 1986, Blackboard system, blackboard application system, blackboard systems from a knowledge engineering perspective. A.I. magazine, August 1986, p. 82-106

Poyet, P., 1986, Discrimination des anomalies géochemiques multi-élémentaires significatives: Thèse de Doctorat d'Etat ès Sciences, Université de Nice — France, 436 p.

Poyet, P., and De La Cruz, P., 1987a, Simulation navale en environment hostile: Proc. Ecole d'Automne sur l'intelligence Artificielle, THOMSON et Ecole Normale Supérieure d'Ulm, Jouy en Josas — France, Sept. 1987, 10 p.

Poyet, P., Haren, P., De La Cruz, P., 1987b, Un système expert de simulation navale: Proc. 6ème Congrès Reconnaissance des Formes et Intelligence Artificielle de l'Association Française de Cybernétique Economique et Technique, Antibes, France, p. 587-592.

Poyet, P., 1987c, La structure de contrôle dans les systèmes experts de simulation. Proc. 6ème Congrès Reconnaissance des Formes et Intelligence Artificielle de l'Association Française de Cybernétique Economique et Technique, Antibes, France, p. 723-738.

Poyet, P., and De La Cruz, P., 1988a, Une nouvelle classe de simulateurs destinée aux aides tactiques et aux systèmes d'armes: Proc. Eighth international workshop on Expert Systems and their Applications, Specialized Conferences devoted to A.I. and Defense, Avignon, France, p. 89-99.

Poyet, P., and Detay, M., 1988b, Aide à l'implantation d'ouvrages d'hydraulique villageoise: Proc. Eighth international workshop on Expert Systems and their Applications, Avignon, France, P. 397-410.

Poyet, P., 1988c, Cours de Modélisation Assistée par Ordinateur: Les problèmes, les langages, les machines et les techniques de l'I.A., Laboratoire de Géologie et Géochimie, Université de Nice, France, 78 p.

Poyet, P., Detay, M., 1988d, HYDROLAB Version V1.4, Un système expert d'aide à l'implantation de forages en hydraulique villageoise, Manuel d'introduction et de référence, INRIA Research Report, Institut National de Recherche en Informatique et en Automatique, Sophia Antipolis, France, 42 p.

Poyet, P., Detay, M., 1988e, Modélisation de la connaissance en hydraulique villageoise, to be published in Comptes Rendus de l'Académie des Sciences de Paris, Série II, France, 4 p.

Poyet, P., Detay, M., 1989a, Enjeux Industriels et Sociaux de l'Intelligence Artificielle en Hydraulique Villageoise, Convention I.A. 88->89, Editions Scientifiques et Techniques Hermès, Janv 1989, 19 p.

Poyet, P., Detay, M., 1989b, HYDROLAB®: An Example of a New Generation of Compact Expert Systems, accepted and to be published in Computers and Geosciences, an International Journal, Pergamon Press, 18 p.

SMECI, 1988, Le manuel de l'utilsateur SMECI V1.3: ILOG S.A. eds., France, 112 p.

Stefik, M., Bobrow, D. G., Mittal S., Conway, L., 1983, "Knowledge programming in LOOPS". A.I. Magazine (Fall 1983) p. 3-13.

Stefik, M. J., Bobrow, D. G., Kahn, K. M., 1986, "Integrated Access Oriented Programming into a Multiparadigm Environment." IEEE Software, January 1986, p. 10-18.

Warren, 1977, Implementing PROLOG, D.A.I. Research Rep. Vol 1 et 2, University of Edinburgh.

PULEC, MIROSLAV, Dionýz Stúr Institute of Geology, Mlynská dol.1, Bratislava, Czechoslovakia

Scheelite-Gold-Bearing and Antimonite-Gold-Bearing Association in Relation to Carbonaceous Formations in the Ore District of the Nízke Tatry Mts.

Within the West Carpathian metallogenetic region of the Carpathian metallogenetic province, from the viewpoint of the presence of deposits suitable economically, the Nízke Tatry ore district is of considerable importance. It is characterised by polymetallic and polycyclic development in the pre-Hercynian, Hercynian and Alpine metallogenetic epoch. The spatial distribution of ores is directly dependent of the complicated geological structure invigorated by several-phases folding, subsequent formation of block structure and considerable tectonic disturbing. The ore deposits and geochemical anomalies have a zonal distribution and gold in them is of regular position. In the central zone situated in crystalline schists and running parallel with the boundary of the granitoid pluton and metamorphic mantle gold is bound to the quartz-scheelite association. Highest concentrations are in quartz-gold-bearing and quartz-scheelite-gold bearing veins. In two peripheral zones, lying northerly /in granitoids/ and southerly /in metamorphites/ from the central zone, gold is developed in the antimonite association. Besides that higher gold contents have been found in Devonian volcanic-sedimentary formations with higher content of carbonaceous substance. Overclark contents are also in migmatites where the average Au is 0,15 g/t. As the result of latest works show, gold accumulations were forming in the process of sedimentation /exhalation-sedimentary origin of Au pyrite/ in anoxyc environment and diagenesis /Proterozoic ?/ and redistribution of gold was taking place during Hercynian metamorphism and granitisation with formation of deposit accumulations. The occurrence of flaky graphite in neubulitic migmatites, gneiss /sillimanite metamorphic subfacies/ and granitoids, isotopic C composition of which indicates an organic origin, confirms this assumtion. Together with graphite pyrrhotite is found in gneisses, which formed by desulphidization of pyrite.

R

RANGIN, CLAUDE, Departement de Geotectonique: T.26-0 Université Pierre et Marie Curie, 4 place Jussieu 75230 PARIS FRANCE

Back-Arc Thrusting and Marginal Basin Opening, A Consequence of Compressive or Extensional Stress Regimes Across a Convergent Plate Boundary: The Case of the Sulu Sea: Philippines.

The Southern margin of the South China Sea extends from Northern Palawan Mindoro Island in the North-East to the Miri-Dangerous Grounds-Luconia Shelf in the South-West.

This continental margin is fringed southward by two parallel volcanic belts encompassing a small oceanic basin (SE Sulu Sea).

— Northwestward, the Cagayan Volcanic ridge is topped by early Middle Miocene carbonates and extends on land into the central Philippines where it was dated as Late Oligocene — Early Miocene in Tablas, Panay and Mindoro Island. This arc terrane was thrust onto the South China Sea Continental margin 15 Ma ago in Panay central Palawan, Mindoro Island and most probably in Sabah (NE Borneo).

— The Sulu Volcanic arc extending from Panay island to Sabah and western Sulawesi is mainly developped from Late Oligocene to late Middle Miocene, and is overthrusting northwestward the Cagayan volcanic arc on both sides of the SE Sulu sea.

Both Island arc terranes are built on a complex assemblage of Cretaceous-Eocene nappes including Early Cretaceous Ophiolites, a Late Cretaceous flysch sequence, a Late Jurassic olistostrome and metamorphic rocks. This pre-Oligocene basement which is also found along the southern continental margin of the South China Sea, belongs to the Eurasian land mass. Consequently the Cagayan and Sulu arc terranes can be considered as built on this continental margin, the SE Sulu Sea is interpreted as a Middle Miocene marginal basin, and the Cagayan ridge as the remnant volcanic arc for the Sulu arc.

If this interpretation is valid, the Cagayan ridge is back-thrusted on the southern continental margin of the South China Sea prior to opening of the SE Sulu Basin.

In Late Miocene time, exotic terranes forming the bulk of the Philippine mobile belt (Luzon, Visayas, Mindanao) collided with the Sulu arc, while in Sulawesi the same exotic terranes impinge upon the southern most branch of the Sulu arc (Western Sulawesi arc).

The complex history of this area documents the successive stress regimes present in late Paleogene-Neogene time at the Eurasian continental margin.

— Northwestward back thrusting of the Cagayan ridge at 15 My is coeval with intensive deformation also registered in the Sunda shelf north of the Sunda arc and testifies a regional compressive stress regime along the Eurasian margin.

— Opening of the Sulu Sea in Middle Miocene time testify a subsequent extensional regime preceding impingement of this active margin by exotic terranes in Late Miocene time.

Tensile stress regimes along a convergent boundary where the predominant driving force is the trench suction, are occassionally punctuated by compressive stress regimes acting on the whole arc area. These short periods of time are apparently independent of collision processes acting along the plate boundary. Such compressive regimes inducing back-arc thrusting are synchronous over wide areas in Asia, and could be coeval with plate kinematic reorganisations, and global sea level changes.

RANGIN, C.*; J.M. AUZENDE; J. AZEMA; R. BLANCHET; J.P. CADET; P. CHOTIN; P. deWEVER; H. FONTAINE; P. HUCHON; L. JOLIVET; M. LATREILLE; J. LETOUZEY; M. MAURY; C. MULLER; M. PUBELLIER; J.P. RAMPNOUX; J.F. STEPHAN; A. BRIAIS; J. TOURNON; H. WHITECHURCH

*Departement de Geologie Structurale T26-0 E1 Université Pierre et Marie Curie, Paris, France

Paleogeodynamic Reconstructions of the Western Pacific Region in Cenozoic Time: A Key to Unravel the Neotethys Pacific Oceans Relationships Through Time and Space

The Western Pacific Region is presently the place where three major plates (Pacific, Eurasian and Indo-Australian) are coverging, inducing collisional belts and opening of marginal basins mainly during Cenozoic time.

On the other hand various Mesozoic ophiolitic belts are present in this region and are interpreted as relics of former oceanic basins such as the Neotethys. With the objective of constraining the reconstruction of this Tethyan ocean and its margins, we propose a paleogeodynamic reconstruction of the Western Pacific region from 65 Ma to the present. Reconstructions are based on the global kinematic data, marginal basins magnetic anomalies, paleomagnetic data and continental geology gathered by our group in this region over the last 10 years. We selected key periods in order to describe major kinematic reorganisations and the beginning or end of marginal basin opening which is probably the best marker for such reconstructions. Marginal basins appear to open in a great variety of tectonic contexts the two end members being the Marianna basin where trench suction is the predominant driving force and the South China Sea where a pull apart mechanism is dominant. This allows us to demonstrate more accurately the presence of trapped marginal basins such as Celebes or the SE Philippines basins, that can be inherited from the Neothethys ocean basin and preserved as rigid crustal fragments in the collisional belts.

These Cenozoic reconstructions also unravel the possible relationships between the Neothethys-Indian Ocean and the Pacific. Early Cenozoic reconstruction of the Philippines Sea Basin indicates a complex intra-oceanic history for this basin and surrounding island arcs and very restricted pathways between The Tethys and the Pacific in late Mesozoic time. However Neothethys can be extended along the northern and eastern coast of the Australian Continent.

The reconstruction presented here will be used as a basis for Mesozoic reconstructions where kinematic constraints are less available.

RE KUHL, Guillermo, Argentina

Mineral Exploration in the Central Andean Range of Argentina by Means of Low Cost Digital Image Processing Techniques

The present applications in geology in Latin America of digital processing of satellite products did not reach the intensity that was expected in previous years based on former experiences in the United States and Europe.

This is due to technical and human inconveniences. The first ones are directly related with the hard economic situation that Latin America is passing through, which makes difficult the acquisition of the necessary hardware and software for this task. The human aspect is refered to the little diffusion that these techniques have had and that only a few specialists have been working in the various countries. This aspect has been undertaken by the Latin America Remote Sensing Specialists Society since 1980.

The challenge of solving the technical-economical inconvenience for the use of digtal image processing techniques was encouraged by the National Institute of Industrial Technology (I.N.T.I. in Argentina) since 1983; as a result of its investigations some programs were developed for conventional alphanumeric terminals that allow to any user to carry on geological studies at low costs.

This software was used in Argentina in the survey of hydrothermal alternation areas in the Andean Range, in the western provinces of San Juan and Mendoza. The results of this case is being discussed in the present paper.

REYES, L. (1), J. C. ROMERO (2)
(1) Eng Ph. D. Dirección Geología, MINBAS, Cuba
(2) Eng. Centro de Investigaciones Minero Metalúrgicas, UÑn

Geological Researches of Zeolites Deposits in Cuba

Introduction

Since the first discovering of outcrops zeolitized tuffs in Cuba (1) the geological investigations began. Geological research methodologies carried out in the beginning by several geologists from different parts of the world (2, 3, 4, 7) were founded in the utilization of low productive equipment. Detailed investigations performed in Central Cuba deposits (5), allow to establish an optimal methodology of researching zeolite ores for industrial uses. Cation exchangeable composition and its dependence of the surfave water chemistry of first was discovered (6). In the course of investigation were observed a new industrial genetical type of zeolitic deposits related with flysh sediments and the presence of clinoptilolite from basaltic glass in opposition with opinions of several researches (2).

Theory and Experiments

Statistical correlation between datum, which directly or indirectly determine the zeolite content: X-ray diffractometry, cation exchangeable capacity, adsorption, calorimetric methods, petrography and other close together with zeolite using properties like animal food adols, greenhouse soils, fertilizer, to prevent pollution, construction raw materials and others.

Results and Discussion

According to datum it is possible to infer that lithological control of zeolite content is demonstrated by the existence of an statistical correlation near numberoone between compared parameters. Cation exchangeable capacity better remarks the using properties.

Conclusions

The above mentioned datum enables us to infer the following:

1. To select the best quality and mining condition places of zeolite deposits.
2. To carry out in a very short time eight deposits per year, which allows us to get a rational distribution of zeolite ores deposits along the country.

BIBLIOGRAPHY

Alexeiev B et al. Las rocas zeolíticas al Sur de la ciudad de Sta. Clara, Las Villas, Geol. yac. minerales útiles de Cuba, No. 3, ACC, 1971.

Chelischev N. et al Zeolite new type of raw material, Moscouw, "Nedra", 1987.

Iijma A. Inter. Conf. on Zeolites, L.V.C. Rees, London, 1980.

Mumpton, F.A. In: Natural Zeolites, Occurrence, Properties. Use. Pergamon Press, 1978.

Reyes, L. et al. Tasajeras — Primer yacimiento con reservas industriales de zeolita de Cuba, XXVII Congreso Mundial de Geología, Moscú, 1984. Abstract.

Reyes L., et al. Ion composition change in the eweathering zone of zeolite tuffs from Tasajeras deposit in Central Cuba. Poster Inter. Conf. "Zeolit '85", Budapest, Hungary, 1985, Abstract.

Utada M. Zeolite zoning of the neogene puroclastic rocks in Japan. Sci. Pa. Coll. Gen. Educ. Univ. Tokyo, 1971.

RIAD*, Samir, Hakam MOUSTAFA**, and Hassan HOUSNEY***
*Geology Department, Assiut University, Assiut, Egypt
**Department of Earth and Environmental Sciences, Yarmouk University, Irbid, Jordan.
***Geophysics Department, Cairo University, Cairo, Egypt.

Recent Earthquake Activity and Neotectonics of the Dead Sea Rift Area, Jordan

The newly established Jordan Seismological Observatory represent an important source of the recent earthquake activity in the Dead Sea area, specially for relatively low magnitude events, that may have direct relations to present-day tectonic activity in this area. Through the net of telemetric stations distributed in the eastern side of the rift, a total number of earthquakes of 1488, with magnitude range (ML) between 1.4 and 5.1 was reported for the period September, 1983 through September 1986. Frequency-magnitude relation have been derived in the following form:

$$\text{Log}\ (N/\text{year}) = 6.85 - 1.375\ ML$$

The obtained relation seems to fit well with the pattern of historical earthquakes in the study area.

The geographical distribution of earthquakes shows, in many cases, a direct relation to geologically known surface faults. Earthquake activity follows, mainly, NW fault trends in the southern parts, whereas, in the central and northern parts it follows NE and E-W patterns and faults.

It is also noticed that the earthquake activity is not uniform with time for the whole area and for selected parts within it as well. This is attribited to local lithological and fault pattern characteristics in the different parts.

The relatively low earthquake activity reported for the western side of the Dead Sea Rift is probably related to major differences in the structure of the crust on both sides. Most probable that both sides of the rift belong to different plates the Sinai and the Arabian plates,with different neotectonic activity.

ROBERTS, D.L., Geological Survey of South Africa, Pretoria, South Africa

Controls on the Distribution of Some Minor Constituents in Certain South African Permian Coals

There is a positive correlation between sulphur concentration and the vitrinite maceral group content in Permian Vryheid Formation coals in South Africa; the reverse is true for the inertinite maceral family. This relationship is evident at intraseam, interseam and intercoalfield levels. Vitrinite precursors formed where the water table in peat-forming environments was high and Eh low. These conditions favoured bacterial sulphate reduction and resulting H_2S reacted with Fe^{2+} and organic peat compounds to form pyrite and organic sulphur components. Inertinite precursors formed when the water table was low and Eh high; aerobic respiration prevailed and little H_2S was produced.

The water table level in Vryheid Formation peat swamps is at least partly dependent on subsidence rates (Tavener-Smith, 1979; Le Blanc Smith, 1980; Roberts, 1986). The peats which formed Orange Free State and South Rand coalfield coals accumulated in the most stable northwestern part of the Karoo Basin and inertinite is the dominant maceral group (Falcon, 1983). These coals are very low in sulphur (±0,6% average). The northern Natal coalfields are situated nearer the Natal and Tugela structural axes in the southeast and subsidence rates were higher; the coals are enriched in vitrinite relative to the coalfields to the north and have relatively high sulphur contents. The Witbank/Highveld coalfields represent a tectonic setting intermediate between the South Rand/O.F.S. coalfields and those of northern Natal, and accordingly have moderately high vitrinite (Falcon 1983) and sulphur contents. Major exceptions are the numbers 1 & 5 Seams which have considerably higher vitrinite contents than the other coals in the Witbank/Highveld coalfields (Falcon, 1983). Significantly, these seams contain markedly higher sulphur concentrations than the intervening coals (average ±1,3% for the No. 5 Seam and 1,7% for the No. 1 Seam). Clearly, sulphur content in Vryheid Formation coals was influenced more by water table levels dictated by tectonic setting and locally by compactional subsidence rates than degree of marine influence.

Certain trace elements, namely Mo, Co, Ni and As are concentrated in syngenetic pyrite in the Vryheid Formation coals studied (Nos. 4,4L and 5 seams). The element As is of particular significance as it is both highly toxic and volatile, reporting in the gaseous phase during coal combustion. Moreover, it may occur at concentrations well in excess of 100 ppm in some coals. The distribution of Mo, Co, Ni and As were examined in different pyrite forms by means of electron microprobe analysis. Framboidal pyrite is most enriched in these elements. Since the controls on pyrite distribution in Vryheid Formation coals has largely been eludicated (Roberts, 1988) it follows that the controls on the distribution of trace elements associated with pyrite are also understood.

Pb, Cu and Zn do not correlate with sulphur content in whole coal analyses, nor are these elements present in significant concentrations in pyrite. Pb correlates strongly with ash content, suggesting a lithophile association. Its concentration in beneficiated coal is drastically reduced. Cu and Zn are only weakly related to ash content and apparantly have both lithophile and organophile tendencies.

Generally, the pyrite content of northern Vryheid Formation coals can be significantly reduced by washing. Not only does beneficiation lower coal sulphur content, but also the concentration of several other environmentally sensitive trace elements.

REFERENCES

FALCON, R.M.S., 1983. A reveiw of coal classification systems in the industrialized countries of the world. Report, Falcon Research Laboratories, 21 Wellington Road, Johannesburg, 115 pp.

LE BLANC SMITH, G., 1980. Genetic stratigraphy and palaeoenvironmental controls on coal distribution in the Witbank Basin Coalfield. Ph.D. thesis (unpubl.), University of the Witwatersrand, Johannesburg, 242 p.

ROBERTS, D.L., 1986. Depositional framework and controls on peat accumulation in the Vryheid Formation of northern Natal. Unpub. Ph.D. thesis, University of Natal, 230 pp.

ROBERTS, D.L., 1988. The relationship between maceral composition and sulphur content of some South African Permian coals. Int. J. Coal Geol. *11*.

TAVENER-SMITH, R., 1979. Regional control on coal seam occurrence in the north-east Karoo basin. S. Afr. J. Sci., 75, 353-355.

RODRIGUEZ, RODOBALDO, Jorge R. SANCHEZ, Sonia TOUCET, Empresa Nacional De Goefisica, Cuba.

The Possibility of Finding Reef Complexes and Salt Diapirs on the Northern Marine Shelf of Cuba

An evaluation of the most recent geologic and geophysical investigations obtained at the northern marine shelf is given, which corresponds to the area between Pinar del Río and Matanzas provinces, Cuba.

The likelihood of finding reef complexes and salt diapirs is based on the following criteria:

a) Pinch-outs and no seismic reflections along the seismic lines.

b) Lithostratigraphic analysis of the geologic profile obtained from the main deep wells drilled at the north coast of Cuba.

c) Correlation of the geologic and geophysical data obtained in Cuba with those of the neighbouring regions as the Gulf of Mexico, Florida and Bahamas.

The total evaluation of the information obtained reveals up to three wave field — anomalies associated to the corresponding geologic interpretation on the southern border of the Northern Cuban Basin.

The anomalies display bands with sublatitudinal striking attained from 30 km up to 120 km in length and from 2 km up to 8 km in width.

It is pointed out that reef complexes and salt diapirs were developed in age — ranging from the Upper Jurassic to the Upper Cretaceous (Turonian) and could represent either the extension to the northeast of the carbonate bank of the Gulf of Mexico or an independent basin.

The existence of oil fields in the Gulf of Mexico as much as in the northwestern area of Cuba supports that these anomalies of the seismic field are related to the likelihood of finding oil accumulations containing important oil and gas reserves.

S

SANCHEZ, JUAN ROMERO, Centro Nacional del Fondo Geologico, Havana, Cuba

Mineral Waters in the Republic of Cuba

The knowledge about mineral waters in Cuba comes from the 18th Century, developing traditions about their use in such manner that at a certain moment there were 25 bottling enterprises and 7 balneological sites.

The Ministry of Basic Industry is recovering suc usefull tradition, by the profound investigation of our potential on mineral waters and their uses.

This paper mainly deals with the work done on this field up to now.

Up to now we are able to classify and describe as the main types: the sulfurous waters, radioactive-radonic waters, bromic and iodine-bromic waters, arsenic rich waters and ferrouginous-siliceous waters with different degree of minelisation (from 0,2 up to 74 gc) temperature from 20 to 51 gc and pH from 2 to 9,5 which are mainly associated with faults and tectonic zones.

According to their chemical and gaseous tenor as well as their content of pharmacologically active micro components, it has been demonstrated how some types of cuban mineral waters are compared with those of France, USSR, Germany, Checoslovaquia, Italy, Bulgaria and other countries where this resources are widely used for bottling therapeutical uses, and health tourism, showing our potentialities on this field.

Today we are working in Cuba to rescue this natural resource for the people increasing our potentiality in national and international tourism (due to our favourable climatic conditions) as well as for the growth of the bottling industry and the cuban medial development on this field.

SEOO, JEONG RYUL, Korea Institute of Energy and Resources

Polymetallic Mineralization at the Dongnam Mine, Republic of Korea.

Dongnam mine is located at 128°46′-128°49′E, 37°16′-37°18′N in Mureong-ri, Nammyun, Jeongseon-gun, Gangweon-do, Korea. Geology of the mine consists of Cambrian slate (Myobong Formation), Ordovician limestone (Poongchon and Hwajeol Formation), Cretaceous (?) granitic intrusives, and Quaternary alluvium deposits. General trend of strike and dip of bedding plane in sedimentary rocks is N10°-45°E, 5°-30° SE, and joints in granitic intrusives strike NS-N30°E, dipping 70°-80°NW. Main lineament of the study area shows a trend of NNE-SSW. Granitic intrusives, diorite and quartz porphyry, are classified depending on the predominent normative minerals and chemical compositions: diorite into diorite-tonalite-monzodiorite-granodiorite, and quartz porphyry into granite-quartz monzonite-granite. The petrochemical nature of source magma inferred from Niggli value is semi-femic, intermediate alkaline and c-normal for diorite, and salic, alkali-rich and c-poor for quartz porphyry. Sedimentary rocks were thermally metamorphosed by the granite intrusives to form mostly albite-epidote hornfels from pelitic Myobong slate and hornblende hornfels from Poongchon limestone.

Contact metasomatic deposits of magnetite and hematite were formed in the country rocks in the border zone with the intrusives, hydrothermal Pb-Zn-Mn deposits along the fault shear zone and a stockwork deposit of molybdenite in the diorite itself. The molybdenite deposit is thought to be formed in the early solidified margin by the residual mineralized solution of the intrusive itself. Molydenite bearing skarn is plotted in the triangular diagram of diopside-hedenbergite-johannesenite within the same compositional region as the copper-bearing skarn of general porphyry copper deposits. The wallrock alteration is potassic in diorite and propylitic in quartz porphyry. As shown in ACF and AKF trangular diagrams quartz porphyry belongs to epidote-chlorite-calcite and kaolinite-muscovite-chlorite facies. The sequence and stage of mineralization is: igneous intrusion → skarnization → Fe oxide → Fe sulphide → rhodochrosite → calcite → supergene enrichment of Mn oxide. The Dongnam mine has 4 types of ore deposits: contact metasomatic deposits at the border zone of limestone with the granitic intrusives, hydrothermal filling and replacement deposits along the fault shear zone, and disseminated hydrothermal stockwork deposit in the diorite itself. Main ore minerals are composed mainly of magnetite and hematite in the contact metasomatic deposits, and Mn carbonates, galena, sphalerite and molybdenite in the hydrothermal deposits. Mn oxides are enriched at surface by supergene oxidation of primary Mn carbonates.

SHAH, S. D. and A. R. KHAN, Dept. of Earth Sciences, Center for Technology Alternative for Rural Areas, Indian Institute of Technology, Bombay

Deccan Trap rock units are exposed in nearly fifty km. wide thin strip between Surat and Ratnagiri on the upper west coast of Indian Peninsula. The complex lithounits of Deccan Traps include fine to medium grained dark grey basalts (both massive and vesicular/amygdular types), diorite and dolerite intrusives, trachytes, occasional rhyolites and varieties of pyroclastic and intertrappean rocks. Pyroclastic rocks include tuffs, breccias and redbole layers, while the intertrappeans are sedimentary shales and siltstones and sandstones. These rocks are mostly compact with random fractures and well defined joint planes. The number of sets in massive crystalline rocks vary from three to four nearly vertical and one or two nearly horizontal sets. Vertical joint planes are mostly range between NW-SE and NE-SW. The rock units have either thin cover (2-3 m) of altered rocks or at many places thick soil cover (> 10 m) followed downwards by weathered rock horizon.

Climate in the region is hot and humid throughout the year where the south-west monsoon between June to October brings the precipitation of more than 2000 mm in this Konkan coastal strip lying between the Western Ghat ridges and the Arabian Sea. The upper weathered rock/soil zone constitutes the upper unconfined groundwater zone in which the water-table level fluctuates from the near surface conditions to absolute dryness in the open dug wells. In the postmonsoon months water is available in plenty in these dug wells but by end of January the water levels reach almost the bottom of the wells (10 m). and by end of February they are absolutely dry. However, the water is generally available in the lower semi-confined or confined aquifers consisting of flow contacts/rocks contacts of different lithounits and fractures in the hard and compact basalts. These deep aquifers occur mainly in the closely spaced or intersecting planes of discontinuity.

Due to the lack of water during the dry months there is always a scarcity of fodder for cattle and vegetation for human beings. The forested area is becoming smaller every year because of tree-felling for fuels and economic survival. Scarcity of water also drives alway the rural population towards the towns and to Bombay. An effective basinwise study of water resources along with appropriate conservation practice for improving the soil moisture conditions and management of surface and subsurface water resources was carried out in a backward Karjat tribal rural development black near Bombay. The investigations consisted of geological studies using geomorphic and hydrogeologic aspects. Soil properties were studied and infiltration tests were carried out. Electrical resistivity soundings were taken at selected sites from which three water-bearing zones were identified in a vertical column of 100 m depth and were compared with the lithologs of the existing bore well data. Groundwater input and monthly soil moisture recharge were predicted using rainfall, evaporation and soil data. Groundwater recharge values computed using (a) infiltration rate, (b) mass-balance concept and (c) water-table hydrograph are found to be 15%, 12% and 10% of the total annual rainfall respectively. From the net rainfall of 3065 mm, taking out 990 mm of evaporation, and 1515 mm of overland and interflows, gave the root zone recharge of 320 mm and groundwater recharge of 240 mm. Static and dynamic groundwater potentials are found to be 146 and 106 million cubic metres respectively. Stanford Watershed Model IV was applied employing input parameters such as impervious area fraction, basin slope, overland flow length, infiltration, interflow, upper and lower zone nominal capacities and flow recession constants. The model was applied to a small sub-tributary of a river basin and the watershed calibration was done through parametric variations. The results how that the major portion of the infiltrated water is lost as interflow which can be checked by light or heavy plantations. The study showed that the catchment treatment in the form of plantation and checkdams does increase the total pondage and groundwater recharge. Also it has been found that the broad valleys and tabletop lands with interconnected fracture systems are the better sites for well installation, and the minor basins with sound foundations are the better sites for checkdam locations.

It is further envisaged that further detailed investigations involving geology, geomorphology and geophysics along with geochemistry and botany would be needed for the hydrologic development of the Deccan Trap region. Hydrogeologic model studies would be held in locating suitable small pondage areas, which in turn would increase the soil moisture, and water recharge capacity in the upper unconfined and middle (20-50 m) zones. Suitable plantation programmes would increase the erosional losses. Demarcation of the deeper equifer zones for their subsurface extent and storage capacity, along with other geomorphic and geophysical data may help in finding easy flowpaths for the uplift of the stored water and transfer to either local checkdams or to the water supplying lakes of the metropolitan areas.

REFERENCES

Crawford, N. M. and Linsley, R. K. (1966): Digital simulation in hydrology, SWM-IV, Stanford University, Dept. of Civil Engg., Tech. Report, No. 39.

Khan, A. R. and Shah S. D. (1988): Hydrogeological Investigation Reports. Centre for Technology Alternatives for Rurual Areas, IIT-Bombay (Unpublished).

SHUCANG, YUE, Lei XINGYOUG, Gong CHANGRUI and Xu XIAOCHUN, Hefei University of Technology, Anhui, P.R. China

Geochemical Studies on the Tin, Tungsten Deposits and Granitoid Along the Coast of Southern China

The Yenshannia igneous rocks in southern China can be divided into two periods with the earlier period being volcanic rocks and admellite, the later period volcanic rocks and biotite granite. Geochemical studies have shown that all the igneous rocks formed by partial melting. The earlier volcanic rocks and admellite were the successive melting extent, and the later volcanic rocks and biotite granite formed in the same mechanism. The source rocks of the earlier volcnic rocks and admellite were mainly sedimentary rocks, and those of the later volcanic rocks and biotite granite were mainly mantle substance, i.e. magmatic rocks of mantle origin which had previously existed in the crust.

The tin mineralizations include cassiterite-quartz stage, cassiterite-sulfide stage and sulfide-silver-gold-carbonate stage. The tungsten mineralizations include wolframite-quartz stage, scheelite-sulfide stage, and carbonate stage. Systematic studies on fluid inclusions, thermodynamics, genetic mineralogy, geochemistry of trace elements and isotopes indicate that the mineralizing fluids and the ore-forming elements in these deposits derived mainly from the granitoid. Tin mineralizations are related to the earlier Yenshannia admellitic magmatic activity, and tungsten mineralization to the later Yenshannia biotite granitic magmatic activity. They all belong to the magmatic hydrothermal deposits.

SHUTIAN, SUO, and You ZHENDONG, China University of Geosciences, Wuhan, China

Tectonic Styly and Deformational Sequence of Dabie Metamorphic Terrane, China

In the Dabieshan area, the late Archaean metamorphic rocks, here termed as the Dabie Group, are complexly deformed in amphibolite facies. Regional mapping and classical mesoscopic structural investigations in the rocks have revealed histories of repeated deformation, metamorphism and magmatism.

Two sets of major shear belts, which played a very important role in the development of the tectonic framework of the Dabie metamorphic massif, are recognized. These shear belts, which can be grouped into NEN- and WNW-trending systems, cross each other to make a complex reticulate pattern. Blocks separated by the shear belts are of rhomb-shaped. They are relatively less deformed.

The deformational characteristics of the metamorphic supracrustal rocks, gneisses and shear belts are analyzed respectively. At least three widespread episodes of deformation can be determined in the metamorphic supracrustal rocks. The late Archaean sialic crust was subjected to intense reworking along the shear belts. Heterogenous simple shear deformation in the shear belts particularly in agmatites is described. The date show that the folds within these belts develop from instabilities in the shear belt or from local components of foliation-parallel shortening. The major crustal shears may be often associated with compressional or extensional flow. Therefore they should not be simply correlated fold to the phase across the massif.

A total deformation sequence of the Dabie Group is established using fold overprinting combined with shear belt and neosomes cutting. The findings suggest that the present tectonic style of Dabie metamorphic terrane has come about as a result of multiple regional deformation and reworking. Consequently this area provides a window for the examination of mid-lower crustal processes.

SINGH, P. and S. ACHARYA, Department of Geology, Utkal University, Bhubaneswar, India

Genetic Modelling of the Manganese Ore Deposits of the Youngest Iron Ore Group, North Orissa, Eastern India

Geological study of the Iron Ore Group of north Orissa has documented manganese ore deposits in relationship with lithounits of unmetamorphosed sedimentary rocks that belong to Proterozoic age. Mapping has revealed that the persistent bedded nature of manganese horizon in lower shale has given rise to Banded Manganese Formation (protore) in the same sense as the Banded Iron Formation. The rocks of the belts have been classified into three distinct but conformable formation, (Jones, 1934). The lower formation of shale-phyllite and upper formation of shale between which is sandwiched a thick Banded Iron Formation. The polyphase deformations have given rise to plunge reversals in the meridional folds resulting in development of zones of culmination and depressions. Faults and shearplane are noticed ending up in brecciation. Regionally the distribution of iron and manganese concentration have a broad space-time relation in the area.

The manganese ore bodies exhibit significantly different association and style of development that can be classified as (i) Stratiform deposits, (ii) Podiform deposits and (iii) Blanket type. (i) Stratiform deposits—These occur as sheet like bodies, inter bedded, consisting of alternate bands of manganese ore and shale. It is established that these persistent manganiferous horizons are essentially confined to lower shale member and exhibit various sedimentogenic features, and cofolded along with the country rocks. (ii) Podiform deposit—Ore bodies of this kind occur as irregular pockets, lenses and veins within the shale and brecciated chert. The ore bodies reveal features such as colloform texture, mammillated and botryoidal structure, crosscutting veins and segregation along the fissure planes within the shale-chert. (iii) Blanket deposits—A characteristic feature in the region is the occurrence of flat topped caps of manganese bearing laterite. The ore occurs as irregular lensoidal pockets. It is usually spongy and cavernous in appearance. These lateritoid deposits occur within autochthonous laterite as blanket (tabular appearance) at times covering the above two types. The manganese ores have shown the presence of cryptomelane, pyrolusite, psilomelane and manganite. Many other minerals like braunite, jacobsite, todorokite, lithiophorite, ramsdelite, nsutite ($YMnO_2$) and minor rhodochrosite are also being reported. The other associated minerals are iron minerals (hematite, limonite), silica mineral (quartz), aluminious hydroxide minerals. The manganese ore deposits of this region are mostly structurally and stratigraphically below the iron formation and also are remobilized product from manganiferous formation along different major and minor lineaments (Acharya, 1987). The primary control of manganese mineralization was stratigraphic as due to perfect banding character of the ore bodies in the absence of metamorphism (type-i) and is also restricted lithologically either to the shale or autochthonous laterite and chert band. That the localization of the ore body in fissure planes and in conformity with the fold pattern of the region is also structurally controlled.

Synsedimentary origin of the manganese ore and their confinement to lower shale and later remobilization along structurally inhomogenous planes (faults) resulting in development of the ore deposits which are necessarily irregular in shape and size broadly suggest the *modus operandi* of the formation of the ore deposits. This is of specific importance at the present time as the older concept of manganese being confined to the younger shales is not accepted in the present paper.

Detailed mapping of the entire region followed by further detailed work on the several individual deposits based on stratigraphic studies indicate the occurrence and confinement of stratiform manganese deposits to the lower shale, lower to the Banded Iron Formation horizon. This is true for the entire southern part of the basin, south of Joda-Kiriburu axis and nowhere in the north any departure is noted. The conclusion, therefore, that the stratiform syn-sedimentary manganese ores from the lower shale gets remobilized and concentrated by supergene processes, is applicable to the entire basin. The economic potential of the deposit is also considered in the light of genetic model which has been valid for several cases.

REFERENCES

ACHARYA, S., (1987) Field relationship of Iron and Manganese Ore deposits in the Iron Ore Basin of Bihar and Orissa. Symposium on Development of Indias mineral and Fuel Resources, I.S.M., Dhanbad, India (in press).

FERMOR, L.L., (1909) Manganese Ore deposits of India. Mem. Geol. Surv. Ind., Vol. 37.

JONES, H.C., (1934) The Iron Ore deposits of Bihar and Orissa. Mem. Geol. Surv. Ind., Vol. 63 pt 2.

NARAYANSWAMY, S., (1961) Manganese Ore. Bull. Geol. Surv. Ind., No. 20, series A.

PARK, C.F., (1956) On the Origin of Manganese, Int. Geol. Cong., XXth session, Mexico.

ROY, S., (1981) Manganese deposit. Academic Press, London.

SINGH, P.P. (1988) Geologica Setting and Paragenesis of Manganese deposits of Bonai-Keonjhor belt, Orissa, India. IAS Inter. Symp. on Sedimentology Related to Mineral Deposits, Beijing, China. (Abstract)

SINGH, V.P. and Daya SHANKER, Department of Geophysics, Faculty of Science, Banaras Hindu University, Varanasi, India

Precursory Swarm and Medium Size Earthquake Occurrences in Pamirs and its Adjoining Regions

A few earthquakes are preceded by swarm activity in the epicentral region. Evison (1977) identified the swarm as a precursor to earthquakes and advocated that causal relationship exists between the swarm and the main shock. Further, he asserted that the swarm can be used as a predictive parameter for estimating the time of occurrence and magnitude of the forthcoming large earthquake in the swarm region. Seven years seismictiy data (1969-75) as reported by the International Seismological Centre(ISC) in the Regional Catalogue of earthquakes have been used. A total of fortythree earthquakes of $m_b \geq 5.3$ occurred in the region 33°-44°N and 70°-80°E during the considered period. The minimum level of magnitude has been taken as 5.3 because an earthquake of this magnitude has been preceded by a swarm. Of these twentythree earthquakes are related with identified seven precursory swarm and major event sequences in the region. The precursory swarm, main event sequences and after shocks have been used to delineate the boundary of the preparation area for the seven main event sequences. The areas have been found to be elliptical wherein some cases main shock sequences have occurred at a corner of the representative source regions. Two regression equations between the largest magnitude in the precursory swarm, and the main shock magnitude, and the precursory gap has been derived. The equations use the largest magnitude in the precursory swarm for estimating the magnitude and time of occurrence of the main shock.

SITTON, DAN, U.S. Forest Service, RRNF, Medford, Oregon

Slope Stability and Hazard Zonation of the Enlarged Mt. Ashland Ski Development Area, Jackson County, Oregon, USA

An area proposed for additional ski area development on Mt. Ashland was studied to determine the general engineering geology and slope stability characteristics. Mt. Asland is located approximately seventeen air miles southeast of Medford, Oregon. The current ski area has eleven ski runs developed on the upper slopes of the 7,500 foot mountain. Four more ski runs, two poma lifts, and two chair lifts are planned in the preferred alternative. The ski area serves primarily southern Oregon and northern California skiers. The existing and proposed ski areas are located within the city of Ashland Watershed. Ashland's drinking water comes from Reeder Reservoir which is situated less than 5 miles downstream from the proposed ski area.

Mt. Ashland is located within the Klamath Mountain Geologic Province. This portion of the province is made up entirely of plutonic rocks. The granitic rocks encompassing the Mt. Ashland area are composed of two different episodes of Jurassic age intrusions. This plutonic rock body was exposed after millions of years of uplift and erosion of overlying metamorphic rocks. Five varieties of grantic rocks were mapped in the Mt. Ashland Ski Area including: diorite, tonalite, syenodioerite, grandodiorite, and quartz monzonite. Contacts between the rock types are gradational with the exception of the syenodiorite outcrops on the uppermost slopes of the mountain. In many areas the rock outcrops are very weathered. As a result of the weathering, much of the rock is partially to completely decomposed. Soils are composed generally of silty sand and sandy loam. This soil is non-cohesive and tends to erode easily. It is these soils that comprised the failure prone mantle over the weathered bedrock.

During field mapping numerous areas of instability were located. Unstable terrain is characterized by: wet meadows/alder glades, severe erosion, landslide scarps, gutted landslide channels, over-steepened stream banks and wet landslide deposits. Several of the debris landslides mapped are currently active. Some of these failures are still largely unvegetated with raw soils on their scarps. The active landslides (shown as activity level "1" on accompanying slope stability map) have occurred within the past ten to fifteen years. Many of these landslides may have been triggered during an intense storm event that occurred January, 1974.

Most of the wet areas mapped are alder glades. The alder glades are groundwater exit points, many of which begin at similar elevations throughout the project area. These sites are interpreted as groundwater conduits which have subsequently collapsed and sediments eroded away. Water was brought to the surface when mass movement occurred as a result of strong groundwater seepage forces in the sapping zone near the upstream origin of groundwater flows. The combination of weak tractive forces, strong seepage and gravity forces, and increased pore pressure on the non-cohesive soils has caused numerous failures to progress upslope in this zone. The evolution of groundwater sapping failures (debris landslides) to alder glade formation is also evidenced by landslide scarps found immediately above the beginning of most alder glades and meadows. Springs usually begin to issue at the base of the landslide head scarps.

Water has also been brought to the surface downstream from these zones via piping holes during peak groundwater flows. The piping channels enlarge over time and partially collapse. Eventually more of the pipe collapses creating gully features. Groundwater pipes are also found underneath landslide deposits well downslope from the zones described above in the upper alder glades. Water was still flowing in the collapsed pipes in late October, 1988.

Several swales and creek channels are deeply incised down to bedrock from surface erosion and debris landsliding. These gutted channels are located both in alder glades and through timbered terrain. Sheet, gully and dry ravel are all erosion processes taking place to develop and deepen these channels after debris landslides originated them.

A slope stability hazard zonation map was created to categorize the ground within the area proposed for further ski developments. Hazard zonation is a method used by the Rogue River National Forest which assigns risk zones to terrain considered for management/development. The Mt. Ashland hazard zonation map is used in conjunction with the one-inch to 500 feet scale slope stability map. The hazard zonation map shows four different hazard zones. Each hazard zone evaluates the risk of landsliding associated with developing ski runs and lifts in each zone. Estimated risk percentages for each zone increase significantly from natural conditions to conditions with development involved.

These hazard zones are stratified according to the risk of activating landslides within each zone of the ski development area. Hazard zone "1" represents the highest risk terrain and is classified as unstable. Hazard zone "2" areas are not as unstable as zone "1", but are still very senstive, (i.e., severe erosion, subactive and potentially active landslides). Hazard zone "2" is classifed as low stability (high risk) terrain. Hazard zone "3" represents dry, somewhat rocky soils with only a few dormant landslides. Zone "3" is classified as moderately stable. The most stable sites are located in hazard zone "4". It generally reflects the rocky ridges and relatively flat, smooth-sloped areas.

Hazard zones are rated for both natural and developed conditions. Risk percentages are used to show how development increases the probability of activating landslides within each zone. The highest risk percentages are for developed conditions in hazard zone "1" and the lowest are in hazard zone "4" for natural conditions. Potential impacts that may be expected from developing the ground within hazard level "1" is higher in terms of cubic yards of sediment and lower in terms of size of the total area than the other hazard zones. The potential impacts are calculated from past landslide volume totals for each hazard level area. During EIS meetings potential impacts were given in size (acres) and volume (cubic yards) expected from landslides. The time frame used for potential risks of natural landsliding conditions is from zero to fifty years. A shorter time frame for developed conditions is used to indicate that the development of the ski area will tend to accelerate landslide frequency. Risk percentages for landsliding are estimated to be much higher for developed conditions, especially in hazard levels "1" and "2".

The final Environmental Impact Statement (FEIS) will use the hazard zonation process as described above for alternative selection. Input from water resources, soils, and geology will then address these hazard zones to describe what mitigation is necessary for each proposed activity for the various alternatives. Since erosion and unstable sites do not always coincide with each other, surface erosion will be assessed and mitigated for in a separate survey.

After the preferred ski development alternative is selected further engineering geologic site investigations will be necessary. These will help confirm boundaries of the most stable zones for the proposed ski runs and lifts. The study to be completed will include an investigation of the source of the groundwater zones. This work will also attempt to determine why the groundwater seems to be structurally controlled (i.e., why groundwater exit points are at similiar elevation). This could influence the preferred alternative if groundwater pipes at similar elevations between the alder glades are also found to be high risk sites. The following lists the investigations needed to properly complete this project:

A. Geophysical Surveys: Seismic refraction and electromagnetic lines controlled with resistivity soundings will be used to locate subsurface geologic features.

B. Drilling: Needed to provide subsurface information to confirm geophysical data and holes for piezometer installation.

C. Sampling and testing of soils to determine the unit weight, cohesion and internal angle of friction.

D. Calculation of factor of safety for proposed development area using the above data collected along cross sections perpendicular to contours in representative areas.

E. Monitoring: Piezometric, photo point and sediment trap monitoring for a full year prior to any site development. If monitoring is done during a dry year then site development activities will wait until the sites are monitored during a year with normal precipitation.

SLEMMONS, D. B., Peizhen ZHANG, and Fengying MAO, Center for the Neotectonic Studies, Mackay School of mines, University of Nevada at Reno, USA

Geometry and Displacement of the Surface Rupture Zone Associated With the 1954 Fairview Peak, Nevada, Earthquake

In 1954, four major earthquakes, the Rainbow Mountain (Ms=6.3 and 6.8), the Fairview Peak (Ms=7.1), and the Dixie Valley (Ms=68), occurred in north-central Nevada. These earthquakes produced the most extensive pattern of surface ruptures in the entire Basin and Range province in historic time, which provide valuable information not only on present tectonic activity in the Basin and Range province but also on detailed deformational process along each individual fault zone.

The Fairview Peak surface rupture zone trends north-northwest, and the right-lateral displacement is predominant component of the movement associated with the earthquake. Geometrically, the rupture zone can be divided into two segments. The southern segment dips to the east, and the hanging wall, the eastern block, down-dropped to the east. On the contrary, the northern segment dips to the west, and the hanging wall, the western block, down-dropped to the west. In the area between two segments the surface rupture shows a complex pattern, and another 12 km long west-facing fault scarp is present 6 km east of the main rupture zone. The two main segments themselves consist of a series of subparallel en echelon scarps, grabens, troughs, and fissures, their patterns exclusively indicate right-lateral movement. South of Bell Flat the zone has a complex rupture pattern, and at about 2.5 km to the southern end, the sense of motion along the main rupture zone changes from strike-slip to normal faulting style. At the northern end there is a minor bifurcation, and the trend of the main rupture zone kinks to about N35E from about N10-15W. Along the kinked rupture zone there is no any evidence of right-lateral movement, and the rupture zone itself consists of troughs, fissures, and normal fault scarps. More than 250 measurements of horizontal and vertical displacement are made along the surface rupture zone, to provide information on how the displacement distributes along the fault, how displacement varies as the geometry of the rupture zone changes, and the style of fault segmentation in this complex fault zone.

SOVADINA, L., I. LANDA, Czechoslovakia

Mass Transfer of Ion-Exchanging Pollutants in Groundwater

Prognostic estimation of ion-exchanging for the case of binary exchange in ground water is based on following systems of differential equations

$$n \frac{\delta C_a}{\delta t} + v \frac{\delta C_a}{\delta x} - D \frac{\delta^2 C_a}{\delta x^2} = 0$$

$$\bar{n} \frac{\delta C_k}{\delta t} + v \frac{\delta C_k}{\delta x} - D \frac{\delta C_k}{\delta x^2} - \frac{C_a}{Ca} B \frac{\delta C_a}{\delta t} = O; \bar{n} = n \frac{B}{Ca}; B = f(Qv, K_2, C_{K1} C_a)$$

where n-porosity, C_a, C_{kj}- aniont and cationt concentration, t-time, B-coefficient, D-coefficient of diffusion. The equations express migration phenomena including both effect, diffusion and ion-exchange.

The exact analytical solution of the equation does not exist at present. Therefore, the methods of mathematical modelling are applied in analyses. Variation methods in modifications of finite elements /F.E./ method or finite differences /F.D./ are usually applied. The method of F.E. can be used when computers with comparatively large operational memory and speed are available. An advantage of the method of F.D. lies in better understanding of the physical analogies of basic equations, while the method of F.E. and even the Galerkin's method are suitable for simple physical interpretations.

Our contribution summarizes our experience with the solution of the equation by the F.D. method using the integral-interpolative Samarskij's approach. We identify and explain some anomalous phenomena in ion-exchange discussed already by A. J. Valocchi et al. /1981/. Algorithm of the solution of the basic equation is verified by several test studies. Conditions of the optimum solution are defined. The conditions improved the method of ion-exchange modelling generally in case of contaminated ground water.

The results are demonstrated by the figures and tables.

REFERENCES

Valocchi, A. J., Street, R. L., Roberts, P. V. /1981/: Transport of ion-exchanging solutes in groundwater: Chromatographic theory and field simulation — Water Resources Research, V. 17, N.5, pp. 1517-1527.

SUBOTOWICZ, WIESLAW, Assistant Professor, Technical University of Gdańsk, Poland

Lithodynamical Model of Cliff Coast in Poland

Introduction

Two types of coast are to be found at the Polish seaside, viz., the cliff coast and the dune coast.

The cliff coast is built of the formations of Pleistocene Age. Those are: moraine clay and intermoraine formations, viz., gravel, sand, silt and loam. Among the pleistocene formations underground waters occur. Those ones appear in form of numerous seepage springs on the cliff slope.

The dune coast is mostly formed of sand of Holocene Age. The sea coasts in Poland are about 500 km long. The length of the cliff coast is 100 km while that one of the dune coast — 400 km.

Prognosing of the development of coast and its protection is based on the estimation of processes shaping the coast zone. It is an estimation both in space and time regard. The coast zone is modelled by the hydrodynamical phenomena and, on the other hand, by the morpho — and geodynamical processes.

The resultant interdependence has been called a lithodynamical process [6]. This process is expressed by the lithodynamical mode of the coast.

The cliff coast is the object of consideration. That is a space which consist of the undercoast, beach and overcoast [4, 5].

In the undercoast a dynamical layer and a matrix bed can be distinguished. Those layers are separated by a surface which is represented by abrasion profile on the vertical projection [6].

The dynamical layer is mostly a sandy one — at the sole with boulders and pebbles while at the top of bed — with shaped banks. Its thickness does not exceed several meters.

The matrix bed is built mostly of moraine clay and silt which are relatively resistant with regard to abrasion. That's why the abrasion profile, which has been shaped in them, is convex [5].

The beach is a continuation of the undercoast during the high levels of the sea.

The sand and boulder rubble in the undercoast and on the beach has mainly a transit character. It is being displaced in the stream of the rubble.

In this case the overcoast is a cliff. A cliff determines not only the shape of the overcoast but also its space. The prognostical range of landslide phenomena is the land and depth border of the cliff.

The height of cliffs in Poland amounts from a few to several tens meters. The maximum rate of cliff abrasion is 1,5 m per year [1].

Program of researches

Lithodynamical researches of cliff coast in Poland have been planned for 1987-1990. They mainly have an expedition character and are performed every year from 15th June till 15th July. They are concentrated on the selected research section on 5 investigative profiles every 100 m especially on the central profile. They comprise: the undercoast, beach and overcoast — cliff.

Undercoast and beach

1. Actual measurements of waving, currents and levels of the sea.

2. Bathymetry of the undercoast and lovelling of the beach.

3. Separation of the dynamical bed and matrix in the undercoast and beach by means of investigative boring and hydrosounder.[x]

4. Morpho — and geodynamical mapping of the undercoast and beach consisting in direct observation of dynamical forms/banks, ripplemarks/ and description of lithological sediments. The samples taken from surface sediments and borings will be put on graulometrical and petrographical tests. To the sedimentary tests will be taken special samples of sediments hardened by means of resin.

Besides, eolin transport measurements are performed on the beach by means of catchers. In the undercoast are used marks i.e. rods driven in the dynamical bed and the matrix to determine the direction of transported sand/measurements of inclination of the rods/.

5. Qualitative estimation of the abrasion of the matrix in the undercoast and on the beach by means of the marks i.e. rods. They were driven into the open clay matrix. They are destined to determine the rate of abrasion and to estimate the volume of abraded subsoil as a source of the rubble feeding the dynamical layer [7].

6. Determination of the abrasive profile in the undercoast and beach of the cliff coast as an effect of hydro — and geodynamical interdependence

between the dynamical bed and the matrix. Formulating the simulation model of transformation of undercoast and beach [2, 6].

ˣHydrosounder — perforated pipe into which water is forced under pressure in. Then it is introduced into dynamical sand bed. The resistance perceptible at its end makes it possible to estimate the occurrence of the matrix. The drill fastened at the end of the hydrosounder makes possible to take a small sample of cohesive sediment.

Overcoast — cliff

1. Morphodynamics, geology, water conditions and geodynamics of the cliff. Dynamical elements will be worked out on the base of comparison of the results of photogrammetrical measurements carried out twice in 1987 and 1990 [3].

2. Granulometrical and mineralogical and petrographical estimation of rocks of which the cliff is built. Their qualitative and quantitative part in feeding the dynamical bed.

3. Dynamics of underground waters in the cliff on the base of stationary measurements of their levels in piezometers.

In the analysis historical and actual data of precipitation will be used.

4. Long-term geodynamical deformations of the subsoil on the base of stationary inclinometrical measurements.

5. Physical and mechanical properties of rocks on the base of laboratory tests of the samples taken from boring places and excavations.

6. Qualitative estimation of the cliff abrasion on the base of photogrammetrical measurements carried out in 1987 and 1990.

7. Analysis of cliff's stability and the formulation of the simulation model of the prognostic geodynamical development of the cliff.

Final remarks

Parallel to the presented range of researches concerning the lithdynamical model of the cliff coast, monitoring measurements on the other sectors of cliff coast in Poland are being performed. On the base of them a lithodynamical model of cliff coast will be verified and formulated as a universal model.

It will be a base for prognosing the development of cliff coast in Poland and, afterwards, for taking decisions concerning its protection [1].

REFERENCES

Cieślak A., Subotowicz W., 1987, Raport o stanie wiedzy o brzegu morskim w Polsce i jego ochronie. Inzynieria Morska, No. 2.

Fox W. T., Davis R. A., Jr., 1973, Simulation model for storm cycles and beach erosion on Lake Michigan. Geological Society of America Bulletin, Vol. 84.

Hutchinson J. N., Chandler M. P., Bromhead E. N., 1981, Cliff recession on the Isle of Wight SW coast. Proceedings of the Tenth International Conference on Soil Mechanics and Foundation Engineering, Stockolm, Vol. 1.

Rudowski S., 1962, Mikroformy strefy brzegowej Baltyku w Polsce. Acta Geologica Polonica, Vol. XII, No. 4.

Subotowicz W., 1982, Litodynamika brzegów klifowych wybrzeza Polski. Ossolineum, Gdańsk.

Subotowicz W., 1988, Litodynamiczny model brzegu klifowego w Polsce. Inzynieria Morska, No. 2.

Szujskij Ju. D., 1986, Problemy issledowanija balansa nanosow w bieregowoj zonie moriej. Gidromieteoizdat, Leningrad.

Asst. prof. de hab. eng. Wieslaw Subotowicz
Technical University of Gdańsk
Faculty of Hydrotechnics
Majakowskiego Str. 11/12, 80-952 Gdańsk, Poland

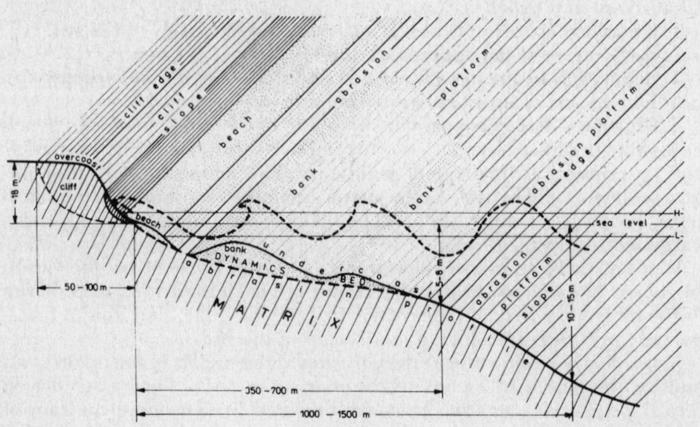

TANKUT, AYLA, Geological Engineering Department, Middle East Technical University, Ankara Turkey

Tertiary lavas of Ankara region comprise two main groups as calcalkaline intermediate-acidic rocks and alkaline basic rocks. The former group includes lava flows of volumetrically dominated andesites and dacites which alternate with agglomerates and tuffs together with local exposures of rhyolites. The basic volcanics are the latest products (U-Pliocene) and consist of basalts and trachy basalts, showing minor magmatic evolution.

The K, Rb, Sr, Zr contents, K/Rb ratios and flat HREE patterns of calcalkaline rocks are similar to those from continental margin volcanics. The REE and HFS contents of the alkaline basaltic rocks indicate primary source magma, probably generated from upper mantle by partial melting of lherzolite. The rocks of the latter group also show some LIL element enrichment which might be due to the modification of the source by the fluids or melts rich in LIL elements.

Trace element contents of the studied rock suggest that two groups are the products of different settings. The calcalkaline activity can be attributed to plate convergence, while alkaline activity is interpreted to reflect more likely a divergent motion.

TAYLOR, W. E. G., Faculty of Applied Sciences, Luton C.H.E., Park Square, Luton, Beds, LU1 3JU, U.K.

Structural controls on the formation of regional Skarns

The division of Skarns into *contact & regional* types is not widely known but a strong case can be made for the viability of such a classification since it can aid considerably in the understanding of the formation of these deposits.

Investigations of Skarns from Western and Central Europe, & North America show that structures control not only the formation position of the rocks but also can be used to predict the location of mineral deposits associated with them.

Structures associated with Skarns vary from simple planar and linear types through to quite complex flats and ramps typical of thrust regions. Temporal studies indicate that in many deposits intrusion of granitic bodies, spacially associated with Skarns, frequently are emplaced **after** the formation of the Skarn.

THUY, NGUYEN NGOC, Institute of Geophysics, National Centre for Scientific research of S.R.V, Hanoi, Vietnam

Geology-Geophysical Criteries of Seismicity in the Territory of Vietnam

In order to find relations between seismicity and geology-geophysical characteristics were used geological map, tectonic maps of many authors, map of earth's crust structure and geophysical fields.

On basis of studying seismological data in the bottom of magnetic and Bougher gravity anomaly maps, their transformations on different heights, Faye gravity anomaly, isostasy, horizontal gradient of Bougher gravity anomaly, gradient of mean velocity of neotectonic motions and parameters of genesis zones.

Were elucidated geology-geophysical criteries of seismicity in Vietnam.

The author has obtained some quantitative relations, allowing to estimate energy of expected maximum earthquakes in this territory.

TORRES, J.A.[1]; M. LAGO[1]; C. BAMBO[1]; A. POCOVI[1]; R. VAQUER[2] & D. ZACHMANN[3]. [1]Department of Earth Sciences. University of Zaragoza. Spain. [2] Department of Petrology and Geochemistry. University of Barcelona. Spain. [3]Institute of Geology and Paleontology. Technical University of Braunschweig. F.R. Germany.

The Calc-Alkaline Magmatism of the Hercynian Belt in the East of the Iberian Massif (Iberian Range, Spain).

Late orogenic magmatism of calcalkaline nature in the Hercynian Belt exhibits the greatest diversity of occurence in the Iberian Chain (Alpine type chain in NE Spain, which "hercynian socle" pertains to Asturoccidental-leonesa zone of the Iberian Massif).

This magmatism is mainly constituted by subvolcanic intrusions (dikes and sills), small-size plutons and even several outcrops of explosive features.

Several chriteria (stratigraphic, tectonic, petrologic, contact metamorphism and radiometric ages) appear to indicate an Stephanian -permian emplacement age, through a preferential fracturation, between the end of a compresive stage and the beginning of the Late - Hercynian distension.

From a petrological point of view there is a complete sequence, from Al-basalts to rhyolites, but the amphibolic and piroxenic andesites are the dominant types (in volume).

Several types of enclaves are enclosed in the andesitic rocks (granitoid, metamorphic xenolithes, etc.). They show reaction relationships with the volcanic wall-rocks.

The results of geochemical analysis indicate the calc-alkaline character of these rocks and a good differentiation by fractionated crystallization. The behavior of compatible and incompatible elements is in good agreement. The REE data verify an orogenic-intraplate environment and indicates the magma genesis by melting, in a thick continental crust enriched in H_2O and K.

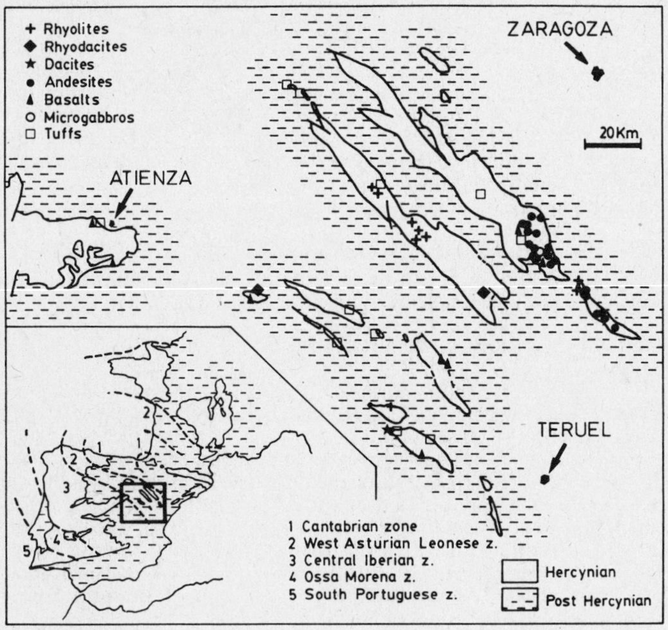

Fig. 1 – Siting of the calc-alkaline magmatism in the Iberian Range

V

VAN BEVER DONKER, J.M., University of the Western Cape, Bellville, South Africa

Regional Structural Constraints on a Major Tectonic Boundary Between a 1000 My Mobile Belt and a 3000 My Craton

At the boundary between the 300 Ma Kaapvaal Craton and the much younger Gordonia sub-province of the Namaqualand Metamorphic Complex developed the Areachap Group, a sequence of amphibolitic and calc-silicate rich rocks. Geochemical evidence has shown that this group is in all probability a volcanic island arc. Investigations into the structural relationship between these adjacent terraines has revealed a complexly deformed area, where at least four phases of deformation can be defined, developing complex interference patterns, displaced by a large number of shearzones.

In order to explain the different orientations of the various phases of deformation, an oblique collison model with a rigid indentor is suggested. The collision between the Kaapvaal Craton and the Namaqua-Natal Belt is thought to have taken place in two stages: the first with the Craton as a pointed indentor, causing early Kheiss thrusting. After the Kheiss had been accreted in this manner, the shape of the indentor changed from pointed to flat, resulting in a differently oriented stress field causing the interference structures so typical for this area.

The thrusting resulting from this collision model, has been interpreted as the lateral ramp of a Northwards directed Cordilleran type collision, causing thrusting to take place obliquely to the North East, so that the NW trending shearzones can be considered frontal ramps. The NNW trending shearzones have consequently been interpreted as oblique ramps.

This thrusting in the early stages of the collision is suggested to have had a prestressing effect, while Sigma-one was sub-perpendicular to the strike-slip zone. Differential transport and the consequent rotation of Sigma-one caused the differential stresses responsible for the development of the enechelon fault pattern with zones of reversed sense shearing.

Finally it is concluded that transpression has been the dominant mechanism during the later stages of the collision.

VAN DER WIEL, A.M., Agricultural University, Dept. of Soil Science and Geology, Wageningen, The Netherlands

Uplift of the Precambrian Garzón Massif (Eastern Cordillera of the Colombian Andes) in relation to fluvial and vocaniclastic sedimentation in the adjacent Neiva Basin.

The purpose of this investigation is to determine the age of the uplift of the southern part of the Eastern Cordillera of Colombia, the Garzón Massif. To that purpose two methods are used: 1) Determination of the uplift ages of the Garzón Massif with the aid of fission-track datings on apatites taken from samples at known heights and 2) study and K-Ar dating of the sedimentary infill of the adjacent Neiva Basin.

The Neiva Basin constitutes a broad tectonic depression situated between the Central and Eastern Cordilleras in the south of Colombia. This basin developed during the Middle to Upper Tertiary as a result of lithospheric loading due to foreland thrusting along the eastern margin of the Central Cordillera. It is filled with 3500m of fluvial deposits resulting from the uplift of the Central Cordillera and with±1000m of volcanic and vocaniclastic sediments derived from the Central Cordillera volcanic arc. These sediments were folded and faulted as a result of uplift of the Garzón Massif.

The fluvial sediments were deposited on an alluvial plain and are subdivided into the Gualanday and Honda Formations. The volcanic and volcaniclastic deposits are grouped into the Gigante Formation. The youngest deposit to be folded and faulted by the uplift of the Garzón Massif is the so-called "Ceibas Conglomerate". This deposit consists of at least 200m of proximal alluvial-fan deposits. These sediments are made up entirely of sand- to boulder-sized fragments of Precambrian granulites and gneisses and Jurassic granites, which come from the crystalline core of the Garzón Massif. The "Ceibas Conglomerate" rests uncomformably on the Gigante Formation, indicating that folding had already begun.

Preliminary results of the fission-track datings on apatites from the Garzón Massif gave an age for the uplift of the massif of 12± to 9±2 Ma.

K-Ar datings were done on biotite separates from samples taken at known depths in the stratigraphical columns. In this way in the middle to upper part of the Gigante Formation was dated at 7.92±0.129 to 6.61±0.250 Ma. Although at present K-Ar datings on the Honda Formation have not been done yet, in the literature fission-track ages are given for part of the formation of 16.1±1.1 to 14.6±1.1 Ma. Unfortunately, the article is unclear about the stratigraphic position of these datings.

The uplift ages of the massif and the age of the Gigante Formation coincide, indicating that uplift was simultaneous to sedimentation of this formation.

Although volcanism of the Central Cordillera volcanic arc started before uplift of the massif – the first indications of volcanism are found at the base of the Honda Formation – the main phase of volcanism, represented by the deposits of the Gigante Formation, coincided with the uplift.

In spite of the fact that the uplift coincided with deposition of the Gigante Formation few indications of this uplift are found in the sedimentary record: paleocurrent directions of the lower part of the Gigante Formation are to the east, while very little material with an eastern provenance is found. Only the upper part of the Gigante Formation shows paleocurrent directions to the north and a somewhat higher percentage of Precambrian pebbles, while the composition of the "Ceibas conglomerate" indicates that unroofing of the Garzón Massif had already occurred.

There is, however, structural evidence that deposition of the Gigante Formation was influenced by the uplift of the massif: the formation show strong lateral changes in thickness and the stratigraphy is very variable, especially in a north-south sense. Apparently, uplift of the Garzón Massif resulted in the development of a fault system in the adjacent basin which caused the creation of local depositional sites due to tilting, uplift or subsidence of local blocks. Perhaps strike-slip movements also played a role: from the literature it is known that subduction of the Nazca Plate below the South-American Plate changed from perpendicular to oblique about 8 Ma. ago.

The lack of material with an eastern provenance in the sedimentary infill of the basin and the predominance of paleocurrent directions to the east make it likely that deposition of erosional material from the Garzón Massif was mainly to the west until deposition of the "Ceibas Conglomerate" took place.

VALDÉS, ANTONIO, Unión de Empresas Geológicas

Some Considerations on the Dynamic Theory of Sea Water Encroachment

At present sea water encroachment into coastal aquifers is a problem affecting most countries. For many island-nations it is a permanent economic burden and for those small island-nations it is a vital problem.

This phenomenon amplies not only surveys cost, but also cost for an extra surveillance, control deficiencies, assessment errors as well as for the restablischment of affected sources. One way or another, the partial knowledge of sea water encroachment behaviour is continuously affecting the conomy of coastal countries.

In order to understand sea water-fresh water relationship, the use of complicated mathematical formulas needed can not be overlooked for interrelating highly changeable factors such as climate, geological environment and sea water and fresh water flows. Variables limiting the implementation of solutions to particular uncertain cases of extrapolations to other coastal areas may not be stablished, because they may lead into some of the above mentioned cost.

The key of sea water encroachment phenomenon depends on the dunamical process of it and for its actual quantification it is necessary firstly to improve the physical conception of its mechanism.

Theoretically the author shows that the dynamic theory is divided into a classic and a modern category according to the approach used for the sea water encroachment interpretation and assessment.

Evidences are given showing that the considerations of rotational flow and the correspondence between sea water and fresh water flows are integrating parts of the modern theory. Based on this, conditions under which Ghyben-Herberg ratio is totally valid is discussed.

Random criteria to be taken into account for coastal aquifers and small islands are shown.

On this theoretical basis, the author establishes maximum and minimum values of sea water wedge intrusion length and ground water drainage flows to the sea.

Finally, metrodological aspects to be used for prospection, assessment and effident control of sea water encroachment into coastal aquifers are explained.

VENKATAPATHI, M., A. V. SUBRAHMANYAM, J.S.R. KRISHNA RAO, M. J. RAO

Genesis of Panchpatmali bauxite deposit, Orissa, India

Panchpatmali bauxite deposit is the largest single deposit in India, located in the hinterlands of East coast bauxite province. The geology of the area is simple and constitutes alternate broad bands of Charnockite and Khondalite, the latter being the dominant rock type, which due to insitu chemical weathering given rise to bauxite deposit. Since this deposit has a very large areal extent (15 sq km) it has been divided into north, south and central blocks. A variety of nodular laterite is noticed in the south and central blocks along the irregular fractures in the bauxite cap. The nodules are showing core (composition ranging from Khondalite to bauxite) and concentric rings (composition, ranges between goethite and gibsite). The nodular laterite is cutting the bauxite and laterite and showing vein like structure in the central block. The nodules are embodied in the matrix of clay. The details of the nodular laterite, its genetic relation, mineralogy also dealt in the paper.

Bore hole log data suggested that this deposit has formed under poor leaching conditions. Landsat imagery studies suggested some post bauxitisation fractures in this deposit along which the nodular laterite is formed. These post bauxitisation fractures (pliestocene) are recorded in this province for the first time.

The Mineralogical studies are carried out on the bauxite samples with special attention on nodular laterite and the textures. Mineral corundum is also noticed in this deposit as an accessory mineral.

The chemistry of the samples revealed that this deposit is having high iron content than other deposits of this province. It is due to post bauxite fractures, and presence of nodular laterite.

Based on the field mineralogical and chemical evidence, it is concluded that the Panchpatmali deposit is less dessected poorly leached with intermittent pockets of laterite showing two phases of lateritisation which includes post bauxitisation movements in pliestocene age.

VILLA, IGOR M., Istituto di Geocronologia, Pisa, Italy

Excess Ar Geochemistry in Potassic Magmas

$^{40}Ar/^{39}Ar$ dating experiments on leucites, sanidines and a kalsilite from the potassic Roman Volcanic Province (Central Italy) reveal a very peculiar excess Ar component released only at high temperature. 34 out of 53 analyzed minerals show asymmetric saddle spectra: step ages increase from typically 0.3 Ma in the first 90% of the ^{39}Ar release to up to 300 Ma in the last 1% (Villa, Rend. Soc. It. Min. Petr., 43, 1988, 95-104). This excess ^{40}Ar ($^{40}Ar_{XS}$) has a highly enriched isotopic composition ($^{40}Ar/^{36}Ar > 6300 = 21\ R_a$); it is not inherited from a restite phase, nor is it related to surface alteration. The latter was controlled on a stratigraphically bracketed analcime, whose excess spectrum is markedly different from the leucite spectra. $^{40}Ar_{XS}$ is common to sanidines from pumices and from the matrix; it is partly removed by deep HF etching.

The origin of the high-temperature $^{40}Ar_{XS}$ must be sought in a process operating at sufficient depth to affect sanidines already formed in pumices. The fluid carrying the ^{40}Ar anomaly was either surface water that had leached old continental crust or, more probably in this case, a magmatic water-poor fluid which derives its Ar from the mantle.

The mechanism to incorporate extraneous Ar into minerals can be simulated in a laboratory experiment. An artificial doping of leucite with ^{38}Ar at 350 C, 2 kb resulted in dissolution of Ar into the lattice; stepwise heating resulted in a high-temperature excess similar to natural age spectra. The activation energy for vacuum release of $^{38}Ar_{XS}$ obtained from an Arrhenius plot is 80 kcal/mole, which should be compared with the 30 kcal/mole obtained from the ^{39}Ar vacuum release. The existence of two sites with two different activation energies explains the camel pattern of the differential ^{40}Ar release vs temperature.

A model which makes use of these two Ar activation energies is able to correctly predict the leucite spectra obtained by stepwise heating in the laboratory. The fact that Ar atoms can occupy two *non*-communicating reservoirs implies that the physical mechanisms of its diffusion are complex. Van der Waals binding energy is one order of magnitude too low; at least three different diffusion mechanisms with different activation energies are operating in the leucite lattice.

The identity of the high-energy site is investigated by means of three-isotope plots: the $^{40}Ar_{XS}$ is not correlated to anion vacancies ($^{38}Ar_{Cl}$).

VRUBLEVSKY, A. A., Institute of Tectonics and Geophysics, Far East Branch of the USSR Academy of Sciences, Khabarovsk

Faults of Fold Systems of the Soviet Far East

The faults determine the cover-fold structure of Mesozoic and Cenozoic fold systems of the Far East and their break into en echelon blocks.

In the modern structure of the fold systems, boundary, intrastructural and through faults and their systems are distinguished. To the first category belong the faults which at their whole length or at most part of it coincide with the boundaries of the fold systems; for example, the Arsenievsky, Alchansky faults and systems of the East-Bureya faults of the Sikhote-Alin fold system. To intrastructural faults belong a number of known faults, in particular, Central Sikhote-Alin, Central Sakhalin, Kamuikotan, and others. Through faults are represented on the surface most often by transregional zones of heightened jointing.

In the fold systems, together with structural-material series are ascertained dislocational vertical and lateral series where heterogeneous rupture dislocations play an important role. Incidentally, the sequence of these faults of this or that genetic variety in these series may be superposed and even repeat depending on the general direction of the region's tectonic evolution. The full vertical and morphogenetic series of faults of the fold systems of the Far East South includes: a) faults of the foundation stage of the fold systems; b) consedimentation nappes accompanied with mixtites; c) scaly-thrust dislocations; d) strike-slip fault late-fold dislocations; e) post-fold systems of riftogenic superimposed basins, normal fault — tension zones and strike-slip faults of marginal-continental volcanic belts, and others. They are most vividly apparent and fully investigated in the Sikhote-Alin fold system. Numerous observations were also made of the Sakhalin-Hokkaido system.

The Jurassic-Cretaceous consedimentation covers and scaly dislocations of the South and East Sikhote-Alin, East Sakhalin are well correlated by the time and character of their manifestation with the close assemblages of the Japanese and Philippine Islands.

The Sikhote-Alin Late Cretaceous marginal-continental strike-slip fault system, the Cenozoic strike-slip faults of Sakhalin, the Amur-Ussuri system of Cenozoic riftogenic basins are related to particular types of deposits of mineral and energy raw materials.

Manifestation of incipient magmatism with specific metallogeny is related to the faults of the initial stages of the fold systems development. Though the consedimentation nappes did not directly serve as canals for ore-bearing fluids, but in many respects they determined special distribution of olistostrome complexes in which polymetallic and tungsten mineralization is often localized. The scaly-fold dislocations formed a complex system of screens for the ore-bearing fluids, that resulted in concentration of minerals, in particular, gold and gold-rare metal ores in the underthrust parts and in the curves of the folds complicated by upthrow faults. The underthrust type of oil- and gas-bearing deposits broadly manifested in similar conditions in the other parts of the Pacific framing, is not excluded in the sedimentary basins of the Far East South. Strike-slip fault dislocations are ore-controlling structures for many syngenetic deposits of the volcanic belts and fold systems at the final stage of their foundation. It is these dislocations that in most cases determine the en echelon structure of the ore fields, location of the ore knots, columns, etc.

W

WITKOWSKI, ANDRZEJ, Institute of Oceanography, University of Gdańsk, Gdańsk, Poland

Aerobic Degradation of Green Coccoid Algae (Chlorophyceae) as a Mechanism of the Proterozoic "Spot" Cells Formation?

Laboratory experiments concerning the influence of aerobic degradation on the coccoid green algae were carried out. Degradational experiments were carried out in aerobic dry conditions (on Petri dishes) and in aerobic wet conditions (in test tubes). The size of single cells of degraded algae varied from 10 to 16 μm. The characteristic feature of the predegraded cells was the presence of a single pyrenoid. In the course of the experiments series of internal "spot-like" bodies were formed, which decreased in size and were rounded in shape; they were mostly placed eccentrically. Two kinds of internal structures were produced depending on the conditions of experiments.

In the dry aerobic conditions they were composed exclusively of opaque starch granules (as indicated staining by tincture of iodine) ca 2.0 μm in size. In contrast in the wet aerobic conditions they divided into two parts, an inner one of ca 2.0 μm in size, opaque (composed of starch) and an outer one a translucent rim of the cellular residuum (Fig. 1). In the second type of experiment among degraded cells containing internal bodies, there were also observed numerous successive stages of shrinking cellular content, and less abundant empty cells.

It was stated that internal structures formed as a result of the degradation of investigated microalgae indicate close resemblance to the "spot structures" occurring in the Proterozoic stromatolites (Schopf, 1968; Hofmann, 1976; Knoll et al., 1978), particularly those of *Glenobotrydion aenigmatis* Schopf and *G. majorinum* Schopf and Blacic. Moreover, these experiments provide an explanation of the origin of some other microfossils from those sediments which are considered as a separate species. It is possible to point out for a particular microfossil an appropriate degradational stage. The initial stage is suitable for a fossil genus *Globophycus.* The process of further degradation decreases significantly the cellular contents inside which the dark "spots" of starch form. This stage is adequate for fossil genus *Caryosphaeroides.* In the next stage "spot-like" structures which resemble those of genus *Glenobotrydion* form (Fig. 2).

The results of these experiments indicate that in natural conditions, the formation of "spot" structures probably resulted from oxygen-mediated decomposition of eukaryotic microalgae in microbial sediments which were periodically devoid of water cover. This phenomenon has been observed in the modern microbial mat of Puck Bay in the Southern Baltic Sea.

Although some evidence indicates that "spot - cells" can be produced as a result of degradation of cyanobacteria (Francis et al, 1978; Knoll et al., 1978), the degradation of eukaryotic microalgae provides a plausible means of producing Proterozoic "spot - cells".

REFERENCES

Francis, S., Barghoorn, E.S. and Margulis, L. 1978. Precambrian Res., 7: 377-383.

Hofmann, H.J. 1976. J. Paleontol., 50: 1040-1073.

Knoll, A.H., Barghoorn, E.S. and Awramik, S.M. 1978. J. Paleontol. 52: 976-992.

Schopf, J.W. 1968. J. Paleontol. 42: 651-688.

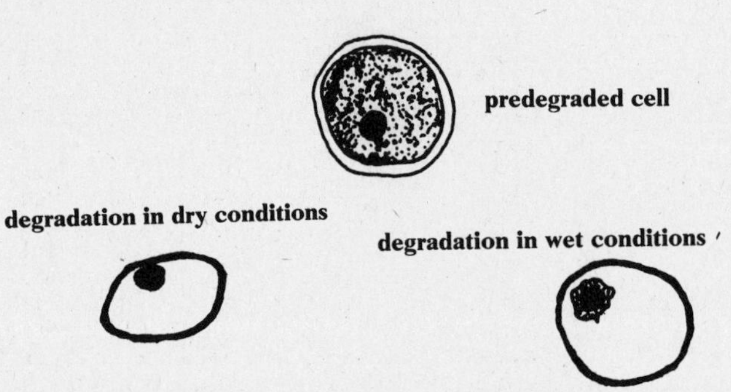

Fig. 1. Influence of degradation on studied microalgae

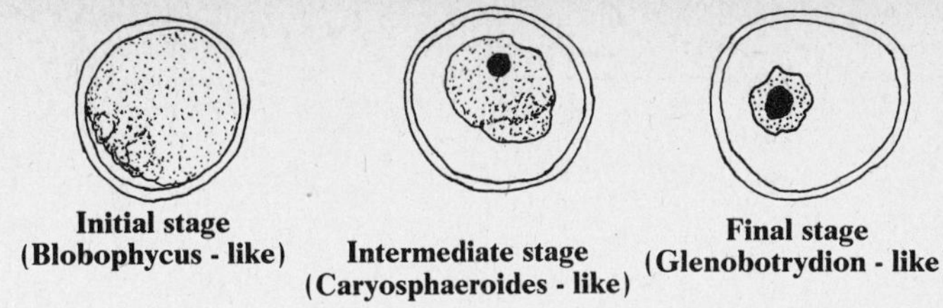

Fig. 2. Successive stages of degradation observed in wet aerobic experiments.

WOOTEN, RICHARD M., U.S. Forest Service, Gifford Pinchot National Forest, Zone II Engineering, Cook, Washington

Geologic and Geotechnical Contributions to Resource Management on the Gifford Pinchot National Forest

The Geology and Geotechnical Group on the Gifford Pinchot National Forest, located in the southern Washington Cascade Mountains, provides a range of skills and services in support of Forest Service activities. Consisting of geologists, engineering geologists, geotechnical engineers and drillers the team focuses on the application of geologic and engineering skills used in risk assessment and the solution of problems in resource management. Involvement in four major areas is presented: geologic resource mapping, slope stability evaluation, rock resource management, and development of visitor facilities at the Mount St. Helens National Volcanic Monument.

A seven year program of resource mapping produced Geologic Resources and Conditions Maps which serve as a geologic data base and are used by several disciplines in timber sale planning. The 1:15,840 maps display soil and rock types with geologic and engineering classifications and properties; slope failures and their classification; common variety mineral material sources and management information; and geologic points of interest.

Slope stability hazard and risk evaluation may be categorized into three levels of intensity of investigation, analysis, and design/recommendations. Recently developed monitoring techniques and computer programs are used in the investigation and analysis. The first level of analysis employs the LISA (**L**evel **I** **S**tability **A**nalysis) computer program which utilizes a Monte Carlo simulation to estimate the probability of slope failure occurrence in the natural and harvested conditions. This information is used in timber sale planning for land allocation and cumulative effects assessment. Level II focuses mainly on cut and fill slope stability of forest roads and uses the SARA (**S**ite **a**nd **R**oad **A**nalysis) computer program. Level III, the most intensive evaluation, may use several analysis methods for critical site stabilization. Subsurface exploration, continuous slope movement and groundwater monitoring using data recording circuitry and down hole pressure transducers aid in the analysis and evaluation of stabilization alternatives.

Rock resource management incorporates geologic expertise in the management of common variety material sources used for forest road and facility construction. Aspects of rock resource management include: maintaining inventories, prospecting, investigating, drilling, and management plan preparation.

Involvement in the planning and preconstruction activities for the development of the Coldwater-Johnston visitor center in the Mount St. Helens National Volcanic Monument includes preparation of environmental analysis documents, facility site evaluation, slope stability assessment, groundwater feasibility studies, and development of aggregate sources.

X

XIANTAO, WU, Mining College of Jiaozuo, China

The Role of Trace Fossils and Their Associated Sedimentary Structures in Determining Marine Environments of Coastal Clastic Deposits

Five types of marine coastal clastic deposits have been recognized in Guanshan Formation of Upper Devonian from Lower reach of Yangtze River, China. Though it has long been treated as terristrial facies since 1919.

Type A is characterized by pebbles deposited directly on the trough of rippled bed without erosion and therefore is interpreted as rip strand gravels on the back shore environment.

Type B deposits recorded the wave-like pebbles distributing vertically in the section and may suggest a beach cusp-like origin in foreshore area.

Type C deposits contains pure white middle-grained sandstone with escape structures above and herringbone cross-bedding below. Thus it is thought to be intertidal bar.

Type D deposits is composed of conglomerate lag, double clay layers and trace fossils of *Arenicolites* and w-shape burrows implying a shallow subtidal channel environment.

Finally type E deposits has lower part of medium to coarse-grained sandstone with pebbles, and middle part of large planar cross-bedded and herringbone cross-beded, as well as thin beded fine to medium grained sandstones alternating with sandy clay layers in which trace fossils of *Cochlichnus, Gordia* and traces of Crustacea have been found. The features of this type probably were of deep subtidal channel.

The sedimentary environments of Guanshan Formation, Wutong Group (Upper Devonian) in South Jiangsu, China, puzzled the geologists for decades because of lacking identifiable body fossils and other depositional features in the sandy deposits. But both trace fossils and sedimentary structures are abundant after carefully looking into the well displayed section in Kongshan near Nanjing, and Taihushan about 80km southeast to Hefee, capital of Anhui province.

Wutong Group is a clastic deposit system of 150-200m in thickness and primaryly was named Wutongshan Mount quartz-sandstone by Ding Wenjiang in 1919. Later works show the Group can be divided into two parts. The upper part is called Leigutai Formation and consists of coloured sandstones, silts and shales including plants of *Leptophloeum rhombicum, Sublepidodendron mirabile, Archaeopteris* sp., and estheritina of *Lioestheria longtanensis.* The thickness is about 90m.

XUYEN, NGUYEN DINH, Institute of Geophysics, National Centre for Scientific Research of S.R.V

A New Method for Determination of Return Period of Large Earthquakes in the Definite Place of Seismic Genesis Zone.

On the basis of concept of seismic cycle the author have established a new formula for determination of return period of large earthquakes in a definite place following the parametres of seismic regime in the focus zones of occured large earthquakes:

$$T = \frac{10^{(3 - \lg S_{k_{max}} + \delta (K_{max} - 10)} - (t_1(A_1 - A_2) - t_3(A_3 - A_2))}{A_2}$$

where : $S_{K_{max}}$ - the surface of focus zone of large earthquakes

K_{max} - energy degree of maximum earthquake

- coefficient of frequency energy relation

$$\lg N = \lg A - \delta (K - K_o)$$

A_1, A_2, A_3 - average seismic activities in the after shock, stabilization and forshock stages of seismic cycle

t_1, t_2 - durations of aftershock and forshock stages.

Y

YAGMURLU, FUZULI, Department of Geology, Engineering Faculty of Akdeniz University, Isparta, Turkey

Depositional Models of the Lignite-Bearing Inter-Mountain Neogene Basins, West Taurus Belt, SW-TURKEY

The Neogene basins are occur in the southwest of Turkey, contain various sedimentary constituent which are deposited in both of marine and continental facies. The marine sediments were deposited during Early Miocene period in southern parts of west Taurus belt. Thus, many large and small size of lignite-bearing inter-mountain basins developed in the west Taurus belt and adjacent northern regions, and were deposited by continental sediments during Middle to Late Miocene. The Neogene sediments of the marine facies consist of, from bottom to top, poor-sorted and coarse-grained conglomerate, platform type allochemical limestone and flysch-like sediments which are in the upper most part of marine sequence. The coarse-grained sediments of the alluvial-fan reach to 1000 m. thick and overlies unconformably the all marine Neogene diments in southern parts of the West Taurus region.

The continental Neogene basins that locate in western region of Antalya-Afyon line of the west Taurus belt, generally are N-NE-trending, although in eastern region of the same line dominantly are NW-trending (Figure 1). The NW-trending continental basins of the region were filled dominantly by Middle Miocene sediments, but the N-NE trending basins deposited with Upper Miocene or younger sediments. These continental basins are limited by the growth faults which were controlled the sedimentary thickness and lithofacies characteristics of the basins. The continental Neogene basins are limited by the N,NE and NW-trending stepwise fault systems which formed under influence of the extension tectonic regime in northern region of Aegean trench, which was developed probably in the Middle Miocene. These stepwise fault systems have total vertical displacement is range of 500 to 2000 meters.

The pre-Neogene basement rocks are classified mainly into two groups, metamorphic and non-metamorphics in the southwest of Turkey. The metamorphic basement rocks of the Sultandag and Menderes Massifs composed of various metamorphic constituents of ranging in age between Precambrian and Upper Paleozoic. The Menderes massif consists of, dominantly bottom to top, gneiss, schist and marble constituent. The Sultandag massif composed mainly of low-degree metamorphic rocks, meta-sediments and recrystallized limestones. The pre-Neogene non-metamorphic basement rocks consist of dominantly platform-type autochthonous carbonate and flysch assemblages and allochthonous ophiolite complex, between Triassic and Eocene in age.

The continental Neogene basins in the West Taurus belt have been filled by the mainly alluvial and fresh-water lacustrine sediments. Both of the sedimentary facies generally have a gradiational contacts in lateral and vertical direction. The sediments within the these basins consist of generally, from bottom to top, conglomerate, crossbedded sandstone, marl, claystone, with lacustrine limestone, and indicate that a transgressive sequence. The alluvial-fan conglomerates in the upper most part of the Neogene sequence, overlies with erosional contact on all older the sediments. Probably, the alluvial-fan conglomerates characterized the regression in the Early Piocene. Therefore, the continental Neogene sediments in the southwest of Turkey occur a complete stratigraphic serious which composed both of transgressive and regressive sedimentary sequence.

The total sedimentary thickness within the basins are range 100 to 1000 meters, there are important lithofacies and thickness differences among the continental basins. Those dissimilaritys are originated from the tectonic and sedimentologic conditions which controlled the sedimentation.

The alluvial-fan deposits in the continental basins consist of blockly horizons, coarse-grained conglomerate and locally sandstone and mudstone interbeds which were derived from pre-Neogene basement rocks in the nearly adjacent areas of basins. The coarsening-upward sequence, seasonal cyclotems, channel-fills, pebble oriantation and locally basement-wash interfingers are main sedimentary structures of the alluvial fan deposits. The alluvial-fan deposits reach maximum thickness to the margin of basin under the influence of growth faults thus, the thickness of the sediments decrease in short distances, and graded into the fluvial and lacustrine sediments in the lateral direction. Therefore, the alluvial-fan deposits developed to the margins of basin, show an ankle-shape geometric feature. These sediments are separated from fluvial deposits with coarse-grained and blockly features, irregular inter-structure and important changes of thickness into short distances. The fluvial sediments within the continental basins generally consist of well-sorted and large-scale cross-bedded channel-fill conglomerate, small-scale cross-bedded and rippled standstone, well-laminated silt-stone and mudstone with locally coal interbeds. The fluvial deposits interfingering with flood-plain and lacustrine sediments to the central parts of basin. Depend on these stratigraphic relations, the dark-coloured mudstone and claystones occur as dominate constituent of the lacustrine sequence, and rich in organic matter.

The lacustrine sediments consist of from bottom to top, mudstone, claystone, porous limestone and volcanoclastic rocks. The varv-like regular lamination gypsiferous intercalations, and clayey interbeds rich in organic matter are wide-spread in the lacustrine sediments. The lacustrine limestones are regular bedded, large porous and contain abundant gastropod and algal constituent. Generally, the lacustrine sediments rest transgressively the under-lying alluvial-fan and fluvial sediments, within the all continental basin in the west Taurus belt (Figure 2).

The main part of economic lignite deposits in the continental Neogene basins occur in the upper most part of the fluvial sequence and overlian concordantly by the lacustrine sediments. Generally, lignites of the west Taurus region, were deposited in the flood-plain marshes of the fluvial system and locally mud-plain of lake margin environment. The dark coloured mudstone and claystone with coaly plant roots are wide-spread in the lower part of the lignite horizon. The palinologic data indicate that the lignites were deposited under the subtropically and moisturaly climatic conditions during the Middle-Late Miocene period. The petrographic results indicate the lignites occuring mainly from the forest-moor plants and small amount of water-plants. The coal rank of the regions are range between dull brown coal and sub-bituminous coal, but the lignites which placed within the active tectonic zone have better quality than other districts of the basin, and its reserves range 5 or 100 million short tons.

The alluvial-fan, fluvial and lacustrine sediments which were filled the continental Neogene basins of the SW-Turkey and deposited under the influence of paleomorphological conditions of basins and were controlled by the growth faults to the margins of basins. Lignite deposits within the region reach the suitable size and thickness in the half-stable basins which have slowly subsidance. But the economic lignite deposits are not occurred in the unstable basins, because the coarse clastic alluvial sedimentation is more dominant than swamp and lacustrine fine sediments in those sediments.

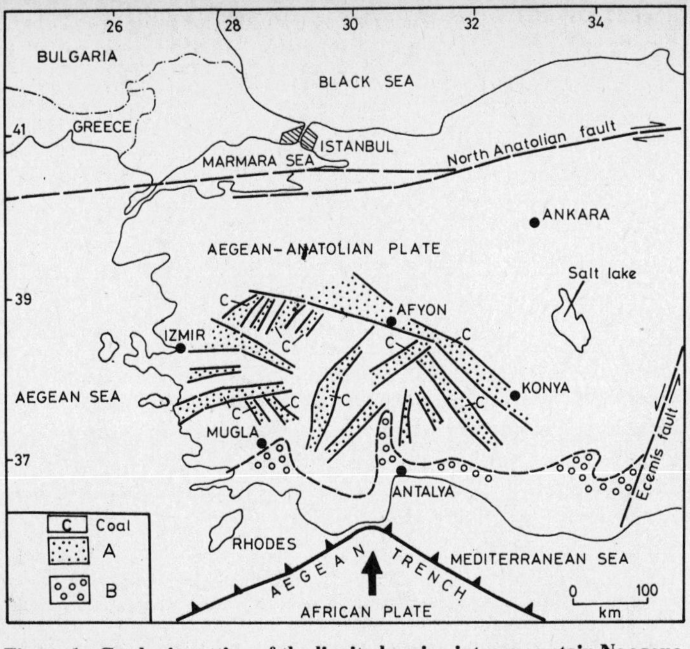

Figure 1 : Geologic setting of the lignite-bearing inter-mountain Neogene basins in Southwestern Turkey. A:Marine Neogene sediments, B:Continental Neogene sediments, C:Location of lignite deposits within the inter-mountain continental basins.

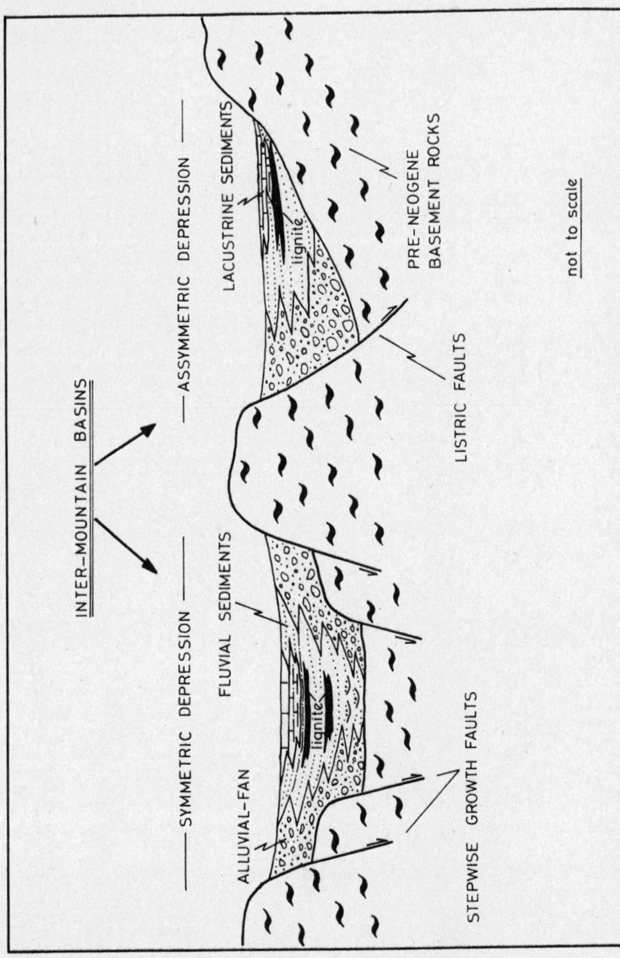

Figure 2 : Depositional model of the lignite-bearing inter-mountain Neogene basins of Southwestern Turkey.

YIM, WYSS WAI-SHU, Department of Geography & Geology, University of Hong Kong, Hong Kong

Electron Spin Resonance Dating of Alluvial Zircons and the Origin of Tin Placers in Northeastern Tasmania, Australia

Electron spin resonance (ESR) is a promising non-destructive method for dating and provenance identification of alluvial zircons The method is based on using the ESR signal to measure the amount of radioactive damage as a result of impurities present in the crystal structure of zircons. Since the amount of damage is approximately proportional to age, by using zircon samples dated previously by the fission track method, it is possible to calibrate the ESR signals and estimate the age of zircons falling within the range of calibration. Furthermore, the ESR signal obtained for zircons with different types of provenance are distinctive.

Two mineralogical types of alluvial zircons have been identified in a heavy mineral provenance study in the placer tin-field of northeastern Tasmania (Yim et al. 1985). The first type is euhedral and fine grained (mostly 0.1 to 0.3 mm). Dating gave a combined age of 367 ± 15 Ma confirming their derivation from Upper Devonian granites. The second type is round and coarser-grained (mostly 2 to 5 mm). They gave a combined age of 46.7 ± 0.6 Ma which agrees with the K-Ar radiometric dates obtained for Tertiary basalts (Sutherland & Wellman 1986).

The presence of both types of zircon in stanniferous placers confirms that they have been subjected to episodic recycling. Because most of the placer reserves occurs within deep leads overlain by Miocene basalt K-Ar dated at ca. 16 Ma (Brown 1977), the deep leads are stratigraphically confined by the 46.7 and 16 Ma ages of the older and younger basalts. Consequently, the cassiterite present in the placers which have derived from the Upper Devonian granites are clearly related to pre-Middle Eocene episodes of recycling.

REFERENCES

Brown, A. V. 1977. Preliminary report on age determination of basalt samples from Ringarooma 1:50 000 sheet. Unpub Rep. Tasmania Dept. of Mines 1977/25.

Sutherland, F. L. & Wellman, P. 1986. Potassium-argon ages of Tertiary volcanic rocks, Tasmania. Papers & Proceedings Roy. Soc. Tasmania 120: 77-86.

Yim, W.W.-S., Gleadow, A.J.W. & Van Moort, J.C. 1985. Fission track dating of alluvial zircons and heavy mineral provenance in northeast Tasmania. J. Geol. Soc. London 142: 351-356.

Z

ZHANG, Peizhen, M. A. ELLIS, D. B. SLEMMONS, and Fengying MAO, Center for Neotectonic Studies, Mackay School of Mines, University of Nevada at Reno, USA

Active Faulting, Low Angle Detachment, and Late Cenozoic Tectonic Evolution of the Death Valley Extensional Region

The Death Valley Extensional Region is the most active area in the entire Basin and Range province. It contains a spectrum of complex Cenozoic extensional structures, such as low angle detachment surfaces, high angle normal faults, grabens and half grabens, tilted blocks, turtleback surfaces, and chaotic megabreccias, all of which are believed to be associated with slip on large scale west-dipping low angle detachment.

Studies on active faults in the region indicate that most recent tectonic activity occurs from the Death Valley fault zone to the west. The minimum Holocene slip rate along the Panamint Valley — Saline Valley fault zone is 2.5 mm/yr. The minimum Quaternary slip rate in the Owens Valley is between 0.7 to 1.1 mm/yr. The exact amount of slip rate is unknown along the Death Valley fault zone, but it must be larger than that of the Panamint Valley fault zone because the spectacular geomorphology suggests much stronger activity. East of the Death Valley, the faults are either no longer active or active at a very slow rate. Tertiary low angle detachment surfaces are broadly exposed, but none are found or reported to be active in the area east of the Death Valley. West of the Death Valley, geological and geophysical evidence suggests active low angle detachment underneath the Panamint Valley. We believe that the pattern of tectonic activity in the Death Valley Extensional Region represents westward migration of tectonic activity which is supported by Cenozoic volcanic history of the same area.

This westward migration can be interpreted to be a result of movement along the major west-dipping low angle detachments. As the hanging wall moves from the footwall, the isostatic rebound causes uplift of footwall. This footwall uplift inactivates the area behind it. New near-surface high angle listric normal faults will develop in front of the uplift and cut the uplifted low angle detachment. We suggest that movement on the major west-dipping low angle detachments control late Cenozoic tectonic evolution of the Death Valley Extensional Region.

ZIMING, YIN, Central South University of Technology, Changsha City, P.R. China

A Study of Genetic Mineralogy of the Late Palaeozoic Bauxite Deposits in China and Formation of the Deposits

With the help of modern research methods (SEM, STEM, EDS, EMP, HOIA etc.), the author studied the origin of bauxite mineralogy.

From the study of the stable isotope geochemistry of hydrogen and oxygen in the AI-minerals and their intergrown kaolinite, it indicates that the coexisting wate which took partin the forming procedure of kaolinite, boehmite, gibbsite and diaspore possessed the characteristics of meteoric water. The changes of δD, $\delta^{18}0$ values of the diaspore samples are connected with the content proportions of clear spar diaspore without Ti and Fe to Ti and Fe containing micrite diaspore. δD and $\delta^{18}0$ values of spar diaspore are much lower than that of micrite diaspore containing Ti and Fe. The former is the product of underground hydrothermal activity. The later is the product of dehydrated recrystallization and its textural water inherited the late Palaeozoic meteoric water.

The study show that diaspore in China is of polygenetic characteristics. The kinds of diaspore include terrigenous diaspore formed in ancient weathering crust, dehydrated recrystallization diaspore formed during diagenesis-epigenesis stage and diaspore formed by middle-low temperature underground hydrothermal activity and by dynamic metamorphism. From the angle of chemical thermodynamics and chemical dynamics the article puts forward “the theory of weathering dewatering”, makes up for “the traditional theory of weathering formation” and “the theory of dewatering formation”. The formation of boehmite, gibbsite and soimonite has been analysed. This paper brings out the general characteristics of the formation of the Al-minerals in the mentioned bauxite.

This paper deals with the protracted nature and complexity of the formation process of diaspore bauxite. The concept of the longitudinal genetic chain of bauxite has been discussed and summarized as the Al-richening model of the laterite soil bauxite—sedimentation—transformation. The author has considered that the diaspore bauxite in China is the polygenetic compound ore deposit.

ZWETKOV, K., P. STOEVA, E. TOZEV, Geol. Ent. for Laboratory Research, Sofia, Bulgaria

Influence of the Diagenetic and Mechanical Heterogenetry of the Salt Diapir on its Geomechanical Properties

The existent salt deposit of Mirovo-Provadia (Bulgaria) present a diapir structure of Upper-Perm age, introduced in the whole profile of Mesozoic and convert with tertiary and quaternary sediments. The internal and external micro- and macrostructures of the diapir are complex due to the intrusion of various viscid salt and salt's solution, the resistence of tectonic forms in the rocks of the roof, the neotectonic processes in the nonsalt sediments, the postformation salt extrusion and the action of the depth creep of the salt's massif.

The study and the exploitation of the stock is complicated by the mezostructure of the salt parts in the vertical and lateral direction. It appears as structural and textural heterogenety and the presence of defects, different classified by size and morphology of the crystals, different ratio salt and nonproductive component with brecciated structure, from the presence of the inner mirror zones.

The laboratory investigations of the strength, deformations and rhéology of the space oriented samples confirm the influence of the local mezostructure and microtectonics as elements for integral estimation of the diapir.

The correlation dependence between the strength properties and the complex heterogenety are obvious for the mechanical conduct of the stock in specific parts as well as the variation values in the whole range of the structure-texture unhomogenety.

The deformative properties correspond to the classified correlation and by means of standarisited dependence it is possible to find the opposite correlation.

The rhéological properties in the different zones of the salt's stock appear in different ways: stage creep between chaotic coarse and gigantic crystals (macrohéological phenomenon), dislocated creep in the oriented structures (mezo- and macrohéological phenomenon), diffusion creep on the basic surfaces or in the defects of the crystals (microrhéological phenomenon).

The complex investigation of the salt stock allows to explain the history of formation and development of the salt deposit; the contemporary deep and shallow deformation of creep inside the massif and the exploiting possibilities and protection of the environment.

Addendum

This section is comprised of abstracts which, due to clerical anomolies, were not complete as printed in the regular section.

AWASTHI, A. K., P. SRIVASTAV and V. KUMAR, University of Roorkee, Roorkee, India

Principal Component Analysis of Common Inorganic Constituents of Waters Associated with Oil Fields, Gas Fields, and Non-Petroliferous Aquifers

Water is intimately associated with petroleum accumulations. It occurs in the pore spaces of rocks as interstitial water and may vary from a maximum of 100% saturation of pore space in an aquifer below the hydrocarbon pool to irreducible limits within the pool. It may, in many cases, be stagnant since its entrapment or may be in continual movement. Although its chemical patterns are complex and vary from area to area, the petroleum pool waters are mostly brines.

Characterization of waters associated with the oil pools, gas pools, and dry wells may be of great value in the correlation of stratigraphic units, in the exploration of petroleum hydrocarbons, and in enhancing of oil recovery. The present investigation explores the possibility of using common inorganic constituents of waters to differentiate between the waters of oil fields and gas fields from waters of nonpetroliferous aquifers, using a multivariate approach.

Chemical analyses of water samples from 18 oil wells, 30 gas wells, and 16 dry wells, obtained mainly from the available published records form the basis of this exploratory investigation. The common inorganic constituents available for the study are NA^+, K^+, Ca^{++}, Mg^{++}, Cl^-, SO_4^{--}, CO_3^{--}, HCO_3^{--}.

Any multivariate data set may be represented in the form of a matrix. In this study, a 64 × 6 order matrix of the data set was constructed. Based on these data, a variance-covariance matrix of order 6 × 6 was restructured. This matrix represents a coordinate of points in six-dimensional space. The points may be thought to lie on the surface of a hyperellipsoid whose center is the origin of the coordinate system. In effect, the hyperellipsoid forms an envelope which just encloses these points. the lengths of each of the six axes, known as principal axes, are equal to six eigen values associated with the variance-covariance matrix; the highest eigen value represents the length of the longest axis, the second highest value represents the second axis, and so on. Each eigen value accounts for the variance of data associated with each axis. The sum of these six eigen values equals the variance of the data set. In general, most of the variance of the data set is found associated with the first few axes. This approach, therefore, attempts to reduce the dimensionality of the data set without losing much of variance and taking into account all variables simultaneously.

In the present investigation, more than 88% of the variability (Table 1) is accounted by first three eigen vectors (Table 2).

The first principal axis, accounting 58.7% of total variance is loaded with almost all constituents, Cl^-, SO_4^{--}, $Na^+ + K^+$, and Mg^{++}. The second axis with second highest eigen value accounts for only 17% of total variance and is loaded mainly with SO_4^{--} and HCO_3^{--} + and CO_3^{--}. The third axis, accounting for only 12.6% of the total variance is loaded with mainly SO_4^{--}.

Considering two axes at a time, bivariate plots of principal scores were prepared using the inorganic constituents of 64 water samples. These plots bring out three clusters of points—characterized by oil field waters, gas field waters, and waters not associated with petroliferous basins, respectively. The plots can be used to allocate a sample of unknown source to one of these three categories and accordingly a decision regarding petroleum exploration may be made. This approach, thus, appears to hold promise in oil/gas exploration. However, the plots need further refinement in terms of the boundaries of various fields, using more and more data for which studies are being done.

Table 1—Eigen values of variance-covariance of inorganic constituents of water samples.

Principal Axes	Eigen values	Percent trace	Cumulative percent trace of variance - covariance matrix
I	3.5227	58.7113	58.7113
II	1.0246	17.0773	75.7886
III	0.7568	12.6138	88.4024
IV	0.4189	6.9822	95.3846
V	0.1485	2.4666	97.8512
VI	0.1289	2.1488	100.0000

Table 2—Principal axis vectors of six-dimensional hyperellipsoid of variance-covariance matrix.

Variables	Principal Axis Vectors I	II	III	IV	V	VI
Cl^-	0.4330	-0.1066	-0.2491	-0.7965	0.3035	-0.1124
SO_4^{--}	0.2871	0.4314	0.8125	-0.1849	-0.1292	-0.1431
$Na^+ + K^+$	0.4954	-0.0827	-0.1841	-0.0031	-0.7531	0.3829
Ca^{--}	0.4730	-0.2594	-0.0680	0.4097	-0.0178	-0.7323
Mg^{--}	0.4867	0.1451	0.1451	0.3714	0.5622	0.5284
$HCO_3^{--} + CO_3^{--}$	0.1554	0.8483	-0.4671	0.1601	0.0874	-0.0692

COHEN, R. E., and M. J. MEHL, Naval Research Laboratory, Washington DC, D. ISAAK, University of California, Los Angeles, CA, and H. KRAKAUER, College of William and Mary, Williamsburg, VA

First Principles Calculations of Elasticity at High Pressures and Temperatures

Elasticity of minerals at high pressures and temperatures is one of the most important quantities for understanding of the deep earth, yet it is also among the most difficult properties to measure. Fortunately, methods have recently been developed that are powerful enough to predict high pressure and temperature elastic properties of complex minerals from first principles. Such studies elucidate the physics behind elastic behavior of minerals in a more rigorous manner than empirical systematics or structural models.

In order to calculate elastic constants for complex minerals, or to calculate thermal effects on elasticity, it is necessary to use a simplified model that allows calculation of the full dynamical matrix and its derivatives. An *ab initio* model known as the potential induced breathing (PIB) model as been demonstrated to give generally reliable predictions for oxides, including the Cauchy violations in simple oxides (Mehl et al., 1986). The PIB model is a Gordon-Kim type model, in which the crystal charge density is approximated by overlapping ionic charge densities, and the local density approximation (LDA) is used to calculate the total energy. The "size" and energy of the O_{2-} ion is a function of the crystal environment, and this has been found to have a strong influence on elastic properties.

The elastic constants of complex minerals can be calculated directly (Cohen, 1987a) by finding the PIB dynamical matrix (Cohen et al., 1987) and taking derivatives with respect to wavevector. Using this approach, elastic constants can be calculated that vary from a few percent from experiment to several tens of percent under room conditions. This technique was used to predict the single crystal elastic constants of orthorhombic $MgSiO_3$ perovskite, a complex structured mineral with 20 atoms per unit cell, before any experimental data were available (Table 1) (Cohea, 1987b). Subsequent experiments verified the accuracy of the calculations (Yeganch-Haeri et al., in press). The pressure derivative of the elastic moduli were also calculated and are most likely reliable; it will be some years before pressure derivatives of $MgSiO_3$ perovskite will be measured!

One disadvantage of the long-wavelength limit approach used above is that it assumes that elastic constants are functions of temperature only due to thermal expansion. In order to go beyond this approximation, we have used a quasiharmonic approximation for the free energy by calculating vibrational frequencies as functions of strain for MgO, and have calculated the high temperature and high pressure elastic constants of MgO (Table 2) (Saak and Cohen, in press). We find that the predicted thermal behavior is in excellent agreement with experiments up to the highest temperatures that experiments have been made.

Different charge densities and functionals can be used in Gordon-Kim models such as PIB, and different schemes can be used to model the stabilization of the O^{2-} ion. We have recently found that similar, but quantitatively different, results are obtained when we use local density (LDA) charge densities, which are closer to those obtained in state-of-the-art electronic structure calculations instead of self-interaction corrected charge densities, and use the Kohn-Sham rather than the Thomas-Fermi self-energy. This technique has the advantage of yielding more reliable lattice dynamics results. This technique was used by Isaak and Cohen (in press) and gives better room condition elastic constants as well, compared with those found by Mehl et al. (1986). Whether this technique will be superior in other materials is not yet clear, and we will present several different techniques for obtaining *ab initio* potentials for oxides.

Ab initio models are by their very nature dependent on the charge density model and on an approximation to the kinetic energy. A better, but much more computationally intensive approach (up to factors of 1 million times!) is to use state-of-the-art electronic structure and total energy methods, such as the linearized augmented plane wave (LAPW) method. Bulk moduli as functions of pressure have been obtained using the LAPW method for MgO, CaO (Mel et al., 1988), and FeO. One advantage of this approach is that elastic constants can be calculated for materials regardless of the complexity of their electronic structure.

REFERENCES

Cohen, R. E., 1987a, Geophysical Research Letters, v. 14, p. 37.

——— 1987b, Geophysical Research Letters, v. 14, p. 1053.

——— L. L. Boyer, and M. J. Mehl, 1987, Phys. Rev., B 35,p. 5749.

Isaak, D. and R. E. Cohen, EOS, in press.

Mehl, M. J., R. J. Hemley, and L. L. Boyer, 1986, Phys. Rev., B 33, p. 8685.

Mehl, M. J., R. E. Cohen, and H. Drakauer, 1988, Journal of Geophysical Research, v. 93, p. 8009.

Yeganeh-Haeri, A., D. J. Weidner and E. Ito, in press.

Table 1—Elastic constants (in GPa) at zero pressure from $MgSiO_3$ perovskite.

	PIB		Exp
	Static	298K	
C_{11}	548	531	520
C_{12}	54	44	114
C_{22}	551	531	510
C_{13}	153	143	118
C_{23}	175	166	139
C_{33}	441	425	437
C_{44}	241	237	181
C_{55}	253	249	202
C_{66}	139	136	176
K	256	249	245
μ	196	192	184

Table 2—Adiabatic elastic constants (GPa) temperature and pressure derivatives, volume, thermal expansivity, and Gruneisen parameter at zero pressure and 300 K for MgO.

	PIB	Exp*
$V_o(Å^3)$	18.66	18.66
K_s	182	163
dK_s/dP	4.12	4.13
dK_s/dT	-0.017	-0.015
$C_{11}-C_{12}$	188	201
$d(C_{11}-C_{12})/dP$	6.12	7.56
$d(C_{11}-C_{12})/dT$	-0.061	-0.060
C_{44}	188	156
dC_{44}/dP	0.95	1.11
dC_{44}/dT	-0.012	-0.012
α	24	31
γ	1.45	1.52

*Jackson and Niesler (1982); Suzuki (1975); Sumino et al. (1983)

D'HEUR, M., Petrofina

The Eldfisk Field

The Eldfisk field is located in the southernmost part of the Norwegian sector of the North Sea, 16 km south of the giant Ikofisk Field. Eldfisk produces oil and gas from chalks of early Paleocene and later Cretaceous age.

The ultimate recovery is expected to be 400 million STB oil plus 1.4 trillion CFT gas given the current production scenario of simple pressure depletion and solution gas drive. The field was discovered in 1971. The reservoir consists of allochthonous (resedimented) biomicrites distributed over a pre-existing relief of half-horst type resulting in drastic thickness variation of the reservoir layers. Chalk porisity often varies between 30 and 40% in the oil zone, which reflects an extremely high degree of porosity preservation for a reservoir depth of more than 9000 feet.

despite its high porosity, the reservoir has an average matrix permeability of less than two millidarcies. The maximum gross pay and net pay thickness are 650 feet and 600 feet respectively. The trap is a 12 km (7.5 mile) NW-SE trending anticline. The seal consists of a thick tertiary claystone cover. The source rock is the prolific black shale of the Mandal formation (upper Jurassic).

The natural fracture intensity is critical for productivity of the chalk reservoir. In areas of moderate fracture intensity, initial production rates of 5000 to 10,000 BOPD per well have bene achieved whereas in areas of high fracture intensity, some wells have produced at initial rates exceeding 20,000 BOPD.

With 40 producing wells, it is expected that a recovery factor of 15 to 17% may be achieved by natural depletion.

Peculiarities of the Eldfisk Chalk reservoir, its anomalously high porosity, the variation in porosity, facies and thickness of the reservoir layers and the hydrocarbon distribution will be discussed.

FERNANDEZ, L. M., and A. DU PLESSIS, Geological Survey of South Africa, Pretoria, Republic of South Africa

Application of Weibull Distribution Function in Analysis of Mine-Induced Seismicity in South Africa

Rockbursts and rock falls are a major hazard to safety in South Africa's deep gold mines and account for 55% of all fatalities. A good overview of the intensive research efforts that have been devoted to an understanding of mine-induced seismicity is given by Gay and Wainwright (1984).

About 1000 mine-induced events of Richter magnitude 2 and above are detected and located yearly by the National Seismological Network of the Geological Survey of South Africa. Shapira et al (1987) have shown this data set has been complete for events above magnitude 3.3 for the past 18 years, since 1971. This seismic history has provided useful insight into the frequency of occurrence and magnitudes of earthquakes that may be expected in the different gold-mining regions. This paper investigates whether, additionally, a relationship exists between an occurrence of an event and the time elapsed since a former, similar event.

In the field of quality control analysis, it has been found that the failure times of products do not fit the classical Gaussian distribution but instead follow a Weibull (1951) distribution function. Hagiwara (1974) argued that where crustal strain increases linearly with time, earthquake occurrences can, in fact, be considered "product failures" and that a Weibull distribution could consequently be used to model a region's temporal pattern of seismicity. He demonstrated the applicability of his model to the South Kanto district southwest of Tokyo, and since then it has been shown that at many other locations in the world and the model also holds. This study investigated whether mine rockbursts too follow a similar pattern. The interest in matching data to a Weibull function is that once the parameters of the best-fitting distribution have been established, a conditional probability of earthquake occurrence can be calculated that is a function of the time elapsed since a former, similar event occurred.

Data from four separate areas of seismicity were considered: the East Rand (ERM), the Far West Range (FWR), the Klerksdorp district (KD), and the Orange Free State gold mines (OFS). The factors that encouraged application of the method were (1) that each of the areas by virtue of small size (20 km × 20 km) constitutes a closed system where the seismicity is controlled by a single set of seismotectonic and mining factors, and (2) that in each area the steady rate of mining results in persistent and even strain; a prerequisite to the application of the method.

The results show that for events of magnitude 3.6 and above in FWR, OFS, and KD the conditional probability increases with the time elapsed since the last event, but for ERM which is not the case. There, activity does not appear to relate to a definite build-up of strain and consequent failure. The gold field that in absolute terms has the highest seismicity is FWR: In Figure 1 it is shown that there, a 75% probability exists for an event of magnitude 3.6 (or above) within 30 days if no event of such magnitude took place in the past 60 days. For the same set of circumstances, the probability at KD is 45%, at OFS 20%, and at ERM 6%. Figure 1 also indicates that the probability of occurrence increases more rapidly at FWR than at other gold fields as the elapsed time increases. There is an 85% probability of an occurrence within 30 days if no event has taken place for 180 days. This probability level is not even reached after 500 days at KD, the second most seismically active area.

In conclusion, a Weibull distribution can be fitted to holding times between rockbursts. This provides support for the notion that seismic hazard in gold mines is not static and depends on past seismicity. In three South African gold fields, the probability of an earthquake increases slightly as the time elapsed since the last occurrence increases.

REFERENCES

Gay, N. C., and E. H. Wainwright, 1984, Rockbursts and seismicity in mines: The South African Institute of Mining and Metallurgy, Symposium Series 6, 363 p.

Hagiwara, Y., 1974, Probability of earthquake occurrences as obtained from a Weibull distribution analysis of crustal strain: Tectonophysics, v. 23, p. 313-318.

Weibull, W., 1951, A statistical distribution function of wide application: Journal of Applied Mechanics, v. 18, p. 293-297.

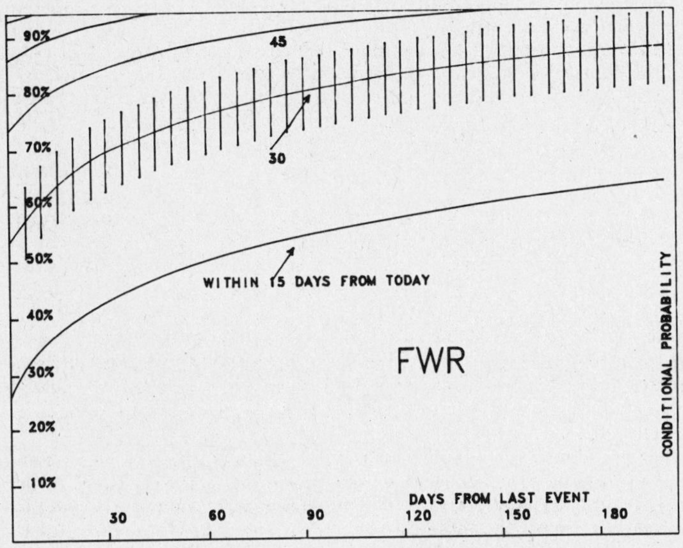

Figure 1.

GORNYI, V.I., A.A. TRONIN, and B.V. SHILIN, All-Union Remote Sensing Research Institute, Leningrad, USSR, and A.G. SALMAN, Academy of Sciences of the USSR, Moscow, USSR

Terrestrial Infrared Radiation as Indicator of Seismic Activity

The 10.5–11.3 μm spectral band of thermal satellite images of the earth has revealed an increase of infrared radiation (IR) intensity appearing from time to time over particular places of some linear structures in the Soviet Central Asia seismic region.

As follows from the retrospective analysis of daily thermal satellite images covering a 10-year period (about 300 earthquakes) in the Soviet Central Asia region, positive IR anomalies appear sporadically always over the same zones of some large faults. Their certain spatial position and duration enable these anomalies to be picked out against the background of other anomalies. The sporadic anomalies originated at the same time the region became seismically alive.

The best demonstration of such seismic activity is the Gazli earthquake dated 19 March 1984, with a magnitude of 7.2. At the junction of Tamdy–Tokrauss and Talass–Ferghana faults (6 and 5, respectively, in Figure 1) on March 11 an extremely intense and large (about 100,000 km^2) positive anomaly of IR flux (7 in Figure 1) was recorded. Again, the aftershocks of that quake (magnitude 4.3–5.3) were also accompanied by sporadic anomalies at those faults' intersection.

Similar positive anomalies of the terrestrial IR flux have been detected in the Eastern Mediterranean region, regarded as another seismic area. The anomalies concentrate over the extending Hellenic system of displacements within the African platform, and over the faults striking in the direction opposite to the Mediterranean trend (between the delta of the Nile and Gulf of Sidra). In this region, the IR anomalies reflect the seismic events genetically related to the Hellenic arch. Both seismic regions have in common the sporadic anomalies that appear quite far from the related quake centers.

A connection established between the sporadic IR anomalies over the active fault zones and the seismic activity of the region gives reason to question such anomalies' nature. Considering their relatively high rate of generation and development (within a day), as well as their intensity of few degrees and the spreading area coming to several scores of thousand square kilometers, we cannot relate these anomalies directly to the conversion of energy from mechanical to thermal in a process leading to a quake. It is not unlikely that these anomalies are caused by a change in a composition and concentrations of gas impurities in the lowest atmospheric layer over the active faults.

A change in gas composition of the atmosphere may produce various effects, and the greenhouse effect is one of them. A gas IR luminescence is also possible in the atmosphere, this luminescence being similar to the light emission of the atmosphere over the epicentrum zones of some quakes.

Another potential factor giving rise to the terrestrial IR flux anomalies over the active fault zones may involve a variation of moisture content in a soil as a result of groundwater level fluctuations. It is obvious that each factor may take some part in the generation of the sporadic IR anomalies.

Summary: The sporadic IR anomalies revealed on the satellite thermal images of Soviet Central Asia and Eastern Mediterranean seismically active regions are connected by seismicity. The sporadic IR anomalies are associated largely with the crustal quakes and can appear quite far (some scores and hundreds of kilometers) from the centers thereof.

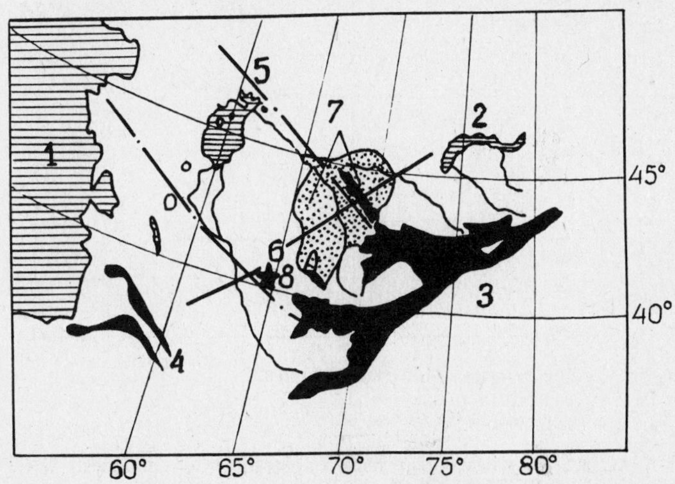

Figure 1—Visual interpretation of NOAA-7 satellite thermal image dated 11 March 1984. 1, Caspian Sea; 2, Balkasch Lake; 3, Tien–Shan; 4. Kopetdag; 5, Talass–Ferghana fault; 6, Tamdy–Tokrauss fault; 7, positive anomaly of IR flux; 8, epicenter of the Gazli earthquake.

KONO, M., Tokyo Institute of Technology, Tokyo, Japan, and Working Group for Japanese GGT

Transects across Trench-Arc-Back-Arc Systems: Japanese Case

We have prepared five routes for the global Geoscience Transect program (Fig. 1). In the Japanese transects, the emphasis is laid on the traverse of the open sea-island arc-back-arc basin-continent system, which will provide a unique opportunity to study the subduction-accretion process associated with the plate boundaries between the Pacific, North American, Eurasian, and Philippine Sea plates.

Four of the transects cross Honshu island while the last one is a long traverse of the Philippine Sea and East China Sea.

1. Hokkaido Transect (Coordinator: Gaku Kimura):
   This transect crosses the Kuril trench, Hokkaido island, and the northern end of the Japan Sea. Hokkaido is thought to have been produced by the collision of two island arcs from east and west. The collisional structure is most well developed near Hidaka mountains, which will be the focus of this transect.
2. Northeast Japan Transect (Coordinator: Kensaku Tamaki):
   This route passes through the Tohoku (northeastern) district of Honshu island at about 40°N latitude, and extends to Shatsky rise in the Pacific basin to the east and Japan Basin in the Japan Sea to the west. This will supply a typical section of the well-developed island arcs.

3. Central Japan Transect (Coordinator: Nobuaki Niitsuma):
This goes through the complex Fossa Magna area, which is the site of the collision between the Philippine Sea plate (Izu, Tanzawa, etc.) and Honshu island. The oceanward extension passes through the TTT triple point of Pacific, Philippine Sea, and Eurasia (North America?) plates.
4. Southwest Japan Transect (Coordinator: Asahiko Taira):
It starts from Shikoku Basin in the Philippine Sea, crosses Nankai Trough, Shikoku island, Seto Inland-Sea, Chugoku district of Honshu island, and extends to the western part of the Japan Sea.
5. Philippine Sea Geotraverse (Coordinator: Seiya Uyeda):
This will be produced by an international cooperation between Japan, USSR, and China. This transect likes on the 18 °N latitude in the Philippine Sea, changes direction to the northwest, crosses Ryukyu Trough and East China Sea, and finally connects with the southeast-northwest transect of China.

The progress made in the preparation of these transects will be reviewed, and some interesting examples of the sections will be shown.

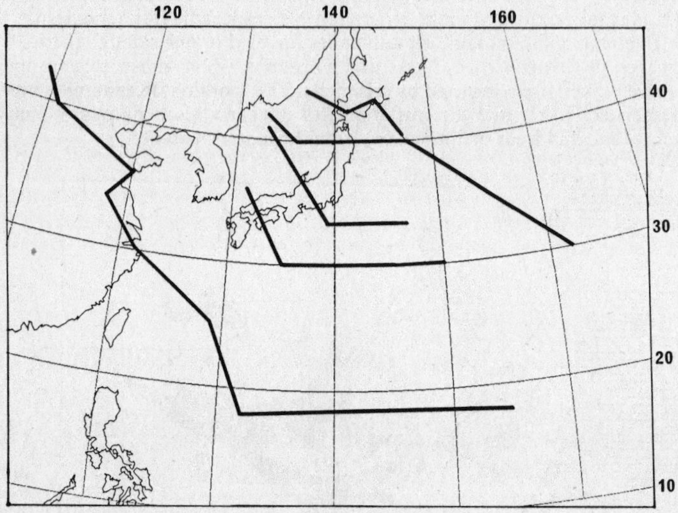

Figure 1.—The Japanese GGT transect routes.

POGACSAS, G., GKV, Geophysical Exploration Co., Budapest, Hungary, and H. P. SEIFERT, ÖMV-AG, Vienna, Austria

Vienna and Pannonian Basin in Central Europe: Comparison of Neogene Sedimentation and Tectonics

This paper is the result of the Austrian–Hungarian scientific geological collaboration.

The Pannonian Basin represents a Neogene–Quaternary system of several deep subbasins or depressions.

The Vienna Basin is an external basin northwest of this above mentioned system (Figure 1). It is situated in the middle of the Mesozoic–Early Tertiary Alpine–Carpathian mountain belt and was formed during the last phase of orogenic movement and the overthrusting of the upper crust northwestwards at the Bohemian Massif. Changing tectonic regimes during the Miocene built the complex present stage (Figure 2). Subduction of the Variscan Massif below the Pannonian microplate along the Pericarpatian lineament in the southeast and compression and thrusting to northwest caused the extension of the upper crust to northeast and the development of the basin (22.5–16.5 m.y.).

A partial uplift, a relief inversion of the basement, and a new tectonic regime took place 16.5 m.y. ago. Strike-slip faulting, striking north-northeast–south-southwest, penetrating down into the crust caused the pull-apart phase and a rapid subsidence. The regional uplift of the Alpine–Carpatian orogene and the crust extension with graben-like faulting since 8 m.y. is still going on now.

The basement of the Pannonian Basin is built by several microplates consisting of Eastern Alpine thrust sheets, Upper Paleozoic–Mesozoic sediments, and Lower Paleozoic metamorphites. Due to the Alpine orogenic movements (Middle Cretaceous–Early Miocene) the sedimentary and crystalline basement was organized into tectonic belts. Along their contacts imbrication and overthrusting can be observed. The detailed analysis of low frequence seismic events within the Neogene basement suggests intensive local compressional folding and thrusting during the Late Cretaceous and extensional fracturing during the Miocene.

The main subsidence phase of the Vienna Basin was the pull-apart phase (16.5–11 m.y. ago). The increasing sediment input resulted in the highest sedimentation rate and the thickest formation at this time (Figure 2). Regional uplift terminated sedimentation 5 m.y. ago.

Neogene deformation of the Pannonian Basin was controlled by strike-slip faulting and local oblique extension and compression.

The rapid subsidence began 11 m.y. ago and was caused by lithospheric extension, combined with outward thrusting of the Carpatian arc, followed by cooling of the lithosphere and massive sedimentation (Figure 2).

After the first phase of the Vienna Basin opening with limnic sedimentation, a marine ingression from the Molasse basin set up a phase of shallow-marine environment (−250 m water depth). During a marine regression phase (17 m.y. ago), the limnic and terrestric sedimentation transgressed 50 km to the south. A great delta complex originated 16.5 m.y. ago using a reactivated west-east graben structure delineated in the underlying Variscan Massif. The delta front area is the most potential hydrocarbon deposit of this region with reserves of 70 million tons of oil and 10 billion m^3 of gas. The shallow-marine environment (−200 m water depth) changed salinity continuously first to brackish and finally to freshwater conditions 8 m.y. ago.

Between 17 and 11.5 m.y. ago, deposition conditions in the Pannonian Basin corresponded to shallow-water, nearshore, marine, brackish environments. During the postrift phase of its evolution (middle Miocene–present), the increasingly subsiding basin was filled in primarily by southward- and eastward-prograding sediment wedges. In some subbasins of the Pannonian Basin, this turbidite sequence represents deep-water conditions between 1000 and 1500 m depth. The interpretation of seismic reflection data combined with the results of magnetostratigraphy in adjacent boreholes proved that, as a consequence of this progradation, the postrift sediments are progressively younger from the basin's margins toward the depocenter.

The Miocene marine faunal assemblages of both basins show varying affinities to neighboring basins; an Atlantic–Boreal affinity between 23 and 18 m.y. ago, a Mediterranean affinity between 18 and 14 m.y. ago and a Caspian affinity between 14 m.y. and the Pliocene.

Sea-level changes in the Vienna Basin are approximately comparable with the global eustatic curve. The geological ages of the several hiatus (e.g., 4.4–5.7; 5.7–6.8; 7.6–7.9 m.y. ago) in the Pannonian Basin, represented by unconformities, checked by paleomagnetic data, correspond approximately with the global eustatic curve (Haq et al, Figure 2).

The whole sediment filling of the Vienna Basin is immature because of the continuous low heat flow that causes a geothermal gradient of 25–32 °C/km. Hydrocarbon generation started 17 m.y. ago, the time of the pull-apart phase of this area, in the underlying upper Jurassic sequence.

The heat flow in the Pannonian Basin reaches 108 mW m^{-2}. The geothermal gradient varies between 42 and 45 °C/km. The hydrocarbon generation in the Upper Miocene turbidite sequence occurred between 5 and 8 m.y. ago, in the Lower Miocene strata between 8 and 11 m.y. ago.

REFERENCES

Horvath, F., and L. Royden, 1981, Mechanism for the Formation of the Intra-Carpatian Basins: A Review: Earth Evol. Sci., v. 1, n. 4, p. 307–316.

Jiricek, R. and P. Seifert, in press, Paleogeography of Neogene in the Vienna Basin and adjacent part of the foredeep.

Pogacsas, G., et al, in press, Correlation of seismo- and magnetostratigraphy in southeast Hungary.

Royden, L., F. Horvath, and J. Rumpler, 1983, Evolution of the Pannonian Basin system: Tectonics, v. 2, n. 1, p. 63–90.

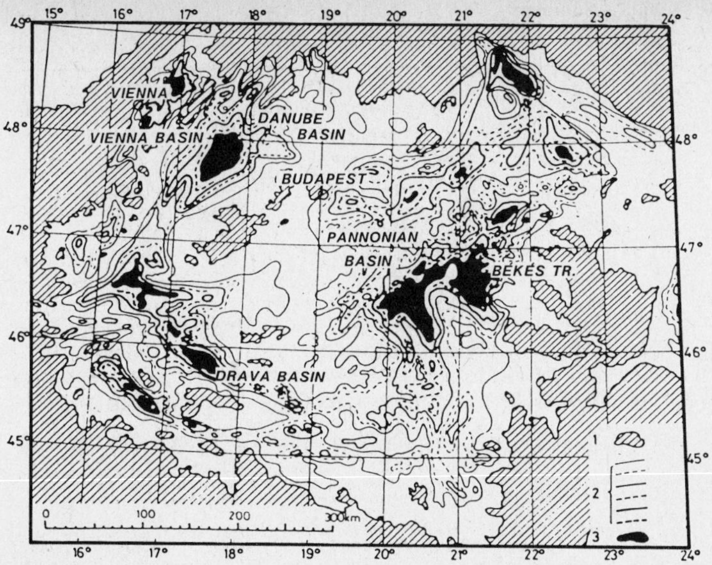

Figure 1—Isopach map of Neogene–Quaternary sedimentary complex of the intracarpatian basins. 1, outcrop of pre-Tertiary complexes; 2, thickness isolines in km; 3, depths over 4.0 km (Cech, 1987).

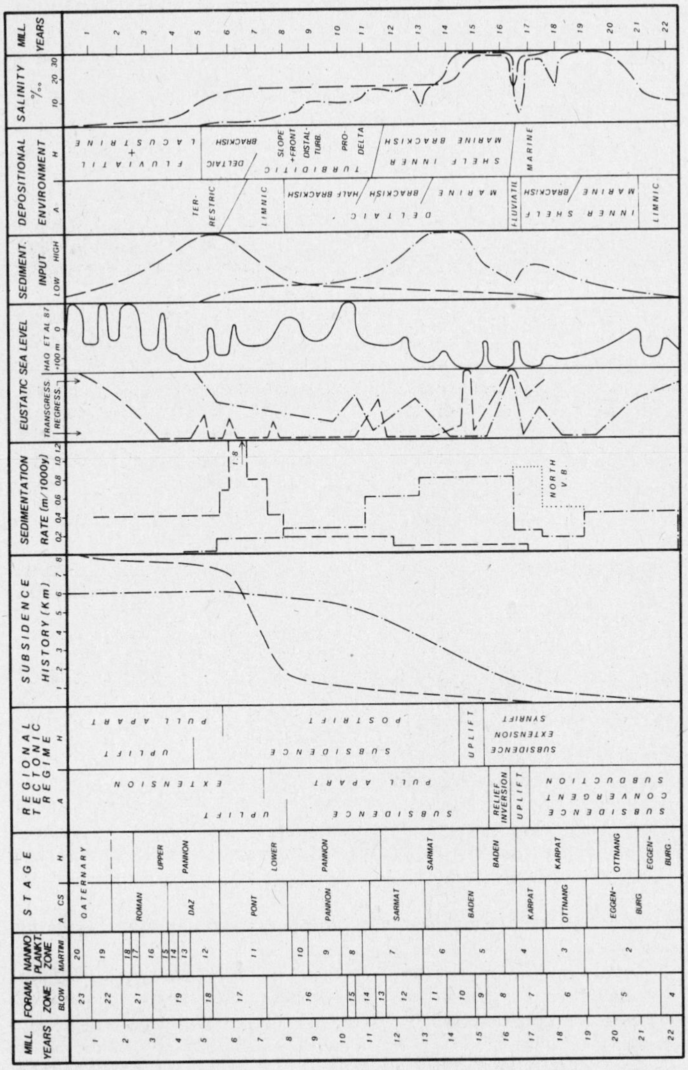

Figure 2—Comparison of the tectonically and sedimentologically development of the Central Vienna Basin (A) and Békés subbasin in the Pannonian Basin (H).

ROSEN, O. M., and V. E. SONYUSHKIN, Academy of Sciences of the USSR, Moscow, USSR

Fluid Inclusions in Polymetamorphic Rocks of Anabar Shield: Differences in Fluid Regime of Metamorphism and Partial Melting

The Anabar shield is the northern window of crystalline basement of the Siberian platform. The Archaean Anabar granulite complex of 2.7 Ga age is one of the most widespread, and the Proterozoic Lamujka complex of 1.9 Ga age occupies about one-fifth of the shield. It is formed by transformation along fault belts of granulites under amphibolite conditions and composed of blastotectonites, hornblende-migmatite gneisses (originated by the retrograde metamorphism of granulites), migmatites, and granites. The methods of thermometry, cryometry, electron microscope, and microprobe analyses were used to study fluid inclusions in quartz of migmatite in enderbite of plagioclase–quartz–hypersthene–magnetite composition in the Anabar complex and in quartz of the Lamujka migmatite composed of potassium feldspar–plagioclase–quartz–biotite–hornblende–ilmenite mineral association.

In the Anabar enderbitic migmatites, the inclusions are described as high-density CO_2 and seem to be of primary and primary-secondary localization (Figure 1a). The Lamujka migmatites contain mostly H_2O inclusions and mixed CO_2 + H_2O inclusions and CO_2 inclusions as well (Figure 1b). NaCl, KCl, and $CaCl_2$ were discovered under cracking in saline residues of H_2O inclusions by electron microscope and microprobe analyses.

In these migmatites, melt inclusions are described to be 10–15 mkm size. The fine-grained devitrificated glass-like material and grains of 1–2 mkm size composed of quartz, potassium feldspar, plagioclase, and biotite were established by microprobe analysis.

Taking into account the temperature of homogenization of melt inclusions and the density of CO_2 inclusions, the conditions of rock-forming processes were estimated as follows: the granulitic facies P = 8.5–10 Kbars, T = 850–900 °C; the amphibolitic facies P = 6.0–6.5 Kbars, T = 780–820 °C. Quartz of the later tectonites was formed under postgranitic H_2O fluids at P = 0.7 Kbar, T = 270 °C.

The data received show that partial melt of granitic substance was generated under granulite facies conditions at CO_2 fluid presence and under amphibolite facies conditions at mostly H_2O fluid invasion. Both of these melting processes are separate at about 1.0 Ga time interval. The important difference consists in change of CO_2 fluid composition in the Arahaean and H_2O fluid composition in the Proterozoic.

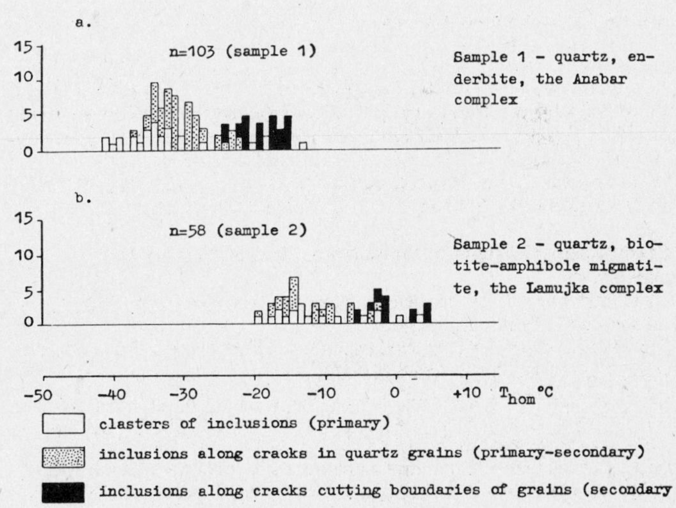

Figure 1—Temperature of homogenization in CO_2 inclusions of different morphology in quartz.

SCOTT SMITH, B. H., North Vancouver, Canada

Lamproites in India

The Majhgawan diamond mine and nearby Hinota pip, Madhya Pradesh are classified here as olivine lamproites and not kimberlites. These bodies, therefore, comprise a newly recognized province of diamondiferous olivine lamproites which are extremely rare worldwide. Other intrusions in India have been proposed as lamproites (e.g., Paul and Sarkar, 1984; Bergman and Baker, 1984; Reddy, 1987). Information for relevant localities is reviewed (Table 1).

Rocks from *Majhgawan* and *Hinota* are lapilli tuffs. The juvenile lapilli are composed of olivine macrocrysts and phenocrysts set in glassy groundmasses containing mica, vesicles, apatite and altered perovskite. The petrographic classification of these rocks is problematic because of their glassy nature and paucity in primary minerals. Various features, however, are atypical of kimberlites and more typical of lamproites. These include the occurrence of the glass, in particular of scoriaceous juvenile lapilli, the complex shapes of the olivine macrocrysts (probably imposed morphology) and phenocrysts (crystal aggregates) and polysynthetic twinning in the phlogopite. Also phlogopite compositions (high TiO_2 contents—5.3–7.3 wt%) and whole rock compositions (e.g., high Ba—3634–27,300 ppm) are more similar to lamproites than kimberlites. These rocks are very similar to those found at Ellendale and Argyle (Western Australia), Prairie Creek (Arkansas) and Kapamba (Zambia). The apparent paucity of mantle-derived xenoliths and xenocrysts at Majhgawan is also similar to that noted in the other lamproites. These data suggest therefore that Majhgawan and Hinota are not kimberlites but should rather be classified as lamproites. The samples examined are hence classified mostly as glassy olivine lamproite lapilli tuffs although some autolithic breccias may be present.

The *Wajrakarur-Lattavaram* bodies (Andra Pradesh) are composed mostly of mica-bearing monticellite kimberlite. Contamination by late stage xenolith digestion produces pectolite-, clinopyroxene- bearing kimberlite. Texturally, the bodies are predominantly hypabyssal but pelletal tuffisitic kimberlite breccia (diatreme- facies) is present in the largest body. They probably represent the root zones of substantially eroded diatremes. Pipes 1, 3, 4, and 6 are typical kimberlites and contain some diamond and relatively common mantle-derived xenoliths and xenocrysts. Pipes 2 and 5 display some petrographic features (e.g., paucity of olivine macrocrysts, probable melilite pseudomorphs) which are not typical of, but do not preclude their classification as, kimberlites. Pipes 2 and 5 also appear to be devoid of diamond, mantle-derived xenoliths and xenocrysts. These differences are not sufficient to suggest that these two pipes represent a different rock type such as lamproite. They are, rather, more extreme varieties of kimberlite.

Lamprophyre dykes both at *Chelima* and in the *Gondwana Coalfields* could include lamproites. Information for *Angor, Banda,* and *Jungel* suggests that they are not kimberlites or lamproites. Insufficient information is available for the intrusions at *Zangamrajupalle, Maddur,* and *Warangal* to comment further but it is doubtful whether they are kimberlites or lamproites.

It is noteworthy that the only confirmed lamproites (Majhgawan) and kimberlites (Wajrakarur–Lattavaram) constitute the only known primary sources of diamond in India. They are Proterozoic and similar in age to other lamproites, kimberlites and related rocks worldwide. The tectonic setting of these provinces may also be similar to those elsewhere. The Wajrakarur–Lattavaram kimberlites occur well within Dharwar craton, while it appears that the lamproites (Majhgawan–Hinota) may occur toward the margin of the Aravalli craton and might even be associated with a rift. Although there are numerous published papers on the "kimberlitic" bodies of India, this review shows the great need for new, detailed petrological studies of many of these localities.

Acknowledgments: The author is indebted to R. B. Hargraves, F. R. Boyd, P. Kresten, and S. C. Bergman for donating samples without which this study would not have been possible.

REFERENCES

Bergman, S. C., and N. R. Baker, 1984, A new look at the Proterozoic dykes from Chelima, Andhra Pradesh, India: diamondiferous lamproite?: GSA Abstracts, v. 16, p. 444.

Paul, D. K., and A. Sarkar, 1984, Petrogenesis of some Indian lamprophyres, *in* Proceedings of SYMPET, Jaipur 1981: Geol. Surv. India Special Publ. v. 12, p. 45–54.

Reddy, T. A. K., 1987, Kimberlite and lamproite rocks of Vajrakarur area, Andhra Pradesh: Geol. Soc. India Journal v. 30, p. 1–12.

Table 1—Summary.

Locality	Name	Ages	Published classifications	Classification this study
1	Majhgawan	1135my	kimberlite	ol lamproite lapilli tuff
1	Hinota		kimberlite	ol lamproite lapilli tuff
2	Angor		kimberlite peridotite-ol-pyroxenite-gabbro	not kimberlite ?peridotite-gabbro
3	Jungel	919my	kimberlite	not kimberlite ?metavolcanics
4	Gondwana Coalfields	110my	lamproites lamprophyres	should be investigated may include lamproite
5	Wajrakarur P1		kimberlite	mica-bearing hypabyssal kimberlite and pelletal tuffisitic kimberlite breccia
5	Wajrakarur P2	840my	lamproite kimberlite	hypabyssal monticellite kimberlite
5	Lattavaram P3	933my 956my	kimberlite	hypabyssal monticellite kimberlite (+\-phlogopite)
5	Lattavaram P4	926my 1023my	kimberlite	hypabyssal mica-bearing monticellite kimberlite contaminated to pectolite-cpx-bearing kimberlite
5	Muligiripalle P5		lamproite kimberlite	hypabyssal melilite- and monticellite-bearing phlogopite kimberlite contaminated to pectolite-cpx-bearing kimberlite
5	Wajrakarur P6		kimberlite	altered hypabyssal kimberlite
6	Chelima	1340my	minette kimberlite/ carbonatite lamproite	minette or lamproite requires further study
7	Zangamrajupalle		kimberlite	requires further investigation
8	Maddur		ol-lamproite	reported data does not support classification as lamproite. further study necessary
9	Warangal		kimberlite/ carbonatite	Reported data unusual for kimberlite or lamproite

Legend : ol - olivine; cpx - clinopyroxene

SUN Tian-xi, Environmental Monitoring Center, Suzhou, China

On Exploiting Drinkable Groundwater Resource under Suzhou

In view of increasing grave pollution of surface water and splendid prospects for exploiting groundwater in Suzhou, the present paper puts forward a technical scheme for the purpose of exploiting drinkable groundwater resource under Suzhou. This is: drinking water for civil use should be drawn from quality groundwater, whereas the existing waterworks could be completely devoted to the uses of industry. As long as 80,000 T of groundwater a day will be exploited, the drinking needs of all residents of Suzhou City could be satisfied.

The bases of the plan are as follows:

1. Examinations confirmed that 179 trace organic compounds have been examined in Suzhou's running water. Of these, the carcinogenic substances (and their concentrations) are benzene (21.1 μg/L), which is 210 times that of USA; cyclohexane (35.80 μg/L);-heptane (5.90 μg/L); tri-toluene (2.89 μg/L); n-hexane (6.70 μg/L); and propyl cyclohexane (4.89 μg/L). Therefore, reinvestigating Suzhou's drinking policy is a task that brooks no delay.

2. Organic pollution in Suzhou's surface water is becoming more serious every day because of industrial development. The ranges of qualified sources of drinking water will be increasingly narrowed. People will have to consider groundwater.

3. According to the standard of drinking water, the quality of the bearing groundwater under Suzhou is fine and occupies fifth place in China (Table 1). In addition, it hasn't worsened through at least the past decade (Tables 2, 3).

4. The allowable exploiting reserves of Suzhou's bearing water, which will not cause continued earth subsidence, is about 30 million tons a year

(about 80,000 T a day). It is enough to satisfy the drinking needs of the 700,000 residents of Suzhou City.

5. The yield of running water will be 540,000 tons a day, whereas the water for Suzhou's industry will be about 500,000 tons per day. It is possible that the existing waterworks will be completely turned to Suzhou's industry.

In brief, the main point of this paper is "quality groundwater for civil use and worse surface water to industry."

Table 1

No.	Cities	Pi	No.	Cities	Pi
1	Hongzhou	24.46	16	Changzhou	0.98
2.	Wenzhou	8.97	17	Shijiazhuang	0.86
3.	Shanghai	7.62	18	Taiyuan	0.78
4.	Benxi	6.26	19	Dukou	0.53
5.	Shashi	6.20	20	Jinzhou	0.45
6.	Zhangjiakou	6.16	21	Luoyang	0.36
7.	Shenyang	4.55	22	Wuxi	0.34
8.	Qiqihaer	4.44	23	Guilin	0.32
9.	Wulumuqi	2.47	24	Hefei	0.25
10.	Beijing	1.86	25	Suzhou	0.06
11.	Yinchuan	1.72	26	Tangshan	0.05
12.	Kunming	1.62	27	Guangzhou	0.01
13.	Handan	1.28	28	Nanning	0.0
14.	Wuhan	1.20	29	Qinhuangdao	0.0
15.	Changchun	1.11			

Table 2—Table of average concentrations of main pollutants in Suzhou's bearing groundwater over the years.

year	PH	H*(Ca,Mg)	TDS	COD	NH_3-N	NO_2^--N	NO_3^--N	Hg
1977	7.8	14.69	---	---	---	---	---	0
1978	7.8	15.38	---	---	1.24	---	---	0
1979	7.8	16.22	---	---	1.44	---	---	0
1980	7.7	11.92	444	---	1.67	0.010	0.15	0
1981	7.6	13.92	332	2.67	2.13	0.016	0.26	0
1982	7.6	9.94	303	2.54	1.06	0.002	0.19	0
1983	7.7	12.12	335	2.40	1.19	0.004	0.47	0
1984	7.4	11.38	433	2.45	1.40	0.005	0.09	0
1985	7.5	11.70	427	2.30	1.30	0.005	0.09	0
1986	7.6	13.66	424	2.34	1.35	0.001	0.06	0
1987	7.7	11.55	462	2.55	1.28	0.001	0.06	0

year	Phenol	F^-	Cl^-	As	Cr^{+6}	Cd	total germ	CN^-
1977	0.029	0.052	---	0.019	0	--	---	0
1978	0	0.20	---	0.009	0	--	---	0
1979	0	0.39	---	0.010	0	--	---	0
1980	0	0.41	24.27	0.017	0	--	---	0
1981	0	0.07	27.83	0.025	0	0	---	0
1982	0	0.25	21.90	0.021	0	0	---	0
1983	0	0.05	23.09	0.024	0	0	---	0
1984	0	0.05	---	0.022	0	0	---	0
1985	0	0.16	25.80	0.024	0	0	---	0
1986	0	0.08	25.12	0.026	0	0	---	0
1987	0.0004	0.08	32.91	0.015	0	0	10-27	0

Remarks : The units in this table are mg/l with the exception of hardness (German degree) and the sum total of germ (several germs/ml) ; H*= hardness .

Table 3—Table of maximum and minimum concentrations of main pollutants in Suzhou's bearing groundwater over the years.

year	PH		H*(Ca,Mg)		TDS		COD		NH_3-N	
	Max.	Min.	Max.	Min.	Max.	Min.	Max.	Min.	Max.	Min.
1980	7.9	6.8	37.9	3.7	536	112	7.49	2.22	13.20	0
1981	8.2	6.9	25.2	6.9	580	48	9.05	1.37	4.72	0.14
1982	10.3	6.4	27.6	4.9	970	30	4.90	0.70	7.20	0.02
1983	8.4	7.0	21.1	5.5	578	46	3.93	1.20	2.60	0.02
1984	7.8	6.5	22.7	4.2	628	105	6.85	1.20	2.72	0.13
1985	8.1	6.8	19.9	4.3	573	92	3.10	1.60	2.26	0.16
1986	8.1	7.0	20.1	8.2	583	104	8.50	1.40	2.71	0.18
1987	8.1	7.2	21.4	6.4	612	289	5.41	1.71	2.79	0.28

year	NO_2^--N		NO_3^--N		Phenol		F^-		Cl^-	
	Max.	Min.	Max.	Min.	Max.	Min.	Max.	Min.	Max.	Min.
1980	0.210	0	0.78	0	0.002	0	0.6	0.2	209.6	42.7
1981	0.410	0.001	1.06	0.02	0.192	0	0.1	0	169.90	5.34
1982	0.158	0.001	13.00	0.01	0.007	0.001	0.6	0.1	193.50	2.90
1983	0.069	0.001	4.28	0.03	0	0	2.0	0.1	40.30	3.30
1984	0.045	0.001	0.73	0.03	0	0	0.5	0.1	---	---
1985	0.024	0	0.90	0	0.002	0	0.8	0	36.30	4.60
1986	0.002	0	1.15	0	0	0	0.29	0.02	35.34	4.80
1987	0.016	0	1.33	0	0.003	0	0.44	0	87.88	2.39

year	CN^-		As		Cr^{+6}		Hg		Cd	
	Max.	Min.	Max.	Min.	Max.	Min.	Max.	Min.	Max.	Min.
1980	0	0	0.09	0	0.02	0	0	0	---	---
1981	0.02	0.006	0.12	0.01	0	0	0	0	0	0
1982	0	0	0.08	0.01	0	0	0	0	0	0
1983	0	0	0.07	0.01	0	0	0	0	0	0
1984	0	0	0.06	0.01	0	0	0	0	0	0
1985	0	0	0.05	0	0	0	0	0	0	0
1986	0	0	0.09	0	0	0	0	0	0	0
1987	0	0	0.04	0	0	0	0	0	0	0

Remarks : H*=Hardness (German degree).

ZHENG, Y-F., Geochemisches Institut der Universitat Gottingen, Gottingen, Federal Republic of Germany, Z-H. ZHANG, and W-Z. SHEN, Nanjing University, Nanjing, China

Genesis of Reddening in the Dabu Uranium Deposit and Its Relationship to the Uranium Mineralization

The Dabu uranium deposit is located in the contact zone of two granitic bodies inside the Taoshan granitic batholith, south China. The uranium mineralization is determined to have a hydrothermal origin. Periore reddening (including hematitization and goethitization) is observed to be widespread in the hydrothermal uranium mineralization. It is superimposed on alkali feldsparization zone. Calcite-pitchblende and fluorite-pitchblende veins are generally distributed in the reddening halos, but dereddening occurs surrounding the veins. The reddening usually shows an antithetical correlation with pyritization. Investigation of the paragenetic association of alteration mineral assemblages reveals that the reddening is formed before deposition of the uranium mineralization, and the visible pitchblende vein is coprecipitated with the pyrite. Physicochemical conditions of the pre-ore alterations are estimated to be of high oxygen fugacity (about 10^{-24} to 10^{-34} atm). The alkali feldspars associated with the reddening have low $\delta^{18}O$ values (+2.9 to +5.6‰), indicating a poor $-{}^{18}O$, meteoric alteration solution (having $\delta^{18}O_{H2O}$ values of about −1.7 to +2.0‰ at alteration temperature of 300-400 °C). Thus, the pre-ore solution dominant in the meteoric water component could have high oxidative potential, resulting in the reddening associated with the alkali feldsparization.

In order to decipher the relationship between the reddening and the uranium mineralization in the Dabu deposit, a thermodynamic calculation is performed at temperatures between 150 and 300 °C using combined thermodynamic data of Naumov (1971) and Robie (1978). The following reactions are taken into account.

$$(1)\quad [UO_2(CO_3)_2]^{2-} + 2Fe^{2+} + 2Ca^{2+} + 3H_2O \longrightarrow UO_2 + Fe_2O_3 + 2CaCO_3 + 6H^+$$

$$(2)\quad [UO_2F_4]^{2-} + 2Fe^{2+} + 2Ca^{2+} + 3H_2O \longrightarrow UO_2 + Fe_2O_3 + 2CaF_2 + 6H^+$$

$$(3)\quad [UO_2(CO_3)_2]^{2-} + 2H_2S + Fe^{2+} + 2Ca^{2+} \longrightarrow UO_2 + FeS_2 + 2CaCO_3 + 4H^+$$

(4) $[UO_2F_4]^{2-} + 2H_2S + Fe^{2+} + 2Ca^{2+} \longrightarrow UO_2 + FeS_2 + 2CaF_2 + 4H^+$

(5) $[UO_2(CO_3)_2]^{2-} + Fe_2O_3 + 4H_2S + 2Ca^{2+}$
$$\longrightarrow UO_2 + 2FeS_2 + 2CaCO_3 + 2H_2O + 2H^+$$

(6) $[UO_2F_4]^{2-} + Fe_2O_3 + 4H_2S + 2Ca^{2+} \longrightarrow UO_2 + 2FeS_2 + 2CaF_2 + 2H_2O + 2H^+$

(7) $4[UO_2(CO_3)_2]^{2-} + H_2S + 8Ca^{2+} + 4H_2O$
$$\longrightarrow 4UO_2 + HSO_4^- + 8CaCO_3 + 9H^+$$

(8) $4[UO_2F_4]^{2-} + H_2S + 8Ca^{2+} + 4H_2O \longrightarrow 4UO_2 + HSO_4^- + 8CaF_2 + 9H^+$

The reactions (1) and (2) do not proceed in the right direction due to the positive Gibbs free energy (i.e., $\Delta G_r > 0$), implying that the hematitization cannot result from redox between UO_2^{2+} and Fe^{2+}. The reactions (3) and (4) also have positive Gibbs free energy, showing that even if H_2S exists in a solution containing Fe^{2+}, the uranyl complexes cannot be reduced to form pitchblende. Therefore, the formation of the hydrothermal uranium mineralization needs a given redox environment. The redox properties of uraniferous solution are too strong or too weak to be favorable for reduction and precipitation of uranium. The reactions (5) and (6) possess the negative Gibbs free energy (i.e., $\Delta G_4 < 0$). This means that when Fe_2O_3 and H_2S coexist with the uranyl complexes, the pitchblende can then be formed. These are consistent with the geological observation. It implies that the hematited granite is favorable for the hydrothermal uranium mineralization. In other words, when a solution containing both the uranyl complexes and H_2S encounters the hematited rock in hydrothermal conduits, the hematite is then reduced to pyrite, concomitant with the reduction of uranium through combined oxidation of UO_2^{2+} and H_2S. If a solution contains only H_2S without Fe_2O_3, the H_2S would not be oxidized by the uranyl complexes before the solution is contacted with the hematited rock, because reactions (7) and (8) have positive Gibbs free energy.

The conclusions are thus drawn that the reddening in the Dabu uranium deposit results from oxidation of the pre-ore solution to the ambient granites. But the reddening rocks provide a suitable redox condition, whereby the uranium in the solution can be reduced to deposit the pitchblende.

Directory of Presenters

S

Sabir, Hussein M.
B.R.G.M.
P.O. Box 1492
Jeddah 21431
Saudi Arabia

Sacchi, Rosalino
Earth Sci. Dept.
Univ. of Torino
5, via Accademia delle Sci.
10123 Torino
Italy

Sacchi, Rosalino
Univ. of Torino
Dept. of Earth Sciences
5, via Accademia delle Scienze
10123 Torino
Italy

Sacks, Laura A.
Environ. Sci. Dept.
Univ. of Virginia
Charlottesville, VA 22903

Sadooni, Fadhil N.
Geol. Dept.
Oil Expl. Co.
P.O. Box 476
Baghdad
Iraq

Sadowski, Georg Robert
Institute de Geociencias da
Univeridade de Sao Paulo
Cx. Post. 20.899
Sao Paulo, SP, 01498
Brazil

Sag, Laszlo
Benczur u. 35/c.
H-1068 Budapest IV
Hungary

Sag, Laszlo
Geol. Res. Inst. of the
Hungarian Academy of Sci.
Pf. 64
Budapest H-1388
Hungary

Sager, William W.
Oceanog. Dept.
Texas A&M Univ.
College Station, TX 77843-314

Sager, William W.
Oceanog. Dept.
Texas A&M Univ.
College Station, TX 77843-314

Sagredo, J..
Dpt. Petrologia
Int. Geol. Econ.,CSIC-UCM
Fac.de Geol. Sci.
28040 Madrid
Spain

Saha, Dilip
Geol. Studies Unit
Indian Statistical Inst.
203 B-T Rd.
Calcutta 700035
India

Sahagian, Dork
Lamont-Doherty Geol. Obs.
Columbia Univ.
Palisades, NY 10964

Sahu, K.C.
Dept. of Earth Sciences
Indian Institute of Technology
Bombay 400076
India

Salda, L. Dalla
Centro Investig. Geologicas
Calle 1 n 644
1900 La Plata
Argentina

Salem, Rafik
3716 Wedgeway
Fort Worth, TX 76133

Saler, Antun
Nuklearna Elektrarna Krsko
P.O. Vrbina 12
Krsko 68270
Yugoslavia

Salinas, Amarilis
Empresa Geologia Pinar del Rio
Sta. Lucia, Matahambre
P. del Rio
Cuba

Salinas, Edgar
Ecole des Mines de Paris
C.G.G.M.-Sedimentologie
35, rue St. Honore
77305 Fontainebleau Cedex
France

Salje, E.
Earth Sci. Dept.
Downing St.
Cambridge CB2 3EQ
England

Salje, Ekhard
Earth Sci. Dept.
Univ.of Cambridge
Downing Street
Cambridge CB2 3EQ
England

Salje, Ekhard
Dept. of Earth Sciences
Univ. of Cambridge
Downing Street
Cambridge CB2 3EQ
England

Salman, Alexander G.
Physics of the Earth Inst.
USSR Academy of Sciences
Bolshaya Gruzinskaya 10
Moscow 123810
U.S.S.R.

Salman, Alexander G.
Physics of the Earth Inst.
USSR Academy of Sciences
Bolshaya Gruzinskaya 10
Moscow 123810
U.S.S.R.

Salminen, Reijo
Geol. Survey of Finland
SF 02150 Espoo 15
Finland

Salters, Vincent J.M.
54-1114
MIT
Cambridge, MA 02139

Salvador, Amos
Univ. of Texas
Dept. of Geological Sciences
P.O. Box 7909
Austin, TX 78713

Samani, Bahram A.
Exploration Division
AEOI
P. O. Box 14155-1339
Tehran
Iran

Samani, Bahram A.
Exploration Division
Atomic Energy Organization of
Iran
P. O. Box 14155-1339
Tehran
Iran

Samaniego, Cesar M.
Geol. Dept.
Univ. of Akron
Akron, OH 44325

Samir, Ahmed
Geol. Dept.
Fac. of Science
Zagazig University
Zagazig,
Egypt

Sanchez Cella, V.
Dept.de Ciencias de la Tierra
Universidad de Zaragoza
50009 Zaragoza
Spain

Sanchez Celle, V.
Dept. de Ciencias de la Tierre
(Petr. & Geoquimica)
Fac. de Ciencias
Univ. of Zaragoza
50009 Zaragoza
Spain

Sanchez-Arango, Jorge R.
Centro de Investigactiones del
Petroleo
Oficios #154
La Habana 1
Cuba

Sandrone, Riccardo
Dipartimento di Georisorse
e Territorio
Politecnico
corso Duca degli Abruzzi 24
I-10129 Torino
Italy

Sandstedt, Hakan
Vattenfall Avd. BEL
S-162 87 Vallingby
Sweden

Sandy, Michael R.
Department of Geology
University of Dayton
300 College Park
Dayton, OH 45469

Sangster, D.F.
Geol. Survey of Canada
601 Booth St.
Ottawa, Ont. K1A 0E8
Canada

Sappal, Krishna K.
Geol. & Geophy. Dept.
Curtain Univ. of Technology
GPO Box U 1987
Perth 6001
Australia

Sardarov, Sard
Inst. of Geothermy Problems
USSR Academy of Sciences
Daghestan Br.
Makhachkala
Gadjieva 45
U.S.S.R.

Sardarov, Sard
Inst. of Geothermy Problems
USSR Academy of Sciences
Daghestan Branch
Makhachkala
Gadjieva 45
U.S.S.R.

Sarfati, Janine
Ctr. Geol. & Geophy.
USTL
Place E. Bataillon
34060 Montpellier Cedex
France

Sarkar, S.C.
Geol. Sci. Dept.
Jadavpur Univ.
Calcutta 700 003
India

Sarkar, S.C.
Geol. Sci. Dept.
Jadavpur Univ.
34 Kasba Rd.
Calcutta 700042
India

Sarkar, Subir
c/o G.C. Bhattacharya, F/6
Staff Quarter
Jadavpur University Campus
Jadavpur University
Calcutta 700 032
India

Sarma, V.V.T.
Geophy. Dept.
Andhara Univ.
Visakhapatnam 530 003
India

Sarmah, Suryya Kanta
Dept. of Environmental Science
Gauhati University
Guwahati 781014
Assam
India

Sarman, Rakesh
Dept. of Geology
Kent State University
Kent, OH 44242

Sarnthein, Michael
Geol.-Paleon. Inst.
Universitat
Olshausenstr. 40
D-2900 Kiel
Fed. Rep. of Germany

Sarti, Massimo
Depart. Di Scienze Della Terra
Univ. Della Calabria
87030 Castiglione
Cosentino Scalo
Cosenza
Italy

Sasada, Masakatsu
Geological Survey of Japan
Tsukuba 305
Japan

Sasowsky, Ira D.
Dept. of Geosciences
Pennsylvania State Univ.
Room 335, Deike Bldg.
University Park, PA 16802

Sassa, Kr.
Disaster Prevention Research
Institute
Kyoto University, Uji
Kyoto 611
Japan

Sassi, Francesco
Dept. Mineralogy & Petrology
University of Padova
Corso Garibaldi 37
Padova I-35100
Italy

Sassi, Francesco P.
Miner. & Petr. Dept.
Univ. of Padova
Corso Garibaldi 37
I 35100 Padova
Italy

Sathyamurthy, K.
Geological Survey of India
Training Institute
3 - 6 - 266, Himayatnagar
Hyderabad 500 029
India

Sato, Hiroaki
Fac.of Intergrated Arts & Sci.
Hiroshima Univ.
Hiroshima 730
Japan

Sato, Hiroki
Geol. Dept.
126 Earth Sci. Bldg.
Univ. of Alberta
Edmonton, Alta T6G 2E3
Canada

Sato, Motoaki
U.S.G.S.
959 National Center
Reston, VA 22092

Sato, Motoaki
U.S.G.S.
MS 959, National Ctr.
Reston, VA 22092

Sato, Tadashi
Geosci. Inst.
Univ. of Tsukuba
Ibaraki 305,
Japan

Sato, Tadashi
Inst. of Geosci.
Univ. of Tsukuba
Ibaraki 305
Japan

Saunders, R. Stephen
Jet Propolsion Laboratory
4800 Oak Grove Drive
Mail Stop 264-316
Pasadena, CA 91109

Sauvagnac, Serge
National Geo. Research Instit.
Hyderabad 500007
India

Saverikko, Matti
(Geological Survey of Finland)
Hakamaki 4 G 97
Espoo SF-02120
Finland

Savostin, Leo A.
Institute of Oceanology
Academy of Sciences of USSR
117218 Moscow
U.S.S.R.

Savostin, Leo A.
Institute of Oceanology
Academy of Sciences of USSR
117218 Moscow
U.S.S.R.

Sawaf, Tarif
c/o Muawia Barazangi, INSTOC
Snee Hall
Cornell University
Ithaca, NY 14853

Sayet, Y. E.
Institute of Minerology,
Geochemistry & Cristallochem.
of Rare Elements
Moscow
U.S.S.R.

Sborshchikov, I.M.
Inst. of Oceanology
Krasikova Str. 23
Moscow 117218
U.S.S.R.

Scanlon, Kathryn M.
U.S. Geological Survey
Woods Hole, MA 02543

Schaeben, Helmut
Dept. of Geology
University of Bonn
Nussallee 8
Bonn 1 5300
Fed. Rep. of Germany

Schaeben, Helmut
Dept. of Geology
University of Bonn
Nussallee 8
Bonn 1 5300
Fed. Rep. of Germany

Schaeffer, Malcolm F.
Duke Power Company
P.O. Box 33189
Charlotte, NC 28242

Schafer, Andreas
Geol. Inst.
Nussallee 8
5300 Bonn
Fed. Rep. of Germany

Schalamuk, Isidoro
Catedra de Geol.de Yacimientos
Univ. Nacional de La Plata
Avenida Paseo del Bosque s/n
La Plata 1900
Argentina

Scheepers, Deon J.
Dept. of Geology
University of Stellenbosch
7600
South Africa

Scheibner, Erwin
NSW Geol. Survey
P.O. Box 5288
Sydney NSW 2001
Australia

Scheibner, Erwin
N.S.W. Geological Survey
P.O. Box 5288
Sydney, N.S.W. 2001
Australia

Scherer, Wolfgang
Intevep S.A.
Apartado 76343
Caracas 1070 A
Venezuela

Scheuing, David F.
Lamont-Doherty Geol. Obs.
Palisades, NY 10964

Schidlowski, M.
Max-Planck-Inst. f. Chermie
Saarstr. 23/Postfach 3060
D-6500 Mainz
Fed. Rep. of Germany

Schlager, Wolfgang
Earth Sci. Dept.
Free Univ.
P.O. Box 7161
Amsterday 1007 MC
Netherlands

Schlager, Wolfgang
Earth Sci. Dept.
Free Univ.
P.O. Box 7161
1007 MC Amsterdam
Netherlands

Schlee, John S.
U.S.G.S.
Woods Hole, MA 02543

Schleicher, Helmut
Miner.-Petro. Inst.
Albertstr. 236
D-7800 Freiburg
Fed. Rep. of Germany

Schleicher, Helmut
Miner-Petr. Inst.
Albentstr. 236
D-7800 Freiburg
Fed. Rep. of Germany

Schleusener, Julie
Geosci. Dept.
201 Deike Bldg.
Pennsylvania State Univ.
University Park, PA 16802

Schmidt, Hannelore
Inst. for Geowissenschaften/
Geol.,Univ. of Mainz
Postfach 3980
6500 Mainz
Fed. Rep. of Germany

Schmidt, P.W.
CSIRO Division of Exploration
Geoscience
P. O. Box 136
North Ryde, NSW 2113
Australia

Schmidt, Susanne T.
Miner.-Petro. Inst.
der Universitat
Im Neuenheimer Feld 236
D-6900 Heidelberg
Fed. Rep. of Germany

Schmincke, H.U.
Inst. f. Mineralogie
Ruhr Univ. Bochum
Postfach 1022 148
D-463 Bochum
Fed. Rep. of Germany

Schmitz, Birger
MS 110A, Bldg. 70
Lawrence Berkeley Lab.
Berkeley, CA 94720

Schneer, Cecil J.
P.O. Box 181
Newfields, NH 03856

Schneider, Jay A.
Geophy Sci. Dept.
Univ.of Chicago
5734 S. Ellis Ave.
Chicago, IL 60637

Scholl, David W.
Br. of Pacific Marine Geol.
U.S.G.S.
345 Middlefield Rd.
Menlo Park, CA 94025

Schot, Paul P.
Environ. Studies Dept.
Univ. of Utrecht
Heidelberglaan 2
3584 CS Utrecht
Netherlands

Schott, Jacques
Lab. de Geochimie
Univ. Paul-Sabatier
38, rue des 36 Ponts
31400 Toulouse
France

Schreiber, Henry D.
Center for Glass Chemistry
Chemistry Dept.
Virginia Military Inst.
Lexington, VA 24450

Schreyer, Werner
Institut fur Mineralogie
Ruhr-Universitat Bochum
Postfach 10 21 48
D-4630 Bochum 1
Fed. Rep. of Germany

Schroder, Niels
Roskilde Univ. Ctr.
Box 260
DK-4000
Roskilde
Denmark

Schroeder, Johannes H.
Institut f. Geologie and
Palaeontologie
Tech. Univ. Berlin,Sekr. EB 10
Ernst-Reuter-Platz 1
D-1000 Berlin 10
Fed. Rep. of Germany

Schubert, Gerald
Earth & Space Sci. Dept.
Univ. of California
405 Hilgard Avenue
Los Angeles, CA 90024

Schubert, Gerald
Earth & Space Sci. Dept.
Univ. of California
405 Hilgard Ave.
Los Angeles, CA 90024

Schultz, Klaus
U.S.G.S.
M/S 954
National Center
Reston, VA 22092

Schulze, Daniel J.
Geol. Sci. Dept.
Queen's Univ.
Kingston, Ont. K7L 3N6
Canada

Schumann, Herbert H.
U.S. Geological Survey, WRD
Suite E
3738 North 16th Street
Phoenix, AX 85016

Schwab, William C.
U.S. Geological Survey
Branch of Atlantic Marine
Geology
Quissett Campus
Woods Hole, MA 02543

Schwarcz, Henry P.
Geol. Dept.
McMaster Univ.
1280 Main St., W.
Hamilton, Ont. L8S 4M1,
Canada

Schwartz, F.W.
Dept. of Geology
University of Alberta
Edmonton Alberta T6G 2E1
Canada

Schwarz, Gerhard
Inst.fur Geophy.Wissenschaften
Freie Univ. Berlin
Rheinbabenalle 49
D-100 Berlin 33
Fed. Rep. of Germany

Sclar, Charles B.
Geol. Sci. Dept.
Williams Hall #31
Lehigh Univ.
Bethlehem, PA 18015-3188

Sclar, Charles B.
Geol. Sci. Dept.
Williams Hall #31
Lehigh Univ.
Bethlehem, PA 18015-3188

Scoon, Roger N.
Geol. Dept.
Rhodes Univ.
Grahamstown
South Africa

Scotese, Christopher R.
Shell Oil Co.
P.O. Box 481
Houston, TX 77001

Scotese, Christopher R.
Shell Oil Co.
P.O. Box 481
Houston, TX 77001

Scott Smith, Barbara H.
2555 Edgemont Blvd.
N. Vancouver, B.C. V7R 2M9
Canada

Scott, Andrew C.
Geol. Dept.
Royal Holloway and Bedford
New College
University of London
Egham, Surrey TW20OEX
England

Scott, R.W.
Amoco Prod. Co.
P.O. Box 3385
Tulsa,, OK 74102

Scott, Steven D.
Geol. Dept.
Univ. of Toronto
Toronto, Ont. M5S 1A1,
Canada

Scribano, Vittorio
Istituto de Scienze della
Terra
Corso Italia
Catania 55-95129
Italy

Seaber, Paul R.
Ground Water Section
Illinois State Geol. Survey
Nat. Res. Bldg.
615 E. Peabody
Champaign, IL 61820

Searle, Michael P.
Geol. Dept.
Univ. of Leicester
Leicester LE1 7RH
England

Sebrier, Michel
5 rue Rataud
75005 Paris
France

Seccombe, Philip K.
Dept. of Geology
Univ. of Newcastle
Newcastle, N.S.W. 2308
Australia

Seccombe, Philip K.
Dept. of Geology
Univ. of Newcastle
Newcastle, N.S.W. 2308
Australia

Seck, H.A.
Miner. Inst.
Univ. Koln
D-5000 Koln 1
Fed. Rep. of Germany

Seeger, C. Ronald
Dept. of Geography & Geology
Western Kentucky University
Bowling Green, KY 42101

Segero, Amram S.
Geoloconsultants Network
P.O. Box 62134
Nairobi
Kenya

Seguin, Maurice K.
Geol. Dept.
Laval Univ.
Quebec City, Que. G1K 7P4,
Canada

Segura, Efren P.
San Miguel de Horcajitas 19
Col. Residencial de Anza
83000 Hermosillo, Sonora
Mexico

Seidel, Eberhard
Mineralogisch-Petrographisches
Institut der Universitat
zu Koln
Zulpicher Str. 49
D-5000 Koln
Fed. Rep. of Germany

Seifert, H.P.
OMV-AG - Expl. Dept.
Gerasdorferstr. 151
A-1211 Vienna
Austria

Seilacher, Adolf
Kline Geology Lab
Yale University
P.O. Box 6666
New Haven, CT 06511-8130

Selden, Paul A.
Extra-Mural Studies Dept.
Univ. of Manchester
Manchester M13 9PL
England

Sempere, Jean-Christophe
School of Oceanog., WB-10
Univ. of Washington
Seattle, WA 98195

Sempere, Jean-Christophe
School of Oceanography WB-10
University of Washington
Seattle, WA 98195

Sempere, Therry
Convenio ORSTOM-YPFB
CC 4875
Santa Cruz
Bolivia

Sempere, Thierry
Convenio ORSTOM-YPFB
CC 4875
Santa Cruz
Bolivia

Semroud, Belkalem
Geol. Dept.
Inst. Sci. della Terra
USTHB
BP 32
Alger
Algeria

Sen Gupta, Barun K.
Geol. & Geophy. Dept.
Louisiana State Univ.
Baton Rouge, LA 70803

Sen, D. K.
Mineral Physics Division
29 Jawaharlai Nehru Road
Geological Survey of India
Calcutta 700 016
India

Sen, S.
166/30 Lake Gardens
Calcutta 700045
India

Sen, S.
166/30 Lake Gardens
Calcutta 700 045
India

Sen, Subhasis
Central Fuel Research Inst.
Post Box 75
Nagpur 440 001
India

Sen, Subhasis
Central Fuel Res. Inst.
Post Box 75
17/C Telenkhedi Area
Nagpur 440001
India

Senderov, Erik
Vernadsky Inst.of Geochemistry
and Analytical Chemistry
Moscow
U.S.S.R.

Sengor, A.M.C.
ITU Maden Fakultesi
Jeologie Bolumu
Tesvikiye
80394 Istanbul
Turkey

Sengupta, Sudipta
Dept. of Geological Sciences
Jadavpur University
Calcutta 700032
India

Seno, T.
Earthquake Research Institute
University of Tokyo
Bunkyo
Tokyo 113
Japan

Seno, T.
Earthquake Research Institute
University of Tokyo
Bunkyo
Tokyo 113
Japan

Seno, T.
Earthquake Res. Inst.
Univ. of Tokyo
Bunkyo-ku
Tokyo 113
Japan

Senske, David
Geol. Sci. Dept.
Brown Univ.
P.O. Box 1846
Providence, RI 02912

Seo, Jeong Ryul
219-5 Garibongdong Gurogu
Seoul
Korea

Seyed-Emami, K.
Faculty of Engineering
University of Tehran
P.O. Box 11365/4563
Tehran
Iran

Shafiqullah, Muhammad
Geosci. Dept.
Univ. of Arizona
Tucson, AZ 85721

Shah, Shirish D.
Dept. of Earth Sciences
Indian Institute of Technology
Powai,Bombay 400076
India

Shahin, Ahmed Nabil
GUPCO Exploration
P.O. Box 2400
Cairo
Egypt

Shankar, R.
Marine Geology Faculty
Mangalore University
Mangalagangotri DK
Karnataka 574199
India

Shanks III, Wayne C. (Pat)
U.S. Geological Survey
Branch of Eastern Mineral Res.
Mail Stop 954, National Center
Reston, VA 22092

Shannon, Robert D.
Central Research Dept.
Experimental Station 356/329
E.I. Dupont De Nemours & Co.
Wilmington, DE 19880-0356

Sharma, Kewal K.
Wadia Instit. of Himalayan
Geology
33, General Mahadeo Singh Road
Dehra Dun - 248 001
India

Sharma, Prabhaker
Room 261
Physical Research Laboratory
Navrangpura, Ahmedabad 380 0
India

Sharma, Ram S.
Department of Geology
Banaras Hindu University
Varanasi 221005
India

Sharp Jr., John M.
Dept. of Geological Sciences
University of Texas
Austin, TX 78713-7909

Shata, Abdu A.
Desent Inst.
Mataria 1 Cairo
Egypt

Shata, Abdu A.
Desent Inst.
Mataria 1 Cairo
Egypt

Shata, Abdu A.
Desert Inst.
Mataria 1
Cairo
Egypt

Shaw, Christopher A.
U.S. Geological Survey
Box 25046, MS939
Lakewood, CO 80225

Shaw, H.F.
Geol. Dept.
Imperial College
London SW7 2BP
England

Shaw, Herbert R.
U.S.G.S.
M/S 910
Menlo Park, CA 94025

Shcheglov, A.D.
Ministry of Geology of USSR
Bolshaya Gruzinskaya, 4/6
Moscow 123812
U.S.S.R.

Shchegolev, I. N.
Geol. Dept.
Voronezh State Univ.
Universitetskaya Pl. 1
394693 Voronezh
U.S.S.R.

Shcherbak, Nikolai P.
Inst. of Geochem. & Physics of
Minerals Acad. Sci.
Ukr. SSR.
Palladin Av, 34
252680 Kiev-142
U.S.S.R.

Shedlock, Robert J.
United States Geological Surv.
Water Resources Division
208 Carroll Bldg.
8600 LaSalle Road
Towson, MD 21204

Shehata, William M.
King Fahd University of
Petroleum & Minerals
P.O. Box 1751
Dhahran
Saudi Arabia

Shemiakin, V.M.
Institute of Precambrian
Geology and Geochronology
USSR Academy of Sciences
Nab Makarova, 2
Leningrad 199034
U.S.S.R.

Shemiakin, V.M.
Institute of Precambrian
Geology and Geochronology
USSR Academy of Sciences
Nab. Makarova 2
Leningrad 199034
U.S.S.R.

Shen, Zhaoli
China University of Geoscience
Xue-yuan Road 29
Beijing
Peo. Rep. of China

Shen, Zhaoli
Geological Society of China
Baiwanzhuang Raod 26
Fuchengmenwai
Beijing
Peo. Rep. of China

Shen, Zhaoli
Hydrogeology & Engr.Geol.Dept.
Wuhan College of Geology
Yujiashan Wuhan
Hubei
Peo. Rep. of China

Shepherd, T.J.
British Geol. Survey
64 Grays Inn Road
London WC1X8NG
England

Sheridan, Robert E.
Dept. of Geological Sciences
Rutgers University
Busch Campus
New Brunswick, NJ 08903

Shi, Dehong
Geological Society of China
Pai Wan Chuang Fuchengmenwai
Beijing
Peo. Rep. of China

Shi, Yaolin
Dept. of Earth Sciences
Graduate School
Academia Sinica
P.O. Box 3908
Beijing
Peo. Rep. of China

Shi, Yaolin
Dept. of Earth Sciences
Graduate School
Academia Sinica
P.O. Box 3908
Beijing
Peo. Rep. of China

Shikazono, Naotatsu
Geological Institute
University of Tokyo
Tokyo 113
Japan

Shimamoto, Toshihiko
Inst. of Geol. & Miner.
Fac. of Sci.
Hiroshima Univ.
1-1-89 Higashi-senda machi
Naka-ku Hiroshims 730
Japan

Shimamoto, Toshihiko
Inst. of Geol. & Miner.
Fac. of Sci.
Hiroshima Univ.
1-1-89 Higashi-senda machi
Naka-ku Hiroshima 730
Japan

Shimp, Robert J.
The Procter & Gamble Co.
5288 Spring Grove Ave.,A100F
Cincinnati,, OH 45217

Shipley, Thomas H.
University of Texas
Institute for Geophysics
8701 Mopac Blvd.
Austin, TX 78759

Shishkin, M.A.
Paleon. Inst. of the Academy
of the Sciences of the USSR
Profsoyuznaya 123
Moscow 117321
U.S.S.R.

Shivakumar, B.S.
Department of Geology
Bangalore University
Bangalore-560 056
India

Shnyukov, E.F.
Inst. of Geological Sciences
55-B Chkalova Ulitsa
Kiev 252054
U.S.S.R.

Short, Nicholas M.
Geog. & Earth Sci. Dept.
Bloomsburg Univ.
Bloomsburg, PA 17815

Shreedhar, Satish K.
Lecturer, Post-Grad.
Geol.Dept.,
Govt. Degree College
Dharamsala-176 215
India

Shringarpure, D.M.
Geol. Dept.
M.S. Univ. of Baroda
Baroda 390 002
India

Shroder, J.F.
Geol. Dept.
Univ. of Nebraska
Omaha, NE 68182

Shukowsky, Wladimir
Caixa Postal 30627
Sao Paulo 01051 SP
Brazil

Sial, Alcides N.
Riverbend Research Labs
Universiyt of Georgia
110 Riverbend Road
Athens, GA 30605

Sial, Alcides Nobrega
Riverbend Research Labs.
University of Georgia
110 Riverbend Road
Athens, GA 30605

Sibrava, Vladimir
Div. des Sci. de la Terre
Secteur des Sciences
UNESCO, Place de Fontenoy
Paris VIIe
France

Sibuet, Jean-Claude
IFREMER, Centre de Brest
B.P. 70
29263 Plouzane
France

Sibuet, Jean-Claude
IFREMER
BP 70
29263 Plousane
France

Sidorov, A.A.
Northeastern Res. Inst.
16 Portovaya St.
Magadan 685010
U.S.S.R.

Siegel, Donald I.
Geol. Dept.
Syracuse Univ.
Syracuse, NY 13210

Signor, Philip W.
Dept. of Geology
University of California
Davis, CA 95616

Sigolo, Joel B.
Geol. Dept.
Inst. of Geosci.
Univ. of Sao Paulo
P.O. Box 20 899
CEP 01498 Sao Paulo
Brazil

Sikka, S.N.
KDMIPE, Room No. 183, ONGC
Kaulagarh Road, Dehradun
Dehradun 248001
India

Silva, Alejandra Cortes
Instituto de Geofisica, UNAM
Ciudad Universitaria
C.P. 04510 Mexico D.F.
Mexico

Silva, Annkarin A.K.
Univ. de Sao Paulo
Inst. de Geosci.
Caixa Postal 20899, CEP
01498 Sao Paulo
Brazil

Silva, Jose Brandao
Departamento De Geologia
Facul. De Ciencias De Lisboa
Campo Grande Edificio C-2
5 Piso
1700 Lisboa
Portugal

Silva, Zenaide C.G.
University of Lisbon-Portugal
Alto da Barra
Bl.B, L-5. 5o E
Oeiras 2780
Portugal

Silver, Eli A.
Earth Sci. Dept.
Univ. of California
Santa Cruz, CA 95064

Silver, Leon T.
Geol. & Planetary Sci. Div.
California Inst. of Tech.
Pasadena, CA 91125

Silver, Leon T.
Division of Geological and
Planetary Sciences
California Institute of Tech.
MC 170-25
Pasadena, CA 188192

Silver, Paul
Carnegie Institution
D.T.M. & Geophy. Lab.
Washington, DC 20015

Silver, Paul
Carnegie Institution
D.T.M. & Geophy. Lab.
Washington, DC 20015

Simkin, Tom
NHB, MS 119
Smithsonian Institution
Washington, DC 20560

Simmons, Marvin D.
U.S. Army Corps of Engineers
P.O. Box 1070
Nashville, TN 37202-1070

Simo, A.
Geol. & Geophy. Dept.
Univ. of Wisconsin
1215 W. Dayton St.
Madison, WS 53706

Simon, Armando A.
Empresa de Geol. de Pinar
de Rio
Km. 1 1/2 Carr. Minas
Santa Lucia
Pinar del Rio 24200
Cuba

Simon, Armando A.
Empresa de Geol. de
Pinar del Rio
Km. 1 1/2 Carr. Minas
Santa Lucia
Pinar del Rio, 24200,
Cuba

Simoneit, Bernd R.T.
College of Oceanog.
Oceanog. Admin. Bldg. 104
Oregon State Univ.
Corvallis,, OR 97331-5503

Simons, Bruno
Mineralogisches Inst.
Univ. Kiel
Olshausenstr. 40
2300-Kiel
Fed. Rep. of Germany

Simpson, Carol
Earth & Planetary Sci. Dept.
The Johns Hopkins Univ.
Baltimore, MD 21218

Simpson, Edward L.
Phys. Sci. Dept.
Kutztown Univ.
Kutztown,, PA 19530

Simpson, Peter R.
Geochem. & Hydrogeol. Dir.
64 Grays Inn Rd.
London, WC1X 8NG
England

Singh, Arun Kumar
H7/80, 228 Quaters
Jawahan Chocok
South T.T. Nagar
Bhopal 462003
India

Singh, Arun Kumar
H7180, 228 Quaters
Jawahar Chowk
South T.T. Nagar
Bhopal, 462003, MP
India

Singh, Indra B.
Geol. Dept.
Lucknow Univ.
Lucknow 226007
India

Singh, Indra Bir
Geol. Dept.
Lucknow Univ.
Lucknow 226007
India

Singh, K.P.
Punjab State Council for
Science & Technology
SCO: 2935-36, Sector 22-C
Chandigarh 160022
India

Singh, Prem
Department of Geology
Utkal University
Bhubaneswar
Orissa 751004
India

Singh, R. N.
National Geophy. Res. Inst.
Hyderabad 500007
India

Singh, R. N.
National Gsophy. Res. Inst.
Hyderabad 5000007
India

Singh, R. P.
Geol. Dept.
Banaras Hindu Univ.
Varanasi-221005
India

Singh, Ramesh P.
Civil Engr. Dept.
Indian Inst. of Tech.
Kanpur 208 016
India

Singh, Ramesh P.
Civil Engr. Dept.
Indian Inst. of Tech.
Kanpur 208 016
India

Singh, V.P.
Reader, Dept. of Geophysics
Baharas Hindu University
Varanasi 221 005
India

Singhal, B.B.S.
Earth Sci. Dept.
Univ. of Roorkee
Roorkee 247667
India

Singhroy, Vernon
Ontario Ctr.for Remote Sensing
40 Sheppard Ave. East
North York, Ont. M2N 3A1,
Canada

Sinha-Roy, S.
Geol. Survey of India
A-3 Sawai Jai Singh Rd.
Jaipur 302006
India

Sinha, A. K.
Wadia Institute of Himalayan
Geology
33, General Mahadeo Singh Road
Dehra Dun-248001
India

Sinha, A.K.
Warden Gokhal Hostel
Maharaja College Campus
Jaipur 302004
India

Sinha, A.K.
Warden Gokhal Hostel
Maharaja College Campus
Jaipur 302004
India

Sinha, A.S.
Exploration Division, Central
Mine Planning & Design
Institute, Ltd.
Gondwana Place
Ranchi - 834008, Bihar
India

Sinha, Anshu K.
Wadia Inst. of Himalayan Geol.
33, General Mahadev Singh Rd.
Dehra Dun 248001
India

Sinha, Ravi Kumar
Geological Survey of India
A/3, Sawai Jai Singh Highway
Bani Park
Jaipur-302016
India

Sinton, John M.
Hawaii Inst. of Geophy.
Univ. of Hawaii
2525 Correa Rd.
Honolulu,, HA 96822

Siskov, Georgi D.
Lab.for Coal Petr.
Univ. "Kliment Ohridski"
bul. Russki 15
1000 Sofia
Bulgaria

Sisodia, M.S.
Geol. Dept.
Univ. of Jodhpur
Jodhpur 342001
India

Sitton, Dan R.
U.S. Forest Saervice
Rogue River National Forest
P.O. Box 520
Medford, OR 97501

Sitton, Dan R.
U.S. Forest Service
Rogue River National Forest
P.O. Box 520
Medford,, Or 97501

Size, William
Department of Geology
Emory University
Atlanta, GA 30322

Skehan, James W.
Weston Observatory
381 Concord Road
Weston, MA 02193

Skinner, Catherine W.
Dept. of Geology & Geophysics
Yale University
210 Whitney Ane.
New Haven, CT 06511-8130

Sklash, Michael
Geological Department
University of Windsor
Windsor, Ontario N9B 3P4
Canada

Skuodis, Vytautas
7053 S. Maplewood Ave.
Chicago, IL 60629

Slack, John F.
USGS
MS 954
1220 Sunrise Valley Dr.
Reston, VA 22092

Sladen, C. P.
Seimentology Branch
BP Britannic House
Moor Lane
London EC2Y 9BU
England

Slaney, V.R.
RADARSAT Project Office
Canada Centre for Remote
Sensing,
110 O'Connor St., Suite 200
Ottawa, Ontario K1P 5M9
Canada

Slemmons, D.B.
Center for Neotectonic Studies
MacKay School of mines
University of Nevada at Reno
Reno, NV 89557

Slosson, James E.
Slosson and Associates
14046 Oxnard Street
Van Nuys, CA 91401

Smale, John L.
Dept. of Geological Sciences
University of South Carolina
Columbia, SC 29208

Smirnoff, Leonid
Smirnoff Minerals Corp.
12620 Forest Canyon
Parker, CO 80134

Smit, C.A.
Dept. of Geology
Rand Afrikaans University
Auckland Park
Johannesburg 2000
South Africa

Smit, Jan
Inst.of Earth Sci.
Free Univ.
P.O. Box 7161
1007 MC Amsterdam
Netherlands

Smith, A.G.
Earth Science Department
University of Cambridge
Downing Street
Cambridge CB2 3EQ
England

Smith, Brian A.
Bechtel National Inc.
P.O. Box 350
Oak Ridge, TN 37831-0350

Smith, Dorian G. W.
Department of Geology
University of Alberta
Edmonton, Alberta T6G 2E3
Canada

Smith, G. Grant
Geological Survey of Canada
Coal Geology Subdivision
3303 - 33rd Street, N.W.
Calgary, Alta. T2L 2A7
Canada

Smith, Leslie
Geol. Sci. Dept.
Univ. of British Columbia
6339 Stores Rd.
Vancouver, BC V6T 2B4
Canada

Smith, Leslie
Geol. Sci. Dept.
Univ. of British Columbia
6339 Stores Rd.
Vancouver, BC V6T 2B4
Canada

Smith, Michael A.
Texaco E & P Tech. Div.
3901 Briarpark
Houston, TX 77042

Smith, Patrick
Physics Dept., Geophy. Lab.
Univ. of Toronto
Toronto, Ont. M5S 1A7
Canada

Smith, Stephen A.
Law/Geoconsult Intl., Inc.
1140 Hammond Drive, N.E.
Building E, Suite 5250
Atlanta, GA 30328

Smoot, Joseph P.
USGS
MS 926
Reston, VA 22092

Snee, Larry
U.S.G.S., MS 905
Denver Federal Center
P.O. Box 25046
Denver, CO 80225

Snelson, S.
Shell Oil Company
P. O. Box 481
Houston, TX 77001

Snelson, S.
Shell Oil Company
P. O. Box 481
Houston, TX 77001

Snodgrass, Gary B.
U.S. Environ. Protection Agen.
Office of Radiation Programs
(ANR-461)
401 M Street S.W.
Washington, DC 20460

Snyder, David B.
B.I.R.P.S., Bullard Lab.
Madingley Rise
Madingley Road
Cambridge CB3 0EZ
England

Snyder, Don
Geol. & Geophy. Dept.
Univ. of California
Berkeley, CA 94720

Snyder, Stephen W.
Marine, Earth, & Atmospheric
Sciences Dept.
North Carolina State Univ.
Raleigh, NC 27695-8208

Snyder, Stephen W.
Marine, Earth & Atmospheric
Sciences Dept.
North Carolina State Univ.
P.O. Box 8208
Raleigh, NC 27695-8208

Soba, D.
P.O. Box 4110
Yaounde
Cameroon

Sobczak, L.W.
Energy, Mines, & Res. Dept.
Geol. Survey of Canada
1 Observatory Crescent
Ottawa, Ont. K1A0Y3
Canada

Sobolev, Alexander
Vernadsky Inst. of Geochem.
Moscow
U.S.S.R.

Sobolev, Alexander
Vernadsky Inst. of Geochem.
Moscow
U.S.S.R.

Sobolev, N.V.
Inst. of Geol. & Geophy.
Siberian Div. of Acad.of Sci.
630090 Novosibirsk-90
U.S.S.R.

Socci, Anthony D.
Geol. Dept.
Ohio Univ.
Athens, OH 45701

Sohn, I.G.
U.S.G.S.
Rm. E-311
Nat'l. Mus. of Natural History
Washington, DC 20560

Sokolov, Boris S.
Paleon. Inst.
USSR Academy of Sciences
Profsoyuznaya ul. 123
Moscow 117321
U.S.S.R.

Soler, Pierre
ORSTOM, C.S.1
213 Rue Lafayette
75010 Paris
France

Soller, David R.
U.S.G.S.
National Center, m/s 926
Reston, VA 22092

Soman, Ganesh, R.
759/83 Shivajinagar
Pune 411 004
India

Soman, K.
Geosci. Div.
Ctr. for Earth Sci. Studies
Trivandrum-695 031
India

Somayajulu, B.L.K.
Geocosmophysics Area
Chemistry Bldg.
Physical Research Lab.
Navrangpura
Ahmedabad 380 009
India

Sonder, Leslie
Dept. of Earth Sciences
Dartmouth College
Hanover, NH 03755

Song, Shuhe
Geological Society of China
Pai Wan Chuang Fuchengmenwai
Beijing
Peo. Rep. of China

Sonnenfeld, Peter
Geol. dept.
Univ.of Windsor
Windsor, Ont. N9B 3P4,
Canada

Sonyushkin, Vladimir E.
Institute of Lithosphere
Staromonetny per 22
Moscow 109180
U.S.S.R.

Sonyushkin, Vladimir E.
Institute of Lithosphere
Staromonetny per 22
Moscow 109180
U.S.S.R.

Sopena, Alfonso
Dpto. Estratigrafia
Facultad de Ciencias Geol.
Univ. Complutense
28040 Madrid
Spain

Sorriso-Valvo, Marino
CNR-IRPI
via G. Verdi 1
87030 Roges di Rende
Italy

Sorriso-Valvo, Marino
CNR-IRPI, via G. Verdi 1
87030 Roges di Rende
Italy

Sosson, Marc
Geol. Structure Lab.
Tour 26-00 1er etage
4, Place Jussieu
75252 Paris Cedex
France

Sotin, Christopher
Lab. de Geol. Dynamique Int.
Univ. de Paris XI Orsay
Bat. 509
91405 Orsay Cedex
France

Sovadina, L.
CESKY
Geologicky' Urad
Kodanska' 10
101 59 Praha 10
Czechoslovakia

Spakman, Wim
Geophy. Dept.
Univ. of Utrecht
Budapestlaan 4
P.O. Box 80.021
3508 TA Utrecht
Netherlands

Spakman, Wim
Geophy. Dept.
Univ. of Utrecht
Budapestlaan 4
P.O. Box 80.021
3508 TA Utrecht
Netherlands

Spamer, Earle E.
The Academy of Natural Science
19th and The Parkway
Philadelphia, PA 19103

Speczik, Stanislaw
Fac. of Geol.
Univ. of Warsaw
Zwirki 1 Wigury 93
02-089 Warsaw
Poland

Speed, R.C.
Geol. Dept.
Northwestern Univ.
Evanston, IL 60201

Speyer, Stephen E.
Geosci. dept.
Univ. of Arizona
Tucson, AZ 85721

Spiers, C.J.
Geol. Dept.
Inst. of Earth Sci.
Univ. of Utrecht
P.O. Box 80.021
3508 TA Utrecht,
Netherlands

Spiridonov, Hernany Borissov
Space Research Institute
Acad.G.Bonchev,STR.BL.3
Sofia, III3
Bulgaria

Spry, Paul G.
Earth Sci. Dept.
253 Science I
Iowa State Univ.
Ames,, IA 50011

Spry, Paul G.
Earth Sci. Dept.
253 Science I
Iowa State University
Ames, IA 50011

Squillace, Paul J.
U.S. Geological Survey WRD
P. O. Box 1230
Iowa City, IA 52244

Squyres, Steven W.
Geol. Dept.
Cornell Univ.
Ithaca, NY 14853

Srecec, I.
Inst. fur Kristallographic
RWTH Aachen
D-5100 Aachen
Fed. Rep. of Germany

Srinivasan, R.
Geochem. Div.
National Geophy. Res. Inst.
Uppal Rd.
Hyderabad 500007
India

Srivastav, Piyush
92 Vigyan Kunj
Univ. of Roorkee
Roorkee 247667
India

Srivastava, H. N.
Deputy Director General (SP)
Indian Meteorological Dept.
Lodi Road
New Delhi 110003
India

Srivastava, H.N.
Deputy Director General (SP)
Indian Metrological Dept.
Lodi Road
New Delhi 110003
India

Srivastava, L. S.
Dept. of Earthquake Eng.
University of Roorkee
Roorkee 247667
India

Srivastava, L.S.
Earthquake Engr. Dept.
Univ. of Roorkee
Roorkee 247667
India

Srivastava, L.S.
Earthquake & Engr. Dept.
Univ. of Roorkee
Roorkee 247667
India

Srivastava, R. Mohan
4106 Staulo Crescent
Vancouver, BC V6N 3S2
Canada

Srivastava, S.P.
Atlantic Geoscience Centre
Geological Survey of Canada
Bedford Institute of Ocean.
P. O. Box 1006
Dartmouth, N.S. B2Y 4A2
Canada

Srivastava, S.P.
Atlantic Geoscience Centre
Geological Survey of Canada
Bedford Inst. of Oceanography
P. O. Box 1006
Dartmouth, N.S., B2Y 4A2
Canada

Stackebrandt, Werner
Central Inst. for Physics
of the Earth
Telegrafenberg A51
Potsdam 1561
Dem. Rep. of Germany

Stalin, Benitez A.
Alborado 8 va Etapa Mz. 810
Villa 4
Casilla 10829
Guayaquil
Ecuador

Stallard, Robert F.
U.S.G.S., WRD
MS 413, DFC
P.O. Box 25046
Denver, CO 80225

Stampfli, Gerard
Institut de Geologie
UNIL - BESH2
CH-1015 Lausanne
Switzerland

Stanley, Daniel J.
Mediterranean Basin Project
Div. Sedimentology
Room E-206 NMNH
Smithsonian Institution
Washington, DC 20560

Stanley, Rolfe S.
20 Overlake Park
Burlington, VT 05401

Stanton, Ron
U.S. Geological Survey
MS956, National Center
Reston, VA 22092

Staritsyna, G. N.
All-Union Research Inst. for
Geology & Mineral Resources
Nabereghraya Moiki 120
Leningrad 190121
U.S.S.R.

Starkel, Leszek
Geomorphology and Hydrol.Dept.
Polish Acad. of Sci.
Inst. of Geog.
31-018 Krakow
Poland

Stattegger, Karl
Institute of Geology and
Paleontology
Karl-Franzens Universitat
Heinrichstrasse 26
A-8010 Graz
Austria

Stattegger, Karl
Institute of Geology and
Paleontology
Karl-Franzens Universitat
Heinrichstrasse 26
A-8010 Graz
Austria

Stein, Ruediger
Inst. fur Geowissenschaften
und Lithospharenforschung
Giessen University
D-6300 Giessen
Fed. Rep. of Germany

Steinberg, Erik
Escuela Centroamericana
de Geol.
Apt. 35
2060 Univ. de Costa Rica
San Jose
Costa Rica

Steiner, Maureen
Geol. & Geophy. Dept.
Univ. of Wyoming
Laramie, WY 82071

Steinhauser, Peter
Institute for Meteorology and
Geophysics
University of Vienna
Hohe Warte 38
A-1190 Vienna
Austria

Steinthorsson, S.
Geosci. Bldg.
Univ. of Iceland
601 Reykjaulk
Iceland

Stemprok, Miroslav
Geol. Survey
Malostraske nam. 19
11821 Praha 1
Czechoslovakia

Stephan, Jean-Francois
GIS Oceanolog. et Geodynamique
Univ. de Bretagne Occidentale
Avenue Le Gorgeu
29287 Brest Cedex
France

Stephan, Jean-Francois
GIS Oceanolog. et Geodynamique
Univ. de Bretagne Occidentale
Avenue Le Gorgeu
29287 Brest Cedex
France

Stephansson, O.
Div. of Rock Mechanics
Lulea Univ of Tech
S-95187 Lulea
Sweden

Stephens, Michael B.
Geol. Survey of Sweden
Box 670
S-751 28 Uppsala
Sweden

Stephens, Robert L.
c/o Tom Novak
U.S. Army Corps of Engineers
Geotechnical Branch
P.O. Box 4970
Jacksonville, FL 32232-0019

Stephenson, Jonathan
Geology Department
Royal Holloway & Beford New
College
University of London
Egham, Surrey TW20 0EX
England

Stern, Robert J.
Programs in Geosci. (FO2.1)
The Univ. of Texas at Dallas
Box 830688
Richardson, TX 75083-0688

Sternberg, Robert S.
Geol. Dept.
Franklin & Marshall College
P.O.Box 3003
Lancaster, PA 17604-3003

Sterner, S. Michael
Geol. Sci. Dept.
Virginia Polytechnic Inst.
and State Univ.
Blacksburg,, VA 24061

Stewart, Alastair J.
Bureau of Mineral Resources
GPO Box 378
Canberra ACT 2601
Australia

Stewart, Brian W.
Earth & Space Sci. Dept.
Univ. of California
Los Angeles, CA 90024

Stewart, David B.
National Center 959
U.S. Geological Survey
Reston, Va 22092

Stewart, David B.
National Center 959
U. S. Geological Survey
Reston, VA 22092

Stewart, David John
K.S.E.P.L. (Shell Research)
Volmerlaan 6, 2288 9D
Rijswijk (Z-H)
Netherlands

Stewart, R.B.
Department of Soil Science
Massey University
Palmerston North
New Zealand

Stillings, Lisa
Geosci. Dept.
204 Deike Bldg.
Pennsylvania State Univ.
University Park, PA 16802

Stix, John
Geol. Dept.
Univ. of Toronto
Toronto, Ont. M5S 1A1
Canada

Stixrude, Lars
Geol. & Geophy. Dept
Univ. of California
Berkeley, CA 94720

Stixrude, Lars
Geol. & Geophy. Dept
Univ. of California
Berkeley, CA 94720

Stock, Joann M.
Harvard University
Dept. of Earth and Planetary
Sciences
24 Oxford Street
Cambridge, MA 02138

Stock, Joann M.
Harvard University
Dept. of Earth and Planetary
Sciences
24 Oxford St.
Cambridge, MA 02138

Stone, David B.
Geophysical Institute
University of Alaska
Fairbanks, AK 99775

Storey, B.C.
British Antarctic Survey
High Cross
Madingley Rd.
Cambridge CB3 OET
England

Storey, Bryan C.
British Antarctic Survey
High Cross
Madingley Rd.
Cambridge CB3 0ET
England

Storey, Michael
Geol. Dept.
Univ. of Leicester
Leicester LE1 7RH
England

Stow, Dorrik A.V.
Dept. of Geology
Univ. of Nottingham
Nottingham NG7 2RD
England

Stow, Stephen H.
Oak Ridge National Laboratory
P.O. Box 2009
Oak Ridge, TN 37831-6038

Stowe, C.W.
Geol. Dept.
Univ. of Cape Town
Private Bag
Rondebusch 7700
South Africa

Stowe, C.W.
Geol. Dept.
Univ. of Cape Town
Private Bag
Rondebusch 7700
South Africa

Strand, Kari Olavi
Department of Geology
University of Oulu
Linnanmaa
SF- 90570 Oulu
Finland

Stribrny, Bernhard
Inst.fur Geochem., Petr. und
Lagerstattenkunde
J.W. Goethe-Univ. Frankfurt
Postfach 111932
D-6000 Frankfurt/M
Fed. Rep. of Germany

Strnad, J.G.
Uranerz Expl.& Mining, Ltd.
1300 Saskatoon Sq.
410-22nd St., East
Saskatoon, Sask. S7K 5T6,
Canada

Subotowicz, Wieslaw
Technical University of Gdansk
Faculty of Hydrotechnics
Majakowskiego Str. 11/12,
80-952 Gdansk
Poland

Subramanyan, V.
Earth Sci. Dept.
Indian Inst. of Tech.
Powai
Bombay 400076
India

Subramanyan, V.
Earth Sci. Dept.
Indian Inst.of Technology
Powai
Bombai 400076
India

Sueno, Shigeho
Inst. of Geosci.
Univ. of Tsukuba
Tsukuba 305
Japan

Suk, Milos
Geofyzika sp.
Jecna 29 a
Brno 612 46
Czechoslovakia

Suk, Milos
c/o Petr Jakes
Geological Survey, Prague
Malostranske nam. 19
118 21 Praha 1
Czechoslovakia

Sukamto, Rab
c/o John A. Katili
Republic of Indonsia
Dept. of Mines & Energy
Jalan Jenderal Gatot Subroto
Kav. 49, Jakarta
Indonesia

Sumbatov, R.A.
USSR Ministry of Geology
4/6 Gruzinskaya
Moscow 123 812
U.S.S.R.

Sun, Dazhong
Geological Society of China
Pai Wan Chuang Fuchengmenwai
Beijing
Peo. Rep. of China

Sun, Freddie Y-Y.
Inst. of Geol.
Academia Sinica
Beijing
Peo. Rep. of China

Sun, Jiacong
Geol. Dept.
Kunming Inst. of Tech.
Kunming, Yunnan
Peo. Rep. of China

Sun, Shen-su
Bureau of Mineral Resoures
G.P.O. Box 378
Canberra 2601
Australia

Sun, Shu
Inst. of Geol., Bur. of Nat.
Res. & Environ. Sci.
Academia Sinica
P.O. Box 634
Beijing
Peo. Rep. of China

Sun, Tian-xi
Environmental Monitoring
Center
Suzhou, Jiangsu
Peo. Rep. of China

Sun, Wucheng
Geophy. Prosp. Brigade
State Seismological Bureau
104 Wennua Rd.
Zhengzhou, Henan Province
Peo. Rep. of China

Sun, Wucheng
Geophy. Prospecting Brigade
State Seismological Bur.
104 Wennua Rd.
Zhengzhou, Henan Province
Peo. Rep. of China

Sundquist, Eric T.
U.S.G.S.
Woods Hole, MA 02543

Sunouchi, Hisao
Japan National Oil Corp.
3-5-5 Midorigaoka
Hamura, Tokyo 190-11
Japan

Suo, Shutian
China University of
Geosciences
Wuhan
Peo. Rep. of China

Suo, Shutian
China University of
Geosciences
Wuhan
Peo. Rep. of China

Surkov, V.S.
NPO SIBGEO
Krashyi prospect 67
Novosibirsk
U.S.S.R.

Suro-Perez, Vinicio
310 Mitchell Bldg.
Stanford Univ.
Stanford, CA 94305

Susse, Peter
Inst. of Mineralogy and
Crystallography
Univ. of Gottingen
V.M. Goldschmidt Str. 1
D-3400 Gottingen
Fed. Rep. of Germany

Suter, John R.
Exxon Production Research
P.O. Box 2189
Houston, TX 77252

Suter, Max
Inst. de Geol.
Univ. Nacional Autonoma
de Mexico
Apt. Postal 70-296
Mexico DF 04510
Mexico

Sutton, Sally J.
Geol. Sci. Dept.
Univ. of Texas
Austin, TX 78713-7909

Sutton, Stephen R.
Building 815
Brookhaven National Laboratory
Upton, NY 11923

Suzuki-Kamata, Keiko
U.S. Geol. Survey
Cascades Volcano Observatory
5400 MacArthur Blvd.
Vancouver,, WA 98661

Swamy, A.S.R.
Geol. Dept.
Andhra Univ.
Visakhapatnam 530003
India

Swamy, A.S.R.
P.O. Box 4110
Yaounde
Cameroon

Swart, Peter K.
Marine Geol. & Geophy Div.
Univ. of Miami
5600 Rickenbacker Cswy.
Miami, FL 33149-1098

Sweeney, Jack
Geol. Survey of Canada
100 W. Pender
Vancouver, B.C. V6B 1R8
Canada

Sweeney, Jack
Geol. Survey of Canada
100 W. Pender
Vancouver, BC
Canada

Sweet, G.
Dept. of Geology
Univ. of Winnepeg
Winnepeg, Manitoba
Canada

Sweet, Michael
Geol. Sci. Dept.
Univ. of Texas
Austin, TX 78713

Swint-Iki, Theresa
U.S. Geological Survey
345 Middlefield Road MS 952
Menlo Park, CA 94025

Swisher, Carl C.
Berkeley Geochronology Ctr.
2453 Ridge Rd.
Berkeley,, CA 94709

Sykes, Martha L.
Earth Sci. Dept.
Univ. of Cambridge
Cambridge
England

Sylvester, Arthur G.
Geol. Sci. Dept.
Univ. of California
Santa Barbara,, CA 93106

Syvitski, James P.M.
Geol. Survey of Canada
Bedford Inst. of Oceanog.
Box 1006
Dartmouth, N.S. B2Y 4A2
Canada

Szabo', Csaba
Department of Petrology and
Geochemistry,
Eotvos University
Muzeum korut 4/A
H-1088 Budapest
Hungary

Szulc, Joachim
Geol. Sci. Inst.
Jagellonian Univ.
Oleandry Str. 2a
30-063 Cracow
Poland

Szulc, Joachim M.
Inst. of Geol. Sci.
Jagellonian Univ.
Oleandry Str. 2a
30-063 Cracow
Poland

T

Tabesh, Elahe
204 Heroy Geol. Lab.
Syracuse Univ.
Syracuse, NY 13244-1070

Taghizadeh, N.
No. 5, Koocheh Larijani
Khiaban Khalili
(Amir Ahmadi Zamany)
Tajrish
Tehran
Iran

Taguchi, Kazuo
Technical Res. Ctr.
Teikoku Oil Co., Ltd.
9-23-30 Kita-Karasuyama
Setagaya-ku
Tokyo 157
Japan

Taira, Asahiko
Ocean Research Institute,
University of Tokyo
1-15-1 Minamidai, Nakano-ku,
Tokyo 164
Japan

Takamitsu, Yamanaka
College of General Education
Osaka Univ.
Toyonaka
Osaka 560
Japan

Takasu, A.
Department of Geology
Faculty of Science
Shimane University
Matsue 690
Japan

Takeda, Hiroshi
Mineralogical Inst.
Fac. of Sci.
Univ. of Tokyo
Hongo
Tokyo 113
Japan

Talukdar, Suhas
INTEVEP, S.A.
Apartado 76343
Caracas 1070-A
Venezuela

Talwani, Manik
Geol. & Geophy. Dept.
Rice Univ.
P.O. Box 1892
Houston,, TX 77251

Talwani, Manik
Geotechnology Res. Inst.
Houston Area, Research Ctr.
4802 Research Forest Dr.
The Woodlands, TX 77381

Talwani, Manik
Houston Area Research Center

, TX

Talwani, Pradeep
Geol. Dept.
Univ. of South Carolina
Columbia, SC 29208

Talwani, Pradeep
Dept. of Geological Sciences
University of South Carolina
Columbia, SC 29208

Tan, Shundao
Geol. Bureau, China Nat'l.
Nonferrous Metals Ind. Corp.
9 Xizhang Hutong
Xizhimennei Dajie
Beijing
Peo. Rep. of China

Tan, Tjong Kie
Institute of Geophysics
Academia Sinica
P. O. Box g28
Beijing
Peo. Rep. of China

Tan, Tjong Kie
Institute of Geophysics
Academia Sinica
P.O. Box G28
Beijing
Peo. Rep. of China

Tanaka, Hisao
Earth Sci. Dept., Fac. of Sci.
Yamagata Univ.
Yamagata 990
Japan

Tanelli, Giuseppe
Dip. Scienze della Terra
della Univ.
Via G. La Pira, 4
I-50121 Firenze
Italy

Tang, Chunfa
20 Daqinglu Harbin 150046

Heilongjiang Province
Peo. Rep. of China

Tang, Tianfu
Nanjing Inst. of Geol.& Paleon
Academia Sinica
No. 39
East Beijing Rd.
Nanjing
Peo. Rep. of China

Tanimoto, Toshiro
Seismological Lab. 252-21
Caltech
Pasadena, CA 91125

Tankut, Ayla
Geological Engineering Dept.
Middle East Technical
University
06531 Ankara
Turkey

Tari, Gabor
Geophysics Department
Eotvos Lorand University
Kun Bela ter 2
Budapest, H-1083
Hungary

Taruna, Maria Flora
Via Trilussa

Padova I-35125
Italy

Tassinari, Colombo Celso Gaeta
Univ. de Sao Paulo
Instituto de Geociencias
Caixa Postal 20.899
01498 - Sao Paulo
Brazil

Tauchid, M.
Div. of Nuclear Fuel Cycle
Intl. Atomic Energy Agency
P.O. Box 100
A-1400 Vienna
Austria

Taylor, Alan E.
Geol. Survey of Canada
1 Observatory Ctr.
Ottawa, Ont. K1A 0Y3
Canada

Taylor, Edith L.
Botany Dept.
Ohio State Univ.
1735 Neil Ave.
Columbus, OH 43210-1293

Taylor, Lawrence A.
Geol. Sci. Dept.
Univ. of Tennessee
Knoxville, TN 37996-1410

Taylor, S.R.
R.S.E.S.
Australian National University
Canberra Act 2601
Australia

Taylor, W.E.G.
Faculty of Applied Sciences
Luton College of Higher
Education
Park Square
Luton Bedfordshire
England

Tazhibaeva, Patchim T.
Institute/Geological Sciences
Kasakh Academy of Sciences
Kalenen Street, 69a
Alma-Ata 480100
U.S.S.R.

Tazioli, Giulio Sergio
Dep. Scienze della Terra,
University of Ancona
Via Brecce Bianche
60123-Ancona
Italy

Tedesco, Lenore P.
RSMAS - MGG
University of Miami
4600 Rickenbacker Causeway
Miami, FL 33149

Tehrani, K. K.
P. O. Box 19615-466

Tehran
Iran

Teichmuller, Marlies
Am Hohen Haus 15

Krefeld 4150
Fed. Rep. of Germany

Teixeira, Wilson
Geosci. Inst.
Univ. of Sao Paulo
Rua do Lago 562
Cx. Postal 20.899
Sao Paulo CEP 01498
Brazil

Telleria, Jose L.
Centro de Investigaciones
Geologicas
P. O. Box 11253
La Paz
Bolivia

Teng, Jiwen W.
Geophy. Inst.
Academia Sinica
Benjing
Peo. Rep. of China

Tera, Fouad
Terrestrial Magnetism Dept.
Carnegie Inst. of Washington
5241 Broad Branch Rd., NW
Washington, DC 20015

Terentiev, V.M.
Ministry of Geology of USSR
Bolshaya Gruzinskaya, 4/6
Moscow 123812
U.S.S.R.

Tessensohn, Franz
BGR
Postfach 510153
3000 Hannover 31
Fed. Rep. of Germany

Teutsch, G.
Inst. fur Wasserbau
Univ. Stuttgart
Pfaffenwaldring 61
D-7000 Stuttgart 70
Fed. Rep. of Germany

Tew, Berry H. .
Geol. Survey of Alabama
P.O. Box 0
Tuscaloosa, AL 35486-9780

Textoris, Daniel A.
Dept. of Geology
University of N. Carolina
Mitchell Hall
Chapel Hill, NC 27599

Thabet, Kanana M.
Jeology Dept.
College of Science
University of Mosul
Mosul
Iraq

Thakur, Vikram C.
Wadia Inst. of Himalayan Geol.
33, General Mahadeo Singh Rd.
Dehra Dun 248 001
India

Thakur, Vikram C.
Wadia Inst. of Himalayan Geol.
33 General Mahadeo Singh Rd.
Dehra Dun 248 001
India

Thiede, Jorn
GEOMAR Research Center for
Marine Geosciences
Wischhofstrasse 1-3,
D-2300 Kiel 14,
Fed. Rep. of Germany

Thigale, S.S.
28, Ashok Nagar

Pune-411007
India

Thigale, S.S.
28, Ashok Nagar

Pune 411007
India

Thiry, Medard
Ecole des Mines de Paris
Centre de Geologie
35, Rue St. Honore
77305 Fontainebleau
France

Thomas, Ellen
Earth & Environ. Sci. Dept.
Wesleyan Univ.
Middletown,, CT 06457-6034

Thomas, Roger E.
U.S. Geological Survey
12201 Sunrise Valley Dr.
MS 956
Reston, VA 22092

Thompson, K.F.M.
Geochemical & Environmental
Research Center
Texas A&M University
College Station, TX 77840

Thompson, R.
Geophy. Dept.
Univ. of Edinburgh
Edinburgh
Scotland

Thomson, M.R.A.
British Antarctic Survey
High Cross
Madingley Road
Cambridge CB2 0ET
England

Thorez, Jacques M.M.
Clay Geology Laboratory
Allee du 6 AOUT, B48
Liege State University
Sart /tilman
Liege B-4000
Belgium

Thorez, Jacques M.M.
Clay Geology Laboratory
Allee du 6 AOUT, B48
Liege State University
Sart /tilman
Liege B-4000
Belgium

Thune, Torsten G.O.
Quaternary Res. Dept.
Odengatan 63
S-113 22 Stockholm
Sweden

Thurow, Juergen
Institute and Museum for
Paleontology & Geology
Universitaet Tuebingen
Sigwartstrasse 10
7400 Tuebingen
Fed. Rep. of Germany

Thurow, Juergen
Institute and Museum for.
Paleon. & Geol.
Univ. Tuebingen
Sigwartstrasse 10
7400 Tuebingen
Fed. Rep. of Germany

Thurston, P.C.
Ontario Geological Survey
77 Grenville St.
Toronto M7A 1W4
Canada

Thurston, P.C.
Ontario Geological Survey
77 Grenville St.
Toronto M7A 1W4
Canada

Thy, Peter
NASA
Johnson Space Center, SN2
Houston,, TX 77058

Tiainen, Markku
Geol. Survey of Finland

SF-02150 Espoo
Finland

Tickshaev, V.
Ministry of Geology, USSR
Saratov, Lenina pr., 70
NVNIIGG 410710
U.S.S.R.

Tidball, Ronald R.
U.S.G.S.
MS 973, Federal Ctr.
P.O. Box 25046
Denver, CO 80225

Tietz, Gerd F.
Geol. Paleon. Dept.
Univ. of Hamburg
Bundesstr. 55
D 2000 Hamburg 13
Fed. Rep. of Germany

Tighe, Stacey A.
Grad. School of Oceanography
Univ. of Rhode Island
So. Ferry
Narragansett, RI 02882

Tijani, Allal
B.R.P.M. Rabat
5-Charii Moulay Hassan
B.P. 99
Rabat
Morroco

Ting, Francis T.C.
Dept. Geology & Geography
West Virginia University
Morgantown, WV 26506

Tomadin, Luciano
Istituto di Miner. & Paleon.
Univ. di Urbino
Via M. Oddi, 14
61029 Urbino
Italy

Tomek, Cestmir
Geofyzika Brno
P.O. Box 62
61246 Brno
Czechoslovakia

Tomek, Cestmir
Geophy. Brno
P.O. Box 62
61246 Brno
Czechoslovakia

Tong, Li
The Geol. Society of China
Pai Wan Chuang Fuchengmenwai
Beijing
Peo. Rep. of China

Toriumi, Mitsuhiro
Dept. of Geology
University of Tokyo
7-3-1 Hongo
Bunkyo-ku, Japan
Japan

Torrens, Hugh S.
Geol. Dept.
Univ. of Keele
Keele ST 5 5 BG
England

Torres Martin, Jose Antonio
Dep. Ciencias De La Tierra
Facultad De Ciencias
Universidad De Zaragoza
50009 Zaragoza
Spain

Torres, Alberto
Jr. Caracas 2410

Lima 11
Peru

Toselli, Alejandro J.
Geol. Dept.
Univ. Nacional de Tucuman
Miguel Lillo 205
Tucuman 4000
Argentina

Toselli, Alejandro J.
Geol. Dept.
Univ. Nacional de Tucuman
Miguel Lillo 205
Tucuman 4000
Argentina

Touret, J.L.R.
Postbus 7161

1007 MD Amsterdam
Netherlands

Touret, J.L.R.
Postbus 7161

1007 MC Amsterdam
Netherlands

Tracy, Robert J.
Geol. Sci. Dept.
Virginia Tech.
Blacksburg, VA 24061

Trayner, P.M.
GECO Expl. Serv.
Merlin House
Boundary Road
Woking Surrey GU21 5BX
England

Treagus, J.E.
Geol. Dept.
Univ. of Manchester
Manchester M13 9PL
England

Treagus, Susan H.
Department of Geology
Manchester University
Manchester M13 9PL
England

Trendall, Alec F.
Geological Survey of Western
Australia
Mineral House
100 Plain Street
Perth 6000
Australia

Trettin, Hans P.
Geol. Survey of Canada
3303-33rd St., NW
Calgary, Alta. T2L 2A7
Canada

Trifinov, V.G.
Geological Institute
USSR Academy of Sciences
7, Pyzhevsky, P.O. Box 17
Moscow 109017
U.S.S.R.

Trifonov, V.G.
Geological Institute
USSR Academy of Sciences
7, Pyzhevsky, P.O. Box 17
Moscow 109017
U.S.S.R.

Trinh, Long
University of Hanoi
Faculty of Geology
S.R.
Vietnam

Triplehorn, Don
Br. of Coal Geol.
U.S.G.S.
MS956
12201 Sunrise Valley Dr.
Reston, VA 22092

Trivedy, A.N.
Geological Survey of India
Operations Nihar
2, Patliputra Colony
Patna 800 013
India

Troester, Joseph W.
U.S.G.S.
GPO Box 4424
San Juan, PR 00936

Trofimov, V.T.
Geological Faculty
Moscow State Univ.
Moscow
U.S.S.R.

Troian, Fabio Luiz
Dept. Geology
University Vale do Rio Dos
Sinos - Unisinos
CP 93000
Sao Leopoldo, RGS
Brazil

Trompette, Roland
Laboratoire de Geologie
Centre de St. Jerome
Universite d'Zix-Marseille III
13397 Marseille Cedex
France

Trotsuk, V.Y.
P.P. Shirshov Inst. for
Oceanology
The USSR Academy of Sciences
Moscow
U.S.S.R.

Trurnit, T.P.
Wacholderweg 6

Burgwedel 1 3006
Fed. Rep. of Germany

Trurnit, T.P.
Wacholderweg 6

Burgwedel 1 3006
Fed. Rep. of Germany

Tsai, Chao-Ming
Radwast Administration/
Atomic Energy Council
65, Lane 144, Keelung Rd.
Section 4
Taipei, Taiwan
Peo. Rep. of China

Tsoar, Haim
Dept. Geography
Gen-Gurion University of Negev
Beer-Sheva IL-84105
Israel

Tsuruta, Haruo
Yokohama Research Institute
for Environmental Science
1-2 Takigashira
Isogo-ku
Yokohama 235
Japan

Tsusue, Akio
Geol. Dept., Fac. of Sci.
Kumamoto Univ.
Kurokami 2-39-1
Kumamoto 860
Japan

Tu, Kan
Dept. of Geological Sciences
University of Illinois
Chicago, IL 60680

Tuezov, I.K.
Inst. of Tectonics and
Geophysics,Far East Branch
Acad. of Sci. of the USSR
65, Kim Yu Chen St.
Khabarovsk 680063
U.S.S.R.

Tuisku, Pekka
University of Oulu
Dept. of Geology
Linnanmaa
SF-90570 Oulu
Finland

Tuisku, Pekka
University of Oulu
Dept. of Geology
Linnanmaa
90570 Oulu
Finland

Turcotte, Donald L.
Geol. Sci. Dept.
Cornell Univ.
Ithaca, NY 14853

Tusak, Boris
INA-PROJEKT
Savska c. 88a
41000 Zagreb
Yugoslavia

Tusak, Boris
Ina-project
Savska 88 A
41000 Zagreb
Yugoslavia

Twichell, David
U.S. Geological Survey
Br. of Atlantic Marine Geology
Quissett Campus
Woods Hole, MA 02543

U

Uchman, Alfred
Institute of Geological Sciences
Jagiellonian University
Cracow,
ul. Oleandry 2a,
Poland

Udintsev, G.B.
Geological Institute of the USSR Academy of Sciences
U.S.S.R.

Ugolini, F.C.
College of Forest Resources
AR - 10
Univ. of Washington
Seattle, WA 98195

Ui, Hirotaka
Earth Sci. Lab.
Fac. of Education
Toyama Univ.
Gofuku
Toyama City 930
Japan

Ulmer, Gene C.
2207 Blackhorse Drive
Warrington, PA 18976

Ulmishek, G.F.
U.S. Geol. Survey
Denver Federal Center
Box 25046, M/S 940
Denver, CO 80225

Underwood, Michael B.
Geol. Dept.
Univ. of Missouri
Columbia,, MO 65211

Unrug, Raphael
Geol. Sci. Dept.
Wright State Univ.
Dayton, OH 45435

Upadhyaya, Rajani
Geol. Dept.
Univ. of Rajasthan
Udaipur 313001 Rajasthan
India

Upadhyaya, Rajani
Geol. Dept.
Univ. of Rajasthan
51 Saraswati Marg
Udaipur (Rajasthan) 313001
India

Upadhyaya, V. S.
Project Geologist
Karnataka Power Corp. Ltd.
Dandeli 581325
India

Upson, Craig
Stellar Computer
920 Hillview CT #180
Milpitas, CA 95035

Urai, Janos L.
Inst. voor Aardwetenschappen
P.O. Box 80021
3508 TA Utrecht
Netherlands

Urbanski, Walter S., Jr.
Intermagnetics General Corp.
P.O. Box 566
Guilderland, NY 12084

Urrutia-Fucugauchi, J.
Lab. de Paleomagnetismo 7
Geofisica Nuclear
Inst. de Geofisica
Unv. Nacional Autonoma de Mex.
D. Coyoacan 04510 D.F.
Mexico

Utada, Minoru
The Univ. Museum
Univ. of Tokyo
Hongo 7-3-1, Bunkyo-ku
Tokyo 113
Japan

V

Vai, Gian Battista
Dipartimento di Scienze Geol.
Viazamboni 67
Bologna I-40127
Italy

Vai, Gian Battista
Dipart. Di Scienze Geologiche
Via Zamboni 67
Bologna I-40127
Italy

Vail, Peter R.
Geol. & Geophy. Dept.
Rice Univ.
Houston,, TX 77005

Vaithiyanathan, P.
School of Environmental Sci.
Jawaharlal Nehru University
New Delhi 110 067
India

Valdes, Antonio
Union de Geologia
Prolongacion Carr. Central
entre Linea FFCC y Via Blanca
San Miguel del Padron
C. Habana
Cuba

Valentine, Page
U.S. Geol. Survey

Woods Hole, MA 02543

Valette-Silver, Nathalie J.
Chemistry Dept.
Univ. of Maryland
College Park, MD 20742

Valette-Silver, Nathalie J.
Chemistry Dept.
Univ. of Maryland
College Park, MD 20742

van den Bogaard, Paul
Inst. fur Mineralogie
Ruhr-Univ. Bochum
Postfach 102148
D-4630 Bochum
Fed. Rep. of Germany

Van der Merwe, S.W.
Department of Geology
Univ. of the Orange Free State
P. O. Box 339
Bloemfontein
South Africa

Van der Voo, Rob
Geol. Sci. Dept.
Univ. of Michigan
Ann Arbor, MI 48109-1063

Van der Voo, Rob
Geol. Sci. Dept.
Univ. of Michigan
Ann Arbor, MI 48109-1063

van der Wiel, A.M.
Agricultural University,
Dept. of Soil Science and
Geology
P.O. Box 37
6700AA Wageningen
Netherlands

van Bever Donker, J.M.
Department of Geology,
University of the Western Cape
Private Bag X17
Bellville, 7530
South Africa

Van Houten, Franklin B.
Geol. & Geophy. Sci. Dept.
Princeton Univ.
Princeton, NJ 08544

Van Schmus, W.R.
Geol. Dept.
Univ. of Kansas
Lawrence, KS 66045

Van Schmus, W.R.
Geol. Dept.
Univ. of Kansas
Lawrence, KS 66045

Van Siclen, DeWitt C.
4909 Bellaire Blvd.

Bellaire, TX 77401

Van Siclen, DeWitt C.
4909 Bellaire Blvd.

Bellaire, TX 77401

Van Tassell, Jay
Eastern Oregon State College
Badgley Hall
La Grande, OR 97850

Van Wagoner, J.C.
Exxon Prod. Res. Co.
P.O.Box 2189
Houston, TX 77252-2189

Van Wagoner, Nancy A.
Geol. Dept.
Arcadia Univ.
Wolfville, NS B0P 1X0,
Canada

Vannier, Michel
Centre de Geologie Generale et
Miniere, IGM
Ecole Nationale Superieure
35 Rue St. Honore
77305 Fontainebleau Cedex
France

Varela, Manuel
Servicio Geologico de Obras
Publicas
Avenida de Portugal, 81
28011 Madrid
Spain

Varentsov, I.M.
Institute of Geology of Ore
Deposits
USSR Academy of Sciences
Moscow
U.S.S.R.

Varentsov, I.M.
Institute of Geology of Ore
USSR Academy of Sciences
Moscow
U.S.S.R.

Vartanyan, Genrich S.
All-Union Research Inst. for
Hydrogeology & Engineering
Geology
142452 Zeleny-village, Noginsk
Moscow region
U.S.S.R.

Vaslet, Denis
B.R.G.M.
P.O. Box 1492
Jeddah
Saudi Arabia

Vasudevan, V.
Marine Sci. Div.
Centre for Earth Sci. Studies
PB 7250
Trivandrum 695031
India

Vebleu, David R.
Earth & Planetary Sci. Dept.
Johns Hopkins Univ.
Baltimore, MD 21218

Veeger, Anne
Dept. of Hydrology
and Water Resources
Univ. of Arizona
Tucson, AZ 85721

Veizer, Jan
Inst. fur Geol.
Ruhr Univ.
4630 Bochum
Fed. Rep. of Germany

Veksler, Ilya
Vernadsky Institute
Kosygin str. 19
Moscow
U.S.S.R.

Velbel, Michael A.
Dept. of Geological Sciences
206 Natural Science Bldg.
Michigan State University
East Lansing, MI 48824-1115

Velbel, Michael A.
Dept. of Geological Sciences
206 Natural Science Bldg.
Michigan State University
East Lansing, MI 48824-1115

Velichko, A.A.
Paleogeog. Dept.
USSR Academy of Sciences
Staromonetny 29
Moscos 109017
U.S.S.R.

Venkatapathi, M.
Dy. Manager (Geology)
National Aluminum Co., Ltd.
Mines & Refinery Complex
Damanjodi-763 008, Dt. Koraput
Orissa
India

Verdon, Andrew J.
American Geol. Inst.
4220 King St.
Alexandria, VA 22302

Vergniolle, Sylvie
Institut de Physique du Globe
Laboratoire de Dynamique des
Systemes Geologiques
4 Place Jussieu
75252 Paris Cedex 05
France

Verhoeven, J.T.A.
Smithsonian and Environmental
Research Center
P.O. Box 28
Edgewater, MD 21037

Verma, Prem N.
International Wing
Geol. Survey of India
27 Jawaharlal Nehru Rd.
Calcutta-700 016
India

Verwoerd, W.J.
Geol. Dept.
Univ. of Stellenbosch
Stellenbosch 7600,
South Africa

Verzhbitsky, E.V.
Inst. of Oceanology
Krasikova Str. 23
Moscow 117218
U.S.S.R.

Verzhbitsky, E.V.
Inst. of Oceanology
Krasikova Str. 23
Moscow 117218
U.S.S.R.

Vessem, E. J. Van
Chevron UK Limited
2 Portman Street
London W1H OAN
England

Vidal, Cesar E.
c/o Don Noble
Mackay School of Mines
Univ. of Nevada-Reno
Reno, NV 89557

Vidic, Natasa
Biotechnical Faculty
Univ. of Ljubljana
Ljubljana
Yugoslavia

Viele, George W.
Geol. Dept.
Univ. of Missouri
Columbia, MO 65211

Vigneresse, Jean L.
Lab. Tectonophysique
Univ. de Nantes
44072 Nantes Cedex 02
France

Vilas, Juan F.
Dto. Diencias Geol.
Fac. de Cs Exactas y Nat.
Univ. de Buenos Aires
Pabellon 2
1428 Buenos Aires
Argentina

Vilas, Juan F.
Dept. de Ciencias Geol.
Ciudad Univ.
Univ.de Buenos Aires y CONICET
Pab. II.
1428 Buenos Aires
Argentina

Villa, I.M.
Istituto di Geocronologia,
CNR, Via Maffi 36
Pisa
Italy

Villasuso-Pino, Miguel
Fac. de Ingenieru
Apdo Postal 1423-B
Merida Yucaran 97002
Mexico

Villeneuve, Michel
Laboratories de Geologie
Faculte Des Sciences
Univ. D'Aix Marseille III
St. Jerome
13397 Marseille Cedex 13
France

Villeneuve, Michel
Laboratories De Geologie
Faculte Des Sciences
Saint Jerome
Univ. D'Aix Marseille III
13397 Marseille Cedex 13
France

Villey, M.
BRGM P.O. Box 6009-65060
Orleans
Cedex 2
France

Villey, Michel
BRGM
BP 6009
Orleans Cedex 2
France

Virdi, Niranjan S.
Wadia Inst. of Himalayan Geol.
33, General Mahadev Singh Rd.
Dehra Dun 248001
India

Virdi, Niranjan Singh
Wadia Inst. of Himalayan Geol.
33-General Mahadev Singh Rd.
Dehra Dun 248001
India

Visser, Johan N.J.
Geol. Dept.
P.O. Box 339
Bloemfontein 9300
South Africa

Visser, Wiekert
Petr. Devel. Oman (XXG/6)
P.O. Box 81
Muscat
Oman

Vistelius, A. B.
Mathematical Geol. Lab.
Inst. of Geol. & Geochronology
of Precambrian,Acad.of Sci.
12 Voinov St.
Leningrad D-187, 11187
U.S.S.R.

Vladimirow, Valentin Draganow
Dep. Economic Geology and
Geochemistry, Faculty of
Geology & Geography, Sofia
University, 1000, Sofia,
Bulgaria

Vlahovic, Milos
Civil Engr. Fac.
Bulevar Revolucije 73
11000 Beograd
Yugoslavia

Vochten, R.
Chem. & Phys. Miner. Lab.
State Univ. Antwerpen
Middelheimlaan 1
2020 Antwerpen
Belgium

Vogel, Koen R.
246 Deike Bldg.
Pennsylvania State Univ.
University Park, PA 16802

Vogelsang, Elke
Geologisch-Palaontologisches
Inst. und Mus. der
Christian-Albrechts Univ.
Olshausenitrasse 40160
D-2300 Kiel 1
Fed. Rep. of Germany

Vokurka, Karel
Geological Survey
Malostranske nam. 79
11821 Praha 7
Czechoslovakia

Vokurka, Karel
Geological Survey,
Malostranske NA'M. 79, 778 27
Praha 7
Czechoslovakia

Volpe, Alan M.
Scripps Inst. of Oceanog.
A-012
Univ. of California
La Jolla, CA 92093-0212

Von Engelhardt, Wolf
Mineralogical Inst.
Univ. of Tuebingen
Wilhelmstr. 56
D-7400 Tuebingen
Fed. Rep. of Germany

von Huene, Roland
U.S.G.S.
345 Middlefield Rd.
Menlo Park, CA 94025

Vorder Bruegge, Richard W.
Geol. Sci. Dept.
Brown Univ.
P.O. Box 1846
Providence, RI 02912

Vorobyov, Yu. K.
Institute of Geology of Ore
Deposits, Mineralogy,
Petrography and Geochemistry
Staromonetnic, 35
Moscow
U.S.S.R.

Vrba, J.
Stavebni
Gorkeho nam. 7
113 09 Praha 1
Czechoslovakia

Vrublevsky, Anatoly A.
Inst. of Tectonics and
Geophysics
65, Kim Yu Chen Street
680063 Khabarovsk
U.S.S.R.

Vrublevsky, Anatoly A.
Institute of Tectonics and
Geophysics
65, Kim Yu Chen Street
Khabarovsk 680063
U.S.S.R.

Vuollo, Jouni
Department of Geology
University of Oulo
Linnanmaa
90570 Oulu
Finland

Vurasinovic, Slobodan
Geoinstitut
Rovinjska 12
11000 Belgrade
Yugoslavia

W

Wada, Hideki
Geosci. Inst.
Shizuoka Univ.
836 Ohya
Shizuoka 422
Japan

Wagner, Jean-Frank
Karlsruhe University
Dept. of Applied Geology
Kaiser Str. 12
Karlsruhe D-7500
Fed. Rep. of Germany

Walker, Patrick H.
CSIRO Div. of Soils
GPO Box 639
Canberra 2601
Australia

Walker, Richard J.
A-21 Physics Building
National Bureau of Standards
Gaithersburg, MD 20899

Wallace, Paul J.
Geol. & Geophy. Dept.
301 Earth Sci. Bldg.
Univ. of California
Berkeley, CA 94720

Walliser, Otto H.
Inst. Geol. Palaeont.
Goldschmidt-Str. 3
D-34 Gottingen
Fed. Rep. of Germany

Walliser, Otto H.
Inst. Geol. Palaeont.
Goldschmidt-Str. 3
D-34 Gottingen
Fed. Rep. of Germany

Walter, Louis S.
Code 620
Goddard Space Flight Center
Greenbelt, MD 20771

Walter, M.R.
Bureau of Mineral Resources
Div. of Continental Geology
P.O. Box 378
Canberra, ACT 499738
Australia

Walter, M.R.
Bureau of Mineral Resources
Div. of Continental Geology
P.O. Box 378
Canberra, ACT 499738
Australia

Waltz, James P.
Dept. of Earth Resources
Colorado State University
Fort Collins, CO 80523

Wan, Erkang
The Geol. Society of China
Pai Wan Chuang Fuchengmenwai
Beijing
Peo. Rep. of China

Wan, Ming-Hao
The Geol. Society of China
Pai Wan Chuang Fuchengmenwai
Beijing
Peo. Rep. of China

Wang, Binhai
Geological Scientific Inst. of
ShengLi Oilfield
Doneying City
Shandong
Peo. Rep. of China

Wang, C.
Sedimentary Institute
Chengdu College of Geology
Chengdu
Peo. Rep. of China

Wang, Chen-yuen
Geol. and Geophy. Dept.
Univ. of California
Berkeley, CA 94720

Wang, Dezi
The Geol. Society of China
Pai Wan Chuang Fuchengmenwai
Beijing
Peo. Rep. of China

Wang, Dongpo
Energy Res. Geol. Dept.
Changchun College of Geol.
Changchun 130026 Jilin
Peo. Rep. of China

Wang, Fangguo
The Geol. Society of China
Pai Wan Chuang Fuchangmenwai
Beijing
Peo. Rep. of China

Wang, H.
Geological Society of China
Baiwanzhuang Road 26
Fuchengmenwai
Beijing
Peo. Rep. of China

Wang, Hongzhen
China Univ. of Geosciences
Xueyuan Road 29
Beijing 100083
Peo. Rep. of China

Wang, Kuiren
Dept. of Earth & Space Science
University of Science and
Technology of China
Hefei, Anhui
Peo. Rep. of China

Wang, Kuiren
Dept. of Earth & Space Science
Univ. of Science & Technology
Hefei, Anhui
Peo. Rep. of China

Wang, Peijun
Geological Society of China
Pai Wan Chuang Fuchengmenwai
Beijing
Peo. Rep. of China

Wang, Shan Shu
Offshore Oil Exploration and
Development Centre
CNOOC
Xin-Chen County
Hebei Prov.
Peo. Rep. of China

Wang, Shigui
Geological Society of China
Pai Wan Chuang Fuchengmenwai
Beijing
Peo. Rep. of China

Wang, Shou Lun
The Geol. Society of China
Pai Wan Chuang Fuchengmenwai
Beijing
Peo. Rep. of China

Wang, Sijing
Institute of Geology
Academia Sinica
P. O. Box 634
Beijing
Peo. Rep. of China

Wang, Yifeng
Geol. Dept.
Indiana Univ.
Bloomington, IN 47405

Wang, Yuan
Department of Geology
National Taiwan University
Taipei 10770
Taiwan

Wang, Zhengbang
Beijing Research Institute of
Uranium Geology
P.O. Box 764
Beijing
Peo. Rep. of China

Wang, Zhitian
Beijing Research Institute of
Mineral Resources & Geology of
CNNC
Beiyuan Andingmenwai
Beijing
Peo. Rep. of China

Wang, Zhitian
Beijing Res. Inst. of Miner.
Res. & Geol. of CNNC
Beiyuan Andingmenwai
Geijing
Peo. Rep. of China

Wanke, Heinrich
Max-Planck-Inst. fur Chemie
Abt. Kosmochemie
Saarstrasse 23
D-6500 Mainz
Fed. Rep. of Germany

Wanke, Heinrich
Max-Planck-Inst. fur Chemie
Abt. Kosmochemie
Saarstrasse 23
D-6500 Mainz
Fed. Rep. of Germany

Wanless, Harold R.
RSMAS-MGG
Univ. of Miami
4600 Rickenbacker Cswy.
Virginia Key, FL 33149

Wardle, Richard J.
Mines Dept.
P.O. Box 4750
St. John's, Nfld. A1C 5T7
Canada

Warren, Paul H.
Institute of Geophysics and
Planetary Physics
University of California
Los Angeles, CA 90024

Warsi, Waris E. K.
Department of Geology
University of Kuwait
P. O. Box 5969
13060 Kuwait
Kuwait

Wartel, Stan
Koninklijk Belgisch Inst.
Voor Natuurwetenschappen
Vautierstraat 29
Brussels 1040
Belgium

Watanabe, Teruo
Geol. & Miner. Dept.
Fac. of Sci.
Hokkaido Univ.
Kita-10, Nishi-8, Kitaku
Sapporo 060
Japan

Watanabe, Teruo
Geol. & Miner. Dept.
Fac. of Science
Hokkaido Univ.
Kita-10, Nishi-8, Kitaku
Sapporo
Japan

Watanabe, Yasushi
Hokkaido Branch
Geological Survey of Japan
Minami 1, Nishi 18
Chuo-ku
Sapporo 060
Japan

Waters, David J.
Earth Sci. Dept.
Univ. of Cambridge
Downing Street
Cambridge CB2 3EQ
England

Waters, David J.
Earth Sci. Dept.
Univ. of Cambridge
Downing St.
Cambridge CB2 3EQ
England

Waters, Johnny A.
Dept. of Geology
West Georgia College
Carrollton, GA 30118

Watkeys, Michael K.
Geochem. Dept.
Univ. of Cape Town
Rondebosch 7700
South Africa

Watney, W. Lynn
Kansas Geol. Survey
Campus West
Univ. of Kansas
1930 Constant Ave.
Lawrence, KS 66046

Watson, E. Bruce
Geol. Dept.
Rensselaer Polytechnic Inst.
Troy, NY 12180-3590

Watson, E. Bruce
Geol. Dept.
Rensselaer Polytechnic Inst.
Troy, NY 12180-3590

Watts, A.B.
Lamont-Doherty Geol. Obs.
Columbia Univ.
Rt. 9W
Palisades, NY 10964

Waychunas, Glenn A.
Center for Material Research
105 McCullough Bldg.
Stanford Univ.
Stanford, CA 94305-4045

Weaver, Jean N.
U.S. Geological Survey
MS972, Box 25046 Federal Ctr.
Denver, CO 80225

Webb Jr., Fred
Geol. Dept.
Appalachian State Univ.
Boone, NC 28608

Webb, John A.
Geol. Dept.
La Trobe Univ.
Bundoora, Vic. 3083
Australia

Webb, John A.
Geol. Dept.
La Trobe Univ.
Bundoora, Vic. 3083
Australia

Webb, Peter N.
Geol. & Miner. Dept.
Mendenhall Lab. 107
Ohio State Univ.
Columbus, OH 43210

Weber, Christian
Bureau de Recherches
Geol. et Minieres
BP 6009
45060 Orleans Cedex
France

Weber, K.
Inst. of Geology & Dynamics
of the Lithosphere
Goldschmidtstr. 3
Gottingen D-3400
Fed. Rep. of Germany

Weeks, Edwin P.
U.S.G.S.
MS413, Box 25046
Denver Federal Center
Denver, CO 80225

Weerasooriya, S.V.R.
Inst. of Fundamental Studies
Nantana Road
Kandy
Sri Lanka

Wei, Jiayong
Geological Society of China
Pai Wan Chuang Fuchengmenwai
Beijing
Peo. Rep. of China

Wei, Siyu
Geophy. Inst.
Academia Sinica
Beijing
Peo. Rep. of China

Weidner, Donald J.
Earth & Space Sci. Dept.
State Univ. of New York
Stony Brook, NY 11794

Weihed, Per A.
Geol. Dept.
Univ. of Gothenburg
Chalmers Univ. of Tech.
S-41296 Gothenburg
Sweden

Weiji, Cheng
Earth & Space Sci. Dept.
Univ.of Sci. & Tech. of China
Hefei, Anhui
Peo. Rep. of China

Weiland, Thomas J.
Geol. Dept.
Georgia Southwestern College
Americus, GA 31709

Weimer, Paul
Mobil Research
13777 Midway Road
Dallas, TX 75244

Weimer, Robert J.
Geol. Dept.
Colorado School of Mines
Golden, CO 80401

Weimer, Robert J.
Geol. Dept.
Colorado School of Mines
Golden, CO 80401

Weissert, Helmut
Geol. Inst.
ETH-Z
CH-8092 Curich
Switzerland

Weng, Shijie
Inst. of Petr. Geol.
Ministry of Geol. & Min. Res.
No. 31, College Road
Beijisip 100083
Peo. Rep. of China

Wenner, David B.
Geol. Dept.
Univ. of Georgia
Athens, GA 30602

Wenrich, Karen
U.S.G.S.
MS 905, Box 25046
Federal Center
Denver, CO 80225

Werner, Bradley T.
Research Dept.
Naval Weapons Center
China Lake, CA 93555

West, Dorene B.
GECO Geophysical Company, Inc.
1325 South Dairy Ashford
Houston, TX 77077

West, Ronald R.
Geol. Dept.
Thompson Hall
Kansas State Univ.
Manhattan, KS 66506

Westerman, Gerd E.G.
Geol. Dept.
McMaster Univ.
Hamilton, Ont. L8S 4M1,
Canada

Westgate, John
Dept. of Geology
University of Toronot
Scarborough Campus
1265 Military Trail
Scarborough, Ontario M1C 1A4
Canada

Wetzel, K.
Central Institute of Isotope
and Radiation Research
Academy of Sciences of G.D.R.
Permoserstr. 15
Leipzig 7050
Dem. Rep. of Germany

Wetzel, K.
Central Institute of Isotope &
Radiation Research
Academy of Sciences of G.D.R.
Permoserstr. 15
Leipzig 7050
Dem. Rep. of Germany

Wetzel, K.
Central Institute of Isotope &
Radiation Research
Academy of Sciences of G.D.R.
Permoserstr. 15
Leipzig 7050
Dem. Rep. of Germany

Wezel, Forese-Carlo
Inst. of Geol.
Univ.
Via S. Chiara, 27
61029 Urbino
Italy

Wezel, Forese-Carlo
Inst. of Geol.
Univ.
Via S. Chiara, 27
61029 Urbino
Italy

White, R.R.
Amoco Production Co.
P.O. Box 800
Denver, CO 80201-0800

Whitehead, John A.
Woods Hole Oceanog. Inst.

Woods Hole, MA 02543

Whitham, A.G.
British Antarctic Survey
High Cross
Madingley Road
Cambridge CB3 0ET
England

Whitham, A.G.
British Antarctic Survey
High Cross
Madingley Road
Cambridge CB3 0ET
England

Whittaker, A.
British Geol. Survey
Keyworth
Nottingham NG12 5GG
England

Wichrowska, Maria
Geol. Survey of Poland
Rakowiecka Str. 4
00-975 Warsaw
Poland

Wickham, Stephen M.
Geophy. Sci. Dept.
Univ. of Chicago
5734 S. Ellis Ave.
Chicago, IL 60637

Wickham, Stephen M.
Geophy. Sci. Dept.
Univ. of Chicago
5734 S. Ellis Ave.
Chicago, IL 60637

Wickham, Stephen M.
Geophy. Sci. Dept.
Univ. of Chicago
5734 S. Ellis Ave.
Chicago, IL 60637

Wilber, R. Jude
Sea Education Association
P.O. Box 6
Woods Hole, MA 02543

Wilde, Pat
Marine Sci. Group
3 Earth Sci. Bldg.
Univ. of California
Berkeley, CA 94720

Wildi, Walter
Geol. & Paleon. Dept.
13, rue des Maraiocher
CH-1211 Geneva
Switzerland

Wildi, Walter
Geol. & Paleon. Dept.
13, rue des Maraiocher
CH-1211 Geneva
Switzerland

Wilding, Martin C.
Grant Inst. of Geology
University of Edinburgh
West Mains Road
Edinburgh EH9 3JW
Scotland

Wilding, Martin C.
Grant Inst. of Geology
University of Edinburgh
West Mains Road
Edinburgh EH9 3JW
Scotland

Williams, George E.
Geol. & Geophy. Dept.
Univ. of Adelaide
G.P.O. Box 498
Adelaide, S.Australia 5001
Australia

Williams, Harold
Earth Sciences
Memorial University of
Newfoundland
St. John's A1B 3X5
Canada

Williams, John W.
Department of Geology
San Jose State University
San Jose, CA 95192-0102

Williams, Patrick J.
Earth Sci. Dept.
Univ. of London
Goldsmith's College
Creek Rd.
London SE8 3BU
England

Williams, Paul F.
Geol. Dept.
Univ. of New Brunswick
P.O. Box 4400
Fredericton, NB E3B 5A3
Canada

Williams, Quentin
Institute of Tectonics
Board of Earth Sciences
University of California
Santa Cruz, CA 95064

Williams, Richard T.
Geol. Sci. Dept.
Univ. of Tennessee
Knoxville, TN 39996

Williams, S.J.
U.S.G.S.
914 National Center
Reston, VA 22092

Williamson, Alex K.
U.S.G.S., WRD
55 N. Interregional Hwy.
Austin, TX 78702-5206

Wilson, A.H.
Geol. Dept.
Univ. of Natal
P.O. Box 375
Pietermaritzburg 3200
South Africa

Wilson, Allan
Geol. Dept.
Univ. of Natal
P.O. Box 375
Pietermaritzburg 3200
South Africa

Wilson, Barbara H.
P. O. Box 1198
Ada, OK 74820

Wilson, J. Richard
Geol. Inst.
C.F. Mollers Alle
8000 Aarhus C
Denmark

Wilson, John L.
Geosci. Dept.
New Mexico Inst. of
Mining & Technology
Socorro, NM 87801

Wilson, John T.
U.S. Environ. Prot. Agency
R.S. Kerr Environ. Res. Lab.
P.O. Box 1198
Ada, OK 74820

Wilson, K.M.
Geol. Sci. Dept.
Univ. of Colorado
Boulder, CO 80309

Wilson, K.M.
Geol. Sci. Dept.
Univ. of Colorado
Boulder, CO 80309

Wilson, William E.
U.S.G.S., Br. of NNWSI
MS 421 Box 25046
Denver Federal Center
Denver, CO 80225

Winston, Don
Geol. Dept.
Univ. of Montana
Missoula, MT 59812

Winterer, Edward L.
Geol. Res. Div.
Univ. of California
Scripps Inst. of Oceanog.
La Jolla, CA 92093

Wintle, Ann G.
Dept. of Geolography
Royal Hollowar & Bedford
Univ. of London
Egham, Surrey TW20 0EX
England

Wise Jr., Sherwood W.
Geol. Dept.
Florida State Univ.
Tallahassee, FL 32306

Wise Jr., Sherwood W.
Geol. Dept.
Flordia State Univ.
Tallahassee, FL 32306

Witherspoon, P.A.
Mat.Sci.&Min.Engr.Dept.
Univ. of California
Berkeley, CA 94720

Witkowski, Andrzej
Institute of Oceanography,
University of Gdansk,
Ozolgistow 46,
81-378 Gdynia
Poland

Wnuk, Christopher
U.S. Geological Survey
956 National Center
Reston, VA 22092

Woermann, E.
Inst. fur Kristallographic
RWTH Aachen
D5100 Aachen
Fed. Rep. of Germany

Wohletz, Kenneth H.
ESS-1, MS D462
Earth & Space Sci. Div.
Los Alamos Natl. Lab.
Los Alamos, NM 87544

Wohnlich, Stefan
Applied Geol. Dept.
Univ. of Karlsruhe
P.O.B. 6980
D7500 Karlsruhe
Fed. Rep. of Germany

Wojcik, Andrzej
Sztygarka-School of Mines
27 Stycznia 69
Dabrowa Gornicza 41 300
Poland

Wojcik, Andrzej
Sztygarka-School of Mines
27 Stycznia 69
Dabrowa Gornicza 41 300
Poland

Wokoma, W.D.
Department of Geography
University of Port Harcourt
P.M.B. 5323
Port Harcourt
Rivers State
Nigeria

Wokoma, W.D.C.
Univ. of Port Harcourt

Port Harcourt
Nigeria

Wold, Christopher N.
Uiktoriastr. 25

8000 Munchen 40
Fed. Rep. of Germany

Wolf, Thomas C.W.
Geol.-Pal. Inst. & Mus. of
Christian-Albrechts Univ.,
Olshausenstrasse 40/60, D-2300
Kiel 1
Fed. Rep. of Germany

Wolf, Thomas C.W.
Geological-Paleontological
Institue and Museum of
Christian-Albrechts University
Olshausenstrasse 40/60
D-2300 Kiel 1
Fed. Rep. of Germany

Wolff, F. Chr.
Dept. of Bedrock Mapping
Geological Survey of Norway
P. O. Box 3006
Trondheim N-7002
Norway

Wong, Ivan G.
Woodward-Clyde Consultants
500 12th Street
Suite 100
Oakland, CA 94607

Woo, Kyung-Sik
Geol. Dept.
College of Natural Science
Kangweon National Univ.
Chuncheon
Kangweon-do 200 092
Korea

Woo, Ming-ko
Dept. of Geography
McMaster University
Hamilton, Ont. L8S 4K1
Canada

Wood, John A.
Harvard-Smithsonian Center for Astrophysics
60 Garden St.
Cambridge, MA 02138

Woodhouse, J.H.
Earth & Planetary Sci. Dept.
Harvard Univ.
Cambridge, MA 02138

Woodward, Robert L.
IGPP, A-025
Scripps Institution of Oceanography, UCSD
La Jolla, CA 92093

Wooten, Richard M.
U.S.D.A. - Forest Service
Gifford Pinchot National Forest Zone II Engineering
Cook, WA 98605

Wortel, Rinus
Inst. of Earth Sci.
Univ. of Utrecht
P.O. Box 80.021
3508 TA Utrecht
Netherlands

Wortel, Rinus
Inst. of Earth Sci.
Univ. of Utrecht
P.O. Box 80.021
3508 TA Utrecht
Netherlands

Wright, Judith
Battelle, Pacific Northwest Lab
P.O. Box 999
Richland, WA 99352

Wright, Thomas L.
U.S. Geological Survey
P. O. Box 51
Hawaii National Park, HI 9671

Wu, Gongjian
Geological Society of China
Pai Wan Chuang Fuchengmenwai
Beijing
Peo. Rep. of China

Wu, Xiantao
Department of Geology
Jiaozuo Mining College
Jiaozuo City
Henan Province
Peo. Rep. of China

Wurzer, F.
Joanneum Research Assoc.
Roseggerstrasse 15
A-8700 Leoben
Austria

Wyllie, Peter J.
Division of Geological and Planetary Sciences
California Institute of Tech.
Pasadena, CA 91125

X Y Z

Xenophontos, Costas
Geological Survey Department
Ministry of Agriculture and
Natural Resources
Nicosix, Cyprus
Turkey

Xiang, Wei
Applied Geol. Dept.
Karsruhe Univ.
7500 Karlsruhe
Fed. Rep. of Germany

Xiang, Wei
Karisruhe Universitye
Dept. of Applied Geology
Kaiserstr. 12
P.O.B. 6980
Karlsruhe D-7500
Fed. Rep. of Germany

Xiao, Xuchang
Geological Society of China
Pai Wan Chuang Fuchengmenwai
Beijing
Peo. Rep. of China

Xie, Douke
534 East Zhongshan Road

Nanjing
Peo. Rep. of China

Xie, Douke
534 East Zhongshan Road

Nanjing
Peo. Rep. of China

Xie, Xuejing
Institute of Geophysical and
Geochemical Exploration
Langfang
Hebei 102801
Peo. Rep. of China

Xie, Xuejing
Inst. of Geophysical and
Geochemical Exploration
Lanfang
Hebei 102801
Peo. Rep. of China

Xin, Xu
Dept. of Geography
Nanjiang University
Nanjian
Peo. Rep. of China

Xing, Yusheng
Institute of Geology
Chinese Acad. of Geol. Sci.
P.O. Box 634
Baiwanzhuang Road
Beijing
Peo. Rep. of China

Xinmin, Chen
Department of Geology
Guizhou Institute of
Technology
Guizhou
Peo. Rep. of China

Xu, Huifang
The Geol. Society of China
Pai Wan Chuang Fuchengmenwai
Beijing
China

Xu, Renchao
Geological Society of China
Baiwanzhuang Road 26
Fuchengmenwai
Beijing
China

Xu, Yanqiang
Geological Society of China
Pia Wan Chuang Fuchengmenwai
Beijing
Peo. Rep. of China

Xu, Yong-Chang
No. 198, Donggangxilu Road
Gansu Province
Lanzhou
Peo. Rep. of China

Xu, Zhigin
Geological Society of China
Pai Wan Chuang Fuchengmenwai
Beijing
Peo. Rep. of China

Xu, Zuyre
The Wuhan Team of Hydrogeology
Hubei Bureau of Geology and
Mineral Resources
Wuhan
Peo. Rep. of China

Xue, Yuxuan
Bur. of Geology
China National Nuclear Corp.
P.O. Box 1436
Beijing 100013
Peo. Rep. of China

Yaghubpur, Abdolmajid
Department of Geology
Tarbiat Moalem University
P.O. Box 19615-116, Tajrish
Tehran
Iran

Yaghubpur, Abdolmajid
P. O. Box 19615-116, Tajrish

Tehran
Iran

Yagi, Takehiko
Inst. for Solid State Physics
Univ. of Tokyo, Roppongi
Minato-ku,
Tokyo 106
Japan

Yagmurlu, Fuzuli
Department of Geology,
Engineering Faculty of Akdeniz
University
32260 Isparta
Turkey

Yair, Aaron
Inst. of Earth Sci.
The Hebrew Univ. of Jerusalem
Jerusalem 91904
Israel

Yamada, Eizo
Geological Survey of Japan
1-1-3 Higashi, Tsukuba
Ibataki 305
Japan

Yamagishi, Hiromitsu
Geol. Survey of Hokkaido
Kita 19
Nishi 12
Kitaku
Sapporo 060
Japan

Yamagishi, Hiromitsu
Geol. Survey of Hokkaido
Kita 19
Nishi 12, Kita-ku
Sapporo 060
Japan

Yang, Bo
Karst Section
Guizhou Academy of Sciences
Guiyang, Huizhou
Peo. Rep. of China

Yang, C.S.
InterGeos BV
Utrechtseweg 62
3704 HE Zeist
Netherlands

Yang, Changshu
Sedimentology Consultants
Utrechtseweg 62
3704 HE Zeist
Netherlands

Yang, Changshu
Sedimentology Consultants
Utrechtsweg 62
3704 HE Zeist,
Netherlands

Yang, Chung Y.
Tianjin Geol. Academy
Ministry of Metallurgical Ind.
Tianjin
Peo. Rep. of China

Yang, Hua
The Geol. Survey of China
Pai Wan Chuang Fuchengmenwai
Beijing,
Peo. Rep. of China

Yang, Jicheng
Ministry of Water Resources
Remote Sensing Appl. Ctr.
Beijing
Peo. Rep. of China

Yang, Jiduan
Inst. of Geology
Chinese Acad. of Geol. Sci
Baiwanzhuang Rd.
Fuchengwai
Beijing
Peo. Rep. of China

Yang, Jingsui
Geol. Dept.
Dalhousie Univ.
Halifax, N.S. B3H 3J5
Canada

Yang, Jingsui
Geol Dept.
Dalhousie University
Halifax, N.S. B3H 3J5
Canada

Yang, Qi
China Unif. of Geosci.
Xueyuan Rd. 29
Beijing
Peo. Rep. of China

Yang, Weiran
Geological Society of China
Baiwanzhuang Road 26
Fuchengmenwai
Beijing
China

Yang, Wenjin
Institute of Geochemistry
Academia Sinica
Guiyang City
China

Yang, Zun-Yi
China Univ. of Geosciences

Beijing 100083
Peo. Rep. of China

Yang, Zuosheng
The Geol. Society of China
Pai Wan Chuang Fuchengmenway
Geijing
Peo. Rep. of China

Yasvin, L. S.
All-Union Research Inst.
Hydrogeology & Engineering
Geology
142452 Zeleny-village, Noginsk
Moscow region
U.S.S.R.

Yates, Marylynn V.
Dept. of Environmental Science
University of California
Riverside, CA 92521

Ye, Lian-jun
Institute of Geology
Academia Sinica
P.O. Box 634
Beijing
Peo. Rep. of China

Yeats, Robert S.
Geol. Dept.
Oregon State Univ.
Corvallis, OR 97331-5506

Yegorkin, A.V.
Ministry of Geology, USSR
4/6 Ulitsa Bolchaya
Gruzinskaya
Moscow 123812
U.S.S.R.

Yilmaz, Y.
I.T.U.
Maden Fakultesi
Jeoloji Bolumu
Ayazuga
Istanbul 80626
Turkey

Yim, Wyss Wai-Shu
Department of Geography and
Geology
University of Hong Kong
Pokflem Road
Hong Kong

Yin, Hanhui
Changsha Inst. of Geotectonics
Academia Sinica
Changsha, Hunan
Peo. Rep. of China

Yochelson, Ellis L.
Dept. of Paleobiology
National Museum of Natural
History
Washington, D.C. 20560

Yogodzinski, Gene
Geol. Sci. Dept.
Cornell Univ.
Ithaca, NY 14853

Yong Ning, Tao
China Off-Shore Oil Expl. and
Development Research Center
Xincheng County
Hebei Province
Peo. Rep. of China

York, Derek
Physics Dept.
Univ. of Toronto
Toronto Ont. M5S 1A7
Canada

York, Derek
Physics Dept.
Univ. of Toronto
Toronto, Ont. M5S 1A7,
Canada

Young, Grant M.
Geol. Dept.
Univ. of Western Ontario
London, Ont. N6A 5B7
Canada

Young, Grant M.
Geol. Dept.
Biological & Geol. Bldg.
The Univ. of Western Ontario
London, Ont. N6A 5B7
Canada

Young, Grant M.
Geol. Dept.
Univ. of Western Ontario
London, Ont. N6A 5Bt
Canada

Young, Kirby D.
Geosci. Dept.
Pennsylvania State Univ.
University Park, PA 16802

Young, Richard A.
Geol. Sci. Dept. .
SUNY College at Geneseo
Geneseo, NY 14454

Young, Richard A.
Geol. Sci. Dept.
SUNY College at Geneseo
Geneseo, NY 14454

Young, Richard E.
Nasa-Ames Research Center
Mail Stop 245-3
Moffett Field, CA 94035

Young, Robert William
Geolography Department
Univ. of Wollongong
Wollongong 2500
Australia

Yousefi, Emamgholi
Geological Survey of Iran
P. O. Box 13185-1494
Tehran
Iran

Yrigoyen, Marcelo R.
Zavalia 2048, 5

1428 Buenos Aires
Argentina

Yu, Guanming
Institute of Sedimentary
Geology & Mineral Resources
Chengdu College of Geology
Chengdu Sichuan
Peo. Rep. of China

Yu, Puzhi
Institute of Oceanology
Academia Sinica
Qingdao
Peo. Rep. of China

Yu, Rongbing
Tianjin Inst. of Geol. and
Miner. Res.
No. 4, 8th Rd., Dazhigu
Tianjin 300170
Peo. Rep. of China

Yu, Shuhu
Geological Society of China
Baiwanzhuang Road 26
Fuchengmenwai
Beijing
Peo. Rep. of China

Yu, Tzuen-dean
21 Minghang Road

Kunming City, Yunnan Province
Peo. Rep. of China

Yu, Yuanchang
Geological Society of China
Baiwanzhuang Road 26
Fuchengmenwai
Beijing
Peo. Rep. of China

Yuan, Daoxian
The Inst. of Karst Geol.

Guilin, Guangxi
Peo. Rep. of China

Yuan, Jun
317 E. Third Street #3

Bloomington, IN 47401

Yuan, Xuecheng
Bur. of Geophy. & Geochemical Exploration
Ministry of Geol. & Mineral Resources
Beijing
Peo. Rep. of China

Yue, Shucang
Department of Resources and Environmental Science
Hefei University of Technology
Hefei, Anhui
Peo. Rep. of China

Yui, Tzen-Fu
Institute of Earth Sciences,
Academic Sinica, R.O.C.
P. O. Box 23-59
Taipei
Taiwan

Yumul, Graciano P.
Geological Institute
University Of Tokyo
7-3-1 Hongo
Bunkyo-Ku
Tokyo 113
Japan

Zahedi, Mostaea
Geological Survey of Iran
P.O. Box 13185-1494
Tehran
Iran

Zainuddin, Syed M.
Sir Syed Nagar

Aligarh 202002
INDIA

Zakharov, Victor A.
Inst. of Geology & Geophysics

Novosibirsk 90
U.S.S.R.

Zakharov, Victor A.
Inst. of Geol. & Geophy.of the Siberian Br. of the U.S.S.R.
Univ. av. 3
Novosibirsk 90
U.S.S.R.

Zambre, M.K.
D.B.F. Dayanand College of Arts & Science
Post
Sulapur 413002
India

Zarayan, Sirous
Department of Geology
University of Tehran
Tehran 19879
Iran

Zen, E-an
U.S. Geological Survey
959 National Center
Reston, VA 22092

Zeng, Xuesi
Geological Society of China
Pai Wan Chuang Fuchengmenwai
Beijing
China

Zentilli, Marcos
Geol. Dept.
Dalhousie Univ.
Halifax, N.S. B3H 3J5
Canada

Zhai, Guangming
Scientific Research Inst. of Petroleum Expl. and Develp.
P.O. Box 910 Xueyuan Road
Beijing
China

Zhan, Bing-shan
The Geol. Society of China
Pai Wan Chuang Fuchengmenway
Beijing
Peo. Rep. of China

Zhang, Fuqin
Earth Sci. Dept.
Nanjing Univ.
Nanjing
Peo. Rep. of China

Zhang, Fuqin
Earth Sci. Dept.
Nanjing Univ.
Nanjing
Peo. Rep. of China

Zhang, Guowei
The Geol. Society of China
Pai Wan Chuang Fuchengmenwai
Beijing
Peo. Rep. of China

Zhang, Huifen
Inst. of Geochemistry
Academia Sinica
Guiyang
Guizhou
Peo. Rep. of China

Zhang, Huifen
Inst. of Geochem.
Academia Sinica
Guiyang
Guizhou Province
Peo. Rep. of China

Zhang, Huimin
Tianjin Inst. of Geology and Mineral Resources
Jintang Hwy. 26
Tianjin 300170
Peo. Rep. of China

Zhang, Ligang
Yichang Institute of Geology and Mineral Resource, CAGS
P. O. Box 502
Yichang City, Hubei Province
Peo. Rep. of China

Zhang, Ligang
Yichang Institute of Geology and Mineral Resource, CAGS
P. O. Box 502
Yichang City, Hubei Province
Peo. Rep. of China

Zhang, Ligang
Yichang Institute of Geology and Mineral Resource, CAGS
P. O. Box 502
Yichang City, Hubei Province
Peo. Rep. of China

Zhang, Renjie
Yichang Institute of Geology
P. O. Box 502
Yichang, Hubei
Peo. Rep. of China

Zhang, Rentie
Yichang Institute of Geology and Mineral Resources
P. O. Box 502
Yichang, Hubei
Peo. Rep. of China

Zhang, RongHua
Geological Society of China
Pai Wan Chuang Fuchengmenwai
Beijing
Peo. Rep. of China

Zhang, Shao Hua
Geol. Co. State Bureau of Bldg. Materials Ind.
No. 11 Shuncheng Street
Beijing
Peo. Rep. of China

Zhang, Shao Nan
Dept. of Petroleum Geology
Chengdu College of Geology
Chengdu Sichuan
Peo. Rep. of China

Zhang, Wei
Ctr.for Analysis & Predictions
State Seismological Bureau
63 Fuxing Avenue
Beijing
China

Zhang, Xianqi
Changchun College of Geology

Changchun
Peo. Rep. of China

Zhang, Xiqi
Geol. Dept.
Shandong Mining College
Taian
Peo. Rep. of China

Zhang, Y.G.
2801 Upton St. N.W.

Washington, DC 20008

Zhang, Yifei
Geological Society of China
Baiwanzhuang Road 26
Fuchengmenwai
Beijing
Peo. Rep. of China

Zhang, YiGang
Geological Society of China
Pai Wan Chuang Fuchengmenwai
Beijing
China

Zhang, Youxue
Lamont-Doherty Geological Observatory
Dept. of Geological Sciences
Columbia University
Palisades, NY 10964

Zhang, Zhongying
The Geol. Society of China
Pai Wan Chuang Fuchengmenwai
Beijing
Peo. Rep. of China

Zhao, Penga
Geological Society of China
Baiwanzhuang Road 26
Fuchengmenwai
Beijing
China

Zheng, Baoshan
Institute of Geochemistry
Academia Sinica
Guiyang, Guizhou
Peo. Rep. of China

Zheng, Jiandong
Institute of Geology
State Seismological Bureau
Beijing 100011
Peo. Rep. of China

Zheng, Jiandong
Institute of Geology
State Seismological Bureau
Beijing 100011
Peo. Rep. of China

Zheng, Y-F.
Geochemisches Inst. der Univ. Gottingen
Goldschmidstr. 1
3400 Gottingen
Fed. Rep. of Germany

Zheng, Y-F.
Geochemisches Inst. der Univ. Gottingen
Goldschmidstr. 1
3400 Gottingen
Fed. Rep. of Germany

Zhou, Huitang
China Unif. of Geosci.
Xueyuan Rd. 29
Beijing
Peo. Rep. of China

Zhou, Lingdi
Institute of Geochemistry
Academia Sinica
Guiyang
Guizhou
Peo. Rep. of China

Zhou, Xiang
Geological Society of China
Baiwanzhuang Road 26
Fuchengmenwai
Beijing
China

Zhou, Xinhua
Inst. of Geology
Academia Sinica
P.O. Box 634
Beijing 100011
Peo. Rep. of China

Zhou, Xinhua
Institute of Geology,
Academia Sinica,
P.O. Box 634
Beijing, 100011
Peo. Rep. of China

Zhou, Zhi-Wen
Dept. of Earth & Space Science
Univ. of Science & Technology
Hefei, Anhui
Peo. Rep. of China

Zhu, Shixing
Tianjin Inst. of Geol. and Mineral Resources
Chinese Academy of Geol. Sci.

Zhu, Xingshan
Dept. of Geology
Shandong Institute of Mining & Technology
Tain
Peo. Rep. of China

Ziegler, Will
Forschungsiastitut Senckenberg
Senckenberganlage 25
6000 Frankfurt 1
Fed. Rep. of Germany

Zierenberg, Robert
U.S. Geological Survey
M.S. 901
345 Middlefield Road
Menlo Park, CA 94025

Zoback, Mark
Geophy. Dept.
Stanford Univ.
Stanford, CA 94305

Zoback, Mark D.
Geophy. Dept.
Stanford Univ.
Stanford, CA 94305

Zoback, Mary Lou
U.S.G.S.
345 Middlefield Rd.
Menlo Park, CA 94025

Zodrow, Erwin L.
University College of Cape Breton
Sydney, N.S. B1P 6L2
Canada

Zolnai, Greg
11. av. des Sayettes

Pau (64.000)
France

Zolotov, Alexei N.
Ministry of Geology, USSR
4/6 Bolshaya Gruzinskaya St.
Moscow 123812
U.S.S.R.

Zonenshain, Lev P.
Institute of Oceanology
USSR Academy of Science
23 Krasikova Str.
Moscow 117218
U.S.S.R.

Zonenshain, Lev. P.
Institute of Oceanology
USSR Academy of Science
23 Krasikova Str.
Moscow 1247942
U.S.S.R.

Zuchiewicz, Witold
Institute of Geological
Sciences
Jagiellonian University
2A Oleandry Str.
Krakow 30-063
Poland

Zuchiewicz, Witold
Institute for Geological
Sciences
Jagiellonian University
2A Oleandry Str.
Kracow 30-063
Poland

Zupan, Martina
Hidrometeoroloski Zavod SRS
61000 Ljubljana
Vojkova 1b
Yugoslavia

Zwetkov, K.
Geol. Ent. for Laboratory
Research
G.A. Nasser str. 16
Sofia 1113
Bulgaria

Zwolinski, Zbigniew
Quaternary Research Institute
Adam Mickiewicz University
Fredry 10,
PL 61-701 Poznan
Poland

0hou, Yingxin
Virginia Division of Mineral
Resources
617 North Main St.
Blacksburg, VA 24060

0raun, Duane D.
Geography and Earth Science
Department
Bloomsburg University
Bloomsburg, PA 17815